T0393812

NUMERICAL METHODS IN GEOTECHNICAL ENGINEERING IX

PROCEEDINGS OF THE 9TH EUROPEAN CONFERENCE ON NUMERICAL METHODS IN GEOTECHNICAL ENGINEERING (NUMGE 2018), 25–27 JUNE 2018, PORTO, PORTUGAL

Numerical Methods in Geotechnical Engineering IX

Editors

António S. Cardoso, José L. Borges, Pedro A. Costa, António T. Gomes, José C. Marques & Castorina S. Vieira

Faculty of Engineering, University of Porto, Porto, Portugal

VOLUME 1

CRC Press is an imprint of the
Taylor & Francis Group, an **informa** business
A BALKEMA BOOK

CRC Press/Balkema is an imprint of the Taylor & Francis Group, an informa business

Typeset by V Publishing Solutions Pvt Ltd., Chennai, India

Published by: CRC Press/Balkema
Schipholweg 107C, 2316 XC Leiden, The Netherlands
e-mail: Pub.NL@taylorandfrancis.com
www.crcpress.com – www.taylorandfrancis.com

ISBN: 978-1-138-54446-8 (set of 2 volumes)
ISBN: 978-1-138-33198-3 (Vol 1)
ISBN: 978-1-138-33203-4 (Vol 2)
ISBN: 978-1-351-00362-9 (eBook set of 2 volumes)
ISBN: 978-0-429-44693-1 (eBook, Vol 1)
ISBN: 978-0-429-44692-4 (eBook, Vol 2)

Numerical Methods in Geotechnical Engineering IX – Cardoso et al. (Eds)
© 2018 Taylor & Francis Group, London, ISBN 978-1-138-33198-3

Table of contents

Earthquake engineering, soil dynamics and soil-structure interaction

Rock mechanics

VOLUME 2

Application of numerical methods in the context of the Eurocodes

Shallow and deep foundations

Preface

The European Regional Technical Committee (ERTC7) of the International Society for Soil Mechanics and Geotechnical Engineering (ISSMGE) and the Organizing Committee welcome all participants of the 9th European Conference on Numerical Methods in Geotechnical Engineering (NUMGE2018), held at the Faculty of Engineering of University of Porto (FEUP), in Porto, Portugal, from 25th to 27th June 2018.

This conference is the ninth in a series of conferences on Numerical Methods in Geotechnical Engineering organized by the ERTC7 under the auspices of the ISSMGE. The first conference was held in 1986 in Stuttgart, Germany, and the series continued every four years (1990 Santander, Spain; 1994 Manchester, United Kingdom; 1998 Udine, Italy; 2002 Paris, France; 2006 Graz, Austria; 2010 Trondheim, Norway; 2014 Delft, The Netherlands).

The conference provides a forum for exchange of ideas and discussion on topics related to numerical modelling in geotechnical engineering. Both senior and young researchers, as well as scientists and engineers from Europe and overseas, attend this conference to share and exchange their knowledge and experiences. Geotechnical engineering researchers and practical engineers submit their papers on scientific achievements, innovations and engineering applications related to or employing numerical methods.

The papers for NUMGE2018 cover topics from emerging research to engineering practice. For the proceedings the contributions are grouped under the following themes:

› Constitutive modelling and numerical implementation
› Finite element, discrete element and other numerical methods. Coupling of diverse methods
› Reliability and probability analysis
› Large deformation – large strain analysis
› Artificial intelligence and neural networks
› Ground flow, thermal and coupled analysis
› Earthquake engineering, soil dynamics and soil-structure interactions
› Rock mechanics
› Application of numerical methods in the context of the Eurocodes
› Shallow and deep foundations
› Slopes and cuts
› Supported excavations and retaining walls
› Embankments and dams
› Tunnels and caverns (and pipelines)
› Ground improvement and reinforcement
› Offshore geotechnical engineering
› Propagation of vibrations

Around 400 abstracts were submitted and the Authors of the approved abstracts were invited to submit full papers for peer review. A total of 204 papers were accepted for inclusion in the conference proceedings. The Editors would like to thank the Scientific and Reviewing Committees for their assistance in the review process.

The Editors are grateful for the support of the Chairman, Core Members and National Representatives of ERTC7, namely for promoting the conference on their respective home countries.

NUMGE2018 is jointly organized by SPG (Portuguese Geotechnical Society) and FEUP. These institutions and conference sponsors are gratefully acknowledged for their generous support.

The Editors want to express their particular thanks to the Authors, for their fundamental contribution to the success of the conference, and to the Participants, wishing that the 3 days of presentations and discussions would be fruitful for their future research and technical work.

On behalf of the Organising Committee and ERTC7, we welcome you to Porto hoping that you enjoy the scientific and technical aspects of the conference, as well as its social programme and the city of Porto.

ERTC7 (ISSMGE) –
Helmut Schweiger (Chairman) &
César Sagaseta (Past-Chairman)
NUMGE 2018 Organizing Committee –
António S. Cardoso, José L. Borges, Pedro A. Costa,
António T. Gomes, José C. Marques & Castorina S. Vieira
April 2018

Committees

SCIENTIFIC COMMITTEE (ERTC7)

Core Members

Helmut Schweiger, *Austria (Chairman)*
César Sagaseta, *Spain (Past-Chairman)*
Philip Mestat, *France*
Steinar Nordal, *Norway*
Manuel Pastor, *Spain*
Juan Pestana, *USA*
David Potts, *United Kingdom*
Scott Sloan, *Australia*

National Representatives

Sergey Aleynikov, *Russia*
Katalin Bagi, *Hungary*
Ronald Brinkgreve, *The Netherlands*
Imre Bojtár, *Hungary*
Albert Bolle, *Belgium*
Harvey Burd, *United Kingdom*
Annamaria Cividini, *Italy*
Pedro Alves Costa, *Portugal*
George Dounias, *Greece*
Torbjörn Edstam, *Sweden*
Pit (Peter) Fritz, *Switzerland*
Maciej Gryczmański, *Poland*
Frands Haahr, *Denmark*
Ivo Herle, *Czech Republic*
Fritz Kopf, *Austria*
Tim Länsivaara, *Finland*
Tom Schanz, *Germany*
Herbert Walter, *Austria*

ORGANIZING COMMITTEE

António Silva Cardoso
José Leitão Borges
Pedro Alves Costa
António Topa Gomes
José Couto Marques
Castorina Vieira

REVIEWING COMMITTEE

Armando Antão, *Portugal*
Imre Bojtár, *Hungary*
Albert Bolle, *Belgium*
José Leitão Borges, *Portugal*
Harvey Burd, *United Kingdom*
António Silva Cardoso, *Portugal*
Annamaria Cividini, *Italy*
David Connolly, *United Kingdom*
Pedro Alves Costa, *Portugal*
Teresa Bodas Freitas, *Portugal*
António Topa Gomes, *Portugal*
Nuno Guerra, *Portugal*
Kianoosh Hatami, *USA*
José Vieira de Lemos, *Portugal*
Fernando Lopez-Caballero, *France*
José Couto Marques, *Portugal*
Arézou Modaressi, *France*
Paulo da Venda Oliveira, *Portugal*
Manuel Pastor, *Spain*
António Pedro, *Portugal*
David Potts, *United Kingdom*
César Sagaseta, *Spain*
Helmut Schweiger, *Austria*
Jorge Almeida e Sousa, *Portugal*
David Taborda, *United Kingdom*
Yiannis Tsompanakis, *Greece*
Ana Vieira, *Portugal*
Castorina Silva Vieira, *Portugal*
Helbert Walter, *Austria*
Lidija Zdravković, *United Kingdom*

 ISBN 978-1-138-33198-3

Institutional support

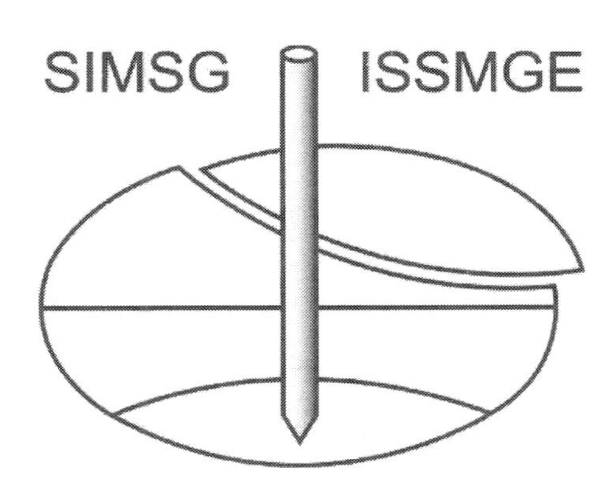

Sociedade Portuguesa de Geotecnia

PORTUGUESE GEOTECHNICAL SOCIETY

Numerical Methods in Geotechnical Engineering IX – Cardoso et al. (Eds)
© 2018 Taylor & Francis Group, London, ISBN 978-1-138-33198-3

Sponsors

Keynote lectures

Numerical Methods in Geotechnical Engineering IX – Cardoso et al. (Eds)
© 2018 Taylor & Francis Group, London, ISBN 978-1-138-33198-3

Numerical modelling of coupled thermo-hydro-mechanical problems: Challenges and pitfalls

D.M. Potts, W. Cui, K.A. Gawecka, D.M.G. Taborda & L. Zdravković
Imperial College London, London, UK

ABSTRACT: Temperature effects in geotechnical engineering are complex as they involve interaction of the soil's constituent phases, such as the soil skeleton and the pore fluid. Accounting for such interaction in the design of geo-thermal infrastructure requires numerical algorithms capable of reproducing thermo-hydro-mechanical (THM) coupling of soil behaviour. However, numerical modelling of transient coupled THM problems may produce erroneous solutions if either the adopted time-step size is too small or highly advective flows are involved. This paper summarises the time-step constraints in both 1D and 2D transient coupled finite element (FE) analysis which prevent the numerical 'shock' problem. Moreover, a coupled thermo-hydraulic boundary condition, as well as a new Petrov-Galerkin finite element method, which are necessary for simulating highly advective flows, are presented and their capabilities are demonstrated in a series of numerical examples.

1 INTRODUCTION

Significant temperature effects have been found in a range of geotechnical engineering problems, such as oil & gas pipelines, pavements, buried power cables, ground energy systems, and the storage of high-level radioactive waste (Gens, 2010). These problems, where *mechanical, hydraulic and thermal systems in soils interact with each other, with the independent solution of any one system being impossible without simultaneous solution of the others*, are defined as coupled thermo-hydro-mechanical (THM) problems, while the interaction between the systems is referred to as the THM coupling (Zienkiewicz 1984).

To simulate adequately a coupled THM problem, it is necessary to first develop the coupled equations based on the governing law of each physical system. A number of studies have been carried out to model the coupled THM behaviour of soils, and the most extensively used numerical tool has been the finite element (FE) method. Most of the exiting coupled FE THM formulations (e.g. Aboustit et al. 1985, Noorishad et al. 1984, Britto et al. 1992, Seneviratne et al. 1994, Thomas & He 1997, Olivella et al. 1996, Vaziri 1996, Gatmiri & Delage 1997, Lewis & Schrefler 1998), are expressed in a full matrix form, in which the diagonal terms govern the main feature of the corresponding system, while the off-diagonal terms describe the coupling effects. As the adopted governing laws are mostly similar, the same diagonal terms have always been obtained in the literature, although various approaches can be employed for deriving the coupled THM formulation. However, the adoption of different assumptions during the derivation procedure may result in differences in the final formulation, especially in the off-diagonal terms representing the couplings. These differences in the coupling terms may significantly influence the results when modelling THM coupled problems.

In an FE analysis of a transient coupled THM problem, it is generally believed that reducing the size of the time-step improves the accuracy of the solutions. However, a lower bound to the time-step size exists, below which the solution may exhibit spatial oscillations at the initial stages of the analysis in the regions where the gradient of the solution is steep. This type of problem is known as a numerical 'shock' problem and may lead to accumulated errors in coupled analyses.

Attention should also be paid in the FE analysis of coupled THM problems involving advective flows, such as those involved in ground source energy systems which depend on a fast flowing fluid to provide heating and cooling. Firstly, accurate solutions may only be obtained if correct boundary conditions are adopted. Secondly, when the conventional Galerkin FE method is applied to simulate flows where advection is the dominant heat transfer mechanism compared to conduction, spatial oscillations of the nodal solution may occur. Although this type of oscillations can be eliminated by refining the finite element mesh, in problems such as those involving heat exchanger pipes, this approach results in an extremely large number of elements, thus becoming computationally expensive.

This paper firstly demonstrates the potential significance of the off-diagonal terms in the coupled FE THM formulation for modelling the thermally induced excess pore water pressure, the behaviour of which is governed by the pore pressure-temperature coupling. Subsequently, the behaviour of two types of numerical oscillations related to the chosen time-step size and the existence of highly advective flows, as well as the corresponding numerical approaches to eliminate these oscillations, are shown through a series of FE analyses. All the analyses presented in this paper were performed using the authors' FE software—the Imperial College Finite Element Program, ICFEP (Potts & Zdravković 1999, 2001), in which the formulation for modelling a coupled THM problem for saturated soils is expressed as:

$$\begin{bmatrix} [K_G] & [L_G] & -[M_G] \\ [L_G]^T & -\beta_1 \Delta t[\Phi_G]-[S_G] & -[Z_G] \\ [Y_G] & -\beta_2 \Delta t[\Omega_G] & \alpha_1 \Delta t[\Gamma_G]+[X_G] \end{bmatrix} \begin{Bmatrix} \{\Delta d\}_{nG} \\ \{\Delta p_f\}_{nG} \\ \{\Delta \theta\}_{nG} \end{Bmatrix} = \begin{Bmatrix} \{\Delta R_G\} \\ \{\Delta F_G\} \\ \{\Delta H_G\} \end{Bmatrix} \quad (1)$$

Details of the coupled THM formulation can be found in Cui et al (2017).

2 TEMPERATURE-PORE PRESSURE COUPLING

2.1 *Numerical formulation*

Under isothermal conditions, the governing equations for pore fluid flow through the soil skeleton can be established by combining the continuity equation with Darcy's law (e.g. Potts & Zdravković 1999). For a soil saturated with a compressible pore fluid, the continuity equation can be formulated based on the volume conservation of the pore fluid, which implies that the net volume of the pore fluid flowing into and out of a compressible element of fully saturated soil is equivalent to the total volumetric change of the soil skeleton. Under non-isothermal conditions, however, the changes in volume of the pore fluid due to the temperature change need to be taken into account, as its coefficient of thermal expansion is different from that of the soil particles. The difference in the two thermal expansion coefficients can generate a volume of pore fluid flowing into or out of the soil element when there is a temperature change. If dissipation of the excess pore fluid is not sufficiently quick, a variation in the pore fluid pressure is induced.

In Equation (1), the effect of temperature change on the volume change of the pore fluid is represented by the matrix $[Z_G]$, which can be expressed as (Cui et al. 2017):

$$[Z_G] = \sum_{i=1}^{N} \left(\int_{Vol} 3[n(\alpha_{T,f} - \alpha_T) + \alpha_T][N_p]^T [N_T] dVol \right)_i \quad (2)$$

where $\alpha_{T,f}$ and α_T are the linear thermal expansion coefficients of the pore fluid and soil particles respectively, and $[N_p]$ and $[N_T]$ are the matrices of pore fluid pressure and temperature interpolation functions (or shape functions) respectively.

Both $\alpha_{T,f}$ and α_T are observed to vary with temperature. It is noted that the dependence of the linear thermal expansion coefficient of the pore water on temperature is particularly significant (i.e. varying between 2.93×10^{-5} m/m.K at 10 °C and 2.51×10^{-4} m/m.K at 100°C, Cengel & Ghajar 2011). However, the variation in the linear thermal expansion coefficient of the soil skeleton has been observed to be substantially smaller, as shown by the linear paths in volumetric strain—temperature space obtained in drained heating tests on overconsolidated clay samples (Abuel-Naga et al. 2007a), and hence can be neglected. In order to model the variation of $\alpha_{T,f}$ with temperature reported by Cengel & Ghajar (2011), a third-order polynomial function is adopted. If the temperature, T, is defined in Celsius, this function can be expressed as:

$$\alpha_{T,f}(T) = 1.48\times10^{-10}T^3 - 3.64\times10^{-8}T^2 + 4.88\times 10^{-6}T - 2.02\times10^{-5} \quad (3)$$

2.2 *Modelling of thermally induced pore fluid pressure in a undrained heating test*

To demonstrate the importance of the coupling term $[Z_G]$, numerical analyses were carried out to reproduce the development of thermally induced excess pore water pressures in an undrained triaxial heating test reported by Abuel-Naga et al. (2007b).

A sample of fully saturated Soft Bangkok clay was used in the test with an initial temperature of 25 °C. The specimen was isotropically consolidated under isothermal conditions to 200 kPa, followed by unloading which resulted in an overconsolidation ratio (OCR) of 4.0. Subsequently, the sample was heated under undrained conditions to 90 °C with increments of 10 °C, and the thermally

Table 1. Material properties for the analysis under undrained heating conditions.

Bulk modulus of soil, K (MPa)	1.98
Initial porosity, n_0 (–)	0.634
α_T $(m/m.K)$	1.0×10^{-5}
Bulk modulus of water, K_f (MPa)	2200

induced pore water pressure was measured after each temperature increment. Laboratory tests of the thermo-mechanical response of Soft Bangkok clay have shown that overconsolidated samples tend to behave elastically (Abuel-Naga 2007a,b), and consequently such an assumption is adopted here.

A single element axi-symmetric fully coupled THM analysis was carried out with the present THM formulation. The relevant properties of Soft Bangkok clay (K, n_o and α_T) were obtained from data provided by Abuel-Naga et al. (2007a) and are listed in Table 1. In the first analysis, a constant value of 8.57 × 10-5 m/m.K is adopted for α_{Tf} which corresponds to a temperature of 25°C, while a temperature dependent value as shown by Equation (3) is employed in the second analysis. In both analyses, all boundaries of the finite element were modelled as impermeable so that an undrained condition was enforced. A total temperature increase of 65 °C was applied to all element nodes in steps of 1°C. A value of 0.8 was adopted for time-marching parameters α_1, β_1 and β_2 (Equation (1)) and the time-step was chosen arbitrarily as 1.0 s. The changes in pore water pressure were monitored in both analyses and good agreement can be observed in Figure 1 when comparing numerical predictions adopting a temperature dependent α_{Tf} against the experimental results. It should be noted that it is the coupling matrix $[Z_G]$ that enables adequate prediction of excess pore water pressures in an undrained heating test, without resorting to a thermo-mechanical constitutive model specially developed to simulate such an effect (Laloui & Francois 2009).

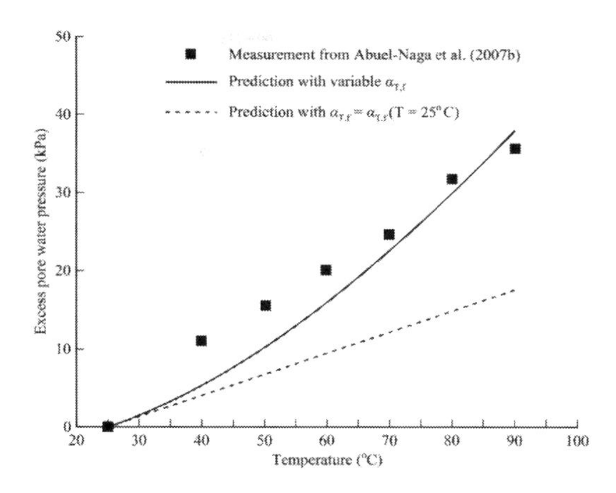

Figure 1. Comparison between numerical and experimental results for an undrained heating test.

3 TIME-STEP CONSTRAINTS

3.1 *Spatial oscillations in 'shock' problems*

The characteristics of the 'shock' problem, existing in the modelling of coupled consolidation and/or heat transfer, can be illustrated by the following example where 1D conductive heat transfer is involved. A plane strain analysis of a 1 m long bar using 4-noded linear elements with an element length of 0.1 m was carried out with ICFEP using the material properties of a typical sandstone (listed in Table 2). The initial temperature was 10 °C and a fixed temperature boundary condition (T = 20 °C) was prescribed on the left-hand side of the mesh. Additionally, a boundary condition imposing no heat flow was prescribed on the other mesh boundaries. The backward difference time marching scheme was applied in which a value of 1.0 was specified for the time marching parameter α_1 and the time-step was chosen arbitrarily as 60 s.

Figure 2 shows the nodal temperature distribution along the bar after one time-step, Δt, of 60 s. The numerical results show spatial oscillations in temperature, deviating from the analytical solution. It can be seen in Figure 3 that these oscillations exist only at the initial stages of the analysis, in the regions where the solution exhibits a steep gradient, and that they decay and finally disappear as the gradient of the solution reduces. It should be noted that, after some increments (in this case 200 increments), the numerical solution matches the analytical solution. Similar behaviour can also be found if the exercise is repeated with quadratic elements as well as in the analyses of coupled consolidation and/or heat advection-conduction (see Cui et al., 2016a).

3.2 *1D time-step constraints*

To avoid oscillations in the simulation of 1D transient problems, the following two non-oscillatory criteria on the nodal solutions should be satisfied:

1. The change of nodal values should not be negative;
2. The variation in the nodal solution should decrease monotonically along the bar.

To fully satisfy the above non-oscillatory criteria, an analytical approach was proposed by Cui

Table 2. Material properties for heat conduction analysis.

Density of solids, ρ_s (t/m^3)	2.5
Density of water, ρ_w (t/m^3)	1.0
Specific heat capacity of solids, C_{ps} ($kJ/t.K$)	880
Specific heat capacity of water, C_{pw} ($kJ/t.K$)	4190
Thermal conductivity, k_T ($kJ/s.m.K$)	0.001
Void ratio, e	0.3

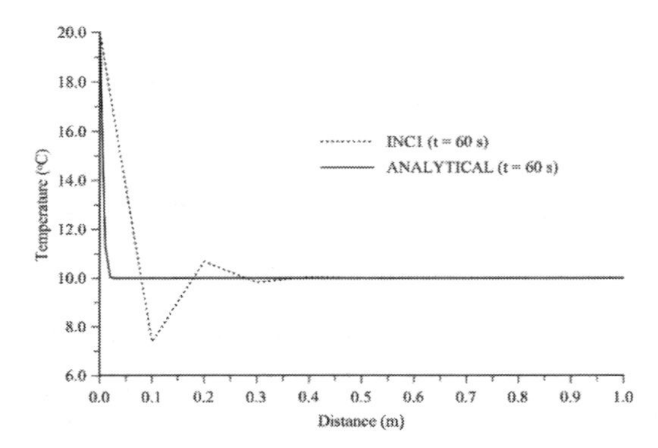

Figure 2. Spatial oscillations in a heat conduction analysis at increment 1 using linear elements with Δt = 60 s, and the corresponding analytical solution.

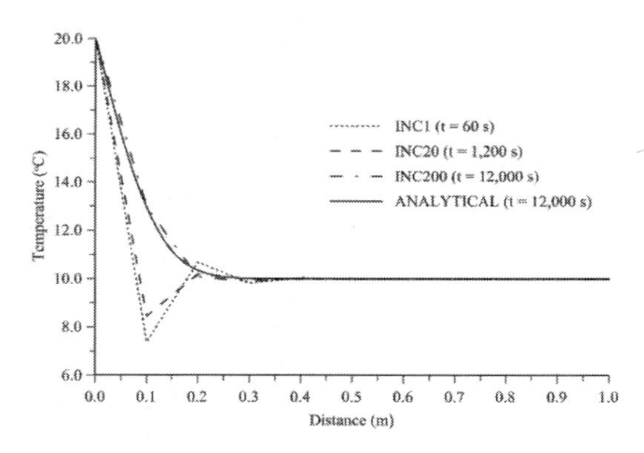

Figure 3. Variation of spatial oscillations with time in a heat conduction analysis using linear elements with Δt = 60 s, and a comparison with the analytical solution.

et al. (2016a), following which the time-step constraints for the FE analysis of problems involving 1D heat conduction (T), heat conduction-advection (TH) and coupled consolidation (HM) have been established and validated for both linear and quadratic elements. Table 3 summarises the time-step constraints corresponding to each type of problem, where ρ and C_p are the density and the specific heat capacity of the soil respectively, h is the element length, k_T is the thermal conductivity, k_w is the permeability, $Pe = \rho_w C_{pw} v_w h/k_T$ is defined as the Péclet number, which represents the ratio between the advective and the conductive transport rates with v_w being the velocity of the pore fluid, M_c is the constrained modulus and α_1 and β_1 are the time marching parameters shown in Equation (1). For a saturated soil, ρC_p can be expressed in terms of the soil's components as $\rho C_p = (1-n)\rho_s C_{ps} + n\rho_w C_{pw}$ where n is the porosity, and the subscripts w and s denote pore water and soil particles respectively. In addition, a time-step constraint of $\Delta t \geq \gamma_f h^2/(6\beta_l M_c k_w)$ is obtained for 1D FE analysis of coupled consolidation with composite elements where pore water pressure varies linearly across the element, whereas displacement varies quadratically.

3.3 *2D time-step constraints*

The non-oscillatory criteria shown above have also been adopted to establish the time-step constraints for the FE analysis of 2D heat conduction problems. However, unlike 1D problems, analytical approaches are not available for 2D problems. Therefore, computational studies were performed by Cui et al. (2018a) based on the numerical example presented by Murti et al. (1989) demonstrating that the time-step constraints for 2D heat conduction problems can be expressed as:

$$\Delta t \geq \beta_T \frac{\rho C_p h_2^2}{3\alpha_1 k_T}(h_2 \geq h_1) \tag{4}$$

where h_1 and h_2 are the lengths of each side of a 2D quadrilateral element respectively, and β_T is a factor which depends on the element type as well as the elemental aspect ratio, i.e. h_2/h_1.

As listed in Table 4, two critical values of α_T have been observed considering the two non-oscillatory criteria proposed above, i.e. one ensures that the temperature increment at any node in the mesh is non-negative, while the other refers to the stricter condition that the temperature variation along each line (horizontal, vertical or diagonal) of a 2D mesh also decreases monotonically. In a 2D FE transient analysis with 4-noded square linear elements ($h_2 = h_1$), both non-oscillatory conditions can be satisfied if a value of $\alpha_T = 3.68$ is adopted in Equation (4), while $\alpha_T = 1$ only ensures the first criterion. When 4-noded rectangular linear elements ($h_2 > h_1$) or 8-noded quadratic elements are employed, only the first criterion can be ensured while the minimum time-step size which satisfies both non-oscillatory conditions does not exist and consequently oscillations are likely for all time steps. For the former case, the critical value

Table 3. Summary of 1D time-step constraints.

Type of problem	Linear element	Quadratic element
T	$\Delta t \geq \dfrac{\rho C_p h^2}{6\alpha_1 k_T}$	$\Delta t \geq \dfrac{\rho C_p h^2}{20\alpha_1 k_T}$
TH	$\Delta t \geq \dfrac{\rho C_p h^2}{3\alpha_1 k_T (2+P_e)}$	$\Delta t \geq \dfrac{3\rho C_p h^2}{20\alpha_1 k_T (3+P_e)}$
HM	$\Delta t \geq \dfrac{\gamma_f h^2}{4\beta_1 M_c k_w}$	$\Delta t \geq \dfrac{\gamma_f h^2}{16\beta_1 M_c k_w}$

of α_T approaches 2 with an increasing elemental aspect ratio, while for the latter case the critical value depends on the Gauss integration order adopted in the analysis (e.g. 2 × 2 integration order or 3 × 3 integration order), as shown in Figure 4.

Table 4. Summary of critical values of β_T for 2D time-step constraints.

	Linear element		Quadratic element	
Criterion	$h_2 = h_1$	$h_2 > h_1$	$h_2 = h_1$	$h_2 > h_1$
(1)	$\beta_T = 1$	1 β_T 2	Figure 4	Figure 4
(1) & (2)	$\beta_T = 3.68$	non existent	non existent	non existent

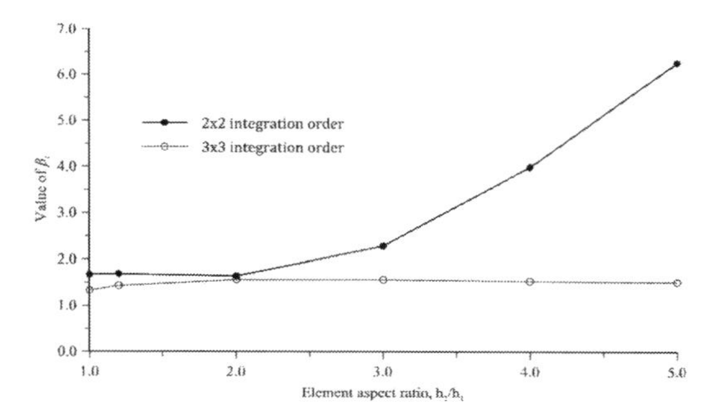

Figure 4. Variation in values of β_t satisfying criterion (1) with aspect ratios using 8-noded quadratic elements.

4 MODELLING OF ADVECTIVE FLOWS

4.1 *Coupled thermo-hydraulic boundary condition*

To demonstrate the necessity of the coupled thermo-hydraulic boundary condition, several analyses on a 1 m bar of soil with 100 elements of length of 0.01 m, subjected to one-dimensional (1D) advective heat transfer, were conducted in ICFEP using 4-noded linear quadrilateral elements. Material properties (see Table 2), boundary conditions (i.e. a fixed temperature boundary condition of $T = 20$ °C prescribed on the left-hand side of the mesh and no heat flow prescribed on the other mesh boundaries) and initial conditions employed here are the same as those adopted in the simulation illustrating the 'shock' problem. In addition, to include the advective flow, a constant pore water pressure gradient was applied over the mesh, inducing a water flow with a constant velocity from left to right. The applied pore water pressure gradient was varied in the analyses resulting in flows with different velocities.

Figure 5 shows the nodal temperature distribution along the bar after 3600s. A sharp increase in temperature can be clearly observed at the right-hand side boundary of the soil bar, even though the heat front has not yet arrived there, as suggested by the fact that the temperature in the middle of the bar is still at its initial value (10 °C). It should also be noted that this unrealistic increase in temperature occurs at the boundary where pore water leaves the mesh, with a higher increase in temperature being found in the analysis with a larger pore water velocity. This scenario is clearly physically impossible, and its characteristics imply that the prescribed boundary condition is incorrect.

To avoid this issue, a boundary condition was proposed and implemented into ICFEP (Cui et al. 2016b), which prescribes the heat flux at the boundary where water leaves or enters the mesh in order to balance the change of the thermal energy associated with the water flow through the boundary. The total amount of the thermal energy associated with the water flow through the boundary can be determined from the following equation:

$$q_{T,b} = \int \rho_w C_{pw} \{v_w\}_b T_b \mathrm{d}' \qquad (5)$$

where $\{v_w\}_b$ represents the velocity of the pore water flowing through the boundary, T_b is the current temperature at the boundary, and Ω denotes either a line, a surface or a volume over which the boundary condition is prescribed. It should be noted that the proposed boundary condition is non-linear, with both $\{v_w\}_b$ and T_b varying over a solution increment, which must be accounted for by the software. Therefore, it is defined as a coupled thermo-hydraulic boundary condition.

The proposed coupled thermo-hydraulic boundary condition was then applied to the previously described numerical analysis of 1D advective flow. As a fixed temperature boundary condition was already prescribed at the left-hand side boundary of the mesh, which automatically balances the change of the thermal energy related to the volume of water flowing into the mesh, the proposed boundary condition was only prescribed at the right-hand side boundary of the bar where water flows out of the mesh. Figure 6 shows the nodal temperature distributions along the bar with the coupled thermo-hydraulic boundary condition. The sharp increase in temperature at the right-hand side boundary of the bar disappears and the numerical results with different values of water velocity obtained using ICFEP agree very well with the analytical results.

4.2 *Modelling of highly advective flows*

In the modelling of heat transfer problems dominated by conduction, the extensively used Galerkin finite element method (GFEM), which assumes that the chosen weighting function is the same as the shape function, is capable of providing bounded numerical solutions, as shown in the previous section. However, if heat transfer is dominated by advection, i.e. if highly advective flow exists, adopting the GFEM may result in spatial oscillations of the nodal temperatures if the Péclet number, *Pe*, is too large (i.e. $Pe > 1$ for linear elements and $Pe > 2$ for quadratic elements). The oscillations become more significant with increasing Péclet number and their behaviour and magnitude also depends on the type of the element and the boundary conditions employed (Cui et al. 2016b). To eliminate the spatial oscillations, a Petrov-Galerkin finite element method (PGFEM) for modelling highly advective flows (Pe > 1) in porous media was developed and investigated by Cui et al. (2018b).

In the conventional GFEM the matrix of weighting functions, $[W]$, is assumed to be the same as the matrix of temperature interpolation functions (or shape functions), $[N]$, while the PGFEM employs weighting functions which differ from the shape functions.

For a 2-noded 1D linear isoparametric element, the continuous PG weighting functions can be defined as:

$$W_{l,1}(s) = N_1(s) + \alpha_{PG} f(s) \tag{6}$$

$$W_{l,2}(s) = N_2(s) - \alpha_{PG} f(s) \tag{7}$$

where the subscripts 1 and 2 denote the nodal number, N_1 and N_2 are the isoparametric nodal shape functions, s represents the natural ordinate of the 1D elements, α_{PG} is the PG weighting factor and the function $f(s)$ is expressed as:

$$f(s) = -\frac{3}{4}(1+s)(1-s) \quad (-1 \le s \le 1) \tag{8}$$

The PG weighting factor controls the level of upwinding and its optimal value can be calculated as:

$$\alpha_{PG,opt} = \coth\left(\frac{Pe}{2}\right) - \frac{2}{Pe} \tag{9}$$

Similarly, the continuous PG weighting functions for 3-noded 1D quadratic elements are given as:

$$W_{q,i}(s) = N_i(s) - \beta_{PG1} g(s) \quad (i = 1,2) \tag{10}$$

$$W_{q,3}(s) = N_3(s) + 4\beta_{PG2} g(s) \tag{11}$$

where nodes 1 and 2 are located at the two ends, node 3 is the mid-side node, β_{PG1} and β_{PG2} are

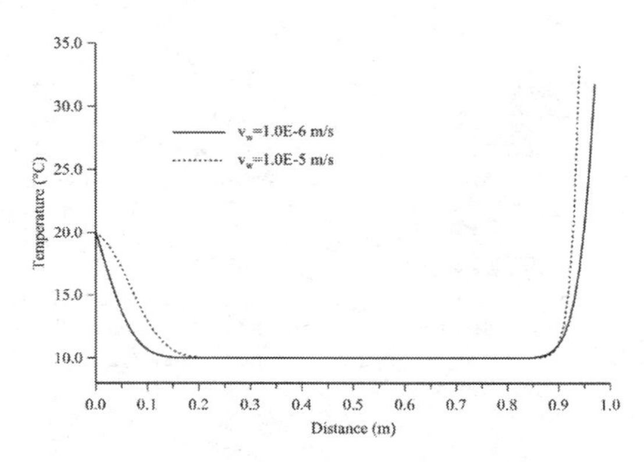

Figure 5. Temperature distribution at t = 3600s in the analyses with no heat flow prescribed at the right-hand mesh boundary.

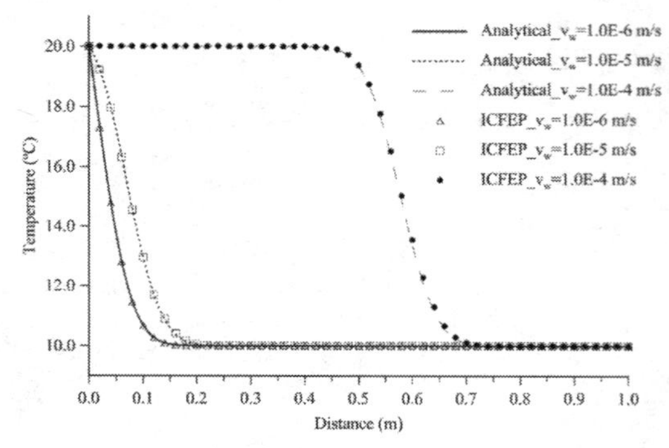

Figure 6. Temperature distribution at t = 3600s in the analyses with coupled thermo-hydraulic boundary condition.

the PG weighting factors and the function $g(s)$ is written as:

$$g(s)=\frac{5}{8}s(s+1)(s-1)\quad(-1\leq s\leq 1) \tag{12}$$

The optimal value of the PG weighting factors β_{PG1} and β_{PG2} can be obtained from:

$$\beta_{PG1,opt}=2\tanh\left(\frac{Pe}{2}\right)\left(1+\frac{3\beta_{PG2,opt}}{Pe}+\frac{12}{Pe^2}\right)\frac{12}{Pe}-\beta_{PG2,opt} \tag{13}$$

$$\beta_{PG2,opt}=\coth\left(\frac{Pe}{4}\right)-\frac{4}{Pe} \tag{14}$$

The PG weighting functions for a 2D 4-noded linear element can be obtained by combining two 1D 2-noded linear element weighting functions. However, the two 1D weighting functions now depend on the S and T natural coordinates, such that:

$$W_{l,1}^{2D}(S,T)=W_{l,1}(S)W_{l,1}(T) \tag{15}$$

$$W_{l,2}^{2D}(S,T)=W_{l,2}(S)W_{l,1}(T) \tag{16}$$

$$W_{l,3}^{2D}(S,T)=W_{l,2}(S)W_{l,2}(T) \tag{17}$$

$$W_{l,4}^{2D}(S,T)=W_{l,1}(S)W_{l,2}(T) \tag{18}$$

The PG weighting functions for a 2D 8-noded quadratic element can be expressed as:

$$W_{q,1}^{2D}(S,T)=W_{l,1}(S)W_{l,1}(T)-\frac{1}{2}W_{q,5}^{2D}(S,T)-\frac{1}{2}W_{q,8}^{2D}(S,T) \tag{19}$$

$$W_{q,2}^{2D}(S,T)=W_{l,2}(S)W_{l,1}(T)-\frac{1}{2}W_{q,5}^{2D}(S,T)-\frac{1}{2}W_{q,6}^{2D}(S,T) \tag{20}$$

$$W_{q,3}^{2D}(S,T)=W_{l,2}(S)W_{l,2}(T)-\frac{1}{2}W_{q,6}^{2D}(S,T)-\frac{1}{2}W_{q,7}^{2D}(S,T) \tag{21}$$

$$W_{q,4}^{2D}(S,T)=W_{l,1}(S)W_{l,2}(T)-\frac{1}{2}W_{q,7}^{2D}(S,T)-\frac{1}{2}W_{q,8}^{2D}(S,T) \tag{22}$$

$$W_{q,5}^{2D}(S,T)=W_{q,3}(S)W_{l,1}(T) \tag{23}$$

$$W_{q,6}^{2D}(S,T)=W_{l,2}(S)W_{q,3}(T) \tag{24}$$

$$W_{q,7}^{2D}(S,T)=W_{q,3}(S)W_{l,2}(T) \tag{25}$$

$$W_{q,8}^{2D}(S,T)=W_{l,1}(S)W_{q,3}(T) \tag{26}$$

where nodes 1–4 are located at the corners and nodes 5–8 are the middle-side nodes. It should be noted that with the new PG weighting functions shown above it is possible to simulate highly advective flow using a mesh which contains a mix of both 8-noded elements and 4-noded elements.

In the proposed 2D PG scheme, the optimal values of the weighting factors are formulated based on the Péclet numbers in the local S and T coordinate directions, $Pe_{,S}$ and $Pe_{,T}$, which are evaluated at each Gauss point. When the derived $Pe_{,S}$ and $Pe_{,T}$ are directly used to calculate the weighting factors, the PG scheme described above leads to accurate results when simulating 1D highly advective flows, with either constant or variable fluid velocities, using a 2D FE mesh with regular (square or rectangular) elements. However, invalid results are found when either multi-dimensional flows are simulated or distorted elements are used. Therefore, an additional numerical scheme is employed here, in which modified average elemental Péclet numbers, as shown by Cui et al. (2018b), are used to determine the weighting factors.

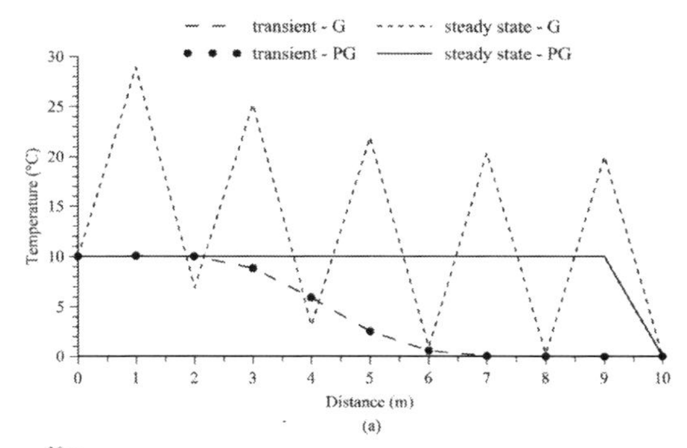

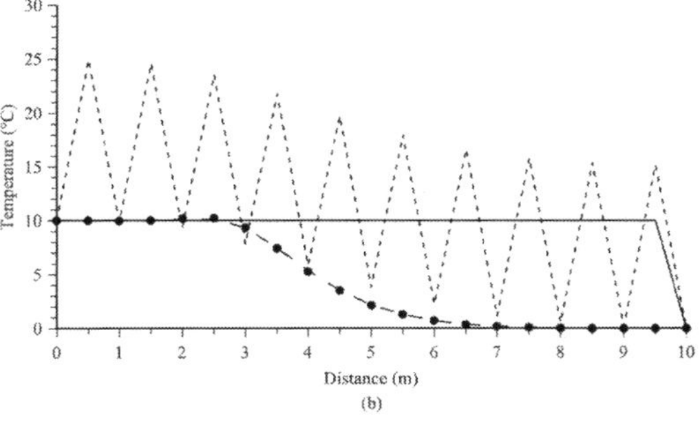

Figure 7. Temperature distribution with fixed temperature BC at the right-hand side boundary and (a) 4-noded linear elements; (b) 8-noded quadratic elements.

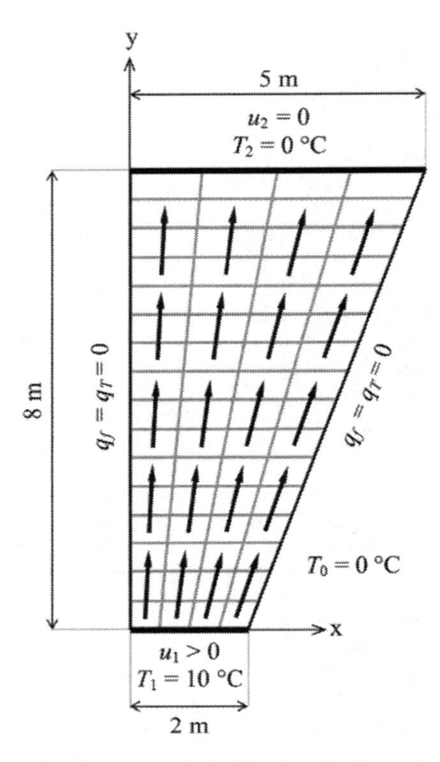

Figure 8. Geometry of the 2D advective flow with initial and boundary conditions and vectors of fluid flow.

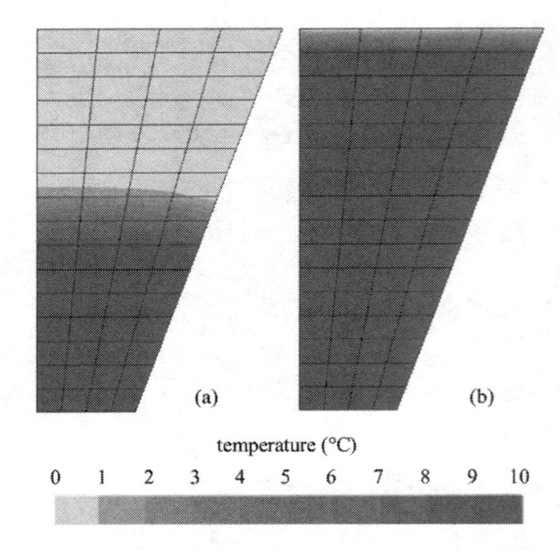

Figure 9. Temperature distribution at (a) transient and (b) steady state stages in the mesh with the proposed PGFEM.

4.3 *Numerical examples of 1D highly advective flow*

In order to validate the formulation for the proposed PGFEM for transient problems, several plane strain analyses of a problem involving a 1D highly advective flow were performed following a similar procedure as described for the simulation demonstrating the coupled thermo-hydraulic boundary condition. A Péclet number of 10000 was imposed by increasing the pore water pressure gradient over the mesh (i.e. flow velocity).

Figures 7(a) and 7(b) show the nodal temperature distribution from left to right in the mesh with 4-noded linear elements and 8-noded quadratic elements, respectively, and a fixed temperature BC at the right-hand side boundary. For comparison, the solution of the same analyses performed with the GFEM are also plotted on the figures. In the case of the GFEM, the oscillations of nodal temperature occur only when the heat front reaches the right-hand side boundary of the mesh. Prior to that stage, the solution obtained using the GFEM is exact, and it can therefore be used to verify the accuracy of the PGFEM. The results of the transient stage shown in Figure 7 indicate that the PG formulation presented in this paper does not lead to any reduction in accuracy, even when the magnitude of the Péclet number is extremely large. Moreover, the oscillations with GFEM increase once the heat front reaches the mesh boundary until a steady state is achieved with maximum amplitude of oscillations. This problem is eliminated by employing the PG formulation. The same conclusions can also be obtained when the coupled thermo-hydraulic boundary condition, instead of the fixed temperature boundary condition, is prescribed at the right-hand side boundary of the bar.

4.4 *Numerical examples of 2D highly advective flows*

A complex case with 2D advective flow and Péclet numbers of around 100 was studied to verify the proposed 2D PG scheme. The 2D mesh using distorted elements shown in Figure 8 was used. Constant pore pressure and temperature boundary conditions were prescribed at the top and bottom boundaries, while no water flow and no heat flux boundary conditions (i.e. boundaries were assumed to be both impermeable and insulated) were prescribed at the two lateral boundaries. This resulted in a 2D highly advective flow with varying direction and magnitude of fluid velocity.

Figure 9 shows the transient and steady state temperature distributions obtained in analyses with the proposed algorithm for 2D flow. Erratic heat transfer with large oscillations (over ±100 °C) was produced either using the GFEM or using the

PGFEM without the additional numerical scheme proposed by Cui et al. (2018b) which modifies the optimal values of the weighting factors.

5 CONCLUSIONS

This paper describes the complexity in numerical modelling of coupled THM problems. The significance of off-diagonal terms in the coupled THM FE formulation, such as the temperature-pore pressure coupling, has been demonstrated. Possible numerical pitfalls, which may result in erroneous solutions such as numerical oscillations, are also shown. Finally, approaches for eliminating these numerical problems are provided and their capabilities are verified through a series of numerical examples.

ACKNOWLEDGEMENTS

The research presented in this paper was funded by the China Scholarship Council (CSC, grant number: 2011602099), the Geotechnical Consulting Group (GCG), UK, and the Engineering and Physical Sciences Research Council (EPSRC, grant number: 1386304), UK.

REFERENCES

Aboustit, B.L., Advani, S.H. & Lee, J.K. 1985. Variation principles and finite element simulation for thermo-elastic consolidation. *International Journal for Numerical and Analytical Methods in Geomechanics*, 9: 49–69.

Abuel-Naga, H.M., Bergado, D.T., Bouazza, A. & Ramana, G.V. 2007a. Volume change behaviour of saturated clays under drained heating conditions: Experimental results and constitutive modeling. *Canadian Geotechnical Journal*, 44(8): 942–956.

Abuel-Naga, H.M., Bergado, D.T. & Bouazza, A. 2007b. Thermally induced volume change and excess pore water pressure of soft Bangkok clay. *Engineering Geology*, 89(1–2): 144–154.

Cengel, Y.A., & Ghajar, A.J. 2011. *Heat and Mass Transfer: Fundamentals and Applications*. 4th ed. New York: McGraw-Hill.

Cui, W., Potts, D.M., Zdravković, L., Gawecka, K.A. & Taborda, D.M.G. 2017. An alternative coupled thermo-hydro-mechanical finite element formulation for fully saturated soils. *Computers and Geotechnics*, 94: 22–30.

Cui W., Gawecka K.A., Taborda D.M.G., Potts D.M., Zdravković L. 2016a. Time-step constraints in transient coupled thermo-hydraulic finite element analysis. *International Journal for Numerical Methods in Engineering*, 106(12): 953–971.

Cui W., Gawecka K.A., Potts D.M., Taborda D.M.G. & Zdravković L. 2016b. Numerical analysis of coupled thermo-hydraulic problems in geotechnical engineering. *Geomechanics for Energy and the Environment*, 6: 22–34.

Cui W., Gawecka K.A., Taborda D.M.G., Potts D.M., Zdravković L. 2018a. Time-step constraints for finite element analysis of 2D transient heat diffusion. (Prepared for submission)

Cui W., Gawecka K.A., Potts D.M., Taborda D.M.G., Zdravković L. 2018b. A Petrov-Galerkin finite element method for 2D transient and steady state highly advective flows in porous media. (under review)

Delage, P., Sultan, N., & Cui, Y.J. 2000. On the thermal consolidation of Boom clay. *Canadian Geotechnical Journal* 37(2): 343–354.

Gatmiri, B. & Delage, P. 1997. A formulation of fully coupled thermal-hydraulic-mechanical behaviour of saturated porous media—numerical approach. *International Journal for Numerical and Analytical Methods in Geomechanics*, 21(3): 199–225.

Gens, A. 2010. Soil-environment interactions in geotechnical engineering. *Géotechnique* 60 (1): 3–74.

Laloui, L. & Francois, B. 2009. ACMEG-T: Soil thermoplasticity model. *Journal of Engineering Mechanics*, 135 (9): 932–944.

Lewis, R.W. & Schrefler, B.A. 1998. *The finite element method in the static and dynamic deformation and consolidation of porous media.* 2nd ed. Wiley.

Noorishad, J., Tsang, C.F. & Witherspoon, P.A. 1984. Coupled thermal-hydraulic-mechanical phenomena in saturated fractured porous rocks: numerical approach. *Journal of Geophysical Research*, 89 (B12): 10365–10373.

Olivella, S., Gens, A., Carrera, J. & Alonso, E.E. 1996. Numerical formulation for a simulator (CODE_BRIGHT) for the coupled analysis of saline media. *Engineering Computations* 13(7): 87–112.

Potts, D.M. & Zdravković, L. 1999. *Finite Element Analysis in Geotechnical Engineering: Theory*. London: Thomas Telford.

Potts, D.M. & Zdravković, L. 2001. *Finite Element Analysis in Geotechnical Engineering: Application.* London: Thomas Telford.

Seneviratne, H.N., Carter, J.P., & Booker, J.R. 1994. Analysis of fully coupled thermomechanical behaviour around a rigid cylindrical heat source buried in clay. *International Journal for Numerical and Analytical Methods in Geomechanics*, 18(3): 177–203.

Thomas, H.R. & He, Y. 1997. A coupled heat–moisture transfer theory for deformable unsaturated soil and its algorithmic implementation. *International Journal for Numerical Methods in Engineering* 40(18): 3421–3441.

Vaziri, H.H. 1996. Theory and application of a fully coupled thermo-hydro-mechanical finite element model. *Computers and structures* 61(1): 131–146.

Zienkiewicz, O.C. 1984. Coupled problems and their numerical solution. In: *Lewis, R.W., Bettess, P. & Hinton, E. (ed.) Numerical Methods in Coupled Systems*. Chichester, Wiley, pp. 35–58.

Numerical Methods in Geotechnical Engineering IX – Cardoso et al. (Eds)
 ISBN 978-1-138-33198-3

Rock failure analysis with discrete elements

J.V. Lemos
National Laboratory for Civil Engineering (LNEC), Lisboa, Portugal

ABSTRACT: Discrete element models are widely used in the representation of discontinuous media, ranging from granular materials to jointed rock masses. These numerical techniques have also become a powerful tool in micro- and meso-scale modeling of fracture processes in rock. A review is presented of the bonded-particle and bonded-block models developed for the analysis of laboratory tests on rock specimens, and their extensions intended for failure analysis of rock engineering works. Key modeling issues are examined and the current research trends are discussed.

1 INTRODUCTION

The designations 'Discrete Element Method' (DEM) or 'Discrete Elements' (DE) encompass a wide class of numerical methods aimed at the representation of the physical behavior of systems composed of particles, grains or blocks. There is a multiplicity of techniques, formulations, terminology and codes that may be included in this class, reflecting the way they evolved, as independent groups of researchers searched solutions for specific applications. The concept of a 'discontinuum' model, which underlies all DE approaches, is also shared by many related numerical techniques, namely Molecular Dynamics, Discrete-Finite Elements, Discontinuous Deformation Analysis, the Manifold Method, and others found in the literature. Underneath the differences in terminology, and the diversity of numerical implementations, there are many common aspects, for example, in the representation of the mechanical contact between blocks, or in their internal discretization to obtain complex deformation patterns. A broad knowledge of the extensive publications in the field is useful to all practitioners.

Cundall (2001) characterized the essential concept underlying discontinuum models: "Assemblies of discrete particles (bonded together to represent rock, and unbonded to represent soil) capture the complicated behavior of actual material with simple assumptions and few parameters at the micro level. Complex overall behavior arises as an emergent property of the assembly". This approach contrasts with one adopted by equivalent continuum modeling, which develops complex constitutive relations to simulate the observed material response.

In geomechanics, DE models have been applied to a wide variety of problems, comprising both frictional and cohesive materials, and ranging in scale from the micro- or meso-level of analysis to the field scale of engineering works. One of the fruitful research areas in recent years has been the study of fracture phenomena in geomaterials by means of bonded particle and block models. The present paper reviews the main results obtained in this field, and examines the evolving research trends. Key aspects of DE modeling are concisely addressed in the next section, followed by the discussion of bonded circular/spherical particle models and polygonal/polyhedral block models.

2 THE EVOLUTION OF DISCRETE ELEMENT MODELS IN ROCK MECHANICS

2.1 *Early developments*

The major motivation for the early development of DE models in rock mechanics was to address safety assessment problems in jointed rock masses. Rock slope stability problems are governed mostly by friction along the discontinuities. Therefore, blocks could be assumed rigid given the low stresses involved, as in the analytical solutions based on limit equilibrium that were used for simple problems. However, failure mechanisms involved large movements and changes in block contact locations that invalidated the small displacement assumptions common in early numerical models. Conceptual models beyond continuum mechanics existed, e.g. the 'clastic mechanics' proposal of Trollope (1968), but the analytical solution procedures limited their practical application. Cundall (1971) devised a general numerical solution technique capable of materializing the block assemblage concept, based on the time integration of the equations of motion of each block assumed to behave as a rigid body. The representation of mechanical contacts between the blocks, and the methods to detect them, completed the novel features of the designated 'Distinct element method'. While the 'Distinct element method' was initially regarded as a sub-set of the 'discrete element' class, the two designations are now commonly used as synonyms. Large displacement analysis became manageable, with the system connectivity automatically updated during a simulation, as some contacts break and new ones are formed. Both polygonal blocks and circular particles were already implemented at this early stage (Fig. 1).

2.2 *Key components*

The key concept underlying the DE approach is the representation of complex material behavior using simple constitutive models for the particles or blocks and for the mechanical interaction between them.

The simplest option is to assume that the particles or blocks behave as rigid bodies, implying that all the system deformability is accounted by the contact compliances. Therefore, the constitutive laws adopted for the contacts govern completely the response of the model. Simple models based on Coulomb friction are often sufficient to analyze systems of loose granular materials or unfilled rock discontinuities. More elaborate contact constitutive models, namely including cohesive bonds, have been proposed to account for more complex modes of interaction between the particles or blocks.

In block models, the consideration of deformability, by means of an internal FE mesh, was found to extend significantly the range of application, namely to address heterogeneous or anisotropic systems. The internal FE elements may obey elastic or nonlinear constitutive relations. Alternatively, a network of potential cracks may be created, allowing fracture propagation inside the block, as discussed in the following sections.

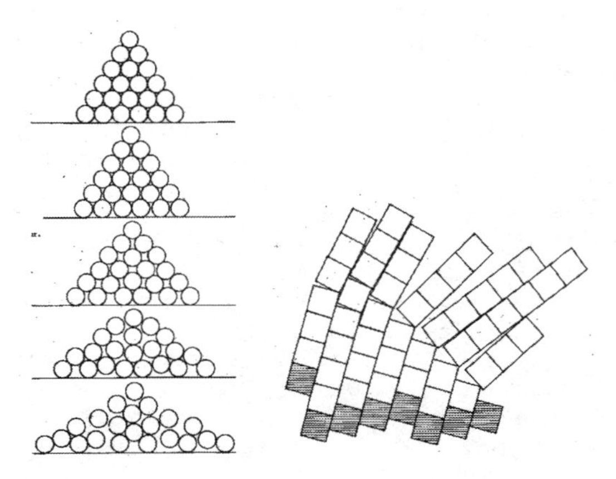

Figure 1. Two examples from Cundall's 1971 paper: collapsing pile of disks and toppling failure of rock slope.

For large systems of blocks or particles, the detection and update of the contacts becomes the dominant task in terms of run time. Therefore, DE development tends to be controlled to a large extent by numerical and programming issues, thus further contributing to the variety of formulations and codes available. The emphasis on numerical issues sometimes obscures physical fundamentals, and it remains important to bear in mind the practical engineering implications of built-in code assumptions.

Since the early days, the solution algorithms of DE codes were required to handle markedly nonlinear behavior, involving large displacements and total separation of blocks. Explicit time-stepping algorithm are naturally suited to follow the system evolution, and thus became the choice in most codes (Lemos 2012a). They are typically employed for both quasi-static problems, as a dynamic relaxation solution procedure, and time domain dynamic analysis.

2.3 *Particle and block models for granular and cohesive materials*

The 2D circular particle code BALL presented by Cundall & Strack (1979) was initially intended to address the micro-mechanics of soils and other granular materials. Since then, extensive research on the analysis of granular systems has produced a variety of approaches and algorithms for DE and related numerical methods well documented in the literature (Pöschel & Schwager 2005, O'Sullivan 2017).

By applying cohesive bonds between the particles, and letting them break in tension or shear, the same numerical formulation became a choice tool to study rock fracture, in the form of the bonded-particle models. The random nature of the particle assemblies simulates the natural arrangement of grains in the rock matrix. Based on elementary constitutive laws governing the interaction between the rigid particles, complex forms of behavior emerge, to be evaluated against experiments (Fig. 2).

Potyondy & Cundall (2004) summarize the results they obtained with these models in rock fracture analysis, using the code PFC (Itasca 2015), following a long research effort in this field, which includes, for example, the study of tuff's behavior by Trent (1988), using Cundall's early 3D particle code TRUBAL. Bonded-particle models have also been applied to study the failure processes in

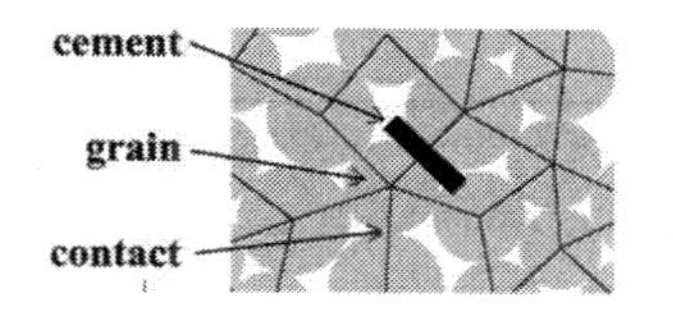

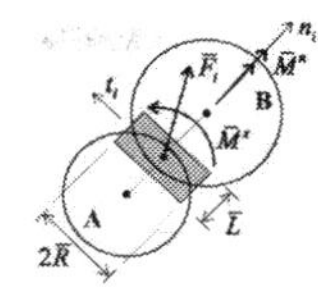

Figure 2. Bonded-particle model concept (Potyondy & Cundall 2004).

other geomaterials, such as concrete (e.g., Donzé 1999, Azevedo & Lemos 2005), or stone masonry (Azevedo et al. 2016).

The use of block models in fracture analysis has been limited by their higher computational costs. Circular and spherical particles models are particularly efficient for contact detection and resolution tasks. Polygonal or polyhedral blocks, as will be discussed in section 4, require more complex operations. Some early experiments were however performed with success. Lorig & Cundall (1987) presented the analysis of the failure of a reinforced concrete beam using the UDEC code (Itasca 2014). The concrete was simulated by a system of rigid blocks, defined by a Voronoi polygon tessellation. The reinforcement was represented by simple spring elements. Christianson et al. (2006) applied the same code to analyze the development of fractures in lithophysal tuff, also resorting to a Voronoi geometry, but now with elastic blocks. Figure 3 shows the development of cracking in a confined compression test, and a detail of the block system. More recently, with better computer resources available, the use of bonded-block models has increased significantly.

Modeling the transition from a continuum to a discontinuum, as fractures develop during the analysis, is the goal of several hybrid techniques, designated as FDEM or FEM/DEM (e.g., Munjiza 2004, Owen & Feng 2001). Several strategies have been used to convert a FE continuum mesh into a DE block system, by progressively detaching elements and creating new interfaces according to given failure criteria. These models have been widely applied at different scales, from lab specimens to underground openings in rock masses.

Many other research groups have developed DE codes based on circular or polygonal particles intended for fracture analysis, a review of the most widely used may be found in Lisjak & Grasselli (2014).

2.4 *Field scale modeling*

The analysis of rock engineering stability problems, including slopes, foundations and underground excavations, was the main early motivation for DE block models. These techniques were seen as a powerful extension of the simple limit

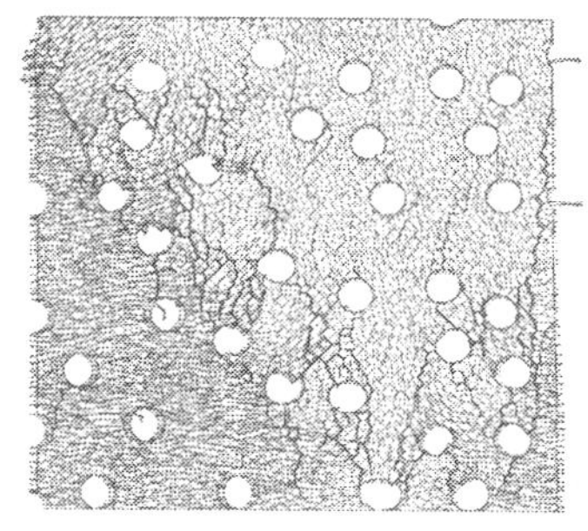
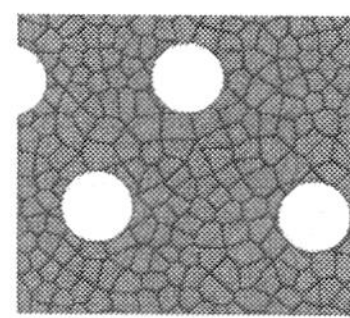

Figure 3. Failure of lithophysal rock under confined compression (left); detail of Voronoi block model (right) (Christianson et al. 2006).

equilibrium methods used in engineering practice. In many cased, only a small number of rock mass discontinuities is required to define the collapse modes. Therefore, the block size may be large, advising more elaborate representation of the intact block behavior, considering their deformability, anisotropy, and possibly non-elastic behavior. This implies discretization of the blocks into internal finite element meshes, leading to some overlap between DE and FE techniques. Many field scale problems in which the rock mass can be represented as a continuum with a small number of major discontinuities may be equally addressed by FE codes with joint or interface elements, or by DE codes with deformable blocks.

Deformable blocks are recommended for most field scale problems. This is particular important when the spatial variation of the rock mass moduli is significant, as in concrete dam foundations, or when wave propagation is analyzed, as in seismic analysis. An example of the failure mode of an arch dam foundation after progressive reduction of the joint friction angle, obtained with the code 3DEC (Cundall 1988, Hart et al. 1988), is shown in Figure 4 (Lemos 2012b).

The use of bonded-particle models at field scale is obviously demanding in computational terms, so there are practical limitations to the size of the problems that can be approached. In addition, it requires the consideration of planar features such as joints and faults. The 'Synthetic Rock Mass' proposal, discussed in section 3.2, is intended to address these issues (Mas Ivars et al. 2011).

3 BONDED-PARTICLE MODELS (BPM)

3.1 *Particles and contacts*

DE models based on circular particles gained wider attention in the geomechanics community following the Cundall & Strack (1979) paper, which presented a complete treatment of the 2D problem. The primary motivation was the detailed examination of the fundamental phenomena governing the failure processes in soils and other granular media. However, changing the contacts laws between particles from Coulomb friction to more elaborate constitutive relations involving breakable bonds allowed the application of these models to cohesive materials such as rock or concrete. It was found that these large, randomly generated systems of particles provided an appropriate geometric framework to simulate fracture initiation and propagation in geomaterials. The computational effort required tends to limit the analysis to lab size specimens, and also to favor 2D rather than 3D simulations, but much larger models are nowadays possible.

Potyondy & Cundall (2004) showed the capabilities of this approach to analyze the fracture of rock specimens in unconfined and confined compression tests. A key component of this model is the representation of the particle interaction by means of 'parallel bonds', capable of transmitting not just forces, but also moments, which respond to the relative rotations between the particles (Fig. 2).

Parallel bonds conceptually correspond to the existence of a cohesive segment on the contact plane tangent to both particles. They break when the maximum tensile stress, which is assumed to vary linearly on the contact plane according to the elementary bending formula, exceeds the tensile strength. A more general approach, the multiple-contact model, was proposed by Azevedo & Lemos (2005). In this case, the interaction between the circular particles is discretized into three or more point contacts, lying along the plane tangent

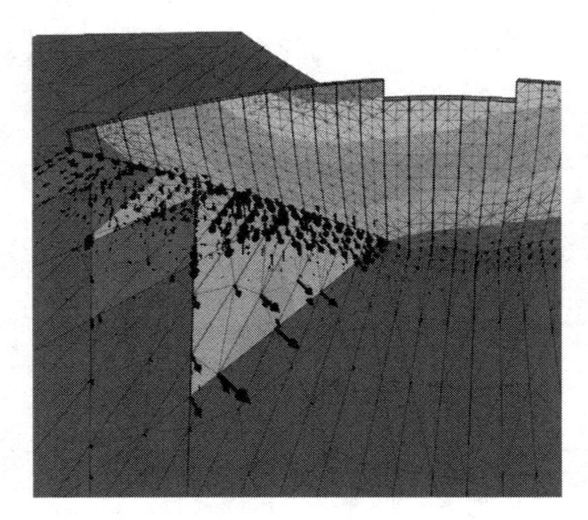

Figure 4. Failure mechanism of rock wedge in arch dam foundation model (Lemos 2012b).

to the particles, each contact representing a segment of the contact area. Forces at each contact point are calculated from the relative displacement between the two particles at that point, allowing general constitutive laws to be used. Additionally, it permits the partial fracture of the bonds, which can break progressively point by point. The multiple-contact model was later extended to 3D, with the contact points lying on concentric circles on the contact plane (Azevedo & Lemos 2013). The 'flat-joint' model proposed by Potyondy (2014) is also based on a discretization of the contact plane into several segments, allowing partial cracking to take place.

3.2 *Modeling planar discontinuities in BPM*

The use of BPM at the rock mass scale requires the representation in the model of planar discontinuity surfaces, such as rock joints and faults. The friction angle of such features has to be assigned by the user, given the field data. As the rock blocks are composed of circular particles, the irregularity of the contact between them implies a high friction value and an excessively dilatant behavior. The 'Smooth joint model' proposed by Cundall is intended to solve this issue, by using a special calculation for the normal vector of the contacts between particles across the disconti-nuity. Instead of being defined by the line connecting the particle centers, the contact normal is set perpendicular to the joint plane, allowing smooth sliding between the two macro-particles.

This scheme makes it possible to extend BPM to simulate a jointed rock medium, as proposed in the 'Synthetic Rock Mass' concept (Mas Ivars et al. 2011), which attempts to bridge the gap from micro-modeling to field scale analysis. The discontinuities are generated according to the parameters characterizing a discrete fracture network, and then they superimposed on a randomly generated circular particle assembly (Fig. 5). In this way, the block structure is defined and the contacts across adjacent blocks are treated by the special logic. The 'Smooth joint model' is the fundamental requirement to convert the response of irregular interfaces into planar joints. Computational costs obviously limit the size of the region that can be analyzed, or impose large circular particle sizes, which can result in an overly coarse definition of the blocks.

3.3 *Model developments*

The early circular particle models displayed several limitations in matching specific features of the observed behavior of rock specimens. First, they were unable to reproduce the ratio of tensile

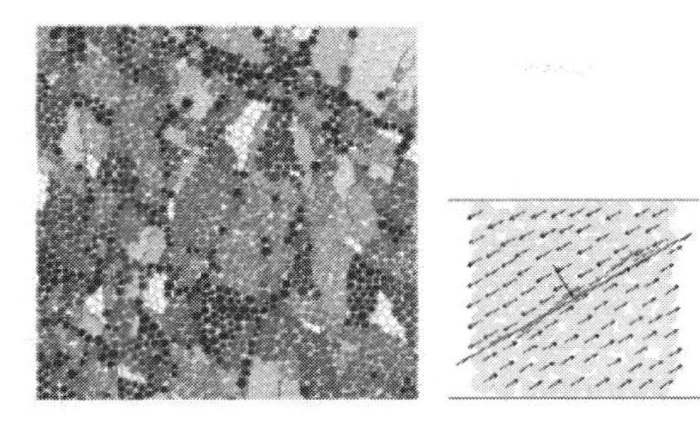

Figure 5. 'Synthetic Rock Mass' model: Planar discontinuities overlaying a particle assembly (left); Contact sliding behavior with 'Smooth joint model' (right) (Mas Ivars et al. 2011).

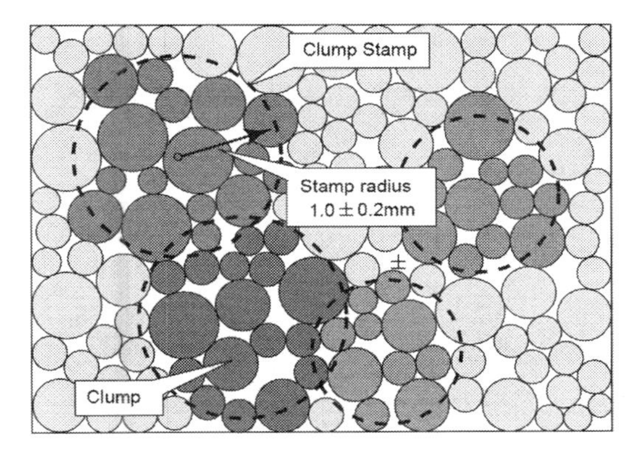

Figure 6. Macro-particles formed by disk 'clumps' (Cho et al. 2007).

to compressive strength of rock, predicting lower values than ones measured in the lab. In addition, they did not fully account for the confining stress effects in triaxial tests, displaying macroscopic friction angles that were too low. This performance is linked to an excessive tendency for rotation of the circular particles once the bonds are broken, which prevents the system from properly mobilizing the frictional effects. Potyondy & Cundall (2004) discussed these issues, showing that creating macro-particles by joining circular particles with strong bonds improved the model performance. Cho et al. (2007) followed this line of development, employing particle 'clumps', i.e., macro-particles formed by rigidly linking groups of circular particles (Fig. 6), and were able to obtain realistic values of compressive to tensile strength ratios for granite, and a good match of the nonlinear failure envelope. Yoon et al. (2011) also applied the clumped-particle logic to uniaxial and biaxial rock tests with good results.

Scholtès & Donzé (2013) discuss the ways to improve the BPM performance by increasing the grain interlocking effects, obtained with a higher bond density. They applied a criterion that creates bonds between particles lying within a given interaction distance, thus allowing connections between disks that are not initially touching, and therefore would not be formed in standard models. The higher number of bonds provided a better match of the experimental failure envelopes. Azevedo & Lemos (2013) proposed an alternative criterion to create bonds between particles, based on the Voronoi generated from the particle centers. A bond is formed if the Voronoi polygons corresponding the two particles have a common edge. This approach, coupled with the use of a multiple-point contact representation, establishes a link between the simpler circular particle models and the more elaborate polygonal block representations, providing a good match of the failure envelopes obtained in lab tests.

Potyondy (2014) discusses the various types of BPM, including a 'grain-based model' using circular particles, also intended to approximate the performance of a polygonal block model. Grains are macro-particles formed by several circular particles that fall inside a given polygon. Along each side of the polygon, the contacts use the 'Smooth joint model' logic described above in section 3.2, so that an equivalent planar contact surface can be simulated.

The run times for large 3D BPM systems are still a key issue, prompting several strategies to speed up the calculations. Cundall (2011) proposed the 'lattice model', in which the finite-sized particles are replaced by point masses, and the contacts between particles are replaced by breakable springs. The model assumes small displacements, in order to achieve high computational efficiency because the interaction geometry (location and apparent stiffness of springs) can be pre-computed, eliminating contact detection as an overhead. The model has been applied in underground hydraulic fracturing problems, and other situation in which the small displacement hypothesis is acceptable.

4 BONDED-BLOCK MODELS (BBM)

4.1 *Contact representations*

Contact calculations are typically responsible for a large fraction of run time in DE simulations. These run times have two main components (e.g., Cundall & Strack 1979). The first is the detection of new contacts, which must be performed initially and when large movements take place. The second component involves the computations that need to

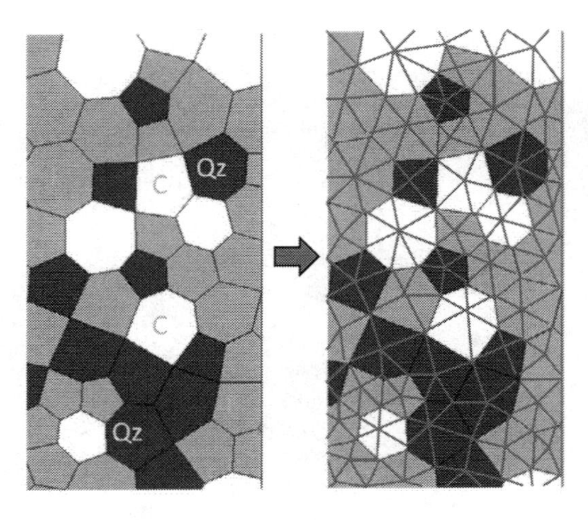

Figure 7. Grain-based model using a triangular block system derived from a Voronoi diagram (Gao & Stead 2014).

be performed at each step: (i) update the contact geometric parameters (location, normal, etc.); (ii) calculate the contact displacements, from the relative movements of the two bodies in contact; (iii) obtain the contact stresses and forces by application of the constitutive law; (iv) apply these forces to the two particles or blocks.

Circular particle models have a great computational advantage in terms of contact calculations, as geometric parameters such as contact orientation or particle overlap can be obtained by fairly straightforward operations. Macro-particles formed by bonded or clumped disks or spheres also rely on the elementary interactions between the sub-particles, possibly corrected by special schemes, such as the smooth joint model. On the other hand, the detection and resolution of contact between polygons or polyhedra involves more intensive computations. For the 2D code UDEC, Cundall developed a ‘rounded corner’ logic, in which the sharp vertices of the polygons were smoothed by tangent circular arcs. This scheme provides a very sound physical representation of contact, since the interaction between two vertices is approximated by the contact between circular arcs, as in the circular particle models, providing a unique normal that rotates progressively as the blocks move. In addition, the transition from a vertex-to-edge to a vertex-to-vertex interaction is gradual and smooth.

For the code 3DEC, Cundall (1988) developed the ‘common-plane’ concept to provide an efficient way of testing the contact between convex polyhedra, and also updating the contact geometric parameters. The common-plane is obtained by an iterative algorithm, which converges fast when

Figure 8. Model for fragmentation analysis in underground mining: tetrahedral deformable block system (left); cross-section showing broken interfaces induced by progressive excavation at the lower boundary (right) (Garza-Cruz and Pierce 2014).

only a minor change is required due to incremental block movements. Sphero-polyehedra provide a generalization of the ‘rounded corner’ concept for 3D systems, with a smooth approximation of edges and vertices of the polyhedron (Pournin & Liebling 2005, Galindo-Torres et al. 2012). This approach is quite appropriate for DE contact problems, particularly for block systems undergoing rapid changes of configuration and modes of interaction.

4.2 *Block system geometry*

Bonded block models are based on a randomly generated network of joints that define the potential fracture paths. As pointed out in section 2.3, there are formulations that allow the progressive fracturing of the continuum elements and the generation of new interfaces. However, these schemes generally add to the complexity and computational cost of the analysis. In most cases, the aim is not to inspect the propagation of fractures at specific locations, but get a broad evaluation of the most likely paths. Therefore, if a more refined analysis is required, the simpler approach it to reduce the block size, so that many more potential cracks are present, making it possible the emergence of other failure modes.

Block systems based on Voronoi diagrams, polygons or polyhedra, have been the most common choice (Herbst et al. 2008). There are algorithms for generation of these diagrams in 2D and 3D, given a random set of points. The tetrahedral network that connects these points is the corresponding Delauney tessellation. For rock specimens, these geometries resemble in some way the granular structure of the material, and the model attempts to reproduce the grain shapes and size distribution (Lan et al. 2010). However, in most cases, the Voronoi diagram is simply a physically attractive way of creating a random network of potential cracks (Figs. 3 and 7).

Ghasvinian et al. (2014) used 3D Voronoi geometries to simulate uniaxial compression tests in anisotropic rocks. Müller et al. (2018) also applied a 3D Voronoi model to the simulation of triaxial tests on rock salt, relating the progression of damage given by bond breakage with the acoustic emissions measured during the experiments.

Other network geometries have been tried. Kazerani & Zhao (2012) used a triangular block geometry in UDEC models, in a study of the cracking of asperities in rock joints. Gao & Stead (2014) also employed a triangular block geometry in UDEC, but generated from a standard Voronoi diagram by splitting each block into triangles formed by connecting each vertex to a central point (Fig. 7). The authors claimed that it produced more realistic fracture patterns, without excessive interlocking and friction. Tan et al. (2015) analyzed a series of Brazilian tests on slate disks, varying the angle of the foliation with the loading direction. The model was formed by small quadrilateral blocks defined by planes parallel and perpendicular to the foliation. In this way, they were able to simulate cracking along the foliation and through the intact material.

Garza-Cruz and Pierce (2014) applied a 3DEC model using tetrahedral blocks generated by a Delauney algorithm to the study of the fragmentation evolution in veined rock. They were able to analyze a very large system, a cube of 8 m size with blocks with an average 0.5 m length, in order to assess the effects of mining for rock masses with different strengths (Fig. 8).

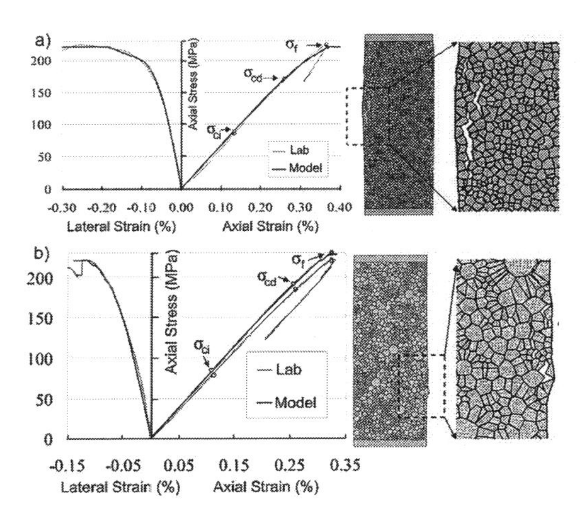

Figure 9. Calibrated stress-strain response with laboratory data for (a) Äspö diorite and (b) Lac du Bonnet granite (Lan et al. 2010).

4.3 *Internal block behavior*

The simplest assumption is to consider that the blocks are rigid, thus concentrating all the deformation at the joints or interfaces, the potential cracks (e.g. Lorig & Cundall 1987, Kazerani & Zhao 2010). This provides an immediate criterion to define the joint stiffnesses, in order to match the global deformability of the rock specimen.

In most models, however, the blocks are assumed elastic, involving a discretization into an internal FE mesh (e.g. Christianson et al. 2006). Lan et al. (2010) used a UDEC model with polygonal deformable blocks representing the rock grain structure, approximating the grain shapes and the elastic moduli of the different minerals. They presented an interesting analysis of the effects of heterogeneity on the stress-strain response and the compressive strength of granite and diorite samples (Fig. 9).

In principle, nonlinear material models can also be assigned to the intact rock material. However, the most common choice of the researchers that wish to study the breakage of grains is to provide potential cracks inside the blocks by dividing them into sub-blocks. Gao et al. (2016) elaborated the model mentioned above, in which a Voronoi polygon is split into triangles, by assigning the contacts between the triangles inside each polygonal grain different micro-properties. Therefore, intra- and inter-granular potential cracks could be simulated. They concluded that the model with unbreakable grains tends to under-estimate the crack initiation threshold, recommending the use of the more elaborate approach for brittle rocks.

Similar conceptual models for grain based analysis of rock fracture are provided by FDEM or FEM/DEM approaches. The main differences

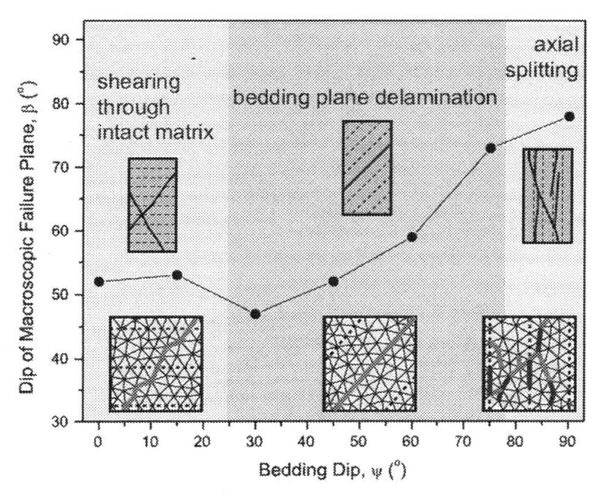

Figure 10. Inclination of macroscopic failure plane versus dip of specimen bedding (Lisjak et al. 2014).

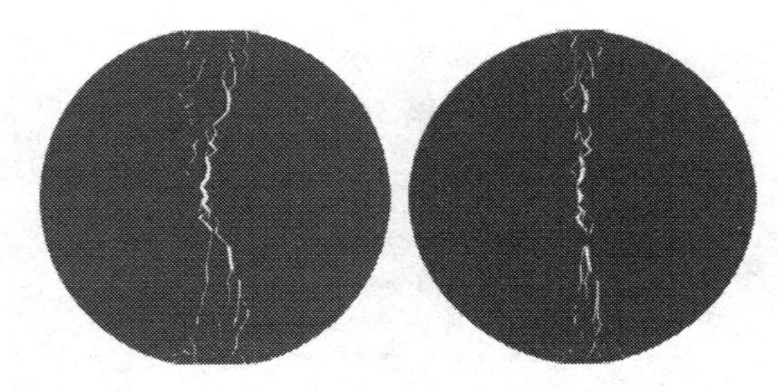

Figure 11. Fracture patterns in Brazilian tests, obtained with a block model (Kazerani 2013).

between these and standard DE codes usually relate to the numerical treatment of contact and interface elements. Analyses with the code Y-Geo (Mahabadi et al. 2012), which extends the work of Munjiza (2004), are often based on an initial network of potential cracks. Lisjak et al. (2014) applied these models to the analysis of underground excavations in clay shales, taking into account the presence of the bedding planes. Figure 10 presents results of compressive test simulations showing the variation of the angle of the macroscopic failure plane as a function of the orientation of the bedding planes, which leads to different failure modes.

5 CONTACT CONSTITUTIVE BEHAVIOR AND MICRO-PARAMETERS

A key issue in bonded-particle and bonded-block models is the assignment of the micro-parameters that govern the contact behavior in order to reproduce the experimentally measured macro-properties. In general, a two-step procedure is followed. First, contact stiffness parameters are calibrated to match the rock deformability. Then, bond strengths, in tension and shear, and frictional angles are determined to approximate the experimental response observed in elementary test setups, namely considering tensile and compressive loads.

Rigid block (or particle) models assume that all the deformability is concentrated at the contacts. This leads to a direct estimation of the best contact stiffness parameters, in the normal and shear directions. For deformable block models, the total deformation is split between the intact material and the contacts. Some authors use large contact stiffnesses, essentially treating them as penalty coefficients, to reduce the relative displacements at the contacts, and assign elastic block moduli similar to the macroscopic values. Other assumptions are possible, since excessively high contact stiffness are not desirable for numerical reasons. For example, in explicit algorithms, they tend to increase run times, often unnecessarily.

The micro-parameters depend on the size selected for the particles or blocks. Often, for computational reasons, this size has to be larger than the real dimension of the micro-structure. The dependence of elastic properties on the particle size can usually be handled in a straightforward manner. Strength parameters require more attention. Potyondy & Cundall (2004) showed that the fracture toughness displayed by the numerical model depends on the particle size, which is natural since the particle dimension controls the numerical discretization of the fractures. Therefore, contact strength parameters have to be calibrated for a given particle size, in order to achieve the macroscopic values of tensile and compressive strength.

The complexity of the constitutive models assigned to the contacts has varied, with a steady tendency towards more elaborate models. It is significant that very good results were achieved employing fairly simple constitutive relations, typically assuming brittle bond breakage when the tensile or shear strengths are reached (e.g., Potyondy 2014, Cho et al. 2007, Lan et al. 2010, and many others). Key features of the nonlinear stress-strain response of rock materials were replicated by these authors with these elementary contact models, the macroscopic complexity arising from the response of the random assemblies of particles or blocks.

The consideration of softening branches after the peak stress is reached, in tension as well as in shear, is driven by the objective of reproducing more closely the fracture energies involved in the bond breakage processes. Ma & Huang (2018) implemented a tensile softening model in PFC in their analysis of the modes of fracture initiation and propagation in Brazilian testing or rock disks. Kazerani (2013) implemented a contact constitutive law in a UDEC block model, which adopted nonlinear stress-strain relations in the pre-peak range and post-peak softening, as a function of the material fracture energies. The underlying rationale of the model also provides the values of the initial normal and shear contact stiffnesses, which progressively decrease until the peak is reached. Figure 11 shows the cracking patterns obtained in two Brazilian tests, using a triangular block model.

Chen et al. (2015), also using a block model, analyzes time-independent and time-dependent damage evolution processes in sandstone, applying a constitutive model with strength reduction as a function of crack growth. The crack generation procedures developed by Munjiza (2004) and Lisjak et al. (2014) are also steered to fulfill the energy requirements of fracture mechanics.

6 CONCLUDING REMARKS

Discrete elements have evolved considerably since the early rigid block models intended for the stability analysis of jointed rock slopes. These idealizations continue to be applied at the scale of engineering works, whether underground excavations or dam foundations. In parallel, the power of the 'discontinuum' concept has created a prime tool for micro- and meso-scale analysis of a wide range of problems, including the flow of granular media, and the damage and fracture processes in rock and other cohesive materials.

An attractive feature of bonded-particle/block models for fracture analysis is their proved capability to simulate the complexity observed in experiments using only simple constitutive assumptions at the contact level. The trend of current research, however, appears to be heading towards more elaborate constitutive relations, as the means to improve the match of test data. This will require more work on the methodologies to calibrate the micro-parameters, which remains a major research need in this field. The lack of reliable and robust methods to define these input parameters is one of the barriers that limits the use of these techniques in engineering practice.

REFERENCES

Azevedo, N.M. & Lemos, J.V. 2005. A generalized rigid particle contact model for fracture analysis. *Int. J. Numer. Analyt. Meth. Geomech.* 29: 269–285.

Azevedo, N.M. & Lemos, J.V. 2013. A 3D generalized rigid particle contact model for rock fracture. *Engineering Computations* 30(2): 277–300.

Azevedo, N.M., Lemos, J.V. & Almeida, J.R. 2016. Discrete Element Particle Modelling of Stone Masonry, In Sarhosis, Bagi, Lemos & Milani (eds), *Computational Modeling of Masonry Structures Using the Discrete Element Method*: 146–169. Hershey, PA: IGI Global.

Chen, W., Konietzky. H. & Abbas, S.M. 2015. Numerical simulation of time-independent and -dependent fracturing in sandstone. *Eng Geol.* 193(2): 118–131.

Cho, N., Martin, C.D. & Sego, D.C. 2007. A clumped particle model for rock. *Int. J. Rock Mech. Min. Sci.* 44: 997–1010.

Christianson, M., Board, M. & Rigby, D. 2006. UDEC simulation of triaxial testing of lithophysal tuff. In *Proc. 41st US Symposium on Rock Mechanics*. Golden, USA: American Rock Mechanics Association.

Cundall, P.A. 1971. A computer model for simulating progressive large scale movements in blocky rock systems. In *Proc. Symp. Rock Fracture (ISRM), Nancy*: vol. 1, paper II–8.

Cundall, P.A. 1988. Formulation of a three-dimensional distinct element model - Part I: A scheme to detect and represent contacts in a system composed of many polyhedral blocks. *Int. J. Rock Mech. Min. Sci.* 25: 107–116.

Cundall, P.A. 2001. A discontinuous future for numerical modelling in geomechanics? *Proc. Inst. Civil Engineers, Geotechnical Engineering* 149(1): 41–47.

Cundall, P.A. 2011. Lattice method for modeling brittle, jointed rock. In Sainsbury, Hart, Detournay & Nelson (eds), *Continuum and Distinct Element Numerical Modeling in Geomechanics - 2011*: Paper 01–02. Minneapolis: Itasca.

Cundall, P.A. & Strack, O.D.L. 1979. A discrete numerical model for granular assemblies. *Geotechnique* 29(1): 47–65.

Donzé, F.V., Magnier, S.-A., Daudeville, L., Mariotti, C. & Davenne, L. 1999. Numerical study of compressive behavior of concrete at high strain rates. *Journal of Engineering Mechanics* 125(10): 1154–1163.

Garza-Cruz, T.V., & Pierce, M. 2014. A 3DEC Model for Heavily Veined Massive Rock Masses. *Proc. 48th US Rock Mechanics/Geomechanics Symposium*: Paper 14–7660. Alexandria: American Rock Mechanics Association.

Galindo-Torres, S., Pedroso, D., Williams, D. & Li, L. 2012. Breaking processes in three-dimensional bonded granular materials with general shapes. *Computer Physics Communications* 83(2): 266–277.

Gao, F.Q. & Stead, D. 2014. The application of a modified Voronoi logic to brittle fracture modelling at the laboratory and field scale. *Int. J. Rock Mech. Min. Sci.* 68: 1–14.

Gao, F.Q., Stead, D. & Elmo, D. 2016. Numerical simulation of microstructure of brittle rock using a grain-breakable distinct element grain-based model. *Computers and Geotechnics* 78: 203–217.

Ghazvinian, E., Diederichs, M.S. & Quey, R. 2014. 3D random Voronoi grain-based models for simulation of brittle rock damage and fabric-guided micro-fracturing. *J. Rock Mech. Geotech. Eng.* 6: 506–21.

Hart, R.D., Cundall, P.A. & Lemos, J.V. 1988. Formulation of a three-dimensional distinct element model - Part II: Mechanical calculations for motion and interaction of a system composed of many polyhedral blocks. *Int. J. Rock Mech. Min. Sci.* 25: 117–125.

Herbst, M., Konietzky, H. & Walter, K. 2008. 3D microstructural modeling. In Hart, Detournay & Cundall (eds) *Continuum and Distinct Element Numerical Modeling in Geo-Engineering - Proc. 1st Int. FLAC/DEM Symposium*: 435–441. Minneapolis, MN: Itasca.

Itasca 2015. *PFC - Particle Flow Code, Version 5.0*. Minneapolis, MN: Itasca.

Itasca 2014. UDEC - Universal Distinct Element Code, Version 6.0. Minneapolis, MN: Itasca.

Kazerani T. 2013. Effect of micromechanical parameters of microstructure on compressive and tensile failure process of rock. *Int. J. Rock Mech. Min. Sci.* 64: 44–55.

Kazerani, T. & Zhao, J. 2010. Micromechanical parameters in bonded particle method for modeling of brittle material failure. *Int. J. Num. Analyt. Meth. Geomech.* 34: 1877–1895.

Kazerani, T., Yang, Z. & Zhao, J. 2012. A discrete element model for predicting shear strength and degradation of rock joint by using compressive and tensile test data. *Rock Mechanics and Rock Engineering* 45(5): 695–709.

Lan, H., Martin, C.D. & Hu, B. 2010. Effect of heterogeneity of brittle rock on micromechanical extensile behavior during compression loading. *J. Geophysical Research* 115: B01202.
Lemos, J.V. 2012a. Explicit codes in geomechanics - FLAC, UDEC and PFC. In Ribeiro e Sousa, L., Vargas Jr., E., Fernandes, M.M. & Azevedo, R. (eds), *Innovative Numerical Modelling in Geomechanics*: 299–315, London: Taylor & Francis.
Lemos, J.V. 2012b. Modelling the failure modes of dams' rock foundations. In Barla, G. (ed.), *MIR 2012 - Nuovi Metodi di Indagine, Monitoraggio e Modellazione Degli Amassi Rocciosi*: 259–278. Torino, Italy: Politecnico di Torino.
Lisjak, A., Grasselli. G. & Vietor, T. 2014. Continuum-discontinuum analysis of failure mechanisms around unsupported circular excavations in anisotropic clay shales. *Int. J. Rock Mech. Min. Sci.* 65: 96–115.
Lisjak, A. & Grasselli, G. 2014. A review of discrete modeling techniques for fracturing processes in discontinuous rock masses. *J. Rock Mech. Geotech. Eng.* 6: 301–314.
Lorig, L.J. & Cundall, P.A. 1987. Modeling of Reinforced Concrete Using the Distinct Element Method. In Shah & Swartz (eds), *Fracture of Concrete and Rock*: 459–471. Bethel: SEM.
Ma, Y. & Huang, H. 2018. DEM analysis of failure mechanisms in the intact Brazilian test. *Int. J. Rock Mech. Min. Sci.* 102: 109–119.
Mahabadi, O.K., Lisjak. A., Munjiza, A. & Grasselli, G. 2012. Y-Geo: new combined finite-discrete element numerical code for geomechanical applications. *Int. Journal of Geomechanics* 12: 676–88.
Mas Ivars, D., Pierce, M.E., Darcel, C., Reyes-Montes, J., Potyondy, D.O., Young, R.P. & Cundall, P.A. 2011. The synthetic rock mass approach for jointed rock mass modelling. *Int. J. Rock Mech. Min. Sci.* 48(2): 219–44.
Müller, C., Frühwirt, T., Haase, D., Schlegel, R. & Konietzky, H. 2018. Modeling deformation and damage of rock salt using the discrete element method. *Int. J. Rock Mech. Min. Sci.* 103: 230–241.
Munjiza, A. 2004. *The combined finite-discrete element method*. Chichester, UK: John Wiley & Sons Ltd.
O'Sullivan, C. 2017. Particulate Discrete Element Modelling: A Geomechanics Perspective. Taylor & Francis.
Owen, D.R.J. & Feng, Y.T. 2001. Parallelised finite/discrete element simulation of multifracturing solids and discrete systems. *Engineering Computations* 18(3/4): 557–76.
Pöschel, T. & Schwager, T. 2005. Computational Granular Dynamics: Models and Algorithms. Berlin: Springer.
Potyondy, D.O. & Cundall, P.A. 2004. A bonded-particle model for rock. *Int. J. Rock Mech. Min. Sci.* 41: 1329–64.
Potyondy, D.O. 2014. The bonded-particle model as a tool for rock mechanics research and application: current trends and future directions. *Geosystem Engineering* 17(6): 342–369.
Pournin, L. & Liebling, T.M. 2005. A generalization of distinct element method to tridimensional particles with complex shapes. In *Powders and Grains 2005*: Vol. 2, 1375–1378. Rotterdam: Balkema.
Scholtès, L., Donzé, F.-V. 2013. A DEM model for soft and hard rocks: role of grain interlocking on strength. *Journal of the Mechanics and Physics of Solids* 61(2): 352–69.
Tan, X., Konietzky, H., Frühwirt, T. & Dinh, Q.D. 2015. Brazilian tests on transversely isotropic rocks: laboratory testing and numerical simulations. *Rock Mechanics and Rock Engineering* 48: 1341–1351.
Trent, B.C. 1988. Microstructural effects in static and dynamic numerical experiments. In Cundall, Sterling & Starfield (eds), *Key Questions in Rock Mechanics - Proc. 29th U.S. Symposium on Rock Mechanics*: 395–402. Rotterdam: Balkema.
Yoon, J.S., Jeon, S., Zang, A. & Stephansson, O. 2011. Bonded particle model simulation of laboratory rock tests for granite using particle clumping and contact unbonding. In Sainsbury, Hart, Detournay & Nelson (eds), *Continuum and Distinct Element Numerical Modeling in Geomechanics - 2011*: Paper 08–05. Minneapolis, MN: Itasca.

Constitutive modelling and numerical implementation

Numerical Methods in Geotechnical Engineering IX – Cardoso et al. (Eds)
© 2018 Taylor & Francis Group, London, ISBN 978-1-138-33198-3

The dilatancy conditions at critical state and its implications on constitutive modelling

E.J. Fern & K. Soga
Department of Civil and Environmental Engineering, GeoSystems
University of California, Berkeley, USA

ABSTRACT: The critical state concept is a central theory of many constitutive models. However, it is modelled in different ways by different models and not always in a straightforward manner. The critical state can be defined by two dilatancy conditions, which allow understanding how constitutive models achieve the critical state. Moreover, it permits understanding some of their limitations. For instance, the Cam-Clay models are not well-suited for dense, or overconsolidated, soils as they cannot distinguish the phase transition point from the critical state. Consequently, they model the peak strength as a yielding point, which is in contraction with the stress-dilatancy theory. This paper explains in a simple manner how the critical state is achieved in three different models - Cam-Clay with and without sub-loading surfaces, and Nor-Sand - and illustrates these models highlighting some of their limitations.

1 INTRODUCTION

The critical state theory (Roscoe et al. 1958) is recognised as being the founding theory of the critical state soil mechanics (Schofield & Wroth 1968) and has permitted the development of stress-strain relationships, also called constitutive models, such as the Cam-Clay models (Roscoe & Schofield 1963, Roscoe & Burland 1968). However, the stress-dilatancy theories (i.e. Taylor 1948) are rarely acknowledged as being a fundamental theory of this framework despite that the critical state is s specific case of the stress-dilatancy theory and that the original Cam-Clay model (Roscoe & Schofield 1963) was derived from it. This paper explains how the critical state, defined from dilatancy conditions, is achieved in different constitutive models and why some models are restricted to 'wet' (contractive) soils.

2 DILATANCY AND STATE

It is well known that density plays an important role in the mechanical behaviour of sand. It increases the dilatancy rate and, thus, the peak strength and the hardening/softening rates. Fig. 1 shows the development of shear strength of (a) loose and (b) dense sand in triaxial compression tests. Loose sand progressively hardens and contracts until reaching the critical state (point A to B'). However, dense sand has a more complex behaviour. It first contracts and hardens from point A to point B, at which point it reaches the *phase transition point*; this state was first identified by Wroth & Bassett (1965) but named as such by Tatsuoka et al. Tatsuoka et al. (1986). It is characterised by a nil dilatancy state ($D = d\varepsilon_v/d\varepsilon_d = 0$) and a stress state equal to the critical state ($\eta' = q/p' = M$), but it is not at the critical state ($e \neq e_{cs}$). The dense sand then dilates and hardens until reaching the *peak state* (point B to C) where the dilatancy rate is minimum $D_{\min}$ and the effective stress ratio maximum $\eta'_{\max}$. It subsequently softens until reaching the same *critical state* as for the loose sand (point C to B'). These different phases and states can be defined solely by their dilatancy characteristics (Eqs. 2–5).

- contractive hardening phase

$$D > 0, \quad \frac{\partial D}{\partial \varepsilon_d} < 0 \tag{1}$$

- phase transition point

$$D = 0, \quad \frac{\partial D}{\partial \varepsilon_d} < 0 \tag{2}$$

- dilative hardening phase

$$D < 0, \quad \frac{\partial D}{\partial \varepsilon_d} < 0 \tag{3}$$

- peak state

$$D < 0, \quad \frac{\partial D}{\partial \varepsilon_d} = 0 \tag{4}$$

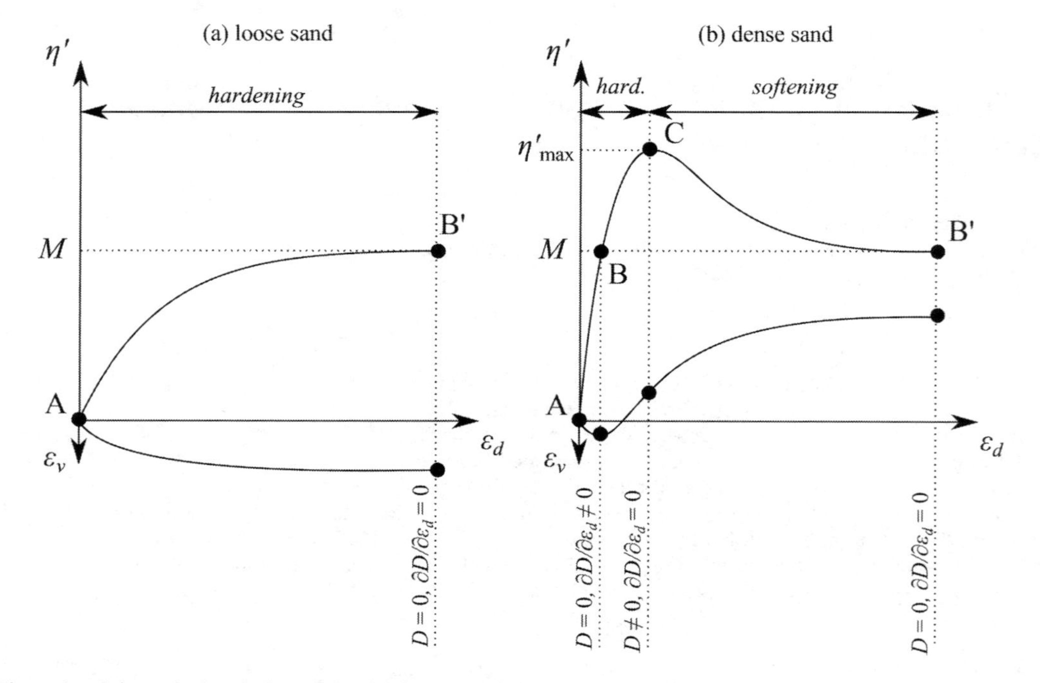

Figure 1. Schematic description of the development of shear strength in triaxial compression tests of a) loose sand and b) dense sand.

- dilative softening phase

$$D < 0, \quad \frac{\partial D}{\partial \varepsilon_d} > 0 \tag{5}$$

- critical state

$$D = 0, \quad \frac{\partial D}{\partial \varepsilon_d} = 0 \tag{6}$$

where $D = d\varepsilon_v/d\varepsilon_d$ is the dilatancy rate, $d\varepsilon_v$ and $d\varepsilon_d$ the incremental volumetric and deviatoric strains, respectively.

2.1 *The critical state*

The critical state theory (Roscoe et al. 1958) suggests that any soil, which is sufficiently sheared, will ultimately reach a unique state called the *critical state* at which point the soil will undergo a continues and monotonic deformation without any volume change. The critical state is defined by the two conditions shown in Eq. 6 from which the stress state and void ratio can be derived.

Roscoe et al. (1958) were driven by the idea of Taylor (1948) that the development of strength was the consequence of inter-granular friction (critical state) and interlocking (dilatancy), and Roscoe & Schofield (1963) suggested a simple stress-dilatancy relationship (Eq. 7). Other relationships have been later proposed but follow the same idea (i.e. Nova 1982, Nova & Wood 1979).

$$\eta' = M - D \tag{7}$$

where $\eta' = q/p'$ is the effective stress ratio, q the deviatoric stress, p' the mean effective stress and M the critical state effective stress ratio.

The critical state is defined by a nil dilatancy and a nil dilatancy change state and, hence, its stress state is uniquely defined and will remain constant (Eqs. 8–9).

$$D = 0 \;\rightarrow\; \eta' = M \tag{8}$$

$$\frac{\partial D}{\partial \varepsilon_d} = 0 \;\rightarrow\; \frac{\partial \eta'}{\partial \varepsilon_d} = 0 \tag{9}$$

The *phase transition point* is also defined by a nil dilatancy state and, hence, the stress state is the same as the critical state ($\eta' = M$). However, the nil dilatancy change is not nil ($\partial D/\partial \varepsilon_d \neq 0$) and the stresses will change as the soil is sheared ($\partial \eta'/\partial \varepsilon_d \neq 0$).

Bolton (1986) suggested a state index called the relative dilatancy index I_R (Eq. 10) as a proxy for dilatancy.

$$I_R = \left(\frac{e_{\max} - e}{e_{\max} - e_{\min}}\right) \cdot \ln\left(\frac{Q}{p'}\right) - R \tag{10}$$

where $e_{\min}$ the minimum void ratio, $e_{\max}$ the maximum void ratio, Q the crushing pressure for which values are given in Bolton (1986), and R a fitting parameter which is by default $R = 1$.

Boulanger (2003), followed by Mitchell & Soga (2005), suggested using the relative dilatancy index and the nil dilatancy conditions of the critical state (Eq. 11) to determine the critical state void ratio and came up with a new non-linear critical state line (Eq. 12)

$$D = \frac{\partial D}{\partial \varepsilon_d} = 0 \quad \rightarrow \quad I_R = \frac{\partial I_R}{\partial \varepsilon_d} = 0 \tag{11}$$

$$\rightarrow \quad e_{cs} = e_{\max} - \frac{e_{\max} - e_{\min}}{\ln(Q/p')} \cdot R = cst \tag{12}$$

2.2 *Peak state*

Jefferies (1993) and Jefferies & Shuttle (2002) followed the idea of Bolton (1986) and suggested using the state parameter ($\psi = e - e_c$) to predict the minimum achievable dilatancy $D_{\min}$ with a dilatancy coefficient χ (Eq. 13), which which is then used to predict the peak stress state with a stress-dilatancy theory (Eq. 14).

$$D_{\min} = \chi \cdot \psi \tag{13}$$

$$\eta'_{\max} = M - D_{\min} \tag{14}$$

As initially suggested by Tatsuoka (1987) for the relative dilatancy index I_R and then by Jefferies and Been (2006, 2016) for the state parameter ψ, the conversion of a state index to a minimum dilatancy is a function of the soil fabric as demonstrated by Fern (2016) with the work of Oda (1972b, 1972a).

3 CONSTITUTIVE MODELLING

Roscoe & Schofield (1963) developed a stress-strain relationship by following the work of Drucker et al. (1957) and derived a yield surface from the stress-dilatancy theory (Eq. 7). The development of the consistency condition and the incremental stress-strain equation gives Eq. 15 for which a hardening term H has to be explicitly given and in which the critical state conditions are embedded.

$$d\vec{\sigma}' = \mathbf{D_e}\left[1 - \frac{\left(\frac{\partial F}{\partial \vec{\sigma}'}\right)^t \mathbf{D_e} \frac{\partial F}{\partial \vec{\sigma}'}}{\left(\frac{\partial F}{\partial \vec{\sigma}'}\right)^t \mathbf{D_e} \frac{\partial F}{\partial \vec{\sigma}'} - H}\right] d\vec{\varepsilon} \tag{15}$$

where F is the yield function, $d\vec{\sigma}'$ is the effective stress increment, $d\vec{\varepsilon}$ the strain increment, $\mathbf{D_e}$ the elastic stiffness tensor, F the yield function and H the hardening term.

The ability of a model to predict the peak strength as a consequence of dilatancy lies in its ability to distinguish the phase transition point from the critical state. This distinction depends on the formulation of the hardening term H and will be illustrated for three different models - Cam-Clay (Roscoe & Schofield 1963) with and without a bounding surface (Hashiguchi & Chen 1998) and Nor-Sand (Jefferies 1993).

3.1 *Cam-Clay*

Roscoe & Schofield (1963) developed a hardening term H (Eq. 16) from the isotropic virgin compression line. It was thus assumed that the hardening of soil was solely driven by the plastic volumetric strains ε_v^p (Eq. 17).

$$H = \left(\frac{\partial F}{\partial \vec{\varepsilon}}\right)^t \frac{\partial F}{\partial \vec{\sigma}'} \tag{16}$$

$$\frac{\partial F}{\partial \vec{\varepsilon}} = \frac{\partial F}{\partial p_c}\left(\frac{\partial p_c}{\partial \varepsilon_v^p}\frac{\partial \varepsilon_v^p}{\partial \vec{\varepsilon}} + \underbrace{\frac{\partial p_c}{\partial \varepsilon_d^p}\frac{\partial \varepsilon_d^p}{\partial \vec{\varepsilon}}}_{=0}\right) \tag{17}$$

where p_c is the consolidation pressure and is the hardening variable, and the superscript p refers to 'plastic'.

Roscoe & Schofield (1963) then assumed that the critical state line was parallel to virgin compression line and a volumetric strain hardening rule was formulated (Eq. 18). These assumptions hold true for clay but not for sand (Been, Jefferies, & Hachey 1991).

$$\frac{\partial p_c}{\partial \varepsilon_v^p} = \frac{\nu}{\lambda - \kappa} p_c \tag{18}$$

where $\nu = 1 + e$ is the specific volume.

Fig. 2 (a,d) shows the stress and strain paths for Original Cam-Clay in drained triaxial compression conditions of dilative soil. It can be seen that the hardening phase (point A to B) is very stiff due to the use of the unloading swelling modulus κ and

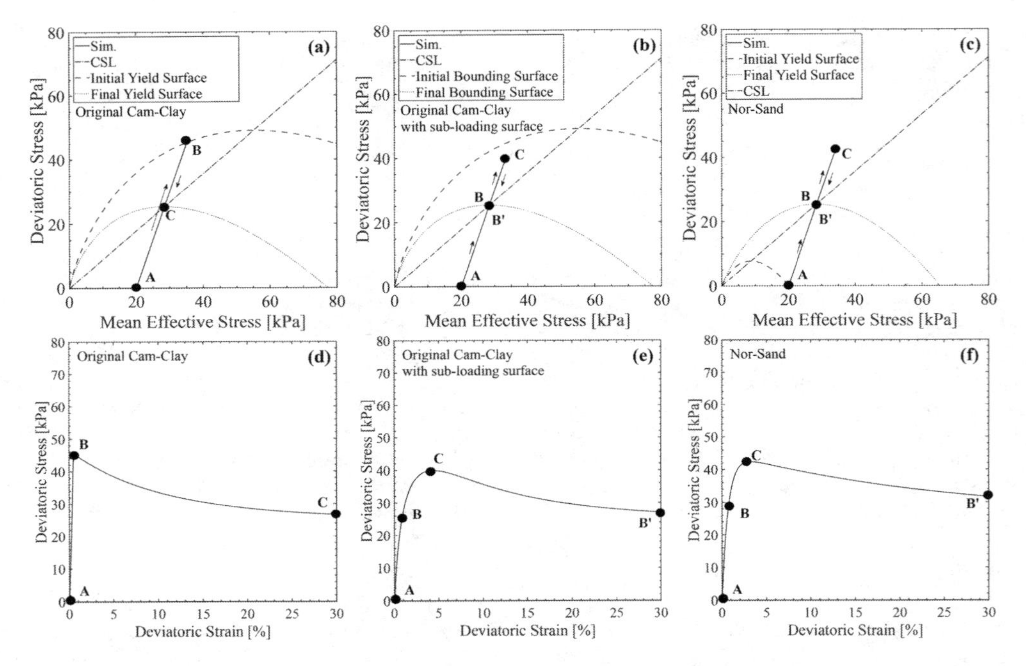

Figure 2. Stress and strains paths of drained 'dry' soil with (a,d) Cam-Clay, (b,e) Cam-Clay with bounding surface and (c,f) Nor-Sand.

that the peak state is a yielding point (point B). Moreover, it is a singularity point, which can cause numerical instabilities when used for numerical analyses. Once the peak state is reached, the yield surface decreases modelling the softening phase (point B and C). Ultimately, the soil reaches the critical state (point C).

Fig. 3 (a,d) shows the stress and strain paths for Original Cam-Clay in undrained triaxial compression conditions of dilative soil. It can be seen that the elastic hardening phase (point A to B) is very stiff with a constant value of mean effective stress p' ; the deviatoric stress q shoots up vertically. For this reason, the elastic properties in Cam-Clay for dilative soils are often curve-fitted for best match and are not representative of the elastic properties of the soil. Once the state reaches the yield surface at point B, the soil hardens with a plastic behaviour until reaching a plateau (point B to C). However, the yield surface decreases although no significant softening is observed in the strain path. The overall stress-strain prediction by Original Cam-Clay was the best estimate for many years but this prediction differed substantial from the behaviour observed in the laboratory.

The hardening rule (Eq. 18) couples the two conditions of the critical state ($D=0 \leftrightarrow \partial D/\partial \varepsilon_d^p = 0$) and, hence, it cannot distinguish the phase transition point from the critical state as the hardening is solely driven by the dilatancy D. Therefore, the hardening and softening rate are nil for any nil dilatancy state (i.e. critical state and phase transition point). For this reason, Roscoe & Schofield (1963), and consequently Roscoe & Burland (1968), restricted the use of the Cam-Clay models to 'wet' (contractive) soils for which the dilatancy D and the changes in dilatancy $\partial D/\partial \varepsilon_d^p$ reach progressively and simultaneously a nil state at the critical state.

In order to overcome this limitation, it has been customary to model the peak state as a yielding point rather than as a consequence of dilatancy. Therefore, the hardening phase becomes solely elastic and the model can by-pass the phase transition point reaching a stress state above the critical state ($\eta' > M$). Hence, the elasto-plastic behaviour starts at the peak state only and when the model predicts a softening phase. This can be viewed as a limitation of the model for which it was not originally developed.

Cam-Clay only possesses one surface - the yield surface. When it is used for 'wet' (contractive) soils, this yield surface defines the elastic domain. However, when it is used for 'dry' (dilative) soil, this yield surface then defines the peak state. The elastic hardening phase and the consolidation

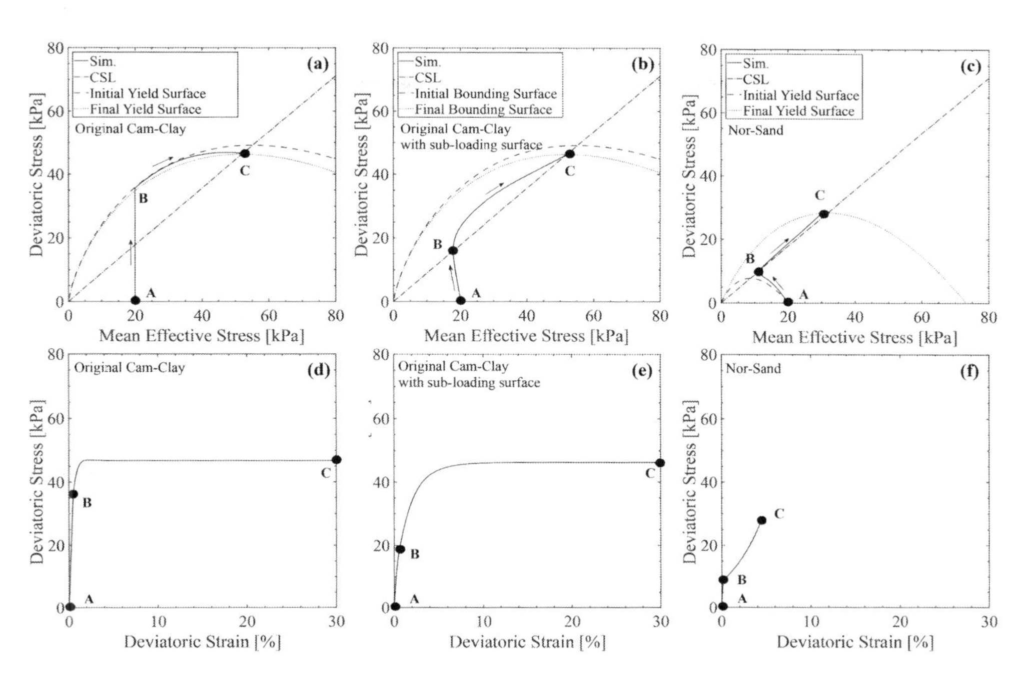

Figure 3. Stress and strains paths of undrained 'dry' soil with (a,d) Cam-Clay, (b,e) Cam-Clay with bounding surface and (c,f) Nor-Sand.

pressure are then modelling assumptions. Fig. 2 (a) illustrates the stress paths and initial yield surface for 'dry' (contractive) soil.

Only one state can be be defined in the Cam-Clay model - the critical state (Eq. 19).

$$p_c = \alpha \cdot p_i \tag{19}$$

where α is the ratio between the critical state, or image pressure p_i and the consolidation pressure p_c (Fig. 4) and is $\alpha = 2.718$ for original Cam-Clay and $\alpha = 2$ for modified Cam-Clay.

Roscoe (1970) had already pointed out the limitations of Cam-Clay and had recommended developing a new hardening rule, especially for sand, which would relate somehow to the critical state.

3.2 *Cam-Clay with bounding surface*

Hashiguchi & Chen (1998) suggested introducing a second surface in order to model both the yielding point and the peak strength, which where coupled in the Cam-Clay model. In the present paper, the yield surface defines the on-set boundary of plastic deformation the bounding surface the peak state. The original terminology of Hashiguchi & Chen (1998) was abandoned in this paper in order to be consistent with the original formulation of the

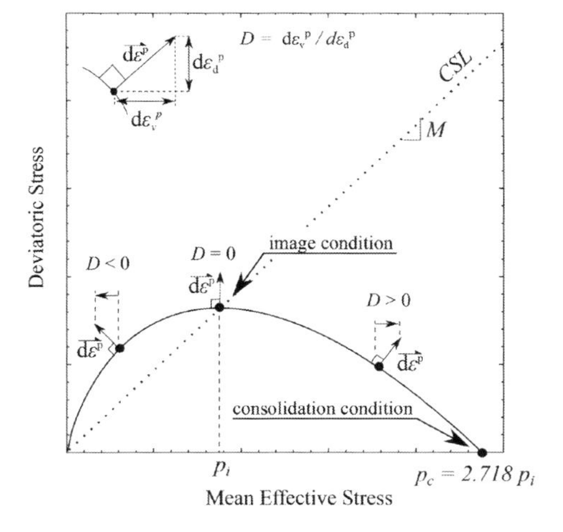

Figure 4. Original Cam-Clay's yield/bounding surface with image pressure and consolidation pressure.

Cam-Clay and Nor-Sand models. The shape of the yield and bounding surfaces are the same and are related to the consolidation ratio R (Eq. 20).

$$p_c = R \cdot p_{c,BS} \tag{20}$$

where p_c is the consolidation pressure of the yield surface, $p_{c,BS}$ the consolidation pressure of the bounding surface and R the consolidation ratio with $0 \leq R \leq 1$. Therefore, the yield surface can only be smaller or the same size as the bounding surface. When the soil is 'wet' (contractive), both surfaces are superimposed and the bounding surface does not play any role; the model behaves as for the original formulation of Cam-Clay. However, when the soil is 'dry' (dilative), the model permits plastic deformation to take place prior to the peak.

The inclusion of a new variable, the consolidation ratio *R*, induces a modification of the consistency condition and a second hardening term appears in the hardening *H* (Eq. 21).

$$H = \underbrace{\left(\frac{\partial F}{\partial \varepsilon^p}\right)^t \frac{\partial F}{\partial \vec{\sigma}'}}_{hard.1} + \underbrace{\frac{\partial F}{\partial R}\frac{\partial F}{\partial \vec{\sigma}'} dR}_{hard.2} \tag{21}$$

The first hardening rule (*hard. 1*) is the hardening rule of Cam-Clay (Eq. 18) and models the nil dilatancy condition ($D = 0$). The second hardening rule (*hard. 2*) models the nil change in dilatancy condition ($\partial D / \partial \varepsilon_d^p = 0$) and has to be explicitly given. Among other expressions, Hashiguchi & Chen (1998) suggested Eq. 22.

$$dR = U \cdot (R^{-m} - 1) \tag{22}$$

where U and m are model parameters.

Therefore, the model permits the soil to harden at the phase transition point as the second hardening rule is not nil (*hard. 2* $\neq 0$) despite the first hardening rule being nil (*hard. 1* $= 0$). Hence, the model is able to distinguish the phase transition point from the critical state, and model the peak strength as a consequence of dilatancy.

Three states can be defined for Cam-Clay with a bounding surface (Hashiguchi & Chen 1998) (Eqs. 23–25).

- phase transition point

$$\alpha \cdot p' = p_c < p_{c,BS} \tag{23}$$

- peak state

$$\alpha \cdot p' < p_c = p_{c,BS} \tag{24}$$

- critical state

$$\alpha \cdot p' = p_c = p_{c,BS} \tag{25}$$

The inclusion of a bounding surface allowed the inclusion of a second hardening term. This approach was driven by mathematical considerations in an attempt to improve the existing models. However, this second hardening term required two additional model parameters which have little physical meaning and are not experimentally quantifiable. Moreover, the consolidation pressure of the bounding surface $p_{c,BS}$ remained a modelling assumption.

3.3 *Nor-Sand*

Jefferies (1993) recognised the necessity to predict the peak state and to distinguish the phase transition point from the critical state. However, unlike Hashiguchi & Chen (1998), Jefferies (1993) followed the path of Roscoe & Schofield (1963) and kept a single hardening rule (Eq. 16). A bounding surface, called the maximum yield surface, was included in order to decouple the two conditions of the critical state ($D = 0$, $\partial D / \partial \varepsilon_d^p = 0$).

Jefferies (1993) argued that the hardening of sand was driven by the plastic deviatoric strains rather than by the plastic volumetric strains (Eq. 26), and that the hardening rate must be proportional to the distance between the current state and the peak state (Eq. 27). Note that other hardening rules have been proposed by Jefferies & Shuttle (2002) and Jefferies & Been (2006, 2016) but followed the same idea.

$$\frac{\partial F}{\partial \varepsilon^p} = \frac{\partial F}{\partial p_i}\left(\underbrace{\frac{\partial p_i}{\partial \varepsilon_v^p}\frac{\partial \varepsilon_v^p}{\partial \vec{\varepsilon}}}_{=0} + \frac{\partial p_i}{\partial lon_d^p}\frac{\partial \varepsilon_d^p}{\partial \vec{\varepsilon}}\right) \tag{26}$$

$$\frac{\partial p_i}{\partial \varepsilon_d^p} = h \cdot (p_{i,max} - p_i) \tag{27}$$

where h is the hardening modulus, p_i the image pressure and $p_{i,max}$ the maximum image pressure.

Fig. 2 (c,f) shows the stress and strain paths of Nor-Sand in triaxial compression conditions for dilative soil. The results in drained conditions are very similar to those obtained for Original Cam-Clay with sub-loading surface (Fig. 2 b,e). The hardening phase is elasto-plastic and the peak strength is a direct consequence of the maximum dilatancy. However, the results for undrained conditions (Fig. 2 c,f) are different than those for Original Cam-Clay with sub-loading surface (Fig. 3 b,e). Both models capture the reduction in mean effective stress p' (point A to B), but then Nor-Sand models an asymptotic hardening behaviour along the critical state line (point B to C). This difference is the consequence of the hardening rule.

Nor-Sand sizes the yield and bounding surfaces with the image pressure p_i and the maximum image pressure $p_{i,max}$, respectively. These are equivalent expressions of the consolidation pressures p_c and $p_{c,BS}$ but relate to the critical state condition of the surfaces as suggested by Roscoe (1970). However, a fundamental difference exists between the bounding surfaces of Hashiguchi & Chen (1998) and Jefferies (1993); the latter can be smaller than the yield surface in which case softening is predicted. Fig. 4 illustrates the consolidation and image pressures. The image pressures offer some modelling advantages as the different states can be solely characterised by them (Eqs. 29–32).

- hardening contractive phase

$$p_i < p' < p_{i,\max} \tag{28}$$

- phase transition point

$$p' = p_i < p_{i,\max} \tag{29}$$

- hardening dilative phase

$$p' < p_i < p_{i,\max} \tag{30}$$

- peak state

$$p' < p_i = p_{i,\max} \tag{31}$$

- dilative softening phase

$$p' < p_{i,\max} < p_i \tag{32}$$

- critical state

$$p' = p_i = p_{i,\max} \tag{33}$$

Additionally, Nor-Sand predicts the maximum image pressure $p_{i,max}$, and hence the bounding surface, from the predicted minimum dilatancy (Eq. 34). Therefore, the bounding surface is no longer a modelling assumption. Moreover, it permits the void ratio e to become a model variable (Eq. 13) and a single set of model parameters is sufficient to model a wide range of initial densities.

$$p_{i,\max} = p' \cdot \exp\left(-\frac{D_{\min}}{M}\right) \tag{34}$$

when using the yield surface of original Cam-Clay.

Both Hashiguchi & Chen (1998) and Jefferies (1993) included a bounding surface in the model in order to decouple the two critical state conditions. However, Hashiguchi & Chen (1998) introduced a second hardening rule whilst Jefferies (1993) replaced the one proposed by Roscoe & Schofield (1963). However, Jefferies (1993) additionally suggested predicting the size of the bounding surface from the dilatancy characteristics of the soil.

4 EXTENSION TO UNSATURATED SOIL

The concepts presented in this paper can be extended to unsaturated soil mechanics and their implications are even more significant. The Barcelona Basic Model (Alonso et al. 1990) is the extension of modified Cam-Clay, in which the peak state and the on-set of a wetting-collapse are modelled as yielding points. Zhou & Sheng (2015) extended this model with sub-loading surfaces and obtained good results whilst using Bishop's effective stress (Bishop 1959). Fern (2016) and Fern et al. (2016) extended Nor-Sand by extending the stress-dilatancy rules and also used Bishop's effective stress. Both approaches were able to model the peak state as a consequence of dilatancy and modelled by a reduction in the bounding surface. However, they explained the wetting-collapse as a loss of dilatancy characteristics resulting in the rearrangement of the grains, giving the wetting-collapse a mechanical explanation. Both models suggest that some criticism towards Bishop's effective stress in constitutive models is due to limitations of the constitutive models rather than the definition of the effective stress.

5 CONCLUSION

It is shown in this paper that mechanical behaviour of soil can be solely defined from dilatancy with the critical state modelled as a nil dilatancy state ($D = 0$) and a nil dilatancy change state ($\partial D/\partial\varepsilon = 0$). It is then shown that these two conditions are coupled in Cam-Clay due to the existence of one surface - the yield surface. This means that the dilatancy rate and the dilatancy change rate must evolve in the same direction, which prevents Cam-Clay from modelling dilative soils correctly. It is shown that these two conditions can be decoupled when an additional surface is included - the bounding surface - and is demonstrated with the sub-loading surface and Nor-Sand. This permits the models to capture the phase transition and model the peak state as a consequence of dilatancy. This is achieved for the case of sub-loading surface by the inclusion of a second hardening rule, and for the case of Nor-Sand by reformulating the hardening rule in terms of current state and predicted peak state. Moreover, it is shown that the yield surface captures the nil dilatancy condition ($D = 0$) and the bounding surface the nil dilatancy change condition ($\partial D/\partial\varepsilon_d = 0$). However, the yield surface is used as a bounding surface in Cam-Clay and with sub-loading surfaces. The decoupling of these two conditions by the inclusion of a bounding surface allows the model to capture the reduction in mean effective stress p' in undrained conditions. The comparison of the mean pressure and the image pressures (p_i, $p_{i,max}$) in Nor-Sand permits defining the mechanical state offering a strong modelling advantage.

REFERENCES

Alonso, E., A. Gens, & A. Josa (1990). A constitutive model for partially saturated soils. *Géotechnique 40*(3), 405–430.

Been, K., M. Jefferies, & J. Hachey (1991). The critical state of sands. *Géotechnique* (3), 365–381.

Bishop, A.W. (1959). The principles of effective stress. *Tecnisk Ukeblad* (8), 859–863.

Bolton, M. (1986). The strength and dilatancy of sands. *Géotechnique 36*(1), 65–78.

Boulanger, R.W. (2003). Relating to relative state parameter index. *Journal of Geotechnical and Geoenvironmental Engineering 129*, 770–773.

Drucker, D., R. Gibson, & D. Henkel (1957). Soil mechanics and work-hardening theories of plasticity. *Journal of Soil Mechanics and Foundation Engineeringl 122*, 338–346.

Fern, E.J. (2016). *The mechanics of unsaturated sand and its application to large deformation modelling*. Ph. D. thesis, University of Cambridge.

Fern, E.J., D. Robert, & K. Soga (2016). Modeling the Stress-Dilatancy Relationship of Unsaturated Silica Sand in Triaxial Compression Tests. *Journal of Geotechnical and Geoenvironmental Engineering 142*(11), 04016055.

Hashiguchi, K. & Z.-P. Chen (1998). Elastoplastic constitutive equation of soils with the subloading surface and the rotational hardening. *International Journal for Numerical and Analytical Methods in Geomechanics 22*(3), 197–227.

Jefferies, M. (1993). Nor-Sand: a simple critical state model for sand. *Géotechnique 43*(1), 91–103.

Jefferies, M. & K. Been (2006). *Soil Liquefaction A Critical State Approach*. London: Taylor & Francis.

Jefferies, M. & K. Been (2016). *Soil Liquefaction A Critical State Apporach* (second ed.). London: CRC Press.

Jefferies, M. & D.A. Shuttle (2002). Dilatancy in general Cambridge-type models. *Géotechnique 52*(9), 625–638.

Mitchell, J. & K. Soga (2005). *Fundamentals of soil behavior* (3rd ed.). Hoboken: John Wiley & Sons.

Nova, R. (1982). A constitutive model for soil under monotonic and cyclic loading. In G.N. Pande and C. Zienkiewicz (Eds.), *Soil mechanics - transient and cyclic loading*, Chichester, pp. 343–373. Wiley.

Nova, R. & D. Wood (1979). A constitutive model for sand in triaxial compression. *International Journal for Numerical and Analytical Methods in Geomechanics 3*(3), 255–278.

Oda, M. (1972a). Initial fabrics and their relations to mechanical properties of granular material. *Soils and Foundations 12*(1), 17–36.

Oda, M. (1972b). The mechanism of fabric changes during compressional deformation of sand. *Soils and Foundations 12*(2), 1–18.

Roscoe, K.H. (1970). The Influence of strains in soil mechanics. *Géotechnique 20*(2), 129–170.

Roscoe, K.H. & J.B. Burland (1968). On the generalised stress-strain behaviour of 'wet' clay. In J. Heyman and F. Leckie (Eds.), *Engineering plasticity*, Cambridge, pp. 535–609. Cambridge University Press.

Roscoe, K.H. & A.N. Schofield (1963). Mechanical behaviour of an idealised wet clay. In *2nd European Conference on Soil Mechanics and Foundation Engineering*, Wiesbaden, pp. 47–54.

Roscoe, K.H., A.N. Schofield, & C.P. Wroth (1958). On the yielding of soils. *Géotechnique 8*(1), 22–53.

Schofield, A.N. & P. Wroth (1968). *Critical State Soil Mechanics*. London: McGraw-Hill.

Tatsuoka, F. (1987). Discussion: The strength and dilatancy of sands. *Géotechnique 37*(2), 219–226.

Tatsuoka, F., M. Sakamoto, T. Kawamura, & S. Fukushiima (1986). Strength and deformation characteristics of sand in plane strain compression at extremely low pressures. *Soils and Foundations 26*(1), 65–84.

Taylor, D. (1948). *Fundamentals of soil mechanics*. New York: Wiley.

Wroth, C.P. & R.H. Bassett (1965). A stress-strain relationship for the shearing behaviour of a sand. *Géotechnique 15*(1), 32–56.

Zhou, A. & D. Sheng (2015). An advanced hydro-mechanical constitutive model for unsaturated soils with different initial densities. *Computers and Geotechnics 63*, 46–66.

Numerical Methods in Geotechnical Engineering IX – Cardoso et al. (Eds)
© 2018 Taylor & Francis Group, London, ISBN 978-1-138-33198-3

An innovative FE approach using soil hyperbolic model for predicting the response of monopiles supporting OWTs in sands

Dj. Amar Bouzid
Department of Civil Engineering, Faculty of Technology, University Saad Dahled of Blida, Algeria

ABSTRACT: An innovative numerical technique called NVSM (Nonlinear Vertical Slices Model) is presented in this paper. Its efficiency is achieved by the ability of the procedure to approximate a full 3D problem into a series of 2D vertical interacting panels which can be solved using a standard 2D FE discretization, while allowing the slices interaction to be accounted for using the finite difference method. The soil nonlinearity is dealt with by the implementation of the hyperbolic model in the new numerical model. A computer code called NAMPULAL written on the basis of nonlinear VSM, has been used to analyze large diameter monopiles supporting Offshore Wind Turbines under combined loading. The computer program is then, assessed against two case studies in which the authors used independently the commercial packages FLAC3D and PLAXIS to study large diameter monopiles embedded in offshore subsoil formed mostly of dense sand. An excellent agreement between NAMPULAL predictions and those of the other numerical codes has been obtained.

1 INTRODUCTION

The Finite Element Method is by far the most powerful and versatile tool for the solution of many engineering problems. It is now a common practice to carry out a finite element analysis on plane strain or geometrically axisymmetric problems routinely. This is not, however, the case for fully 3D nonlinear problems which require a tremendous computational effort for processing data.

Most soil/structure interaction problems are geometrically complex to be dealt with a simple 2D Finite element analysis. In facing such situations where huge amounts of computer resources are required, the engineer is always tempted to find ways to simplify his problem into a two-dimensional model. In this context, many authors developed procedures to introduce a certain "three-dimensional effect" into their two-dimensional model allowing thus for a reduction in computer resources.

The present paper describes first, the equations constituting the theory of the elastic version of the vertical slices model, and then gives the equations formulating Duncan and Chang's hyperbolic model used for modelling cohesionless soils. The combination of both sets of equations formulates the innovative nonlinear numerical tool proposed in this paper.

To assess the capabilities of the proposed numerical model, two case studies dealing with the lateral behavior of monopiles supporting Offshore Wind Turbines, are considered.

2 THE FE VERTICAL SLICES MODEL: MAIN EQUATIONS GOVERNING THE PROCEDURE

The concept of this model is based on the slicing of the medium into a finite number of vertical panels in which the finite element and finite differences are combined to analyze the soil-structure interaction problem (For more details, the reader is referred to Amar Bouzid *et al.* (2005b). The equilibrium equations in each panel are given by:

$$\frac{\partial \sigma_x}{\partial x} + \frac{\partial \tau_{xy}}{\partial y} + b_x = 0, \frac{\partial \tau_{xy}}{\partial x} + \frac{\partial \sigma_y}{\partial y} + b_y = 0 \quad (1)$$

The terms b_x and b_y appearing in the equations (1) are considered to be the key elements in the development of the FE vertical slices model. These physical parameters termed body forces in this model have been interpreted as fictitious forces transmitted to the slice under consideration by shear forces applied from slices at the left and at the right. For a given slice i, this can be mathematically expressed as:

$$b_x = \frac{\tau^l_{zx_i} - \tau^r_{zx_i}}{t_i}, b_y = \frac{\tau^l_{zy_i} - \tau^r_{zy_i}}{t_i} \quad (2)$$

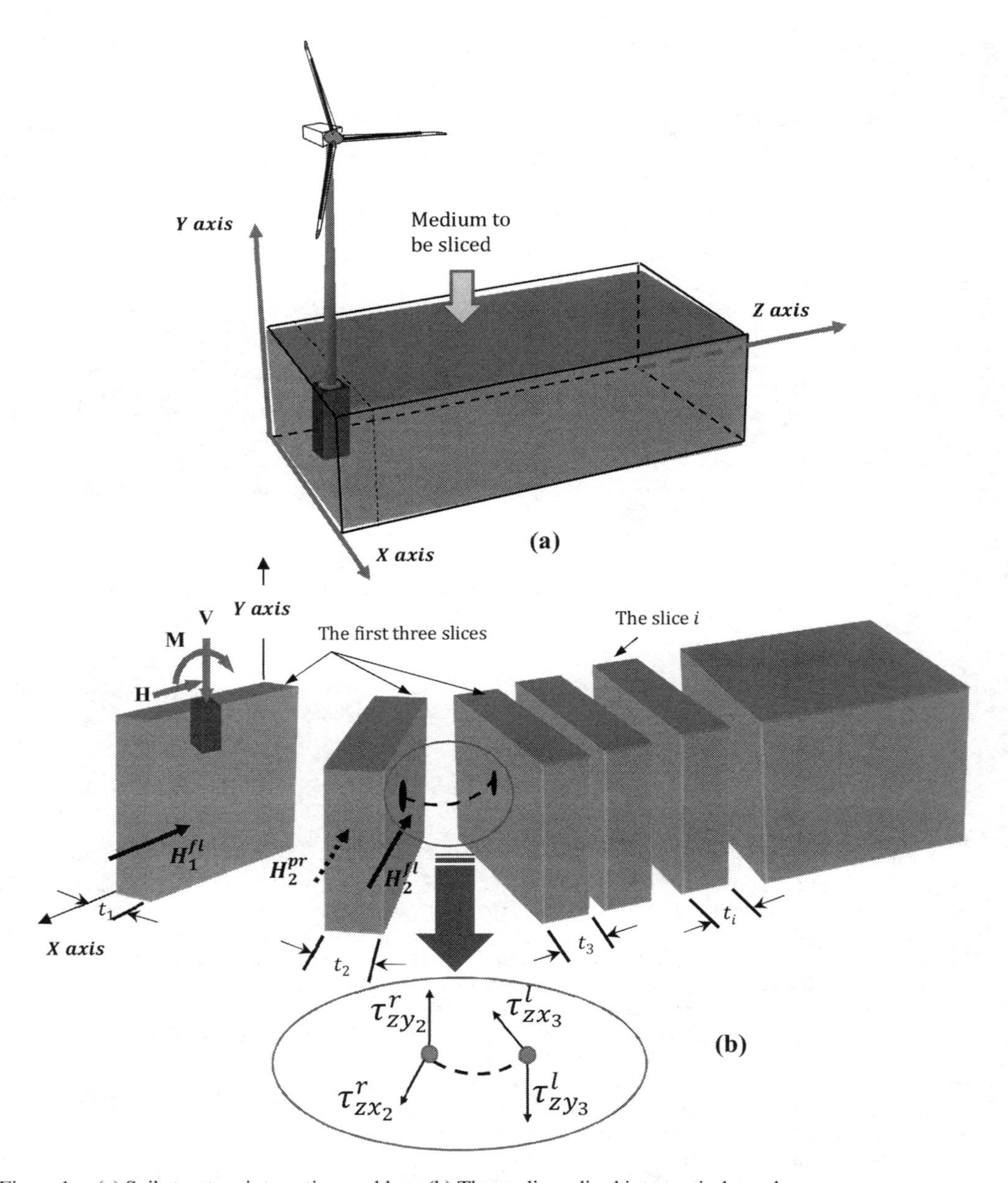

Figure 1. (a) Soil structure interaction problem, (b) The medium sliced into vertical panels.

where $\tau^l_{zx_i}$ and $\tau^l_{zy_i}$ are shear stresses acting at the left interface of slice i, whereas $\tau^r_{zx_i}$ and $\tau^r_{zy_i}$ are shear stresses acting at the right interface of slice i. t_i is the slice thickness. The stress and deformation analyses in each slice are conducted by the conventional finite element method using 2D finite elements.

According to the standard formulation in the displacement based finite element method, the stiffness matrix of an element belonging to the slice i can be written:

$$\int_v \boldsymbol{B}^t \boldsymbol{D} \boldsymbol{B} \boldsymbol{a}_i dv = \int_v \boldsymbol{N}^t \boldsymbol{b}_i dv + \boldsymbol{p}_i \tag{3}$$

$\boldsymbol{p}_i$ in equation (3), is the external forces vector to which the slice i is subjected, $\boldsymbol{D}$ is the matrix corresponding to a problem of plane stresses and $\boldsymbol{b}_i$

is the body forces vector which has the following compact form:

$$b_i = b_i^{pr} - b_i^{pc} + b_i^{fl} \tag{4}$$

$$b_i^{pc} = L^{pc} N a_i ,\ b_i^{pr} = L^{pr} N a_{i-1} ,\ b_i^{fl} = L^{fl} N a_{i+1} \tag{5}$$

$$\text{where } L^{pc} = l_i^{pc} I ,\ L^{pr} = l_i^{pr} I \text{ and } L^{fl} = l_i^{fl} I \tag{6}$$

l_i^{pr}, l_i^{fl} and l_i^{pc} are the slices interaction factors and a_{i-1}, a_i and a_{i+1} are the element nodal displacement vectors of slices i–1, i and i+1 respectively. I is the identity matrix.

It is clear from equation (4) that the fictitious body forces applied to a slice i depend essentially on its own nodal displacements and on those of slices sandwiching it. The resulted equations are those of a pseudo plane stress problem with body forces representing the interaction between the slices forming the structure and its surrounding medium. In a more detailed development where the equations (4) and (5) are substituted into equation (3), the latter becomes:

$$\int_v \left(B^t D\, B + N^t L^{pc} N\right) a_i dv = \int_v \left(N^t L^{pr} N\right) a_{i-1} dv + \int_v \left(N^t L^{fl} N\right) a_{i+1} dv + p_i \tag{7}$$

expression (7) may be re-written in a compact form as:

$$S_i a_i = H_i^{pr} + H_i^{fl} + p_i \tag{8}$$

This equation cannot be resolved straight-fully, since the right hand terms are not available explicitly at the same time. Thus, it must be resolved according to an updating iterative process:

$$S_i^j a_i^j = H_i^{pr^j} + H_i^{fl^{j-1}} + p_i \text{ for } j = 1,2,\ldots, j_{\max} \tag{9}$$

where, j denotes the iteration number and j_{max} is determined by a certain convergence criterion. It is clearly shown from equation (9) that even for elastic problems the resolution process is naturally iterative. Hence, this expression is the key element of the Vertical Slices Model. Numerical results obtained in this paper are fundamentally based on its encoding in a computer program.

3 DUNCAN-CHANG HYPERBOLIC MODEL FOR MODELLING SOILS

In practice, geotechnical engineers often prefer relatively simple constitutive models that do not require a high number of behavior parameters and

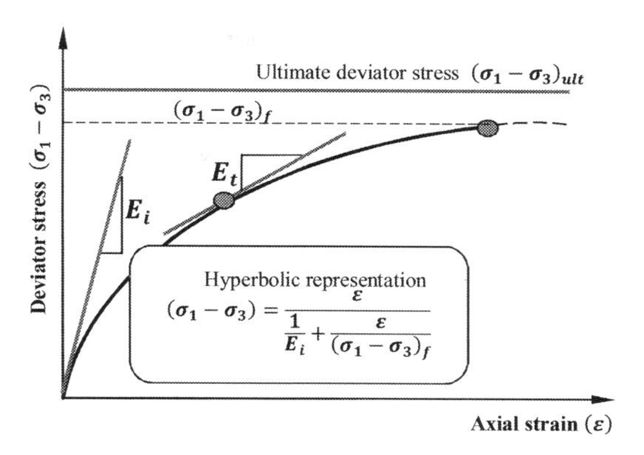

Figure 2. Hyperbolic representation for stress-strain relation.

consequently easy to implement in a computer programs. More than five decades ago Kondner (1963) and Kondner & Zelako (1963) approximated the stress-strain curves for both clays and sands according to the shape illustrated in Figure 2. Analytically, such curves are represented by the following hyperbolic relationship:

$$(\sigma_1 - \sigma_3) = \frac{\varepsilon}{\frac{1}{E_i} + \frac{\varepsilon}{(\sigma_1 - \sigma_3)_{ult}}} \tag{10}$$

where $(\sigma_1 - \sigma_3)$ is the principal stress difference, $(\sigma_1 - \sigma_3)_{ult}$ is the asymptotic value of the principal stress difference at large axial strain, ε is the axial strain and E_i is the initial tangent modulus.

Duncan and Chang (1970) found out that both tangential modulus E_i and ultimate stress deviator $(\sigma_1 - \sigma_3)_{ult}$ present in equation (10) should rely on the minor principal stress σ_3. More precisely, they suggested for the initial tangent modulus the following formula:

$$E_i = KP_a \left(\frac{\sigma_3}{P_a}\right)^n \tag{11}$$

where K and n are respectively the modulus number and the modulus exponent. P_a is the atmospheric pressure used to make K and n non-dimensional. The failure ratio R_f is given by:

$$R_f = \frac{(\sigma_1 - \sigma_3)_f}{(\sigma_1 - \sigma_3)_{ult}} \tag{12}$$

The principal stress difference at failure is related to the confining pressure σ_3 as:

$$(\sigma_1 - \sigma_3)_f = \frac{2C\cos\phi + 2\sigma_3 \sin\phi}{1-\sin\phi} \tag{13}$$

where C is the cohesion intercept and ϕ the friction angle.

The tangent modulus E_t in Duncan-Chang's formulation is given by:

$$E_t = \left[1 - \frac{R_f(1-\sin\phi)(\sigma_1 - \sigma_3)}{2c\cos\phi + 2\sigma_3\sin\phi}\right]^2 KP_a\left(\frac{\sigma_3}{P_a}\right)^n \tag{14}$$

For unloading/reloading conditions, Duncan and Chang proposed the following expression:

$$E_{ur} = K_{ur}P_a\left(\frac{\sigma_3}{P_a}\right)^n \tag{15}$$

where, E_{ur} is the loading-unloading modulus and K_{ur} is its corresponding modulus number.

For modeling most cohesionless soils a total of five (05) independent model parameters (ϕ, K, K_{ur}, n, R_f) are required in the hyperbolic model to accurately describe the non-linear, stress-dependent stress-strain relationship. These soil and model parameters should be determined directly from triaxial test equipment using carefully established curve fitting. However, in many situations especially those relevant to marine subsoils, triaxial data may be unavailable and soil strength in terms of ϕ and soil deformation in terms of E_s and v_s are only the unique papramenters at hand for design purposes.

In order to keep the use of hyperbolic model sufficiently accurate and to make its use practical for solving soil structure interaction problems, a thorough literature investigation performed by Otsmane and Amar Bouzid (2018) who examined a large number of well-established correlations between soil physical parameters especially those of sandy deposits whose behaviors are mainly governed by their internal friction angles. They established some relationships while others have been set on the basis of recommendations made by well-known researchers who carried out an overwhelming number of careful experiments. These parameters are listed in Tables 1 and 2 along with references of their origin. It has been found, besides the friction angle which is the crucial physical parameter, most parameters governing the behavior of sands in the hyperbolic model depend on the sand relative density and the confining pressure σ_3.

Expressions of both Tables 1 and 2 allow not only the successful implementation of the hyperbolic model in the finite element code but reducing the relatively high number of parameters required to model both sand deformation and strength to only the internal friction angle using well established correlations and recommendations from well-known research investigations. These expressions have been implemented in the FE computer code NAMPULAL which will be described in the next subsection.

Table 1. Parameters governing the hyperbolic model according to correlations and recommendations.

Parameter	Relevant expression or value	Reference
Exponent $\boldsymbol{n}$	0.51	Otsmane and Amar Bouzid (2018)
Failure ratio $\boldsymbol{R_f}$	0.7	Wong and Duncan (1974)
$\boldsymbol{K_{ur}}$	$1025e^{2.93D_r}\left(\frac{1+2k_0}{3k_0}\right)^{0.51} p_a^{-0.49}$	Otsmane and Amar Bouzid (2018)
$\boldsymbol{K}$	$0.667\ K_{ur}$	Duncan and Wong (1999)

Table 2. Soil stiffness parameters in terms of soil friction angle and confining pressure.

Parameter	Relevant expression or value	Reference
Sand relative density $\boldsymbol{D_r}$	$\frac{\phi}{15} - 1.8\sigma_3^{0.28} - 1.53[°]$	Ibsen et al. (2009)
Sand modulus of elasticity $\boldsymbol{E_s}$	$1025e^{2.93D_r}\left(\frac{1+2k_0}{3}\sigma_{v0}\right)^{0.51}$	Otsmane and Amar Bouzid (2018)
Earth pressure coefficient at rest $\boldsymbol{k_0}$	$1-\sin\phi$	Jaky (1944)

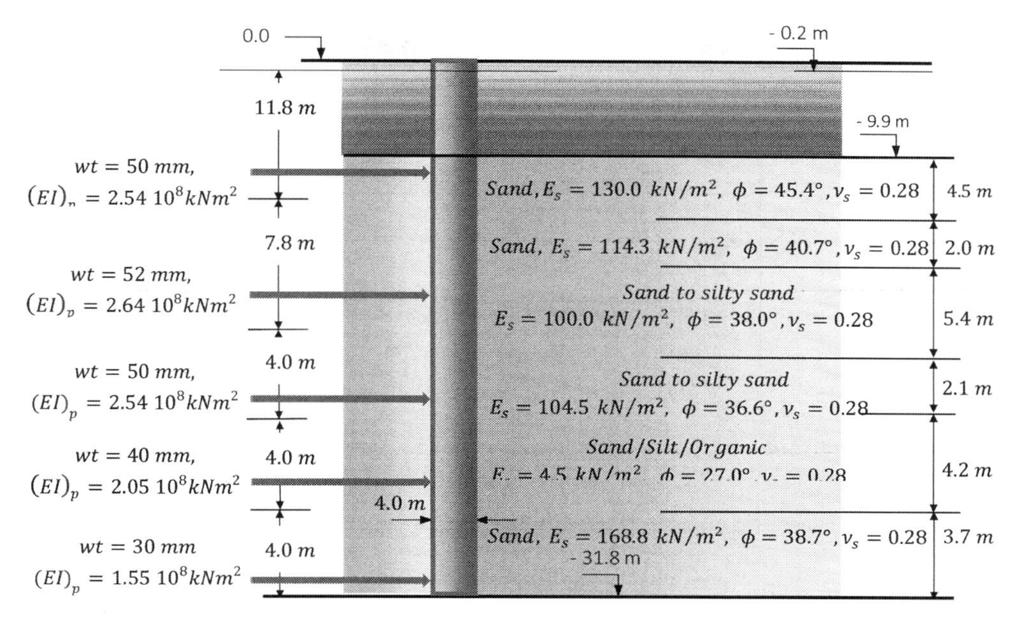

Figure 3. Details of monopile and soil strata at Horns Rev.

4 NAMPULAL: THE COMPUTER PROGRAM USED

A Fortran computer program called **NAMPULAL** (**N**onlinear **A**nalysis of **M**ono**P**iles **U**nder **L**ateral and **A**xial **L**oading) for the analysis of axially and laterally loaded single monopiles has been written on the basis of the finite element vertical slices equations. For further details, the reader is referred to Otsmane and Amar Bouzid (2018), only the features of this computer program are given here.

1. A number of twenty slices has been implemented in NAMPULAL. This number has been found to be sufficient to model accurately many soil/structure interaction problems (Amar Bouzid *et al.* 2005(a) and Otsmane and Amar Bouzid 2018).
2. NAMPULAL requires only two iterations. This fact, alleviates considerably the whole process of solution and makes it easier to find fast solutions even for the most complex soil/structure interaction problems.
3. In addition to rectangular cross-section shaped monopiles that are automatically taken into account due to the shape of the vertical slice, the solid circular or tubular cross-section monopiles are easily dealt with by prescribing the effective bending stiffness. Hence, an equivalent Young's modulus is extracted according to the following formula:

Table 3. Monopile properties.

Monopile properties			
$E_p (kN/m^2)$	(L_p/D_p)	$(L_p/D_p)_{cr}$	v_p
21.0×10^7	7.9	13.13	0.30

Table 4. Hyperbolic model parameters used in NAMPULAL.

Soil hyperbolic model parameters			
R_f	K	K_{ur}	n
Soil layer 1			
0.7	1029.326	1543.989	0.51
Soil layer 2			
0.7	545.598	818.397	0.51
Soil layer 3			
0.7	390.883	586.325	0.51
Soil layer 4			
0.7	297.473	446.209	0.51
Soil layer 5			
0.7	136.693	205.039	0.51
Soil layer 6			
0.7	404.632	606.947	0.51

$$E_{P_{eq}} = \frac{192(EI)_{act}}{\pi^2 (D_p)^4} \tag{16}$$

Table 5. PLAXIS parameters.

PLAXIS parameters						
E_{50}^{ref} (kPa)	E_{ur}^{ref} (kPa)	E_{oed}^{ref} (kPa)	$\psi(°)$	m	R_f	σ'_{pmax} (kPa)
44000.0	155000.0	25000.0	6.6	0.4	0.8	0.55

Table 6. Hyperbolic model parameters.

NAMPULAL parameters			
N	R_f	K_{ur}	K
0.51	0.7	2614.123	1742.75

This expression has been set on the assumption that the square cross-section monopile under consideration in NAMPULAL has the same cross-section area as the effective circular cross-section monopile.

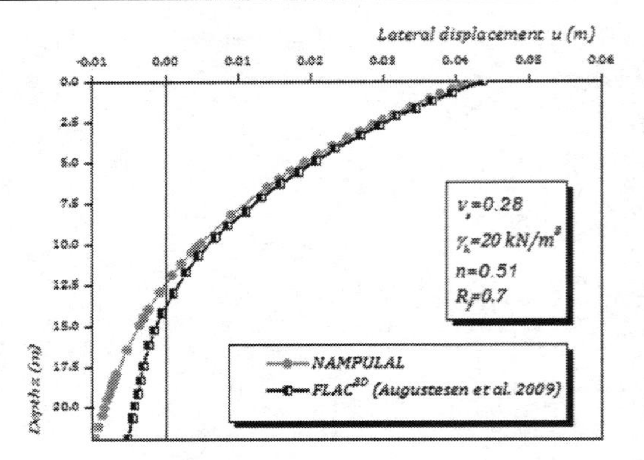

Figure 4. Lateral displacement profiles.

5 NAMPULAL PERFORMANCE ASSESSMENT AND VERIFICATION

To substantiate the performances of NAMPULAL results in problems involving laterally loaded monopiles embedded in sands, the works by Augustesen *et al.* (2009) and Doherty *et al.* (2012) were investigated. Indeed, Augustesen *et al.* (2009) studied monopiles at Horns Rev Offshore Wind Farm, built during 2003 and located in the North Sea west of Esbjerg in Denmark, whereas Doherty *et al.* (2012) performed numerical and field test investigations on OWT monopile installed in a research area, Ireland. Augustesen *et al.* (2009) employed FLAC^{3D} and Doherty *et al.* (2012) used PLAXIS which are both powerful numerical tools that can provide reliable results if accurate soil properties are provided as input data.

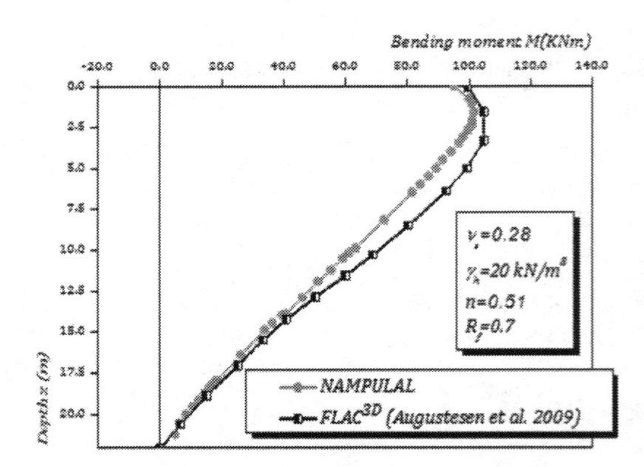

Figure 5. Evolution of bending moment with depth.

5.1 *NAMPULAL results versus those of FLAC^{3D}*

In addition to a numerical investigation, Augustesen *et al.* (2009) carried out a subsoil in-situ program in order to identify the different subsoil strata at the site of Horns Rev (Danmark). The monopile which has been the subject of research of Augustesen *et al.* has an outer diameter of 4 *m*, a length of 31.6 *m* and a variable wall thickness *wt*. Furthermore, the monopile has a varying flexural stiffness $(EI)_p$ with depth and has been driven to 31.8 *m* below the mean sea level (MSL) which results in an embedded depth equal to 21.9 *m* (Fig 3).

The monopile is subjected to a static horizontal load $H = 4.6$ *MN* and a bending moment $M = 95$ *MN.m*, both acting at seabed level.

By means of FLAC^{3D}, Augustesen *et al.* performed numerical computations, in which the soil has been considered as an elasto-plastic material obeying to Mohr-Coulomb failure criterion, and the monopile is assumed to be linear elastic steel with monopile Young's modulus $E_p = 210$ *GPa* and monopile Poisson's ratio $\nu_p = 0.3$. The FLAC^{3D} mesh has an outer diameter of $40D_p = 160$ *m* and the bottom boundary is placed approximately 18 *m* below the monopile tip.

In order to get a reliable comparative study where the mesh effects can be excluded, the same mesh has been adopted in the present computations carried out by the computer code NAMPULAL.

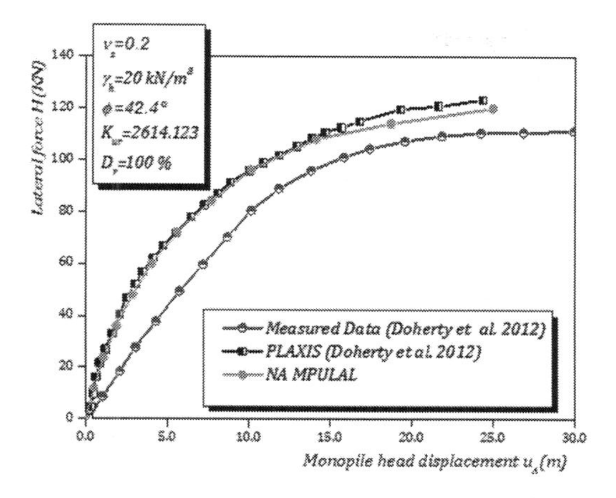

Figure 6. Monopile head lateral displacement in function of applied loading.

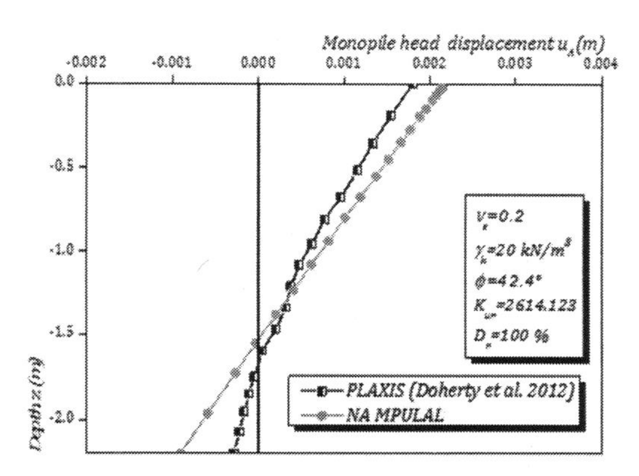

Figure 7. Displacement profiles for $H = 40kN$.

Indeed, a lateral extent of the mesh of a magnitude of 20 times the monopile diameter has been prescribed on both sides of the monopile. The bottom boundary in NAMPULAL has been kept at the same distance as that in FLAC^{3D}. However, the monopile in NAMPULAL is modeled as a square cross-section rather than an open tubular monopile with an internal soil plug. Poisson's ratio is unaltered, since the value for the soil is close to that of steel.

The monopile and soil hyperbolic model parameters used in NAMPULAL are listed in Tables 3 and 4.

The critical slenderness ratio $(L_p/D_p)_{cr}$ appearing in Table 3, has been computed using $\phi = 39.9°$ as an average value from all sand layers friction angles, excluding the organic layer which exhibited poor stiffness characteristics (Fig. 3).

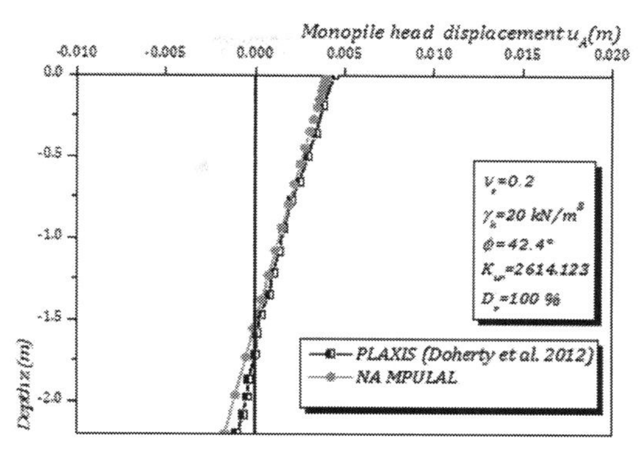

Figure 8. Displacement profiles for $H = 60kN$.

The comparison between the results of NAMPULAL and those of FLAC^{3D} have been performed through the most important design parameters which are the lateral displacement profile and the distribution of the bending moment with depth. Figures 4 and 5 illustrate respectively the evolution with depth of the monopile lateral displacement and bending moment. Firstly, results of both NAMPULAL and FLAC^{3D} are in perfect agreement, especially at the part of the monopile above the point of zero displacement. Secondly, the monopile behaves, in both methods, as relatively stiff monopile, exhibiting thus, a 'toe kick' shape, which corresponds exactly to the behavior of monopiles when their slenderness ratios are less than the critical ones $((L_p/D_p) = 7.9 < (L_p/D_p)_{cr} = 13.13)$.

The examination of Figure 5, which illustrates the evolution of bending moment with depth, confirms again the excellent agreement obtained between NAMPULAL and FLAC^{3D}.

5.2 *NAMPULAL predictions against PLAXIS*

Doherty *et al.* (2012), in addition to numerical analyses dealing with finite element method using PLAXIS, performed field test investigations on instrumented monopiles installed in a dense sand. The tested open ended steel monopiles have an outer diameter D_p = 340 *mm*, a wall thickness of t_p = 13 *mm*. The monopile was driven to an embedded length of L_p = 2.2 *mm*, which results in a slenderness ratio L_p/D_p = 6.47. These authors studied the evolution of monopile head displacement in function of the applied horizontal load, on one hand and the displacement profiles under two different values of the applied lateral load on the other hand. The latter was applied at a location situated 0.4 *m* above the mudline. PLAXIS input parameters and those of NAMPULAL employed in this case study are summarized in Tables 5 and 6.

Results of comparisons are presented in Figures 6–8. Although NAMPULAL necessitates only strength parameters as input data, in comparison to PLAXIS which requires a high number of parameters, the written code proved to be a powerful tool to capture the lateral response of monopiles in dense sand deposits.

6 CONCLUSIONS

A Fortran computer program bearing the acronym NAMPULAL has been written for analyzing horizontally and vertically loaded single monopiles. Monopiles which are very stiff deep foundations have proven to be the best solution for Offshore Wind Turbines (OWT) in water depths ranging between 15 m and 35 m, due to the ease and speed on installation

The very close agreement obtained between the results of the pseudo 3D Vertical Slices model (NAMPULAL code) and those of the finite differences model based code FLAC on one hand and PLAXIS on the other hand has no explanation than the robustness of the proposed approach. In fact, NAMPULAL has been written on the basis of two powerful numerical procedures which are the finite difference method and the finite element method. Furthermore, it encompasses the hyperbolic model for modeling soils, which is an excellent soil criterion in the stage prior to failure. Although, the hyperbolic model parameters for the two case studies investigated were unavailable, the formulation used to assess soil stiffness confirmed its validity in the accurate estimation of the monopile response in a medium composed mostly of dense sand. The obtained predictions were in reasonable agreement with those of the other well-known packages in the field of computational geomechanics.

Considering that the nonlinear finite element vertical slices model is comparatively less demanding and more efficient than conventional 3D finite element analysis, the proposed model, is seen to be a powerful tool for the analysis of complex 3D soil/structure interaction problems involving nonlinear soil behavior.

REFERENCES

Amar Bouzid Dj, Vermeer PA, Tiliouine B, Mir M. (2005a). An efficient pseudo three-dimensional FE model: Presentation and analysis of two soil/foundation interaction problems. International journal of Computational methods; 2(2): 231–253.

Amar Bouzid Dj, Vermeer PA, Tiliouine B. (2005 b). Finite element vertical slices model: Validation and application to an embedded square footing under combined loading. Computers and Geotechnics; 32: 72–91.

Augustesen AH, Brodbaek KT, Moller M, Sorensen SPH, Ibsen LB, Pedersen TS, Andersen L. (2009). Numerical modelling of large-diameter steel piles at Horns Rev. Proceedings of the Twelfth International Conference on Civil, Structural and Environmental Engineering Computing. Paper 239, Civil-Comp Press, Stirlingshire, Scotland.

Doherty P, Li W, Gavin K. (2012). Field lateral load test on monopile in dense sand. Offshore Site Investigation and Geotechnics: Integrated Technologies-Present and Future, London, UK.

Duncan JM, Chang C-Y. (1970). Nonlinear analysis of stress and strain in soils. Journal of Soil Mechanics and Foundation Division, ASCE; 96 (SM5):1629–1653.

Duncan JM, Wong KS (1999). User's Manual for SAGE. Volume II- "Soil Properties Manual." Centre for Geotechnical Practice and Research, Virginia Polytechnic Institute and State University.

Ibsen, L.B., Hansen, M., Hjort, T.H. and Thaarup, M. (2009). "MC-parameter calibration of Baskarp sand no.15.", *DCE Technical report no. 62*, Department of Civil Engineering, University of Aalborg, Danmark.

Jâky J. (1944). A nyugalmi nyomâs tényezöje (The coefficient of earth pressure at rest)," Magyar Mérnok és Epitész Egylet Közlönye (Journal for Society of Hungarian Architects and Engineers); 355–358.

Kondner RL (1963). Hyperbolic stress-strain response: cohesive soils. Journal of Soil Mechanics and Foundation Division, ASCE; 89(SM1):115–143.

Kondner RL, Zelako JS (1963). A hyperbolic stress-strain formulation for sands. Proceedings of the 2nd Pan American Conference on Soil Mechanics and Foundation Engineering; 1: 289–324.

Otsmane and Amar Bouzid (2018). An Efficient FE model for SSI: Theoretical background and assessment by predicting the response of large diameter monopiles supporting OWECs. Computers and Geotechnics; 97: 155–166.

Wong KS, Duncan JM (1974). Hyperbolic stress-strain parameters for non-linear finite element analyses of stresses and movements in soil masses. Report, no.TE-74-3; University of California, Berkeley, USA.

Numerical Methods in Geotechnical Engineering IX – Cardoso et al. (Eds)
© 2018 Taylor & Francis Group, London, ISBN 978-1-138-33198-3

Numerical modeling of the creep behavior of a stabilized soft soil

P.J. Venda Oliveira
ISISE, Department of Civil Engineering, University of Coimbra, Portugal

A.A.S. Correia
CIEPQPF, Department of Civil Engineering, University of Coimbra, Portugal

L.J.L. Lemos
Department of Civil Engineering, University of Coimbra, Portugal

ABSTRACT: This paper examines the ability of two creep laws (volumetric and deviatoric), associated with a constitutive model composed by the Modified Cam Clay yield surface combined with the Von Mises yield surface, to simulate the creep behavior of a soft soil chemically stabilized with binders. The constitutive models associated with the two creep laws are validated by oedometer and triaxial creep tests done for the stabilized soil. A 2-D finite element code with these constitutive models and creep laws is used in these analyses.

1 INTRODUCTION

Over the last decades, numerous embankments have been built on soft soils, which, in general, contain a high organic matter content. In these conditions significant creep deformations are expected (Venda Oliveira et al. 2012) to occur in the long term. The creep deformations can be mitigated by installing rigid vertical inclusions such as: concrete piles, stone columns and deep soil mixing columns (DMCs). However, due to the arching effect there is a transfer of stresses from the soft soil to the DMCs, which increases the creep settlements of the DMCs (Venda Oliveira et al. 2009, Correia et al. 2009). The consequences of these creep deformations may be quite significant, for instance in high-speed train embankments and crane foundations, that is, in structures where the allowable deformation is very low. Furthermore, an accurate evaluation of the creep deformations is important to minimize the works and the costs during the service life of the embankment (Venda Oliveira et al., 2017a).

Few research works concerning the numerical prediction of creep deformations of stabilized soils have been published (Correia et al., 2009; Venda Oliveira et al. 2009, 2017a, 2017b), since it is usually considered that DMCs suffer negligible creep deformations.

Indeed, the field data of some embankments seem to indicate the existence of creep deformations, due to a continuous increase of the settlement (Voottipruex et al. 2011, Chai et al. 2015, Bergado et al. 1999) and the measurement of settlements higher (about 100%) than those predicted numerically (Yapage et al. 2014).

This work evaluates the ability of two creep laws to simulate the creep behavior of a soft soil chemically stabilized with binders, based on the experimental results of eodometer and triaxial creep tests.

A 2-D finite element program with several constitutive models, upgraded at the University of Coimbra (Venda Oliveira 2000) and capable of carrying out elastoplastic analyses with coupled consolidation and creep, is used in these analyses. The constitutive model used is the Modified Cam Clay combined with the Von Mises (MCC/VM) model. These two models are associated to two creep laws, composed by a volumetric and a deviatoric component.

2 DESCRIPTION OF THE MODELS

2.1 *Creep laws*

The creep deformations in soils are composed by a volumetric and a deviatoric component. Creep volumetric strains (1) are calculated with the equation of the secondary consolidation (Taylor 1948):

$$\varepsilon_v^c = \int \frac{C_{\alpha e}}{2.3\,(1+e)\,t_v}\,dt \qquad (1)$$

where $C_{\alpha e}$ (= $\Delta e/\Delta \log t$) is the index of secondary consolidation, e the void ratio and t_v the volumetric age of the soil in relation to the reference volumetric time (t_{vi}). Considering Kavazanjian & Mitchell's (1980) proposal, $C_{\alpha e}$ is constant and independent of the shear stress level.

Creep deviatoric strains (2) are calculated based on Singh-Mitchell's law (Singh & Mitchell 1968) derived from triaxial creep tests:

$$\varepsilon_{ax}^{c} = \int A\, e^{\bar{\alpha}\bar{D}} \left(\frac{t_{di}}{t_d}\right)^{m} dt \tag{2}$$

where A, α and m are soil creep parameters, $\bar{D} = (q/q_{ult})$ is the deviatoric stress level and t_d the deviatoric age of the soil in relation to the deviatoric reference time (t_{di}). The influence of the composition, structure and stress history is shown by parameter A (Morsy et al. 1995a, 1995b); parameter α reproduces the effect of the shear stress level (Kuhn & Mitchell 1993), while m controls the decrease of the creep strain rate with time (Singh & Mitchell 1968). Equation (2) is valid for $0.2 < \bar{D} < 0.9$ (Borja & Kavazanjian 1985), i.e., for a steady state creep phase.

2.2 *Constitutive models associated with creep laws*

A constitutive model composed by two yield surfaces (MCC/VM) is used to simulate the behavior of the stabilized soil.

The MCC model is an elastoplastic soil model based on isotropic conditions, described by a yield function represented by an ellipse oriented in line with the p' axis (Figure 1a). The yield function of the MCC model with creep can be described by:

$$F(\sigma'_{ij}, e_k, t_v) = (\lambda - \kappa).\ln\left[p'\left(1 + \frac{\eta^2}{M^2}\right)\right] - h(e_k, t_v) \tag{3}$$

where λ and κ are, respectively, the slope of the virgin consolidation line (VCL) and the slope of the overconsolidation line in the plot (e-lnp'), M is the slope of the critical state line (CSL) and η is the stress ratio q/p'. This model uses an isotropic hardening rule, $h(e_k, t_v)$, which is composed by a plastic deformation, $h(e_\kappa)$ (Figure 1b), and a volumetric creep law, $\Delta h(t_v)$, described by:

$$h(e_k, t_v) = e_{\lambda o} - \underbrace{\underbrace{(e + \kappa.\ln p')}_{e_\kappa}}_{h(e_\kappa)} + \underbrace{\frac{C_{\alpha e}}{2.3}.\ln\left(\frac{t_v}{t_{vi}}\right)}_{\Delta h(t_v)} \tag{4}$$

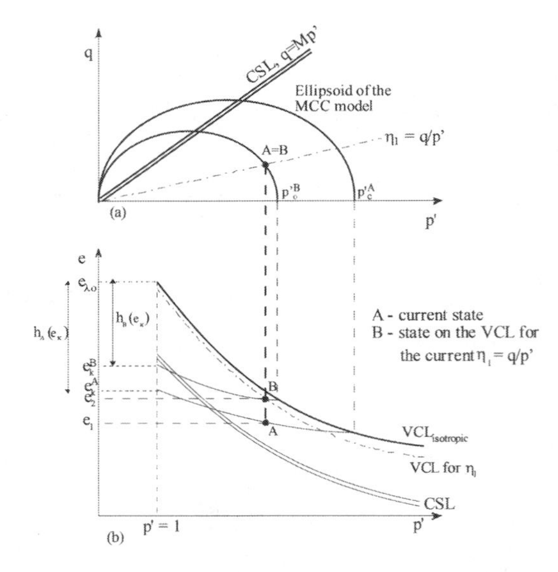

Figure 1. MCC model: a) yield surface; b) hardening rule.

where $e_{\lambda o}$ is the void ratio for p' = 1. The creep phenomenon promotes the change of the $h(e_k, t_v)$, which induces the change of the isotropic preconsolidation stress (p'_c) (Figure 1).

The VM model reproduces the non-linearity within the yield surface of the MCC model. The MCC/VM model is expressed by two yield functions, described by equations (3), (4) for the MCC model, and equation (5) for the VM model:

$$G(\sigma'_{ij}, \gamma^p, \varepsilon_v^p, t_v, t_d) = \underbrace{q}_{g(\sigma'_{ij})} - \underbrace{\frac{\gamma\, p'_c}{a + b\gamma} R_f}_{h(\gamma^p, \varepsilon_v^p, t_v, t_d)} \tag{5}$$

where $g(\sigma'_{ij})$ is the load function and $h(\gamma^p, \varepsilon_v^p, t_v, t_d)$ the hardening rule. The parameter $h(\gamma^p, \varepsilon_v^p, t_v, t_d)$ depends on the plastic deformations and the volumetric and deviatoric creep time. This is calculated assuming that the trace of the yield surface on plane q-γ is a hyperbola where a and b are normalized hyperbolic parameters, and R_f (= $q_{failure}/q_{ult}$) is the failure ratio (Duncan & Chang 1970).

Both models (MCC and VM) use an associated plastic flow rule.

2.3 *Creep time (t_v, t_d)*

The volumetric age, t_v, is stated based on the difference between the void ratio of the current state (e_1) and that of the boundary surface state (e_2) (Figure 1):

$$t_v = t_{vi} \cdot e^{\left(\frac{e_2 - e_1}{C_{\alpha e}/2.3}\right)} \quad (6)$$

The deviatoric age, t_d, is stated from the difference between the actual deviatoric strain (γ_1) and that of the normally consolidated state (γ_2) (equation 7), which is evaluated considering the trace of the MCC ellipsoid on the q-γ plane (equation 8) (Konder 1963):

$$t_d = \left[\frac{(\gamma_1 - \gamma_2).(1-\mathrm{m})}{\mathrm{A}.e^{\bar{\alpha}.\bar{D}}.(\mathrm{t_{di}})^{\mathrm{m}}}\right]^{\frac{1}{1-\mathrm{m}}} \quad (\mathrm{m} \neq 1) \quad (7)$$

$$\gamma_2 = \frac{\mathrm{q.a}}{\mathrm{p}'_c.R_f - \mathrm{q_c.b}} \quad (8)$$

A detailed description of the creep laws associated with the MCC and MCC/VM models are presented in some research studies (Hsieh et al. 1990, Borja 1992, Borja & Kavazanjian 1985, Morsy et al. 1995a, 1995b, Venda Oliveira 2000, Venda Oliveira et al. 2017a, 2017b).

3 PROPERTIES/PARAMETERS OF THE STABILIZED SOIL

The stabilized soil used in this study is based on a Portuguese soft soil, from the "Baixo Mondego" area.

The natural soft soil is a silty soil, classified by the Unified Soil Classification System as OH (ASTM D2487 1998), due to its high organic matter content (9.3%), which affects the main properties of the soil. Thus, it exhibits low unit weight, high void ratio and plasticity, low undrained shear strength and high compressibility (C_c, C_r and $C_{\alpha e}$) (Venda Oliveira et al., 2010).

The soft soil was chemically stabilized with a binder mixture composed by Portland cement Type I 42.5 R (EN 197–1 2000) and blast furnace granulated slag, with a dry weight proportion of 75/25, and a binder quantity of 125 kg/m^3 (Correia 2011). Thus, the Portland cement behaves as a hydraulic material (i.e., reacts spontaneously with water) while the slag promotes pozzolanic reactions, which induce the long-term strengthening (Janz & Johansson 2002).

Table 1 presents the main characteristics of the stabilized soil, which has a high unconfined compressive strength (q_u = 1246 kPa), high vertical yield stress (σ'_y = 2500 kPa), low pre-yield compression index and a low secondary compression index.

Table 2 shows the parameters of the MCC model and both creep laws used (volumetric and deviatoric) to reproduce the behavior of the soft soil, evaluated based on the results of oedometer and triaxial tests (Venda Oliveira et al. 2010, Coelho 2000, Correia 2011).

4 NUMERICAL PREDICTION OF THE BEHAVIOR

4.1 *CIU triaxial tests*

Figure 2 compares the experimental results of the CIU triaxial tests carried out with samples isotropically consolidated with confining pressures (p'_0) of 50 and 100 kPa, with the numerical predictions.

Table 1. Properties of the stabilized soil.

Property	Value
Binder quantity (kg/m^3)	125
Binder content, a_w (%)	15.0
Water-binder ratio (w/a_w)	5.4
After curing void ratio, $e_{0\text{-}ac}$	1.72
Unit weight, γ(kN/m3)	15.3
Post-yield compression index, C_c	1.00
Pre-yield compression index, C_r	0.032
Vertical yield stress, σ'_y (kPa)	2500
Secondary compression index, $C_{\alpha e}$	0.00075
Unconfined compressive strength, q_u (kPa)	1246

Table 2. Parameters used to simulate the behavior of the stabilized soil.

Parameter		Value
Constitutive model		MCC/VM
Elastic parameters	E'	164.7 MPa
	v	0.3
MCC model	$e_{\lambda 0}$	5.07
	e_0	(*)
	λ	0.435
	κ	0.0074
	M	1.50
VM model	a	0.0013
	b	1.683
	R_f	1.0
Volumetric creep law	C_α	(*)
	t_{vi}	(**)
Deviatoric creep law	A	0.0000244%/min
	m	0.28
	$\propto$	0.671
	t_{di}	(**)
	a	0.0013
	b	1.683
	R_f	1.0

(*) Variable, depends on the stress level,
(**) The creep time is equal to the current time.

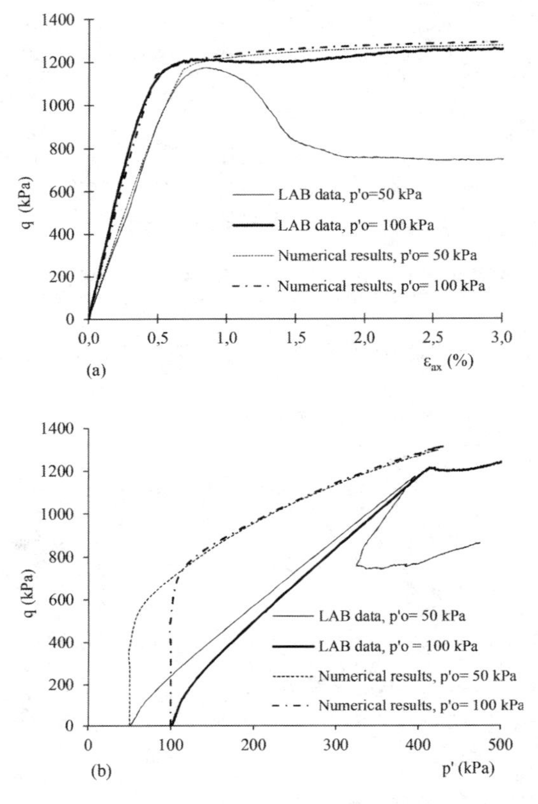

Figure 2. Numerical prediction of the CIU triaxial tests of the stabilized soil: a) q-ε_{ax} curves, b) stress path curves.

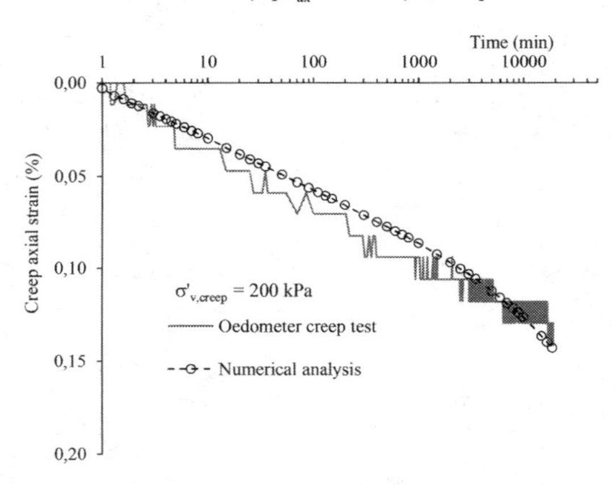

Figure 3. Numerical prediction of the eodometer creep test.

The elastic parameters used in the numerical predictions were obtained from the experimental results: $\nu' = 0{,}3$ and E' of 164,7 and 220 MPa for p'_0 equal to 50 and 100 kPa, respectively. The stress-strain

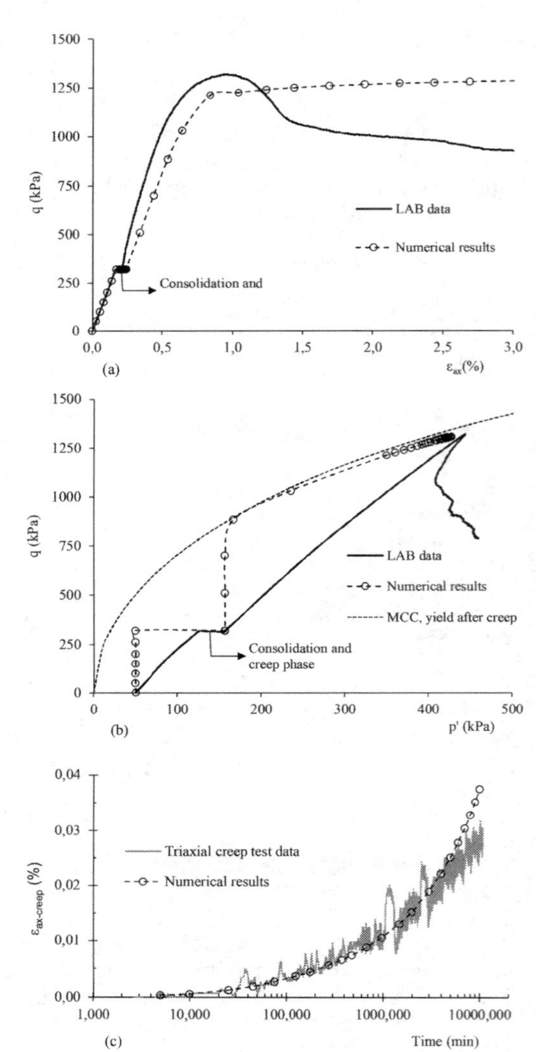

Figure 4. Numerical prediction of the multi-stage triaxial test (with creep phase: a) q-ε_{ax} curves, b) stress path curves; c) evolution of the axial strain during creep phase.

behavior (Figure 2a) shows very good numerical predictions before the peak strength. Naturally, after the peak failure, the MCC/VM model is not able to predict the softening observed in the sample consolidated for p'_0 of 50 kPa, since this model does not incorporate the effect of the structure of the soil. In terms of stress paths (Figure 2b), a similar final state is obtained. However the differences between the experimental and numerically predicted results are significant. This is fundamentally due to the consideration of an unrealistic linear elastic behavior inside the yield surface by the MCC model, which induces vertical effective stress paths inside the yield surface.

4.2 *Creep tests*

The numerical simulation of the behavior of the stabilized soil under creep conditions is studied with the parameters for the constitutive models (MCC and MCC/VM) and the creep laws (Table 2). For this purpose, the results of oedometer creep tests and multi-stage triaxial tests are used. The multi-stage triaxial tests are composed by an undrained initial stage, followed by a consolidation and creep stage and by a final undrained stage.

The numerical predictions, were carried out according to the following conditions: i) axisymmetric FE analyses; (ii) the oedometer test was simulated by an FEM mesh with 13 FE elements and 68 nodes; (iii) multi-stage triaxial tests were simulated with only 1 FE element and 8 nodes; (iv) eight-noded isoparametric quadrilateral elements with 20 nodal degrees of freedom are used in both analyses (thus, the displacement is evaluated at eight nodes and the excess pore pressure at the four corner nodes).

Figure 3 compares the creep axial strain ($\varepsilon_{ax,creep}$) obtained in the oedometer tests with the numerical predictions under a creep vertical pressure of 200 kPa. The results show that the $\varepsilon_{ax,creep}$ obtained is very reduced and the numerical model used is able to predict the creep behavior under oedometer conditions, since the numerical calculations match the experimental results very well.

Figure 4 illustrates the experimental and numerical results of a multi-stage triaxial test carried out with a sample of stabilized soil isotropically consolidated ($p'_0 = 50$ kPa). The numerical prediction of the first stress-strain undrained stage shows an excellent agreement with the experimental results (Figure 4a), since the behavior is, approximately, linear elastic in this zone. After the creep stage, the behavior observed experimentally seems to show a slight increase in the stiffness, which is not replicated by the numerical results, since constant elastic parameters (evaluated for the first stage) were used. As observed in Figure 2, the numerical prediction is not able to reproduce the softening observed after peak strength. As expected, the stress paths (Figure 4b) obtained numerically show a vertical path inside the yield surface while the stress paths obtained experimentally are more flattened. Despite these differences, the point correspondent to the peak strength is predicted reasonably well numerically. In terms of the evolution of the axial strain during the creep phase (Figure 4c), the numerical model predicts the variation of the $\varepsilon_{ax,creep}$ with time very well.

2 CONCLUSIONS

This work studies the ability of two constitutive models associated with volumetric and deviatoric creep laws to replicate the creep behavior of a chemically stabilized Portuguese soft soil. The behavior of this material is predicted by the Modified Cam Clay model associated with the Von Mises (MCC/VM). Results of oedometer creep tests and multi-stage triaxial tests were used to validate the constitutive models and creep laws. Some conclusions can be drawn from this study:

- The MCC/VM model is able to replicate the stress-strain behavior until failure and the state at the peak strength reasonably well.
- However, the MCC/VM model does not reproduce the stress path paths inside the yield surface of the MCC model, since a vertical effective stress path is predicted by the numerical model, which is different from those obtained in the CIU triaxial tests.

- The MCC/VM model associated with the creep laws predicts a creep axial strain similar to those observed in the oedometer creep tests and multi-stage triaxial tests.
- The numerical predictions of the stress path of the multi-stage triaxial tests (with an intermediate creep stage) is coherent with the theoretical framework of the MCC model, although they exhibit some differences in relation to the stress paths obtained experimentally.

ACKNOWLEDGMENTS

The authors would like to express their thanks to the institutions that financially supported the research: ISISE (project UID/ECI/04029/2013) and CIEPQPF (project EQB/UI0102/2014).

REFERENCES

ASTM D2487 1998. *Standard Classification of Soils for Engineering Purposes (Unified Soil Classification System)*. West Conshohoken, PA.

Bergado D.T., Ruenkrairergsa, T., Taesiri, Y. & Balasubramaniam, A.S. 1999. Deep soil mixing used to reduce embankment settlement. *Ground improvement* 3(4):145–162.

Borja, R.I. & Kavazanjian Jr, E. 1985. A constitutive model for the stress-strain-time behavior for "wet" clays. *Géotechnique* 35(3): 183–198.

Borja, R.I. 1992. Generalized creep and stress relaxation model for clays. *Journal of the Geotechnical Engineering (ASCE)* 118(11): 1765–1786.

Chai, J.C., Shrestha, S., Hino, T., Ding, W.Q., Kamo, Y. & Carter, J. 2015. 2D and 3D analyses of an embankment on clay improved by soil-cement columns. *Computers and Geotechnics* 68: 28–37.

Coelho, P.A.L.F. 2000. Geotechnical characterization of soft soils. *Study of the experimental site of Quinta*

do Foja, MSc Dissertation, University of Coimbra, Portugal (in Portuguese).
Correia, A.A.S. 2011. *Applicability of deep mixing technique to the soft soil of Baixo Mondego.* Ph.D. dissertation, University of Coimbra, Coimbra, Portugal (in Portuguese).
Correia, A.A.S., Venda Oliveira, P.J., Lemos, L.J.L. & Mira, E.S.P. 2009. Creep deformations of deep mixing columns. Evaluation from laboratorial tests. *Proc., Int. Symp. of Deep Mixing and Admixture Stabilization*, Port and Airport Research Institute, Yokosuka City, Japan.
Duncan, J.M. & Chang, C.Y. 1970. Nonlinear analysis of stress and strain in soils. *Journal of the Soil Mechanics and Foundations Division (ASCE)* 96(SM 5): 1629–1653.
EN 197–1 2000. *Cement e Part 1: Composition, Specifications and Conformity Criteria for Common Cements.* European Committee for Standardization, June 2010.
Hsieh, H.S., Kavazanjian Jr, E. & Borja, R.I. 1990. Double yield surface Cam-Clay plasticity model I: Theory. *Journal of Geotechnical Engineering (ASCE)* 116(9): 1381–1401.
Janz, M. & Johansson, S.-E. 2002. *The function of different binding agents in deep stabilization.* Swedish Deep Stabilization Research Centre, Report 9, Linköping, Sweden.
Kavazanjian, E. & Mitchell, J.K. 1980. Time-dependent deformation behaviour of clays. *Journal of the Geotechnical Engineering Division (ASCE)* 106(GT6): 611–630.
Konder, R.L. 1963. Hyperbolic stress-strain response: Cohesive soils. *Journal of the Soil Mechanics and Foundation Division(ASCE)* 106(6): 611–630.
Kuhn, M.R. & Mitchell, J.K. 1993. New perspectives on soil creep. *Journal of Geotechnical Engineering* 119(3): 507–523.
Morsy M.M., Chan, D.H. & Morgenstern, N.R. 1995a. An effective stress model for creep of clay. *Canadian Geotechnical Journal* 32(5): 819–834.
Morsy, M.M., Morgenstern, N.R. & Chan, D.H. 1995b. Simulation of creep deformation in the foundation of Tar Island Dyke. *Canadian Geotechnical Journal* 32(6): 1002–10023.
Singh, A. & Mitchell, J.K. 1968. General stress-strain-time functions for soils. *Journal of the Soil Mechanics and Foundation Division (ASCE)* 94(SM1): 21–46.
Taylor D.W. 1948. *Fundamentals of Soil Mechanics.* John Wiley and Sons, Inc., New York.
Venda Oliveira P.J. 2000. *Embankments on soft clays—Numeric analysis.* Ph.D. Dissertation, University of Coimbra, Portugal (in Portuguese).
Venda Oliveira P.J., Lemos L.J.L. & Coelho P.A.L.P. 2010. Behavior of an atypical embankment on soft soil: field observations and numerical simulation. *Journal of Geotechnical and Geoenvironmental Engineering* 136(1): 35–47.
Venda Oliveira, P.J., Correia, A.A.S. & Garcia, M.R. 2012. Effect of organic matter content and curing conditions on the creep behavior of an artificially stabilized soil. *Journal of Materials in Civil Engineering* 24(7): 868–875.
Venda Oliveira, P.J., Correia, A.A.S. & Lemos, L.J.L. 2009. Numerical prediction of creep of an embankment over deep soil mixing columns. *Proc., Int. Symp. of Deep Mixing and Admixture Stabilization*, Port and Airport Research Institute, Yokosuka City, Japan.
Venda Oliveira, P.J., Correia, A.A.S. & Lemos, L.J.L. 2017a. Numerical predictions of the creep behaviour of an unstabilised and a chemically stabilized soft soil. *Computers and Geotechnics* 87: 20–31.
Venda Oliveira, P.J., Correia, A.A.S. & Lemos, L.J.L. 2017b. Numerical modelling of the effect of curing time on the creep behaviour of a chemically stabilised soft soil. *Computers and Geotechnics* 91: 117–130.
Voottipruex, P., Bergado, D.T., Suksawat, T., Jamsawang, P. & Cheng, W. 2011. Behavior and simulation of deep cement mixing (DCM) and stiffened deep cement mixing (SDCM) piles under full scale loading. *Soils and Foundations* 51(2): 307–320.
Yapage, N.N.S., Liyanapathirana, D.S., Kelly, R.B., Poulos, H.G. & Leo, C.J. 2014. Numerical modeling of an embankment over soft griund improved with deep cement mixed columns: case history. *Journal of Geotechnical and Geoenvironmental Engineering* 140(11): 04014062.

Numerical Methods in Geotechnical Engineering IX – Cardoso et al. (Eds)
© 2018 Taylor & Francis Group, London, ISBN 978-1-138-33198-3

On the use of NURBS plasticity for geomaterials

W.M. Coombs
Department of Engineering, Durham University, Durham, UK

Y. Ghaffari Motlagh
School of Chemical and Process Engineering, University of Leeds, Leeds, UK

ABSTRACT: There are a huge number of plasticity models available in the literature to represent the behaviour of geomaterials and the majority include the concept of a yield surface which bounds the allowable stress in the material. However, each surface is distinct and requires a specific equation describing the shape of the surface to be formulated in each case. Recently Coombs *et al.* (2016) proposed a non-uniform rational basis spline (NURBS) plasticity framework where any isotropic yield surface can be represented and integrated using the same numerical algorithm. However, the NURBS plasticity framework is currently restricted the to case of associated plastic flow which will lead to an overly dilative response for most geomaterials. This paper extends the approach to include a non-associated plastic flow rule whilst guaranteeing the recovery of associated plastic flow if required. The algorithm's performance is demonstrated at both material point (stress-strain) and boundary value problem levels.

1 INTRODUCTION

Numerical analysis of engineering problems rely on robust and efficient constitutive models that provide the link between stress and strain in the material under consideration. Within this, conventional plasticity models use the concept of a yield surface to distinguish between elastic (inside the surface) and elastoplastic (on the surface) behaviour. Depending on the material analysed the shape of this surface will change and this impacts on the stress integration algorithm[1] (which requires changes in the numerics) for each implemented yield surface. The non-uniform rational basis spline (NURBS) plasticity framework was first proposed by Coombs et al. (2016) and extended to include linear isotropic hardening by Coombs and Ghaffari Motlagh (2017). The key idea of the framework is to represent the yield surface of an isotropic plasticity model using a NURBS surface. This allows different yield criteria to be included within the same numerical algorithm by only changing information associated with the NURBS surface (control point positions, spline order, etc.).

This paper adopts the NURBS plasticity framework and applies it to yield surfaces typically used for the behaviour of geomaterials, such as Mohr-Coulomb and Drucker-Prager. However, it is well known that associated plastic flow frictional yield surfaces overestimate dilative behaviour. Therefore this paper extends the framework to include non-associated plastic flow whilst maintaining that all information associated with the yielding of the material is held at control points.

The layout of the paper is as follows. After this introduction, Section 2 describes the NURBS plasticity framework and its extension to include nonassociated flow. Section 3 discusses the implementation of the model and Section 4 presents a series of material point and boundary value simulations to demonstrate the capabilities of the non-associated plastic flow NURBS framework. Finally, conclusions are drawn in Section 5.

The majority of this paper is presented in terms of principal stresses with a tension positive notation and the conventional ordering of the principal stresses

$$\sigma_1 \geq \sigma_2 \geq \sigma_3,$$

which restricts the principal stress state to a single sextant of stress space. Adopting a principal stress notation is common in other isotropic plasticity models, for example see the work of Coombs and Crouch (2011a), Coombs et al. (2010), Coombs and Crouch (2011b) and Clausen et al. (2007) amongst

1. A stress integration algorithm is required to convert the constitutive equations, which are typically developed in rate form, into an incremental relationship that can be used within a boundary value solver (finite elements, for example).

others, and does not change the generality of the algorithm (it is suitable for 1D, 2D and 3D analysis). The principal values are simply mapped back to conventional six-component stress space at the end of the constitutive algorithm.

2 NURBS PLASTICITY

The following section outlines the NURBS plasticity framework and extends the approach of Coombs et al. (2016) to include non-associated plastic flow.

A general NURBS surface can be expressed as

$$S_k(\xi,\eta)=\sum_{i=0}^{n}\sum_{j=0}^{m} R_{i,j}(\xi,\eta)(C_k)_{i,j} \tag{1}$$

where k is the physical index, C_k are the control point positions and n and m are the number of control points in the ξ and η directions.[2] The NURBS basis functions, $R_{i,j}$, are given by

$$R_{i,j}(\xi,\eta)=\frac{N_{i,p}(\xi)N_{j,q}(\eta)w_{i,j}}{\sum_{k=0}^{n}\sum_{l=0}^{m} N_{k,p}(\xi)N_{l,q}(\eta)w_{k,l}} \tag{2}$$

where $N_{i,p}$ and $N_{j,q}$ are the pth and q^{th}-degree B-spline basis functions, ξ and η are the local positions within the Knot vectors and $w_{i,j}$ are the weights associated with the control points.

2.1 *Yield surfaces & associated plastic flow*

Within the framework of NURBS plasticity (Coombs et al. 2016), the yield surface can be expressed as

$$f=(\sigma_i - S_i(\xi,\eta))(S,_\sigma)_i = 0, \tag{3}$$

where σ_i is the principal stress state and $(S,_\sigma)_i$ is the partial derivative of (1) with respect to stress which is the same as the outward normal to the surface. This can be obtained through the cross product of the two local derivatives

$$(S,_\sigma)_i=(S,_\eta \times S,_\xi)_i=\varepsilon_{ijk}(S,_\eta)_j(S,_\xi)_k, \tag{4}$$

where ε_{ijk} is the Levi-Civita tensor.[3]

In associated flow plasticity theory the plastic strains evolve according to

$$\dot{\varepsilon}_i^{p}=\dot{\gamma}(S,_\sigma)_i, \tag{5}$$

where $\dot{\gamma}$ is the scalar plastic multiplier (or consistency parameter). This plastic multiplier must satisfy the Kuhn-Tucker-Karush consistency conditions

$$f(\sigma_i)\le 0, \quad \dot{\gamma}\ge 0 \quad \text{and} \quad f(\sigma_i)\dot{\gamma}=0. \tag{6}$$

These conditions enforce that the material must either be on the yield surface undergoing elasto-plastic deformation ($f=0$ and $\dot{\gamma}\ge 0$) or inside the yield surface with purely elastic behaviour ($f\le 0$ and $\dot{\gamma}=0$).

Figure 1 shows a spherical NURBS surface and associated control points (shown by the red points). Different surfaces can be obtained by moving the positions of the control points and/or modifying the basis functions, $R_{i,j}$.

2.2 *Non-associated plastic flow*

In the case of non-associated flow the evolution of plastic strains is decoupled from the spatial gradient of the yield envelope and the plastic strains evolve according to

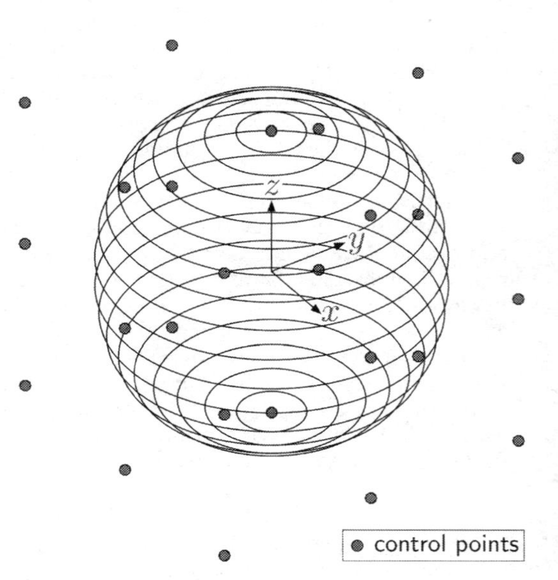

Figure 1. Spherical NURBS surface and control point net where the control points are shown by the red-shaded circles.

2. Note that the total number of control points used to define the surface is $n \times m$.
3. $\varepsilon_{ijk}=0$ if $i=j$, $j=k$ or $k=i$, $\varepsilon_{ijk}=1$ for even permutations of i, j and k and $\varepsilon_{ijk}=-1$ for odd permutations of i, j and k.

$$\dot{\varepsilon}_i^{\rm p} = \dot{\gamma}(g_{,\sigma})_i, \tag{7}$$

where $(g_{,\sigma})_i$ is the gradient of the plastic potential surface.

In this NURBS plasticity approach the gradient of the plastic potential surface is given by

$$(g_{,\sigma})_i = (S^g{}_{,\eta} \times S^g{}_{,\xi})_i = \varepsilon_{ijk}(S^g{}_{,\eta})_j(S^g{}_{,\xi})_k, \tag{8}$$

where S^g is the plastic potential surface

$$S_k^g(\xi,\eta) = \sum_{i=0}^{n}\sum_{j=0}^{m} R_{i,j}^g(\xi,\eta)(G_k)_{i,j} \tag{9}$$

and G_k are control point coordinates that control the shape of the surface. The NURBS basis functions $R_{i,j}^g(\xi,\eta)$ are calculated in the same way as (2). Note that it is not necessary to have the same basis functions from the direction of plastic flow as used to describe the geometry of the yield surface however it is assumed that a single set of control points control the form of S_k and S_k^g.

Previous attempts to extend the NURBS plasticity framework to include non-associated flow have directly specified the flow direction at the control points rather than specifying the geometry of a plastic potential surface (see Coombs (2017)). However, due to the presence of the cross product in (4) it is not possible to specify the flow direction at the control points and guarantee the recovery of associated flow. With this approach it is only possible to recover associated flow over the entire yield surface if there is no coupling between the local knot coordinates (ξ,η); the von Mises yield surface is one example. In this paper we adopt a more conventional approach, in terms of plasticity theory, and specify the geometry of the plastic potential surface using control points. In the case of associated plastic flow the plastic potential control point positions coincide with those used to define the geometry of the yield surface.

2.3 *Stress integration*

A robust and efficient stress integration routine is essential if a constitutive model is to be used to analyse engineering problems. These routines convert the rate relationships presented in the previous section into an incremental algorithm that, given the previous stress state and a strain increment, will provide an updated stress state.

In this work we use an implicit elastic predictor, plastic corrector scheme (Wilkins 1964), where the elastic trial stress is given by

$$\sigma_i^t = \sigma_i^n + \Delta\sigma_i. \tag{10}$$

The stress increment and the previous convergence stress state are given by

$$\Delta\sigma_i = D_{ij}^{\rm e}\Delta\varepsilon_j \quad \text{and} \quad \sigma_i^n = D_{ij}^{\rm e}(\varepsilon_n^{\rm e})_j. \tag{11}$$

$(\varepsilon_n^{\rm e})_j$ is the elastic strain state from the previous load (or time) step in the global solution algorithm, $\Delta\varepsilon_i$ is the strain increment associated with the global boundary value displacement and $D_{ij}^{\rm e}$ contains the principal components of the linear elastic stiffness matrix.

If the trial elastic stress state exceeds the yield envelope ($f > 0$) then it must be corrected back onto the yield surface using a plastic stress increment, that is

$$\sigma_i^r = \sigma_i^t - \Delta\sigma_i^{\rm p}, \quad \text{where} \quad \Delta\sigma_i^{\rm p} = D_{ij}^{\rm e}\Delta\varepsilon_j^{\rm p}. \tag{12}$$

σ_i^r is the *returned* stress state on the yield surface and $\Delta\varepsilon_j^{\rm p}$ is the plastic strain increment obtained from the incremental form of (7). Once this correction has been applied the updated elastic strain can be obtained from

$$(\varepsilon_{n+1}^{\rm e})_i = (\varepsilon_n^{\rm e})_i + \Delta\varepsilon_i - \Delta\varepsilon_i^{\rm p}. \tag{13}$$

3 IMPLEMENTATION

Consistent with the perfect plasticity implementation of Coombs et al. (2016), here we use a coarse initial subdivision algorithm to provide the initial starting point for a backward Euler (bE) implicit stress integration process. This is to provide an initial estimate for the local positions within the Knot vectors, ξ and η in (3) that act as the primary unknowns in the implicit bE closest point projection (CPP) problem. However, despite this process being referred to as a CPP, the return stress is not generally the closest point geometrically in standard stress space.

In this paper we make use of energy-mapped space (Crouch et al. 2009) to convert this CPP minimisation into a problem of finding the point on the yield envelope that the normal to the plastic potential surface passes through when intersecting with a trial point outside of the surface. Once the closest point solution in energy-mapped stress space has been found, the solution can be transformed back to conventional stress space. For a NURBS yield surface we only need to map the control point coordinates for both the yield and plastic potential surfaces into energy-mapped space, the rest of the NURBS information remains unchanged.

As with the algorithm for associated flow perfect plasticity, see Coombs et al. (2016), the stress return path for bE procedure described in this paper starts and remains in the yield envelope and thereby satisfies the consistency conditions not only at the final state but also during the stress updating algorithm. This removes an issue associated with bE stress integration algorithms where they can become trapped in a local minimum, or converge to a spurious auxiliary surface, outside of the true yield surface. Introducing non-associated flow does not increase the number of unknowns in the stress integration algorithm; the key unknowns are the local coordinates on the NURBS surface, (ξ,η). The adopted implicit stress integration algorithm also allows the formulation of the algorithmic consistent tangent (Simo and Taylor 1985) which ensures optimum convergence of the global equilibrium equations when implemented within an implicit boundary value solver.

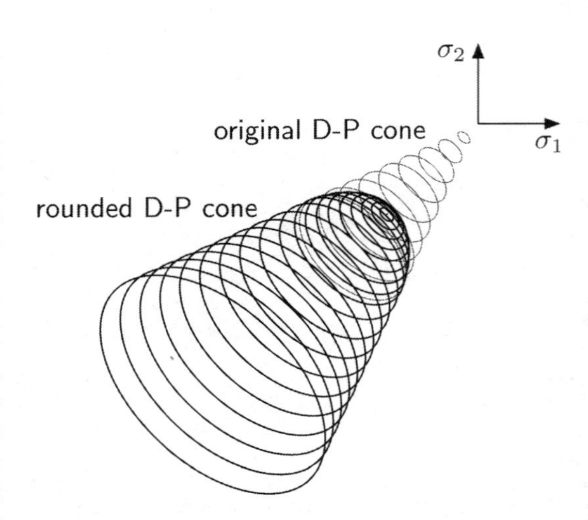

Figure 2. Drucker-Prager yield surface with (exaggerated) apex rounding (reproduced from Coombs and Ghaffari Motlagh (2017)).

4 NUMERICAL EXAMPLES

This section presents a series of numerical examples to demonstrate the performance of the non-associated flow NURBS plasticity formulation.

All of the analyses use a Drucker-Prager (D-P) yield surface (Drucker and Prager 1952) which can be expressed as

$$f = \rho + \beta(\zeta - \zeta_a) = 0, \tag{14}$$

where the deviatoric stress is $\rho = \sqrt{2J_2}$ with $J_2 = \frac{1}{2}\mathrm{tr}(s_{ij}s_{jk})$, $s_{ij} = \sigma_{ij} - \frac{1}{3}\sigma_{kk}\delta_{ij}$ and δ_{ij} the Kronecker delta tensor. $\zeta = \sigma_{ii}/\sqrt{3}$ is the hydrostatic stress, $\beta = \tan(\phi)$ is the opening angle of the cone, $\zeta_a = c\sqrt{3}\cot(\phi)$ is the location of the cone's tensile apex, φ is the friction angle and c the cohesion. The tensile apex of the yield surface poses an issue for the stress return algorithm presented in this paper as the derivatives of the NURBS surface are undefined at this point. Here we follow the same approach as Coombs and Ghaffari Motlagh (2017) and locally round the apex, as shown in Figure 2 with $\zeta_a = 0$. Both the original (fine lines) and rounded (thick lines) surfaces are shown in principal stress space. The plastic potential surface is similar to (14) but with β replaced with $\beta_g = \tan(\psi)$ where $\psi \in [0,\phi]$ is the dilation angle. For associated plastic flow $\psi = \phi$.

Within the framework of NURBS plasticity, the yield envelope is defined in a single sextant of stress space using a bi-quadratic NURBS surface with 15 control points (3 in the deviatoric direction by 5 in the hydrostatic direction). All of the examples presented in this section use a friction angle of $\phi = \pi/9$ (20° degrees).

4.1 *Material point investigations*

This section analyses the errors associated with the implicit stress integration algorithm. The material had a Young's modulus of 100 kPa and a Poisson's ratio of 0.2. The cohesion was set to 0.49 kPa and the final 0.1 kPa of the yield surface apex was rounded.[4] The errors associated with the stress return algorithm were evaluated using dilation angles of $\pi/18$ and $\pi/36$ (10° and 5° degrees).

The stress state was initially located on the shear meridian in one of the sextants of stress space with a hydrostatic stress of $\zeta = 0$ kPa. This point was then subjected to a stress increment that took the trial stress state outside of the yield envelope. The space of trial states explored was $\rho_t/\rho_n \in [1,6]$, where the t and n subscripts denote the trial and starting locations.

The errors associated with the trial state are shown in Figure 3, using the following normalised error measure

$$\text{error} = \frac{\| \{\sigma_{\text{NURBS}}\} - \{\sigma_e\} \|}{\| \{\sigma_e\} \|}, \tag{15}$$

4. Note that in all cases associated flow was imposed on the rounded part of the yield surface.

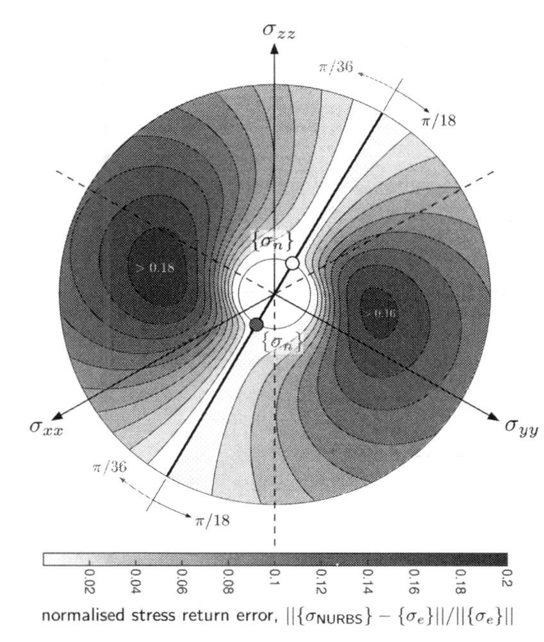

Figure 3. Stress return error analysis for D-P with non-associated flow with $\psi = \pi/18$ and $\psi = \pi/36$. The inner circle shows a deviatoric section through the yield surface and the white and red filled circles the starting points for the $\psi = \pi/18$ and $\psi = \pi/36$ error analyses, respectively.

where $\{\sigma_{\mathrm{NURBS}}\}$ is the stress return location associated with the NURBS model and $\{\sigma_e\}$ is the *exact* stress return.[5] The errors associated with $\psi = \pi/18$ and $\psi = \pi/36$ are shown on the right and left of the thick black line, respectively. The starting point for $\psi = \pi/18$ is shown by the white-shaded circle whereas the red-shaded circle is the starting point for the $\psi = \pi/36$ analysis.

Although errors of almost 20% are present in the model, exactly the same level of errors are observed in the D-P yield surface integrated with a conventional implicit stress integration procedure. As expected with any predictor-corrector stress integration algorithm, the error increases as the tangential proportion of the stress increment increases. Errors also increase with increasing non-associativity, with $\psi = \pi/18$ having a maximum error of 1.66×10^{-1} whereas for the $\psi = \pi/36$ the maximum error was 1.92×10^{-1}, again this is due to the return path having a larger tangential component relative to the yield surface normal direction.

5. The exact stress state was approximated by using a conventional implicit stress return algorithm for the D-P model with the stress increment applied in 1000 sub steps.

4.2 *Rigid footing*

The final example is that of a 1 m wide plane strain rigid smooth footing[6] displacing into a weightless 10 m by 5 m domain with a Young's modulus of $E = 1 \times 10^7$ kPa and a Poisson's ratio of $\nu = 0.48$. Yielding of the material was governed by a D-P yield envelope with cohesion of $c = 490$ kPa and a friction angle of $\theta = \pi/9$ (20° degrees).

The problem was analysed using a mesh comprising 135 eight-noded bi-quadratic quadrilateral elements integrated using reduced four-point quadrature. Due to symmetry only half of the problem was modelled. The mesh is the same as that used by de Souza Neto et al. (2008) (amongst

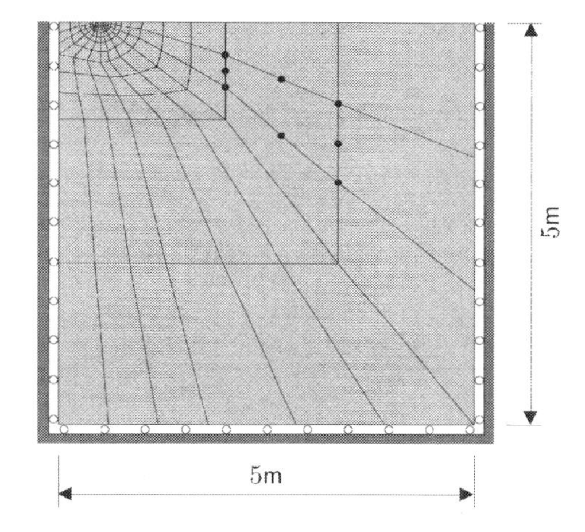

Figure 4. Rigid footing: finite element mesh. The mesh detail around the footing is shown in Figure 5.

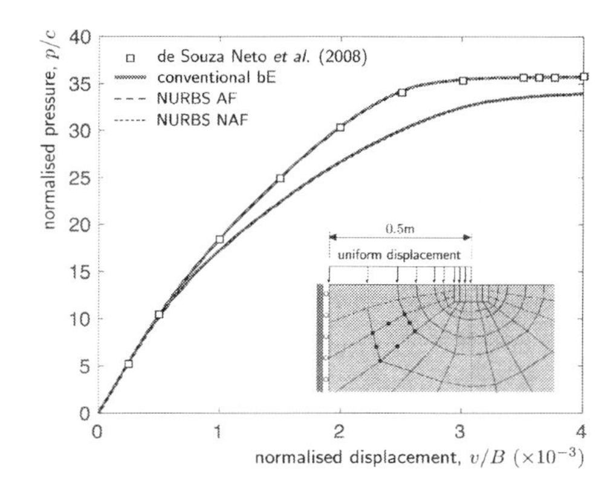

Figure 5. Rigid footing: normalised pressure versus displacement response where $B = 1$ m is the footing width.

6. The term *smooth* denotes that the nodes underneath the footing are free to displace in horizontal direction.

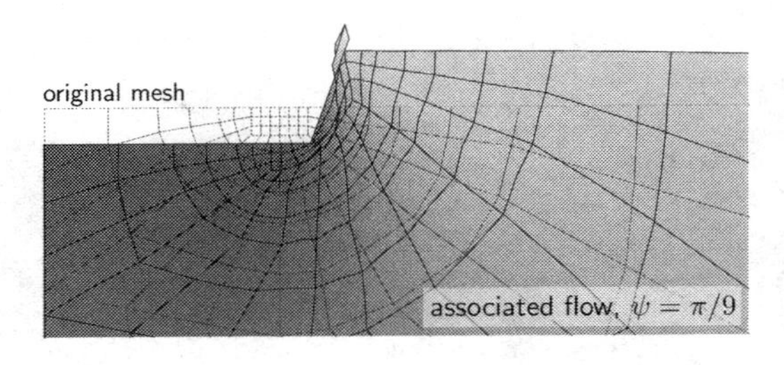

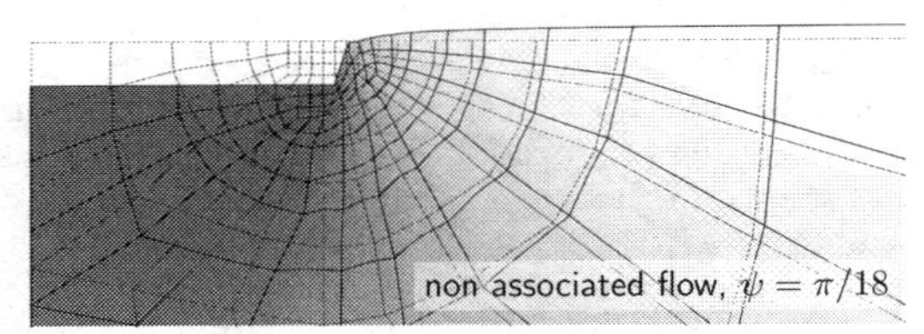

Figure 6. Rigid footing: deformed mesh for associated (top) and non associated flow (bottom) with × 20 displacement magnification.

others) and is shown in Figure 4. A vertical displacement of 4 mm was applied to the footing over 20 equal displacement-controlled loadsteps.

The normalised pressure versus displacement response is shown in Figure 5 along with an inset figure showing the mesh detail around the rigid footing. The associate flow (AF) NURBS plasticity response (long dashed line) is compared with the result of de Souza Neto et al. (2008) (discrete points) and that of a conventional backward Euler (bE) closest point projection implementation of the D-P yield surface (thick grey line). Excellent agreement is seen between the three results.

The non-associated flow (NAF) result with $\psi = \pi/18$ (10° degrees) is also presented in Figure 5 for both the conventional bE (thick grey line) and the NURBS (short dashed line) models. As with the associated flow results, excellent agreement is seen between the two models. Combined with the material point investigation presented in the previous section, this verifies the implementation of non-associated flow within the NURBS plasticity framework.

Figure 6 shows the deformed mesh around the footing for associated (top) and non-associated (bottom) plastic flow where the mesh has been shaded by the vertical displacement with dark grey being the maximum downwards displacement. The original mesh is shown by the fine dashed line and the displacements have been exaggerated by × 20. The associated flow case exhibits excessive volumetric dilation in the region of soil adjacent to the footing leading to unrealistic heaving of the ground surface. Reducing the dilation angle from $\pi/9$ to $\pi/18$ significantly reduces the heave leading to a more realistic deformed surface profile.

5 CONCLUSION

This paper has applied the NURBS plasticity framework to frictional plasticity models and extended the approach to include non-associated plastic flow. The approach was demonstrated at both a stress-strain and boundary value simulation level and the results validated against published data and conventional constitutive models. Extending the NURBS plasticity approach to include non-associated plastic flow allows the framework to model the frictional response of geomaterials in a more realistic way, specifically avoiding the excessive volumetric dilation observed in associated flow plasticity models.

REFERENCES

Clausen, J., L. Damkilde, & L. Andersen (2007). An efficient return algorithm for non-associated plasticity with linear yield criteria in principal stress space. *Computers & Structures 85*(23–24), 1795–1807.

Coombs, W.M. (2017). Hardening and non-associated flow NURBS plasticity. In E. Oñate, D. Owen, D. Peric, and M. Chiumenti (Eds.), *COMPLAS 2017: XIV International Conference on Computational Plasticity: Fundamentals and Applications*, Barcelona, Spain, pp. 363–372.

Coombs, W.M. & R.S. Crouch (2011a). Algorithmic issues for three-invariant hyperplastic critical state models. *Comput. Methods Appl. Mech. Engrg. 200*(25–28), 2297–2318.

Coombs, W.M. & R.S. Crouch (2011b). Non-associated Reuleaux plasticity: analytical stress integration and consistent tangent for finite deformation mechanics. *Comput. Methods Appl. Mech. Engrg. 200*(9–12), 1021–1037.

Coombs, W.M., R.S. Crouch, & C.E. Augarde (2010). Reuleaux plasticity: analytical backward Euler stress integration and consistent tangent. *Comput. Methods Appl. Mech. Engrg. 199*(25–28), 1733–1743.

Coombs, W.M. & Y. Ghaffari Motlagh (2017). NURBS plasticity: yield surface evolution and implicit stress integration for isotropic hardening. *Comput. Methods Appl. Mech. Engrg. 324*, 204–220.

Coombs, W.M., O.A. Petit, & Y. Ghaffari Motlagh (2016). NURBS plasticity: Yield surface representation and implicit stress integration for isotropic inelasticity. *Comput. Methods Appl. Mech. Engrg. 304*, 342–358.

Crouch, R.S., H. Askes, & T. Li (2009). Analytical CPP in energy-mapped stress space: application to a modified Drucker-Prager yield surface. *Comput. Methods Appl. Mech. Engrg. 198*(5–8), 853–859.

de Souza Neto, E.A., D. Perić, & D.R.J. Owen (2008). *Computational methods for plasticity: Theory and applications*. John Wiley & Sons Ltd.

Drucker, D. & W. Prager (1952). Soil mechanics and plastic analysis or limit design. *Quart. Appl. Math 10*(2), 157–164.

Simo, J.C. & R.L. Taylor (1985). Consistent tangent operators for rate-independent elastoplasticity. *Comput. Meth. Appl. Mech. Eng. 48*(1), 101–118.

Wilkins, M.L. (1964). Calculation of elastic-plastic flow. Technical report, California Univ Livermore Radiation Lab.

Numerical Methods in Geotechnical Engineering IX – Cardoso et al. (Eds)
© 2018 Taylor & Francis Group, London, ISBN 978-1-138-33198-3

Numerical simulation of the behavior of collapsible loess

R. Schiava
National University of Santiago del Estero, Argentina

ABSTRACT: The behavior of partially saturated soils is determined by the characteristics of their macro porous structure. The matrix suction is an additional fundamental variable in its response to deformation under variable moisture conditions. In the northwest of Argentina, there are deposits of loess of predominantly Aeolian origin and some of them have collapsible characteristics. The aim of the study is propose solutions to the geotechnical problems that affect these types of soils. This work shows the numerical implementation of a flow-mechanical coupled model. The extended MRS-Lade elastoplastic constitutive model was applied. Finally, numerical simulations were performed to reproduce the behavior of this unsaturated loess under conditions of infiltration and variable levels of load, after the calibration of the parameters of the model through experimental tests.

1 INTRODUCTION

1.1 *Collapsible loess*

The behavior of partially saturated soils is determined by the characteristics of its macro porous structure. In the northwestern of Argentina the loess deposits have a macro porous structure in which the grains of fine sand and silt are joined by bridges of clay and / or salts that give it rigidity and resistance with reduced moisture content. When the moisture content increases, the salts dissolve and the clays tend to expand, so destroying the links between the particles causing a rapid decrease of volume. This phenomenon is called "collapse" and implies an important volumetric strain. At the engineering works involving this type of soil, such as the roads or hydraulic constructions, this condition or the magnitude of the strain can be reduced by increasing the compactness of the soil. In such cases, partially saturated soils and their mechanical properties are strongly influenced by their initial moisture content and their variation over time due to changes in environmental conditions, the surface flow or infiltration by various causes. The application of constitutive models to simulate the unsaturated soil's behavior and to predict the stress strain response is very important. In this work an elastoplastic constitutive theory for unsaturated soils is presented. The proposed material model is formulated in the general framework of the porous media theory and of the flow theory of plasticity. The model is based on an extension of the well-known MRS-Lade model whereby the suction and the effective stress tensor are introduced as additional independent and dependent stress components, respectively.

The numerical implementation of a flow-mechanical coupled model is presented. For the resolution of the flow on the partially saturated porous medium is used the equation proposed by Richards (1992).

1.2 *Flow in porous media*

To represent the behavior of the partially saturated porous medium the equation proposed by Richards is used. This equation derived from the combination of the mass conservation equation or continuity equation and Darcy's law, assuming that the effects of the occluded air in the water and the compressibility of the solid matrix are negligible. The porosity of the soil is described based on the water retention curve and the hydraulic conductivity or permeability tensor. This relations can be expressed as functions of the suction, according to Sheng et al. (2003).

The continuity equation takes the form

$$\frac{\partial \theta(s)}{\partial t} = -\nabla \mathbf{q} + \mathrm{G}. \tag{1}$$

$$\mathbf{q} = -\mathbf{K}(\mathrm{s})\nabla p. \tag{2}$$

where, $\mathbf{q}$ = volumetric flow given by Darcy's law, G = the rate at which flow is generated or lost per volume unit in steady or permanent regime, s = suction matrix, t = time, $\theta(s)$ = volumetric water content, p = potential of water in pore of soil and ∇ the mathematical operator $(\partial/\partial x, \partial/\partial y)$.

Generally, the parameters of soil used for these functions are assumed to be invariant over time, following the Van Genuchten`s formulation.

The total pressure can be defined as the equivalent suction at the point considered relative to the free water pressure according to Gräsle et al. (1995) and Richards et al. (1995).

$$\frac{\partial\theta}{\partial s}\frac{\partial s}{\partial t} = \nabla[\ \mathbf{K}(s)\nabla p\ ] + G. \tag{3}$$

$$\frac{\partial\theta}{\partial s} = c(s) \tag{4}$$

where, $c(s)$ = the capillary capacity or slope of the suction-volumetric water content curve. This formulation uses the function proposed by van Genuchten and Hillel and which was used by Sheng et al. (2003).

1.3 *Elastoplastic model for unsaturated soil*

For numerical simulation of the stability of slopes, the extended MRS-Lade model for partially saturated soils are used, Schiava R. (2009). The main features are:

- Formulation of two surfaces of failure, a curved surface for the cone that intersects with another curved surface of cap in the meridian plane.
- Definition of the yield surface load-collapse (LC).
- The variable-hardening and softening of both surfaces are based on the work of plastic dissipation.
- Rule of non-associatively on the meridian plane and the associated in the deviatory plane in the cone`s region and associatively rule in the cap`s region.

The area of influence is shaped like an asymmetrical cone with the vertex located on the left of the origin of the space of stress, depending on the cohesive characteristics of the material involved and represented by the equation

$$F_{cone} = f\{q,\theta,s\} - \eta_{cone}(\kappa_{cone})(p + s - p_c) = 0 \tag{5}$$

$$f(q,\theta,s) = q\left(1 + {}^{q}\!/_{q_a}\right)^{m} g(\theta) \tag{6}$$

$$p = -\frac{I_1}{3} q = \sqrt{3J_{2D}}\cos 3\theta = \frac{3\sqrt{3}}{2}\frac{J_{3D}}{\sqrt{(J_{2D})^3}} \tag{7}$$

where I_1 is the first invariant of the net stress tensor, J_{2D} and J_{3D} the second and third invariants of the deviatoric stress tensor, q_a, m, η_{cone} are model parameters and κ the hardening variable, calculated based on the accumulated plastic work. The cohesion of the material p_c varies with the suction according to the equation

$$p_c = r_{pc}\ s \tag{8}$$

with r_{pc} material parameter.

The suction $s = p_a - p_w$ is defined as the difference between air pressure and water pressure in the voids of the soil. The cap surface, which involves the volumetric response, is given by an elliptic surface in the meridian plane in terms of stress invariants

$$F_{cap}(p,q,\theta,s,\kappa_{cap}) = \left(\frac{p - p_m}{p_r}\right)^2 + \left(\frac{f}{f_r}\right)^2 - 1 = 0 \tag{9}$$

The parameters of hardening and softening κ are defined in terms of accumulated plastic work

$$\begin{aligned}\dot{\kappa}_{cone} &= \frac{1}{c_{cone}\ p_a}\left(\frac{p + s - p_c}{p_a}\right)^{-l} \dot{w}^p\,\dot{\kappa}_{cap} \\ &= \frac{1}{c_{cap}\ p_a}\left(\frac{p_{cap,0}}{p_a}\right)^{-r} \dot{w}^p\end{aligned} \tag{10}$$

where: $p_{cap,0}$ is the pre consolidate pressure depending on the value of the suction, which is defined as the expression of Schrefler (Schrefler et al 1997), and used as an additional failure surface known as "load-collapse"

$$p_{cap,0} = p_0^* + i\ s \tag{11}$$

p_0^* = the pre-consolidation pressure for saturated condition and i a constant of material. The parameters of the model extended Lade-MRS used for the simulation are shown in Table 1.

Table 1. Parameters of extended MRS-Lade model.

Parameter	Layer 1	Layer 2
E (kPa)	3000	9000
q_a (kPa)	1.00	1.00
Poisson ratio	0.25	0.25
m	0.0387	0.0387
η_{cone}	1.64	1.87
r_{pc}	–0.10	-0.55
l	1.065	1.065
r	1.102	1.102
p_0	10	320
i	1.00	1.50
c	0.10	0.12
γ(kN/m3)	13,1	14.7

The model proposed was implemented in FEAP, finite element program of Taylor (2000). For further details see Schiava & Etse (2014)

2 SOIL CHARACTERISTIC

2.1 *Type of soil*

The soil is silty clay of unified classification (SUCS = CL-ML). They have a wind origin and collapsible characteristics when increasing its humidity to critical values.

The dry density of soil layer is of 13.10 kN/m³. The average parameters of soil layer are summarized in Table 2.

2.2 *Simulation of material behavior*

In order to determine the stress strain behavior of the soils, a confined compression tests were carried out with different moisture contents. The moisture content was correlated with the suction value by means of the characteristic curves defined by Xie et al. (1995) for soils of identical characteristics and adjusted with own tests.

To reproduce the behavior of the unaltered soil obtained in this experimental test the parameters of the model were calibrated and detailed in Table 1. The results obtained can be seen in Figure 1.

Table 2. Soil parameters.

SUCS	% 200	L.L.	I.P.	w%
ML-CL	78	22.60	6.50	10.20

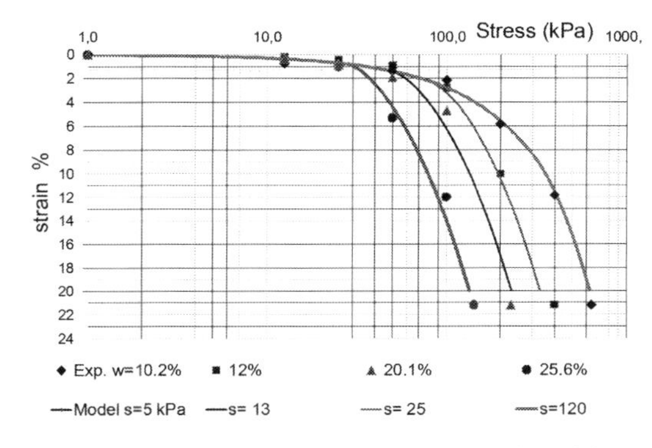

Figure 1. Oedometric test: experimental and model.

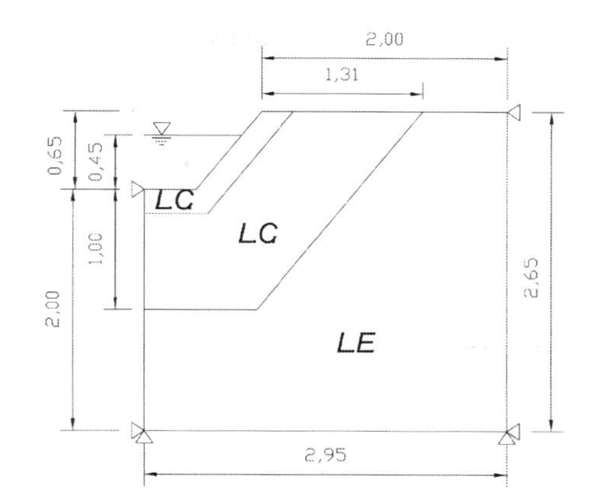

Figure 2. Case of study.

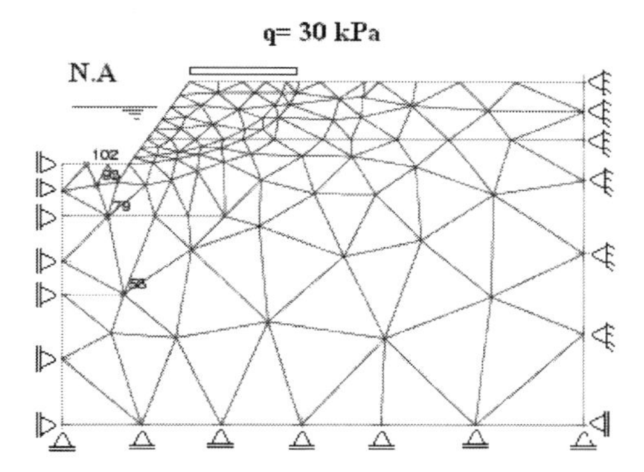

Figure 3. Element finite mesh.

3 NUMERICAL SIMULATION

3.1 *Description*

The developed model is used to reproduce and analyze the behavior of drains of a road implanted in the field of collapsible loess.

The process of filtration through the soil with the distribution of equivalent pore pressures, strains and state of vertical stresses is performed using the partially coupled flow-mechanical analysis described above, Lu et al. (2012)

The profile is composed of two types of material, one corresponding to the compacted soil (L.C.) and the second one to the original soil (L.E.), see Figure 2. For all the cases analyzed, a water load of 0.45 m was considered the maximum in the design and at the upper edge a uniform pressure of q = 30 kPa was applied in representation of the lateral transit load. The conceptual scheme analyzed, the element finite mesh with imposed boundary conditions, is shown in Figure 3. Four cases were considered for study according to the thickness of the treatment with compacted soil, to evaluate the response in each of them. Sc. 1 corresponds to the soil in natural condition, that is unchanged and the other cases 2, 3 and 4, by treatment with compacted soil at 98% of the compaction test with a variable thickness of 0.20 m, 0.40 m and 1.00 m, respectively. The initial suction (kPa) condition is shown in Figure 4.

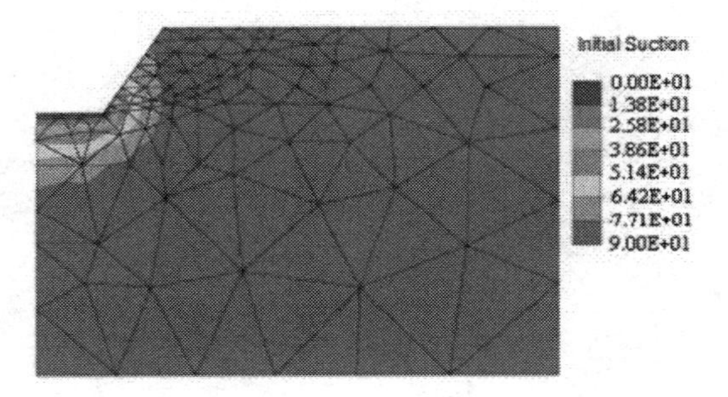

Figure 4. Initial suction.

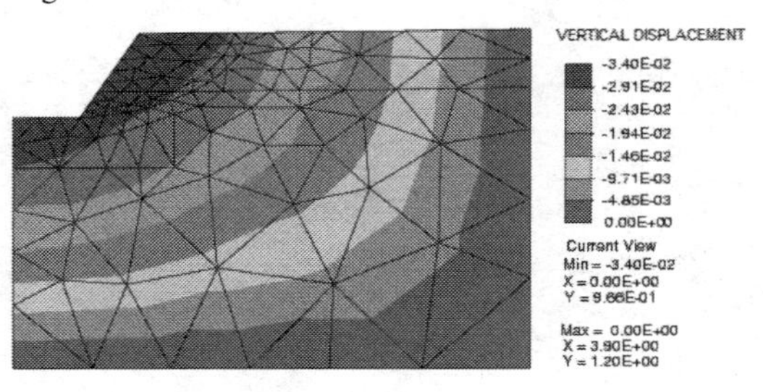

Figure 5. Vertical displacement for Sc 1.

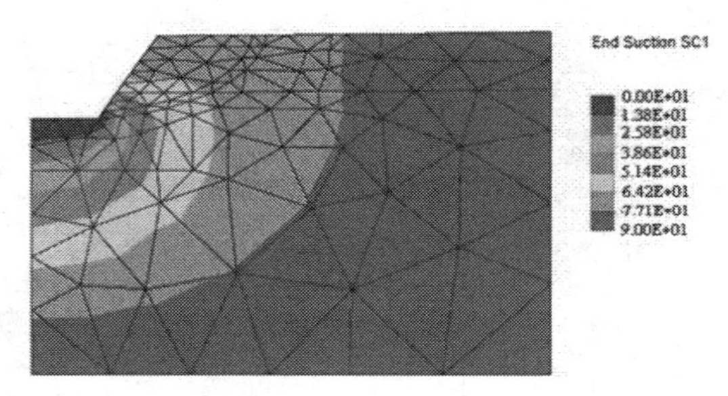

Figure 6. End suction for Sc 1.

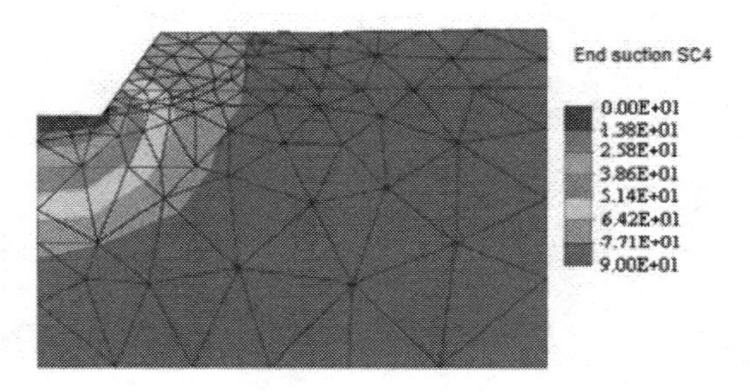

Figure 7. End suction for Sc 4.

4 RESULTS

The values of the soil's deformations for the cases analyzed in different points of interest of the model were determined. Figure 5 shows the vertical displacement, and the evolution to the steady state. The final suction state, for Sc. 1 and Sc. 4, are shown in Figure 6 and 7, respectively.

The displacements of the nodes 102 (bottom) and the node 108 (upper edge of the node), for Sc. 1 and Sc. 4 are shown in Figure 8 as a function of the time expressed in days. Figure 9 shows the displacement of the soil in depth below of bottom as displacements of nodes 102, 93, 79 and 58. Shows the effect of the thickness of the treatment with compacted soil.

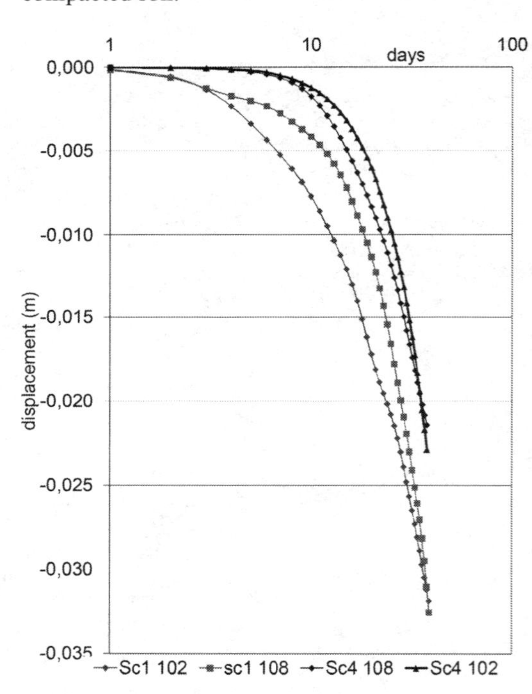

Figure 8. Nodal displacement.

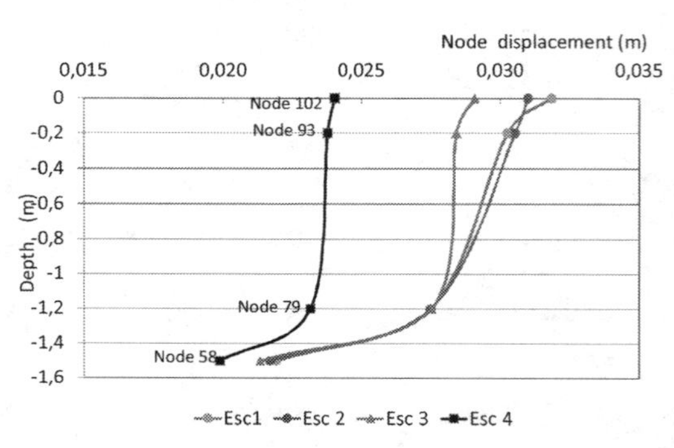

Figure 9. Deep displacement.

5 CONCLUSIONS

The numerical simulations obtain the evolution of the infiltration over time, with the resulting suction values and the soil's deformation. The model allows the study of the behavior of clay silty with collapsible characteristics and the analysis of the thicknesses of treatment with compacted soil to control the infiltration and reduce the settlements.

The influence of the lateral overload due to the traffic and the comparison with deformation due to the weight of the soil and the water load was also determined. It is concluded that in order to achieve significant results in response change, to limit infiltration and to reduce large volumetric strain, the unaltered loess must be replaced with compacted soil in a thickness of not less than one meter.

REFERENCES

Gräsle W., Baumgardtl T., Horn R. & Richards B.G. 1995. Interaction between soil mechanical properties of structured soils and hydraulic process. Theorical fundamentals of a model. Unsaturated Soils, 2, 719–724. Balkema,.

Lu N, Kaya B.S, Wayllace A., Godt J.W. 2012. Analysis of Rainfall-induced slope instability using a field of local factor of safety, *Water Resources Research,* **48**,(9), 1–14.

Richards, B.G. 1992. Modelling interactive load deformation and flow procesess in soils, including unsaturated and swelling soils. Proceedings of the 6th Asut-NZ Conference Geomechanics, 18–37. New Zealand.

Schiava, R. 2009. Modelation and computational analysis of unsaturated soils. Doctoral thesis. National University of Tucuman. Argentina

Schiava, R. & Etse, G. 2014. Analysis of Failure of Unsaturated Solis. Journal of Geological Resource and Engineering. David Publishing Company, New York, V. 2.2. 99–106.

Schrefler B. & Bolzon G. 1997. Compaction in gas reservoirs due to capillary effects. Computational Plasticity, Section Plasticity approaches geotechnics, CIMNE, 1625–1630,

Sheng D., Smith D., Sloan S. & Gens A. 2003. Finite Element Formulation and algorithms for unsaturated soils. Part I: Theory. *International Journal for Numerical and Analytical Methods in Geomechanics.* 27(9), 745–765.

Taylor R.L. 2000. *A Finite Element Analysis Program-FEAP,* Theory Manual, University of California at Berkeley.

Xie D.Y., Liu F.Y., Han X.L. & Wu L.Y. 1995. A new type of triaxial apparatus for unsaturated soils. Unsaturated Soils, 3, 1551–1558. Balkema,

Zur, A. & Wiseman, G. 1973. A study of collapse phenomena of an undisturbed loess. *8 International Conference on soil Mechanics and Foundation Engineering*, 2.2, 256–268.

Numerical Methods in Geotechnical Engineering IX – Cardoso et al. (Eds)
© 2018 Taylor & Francis Group, London, ISBN 978-1-138-33198-3

The influence of non-coaxial plasticity in numerical modelling of soil-pipe interaction

H.E. Mallikarachchi
Department of Engineering, University of Cambridge, UK

K. Soga
Department of Civil and Environmental Engineering, University of California, Berkeley, USA

ABSTRACT: Practical boundary value problems with large deformations often involve stress paths which deviate from proportional loading. This leads to plastic instabilities inducing shear bands or diffusive modes of failure. Traditional coaxial plasticity theories are not sufficient to predict the inelastic stretching caused by principal axis rotations which occur during the formation of shear bands. This paper highlights the influence of the non-coaxial plasticity on granular soil followed by its influence on the pipe-soil interaction under large deformations. The tangential plasticity concept proposed by Hashiguchi (1993) is integrated into the three invariant Nor-Sand constitutive model which is enriched with critical state and state parameter theories. This model is numerically implemented into the finite element software ABAQUS and is validated for simple shear simulations of drained sand.

The peak mobilised force and the corresponding displacement are main design parameters of the pipe-soil interaction problems. The predictions of these design parameters for the upheaval buckling of buried pipelines using the non-coaxial model are compared with those of conventional coaxial models. The effect of soil friction and dilation along with the pipe depth to diameter ratio are discussed. It is observed that the non-coaxial model predicts a softer response leading to higher mobilised displacements at the peak uplift force. Ultimate pulling capacities of the non-coaxial model are close to those of the coaxial model. The discrepancy between coaxial and non-coaxial models increases with the greater rotation of principal stress axis inside the shear bands formed around the pipe. This study sheds light on the fundamental understanding of the phenomenon of principle axis rotation during strain localisation and its effect on the soil-pipe interaction.

Keywords: Non-coaxial, simple shear, soil-pipe interaction, principle axis rotation, large deformations

1 INTRODUCTION

1.1 *Principal axis rotation and non-coaxial plasticity*

Soil constitutive models play a prominent role in finite element predictions of geotechnical structures. The accuracy of these material models dictate the precision of the finite element predictions. Conventional constitutive models assume that orientation of principal stresses coincide with that of principal plastic strain rate which is termed as coaxility. Although this assumption is valid for proportional loading paths such as triaxial and biaxial tests, practical geotechnical problems often deviates from this idealistic behaviour. Hence lab measured soil properties may not be appropriate for the design of soil structure interaction.

The progressive failure of buried pipelines often accompanies with diffused or localised modes of instabilities. Both experimental and numerical studies indicate that multiple shear bands tend to propagate from the shoulder of pipelines during upheaval or lateral movements (Robert and Thusyanthan 2014,Wang 2012, Cheuk et al. 2008). A considerable amount of principal axis rotations take place within these shear band under large shear deformations. As a result the orientation of principal stresses does not coincide with that of principal plastic strain rates as shown in Figure 1. In essence, conventional flow theories are proved to be not valid under principal stress rotations.

This phenomenon is verified by various laboratory experiments such as stress probe tests (Gutierrez and Ishihara 2000, Gutierrez et al. 1991, Gutierrez et al. 1993), simple shear tests (Roscoe 1970), hollow cylinder tests (Miura et al. 1986, Cai 2010) etc. These laboratory experiments report that magnitude and direction of plastic deforma-

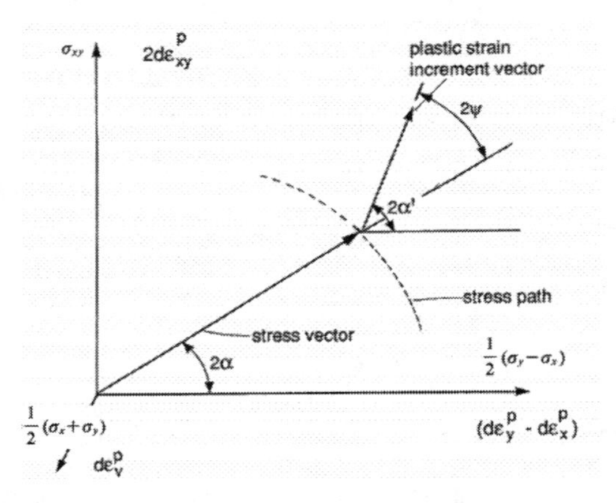

Figure 1. Representation of stress path and plastic strain increment path in deviatoric plane, principal stress direction = α, principal plastic strain increment direction = α', angle of non-coaxility = ψ, (Gutierrez and Ishihara 2000).

tion is dependent on the direction of stress rates. The principal stress direction lags behind principal plastic strain increment direction during shear deformations or non proportional loading. The degree of non-coaxility is greater in the beginning, but principal stress and principal plastic strain rate directions coincide near the critical state.

Theoretical studies on micro-mechanical point of view of granular materials also support the concept of non-coaxility (Oda & Konishi 1974a, Oda & Konishi 1974b). Based on the kinematics of inter-granular contact and friction angles, DEM simulations also have reported evidence on the non-coaxial plastic flow (Yu 2008).

Within the context of continuum plasticity, various theories have been put forward to describe the principal axis rotation and non-coaxility of granular material. The yield vertex theory initially introduced by Rudnicki and Rice (1975) has been applied by many researchers to successfully model the non-coaxial behaviour. This theory recognises the plastic stretching caused by stress rates tangential to the yield surface in the deviatoric plane. Over the past years, various groups of researchers implemented yield vertex theory to reproduce simple shear and hallow cylinder tests (Yu and Yuan 2006, Yang and Yu 2006a, Yang and Yu 2006b, Yang and Yu 2010, Yang et al. 2011, Hashiguchi et al. 1998, Tsutsumi and Hashiguchi 2005). Another way to integrate the effects of non-coaxility into plastic deformation is modifying the expression for energy dissipation. Gutierrez and Wang (2009) derived a novel stress-dilatancy correlation including the non-coaxial angle, under plane strain conditions. The double shearing theory of Spencer (1982) has also been combined with conventional plastic theory to replicate the non-coaxial behaviour by Yu and Yuan (2006). Recent numerical investigations by Zhao and Gao (2015),Gao and Zhao (2013) and Gao and Zhao (2017) declare that accounting soil fabric and its evolution leads to capture non-coaxial behaviour of soil naturally. However, Tsutsumi and Hashiguchi (2005) points out that anisotropic plastic potential itself is not sufficient to predict the observed non-coaxial behaviour.

Rotation of principal axes in the continuum is a result of soil grains establishing a compatible internal structure with regards to particle contacts and orientation under external loading. In author's point of view, all aforementioned theories are different mathematical artefacts to replicate the microstructure of granular particles by a continuum.

Although abundant literature is available on numerical validation of non-coaxial theories with laboratory experiments, only handful of investigators have used non-coaxial material models to study soil-structure interaction problems. Yang and Yu (2006b),Yang and Yu (2010) and Yang et al. (2011) utilised the Drucker-Prager model with a yield vertex based non-coaxial component to analyse shallow foundations. They found out that ultimate bearing capacities are not significantly affected by the non-coaxial theory although settlement of the footings at peak load are increased considerably. Analogous results were observed by Yang and Yu (2010) using the same model on the uplift resistance of horizontal and vertical anchor plates buried at various embedments. Yuan (2015) investigated the effect of non coaxility on tunnel subsurface displacements using an anisotropic Mohr-Coulomb model. Both vertical settlement and horizontal displacement were increased when large magnitudes of anisotropic and non-coaxial coefficients are used.

This study amalgamates yield vertex theory with a critical state soil constitutive model to predict the load displacement response during the upheaval buckling of buried pipelines. The influence of non-coaxility on main design parameters such as peak uplift force and the corresponding displacement are scrutinised.

2 FINITE ELEMENT IMPLEMENTATION

In conventional elasto-plastic theory, total strain rate is composed of plastic and elastic components (Equation 1*a*). The traditional plastic flow theory assumes that plastic flow depends only on the current state of stress; the direction of the plastic flow is normal to the plastic potential function as given in equation $1b$. This assumption fails to describe the phenomenon observed in stress probes, hallow cylinder or simple shear tests.

$$d\varepsilon_{ij} = d\varepsilon^{e}_{ij} + d\varepsilon^{p}_{ij} \quad (1a)$$

$$d\varepsilon^{p}_{ij} = d\lambda \frac{\partial g}{\partial \sigma_{ij}} \quad (1b)$$

Yield vertex theory modifies the plastic strain rate in Equation $1a$ to include coaxial $d\varepsilon^{pc}_{ij}$ and non-coaxial $d\varepsilon^{pn}_{ij}$ parts.

$$d\varepsilon_{ij} = d\varepsilon^{e}_{ij} + d\varepsilon^{pc}_{ij} + d\varepsilon^{pn}_{ij} \quad (2)$$

Coaxial $d\varepsilon^{pc}_{ij}$ part can be determined by conventional plastic flow theory in Equation $1b$ whereas non-coaxial non-coaxial plastic flow rate $d\varepsilon^{pn}_{ij}$ is mathematically determined by equation $3a$ and $3b$ in which h_{nc} is the non-coaxial plastic modulus and n_{ij} represent the vector of plastic flow direction. N_{ijkl} in Equation $3b$ denotes the flow direction matrix.

$$d\varepsilon^{pn}_{ij} = \frac{1}{h_{nc}} n_{ij} \quad (3a)$$

$$d\varepsilon^{pn}_{ij} = \frac{1}{h_{nc}} N_{ijkl} d\sigma_{ij} = \frac{2G}{h_{nc} + 2G} N_{ijkl} d\varepsilon_{ij} \quad (3b)$$

Original yield vertex theory by Rudnicki & Rice (1975) suggests that non coaxial part n_{ij} is normal to the unit vector along the deviatoric stress (Equation 4 and Figure 2). This has been adopted by Yu & Yuan (2006),Yang and Yu (2006a),Yang and Yu (2006b),Yang and Yu (2010) and Yang et al. (2011) for simple constitutive models with circular yield surfaces such as Drucker-Prager and Mohr-Coulomb with only isotropic hardening.

$$n_{ij} = ds_{ij} - \frac{s_{ij}s_{kl}}{2\tau^2} ds_{kl} \quad (4)$$

where $s_{ij} = \sigma_{ij} - \frac{\sigma_{kk}}{3}$ is the deviatoric stress tensor and $\tau = \sqrt{0.5 s_{ij} s_{ij}}$.

In this paper, a generalised yield vertex theory (Equation 4 and Figure 3) developed by Hashiguchi (1993),Hashiguchi (1998),Hashiguchi et al. (1998),Tsutsumi and Hashiguchi (2005) is utilised. This is valid for non-circular yield surfaces with any type of hardening. Equation 5 coincides with the original yield vertex theory when the yield surface is circular in the deviatoric plane and only isotropic hardening is adopted.

$$n_{ij} = ds_{ij} - \bar{n}^{*}_{ij}\bar{n}^{*}_{kl} ds_{kl} \quad (5a)$$

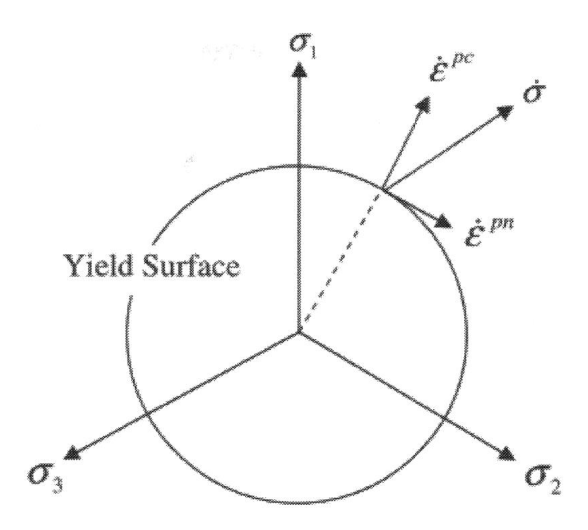

Figure 2. Illustration of coaxial and non-coaxial plastic strain rate for a circular yield surface assuming associative plasticity (Yang et al. 2011)

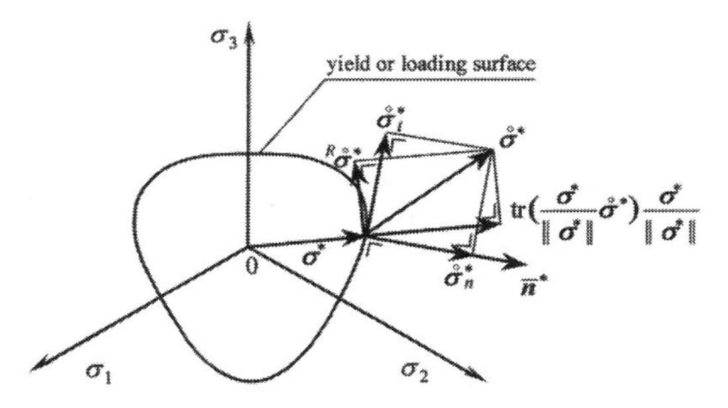

Figuere 3. Illustration of coaxial and non-coaxial plastic strain rate for non-circular yield surface assuming associative plasticity (Hashiguchi and Tsutsumi 2001).

$$\bar{n}^{*}_{ij} = \frac{\left(\frac{\partial f}{\partial \sigma_{ij}}\right)^{*}}{\left\| \left(\frac{\partial f}{\partial \sigma_{ij}}\right)^{*} \right\|} \quad (5b)$$

$$\left(\frac{\partial f}{\partial \sigma_{ij}}\right)^{*} = \left(\frac{\partial f}{\partial \sigma_{ij}}\right) - \left(\frac{\partial f}{\partial \sigma}\right)_{ii} \quad (5c)$$

$$N_{ijkl} = \frac{1}{2}\left(\delta_{ik}\delta_{jl} + \delta_{il}\delta_{jk} - \frac{2}{3}\delta_{ij}\delta_{kl} - \bar{n}^{*}_{ij}\bar{n}^{*}_{kl} \right) \quad (6)$$

The final constitutive relationship is given by,

$$d\sigma_{ij} = D^{ep} d\varepsilon_{ij} \quad (7a)$$

$$D^{ep} = \left[D^{e} - \frac{D^{e}\left(\frac{\partial f}{\partial \sigma_{ij}}\right) D^{e}\left(\frac{\partial g}{\partial \sigma_{ij}}\right)}{K_p + \left(\frac{\partial f}{\partial \sigma_{ij}}\right) D^{e}\left(\frac{\partial g}{\partial \sigma_{ij}}\right)} - \frac{4G^2}{h_{nc} + 2G} N_{ijkl} \right] \quad (7b)$$

where D^e is elastic stiffness matrix. Initial two terms indicates the coaxial effect and last term deflects the non-coaxial effect. Smaller non-coaxial modulus leads to higher non-coaxial influence.

Aforementioned yield vertex theory can be utilised with any constitutive model. In this paper Nor-Sand model by Jefferies (1993) based on fundamental axioms of critical state theory has been used. It is an elasto-plastic model which incooperates associative flow rule with a bounding surface concept. The state parameter concept integrates the effect of both void ratio and the mean pressure. Nor-Sand yield surface is given by,

$$F = \frac{q}{p} - \frac{M_\theta}{N}\left[1 + (N-1)\left(\frac{p}{p_i}\right)^{\frac{N}{N-1}}\right] \tag{8}$$

where p, q and M_{teta} are three stress invariants namely mean pressure, deviatoric stress and critical stress ratio which is a function of lode angle θ and calculated by Matsuoka-Nakai failure criteria Cheong (2006). N is the dilatancy parameter. The size of the bounding yield surface is determined through the image pressure p_i. Nor-Sand assumes an isotropic hardening rule driven solely by the plastic deviatoric strain increment. In line with Hashiguchi et al. (1998)'s concept of sub-loading surface, current Nor-Sand model is elasto-plastic from the beginning. Further information about the implementation of the Nor-Sand model can be found in Cheong (2006). Constitutive formulations are implemented in ABAQUS finite element software as a user defined material model.

Following sections presents three types of numerical simulations to depict the influence on non-coaxility in granular soil.

3 NUMERICAL RESULTS

3.1 *Simulations of simple shear*

Aforementioned model is validated with direct simple shear experimental results of dry Leighton-Buzzard sand Cole (1968). Dense, medium dense and loose sand had initial void ratios of 0.53, 0.64 and 0.75 respectively. Tests were conducted at confining pressure between $40 - 48kPa$.

To display the non-coaxial influence, a plane strain four node element with reduce integration is subjected to simple shear which causes the rotation of principle axes. The numerical model consisted of two steps; the application of confining pressure and the shear loading. Throughout the shearing, both pairs of horizontal and vertical sides are maintained linear and parallel. Horizontal linear strain is constrained be zero and volumetric strain is caused only by vertical strain. Table 1 lists the model parameters calibrated for Leighton-Buzzard sand. Initial lateral earth pressure coefficient is assumed to be 0.4. Numerical vs experimental comparison of stress strain and volumetric relationships of are shown in Figure 4 and 5 respectively.

It is observed that non-coaxial model closely approximates the experimental behaviour of dry Leighton-Buzzard sand. The difference between coaxial and non-coaxial model is well displayed in stress strain relationship, even though it is barely visible in volumetric relationships. Non-coaxiality reduces the initial stiffness and softens the stress strain relationship. The effect of non-coaxility is greater for smaller void ratios. Compared to previously implemented non-coaxial models by Yuan (2006),Yang and Yu (2006a), Yang and Yu (2006b),Yang and Yu (2010) and Yang et al. (2011) in-which the influence of non-coaxility is only visible in the plastic regime, the current Nor-Sand model incorporate the non-coaxility from the beginning.

As observed in Figure 4, numerical simulations of dense and medium dense match test data more closely where as those on loose sample over predict the peak stress ratio. However, it should be mentioned that "loose" sample in Cole (1968) also shows a dilative volumetric behaviour which inherits it a denser response. This is because the initial state parameter is negative for dilative sand. Practically it is difficult to maintain a constant maximum and minimum void ratio between different batches even for the same sand. Hence the calibrated maximum void ratio is changed based on the initial density. When sand reaches the critical state, the disparity between coaxial and

Table 1. Material parameters of Nor-Sand.

M_{tc}	e_{max}	e_{max}	N	Hardening modulus	A	n	v	X_{tc}	Non-coaxial modulus
1.27	$0.5e_0 + 0.48$	5.1	0.3	400	1500	0.5	0.3	3.5	$0.7G$

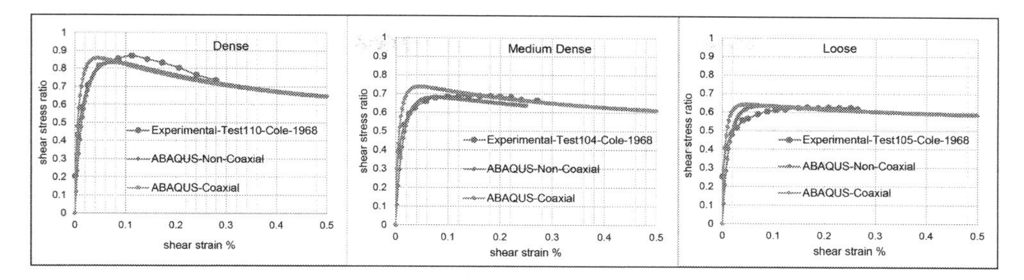

Figure 4. Comparison of simple shear simulations with experimental data-Stress strain relationships.

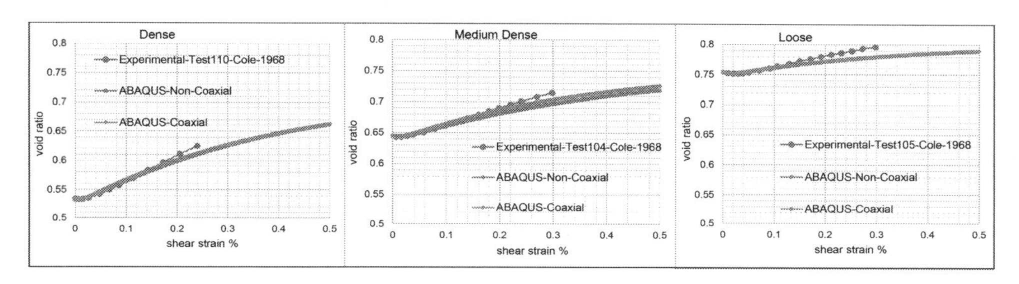

Figure 5. Comparison of simple shear simulations with experimental data-Volumetric relationships.

non-coaxial predictions disappears. Theoretically if maximum void ratio and confining pressure are similar for three samples, same critical void ratio should be achieved in Figure 5 and shear stress ratio must reach a same ultimate value in Figure 4.

Rotation angles of principle stress and plastic strain increment directions for dense sand are shown in Figure 6. Confirming the experimental evidence, the orientation of principal plastic strain rate is ahead of the direction of pricipal stress during early stage of loading and coincide at the critical state.

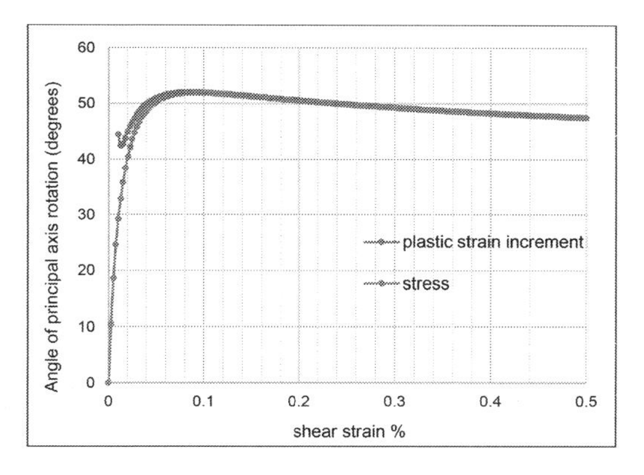

Figure 6. Rotation angles of principal stress and plastic strain increments-dense sand.

3.2 *Simulations of biaxial compression*

In this section, coaxial vs non-coaxial simulations are compared for a plain strain biaxial compression test. This simulation depicts the influence of non-coaxility during formation of a shear band. In contrast to simple shear simulations in the previous section, here the principal axis rotation is confined only to the material points inside the shear band.

Material parameters of dense Leighton-Buzzard sand are used for this simulation. Initial lateral earth pressure co-efficient is assumed to be 1 in this case. Plane strain bi-quadratic (8-node) elements with reduced integration are utilised in the analysis. A soil sample with aspect ratio of 2 is pinned at the bottom left node and other bottom nodes are roller supported to allow the formation of shear band. After the application of confining pressure, top nodes are vertically displaced downwards. Figure 7 depicts the comparison of coaxial and non-coaxial predictions of total reaction force vs displacement.

The shear band is responsible for a drastic reduction in force after the peak in the force dis-

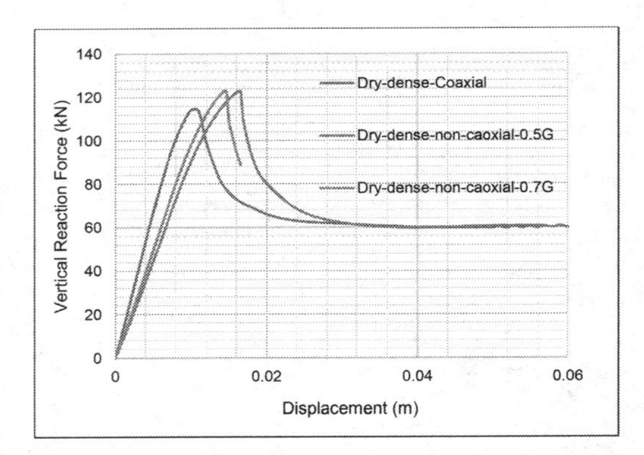

Figure 7. Force displacement response of biaxial compression test-dense sand.

placement relationship. Evidently the non-coaxial model softens the response. Smaller the non-coaxial modulus, the initial stiffness gets lower. The peak force as well as the displacement to reach the peak are enhanced by the non-coaxility. However inclusion of tangential or non-coaxial plastic strain rate cause numerical problems and premature failure as discussed by Yang and Yu (2010) and Yang and Yu (2006b). This is mainly attributed to total plastic strain rate not being normal to the yield surface even with an associative flow rule.

3.3 *Simulations of soil-pipe interaction*

A pipeline buried in sand backfill is simulated in plane strain condition. The pipe is 600*mm* in diameter and assumed to be rigid elastic. Material properties of both dense (unit weight $17.7kNm^3$) and medium dense (unit weight $16.4kNm^3$) Leighton-Buzzard sand are used for the analysis. Deep and shallow embedments with depth to diameter ratios of $H/D = 2,4$ are considered. The lateral coefficient of earth pressure is assumed to be 0.6. Eight node bi-quadratic elements with reduced integration are utilised for the analysis. The boundaries are sufficiently large to eliminate their effect on the results.

The interaction between the pipe and soil is treated by a GAP contact method, which allows the soil to separate from the pipe. Normal (hard contact) and tangential interaction properties are specified for the interface with a friction coefficient of 0.3. A surface to surface contact algorithm was prescribed, naming more rigid pipe as the master surface and the soil boundary as the slave surface.

The analysis consists of three steps, the initial stress condition, geostatic stress state and loading step. All boundary and initial conditions are applied at the initial state where as gravity load is applied at the geostatic state. Vertical upward displacement of $0.1m$ is applied to the pipe at the loading stage. The updated Lagrangian algorithm is utilised to take account of the geometrical non-linearities during large deformations.

Predicted load displacement curves for two densities at different depths are shown in Figures 8 and 9. Generally, deep burial depths and denser backfills contribute for larger reaction forces. For both soil densities, non-coaxiality has influenced theload displacement curves of deep embedment scenario. As shown in Figure 8(*a*) and 9(*a*), the non-coaxial model predict a reduced stiffness showing a softer response. Opposed to biaxial simulations, the peak load exerted on the pipe is not significantly affected, although the displacement to reach the peak is delayed by the non-coaxility. Both predictions reach the same ultimate value at the latter stage of shearing. For shallow depths, non-coaxility has not significantly influenced the force exerted onthe pipe, except for a slight softening. This is attributed to the fact that extent of non-coaxility depends on the magnitude of shear deformation. Figure 10 illustrates the shear strain contours at a displacement of $0.056m$. When a buried pipeline is subjected to the upheaval buckling, two shear bands tends to generate from the shoulder of pipe and extends towards the ground surface. Considerable amount of shear deformation concentrates within these shear bands which leads to principal axis rotation. Figure 11 compares the direction of minor principal stress (in plane) for both shallow and deep embedments for the dense sand case. Evidently in deep embedments a larger area of soil mass is subjected to principal stress rotations than in shallow depths. Hence the non-coaxial effect is greater. Moreover, pipes closer to the ground surface quickly approach failure after start of loading.Thus the plastic deformation required to reach the ultimate bearing capacities are comparatively smaller. These findings are in agreement with non-coaxial simulations of shallow foundations (Yang and Yu 2006b) and buried anchors (Yang and Yu 2010) using Drucker-Prager model.

Also when comparing Figures 8 with 9, it is evident that the denser sand case creates a larger non-coaxial response. This is because dense sand is susceptible to more localised shear deformations. Also since sand is stronger larger plastic strain is required to reach ultimate bearing capacities which encourage the non-coaxial influence. This observation however contradicts Yang and Yu (2010)'s claim that larger friction angles lead to lower non-coaxial response. This is based on the fact that large friction delays the yielding and most of principal axis rotations take place in elastic regime which

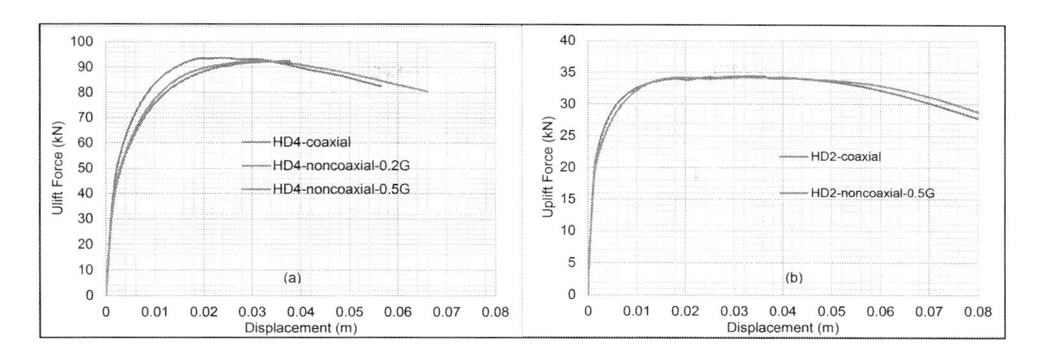

Figure 8. Force displacement relationship for dense sand $(a) H/D = 4 \, and \, (b) H/D = 2$.

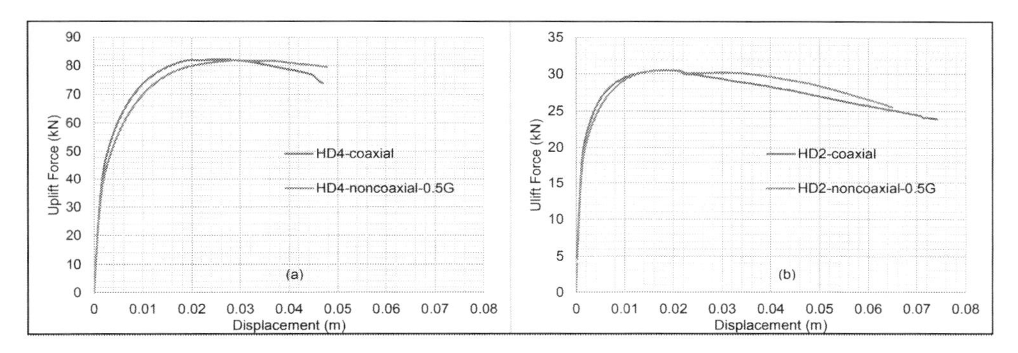

Figure 9. Force displacement relationship for medium dense sand $(a) H/D = 4 \, and \, (b) H/D = 2$.

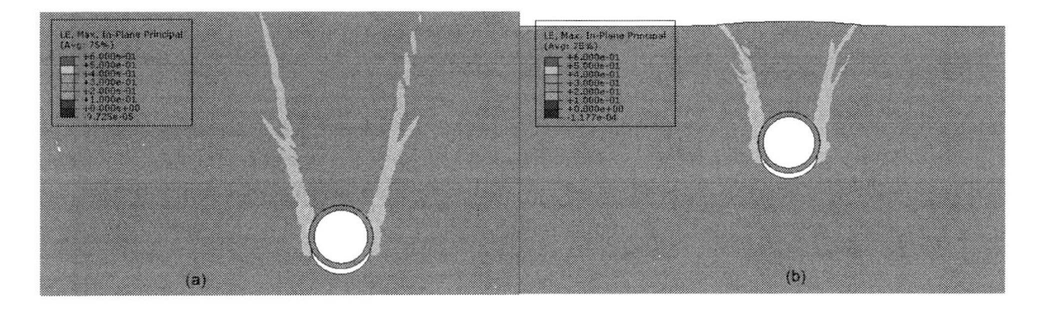

Figure 10. Contours of shear strain for dense sand $(a) H/D = 4 \, and \, (b) H/D = 2$.

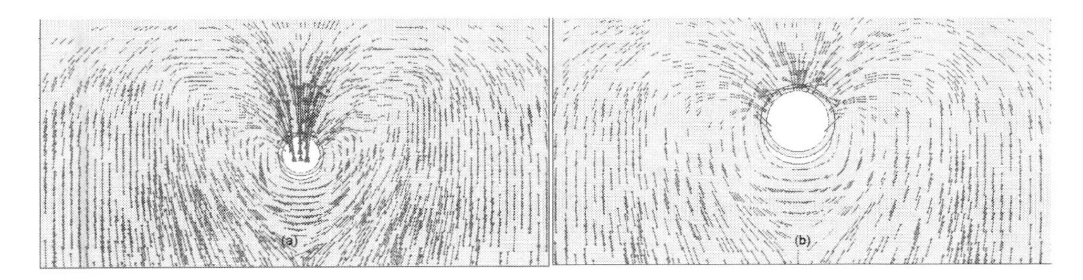

Figure 11. Directions of minor in plane principal stress for dense sand $(a) H/D = 4 \, and \, (b) H/D = 2$.

can not be captured if non-coaxility is confined to plastic regime. Since current Nor-Sand model is elasto-plastic from the beginning, it is capable of taking account of principal axis rotation from the start of the loading.

4 DISCUSSION

From the three types of analysis mentioned above, it is evident that the influence of non-coaxiality softens the stress strain response without changing the ultimate residual force. This is most pronounced in the simple shear test where soil is subjected to direct rotation of principal stress. The impact on the peak load seems to be not significant. The plane strain biaxial compression test also comply with non-coaxial theory because a significant shear deformation occur inside the shear band. The more stress paths deviates from the proportional loading, greater the non-coaxial influence. The observed increase of peak load predicted by non-coaxial model can be attributed to progressive shear failure inside the shear band.

Although apparent changes in the load displacement response of soil-pipe interaction introduced by non-coaxility is marginal, it can have an impact on main design parameters. The soil stiffness is one of the most critical inputs required for soil-spring based pipe designs which is still adopted by current industrial guidelines such as ASCE (1984),American Lifelines Alliance (2005),DNV (2007) and Pipeline Research Council International (2009). In these, the soil is assumed to follow either bilinear(ASCE 1984, American Lifelines Alliance 2005, Pipeline Research Council International 2009) or tri-linear (DNV 2007) elastic-perfectly plastic force displacement relationship. Hence the displacement mobilised at the peak force is equally important as the peak uplift load itself, to decide the minimum cover depth against the upheaval buckling of buried pipelines.

The uplift resistance is assumed to be fully mobilised at the vertical uplift displacement of (0.010.015) of burial depth according to ASCE (1984),American Lifelines Alliance (2005) and of (0.0050.008) of cover depth according to DNV (2007). These empirical correlations are based on experimental observations on limited H/D ratios and must be dependent on properties of the soil considered. While the finite element method provides a convenient tool to estimate the peak mobilised load and displacement specific to the considered soil properties, the accuracy hangs in the assumptions used in constitutive models. Numerical predictions based on coaxial models underestimates the mobilisation distance resulting in a stiffer response than reality.

However, there are some drawbacks in the finite element method associated with modelling localisations. The properties of shear bands formed during the uplift of buried pipelines depend on burial depth, diameter and roughness of the pipe and soil properties such as friction, dilation, grain size etc. These geometrical and material properties influence the degree of non-coaxility. Furthermore, the width of the shear band is affiliated physically to the grain diameter and numerically to mesh size. Hence then on-coaxial effect is mesh sensitive as well. A regularisation technique will be applied in a future study to minimise mesh dependency of shear localisation analysis.

5 CONCLUSION

In this paper, force displacement response of pipelines buried in sand is numerically investigated using a non-coaxial elasto-plastic model. The modified yield vertex theory by Hashiguchi (1993) is utilised to introduce a non-coaxial plastic component to the Nor-Sand constitutive model. It is validated by simple shear simulation, followed by the numerical evaluation of biaxial compression test and soil-pipe interaction in plane strain condition. Results are compared with conventional coaxial plasticity.

It is observed that non-coaxiality softens the load-displacement relationship during non-proportional loading. Principal axis rotations which take place during the shear deformation is responsible for the reduced stiffness. Hence the degree of non-coaxility depends on the extent of shear localisation in the soil mass. Deep burial depths and denser backfills promotes the degree of non-coaxility. The estimation of mobilised distance based on non-coaxial models can potentially lead to better pipeline design.

REFERENCES

American Lifelines Alliance (2005). Guidelines for the design of buried steel pipe.

ASCE (1984). Guidelines for the seismic design of oil and gas pipeline systems.

Cai, Y. (2010). *An experimental study of non—coaxial soil behavior using hollow cylinder testing.* Ph. D. thesis, University of Nottingham.

Cheong, T. (2006). *Numerical modelling of soil—pipeline interaction.* Ph. D. thesis, University of Cambridge.

Cheuk, C.Y., D.J. White, & M.D. Bolton (2008). Uplift Mechanisms of Pipes Buried in Sand. *Journal of Geotechnical and Geoenvironmental Engineering 134*(February), 154–163.

Cole, E. (1968). *The Behaviour of Soil in Simple Shear Apparatus.* Ph. D. thesis, University of Cambridge. DNV (2007). OFFSHORE STANDARD, DNV-OS-F101.

Gao, Z. & J. Zhao (2013). Strain localization and fabric evolution in sand. *International Journal of Solids and Structures 50*(22–23), 3634–3648.
Gao, Z. & J. Zhao (2017). A non-coaxial critical-state model for sand accounting for fabric anisotropy and fabric evolution. *International Journal of Solids and Structures 106–107,* 200–212.
Gutierrez, M. & J. Wang (2009). Non-coaxial version of Rowe's stress-dilatancy relation. *Granular Matter 11*(2), 129–137.
Gutierrez, M. & K. Ishihara (2000). Non coaxility and energy dissipation in granular materials. *Soils and Foundations 40*(2), 49–59.
Gutierrez, M., K. Ishihara, & T. Ikuo (1991). Flow theory for sand during rotation of principal stress direction. *Soil mechanics—transient and cyclic loads 31*(4), 121–132.
Gutierrez, M., K. Ishihara, & T. Ikuo (1993). Model for the deformation of sand during rotation of principal stress sirections. *Soils and Foundations 33*(3), 105–117.
Hashiguchi, K. & S. Tsutsumi (2001). Elastoplastic constitutive equation with tangential stress rate effect. *International journal of plasticity 17*(1), 117–145.
Hashiguchi, K. (1993). Fundamental requirements and formulation of elastoplastic constitutive equations with tangential plasticity. *International Journal of Plasticity 9*(5), 525–549.
Hashiguchi, K. (1998). The Tangential Plasticity. *Metals and Materials International 4*(4), 652–656.
Hashiguchi, K., M. Ueno, & Z.-P. Chen (1998). Elastoplastic Constitutive Equation of Soils Based on the Concepts of Subloading Surface and Rotational Hardening. *International Journal of Analtical Methods of Geomechanics 22*, 197–227.
Jefferies, M.G. (1993). Nor-Sand: a simple critical state model for sand. *Géotechnique 43*(1), 91–103.
Miura, K., S. Miura, & S. Toki (1986). Deformation behavior of anisotropic dense sand under principal stress axes rotation. *Soils and Foundations 26*(1), 36–52.
Oda, M. & J. Konishi (1974a). Microscopic Deformation Mechansim of Granualr Material in Simple Shear. *Soils and Foundations 14*(4), 25–38.
Oda, M. & J. Konishi (1974b). Rotation of Principal Stresses in Granular Materials during Simple Shear. *Soils and Foundations 14*(4), 39–53.
Pipeline Research Council International (2009). Guidelines for Constructing Natural Gas and Liquid Hydrocarbon Pipelines through Areas Prone to Landslidesa and Subsidence Hazards. Technical report, C-CORE,D.G. Honegger Consulting.
Robert, D.J. & N.I. Thusyanthan (2014). Numerical and experimental study of uplift mobilization of buried pipelines in sands. *Journal of Pipeline Systems Engineering and Practice 6*(1), 4014009.
Roscoe, K. (1970). The influence in strains in soil mechanics. *Geotechnique 20*(2), 129–170.
Rudnicki, J. & J. Rice (1975). Conditions for the localization of deformation in pressure-sensitive dilatant materials. *Journal of the Mechanics and Physics of Solids 23*(6), 371–394.
Spencer, A.J.M. (1982). Deformation of ideal granular materials. In Hopkins and M.J. Sewell (Eds.), *Mechanics of solids,* pp. 607–652. Oxford University Press.
Tsutsumi, S. & K. Hashiguchi (2005). General non-proportional loading behavior of soils. *International Journal of Plasticity 21*(10), 1941–1969.
Wang, J. (2012). *Monotonic and Cyclic uplift resistance of buried pipelines in cohesionless soils.* Ph. D. thesis, University of Cambridge.
Yang, Y. & H.S. Yu (2006a). Numerical simulations of simple shear with non-coaxial soil models. *International Journal for Numerical and Analytical Methods in Geomechanics 30*(1), 1–19.
Yang, Y. & H.S. Yu (2010). Numerical aspects of non-coaxial model implementations. *Computers and Geotechnics 37*(1–2), 93–102.
Yang, Y. & S. Yu (2006b). Application of a non-coaxial soil model in shallow foundation. *Geomechanics and Geoengineering: An International Journal 1*(2), 139–150.
Yang, Y., H.S. Yu, & L. Kong (2011). Implicit and explicit procedures for the yield vertex non-coaxial theory. *Computers and Geotechnics 38*(5), 751–755.
Yu, H.S. & X. Yuan (2006). On a class of non-coaxial plasticity models for granular soils. *Proceedings of the Royal Society of London A: Mathematical, Physical and Engineering Sciences 462*(2067), 725–748.
Yu, H.S. (2008). Non-coaxial theories of plasticity for granular materials. *Proceedings of the 12th International Conference of IACMAG* (1988), 1–6.
Yuan, R. (2015). *A non-coaxial theory of plasticity for soils with an anisotropic yield criterion.* Ph. D. thesis, University of Nottingham, Nottingham.
Zhao, J. & J. Gao (2015). Modelling non-coaxiality and strain localisation in sand: The role of fabric and its evolution. In S.K.F. Oka, A. Murakami, R. Uzuoka (Ed.), *Computer Methods and Recent Advances in Geomechanics,* Chapter Chapter 22, pp. 1367–1372. CRC Press, Taylor & Francis.

Numerical Methods in Geotechnical Engineering IX – Cardoso et al. (Eds)

Hypoplastic model and inverse analysis for simulation of triaxial tests

S. Cuomo, P. Ghasemi & M. Calvello
Department of Civil Engineering, University of Salerno, Fisciano, Italy

V. Hosseinezhad
Department of Industrial Engineering, University of Salerno, Fisciano, Italy

ABSTRACT: The paper deals with the use of an optimization procedure to estimate the parameters of an hypoplasticity constitutive model. The latter is used for simulating the soil mechanical response of a literature sand, for which a well-reported data-set is available. Standard saturated drained triaxial tests were simulated. The calibration procedure was based on inverse analysis herein used to optimize the values of the model parameters. A Quantum Particle Swarm Optimization algorithm, specifically designed to calibrate a large number of parameters, was applied. The capabilities of the calibrated model to accurately simulate the stiffness, shear strength and mechanical behaviour of sand are outlined. The advantages deriving from the use of inverse analysis for model calibration are also discussed.

1 INTRODUCTION

Boundary value problems are often associated to complex evolutions of stress and strain, depending on external loads, changes in pore water pressure and geometric constraints. Thus, general formulations are needed to properly reproduce stiffness variation with stress level, contractive or dilative soil behaviour, shear strength and mechanical instabilities including strain softening and static liquefaction.

Since the pioneering works on Classic Plasticity and Critical State Theory, a number of constitutive models have been developed to simulate unsaturated soil behaviour (Alonso et al., 1990), small strain stiffness, and other peculiar features of soils, which are complex stress-dependent multiphase geomaterials (Tamagnini and Viggiani, 2002). Alternative to Critical State, other models have been proposed like Hypoplasticity (Kolymbas and Wu, 1985) or Generalized Plasticity (Pastor et al, 1990), among others.

Nowadays, very few of these models are implemented in standard engineering codes, because a large number of parameters must be calibrated (Cuomo et al., 2015). This paper shows a procedure for model calibration based on inverse analysis considering the results of standard triaxial tests. The type and number of tests to be taken into the account in estimation procedures is not a trivial issue. In this paper, an attempt has been carried out to find the minimum required data, which leads to appropriate parameters estimation via inverse analysis.

The selected constitutive model belongs to the class of hypoplasticity. Previous works showed that hypoplastic model can describe the dependence of material behaviour on the pressure level and on the density with a single set of material parameters, also considering small-strain stiffness effects. A currently-considered standard hypoplastic model for granular materials is that of Gudehus (2004). It includes 8 main material parameters, plus other 5 so-called intergranular parameters introduced by Niemus and Herle (1997) to eliminate ratcheting (excessive accumulation of deformation predicted for small stress cycles) and improve the model performance in cyclic loading. The definition of the considered hypoplastic model parameters and their range of values are reported in von Wolffersdorff (1996).

A model calibration procedure was proposed by Herle and Gudehus (1999) and applied to a variety of coarse-grained soils, such as Hochestten sand, Karsluhe sand, and Toyora Sand. Similarly, Anaraki (2008) obtained the model parameters for Baskarp sand, and Masin (2017) did for Geba sand.

In this study, triaxial tests on Baskarp sand (Anaraki, 2008) were selected to verify the performance of an inverse analysis procedure, based on a species-based Quantum Particle Swarm Optimization (SQPSO) algorithm.

2 MATERIALS AND METHODS

2.1 *SQPSO algorithm for inverse analysis*

The adopted SQPSO algorithm was proposed by Hosseinnezhad et al. (2014). SQPSO is a population-based evolutionary technique created by combining a Quantum Particle Swarm Optimization (QPSO) algorithm and the speciation concept. QPSO is the quantum model of PSO (Sun et al 2004), in which the particle dynamics follow quantum mechanics rules. By applying the notion of species in QPSO, the population (solution) is classified into some groups based on the Euclidean distance. Group search improves the performance of SQPSO against the potential problem of being trapped in local optimum points when large search spaces are used.

A detailed description of the algorithm is available in Hosseinnezhad et al. (2014), as well as a comprehensive analysis of the exploration ability and solution quality of SQPSO in relation to several standard functions and a practical, power-engineering-related task. Comparing the results of the SQPSO with those obtained using other methodologies, Hosseinnezhad et al. (2014) outlined the convenience of adopting a SQPSO algorithm for stability, convergence and accuracy. The results obtained by Hosseinnezhad et al. (2016, 2018) with different operational limitations and variables confirmed that SQPSO is a very reliable tool for solving complex practical problems.

2.2 *Error functions*

The calibration of the input parameters of the hypoplastic constitutive model was performed using results of standard triaxial compression tests.

The experimental curves used for determining the error function were: deviatoric stress vs. axial strain (q, ε_a) and volumetric strain vs. axial strain (ε_v, ε_a).

The fit between the observed and the simulated values was quantified by defining an error of the numerical simulation, i.e. the objective function to minimise by inverse analysis. The error function (EF) was defined as the sum of the error functions of each considered experimental curve, $EF(k)$:

$$EF = \sum_{k=1}^{N} EF(k) \tag{1}$$

$$EF(k) = \sum_{k=1}^{m_k} e_k^2(i) \tag{2}$$

$$e_k(i) = [(y_k(i) - y'_k(i)]w_k(i) \tag{3}$$

where: N is the number of experimental curves considered, m_k is the number of observations adopted to define the k-th experimental curve, $e_k(i)$ is the weighted residual related to the i-th observation of the k-th experimental curve, $y_k(i)$ is the value of the i-th observation of the k-th experimental curve, $y'_k(i)$ is the value computed by the model which corresponds to the i-th observation of the k-th experimental curve, $w_k(i)$ is the weight assigned to the i-th observation of the k-th experimental curve.

The weights were assigned to produce dimensionless weighed residuals, $e_k(i)$, so that the error functions of the considered experimental curves, $EF(k)$, can be summed to produce a global dimensionless error function, EF. The weight assigned to the i-th observation of the k-th experimental curve is thus defined as follows:

$$w_k(i) = \frac{1}{s_k(i)} \tag{4}$$

where: $s_k(i)$ is the acceptable error related to the i-th observation of the k-th experimental curve.

The acceptable error, $s_k(i)$, has the units of measure of the observation to which it refers. It defines an acceptable range for the difference between $y_k(i)$ and $y'_k(i)$, and produces dimensionless weighted residuals lower than 1.00 when the difference falls within that range.

Herein, the weights assigned to all the observations of a given experimental curve are always equal, i.e. they do not depend on the value of the i-th observation of that curve. Indeed, the expression used to quantify the acceptable error of the k-th experimental curve, s_k, is the following:

$$s_k(i) = s_k = r_k \max(y_k) \tag{5}$$

where: r_k is the dimensionless tolerance coefficient of the k-th experimental curve, $max(y_k)$ is the maximum observed value of the k-th experimental curve.

Considering the above relationships, Eq. 1 can be expressed as follows:

$$\boldsymbol{EF} = \sum_{K=1}^{N} \frac{1}{[r_k \max(y_k)]^3} \sum_{i=1}^{m_k} [y_k(i) - y'_k(i)]^2 \tag{6}$$

2.3 *Results of laboratory soil testing*

We referred to the Baskarp sand, which is a yellow-orange fine-grained sand deposited a few miles north of Jonkoping (Sweden) under melting of the icecap ca. 10'000 years ago. The grains are classified as angular to sub-angular. It is a uniform sand with a D_{50} approximately equal to 140 μm. Anaraki (2008) conducted a comprehensive series of laboratory tests on samples of Baskarp sand,

Table 1. Triaxial tests used for model calibration.

Test	σ' (kPa)	$e_{initial}$(-)	s_k for q (kPa)	s_k for ε_v(-)
1	50	0.59	2.3	6.4 e-4
2	100	0.60	4.2	6.4 e-4
3	200	0.60	8.3	6.0 e-4
4	50	0.70	1.5	2.5 e-4
5	100	0.84	2.7	1.3 e-4
6	200	0.81	5.1	0.4 e-4

reporting: minimum and maximum void ratios equal to 0.58 and 1.08, respectively; critical friction angle of 30°.

Six drained triaxial tests were considered for model calibration (Tab. 1). Three tests were performed on dense specimens and three tests on loose specimens, characterized by different initial void ratio and three different confining pressures (50, 100 and 200 kPa). For each test the two (q, ε_a) and (ε_v, ε_a) experimental curves were considered. Each curve was discretized considering 50 points, thus the observations were 600.

As already mentioned, the weights assigned to the observations were defined as the inverse of an acceptable error, s_k, which depends on a tolerance coefficient, r_k, and on the maximum observed value of the k-th experimental curve, $max(y_k)$ (Eq. 5). The values of r_k adopted herein are always equal to 0.01. The resulting acceptable errors range from 1.5 to 8.3 kPa for the deviator stress values and from 0.4 to 6.4 10^{-4} for the volumetric strain values (Table 1).

3 RESULTS

3.1 *Model calibration*

The aim of the calibration was to estimate the following 5 parameters of the adopted hypoplastic constitutive model: h_s, n, e_{crit}, α and β. Parameters h_s, n *and* β influence the stiffness of the material. Parameter α influences the frictional angle mobilized upon shearing. Parameter e_{crit} plays a role in the model equation which takes into account the mean skeleton pressure (Herle and Gudehus, 1999). The detailed definition and function of these parameters are also well explained in Von Wolffersdorff (1996).

The other model parameters were assumed fixed and their values were assigned as follows: e_{d0} and e_{i0}, i.e. maximum and minimum void ratio at zero pressure, respectively, were set equal to 0.58 and 1.23, considering the values reported in the literature, the critical friction angle was set equal to 30°. The intergranular parameters were out of the scope of this study, since their main effect is related to the cyclic behavior. The range of values assigned to the parameters being calibrated was assumed based on suggestions from Von Wolffersdorff (1996). Accordingly, h_s was calibrated in the range 1000 to 6'000'000, n ranged from 0 to 0.67, e_{crit} varied between the minimum and maximum adopted void ratios, α from 0 to 0.3, and β from 0 to 2. It should be underlined that the starting values adopted by the inverse analysis algorithm were the lowest values of the defined range of the parameters.

Table 2. Inverse analyses performed.

Inverse analysis ID	Tests	# of observations
INV 01	4, 6	200
INV 02	1, 2, 3	300
INV 03	4, 5, 6	300
INV 04	3, 6	200
INV 05	1, 4	200

Table 3. Calibrated parameters compared to Anaraki (2008).

Parameters	α	β	hs	n	$_{ec}$
INV 01	0.13	2.48	5.33e+6	0.29	0.99
INV 02	0.15	1.57	3.39e+6	0.37	0.98
INV 03	0.10	0.21	5.29e+6	0.39	1.02
INV 04	0.15	1.31	3.06e+6	0.41	0.91
INV 05	0.17	1.88	4.90e+6	0.33	0.98
Average Values	0.15	1.55	3.37e+6	0.358	0.96
Anaraki (2008)	0.12	0.96	4.00e+6	0.42	0.93

Inverse analysis was conducted for the triaxial tests of loose and dense specimens, separately. Table 2 shows the tests adopted for calibration in the five inverse analyses performed (INV 01 to INV 05). Each calibration was conducted via 80 iterations of SQPSO, adopting a population number equal to 40.

The aim of INV 01 was to check the performance of the procedure while using two tests which differs in both initial void ratio and confining pressure. INV 02 aimed at checking the performance of the procedure in the case of three triaxial tests on dense material with different confining pressures. INV 03 was based on the same idea of INV 02, considering the loose sand specimens. INV 04 and INV 05 were conducted to see the outcomes of the inverse analysis when the objective function includes two tests with similar confining pressures and different initial void ratios.

As reported in Table 2, the number of employed observations differs for each analysis. Therefore, the direct comparison of the values of the error functions is not possible. The obtained parameters are reported in Tab. 3, in addition to those estimated by Anaraki (2008).

The comparison of the parameters values estimated in the different inverse analyses shows that observations from different tests lead to relatively different optimal parameters values, although the values are often close to those reported in the literature. The value obtained in INV 03 for α and β are significantly different from those obtained with the other analyses. Similarly, α calculated by INV 05 is relatively higher than the other three estimates. The estimated values for n ranged from 0.29 to 0.41 with an average value of 0.35. The values of h_s are scattered around a mean value of 4.9 10^6. In all the analyses the estimated values for β are not close to the values reported by Anaraki (2008), who has obtained this parameter by using experimental results from oedometer tests.

3.2 *Model results versus experimental data*

The numerical results obtained using the model parameters calibrated in the five performed analyses, from INV 01 to INV 05, are illustrated in Figures 1 to 5, respectively. The results of the model simulations adopting the parameters values indicated by Anaraki (2008) are also reported in each figure.

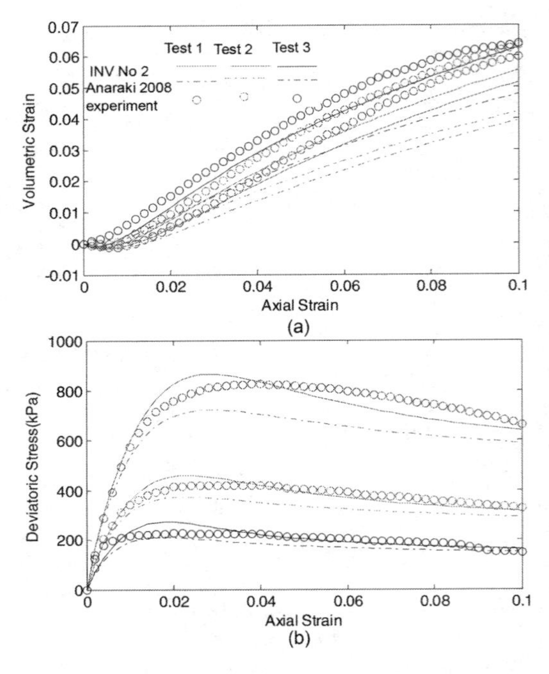

Figure 2. Model results using parameters calibrated in INV 02.

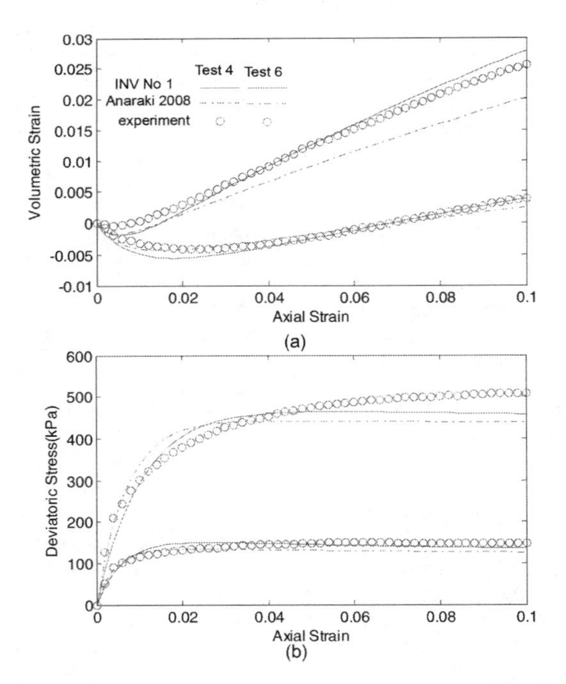

Figure 1. Model results using parameters calibrated in INV 01.

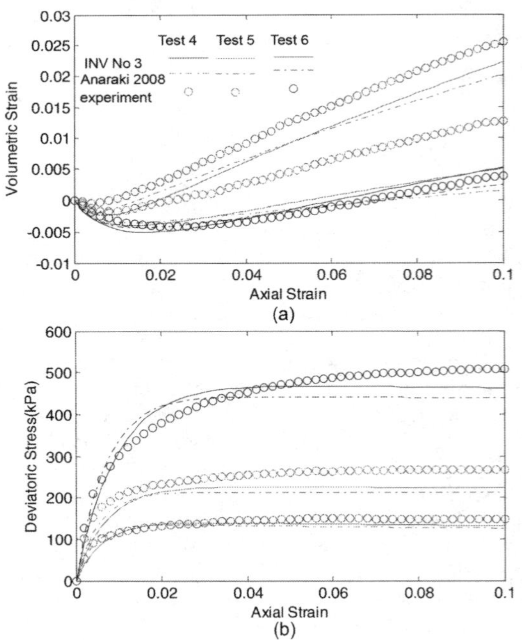

Figure 3. Model results using parameters calibrated in INV 03.

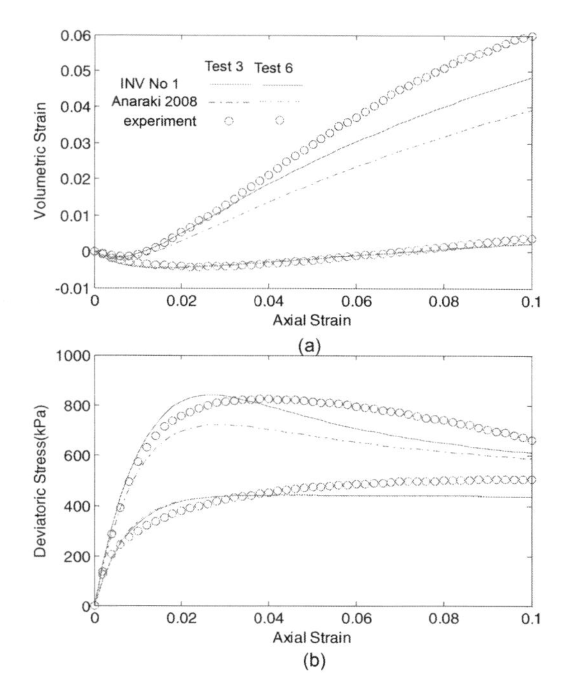

Figure 4. Model results using parameters calibrated in INV 04.

INV 01 and INV 03 only consider tests on loose soil samples. In the first analysis, two tests were taken into the account (tests 4 and 6). In the latter, also test 5 was considered in addition to the previous ones. The experimental data from the two tests employed in INV 01 well match the numerical simulations calibrated using those tests (Fig. 1). On the contrary, INV 03 leads to an unsatisfactory simulated behaviour in the (ε_v, ε_a) space for all the three tests (Fig. 3). This unsatisfactory performance of the inverse analysis highlights the role played by test 5 in worsening the calibration of the model also in relation to the simulations for the other two tests. The improper calibration of this analysis could also be perceived by looking at the values of the calibrated parameters, most of which are quite different from the average values (Tab. 3). Indeed, even adopting the parameters values suggested by Anaraki (2008) to model results do not reproduce the experimental volumetric strain behaviour recorded in test 5. Experimental results from test 5 should thus be considered peculiar in relation to the other test results, thus unfit as a source of observations for model calibration purposes.

The outcome of the calibration performed in INV 02 resulted in a satisfactory fit between simulated and measured values for both (q, ε_a) and (ε_v, ε_a) plots. However, the final predicted volumetric strains are slightly underestimated for all three tests. Nevertheless, the predicted curves are closer to the experimental data than the curves produced using the parameters values from the literature.

As mentioned before, INV 04 and INV05 were conducted to check the performance of the inverse analysis when the observations come from two tests on specimens with similar confining pressures and different initial void ratios. The curves from INV 04 (tests with a confining pressure equal to 200 kPa) are in good agreement with the experimental data, as shown in Fig 4. The values of the predicted peak shear strengths are consistent with the observations. The simulated residual shear strength for test 6 is slightly lower than the value recorded in the experiment. The results of INV 05 (tests with a confining pressure equal to 50 kPa) are illustrated in Fig 5. The results show proper matches for the loose soil sample. For the dense sample the peak deviatoric stress is overestimated while the residual shear strength is adequately predicted.

The inverse analyses conducted adopting observations from specimens confined at lower stresses generally lead to lower estimates for parameters α and h_s. The analyses conducted adopting observations from specimens confined at higher stresses also produce higher values of the critical void ratio.

The values of α obtained in the latter cases are also relatively higher than the value reported by Anaraki (2008). These differences are in accordance with the dissimilarities observed in the simulation of the dilative behaviour of the three tests on dense specimens.

As shown in the Figures, lower values of α produce lower dilation. The best fit between experimental data and numerical results occurs when this parameter is equal to its maximum values (0.17).

As a summary, the outcomes of all the analyses except INV 03 appear to be acceptable, although the calibrated values of β are always very different from the ones reported in the literature. In all these four cases, the modelled volumetric strains match the experimental data better than when the model is adopting the set of input parameters from the literature. The analyses also highlight that using observations from tests with similar initial void ratios and different confining pressures (INV 01 and INV 02) or similar confining pressures and different initial void ratios (INV 04 and INV 05), works similarly well. This means that all the 4 sets represent 4 solutions of the optimization algorithm, i.e. they are local minima of the objective function (error) that we are trying to minimize.

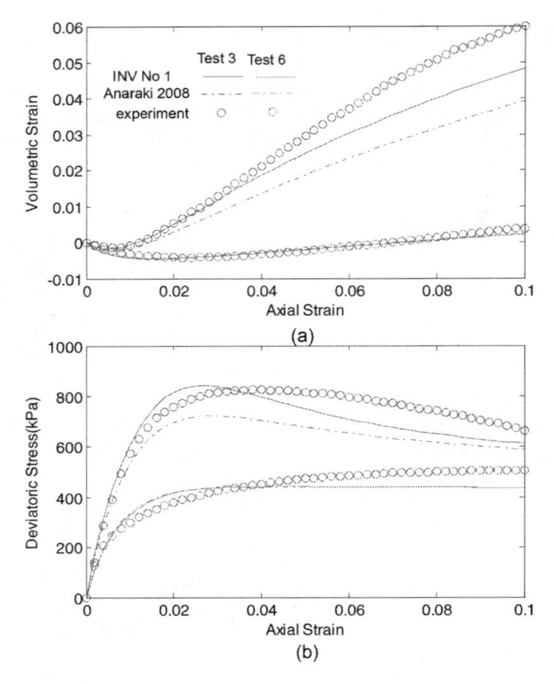

Figure 5. Model results using parameters calibrated in INV 05.

3.3 *Validation towards an oedometer test*

The four sets of input parameters of the hypoplastic constitutive model previously calibrated should be able to properly simulate trixial test results performed on the same material also for tests on specimens that are different in both initial void ratio and confining stress level. In addition, the same set of parameters should also allow reproducing the soil behaviour along other stress paths. Therefore, the performance of achieved calibrated sets of input parameters was rechecked in the simulation of the experimental results of an oedometer test.

Figure 6 shows the oedometric response of a specimen with initial void ratio equal to 0.825 and loaded up to 5 MPa, as well as the simulated responses adopting the four good-performing sets of calibrated.

According to Bauer (1996), the compressive behaviour of the model is related to the parameters h_s n and β. As shown in Fig. 6, the parameters set estimated with the analyses INV 01 and INV 05 were not able to reproduce the measured curve correctly. Indeed, the above sets of parameters underestimate the soil stiffness. Adopting INV 04 and INV 02, the results provide a good simulation of the oedometer test. However, some small differences between the simulated and the experimental response still exist.

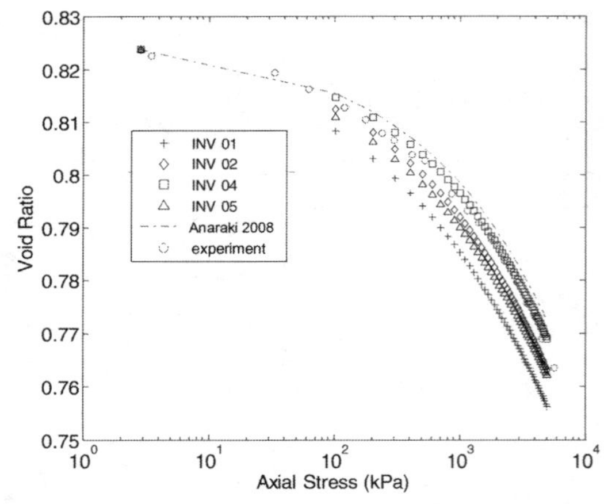

Figure 6. Performance of obtained parameter sets in oedometer test.

In fact, INV04 overestimates the oedometric test results and INV02 underestimates them. It is not straightforward to individuate the best optimal parameter set. Therefore, to check the accuracy of the obtained values and to individuate one final set of parameters, other two inverse analysis calculations were carried out in which the error function was defined considering the observations from the oedometer test and, respectively, INV02 and INV04. In the former case, the obtained parameters were almost the same obtained before; in the latter case, the optimal parameters differed from the previous values. Therefore, it is herein assumed that the values obtained in INV02 are more prone to simulate the material behaviour properly.

4 CONCLUSIONS

This paper presented a robust curve-fitting (inverse analysis) algorithm applied to determine the parameters of an hypoplastic constitutive model.

Firstly, the proposed curve fitting algorithm was able to determine a set of parameters by which the simulated soil behavior was closer to the experimental result than the response computed using parameters values reported in the literature. Secondly, the paper showed the importance of the right selection of the type of tests used to provide the observations for the inverse analysis algorithm.

The obtained set of calibrated parameters significantly depends on the observations used to calibrate them. For example, it appears that conducting the inverse analysis employing triaxial tests conducted on specimens subjected to higher confining pressures more likely leads to a correct estimate of the parameters. However, it should also be stated that a set of parameters providing good predictions of triaxial tests does not necessarily allow simulating oedometer tests properly. It entails that combinations of different types of laboratory tests should be used for a general calibration of constitutive model along different stress paths.

ACKNOWLEDGMENTS

The simulations of triaxial tests have been conducted by the element test module of Anura 3D software (http://www.mpm-dredge.eu/). The authors also thank Dr. Mario Martinelli and Dr. Vahid Galavi (Deltares, Netherlands), for their consulting on the hypoplastic model.

REFERENCES

Alonso, E.E., Gens, A., & Josa, A. 1990. A constitutive model for partially saturated soils. *Géotechnique*, *40*(3): 405–430.

Anaraki, K.E. 2008. Hypoplasticity investigated: parameter determination and numerical simulation. M.Sc. thesis, TU Delft, The Netherland

Bauer, E. 1996. Calibration of a comprehensive hypoplastic model for granular materials. *Soils and foundations*, *36*(1): 13–26.

Cuomo S., Manzanal D., Moscariello M., Pastor M., Foresta V. 2015. Application of a generalized plasticity constitutive model to a saturated pyroclastic soil of Southern Italy. In: Proc. of "Soil Mechanichs 2015 – Panamerican Conf. on Soil Mechanics and Geotechnical Engineering", Buenos Aires, Argentina, 15–18 Nov. 2015, D. Manzanal and A.O. Sfriso (Eds.) ISBN: 978–1-61499–600–2, pp. 1215–1222.

Gudehus, G. 2004. A visco-hypoplastic constitutive relation for soft soils. *Soils and Foundations*, 44(4): 11–25.

Herle, I., & Gudehus, G. 1999. Determination of parameters of a hypoplastic constitutive model from properties of grain assemblies. *Mechanics of Cohesive-frictional Materials*, *4*(5): 461–486.

Hosseinnezhad, V., Rafiee, M., Ahmadian, M., & Ameli, M.T. 2014. Species-based quantum particle swarm optimization for economic load dispatch. *International Journal of Electrical Power & Energy Systems*, *63*: 311–322.

Hosseinnezhad, V., Rafiee, M., Ahmadian, M., & Siano, P. 2016. Optimal day-ahead operational planning of microgrids. *Energy Conversion and Management*, *126*: 142–157.

Hosseinnezhad, V., Rafiee, M., Ahmadian, M., & Siano, P. 2018. Optimal island partitioning of smart distribution systems to improve system restoration under emergency conditions. *International Journal of Electrical Power & Energy Systems*, *97*: 155–164.

Kolymbas, D., & Wu, W. 1993. Introduction to hypoplasticity. *Modern approaches to plasticity*, 213–223.

Masin, D. 2017, Report on the triaxial tests on GEBA sand carried out for client: *IHC IQIP, NL*. Technical report.

Niemunis, A., & Herle, I. 1997. Hypoplastic model for cohesionless soils with elastic strain range. *Mechanics of Cohesive-frictional Materials*, *2*(4): 279–299.

Pastor, M., Zienkiewicz, O.C. and Chan, A.H.C. 1990. Generalized plasticity and the modelling of soil behaviour. *International Journal for Numerical and Analytical Methods in Geomechanics*, 14: 151–190.

Sun, J., Feng, B., & Xu, W. 2004, June. Particle swarm optimization with particles having quantum behavior. In *Evolutionary Computation, 2004. CEC2004. Congress on* (Vol. 1, pp. 325–331). IEEE.

Tamagnini, C., & Viggiani, G. 2002. On the incremental non-linearity of soils. Part I: theoretical aspects. *Rivista italiana di geotecnica*, *36*(1): 44–61.

von Wolffersdorff, P.A. 1996. A hypoplastic relation for granular materials with a predefined limit state surface. *Mechanics of Cohesive-frictional Materials*, *1*(3): 251–271.

Numerical Methods in Geotechnical Engineering IX – Cardoso et al. (Eds)
© 2018 Taylor & Francis Group, London, ISBN 978-1-138-33198-3

Application of a generalized continuous Mohr–Coulomb criterion

Gustav Grimstad, Jon A. Rønningen & Steinar Nordal
Norwegian University of Science and Technology, Trondheim, Norway

ABSTRACT: This paper presents an application of a generalized failure criterion for soils. The presented formulation was originally adopted by the authors to round the corners of the Mohr–Coulomb (MC) yield surface in principals stress space. The formulation was later extended to a generalized yield criterion that, in addition to the rounded MC, includes the Matsuoka-Nakai, the Lade-Duncan yield criterion and other well-known criteria. Two additional material parameters controls the alternative Lode angle dependencies. It has certain advantages over already existing generalized yield criteria: It is general, but still mathematically rather simple. It is infinitely differentiable, meaning that derivatives of any order may be calculated anywhere in stress space. The criterion defines a unique, single yield surface and does not suffer from false solutions in other octants in principal stress space (due to unwanted secondary surfaces that is present in many other formulations). Convexity may be analytically checked. Analytical relations between parameters for the formulation and existing criteria have been derived and some are presented in the paper as a function of the friction angle in triaxial compression. Because of its general formulation and beneficial properties, the criterion offers a convenient alternative for implementation in numerical methods using elasto-plasticity. One unified implementation covers several failure criteria by selecting appropriate parameters. The uniqueness of the criterion and its derivatives contribute to numerical stability and robustness. The paper compares numerical result with the proposed failure criterion in a FEA of a plane strain passive earth pressure problem. Different parameters, such that the criterion corresponds to the different criteria mentioned above, are used.

1 INTRODUCTION

An often-used criterion in geotechnical engineering is the Mohr-Coulomb (MC) criterion. MC criterion has a disadvantage, because of the discontinuities in the first differentials in corners. Continuous surfaces like the Drucker-Prager (DP) criterion (Drucker and Prager, 1952), Matsuoka-Nakai (MN) (i.e. the Spatial Mobilized Plane, SMP, criterion) (Matsuoka, 1976, Matsuoka and Nakai, 1974, Matsuoka and Nakai, 1985), Lade-Duncan (LD) criterion (Lade and Duncan, 1975), Bardet (LMN) criterion (Bardet, 1990) overcomes this corner problem. However, the 'cost' is higher friction angles for other Lode angles and for some of the formulations branches of F equal to zero ($F = 0$ means that the stress state is at yield) appearing in other octants, see Figure 1. Some of the criteria are concave for trial stress states, i.e. for values of $F > 0$. Which could be the case for the criterion of e.g. Argyris et al. (1974). For one or both of these two reasons some of the continuous surfaces are not well suited for numerical implementations following an implicit scheme with large increments.

2 GENERALIZATION OF THE MOHR–COULOMB FAILURE CRITERION

Equation (1) expresses the Mohr-Coulomb criterion in invariant form as a variant of the Drucker–Prager criterion with Lode angle dependency (compression is positive).

$$F = \sqrt{3 \cdot J_2} - \frac{\sin\varphi}{\sqrt{3} \cdot \cos\theta + \sin\theta \cdot \sin\varphi} \cdot (I_1 + 3 \cdot a) = 0 \tag{1}$$

where: $I_1 = \sigma_{ii}$, $J_2 = -s_1 \cdot s_2 - s_2 \cdot s_3 - s_3 \cdot s_1$, ϕ is the friction angle, $a = c \cdot \cot\phi$ (attraction), c is the cohesion, θ is the Lode angle calculated from eq. (2), σ_{ij} are Cauchy stresses and s_i are the principal deviatoric stresses calculated from $s_{ij} = \sigma_{ij} - I_1/3 \cdot \delta_{ij}$.

$$\sin 3\theta = -\frac{3\sqrt{3} \cdot J_3}{2 \cdot \sqrt{J_2^{\,3}}} \tag{2}$$

where $J_3 = \det(\mathbf{s}) = s_1 \cdot s_2 \cdot s_3$. At a value of $\theta = -\pi/6$ (Triaxial compression, TXC) or for $\theta = \pi/6$ (Triaxial

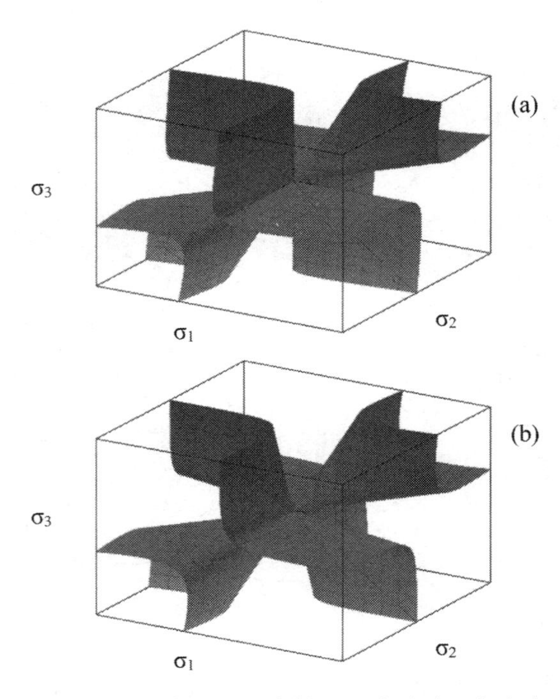

Figure 1. (a) the MN and (b) LD criteria in principal stress space, showing the complementary surfaces in other octants.

extension, TXE) differentiates of equation (1) with respect to stress are not uniquely defined in the corners. Therefore a simple way to overcome this is sought by replace the trigonometric functions by some 'approximations'. Billington (1988) proposed an approximation to the Tresca criterion (i.e. rounded corners). This formulation is C^{∞} continuous in the π-plane (i.e. an analytic function); meaning that it has derivatives of all orders and it equals its Taylor series expansion around any point in its domain. The formulation of Billington is previously used in soil modelling by e.g. Grimstad et al. (2012). Extending the Tresca approximation to the Mohr–Coulomb criterion results in the following form for the yield surface:

$$F = \sqrt{3 \cdot J_2} - \frac{\sin\varphi_0}{\sqrt{3}\cdot c_\theta + s_\theta \cdot \sin\varphi_0} \cdot (I_1 + 3\cdot a) = 0 \qquad (3)$$

$$c_\theta = \cos\left(\frac{1}{3}\cdot \arcsin(a_1 \cdot \sin 3\theta)\right) \qquad (4)$$

$$s_\theta = \sin\left(\frac{1}{3}\cdot \arcsin(a_1 \cdot \sin 3\theta)\right) \qquad (5)$$

Equation (4) and (5) give c_θ and s_θ as a function of the Lode angle. This means that the 'approximations' simply replace the sinus and cosine term in equation (1). a_1 is an input parameter between 0.0 and 1.0. Equation (4) is rewritten form of Billington to avoid use of the 'sign' function in the s_θ term, and is necessary for C^{ω} continuity. In this article the notation ϕ_0 means the friction angle given for $\theta = 0°$.

For $a_1 = 0$ equation (3) gives the DP criterion while for $a_1 = 1$ the equation gives the standard MC criterion. For all $0 < a_1 < 1$ the criterion is smooth, see Figure 2.

The proposed approximation to the MC develops further by introducing one more parameters, a_2, into the criterion. The proposed 'generalized Mohr-Coulomb yield/failure criterion (GMC criterion) will then take the form of eq. (6).

$$F = \sqrt{3 \cdot J_2} - \frac{\sin\varphi_0}{\sqrt{3}\cdot c_\theta + s_\theta \cdot \frac{\sin\varphi_0}{a_2}} \cdot (I_1 + 3\cdot a) = 0 \qquad (6)$$

Again, for θ equal to zero, the subscript 0 is used.

From Figure 2 it becomes apparent that by smoothing the corners of MC using $a_1 < 1$ in equation (3) the strength in triaxial compression varies with a_1 even for a constant input friction angle ϕ_0. From equation (6) and Figure 2 we observe that all surfaces intersect at Lode angle $\theta = 0$. Alternatively they could intersect at triaxial compression (TXC) by expanding the surfaces for $a_1 < 1$. This is easily achieved by using input friction angle ϕ_0 as a function of a_1, a_2 and ϕ_{TXC}, equation (7).

$$\sin\varphi_0 = \frac{2\cdot\sqrt{3}\cdot a_2 \cdot \cos\beta \cdot \sin\varphi_{TXC}}{3\cdot a_2 - (a_2 - 2\cdot\sin\beta)\cdot\sin\varphi_{TXC}} \qquad (7)$$

where: $\beta = \frac{1}{3}\cdot\arcsin(a_1)$

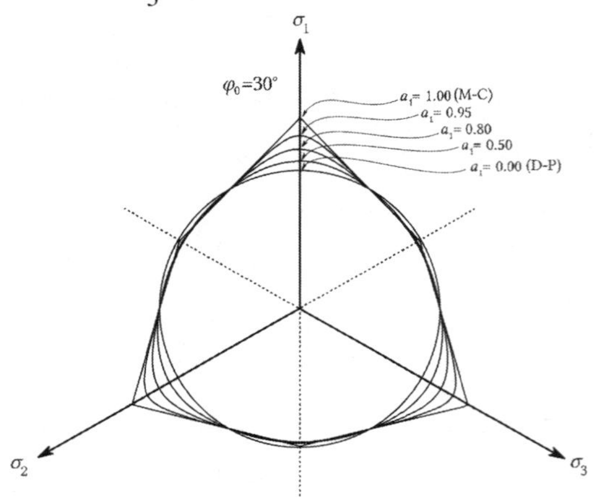

Figure 2. Mohr–Coulomb approximation for different values of a_1, $\phi_0 = 30°$, in the π-plane.

Further, it is easy to show that the proposed criterion is equal to the MN criterion (as a function of the friction angle in triaxial compression, ϕ_{TXC}) when:

$$\sin\varphi_0 = a_2 = \frac{2\cdot\sin\varphi_{TXC}}{\sqrt{3+\sin^2\varphi_{TXC}}} \tag{8}$$

$$a_1 = \sin\varphi_0 \cdot \frac{3-\sin^2\varphi_0}{2} \tag{9}$$

For the LD criterion:

$$\sin\varphi_0 = a_1 = a_2 = \sqrt{\frac{9-7\cdot\sin\varphi_{TXC}}{3-\sin\varphi_{TXC}}\cdot\frac{2\cdot\sin\varphi_{TXC}}{3-\sin\varphi_{TXC}}} \tag{10}$$

And, for the generalized LMN criterion:

$$\sin\varphi_0 = a_2 = \sqrt{\frac{3}{\rho_{TXE}^2 - \rho_{TXE}\cdot\rho_{TXC} + \rho_{TXC}^2}} \tag{11}$$

$$a_1 = \frac{1}{2}\cdot\rho_{TXE}\cdot\rho_{TXC}\cdot(\rho_{TXE}-\rho_{TXC})\cdot\sin^3\varphi_0 \tag{12}$$

where: $\rho_{TXE} = \frac{3+\sin\varphi_{TXE}}{2\cdot\sin\varphi_{TXE}}$ and $\rho_{TXC} = \frac{3-\sin\varphi_{TXC}}{2\cdot\sin\varphi_{TXC}}$

Figure 3 shows the input according to LMN (ϕ_{TXC} = 30° and varying ϕ_{TXE}).

For the DP criterion:

$$\sin\varphi_0 = \frac{2\cdot\sqrt{3}\cdot\sin\varphi_{TXC}}{3-\sin\varphi_{TXC}}, a_1 = 0, a_2 > 0 \tag{13}$$

Similar procedure can also be used with other criteria in literature e.g. Argyris et al. (1974). Then:

$$\sin\varphi_0 = \frac{4}{\sqrt{3}}\cdot\frac{\sin\varphi_{TXE}\cdot\sin\varphi_{TXC}}{\sin\varphi_{TXE}+\sin\varphi_{TXC}} \tag{14}$$

Further, for the other constants only a close approximation of the Argyris criterion is found, assuming $\sin(\varepsilon) \approx \varepsilon$, results in the following constants after selecting a small value for a_2.

$$a_1 = \frac{3}{2}\cdot a_2\cdot(\rho_{TXE}-\rho_{TXC}) \tag{15}$$

The extended Tresca criterion (Van Eekelen, 1980) can be achieved by setting $1/a_2 = 0$ and letting $a_1 \to 1$ then using eq. (16) for calculation of $\sin\phi_0$.

Note that exact original Tresca can easily be obtained by removing the I_1 dependency from eq. (6).

$$\sin\varphi_0 = \frac{3\cdot\sin\varphi_{TXC}}{3-\sin\varphi_{TXC}} \tag{16}$$

Table 1. Input for Generalized Mohr-Coulomb (ϕ_{TXC} = 30°).

	$\sin\phi_0$	a_1	a_2
GMC	0.5201	0.9750	0.8680

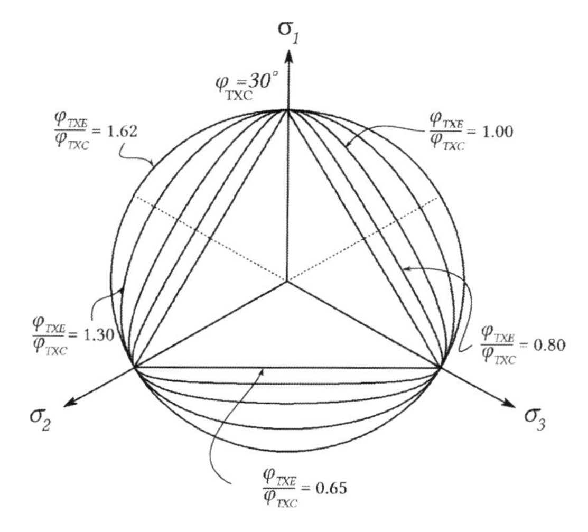

Figure 3. Generalized failure criteria in the π-plane for input according to the LMN criterion.

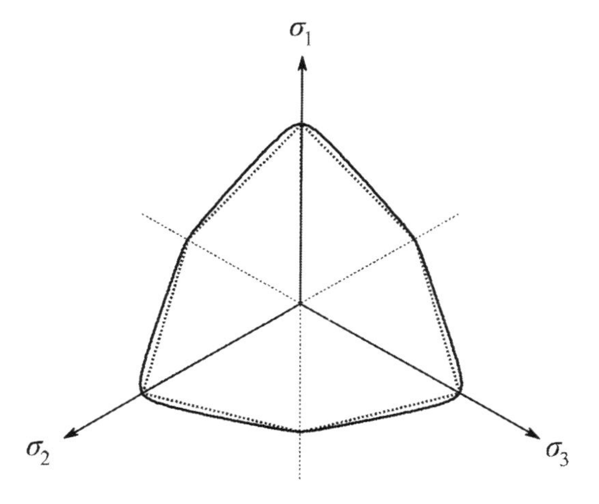

Figure 4. Generalized Mohr-Coulomb (GMC) in the π-plane with input according to Table 1.

Figure 4 shows how the GMC criterion looks like for ϕ_{TXC} = 30° and a_1 and a_2 from Table 1. While Figure 5 shows the GMC criterion in principal stress space. Table 4 at the end of the paper gives an overview of the different relationships between the Generalized Mohr Coulomb criterion

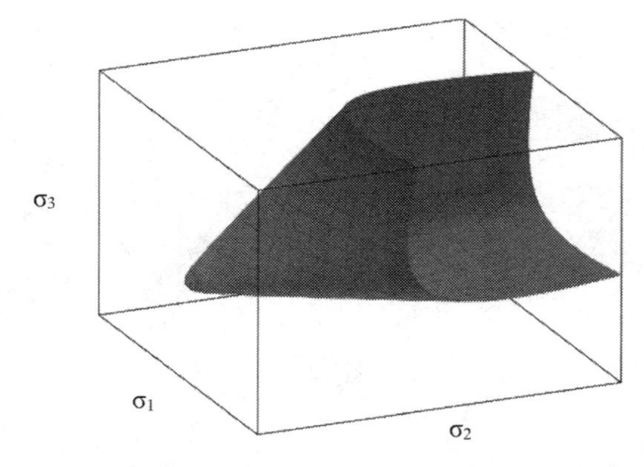

Figure 5. The GMC criterion in principal stress space.

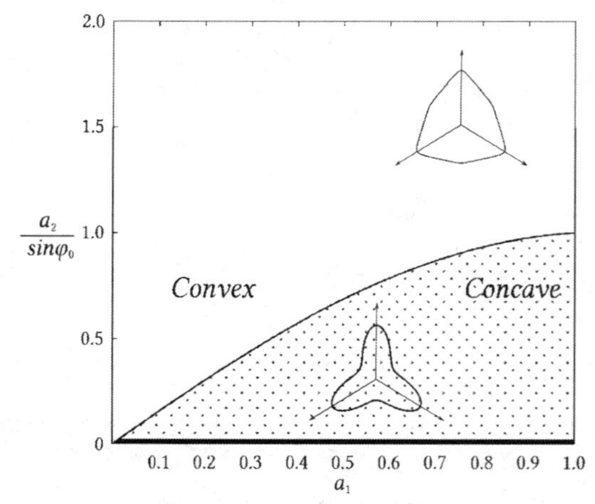

Figure 6. Visualization of the convexity limit of the parameters.

and some different criteria from literature for $\phi_{TXC} = 30°$.

It should be noted that a similar, but less versatile, generalized criterion is found in Panteghini and Lagioia (2014). They proposed a fully convex reformulation of the original Matsuoka–Nakai failure criterion that avoids the branches of $F = 0$ appearing in other octants. Recently Lagioia and Panteghini (2016) extended their work from 2014 by using a phase change in their MN description, giving the capability to describe the same class of surfaces as the formulation presented here, using three additional parameters to the MC parameters. Note also that even though Lagioia and Panteghini (2016) state that their formulation ensures convexity of the failure surface, this is not rigorously demonstrated in their paper. In fact, restrictions on parameter selection is necessary to ensure convexity.

3 CONVEXITY

To ensure a convex yield/potential surface there are limits to the ratio between a_1 and $a_2/\sin\phi_0$. Satisfying eq. (17) for all values of θ between $-\pi/6$ and $\pi/6$ ensures convexity. In reality it will be sufficient to check for TXE ($\theta = \pi/6$) since $h'(\theta) < 0$ for the chosen function when convexity is satisfied. Eq. (18) gives the limits between a_1 and $a_2/\sin\phi_0$.

Figure 6 visualizes this limit in form of a convex and a concave region. From the figure it is clear that the LMN criterion always is convex since $a_2/\sin\phi_0 = 1.0$, while the Argyris criterion is not necessarily convex.

$$h''(\theta) + h(\theta) \geq 0 \tag{17}$$

where: $h(\theta) = \sqrt{3} \cdot c_\theta + s_\theta \cdot \dfrac{\sin\varphi_0}{a_2}$

$$\frac{a_2}{\sin\varphi_0} \geq \frac{1}{\sqrt{3}} \cdot \frac{2\cdot\sin(3\beta)\cdot\cos(\beta)+\sin(2\beta)}{2\cdot\sin(3\beta)\cdot\sin(\beta)+\cos(2\beta)} \tag{18}$$

where: $\beta = \dfrac{1}{3}\cdot\arcsin(a_1)$

For reasonably small values of a_1 (as $a_1 \to 0$) eq. (18) is equal to $a_2/\sin\phi_0 \geq (4/3)^{3/2} \cdot a_1$. This means that for the Argyris criterion convexity is ensured by the simple ratio of minimum $(4/3)^{3/2} \cdot \sin\phi_0$ between a_2 and a_1. Inserting this requirement back into Eq. (15) and combining this with equation (14) gives then the limit for $\sin\phi_{TXE}$ as a function of $\sin\phi_{TXC}$ (Eq. (19)), this is the same limit as given by Lin and Bažant (1986).

$$\sin\varphi_{TXE} \geq \frac{21\cdot\sin\varphi_{TXC}}{27-16\cdot\sin\varphi_{TXC}} \tag{19}$$

Eq. (18) holds also for convexity of the potential surface. Therefore, for $0 \leq \psi_0 \leq \phi_0$ convexity of the potential surface is automatically satisfied when the yield surface is convex. For negative values of the dilatancy angle, then TXC ($\theta = -\pi/6$) is critical, rather the TXE. Now the criterion for convexity is simply a mirror of the curve in Figure 6 around the horizontal axis. This means that one ensures a convex potential surface also for the interval $-\phi_0 \leq \psi_0 < 0$, again this is only valid as long as the yield surface is convex. For values of $F > 0$ (i.e. for a trial stress state outside $F = 0$) convexity is ensured, when the surface for $F = 0$ is convex, since the surfaces for $F =$ *const.* are *semi*-conform in the π-plane, meaning that the surface is more and more circular for higher value of F. When $F \to \infty$ the shape in the π-plane converges towards a circle with infinite radius, see Figure 7.

Table 2. Input for Generalized Mohr-Coulomb (ϕ_{TXC} = 30°).

	$\sin\phi_0$	a_1	a_2
LD	0.5933	0.5933	0.5933
MN	0.5547	0.7467	0.5547

4 PLASTIC POTENTIAL

An implicit integration scheme is used in the implementation of the proposed generalized MC failure criterion. The model is implemented as a user defined soil model in the finite element program PLAXIS (www.plaxis.nl). The scheme requires double differentiates of the plastic potential function with respect to stresses; hence, a criterion of minimum class C^2 continuity is preferred. The plastic potential can be expressed (using non-associated flow) as:

$$Q_1 = \sqrt{3\cdot J_2} - \frac{\sin\psi_0}{\sqrt{3}\cdot c_\theta + s_\theta\cdot\frac{\sin\psi_0}{a_2}}\cdot(I_1 + 3\cdot a) - k_\psi \tag{20}$$

where ψ is the dilatancy angle and the 'apparent cohesion' k_ψ is found as a 'constant' for the given stress state, as determined by:

$$\frac{k_\psi}{I_1 + 3\cdot a} = \left(\frac{\sin\varphi_0}{\sqrt{3}\cdot c_\theta + s_\theta\cdot\frac{\sin\varphi_0}{a_2}} - \frac{\sin\psi_0}{\sqrt{3}\cdot c_\theta + s_\theta\cdot\frac{\sin\psi_0}{a_2}}\right) \tag{21}$$

This formulation follows the apparent cohesion approach used by Krabbenhoft et al. (2012).

As an alternative the potential surface can be expressed of Cam Clay type of formulation equal to:

$$Q_2 = \frac{3\cdot J_2\cdot\left(\sqrt{3}\cdot a_2\cdot c_\theta + s_\theta\cdot\sin\varphi_{0,cs}\right)^2}{(I_1 + 3\cdot a)\cdot a_2^{\,2}\cdot\sin^2\varphi_{0,cs}} + I_1 - \kappa = 0 \tag{22}$$

where $\phi_{0,cs}$ is the critical state friction angle, and the state parameter κ is only of interest for associated flow ($F = Q$), then it will be the hardening parameter defining the size of the yield surface. The actual value of κ has no influence on the flow direction, but is needed for an apparent pre-consolidation approach (to be used in a scheme similar to the apparent cohesion scheme). The critical state friction angle can be found by comparing flow directions from the different potential functions above and at the same time satisfying the yield criterion. An equation of the following form can be found: $\sin\phi_{cs} = f(\sin\phi_0, \sin\psi_0, \theta)$. For $\theta = 0°$ the relationship between the critical state friction angle and the dilatancy angle is:

$$sin\varphi_{0,cs} = \sqrt{\sin^2\varphi_0 - 2\cdot\sin\varphi_0\cdot\sin\psi_0} \tag{23}$$

For other Lode angles, the relationship is quite complicated. One sees that the limit for maximum value is $\sin\psi_0 < 0.5\cdot\sin\phi_0$. For other values of a_1 and a_2 this is limited by $\sin\psi_0 < f(a_1, a_2/\sin\phi_0)\cdot\sin\phi_0$. Figure 8 gives the range for limit of the normalized dilatancy angle as a function of a_1 and the ratio $a_2/\sin\phi_0$.

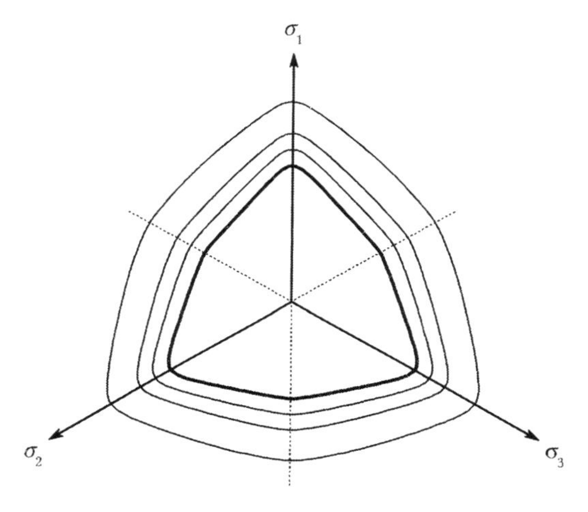

Figure 7. Semi-conformity for F = *const.* for convex yield surfaces shown in the π-plane.

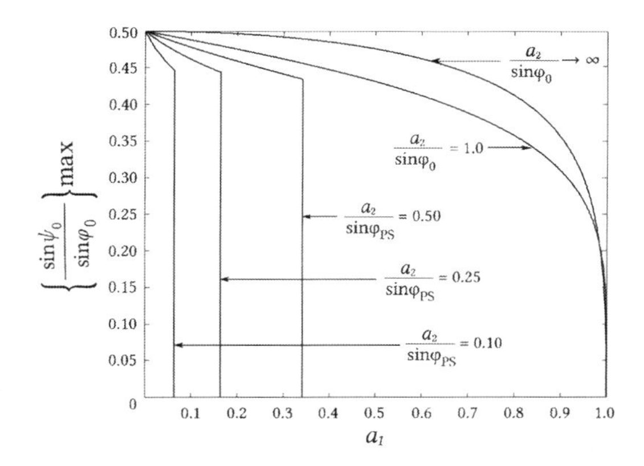

Figure 8. Limit of dilatancy angle for Cam-Clay type of plastic potential function.

5 NUMERICAL SIMULATIONS

The generalized failure criterion is implemented as a user defined soil model (UDSM) in the finite element package PLAXIS, using an implicit calculation scheme. The elastoplastic model is in this study a linear elastic perfectly plastic model. The numerical scheme uses the Newton-Raphson method for iteration with the unknowns being the plastic multiplier $\Delta\lambda$ and the new stress state, $\boldsymbol{\sigma}^{n+1}$. The residuals used are that the sum of elastic strain increment, $\Delta\boldsymbol{\varepsilon}^e$, and the plastic strain increment, $\Delta\boldsymbol{\varepsilon}^p = \Delta\lambda \cdot \partial Q/\partial\boldsymbol{\sigma}$ equals the total strain increment, as well as satisfying $F = 0$. The value for F is normalized such that the error is independent on dimension of the stress measure.

Table 3. Calculated plane strain friction angle and passive earth pressure coefficients from the analytical solution.

	$\sin\phi_0$	$\sin\phi_{PS}$	K_P [Analytical]
LD	0.5933	0.5933	8.10
MN	0.5547	0.5608	6.77
GMC	0.5201	0.5240	5.62

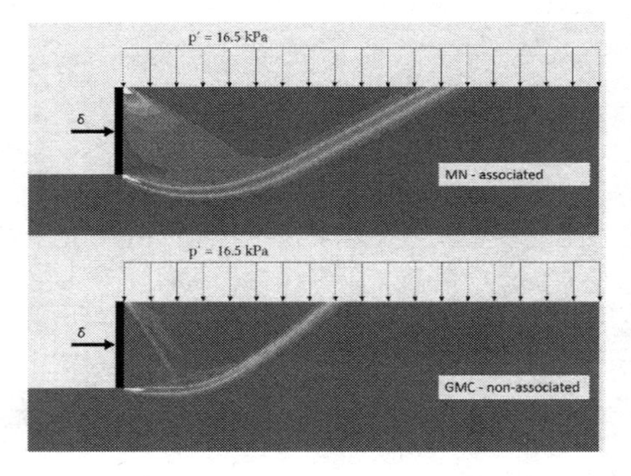

Figure 9. Geometry of the passive earth pressure problem and simulation results in terms of shear strain contours.

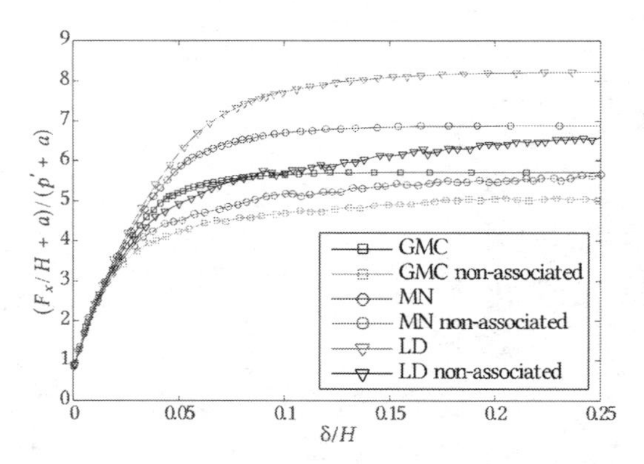

Figure 10. Load-displacement curves for the different simulations.

Note that when the failure criterion is used as a yield criterion, the apex in the meridional plane is not continuously differentiable; this may be solved by using e.g. a hyperbolic shape for the yield criterion (this is done in Figure 5). Continuity for the potential surface in vicinity of the apex may be solved by using an elliptically shaped potential surface (eq. (22)), which in the apex will be tangent to the hyperbola.

A plane strain passive earth pressure problem with rough contact is simulated in order to tests the model performance and verify the numerical scheme. The problem is calculated with the GMC, LD and MN parameters from Table 1 and Table 2 and is combined with shear stiffness, G, of 4 MPa and bulk stiffness, K, of 10 MPa. All three types of failure criteria use associated and non-associated flow. In case of non-associated flow, the apparent cohesion approached is used (Krabbenhoft et al., 2012), with $\psi_0 = 10°$. A surface load of 16.5 kPa and attraction of 25 kPa is applied and the soil is weightless. Nordal (2008) has previously simulated the same problem using the Mohr-Coulomb model.

Nordal showed large difference in capacity dependent on dilatancy angle. Similar differences are expected in this study and is not considered as part of the main topic for this article. Capacities for non-associated flow is a large subject of its own.

Figure 9 shows the geometry of the problem and results in terms of shear strain at failure for two of the simulations. The analyses uses a number of 2019 [15 noded] triangular elements. Figure 10 gives the normalized load $[(F_x/H + a)/(p' + a) = K_P]$ – displacement $[\delta/H]$ curves for the six analyses. Again considering that all the models will predict the same drained triaxial compression strengths, the variation in the earth pressure is significant. Observe that the associated GMC and the non-associated MN predicts similar capacities. This means that, if the MN failure criteria with non-associated flow best captures "true" soil behavior, the associated GMC model can replace it for this particular case. Said in other words, the additional capacity given by MN criteria is "eaten" up by the non-associated flow.

Plasticity theory gives a Passive Earth pressure coefficient, K_p, for a rough wall of:

$$K_P = \exp\left(\left(\frac{\pi}{2} + \varphi_{PS}\right) \cdot \tan\varphi_{PS}\right) \cdot \left(1 + \sin\varphi_{PS}\right) \quad (24)$$

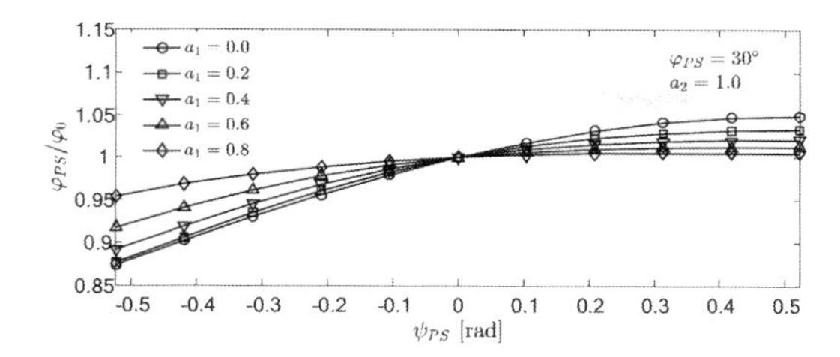

Figure 11. Plane strain friction angle for different dilatancy angles and a_1 (for $a_2 = 1.0$).

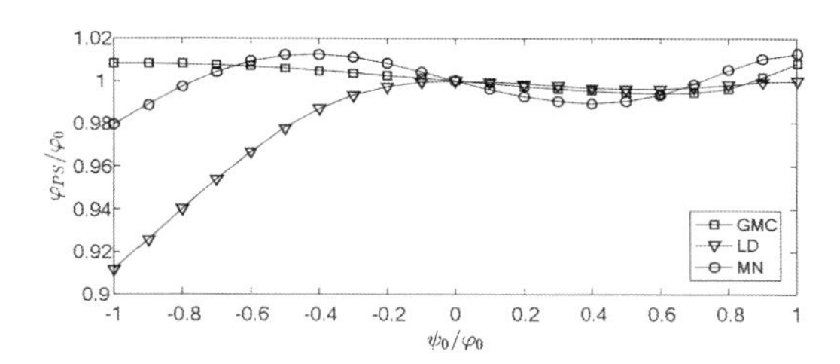

Figure 12. Plane strain friction angle for different dilatancy angles for the three parameter sets used, $\phi_{TXC} = 30°$.

where ϕ_{PS} is the friction angle under plane strain condition (in radians). If one takes into account, the actual predicted friction angle in plane strain condition, for the different combination of parameters, the calculations predicts reasonable capacities for the cases with associated flow, see Table 3. Figure 11 gives an example of the plane strain friction angle as a function of dilatancy angle for different a_1 parameter, for $a_2 = 1.0$ and $\phi_0 = 30°$. Similar curves can be established for other values of ϕ_0 and a_2 by solving the condition for $\partial Q/\partial \sigma_2 = 0$ satisfying $F = 0$. Figure 12 gives the solution for the parameters used in this article. Interestingly for LD ϕ_{PS} is equal to ϕ_0 for associated flow.

6 CONCLUSION

A generalized frictional failure criterion for soil is successfully formulated, implemented and verified. The criterion gives C^ω continuity for a generalized Mohr-Coulomb criterion (i.e. slightly rounded shape). The rounded shape beneficial for implicit schemes where C^2 continuity is required. The formulation also incorporates existing failure criteria as e.g. Lade-Duncan and Matsuoka-Nakai. In addition, it avoids branches of $F = 0$ appearing in other octants. It is worth mentioning that the original formulation of Lade-Duncan and Matsouka-Nakai suffers with this problem. Due to the general form of the criterion, it can describe concave surfaces. However, for reasonable combinations of parameters convexity is ensured, both for surfaces corresponding to $F = 0$ and $F > 0$. For $F \to \infty$ the shape in the π-plane converges towards a circle with infinite radius. This means that it gives additional robustness, when used in an implicit integration scheme, especially for cases of a trial stress state far outside yield. Stress states that may even be in other octants, giving $F < 0$ for the LD criterion or the MN criterion. The proposed criterion can easily be incorporated into more advanced constitutive models with strain hardening/softening. As an example, it may be used for Lode angle dependency of the Critical State line in different versions of the Modified Cam Clay model.

The numerical analysis demonstrates that the criterion implements well into a FE code. The passive earth pressure problem shows the large differences in predicted capacity for the different combinations of parameters, even though all combinations gives the same shear strength in triaxial compression. For the case of the cases with associated flow, the calculated value is according to plasticity theory.

REFERENCES

Argyris, J.H., Faust, G., Szimmat, J., Warnke, E.P. & Willam, K.J. 1974. Recent developments in the finite element analysis of prestressed concrete reactor vessels. *Nuclear Engineering and Design,* 28, 42–75.

Bardet, J.P. 1990. Lode Dependences for Isotropic Pressure-Sensitive Elastoplastic Materials. *Journal of Applied Mechanics,* 57, 498–506.

Billington, E.W. 1988. Generalized isotropic yield criterion for incompressible materials. *Acta Mechanica,* 72, 1–20.

Drucker, D.C. & Prager, W. 1952. Soil Mechanics and Plastic Analysis or Limit Design. *Quarterly of Applied Mathematics,* 10, 157–165.

Grimstad, G., Andresen, L. & Jostad, H.P. 2012. NGI-ADP: Anisotropic shear strength model for clay. *International Journal for Numerical and Analytical Methods in Geomechanics,* 36, 483–497.

Krabbenhoft, K., Karim, M.R., Lyamin, A.V. & Sloan, S.W. 2012. Associated computational plasticity schemes for nonassociated frictional materials. *International Journal for Numerical Methods in Engineering,* 90, 1089–1117.

Lade, P.V. & Duncan, J.M. 1975. Elastoplastic stress-strain theory for cohesionless soil. *ASCE Journal of the Geotechnical Engineering Division,* 101, 1037–1053.

Lagioia, R. & Panteghini, A. 2016. On the existence of a unique class of yield and failure criteria comprising Tresca, von Mises, Drucker–Prager, Mohr–Coulomb, Galileo–Rankine, Matsuoka–Nakai and Lade–Duncan. *Proceedings of the Royal Society A: Mathematical, Physical and Engineering Science,* 472.

Lin, F. & Bažant, Z. 1986. Convexity of Smooth Yield Surface of Frictional Material. *Journal of Engineering Mechanics,* 112, 1259–1262.

Matsuoka, H. 1976. On the significance of the spatial mobilized plane. *Soils and Foundations,* 16, 91–100.

Matsuoka, H. & Nakai, T. Stress-deformation and strength characteristics of soil under three different principal stresses. Proceedings of JSCE, 1974. 59–70.

Matsuoka, H. & Nakai, T. 1985. Relationship among tresca, mises, mohr-coulomb and matsuoka-nakai failure criteria. *Soils and foundations,* 25, 123–128.

Nordal, S. 2008. Can we trust numerical collapse load simulations using nonassociated flow rules? *12th Int Conference of International Assoc for Computer Methods and Advances in Geomechanics.* Goa.

Panteghini, A. & Lagioia, R. 2014. A fully convex reformulation of the original Matsuoka–Nakai failure criterion and its implicit numerically efficient integration algorithm. *International Journal for Numerical and Analytical Methods in Geomechanics,* 38, 593–614.

Van Eekelen, H.A.M. 1980. Isotropic yield surfaces in three dimensions for use in soil mechanics. *International Journal for Numerical and Analytical Methods in Geomechanics,* 4, 89–101.

APPENDIX

Deriving expression for $\sin\phi_{cs}$:

$\mathrm{Sin}\phi_{cs}$ can be expressed as a function of $\sin\phi_0$ and $\sin\psi_0$, First derivatives for the two potentials with respect to I_1 and J_2 at a stress state that satisfies $F = 0$ must be giving equal flow direction:

$$\frac{\partial Q_1}{\partial I_1}\cdot\frac{\partial Q_2}{\partial J_2} = \frac{\partial Q_1}{\partial J_2}\cdot\frac{\partial Q_2}{\partial I_1}$$

Resulting in:

$$-\frac{\left(\sqrt{3}+t_\theta\cdot\dfrac{\sin\varphi_{0,cs}}{a_2}\right)^2}{\left(\sqrt{3}+t_\theta\cdot\dfrac{\sin\psi_0}{a_2}\right)\cdot\left(\sqrt{3}+t_\theta\cdot\dfrac{\sin\varphi_0}{a_2}\right)}\cdot\left(1-\frac{\dfrac{\sin\varphi_{0,cs}}{a_2}-\sqrt{3}\cdot t_\theta}{\sqrt{3}+t_\theta\cdot\dfrac{\sin\varphi_{0,cs}}{a_2}}\cdot\frac{a_1\cdot\sin 3\theta}{\sqrt{1-a_1^2\cdot\sin^2 3\theta}}\right)\cdot\frac{\sin\psi_0\cdot\sin\varphi_0}{\sin^2\varphi_{0,cs}}$$

$$=\left(1-\frac{\left(\dfrac{\sin\psi_0}{a_2}-\sqrt{3}\cdot t_\theta\right)\cdot\left(\sqrt{3}+t_\theta\cdot\dfrac{\sin\varphi_0}{a_2}\right)}{\left(\sqrt{3}+t_\theta\cdot\dfrac{\sin\psi_0}{a_2}\right)^2}\cdot\frac{a_1\cdot\sin 3\theta}{\sqrt{1-a_1^2\cdot\sin^2 3\theta}}\cdot\frac{\sin\psi_0}{\sin\varphi_0}\right)\cdot\frac{1}{2}\cdot\left(1-\frac{\left(\sqrt{3}+t_\theta\cdot\dfrac{\sin\varphi_{0,cs}}{a_2}\right)^2}{\left(\sqrt{3}+t_\theta\cdot\dfrac{\sin\varphi_0}{a_2}\right)^2}\cdot\frac{\sin^2\varphi_0}{\sin^2\varphi_{0,cs}}\right)$$

where: $\dfrac{s_\theta}{c_\theta}=t_\theta$

This equation can be solved for different values of θ, for $\theta = 0°$ (or $a_1 = 0$) the expression becomes rather simple as the equation given in the paper.

Table 4. Generalized Mohr-Coulomb ($\phi_{TXC} = 30°$) for different shapes.

	$\sin\phi_0$	a_1	a_2
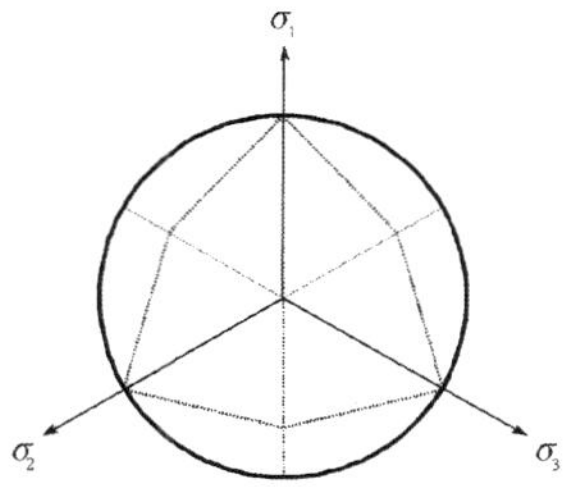 Drucker-Prager	$\frac{2\cdot\sqrt{3}\cdot\sin\varphi_{TXC}}{3-\sin\varphi_{TXC}} = 0.6930$	0	1
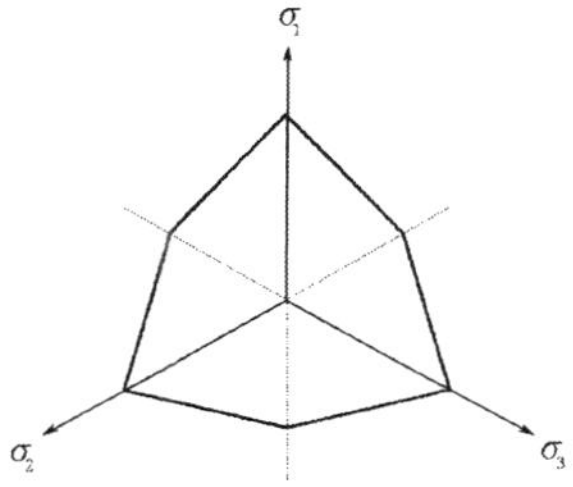 Mohr-Coulomb	$\sin\phi_{TXC} = 0.5000$	$a_1 \to 1$	1
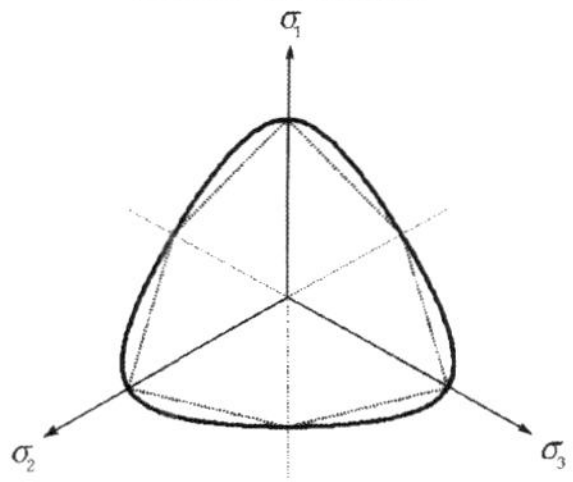 Matsuoka-Nakai	$\frac{2\cdot\sin\varphi_{TXC}}{\sqrt{3+\sin^2\varphi_{TXC}}} = 0.5547$	$\sin\varphi_0\cdot\frac{3-\sin^2\varphi_0}{2} = 0.7467$	$\sin\varphi_0 = 0.5547$
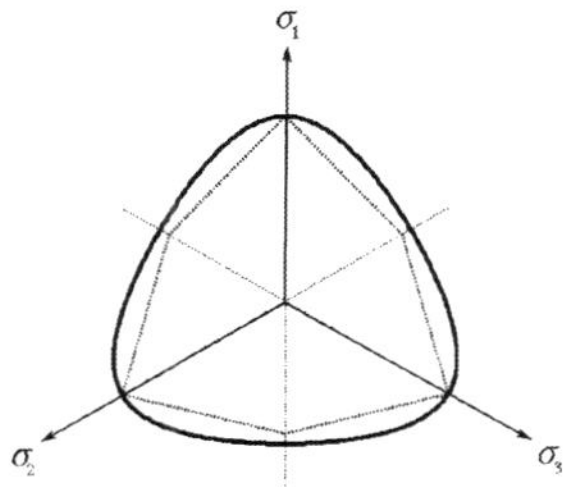 Lade-Duncan	$\sqrt{\frac{9-7\cdot\sin\varphi_{TXC}}{3-\sin\varphi_{TXC}}}\cdot\frac{2\cdot\sin\varphi_{TXC}}{3-\sin\varphi_{TXC}} = 0.5933$	$\sin\phi_0 = 0.5933$	$\sin\phi_0 = 0.5933$

(*Continued*)

Table 4. (*Continued*).

	$\sin\phi_0$	a_1	a_2
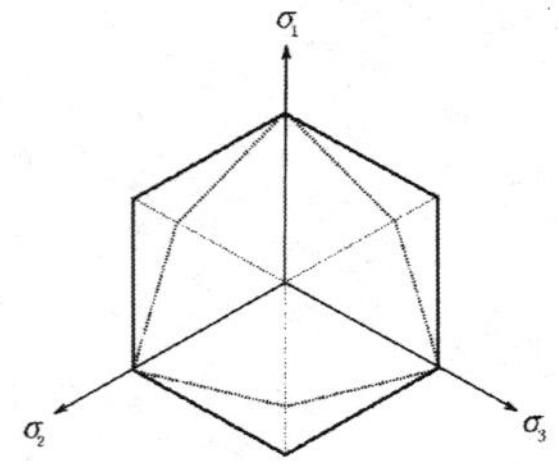 σ_1 σ_2 σ_3 Extended Tresca	$\frac{3 \cdot \sin\varphi_{TXC}}{3 - \sin\varphi_{TXC}} = 0.6000$	$a_1 \to 1$	$a_2 \to ($
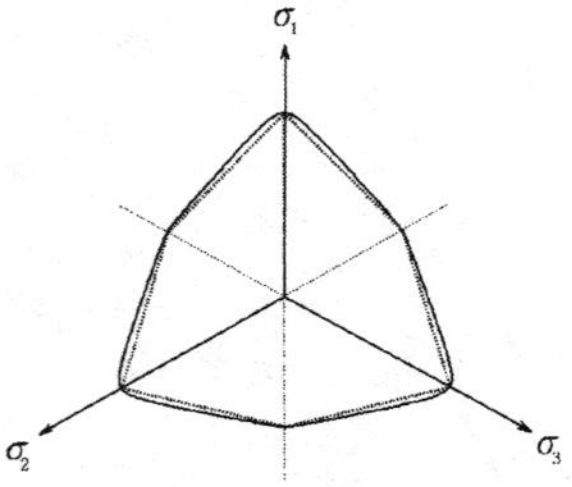 σ_1 σ_2 σ_3 Generalized Mohr-Coulomb	$\frac{2}{\sqrt{3}} \cdot \cos\beta \cdot \sin\varphi_{TXC} = 0.5201$	0.9750	$2 .\sin\beta = 0.8680$

Numerical Methods in Geotechnical Engineering IX – Cardoso et al. (Eds)
© 2018 Taylor & Francis Group, London, ISBN 978-1-138-33198-3

Enhanced plasticity modelling of high-cyclic ratcheting and pore pressure accumulation in sands

H.Y. Liu & F. Zygounas
Section of Geo-Engineering, Department of Geoscience and Engineering, Delft University of Technology, Delft, The Netherlands

A. Diambra
Department of Civil Engineering, Faculty of Engineering, University of Bristol, Bristol, UK

F. Pisanò
Section of Geo-Engineering, Department of Geoscience and Engineering, Section of Offshore Engineering, Department of Hydraulic Engineering, Delft University of Technology, Delft, The Netherlands

ABSTRACT: Predicting accurately the response of sands to cyclic loads is as relevant as still challenging when many loading cycles are involved, for instance, in relation to offshore or railway geo-engineering applications. Despite the remarkable achievements in the field of soil constitutive modelling, most existing models do not yet capture satisfactorily strain accumulation under high-cyclic drained loading, nor the the build-up of pore pressures under high-cyclic undrained conditions. Recently, bounding surface plasticity enhanced with the concept of memory surface has proven promising to improve sand ratcheting simulations under drained loading conditions (Corti et al. 2016). This paper presents a new model built by combining the memory surface concept by (Corti et al. 2016) with the well-known SANISAND04 bounding surface formulation proposed by Dafalias and Manzari (2004). The outcome is a new sand model that can reproduce phenomenologically the fabric evolution mechanisms governing strain accumulation under long-lasting loading histories (here up to 10^4 loading cycles). In undrained test simulations, the model proves capable of correctly capturing the rate of pore pressure accumulation, preventing precocious occurrence of cyclic liquefaction.

1 INTRODUCTION

Predicting accurately the response of sands to cyclic loads is still challenging when many loading cycles are involved. More specifically, cyclic accumulation of permanent strain and pore water pressure may lead to reduction of capacity, serviceability and fatigue resistance (Andersen 2015). The term 'ratcheting' is adopted to denote the gradual accumulation of plastic strains under loading cycles (Houlsby et al. 2017). In engineering practice, soil ratcheting is often described by using empirical equations derived from experimental measurements (Pasten et al. 2013). Empirical formulations of this kind may be found for long-term strain accumulation phenomena (Sweere 1990, Lekarp and Dawson 1998, Wichtmann 2005) and short-term pore pressure build-up during earthquakes or storms (Seed et al. 1975, Green et al. 2000, Idriss and Boulanger 2006). Although empirical equations have proven efficient to use, some limitations stand out clearly: (1) high-cyclic tests for calibrating empirical relations are usually costly and time-consuming; (2) the use of reconstituted sand specimens goes beyond the scope of standard soil characterisation. An alternative approach is to set up a reliable advanced constitutive model that can capture satisfactorily sand ratcheting and related strain/pore-water-pressure accumulation trends under different cyclic loading conditions and drainage scenarios.

In this work, the memory surface concept is combined with the structure of the model prosed by Dafalias & Manzari (2004) (SANISAND04 model). The reference SANISAND04 model, which is build upon a critical state and bounding surface plasticity framework, includes a fabric-dilatancy tensor to reproduce soil fabric effects. The suitability of combining bounding surface theory and memory surface concept has been proven by the work of Corti et al. Soil fabric and its evolution are recorded by a newly introduced memory surface, enclosing a stress region which the soil feels to have already experienced and thus characterised by high stiffness.

The main purpose of the paper is to provide a reliable constitutive model to: (1) complement/ replace demanding laboratory sand testing as input to displacement/rotation accumulation procedures; (2) enhance the real time-domain simulation during shorter cyclic loading histories. The proposed model is validated by simulating results of experimental tests performed on a quartz sand and the Karlsruhe sand under cyclic loading conditions. The experimental data regard both drained (Wichtmann 2005) and undrained (Wichtmann and Triantafyllidis 2016) triaxial tests investigating the influence of varying the mean confining pressure, cyclic stress amplitude and void ratio.

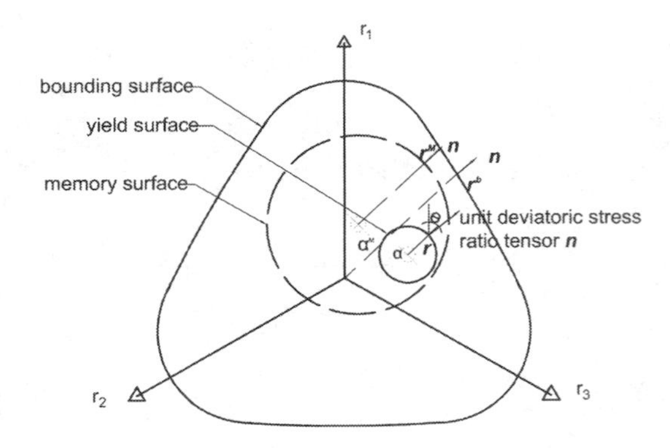

Figure 1. Visualisation in the π plane of the three-surface model formulation.

2 MODEL FORMULATION

2.1 *General modelling strategy*

The proposed model improves the cyclic performance of SANISAND04 by introducing a memory surface to keep track of relevant fabric effects and to simulate realistic sand cyclic behaviour under both drained and undrained loading conditions. The selected backbone bounding surface model (SANISAND04 model) was developed by Dafalias & Manzari (2004), with a fabric-dilatancy tensor defined to take into account the effects of increased dilation following cyclic contractive behaviour. In SANISAND04, soil stiffness is determined by the distance between current stress state and its conjugate point on the bounding surface. However, the fabric-dilatancy tensor, which only activates if the stress path crosses the phase transformation line (PTL), is not sufficient to fully describe the influence of soil fabric especially when the soil is responding within its contractive regime. Therefore, SANISAND04 is less suitable for simulating soil progressive stiffening during cyclic loading, since it will overpredict strain accumulation and the rate of pore pressure build-up. The proposed model introduces a circular-shape memory surface as a third model surface in the normalised π plane, as illustrated in Figure 1. The memory surface evolves during the loading process following three main rules: (1) changes size and position with plastic strains to simulate gradual evolution of the soil fabric; (2) always encloses the current stress state and (3) always enclose the yield surface. The bounding, critical and dilatancy surfaces are defined using an Argyris-type shape to capture the response of the soil at varying load angle. The adoption of state parameter p'_{in} (Been and Jefferies 1985), defined as the difference between the current void ratio and the critical void ratio ($p'_{in} = 100\ kPa, e_{in} = 0.825;$) at the same mean stress level, allows to reproduce the influence of the void ratio over the whole loose-to-dense range. For the critical state line, a power relationship is adopted as suggested by (Li and Wang 1998). The complete model formulation is presented in the work of Liu et al. (2018), following the previous developments by Corti et al. (2016).

2.2 *Implementation of the memory surface*

Introduction of the memory surface is linked to a modification of the hardening coefficient h, see Equation 1:

$$p'_{in} = 100\ kPa, e_{in} = 0.759; \tag{1}$$

Here, $p'_{in} = 300\ kPa, ein = 0.744.$ denotes the distance between current stress ratio point $p'_{in} = 200\ kPa, (a) = e_{in} = 0.842, \varsigma = 0.2; (b)\ e_{in} = 0.813, \varsigma = 0.25.$ and its image point on memory surface $\mathbf{r}^M$, projecting along the direction of unit normal to the surface $\mathbf{n}$. The dependency of h on the current pressure p through the term $(p/p_{atm})^{0.5}$ is introduced to improve the original formulation of Corti et al. (2016) and follows some findings from Corti et al. (2017). Definitions for b_0 and $\mathbf{r}_{in}$ are the same as in Dafalias & Manzari (2004). p_{atm} is the atmospheric pressure. b_{ref} , which is adopted for normalisation, is the reference distance indicates the size of bounding locus along the current load angle. The memory-surface-related parameter μ_0 quantifies the influence of fabric effects on soil stiffness, especially in the transition from ratcheting to shakedown behaviour. In this case, soil stiffness depends not only on the relative distance between current stress state ($\mathbf{r}$) and its image point on bounding surface ($\mathbf{r}^b$), but also on the distance between $\mathbf{r}$ and its image point on memory surface ($\mathbf{r}^M$). This is reflected in the expression of the hardening modulus K_p in Equation 2.

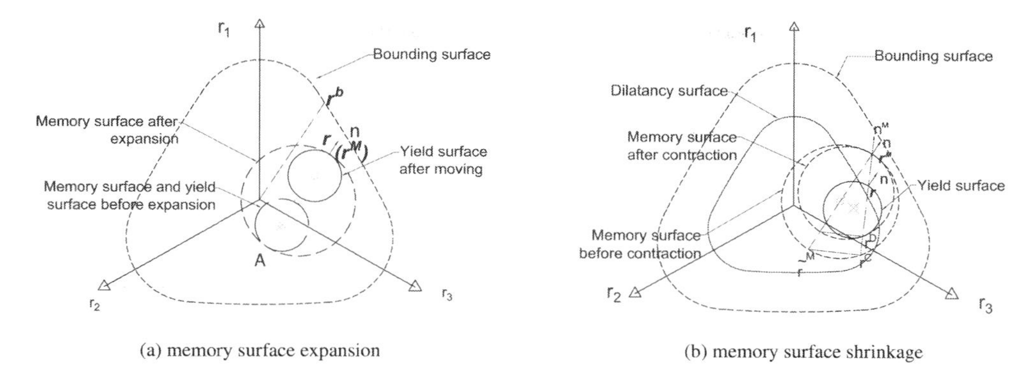

(a) memory surface expansion (b) memory surface shrinkage

Figure 2. Evolution of the memory surface size: (a) memory surface expansion during virgin loading conditions; (b) memory surface shrinkage during dilative straining.

$$K_p = \frac{2}{3} ph(\mathbf{r}^b - \mathbf{r}) : \mathbf{n} \quad (2)$$

The evolution of memory surface center α^M is assumed to be along the direction $\mathbf{r}^b - \mathbf{r}^M$ - see Equation 3:

$$d\alpha^M = \frac{2}{3} \langle L \rangle h^M (r^b - r^M) \quad (3)$$

The evolution of memory surface size is shown mathematically with Equation 4. From experimental observations, contractive soil behaviour (positive volumetric strain by geotechnical convention) leads to more stable soil fabric configuration and therefore, stiffer soil behaviour. Thus, it is reasonable to link it with an expansion of the memory surface. The expansion of the memory surface mainly take places during virgin loading (see Figure 3), which is defined as the state when the yield and memory surfaces are tangential to each other at the current stress point. The size of the memory surface is linked to the variable m^M and its expansion is linked to the evolution of memory surface center, as shown in the first term of the Equation 4 (i.e., $\sqrt{2/3} d\boldsymbol{\alpha}^M : \boldsymbol{n}$):

$$dm^M = \sqrt{\frac{3}{2}} d\boldsymbol{\alpha}^M : \boldsymbol{n} - \frac{m^M}{\zeta} \left(1 - \frac{x_1 + x_2}{x_3}\right) \langle -d\varepsilon_v^p \rangle \quad (4)$$

Here, $x_3 = (\mathbf{r}^M - \mathbf{r}^C) : \mathbf{n}^M$, $x_1 + x_2 = (\mathbf{r}^M - \mathbf{r}^D) : \mathbf{n}^M$ (see Figure 3). The unit tensor $\mathbf{n}^M$ represents the direction of $\mathbf{r}^M - \mathbf{r}$ .During virgin loading conditions, soil stiffness is governed only by the distance between the current stress and its image on the bounding surface. Under non-virgin conditions (for example after load reversal), the memory surface acts as an additional bounding surface. Soil stiffness is increased by the non-zero distance b^M (see Equation 1) between yield surface (f) and memory surface (f^M). Stiffer soil behaviour is captured in this manner.

By contrast, dilative soil behaviour leads to a weaker granular arrangement (also known as 'fabric damage'), and the soil loses part of its stiffness. This situation can therefore be linked to the shrinkage of the memory surface size. In the model, the memory surface contraction mechanism, or fabric damage mechanism, only activates when negative plastic volumetric strains are generated. As presented by the second term of Equation 4, if positive volumetric strains are generated, $\langle -d\varepsilon_v^p \rangle = 0$ because of the MacCauley brackets. The reduction in memory surface size, which is schematically described in Figure 3, happens on the opposite side of $\mathbf{r}^M$, along thedirection ($\mathbf{r}^M \mathbf{r}^C$). To guarantee that the yield surface always lies inside the memory surface, the size reduction ends if the memory surface and the yield surface become tangential to each other at the contraction point (in other words, when $1-(x_1+x_2)/x_3 = 0$), even if the soil is still experiencing negative volumetric strain. The reduction rate of memory surface size is controlled by the parameter ζ.

2.3 *Enhanced dilatancy coefficient*

In SANISAND04, a fabric-dilatancy tensor is introduced into the flow rule, or to be more specific, into the dilatancy coefficient D. The feature is introduced to reproduce the experimental observation that when stress path crosses the PTL and load increment reversal is imposed, there is an obvious increase in pore water pressure build-up with respect to the number of cycles under undrained cyclic loading conditions. The underlying mechanism is the change of fabric orientation. Instead of using a single tensor, the proposed model accounts for fabric effects through the relative position and distance between the memory surface and the dilatancy surface, according to Equation 5:

$$D = A_d(\mathbf{r}^d - \mathbf{r}) : \mathbf{n} \qquad A_d = A_0 \exp\left(\beta \frac{\langle \tilde{b}_d^M \rangle}{b_{ref}}\right) \qquad (5)$$

In this work, whether the sand is more prone to dilation or contraction is determined by the term $\tilde{b}_d^M = (\tilde{\mathbf{r}}^d - \tilde{\mathbf{r}}^M) : \mathbf{r}$, which basically modulates the magnitude of D (not the sign) depending on the occurrence of previous dilation or contraction. The graphical representation of this mechanism is provided in Figure 3. If $\tilde{b}_d^M > 0$, the soil has experienced dilation during the previous loading process, implying a sort of 'contraction bias' under subsequent unloading. The dilatancy D is enlarged because of a larger dilation coefficient A_d by noticing that the term $\exp\left(\beta\langle \tilde{b}_d^M \rangle / b_{ref}\right) > 1$ in this situation. In this way, the model can simulate more significant pore water pressure build-up under undrained loading. Conversely, if $\tilde{b}_d^M < 0$, soil fabric orientation is biased toward dilation, $\exp\left(\beta\langle \tilde{b}_d^M \rangle / b_{ref}ht\right) = 1$. It should be noted that the dilatancy coefficient in Equation 5 substitutes the fabric-tensor concept used by Dafalias and Manzari (2004), and the two associated constitutive parameters.

3 MODEL CALIBRATION

The model parameters can be divided into two sets. The first set can be entirely derived from monotonic tests, although they influence the cyclic performance as well (from G_0 to n^d in Table 1and Table 2.

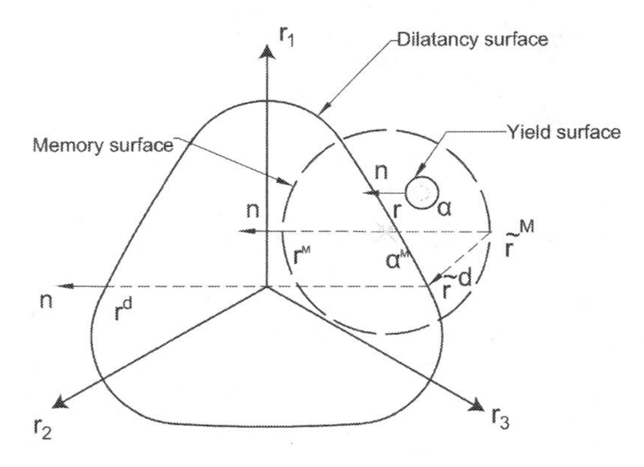

Figure 3. Illustration of image points for modifying current dilatancy coefficient.

Their calibration rely on drained and/or undrained monotonic triaxial tests (Dafalias and Manzari 2004, Taiebat and Dafalias 2008) The second set contains the memory surface related parameters i.e., μ_0, ζ and β in the Table 1 which can be estimated by fitting procedures (Liu et al. 2018). Calibration of μ_0 can be more easily achieved through cyclic tests where the soil is experiencing contractive behaviour. In this condition, ζ and β, have no influence on the model prediction because of the absence of dilative soil behaviour. Conversely, calibration of ζ and β requires tests with (at least part of) loading paths crossing the PTL. In particular, the parameter β controls the reduction rate of the mean effective stress during undrained loading after stages of dilative deformation (after which more pronounced effective stress reduction is observed experimentally).

In Table 1 model parameters for a quartz sand are calibrated based on the experimental work conducted by Wichtmann (2005). Six sets of drained monotonic triaxial tests are used for calibrating the monotonic parameters, while memory surface-related parameters are calibrated based on the strain accumulation curve with loading cycles up to 10^4. Table 2 lists model parameters for the Karlsruhe fine sand-experimental data published by Wichtmann and Triantafyllidis. In total, 10 sets of drained and undrained monotonic triaxial tests are simulated to identify the monotonic parameters. For the memory surface parameters, the calibration relies on isotropic cyclic triaxial undrained tests, using the observed trends of accumulated pore water pressure ratio against number of cycles.

4 MODEL VALIDATION

4.1 *Drained tests*

For the drained case, model parameters are calibrated against experimental results from Wichtmann (2005) on a quartz sand, as listed in Table 3. For memory surface-related parameters, the calibration procedure relies on the drained cyclic test with $e_{in} = 0.674$, with other load conditions the same as described in Figure 4.1.1. Performance of the proposed model is validated through the comparison between triaxial experimental results and simulation results under cyclic loading conditions, with focus on different initial void ratio and different cyclic stress amplitude q^{ampl}. Wichtmann's experiments concern one-way asymmetric cyclic loading performed in two stages: after the initial isotropic consolidation up to $p = p_{in}$, p- constant shearing is first performed to reach the target average stress ratio $\eta^{ave} = q^{ave}/p_{in}$; then, cyclic axial loading at constant radial stress is applied to obtain

Table 1. Model parameters for the quartz sand for drained cyclic simulations.

Elasticity		Critical state					Yield surface	Plastic modulus			Dilatancy		Memory surface		
G_0	v	M	c	λ_c	e_0	ξ	m	h_0	c_h	n^b	A_0	n^d	u_0	ζ	β
130	0.05	1.25	0.702	0.015	0.81	0.7	0.01	5.05	1.05	2.25	1.06	1	270	0.0005	0.2

Table 2. Model parameters for the Karlsruhe fine sand for undrained cyclic simulations.

Elasticity		Critical state					Yield surface	Plastic modulus			Dilatancy		Memory surface		
G_0	v	M	c	λ_c	e_0	ξ	m	h_0	c_h	n^b	A_0	n^d	u_0	ζ	β
95	0.05	1.35	0.85	0.056	1.038	0.28	0.01	7.6	1.015	1.2	0.56	2.15	85	0.0005	6

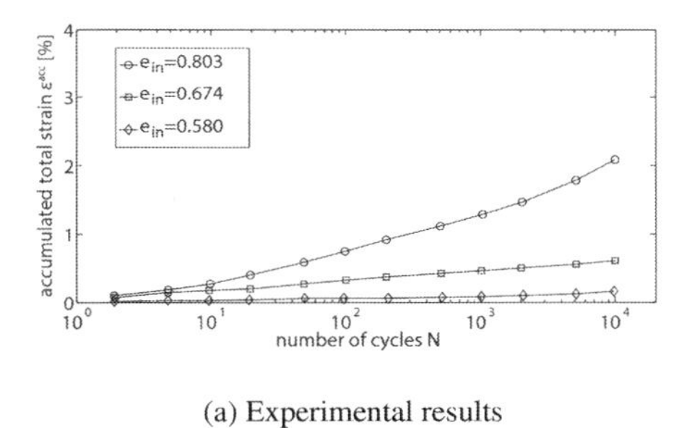

(a) Experimental results

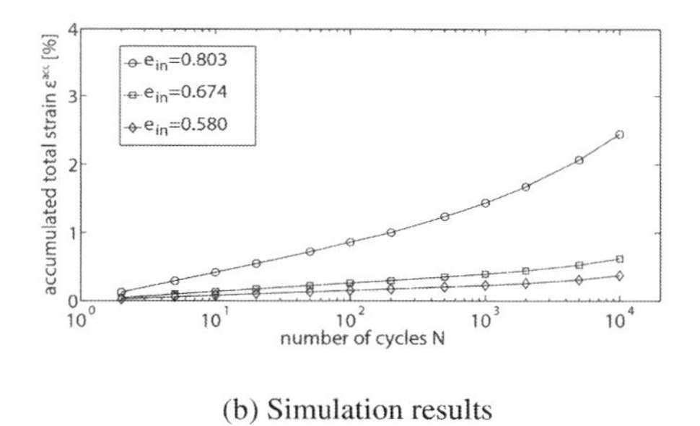

(b) Simulation results

Figure 4. Influence of soil density on strain accumulation under drained conditions (a) experimental results; (b) simulation results. Confining pressure $p_{in} = 200\ kPa$, stress obliquity $\eta^{ave} = 0{:}75$, stress amplitude $q^{ampl} = 60\ kPa$.

cyclic variations in deviatoric stress q about the average value q^{ave}, i.e. $q = q^{ave} \pm q^{ampl}$. High-cyclic sand parameters are tuned to match the evolution during regular cycles of the accumulated strain norm ε^{acc} defined through the accumulated axial strain ε_a^{acc} and accumulated radial strain ε_a^{acc}, as:

$$\varepsilon^{acc} = \sqrt{\left(\varepsilon_a^{acc}\right)^2 + 2\left(\varepsilon_r^{acc}\right)^2} \tag{6}$$

4.1.1 *Influence of initial void ratio*

The influence of the soil initial void ratio on soil drained cyclic behaviour is studied by performing drained triaxial tests on soil samples with initial void ratios $e_{in} = 0.803$, 0.674 and 0.580. Other loading parameters are kept unaltered: pin = 200 kPa, $\eta_{ave} = 0.75$ and $q_{ampl} = 60\ kPa$. Comparison between experimental results and model simulations are presented in Figure 4.1.1. Based on the experimental observations, denser sand specimens accumulate less strains. The model seems to capture quantitatively well all relevant trends. The model also accurately predicts the accumulated strain for the cases of $e_{in} = 0.803$ and 0.674, the loose and medium dense sand. However, for the dense sample $e_{in} = 0.580$, the model slightly overestimates the accumulated strain.

4.1.2 *Influence of cyclic stress amplitude*

The impact of the cyclic stress amplitude on the accumulated permanent stain under drained cyclic conditions is studied by performing three different stress amplitudes $q^{ampl} = 80\ kPa$, $60\ kPa$ and $31\ kPa$ on three soil samples. In the tests, confining pressure $p_{in} = 200\ kPa$, stress obliquity $\eta_{ave} = 0.75$, initial void ratio $e_{in} = 0.702$ are considered. Comparison between experimental results and model simulations are presented in Figure 5. The experimental results show that increasing cyclic stress amplitude leads to larger accumulated strain, corroborating findings of other experimental works (Escribano

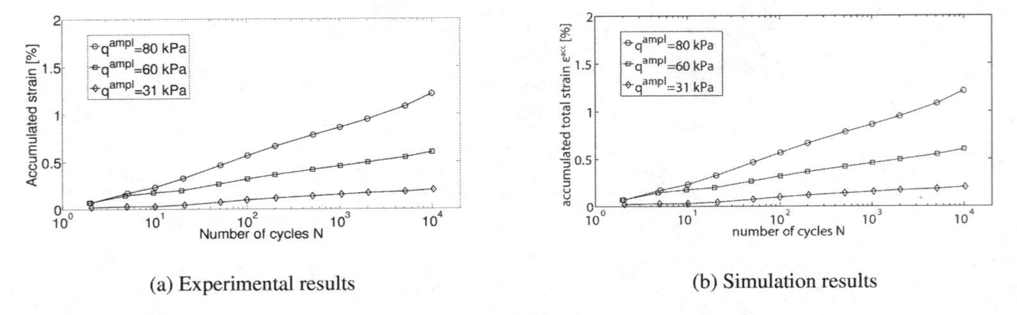

(a) Experimental results

(b) Simulation results

Figure 5. Influence of cyclic stress amplitude on strain accumulation under drained conditions (a) experimental results; (b) simulation results. Confining pressure $p_{in} = 200\ kPa$, stress obliquity $\eta^{ave} = 0{:}75$, initial void ratio $e_{in} = 0{:}702$.

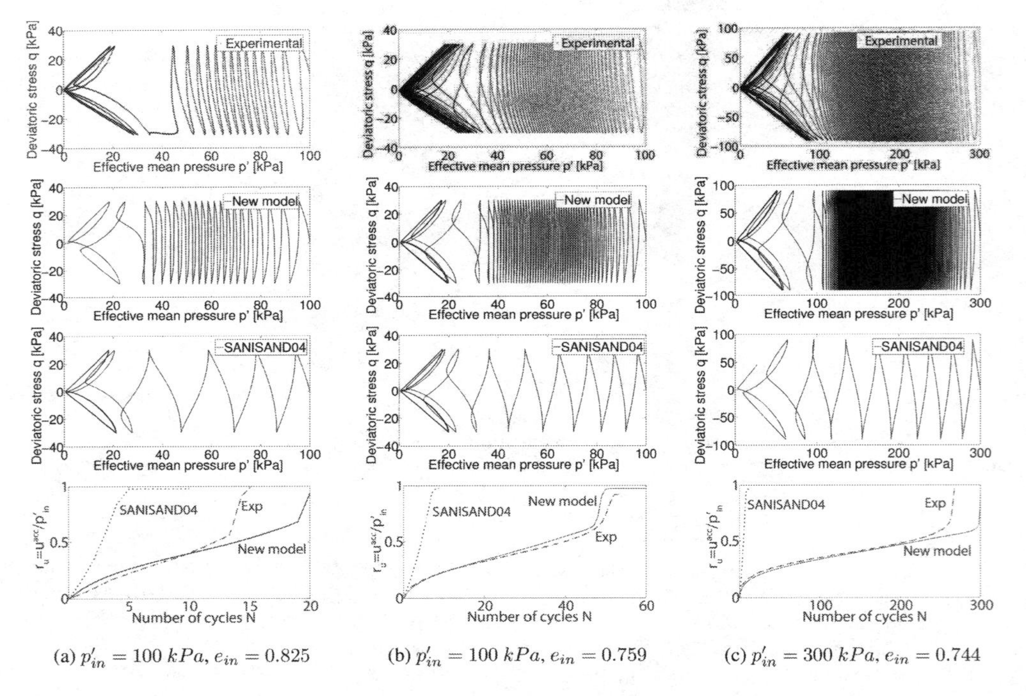

(a) $p'_{in} = 100\ kPa$, $e_{in} = 0.825$

(b) $p'_{in} = 100\ kPa$, $e_{in} = 0.759$

(c) $p'_{in} = 300\ kPa$, $e_{in} = 0.744$

Figure 6. Influence of initial effective confining pressure p'_{in} in and initial void ratio ein on soil undrained cyclic behaviour, with a drained cycle be applied prior to the undrained cycles. (a) $p'_{in} = 100\ kPa$, $e_{in} = 0.825$; (b) $p'_{in} = 100\ kPa$, $e_{in} = 0.759$; (c) $p'_{in} = 300\ kPa$, $e_{in} = 0.744$. All tests with $\varsigma = 0.3$.

et al. 2018). This is successfully predicted by the model, see Figure 5.

4.2 *Undrained tests*

The undrained perfromance of the model is validated by comparing model simulations with five sets of experimental results by Wichtmann and Triantafyllidis (2016) on Karlsruhe fine sand. Model parameters are listed in Table 2. The results are presented in terms of accumulated pore water pressure ratio (r_u, which represents the accumulated pore water pressure u^{acc} normalised by the initial effective confining pressure, i.e. $[\ r_u = u^{acc}/p'_{in}\ ]$). Parameters μ_0,ζ and β are calibrated by fitting the experimental data presented in Figure 6 (b).

4.2.1 *Influence of initial effective confining pressure and void ratio*

In Figure 6, three sets of results are presented with simulation results from both the new model and the original SANISAND04 model. For all three sets of results, the simulation of the SANISAND04 model show clear liquefaction with accurate ultimate r^u level. However, the predicted number of cycles to liquefaction is in all instances too low. In other words, SANISAND04 underestimates significantly the resistance to liquefaction, especially as the void ratio deviates from e_{max}.

The impact of the initial void ratio is studied by combing the test results on medium dense sand (Figure 6(a), the initial void ratio $e_{in} = 0.825$) with the dense sand results (Figure 6(b), the initial void ratio $e_{in} = 0.759$). For both tests, the initial effective pressure $p'_{in} = 100\ kPa$, the cyclic stress ratio ς, defined as $\varsigma = q^{ampl} / p'_{in}$, equals to 0.3. A slower increasing of r_u is observed compared to that of for densersand. It indicates that under the same loading condition, the resistance to liquefaction for dense sands is higher than for loose sands. The new model predicts the ultimate r_u level and number of loading cycles in a reasonable accurate magnitude with a single set of parameters, especially for the dense sand.

The influence of the initial effective confining pressure p'_{in} is shown in Figure 6(a) and Figure 6(c). For both tests, ς. In Figure 6(b), the initial effective confining pressure $p'_{in} = 100\ kPa$ and the number of cycles N to liquefaction is 54. For Figure 6c, $p'_{in} = 300\ kPa$ and $N = 269$. Overall, the new model predicts satisfactory the undrained behaviour at both qualitative and quantitative levels.

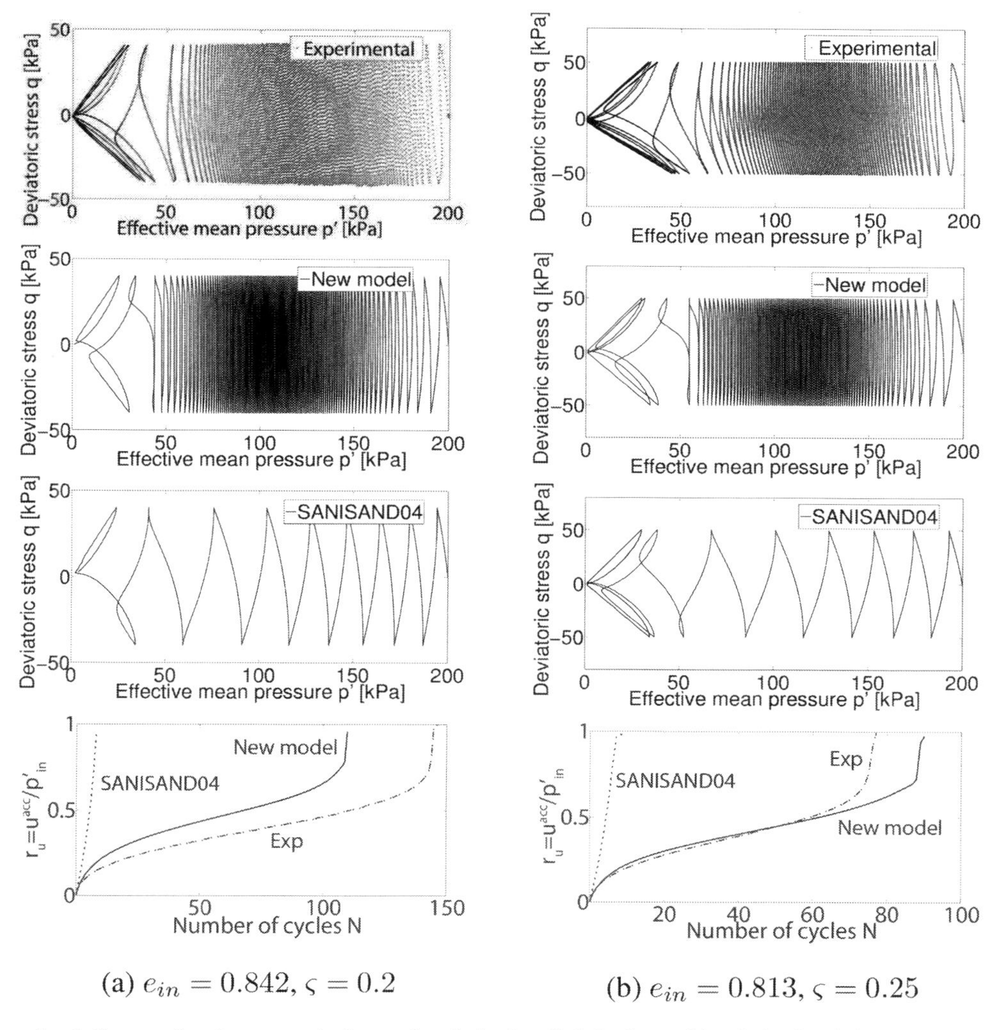

Figure 7. Influence of cyclic stress ratio & on soil undrained cyclic behaviour, with a drained cycle be applied prior to the undrained cycles. Effective confining pressure $p'_{in} = 200\ kPa$, (a) = $e_{in} = 0.842$, (b) $e_{in} = 0.813$, $\varsigma = 0.25$.

4.2.2 *Influence of cyclic stress ratio*
Cyclic triaxial tests are conducted on two soil samples at $\varsigma = 0.2$ and $\varsigma = 0.25$, respectively. For both tests, the effective confining pressure is p'_{in} The results are shown in Figure 7. The experimental results in Figure 7(a) indicate that for the soil sample with $e_{in} = 0.842$ and $\varsigma = 0.2$, liquefaction occurs after 146 loading cycles. The other soil sample, with $e_{in} = 0.813$ and $\varsigma = 0.25$, undergoes 77 loading cycles before liquefaction under the same loading conditions, see Figure 7(b). It indicates that the looser sample subjected to smaller cyclic stress amplitude resists more to liquefaction compared with the denser sample under larger cyclic stress amplitude. Again, SANISAND04 predicts higher pore water pressure generation for each cycle, while the new model performs substantially better.

4.2.3 *Best-fit simulations*
For all undrained tests above, only one single set of parameters has been adopted, including an average value of μ_0 equal to 85. The model performs well in (both qualitatively and quantitatively) capturing the undrained cyclic behaviour under different loading conditions and relative densities. This demonstrates that the strategy of combining memory surface concept with bounding surface hardening theory is suitable to analyse the influence of fabric effects in cyclic loading problems. However, better simulation results are possible if different values of memory surface parameters are adopted for different tests. In Figure 8, five different μ_0 values are adopted to simulate the aforementioned five undrained cyclic triaxial tests (Wichtmann and Triantafyllidis 2016). Simulation results are presented in terms of $u_r - N$. Simulation results with different μ_0 values fit the experimental results better than that of the simulation results obtained using $\mu_0 = 85$ for all tests. This indicates that even if the modelling strategy is suitable, better memory surface evolution law and more suitable flow rule can still be devised.

5 CONCLUSIONS

In this work, a new model was proposed by introducing the memory surface hardening theory into SANISAND04 model to properly simulate sand ratcheting and pore pressure accumulation under high-cyclic loading. The memory surface, which is allowed to evolve both in size and position in this model, represents phenomenologically the evolution of sand fabric during repeated loading. The model predictions agrees well with both drained and undrained cyclic triaxial experimental results under different loading conditions and different soil densities. The validation of the model allowed to check its accuracy with respect to the following aspects: (1) progressive soil stiffening at increasing number of cycles; (2) larger accumulated strain for looser soil samples under drained loading conditions at given cycles; (3) larger accumulated strain for larger stress amplitude under drained loading conditions at given number of cycles; (4) prediction of pore water pressure accumulation under different undrained cyclic loading conditions and soil densities. Overall, the model has promising potential to complement

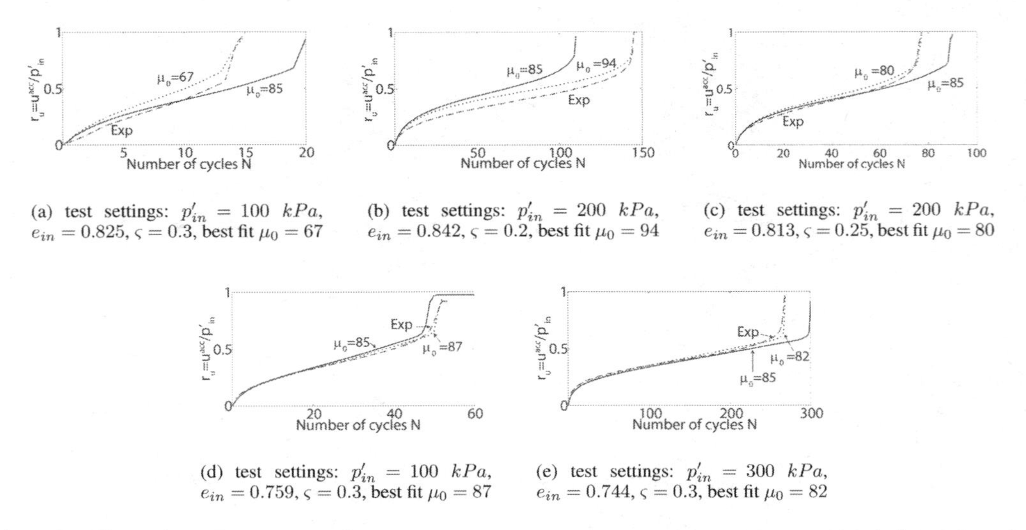

(a) test settings: $p'_{in} = 100\ kPa$, $e_{in} = 0.825$, $\varsigma = 0.3$, best fit $\mu_0 = 67$

(b) test settings: $p'_{in} = 200\ kPa$, $e_{in} = 0.842$, $\varsigma = 0.2$, best fit $\mu_0 = 94$

(c) test settings: $p'_{in} = 200\ kPa$, $e_{in} = 0.813$, $\varsigma = 0.25$, best fit $\mu_0 = 80$

(d) test settings: $p'_{in} = 100\ kPa$, $e_{in} = 0.759$, $\varsigma = 0.3$, best fit $\mu_0 = 87$

(e) test settings: $p'_{in} = 300\ kPa$, $e_{in} = 0.744$, $\varsigma = 0.3$, best fit $\mu_0 = 82$

Figure 8. Illustration of memory surface parameter μ_0 on pore water pressure pressure accumulation ratio under undrained conditions. Experimental results compared with new-model simulation results with average $\mu_0 = 85$ for all five sets and with simulation results with best-fit μ_0 value for each set.

costly high-cyclic laboratory tests, as well as to be employed in time-domain simulations of cyclic/dynamic soil-structure interaction problems.

REFERENCES

Andersen, K. (2015). Cyclic soil parameters for offshore foundation design. *Frontiers in Offshore Geotechnics III 5.*

Been, K. & M. G. Jefferies (1985). A state parameter for sands. *Géotechnique 35*(2), 99–112.

Corti, R., A. Diambra, D. M. Wood, D. E. Escribano, & D. F. Nash (2016). Memory surface hardening model for granular soils under repeated loading conditions. *Journal of Engineering Mechanics 142*(12), 04016102.

Corti, R., S. M. Gourvenec, M. F. Randolph, & A. Diambra (2017). Application of a memory surface model to predict whole-life settlements of a sliding foundation. *Computers and Geotechnics 88*, 152–163.

Dafalias, Y. F. & M. T. Manzari (2004). Simple plasticity sand model accounting for fabric change effects. *Journal of Engineering mechanics 130*(6), 622–634.

Escribano, D., D. Nash, & A. Diambra (2018). Local and global volumetric strain comparison in sand specimens subjected to drained cyclic and monotonic triaxial compression loading. *Geotechnical Testing Journal submitted for publication.*

Green, R., J. Mitchell, & C. Polito (2000). An energy-based excess pore pressure generation model for cohesionless soils. In *Proc.: Developments in Theoretical Geomechanics–The John Booker Memorial Symposium*, pp. 16–17. Sydney, New South Wales, Australia, Nov.

Houlsby, G., C. Abadie, W. Beuckelaers, & B. Byrne (2017). A model for nonlinear hysteretic and ratcheting behaviour. *International Journal of Solids and Structures 120*, 67–80.

Idriss, I. & R. Boulanger (2006). Semi-empirical procedures for evaluating liquefaction potential during earthquakes. *Soil Dynamics and Earthquake Engineering 26*(2), 115–130.

Lekarp, F. & A. Dawson (1998). Modelling permanent deformation behaviour of unbound granular materials. *Construction and building materials 12*(1), 9–18.

Li, X.-S. & Y. Wang (1998). Linear representation of steadystate line for sand. *Journal of geotechnical and geoenvironmental engineering 124*(12), 1215–1217.

Liu, H. Y., J. A. Abell, A. Diambra, & F. Pisan`o (2018). A three-surface plasticity model capturing cyclic sand ratcheting. *Géotechnique submitted for publication.*

Pasten, C., H. Shin, & J. C. Santamarina (2013). Long-term foundation response to repetitive loading. *Journal of Geotechnical and Geoenvironmental Engineering 140*(4), 04013036.

Seed, H. B., P. P. Martin, & J. Lysmer (1975). *The generation and dissipation of pore water pressures during soil liquefaction.* College of Engineering, University of California.

Sweere, G. T. (1990). *Unbound granular bases for roads.* Ph. D. thesis.

Taiebat, M. & Y. F. Dafalias (2008). Sanisand: Simple anisotropic sand plasticity model. *International Journal for Numerical and Analytical Methods in Geomechanics 32*(8), 915–948.

Wichtmann, T. (2005). *Explicit accumulation model for noncohesive soils under cyclic loading.* Ph. D. thesis, Inst. f¨ur Grundbau und Bodenmechanik Bochum University, Germany.

Wichtmann, T. & T. Triantafyllidis (2016). An experimental database for the development, calibration and verification of constitutive models for sand with focus to cyclic loading: part itests with monotonic loading and stress cycles. *Acta Geotechnica 11*(4), 739–761.

Numerical Methods in Geotechnical Engineering IX – Cardoso et al. (Eds)
© 2018 Taylor & Francis Group, London, ISBN 978-1-138-33198-3

A generalized plasticity model adapted for shearing interface problems

B. Kullolli
Federal Institute for Material Testing and Research, Berlin, Germany

H.H. Stutz
Department of Engineering, Aarhus University, Aarhus, Denmark

P. Cuéllar & M. Baeßler
Federal Institute for Material Testing and Research, Berlin, Germany

F. Rackwitz
Chair of Soil Mechanics and Geotechnical Engineering, Technical University of Berlin, Berlin, Germany

ABSTRACT: The response of many geotechnical systems, whose structural behavior depends on shearing effect, is closely related to soil structure interaction phenomenon. Experimentally it is found that the localisation of these effect happens at a narrow soil layer next to the structure. Numerically, this behavior can be modelled through interface elements and adequate constitutive models. In this work, a constitutive model in the framework of Generalized Plasticity for sandy soils has been chosen to be adapted for the interface zone. From the direct shear experiments a sandy soil at loose and dense states under different normal pressures is considered. The adapted constitutive model is able to reproduce contraction and dilatation of the soil according to its relative density and it shows a good agreement with the experimental data.

1 INTRODUCTION

The performance of many geotechnical systems such as piles, gravity foundations and anchors, depends on the soil-structure interaction behavior. For pile foundations under axial loading, important tangential forces develop in a small volume of soil next to the pile as shown by Yoshimi & Kishida (1981) and Boulon (1989). This phenomenon on itself is very complex and is related to different factors such as: roughness of the contact body (Kishida & Uesugi (1987)), normal pressure (Boulon (1989)), soil density (Dejong et al. (2006)) and grading of soils (Uesugi et al. (1988)). According to (Kishida & Uesugi (1987)) the thin layer of soil close to the structural part has a thickness approx. 5–10 times the soil mean diameter D_{50}.

Numerically the soil-pile interface is often modelled by zero or thin-layer interface elements Goodman et al. (1968), Beer (1985) and Desai et al. (1984) which need to reproduce the important shearing effects localized in this particular area. More advanced interface elements were developed recently. Cerfontaine et al. (2015) proposed a 3D hydro-mechanical coupled element. The element belongs to the zero-thickness formulation and the contact constraint is ensured by the penalty method. Fluid flow is modelled through a three-node scheme discretizing the inner flow by additional nodes. Stutz et al. (2014) enhanced an interface element with a cyclic gap opening function to reproduce the cyclic behavior of gap openings in cohesive soils.

An important aspect of modelling interfaces, is the choice of constitutive model for it. One of the most common models is the elastic-perfectly plastic model of Mohr-Coulomb ((De Coulomb (1821), Van Langen & Vermeer (1991)). More recent ones, e.g. Lashkari (2013) and Mortara (2001) have been introduced in the framework of elasto-plasticity, by incorporating important features of soil behavior such as hardening and dilatancy. Beside the elasto-plastic models, generalized plasticity models Liu et al. (2014) or hypoplastic models by Stutz & Mašín (2017) have also been proposed.

In this paper the Pastor et al. (1990) model in the frame of Generalized Plasticity is adapted for interface modelling. Its formulation for the soil continuum has not only the advantage of reproducing important features of soil behavior such as compaction or dilatation but also includes cyclic loading effects. Based on the experimental results from Boulon & Nova (1990) and Feda (1976) it is found that the governing shear mechanisms of interfaces are similar to simple triaxial shearing of the soil continuum. The interface which is a small part of the continuum has the same deformation pattern as the continuum itself (e.g. Arnold & Herle (2006) and Stutz et al. (2017)). In this work, the Pastor et al. (1990) model adapted through a tensor reduction formulation is applied in 2D shear test problems to reproduce the shear behavior of an interface.

2 CONSTITUTIVE MODEL

2.1 *Generalized plasticity formulation*

The Generalized-Plasticity framework was introduced by Zienkiewicz & Mroz (1984) and extended by Pastor et al. (1985), Pastor & Zienkiewicz (1986) and Pastor et al. (1990). The plastic material behavior is defined by means of two tensorial directions in the space of stress invariants namely: the loading direction vector n and the plastic flow direction vector n_g, instead of yield and plastic potential surfaces. The incremental stress tensor can be expressed as:

$$d\sigma = D^{ep} \cdot d\varepsilon = D^e - \frac{D^e : n_g n : D^e}{H_L + n^T \cdot D^e \cdot n_g} \quad (1)$$

where D^{ep} is the elasto-plastic material matrix; D^e is the elastic material matrix and H_L is the plastic modulus. The elastic matrix is assumed uncoupled in its volumetric and shear components.

$$D^e = \begin{bmatrix} K^e & 0 \\ 0 & G^e \end{bmatrix} \quad (2)$$

The elastic shear moduli G^e and bulk moduli K^e depend on the confining pressure p as follows:

$$K^e = K_0^e \frac{p}{p_0} \quad (3)$$

$$G^e = G_0^e \frac{p}{p_0} \quad (4)$$

where p_0 is the reference pressure; K_e^0 is the bulk modulus and G_e^0 is the shear modulus at reference pressure. The plastic flow direction vector n_g can be decomposed in volumetric $n_{g,v}$ and shear component $n_{g,s}$:

$$n_g = \begin{Bmatrix} n_{g,v} \\ n_{g,s} \end{Bmatrix} \quad (5)$$

$$n_{g,v} = \frac{d_g}{\sqrt{1+d_g^2}} \quad (6)$$

$$n_{g,s} = \frac{1}{\sqrt{1+d_g^2}} \quad (7)$$

Based on the experimental results obtained by Frossard (1983), Pastor & Zienkiewicz (1986) proposed an expression for the dilatancy d_g as follows:.

$$d_g = (1+\alpha_g)\cdot(M_g - \eta) \quad (8)$$

The dilatancy depends on the stress ratio $\eta = q/p$ between deviatoric q and confining pressure p and the slope M_g of the CSL in the p-q plane. The parameter α_g is a material constant that can be obtained from experimental diagrams of dilatancy versus stress ratio.

$$M_g = \frac{6\sin\varphi}{3 - \sin\varphi \sin 3\theta} \quad (9)$$

where φ is the friction angle and θ is the Lode angle. Similarly, as for n_g the load vector n is defined by M_f and α_f which are constitutive parameters. As there is no explicit definition of a yield surface f, these parameters tend to define a similar expression in an implicit manner.

$$n_g = \begin{Bmatrix} n_v \\ n_s \end{Bmatrix} \quad (10)$$

$$n_v = \frac{d_f}{\sqrt{1+d_f^2}} \quad (11)$$

$$n_s = \frac{1}{\sqrt{1+d_f^2}} \quad (12)$$

Where:.

$$d_f = (1+\alpha_f)(M_f - \eta) \quad (13)$$

The volumetric and deviatoric plastic strain increments can be defined as:.

$$d\varepsilon_v^p = \frac{1}{H_L} n_{g,v} (n_v dp + n_s dq) \quad (14)$$

$$d\varepsilon_s^p = \frac{1}{H_L} n_{g,s} (n_v dp + n_s dq) \quad (15)$$

The plastic modulus H_L, permits the consideration of different aspects of sand behavior like the existence of a critical state line where all the residual stress states lie.

$$H_L = H_0 \cdot p \cdot H_f \cdot (H_v + H_s) \cdot H_{DM} \quad (16)$$

This way H_L incorporates the following ingredients: (i) a pressure dependence through the effective confining stress p, (ii) an isotropic plastic modulus H_0, (iii) a frictional factor H_f that limits the possible stress states within the sand, (iv) a volumetric strain hardening function H_v with a dependence on the mobilized stress ratio which makes it zero at the critical stress line, (v) a deviatoric strain hardening H_S, which models the material degradation by accumulated strains, (vi) a discrete memory factor H_{DM} that accounts for the previous loading.

Pastor & Zienkiewicz (1986) proposed the following expressions for the components of the plastic modulus:.

$$H_f = \left(\frac{1-\eta}{\eta_f}\right)^4 ; \eta_f = \left(\frac{1-\eta}{\alpha_f}\right) M_g \quad (17)$$

$$H_v = \left(\frac{1-\eta}{M_g}\right) \quad (18)$$

$$H_s = \beta_0 \beta_1 \exp(\beta_0 \xi) \quad (19)$$

$$H_{DM} = \left(\frac{\zeta_{max}}{\zeta}\right)^{\gamma} \quad (20)$$

Altogether a set of 10 parameters needs to be determined to fully characterize the sand for monotonic loading at a given initial pressure and density and is summarized in Table 1.

Table 1. Model parameters.

Parameter	Description	Units
K_0^e	Elastic volumetric modulus	kPa
G_0^e	Elastic Shear modulus	kPa
M_g	Slope of the Critical State Line	-
α_g	Slope of the dilatancy vs. stress ratio	-
M_f	Shape parameter, function of CSL	-
α_f	Shape parameter, usually equal to α_g	-
H_0	Plastic modulus of isotropic compression	kPa
β_0	Deviatoric strain hardening parameter	
β_1	Deviatoric strain hardening parameter	
γ	Discrete memory interpolation parameter	

Applications for 2D problems were simulated by Mira et al. (2003) and Mira et al. (2004). An implementation of the 3D formulation in a finite element code (GehoMadrid) is shown by Cuellar (2011), where the influence of pore pressure around a cyclically loaded pile is studied.

2.2 *Reduced stress and strain tensors for modelling an interface*

As it is shown by Stutz et al. (2016) and Stutz et al. (2017) continuum models can be used to simulate the interface shear response. The proposed methodology is to use reduced stress and strain tensors (σ_r, ε_r) instead of the full stress and strain tensor (σ_t, ε_t) for a 2D case.

$$\sigma_t = \begin{bmatrix} \sigma_{11} & \tau_{12} & \tau_{13} \\ \tau_{21} & \sigma_{22} & \tau_{23} \\ \tau_{31} & \tau_{32} & \sigma_{33} \end{bmatrix} \quad (21)$$

$$\sigma_r = \begin{bmatrix} \sigma_n & \tau_x \\ \tau_x & 0 \end{bmatrix} \quad (22)$$

where σ_n is the normal stress and τ_x is the shear stress component. The full and reduced strain tensor is expressed in 2D as:.

$$\varepsilon_t = \begin{bmatrix} \varepsilon_{11} & \varepsilon_{12} & \varepsilon_{13} \\ \varepsilon_{21} & \varepsilon_{22} & \varepsilon_{23} \\ \varepsilon_{31} & \varepsilon_{32} & \varepsilon_{33} \end{bmatrix} \quad (23)$$

$$\sigma_r = \begin{bmatrix} \varepsilon_n & \varepsilon_x \\ \varepsilon_x & 0 \end{bmatrix} \quad (24)$$

where ε_n is the normal strain, and ε_x is the shear strain. For a 2D problem, the in-plane stress and strain for its contact is considered zero here.

Using these reduced stress and strain tensors, the continuum model considers simple shear conditions. This is the deformation mode which is undergone at the interface under shearing. The interested reader can refer to Stutz et al. (2016), Stutz et al. (2017) and Stutz & Mašín (2017). Using such an approach, usually different interface model developments can be simplified.

3 MODEL DESCRIPTION

3.1 *Zero thickness interface element*

The interface element used here for the contact between soil and solid is a zero-thickness one as in Beer (1985). For each node, there are two displacement components: one in the normal direction and the other one in the tangential direction (Fig. 1). It has 4 nodes and 8 displacement degrees of freedoms in total. The formulation is derived based on two relative displacements of the continuum element on both sides of the interface (e.g. nodes (1,4) and nodes (2,3)).

$$\{w\} = \begin{Bmatrix} w_s \\ w_n \end{Bmatrix} = \begin{Bmatrix} u_3 - u_2 \\ v_3 - v_2 \end{Bmatrix} \tag{25}$$

where w_s, w_n are the tangential and normal relative displacements; u, v are the displacements along x and y axis (Fig. 1).

3.2 *Boundary value problem*

The reduced constitutive model (subsection 2.2) is considered to represent the interface behavior of a direct shear test (Fig. 2). The experimental results are taken from Shahrour & Rezaie (1997). They tested the behavior of Hostun Sand at two different relative densities D_r (15% and 90%). Three different normal pressures p_n (100 kPa, 200 kPa, 300 kPa) are applied for each relative density of the soil on top of the device. A shear displacement u_x was imposed on the lateral sides.

Numerically the direct shear test is treated as 2D plane strain problem (Fig. 3). The model consists of two different domains: soil (upper part) and steel plate (lower part). The dimensions of each block were 0.25 m × 0.05 m. The model has in total 20 elements. Each block was divided in 8 quadrilateral elements with 4 nodes and the contact area has 4 zero thickness elements as in Beer (1985) with 4 nodes and 2 integration

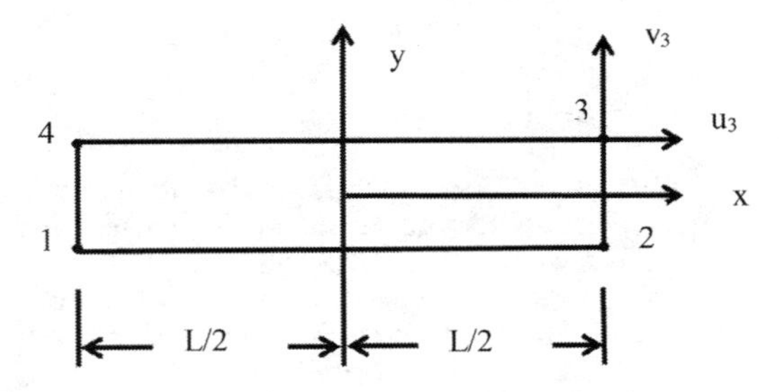

Figure 1. Zero thickness element geometry.

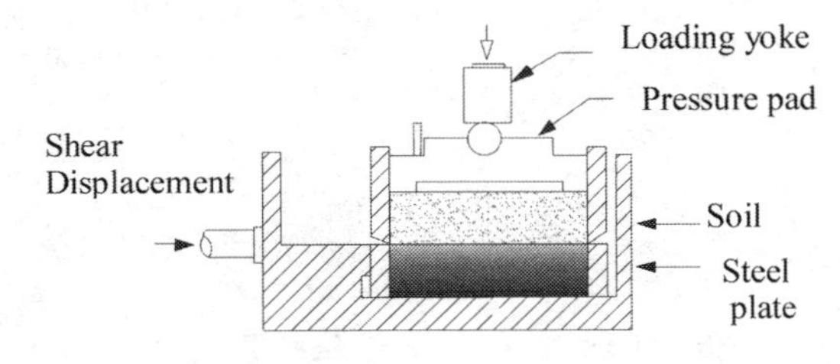

Figure 2. Direct shear test device.

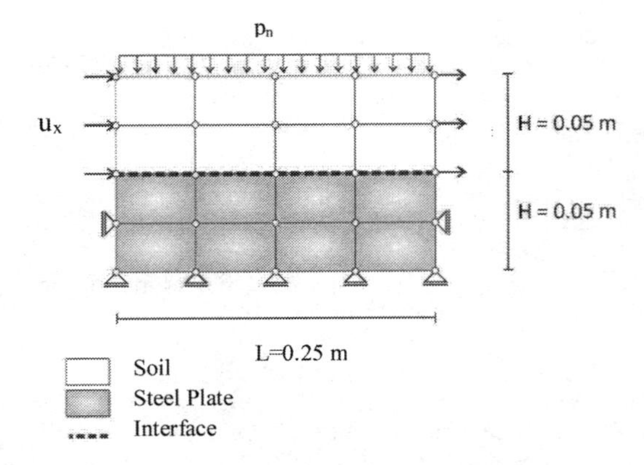

Figure 3. Boundary value problem of direct shear test.

Table 2. Loose sand case.

	Dr = 15%		
Param.	p_n = 100 kPa	p_n = 200 kPa	p_n = 300 kPa
K_0^e (kPa)	15500	17500	35500
G_0^e (kPa)	10000	21000	25000
M_g (–)	1.05	1.05	1.05
α_g (–)	0.40	0.40	0.40
M_f (–)	1.29	1.19	1.19
α_f (–)	0.40	0.40	0.40
H_0 (kPa)	65	65	65
β_0 (–)	0.12	0.12	0.12
β_1 (–)	0.05	0.05	0.45

Table 3. Very dense sand case.

	Dr = 90%		
Param.	p_n = 100 kPa	p_n = 200 kPa	p_n = 300 kPa
K_0^e (kPa)	5000	6000	9000
G_0^e (kPa)	9500	9500	15500
M_g (–)	1.85	1.68	1.55
α_g (–)	0.45	0.45	0.45
M_f (–)	0.77	0.66	0.70
α_f (–)	0.45	0.45	0.45
H_0 (kPa)	150	150	150
β_0 (–)	20.2	6.2	7.2
β_1 (–)	0.18	0.2	0.15

Gauss-points. It is important to mention that the continuum behavior of the steel plate and the soil elements is simulated as purely elastic. The steel plate is characterized by a Young's modulus E = 1 GPa and a Poisson's ratio $\nu = 0.35$ and the soil has E = 11 MPa and $\nu = 0.30$.

The parameters of Pastor et al. (1990) model (Table 2 and 3) are calibrated according to the shear stress development of experimental test for monotonic loading of Shahrour & Rezaie (1997).

4 RESULTS AND DISCUSSION

From the numerical results, the adapted Pastor et al. (1990) model for sands is able to model not only the plastic behavior of the interface, but also the post peak behavior for loose and very dense sand as can be seen in Figure 4 and 5.

As given by the experimental data, loose sands tend to contract during all the loading process and no softening is noticed. This effect comes due to the rearrangement of particles and void fill during shearing. Numerically this behavior was achieved by changing the shape parameter β_1 and M_f as the normal pressure applied is increased from 100 kPa to 300 kPa. The initial slope is captured by increasing shear modulus G_0^e.

Under shearing, dense sands behave differently from loose ones. After maximal compaction under shearing is reached, a lever motion occurs between neighboring grains. In Figure 5, initially a peak behavior is observed, which implies a phase change from contraction to dilatation. Numerically from p_n = 100kPa to p_n = 300kPa, the behavior is captured by changing both shear and bulk modulus, and by varying β_l and β_0 which are responsible for the post-peak behavior.

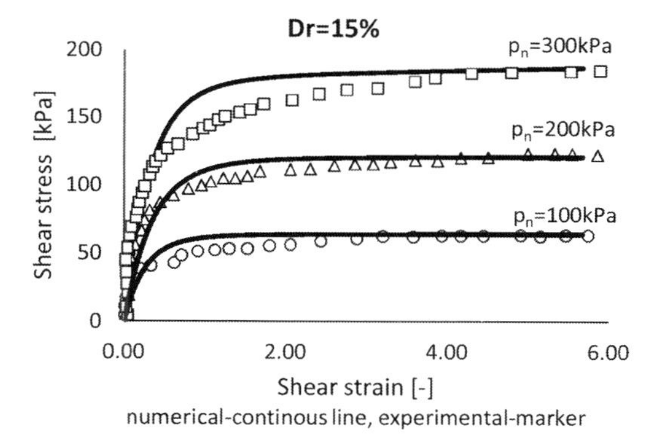

Figure 4. Shear stress vs shear strain behavior of loose sand.

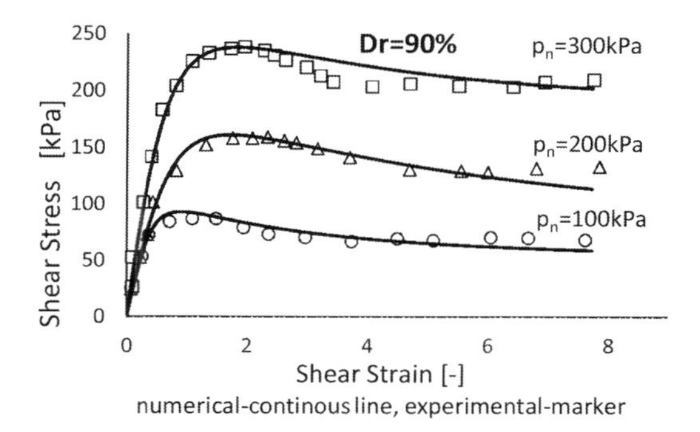

Figure 5. Normal stress vs shear strain behavior of very dense sand.

5 CONCLUSION AND OUTLOOK

In the present study, the full formulation of the Pastor et al. (1990) model was adapted to interface conditions using a method proposed by Stutz et al. (2016). The reduced model was used in a zero-thickness interface element. The application was done in a 2D problem of direct shear test. So far, the model can reproduce the shear behavior of dense and loose sands under monotonic loading. A drawback of the model is, that it needs to be calibrated for every pressure and relative density. A solution would be to calibrate the Generalized Plasticity model which includes state parameters as in Manzanal et al. (2011) based on the work of Been & Jefferies (1985). Further investigation is needed to properly model the normal stress and strain behavior. Another aspect to consider is to calibrate and validate this method for the behavior of interfaces under cyclic loading.

REFERENCES

Arnold, M. & Herle, I. 2006. Hypoplastic description of the frictional behaviour of contacts. *Numerical Methods in Geotechnical Engineering*, 101–106.

Been, K. & Jefferies, M.G. 1985. A state parameter for sands. *Géotechnique*, 35, 99–112.

Beer, G. 1985. An isoparametric joint/interface element for finite element analysis. *International journal for numerical methods in engineering*, 21, 585–600.

Boulon, M. 1989. Basic features of soil structure interface behaviour. *Computers and Geotechnics*, 7, 115–131.

Boulon, M. & Nova, R. 1990. Modelling of soil-structure interface behaviour a comparison between elastoplastic and rate type laws. *Computers and Geotechnics*, 9, 21–46.

Cerfontaine, B., Dieudonné, A.-C., Radu, J.-P., Collin, F. & Charlier, R. 2015. 3d zero-thickness coupled interface finite element: Formulation and application. *Computers and Geotechnics*, 69, 124–140.

Cuellar, P. 2011. Pile foundations for offshore wind turbines: Numerical and experimental investigations on the behavior under short term and long term cyclic loading. TU Berlin.

De Coulomb, C.A. 1821. *Théorie des machines simples: En ayant égard au frottement de leurs parties et à la roideur des cordages*, Bachelier.

Dejong, J., White, D. & Randolph, M. 2006. Microscale observation and modeling of soil-structure interface behavior using particle image velocimetry. *Soils and Foundations,* 46, 15–28.

Desai, C., Zaman, M., Lightner, J. & Siriwardane, H. 1984. Thin-layer element for interfaces and joints. *International Journal for numerical and analytical methods in geomechanics*, 8, 19–43.

Feda, J. 1976. Skin friction of piles. *Proc. VI ECSMFE*, 423–428.

Frossard, E. 1983. Uneéquation d'éoulement simple pour les matériaux granulaires. *Géotechnique,* 33, 21–29.

Goodman, R.E., Taylor, R.L. & Brekke, T.L. 1968. A model for the mechanics of jointed rocks. *Journal of Soil Mechanics & Foundations Div.*

Kishida, H. & Uesugi, M. 1987. Tests of the interface between sand and steel in the simple shear apparatus. *Géotechnique,* 37, 45–52.

Lashkari, A. 2013. Prediction of the shaft resistance of nondisplacement piles in sand. *International Journal for numerical and analytical methods in geomechanics,* 37, 904–931.

Liu, J., Zou, D. & Kong, X. 2014. A three-dimensional state-dependent model of soil–structure interface for monotonic and cyclic loadings. *Computers and Geotechnics,* 61, 166–177.

Manzanal, D., Fernández Merodo, J.A. & Pastor, M. 2011. Generalized plasticity state parameter-based model for saturated and unsaturated soils. Part 1: Saturated state. *International Journal for numerical and analytical methods in geomechanics,* 35, 1347–1362.

Mira, P., Pastor, M., Li, T. & Liu, X. 2003. A new stabilized enhanced strain element with equal order of interpolation for soil consolidation problems. *Computer methods in applied mechanics and engineering,* 192, 4257–4277.

Mira, P., Pastor, M., Li, T. & Liu, X. 2004. Failure problems in soils: An enhanced strain coupled formulation with application to localization problems. *Revue française de génie civil,* 8, 735–759.

Mortara, G. 2001. An elastoplastic model for sand-structure interface behaviour under monotonic and cyclic loading. Ph. D. Thesis. Technical University of Torino.

Pastor, M. & Zienkiewicz, O. 1986. A generalized plasticity, hierarchical model for sand under monotonic and cyclic loading. *Proc., 2nd Int. Symp. on Numerical Models in Geomechanics.* Jackson and Son, Ghent, Belgium.

Pastor, M., Zienkiewicz, O. & Chan, A.1990. Generalized plasticity and the modelling of soil behaviour. *International Journal for numerical and analytical methods in geomechanics,* 14, 151–190.

Pastor, M., Zienkiewicz, O. & Leung, K. 1985. Simple model for transient soil loading in earthquake analysis. Ii. Non-associative models for sands. *International Journal for numerical and analytical methods in geomechanics*, 9, 477–498.

Shahrour, I. & Rezaie, F. 1997. An elastoplastic constitutive relation for the soil-structure interface under cyclic loading. *Computers and Geotechnics,* 21, 21–39.

Stutz, H. & Mašín, D. 2017. Hypoplastic interface models for fine-grained soils. *International Journal for numerical and analytical methods in geomechanics*, 41, 284–303.

Stutz, H., Wuttke, F. & Benz, T. 2014. Extended zero-thickness interface element for accurate soil–pile interaction modelling. Numerical Methods in Geotechnical Engineering, Proceedings of the 8th European Conference on Numerical Methods in Geotechnical Engi-neering, NUMGE 2014, 1, 283–285.

Stutz, H., Mašín, D. & Wuttke, F. 2016. Enhancement of a hypoplastic model for granular soil–structure interface behaviour. *Acta Geotechnica,* 11, 1249–1261.

Stutz, H., Mašín, D., Sattari, A. & Wuttke, F. 2017. A general approach to model interfaces using existing soil constitutive models application to hypoplasticity. *Computers and Geotechnics,* 87, 115–127.

Uesugi, M., Kishida, H. & Tsubakihara, Y. 1988. Behavior of sand particles in sand-steel friction. *Soils and Foundations,* 28, 107–118.

Van Langen, H. & Vermeer, P. 1991. Interface elements for singular plasticity points. *International Journal for numerical and analytical methods in geomechanics,* 15, 301–315.

Yoshimi, Y. & Kishida, T. 1981. A ring torsion apparatus for evaluating friction between soil and metal surfaces. *Geotechnical Testing Journal,* Vol. 4, 145–152.

Zienkiewicz, O. & Mroz, Z. 1984. Generalized plasticity formulation and applications to geomechanics. *Mechanics of engineering materials,* 44, 655–680.

Numerical Methods in Geotechnical Engineering IX – Cardoso et al. (Eds)
© 2018 Taylor & Francis Group, London, ISBN 978-1-138-33198-3

Comparative investigation of constitutive models for shotcrete based on numerical simulations of deep tunnel advance

M. Neuner, M. Schreter & G. Hofstetter
Unit for Strength of Materials and Structural Analysis, Institute of Basic Sciences in Engineering Sciences, University of Innsbruck, Austria

ABSTRACT: According to the New Austrian Tunneling Method (NATM), subsequent to each excavation step shotcrete is directly applied onto the surrounding rock mass, forming a supporting structure which is usually loaded already several hours after application due to further advance of the tunnel. Only a few advanced material models for shotcrete for representing the highly nonlinear, time-dependent material behavior have been proposed in the literature so far. Recently, in (Neuner, Gamnitzer, & Hofstetter 2017) a new constitutive model for shotcrete, denoted as the SCDP model, based on a combination of the theory of plasticity, the theory of continuum damage mechanics, and a modified version of the solidification theory (Bažant & Prasannan 1989) was presented. The model represents time-dependent material behavior, as well as hardening and softening material behavior of shotcrete. In the present contribution, an extension of the SCDP model, and a comparison with two selected, frequently used shotcrete models, i.e., the models proposed by Meschke (1996), and by Schädlich and Schweiger (2014), are presented. The comparison is based on a benchmark example of a deep tunnel driven by the NATM, which is derived from a stretch of the Brenner Basetunnel.

1 INTRODUCTION

Shotcrete linings, installed immediately after an excavation step, serve as a securing measure during tunneling according to the New Austrian Tunneling Method. Since these shotcrete shells are loaded already at a very early age of the material due to further tunnel advance, i.e., usually a few hours after application of the shell, the representation of the early age material behavior is an important aspect in numerical simulations of tunnel advance. In addition, shotcrete exhibits nonlinear material behavior like hardening and softening, creep and shrinkage, which have to be taken into account properly.

Only a few advanced material models for representing these nonlinear, time-dependent material phenomena were proposed in the literature so far. Among them are the viscoplastic model by Meschke (1996), the viscoelastic-plastic model by Schädlich and Schweiger (2014), and a recently proposed damage-plasticity model presented in (Neuner, Gamnitzer, & Hofstetter 2017), denoted as the SCDP model. In (Neuner, Schreter, Unteregger, & Hofstetter 2017), the three models were evaluated and compared by means of a benchmark example of deep tunnel advance. For calibrating the models, experimental data by Huber (1991) and Müller (2001) was used. However, since these experimental results were obtained from dry-mix shotcrete compositions which do not represent the state-of-the-art of modern shotcrete technology, a new experimental program on a modern wet-mix shotcrete composition was designed and presented in (Neuner, Cordes, Drexel, & Hofstetter 2017).

In the present contribution, the experimental program and the investigated material models are summarized briefly, and an extension of the SCDP model is presented. Finally, the application of the three models to the benchmark example of deep tunnel advance, supported by the novel experimental test results is presented.

2 EXPERIMENTAL PROGRAM

The new experimental program on the early age material behavior of a modern wet-mix shotcrete composition focused on the evolution of the Young's modulus and the evolution of the uniaxial compressive strength, creep and shrinkage, with the latter two tested on sealed specimens to exclude drying effects. The specimens of the investigated composition (manufacturer designation *Sp C25/30/(56)/ÜK3/J2/XC4/XF3/GK8* according to the Austrian guideline for shotcrete (ÖVBB 2009)) were sampled directly at a construction site of the Brenner Base Tunnel. To this end, two different techniques were employed: For tests on shotcrete younger than 24 hours, *sprayed* specimens directly sampled from tubular molds were used, while for tests on shotcrete older than 24 hours both sprayed specimens as well as drill cores sampled from spray boxes were used. Sprayed specimens could already be unmolded safely 6 hours after casting, and experimental tests on the evolution of the uniaxial compressive strength were started immediately afterwards. Although sprayed specimens are a suitable alternative to drill cores, special care has to be taken of the spraying direction, which must be aligned coaxial with the axis of the tubular molds in order to minimize the number of faulty specimens due to shotcrete rebound and air pockets.

The evolution of the Young's modulus was investigated between a material age of 24 hours and 28 days, and the uniaxial compressive strength was tested from 6 hours to 28 days. The mean values of both the Young's modulus and the uniaxial compressive strength, as well as the respective standard deviations (SD) are shown in Figure 1. Compared to the uniaxial compressive strength, the ultimate Young's modulus is already attained at a material age of 7 days, while a substantial increase of material strength still can be observed between 7 days and 28 days.

Shrinkage and creep tests were started simultaneously at three different ages of the material, i.e., 8 hours, 24 hours and 27 hours, and lasted for 56 days each. During the shrinkage and creep tests, the time-dependent strain at the center of the specimens was determined by measuring the time-dependent displacements over a distance of 200 mm, using three displacement transducers which were arranged along the perimeter of each specimen. Creep tests were conducted on a hydraulic creep test bench (Figure 2), and the magnitude of the applied compressive load was chosen to ensure linear viscoelastic material behavior. Accordingly, the specimens tested at 8 hours, 24 hours and 27 hours were loaded with compressive stresses of 1.9 MPa, 2.9 MPa and 2.7 MPa. After application of the load, it was held constant throughout the complete testing period of 56 days.

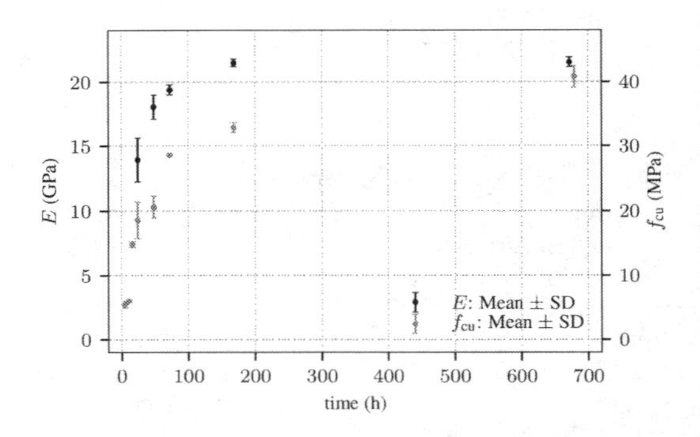

Figure 1. Evolution of the Young's modulus (E) and the uniaxial compressive strength (f_{cu}): Mean values and standard deviations (SD) form shotcrete ages of 6 hours to 672 hours (28 days).

3 MATERIAL MODELS

In the following, the assessed shotcrete models are described briefly, and an extension of the SCDP model is presented.

3.1 *Viscoplastic model by Meschke*

In the viscoplastic model by Meschke (1996), nonlinear mechanical behavior of shotcrete is described on the basis of multisurface viscoplasticity theory. A hardening Drucker-Prager criterion is used for predominantly compressive stress states and mixed stress states, and a softening Rankine criterion for tensile stress states. Creep of shotcrete is described by the evolution of viscoplastic strains, employing associated flow rules. Aging of shotcrete is considered by empirical time functions for the Young's modulus, the uniaxial compressive strength and the uniaxial tensile strength. Shrinkage of shotcrete is taken into account on the basis of the model proposed by Bažant and Panula (1978).

Figure 2. Sealed specimen during a creep test in the hydraulic creep test bench.

3.2 *Viscoelastic-plastic model by Schädlich and Schweiger*

In the shotcrete model by Schädlich and Schweiger (2014) hardening and softening material behavior of shotcrete is described on the basis of non-associated multisurface plasticity. To delimit the elastic domain, a hardening and softening Mohr-Coulomb criterion for predominantly compressive stress states and mixed stress states, and a softening Rankine criterion for tensile stress states is employed. Aging of shotcrete is considered by empirical time functions for stiffness, strength and ductility. Shrinkage of shotcrete is taken into account on the basis of the model proposed by the ACI committee 209 (1992), and nonlinear creep of shotcrete is modeled on the basis of the theory of viscoelasticity, derived from the model proposed in the Eurocode 2 guidelines (2004).

3.3 *SCDP model by Neuner et al*

The SCDP model for shotcrete presented in (Neuner, Gamnitzer, & Hofstetter 2017) is based on three material models for concrete, which are the damage plasticity model for concrete by Grassl and Jirásek (2006), the shrinkage model by Bažant and Panula (1978) and the solidification theory by Bažant and Prasannan (1989). The latter model is used to describe the evolution of material stiffness and nonlinear creep. To model the evolution of material strength and ductility, empirical time functions are employed.

The stress–strain relation in total form is expressed as

$$\sigma = (1-\omega)\mathbb{C} : (\varepsilon - \varepsilon^{p} - \varepsilon^{ve} - \varepsilon^{f} - \varepsilon^{shr}), \tag{1}$$

in which σ denotes the nominal stress (force per total area), ω is the isotropic scalar damage parameter and $\mathbb{C}$ is the fourth order stiffness tensor. The total strain ε is decomposed into the instantaneous elastic strain $\varepsilon^{el} = \varepsilon - \varepsilon^{p} - \varepsilon^{ve} - \varepsilon^{f} - \varepsilon^{shr}$, the plastic strain ε^{p}, the viscoelastic strain ε^{ve}, the flow (viscous) strain ε^{f} and the shrinkage strain ε^{shr}.

The evolution of ε^{el}, ε^{ve} and ε^{f} is described according to the solidification theory and formulated within the framework of coupled continuum damage mechanics and plasticity theory as

$$\begin{aligned} \dot{\varepsilon}^{el}(t) &= q_1 \mathbb{C}_{\nu}^{-1} : \dot{\bar{\sigma}}(t), \\ \dot{\varepsilon}^{ve}(t) &= \frac{F(\bar{\sigma}(t))}{v(t,q_2,q_3)} \int_0^t \dot{\Phi}(t-t')\mathbb{C}_{\nu}^{-1} : d\bar{\sigma}(t'), \\ \dot{\varepsilon}^{f}(t) &= \frac{q_4 F(\bar{\sigma}(t))}{t} \mathbb{C}_{\nu}^{-1} : \bar{\sigma}(t). \end{aligned} \tag{2}$$

Therein, $\bar{\sigma}$ is the effective stress (force per undamaged area), $\mathbb{C}_{\nu}$ the fourth order unit stiffness tensor, $v(t,q_2,q_3)$ denotes the solidified volume at time t (in days), $\dot{\Phi}(t-t')$ is the unit viscoelastic compliance rate with its antiderivative given as

$$\Phi(t-t') = q_2 \ln(1+(t-t')^{0.1}), \tag{3}$$

q_1 to q_4 are compliance parameters, and $F(\bar{\sigma}(t)) = 1 + s(\bar{\sigma})^2$ is an amplifying function dependent on the degree of loading $s(\bar{\sigma})$ to account for nonlinear creep. To determine the degree of loading for multiaxial stress states in $F(\bar{\sigma}(t))$, the invariant J_2 of the effective stress tensor and the uniaxial compressive strength are used for the SCDP model. The volume function $v(t,q_2,q_3)$ of the solidification theory is given as

$$v(t,q_2,q_3) = \frac{1}{t^{-0.5} + \frac{q_3}{q_2}}. \tag{4}$$

Based on these relations, the compliance function of the solidification theory for a constant unit stress $\bar{\sigma}$ applied at time t' is expressed as

$$J(t,t') = q_1 + F(\bar{\sigma}) \int_{t'}^{t} \frac{\dot{\Phi}(\tau - t')}{v(\tau,q_2,q_3)} d\tau + F(\bar{\sigma}) q_4 \ln\left(\frac{t}{t'}\right). \tag{5}$$

In (Neuner, Gamnitzer, & Hofstetter 2017), a time transformation function $\tau(t)$ is introduced to account for the hydration behavior of shotcrete, and $v(t,q_2,q_3)$ in (2) is replaced accordingly by $v(\tau(t),q_2,q_3)$. Time transformation function $\tau(t)$ is calibrated by the experimentally determined value of the Young's modulus at the age of 1 day, $E^{(1)}$, to represent the early age material behavior. For matured material stages, it is characterized by $\tau(t) \to t$ and thus, the original formulation is recovered.

Although the SCDP model is in good agreement with the experimentally observed material behavior, two shortcomings can be identified for the present approach: While $\tau(t)$ is formulated by means of $E^{(1)}$, the respective 28 days value $E^{(28)}$ does not enter the model, and hence compliance parameters q_1, q_2 and q_3 must be calibrated to represent the evolution of material stiffness properly. The second shortcoming is related to q_3, which governs that part of the viscoelastic creep strain rate which is independent of the material age. Since calibration of q_3 is rather difficult and requires comprehensive experimental results which are usually not available, it is proposed to eliminate this parameter from the formulation, and replace the formulation by a simplified one which can be parameterized more easily by $E^{(1)}$ and $E^{(28)}$. Accordingly, the modified volume function $v(\tau(t),q_2,q_3)$ is substituted by a function $v_{sc}(t)$, specifically designed to represent the hydration behavior of shotcrete.

The relation between $v_{sc}(t)$ and $E^{(1)}$ and $E^{(28)}$ is introduced as follows: As described in (Bažant & Baweja 1995), the experimentally determined Young's modulus can be approximated by computing the effective stiffness for a short time period Δt as $E(t) = J(t+\Delta t,t)^{-1}$. Approximating the integral in (5) for a short Δt as

$$\int_t^{t+\Delta t} \frac{\dot{\Phi}(\tau - t)}{v_{sc}(\tau)} d\tau \approx \frac{\Phi(\Delta t)}{v_{sc}(t)}, \tag{6}$$

assuming linear viscoelastic behavior only, i.e., $F(\sigma(t)) = 1$, and neglecting the effect of flow creep for short Δt, $E(t)$ can be approximated as

$$E(t) = \left(q_1 + q_2 v_{sc}(t)^{-1} \ln\left(1+\Delta t^{0.1}\right)\right)^{-1}. \tag{7}$$

Based on this approximation and assuming that $v_{sc}(28\text{d}) = 1$, $E^{(1)}$ and $E^{(28)}$ are given as

$$\begin{aligned} E^{(1)} &= \left(q_1 + q_2 v_{sc}(1\text{d})^{-1} \chi\right)^{-1}, \\ E^{(28)} &= \left(q_1 + q_2 \chi\right)^{-1}, \end{aligned} \tag{8}$$

with $\chi = \ln(1+\Delta t^{0.1})$.

Accordingly, q_2 and $v_{sc}(1\text{d})$ are computed as

$$\begin{aligned} q_2 &= \frac{1/E^{(28)} - q_1}{\chi}, \\ v_{sc}(1\text{d}) &= \frac{q_2 \chi}{1/E^{(1)} - q_1}. \end{aligned} \tag{9}$$

The remaining unknown compliance parameter q_1 is estimated as $q_1 = 0.6/E^{(28)}$ and Δt is assumed as 10^{-3} d, as proposed by Bažant and Baweja (1995).

Derived from the evolution law for the uniaxial compressive strength in (Neuner, Gamnitzer, & Hofstetter 2017), the improved solidified volume function for shotcrete $v_{sc}(t)$ is proposed as

$$v_{sc}(t) = \begin{cases} v_{sc}^{\text{I}}(t) = r_v + a_v t + b_v t^2 & \text{if } t < t_T, \\ v_{sc}^{\text{II}}(t) = \dfrac{c_v}{(t - t_D)^{e_v} + d_v} & \text{otherwise,} \end{cases} \tag{10}$$

in which $t - t_D$ is a time period to model the delayed start of hydration, t_T is the time of transition between the two expressions v_{sc}^{I} and v_{sc}^{II}, r_v is a residual value at $t = 0$, and a_v, b_v, c_v, d_v, and e_v are model parameters.

To ensure a smooth transition at $t = t_T$, parameters a_v and b_v are computed as

$$a_v = \left.\frac{dv_{sc}^{\text{II}}}{dt}\right|_{t=t_T} - 2b_v t_T \tag{11}$$

and

$$b_v = \left(r_v + t_T \left.\frac{dv_{sc}^{\text{II}}}{dt}\right|_{t=t_T} - \left.v_{sc}^{\text{II}}\right|_{t=t_T} \right) / t_T^2. \tag{12}$$

To satisfy $v_{sc}(28\text{ d}) = 1$ in conjunction with the prescribed value of $v_{sc}(1\text{ d})$, parameters c_v and d_v follow as

$$c_v = (28\text{d} - t_D)^{e_v} + d_v \tag{13}$$

and

$$d_v = \frac{v_{sc}(1\text{d})(1\text{d} - t_D)^{e_v} - (28\text{d} - t_D)^{e_v}}{1 - v_{sc}(1\text{d})}. \tag{14}$$

The remaining model parameters t_D, t_T, r_v and e_v are proposed as $t_D = 0.3$ d, $t_T = 0.5$ d, $r_v = 0.01$, and $e_v = -0.65$, calibrated based on the experimentally observed evolution of the Young's modulus. While the proposed improved formulation predicts a very similar evolution of the material stiffness compared to the original formulation for shotcrete in (Neuner, Gamnitzer, & Hofstetter 2017), it can be calibrated more easily by means of the commonly experimentally determined values of $E^{(1)}$ and $E^{(28)}$.

A further extension of the SCDP model is proposed for the evolution law of the flow strain, based on the experimental results from the long-term creep tests: By introducing an additional exponential material parameter e_{flow}, the prediction of the long-term creep behavior can be improved slightly. Departing from (2), the extended expression for the flow creep strain rate is given as

$$\dot{\varepsilon}^f(t) = \frac{q_4 F(\bar{\sigma}(t))}{t^{e_{\text{flow}}}} \mathbb{C}_v^{-1} : \bar{\sigma}(t). \tag{15}$$

A least square optimization based on the experimental results from the three long term creep tests yields $e_{\text{flow}} = 1.22$, together with $q_4 = 34 \times 10^{-6}\ \text{MPa}^{-1}$.

Finally, concerning the evolution of the shrinkage strain, in (Neuner, Cordes, Drexel, & Hofstetter 2017) based on the experimental results from the shrinkage tests it was concluded that the time function proposed by the ACI committee 209 (1992) is in better agreement with the observed material behavior than the one of the Bažant and Panula model. For this reason, for the subsequently presented numerical results, the time function proposed by the ACI committee 209 is employed for the SCDP model to represent the evolution of the shrinkage strain, identical to the Schädlich model.

3.4 *Calibration of the material models*

The parameter identification procedure for the three shotcrete models is described in (Neuner, Gamnitzer, & Hofstetter 2017), and the identification from the data of the present experimental program is reported in (Neuner, Cordes, Drexel, & Hofstetter 2017). The resulting material parameters are listed in Tables 1, 2 and 3. For the sake of brevity, a detailed description of the parameters is omitted here.

The evolution of the Young's modulus (Figure 3) is predicted very well by all models, although it is slightly underestimated between a material age of 1 day and 28 days by all models.

The evolution of the uniaxial compressive strength and the computed values are shown in Figure 4. Both the Schädlich model and the SCDP model are in good agreement with the experimental results, while the compressive strength is overestimated by the Meschke model between 1 day and 28 days.

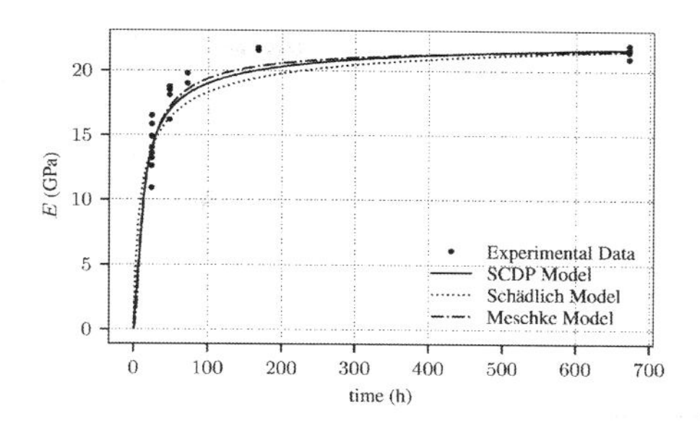

Figure 3. Measured and computed evolution of the Young's modulus E.

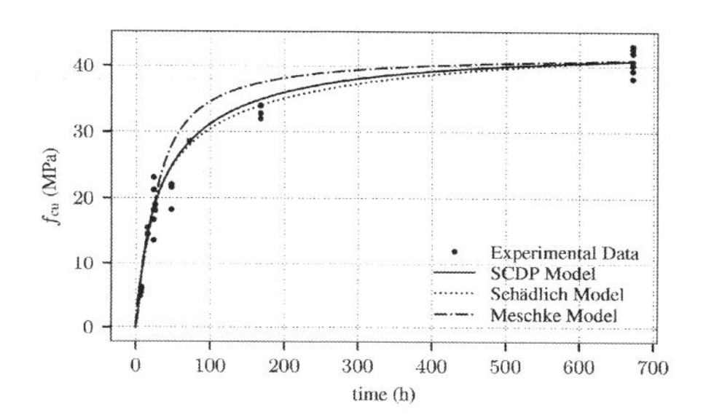

Figure 4. Measured and computed evolution of the uniaxial compressive strength f_{cu}.

Table 1. Material parameters for the SCDP model.

$E^{(1)}$ (MPa)	$E^{(28)}$ (MPa)	q_4 (MPa^{-1})	ν (–)	e^{flow} (–)	$f^{(1)}_{cu}$ (MPa)	$f^{(28)}_{cu}$ (MPa)	f_{cy}/f_{cu} (–)
13 943	21 537	34 (10^{-6}	0.21	1.22	18.56	40.85	0.1
f_{cb}/f_{cu} (–)	f_{tu}/f_{cu} (–)	ε^{shr} (–)	t^{shr}_{50} (h)	$\varepsilon^{p(1)}_{cpu}$ (–)	$\varepsilon^{p(8)}_{cpu}$ (–)	$\varepsilon^{p(24)}_{cpu}$ (–)	$G^{(28)}_{f1}$ (N/mm)
1.16	0.1	–0.0019	8645	–0.03	–0.0007	–0.0007	0.1

Table 2. Material parameters for the Meschke model.

$E^{(1)}$ (MPa)	$E^{(28)}$ (MPa)	ν (–)	$f^{(1)}_{cu}$ (MPa)	$f^{(28)}_{cu}$ (MPa)	f_{cy}/f_{cu} (–)	f_{cb}/f_{cu} (–)	f_{tu}/f_{cu} (–)
13 943	21 537	0.21	18.56	40.85	0.1	1.16	0.1
η (h)	ε^{shr}_{50} (–)	k_h (–)	τ_{shr} (d)	Δt_E (h)	t_E (h)	$G^{(28)}_{f1}$ (N/mm)	
5	–0.002	1.0	4082	6	8	0.1	

Table 3. Material parameters for the Schädlich model.

$E^{(1)}$ (MPa)	$E^{(28)}$ (MPa)	ν (–)	ψ (°)	t^{shr}_{50} (h)	φ^{cr} (–)	$f^{(1)}_{cu}$ (MPa)	$f^{(28)}_{cu}$ (MPa)	$f^{(28)}_{cu}$ (MPa)	f_{cy}/f_{cu} (–)
13 943	21 537	0.21	0	36	2.62	18.56	40.85	4.08	0.1
f_{cfn} (–)	f_{cun} (–)	f_{tun} (–)	ε^{shr}_{50} (–)	t^{shr}_{50} (h)	$\varepsilon^{p(1)}_{cpu}$ (–)	$\varepsilon^{p(8)}_{cpu}$ (–)	$\varepsilon^{p(24)}_{cpu}$ (–)	$G^{(28)}_{f1}$ (N/mm)	$G^{(28)}_{c}$ (N/mm)
0.1	0.1	0.1	–0.0019	8645	–0.03	–0.0007	–0.0007	0.1	30

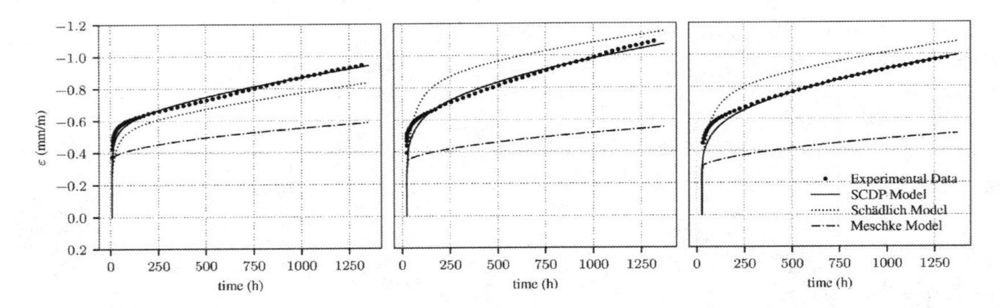

Figure 5. Measured and computed evolution of the total strain during the creep tests on sealed specimens loaded at material ages of 8 hours (left), 24 hours (center) and 27 hours (right).

The experimental results from the creep tests on sealed specimens loaded at material ages of 8 hours, 24 hours and 27 hours with compressive stresses of 1.9 MPa, 2.9 MPa and 2.7 MPa are shown together with the predicted results in Figure 5. It can be concluded that compared to the Meschke model the creep behavior is represented better by the Schädlich model and the SCDP model. As pointed out in (Neuner, Cordes, Drexel, & Hofstetter 2017), this is a consequence of the applied low stress levels in combination with the viscoplastic formulation of the Meschke model. Since calibrating viscosity parameter η on the basis of the present creep test results is not reasonable, the default viscosity parameter $\eta = 5$ h as proposed in (Meschke, Kropik, & Mang 1996) is assumed for the subsequent numerical simulations.

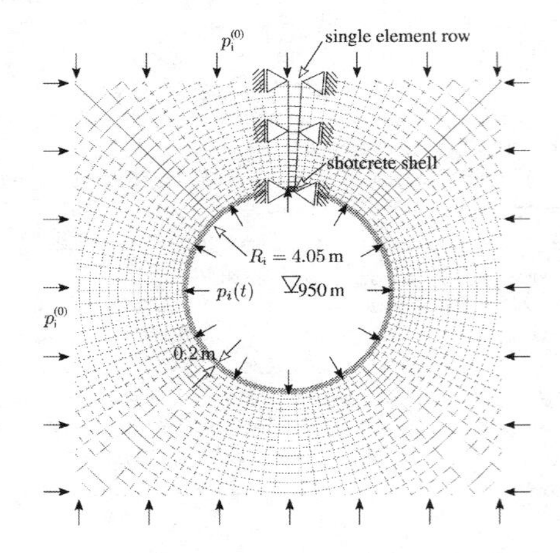

Figure 6. Schematic view of the benchmark example together with the 2D axisymmetric finite element model.

4 APPLICATION

The structural benchmark example for comparing the shotcrete models was presented in (Neuner, Schreter, Unteregger, & Hofstetter 2017). It is derived from a stretch of the Brenner Basetunnel, for which in-situ measurement results of deformations in the surrounding rock mass during tunnel advance are available. The investigated stretch is characterized by a circular full face excavation profile with a diameter of 8.5 m and an overburden of 950 m measured from the tunnel axis. The installed shotcrete shell has a thickness of 0.2 m.

The problem is analyized by means of a simplified 2D finite element model (Figure 6), assuming plane strain conditions. For representing the surrounding rock mass, the linear-elastic perfectly-plastic Hoek-Brown criterion (Hoek & Brown 1980), calibrated for representing the material behavior of Innsbruck Quartz-phyllite (Schreter, Neuner, Unteregger, Hofstetter, Reinhold, Cordes, & Bergmeister 2017), characteristic for this specific stretch, is employed. The employed material parameters for the Hoek-Brown model are summarized in Table 4.

Due to the geometry, boundary conditions, an assumed initial hydrostatic geostatic stress state, and full face excavation, axial symmetry can be exploited, leading to a reduced finite element model consisting of a single element row. Accordingly, homogeneous Dirichlet boundary conditions are assumed perpendicular to the remaining part of the boundary of the single row of finite elements. For discretizing the rock mass and the shotcrete shell, 8-node quadrilateral continuum elements are used, with the shotcrete shell modeled by four elements through its thickness. Figure 6 shows the full and the reduced (highlighted in black) 2D finite element model. As a consequence of the employed geometry and boundary conditions, in the numerical simulations only radial displacements of the shotcrete shell are arising, and bending moments

Table 4. Material parameters for the Hoek-Brown model representing the rock mass.

E (MPa)	ν (–)	f_{cu} (MPa)	m_0 (–)
56 670	0.21	42	12
ψ (°)	e (–)	GSI (–)	D (–)
11.6	0.51	40	0

in the shotcrete shell are excluded. Obviously, this is a simplification of the actual deformations of the shotcrete shell during tunneling, however it allows for an easy comparison of the shotcrete models.

In the numerical simulations, an initial hydrostatic stress state in the rock mass of $p_i^{(0)} = 25.7\,\text{MPa}$ is assumed. Within the convergence-confinement method, the excavation of the investigated cross section is modeled by reducing an equivalent fictitious internal pressure $p_i(t)$ acting on the tunnel lining by means of a stress release ratio $\lambda(t)$ ranging from 0% (no reduction) to 100% (full removal), expressed by the relation

$$p_i(t) = (1 - \lambda(t))\, p_i^{(0)}. \tag{16}$$

Before installing the shotcrete shell in the numerical simulations, the stress is released according to an initial stress release ratio. Subsequently, the shotcrete shell is installed and the remaining fictitious internal pressure is removed employing a stepwise, timedependent stress release function representing the sequence of *drill, blast and idle periods* during further tunnel advance. This stepwise stress release is assumed according to a parabolic decline, as suggested by Pöttler (1990). In total, 9 excavation steps are considered, separated by idle periods of 8 hours. After the last excavation step, an additional time period of two weeks, i.e. 336 hours, is investigated for observing the relaxation behavior of the shotcrete shell.

Two different initial stress release ratios are considered: 85% and 95%. They were derived according to the in-situ measured range of predeformations in the rock mass, as reported in (Schreter, Neuner, Unteregger, Hofstetter, Reinhold, Cordes, & Bergmeister 2017) and (Neuner, Schreter, Unteregger, & Hofstetter 2017).

The time-dependent mechanical response of the shotcrete shell is assessed by comparing the time-dependent evolution of the radial displacement of an arbitrary node at the inner surface of the shotcrete shell, as well as the time-dependent evolution of the stresses in circumferential and longitudinal direction. The latter are taken at an integration point close to the inner surface of the shotcrete shell. Since the stress state in the shotcrete shell can be classified as nearly biaxial, the radial stress in thickness direction of the shotcrete shell is neglected in the present assessment. This observation was already reported in (Meschke 1996), and confirmed in (Neuner, Schreter, Unteregger, & Hofstetter 2017).

Figure 7 shows the predicted time-dependent evolution of the radial displacement, the circumferential stress and the longitudinal stress for both initial stress release ratios. Therein, the time-dependent radial displacement consists of the instantaneous displacement of the unsupported rock mass due to initial stress release, the instantaneous displacements due to the 9 excavation steps after installation of the shotcrete shell, and the displacements due to the timedependent deformations due to shrinkage and creep of shotcrete. Comparing the obtained radial displacements for the initial stress release ratio of 85%, it can be seen that during the first excavation steps the predicted responses by the Schädlich model and the SCDP model are very similar, with larger

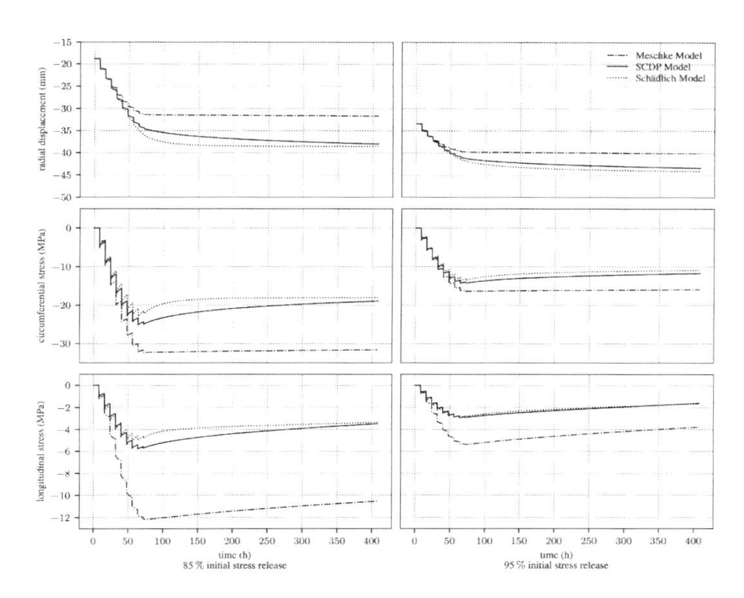

Figure 7. Evolution of the radial displacement (top), circumferential stress in the shotcrete shell (center) and longitudinal stress in the shotcrete shell (bottom), for two different initial stress releases 85% (left column) and 95% (right column).

displacements predicted by the Schädlich model during the last excavations steps. However, due to the stronger relaxation predicted by the SCDP model, the radial displacement and stresses attain those predicted by the Schädlich model at the end of the investigated time period. A stiffer material behavior, and thus smaller radial displacements are predicted by the Meschke model throughout the complete investigated time period. As explained in (Neuner, Schreter, Unteregger, & Hofstetter 2017), this is attributed to the viscoplastic formulation of the Meschke model. Both the circumferential and longitudinal stress predicted by the Meschke model are considerably higher, which again is a consequence of the employed viscoplastic formulation, neglecting creep in the elastic domain.

For the initial stress release of 95%, nearly identical displacements and stresses are predicted by the Schädlich model and the SCDP model, and again, a stiffer response is predicted by the Meschke model due to the viscoplastic formulation.

An interesting detail can be observed for the Meschke model: In contrast to the SCDP model and the Schädlich model, no relaxation of the longitudinal stress during the idle periods occurs, but conversely, it is increasing. This is a consequence of the employed associated viscoplastic formulation, assuming dilatant material behavior during creep.

5 CONCLUSIONS

Three different constitutive models for shotcrete, i.e., the Meschke model, the Schädlich model and the SCDP model were described briefly, and an extension of the SCDP model was presented. The proposed improvements of the SCDP model comprise the reformulation of the evolution law of the viscoelastic strain aiming at an easier calibration procedure on the basis of experimental data, as well as an additional material parameter for the evolution law of the flow creep strain was introduced for an improved agreement with the experimentally observed material behavior.

Finally, a comparison of the influence of the different shotcrete models on the evolution of displacements and stresses by means of a benchmark example of deep tunnel advance was presented. Thereby, it was shown that despite of the different formulations for the evolution of stiffness and strength, hardening material behavior, shrinkage and creep, similar displacements and stresses are predicted by the Schädlich model and the SCDP model.

REFERENCES

ACI Committee 209 (1992). 209r-92: Prediction of creep, shrinkage, and temperature effects in concrete structures.

Bažant, Z. & S. Baweja (1995). Creep and shrinkage predicition model for analysis and design of concrete structures – model B3. *Mater. Struct. 28*, 357–365.

Bažant, Z. & L. Panula (1978). Practical prediction of timedependent deformations of concrete. *Mater. Struct. 11*, 307–328.

Bažant, Z. & S. Prasannan (1989). Solidification theory for concrete creep. i: Formulation. *J. Eng. Mech. 115*(8), 1691–1703.

EN 1992-1-1 (2004). *Eurocode 2: Design of concrete structures*. European Committee for Standardization.

Grassl, P. & M. Jirásek (2006). Damage-plastic model for concrete failure. *Int. J. Solids Struct. 43*(22–23), 7166–7196.

Hoek, E. & E.T. Brown (1980). Empirical strength criterion for rock masses. *J. Geotech. Eng. Div. ASCE 106*(GT9), 1013–1035.

Huber, H.G. (1991). Untersuchungen zum Verformungsverhalten von jungem Spritzbeton im Tunnelbau. Diplomarbeit, Leopold-Franzens-Universität Innsbruck.

Meschke, G. (1996). Consideration of aging of shotcrete in the context of a 3d-viscoplastic material model. *Int. J. Num. Meth. Eng. 39*(18), 3123–3143.

Meschke, G., C. Kropik, & H. Mang (1996). Numerical analysis of tunnel linings by means of a viscoplastic material model for shotcrete. *Int. J. Num. Meth. Eng. 39*(18), 3145–3162.

Müller, M. (2001). Kriechversuche an jungen Spritzbetonen zur Ermittlung der Parameter für Materialgesetze. Diplomarbeit, Montanuniversität Leoben.

Neuner, M., T. Cordes, M. Drexel, & G. Hofstetter (2017). Time-Dependent Material Properties of Shotcrete: Experimental and Numerical Study. *Materials 10*(9), 1067.

Neuner, M., P. Gamnitzer, & G. Hofstetter (2017). An extended damage plasticity model for shotcrete: Formulation and comparison with other shotcrete models. *Materials 10*(1), 82.

Neuner, M., M. Schreter, D. Unteregger, & G. Hofstetter (2017). Influence of the Constitutive Model for Shotcrete on the Predicted Structural Behavior of the Shotcrete Shell of a Deep Tunnel. *Materials 10*(6), 577.

ÖVBB (2009). *Richtlinie Spritzbeton - 2009*. Österreichische Vereinigung für Beton und Bautechnik.

Pöttler, R. (1990). Time-dependent rock-shotcrete interaction – a numerical shortcut. *Comput. Geotech. 9*, 149–169.

Schreter, M., M. Neuner, D. Unteregger, G. Hofstetter, C. Reinhold, T. Cordes, & K. Bergmeister (2017). Application of a damage plasticity model for rock mass to the numerical simulation of tunneling. In *Proceedings of the 4th International Conference on Computational Methods in Tunneling and Subsurface Engineering (EURO:TUN 2017)*, pp. 549–556.

Schädlich, B. & H. Schweiger (2014). A new constitutive model for shotcrete. In *Numerical Methods in Geotechnical Engineering: 8th European Conference on Numerical Methods in Geotechnical Engineering*, pp. 103–108. CRC Press Taylor & Francis Group.

Numerical Methods in Geotechnical Engineering IX – Cardoso et al. (Eds)
 ISBN 978-1-138-33198-3

On modelling of anisotropic undrained strength for non-horizontal terrain

S. Nordal & G. Grimstad
Norwegian University of Science and Technology, Norway

T. Jordbakke
Sweco AS, Norway

K. Rabstad
Løvlien Georåd AS, Norway

M. Isachsen
Multiconsult, Norway

ABSTRACT: Low plasticity soft clays show pronounced variation in undrained shear strength with the direction of loading. The active undrained shear strength (A) is significantly larger than the direct shear strength (D), which again is significantly larger than the passive shear strength (P). The total stress based NGI-ADP model, available in Plaxis, captures such shear strength anisotropy and works well when applied to embankments on or excavations from a horizontal or almost horizontal terrain. For non-horizontal terrain the direction of the in-situ principal stresses is inclined. This paper presents a simple linear elastic, perfectly plastic ADP model that adds anisotropy induced by initial shear stresses on horizontal and vertical planes to an ADP framework. One model parameter controls the conventional anisotropy related to compression versus extension, while another parameter controls the anisotropy caused by the initial shear stress on horizontal and vertical planes. The model is using total stresses. A plane strain version is presented herein. The formulation is inspired by results from DSS laboratory testing where samples were consolidated under inclined effective stresses before shearing in the same or the opposite direction of the initial shear stress. As expected, the extended model called ADPX shows higher factors of safety when applied to a slope than a conventional ADP model. The paper discusses to what extent this represents a real safety margin that has previously been neglected.

1 ACTIVE, DIRECT AND PASSIVE STRENGTH

1.1 *Background*

Soft clays, and in particular lean soft clays have different undrained shear strengths when sheared on differently oriented planes, Soydemir (1976). Eide and Bjerrum (1973) present early results illustrating the significant difference between undrained strength obtained from triaxial compression and extension tests. Pragmatically the triaxial compression tests provides the plane strain active strength while the extension test provides the plane strain passive strength, Figure 1. The passive strength is for lean Drammen clay about 1/3 of the active strength.

1.2 *Soil models for anisotropic undrained strength*

Several well known effective stress based soil models like the MIT-E3, Whittle (1993), S-CLAY1-S, (Karstunen et al. (2005), and SaniClay, Dafalias (2006) provide an anisotropic undrained shear strength, lower for triaxial extension than for triaxial compression. One challenge with these models is that they require careful calibration of several effective stress based input parameters to provide a specific undrained strength design profile.

The NGI-ADP model in Plaxis, Grimstad et al. (2012), is a conceptually simple, total stress model where the ADP strength is direct input. This is convenient for practical applications since in practice the undrained strength (active or direct)

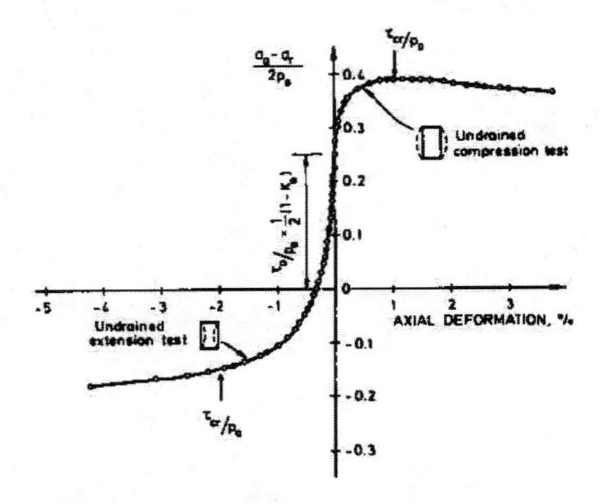

Figure 1. Anisotropic undrained strength Drammen clay (Eide and Bjerrum, 1973).

may be the measured and available strength parameter.

1.3 *The NGI-ADP model in plain strain*

The model is available in Plaxis for a full 3D stress state, Grimstad et. al (2012). A plane strain version of the NGI-ADP model with nested yield surfaces is illustrated in Figure 2. A vertical y-axis and a horizontal x – axis is used. In the model s_u^A, s_u^P, s_u^{DSS} are the active, passive and direct shear strengths respectively. The initial, vertical and horizontal effective stresses, σ'_{yy0} and σ'_{xx0}, define the starting point for loading by $\tau_0 = (\sigma'_{yy0} - \sigma'_{xx0})/2$. The initial value of $\tau_{xy} = \tau_{xy0}$ is assumed to be zero.

1.4 *Initially inclined principal stresses*

Anisotropy in undrained shear strength may originate from the direction of sedimentation. Anisotropy may also be stress induced. The NGI-ADP has a "K_0 – related" anisotropy, related to τ_0, and implicitly assumes that the initial, major principal stress is vertical. However, in slopes the initial principal stresses will be inclined, with $\tau_{xy0} \neq 0$, Figure 3. The ADPX model opens for "shear induced" anisotropy caused by $\tau_{xy0} \neq 0$.

The "τ_{xy0} - shear induced" anisotropy has been investigated by direct simple shear tests at NGI, where in addition to a vertical effective stress, σ'_{vc}, the sample was left to consolidate (drained) with a horizontal shear stress, τ_c, before sheared undrained to failure in the direction of τ_c, Figure 4. It is demonstrated that the undrained strength is increasing for increasing τ_c, for further loading in the "direction" of τ_c. The direct, undrained, simple shear stress increases by almost

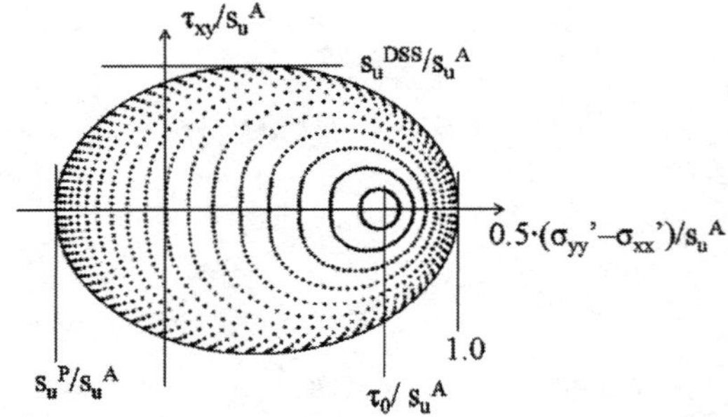

Figure 2. The NGI-ADP yield surfaces, Grimstad et al. (2012).

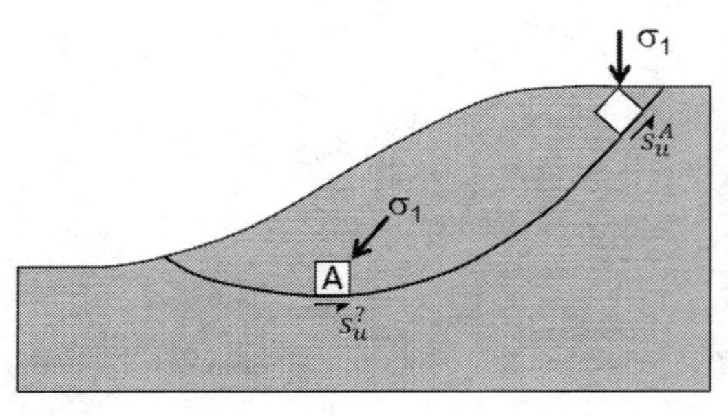

Figure 3. In soli element A the major principal stress is inclined and the shear strength may be closer to s_u^a than to s_u^{DSS}.

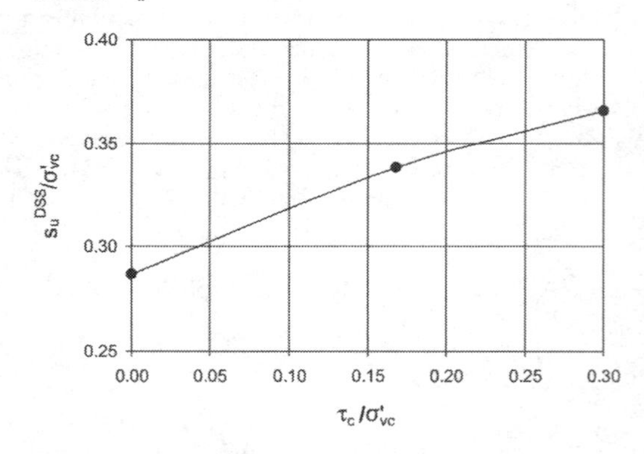

Figure 4. The DSS undrained shear strength increase when consolidated for an inclined, initial effective stress in the direction of undrained loading, Andersen (2009).

30% for $\tau_c/\sigma'_{vc} = 0.30$ compared to $\tau_c/\sigma'_{vc} = 0$, Andersen (2009).

2 AN EXTENDED ADP MODEL, ADPX

2.1 *The anisotropic yield surface*

This paper reports on work done by MSc students at NTNU aiming to study the effect of an anisotropy induced by inclined principal, initial effective stresses. A simple linear elastic, perfectly plastic soil model is adopted for this purpose. It is often found that for vertical σ'_1 then s_u^{DSS} is almost $s_u^{DSS} \approx (s_u^A + s_u^P)/2$. Thus a circular yield surface in a $(\sigma_{yy} - \sigma_{xx})/2$ versus τ_{xy} stress space is selected for this study, Equation 1 and Figure 5.

$$F = \sqrt{\left(\frac{\sigma_{yy} - \sigma_{xx}}{2} - \xi \cdot \tau_0\right)^2 + \left(\tau_{xy} - \eta \cdot \tau_{xy0}\right)^2} - \bar{s}_u \quad (1)$$

For isotropic conditions, the circular yield surface has its origo in the center. For combined "K_0 – induced" and "τ_{xy0} - induced" anisotropy the center of the circle is suggested to move in the direction of the initial effective stress point, (τ_0, τ_{xy0}). This movement is in the formulation controlled by two dimensionless parameters, ξ and η, both in the interval [0, 1]. The circle center is at ($\xi \cdot \tau_0$, $\eta \cdot \tau_{xy0}$). Isotropic conditions are given by $\xi = \eta = 0$, while maximum anisotropy is given by $\xi = \eta = 1$. The consequence of the formulation is the anisotropic strengths expressed by Equations 2–5:

$$s_u^A = \bar{s}_u + \xi \cdot \tau_0 \quad (2)$$

$$s_u^P = \bar{s}_u - \xi \cdot \tau_0 \quad (3)$$

$$s_u^{DSS1} = \bar{s}_u + \eta \cdot \tau_{xy0} \quad (4)$$

$$s_u^{DSS2} = \bar{s}_u - \eta \cdot \tau_{xy0} \quad (5)$$

Where $\bar{s}_u = (s_u^A + s_u^P)/2$. The $\bar{s}_u$ may be given proportional with depth or with $(\sigma'_{yy0} + \sigma'_{xx0})/2$ for plane strain. The model is linear elastic, perfectly plastic with an associated flow rule. The ADPX is implemented as a user defined soil model in Plaxis.

2.2 *Initial effective stresses and model parameters*

Application of the model requires a known initial effective stress as a starting stress for all integration points. These initial stresses may be computed using an effective stress based model under drained conditions in an initial computational phase. Adding soil weight is one possible procedure. In Plaxis a Hardening Soil (HS) model may be used. A challenge of consistency may occur when the ADPX model is applied in the next computational phase: The ADPX parameters, s_u, ξ and η must be selected so that the undrained shear strength is consistent with the initial effective stresses. In an adjusted version the ADPX-model it is pragmatically suggested that for normally consolidated clays the resulting shear strength could be limited

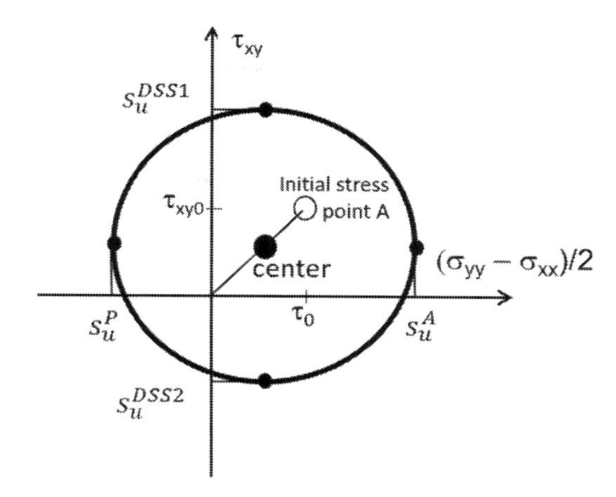

Figure 5. The ADPX yield surface is translated towards point A representing the initial deviatoric stresses. Here: $\xi = \eta = 05$.

by a maximum undrained shear strength and a minimum undrained shear strength, Figure 6. For simplicity, the limits may be given by the Mohr Coulomb parameters used during stress initiation, and a parameter B, Equations 6 and 7.

$$s_u^{max} = c \cdot cos\varphi + \left(\frac{\sigma'_{yy0} + \sigma'_{xx0}}{2}\right) \cdot sin\varphi \quad (6)$$

$$s_u^{min} = s_u^{max} / B \quad (7)$$

When implemented, these restrictions will to a large degree limit and overrun the parameters s_u and ξ. Note that it will indirectly also affect η, but η is given from the input ratio of η/ξ. This restriction (eq. 6) means that effectively the effective friction and effective cohesion become input for undrained strength in the ADPX model. In addition, the ratio η/ξ and the value of B position the yield surface within the limiting undrained maximum and minimum strengths, see Figure 6.

In application it is necessary to evaluate the resulting shear strength in key points to ensure that the model provides realistic values for s_u^A, s_u^P, s_u^{DSS1} and s_u^{DSS2}. The friction angle may have to be a bit low to provide measured undrained strengths. Further, current experience indicate that B around 3 and η/ξ between 0.5 and 1 may be realistic.

3 EXAMPLES OF APPLICATION

3.1 *Stress driver testing*

The implemented formulation has been tested by applying strains and studying the resulting stress

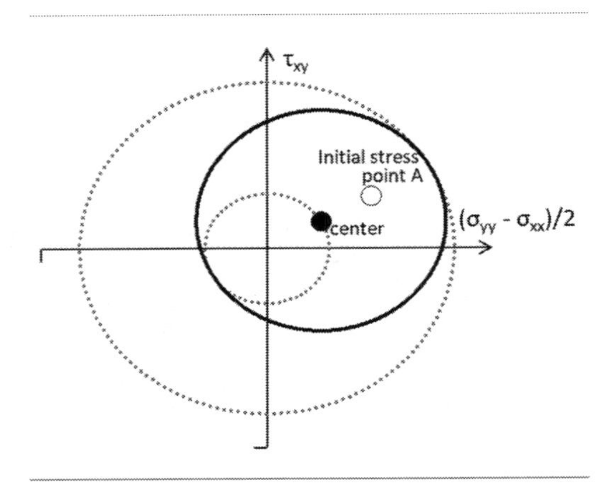

Figure 6. The ADPX yield surface as limited by the maximum and the minimum undrained strength circles around origo. The initial stress is point x. Here: B = 3 and $\xi/\eta = 1$.

paths. One such example is given in Figure 7. A plane strain soil element is tested with initial stresses $\left(\sigma'_{yy0} + \sigma'_{xx0}\right)/2$ = 17,5 kPa and τ_{xy0} = 19 kPa.

The shear stiffness is 5 MPa, $\overline{s_u} = 35kPa$ and $\xi = \eta = 1$.

First, pure shear strain is applied with $\Delta\varepsilon_y = \Delta\varepsilon_x = 0, \Delta\gamma_{xy} = 0{,}5\%$.

Next, $\Delta\varepsilon_y = 2\%, \Delta\varepsilon_x = -2\%, \Delta\gamma_{xy} = 0$ is applied. The resulting stress will slide along the yield surface until a final position is reached where the applied strains are all plastic as given by normality to the yield surface (associated flow). In step 2 the elastic shear strains γ_{xy} are unloaded and replaced by plastic shear strains since the total shear strain is kept constant. Step 2 involves a rotation of principal stresses during pure normal compression.

3.2 *Application to slope stability*

A slope at Vestfossen in Norway failed in 1984 during construction of an embankment, a fill, in the slope. The case has been studied for investigating the effect of anisotropy related to inclined, initial effective stresses, i.e. $\tau_{xy0} \neq 0$. Figure 8 shows the slope geometry with sensitive NC clay under a dry crust. The figure also shows the fill that was placed when the slide took place. The failure surface indicated is the one resulting from a simulation. The ADPX model was used for the NC clay. Drained conditions and a Mohr Coulomb model was used for both the fill material and the dry crust. The active undrained shear strength of the NC clay was 15 kPa, increasing in depth by 2.5 kPa pr. meter. This was fitted to Equation 6 using the effective

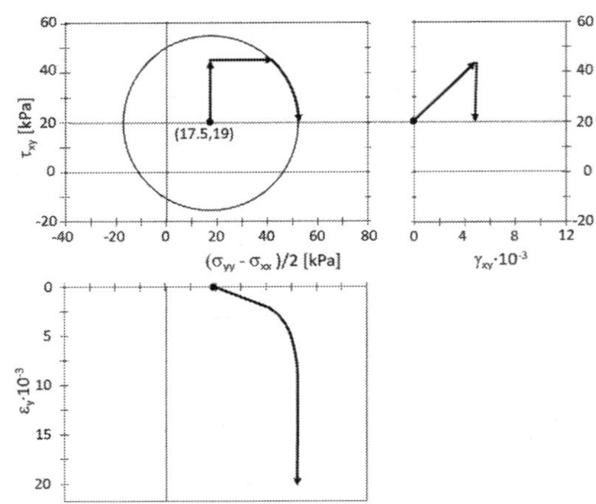

Figure 7. The stress path and stress strain paths obtained from applying shear followed by normal compression.

Table 1. Values of ΣM_{stage} obtained when constructing the embankment in the slope at Vestfossen. A value less than 1 shows that too low strength is used in that particular simulation.

B	η/ξ				
	0	.25	.5	.75	1.0
2	.9995	≥ 1.0	≥ 1.0	≥ 1.0	≥ 1.0
3	.9445	.9886	.9994	.9999	≥ 1.0
4	.9140	.9640	.9935	.9992	.9997

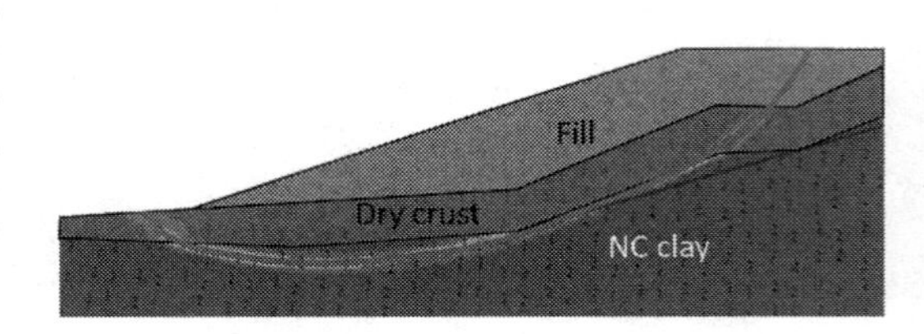

Figure 8. The Vestfossen slope were a fill added in the slope caused failure in the NC clay. Here modelled by using ADPX.

stress parameters $\varphi = 25^0$ and $c' = 2.3$ kPa. For the ADPX model, G = 5 MPa and an undrained Poisson ratio of 0.495 were used. A series of simulations were performed varying B and the ratio η/ξ. Table 1 shows the results in terms of ΣM_{stage} for adding the fill. A value very close to 1.00 should be obtained for a realistic simulation. It is observed that a standard ADP approach with $\eta/\xi = 0$ underestimates the capacity of the slope with respect to carrying the fill. B = 3 and $\eta/\xi = 0{,}75$ is a parameter set that correctly predicts pending failure. The numbers illustrate that taking the shear induced anisotropy due to $\tau_{xy0} \neq 0$ into account as in ADPX, lifts the calculated safety margin by about 10%. This contribution is neglected or ignored in a conventional ADP stability analysis, which corresponds to $\eta/\xi = 0$.

4 DISCUSSION AND CONCLUSION

It is believed that for a slope the initial inclination of the principal stresses increases the undrained shear strength compared to the strength predicted by standard ADP. The ADPX model includes this by a translation of the failure surface in the direction of the initial shear stress on horizontal and vertical planes in stress space. Preliminary results indicate that the calculated safety factor may increase by about 10% using ADPX compared to ADP.

Many aspects affect calculation of slope stability in soft clays using alternative soil models. Anisotropy in strength, sample disturbance, rate dependency, partial drainage and softening are important features that complicates the problem. Still, since the aim is to make realistic soil models and use relevant simulations, it makes sense to include the ADPX type anisotropy in stability evaluations. Whether a 10% strength increase can allow higher slopes, must be discussed in terms of safety levels and required partial material factors calibrated to avoid failure. The factors are given without taking the ADPX contribution to an "upgraded" strength into account.

REFERENCES

Andersen, K.H. 2009. Bearing capacity under cyclic loading - offshore, along the coast, and on land. *The 21st Bjerrum lecture presented in Oslo, 23 November 2007. Canadian Geotechnical Journal* 46: 513–535.

Berre T., L.B. 1973. Shear strength of normally consolidated clays. In T.U.N.S. for Soil Mechanics and F. Engineering (Eds.), Proceedings of the eigth international conference on soil mechanics and foundation engineering: 39–49.

Dafalias, Yannis F., Manzari, Majid T. Papadimitriou, Achilleas G. 2006. SANICLAY: simple anisotropic clay plasticity model. International Journal for Numerical and Analytical Methods in Geomechanics, Vol 30 Issue 12: 1231–1257.

Grimstad, G., L. Andresen, and H.P. Jostad 2012. NGI-ADP: Anisotriopc shear strength model for clay. International journal for numerical and analytical methods in geomechanics. 36(4): 483–497.

Karstunen M, Krenn H, Wheeler SJ, Koskinen M, Zentar R. 2005. The effect of anisotropy and destructuration on the behaviour of Murro test embankment. International Journal of Geomechanics (ASCE) 2005, 5(2):87–97.

Soydemir, C. 1976. Strength anisotropy observed through simple shear tests. Laurits Bjerrum Memorial Volume Laurits Bjerrum Memorial Volume: 99–112.

Whittle AJ. 1993. Evaluation of a constitutive model for overconsolidated clays. Géotechnique; 43(2):289–313.

Numerical Methods in Geotechnical Engineering IX – Cardoso et al. (Eds)
© 2018 Taylor & Francis Group, London, ISBN 978-1-138-33198-3

Elastic and plastic anisotropy in soft clays: A constitutive model

J. Castro & J. Justo
University of Cantabria, Santander, Spain

N. Sivasithamparam
Norwegian Geotechnical Institute, Oslo, Norway

ABSTRACT: This paper presents the application to a boundary value problem, namely a benchmark embankment, of a novel constitutive model for soft structured clays that includes anisotropic behavior both of elastic and plastic nature. The model was developed based on an existing model called S-CLAY1S, which is a Cam clay type model that accounts for plastic anisotropy and destructuration. The new model incorporates stress-dependent cross-anisotropic elastic behavior within the yield surface using just 3 independent elastic parameters. Only an additional parameter that relates the vertical and horizontal elastic stiffnesses was introduced for the sake of simplicity. The model does not consider evolution of elastic anisotropy, but laboratory results show that large strains are necessary to cause noticeable changes in elastic anisotropic behavior. Model predictions show a good agreement with laboratory tests available in the literature. Finally, the application of the constitutive model to a benchmark embankment on a soft soil, is presented. A parametric analysis shows the influence of the anisotropic behavior within the elastic range.

1 INTRODUCTION

1.1 *Soft soil anisotropic behavior*

Natural clays exhibit a significant degree of anisotropy in their fabric, which initially is derived from the shape of the clay platelets, deposition process and one-dimensional consolidation. Neglecting this anisotropy of natural clay behavior may lead to highly inaccurate predictions of material response under loading (see, for example, Zdravkovic et al. 2002). Due to the success of critical state models, the most common way of introducing anisotropic behavior is through an inclined yield surface (e.g., Dafalias 1986, Whittle et al. 1994, Sekiguchi & Otha 1977). This accounts for anisotropic behavior of plastic nature, but isotropic elastic behavior is usually adopted within the inclined yield surface to keep the model simple (e.g., Wheeler et al. 2003).

Yet plastic deformations are likely to dominate for many problems of practical interest, anisotropic elastic strains should be considered for a more accurate description of the soft clay behavior (e.g. Mouratidis & Magnan 1983). Furthermore, it is not necessary to consider the development or erasure of elastic anisotropy because it requires important plastic strains; contrary to plastic anisotropy, which evolves at a much faster rate; for example, Mitchel (1972) and Huekel & Tutumluer (1994) showed that the rate of plastic anisotropy demise is notably greater than that of elastic anisotropy after isotropic loading and unloading.

The MELANIE model is a constitutive model for soft soils that incorporates both elastic and plastic anisotropy and has been used to model real boundary value problems, namely several experimental embankments in Cubzac-les-Ponts (France) (Mouratidis & Magnan 1983). However, the MELANIE model does not consider two important features of soft clay behavior: evolution of plastic anisotropy with plastic straining and stress-dependent elastic stiffness.

The authors (Castro & Sivasithamparam 2017) have recently extended an existing elasto-plastic model that considers plastic anisotropy and its evolution with plastic straining, namely S-CLAY1S model (Karstunen et al. 2005), to incorporate elastic cross-anisotropy, also known as transverse isotropy. To keep the model simple and for practical usage, only an additional parameter was added to describe the elastic stress-dependent cross-anisotropy. Furthermore, evolution of elastic anisotropy was neglected. The model was successfully validated against triaxial tests in Bothkennar clay available in the literature (McGinty 2006)

Here, this new constitutive model is applied to a benchmark embankment in Bothkennar clay and a parametric analysis is performed. For the sake of completeness, the main features and capabilities of the new constitutive model are briefly presented first.

1.2 *S-CLAY1S*

S-CLAY1S (Karstunen et al. 2005) is a Modified Cam Clay (MCC) (Roscoe & Burland 1968) type model that accounts for plastic anisotropy and destructuration. Anisotropy of plastic behaviour is represented through an inclined and distorted elliptical yield surface and a rotational hardening law to model the development or erasure of fabric anisotropy during plastic straining; while interparticle bonding and degradation of bonds (structure) is reproduced using intrinsic and natural yield surfaces (Gens & Nova 1993) and a hardening law describing destructuration as a function of plastic straining. The last letter of the model ("S") refers to the soil structure. So, when the hierarchical version of the model without destructuration is used, the model is simply called S-CLAY1 (Wheeler et al. 2003).

For the sake of simplicity, the mathematical formulation is presented in the following in triaxial stress space, which can be used only to model the response of cross-anisotropic samples (cut vertically from the soil deposit) subjected to oedometer or triaxial loading. The original inclined yield surface of the S-CLAY1 model is elliptical (Dafalias 1986) (see Figure 1):

$$f_y = 1 + \frac{(\eta - \alpha)^2}{M^2 - \alpha^2} - \frac{p'_{mi}}{p'} = 0 \quad (1)$$

Where η is the stress ratio, M is the stress ratio at critical state, p' is the mean effective stress, p'_m is the size of the yield surface related to the soil's pre-consolidation pressure and α is the plastic anisotropy parameter that gives the inclination of the yield surface. In triaxial stress space, α is a scalar value but in the full three-dimensional formulation, $\boldsymbol{\alpha}$ is a fabric tensor (3×3).

The effect of bonding in the S-CLAY1S model (Karstunen et al. 2005) is described by an intrinsic yield surface (Gens & Nova 1993) that has the same shape and inclination of the natural yield surface but with a smaller size (see Figure 1). The size of the intrinsic yield surface is specified by the parameter p'_{mi}, which is related to the size p'_m of the natural yield surface by parameter χ as the current amount of bonding:

$$p'_m = (1 + \chi)\, p'_{mi} \quad (2)$$

S-CLAY1S model incorporates three hardening laws: isotropic hardening, evolution of anisotropy and destructuration. For isotropic hardening, the law is similar to that of Modified Cam-Clay (MCC) model (Roscoe & Burland 1968). It controls the expansion or contraction of the intrinsic yield surface as a function of the increments of plastic volumetric strains $(d\varepsilon_v^p)$

$$dp^{\boxtimes}_{mi} = \frac{v p^{\boxtimes}_{mi}}{\lambda_i - \kappa} d\varepsilon_v^p \quad (3)$$

Where v is the specific volume, λ_i is the gradient of the intrinsic normal compression line in the compression plane ($\ln p' - v$ space), and κ is the slope of the swelling line in the compression plane.

The second hardening law is the rotational hardening law, which describes the rotation of the yield surface with plastic straining (Wheeler et al. 2003)

$$d\alpha = \omega \left[\left(\frac{3\eta}{4} - \alpha_d \right) d\varepsilon_v^p + \omega_d \left(\frac{\eta}{3} - \alpha_d \right) \left| d\varepsilon_d^p \right| \right] \quad (4)$$

Where $d\varepsilon_d^p$ is the increment of plastic deviatoric strain, and ω and ω_d are additional soil constants that control, respectively, the absolute rate of rotation of the yield surface toward its current target value, and the relative effectiveness of plastic deviatoric strains and plastic volumetric strains in rotating the yield surface.

The third hardening law in S-CLAY1S model considers destructuration, which describes the degradation of bonding with plastic straining. The destructuration law is formulated in such a way that both plastic volumetric strains and plastic shear strains tend to decrease the value of the bonding parameter χ towards a target value of zero (Karstunen et al. 2005), it is defined as

$$d\chi = -\xi\chi \left(\left| d\varepsilon_v^p \right| + \xi_d \left| d\varepsilon_d^p \right| \right) \quad (5)$$

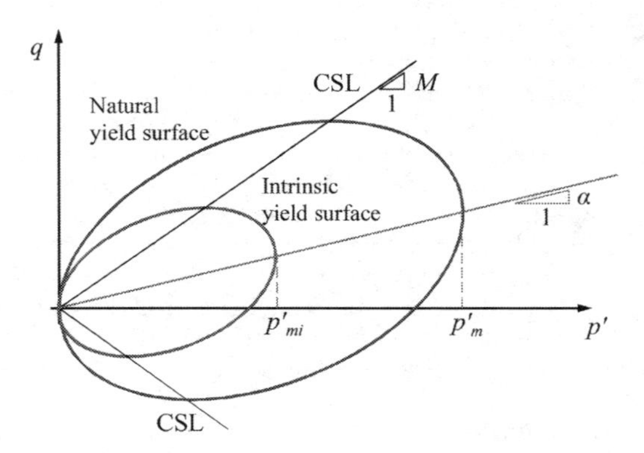

Figure 1. Yield surfaces of the S-CLAY1S model (Karstunen et al. 2005).

where ξ and ξ_d are additional soil constants. Parameter ξ controls the absolute rate of destructuration, and parameter ξ_d controls the relative effectiveness of plastic deviatoric strains and plastic volumetric strains in destroying the inter-particle bonding (Koskinen et al. 2002).

The elastic behaviour in the model is formulated as isotropic, with elastic increments of volumetric and deviatoric strains given by the same expressions as in MCC:

$$\begin{bmatrix} d\varepsilon_v^e \\ d\varepsilon_d^e \end{bmatrix} = \begin{bmatrix} 1/K & 0 \\ 0 & 1/3G \end{bmatrix} \begin{bmatrix} dp' \\ dq \end{bmatrix} \tag{6}$$

Where K is a stress-dependent elastic bulk modulus $(K = vp'/\kappa)$. The model has traditionally been implemented assuming a constant Poisson's ratio, ν (e.g., Wheeler et al. 2003, Sivasithamparam & Castro 2016a). So, the shear modulus G is related to the elastic bulk modulus K and, then, it is also stress dependent, instead of being constant.

1.3 *Linear elastic cross-anisotropy*

Fully generalized elastic anisotropy involves 21 independent elastic parameters due to the symmetry of the 6×6 stiffness matrix that relates the six independent stress increment components with the six independent strain increment components. However, most materials present any kind of symmetry that allows to reduce the number of independent elastic parameters. Soils usually present a cross-anisotropic or transversely isotropic behavior, i.e. both horizontal axes are interchangeable. That reduces the number of independent elastic parameters to 5:

$$\begin{bmatrix} d\varepsilon_{xx}^e \\ d\varepsilon_{yy}^e \\ d\varepsilon_{zz}^e \\ d\gamma_{xy}^e \\ d\gamma_{yz}^e \\ d\gamma_{zx}^e \end{bmatrix} = \begin{bmatrix} a & b & d & 0 & 0 & 0 \\ b & a & d & 0 & 0 & 0 \\ d & d & c & 0 & 0 & 0 \\ 0 & 0 & 0 & f & 0 & 0 \\ 0 & 0 & 0 & 0 & f & 0 \\ 0 & 0 & 0 & 0 & 0 & 2(a-b) \end{bmatrix} \begin{bmatrix} d\sigma'_{xx} \\ d\sigma'_{yy} \\ d\sigma'_{zz} \\ d\tau_{xy} \\ d\tau_{yz} \\ d\tau_{zx} \end{bmatrix} \tag{7}$$

where a, b, c, d and f may be expressed using E_V, E_H, ν_{VV}, ν_{VH} and G_{VH} (V and H subscripts refer to vertical and horizontal directions, respectively).

$$a = 1/E_H; b = -\nu_{HH}/E_H; c = 1/E_V;$$
$$d = -\nu_{HV}/E_V; f = 1/G_{HV}$$

Graham and Houlsby (1983) showed that only three parameters can be deduced from triaxial tests on vertically cut specimens and proposed a simplified three parameter model (E^*, ν^*, α_e). That is also called one-parameter anisotropy because only one parameter is added to describe anisotropy. Here, the subscript "e" is introduced in the α parameter to indicate that it refers to the elastic part and to distinguish it from the α parameter of the S-CLAY1S model that controls anisotropy in the plastic part. The 5 parameters (E_V, E_H, ν_{VV}, ν_{VH} and G_{VH}) of the general compliance matrix for cross-anisotropic behavior (Eq. 7) may now be expressed as a function of the above mentioned 3 parameters:

$$E_V = E^*; E_H = \alpha_e^2 E^*; \nu_{HH} = \nu^*; \nu_{HV} = \nu^*/\alpha_e;$$
$$2G_{HV} = \alpha_e E^*/(1+\nu^*)$$

2 NOVEL CONSTITUTIVE MODEL

2.1 *Introduction*

The novel constitutive model is described in detail in Castro and Sivasithamparam (2017). The stress-dependent cross-anisotropic elastic stiffness of the proposed model is based on the Graham & Houlsby (1983) model because it allows to keep E^* and ν^* with a similar meaning and the same values as the traditional E and ν parameters. Besides, the only additional parameter (elastic anisotropic parameter, α_e) has a clear physical meaning $(\alpha_e^2 = E_H/E_V)$ and may be analytically obtained from conventional triaxial test, either from the undrained stress path during shearing or from the shear-volumetric strain ratio during drained isotropic compression.

2.2 *Stress-dependent elastic cross-anisotropy*

The stress-dependent stiffness of the present model has been introduced relating the E^* parameter with κ through a linear dependency on the effective mean pressure p'. To preserve the original meaning of κ as the slope of the swelling line in the compression plane $(\ln p' - v \text{ space})$, the corresponding one-dimensional (confined) compression modulus has been used $(dp'/d\varepsilon_{zz})$

$$\frac{dp'}{d\varepsilon_{zz}^e} = \frac{E^*(1-{}^* + 2\alpha_e{}^*)}{3(1+{}^*)(1-2{}^*)} \tag{8}$$

For isotropic elasticity ($\alpha_e = 1$), Eq.(8) reduces to the bulk modulus K

$$\frac{dp'}{d\varepsilon_{zz}^e} = \frac{E}{3(1-2\nu)} = K \tag{9}$$

The stress-dependent stiffness of a soil under unloading or reloading in the compression plane is given by

$$\frac{de^e}{dp'} = -\frac{\kappa}{p'} \tag{10}$$

where the variation of the void ratio that corresponds to the elastic part (de^e) may be related to the increment of the vertical elastic strain $d\varepsilon^e_{zz}$

$$d\varepsilon^e_{zz} = \frac{de^e}{1+e} \tag{11}$$

So, using Eqs. (8, 10, 11), the stress-dependent expression for the E^* parameter is obtained

$$E^* = \frac{3(1+^*)(1-2^*)}{1-^*+2\alpha_e^*} \frac{(1+e)}{\kappa} p' \tag{12}$$

The present formulation of elastic anisotropy is hierarchical because if $\alpha_e = 1$, the model reduces to S-CLAY1S (Karstunen et al. 2005) with isotropic elasticity. The current formulation of elasticity may be considered as hypoelastic, as it is defined in terms of stress and strain increments (Eq. 7). The authors are aware that this model is non-conservative (Zytynski et al. 1978), but assuming a constant Poisson's ratio is widely used (e.g., Wheeler et al. 2003) and the differences for non-cyclic loads are small for practical purposes (Borja et al. 1997). Application of a hyperelastic formulation (e.g., Puzrin 2012) can solve the problem. However, most hyperelastic models are fundamentally isotropic and only predict some stress-induced anisotropy (e.g., Castro, in prep.).

2.3 *Parameter determination*

The additional parameter (elastic anisotropic parameter, α_e) may be analytically obtained from conventional triaxial test, either from the undrained stress path during shearing

$$\frac{dq}{dp'} = \frac{3(2-2\nu^* - 4\alpha_e\nu^* + \alpha_e^2)}{2(1-\nu^* + \alpha_e\nu^* - \alpha_e^2)} \tag{13}$$

or from the shear-volumetric strain ratio during drained isotropic compression

$$\frac{d\varepsilon^e_d}{d\varepsilon^e_v} = \frac{-2(1-\nu^* + \alpha_e\nu^* - \alpha_e^2)}{3(2-2\nu^* - 4\alpha_e\nu^* + \alpha_e^2)} \tag{14}$$

3 COMPARISON WITH LABORATORY RESULTS

3.1 *Bothkennar clay*

The novel constitute model was used to represent the behavior of a well-documented soft natural clay, namely Bothkennar clay (e.g., Atkinson et al. 1992, McGinty 2006). The results were compared with the results of S-CLAY1S ($\alpha_e = 1$) to highlight the influence of anisotropic elasticity. Parameter calibration for S-CLAY1S was based on the methods proposed by Wheeler et al. (2003) and Karstunen et al. (2005). The additional parameter of the proposed model ($\alpha_e = 1.3$) was analytically obtained using Eq. (14) from the ratio of deviatoric to volumetric strain increments $(d\varepsilon^e_d / d\varepsilon^e_v = 0.2)$ measured in the laboratory (McGinty 2006) during drained isotropic compression at pressures lower than the preconsolidation one, i.e. initial straight part in Figure 2. $\alpha_e = 1.3$ means that the ratio of horizontal to vertical elastic stiffness is equal to $E_H/E_V = 1.7$. Those values are rounded to one decimal due to the lack of higher precision. Tables 1 and 2 show the adopted parameters for the proposed model.

3.2 *Isotropic compression*

Anisotropic behavior is visible under drained isotropic compression when deviatoric strains appear. Figure 2 shows how the proposed model reproduces the deviatoric strains during isotropic compression. Initially, during the elastic part, positive deviatoric strains $(d\varepsilon^e_d > 0)$ are predicted because of the higher horizontal elastic stiffness ($\alpha_e > 1$). After yielding, the deviatoric strain increments are mainly plastic and negative $(d\varepsilon^p_d < 0)$. Negative plastic deviatoric strain increments are predicted by the model because the yield surface is rotated and an associated plastic flow rule is adopted (see Fig. 1). Obviously, yielding is sharply represented

Table 1. Model parameters for Bothkennar clay.

κ	ν'	λ_i	M	ω	ω_d	ξ	ξ_d	α_e
0.025	0.2	0.2	1.4	50	1	9	0.2	1.3

Table 2. Model initial state variables.

p'_0* (kPa)	K_0*	σ'_{Vc}* (kPa)	e_0	α_0	χ_0
15	1	90	1.55	0.3	8

* Only for drained isotropic compression tests

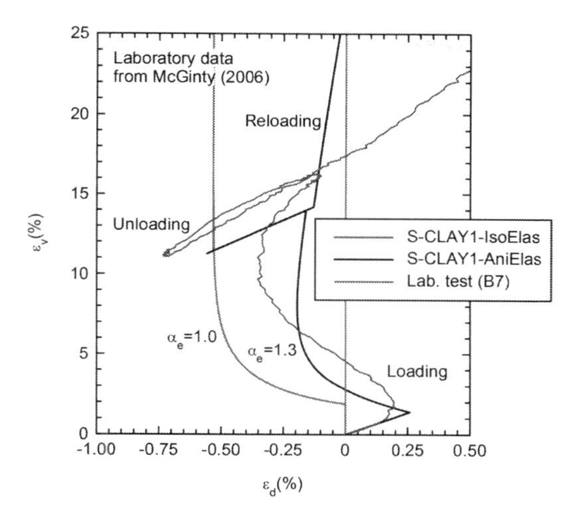

Figure 2. Influence of elastic anisotropy during isotropic compression and for an isotropic unloading-reloading loop.

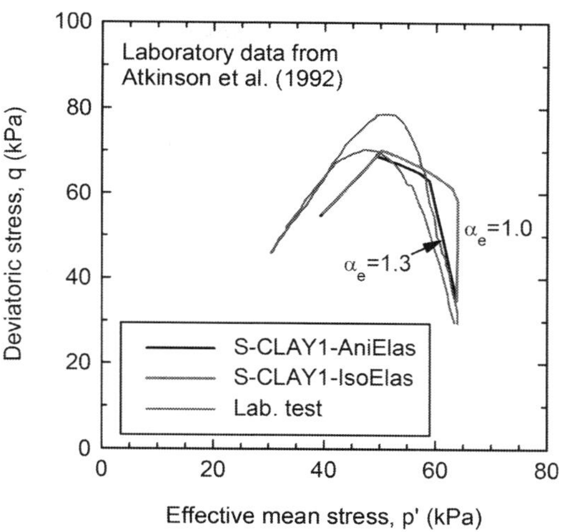

Figure 3. Influence of elastic anisotropy in effective stress path for undrained triaxial compression.

by the constitutive model, while the laboratory results (McGinty 2006) show a smooth transition. As plastic strains develop during isotropic loading, plastic anisotropy demises, the yield surface rotates toward the hydrostatic axis and plastic deviatoric strains tend to be 0 $(d\varepsilon_d^p = 0)$. For the unloading-loading loop, the constitutive model with isotropic elasticity predicts a vertical straight line ($(d\varepsilon_d^e = 0)$, while an inclined straight line is predicted when anisotropic elasticity is included. The strain path direction is the same as that at the beginning of the test $(d\varepsilon_d^e / d\varepsilon_v^e = 0.2)$ for the proposed constitutive model because a constant elastic anisotropy is adopted, while McGinty (2006) data show a demise of elastic anisotropy $(d\varepsilon_d^e / d\varepsilon_v^e = 0.12)$. An anomaly in McGinty (2006) data is that no detectable change in strain path direction associated with the yield point is observed after reloading. This anomaly may be attributed to testing imperfections at high strains (>15%).

3.3 *Undrained shearing in triaxial tests*

The effect of elastic anisotropy is also visible under undrained triaxial compression when the initial (elastic) effective stress path is no longer vertical (Fig. 3). Atkinson et al. (1992) present results of undrained triaxial compression tests carried out on samples of intact Bothkennar clay recovered using a Laval sampler. The fitted value α_e = 1.3 analytically obtained using Eq. (14) and McGinty (2006) data agrees with that obtained using Eq. (13) and Atkinson et al. (1992) data (Fig. 3).

4 NUMERICAL ANALYSES OF A BENCHMARK EMBANKMENT

The performance of the recently proposed model, which includes anisotropic elasticity, was analyzed in a finite element benchmark problem using PLAXIS 2D 2016 code (Brinkgreve et al. 2016) and their results are presented in this section. The proposed model has been implemented into the finite element code as user-defined soil model using an implicit (stress-strain integration) algorithm presented in Sivasithamparam & Castro (2016a).

A benchmark embankment problem (Sivasithamparam & Castro 2016b) was selected to represent a typical geotechnical engineering problem on Bothkennar clay where elastic and plastic anisotropy have a role. An embankment constructed on soft soil is assumed 2 m high, with a width at the top of 8 m and the side slopes with a gradient of 1:4. The soft soil Bothkennar clay below embankment extends for 20 m depth. At the surface, there is a 1 m thick over-consolidated dry crust. The geometry of the embankment and the finite element mesh used are shown in Figure 4. The groundwater table is assumed to be located at 1 m below the ground surface.

Tables 1–3 give the values for the initial state variables as well as the conventional soil constants for Bothkennar clay as required for the recently proposed model. The permeability is assumed to be the same in the vertical and horizontal direction for the sake of simplicity. The embankment, which is assumed to be made of granular material, was modelled using a simple linear elastic perfectly plastic model

(Mohr-Coulomb model). Its model parameters are depicted in Table 4. In order to make the results fully comparable, the over-consolidated dry crust layer is also modelled with a linear elastic perfectly plastic model (see Table 4 for material parameters). This embankment problem is hence expected to be dominated by the soft soil response and is not very sensitive to the embankment and dry crust parameters.

The analyses were performed using small deformation assumption as the idea is just to compare the influence of the elastic anisotropic parameter (α_e) at a boundary value level. The construction of the embankment was simulated by two undrained phases of 5 days each. In all analyses, drained conditions and zero initial pore pressures have been assumed above the water table.

For the initial condition, the in-situ K_0 value was assumed to be K_0 = 0.5 due to the slight overconsolidation of Bothkennar clay (Table 3). The first construction phase, in which the first layer of the embankment was built, was followed by a 5-day consolidation stage. After the completion of the second layer of the embankment, the final consolidation was simulated until excess pore pressures (EPP) are lower than one kPa.

Table 3. Model initial state variables for benchmark embankment simulation.

depth (m)	γ (kN/m³)	$k_V = k_H$ (m/day)	OCR	POP (kPa)	K_0
0–1	dry crust				
1–20	16.5	2.5E-4	1.5	–	0.5

Table 4. Embankment and dry crust parameters.

	γ/γ_{sat} (kN/m³)	c (kPa)	ϕ (°)	ψ (°)	ν	E (MPa)
Dry crust	19	2.0	37.0	0	0.2	3
Embank.	20	0.0	40.0	0	0.35	40

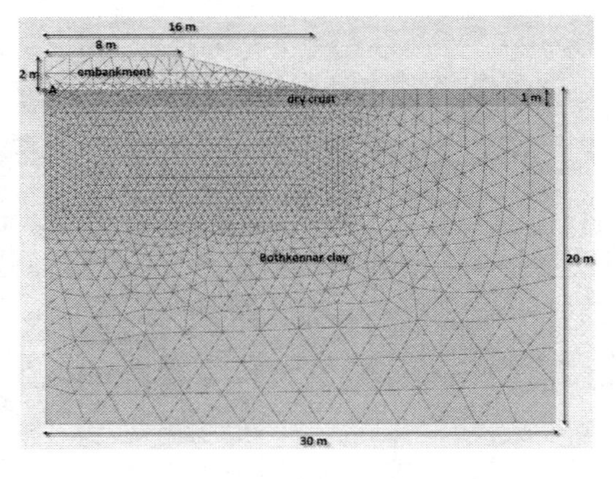

Figure 4. Finite element mesh and geometry of the benchmark embankment.

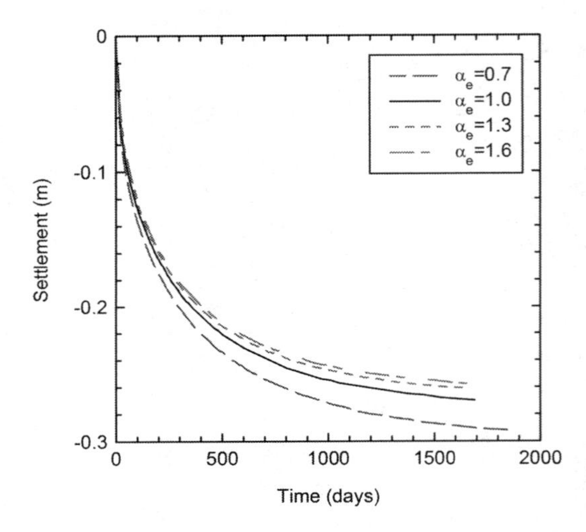

Figure 5. Settlement at the center point beneath the embankment with time for different elastic anisotropic parameters.

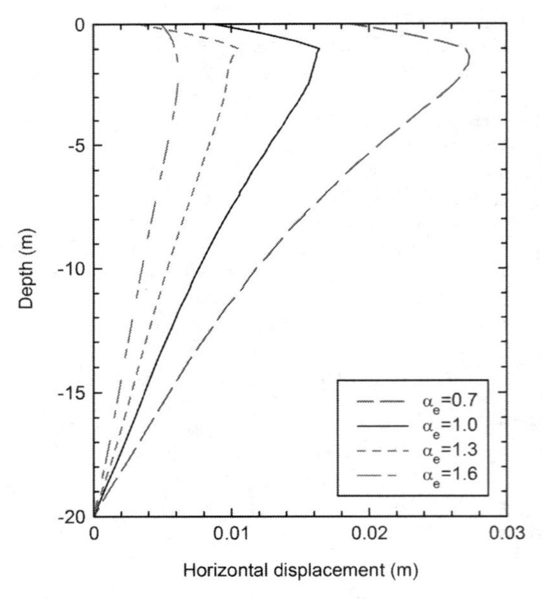

Figure 6. Horizontal displacements at the toe of the embankment at the end of analysis (i.e. EPP < 1 kPa) for different elastic anisotropic parameters.

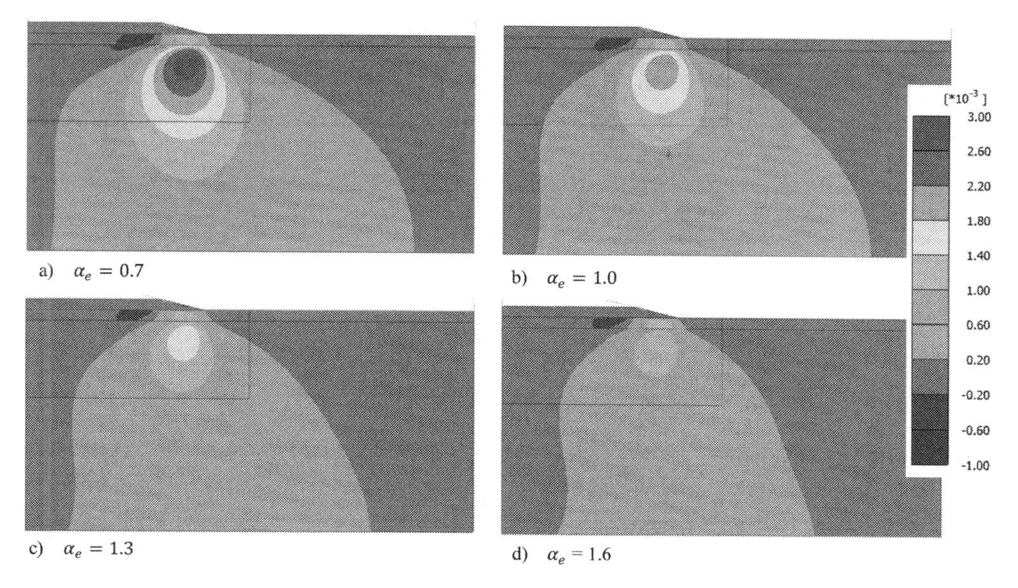

Figure 7. Shear strain contours at the end of the first layer embankment construction

The comparison of vertical settlement for varying elastic anisotropic parameter (α_e) of the finite element simulations is presented in Figure 5, which shows the settlement at point A (Figure 4) (center point beneath the embankment) with time. The settlement decreases with increasing values of α_e, because it means higher values of the horizontal stiffness $\left(\alpha_e^2 = E_H / E_V\right)$ and the vertical stiffness is the same. The final settlements for $\alpha_e = 0.7$ and 1.6 are 29.2 and 25.8 cm, respectively, what means a difference of 12%.

The assumed overconsolidation ratio is low (OCR = 1.5) (Table 3) because Bothkennar clay may be considered as a nearly normally consolidated clay. Although the overconsolidation is low and the vertical direction dominates for the chosen boundary value problem, namely an embankment, the differences are not negligible and elastic strains play a role as Figure 5 shows.

The calculated displacements concentrate underneath the slopes approximately 1 meter below the ground surface (at the contact between the dry crust and the soft soil). Therefore, the vertical cross section at the toe of the embankment was chosen to study the horizontal displacements. Figure 6 shows those horizontal displacements at the end of the analyses (EPP < 1 kPa). The parametric study of α_e shows that horizontal displacements increase with decreasing α_e values because it means a lower value of the elastic horizontal stiffness $\left(\alpha_e^2 = E_H / E_V\right)$, being the elastic vertical stiffness the same. The maximum horizontal displacements for $\alpha_e = 0.7$ and 1.6 are 27 and 6 mm, respectively, what means a difference of 77%. So, the α_e parameter has obviously a greater effect on horizontal deformations of subsoil beneath the toe of the embankment compared to the surface settlement at the center of the embankment (Figs. 5 and 6).

Figure 7 presents the contour of shear strains at the end of the first layer construction. The comparison is done for the first layer construction because the effect of elastic strains is more important in this first phase. The shear strains are concentrated beneath the toe of the embankment and approximately 2 m below the ground surface. It is clear from Figure 7 than the shear strains decrease with an increasing α_e value because the elastic horizontal stiffness is higher.

5 CONCLUSIONS

The application to a boundary value problem, namely a benchmark embankment, of a novel constitutive model for soft structured clays that includes anisotropic behavior both of elastic and plastic nature shows that the elastic anisotropic behavior has a non-negligible influence on the simulated displacements, particularly the horizontal displacements.

The novel constitutive model was developed based on an existing model that accounts for plastic anisotropy and destructuration (S-CLAY1S).

The new model incorporates stress-dependent cross-anisotropic elastic behavior within the yield surface using an additional parameter that relates the vertical and horizontal elastic stiffnesses and may be analytically obtained from conventional triaxial test, either from the undrained stress path during shearing or from the shear-volumetric strain ratio during drained isotropic compression.

Model parameters for Bothkennar clay have been presented and finite element simulations of a benchmark embankment on this soil show the influence of the elastic anisotropy parameter on horizontal displacements and settlements.

REFERENCES

Atkinson, J.H., Allman, M.A. & Böese, R.J. 1992. Influence of laboratory sample preparation procedures on the strength and stiffness of intact Bothkennar soil recovered using the Laval sampler. *Géotechnique* 42: 349–354.

Brinkgreve, R.B.J., Kumarswamy, S., Swolfs, W.M. 2016. *Plaxis 2D 2016 Manual*. Delft: Plaxis bv.

Borja, R.I., Tamagnini, C., Amorosi, A. 1997. Coupling plasticity and energy-conserving elasticity models for clays. *J. Geotech. Engng.* 123: 948–956.

Castro, J. E-potentials as yield surfaces for isotropic soils (in prep.).

Castro, J. & Sivasithamparam, N. 2017. A Constitutive Model for Soft Clays Incorporating Elastic and Plastic Cross-Anisotropy. *Materials* 10(6): 584.

Dafalias, Y.F. 1986. An anisotropic critical state soil plasticity model. *Mech. Res. Commun.* 13: 341–347.

Gens, A. & Nova, R. 1993. Conceptual bases for a constitutive model for bonded soils and weak rocks. In A. Anagnostopoulos, F. Schlosser, N. Kaltesiotis, & R. Frank (eds) *Geomechanical engineering of hard soils and soft rocks* Vol. 1: 485–494. Rotterdam: Balkema.

Graham, J. & Houlsby, G.T. 1983. Anisotropic elasticity of natural clay. *Géotechnique* 33: 165–180.

Hueckel, T. & Tutumluer, E. 1994. Modeling of elastic anisotropy due to one-dimensional plastic consolidation of clays. *Comput. Geotech. 16*: 311–349.

Karstunen, M., Krenn, H., Wheeler, S.J., Koskinen, M. & Zentar, R. 2005. Effect of anisotropy and destructuration on the behaviour of Murro test embankment. *Int. J. Geomech. 5*: 87–97.

Koskinen, M., Karstunen, M. & Wheeler, S. 2002. Modelling destructuration and anisotropy of a soft natural clay. In *5th Eur. Conf. Numer. Methods Geotech. Engng.*: 11–20. Paris: Presses de l'ENPC.

McGinty, K. 2006. *The stress-strain behaviour of Bothkennar clay*. PhD thesis, University of Glasgow.

Mitchell, R.J. 1972. Some deviations from isotropy in a lightly overconsolidated clay. *Géotechnique* 22: 459–467.

Mouratidis, A. & Magnan, J.-P. 1983. *Modèle élastoplastique anisotrope avec écrouissage pour le calcul des ouvrages sur sols compressibles*. Rapport de recherché LPC 121. Paris: Laboratoire central des Ponts et Chaussées.

Puzrin, A.M. 2012. Constitutive Modelling in Geomechanics – Introduction. Heidelberg: Springer-Verlag.

Roscoe, K.H. & Burland, J.B. 1968. On the generalised stress-strain behaviour of 'wet' clay. In J. Heyman, & F.A. Leckie (eds), *Engineering Plasticity*: 563–609. Cambridge: Cambridge University Press.

Sekiguchi, H., Ohta, H. 1977. Induced anisotropy and time dependency in clays. In Proc. specialty session 9, Constitutive equation of soils, *9th Int. Conf. Soil Mech. Found. Eng.*, Tokyo: 306–315.

Sivasithamparam, N., Castro, J. 2016a. An anisotropic elastoplastic model for soft clays based on logarithmic contractancy. *Int. J. Numer. Anal. Methods Geomech. 4*0: 596–621.

Sivasithamparam, N., Castro, J. 2016b. A framework for versatile shape yield surfaces for structured anisotropic soft soils. In J. Medzvieckas (ed) *The 13th Baltic Sea Region Geotechnical Conference Historical Experience and Challenges of Geotechnical Problems in Baltic Sea Region*: 154–158. Vilnius Gediminas Technical University Press.

Wheeler, S.J., Naatanen, A., Karstunen, K. & Lojander, M. 2003. An anisotropic elastoplastic model for soft clays. *Can. Geotech. J.* 40: 403–418.

Whittle, A.J., Kavvadas, M.J. 1994. Formulation of MIT-E3 constitutive model for overconsolidated clays. *J. Geotech Eng., ASCE* 120: 173–198.

Zdravkovic, L., Potts, D.M. & Hight, D.W. 2002. The effect of strength anisotropy on the behaviour of embankments on soft ground. *Géotechnique* 52: 447–457.

Zytynski, M., Randolph, M.K., Nova, R. & Wroth, C.P. 1978. On modelling the unloading-reloading behaviour of soils. *Int. J. Numer. Anal. Methods Geomech. 2*: 87–93.

Numerical Methods in Geotechnical Engineering IX – Cardoso et al. (Eds)
© 2018 Taylor & Francis Group, London, ISBN 978-1-138-33198-3

Fe-analysis of anchor pull out tests using advanced constitutive models

C. Fabris, H.F. Schweiger & F. Tschuchnigg
Graz University of Technology, Graz, Austria

ABSTRACT: The bearing capacity of ground anchors is significantly influenced by the surrounding soil and by the mechanical behaviour of the grout in the bonded length. In stiff overconsolidated clays, peak strength and strain softening leads to a non-uniform distribution of the mobilised shear strength. In order to reproduce an in-situ anchor pull out test, advanced constitutive models are employed in numerical analyses, namely a multilaminate model developed for overconsolidated stiff soils and a constitutive model accounting for strain softening for the grout. The numerical results obtained with the Multilaminate model are compared to the results obtained with the Hardening Soil Small model and to in-situ fibre optics measurements. The Multilaminate soil model was capable of reproduce the softening in the stiff overconsolidated soil and the cracks in the grout could be captured reasonably well. The strain measurements and the load-displacement behaviour were well reproduced.

1 INTRODUCTION

In numerical simulations the constitutive model describing the mechanical behaviour of soils plays a crucial role. Whereas it is clear that advanced constitutive models are appropriate for modelling the nonlinear behaviour of soils, it depends on the geotechnical problem what level of complexity is actually required with respect to the constitutive model in order to capture the main features of the problem.

Stiff overconsolidated soils show a very different behaviour in comparison to normally consolidated soils. Shearing of overconsolidated samples results in expansion of the soil and the maximum shear strength is followed by a reduction of the shear stress with continuous displacements. This behaviour has some implications such as the development of a progressive debonding at the interface soil/grout when an anchor tendon is submitted to a load test and it is is commonly referred to as a progressive failure mechanism.

The progressive failure mechanism causes a non-uniform distribution of load along the fixed length. This mechanism is usually observed at the interface soil/grout and was described by Ostermayer & Scheele (1977) during the performance of pull out tests in different soil conditions. According to Ostermeyer & Barley (2003), during an anchor pull out test in a stiff soil, the maximum bond stress moves towards the fixed length and, as the load increases, the ultimate stress at the interface tendon/grout or grout/soil is exceeded and reaches its residual value. With further increase of the load, the bond stress concentration reaches the bottom end of the fixed length, as shown schematically in Figure 1.

A Multilaminate model (MLSM) is employed in this work to simulate the progressive failure mechanism (Schädlich & Schweiger 2014b) and the results are compared with the Hardening Soil Small model (HSS model), described in Schanz et al. (1999) and Benz (2007). The latter is implemented in the commercial finite element code Plaxis 2D 2016 (Brinkgreve et al. 2016), which is used throughout this study, and has gained some popularity for solving practical geotechnical problems. However, it is not capable of modelling strain softening behaviour. The Multilaminate model is implemented via the so-called user defined soil model option. The grout is modelled with the Shotcrete model, originally developed for modelling the mechanical behaviour of shotcrete (Schädlich & Schweiger 2014a), which is capable of simulating the development of cracks in the grout. The results are compared with in-situ measurements of strains, along the tendon and along the grout.

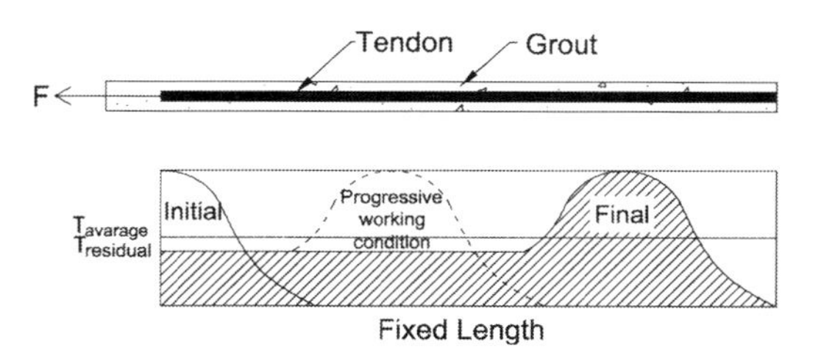

Figure 1. Progressive failure mechanism.

2 ADVANCED CONSTITUTIVE MODELS

2.1 *Multilaminate model*

The development of the multilaminate framework relates macro-mechanical behaviour of materials to the micro-mechanical scale and dates back to the slip theory of Taylor (1938), describing the sliding phenomena in metals with contact planes of different orientations. The Multilaminate material models are based on the concept that the material behaviour can be formulated on a distinct number of local planes with varying orientation. The stress–strain state varies from plane to plane, and the global behaviour is obtained by integration of the contributions of all planes (Bažant & Oh 1986).

Local stress is obtained by projection of the global stress vector into the integration planes (Fig. 2) and yield surfaces are defined in terms of local shear stress τ and normal stress σ'_n.

The original Multilaminate model for clay as proposed by Pande & Sharma (1983) has been extended by Schweiger et al. (2009) for modelling the behaviour of loose to medium dense sand or normally to slightly overconsolidated clays. The model features an elliptical volumetric hardening surface f_{cap} for loading in compression, a linear shear hardening surface f_{cone} for deviatoric loading and a tension cut-off surface f_{tens} and takes into account small strain stiffness.

This Multilaminate model has been extended further by Schädlich & Schweiger (2014b) in order to reproduce specific features of overconsolidated clays. A Hvorslev failure envelope is introduced in order to describe peak shear strength of heavily overconsolidated clays. Therefore, the Hvorslev surface definition was adapted to the integration plane level by normalizing the local stresses with the equivalent normal stress σ'_{ne}, as shown in Figure 3.

Because σ'_{ne} decreases with reduction of the normal stress σ'_n, the Hvorslev yield surface is a curved line in the non-normalized τ-σ'_n plot of the yield surfaces (Fig. 4) and only gets activated when the local stress path reaches the Hvorslev surface. Plastic strains are obtained also from the strain hardening deviatoric yield surface (Schädlich 2012). A regularization technique, namely a non-local formulation (Galavi & Schweiger 2010) is introduced in order to avoid the well known mesh dependency when modelling strain softening behaviour by means of standard finite element formulations.

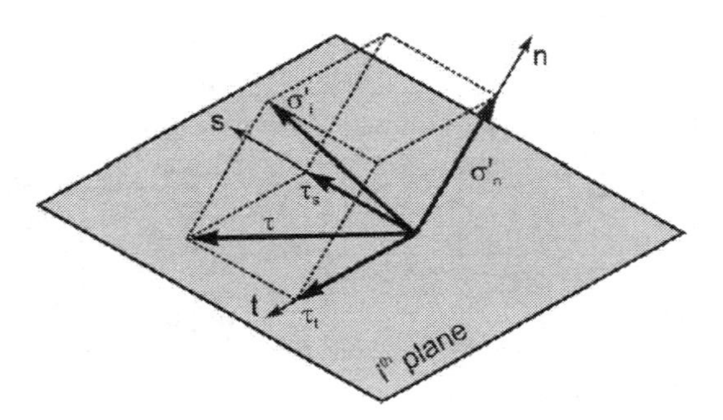

Figure 2. Local stress components (Schädlich 2012).

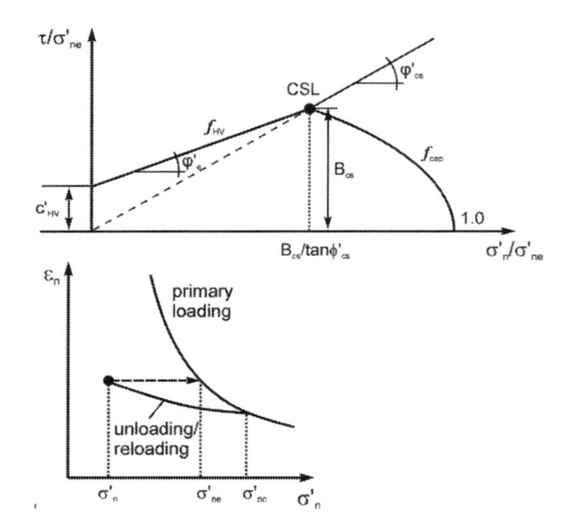

Figure 3. Normalized Hvorslev surface and equivalent stress definitions (Schädlich 2012).

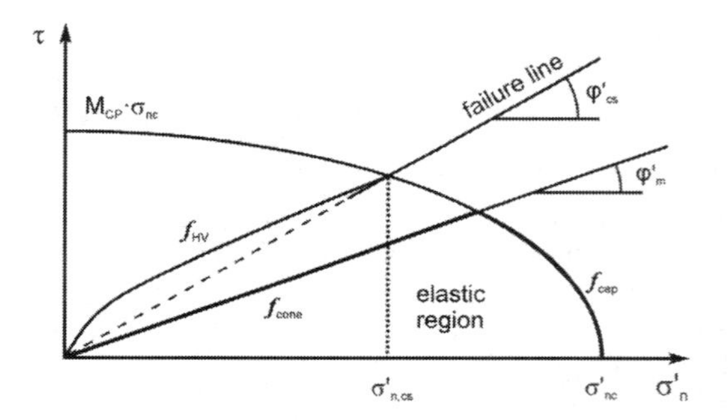

Figure 4. Local yield surface (Schädlich 2012).

2.2 *Material model for the grout*

The shotcrete model is employed for the grout and is capable to consider the time dependent strength and stiffness development, strain hardening and softening, creep and shrinkage. However, in this study, no time dependency is introduced and the grout is assumed cured when it is loaded.

The model decomposes the total strain ε into elastic strains ε_e, plastic strains ε_p, creep strains ε_{cr} and shrinkage strains ε_{shr}. Plastic strains are calculated according to strain hardening/softening elastoplasticity. Creep and shrinkage are not taken into account herein.

For deviatoric loading, a Mohr-Coulomb yield surface F_c is employed and, in the tension regime, a Rankine yield surface F_t is assumed (Fig. 5).

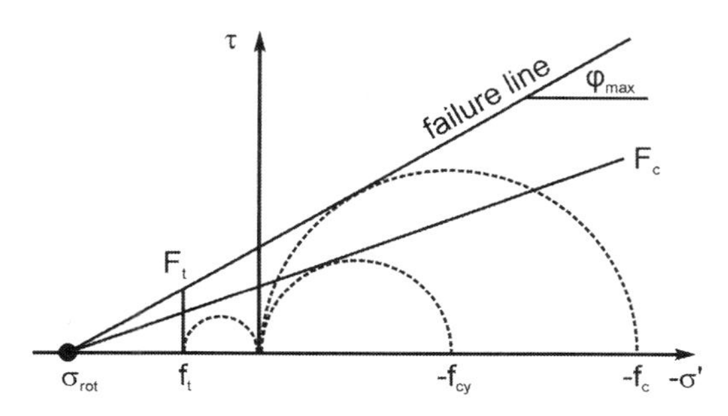

Figure 5. Yield surfaces and failure envelope of the model (Schädlich & Schweiger 2014a).

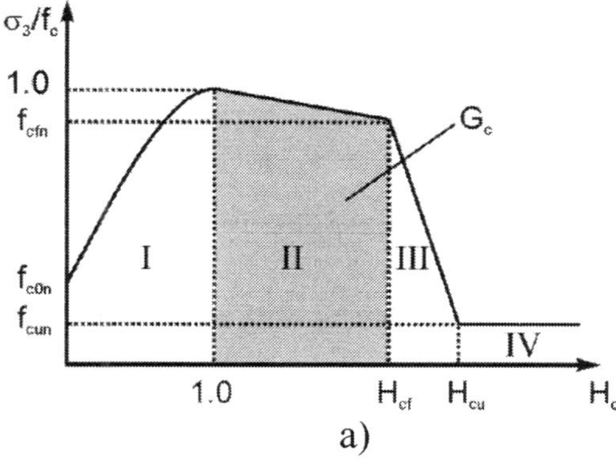

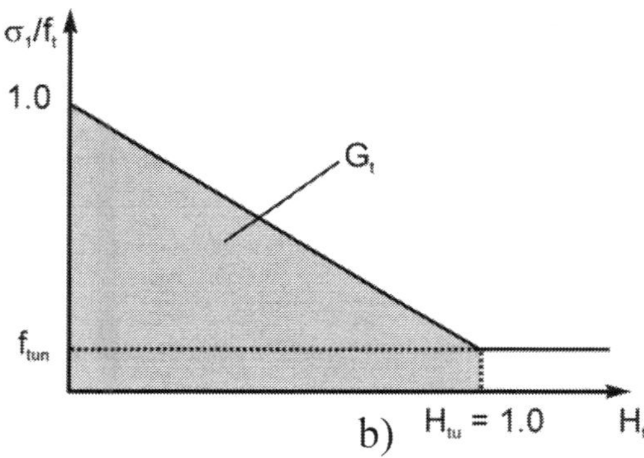

Figure 6. Stress-strain curve—a) in compression and b) in tension (Schädlich & Schweiger 2014a).

The behaviour in compression and in tension is shown in Figure 6a and Figure 6b. In compression, it is divided in four parts, according to Figure 6a:

- Quadratic strain hardening: comprised between the normalized initial strength f_{c0n} and the uniaxial compressive strength of cured shotcrete f_{c28};
- Linear strain softening in between the uniaxial compressive strength and the normalized failure strength f_{cfn};
- Linear strain softening in between the normalized failure strength and the normalized residual strength f_{cun}.

H_c is a normalized hardening/softening parameter equal to $\varepsilon_3^p / \varepsilon_c^p$, where ε_3^p is the minor principal plastic strain and ε_c^p is the plastic peak strain in uniaxial compression. Full mobilisation of f_c coincides with $H_c = 1$ and failure strength is reached at H_{cf}. The fracture energy in compression is an input parameter provided by G_c.

The model behaviour in tension is linear elastic up to the tensile strength f_t, followed by linear strain softening (Fig. 6b). The fracture energy in tension is given by G_t and f_t is governed by a softening parameter H_t, equal to $\varepsilon_1^p / \varepsilon_{tu}^p$, where ε_1^p is the major principal plastic strain and ε_{tu}^p is the plastic ultimate strain in uniaxial tension. Softening takes place until the residual strength $f_{tu} = f_{tun}$. f_t is reached.

More details on the model can be found in Schädlich & Schweiger (2014a) and Schädlich et al. (2014).

3 NUMERICAL SIMULATIONS

The finite element model for the in-situ anchor load test is axisymmetric with the axis of symmetry being vertical reflecting the test arrangement for this particular case. The grout is introduced at the free length, for corrosion protection, and at the fixed length. In order to ensure no load transfer between the tendon and the grout at the free length, the tendon is modelled only at the fixed length and a gap is introduced between the grout and the vertical axis of symmetry at the section presenting the free length. The load is applied by means of prescribed displacements on top of the tendon (Fig. 7a).

Four different soil layers are considered (Fig. 7b). The first two layers are modelled with the Mohr-Coulomb model and consist of a fill material of 5.5 m thickness and a 2 m thick layer of overconsolidated soil. The third layer is also an overconsolidated soil and is modelled with the Hardening Soil Small model (HSS model) or with both, the Hardening Soil Small model and the Multilaminate soil model (MLSM). A fourth layer

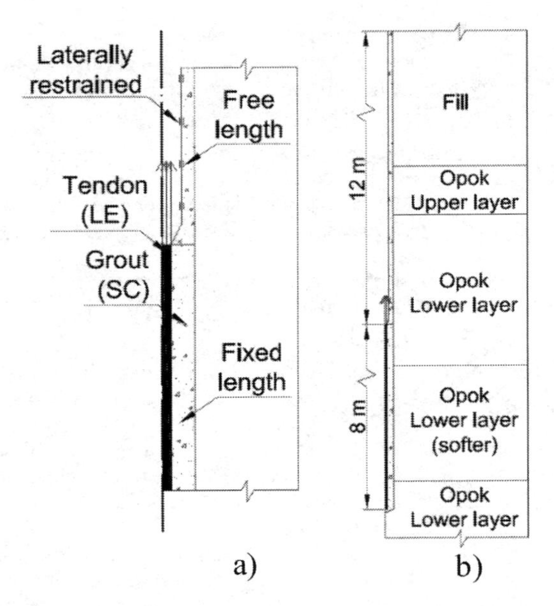

Figure 7. Anchor geometry—a) anchor detail and b) soil layers.

of 3 m thick is assumed between the third one, with the same strength properties but with lower stiffness parameters. This assumption was made after the evaluation of the in-situ strain measurements along the fixed length and the inspection of the borehole. The overconsolidated soil is known as "Opok" and consists of a clayey, sandy silt. The parameters of the Opok were obtained from CU triaxial tests.

Two simulations are performed: one with the entire Opok lower layer modelled with the Hardening Soil model and another one employing the Multilaminate soil model in the vicinity of the bonded length by assigning this model to a 50 cm wide cluster next to the grouted body. This enables the modelling of the softening behaviour at the contact grout/soil but avoids excessive computational time when the Multilaminate model is assigned to the entire domain of the Opok. The material of the tendon is linear elastic and the grout is modelled with the Shotcrete model.

3.1 *Material parameters*

The material parameters are presented in Tables 1–6. A detailed description of the parameters of the Multilaminate model is provided in Schädlich & Schweiger 2014b and Schädlich 2012.

The grout at the free length is assumed to have a lower stiffness in comparison with the grout at the fixed length. This assumption arises from the fact that, after performing numerical simulation applying the same grout properties at the free and fixed length, the compressive stresses in the grout between the free and the fixed length were higher than what followed from the measurements. This assumptions can be justified considering that the grout in this part of the anchor is supposed to work mainly as corrosion protection. However, additional studies are currently undertaken to investigate this effect in more detail.

Table 1. Material parameters for the tendon.

Material Model: Linear Elastic			Tendon
Symbol	Description	Unit	Value
γ	Unit weight unsat	kN/m³	78.5
E_{tendon}	Young's Modulus	kN/m²	16 3900 000
ν	Poisson's ratio	-	0.3

Table 2. Material parameters for the grout.

Material Model: Shotcrete Model			Grout
Symbol	Description	Unit	Value
E_{28}	Young's modulus of cured grout	kPa	16 260 000
v	Poisson's ratio	-	0.2
$f_{c,28}$	Uniaxial compressive strength	kPa	32,120
$f_{t,28}$	Uniaxial tensile strength	kPa	2000
f_{c0n}	Normalized initially mobilised strength	-	0.15
f_{cfn}	Normalized failure strength (compression)	-	0.95
f_{cun}	Normalized residual stength (compression)	-	0.1
$G_{c,28}$	Compressive fracture	kN/m	50
f_{tun}	Ratio of residual vs. Peak tensile strength	-	0.05
$G_{t,28}$	Tensile fracture energy	kN/m	0.15
ϕ_{max}	Maximum friction angle	°	40

Table 3. Material parameters for the Fill.

Material Model: Mohr-Coulomb Model			Fill
Symbol	Description	Unit	Value
E'	Stiffness	kN/m²	15,000
ν	Poisson's ratio	-	0.3
c'_{ref}	Cohesion	kN/m²	2
ϕ'	Friction angle	°	27.5
Ψ	Dilatancy angle	°	0
K_0	Earth pressure coefficient at rest	-	0.54

Table 4. Material parameters for the Opok upper layer.

Material Model: Mohr-Coulomb Model			Opok upper
Symbol	Description	Unit	Value
E'	Stiffness	kN/m²	40,000
ν	Poisson's ratio	-	0.35
c'_{ref}	Cohesion	kN/m²	100
ϕ'	Friction angle	°	35
Ψ	Dilatancy angle	°	0
K_0	Earth pressure coefficient at rest	-	1.5

Table 5. Material parameters for the Opok lower layer.

Material Model: Hardening Soil Small Model			Opok lower
Symbol	Description	Unit	Value
$E_{50,ref}$	Primary loading stiffness (reference)	kN/m²	6000/600*
$E_{oed,ref}$	Oedometric stiffness (reference)	kN/m²	6000/600*
$E_{ur,ref}$	Un/reloading stiffness (reference)	kN/m²	29,000/2900*
ν_{ur}	Poisson's ratio un/ reloading	-	0.2
c'_{ref}	Effective cohesion	kN/m²	200
ϕ'	Effective friction angle	°	35
M	Stress dependency index	-	0.85
Ψ	Dilatancy angle	°	0
$K_{0,nc}$	Earth pressure coefficient in normal compression	-	0.426
POP	Pre-overburden pressure	kN/m²	3500
$\gamma_{0.7}$	Shear strain at 70% G0ref	-	1.50E-04
G_{0ref}	Small strain shear modulus	kN/m²	48,300/4830*
K_0	Earth pressure coefficient at rest	-	1.5

*Values assumed for the softer soil layer

Table 6. Material parameters for the Opok lower layer.

Material Model: Multilaminate Model			Opok lower
Symbol	Description	Unit	Value
$E_{oed,ref}$	Oedometric stiffness (reference)	kN/m²	6000
$E_{ur,ref}$	Un/reloading stiffness (reference)	kN/m²	29,000
m	Power	-	0.85
v'_{ur}	Poisson's ratio in un-/reloading	-	0.2
A_{mat}	Shear hardening parameter	10^{-3}	2.5
R_f	Failure ratio	-	0.95
$K_{0,nc}$	Earth pressure coefficient in normal compression	-	0.47
n_{cp}	Number of integration planes	-	21
φ_e	Hvorslev surface inclination	o	24
σ_{nc0}	Initial value of σ_{nc}	kN/m²	–3500
φ_{cs}	Critical state friction angle	o	32
h_{soft}/L_{calc}	Ratio h_{soft}/L_{cal}	m⁻¹	10,000
K_0	Earth pressure coefficient at rest	-	1.5

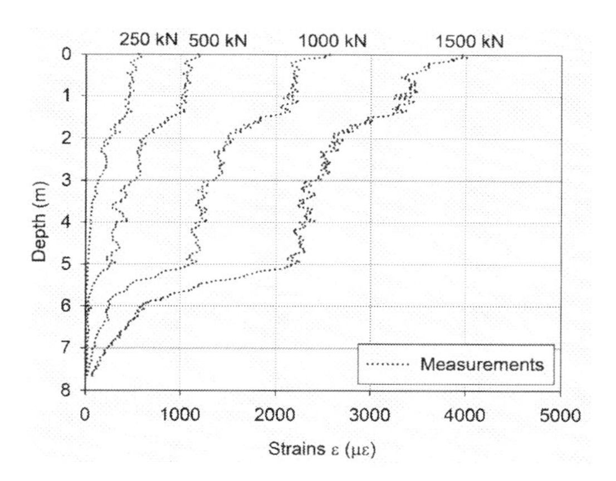

Figure 8. Strain measurements along the tendon at the fixed length.

3.2 *Numerical versus in-situ measurements*

The pull out test was performed in 2015 by Keller Grundbau, in Söding, Austria. During the test, the strains along the tendon and the grout were monitored with a fibre optic sensing system and results are presented in more detail in Racansky et al. (2016) and Monsberger et al. (2017).

The anchor was vertically installed and the borehole diameter was 178 mm. The fixed length was 8 m and the free length 12 m. The grout at the free and fixed length was only gravity grouted and 11strands were used. The strain measurements in the tendon and in the grout at the fixed length are presented in Figures 7 and 8.

3.2.1 *Load-displacement curve*

The load-displacement curve obtained in-situ and the curves obtained with the HSS and the MLSM are presented in Figure 9. The displacement and

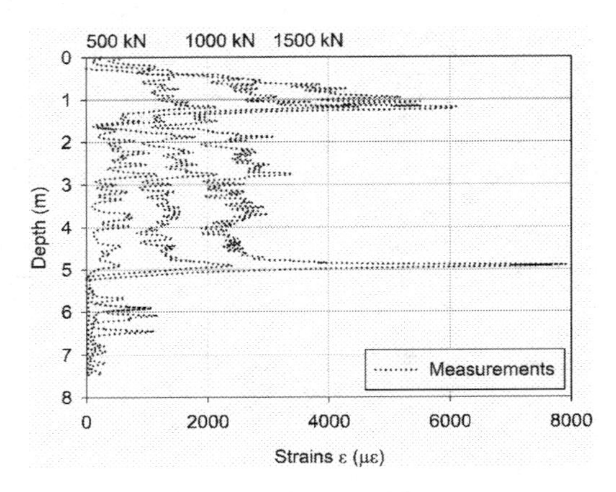

Figure 9. Strain measurements along the grout at the fixed length.

stress points selected for the comparison are located on top of the tendon.

Regarding the first 4 mm of displacement, the prediction obtained with the HSS model agreed almost perfectly with the measurements, whilst the MLSM curve was stiffer. This is due to the stiffness definition in the MLSM, which in general results in a stiffer behaviour in comparison to the HSS model. The kink in the MLSM curve at 4 mm is due to the reduction of the shear stress in the soil after softening starts.

From 4 mm onwards, the best agreement was achieved with the MLSM, and the curve is above the one obtained with the HSS model. This is a result of peak strength in the MLSM, which is an output of the model and is higher than the friction angle defined in the HSS model.

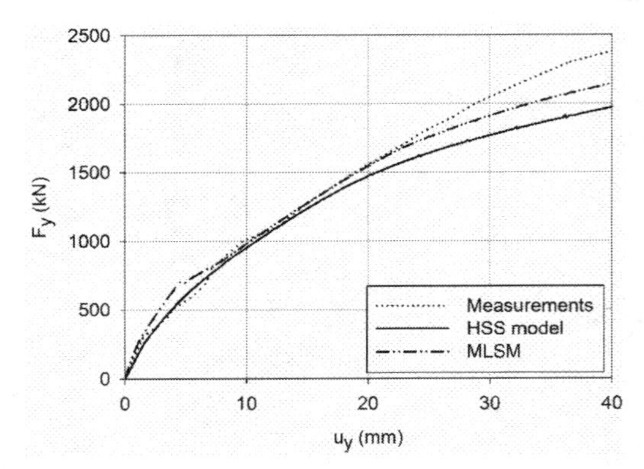

Figure 10. Load-displacement curves.

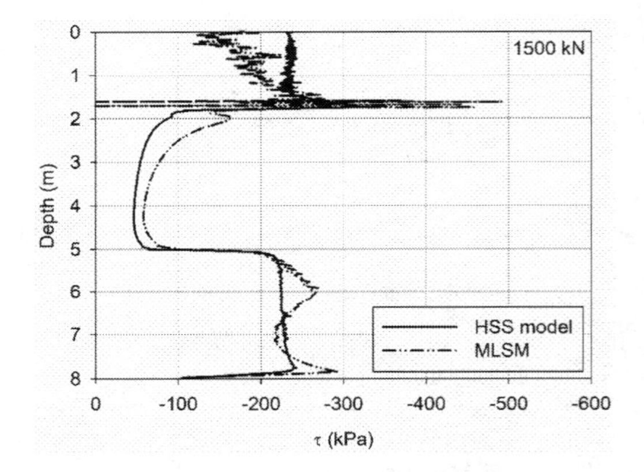

Figure 11. Shear stress distribution in the soil.

3.2.2 *Mobilised shear stress*

The shear stress distribution in the soil, at a distance of 6 cm from the grout at the fixed length, is shown in Figure 10 for the load level of 1500 kN. Within the first 2 m, the shear stress for the HSS model has reached its maximum values. Considering the MLSM, within this depth softening has started in the soil, decreasing the shear stress. Between 2 m and 5 m, the mobilised shear stress is considerably lower, due to the softer soil layer in this region.

It is also notable that, for the MLSM, the peak shear stress is located at about 6 m and its value is considerably higher than the shear stress obtained with the HSS model. This is because dilatant behaviour is involved when the Hvorslev surface is activated, which in turn leads to an increase in normal stress at the grout/soil interface.

3.2.3 *Softening parameter in tension*

The softening parameter H_t is an output of the Shotcrete model and indicates if the grout has reached the peak tensile strength or the residual strength respectively. A value of $H_t > 0$ indicates that softening in tension has started and $H_t > 1$ indicates residual level. The development of cracks in the grout starts as soon as $H_t > 0$.

The H_t distribution and the strain measurement along the grout at the fixed length are presented in Figure 11 for 1500 kN. The distribution is similar for the HSS and the MLSM, but the oscillations are more pronounced for the MLSM. These oscillations obtained with the numerical simulations indicate that the grout is cracked. If the interface soil/grout is softer, that is, between 2 m and 5 m, fewer cracks are observed. At the transition between the soft and the stiff soil layers, cracking is more pronounced. The same is observed for the strain measurement, indicating that the shotcrete model reproduces the cracking pattern reasonably well. The H_t shadings allows a better visualisation of the cracking pattern and it is presented in Figure 12 for the same load level of 1500 kN, considering the first 2 m of the fixed length.

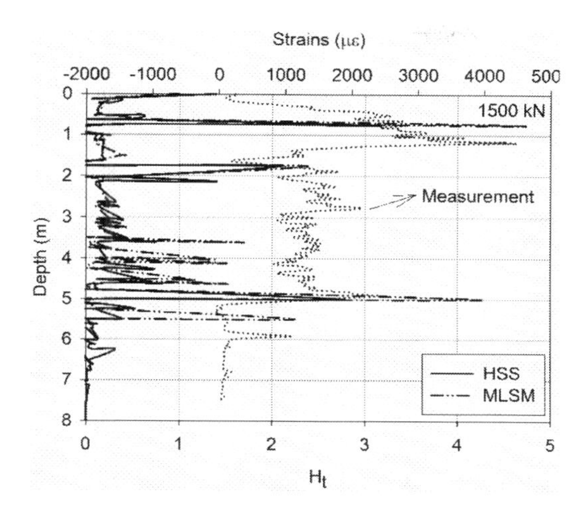

Figure 12. H_t distribution in the grout.

Although the patterns in Figure 13 are very similar, the H_t values are slightly higher for the MLSM. In this case, the crack development in the grout is more pronounced because after softening behaviour in the soil is initiated, the load transfer mechanism changes significantly if the Multilaminate model is employed.

3.2.4 *Strains in the tendon*

The strains in the tendon, at the fixed length, obtained from the fibres optic measurements and from the numerical analyses are presented in Figure 13.

Strains start to decrease at the top of the fixed length due to the load transfer in the interfaces tendon/grout and grout/soil respectively. The slope of the strains indicates the rate of the load transfer. Within the first 2 m, the slope in relation to the vertical axis is higher because the transfer in this part is higher. However, the slope decreases between 2 m and 5 m, where the soil is softer.

The strain distributions are similar for the HSS model and the MLSM. Up to 5 m depth a very good agreement is obtained with the MLSM. From 5 m below the top to the end of the fixed length, differences between simulation and measurement are obvious but this can be most probably attributed to inhomogeneities in the soil.

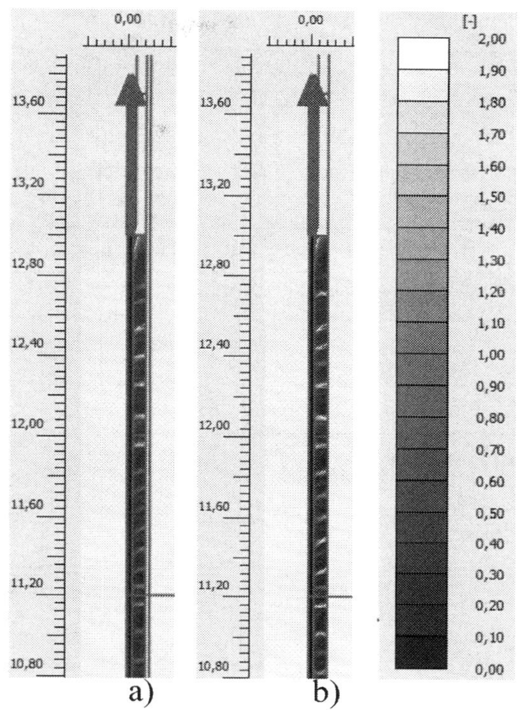

Figure 13. Cracking pattern—a) HSS model and b) MLSM.

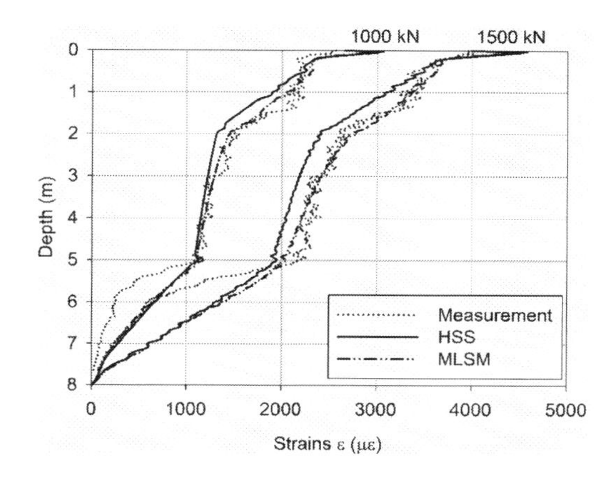

Figure 14. Predicted strains along the tendon at the fixed length.

4 CONCLUSION

The load-displacement curve obtained with both, the MLSM and the HSS model showed good agreement with the in-situ measurements. The HSS model presented a better fit within the first 4 mm of displacement and, from 4 mm onwards, the MLSM showed a better response. The Shotcrete model was able to simulate the crack pattern in the grout.

The main difference between the MLSM and the HSS model is the decrease of the shear stress at the interface soil/grout after softening starts in the MLSM. The Multilaminate constitutive model was capable of reproduce the progressive failure mechanism along the interface soil/grout by considering softening after the peak strength was reached.

Softening in the soil had a minor effect on the strains distribution in the tendon along the fixed length and, for that reason, both simulations showed very similar results. The measured strain distribution could be captured reasonably well.

REFERENCES

Bažant, Z.P. & Oh, B.H. 1986. Efficient Numerical Integration on the Surface of a Sphere, *Zeitschrift für angewandte Mathematik und Mechanik* 66: 37–49.

Benz, T. 2007. *Small-Strain Stiffness of Soils and its Numerical Consequences*. Publication No. 55, Institute for Geotechnical Engineering, University of Stuttgart.

Brinkgreve, R.B.J., Kumarswamy, S. & Swolfs, W.M. 2016. PLAXIS 2016. *Finite element code for soil and rock analyses*, User Manual. Delft: Plaxis bv.

Galavi, V. & Schweiger, H.F. 2010. Nonlocal multilaminate model for strain softening analysis. *International Journal of Geomechanics*, Vol. 10, No. 1, 30–44.

Monsberger, C., Woschitz, H., Lienhart, W., Racansky, V. & Hayden, M. 2017. Performance assessment of geotechnical structural elements using distributed fiber optic sensing, *Sensors and Smart Structures Technologies for Civil, Mechanical, and Aerospace Systems; Proc.*, 2017. DOI: 10.1117/12.2256711.

Ostermayer, H. & Scheele, F. 1977. Research on ground anchors in non-cohesive soils. *Revue Française de Géotechnique* 3: 92–97.

Ostermeyer, H. & Barley, T. 2003. *Fixed Anchor Design Guidelines*. Geotechnical Engineering Handbook, 2.

Pande, G.N. & Sharma, K.G. 1983. Multilaminate model of clays – a numerical evaluation of the influence of rotation of principal stress axes. *International Journal for Numerical and Analytical Methods in Geomechanics*, Vol. 7, No. 4, 397–418.

Racansky, V., Weidacher, R., Lienhart, W., Monsberger, C., Woschitz, H. & Schweiger, H.F. 2016. Überwachung eines Ankerausziehversuches mittels Glasfasersensoren, *Deutsche Baugrundtagung; Proc., Bielefeld, 14–17 September 2016*.

Schanz, T., Vermeer, P.A. & Bonnier, P.G. 1999. The Hardening-Soil Model: formulation and verification. *Beyond 2000 in Computational Geotechnics*, Brinkgreve (ed.), Rotterdam, Balkema, 281–290.

Schädlich, B. & Schweiger, H.F. 2014a. A new constitutive model for shotcrete. *Numerical Methods in Geotechnical Engineering*, 103–108.

Schädlich, B. & Schweiger, H.F. 2014b. Modelling the shear strength of overconsolidated clays with a Hvorslev surface. *Geotechnik* 37: 47–56.

Schädlich, B., Schweiger, H.F., Marcher, T. & Saurer, E. 2014. Application of a novel constitutive shotcrete model to tunnelling. *Rock Engineering and Rock Mechanics: Structures in and on Rock Masses*, 799–804.

Schädlich, B. 2012. *A Multilaminate Constitutive Model for Stiff Soils*. Ph.D. thesis. Gruppe Geotechnik Graz, Graz University of Technology, Austria, Heft 57.

Schweiger, H.F., Wiltafsky, C., Scharinger, F. & Galavi, V. 2009. A multilaminate framework for modelling induced and inherent anisotropy of soils. *Géotechnique* 59 (2): 87–101.

Taylor, G.I. 1938. Plastic strain in metals. *Journal of the Institute of Metals* 62: 307–324.

Numerical Methods in Geotechnical Engineering IX – Cardoso et al. (Eds)
 ISBN 978-1-138-33198-3

On the convexity of yield and potential surfaces in rotational hardening critical state models

Jon A. Rønningen, Gustav Grimstad & Steinar Nordal
Norwegian University of Science and Technology, Trondheim, Norway

ABSTRACT: This article shows that a common formulation often used for an anisotropic modified cam clay yield and potential surface with a modified Lode angle dependency may become concave for high values of anisotropy. Concave surfaces are undesirable in plasticity theory and could lead to numerical problems. To remediate this problem the article suggests a formulation for the Lode angle dependency that will not suffer from concavity. The suggested formulation is discussed. The formulation does not introduce any additional parameters for Lode angle dependency than that used to describe Lode angle dependency of an isotropic yield surface. In this paper, a generalized continuous Mohr–Coulomb criterion is used that allows a π-plane cross-section to take the shape of several criteria including Mohr-Coulomb, Matsuoka-Nakai and Lade-Duncan.

1 INTRODUCTION

Many of the state-of-the-art soil models today, in particular for clays, are based on the critical state (CS) concept (Schofield and Wroth, 1968) with the Modified Cam-Clay (MCC) model (Roscoe and Burland, 1968) as formulation basis. New features have been added such as rotational hardening. A common formulation defines a yield and potential surface which when rotated changes its shape and become "sheared", as first proposed by Dafalias (Dafalias, 1986) (Figure 1). This model is often referred to as the Anisotropic Modified Cam Clay Model (AMCCM). The formulation was originally derived considering plastic dissipation in triaxial coordinates. In general, stress space the yield surface may be given by:

$$p_{eq} = p' \cdot \left(1 + \frac{3}{2} \cdot \left|\frac{\sigma_d}{p'} - \alpha_d\right|^2 \cdot g^2(\theta^\alpha, \alpha)\right) \qquad (1)$$

$$F = p_{eq} - p_{mi} = 0$$

where $\boldsymbol{\sigma}_d$ and $\boldsymbol{\alpha}_d$ are the deviatoric stress and rotational vector respectively (see appendix), p' is the mean effective stress and p_{mi} is the size of the surface and represents the pre-consolidation pressure. The function $g(\theta^\alpha,\alpha)$ is the subject of this paper. The modified Lode angle, θ^α, is defined as:

$$\sin 3\theta^\alpha = -\frac{3\sqrt{3} \cdot J_3^\alpha}{2 \cdot (J_2^\alpha)^{3/2}} \qquad (2)$$

where J_2^α and J_3^α is the second and third invariant of the modified deviatoric stress vector, $\boldsymbol{\sigma}_d - p' \cdot \boldsymbol{\alpha}_d$. The function, $g(\theta^\alpha, \alpha)$, represents the shape of the surface in a plane normal to the α-line (Figure 2, Eq. (3)) and includes a dependency of a critical state ratio, M. If M is a constant then the shape will be circular, but most often it varies in accordance with a failure criteria such as Mohr-Coulomb, Matsuoka-Nakai (Matsuoka and Nakai, 1974), Lade-Duncan (Lade and Duncan, 1975), or a generalized form (Grimstad et al., 2018). In general M must be a function of the modified Lode angle, $M = M(\theta^\alpha)$.

$$\alpha = \sqrt{\frac{3}{2} \cdot \alpha_d^T \cdot \alpha_d} \qquad (3)$$

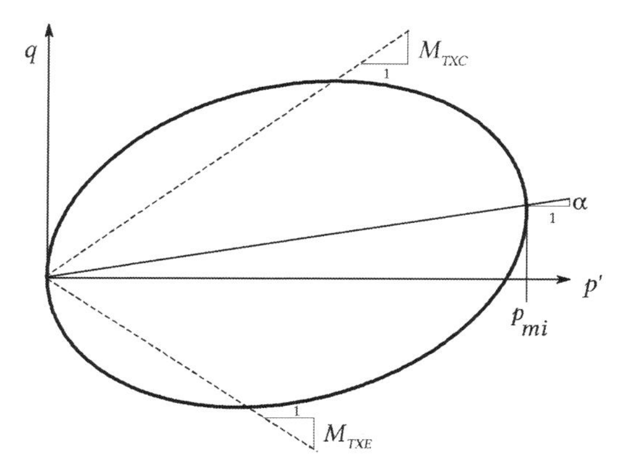

Figure 1. Yield surface for AMCCM in p'-q space.

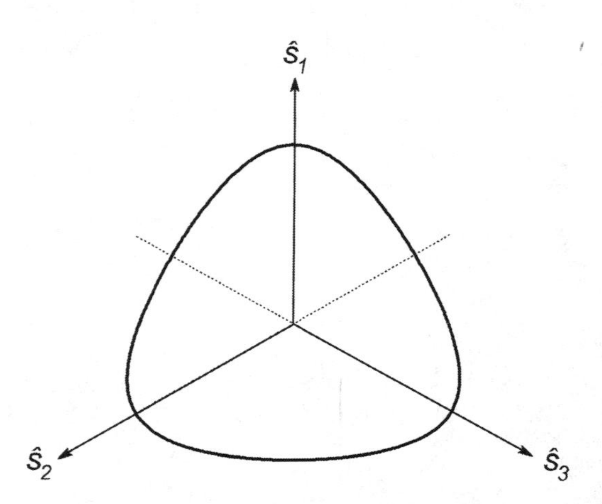

Figure 2. Typical projected cross section of the yield surface in the translated π-plane. Here $\hat{s}_i = \sigma_i - \alpha_i \cdot p$.

2 PROBLEM DESCRIPTION

In many formulations found in the literature the function $g^2(\theta^\alpha, \alpha)$, from eq. (1), is defined as:

$$g^2(\theta^\alpha, \alpha) = \frac{1}{M^2(\theta^\alpha) - \alpha^2} \tag{4}$$

In this expression, M is directly a function of the modified Lode angle, while α might change due to kinematic hardening. It can be seen in the equation above that whenever α approaches M (i.e. due to kinematic hardening) the expression $g(\theta^\alpha, \alpha)$ becomes very large and approaches infinity as a limit. As stated in (Crouch and Wolf, 1995) the introduction of a dependency of the anisotropic bounding surface on the Lode angle is not straightforward. In the work on this subject it is discovered that due to the specific form of eq. (4) and (1), the shape of the surface could actually become concave as it is "sheared" (Figure 3 and Figure 4). This may happen even if the shape is convex for α equal to zero. In general, concave surfaces are undesirable in plasticity theory and could lead to numerical problems.

It can be shown that convexity is ensured if eq. (5) is satisfied.

$$g''(\theta^\alpha) + g(\theta^\alpha) \geq 0 \tag{5}$$

Solving this equation at the limit, for any θ^α, gives the maximum rotation, i.e. the limit for α, for which the surface remains convex. For the Lade-Duncan criterion the solution will follow the line LD in Figure 5. For other criteria such as generalized Mohr-Coulomb, the curve may be lower or higher up, this is dependent on the shape parameters used. Typically, for sharper corners and straighter segments the curve is lower than the one for LD.

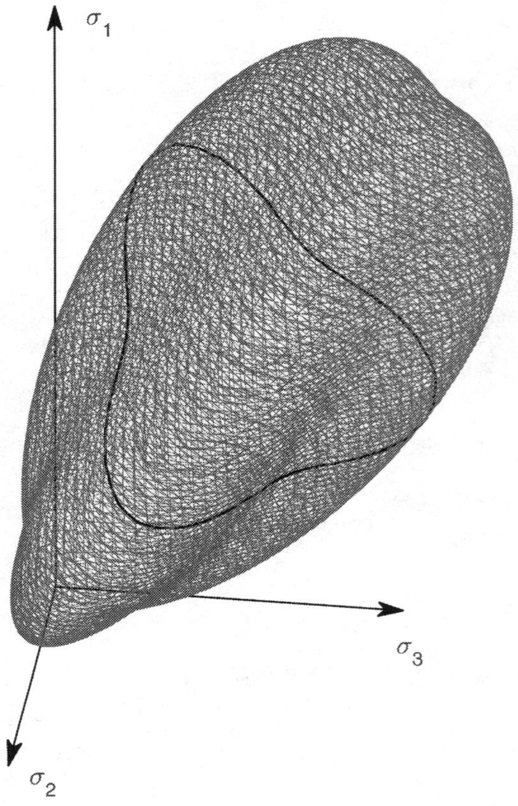

Figure 3. Yield surface for AMCCM in principal stress space for high values of α.

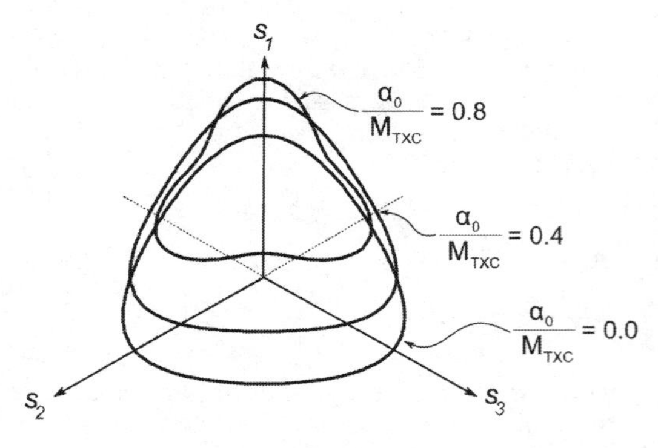

Figure 4. Cross section of the Modified Lode angle dependent AMCCM yield surface in the π-plane.

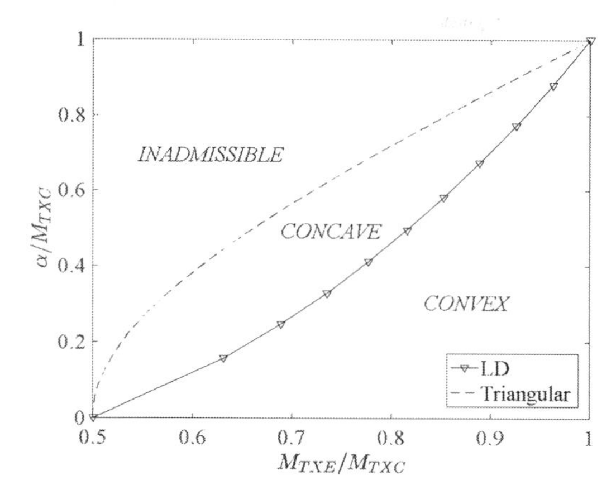

Figure 5. Limit value for α using the LD criterion for critical state description, the line for "Triangular" to refer to eq. (10), restricting the required shape to not be "beyond" triangular.

In order to investigate if the concavity is a problem in practice one need to see if typical material parameters for a typical clay material could lead to concavity.

It is common to use the earth pressure coefficient at rest under virgin loading, K_0^{NC}, to define the initial value of α. Dafalias provided an exact expression for the initial value α_0 corresponding to virgin compression, which, when disregarding elastic strain contributions, simplifies to:

$$\alpha_0 \approx \frac{\eta_0^2 + 3 \cdot \eta_0 - M_{TXC}^2}{3}$$
$$\eta_0 = \frac{3 \cdot (1 - K_0^{NC})}{1 + 2 \cdot K_0^{NC}} \tag{6}$$

For soft Scandinavian clays typical values for K_0^{NC} is in the range 0.55 to 0.65, which for an effective friction angle of 25° to 32° might, in a worst case combination, give a value of $\alpha_0 \sim 0.465 \cdot M_{TXC}$, where M_{TXC} is the value of M for $\theta^\alpha = -\pi/6$ (i.e. for triaxial compression). For such an extreme case, concavity of the surfaces may develop initially. However, a low effective friction angle will normally results in higher value for K_0^{NC}, and a value of $\alpha_0 \sim 0.2$ will be a typical case. Then concavity does not have to be a problem. If this is the case or not will depend on the type of critical state (failure) criterion used. For boundary value problems where soil is to failure in shear, concavity will normally increase as loading is applied since α increases due to kinematic hardening. Concavity may be totally avoided, or eventually become negligible for volumetrically dominated deformation, i.e. a state resulting in $\alpha \to 0$. One should notice that concavity would often not be experienced for K_0 loading since the rotational hardening is modest in this case, $\alpha \approx \alpha_0$ (if this is the case or not is again dependent on the failure criterion deployed).

3 PROPOSED SOLUTION

Some simple adjustments are suggested to avoid concavity. Eq. (7) is proposed as an alternative to eq. (4) . In Eq. (7) the triaxial compression value, M_{TXC}, is used instead of the function $M(\theta^\alpha)$ (in eq. (4)). The function $h(\theta^\alpha,\alpha)$ (eq. (8)) can ensure that the ratio between the critical state line for compression and extension states remains constant when the function $M(\theta^\alpha,\alpha)$ is carefully selected.

$$g^2(\theta^\alpha,\alpha) = \frac{h^2(\theta^\alpha,\alpha)}{M_{TXC}^2 - \alpha^2} \tag{7}$$

where:

$$h(\theta^\alpha,\alpha) = \frac{M_{TXC}}{M(\theta^\alpha,\alpha)} \tag{8}$$

In order to find a proper function $h(\theta^\alpha,\alpha)$ (i.e. $M(\theta^\alpha,\alpha)$), that is consistent with critical state in triaxial extension, one need to establish a relation between the input value of M_{TXE} and the value for M at $\theta^\alpha = \pi/6$ (mod. Lode angle equivalent to triaxial extension). Eq. (9) gives the required value of $M(\theta^\alpha = \pi/6, \alpha)$ such that the critical state line in triaxial extension is equal to the desired value M_{TXE}. If limiting oneself to a shape of the failure criterion reproducible in six sectors (i.e. defined in the sector $-\pi/6 < \theta^\alpha < \pi/6$), the ratio between triaxial compression (TXC) and triaxial extension (TXE) strengths are limited by a triangular shape in order not to become concave. This means that it will be impossible to maintain a convex surface if the ratio between $M(\theta^\alpha = \pi/6)$ and $M(\theta^\alpha = -\pi/6)$ is less than ½. Eq. (10) gives the limit for M_{TXE} in this case. Figure 5 gives a graphical representation of eq. (10), separating admissible from inadmissible (dotted line).

$$M\left(\theta^\alpha = \frac{\pi}{6}, \alpha\right) = M_{TXC} \cdot \sqrt{\frac{M_{TXE}^2 - \alpha^2}{M_{TXC}^2 - \alpha^2}} \tag{9}$$

$$M_{TXE} > \sqrt{\frac{M_{TXC}^2 + 3 \cdot \alpha^2}{4}} \tag{10}$$

For moderate values of α there is moderate difference between the necessary $M(\theta^\alpha = \pi/6)$ and the value for M_{TXE}. Hence, it might be practical to leave the formulation without this correction, and instead directly apply the Lode angle dependency to $M(\theta^{\alpha}, [\alpha = 0])$, eq. (11). The great benefit in doing this is that the yield surface will never become concave as long as it is not concave for the isotropic MCCM ($\alpha = 0$).

$$h(\theta^{\alpha}) = \frac{M_{TXC}}{M(\theta^{\alpha})} \tag{11}$$

By using this function the maximum ratio between the input values M in extension and compression, M_{TXE}/M_{TXC}, is as "normal" the factor of ½. The downside of eq. (11) is an over-prediction of the experienced critical state line in extension (when compared to the input value of M_{TXE}). Eq. (12) gives the ratio between the experienced triaxial extension critical state, M_{TXE}^{exp}, compared to the input value, M_{TXE}^{inp}, when eq. (11) is used for the function h.

$$\frac{M_{TXE}^{\exp}}{M_{TXE}^{inp}} = \sqrt{1 + \left(\frac{1}{\left(\frac{M_{TXE}^{inp}}{M_{TXC}} \right)^2} - 1 \right) \cdot \left(\frac{\alpha}{M_{TXC}} \right)^2} \tag{12}$$

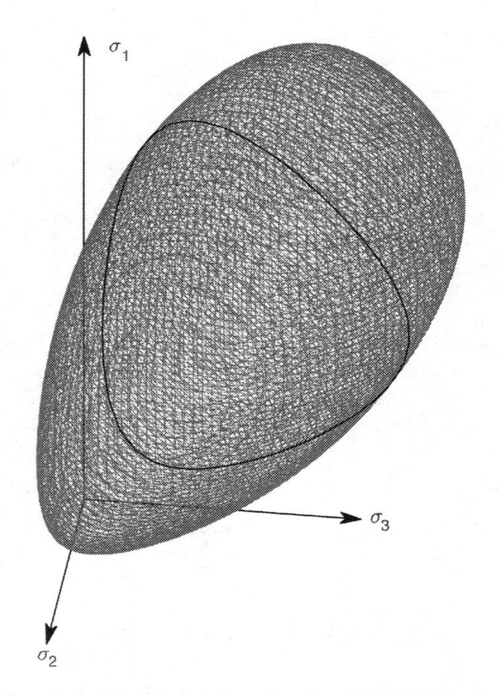

Figure 6. Modified yield surface for AMCCM in principal stress space.

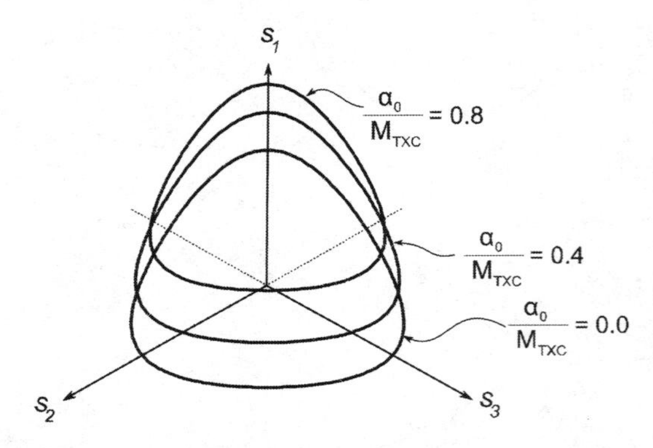

Figure 7. Cross section of the Modified Lode angle dependent AMCCM yield surface in the π-plane with proposed modification.

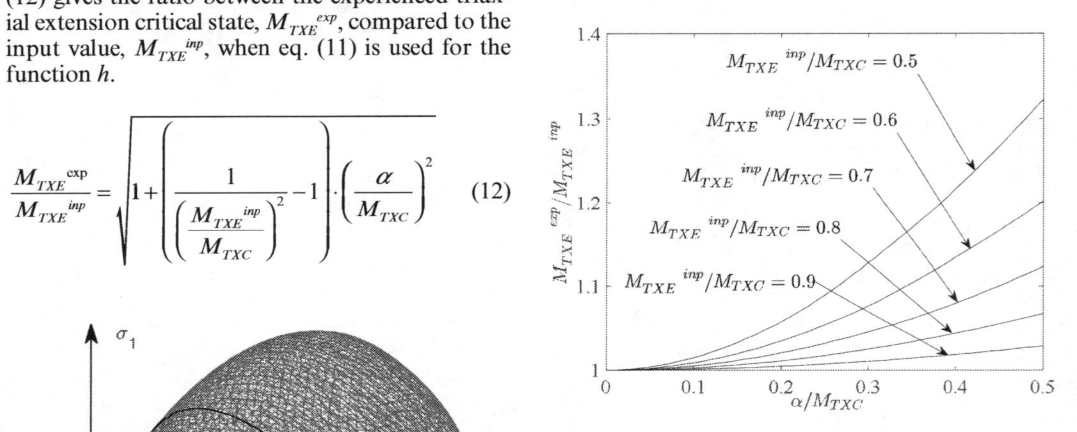

Figure 8. Graphical representation of eq. (12), showing the experienced value of critical state in extension compared to the input value as a function of α.

The equation for $M(\theta^{\alpha})$ can be selected to provide surfaces like Mohr-Coulomb (MC), Lade-Duncan (LD) or Matsouka-Nakai (MN), or other surfaces.

Graphical illustrations of eq. (7) with use of eq. (11) are given in Figure 6 and Figure 7. Here parameters in the generalized criterion (Grimstad et al., 2018) equivalent to the Lade-Duncan criterion are used.

Figure 8 shows a graphical representation of eq. (12). As seen in the figure, for "normal" input, the difference is typically less than 10% between M_{TXE}^{exp} and M_{TXE}^{inp}.

4 FINAL FORMULATION

As mentioned above the equation for $M(\theta^{\alpha})$ can be selected to provide surfaces like Mohr-Coulomb

(MC), Lade-Duncan (LD) or Matsouka-Nakai (MN), or other surfaces. In the following the GMC criterion from (Grimstad et al., 2018) is used to provide a description of the $h(\theta^\alpha)$. Eq. (13) to (15) is taken from Grimstad et al. where the terms c_θ ans s_θ is modified to be a function of the Modified Lode angle, θ^α.

$$M(\theta^\alpha) = \frac{3 \cdot \sin\varphi_0}{\sqrt{3} \cdot c_{\theta\alpha} + s_{\theta\alpha} \cdot \dfrac{\sin\varphi_0}{a_2}} \tag{13}$$

$$c_{\theta\alpha} = \cos\left(\frac{1}{3} \cdot \arcsin\left(a_1 \cdot \sin 3\theta^\alpha\right)\right) \tag{14}$$

$$s_{\theta\alpha} = \sin\left(\frac{1}{3} \cdot \arcsin\left(a_1 \cdot \sin 3\theta^\alpha\right)\right) \tag{15}$$

where $\sin\phi_0$ is the friction angle for modified Lode angle of 0° and the two new parameters a_1 and a_2 are used to control the shape of the cross section in the π-plane (the Lode angle dependency).

In order to use the criterion the parameter $\sin\phi_0$ must be linked to the value of M_{TXC} used in Eq. (11). By inserting the lode angle for triaxial compression into eq. (13) the following relationship is obtained:

$$\sin\varphi_0 = \frac{\sqrt{3} \cdot \cos\beta \cdot M_{TXC}}{3 + \dfrac{\sin\beta \cdot M_{TXC}}{a_2}} \tag{16}$$

where: $\beta = \frac{1}{3} \cdot \arcsin(a_1)$

Then inserting eq. (16) into eq. (13) gives:

$$M(\theta^\alpha) = \frac{M_{TXC}}{\dfrac{c_{\theta\alpha}}{\cos\beta} + \left(c_{\theta\alpha} \cdot \tan\beta + s_{\theta\alpha}\right) \cdot \dfrac{M_{TXC}}{3 \cdot a_2}} \tag{17}$$

Finally inserting eq. (17) into eq. (11) gives:

$$h(\theta^\alpha) = \frac{c_{\theta\alpha}}{\cos\beta} + \left(c_{\theta\alpha} \cdot \tan\beta + s_{\theta\alpha}\right) \cdot \frac{M_{TXC}}{3 \cdot a_2} \tag{18}$$

The parameters a_1 and a_2 can be selected based on knowledge of the actual Lode angle dependency. However, in most cases such information is not available for other states than triaxial compression and triaxial extension. In cases where only information on M_{TXC} is available e.g. the Matsouka-Nakai criterion could be employed. Then eq. (19) and (20) gives the parameters.

$$a_1 = \frac{27}{2 \cdot \sqrt{3}} \cdot \frac{M_{TXC} \cdot (3 + M_{TXC})}{\sqrt{\left(9 + 3 \cdot M_{TXC} + M_{TXC}^2\right)^3}} \tag{19}$$

$$a_2 = \frac{3 \cdot M_{TXC}}{\sqrt{3 \cdot \left(9 + 3 \cdot M_{TXC} + M_{TXC}^2\right)}} \tag{20}$$

$$\sin\varphi_0 = a_2 \tag{21}$$

One may also express the relation between the parameters as:

$$a_1 \in \langle 0,1 \rangle \tag{22}$$

$$a_2 = \frac{2}{3} \cdot \left(\frac{1}{M_{TXE}} - \frac{1}{M_{TXC}}\right)^{-1} \cdot \sin\beta \tag{23}$$

If one decides to use the GMC criterion with parameters equal to a rounded Mohr-Coulomb (instead of the MN formulation), then:

$$a_1 \to 1 \tag{24}$$

$$a_2 = 2 \cdot \sin\beta \tag{25}$$

Giving:

$$h(\theta^\alpha) = \frac{c_{\theta\alpha}}{\cos\beta} + \left(\frac{c_{\theta\alpha}}{\cos\beta} + \frac{s_{\theta\alpha}}{\sin\beta}\right) \cdot \frac{M_{TXC}}{6} \tag{26}$$

5 CONCLUSION

This paper shows that when considering the anisotropic Modified Cam Clay (AMCC) type of yield surface and introducing a modified Lode angle dependency for critical state, the yield/potential surface might become concave for high values of anisotropy. Therefore, this paper proposes a function, $g(\theta^\alpha, \alpha)$, that ensures convexity, for the Lode angle dependent AMCCM, for any $\alpha < M$, as long as the surface itself is convex for $\alpha = 0$. Avoiding concave yield/potential surfaces is important and the proposed remedy offers a simple solution to solve this problem. The simplified modification proposed will match the input value for the critical state line in triaxial compression. However, it will result in higher experienced values for critical state in extension that the value given for $\alpha = 0$. Typically, the difference is in the order of up to 10%.

Finally a generalized form of the function $h(\theta^\alpha)$ is proposed that allow a variety of shapes of the surfaces. The formulation is shown to depend on the ratio between M_{TXC} and M_{TXE}. If the GMC criterion is used to give a rounded Mohr-Coulomb, then one additional parameter (that is just below 1.0) is needed.

REFERENCES

Crouch, R.S. & Wolf, J.P. 1995. On a three-dimensional anisotropic plasticity model for soil. *Géotechnique,* 45, 301–305.

Dafalias, Y.F. 1986. An anisotropic critical state soil plasticity model. *Mechanics Research Communications,* 13, 341–347.

Grimstad, G., Rønningen, J.A. & Nordal, S. Application of a generalized continuous Mohr–Coulomb criterion 9th Conference on Numerical Methods in Geotechnical Engineering (NUMGE), 25–27 June 2018 2018 Porto, Portugal.

Lade, P.V. & Duncan, J.M. 1975. Elastoplastic stress-strain theory for cohesionless soil. *ASCE Journal of the Geotechnical Engineering Division,* 101, 1037–1053.

Matsuoka, H. & Nakai, T. Stress-deformation and strength characteristics of soil under three different principal stresses. Proceedings of JSCE, 1974. 59–70.

Roscoe, K.H. & Burland, J. 1968. On the generalized stress-strain behaviour of wet clay.

Schofield, A. & Wroth, P. 1968. Critical state soil mechanics.

APPENDIX

Mean effective stress:

$$p' = \frac{1}{3}\left(\sigma'_{11} + \sigma'_{22} + \sigma'_{33}\right) \tag{27}$$

Deviatoric stress vector:

$$\sigma_d = \begin{Bmatrix} \sigma'_{11} - p' \\ \sigma'_{22} - p' \\ \sigma'_{33} - p' \\ \sqrt{2}\sigma'_{12} \\ \sqrt{2}\sigma'_{23} \\ \sqrt{2}\sigma'_{13} \end{Bmatrix} \tag{28}$$

Rotational vector:

$$\alpha_d = \begin{Bmatrix} \alpha_{11} - 1 \\ \alpha_{22} - 1 \\ \alpha_{33} - 1 \\ \sqrt{2}\alpha_{12} \\ \sqrt{2}\alpha_{23} \\ \sqrt{2}\alpha_{13} \end{Bmatrix} \tag{29}$$

For cross ansiotropic condition (major direction as z-direction):

$$\sigma_d = \begin{Bmatrix} \sigma'_{xx} - p' \\ \sigma'_{yy} - p' \\ \sigma'_{zz} - p' \\ 0 \\ 0 \\ 0 \end{Bmatrix} = \begin{Bmatrix} \sigma'_2 - p' \\ \sigma'_3 - p' \\ \sigma'_1 - p' \\ 0 \\ 0 \\ 0 \end{Bmatrix} \tag{30}$$

$$\alpha_d = \begin{Bmatrix} -\frac{1}{3}\cdot\alpha \\ -\frac{1}{3}\cdot\alpha \\ \frac{2}{3}\cdot\alpha \\ 0 \\ 0 \\ 0 \end{Bmatrix} \tag{31}$$

Numerical Methods in Geotechnical Engineering IX – Cardoso et al. (Eds)
© 2018 Taylor & Francis Group, London, ISBN 978-1-138-33198-3

On constitutive modelling of anisotropic viscous and non-viscous soft soils

M. Tafili & Th. Triantafyllidis
Institute of Soil Mechanics and Rock Mechanics, Karlsruhe Institute of Technology KIT, Karlsruhe, Germany

ABSTRACT: The inherent anisotropy showed by most natural clays, along with the time dependent phenomena of soft soils have practical consequences for i.e. the passive lateral thrust on piles.

This paper examines the anisotropic material response of soft soils resulting in transverse isotropy. Only one additional parameter is needed to describe this type of inherent elastic anisotropy. Among all strain amplitudes the strain rate dependency of clays is experimentally investigated. The effects are incorporated into the proposed model by an extension of the (hypo)elastic stiffness tensor and the incorporation of an additional strain mechanism depending on the material viscosity. The model is proposed under the platform of hypoplasticity, well-established for monotonic loading.

The proposed model is able to describe the material behavior of viscous and non-viscous soft soils, capturing also the inherent anisotropy of clays without imposing any restriction on the overconsolidation ratio. This is very interesting because the proposed model is able to simulate each effect without changing the model and keeping the same set of material parameters.

1 INTRODUCTION

The simulation of geotechnical problems dealing with soft soils require a realistic description of the soil behavior (Avgerinos et al. 2016, Burland et al. 1979, Chatterjee et al. 2012, Vlahos et al. 2011). Thereby, the first concern is mostly about the time-dependent behavior of clays (Karstunen & Yin 2010) and the second important aspect is the anisotropic behaviour of soft soils Teng et al. (2014). On the other side, there are also non-plastic cohesive materials e.g. silt or low-plasticity clays (Wichtmann 2016, Nocilla et al.2006), whose behavior is time independent. On the other side, most natural clays show anisotropic behavior because of their mode of deposition and the elongated shape of the particles. This anisotropy, known as inherent anisotropy, along with the time dependent phenomena of soft soils have practical consequences for i.e. the passive lateral thrust on piles, which has been the target of many researchers in the last decades. By creeping slopes or construction of an embankment on such formations, passive time-dependent lateral pressure may occur to the pile shaft. This may lead to deformations of the pile foundation and of the superstructure.

A single model capturing all these effects is rare, and therefore, researchers recommend the usage of an "'appropriated"' model for each particular problem. Thus, there are models only able to describe undrained conditions and non-viscous behavior of clays e.g. (Hsieh & Ou 2011) and other more general models which describe the behavior of non-viscous clays unter monotonic loading e.g. (Manzari et al. 2006). To consider the influence of the loading velocity of plastic clays, viscous constitutive models are required e.g. (Niemunis 2003, Olszak and Perzyna 1970, Olszak and Perzyna 1966, Madaschi and Gajo 2017, Rezania et al. 2016). Most of them were fairly reviewed in the works of (Liingaard et al. 2004, Tafili et al. 2018), so there is no need for a repetition here. Thus, as one may see, the type of constitutive model varies substantially depending on the problem type and one of the challenges for geotechnical engineers and researchers is to find a model capturing all these observations.

This challenge is accepted in this work: the proposed constitutive approach distinguishes between short term and long term plastic responses. In detail a pure plastic strain rate as well as a visco-plastic strain rate are introduced. The former one follows the hypoplastic framework (Kolymbas 1991), but is completely reformulated to account for soft soils. It is responsible for the description of rate and time independent behavior of soft soils. The viscous strain rate is formulated using the Norton's powerlaw (Norton 1929), without introducing any reference time, avoiding so the difficulties arising with the definition of an equivalent time. This part of the constitutive approach consistently simulates the rate and the time dependence of clays includingstress relaxation, strain creep and rate dependence, capturing also the isotache concept (unique stress-strain curves for each strain rate). The behavior of both normalconsolidated, as well as slightly and heavily overconsolidated soft soils is reliably described. Avoiding the index discontinuity of the viscous strain rate, arising in other viscous models e.g. (Niemunis 2003, Niemunis et al. 2009), the model is able to describe also the behavior of non-viscous soft soils.

The second important aspect regarding soft soils is their anisotropic behavior due to their mode of deposition and the elongated shape of the particles. This anisotropy, known as inherent anisotropy or fabric, tends to induce a horizontal bedding plane in the soil layer, is transverse isotropic and effects the elastic stiffness of the constitutive approach. The elastic stiffness E of the proposed model is extended and rotated to account for the fabric of clays. The model is in that sense able to describe the monotonic behavior of anisotropic as well as isotropic soft soils.

The structure of this article is as follows: at the beginning, the formulation of the proposed model is explained. Subsequently, some information about the inherent anisotropy is provided and the manner to introduce it into the model is described. Then, some details about the numerical implementation and the material parameters are given. Finally some simulations are carefully analyzed to discuss the models performance.

2 MECHANICAL MODEL FORMULATION

In this section the mechanical model formulation of the proposed visco-hypoplastic model is explained briefly. The focus lies on the mobilized state, thus the intergranular strain model described in (Tafili and Triantafyllidis 2017) is omitted. In order to capture the mechanical behavior of both viscous and non-viscous clays, the strain rate is decomposed into three parts connected in series: elastic $\dot{\varepsilon}^e$, hypoplastic $\dot{\varepsilon}^{hp}$ and viscoplastic $\dot{\varepsilon}^{vp}$:

$$\dot{\varepsilon} = \dot{\varepsilon}^e + \dot{\varepsilon}^{hp} + \dot{\varepsilon}^{vp}. \tag{1}$$

In the following, the classical hypoplastic strain rate relation (Niemunis 2003):

$$\dot{\varepsilon}^{hp} = \mathrm{Y}\mathbf{m}\|\dot{\varepsilon}\|. \tag{2}$$

has been employed, whereby Y is the degree of nonlinearity and **m** the plastic flow rule. The degree of nonlinearity is introduced with a similar interpolation function as in (Niemunis 2003, Masin 2006) and previously analyzed in (Fuentes et al. 2017):

$$\mathrm{Y} = \mathrm{Y}_i + (1-\mathrm{Y}_i)\left(\frac{\|\mathbf{r}\|}{\|\mathbf{r}b\|}\right)^{nY} \tag{3}$$

with $\mathrm{Y}_i = \mathrm{Y}_{im}(p/p_e)^2$, described in detail in (Fuentes and Triantafyllidis 2013).

The direction of the plastic flow **m** is adopted to account for the behavior of soft soils:

$$\mathbf{m} = \left(-1/2F\mathbf{1} + \mathbf{r}\right)^{\rightarrow} \tag{4}$$

with F as a scalar function of the critical state and the stress ratio tensor **r**: $F = F(g(\theta), \mathbf{r})$ (Fuentes et al. 2017).

Let us now consider the viscous strain rate formulation. It introduces the time-dependent phenomena clays may exhibit including strain creep, stress relaxation and rate dependence into the model. Niemunis and Krieg showed in (Niemunis and Krieg 1996) that the viscous strain rate is only dependent on the mean stress and the void ratio, both presented through the overconsolidation ratio. In this work, the intensity of the viscous strain rate is described analogously to the Norton's power law (Norton 1929) and for its direction the plastic flow rule **m** is adopted:

$$\dot{\varepsilon}^{vp} = -I_v\lambda(1\mathrm{OCR})^{1/I_v}\,\mathbf{m} \tag{5}$$

with Leinenkugels (Leinenkugel 1976) index of viscosity I_v and the compression index λ. The overconsolidation ratio OCR introduces the stress and void ratio dependence of the viscous strain rate through the relation (Fuentes et al. 2017):

$$\mathrm{OCR} = \frac{p_i}{p} + \left(1-\frac{p_i}{p}\right)\left(\frac{\|\mathbf{r}\|}{\|\mathbf{r}_b\|}\right)^{n_{OCR}} \tag{6}$$

whereby n_{OCR} is a material parameter defining the shape of the OCR surface, $\mathbf{r}_b$ the deviator stress ratio at the bounding surface and p_i is the Hvorslev mean stress. For further informations about the critical and bounding surface introduced into the model, the interested reader is referred to (Fuentes et al. 2017).

After the definition of each plastic strain rate, we proceed with the evolution equation of the stress. It relates the stress rate with the elastic strain rate $\dot{\varepsilon}^e$, see Eq. 1, through the elastic stiffness E:

$$\dot{\sigma} = \mathrm{E} : \left(\dot{\varepsilon} - \dot{\varepsilon}^{hp} - \dot{\varepsilon}^{vp}\right). \tag{7}$$

The elastic stiffness remains to define. It obeys the relation proposed in (Fuentes et al. 2015):

$$\mathsf{E} = K\mathbf{1}\otimes\mathbf{1} + 2GI^{dev} - K\sqrt{3}M_c(\mathbf{1}\otimes\mathbf{r} + \mathbf{r}\otimes\mathbf{1}) \tag{8}$$

with the slope of the critical state line denoted as M_c. For the isotropic case, the bulk K and the shear G modulus are adjusted to account for the behaviour of isotropic clays:

$$K = \frac{p(1+e)}{\lambda(1-\Upsilon_{im})}, \quad G = K\frac{(1-2\nu)}{2(1+\nu)} \tag{9}$$

with the Poisson ratio ν as material parameter to be calibrated. Υ_{im} introduces the swelling index κ into the model, see (Fuentes and Triantafyllidis 2013).

This manageable number of equations governs the mechanical model formulation for isotropic viscous clays. The model is not only able to describe viscous behavior of clays, but also non-viscous behavior by just setting the viscosity index to $I_v = 0$. Therefore, the viscous part of the strain rate vanishes and the models response is hypoplastic. This feature of the model is particulary attractive for modelling non-viscous clays. The developed 'visco-hypoplastic models' in literature (Niemunis 2003, Niemunis et al. 2009) do not allow for the viscous strain rate to vanish, due to their I_v – exponent discontinuity. Furthermore, the visco-hypoplastic model in (Niemunis 2003) incorporates only the viscous strain rate, not the pure hypoplastic strain rate. Thus, its mechanical structure coincides rather with a viscoplastic definition. These models can only describe the behavior of viscous clays in contrast to the model described in this paper.

Despite this, neither the models proposed in (Niemunis 2003, Niemunis et al. 2009) nor this model are able to capture the behavior of clayey soils in terms of their fabric and structure.

3 INTRODUCING THE INHERENT ANISOTROPY

Most natural clays show anisotropic behavior because of their mode of deposition and the elongated shape of the particles. This anisotropy, known as inherent anisotropy or fabric, resulting from the deposition process, which tends to induce a horizontal bedding plane in the soil layer, can be regarded as transverse isotropy, see also Fig. 1.

It effects the elastic stiffness of the material. Eventhough five elastic parameters are needed to describe transverse isotropy, Graham and Houlsby (Graham and Houlsby 1983) deduced a simplified transversal isotropic elasticity for the special case of anisotropic clays, whereby only three parameters are needed: the Young's modulus $E = E_v$,

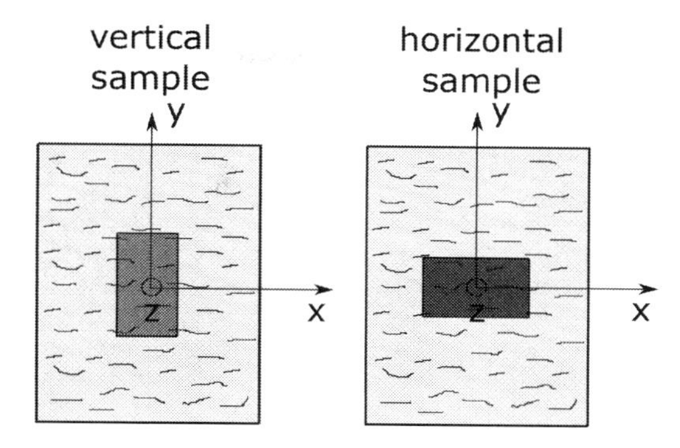

Figure 1. Fabric and cutting direction of the samples for experiments.

the Poisson ratio $\nu = \nu_h$ and the scalar factor α governing the following relations between the horizontal and vertical stiffness:

$$G_h/G_v = (E_h/E_v)^{1/2} = \nu_h/\nu_{vh} = \alpha \tag{10}$$

With the y-axis as the vertical direction representing the direction of the anisotropy and the x, z-plane as the plans of isotropy, see Fig. 1, the stress-strain increment equation for a transverse isotropic material considering the simplifications provided by (Graham and Houlsby 1983) can be expressed through the relation given in Eq. 26, whereby H is the transverse isotropic stiffness.

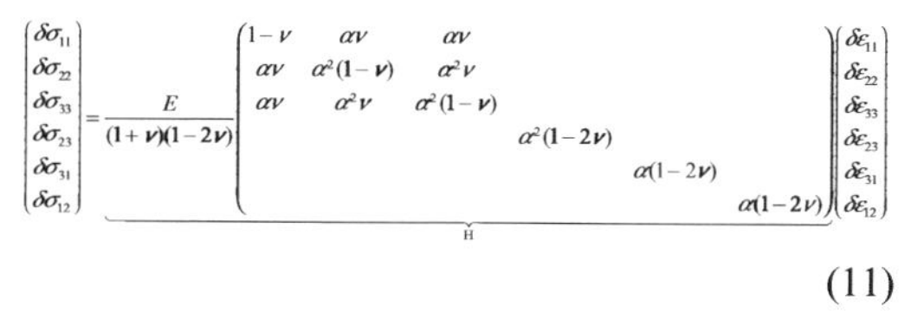

$$\begin{Bmatrix}\delta\sigma_{11}\\ \delta\sigma_{22}\\ \delta\sigma_{33}\\ \delta\sigma_{23}\\ \delta\sigma_{31}\\ \delta\sigma_{12}\end{Bmatrix} = \underbrace{\frac{E}{(1+\nu)(1-2\nu)}\begin{pmatrix}1-\nu & \alpha\nu & \alpha\nu & & & \\ \alpha\nu & \alpha^2(1-\nu) & \alpha^2\nu & & & \\ \alpha\nu & \alpha^2\nu & \alpha^2(1-\nu) & & & \\ & & & \alpha^2(1-2\nu) & & \\ & & & & \alpha(1-2\nu) & \\ & & & & & \alpha(1-2\nu)\end{pmatrix}}_{\mathrm{H}}\begin{Bmatrix}\delta\varepsilon_{11}\\ \delta\varepsilon_{22}\\ \delta\varepsilon_{33}\\ \delta\varepsilon_{23}\\ \delta\varepsilon_{31}\\ \delta\varepsilon_{12}\end{Bmatrix} \tag{11}$$

To incorporate this stiffness into a constitutive model, the present stiffness needs to be adjusted, scaled or rotated. After some mathematical manipulations one may notice that any hyperelastic stiffness E renders a transverse isotropic stiffness like H by a special scaling transformation as denoted in (Niemunis et al. 2015):

$$\begin{aligned} &\mathsf{H} = \mathsf{Q} : \mathsf{E} : \mathsf{Q} \\ &\text{with} \mathsf{Q} = \mu\otimes\mu \\ \text{and } &\mu = \sqrt{\alpha}\delta_{ij} + (1-\sqrt{a})\mathbf{m}\otimes\mathbf{m}, \end{aligned} \tag{12}$$

whereby $\mathbf{m}$ is the unit vector along the sedimentation axis e.g. if the sedimentation axis is vertical than $\mathbf{m} = \{0,0,1\}$.

Some recent research works have shown the importance of the inherent anisotropy on the behavior of clays. According to the investigations taken on Kaolin clay by Wichtmann (Wichtmann 2016) the samples cut out in the horizontal direction, Fig. 1, show a more dilative response and a higher undrained shear strength than those taken vertically. Furthermore, depending on the cutting direction—if all other conditions remain the same—the horizontal samples can withstand a much larger number of stress cycles to failure than the vertical samples (Wichtmann 2016). This postulates the importance of simulating the different elastic stiffness for different cutting directions.

4 NUMERICAL IMPLEMENTATION AND MATERIAL PARAMETERS

The proposed model has been implemented with the programming language FORTRAN, as a material subroutine for the commercial finite element software ABAQUS standard. A substepping scheme with small strain increments has been implemented to guarantee numerical convergence. All the terms except of the viscous strain rate were explicitly implemented (evaluated at the beginning of the subincrement). For the viscous strain rate, a semi-implicit scheme has been implemented following a similar integration scheme as by Niemunis (Niemunis 2003).

The proposed model requires the calibration of 9 parameters. They are subdivided in different groups according to their role within the model, namely "Elasticity" (ν, κ, λ, α), "Plasticity" (e_{i0}, M_c, f_{b0}) and "Viscosity" (I_v, n_{ocr}). All parameters exept α have been already used in former works, e.g. (Fuentes et al. 2015, Fuentes et al. 2017, Niemunis and Krieg 1996) and detailed in (Hadzibeti 2016), but their calibration is once more summarized in Table 1.

The compression λ and swelling index κ, are calibrated with isotropic or oedometric compression tests in the e vs. $\ln(p)$ space upon loading and unloading paths, respectively. The Poisson ratio ν and the anisotropic coefficient *pha* can be determined by measuring the shear modulus G for small strain amplitudes and using the relations given in Eq. 9 and the transverse isotropic stiffness H from Eq. 11. The parameter M_c is adjusted to the slope of the critical state line CSL within the p vs. q space under triaxial compression. The parameter f_{b0} controls approximately the maximum stress ratio for triaxial compression $f_{b0} = \eta_{max}/M_c$ and can therefore be adjusted to highly overconsolidated samples OCR > 2. When data is scarce, a recommended value of $f_{b0} = 1.3$ may be carefully used according to our experience with some soft soils. The reference void ratio e_{i0} corresponds to the maximum void ratio at the reference pressure p_{ref} = 1 kPa. Due to the decomposition of the strain rate into three parts, what differs this model to other models, the e_{i0} should be calibrated at the isotach with infinite strain rate $\| \dot{\varepsilon} \| = \infty$. To simplify the calibration (Fuentes et al. 2017) deduced the relation $e_{i0} = e_{ri} - \lambda \ln(\text{OCR}_{ri})$. For the viscosity index two isotropic compression tests with different strain rates, $\| \dot{\varepsilon}_a \|$ and $\| \dot{\varepsilon}_b \|$, respectively, are required. I_v can then be computed through the

Table 1. Material parameters of the proposed model.

Description		Units	Approx. range	Required test	Lower Rhine clay	Kaolin clay	Cons. Lac. Clay	Berg. silt
Elasticity								
λ	Compression index	[-]	10^{-6}–1	ICT*	0.18	0.13	0.04	0.026
κ	Swelling index	[-]	10^{-6}–1	ICT*	0.04	0.05	0.007	0.005
ν_h	Poisson ratio	[-]	0–0.5	TCT**	0.25	0.33	0.2	0.2
α	Anisotropy coefficient	[-]	0–4	TCT**	2.0	2.0	0.8	1.0
Plasticity								
e_{i0}	Maximum void ratio at $p = p_{ref}$	[-]	0.5–2	ICT*	2.03	1.76	0.675	0.708
M_c	CS slope	[-]	0.5–1.5	TCT**	0.95	1.0	1.37	1.5
f_{b0}	Bounding surface factor	[-]	1–2	TCT**	1.45	1.5	1.5	1.5
Viscosity								
I_v	Viscosity index	[-]	1–2	$2 \times \varepsilon_v$ ICT*	0.025	0.015	0	0
n_{ocr}	Viscous exponent	[-]	0.4–2.0	TCT**	0.5	0.4	0	0

* = isotropic compression test, ** = undrained triaxial compression test

Table 2. Index properties of Kaolin clay, Constance lacustrine clay and Berghausen silt.

Material	W_L[%]	W_P[%]	I_P[%]	$\rho_s[g/cm^3]$
Lower Rhine clay	56.1	22.0	34.1	2.590
Kaolin clay	47.2	35.0	12.2	2.675
Cons. lac. clay	22.5	17.2	5.3	2.708
Berg. silt	21.1	21.1	0	2.715

relation $I_v = \ln(\text{OCR}_b/\text{OCR}_a)/\ln(\| \dot{\varepsilon}_a \| / \| \dot{\varepsilon}_b \|)$ (Niemunis 2003). The parameter n_{ocr} controls the shape of the OCR surface. We recommend to calibrate this parameter by trial and error given some undrained tests.

5 SIMULATIONS WITH EXPERIMENTS

In order to evaluate the performance of the proposed model, this section presents some simulations of the experimental results of three clays with different plasticity and one silt, namely the Lower Rhine clay (high plasticity clay), Kaolin clay (medium plasticity clay), Constance lacustrine clay (low plasticity clay) and Berghausen silt (non-plastic material). Table 2 provides an overview of the index properties of these materials, including the liquid limit W_L, plastic limit W_P, plasticity index I_P and grain density ρ_s (Wichtmann 2016). Table 1 lists the calibrated parameters for these soft soils.

5.1 *Simulations with Lower Rhine clay*

The Lower Rhine clay is a high plasticity clay with a viscosity index equal to $I_v = 0.025$. The calibrated parameters are listed in Table 1, whereby the anisotropy coefficient is $\alpha = 2$.

The rate dependent one-dimensional compression behavior is studied with the use of three oedometer tests conducted for this work with different axial strain rates $\dot{\varepsilon}_1 = \{0.0005; 0.01; 0.02\}$ 1/s. The three axial strain rates generate different compression paths, see Figs. 2. The simulations depicted in the same Figures show a good agreement with the experimental results.

5.2 *Simulations with Kaolin clay*

Kaolin is a medium plasticity clay with a viscosity index equal to $I_v = 0.015$. Reconstituted samples of height and diameter equal to 50 mm were used for the experiments.

Similar as done with the Lower Rhine clay, the rate dependent one-dimensional compression behavior of the Kaolin clay is studied with the use of three oedometer tests conducted for this work with different axial strain rates $\dot{\varepsilon}_1 = \{0.0005; 0.01; 0.02\}$ 1/s. Therefore three different compression paths are generated. The experiments and simulations are in good agreement, which is shown in Figs. 3. Note, that the one-dimensional compression behavior of Kaolin clay is in comparison with Lower Rhine clay more nonlinear over the void ratio-logarithm of effective stress range. This is a cause of the lower plasticity.

All experiments in the following are conducted by (Wichtmann 2016) with a constant axial strain rate equal to $\dot{\varepsilon}_1 = 10^{-8}[1/s]$. Figs. 4 and 5 show the stress-strain relationships and the effective stress paths measured in undrained monotonic triaxial tests of three normal consolidated samples with different confining pressures $p_0 = \{100; 200; 300\}$ kPa. The simulation in Fig. 4 captures fairly well most of the observed behavior. For comparison purposes a simulation with the isotropic model is presented in Fig. 5. In order to obtain isotropic behavior the anisotropic factor α is set equal to $\alpha = 1$. Especially, in the p vs. q space a fundamental difference can be observed. The isotropic model does not provide the same decrease in the mean preassure as the experiment. The anisotropic model, in contrast, fits very well the experiments, see Fig. 4.

5.3 *Simulations with Constance lacustrine clay*

The Constance lacustrine clay is a low plasticity clay without viscosity i.e. its viscosity index is equal to $I_v = 0$. The calibrated parameters are listed in Table 1, whereby the anisotropy coefficient is $\alpha = 0.8$.

As pointed out in Sec. 2, viscous models for clays in literature as the viscohypoplastic model from (Niemunis 2003, Niemunis et al. 2009) and the viscoplastic model with overstress (Olszak and Perzyna 1970) are due to their index discontinuity of the viscosity index and the absence of a pure (hypo)plastic strain rate restricted to viscous clays only.

With its general formulation, the proposed model can describe also the behavior of non-viscous clays. This is shown through the simulations in Figs. 6 and 7. Fig. 6 shows the first experiment and corresponds to an oedometer test showing time-independent effects. The experiment includes some unloading-reloading cycles showing hysteretic behavior. All these observations were captured by the proposed model which is also shown in the same figure.

Fig. 7 shows the effective stress paths measured in undrained monotonic triaxial tests of four normal consolidated samples with different confining pressures $p_0 = \{50; 100; 200; 300\}$ kPa. The initial void ratios of these experiments were equal to

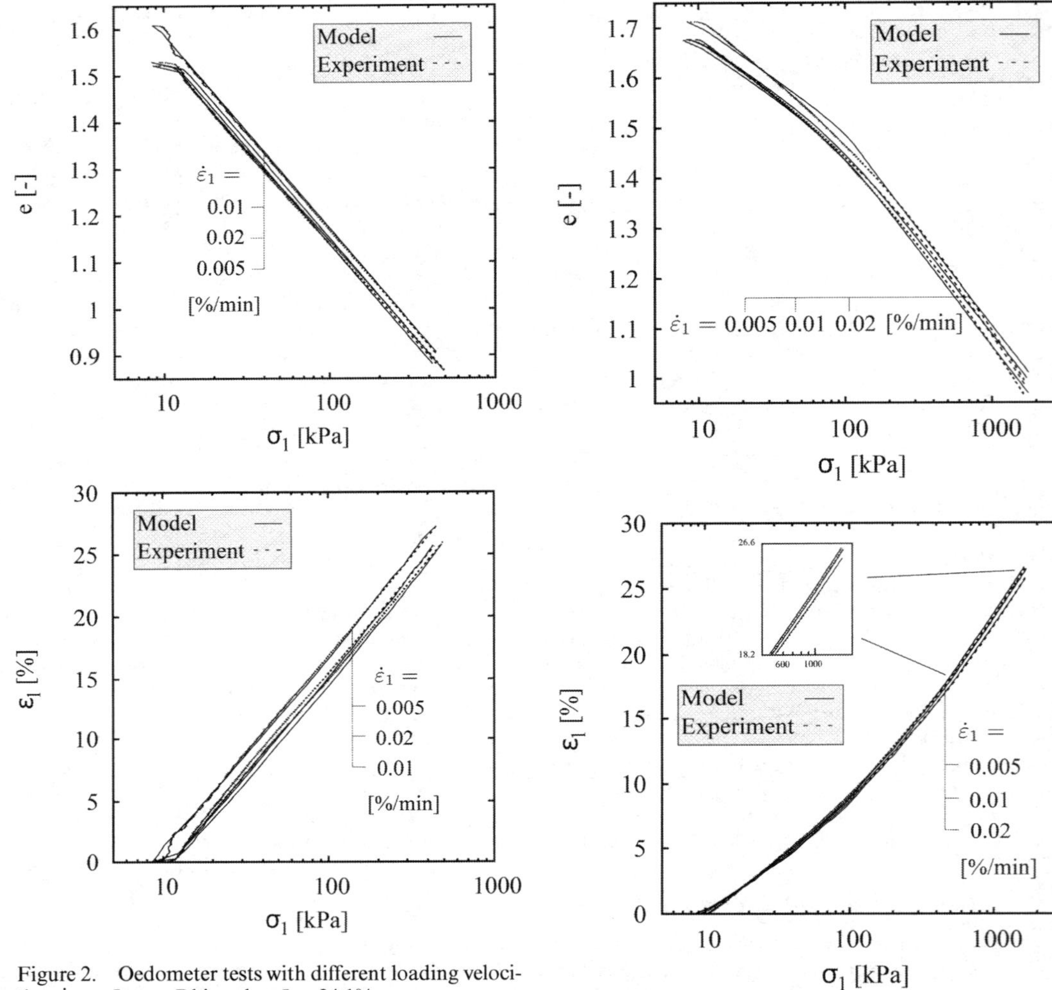

Figure 2. Oedometer tests with different loading velocities $\dot{\varepsilon}_1$ on Lower Rhine clay, $I_P = 34.1\%$.

Figure 3. Oedometer tests with different loading velocities $\dot{\varepsilon}_1$ on Kaolin clay, $I_P = 12.2\%$.

e_0 = {0.604; 0.520; 0.508; 0.447}, respectively for each confining pressure. The simulations showed some discrepancies with respect to the dilatancy at larger strains and especially higher confining pressures e.g. $p_0 = 300$ kPa.

5.4 *Simulations with Berghausen silt*

Berghausen silt is a non-plastic material i.e. silt with $I_P = 0\%$. Hence, its viscosity index is equal to $I_v = 0$. The calibrated parameters are listed in Table 1, whereby $\alpha = 10$ describing an isotropic material. Reconstituted samples of height and diameter equal to 50 mm were used for the experiments.

Fig. 8 presents an oedometer test showing time-independent effects. The experiment includes some unloading-reloading cycles showing hysteretic behavior. All these observations were captured by the proposed model which is also shown in the samefigure.

Fig. 9 illustrates the effective stress paths measured in undrained monotonic triaxial tests of four normal consolidated samples with different confining pressures p_0 = {50; 100; 200; 300} kPa. The simulations showed again some discrepancies with respect to the dilatancy at larger strains. Comparing these experiments with those conducted on Constance lacustrine clay ($I_P = 5.3\%$), shown in Fig. 7, it can be concluded that with decreasing plasticity of the test material, the dila-

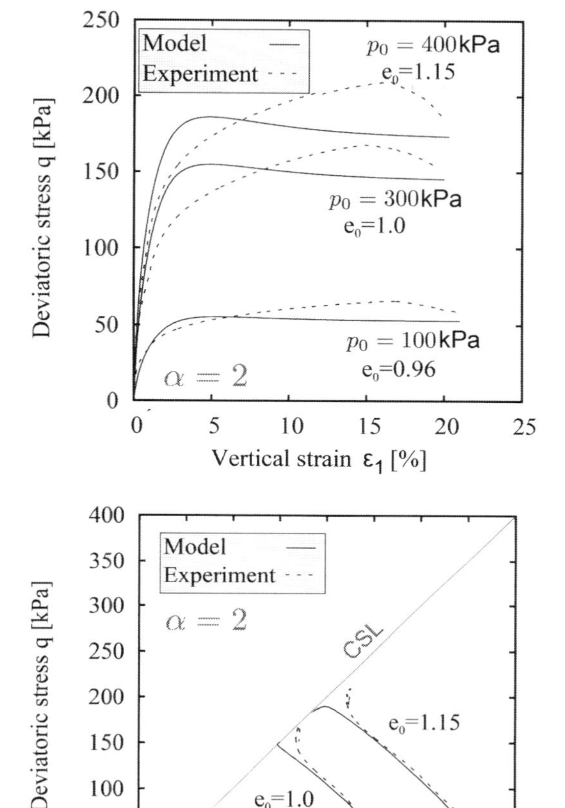

Figure 4. Undrained triaxial tests with $\alpha = 2$ on Kaolin Clay, $I_P = 12.2\%$.

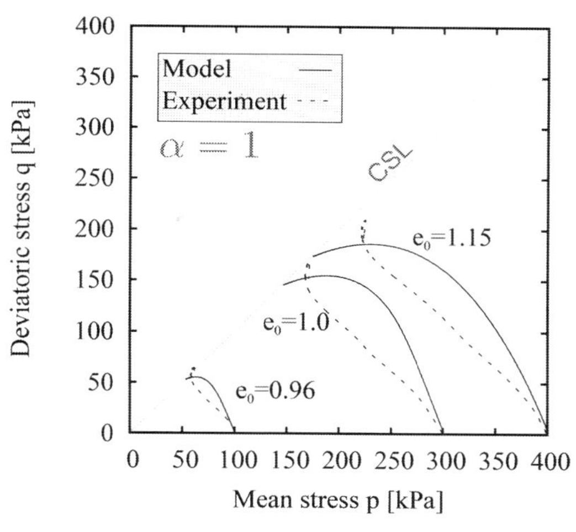

Figure 5. Undrained triaxial tests with $\alpha = 1$ on Kaolin clay, $I_P = 12.2\%$.

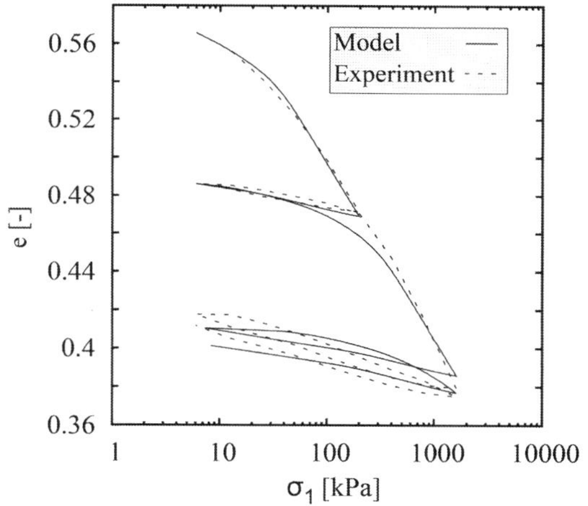

Figure 6. Oedometer tests with loading, unloading and reloading on Cons. lac. clay., $I_P = 5.3\%$.

tancy observed at larger strains increases. Hence, the Berghausen silt ($I_P = 0.0\%$) shows a material response similar to a medium dense air-pluviated sand (Wichtmann 2016).

It is well-known that overconsolidated clays exhibit positive dilatancy tendencies and greater shear strengths than normal consolidated clays. This behavior is described by the proposed model and illustrated in Fig. 10. In order to draw conclusions about the dilatancy of low-plasticity normalconsolidated clays more experimental research is required. Subsequently, an extension or adjustment of the degree of nonlinearity and the critical void ratio, will be required. This extension will improve the simulations depicted in Figs. 7 and 9.

6 CONCLUSIONS

In the present work a constitutive model for the simulation of saturated soft soils has been proposed. The model presents some interesting capabilities, such as the simulation of the strain rate dependency and the incorporation of the inherent anisotropy resulting in transverse isotropy. From the point of constitutive modeling, this is very interesting because it permits the user to evaluate the influence of each effect on a simulation, such as viscosity or anisotropy, without changing

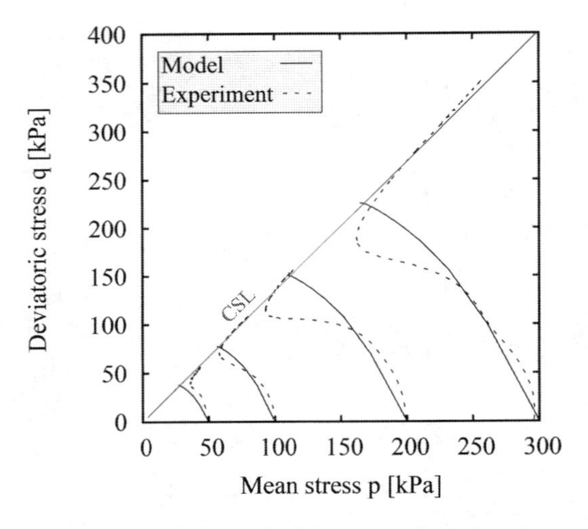

Figure 7. Undrained triaxial tests on Cons. lac. clay., $I_P = 5.3\%$.

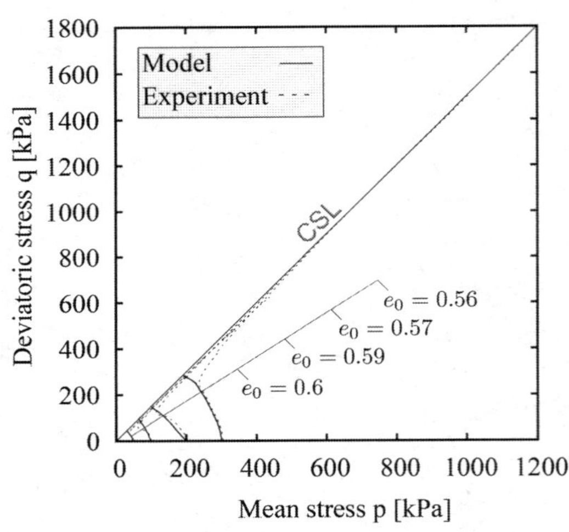

Figure 9. Undrained triaxial tests on Berghausen silt, $I_P = 0\%$.

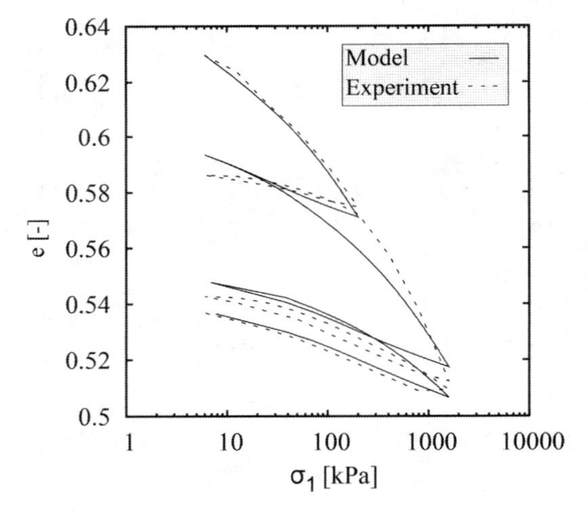

Figure 8. Oedometer tests with loading, unloading and reloading on Berghausen silt, $I_P = 0\%$.

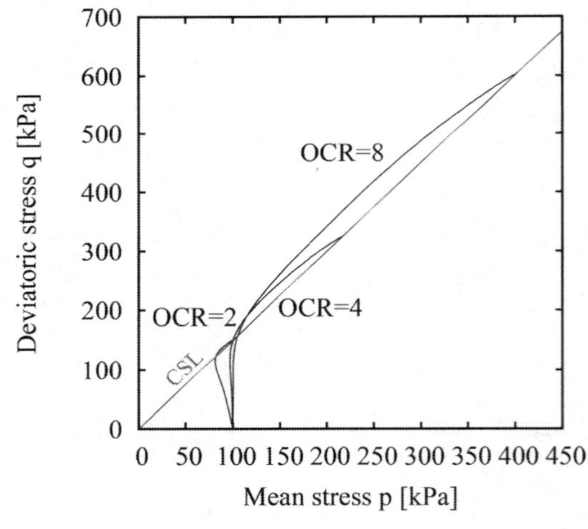

Figuer 10. Qualitative undrained triaxial tests on overconsolidated samples. Parameters of Berghausen Silt were borrowed, $I_P = 0.0\%$, see Table 1.

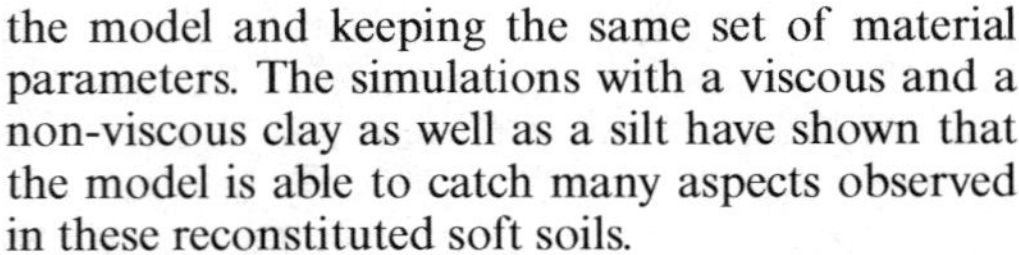
the model and keeping the same set of material parameters. The simulations with a viscous and a non-viscous clay as well as a silt have shown that the model is able to catch many aspects observed in these reconstituted soft soils.

Although not shown in this paper, the models formulation is so general that it can be adjusted (just by reformulating the critical void ratio and the degree of nonlinearity) to account also for the behaviour of grained soils as for e.g. sands, sand-silt or sand-clay mixtures. For the latter one the viscous effects may be of a great importance.

Some experimental observations are left out for improvement in future works, such as the cementation, partially saturation and the behavior of soft soils under cyclic loading. Moreover, the dilatancy that show non-plastic materials is not captured by the proposed model. The adjustment of the model to simulate the behavior of sands can be the first step towards reliably describing also the dilatancy of other non-plastic materials.

In order to draw conclusions about the evolution and the development of the inherent anisotropy during large shearing, the material behavior under cyclic loading should be examined also for large strain amplitudes towards to failure.

REFERENCES

Avgerinos, V., D. Potts, & J. Standing (2016). The use of kinematic hardening models for predicting tunnelling-induced ground movements in London Clay. *Géotechnique 66*(2), 106–120.

Burland, J., B. Simpson, & H. St. John (1979). Movements around excavations in London Clay. in Design parameters in Geotechnical Engineering. In B.G.S. BGS (Ed.), *VII ECSMFE*, Volume 1, London, UK, pp. 13–29.

Chatterjee, S., D. White, & M. Randolph (2012). Numerical simulations of pipe-soil interaction during large lateral movements on clay. Géotechnique 62(8), 693–705.

Fuentes, W., M. Hadzibeti, & T. Triantafyllidis (2015). Constitutive model for clays under the isa framework. In T. Triantafyllidis (Ed.), *Holistic Simulation of Geotechnical Installation Processes - Benchmarks and Simulations*, pp. 115–130. Springer.

Fuentes, W., M. Tafili, & T. Triantafyllidis (2017). An isa-plasticity-based model for viscous and non-viscous clays. *Acta Geotechnica*, https://doi.org/10.1007/s11440–017–0548–y.

Fuentes, W. & T. Triantafyllidis (2013). Hydro-mechanical hypoplastic models for unsaturated soils under isotropic stress conditions. *Computers and Geotechnics 51*, 72–82.

Graham, J. & G. Houlsby (1983). Anisotropic elasticity of a natural clay. *Géotechnique 33*(2), 165–180.

Hadzibeti, M. (2016). Formulation and calibration of a viscous ISA model for clays. Master Thesis. Karlsruhe Institute of Technology KIT. Institute of Soil Mechanics and Rock Mechanics IBF.

Hsieh, P. & C. Ou (2011). Analysis of nonlinear stress and strain in clay under the undrained condition. *Journal of Mechanics 27*(2), 201–213.

Karstunen, M. & Z. Yin (2010). Modelling time-dependent behavior of murro test embankment. *Géotechnique 60*(10), 735–749.

Kolymbas, D. (1991). An outline of hypoplasticity. *Arch Appl Mech 61*(3), 143151.

Leinenkugel, H. (1976). Deformation and strength behaviour of cohesive soils experiments and their physical meaning. PhD Dissertation. University of Karlsruhe. Publications of Institute of Soil and Rock Mechanic. Germany. Heft 66.

Liingaard, M., A. Augustesen, & P. Lade V. (2004). Characterization of models for time-dependent behavior of soils. *International Journal of Geomechanics 4*(3), 157–177.

Madaschi, A. & A. Gajo (2017). A one-dimensional viscoelastic and viscoplastic constitutive approach to modeling the delayed behavior of clay and organic soils. *Acta Geotechnica 12*, 827–847.

Manzari, M., A. Papadimitriou, & Y. Dafalias (2006). Saniclay: Simple Anisotropic Clay Plasticity Model. *Int. J. Num. Anal. Meth. Geomech. 30*(4), 1231–1257.

Masin, D. (2006). Hypoplastic models for fine-grained soils. PhD Thesis. Charles University, Prague.

Niemunis, A. (2003). Extended hypoplastic models for soils. Schriftenreihe des Institutes für Grundbau und Bodenmechanik der Ruhr-Universität Bochum. Habilitation. Germany. Heft 34.

Niemunis, A., C.E. Grandas-Tavera, & L.F. Prada-Sarmiento (2009). Anisotropic visco-hypoplasticity. *Acta Geotechnica 4*(4), 293–314.

Niemunis, A., C.E. Grandas-Tavera, & T. Wichtmann (2015). Peak stress obliquity in drained and undrained sands. Smulations with neohypoplasticity. In T. Triantafyllidis (Ed.), *Holistic Simulation of Geotechnical Installation Processes - Benchmarks and Simulations*, pp. 85–114. Springer.

Niemunis, A. & S. Krieg (1996, January). Viscous behaviour of soils under oedometric conditions. *Can. Geot. &Journal 33*, 159–168.

Nocilla, A., M.R. Coop, & F. Colleselli (2006). The mechanics of an italian silt: an example of 'transitional' behaviour. *Géotechnique 56*(4), 261271.

Norton, F. (1929). *The Creep of Steel at High Temperatures*. New York: Mc Graw Hill Book Company, Inc.

Olszak, W. & P. Perzyna (1966). The constitutive equations of the flow theory for a nonstationary yield condition. *Proc., 11th Int. Congress of Applied Mechanics*, 545–553.

Olszak, W. & P. Perzyna (1970). Stationary and non-stationary viscoplasticity. *McGraw-Hill, New York*, 53–75.

Rezania, M.,M. Taiebat, & E. Poletti (2016). A viscoplastic saniclay model for natural soft soils. *Computers and Geotechnics 73*, 128–141.

Tafili, M., W. Fuentes, & T. Triantafyllidis (2018). A comparative study of different constitutive frameworks on simulating the viscous behavior of clays. *Submitted for Géotechnique*.

Tafili, M. & T. Triantafyllidis (2017). Constitutive model for viscous clays under the isa framework. In T. Triantafyllidis (Ed.), *Holistic Simulation of Geotechnical Installation Processes - Theoretical Results and Applications*, pp. 324–340. Springer.

Teng, F.-C., C.-Y. Ou, & P.-G. Hsieh (2014). Measurements and numerical simulations of inherent stiffness anisotropy in soft taipei clay. *Journal of Geotechnical and Geoenvironmental Engineering 140*(1), 237–250.

Vlahos, G., M. Cassidy, & M. Martin (2011). Numerical simulation of pushover tests on a model jack-up platform on clay. *Géotechnique 61*(11), 947–960.

Wichtmann, T. (2016). Soil behaviour under cyclic loading - experimental observations, constitutive description and applications. Veröffentlichungen des Instituts für Bodenmechanik und Felsmechanik am KIT. Habilitation. Germany. Heft 181.

Numerical Methods in Geotechnical Engineering IX – Cardoso et al. (Eds)
© 2018 Taylor & Francis Group, London, ISBN 978-1-138-33198-3

Measurement and analysis of ground contacts during rockfall events

W. Gerber
Swiss Federal Institute for Forest, Snow and Landscape Research WSL, Birmensdorf, Switzerland

A. Caviezel
WSL Institute for Snow and Avalanche Research SLF, Davos, Switzerland

ABSTRACT: Rockfall trajectories are mainly influenced by ground contacts, which cause deceleration and changes in the rocks' rotation. The duration of contacts, resulting deceleration and rotational changes are highly variable and generally unknown. Appropriate data are important for improving our understanding of the rockfall process. So we placed three-axis gyroscopes and accelerometers in various rocks and concrete blocks to measure such values. For the first time, we succeeded in recording high-quality data with our newly developed sensors. The measurements detail the path of deceleration while simultaneously recording rotational changes. During contacts with the ground, 400–500 deceleration and rotation values per second are recorded. Very brief ground contacts range between 8 and 15 milliseconds (ms); longer contacts last between 50 and 70 ms.

1 INTRODUCTION

Rockfalls are natural hazards that usually cause several rocks to come crashing down, leaving traces on trees and in the ground. If infrastructure or buildings are damaged, protective measures are required, prompting the question as which design parameters need to be taken into account. Typical questions are: how high does a protective structure need to be, and what energy should it be designed to cater for? Today, answers to these questions are often provided by corresponding simulation programs (Volkwein et al. 2011). The shape of the stones is usually treated as a mass point, sphere or quboid block. More recent programs can even take the actual shape of rocks into account (Leine et al. 2014). The calculated velocities and jump heights serve as design parameters for the measures. First, however, the results from simulations have to be critically assessed and verified. Traces left by falling rocks provide important clues about velocities. For example, if trajectories can be matched to ensuing ground contacts, this provides physical bases for calculating flight parabolas. Two of the three measures required to calculate the flight parabola are known: the jump distance and the slope angle between the ground contacts. The unknown factor, as yet, is the rock's jump height. For an initial estimation it can be assumed that the jump height will be about a tenth of the jump distance on the slope. Based on such an assumption, a possible flight parabola is determined and the relevant velocities can be calculated (Gerber 2015). If other jump heights are assumed, for example an eighth or a twelfth, this only changes the maximum velocities shortly before impact with the surface by less than 10%. In many cases, this accuracy suffices to calculate a rough velocity. By contrast, the height of a jump must not be verified solely on the basis of the distance covered, but also compared with any traces left on trees.

In this report, flight parabolas are not verified based on traces, but on accelerations and rotations measured in rocks. Using newly developed sensors, these values are taken and recorded on 3 axes at a frequency of 400–500 measurements per second. The resulting measurements yield extremely detailed insights into how rocks behave, both in flight and upon contact with the ground. Before any conclusions can be reached or any further calculations made based on the results, the measurements must be subjected to a quality check and verified. For example, the acceleration measurements at rest must correspond to the value of gravitational acceleration and indicate a value of zero in free flight.

2 METHODS OF MEASUREMENT AND EVALUATION

2.1 *Field studies*

The test site, near Tschamut in the Canton of Grisons, is a slope 50 m high with a maximum inclination of 42° at the top, running down to a horizontal surface. The vegetation there consists mainly of grass, with a few scattered shrubs in the steeper part. This makes it ideal for conducting rockfall experiments and filming the rocks' movements. The rocks used were all fitted with sensors that recorded their acceleration and rotation.

The results presented in this article are taken from just one out of more than 50 runs in a series of experiments carried out at the test site (Caviezel et. al. 2017a). In the measured run, an artificially manufactured concrete block with an 0.3 m edge length and a mass of 44 kg was used. Its edges were pared back a quarter to the blocks more dice-shaped. A hole 68 mm in diameter was drilled through the block to accommodate the sensor. The block's mass and volume (0.019 m^3) make it equivalent to a sphere with a radius of 0.165 m and a circumference of 1.04 m.

2.2 *Sensor*

In view of developments in consumer electronics for devices including tablets, mobile phones and unmanned aerial vehicles (UAVs), the measurement ranges and performances of available miniaturised motion sensors are steadily increasing. In-situ data were recorded using a dedicated low-power sensor node, dubbed StoneNode (Figure 1). The main components of StoneNode v1.0, which was used to record the data presented here, are an tri-axial accelerometer with a measurement range of 400 *g* and an InvenSense 3-axis gyroscope recording up to 4,000 dps.

An microcontroller manufactured by Texas Instruments hosts the sensors and was selected for its low power consumption (roughly 3.6 mW at 3 V). Thus, an 1,100-mAh LiPo battery can gather 56 hours of data. Efficient data retrieval is ensured using a plug-and-play USB device. For detailed information on the sensors used, see the References below. P. Niklaus et. al. 2017 and A. Caviezel et. al. 2017.

2.3 *Quality analysis*

Before the measurements can be processed, the raw data need to be verified. Assuming that the sensors are functioning properly, the raw data should be checked for the following criteria:

- the measuring range of each individual sensor should not be exceeded;
- when at rest, the rotational velocity should equal zero and the acceleration values should equal one, corresponding to gravitational acceleration;
- during freefall, the rotational velocity should remain constant, with zero absolute acceleration; this analysis must be performed when there is relatively little rotation, because the influence of central acceleration will grow at higher rotational velocities;
- theoretically, when rotational velocity is constant, the phenomenon of central acceleration should result in the measurement of higher values.

Figure 1. An exposed microcontroller board hosting all the MEMS sensors, microSD card and a USB connector powered by an 1,100 mAh battery (both covered by the board).

The rotational velocities and acceleration values can then be correlated, representing the eccentrically fitted sensor as the rock's centre of gravity. In physical terms, this relationship can be expressed by the formula (1):

$$R_e = \frac{a_z}{\omega^2} \tag{1}$$

whereby R_e = the sensor's eccentricity (m); a_z = central acceleration (m/s²); ω = rotation (rad/s).

In theory, rotational differences should result in the same eccentricities. However, this only holds true for horizontal rotation: for vertical rotation, gravitational acceleration g = 9.81 m/s ^{2} should also be taken into account.

2.4 *Fall trajectory's slope length*

Video analysis of the fall trajectory and test site model were used to determine a longitudinal profile. In this connection, the corresponding height was derived (horizontally) from the site model every 3 m on steeper terrain and every 5 m in the flatter part. Local slope distances could then be calculated enabling a longitudinal profile of the effective fall trajectory with slope distances.

3 RESULTS

3.1 *General*

The raw data comprise measurements starting from when the sensor was switched on until the block's deposition some 74 s later. The effective start of the rockfall occurred after around 54 s. During this period, 8,000 acceleration values for all three axes (x, y and z) and 9,750 rotational

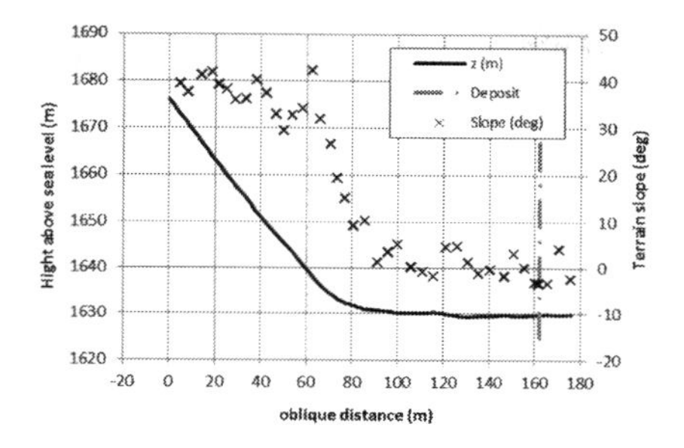

Figure 2. Slope distance of the projected trajectory of the stone with its location and slopes.

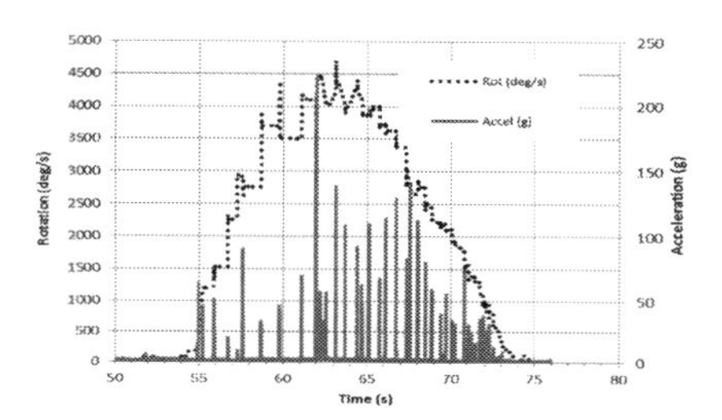

Figure 3. Absolute rotational velocities and acceleration values during the 20-s movement phase (from 54 to 74 s).

values were measured. Analysis of frequency measurements yielded values of 400 Hz during acceleration and 487.5 Hz for rotation. During the 20-second rockfall (from initiation to deposition), the block covered a horizontal distance of 147 m and negotiated a height difference of 49 m. The maximum inclination on site was –42°, dropping to zero and even +4° on the upslope of the depositional area. The effective fall trajectory's slope length was 162 m (Figure 2).

Ground contacts are very clearly indicated by large discrepancies in the values of acceleration measurements and rotational velocities. In steeper terrain, significantly fewer ground contacts were registered than in the shallower depositional area. Whereas ground contacts occur every two seconds on steep slopes, on shallower inclines there are between twice and three times as many. Absolute rotation increases from an initial value of zero to 4,500 degrees per second (deg/s) before falling back to zero.

The maximum absolute acceleration value measured was 225 g (at second 62.0). All other values lay below 140 g, and many were even less than 50 g (Figure 3).

3.2 *Quality analysis*

In the previous section, the absolute rotational velocities and acceleration values were calculated and presented. Prior to that, the peak values of the individual measurements were checked, and it was confirmed that none of the sensors' axial capacity limit values had been reached.

The used to analyse measurements of the block at rest were taken from the period 53.0–53.1 s, when it was still on the ground. The acceleration values ranged from 2.0 g to –1.0 g. If absolute values are calculated, the range of variation only slightly decreases, still yielding values of 0.3 g to 2.3 g, with a mean of 1.37 g. This value should equal 1.0 g, but is 0.37 g higher than that (Figure 4). The rotational measurements indicate the zero-point distance along each axis. Whereas this distance is very well maintained along the x axis, along the y and z axes it is off by about 3 deg/s. The absolute values range from 4.3–4.8 deg/s, at an average of 4.52 deg/s (Figure 5).

The analysis of measurements taken when the block was in freefall began immediately its motion began, at 54.5 s. At this time its rotation was relatively low (180.5 deg/s) and its influence on the acceleration value was small, as the central acceleration value calculated using the transformed formula 1 shows:

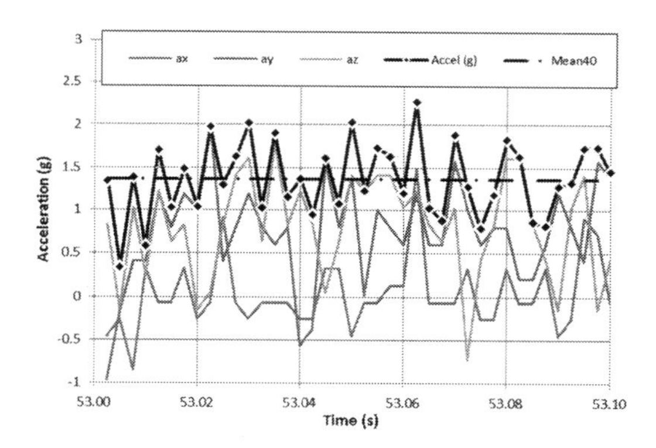

Figure 4. Measurements, absolute values, and mean acceleration during a rest period lasting 0.1 s.

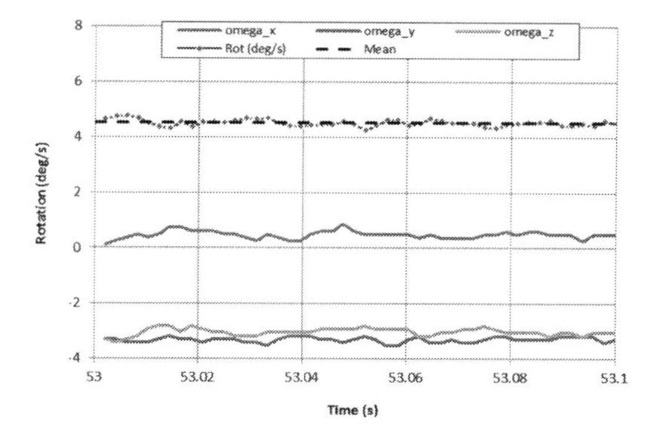

Figure 5. Measurements, absolute values and rotational mean during a rest period lasting 0.1 s.

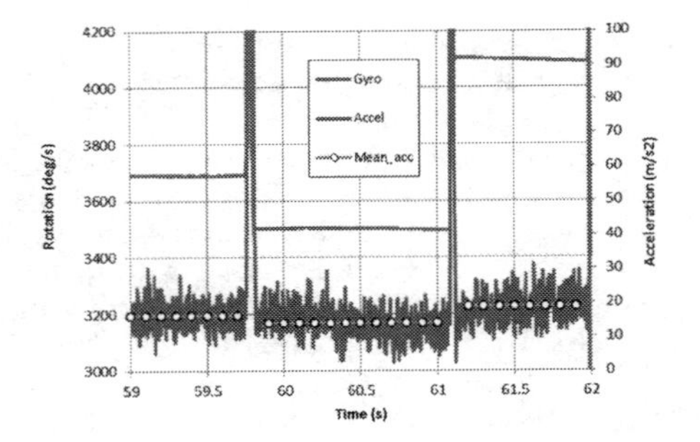

Figure 6. Absolute rotational velocities and acceleration values, plus mean acceleration values for calculating eccentricity.

Table 1. Mean values of absolute rotational and acceleration data for calculating eccentricities.

Time (s)	59–59.7	59.9–61	61.2–91.9
Rotation (deg/s)	3,690	3,501	4,098
Rotation (rad/s)	64.39	61.11	71.52
Acceleration (g)	1.651	1.428	1.927
Acceleration (m/s²)	16.20	14.01	18.9
Eccentricity (m)	0.004	0.004	0.004
Eccentricity (mm)	3.91	3.75	3.70

$$a_z = R_e \cdot \omega^2 = 0.165 \cdot \left(\frac{180.5}{360} \cdot \pi\right)^2 = 0.41 \, {}^{m}\!/_{s^2}$$

The acceleration measurements for each axis, with maxima of 1.3 g and minima of -1 g, indicate a good mean absolute value, albeit one of 0.83 g, making it this amount off.

Eccentricity was analysed between 59.0 and 61.9 s, except for data from the two intervening ground contacts (Figure 6).

This results in three time intervals in which the mean values of the absolute data could be calculated. Based on these, formula 1 was used to determine the eccentricity radius, which is virtually identical for all three time intervals, equalling 0.004 m (Table 1).

3.3 *Duration of ground contacts*

As already mentioned at the beginning of the article, ground contacts are clearly recognisable from the measured rotational and acceleration values. Very short ground contacts last 8–15 ms, medium-length contacts 20–40 ms and lengthy contacts 50–75 ms. During the first 2–3 seconds after the rock has been set in motion, the duration of ground contacts increases very quickly to the peak values and then drops back to values of 10–30 ms, remaining at this level on flat terrain (Figure 7).

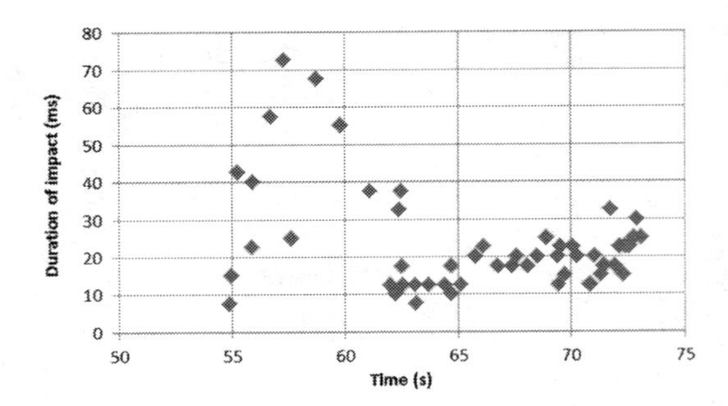

Figure 7. Duration of ground contacts.

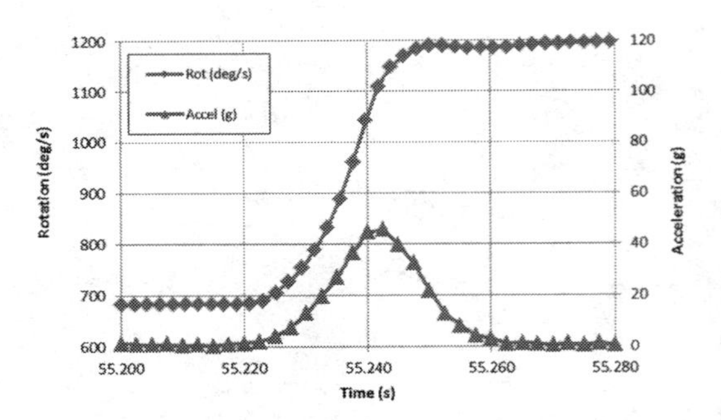

Figure 8. Absolute rotational and acceleration values for the ground contact lasting 80 ms at 55.24 s.

3.4 *Details of individual ground contacts*

Individual results on absolute rotational and acceleration values during ground contacts are presented below. At the start of the fall trajectory, rotation increases with almost every ground contact, with a relatively short ground contact increasing rotation, as at 55.24 s. This contact lasted 42 ms at a maximum acceleration of 45.6 g and increased rotation from 683 to 1,087 deg/s (Figure 8).

For the next ground contact (which lasted 28 ms, starting at 57.62 s), larger accelerations (90.0 g) were measured while instantaneous rotation dropped from 2,921 deg/s to 2,766 deg/s (Figure 9).

Both the ground contacts shown above have clear maxima during the acceleration phase. However, some contacts with two or even more maxima were also recorded. A relatively long ground contact occurred at 58.72 s, lasting 68 ms. During this time, two main acceleration maxima were measured: 33.4 g and 30.4 g respectively. During this ground contact, rotation increased steadily from 2,758 deg/s to 3,696 deg/s (Figure 10).

If rotation between two peak acceleration values remains constant, neither steadily rising nor falling, this indicates two separate ground contacts, like those occurring at 63.13 and 63.15 s. Here, the contact times are very short (lasting

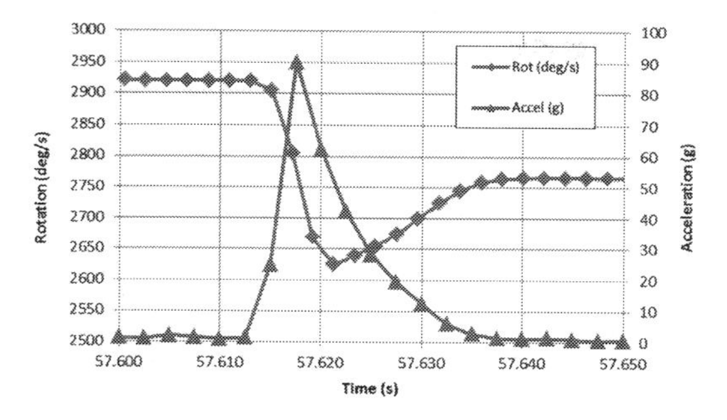

Figure 9. Absolute rotational and acceleration values for the ground contact at 57.62 s over a 50 ms time frame.

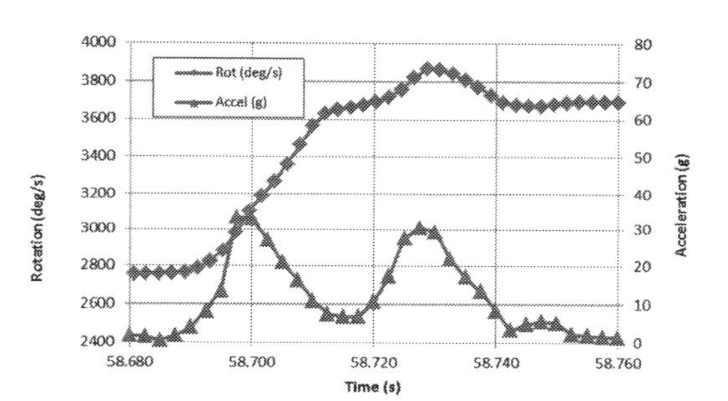

Figure 10. Absolute rotational and acceleration values for the ground contact at 58.72 s, which lasted 80 ms.

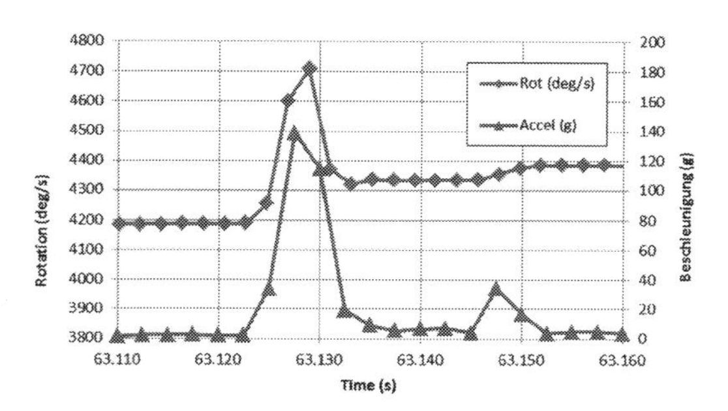

Figure 11. Absolute rotational and acceleration values for the ground contacts at 63.13 and 63.15 s.

13 ms and 8 ms) and the acceleration maxima differ (138.6 g and 34.3 g). During these two contacts, rotation increased from 4,186 deg/s to 4,387 deg/s, with a constant intermediate value to 4,334 deg/s. Interestingly, the maximum rotation of 4,709 deg/s occurred during the first contact, subsequently decreasing to the intermediate value (Figure 11).

Towards the end of the trajectory, the decrease in rotation occurred at much shorter time intervals than the increase on steeper terrain. A typical

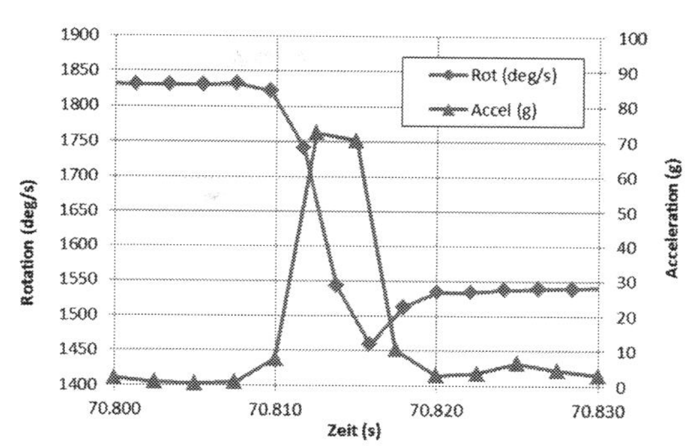

Figure 12. Absolute rotational and acceleration values for the ground contact at 70.81 s, lasting 30 ms.

example of presented here is a relatively short ground contact at 70.81 s, lasting 13 ms. During this time, rotation decreased from 1,831 deg/s to 1,539 deg/s, reaching a local minimum of 1,458 deg/s in between. The maximum acceleration for this ground contact was 72.5 g (Figure 12).

4 DISCUSSION

4.1 *General results*

For the measurements and analyses shown, mainly temporal information was gathered and processed. A connection to the spatial extent of the trajectory is still missing. An approach involving the projected longitudinal profile is available, but the exact connection to the inclination of the terrain or the assignment of slopes to each ground contact is not yet possible. This information would provide a better explanation of the general increases and decreases in rotation.

4.2 *Quality analysis*

The measurements were of a very high quality. Almost every sensor, has a very small zero-point distance, for both rotation and acceleration, so the raw data did not need to be corrected. For measuring rotation, with the capacity of each individual sensor equalling 4,000 deg/s, the zero point is about 4.5 deg/s off, corresponding to an average error of 0.1%. For acceleration measurements, the raw data showed a mixed picture. While an error of 0.37 g was determined when the block was at rest, the corresponding figure during free fall was 0.83 g, whereby the rotational influence was 0.41 m/s^2, corresponding to a figure of 0.04 g. However, the time interval, a duration of 0.1 s, was relatively short. Nonetheless, it was deliberately chosen so that the individual measurement data would still be visible on the graph. In principle, several such calculations should be carried out over a lengthier interval, to form a corresponding mean value, for example, before the rock is set in motion and after its deposition.

An evaluation of the sensor's centric installation in the block indicated a very small eccentricity of 4 mm. This most probably happened accidentally, but should be checked every time and experiment is carried out.

4.3 *Duration of ground contacts*

In this experiment, ground contact duration was shown to vary considerably, ranging from a minimum of 8 ms to a maximum of 75 ms. These measurements show that longer contacts occurred on steeper terrain and shorter ones on flatter terrain. However, no precise categorisation is possible, because the spatial data cannot be linked to the temporal data.

Temporal information on the block's flight duration between ground contacts can be used to calculate the jump height of the flight parabola (Gerber 2015). A temporal and/or spatial link could be used to calculate the jump distance on the slope, but no such link has been established yet.

4.4 *Details on ground contacts*

The measurements suggest very different forms of contact, in terms of both acceleration and rotation. For very short contacts of less than 10 ms, the individual measurements are not quite as reliable as the quality control purports. To measure such short contacts, the measurement frequency would have to be increased, which would in turn increase the total volume of data. The selected frequency of 400–500 Hz was a good compromise, since the duration of virtually all ground contacts exceeds 10 ms. The sampling rate issue was tackled by updating to StoneNode v1.1, which has an increased sampling rate of 1 kH for the accelerometer and gyroscope.

5 CONCLUSIONS

These measurements show that high-quality, detailed and reliable analyses of rotational and acceleration data for rocks hitting the ground are possible. The sensors used provide a logistically rather simple, but effective measurement tool for sampling kinematic data during rockfall events. The results obtained provide fresh insights, but also raise new questions, primarily with respect to the spatial relation of the rock to the surface of the terrain at the start and end of ground contacts. For example, the angle of the rock's trajectory at the onset of contact and the slope of the surface should be known. The device's robustness and updatability will allow these unanswered questions to be taken on board in any further developments of the sensor node.

However, even now the data are already extremely valuable as real-world references for calibrating changes and updates to the code kernel of the RAMMS::rockfall software module (Caviezel 2017a). Being based on a rigid body model that calculates the forces in play at any given time for every simulated rockfall trajectory, it provides in-situ rotational and acceleration values. These simulated results can now be calibrated to measured data, to provide an unprecedented calibration methodology for rockfall simulation code.

The results show also that it is not possible with restitution coefficients to describe the complicated rock-soil interaction process. The interaction is more accurately described by the local orientation and extremely short contact times. Without a methodology to consider rock geometry, surface roughness, and soil scarring of the rock-soil interaction process is overly simplified and can not be effectively used to make consistent runout or jump altitude forecasts.

REFERENCES

Caviezel, A., Christen, M., Bühler, Y., Bartelt, P., Corominas, J., Moya, J., & Janeras, M., 2017a: Calibration methods for numerical rockfall models based on experimental data. RocExs 2017. 6th Interdisciplinary Workshop on Rockfall Protection, 59–62.

Caviezel, A., Schaffner, M., Cavigelli, L., Niklaus, P., Bühler, Y., Bartelt, P., Magno, M., Benini, L., 2017b: Design and evaluation of a low-power sensor device for induced rockfall experiments. IEEE transactions on instrumentation and measurement, vol. PP, no. 99, pp. 1–13.

Gerber, W., 2015: Geschwindigkeit und Energie aus der Anlayse von Steinschlagspuren - Velocity and kinetic energy from the analysis of rockfall trajectories. *Österr. Ing.- Archit.-Z.* 160, 1–12: 171–175.

Leine, R.I., Schweizer, A., Christen, M., Glover, J., Bartelt, P., Gerber, W., 2014: Simulation of rockfall trajectories with consideration of rock shape. *Multibody System Dynamics* 32, 2: 241–271.

Niklaus, P. et al., StoneNode: A low-power sensor device for induced rockfall experiments, in Proc. IEEE SAS, Mar. 2017, pp. 1–6.

Volkwein, A., Schellenberg, K., Labiouse, v., Agliardi, F., Berger, F., Bourrier, F., Dorren, L., Gerber, W., Jaboyedoff, M., 2011: Rockfall characterisation and structural protection – a review. *Nat. Hazards Earth Syst. Sci.* 11: 12617–2651.

Numerical Methods in Geotechnical Engineering IX – Cardoso et al. (Eds)
© 2018 Taylor & Francis Group, London, ISBN 978-1-138-33198-3

The role of evolutionary algorithms in soil constitutive models

C. Pereira & J.R. Maranha
Geotechnical Department, LNEC (National Laboratory for Civil Engineering), Lisbon, Portugal

ABSTRACT: The complex nature of soil behaviour must be modelled with advanced constitutive models. However, some difficulties can be highlighted that hinder the use and the development of advanced constitutive soil models in geotechnical research and practice: the objective evaluation of the performance of model variants, the objective comparison between different models, the information on the effect of individual parameters (material model constants and initial values of internal variables) on the model's response and the calibration of the model's parameters. Evolutionary algorithms appear to play an important role in the development, testing and application of advanced constitutive models. Several papers have been published concerning only the determination of the constitutive models' parameters by means of evolutionary algorithms. Despite their success when applied to model calibration, this paper illustrates that they play an important role, not only in determining the parameters values, but also in the development, improvement, testing, comparison and application of advanced constitutive models.

1 INTRODUCTION

The complex nature of soil behaviour must be modelled with advanced constitutive models which have the disadvantage of requiring a significant amount of parameters that must be determined. Constitutive model's parameters are considered here as the model's constants and the initial values of internal variables. In most cases, some of the parameters cannot be obtained directly from the results of specific laboratory tests. This type of constitutive models is nonlinear and some parameters have coupled effects which makes the determination of their values difficult by means of the traditional approaches. Instead, they need to be obtained through the fitting of the computed results to the available data, usually from several distinct laboratory tests. Consequently, the determination of the parameters can be an arduous task that hinders the use of advanced constitutive models in engineering practice. The application of evolutionary algorithms to calibrate advanced constitutive models appears to be a useful technique to solve this issue, as shown by Pereira et al. (2014), Zhao et al. (2015), Jin et al. (2016) and Yin et al. (2017), among many others.

As explained in the present paper, the application of evolutionary algorithms only to the determination of constitutive model's parameters underuses all the potential that they have in the development of advanced constitutive models. The evolutionary algorithms appear to play an important role, not only in the calibration of constitutive models but also in their development, improvement and testing. With the use of these algorithms, the complexity of constitutive models is no longer such a relevant obstacle. The parameters are not dependent on a specific laboratory test to be determined. Instead, these algorithms may be useful in accessing the physical meaning of some of the constants used or how they affect the model's response. In the end, the development of a constitutive model is limited by the quantity and quality of the information available.

This paper illustrates how evolutionary algorithms were essential to the development or improvement of four advanced constitutive models: the ratedependency (section 3), soil-rockfill mixtures (section 4), cemented soils (section 5) and thermomechanical response (section 6). The evolutionary algorithms used are the Genetic Algorithm, AG, and the Differential Evolution method, DE. In section 2, both methods are briefly described.

2 EVOLUTIONARY ALGORITHMS

Evolutionary algorithms are heuristic algorithms used in optimisation problems that draw inspiration from the scientific theory of evolution by natural selection. Differently from traditional optimisation techniques, evolutionary algorithms usually evolve a population of solutions (individuals) in the search space of decision variables (models' parameters). Each iteration generates a new set of solutions and performs a competitive selection that weeds out weak solutions. In comparison with traditional optimisation techniques, evolutionary algorithms are usually more robust and more generally applicable.

This section briefly describes two distinct evolutionary algorithms, the Genetic Algorithm, GA, and the Differential Evolution method, DE, used in the development or improvement of constitutive models and the determination of their parameters using the experimental data available. This paper does not make a comparison between both methods. More important than the specific evolutionary algorithm used, is this new concept that takes advantage of them in constitutive model development.

2.1 *Genetic algorithm*

GAs employ analogies to some of the concepts of the Darwinian theory of the evolution of species, such as selection, mutation, inheritance and crossover. In general, GAs tend to be able to find accurate global solutions, but are inefficient at finding the absolute optimum. In this technique, possible solutions to the problem to be solved (minimised or maximised) are interpreted as individuals in a population. These individuals have a chance of survival related to their fitness or error criterion measure (the higher the fitness, the higher their probability of mating and survival). GAs require a significant number of generations, with the associated computational cost, to obtain good solutions. The implementation of the GA presented by Pereira et al. (2014) followed the work presented by Azeiteiro (2008).

To improve the capability of the GA in the local search, the Hill Climbing procedure, HC, was combined with the GA. HC is a local search optimisation technique which, starting from a given initial solution, attempts to improve it by randomly changing its parameters.

If the change produces a better solution, the new parameters are used in the next iteration. The implementation of HC and the linkage between GA and HC was done taking into account the recommendations presented in Taborda et al. (2011).

Figure 1 presents a simplified diagram illustrating the implementation process. The algorithm was implemented in Python and the numerical simulation of the constitutive model was performed with the program FLAC. The combination of Python and FLAC provides significant flexibility in the determination of parameters of any constitutive model with the experimental test available. However, to speed up the algorithm other types of computer implementations were considered, such as the use of compiled programming languages with parallel implementations.

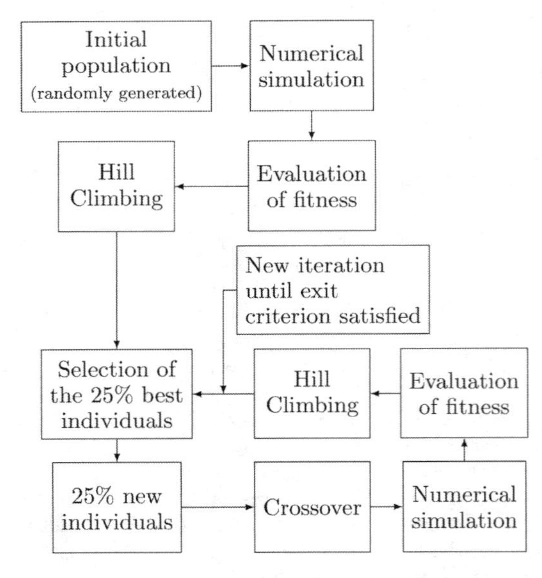

Figure 1. Simplified diagram of the Genetic Algorithm with the Hill Climbing procedure (adapted from Pereira et al. (2014)).

2.2 *Differential evolution method*

Like the previous method, the Differential Evolution method, DE, presented by Storn and Price (1997), is a population-based, derivative-free function optimiser. Like other evolutionary algorithms, DE relies on the concept of a population of individuals, which undergo such probabilistic operations as mutation, crossover and selection in a loop of successive generations to evolve towards solutions of better fitness in the search space of decision variables. The algorithm executes successive generations until the best current solution is deemed satisfactory. Figure 2 represents a simplified diagram of DE.

Briefly, consider that in each generation g there is a population of n_p individuals or solutions and each has nv parameters to be determined. The initial population $x^0 = x^0_{ij}$, for $i = 1,2,\ldots,n_p$ and $j = 1,2,\ldots,n_v$, is randomly generated according to a uniform distribution $l_j^{low} \le x_{ij}^0 \le l_j^{up}$, where l_j^{low} are the upper and lower limits of the *j*-th component of the solutions vectors of individual i. After the evaluation of the initial population, DE enters in a loop of evolutionary operations: mutation, crossover and selection.

At each generation g, mutant solutions v_i^g based on the current parent population x^{g-1} are created. Parent population designates the population that came from the last generation, g − 1. The mutation introduces new information into the population by randomly generating variations to

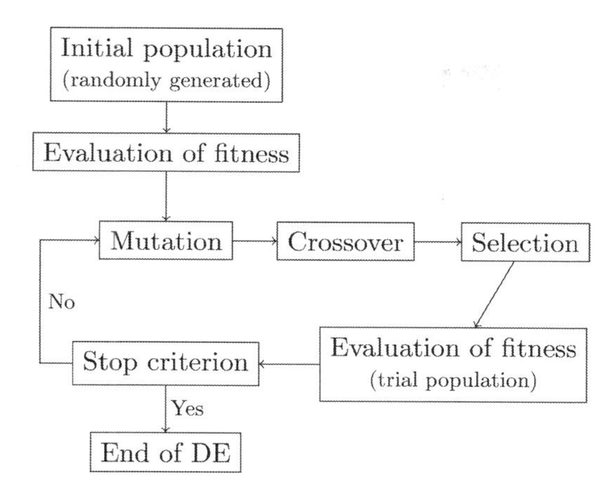

Figure 2. Simplified diagram of Differential Evolution method.

existing individuals. The following mutation strategy, namely DE/rand/1,

$$v_i^g = x_{r0}^{g-1} + f_m\left(x_{r1}^{g-1} - x_{r2}^{g-1}\right), \tag{1}$$

was the first to be proposed and one of the most used. The indices r_0, r_1 and r_2 are distinct integers randomly chosen from the set $\{1,2,\ldots,n_p\}/\{i\}$ and f_m the mutation factor which usually ranges between 0 and 1. The mutation operation may generate trial vectors vg i whose components violate the boundary constraints, predefined by vectors $\boldsymbol{l}_{low}$ and $\boldsymbol{l}^{up}$. A simple way to correct that is to set the violating component to the middle value between the violated bound and the corresponding components of the parent individual, i.e.,

$$v_{ij}^g = \frac{l_j^{low} + x_{ij}^{g-1}}{2} \text{ if } v_{ij}^g < l_j^{low} \tag{2}$$

and

$$v_{ij}^g = \frac{l_j^{up} + x_{ij}^{g-1}}{2} \text{ if } v_{ij}^g > l_j^{up}, \tag{3}$$

where v_{ij}^g and x_{ij}^{g-1} are the j-th components of the nutation vector v_i^g and the parent vector x_i^{g-1} at the eneration g, respectively.

After mutation, a crossover operation is performed between mutated and parent solutions in order to increase the diversity of the perturbed parameter's vectors. The new trial solutions $\boldsymbol{u}_i^g$ are obtained by

$$u_{ij}^g = \begin{cases} v_{ij}^g & \textbf{if rand}_j\,(0,1) \le f_c \vee j = j_{rand} \\ x_{ij}^{g-1} & \text{otherwise} \end{cases}, \tag{4}$$

where rand$_j$ (0,1) is a uniform random number on the interval (0,1) and newly generated for each j component, j_{rand} an integer randomly chosen from 1 to n_v and newly generated for each i individual, and $f_c \in [0,1]$ the crossover probability. The crossover typically performs an information exchange between the parent solution x_i^{g-1} and the mutant vector v_i^g. The crossover probability is a user-defined value that controls the fraction of parameters values that are copied from the mutant solution. In all situations, the j-th parameter is obtained from the mutant vector. Then, the trial solutions $\mathbf{u}^g$ is evaluated.

The selection operation selects the better one from the parent vector x_i^{g-1} and the trial vector u_i^g, comparing their objective function values f (·). In a minimisation problem, the selected vector is given by

$$x_i^g = \begin{cases} u_i^g & \text{if } f(u_i^g) < f(x_i^g) \\ x_i^{g-1} & \text{otherwise} \end{cases} \tag{5}$$

and used as the parent vector in the next generation $g + 1$. The selection imposes a driving force towards the optimum by preferring individuals with better evaluation result. The loop is repeated until a maximum number of generation was reached, the minimum objective value was less than a given error (in a minimisation problem), or all the solutions became equal.

When dealing with constitutive models, some combinations of parameters, even with all the values inside the search space, are not admissible regarding the models' response. The imposition of some restrictions has to be considered. An example of that is k \$ λ in the Cam Clay Model. Therefore, the generation of the initial population, mutation and crossover stages must satisfy the constitutive model's restrictions.

2.3 *Objective function to the determination of constitutive models' parameters*

The objective function is a measurement of the quality of the solutions. When applied to the determination of the parameters of advanced constitutive models using experimental tests data, the objective function can be a measurement of the difference between the experimental and the numerical polygonal curves, also named the error measure. Numerical polygonal curves derive from the results computed with the constitutive models when simulating the experimental tests under the same conditions.

As pointed out by Pereira et al. (2014), one way to evaluate the error measure between polygonal curves is by the average distances between points equivalently positioned in both curves, according to the corresponding cumulative normalised length. This type of error measure may not produce acceptable results under certain circumstances, as illustrated in Figure 3.

The mathematical concept known as Discrete Fréchet Distance, DFD, an approximation of the Fréchet Distance presented by Fréchet (1906) (Altand Godau 1995) was adopted to solve this problem. The Fréchet Distance is a measure of similarity between curves that takes into account the location and ordering of the points along the curves. An intuitive illustration of DFD was presented by Agarwal et al. (2014). Informally, think of $A = (a_1, a_2, \ldots, a_m)$ and $B = (b_1, b_2, \ldots, b_n)$ as two sequences of stepping stones and of two frogs, the A-frog and the B-frog, where the A-frog has to visit all the A-stones in order, and the B-frog has to visit all the B-stones in order. The frogs are connected to each other by a rope of length δ, and are initially placed at a_1 and b_1, respectively. At each move, one or both frogs can jump from its current stone to the next one, but backtracking is not allowed. In DFD, δ_{dF}, is the length of the shortest rope that is enough to allow the frogs A and B to go from a_1 and b_1 to a_m and b_n, respectively. When dealing with multiple polygonal curves, the objective function to be minimised can be the sum of the DFD between the several experimental and numerical data polygonal curves.

3 VISCOPLASTIC SUBLOADING CONSTITUTIVE MODEL

A new purely viscoplastic soil model, based on the subloading surface concept, with a mobile centre of homothety, was initially presented by Maranha et al. (2014). The model enables the occurrence of viscoplastic strains inside the yield surface and avoids the abrupt change in stiffness of the traditional overstress viscoplastic models (a requirement to describe the behaviour of overconsolidated soils). The model was formulated to reproduce the soil rate-dependent behaviour under

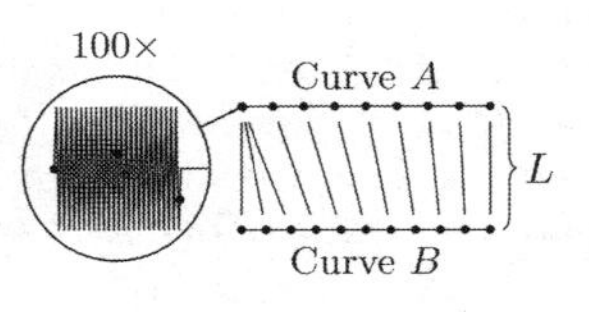

Figure 3. Schematic representation of an example where the fitness of curves, performed according to the mean of distances between points equivalently positioned along the length of both curves, fails (Pereira et al. 2014).

cyclic loading (changes in loading direction) and incorporates both initial and induced anisotropy. This model also has the advantage that, for sufficiently small values of the viscosity, it becomes identical to the corresponding elastoplastic (inviscid) subloading model. Figure 4 schematically represents, in the effective stress space, σ, the several homothetic surfaces that define the viscoplastic subloading overstress model with a mobile centre of homothety, a, where p is the mean stress and s the deviatoric stress. _σ$_y$, given by

$$\sigma_y = \mathbf{a} + \frac{1}{R}(\sigma - \mathbf{a}) = \mathbf{a} + \frac{1}{R_s}(\sigma_s - \mathbf{a}), \quad (6)$$

is the homothetic image of both σ and σ_s on the yield surface relative to a, R and R_s are the homothety ratios of the dynamic loading and subloading surfaces relative to the yield surface, respectively. The stress state is always on the current dynamic loading surface. Additionally, it was assumed that ⊠$_m = \bar{R}_m \sigma_y$ is the homothetic image of _y on the infinite strain rate yield surface, ISRYS, with the centre of homothety at the origin of the stress space, where $\bar{R}_m$ is the homothety ratio of the ISRYS relative to the yield surface.

Viscoplastic strain rate, $\dot{⊠}^{vp}$, occurs when σ is outside the subloading surface ($R > R_s$), even if it is still inside the yield surface. The ISRYS defines a limit to the evolution of σ.

According to Figure 5, the model was only able to reproduce accurately the stress-strain response from cyclic undrained triaxial creep tests performed on an overconsolidated stiff clay from the Lisbon region (Formąção de Benfica) (Maranha and Vieira 2011). The model's parameters were

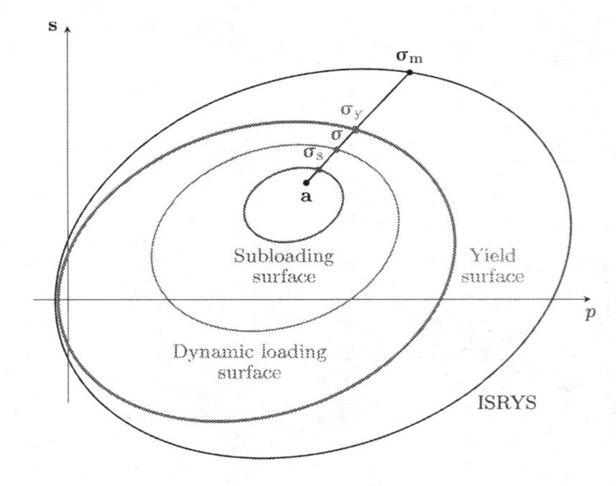

Figure 4. Schematic representation, in _ space, of the homothetic surfaces that define the viscoplastic subloading overstress model with a mobile centre of homothety, a (adapted from (Maranha et al. 2014)).

determined using the GA (Pereira et al. 2014). The laboratory test was performed in undrained triaxial compression conditions, which applies both to the change of deviatoric stress and creep stages. In the first stage of the test, the sample was gradually loaded from the isotropic stress state, at constant total mean stress, $q = 200$ kPa, to deviatoric stress $= 250$ kPa, during 3 days. The loading sequence modelled, consisted in a series of unloading steps followed by a series of reloading steps, defining a complete load cycle, with each step consisting in a relatively fast change in deviatoric stress at $\dot{\varepsilon}_a = 6 \times 10^{-6} s^{-1}$, followed by 24 h creep at constant deviatoric stress and constant total mean stress equal to 200 kPa.

In Maranha et al. (2016b), an improved version of the model was presented that reproduces well both the stress-strain response and the evolution of the pore pressure (Figure 5). As explained in Maranha et al. (2016b), all the hardening laws, $\dot{p}_c$, ȧ, ⊠ and $\dot{R}_s$, were modified. The study on how each hardening law influences the model's response and how it could be improved was greatly facilitated by the use of the evolutionary algorithm in each stage of the development process.

GA was also essential to verify the capability of the same constitutive model to reproduce the strain rate change effects in soils. As described in Maranha et al. (2015) and Maranha et al. (2016a), an undrained triaxial compression test on a reconstituted London Clay sample (Sorensen et al. 2007) was simulated. The sample was isotropically consolidated to a mean effective stress $p = 300$ kPa, corresponding to a normally consolidated state. Then, the sample was sheared in undrained triaxial compression conditions from the initial isotropic stress state. The viscous behaviour of the soil sample was characterised by its response to the change in a stepwise manner of the axial strain rate, $\dot{\varepsilon}_a$, between three distinct values, 0:05, 0:2 and 0:9% = h. Figure 6 shows the stress-strain experimental response of a normal consolidated sample of the reconstituted London Clay (Sorensen et al. 2007). Figure 6 also represents the numerical simulation results obtained with the viscoplastic subloading model. The model's parameters were obtained through the application of the GA (Pereira et al. 2014). This result was achieved without any specific modification of the model regarding this type of behaviour, showing the good performance of the model to reproduce the soil's viscous behaviour. Having determined the parameters, a numerical study to find the reason behind the temporary effects was conducted. The temporary effects produced by the viscoplastic subloading model were found to be influenced by the strain acceleration value. However, the temporary effects seem to be insensitive to the higher values of strain acceleration comprising

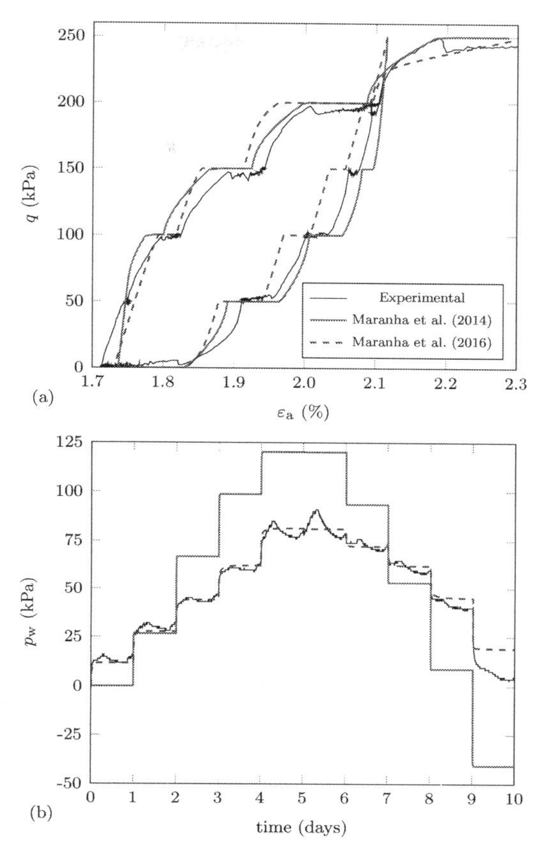

Figure 5. Comparison between the experimental and both numerical results (Maranha et al. (2014) and Maranha et al. (2016b)). (a) Results in (ε_a, q) representation space. (b) Results in (p_w, time) representation space.

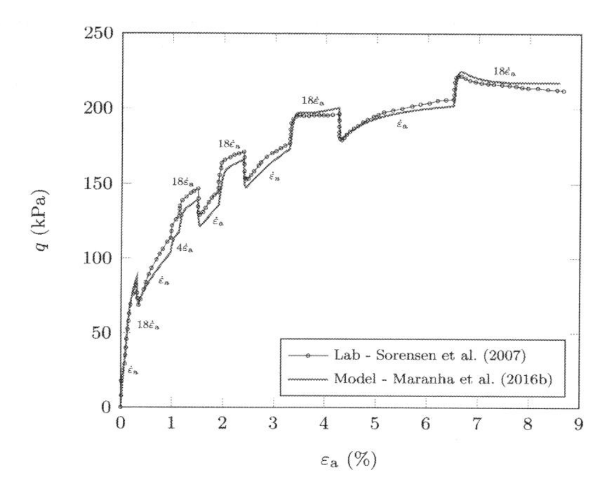

Figure 6. Modelling and experimental results of the undrained triaxial test (adapted from Maranha et al. (2015)).

several orders of magnitude, which includes the values typically used in the laboratory tests.

4 CONSTITUTIVE MODEL FOR SOIL-ROCKFILL MIXTURES

Brito et al. (2017) presented a new constitutive model, namely the Subloading Surface Rockfill Model, SSRM, developed and tested to reproduce the behaviour of soil-rockfill mixtures. The SSRM is an extension of the Modified Cam Clay Model that incorporates the subloading surface concept. Other characteristics of the model are tensile strength, a nonassociated flow rule and a curved critical state line. The SSRM allows the occurrence of plastic strains inside the yield surface, including a smooth transition between the elastic and plastic behaviours. The study of these, and others characteristics which were not adopted due to the lack of improvement in reproducing the laboratory results, such as, anisotropic hardening and the λ dependence on the volumetric plastic strain, was supported by the use of the GA (Pereira et al. 2014). According to Brito et al. (2017), the evolutionary algorithm was essential to develop, test and validate the different parts of the model and determine all the constants and initial values of internal variables of each soil-rockfill mixture.

As an example of the application of the GA to the determination of the model's parameters and the definition of the model's characteristics that should be adopted, the GA had to simultaneously adjust several experimental polygonal curves of the soil-rockfill mixtures. For the samples without rockfill material, ten experimental tests were available: four undrained triaxial tests with isotropic consolidation equal to 190, 376, 760 and 1520 kPa, and the K_0 test. Two representation spaces were considered for each laboratory test: the axial strain vs the deviatoric stress (ε_a, q) and the effective stress path (p, q). In the end, the SSRM was able to reproduce reasonably well the response of the several soil-rockfill mixtures considering the intrinsic variability of the tested specimens and the number of experimental curves available for each mixture (Brito et al. 2017). The analyses of the soil's parameter values of each mixture permitted the establishment of a relation between the values of some parameters and the proportion between soil and rockfill in the samples.

5 CEMENTED SUBLOADING CONSTITUTIVE MODEL

Ribeiro et al. (2017) developed a new Cemented Modified Cam Clay Model, CMCCM, using the subloading surface concept (Ribeiro 2017). An isotropic hardening law controls the hardening of the yield surface. The model enables continuous plasticity which was found suitable to handle plastic strains inside the yield surface region, which was fundamental to the reproduction of the experimental data. The effective stress concept and the pore water pressure follow Biot's theories.

The GA (Pereira et al. 2014) was used to search for combinations of the model's parameters that better reproduce the experimental tests for each cement content, to evaluate the performance of some discarded features of the model, such as, an anisotropic hardening law and a curved critical state line, and to adopt some model characteristics that improve the adjustment, including a non-associative flow rule. The model was able to adjust well the stress-strain curves and reasonably well the pore pressure evolution for each dosage of cement (50, 150, 200, and 250 kg/m^3). In the process of development of the constitutive model, the evolutionary algorithm enable the evaluation of the impact in the performance of each attempt to improve the model.

6 SUBLOADING THERMO-VISCOPLASTIC CONSTITUTIVE MODEL

Maranha et al. (2017) presented a new thermo-viscoplastic subloading soil model. The constitutive model was formulated to represent the non-isothermal behaviour of soils under strictly isotropic stress and strain conditions. This model was an extension of the viscoplastic subloading soil model presented by Maranha et al. (2016b), restricted to isotropic stress and strain conditions. Figure 7 schematically represents, in the isotropic effective stress space, the different homothetic surfaces which define the thermoviscoplastic subloading model, with a mobile centre of homothety, p_a.

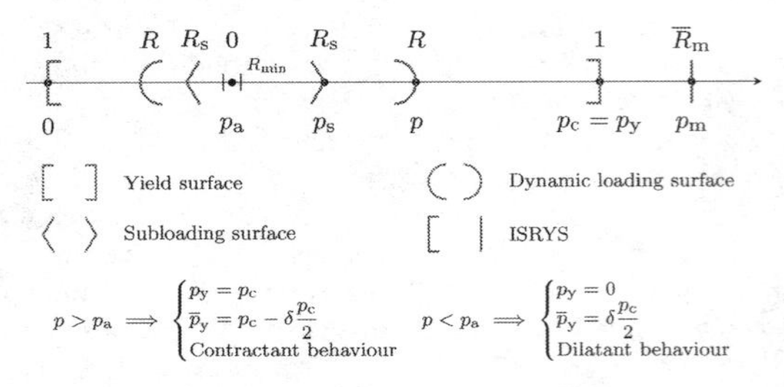

Figure 7. Schematic representation, in the isotropic stress space, of the several homothetic surfaces which define the thermalviscoplastic subloading model, with a mobile centre of homothety (adapted from Maranha et al. (2017)).

In one-dimensional isotropic effective stress space, the surfaces are represented by two points.

The model was developed to represent the isotropic thermo-mechanical behaviour of soil. Three laboratory tests performed on Kaolin clay by Cekerevac (2003) were used to support the development of the constitutive model. Three samples of saturated Kaolin clay, HT-T21 ($OCR = 1$), ISO-T1 ($OCR = 6$) and HT-T17 ($OCR = 12$), were isotropically consolidated to 600 kPa, followed by, in some cases (ISO-T1 and HT-T17), isotropic unloading to achieve the OCR values of each test. Then, the samples were heated in drained conditions from 22° to 90°C, with a heating rate of 10°C per 3 h under constant mean effective stress p (Cekerevac and Laloui 2004).

An excellent adjustment to the experimental results was achieved (see Figure 8). The viscosity (ratedependent) properties of the model were responsible for the adjustment by means of the assumption that the unloading stage, which precedes the thermal loading, was not slow enough to extinguish the viscoplastic dilatant volumetric strains associated with unloading, i.e. assuming at the beginning of the heating path $R > R_s$ in tests with OCR = 6 and 12. This assumption was made due to the lack of information regarding the tests' conditions and supported by the significant improvement in error measure obtained by the DE.

At this time, with the support of the DE, this model has been improved to have the capability to reproduce the eating and cooling cyclic stages (Maranha et al. 2018).

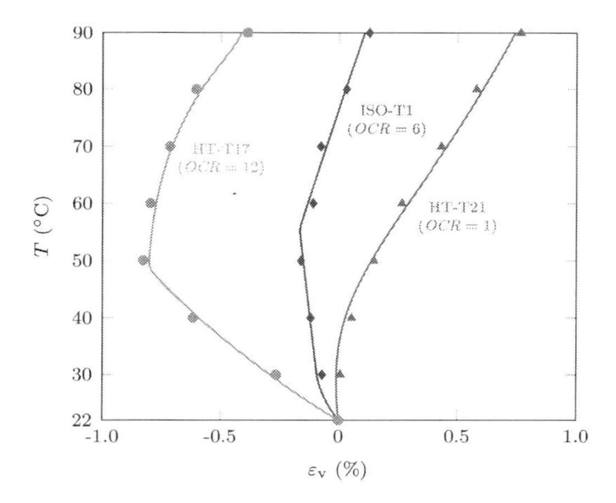

Figure 8. Volumetric strains of Kaolin clay under isotropic drained heating from 22° to 90°C for three different OCR values. Comparison between the laboratory results obtained by Cekerevac (2003) and published in Cekerevac and Laloui (2004) (marks) and the numerical simulation results obtained by Maranha et al. (2017) (lines) (adapted from Maranha et al. (2017)).

7 CONCLUSIONS

This paper shows how evolutionary algorithms are fundamental, not only to the determination of the parameters of constitutive models, but as an essential key in the development and improvement of advanced constitutive models taking into account the information available.

By itself, complex constitutive models, characterised by a significant number of parameters, are no longer such a problematic issue. Beyond the determination of the models' parameters, this type of optimisation algorithms is an expedite way of testing the effectiveness of small improvements of parts of the models to reproduce the soils' behaviour. The accumulated experience in the development and improvement of constitutive models, briefly described in the previous sections, shows that the progress achieved in such a short amount of time was possible due to the use of this type of mathematical tools.

ACKNOWLEDGEMENTS

The financial support provided by FCT (the Portuguese Foundation for Science and Technology) to the first author with the PhD Grant No. SFRH/BD/120336/2016 is gratefully acknowledged.

REFERENCES

Agarwal, P.K., R.B. Avraham, H. Kaplan, & M. Sharir (2014). Computing the discrete Fréchet distance in subquadratic time. *SIAM Journal on Computing 43*(2), 429–449.

Alt, H. & M. Godau (1995). Computing the Fréchet distance between two polygonal curves. *International Journal of Computational Geometry & Applications 5*(1 & 2), 75–91.

Azeiteiro, R. (2008). Aplicação de Algoritmos Genéticos na calibração de modelos de comportamento de solos. Master's thesis, Universidade de Coimbra (in Portuguese).

Brito, A., J. ao Ribas Maranha, & L.M.M.S. Caldeira (2017). A constitutive model for soil-rockfill mixtures. *Computers and Geotechnics.*

Cekerevac, C. (2003). *Thermal effects on the mechanical behaviou of saturated clays: an experimental and constitutive study*. Ph. D. thesis, École Polytechnique Fédérale de Lausanne.

Cekerevac, C. & L. Laloui (2004). Experimental study of thermal effects on the mechanical behaviour of a clay. *International Journal for Numerical and Analytical Methods in Geomechanics 28*(3), 209–228.

Fréchet, M.M. (1906). Sur quelques points du calcul fonctionnel. *Rendiconti del Circolo Matematico di Palermo 22*(1), 1–72.

Jin, Y.F., Z.Y. Yin, S.L. Shen, & P.-Y. Hicher (2016). Selection of sand models and identification of parameters using an enhanced genetic algorithm. *International Journal for Numerical and Analytical Methods in Geomechanics 40*(8), 1219–1240.

Maranha, J.R., C. Pereira, & A. Vieira (2014). A viscoplastic subloading overstress model with a moving centre of homothety. In F. Oka, A. Murakami, R. Uzuoka, and S. Kimoto (Eds.), Proc. *The 14th International Conference of the International Association for Computer Methods and Advances in Geomechanics, 22–25 September 2014, Kyoto, Japan*, Boca Raton, pp. 243–248. CRC Press.

Maranha, J.R., C. Pereira, & A. Vieira (2015). Viscoplastic subloading modelling of strain rate shange effects in London clay. In J. Dijkstra, M. Karstunen, J. Gras, and M. Karlsson (Eds.), *Proceedings of the International Conference on creep and deformation characteristics in geomaterials, Gothenburg, Sweden*, Gothenburg: Chalmers University of Technology., pp. 55–58.

Maranha, J.R., C. Pereira, & A. Vieira (2016a). Strain-rate change effects in reconstituted London clay using a viscoplastic subloading model. *European Journal of Environmental and Civil Engineering*, 1–12.

Maranha, J.R., C. Pereira, & A. Vieira (2016b). A viscoplastic subloading soil model for rate-dependent cyclic anisotropic structured behaviour. *International Journal for Numerical and Analytical Methods in Geomechanics 40*(11), 1531–1555.

Maranha, J.R., C. Pereira, & A. Vieira (2017). Thermoviscoplastic subloading soil model for isotropic stress and strain conditions. In A. Ferrari and L. Laloui (Eds.), *Advances in Laboratory Testing and Modelling of Soils and Shales (ATMSS)*, pp. 479–485. Cham: Springer International Publishing.

Maranha, J.R., C. Pereira, & A. Vieira (2018). Improved subloading thermo-viscoplastic model for soil under strictly isotropic conditions. *Geomechanics For Energy and the Environment.*

Maranha, J.R. & A. Vieira (2011). An elastoplasticviscoplastic soil model for cyclic loading. In E. O˜nate, D.R.J. Owen, D. Peric, and Su´arez (Eds.), *Proceedings of the XI International Conference on Computational Plasticity. Fundamentals and Applications, COMPLAS XI, 7–9 September 2011, Barcelona*, CIMNE: Barcelona, pp. 1201–1211.

Pereira, C., J.R. Maranha, & A. Brito (2014). Advanced constitutive model calibration using genetic algorithms: some aspects. In M.A. Hicks, R.B.J. Brinkgreve, and A. Rohe (Eds.), *Proc. 8th European Conference on Numerical Methods in Geotechnical Engineering, 18–20 June 2014, Delft, The Netherlands*, pp. 485–490. Boca Raton: CRC Press.

Ribeiro, D., J.R. Maranha,&R. Cardoso (2017). Modeling the mechanical behaviour of sand-cement mixtures. In *Proc. 6th International Young Geotechnical Engineers Conference, Seul, South Korea.*

Ribeiro, D.C. (2017). *Prediction of the hydro-mechanical behavior of jet-grouting columns*. Ph. D. thesis, Instituto Superior Técnico.

Sorensen, K.K., B.A. Baudet, & B. Simpson (2007). Influence of structure on the time-dependent behaviour of a stiff sedimentary clay. *Géotechnique 57*(1), 113–124.

Storn, R. & K. Price (1997). Differential evolution a simple and efficient heuristic for global optimization over continuous spaces. *Journal of Global Optimization 11*(4), 341–359.

Taborda, D.M.G., A. Pedro, P.A.L.F. Coelho, & D. Antunes (2011). Impact of the integration of a Hill Climbing procedure on the performance of a genetic algorithms-based software. In S. Pietruszczak and G.N. Pande (Eds.), *Proc. 2nd International Symposium on Computational Geomechanics COMGEO II*, International Centre for Computational Engineering, Cavtat-Dubrovnik, Croatia.

Yin, Z.Y., Y.F. Jin, J.S. Shen, & P.Y. Hicher (2017). Optimization techniques for identifying soil parameters in geotechnical engineering: Comparative study and enhancement. *International Journal for Numerical and Analytical Methods in Geomechanics.*

Zhao, B.D., L.L. Zhang, D.S. Jeng, J.H. Wang, & J.J. Chen (2015). Inverse analysis of deep excavation using differential evolution algorithm. *International Journal for Numerical and Analytical Methods in Geomechanics 39*(2), 115–134.

Numerical Methods in Geotechnical Engineering IX – Cardoso et al. (Eds)
© 2018 Taylor & Francis Group, London, ISBN 978-1-138-33198-3

Estimation and calibration of input parameters for Lake Texcoco Clays, Mexico City

N. O'Riordan, S. Kumar, F. Ciruela-Ochoa & A. Canavate-Grimal
Arup Geotechnics, London, UK

ABSTRACT: The soils of the former Texcoco Lake located in the Mexico Basin are structured soft lacustrine sediments, comprising a variable mixture of high plasticity clay minerals and siliceous diatoms. These soils present an unusual combination of high void ratio coupled with high friction angles. Other properties like the ability to sustain large shear deformations without generation of any significant excess pore pressure, thixotropic behaviour and high natural period of vibrations makes it a uniquely challenging material to work with. The properties of this soft lake deposit are a moving target due to ongoing consolidation caused by the groundwater abstraction from the deep aquifer. The former lake is effectively undeveloped, but importantly site investigations including excavation trials were carried out in late 1960 demonstrating the difficulties of working with these materials fifty years ago when ground levels were about 8 to 10 m higher than they are today. As a part of the development of the New International Airport of Mexico City (NAICM), extensive ground investigations were carried out including a large-scale trial excavation in 2016. This paper presents the key results from the ground investigation and the detailed calibration work carried out to develop input parameters for the Hardening Soil model with Small Strain stiffness (HSsmall) constitutive model. This paper also shows how the Finite Element analyses were successfully used to reproduce the trial excavation behaviour.

1 INTRODUCTION

1.1 *Mexico City Basin*

The Basin of Mexico is a predominately flat lacustrine plain, approximately 2228 m above sea level, and occupies an area of 10,000 km^2. The Basin was open until about 700,000 years ago, during the Pleistocene epoch, when volcanic activity caused the creation of the Chichinutzin Range, which acted to close off the Basin, surrounded on all sides by the southern Mexico plateau. At the point projections of the ranges surrounding the Basin are alluvial fans and debris flows (lahars) interbedded with volcanic pumice and ash. This volcanic-sedimentary complex is known as the Tarango Formation.

Several large lakes, including Lake Texcoco, were formed within the Basin by periods of glaciation and persistent rains within the last 100,000 years. As the surrounding mountains gradually eroded, the fine and ultra-fine volcanic ash particles were transported within the basin. Subsequent volcanic eruptions created dense ash clouds that settled onto the lake surface by rainfall.

The basin remained closed until 1789 when Nochistongo Cut first drained the lake water. The lake was further drained during the early 20th century and much of the city remains built on lacustrine sediments (Auvinet 2009). Figure 1 shows a cross-section through the basin illustrating the infilling with Mexico City clay (Marsal and Mazari 1959).

1.2 *The project site*

The Lake Texcoco region on the east of Mexico City was selected to build the New Mexico City International Airport (NAICM). The project site

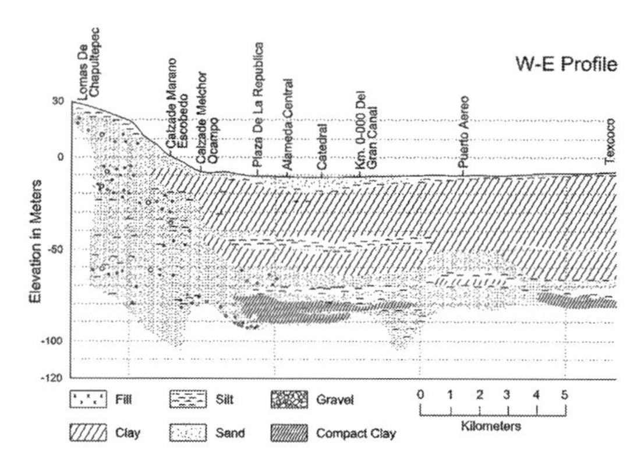

Figure 1. Generalized east-west geological profile. (note that elevation 0 is approximately +2240 m above sea level at 2015).

Table 1. Geology of Lake Texcoco (Marsal & Graue, 1969).

Geology of Lake Texcoco				
Depth (m)	Period	Epoch	Formation	Approximate Age (millions of years)
0–53		Holocene	Lacustrine deposits. Clay	0.000–0.008
53–59		Late Pleistocene		0.008–0.012
59–64		Late Pleistocene		0.012–0.013
64–180	Quaternary	Late Pleistocene		0.013–0.046
180–505		Middle Pleistocene to Late Pliocene	Tarango formation	0.046–0.800
505		Refractor A		
505–814		Middle/Early Pliocene		8.0–13.0
814–1030		Miocene	Huatepec	13.0–21.0
1030–1125		Early Miocene - Late Oligocene		21.0–24.0
1125–1437	Tertiary	Late Oligocene		24.0–29.0
1450		Refractor B		
1437–2065		Middle Oligocene - Middle Eocene	Balsas formation	29.0–(?)

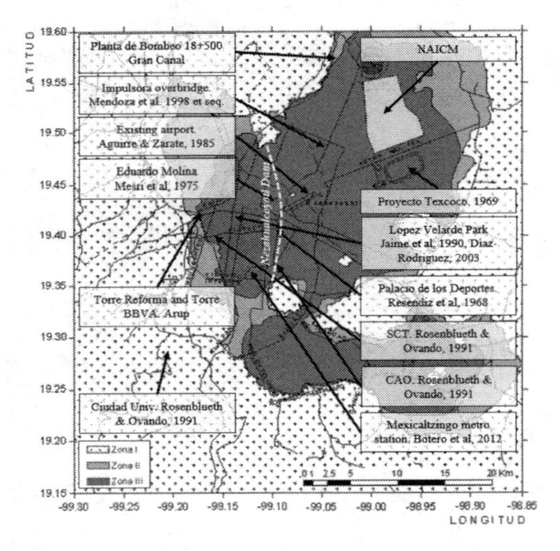

Figure 2. Approximate location of the new airport.

is located to the east of Nezahualcoyotl dam built by the Aztecs to separate freshwater lake from saltwater. Figure 2 shows the approximate location of the project in relation to the Aztec dam, seismic zonation and published case histories.

As the selected site on the saltwater side of the Lake Texcoco was largely undeveloped, the existing information on the in-situ ground conditions was limited. The Mexico City basin has on-going subsidence due to deep aquifer pumping. At the Airport site, the compressible aquifer from which groundwater is extracted is considered to be the strata between 120 m and 600 m depth. The site was also subjected to brine pumping from wells (~1 l/s yield) of between 30 and 60 m depth by Sosa Texcoco Co. (Santoyo Villa et al.2005) that took place between 1945 and 1995.

2 LAKE TEXCOCO GROUND CONDITIONS

Although the lakebed soils of downtown Mexico City have been studied extensively (Santoyo et al, 2005, Diaz-Rodriguez, 2003, Auvinet, 2002), the saltwater Lake Texcoco area was studied only under Proyecto Texcoco (Carillo, 1969) a volume that was produced as a memorial to Nabor Carillo, an important Mexican geotechnical engineer at UNAM.

The lake bed comprises interbedded clays, sands and volcanic glass deposits extending to a depth of about 200 m. The general geology of the area is summarized in Table 1.

The Holocene deposits comprise the softest clay unit Formación Arcillosa Superior (FAS) in the upper 25 to 30 m of the sequence, followed by 2 to 5 m thick dense sands/volcanic glass knows as Capa Dura (CD). The CD overlays lower clay units known as Formación Arcillosa Inferior (FAI), followed by alternate layers of sand and lower clay deposits.

All structures for NAICM are founded in the FAS unit. As the foundation design for the project is mainly governed by the physical and mechanical properties of the FAS layer, only this unit has been considered for detailed description in this paper.

The paper presents an estimation of geotechnical and geo-seismic parameters for the FAS layer. The selection and calibration of the parameters for the non-linear soil constitutive model is presented. The estimation of the regional subsidence carried out as a part of the calibration process is discussed. Forward predictions made for a trial excavation and comparison with monitoring results are also presented.

3 CHARACTERISATION OF FAS

3.1 *Impact of regional subsidence*

The ongoing regional subsidence due to deep aquifer pumping has a significant effect on the soil properties. In the Lake Texcoco area, the groundwater abstraction from a depth of below 120 m has caused continuous compression in the Formacion Tarango and above has led to regional subsidence as high as 35 cm/year (Carrillo, 1948). The other effect of pumping is drop in the piezometric profile to below hydrostatic in and around the pumping zone.

With the continual process of aquifer compaction and aquitard consolidation, the stiffness of the FAS is increasing over time and therefore the geotechnical properties are a 'moving target'. The effect is clearly shown in Figure 3, which presents the results of consolidation tests performed on Mexico City clays at the same location between 1950 and 2001 (Ovando, 2011).

3.2 *Ground investigations*

The ground investigation took place in several phases starting with pre-master plan investigations, followed by various campaigns carried out by Comisión Federal de Electricidad (CFE) and Ingeniería Experimental (IE).

The ground investigation for the site comprises of drilling boreholes to a depth of 50 to 75 m using rotary wash drilling fluid to minimise sample disturbance. Continuous sampling was performed in each boring and samples in the soft clays were recovered using the TGC thin walled tube sampler (Santoyo Villa 2010) 10 cm in diameter and 1 m in length split in 5 aluminium tubes of 1.2 mm thickness and 20 cm long. CPTu were performed to a depth of 50 m in accordance with ASTM D 5778-12.

Measurement of shear wave velocity with depth was performed using P-S suspension logging.

An extensive laboratory testing program included strength and stiffness from consolidation testing, consolidated-undrained (CU) tests, isotropically and K_0 consolidated triaxial tests (TXCU and K_0TXCU) following ASTM D4767. The specialist testing, including cyclic triaxial tests plus bender elements (ASTM D5311) and resonant column tests was carried out to establish behaviour under cyclic loading including shear modulus and strength degradation profile.

3.3 *Current index properties*

The FAS unit is characterized as silty clays with thin interbedded sand lenses. The clay fraction is typically between 50 to 60% and nearly half of those are smectite mineral. The presence of siliceous diatoms, typically 60 to 80% by weight, promotes an open structure giving unusual characteristics like high void ratio and ability to sustain high moisture contents. The pore water salinity is typically of the order of 45 g/L that has a flocculating effect. The variation of strength properties that result from variable diatom and salt contents are considered in later sections.

The total unit weight was found to vary between 11.5 and 12.5 kN/m^3. The measured specific gravity ranged between 2.2 and 3.0. The average moisture content of FAS is 200%, measurements lay mostly between 150% and 350%, which is close to the liquid limit. FAS layer presents some thin sandy layers with water contents below 150% as lower moisture contents are not regarded a representative of the FAS unit. Plastic limit values typically vary between 30 and 75%, but do not show a noticeable change with depth or soil units. The void ratio e_0 was found to reduce with depth and could be as high as about 12 near the top but varies typically between 8 and 4. An average of the measured index properties are summarised in Table 2.

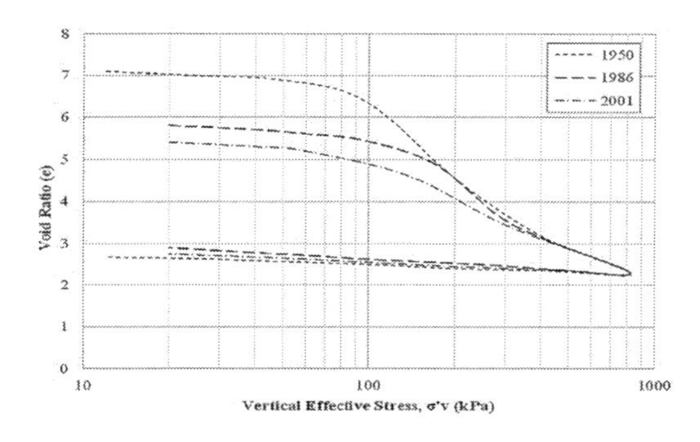

Figure 3. Consolidation tests performed at same site (Ovando, 2011).

Table 2. Typical FAS index properties.

Index properties						
Specific Gravity, Gs	Void Ratio, e	Moisture Content (%)	Liquid Limit, w_l (%)	Plastic Limit, w_p (%)	Plasticity Index, I_p (%)	Liquidity Index, I_L
2.7	6	200	212	60	154	177

3.4 *Compressibility and yield stress ratio*

Results of a large number of oedometer tests indicate that the virgin line compression index of FAS ranges between 4 and 10 with an average of 5.6.

Recompression index, Cr, varies mostly between 0.1 and 0.4. Cr/Cc ratio is generally greater than 10 and in many cases in excess of 20. The high recompression stiffness is also considered an attribute of the high diatom content because for non-microstructured clays, the ratio Cr/Cc is typically between 4 and 6 (Leroueil and Hight 2003). The very stiff response in the unload-reload state has proved to be very useful for the construction of compensated foundations for NAICM.

The coefficient of consolidation is derived from oedometer test at the reference stress level close to the in situ vertical effective stress plus 20 kPa. A typical consolidation coefficient of FAS is 24 m^2/year.

3.4.1 *Yield stress ratio*

The yield stress ratio (YSR) for natural clays is defined as:

$$YSR = \frac{\sigma'_{vy}}{\sigma'_{v0}}$$

where σ'_{vy} is the effective vertical yield stress and σ'_{v0} is the in situ effective vertical stress.

The oedometer test results indicate an average *YSR* equal to 1.8, decreasing to 1.0 about 20 mbgl due to the effects of underdrainage from Capa Dura. The larger YSR values in upper FAS are a combined effect of brine pumping and the rebound of piezometric profile, and thereafter, structuration due to the presence of diatoms, cementation and ageing effects.

3.5 *Strength and stiffness parameters*

3.5.1 *Effective strength*

Diatom content has a significant effect on the strength and stiffness properties of these clays. Figure 4 shows the normalized plot of the stress-strain path from a series of isotropic TXCU. The measured friction angle is comparable in magnitude to those in sands and it can be explained by the fact that FAS contains glass particles and diatom shell fragments generating friction angles well in excess of 40 degrees (Leroueil and Hight, 2003).

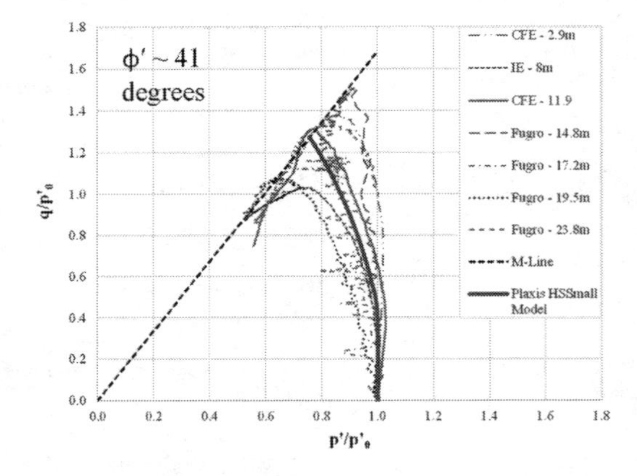

Figure 4. Typical normalized stress-strain path TXCU.

The friction angle determined from the triaxial CU testing ranges typically between 27 and 52 degrees. Díaz Rodríguez (2011) also shows for kaolin and diatoms mixtures that the effective friction angle increases with the diatom content.

3.5.2 *Undrained shear strength*

Obtaining a representative undrained strength from laboratory testing is quite challenging due to difficulties in obtaining an undisturbed sample and being highly sensitive to the rate of shearing. The measured peak undrained strength could vary by as much as 15% for a tenfold increase in the strain rate (Díaz Rodríguez et al. 2009). The stress-strain behaviour of the soil also appears to vary with diatom and salt content, which is also reflected in the wide bandwidth obtained from CPTu testing, refer Figure 12. The variable structure of the soil, sometimes brittle, but often not, is seen in the appearance of peak shear strengths within an axial strain of 5% in triaxial tests. Salt and diatom content cannot be measured accurately and so, to avoid progressive failure and shear strain concentration effects in design, lower bound strength and stiffness parameters were derived.

The CPTu tests were correlated with the PS logging by use of a modified form of a derivation by Ovando & Romo (1991):

$$v_s = \xi\left(\frac{q_t}{\gamma_s}\right)^{0.5}$$

where v_s is the shear wave velocity, *qt* is the pore water pressure corrected cone resistance, ξ is an empirical parameter and γ_s is the soil unit weight.

A strong correlation was found between the measured and back-calculated v_s profile. Therefore, it is considered that the CPTu can describes the FAS adequately and could be used to properly determine other parameters such as Su. The Su profile was obtained using correlation Su = q_t/13.2 (Ovando, 2011), coupled with, generally, the lower values of q_t.

The CPTu, and laboratory testing show that the FAS display stiffness and strength characteristics that generally increase linearly with depth and thus can be described in terms of the current vertical effective stress (σ'_v). The precision with which that variation with current vertical effective stress can be characterized is however compromised by the variable presence of diatoms within these soils. Díaz Rodríguez (2011, 2017) shows the sensitivity to diatom content of stiffness and normalized strength behaviour in remoulded kaolinite-diatom mixtures. Rangel Núñez et al (2015) shows that the presence of calcium montmorillonite and diatoms in Lake Texcoco clays had a much stronger control on stiffness and strength than, for example, water content and void ratio.

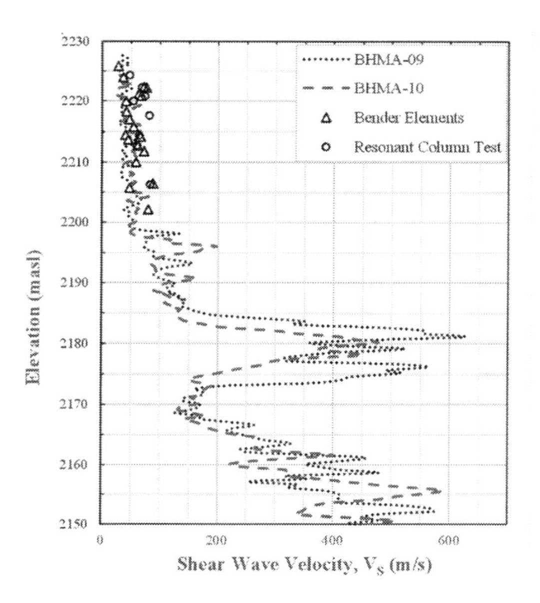

Figure 5. PS suspension logging, benders element and resonant column test results.

A lower bound undrained shear strength of the upper FAS can be conveniently described as Su ≈ 7 + 1.05z kPa, where z is the depth in meter up to 20 m depth. Using this linear approximation, a ratio of Su/σ'_v = 0.7 at a YSR of about 1.8 is obtained.

3.6 *Stiffness*

The large strain secant moduli and volumetric stiffness were derived using laboratory TXCU and Oedometer test results. Using a range of trial test data, the average compression stiffness for FAS was found to be 300 kPa for a reference stress of 100 kPa.

The small strain modulus (Gmax) was obtained using the shear wave velocity measured from PS suspension logging and bender elements tests (Figure 5). A best-fit trend of Gmax = 0.5+0.15z MPa was adopted for FAS. Based on this value the ratio Gmax/Su is found to vary from 70 to 150 within the upper 20 m of FAS. The decay of shear modulus with increasing shear strain is obtained from resonant column testing, modified for monotonic undrained strength at large strains. The behaviour of FAS was found to be elastic for large strains up to 0.1%, which means it can sustain large cyclic loading without much degradation. This also means unusually low hysteretic damping even when subject to high shear stress.

3.7 *Artificial bentonite-diatom mix testing*

In order to gain more understanding on the influence of diatoms and salt content on FAS properties, tests have been made using artificial mixtures of bentonite, diatoms and salt. Figure 6 illustrates the flocculating effect of salt and shows how strength (using PI as a proxy for strength) can vary widely at given water content and void ratio,

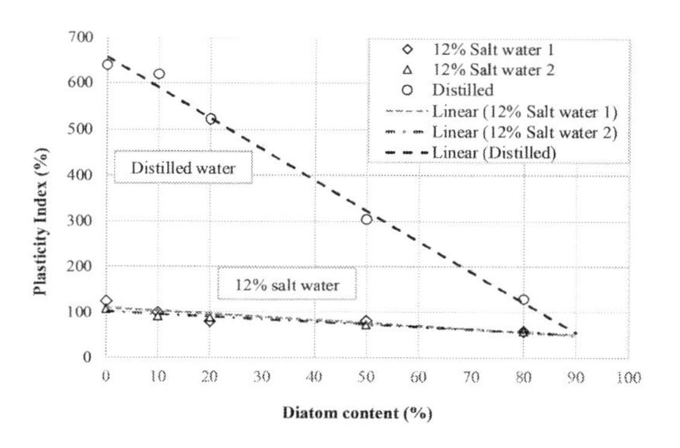

Figure 6. Plasticity index variation of diatom/bentonite mixtures.

Table 3. Plaxis ground model input parameters.

HSSmall model properties

Soil unit	Layer thick.	Unit weight γ_t (kN/m³)	Coeff of permeability k_v (m/s)	E_{oed}^{ref} (kN/m²)	E_{ur}^{ref} (kN/m²)	m	c^{ref} (kN/m²)	Friction Angle (deg)	γ_{70}	G_0^{ref} (kN/m²)
1 Crust	1.5	14	1.00E-07	10,000	10,000	–	10	30	–	–
2 FAS	28.5	12	4.00E-09	300	6,000	0.8	0.1	41	1.00E-03	1.25E+04
3 CD	1.5	17	5.90E-05	1,250	6,250	0.8	0.1	38	1.00E-04	2.60E+04
4 FAI	11.5	13.1	2.00E-09	375	3,000	1	0.1	36	1.00E-03	1.25E+04
5 SES	11	14	1.00E-06	1,000	5,000	1	0.1	35	1.00E-03	2.05E+04
6 FAP	13	13.5	4.00E-09	800	5,000	1	0.1	35	1.00E-03	2.00E+04
7 SEI	23	16	1.00E-06	1,850	10,000	1	0.1	35	1.00E-03	4.00E+04
8 FAP 2	15	13.5	4.00E-09	900	5,450	1	0.1	35	1.00E-03	2.25E+04
9 SEI 2	15	16	1.00E-06	2,000	10,000	1	0.1	35	1.00E-03	4.10E+04

depending upon the proportions of diatom, bentonite and salt content.

It is interesting to note that excess pore pressure generated during undrained shearing for both monotonic and cyclic loading reduces with diatom content.

Figure 7 shows a pronounced increase in stiffness for both primary loading and unload-reload with diatom content.

4 NUMERICAL ANALYSIS

Table 3 shows the HSsmall parameters used for various Plaxis 2D and 3D analyses carried out for the foundation design and assessment of short and long-term behaviour in terms of total and differential settlements.

The input parameters were developed based on extensive calibrations carried out using TXCU stress path, oedometer test results, CPTu interpreted shear strength and PS suspension logging test results. The stiffness parameters were further adjusted using back-analysis of the regional subsidence, together with brine pumping and historical field trials. A lower bound stiffness was adopted to account for the creep effects to some extent, as HSS model could not consider it explicitly.

A comparison of the HSsmall model predictions against laboratory TXCU and oedometer tests is presented in Figure 8 and Figure 9. As shown, the HSS model is able to mimic the stress-strain and 1D compression/unload-reload behaviour reasonably well.

As mentioned earlier, the model parameters were also calibrated by comparing computed settlements in the site for a period between 1930 and 2015 with

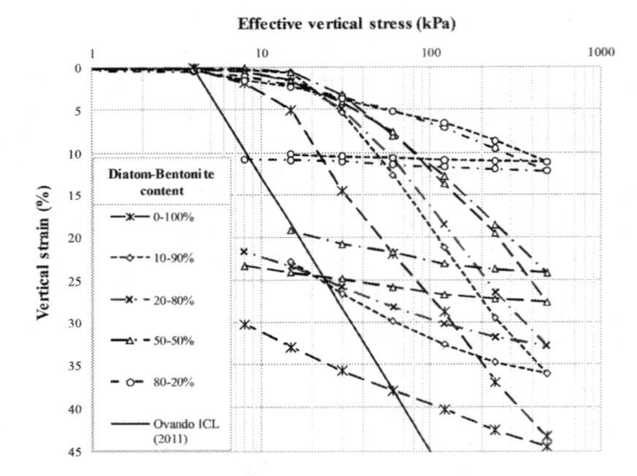

Figure 7. Variation of oedometer behaviour: diatom-bentonite mixtures compared to the intrinsic compression line (ICL) of the native soils suggested by Ovando (2011).

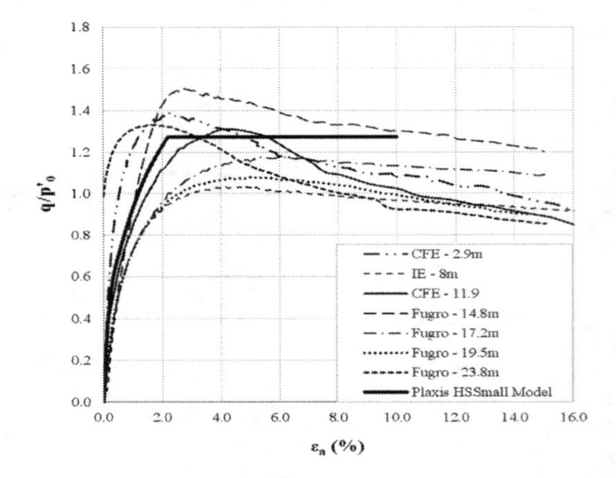

Figure 8. Normalized stress path TXCU.

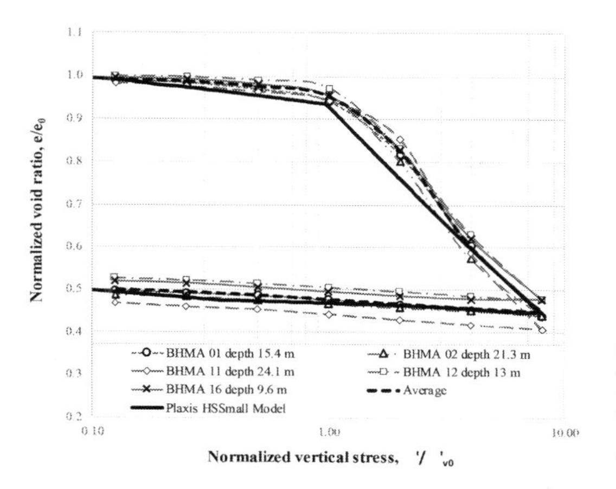

Figure 9. Normalized oedometer tests.

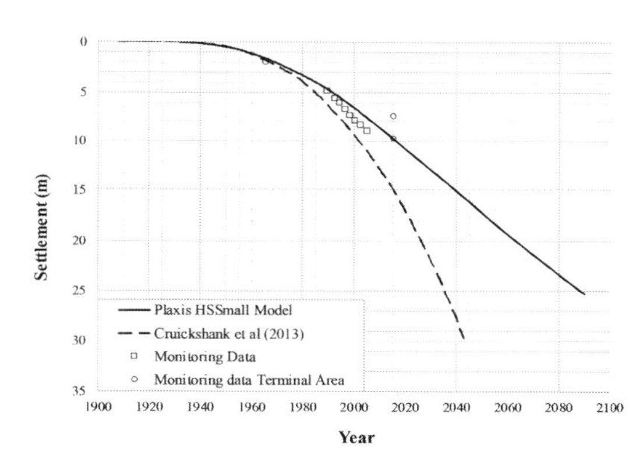

Figure 10. Comparison of measured and predicted settlements.

recorded regional subsidence in the surrounding area as shown in Figure 10 (II UNAM 2015; Morales, Murillo Fernández, and Hernández Rubio 1991). The settlement predicted for year 2015 were found to be in good agreement with the measurements. Other successful calibrations carried out includes back analysis of a trial excavation carried out in 1968 (Comité Técnico Proyecto Texcoco 1969b; Hiriart and Graue 1969) and pumping trials to induce surface subsidence (Hiriart and Marsal 1969).

4.1 *Stress-history of site*

The airport site stress history was simulated by considering the brine pumping between 1945 and 1995 by Sosa Texcoco, in which wells were installed across the site at approximately 200 m centres to depths between 30 and 60 m and pumped at 1 l/s (Santoyo et al, 2005). The evolution of pore water pressure starting from year

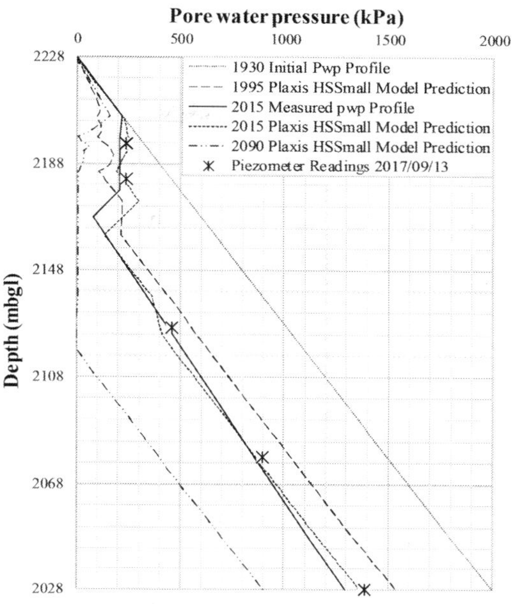

Figure 11. Evolution of water pressure.

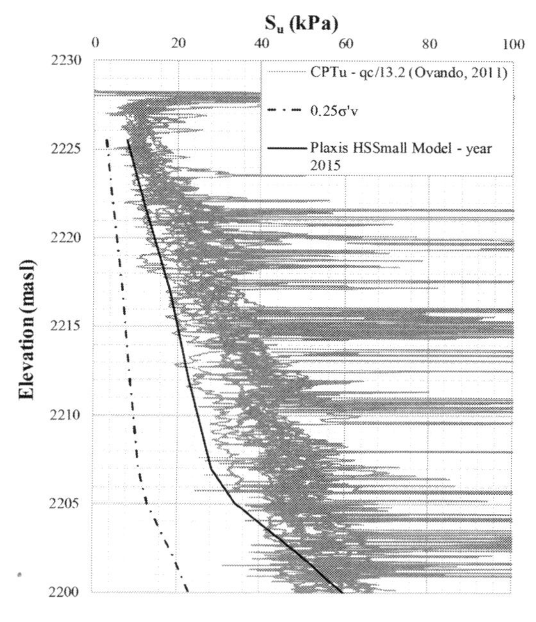

Figure 12. Undrained shear strength profile.

1930 (start of deep aquifer pumping) was carried out by including shallow brine pumping in FAS (Figure 11).

The water pressure recovery in top 20 m FAS post decommission of brine pumping was considered and prediction on future water pressure profile was made. A good match was found between the predictions and measured piezometric profile for both shallow and deeps measurements. This process allowed checking of the current state of in-situ stress, strength and stiffness properties using adopted soil parameters.

A comparison of the current undrained shear strength profile predicted by the model for the current state of stress and CPTu interpreted profile is shown in Figure 12

The projection on the future settlement was also made (Figure 10) and was compared with the predictions made by Cruickshank et al (2013) using linear elastic model.

5 TRIAL EXCAVATION

5.1 *Purpose*

As compensated foundation solutions were to be used for all main structures, it was imperative to understand the behaviour of a large excavation in these difficult ground conditions. Groundwater controls, excavation duration and behaviour of two adjacent cast raft elements were also studied in the trial.

5.2 *Trial excavation details*

A 96 m × 96 m × 6 m deep fully instrumented trial excavation was proposed as shown in Figure 13. Side slopes were 1 in 4 and eductor wells were installed to control the groundwater ingress from sand lenses and to avoid base instability.

5.3 *Finite element analysis*

The FE analyses of the trial excavation were undertaken using Plaxis 2D (version 2016) software. The 2D Plaxis model was 300 wide and almost 120 m deep to minimise boundary effects. The mesh and geometrical arrangement from a typical model is shown in Figure 14.

The stratigraphy and soil parameters adopted in the model are presented in Table 3. The stress initilisation in the model was done using K_0 method with the direct input of OCR to obtain the current state of stress.

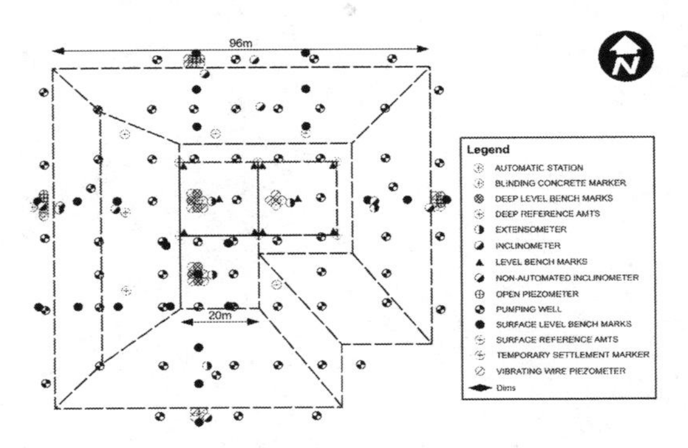

Figure 13. Trial excavation plan with monitoring instruments.

5.4 *Comparison of FE predictions with monitoring results*

The measured excavation-induced pore pressure and lateral displacements were found to be in a close agreement with the predictions. Figure 15 and Figure 16 show a comparison of lateral movements at the crest of the slope and excavation induced heave at the bottom of the excavation.

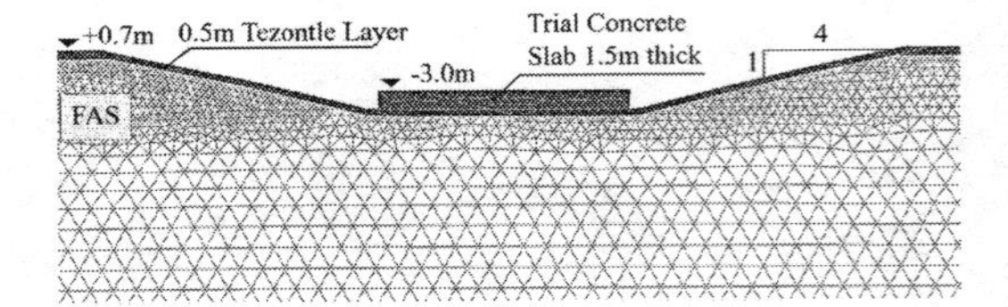

Figure 14. FE model of the trial excavation.

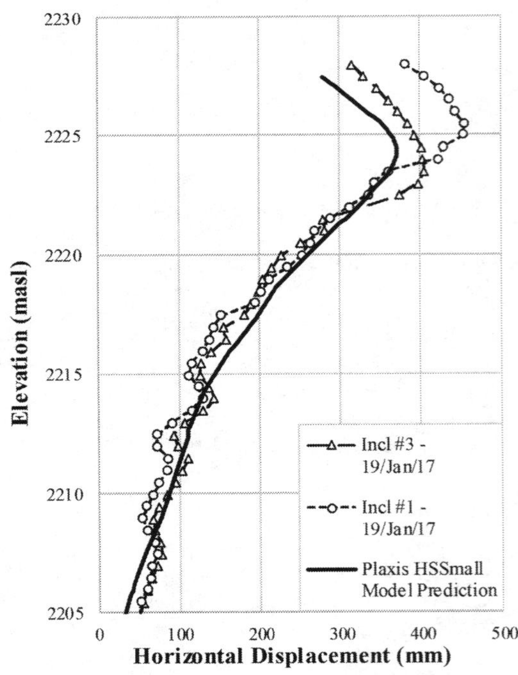

Figure 15. Comparison of lateral movements.

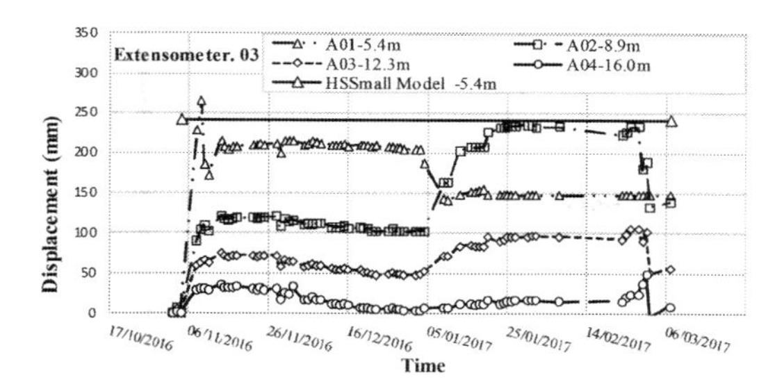

Figure 16. Comparison of excavation heave.

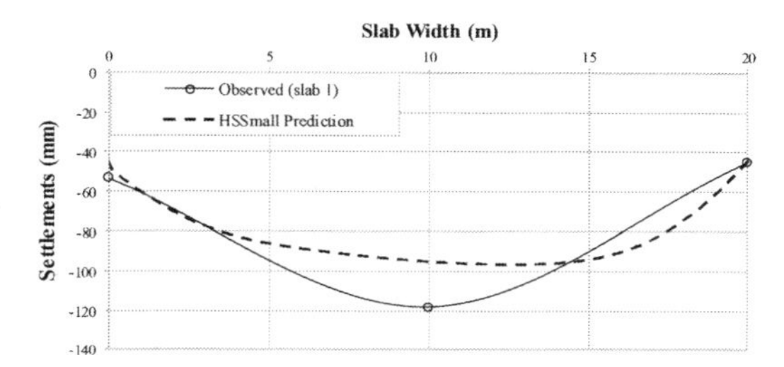

Figure 17. Comparison of settlements under the raft weight.

A comparison of undrained distortion due to reloading is shown in Figure 17.

The pore pressure measurement confirmed the soils to behave as undrained with total stress change reflected closely in the pore pressure changes with both excavation and reloading. No significant increase in pore pressure was noted for a period of more than a month between the final excavation and the raft construction.

6 CONCLUSIONS

The development of advanced numerical models has enabled the difficult, very soft clays of Lake Texcoco to be placed into a predictive framework for numerical analysis and design of NAICM. This has been achieved by a careful and extensive investigation of the site's stress history, the mineralogy and salinity of the soils, behaviour in laboratory tests and field trials. Tests on artificial soil mixtures have helped to understand the natural variability of the FAS deposit whose mechanical properties depend strongly upon salt, diatom and smectite mineral content.

ACKNOWLEDGMENTS

We thank GACM. FR+EE and Foster & Partners for permission to publish some of the results of ground investigations for NAICM.

REFERENCES

Auvinet G & Juarez M (2009). Geotechnical engineering inurban areas affected by land subsidence. ISSMGE36 TC36#

Botero E, Ovando E, Ossa A, Grialdo V & Sierra L (2012) Model for the determination of seismic interaction between tunnel and underground station.*15th World Conf Earthquake Eng, Lisboa*. 10pp.

Carrillo N (1969) El Hundimiento de la ciudad de México y Proyecto Texcoco. *Memorial volume*. SHCP 340 pp.

CONAGUA (2015) Well construction details, private communication.

Cruickshank et al (2013) Cruickshank, C. González, F. Palma, A. (2013) Estudios para la primera etapa de la actualización del modelo numérico del acuífero del Valle de México, CONAGUA-UNAM, Mexico.

Diaz-Rodriguez J A (2003) Characterization and engineering properties of Mexico City lacustrine soils. In Vol 1 of *Characterization and engineering properties of Natural Soils* (eds Tan et al) Swets & Zeitlinger, Lisse, pp 725 to 755.

Diaz-RodriguezJ A(2011)Diatomaceoussoils:monotonic behavior. *Proc. Int. Symp Deformation Characteristics of Geomaterials*. Seoul. 8pp.

Diaz-Rodriguez J A & Santamarina J C (2009) Strain-rate effects in Mexico City soil. *Jnl Geotechnical and Geoenvironmental Engineering* Vol 135 No 2,Tech Note. ASCE.

Fugro (2016) Informe de pruebas de laboratorio en muestras de suelo en el área destinada al nuevo Aeropuerto de la Ciudad de México.

González Blandón, C.M., & Romo Organista, M.P. (2011). Estimación de propiedades dinámicas de arcillas. *Ingeniería sísmica*, (84), 1–23.

Ingeniería Experimental (2015) Informe de investigación Geotécnica del subsuelo del Ex-Lago de Texcoco para el Nuevo Aeropuerto Internacional de la Ciudad de México.

Leroueil S and Hight D (2003) Behaviour and properties of natural soils and soft rocks. *Characterisation of natural soils and soft rocks* Tan et al (eds.) Swets&Zeitlinger pp 29 to 254.

Jaime A, Romo M & Resendiz D (1990) Behavior of friction piles in Mexico City Clay. *Jnl Geotechnical Eng* Vol 116 No 6 pp 915 to 931.

Marsal (1969) Development of a lake by pumping-induced consolidation of soft clays. In Carillo(1969) pp229 to 267.

Mendoza M, Romo M, Orozco M & Dominguez L (1998 to 2000) Static and seismic behavior of a friction pile-box foundation in Mexico City Clay. *Soils and Foundations* Vol 40 No4 pp 143 to 154.

Mesri G, Rokhsar A & Bohor B (1975) Composition and compressibility of typical samples of Mexico City clay. *Geotechnique* Vol 25 No 3 pp 527 to 554.

Ovando E & Romo M (1991) Estimacion de la velocidad de ondas S en la arcilla de la ciudad de Mexico con ansayos de cono. Revista Sismodinamica, 2, pp 107–123.

Ovando E. (2011). Some geotechnical properties to characterize Mexico City clay. *Pan-Am CGS Geotechnical Conference*, October 2011, Toronto, Canada.

Resendiz D, Auvinet G & Silva C (1968) Conception des fondations du Palais des Sports de la Ville de Mexico en presence de frottement negatif. Specialty session at 7th ICSMFE Mexico City, 20 pp.

Santoyo E S, Ovando-Shelley E, Mooser F & Plata E L (2007) Sintesis Geotecnica del la Cuenca del Valle de Mexico. TGC Geotecnica de CV. Mexico DF.

UNAM (2015) Investigaciones y estudios especiales relacionados con aspectos geotécnicos del nuevo aeropuerto de la Ciudad de México, en el vaso del ex Lago de Texcoco, Zona Federal. Nota Técnica Nº GEO 12 Informe final. Ciudad Universitaria DF.

Vardanega P J & Bolton M D (2013) Stiffness of clays and silts: Normalizing shear modulus and shear strain. Jnl of Geotechnical and Geoenvironmental Engineering, 139 (9), 1575–1589.

Numerical Methods in Geotechnical Engineering IX – Cardoso et al. (Eds)
© 2018 Taylor & Francis Group, London, ISBN 978-1-138-33198-3

A comparison of the series and parallel Masing-Iwan model in 2D

W.J.A.P. Beuckelaers, G.T. Houlsby & H.J. Burd
Department of Engineering Science, University of Oxford, UK

ABSTRACT: For a material or structural model, which satisfies the extended Masing rules under cyclic loading, a unique parallel or series Iwan-model exists which characterises the stress-strain response. Both the parallel and series version of the model result in the same stress-strain response when loaded in one dimension. In numerical implementation, the series approach is particularly suited for a stress controlled algorithm since the same stress applies to each of the spring-slider elements. For the parallel approach, the same strain applies to each of the spring-slider elements, making it suitable for a strain controlled algorithm. The hyperplasticity approach provides a rigorous framework to extend both the series and parallel version to higher dimensions and derive the incremental response. Though both models can be calibrated to simulate the same response in 1D, the extensions to 2D illustrate that there is a small difference between the stress-strain behaviour of these models in higher dimensions. This paper illustrates the similarities and differences between the behaviour of the series and parallel Iwan-models.

1 INTRODUCTION

Masing type hysteretic behaviour has proven to be a realistic approximation for the cyclic response of structures and materials in many applications (Houlsby and Puzrin 2006). The method originates from the modelling of brass behaviour (Masing 1926) and has been adopted in various geotechnical applications including: macro-element modelling for suction buckets (Byrne and Houlsby 2004), integration into a Winkler model for monopile foundations (Beuckelaers 2015, Beuckelaers 2017) and integrated into the constitutive behaviour for 3D finite element analyses (Houlsby and Mortara 2004).

Masing type hysteretic behaviour is often referred to as a kinematic hardening material model, since it is modelled using yield surfaces which translate but do not expand or contract on yielding. The original publication by Masing (1926) suggests that a kinematic hardening material behaves according to the following rules:

1. The tangent stiffness at the start of each loading reversal assumes a value equal to the initial tangent stiffness for the initial loading curve.
2. The shape of the unloading and reloading curves is the same as that of the initial loading curve, except that the scales of both the stress and strain axes are enlarged by a factor of two.

As discussed by Byrne & Houlsby (2004), the first rule follows directly from the second. The stiffness upon unloading is defined by the initial stiffness of the loading curve (usually called the backbone curve) resulting in the same initial tangent stiffness. To define fully the material behaviour for cyclic loading, Pyke (1979) suggested the following additional rules, known as the extended Masing rules:

3. The unloading and reloading curves should follow the initial loading curve (backbone curve) if the previous maximum shear strain is exceeded.
4. If the current loading or unloading curve intersects the curve described by a previous loading or unloading curve, the stressâ€“strain relationship follows the previous curve.

The material behaviour in 1D loading is fully defined from the backbone curve using the four rules above. However, the extension to multiple dimensions is not fully defined. This paper identifies the similarities and differences between two common implementations for the kinematic hardening model: the series and the parallel approach. The hyperplasticity framework allows for a straightforward extension of these models to multiple dimensions. This approach is used to compare the models in 2D.

2 SERIES AND PARALLEL IWAN MODELS

Series and parallel versions of the kinematic hardening model are illustrated in Figure 1 and Figure 2. Both models use a number of linear elastic springs with stiffnesses H_i and slider elements with strengths k_i and associated plastic strains α_i. These

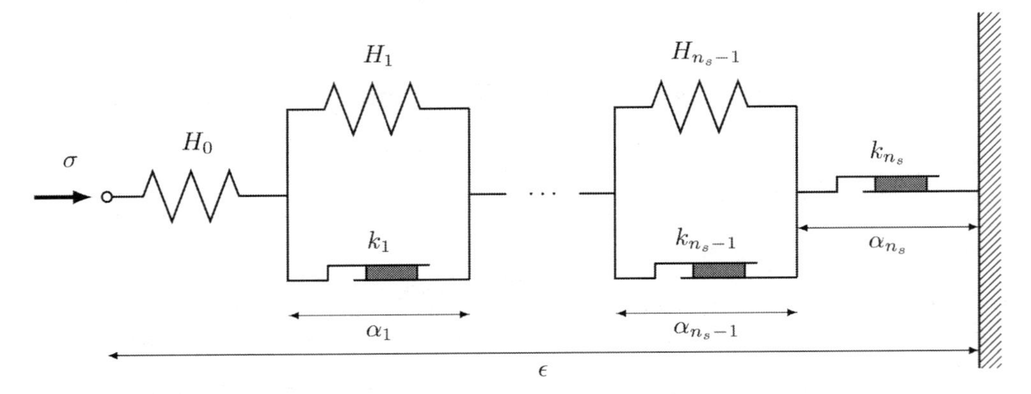

Figure 1. Illustration of the series Iwan model.

models were originally developed by Iwan (1967), who also introduced the extensions of these models to multiple dimensions. For these higher dimension cases, the behaviour was not examined in furtherdetail.

The series and parallel Iwan model have been formulated in the hyperplasticity framework by Houlsby & Puzrin (2000) and Einav (2004). Models derived in the hyperplasticity framework are fully defined using a combination of an energy potential and a dissipative potential. The energy potential can either be the internal energy (u), the Helmholtz free energy (f), the enthalpy (h) or the Gibbs free energy (g). These functions are related by Legendre-Fenchel transforms, so that models developed with any of these functions are equivalent. For rate independent models, the dissipative potential can either be the dissipation potential (d) or the yield function (y), preferably but not necessarily in canonical form $(\bar{y})$. The interested reader may find an overview of these formulations in Houlsby & Puzrin (2006).

The Gibbs free energy, yield functions (not in canonical form) and dissipation potential for the series and parallel Iwan models are compared in Table 1. This table shows that the yield functions and the dissipation potentials for both models take the same form, although importantly it should be noted that the physical interpretation of α_i and the numerical values of k_i and H_i differ between the two models, as is clear from comparing Figure 1 to Figure 2. In the series model, the same stress applies to each of the spring-slider elemenets; whereas in the parallel model, the same strain applies to each of the spring-slider elements.

The parameters k_i and H_i are uniquely defined for a given backbone curve (Figure 3). The equations to calculate these parameters are given in Table 2.

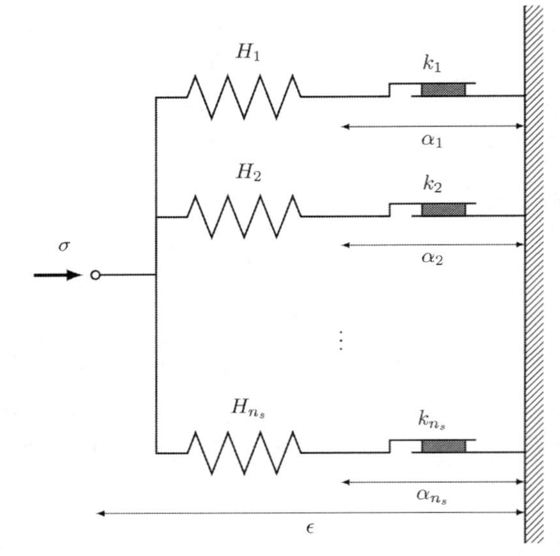

Figure 2. Illustration of the parallel Iwan model.

From the Gibbs free energy, the strain can be derived: $\varepsilon = -\partial g / \partial \sigma$. The plastic behaviour of both models can be derived using the generalised stress $\bar{\mathcal{X}}$ and the dissipative generalised stress $\mathcal{X}$:

$$\bar{\mathcal{X}}_i = -\frac{\partial g}{\partial \alpha_i} \qquad \mathcal{X}_i = \frac{\partial d}{\partial \dot{\alpha}_i} \tag{1}$$

The second law of thermodynamics implies that: $\sum(\bar{\mathcal{X}}_i - \mathcal{X}_i)\alpha_i = 0$. The hyperplasticity framework presented in Houlsby & Puzrin (2006) makes use of a stronger assumption:

$$\mathcal{X}_i = \bar{\mathcal{X}}_i \tag{2}$$

Table 1. Potential functions for the series and parallel Iwan model.

Series Iwan model

$$g = -\frac{\sigma^2}{2H_0} - \sigma\sum_{i=1}^{n_s}\alpha_i + \sum_{i=1}^{n_s-1}\frac{H_i}{2}\alpha_i^2$$
$$y_i = |\mathcal{X}_i| - k_i$$
$$d = \sum_{i=1}^{n_s} k_i |\dot{\alpha}_i|$$

Parallel Iwan model

$$g = -\frac{\left(\sigma + \sum_{i=1}^{n_s} H_i\alpha_i\right)^2}{2\sum_{i=1}^{n_s} H_i} + \sum_{i=1}^{n_s}\frac{H_i}{2}\alpha_i^2$$
$$y_i = |\mathcal{X}_i| - k_i$$
$$d = \sum_{i=1}^{n_s} k_i |\dot{\alpha}_i|$$

Table 2. Formula for the calculation of H_i and k_i based on the backbone curve.

Series Iwan model

$$\sigma_i = k_i, i = 1 \ldots n_s$$
$$\epsilon_i = \sum_{j=0}^{i-1}\frac{k_i - k_j}{H_j}, i = 1 \ldots n_s \text{ with } k_0 = 0$$
$$\frac{1}{E_i} = \sum_{j=0}^{i}\frac{1}{H_j},\ i = 0 \ldots n_s - 1$$

Parallel Iwan model

$$\sigma_i = \sum_{j=1}^{i-1} k_j + \sum_{j=i+1}^{n_s} H_j \epsilon_i, i = 1 \ldots n_s$$
$$\epsilon_i = \frac{k_i}{H_i}, i = 1 \ldots n_s$$
$$E_i = \sum_{j=i+1}^{n_s} H_j, i = 0 \ldots n_s - 1$$

which is also known as Ziegler's orthogonality principle (Ziegler 1983). The strain and the generalised stresses for the series and parallel versions of the Iwan model are given in Table 3, where $S(\cdot)$ is the generalised signum function $\left(S(x>0)=1, S(x<0)=-1 \text{ and } S(0)\in[-1,1]\right)$.

The incremental response of both models follows from a standardised approach (Houlsby and

Table 3. Strain and generalised stresses for the series and parallel Iwan model.

Series Iwan model

$$\epsilon = \frac{\sigma}{H_0} + \sum_{i=1}^{n_s}\alpha_i$$
$$\bar{\mathcal{X}}_i = \sigma - H_i\alpha_i$$
$$\mathcal{X}_i = S(\dot{\alpha}_i)k_i$$

Parallel Iwan model

$$\epsilon = \frac{\sigma + \sum_{i=1}^{n_s} H_i\alpha_i}{E_0}$$
$$\bar{\mathcal{X}}_i = H_i\left(\frac{\sigma + \sum_{j=1}^{n_s} H_j\alpha_j}{E_0} - \alpha_i\right) = H_i(\epsilon - \alpha_i)$$
$$\mathcal{X}_i = S(\dot{\alpha}_i)k_i$$

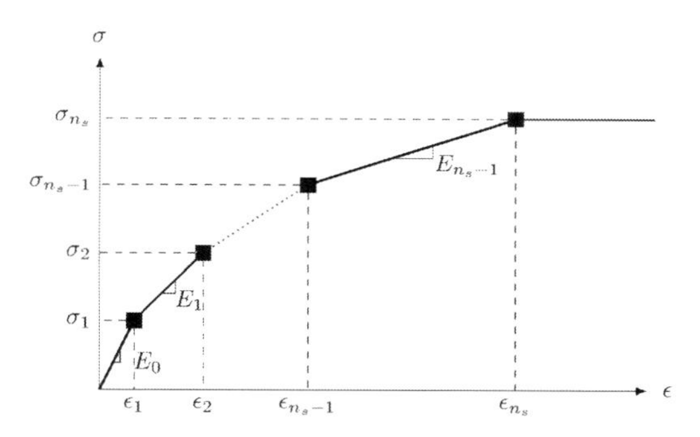

Figure 3. Discretisation of the backbone curve for the calculation of the strength k_i and stiffness H_i parameters.

Puzrin 2006), which can be written in the following form:

$$\begin{Bmatrix} -\dot{\epsilon} \\ -\dot{\bar{\mathcal{X}}}_i \end{Bmatrix} = \begin{bmatrix} \dfrac{\partial^2 g}{\partial\sigma^2} & \dfrac{\partial^2 g}{\partial\sigma\partial\alpha_j} \\ \dfrac{\partial^2 g}{\partial\alpha_i\partial\sigma} & \dfrac{\partial^2 g}{\partial\alpha_i\partial\alpha_j} \end{bmatrix} \begin{Bmatrix} \dot{\sigma} \\ \dot{\alpha}_j \end{Bmatrix} \tag{3}$$

When using the yield functions y_i rather than the dissipation potential, the explicit stress-strain response can be obtained as follows. The rate of hardening variables are defined by:

$$\dot{\alpha}_j = \sum_{i=1}^{n_s}\lambda_i\frac{\partial y_i}{\partial\mathcal{X}_j} \tag{4}$$

where the multipliers λ_i are derived from the consistency condition, which is obtained from differentiating the yield function:

$$\dot{y}_i = \frac{\partial y_i}{\partial \sigma}\dot{\sigma} + \sum_{j=1}^{n_s}\frac{\partial y_i}{\partial \alpha_j}\dot{\alpha}_j + \sum_{j=1}^{n_s}\frac{\partial y_i}{\partial \chi_j}\dot{\chi}_j = 0 \tag{5}$$

Combining these equations with the orthogonally condition, and substituting $\frac{\partial y_i}{\partial \alpha_j} = 0$, $\frac{\partial y_i}{\partial \sigma} = 0$ and $\left(\frac{\partial y_i}{\partial \chi_j}\right)_{i \neq j} = 0$ (which holds for both models), gives:

$$0 = \dot{y}_i = \frac{\partial y_i}{\partial \chi_i}\left(\frac{\partial^2 g}{\partial \alpha_i \partial \sigma}\dot{\sigma} + \sum_{j=1}^{n_s}\frac{\partial^2 g}{\partial \alpha_i \partial \alpha_j}\lambda_j\frac{\partial y_j}{\partial \chi_j}\right) \tag{6}$$

The multipliers λ_j can be determined by solving the above set of equations.

The hyperplasticity approach allows a straightforward extension of the potential functions to multiple dimensions. The extensions to two dimensions are given in Table 4.

The model extensions to higher dimensions can be used for e.g. macro-element models with multi degree-of-freedom loading, 3D constitutive models or soil reaction models for Winkler foundations in 2D loading. The incremental response of the extensions can be derived using the procedure described above.

3 COMPARISON OF THE BEHAVIOUR FOR THE TWO MODELS

The stiffness and strength parameters of the series and parallel Iwan models are calibrated to a a conic backbone curve:

$$-n\left(\frac{\sigma}{\sigma_u} - \frac{\epsilon}{\epsilon_u}\right)^2 + (1-n)\left(\frac{\sigma}{\sigma_u} - \frac{\epsilon E_{ini}}{\epsilon_u}\right)\left(\frac{\sigma}{\sigma_u} - 1\right) = 0 \tag{7}$$

with a curvature parameter $n = 0.5$, initial stiffness $E_{ini} = 1$, ultimate stress $\sigma_u = 1$ and ultimate strain $\epsilon_u = 10$, using a discretisation with $n_s = 20$. This formulation has been used *e.g.* to define the soil reaction curves for an extended Winkler model (Byrne et al. 2017). For generality, the behaviour is presented using non-dimensional variables. The dimensions for a specific model depend on the application, for example in a macro-element foundation model, the stress and strain variables

Table 4. Potential functions for the series and parallel Iwan models in 2D.

Series Iwan model
$g = -\frac{\sigma_x^2 + \sigma_y^2}{2H_0} - \sigma_x\sum_{i=1}^{n_s}\alpha_{ix} - \sigma_y\sum_{i=1}^{n_s}\alpha_{iy} + \sum_{i=1}^{n_s-1}\frac{H_i}{2}\left(\alpha_{ix}^2 + \alpha_{iy}^2\right)$ $y_i = \sqrt{\chi_{ix}^2 - \chi_{iy}^2} - k_i$ $d = \sum_{i=1}^{n_s} k_i\sqrt{\dot{\alpha}_x^2 + \dot{\alpha}_y^2}$
Parallel Iwan model
$g = -\frac{\left(\sigma_x + \sum_{i=1}^{n_s} H_i\alpha_{ix}\right)^2}{2E_0} - \frac{\left(\sigma_y + \sum_{i=1}^{n_s} H_i\alpha_{iy}\right)^2}{2E_0} + \sum_{i=1}^{n_s}\frac{H_i}{2}\left(\alpha_{ix}^2 + \alpha_{iy}^2\right)$ $y_i = \sqrt{\chi_{ix}^2 - \chi_{iy}^2} - k_i$ $d = \sum_{i=1}^{n_s} k_i\sqrt{\dot{\alpha}_x^2 + \dot{\alpha}_y^2}$

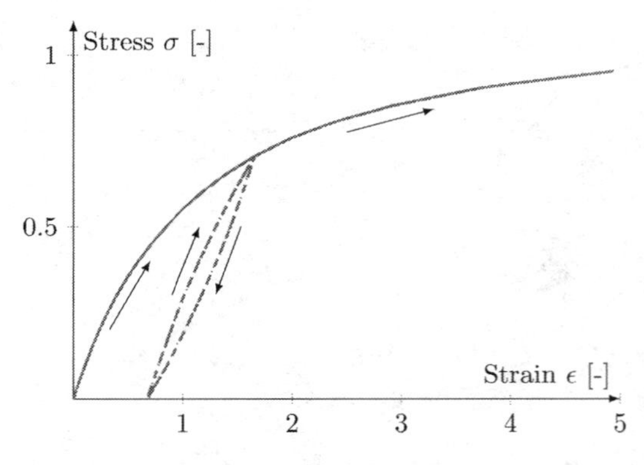

Figure 4. Backbone curve (grey) and the simulated response in 1D using the series (blue dashed) and the parallel (red dashdotted) Iwan model.

could represent the load and displacement variables of the foundation response. The backbone curve is illustrated in Figure 4 along with the response using both models for a 1D stress path of $\sigma = 0 \to 0.7 \to 0 \to 0.9$. This figure illustrates that the extended Masing rules are satisfied. As expected, the two models result in an identical response throughout the stress path.

The 2D extensions of the Iwan models are used to simulate the response under a rectangular spiralling stress path and a cross shaped stress path. The responses shown in Figure 5 and 6, illustrate

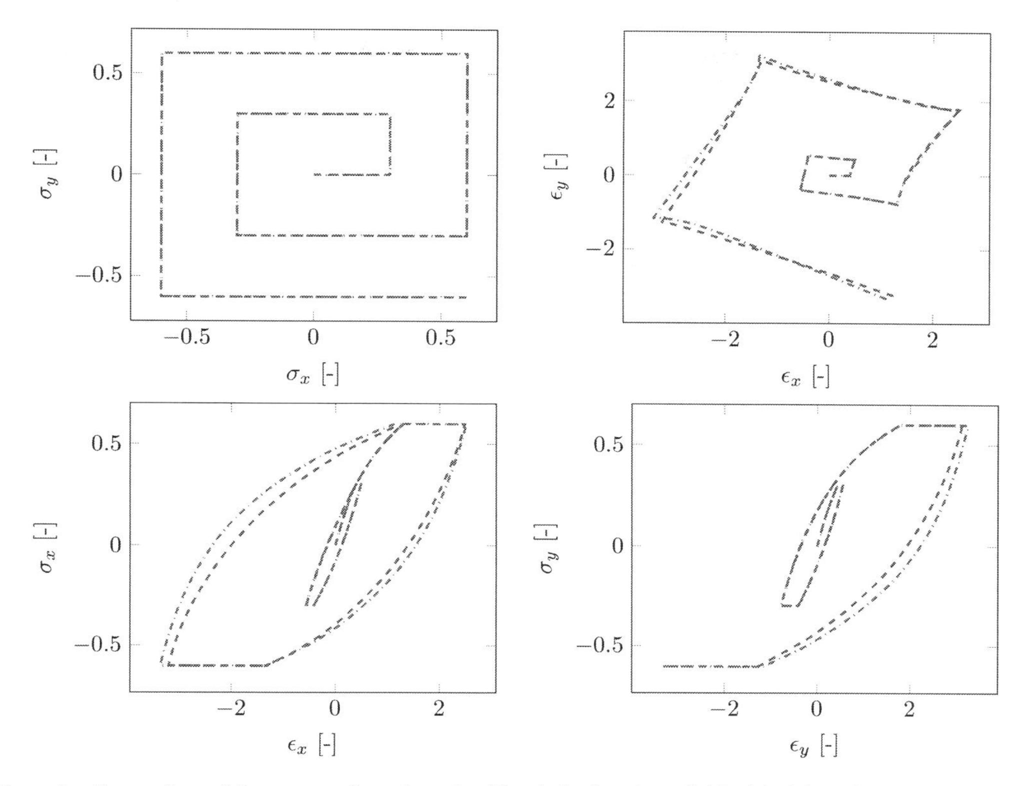

Figure 5. Comparison of the response from the series (blue dashed) and parallel (red dashdotted) Iwan models under a rectangular spiralling 2D loading.

the difference between the behaviour of the two models when a 2D stress path is applied.

Experimental data could be used to select between the models for the behaviour under 2D loading, provided that the model behaviour can be reasonably approximated with the Masing rules under 1D loading. This exercise would require extensive experimental data, and the relatively small difference between the predictions of the two models may be within the experimental error of a test setup. In the absence of such data, the choice of between the models could be based pragmatically on the practicality of their implementation rather than on theoretical grounds. In the series model, the same stress is applied to each of the spring-slider elements, which makes it suitable for stress driven algorithms. In the parallel model, the same strain applies to each of the spring-slider elements and the implementation of this model may be more suitable for strain-driven algorithms (such as those that are usually employed in the FE-method).

4 CONCLUSIONS

This paper compares the series and parallel versions of the Iwan model in 1D and 2D loading. The responses of the models are derived in the hyperplasticity framework, which allows a straightforward extension from 1D models to multi-dimensional models. When calibrated to the same backbone curve, the series and parallel models show the same behaviour on cyclic loading in 1D. However, when 2D loading is applied, a noticeable difference occurs between the models. Calibration of the 2D models to experimental tests (assuming that they respond according to the Masing rules in 1D) could be used to select between the models for a specific application. In the absence of extensive calibration data, the limited difference between these models shows that the choice can be based on the practicality of numerical implementation rather than on theoretical grounds.

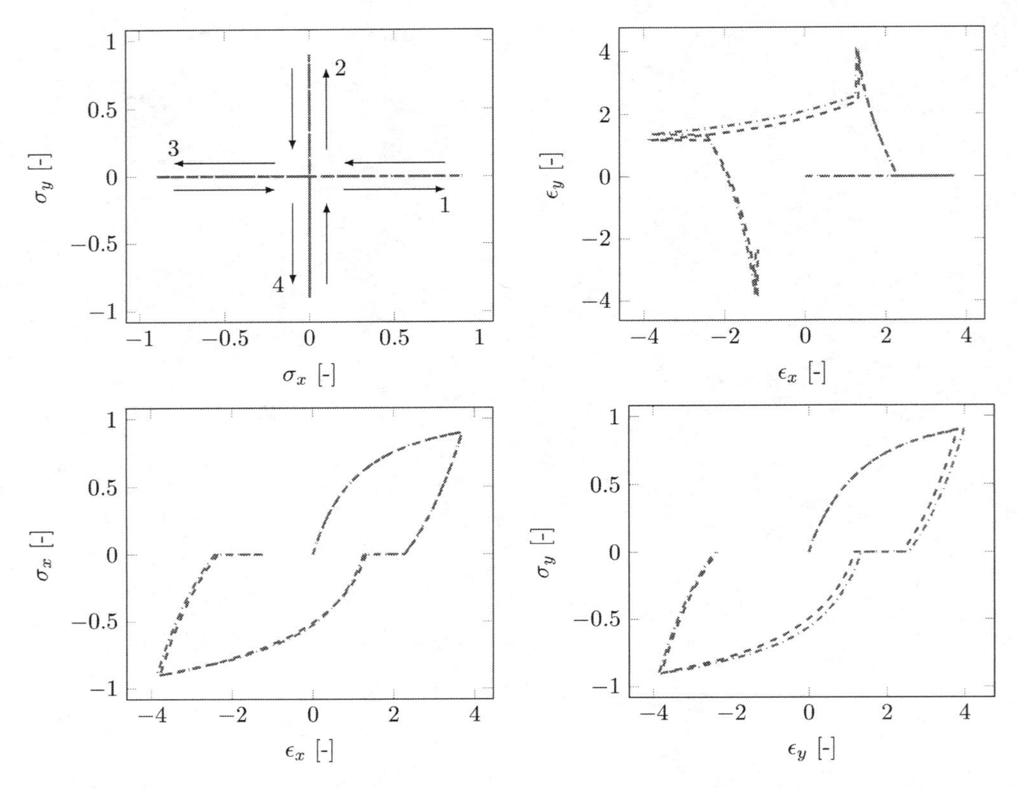

Figure 6. Comparison of the response from the series (blue dashed) and parallel (red dashdotted) Iwan models under a cross shaped 2D loading.

REFERENCES

Beuckelaers, W.J.A.P. (2015). Fatigue life calculation of monopiles for offshore wind turbines using a kinematic hardening soil model. In *Proceedings of the 24th European Young Geotechnical Engineering Conference*, Durham, U.K.

Beuckelaers, W.J.A.P. (2017). *Numerical Modelling of Laterally Loaded Piles for Offshore Wind Turbines*. DPhil thesis, University of Oxford.

Byrne, B.W. & G.T. Houlsby (2004). Experimental investigations of the response of suction caissons to transient combined loading. *Journal of Geotechnical and Geoenvironmental Engineering 130*(3), 240–253.

Byrne, B.W., R. McAdam, H.J. Burd, G.T. Houlsby, C.M. Martin,W.J.A.P. Beuckelaers, L. Zdravkovi´c, D.M.G. Tarboda, D.M. Potts, R.J. Jardine, E. Ushev, T. Liu, D. Abadias, K. Gavin, D. Igoe, P. Doherty, J. Skov Gretlund, M. Pacheco Andrade, A. Muir Wood, F.C. Schroeder, S. Turner, & M.A.L. Plummer (2017). PISA: new design methods for offshore wind turbine monopiles. In *Proceedings of the 8th Offshore Site Invesigation & Geotechnics (OSIG) International Conference*, London, U.K.

Einav, I. (2004). Thermomechanical relations between stressspace and strain-space models. *Géotechnique 54*(5), 315–318.

Houlsby, G.T. & G. Mortara (2004). Continuous hyperplasticity model for sands under cyclic loading. In *Proceedings of the International Confernce on Cyclic Behaviour of Soils and Liquefaction Phenomena)*, Volume 31, Bochum, Germany, pp. 21–26.

Houlsby, G.T. & A.M. Puzrin (2000). Thermomechanical framework for constitutive models for rate-independent dissipative materials. *International Journal of Plasticity 16*(9), 1017–1047.

Houlsby, G.T. & A.M. Puzrin (2006). *Principles of Hyperplasticity: an Approach to Plasticity Theory Based on Thermodynamic Principles*. Springer.

Iwan,W.D. (1967). On a class of models for the yielding behavior of continuous and composite systems. *Journal of Applied Mechanics 34*(3), 612–617.

Masing, G. (1926). Eigenspannungen und Verfestigung beim Messing. In *Proceedings of the 2nd International Congress of Applied Mechanics*, pp. 332–335.

Pyke, R. (1979). Nonlinear soil models for irregular cyclic loadings. *Journal of the Geotechnical Engineering Division (ASCE) 105*(6), 715–725.

Ziegler, H. (1977,1983). *An Introduction to Thermomechanics*. North-Holland, Amsterdam. (2nd edition 1983).

Numerical Methods in Geotechnical Engineering IX – Cardoso et al. (Eds)
© 2018 Taylor & Francis Group, London, ISBN 978-1-138-33198-3

Effective stress based model for natural soft clays incorporating restructuration

Jesper Bjerre
Multiconsult Norge AS, Bergen, Norway

Jon A. Rønningen, Gustav Grimstad & Steinar Nordal
Norwegian University of Science and Technology, Trondheim, Norway

ABSTRACT: Several advanced constitutive models have been developed over the last decades to reproduce the mechanical response of natural soft clays observed during laboratory testing and in the field. Features such as anisotropy, rate-dependence (creep) and the effect of structure (destructuration) are typically implemented in constitutive models using various approaches. This paper describes the development of an effective stress based model for natural soft clays, which incorporates the features earlier mentioned, as well as a new concept for regaining of structure (bondings) over time. The regaining of structure introduces a link between the development of creep strains and the amount of regained structure through the intrinsic reference surface and the apparent reference surface. The constitutive model is based on the assumption of associated flow and utilizes three hardening laws, which introduces eight hardening parameters in total. The model has been implemented in computer code for use in the finite element application PLAXIS as a user defined soil model. The user defined soil model is called SCA-R "Structure, Creep, Anisotropy—Restructuration". Examples of application are presented and the results are discussed.

1 INTRODUCTION

Presently, the main topics of research related to behavior of natural soft clays are associated with fabric (giving anisotropy in strength and stiffness), viscous behavior (time dependence) and aging effects (bonding). Some of these characteristics are also known as secondary compression (creep), relaxation, rate-dependence and the influence of structure (bonding/debonding). All these aspects need to be taken into consideration if higher accuracy in estimating the mechanical response of natural soft clay is desired.

This paper describes the development of a constitutive model for natural soft clays that includes features such as anisotropy in the elastic and plastic region, time-dependence and the effect of loss of structure (destructuration). As a new concept, a method to simulate the regaining of structure as a function of time is implemented to capture one of the ageing effects (bonding) observed during laboratory testing or in the field.

The constitutive model is based on Critical State Soil Mechanics (CSSM) Schofield & Wroth (1968) and can simplify to the Modified Cam Clay (MCC) model Roscoe & Burland (1968) if certain features are omitted. The model assumes associated flow and utilizes three hardening laws, which introduce eight state parameters in total, and can be thought of as an extension of the S-CLAY1S model proposed by Koskinen et al. (2002). The assumption of associated flow, i.e. yield and plastic potential surfaces being the same, is seen as sufficient to simulate natural soft clays since experimental data have shown minor deviations between the normality of the plastic strain increment directions and the yield surface for this particular soil.

The user defined soil model is called SCA-R "Structure, Creep, Anisotropy—Restructuration".

2 FEATURES

The MCC model is based on the assumption of an isotropic reconstituted soil with elasto-plastic mechanical response that deforms as a continuum. The MCC model utilizes isotropic hardening of the yield surface (i.e. a change in size) where only plastic volumetric strains contribute.

The concept of CSSM states that a soil, if continuously sheared until it flows as a frictional fluid, will come into a well-defined Critical State (CS) Wood (1990). At this state, shear distortions can occur without further change in effective stress or specific volume.

The path between the initial stress condition and the CS can be rather complex and it may be necessary to introduce several state parameters that evolve with the state of the soil. This will necessarily lead to a more complicated model formulation, but one may obtain a more accurate mechanical response of the soil.

In the sections which follow, several ways to extend the MCC formulation with more features are described to better reproduce the mechanical response of natural soft clays.

2.1 *Anisotropy*

2.1.1 *Plastic region*

Dafalias (1986) proposed an extension of the MCC model from isotropic to anisotropic response in the plastic region, where the energy dissipation was related not only to plastic volumetric and deviatoric strain increments alone (as in MCC), but was modified to include a coupling between the two as well. This gave rise to a yield surface with a rotational degree of freedom. With respect to general 3D space the equivalent stress condition is given as:

$$p^{eq} = p' + \frac{\frac{3}{2}\{\sigma_d - p'\alpha_d\}^T\{\sigma_d - p'\alpha_d\}}{\left(M^2 - \frac{3}{2}\alpha_d^T\alpha_d\right)p'} \tag{1}$$

where p' = the effective mean stress; M = the critical state line; α_d = the deviatoric fabric (rotational) tensor; and σ_d = the deviatoric stress tensor.

The equation above describes a rotated and distorted ellipsoid in general stress space, corresponding to the MCC yield surface if the rotation is zero.

Rotational hardening (evolution of the fabric tensor) is in general driven by both volumetric and deviatoric plastic strains, and may vary among existing models found in literature.

The expression suggested by Wheeler et al. (2003) is highlighted and is given as:

$$\frac{d\alpha_d}{d\lambda} = \mu\left[\left(\frac{3\sigma_d}{4p'} - \alpha_d\right)\frac{\partial p^{eq}}{\partial p'} + \beta\left(\frac{\sigma_d}{3p'} - \alpha_d\right)\sqrt{\frac{2}{3}\left\{\frac{\partial p^{eq}}{\partial \sigma_d}\right\}^T\frac{\partial p^{eq}}{\partial \sigma_d}}\right] \tag{2}$$

where β = the relative effectiveness of deviatioric and volumetric plastic strains; μ = the absolute rate at which α approaches its boundary value; and $d\lambda$ = the change in the plastic multiplier. The brackets $\langle\rangle$ are Macaulay brackets.

For the critical state surface in stress space, given through the parameter M, the constitutive formulation uses the LMN dependence proposed by Bardet (1990), i.e. making M dependent upon the modified lode angle, and is based on the failure criteria suggested by Lade & Duncan (1975) and Matsuoka & Nakai (1974). A three dimensional view of the LMN dependency with respect to the isotropic line is presented in Figure 1.

2.1.2 *Elastic region*

A simple approach to describe an anisotropic response in the elastic region can be done by assuming transverse anisotropy proposed by Graham & Houlsby (1983). Hence, the influence of fabric in the soil can be introduced by one anisotropy parameter (α_e):

$$E^* = \frac{27}{4}\frac{(1+e_0)p'}{\kappa}\frac{(1+\nu)(2\nu-1)}{4\alpha_e^2\nu - 2\alpha_e^2 - 9(\nu+1)} \tag{3}$$

where κ = the slope of the swelling line; ν = the Poisson's ratio; e_0 = the initial void ratio; and p' = the effective mean stress. Notice that $\alpha_e = 0$ yields isotropic conditions in the elastic region.

2.2 *Preconsolidation*

One of the most important parameters in geotechnical engineering is the preconsolidation pressure. In CSSM plasticity theory this measure separates

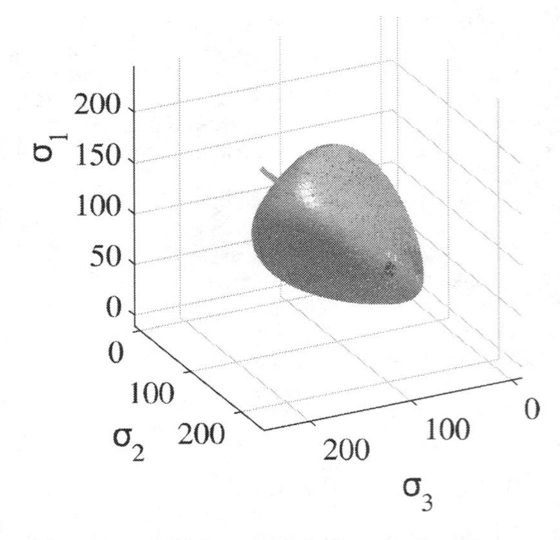

Figure 1. LMN dependency in the general stress space.

the elastic mechanical response (reversible deformations) from the plastic mechanical response (irreversible deformations).

In this article, the preconsolidation pressure is defined as the measured apparent yield stress in an oedometer test. The apparent yield stress is associated with the highest stress condition the soil has ever experienced plus the effect of bonding between soil particles. The over consolidation ratio (*OCR*) is the ratio between the apparent yield surface (p_m) and the in-situ stress reference surface (p^{eq}), i.e. using equivalent stress measures rather than the vertical stress.

The intrinsic yield surface, apparent yield surface and the reference surface which represent the in-situ stresses are shown in Figure 2. A coupling between the apparent yield surface and intrinsic yield surface (reconstituted material) is presented in eq. (5).

An evolution rule for the pre-consolidation stress can be determined by linking the plastic volumetric strain increment to a change in preconsolidation:

$$\frac{dp'_{mi}}{d\lambda} = \frac{1+e_0}{\lambda_i - \kappa} \cdot p'_{mi} \frac{\partial p^{eq}}{\partial p'}$$
$$dp'_m = dp'_{mi}(1+x) + p'_{mi} \cdot dx \qquad (4)$$

where λ_i = the slope of the intrinsic normal compression line; κ = the slope of the swelling line; and e_0 = the initial void ratio. The parameter x is the amount of structure that can be changed.

2.3 *Time dependent behavior*

According to Sorensen (2006) the time dependent behavior can be divided into two general aspects: ageing and viscous effects as presented in Figure 3.

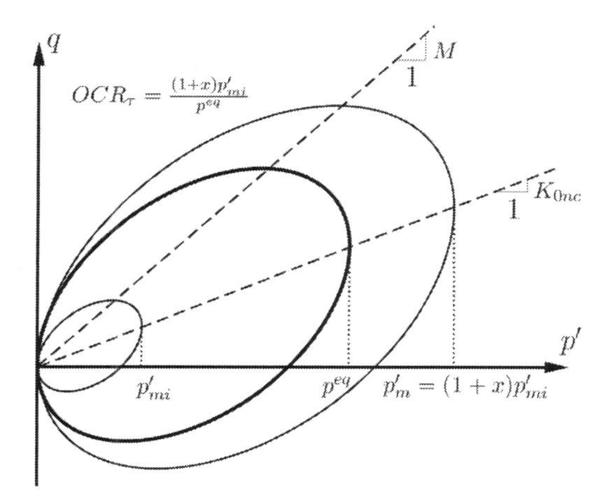

Figure 2. The intrinsic yield surface, its associated surface through the structure and the reference surface representing the stress condition. Note that $q = \sigma'_1 - \sigma'_3$; K_{0nc} = oedometer condition; and τ = rate dependency.

The effect from bondings and creep are shown in Figure 4 with respect to oedometer testing, and will be described in the following paragraphs.

2.3.1 *Ageing effects—bondings*

The term bonding includes all the inter-particle forces which are not generated by pure friction between the grains. These forces may consist of electrostatic or electromagnetic nature, Van der Wall forces and viscous stresses within the absorbed water layers Gasparre (2005). Additional forces from for instance organic content or non-clay minerals may be defined within the term cementation. Several of these effects are associated to the tiny dimension and large surface area which characterize the clay particles. The bonding (structure) between the particles will break during straining, i.e. the soil will progressively lose structure until it reaches a state where all the original bonding is gone.

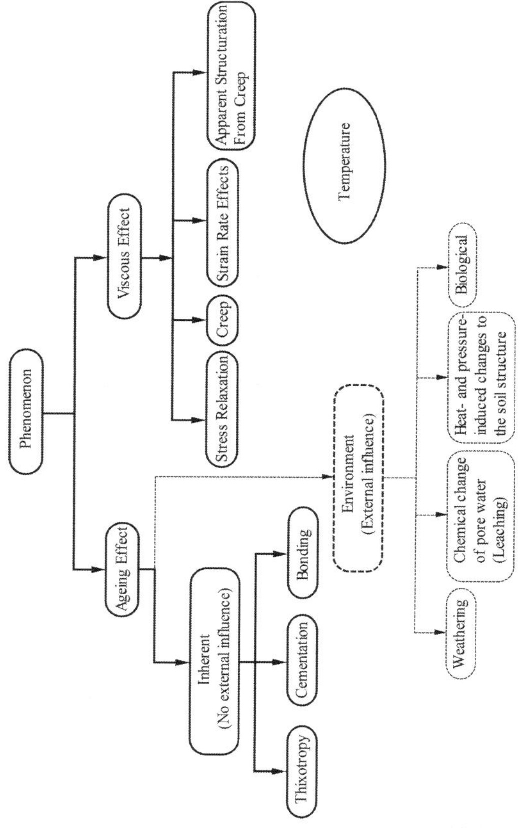

Figure 3. Classification of common time effects in natural clays.

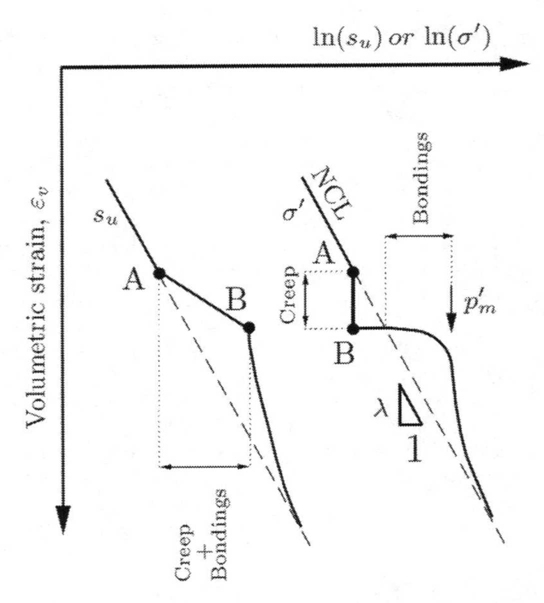

Figure 4. Combined effect of creep and bondings. NCL = Normal Compression Line.

A relation between the intrinsic yield surface and the apparent yield surface associated to the amount of bondings can be formulated as proposed by Gens & Nova (1993):

$$p'_m = (1 + x)\, p'_{mi} \tag{5}$$

where x refers to amount of structure (bonding) as the ratio between the yield surface of the natural soft clay and the yield surface of the reconstituted clays (notice that i denotes intrinsic).

The change of the state parameter, x, is generally dependent on both the volumetric and deviatoric plastic strain increment. Koskinen et el. (2002) proposed a hardening rule related to the degradation of structure which can be rewritten into the general stress space as:

$$\frac{dx}{d\lambda} = -ax\left(\left|\frac{\partial p^{eq}}{\partial p'}\right| + b\sqrt{\frac{2}{3}\left\{\frac{\partial p^{eq}}{\partial \sigma_d}\right\}^T \frac{\partial p^{eq}}{\partial \sigma_d}}\right) \tag{6}$$

where a = the absolute rate of destructuration; and b = the relative effectiveness of plastic volumetric and deviatoric strains in destroying the bondings.

2.3.2 *Viscous effects*

The term, viscosity, will normally be associated with liquids and gasses by symbolizing the measure of shear resistance to an applied shear force. The relationship between resistance and shear force may describe the characteristics of the matter.

The general understanding of viscous effects are related to the consequence of sliding at inter-particle contacts and thereby associated to a rearrangement of particles in the soil medium.

2.3.2.1 Creep and strain rate effects

One of the first to encounter long-term deformation was Bjerrum (1967) during a research project regarding the soft clay found nearby Drammen in Norway. Bjerrum (1967) suggested a concept of isochrones or time lines in the one-dimensional compression plane to represent the increase in the apparent yielding point caused by the changed void ratio.

Laboratory experiments have shown that the chosen strain rate influences the soil response. During undrained triaxial testing, the strain rate will influence the stress path Lunne et al (2006). The general tendency may be concluded as: An increase in strain rate may be associated to an increase in the undrained shear strength. Several studies have shown a unique relationship between void ratio and effective stresses for a given strain rate for clays. Thereby, the concept of isochrones suggested by Bjerrum (1967) may be replaced by the concept of isotach as suggested by Šuklje (1957) (same rate). Hence, by increasing the strain rate the apparent yielding point increase. However, other materials such as sand and gravel may not show isotach tendency.

One approach to incorporate these effects is to define a plastic multiplier as a function of time proposed by Grimstad (2009). Utilizing the time resistance concept of Janbu (1969) which uses a single creep parameter it can be shown that the change in multiplier per time is given as:

$$\frac{d\lambda}{dt} = \dot{\lambda} = \frac{1}{r_{si}\tau}\left(\frac{p^{eq}}{(1+x)\,p'_{mi}}\right)^{r_{si}\zeta_i} m_{K_{0nc}}$$
$$\zeta_i = (\lambda_i - \kappa)/(1 + e_0) \tag{7}$$

where ζ_i = the intrinsic viscoplastic compressibility coefficient; r_{si} = the intrinsic time resistance number associated by no structural effects; and τ = the reference time, typically 1 day. The parameter m_{K0nc} defines the plastic flow during one-dimension compression:

$$m_{K_{0nc}} = \frac{M_c^2 - \alpha_{K0nc}^2}{M_c^2 - \eta_{K0nc}^2} \tag{8}$$

where M_c = the critical state line in compression; α_{K0nc} = the rotation in a K_{0nc} type of loading; and $\eta_{K0nc} = 3(1 - K_{0nc})/(1 + 2K_{0nc})$. The formulation for

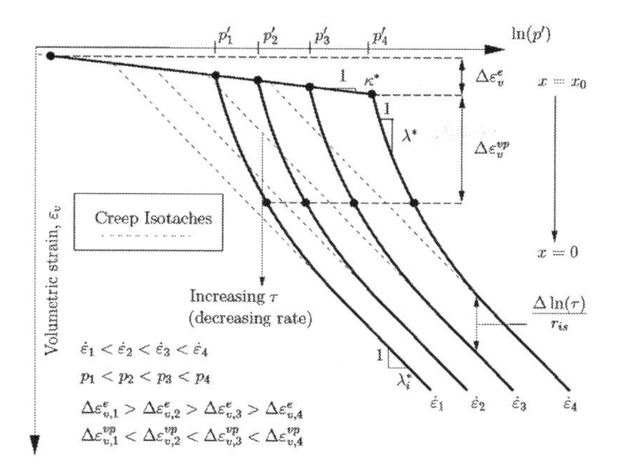

Figure 5. Concept of isotaches effected by structure. Note that e = elastic; vp = viscoplastic; and * = modified compression indies.

the plastic multiplier is not directly dependent on time itself but rather a change in preconsolidation. The derivation of this expression involves finding the plastic volumetric change due to creep from a reference time τ to time t and linking this change in volume to a change in preconsolidation, thus eliminating time.

The concept of isotaches including the effect of structure is shown in Figure 5.

2.4 *Recovery of structure*

It is expected that various factors such as: plastic index, overburden stress, temperature, chemical composition etc. would influence the recovery process. However, for the sake of simplicity a simple procedure has been chosen which requires two additional input parameters. The recovery is implemented as a differential equation describing a simple exponential transition towards a prescribed value of structure. This asymptotic value is controlled by the initial amount of structure and an input parameter determining how much recovery is allowed, as a fraction of the initial amount. Such a formulation may be given as:

$$\frac{dx}{dt} = R_t \cdot Rx_0 - x \tag{9}$$

where R = the amount of recovery, from 0 to 1; and R_t = the pace of recovery. To ensure that the recovery does not initiate before the amount of structure is below the initial value (Rx_0), Macaulay brackets are used. The recovery process will influence the creep rate caused by the increase in x which yields a lower creep rate.

3 NUMERICAL IMPLEMENTATION

The relationship between stresses and strains can be assessed through a constitutive model. With respect to a typical computer implementation it would involve that a strain and time increment are given as input and the model provides the corresponding stress state as output.

PLAXIS is a commercial finite element software for 2D and 3D soil calculations, which offers the possibility of creating user defined soil models by defining a subroutine with specific arguments.
The constitutive model is implemented in PLAXIS through the interface between PLAXIS and FORTRAN. The FORTRAN code is supported by additional MATLAB coding.

3.1 *Local iterative procedure—stress update*

The local iterative process is related to the iterative procedure within the constitutive model itself. The constitutive model utilizes the concept of explicit integration, which represents a direct integration procedure. The explicit approach utilizes a linear incremental stress-strain-time relationship yielding an error due to non-linearity of the constitutive equations, because each step is finite and not infinitesimal. However by utilizing a small (controlled) incremental step the error may be minimized to an acceptable limit. The simplest explicit method is known as the forward Euler scheme, where the stress increment is evaluated by the finite strain and time increment:

$$\Delta\sigma_{n+1} = \boldsymbol{D}(\sigma_n, \kappa_n)\Delta\varepsilon_{n+1} \tag{10}$$

where $\boldsymbol{D}(\sigma_n, \kappa_n)$ = the tangential stiffness determined by the previous iteration; n = the previous stage; and $n+1$ = the current stage. κ_n is a vector containing the eight state parameters at the previous step, as defined in eq. (14).

3.2 *Numerical scheme*

The computational framework is presented in Figure 6.

The elastic stiffness matrix (***D***) and the differential equations determining the constitutive behavior of the soil model (***v***) are assembled in MATLAB using symbolic differentiation in an approach suggested by Roenningen (2014). The code is afterwards transformed automatically into FORTRAN code. The initializing of the state parameters are defined in the FORTRAN code and the code is compiled into a DLL-file containing the soil model.

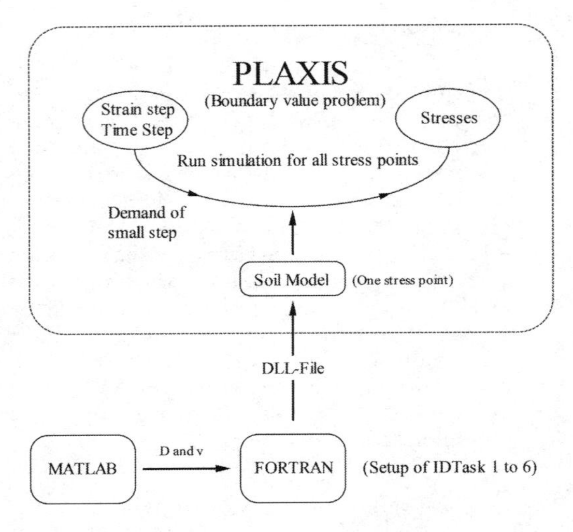

Figure 6. Principle of the computational framework.

The vector ($\boldsymbol{v}$) contains the six Cartesian stresses, the eight state parameters and the plastic multiplier and is assembled as:

$$\boldsymbol{v} = \left[\sigma \quad \lambda \quad \alpha_d \quad p'_{mi} \quad x \right]^T \tag{11}$$

The local integration in the stress points is determined through a forward Euler scheme which may be formulated as:

$$\boldsymbol{v}_{n+1} = \boldsymbol{v}_n + d\boldsymbol{v} \tag{12}$$

where $d\boldsymbol{v}$ = the change in $\boldsymbol{v}$. The vector $\boldsymbol{v}$ is given by:

$$d\boldsymbol{v} = \begin{bmatrix} \boldsymbol{D}\left(\Delta\varepsilon - \Delta\lambda \dfrac{\partial Q}{\partial \sigma} \right) \\ \Delta\lambda \\ \Delta\lambda \dfrac{\partial \kappa}{\partial \lambda} + \Delta t \dfrac{\partial \kappa}{\partial t} \end{bmatrix} \tag{13}$$

where

$$\frac{\partial \kappa}{\partial \lambda} = \left[\frac{\partial \alpha_d}{\partial \lambda} \quad \frac{\partial p_{mi}}{\partial \lambda} \quad \frac{\partial x}{\partial \lambda} \right]^T \tag{14}$$

$$\frac{\partial \kappa}{\partial t} = \left[0 \quad 0 \quad \frac{\partial x}{\partial t} \right]^T \tag{15}$$

Eq. (14) and (15) are vectors containing the derivate of the state parameters with respect to the plastic multiplier and time, respectively. The change $d\boldsymbol{v}$ is obtained using known quantities from iteration n.

The change in the plastic multiplier may be approximated as suggested by Grimstad & Benz (2014):

$$\Delta\lambda \approx \frac{\dot{\lambda} + \left\{ \dfrac{\partial \dot{\lambda}}{\partial \sigma} \right\}^T \boldsymbol{D}\Delta\varepsilon + \left\{ \dfrac{\partial \dot{\lambda}}{\partial \kappa} \right\}^T \dfrac{\partial \kappa}{\partial t} \Delta t}{\dfrac{1}{\Delta t} - \left\{ \dfrac{\partial \dot{\lambda}}{\partial \kappa} \right\}^T \dfrac{\partial \kappa}{\partial \lambda} + \left\{ \dfrac{\partial \dot{\lambda}}{\partial \sigma} \right\}^T \boldsymbol{D} \left\{ \dfrac{\partial Q}{\partial \sigma} \right\}} \tag{16}$$

where $\boldsymbol{D}$ = the elastic stiffness matrix; and $\dot{\lambda} = dy/dt$ is as given by Eq. (7).

4 RESULT AND DISCUSSION

The performance of the constitutive model has been investigated by comparing the mechanical response with laboratory tests on block samples from the Onsøy site, Norway. Soil parameters have been interpreted from the block samples and afterwards modified with respect to "best fit" in back-calculations. The soil parameters are presented in Table 4–1.

Oedometer test on block sample "Block-26-B1-20-50" from 10.8 m depth Berre (2013) along with the performance of the model is presented in Figure 7.

The mechanical response deviate in the preloading of the soil sample until the preconsolidation pressure is reached. The constitutive model utilizes the same stiffness in the preloading region and the unloading-reloading region and it has be chosen to fit the unloading-reloading region for this paper.

Triaxial test on block sample "Block-26-A-2" from 10.0m depth Berre (2013) along with the

Table 4–1. Input data—Interpreted/back-calculations.

Parameter	Value	Unit
λ_i	0.219	-
κ	0.034	-
e_0	1.95	-
OCR_τ	1.18	-
r_{si}	600	-
$r_{si,min}$	104	-
a	12	-
b	8.0	-
K_{0nc}	0.53	-
K'_0	0.60	-
ϕ'	36	°
ν_{ur}	0.15	-
R, R_t	0.0	-
μ	8.0	-
α_e	–0.17	-

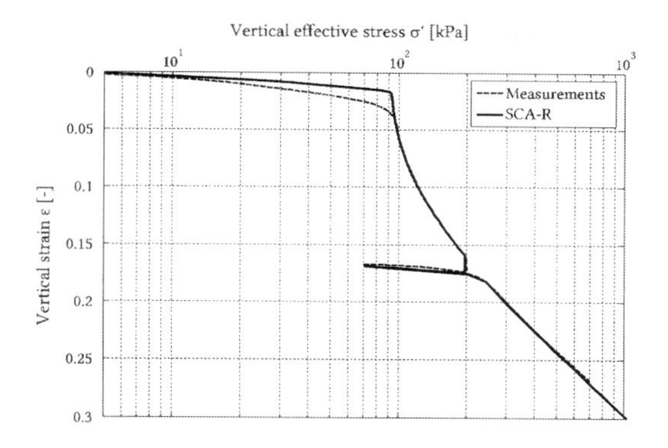

Figure 7. Back-calculation of CRS response forBlock-26-B1-20-50.

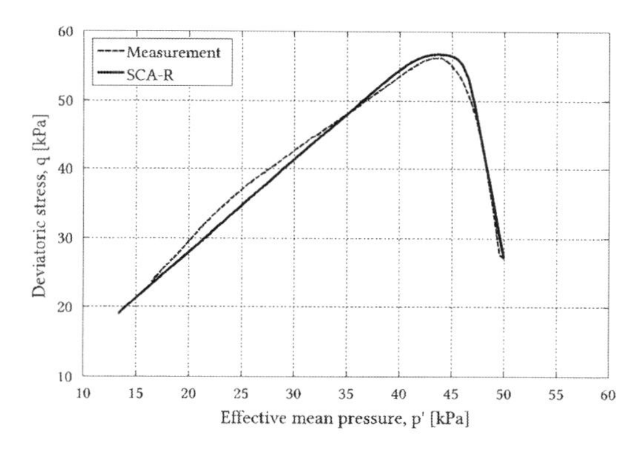

Figure 8. Effective stress path.

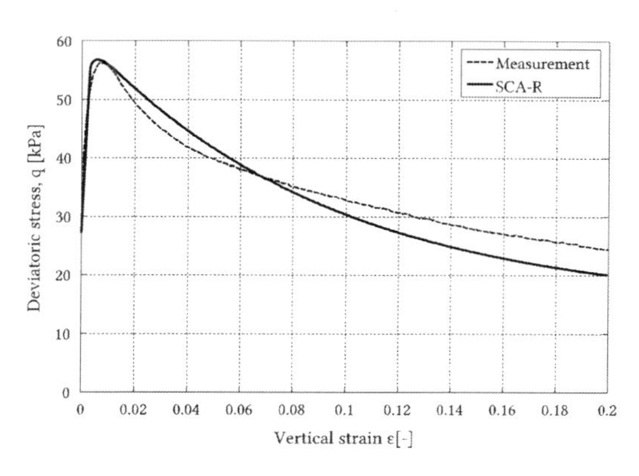

Figure 9. Stress-strain relationship.

performance of the model are presented in Figure 8 to Figure 10. Both block samples are taken from the same site. It can be observed that the model predicts the soil response in a relatively accurate manner in both the oedometer and the triaxial test.

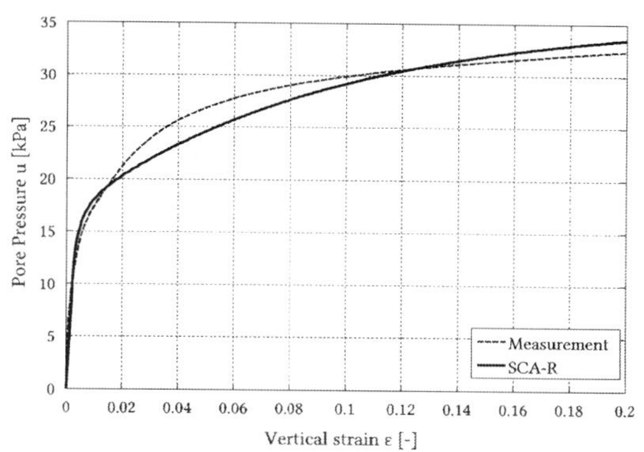

Figure 10. Development of excess pore pressure.

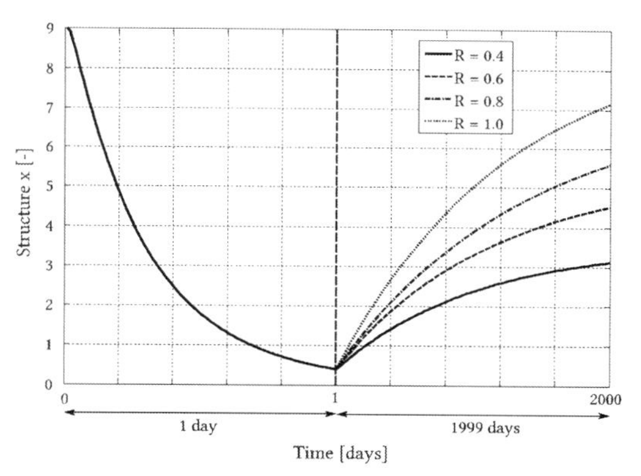

Figure 11. Recovering of structure (x) over time. 1 day shearing, i.e. destroying the structure. Second part symbolize constant effective stresses over a period of 1999 days. $R_t = 0.001$ for all cases.

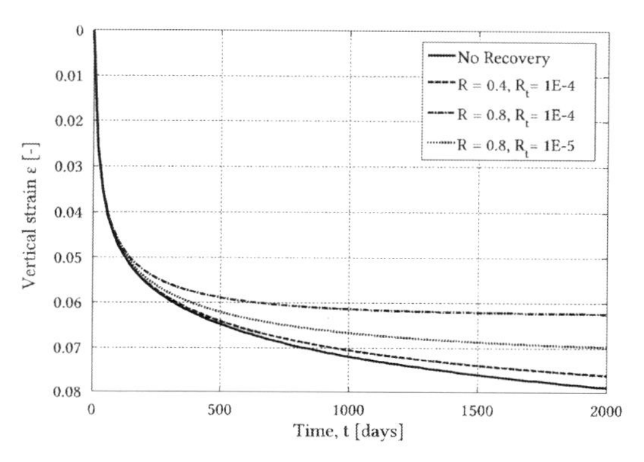

Figure 12. Influence of recovering of structure with respect to creep under constant stress conditions. The initial structure of the test is equal to $x_0 = 1$ going towards the boundary value of $x_b = 9R$.

Figure 11 presents four scenarios with the same shearing phase (triaxial) and four different magnitudes of allowed structural recovering with the recovery pace $R_t = 0.001$. The first part of the curves symbolize a shearing phase where destroying the initial structure ($x_0 = 9$) occurs during one day. The other part day 1 to 2000 illustrates the recovering of structure during constant effective stresses over a period of 1999 days.

By observing Figure 11 the evolution of the structure is shown and the regaining of structure is shown as an exponential rate as implemented. Note that the boundary value, R, indirectly affect R_t, which is associated with the pace of recovery.

The recovering of structure induces a coupling term between the recovering of structure and implemented time evolution of the visco-plastic multiplier. Figure 12 shows four scenarios illustrating the coupling effect for a drained creep test at constant effective stresses.

It may be observed on Figure 12 that the initial part of the curves are unaffected by the recovering. After approximately 5% of vertical strain the evolution in creep strains deviate and the effect of recovery of structure is shown. The recovering of structure indirectly influence the creep evaluation.

5 CONCLUSION

This paper describes the development of a constitutive model for natural soft clays, which includes features such as anisotropy, rate-dependence (creep) and the effect of structure (destructuration and restructuration). The model predictions have been compared to laboratory tests, and fits well for both oedometer and undrained traxial compression tests. The model introduces a new concept of regaining structure with time. The implementation yields a coupling to the development of creep strains over time through the intrinsic reference surface and the apparent reference surface and may be used to "slow down" creep.

ACKNOWLEDGEMENTS

The work presented in this article is a product of a Master Thesis carried out at NTNU, spring 2015, at the Geotechnical department. The Master Thesis is related to Jon A. Rønningen's research on the going GeoFuture project, see www.GeoFuture.no for more information.

REFERENCES

Bardet, J.P. (1990). Lode dependences for isotropic pressure-sensitive elastoplastic materials. Journal of Applied Mechanics, Transactions ASME, 57(3), 498–506.

Berre, T. (2013): Test fill on soft plastic marine clay at Onsøy, Norway. Canadian Geotechnical Journal. 1–22.

Bjerrum L (1967) Engineering geology of Norwegian normallyconsolidated marine clays as related to settlements of buildings. Géotechnique 17(2):81–118

Dafalias, Y.F. (1986). An anisotropic critical state soil plasticity model. Mechanics Research Communications 13(6), 341–347.

Gasparre A (2005). Advanced Laboratory of London Clay. PhD thesis, Imperial College, London.

Gens, A., & Nova, R. (1993). Conceptual bases for a constitutive model for bonded soils and weak rocks. Geotechnical engineering of hard soils-soft rocks, 1(1), 485–494.

Grimstad, G. (2009). Development of effective stress based anisotropic models for soft clays.

Grimstad G. & Benz T. (2014). Lecture Notes, Soil Modeling (BA8104) - Implementation of Soil Models. NTNU, Geotechnical division, Trondheim, 1. Edition.

Graham T, Houlsby GT (1983) Anisotropic elasticity of a natural clay. Géotechnique 33(2):165–180

Janbu, N. (1969). The resistance concept applied to deformations of soils. Proc. 7th Int. Conf. on Soil Mech. & Found. Eng, Mexico city: 1 191–196.

Koskinen, M., Karstunen, M., & Wheeler, S.J. (2002). Modelling destructuration and anisotropy of a natural soft clay, in: Proceedings of the Fifth European Conference on Numerical Methods in Geotechnical Engineering, Paris, France, 2002, pp.11–20.

Lade, P.V., & Duncan, J.M. (1975), "Elastoplastic Stress-Strain Theory for Cohesionless Soil," J. Geotechnical Engineering Division, ASCE, New York, Vol. 101, pp. 1037–1053.

Lunne T, Berre T, Andersen KH, Strandvik S, Sjursen M. (2006). Effects of sample disturbance and consolidation procedures on measured shear strength of soft marine Norwegian clays. Canadian Geotechnical Journal 43(7): 726–750

Matsuoka, H., & Nakai, T., 1974, "Stress-deformation and Strength Characteristics of Soil Under Three Different Principal Stress," Proceedings Japanese Society of Civil Engineers, Vol. 232, pp. 59–70.

Roenningen J.A., Gavel-Solberg V., & Grimstad G (2014). Effective Stress Model for Soft Scandinavian Clays. In European Young Geotechnical Engineers Conference, pages 1–4, Barcelona, Spain.

Roscoe KH & Burland JB (1968) On the generalized stress-strain behavior of wet clay. In: Engineering plasticity, Hetman Leckie,

Cambridge, pp. 535–609

Schofield, A., & Wroth, P. (1968). Critical state soil mechanics.

Šuklje L (1957). The analysis of the consolidation process by the isotaches method. Proceedings 4th International Conference Soil Mechanics Foundation Engineering London 1:200–206

Sorensen K.K. (2006). Influence of Viscosity and Aging on the Behaviour of Clays. PhD thesis, University College London.

Wheeler, S.J., Näätänen, A., Karstunen, M., & Lojander, M. (2003). An anisotropic elastoplastic model for soft clays. Canadian Geotechnical Journal, 40(2), 403–418.

Wood D.M. Soil Behaviour and Critical State Soil Mechanics (1990). Cambridge University Press.

Numerical Methods in Geotechnical Engineering IX – Cardoso et al. (Eds)
© 2018 Taylor & Francis Group, London, ISBN 978-1-138-33198-3

Interpretation of the cyclic behaviour of a saturated dense sand within an elasto-plastic framework

B.M. Dahl, M.S. Løyland & H.P. Jostad
Norwegian Geotechnical Institute (NGI), Oslo, Norway

ABSTRACT: Offshore structures, as for instance offshore wind turbines, are subjected to combined static and cyclic loading from wind, waves and current. To be able to perform an optimised design of the foundation of these structures it is important to account for the characteristic behaviour of the soil under these irregular load conditions. This paper presents detailed interpretation of undrained, partly drained and drained monotonic and cyclic triaxial tests on a typical dense North Sea sand within an elasto-plastic framework. The incremental shear (or deviatoric) strain is given by a plastic hardening modulus and the ratio between the incremental plastic volumetric strain and shear strain by a dilatancy parameter. The plastic strain increments are derived from the total strains by first subtracting the elastic strains from the measured total strains. The variations of the hardening and dilatancy parameters are presented as function of mobilised friction. In addition, the rate of shear strain and pore pressure accumulations are presented. Then, the interpreted results are compared with values obtained from the equations used in the SANISAND model. Based on this, limitations in the existing formulation of the SANISAND model are discussed. The presented results may further be used to improve existing models for cyclic behaviour of water saturated dense sands.

1 INTRODUCTION

1.1 *Background*

Offshore structures, as for instance offshore wind turbines, are subjected to combined static and cyclic loading from wind, waves and current. To be able to perform an optimised design of the foundation of these structures it is important to account for the characteristic behaviour of the soil under these irregular load conditions (Andersen 2009). One approach to represent the change in soil properties as a result of cyclic loading is to establish pore water pressure and strain contour diagrams, as described in Andersen (2015). A large number of undrained cyclic laboratory tests have to be performed to establish these diagrams. These diagrams are then applied to the Partially Drained Accumulation Model in a finite element analysis in order to assess displacements, stiffness and capacity of foundations (Jostad et al. 2015).

In order to verify and improve existing calculation tools, the Norwegian Geotechnical Institute (NGI) has initiated a three years strategic research project. As part of this project, undrained, partly drained and drained monotonic and cyclic triaxial tests on a typical dense North Sea sand are carried out.

1.2 *Formulation of the problem*

A detailed interpretation of the triaxial test data is desirable to get a better understanding of the behaviour of dense saturated sand. The interpretation is carried out within an elasto-plastic constitutive framework, and characteristic behaviour is determined. Incremental shear strain is given by a plastic hardening modulus, H, and the ratio between the incremental plastic volumetric strain and shear strain by a dilatancy parameter, d. The variations of H and d are presented as function of mobilised friction. In addition, the rate of shear strain and pore pressure accumulations are presented. The triaxial test data is further used to evaluate the Simple ANIsotropic SAND (SANISAND) constitutive material model (Manzari and Dafalias 1997, Dafalias and Manzari 2004, Taiebat and Dafalias 2007) describing static and cyclic behaviour of sand. Based on the results, limitations in the existing formulation of the SANISAND model are discussed. The undrained cyclic behaviour of sand is the main focus.

2 SANISAND CONSTITUTIVE MODEL

SANISAND is the name used for a family of Simple ANIsotropic SAND constitutive models (Taiebat & Dafalias 2007). SANISAND is developed to simulate the stress-strain behaviour of sands under monotonic and cyclic loading conditions. Manzari & Dafalias (1997) introduced a critical state two-surface plasticity model for sands, a model in the triaxial $p' - q$ space based on the critical state soil mechanics framework (Schofield and Wroth 1968,Wood 1990) and the bounding surface plasticity model (Dafalias & Popov 1975, Bardet 1986, Dafalias 1986). The presented interpretations are based on a SANISAND model which is extended to account for the effect of fabric changes during loading (Dafalias and Manzari 2004).

The SANISAND model is based on the assumption that only changes of stress ratio, $\eta = q/p'$, can cause plastic deformation. Increased stress with a constant stress ratio is assumed to only cause elastic deformations, assuming there is no crushing of grains.

2.1 *Elastic and plastic strains*

The SANISAND equations are formulated in the triaxial stress space, with all stress components considered as effective stresses. The principal stresses and strains are defined in the axisymmetric triaxial space as $\sigma'_1, \sigma'_2 = \sigma'_3$ and $\varepsilon_1, \varepsilon_2 = \varepsilon_3$.

The deviatoric stress and mean stress is defined as $q = \sigma_1 - \sigma_3$ and $p = \frac{1}{3}(\sigma_1 + 2\sigma_3)$, and the deviatoric strain and volumetric strain are defined as $\varepsilon_q = \frac{2}{3}(\varepsilon_1 - \varepsilon_3)$ and $\varepsilon_v = \varepsilon_1 + 2\varepsilon_3$ respectively.

The strains are divided into an elastic and a plastic part. The incremental stress-strain relations for the elastic and plastic parts are given in Equation 1 and 2 respectively.

$$d\varepsilon_q^e = \frac{dq}{3G}; d\varepsilon_v^e = \frac{dp}{K} \quad (1)$$

$$d\varepsilon_q^p = \frac{d\eta}{H}; d\varepsilon_v^p = d\left|d\varepsilon_q^p\right| \quad (2)$$

Where η is the stress ratio, G is the elastic shear modulus, K is the elastic bulk modulus, H is the plastic hardening modulus, and d is the dilatancy parameter.

2.2 *Plastic modulus*

The stress ratio, η, will increase under monotonic drained triaxial compression loading, but is bounded by the bounding stress ratio, M^b. The bounding stress ratio varies with the material state and is related to the hardening modulus, H, Equation 3. The magnitude of H depends on the distance from the current stress ratio to the bounding line.

$$H = h\left(M^b - \eta\right) \quad (3)$$

Where h is a function of the state variables, and varies with b_0. The equations of h and b_0 are shown in Equation 4.

$$h = \frac{b_0}{|\eta - \eta_{in}|}; b_0 = G_0 h_0 \left(1 - c_h e\right)\left(\frac{p}{p_{atm}}\right)^{-1/2} \quad (4)$$

h_0 and c_h are scalar parameters and η_{in} is the initial stress ratio at the initiation of a loading process, and is updated when the loading is reversed, according to Dafalias (1986).

2.3 *Dilatancy*

The dilatancy parameter, d, given in Equation 5, is proportional to the difference of current stress ratio, η, from dilatancy stress ratio, M^d. The dilatancy stress ratio, M^d is also called the phase transformation line.

$$d = A_d\left(M^d - \eta\right) \quad (5)$$

Where A_d is a function of the state. A contractant behaviour is obtained when $d > 0$, and for $d < 0$ there is a dilatant behaviour of the material.

2.4 *Critical state*

The critical state stress ratio, M, is defined from the critical deviatoric stress, q_c, and the critical mean stress, p_c. While M is related to the friction angle, the critical void ratio, e_c, and the critical mean stress, p_c, vary depending on the soil and for the range of pressure considered. A relation between e_c and p_c is suggested by Wang & Li (1998) in Equation 6.

$$e_c = e_0 - \lambda_c\left(\frac{p_c}{p_{atm}}\right)^{\xi} \quad (6)$$

Where e_0 is the void ratio at $p_c = 0$, and λ_c and ξ are constants.

The bounding and dilatancy lines vary with the material state, and the lines are defined in such a way that when $e = e_c$ and $p = p_c$ then $M^b = M^d = M$.

$$M^b = M\exp(-n^b\Psi); M^d = M\exp(n^d\Psi) \quad (7)$$

Where n^b and n^d are positive material constants, and Ψ is called the state parameter by Been & Jefferies (1985) defined as $\Psi = e - e_c$.

2.5 *Fabric change*

The fabric-dilatancy internal variable, z, will influence the dilatancy, d, to give a more realistic interpretation of the soil behaviour under cyclic loading. Initially $z=0$, and Equation 8 gives the development in z. For a contractant behaviour there is no development in z, $dz=0$, because $\langle -d\varepsilon_v^p \rangle = 0$, and for a dilatant behaviour a negative z will develop in compression and a positive z will develop in triaxial extension.

$$dz = -c_z \langle -d\varepsilon_v^p \rangle (s z_{\max} + z) \tag{8}$$

$z_{\max}$ is the maximum value z can attain and c_z controls the speed of the change in z. The parameter A_d of the dilatancy is a function of z given in Equation 9.

$$A_d = A_0 (1 + \langle sz \rangle) \tag{9}$$

3 TRIAXIAL TESTS ON SAND

NGI has performed 10 monotonic and 9 cyclic triaxial tests on dense saturated sand from the offshore field Siri in the North Sea. The triaxial test specimens are reconstituted to the desired relative density by the moist tamping method according to NGI standard procedures. The average minimum and maximum void ratios of the tests are $e_{\min} = 0.502$ and $e_{\max} = 0.787$. The grain size distribution is shown in Figure 1, plotted together with other typical North Sea sands.

3.1 *Description of monotonic tests*

The tests are carried out with a relative density of about 80% and consolidated to an axial consolidation stress of $\sigma'_{ac} = 200$ kPa and a radial consolidation stress of $\sigma'_{rc} = 90$ kPa. Both drained, undrained, compression and extension monotonic tests are performed.

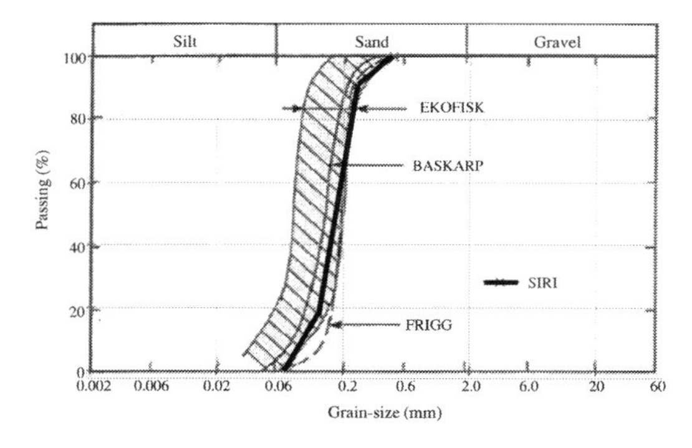

Figure 1. Grain size distribution of Siri sand (Carotenuto & Suzuki 2016).

3.2 *Description of cyclic tests*

The cyclic tests are loaded to a state where the average deviatoric stress, $q = 110$ kPa, and cyclically loaded with a $\Delta q = \pm 200$ kPa such that $q_{\max} = 310$ kPa and $q_{\min} = -90$ kPa. For the cyclic samples, drained preshearing of $\tau_{cy} = 0.06 \cdot \sigma'_{ac}$ for 400 cycles is performed. Table 1 summarises the maximum shear modulus and relative densities of the cyclic tests.

Two tests (Test 12 and 12b) are identical standard undrained cyclic triaxial tests consolidated to $\sigma_{ac} = 200$ kPa and $\sigma_{rc} = 90$ kPa. One test (Test 17) is also a standard undrained cyclic triaxial test which is consolidated to $\sigma_{ac} = 200$ kPa and $\sigma_{rc} = 90$ kPa, but unlike Test 12 the sample is unloaded to $\sigma_{ac} = 170$ kPa before the test is run.

What happens to the sand properties after partial drainage is of interest. Three tests (Test 13, 14 and 15) have partial drainage. The tests are standard undrained tests before drainage of the pore water is applied. One test (Test 13) includes packages of19–20 cycles with full drainage in between, while two tests (Test 14 and 15) include packages of 19–20 cycles with dissipation of 50% of the excess pore water pressure after each package. Test 13 has 20 undrained cycles before the applied drainage, Test14 has 41 undrained cycles, and Test 15 has 21 undrained cycles.

3.3 *Test results*

The results are interpreted within the framework of the SANISAND model. The raw data from the triaxial tests are filtered using the *filtfilt* function in MATLAB, reducing high-frequency fluctuations in the data (MathWorks 2017). Stress paths of the tests are studied from plots of effective mean stress, p', versus deviatoric stress, q, and the sand's tendency of contraction and dilation is evaluated from plots of dilatancy, d, versus stress ratio, η Incremental plastic shear and volume strains are plotted for evaluation of the hardening and dilatancy trends respectively, and the total plastic shear and volume strains are included to understand the accumulation of strains.

Table 1. Maximum shear modulus and relative density of the cyclic tests.

Test no.	G_{max} (MPa)	D_r (%)
Test 12	142.6	–
Test 12b	118.9	77.9
Test 13	146.3	81.3
Test 14	135.1	76.1
Test 15	135.2	80.5
Test 17	–	–

3.3.1 *Spread in data*

All undrained and partly drained cyclic tests started with at least 20 undrained cycles. The behaviour of the sand in the different tests is expected to be similar, but deviations are observed in pore water pressure generation among the first 20 cycles in the different tests. This can be seen in Figure 2, where the generated pore pressure is the difference between $p'_{initial}$ kPa for the first cycle and the $p'_{initial}$ for the cycle of interest. It is also seen that the rate of pore pressure generation, i. e. the slope of the curves, is different at a given initial effective stress, $p'_{initial}$. This especially applies for the initial part of the cyclic loading history. There is no obvious reason for the deviation in the data, all the tests are prepared by the same method, they are anisotropically consolidated to the same stress state, and loaded with the same cyclic amplitude, $q = 110 \pm 200$ kPa. The build-in relative density, D_r, ranges from 76% to 81% and the G_{max} measurements vary from 118 MPa to 147 MPa, but there is no distinct correlation with G_{max}. There is tendencies of higher rate of pore water pressure generation in the samples with higher D_r.

3.3.2 *Effect of part drainage*

Three of the tests, Test 13, Test 14 and Test 15, are partly drained cyclic triaxial tests. The pore water dissipation for the first 100 cycles is illustrated in Figure 2.

The first and second packages for Test 13 are plotted in Figure 3. The response of the soil is similar. The $p' - q$ plot shows the same tendencies in the stress paths, more pore water pressure is generated in the first cycle of the test, and the first cycle after drainage. The phase transformation line ($d = 0$) is similar for the cycles in the first and the second package, and shows that the dilative behaviour of the soil is not affected by the drainage.

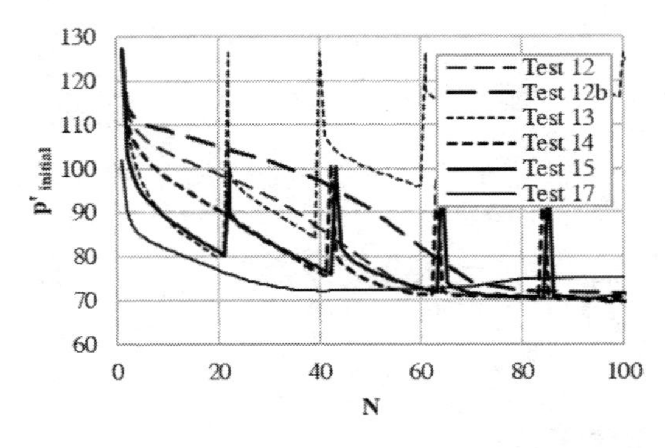

Figure 2. $p'_{initial}$ versus number of cycles, N.

3.3.3 *Mean stress condition*

The mean stress condition seems to be a state parameter. How the material got to the specific stress condition does not seem to have a significant impact on the behaviour of the material. Different cycles with a similar $p'_{initial}$ are compared, and the material behaves in a similar way in the different tests performed. Two examples are given in Figures 4 and 5. In Figure 4 all tests are represented with a cycle from the undrained part of the triaxial tests, Test 13 is in addition presented with a cycle after the first drainage. Despite deviation in pore water pressure generation, the different tests behave very similar when considering cycles starting at the same $p'_{initial}$.

The phase transformation line ($d = 0$) is the same for all tests in Figure 4, but Test 12 has a behaviour that is less dilative compared to the other tests seen in Figure 4b. The $p' - q$ plot verifies the less dilative behaviour as the butterfly-shape is not as prominent as in the other tests.

When studying the tests exposed to dissipation of water, the material behaviour does not seem to have changed after the drainage. In Figure 5, Test 14 and Test 15 have been exposed to dissipation of pore water pressure twice. The responses are similar to the tests run continuously undrained. There is deviation in the $\varepsilon_q^p - \eta$ plot in Figure 5, where the amplitude of the plastic shear strain is different. Test 14 and Test 15 with partial drainage are extremes, Test 15 with the smallest amplitude and Test 14 with the greatest amplitude. The same can also be seen in the $d\varepsilon_q^p - \eta$ plot in Figure 5. The difference in incremental strain and the amplitude for the two partly drained tests imply that there is no direct correlation between the dissipation of pore water pressure and change in incremental strains. However, the strains are rather small and the deviation is not distinct.

4 IMPLEMENTATION OF THE SANISAND CONSTITUTIVE MODEL

4.1 *Input parameters*

The SANISAND model is implemented in Excel Worksheet to simulate both monotonic and cyclic behaviour. Figures 6 and 7 illustrate how SANISAND reproduce the monotonic and cyclic triaxial tests on Siri sand respectively. The SANISAND input parameters have been fitted to the monotonic tests, and then developed further for the cyclic simulations. This is done by evaluating the different SANISAND formulations and how the different parameters affect the behaviour. The input parameters are separated into one set for compression and one for extension, and the same set of parameters has been applied in all

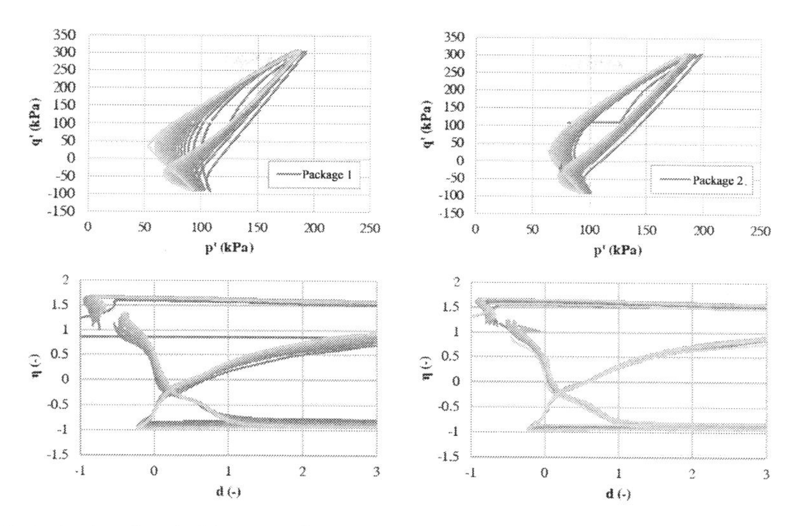

Figure 3. Partly drained cyclic triaxial test, package 1 and 2 in Test 13. The colour fades throughout the cycles. The material is contractant for $d > 0$, dilative for $d < 0$ and the phase transformation line is found in $d = 0$.

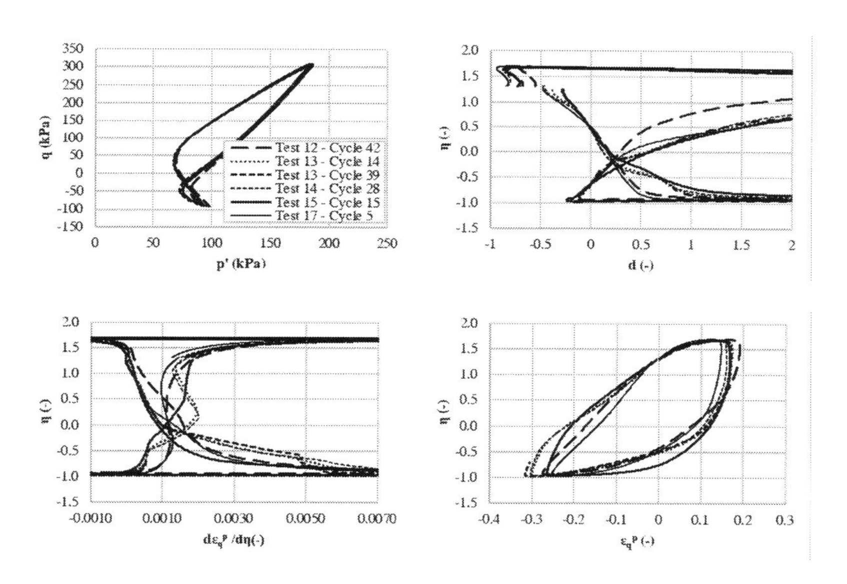

Figure 4. Comparison of cycles from various tests starting at p⊠ = 84 kPa.

cyclic simulations of the Siri sand. It has been observed in the results that due to some variation in stress condition and relative density of the tests, small modifications of the parameters are required to represent the behaviour of each test in an optimal manner. Table 2 gives a summary of parameters for monotonic simulations. Some parameters are changed to improve simulations of cyclic behaviour. These parameters, in addition to parameters describing fabric change, are listed in Table 3.

4.2 *Limitations of the SANISAND constitutive model*

Figure 6 shows the monotonic SANISAND simulation of Siri sand. The results show that the constitutive model simulates the response well with one unique set of parameters for drained and undrained behaviour in both compression and extension. However, the SANISAND model is significantly underestimating the initial dilatation/contraction in compression/extension. The model

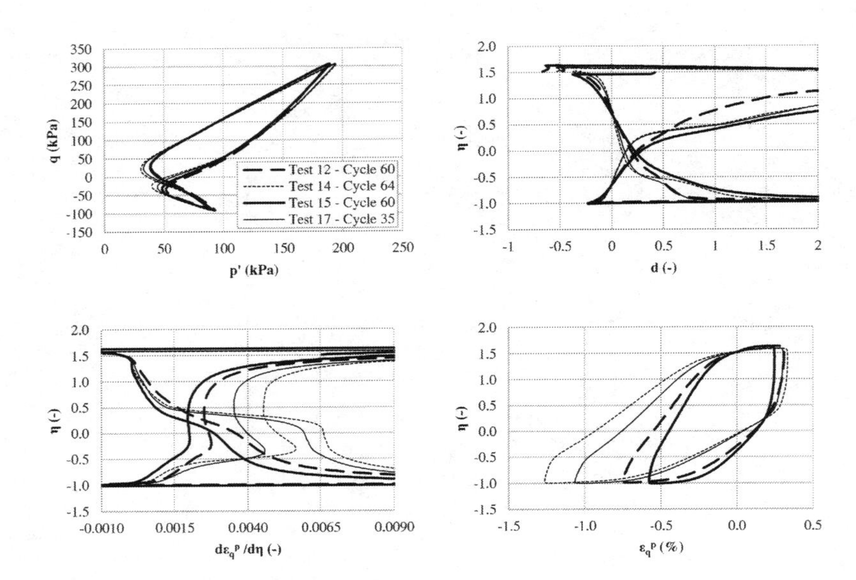

Figure 5. Comparison of cycles from various tests starting at p′ = 72 kPa.

is only able to simulate a linear trend in dilatancy behaviour, and it has been emphasised that the phase transformation line is described correctly. This has been successful by varying the material constant n^d in compression and extension.

Figure 7 shows the cyclic SANISAND simulation of Siri sand, and there are observed several limitations in the SANISAND model. The model only simulates linear behaviour of dilatancy, but predicts behaviour fairly well with similar trends and phase transformation lines that coincide. It is however noticed that the contraction behaviour after load reversal is significantly underestimated.

The cyclic triaxial tests on Siri sand show increasing incremental plastic shear strains throughout the cycles, in contrast to the SANISAND model which has an insignificant increase. As a consequence of this, SANISAND describes the total strains inaccurately. The SANISAND model predicts a constant amplitude in total strains for the chosen input parameters, but the triaxial test results display an increasing amplitude throughout the cycles.

To get a deeper understanding of how the different parameters in the SANISAND formulations affect the representation of the Siri sand, a comparison with the triaxial tests is carried out. The SANISAND formulations are applied to the triaxial data to back calculate the parameters h and h_0. Figure 8 shows the back calculation of the hardening parameter h from the triaxial data together with the development of h in the SANISAND model. The development of h in the triaxial data and the SANISAND model seem to have similar trends, but they do not coincide accurately.

Table 2. Summary of monotonic SANISAND parameters.

Input parameters		Compression	Extension
Elasticity	G_0	250	250
	ν	0.05	0.05
Critical state	M	1.49	0.90
	λ_c	0.013	0.013
	e_0	0.71	0.71
	ξ	0.67	0.67
Yield surface	m	0.0	0.0
Plastic modulus	h_0	5.00	8.00
	c_h	1.10	1.10
	n^b	1.30	2.00
Dilatancy	A_0	0.6	0.8
	n^d	5.00	7.00

Table 3. Cyclic SANISAND parameters.

Input parameters		Compression	Extension
Plastic modulus	h_0	7.00	8.00
	n^b	6.00	6.00
Dilatancy	A_0	0.6	0.5
	n^d	8.00	3.00
Fabric dilatancy	z_{max}	3.0	3.0
	c_z	100	100

The deviation in h explains the difference in incremental shear strains as pointed out above.

The parameter h_0 is a constant input parameter in the SANISAND model. However, when the

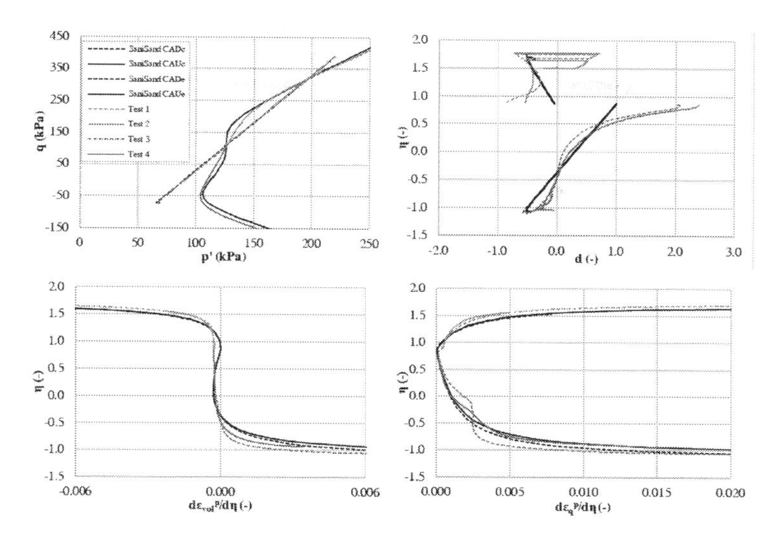

Figure 6. Monotonic SANISAND simulation.

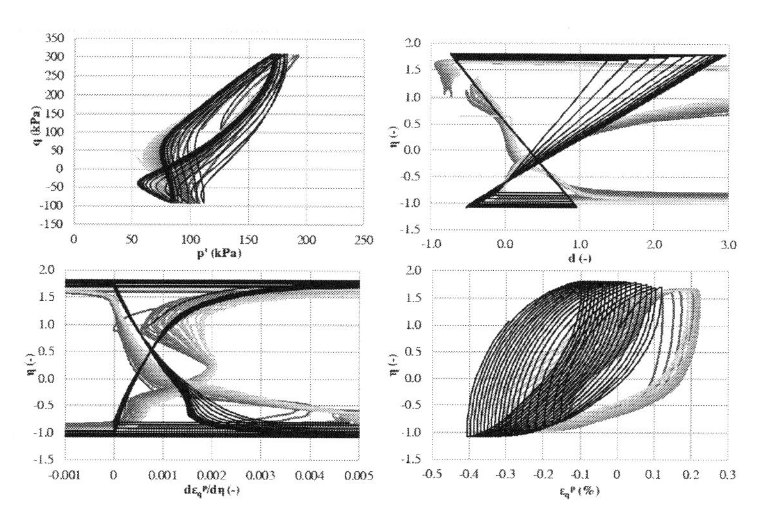

Figure 7. Undrained cyclic SANISAND simulation.

triaxial data is back calculated, the tests show a decreasing h_0 throughout the cycles. Figure 9 illustrates how the parameter h_0 develops with cycles at a specific stress ratio for the first 20 cycles of Test 13 and Test 15 and first 40 cycles of Test 12, Test 14 and Test 17.

To improve the prediction of the hardening of the sand, a equation of the hardening parameter, h_0, could be implemented. The function should decrease throughout the cycles, and this may give a better representation of the incremental and total shear strains.

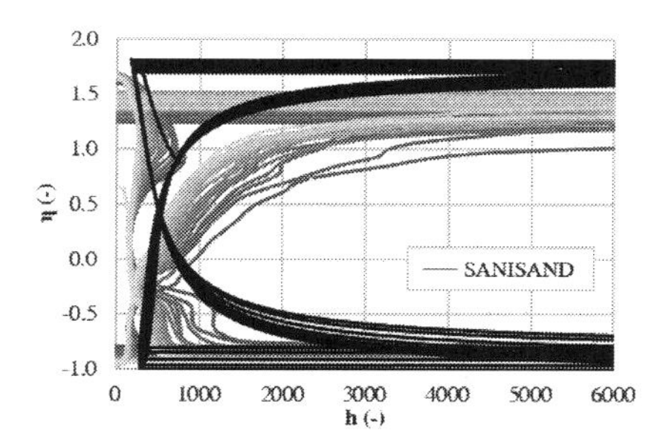

Figure 8. Comparison of h from the SANISAND formulation and triaxial test data.

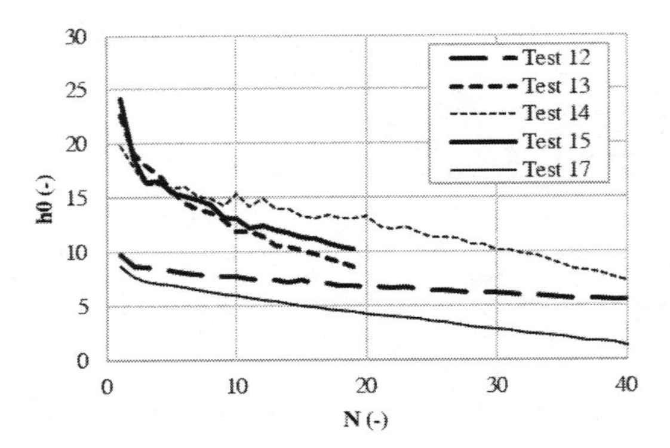

Figure 9. Back calculation of h_0 from different tests.

5 CONCLUSIONS

Interpretations of monotonic and cyclic triaxial tests within the framework of SANISAND have been performed. Deviations have been observed in the pore water pressure generation in the different cyclic triaxial tests. Sample preparation affects the properties of the sand, and is therefore most likely the reason for the deviations. It would be advantageous to develop a standardised sample preparation method that reduces the uncertainties in the results.

Cycles with the same initial mean stress from different tests have been compared, and there is a striking resemblance of the response during the individual cycles. This indicates that the dense sand material considers the mean stress condition as a state parameter. However, the rate of pore pressure accumulation is sensitive to the stress history.

This study indicates that the SANISAND model has some limitations. It has difficulties describing the Siri sand, and to describe the sand in an optimal manner it is necessary to use different properties in compression and extension. The representations of incremental plastic and volume strains are inaccurate, which result in inaccurate simulations of total shear and volume strains. It is suggested that the hardening parameter, h_0, is implemented as a decreasing function throughout the cycles. The intention is to obtain a better representation of the hardening of the sand, in addition to the incremental shear strains. It is also found that the contractant behaviour after reversing the load is underpredicted by the SANISAND model.

ACKNOWLEDGEMENTS

The authors wish to acknowledge the NGI internal research project, SP9 NGI foundations. A thank you is extended to Pasquale Carotenuto for organising and conducting the triaxial tests.

REFERENCES

Andersen, K.H. (2009). Bearing capacity under cyclic loadingoffshore, along the coast, and on land. *Canadian Geotechnical Journal 46*(5), 513–535.

Andersen, K.H. (2015). Cyclic soil parameters for offshore foundation design. *Frontiers in Offshore Geotechnics III 5.*

Bardet, J.P. (1986). Bounding surface plasticity model for sands. *Journal of Engineering Mechanics 112*(11), 1198–1217.

Been, K. & M.G. Jefferies (1985). A state parameter for sands. *Geotechnique 35*(2), 99–112.

Carotenuto, P. & Y. Suzuki (2016, 12). WP3 - Technical Note—Summary 2016. Technical report, NGI.

Dafalias, Y.F. (1986). Bounding surface plasticity. i: Mathematical foundation and hypoplasticity. *Journal of Engineering Mechanics 112*(9), 966–987.

Dafalias, Y.F. & M.T. Manzari (2004). Simple plasticity sand model accounting for fabric change effects. *Journal of Engineering Mechanics 130*(6), 622–634.

Dafalias, Y.F. & E.P. Popov (1975). A model of nonlinearly hardening materials for complex loading. *Acta Mechanica 21*(3), 173–192.

Jostad, H., G. Grimstad, K. Andersen, & N. Sivasithamparam (2015). A fe procedure for calculation of cyclic behaviour of offshore foundations under partly drained conditions. *Frontiers in Offshore Geotechnics III*, 153–172.

Manzari, M.T. & Y.F. Dafalias (1997). A critical state twosurface plasticity model for sands. *Geotechnique 47*(2), 255–272.

MathWorks (2017). *Signal Processing Toolbox™, User's Guide.* MathWorks.

Schofield, A.N. & C.P. Wroth (1968). *Critical State Soil Mechanics.* McGraw-Hill.

Taiebat, M. & Y.F. Dafalias (2007). Sanisand: Simple anisotropic sand plasticity model. *International journal for numerical and analytical methods in geomechanics 32*, 915–948.

Wang, Y. & X.S. Li (1998). Linear representation of steady-state line for sand. *Journal of Geotechnical and Geoenvironmental Engineering, ASCE 123*(7), 1215–1217.

Wood, D.M. (1990). *Soil Behaviour and Critical State Soil Mechanics.* Cambridge: Cambridge University Press.

Numerical Methods in Geotechnical Engineering IX – Cardoso et al. (Eds)
© 2018 Taylor & Francis Group, London, ISBN 978-1-138-33198-3

Numerical implementation of hardening soil model

L.J. Cocco & M.E. Ruiz
Structural Department of Structures, Faculty of Exact, Physical and Natural Sciences of the National University of Córdoba, Spain

ABSTRACT: The "hardening soil model" (HSM) is a very recognized constitutive model and has been proved to produce reasonable accurate results when applied to different geotechnical problems. The model incorporates in its formulation different aspects related to the complex non-linear behavior of soils, such as dependency of elastic modulus on the confined pressure, hyperbolic elastic behavior, two yield surfaces with non-associated, and associated plasticity and friction and dilatancy angles of the material. In this article the HSM is implemented based on a new plastic potential and the Matsuoka-Nakai failure criterion as proposed by Benz (Benz, 2007). The numerical implementation is performed through a user subroutine, and some minor modifications were performed to the original model. The results obtained are compared to those produced by Plaxis. A very good agreement between the models was found. Finally, the numerical implementation was used to model a boundary value problem of a spread footing.

1 INTRODUCTION

1.1 *The hardening soil model*

The numerical simulation of soil behavior through its path from rest to failure is a very complex task and involves different concepts present on soil mechanical behavior. Constitutive models try to establish relations between stresses and strains that represent soil behavior as closely as possible, and also bring with them suitable stiffness estimations. There are several constitutive models available in the literature. Some of them are based on critical states mechanics and others are based on experimental data or combine different kind of concepts in order to achieve accurate results. Among the available constitutive models, the Hardening Soil Model (HSM) has proved to produce reasonable accurate results by several authors when applied to different geotechnical problems. In addition to this appropriate performance, the HSM has another advantage: its user-friendliness that comes from the input parameters required to characterize the soil being analyzed. In order to be user-friendly, the parameters should be limited in their number, easy to understand in their physical meaning, and easy to quantify using common soils test data or the experience. The HSM fulfill appropriately these aspects, or at least performs better than other constitutive models.

The original HSM was developed by Schanz (Schanz, 1998; Schanz et al, 1999) based on the Double Hardening model by Vermeer (Vermeer, 1978). Consequently, the HSM also comprises ideas by Kondner (Kondner & Zelasko, 1963), Duncan & Chang (Duncan & Chang, 1970), Janbu (Janbu, 1963), and Rowe (Rowe, 1962). The model parameters are characterized by standard laboratory test, such as triaxial and oedometer tests. In its formulation incorporates different aspects based on the formerly mentioned ideas in which the HSM model stands on, such as dependency of elastic modulus on the confined pressure, hyperbolic elastic behavior, two yield surfaces with non-associated and associated plasticity and friction and dilatancy angles of the material. All these features make of the HSM a constitutive model suitable to simulate soil behavior in many boundary value problems with significant importance in geotechnical engineering.

1.2 *Numerical implementation and validation*

The numerical implementation of the HSM in this study is based on the version developed by Benz (Benz, 2007), in which a new plastic potential and the Matsuoka-Nakai failure criterion are proposed and implemented in order to enhance and update the original version of the HSM. In the numerical implementation of a constitutive model some issues might arrive and sometimes enforce to perform modifications to the original algorithm. In this study, some minor modifications were performed to the original model proposed by Benz in order to solve numerical complications encountered in the simulations that were carried out. The numerical implementation is performed using an implicit Euler scheme to integrate the constitutive equations. The consistent algorithmic tangent stiffness tensor for one and two yield surfaces active are determined, and finally a two surface return strategy proposed by Bonnier (Bonnier, 2000) is implemented.

The numerical implementation is performed through a user subroutine implemented in a finite element program and then validated for several stress paths. Triaxial tests (both in compression and in extension) along with an oedometric test were simulated in order to validate the new subroutine. Results obtained are compared with those produced by a Plaxis finite element model that used the built-in HSM constitutive law. A very good agreement of the results for both approaches was found and is shown in this paper. Finally, the numerical implementation was used to model a boundary value problem represented by the study of the bearing capacity of a spread footing.

2 HARDENING SOIL MODEL FORMULATION

2.1 *Yield functions and plastic potentials*

The HSM implemented in this study is based on the work performed by Benz (Benz, 2007), and therefore some changes in the original formulation of the HSM were made. The HSM has two yield functions, one controlling the deviatoric behavior (cone-type yield surface) and the other one controlling the volumetric behavior (cap-type yield surface). Two different types of plastic potentials are used for the yield surfaces. A non-associated plastic potential is used for the cone-type yield surface, and for the cap-type yield function an associated plastic potential is adopted. Precisely, in the two yield surfaces and plastic potentials modifications were made with respect to the original formulation of the HSM and a different failure criterion was adopted in this implementation. The classical Mohr-Coulomb failure criterion (Mohr, 1900) was replaced by the Matsuoka-Nakai criterion (Matsuoka & Nakai, 1982).

The cone-type yield function of the original version of the HSM needs to be modified in order to incorporate the new failure criterion proposed in this implementation. The original version of the yield function is presented in Equation 1, in which its relation to triaxial stress conditions is clearly shown.

$$f^s = \frac{q_a}{E_{50}} \frac{q}{q_a - q} - \frac{2q}{E_{ur}} - \gamma^{ps} \tag{1}$$

Where γ^{ps} is an internal material variable for the accumulated plastic deviatoric strain, $q = \sigma_1 - \sigma_3$ is defined for triaxial loading, q_a is the asymptotic deviatoric stress as defined in the original Duncan & Chang model (Duncan & Chang, 1970), and E_{50} is the stiffness modulus for primary loading. The stress-strain relation of soils in unloading and reloading can be approximated by a linear function, and also inside the yield function isotropic elasticity is assumed. The elastic unloading-reloading stiffness is characterized by E_{ur}.

To apply the Matsuoka-Nakai criterion the yield function is reformulated in terms of mobilized friction angle φ_m (Benz, 2007). The Mohr-Coulomb failure criterion in triaxial conditions yields:

$$\sin\varphi_m = \frac{\sigma_1 - \sigma_3}{\sigma_1 + \sigma_3 + 2c\cot\varphi} \tag{2}$$
$$q = \sin\varphi_m(\sigma_1 + \sigma_3 + 2c\cot\varphi)$$

and

$$q_f = \frac{2\sin\varphi}{1-\sin\varphi}(\sigma_3 + c\cot\varphi) \tag{3}$$

consequently

$$\frac{q}{q_a} = R_f\left(\frac{1-\sin\varphi}{\sin\varphi}\right)\left(\frac{\sin\varphi_m}{1-\sin\varphi_m}\right) \tag{4}$$

With the objective shear strain measure:

$$\gamma_s = \sqrt{\frac{1}{2}\left((\varepsilon_1-\varepsilon_2)^2 + (\varepsilon_2-\varepsilon_3)^2 + (\varepsilon_3-\varepsilon_1)^2\right)} \tag{5}$$

which reduces to $\gamma_s = \varepsilon_1 - \varepsilon_3$ in triaxial conditions, one obtains $\gamma_s = 3/2\varepsilon_1$ for zero volumetric strain. Adapting the original yield function of HSM to shear strain γ_s and considering Equation (4) results in Equation (6):

$$f_s = \frac{3}{2}\frac{q}{E_i}\frac{\left(\frac{1-\sin\varphi_m}{\sin\varphi_m}\right)}{\left(\frac{1-\sin\varphi_m}{\sin\varphi_m}\right) - R_f\left(\frac{1-\sin\varphi}{\sin\varphi}\right)} - \frac{3}{2}\frac{q}{E_{ur}} - \gamma_s^{ps} \tag{6}$$

where φ_m is the mobilized frictional angle in triaxial compression. In mobilized friction, the Matsuoka-Nakai yield criterion can be written as:

$$\sin^2\varphi_m = \frac{9 - \frac{I_1 I_2}{I_3}}{1 - \frac{I_1 I_2}{I_3}}. \tag{7}$$

Alternatively, Equation (7) can be expressed in terms of the Lode angle (θ) as:

$$\sin\varphi_m = \frac{3q}{6\chi(p + c\cot\varphi) + q} \tag{8}$$

where $\chi = \chi(\theta)$ and θ is defined in this article as:

$$\theta = \frac{1}{3}\arcsin\left(\frac{3\sqrt{3}J_3}{2J_2^{3/2}}\right) \tag{9}$$

where J_2 and J_3 are the second and third deviatoric stress invariants, and $\chi(\theta)$ is defined as follow:

$$\chi(\theta) = \frac{\sqrt{3}\delta}{2\sqrt{\delta^2 - \delta + 1}}\frac{1}{\cos\vartheta}$$
$$with\begin{cases} \vartheta = \frac{1}{6}\arccos\left(-1 + \frac{27\delta^2(1-\delta)^2}{2(\delta^2 - \delta + 1)^3}\sin^2(3\theta)\right) for\theta \leq 0 \\ \vartheta = \frac{\pi}{3} - \frac{1}{6}\arccos\left(-1 + \frac{27\delta^2(1-\delta)^2}{2(\delta^2 - \delta + 1)^3}\sin^2(3\theta)\right) for\theta > 0 \end{cases}$$
$$\delta = \frac{3 - \sin\varphi}{3 + \sin\varphi} \tag{10}$$

The plastic potential to the cone-type yield surface is defined as:

$$g^s = (p + c\cot\varphi)\frac{6\sin\psi_m}{3 - \sin\psi_m} \tag{11}$$

in which,

$$\sin\psi_m = \begin{cases} \frac{\sin\varphi_m - \sin\varphi_{cs}}{1 - \sin\varphi_m\sin\varphi_{cs}} for\sin\varphi_m - \sin\varphi_{cs} \geq 0 \\ \frac{1}{10}\left(Me^{\frac{1}{15}\ln(p_{cs}/p)} - \eta\right) for\sin\varphi_m - \sin\varphi_{cs} < 0 \end{cases}$$
$$\sin\varphi_{cs} = \frac{\sin\varphi - \sin\psi}{1 - \sin\varphi\sin\psi} \tag{12}$$
$$\frac{p_{cs}}{p} = \frac{\eta\sin\varphi_{cs}(1 - \sin\varphi_m)}{M\sin\varphi_m(1 - \sin\varphi_{cs})}$$

As the cone-type yield surface was modified, the cap-yield surface has to be reformulated . The Roscoe invariant $q = \sqrt{3}J_2$ is used in its formulation, and also the Lode angle dependency of the cone-type yield surface is translated to the cap by scaling its steepness. Consequently, the cap-yield surface could be written as:

$$f^c = \frac{q^2}{(\chi\alpha)^2} - p^2 - p_p^2. \tag{13}$$

The cap-type yield surface uses the associated plastic potential.

$$g^c = \frac{q^2}{(\chi\alpha)^2} - p^2 - p_p^2. \tag{14}$$

The cap's deviatoric plastic flow direction is thus consistent with the cone's radial Drucker-Prager potential. Finally, the evolution laws for the objective shear strain measure are as follows:

$$d\gamma_s^{ps} = d\lambda^s h_{\gamma_s^{ps}}$$
$$h_{\gamma_s^{ps}} = \sqrt{\frac{1}{2}\left(\left(\frac{\partial g}{\partial\sigma_1} - \frac{\partial g}{\partial\sigma_2}\right)^2 + \left(\frac{\partial g}{\partial\sigma_2} - \frac{\partial g}{\partial\sigma_3}\right)^2 + \left(\frac{\partial g}{\partial\sigma_3} - \frac{\partial g}{\partial\sigma_1}\right)^2\right)} = \frac{3}{2}$$
$$dp_p = d\lambda^c h_{p_p}$$
$$h_{p_p} = 2H\left(\frac{\sigma_3 + c\cot\varphi}{p^{ref} + c\cot\varphi}\right)^m p \tag{15}$$

For further explanations and a complete detail of developments of the HSM formulation implemented in this article, the reader is referred to Benz (2007).

2.2 *HSM parameters*

The HSM, as most of the isotropic models based on critical states, has a number of parameters in between the simplest and complex constitutive models. For this reason, the HSM has an enormous applicability and also because most of its parameters might be determined through laboratory tests. The HSM, as was implemented in this study, has eleven user-defined parameters and three internal parameters. The HSM parameters are summarized in Table 1. The internal parameters are determined in an optimization scheme that will be explained in the next section.

The model presents stress dependent stiffness, which allows to characterize the soil independently of the confining pressure. Consequently, the stiffness is defined for a reference pressure and is updated throughout the analysis for each integration point. For this reason, in Table 1 the parameters related to stiffness are referred to as "reference". The power law used to consider the confining pressure is the following:

Table 1. Hardening soil model parameters.

User-defined parameters	Symbol	Units
Triaxial secant stiffness	E_{50}^{ref}	[kN/m²]
Oedometric tangent stiffness	E_{oed}^{ref}	[kN/m²]
Unloading/reloading stiffness	E_{ur}^{ref}	[kN/m²]
Power of stress dependency	m	[-]
Cohesion (effective)	c	[kN/m²]
Friction angle (effective)	φ	[°]
Dilatancy angle	ψ	[°]
Poisson's ratio	ν_{ur}	[-]
Reference stress for stiffness	p_{ref}	[kN/m²]
K_0-value (normal consolidation)	K_0^{nc}	[-]
Failure ratio	R_f	[-]
Internal parameters	**Symbol**	**Units**
Initial secant stiffness	E_i^{ref}	[kN/m²]
Cap parameter (steepness)	α	[-]
Cap parameter (stiffness ratio)	K_s/K_c	[-]

$$E_i = E_i^{ref}\left(\frac{\sigma_3 + c\cot\varphi}{p^{ref} + c\cot\varphi}\right)^m$$
$$E_{ur} = E_{ur}^{ref}\left(\frac{\sigma_3 + c\cot\varphi}{p^{ref} + c\cot\varphi}\right)^m \quad (16)$$

where σ_3 is the minor principal stress.

3 NUMERICAL IMPLEMENTATION

3.1 *General aspects*

The numerical implementation of the HSM performed in this study is based on the concepts of classical plasticity and therefore uses a return mapping algorithm to integrate the constitutive equations. In this implementation an implicit backward Euler scheme was used in the integration mentioned. The strategy proposed by Bonnier (Bonnier, 2000) was used in order to determine which yield surface is active and, therefore, the trial stress should returned to.

Some issues were found during the numerical implementation of the model and solutions adopted to overcome those issues are considered interesting to remark, as describe in the next paragraphs, where each one of the problems found and its solution adopted are described.

3.2 *Internal parameters optimization*

As was mentioned in the former section, the model has three internal parameters, two of them related to the shape of the cap-yield surface and one related to the initial stiffness of the material. The determination of the three internal parameters is performed through an optimization process. This process requires that, in oedometric conditions, the K_0^{nc} and the E_{oed} predicted by the model should be the same as those specified by the user (this condition is strongly dominated by α and K_s/K_c); simultaneously the triaxial compression curve predicted by the model must fit the experimental curve (this condition is strongly dominated by E_i^{ref}).

The conditions required to be accomplished in the oedometric test can be numerically modeled in a single integration gauss point, in order to simplify the problem and make the process faster. A strain driven oedometric test is conducted with a very small vertical strain amplitude ($\Delta\varepsilon = 1 \times 10^{-6}$) and the tangent oedometric modulus is estimated as $E_{oed} = \delta\sigma/\delta\varepsilon = \Delta\sigma/\Delta\varepsilon$ and K_0^{nc} is calculated as $K_0^{nc} = \Delta\sigma_h/\Delta\sigma_v$. The test continues until the vertical stress reaches the reference stress. The so explained calculated parameters are used to estimate the numerical derivatives with respect to the internal parameters α and K_s/K_c, which are needed to perform the optimization process required to determine appropriate internal parameters.

Simultaneously, a triaxial test is modeled numerically and compared to an actual test. Thus, the third internal parameter can be determined (E_i^{ref}) through the optimization process. This parameter controls the initial stiffness of the model in triaxial conditions and is one of the stiffness parameters used in the cone-type yield function. To avoid excessive iterations, a suitable value of the parameter to start the optimization process can be determined using the following expression (Plaxis, 2014):

$$E_i^{ref} = \frac{2E_{50}^{ref}}{\left(2 - R_f\right)} \quad (17)$$

3.3 *Consistent algorithmic tangent stiffness tensor*

Most of the commercial finite element programs use a Newton-Raphson approach to solve the global equilibrium equations. Consequently, to preserve its quadratic convergence it is needed a consistent algorithmic tangent stiffness tensor. The consistent elastoplastic algorithmic tangent stiffness was originally introduced by Hughes and Taylor (Hughes & Taylor, 1978) for viscoplastic materials. Then, Simo and Taylor (Simo & Taylor, 1985) used this concept and applied to elastoplastic materials.

For the HSM implemented in this study, three different consistent algorithmic tangent stiffness tensors were determined, one for the cone-type yield surface, another one for the cap-type yield surface when these surfaces are active individually, and finally, one when the two yield surfaces are active simultaneously. In last case, the approach proposed by Koiter (Koiter, 1960) was applied to additively decompose the plastic strain rates when two surfaces are active.

The expression for the consistent algorithmic tangent stiffness tensor when one yield surface is active can be derived enforcing the consistency condition for the yield surface active, as follows:

$$\begin{aligned} d^{n+1}f &= \left.\frac{\partial f}{\partial \sigma_{ij}}\right|_{n+1} d\sigma_{ij} + \left.\frac{\partial f}{\partial q_*}\right|_{n+1} dq_* \\ &= {}^{n+1}n_{ij} d\sigma_{ij} + {}^{n+1}\xi_* dq_* = 0. \end{aligned} \tag{18}$$

The increment of stress $\delta\sigma_{ij}$ is obtained by linearization or differentiation of the Equation (19).

$${}^{n+1}\sigma_{ij} = {}^{Trial}\sigma_{ij} - \lambda D_{ijkl}\,{}^{n+1}m_{kl} \tag{19}$$

yielding to:

$$\begin{aligned} d\sigma_{ij} &= D_{ijkl} d\varepsilon_{kl} - d\lambda D_{ijkl}\,{}^{n+1}m_{kl} \\ &-\lambda D_{ijkl} \left.\frac{\partial m_{kl}}{\partial \sigma_{mn}}\right|_{n+1} d\sigma_{mn} - \lambda D_{ijkl} \left.\frac{\partial m_{kl}}{\partial q_*}\right|_{n+1} dq_* \end{aligned} \tag{20}$$

isolating the increment of stress:

$$d\sigma_{mn} = \Xi_{ijmn} D_{ijkl} \left(d\varepsilon_{kl} - d\lambda \Theta_{kl} \right) \tag{21}$$

where

$$\begin{aligned} \Xi_{ijmn} &= \left(\delta_{im}\delta_{nj} + \lambda D_{ijkl} \left.\frac{\partial m_{kl}}{\partial \sigma_{mn}}\right|_{n+1} \right)^{-1} \\ \Theta_{kl} &= {}^{n+1}m_{kl} + \lambda \left.\frac{\partial m_{kl}}{\partial q_*}\right|_{n+1} h_*. \end{aligned} \tag{22}$$

Finally, substituting Equation (21) in Equation (18) the expression for the consistent algorithmic tangent stiffness tensor is obtained.

$$\begin{aligned} d\sigma_{pq} &= D^{ep}_{pqmn} d\varepsilon_{mn} \\ D^{ep}_{pqmn} &= \left(\Xi_{rspq} D_{rsmn} \right) - \frac{\left(\Xi_{rspq} D_{rskl} \right) \Theta_{kl}\,{}^{n+1}n_{ij} \left(\Xi_{rsij} D_{rsmn} \right)}{{}^{n+1}n_{ot} \left(\Xi_{rsot} D_{rspq} \right) \Theta_{pq} + {}^{n+1}\xi_* h_*} \end{aligned} \tag{23}$$

When two yield surfaces are active at the same time, is necessary to use the approach proposed by Koiter, which is described in detail by De Borst (De Borst, 1987). The approach additively decomposes plastic strain rates as follows:

$$d\varepsilon^{p}_{ij} = d\lambda_1 \frac{\partial g_1}{\partial \sigma_{ij}} + d\lambda_2 \frac{\partial g_2}{\partial \sigma_{ij}} \tag{24}$$

where g_1 and g_2 are the plastic potentials of the two yield surfaces active; and $d\lambda_1$ and $d\lambda_2$ are the corresponding plastic multipliers. To obtain the consistent algorithmic tangent stiffness tensor, as was done for the case when one yield surface is active, is necessary to enforce the consistency condition. In this case, the consistency condition must be fulfilled by the two yield surfaces active.

$$\begin{aligned} d^{n+1}f_1 &= \left.\frac{\partial f_1}{\partial \sigma_{ij}}\right|_{n+1} d\sigma_{ij} + \left.\frac{\partial f_1}{\partial q_1}\right|_{n+1} dq_1 = 0 \\ d^{n+1}f_2 &= \left.\frac{\partial f_2}{\partial \sigma_{ij}}\right|_{n+1} d\sigma_{ij} + \left.\frac{\partial f_2}{\partial q_2}\right|_{n+1} dq_2 = 0 \end{aligned} \tag{25}$$

The increment of stress $\delta\sigma_{ij}$ is, in this case, by linearization of differentiation of the following expression:

$${}^{n+1}\sigma_{ij} = {}^{Trial}\sigma_{ij} - \lambda_1 D_{ijkl}\,{}^{n+1}m_{kl} - \lambda_2 D_{ijkl}\,{}^{n+1}w_{kl} \tag{26}$$

Following a process similar to that presented for a single yield surface active, the consistent algorithmic tangent stiffness tensor is obtained. This expression is shown in Equation (28) with the following definitions:

$$\begin{aligned} \mu_1 &= {}^{n+1}n_{ij} \Xi_{ijmn} D_{ijkl} \Theta^1_{kl} - \xi_1 h_1 \\ \mu_2 &= {}^{n+1}n_{ij} \Xi_{ijmn} D_{ijkl} \Theta^2_{kl} \\ \mu_3 &= {}^{n+1}u_{ij} \Xi_{ijmn} D_{ijkl} \Theta^1_{kl} \\ \mu_4 &= {}^{n+1}u_{ij} \Xi_{ijmn} D_{ijkl} \Theta^2_{kl} - \xi_2 h_2 \end{aligned} \tag{27}$$

$$\begin{aligned} D^{ep}_{pqmn} &= \left(\Xi_{rspq} D_{rsmn} \right) \\ &- \left[\frac{\left(\Xi_{rspq} D_{rskl} \Theta^1_{kl}\,{}^{n+1}n_{ij} \Xi_{rsij} D_{rsmn} \right) \mu_4}{\mu_1\mu_4 - \mu_3\mu_2} - \frac{\left(\Xi_{rspq} D_{rskl} \Theta^1_{kl}\,{}^{n+1}u_{ij} \Xi_{rsij} D_{rsmn} \right) \mu_2}{\mu_1\mu_4 - \mu_3\mu_2} \right] \\ &- \left[\frac{\left(\Xi_{rspq} D_{rskl} \Theta^2_{kl}\,{}^{n+1}u_{ij} \Xi_{rsij} D_{rsmn} \right) \mu_1}{\mu_1\mu_4 - \mu_3\mu_2} - \frac{\left(\Xi_{rspq} D_{rskl} \Theta^2_{kl}\,{}^{n+1}n_{ij} \Xi_{rsij} D_{rsmn} \right) \mu_3}{\mu_1\mu_4 - \mu_3\mu_2} \right] \end{aligned} \tag{28}$$

When the consistent algorithmic tangent stiffness tensors were applied in the current implementation of HSM, numerical instabilities were found in triaxial tests, especially when failure approaches. These problems encountered, forced to disregard the consistent algorithmic tangent stiffness tensors and use the elastic stiffness tensor. The use of the elastic stiffness tensor gives a very robust iterative procedure as long as the material stiffness does not increase, even when using non-associated plasticity models (Brinkgreve, 1994; Sloan et al., 2000).

3.4 *Cone-type yield surface*

The implementation of the renewed HSM as it was described in its formulation, presented some numerical inconsistencies. Particularly, in the evolution of the mobilized friction angle in a triaxial test from zero to its maximum value at failure, some numerical problems were encountered in the cone-type yield surface. As a triaxial test is an increasing monotonic loading, the expected behavior of the numerical model is to undergo plastic behavior from start to end. However, results showed an initially plastic behavior and then, suddenly, an elastic response near to failure friction angle values.

A close analysis of this issue showed that the problem was due to the presence of a factor in the cone-type yield surface, involving the mobilized friction angle, the failure ratio, and the friction angle. Equation (29) shows the expression of the cone-type yield surface as defined by Benz (Benz, 2007).

$$f_s = \frac{3}{2}\frac{q}{E_i}\frac{\left(\dfrac{1-\sin\varphi_m}{\sin\varphi_m}\right)}{\left(\dfrac{1-\sin\varphi_m}{\sin\varphi_m}\right) - R_f\left(\dfrac{1-\sin\varphi}{\sin\varphi}\right)} - \frac{3}{2}\frac{q}{E_{ur}} - \gamma_s^{ps} \tag{29}$$

where the factor responsible for the inconsistencies is:

$$\frac{\left(\dfrac{1-\sin\varphi_m}{\sin\varphi_m}\right)}{\left(\dfrac{1-\sin\varphi_m}{\sin\varphi_m}\right) - R_f\left(\dfrac{1-\sin\varphi}{\sin\varphi}\right)} \tag{30}$$

This factor entails a singularity in its definition that is evident when represented against the mobilized friction angle, as shown in Figure 1, in which the failure ratio and the friction angle are kept constant.

In order to overcome this issue a change in the cone-type yield surface was introduced. The

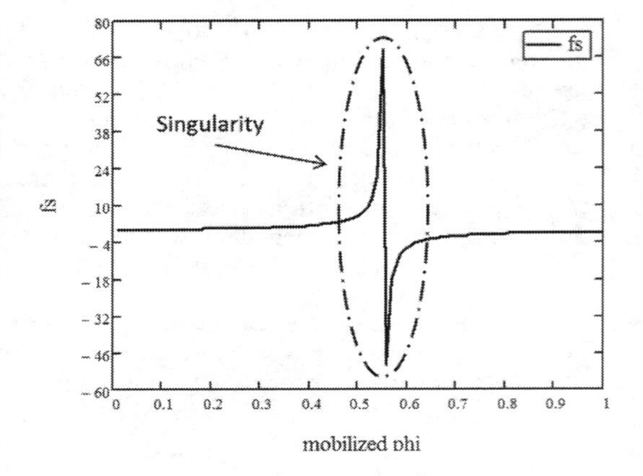

Figure 1. Factor present on cone-type yield surface versus mobilized friction angle.

proposed modification was to multiply the whole yield function by the inverse of the factor analyzed.

This solution has proved to be efficient and robust in solving the numerical inconsistencies and problems encountered. Although, the modified factor presents rather different values than the original factor, in the whole range of mobilized friction angle, this does not modify the main function of the cone-type yield surface, i.e. determine stress states which involve plasticity. It is also important to note that neither numerical results nor the accuracy of the model have been affected, even though the absolute value of the factor has been significantly modified for a wide range of the mobilized friction angle, as shown later in this paper.

3.5 *Apex return algorithm and sub-stepping*

Another singularity in the HSM was found at the apex of the yield surface in its intersection with the hydrostatic axis. Because of this, the same return algorithm to the cone-type yield surface cannot be applied to the apex region. To overcome this issue, the apex region must be determined and a different return strategy should be applied. In this study this was achieved limiting the apex region by the gradient of the cone-type plastic potential. All the stress points that lay on the left of this limit return to the apex, with a new strategy, and the stress points that lay on the right of the limit return to cone type yield surface, with the return algorithm presented previously.

The return algorithm used in the apex region simply consist in enforcing the stress point to return to the apex. This means that the stress point, once that has returned to the apex, has a zero deviatoric component and the mean stress is equal to

$c \cdot cot(\varphi)$. In this situation the whole deformation is considered plastic and there is no hardening in the yield surfaces.

In order to appropriately represent the approach to failure of the material, the numerical implementation of the HSM needed a sub-stepping scheme. Precisely, it was needed to capture the evolution of the mobilized friction angle from its initial value to its failure value. There are different sub-stepping schemes proposed in the literature (Sloan, 1987), however the simplest one is just to divide the increment analyzed into a number of smaller increments. This scheme was selected in this implementation and each increment is divided into one hundred increments, when the trial mobilized friction angle is higher than the friction angle specified by the user. This scheme has proved to be robust and suitable for the cases analyzed.

4 RESULTS

The validation of the numerical implementation of the HSM was performed through the numerical modeling of four triaxial tests and one oedometric test. In these tests, different stress path were followed to evaluate the performance of the implementation. The four triaxial test performed were: i) Triaxial Compression Loading (TCL), ii) Triaxial Compression Unloading (TCU), iii) Triaxial Extension Loading (TEL), iv) Triaxial Extension Unloading (TEU).

The model parameters used to perform the triaxial tests and the oedometric test are summarized in Table 2 and correspond to a sample of dense Hostun sand at a reference pressure of 100 kPa.

Table 2. Model parameters of dense hostun sand.

User-defined parameters	Values	Units
E_{50}^{ref}	30000	[kN/m^2]
E_{oed}^{ref}	30000	[kN/m^2]
E_{ur}^{ref}	90000	[kN/m^2]
m	0.55	[-]
c	0	[kN/m^2]
φ	42	[°]
ψ	16	[°]
ν_{ur}	0.25	[-]
p_{ref}	100	[kN/m^2]
K_0^{nc}	0.40	[-]
R_f	0.90	[-]
Internal parameters	**Values**	**Units**
E_i^{ref}	65000	[kN/m^2]
α	1.46	[-]
H	72028	[-]

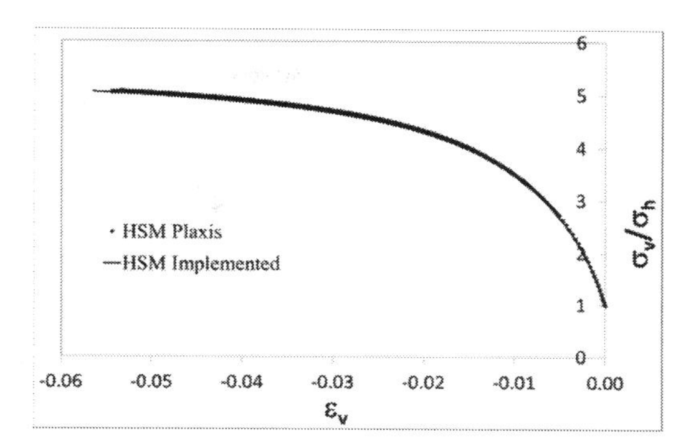

Figure 2. Results of stress-strain TCL test.

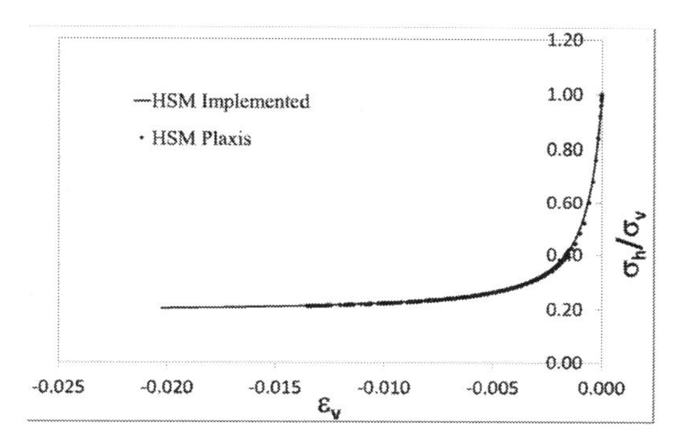

Figure 3. Results of stress-strain TCU test.

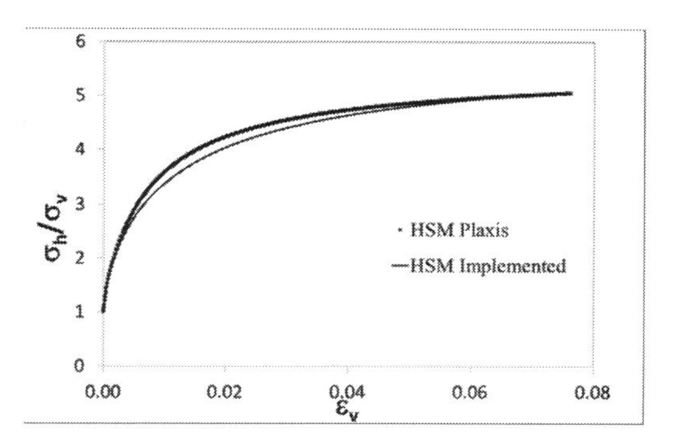

Figure 4. Results of stress-strain TEL test.

The results of the triaxial and oedometric tests obtained by the numerical implementation are compared to those obtained by a numerical model analyzed using the HSM constitutive law built-in the commercial finite element software, Plaxis. From Figure 2 through Figure 6 the four triaxial tests and the oedometric test are depicted.

The results obtained by the numerical implementation proposed in this study are considered,

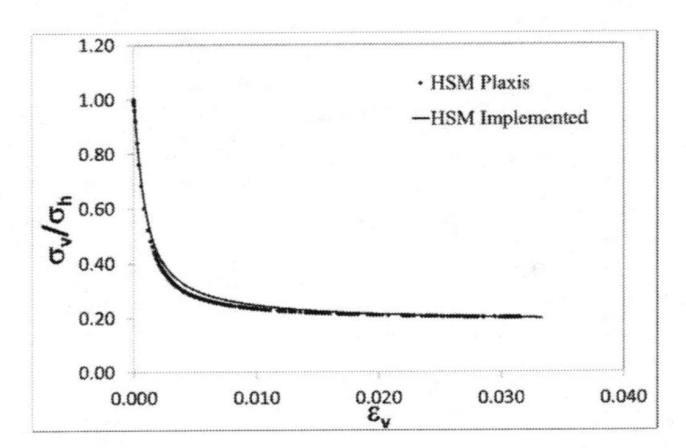

Figure 5. Results of stress-strain TEU test.

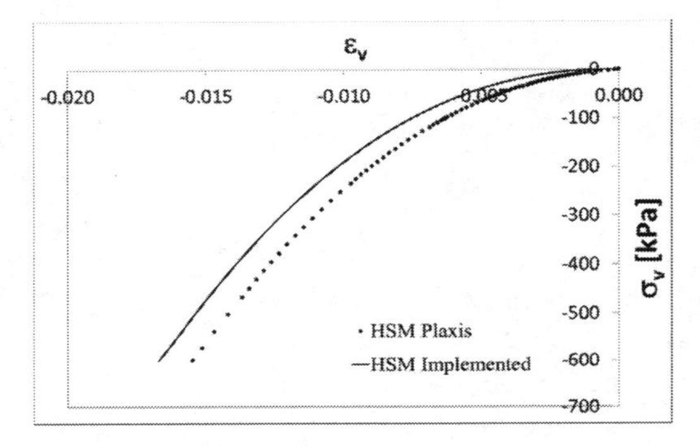

Figure 6. Results of oedometric test.

Table 3. Model parameters of BVP.

User-defined parameters	Values	Units
E_{50}^{ref}	10000	[kN/m^2]
E_{oed}^{ref}	10000	[kN/m^2]
E_{ur}^{ref}	30000	[kN/m^2]
m	0.50	[–]
c	10	[kN/m^2]
φ	30	[°]
ψ	5	[°]
ν_{ur}	0.20	[–]
p_{ref}	100	[kN/m^2]
K_0^{nc}	0.50	[–]
R_f	0.90	[–]
Internal parameters	**Values**	**Units**
E_i^{ref}	21000	[kN/m^2]
α	1.04	[–]
H	25836	[–]

in general terms, accurate enough in order to capture the soil behavior and, therefore, validate the numerical implementation of the model.

Table 4. Results of BVP.

Model	Bearing Capacity [kPa]	Ultimate vertical displacement [m]
HSM plaxis	660	–0.49
HSM implemented	660	–0.36

Once the numerical implementation was validated, a boundary value problem (BVP), which consisted on the simulation of a 2.0 meters wide spread foundation, was simulated to evaluate the performance of the implementation of the HSM study in this article. The parameters used to perform this analysis are summarized in Table 3.

The bearing capacity of the spread foundation and the ultimate vertical displacement are listed in Table 4. The results obtained show a stiffer response produced by the HSM implemented in this study, when compared to the one obtained with Plaxis.

One of the causes of this stiffer response might be the different type of elements used in the two models (for the HSM Plaxis quadratic triangular elements were used and for the HSM implemented bilinear quadrilateral elements were used). In addition to this, some differences might be due to different numerical implementation of the constitutive model, integration scheme of the equation, general solver algorithm, etc.

5 CONCLUSIONS

A renewed version of the HSM based on the work by Benz (Benz, 2007) was numerically implemented in a finite element program through a user-subroutine. Minor modifications were performed to the original model proposed by Benz in order to solve numerical problems encountered in the simulations that were carried out. A very good agreement of the results from the two models (HSM implemented and Plaxis) was found, which allows to conclude that the modified HSM implemented as an user material is capable of adequately capture the main characteristics of soil behavior.

The BVP of a spread foundation was modeled in order to obtain its bearing capacity and its associated ultimate displacement. For this case the results show a good agreement in the prediction of the ultimate bearing capacity of the footing, although the ultimate displacement predicted by the modified HSM model was 36% smaller than the displacement estimated by Plaxis. Despite of these differences encountered in the BVP, the numerical implementation of the modified HSM has brought an insight in the constitutive model itself and also in the general plasticity framework and in its numerical application. These aspects are crucial to enhance the implementation performed and to overcome the problems encountered in future developments.

REFERENCES

Benz, T. 2007. Small-Strain Stiffness of Soils and its Numerical Consequences. PhD Thesis Dissertation. University of Stuttgart. Stuttgart, Germany.

Bonnier, P.G. 2000. Implementational aspects of constitutive modelling. SCMEP Workshop N° 1 at NTNU. Trondheim, Norway.

Brinkgreve, R.B.J. 1994. Geomaterial Models and Numerical Analysis of Softening. PhD Thesis Dissertation. Delft University of Technology. Delft, The Netherlands.

De Borst, R. 1987. Integration of plasticity equations for singular yield functions. Computers & Structures. Vol. 26, N° 5: 823–829.

Duncan, J.M. & Chang, C.Y. 1970. Nonlinear analysis of stress and strain in soil. Proc. ASCE: Journal of the Soil Mechanics and Foundation Division. 96: 1629–1653.

Hughes, T.J.R. & Taylor, R.L. 1978. Unconditionally stable algorithms for quasi-static elasto/viscoplastic finite element analysis. Computers and Structures. 8: 169–173.

Janbu, N. 1963. Soil compressibility as determined by oedometer and triaxial tests. Proc. 3rd ECSMFE. Vol. 1: 19–25. Wiesbaden.

Koiter, W.T. 1960. General theorems for elastic-plastic solids. Progress in Solid Mechanics. Vol. 1: 165–221. North-Holland, Amsterdam.

Kondner, R.L. & Zelasko, J.S. 1963. A hyperbolic stress-strain formulation for sands. 2nd Pan. Am. Conference in Soil Mechanics Foundation Engineering. Vol. 1: 289–394. Brazil.

Matsuoka, H. & Nakai, T. 1982. A new failure criterion for soils in three dimensional stresses. IUTAM Conference on Deformation and Failure of Granular Material. pp.: 289–394. Delft, The Netherlands.

Mohr, O. 1900. Welche Umstände bedingen die Elastizitätsgrenze und den Bruch eines Materials?. VDI-Zeitschrift. 44: 1524.

Plaxis, 2014. Material Models Manual. Plaxis Company (Plaxis bv), Delft, The Netherlands.

Rowe, P.W. 1962. The stress-dilatancy relation for static equilibrium of an assembly of particles in contact. Proc. of the Royal Society of London. Series A, Mathematical and Physical Science. Vol. 269: 500–527.

Schanz, T. 1998. Zur Modellierung des mechanischen Verhaltens von Reibungsmaterialien. Mitt. Inst. für Geotechnik 45. University ofStuttgart. Stuttgart, Germany.

Schanz, T., Vermeer, P.A. & Bonnier, P.G. 2000. The Hardening Soil Model: Formulation and Verification. Beyond 2000 in Computational Geotechnics. pp.: 281–290. Rotterdam: Balkema.

Simo, J.C. & Taylor, R.L. 1985. Consistent tangent operators for rate-independent elastoplasticity. Computer Methods in Applied Mechanics and Engineering. 48: 101–118.

Sloan, S.W. 1987. Substepping schemes for the numerical integration of elastoplastic stress-strain relations. International Journal for Numerical Methods in Engineering. Vol. 24, Issue 5: 893–911.

Sloan, S.W., Sheng, D. & Abbo, A.J. 2000. Accelerated initial stiffness schemes for elastoplasticity. International Journal for Numerical Methods and Analytical Methods in Geomechanics. 24: 579–599.

Vermeer, P.A. 1978. A double hardening model for sand. Géotechnique. Vol. 28, Issue 4: 413–433.

Numerical Methods in Geotechnical Engineering IX – Cardoso et al. (Eds)
© 2018 Taylor & Francis Group, London, ISBN 978-1-138-33198-3

A hypoplastic model for soft clays incorporating strength anisotropy

J. Jerman & D. Mašín
Faculty of Science, Charles University, Prague, Czech Republic

ABSTRACT: The paper presents a new hypoplastic model for soft clays accounting for their typical feature—strength anisotropy. The model is based on the latest version of the hypoplastic model for clays (Mašín 2014), enhanced by the anisotropic shape of the state boundary surface. It has been shown that by skewing of the asymptotic state boundary surface the model predicts different ultimate strength of the material. Additionally, the tensor L is made bilinear in D to more realistically predict the stress path. The new model has been evaluated by simulating laboratory experiments on soft marine clays (Corral & Whittle 2010) involved in the Nicoll highway collapse in Singapore, a prominent case of geotechnical failure. Furthermore, other typical soft clays were used for calibration, such as Bangkok clay. The significant advantage of the model lies in its easy calibration using only simple undrained triaxial and oedometer tests.

1 INTRODUCTION

The construction on soft soil deposits has become increasingly important in past decades, since urban areas at many places around the world become congested due to increasing population, it is often necessary to reclaim land. For land reclamation local soils such as clay slurry and soft clay from the seabed are usually used. Soft clay seabeds are typical in locations of hydrocarbon and renewable energy sources, which provides a significant economic motivation for better developing our understanding of their behavior. Very soft clays occur in many coastal regions (such as in Australia and southeast Asia), Mexico City can be named as another typical example. Problems in the construction of deep excavations and tunnels can arise due to occurrence of deep deposits of normally or nearly normally consolidated clay. These soils are difficult to characterize and model; their stress-strain behavior is complex.

The two most important characteristics of a constitutive model is firstly the accuracy in reproducing the soil behavior, and secondly the model used for this purpose should be easy to calibrate on the basis of laboratory experiments performed in a standard experimental equipment available in practice. The developed model presents an advanced constitutive model for soft natural clays that fulfills these requirements.

Non-linearity and stiffness reduction from very small strain range to large strains are very well-known characteristics of soil behavior—however, these features are not predicted by typical elasto-plastic models such as Modified Cam clay model and the behavior inside and on state boundary surface are strictly distinct, thus not allowing to model soil non-linearity. The fundamental difference between group of conventional elasto-plastic models and hypoplastic models is in the separation of elastic and plastic deformations in elasto-plastic models, while hypoplastic models allow for irreversible strains inside state boundary surface and do not decompose the strain rate into elastic and plastic parts. In case of elasto-plastic models, incorporation of non-linearity brings more complex constitutive models (for example kinematic hardening models such as the model by Rouainia & Muir Wood (2000)) with large number of not-easily obtainable material parameters, which discourages practicing engineers from using them in application. However, hypoplasticity offers completely different approach to model non-linearity characterized by a single equation non-linear in stretching $\boldsymbol{D}$, thus allowing to model non-linearity in more straightforward way compared to elasto-plasticity (Kolymbas 1991). Hypoplastic models can predict such features of soil behavior as critical state, non-linearity in large and small strains, stiffness dependency on loading direction, etc.

2 MODEL FORMULATION

2.1 *Reference hypoplastic model*

The reference hypoplastic model clays for the present improvements was proposed by Mašín (2014) and is based on hypoplastic model for clays by Mašín (2005). This approach to hypoplasticity was developed at the University of Karlsruhe and is represented by the work of Kolymbas (1991). The basic rate formulation of the hypoplastic model is characterized by a single equation (Gudehus 1996)

$$\dot{\boldsymbol{\sigma}} = f_s \boldsymbol{\mathcal{L}} : \boldsymbol{D} + f_s f_d \boldsymbol{N} \|\boldsymbol{D}\| \tag{1}$$

where $\dot{\boldsymbol{\sigma}}$ is the objective (Zaremba—Jaumann) stress-rate tensor, $\boldsymbol{D}$ is the Euler stretching tensor, Λ and $\boldsymbol{N}$ are the fourth and second order constitutive tensors, respectively. f_s is the barotropy factor (function of mean stress p) and f_d is the pyknotropy factor (function of void ratio e and mean stress p). The isotropic normal compression line is defined by Butterfield (1979) such as

$$\ln(1+e) = N - \lambda^* \ln\left(\frac{p}{p_r}\right) \tag{2}$$

where parameters N and $\lambda*$ define the position and shape of the isotropic normal compression line and pr is the reference stress 1 kPa.

For detailed mathematical formulation of the original and reference hypoplastic models for clays the readers are referred to relevant publications (Mašín 2005, Mašín 2014).

2.2 *Hypoplastic model for soft clays*

The motivation for enhancement of the hypoplastic model for clays (Mašín, 2014) was driven by a need for realistic prediction of experimental results, in particular of extension undrained triaxial tests on normally consolidated or slightly overconsolidated soft clays. Fabianová (2014) performed a comparison of undrained triaxial test results on Singapore soft clays carried out by Corral (2010) with the hypoplastic model for clays by Mašín (2014). The material in concern were soft marine clays involved in the collapse of a deep excavation which occurred at Nicoll highway in the southeast part of Singapore. The hypoplastic model has displayed an inability to predict undrained triaxial tests (in extension in particular) realistically, which resulted in insufficiently accurate predictions of the excavation deformations and collapse. Two main drawbacks of the model were identified:

- Prediction of soil strength, especially in extension and inability to predict strength anisotropy, a feature typical for soft normally consolidated clays.
- Incorrect shape of stress paths in K_0 consolidated samples.

Two distinct improvements to the original hypoplastic model (Mašín, 2014) aiming to enhance model performance were developed and are described in Chapter 2.2.1 and 2.2.2 – denoted as strength anisotropy and stress path correction (D-dependent L approach, abbreviated as DdepL hereafter), respectively. The new model with both enhancements combined is entitled soft clay model hereafter. By setting the additional soft clay model parameters to reference values the model reduces to the original hypoplastic model for clays by Mašín (2014).

2.2.1 *Strength anisotropy*

Significant degree of anisotropy is common in natural clays, developed by the mechanism of deposition, the shape of the particles and any subsequent loading, which is mostly one-dimensional consolidation under the soil's own weight (Leoni et al, 2008).

To account for this property of soil, typically models with rotational hardening are used. Rotational hardening is means rotation of the yield surface (and plastic potential surface) in stress space, which was first introduced by Ohta (1979) for clay plasticity. First suggestions to consider both initial and plastic strain induced anisotropy of clays using the well-known elastoplastic framework were published by Nova (1985) and Dafalias (1986). These constitutive models incorporate hardening laws describing (as a function of plastic strains) either the change in orientation of the yield surface—rotation (Dafalias, 1986), or a shift of the centre of the yield surface via translational kinematic hardening laws (Di Prisco et al, 1993). Various authors have proposed anisotropic elastoplastic soil models with an inclined yield surface with the possibility to change an inclination of the yield curve during plastic straining as fabric anisotropy is erased or developed (Wheeler et al. 2003). All these models include rotational hardening law relating the change of yield curve inclination to the increments of plastic strains and to the current soil state. The hardening is expressed in terms of a scalar valued constitutive variable in triaxial space or in multiaxial stress space in a tensor valued, deviatoric variable (Taiebat & Dafalias 2013). S-CLAY1 (Wheeler et al. 2003) is a good example of typical elastoplastic anisotropic model for soft clays, it is an extension of the critical state models as Modified Cam clay

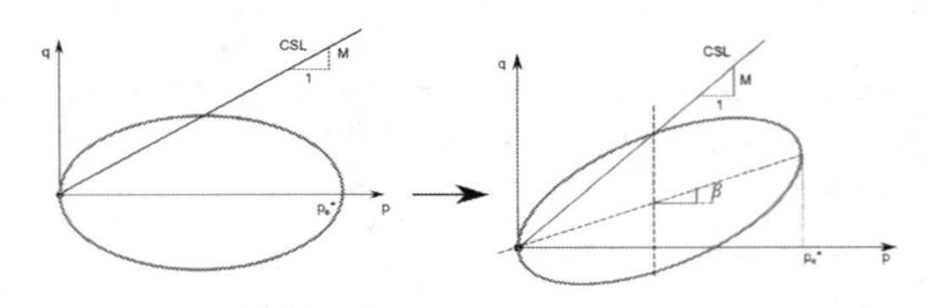

Figure 1. Sketch of ASBS rotation represented by angle β; shown in triaxial q-p plane.

model (Roscoe & Burland, 1968). S-CLAY1 incorporates two hardening laws—the first one describes the change of the yield curve, while the second one is the rotational hardening law (Wheeler et al. 2003). It should be pointed out, that S-CLAY1 accounts only for plastic anisotropy and not for elastic anisotropy and cyclic loading.

Mašín (2012) developed a method, which enables incorporating an explicit formulation of the asymptotic state boundary surface (ASBS) into hypoplasticity independently of hypoelastic tensor Λ. Using this method, apparent rotation of ASBS has in the model proposed in this paper been done by skewing the stress space similarly to Gajo and Muir Wood (2001), while defining a "skewed" and "normalized" stress space as sketched in Fig. 1.

Gajo & Muir Wood (2001) in their elastoplastic model related skewed ("normalised") stress space to a real one such that:

$$\boldsymbol{T}_{sk} = \boldsymbol{T} + \frac{1}{3}tr(\boldsymbol{T})\boldsymbol{\beta} \tag{3}$$

where $\boldsymbol{\beta}$ is a deviatoric second-order symmetric tensor defining skewing of the stress space and thus implying the rotation of the ASBS in the real stress space. It is acquired by decreasing all deviatoric components by the deviatoric tensor $\boldsymbol{\beta}$. When the surface is not rotated, $\boldsymbol{\beta}$ is null and the model reduces to the original hypoplastic model for clays (Mašín, 2014). The tensor $\boldsymbol{\beta}$ is defined as:

$$\boldsymbol{\beta} = \left[-\frac{2}{3}tg(\beta), \frac{1}{3}tg(\beta), \frac{1}{3}tg(\beta), 0, 0, 0\right]^T \tag{4}$$

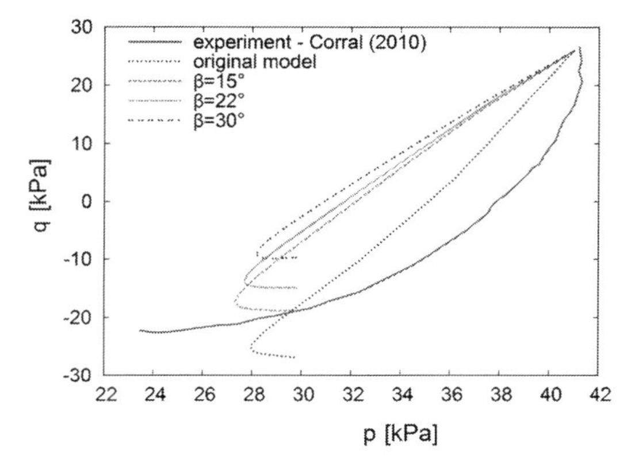

Figure 2. Stress paths of extension triaxial tests on Singapore clay performed by Corral & Whittle (2010) compared to results of original and modified hypoplastic model.

where the scalar β stands for inclination of the surface with respect to p-axis in q-p plane. The predicted shape of the ASBS is shown in Fig. 3.

The effect of stress rotation on stress paths in q-p plane is demonstrated on extension triaxial tests on Singapore clay. The experiments performed by Corral & Whittle (2010) are plotted against original hypoplastic model by Mašín (2014) and three β rotations (15°, 22° and 30°), see Figure 2. For simulations basic hypoplastic model parameters from Table 1 were used, DdepL (see Sec. 2.2.2) parameters were set to 0 and only value of β rotations are changed. Figure 3a shows normalized stress paths starting from two arbitrary states inside of ASBS. ASBS is formed by connecting all asymptotic states. Figure 3b depicts ASBS for various values of rotation, ranging from $\beta = 0°$ (equivalent to model by Mašín (2014)) to $\beta = 38°$.

a)

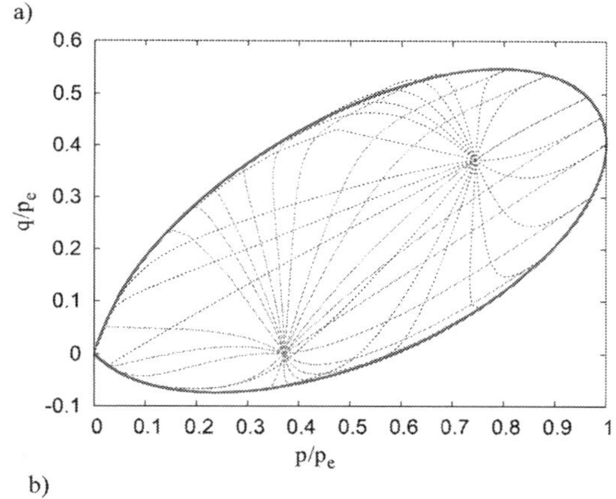

b)

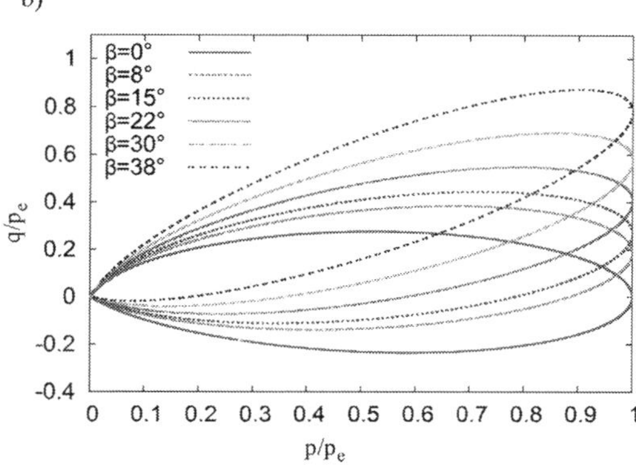

Figure 3. Results of the model enhanced by strength anisotropy: a) Normalized stress paths starting from two arbitrary states leading to an asymptotic state bound by ASBS, b) ASBS for various values of rotation (values of scalar parameter β are shown).

Table 1. Material parameters for two clays used in calibrations—Singapore upper marine clay and Bangkok clay.

Basic hypoplastic model for clays					
	CS friction angle (ϕ)	NCL slope (λ)	Slope of URL (κ)	NCL for p = 1 kPa (N)	Shear modulus parameter (ν)
Singapore clay	32	0.165	0.005	1.76	0.21
Bangkok clay	20.5	0.227	0.015	2.36	0.27
Soft clay model					
	Stress path factor (ξ)	Stress space rotation for DdepL (γ)	ASBS rotation (β)		
Singapore clay	1	31	8		
Bangkok clay	1	9	0		

2.2.2 *Stress path correction*

In hypoplasticity, asymptotic behavior and strength are governed by a combination of L and N tensors, while stiffness is controlled by the tensor Λ. In Sec. 2.1.1, N was changed to imply the pre-defined rotated shape of ASBS, which did not improve

of extension test satisfactorily (see Fig. 2). The initial stiffness has especially been significantly underestimated. We thus introduced further modification of the model, which is based on procedure proposed by Mašín & Herle (2007). In their approach, the tensor Λ was made bilinear in $\boldsymbol{D}$, so the predictions resembled an elasto-plastic model with different tangent stiffness in isotropic loading and unloading, which also corrected the shape of undrained stress paths. In this paper, we adopted similar approach, but instead of defining different stiffnesses in loading and unloading, the stiffnesses were defined different in shear in compression and extension. In this way, we were able to predict smaller shear stiffness in extension shear tests without modifying stiffness in compression. Thanks to the use of the explicit ASBS approach, this method did not affect the ASBS shape.

The rate form of the DdepL model is for $\beta = 0°$ defined as:

$$\dot{\boldsymbol{T}} = f_s \boldsymbol{\mathcal{L}} : \boldsymbol{D} + f_s f_d \boldsymbol{N} D_n \tag{5}$$

where $D_n = \|\boldsymbol{D}\|$ of the original model is replaced by

$$D_n = w_y \boldsymbol{D} + \left(1 - w_y\right) |\boldsymbol{d} : \boldsymbol{D}| \tag{6}$$

The weighting factor is defined as

$$w_y = F_{msk}^{\,\xi} \tag{7}$$

where $\boldsymbol{\xi}$ is a new model parameter. F_{msk} is a Matsuoka-Nakai factor calculated using invariants from skewed stress space:

$$F_{msk} = \frac{9I_{3sk} + I_{1sk} I_{2sk}}{I_{3sk} + I_{1sk} I_{2sk}} \tag{8}$$

The changes described above are sufficient for altering the stress paths for isotopically consolidated soils, however, the procedure is not adequate typical for soft marine clays of interest in this paper, which are K0 consolidated. For this purpose, Λ wy and $\boldsymbol{N}$ (used to find Λ^{D}) as well as stress invariants used in Equation (8) needs to be calculated from a skewed stress space defined by Gajo & Muir Wood (2001) as:

$$\boldsymbol{T}_{sk} = \boldsymbol{T} + \frac{1}{3} tr(\boldsymbol{T}) \gamma \tag{9}$$

where $\boldsymbol{\gamma}$ is a deviatoric second-order symmetric tensor defining skewing. When $\boldsymbol{\gamma}$ is null, the model reduces to model by Mašín (2014). The tensor $\boldsymbol{\gamma}$ is defined as

$$\gamma = \left[-\frac{2}{3} tg\, \gamma, \frac{1}{3} tg\, \gamma, \frac{1}{3} tg\, \gamma, 0, 0, 0 \right]^{T} \tag{10}$$

where scalar γ stands for inclination of the skewed stress space with respect to p-axis in q-p plane. By skewing of the stress space and using the skewed stress for stress invariants and components of stiffness matrix, the stress path becomes perpendicular to p-axis at the stress deviator specified by the parameter γ.

The model can be rewritten in the form

$$\dot{\boldsymbol{T}} = f_s \left(\boldsymbol{\mathcal{L}}^{D} : \boldsymbol{D} + w_y f_d \boldsymbol{N} \boldsymbol{D} \right) \tag{11}$$

where the tensor Λ^{D} is dependent on the direction of stretching with respect to $\mathbf{d}_{sk}$ via the equation

$$\mathcal{L}^D = \begin{cases} L + f_d\left(1-w_y\right)\left(\boldsymbol{N}_{sk} \otimes \mathbf{d}_{sk}\right) for\left(\mathbf{d}_{sk} : \mathbf{D}\right) > 0 \\ L - f_d\left(1-w_y\right)\left(\boldsymbol{N}_{sk} \otimes \mathbf{d}_{sk}\right) for\left(\mathbf{d}_{sk} : \mathbf{D}\right) \leq 0 \end{cases} \tag{12}$$

where $\mathbf{d}_{sk}$ is the direction of stretching at the asymptotic state

$$\mathbf{d}_{sk} = \frac{\boldsymbol{A}_{sk}^{-1} : \boldsymbol{N}_{sk}}{\left\|\boldsymbol{A}_{sk}^{-1} : \boldsymbol{N}_{sk}\right\|} \tag{13}$$

with fourth order tensor $\mathbf{A}_{sk}$ defined as

$$\mathrm{A}_{sk} = f_s \mathcal{L} + \frac{1}{\lambda^*} \boldsymbol{T}_{sk} \otimes 1 \tag{14}$$

and

$$\boldsymbol{N}_{sk} = -\frac{\boldsymbol{A}_{sk} : \boldsymbol{d}_{sk}^A}{f_s f_d^A} \tag{15}$$

with

$$\mathrm{d}_{sk}^A = -\hat{T}_{sk}^* + \mathbf{1}\left[\frac{2}{3} - \frac{cos3\theta + 1}{4} \boldsymbol{F}_{m,sk}^{1/4}\right] \frac{F_{m,sk}^{\xi/2} - sin^{\xi}\varphi_c}{1 - sin^{\xi}\varphi_c} \tag{16}$$

For further inspection into implementation of Λ^D tensor and DdepL the reader is referred to the model by Mašín & Herle (2007).

2.3 *Model parameters*

Parameters used in the model can be separated into two groups: 1) Parameters used in basic hypoplastic model for clays by Mašín (2014); 2) Parameters of the proposed soft clay model. In the following, the complete list of model parameters of the first group is described, starting with parameters from model by Mašín (2014)

- ϕ_{cr} – Critical state friction angle.
- λ^* – Slope of normal compression line in the ln p/p_r vs ln$(1+e)$ plane, where p_r is reference stress of 1 kPa.
- N – Position of normal compression line in the ln p/p_r vs ln$(1+e)$ plane for $p = p_r$.
- κ^* – Parameter controlling unloading-reloading line.
- ν – Parameter controlling stiffness in shear.

The basic clay hypoplastic model requires 5 material parameters equivalent to the Modified Cam clay model. The parameters are obtainable from standard element tests—oedometric and triaxial tests. However, the definitions of individual parameters are not identical to the Modified Cam clay model due to non-linear nature of hypoplasticity, even though their physical meaning is similar.

The enhanced model adds three parameters to the original model:

- β – Defines inclination of the ASBS with respect to p-axis in q-p plane.
- γ – Defines skewing of the stress space for calculation of all variables denoted with $_{sk}$ for use in DdepL approach.
- ξ – Parameter for control of the stress path shape in DdepL approach.

For calibration of parameter β, extension triaxial test is necessary to evaluate strength difference in compression and extension. For parameters γ and ξ anisotropically (ideally K_0) consolidated triaxial tests are required in compression and extension to find the best agreement between experiments and stress paths predicted by the model. In the most efficient approach, the hypoplastic model for soft clays can be calibrated using three separate tests:

- Compression anisotropically consolidated triaxial test.
- Extension anisotropically consolidated triaxial test.
- Oedometric test.

3 EVALUATION OF THE MODEL PREDICTIONS

The model has been evaluated with respect to experimental data on anisotropically consolidated Singapore (Corral & Whittle 2010) and Bangkok clays. For details on calibration of the original hypoplastic model the reader is referred to Mašín (2014). Compression and extension triaxial tests are used for model calibration, together with oedometric tests. The parameters from Table 1 were used for simulation of undrained compression and extension tests on studied clays.

The parameter γ was obtained by following assumptions: from the ϕ_{cr} of the soil of interest the equivalent value of K_0 was determined using Jaky formula 1 – sin (φ). Then, from γ, K_0 the rotation in q/p space was determined. Parameter β was obtained by comparison of strength in compression and extension. For determination of ξ, a parametric study needs to be performed to find the best agreement in stress path shape between experimental results and the simulations.

3.1 *Model predictions*

The samples were normally consolidated under K_0 conditions and then tested in triaxial undrained compression and extension. The predictions of

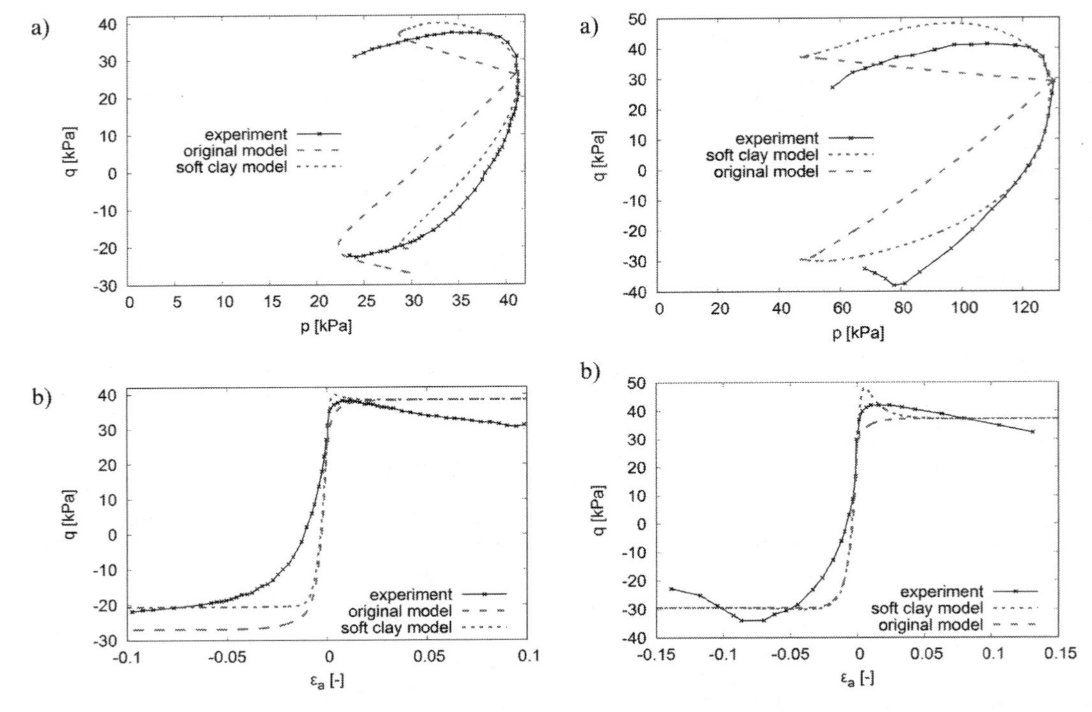

Figure 4. Soft clay and original model predictions compared with experimental results of Singapore clay, compression and extension triaxial tests: a) stress paths, b) q vs ε_a.

Figure 5. Soft clay and original model predictions compared with experimental results of Bangkok clay, compression and extension triaxial tests: a) stress paths, b) q vs ε_a.

the modified model for tests on anisotropically consolidated soils starting from the isotropic stress states were significantly different as compared to the original model.

Model predictions (abbreviated "soft clay model" for enhanced model and "original model" for model by Mašín (2014)) are shown in Figure 4 and 5. Experiments on Singapore and Bangkok clays, respectively, are abbreviated "experiment". Parameters used for the original model are shown in Table 1 under heading "Basic hypoplastic model for clays". As discussed earlier, 5 parameters are needed only for the calibration of the model together with initial conditions (stress state and void ratio). For further inspection into calibration of the original model, the reader is referred to Mašín (2014). In the enhanced soft clay model simulations, the original model parameters were left similar to the original model simulations and only three soft clay model parameters were calibrated.

Figure 4a, b depicts comparison of model predictions and experimental results for Singapore upper marine clay. Soft clay model parameters are shown in Table 1. Both enhancements were used—ASBS rotation as well as DdepL, meaning that for the full calibration of the soft clay model 8 parameters must be identified. The advantage of the parameter β causing rotation of the ASBS is apparent in the q vs ε_a space. By setting β to 8° the ultimate strength in compression remains the same, however, the strength in extension is lowered, thus the strength predicted in the simulation becomes closer to experiment than for the original model. It should be pointed out that parameter β influences only ultimate strength, not the stress path, as can be seen in Figure 2. The second enhancement comprises the alteration of the stress path shape for anisotropically consolidated samples as described in Chapter 2.2.2, however, the asymptotic state remains unchanged. This characteristic is evident in the Figure 4a and 5a, where the first part of the stress path captures key features of clay soil's response in compression and extension. The influence of tensor Λ^D is governed by parameter ξ – for $\xi = 0$ model reduces to the original hypoplastic model and for very high values of ξ the very beginning of the stress path becomes parallel to y axis. As seen from Figure 4a and 5a it turned out to be necessary to use $\xi = 1$ to capture the soil behavior correctly.

This solution leads to a slight overestimation of peak value of q for compression samples, as seen especially from Figure 5a, on the other hand the stiffness at the beginning is showing better agreement compared to the original model, especially in compression.

The asymptotic response is strictly dependent on the prediction of original hypoplastic model and remains unchanged with exception of strength reduction in extension due to ASBS rotation. This can be illustrated on example in Figure 5a, since for simulation of Bangkok clay the value of ASBS rotation was set to $\beta = 0°$ for the best fit of experimental curve. The model predicts the same friction angles in compression and extension, thus the DdepL and anisotropic ASBS enhancements may be seen as independent. The shape of the undrained stress paths may be considered as satisfactory, especially while compared to the original model predictions, which show significantly worse performance, both in compression and extension.

4 CONCLUSIONS

An enhanced hypoplastic model for clays has been proposed with the aim to improve performance of the model for natural K_0 consolidated soft soils. Two enhancements to the model were described: anisotropic state boundary surface was implemented into the hypoplastic model, accounting for strength anisotropy in compression and extension; and the tensor $\boldsymbol{L}$ was made bilinear in $\boldsymbol{D}$ to more realistically predict the stress paths. The model contains five material parameters from the original hypoplastic model for clays by Mašín (2014), in addition three new parameters were defined for soft clay model. Fundamental features of the model formulation were described, followed by evaluation of performance of the model in element tests. All parameters can be determined from an oedometric test and anisotropically consolidated undrained triaxial compression and extension tests.

It was demonstrated that the predictions of the proposed model are showing better agreement with experimental results as compared to the original model for tests on anisotropically consolidated soils. The proposed modifications improve applicability of the hypoplastic model by improving its predictions of soil initially in K_0 states, which are typical for many naturally occurring soils used for construction. No additional tests are needed for calibration except conventional triaxial and oedometer tests.

The DdepL model has been validated using large scale boundary value problem, in this case retrospective simulations of centrifuge test of spudcan installation incorporating very large deformations (Jerman & Mašín 2018). The model implementation within an element test driver can be found at soilmodels.com/triax.

ACKNOWLEDGMENT

The first author has been supported by grant No. 1075516 of the Charles University Grant Agency. The second author received funding from grant No. 15-059355 of the Czech Science Foundation.

REFERENCES

Butterfield, R. 1979. A natural compression law for soils. *Géotechnique* 29(4): 469–480.

Corral, G., & Whittle, A. J. 2010. Re-analysis of deep excavation collapse using a generalized effective stress soil model. In *Earth Retention Conference* 3: 720–731.

Dafalias, Y. F. 1986. Bounding surface plasticity. I: Mathematical foundation and hypoplasticity. *Journal of Engineering Mechanics* 112(9): 966–987.

Di Prisco, C., Nova, R., & Lanier, J. 1993. *A mixed isotropic-kinematic hardening constitutive law for sand.* Elsevier App Sci. 83–124.

Gajo, A., & Muir Wood, D. 2001. A new approach to anisotropic, bounding surface plasticity: general formulation and simulations of natural and reconstituted clay behaviour. *International journal for numerical and analytical methods in geomechanics* 25(3): 207–241.

Gudehus, G. 1996. A comprehensive constitutive equation for granular materials. *Soils and Foundations* 36(1): 1–12.

Jerman, J. & Mašín, D. 2018. Modelling of spudcan foundation penetrations using an improved hypoplastic model for soft clays. In *4th GeoShanghai International Conference*.

Kolymbas, D. 1991. An outline of hypoplasticity. *Archive of applied mechanics* 61(3): 143–151.

Mašín, D. 2005. A hypoplastic constitutive model for clays. International Journal for Numerical and Analytical Methods in Geomechanics 29(4): 311–336.

Mašín, D. 2007. A hypoplastic constitutive model for clays with meta-stable structure. *Canadian Geotechnical Journal* 44(3): 363–375.

Mašín, D. 2013. Clay hypoplasticity with explicitly defined asymptotic states. *Acta Geotechnica* 8(5): 481–496.

Mašín, D. 2014. Clay hypoplasticity model including stiffness anisotropy. *Géotechnique* 64(3): 232–238.

Mašín, D. & Herle, I. 2007. Improvement of a hypoplastic model to predict clay behaviour under undrained conditions. *Acta Geotechnica* 2(4): 261–268.

Nova, R. 1985. Mathematical modelling of anisotropy of clays. *Proceedings of the 11th ICSMFE, San Francisco* 1, 607–661.

Ohta, H. 1979. Constitutive equations considering anisotropy and stress reorientation in clay. *In Proc. 3td Int. Conf. on Numerical Methods in Geomechanics* 1: 475–484.

Roscoe, K., & Burland, J. B. 1968. *On the generalized stress-strain behaviour of wet clay.*
Rouainia, M., & Muir Wood, D. 2000. A kinematic hardening constitutive model for natural clays with loss of structure. *Géotechnique* 50(2): 153–164.
Taiebat, M., & Dafalias, Y. F. 2013. Rotational hardening and uniqueness of critical state line in clay plasticity. *In Constitutive Modeling of Geomaterials* 223–230. Berlin, Heidelberg: Springer.
Wheeler, S. J., Näätänen, A., Karstunen, M., & Lojander, M. (2003). An anisotropic elastoplastic model for soft clays. *Canadian Geotechnical Journal* 40(2), 403–418.

Numerical Methods in Geotechnical Engineering IX – Cardoso et al. (Eds)
© 2018 Taylor & Francis Group, London, ISBN 978-1-138-33198-3

A modified bounding surface plasticity model for sand

A. Amorosi & F. Rollo
Department of Structural and Geotechnical Engineering, Sapienza University of Rome, Rome, Italy

D. Boldini
Department of Civil, Chemical, Environmental, and Materials Engineering, University of Bologna, Bologna, Italy

ABSTRACT: A modified bounding surface plasticity model developed within the critical state soil mechanics framework is presented. Herein a modified version of the Dafalias and Manzari (2004) model is proposed, specifically developed to improve its capability to reproduce the cyclic behaviour of sands under different strain levels adopting a unique set of constitutive parameters. The proposed modified version of the model includes a thermodynamically consistent isotropic hyperelastic formulation for the very small strain range, which realistically reproduces the nonlinear stress dependence of the elastic stiffness. Furthermore, modified versions of the plastic modulus and dilatancy laws are introduced leading to a more realistic small to medium strain range cyclic behaviour, while preserving the original good predictive capability for monotonic and cyclic loading at large strains. The model response is illustrated by comparison with experimental data over a wide range of void ratios and stress states.

1 INTRODUCTION

The simulation of a seismic wave propagation process in saturated non-cohesive soil deposits requires the adoption of sufficiently advanced constitutive models, capable of capturing the complex dynamic phenomena typically observed in porous media. Those latter range from stress-dependency of the elastic response at very small strain amplitudes to the large strain plasticity-driven soil liquefaction. In detail, an accurate prediction of coarse-grained soil response under cyclic loading should realistically account for the accumulation of deformation, the degradation of the shear stiffness, the increase of damping and the build-up of the excess pore water pressures, possibly in a multidirectional context.

Most of the single surface plasticity-based constitutive models proposed in the literature to model the monotonic response of sands are, generally, not able to capture the above essential features of their cyclic response. On the contrary, more advanced elasto-plastic constitutive models reveal to be more appropriate at this scope. Among them the so called Simple ANIsotropic SAND constitutive models (SANISAND), originally developed by Manzari and Dafalias (1997) and then further modified by others (Papadimitriou et al. 2001, Papadimitriou and Bouckovalas 2002, Dafalias and Manzari 2004, Taiebat and Dafalias 2008, Taborda et al. 2014, Dafalias and Taiebat 2016), have received much attention in the research community, as they explicitly account for some of the above mentioned key features of the cyclic response of sands.

In this paper the 2004 version of the model is first summarised and its performance is analysed through the simulation of strain-and stress-controlled cyclic triaxial tests, carried out under undrained conditions. A specific calibration strategy of the model parameters is then proposed to reproduce the soil behaviour at small-to-medium strain range, relevant to seismic site response problems, highlighting limits and advantages emerging form the results.

Based on these evidences, a modified version of the model is finally presented, specifically arranged to improve the predictive capabilities of the constitutive law in a wide range of strain amplitude while adopting a unique set of parameters. The formulation includes a hyperelastic law for the reversible behaviour and new state-dependent plastic modulus and dilatancy laws.

2 SANISAND CONSTITUTIVE MODEL

The model formulation is based on the bounding surface plasticity and critical state theory. Basically, the constitutive law includes four surfaces in the stress-ratio space. A conical yield surface with the apex at the origin of stress space, capable of

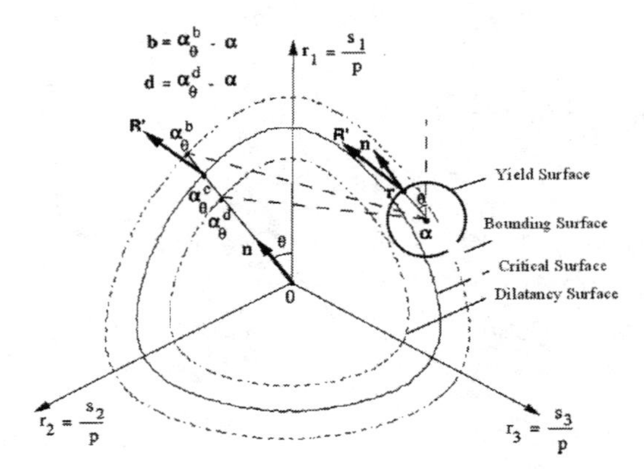

Figure 1. Representation of the yield, critical, dilatancy and bounding surfaces on the stress ratio π plane (modified from Dafalias and Manzari 2004).

capturing the plastic response for variable stress ratios. The critical state surface, characterised by a noncircular shape, which determines the failure of the material, i.e. the state at which shear deformations occur for fixed stresses and zero increments of volumetric strain. The material softening for dense states is governed by the bounding surface through the peak stress-ratio, while the dilatancy surface is adopted to simulate the plastic volumetric response during loading (Figure 1).

The bounding, dilatancy and critical surfaces are described as a function of the state parameter $\Psi = e - e_c$, used to incorporate the effect of the state of the material in the critical state stress-strain behaviour, where e is the current void ratio and e_c is the critical void ratio corresponding at the current confining pressure (Been and Jefferies 1985). The dependence on the state parameter Ψ enables the simulation of sand response with a single set of model parameters, irrespective of the initial conditions (i.e. loose or dense).

The model formulation in triaxial stress space and its multiaxial stress-space generalisation are presented in detail in Dafalias and Manzari (2004). Herein, for the sake of brevity, only some basic constitutive equations in the multiaxial generalisation are summarised, to ease the understanding of the proposed modifications discussed in section 4. All stress components are expressed as effective stress. The stress and strain tensors are denoted by bold-faced symbols; superposed dot denotes the increment, while the superscript e and p indicate the elastic and plastic parts of strain.

2.1 *Multiaxial stress-space formulation*

The elastic part of the SANISAND model is assumed to be hypoelastic:

$$\dot{\varepsilon}_v^e = \frac{\dot{p}}{K} \quad \dot{\mathbf{e}}^e = \frac{\dot{\mathbf{s}}}{2G} \tag{1}$$

where ε_v^e is the elastic volumetric strain, $\mathbf{e}^e$ is the elastic deviatoric strain tensor, $\mathbf{s}$ is the elastic deviatoric stress tensor and p is the mean effective stress. G and K are the elastic shear and bulk moduli, respectively, function of the mean effective pressure p and the current void ratio e:

$$G = G_0 p_{at} \frac{(2.97 - e)^2}{1+e}\left(\frac{p}{p_{at}}\right)^{0.5} \quad K = \frac{2(1+\nu)}{3(1-2\nu)} G \tag{2}$$

where G_0 is a dimensionless material constant, ν is the Poisson's ratio and p_{at} is the atmospheric pressure.

The elastic domain is limited in the stress space by a conical yield surface with circular cross section in π plane, with centre α and radius $\sqrt{2/3}m$ (Figure 1):

$$f = \left[(\mathbf{s} - p\boldsymbol{\alpha}) : (\mathbf{s} - p\boldsymbol{\alpha})\right]^{1/2} - \sqrt{2/3}pm = 0 \tag{3}$$

where m is a model constant defining the size of the yield surface and $\boldsymbol{\alpha}$ is the back-stress ratio deviatoric tensor, indicating the location of the axis of the cone. When plastic deformation occur, the back-stress ratio $\boldsymbol{\alpha}$ changes, inducing a translation of the yield surface (kinematic hardening), as well as the size m (isotropic hardening), involving an enlargement of the elastic domain.

The normal to the yield surface defines the loading direction:

$$\frac{\partial f}{\partial \sigma} = \mathbf{n} - \frac{1}{3}(\mathbf{n} : \mathbf{r})\mathbf{I} \quad \mathbf{n} = \frac{\mathbf{r} - \alpha}{\sqrt{2/3}m} \tag{4}$$

where $\mathbf{r} = \mathbf{s}/p$ is the deviatoric stress-ratio tensor, $\mathbf{I}$ is the second rank identity tensor and $\mathbf{n}$ is the unit tensor oriented outward along the radius $\mathbf{r} - \boldsymbol{\alpha}$ of the yield surface, satisfying the relations tr $\mathbf{n} = 0$ and tr $\mathbf{n}^2 = 1$.

The three concentric non-circular surfaces, the bounding, dilatancy and critical surfaces, are analytically defined in terms of image back-stress ratio tensors α_θ^b, α_θ^c and α_θ^d, respectively:

$$\alpha_\theta^a = \sqrt{2/3}\left[g(\theta,c) M_c \exp(\mp n^a \Psi) - m\right]\mathbf{n} \tag{5}$$

$$g(\theta,c) = \frac{2c}{(1+c) - (1-c)\cos 3\theta}; \quad c = \frac{M_e}{M_c} \tag{6}$$

where a is equal to b, c or d, $n^c = 0$, $\mp$ refers to b and c, respectively. M_c and M_e are the stress ratios at critical state in compression and extension respectively, n^b and n^d are positive material constants and θ is the Lode angle, defined as

$$\cos 3\theta = \sqrt{6}\,\mathrm{tr}\,\mathbf{n}^3 \tag{7}$$

The critical state behaviour is described by the following power relation

$$e_c = e_0 - \lambda_c \left(p_c / p_{at}\right)^{\xi} \tag{8}$$

where λ_c, ξ and e_0 are material constants, the latter representing the void ratio at $p_c = 0$. It is worth observing that when $\Psi = 0$ (i.e. at critical state), the bounding and dilatancy surfaces collapse onto the critical one.

The plastic strain evolution $\dot{\varepsilon}^p$ is provided by a non-associative flow rule

$$\dot{\varepsilon}^p = \langle L \rangle \left(\mathbf{R}' - \frac{1}{3} D\mathbf{I} \right) \quad \mathbf{R}' = B\mathbf{n} - C\left(\mathbf{n}^2 - \frac{1}{3}\mathbf{I} \right) \tag{9}$$

$$B = 1 + \frac{3}{2}\frac{1-c}{c} g \cos 3\theta \quad C = 3\sqrt{\frac{3}{2}}\frac{1-c}{c} g \tag{10}$$

where $\mathbf{R}'$ is the deviatoric part of the plastic strain rate direction (normal to the critical surface at image point α_θ^c), L is the plastic multiplier, defined as a function of plastic modulus K_p, and D is a dilatancy function:

$$L = \frac{1}{K_p}\left(\frac{\partial f}{\partial \sigma} : \dot{\sigma} \right) = \frac{2G\mathbf{n} : \dot{\mathbf{e}} - \mathbf{n} : \mathbf{r}\dot{\varepsilon}_v}{K_p + 2G(B - C\,tr\,\mathbf{n}^3) - KD\mathbf{n} : \mathbf{r}} \tag{11}$$

$$D = \mathrm{A}_d \left(\alpha^d - \alpha \right) \cdot \mathbf{n} \quad \text{with} \quad A_d = A_0 \left(1 + \langle \mathbf{z} : \mathbf{n} \rangle \right) \tag{12}$$

with A_0 a material constant and $\mathbf{z}$ the so-called fabric-dilatancy internal variable, introduced to account the effect of fabric change on dilatancy. Note that for states $(\alpha^d - \alpha) : \mathbf{n} > 0$ (i.e. $\boldsymbol{\alpha}$ is inside the dilatancy surface), D is positive and plastic contraction is expected; if the back-stress ratio $\boldsymbol{\alpha}$ is outside the dilatancy surface, plastic dilation will occur ($D < 0$).

The model formulation is completed by the kinematic hardening law that governs the evolution of the position $\boldsymbol{\alpha}$ of the yield surface. Assuming $\dot{m} = 0$ (no isotropic hardening) it is possible to express the plastic modulus K_p as follows:

$$\dot{\alpha} = \langle L \rangle \frac{2}{3} h \left(\alpha_\theta^b - \alpha \right) \tag{13}$$

$$K_p = \frac{2}{3} p h \left(\alpha_\theta^b - \alpha \right) : \mathbf{n} \tag{14}$$

with h being a hardening function defined as:

$$h = \frac{G_0 h_0 \left(1 - c_h e\right)\left(p / p_{at}\right)^{-1/2}}{\left(\alpha - \alpha_{in}\right) : \mathbf{n}} \tag{15}$$

where h_0, c_h are model constants and $\boldsymbol{\alpha}_{in}$ the initial value $\boldsymbol{\alpha}$ at the beginning of a new loading process.

2.2 *Calibration of model parameters*

The first set of simulations are run with the original model parameters calibrated by Dafalias and Manzari (2004) with reference to Toyoura sand, based on experimental data from standard laboratory tests, e.g. drained and undrained triaxial compression tests (Verdugo and Ishihara 1996). The related results are then compared to those obtained adopting a new calibration procedure, specifically aimed at differently selecting three fundamental model parameters: G_0 (related to the elastic response), A_0 (related to the dilatancy) and h_0 (related to distance from bounding surface). The new calibration is proposed to better match the non-linear cyclic response at small-to-medium strain range, which was not considered in the original paper of 2004.

In particular, the new calibration is performed against a set of experimental data consisting of undrained cyclic triaxial tests carried out by Kokusho (1980) at small-to-medium strain levels on the same Toyoura sand originally considered by Dafalias and Manzari (2004). The experimental data are illustrated in terms of shear modulus reduction curves and variation of damping ratio with shear strain (Figure 2a, b and c).

To the purpose, a series of strain-controlled cyclic triaxial simulations are performed under undrained conditions on relatively dense Toyoura sand at initial void ratio $e_{in} = 0.735$. After the isotropic consolidation stage at a mean effective stress of 100 kPa, the ideal soil specimen is subjected to a vertical displacement varying with time by a sinusoidal law, so that its maximum value ranges from very small strain level (0.001%) to relatively large one (0.3%) for a number of cycles equal to 10 for each strain amplitude. The deviatoric strain is directly related to the shear strain γ by means of the relation $\gamma = \sqrt{3}\varepsilon_s$ (Georgiannou et al. 1991). For each investigated strain level, the secant shear modulus and the damping ratio are evaluated stemming from the stress-strain curves with reference to the tenth cycle.

Based on the experimental curve of secant shear modulus (Figure 2a), the elastic parameter G_0 is calibrated in order to attain the same maximum shear modulus G_{max} observed in the experiments for such a dense state of Toyoura sand.

Table 1. Summary of material constants of the SANISAND model for Toyoura sand.

Constant	Variable	Original set	Modified set
Elasticity	G_0	125	375
	ν	0.05	0.05
Critical state	M_c	1.25	1.25
	c	0.712	0.712
	λ_c	0.019	0.019
	e_0	0.934	0.934
	ξ	0.7	0.7
Yield surface	m	0.01	0.01
Plastic modulus	h_0	7.05	20
	c_h	0.968	0.968
	n^b	1.1	1.1
Dilatancy	A_0	0.704	0.05
	n^d	3.5	3.5
Fabric-dilatancy tensor	z_{max}	4	4
	c_z	600	600

To best fit the normalised secant shear modulus reduction curve, the dilatancy parameter A_0 is reduced, while the hardening constant h_0 is increased (Figure 2b). Both model parameters have a significant effect on the plastic response of the constitutive model, as described in detail in the next section.

The original set of model parameters by Dafalias and Manzari (2004) and the modified one are listed in Table 1.

3 MODEL PERFORMANCE UNDER CYCLIC TRIAXIAL TEST CONDITIONS

The influence of the material parameters on the cyclic constitutive response is assessed by performing the same numerical simulations of undrained cyclic triaxial tests, adopting both the original and updated sets of material constant (Dafalias and Manzari 2004).

Figure 3 shows the cyclic response as obtained based on the two sets of parameters for a cyclic deviatoric strain amplitude equal to 0.1%.

The response based on the original set of constants (Figure 3a and b) is characterised by a pronounced development of excess pore water pressure, leading to a rapid decrease of the mean effective stress and a consequent loss of shear strength. This is not the case for the simulation carried out adopting the modified set (Figure 3c and d), which shows a stable nonlinear hysteretic stress-strain curve, characterised by negligible accumulation of excess pore water pressure and a related almost constant value of the mean effective stress.

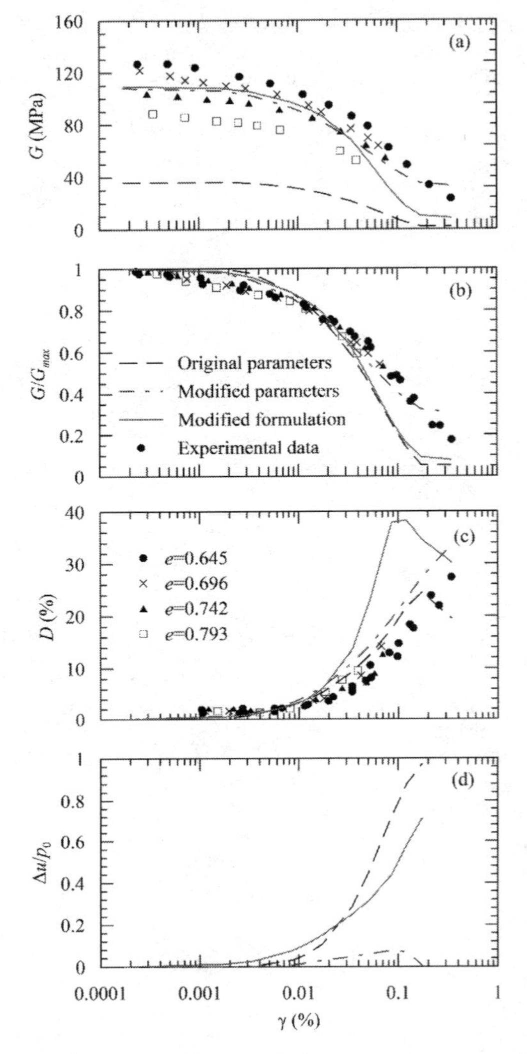

Figure 2. Comparison of model simulations to experimental data (Kokusho 1980) for Toyoura sand: (a) secant shear modulus reduction curves, (b) normalised secant shear modulus reduction curves, (c) variation of damping ratio with shear strain and (d) normalised excess pore water pressure.

The above typical cyclic behaviour directly reflects on the shear modulus reduction curves illustrated in Figure 2b: the one based on the original set of parameters shows a more abrupt decrease for increasing strain level as compared to what exhibited by both the modified parameters simulations and the experimental data. Furthermore, the modified set of model parameters leads to a greater dissipative capacity as compared to the original one, as it can be observed in Figure 2c illustrating the variation of damping ratio with the cyclic strain level.

Figure 2d clearly indicates that the reduction of the dilatancy parameter A_0 and the increment of the hardening constant h_0 adopted in the new calibration lead to a significant reduction of the build-up of excess pore water pressure with the shear strain level, which attains much lower values than those typically observed in laboratory tests and in the simulation based on the original set of parameters.

A further check of the predictive capability of the model is carried out with reference to the experimental results obtained by Zhang et al. (2010). The cyclic test here selected was carried out on Toyoura sand under undrained stress-controlled cyclic triaxial conditions at an initial void ratio $e_{in} = 0.77$ (Figure 4).

After the usual isotropic consolidation stage at a mean effective stress of 100 kPa, the sample is subjected to a deviatoric cyclic loading stage characterised by a sinusoidal law of amplitude ±50 kPa.

The simulation of the model is depicted in Figure 5 for ten cycles of loading. Because of the pronounced development of excess pore water pressures, the response for the original set of parameters is characterised by the well-known butterfly shape path in the *p-q* plane with a rapid loss of shear strength (Figure 5a).

Another expected result, typical of the whole class of bounding surface plasticity models, is the unlimited accumulation of plastic strains during loading commonly known as ratcheting (Figure 5b). On the contrary, the simulation conducted with the modified set of parameters (Figure 5c and d) is characterised by the absence of detectable pore water pressure build-up, which makes the model incapable of reproducing the phenomenon of cyclic mobility for medium to dense sands.

It should also be remarked that the initial response of the model with the original set of parameters underestimates of about 70% the value of the maximum shear modulus G_{max} as observed by the cyclic experimental data (see Figure 2a). This discrepancy is associated to the calibration

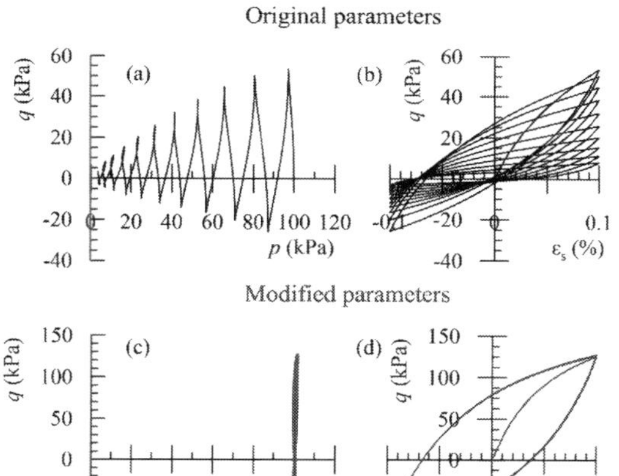

Original parameters

Modified parameters

Modified formulation

Figure 3. Model performance under undrained strain-controlled cyclic triaxial compression tests carried out with the original and the modified sets of material constants and with the modified formulation ($e_{in} = 0.735$, $p_0 = 100$ kPa, $N_{cycles} = 10$).

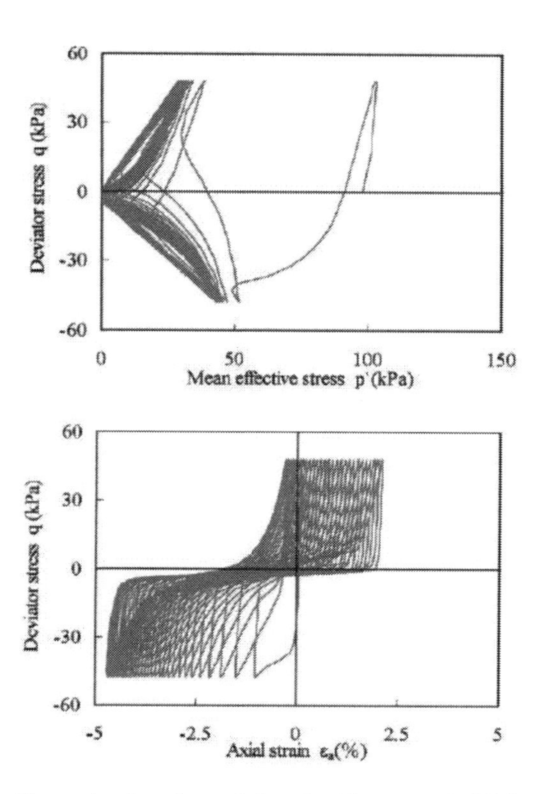

Figure 4. Experimental data by Zhang *et al.* (2010) in undrained stress-controlled cyclic triaxial test on Toyoura Sand ($e_{in} = 0.77$, $p_0 = 100$ kPa, $q = \pm 50$ kPa).

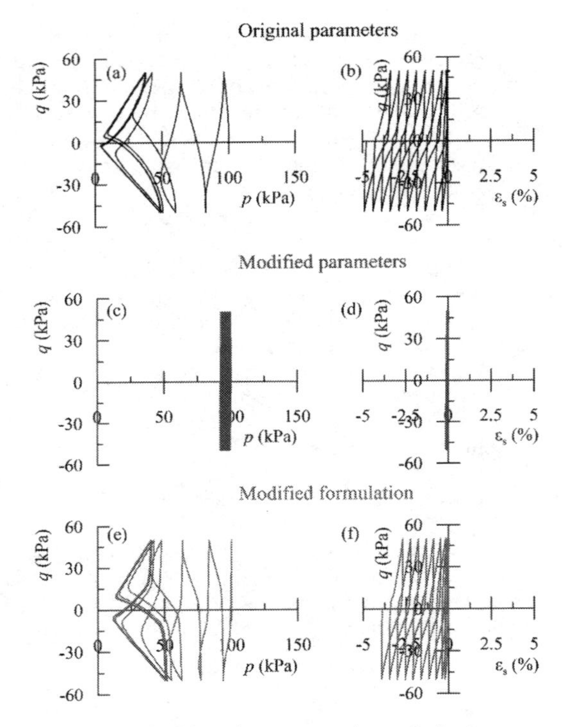

Figure 5. Model performance under undrained stress-controlled cyclic triaxial compression tests carried out with the original and the modified sets of material constants and with the modified formulation (e_{in} = 0.735, p_0 = 100 kPa, q = ±50 kPa, N_{cycles} = 10).

strategy adopted by Dafalias and Manzari (2004), which were focused on the best fitting the overall stress-strain curves of the triaxial tests by Verdugo and Ishihara (1996), without any emphasis on the small to medium strain levels: in fact those latter, which play a major role in any seismic-related problem, require the adoption of specific testing procedure, as those implemented by Kokusho (1980), that were out of the scope in the experimental work assumed as a reference by Dafalias and Manzari.

It is interesting to analyse the back-predictions of the large stress-strain behaviour obtained by Verdugo and Ishihara and illustrated in Figure 6. In this case, the triaxial response predicted by the modified set of constants is absolutely unsatisfactory, both in terms of effective stress path and stress-strain behaviour. In particular, the negative excess pore water pressures increase more slowly with the strain level as compared to those exhibited by the original parameters (Figure 6c). This is clearly an effect of the reduction of the dilatancy parameter A_0. Related to this is the different stress-strain behaviour (Figure 6b), characterised by a

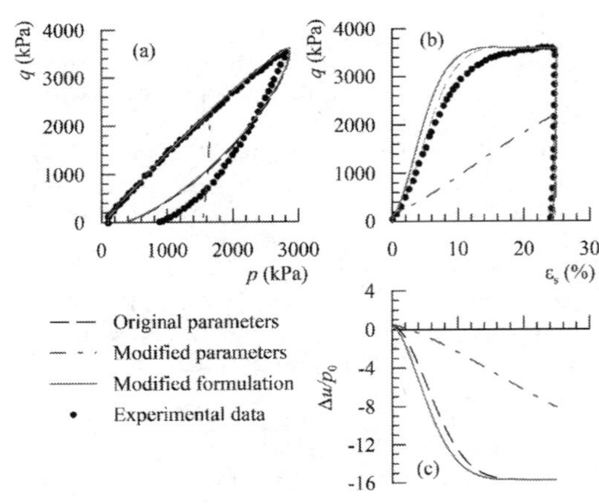

Figure 6. Comparison of model simulations to experimental data by Verdugo and Ishihara (1996) in undrained loading-unloading triaxial compression test (e_{in} = 0.735, p_0 = 100 kPa).

less stiff response at large strain, due to the lower mean effective pressure attained at the end of the simulation (Figure 6a).

It can be concluded that the original SANISAND model requires two different sets of parameters to reproduce the small to medium strain behaviour and the large strain (up to failure) one observed for Toyoura sand: each set does not perform well in the other strain level range.

4 MODIFIED FORMULATION

In this section, a modified formulation of the model is presented, in order to improve its overall performance at different strain levels.

First, the nonlinear isotropic hyperelastic model proposed by Houlsby et al. (2005) is adopted for the reversible response. Contrary to the original elasticity, this latter formulation guarantees a thermodynamically consistent response. The definition of the free energy functions allows to directly obtain the nonlinear constitutive equations in both stiffness and compliance forms given the current state of stress or strain, avoiding any numerical integration. The elastic formulation is described by the parameters g, k, n instead of G_0 and the Poisson ratio ν. Furthermore, the constant G_0 entering in the Equation (15) is substituted by the hyperelastic parameter g.

As described above, the improvement of the small to medium strain cyclic response, obtained by increasing the parameter h_0 and decreasing the dilatancy parameter A_0, does not produce a satisfactory performance of the model at large strains. The modified version of the model here proposed considers h_0 and A_0 as no longer constants but functions of the current state of the material. This latter is characterised by the Been & Jefferies parameter Ψ and by the variable d, representing the distance along the tensor **n** direction between the current value of $\boldsymbol{\alpha}$ and its initial value $\boldsymbol{\alpha}_{in}$ evaluated at the beginning of any new loading process. The basic idea is that at the beginning of a reverse loading the material is stiffer, getting gradually softer as the loading monotonically increases. Therefore, the parameter h_0 attains its maximum value for small d and decreases as this latter increases.

For $d \le d_b$, a bounding value for d, and $\Psi \le 0$ it reads:

$$h_0 = h_{0,min} + \bar{h}_0 - h_{0,min} \exp\left[\ln\left(\frac{\bar{h}_0}{h_{0,min}}\right)\frac{d}{d_b}\right] \tag{16}$$

where $h_{0,min}$ is the minimum value for h_0 and $\bar{h}_0$ is the value of h_0 when $d = 0$. The dependence of the state parameter Ψ enters as it follows:

$$\bar{h}_0 = h_{0,min} + \left(h_{0,max} - h_{0,min}\right)\cos\left(\frac{\Psi_{min} - \Psi}{\Psi_{min}}\frac{\pi}{2}\right) \tag{17}$$

where $h_{0,max}$ is the maximum value for h_0 and Ψ_{min} is the minimum value for the state parameter. The bounding distance d_b varies as well with the Been & Jefferies parameter, following the linear law:

$$d_b = \frac{d_{max}}{\Psi_{min}}\Psi \quad \text{with} \quad d_{max} = \sqrt{\frac{2}{3}}\left(M_e - m\right) \tag{18}$$

The variation of the distance d is controlled by the maximum value d_{max}, as reported in Equation (18).

Whenever $d > d_b$ or $\Psi > 0$, h_0 attains its minimum, constant value.

Following the same philosophy, the dilatancy parameter A_0 varies with d and attains a lower value at the beginning of any reversal loading and increases as the distance increases.

For $d \le d_b$ and $\Psi \le 0$ it follows:

$$A_0 = \bar{A}_0 \exp\left[\ln\left(\frac{A_{0,\max}}{\bar{A}_0}\right)\frac{d}{d_b}\right] \tag{19}$$

with $A_{0,max}$ the maximum value of A_0 and $\bar{A}_0$ the value of A_0 for $d = 0$, defined as:

$$\bar{A}_0 = A_{0,\min} \exp\left[\ln\left(\frac{A_{0,\min}}{A_{0,\max}}\right)\frac{\Psi - \Psi_{\min}}{\Psi_{\min}}\right] \tag{20}$$

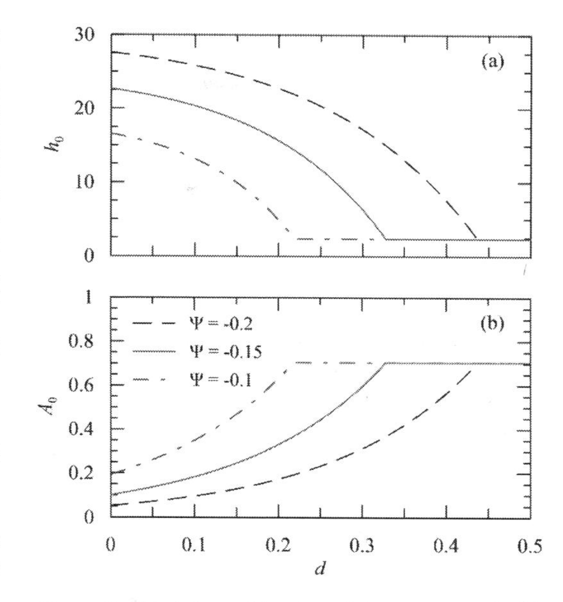

Figure 7. Variation of the two state parameters h_0 (a) and A_0 (b) with the distance d for different values of Ψ.

Table 2. Summary of material constants of the modified SANISAND model for Toyoura sand.

Constant	Variable	New formulation
Elasticity	g	375
	n	0.5
	k	97.2
Critical state	M_c	1.25
	c	0.712
	λ_c	0.019
	e_0	0.934
	ξ	0.7
Yield surface	m	0.01
Minimum Ψ	Ψ_{min}	−0.329
Plastic modulus	$h_{0,min}$	2.35
	$h_{0,max}$	33.3
	c_h	0.968
	n^b	1.1
Dilatancy	$A_{0,max}$	0.704
	$A_{0,min}$	0.01
	n^d	3.5
Fabric-dilatancy tensor	z_{max}	4
	c_z	600

with $A_{0,min}$ the minimum value of A_0. Similarly as above, for $d > d_b$ or $\Psi > 0$, A_0 attains its maximum, constant value.

Equations (16) and (19) are plotted in Figure 7 for three different values of the state parameter Ψ.

The calibration procedure of the new state parameters can be summarised as follows.

The constant Ψ_{min} is the minimum possible value for the state parameter, therefore it can be calculated in correspondence to the zero mean effective pressure knowing the minimum void ratio for the material. The values of $h_{0,min}$ and $A_{0,max}$ assume the same meaning of the original parameters h_0 and A_0 of the model and, as such, following Dafalias and Manzari (2004) can be calibrated by a trial-and-error procedure based on the experimental results of monotonic triaxial compression and extension tests. The additional parameters $h_{0,max}$ and $A_{0,min}$ should be calibrated on the undrained cyclic response of sand at small strains as observed in resonant column, torsional shear, simple shear or cyclic triaxial tests.

In Table 2 the values of the new parameters are reported as calibrated against the experimental data of monotonic triaxial compression tests (Figure 6) obtained by Verdugo & Ishihara (1996) and of undrained strain-controlled cyclic triaxial tests (Figures 2 and 3) by Kokusho (1980) for Toyoura sand.

It can be observed that, in all the investigated test conditions, the new version of the model is capable to reproduce the experimental response with sufficient accuracy, capturing both the initial stiffness and stiffness decay at low to medium strain levels and the cycling mobility of sands at large strain amplitudes, characterised by a significant amount of pore pressure build-up, stiffness degradation in strain-controlled cyclic tests and the classical butterfly shape of the stress-strain cycles in the stress-controlled cyclic tests.

5 CONCLUSIONS

The paper investigates the performance of a well-known SANISAND model when simulating the cyclic soil behaviour, highlighting the limitations of its original formulation when attempting to reproduce the soil response at small and large strain levels with a unique set of material parameters.

An alternative parameters calibration procedure is initially proposed in this paper aiming at better capturing the cyclic soil response at small-to-medium strain range. However, the numerical simulations showed that this approach does not allow to satisfactory simulate the cyclic mobility phenomenon and pore pressure build-up observed at large strains.

A satisfactorily compromise in reproducing both small and large strain test conditions is finally achieved by reformulating the model, introducing state-dependent plastic modulus and dilatancy laws and, with a lesser extent, by implementing a hyperelastic formulation for the reversible response. In this way the constitutive law, fed by a unique set of material constants, proves to reproduce the soil response at rather different strain levels with a sufficient degree of accuracy.

REFERENCES

Been, K. & Jefferies, M.G. 1985. A state parameter for sands. *Géotechnique* 35(2): 99–112.

Dafalias, Y.F. & Manzari, M.T. 2004. Simple Plasticity Sand Model Accounting for Fabric Change Effects. *Journal of Engineering Mechanics* 130(6): 622–634.

Dafalias, Y.F., Papadimitriou, A.G. & Li, X.S. 2004. Sand Plasticity Model Accounting for Inherent Fabric Anisotropy. *Journal of Engineering Mechanics* 130(11): 1319–1333.

Dafalias, Y.F. & Taiebat, M. 2016. SANISAND-Z: zero elastic range sand plasticity model. *Géotechnique* 66(12): 999–1013.

Georgiannou, V.N., Rampello, S. & Silvestri, F. 1991. Static and dynamic measurements of undrained stiffness on natural overconsolidated clays. *10th European Conference on Soil Mechanics and Foundation Engineering, Firenze,* Vol. 1: 91–95.

Kokusho, T. 1980. Cyclic triaxial test of dynamic soil properties for wide strain range. *Soils and Foundations* 20(2): 45–60.

Papadimitriou, A.G. & Bouckovalas, G.D. 2002. Plasticity model for sand under small and large cyclic strains: a multiaxial formulation. *Soil Dynamics and Earthquake Engineering* 22(3): 191–204.

Papadimitriou, A.G., Bouckovalas, G.D. & Dafalias, Y.F. 2001. Plasticity Model for Sand under Small and Large Cyclic Strains. *Journal of Geotechnical and Geoenvironmental Engineering* 127(11): 973–983.

Taborda, D.M.G., Zdravković, L., Kontoe, S. & Potts, D.M. 2014. Computational study on the modification of a bounding surface plasticity model for sands. *Computers and Geotechnics* 59: 145–160.

Taiebat, M. & Dafalias, Y.F. 2008. SANISAND: Simple anisotropic sand plasticity model. *International Journal for Numerical and Analytical Methods in Geomechanics* 32(8): 915–948.

Verdugo, R. & Ishihara, K. 1996. The steady state of sandy soils. *Soils and Foundations* 36(2): 81–91.

Zhang, F., Jin, Y., & Ye, B. 2010. A try to give a unified description of Toyoura sand. *Soils and Foundations,* 50(5): 679–693.

Numerical Methods in Geotechnical Engineering IX – Cardoso et al. (Eds)
 ISBN 978-1-138-33198-3

Evaluating the effects of noise on full field displacement data used for the identification of soil stress-strain response

J.A. Charles, C.C. Smith & J.A. Black
Department of Civil and Structural Engineering, University of Sheffield, UK

ABSTRACT: Plane-strain physical modelling tests can provide large quantities of often unused data in the form of Particle Image Velocimetry derived displacement fields. Using an Identification Method, it is possible to use this full-field displacement data, along with external force data, such that a stress-strain response can be reconstructed using optimisation by minimising the energy gap between internal and external work. Previous work indicates that the method is viable with perfect artificial data, but noise, discretisation, and other issues, can cause a reduction in accuracy of the recovered curve. This contribution demonstrates the applicability of this method to soils with non-linear stress-strain responses undergoing non-uniform deformation. Finite element modelling was used to obtain "perfect" strain fields which were artificially degraded to simulate noisy experimentally obtained data, with higher levels of noise increasingly reducing the quality of the recovered stress-strain curve, particularly for higher strain values in which less data points are available.

1 INTRODUCTION

An Identification method (such methods are described extensively by Avril et al. 2008), is a back analysis method developed for testing solids in order to recover the constitutive parameters of the material based on recorded loading data and image derived displacement data. Although there are multiple algorithms to achieve this, the basic principle is to find a set of parameters such that a numerically modelled response of an object is as similar to the physically modelled response as possible.

The application of such a method to geotechnics would potentially allow for the utilisation of often underused physical full-field displacement data. Although this contribution details work in progress and preliminary validation tests, the method could allow for the recovery of constitutive parameters for different sections a highly non uniform soil sample, facilitating more accurate numerical modelling, or potentially automated anomaly identification in cases where the physical soil body is taken to be uniform.

Identification methods were originally developed for use in materials science, however with the adaptation of Particle imaging velocimetry (PIV) by White et al. (2001) for geotechnical usage, it is now possible to obtain the full field displacement data for soil samples undergoing testing. PIV is a non-intrusive method that provides no further disturbance to the soil sample.

Although some work has been done on incorporating non-linear responses such as plastic deformation into Identification Methods, (Grédiac & Pierron 2006), the difficulties of dealing with deformations within a soil, such as noise, discretisation, and irregular particle movement, will require novel approaches. This paper reports work in progress towards this goal.

The identification method chosen for adaption for geotechnical problems is based on the Virtual fields method (VFM), described by Grédiac & Pierron (1998), which utilises full field displacement measurements, along with the principles of the conservation of energy and of virtual work to recover the parameters of a chosen constitutive model.

Work by Gueguin et al. (2015) demonstrated an identification method based on a simplification of VFM developed with the goal of recovering and reconstructing the stress-strain response of a soil sample undergoing loading. In that work artificial datasets were generated using FEA (Finite element analysis) software based on a linear elastic, perfectly plastic material response

This contribution builds on the work by Gueguin et al. by investigating issues relating to noisy data. Artificial displacement fields and load-displacement data were generated using Abaqus FEA software to model several simple cases: a simple shear case, in which soil will be uniformly sheared, and a case modelling the soil at the base of a rotating retaining wall. In each case, non-linear soil properties were adopted.

For simplicity, in this initial study, soil will be modelled as weightless. As such, no energy will be expended due to moving soil against gravity and will not need to be considered in the identification method.

The obtained data was then input into the identification method to reconstruct the soil stress-strain response. Noise was artificially added to the FEA data to investigate the degree to which the recovered curve is degraded.

2 AN ENERGY BASED IDENTIFICATION METHOD FOR UNDRAINED SOILS

2.1 *Principles of the identification method*

The principle of this energy based identification method is that for any valid physical system the stress field σ and the strain field ε must satisfy the following equation:

$$W_{\text{int}}(\sigma,\varepsilon) - W_{\text{ext}=0} \quad (1)$$

Where W_{int} and W_{ext} represent the internal work (i.e. energy expended by deformation etc.) and external work (i.e. energy imparted by loading etc.)

For a simple model test, in which the effects of gravity are neglected and plane strain is assumed, external work, where u is load displacement $F_{(u)}$ is applied force corresponding to displacement u, is defined as:

$$W_{\text{ext}} = \int F\,\text{d}u \quad (2)$$

Internal work W_{int} can be written as an integration of local energy across the whole field area A with the following equation:

$$W_{\text{int}} = \int_{\text{A}} \left[s\varepsilon_{\text{v}} + 2t\varepsilon_{\text{s}}\left(1 - 2\sin^2(\theta_\sigma - \theta_\varepsilon)\right)\right] \text{d}A \quad (3)$$

In which θ_σ and θ_ε are the orientations of the principle stress and principle strain respectively, and in which the other terms, calculated from eigenvalues of the stress field (σ_1, σ_3) and the strain field $(\varepsilon_1, \varepsilon_3)$, are: mean stress s, deviatoric stress t, volumetric strain ε_v, and shear strain ε_s.

In the current work, application of the method will currently be limited to monotonically loaded, plane strain tests on undrained soils, allowing a number of simplifications to be made. Volumetric strain will be taken as zero and associative flow will be assumed, allowing equation 3 to be reformulated as:

$$W_{\text{int}} = \int_{\text{A}} 2t\varepsilon_{\text{s}}\,\text{d}A \quad (4)$$

2.2 *Numerical implementation of the identification method*

Several issues adversely affect the ability to ensure equality in equation 1. Discretisation of the displacement field and hence strain field into elements, low quality imaging data due to noise or resolution, and complexities in the soil response, require the reformulation of the equilibrium equation into an optimization problem, with internal and external energies for each of n images, or time steps, measured in place of global values. Thus equation 1 can be written as:

$$W_{\text{int}}^{(j)} + \textbf{localgap}^{(j)} = W_{\text{ext}}^{(j)} \quad (5)$$

The external work imparted during a time step, in which there is a known force value corresponding to each displacement value is:

$$W_{\text{ext}}^{(j)} = \int_{u^{j-1}}^{u^j} F\,\text{d}u \quad (6)$$

The internal work done during a time step will be the sum of the work done by each of E elements within the strain field, each of which may be experiencing a shear stress at a different point on the stress-strain curve and a corresponding and unknown deviatoric stress $t_{(\varepsilon_s)}$. As such, the incremental internal work for a time step j is defined as follows:

$$W_{\text{int}}^{(j)} = \sum_{e=1}^{E\text{ elements}} \left(\int_{\varepsilon_s^{j-1,e}}^{\varepsilon_s^{j,e}} 2t\,\text{d}\varepsilon_{\text{s}} \right) \quad (7)$$

Although each element will be at a different stage of the stress-strain response, it is, as the soil is uniform, assumed that every element is at some point on a global curve and that all elements start at the same point.

The stress-strain response to be reconstructed will be split into a number of intervals where each interval may chosen and distributed arbitrarily by the user, unrelated to both the number of images and the number of elements. Figure 1 demonstrates the relationship between the piecewise stress-strain curve that will be obtained, and the continuous curve. $W_{\text{int}}^{(j)}$ will be formulated in terms of these pre-chosen strain increments and the corresponding t values will be optimised for.

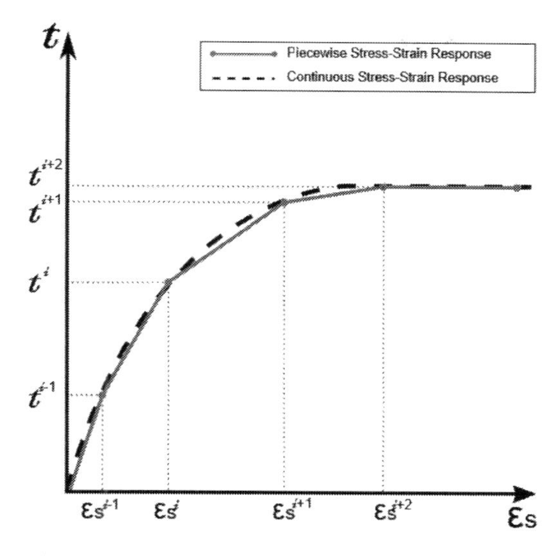

Figure 1. Continuous and piecewise stress-strain response.

The optimization problem is defined as follows:

$$\text{Minimize} \sqrt{\sum_{j=1}^{n_{\text{images}}} (\text{localgap}^{(j)})^2} \tag{8}$$

Subject to:

$$W_{\text{int}}^{(j)} + \textbf{localgap}^{(j)} = W_{\text{ext}}^{(j)} \tag{9}$$

$$G^{(i)} - G^{(i-1)} \geq 0 \tag{10}$$

$$G^{(i)} \geq 0 \tag{11}$$

where G is the shear modulus at any point. Equation 8 is the objective function, with the optimal curve being the one that results in the smallest gap between internal and external energies. Equation 9 defines the incremental energy gap, equation 10 enforces strain hardening by requiring a monotonically reducing shear modulus G and equation 11 requires the shear moduli to be positive.

The method as described was implemented using the Mosek (2017) mathematical optimiser in the Matlab environment.

3 FEA MODEL DESCRIPTION

In order to generate high quality displacement data, Abaqus CAE (Dassault Systems 2017) finite element software was used to simulate two simple model tests on clay. FEA obtained data was used in place of physical modelled data in order to avoid experimental sources of noise. Instead, this was applied artificially in a controlled way for testing purposes.

The soil was simulated as a clay undergoing strain hardening. The stress-strain response of the clay was a non-linear curve generated using a power law such as described by Vardanega & Bolton (2011). A horizontal plateau was added to the curve to simulate a perfect plastic response once the yield shear strength has been reached. The chosen curve is shown in Figure 2.

The soil was modelled with a Poisson's ratio of 0.49999. This value, the highest allowable in the FEA simulation used, was chosen to minimise volumetric strain in line with the assumptions made. As the FEA modelling was used to simulate PIV data, the nodal displacement data was used to calculate strains rather than using the strain values output by ABAQUS. Although negligible in relation to shear strain, some volumetric strains were found using this approach, potentially reducing the quality of results.

Figure 3 and Figure 4 shows the problem layout for the two chosen models.

Model 1 is a simple shear test, the base fixed and loading distributed uniformly along the top. A grid of 64 4 node square elements was used, with a total of 81 nodes at which displacement will be recorded.

Model 2 simulates a section of soil at the base of a rotating wall. A point load is applied to the top of the rigid wall. A grid of 512 6 node triangular elements was used, with a total of 1089 nodes.

In both cases the soil was assumed to be weightless and isotropic.

Each FEA model was run over a time period of 5 seconds, with a total of 10 uniformly distributed frames. Loading was increased linearly during this

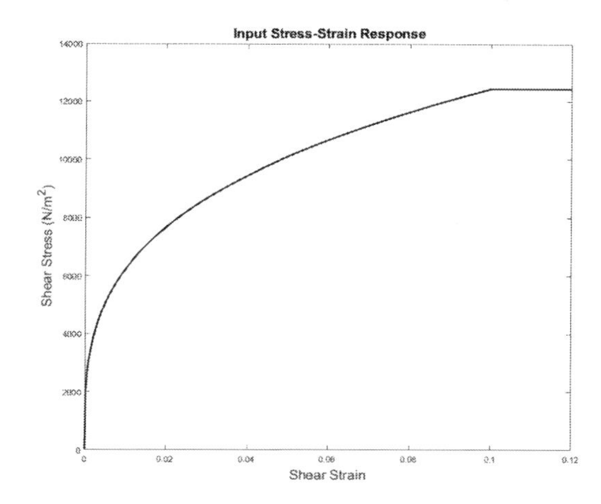

Figure 2. The stress-strain curve input into the FEA software to define material properties of the clay.

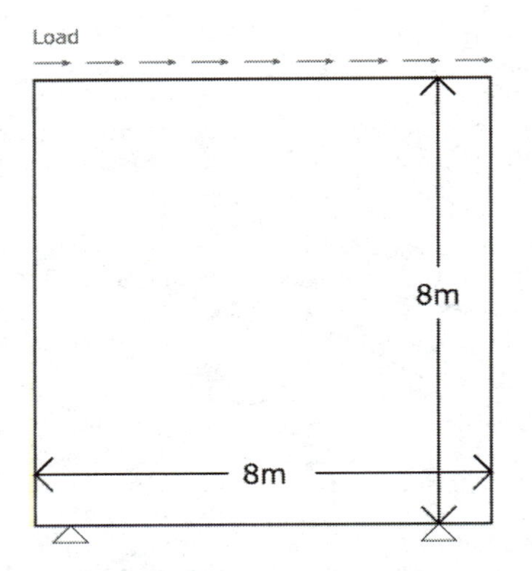

Figure 3. A not to scale representation of the layout of Model 1, a block of soil undergoing simple shear.

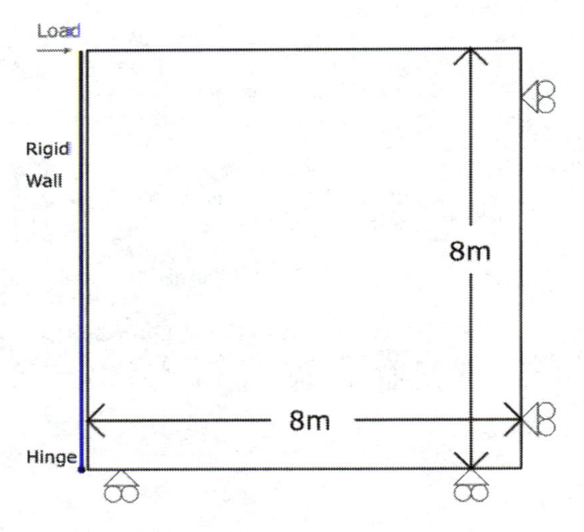

Figure 4. A not to scale representation of the layout of Model 2, A block of soil at the base of a rotating wall.

time from 0 to 75 kN for model 1 (distributed uniformly along the top of the soil) and 0 to 120 kN for model 2 (applied as a point load). Figure 5 shows a plot with the force-displacement curves for each of the models.

The ABAQUS generated shear strain fields for each of the models can be seen in Figure 6 and Figure 7. Max in-plane principle strain refers is the maximum shear strain ε_s as defined in the derivation in Section 2.1. The plots show the strain and displacement of each model as of the final time frame.

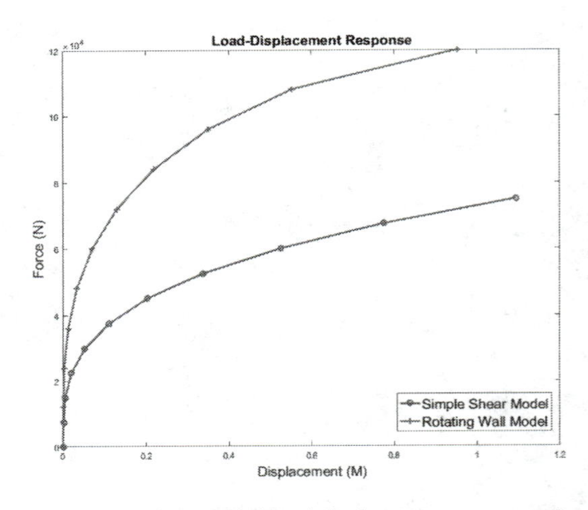

Figure 5. The recorded load-displacement responses of the FEA models. Each marker represents a 0.5 second frame.

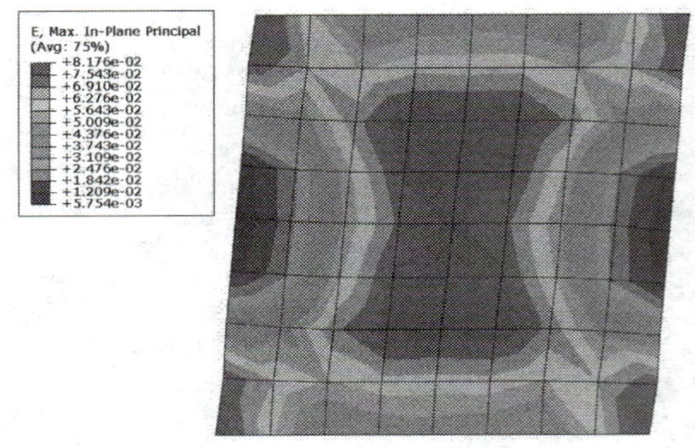

Figure 6. The shear strain field as output by ABAQUS for the simple shear model.

4 RECOVERY OF STRESS STRAIN-RESPONSE FROM FEA DISPLACEMENT FIELD

To simulate the recovery of strain data from physical PIV data, the nodal displacements were taken from the FEA data and used in place of PIV patch movement. The nodes were assigned to a triangular grid with strains calculated based on movement relative to their initial locations.

The obtained strain fields along with the force-displacement histories were input into the Identification Method software. The tests were rerun with noise artificially added to the shear strain values of each element in order to evaluate the methods resilience to noise. Each element was adjusted by a random amount of up to either 2.5, 5, or 7.5% of its value.

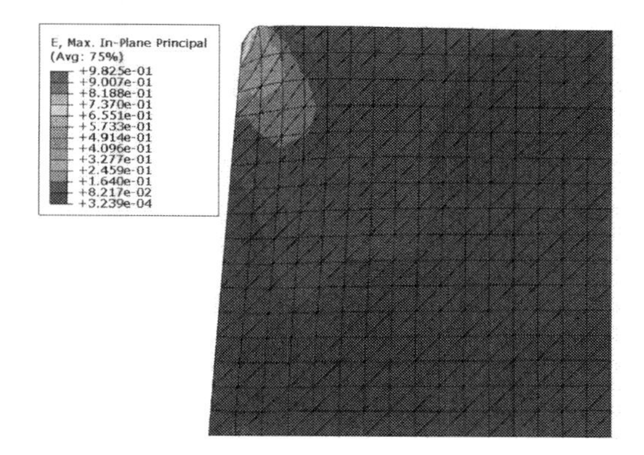

Figure 7. The shear strain field as output by ABAQUS for the rotating wall model.

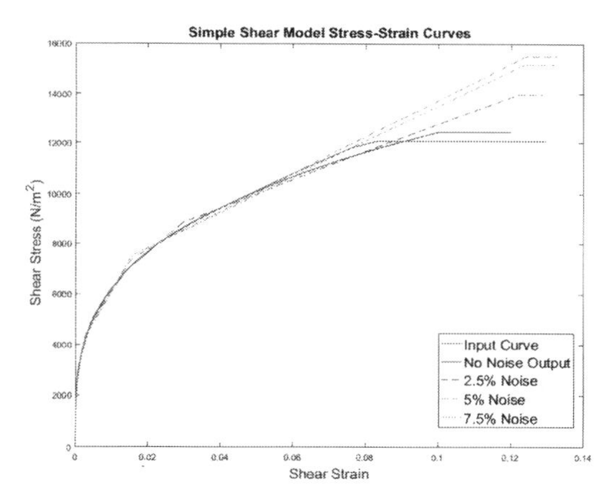

Figure 8. A comparison between the input stress-strain curve and the output curves with varying levels of added noise for the simple shear model.

Noise was added to the shear strain field as opposed to the displacement field (which would be a more realistic simulation of noisy PIV images) as adding noise to the latter causes large amounts of volumetric strain, something that the method is not currently equipped to handle. Thus, adding noise only to the shear strain field allows the assumptions regarding volumetric strain to reman valid.

Figure 8 shows the output stress strain curves for the simple shear model and the global energy gap data is given in Table 1. Of note is the similarity between the actual input curve and the output with no added noise. Particularly for low strains the curves are almost identical with differences only becoming apparent for higher strain values. The difference between recovered and actual c_u is approximately 3%.

A possible reason for the slight discrepancy between the curves is that due to the derivation of strains from the displacement data, some volumetric strain was interpreted, with the sum of observed absolute volumetric strains being around 5% of the sum of observed shear strains. As work done by volumetric straining is currently discounted, this would cause a slight overestimate in work done by shear, resulting in a "higher" stress-strain curve.

Predictably, adding noise reduces the quality of the recovered curve. Of note is the tendency of lower quality curves to be less smooth and made up of line segments. The addition of noise does not appear to affect the ability of the optimiser to find a suitable curve for the data it is given. For the curve with 7.5% noise, the global energy gap is only 0.35%.

Figure 9 shows the output curves for the rotating wall model. Again, the recovered curve is very similar when no noise is artificially added, with negligible difference between the c_u values. Again, it is likely that the slight discrepancy between the curves is due to unaccounted for volumetric strain.

The rotating wall model appears to be less affected by noise than the simple shear model, with the curve representing 7.5% added noise having a c_u value within 5% of the input curve. Table 1 additionally shows that the quality of fit is better for this model with lower global energy gaps observed even in cases with added noise.

It should be noted that the presented plots are typical recovered curves. Randomly added noise can potentially produce a range of effects, or no effects at all. However, due to the large number of elements, little change was observed across multiple runs.

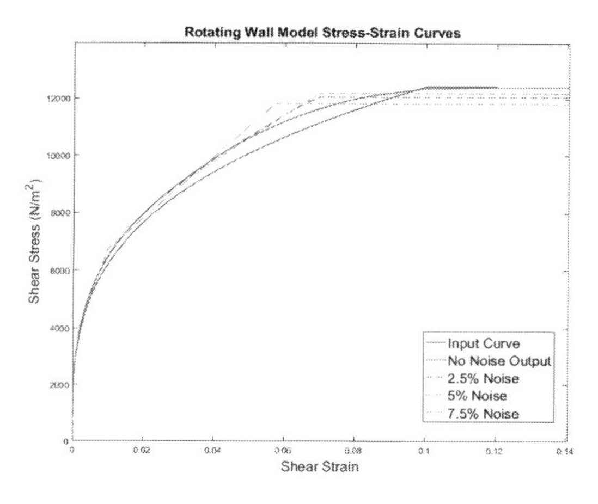

Figure 9. A comparison between the input stress-strain curve and the output curves with varying levels of added noise for the rotating wall model.

Table 1. Table to show variance in global error values measured in joules. Total external work is 6.3e+4 J for the Simple Shear case and 9.2e+4 J for the Rotating Wall case.

Noise %	Simple Shear Global Error	Rotating Wall Global Error
0	1.33e-12	1.99e-14
1	3.64e-12	6.51e-14
2.5	153	8.18e-14
5	201	2.90e-13
7.5	222	38.2

Whereas the simple shear model features a more uniform distribution of strain values, the rotating wall model has large regions of the field in which little to no strain takes place. As noise was applied as a percentage, these regions will be less affected. And as proportionally more of the elements are undergoing low strains, the additional weighting to the earlier sections of the stress strain curve results in a better fit.

Due to the lower number of elements experiencing high strain, and the disproportionate effect of noise on them, it is likely to be possible to automatically identify areas of the curve in which the user should have lower confidence. Knowing, for example, that the first half of the recovered stress-strain curve has a high degree of validity would be of use even if the latter half of the curve was ultimately discarded.

5 CONCLUSIONS

An identification method has been described and demonstrated to function with a good degree of precision for soils with non-linear stress-strain responses undergoing non-uniform deformation.

The effects of noise on both the ability of the optimiser to find a suitable stress-strain response and the quality of the recovered curves has been discussed, with increasing levels of noise shown to adversely effect the quality of the recovered stress-strain curves.

Future work will include evaluating the effects of noise in increasingly complex artificial and physical experiment scenarios and and developing techniques to minimise or automatically identify its effects. Developing an algorithm to assess confidence in different sections of the output curve, will also increase the usability of the method with lower quality data.

The method will also be extended to handle frictional soils and self weight effects, removing for example the assumptions regarding volumetric strain and flow rules.

ACKNOWLEDGEMENTS

The primary author acknowledges the PhD Studentship provided by the UK Engineering and Physical Sciences Research Council (EPSRC).

REFERENCES

Avril, S., M. Bonnet, A.S. Bretelle, M. Grédiac, F. Hild, P. Ienny, F. Latourte, D. Lemosse, S. Pagano, E. Pagnacco, & F. Pierron (2008). Overview of identification methods of mechanical parameters based on full-field measurements. *Experimental Mechanics 48*(4), 381–402.

Dassault Systems (2017). Abaqus cae finite element analysis software. https://www.3ds.com/productsservices/simulia/products/abaqus/abaquscae/.

Grédiac, M. & F. Pierron (1998). A T-shaped specimen for the direct characterization of orthotropic materials. *International Journal for Numerical Methods in Engineering 41*(September 1996), 293–309.

Grédiac, M. & F. Pierron (2006). Applying the Virtual Fields Method to the identification of elasto-plastic constitutive parameters. *International Journal of Plasticity 22*(4), 602–627.

Gueguin, M., C. Smith, & M. Gilbert (2015). *Use of digital image correlation to directly derive soil stress-strain response from physical model test data*, pp. 3881–3886.

Mosek (2017). Mosek optimisation software. https://www.mosek.com/.

Vardanega, P. & M. Bolton (2011). Practical methods to estimate the non-linear shear stiffness of fine grained soils. *International Symposium on Deformation Charateristics of Geomaterials* (1972), 372–379.

White, D.J., W.A. Take, & M. Bolton (2001). Measuring soil deformation in geotechnical models using digital images and PIV analysis. *10th International Conference on Computer Methods and Advances in Geomechanics*, 997–1002.

Numerical Methods in Geotechnical Engineering IX – Cardoso et al. (Eds)
© 2018 Taylor & Francis Group, London, ISBN 978-1-138-33198-3

Analysis of the bearing capacity of strip footing on crushable soil

Vu Pham Quang Nguyen, Mamoru Kikumoto & Keita Nakamura
Department of Civil Engineering, Yokohama University, Japan

ABSTRACT: The analysis of particle crushing effect on the bearing capacity of strip footing is the main focus of this paper. Firstly, an elastoplastic constitutive model exhibiting particle crushing is developed by extending the critical state theory with the evolution of particle size distribution. Then, the performance of the proposed model is validated by experimental results. Finally, the Finite element analysis of the bearing capacity of strip footing is performed using the proposed model to study the bearing capacity of strip footing on crushable soil.

1 INTRODUCTION

The bearing capacity analysis of shallow foundation is one of the most fundamental problems in geotechnical engineering. The evaluation of the bearing strength of a shallow foundation is obtained from the mobilization of the shear strength of soil beneath and adjacent to the foundation. The famous equation for estimating the ultimate bearing capacity of shallow strip foundation by Tezaghi (1943) has been widely used among geotechnical engineers. His estimation is the summation of three distinct components: the cohesive component of shear strength; the component for the surcharge pressure adjacent to the foundation; and the component accounting for frictional resistance of the soil.

In the context of bearing capacity of shallow foundation, the combination of the applied pressures can cause the particle crushing phenomenon in soil particles. The important role of particle crushing in the mechanical behavior of soil has been long recognized. This crushing will reduce the strength of the soil which leads to the reduction of the footing's bearing capacity. This reduction is significant in case of very crushing sensitive soil. Thus, geotechnical engineers may jeopardize the safety of strip foundation in crushable soil areas. Nevertheless, there has been no research considering particle crushing effect on the bearing capacity of shallow foundation.

Even though experimental testing is an attractive approach to study the effect of particle crushing to strip footing's bearing capacity, experimental validation is not a straightforward because it does not provide a simple way to visualize the evolution of this phenomenon. Therefore, the aim of this study is to perform the numerical analysis of the particle crushing effect on bearing capacity of strip footing by utilizing an elastoplastic constitutive model for crushable soils.

In simulating the bearing capacity problems of shallow foundation, Finite Element Method (FEM) is applied in this study. However, it is usually known that the so-called volumetric locking behavior occurs and causes unrealistically stiff solution in FE simulation especially when distortional deformation is exhibited without significant change in the volume. In order to overcome this issue, B-bar method (Hughes, 1987) is applied in our study.

In this paper, a constitutive model for particle crushing is firstly introduced. Then, the FEM in which B-bar method is implemented is validated by comparing the simulation with analytical solution of the elastic beam. Finally, the simulation result of strip footing's bearing capacity using the proposed constitutive model is discussed.

2 GRADING INDEX AND ITS EVOLUTION DUE TO PARTICLE CRUSHING

In order to describe the evolving grading due to particle crushing, it is convenient to define a simple index which can indicate where the current changing grading of the soil. The relative breakage B_r (Hardin, 1985), which is defined as the ratio of areas ABC and ABD in Figure 1a and which takes a value between 0 and 1, is usually used as an index of the changing grading of soils. However, this index is not convenient in describing the particle breakage in a constitutive model as any grading can be selected as the initial grading and as it does not incorporate the critical grading at which any further crushing will not occur.

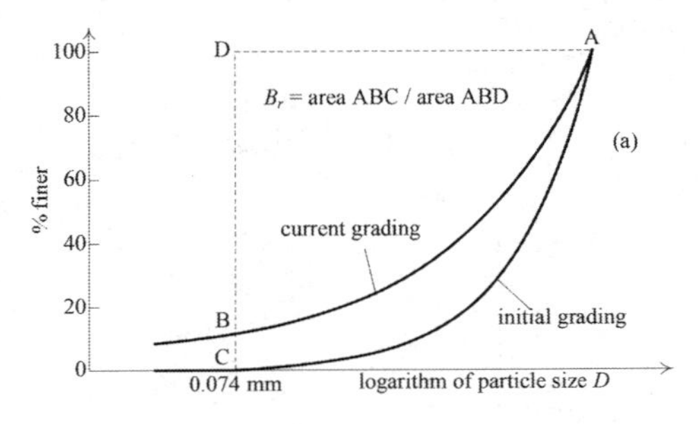

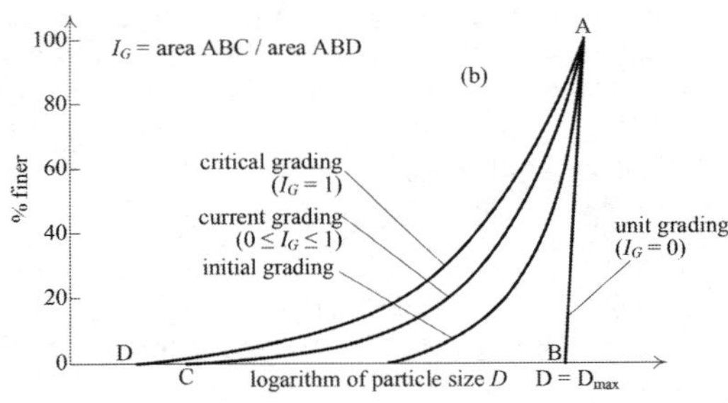

Figure 1. Definition of grading indices (a) breakage parameter B_r (b) grading index I_G.

On the other hand, Muir Wood (2007) proposed a grading index I_G which is defined in Figure 1b as the ratio of areas ABC and ABD. I_G takes the values of 0 and 1 at the single size grading I_G (AB) and at the critical grading (AD), respectively. In this study, I_G is selected as a scalar index of changing grading and is assumed to increase monotonically from 0 to 1 due to particle breakage. The evolution of I_G is linked with a stress variable p_x which reflects the maximum stress level that the soil has experienced.

$$I_G = 1 - \exp\left(-\frac{p_x - p_{x0}}{p_r}\right) \tag{1}$$

Here, p_{x0} is a constant representing the stress level at which the crushing of particle is initiated in the soil having unit grading. p_r is a material constant which controls the rate of particle crushing (Fig. 2). The crushing behavior of soil particles obviously depends on the properties of grains (such as mineralogy, hardness, shape and size) and other conditions (such as packing density, particle size distribution, stress level and mobilized friction), which would be reflected in soil parameters p_r and p_{x0}. Equation 1 ensures that I_G monotonically increases from 0 to 1 with the increase of p_x to infinity (Fig. 2). Rate form of the evolution of I_G is derived as follow:

$$\dot{I}_G = \frac{1}{p_r}\exp\left(-\frac{p_x - p_{x0}}{p_r}\right)\dot{p}_{x0} = \frac{1 - I_G}{p_r}\dot{p}_x \tag{2}$$

In isotropic compression, particle crushing is initiated when the mean effective stress p' reaches p_x.

As particle crushing is also exhibited during shearing, combined effect of compression and shearing is is considered by linking I_G with mean stress p and deviator stress q. For this, we employ a non-positive function f_x defined in p-q plane (Fig. 3) as:

$$f_x = \ln p + \frac{2}{\alpha}\ln\left\{1 + \left(\frac{\eta}{M_x}\right)^{\alpha}\right\} - \ln p_x \le 0 \tag{3}$$

where η (= q/p') is stress ratio, q is deviator stress and M_x and α are material parameters that controls the effect of shearing on the magnitude of particle crushing. When $f_x = 0$ and $(\partial f_x / \partial \sigma_{ij})\dot{\sigma}_{ij} > 0$, particle crushing occurs and I_G increases satisfying consistency condition of f_x. The rate of the crushing stress p_x is given by:

$$\dot{p}_x = p_x \frac{\partial f_x}{\partial \sigma_{ij}} \dot{\boldsymbol{\sigma}}_{ij} \tag{4}$$

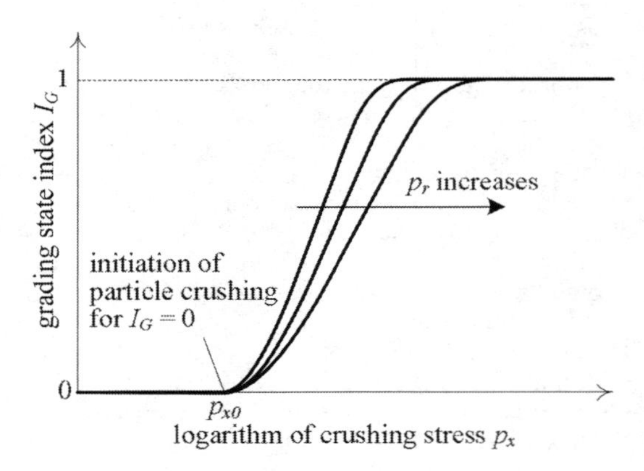

Figure 2. Variation of the grading index I_G with the increase of crushing.

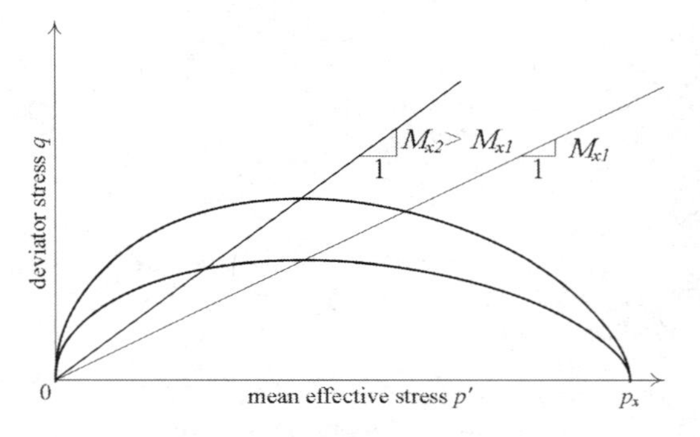

Figure 3. M_x parameter and crushing surface f_x.

3 ELASTOPLASTIC MODEL FOR SOILS CONSIDERING PARTICLE CRUSHING

An elastoplastic constitutive model considering crushing effect is developed based on continuum mechanics. The proposed model is formulated based on the extension of critical state model (Roscoe *et al.*, 1963; Wood, 2007) to incorporate the density effect on stress-strain characteristics by utilizing subloading surface concept (Hashiguchi & Ueno, 1977). Then, the effect of particle crushing is further implemented by considering the evolution of particle size distribution due to the change of crushing stress and its effect on the constitutive behavior.

Critical state is an ultimate state towards which all states of soil finally approach when the soil is sheared. The critical state line (CSL) has been usually chosen as linear in a semi logarithmic compression plane, which is the specific volume, $v(=1+e)$, versus the logarithm of mean effective stress, $ln\,p'$. Similar to CSL, limiting isotropic compression line (LICL) is a reference line in the v-$ln\,p'$ plane where any stress state finally approaches under isotropic compression.

It is customary to utilize the State boundary surface (SBS) containing CSL and LICL (Fig. 4) in the formation of critical state model. Specific volume on the SBS v_{sbs}, which defines the loosest state of soil at the current stress (p, η) is given by considering the combined effects of compression and dilation as:

$$v_{sbs} = \mathrm{N} - \lambda \ln \frac{p'}{p_a} + (\Gamma - \mathrm{N})\zeta(\eta) \tag{5}$$

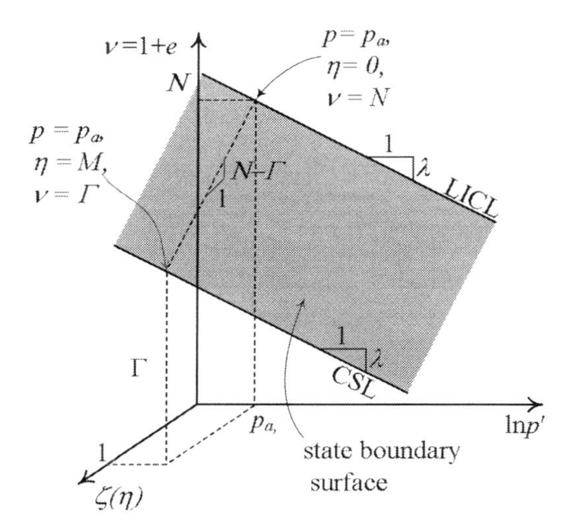

Figure 4. Specific volume in the loosest state.

where p_a (= 98 kPa) denotes atmospheric pressure, λ is compression index, $\zeta(\eta)$ is a monotonic increasing function of stress ratio η satisfying $\eta(0) = 0$ on NCL and $\eta(M) = 1$ on CSL. Γ and N represent specific volumes on LICL ($\eta = 0$) and CSL ($\eta = M$) at $p' = p_a$ respectively. It is postulated that different functions of $\zeta(\eta)$ are used for different versions of critical state models. In the current model, Equation 6 is employed in accordance with the model of Roscoe and Burland (1968) when $\alpha = 2$.

$$\zeta = \frac{\ln\left\{1+\left(\frac{\eta}{\mathrm{M}}\right)^{\alpha}\right\}}{\ln 2} \tag{6}$$

where M is critical state stress ratio ($= \eta_{cs}$).

The effect of particle crushing on soil behavior is then incorporated by extending the critical state concept. As noticed in the past researches (Daouadji *et al.*, 2001, Kikumoto *et al.*, 2010, Ghafghazi *et al.*, 2014), breakage caused a downward shift in the CSL in e-lnp' plane. Therefore, our key concept in the formulation of the model for particle crushing is that the particle crushing effect of soils is considered by the downward parallel movement of the SBS in the volumetric plane of p' and v. For this purpose, a state variable Ψ is newly introduced to represent the downward shift of the SBS in the p'–$\zeta(\eta)$–v space as indicated in Figure 5. From this, the state parameter Ψ is a non-negative variable defined as the volumetric distance between the SBSs for soil with and without density effect. Ψ works as a state variable controlling the elastoplastic response in the constitutive model. The specific volume on the SBS of soil exhibiting particle breakage, v^b_{sbs}, is thus given in a similar way as Equation 5.

$$\begin{aligned} v^b_{sbs} &= v_{sbs} + \Psi \\ &= N - \lambda \ln \frac{p'}{p_a} - (\mathrm{N}-\Gamma)\zeta(\eta) + \Psi \end{aligned} \tag{7}$$

Due to the fact that the soils whose states lie under the SBS do exhibit plastic strain together with elastic strain, subloading surface concept (Hashiguchi & Ueno, 1977) is further introduced to portray this behavior. A state parameter Ω, which is a combination of specific volume and mean effective stress to describe the changing strength and stiffness is incorporated in this model. As all states of soil locate on or below the SBS in Figure 4, the SBS defines the loosest, upper limit of specific volume of soils. Consequently, the state parameter $\Omega(\geq 0)$ is thus defined as the specific volume difference between the current state and the loosest state under the same stress (p', η) on the

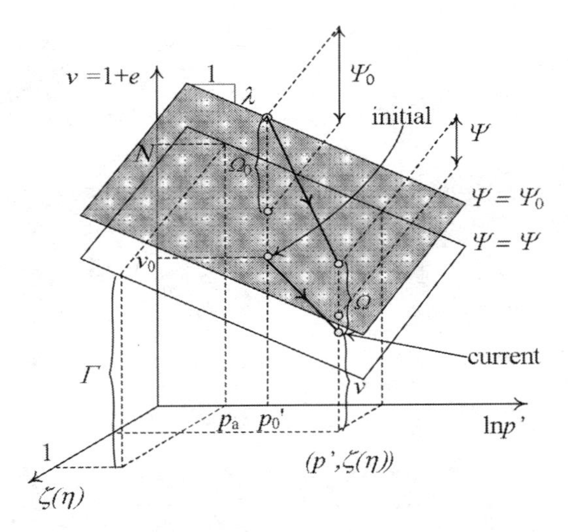

Figure 5. Modeling of volumetric behavior of soil considering particle crushing and density effect.

SBS as shown in Figure 5. According to this concept, soil exhibits irreversible deformation below the SBS and gradually approaches the SBS with loading. Taking a state variable Ω which is the difference between the specific volume of the current state and that on the SBS under the same stress (p', η), we can represent an arbitrary specific volume v:

$$\begin{aligned} v &= v^{\mathrm{b}}_{\mathrm{sbs}} - \Omega \\ &= N - \lambda \ln \frac{p'}{p_a} + (\Gamma - \mathrm{N})\zeta(\eta) + \Psi - \Omega \end{aligned} \tag{8}$$

Ω decreases gradually with the development of plastic deformation and finally converges to zero. Therefore, an evolution law of Ω can be chosen as:

$$\frac{\dot{\Omega}}{v_0} = -Q(\Omega)\|\dot{\varepsilon}^p\| \tag{9}$$

where $\dot{\varepsilon}^p$ is plastic strain rate tensor and $Q(\Omega)$ is a function of Ω given as:

$$Q(\ ) = \omega\Omega|\Omega| \tag{10}$$

ω is a parameter controlling the effect of density.

From the current specific volume v given by Equation 8, and we can also calculate the initial specific volume v_0 by substituting the initial states: $v = v_0$, $\Psi = \Psi_0$, $\Omega = \Omega_0$, $p' = p'_0$ and $q = 0$ as:

$$v_0 = \mathrm{N} - \lambda \ln \frac{p'_0}{p_a} + \Psi_0 - \Omega_0 \tag{11}$$

Total volumetric strain (compression is taken to be positive) generated from the initial state to the current state is given by:

$$\varepsilon_v = -\frac{\mathrm{d}v}{v_0} = \frac{v_0 - v}{v_0} \tag{12}$$

By substituting Equations 8 and 11 for Equation 12, we obtain:

$$\varepsilon_v = \frac{1}{v_0}\left\{\lambda \ln \frac{p'}{p'_0} + (\mathrm{N} - \Gamma)\zeta(\eta) - (\Psi - \Psi_0) + (\Omega - \Omega_0)\right\} \tag{13}$$

Then, the plastic volumetric strain can be determined by taking a difference between the total volumetric strain given by Equation 13 and the elastic volumetric strain:

$$\begin{aligned} \varepsilon_v^p = \varepsilon_v - \varepsilon_v^e = \frac{1}{v_0}\Big\{&(\lambda - \kappa)\ln \frac{p'}{p'_0} + (\mathrm{N} - \Gamma)\zeta(\eta) \\ &- (\Psi - \Psi_0) + (\Omega - \Omega_0)\Big\} \end{aligned} \tag{14}$$

From Equation 14, yield function f for soil considering the effect of crushing phenomenon can be written as follows:

$$\begin{aligned} f = \frac{1}{v_0}\Big\{&(\lambda - \kappa)\ln \frac{p'}{p'_0} + (\mathrm{N} - \Gamma)\zeta(\eta) \\ &- (\Psi - \Psi_0) + (\Omega - \Omega_0)\Big\} - \varepsilon_v^p \end{aligned} \tag{15}$$

Assuming associated flow in the proposed model, plastic strain rate tensor is derived as:

$$\dot{\varepsilon}^p = \langle \dot{\Lambda} \rangle \frac{\partial f}{\partial \sigma'} \tag{16}$$

where $\dot{\Lambda}$ is the rate of the plastic multiplier. The loading criterion is thus given by $\dot{\Lambda} > 0$. As unlimited distortional strain is exhibited at critical state without any change in stress or volume, $tr(\partial f / \partial \sigma)$ becomes zero when η is equal to M. Furthermore, $(N - \Gamma)$ is equal to $2\ln 2(\lambda - \kappa)/\alpha$ in case Equation 6 is applied. The yield function is finally given as follows.

$$\begin{aligned} f = \frac{\lambda - \kappa}{v_0}&\left[\ln \frac{p'}{p'_0} + \frac{2}{\alpha}\ln\left\{1 + \left(\frac{\eta}{\mathrm{M}}\right)^{\alpha}\right\}\right] \\ &- \frac{\Psi - \Psi_0}{v_0} + \frac{\Omega - \Omega_0}{v_0} - \varepsilon_v^p \end{aligned} \tag{17}$$

Now that we have known I_G accounts for the level of particle crushing and Ψ is responsible for

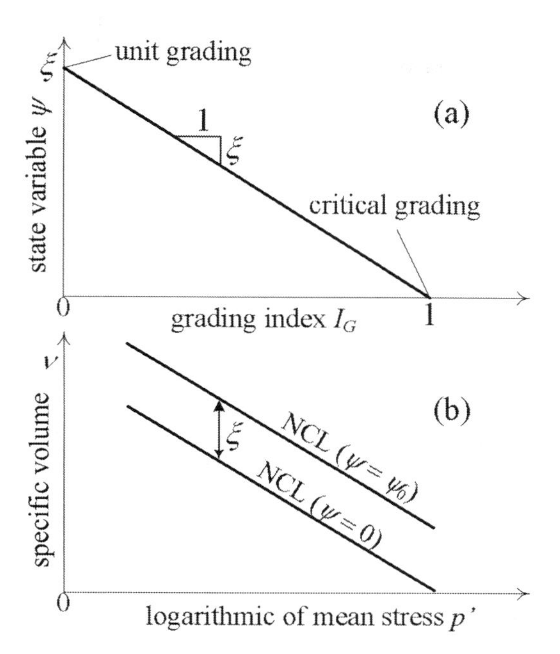

Figure 6. Relationship between I_G and Ψ.

the packing & density effect on soil. Apparently, particle crushing will increase the soil's density, thus I_G and Ψ has a close relationship. Therefore, it is possible to relate I_G and Ψ as follow (Fig. 6):

$$\Psi = \xi(1 - I_G) \tag{18}$$

From Equations 16, 18 and 20, the time derivative of ψ is derived:

$$\dot{\Psi} = \frac{\partial \psi}{\partial I_G}\frac{\partial I_G}{\partial p_x}\frac{\partial p_x}{\partial \sigma_{ij}}\dot{\sigma}_{ij} = -\frac{\xi}{p_r}(1 - I_G)p_x\frac{\partial f_x}{\partial \sigma_{ij}}\dot{\sigma}_{ij} \tag{19}$$

The performance of the constitutive model for particle crushing has been validated via triaxial elementary tests (Nguyen & Kikumoto, 2017) by the comparison between simulation results and experimental results of Dogs Bay Sand conducted by Coop (1993)

4 VOLUMETRIC LOCKING AND ITS COUNTERMEASURE

When dealing with distortional behavior of incompressible material, FEM may exhibit an unrealistically stiff behavior. In the similar manner, elastoplastic soil model in which constant volume is predicted at critical state, also witnesses this unrealistically high stiffness. This is usually known

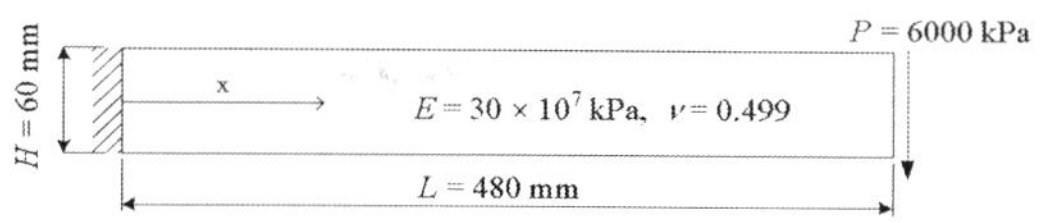

Figure 7. Coordinate system for the cantilever beam.

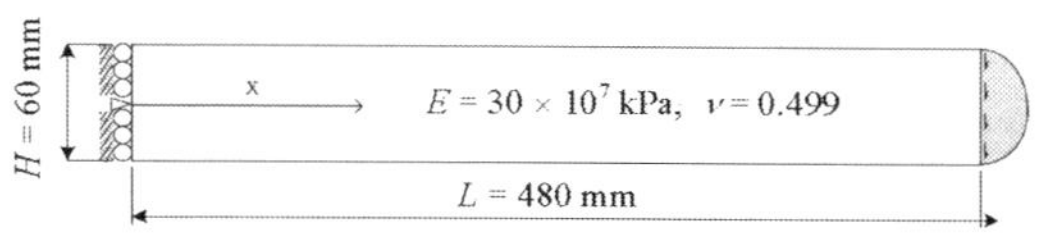

Figure 8. Cantilever problem analyzed.

as "volumetric locking" phenomenon. This volumetric locking occurs when a finite element mesh use low order full integration element, such as constant strain triangle elements, or 4-node isoparametric quadrilateral elements. In this study, the modified version of the B-bar approach developed by Hughes (1987) is utilized to overcome volumetric locking for elastoplastic materials. The following section will present this modified B-bar method in detail.

4.1 *B-bar method*

The idea of B-bar method is to evaluate separately shear and volumetric contribution of strain to element stiffness. Shear strain is calculated with full integration as the standard FEM. However, volumetric strain is calculated with one order lower than the standard FEM.

The B-bar method in plane strain condition separates total strain into "shear strain" and "mean strain":

$$\varepsilon_{ij} = e_{ij} + \frac{\varepsilon_{kk}}{2}\delta_{ij} \tag{20}$$

in which e_{ij} is shear strain, and "mean strain" is denoted by $1/2(\varepsilon_{kk}\delta_{ij})$. We replace this mean strain by:

$$\omega = \frac{1}{V^{el}}\int_{V^{el}} \varepsilon_{kk} dV \tag{21}$$

As such, the total strain is rewritten as:

$$\bar{\varepsilon}_{ij} = e_{ij} + \frac{\omega}{2}\delta_{ij} \tag{22}$$

To implement this to FEM framework, we defined a new $\bar{B}$ matrix so that: $\bar{\varepsilon} = [\bar{B}]\boldsymbol{u}^{el}$

$$\frac{\partial \overline{N^a}}{\partial X_j}=\frac{1}{V^{el}}\int_{V^{el}}\frac{\partial N^a}{\partial X_j}dV \tag{23}$$

$$\bar{B}=\begin{bmatrix}\frac{\partial N_1}{\partial x_1} & 0 & \frac{\partial N_2}{\partial x_1} & 0 & \frac{\partial N_3}{\partial x_1} & 0\\ 0 & \frac{\partial N_1}{\partial x_2} & 0 & \frac{\partial N_1}{\partial x_2} & 0 & \frac{\partial N_3}{\partial x_2}\\ \frac{\partial N_1}{\partial x_2} & \frac{\partial N_1}{\partial x_1} & \frac{\partial N_2}{\partial x_2} & \frac{\partial N_2}{\partial x_1} & \frac{\partial N_3}{\partial x_1} & \frac{\partial N_3}{\partial x_2}\end{bmatrix} -\frac{1}{2}\begin{bmatrix}\frac{\partial N_1}{\partial x_1} & \frac{\partial N_1}{\partial x_2} & \frac{\partial N_2}{\partial x_1} & \frac{\partial N_2}{\partial x_2} & \frac{\partial N_3}{\partial x_1} & \frac{\partial N_3}{\partial x_2}\\ \frac{\partial N_1}{\partial x_1} & \frac{\partial N_1}{\partial x_2} & \frac{\partial N_2}{\partial x_1} & \frac{\partial N_2}{\partial x_2} & \frac{\partial N_3}{\partial x_1} & \frac{\partial N_3}{\partial x_2}\\ 0 & 0 & 0 & 0 & 0 & 0\end{bmatrix} +\frac{1}{2}\begin{bmatrix}\frac{\partial N_1}{\partial x_1} & \frac{\partial N_1}{\partial x_2} & \frac{\partial N_2}{\partial x_1} & \frac{\partial N_2}{\partial x_2} & \frac{\partial N_3}{\partial x_1} & \frac{\partial N_3}{\partial x_2}\\ \frac{\partial N_1}{\partial x_1} & \frac{\partial N_1}{\partial x_2} & \frac{\partial N_2}{\partial x_1} & \frac{\partial N_2}{\partial x_2} & \frac{\partial N_3}{\partial x_1} & \frac{\partial N_3}{\partial x_2}\\ 0 & 0 & 0 & 0 & 0 & 0\end{bmatrix} \tag{24}$$

Finally, stiffness matrix is calculated as follow:

$$[K^{el}]=\int_{V^{el}}[\bar{B}]^T[D][\bar{B}]dV \tag{25}$$

Noticed that volumetric strain in the element is everywhere equal to its mean value in B-bar method. As for 4-nodes isoparametric element, volumetric strain is integrated with only one Gauss point. In contrast, shear strain is fully integrated with four Gauss points.

4.2 *Validation of the B-bar method*

As an example to illustrate the performance of B-bar method to incompressible material, the behavior of a cantilever shown in Figure 7 has been studied. The boundary condition of cantilever problem is analyzed as in Figure 8. The cantilever is of dimension H = 60 mm, L = 480 mm, and the applied end load is equivalent to a uniform stress of 1 unit per unit area (P = 6000 kPa). The material properties used are Young modulus $E = 3 \times 10^7$ kPa, and Poisson ratio ν = 0.499 (nearly incompressible). The 4-node isoparametric element is used in this simulation. Different mesh densities are conducted. The Cartesian coordinate (x, y) are set along the parallel and normal directions to the beam axis, respectively.

The theoretical solution of the vertical deflection is expressed as:

$$u_y=\frac{P}{6\bar{E}I}\left\{3\bar{\nu}y^2(L-x)+\frac{1}{4}(4+5\bar{\nu})H^2x+(3L-x)x^2\right\} \tag{26}$$

in which:

$$I=\frac{1}{12}H^3,\bar{E}=\frac{E}{1-\nu^2},\bar{\nu}=\frac{\nu}{1-\nu} \tag{27}$$

Here, the deflection is positive in the y direction, and I is the second moment of area when the unit depth is assumed.

It is clearly seen (Figs 9–12) that without B-bar approach, the beam's displacement is far different from analytical solution. On the other hand, B-bar method has a really good effect on the behavior of nearly incompressible bar as long as the number of DOF is large enough. For a fine mesh of DOF = 3782, the simulation result agrees well with the theoretical solution. Therefore, the results demonstrate that the B-bar approach improved the performance of FEM when dealing with incompressible material or critical state model. Thus this B-bar method will be applied in the analysis of bearing capacity of strip footing.

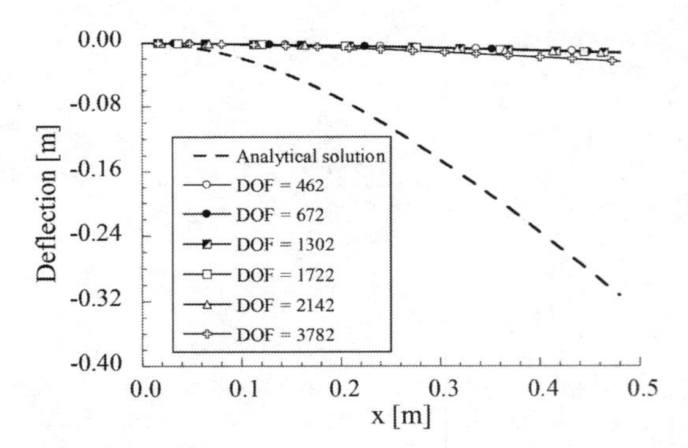

Figure 9. Distribution of deflection (traditional FEM).

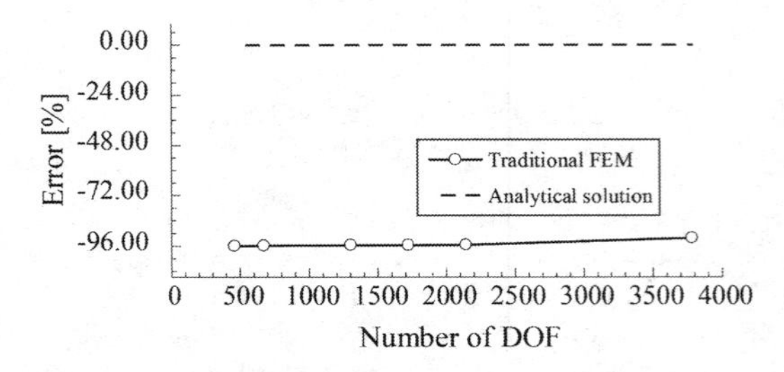

Figure 10. Error in tip deflection with (traditional FEM).

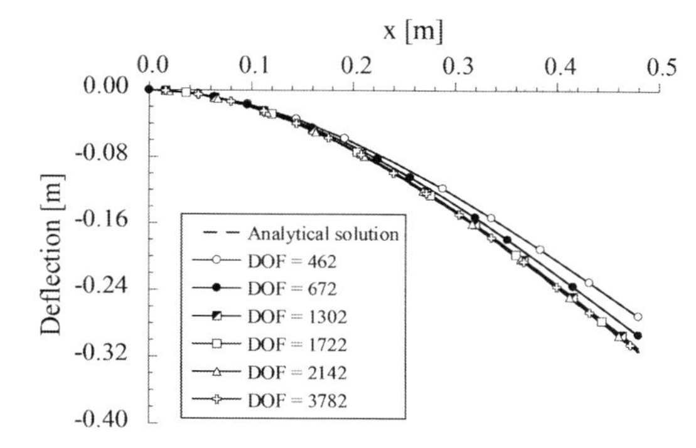

Figure 11. Distribution of deflection (FEM with B-bar).

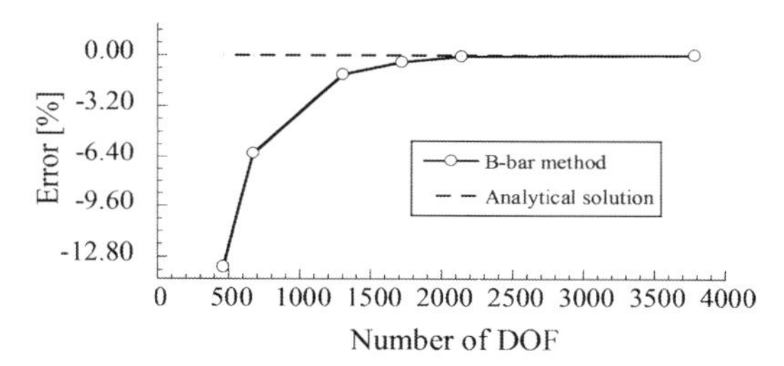

Figure 12. Error in tip deflection (FEM with B-bar method).

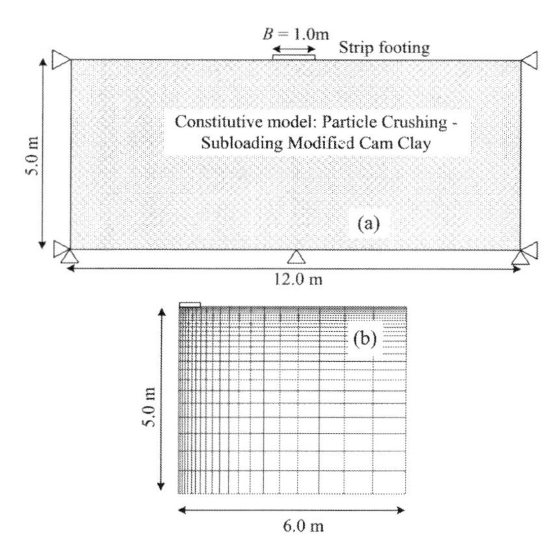

Figure 13. (a) Analytical domain and boundary conditions for the simulation of bearing capacity of strip footing.(b) Grid Mesh (20 ×24) of a half of the domain.

5 SIMULATION OF BEARING CAPACITY OF STRIP FOOTING

The bearing capacity analysis are carried out by FEM which incorporates our constitutive model for crushable soils. The analytical domain and boundary conditions are shown in Figure 13a. Finite element mesh used is shown in Figure 13b. The area near the footing is finely mesh while the coarser mesh is applied at the area far away from the footing. The soil material parameters are assumed as in Table 1. The soil medium is discretized by isoparametric plane strain elements. The nodes representing the footing width are incrementally displaced by an equal amount in the vertical direction, simulating a rigid footing condition with a uniformed vertical settlement but without any rotation. Smooth footing conditions are simulated by not restraining horizontal movement of these nodes. The footing load for each increment is the summation of the nodal forces back-calculated form the conversed stress field after each incremental step.

5.1 *Bearing capacity of strip footing of several widths on crushable soil and non-crushable soil*

The bearing capacity of crushable soil on a variety of width of foundation has been simulated. The same set of soil parameters but without crushing phenomenon has been utilized to compare the bearing capacity with that of crushable soil. As seen in Figure 14, for the same width of strip footing, when particle crushing occurs, the bearing capacity of foundation is reduced significantly in comparison with the bearing capacity of uncrushable soil.

5.2 *Distribution of grading index I_G*

In the simulation of bearing capacity of strip footing with B = 1.0 m, we can observe the changing of grading index I_G in the ground as shown in Figure 15. The area near the footing exhibits more crushing than the area far from the footing. The more settlement the footing is, the large value I_G is and the more serious crushing is observed. Thus, we can both observe the settlement-load relationship and the evolution of grading index by using our proposed constitutive model.

5.3 *Parametric studies*

In the parametric studies that follow, the material parameters for Modified Cam clay, Subloading effect, unit-weight are kept constant. Only the parameters for partical crushing mechanism are varied systematically.

Table 1. Material parameters (*).

Parameter	Description	Classification	Value
λ	Compression index		0.3
κ	Recompression index		0.005
N	Specific volume on NCL at $p = 98$ kPa	Modified Cam Clay	1.8
M	Critical stress ratio		1.65
ν	Poisson ratio		0.2
a	Parameter controlling density effect	Subloading	500
p_r	Parameter controlling crushing resistance		70
M_x	Parameter controlling the shape of crushing surface	Particle Crushing	1.0
ξ	Volumetric distance between NCL of $I_G = 0$ & $I_G = 1$		0.5
p_{x0}	Crushing stress when $I_G = 0$: (kN/m^2)		50

* The unit weight ($\gamma = 18$ kN/m^3) is applied by the gravity load with the initial soil state ($e_0 = 2.5$, $p_0 = 0.01$ kPa, $I_{G0} = 0$) before applying the loading process

5.3.1 p_{x0} parameter

p_{x0}, which controls the stress level at which crushing start to occur at the unit grading, is varied to study its effect to bearing capacity of strip footing. As mention in the formulation of the model, the crushing surface controls the occurrence of crushing phenomenon. When the applied stress reaches the crushing surface, crushing occurs. Whether stress required for crushing is small or large depends on the parameter p_{x0}, which is the crushing stress corresponding to grading index $I_G = 0$. With a small value of p_{x0}, the crushing phenomenon occurs faster. This leads to the sooner reduction of the strength of the soil. Thus, a highly crushable soil will possess a small value of p_{x0} and a smaller bearing capacity. As expected, Figure 16 shows that when p_{x0} increase, bearing capacity increases. When p_{x0} is larger and larger, the bearing capacity will convergence to that of uncrushable soil.

5.3.2 p_r parameter

Whenever the crushing phenomenon has been initiated the resistance to crushing parameter p_r controls the speed of crushing for one paticular material. The evolution rate of PSD to the limiting PSD is governed by this p_r parameter. Thus, the larger the value of p_r is, the slower rate of crushing will be and the larger bearing capacity will be obtained. As seen in Figure 17, when p_r is larger and larger, the bearing capacity of crushable soil will convergence to that of uncrushable soil.

5.3.3 M_x parameter

The combination of the effect of compression stress and shearing stress on the initialization of crushing is conveniently expressed via the crushing surface in our model. When the slope M_x of this crushing surface is small, under the same confining pressure, the smaller shearing stress is needed to initiate the crushing phenomenon. Therefore, the soil with a

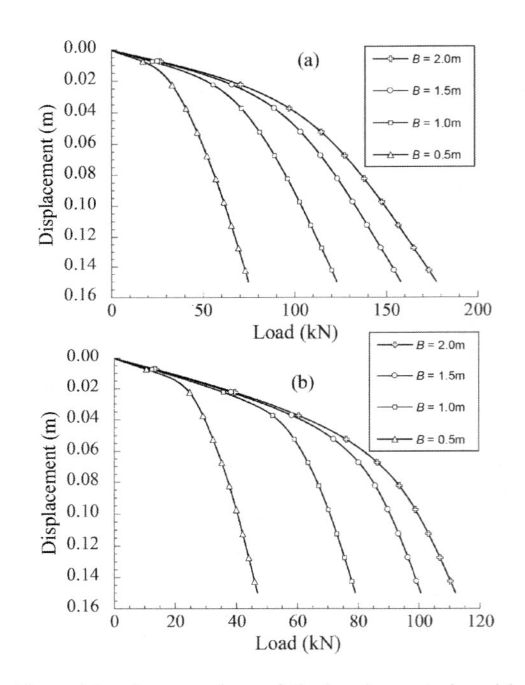

Figure 14. A comparison of the bearing capacity with different sizes of strip footing on (a) uncrushable soil (b) crushable soil.

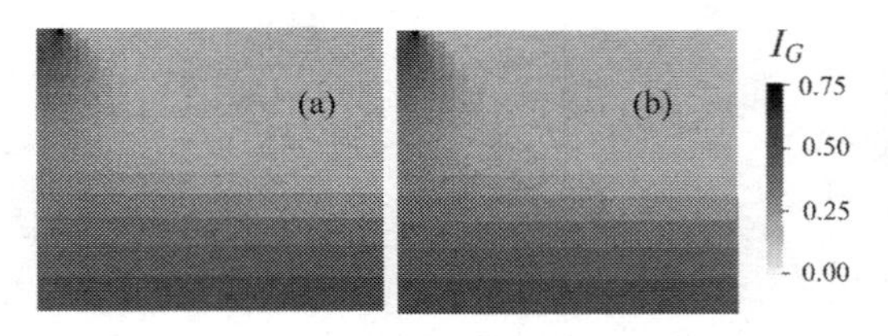

Figure 15. I_G distribution with the variation of settlement d (a) $d = 0.1$ m (b) $d = 0.2$ m.

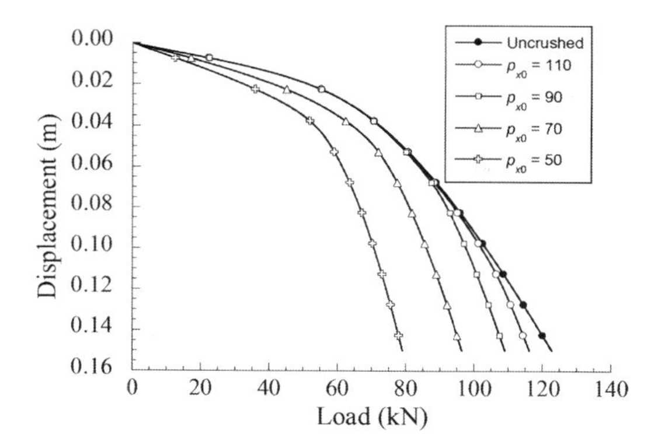

Figure 16. Effect of the initial crushing stress p_{x0} corresponding to $I_G = 0$.

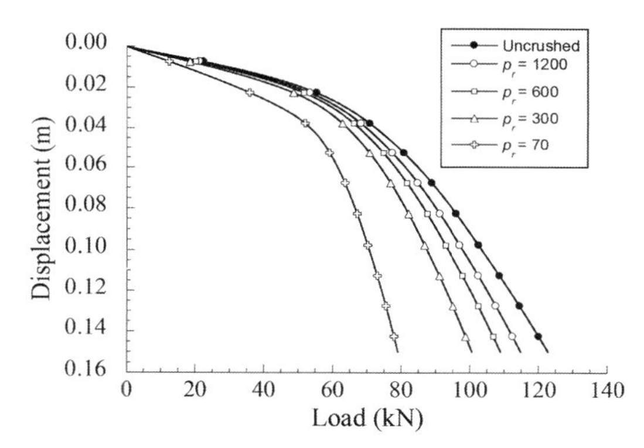

Figure 17. Effect of crushing resistance p_r on the bearing capacity of strip footing.

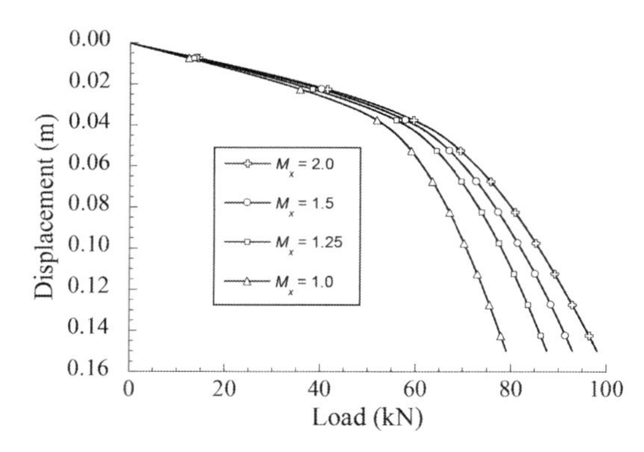

Figure 18. Effect of M_x parameter on the bearing capacity of strip footing.

large value of M_x will be stronger to crushing in term of shearing stress than the soil with a smaller value of M_x. This leads to the larger bearing capacity with the soil having a larger value of M_x (Fig. 18). It it worth noticing that even if M_x is increased to infinitive value, the bearing capacity of uncrushable soil is also not be able to reach because of the effect of compression stress in the yield surface.

6 CONCLUSION

An elastoplastic model considering the particle crushing phenomenon for soil has been developed and validated at elementary level (Nguyen & Kikumoto, 2017). In this research, B-bar approach in FEM to overcome volumetric locking problem has been successfully implemented to analyze the bearing capacity of strip footing using our particle crushing constitutive model. The crushing effect reduced the soil strength; therefore, the bearing capacity is reduced. Not only bearing load - settlement but also the variation of grading index in the soil can also be observed by using our constitutive model. The presented constitutive model and B-bar approach to FEM is a simple and useful techniques for estimating the bearing capacity of crushable soil.

REFERENCES

Coop, M.R. and Lee, I.K. 1993. The behavior of granular soils at elevated stresses. Predictive Soil Mechanics, *Proc. C.P. Worth Memorial Sym., Thomas Telford, London*, 186–198.

Daouadji, A., Hicher, P.Y. and Rahma, A. 2001. Elastoplastic model for granular materials taking into account grain breakage. *European Journal of Mechanics, A/Solids*, 20 (1), 113–137.

Gajo, A. and Muir Wood. 1999. Severn–Trent sand: a kinematic-hardening constitutive model: the q–p formulation. Géotechnique 49.5, 595–614.

Ghafghazi, M., Shuttle, D.A. and DeJong, J.T. 2014. Particle breakage and the critical state of sand. *Soils Found*, 54 (3), 451–461.

Hardin, B.O. 1985. Crushing of soil particles. *J. Geotech. Eng.*, 111(10), 1170–1192.

Hashiguchi, K. and Ueno, M. 1977. Elastoplastic constitutive laws of granular media. *Proc. JSCE*, 227, 45–60.

Hughes, T.J. 2012. *The Finite Element Method, Linear Static and Dynamic Finite Element Analysis*. Courier Corp.

Kikumoto, M., Muir Wood, D., & Russell, A. 2010. Particle Crushing and Deformation Behaviour. *Soils Found*, 50(4), 547–563, 2010

Nguyen, V.P.Q. and Kikumoto, M. 2017. An elastoplastic model for soils exhibiting particle breakage. *Proc. 4th Congres International de Geotechnique-Ouvrages-Structures*. Ho Chi Minh, 644–655

Roscoe, K.H., and Burland, J.B. 1968. On the generalized stress-strain behaviour of wet clay. *Engineering Plasticity*. 535–609.

Rosco, K.H., Schofileld, A.N. and Thurairajah A. 1963. Yielding of clays in states wetter than critical. *Géotechnique*, 13, 211–140.

Terzaghi, K. 1943. Theoretical Soil Mechanics, *Wiley, New York*.

Muir Wood, D. 2007. The magic of sand. *Can. Geotech. J.*, 44 (11), 1329–1350.

Numerical Methods in Geotechnical Engineering IX – Cardoso et al. (Eds)
© 2018 Taylor & Francis Group, London, ISBN 978-1-138-33198-3

Numerical simulations of the dynamic soil behaviour in true triaxial conditions

C. Ferreira
CONSTRUCT-GEO, Faculty of Engineering of the University of Porto, Portugal

A.R. Silva
Formerly MSc research student, Faculty of Engineering of the University of Porto, Portugal

J. Rio
Formerly CONSTRUCT-GEO, Faculty of Engineering of the University of Porto, Portugal

ABSTRACT: The dynamic behaviour of a soil depends of its inherent properties, namely stiffness, density, stress state, anisotropy, structure, among others. In this research, a detailed three-dimensional numerical model was implemented in the finite differences program $FLAC^{3D}$, in order to simulate the dynamic behaviour of a soil in true triaxial conditions. A series of true triaxial tests with shear wave measurements was modelled, which required the implementation of bender elements in the model. The influence of the cubical geometry and boundary conditions (reflective and absorbent) was investigated, by means of parametric and sensitivity studies, considering the linear elastic constitutive model. After these studies, a comparison was established, based on the extensive laboratory data provided by Ferreira (2009) on reconstituted specimens of residual soil from Porto granite. In the present paper, only test results on dry (w 0%) specimens will be used and discussed for the purpose of comparison.

1 INTRODUCTION

1.1 *Background*

The understanding of the dynamic properties of soils is fundamental in geotechnical design, since these influence soil behaviour and can be linked to other relevant soil properties, such as stiffness, void ratio, stress state, anisotropy, natural structure and fabric, saturation conditions, among others. The correct determination of these properties is especially important due to the highly non-linear response of practically all soils, given its particulate nature and heterogeneity. The discovery of the non-linear behaviour led to the development of several methods to better comprehend and evaluate the dynamic response and particulate nature of soils. For this purpose, the measurement of seismic wave velocities is particularly appropriate, since it enables to study its nature without affecting the structural equilibrium, fabric or inherent mechanical properties (Fam and Santamarina 1995; Atkinson 2000).

Wave measurements are a practical, non-destructive, cost-effective and usually non-invasive means of determining small-strain stiffness of soils and can be applied to estimate soil properties and assess soil conditions, from the stress state to anisotropy. Shear and compression wave velocities can be measured both in situ, in geophysical surface or borehole tests and in the laboratory, with ultrasonic or acoustic transducers installed in testing devices, namely in the resonant-column, hollow cylinder, oedometer, triaxial and true triaxial apparatuses.

This work focuses on the validation of the elastic, small-strain stiffness parameters from local strain and seismic wave measurements in a true triaxial apparatus, from previous studies developed by Ferreira (2009), through numerical modelling in a finite differences software named $FLAC^{3D}$. The selection of a three-dimensional numerical model to simulate this apparatus was made simply to overcome some limitations of more standard devices and to better understand wave propagation phenomena and the influence of several non-linearity aspects in the results. The use of a numerical model is also important to evaluate the influence of the cubical geometry and reflective conditions of the boundaries of the equipment in the dynamic behaviour of this soil.

2 METHODOLOGY

2.1 $FLAC^{3D}$ *program*

The numerical code $FLAC^{3D}$ (Fast Lagrangian Analysis of Continua) is a three-dimensional finite difference program developed primarily by *Itasca Consulting Group* in 1994, for geotechnical engineering calculations, making it the ideal choice for this study. Particularly, this program is suited to simulate the behaviour of three-dimensional structures of soil, rock or other materials where a continuum analysis is required. The main characteristics of this program are its explicit calculation, by means of a time-stepping procedure to solve a problem without forming the stiffness matrix (Neiva 2011), providing stable solutions to unstable physical processes and ensuring that plastic collapse and flow are accurately modelled. It integrates several constitutive models, such as elastic, plastic, Mohr-Coulomb, hardening-soil, among others. In the dynamic module, it considers two types of hysteretic damping, namely Rayleigh damping (used in time-domain programs) and local damping (frequency independent). It enables an adequate specification of boundary conditions; possible definition of water table for effective stress calculations; graphical output in a variety of formats; built-in text corrector that offers command syntax error checking.

FLAC^{3D} is robust in the sense that it is especially suited for dynamic calculations in the time-domain, enabling the direct application to the grid points and the consideration in the consecutive calculation cycles, keeping the same computational effort, regardless of the alterations made to the properties of the model during calculation. Therefore, this software is capable to simulate situations varying with time, namely in this case for the simulation of bender elements and the dynamic behaviour of the soil.

2.2 *Experimental setup*

The numerical model was developed with the aim of reproducing previously obtained experimental results from Ferreira (2009), where a rigid-boundary true triaxial apparatus embedded with bender elements was employed for testing intact and reconstituted specimens of residual soil from Porto granite.

A true triaxial apparatus provides the capability of applying independent stresses or strains in three orthogonal directions to cubical specimens of soil and thus provides one more degree of freedom that the conventional triaxial apparatus. Several models of true triaxial apparatuses have appeared over the years, all of which fall with three main categories: a) strain-controlled rigid boundaries; b) stress-controlled flexible boundaries; c) mixed boundary apparatuses.

In the present case, the true triaxial apparatus was developed as result of the cooperation between the Faculty of Engineering of the University of Porto (FEUP) and the University of Western Australia (UWA), as part of the research of Ferreira (2009). This apparatus is made of anodized aluminium and consists of six square platens mounted in a cubical frame of 250 mm internal side length. Uniform strains are applied in the three orthogonal directions by means of rigid platens (Figure 1). A thick rubber diaphragm, between each piston and an outer plate, is pressurized using a water pressure controller. For the application of independent stresses in the three directions (X, Y and Z), three water pressure controllers were used, distributing the pressure to each pair of platens in the same direction.

For the measurement of seismic wave velocities, piezoelectric transducers have been fitted in each platen of the device. Piezoelectric transducers have the ability of converting electrical energy into mechanical energy or vice-versa, generating an electric potential during deformation, therefore allowing the transducers to act as sensors, actuators or both. The most common piezoelectric transducers currently use in geotechnical laboratories are bender elements (Viana da Fonseca et al., 2009). Bender elements (BE) function as a cantilever beam and consist of two piezoceramic plates rigidly attached to an intermediate metallic shim and electrodes on its exterior surfaces. The piezoceramic plates are very thin, which enable the production and detection of transversal motion, from a bending deformation. The approximate size is about 12 mm long, 8 mm wide and 0.6 mm thick. In this setup, a probe containing a pair of T-shaped bender elements has been installed in each platen, to enable seismic wave measurements in two orthogonal directions.

From the available database of laboratory true triaxial tests results, the preliminary selection of tests considered simple stress-paths, allowing the anticipation of the behaviour and the quick interpretation of the differences between tests, since these could only be attributed to soil conditions, state and stress field parameters. From those tests,

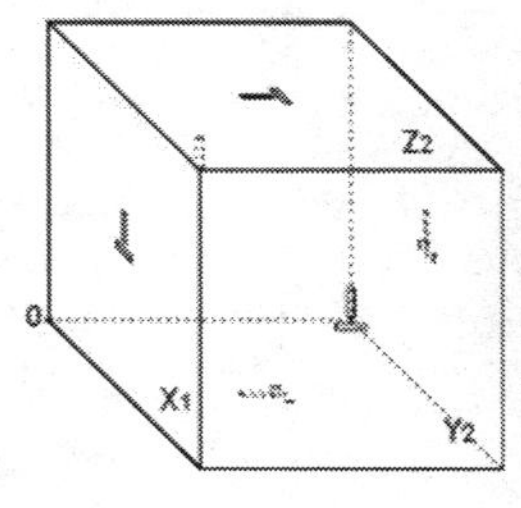

Figure 1. True triaxial apparatus: a) Nomenclature for the platens and shape of the bender elements; b) the apparatus at LabGEO (Ferreira, 2009).

the final selection focused on tests with an isotropic or anisotropic (biaxial) loading on reconstituted residual soil samples from Porto granite, with consideration of the inherent and induced anisotropy.

2.3 *Description of the model simulation program*

2.3.1 *Mesh generation*

The characteristics of the mesh, namely the number and distribution of elements, is usually trusted to the judgement of the engineer, considering that its choice will have a major influence on the accuracy of the results (Hardy et al. 2002). Since this study regards the numerical modelling of a true triaxial apparatus, the desired shape of the mesh is a cube. In the program *FLAC^{3D}*, the corresponding

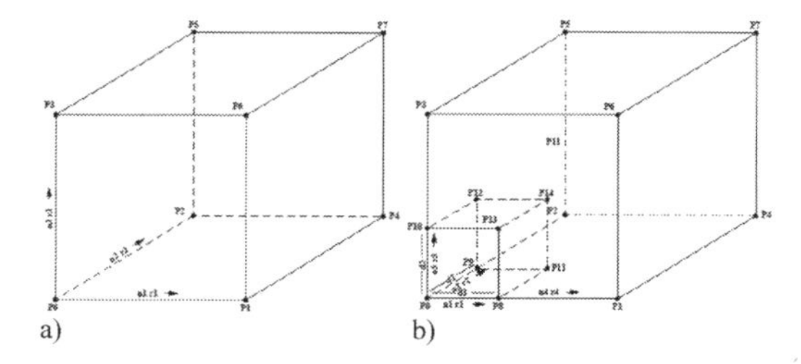

Figure 2. a) Brick shaped mesh; b) Radially graded mesh around brick (Itasca 2002).

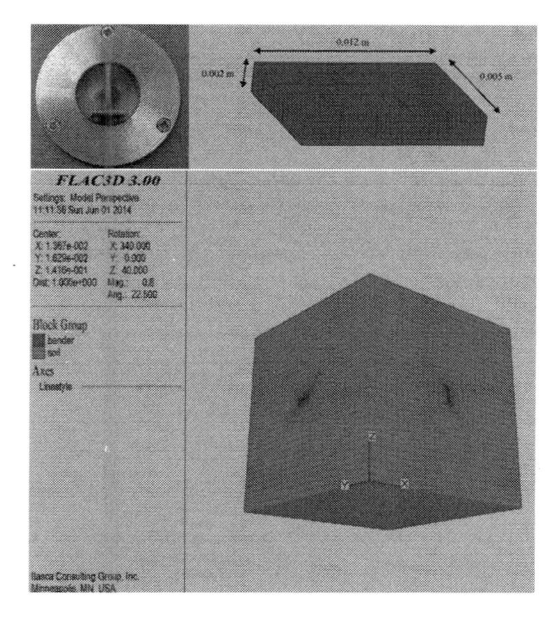

Figure 3. Details of the BE geometry and full representation of the true triaxial apparatus with embedded bender elements.

primitive mesh shape is a brick, as can be observed in Figure 2a. On the other hand, since the true triaxial apparatus is embedded with bender elements, its insertion in the brick is fundamental. Therefore, another shaped mesh is necessary, namely a radbrick, so that the different and stiffer properties of the BE are considered, as shown in Figure 2b. The soil and the BE are necessarily modelled as two different media.

With this information in mind and considering the usual position of the BE in the centre of each platen of the true triaxial apparatus, it was decided to use the combination of the two shapes presented in Figure 3. Since it was not feasible to create T-shaped BE in each of the 6 faces of the model (as in the apparatus), it was decided to consider separate vertical and horizontal bender elements in different vertical faces, which would lead to nearly identical results for a much simpler mesh. This hypothesis was further investigated and discussed in Silva (2014).

The dimensions used for the brick were $0.15 \times 0.15 \times 0.15$ m with five smaller sized square dimensions or zones of the bender elements as illustrated in Figure 3. The number of zones was carefully selected, since a balance is needed among the accuracy of the model and the calculation time. Nevertheless, when constructing the radbrick, a geometric ratio is in order, given that a finer mesh with a gradual change in size, from the smallest to the largest, can lead to significantly improved results. For this, it was necessary to "cut" the brick and the radbrick, then through simple brick shapes, the rest of the upper half of the cube was made and the final mesh was achieved by a series of reflections of the YZ and XZ planes, resulting in Figure 3.

2.3.2 *Constitutive models and properties*

The soil samples tested and analysed in this study originate from a residual soil from Porto granite, used in remoulded or reconstituted conditions. This residual soil results from the weathering of Porto granite, a leucocratic, alkaline rock, with two micas and medium-to-coarse grains (Viana da Fonseca 2003). Despite the complex characteristics of the intact soil, which are a consequence of the variability and heterogeneity of the parent rock, as well as the distribution and spatial arrangement of the particles, reconstituted specimens are usually classified as silty sands.

Once the mesh is generated and the properties decided, it is time to associate them to a constitutive model. There is not (yet) a single and unique constitutive model capable of capturing all aspects of the real behaviour of a soil. For that reason, it is necessary to make decisions concerning the most relevant features for a specific problem and then

Table 1. Material properties for the soil and the BE.

Material properties	Soil	Bender element
Shear modulus G	80 MPa	1000 MPa
Poisson's ratio ν	0.10	0.25
Young's modulus E	176 MPa	2500 MPa
Bulk modulus K	73.33 MPa	1700 MPa
Density ρ	2000 kg/m^3	3000 kg/m^3

choose the most suitable constitutive model (Potts & Zdravkovic, 1999). One of the models chosen for this study was the linear elastic model, given that in dynamic conditions, which is the case of this study, the soil is expected to exhibit elastic behaviour. In addition, this model provides a straightforward and reasonably accurate response. The elastic model can be characterized by its linear stress-strain behaviour with no hysteresis on unloading, leading to a coincidence of the directions of the main incremental stresses and strains.

In this model, the materials are considered continuous and homogeneous, and its application only extends to stress states that do not produce yielding, independently of the point of departure for most parameters. This model is also characterized by an isotropic behaviour. This description, along with the existence of only two material constants, E, μ, K or G, distances itself from the non-linear elastic constitutive model, because in the latter, the elastic parameters (five in this case) vary with both stress and/or strain level. The non-linear models can assume an isotropic or anisotropic behaviour and are unable to reproduce the tendency to change volume when sheared (Potts & Zdravkovic 1999). Besides, they strongly depend on the constitutive formulation along with its undergoing of large deformations and disability of accurately reproduce failure mechanisms.

To guarantee an adequate stiffness of the BE, its elastic properties were considered approximately ten times higher than those of the soil (Table 1).

The properties of the bender elements were experimentally determined by Rio (2006) and although some of the values refer to specific dimensions and materials of particular bender elements, which may be different from the ones in this study, these were found to not affect significantly the performance of the model. In terms of the constitutive model for BE, the adopted model was also the linear elastic, in order to adequately reproduce its behaviour.

The elastic constants for the soil were determined based on the small strain shear modulus, G, parameter experimentally measured. The value of 80 MPa for this particular soil was attained, having in consideration a reference value of the

Table 2. Transversely isotropic elastic properties of the residual soil from Porto granite.

Soil Properties	
Young's Modulus (E_1) in the plane of isotropy	176 MPa
Young's Modulus (E_3) normal to the plane of isotropy ($E_3 = 0.70xE_1$)	123.2 MPa
Shear modulus (G_{13}) for any plane normal to the plane of isotropy	80 MPa
Poisson's ratio (ν_{12}) charactering lateral contraction in the plane of isotropy when tension is applied in the plane	0.10
Poisson's ratio (ν_{13}) charactering lateral contraction in the plane of isotropy when tension is applied normal to the plane	0.10

consolidation isotropic stress of 100 kPa. Poisson's ratio, ν, at low strain levels has been found to assume very low values. Cascante & Santamarina (1997) suggested values as low as 0.07, in agreement with other analytical estimates (Chang et al. 1990). More recently, Rio (2006) also observed that the best suited value of Poisson's ratio for modelling the dynamic behaviour of soils was very low, around 0.1. A specific parametric study was carried in this research, confirming the value of 0.1 for Poisson's ratio.

The other model relevant to this study, only as far as the soil is concerned, was the transversely isotropic or cross-anisotropic model. Anisotropy is a significant parameter, which varies with direction, providing information about the physical and mechanical properties of a material or soil. There can be distinguished two components of soil anisotropy, namely inherent and induced anisotropy, which may occur in most field situations. Anisotropy is usually expressed in terms of stiffness, given the dependency of the latter with effective stress. This dependency leads to an increase of stiffness with stress and to an expectancy of stress anisotropy due to frictional resistance. Moreover, under isotropic loading, soils can exhibit distinct small-strain stiffness, direction dependent (Stokoe and Santamarina 2000).

Regarding laboratory testing, the use of seismic wave velocities enables the observation of several degrees of anisotropy. The direct measurement, whether propagated (wave direction) or polarised (particle movement direction) in different directions allows the identification of distinct values of stiffness from which the level of anisotropy can be determined. However, even though the strength and deformation characteristics of residual soils are dependent of the shearing loading direction, due to anisotropy, these direct shear wave measurements are only preliminary, given the limited amount of samples that were carried out (Ferreira 2009).

The assessment of anisotropy, namely the inherent anisotropy was also carried out in $FLAC^{3D}$, by a series of simulations, considering the cross-anisotropic or transversely isotropic properties of the residual soil presented in Table 2. These properties are based on those in Table 1, the only differences being the Young's modulus normal to the plane of isotropy, E_3, which is about 70% of the value of the Young's modulus in the plane of isotropy. Poisson's ratios in different directions were assumed identical.

3 PARAMETRIC AND SENSITIVITY STUDIES

In order to comprehend the degree of influence of a number of parameters of the numerical model such as geometry, time step, amplitude, frequency, damping, Poisson's ratio, boundary conditions, anisotropy, a series of parametric and sensitivity studies were made, from which the parameters of the final model were defined. On one hand, there are the parametric studies where only one of the parameters was varied, in order to facilitate the interpretation of the results and decide which features are best suited for the model. This process consisted of a trial and error approach. On the other hand, the sensitivity studies are associated to the uncertainty regarding the output of a certain system, being therefore useful to test the relationship between the parameters and the validity of this particular model. Since these studies belong to a numerical model, all the parameters are controlled by the user, contrary to what happens in laboratory testing. For this purpose, the model was simplified and consider without the inclusion of the BE, so that the decision of the most appropriate features could be made as efficiently as possible. Nevertheless, in this simple model, the bender elements are in fact simulated, by the application of a velocity starting in the midpoints of faces A (horizontal bender element) and B (vertical bender element) and finishing in faces C and D, respectively.

In terms of the results of the parametric and sensitivity studies, only the most relevant parameters are discussed below, namely time step, damping and boundary conditions. Detailed information on these studies can be found in Silva (2014). The parametric studies were performed only in the time domain approach, for identification of the first arrival of the received wave. These studies comprised two main stages: one where gravity and the stress state is applied and the other where wave propagation is induced.

3.1 *Time step*

In $FLAC^{3D}$, time step adopts an explicit method, where no iterations are needed to find equilibrium between the elements, considering that the time step is small enough to guarantee no variation within it. The time step can be automatically calculated and there is not a need of manual input. Nevertheless, since information can propagate, depending on the maximum speed characteristic of each material, a series of experiments were conducted, with different time step values, in order to achieve the appropriate balance between the results and the computer processing capacity. It was found that the natural time step of 1.783×10^{-5} was not the most appropriate, given the lack of sufficient data to correctly translate the shape of the output signal. The value chosen for the model was 2.5×10^{-7}, as it provided the right balance between the number of data points, signal quality and computer processing capacity.

3.2 *Damping*

The most common models of damping are the viscous damping more suitable for analytical problems and the hysteretic damping, as used by $FLAC^{3D}$. The latter can be divided into two forms of damping, namely local and Rayleigh damping.

Local damping is frequency dependent for only simpler cases, besides requiring less computational processing capacity than Rayleigh damping. Also, its modelling can be done by the addition or removal of mass proportional to both the mass and acceleration of the element in question, hence conserving the total mass of the system, to oscillating elements.

On the other hand, Rayleigh damping is frequently used in time-domain programs, and is frequency-independent in a specific range of frequencies, being frequency-dependent outside that range (Wilson and Bathe 1976). However, contrary to the local damping, this kind of damping requires two constants, since it is independently proportional to the stiffness and mass of the system, making its estimate more intuitive (Rio 2006).

Both types of damping were applied in $FLAC^{3D}$, in order to better understand the influence that frequency has in each one. Based on these studies, the value chosen for local damping for use in the model was 0.05, given its slow and steady approximation to equilibrium. With this value, it was possible to obtain a high variation of amplitudes and thus more detailed results in the numerical simulations of this type of soil. On the other hand, for Rayleigh damping, the most appropriate value was found to correspond to 10.0 kHz.

3.3 *Boundary conditions*

For the type of apparatus used in the laboratory, it is possible to distinguish two types of boundary conditions, namely reflective and absorbent boundaries. Even though it is possible to demonstrate the reflective boundary condition in the true triaxial apparatus, given its metallic platens, which favour the reflection of waves inside the specimen, it is not possible to study a specimen in this apparatus considering an absorbent boundary condition, as it would be unrealistic. It is only possible to study an absorbent boundary in a cubical cell, with flexible boundaries, usually made of rubber. Therefore, although the study considers a true triaxial apparatus as the source of the experiments, this study is important to confirm the complexity of the reflective boundary in comparison with the absorbing one.

The constructed model, including the bender elements, have the bottom end vertically fixed, simulating an infinite medium downwards, but with free horizontal movement, along with null initial displacement and velocity conditions in all directions.

Due to the existence of a free boundary or reflective boundary in the vertical planes of the model, there is no resistance to wave propagation, causing it to be reflected back into the system with the same signal, enabling the existence of a more complex behaviour, as opposed to an absorbent boundary.

Figure 4 illustrates the results obtained in the simulations for reflective (R) and absorbent (A) boundaries at three distinct points in the model: at the transmitter, at midpoint and at the receiver. The figure also includes the identification of the arrival times of P and S-waves, based on the defined soil properties.

As expected, no significant differences can be observed with respect to the input signals at the transmitter (face A), since the nature of the boundaries has no influence at transmission. At midpoint, the differences are still very subtle. In fact, the reflective and absorbent boundary simulations show similar signals approximately half the time (or half the travel length), as evidenced in the figure.

On the contrary, the output signals at the receiver (face C) are clearly distinct and strongly depend on the type of boundary conditions.

Since the results at midpoint are similar, regardless of the boundary conditions, it can be argued that at that point the response corresponds to that of an infinite medium. The further from the transmitter, hence the closer to the edge of the model at the receiver, the more the difference among first arrival times of P and S waves, which is in compliance to its actual behaviour. In effect, due to the higher reflection present near the edge of the model, it can be concluded that the higher the reflection, the higher the energy of the wave and the lower the absorbency.

Additionally, it is important to mention that the midpoint is more affected by the posterior reflections of the wave signal, than the signal of the receiver, although with far less energy. The reflection velocity of the receiver was calculated considering three time the total length of the cube (3 x 0.30 m = 0.90 m), contrary to the midpoint where it was considered the total length of the cube plus the distance to the middle point (0.15 m), making a total of 0.45 m. These reflections are not usually easily visualized in the output of the waves; however, there is a higher approximation between the midpoints and the receivers of the reflected and absorbent boundaries near the values theoretically calculated. Figure 4 shows that, for the midpoint of the model, there is clear evidence of the second passage of the S wave for the case of reflective boundaries, confirming that existence of important reflections at the faces of the model. On the other hand, at the receiver, the differences between reflective and absorbent boundaries are mainly on the longer oscillation of the first arrival wave, which is actually mixed with the second arrival of the wave due to its reflection.

There is a correlation between the presence of boundary conditions and the complexity of wave propagation. Comparing the reflective and the absorbent boundaries, it is possible to conclude that the absorbent boundaries do not introduce additional components regarding reflected waves back to the system, thus reducing the complexity of the signals and easing its interpretation.

Since the present study focuses on a true triaxial apparatus with reflective boundaries, reflected waves are expected to appear, as confirmed by these numerical results and as observed experimentally.

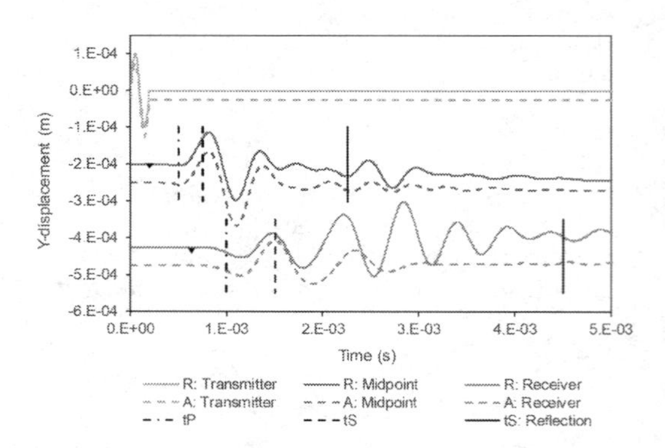

Figure 4. Reflective (R) and absorbent (A) boundary conditions.

4 SIMULATION RESULTS AND DISCUSSION

Among the true triaxial tests made by Ferreira (2009), a number of tests consisted of dry (w ≈ 0%) and wet (w ≈ 30%) reconstituted residual soil specimens, intended to reproduce the behaviour of a non-structured soil. The dry condition allows a complete elimination of suction effects, despite its difficult preparation and reconstitution process, causing an unwanted but inevitable heterogeneity and layering. On the contrary, the wet condition is far easier to prepare at a very high void ratio, leading to a much more homogeneous sample. Nevertheless, given the probability of entrapment of air bubbles among the soil particles, there is a possibility of an unmeasured quantity of capillary negative pressure (Ferreira 2009). For this paper, only two dry samples have been considered, as summarily identified in Table 3.

Here, a comparison is established between the results from these specific laboratory tests, for validation of the simulation model.

4.1 *R8D-TT: dry soil, isotropic conditions*

The first test to be analysed refers to specimen R8D-TT, which included an array of unload-reload cycles, providing the opportunity to measure shear wave velocities repeatedly at the same stress levels. This stress path led to significant soil deformations during isotropic loading, thus varying the values of the void ratio. After these isotropic cycles were finalized, a series of three-dimensional stress conditions were imposed, taking into account the same mean effective stress, *p'* (Ferreira 2009). For this comparison, only the isotropic stages will be analysed. For a realistic comparison, several values of stresses were considered in the numerical model, namely 100 kPa, 200 kPa and 500 kPa, under isotropic conditions. Considering these stresses, a study of the displacements induced by the loading process was made as shown three-dimensionally in Figure 5.

Considering this figure, it is possible to verify that the deformation of the model is concomitant with the applied stresses, both in the laboratory and in the simulations. The displacements

Table 3. Physical properties of the reconstituted specimens tested in the true triaxial apparatus.

Specimen	γ[kN/m^3]	w_0 [%]	e_0
R4D-K_0TT	14.1	1.2	0.883
R8D-TT	12.7	1.0	1.067

measured in the laboratory test are significantly higher than those of the model. However, the comparison between numerical and experimental results is valid. The main reason why these values are so different is associated with the compliance and bedding errors of the laboratory equipment. Since the platens of the true triaxial apparatus are rigid and the measurement is made on its outside, it is not possible to accurately measure such small displacements. Additionally, the capacity of the true triaxial apparatus to measure small strains comes with a slight margin of error, given its compliance. Therefore, it was expected that the degree of magnitude of the displacement values of the experimental tests were higher than the numerical.

Finally, it is worth understanding how the seismic waves behaved by varying the stress applied, as illustrated in Figure 6. It is possible to notice that the shape and configuration of the waves is not considerably affected by the different stresses, what leads to conclude that these are independent from the applied loading.

This result can be justified by the characteristics of the constitutive model, which is linear elastic, with imposed stiffness moduli. On the other hand, the soil responds with volumetric deformations, thus increase the stiffness and necessarily the seismic wave velocities of the specimen with increasing loads. This non-linearity is naturally included in the laboratory results, since the tests were developed in residual soil specimens that incorporate such behaviour.

4.2 *R4D-K_0TT: dry soil, anisotropic K_0 conditions*

Since there was only a slight distinction among the shear wave velocities regarding the different stress components applied, in the previous specimen, a second analysis was carried out under anisotropic stress conditions. The laboratory test was also carried out in order to investigate if the true triaxial apparatus would be able to impose a K_0

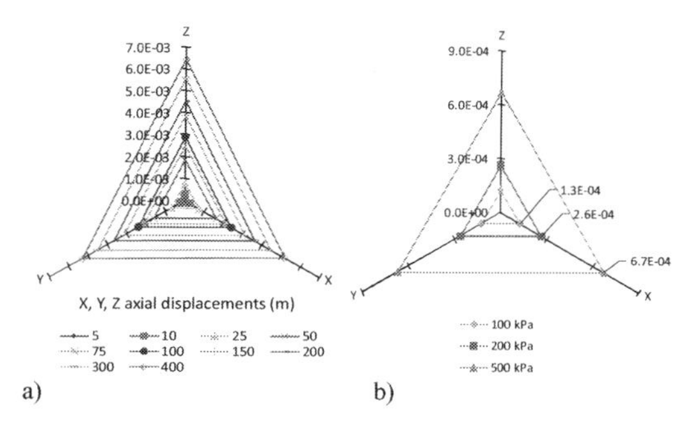

Figure 5. Three-dimensional view of the strains in isotropic loading: a) experimentally; b) numerically.

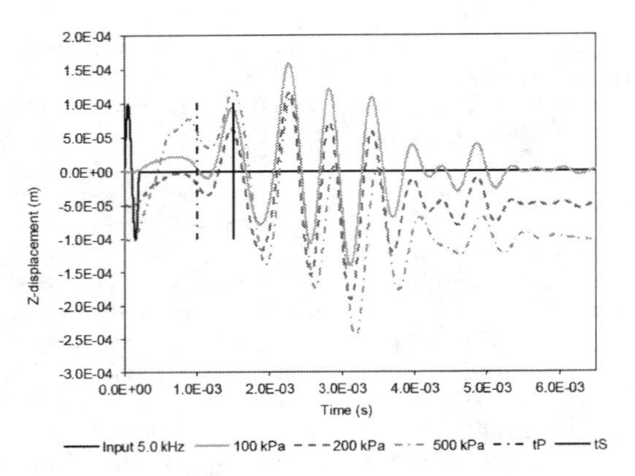

Figure 6. Simulations of seismic wave propagation for several stress values.

stress state condition. If this was possible, the at rest stress parameters could be achieved (Ferreira 2009). Two different loading conditions were considered, one with the horizontal stress equal to 35 kPa and the vertical stress at 100 kPa and another with the double, that is, a horizontal stress of 70 kPa and a vertical stress of 200 kPa. This provided an opportunity for comparison not only between isotropic and anisotropic constitutive models, but also for two different values of anisotropic ratio, in order to evaluate the anisotropic behaviour of the model. The obtained numerical results in terms of horizontal and vertical displacements are provided in Figure 7.

From Figure 7, it is possible to conclude that the higher the stress applied, the higher the displacement, as expected. Nevertheless, when the soil is loaded under isotropic conditions, it can be verified that the displacements should have been higher than they actually are, especially since the same loading values are applied in the three principal directions and considering the values of the displacements of both the anisotropic models. Furthermore, when considering the anisotropic model with the higher vertical Young's modulus in this figure, it can be observed that the horizontal displacements are slightly higher than the vertical ones. In other words, even though the soil is more rigid in the vertical direction, it exhibits higher strains in the horizontal direction.

5 CONCLUSIONS

The main objective of this research consists on the study of the small-strain behavior of a reconstituted residual soil from Porto granite in true triaxial

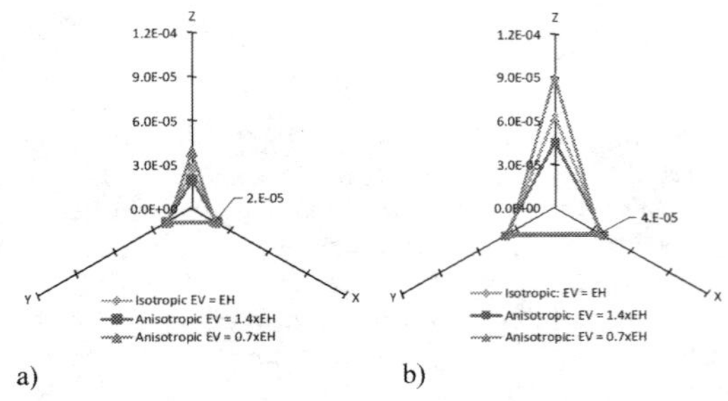

Figure 7. Horizontal and vertical displacements, correspondent to horizontal and vertical stresses of: a) 35 kPa and 100 kPa; b) 70 kPa and 200 kPa.

conditions. For this purpose, a numerical model of the true triaxial apparatus was developed, in order to simulate soil response, namely the propagation of seismic wave velocities by means of bender elements. For comparison, a large set of experimental results was available from Ferreira (2009). However, even though a correctly designed model was achieved, a simplified model was used instead, with benefits in terms of simplicity of the program itself and computation time, without affecting the quality and representativeness of the results. The model used to assess the various parameters that could influence the behavior of the soil, considered a cubical shape and applied velocities in the lateral boundaries in order to simulate the existence of both vertical and horizontal bender elements. The parametric and sensitivity studies proved that this model was able to adequately characterize soil response. After the validation of the simulated model by these studies, a comparison between numerical and experimental tests was made, in order to have a more realistic view of the results.

In this paper, the numerical results obtained with the dry reconstituted residual soil specimens are compared, namely one under isotropic loading conditions and other under anisotropic stresses. Having in consideration the differences between numerical analyses and experimental data, namely the linear elastic behaviour of the first versus the highly non-linear soil response of the specimens, it is possible to conclude that the comparison was successful and the measurement of the stiffness parameters achieved in the laboratory was verified. This work represents a first approach to the numerical modelling of this particular apparatus, as well as of the seismic wave propagation conditions, from which many other analyses will follow. In this paper, the successful implementation of the model to simulate small-strain soil response has been demonstrated by comparison with experimental data.

ACKNOWLEDGEMENTS

The first author acknowledges the support of FCT through the grant SFRH/BPD/120470/2016.

REFERENCES

Atkinson, J. 2000. Non-linear soil stiffness in routine design. Geotéchnique no. 50 (5): 487–508.

Cascante, G. and Santamarina, C. 1997. Low strain measurements using random noise excitation. Geotechnical Testing Journal, 20 (I): 29–39.

Chang, T., Misra, A., and Sundaram, S.S. 1990, Micromechanical modeling of cemented sands under low amplitude oscillations, Geotéchnique, 40 (2): 251–263.

Fam, M. & Santamarina, C. 1995. Study of geoprocesses with complementary mechanical and electromagnetic wave measurements in an oedometer. Geotechnical Testing Journal, 18 (3): 307–314.

Ferreira, C. 2009. The use of seismic wave velocities in the measurement of stiffness of a residual soil. PhD thesis in Civil Engineering, University of Porto, Portugal.

Hardy, S., Zdravkovic, L. & Potts, D. 2002. Numerical interpretation of continuously cycled bender element tests. Numerical Models in Geomechanics (NUMOG VII). Pande & Pietruszczak. Swets & Zeitlinger, Lisse.

Itasca, 2002. FLAC3D, Online Manual.

Neiva, A.A. S. 2011. Considerations of wave-transmission from soil into structure based on numerical calculations. Porto.

Potts, D.M. and Zdravkovic, L. 1999. Finite element analysis in geotechnical engineering. London: Thomas Telford.

Rio, J.F. M.E. 2006. Advances in laboratory geophysics using bender elements. University College of London.

Silva, A.R. 2014. Numerical modelling of the dynamic behaviour of a soil in true triaxial tests with bender elements. Dissertation in Civil Engineering, University of Porto, Portugal.

Stokoe, K.H. and Santamarina, J.C. 2000. Seismic-wave-based testing in geotechnical engineering. Proceedings of the International Conference on Geotechnical and Geological Engineering (GeoEng 2000), Melbourne, Australia: 1490–1536.

Viana da Fonseca, A. 2003. Characterising and deriving engineering properties of a saprolitic soil from granite. Characterisation and Engineering Properties of Natural Soils. Swets & Zeitlinger, Lisse: 1341–1378.

Viana da Fonseca, A., Ferreira, C. and Fahey, M. 2009. A framework interpreting bender element tests, combining time-domain and frequency-domain methods. Geotechnical testing Journal, 32 (2): 1–17.

Wilson, E.L. & Bathe, K.-J. 1976. Numerical methods in finite element analysis.

Numerical Methods in Geotechnical Engineering IX – Cardoso et al. (Eds)
© 2018 Taylor & Francis Group, London, ISBN 978-1-138-33198-3

The role of soil fabric anisotropy for reaching and maintaining critical state

A.I. Theocharis & E. Vairaktaris
Department of Mechanics, National Technical University of Athens, Athens, Greece

Y.F. Dafalias
Department of Mechanics, National Technical University of Athens, Athens, Greece
Department of Civil and Environmental Engineering, University of California Davis, Davis CA, USA

A.G. Papadimitriou
Department of Geotechnical Engineering, National Technical University of Athens, Athens, Greece

ABSTRACT: According to Critical State Theory (CST) for granular media, two conditions on the stress ratio and void ratio are considered to be necessary and sufficient for reaching and maintaining the critical state (CS). A two-dimensional discrete element method experiment questions whether these two conditions, that do not consider soil fabric, are sufficient for CS. For this purpose, a virtual sample is first brought to CS and then rotation of the stress principal axes is imposed while keeping stress principal values fixed, hence maintaining the satisfaction of the aforementioned two CST conditions. The rotation induces a void ratio reduction and thus, abandonment of CS, proving incompleteness of classical CST. The recently proposed Anisotropic Critical State Theory (ACST) remedies this incompleteness by enhancing the two foregoing conditions by a third, related to the critical state value of a fabric anisotropy variable. The ACST can also explain various other soil response characteristics that cannot be addressed by classical CST with no fabric anisotropy consideration, as depicted by simulations performed with a constitutive model developed within ACST.

1 INTRODUCTION

Critical state (CS) is a physically observed phenomenon that defines the state of ultimate failure for geomaterials and was initially described by Schofield and Wroth (1968). During CS the material keeps deforming in shear under constant stress and volume, which in terms of triaxial variables is expressed by:

$$\dot{p}=0,\ \dot{q}=0,\ \dot{\varepsilon}_v=0,\ \dot{\varepsilon}_q\neq 0 \tag{1}$$

where p is the hydrostatic pressure, q is the deviatoric stress tensor, ε_v is the volumetric strain and ε_q is the deviatoric strain; a superposed dot indicates the material time derivative.

The classical critical state theory (CST) was defined based on CS, in order to analytically describe the necessary and sufficient conditions to reach and maintain CS. These two conditions of CST have been explicitly defined by Schofield and Wroth (1968):

$$\eta=\eta_c=(q/p)_c=M,\ e=e_c=e_c(p) \tag{2}$$

It is crucial to distinguish between the physical phenomenon, the CS described by Eq. (1), and the theory that is proposed in order to describe analytically the conditions of the CS phenomenon, namely the CST, whose conditions are provided in Eq. (2). As such, the CST provides the tools for understanding the response of granular materials as they approach CS failure. In addition, the CST has been a hugely successful framework for the constitutive modeling of such materials. For the past 60 years, soil constitutive modeling has flourished within CST, and constitutive models that do not fall into this framework are rarely used for advanced applications.

However, issues have been raised lately concerning the accuracy and completeness of the CST. Specifically, Dafalias (2016) underlined that CST essentially neglects fabric effects, thus rendering CST an incomplete theory, and outlined an experiment that could prove this incompleteness. On the other hand, Li and Dafalias (2012) introduced the anisotropic critical state theory (ACST), which incorporates fabric anisotropy of soils in CST, and adds a new third condition to the 2 conditions of

Eq. (2). However, neither the incompleteness of the CST, nor the necessity for this third condition had been proven until recently by Theocharis et al. (2017).

Along the lines of the aforementioned studies, this paper presents a virtual experiment with the use of the discrete element method (DEM), which proves the incompleteness of CST. In addition, it provides evidence that the third condition of ACST is an indispensable addition to Eq. (2), thus underlining the necessity for ACST as a constitutive modeling framework that also includes fabric anisotropy as an indispensable ingredient. Then, it shows simulations from a constitutive model for sands (Papadimitriou et al., 2015) that is developed within ACST. These simulations underline the role of fabric anisotropy in altering the sand response, despite that the critical state line (CSL) in the e – p space is unique, a thermodynamic requirement (Li and Dafalias, 2012). In the presentation that follows scalar valued quantities are presented in italics, while tensors are presented in bold face characters.

2 ELEMENTS OF ANISOTROPIC CRITICAL STATE THEORY

In order to address the missing link of fabric anisotropy in CST, Li and Dafalias (2012) introduced the *fabric anisotropy variable A*, defined as:

$$A = \mathbf{F} : \mathbf{n} = F\mathbf{n}_F : \mathbf{n} = FN \tag{3}$$

where symbol ":" denotes the double inner product of 2 tensors, $\mathbf{F}$ is the evolving deviatoric fabric tensor that can be decomposed into its (non-negative) norm $F = (\mathbf{F}{:}\mathbf{F})^{0.5}$ and its unit-norm deviatoric tensor-valued direction $\mathbf{n}_F$, while $\mathbf{n}$ is the unit-norm deviatoric tensor-valued plastic strain rate direction. Note that norm $F = 0$ corresponds to isotropy, while any value $F \neq 0$ depicts fabric anisotropy. Scalar $N = \mathbf{n}_F : \mathbf{n}$ quantifies the relative orientation of the fabric direction $\mathbf{n}_F$ with respect to the plastic strain rate direction $\mathbf{n}$.

According to ACST, when plastic loading occurs the $\mathbf{F}$ is expected to evolve during loading towards $\mathbf{n}$, and as such the relative orientation N of the two tensors is expected to evolve towards 1. The $\mathbf{F}$ is considered normalized with respect to its norm at critical state, therefore the value of the (always positive) norm F has a value of 1 at CS. For loose samples norm $F = 1$ is the maximum value, whereas for sufficiently dense samples the norm F may increase above 1 before falling back to 1 at CS. Hence, at critical state both scalars F and N become 1, and so does the fabric anisotropy variable A, based on Eq. (3). According to ACST, this condition of $A = 1$ is the 3rd necessary and sufficient condition for CS to be reached and maintained, additionally to the 2 conditions of Eq. (2).

A simple rate equation that describes the fabric tensor evolution from deposition to the critical state is the following:

$$\dot{\mathbf{F}} = \langle L \rangle c(\mathbf{n} - r\mathbf{F}) \tag{4}$$

where the superposed dot on the fabric tensor depicts its rate, the scalar c quantifies the intensity of fabric evolution (and requires calibration) and r is a scalar variable that is smaller than 1 before CS, but equals 1 at CS and allows the norm F to reach values greater than 1 before $F = 1$ at CS. L is the loading index of any constitutive model that gives $L > 0$ when plastic loading occurs. Setting the L in Macauley brackets $< >$ portrays that fabric evolves only when plastic loading occurs, since by definition $<L> = L$ when $L > 0$ and $<L> = 0$ when $L \leq 0$. For the usual gravitational deposition of sands, the initial fabric is anisotropic (initial value of norm $F_{in} > 0$) and axisymmetric, with the major axis aligned vertically. Loading in triaxial compression (TC) is also axisymmetric with its major axis being vertical. Based on Eq. (4), if TC loading is applied, $N = 1$ and the norm F increases in value initiating from F_{in} (a parameter requiring calibration). If the TC loading is sustained, the norm F reaches value of 1 at CS, which corresponds to a fabric $\mathbf{F} = \mathbf{n}$. Based on Eq. (3), this means that for loading in TC the A initiates from $A = F_{in} > 0$ and then increases until $A = 1$, i.e. until the 3rd condition for CS is met, when also the fabric stops evolving based on Eq. (4), despite that plastic loading continues.

On the other hand, loading of a sand with gravitational fabric towards triaxial extension (TE) would lead initially to a decrease of norm F from its initial value F_{in} due to $N = -1$ (since $\mathbf{n} = -\mathbf{n}_F$). This decrease would continue until $F = 0$ is reached, i.e. until an isotropic fabric is temporarily attained. If loading in TE is sustained, then the fabric becomes re-oriented along $\mathbf{n}$, i.e. $\mathbf{n}_F = \mathbf{n}$, and $N = 1$ holds. Hence, based on Eq. (4), the norm F will increase in value initiating from zero and this increase continues until the norm F reaches a value of 1, which corresponds to $\mathbf{F} = \mathbf{n}$, an axisymmetric fabric with the minor axis being vertical but now in TE format. Based on Eq. (3), this means that for sustained loading in TE the A initiates from $A = -F_{in} < 0$ (due to $N = -1$) and increases algebraically. It passes through $A = 0$ (due to $F = 0$) and then increases until $A = 1$, i.e. until the 3rd condition for CS is met, according to ACST. Of course this triaxial stress space evolution of fabric and A is only a particular case of the general stress space evolution, where the $\mathbf{n}$ is not co-axial (not same eigenvectors) with $\mathbf{n}_F$, the latter being aligned eventually with $\mathbf{n}$ and becoming identical to it at CS.

Additionally, the ACST uses the A to introduce the evolving fabric anisotropy effect on the basic premises of critical state soil mechanics. In the ACST of Li and Dafalias (2012), the A is used to define a Dilatancy State Line (DSL) in the e–p–A space, which delineates the dilative or contractive state of the soil, as does the Critical State Line (CSL) in isotropic critical state theory. The distance of the current void ratio e from the one on the DSL at the same p, denoted by e_d, is named the *Dilatancy State Parameter* ζ and is defined by:

$$\zeta = e - e_d = \psi - e_A(A-1) \tag{5}$$

where e_c is the critical void ratio on the CSL at current p, $\psi = (e - e_c)$ is the state parameter of Been and Jefferies (1985), A is defined in Eq. (3), and the quantity $-e_A(A-1)$ depicts the evercurrent parallel translation of the DSL in comparison to the constant CSL (see Fig. 1). The parallel translation is downward when $A < 1$ and upward when $A > 1$, the latter when the norm $F > 1$ (due to $r < 1$, as explained in the text following Eq. (4)).

The e_A is a parameter requiring calibration that quantifies this translation, as that when the fabric is isotropic and $A = 0$. Note that states with $\zeta > 0$ and $\zeta < 0$ correspond to contractive and dilative soil response, respectively. This is exactly similar to how states with $\psi > 0$ and $\psi < 0$ correspond to contractive and dilative response, respectively, in isotropic critical state theory. Observe, that the DSL becomes identical to the CSL when $A = 1$, i.e. at critical state, and consequently the $\zeta = \psi$.

As described above, for the usual gravitational fabric structures, the A values for loading in TE are initially negative, while the opposite holds for loading in TC. Moreover, given Eqs. (3) and (4), the A values in TE are generally expected to be

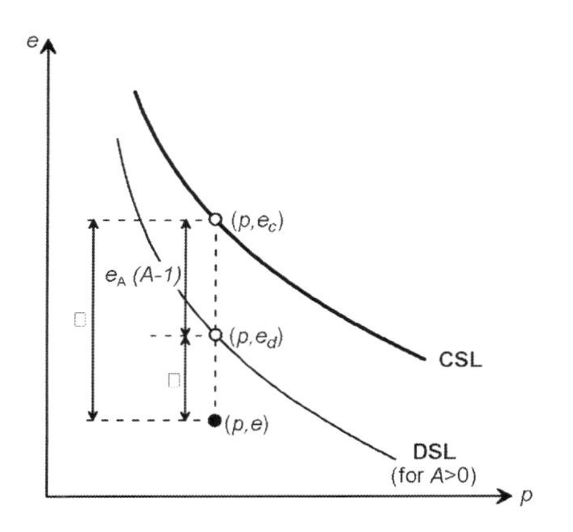

Figure 1. Illustration of state parameters ζ and ψ and definition of DSL and CSL in the e—p space (Li and Dafalias, 2012).

algebraically smaller than in TC. Based on Eq. (5), and given that $e_A > 0$, this translates to algebraically larger ζ values in TE than in TC. Hence, according to the ACST of Li and Dafalias (2012), loading in TE is expected to be more contractive than in TC for the usual gravitational fabric structures, an event that is repeatedly confirmed by experiments (e.g. Yoshimine et al., 1998).

3 DEM PROOF OF ACST NECESSITY

The ACST introduced one extra condition for reaching and maintaining CS with regard to CST. This conceptual framework difference constitutes the basis of comparison for the two theories. Thus, one can vividly challenge CST, and ACST, by reaching CS on a granular sample and then violate only the third condition of ACST. If CS is maintained, then CST is supported and appears complete; while if CS is violated, then CST is incomplete to maintain CS, and ACST could be its remedy.

In order to realize the concept mentioned above, a granular sample should be loaded until reaching CS (phase 1) and then loaded under a particular loading stress path to fulfill the violation of ACST's third condition only. This loading is the stress principal axes (PA) rotation while keeping the stress principal values constant (phase 2).

Before we proceed with further elaboration on this stress PA rotation loading, a few crucial observations are necessary. Such stress PA rotation will violate the physical event of CS that presumes fixed stress and plastic strain rate directions, but will not violate the two conditions of the CST, Eq. (2), that depend only on the principal stress values that are kept fixed. Thus, according to CS definition as a physical phenomenon this stress PA rotation may cause abandonment of CS, while according to CST the CS must be maintained during the stress PA rotation. The only way these two opposing conclusions can be compatible, is for the plastic strain rate to continuously rotate with the stress and in parallel maintain the fixity of volume. It is important to distinguish between the CS (physical phenomenon) and the CST (relevant theory); in this work it is the latter that is being challenged.

The difficulties of the implementation of this 2-phase procedure are profound; the sample must be initially homogeneous and also remain homogenous while reaching CS. A localized failure mode, i.e. a shear band, during phase 1 will make the sample uncontrollable and the second phase, which is crucial, impossible to perform. As a result, the realization of this experimental procedure in physical apparatus appears practically impossible, and the discrete element method (DEM) was employed in order to implement it. The exceptional conditions of the loading path present significant issues even on the numerical implementation in DEM, which are discussed in the sequel.

The virtual experiment consists of one preliminary stage and three loading phases. During the preliminary stage, the particles were created using the radius expansion method, inside the sample's specified area. Then an isotropic compression loading was applied in order for the sample to initiate at a hydrostatic pressure $p = 200$kPa.

During the first phase, a biaxial loading took place; the vertical stress (which is also the major principal stress) was increased while the horizontal stress (minor principal stress) was decreased, keeping the hydrostatic pressure p constant. The sample was loaded until undoubtedly reaching CS. While being in CS, the second phase was applied. During this phase, the principal stress values are kept constant, and the stress principal axes are continuously rotated (Figure 2). This second phase is of primary importance, as it is the one that can validate or not the completeness of CST and ACST. After the end of phase 2, a third loading phase was applied. During this third phase a biaxial load, similar to the initial one of phase 1 was implemented, for reasons explained below.

3.1 *Implementation issues*

DEM is a numerical method that can simulate the behavior of a granular material at its grain scale. The software PFC 2D v4.0 (Cundall & Strack, 1979; Itasca, 2013) was used for the virtual experiments presented herein. Boundary conditions are essential for the procedure as shear bands must be avoided, and the homogeneity and uniformity of the sample must be retained throughout the whole 3-phase loading. Thus, particles at the outer limits of the sample were used as rigid-type boundaries, and a proper velocity field was implemented on them for the application of the desired loading paths. In addition, a loose sample was created to avoid non-uniformities that usually accompany dense samples. The fabric was initially isotropic, due to the particles' creation method and the sample was checked thoroughly to be homogeneous.

Stress, fabric and strain rate tensors were measured inside a circular representative volume of the sample and were checked for their homogeneity on the whole of the sample. The stress tensor was calculated based on its typical DEM formula that uses branch vectors and contact forces (Love, 1927). For the strain rate tensor, a best fit procedure was used (Liao 1997; Itasca, 2008). Finally, the fabric tensor was calculated based on the contact normal vectors, which also constitute the typical way for quantifying fabric on granular DEM samples (Satake, 1982). More details on the implementation issues can be found in Theocharis et al. (2017).

3.2 *Macroscopic results*

The macroscopic results of the virtual experiment are presented in Figure 3. On Figure 3(a) the evolution of the stress is displayed on a stress space as in Figure 2; during phase 1 (light grey) the shear stress is practically zero while the difference between the horizontal and the vertical stress increases. This phase consists a standard biaxial test of a loose sample, with constant hydrostatic pressure p. During phase 2 (black) the stress circle expected on this stress space is presented, as sketched out in Figure 2. The center of this circle remains at the axes origin, as the loading initiated after isotropic compression, and the radius of the circle is equal to $Mp/2$, where M is the deviatoric stress ratio η at CS. The stress PA rotation of the phase 2 was applied for ten cycles of rotation.

On Figure 3(b) the evolution of the stress ratio is presented with respect to the cumulative plastic strain. This equivalent strain measure is calculated through the strain rate and is used for the uniformity of the presentation of the results of all three phases. During phase 1 the stress ratio η increases until it reaches its CS value. Then, during phase 2 of stress PA rotation, the η remains constant, equal to the critical value M. Based on Figure 3(a) and (b) it becomes apparent that the loading of the phase 2 is the desirable one, meaning the principal stress values remained constant and the principal stress axes were constantly rotating.

Figure 3(c) is a crucial figure. During phase 1, the void ratio e presents predominantly compression, as expected from a very loose sample, until it reaches its CS value where it remains fixed. When phase 2 initiates, the void ratio almost immediately abandons its CS value in a sharp way. This result alone proves the incompleteness of CST. During phase 2, because the stress ratio η remains

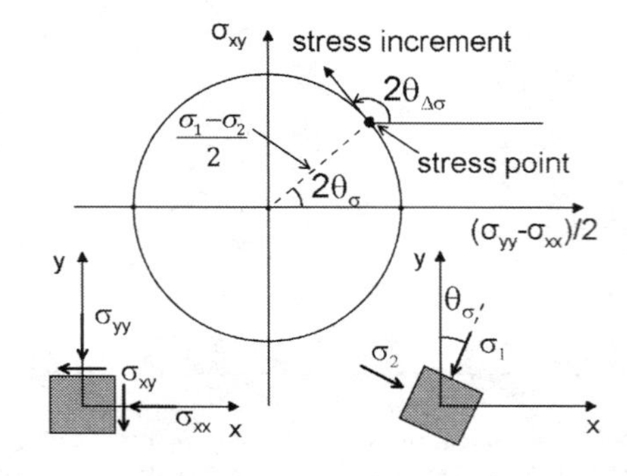

Figure 2. Stress principal axis (PA) rotation represented on stress space.

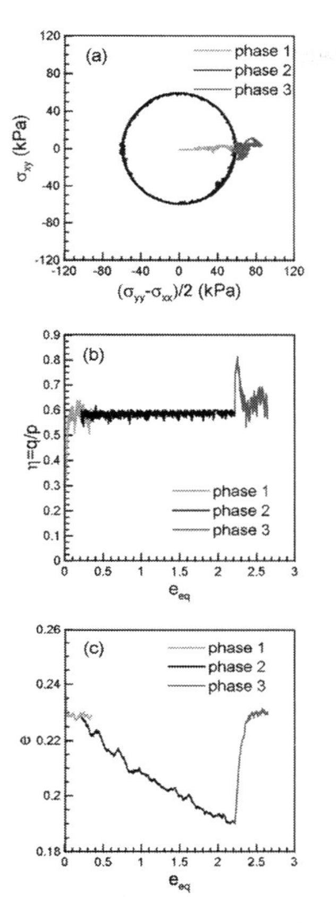

Figure 3. Macroscopic DEM results of the 3-phase loading.

constant (see Figure 3(b)) and the hydrostatic pressure p remains also constant, CST would predict constant void ratio, which apparently is not the case here. On the contrary, the void ratio decreases continuously during the ten cycles of stress PA rotation. It appears to be reaching another limit state value, which corresponds to a very dense sample. During phase 2, a constant densification of the sample took place, underlining that the sample is moving away from its CS void ratio value and towards another state.

It is then intriguing to reload the sample, as in phase 1, to monitor its response. During this phase 3, a biaxial loading was applied, similarly as in phase 1, but after the end of the 2 preceding phases. On Figure 3(a) phase 3 (dark grey) implies that the stress increases outside the circle, which represents the CS stress limits. The same result is presented in Figure 3(b) where the stress ratio of phase 3 increases over its CS value very rapidly, and then falls back to CS. This abrupt increase of CS is due to the densification of the sample during the preceding phase 2.

The increase of the stress ratio is accompanied by an astounding dilation for the sample (Figure 3(c)). Its void ratio very rapidly and monotonically rebounds to its CS value reached after the biaxial loading (with fixed axes) of phase 1. The fact that the void ratio of rebound is equal to the CS value of phase 1 is due to fact that the hydrostatic pressure p of phases 1, 2 and 3 is the same. This is also another verification of the homogeneous deformation for the sample at all stages

3.3 *Fabric results*

So far, we have proved the incompleteness of CST in reference to not accounting for the fabric effects in stating the conditions for CS, but it would also be essential to check whether the ACST of Li and Dafalias (2012), which accounts for fabric effects, is a viable substitution. The check will be performed on the basis of the results of the DEM experiment, the most vital part of which is that during the stress PA rotation (phase 2), CS is abandoned. In order to address this issue, the notion of fabric as a descriptor of the sample's orientation comes into place.

For ACST the crucial ingredient is the fabric anisotropy variable A, which is a function of the fabric tensor and the (plastic) strain rate direction tensor. These two tensors can be directly calculated in DEM, since all the information regarding fabric and strain rate can be measured from the grain scale. Given their calculation, other ingredients of ACST like the scalar norm F of the fabric tensor and the scalar N depicting the relative orientation of the 2 tensors follow.

The evolution of A during the 3-phase experiment, as well as the evolution of the two scalars F and N constituting A as per Eq. (3), are presented in Figure 4 versus the equivalent plastic strain e_{eq} measure. During phase 1, the relative orientation of fabric and plastic strain rate N becomes rapidly equal to 1.

The fabric norm F, which accounts for the intensity of the sample's fabric anisotropy, initiates from zero (in this sample) and increases until it reaches its critical state value of 1. Thus, at critical state $F = N = 1$ and $A = FN = 1$, and the 3rd ACST condition is fulfilled.

During phase 2, the fabric norm F remains equal to its critical state value of 1, but the N immediately after the initiation of the stress PA rotation loading becomes less than 1, and on average equal to 0.7. As a result, $A = FN < 1$ and based on ACST the sample should abandon the critical state, since the 3rd ACST condition is no longer valid. This observation is of profound importance as it validates the ACST, which provides the abandonment of critical state during phase 2, in contrast to CST that prescribes that the sample remains at critical state since the 2 conditions of Eq. (2) remain valid (see Figure 3).

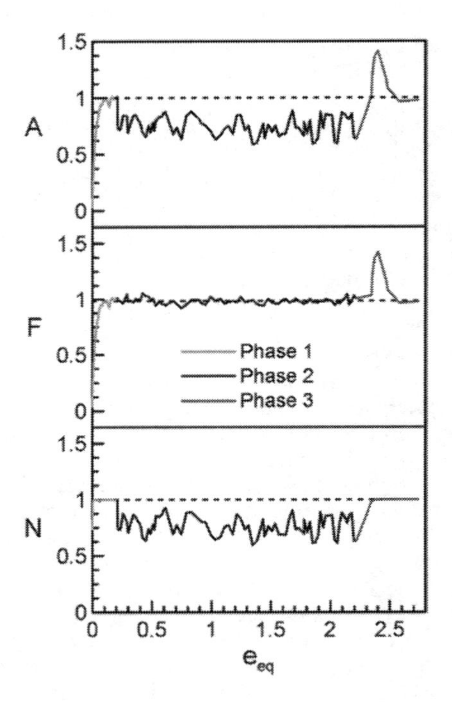

Figure 4. Evolution of fabric anisotropy variable A, fabric norm F and relative orientation of fabric and strain rate N during the 3-phase loading with DEM.

Finally, during phase 3, strain rate and fabric become again coaxial very quickly. The fabric norm F, following the evolution of the stress ratio η, becomes greater than 1 and then falls back to its critical state value of 1. Notice that during phases 1 and 3 the fabric norm F is the part providing the evolution of A, because the strain rate is coaxial with the fabric practically from the beginning of each of these phases. On the other hand, during phase 2 the fabric norm remains constant at its critical state value of 1, while the relative orientation of fabric and strain rate is providing the evolution of N and A.

4 IMPORTANCE OF FABRIC ON SIMULATED RESPONSE

The ACST, outlined in section 2 and proved accurate in section 3, has generic value because it can be combined with any constitutive model platform constructed within isotropic CST and produce a complete model within ACST. Papadimitriou et al. (2015) proposed such a model, by adopting the SANISAND model platform. Particularly, it is a constitutive model for sands developed within Bounding Surface plasticity, that adopts a dependence of the peak M^b and dilatancy M^d stress ratios on the dilatancy state parameter ζ, Eq. (5), according to the proposition of Li and Dafalias (2012). This dependence is expressed via:

$$M^b = M\exp\left(n^b\langle -\zeta\rangle\right) \tag{6}$$

$$M^d = M\exp\left(n^d\zeta\right) \tag{7}$$

These equations follow the spirit of the original propositions of the two-surface model of Manzari and Dafalias (1997) and the subsequent modification by Li and Dafalias (2000), in which Eqs. (6) and (7) were stated with ψ of Been and Jefferies (1985) in lieu of ζ, when no fabric effects were considered.

The thus-formulated M^b and M^d enter the definitions of the plastic modulus and the dilatancy of the model. In more detail, based on Eq. (5) the higher the value of A, the lesser the value of ζ. This decrease of ζ corresponds to an increase of M^b and a decrease of M^d, which lead to stiffer and more dilative response, respectively. In addition, the use of ζ in Eqs. (6) and (7), or of ψ in Manzari and Dafalias (1997) and Li and Dafalias (2000), enables successful simulations for very different densities and stress levels without change in the values of model constants. The ACST framework also enters in other constitutive ingredients of the model via the fabric anisotropy variable A. Namely, the plastic modulus is an increasing function of A, as is the scalar c that quantifies the rate of fabric evolution in Eq. (4).

All the foregoing ACST-related interventions in the constitutive model aim at simulating experimental results that show more dilative and stiff response when the loading is applied along the direction of the fabric, i.e. when the A takes relatively large values. Figure 5 presents examples of the simulative performance of this new anisotropic model.

The emphasis is placed on triaxial compression (TC) and extension (TE) tests performed by Yoshimine et al. (1998). These tests explore the monotonic undrained triaxial response of Toyoura sand for a relatively narrow void ratio range e = 0.86 to 0.89, but a relatively large range of confining pressures p = 50 to 500kPa. Subplot (a) compares data to simulations for the effective stress paths, while subplot (b) does the same for the shear stress-strain relations. Data and simulations pertaining to positive values of q and γ correspond to the TC paths, while negative values of these parameters correspond to the TE paths. Observe that the data highlight the much more contractive and soft response in TE in comparison to TC loading paths with similar void ratio e and identical mean effective stress p values. The model simulates this intense difference in the response very well.

The samples of Toyoura sand in Figure 5 were prepared with a gravitational (dry) deposition method. Therefore, their fabric was axisymmetric, with the major fabric axis being vertical and

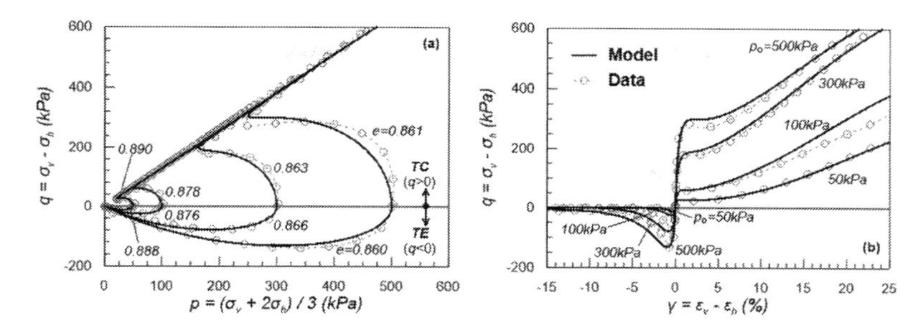

Figure 5. Experimental results versus model simulations of undrained triaxial compression (*TC*) and extension (TE) tests. Data on dry-deposited Toyoura sand after Yoshimine et al (1998).

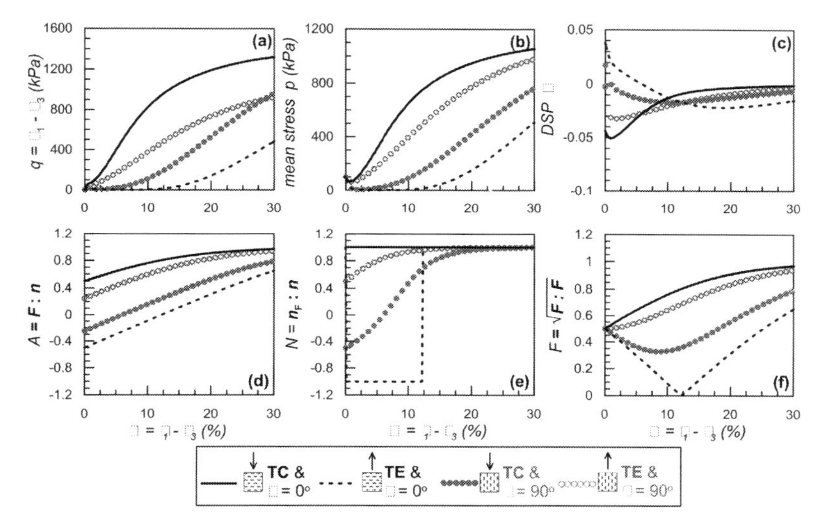

Figure 6. Model simulations demonstrating the effects of loading direction (triaxial compression, triaxial extension) and deposition plane orientation (horizontal, vertical) under undrained testing starting from the same initial conditions (p_o = 100kPa, e_o = 0.83).

the two minor fabric axes horizontal, as was the deposition plane. If one rotates the samples by 90° around one of the two minor fabric axes, then the deposition plane becomes vertical and the major fabric axis horizontal (and perpendicular to the axis of sample rotation). This change in fabric orientation cannot be taken into account by any model developed within CST, however it can be considered by models developed within ACST. Hence, Figure 6 compares undrained TC and TE paths predicted by the model of Papadimitriou et al. (2015) for samples with horizontal (δ = 0°) and vertical (δ = 90°) deposition planes, where δ is the angle of the deposition plane with respect to the horizontal plane. The initial conditions of all 4 simulations are identical: p = 100kPa and e = 0.83. On top of the macroscopic response in terms of p and q evolution, this figure also includes the evolution of fabric entities ζ, A, N and F for these 4 tests.

Observe firstly the huge difference in macroscopic response in TC and TE for the usual case of δ = 0°, which are presented in black lines (solid and dashed, respectively) without symbols in Figure 6. For example, note the difference in mobilized q for γ = 10%, which is less than 5kPa in TE and more than 800kPa in TC. This huge difference is governed by the values of p, which is almost zero for TE (due to excessive contraction) as opposed to a few hundreds for TC (due to significant dilation). This figure illustrates that this difference in response originates from the very different values of A, that initiate from 0.5 and –0.5 for TC and TE, respectively (since F_{in} = 0.5 in the model calibration). These widely different values of initial A make the ζ change sign, and become approximately –0.04 and 0.04 for TC and TE, thus dictating the dilative and contractive response, respectively. Note also how the norm F increases monotonically from 0.5 to 1.0 for the case of TC, while for TE the norm F first becomes zero and then starts increasing towards 1.0, due to form of Eq. (4). This non-monotonic evolution of norm F, which is always positive or zero, is due to the evolution of N that jumps from −1.0 to 1.0 exactly when $F = 0$ momentarily.

The predicted TC and TE paths after sample rotation by 90° are presented by curves with solid and hollow gray symbols in Figure 6. Observe in this figure that upon sample rotation (for $\delta = 90^o$) the relative orientation scalar N becomes approximately −0.5 and 0.5 for TC and TE, respectively, and not −1.0 and 1.0 that one would get for a TE-like fabric. These intermediate values of N lead to values of A of −0.25 and 0.25 (given the $F_{in} = 0.5$) for TC and TE, respectively. Given these values, the TE response for $\delta = 90^o$ is stiffer and more dilative than that for TC, and this again is dictated by the ζ values that have opposite signs ($\zeta < 0$ for TE and $\zeta > 0$ for TC). Overall, the response for TC and TE under the usual $\delta = 0^o$ sets the upper and lower limits of macroscopic response for any given initial conditions. On the other hand, the response for $\delta = 90^o$ is less affected by loading direction and lies in between the upper and lower limits, with the TE response being stiffer and more dilative than that in TC. The reader can refer to Li and Dafalias (2012) for the response under various orientations of stress with respect to fabric, and comparison with experimental data, but without the detailed interpretation presented here for the TC and TC in regards to $\delta = 0^o$ and 90° orientations of the deposition plane.

5 CONCLUSIONS

This paper provides evidence of the incompleteness of the critical state theory (CST) of Schofield and Wroth (1968) for granular media in regards to not accounting for fabric effects. A virtual DEM experiment showed that the two conditions in terms of stress ratio and void ratio, Eq. (2), are necessary for reaching critical state but are not sufficient for maintaining it. The anisotropic critical state theory (ACST) of Li and Dafalias (2012) introduces fabric effects in CST and adds a 3rd condition in terms of fabric reaching is critical value. The triplet of conditions prove necessary and sufficient for reaching and maintaining critical state, thus making ACST a necessary enhancement of CST by accounting for the effect of fabric.

The ACST poses as an integrated framework for developing constitutive models that take into account fabric anisotropy. One such model is presented here, and it is shown capable to predict both the stiff and dilative triaxial compression (TC) paths, as well as the soft and contractive paths in triaxial extension (TE). More importantly, this model is able to predict what would happen if the sand samples that were prepared by gravitational deposition are rotated by 90°. The simulations show that in such a case the TE loading leads to stiffer and more dilative response, as compared to the TC loading initiating from the same initial stresses and density.

ACKNOWLEDGMENT

The research leading to these results has received funding from the European Research Council under the European Union's Seventh Framework Program FP7-ERC-IDEAS Advanced Grant Agreement no. 290963 (SOMEF).

REFERENCES

Been, K., & Jefferies, M.G. 1985. A state parameter for sands. *Géotechnique*, 35(2): 99–112.

Cundall, P.A., & Strack, O.D. 1979. A discrete numerical model for granular assemblies. *Géotechnique*, 29(1): 47–65.

Dafalias, Y.F. 2016. Must Critical State Theory be revisited to include fabric effects?. *Acta Geotechnica,* 11(3): 479–491.

Itasca (2008). Particle Flow Code in Two Dimensions. *Users' Manual: Theory and Background*. Minneapolis, MN, 3.13–3.16.

Itasca Consulting Group, Inc. (2013). *PFC— Particle Flow Code, Ver. 4.0*. Minneapolis, MN.

Li, X.S., & Dafalias, Y.F. 2000. Dilatancy for cohesionless soils. *Géotechnique*, 50(4): 449–460.

Li, X.S., & Dafalias, Y.F. 2012. Anisotropic critical state theory: role of fabric. *Journal of Engineering Mechanics*, 138(3): 263–275.

Liao, C. 1997. Stress-strain relationship for granular materials based on the hypothesis of best fit. *Int. J. Solids Structures,* 34 (31,32): 4087–4100.

Love, A.E.H. 1927. *A Treatise of Mathematical Theory of Elasticity*. Cambridge: Cambridge University Press

Manzari, M.T., & Dafalias, Y.F. 1997. A critical state two-surface model for sands. *Géotechnique*, 47(2): 255–272

Papadimitriou, A.G., Dafalias, Y.F., & Li., X.S. 2015. Sand model within anisotropic critical state theory with evolving fabric. In K. Soga. K. Kumar, G. Biscontin, M. Kuo (eds.), *Geomechanics from Micro to Macro*, 2, p. 627–632, Taylor & Francis Group, London

Satake, M. 1982. Fabric tensor in granular materials. In L.H. Lugar, P.A. Vermeer (eds), *IUTAM Symposium on Deformation and Failure of Granular Materials*. Amsterdam: Delft (pp. 63–68).

Schofield, A.N., & Wroth, C.P. 1968. *Critical State Soil Mechanics*. London: McGraw-Hill.

Theocharis, A.I., Vairaktaris, E., Dafalias, Y.F., & Papadimitriou, A.G. 2017. Proof of Incompleteness of Critical State Theory in Granular Mechanics and Its Remedy. *Journal of Engineering Mechanics*, 143(2): 04016117.

Yoshimine, M., Ishihara K., & Vargas W. 1998. Effects of principal stress direction and intermediate principal stress on undrained shear behavior of sand. *Soils and Foundations*, 38(3): 177–186.

Numerical Methods in Geotechnical Engineering IX – Cardoso et al. (Eds)

A method to consider the electrical/chemical interaction of clay crystal in general constitutive model

H. Kyokawa
The University of Tokyo, Tokyo, Japan

S. Ohno & I. Kobayashi
Kajima Corporation, Tokyo, Japan

ABSTRACT: A mechanical behavior of expansive soil like bentonite variously changes depending on the pore fluid condition. It is well known that the electrical/chemical phenomena on the surface of the clay mineral crystal affects the macroscopic behavior. However, a mechanical model considering such surface phenomena is few in geotechnical engineering. In this study, a method considering the interlaminar behavior of clay mineral, which is governed by the electrical/chemical force and mechanical force, in a general constitutive model is proposed. Several element tests on expansive clay are simulated by the proposed model. Simulations show the proposed model can reasonably and uniformly represent a typical behavior of expansive clay.

1 INTRODUCTION

Bentonite whose main constituent is clay mineral montmorillonite shows highly expansive behavior and the self-healing property. Because of such special mechanical characteristics, low permeability and the absorptivity of radioactive nuclide, bentonite has been examined to use it as the buffer material surrounding the radioactive waste in a waste disposal facility. In particular, the swelling behavior by water absorption of a compacted "unsaturated" bentonite has been widely investigated to consider the reflooding process of a disposal pit after closing it. Since unsaturated general soils show volume contraction due to soaking. Moreover, by 100,000 years later when the radiological does sufficiently reduces, the buffer bentonite would be exposed to several environments: groundwater and/or seawater inflow, Ca^{2+} ion leached from the cement material, and Fe^{2+}/Fe^{3+} ion from the frame structure. It is concerned that the expansive characteristic of bentonite intricately changes due to such ion conditions. Because of the large swelling behavior by water absorption and the change of mechanical characteristics due to several ion conditions, an expansive soil such as bentonite is classified as a special/problem soil of which the mechanical theory is not established in soil mechanics and geotechnical engineering.

Meanwhile, the composition of montmorillonite, which is the negative charged clay mineral crystal (aluminum silicate) and the hydrated cation surrounding the crystal (see Figure 1), is broadly known since it is explained in a general text book of soil mechanics (e.g., Mitchell and Soga, 2005). Moreover, in surface chemistry and molecular dynamics considering the microscopic structure of clay minerals, it is generally known that the characteristics of clay material are affected by the electrical/chemical phenomena on the surface of the clay mineral crystal, since the specific surface area of clay materials is considerably larger than that of granular materials like sand. Such surface phenomena consequently yield the peculiar characteristics of clay, i.e., "plasticity", "expansion", "ion exchange" and "suspension/aggregation".

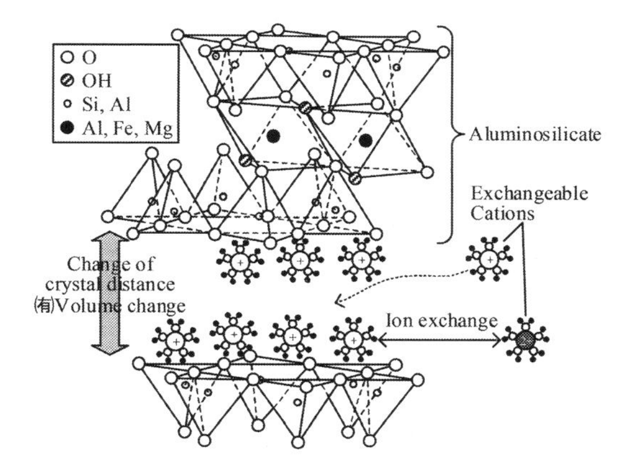

Figure 1. The composition of montmorillonite.

From the point of view of the surface electrical/chemical phenomena of clay minerals, an expansive soil which is a special/problem soil in geotechnical engineering can be regarded as just one of the clay materials. However, a mechanical model considering such surface electrical/chemical phenomena, which obviously affects the mechanical characteristic of clay, is few in geotechnical engineering. In this study, a method to consider the interlaminar behavior of clay crystal affected by the electrical/chemical phenomena in a general constitutive model is proposed. Considering the effective stress and the hydration force in the DLVO theory, which explains the stability of colloid based on the interaction of the electrical double layer between the negatively charged clay crystals, the interlaminar behavior of clay crystal is described (the micro-pore). On the other hand, the mechanical behavior of void except the interlayer of crystal is given by a general constitutive model as the relationship between the effective stress and the macro-strain (the macro-pore). Both micro and macro pore behaviors are simultaneously solved regarding the effective stress. The proposed method is applied to the modified Cam clay model in this study. Several element tests on expansive clay are simulated by the proposed model. Simulations show the proposed model can reasonably and uniformly represent a typical behavior of expansive clay.

2 THE ELECTRICAL/CHEMICAL BEHAVIOR OF SATURATED EXPANSIVE CLAY—OSMOTIC CONSOLIDATION

Figure 2 shows the volume change of the saturated bentonite by replacing the water in cell to the saturated solution: NaCl (Di Maio, 1996). The bentonite specimen saturated by the distilled water was firstly mechanically consolidated to a certain stress level in a conventional oedometer, and then repeatedly exposed to the saturated solution and the distilled water under constant stress. It can be observed from figure that the volume contraction occurs due to exposure to the solution and volume expansion occurs due to the exposure to the distilled water vice versa. Moreover, the repeated substitution of cell fluid causes the repeated volume changes. These volume changes due to the change of fluid are known as the osmotic consolidation and the osmotic swelling respectively.

Figure 3 shows the oedometer tests with replacing the cell fluid (Di Maio, 1996). The dashed line represents the consolidation behavior of the bentonite specimen saturated by the distilled water. The result shows the typical consolidation behavior of bentonite, namely highly recoverable volume change. On the other hand, the replacing cell fluid: the distilled water to the solution and the solution to the distilled water were conducted in Cases 1 and 2. Both cases shows that the volume change of osmotic consolidation and the osmotic swelling highly depends on confining pressure. The consolidation behaviors of both cases after the osmotic consolidation are similar and finally approach to the swelling line of the water saturated sample due to exposure to the distilled water. Moreover, the swelling index during unloading pass is quite smaller than that of the water saturated bentonite. Similar laboratory tests on expansive soil considering the chemical osmotic effects were conducted

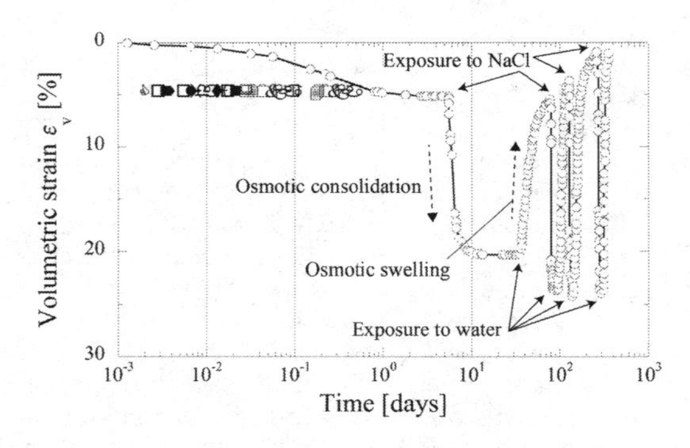

Figure 2. Osmotic volume change by the repeated replacing cell fluid (after Di Maio, 1996).

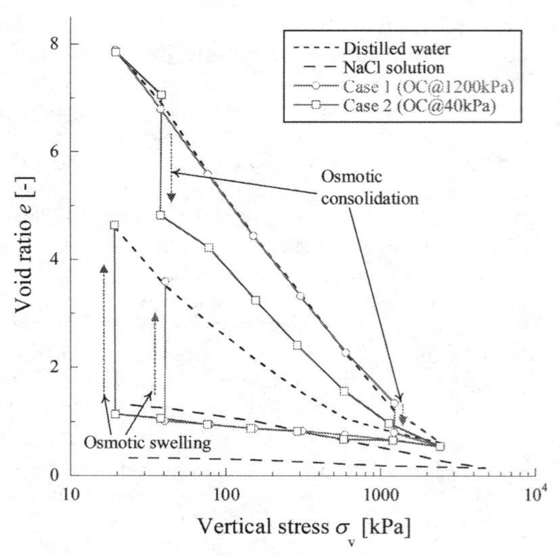

Figure 3. Osmotic consolidation / swelling tests in oedometer test (after Di Maio, 1996).

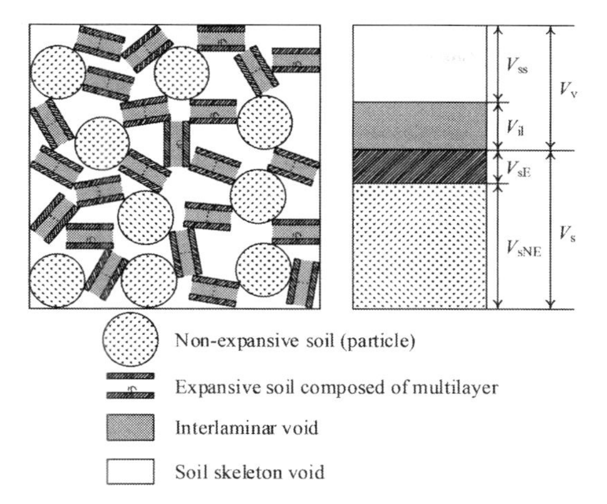

Figure 4. Schematic figure of expansive soil and the volume components.

also by Musso et al. (2003), and Thyagraj and Rao (2013).

The investigation of swelling behavior on the unsaturated bentonite has been actively conducted. Numerical models based on unsaturated soil mechanics also has been proposed in the past (e.g., Gens and Alonso, 1992, Alonso et al., 1999 and Takayama, 2017). However, the saturated bentonite also shows the expansive behavior due to the solution change as shown in Figures 2 and 3. It can be interpreted that the solution change, precisely the decreasing in concentration, occurs inside of the unsaturated soil due to water supply; this process causes the swelling behavior as a result. Therefore, the proposed method considering the electrical/chemical interlaminar behavior is established based on saturated soil mechanics. It would be possible to extend to unsaturated soil mechanics.

3 CONSTITUTIVE MODELING OF ELECTIRCAL/CHEMICAL BEHAVIOR OF EXPANSIVE CLAY

3.1 *Framework of the proposed method*

Figure 4 shows the schematic figure of expansive soil including the multilayer clay crystals and its volume fraction. V_s is total volume of soil particles; V_{sE} is volume of the expansive soil minerals; V_{sNE} is volume of the non-expansive soil; V_v is volume of void; V_l is volume of interlaminar layers; V_{ss} is volume of the soil skeleton void. The soil skeleton void is composed of the granular structure (the soil fabric). Each void ratio variables are given as

$$\begin{cases} e = V_v / V_s \\ e^{il} = V_{il} / V_s \\ e^{ss} = V_{ss} / V_s \end{cases} \tag{1}$$

A volume change of expansive soil is schematically represented as Figure 5. Assuming small deformation theory, the volumetric strain can be given as follows.

$$\varepsilon_v = \frac{V - V_0}{V_0} = \frac{\Delta V_{ss} + \Delta V_{il}}{V_{s0} + V_{ss0} + V_{il0}} \tag{2}$$

Referencing to the montmorillonite mineral model proposed by Komine and Ogata (1996), the volume change of interlayers is described by

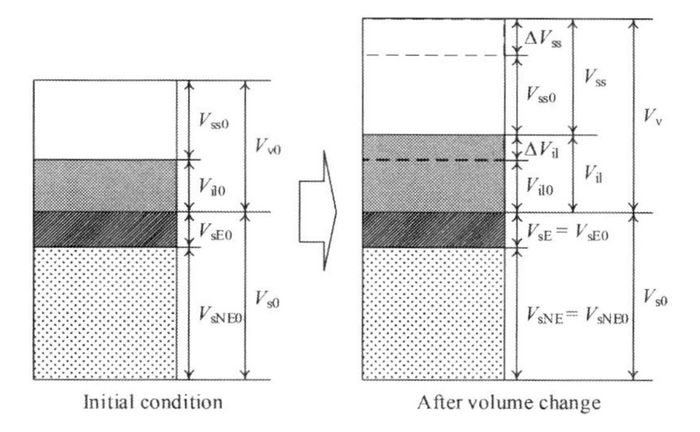

Figure 5. Volume change of expansive soil.

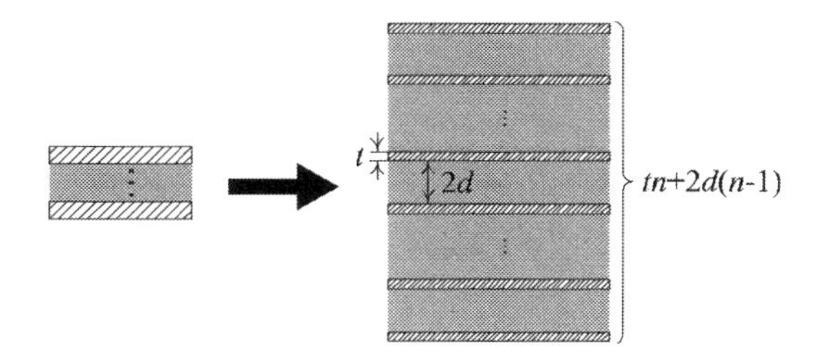

Figure 6. Multilayer structure of clay crystal.

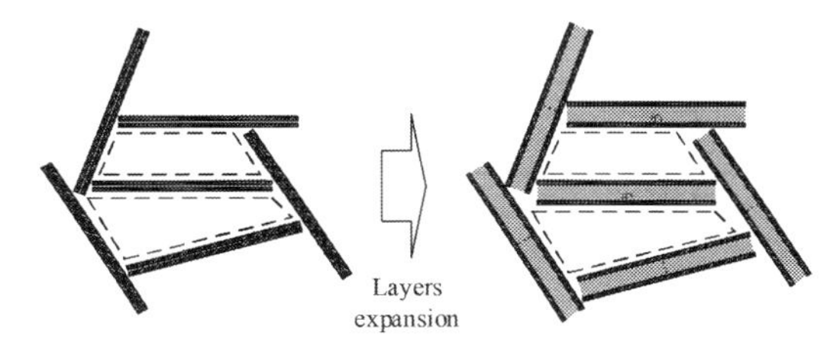

Figure 7. Relation between the soil skeleton void and interlaminar void under non-constrained condition.

upscaling the interlaminar behavior into the macroscopic value (Figure 6) and that relation is given by

$$\frac{\Delta V_{il0}}{V_{il0}} = \frac{2(d-d_0)S(n-1)N}{2d_0 S(n-1)N} = -\frac{d-d_0}{d_0} \quad (3)$$

Here, d is the average distance of interlayer and d_0 is its initial value. S is a specific surface area of montmorillonite. n is number of layers of mineral. N is number of mineral. Assuming that the soil skeleton does not change when the interlayer distance changes under non-constrained condition (Figure 7), the change in the soil skeleton void is given as follows.

$$\Delta V_{ss} = (V_{s0} + V_{ss0} + V_{il0})\,\varepsilon_v^{ss} \quad (4)$$

Substituting Equations (3) and (4) into Equation (2), the volume change in expansive soil is given as

$$\varepsilon_v = \frac{V-V_0}{V_0} = \varepsilon_v^{ss} - \theta^* \frac{d-d_0}{d_0} \quad (5)$$

$$\theta^* = \frac{V_{il0}}{V_{s0}+V_{ss0}+V_{il0}} = \frac{e_0^{il}}{1+e_0^{ss}+e_0^{il}} \quad (6)$$

Equation (6) is extended to ordinary strain by assuming the interlaminar behavior only causes volume change.

$$\varepsilon_{ij} = \varepsilon_{ij}^{ss} - \frac{1}{3}\theta^* \frac{d-d_0}{d_0}\delta_{ij} \quad (7)$$

The details of each value, namely the interlayer distance d and the soil skeleton strain ε_{ij}^{ss} are explained in a later chapter. The outline and assumptions of the proposed model are summarized below.

- The interlaminar behavior of clay minerals obeys the equilibrium equation based on the DLVO theory in which the electrical/chemical phenomenon is considered
- A mechanical behavior of the soil skeleton is described by the general constitutive equation
- The interlaminar behavior and the soil skeleton behavior are related through the effective stress
- Based on saturated soil mechanics
- Assuming small deformation

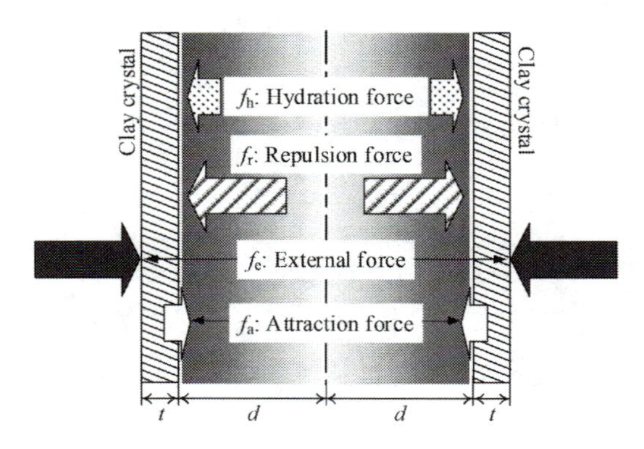

Figure 8. Interlaminar equilibrium (forces acting on interlayer).

3.2 *Interlaminar behavior—micro-pore behavior*

The interlaminar behavior is described by referring to study of surface chemistry. The Gouy-Chapman theory (Chapman, 1913) describes the parallel two layers of charge surrounding surface of an object which might be a solid particle, a bubble or a liquid droplet. The charged surface of an object is adsorbed by the oppositely charged ions in the fluid due to chemical interactions to keep an electrical neutrality. The concentration of the oppositely charged ions decreases with distance from the surface due to the influence of electric attraction and thermal motion. It is thus called the double diffuse layer. When two objects approach each other, the electrostatic repulsion increases and the interference between their electrical double layers increases. It results in the osmotic force due to overlapping the double layer.

Meanwhile, van der Waals force (London, 1937) also acts on between two objects and it increases as they get closer. Considering the combination of these forces, the DLVO theory explains (Derjaguin and Landau, 1941, and Verwey and Overbeek, 1948) the suspension and aggregation phenomena of the colloidal particles in a liquid phase.

Regarding the expansive clay materials, the surface of clay crystal is negatively charged and the exchangeable cations are adsorbed onto the surface and distribute in the fluid. This electrical double layer causes the repulsion force between crystals and results in the osmotic swelling behavior. Of course, van der Waals force also acts simultaneously. Additionally, it would be appropriate to consider the effective stress in the interlaminar equilibrium especially for geomaterials. Figure 8 is the schematic figure representing forces acting on the interlayer of clay crystals. The equilibrium condition is given by

$$F^*(c,d,\sigma_{ij}) = f_a - f_r - f_h + f_e = 0 \quad (8)$$

c is the concentration of solution in the skeleton void. Referring to the past study (Komine and Ogata, 1996 and Mitchell and Soga, 2005), each force are defined. The osmotic repulsion force is given by the following equations.

$$f_r(c,d) = 2ckT\{\cosh(u)-1\}\times 10^{-6}\,[MPa] \quad (9)$$

$$u = 8\tanh^{-1}\{\exp(-\omega d)\cdot\tanh(z/4)\} \quad (10)$$

$$\omega = \sqrt{2cv^2e'^2/\epsilon kT} \quad (11)$$

$$z = \sinh^{-1}\left(96.5\frac{\mathrm{CEC}}{\mathrm{S}}\sqrt{\frac{1}{8\epsilon\mathrm{ckT}}}\right) \quad (12)$$

Here, T is absolute temperature, k is Boltzmann constant, v is a valence of exchangeable cation, e' is electronic charge, ε is an electronic permittivity of pore water, CEC is a cation exchange capacity, and S is a specific surface area of crystal. On the other hand, the attractive van del Waals force is give as

$$f_a(d) = \frac{A_h}{24\pi}\left\{\frac{1}{d^3}+\frac{1}{(d+t)^3}-\frac{1}{(d-t/2)^3}\right\}\times 10^{-6}\,[\mathrm{MPa}] \quad (13)$$

Here, A_h is Hamaker constant and t is a thickness of clay crystal. Additionally, the hydraulic force, which is one of the non-DLVO forces, has to be considered in particular for clay minerals (Pashley and Israelachvili, 1983). This repulsive force monotonically decreases with distance and it is empirically given by follows (Afzal et al., 1994).

$$f_h(d) = \frac{W_0}{\chi_0}\exp\left(-\frac{d}{\chi_0}\right)\times 10^{-3}\,[\mathrm{MPa}] \quad (14)$$

Here, W_0 and χ_0 is material parameter. In this study, f_e is assume to be equal to mean effective stress p for simplicity. Namely, the effective stress isotropically affect the interlaminar behavior.

$$f_e(\sigma_{ij}) = p\,[\mathrm{MPa}] \quad (15)$$

It can be noted from these equations that the interlaminar behavior is assumed as the reversible relation. Other non-DLVO forces, e.g., steric effect, should be considered in the equilibrium of the interlaminar behavior as necessary for high precision.

3.3 *Free void behavior—macro-pore behavior*

The soil skeleton strain in Equation (7) is equal to the deformation of granular structure. Thus, it can be given by the ordinary constitutive model for non-expansive soil. The modified Cam clay model (Roscoe and Burland, 1968) is applied in this study. The yield function, the plastic and elastic strain relations are summarized as below.

$$\varepsilon_{\mathrm{ij}} = \varepsilon_{\mathrm{ij}}^{\mathrm{e}} + \varepsilon_{\mathrm{ij}}^{\mathrm{p}} \quad (16)$$

$$f(\sigma_{\mathrm{ij}},\varepsilon_{\mathrm{ij}}^{\mathrm{p}}) = \ln\frac{p}{p_0} + \ln\left(\frac{M^2+\eta^2}{M^2}\right) - \frac{1+e_0}{\lambda-\kappa}\varepsilon_{\mathrm{v}}^{\mathrm{p}} = 0 \quad (17)$$

$$d\varepsilon_{\mathrm{ij}}^{\mathrm{p}} = \Lambda\frac{\partial f}{\partial\sigma_{\mathrm{ij}}}\left(\Lambda = \frac{C_{\mathrm{p}}df_\sigma}{\partial f/\partial\sigma_{\mathrm{kk}}}, C_{\mathrm{p}} = \frac{\lambda-\kappa}{1+e_0}\right) \quad (18)$$

$$\mathrm{d}\varepsilon_{\mathrm{ij}}^{\mathrm{e}} = \frac{1+\nu_{\mathrm{e}}}{E}d\sigma_{\mathrm{ij}} - \frac{\nu_{\mathrm{e}}}{E}d\sigma_{\mathrm{kk}}\delta_{\mathrm{ij}} \quad \text{where} \quad E = \frac{3(1-2\nu_{\mathrm{e}})(1+e_0)}{\kappa}p \quad (19)$$

It should be noted that any general constitutive model is applicable to the proposed method.

3.4 *Coupled behavior of the proposed method*

Figure 9 is the schematic figure of coupling behavior (unloading pass) described by the proposed model. The electrical/chemical equilibrium of the interlaminar behavior of clay crystal (Equation (8)) and the soil skeleton behavior given by a general constitutive model (Equations (16)–(19) in this study) are related to each other through the effective stress, thus both equations are solved simultaneously. The points of the coupling behavior of the proposed model are simply summarized below.

- Increasing/decreasing in the effective stress causes volume contraction/expansion on both interlaminar and soil skeleton relations.
- Increasing/decreasing in ionic strength, which is consist of "ion concentration" and "ion valence" of solution causes volume contraction/expansion respectively on the interlaminar behavior through the repulsion force (Equation (9)).
- The electro/chemical action at the interlaminar immediately occurs while the concentration of fluid (solution) changes.

Therefore, the influence of various exchangeable cations, which are difference in ion valence and ion exchange capacity, on mechanical behavior is suitably considered in the proposed method. Actually, Komine (2008) has uniformly evaluated the expansive pressures of the different type of bentonites using Equations (9) and (13).

Unlike the existing double porosity model, e.g., Alonso (1999), the ratio of deformations of micro and macro structures doesn't have to be arbitrarily

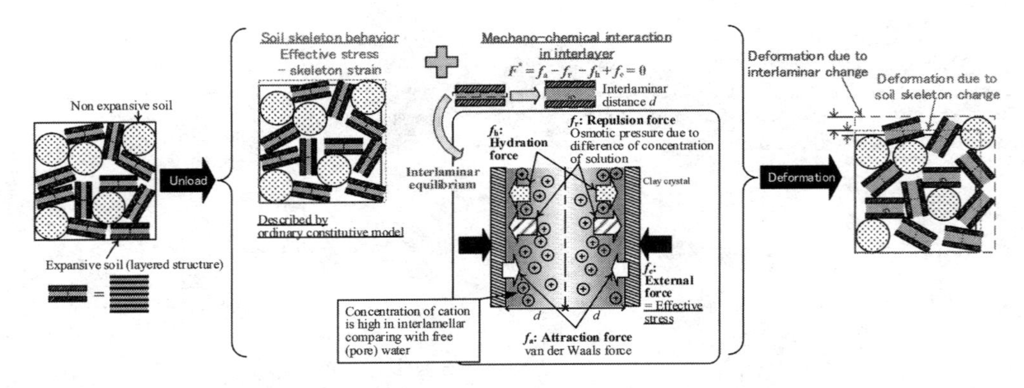

Figure 9. Soil skeleton and interlayer coupling behavior of the proposed method.

Table 1. Constitutive parameters.

Parameters of consitutitive model (soil skeleton deformation)			
λ	Compression index	–	0.90
κ	Swelling index	–	0.30
M	Critical stress ratio	–	0.435
ν_e	Poisson ratio	–	0.43
e_{NC}	Reference void ratio on NCL at p = 20kPa	–	4.5
Material specification			
e_0	Initial void ratio	–	8.0
e_0^{ss}	Initial soil skeleton ratio	–	4.5
e_0^{il}	Initial interlaminar void ratio	–	3.5
Physical constants			
A_h	Hamaker constant	J	2.200E-20
K	Boltzmann constant	J/K	1.380E-23
e'	Electronic charge	C	1.609E-19
Bentonite specification			
t	Thickness of crystal	m	9.60E-10
S	Specific surface area of crystal	m²/g	810
CEC	Cation exchange capacity	mEq/g	0.405
ν	Valence of exchangeable cation	–	1
W_0	Parameter for the hydration force	MJ/m²	1.00E-02
χ_0	Parameter for the hydration force	m	2.50E-10
Pore water specification			
T	Absolute temperature	K	293.15
ε	Electronic permittivity of pore water	C²/(J·m)	7.083E-10

given in the proposed model since that ratio can be obtained as a result of simulation and changes according to the boundary condition. For instance, in the expansive pressure test in which the displacement is fixed in any direction, the deformations of micro and macro structures are cancelled out, namely the micro structure expands and the macro structure contracts vice versa.

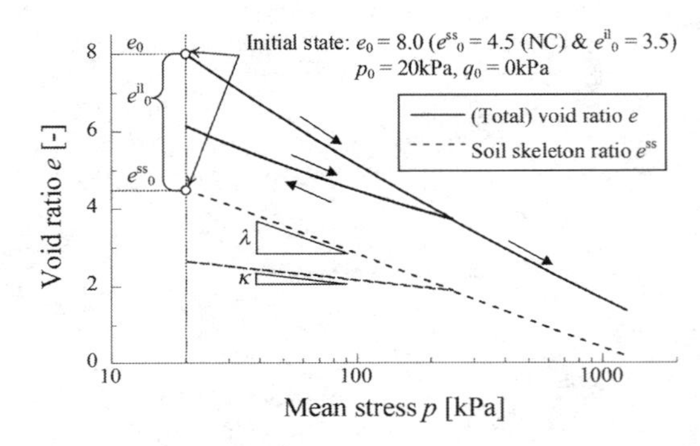

Figure 10. Calculated result of isotropic consolidation.

4 MODEL PERFORMANCE

4.1 *Constitutive parameters*

The series of simulation are carried out by the proposed model here. The analyses were carried out using parameters for Ponza bentonite listed in Table 1. The bentonite and pore water specifications are set by referring to the past work (Komine and Ogata, 1996). Only sodium ion: Na^+ is considered in this study. On the other hand, the parameters for soil skeleton behavior, i.e., the modified Cam clay model, are given by fitting the observed results of the conventional oedometer test on the distilled water saturated bentonite (Figure 2). e_{NC} is the reference void ratio on the *NCL* of the soil skeleton void: and the initial condition of simulations was the normally consolidated condition.

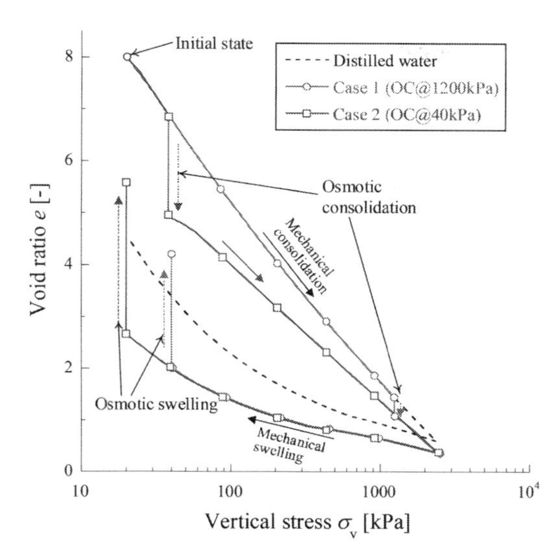

Figure 11. Calculated results of the osmotic consolidation and swelling test in oedometer test.

4.2 *Simulation results*

Firstly, the simulation of the isotropic loading/unloading behavior on the water-saturated clay is shown in Figure 10. The concentration of Na^+ of the water-saturated clay was assumed 0.01mol/l which is sufficiently small in this study. In the calculation, the mean effective stress increased to 240kPa, and then decreased to 20kPa and finally increased again to 1250kPa while the concentration of Na^+ solution was kept constant. As explained above, total behavior of the proposed model consists of the interlaminar behavior and the soil skeleton behavior. In Figure 10, the soil skeleton behavior shown by the dashed line represents a familiar compression behavior of general clay materials, namely the linear relation in logarithmic scale. On the other hand, total void ratio represents a non-linear compression line even in logarithmic scale because of the interlaminar behavior. The difference between total void ratio and the soil skeleton void ratio represents the interlaminar void ratio and it decreases with mean stress.

Next, the simulations of the oedometer tests with replacing the cell fluid are shown in Figure 11 (the corresponding observed results is Figure 2). Initial condition was same as the isotropic consolidation test in Figure 10. The dashed line is the loading-unloading behavior of the water saturated sample in oedometer test. The material parameters in Table 1 were set through the calibration of this result. From result, the highly swelling behavior during unloading path can be observed regardless of the same parameter with the isotropic consolidation in Figure 10. Not only the interlaminar behavior but also dilatancy affect this relatively large swelling behavior during unloading path in the oedometer test.

Cases 1 and 2 are the oedometer test with replacing the cell fluid. In the analysis, the replacing the cell fluid was simulated by gradually changing the concentration of solution (the replacing water to solution: $c_0^{Na+} = 0.01$mol/l $\rightarrow$ 1.0mol/l; the replacing solution to water: $c_0^{Na+} = 1.0$mol/l $\rightarrow$ 0.01mol/l). The simulation results show that the proposed model can suitably describe both osmotic consolidation and swelling behavior. Moreover, after the osmotic consolidation, the compression line and swelling line are difference from those of the water-saturated clay. However, there is the residual issue for the quantitativity.

Regarding the osmotic swelling behavior, the final states of observed results (Cases 1 and 2 in Figure 2) coincide with the unloading swelling line of the water-saturated clay. On the other hand, the simulation result of the osmotic swelling does not necessarily approach to the same void ratio of the water-saturated clay. The interlaminar behavior is reversible as explained above; however, the soil skeleton behavior is affected by the stress state and stress history since it is described by the ordinary elastoplastic constitutive model in this study. Therefore, the simulated final state after the osmotic swelling would not coincide with the swelling line of the water-saturated clay.

Figure 11 shows that the magnitude of the osmotic consolidation and swelling depends on confining pressure as seen in the observed results in Figure 2. Figures 12 and 13 summarize the confining pressure dependency on the osmotic behavior. The dashed line is the compression/swelling behavior of the distilled-water saturated clay. In particular, the results of the osmotic consolidation in Figure 12 is similar to the observed results shown (Figure 14: Di Maio, 1996) in which the magnitude of contraction gradually decrease with increasing in confining pressure.

Figures 15 and 16 respectively show the simulation results of the swelling deformation and the swelling pressure tests on several density clays in the oedometer test. The initial water saturated clays (p = 20 kPa, $c_0^{Na+} = 0.01$ mol/l) were firstly replaced its fluid to 1.0 mol/l solution in the oedometer test and then they were over-consolidated by specific vertical stresses (σ_v = 20 kPa, 245 kPa and 2450 kPa) and unloaded to σ_v = 20 kPa under constant 1.0mol/l concentration. As a result, the several density clays (e_{ini} = 5.69, 3.80 and 2.90; $c_0^{Na+} = 1.0$ mol/l) were made. Subsequently, those clays were replaced its solution to the water ($c_0^{Na+} = 0.01$mol/l) with constant 20 kPa vertical

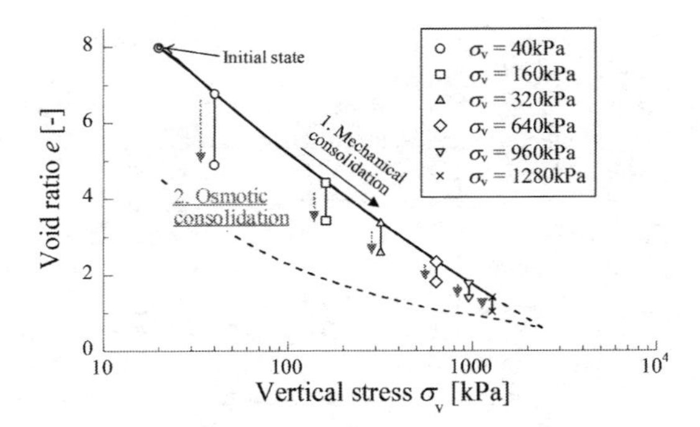

Figure 12. Calculated results of osmotic consolidation under several vertical stresses.

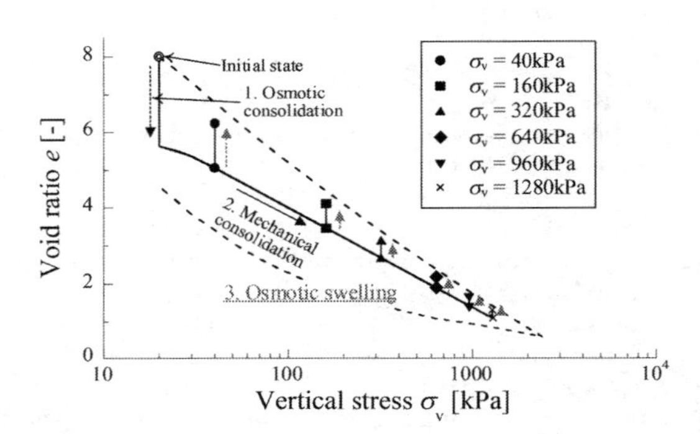

Figure 13. Calculated results of osmotic swelling under several vertical stresses.

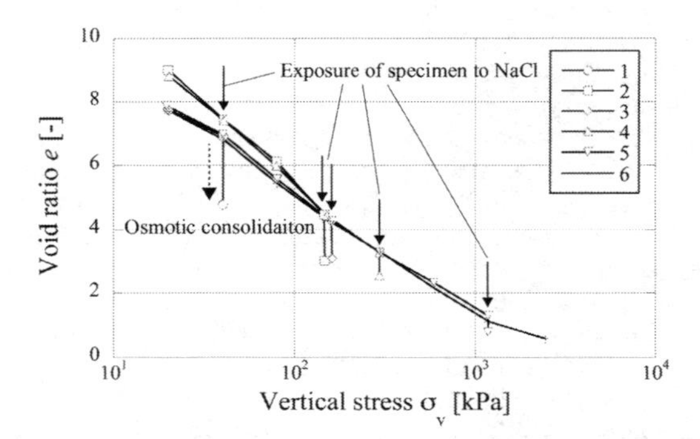

Figure 14. Observed results of osmotic consolidation under several vertical stresses (after Di Maio, 1996).

stress (the swelling deformation test) or constraining deformation (the swelling pressure test). It can be observed from figures that both the vertical strain and the vertical stress increase with decreasing in the concentration. As well as the swelling behavior of unsaturated bentonite due to water absorption, the proposed mode can describe that the higher density soil shows the larger swelling

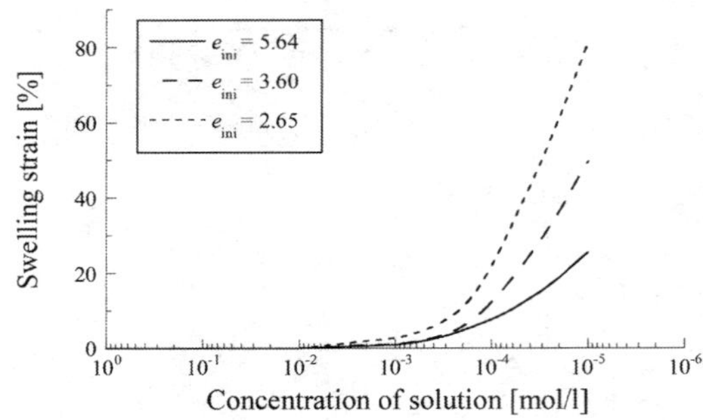

Figure 15. Calculated results of swelling deformation test.

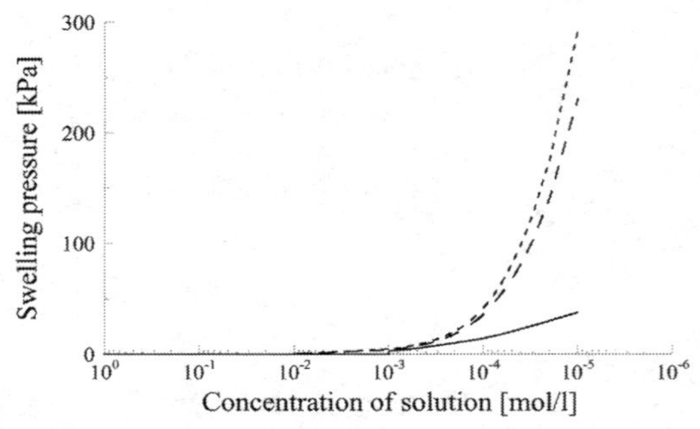

Figure 16. Calculated results of swelling pressure test.

characteristics, namely larger swelling deformation and larger expansive pressure.

5 CONCLUSION

In this paper, the modeling of the electrical/chemical phenomena on the surface of the clay mineral crystal affecting the macroscopic mechanical behavior was proposed. The interlaminar behavior (the micro-pore behavior) of clay mineral was described by the modified DLVO theory, i.e., the equilibrium governed by the osmotic repulsion force, van der Walls attraction force, hydration force and the effective stress. On the other hand, the soil skeleton behavior (the macro-pore) was modeled by the ordinary constitutive model describing the stress-strain relation, and the modified Cam clay model was applied in this study. The electrical/chemical and mechanical coupling behavior was represented by simultaneously solving both micro and macro behaviors through the effective stress. It was shown from the simulation results that the proposed model can suitably describe the typical mechanical behaviors of expansive soil depending on the fluid condition, namely the highly nonlinear swelling behavior in oedometer test, the osmotic consolidation and swelling, and its confining pressure dependency.

ACKNOWLEDGEMENT

We are grateful to the Japan Society for the Promotion of Science for its financial support with Grant-in-Aid for Research Activity Start-up.

REFERENCES

Afzal, S., Tesler, W.J., Blessing, S.K., Collins, J.M. and Lis, L.J.., 1983: Hydration force between phosphatidylcholine surfaces in aqueous electrolyte solutions, *Colloid and Interface Science*, 97(2): 303–307.

Alonso, E.E., Vaunat, J. and Gens, A., 1999: Modelling the mechanical behaviour of expansive clays. *Engineering Geology*, 54: 173–183.

Chapman, D.L., 1913: A contribution to the theory of electro capillarity, *Philosophical Magazines*, 25(6): 475–481.

Derjaguin, B. and Landau, L., 1941: Theory of the stability of strongly charged lyophobic sols and of the adhesion of strongly charged particules in solution of electrolytes, *Acta Physicochim*, URSS 14: 633–662.

Di Maio, C., 1996. Exposure of bentonite to salt solution: osmotic and mechanical effects. *Geotechnique*. 46(4): 695–707.

Gens, A. and Alonso E.E., 1992: A framework for the behaviour of unsaturated expansive clays. *Canadian Geotechnical Journal*, 14(1): 1013–1032.

Komine, H. and Ogata, N., 1996: Prediction for swelling characteristics of compacted bentonite. *Canadian Geotechnical Journal*, 33: 11–22.

Komine, H., 2008: Theoretical equations on hydraulic conductivities of bentonite-based buffer and backfill for underground disposal of radioactive wastes, *Geotechnical and Geoenvironmental Engineering*, 134(4): 497–508.

London, F., 1937: The general theory of molecular forces, *Transactions of the Faraday Society*, 33(8).

Mitchell, J.M. and Soga, K., 2005. Fundamentals of soil behavior. 3rd ed. *Joh Wiley & Sons, Inc*.

Musso, G., Romero, E., Gens, A. and Castellanos, E., 2003: The role of structure in the chemically induced deformations of FEBEX bentonite. *Applied Clay Science*. 23: 229–237.

Pashley, R.M and Israelachvili, J.N., 1984: Molecular layering of water in thin films between mica surfacdes and its relation to hydration forces, *Colloid and Interface Science*, 101(2): 511–523.

Roscoe K.H. and Burland J.B., 1968: On the generalised stress-strain behaviour of 'wet' clay, Eng. plasticity, Cambridge Univ. Press, 535–609.

Takayama, Y., Tachibana, S., Iizuka, A., Kawai, K. and Kobayashi, I., 2017: Constitutive modeling for compacted bentonite buffer materials as unsaturated and saturated porous media. *Soils and Foundations*, 57(1): 80–91.

Thyagaraj, T. and Rao, S.M., 2013: Osmotic swelling and osmotic consolidation behaviour of compacted expansive clay. *Geotechnical and Geological Engineering*, 31: 435–445.

Verwey, E.J.W. and Overbeek, J.T.G., 1948: Theory of the Stability of Lyophobic Colloids (The interaction of so particles having an electric double layer), Elsevier Publishing Company.

Numerical Methods in Geotechnical Engineering IX – Cardoso et al. (Eds)
© 2018 Taylor & Francis Group, London, ISBN 978-1-138-33198-3

Extended bounding surface model for general stress paths in practical applications

K. Bergholz
Department of Geotechnical Engineering, Federal Waterways Engineering and Research Institute (BAW), Karlsruhe, Germany

ABSTRACT: A new elastoplastic soil model has been developed for non-cohesive soils, meeting both the challenges of geotechnical problems and the demands of engineering practice. Based on the concept of bounding surface plasticity (Manzari and Dafalias 1997), it captures the state-dependence of basic soil behavioural patterns. For improving the model's reliability with regard to fundamental and more complex stress paths, the original formulation has been extended with optional features. The increased shear stiffness after load reversals as well as its strain dependent degradation within the small strain range are taken into consideration. A corresponding state variable scales the plastic hardening modulus of the kinematically hardening yield surface. An additional cap yield surface accounts for the evolution of plastic strains in constant stress ratio loading. In order to ease the model's application in routine design, a parameter calibration procedure internally determines selected bounding surface parameters. The implementation in an open source environment ensures public availability with potential for further development.

1 INTRODUCTION

The prediction of life-time settlements of infrastructural constructions puts high demands on the soil model to be applied in numerical simulations: the complex installation process and the repetitive character of live loads pose considerable challenges. Concentrating primarily on the analytical requirements of the geotechnical problem, the demands of engineering practice should not be neglected in constitutive modelling. Along these lines, a new soil model for non-cohesive soils has been developed in the theoretical framework of elastoplasticity. The development work was focused on correctly reproducing geotechnically relevant stress paths of low and higher complexity with a material model applicable in routine design. The former aspect comprises modelling of elementary behavioural patterns of soil, including for example shear related phenomena such as hardening/softening, contraction/dilation and attainment of critical state (constant volume shear strength). In addition, capturing barotropy and pycnotropy—the soil's state dependence with respect to strength, stiffness and dilatancy—enables unified modelling. Hence, with only one set of material parameters, the mechanical behaviour of a wide range of initial soil states can be simulated. Moreover, when it comes to stress paths of unconventional orientation, to load reversals or composed stress paths with changes in loading direction, additional features of soil behaviour become important. There is kinematic hardening to allow for stress/strain accumulation, an increased stiffness at small strains or an additional cap yield surface for plastic straining in constant stress ratio loading, to name only a few.

Bounding surface plasticity was identified to offer an appropriate modelling framework for the defined needs, with the potential for extensions. In view of its suitability for practical use, the model formulation should be straightforward and well documented. Parameters should be manageable in amount and determinable by means of conventional laboratory testing. And finally, its implementation should be freely accessible. The model's constitutive framework, particular advancements and measures taken for meeting practical demands are described in the following. Emphasis is put on the newly introduced features and their potential for adequately simulating soil behaviour.

2 CONSTITUTIVE FRAMEWORK

The model's constitutive framework is based on the concept of bounding surface plasticity, which has been introduced by Dafalias and Popov (1975) and Krieg (1975). Their original two surface formulation has been advanced over the years (Mróz et al. 1979, Dafalias and Herrmann 1980, Bardet 1984, among others) and obtained its four surface shape given by Manzari and Dafalias (1997). The fundamental idea consists in relating the distance between the actual stress state (on the yield surface) and a certain limit stress (on the bounding surface) to the plastic modulus in order to control the soil stiffness evolution. This principle was extended to volumetric behaviour and ultimate strength by adding two surfaces: dilatancy and critical state surface. The critical state line (CSL) is incorporated in stress (p, q) as well as stress—void ratio (p, e) space as an ultimate limit, which corresponds to the intention of setting up the constitutive relations in a critical state soil mechanics framework (Roscoe et al. 1958, Schofield and Wroth 1968). Furthermore, by introducing the state parameter according to Been and Jefferies (1985) into the model formulation, Manzari and Dafalias (1997) enhanced the bounding surface theory by the concept of state dependence with respect to density and stress. The state parameter ψ quantifies the distance of the actual state in p, e – space from its image on the CSL in terms of void ratio:

$$\psi = e - e_{cs} \tag{1}$$

Hence, its sign indicates a loose and contractive ($\psi > 0$) or a dense and dilative state ($\psi < 0$). Entering the mathematical expressions of bounding and dilatancy surface, the state parameter determines their sizes, defined by M_b and M_d. These deviatoric stress ratios quantify the inclinations of the surfaces' linear representations in the **c**ompression and **e**xtension range of p, q- space (see Fig. 1):

$$M_{c,e}^{b,d} = f\left(\psi, M_{c,e}^{c}\right) \tag{2}$$

Regarding the compression case for simplicity, the generalized expression in Eq. (2) ensures that $M_b \leq M_c$ and also that $M_d \leq$ Mc for dense and M_d \$ M_c for loose sand. Furthermore, when approaching critical state ($\psi \rightarrow 0$), $M^{b,d} \rightarrow M_c$, so that all surfaces ultimately fall onto the critical state surface. The difference between the surface's inclination $M^{b,d}$ and the current stress ratio η finally enters the formulations of the yield surface's hardening modulus H^{cone} and the dilatancy ratio, $D = \frac{\dot{\varepsilon}_p}{\dot{\varepsilon}_q}$,

$$H^{cone} = f\left(M^b - \eta\right) \text{and } D = f\left(M^d - \eta\right) \tag{3}$$

This way, it controls stiffness evolution and volumetric behaviour accordingly via hardening law $\left(\dot{\alpha} = H^{cone} \cdot \dot{\varepsilon}^{pl}\right)$ and flow rule, respectively. It is important to note that at the event of a load reversal, the reference surfaces for the calculation of the distance measures change from compression to extension (and vice versa). Consequently, the plastic stiffness increases instantaneously (due to the larger distance) and the plastic volumetric response becomes contractive, irrespective of the current soil state.

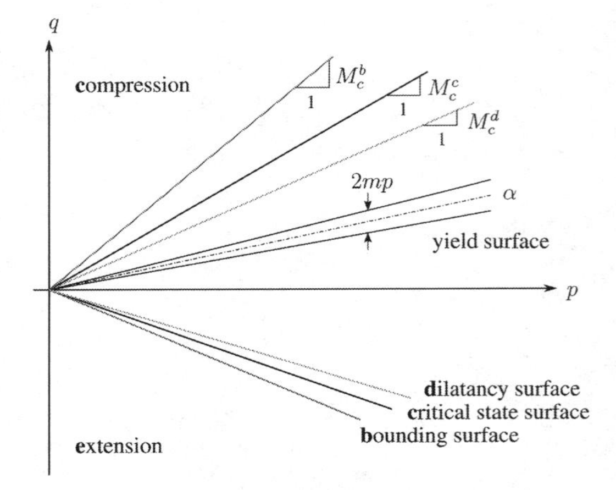

Figure 1. Model surfaces in triaxial stress space.

Due to limitation in space, Eqs. (2) and (3) will not be expanded on in more detail here; the interested reader is referred to the publication by Manzari and Dafalias (1997).

The elastic limit is described by the fourth surface, a conical yield surface with an opening m, whose centreline marks the current back stress ratio α. In triaxial stress space it is defined by the following expression:

$$f^{cone} = \eta - \alpha \pm m = 0 \tag{4}$$

The model hardens purely kinematically as a function of the back stress ratio α, which controls the rotation of the cone around its apex at the origin. The size of the cone remains constant and determines the stress range of elastic behaviour. This is particularly important, if loading is reversed, because the stress state needs to cross the entire elastic domain before plastic strains occur anew due to the violation of the yield criterion.

Since the publication of the model by Manzari and Dafalias (1997), numerous modifications followed. The original constitutive equations were reformulated in order to improve the model response with respect to experimental evidence. Likewise, this bounding surface model served as starting point for the present development, because it fulfils the initially listed elementary key requirements, though leaving a few aspects open (such as constant stress ratio loading or small strain stiffness). The following sections give information on the implemented major advancements.

For the sake of simplicity, the model by Manzari and Dafalias (1997) will be referred to as the öriginal" model.

3 DILATANCY FORMULATION

In contrast to well-known stress-dilatancy theories (Rowe 1962, Roscoe et al. 1963, Schofield and Wroth, 1968) which are mostly based on energy considerations and relate the volumetric behaviour to the stress ratio only, the bounding surface approach enriches this dependence by the soil's density. The dilatancy ratio D as implemented in the Cam Clay dilatancy formulation (Roscoe et al. 1963) used the (constant) critical state stress ratio M^c as reference value. Its substitution by the state dependent dilatancy stress ratio M^d (see Eq. (3)) allows for different volumetric responses with respect to the soil's initial void ratio.

In the original bounding surface model by Manzari and Dafalias (1997) the general expression for the dilatancy surface parameter M^d in Eq. (2) has been chosen to be a linear function of the state parameter ψ. Li and Dafalias (2000) assumed an exponential modification of the critical state stress ratio. A new proposal for the dilatancy stress ratio by Tsegaye (2014) is based on a more general approach. The stress ratio M^d is converted into a function of the criticalstate friction angle and made state dependent by multiplication with a function f_{sd}:

$$M^d_{c,e} = \frac{6 f_{sd} \sin\varphi_{cs}}{3 \mp f_{sd} \sin\varphi_{cs}} \quad (5)$$

The so called state function f_{sd} is dependent on stress, void ratio and potentially other state variables. It scales the critical state friction angle according to the actual state, ensuring that f_{sd} = 1 and hence $M^d = M^c$ at critical state ($\psi = 0$). Consequently, the two constant volume states ($D = 0$) – phase transformation and critical state—can occur at different stress ratios as documented by experimental observations (e. g. Been and Jefferies 2004).

In the present model a variation of an exponential function proposed by Li et al. (1999) is implemented, increasing its non-linearity by taking the square root of ψ.

$$f_{sd}^{new} = \exp\left(m_d \cdot \operatorname{sgn}\psi \cdot \sqrt{|\psi|}\right) \quad (6)$$

The effect of this modification with respect to the formulation by Li and Dafalias (2000), $f_{sd}^{LD} = \exp(m_d \cdot \psi)$, can be seen by looking at Fig. 2: the stronger deflection of M^d the curve towards critical state ($\psi \to 0$) causes a larger difference (a) and hence more intense dilation and contraction, respectively (b).

4 CAP YIELD SURFACE

The constitutive formulation of the original model was built on the premise that sand primarily deforms under shear; plastic strains resulting from loadings along constant stress ratio stress paths such as oedometric or isotropic compression were considered negligible. The choice of an open conical yield surface accommodated this simplification, producing merely elastic strains in constant stress ratio loading. However, in case of very loose sands and high pressures leading to grain crushing, irreversible deformations do occur. In order to take account of these loading conditions, a second yield surface capping the open cone of the original bounding surface model is introduced. Although the numerical treatment of two yield surfaces is more intricate than for example using a closed form, as proposed by Taiebat and Dafalias (2008), a separate loading mechanism for the cap has a decisive advantage: it can be calibrated independently from the conical yield surface using 1D or isotropic compression test data, where the hardening mechanisms are decoupled.

In triaxial stress space the cap yield surface is expressed by an ellipse of the form:

$$f^{cap} = \frac{q^2}{M_{cap}^2} + p^2 - p_c^2 = 0 \quad (7)$$

It is centred at the origin, intersects the p-axis at preconsolidation stress p_c and its steepness is controlled by the parameter M_{cap}. Its multiaxial equivalent is shown in Fig. 3a. In contrast to the cone, the cap is associated, so that $f^{cap} = g^{cap}$. The shared parameter M_{cap} is not a constant, but a function of earth pressure coefficient at rest K_0, oedometer stiffness E_{oed} and elastic properties (E, ν). This way it ensures a stress path inclination in p, q-space corresponding to K_0, if one-dimensional compression conditions apply. The hardening rule is adopted accordingly as to realize a stress-strain evolution in accordance with the user defined stress and void ratio dependent E_{oed}.

The challenge of handling two interacting yield surfaces lies in treating loading cases, where both cone and cap yield conditions are violated. Numerically, this corner problem is solved by applying an integration algorithm for singular yield surfaces according to de Borst (1987). It is based on Koiter's rule Koiter (1987), stating that

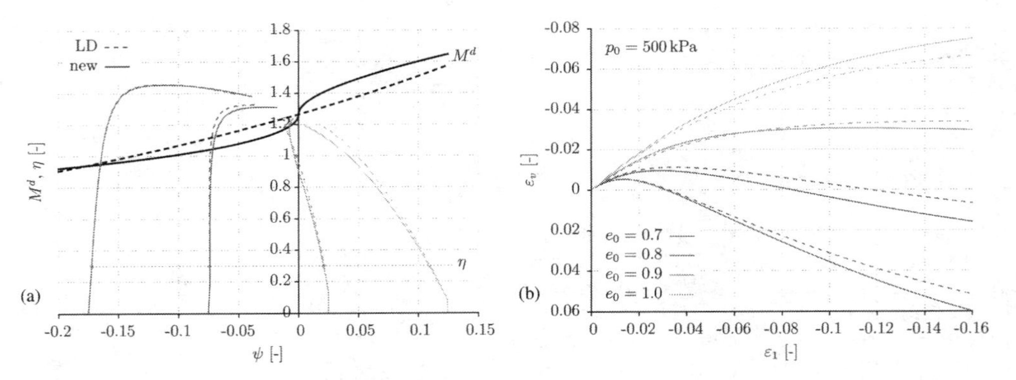

Figure 2. Triaxial compression tests (p_0 = 500 kPa) of different initial void ratios e_0 applying two different state functions f_{sd}^{new}, f_{sd}^{LD}): (a) evolution of M^d (bold black) and η (coloured lines) with ψ, (b) strain evolution ε_1-ε_v.

the plastic strain rate in case of two active yield surfaces is composed of both contributions.

However, simulations of conventional triaxial tests starting from a normally consolidated isotropic stress state (cone and cap active) have shown that the summation of the irreversible strain portions results in a soil response that is distinctly softer than experiments on sand might justify. This observation gave rise to modifying the flow rule of the cap yield surface in a way that it remains very stiff when hardening under shear. Consequently, the cap does not contribute to the plastic deformation in non-constant stress ratio loadings. The applied constitutive measure is to make the hardening modulus dependent of stress ratio rate $\dot{\eta}$ as well as the strain rate direction in terms of volumetric and deviatoric portions $\dot{\varepsilon}_p$ and $\dot{\varepsilon}_q$:

$$H^{cap} = h\left(\dot{\eta}, \dot{\varepsilon}_p, \dot{\varepsilon}_q\right) \cdot H_0^{cap} \tag{8}$$

H_0^{cap} corresponds to the basis cap hardening modulus resulting from the considerations discussed above. The latter dependence is required for ensuring that aside from constant stress ratio paths, proportional strain paths, featuring constant rates of principal strain values ε_1: ε_2: ε_3, can be modelled the same way. The simulation of an oedometric compression test (1: 0: 0) for example, where $\dot{\eta}$ approximates zero only $(\dot{\eta} \approx 0)$, would cause problems otherwise. The auxiliary function $h\left(\dot{\eta}, \dot{\varepsilon}_p, \dot{\varepsilon}_q\right)$ in Eq. (8) is chosen to be an exponential relation of the form

$$h\left(\dot{\eta}, \dot{\varepsilon}_p, \dot{\varepsilon}_q\right) = C^{f\left(\dot{\eta}, \dot{\varepsilon}_p, \dot{\varepsilon}_q\right)} \tag{9}$$

which acts as some kind of switching function without discontinuities. C is a large number (e. g. 1000) and the exponent takes values of ≈ 1 for stress ratio changes or ≈ 0 for proportional stress and strain paths, resulting in $h\left(\dot{\eta}, \dot{\varepsilon}_p, \dot{\varepsilon}_q\right) = C$ or $h\left(\dot{\eta}, \dot{\varepsilon}_p, \dot{\varepsilon}_q\right) \approx 1$. Consequently, the hardening modulus is either artificially increased in order to produce very little plastic deformation, or remains unchanged at its basis value H_0^{cap}.

5 SMALL STRAIN STIFFNESS

According to the bounding surface concept, it is the soil state in terms of stress and void ratio that controls the stiffness evolution in elastoplastic straining, which is reflected in the state dependent formulation of the constitutive equations (Eqs. (1)-(3)). However, besides pressure and density, it is also the level of strain that has an influence on the soil's stiffness, particularly in the smaller strain range (Burland 1989, Atkinson and Sällfors 1991, Mair 1993, among others). A large quasi elastic shear stiffness at very small strains (e. g. straight after a load reversal, $< 10^{-6}$) is followed by a steady stiffness decay at small strains, down to a certain minimum within the range of medium to large strains ($> 10^{-3}$) – a behaviour described by the shear stiffness degradation curve. Since many geotechnical problems involve deformations within the small strain domain—e. g. retaining walls, foundations and tunnels—small strain stiffness is a feature of soil behaviour worth considering, if the realistic prediction of displacements is of importance.

In the original model the transition from elastic to elastoplastic behaviour is accompanied by a sudden stiffness change. Its point of appearance is primarily dependent on stress quantities, since the elastic domain is defined in stress space (see broken line in Fig. 4a). In order to adopt the described degradation characteristic, it is hence desirable to smoothen the abrupt reduction of the

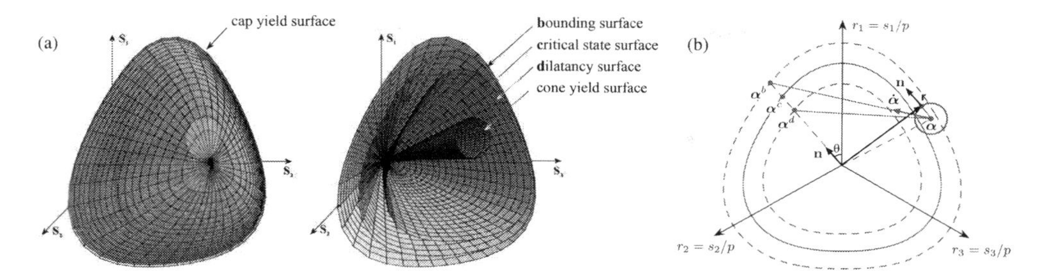

Figure 3. Model surfaces of the bounding surface model with cap extension: (a) 3D and (b) 2D (after Taiebat 2009) representation in multiaxial stress space.

overall stiffness at the onset of plastic straining and to incorporate a strain dependence. In contrast to previous suggestions following the strategy of adjusting the elastic stiffness (Papadimitriou and Bouckovalas 2002, Benz 2007), the hardening law of the conical yield surface is altered. An additional factor is introduced that scales the hardening modulus as a function of shear strain:

$$H^{cone} = h_{ss}(\gamma) \cdot \mathrm{H}_0^{cone} \quad (10)$$

H_0^{cone} corresponds to the bounding surface related hardening modulus presented in Eq. (3).

The key role of h_{ss} is to trace the shear strain evolution and modify the hardening modulus accordingly (Fig. 4). It takes high values at very small strains for enlarging H^{cone} in order to prevent the overall stiffness from dropping too quickly from the quasi-elastic level of the secant shear modulus. Towards higher strains, $h_{ss} \to 1$, so that the stiffness is progressively controlled by the bounding surface hardening mechanism only. Furthermore, it also needs to identify load reversals as indicator for resetting the stiffness degradation process.

Different approaches have been developed for tracking the shear strain evolution including load reversals. There are stress based methods applied by elastoplastic models (mostly kinematically hardening), which follow loading history in stress space with some sort of memory surface(s), e. g. Al-Tabbaa and Muir Wood (1989), Puzrin and Burland (1998) and Papadimitriou and Bouckovalas (2002). Alternatively, loading can be memorized in terms of strain, as for example proposed by Simpson (1992) in his brick model (implemented in an elastoplastic constitutive framework by Länsivaara 1999), the small strain overlay by Benz (2007) or the hypoplastic concept of intergranular strain (Niemunis and Herle 1997). Both strategies have their justification. Since stiffness degradation in the range of small deformations is a strain dependent phenomenon,it seems evident to remain in strain space. On the other

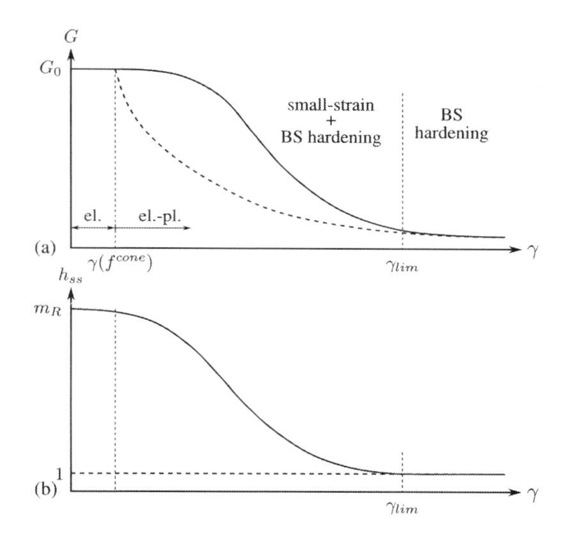

Figure 4. Degradation of (a) the secant shear modulus G and (b) the small strain stiffness factor h_{ss} with () and without () small strain hardening option.

hand, the (extended) Masing rules (Masing 1926, Pyke 1979, among others), which are considered to describe important features of cyclic soil behaviour, are formulated with reference to stress quantities. Different models have been implemented and tested; the strain based contour model is exemplarily presented in the following.

The strain tracing mechanism is reduced to the deviatoric component of strain, more precisely: its change with respect to the last shear reversal. The history is memorized by nested contours in the shape of circles. These expand and shrink in diameter (representing experienced shear strain limits) and are defined by their centre coordinates and the coordinates of their fixed end. The latter point corresponds to the last shear reversal, the moving point on the opposite side of the contour represents the current shear strain and consequently, the centre can be located and stored in the memory as auxiliary point. Crucial for tracing strain history is the identification of reversals in loading direction. They may be found by continuously checking whether the increment $\dot{\mathcal{X}}_e$, quantifying the change in shear strain (with respect to the last shear reversal) from one calculation step to the next, changes its sign:

$$\dot{\mathcal{X}}e = \dot{\mathcal{X}}_e^i - \dot{\mathcal{X}}_e^{i-1} \quad \textbf{with} \quad \mathcal{X}_e = \Delta\gamma^{SR} = \gamma - \gamma^{SR} \tag{11}$$

The evolution of contours is displayed in Fig. 5: the actual state is traced by the active strain contour (bold red line). In primary loading, a particular case, loading starts with one sphere being centred at zero (0), expanding equally to both sides (centre point remains fixed). As soon as a shear reversal is detected (1), a new sphere is created, becoming the new active contour that is fixed to the updated reversal point and starts expanding in the opposite direction. The two final coordinates of the previous deactivated contour are shifted one level down in the memory stack. At the next load reversal (2) the same procedure recurs (Fig. 5a): the previous contour becomes inactive, is stored at the top position of the memory vector and a newly created sphere traces the current strain evolution (3). Now imagine that the innermost contour expands until it reaches the closest inactive sphere belonging to the previous strain cycle. In this case the actual circle is erased from the active strain history and the most recent contour from the memory stack is reactivated (4, Fig. 5b). It is reloaded into the active contour vector and continues to expand (5, Fig. 5c). This way a previous strain path can be intersected and resumed, which may also apply to the primary loading path.

The radius of the current contour, which can be easily derived from its coordinates, reflects the actual shear strain and makes the link to the plastic stiffness. By relating the current shear strain with reference to the last shear reversal, $\Delta\gamma^{SR}$, to the shear strain limit γ_{lim}, the small strain factor h_{ss} is interpolated between 1 and a maximum value m_R:

$$h_{ss} = 1 + (m_R - 1)\left(1 - \min\left(\frac{\Delta\gamma^{SR}}{\gamma_{lim}}, 1\right)\right)^{\zeta} \tag{12}$$

m_R corresponds to the maximum multiplier of the plastic stiffness applying straight after shear reversal. The exponent ζ controls the curvature of the stiffness degradation function.

Applying the described stiffness degradation approach to the simulation of triaxial compression from very small to large strains, results in a secant shear stiffness reduction as depicted in Fig. 6. The comparison to experimental data on Toyoura sand by Kokusho (1980) gives a satisfactory fit. It has to be noted that depending on the choice of the yield surface's parameter m, the transition from elastic to elastoplastic can be recognized as a marked drop in stiffness. Due to the linearly increasing opening of the elastic cone with mean stress, this salient point is shifted to larger strains for growing levels of confining stress. This phenomenon can be remedied by shrinking the yield surface to a line with a very small value for m and adapting m_R and ζ accordingly.

The bounding surface concept in combination with the tracking mechanism of shear history (by creation, erasement and resumption of contours) and the degradation rule of plastic stiffness in Eq. (12), allows to reproduce the characteristics of the stress-strain hysteresis due to reversed loadings as described by the extended Masing rules. These include a quasi-elastic tangent shear modulus after each shear reversal, a twice faster stiffness decay in primary loading (compared to reloadings) and the continuation of formerly abandoned stress-strain paths at intersections.

In its strict sense, particularly the latter observation refers to stress space (intersection with respect to stress or stress ratio). If the exact compliance with the extended Masing rules is of importance, the strain contours could be transferred into stress contours by tracking changes of the deviatoric stress ratio η instead of the shear strain γ. The limit shear strain γ_{lim} needs to be converted into a limit stress ratio η_{lim}, restricting the change of deviatoric stress ratio with reference to the last shear reversal. The deviatoric stress component is approximated by the product of the elastic shear modulus at shear reversal and the limit shear strain:

$$\eta_{lim} \approx a_{lim} = \frac{2G^{SR}\gamma_{lim}}{p^{SR}} \tag{13}$$

a_{lim} is an auxiliary mean to assess the non-linearity of stiffness degradation and can be approximated by $\frac{1}{3}$ (following the proposition of Papadimitriou and Bouckovalas 2002).

6 FURTHER NOTES ON IMPLEMENTATION

Having given information on theoretical aspects of the new model in the previous sections, it should be noted that only a limited selection of extensions could be presented. Another interesting feature should only be mentioned here. In analogy to the small strain stiffness mechanism, a function based on the total dissipated energy adjusts the

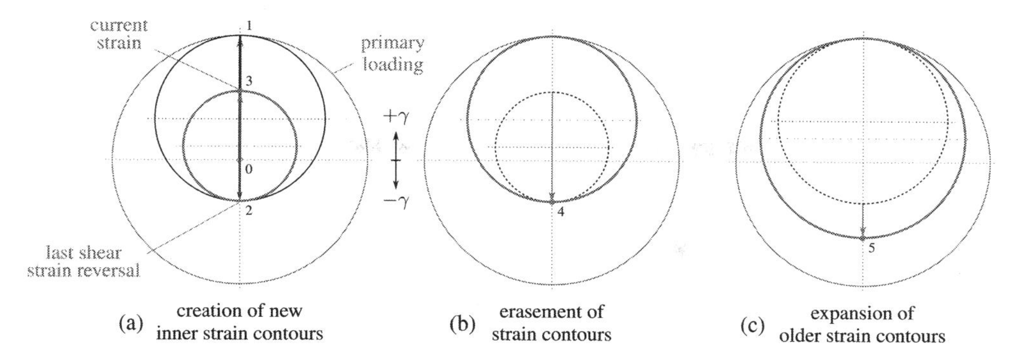

Figure 5. Evolution of strain contours with shear loading including shear reversals.

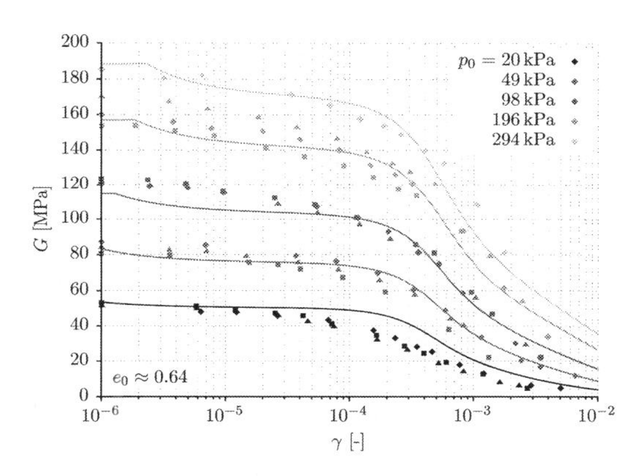

Figure 6. Small strain stiffness degradation for Toyoura sand with an initial density of $e_0 \approx 0.64$ and different levels of initial stress: experimental data by Kokusho (1980), simulation with m = 0.03, $\gamma_{lim} = 2 \cdot 10^{-3}$, $m_R = 17$, $\zeta = 12$.

hardening rule with respect to peculiarities in the accumulation behaviour (such as the shakedown phenomenon or cyclic mobility).

In view of the requirements in engineering routine, it should be noted that the model implementation is structured in a modular way. This allows to either use the basic bounding surface formulation or to additionally activate particular extensions, if special features of soil behaviour need to be taken into account. Consequently, the complexity of the model and the amount of parameters can be adapted to the intricacy of the geotechnical problem.

Most material parameters are determinable by means of conventional laboratory testing, only few require trial and error calibration. In order to ease the determination of parameters related to the bounding surfaces concept, which are only indirectly correlated to experimental data, an auxiliary routine has been developed for supporting the calibration process. By means of other established material constants and alternative measures from laboratory tests, these parameters are internally calibrated with a derivative free evolutionary optimization algorithm.

The extended bounding surface model is implemented in the standardized user defined soil model format (UMAT/UDSM) and will be freely available in an open source context.

7 CONCLUSIONS

A newly developed material model based on bounding surface plasticity according to Manzari and Dafalias (1997) has been presented. Due to the state dependent formulation of the constitutive equations, elementary behavioural patterns such as hardening/softening or contraction/dilation can be reproduced with one set of material parameters, disregarding the soil's initial state. A selection of implemented extensions enriching the original model has been described, including:

- an improved dilatancy formulation,
- a cap yield surface for irreversible deformations along constant stress ratio paths, and
- a strain dependent degradation mechanism for the shear stiffness at small strains, combined with a contour based memory for strain history.

The calibration of bounding surface specific parameters is facilitated by an auxiliary internal routine using alternative user input with a stronger link to experimental data. The implementation in an open source environment ensures public availability with potential for further development.

REFERENCES

Al-Tabbaa, A. & D. Muir Wood (1989). An experimentally based 'bubble' model for clay. In S. Pietruszczak and G.N. Pande (Eds.), *3rd International Symposium on Numerical*

Models in Geomechanics: Proceedings, London and New York, pp. 91–99. Elsevier Applied Science.

Atkinson, J.H. & G. S¨allfors (1991). Experimental determination of soil properties. In *10th European Conference on Soil Mechanics and Foundation Engineering: Proceedings*, Volume 3, pp. 915–956.

Bardet, J.P. (1984). *Application of plasticity theory to soil behavior: a new sand model*. PhD thesis, California Institute of Technology, Pasadena.

Been, K. & M. Jefferies (2004). Stress dilatancy in very loose sand. *Canadian Geotechnical Journal 41*(5), 972–989.

Been, K. & M.G. Jefferies (1985). A state parameter for sands. *Géotechnique 35*(2), 99–112.

Benz, T. (2007). *Small-strain stiffness of soils and its numerical consequences*. PhD thesis, University of Stuttgart, Stuttgart.

Burland, J.B. (1989). Ninth Laurits Bjerrum Memorial Lecture: Small is beautiful - the stiffness of soils at small strains. *Canadian Geotechnical Journal 26*(4), 499–516.

Dafalias, Y.F. & H.J. Herrmann (1980). A bounding surface soil plasticity model. In G.N. Pande and O.C. Zienkiewicz (Eds.), *Soils under cyclic and transient loading: Proceedings*, Rotterdam, pp. 335–345. Balkema.

Dafalias, Y.F. & E.P. Popov (1975). A model of nonlinearly hardening materials for complex loading. *Acta Mechanica 21*(3), 173–192.

de Borst, R. (1987). Integration of plasticity equations for singular yield functions. *Computers & Structures 26*(5), 823–829.

Koiter,W.T. (1953). Stress-strain relations, uniqueness and variational theorems for elastic-plastic materials with a singular yield surface. *Quarterly of Applied Mathematics 11*(3), 350–354.

Kokusho, T. (1980). Cyclic triaxial test of dynamic soil properties for wide strain range. *Soils and Foundations 20*(2), 45–60.

Krieg, R.D. (1975). A practical two surface plasticity theory. *Journal of Applied Mechanics 42*(3), 641.

L¨ansivaara, T. (1999). *A study of the mechanical behavior of soft clay*. PhD thesis, Norwegian University of Science and Technology, Trondheim.

Li, X.S. & Y.F. Dafalias (2000). Dilatancy for cohesionless soils. *Géotechnique 50*(4), 449–460.

Li, X.-S., Y.F. Dafalias, & Z.-L. Wang (1999). State-dependant dilatancy in critical-state constitutive modelling of sand. *Canadian Geotechnical Journal 36*(4), 599–611.

Mair, R.J. (1993). Developments in geotechnical engineering research: Application to tunnels and deep excavations: Unwin memorial lecture. *Proceedings of the Institution of Civil Engineers - Civil Engineering 97*(1), 27–41.

Manzari, M.T. & Y.F. Dafalias (1997). A critical state twosurface plasticity model for sands. *Géotechnique 47*(2), 255–272.

Masing, G. (1926). Eigenspannungen und Verfestigung beim Messing. In *2nd International Congress of Applied Mechanics: Proceedings*, pp. 332–335.

Mr´oz, Z., V.A. Norris, & O.C. Zienkiewicz (1979). Application of an anisotropic hardening model in the analysis of elasto–plastic deformation of soils. *Géotechnique 29*(1), 1–34.

Niemunis, A. & I. Herle (1997). Hypoplastic model for cohesionless soils with elastic strain range. *Mechanics of Cohesive-frictional Materials 2*(4), 279–299.

Papadimitriou, A.G. & G.D. Bouckovalas (2002). Plasticity model for sand under small and large cyclic strains: a multiaxial formulation. *Soil Dynamics and Earthquake Engineering 22*(3), 191–204.

Puzrin, A.M. & J.B. Burland (1998). Non-linear model of small-strain behaviour of soils. *Géotechnique 48*(2), 217–213.

Pyke, R.M. (1979). Nonlinear soil models for irregular cyclic loadings. *Journal of the Geotechnical Engineering Division 105*(6), 715–726.

Roscoe, K.H., A.N. Schofield, & A. Thurairajah (1963). Yielding of clays in states wetter than critical. *Géotechnique 13*(3), 211–240.

Roscoe, K.H., A.N. Schofield, & C.P. Wroth (1958). On the yielding of soils. *Géotechnique 8*(1), 22–53.

Rowe, P.W. (1962). The stress-dilatancy relation for static equilibrium of an assembly of particles in contact. *Proceedings of the Royal Society A: Mathematical, Physical and Engineering Sciences 269*(1339), 500–527.

Schofield, A.N. & P. Wroth (1968). *Critical state soil mechanics*. European civil engineering series. London and New York: McGraw-Hill.

Simpson, B. (1992). Retaining structures: Displacement and design. *Géotechnique 42*(4), 541–576.

Taiebat, M. (2009). *Advanced elastic-plastic constitutive and numerical modeling in geomechanics*. PhD thesis, University of California Davis, Davis.

Taiebat, M. & Y.F. Dafalias (2008). Sanisand: Simple anisotropic sand plasticity model. *International Journal for Numerical and Analytical Methods in Geomechanics 32*(8), 915–948.

Tsegaye, A.B. (2014). *On the modelling of state-dilatancy and mechanical behaviour of frictional material*. PhD thesis, Norwegian University of Science and Technology, Trondheim.

Numerical Methods in Geotechnical Engineering IX – Cardoso et al. (Eds)
 ISBN 978-1-138-33198-3

A time dependent constitutive model for soft clay based on nonstationary flow surface theory

M. Rezania & M. Mousavi Nezhad
School of Engineering, The University of Warwick, Coventry, UK

H. Nguyen
Department of Civil Engineering, The University of Nottingham, Nottingham, UK

ABSTRACT: This paper presents a time dependent constitutive model, namely NSFS-MCC, developed based on nonstationary flow surface theory and in the framework of critical state soil mechanics. The model is capable of capturing the effect of time and loading rate on the long term behaviour of soft soils. Model description is presented before some qualitative and quantitative simulations to verify the model's capacity. Determination of model parameter values via standard laboratory test data is also introduced. The new model has six parameters in total, among which two parameters refer to the creep effects and the remaining four are associated with the Modified Cam-Clay (MCC) model. Three different loading scenarios, over an arbitrary soil type, are simulated to verify the model's performance in capturing different aspects of time and rate effects. Furthermore, experimental data from tests on two different clays, namely Haney clay and Osaka clay, are used to validate the prediction capability of the NSFS-MCC model at element test level. The numerical modelling results show that the new model can capture well the time and rate dependent nature of clayey soils.

1 INTRODUCTION

Long-term deformation of soft soils can potentially cause excessive damages to geo-structures. Developing a practical model, which can provide accurate prediction of natural soil behaviour, has been the focus of many geotechnical engineers and researchers. Several constitutive models have been developed since the framework of critical state soil mechanics was introduced by Roscoe and Schofield (1963). While a number of advanced models has attempted to capture different important features of soft soil behaviour, in particular regarding fabric anisotropy and inter-particles bonding (e.g. Gens & Nova 1993, Koskinen et al. 2002, Taiebat et al. 2010), they cannot describe the time dependent aspects of soft soil behaviour. Over the past 50 years the time dependency of soil response has been studied via various approaches; the resulting developed models can be categorised into different groups, most notably empirical models (e.g., Bjerrum 1967, Garlanger 1972, Sing & Mitchell 1968, Tavenas et al. 1978)), or overstress-based models (e.g., Yin & Karstunen, 2008, Rezania et al. 2016). The empirical time dependent models often require some restricting conditions and hence are not favoured for application in many practical problems; the overstress type models on the other hand, despite the simplicity in their mathematical framework, are only able to capture the strain rate effects and not the true creep response of the soils. Creep deformation capacity of natural clays is known as one of the most important and influential features of their behaviour. This paper proposes a time dependent model based on the nonstationary flow surface (NSFS) theory (Olszak & Perzyna, 1964), namely NSFS-MCC, which is capable of capturing the strain rate effects and creep response of soft soils. In particular the developed model is able to simulate tertiary creep response, which is an important phase of creep behaviour leading to creep failure of soft soil deposits. In the following, a description of model is presented before some qualitative and quantitative simulation are illustrated to show the model's capacity in capturing strain rate effects on strength and yield stress, as well as simulating creep, stress relaxation and creep rupture. The determination of model parameters using standard laboratory test data is also presented before model validation at element level.

2 MODEL DESCRIPTION

The proposed model is developed based on the NSFS theory (Olszak and Perzyna, 1964). This theory is a further development of inviscid elasto-plastic theory in which a time variable parameter t is introduced in the yield surface equation to obtain the simultaneous description of strain hardening and the effect of time. It means that the yield surface changes in every moment; the so called nonstationary yield surface is expressed in the following form

$$F\left(\sigma_{ij}^{'},\varepsilon_{ij}^{vp},t\right)=0 \quad (1)$$

where F depends on the current stress state $\sigma_{ij}^{\prime}$, viscoplastic strain ε_{ij}^{vp} and a time variable parameter.

According to classical elasto-viscoplasticy theory, the total strain increment may be additively decomposed into an elastic strain increment $d\varepsilon_{ij}^{e}$ and a viscoplastic increment $d\varepsilon_{ij}^{vp}$

$$d\varepsilon_{ij}=d\varepsilon_{ij}^{e}+d\varepsilon_{ij}^{vp} \quad (2)$$

where the superscript e and vp denotes the elastic component and viscoplastic component, respectively. The elastic strain increment can be determined, following the modified Cam-Clay (MCC) model formulation, as

$$d\varepsilon_{ij}^{e}=D_{ijkl}^{-1}d\sigma_{kl}^{'} \quad (3)$$

where D_{ijkl} is the elastic tensor.

The viscoplastic strain increment is obtained via a flow rule

$$d\varepsilon_{ij}^{vp}=\lambda\frac{\partial G}{\partial\sigma_{ij}^{'}} \quad (4)$$

where G is viscoplastic potential or flow surface and λ is a non-negative viscoplastic multiplier. The proposed model employs the associated flow rule, i.e. G = F where F is yield surface, or a nonstationary flow surface in this case. The nonstationary flow surface in NSFS-MCC model is proposed as

$$F=\mu\ln\{1+\frac{\dot{v}_{0}.t}{\mu}\exp(\frac{f}{\mu})\}-\varepsilon_{v}^{vp}=0 \quad (5)$$

where μ is a creep parameter, related to coefficient of secondary compression $C_{\alpha e}$ via $\mu=\left\{C_{\alpha e}/\left[ln10\left(1+e_{0}\right)\right]\right\}$ with e_0 being the initial void ratio, t is time parameter, $\dot{v}_0$ is the initial volumetric strain rate, ε_{v}^{vp} is the volumetric viscoplastic strain and f is a function of current stress state, derived from yield surface of the MCC model

$$f=\frac{\lambda-\kappa}{1+e_{0}}\ln(\frac{p'}{p'_{0}})+\frac{\lambda-\kappa}{1+e_{0}}\ln\left(\frac{M^{2}+\eta^{2}}{M^{2}}\right) \quad (6)$$

where λ and κ are the compression and swelling indices from isotropic compression test, respectively; p' is the mean effective stress, η is stress ratio, M is critical state value of the stress ratio and p_0' is the initial size of the yield surface.

Similar to elasto-plasticity, viscoplastic behaviour is trigged until the stress state reaches the initial yield surface and this leads to the continuous expansion of the yield surface. This consistency condition yields an expression

$$dF=\frac{\partial F}{\partial\sigma_{ij}^{'}}d\sigma_{ij}^{'}+\frac{\partial F}{\partial\varepsilon_{v}^{vp}}d\varepsilon_{v}^{vp}+\frac{\partial F}{\partial t}=0 \quad (7)$$

Combining (4) and (7), the viscoplastic multiplier can be derived as

$$\lambda=-\frac{\frac{\partial F}{\partial\sigma_{ij}^{'}}d\sigma_{ij}^{'}+\frac{\partial F}{\partial t}}{\frac{\partial F}{\partial\varepsilon_{v}^{vp}}\frac{\partial F}{\partial\sigma_{ij}^{'}}} \quad (8)$$

Equation (8) has two terms, one represents the stress loading and the other one refers to the time effect. It implies that the viscoplastic strain occurs even when the stress or strain is constant. That is why the model is capable of describing the creep behaviour and stress relaxation response of natural clays.

The NSFS-MCC model requires a total of six parameters. They can be categorised into the following two groups:

- Isotropic parameters which are similar to those of the MCC model.
- Creep parameters.

The MCC-based parameters include Poisson's ratio (ν'), M, λ, and κ. In terms of the creep parameters, the creep index, μ, is determined based on the coefficient of secondary compression, $C_{\alpha e}$, which is the slope of secondary compression line (assumed constant) plotted in void ratio and logarithm of time space. The other creep parameter is called the initial volumetric strain rate, $\dot{v}_0$, at the end of primary consolidation and it can be determined from $C_{\alpha e}$ as

$$\dot{v}_{0}=\frac{C_{\alpha e}}{t_{0}} \quad (9)$$

where t_0 can be estimated by Taylor's square root method from Terzaghi's theory of consolidation

$$t_{0}\approx t_{90}=\frac{H^{2}T_{v}(90\%)}{C_{v}} \quad (10)$$

Table 1. Parameters values of NSFS-MCC model adopted for three types of clays.

Model constant		Hypothetical	Osaka	Haney
Elasticity	κ	0.02	0.02	0.05
	ν	0.2	0.3	0.25
Critical state	M	1.51	1.43	1.28
	λ	0.3	0.25	0.32
Creep	μ	0.0115	3.95e-2	4e-3
	$\dot{\upsilon}_0$ (sec^{-1})	3.5e-6	4.63e-7	2.78e-7
Initial void ratio	e_0	2	1.4	2
Over consolidation ratio		1	1	1

in which C_v is coefficient of consolidation, H is the drainage distance, and T_v (90%) is a constant.

3 MODEL PERFORMANCE

To illustrate the capabilities of the model in capturing the rate and time dependency of soil behaviour, in the following, three different hypothetical loading scenarios are considered and the model performance in each scenario is numerically examined. Furthermore, the experimental data from tests over two different clays, namely Haney clay (Vaid & Campanella 1977) and Osaka clay (Nakano et al. 2000) are used to verify the model performance at element level. The values of model constants and state variables used for the different soil types analysed in this paper are summarised in Table 1.

3.1 *Strain rate effect on undrained shear strength*

Simulations of the model prediction for undrained compression of a hypothetical clay at four different strain rates of 1%/min, 0/1%/min, 0.01%/min and 0.001%/min have been carried out. The results, presented in Figure 1, show that the model is able to capture the effect of different strain rates and predict different undrained shear strengths for the soil. With increasing strain rates, the soil shows higher strength values, which is an expected soil response backed up by many experimental observation.

3.2 *Strain rate effect on preconsolidation pressure*

To study the effect of strain rate on the yield stress value, simulations of three constant strain rate tests under one dimensional compression condition have been carried out. The results are displayed in Figure 2. It can be seen that the higher the strain rate is, the greater the value of apparent preconsolidation pressure becomes. For example in these simulations, the yield stress of a sample

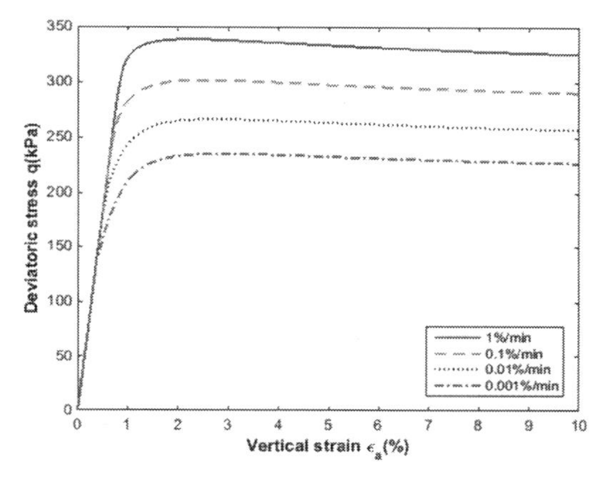

Figure 1. Undrained compression tests at different strain rates.

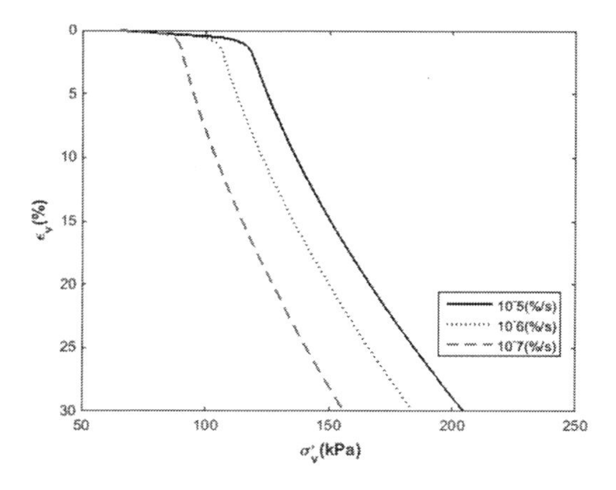

Figure 2. Strain rate effect on yield stress.

which is compressed at a constant strain rate of 1×10^{-5} (%/s) is nearly 125 kPa; whereas, the sample compressed at a constant strain rate of 1×10^{-7} (%/s) yields at a stress of about 78 kPa.

3.3 *Creep rupture*

Creep rupture, which occurs at the tertiary creep stage, is the failure of soil as creep rate increases at a constant deviatoric stress. Capturing this phenomenon is vital for modelling the long-term stability of embankment and slopes. Simulations of creep deformation at three different constant deviatoric stresses have been conducted to study the creep rupture occurrence of a hypothetical clay. The results are shown in Figure 3. It can be seen that at a low deviatoric stress (193 kPa), creep rate has a slight increase and then tends to a rather stable state. However, at higher deviatoric stress values of 267 kPa or 329 kPa at some stage accelerating strain rate occurs signalling the tertiary creep phase. This confirms the elegant capacity of the model in capturing the eventual long term rupture failure of natural clay deposits.

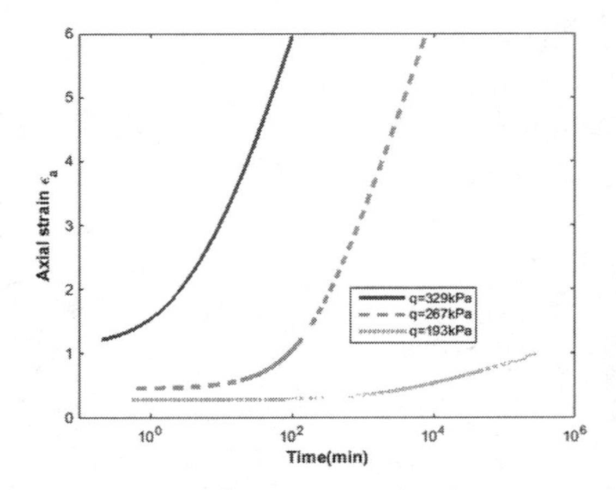

Figure 3. Creep rupture simulations at different constant deviatoric stresses.

3.4 *Validation at element test level—undrained traixial shearing tests on Hanney clay*

Vaid and Campanella (1977) conducted a series of undrained triaxial tests on undisturbed saturated sensitive marine clay named as Haney clay. It is a grey silty clay with the liquid limit $w_L = 40\%$, the plastic limit $w_P = 26\%$, the maximum past pressure of about 340 kPa, and a sensitivity between 6 and 10. All test samples were normally consolidated under an effective confining pressure of 515 kPa for 36 hours, and then left undrained for 12 hours under the consolidation stresses prior to shear loading. For investigation of rate dependency of undisturbed clay response, the undrained triaxial shearing tests were carried out at different constant strain rates, varying from 10^{-3} to 1.1% per min.

Values of conventional parameters e_0, ν', M, λ, and κ (summarised in Table 1) were taken from Vermeer and Neher (1999). Thee different strain rates of 1.1%/min, 0.15%/min and 0.00094%/min, which were conducted in undrained triaxial tests by Vaid and Campella (1997), were selected to verify the model performance in realistic simulation of strain rate effects. The simulation results together with the experimental data are shown in Figure 4.

Figure 4 shows a very good agreement between model simulations and experimental observations. Both low strain rate and high strain rate strengths of the soil are well captured by the model. However, post-peak strain softening part of the response is not well simulated, particularly at high strain rates. This is mainly because NSFS-MCC is an isotropic model that does not take into consideration the anisotropy and destructuration features of sensitive behaviour.

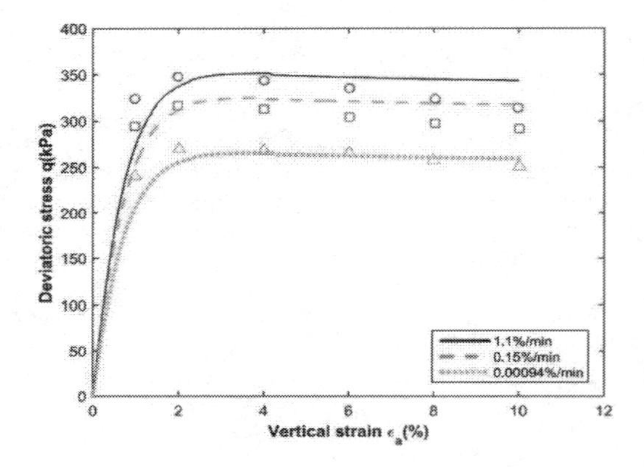

Figure 4. Prediction of undrained shear strength of Haney clay at different strain rates.

3.5 *Reproduction of stress path of Osaka clay under an undrained compression test*

Osaka Bay clay is a Pleistocene clay, which is deposited in Osaka Bay, Japan (Mimura & Jang 2004). A series of undrained triaxial shear tests were conducted on undisturbed samples of Osaka clay by Nakano et al. (2000). The general properties of the test samples were Gs (specific gravity) = 2.68, $w_L(\%) = 98.5$ and $w_P(\%) = 32.8$. The experimental data over an arbitrary sample were chosen for simulation in this paper. For the selected sample, the soil was consolidated at an isotropic consolidation stress of 490 kPa and then subjected to an undrained triaxial compression test at a constant cell pressure and under a controlled axial strain rate of 0.007mm/min. Model parameter values (summarised in Table 1) have been taken from Nakano et al. (2000).

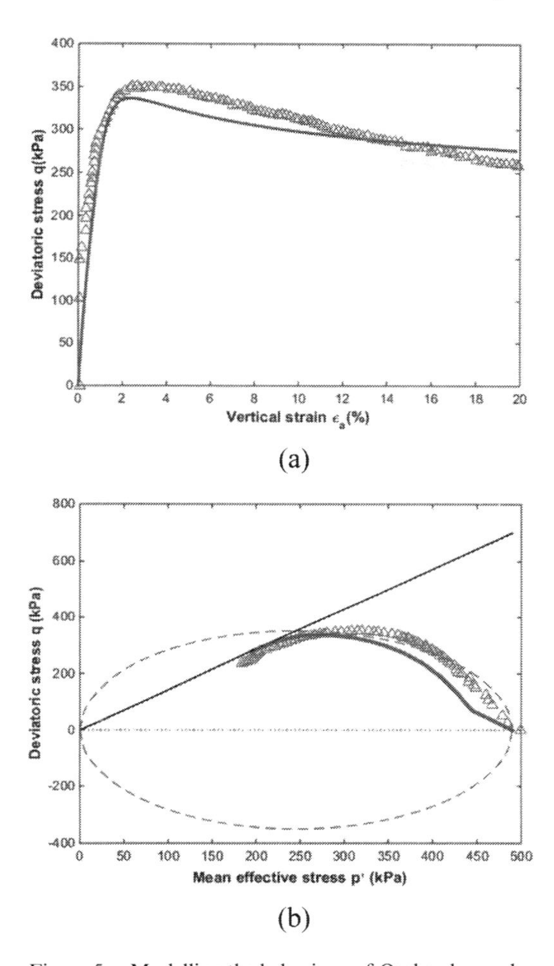

Figure 5. Modelling the behaviour of Osaka clay under an undrained triaxial compression test: (a) deviator stress versus axial strain, (b) effective stress path.

Figure 5 shows the comparison between model simulations and data from the undrained triaxial shear test. It is clear that the simulations fit very well with the experimental data. Looking at modelling results for deviatoric stresses versus vertical strains (Figure 5a) the end slope of the simulation curve is not the same as that of the softening plot in the experimentally observed data. This is likely to be due to the lack of the description of anisotropy and destructuration features in the model.

4 CONCLUSION

In this paper a time dependent constitutive model has been proposes based on the nonstationary flow surface theory. MCC model has been employed as the base model and it was extended to an isotropic creep model. With a modification in yield condition, so that it can describe the time effects, the model took advantage of classical elasto-plasticity framework for numerical implementation. The simulations exhibited that the new model is capable of describing the time and rate effects on the behaviour of natural clays. Validations at element level also confirmed the model's prediction capability particularly for normally consolidated soils. It is shown that the model provides a promising solution for the simulation of the long-term creep deformations in soft ground. However, some additional developments are required to enable the model to describe other fundamental features of natural clays' behaviour, in particular with respect to anisotropy and structure, so that a more accurate prediction can be obtained.

REFERENCES

Bjerrum, L., 1967. Engineering Geology of normally-consolidated Norwegian Marine Clays as related so settlement of buildings. Geotechnique 17, 81–118.

Gens, A.& Nova, R., 1993. Conceptual bases for a constitutive model for bonded soils and weak rocks. Geotechnical engineering of hard soils-soft rocks 1, 485–494.

John E. Garlanger, 1972. The consolidation of soils exhibiting creep under constant effective stress. Geotechnique 22, 71–78.

Koskinen, M., Karstunen, M.& Wheeler, S., 2002. Modelling destructuration and anisotropy of a soft natural clay, in: 5th European Conference on Numerical Methods in Geotechnical Engineering. Presses de l'ENPC, Paris, pp. 11–20.

Mimura, M.& Jang, W.Y., 2004. Description of time dependent behaviour of quasi-overconsolidated osaka pleistocene clays using elasto-viscoplastic finite element analyses. Soils and Foundations 44, 41–52.

Nakano, M., Asaoka, A., Noda, T., Hayashi, N.& Yamada, T., 2000. Difference between sand and clay based on decay of structure and overconsolidation, in: Proc. of the 35th JNCGE. pp. 573–574.

Olszak, W.& Perzyna, P., 1964. Olszak-Perzyna 1964 - Constitutive equation of the flow theory for Non-stationary Yiel Condition.pdf, in: Proceedings of the International Congress for Applied Mechanics. pp. 545–553.

Rezania, M., Taiebat, M.& Poletti, E., 2016. A viscoplastic SANICLAY model for natural soft soils. Computers and Geotechnics 73, 128–141. doi:10.1016/j.compgeo.2015.11.023

Roscoe, K.H.& Schofield, A., 1963. Mechanical behaviour of an idealized'wet'clay, in: Proc. 3rd European Conference on Soil Mechanics and Foundation Engineering. pp. 47–54.

Sing, A.& Mitchell, J.K., 1968. General stress–strain-time functions for soils. J Soil Mech Found Div ASCE 94, 21–46.

Taiebat, M., Dafalias, Y.F.& Peek, R., 2010. A destructuration theory and its application to SANICLAY model. International Journal for Numerical and Analytical Methods in Geomechanics 34, 1009–1040. doi:10.1002/nag.841

Tavenas, F., Leroueil, S.& Rochelle, P.L., Roy, M., 1978. Creep behaviour of an undisturbed lightly overconsolidated clay. Canadian Geotechnical Journal 15, 402–423.

Vaid, Y.P.& Campanella, R.G., 1977. Time-dependent behaviour of undisturbed clay. Journal of Geotechnical Engineering 103, 693–709.

Vermeer, P.& Neher, H., 1999. A soft soil model that accounts for creep. Proceedings of the International … 1–13.

Yin, Z.Y.& Karstunen, M., 2008. Influence of Anisotropy, Destructuration and Viscosity on the Behavior of an Embankment on Soft Clay. 12th International Conference of IACMAC 1–6.

Numerical Methods in Geotechnical Engineering IX – Cardoso et al. (Eds)
© 2018 Taylor & Francis Group, London, ISBN 978-1-138-33198-3

Simple constitutive models to represent the effect of mechanical damage and abrasion on the short-term load-strain response of geosynthetics

A.M. Paula
Polytechnic Institute of Bragança, Bragança, Portugal

M. Pinho-Lopes
University of Southampton, Southampton, UK

ABSTRACT: This paper discusses simple constitutive models to represent the tensile response of two geosynthetics (geotextile GTX and reinforcement geocomposite GCR) and the influence of two endurance durability agents on that response: mechanical and abrasion damage, acting independently and combined. The damage was induced in laboratory under standard conditions. The polynomial models approximated the short-term tensile experimental data very well and better than the hyperbolic-based models. The polynomial model parameters have no physical meaning. The equations from the literature did not always represent the parameters of hyperbolic-based models and their physical meaning. The secant stiffness for 5% strain obtained from the models was conservative for GTX and too optimistic for GCR. No clear relationship was found between the model parameters of the undamaged samples and the samples submitted to mechanical damage (MEC), abrasion damage (ABR) or mechanical and abrasion damage (MEC+ABR). The model parameters were normalised to the reduction factors for damage and to the tensile strength of each sample. A unique trend between parameters was not found. For GTX the normalised parameter $b/T_{ult,sample}$ was reduced after damage and all damaged samples exhibited a similar value.

1 INTRODUCTION

Many commercially available numerical codes tend to represent the tensile response of geosynthetics using linear elastic models or elastoplastic models (i.e., using a stiffness and a tensile strength). However, the accuracy of results and their realism is likely to be improved significantly by incorporating non-linear constitutive models for geosynthetics. Additionally, such constitutive models should be able to represent the response of geosynthetics allowing for influence of mechanisms and agents affecting the durability of geosynthetics. This paper discusses simple constitutive models to represent the tensile response of two geosynthetics and the influence of two endurance durability agents on that response: mechanical and abrasion damage, acting independently and combined.

2 BACKGROUND

For most applications of geosynthetics, and in particular for soil reinforcement, the tensile properties of geosynthetics are primary functional properties. Thus, the tensile strength and corresponding strain are key parameters in design. For some structures and limit states, the tensile stiffness of the geosynthetics can also be a relevant design parameter. However, the durability of geosynthetics also needs to be allowed for in the design, as the tensile response of geosynthetics can be affected significantly. Depending on the project, some durability agents and mechanisms will be more critical than others; among these, installation damage and abrasion damage stand out.

Mechanical damage associated with installation is common to all applications of geosynthetics, as handling and placing the geosynthetics on site and compaction actions of adjacent fill material can be severe. Additionally, for particular applications of geosynthetics abrasion damage is also key (e.g., railways, temporary roads, canal revetments, sea shores with sediments and sliding masses washing up and down). The abrasion damage occurs mostly during service, particularly when cyclic relative motion (friction) between a geosynthetic and contact soil occurs.

In design, changes in tensile response, namely in tensile strength, are represented by independent reduction factors, each representing the changes in tensile strength due to a single durability agent or mechanism. Such reduction factors are then multiplied. For mechanical and abrasion damage there can be some interaction, positive or negative synergy (as shown by Rosete et al. 2013), resulting in a combined reduction factor different than the product between the two independent reduction factors. This synergy can also influence the stiffness of geosynthetics.

Numerical models, particularly using the finite element method, are becoming more and more popular for example to assist the design of reinforced soil structures and/or analyse their performance. However, as mentioned before, often such models are simplistic and represent geosynthetics by a stiffness and a tensile strength. As durability is key for the design and performance of geosynthetics, representing their response using non-linear constitutive models allowing for durability is essential. Thus, it is important to understand how damage may influence the load-strain response of geosynthetics and how such changes can influence the parameters used or defining simple non-linear constitutive models.

3 SIMPLE CONSTITUTIVE MODELS

The short-time tensile response of geosynthetics can be represented by simple constitutive models, such as polynomial models and hyperbolic-based models. The model parameters can be obtained by curve fitting experimental data, in particular using the least-squares method (Bathurst and Kaliakin 2005).

Equation 1 represents a generic polynomial model, where T is the load per unit width, ε is the axial tensile strain, a_i is the polynomial coefficient of order i and n is the order of the polynomial. Equation 2 represents the tangent stiffness ($J_{t\varepsilon\%}$) of the geosynthetic for a strain ε (in%), obtained by derivation of Equation 1.

$$T = \sum_{i=0}^{n} a_i \varepsilon^i \tag{1}$$

$$J_{t\varepsilon\%} = \frac{dT}{d\varepsilon} = \sum_{i=1}^{n} a_i \varepsilon^{i-1} \tag{2}$$

Depending on the type of tensile response observed, herein identified as types A and B (Figure 1), two families of hyperbolic-based models can be used to model the tensile response of geosynthetics.

Equation 3 represents the hyperbolic-based model for type A response (Liu and Ling 2006). Equation 4 represents the tangent stiffness for this model. According to Liu and Ling 2006, a and b

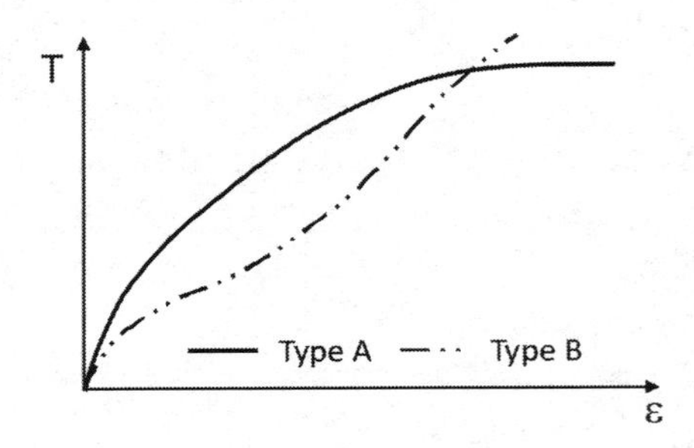

Figure 1. Typical tensile response of geosynthetics, type A and B (not to scale).

are model parameters representing the inverse of the initial stiffness, J_i, and the inverse of the tensile strength of the geosynthetic, T_{max} (Equations 5 and 6, respectively).

$$T = \frac{\varepsilon}{a + b\varepsilon} \tag{3}$$

$$J_{t\varepsilon\%} = \frac{dT}{d\varepsilon} = \frac{a}{(a + b\varepsilon)^2} \tag{4}$$

$$a = \frac{1}{J_i} \tag{5}$$

$$b = \frac{1}{T_{max}} \tag{6}$$

Type B tensile response is more complex and can be modelled by combining a hyperbola (for low strains) with an exponential function (for high strains), as in Equation 7 (Liu and Ling 2006). Equation 8 represents the corresponding tangent stiffness.

$$T = \frac{\varepsilon}{a + 2b\varepsilon} + \frac{1}{2b} \cdot e^{-\alpha(\varepsilon - \varepsilon_{max})^2} \tag{7}$$

$$J_{t\varepsilon\%} = \frac{dT}{d\varepsilon} = \frac{a}{a + 2b\varepsilon} + \frac{-\alpha(\varepsilon - \varepsilon_{max})}{b} \cdot e^{-\alpha(\varepsilon - \varepsilon_{max})^2} \tag{8}$$

where ε_{max} is the rupture tensile strain of the geosynthetic and α is a material constant (Liu and Ling 2006).

4 MATERIALS AND TEST CONDITIONS

4.1 *Geosynthetics*

This paper includes results for two geosynthetics (Table 1 and Figure 2): a nonwoven geotextile, consisting of continuous mechanically bonded polypropylene (PP) filaments (GTX); and a uniaxial

Table 1. Nominal properties of the geosynthetics studied (GTX and GCR): nominal tensile strength, T_{nom}; corresponding strain, ε_{nom}; mass per unit area, μ; nominal thickness, d_{nom}.

	T_{nom}	ε_{nom}	μ	d_{nom}
Geosynthetic	kN/m	%	g/m^2	mm
GTX	50	65	800	6
GCR	75	10	362	2.2

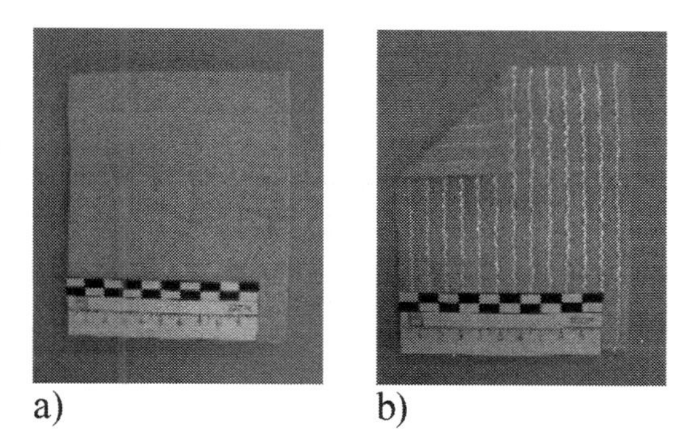

a) b)

Figure 2. Geosynthetics studied: a) GTX; b) GCR.

reinforcement geocomposite, composed of high modulus PET fibres attached to a nonwoven continuous filament PP geotextile backing (GCR).

The tensile response of the geosynthetics was characterized using wide-width tensile tests (BSI 2008, EN ISO 10319), with five valid specimens per sample tested using hydraulic jaws and measuring strains using a video-extensometer.

4.2 *Damage simulated*

The mechanical damage associated with installation was simulated in laboratory using standardised procedures (ENV ISO 10722-1:1998), in which a specimen of geosynthetic is placed between two layers of a synthetic aggregate (sintered aluminium oxide, particle sizes between 5 mm and 10 mm) and submitted to cyclic loading (5 kPa to 900 kPa) at a frequency of 1 Hz for 200 loading cycles (Figure 3a).

The abrasion damage was simulated using the procedures in EN ISO 13427:1998. In this test, a geosynthetic specimen is placed on an upper plate of a stationary platform and rubbed by a P100 abrasive, placed on the lower plate and moved along a horizontal axis under controlled pressure (Figure 3b). Some adjustments to the test procedure were necessary (described by Rosete et al. 2013): for GTX, a P24 abrasive film was placed between the specimen and the upper plate, to

a)

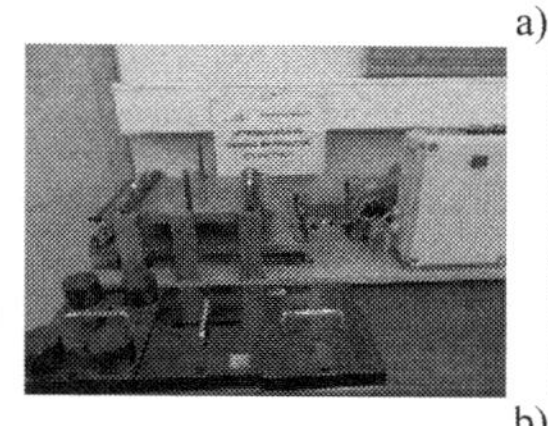

b)

Figure 3. Equipment used to simulate damage in laboratory: a) mechanical damage, box, compaction plate and synthetic aggregate; b) abrasion damage.

ensure that during the test the specimen did not adhere to the abrasive film placed on the lower plate (no additional damage was observed).

5 RESULTS AND DISCUSSION

The experimental data were approximated using polynomial and hyperbolic-based models (described in section 3), excluding data points after the failure of the specimens. For some samples, particularly of GCR after abrasion damage, the tensile response varied and often exhibited two peaks, depending on where the failure occurred (filament or geotextile backing). For defining the constitutive models, only data up to the first peak of the experimental load-strain curves was used. The constitutive models parameters were analysed to assess the changes associated to the damage induced, isolated and combined.

5.1 *Base experimental data*

The tensile tests results are summarised in Table 2: tensile strength, T_{max} (herein defined as the maximum tensile force mobilised during the test); strain for maximum load, ε_{max}; secant stiffness for 5% strain, $J_{s5\%}$. The results refer to five valid specimens per sample and include the corresponding coefficient of variation (CV). The secant stiffness for 5% strain, $J_{s5\%}$, was determined using Equation 9,

Table 2. Summary of tensile tests results for different types of samples of GTX and GCR: undamaged (UND) and submitted to damage in laboratory – mechanical (MEC), abrasion (ABR) and sequential mechanical and abrasion damage (MEC+ABR).

	GTX			
Property	UND	MEC	ABR	MEC+ ABR
Tensile strength, T_{max} (kN/m)				
Mean	42.34	33.69	40.80	30.42
CV (%)	5.2	9.8	13.1	4.8
Strain for maximum load, ε_f (%)				
Mean	89.28	84.64	113.20	96.87
CV (%)	5.4	7.9	5.3	8.1
Secant stiffness for 5% strain, $J_{s5\%}$ (kN/m)				
Mean	120.86	80.55	69.97	60.51
CV (%)	13.1	36.6	21.9	19.0
	GCR			
Property	UND	MEC	ABR	MEC+ ABR
Tensile strength, T_{max} (kN/m)				
Mean	83.46	41.82	17.86	15.37
CV (%)	5.1	4.0	10.4	2.7
Strain for maximum load, ε_f (%)				
Mean	14.24	17.41	72.34	8.53
CV (%)	5.9	23.4	62.7	12.3
Secant stiffness for 5% strain, $J_{s5\%}$ (kN/m)				
Mean	380.31	247.11	145.80	210.13
CV (%)	18.3	26.5	20.2	11.9

where $T_{5\%}$ is the force for 5% strain measured during the tests.

$$J_{s5\%} = \frac{T_{5\%} \times 100}{5} \tag{9}$$

Some experimental load-strain responses were summarised in Figure 4, including data for representative specimens per sample. For the undamaged samples two different types of responses were observed: type A, for GTX, and type B for GC. The range of strains represented Figure 4 include the full load-strain response observed in the tests. However, it is not likely that such high strains are achieved within a structure.

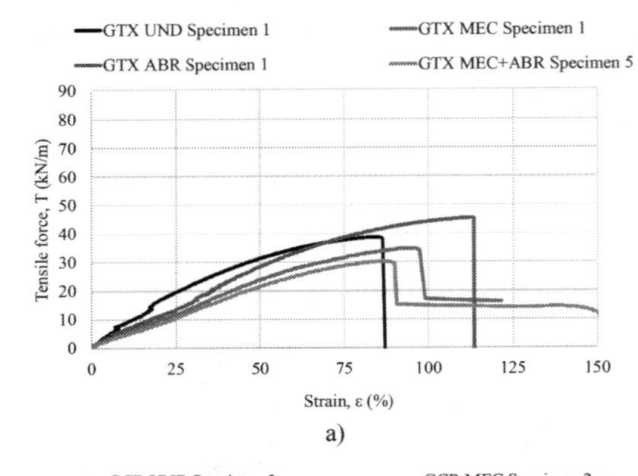

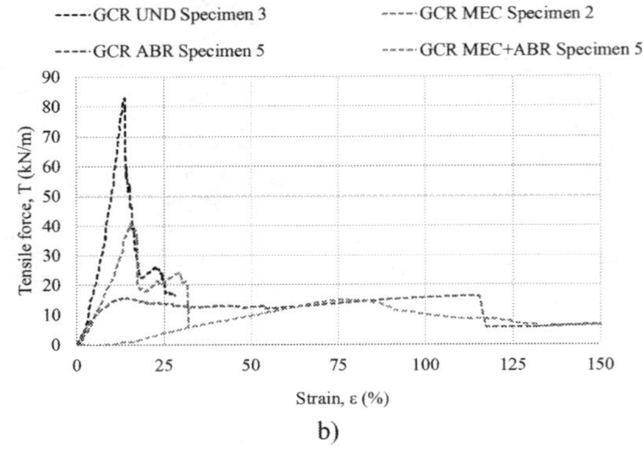

Figure 4. Load-strain response of the geosynthetics for undamaged samples and submitted to mechanical damage (MEC), abrasion damage (ABR) and sequential mechanical and abrasion damage (MEC+ABR) for: a) GTX; b) GCR.

5.2 *Simple constitutive models*

The experimental data was fitted with two simple constitutive models: an order 6 polynomial model and a hyperbolic-based model. The model parameters were obtained via curve fitting of the experimental data for each specimen; then, average parameters were calculated. Tables 3 and 4 summarise the average model parameters for different samples of GTX and GCR, respectively, undamaged (UND) and submitted to laboratory damage: mechanical (MEC), abrasion (ABR) and sequential mechanical and abrasion (MEC+ABR). Tables 3 and 4 also include values for the coefficient of determination (R^2) for some of those model parameters. Figure 5 includes load-strain curves for selected specimens of each sample obtained experimentally and for the models analysed.

The polynomial models fit the experimental data very well (Figure 5 and Tables 3 and 4), with R^2 ranging between 0.994 and 1.000 and coefficient of variation between 0.01% and 0.50%. Nevertheless, these models are simple mathematical functions fitted to the experimental data, without physical meaning.

On the contrary, the hyperbolic-based models are believed to be associated with tensile parameters of the materials, as represented in Equations 5 and 6.

The hyperbolic model (type A) fitted well the experimental data for GTX. For GCR, the

Table 3. Model parameters for GTX for different types of samples: undamaged (UND) and submitted to damage in laboratory—mechanical (MEC), abrasion (ABR) and sequential mechanical and abrasion damage (MEC+ABR).

	GTX			
Model	UND	MEC	ABR	MEC+ ABR
Polinomial				
a_0	1.32	$4.8(10^{-1}$	$4.8(10^{-1}$	$5.8(10^{-1}$
a_1	$9.5(10^{-1}$	$8.4(10^{-1}$	$6.6(10^{-1}$	$5.6(10^{-1}$
a_2	$-1.6(10^{-2}$	$-3.9(10^{-2}$	$-1.9(10^{-2}$	$-2.1(10^{-2}$
a_3	$5.7(10^{-4}$	$1.8(10^{-3}$	$7.5(10^{-4}$	$9.6(10^{-4}$
a_4	$-1.3(10^{-5}$	$-4.0(10^{-5}$	$-1.3(10^{-5}$	$-1.9(10^{-5}$
a_5	$1.4(10^{-7}$	$4.0(10^{-7}$	$9.6(10^{-8}$	$1.7(10^{-7}$
a_6	$-5.2(10^{-10}$	$-1.6(10^{-9}$	$-2.7(10^{-10}$	$-5.6(10^{-10}$
R^2	0.999	1.000	1.000	1.000
CV, R^2	0.06	0.05	0.01	0.04
Hyperbolic-based type A				
a	0.882	1.480	1.592	1.921
CV, a	6.12	20.94	18.87	8.05
b	0.013	0.011	0.009	0.011
CV, b	3.96	30.58	18.07	19.01
R^2	0.996	0.994	0.993	0.993
CV, R^2	0.22	0.27	0.12	0.32

Table 4. Model parameters for GCR for different types of samples: undamaged (UND) and submitted to damage in laboratory—mechanical (MEC), abrasion (ABR) and sequential mechanical and abrasion damage (MEC+ABR).

	GCR			
Model	UND	MEC	ABR	MEC+ ABR
Polinomial				
a_0	$4.4(10^{-1}$	$6.8(10^{-1}$	$3.2(10^{-1}$	$4.1(10^{-1}$
a_1	-1.30	1.37	1.27	1.58
a_2	2.71	$1.8(10^{-1}$	$2.5(10^{-1}$	1.40
a_3	$-6.0(10^{-1}$	$-3.4(10^{-1}$	$-8.8(10^{-2}$	$-8.2(10^{-1}$
a_4	$6.4(10^{-2}$	$-6.9(10^{-3}$	$1.2(10^{-2}$	$2.0(10^{-1}$
a_5	$-3.0(10^{-3}$	$7.0(10^{-4}$	$-7.2(10^{-4}$	$-2.2(10^{-2}$
a_6	$4.7(10^{-5}$	$-2.2(10^{-5}$	$1.6(10^{-5}$	$9.1(10^{-4}$
R^2	0.999	0.999	0.999	0.994
CV, R^2	0.02	0.05	0.04	0.50
Hyperbolic-based type B				
a	0.253	0.310	0.709	0.458
CV, a	23.19	56.36	56.55	30.58
b	0.009	0.018	0.058	0.056
CV, b	5.40	31.99	12.84	6.46
α	0.035	5.584	0.014	0.035
CV, α	13.71	197.76	30.94	33.36
R^2	0.996	0.937	0.993	0.990
CV, R^2	0.19	11.96	0.48	0.66

hyperbolic-based model type B was used. In some cases, particularly after damage, where the specimens exhibited some scatter in response, the model could not always reproduce the observed responses well. For example, Figure 5b includes two specimens for the sample submitted to mechanical damage (MEC) and to abrasion damage (ABR), as different types of load-strain response and/or a wider range of responses were observed.

Figure 6 illustrates the relationship between the inverse of the tensile strength of the specimens tested and the parameter b obtained from the curve fitting exercise (as per Equation 6).

For GTX (Figure 6a) there seems to be no relationship between the model parameter b and the tensile strength of the specimens, showing that Equation 6 is not applicable to the load-strain response of this geosynthetic, both undamaged and after damaged in laboratory. For GCR (Figure 6b), Equation 6 is a good approximation for the model parameter b obtained for the undamaged specimens tested; however, after damage in laboratory (regardless of the type of

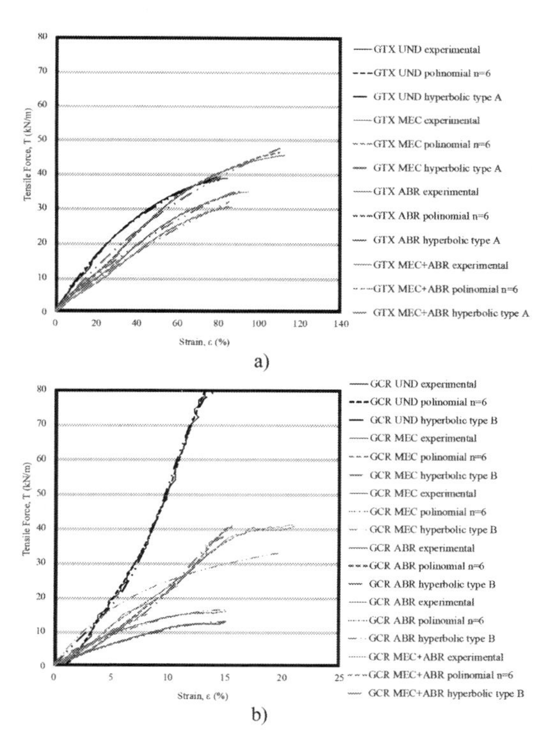

Figure 5. Tensile response of: a) GTX and b) GCR, undamaged (UND) and after laboratory mechanical (MEC) damage, abrasion (ABR) damage and mechanical followed by abrasion (MEC+ABR) damage [Note: The scale for the strains is different in the two subparts of the figure].

damage induced), that relationship was no longer applicable. This could be due to the consequences of the damage induced in this geosynthetic. As reported by Rosete et al. (2013), after mechanical damage GCR exhibited detachment and cuts of some of the PET yarns and incrustation of fines in the geotextile backing; after abrasion, there was partial detachment and damage of the PET yarns and some accumulation of fibres perpendicular to the abrasive movement; after mechanical damage followed by abrasion, there was detachment and superficial disintegration of the PET yearns, filament cutting and reorientation, causing the accumulation of fibres perpendicular to the direction of the abrasive movement. After damage the tensile failure of GCR occurred in different parts of the geocomposite, particularly after abrasion damage (filaments or geotextile backing). For GTX the observed effects of damage were similar to those observed for the geotextile backing of GCR.

For the hyperbolic-based type B mode, the values α obtained for the undamaged (UND) specimens of GCR, 0.035, exhibited some scatter (coefficient of variation of ~14%). After mechanical damage (MEC) the value of α changed and varied significantly (average value of 5.584 and CV of 198%). After abrasion damage (ABR), α was reduced relatively to the undamaged sample ($\alpha = 0.014$; CV~31%). After mechanical and abrasion damage (MEC+ABR), the parameter α was the same as the undamaged sample, but its scatter increased (CV ~33%). This seems to indicate that α may not be a material constant and it can be significantly altered by damage. In fact, as the damage can change GCR significantly, it is likely that the load-strain responses after damage will have different model parameters. Thus, it seems that, contrary to what has been reported in the literature, for the geosynthetics and test conditions reported herein, the model parameter α is not a material constant.

Table 5 summarises the values of the tangent and secant stiffness for 5% strain ($J_{t5\%}$ and $J_{s5\%}$) obtained for GTX and GCR using the constitutive models. $J_{t5\%}$ was obtained from Equations 2, 4 and 8, respectively, for the polynomial, hyperbolic-based type A and type B models; $J_{s5\%}$ was obtained from Equation 9. For GTX the hyperbolic-based models overestimated the tangent stiffness for 5% strain from the polynomial models (that fit the experimental data very well). For GCR an opposite trend was observed (except after MEC+ABR). These trends seem to indicate that the hyperbolic-based models are not able to represent well the initial part of the load-strain curves of these geosynthetics. For GTX, the secant stiffness for 5% strain obtained from the simple constitutive models was lower than the experimental data (Table 2) independently of the model

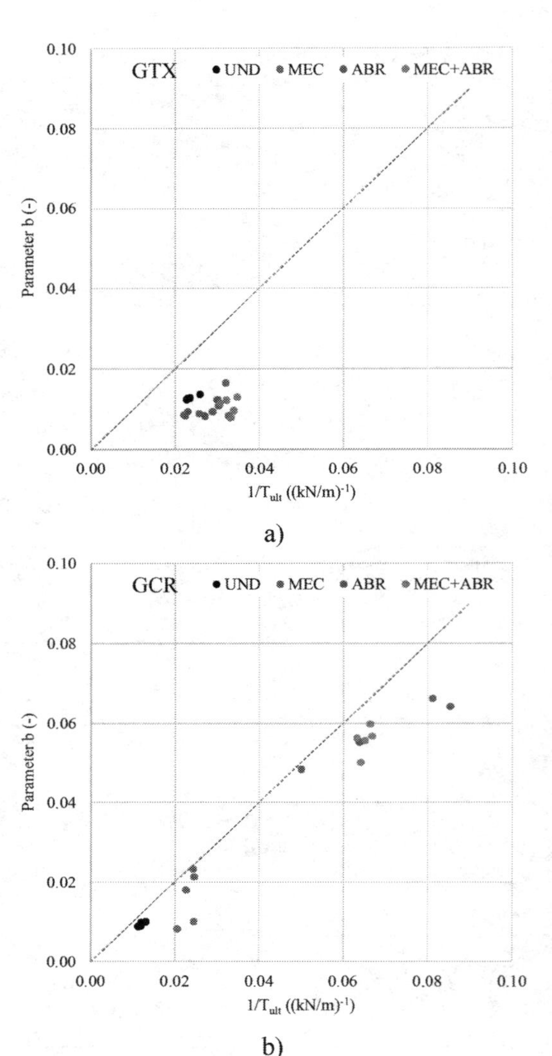

Figure 6. Parameter b for the hyperbolic-based models versus $1/T_{ult}$ for the four types of samples tested (UND, MEC, ABR, MEC+ABR): a) GTX; b) GCR.

analysed, with variations ranging between -19% (hyperbolic-based type A) to -2% (polynomial), for samples submitted to mechanical damage. On the contrary, for GCR, all models were too optimistic, with estimated $J_{s5\%}$ 1% (polynomial, ABR) to 54% (hyperbolic-based type B, MEC) higher than the corresponding experimental data.

5.3 *Influence of mechanical and abrasion damage on the model parameters*

One of the aims of this paper was to assess if the model parameters of the damaged geosynthetics

Table 5. Tangent stiffness ($J_{t5\%}$) and secant stiffness ($J_{s5\%}$) for 5% strain obtained from the simple constitutive models for GTX and GCR: undamaged (UND) and submitted to damage in laboratory—mechanical (MEC), abrasion (ABR) and sequential mechanical and abrasion damage (MEC+ABR).

Stiffness modulus (kN/m)	GTX			
	UND	MEC	ABR	MEC+ ABR
Polynomial				
$J_{t5\%}$	82.72	57.28	52.03	40.88
$J_{s5\%}$	114.58	78.70	68.10	58.93
Hyperbolic type A				
$J_{t5\%}$	98.44	62.87	59.25	49.22
$J_{s5\%}$	105.64	65.18	61.01	50.62
Stiffness modulus (kN/m)	**GCR**			
	UND	MEC	ABR	MEC+ ABR
Polynomial				
$J_{t5\%}$	469.80	245.23	112.32	175.93
$J_{s5\%}$	374.10	223.39	145.24	206.20
Hyperbolic type B				
$J_{t5\%}$	381.53	128.42	108.94	187.24
$J_{s5\%}$	341.20	203.62	127.05	198.06

could be estimated from the parameters obtained for the undamaged sample. For that, the model parameters of the damaged samples of GTX and GCR were normalised to the reduction factor for damage for each type of damage induced in laboratory (MEC, ABR, MEC+ABR). Those reduction factors were defined using Equation 10, as the ratio of the tensile strength of the undamaged sample ($T_{ult,UND}$) to that of the damaged sample ($T_{ult,DAM}$).

$$RF_{dam} = \frac{T_{ult,UND}}{T_{ult,DAM}} \quad (10)$$

Figure 7 summarises the parameters for the polynomial models for GTX and GCR, respectively, normalised to the reduction factor for damage. Some of these normalized parameters are very small and not quite visible in Figure 7. Nevertheless, the values of the normalized parameters exhibited significant variation and there is no clear relationship between the polynomial model parameters and the corresponding reduction factors for damage.

For the hyperbolic-based models, the model parameters normalised to the RF_{dam} are summarised in Figure 8. For GTX, a/RF_{dam} increased after damage, particularly after abrasion damage; the normalised parameter b/RF_{dam} decreased after damage and the values for the damaged samples are similar (~0.009, Figure 8b), regardless of the type of damage induced. For GCR different trends were observed. The normalised parameter a/RF_{dam} decreased after damage; the specimens submitted to MEC and ABR exhibited similar values of a/RF_{dam} (a/RF_{dam}~0.16), with a reduction to half of that value for specimens submitted to MEC+ABR (a/RF_{dam} = 0.08). The parameter b/RF_{dam} is practically constant after mechanical damage (MEC)

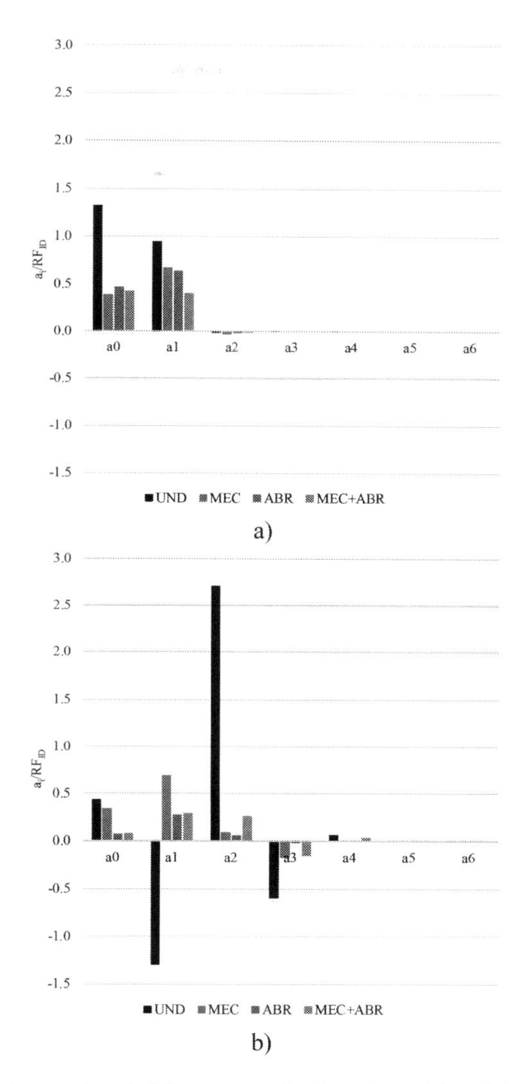

Figure 7. Model parameters for the polynomial models normalised relatively to the reduction factor for damage (a_i/RF_{dam}) for the four types of samples tested (UND, MEC, ABR, MEC+ABR) for: a) GTX; b) GCR.

and increased after abrasion damage, isolated or induced after mechanical damage. The normalised parameter α/RF_{dam} (Figure 8c) had a very important increase after mechanical damage (MEC), showing that the response of the material was clearly altered after this type of damage; after abrasion damage and mechanical and abrasion damage this parameter reduced. As before, there is no indication that α is a material constant.

The data seems to indicate that there is no unique relationship between the models parameters before and after damage, even for the normalized parameters with the exception of b/RF_{dam} for GTX. Thus, these results indicate that, for the conditions considered herein and for the simple constitutive models used, there is no obvious link between the model parameters and the reduction factor for damage (as usually defined in design). As discussed in the previous section, the model parameters cannot be estimated using the equations from the literature, for both undamaged and damaged samples.

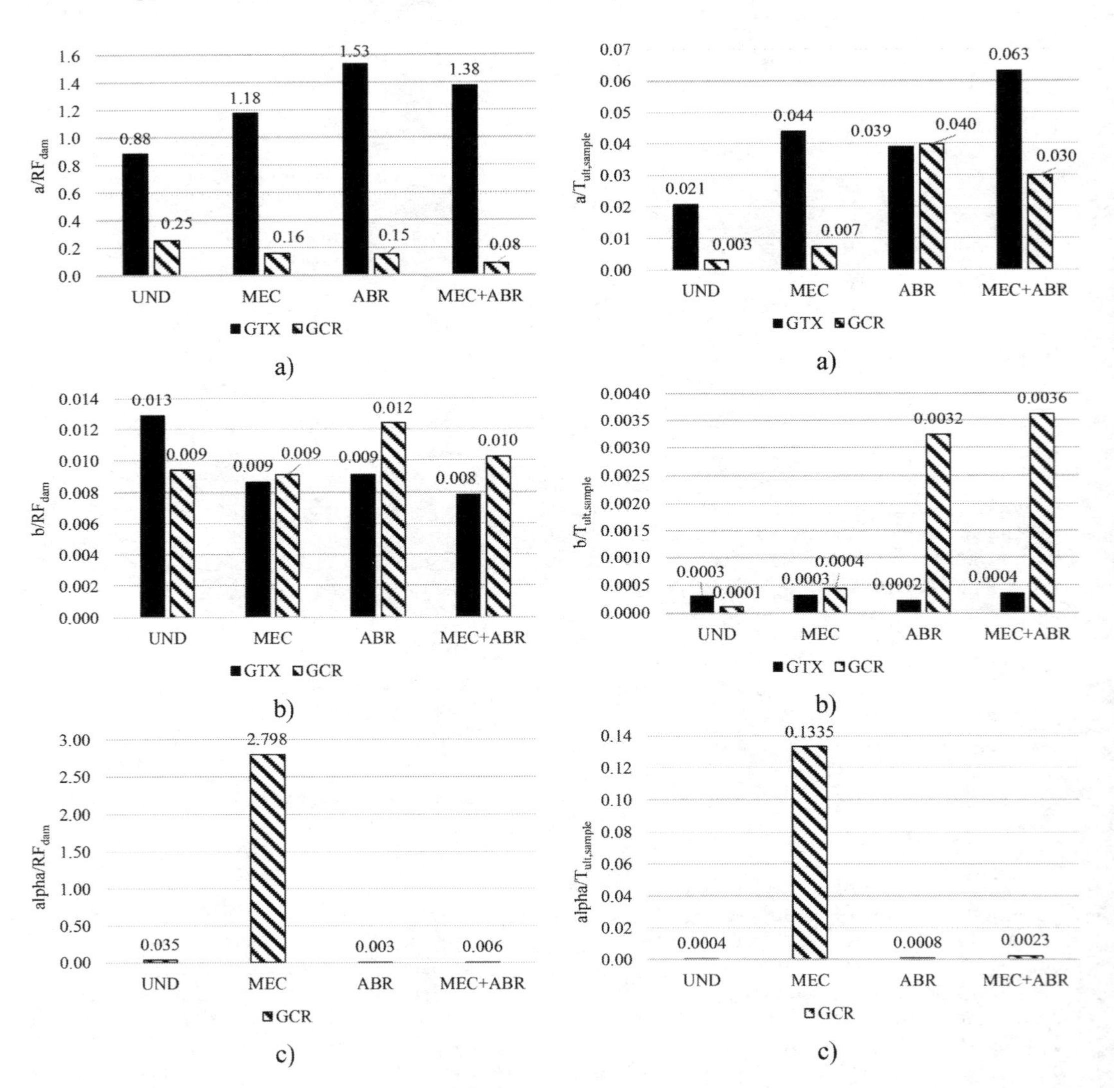

Figure 8. Hyperbolic-based model parameters normalised relatively to the reduction factor for damage for the four types of samples tested (UND, MEC, ABR, MEC+ABR) for GTX and GCR: a) a/RF_{dam}; b) b/RF_{dam}; c) α/RF_{dam} (for GCR only).

Figure 9. Hyperbolic-based model parameters normalised relatively to the tensile strength of each sample for the four types of samples tested (UND, MEC, ABR, MEC+ABR) for GTX and GCR: a) $a/T_{ult,sample}$; b) $b/T_{ult,sample}$; c) $\alpha/T_{ult,sample}$ (for GCR only).

The parameters of the hyperbolic-based models were also normalised to the tensile strength of each sample ($T_{ult,sample}$), to assess if they could indicated alternative physical meaning for the model parameters. Figure 9 summarises those values.

For both GTX and GCR, the normalised parameter $a/T_{ult,sample}$ increased after damage, although differently for the three types of damaged samples studied and for each geosynthetic. For GTX the parameter $b/T_{ult,sample}$ did not change significantly after mechanical damage (MEC), decreased after abrasion damage (ABR) and increased after mechanical and abrasion damage (MEC+ABR). For GCR, the parameter $b/T_{ult,sample}$ increased after damage, mainly after abrasion damage, isolated (ABR) or induced after mechanical damage (MEC+ABR). Lastly, the normalised parameter $\alpha/T_{ult,sample}$ increased after damage, particularly after mechanical damage (MEC).

6 CONCLUSIONS

In this paper, the short-term tensile response of two geosynthetics was approximated using simple constitutive models (polynomial and hyperbolic-based). The influence of two endurance durability agents on that response was analysed, namely mechanical and abrasion damage, acting independently and combined. From the results the main conclusions can be summarised as:

- The polynomial models (order 6) approximated the short-term tensile experimental data very well and better than the hyperbolic-based models. However, the polynomial model parameters have no physical meaning.
- The hyperbolic-based models were not able to replicate the tangent stiffness for 5% strain obtained experimentally, indicating that such models may not represent well the responses of these geosynthetics for low strains.
- The secant stiffness for 5% strain obtained from the models exhibited different trends, depending on the geosynthetic. While for GTX the models underestimated the experimental data, independently of the model analysed, for GCR, all models were too optimistic when estimating the secant stiffness for 5% strain.
- The equations from the literature did not always represent the parameters of hyperbolic-based models and their physical meaning. For GTX, contrary to what has been reported in the literature, for the test conditions presented herein, the parameter b of the hyperbolic-based models could not be estimated as the inverse of the materials' tensile strength. For GCR, while for the undamaged specimens the relationship from the literature was applicable, after damage in laboratory (regardless of the type of damage induced), that relationship was no longer applicable.
- No clear relationship was found between the model parameters of the undamaged samples and the samples submitted to mechanical damage (MEC), abrasion damage (ABR) or mechanical and abrasion damage (MEC+ABR).
- The model parameters were normalised to the reduction factors for damage and to the tensile strength of each sample. A unique trend between parameters was not found.
- For GTX the normalised parameter $b/T_{ult,sample}$ was reduced after damage and all damaged samples exhibited a similar value.

The results presented herein seem to indicate that the hyperbolic-based models from the literature can reproduce well the overall load-strain response of the geosynthetics studied. However, the model parameters do not have the physical meaning commonly attributed in the literature. The damage induced under standardised and repeatable conditions caused the model parameters to change, in some cases, significantly. Although the simple constitutive models used herein could capture the load-strain response after damage, for most cases no obvious relation between model parameters and the reduction factor for damage nor the tensile strength after damage was found. The only exception was the parameter b for the hyperbolic-based model for GTX when normalised to the tensile strength of each sample—after damage all samples exhibited a similar value.

REFERENCES

Bathurst, R.J. & Kaliakin, V.N. 2005. Review of numerical models for geosynthetics in reinforcement applications. Computer Methods and Advances in Geomechanics: 11th Intern. Conf. of the International Association for Computer Methods and Advances in Geomechanics, Torino, 19–24 June 2005, 4: 407–416.

BSI 1998. EN ISO 13427. Geotextiles and geotextile-related products. Abrasion damage simulation (sliding block test). BSI, London, UK.

BSI 1998. ENV ISO 10722–1. Geotextiles and geotextile-related products - Procedure for simulating damage during installation – Part 1: Installation in granular materials. BSI, London, UK.

BSI 2008. EN ISO 10319. Geosynthetics. Wide-width tensile test. BSI, London, UK.

Liu, H. & Ling, H.I. 2007. Unified Elastoplastic-Viscoplastic Bounding Surface Model of Geosynthetics and Its Applications to Geosynthetic Reinforced Soil-Retaining Wall Analysis. *Journal of Engineering Mechanics* 133(7): 801–815.

Rosete, A., Mendonça Lopes, P., Pinho-Lopes, M. & Lopes, M.L. 2013. Tensile and hydraulic properties of geosynthetics after mechanical damage and abrasion laboratory tests. *Geosynthetics International* 20(5): 358–374.

Numerical Methods in Geotechnical Engineering IX – Cardoso et al. (Eds)
© 2018 Taylor & Francis Group, London, ISBN 978-1-138-33198-3

Modelling the small strain behaviour of a cemented silty sand with bounding plasticity

F. Panico & A. Viana da Fonseca
CONSTRUCT-GEO, Faculty of Engineering (FEUP), University of Porto, Portugal

J. Vaunat
Department of Geotechnical Engineering and Geosciences, Technical University of Catalonia (UPC), Barcelona, Spain

ABSTRACT: Modelling the pre-yield behaviour of soils using isotropic hardening models is a difficult task. One-surface elasto-plastic models can predict the post-yield behaviour but fail to reproduce the pre-yield behaviour of soils, even if a non-linear elastic law is adopted. The complexity is increased when artificial cement is added to the soil, as the level of cementation influences the characteristics of the specific mixture. Models based on bounding surface plasticity are able to reproduce the experimentally observed smooth transition from purely elastic to fully plastic state. The model proposed in the present paper represents the extension of an existing model for cemented soil, reformulated in a two-surface bounding plasticity framework. The model is calibrated against a set of triaxial tests performed over an artificially cemented soil. The results of the calibration are compared against the corresponding one-surface model, emphasising the improved performance of the new model in the small strain domain.

1 INTRODUCTION

Artificially or naturally cemented soils are found in many geotechnical structures and infrastructures, onshore and offshore. Some examples are embankments, wind turbines subjected to the action of wind and sea waves, foundations layers of structures subjected to variable loads (e.g. reservoirs or deposits). Modelling the constitutive behaviour of these soils is a difficult task, as cementation modifies the response of the soil at small and large strain.

Advanced constitutive models were developed to deal with this increased complexity. Such model were developed from the usual elasto-plasticity with isotropic hardening, and in the framework of critical state soil mechanics. To this common framework, extra features were added to represent the increased strength and stiffness due to cementation, and the progressive loss of bonding associated with the development of plastic strain. Gens & Nova (1993) proposed a constitutive model for bonded soils and weak rocks, focused on the destructuration process of a bonded material during yielding. The model is based on the model proposed by Nova (1988) for sand but can be extended to any elastoplastic model for unstructured soil. It is equipped with a bonding variable taking into account the level of cementation and a destructuration law, which regulates the loss of bonding with the progressive accumulated strain. Vaunat & Gens (2003) model considers the cemented soil as a system composed of two materials working in parallel: a granular matrix modelled through an elastic-plastic constitutive law, typical of soils, and the bonds, modelled through a damage elastic law (Carol et al., 2001) typical of quasi-brittle materials.

These approaches were both originally conceived for cohesive material. CASM-n model (Yu et al., 2007) combines Gens & Nova (1993) framework for cemented soils with an existing model for unbonded cohesive or granular material (CASM model—Yu, 1998). Such model was implemented by Rios et al. (2016) and used to reproduce a set of drained and undrained monotonic compression triaxial tests (Rios et al., 2014). These tests were performed over a well graded silty-sand originated by the weathered horizons of a Porto granite residual soil (Viana da Fonseca, 2003; Viana da Fonseca et al., 2006), remoulded in laboratory and artificially cemented with Portland cement. While the output showed a good agreement in the large-strain behaviour of the cemented soil, the small-strain behaviour does not seem to be equally well reproduced. This is because the bonding component has the effect of enlarging the purely elastic domain of the material, corresponding to a large elastic response of the material, followed by a sudden loss of stiffness when yielding is reached. Such outcome is undesirable as it contradicts the experimental evidence, and it leads to large discrepancies when small-stress repeated loading is considered. Nevertheless, it is difficult to avoid if isotropic hardening frameworks are adopted, even if equipped with sophisticate elastic laws.

To overcome these limitations, a model based on bounding surface plasticity (Dafalias & Popov, 1975; Krieg, 1975) with a kinematic hardening rule can be adopted. According to this framework, the conventional single yield surface approach is replaced by a two-surface option, which is capable of reproducing the experimentally observed smooth transition from elastic to fully plastic state and the effect of loading reversal during cycles. The framework uses an inner yield surface (f) free to move inside an outer reference (or bounding) surface (f_{ref}).

In the present implementation, a bounding surface extension based on Al Tabbaa & Wood (1989) model is added to the bonded elasto-plastic CASM-n model presented by Rios et al. (2016). In the spirit of hierarchical modelling (Desai et al., 1986), increasingly complex models are developed by superimposing various characteristics as corrections on the basic and simplest model. The added features will not alter the previous framework of the model, which can be restored by setting specific parameters for the additional characteristics. The presented model is compared with Rios et al. (2016) model, by modelling the same tests and comparing the two outputs.

2 MODEL DESCRIPTION AND IMPLEMENTATION

2.1 *Yield and reference surface*

The new formulation is presented schematically in Figure 1. Reference surface has the following expression:

$$f_{ref} = \left(\frac{|q|}{M(p' + p'_t)} \right)^n + \frac{1}{\ln r} \ln \frac{p' + p'_t}{p'_0} \tag{1}$$

where p'_t is the tensile strength, p'_0 is the horizontal dimension of the surface, n and r are parameters defining the shape of the surface, and M is the effective stress ratio (q/p') at critical state.

Inside this surface, a smaller yield surface of same shape is defined:

$$f = \left(\frac{|q - q_\alpha|}{M(p' - p'_\alpha)} \right)^n + \frac{1}{\ln r} \ln \frac{p' - p'_\alpha}{Rp'_0} \tag{2}$$

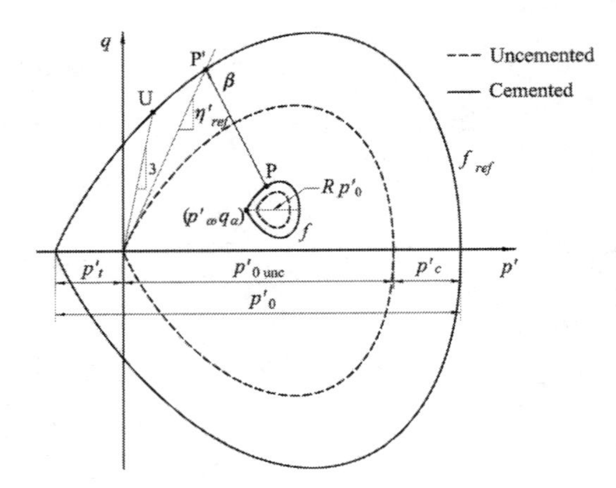

Figure 1. Scheme of the presented model.

This expression represents a kinematic hardening surface of apex (p'_α, q_α) and horizontal axis (Rp'_0). Parameter R represents the scale ratio between the two surfaces.

2.1 *Bonding and damage law*

According to Gens & Nova (1993) framework, in order to define the initial magnitude and the rate of destructuration of the defined surfaces, a non-dimensional variable b is introduced, representing the amount of bonding at a certain strain level. The expression of p'_0 is:

$$p'_0 = p'_{0unc}(1 + b) \tag{3}$$

and the tensile strength is:

$$p'_t = \alpha_t b \tag{4}$$

where α_t is an additional model parameter that can be determined from the initial value of the tensile strength. While p'_0 depends on p'_{0unc}, p'_t is only dependent on the level of bonding. Otherwise, a test in which the contribution of soil densification were predominant over a small loss of structure would result in an increase of tensile strength. With the adopted formulation, p'_t is monotonically decreasing and tends to zero for large bonding damage.

Bonding variable b is dependent on an initial value b_0 (referred to the undamaged material) and a damage variable X (initially nil), measuring the progressive loss of bonding with accumulation of plastic strain:

$$b = b_0 \, e^{-X} \tag{5}$$

The degree of bonding tends towards zero as the damage variable increases. Parameter b_0 depends on the cement content and the soil compaction and can be determined from the results of isotropic compression tests, as described in the following sections.

Substituting (5) into (3) and (4), one obtains:

$$p_0' = p_{0unc}'\left(1 + b_0 e^{-X}\right) \quad (6)$$

$$p_t' = \alpha_t b_0 e^{-X} = p_{ti}' e^{-X} \quad (7)$$

In the last expression, α_t and b_0 are two constant parameters that can be unified in a single parameter, corresponding to the tensile strength of the undamaged material p'_{ti}.

Damage variable X increases monotonically and linearly with the accumulation of volumetric (ε_v^p) and deviatoric (ε_s^p) plastic strain:

$$X = b_1\left|\dot{\varepsilon}_v^p\right| + b_2\left|\dot{\varepsilon}_s^p\right| \quad (8)$$

Where b_1 and b_2 are two model parameters and the dot superscript indicates an increment of the variable.

2.2 *Translation rule*

Translation of the apex of the yield surface (vector $\boldsymbol{\alpha}$) is defined in a similar way as in Al-Tabbaa & Wood (1989):

$$\dot{\alpha} = S\boldsymbol{\beta} \quad (9)$$

where S is a positive scalar (derived applying consistency condition—Prager, 1955) and $\boldsymbol{\beta}$ is the vector joining the current stress point P with the corresponding conjugate point P' on the reference surface (Figure 1).

2.3 *Hardening modulus and flow rule*

Hardening modulus (H) results from the sum of three components, referring to the isotropic hardening (H_0), the evolution of bonding (H_b), and the kinematic hardening (H'). The usual incremental relation between the uncemented component of p'_0 and the volumetric plastic strain regulates isotropic hardening:

$$\frac{\dot{p}_{0unc}'}{p_{0unc}'} = \frac{v_i}{\lambda - \kappa}\dot{\varepsilon}_v^p \quad (10)$$

Where v_i is the initial specific volume and λ and κ are the slopes of the normal compression line (NCL) and of a swelling line, respectively, in the isotropic compression plane. Isotropic modulus H_0 is determined using the precedent relationship; bonding modulus H_b is calculated combining the relations expressed in section 2.1 and applying consistency condition. For kinematic hardening modulus, a convenient expression shall be adopted, consistent with experimental data. The expression adopted is similar to Rouainia & Wood (2001) expression:

$$H' = \frac{v_i}{\lambda - \kappa}\left(B\frac{b}{b_{max}}\right)^{\psi} p_0'^3 \quad (11)$$

B and ψ are interpolating parameters; b is the component of vector $\boldsymbol{\beta}$ in the direction of the normal to the yield surface (Hashiguchi, 1985); b_{max} is the maximum value of b. Such expression is always positive and decreases as the yield surface approaches the reference surface, being nil when the surfaces are in contact.

Several flow rules can be implemented in the model. Rowe's flow rule was selected, as this same stress-dilatancy relation is adopted in Rios et al. (2016). The expression is adapted to the kinematic yield surface:

$$d = \frac{\dot{\varepsilon}_v^p}{\dot{\varepsilon}_s^p} = \frac{9\left(M - \left|\eta_{ref}'\right|\right)}{9 + 3M - 2M\left|\eta_{ref}'\right|}\operatorname{sgn}\left(\eta_{ref}'\right) \quad (12)$$

η'_{ref} is the effective stress ratio of the point on the reference surface corresponding to the actual stress point on the yield surface (Figure 1), and sng(•) is the signum function.

2.4 *Elastic behaviour*

Purely elastic domain is much smaller than in a single-surface model because of the reduced size of the yield surface. Thus, for the uncemented soil, a simple linear elastic relationship can be defined, with constant (independent from stress level) bulk (K'_{unc}) and shear (G_{unc}) moduli. The observed non-linearity is modelled through the elasto-plastic relationship. The relation between elastic moduli follows the usual relation:

$$G_{unc} = \frac{3(1-2\nu)}{2(1+\nu)}K_{unc}' \quad (13)$$

where Poisson's ratio ν assumes the usual value of 0.3. Thus, shear modulus is determined once a value for bulk modulus is selected. For cemented soils, bulk (and shear) modulus increases for increasing cementation. Yu et al. (2007) expression is adopted:

$$K' = K_{unc}'\left(1 + \sqrt{\frac{p_{ci}'}{\sigma_c'}}\right) \quad (14)$$

Effective confining pressure σ'_c is considered instead of the mean effective stress p' in order to have a constant modulus for each test; p'_c is the extra strength in isotropic compression induced by bonding (Figure 1 - suffix i refers to the value at the beginning of the test).

2.5 *Numerical implementation*

The solution of the constitutive equations in the purely elastic domain is evident. The closed-form solution of the boundary value problem constituted by the constitutive equations and the respective boundary conditions is not known in the elasto-plastic domain. Thus, the problem is divided in a set of finite increments and an approximate solution for the system of differential equations is calculated in each interval along an incremental strain path. An explicit stress point algorithm is applied to find an approximate solution of the problem, namely a substepping algorithm with second order Modified Euler integration method with error control. Modified Euler method is a simple, reliable, method for numerical resolution of differential equations commonly used in literature (e.g. Couto Marques, 1984; Sloan, 1987; Potts & Ganendra, 1994; Ding et al., 2015; Ghorbani et al., 2016).

Constitutive equations are rewritten in finite form:

$$\Delta\sigma'_s = \mathbf{D}\Delta\varepsilon_s \tag{15}$$

where $\boldsymbol{D}$ is the elasto-plastic stiffness matrix and suffix s indicates the generic increment. This generic system of equations can be simplified by applying the specific drained or undrained boundary conditions of the particular triaxial compression test performed. For drained conditions, equations (15) become:

$$\begin{pmatrix} \Delta q_s \\ \Delta\varepsilon_{vs} \end{pmatrix} = \frac{3}{9D_{11} - 3D_{12} - 3D_{21} + D_{22}} \begin{pmatrix} 3\det(\mathbf{D}) \\ D_{22} - 3D_{12} \end{pmatrix} \Delta\varepsilon_{as} \tag{16}$$

Terms D_{ij} represent the elements of matrix $\boldsymbol{D}$. ε_a is the (imposed) axial strain. Since the test is drained, the effective stress path is completely determined:

$$\Delta p'_s = \Delta q_s / 3 \tag{17}$$

For undrained conditions, it is:

$$\begin{pmatrix} \Delta p'_s \\ \Delta q_s \end{pmatrix} = \begin{pmatrix} D_{12} \\ D_{22} \end{pmatrix} \Delta\varepsilon_{ai} \tag{18}$$

Pore pressure variation is calculated as the difference between the total and the effective mean stress increment:

$$\Delta u_s = \Delta q_s / 3 - \Delta p'_s \tag{19}$$

In both cases, the algorithm divide the elasto-plastic domain in small axial strain increments and performs two approximate integrations on the interval, considering the stiffness matrix at the beginning and at the end of the interval, respectively. The difference of the two calculations gives an estimate relative error associated to the interval, which is checked against a set tolerance. If the tolerance criterion is not met, the interval length is reduced and the calculation is repeated. The tolerance was set equal to 10^{-9}, as this value is a good compromise between accuracy and computation time.

3 MODEL CALIBRATION

The presented model is calibrated using the monotonic drained and undrained triaxial tests results presented by Rios et al. (2014). Some of the model parameters are shared with Rios et al. (2016) model and can be initially taken as equal. It is expected that the large-strain behaviour (i.e. the configuration with the surfaces in contact) is the same in the two models; conversely, the small-strain behaviour should show different responses. In fact, in the CASM-n model such domain is elastic, while in the present model the kinematic yield law produces plastic strain since an early stage of deformation. A staged approach is followed for the initialization of the five state variables and the calibration of the 13 model parameters.

3.1 *Initialization of surface size in uncemented soil*

The initial value of the horizontal axis of the uncemented surface, p'_{0unc_i}, is calculated as the intersection between the NCL of the uncemented soil and a swelling line passing through the point defining the initial state of the specimen (v_i, p'_i). The usual expression is used:

$$p'_{0unc\_i} = \exp\left(\frac{N - v_i}{\lambda - \kappa}\right) \tag{20}$$

N is the specific volume of the NCL at $p' = 1$ kPa.

3.2 *CASM parameters for uncemented specimens*

These parameters are those shared with CASM model and were determined by Rios et al. (2016).

Table 1. Basic CASM model parameters.

κ	N	λ	M	ν	n	r	K'_{0unc} (kPa)
0.0097	2.35	0.112	1.4	0.3	2.2	3.7	9.75e4

Table 2. Kinematic hardening parameters.

R	B	ψ
0.08	1e-6	1

From this study, the values reported in Table 1 were retrieved. In addition, the uncemented small-strain bulk modulus K'_{0unc} shall be determined. This was determined with a heuristic process by comparing the initial slope of the stress-strain curves calculated in simulated monotonic triaxial tests with the experimental results.

3.3 *Kinematic hardening parameters*

These parameters control the size of the kinematic yield surface (through ratio R) and the interpolating function regulating the kinematic hardening modulus (B and ψ). It is difficult to establish a direct relation between these parameters and one or more physical properties. Thus, their value was established by assigning tentative values and then comparing the results of the simulated test with the experimental results. Attention was focused on the elasto-plastic stiffness in the region inside the reference surface, namely its initial value and decrease with progressive yielding. The three parameters were modified one at a time until a reasonable fit was obtained. The parameters reported in Table 2 were established at the end of this process.

3.4 *Initial position of the yield surface*

The initial coordinates of the yield surface apex ($p'_{\alpha i}$, $q_{\alpha i}$) shall be selected. Before the beginning of the triaxial tests, all the specimens were isotropically consolidated to a specific effective confining pressure, σ'_c. Thus, the stress point in the (p', q) plane has coordinates (σ'_c, 0) at the beginning of triaxial shearing. Since the yield surface was dragged along with the stress state during isotropic compression, one can assume that the initial values of the apex coordinates are:

$$p'_{\alpha i} = \sigma'_c - Rp'_{0i}; \quad q_{\alpha i} = 0 \tag{21}$$

In such configuration, the stress point is in contact with the yield surface from the beginning of the triaxial test. Thus, there is no purely elastic domain and the model response is elasto-plastic from the beginning of shearing.

3.5 *Initialization of bonding state parameters*

The initial bonding parameters are the initial amount of bonding b_0 and the initial tensile strength p'_{ti}. The first parameter can be found by inverting equation (6). Two terms p'_{0i} and p'_{0unc_i} shall be determined. The last term refers to an equivalent uncemented specimen and can be determined using relation (20). The first is composed by three components (Figure 1). Rios et al. (2016) showed that the initial value of p'_c depends on the level of cementation and compaction, or, more specifically, on an adjusted porosity/cement ratio, $n/C_{iv}^{0.21}$ (Rios et al., 2012). The authors present the following correlation:

$$p'_{ci} = 9.95\text{e}9\left(n/C_{iv}^{0.21}\right)^{-4.265} \tag{22}$$

Porosity/cement ratio is a known moulding variable of each specimen (dependent on initial specific volume and cement content). Thus, p'_{ci} is readily calculated from the expression.

Initial tensile strength p'_{ti} was determined indirectly. A series unconfined compression tests (UCT) were performed on the same cemented material, and a similar law to expression (22) was determined relating the unconfined compression strength (q_{UCT}) and $n/C_{iv}^{0.21}$:

$$q_{UCT} = 1.3\text{e}10\left(n/C_{iv}^{0.21}\right)^{-4.527} \tag{23}$$

Assuming that an UCT follows a drained stress path, and in the hypothesis that the ultimate value lies on the reference surface, the value of q_{UCT} corresponds to point U in Figure 1, of coordinates ($q_{UCT}/3$, q_{UCT}).

Substituting the coordinates of point U in equation (1) (solving for $f_{ref} = 0$) and having previously determined p'_{ci} and p'_{0unc_i}, tensile strength p'_{ti} is the only unknown term. This can be determined by solving the implicit equation with an iterative method. Finally, p'_{0i} is determined as the sum of the three calculated terms, and b_0 is determined from equation (6). This procedure shall be performed for each specimen. Cemented bulk modulus K' can be calculated at this stage using expression (14).

3.6 *Calibration of bond damage parameters*

Damage parameters b_1 and b_2 regulate the accumulation of damage with increasing volumetric and distortional plastic strain, respectively. For

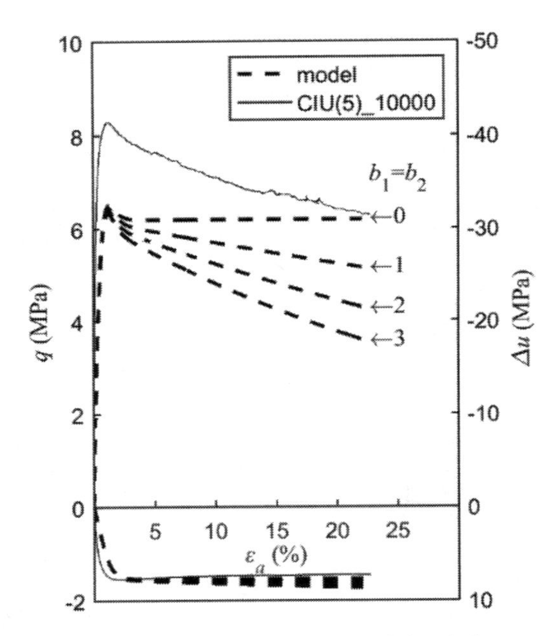

Figure 2. Example of calibration of bond damage parameters.

simplicity, the same value is considered for both parameters.

The calibration process was performed by comparison with the experimental results. A first attempt value was selected equal to zero for both the parameters and the model output compared with the experimental results. Then, the parameter value was successively increased until a good agreement was attained.

An example of such process is shown in Figure 2, where a high confined test (10 MPa) is presented. The influence of damage parameters is more relevant for large strain (especially after critical state has been reached), because the damage increases linearly with the accumulation of plastic strain. A value of $b_1 = b_2 = 3$ was selected as the best fitting value for the damage parameters.

4 PERFORMANCE OF THE CALIBRATED MODEL

4.1 *Uncemented specimens*

In this case it is $b_0 = p'_{ti} = 0$. Thus, bonding hardening modulus is nil.

One drained (CV) and two undrained (CIU) triaxial tests have been modelled (initial characteristics are reported in Table 3, where the denomination used in Rios et al., 2016, is maintained).

Table 3. Initial state variables and index properties for uncemented triaxial tests.

Name	v_i	σ'_c (kPa)	p'_{0unc_i} (kPa)	$p'_{\alpha i}$ (kPa)
CV90(0)_100	1.6	100	1558	–24.6
CIU(0)_30	1.75	30	359.5	1.2
CIU(0)_250	1.75	250	359.5	221.2

Model output is plotted along with the respective experimental results in Figure 3 and Figure 4.

Due to the different confining pressures, the two undrained tests show a dilative and a compressive behaviour, respectively. In general, it can be observed a reasonable agreement between the model and the experimental results. These results can be compared with the corresponding results obtained in the single-surface CASM-n model used by Rios et al. (2016). In the large strain domain (i.e. the post-yield domain in CASM-n model, corresponding to the configuration with the surfaces in contact in the present model) the two models give similar results, confirming the validity of the present implementation.

In the small-strain domain, Rios et al. (2016) adopted a non-linear elastic law. Nevertheless, the authors state that this law cannot reproduce properly the non-linearity of elastic domain (especially the stress path of undrained tests) and that a more complex elastic model would be needed. In the present approach, the simpler linear elastic law, combined with kinematic hardening, is sufficient to improve the response in the initial non-linear domain. In Rios et al. (2016), test CIU(0)_30 shows a vertical effective stress path for deviatoric stress up to 60 kPa. In the present model, a non-linear behaviour is observed from the beginning of the test.

The model is able to reproduce the experimentally observed change in volumetric behaviour from a compressive to dilative tendency. This outcome is obtained thanks to the particular translation rule adopted. In Figure 5, the initial and final configurations of the surfaces are shown for the drained test, along with the stress path and the apex translation. The initial position of the surface results in a positive value of flow rule (12), corresponding to an initial compressive tendency of the material. This behaviour is observed in all the tests, regardless the initial confining pressure. As the stress increases, the yield surface is dragged along with the stress state, following the translation rule. During this process, the value of the flow rule switches from positive to negative values in the tests confined at 30 and 100 kPa. Thus, volumetric

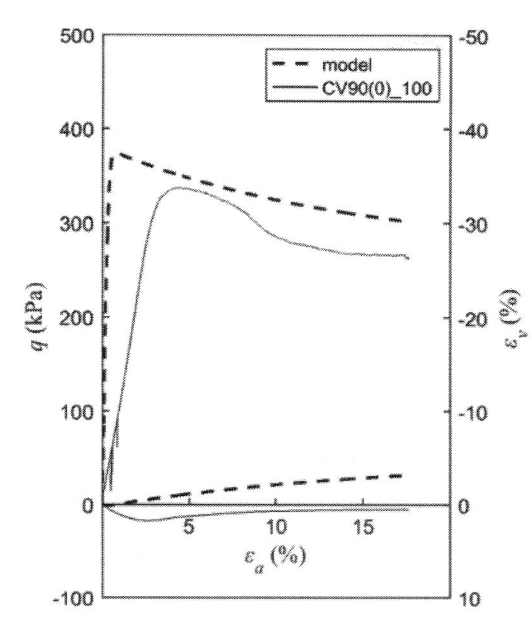

Figure 3. Performance of the model on uncemented drained triaxial tests.

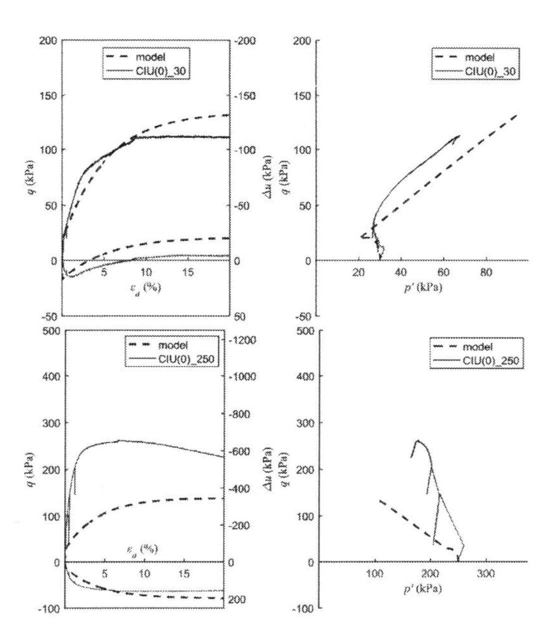

Figure 4. Performance of the model on uncemented undrained triaxial tests.

tendency changes from compressive to dilative. This is due to the relatively low confining pressure applied. In the test confined at 250 kPa, dilatancy

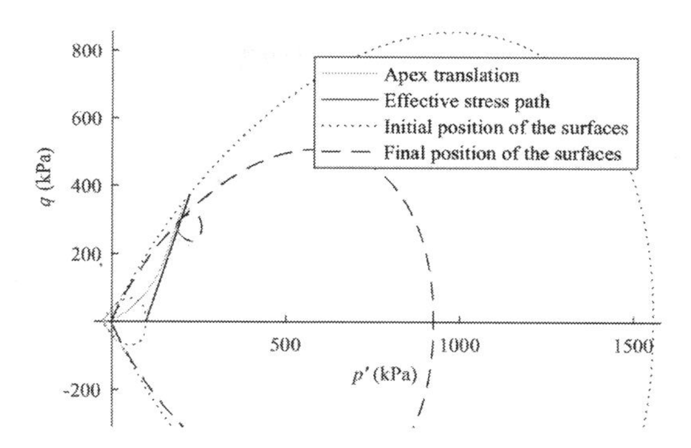

Figure 5. Kinematic and isotropic hardening in drained test CV90(0)_100.

is positive and volumetric tendency is compressive throughout the test.

When yield and reference surfaces are in contact, the two surfaces shrink (or expand) simultaneously, until the stress point reaches the critical stress ratio $\boldsymbol{M}$.

4.2 *Cemented specimens*

Four triaxial tests were selected, two drained (CV) and two undrained (CIU) (Table 4). The model output is plotted along with the respective experimental result in Figure 6 and Figure 7.

Also in this case a similar output at large strain is observed for the two models. In the small strain domain, the present model is more capable of representing the initial non-linearity than the corresponding single-surface model. Drained test CV(5)_250 shows practically the same output in the two implementations. The estimated peak strength in test CV(2)_30, although higher than the experimental result, is more accurate in the present implementation (around 1200 kPa) than in Rios et al. (2016) (~1500 kPa). This is because the dilative tendency of the test causes a shrinkage of the surfaces before they are in contact, while in a single-surface model the yield surface remains unchanged until the stress point is in contact with it. Hence, the progressive yielding observed in two-surface model leads to more accurate results than the sudden yield in the single-surface approach.

The most evident improvement is observed in the undrained effective stress paths presented in Figure 7. Soil non-linearity is well represented from an early stage of deformation, as well as the change in volumetric behaviour from an initial compressive tendency to dilation. In Figure 8, the initial and final configurations of the surfaces and the path followed by the yield surface are shown for an undrained test. As expected, as stress increases, reference surface shrinks as an effect of the

Table 4. Initial state variables and index properties for cemented triaxial tests.

Name	v_i	C	$n/C_{iv}^{0.21}$	σ'_c (kPa)	p'_{0i} (kPa)	b_0	p'_{ti} (kPa)	p'_{ci} (kPa)
CV(2)_30	1.61	2	36	30	4057	1.87	353	-295
CV(5)_250	1.58	5	29	250	8775	3.63	1117	-452
CIU(2)_250	1.60	2	36	250	4194	1.69	344	-85
CIU(5)_250	1.58	5	29	250	8775	3.96	1117	-452

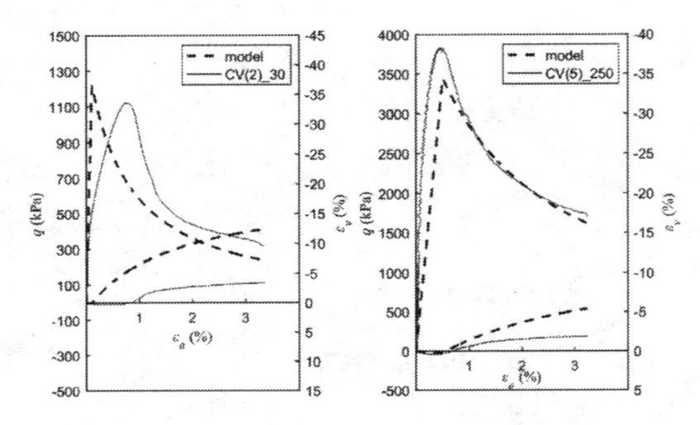

Figure 6. Performance of the model on cemented drained triaxial tests.

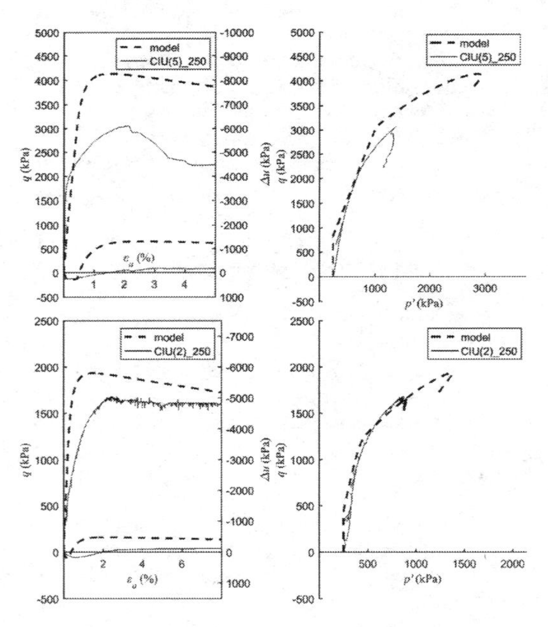

Figure 7. Performance of the model on cemented undrained triaxial tests.

reduction in p'_t and p'_c. Yield surface approaches the reference surface and translates in contact with it afterwards, tending towards the critical state

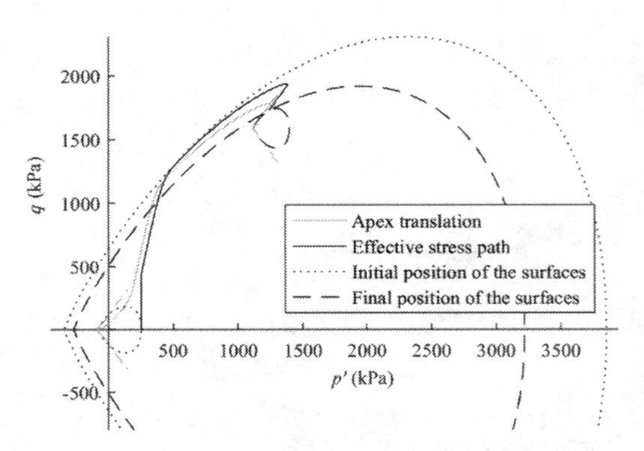

Figure 8. Kinematic and isotropic hardening in undrained triaxial test CIU(2)_250.

stress ratio. A further shrinkage is observed after this state is reached, because of the progressive loss of cementation due to the incremental distortional plastic strain.

5 CONCLUSIONS

A novel constitutive model has been presented, based on boundary surface plasticity and kinematic hardening, equipped with a framework for bonded soils.

The comparison performed between the proposed model and an existing implementation of a single-surface, isotropic hardening model showed the advantages of the present implementation for modelling uncemented and cemented soil. The large-strain output is very similar for the two models, while in the small-strain domain a better performance of the kinematic hardening approach is observed, due to the specific capacity of this framework to model non-linear behaviour since an early stage of deformation.

The shortcoming of the improved accuracy is an increased number of parameters to be calibrated. Nevertheless, the type and number of experimental tests needed for calibration is the same as in CASM-n model.

ACKNOWLEDGEMENTS

The authors would like to acknowledge to the Portuguese Science and Technology Foundation (FCT) through SFRH/BD/92810/2013scholarship, which is co-funded by the Portuguese Ministry of Science and Technology (MCTES).

REFERENCES

Al-Tabbaa, A. & Wood, D. 1989. An experimentally based 'bubble' model for clay. In S. Pietruszczak & G.N. Pande (eds), *Proc. Numerical models in geomechanics (NUMOG 3), Niagara Falls, Canada, 8–11 May 1989*: 91–99. London: Elsevier Applied Science.

Carol, I., Rizzi, E., & Willam, K. 2001. On the formulation of anisotropic elastic degradation. I. Theory based on a pseudologarithmic damage tensor rate. *Int. J. Solids Struct.* 38(4): 491–518.

Couto Marques, J.M.M. 1984. Stress computation in elastoplasticity. *Engineering Computations* 1(1): 42–51.

Dafalias, Y.F. & Popov, E.P. 1975. A model of nonlinearly hardening materials for complex loading. *Acta Mechanica* 21(3): 173–192.

Desai, C.S., Somasundaram, S., & Frantziskonis, G. 1986. A hierarchical approach for constitutive modelling of geologic materials. *International Journal for Numerical and Analytical Methods in Geomechanics* 10(3): 225–257.

Ding, Y., Huang, W., Sheng, D., & Sloan, S.W. 2015. Numerical study on finite element implementation of hypoplastic models. *Computers and Geotechnics* 68: 78–90.

Gens, A. & Nova, R. 1993. Conceptual bases for a constitutive model for bonded soils and weak rocks. In Anagnostopoulos et al. (eds), *Geotechnical Engineering of Hard soils—Soft Rocks; Proc. intern. symp. under the auspices of the ISSMFE, Athens, 20–23 September 1993*: 485–494. Rotterdam: Balkema.

Ghorbani, J., Nazem, M., & Carter, J.P. 2016. Numerical modelling of multiphase flow in unsaturated deforming porous media. *Computers and Geotechnics* 71: 195–206.

Hashiguchi, K. 1985. Two and three surface models of plasticity. In Kawamoto & Itikawa (eds), *Proc. 5th Int. Conf. on Numerical Methods in Geomechanics, Nagoya, 1–5 April 1985*: 125–134. Balkema, Rotterdam.

Krieg, R.D. 1975. A practical two-surface plasticity theory. *Journal of Applied Mechanics* 42: 641–646.

Nova, R. 1988. Sinfonietta classica: an exercise on classical soil modelling. In Saada & Bianchini (eds), *Proc. Symp. Constitutive Eq. for Granular Non-Cohesive Soils, Cleveland, USA, 23 July 1987*: 501–520. Rotterdam: Balkema.

Potts, D.M. & Ganendra, D. 1994. Evaluation of substepping and implicit stress point algorithms. *Computer Methods in Applied Mechanics and Engineering* 119: 341–354.

Prager, W 1955. The theory of plasticity—a survey of recent achievements. In: *Proc. Inst. Mech. Eng.*: 3–19. London.

Rios, S., Viana da Fonseca, A., & Baudet, B.A. 2012. The Effect of the Porosity/Cement Ratio on the Compression of Cemented Soil. *Journal of Geotechnical and Geoenvironmental Engineering* 138(11): 1422–1426.

Rios, S., Viana da Fonseca, A., & Baudet, B.A. 2014. On the shearing behaviour of an artificially cemented soil. *Acta Geotechnica* 9(2): 215–226.

Rios, S., Ciantia, M., Gonzalez, N., Arroyo, M., & Viana da Fonseca, A. 2016. Simplifying calibration of bonded elasto-plastic models. *Computers and Geotechnics* 73: 100–108.

Rouainia, M. & Wood, D. 2001. Implicit numerical integration for a kinematic hardening soil plasticity model. *International Journal for Numerical and Analytical Methods in Geomechanics* 25: 1305–1325.

Sloan, S.W. 1987. Substepping schemes for numerical integration of elasto-plastic stress-strain relations. *International Journal for Numerical Methods in Engineering* 24: 893–911.

Vaunat, J. & Gens, A. 2003. Numerical modelling of an excavation in a hard soil/soft rock formation using a coupled damage/plasticity model. In *Proc. 7th Int. Conf. on Computational Plasticity (COMPLAS 2003), Barcelona, 7–10 April 2003*. CD-ROM.

Viana da Fonseca, A. 2003. Characterising and deriving engineering properties of a saprolitic soil from granite, in Porto. In Tan et al. (eds.), *Characterization and Engineering Properties of Natural Soils*: 1341–1378. Swets and Zeitlinger, Lisse.

Viana da Fonseca, A., Carvalho, J., Ferreira, C., Santos, J.A., Almeida, F., Pereira, E., Feliciano, J., Grade, J., & Oliveira, A. 2006. *Characterization of a profile of residual soil from granite combining geological, geophysical and mechanical testing techniques*. Geotechnical and Geological Engineering 24(5): 1307–1348.

Yu, H.S. 1998. CASM: A Unified state parameter model for clay and sand. *International Journal for Numerical and Analytical Methods in Geomechanics* 22: 621–653.

Yu, H.S., Tan, S.M., & Schnaid, F. 2007. A critical state framework for modelling bonded geomaterials. *Geomechanics and Geoengineering* 2(1): 61–74.

Numerical Methods in Geotechnical Engineering IX – Cardoso et al. (Eds)

Incorporation of creep into an elasto-plastic soil model for time-dependent analysis of a high rockfill dam

P. Pramthawee & P. Jongpradist
King Mongkut's University of Technology Thonburi, Bangkok, Thailand

ABSTRACT: This article proposes an approach to extend an elasto-plastic soil model into the time-dependent analysis of high rockfill dam. The key feature of the model is the incorporation of a creep model and the double yield surface elasto-plastic model, via modifying the hardening functions. The incorporated model is implemented into ABAQUS, an FEM program through user defined subroutine. The model and its implementation are validated by a laboratory multistage creep tests of both sand and rockfill.

1 INTRODUCTION

The finite element method (FEM) has been broadly used as a tool for the stress and deformation analysis of several rockfill dams modelled under various essential situations. A crucial element of this approach is a constitutive model, providing the essential relation between the strains and the stresses. It is widely known that rockfills are time-dependent materials of increasing interest to many researchers. The time-dependent analysis of rockfill dams has mostly received attention regarding long-term deformation affecting the dam safety during the operational stage (Arici, 2011; Dolezalova and Hladik, 2011; Zhou, et al., 2011) and has been omitted for the stage of construction and first impounding. The processes of the construction and impoundment stages for high concrete faced rockfill dams (CFRDs), nevertheless, are time-consuming, taking approximately 3–6 years. The rockfill creep of a high CFRD mostly occurs during the construction process and clearly influences the stresses and deformations of the concrete face as well (Zhou, et al., 2010) Moreover, dam deformation analyses without consideration of time dependence in the past (Lollino, et al., 2005; Xu, et al., 2012; Sukkarak, et al., 2017) have commonly provided underestimated values of the predicted settlement. Therefore, it is more reasonable to consider the rockfill creep in the dam simulation during the construction and first impoundment stages (Kim, et al., 2014).

To reasonably analyze the deformation in the case of a high rockfill dam, the complexity of construction sequences, a long construction period and a long impoundment period make it necessary to consider many loading increments in the analysis together with the creep between them. For this case, the strategy currently used in FEM is to perform a time-independent analysis for each incremental loading step and a time-dependent analysis for each constant stress state, alternately. The essential elements are thus (i) reasonable and rational constitutive models for both time-independent and time-dependent behaviors and (ii) the appropriate incorporation of those two models.

This article focuses on the second element; the incorporation between the creep and elasto-plastic models. The hardening functions are modified in order to link an interaction between the creep model and modified HS model. Then, the incorporated model is applied to the FEM computer program ABAQUS via the defining subroutine UMAT. The validation is made by comparison of the stress-strain curves between the simulated results and the testing results of Virginia Beach sand and rock fill of Nam Ngum 2 dam.

2 INCOPORATION OF THE CREEP MODEL AND THE DOUBLE YIELD SURFACE ELASTO-PLASTIC MODEL

2.1 *Concept*

The modified hardening soil (MHS) model (Sukkarak, et al., 2017), an elasto-plastic model is selected as the basis model to extend for time-dependent analysis in this study, and the approach to extend the model for time-dependent analysis will be described in this section.

Lade (1994) concluded that granular materials become stiffer and the yield surface moves out as a result of creep deformation. He also predicted the movement and new location of the yield surface during creep. Correspondingly, Lade and Lui (1998) have indicated the yield surface and the plastic potential surface of granular material moving out together during creep. They also stated that the nature of creep strain is similar to that of plastic strain, which may be predicted from the same framework of the hardening plasticity theory. These indicate that the creep strain, which involves inelastic strain, occurs inside the plastic yield surface that is moving out to higher stresses. Therefore, the stress state of that creep, which is assumed to be constant, becomes within the elastic region before further loading. The creep model and the MHS model are incorporated by assuming that the yield surfaces are constant in shape and move away from the given stress state for creep.

2.2 *Incorporation of the models*

The assumption is the yield surfaces expand (during creep) as increasing creep strains. These creep strains are the plastic strains defined in elasto-plasticity theory. Accordingly, the MHS model will be the function of time via modifying the hardening functions with creep strain.

Figure 1 illustrates the assumption of the yield surface and plastic potential movement during the creep strain change. Point A and B are the yield points for a particular stress state on a monotonic stress-strain curve of each imposed confining pressure obtained from the general triaxial tests of rockfill material. The curve is presumed as a reference curve for the movement prediction of the yield surface during creep. If q_A is constant during given creep time t_c, an increase of the creep strain from point A to B at each time interval Δt_c will move the yield point A to C. Hence, the rockfill becomes stiffer, and additional loading (B to C) occurs under the elasticity regime. After a given loading rejoins at point C, the stress state will continually place on the reference curve, where the elasto-plasticity theory will be employed. By this concept, the creep strains, which are the plastic strains varying with time, are imposed as a function of the hardening parameter to predict the new location of the yield surface for the time-dependent approach. Therefore, the modified hardening laws $h(\varepsilon^p, t_c)$ for time-dependent analysis are written as a correlation of plastic strain and time, for Shear and Cap surface respective, as:

$$\gamma^p = 2\varepsilon_1^c(t_c) \tag{1}$$

$$p_p = \beta_c \left(\frac{\sigma_3' + c \cot\varphi_p'}{p^{ref} + c \cot\varphi_p'} \right)^m \varepsilon_v^c(t_c) \tag{2}$$

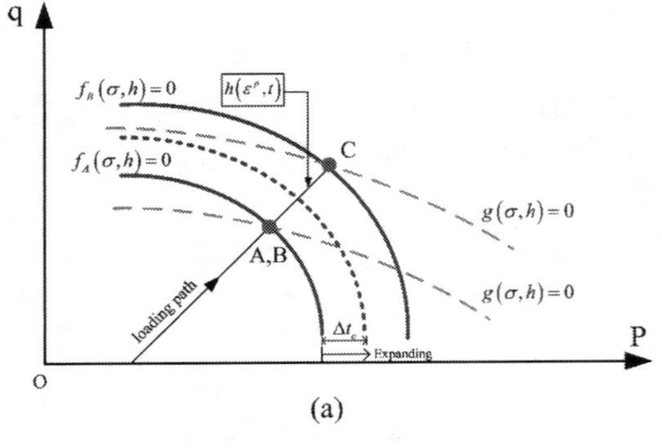

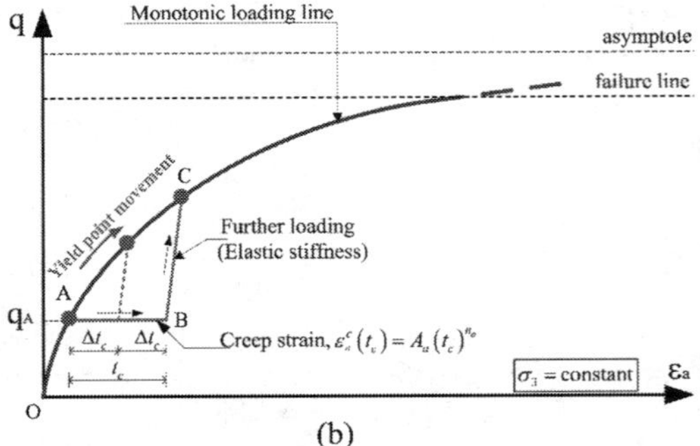

Figure 1. Idealized movements of the yield surface and the plastic potential during creep strain change (Source: Pramthawee, et al., 2017).

where:

T_c = creep time;
ε^p = plastic strain;
γ^p = hardening parameter for shear surface;
ε_s^{ep} = plastic shear strain obtained from shear surface;
ε_s^c = plastic shear strain from the creep model;
P_p = hardening parameter for cap surface;
ε_v^{pc} = plastic volumetric strain from the cap model;
ε_v^c = plastic volumetric strain from the creep model.

The creep strains in Eqs. (1) and (2) can be obtained from:

$$\varepsilon_a^c(t) = A_a(t)^{n_a} \tag{3}$$

$$\varepsilon_v^c(t) = A_v(t)^{n_v} \tag{4}$$

The initial creep parameter $A_a \& A_v$ and the creep rate parameter $(n_a \& n_v)$ can be determined from the best fitting curve of correlation between strain and time (Pramthawee et al., 2016). The Prandtl–Reuss flow law is employed for general stress space as:

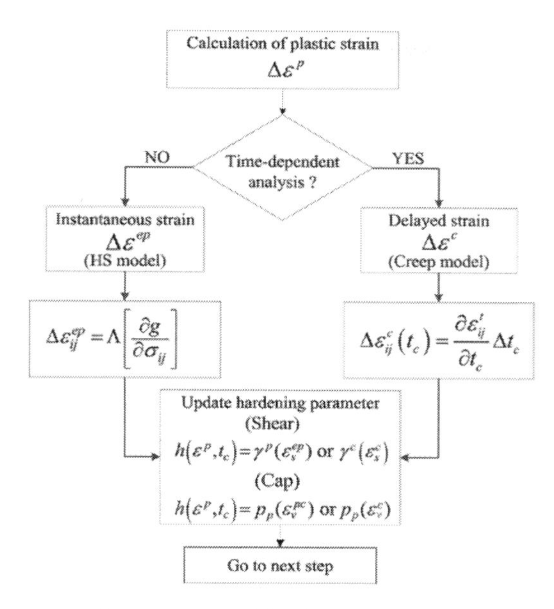

Figure 2. Schematic of hardening parameter calculation (Source: Pramthawee, et al., 2017).

$$[\dot{\varepsilon}^{c}] = \frac{1}{3}\dot{\varepsilon}^{c}_{v}[I] + \frac{1}{2q}\dot{\varepsilon}^{c}_{s}[S] \tag{5}$$

where $[I]$ is a unit tensor and $[S]$ is a deviatoric stress tensor. The shear creep strain rate $\dot{\varepsilon}^{c}_{s}$ can be calculated as follows:

$$\dot{\varepsilon}^{c}_{s} = \frac{\sqrt{2}}{3}(3\dot{\varepsilon}^{c}_{a} - \dot{\varepsilon}^{c}_{v}) \tag{6}$$

Another assumption of the model is the categorization of the plastic strains into two groups, i.e., instantaneous strain and delayed strain. The instantaneous strains can be derived from the MHS model while the delayed strains can be obtained from the creep model. Both groups interact with each other via the hardening parameter, and the algorithm is illustrated by the schematic in Figure 2. When dealing with time-independent analysis the plastic strains in Eqs. (1) and (2) are derived from the MHS model. The instantaneous strain and the delayed strain are computed at each step of calculation, alternately. It is thus necessary to determine if which one the current step is.

3 IMPLEMENTATION INTO THE FEM COMPUTER PROGRAM

The incorporated model is implemented in the FEM program ABAQUS/Standard via a user-defined material subroutine known as 'UMAT'. With this module, there are three tasks must be executed namely:

i. Update the stresses at the end of time increment;
ii. Determine the Jacobian (DDSDDE), or tangent stiffness;
iii. Update the state variables, which are the hardening parameters for this study, at the end of the time increment.

In order to achieve these tasks, it is necessary for the user to specify a time condition, i.e., time-dependent or time-independent, and the creep

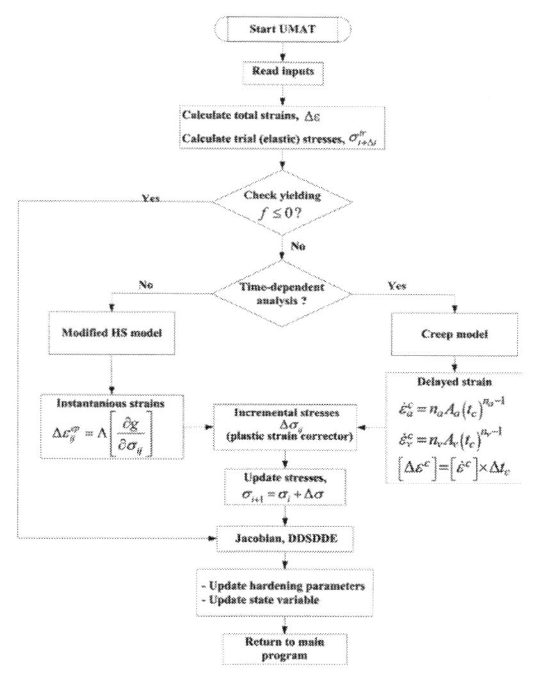

Figure 3. Algorithm of UMAT in ABAQUS (Source: Pramthawee, et al., 2017).

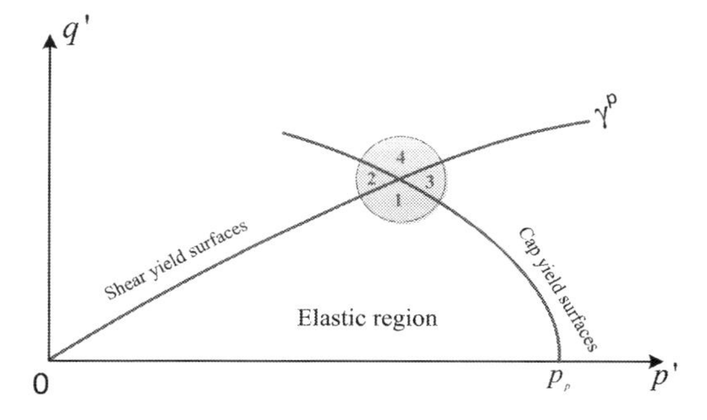

Figure 4. Classification of the model responses.

time at each step of the calculation. The algorithm of the model in UMAT code implemented in ABAQUS is demonstrated schematically in Figure 3.

Since the model is based on double hardening yield surfaces, its character is yet similar to general that type of model. The model response, which depends on the direction of the stress path, can be classified into four cases as shown in Figure 4.

If the stress path is enclosed within both yields surfaces (i.e. zone 1), the nonlinear elastic response is active. If the stress state is in zone 2, plastic shear strain hardening is active only. The shear surface expands following its hardening law while the cap surface is dormant. Correspondingly, if the stress path directs into zone 3, the shear surface is dormant and the cap surface expands according to its hardening law. For the last one, both yield surfaces are activated simultaneously if the stress path is reached zone 4. Both yield surfaces expand following their laws. This case is the hardest one regarding the corner problem for implementation.

4 VALIDATION OF HSC

To validate the HSC model, creep tests carried out on Virginia Beach sand by Karimpour (2012) and rock fill from construction of NamNgum 2 (NN2) concrete faced rockfill dam (CFRD) in Lao People's Democratic Republic, are adopted in this research. Stress-strain curves and axial and volumetric strain curve results of a set of creep tests performed under multiple loading on a sample are used to compare with those results simulated by the incorporated model. The interaction between creep stage and further shearing at each desired deviator stress level is a main point of the verification. Strain hardening type of integration method for time-dependent analysis is applied to model the behaviors of both materials. The objective is to assess the performance of the incorporated model.

4.1 *Experimental data*

Multiple creep tests on Virginia Beach sand have been performed on triaxial apparatus under fully drained condition by Karimpour (2012). The nominal diameter and height of specimens are 38.1 and 108.0 mm, respectively. The sample was sheared at a strain rate of 0.0416%/min under confining pressure of 8,000 kPa. It was allowed to creep for 1 day once the desired deviator stresses had been reached at 6,210, 11,140, and 13,550 kPa.

The rockfill materials for NN2 CFRD project has been performed creep tests on a large oedometer apparatus. The confining cylinder is 1.24 m high having an outside diameter of 0.762 m and the inner diameter of 0.726 m. The rockfill material was placed and compacted in layers of 0.25 m thick by a vibrating plate. The final height of rockfill specimen was 1.05 m. Five steps of loading were applied to the specimen with the axial stress from 0 MPa to 1.80 MPa, namely the axial stresses of 0.00, 0.15, 0.30, 0.60, 1.20 and 1.80 MPa. These axial stresses were maintained constantly for a period 3 to 6 days in each step while the change in the axial stress was monitored. At the maximum axial stress of 1.80 MPa, the load was maintained for two months after that which the specimen was unloaded stepwise from 1.80 MPa to 0.15 MPa, each step 1.20, 0.60, 0.30 and 0.15 MPa. Each unloaded step was maintained for one day. Finally, the specimen was reloaded stepwise from 0.15 MPa to 1.80 MPa. Each reloaded step of 0.15 MPa to 1.20 MPa was maintained for one day while at the maximum axial stress of 1.80 MPa was maintained for ten days. Due to this thesis only focus on the behaviours of rockfill dam during construction and first impounding stage, the unloaded and reloaded steps of simulation for one-dimensional creep test are neglected. Consequently, the parameter determination in this case of simulation employs the creep testing results of the loading steps only.

4.2 *Parameter determination*

Two sets of parameters are required for the HSC model. One is of time-independent analysis which the parameters can be obtained in the same way as the MHS model. For this study, time-independent parameters of both materials are derived by following Sukkarak et al., (2017).

4.2.1 *Virginia Beach sand*

The HSM parameters are shown in Table 1. According to the regression analysis approach suggested by Pramthawee et al. (2016), the set of time-dependent parameters can be obtained from the experimental results of multiple creep tests. These parameters are summarized in Table 2.

4.2.2 *NN2 Dam rockfill*

The HSM parameters are shown in Table 3. The creep parameters have been derived from one-dimensional creep test results of rockfill material. Based on the recommended procedure for parameter determination, the computed results of strain-time curves for various stress states using a single set of parameters are in good agreement with the creep test. This set of parameters presented in Table 4 is used as the time-dependent parameters to simulate one-dimensional multiple creep tests.

Table 1. Time-independent parameters of Virginia Beach sand.

c	φ(	ψ(	m	ν_{ur}	p^{ref}	R_f	OCR	E_{ur}^{ref}	E_{oed}^{ref}	E_{50}^{ref}
(kPa)					(kPa)			(MPa)	(MPa)	(MPa)
1	37	0	0.08	0.3	8000	0.9	1	1560	520	520

Table 3. Time-independent parameters of NN2 rockfill.

c	φ(	ψ(	m	ν_{ur}	p^{ref}	R_f	OCR	E_{ur}^{ref}	E_{oed}^{ref}	E_{50}^{ref}
(kPa)					(kPa)			(MPa)	(MPa)	(MPa)
1	42	-5	0.68	0.3	100	0.68	1	60	17	20

Table 2. Time-independent parameters of Virginia Beach sand.

Parameters	α	β	η	m
$\varepsilon_a^c(t)$	0.000004	0.556	0.0227	0.6510
$\varepsilon_v^c(t)$	0.000002	0.621	0.0370	0.5171

Table 4. Time-independent parameters of NN2 rockfill.

Parameters	α	β	η	m
$\varepsilon_a^c(t)=\varepsilon_v^c(t)$	0.000005	5.413	0.0103	0.4473

4.3 *Finite element mesh and condition*

The multiple creep tests performed on triaxial apparatus are simplified in FEM program ABAQUS. By taking advantage of the symmetry of the sample, a quarter of soil sample is modelled as a single element with four nodes. The finite element mesh and boundary conditions are illustrated in Figure 5. The material subroutine coded as UMAT will be called by ABAQUS at each integration point.

In case of the multiple creep tests, the simulation has been performed under stress control condition. It consists of eight steps following 4 sets of loading and creep steps as did in the test. The creep steps were operated until given creep time t_c was reached at 1,440 minutes. Time intervals or incremental sizes of creep time Δt_c are 1 minute.

In the computation of each step, time period might be divided into many sub-steps related to an incremental size of applied load or displacement. The magnitude of sub-step very influences on the performance of calculation. Note that the setting time period of each step is very important. Therefore, timescale of shearing step and creep step must be identical. In this study, the time period of shearing is a minute which is equal to time increments Δt_c of the creep or stress relaxation step. The time period of the creep or stress relaxation step is set as 1440 minutes. These are made to assume that the instantaneous strain occurs within 1 minute.

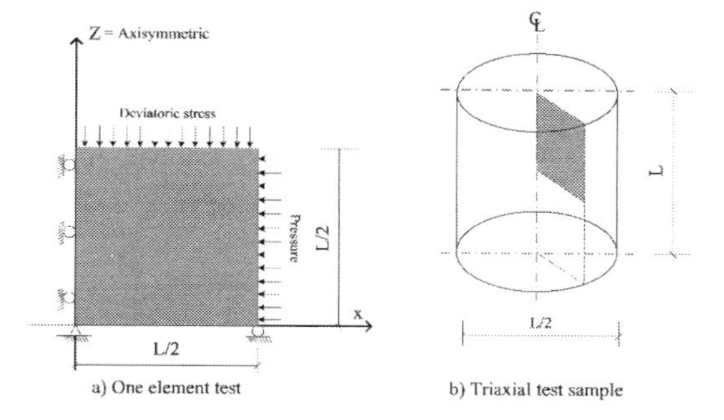

Figure 5. Finite element mesh and boundary conditions for simulation of the multiple creep tests Virginia Beach sand.

For NN2 CFRD rock fill, similar methodology is applied.

5 COMPARISON OF RESULTS

5.1 *Virginia Beach sand*

Comparison between experimental data and simulation results for the multiple creep tests of Virginia Beach sands is presented in Figure 6. It can be seen

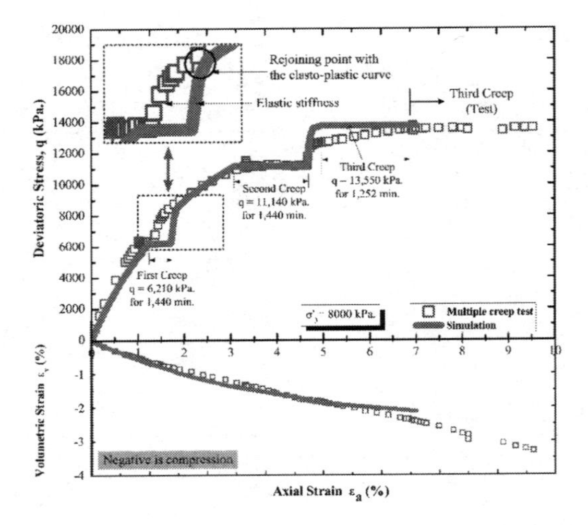

Figure 6. Comparison between experimental data of multiple creep tests and simulation results by using strain hardening integration method.

that the model simulations can reproduce the key characteristics of the sample under shearing and creep stages pleasingly. After creep, the result shows stiffer behaviour (supposing as elastic stiffness) than the previous shearing stage. Then, the behaviour softens to becoming elasto-plastic state at somewhere in which the stress state reaches the yielding state. These are an interaction between shearing and creep stages affecting the movement of yielding state which can be described by the incorporated model. For correlation between volumetric strain and axial strain, the incorporated model gives the closely similar result to measured data for both cases.

Yet, there are some deviations between measured data and model simulation. Mostly, these are about the difference in quantitative prediction. This error will be accumulated and then will be obviously seen in a range of high axial strain.

5.2 *Nam Ngum Dam rockfill*

Figure 7 shows the comparison between the testing results and the simulated results of multiple creep tests performed on large oedometer apparatus. The test program includes 5 loading steps and 4 unloading steps. In each step, the axial stress was maintained constant during the creep period. The prediction result displays the similar tendency with the testing data but some deviations appear. Mostly, the deviations occur during loading stages. Perhaps, the difference of experimental data (large triaxial tests) used to obtain time-independent parameters for this simulation is the cause for this

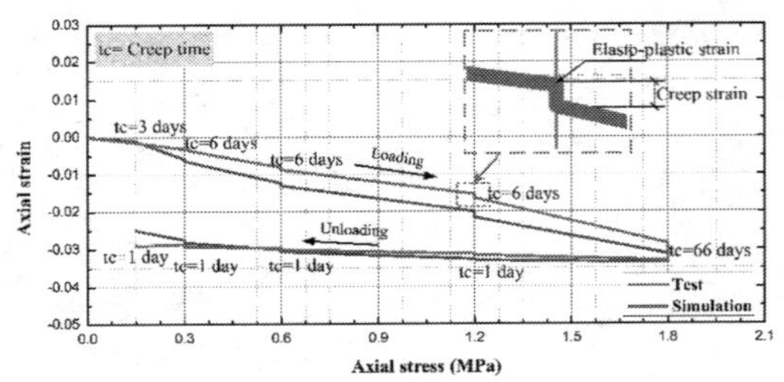

Figure 7. Relationship between axial stress and axial strain at the end of the creep period of each loading step.

error. The magnitude of axial creep strain of each step is largely similar to the measured data except at the maximum axial stress. For unloading stages, the HSC model can reproduce the measured data for both unloading and creep stages pleasingly. These demonstrate that the HSC model is moderately able to simulate the combination of elatoplastic and creep behavior of the rockfill material under one-dimensional loading and unloading stages. It can be noticed that the creep strain under the elastic condition (unloading stage) can be predicted by the HSC model. Nevertheless, an interaction between this creep and further loading under the elastic condition is unclear. For this study, it is assumed that the creep stain can occur under elastic state but not affect the yielding state.

6 CONCLUSIONS

In this article, the performance of incorporated model is validated. The multiple creep of Virginia Beach sands performed on triaxial tests have been simulated in finite element program ABAQUS with employing the incorporated model to represent the material behaviours. In the multiple creep case, the strain hardening type of integration methods have been taken into account. From the comparison, the simulated results can capture the key characteristics of the sample under shearing and creep stages pleasingly.

Furthermore, the simulation on the multiple creep tests of rockfill material in ABAQUS with employing the incorporated model has been carried out in order to evaluate the capability of the HSC model on rockfill materials under one-dimensional condition. The comparison between the experimental results and the modelling results proves that the incorporated model is moderately able to capture the instantaneous and delayed strains of the rockfill material under one-dimensional loading and unloading stages. These demonstrate that the incorporated model can represent the key characteristics under time-dependent analysis of granular materials.

REFERENCES

Arici, Y.L. 2011. Investigation of the cracking of CFRD face plates. *Computers and Geotechnics* 38: 905–916.

Dolezalova, M.& Hladik, I. 2011. Constitutive models for simulation of field performance of dams. *International Journal of Geomechanics*. 11(6): 477–489.

Karimpour, H. 2012. Time effects in relation to crushing in Sand. Ph.D. dissertation, Department of Civil Engineering, The Catholic University of America, Washington, D.C.

Kim, Y.S., Seo, M.W., Lee, C.W. & Kang, G.C. 2014. Deformation characteristics during construction and after impoundment of the CFRD-type Deagok Dam. Korea. *Engineering Geology* 178: 1–14.

Lade, P.V. 1994. Creep effects on static and cyclic instability of granular soils. *Journal of Geotechnical Engineering* 120(2): 404–419.

Lade, P.V. & Liu, C.T. 1998. 'Experimental study of drained creep behavior of sand. *Journal of Engineering Mechanics* 124(8): 912–920.

Lollino, P., Cotecchia, F., Zdravkovic, L. & Potts, D.M. 2005. Numerical analysis and monitoring of Pappadai dam. *Canadian geotechnical journal* 42(6): 1631–1643.

Pramthawee, P., Jongpradist, P. & Sukkarak R., 2017. Integration of creep into a modified hardening soil model for time-dependent analysis of a high rockfill dam. *Computers and Geotechnics* 91: 104–116.

Pramthawee P., Jongpradist P., Phutthananon C. & Sukkarak, R. 2016. Incorporation of creep model with hardening soil model for deformation analysis of rockfill dam. *The 2016 World Congress on Advances in Civil, Environmental, and Materials Research, Jeju Island, Korea, August 28-September 2016*:1–16.

Sukkarak, R., Pramthawee, P. & Jongpradist, P. 2017. A modified elasto-plastic model with double yield surfaces and considering particle breakage for the settlement analysis of high rockfill dams. *KSCE Journal of Civil Engineering* 21(3): 734–745.

Xu, B., Zou, D. & Liu, H. 2012. Three-dimensional simulation of the construction process of the Zipingpu concrete face rockfill dam based on a generalized plasticity model. *Computers and Geotechnics* 43:143–154.

Zhou, W., Chang, X.L., Zhou, C.B. & Liu X.H. 2010. Creep analysis of high concrete-faced rockfill dam. *International Journal for Numerical Methods in Biomedical Engineering* 26: 1477–1492.

Zhou, W., Hua, J.J., Chang, X.L. and Zhou, C.B., 2011. Settlement analysis of the Shuibuya concrete-face rockfill dam. *Computers and Geotechnics* 38: 269–280.

Numerical Methods in Geotechnical Engineering IX – Cardoso et al. (Eds)
© 2018 Taylor & Francis Group, London, ISBN 978-1-138-33198-3

Numerical simulation of a SHTB system for a constant-resistance large-deformation bolt

He Manchao, Gong Weili & Li Chen
SKL-GDUE, China University of Mining and Technology, Beijing, China

Luis R. Sousa
SKL-GDUE, China University of Mining and Technology, Beijing, China
Construct, University of Porto, Portugal

ABSTRACT: The paper introduces a Constant-Resistance-Large-Deformation (CRLD) bolt which has been developed at State Key Laboratory for Geomechanics and Deep Underground Engineering (GDUE), in Beijing. Advancements in Split-Hopkinson Tension Bar (SHTB) tests were recently developed for CRLD bolts and experimental test results for these bolts were obtained. For a better interpretation, complex numerical simulations using several softwares were done for the case of one bolt, which had the dual goal of verifying the experimental and numerical results. In addition, the numerical predictions permitted to do a deep analysis of the behavior of the bolt focusing the development of large deformations.

1 INTRODUCTION

Bolts and anchors are efficient measures for the control of rockburst in underground excavations and also for large deformations that can occur during excavations (Camiro, 1995; Tang et al., 2010; He et al., 2011; He et al., 2015a; He et al., 2017). They were widely investigated by many researchers worldwide and different types are in existence, like the Cone bolt type of anchor bolt and the rock bolt Roofex (He and Sousa, 2014). Nevertheless, given that mining depths advance deeper, the demands for anchors or bolts with larger extension and higher loading capacity are growing.

This paper introduces the state-of-the-art advancements in anchor and bolt technology that resulted in CRLD bolts or anchors developed by GDUE, at CUMTB. The developed CRLD bolt or anchor has the ability to accommodate larger deformations of the adjoining rock masses at great depths in response to external forces (He et al., 2015b). A large number of tests have been conducted and developed for this bolt and test results showed that its mechanical properties are unique and can keep the constant resistance during elongation in a very good performance. Also in situ tests were performed in order to control dynamic loads introducing by rockbursts or by blasting. The feasibility of the new CRLD bolt or anchor was successfully verified, which is expected to have a significant role in the control and prevention of the occurrence of rockburst and large deformations (He and Sousa, 2014). Numerical simulations of the experimental studies were developed to verify and complement the experimental results (Castro et al., 2012; He et al., 2017).

2 CRLD BOLT

2.1 *Concepts*

Bolts are a major method used in rock support in many practical situations. However particularly in high stressed rock masses and for large deformation situations, bolts can break if they are not able to adapt to the induced large deformations. A new bolt or anchor with constant working resistance under large deformations and impact loads has been developed. The bolt device consists of two parts, the constant resistance element and the bolt rod. Figure 1 illustrates the layout of the bolt, the constant resistance element is composed of a slide track sleeve and a constant-resistance body.

The CRLD bolt can withstand pulse-type energy impact for many times while retaining its supporting performance and thus is suitable for rockburst support. In order to evaluate the anti-impact properties of this anchor (bolt), dynamic load impact tests were carried out (He et al., 2007). The development of CRLD bolt has been under testing both in situ and laboratory. The maximum extension of rock bolt for CRLD is about

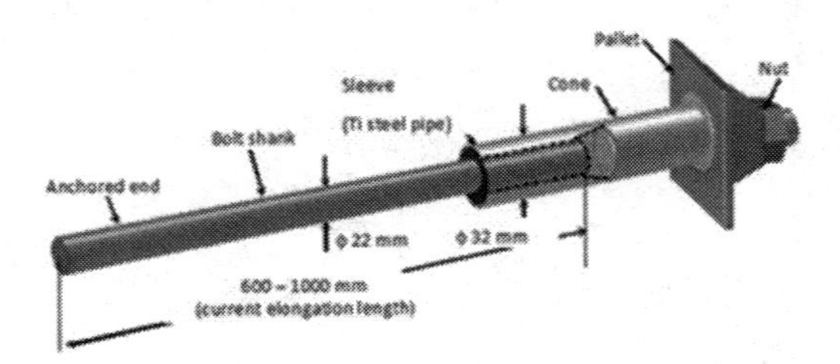

Figure 1. Layout of the CRLD bolt.

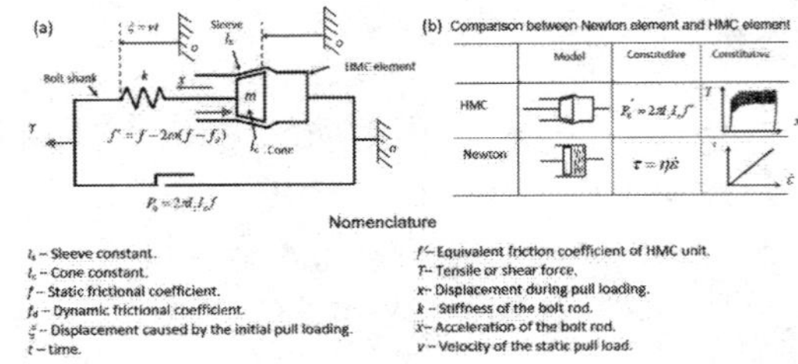

Figure 2. Schematic illustration of the lumped-mass model of the CRLD bolt; (a) the lumped-mass model, and (b) a comparison of the HMC unit with the conventional Newton unit.

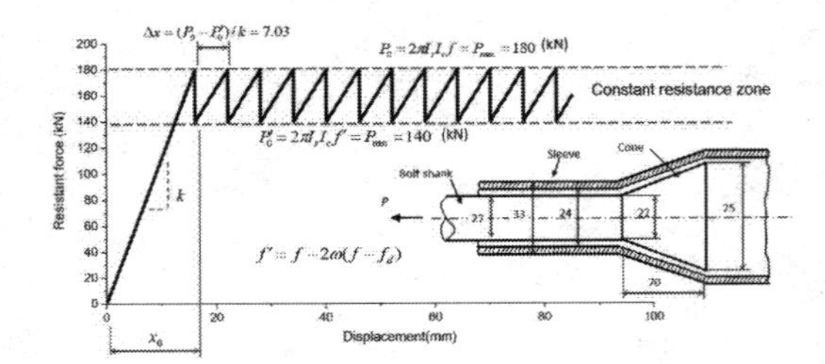

Figure 3. Analytical load-elongation curve of CRLD bolt (He et al., 2014).

1,000 mm which can fully meet the demands for accommodating the displacement extent of the rock mass adjacent to deep underground excavations. Compared to the current existing large-deformation bolts and anchors, this bolt has much longer extension length under the same external pulling force than those current bolt technologies, and at the same time its maximum load-carrying capacity is much larger.

A mathematical model was developed for the CRLD bolt (He et al., 2014). The constitutive model for the bolt is schematically represented in Figure 2a and consists of the following units: a mass unit m representing the cone and shank of the bolt; a spring k representing the stiffness of the shank; an element with a stick-slip behavior of the cone friction (HMC); and a plastic unit representing the large-deformation nature at a constant resistance P_0. A comparison between HMC element and the Newton element is given in Figure 2b. An analytical load-elongation curve of the CRLD bolt is illustrated in Figure 3, showing different cycles computed in accordance to the analytical formulation developed (He et al., 2014).

2.2 *Experimental results of SHTB tests*

The device developed for the CRLD bolt consists of the following (Figure 4): a SHTB dynamic loading system; an impact and elongation system for the bolt; and a data acquisition system. The SHTB dynamic loading system includes a hydraulic-loading component where it is possible to control nitrogen gas with different pressures and push a bullet with different velocities. Bar components include a bullet and input bar of high strength cast steel. Also an automatic control device permits to control the initial gas pressure by using a customized software. The length of the bullets can be achieved for different impact velocities and energies under different gas pressure.

In the conducted research, the CRLD bolts where considered isolated (Figure 5a) or in groups from a minimum of two to a maximum of six (Figure 5b). This permitted to compare, to analyze and to verify the mechanical properties of single bolt, double bolts and bolt groups. For the single bolt system, the impact bar was supported by the guiding frame, in order to be coaxial with the SHTB bars. The CRLD bolt specimen was fixed and guided by the specimen placement site, ensuring coaxial with the impact bar. The specimen placement site, whose material had high strength, was fixed on the working bench by high strength screws. Three force sensors were placed inside the pallet and a displacement sensor was connected with screws. Also a high speed camera was installed in parallel with the composite bearing tray while leaving enough shooting space to ensure comprehensiveness.

From the Figure 5b, it can be shown that the working principle of elongation and impacting system for group of CRLD bolts is similar to that of the single bolt system. The differences are as follows: the specimens are not aligned with the center of the composite bearing tray, but in the directions of the holes of specimen placement site and fixed with screws to ensure free sliding along the guiding plate when impacting. The displacement sensor was connected with the composite bearing tray, rather than the screw in the single bolt system.

Before starting the experiments, a segmented calibration was performed for each bolt specimens in the axial and radial directions, including the

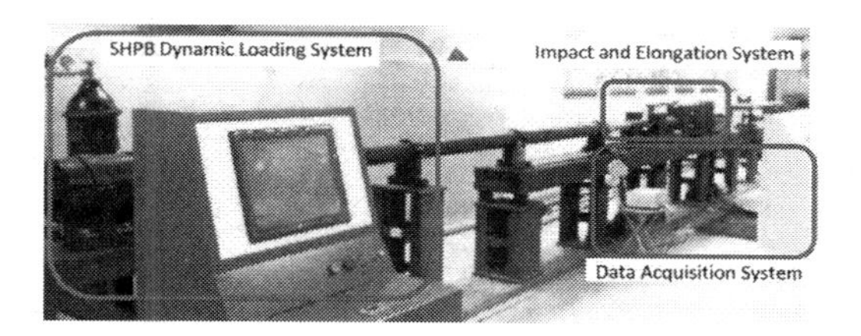

Figure 4. Elongation and impact experiment system for CRLD.

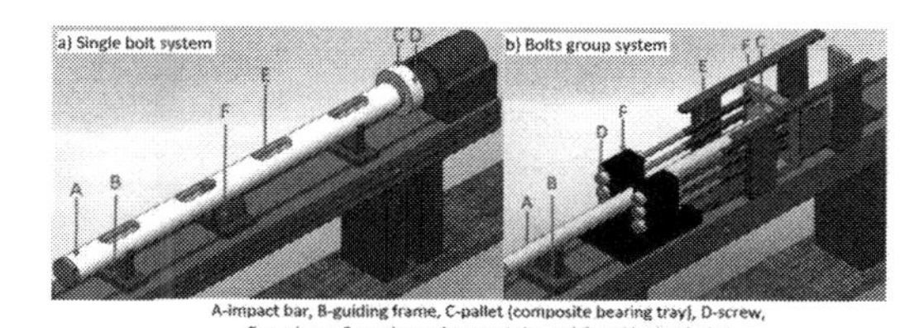

Figure 5. Schematic of elongation and impacting system for single CRLD bolt and bolts group.

total length of bolts, the constant resistance device and the sleeve pipe, diameters of the sleeve pipe and the rod. 9 points were considered (after each 5 rings of the sleeve pipe) on axial direction, and 2 diameters were calibrated on the radial direction. After the experiments, the same calibration procedure was performed.

When starting the single bolt experiment, four levels of gas pressure were applied (1.5, 2.0, 2.5 and 3.0 MPa), and in each level 20 impacts were considered. If the rod is not pushed out of the sleeve pipe after the total 80 impacts, the experiment was continued from the first level. Velocities, forces and displacements were measured after each impact process. When starting the double bolt experiment, four other levels of gas pressures were applied (1.0, 1.5, 1.8 and 2.0 MPa), and the same procedure was followed. The experiments finished with the pushing out of the sleeve pipe.

The basic geometrical and material parameters of CRLD bolt experiment systems are shown in Table 1. The specimens of the single bolt experiment were designated by MG-15-1, MG-15-4, MG-15-5 and MG-15-6. For the bolt group system only two specimens were considered and designated by MG-15-2 and MG-15-3.

Expansion changes on the sleeve pipe were measured in the nine points at the convex and concave zones of outer thread. The curves for the specimen MG-15-1 are shown in Figure 6 for a-a′ diameter direction. From the figure one can observe that on the a-a’ directions of outer thread of sample MG-15-1, the average radial measured diameter of the convex thread before and after experiment are 32.40 mm and 33.10 mm, respectively, and the expansion of convex thread can reach up to 0.60 mm; the average radial diameter of the concave thread before and after experiment are 29.60 and 31.10 mm, respectively and the expansion of concave thread can reach up to 1.50 mm. The expansion relationships of the specimens in single bolt experiment as well for the double tests are shown in He et al. (2017).

As an example, the elongation curve of the specimen MG-15-1 is shown in Figure 7. It is shown that the impact force reached peak value of 538.9 kN under 2.0 MPa in 1 ms, and the force was stable in about 5 ms which showed that the CRLD bolt had a good capacity to adapt to dynamic impacts. It is worth noting that in double bolt experiments an un-synchronization motion phenomena was observed in the two specimens when impact happened, which was due to manufacture deviation and uneven force. In this experiment, 77 impact times were considered, including gas pressures of 1.0, 1.5 and 1.8 MPa per 20 times and pressure of

Table 1. Basic geometrical and material parameters of CRLD bolt experiment systems.

Component	Length (mm)	Outer diameter (mm)	Inner diameter (mm)	Density (kg/m^3)	Yield strength (MPa)	Wave velocity (m/s)	Wave impedance (10^7 kg/m^{2}s)
Bullet	800	75	-	7,800	835	5,190	4.05
Input bar	4,000	115	75	7,800	835	5,190	4.05
Impacting pipe (single boltsystem)	2,000	140	107	7,800	835	5,190	4.05
Impacting bar (bolt group system)	1,500	75	-	7,800	835	5,190	4.05
Sleeve pipe	460	33	25	7,800	245	5,820	4.54
Rod	750	28	25	7,850	355	5,935	4.66
Pallet	50	200	-	7,850	600	5,977	4.69

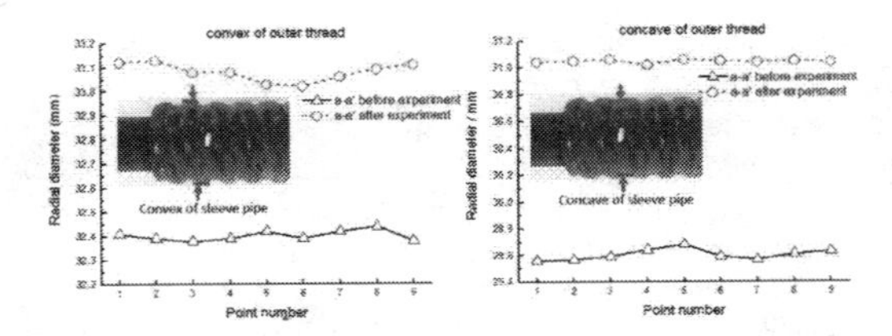

Figure 6. Changes in the expansion of MG-15-1 specimen.

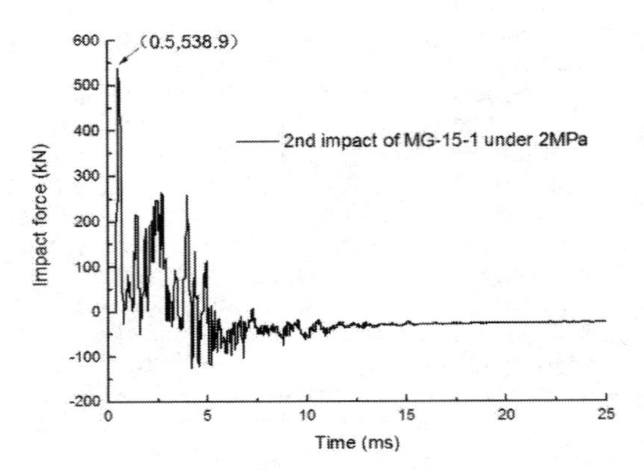

Figure 7. Impact force vs time curve of MG-15-1 bolt.

2.0 MPa in 17 times. The MG-15-2 failed when the 76th impact occurred, and the MG-15-3 thrust to failure with the 77th impacting.

3 NUMERICAL SIMULATION OF THE SHTB TEST

3.1 *Discretization of one bolt*

Numerical simulations of the SHTB impact test with CRLD bolts, by using the Finite Element (FE) method, are an adequate tool for evaluating of the complex tests performed with these large deformations bolts, permitting to determine the soundness and the accuracy of the experiment data obtained. These numerical models permitted to reduce the experiments performed once calibrated the simulations and to better analyze the behavior associated to the deformation process. Complex numerical dynamic 3D simulations were only possible in recent years. To evaluate the test setup, laboratory measurements and the sliding inside the bolt, explicit FE models of the SHTB tests were constructed using different software with the mechanical calculations being performed by LS-DYNA.

The 3D model was developed only for the one bolt test. A CAD (Computer-Aided Design) software was used for the geometric modeling; and then a FE mesh was setup, as well as the boundary conditions and mechanical properties. After this, the data obtained was converted to be used by a FE calculation software that permitted to optimize the errors of the mesh and adjustment of parameters if needed. Then, when necessary, a new mesh was generated being the process repeated several times until an optimal mesh was obtained. Once an iterative process was finalized, the calculations by LS-DINA were done. Finally the results which included displacements, stresses, strains and energy, were analyzed. The flowchart of the calculation process is illustrated in Figure 8.

The main components of the system in order to get the 3D geometric model are: bullet, impact bar, sleeve pipe, rod, pallet and screw. The sleeve pipe and the pallet could be considered as rigid body in one component. A schematic of 3D geometric model is shown in Figure 9. The 3D model was simplified to a quarter of the all body in order to reduce the nodal points and elements of the FE mesh. The size of the components are illustrated in Table 2. The FE mesh, which uses 8-node solid elements, is illustrated in Figure 10. The whole computational model includes a total of 116,526 8-node elements and a total of 139,643 nodal points. The main parameters adopted are also shown in Table 2. The non-linear model adopted followed the Von Mises yield criterion. In the impact test system, bullets, impact bars, screw, pallet and bolts were considered not to have deformation.

The FE mesh employed is illustrated in Figures 10 and 11. The main material parameters adopted are also shown in Table 2.

The surface contacts were set in an automatic manner without permitting their penetration. Also the definition of the initial conditions for the model considering the impact tensile system of gas source velocity relationship were introduced in the numerical simulations, as well as the coefficient of kinematical friction. The friction coefficient between bodies was equal to 0.18. The LS-DYNA uses Newmark time integration method and it is based in an explicit formulation on the fundamental differential equations of dynamics. For this integration time increments should be very small taking into consideration the characteristic element dimension and smallest transit time of dilatancy waves to cross any element of the FE mesh (Tasneen, 2002). In the calculations, the time step was set equal to 9.0×10^{-7} sec in order to guarantee the accuracy of the results following the guidelines recommended by LS-DYNA software. Finally the post-processing calculations were done in order to analyze the results including common results in displacements, stresses and strains.

Table 2. Size and parameters of components of the model.

Components	Density (kg/m³)	Modulus of elasticity (GPa)	Poisson's ratio	Yield strength (MPa)	Length (mm)	Outer diameter (mm)	Inner diameter (mm)
Impact Bar	7,850	206	0.300	400	1500	135	-
Pallet	7,850	206	0.300	400	50	200	-
Screw	7,850	206	0.300	400	50	40	33
Sleeve Pipe	7,800	210	0.269	255	450	33	25
Rod	7,850	206	0.300	400	460	28	25
Bullet	7,850	206	0.300	400	450	75	-

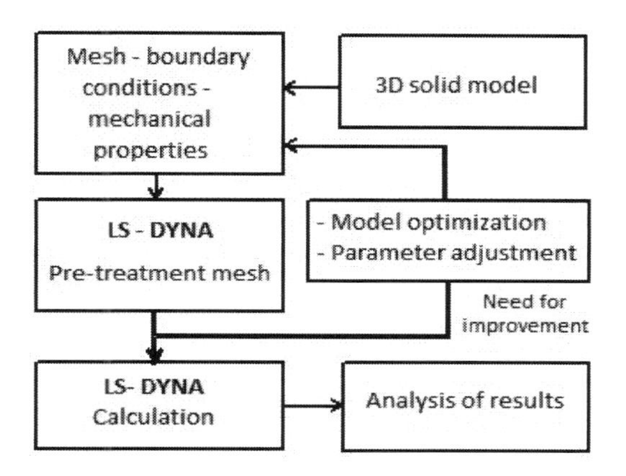

Figure 8. Flowchart of the FE analysis.

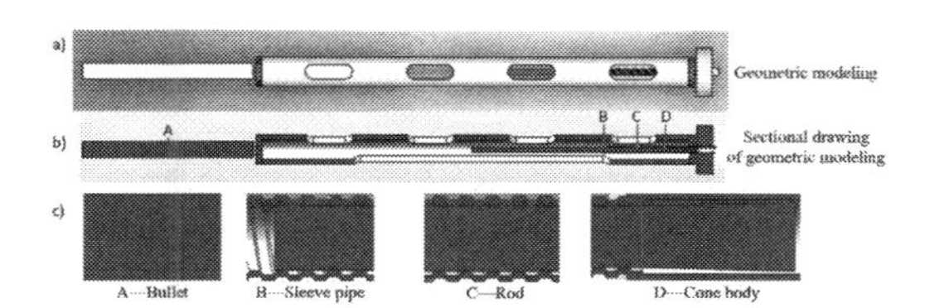

Figure 9. 3D geometric model of the SHTB for a single CRLD bolt.

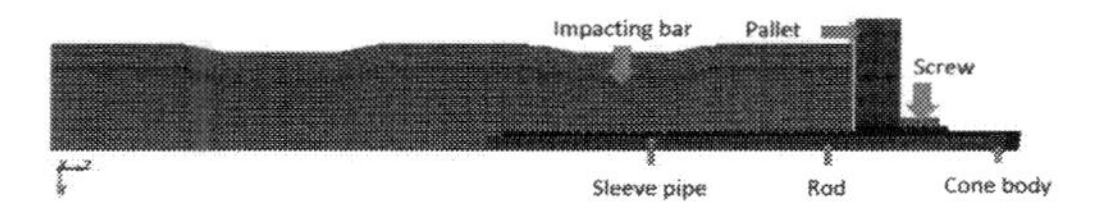

Figure 10. Numerical model developed for the impact process with CRLD bolt.

3.2 *Numercal calculations*

Different numerical calculations were considered adopting increasing gas pressure from 0.5 MPa until 3.0 MPa with constant increments of 0.5 MPa. Six cases were adopted as indicated in

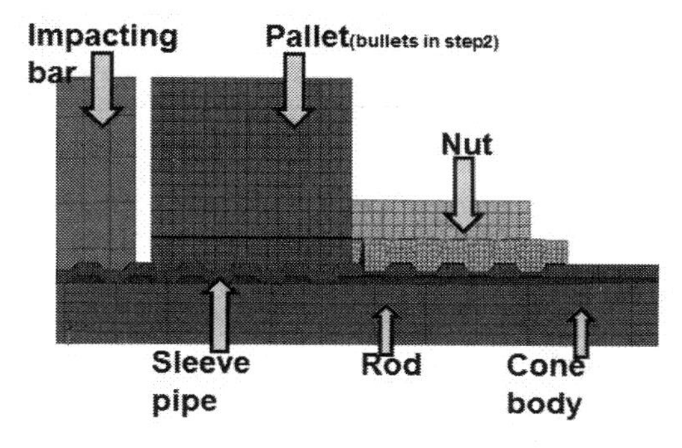

Figure 11. Schematic diagram of part of the mesh.

Figure 12. To each pressure correspond values for the velocity of the bullet (He et al., 2017). For the each calculation five equal impacts were adopted and the analysis of the results will be presented in the next section.

Calculation C_3 was adopted as representative of the mechanical properties of single CRLD bolt under a bullet velocity of 12 m/s, and a detailed analysis of the expansion and tension of the sleeve pipe, and the impact forces of the bolt was performed. These results were compared with the experimental data. For the other calculations, with different bullet velocities, the relationships between impact times and deformation of sleeve pipe and forces and time curves were also analyzed.

Figure 13 illustrates a comparison of axial displacements obtained by the reference numerical simulation (i.e the C_3) for the bolt MG-15-1 and the experimental data obtained by the wired sensor. As it can be seen, the maximum values are similar and around 2.5 mm.

Also the impact forces were validated with experimental results for bolt MG 15-1 that were measured by a cell force as illustrated in Figure 14. The comparison of results is shown by Figure 15. The maximum measured force was about 455 kN, while the calculated value was 496 kN. The deviation is about +9%, which can be justified by the fact that the contact surface of the cell was about 82 cm^2.

4 ANALYSIS OF RESULTS OF COMPUTATIONAL MODELS

An analysis of the numerical results of the FE simulations was performed. This included the data analysis of expansion, tension and impact force of the reference calculation C_3 and the data comparison of deformation and impact force of the whole six calculations. For each analysis, the results of the simulation and the experimental results are also compared.

For the reference simulation C_3, the deformation of selected areas and elements of sleeve pipe in the whole five impacts are shown in Figure 16. Taking the 3rd impact as an example, 3 areas were selected including the concave area (A) and the convex areas (B and C), and 2 elements, concave element (a) and the convex element (b). For the selected areas are illustrated the deformation of the sleeve pipe due to the axial displacements of the stiff rod after the impacts and it can be clearly seen that: i) The range of deformation of area A was significantly greater than that of areas B and C; the thickness of area A became thinner and the thickness changed while deformation in areas B and C were insignificant. The phenomena is in conformity with the experimental results, which also showed that the concave thickness of the sleeve pipe became thinner and expanded in radial direction; ii) Before this impact, area C had an initial deformation because of the impact effect of the previous ones. And after this impact, area C continued to be squeezed; and iii) The expansion of CRLD bolt came from the elastic and plastic deformation of elements, which can verify the NPR (Negative Poisson Ratio) effect of CRLD bolt caused by the structure deformation.

Figure 17 shows the radial displacement of selected elements (a and b) of sleeve pipe after the 3th impact. Both curves showed elastic rebound phenomena after impact. The convex one is the most obvious because there is no space to release elastic energy for the concave elements of sleeve pipe. After this impact, the radial displacement of concave element **a** was 1.39 mm and the convex one **b** was 0.59 mm. The experimental results were 1.58 mm and 0.63 mm, respectively, an acceptable deviation from the results of the numerical simulations.

Figure 18 illustrates the impact forces calculated at the 3rd impact at the area of the sensor showed in Figure 13. Also the experimental data is shown for an equivalent condition. In the simulation of 3rd impact, the peak impact force is 478.68 kN, and the peaking impact force measured by force sensors is 436.67kN, with a small deviation. Both impact forces returned to a stable state in 6 ms, which means the whole process is an instantaneous one. The above analysis shows that the simulation result is in conformity with the experimental data, and the reason of the peaking force of simulation bigger than experimental result is that there is unforeseen and unavoidable energy loss and other factors of the experiment system.

The results of the different numerical calculations were analyzed in terms of radial and axial deformations as well as impact forces. Figure 18 illustrated the deformation of sleeve pipe after 30 times impact from calculations C_1 to C_6 in sequence way as shown in the graph. For axial displacements, Figure 19 showed for initial bullet velocity equal to 5.5 m/s (C_1), there are barely relative small variations between sleeve pipe and rod (average value is 0.06 mm per impact, even smaller than the mesh size of 1 mm). When the

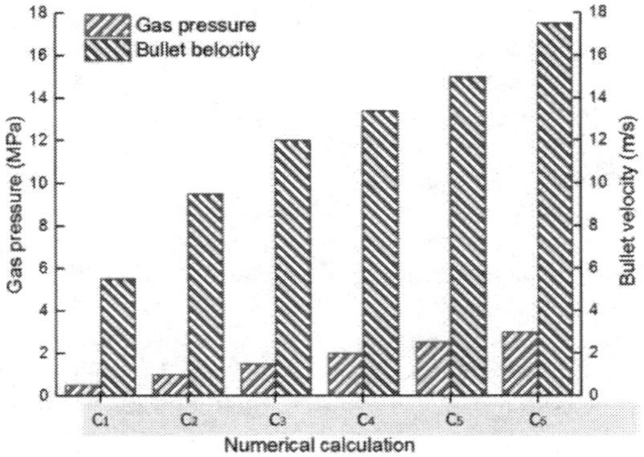

Figure 12. Relationship of gas pressure and bullet velocity in numerical calculation.

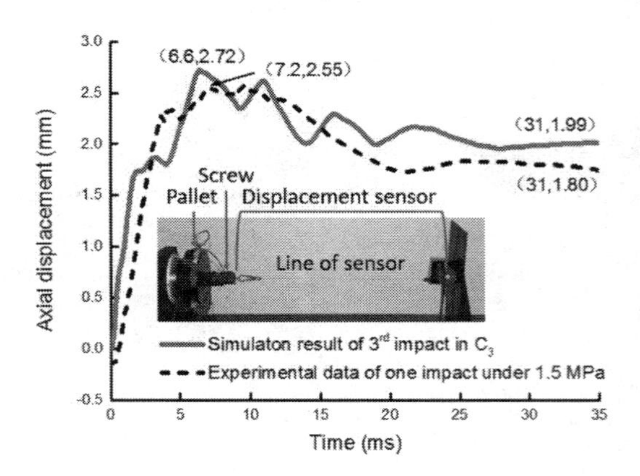

Figure 13. Comparison of axial displacements obtained by numerical calculation and by the sensor of the screw.

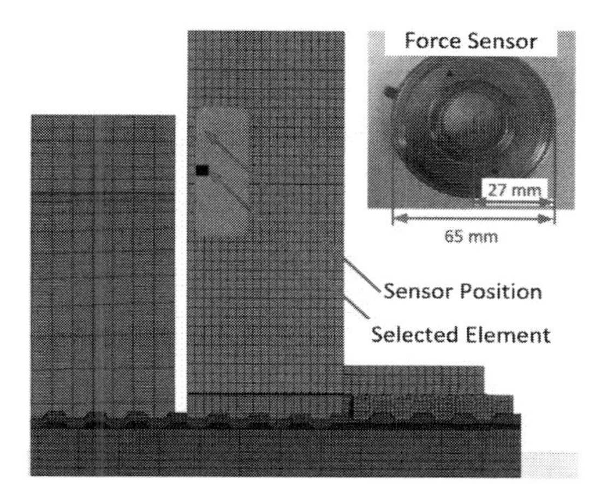

Figure 14. FE reference and installation position of force sensor in pallet.

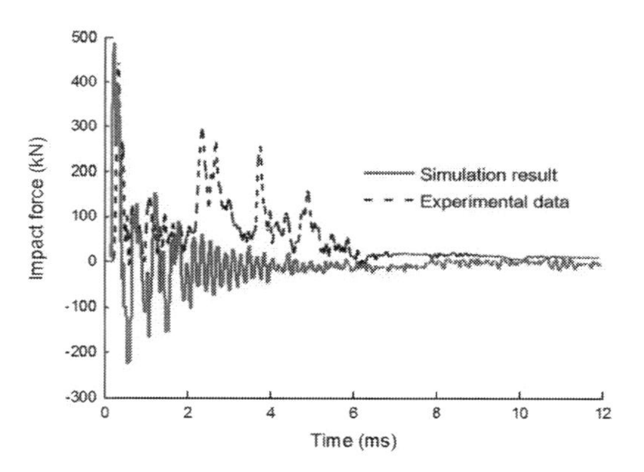

Figure 15. Impact forces obtained by the sensor and comparison with the numerical simulation.

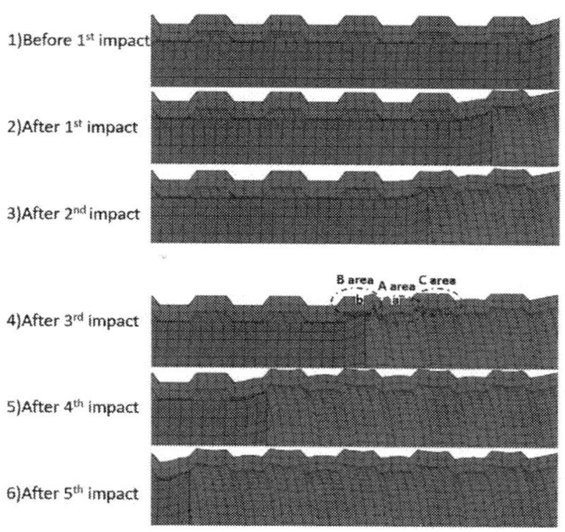

Figure 16. Deformation of selected areas and elements of the sleeve pipe for simulation C_3.

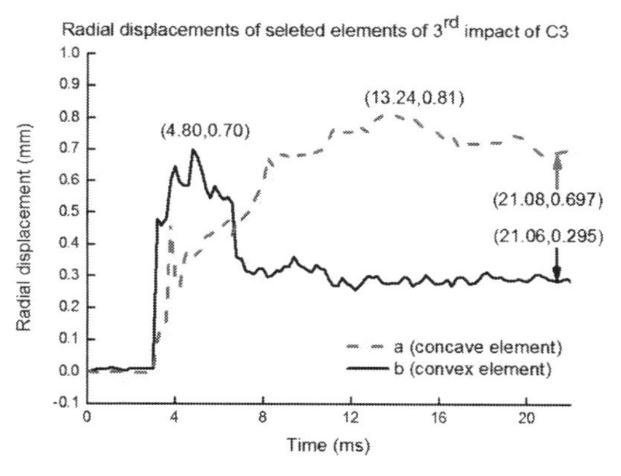

Figure 17. Radial displacement of selected elements after the 3rd impact.

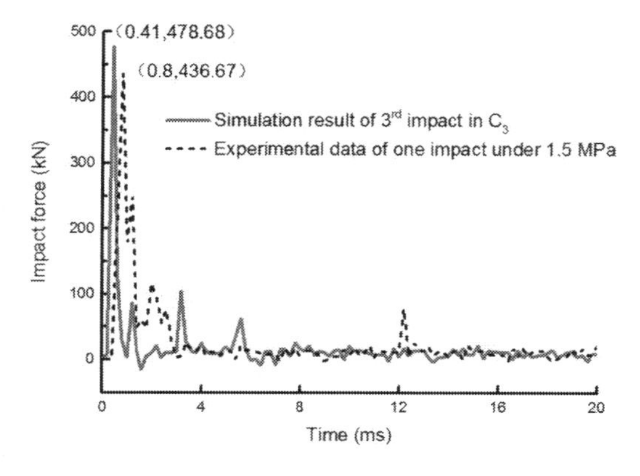

Figure 18. The relationship of axial and radial displacement of sleeve pipe of total 30 times impacts.

initial bullet velocity equals to 9.5 m/s (C_2), the relative axial displacements had a total of 5.1 mm after the 5 impacts. The applied distance can only make the cone body pass through one set of concave and convex elements of sleeve pipe. For C_3, where the initial velocity was 12.0 m/s, there is an obvious increase in the relative axial displacement of about 11.8 mm, that means in each impact the yield strength was exceeded. In the following cases (C_4 to C_6) there was a considerable increase in the displacements following an exponential curve.

Radial displacement on situation C_1 increased in an almost linear relation, with small values, 0.5 mm in the total 5 impacts. The whole process could be considered in the elastic behavior range. For C_2, the deformation value of concave elements of sleeve pipe rebound elastically to 1.3 mm. The radial deformation process of sleeve pipe entered

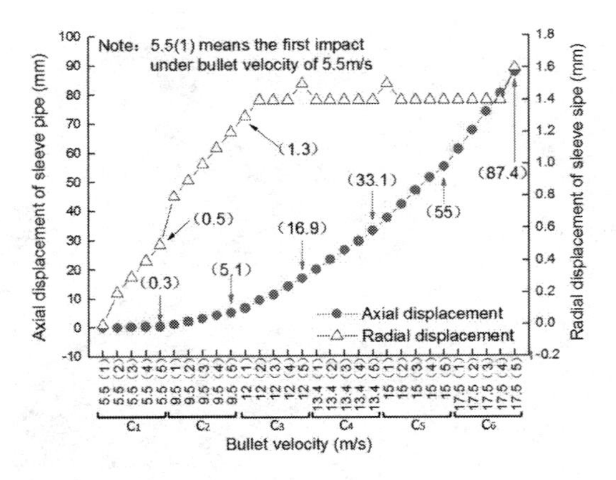

Figure 19. Relationship of axial and radial displacements of sleeve pipe of total 30 times impacts.

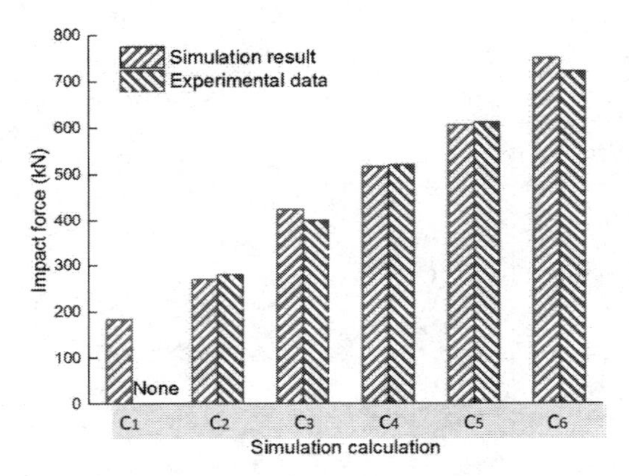

Figure 20. Average impact forces in the simulations and experiments.

a plastic behavior domain. For C_3 to C_6, the radial displacements were almost constant and equal to 1.4 mm, which is due to the fact that the maximum displacement is equal to 1.6 mm, followed by an elastic rebound in each impact.

Figure 20 illustrated the average forces obtained during the calculations. The experimental values are also added, except for C_1, because no experiences were performed. Again, the feasibility of the FE numerical solutions and the potential of these methods were emphasized.

5 CONCLUSIONS

Numerical simulations were used to verify the accuracy and complement the results of complex experimental tests developed for CRLD bolts, using the FE method. In addition the numerical simulations of the experimental tests were used to better understand the behavior of the bolts during the elongation process. A 3D model was settled due to computation difficulties for one bolt case obtained after following an iterative process for generation of the mesh optimizing the committed errors. The main components were considered, von Mises yield criterion was adopted and the surface contacts were set in an automatic way. Six calculations were done for different velocities of the bullet and for each calculation five equal impacts were considered. One scenario was adopted as reference under (bullet velocity of 12 m/s) and a deep analysis of the numerical analysis the expansion and tension of the sleeve pipe and impact forces results for the reference case was done. For the other cases of different bullet velocities the relations between impacting times and deformation of sleeve pipes as well as forces vs, time curves were analyzed.

The findings of the analysis of the numerical simulations on CRLD bolts, led to the following main conclusions:

- Control of rockburst is a very important issue in mining engineering and for this purpose a CRLD bolt was developed at GDUE, which permits constant resistance and large deformations. The performance was experimentally verified through the development of special static and dynamic equipments. The dynamic tests performed consisted of weight-dropping tests and SHTB Hopkinson dynamic tests.
- Numerical simulations of one bolt test were setup by using FE method in order to well evaluate the complex SHTB tests giving a mutual verification and permitted to determine the accuracy of the experimental data. Also the numerical simulations permitted to reduce the tests performed and once calibrated they better analyzed the behavior of the CRLD bolts associated to the deformation and rupture process.
- The results of the FE simulations (namely the data concerning expansion, tension and impact force for the different calculations), were compared with the experimental results, and showed a good agreement. The feasibility of the FE numerical solutions and the potentiality of these methods were emphasized. In the future suggestions for the creation of more complex numerical for the cases of bolts group are addressed in practical situations in the geotechnical field.

ACKNOWLEDGEMENTS

The authors would like to express their gratitude for the financial support from the Special Funds for the National Natural Science Foundation of China (51574248) and National Key Research and Development Program (2016YFC0600901).

REFERENCES

Camiro 1995. Rockburst research handbook. CAMIRO (Canadian Mining Industry Research Organization) Mining division, Canadian Rockburst Research Program, 1990–1995. Sudbury, 6 vol., 977.

Castro, L.M.; Bewick, R.P. & Carter, T.G. 2012. An overview of numerical modelling applied to deep mining. In Innovative Numerical Modeling in Geomechanics, Eds. Sousa, Vargas, Fernandes and Azevedo, CRC Press, London, 393–414.

He M.C.; Miao J.L.; Li D.J. (2007). Experimental study on rockburst processes of granite specimen at great depth. Chinese Journal of Rock Mechanics and Engineering, 26(5): 865–876 (in Chinese).

He M.C., Gong W., Wang J., Qi P., Tao Z. & Du S. 2014. Development of a novel energy-absorbing bolt with extraordinarily large elongation and constant resistance. Int. J. Rock Mechanics Min. Science, 67, 29–42.

He M.C.; Jia X.N.; Gong W.L.; Liu G.J. & Zhao F. 2011. A modified true triaxial test system that allows a specimen to be unloaded on one surface. In True Triaxial Testing of Rocks, Eds. Kwasniewski, Li and Takahashi, Chapter 19, pp. 251–206.

He M. & Sousa, L.R. 2014. Experiments on rock burst and its control. AusRock 2014: Third Australasian Ground Control in Mining Conference, Sydney, 19–31.

He M.C.; Sousa, L.R.; Miranda, T. & Zhu G. 2015a. Rockburst laboratory tests database—Application of Data Mining techniques. Journal of Engineering Geology for Geological and Geotechnical Hazards, 185, 190–202.

He M.C.; Sousa, R.L.; Muller, A.; Vargas Jr., E.; Sousa, L.R. & Chen X. 2015b. Analysis of excessive deformations in tunnels for safety evaluation. J. of Tunneling and Underground Space, 45, 190–202.

He M.; Li C.; Gong G.; L.R. Sousa & Li S.2017. Dynamic tests for a Constant-Resistance-Large-Deformation bolt using a modified SHTB system. Tunnelling and Underground Space Technology, 64, 103–116.

Tang C.A.; Wang J.M. & Zhang J.J. 2010. Preliminary engineering application of microseismic monitoring technique to rockburst prediction in tunneling of Jinping II project. Journal of Rock Mechanics and Geotechnical Engineering, 2(3), 193–208.

Tasneen, N. 2002. Study of wave shaping techniques of split Hopkinson pressure bar using finite element analysis. Wishita State University, 77.

Numerical Methods in Geotechnical Engineering IX – Cardoso et al. (Eds)
 ISBN 978-1-138-33198-3

SHANSEP approach for slope stability assessments of river dikes in The Netherlands

T.D.Y.F. Simanjuntak, D.G. Goeman, M. de Koning & J.K. Haasnoot
CRUX Engineering BV, Amsterdam, The Netherlands

ABSTRACT: The sliding of an inner slope, referred to as macro-instability, is one of the main failure modes for dikes. This failure mostly occurs when the outside water level elevates, resulting in an increase of pore water pressures in and beneath the dike and ultimately in a decrease of undrained shear strength. In this study, the inner slope stability or macro-stability at one of the dike cross-sections between Krimpen aan den IJssel and Gouderak in the province of Zuid-Holland in the Netherlands, is evaluated by using the constitutive model SHANSEP Mohr-Coulomb (MC) in PLAXIS. The safety factor against macro-instability is assessed based on the SHANSEP approach. To validate the model, the calculated safety factor and the predicted slip circle are compared with those obtained using the Uplift-Van model using Deltares software, D-Geo Stability. The discrepancy of the results between the SHANSEP MC model and the Uplift-Van model is discussed and the modelling aspects for macro-stability assessments for river dikes are outlined.

1 INTRODUCTION

1.1 *WBI2017*

Approximately two-thirds of the Netherlands is at risk of flooding (see Fig. 1). Flood defences, such as primary dikes, have long been used to protect the people and the country from flooding.

To offer protection against floods, the Netherlands is divided into 53 dike rings. Together with natural sand dunes, dams and locks, primary dikes create protection against water from the sea and large rivers. The attribute "primary" has been used to distinguish these dikes from the many regional dikes in the polders. Regional dikes defend the country from flooding from canals and small rivers.

The safety of primary dikes in the Netherlands has to comply with the flood protection standards determined by the Dutch law (Water Act). Since January 1st 2017 the flood protection standards are defined based on maximum allowable probabilities of flooding. Accordingly, the dike assessment rules in the Netherlands, called Wettelijk Beoordelings-instrumentarium (WBI) need to be reviewed in order to assess whether or not a primary dike meets the flood protection standards of the Water Act.

One of the new assessment rules according to the WBI2017 is the implementation of the Critical State Soil Mechanics (CSSM) (Schofield and Wroth, 1968), using the Stress History and Normalised Soil Engineering Properties (SHANSEP) method (Ladd and Foott, 1974).

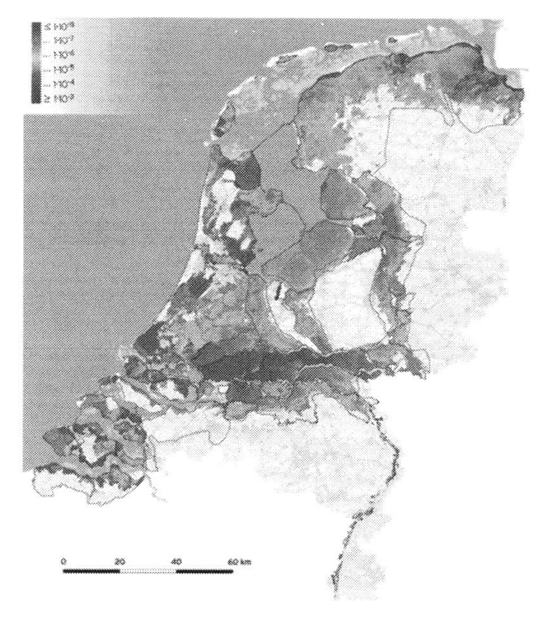

Figure 1. Individual risk in The Netherlands in 2015, called VNK2 (Rijkswaterstaat, 2017).

Hence, the stress history and stress path of the soil can be considered in characterising soil strength and predicting field behaviour. In the past, the macro-stability of dikes in the Netherlands was assessed based on the Mohr-Coulomb model.

1.2 *Failure mechanism model*

When the inner slope of a dike becomes unstable as a result of an increased outside (river) water level, the sliding of the slope towards the protected area may occur. In the Netherlands, this failure mechanism is known as macro-instability and is different from micro-instability, which is the instability of relatively thin layers at the surface of an inner slope due to seepage (TAW, 2001).

According to the WBI2017, the default model to assess the slope stability of a river dike is the Uplift-Van model developed by Van (2001). This model is similar to the Bishop's model, but it allows for a non-circular slip circle and can accommodate uplift conditions, which typically occur when a thin blanket layer is located on top of an aquifer connected to the outside water.

The failure mechanism based on the Uplift-Van model is described by two circular slip circles: one on the active zone and another on the passive zone, bound by a horizontal slip line. This horizontal line, which is part of the passive zone, usually lies along the bottom of a weak soil layer (Fig. 2). The safety factor against macro-instability is expressed as the ratio of the resisting moment to the driving moment. The resistance against sliding is governed by the shear strength of soils.

For cohesive soils with low permeability, such as clay and peat, the undrained shear strengths along the slip circle are determined according to the CSSM and more specifically the SHANSEP approach. For non-cohesive soils with high permeability like sand, the drained material model based on a CSSM effective friction angle is used. The lowest safety factor of the numerous possible slip circles determines the safety factor of the dike cross-section under consideration.

One approach, which considers the influence of stress history on the shear strength, is the SHANSEP approach. The general idea behind this approach is to perform a series of laboratory tests that carefully control the stress conditions during consolidation and the stress path during undrained shear.

These tests are performed over a range of stress histories and stress paths. The in-situ stress history of the soil is evaluated, and the stress path to which the soil is imposed, is determined. The soil strengths from the laboratory tests that most closely replicate the in-situ conditions are then used to predict the soil behaviour.

Based on the SHANSEP approach, the undrained shear strength, s_u, of a soil subjected to a particular stress path is calculated as:

$$s_u = \sigma'_v \, S \, (OCR)^m \tag{1}$$

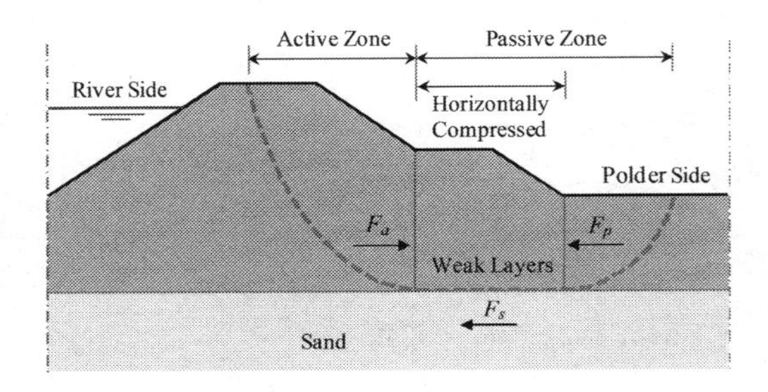

Figure 2. The Uplift-Van model (Rijkswaterstaat, 2017).

with

$$OCR = \frac{\sigma'_y}{\sigma'_v} = \frac{\sigma'_v + POP}{\sigma'_v} \tag{2}$$

where s_u is the undrained shear strength (kPa), σ'_v and σ'_y represent the in-situ effective vertical stress (kPa) and the vertical yield stress (kPa) respectively, OCR denotes the over-consolidation ratio (−), POP is the pre-overburden pressure (kPa), S represents the undrained shear strength ratio (–) and m indicates the strength increase exponent (−).

2 THE SHANSEP MOHR-COULOMB (MC) CONSTITUTIVE MODEL IN PLAXIS

2.1 *Model application and the 'switch' to the SHANSEP approach*

Besides the Uplift-Van model, the SHANSEP Mohr-Coulomb (MC) model developed by PLAXIS may be used to model isotropic undrained shear strength of soils. This model is based on the Mohr-Coulomb model, but is modified such that it is able to simulate potential changes of undrained shear strength based on the effective stress of the soil (Panagoulias et al., 2016). The SHANSEP MC model behaves initially as the Mohr-Coulomb model until it is switched to the SHANSEP approach by the user. It will only take effect for the activated SHANSEP MC soil clusters. The material model for the SHANSEP MC is facilitated through a user-defined soil model (by means of DLL files provided by Plaxis).

It has to be mentioned that in the SHANSEP MC model, unlike in the Uplift-Van model, the effective major principal stress, σ'_1, is used for calculating the over-consolidation ratio OCR, and the undrained shear strength, s_u. In the SHANSEP MC model, the undrained shear strength is determined by:

$$s_u = \max\left[\, \sigma'_1 \, S \, (OCR)^m, s_{u,\min} \right] \tag{3}$$

with

$$OCR = \max\left(\frac{\sigma'_{1,\max}}{\sigma'_1}, OCR_{\min}\right) \tag{4}$$

in which $\sigma'_{1,max}$ represents the effective yield stress (kPa) that is independent on the Cartesian system of axes, $s_{u,min}$ denotes the minimum undrained shear strength (kPa), and OCR_{min} is the minimum over-consolidation ratio (–).

Assuming horizontal soil layering, the calculated OCR and s_u according to the SHANSEP MC model should be equal to that calculated using the Uplift-Van model.

2.2 *Model parameters*

The model parameters in the SHANSEP MC model consist of the Mohr-Coulomb model parameters and the SHANSEP model parameters. These parameters are listed in Table 1.

2.3 *Model outputs*

The SHANSEP MC model in PLAXIS delivers two outputs, namely the State Parameter 1, which corresponds to the effective yield stress $\sigma'_{1,max}$ and the State Parameter 2, which is the calculated undrained shear strength, s_u.

2.4 *Safety factor*

Usually, the safety of a slope is assessed in terms of a safety factor. In a finite element model, the phi-c reduction or strength reduction method (Brinkgreve and Bakker, 1991; Griffiths and Lane, 1999) has been used for many years as the way to calculate the safety factor.

In the SHANSEP MC model, the safety factor is defined as the ratio between the actual shear strength of the soil to the shear strength needed to maintain the equilibrium.

Table 1. SHANSEP MC model parameters.

Parameters		Symbol	Unit
Mohr-Coulomb	Shear Modulus	G	kPa
	Poisson's Ratio	v'	-
	Cohesion	c'	kPa
	Friction Angle	ϕ'	o
	Dilatancy Angle	ψ	o
	Tensile Strength	*Tens*	kPa
SHANSEP	Undrained Shear Strength Ratio	S	-
	Strength Increase Exponent	m	-
	Stiffness and Strength Ratio	G/s_u	-
	Minimum Shear Strength	$s_{u,min}$	kPa
	Min. Over Consolidation Ratio	OCR_{min}	-

3 COMPUTATIONAL EXAMPLE

3.1 *Study area*

This study investigates the applicability of the SHANSEP MC model to assess the macro-stability of a river dike in the Netherlands. It deals with the macro-stability assessment at one of the dike cross-sections located between Krimpen aan den IJssel and Gouderak in the province of Zuid-Holland.

The river dike itself has a length of about 10 km (Fig. 3), and has an important role in protecting the dike ring 15 Lopiker- en Krimpenerwaard.

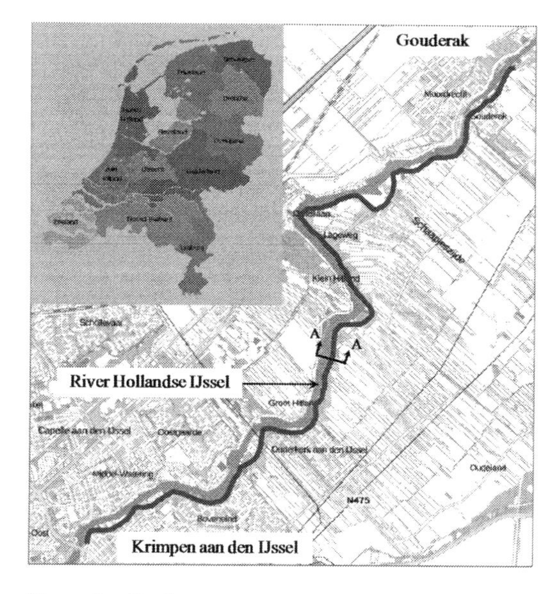

Figure 3. Study area

Table 2. Loading scenarios.

Loading Scenario	Traffic Load (kPa)	High Water Level (m NAP)
I	5	+3.32
II	12	+2.60
III	15	+1.20

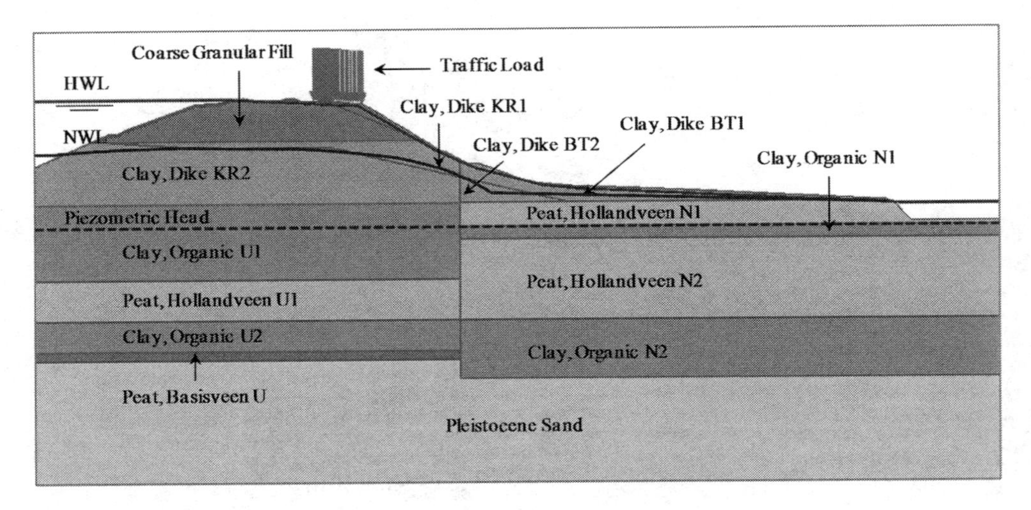

Figure 4. Model geometry (Cross section A-A).

3.1.1 *Model geometry*

Figure 4 shows the model geometry, which portrays the soil deposit, the water levels in the dike and the traffic load on the dike crest.

3.1.2 *Hydraulic head and water levels*

The piezometric (hydraulic) head in the Pleistocene Sand is located at NAP –3.5 m. The normal water level (NWL) of the river Hollandse IJssel is at NAP +0.29 m, and the maximum high water level (HWL) according to the 3000-year flood (probability of flooding of 1/3000 per year) is at NAP +3.32 m.

3.1.3 *Traffic load*

Depending on the high water level, the traffic load with a width of 2.5 m is schematised at the inner side of the dike crest. When the high water level is at NAP +3.32 m, the traffic load is 5 kPa. When the water level decreases to NAP +2.60 m and further to NAP +1.20 m, the traffic load becomes 12 kPa and 15 kPa, respectively. For clarity, these combinations of loading are listed in Table 2.

3.1.4 *Model parameters and POP-values*

It has to be acknowledged that it is not possible to directly convey the stress history of the soils defined through either *OCR* or *POP*-values in the SHANSEP MC model. Nevertheless, this gap can be bridged by combining the SHANSEP MC model with another PLAXIS advanced soil model, which requires either the *OCR* or *POP*-value as one of the input parameters, such as the Hardening Soil (HS) model.

The adopted SHANSEP MC model parameters are listed in Table 3, whereas the relevant HS model parameters including the *POP*-values for Peat and Clay layers are summarised in Table 4.

For both the Coarse Granular Fill and Pleistocene Sand layers, the adopted HS model parameters are presented in Table 5.

3.1.5 *Calculation phases*

The applied calculation phases in PLAXIS for each loading scenario are presented in Table 6.

For the first loading scenario, i.e. when the river water level is at NAP + 3.32 m and the traffic load is 5 kPa, the *POP* for the Peat and Clay layers was generated in the Initial Phase using the HS model with K0-procedure. The water level was raised from the normal level (NWL) at NAP +0.29 m to the high level (HWL) at NAP +3.32 m in Phase 2 with beforehand replacing the HS soil clusters for the Peat and Clay layers with the SHANSEP MC clusters in Phase 1.

To allow for different hydraulic levels for each soil layer, the linear interpolation of pore pressures was set to the soil cluster Peat U1 and Peat N2 in Phase 2. The hydraulic head of the soil layers below these two clusters were set to NAP –3.5 m.

A consolidation analysis was performed in Phase 3, followed by the update of the shear strength in Phase 4. The traffic load of 5 kPa was activated in Phase 5, followed by the update of the shear strength in Phase 6. The safety calculation as a result of the increase of water level and the activation of traffic load was carried out in Phase 7. In order to make sure that no additional deformations will occur in the safety calculation, the reset displacements to zero option in PLAXIS was selected.

Table 3. Adopted SHANSEP MC model parameters.

	Parameters	Clay KR1	Clay KR2	Clay U1	Peat U1	Clay U2	Peat U	Clay BT1	Clay BT2	Peat N1	Clay N1	Peat N2	Clay N2	Unit
Mohr-Coulomb	G	4000	4000	2350	2000	2500	2200	2480	2480	1350	1600	1500	1630	kPa
	ν'	0.15	0.15	0.15	0.15	0.15	0.15	0.15	0.15	0.15	0.15	0.15	0.15	-
	c'	15	15	7	5	7	6	15	15	5	8	5	8	kPa
	ϕ'	34	34	33	20	33	21	34	34	20	34	20	34	°
	ψ	4	4	3	0	3	0	4	4	0	4	0	4	°
	Tens	0	0	0	0	0	0	0	0	0	0	0	0	kPa
SHANSEP	S	0.37	0.37	0.32	0.39	0.32	0.39	0.37	0.37	0.39	0.32	0.39	0.32	-
	m	0.91	0.91	0.88	0.85	0.88	0.85	0.91	0.91	0.85	0.88	0.85	0.88	-
	G/s_u	95	95	65	55	65	55	185	185	70	200	50	80	-
	$s_{u,min}$	1	1	1	1	1	1	1	1	1	1	1	1	kPa
	OCR_{min}	1	1	1	1	1	1	1	1	1	1	1	1	-

Table 4. Adopted HS model parameters for peat and clay.

Parameters	Clay R1	Clay R2	Clay 1	Peat 1	Clay 2	Peat	Clay T1	Clay T2	Peat 1	Clay 1	Peat 2	Clay 2	Unit
γ	18.5	18.5	16.1	10.9	16.2	11.8	18.5	18.5	10.5	14.2	10.5	14.8	kN/m^3
γ_{sat}	18.5	18.5	16.1	10.9	16.2	11.8	18.5	18.5	10.5	14.2	10.5	14.8	kN/m^3
E_{50}^{ref}	9	9	9	2	9	2	9	9	2	9	2	9	MPa
E_{oed}^{ref}	4	4	5	1	5	1	4	4	1	5	1	5	MPa
E_{ur}^{ref}	60	60	25	10	25	10	60	60	10	18	10	18	MPa
m	1	1	1	1	1	1	1	1	1	1	1	1	-
c'_{ref}	15	15	7	5	7	6	15	15	5	8	5	8	kPa
ϕ'	34	34	33	20	33	21	34	34	20	34	20	34	°
ψ	4	4	3	0	3	0	4	4	0	4	0	4	°
ν'	0.15	0.15	0.15	0.15	0.15	0.15	0.15	0.15	0.15	0.15	0.15	0.15	-
K_0^{nc}	0.45	0.45	0.46	0.65	0.46	0.65	0.45	0.45	0.65	0.45	0.65	0.45	-
POP	21.1	59.8	44.9	127.7	14.1	16.2	22.0	35.1	64.0	57.0	34.2	17.2	kPa

Table 5. Adopted HS model parameters for coarse granular fill and pleistocene sand.

Parameters	Coarse Granular Fill	Pleistocene Sand	Unit
γ	19	18	kN/m³
γ_{sat}	20	20	kN/m³
E_{50}^{ref}	25	35	MPa
E_{oed}^{ref}	24	20	MPa
E_{ur}^{ref}	75	100	MPa
m	0.5	0.5	-
c'_{ref}	1	0	kPa
ϕ'	32	35	°
ψ	2	5	°
v'	0.2	0.2	-
K_0^{nc}	0.4	0.4	-
POP	0	0	kPa

Table 6. Applied calculation phases in PLAXIS (Case 2).

Calculation Phases	Remarks
Initial Phase	Stress Initialisation, K0-Procedure
Phase 1 (Plastic)	SHANSEP MC Clusters
Phase 2 (Plastic)	Water Level Increase (HWL)
Phase 3 (Consolidation)	Consolidation Analysis
Phase 4 (Plastic)	Update on Shear Strength
Phase 5 (Plastic)	Activation of Traffic Load
Phase 6 (Plastic)	Update on Shear Strength
Phase 7 (Safety)	Safety Calculation

The same calculation phases were applied for the other two loading scenarios, i.e. when the river water level is at NAP +2.60 m and the traffic load is 12 kPa, and when the river water level is at NAP +1.20 m and the traffic load is 15 kPa.

The switch to the SHANSEP approach for each loading scenario occurs in Phases 1, 4, and 6.

3.1.6 *Failure mechanisms and safety factors*

Figures 5a, 6a, and 7a illustrate the obtained failure mechanisms according to the SHANSEP MC model, which are in accordance with those obtained using the Uplift-Van model in D-Geo Stability (Figs. 5b, 6b, and 7b). The good agreement between the safety factors obtained using the SHANSEP MC model and those calculated using the Uplift-Van model is also evident (Table 7). This

Table 7. The comparison of calculated safety factors.

		Safety Factor (SF)	
Traffic Load (kPa)	HWL (m NAP)	SHANSEP MC Model	Uplift-Van Model
5	+ 3.32	1.44	1.44
12	+ 2.60	1.43	1.43
15	+ 1.20	1.42	1.42

means that there is a global coherence between the SHANSEP MC model and the Uplift-Van model even if they are methodologically different in defining the *OCR*.

For completeness, the data used for the safety factor calculation in D-Geo Stability program are summarised in Tables 8 and 9. Unlike in PLAXIS, the stress history of the soils in D-Geo Stability is represented through a yield stress value, σ'_y, which can be calculated as:

$$\sigma'_y = \sigma'_v \times OCR \tag{5}$$

$$\sigma'_y = \sigma'_v + POP \tag{6}$$

The relation between *POP* and *OCR* can therefore be written as:

$$POP = \sigma'_v (OCR - 1) \tag{7}$$

Once the yield stress for the Peat and Clay layers has been determined, in D-Geo Stability this value needs to be specified in the middle level of the layer (see Tables 8 and 9). Taking the dike surface as a reference level for the ratio *S*, the undrained shear strength s_u can be calculated (Deltares, 2016).

In accordance with the model created in PLAXIS, the traffic load was also schematised at the inside of the dike crest. It was positioned as temporary load on the dike surface and has a width of 2.5 m. The load distribution of 30° was adopted. For the Peat and Clay layers, the degree of consolidation for the load was taken as 0%, while for the Coarse Granular Fill and Pleistocene Sand was 100%.

Table 10 provides the *X* and *Y* co-ordinates of the water levels, namely the NWL and HWL. The *X* and *Y* co-ordinates for the dike surface, which are used as the reference level for the ratio *S* in the undrained shear strength calculations, are listed in Table 11. In D-Geo Stability, an infinite amount of slip circles occur during the computation; but, the minimum safety factor that ensures the equilibrium determines the safety factor.

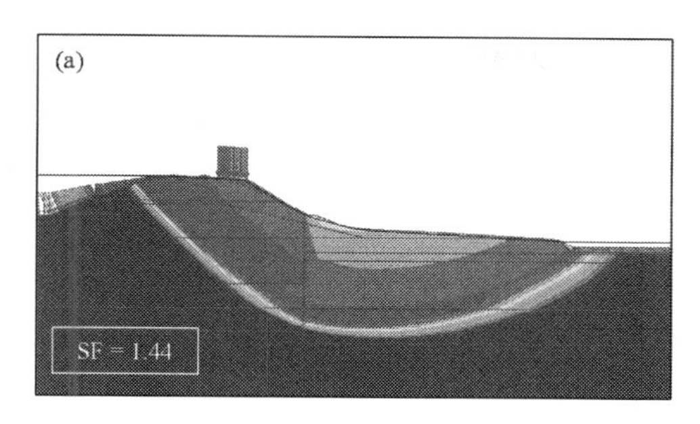

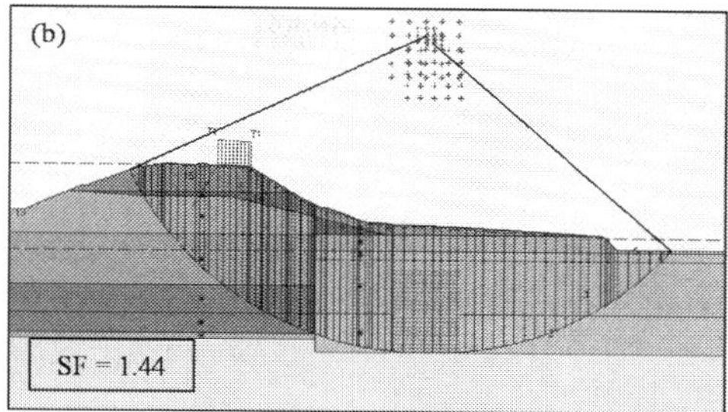

Figure 5. The predicted slip circle and safety factor (Scenario I).

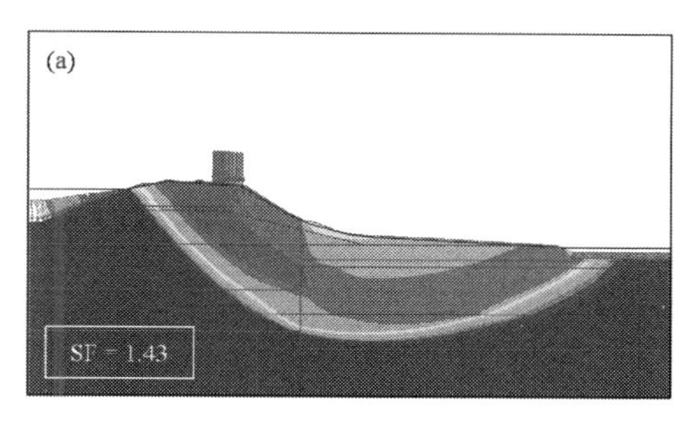

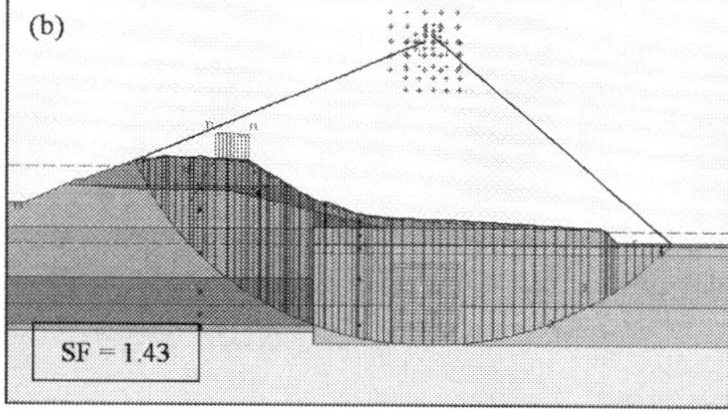

Figure 6. The predicted slip circle and safety factor (Scenario II).

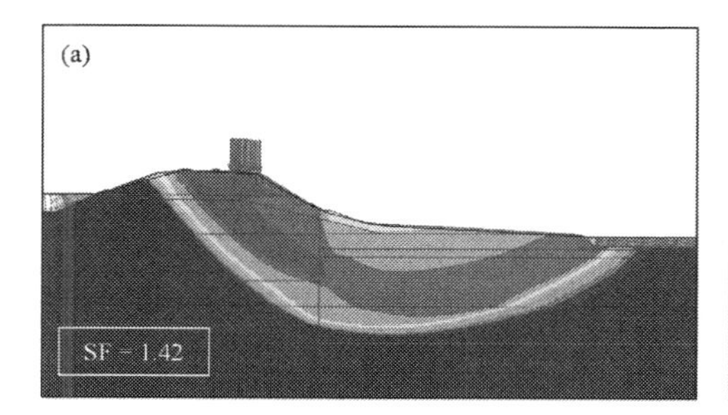

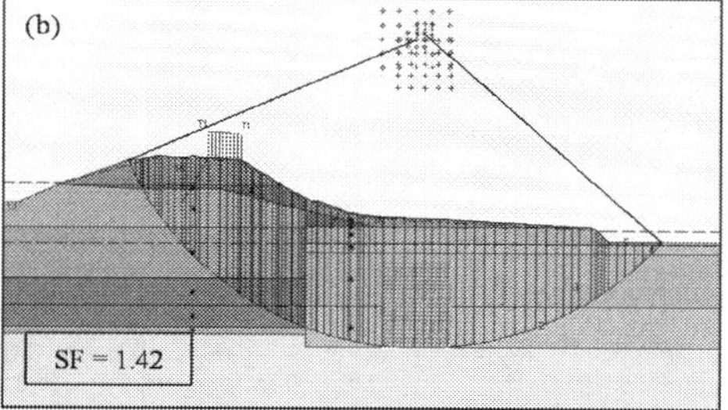

Figure 7. The predicted slip circle and safety factor (Scenario III).

Table 8. Adopted soil parameters for the uplift-van model (Dike Crest at $X = 198.4$ m)

Soil Layers	Top Level (m NAP)	Bottom Level (m NAP)	Middle Level (m NAP)	γ(kN/m3)	γ_{sat} (kN/m3)	S (-)	m (-)	c' (kPa)	ϕ' (°)	σ'_y (kPa)
Coarse Granular Fill	+3.38	+1.12	-	19.0	20.0	-	-	0	32	-
Clay, Dike KR1	+1.12	+0.76	+0.83	18.5	18.5	0.37	0.91	-	-	69.4
Clay, Dike KR2	+0.76	–2.22	–0.82	18.5	18.5	0.37	0.91	-	-	124.4
Clay, Organic U1	–2.22	–6.22	–4.22	16.1	16.1	0.32	0.88	-	-	134.1
Peat, Hollandveen U1	–6.22	–8.40	–7.31	10.9	10.9	0.39	0.85	-	-	230.5
Clay, Organic U2	–8.40	–10.00	–9.20	16.2	16.2	0.32	0.88	-	-	123.1
Peat, Basisveen U	–10.00	–10.50	–10.25	11.8	11.8	0.39	0.85	-	-	130.7
Pleistocene Sand	–10.50	–40.00	-	18.0	20.0	-	-	0	35	-

Table 9. Adopted soil parameters for the uplift-van model (Dike Toe at $X = 210.6$ m).

Soil Layers	Top Level (m NAP)	Bottom Level (m NAP)	Middle Level (m NAP)	γ(kN/m^3)	γ_{sat} (kN/m^3)	S (-)	m (-)	c' (kPa)	ϕ' (°)	σ'_y (kPa)
Clay, Dike BT1	−1.06	−1.70	−1.63	18.5	18.5	0.37	0.91	-	-	32.6
Clay, Dike BT2	−1.70	−2.16	−2.11	18.5	18.5	0.37	0.91	-	-	50.2
Peat, Hollandveen N1	−2.16	−3.46	−2.78	10.5	10.5	0.39	0.85	-	-	79.9
Clay, Organic N1	−3.46	−4.16	−3.75	14.2	14.2	0.32	0.88	-	-	74.6
Peat, Hollandveen N2	−4.16	−8.40	−6.25	10.5	10.5	0.39	0.85	-	-	54.9
Clay, Organic N2	−8.40	−11.50	−9.95	14.8	14.8	0.32	0.88	-	-	47.1
Pleistocene Sand	−11.50	−40.00	-	18.0	20.0	-	-	0	35	-

Table 10. Water levels (NWL and HWL).

Phreatic Line in the Dike during the Normal Water Level (NWL)													
X (m)	169.5	186.2	186.7	189.9	193.1	196.3	198.4	202.8	206.0	208.8	210.6	230.0	260.5
NWL (m NAP)	+0.29	+0.29	+0.34	+0.62	+0.73	+0.72	+0.64	+0.24	−0.33	−1.31	−1.67	−2.35	−2.35
Phreatic Line in the Dike during the High Water Level (HWL)													
X (m)	169.5	186.2	188.4	192.2	195.3	195.8	201.9	205.9	207.9	211.3	213.1	230.0	260.5
HWL_1 (m NAP)	+3.32	+3.32	+3.32	+3.32	+3.32	+3.29	+2.96	+0.54	−0.36	−1.28	−1.48	−2.35	−2.35
HWL_2 (m NAP)	+2.60	+2.60	+2.60	+2.60	+3.20	+3.29	+2.96	+0.54	−0.36	−1.28	−1.48	−2.35	−2.35
HWL_3 (m NAP)	+1.20	+1.20	+1.20	+2.26	+3.16	+3.29	+2.96	+0.54	−0.36	−1.28	−1.48	−2.35	−2.35

Table 11. Dike surface (Reference Level for Ratio S).

X (m)	Y (m NAP)	X (m)	Y (m NAP)	X (m)	Y (m NAP)	X (m)	Y (m NAP)	X (m)	Y (m NAP)	X (m)	Y (m NAP)
170.0	−1.64	188.5	+1.30	197.9	+3.26	204.5	+1.61	208.0	−0.13	229.0	−2.16
177.9	−1.18	188.7	+1.40	198.2	+3.38	204.8	+1.40	208.5	−0.28	229.1	−2.16
179.7	−1.08	189.0	+1.53	198.6	+3.38	204.9	+1.30	209.0	−0.49	229.5	−2.39
180.5	−1.03	189.5	+1.53	198.9	+3.23	205.0	+1.26	209.5	−0.67	229.8	−2.76
181.2	−0.75	190.0	+1.79	200.5	+3.20	205.2	+1.12	210.0	−0.91	230.3	−3.29
182.0	−0.39	192.2	+2.60	201.0	+3.16	205.5	+0.97	210.4	−1.00	245.4	−3.29
183.2	−0.35	193.5	+3.07	201.5	+3.09	206.0	+0.62	210.5	−1.04	245.9	−2.76
185.0	−0.28	194.0	+3.10	202.0	+3.09	206.5	+0.43	211.0	−1.16	246.5	−2.16
185.8	+0.09	194.5	+3.11	202.5	+2.84	206.8	+0.28	211.5	−1.22	246.6	−2.13
186.0	+0.18	195.0	+3.21	202.9	+2.60	206.9	+0.18	212.0	−1.27	247.0	−1.92
186.2	+0.28	195.2	+3.38	203.0	+2.53	207.0	+0.15	212.5	−1.31	247.5	−1.76
188.1	+1.12	196.0	+3.38	203.5	+2.19	207.1	+0.10	213.0	−1.36	248.0	−1.71
188.4	+1.26	196.5	+3.29	204.0	+1.91	207.5	−0.13	218.0	−1.55	260.0	−1.70

Table 12. The calculated yield stress at $X = 198.4$ m.

X (m)	Y (m NAP)	σ'_y (kPa) SHANSEP MC Model	Uplift-Van Model
198.4	+0.83	68.9	69.4
198.4	−0.82	118.2	124.4
198.4	−4.22	136.7	134.1
198.4	−7.31	205.9	230.5
198.4	−9.20	121.1	123.1
198.4	−10.25	127.4	130.7

Table 13. The calculated yield stress at $X = 210.6$ m.

X (m)	Y (m NAP)	σ'_y (kPa) SHANSEP MC Model	Uplift-Van Model
210.6	−1.63	27.0	32.6
210.6	−2.11	41.9	50.2
210.6	−2.78	55.1	79.9
210.6	−3.75	74.1	74.6
210.6	−6.25	47.5	54.9
210.6	−9.95	52.5	47.1

In order to verify whether or not the stress history of the soils was appropriately transferred in the SHANSEP MC model, the calculated yield stresses (State Parameter 1) in Phase 1 from PLAXIS Output are compared with the data used in D-Geo Stability program (Tables 8 and 9). The values of yield stress are summarised in Tables 12 and 13.

From Tables 12 and 13 it is seen that there is a slight discrepancy between the yield stress from the SHANSEP MC model and the applied yield stress in the Uplift-Van model. This is due to the fact that the input *POP*-values in PLAXIS were determined by subtracting the vertical yield stress from the in-situ vertical stress that conform the Uplift-Van model in D-Geo Stability, whereas the yield stress in PLAXIS (State Parameter 1) is not necessarily vertical and is independent of the Cartesian system of axes.

4 REMARKS ON THE SHANSEP MC MODEL

The assessment of macro-stability for dikes using the SHANSEP approach is new to the Netherlands. An attempt has been made in this study to describe the knowledge gained so far in the field of numerical models and thus to make this knowledge available to a wider public.

Based on this study, some remarks on the use of the SHANSEP MC model are outlined as follows:

1. The transfer of stress history through *POP*-values works properly by using an advance soil model in PLAXIS with K0-procedure. Whenever possible, it is highly recommended to verify the results of State Parameter 1 in PLAXIS (yield stress), after the model has been switched to the SHANSEP approach.
2. An update for undrained shear strength has to be done prior to performing a safety calculation so as to obtain a correct safety factor against macro-instability.
3. For cases of dikes with shallow slip circles, the calculated safety factor against macro-instability may be affected by the rotation of principal axes near the slope. The discrepancy of results is due to the different definition of the *OCR* and thus the undrained shear strength.
4. For cases of dikes with non-cohesive soils at small depths and with water level near to the dike surface, where the soil strength is determined by the effective friction angle before switching the model to the SHANSEP approach, the instability of relatively thin layers at the surface of an inner slope, called micro-instability, may occur. Such failure mechanisms can be avoided in the model by providing the non-cohesive soils adjacent to the slope with a low value of cohesion.

5 CONCLUSIONS

In this paper, the applicability of the SHANSEP MC model for assessing the macro-stability of a primary dike in the Netherlands is presented. The study deals with the assessment at one of the river dike cross-sections located between Krimpen aan den IJssel and Gouderak, in the province of Zuid-Holland.

To assess the macro-stability of a dike according to the WBI2017, the required parameters for cohesive soils are the undrained strength ratio *S*, and the stress increase exponent *m*. The undrained shear strength s_u is depending on the stress history in the long-term equilibrium situation, which can be taken into account through either the overconsolidation ratio *OCR*, or the in-situ effective vertical stress of the soils in combination with the pre-overburden pressure (*POP*).

The shear strength of non-cohesive soils is determined based on the drained behaviour and the required parameter is the friction angle ϕ' and without cohesion c'.

The study shows that the stress history of the soils is conveyed properly in the SHANSEP MC model by combining the SHANSEP MC model with an advanced soil model in PLAXIS, such as the use of the Hardening Soil Model. The good agreement between the predicted slip circles as well as the safety factors using the SHANSEP MC model in PLAXIS and those obtained using the Uplift-Van model in D-Geo Stability program is also obvious. Therefore, the SHANSEP MC model can be used in addition to the default Uplift-Van model to assess the macro-stability of primary dikes (flood defences) in the Netherlands, which comply with the new Dutch WBI2017 standards.

ACKNOWLEDGEMENTS

The work presented in this paper is part of the KIJK dike strengthening project in the Netherlands. The authors are grateful to Het Hoogheemraadschap van Schieland en de Krimpenerwaard for providing the data, and would also like to thank the anonymous reviewers for their comments to improve the quality of the paper.

REFERENCES

Brinkgreve, R.B.J. & Bakker, H.L. 1991. Non-Linear Finite Element Analysis of Safety Factors, *Proc. 7th International Conference on Computer Methods and Advances in Geomechanics*. Balkema Rotterdam, pp. 1117–1122.

Deltares, 2016. *D-Geo Stability User Manual Version 16.2*.

Griffiths, D.V. & Lane, P.A. 1999. Slope Stability Analysis by Finite Elements. *Geotechnique, 49*(3): 387–403.

Ladd, C.C. & Foott, R. 1974. New Design Procedure for Sta-bility of Soft Clays. *Journal of Geotechnical Engineering Division. ASCE, 100*(7): 763–786.

Panagoulias, S., Palmieri, F. & Brinkgreve, R.B.J. 2016. *The SHANSEP MC Model*. Plaxis BV.

Rijkswaterstaat, 2017. *WBI2017 Code Calibration*.

Schofield, A.N. & Wroth, C.P. 1968. *Critical State Soil Mechanics*. McGraw-Hill.

TAW, 2001. *Technisch Rapport Waterkerende Grondconstructies; Geotechnische Aspecten van Dijken, Dammen en Boezemkaden (in Dutch)*.

Van, M.A. 2001. New Approach for Uplift Induced Slope Failure, *Proceedings of the International Conference on Soil Mechanics and Geotechnical Engineering*. AA Balkema Publishers, pp. 2285–2288.

Numerical Methods in Geotechnical Engineering IX – Cardoso et al. (Eds)
© 2018 Taylor & Francis Group, London, ISBN 978-1-138-33198-3

Incorporating the state parameter into a simple constitutive model for sand

D.M.G. Taborda, D.M. Potts & L. Zdravkovic
Imperial College London, London, UK

A.M.G. Pedro
ISISE, University of Coimbra, Coimbra, Portugal
Formerly Imperial College London, London, UK

ABSTRACT: Sand behaviour is complex and requires the adoption of predictive frameworks capable of capturing the effects of stress level and void ratio on the strength and dilatancy of the material. The state parameter has been shown to be an effective form of explaining sand behaviour and, as a result, has been incorporated into the formulation of various constitutive models. This paper presents a constitutive model for sands, which enhances the Mohr-Coulomb failure criterion in order to include explicitly the influence of the state parameter on the material's strength and dilatancy. The calibration process is illustrated for Fraser River sand and the performance of the model is demonstrated by simulating the response of a monopile embedded within sand deposits of different density. The results show that the constitutive model successfully reproduces the expected influence of the relative density, making it suitable to be employed in a wide range of geotechnical problems.

1 INTRODUCTION

The accurate modelling of sand is a complex task due to the pronounced effect of both the stress level and void ratio on its behaviour. As a result, simple constitutive models are often unable to reproduce the observed response of geotechnical structures founded on sand deposits. In addition, these models typically require different sets of parameters for each relative density, thus increasing the uncertainty surrounding their use in practical design. In order to address this issue, more complex constitutive models, either based on general plasticity (Jefferies & Been, 2006) or bounding surface plasticity (e.g. Manzari & Dafalias, 1997; Loukidis & Salgado, 2009; Taborda et al., 2014), have been proposed. At the centre of their formulation is the concept of state parameter, ψ (Been & Jefferies, 1985), which is schematically defined in Figure 1. If a power law is used to describe the CSL (Li & Wang, 1998), the state parameter can be calculated using:

$$\psi = e - e_{CS} = e - \left[\left(e_{CS}\right)_{ref} - \lambda \left(\frac{p'}{p'_{ref}} \right)^{\xi} \right] \qquad (1)$$

where p'_{ref} is a reference pressure and $\left(e_{CS}\right)_{ref}$, λ and ξ are model parameters. Clearly, as the state parameter has been shown to be a competent predictor of sand response (Jefferies & Been, 2006), it is unsurprising these models generally perform well. However, their formulation is complex and many of their components lack physical meaning, often requiring a convoluted calibration procedure. Moreover, it is often difficult to establish directly the characteristics of sands that are commonly used to describe them: peak angle of shearing resistance and angle of dilatancy.

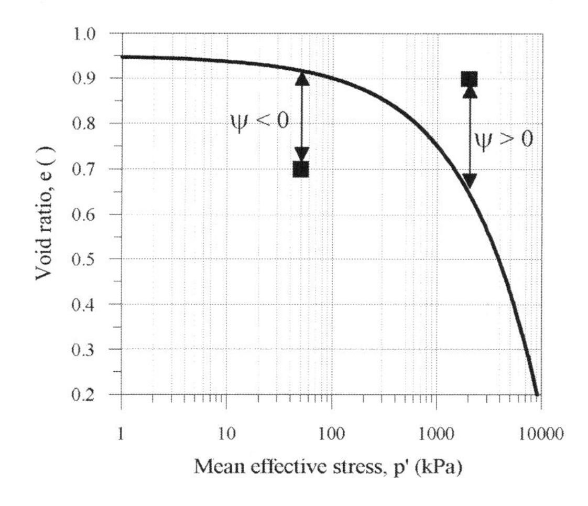

Figure 1. Definition of state parameter.

This paper bridges the gap between the two aforementioned types of constitutive model, retaining the simplicity and familiarity of the Mohr-Coulomb failure criterion, but enhancing it by introducing a direct link between the state parameter and the material's strength and dilatancy. As a result, the proposed constitutive model is capable of simulating the combined effect of stress level and void ratio.

2 CONSTITUTIVE MODEL

2.1 *Fundamental concepts*

The constitutive model assumes that the strength of the material can be described by the Mohr-Coulomb failure criterion, with an angle of shearing resistance which depends directly on the current value of the state parameter. As proposed by Wood et al. (1994), the stress ratio at failure for triaxial compression is given by:

$$M_S = M_{CS} + k \cdot \langle -\psi \rangle \tag{2}$$

where M_{CS} is the stress ratio at critical state, k is a parameter and denotes the Macauley brackets, according to which $A = A$ if $A \geq 0$ or $A = 0$ if $A < 0$. Clearly, as a consequence of Equation 2 and as illustrated schematically in Figure 2 in the $p' - q$ plane, the strength of looser-than-critical samples

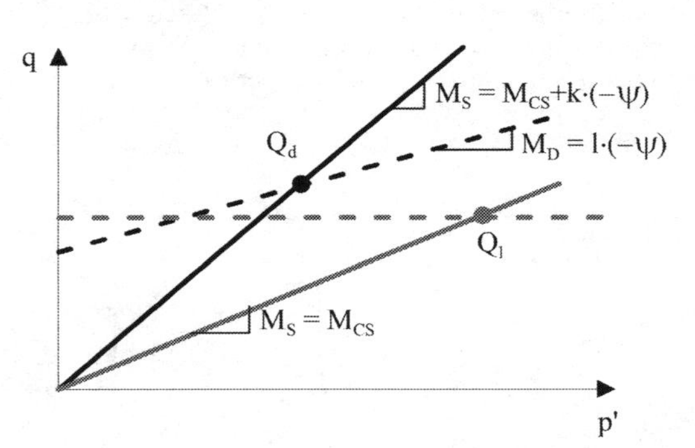

Looser-than-critical samples (ψ≥0)

Yield surface F(σ,ψ)=0

Plastic potential surface P(σ,ψ)=0

Denser-than-critical samples (ψ≥0)

Yield surface F(σ,ψ)=0

Plastic potential surface P(σ,ψ)=0

Figure 2. Yield and plastic potential surfaces adopted by the model for a purely frictional material.

(i.e. $\psi > 0$, point Q_l) is limited to that at critical state. Conversely, for denser-than-critical samples (i.e. $\psi < 0$, point Q_d), the mobilisable strength is larger than that at critical state, with parameter k controlling the relationship between the state parameter and the gain in available strength.

The plastic response is assumed to be non-associated, with the stress ratio defining the inclination of the plastic potential, M_D, depending on the current value of the state parameter according to:

$$M_D = l \cdot \langle -\psi \rangle \tag{3}$$

where l is a model parameter. As a result of the form adopted for Equation 3, looser-than-critical samples ($\psi > 0$, point Q_l in Figure 2) or those at critical state ($\psi = 0$) do not experience any plastic volumetric changes. Conversely, as the absolute value of ψ increases for denser-than-critical samples, the tendency to dilate increases. Despite having a different use, it is interesting to note that this is a similar form to that employed by some bounding surface plasticity models (Manzari & Dafalias, 1997; Loukidis & Salgado, 2009; Taborda et al., 2014) to describe the position of the dilatancy surface, the distance to which determines the volumetric component of the plastic potential.

2.2 *Formulation*

The expression for the Mohr-Coulomb failure criterion in three-dimensional stress space (Potts & Zdravkovic, 1999) is used in this model, adding the state parameter as a variable, since it now controls the strength of the material. This is accomplished by using the current angle of shearing resistance, ϕ_c, in the expression for the yield surface:

$$F(\sigma,\psi) = J - \left(\frac{c}{\tan\phi_c} + p' \right) \cdot g(\theta) = 0 \tag{4}$$

where c is the cohesion, J is the deviatoric stress, θ is the Lode's angle and $g(\theta)$ is the expression defining the opening of the yield surface and its shape in the deviatoric plane, which for the Mohr-Coulomb failure criterion assumes the following form (Potts & Zdravkovic, 1999):

$$g(\theta) = \frac{\sin\phi_c}{\cos\theta + \frac{\sin\theta \cdot \sin\phi_c}{\sqrt{3}}} \tag{5}$$

with the current angle of shearing resistance being determined based on the adopted relationship between strength and the state parameter (Equation 2) using:

$$\phi_c = \arcsin\left(\frac{3 \cdot M_S}{6 + M_S}\right) \tag{6}$$

A similar process is used for the plastic potential, for which the expression in general stress space (Potts & Zdravkovic, 1999) is used:

$$P(\sigma, \psi) = J - [a_P + p'] \cdot g_P(\theta) = 0 \tag{7}$$

where $g_P(\theta)$ controls the opening of the plastic potential surface and its shape in the deviatoric plane, which in this case is assumed to be identical to that employed for the yield surface:

$$g_P(\theta) = \frac{\sin \nu_c}{\cos\theta + \frac{\sin\theta \cdot \sin \nu_c}{\sqrt{3}}} \tag{8}$$

with ν_c being the current value of the dilatancy angle. Moreover, in Equation 7, a_P is the distance from the origin of the principal stress space to the apex of the plastic potential surface (Potts & Zdravkovic, 1999):

$$a_P = \left(\frac{c}{\tan\phi_c} + p'_c\right) \cdot \frac{g(\theta_c)}{g_P(\theta_c)} - p'_c \tag{9}$$

The opening of the plastic potential surface, which according to Equation 8 depends on the current angle of dilatancy, is calculated based on Equation 3 using:

$$\nu_c = \arcsin\left(\frac{3 \cdot M_D}{6 + M_D}\right) \tag{10}$$

Clearly, the consistency condition needs to be updated in order to account for the fact that, in the adopted formulation, the state parameter is a hardening parameter:

$$dF = \frac{\partial F}{\partial \sigma} d\sigma + \frac{\partial F}{\partial \psi} d\psi = 0 \tag{11}$$

However, the state parameter evolves both with changes in mean effective stress level and void ratio (Eq. 1). The latter presents a considerable challenge, as it renders the hardening parameter dependent on the total volumetric strain and not just on the plastic volumetric strain. To overcome this obstacle, and taking into account that in sands the elastic volumetric strains are typically much smaller than those of plastic origin, the concept of "plastic" state parameter, ψ^p, is introduced:

$$\psi \approx \psi^p = e^p - e_{CS} \tag{12}$$

where e^p is the "plastic" void ratio which evolves solely based on plastic volumetric strains, $\Delta\varepsilon^p_{vol}$:

$$\Delta e^p = -(1 + e_0) \cdot \Delta\varepsilon^p_{vol} \tag{13}$$

with e_0 being the initial void ratio of the material. Consequently, the relationships defined by Equations 2 and 3 must also be rewritten in terms of ψ^p.

2.3 *Calibration for Fraser River sand*

To demonstrate the ability of the proposed model to represent soil behaviour, a calibration based on the results of a set of tests on Fraser River sand is performed. Fraser River sand is a material found around the delta of the river Fraser in British Columbia, Canada. It is a uniform, clear sand with particle shapes ranging from angular to sub-angular and characterised by maximum and minimum void ratios of 0.989 and 0.627, respectively (Williams, 2014; Klokidi, 2015). Three isotropically consolidated drained triaxial compression tests performed by Ghafgazi (2011) were chosen for the calibration of the model. As can be seen in Table 1, the initial conditions included a considerable range of mean effective stresses (50 kPa to 515 kPa) and relative densities (65% to 89%). More importantly, given the framework upon which the proposed model is built, it is possible to establish the initial value of the state parameter for each sample by adopting the critical state line in $e - p'$ space estimated by Klokidi (2015), which is defined by the parameters listed in Table 2. Clearly, the chosen tests are characterised by a substantial interval of values of the initial state parameter (−0.228 to −0.127), despite the fact that all samples are initially considered as being denser-than-critical.

To calibrate the function defining the strength of the material (Eq. 2), the mobilised stress ratio must be plotted against the corresponding value of the state parameter, as shown in Figure 3a. Note that only the post-peak portion of the tests are represented in this graph since the constitutive model is unable to simulate the pre-peak behaviour of sand. Moreover, while in effect the plastic state parameter should be used in this calibration, the scatter observed in the data is likely to be of similar significance as the difference between ψ^p and ψ, with the added factor that determining ψ^p would require the introduction of further approximations inherent to the estimation of plastic strains (Loukidis & Salgado, 2009; Taborda, 2011). Based on the observed behaviour,

Table 1. Details of the drained triaxial compression tests considered in the calibration of the proposed plastic model.

	p_0'	e_0	D_r	ψ_0
Designation	(kPa)	(%)		
CID-D-50	50.3	0.753	65.2	−0.164
CID-D-115	113.9	0.668	88.7	−0.228
CID-D-515	514.5	0.689	82.9	−0.127

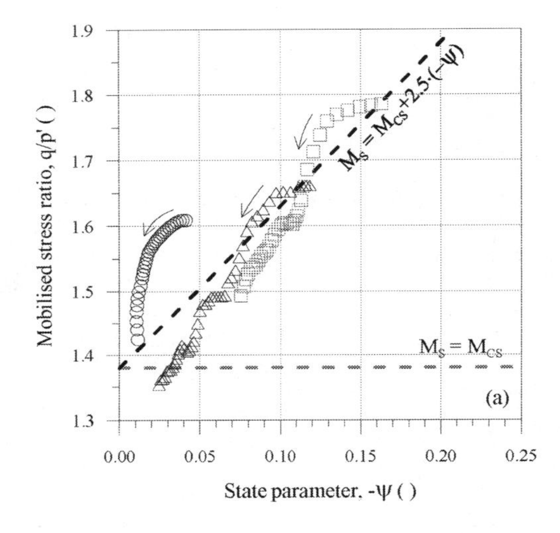

△ CID-D-50 □ CID-D-115 ○ CID-D-515

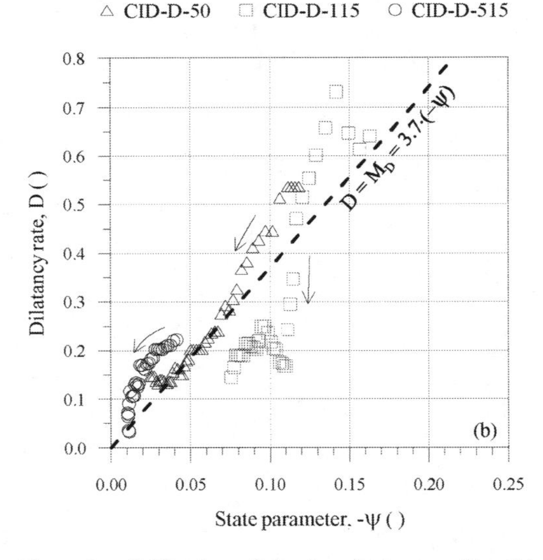

Figure 3. Calibration of the functions controlling (a) the material strength and (b) the simulated dilatancy rate.

Table 2. Summary of the calibrated model parameters.

Proposed plastic model		Nonlinear elastic model (Taborda et al., 2016)	
$e_{CS,ref}$	0.95	$G_{0,ref}\,(kPa)$	42200.0
λ	0.05	μ	0.2
ξ	0.60	$p'_{ref}\,(kPa)$	100.0
$\varphi_{CS}(°)$	34.1	m_G	0.5
$c(kPa)$	0.0	a(m/m)	2.0E-5
k	2.5	b	1.4
l	3.7	R_G	0.4

it is clear that as shearing progresses (indicated by the arrows in the figure), the value of the state parameter approaches zero, for which the mobilised strength must approach that at critical state, estimated by Klokidi (2015) as being defined by M_{CS} = 1.38 (corresponding to an angle of shearing resistance of approximately 34.1°). Given the determined data points, a value of k = 2.5 is considered adequate for this material.

A similar process is carried out to calibrate parameter l, which controls the opening of the plastic potential (Eq. 3). For this aspect of the constitutive model, the plastic dilatancy rate should be plotted against the plastic state parameter. However, invoking the considerably larger magnitude of the plastic strains when compared to those of elastic origin, the total dilatancy rate is used:

$$D^p \approx D = \frac{\Delta\varepsilon_{vol}}{\Delta\varepsilon_s} \tag{14}$$

where $\Delta\varepsilon_s$ is the incremental deviatoric strain defined as $\Delta\varepsilon_s = 2/3\cdot(\Delta\varepsilon_{ax} - \Delta\varepsilon_{rad})$, with $\Delta\varepsilon_{ax}$ and $\Delta\varepsilon_{rad}$ being the incremental axial and radial strains, respectively. As expected, as the material is sheared, the observed dilatancy ratio reduces, approaching zero, which would correspond to the occurrence of critical state conditions. Equation 3 with a value of l = 3.7 is seen to reproduce adequately the observed response.

The last parameters to be determined are those controlling the elastic response. Since the proposed model only defines the post-peak plastic behaviour of the material, the coupling with any elastic model is possible. In the present case, the Imperial College Generalised Small Strain Stiffness model (IC.G3S, Taborda & Zdravkovic, 2012; Taborda et al., 2016) is used. Briefly, this versatile model is capable of taking into account the effects

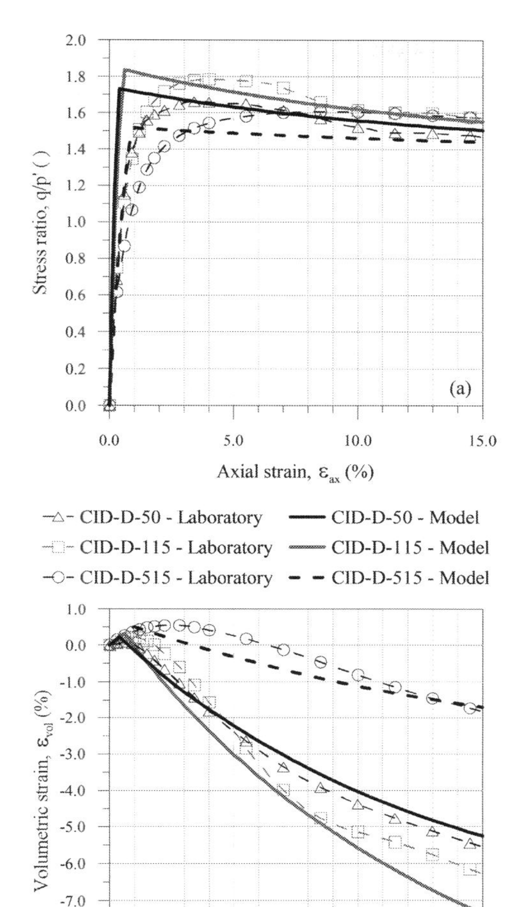

Figure 4. Comparison between the measured and predicted behaviour in terms of (a) axial strain—stress ratio and (b) axial strain—volumetric strain.

of confining pressure, deformation level, void ratio and loading direction on the tangent shear modulus and bulk modulus. For Fraser River sand, a constant Poisson's ratio of 0.2 is assumed (Williams, 2014), while the elastic shear modulus is defined according to:

$$G_{tan} = G_0 \cdot \left[R_G + \frac{1 - R_G}{1 + \left(\frac{E_d}{a} \right)^b} \right] \tag{15}$$

where E_d is the second invariant of the strain tensor (Potts & Zdravkovic, 1999), a, b and R_G are model parameters and G_0 is given by:

$$G_0 = \frac{G_{0,ref}}{0.3 + 0.7e^2} \cdot \left(\frac{p'}{p'_{ref}} \right)^{m_G} \tag{16}$$

with m_G controlling the nonlinearity of the dependency of the elastic shear modulus on the mean effective stress. The model parameters controlling the maximum shear stiffness ($G_{0,ref}$ and m_G) were calibrated by Williams (2014) based on bender element tests presented by Chillarige et al. (1997). However, for the calibration of the parameters defining the reduction of stiffness with deformation level (a, b and R_G), insufficient data obtained at low to medium strains were found and reasonable values are adopted.

The complete set of parameters is listed in Table 2, while the modelled behaviour is compared to the experimental data in terms of axial strain—deviatoric stress and axial strain—volumetric strain in Figures 4a and 4b, respectively. Clearly, the model is capable of reproducing the observed volumetric response with impressive accuracy, matching well both the peak dilatancy rate and the final volumetric strain. Moreover, the mobilised strength is adequately captured, although the inability to simulate pre-peak plastic hardening leads to the proposed model predicting the peak strength and the consequent onset of strain-softening at slightly lower strain levels than those measured in the laboratory. Overall, the model performs remarkably well, particularly considering the simplicity of the adopted formulation.

3 APPLICATION TO THE DESIGN OF MONOPILE FOUNDATIONS

3.1 *General considerations*

To demonstrate the application of the proposed constitutive model to boundary value problems, the case of a monopile foundation for an offshore wind turbine was selected. A geometry representative of current monopile design (Zdravkovic et al., 2015) was chosen, with a diameter of D = 7.5 m, an embedment depth of L = 30 m and a stick-up height of h = 75 m. This results in a slenderness ratio of L/D = 4 and in a normalised stick-up height of h/D = 10. These geometric characteristics, illustrated in Figure 5, suggest that pile response to lateral loading should result in a rigid mode of deformation.

As the problem is symmetric with respect to the plane containing the applied load (x-z, see

Table 3. Initial conditions for the two analysed cases.

D_r (%)	γ_{sat} (kN/m^3)	K_0 (–)	e_0 (–)
50	18.8	0.5	0.81
80	19.3	0.5	0.70

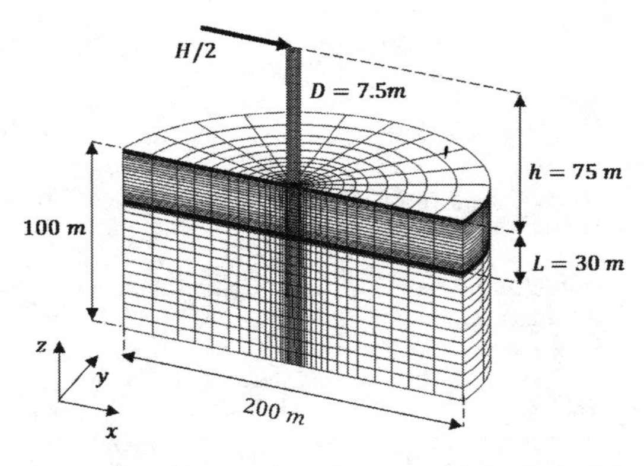

Figure 5. Dimensions of the finite element mesh and the geometry of the analysed pile.

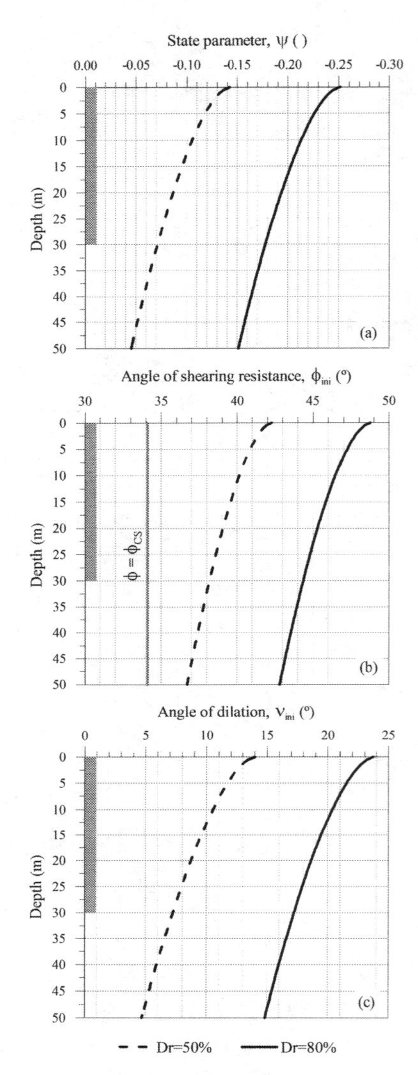

Figure 6. Variations in depth of (a) the state parameter, (b) the initial angle of shearing resistance and (c) the initial angle of dilation for the two values of relative density.

Figure 5), only half of the domain needs to be considered. The soil is discretised using 20-noded hexahedral elements, the pile is represented by 8-noded shell elements (Schroeder et al., 2007), while the interface between the structure and the surrounding soil mass is modelled using 16-noded zero-thickness interface elements (Day & Potts, 1994). The latter were assumed to be purely frictional, with their strength being characterised by the angle of shearing resistance at critical state of the surrounding soil (34.1° for Fraser River sand). The model parameters adopted for the soil are those calibrated in the previous section (Table 2), while steel was assumed to behave elastically with a Young's modulus of 200 GPa and a Poisson's ratio of 0.3.

As shown in Figure 5, the far vertical boundary of the mesh is placed at a radial distance of 100 m in order to guarantee that the pile response is not affected by any boundary effects. Similar considerations led to the adoption of a large depth (100 m) for the soil deposit. The displacements along the three directions (x, y, z) are set to zero over the bottom boundary of the mesh, ensuring that no rigid body movements of the mesh are possible. Furthermore, along the vertical cylindrical boundary, the displacements in the radial direction (i.e. normal to the boundary) are set to zero. Given that x-z is a plane of symmetry, the displacements at the nodes on this plane are restricted in the y direction. Additionally, no rotation about the x— and z-axes are allowed at the nodes on the shell elements contained in this plane. Lastly, horizontal loading is carried out by incrementally applying uniform displacements along the x-direction at the pile top (h = 75 m).

To highlight the effects of the initial state of the sand on the simulated pile response, two different analyses with distinct values of relative density (50% and 80%) were carried out. Moreover, while it is assumed that the value of K_0 is related to the angle of shearing resistance at critical state and, hence, is independent of the void ratio of the soil deposit, slightly different values of bulk unit weight are used in the two analyses (see Table 3). The distribution of the pore water pressures is assumed to be hydrostatic.

Similar to other state parameter-based constitutive models (Manzari & Dafalias, 1997; Loukidis & Salgado, 2009; Taborda et al., 2014), it is not possible to determine the simulated peak strength and dilatancy angle directly from the adopted material properties. To establish these quantities, it is first necessary to determine the profiles of the state parameter (Eq. 1), shown in Figure 6a, where the shaded grey rectangle denotes the depth of the pile. Clearly, both soil deposits are characterised as denser-than-critical ($\psi < 0$) for the entire depth of the pile, with larger absolute values of ψ being determined for the case with the larger relative density. It is interesting to note that the two lines are almost parallel, suggesting that the adjustment of the bulk unit weight with relative density has a very small effect.

Using Equations 2 and 3, it is possible to estimate the peak angle of shearing resistance and angle of dilation, respectively, for both analyses. The obtained variations with depth of these soil characteristics are illustrated in Figures 6b and 6c, with the material closer to the ground surface being characterised by higher strength and higher tendency to dilate. Indeed, for the case with a relative density of 80%, the strength rises to about 49° and the angle of dilatancy to 24°, suggesting that for even denser deposits a cut-off may be needed to prevent unrealistic values from being simulated. At the other extreme, the looser sand deposit at the depth of the pile toe (30 m) still has a peak strength about 4° higher than that at critical state, associated to an angle of dilatancy of around 7°.

3.2 *Results*

The load-displacement curves at mudline level for both values of relative density are shown in Figure 7, highlighting both the initial response of the piles, here assumed to correspond to displacements up to D/1000, and their ultimate capacity, for which a criterion of D/10 was chosen.

The initial response of the piles, which controls their response in the operational range and is therefore of importance to fatigue analysis, is very similar for the two values of relative density, with a slightly stiffer response being predicted for the denser deposit. This arises from the difference in elastic behaviour which, in the present case, was assumed to be void ratio dependent (Eq. 16). There is also a slight increase in stiffness due to the larger unit weight of the denser deposit, though this difference is marginal.

The ultimate capacity of the pile installed in the denser material is about 50% higher than that in the looser sand. This very large difference arises not only from the higher strength the former can mobilise (Fig. 6b), but also from its considerable higher tendency to dilate. Clearly, these results suggest the need to employ a constitutive model capable of simulating the response of sand under a variety of initial stress states using a single set of parameters.

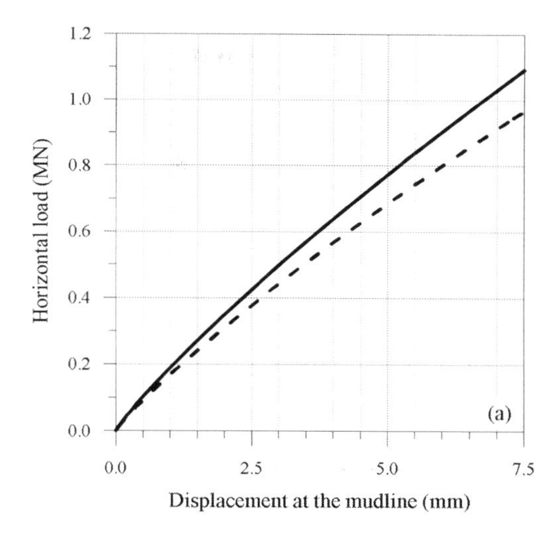

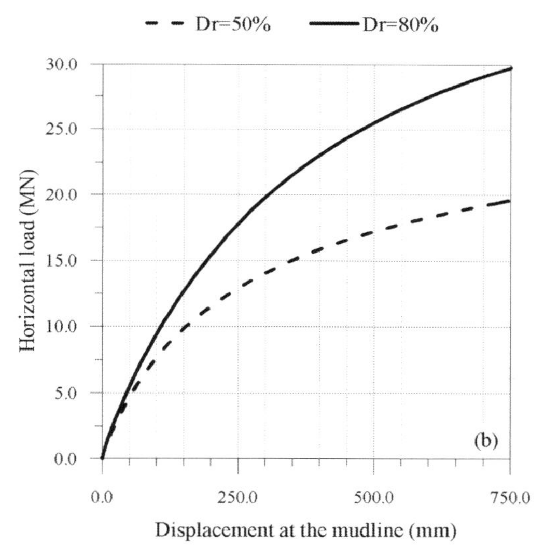

Figure 7. Load-displacement curves focusing on (a) the response at small displacements (D/1000) and (b) the behaviour at large displacements (D/10).

4 CONCLUSIONS

A simple state parameter-based constitutive model is presented for sands and its calibration is performed for Fraser River sand. It is shown that, despite the simplicity of the adopted formulation and the small number of required parameters (only two more parameters than a standard Mohr-Coulomb failure criterion), the model is capable of reproducing accurately the post-peak plastic response of sands, both in terms of mobilised strength and volumetric behaviour. The model is subsequently used in the simulation of two monopiles installed in deposits of sands of different density. The obtained results illustrate the impact of the model's ability to successfully reproduce the effect of the initial state of the soil on the response of these geotechnical structures using a single set of parameters.

REFERENCES

Been, K. & Jefferies, M.G. 1985. A state parameter for sands. *Geotechnique* 35(2): 99–112.

Chillarige, A., Robertson, P., Morgenstern, N. & Christian, H. 1997. Evaluation of the in situ state of Fraser River sand. *Canadian Geotechnical Journal* 34(4): 510–519.

Day, R.A. & Potts, D.M. 1994. Zero thickness interface elements—numerical stability and application. *International Journal of Numerical and Analytical Methods in Geomechanics* 18: 689–708.

Ghafghazi, M. 2011. *Towards comprehensive interpretation of the state parameter from cone penetration testing in cohesionless soils.* DPhil thesis. University of British Columbia.

Jefferies, M.G. & Been, K. 2006. *Soil liquefaction: a critical state approach.* Boca Raton: CRC Press.

Klokidi, M.E.P. 2015. *Numerical simulation of liquefaction behaviour with focus on the re-liquefaction of sand deposits.* MSc thesis. Imperial College London, London.

Li, X.S. & Wang, Y. 1998. Linear representation of steady-state line for sand. *Journal of Geotechnical and Geoenvironmental Engineering* 124(12): 1215–1217.

Loukidis, D. & Salgado, R. 2009. Modeling sand response using two-surface plasticity. *Computers & Geotechnics* 36(1–2): 166–186.

Manzari, M.T. & Dafalias, Y.F. 1997. A critical state two-surface plasticity model for sands. *Geotechnique* 47(2): 255–272.

Muir Wood, D., Belkheir, K. & Liu, D.F. 1994. Strain softening and state parameter for sand modelling. *Geotechnique* 44(2): 335–339.

Potts, D.M. & Zdravkovic, L. 1999. *Finite element analysis in geotechnical engineering: theory.* London: Thomas Telford.

Schroeder, F.C., Day, R.A., Potts, D.M. & Addenbrooke, T.I. 2007. An 8-noded isoparametric shear deformable shell element. *International Journal of Geotemechanics* 7(1): 44–52.

Taborda, D.M.G. 2011. *Development of constitutive models for application in soil dynamics.* PhD Thesis, Imperial College, University of London.

Taborda, D.M.G., Potts, D.M. & Zdravkovic, L. 2016. On the assessment of energy dissipated through hysteresis in finite element analysis. *Computers & Geotechnics* 71: 180–194.

Taborda, D.M.G. & Zdravkovic, L. 2012. Application of a Hill-Climbing technique to the formulation of a new cyclic nonlinear constitutive model. *Computers & Geotechnics* 42: 80–91.

Taborda, D.M.G., Zdravkovic, L., Kontoe, S. & Potts, D.M. 2014. Computational study on the modification of a bounding surface plasticity model for sands. *Computers & Geotechnics* 59: 145–160.

Williams, J.D. 2014. *Modelling of anisotropic sand behaviour under generalised loading conditions.* MSc thesis. Imperial College London, London.

Zdravkovic, L., Taborda, D.M.G., Potts, D.M., Jardine, R.J., Sideri, M., Schroeder, F.C., Byrne, B.W., McAdam, R., Burd, H.J., Houlsby, G.T., Martin, C.M., Gavin, K., Doherty, P., Igoe, D., Muir Wood, A., Kallehave, D. & Skov Gretlund, J. 2015. Numerical modelling of large diameter piles under lateral loading for offshore wind applications. *Proceedings of the 3rd international symposium on frontiers in offshore geotechnics, Norway.*

Numerical Methods in Geotechnical Engineering IX – Cardoso et al. (Eds)
© 2018 Taylor & Francis Group, London, ISBN 978-1-138-33198-3

Governing parameter method for numerical integration of constitutive models for clays

M. Vukićević & S. Jocković
Department of Geotechnical Engineering, Faculty of Civil Engineering, University of Belgrade, Serbia

ABSTRACT: Governing Parameter Method (GPM) is robust, accurate and efficient method for stresses calculation from given strains within incremental inelastic analysis. GPM is implicit stress point algorithm and basic step is to express all unknown variables in the integration procedure in terms of one governing parameter. Within the return mapping algorithm, stress integration is achieved by solving one nonlinear equation with respect to the governing parameter. Two GPM stress point algorithms are shown in the paper, for the Modified Cam Clay model and for the HASP constitutive model for overconsolidated clays developed by authors. Constitutive models were implemented in Abaqus. Verification of the numerical procedure was performed.

1 INTRODUCTION

In order for the nonlinear material model to be used in the analysis of boundary value problems, it is necessary to implement the material model into a software package. The implementation implies the formulation of the numerical procedure for the integration of stresses for the known strain increment. The numerical procedure must be stable and accurate, because the accuracy of the solution of the boundary value problem depends on the integration algorithm for constitutive equations. The paper presents numerical integration algorithm for two constitutive models for clays: Modified Cam Clay—MCC model (Roscoe & Burland 1968) and HASP model (Jocković & Vukićević 2016), called Governing Parameter Method—GPM. Governing parameter method was used in the plasticity theory for metals and it was later successfully expanded to soil constitutive models. Constitutive models are implemented in Abaqus using UMAT user subroutine.

2 CONSTITUTIVE MODELS FOR CLAYS

2.1 *MCC model*

The MCC model is based on a great amount of experimental research on normally consolidated and over consolidated clays, so that it can be reliably used with these materials with monotonic load. The MCC model is developed in triaxial p'-q plane. Complete constitutive relations of the MCC model can be presented as follows:

$$d\varepsilon_v^p = \frac{\lambda-\kappa}{v}\frac{1}{p'}\left(\frac{M^2-\eta^2}{M^2+\eta^2}dp' + \frac{2\eta}{M^2+\eta^2}dq\right) \tag{1}$$

$$d\varepsilon_q^p = \frac{\lambda-\kappa}{v}\frac{1}{p'}\left(\frac{2\eta}{M^2+\eta^2}dp' + \frac{4\eta^2}{(M^2+\eta^2)(M^2-\eta^2)}dq\right) \tag{2}$$

$$d\varepsilon_v^e = \kappa\frac{dp'}{vp'} = \frac{1}{K}dp' \tag{3}$$

$$d\varepsilon_q^e = \frac{1}{3G}dq \tag{4}$$

where v is specific volume, λ is a slope of the virgin compression line, κ is a slope of an unloading/reloading (swelling) line in v-lnp' plane, M is slope of critical state line, η is stress ratio and G is shear moduli.

2.2 *HASP model*

As MCC model does not provide satisfactory results for overconsolidated clays, some modifications are needed. This has led to the development of the HASP (HArdening State Parameter) model (Jocković & Vukićević 2016) for the description of the mechanical behavior of overconsolidated clays. HASP model is developed within the concept of bounding surface plasticity (Dafalias & Herrmann 1982). The HASP model overcomes many deficiencies of the MCC model and at the same time it retains the same simplicity and the same set of parameters.

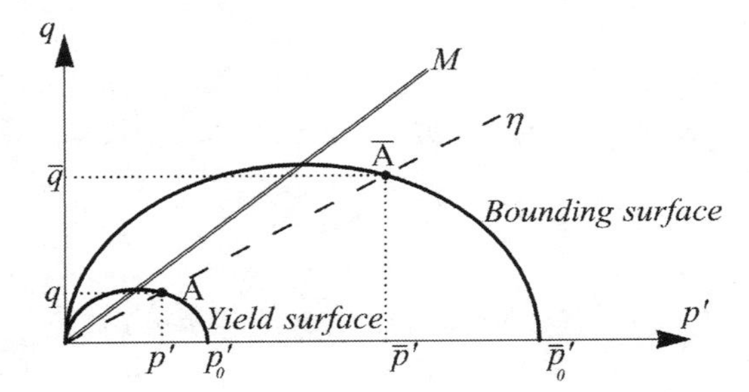

Figure 1. Bounding surface concept.

With the state parameter (Been & Jefferies, 1985) and the degree of overconsolidation as the state variables a novel form of the hardening rule was proposed and it is possible to describe a number of elements of mechanical behavior of overconsolidated clays. Relationships in the model are based on the principle that plastic strains develop from the very beginning of loading. Thus, point A (Figure 1) representing the current stress state is always on the inner loading surface—yield surface, the size of which is defined by the value of the mean effective stress p_0':

$$\frac{p'}{p_0'} = \frac{M^2}{M^2 + \eta^2} \tag{5}$$

Expressions for plastic strains of the HASP model can be written as:

$$d\varepsilon_v^p = \frac{\lambda - \kappa}{v} \frac{1}{p'} \frac{1}{\omega} \left(\frac{M^2 - \eta^2}{M^2 + \eta^2} dp' + \frac{2\eta}{M^2 + \eta^2} dq \right) \tag{6}$$

$$d\varepsilon_q^p = \frac{\lambda - \kappa}{v} \frac{1}{p'} \frac{1}{\omega} \left(\frac{2\eta}{M^2 + \eta^2} dp' + \frac{4\eta^2}{(M^2 + \eta^2)(M^2 - \eta^2)} dq \right) \tag{7}$$

where ω is hardening coefficient. It is noticeable that the hardening coefficient is at the same time the re-duction coefficient for plastic strains. Consequently, plastic strains of an overconsolidated clay in the initial loading stage, when the MCC model predicts only elastic strains, can be significantly reduced with adequate formulation of the hardening coefficient ω. It is then possible to assume that soil deforms plastically from the very beginning of loading. Expressing the hardening parameter through state parameter of stress point on current yield surface Ψ, state parameter of image stress point on bounding surface $\bar{\Psi}$ and overconsolidation ratio R:

$$\omega = \left(1 + \frac{\bar{\Psi} - \Psi}{\bar{\Psi}}\right) R \tag{8}$$

enables the describing of strain hardening, strain softening, as well as negative dilatancy of overconsolidated clays. In undrained conditions, the general form of the effective stress paths depending on the overconsolidation ratio is well predicted, as well as pore water pressure. In drained conditions, HASP model predicts the smooth transition from contractive to dilatant behaviour before the peak strength is reached and a smooth transition from hardening to softening, without mathematical description.

3 GOVERNING PARAMETER METHOD

3.1 *General*

The implicit scheme of stress integration named Governing parameter method was developed by Kojić (Kojić 1996, Kojić & Bathe 2003). It is a generalization of radial return method presented by Wilkins (1964). The basic principle is to present all unknown variables in the function of one parameter (governing parameter), so that the problem is reduced to solving one nonlinear equation per one unknown governing parameter. The initial assumption is that the stress-strain state in the material point at the start of the increment, at time t, is known. The known variables are:

$${}^{t}\boldsymbol{\sigma}, {}^{t}\varepsilon, {}^{t}\varepsilon^{p}, {}^{t}\boldsymbol{\beta}, {}^{t+\Delta t}\varepsilon \tag{9}$$

where ${}^{t}\boldsymbol{\sigma}, {}^{t}\varepsilon, {}^{t}\varepsilon^{p}$ are stresses, total strains and inelastic strains at time t, ${}^{t}\boldsymbol{\beta}$ is the total of internal variables presenting the history of inelastic deformations depending on the type of constitutive model and ${}^{t+\Delta t}\boldsymbol{\varepsilon}$ presents the total deformation at time $t+\Delta t$.

The task of the numerical integration is to define the stresses ${}^{t+\Delta t}\boldsymbol{\sigma}$, total strains ${}^{t+\Delta t}\boldsymbol{\varepsilon}^{p}$ and the total of internal variables ${}^{t+\Delta t}\boldsymbol{\beta}$ at time $t+\Delta t$. The main steps in the implicit GPM are:

1. express all unknown variables in the function of one parameter p – governing parameter;
2. form a governing function $f(p)$ and find the zero of the function. The zero of the function presents the value of the governing parameter at time $t+\Delta t$;
3. calculate all the unknown variables using the value of the governing parameter at time $t+\Delta t$.

The governing function $f(p)$ is most frequently the yield function which, at time $t+\Delta t$, can be presented:

$$ {}^{t+\Delta t}f_y\left({}^{t+\Delta t}\boldsymbol{\sigma}, {}^{t+\Delta t}\boldsymbol{\beta}\right) = {}^{t+\Delta t}f_y\left({}^{t+\Delta t}q, {}^{t+\Delta t}p', {}^{t+\Delta t}\boldsymbol{\beta}\right) = 0 \tag{10} $$

where p' is the mean effective stress and q is deviatoric stress. GPM uses the return mapping concept (elastic predictor—plastic corrector method), where in the first step it is assumed that strains are elastic so that we have:

$$ {}^{t+\Delta t}\boldsymbol{\sigma}^e = \boldsymbol{D}^e\,{}^{t+\Delta t}\boldsymbol{\varepsilon}^e = \boldsymbol{D}^e\left({}^{t+\Delta t}\boldsymbol{\varepsilon} - {}^t\boldsymbol{\varepsilon}^p\right) \tag{11} $$

where $\boldsymbol{D}^e$ is elasticity matrix, ${}^{t+\Delta t}\boldsymbol{\varepsilon}^e$ is the elastic strain vector at the end of the time step (assuming there is no plastic flow for given time/load increment).

The yield function is then:

$$ {}^{t+\Delta t}f_y^e = {}^{t+\Delta t}f_y\left({}^{t+\Delta t}\boldsymbol{\sigma}^e, {}^t\boldsymbol{\beta}\right) \tag{12} $$

If:

$$ {}^{t+\Delta t}f_y^e > {}^t f_y \tag{13} $$

there has been plastic straining. In the further integration procedure the total stresses are:

$$ {}^{t+\Delta t}\boldsymbol{\sigma} = {}^{t+\Delta t}\boldsymbol{\sigma}^e - \boldsymbol{D}^e\Delta\boldsymbol{\varepsilon}^p \tag{14} $$

and they must fulfill the yield condition (10).

3.2 *Selection of the governing parameter*

The crucial step in GPM is the selection of the governing parameter. The complete GPM algorithm where the increment of plastic volumetric strain was taken as the governing parameter can be found in Kojić & Bathe (2003). In the paper by Vukićević (2010), the mean effective stress p' was taken as the governing parameter for the MCC model, as a much more convenient parameter with clear physical meaning. For the HASP model, the mean effective stress was also taken as the governing parameter. The main reason for that selection is the fact that the increment of plastic volumetric strain is often difficult to predict, particularly with hard overconsolidated clays where the effects of dilatancy are very pronounced. On the other hand, the boundaries of the possible stress states are always known in each time/load increment and all variables of the HASP model can be expressed in the function of the mean effective stress.

3.3 *α - method*

In accordance with the standard α method, the integral of a function $f(t)$ at the interval Δt can be approximated as:

$$ \int_t^{t+\Delta t} f(t)dt = \left[(1-\alpha)\,{}^t f + \alpha\,{}^{t+\Delta t}f\right]\Delta t \tag{15} $$

where α is the integration parameter. Value $\alpha = 0$ corresponds to Euler forward integration method, while value $\alpha = 1$ corresponds to Euler backward integration method. For $\alpha = 0.5$ we obtain the trapezoidal rule. In the implicit GPM $\alpha = 1$ applies and the following approximation can be taken for the increment of plastic strain:

$$ \Delta\boldsymbol{\varepsilon}^p = \int_t^{t+\Delta t}\left(\frac{\partial f}{\partial\boldsymbol{\sigma}}d\chi\right)dt = \Delta\chi\frac{\partial\,{}^{t+\Delta t}f}{\partial\,{}^{t+\Delta t}\boldsymbol{\sigma}} \tag{16} $$

where $\Delta\chi$ is a positive scalar (scalar multiplicator) corresponding with interval Δt. This is one of the main approximations in the integration algorithm. From the normality rule, the expression for the increment of plastic volumetric strain can be obtained:

$$ d\varepsilon_v^p = d\chi\frac{\partial f}{\partial p'} = d\chi(2p' - p_0') \tag{17} $$

Based on the expression (16) it can be concluded that:

$$ \Delta\varepsilon_v^p = \Delta\chi(2\,{}^{t+\Delta t}p' - {}^{t+\Delta t}p_0') \tag{18} $$

In addition, from normality rule, the following expression for the increment of plastic shear strain can be obtained:

$$ d\varepsilon_q^p = d\chi\frac{\partial f}{\partial q} = d\chi\frac{2q}{M^2} \tag{19} $$

Also, based on the expression (16) it can be concluded that:

$$ \Delta\varepsilon_q^p = \Delta\chi\frac{2\,{}^{t+\Delta t}q}{M^2} \tag{20} $$

or:

$$ \Delta\varepsilon_{ij}'^p = 3\Delta\chi\frac{{}^{t+\Delta t}S_{ij}}{M^2} \tag{21} $$

where ${}^{t+\Delta t}S_{ij}$ is deviatoric component of stress tensor.

4 NUMERICAL INTEGRATION OF THE HASP MODEL

4.1 *Relation of the HASP model at time t+Δt*

For the known governing parameter ${}^{t+\Delta t}p'$, void ratio in configuration $t+\Delta t$ can be expressed as:

$$ {}^{t+\Delta t}e = \left(1+e_0\right)\exp\left(-{}^{t+\Delta t}\varepsilon_v\right)-1 \tag{22}$$

The expression for the hardening parameter at time $t+\Delta t$ can be obtained using additive decomposition of volumetric strain:

$$ {}^{t+\Delta t}p_0' = {}^{t}p_0'\exp\left(\frac{\left({}^{t}e-{}^{t+\Delta t}e\right){}^{t}\omega}{\lambda-\kappa}\right)\left(\frac{{}^{t}p'}{{}^{t+\Delta t}p'}\right)^{\frac{\kappa\,{}^{t}\omega}{\lambda-\kappa}} \tag{23}$$

If the stress ratio $\eta=M$ is achieved, the following is true:

$$ {}^{t+\Delta t}q = M\,{}^{t+\Delta t}p' \tag{24}$$

$$ \Delta q = {}^{t+\Delta t}q-{}^{t}q \tag{25}$$

$$ \Delta\varepsilon_q^p = \left(\Delta\varepsilon_q-\frac{\Delta q}{3\,{}^{t}G}\right)\frac{1}{{}^{t}\omega} \tag{26}$$

$$ \Delta\chi = \frac{M\Delta\varepsilon_q^p}{2\,{}^{t+\Delta t}p'} \tag{27}$$

If the stress ratio $\eta=M$ is not achieved then:

$$ \Delta\varepsilon_v^p = \frac{\lambda-\kappa}{{}^{t}v}\ln\left(\frac{{}^{t+\Delta t}p_0'}{{}^{t}p_0'}\right)\frac{1}{{}^{t}\omega} \tag{28}$$

$$ \Delta\chi = \frac{\Delta\varepsilon_v^p}{2\,{}^{t+\Delta t}p'-{}^{t+\Delta t}p_0'} \tag{29}$$

Deviatoric component of stress tensor is:

$$ {}^{t+\Delta t}S_{ij} = \frac{\Delta\varepsilon_{ij}'+\dfrac{{}^{t}S_{ij}}{2\,{}^{t}G}}{\dfrac{3\Delta\chi}{M^2}+\dfrac{1}{2\,{}^{t}G}} \tag{30}$$

Finally, stress tensor in configuration $t+\Delta t$ can be obtained:

$$ {}^{t+\Delta t}\sigma_{ij} = {}^{t+\Delta t}p'\delta_{ij}+{}^{t+\Delta t}S_{ij} \tag{31}$$

Plastic shear strains:

$$ \Delta\varepsilon_{ij}'^{p} = 3\Delta\chi\frac{{}^{t+\Delta t}S_{ij}}{M^2} \tag{32}$$

Yield function can be written as:

$$ {}^{t+\Delta t}f = \frac{{}^{t+\Delta t}q^2}{M^2}+{}^{t+\Delta t}p'\left({}^{t+\Delta t}p'-{}^{t+\Delta t}p_0'\right) \tag{33}$$

4.2 *Procedure of numerical integration*

Considering the fact that, according to the HASP model, the stress point is always on the yield surface and that there is no pure elastic region, in the integration algorithm which will be presented, there is no elastic prediction and plastic correction in case of plastic deformation, as was described by Kojić & Bathe (2003). For the value of the mean effective stress as the governing parameter, the following condition must be met:

$$ {}^{t+\Delta t}f = \left|{}^{t+\Delta t}f\left({}^{t+\Delta t}p'\right)\right| \le \mathrm{TOL} \tag{34}$$

I.e. ${}^{t+\Delta t}p'$ represents the zero of the governing function. For the calculation of the zero of the function, bisection method can be applied, i.e. halving the interval where the function is monotonic and has the opposite sign. The interval of the mean effective stresses which meets the aforesaid conditions can be found in the vicinity of the known stress state ${}^{t}p'$ (bracketing the root). For each of the boundaries of an interval, the value of the governing functions ${}^{t+\Delta t}f_1$ and ${}^{t+\Delta t}f_2$ should be found. For the solution (zero of the function) to be in the assumed interval, the following condition must be met:

$$ {}^{t+\Delta t}f_1\cdot{}^{t+\Delta t}f_2 < 0 \tag{35}$$

The boundaries of the intervals gradually expand in small increments until the condition (35) is met. When the boundaries of the intervals of the mean effective stress are known, the zero of the function is easily found using bisection method. For known governing parameter, using the relations presented in section 4.1, unknown variables at time $t+\Delta t$ can be calculated.

5 NUMERICAL INTEGRATION OF THE MCC MODEL

5.1 *Relation of the MCC model at time t+Δt*

There is no considerable difference between the numerical integration algorithm for the MCC model and for the HASP model. For the hardening coefficient $\omega = 1$, the HASP model is transformed into the MCC model. In the numerical integration algorithm for the MCC model, all the relations presented in section 4.1 are valid for value $\omega = 1$.

5.2 *Procedure of numerical integration*

As the MCC model predicts elastic behavior within the initial yield surface, the integration algorithm contains elastic predictor and plastic corrector, i.e. the procedure described by expressions (11) to (14). The interval of the mean effective stresses where the governing function is monotonic and has the opposite sign may be found around a trial elastic solution. The boundaries of the interval gradually expand in small increments until condition (35) is met, and the zero of the function is obtained by bisection method.

6 IMPLEMENTATION IN ABAQUS

The HASP model and the MCC model are implemented in Abaqus/Standard using UMAT user subroutine and the numerical procedure for stress integration previously shown. It should be noted that in UMAT, Jacobian matrix is presented by an elastic matrix for both models. The verification of the developed GPM algorithm was done in two ways:

1. by comparing the results of the finite element simulation of triaxial test using the MCC model already built in Abaqus (Abaqus algorithm—Critical state (clay) plasticity model, Abaqus Analysis User's Manual 2011) and the MCC model implemented through UMAT (GPM algorithm);
2. by comparing the results of the finite element simulation of triaxial test using the HASP model implemented in Abaqus (GPM algorithm) and using Euler deformation integration along the known stress path (procedure programmed in Excel) according to the HASP model.

For the verification procedure, two overconsolidated clays were selected with material parameters presented in Table 1:

Table 1. Parameters of MCC and HASP models.

	λ	κ	M	Γ	μ
Cardiff clay CU tests	0.140	0.050	1.05	2.63	0.2
Bangkok clay CD tests	0.100	0.020	1.13	2.85	0.2

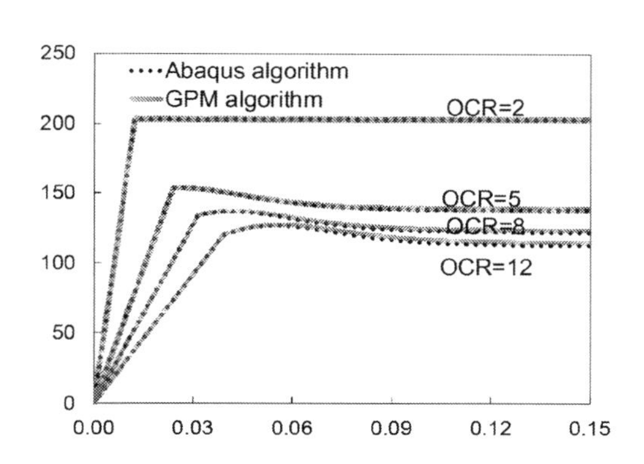

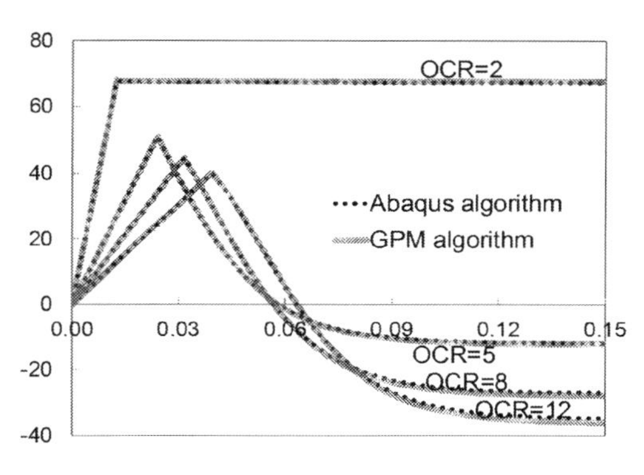

Figure 2. Comparison of results of CU tests, MCC model (Abaqus) and implemented MCC model (GPM), Cardiff clay.

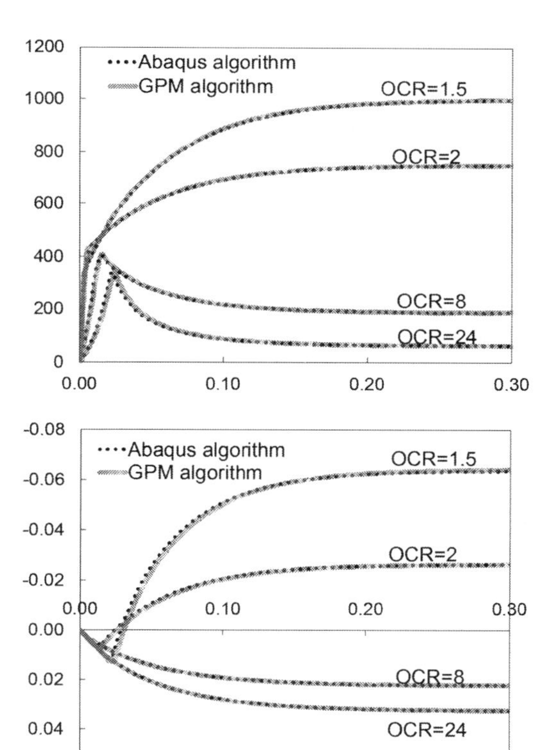

Figure 3. Comparison of results of CD tests, MCC model (Abaqus) and implemented MCC model (GPM), Bangkok clay.

Table 2. CPU user time, MCC model.

	OCR	Abaqus algorithm	GPM algorithm
Cardiff clay CU tests	12	2.3	2.2
	8	2.3	2.3
	5	2.4	2.5
	2	1.3	1.2
Bangkok clay CD tests	24	1.2	2.4
	8	1.3	1.7
	2	1.6	1.6
	1.5	1.6	1.6

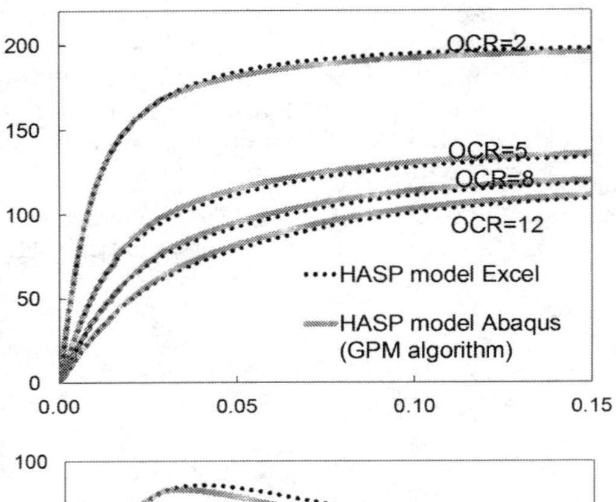

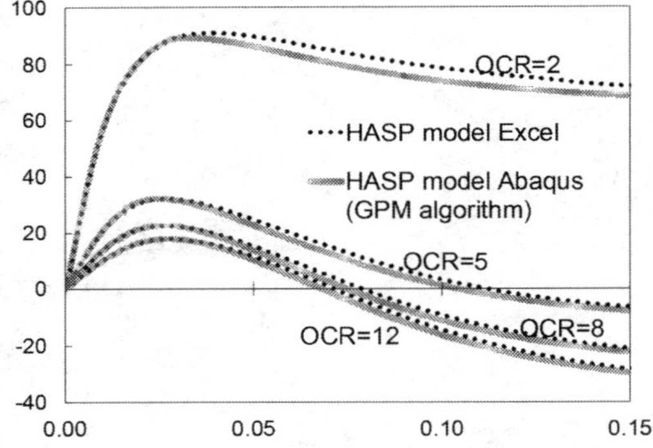

Figure 4. Comparison of results of CU tests, implemented HASP model—GPM and HASP model—Excel, Cardiff clay.

1. Cardiff clay (Banerjee & Stipho 1979) with overconsolidation ratios OCR = 12, 8, 5, 2;
2. Bangkok clay (Hassan 1976) with overconsolidation ratios OCR = 24, 8, 2, 1.5.

Simulation results of undrained triaxial compression tests and drained triaxial compression tests are given in Figure 2 and Figure 3, respectively. Comparison of results is presented using MCC model from Abaqus and the implemented MCC model using GPM. In order to determine

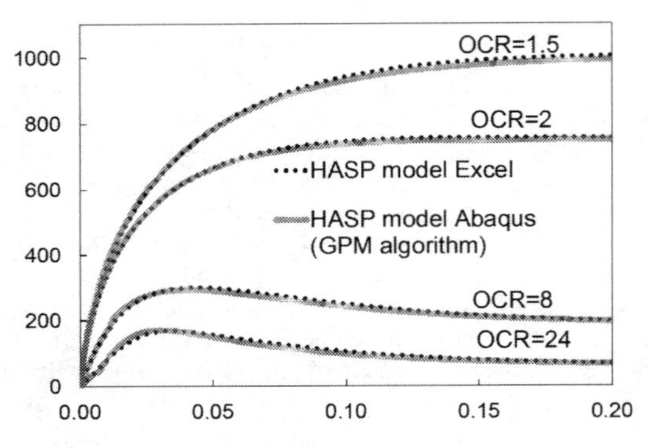

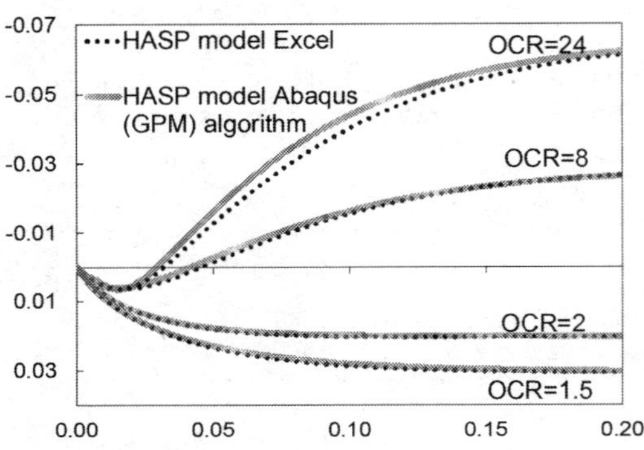

Figure 5. Comparison of results of CD test, implemented HASP model—GPM and HASP model—Excel, Bangkok clay.

the speed of convergence, in Table 2, CPU user times are given for each simulated test, which show that there are no significant discrepancies although elastic constitutive matrix was used in the numerical GPM procedure.

This is followed by the comparison of the results of CU tests (Figure 4) and CD tests (Figure 5) using HASP model implemented in Abaqus applying GPM method and using Euler deformation integration along the known stress path. Euler integration procedure was programmed in Excel and the computational steps for drained and undrained triaxial shear test were presented in the paper by Jocković & Vukićević (2016).

From comparison of the results shown in figures, it can be concluded that there is an excellent match for all degrees of overconsolidation and for both types of tests.

7 CONCLUSIONS

Governing parameter method has proved very effective as stress point algorithm for soil constitutive models. One of the key steps in the method is the selection of the governing parameter. In the presented numerical integration procedure, the mean effective stress p' was taken as the governing parameter, because the boundaries of the possible stress states are always known in every strain increment. Also, all the relevant variables of the HASP model and of the MCC model can be expressed in the function of the mean effective stress. Constitutive models were implemented in Abaqus and the verification of the implementation procedure was done. Results of finite element simulation of triaxial tests were compared with Euler deformation integration for known stress path for different degrees of overconsolidation. For both presented types of triaxial tests there was an excellent match of the results. For the MCC model, the speed of convergence was analyzed. All presented results indicate the accuracy and efficacy of GPM method.

REFERENCES

Abaqus 2011. *User manual,* Hibbit, Karlsson & Sorensen, Inc.

Banerjee, P.K. & Stipho, A.S. 1979. Elastoplastic model for undrained behavior of heavily overconsolidated clays. *Int J Numer Anal Mech Geomech* 3(1): 97–103.

Been, K. & Jefferies, M.G. 1985. A state parameter for sands. *Géotechnique* 35(2): 99–112.

Dafalias, Y.F. & Herrmann, L.R. 1982. Bounding surface formulation of soil plasticity. *Soil Mechanics—Transient and Cyclic Loads*, eds. G.N. Pande, & O.C. Zienkiewicz, pp. 253–282, John Wiley & Sons Ltd.

Hassan, Z. 1976. *Stress-strain behaviour and shear strength characteristics of stiff Bangkok Clays.* Master Thesis, Asian Institute of Technology, Thailand.

Jocković, S. & Vukićević, M. 2016. Bounding surface model for overconsolidated clays with new state parameter formulation of hardening rule. *Comput Geotech* 83:16–29. doi:10.1016/j.compgeo.2016.10.013.

Kojić, M. 1996. The governing parameter method for implicit integration of viscoplastic constitutive relations for isotropic and orthotropic metals, *Computational Mechanics*, Vol. 19, pp. 49–57.

Kojić, M. & Bathe, K.J. 2003. *Inelastic analysis of solids and structures*, Springler-Verlag, Berlin.

Roscoe, K.H. & Burland J.B. 1968. On the generalised stress-strain behaviour of 'wet' clay. In: Heyman J, Leckie FA, editors. *Engineering plasticity*. Cambridge, UK: Cambridge University Press pp. 535–609.

Vukićević, M. 2010. Governing parameter method for implicit stress integration of Modified Cam Clay model, using the mean stress as the governing parameter. *Proceedings of XIV Danube—European Conference on Geotechnical Engineering*, DECGE 2010, Bratislava, 274–275.

Wilkins, M.L. 1964. Calculation of elastic-plastic flow. *Methods of Computational Physics*. Academic Press, N.Y. Vol 3.

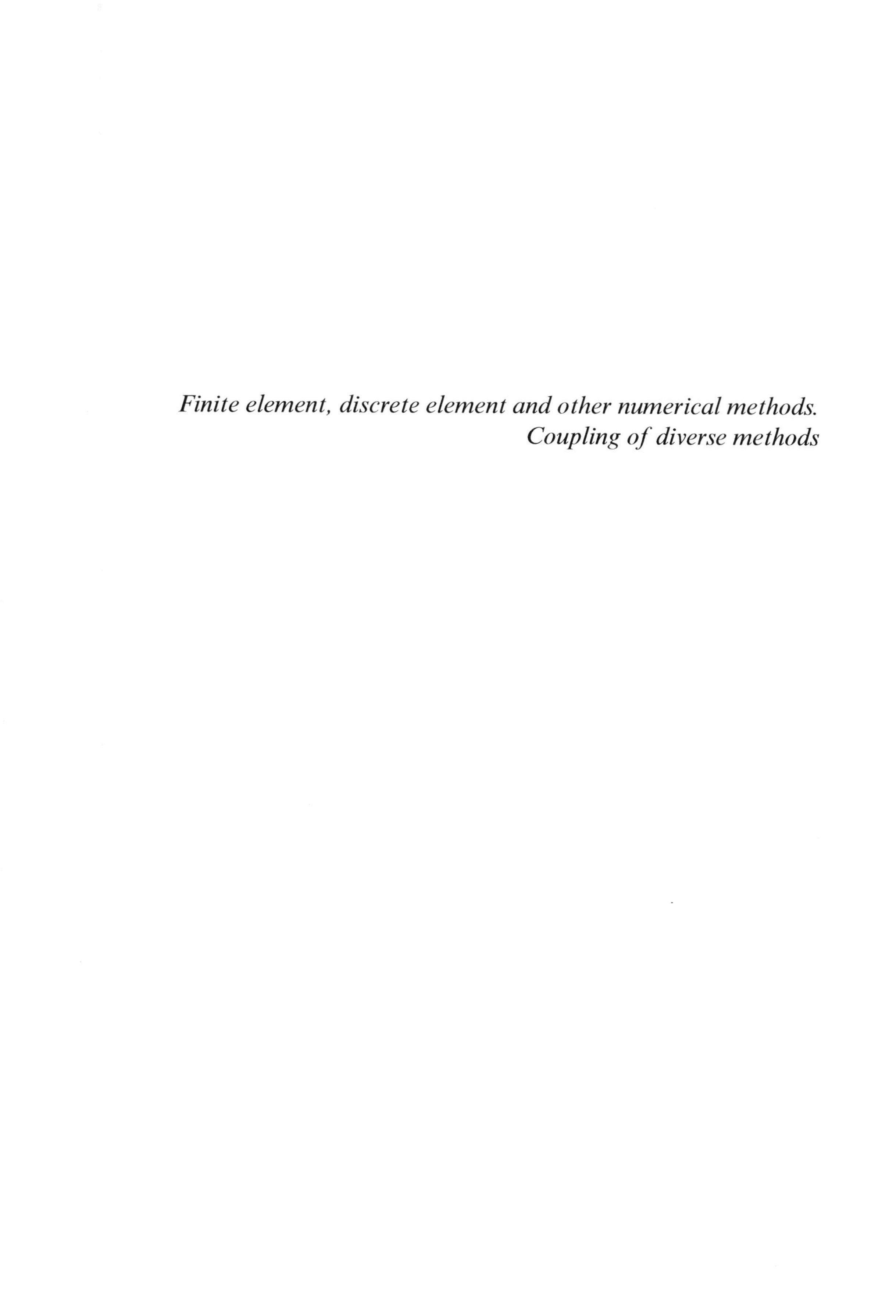

Finite element, discrete element and other numerical methods.
Coupling of diverse methods

Numerical Methods in Geotechnical Engineering IX – Cardoso et al. (Eds)
© 2018 Taylor & Francis Group, London, ISBN 978-1-138-33198-3

DEM simulation of the mechanism of particle dissolution on the behavior of collapsible soils

H. Bayesteh & T. Ghasempour
Department of Civil Engineering, University of Qom, Qom, Iran

M.R. Nabizadeh Shahrbabak
School of Civil Engineering, Iran University of Science and Technology, Tehran, Iran

ABSTRACT: A lot of experimental studies have been done to understand the real behavior of Collapsible soil. But there is some limitation to understand their behavior with experimental methods. DEM is a numerical method that can be used to address the behavior of collapsive soils. In this paper, a 2D DEM code was developed to simulate the collapse behavior of granular soil during isotropic compression by focusing on changes in the physical properties of soluble mixtures during dissolution. The effect of percentage of dissoluble particles are simulated on the load-deformation behavior of Collapsible soil. The results show that the location of soluble particle is very important parameter on the deformation. If the dissolution particle not located in the main chain forces, no settlement takes place. Also, the well graded particles reaches to higher density during confining pressure and the effect of local dissolution has lower impact on their load-deformation behavior.

1 INTRODUCTION

Collapsible soil is one of the problematic soils which are widely dispersed in the various areas with high geotechnical problems (Das, 2015). The main problems in the description of their properties are the assessment of the degradation of mechanical characteristics due to the presence of water and particle dissolution (Gaaver, 2012). Collapsibility can be described as the sudden settlement of the soil due to loss of the contacts between the soil particles. The degree of the collapsibility depends on the initial void ratio and also any change in the percentage of the soluble particles in the soil formations (Borja, 2006).

Sudden volume change behavior of the collapsible soils is one of the challenging tasks in soil engineering. Changing in the environmental condition, water table and stress level after loading on the soil are the important factors that cause settlement in collapsible soils (Barden et al., 1973). Collapsible soil is an unsaturated porous media usually consist of poorly graded soil which some soluble parts such as carbonate particles joint them together and have high porosity. During saturation, these carbonate will be dissolve and the soil particles collapse during saturation with low pressure (Gao et al., 2007, Reznik, 2007). In contrast to ordinary granular materials, volume change behavior of the collapsible soils depend not only on the pore size distribution, but also on the interaction between soluble particle in the soil structure (Mitchell and Soga, 2005).

In the past decades several experimental studied have been reported on the influence of particle dissolution on the compressibility behavior of collapsible soil (Barden et al., 1973, Alawaji, 2001, Gaaver, 2012, Fattah et al., 2014, Li et al., 2016). In these studies, laboratory odeometer test commonly has been used to model one-dimensional compressibility behavior of collapsible soils (Fattah et al., 2014). Studies have been indicated that pore fluid chemistry, carbonate content, stress state, and particle cementations are the most important factors which are controls the volume change behavior of collapsible soils (Li et al., 2016).

Although the experimental methods lead to understanding the real collapsible soil behavior, they have some limitations to control all parameters influencing the results. Therefore the effect of important parameters may not be considered. In particular, the number of physical contacts between particles and change in void ratio during dissolution are main parameters controls volume change behavior of collapsible soils (Barden et al., 1973). These processes are very difficult and sensitive in some cases and sometimes inaccessible. For example, investigation the role of contact reduction on the one-dimensional behavior of collapsible soils cannot be catches in experimental tests.

Based on above limitations to control and monitor all parameters affecting on material behavior in laboratory tests, in the past decade, researchers have been urged to use proper numerical methods instead of experimental method (Zdravkovic and Carter, 2008, Simpson and Tatsuoka, 2008).

The common numerical method which can simulate material behaviors at particle-level (microscopic) has been known as discrete element method (Cundall and Strack, 1979, Zhu et al., 2007) which recently has been used as a computer virtual laboratory (Munjiza, 2004). According to the discontinuous nature of the granular soils, many researchers have been interested in the use of DEM in the past decades for modeling non-cohesive soils and clays behaviors (Zhu et al., 2008, Ng, 2006, Mirghasemi et al., 1997, Bayesteh and Mirghasemi, 2013a, Bayesteh and Mirghasemi, 2013b, Bayesteh and Mirghasemi, 2015).

In order to simulate collapsible soils behaviors by DEM, some studies have been performed (Liu et al., 2003, Lee et al., 2012, Shire et al., 2016). According to the acceptable agreement between DEM simulations and experimental data in these researches, these studies have been indicated that DEM can model dissolution mechanism in collapsive soil behaviors (Lee et al., 2012).

Most results reported in the past studies refer to particular aspects of collapsive behavior with some assumptions. The main aim of this paper is to simulate the effect of particle size distribution and the particle solution percentage on the load-deformation behavior in collapsible soils at different stress levels. For this purpose, the simulations have been performed on uniformly particle size distribution with different coefficient of uniformity (C_u). Various percentage of dissolution are simulated. In order to investigate micromechanical changes during loading process, the number of mechanical contact are calculated in each step.

2 THEORY

The DEM assumes an assembly of the particles and discretize its particles to the rigid elements that are in contact with neighboring elements based on the existing inter-particle contact law. Evaluation of the inter-particle forces, contact detection and integration of the Newton's second law by central difference method are three important computational steps in DEM. In a DEM analysis, from the known particle velocity (boundary loading), integrate to find the relative displacement between tow particles. Next, the inter-particle forces are determined according to the suitable contact laws. Then, Newton's second law is applied to find acceleration of the particles. Finally, integrate two times from acceleration to find the new relative displacement on location of the particles. This cycle is repeated for all particles to reach the needed cycles (Cundall and Strack, 1979). In this study, the particle shape assumed as a circle which is similar to the ideal granular soil.

When two near particles reach each other, virtual overlap (Δ) produced between them. So, the normal (F_N) and shear(F_S) (F_S).forces are developed as a function of the penetration distance "(Δ)" illustrated in the Equation 1 where K is the stiffness coefficient. The normal and shear penetration length between two particles are calculated based on the magnitude of their overlap. The overlap length is calculated according to the position of the cross point between two particles. Firstly, the equation line of the each particle is found according to its geometry.

The used contact law is shown schematically in Figure 1 where the upper bound of the shear force demonstrated by Mohr-Coulomb failure criteria. In this model, contacts cannot carry tensile force so that when the normal force is negative, contact will be deleted. When two particles closed each other less than the defined mechanical cut off, mechanical force will be developed. Also, the resultant mechanical repulsion (F_m) induces the moment (M_m) which is calculated according to the distance between center of the particle and the cross point as Equation 2.

$$\begin{bmatrix} F_N \\ F_S \end{bmatrix} = \begin{bmatrix} K_N & 0 \\ 0 & K_S \end{bmatrix} \times \begin{bmatrix} \Delta_N \\ \Delta_S \end{bmatrix} \tag{1}$$

$$\overrightarrow{M_m} = \overrightarrow{r_m} \times \overrightarrow{F_m} = \overrightarrow{r_m} \times \overrightarrow{F_N} + \overrightarrow{r_m} \times \overrightarrow{F_S} \tag{2}$$

In the simulations carried out in this research, KN and KS are assumed to be $1.5 \times 10^{-7}\,\mathbf{N/m}$ and $5 \times 10^{-6}\,\mathrm{N/m}$ respectively.

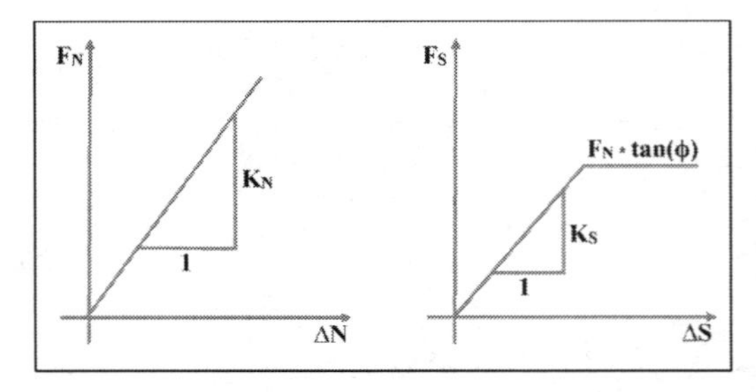

Figure 1. Contact law (Bayesteh and Mirghasemi, 2013a).

3 SIMULATION METHOD

A 2D DEM computer program has been developed in this study for simulating load-deformation behavior of collapsible soil assembly. For this propose, a circular area with 1300 m diameter is selected and 1200 particles with random location generated within this area by using a computer random-generator algorithm. The prepared code tracks the center of the particle and this location is randomly generated. Grain size distribution of spherical particles are uniform (R_{min} = 0.8 mm, R_{max} = 1.5 mm). The sampled are prepared which various characteristics. The coefficient of uniformity (C_u) of particle size distribution, particle dissolution percent, assembles density and confining pressure is change in order to do a suitable parametric study. In the Table 1, the test program is presented.

After particle generation, the boundary particles are moved with a suitable strain rate to reach a compact assembly boundary particles are fixed in order to redistribute the forces between particles. This stage is named relaxation. To simulate biaxial compression test, samples are confined with different confining pressures between 200 KPa to 700 KPa. Then, biaxial test are simulated by using vertical strain rate in boundary particles. Figure 2 shows the particle assembles after biaxial test in two different assembles. In the zoomed area, it is cleared that the void between particles in the assembly with dissolution is more than ordinary condition. The dissolution processes are simulated by reducing the size of the soluble particles at each cycle. The particle size reduction was simulated by multiplying a factor (0.99) to the radii of the soluble particles.

Table 1. Test program.

Test ID	C_u	Density	Confining Pressure	Dissolution Percent
C1.20 D70 S7	1.2	0.7	700 KPa	0
C1.20 D75 S7	1.2	0.75	700 KPa	0
C1.20 D80 S7	1.2	0.8	700 KPa	0
C1.35 D84 S4–3%	1.35	0.84	400 KPa	0.03
C1.35 D84 S4–5%	1.35	0.84	400 KPa	0.05
C1.35 D84 S4	1.35	0.84	400 KPa	0
C1.35 D84 S7–3%	1.35	0.84	700 KPa	0.03
C1.20 D84 S7–3%	1.2	0.84	700 KPa	0.03
C1.20 S2	1.2	0.76	200 KPa	0
C1.20 S4	1.2	0.78	400 KPa	0
C1.20 S7	1.2	0.79	700 KPa	0
C1.35 D84 S2–3%	1.35	0.84	200 KPa	0.03
C1.35 D84 S4–3%	1.35	0.84	400 KPa	0.03
C1.35 D84 S7–3%	1.35	0.84	700 KPa	0.03

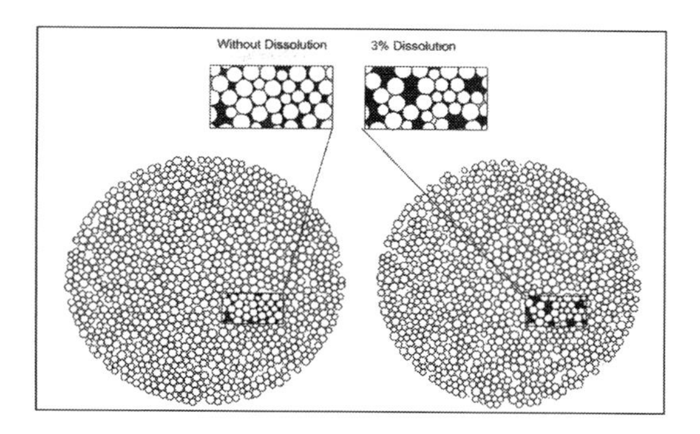

Figure 2. Particle assembly with and without dissolution.

Based on selected time step, the number of computation cycles should be chosen in such a way to reach a static equilibrium state. The number of cycles is calculated by Equation 3. In order to prevent any numerical errors, after each loading step, the relaxation mode is activated without any wall movement. This assembly has been loaded with 10^7 cycles in which 10% strain is met.

$$\dot{\varepsilon} \times N_{cycle} \times \Delta t = \varepsilon\% \tag{3}$$

A critical time step is chosen as Equation 4 and further reduced by a factor of 10, where "m" is the mass of a particle and "k" is the stiffness coefficient (Cundall and Strack, 1979). The particle density is assumed = $2.7(kN.S^2/m^4)$.

$$\Delta t_{cr} = \frac{2\pi}{10}\sqrt{\frac{m}{K}} \tag{4}$$

4 RESULTS

First, the effect of confining pressure on the load deformation behavior is depicted in Figure 3 and Figure4. It is clear that higher confining pressure (700 KPa) lead to higher shear strength (q) and dilative behavior.

The evolutions of the shear strength and volumetric strain versus axis strain generated by the simulations of the biaxial tests for all simulated soluble particle mixture. Figures 5 shows the reduction of the peak shear strength based on the increase in the soluble particle percentage. The volumetric strain versus axis strain are depicted in Figure 6. Clearly, the volumetric strain increase with an increase in the axis strain for all specimens and the dissolution causes the assembly to reach more contractive and the specimen containing

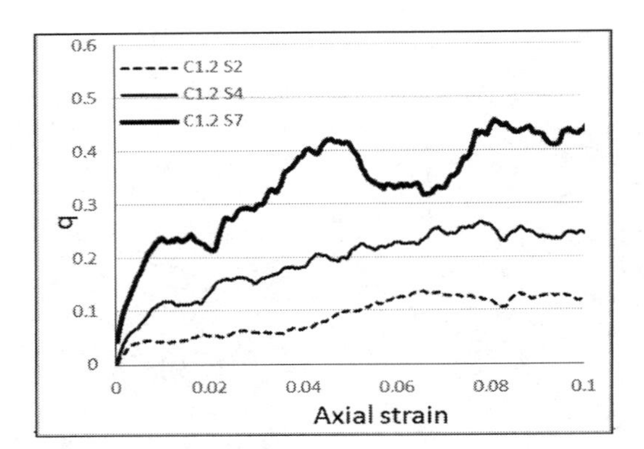

Figure 3. Shear Stress vs. Axial Strain for assembly without dissolution.

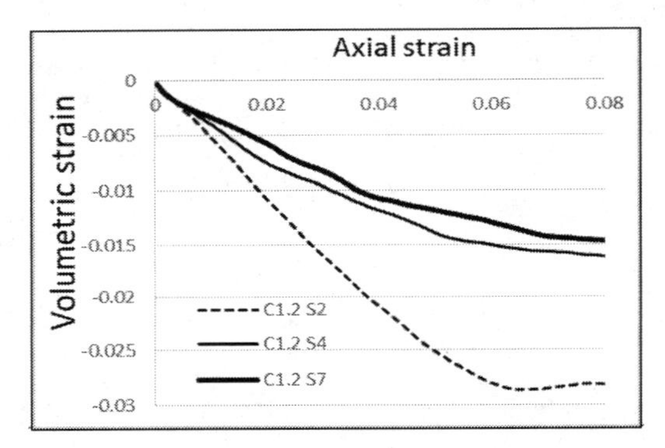

Figure 4. Volumetric Strain vs. Axial Strain for assembly without dissolution.

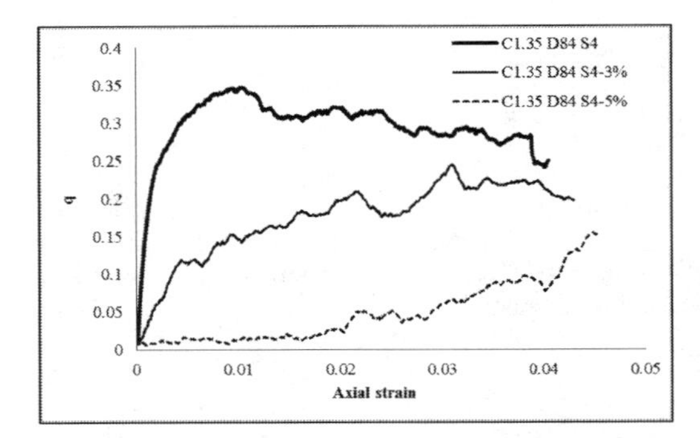

Figure 5. Shear Stress vs. Axial Strain for assembly with various dissolution percentage.

lower dissolution percentage have more dilative behaviors.

In order to describe the reason on the above behavior, the number of mechanical contact

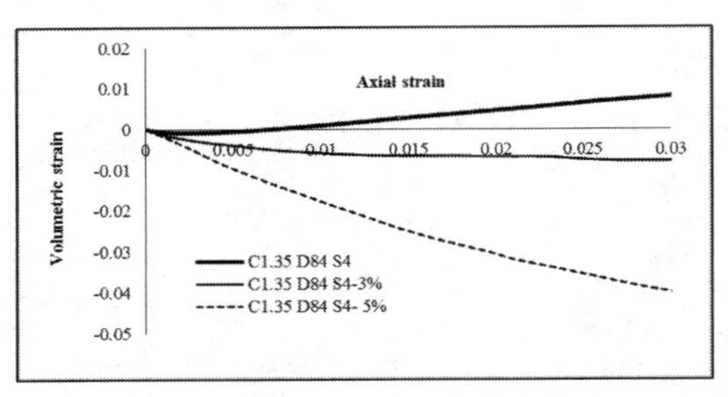

Figure 6. Volumetric Strain vs. Axial Strain for assembly with various dissolution percentage.

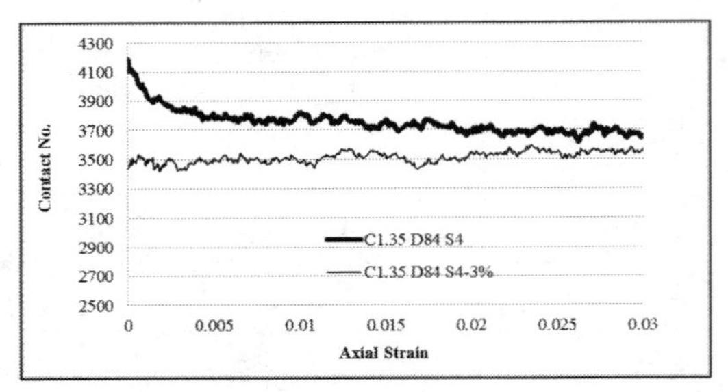

Figure 7. Mechanical Contact vs. Axial Strain for assembly with various dissolution percentage.

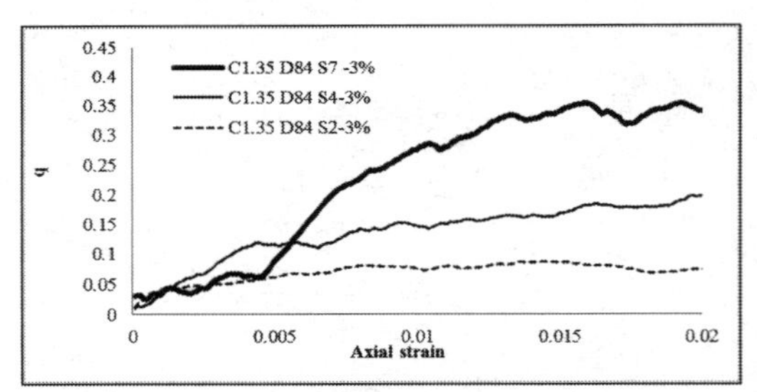

Figure 8. Shear Stress vs. Axial Strain for assembly with various confining pressure in 3% dissolution.

between particles is plotted versus increasing in the axial strain in Figure 7. It is shown that the pick number of the mechanical contact between particles without dissolution is about 4100 contacts. But in the sample with 3% dissolution, the pick number of the mechanical contact between particles is about 3500 contacts. This fact shows that during dissolution, some mechanical contact is lost and leads to more volumetric strain and contractive behavior. The loss of the mechanical contact leads to reduction in the shear strength. But after large deformation, the number of mechanical contact in two particle assemblies, reach to a

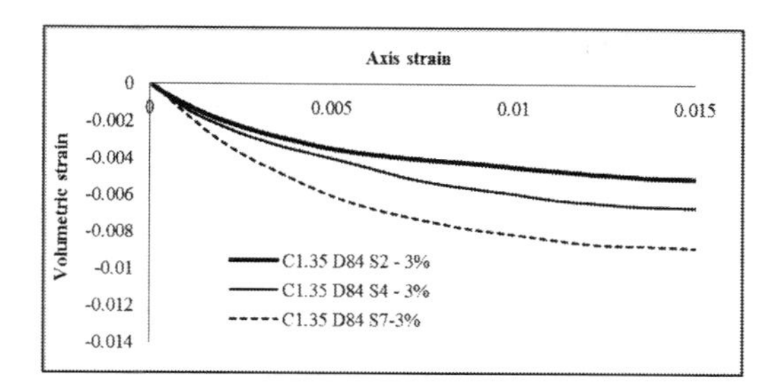

Figure 9. Volumetric Strain vs. Axial Strain for assembly with various confining pressure in 3% dissolution.

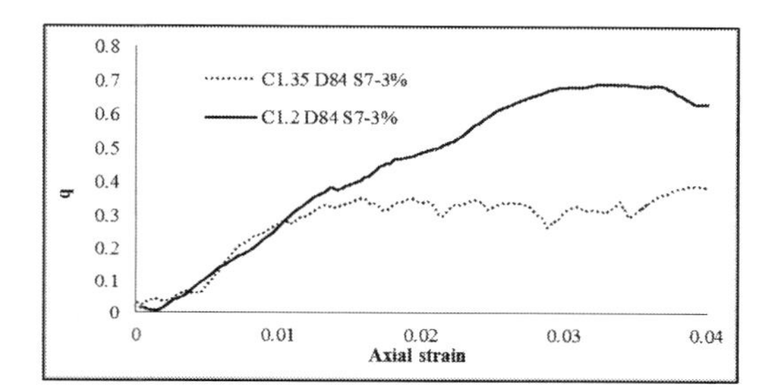

Figure 10. Shear Stress vs. Axial Strain for assembly with various C_u in 3% dissolution.

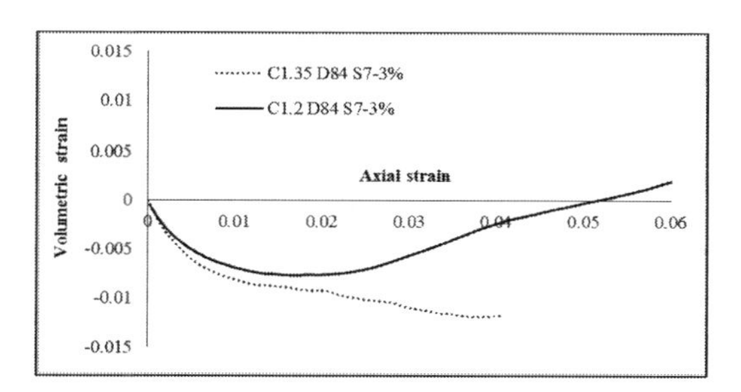

Figure 11. Volumetric Strain vs. Axial Strain for assembly with various C_u in 3% dissolution.

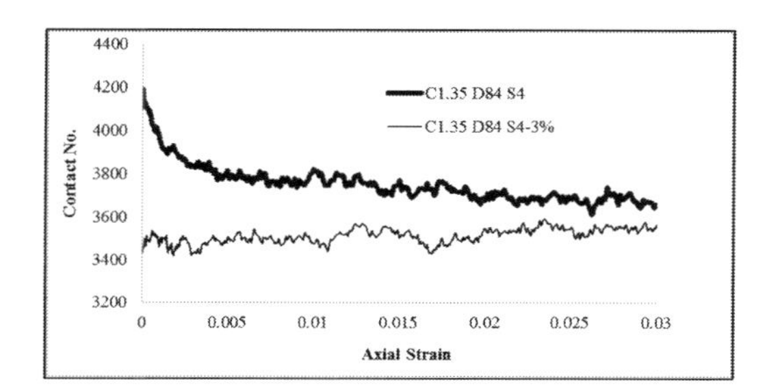

Figure 12. Mechanical contact Vs. Axial Strain for assembly with various C_u in 3% dissolution.

constant number according to the re-arrangement of particles.

Figure 8 shows shear Stress versus axial Strain for assembly with various confining pressure in 3% dissolution. Obviously, higher confining pressure (700 KPa) lead to reduction in shear stress at low axial strain based on the collapse in assembly. But with increasing in the axial strain, more confining pressure leads higher pick shear increase.

Figure 9 shows volumetric strain versus axial strain for assembly with various confining pressure in 3%dissolution. It is clear that higher confining pressure lead more contractive behavior. Because when dissolution happens, higher confining pressure causes more reduction in voids and re-arrangements of the particles.

In order to investigate the effect of particle size distribution (PSD) on the behavior of collapsible soil, to simulation teste are done with various coefficient of uniformity (C_u). Figure 10 shows shear stress versus axial strain for assembly with various C_u in 3% dissolution. Also, Figure 11 shows volumetric strain versus axial strain for assembly with various C_u in 3%dissolution. It can be found that type of PSD directly change the load-deformation behavior of collapsible soils. More C_u, indicated that the widely range of particle exists and leads more well graded PSD produced.

Figure 12 shows number of mechanical contact versus axial strain for assembly with various C_u in 3%dissolution. In the test with $C_u = 1.35$, the number of mechanical contacts is higher than the test with $C_u = 1.2$. Because wider range of particle cause more mechanical contact due to reduction in the void ratio.

The contractive behavior in sample with $C_u = 1.35$ is due to the re-arrangement of particle and reduction in the void ratio. But in sample with $C_u = 1.2$, uniformity lead dilative behavior.

5 CONCLUSION

In this paper, a 2D DEM code was developed to simulate the behavior of collapsible soil during biaxial compression. This study focuses on changes in the physical properties of soluble mixtures during dissolution. The effect of initial confining pressure, density, particle size is simulated on the load-deformation behavior of Collapsible soil. Numerical results obtained in this study show the following results.

Grading curves influences the mechanical behavior of granular materials such as the shear strength and the volumetric strain. The well graded particles reaches to more number of mechanical contacts and more contractive behavior. In a constant dissolution, volumetric strain of specimens increase with an increase in confining pressure. In this process, the sample with higher confining pressure and stress state, more contractive behavior takes place during dissolution. Confining pressure influences the shear strength and the volumetric strain of granular materials and shear strength for the specimen containing a more confining pressure (S = 700 KPa), is greater than others. Also, in the constant confining pressure, the pick shear strength reduces with increasing dissolution percentage. This fact leads to more dilative behavior in samples with lower dissolution percentage.

REFERENCES

Alawaji, H. 2001. Settlement and bearing capacity of geogrid-reinforced sand over collapsible soil. *Geotextiles and Geomembranes,* 19, 75–88.

Barden, L., Mcgown, A. & Collins, K. 1973. The collapse mechanism in partly saturated soil. *Engineering Geology,* 7, 49–60.

Bayesteh, H. & Mirghasemi, A. 2013a. Numerical simulation of pore fluid characteristic effect on the volume change behavior of montmorillonite clays. *Computers and Geotechnics,* 48, 146–155.

Bayesteh, H. & Mirghasemi, A. 2013b. Procedure to detect the contact of platy cohesive particles in discrete element analysis. *Powder technology,* 244, 75–84.

Bayesteh, H. & Mirghasemi, A. 2015. Numerical simulation of porosity and tortuosity effect on the permeability in clay: Microstructural approach. *Soils and Foundations,* 55, 1158–1170.

Borja, R.I. 2006. Conditions for instabilities in collapsible solids including volume implosion and compaction banding. *Acta Geotechnica,* 1, 107–122.

Cundall, P.A. & Strack, O.D. 1979. A discrete numerical model for granular assemblies. *geotechnique,* 29, 47–65.

Das, B.M. 2015. *Principles of foundation engineering,* Cengage learning.

Fattah, M.Y., Al-Ani, M.M. & Al-Lamy, M.T. 2014. Studying collapse potential of gypseous soil treated by grouting. *Soils and Foundations,* 54, 396–404.

Gaaver, K.E. 2012. Geotechnical properties of Egyptian collapsible soils. *Alexandria Engineering Journal,* 51, 205–210.

Gao, X.-J., Wang, J.-C. & Zhu, X.-R. 2007. Static load test and load transfer mechanism study of squeezed branch and plate pile in collapsible loess foundation. *Journal of Zhejiang University-Science A,* 8, 1110–1117.

Lee, J.-S., Tran, M.K. & Lee, C. 2012. Evolution of layered physical properties in soluble mixture: Experimental and numerical approaches. *Engineering geology,* 143, 37–42.

Li, P., Vanapalli, S. & Li, T. 2016. Review of collapse triggering mechanism of collapsible soils due to wetting. *Journal of Rock Mechanics and Geotechnical Engineering,* 8, 256–274.

Liu, S., Sun, D. & Wang, Y. 2003. Numerical study of soil collapse behavior by discrete element modelling. *Computers and Geotechnics,* 30, 399–408.

Mirghasemi, A., Rothenburg, L. & Matyas, E. 1997. Numerical simulations of assemblies of two-dimensional polygon-shaped particles and effects of confining pressure on shear strength. *Soils and Foundations,* 37, 43–52.

Mitchell, J.K. & Soga, K. 2005. Fundamentals of soil behavior.

Munjiza, A.A. 2004. *The combined finite-discrete element method,* John Wiley & Sons.

Ng, T.-T. 2006. Input parameters of discrete element methods. *Journal of Engineering Mechanics,* 132, 723–729.

Reznik, Y.M. 2007. Influence of physical properties on deformation characteristics of collapsible soils. *Engineering Geology,* 92, 27–37.

Shire, T., O'sullivan, C. & Hanley, K. 2016. The influence of fines content and size-ratio on the micro-scale properties of dense bimodal materials. *Granular Matter,* 18, 1–10.

Simpson, B. & Tatsuoka, F. 2008. Geotechnics: the next 60 years. *The Essence of Geotechnical Engineering: 60 years of Géotechnique.* Thomas Telford Publishing.

Zzravkovic, L. & Carter, J. 2008. Constitutive and numerical modelling. *GEOTECHNIQUE-LONDON-,* 58, 405.

Zhu, H., Zhou, Z., YANG, R. & YU, A. 2007. Discrete particle simulation of particulate systems: theoretical developments. *Chemical Engineering Science,* 62, 3378–3396.

Zhu, H., Zhou, Z., Yang, R. & Yu, A. 2008. Discrete particle simulation of particulate systems: a review of major applications and findings. *Chemical Engineering Science,* 63, 5728–5770.

Numerical Methods in Geotechnical Engineering IX – Cardoso et al. (Eds)
 ISBN 978-1-138-33198-3

Modelling soil-water interaction with the material point method. Evaluation of single-point and double-point formulations

F. Ceccato
University of Padua, Italy

A. Yerro
Virginia Tech, Blacksburg, USA

M. Martinelli
Deltares, Delft, The Netherlands

ABSTRACT: Many problems in geotechnical engineering involve large deformations and soil-water interactions, which pose challenging issues in computational geomechanics. In the last decade, the Material Point Method (MPM) has been successfully applied in a number of large-deformation geotechnical problems and multiphase MPM formulations have been recently proposed. In particular, there exist two advanced coupled hydro-mechanical MPM approaches to model the interaction between solid grains and pore fluids: the single-point and the double-point formulation. The first discretizes the soil-water mixture with a single set of Material Points (MP) which moves according to the solid velocity field. The latter uses two sets of MP one for the fluid phase and the other for the solid phase and they move according to the respective velocity field. The aims of this work is to present and compare the two theories, to emphasize their limitations and potentialities, and to discuss their applicability in the geotechnical field. To this end, the results of two numerical examples carried out by using both formulations are presented: a 1D-consolidation problem and a saturated column collapse problem.

1 INTRODUCTION

Soil-water interaction problems are of great interest in the field of geotechnical engineering. Underground excavations, pile installations, seepage failures, slope instabilities and landslides are just a few examples. In many cases, the material involved can experience large deformations, which might lead to dramatic events endangering human lives.

The numerical simulation of these problems is challenging because the treatment of large deformations, interactions between solid and fluid, and fluidization and sedimentation processes is not straight-forward.

Large deformations can be effectively simulated with the Material Point Method (MPM) (Sulsky et al., 1994), which is a continuum-based technique that discretizes the media into a set of Lagrangian material points (MP), which move attached to the material and carry all the updated information such as velocities, strain, stresses, and history variables. Large deformations are simulated by MPs moving through a computational nodal grid that covers the full problem domain. The main governing equations are solved incrementally at the nodes of this grid that typically remains fixed throughout the calculation. Variables required at the mesh to solve the governing equations are transferred from MPs to the nodes using mapping functions. The same mapping functions are used to update the quantities carried by the MPs by interpolation of the mesh results. This dual description of the media, i.e. MPs and nodal grid, prevents mesh distortion problems hence re-meshing techniques are not required. In addition, the original 1-phase MPM-formulation has the advantage that the mass is automatically conserved because the total mass of each MP remains constant through the calculation.

The need of taking into account the interaction between soil and pore fluids brought to the development of multi-phase MPM-formulations. Within the recent years, several approaches were presented in the literature to model coupled hydro-mechanical 2-phase (e.g., saturated soils) (Zabala & Alonso, 2011; Jassim et al. 2013, Abe et al., 2013, Bandara & Soga, 2015, Martinelli, 2016) and 3-phase (e.g., partially saturated soils) (Yerro et al., 2015) problems (Figure 1).

The interaction between two phases is formulated essentially in two different manners: adopting either one set of MPs (i.e., single-point approach, see Section 2) or two separate sets of MPs (i.e., double-point approach, see Section 3).

The aim of this paper is to compare the two formulations emphasizing their limitations and potentialities in different geotechnical applications. Two numerical examples are presented in order to validate and compare the results of both theories. The first example (Sec. 4.1) considers a 1D consolidation problem, while the second (Sec. 4.2) simulates the collapse of a water saturated column.

2 2-PHASE, SINGLE-POINT FORMULATION

The single-point formulation (Zabala & Alonso, 2011; Jassim et al. 2013) considers only one set of MPs. Each MP represents a portion of saturated porous media and carries the information of the solid (solid skeleton) and liquid in the pores (e.g. water). In this case, the MPs remain attached to the solid skeleton giving a Lagrangian description of the solid-phase movement, while the liquid-phase behaviour is described with respect the MPs by means of an Eulerian approach.

This formulation has been successfully applied to the simulation of CPT in partially drained conditions (Ceccato et al. 2016a,b; Galavi et al., 2017), and to model landslides and slope failures (Alonso et al, 2014, Soga et al. 2015; Yerro et al. 2016).

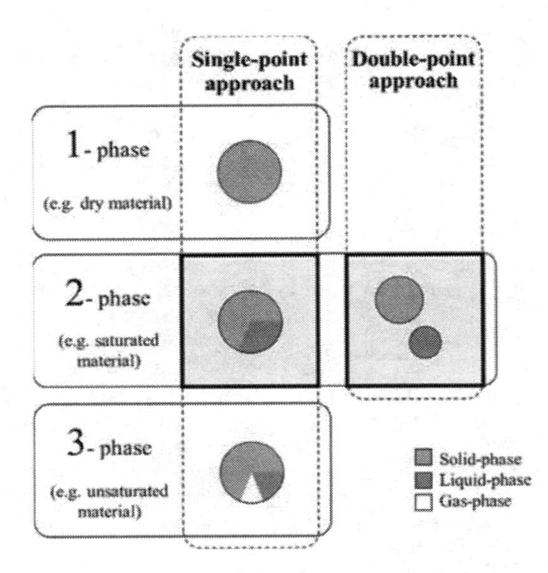

Figure 1. Scheme of multi-phase MPM-formulations (modified from Yerro et al. (2015)).

This work uses the approach proposed by Jassim et al. (2013), in which dynamic momentum balances of the liquid phase (Eq.1) and the mixture (Eq.2) are the governing equations posed at the nodes. In this case, all dynamic terms are taken into account, and $\boldsymbol{a}_S$ and $\boldsymbol{a}_L$ are the primary unknowns. Solid and total mass balances (Eq.3 and 4), as well as constitutive relationships are solved at the MPs.

$$\rho_L \boldsymbol{a}_L = \nabla_p - \boldsymbol{f}^d + \rho_L \boldsymbol{b} \tag{1}$$

$$n_S \rho_S \boldsymbol{a}_S + n_L \rho_L \boldsymbol{a}_L = \nabla \cdot \boldsymbol{\sigma} + \rho_m \boldsymbol{b} \tag{2}$$

$$\frac{D^S n_L}{Dt} = n_s \nabla \cdot \boldsymbol{v}_s \tag{3}$$

$$\frac{D^S \varepsilon_{vol,L}}{Dt} = \frac{1}{n_s}[n_s \nabla \cdot \boldsymbol{v}_s + n_L \nabla \cdot \boldsymbol{v}_L] \tag{4}$$

where n_S =volumetric concentration ratio of solid, n_L = volumetric concentration ratio of liquid (equivalent to porosity in saturated soils), ρ_S= solid density, ρ_L =liquid density, ρ_m =density of the mixture, v_S and v_L are the solid and liquid velocities, $\boldsymbol{\sigma}$ =Cauchy total stress tensor, $\boldsymbol{f}^d$ =drag force, and $\varepsilon_{vol,L}$ =volumetric strain of liquid-phase. $D^S(\cdot)/Dt$ denotes the material time derivative with respect to the solid phase.

In this approach, the flow is considered laminar and stationary in slow velocity regime, hence the interaction force between solid and liquid phases (i.e. drag force $\boldsymbol{f}^d$, term in Eq.1) is governed by Darcy's law (Eq.5). This hypothesis can be controversial in high velocity flows where drag forces may become nonlinear as better explained in the following.

$$\boldsymbol{f}^d = \frac{n_L \mu_L}{\kappa_L}(\boldsymbol{v}_L - \boldsymbol{v}_S) \tag{5}$$

In Equation 5 μ_L is the dynamic viscosity of the liquid and κ_L the liquid intrinsic permeability, which are assumed constant throughout the simulation.

Equation 3 is the expression for the mass balance of the solid and is used to update the porosity according to volumetric deformation of the solid skeleton.

In the framework of the 2-phase single-point approach, solid mass conservation is automatically fulfilled because the solid mass remains constant in each MP. However, this condition is not naturally satisfied for the liquid, because liquid can move apart from the solid skeleton depending on solid volumetric strain changes (porosity changes). Consequently, liquid mass in MPs can change and the conservation of the liquid mass is totally controlled by the accuracy in which the liquid mass balance is solved. Fluxes due to spatial variations of liquid mass are neglected in the 2-phase single-point formulation ($\nabla n_L \rho_L \approx 0$), hence the total mass balance results in Equation 4, and describes the volumetric strain rate of the liquid-phase. This hypothesis is reasonable when gradients of porosity are relatively small, but can induce errors when two materials with very different porosity are in contact. In addition, to obtain Equation 4, liquid is assumed to be weakly compressible.

Finally, constitutive relationships for solid and liquid are solved at the MPs to update stresses and pore pressure. The water is assumed linearly compressible via the bulk modulus of the fluid K_L and shear stresses in the liquid phase are neglected.

As usual in MPM, Equations 1 and 2 are discretized in space by means of the Galerking method and solved in time with a semi-explicit time discretization scheme.

The MPM solution scheme for each time step can be summarized as follow:

1. Liquid nodal acceleration $\boldsymbol{a}_L$ is calculated by solving the discretized form of Equation 1.
2. $\boldsymbol{a}_L$ is subsequently used to obtain the nodal acceleration of the solid $\boldsymbol{a}_S$ from the discretized form of Equation 2.
3. Velocities and momentum of the MPs are updated from nodal accelerations of each phase.
4. Nodal velocities are then calculated from nodal momentum and used to compute the strain rate at the MP location.
5. Liquid and soil constitutive laws give the increment of excess pore pressure and effective stress respectively.
6. Displacement and position of each MP is updated according to the velocity of the solid phase.

3 2-PHASE, DOUBLE-POINT FORMULATION

The 2-phase double-point formulation was initially presented by Bandara (2013) and Wieckowski (2013), and later extended by Abe et al. (2013) and Martinelli (2016). It assumes that the saturated porous media consist of a superposition of two independent continuum media, hence the solid skeleton and the liquid phase are represented separately by two sets of Lagrangian MPs: solid material points (SMPs) and liquid material points (LMPs). While SMPs moves attached to the solid skeleton, LMPs follow the liquid motion, both carrying properties of respective phases. As a result, the required number of MPs to discretize a saturated porous domain increases substantially (at least it doubles) compared to the single-point formulation.

An important advantage of this approach compared to the single-point formulation is that the mass of all MPs remains constant. Therefore, the conservation of both solid and liquid mass is fulfilled through the calculation.

Another important feature of the double-point formulation is that LMPs embodies either liquid within the pores or free liquid. According to this framework, Martinelli (2016) describes three possible domains:

i) porous media in saturated conditions, when SMPs and LMPs share the same grid element,
ii) porous media in dry conditions, when only SMPs are located in the grid element,
iii) free liquid, when only LMPs are located in the grid element.

In any case, the dynamic behaviour of the continuum can be described with the solid and liquid dynamic momentum balances (Eq.6 and 7 respectively) which are solved at the nodes of the grid, being $\boldsymbol{a}_S$ and $\boldsymbol{a}_L$ the primary unknowns. Note that solid momentum balance is considered for convenience instead of momentum balance of the mixture. Solid and total mass balances (Eq.4 and 8 respectively) and constitutive relationships are posed at the corresponding MPs in order to update secondary variables.

$$n_S \rho_S \boldsymbol{a}_S = \nabla \cdot \bar{\sigma}_S + \boldsymbol{f}^d + n_S \rho_S \boldsymbol{b} \tag{6}$$

$$n_L \rho_L \boldsymbol{a}_L = \nabla \cdot \bar{\sigma}_L - \boldsymbol{f}^d + n_L \rho_L \boldsymbol{b} \tag{7}$$

$$\frac{D^L \varepsilon_{vol,\mathrm{L}}}{Dt} = \frac{1}{nL}[n_S \nabla \cdot \boldsymbol{v}_S + n_L \nabla \cdot \boldsymbol{v}_L + (\boldsymbol{v}_L - \boldsymbol{v}_S)\nabla \cdot n_L] \tag{8}$$

In the previous expressions, $\bar{\boldsymbol{\sigma}}_S = \boldsymbol{\sigma}' - n_S \boldsymbol{\sigma}_L$ and $\bar{\boldsymbol{\sigma}}_L = n_L \boldsymbol{\sigma}_L$ correspond to the partial stresses for solid and liquid phases respectively, σ'= effective stress tensor, and σ_L =stress tensor of the liquid phase (equivalent to pore pressure p in saturated porous media). $D^L(\cdot)/Dt$ denotes the material time derivative with respect to the liquid phase.

This formulation presents two additional differences compared to the single-phase approach. Both are related to the fact that the behaviour of the continuum described in the double-point framework can vary from dry porous media to pure fluid. This leads to extreme changes in flow regime and huge gradients of volumetric concentration ratios in transition zones.

The first one is that the drag force $\boldsymbol{f}^d$ is generalized and Equation 5 is extended in order to account for laminar and steady flow in high velocity regime (Forchheimer, 1901), leading to Equation 9 where

β is the non-Darcy flow coefficient (Ergun, 1952) and can be computed with Equation 10

$$f^d = \frac{n_L{}^2 \mu_L}{\kappa_L}(\boldsymbol{v}_L - \boldsymbol{v}_S) + \beta n_L{}^3 \rho_L \,|\boldsymbol{v}_L - \boldsymbol{v}_S|(\boldsymbol{v}_L - \boldsymbol{v}_S) + \boldsymbol{\sigma}_L \cdot \nabla n_L \tag{9}$$

$$\beta = B / \sqrt{\kappa_L A n_L{}^3} \tag{10}$$

Moreover, the intrinsic permeability κ_L is computed and updated as a function of the effective porosity n_L with the Kozeny-Carman formula (Bear, 1972):

$$\kappa_L = \frac{D^2}{A} n_L{}^3 / (1 - n_L)^2 \tag{11}$$

The second difference recalls in the liquid mass balance (Eq.8). Now, all convective terms are accounted including the spatial variations of liquid volumetric concentration ratio ($(\boldsymbol{v}_L - \boldsymbol{v}_S) \cdot \nabla n_L$), hence liquid fluxes due to changes in porosity are accounted. Liquid is also considered weakly compressible.

This formulation can distinguish between mixtures characterized by low and high porosities (see Figure 2). Figure 2a shows a low-porosity mixture, where the grains of the solid skeleton are in contact and the behaviour can be described by constitutive models developed for granular materials (solid-like response). Conversely, as shown in Figure 2b, in a high-porosity mixture the grains are not in contact and float together with the liquid phase. In this case, the effective stresses are equal to zero and the response of the mixture is described by the Navier-Stokes equation (liquid-like response).

In the current formulation, the two aforementioned states are distinguished through the maximum porosity n_{max} of the SMP, which is the maximum value of the porosity for a given soil in its loosest state. During the fluidization process, when the mixture porosity is lower than the maximum porosity ($n_L = 1 - n_S < n_{max}$), the decrease in the mean effective stress results in increase in the porosity. When the contact forces between the grains vanish, the mean effective stress becomes nil. However, the fluidization occurs only if the grains are significantly separated, so that the porosity of the SMP is larger than n_{max}. In the reverse process, i.e. the sedimentation of a fluidized mixture, the porosity decreases due to the fact that the solid grains get closer to each other. However, the effective stresses recur only if the porosity is smaller than n_{max}, i.e. the grains are close enough to be in contact.

Equation 7 is used to describe the behavior of water in the soil-water mixture. In case of liquid-like response, the deviatoric part of the stress tensor of the liquid is computed using the liquid strain rate tensor and a viscosity which takes into account the solid concentration ratio of the mixture. In case of solid-like behavior the deviatoric stress tensor is set to zero.

In this formulation, all LMPs belonging to the liquid free surface are detected and the liquid stress is set to zero to these material points.

The MPM solution scheme for each time step can be summarized as follow:

1. Nodal acceleration of the liquid $\boldsymbol{a}_L$ is calculated by solving the discretized form of Equation 7.
2. Nodal acceleration of the solid $\boldsymbol{a}_S$ is calculated by solving the discretized form of Equation 6.
3. Velocities and momentum of the MPs are updated from nodal accelerations of each phase.
4. Nodal velocities are then calculated from nodal momentum and used to compute the strain rate at the MP location.
5. Liquid and soil constitutive laws give the increment of liquid stress and effective stress respectively in LMP and SMP.
6. All LMPs that belong to the liquid free surface are detected.
7. Displacement and position of each MP is updated according to the corresponding velocity field.

The double-point formulation has been used to simulate the submerged column collapse (Martinelli and Rohe, 2015), the fluidization of a vertical column test (Bolognini et al., 2017), the interaction between water jet and soil bed (Liang et al. 2017), the simulation of the crater development around a damaged pipeline (Martinelli et al. 2017a), and a dike failure (Martinelli et al., 2017b). Other applications are also presented in Martinelli (2016).

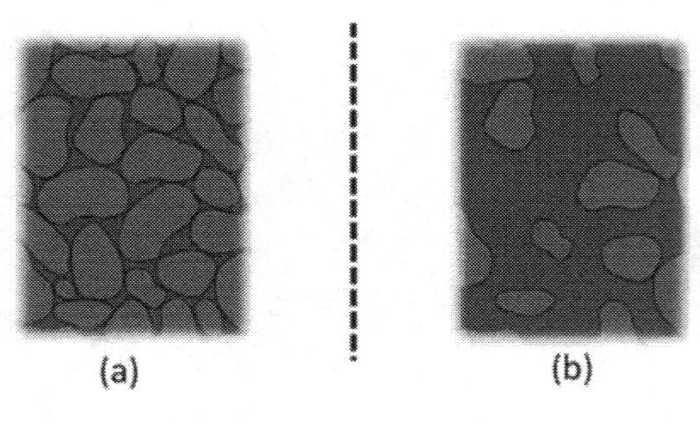

Figure 2. Solid-liquid mixture with (a) low porosity (solid-like response) and (b) high porosity (liquid-like response).

4 EXAMPLES

Two numerical examples are presented in this section. The first one is a 1D consolidation problem, and the aim of it is to validate both theories with an analytical solution. The second example considers the collapse of a saturated sand column and shows the importance of considering the mobility of the two phases separately and the spatial and temporal variation of the volumetric concentration ratios n_L and n_S.

4.1 *1D consolidation*

The two 2-phase formulations are validated by means of the problem of one dimensional consolidation for which an analytical solution by Terzaghi exists for the case of small deformations. In this case, the assumptions of laminar liquid flow through the pores, constant volumetric concentration and permeability are well satisfied.

A 1 m-column of saturated weightless, linear-elastic material is considered (Tab. 1). The column is discretised with 40 rows of 6 tetrahedral elements. Standard oedometric boundary conditions are applied, the base of the column is impermeable and the top is permeable. Each element contains 4 MPs (single-point formulation) or 4 LMPs and 4 SMPs (double-point formulation). Note that the number of MPs required to discretize the problem doubles in the double-point analysis.

A total load of 10 kPa is applied at the top of the column. The initial pore pressure is $p_0 = 10$ kPa and the initial effective stress is 0. Subsequently, the water is allowed to drain out of the top surface. Gradually, the load redistributes from the pore pressure to the soil skeleton.

This example considers a seepage problem in a homogeneous media at small deformations, thus the intrinsic permeability is assumed constant throughout the simulation and the second term of Equation 9 can be neglected.

Figure 3 shows the change of normalized pore pressure over the normalized height of the column with time for selected MPs. The dimensionless time factor T is defined as:

$$T = \frac{c_v t}{h^2} \tag{12}$$

with c_v = consolidation coefficient, h = height of the column.

The results of both formulations are in excellent agreement with the analytical solution, thus validating the implementation.

In this example the single-point and the double-point formulation give the same results, i.e. they are both well applicable when the fluid flow is laminar and the spatial variability of the porosity is negligible, which is the case of many seepage problems in engineering.

4.2 *Column collapse*

In this section, we consider the collapse of a column of saturated soil in air. This problem is well suited to highlight the differences between the two approaches considered in this study.

The geometry is shown in Figure 4, a 1 m-wide 2 m-high column, subjected to gravity is allow to collapse on a flat surface. All the boundaries of the model are impermeable. The bottom boundary is fully fixed, while roller boundary conditions are applied at the remaining surfaces. The width of the model perpendicular to the xy-plane of Figure 4 is 0.2 m and it is discretized with only one row of ele-

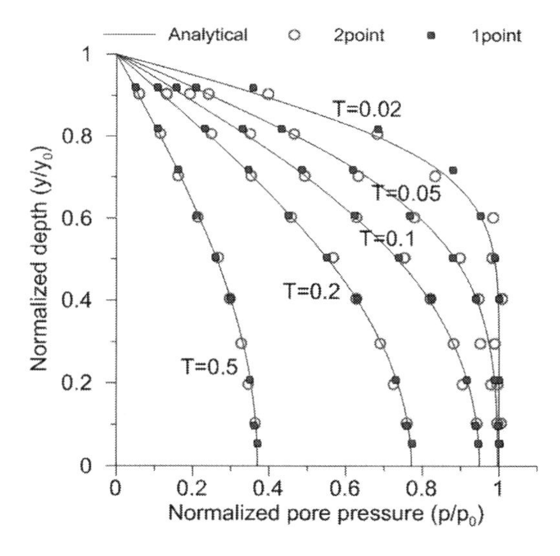

Figure 3. Results of 1D consolidation problem.

Table 1. Material parameters.

Parameter		Value
Initial porosity [-]	n_L	0.4
Grain density [kg/m³]	ρ_S	2650
Liquid density [kg/m³]	ρ_L	1000
Intrinsic permeability [m²]	κ_L	1.0214e-10
Dynamic viscosity [kPa s]	μ_L	1.002e-6
Young modulus [kPa]	E	10000
Poisson ratio [-]	ν	0.2
Fluid bulk modulus [kPa]	K_L	21500
Consolidation coefficient [m²/s]	c_v	1.1

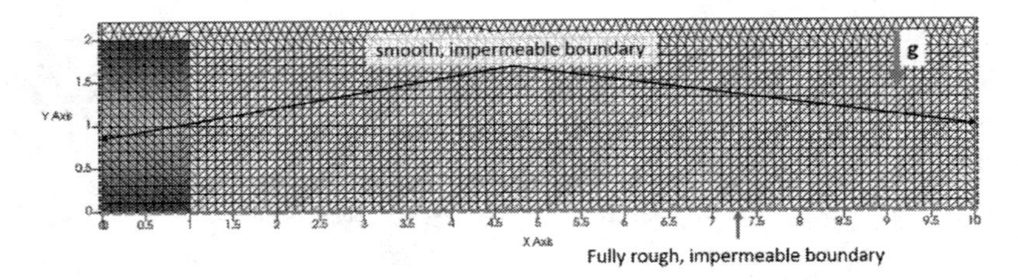

Figure 4. Geometry and discretization of the column collapse problem.

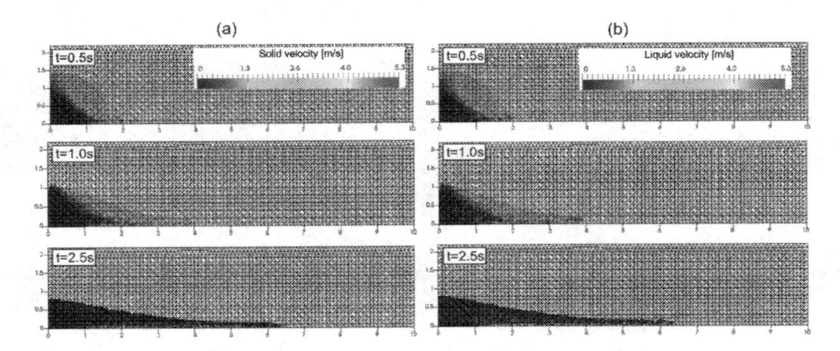

Figure 5. 2-phase single-point formulation. (a) solid velocity, (b) liquid velocity.

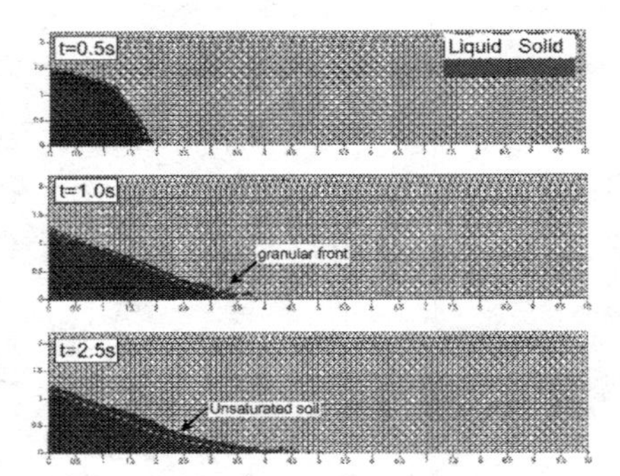

Figure 6. SMP and LMP with the double-point formulation.

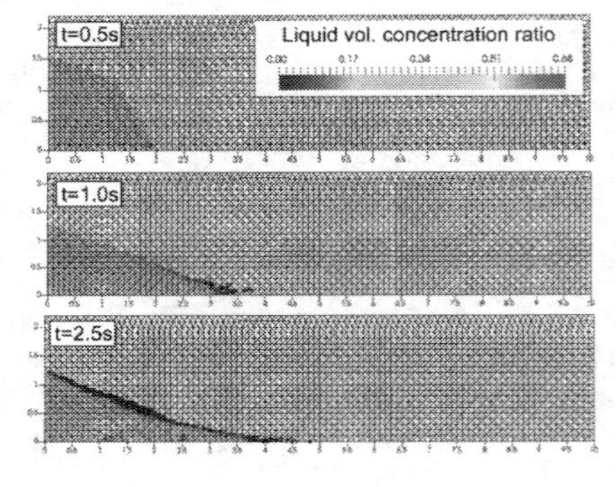

Figure 7. Liquid volumetric concentration ratio at the LMP.

ments to simulate quasi-2D conditions with a full 3D code.

A standard linear elastic perfectly plastic constitutive model with a Mohr-Coulomb failure criterion is used for the solid skeleton and a standard Newtonian compressible constitutive model is used for water. The material parameters are listed in Table 2. Initial effective stresses are generated via K0 procedure and the pressure distribution is assumed hydrostatic.

In both formulations, tensile stress is not allowed in the liquid by setting a cavitation threshold to 0 kPa, this prevents numerical problems with traction stresses in the double-point formulation which will be further investigated in the future.

Figure 5 shows the results obtained with the single-point formulation. The motion is driven by gravity: a shear surface develops and part of the soil accelerates flowing on the flat surface; kinetic energy is dissipated by friction inside the soil mass and at the base and by the drag force. Finally, the material decelerates and stops.

With the single-point formulation the MPs carry the information of both solid and liquid, thus the material is assumed to be fully saturated throughout the computation. In contrast, the double-point

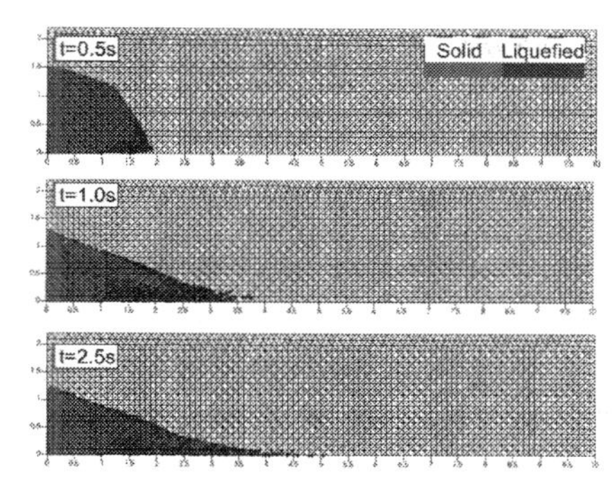

Figure 8. Phase status of SMP during column collapse.

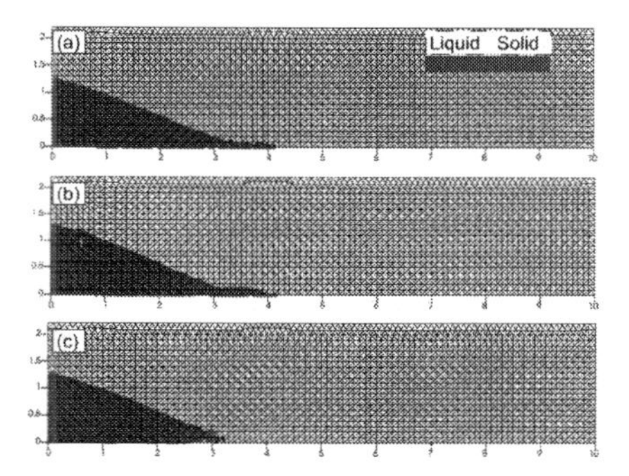

Figure 9. Effect of drag force equation on the SMP and LMP position at t=0.85 s: (a) linear term only, KL =const.; (b) linear term only, KL≠const; (c) linear and quadratic term, KL≠const.

formulation allows large relative movements between SMPs and LMPs and thus the separations between the phases (Fig. 6).

During the column collapse using the double-point approach, SMPs moves ahead with respect to LMPs developing a granular front and a small layer of dry material can be recognized at the surface (Fig. 6). This phenomenon is often recognized in debris flow (Gray et al. 2009, Johnson et al. 2012, Pudasaini 2012). Recently, it has been shown that the formation of a fluid front, i.e. the liquid moves ahead of the solid, or a granular front depends on the shear rate of the moving mass, the characteristics of the grain assembly (e.g. particle concentration) and the viscosity of the fluid (Leonardi et al. 2015).

In the implemented double-point formulation, the effect of partial saturation, e.g. suction, in the behaviour of elements filled only with solid MPs is not considered, thus the soil is assumed to be dry.

Figure 7 shows the liquid volumetric concentration ratio (n_L) at different time instants. It is null in the dry part of the soil; moreover, it should be noted that it varies significantly in space and time

Table 2. Material parameters (** applicable only for the single-point formulation, *applicable.

Parameter		Value
Initial porosity [-]	n_L	0.4
Grain density [kg/m³]	ρ_S	2650
Liquid density [kg/m³]	ρ_L	1000
Intrinsic permeability [m²]	κ_L	1.021e-10**
Dynamic viscosity [kPa s]	μ_L	1.002e-6
Young modulus [kPa]	E	10000
Poisson ratio [-]	ν	0.2
Fluid bulk modulus [kPa]	K_L	21500
K_0 coefficient [-]	K_0	0.5
Ref. grain size diameter [mm]	d	2*
Ergun parameter	A	150*
Ergun parameter	B	1.75*
Maximum porosity	n_{max}	0.5*

(Fig. 7). For this reason, the term ∇n_L should not be neglected in the governing equation of motion (Eq. 8).

During column collapse, the solid concentration decreases along the superficial and front part of the moving mass, because solid grains tend to separate. When they are no longer in contact between each other the effective stresses nullify and the soil is in a liquefied state. The transition between the solid and the liquefied state is controlled by a maximum porosity of $n_{max} = 0.5$. Figure 8 shows with blue dots the solid MP which are in a liquefied state, i.e. for which the effective stress is null.

Comparing Figures 5 and 6 it can be noted that the simulations with the single-point formulation predicts longer runout of the soil mass. This is probably due to the underestimation of the drag force (Eq. 5) because the assumption of the validity of Darcy's law is not satisfied in this case.

The definition of the drag force is a key issue in this type of problems. Figure 9 compares the results obtained considering a material with a reference solid particle diameter d=7 mm and using a drag force computed as follows:

a. Neglecting the quadratic term in Equation 9, i.e. the second addend, and assuming κ_L = constant (Fig. 9a)
b. Neglecting the quadratic term in Equation 9, and updating the intrinsic permeability using Equation 11 (Fig. 9b);
c. Considering the full form of Equation 9, and updating the intrinsic permeability using Equation 11 (Fig. 9c).

It can be seen that the movement of the liquid phase differs significantly in the considered cases.

5 DISCUSSION AND CONCLUSIONS

This paper presents and compares two recently proposed approaches to simulate multiphase problems with MPM, i.e. 2-phase single-point formulation and 2-phase double-point formulation.

2-phase granular-fluid mixture flows are characterized primarily by the relative motion and interaction between the solid and fluid phases. Drag is one of the very basic and important mechanisms of two-phase flow as it incorporates coupling between the phases. The drag force used in the applied double-point formulation considers the gradient of the volumetric phase concentration, a linear (laminar-type, at low velocity) and quadratic (nonlinear-type, at high velocity) contribution. In contrast, the drag force used in the single-point formulation assumes the validity of the Darcy law, thus it is only valid when the fluid motion inside the pores is laminar.

The single-point and the double-point formulations are equivalent and both are well applicable to seepage problems when the fluid velocity is low and the spatial variability of solid concentration is negligible. However, because the number of MPs required to discretize the saturated media is much larger in the double-point formulation, the single-point formulation can be slightly more efficient.

In contrast, the use of double-point formulation is necessary when flow velocity and variability of concentrations are relevant. Moreover, interaction between porous media and free liquid can be captured. This is an important feature in many problems such as the study of debris flow propagation, dike stability, erosion and scouring and other coastal applications.

REFERENCES

Abe, K., Soga, K., Bandara, S. 2014. Material point method for coupled hydromechanical problems. *Journal of Geotechnical and Geoenvironmental Engineering* 140 (3): 1–16.

Alonso, E.E., Pinyol, N.M. & Yerro A. 2014. Mathematical Modelling of Slopes. *Procedia Earth and Planetary Science* 9: 64–73.

Bandara, S. 2013. Material point method to simulate large deformation problems in fluid-saturated granular medium. *PhD thesis, University of Cambridge, Cambridge, UK.*

Bandara, S. & Soga, K. 2015. Coupling of soil deformation and pore fluid flow using material point method. *Computers and Geotechnics* 63(1): 199–214.

Bear, J. 1972. Dynamics of fluids in porous media. Elsevier.

Bolognini, M., Martinelli, M., Bakker, K.J., Jonkman, S.N. 2017. Validation of material point method for soil fluidisation analysis. *Journal of Hydrodynamics* 29(3): 431–437.

Ceccato, F., Beuth, L., Simonini, P. 2016a. Analysis of piezocone penetration under different drainage conditions with the two-phase material point method. *Journal of Geotechnical and Geoenvironmental Engineering* 142(12).

Ceccato, F. & Simonini, P. 2016b. Numerical study of partially drained penetration and pore pressure dissipation in piezocone test. *Acta Geotechnica* (published online).

Ergun, S. 1952. Fluid flow through packed column. *In Chemical Engineering Progress.*

Forchheimer, P. 1901. Wasserbewegung durch Boden. *Z Ver Deutsch Ing* 45: 1782–1788.

Galavi, V., Beuth, L., Coelho, B.Z., Tehrani, F.S., Hölscher P. & Van Tol, F. 2017. Numerical Simulation of Pile Installation in Saturated Sand Using Material Point Method. *Procedia Engineering* 175: 72–79.

Gray, J. & Ancey, C. 2009 Segregation, recirculation and deposition of coarse particles near two-dimensional avalanche fronts. *Journal of Fluid Mechanics* 629: 387–423.

Jassim, I., Stolle, D., Vermeer, P.A. 2013. Two-phase dynamic analysis by material point method. *International Journal for Numerical and Analytical Methods in Geomechanics* 37(15): 2502–2522.

Johnson, C.G, Kokelaar, B.P, Iverson, R.M., Logan, M., LaHusen, R.G. Gray, J.M.N.T. 2012. Grain-size segregation and levee formation in geophysical mass flows, *Journal of Geophysical Research* 117, F01032.

Leonardi, A., Cabrera, M., Wittel, F.K., Kaitna, R., Mendoza, M., Wu, W., Herrmann, H.J. 2015. Granular-front formation in free-surface flow of concentrated suspensions. *Physical Review E* 92(5), 052204.

Liang, D., Zhao, W., Martinelli, M. 2017. MPM simulations of the interaction between water jet and soil bed. *Procedia Engineering* 175: 242–249.

Martinelli, M. & Rohe, A. 2015. Modelling fluidisation and sedimentation using material point method. *1st Pan-American Congress on Computational Mechanics.*

Martinelli, M. 2016. Soil-water interaction with Material Point Method. Double-Point Formulation. *Report on EU-FP7 research project MPM-Dredge PIAP-GA-2012–324522.*

Martinelli, M., Tehrani, F.S., Galavi, V. 2017a. Analysis of crater development around damaged pipelines using the material point method, *Procedia Engineering* 175: 204–211.

Martinelli, M., Rohe, A., Soga, K. 2017b. Modeling dike failure using the material point method. *Procedia Engineering* 175: 341–348.

Pudasaini, S.P. 2012. A general two-phase debris flow model, *Journal of Geophysical Research* 117, F03010.

Soga, K., Alonso, E., Yerro, A., Kumar, K., Bandara, S. 2016. Trends in large-deformation analysis of landslide mass movements with particular emphasis on the material point method. *Géotechnique* 66 (3): 248–273.

Sulsky, D., Chen, Z., Schreyer, H.L. 1994. A particle method for history-dependent materials. *Computer Methods in Applied Mechanics and Engineering* 118(1–2): 179–196.

Wieckowski, Z. 2013. Two-phase numerical model for soil-fluid interaction problems. In *Proceedings of ComGeoIII* pp.410–41.

Yerro, A., Alonso, E., Pinyol, N. 2015. The material point method for unsaturated soils. *Géotechnique* 65(3), 201–217. Yerro, A., Alonso, E.E., Pinyol, N.M. 2016. Run-out of land-slides in brittle soils. *Computers and Geotechnics* 80: 427–439.

Zabala, F. & Alonso, E.E. 2011. Progressive failure of Aznalcóllar dam using the material point method. *Géotechnique* 61(9): 795–808.

Numerical Methods in Geotechnical Engineering IX – Cardoso et al. (Eds)
© 2018 Taylor & Francis Group, London, ISBN 978-1-138-33198-3

Non-Euclidian discrete geometric modeling of granular soils

Y. Larom
Civil and Environmental Engineering Faculty, Technion - Israel Institute of Technology, Haifa, Israel

S. Pinkert
Structural Engineering Department, Faculty of Engineering Sciences, Ben-Gurion University of the Negev, Israel

ABSTRACT: This paper suggests a new geometric modeling approach for discrete representation of a realistic pore-scale morphology using an amorphous non-Euclidian geometry of soil grains' shape. The soil skeleton composition process follows a statistically-based consistent algorithm, which applies random non-repetitive soil skeleton morphology while keeping the same macro-scale geotechnical properties. The model parameters are determined such that both micro-scale indexes (as angularity, roundness and roughness) and macro-scale outcomes (as phase and grading parameters) are correlated. The stability and the convergence of the solution is demonstrated in this paper upon 2D model. In addition, the model allows a direct representation of the amorphous pore space and can be used for further issues related to a non-homogeneous distribution of various materials (liquids or solids) in the porous medium.

1 INTRODUCTION

In geotechnical engineering, a soil can be described either by continuum or discrete approaches. Continuum soil-mechanics models describe the general soil behavior through stress-strain relationships and consider averaged phased parameters. Conventional soil parameters (such as friction, cohesion, dilatation, porosity, grading curve etc.) are utilized for geotechnical simulation, which are commonly studied in centimeter to kilometer scales, and not in a soil-grain resolution. To account for effects which are studied in context of soil-grains resolution, and for describing the soil as an assembly of discrete soil particles, one can use discrete element methods (DEM). DEM enable investigation of the whole specimen mechanical response through known interaction relationships between soil grains (Cundall & Strack 1979). Theoretical and experimental studies based on discrete description have been performed using spherical grain shape; using laboratory testing on tiny balls assembles (e.g., Rowe 1962, Skempton 1961) or DEM numerical simulations (e.g., Ting et al. 1989). The use in spherical grains shape is widely spread and modification for other elliptical shapes (e.g., Ng 1994) or few particles attachment (e.g., Jensen et al. 1999) have been presented in the literature. These geometries can be categorized under the family of Euclidian shapes, which do not represent a natural (realistic), non-repetitive and amorphous grain morphology.

The natural structure of soil skeleton has been investigated by various micro-scale imaging approaches (Rodriguez et al. 2012). Due to the challenge in describing natural amorphous geometries of soil grains, a common simplification was suggested, using categorization of the grain shape into three major classes (Mitchell & Soga 2005); [1] "form", which describes the variance of grain length in different directions, [2] "angularity" or "roundness", describing irregularities on grain edges, and [3] "roughness", which describe the surface texture. One should note, that these parameters can only describe a given data but cannot be used in a straightforward modeling.

A more representative natural morphology can be described based on fractal geometry principles. Fractal geometry is a branch of mathematics that describes a non-Euclidian amorphous geometry using a random but systematic calculation procedure (Perrier et al. 1999). This paper implements such principles to generate soil skeleton as an assembly of discrete amorphous particles.

2 MODEL

2.1 *Model fundamentals*

In the suggested model, the soil is represented as a collection of unique individual grains, such that each of which is composed of a collection of small numerical elements. Each element in the numerical domain can be related either to a specific soil-grain ("full") or a void space unit ("empty"). The soil-skeleton development is a non-physical grain crystallization process, which based on random selections and statistical fundamentals in a consistent framework.

The main model fundamentals and the crystallization process are demonstrated upon a typical model output, given in Fig. 1. Figs. 1a to 1f illustrate the crystallization progress, starting at a fully voided space (Fig. 1a), related to a porosity equals 1.0 and converges into a soil skeleton morphology that is related to the required porosity, n_f, (Fig. 1f). This is one directional process, in which only "empty" elements may become "full", but not the opposite. The crystallization process is such that once a numerical element turns "full" it increases the probability of the grain to inflate at its nearby "empty" elements. The binary decision, of turning "empty" element into "full", is based on a comparison between adjusted probability values and random values. This random selection process ensures an amorphous morphology of each grain. In addition to each unique soil-grain shape, the model satisfies realistic boundary conditions and grain contact morphologies.

2.2 *Crystallization process*

In the crystallization process, a local statistical value is calculated for each "empty" element. The ratio of this value to all other statistical values in the grid dominates the probability of the cell to become "full" in the next calculation cycle and may later be updated. This process is demonstrated qualitatively in Fig. 2, showing a numerical grid, crystallized grains and attributed probabilities (which are sketched along a given cross-section at the upper part of the figure), for three following stages. Fig. 2a shows an initial state, where all elements are "empty" (i.e., porosity equals 1.0). As can be seen, at this stage all elements have the same a priori-determined filling probability, ε. Fig. 2b demonstrates a possible filling result of Fig. 2a's random selection process, in which one element turned "full" and taged #1 (permanently). Then, this element increases the probability to inflate itself to the nearby elements through a radial influence, R. This probability is degraded with increasing distance from the center of the filled element, using a monotonic decreasing bell-shaped function, $w(r)$. One should note, that the model still keeps the probability to generate random new grains, as demonstrated in Fig. 2c, for which grain #2 is formed, in parallel to grain #1's expansion. For the calculation of next cycle probabilities, four cases can be recognized among all "empty" ele-

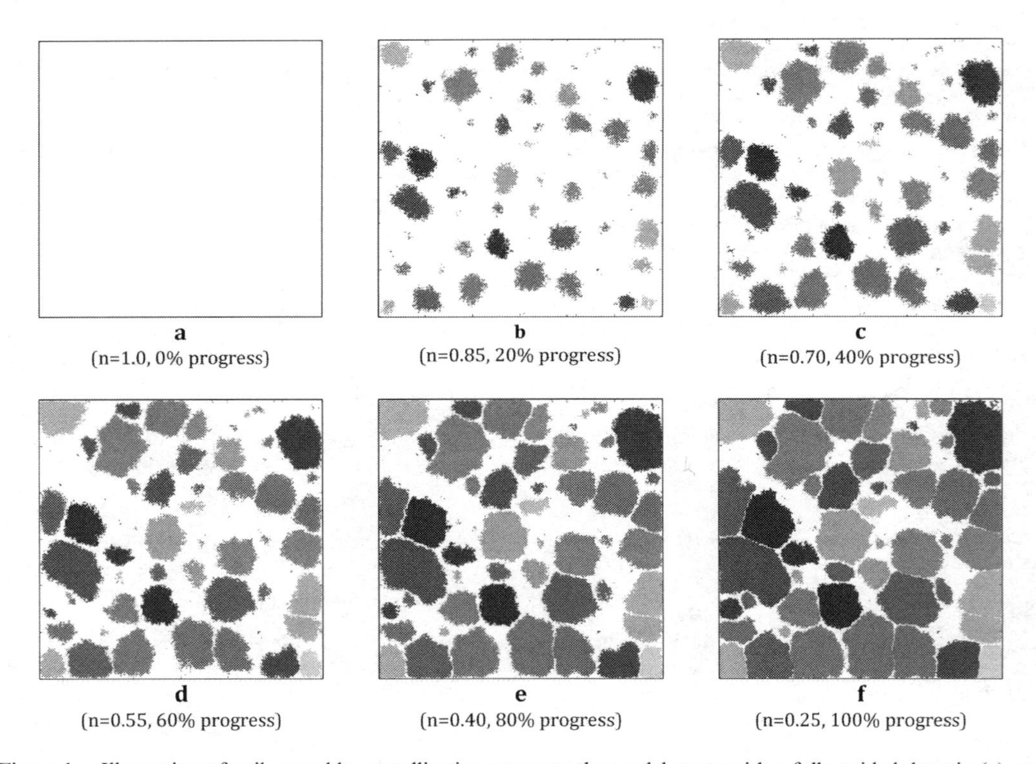

Figure 1. Illustration of soil assembly crystallization progress; the model starts with a fully voided domain **(a)** and operates iterative numerical crystallization **(b, c, d & e)** until a convergence to a pre-defined porosity is achieved **(f)**.

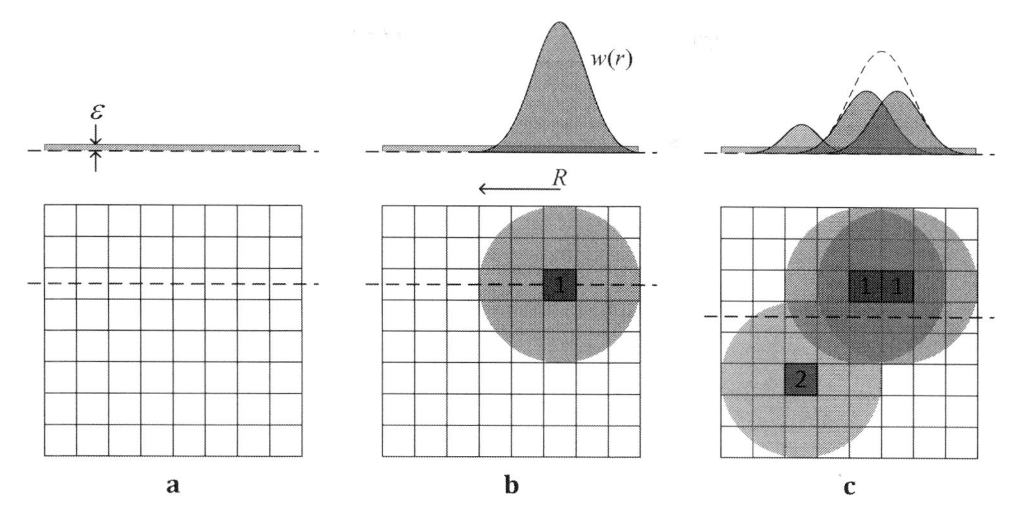

Figure 2. Demonstration of attributing probabilities and random crystallization at two calculation steps.

ments: [1] elements that are not under the influence of grains #1 and #2; i.e., uncolored cells, [2] elements which are affected by cumulative influence of grain #1's related elements (dashed curve), [3] those who influenced by grain #2, and [4] element under a joint influence.

2.3 *Probability calculation*

In the crystallization process, once an element becomes "full" it influences the probability for further crystallization of other nearby elements using the function $w(r)$. r is the radial distance between the center of the new "full" element and its influenced cells (ranging between 0 and R). The crystallization probability of each "empty" cell i is associated with a list of j nearby grains, $i_1, i_2, \ldots, i_j$, and their cumulative probability values:

$$\Psi_i^j = \sum_{k=1}^{N_i^j} w(r_k) \tag{1}$$

where N_i^j is the number of "full" type j numerical elements which cell i is affected by, as shown in Fig. 3a. In this figure, only dark elements of grain i_j with $r < R$ are considered in the calculation of Ψ_i^j. The principle that "empty" cell can be affected by several types of j elements is illustrated in Fig. 3b upon three "empty" cells, following Fig. 2c, where each arrow refers to r and its influencing element.

This data is accumulated and used for calculation of both probability and tagging for the following calculation cycles throughout the whole cyclic

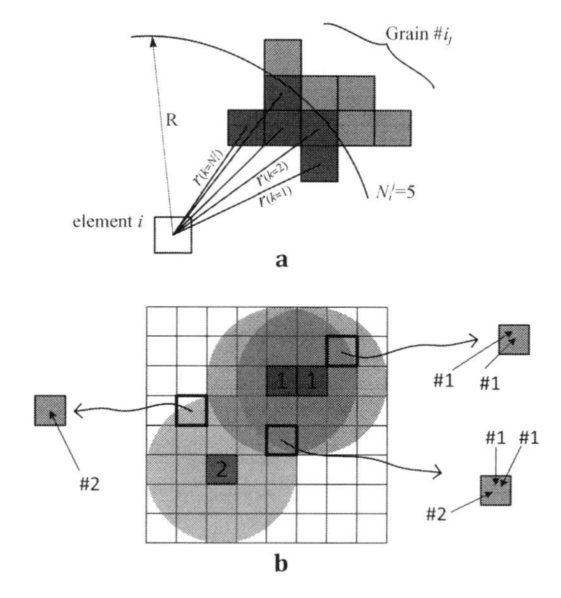

Figure 3. Demonstration of (a) number of "full" type j elements which cell i is affected by and (b) the accumulated data within each "empty" cell.

procedure to the point that the cell is turned "full" (or remains "empty" if a model convergence is achieved). The element will be examined as related to the most influencing nearby grain, according to a choosing function which prefers the maximum cumulative Ψ_i^j value:

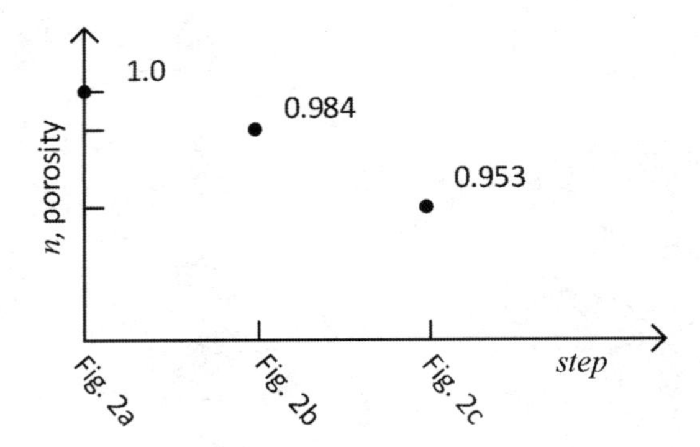

Figure 4. Porosity reduction related to Fig. 2.

$$\#i = \begin{cases} 1 & \Psi_i^1 = \max\left(\varepsilon, \Psi_i^1, \ldots, \Psi_i^{Nj}\right) \\ \vdots & \vdots \\ N^j & \Psi_i^{Nj} = \max\left(\varepsilon, \Psi_i^1, \ldots, \Psi_i^{Nj}\right) \\ N^{j+1} & \varepsilon = \max\left(\varepsilon, \Psi_i^1, \ldots, \Psi_i^{Nj}\right) \end{cases} \tag{2}$$

One should note that in addition to grain expansion, the function allows generation of new grains, $\#N^{j+1}$, under a pre-defined value, ε (as demonstrated in Fig. 2c).

The probability value for an element i, P_i, is calculated in two steps, in which a local value, P_{0i}, is calculated for all "empty" cells and then normalized to adjust the required soil-skeleton generation rate. P_{0i} is calculated by multiplication of three sub-functions; f_{1i}, f_{2i} and f_{3i}, as follows.

$$P_{0i} = f_{1i} f_{2i} f_{3i} \tag{3}$$

where f_{1i} is a crystallization function which gives rise to the most effective Ψ_i^j (or ε, according to Eq. 2), f_{2i} is a decay function for non-realistic grain attachments (the model prefers nodal attachments rather than a whole interface attachment) and f_{3i} is a geometric boundary constraining function, for which grains will not be cut at the edges of the domain. The adjusted probability for each element, P_i, is calculated by:

$$P_i = \Delta n(step) N_0 \frac{P_{0i}}{\sum_{m=1}^{N_1} P_{0m}} \tag{4}$$

where $\Delta n(step)$ is the porosity difference between the current cycle porosity and the required one for the next cycle, N_0 is the total number of all numerical elements in the grid and N_1 is the total number of "empty" elements. This way, $\Delta n(step)N_0$ is the number of "empty" element which will optimally become "full" at the next cycle. The normalization component ensures:

$$\sum_{i=1}^{N_1} \frac{P_{0i}}{\sum_{m=1}^{N_1} P_{0m}} = 1.0 \tag{5}$$

and thus enable using a systematic random selection between 0 and 1 for comparison with Eq. 4's product.

2.4 *Convergence*

The actual porosity of the model is calculated by the ratio between a number of "empty" elements and a total number of elements. Fig. 4 demonstrates the porosity reduction related to Fig. 2 (64/64, 63/64 and 61/64, where 64 is the total number of elements). In the model, a pre-defined porosity monotonic reduction function is used, $n_r(n,step)$, which is interactively updated in accordance with the random crystallization process, where actual porosity is not necessarily be equal the originally required one. In other words, the function may be updated at each *step* according to the current (actual) porosity, n, in a self-curing process. Fig. 5 illustrates the interactive $n_r(n,step)$ function, where filled symbols denote actual porosities calculated at a current *step* and empty symbols refer

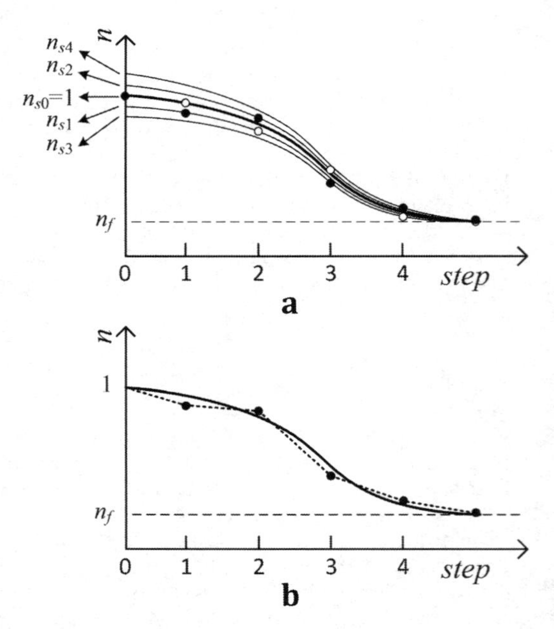

Figure 5. Illustration (**a**) the interactive updated porosity reduction function and (**b**) actual and theoretical convergence curves.

to demanded porosities for *step* + 1. As can be seen at Fig. 5a, the convergence process starts at $n = 1.0$. After the first calculation cycle (*step* = 1), the actual porosity is obtained lower than the required value. Then, the function is updated to begin from a lower initial value (n_{s1}) which ensures the same shape function that crosses the actual porosity at *step* = 1 and converges into the same final porosity, n_f. This interactive function-updating process is performed at the beginning of every *step* till a convergence to n_f is achieved. The actual convergence process (dashed line) and the theoretical function (solid line) are both given in Fig. 5b. As can be seen, the actual porosity may be above or below the theoretical curve, but cannot be higher than the actual porosity at the previous step (i.e., "filled" elements cannot turn back to be "emptied"). Fig. 5's illustrations consider a very general n_r shape function. In the model, any shape function can be considered as long as it satisfies monotonic decreasing from 1.0 at *step* = 0 and converges into n_f at a pre-determined step, $step_f$. Variations of n_f value and n_r function yield different end-products of soil skeleton morphologies.

2.5 *Model algorithm*

The model algorithm implements all the above-mentioned principles in a calculation sequence that described in Fig. 6, and detailed as follows. First, the model is initialized with input parameters related to: [1] global dimensions and resolution, [2] direct geotechnical phase relationships, as porosity, and [3] independent parameters (as R and ε_0) and functions parameters (for the functions n_r, $w(r)$, f_{1i}, f_{2i} and f_{3i}). Then, a cyclic procedure begins, in which at each cycle the model examines convergence by comparison of the actual porosity with n_f. If convergence has not been achieved, a probability for crystallization is adjusted (for all "empty" cells, as described in section 2.3) and a crystallization process is carried out through comparison of the probability values with random values. Once a convergence is achieved, the soil skeleton end-product is examined in light of macro-scale geotechnical characteristics (as grading curve and related coefficients) and micro-scale grain morphology effects (as angularity, roundness and roughness).

3 RESULTS

The suggested model has been presented in Section 2 in its most general form. The results shown in this paper are associated with the following relationships (functions) and specific model parameters.

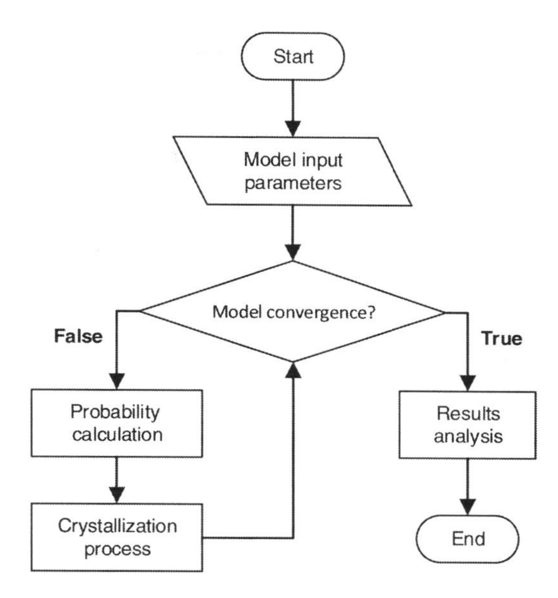

Figure 6. Flowchart of the calculation process.

3.1 *Governing functions*

The interactive convergence function, n_r, describes monotonic reduction in porosity from $n_r(0) = 1.0$ to $n_r(\infty) = n_f$, given by:

$$n_r(step+1) = n_f + \left[n_s(step) - n_f\right]e^{-K\left(\frac{step+1}{step_f}\right)^{\beta}} \tag{6}$$

initialized by $n_s(0) = 1$ (bold curve in Fig. 5) and updated at each cycle according to the actual porosity obtained at the previous step (thin curves in Fig. 5a) by:

$$n_s(step) = n_f + \left[n(step) - n_f\right]e^{K\left(\frac{step}{step_f}\right)^{\beta}} \tag{7}$$

where β is a model parameter which governs the degradation slope and $K = 4.6$ for 99% convergence at $step = step_f$ ($e^{-4.6} \approx 1\%$), in which $step_f$ is an input model parameter.

The influence degradation function, $w(r)$, is given by:

$$w(r) = e^{-K\left(\frac{r}{R_e}\right)^{\alpha}} \tag{8}$$

where R_e is a reference distance within the influence circle, $0 < R_e \le R$, and α is a model parameter.

The value of ε dominates the probability to generate new grains rather than expansion of existing grains (Eq. 2), and is therefore set as proportional to the actual porosity reduction, $n(step)$:

$$\varepsilon = \frac{n(step) - n_f}{1 - n_f} \varepsilon_0 \quad (9)$$

For $n(0) = 1.0$, at the start of the process, $\varepsilon = \varepsilon_0$, and equales 0 when $n(step)$ converges into n_f.

The probability function, Eq. 3, is given by multiplication of the following three sub-functions. The crystallization function, f_{1i}, is given by:

$$f_{1i} = \left[max(\varepsilon, \Psi_i^1, \Psi_i^2, ..., \Psi_i^{Nj})\right]^{\lambda_1} \quad (10)$$

where λ_1 is an amplification factor. The grain attachment function, f_{2i} decays the probability for new crystallization by reducing Eq. 3's value in a factor ranges between 0 and 1, given by:

$$f_{2i} = e^{\ln(vt)c_i^{\lambda_2}} \quad (11)$$

where λ_2 is a model parameter which dominates f_{2i}'s degradation slope, and v_t is a threshold value for f_{2i} when c_i approaches 1.0. c_i function is given by:

$$c_i = \left\{ \sum_{k=2}^{N_i^j} \left[\frac{s_i^k}{s_i^1} \ln\left(1 + \frac{\Psi_i^1}{\Psi_i^k}\right) \right]^{-\delta} \right\}^{1/\delta} \quad (12)$$

where $k = 1$ refers to the nearby grain with the highest Ψ_i value, s_i^k is the shortest distance between element i and grain k, and δ is a model parameter. f_{3i}, is a boundary constraining function, which decays the probability at a distance R_e from the domain boundaries in a similar manner as used in f_{2i} (Eq. 11), given by:

$$f_{3i} = \begin{cases} 1.0 & d_i \geq R_e \\ e^{\ln(vt)d_i^{\lambda_2}} & d_i < R_e \end{cases} \quad (13)$$

where d_i is the shortest distance from the center of cell i to the domain boundaries.

3.2 *Model parameters*

The results shown in this paper refers to the following input parameters detailed in Table 1.

3.3 *Model output*

The model was implemented in MATLAB R2016b program. Fig. 7 shows model outputs of three 5 mm × 5 mm 2D samples which were produced with the same input parameters (given in Table 1). The clear visual difference between the three obtained samples demonstrates random soil skeleton crystallization, which is a main model fundamental. In spite of the difference between the soil skeletons, they describe similar geotechnical features. Fig. 8 shows the three related grading curves. As can be seen, the difference between the three is minor, such that same geotechnical properties may be extracted; as average grain size (D_{50}) and curvature and unity of the curve (C_c and C_u respectively). One should note that similarity in results has been obtained although different number of grains were generated (80, 74 and 71). For the grading curves generation, the grain size is taken as the minimal diameter of an ellipse that has a same second-moments as the grain. Fig. 9 shows a zoom-in of one grain and its related ellipse. The figure also shows the amorphous nature of the grain and its resolution by plotted upon the numerical grid.

The porosity reduction function, n_r (Eq. 6), related to Fig. 7a, is given in Fig. 10. This figure demonstrates the robustness of the developed interactive convergence algorithm (Eqs. 6 and 7), in which an almost perfect matching between the theoretical and the actual degradations is obtained. At $step = step_f$ (300 in this case) 99% convergence is achieved, and the model continues a smooth convergence.

4 CONCLUSIONS

The paper presents the development of a new soil-skeleton geometric model, for which the soil is represented as an assembly of discrete non-Euclidian non-repetitive amorphous particles. Each soil particle in the model is unique and composed of numerous numerical elements. The soil skeleton generation method is operated in a cyclic crystallization process. In each calculation cycle, set of equations are utilized to define crystallization probabilities for the next calculation step

Table 1. Model input parameters.

Parameter	Value	Related Eq.	Remark
Width	5 mm	–	Sample width
Height	5 mm	–	Sample height
N. elements	90,000	–	–
n_f	0.25	6, 7, 9	Required porosity
$step_f$	300	6, 7	Required steps
R	0.50	8 (not directly)	Radial influence
R_e	0.15	8	Reference distance
α	1.50	8	Model parameter
β	0.03	6, 7	Model parameter
λ_1	1.00	10	Model parameter
λ_2	0.50	11, 13	Model parameter
δ	0.30	12	Model parameter
v_t	0.03	11, 13	Model parameter
ε_0	0.01	9	Model parameter

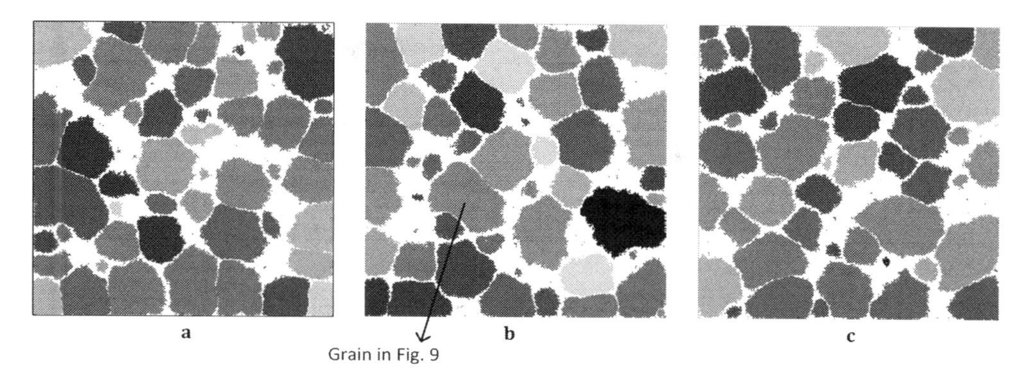

Figure 7. Three model outputs of 5 mm × 5 mm samples, which were all obtained for the same given set of parameters (Table 1). Fig. 7a is sample presented in Fig. 1f.

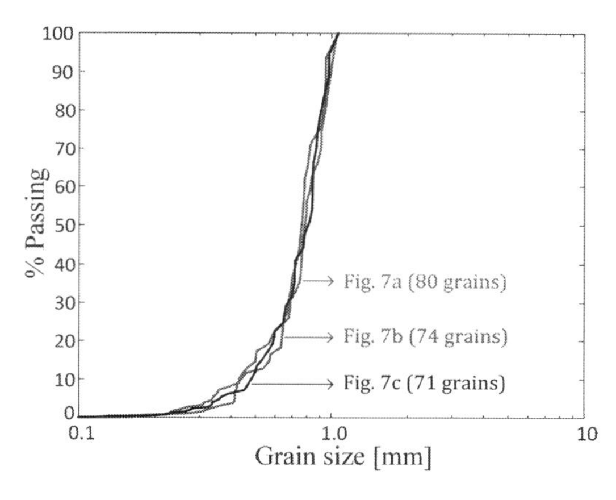

Figure 8. Grading curves related to the three soil samples presented in Fig. 7.

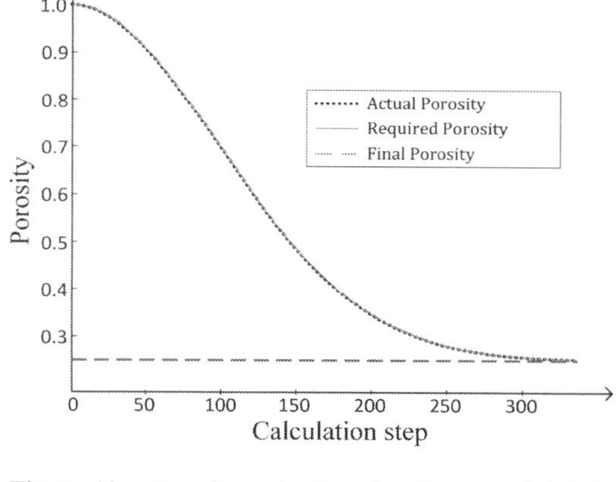

Figure 10. Porosity reduction function n_r related to Fig. 7a.

Figure 9. A specific grain from Fig. 7b and its related ellipse (for grading curve calculation), upon the numerical grid.

in all numerical elements, and so on to the point where the structured sample reaches a pre-defined porosity. The model-user controls the soil-skeleton product through parameters related to the crystallization rate, the probability to generate new grains throughout the cyclic process, and parameters related to natural rearrangement of granular assembly; i.e., small contact areas and general convex grain morphologies. Although the model applies random selections within the crystallization process, such that a unique micro-scale morphology is obtained in each run, it ensures similar geotechnical (macro-scale) characteristics for the same set of input parameters. This principle is demonstrated in the paper by showing similar grading curves calculated for three different samples created by the model for the same input parameters.

The model principles are presented upon a 2D domain and will be further developed for a 3D model. Through direct correlation between standard geotechnical testing and model input parameters, the model could further be used as a numerical laboratory for more advanced issues, where the pore space is filled with other liquids or solids.

REFERENCES

Cundall, P.A. & O.D. Strack (1979). A discrete numerical model for granular assemblies. *geotechnique 29*(1), 47–65.

Jensen, R.P., P.J. Bosscher, M.E. Plesha, & T.B. Edil (1999). Dem simulation of granular mediastructure interface: effects of surface roughness and particle shape. *International Journal for Numerical and Analytical Methods in Geomechanics 23*(6), 531–547.

Mitchell, J.K. & K. Soga (2005). *Fundamentals of soil behavior*. New York; Chichester: Wiley.

Ng, T.-T. (1994). Numerical simulations of granular soil using elliptical particles. *Computers and geotechnics 16*(2), 153–169.

Perrier, E., N. Bird, & M. Rieu (1999). Generalizing the fractal model of soil structure: the pore–solid fractal approach. *Geoderma 88*(3), 137–164.

Rodriguez, J., J. Johansson, & T. Edeskär (2012). Particle shape determination by two-dimensional image analysis in geotechnical engineering. In *Nordic Geotechnical Meeting: 09/05/2012–12/05/2012*, pp. 207–218. Danish Geotechnical Society.

Rowe, P. (1962). The stress-dilatancy relation for static equilibrium of an assembly of particles in contact. *269*(1339), 500–527.

Skempton, A. (1961). Effective stress in soils, concrete and rocks. *Selected papers on soil mechanics*, 106–118.

Ting, J.M., B.T. Corkum, C.R. Kauffman, & C. Greco (1989). Discrete numerical model for soil mechanics. *Journal of Geotechnical Engineering 115*(3).

Numerical Methods in Geotechnical Engineering IX – Cardoso et al. (Eds)
© 2018 Taylor & Francis Group, London, ISBN 978-1-138-33198-3

Numerical investigations on the liquid-solid transition of a soil bed with coupled CFD-DEM

M. Kanitz, E. Denecke & J. Grabe
Institute of Geotechnical Engineering and Construction Management, Hamburg University of Technology, Hamburg, Germany

ABSTRACT: Intense pore water flow through a stable packed soil bed can lead to a liquefaction and hence to failure of structures like submarine slopes. While in geomechanics the soil is treated as a stable soil skeleton subjected to effective stresses, the liquefied soil has to be dealed with in a different manner. With the transition of the soil bed to a liquid state, it has to be treated as a viscous liquid with different properties compared to its stable condition. To gain a deeper understanding of the micromechanic actions taking place at the transition from a solid to a liquid state and vice versa, a multiscale approach, namely a combination of the Computational Fluid Dynamics (CFD) and the Discrete Element Method (DEM), is used. In the coupled CFD-DEM, the soil particles are tracked in a Lagrangian way at a microscale level. The fluid phase, e.g. the pore water, is modelled as a continuum in CFD, solving the pore water flow with the Reynolds averaged Navier-Stokes equations. In a first step, the saturated soil of the seabed will be disturbed by a pressurized flow leading to liquefaction. The second step consists of the sedimentation of the soil grains. These investigations will help to draw conclusions about the void ratio and the stress state in the soil during liquid-solid transition.

1 INTRODUCTION

The transition from a solid to a liquid state of soil is of critical importance in various cases, especially when it occurs abruptly. In this way, submarine landslides can suddenly liquefy due to an external trigger and turn to debris flows (Friedman et al. 1992), clayey slopes can collapse and flow over large distances (Mitchell and Markell 1974), (Evans and Brooks 1994). Hence a major danger is imposed to the natural and rural environment. This sudden change within the solid phase can be triggered by earthquakes or a rapid pore flow through the soil skeleton. From a mechanical point of view, soil can be considered as a yield stress fluid. It remains at rest until a force is applied to it which causes the flow. Granular flows arise the interest of research as they could constitute a fourth state of matter besides gas, liquid and solid. This state is called a jammed system (Liu and S. R. Nagel 1998). A jammed system is characterized by the fact, that the particles of the granular material in a confined volume form a continuous network of interactions throughout the sample. In order for a flow to occur, these interactions need to be separated. Hence, investigations of the transition of granular material from a solid to a liquid state needs extensive consideration of particle-particle and particle-fluid interaction mechanisms. A multiscale approach allows to depict the particle movement and the particle-particle interactions at a microscopic scale while solving the fluid flow and the particle-fluid interactions at macroscopic scale. The combination of the Computational Fluid Dynamics (CFD) and the Discrete Element Method (DEM) represents such an approach. This contribution will present numerical investigations with the coupled CFD-DEM analyzing the transition from a solid to a liquid state due to an intense flow trough a stable packed soil bed in a water column. Small-scale experiments with glass beads are carried out in order to compare the numerical results with the experimental ones in order to validate the numerical model and the numerical method for this kind of application. The calculations are carried out using the open source software package CFDEMcoupling®, which combines the discrete element code LIGGGHTS® with CFD solvers based on OpenFOAM®.

2 CFD-DEM METHOD

The combination of the Computational Fluid Dynamics (CFD) and the Discrete Element Method (DEM) is known as the coupled CFD-DEM method, which was firstly proposed by

Tsuji et al. (1993). The fluid flow is thereby determined using a cell-based continuum approach and the movement of the solid phase is represented on an individual particle level. This method is hence able to take into account solid-fluid interaction forces and to deal with large deformation problems. The following section will give a brief introduction of the coupled CFD-DEM method.

2.1 *Discrete element method*

The DEM tracks the particle position and movement explicitely in a Lagrangian way. It was originally developed by Cundall and Strack (1979), who investigated the motion of rock masses. Its governing equations are formed by Newton's second law of motion to determine the translational and rotational movement of a particle. Hence, the motion of a particle i with a mass m is calculated using the following equations:

$$m_i \frac{\partial v_i}{\partial t} = \sum_j \mathbf{F}_{ij}^c + \mathbf{F}_i^f + \mathbf{F}_i^g, \tag{1}$$

$$I_i \frac{\partial \omega_i}{\partial t} = \sum_j \mathbf{M}_{ij}, \tag{2}$$

where v_i and ω_i describe the translational and angular velocity of the particle i. $\mathbf{F}_{ij}^c$ and $\mathbf{M}_{ij}$ represent the contact force and torque acting on particle i by particle j or the walls. The influence of the interaction with the fluid and the gravitation are included in Eqn. 1 by F_i^f and F_i^g. The contact force is determined using the so-called soft sphere approach proposed by Cundall and Hart (1992). Thereby, the particles in contact are allowed to overlap slightly. Elastic, plastic and frictional forces are then calculated by their spatial overlap. The accurate description of the contact traction distribution in normal and tangential direction is quite complex as it is related to the shape of the particle, its material properties and its state of motion. The simplest model to determine the contact force distribution is based on a linear spring-dashpot model proposed by Cundall and Strack (1979), see Fig. 1. The elastic deformation is represented by the spring while the dashpot accounts for the viscous dissipation. The normal and tangential forces are then determined by the spatial overlap $\delta\mathbf{x}_p$ through

$$\mathbf{F}_n = k_n \delta\mathbf{x}_p - c_n \delta\mathbf{v}_{p,n} \tag{3}$$

and

$$\mathbf{F}_t = k_t \delta\mathbf{x}_p - c_t \delta\mathbf{v}_{p,t}. \tag{4}$$

where k_n and k_t are the normal and tangential spring stiffness and c_n and c_t the damping coefficients in normal and tangential direction. A more accurate description of the contact forces, which is used in the numerical simulations presented in this paper, is given by the simplified Hertz-Mindlin contact model porposed by Langston et al. (1994). It is based on the Hertz-Mindlin contact model, which includes a non-linear relationship between normal force and normal displacements (Hertz 1881) and a force-displacement relationship depending on the whole loading history and the rate of change of normal and tangential displacement (R.W. Mindlin, H. Deresiewicz 1953). The original Hertz-Mindlin contact model is quite time-consuming in DEM simulations, so Langston et al. (1994) adopted a direct force-displacement relation for the tangential force. Traditionally, the particles are simplified as perfect spheres in DEM simulations. Hence, to account for shape effects and interparticle forces that are not acting at the mass centre of the particle and generate a torque, a so-called rolling friction is introduced in DEM simulations by Iwashita (1998). Therefore, an additional set of a spring, a dashpot and a slider is located at each contact point, see Fig. 1. The spring and the dashpot act against the rolling and theslider is activated when the torque is exceeding the rolling resistance.

2.2 *Computational fluid dynamics*

The basis for the determination of the fluid flow with CFD is formed by the Navier-Stokes equations. In this contribution, the Finite Volume Method (FVM) is used, which solves the locally-averaged

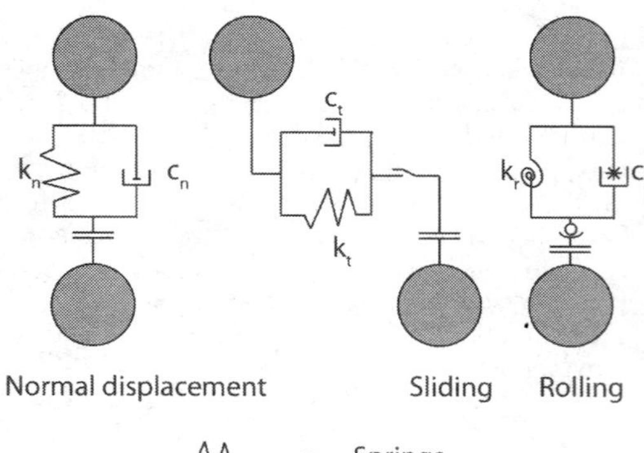

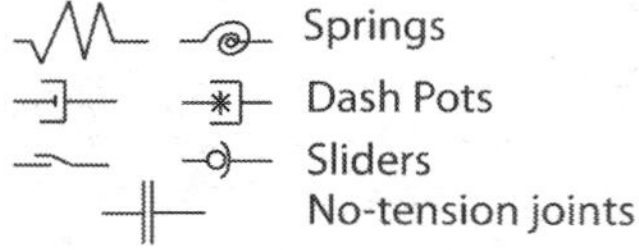

Figure 1. Implemented contact model (Iwashita 1998).

Navier-Stokes equations for every cell in the fluid domain. The governing equations for the conservation of mass, Eqn. 5, and momentum, Eqn. 6, are expressed in terms of locally averaged variables (Anderson and Jackson 1967) through.

$$\frac{\partial}{\partial t}\alpha_f + \nabla \cdot (\alpha_f \mathbf{u}_f) = 0, \tag{5}$$

$$\frac{\partial}{\partial t}(\rho_f \alpha_f \mathbf{u}_f) + \nabla \cdot (\rho_f \alpha_f \mathbf{u}_f \mathbf{u}_f) = -\alpha_f \nabla p + \nabla \cdot (\alpha_f \tau_f) + \alpha_f \rho_f \mathbf{g} + R_{pf}, \tag{6}$$

where $\mathbf{u}_f$ and p describe the fluid velocity and pressure. τ_f and ρ_f are the fluid viscous stress tensor and the density of the fluid. In these equations, there are two major differences when compared to the classic locally averaged Navier-Stokes equations. In Eqn. 5 and Eqn. 6 two terms, α_f and R_{pf}, are added to the equations, which include the influence of the solid phase on the fluid flow. The factor α_f stands forthe porosity, referring to the occupied volume of fluid in a cell in the presence of particles. R_{pf} describes the momentum exchange between the fluid and the particles. It stores the particle-fluid interaction forces and is permanently exchanged between the CFD- and the DEM-solver. R_{pf} is determined through

$$R_{pf} = K_{pf}(\mathbf{u}_f - \mathbf{v}), \tag{7}$$

including the interaction force coefficient K_{pf}. It is defined as the superposition of the particle-fluid interaction forces.

2.3 *Coupling scheme*

The CFD solves its governing equations on a cell level and the DEM on the individual particle level. Hence, one has to take care of the scale level between these two methods when it comes to the coupling. Basically, two different methods exist to couple the CFD and the DEM: the resolved and unresolved method CFD-DEM. In the resolved CFD-DEM, one particle is covered by several CFD-cells, see Fig. 2a. Using this method, the fluid flow around the particles can be solved quite accurately, but it is only feasible for systems with a few particles (some hundred) only. Larger problems, with several millions of particles can be solved using the unresolved CFD-DEM, see Fig. 2b. Within this approach, the CFD cells are notably larger than the particles, causing several particles to be in one CFD cell. Using this method, one has to deal with a loss of information concerning the fluid flow and the fluid-particle interaction forces.

In the numerical simulations presented in this contribution about 9 500 000 particles are computed to model the submarine slope. For this reason, the unresolved CFD-DEM is used.

The coupling routine, described by Goniva et al. (2010), consists of several steps. Generally, the CFD and the DEM solve its governing equations independently. The DEM solver calculates the positions and velocities of the particles and determines the corresponding CFD cell. Afterwards, for each CFD cell, the porosity α_f and the mean particle velocity is caluclated. Based on the calculated flow field and the porosity, the fluid forces acting on each particle are determined on the CFD side. After that, applying the unresolved method, the particle-fluid momentum exchange term R_{pf} is determined by averaging over all particles in the CFD cells. In the next step, the fluid forces acting on the particle are handed to the DEM solver, which calculates the new positions and velocities of the particles. Meanwhile, the CFD solver computes the fluid velocity taking into account the porosity and the momentum exchange and the coupling routine starts again.

The momentum exchange term R_{pf} for the CFD side is solved using Eqn. 7 (Goniva et al. 2010). The interaction forces for the determination of K_{pf} are formed by the drag force, buoyancy force, pressure gradient force and the virtual mass force. Concerning the DEM side, the interaction forces are summed up to the interaction forces F_i^f between fluid and particle and hence added to the sum of forces and torques in Newton's laws of motion (Eqn. 1 and Eqn. 2).

2.4 *Particle-fluid interaction forces*

In the unresolved method, where the fluid flow around the particles is not solved directly, the determination of the particle-fluid forces signifies a crucial point in the coupled CFD-DEM simulations. Therefore, force models are used to

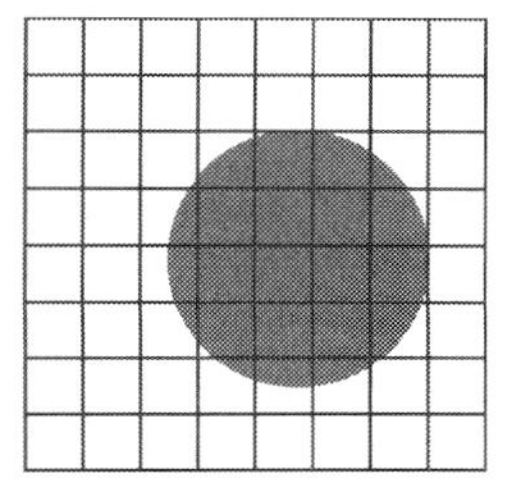

a) Resolved CFD-DEM

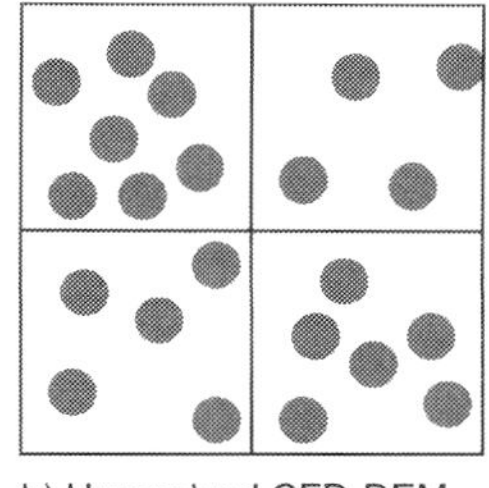

b) Unresolved CFD-DEM

Figure 2. Treatment of the scaling between cell-based CFD and particle-based DEM (Hager 2014).

account for the buoyancy force, the drag force, the pressure gradient force and unsteady forces as the Basset history force, the virtual mass force and lift forces (Li et al. 1999), (Xiong et al. 2005). Concerning the drag force, well established correlations for the drag around a single particle exist (Zhu and Yu 2006). The fluid flow around the particle generates three regions: the Stoke's Law region, the transition region and the Newton's law region. For each of these regions, the drag coefficient C_D can be determined by well-established correlations. Nevertheless, in particulate problems, the determination is much more complex, as the space for the fluid to flow is reduced. This generates sharp velocity gradients, which leads to an increased shear stress on the particle surface.

Table 1. Particle-fluid interaction forces.

Forces	Correlations
Drag force	$F_D = F_0(\alpha_f) + F_1(\alpha_f)\ Re_p^2\ \ 20$ for $Re_p\ \ 20$
	$F_D = F_0(\alpha_f) + F_3(\alpha_f)\ Re_p^2$ for $Re_p > 20$
	$F_0(\alpha_f) = \dfrac{1 + 3(\alpha_f/2)^{1/2} + (135/64)\alpha_f \ln\alpha_f + 16.14\,\alpha_f}{1 + 0.681\alpha_f - 8.48\alpha_f^2 + 8.16\alpha_f^3}$ for $\alpha_f\ \ 0.4$
	$F_0(\alpha_f) = 10\alpha_f/(1-\alpha_f)^3$ for $\alpha_f > 0.4$
	$F_1(\alpha_f) = 0.110 + 5.10 \cdot 10^{-4}\ \exp^{11.6\alpha_f}$
	$F_3(\alpha_f) = 0.0673 + 0{:}212\alpha_f + 0.0232(1-\alpha_f)^5$
Pressure gradient force	$F_p = -V_p\partial_p/\partial\mathbf{x} = -V_p(\rho_f\mathbf{g} + \rho_f u_f\partial\mathbf{u}_f/\partial\mathbf{x})$

In the presented simulation, the particle based drag force is calculated following the correlation of Koch and Hill (2001) and Koch and Sangani (1999). It is based on results obtained from numerical simulations of fluid flow in a randomly dispersed particle system in a bubbling fluidized bed using Lattice-Boltzmann equations. Koch and Hill (2001) and Koch and Sangani (1999) established a correlation depending on the Reynolds particle number Re_p and the porosity α_f. Besides the drag force, the pressure gradient force and the viscous forces are included in the numerical simulations presented in this contribution. The pressure gradient force is derived by Anderson and Jackson (1967). It describes the phenomenon of a particle immersed in a flow field with slowly diverging streamlines. The additional pressure gradient will provoke a supplementary component of force acting on the particle. The pressure gradient force model of Anderson and Jackson (1967) includes the buoyancy force due to gravity and the effects resulting from the accelerated pressure gradients in the fluid. The force models used in the presented numerical simulations are listed in Table 1.

Table 2. Mechanical properties of the glass spheres.

Parameter	Symbol	Value	Unit
Young's modulus	Y	2.15	GPa
Poisson ratio	ν	0.3	-
Restitution coefficient	μ	0.98	-
Friction coefficient	φ	0.454	-
Rolling friction coefficient	μ_R	0.057	-
Density	ρ	2500	kg/m^3

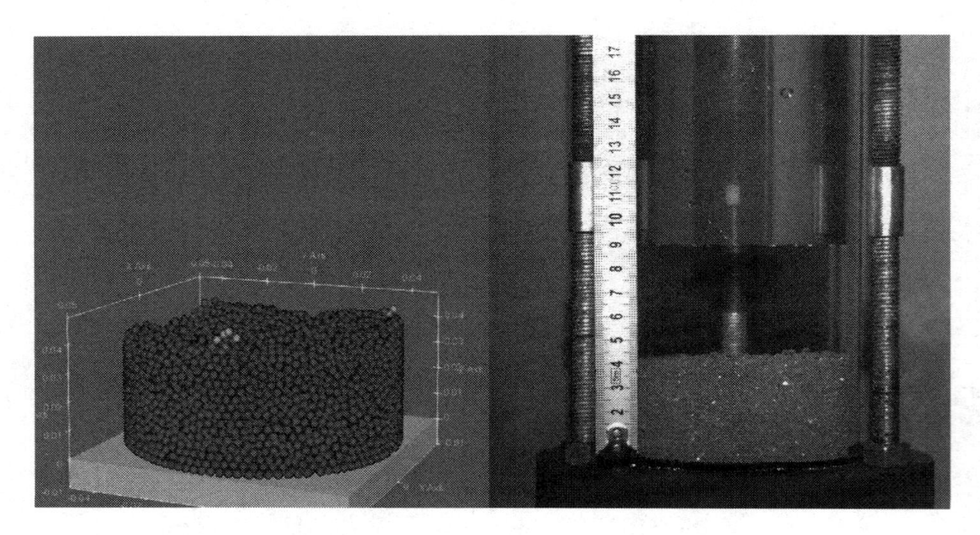

Figure 3. Left: simulation after initial fluidisation, right: initial state of the experiment.

3 EXPERIMENTAL AND NUMERICAL SETUP

The numerical model is designed according to experiments carried out at the Institute of Geotechnical Engineering and Construction Management as part of a master thesis. The setup of the experiments is inspired by a common fluidized bed. The fluid flows vertically from the bottom to the top through an acrylic glass tube which has an inner diameter of 10 cm. The particles at the bottom of the tube are fluidized by the forces of the fluid flow. The whole process is recorded with a high-speed camera. The recordable frame is limited to a height of 20 to 30 cm to compensate the low recording resolution of 1024 × 1024 pixels. To resemble the natural conditions of submerged soil being fluidized the fluid used for the experiments is water at room temperature. Instead of using actual soil samples like sand, small glass spheres with a diameter of 3 mm are used. Therefore, the properties of the used material are defined in a more definite way, are faster to simulate and easier to match with the simulation. Those glass spheres are well analyzed by previous experiments carried out in order to calibrate their mechanical properties (see Table 2).

The installation of the glass spheres in the acrylic glass tube is done by filling the particles into the open top of the tube. They are prevented from dropping out the bottom by a thin metal grid. After installing the spheres the top of the tube is sealed air tight. The top seal is equipped with a ball valve which is opened to start the experiment after anchoring the acrylic cylinder under water. The sealed container enables a set difference in water levels between the inner tube and the outer environment. By opening the ball valve, the water level difference outside and inside the tube starts leveling itself. With this procedure a high velocity flow can be achieved. Tests with an open tube and the fluid being pumped actively through the acrylic tube from water sources like the local water tap results in a water flow too slow to fluidize the particles. Fig. 3 shows the initial state of the experiment. The starting conditions in the simulation are given by the packed installation of the particles achieved by growing the particles into place. They are inserted with a diameter of 2.44 mm and grow up to the before mentioned diameter of 3 mm. With a targeted volume displacement of 0.65 within the 4 cm high space, around 14 300 particles are needed to resemble the approximated number of particles used in the experiments. For the numerical simulation of the particles the Hertz-Mindlin model is used as described in chapter 2.1. Additionally a simplified elastic-plastic spring-dashpot model is used to account for the rolling friction. The high value of the Youngs modulus requires the DEM simulation to run at a low time step. The time step for the DEM analysis is 5 microseconds while the CFD simulation can run at faster time step of 1 millisecond. For every 200 DEM calculations one CFD calculation is performed. If the particle size needs to be changed to a smaller diameter the time step must be lowered as well. The simulation of the metal grid is conducted by implementing a barrier for the particles in LIGGGHTS® which does not interfere with the CFD simulation but stops the particles from sitting directly on top of the inlet. This barrier is pictured on the left side in Fig. 3. The experimental setup allows flow velocities up to 0.3 m which corresponds to a flow of 9.42 l through the cross section of the tube. This highly turbulent flow is simulated using a k-ϵ model. The drag force acting on the particles is calculated by a Koch Hill model as described in chapter 2.4. The additional acting forces are simulated by the pressure gradient model and a viscous force model as described by Zhou et al. (2010).

The fluidization of the glass spheres in the experiments is finished after two seconds. To reach a comparable behavior in the simulation the packed particles must be transferred into a less dense packing. The installation of the particles in the experiments results in a loose packing of the spheres. Therefore, the tightly packed spheres in the simulation are fluidized once and settle into a loose packing similar to the initial condition in the experiment. Contrary to the particle-particle interaction described by the parameters in Tab. 2 the particle-wall interaction must be reduced to a minimum. With a friction coefficient of $\varphi = 0.01$ and a rolling friction coefficient of $\mu_R = 0.005$ the influence on the particle movement is minimized. With these improvements the uprising particle front does not slow down near the walls and brings the deciding improvement in simulating the transition between the start phase and the turbulent fluidized phase.

To improve the capabilities of comparing the experiments with the simulation, the recordings of the highspeed camera are analyzed with particle image velocimetry (PIV). The frame rate of 250 frames per second offers sharp pictures in most frames and eases the lighting conditions. With an increased frame rate a sharper image can be achieved but the quality of the picture analysis suffers from poor lighting. The transparent property of the glass spheres lets single spheres become almost invisible if they are separated and can cause reflections if the light source is too bright. A compromise between the recording frame rate and the number of used particles must be found in order to obtain a usable recording. The best result was achieved with about 14 300 particles which reached a height of 20 cm during fluidization and a frame rate of 250 frames per second.

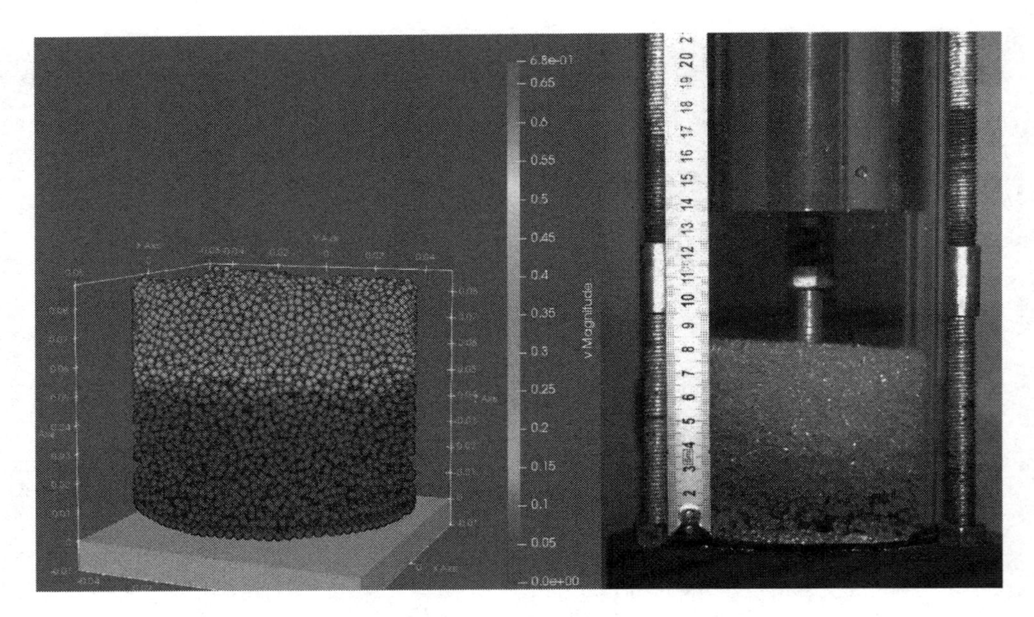

Figure 4. Comparison between simulation (left) and experiment (right) at t = 0,228.

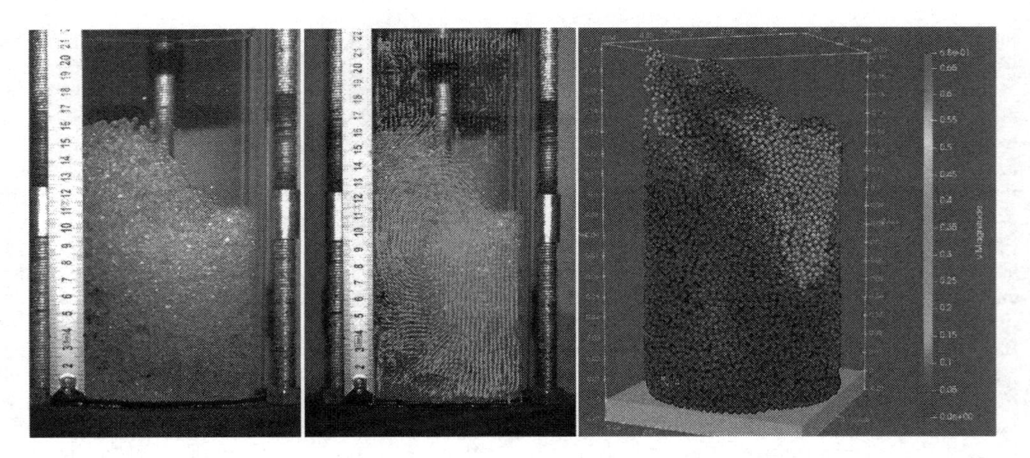

Figure 5. Experimental results (right), PIVlab analysis (middle) and simulation (right) at t = 0.624 s.

4 RESULTS

The evaluation of the particle movement in the experiments is done with PIVlab. PIVlab is an open source time resolved picture image velocimetry tool for MATLAB (Thielicke and Stamhuis 2014). By comparing two timely close pictures it can calculate the velocity distributions within the region of interest. The experiment can be split up into three different phases. The first includes the initial lifting of the particles. The glass spheres stay as a bulk with a straight surface and completely lift of the ground. Only a few particles start settling instantly after the lifting. The simulation in comparison forms a uniform layer at the bottom of the particle formation shown in Fig. 4. As it can be seen, the simulation catches quite accurately the behavior of the glass spheres in the experiment. As in the experiment, the particles are lifted to a heigth of 9 cm.

The second phase consists of the turbulent mixing of the spheres as can be depicted in Fig. 5. As described in section 3 the interaction between wall and particle must be reduced to a minimum. The

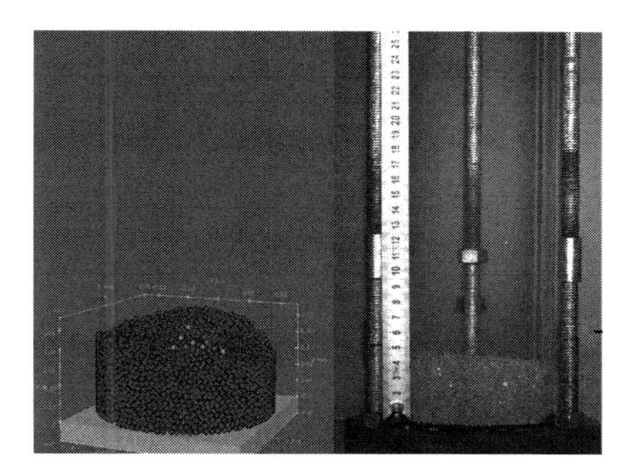

Figure 6. Comparison between simulation (left) and experiment (right) at the end.

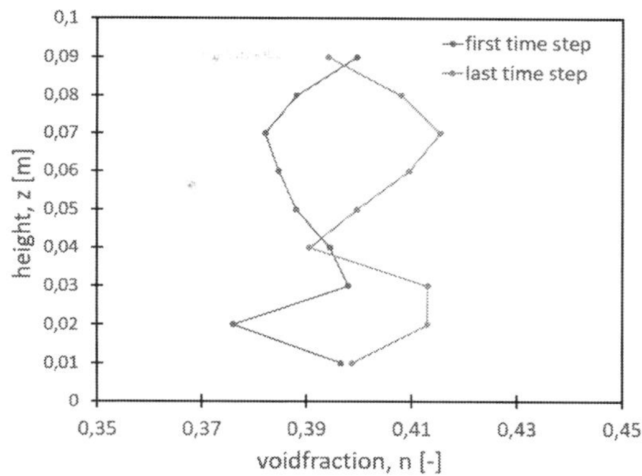

Figure 7. Measured void fraction by the probes over the height of the sphere bulk for the first (blue) and last (orange) time step.

asymmetric deformation within the second phase is only possible, if the outer particles are not slowed down by the contact with the acrylic glass walls.

The results of the PIVLab analysis show the velocity distribution in the center of Fig. 5. The movement of single glass spheres is difficult to follow. PIVlab is able to determine the overall movement but the velocity of every particle cannot be tracked accurately. Nevertheless, the PIV analysis visualizes the turbulent vortex that is created. The water flow is not able to break through the middle of the particle formation but pushes through the sides resulting in the observed vortex. This can equally be identified in the numerical analysis with coupled CFD-DEM. The break pattern of the particle formation results in a circular motion of the glass spheres starting at the highest point on the left site. This circulation slows down and starts showing a symmetrical behavior similar to a water fountain with the particles rising in the center of the tube and slowing down when drifting towards the walls. The calculated particle velocities show a peak at around $0.3\,\frac{m}{s}$. The simulation shows fast velocities in small areas but the comparison of the particle movements reveals an adequate match. The DEM analysis is able to determine the velocity of each particle but the PIV analysis calculates the velocity of the visible spheres.

The last phase covers the sedimentation of the particles. The flow rate slows down and is no longer able to lift the glass spheres until the flow rate completely stops. As already mentioned the capacities of the experimental setup are limited to a couple of seconds of water flow. A prolonged second phase is not possible. Fig. 6 shows the glass spheres at the end of the experiment. In this third phase of the setup, the outcome of the numerical simulation is similar to the experimental results as in both cases the end height of the particle bed is around 5 cm.

Despite minor differences in the behavior of the glass spheres, especially at the lower end near the water inlet, the overall resemblance between simulation and experiment is promising and enables the simulation of more complex problems with the developed numerical model. Based on this model, a variation of the particle diameter is the first step towards an in-depth analysis of fluidized soil. The diameters of the simulated particles are varied to a grain size distribution of Hamburger Sand but upscaled by a factor of 5 to reduce the amount of particles needed and still include the already used 3 mm glass spheres. The spheres are packed tightly to get a good comparison between the porosity before and after fluidization. Besides the visual segregation of the particles where the smallest spheres are located on top and the biggest particles are at the bottom a difference in the void fraction can be measured. Fig. 7 shows the estimated void fraction over the height of the particle bed. The void ratio changes are computed in the center of the packed bed and measured by five probes every centimeter in vertical direction. Overall the void fraction increases by 0.01 which results in a 0.04 change in the void ratio. This result confirms the anticipated dispersal of the particle distribution and the loosening of the particle bed due to the fluidization.

5 CONCLUSION

The physical small scale fluidized bed experiments with glass spheres with a diameter of 3 mm were successfully recreated with a coupled CFD-DEM simulation. CFDEMCOUPLING® offers extensive possibilities by combining the

DEM tool LIGGGHTS® and the CFD program OPENFOAM®. Despite small differences between experiment and simulation an overall good resemblance of the particle movement during the fluidization was achieved. The numerical model requires well investigated mechanical properties of the fluidized material in order for the interacting models between DEM and CFD to be able to transform the input into a respectable simulation. Even though the experiments are done with simplified monodisperse glass spheres, the simulation offers the capabilities to build up on those results. The PIV analysis leaves room for improvement considering the choice of particles to get a better analytical comparison between experiment and simulation. A first attempt based on the accomplished numerical model is made by varying the particle sizes which shows promising results, but still needs validation and does not include grain shape effects, which will play a major role when simulating actual sand.

AKNOWLEDGMENT

The authors thank the German Research Foundation (DFG) for funding this project.

REFERENCES

Anderson, T.B. & R. Jackson (1967). A fluid mechanical description of fluidized beds. *Industrial and Engineering Chemistry Fundamentals 6*(4), 527–538.

Cundall, P.A. & R.D. Hart (1992). Numerical Modelling of Discontinua. *Engineering Computations 9*(2), 101–113.

Cundall, P.A. & O.D.L. Strack (1979). A discrete numerical model for granular assemblies. *Géotechnique (29)*, 47–65.

Evans, S.G. & G.R. Brooks (1994). An earthflow in sensitive Champlain Sea sediments at Lemieux, Ontario, June 20, 1993, and its impact on the South Nation River. *Canadian Geotechnical Journal 31*(3), 384–394.

Friedman, G.M., J.E. Sanders, & D.C. Kopaska-Merkel (1992). *Principles of sedimentary deposits: Stratigraphy and sedimentology*. New York, NY: Macmillan.

Goniva, C., C. Kloss, A. Hager, & S. Pirker (2010). An Open Source CFD-DEM Perspective. In *Proceeding of OpenFOAM Workshop*. Göteborg.

Hager, A. (2014). *CFD-DEM on Mulitple Scales – An Extensive Investigation of Particle-Fluid Interactions*. Dissertation, Johannes Kepler Universit¨at Linz, Linz.

Hertz, H. (1881). Über die Berührung fester elastischer Körper. *Journal für die reine und angewandte Mathematik 92*, 156–171.

Iwashita, K. (1998). Rolling resistance at contacts in simulation of shear band development by DEM. *Journal of Engineering Mechanics (124)*, 285–292.

Koch, D.L. & R.J. Hill (2001). Inertial effects in suspension and porous-media flows. *Annual Review of Fluid Mechanics 33*(1), 619–647.

Koch, D.L. & A.S. Sangani (1999). Particle pressure and marginal stability limits for a homogeneous monodisperse gas-fluidized bed: Kinetic theory and numerical simulations. *Journal of Fluid Mechanics 400*, 229–263.

Langston, P.A., U. Tüzün, & D.M. Heyes (1994). Continuous potential discrete particle simulations of stress and velocity fields in hoppers: Transition from fluid to granular flow. *Chemical Engineering Science 49*(8), 1259–1275.

Li, Y., J. Zhang, & L.-S. Fan (1999). Numerical simulation of gas–liquid–solid fluidization systems using a combined CFD-VOF-DPM method: Bubble wake behavior. *Chemical Engineering Science 54*(21), 5101–5107.

Liu, A.J. & S.R. Nagel (1998). Jamming is not just cool any more. *Nature 396*(21).

Mitchell, R.J. & A.R. Markell (1974). Flowsliding in Sensitive Soils. *Canadian Geotechnical Journal 11*(1), 11–31.

R.W. Mindlin, H. Deresiewicz (1953). Elastic spheres in contact under varying oblique forces. *Journal of Applied Mechanics 20*, 327–344.

Thielicke, W. & E.J. Stamhuis (2014). PIVlab – Towards User-friendly, Affordable and Accurate Digital Particle Image Velocimetry in MATLAB. *Journal of Open Research Software 2*, 1202.

Tsuji, Y., T. Kawaguchi, & T. Tanaka (1993). Discrete particle simulation of two-dimensional fluidized bed. *Powder Technology 77*(1), 79–87.

Xiong, Y., M. Zhang, & Z. Yuan (2005). Threedimensional numerical simulation method for gas–solid injector. *Powder Technology 160*(3), 180–189.

Zhou, Z.Y., S.B. Kuang, K.W. Chu, & A.B. Yu (2010). Discrete particle simulation of particle–fluid flow: Model formulations and their applicability. *Journal of Fluid Mechanics 661*, 482–510.

Zhu, H.P. & A.B. Yu (2006). A theoretical analysis of the force models in discrete element method. *Powder Technology 161*(2), 122–129.

Numerical Methods in Geotechnical Engineering IX – Cardoso et al. (Eds)
© 2018 Taylor & Francis Group, London, ISBN 978-1-138-33198-3

Free vibration analysis of piled raft foundation by FE-BE coupling method

Jagat Jyoti Mandal
Department of Civil Engineering, N.I.T.T.T.R, Kolkata, India

Sayandip Ganguly
Department of Civil Engineering, G.I.M.T, Nadia, India

ABSTRACT: Finite Element Method has been applied by various researchers to analyse many complex structures with greater accuracy. As an alternative to overcome many constraints of FEM a FE-BE coupling technique is applied taking advantages of both Finite Element and Boundary Element method to determine the natural frequency of rectangular piled-raft foundation. The rectangular raft is idealised as thick plate and the soil as isotropic, homogeneous, semi infinite elastic half-space. The plate and half-space is modelled separately with a unilateral and frictional contact at their interface. The stiffness of pile is calculated based on elastic theory using simplified approach. A computer code is developed in which discretisation is automatic and very nominal useful data is required. Effect of various pile parameters and the pile configuration on the natural frequency is studied to demonstrate the effect of pile on natural frequency of the pile raft system.

1 INTRODUCTION

Strong superstructure built on a weak sub structure is similar to human with weak legs. Hence proper choice and accurate analysis of sub structure is very important which mainly depends on the nature of supporting soil. Pile foundation is considered as good solution where bearing capacity of soil is low and significant settlement is to be resisted. Generally in case of pile foundation a group of pile is constructed and tied together using a pile cap, The main purpose of the pile cap is to distribute the load from structure to piles under the cap. This pile cap is designed for structural capacity only. But the contribution of pile cap is significant when it is in direct contact with soil. Pile raft foundation is the one in which pile as well as pile cap takes part in the load distribution mechanism. The piled-raft foundation needs evaluation of number of factors to come up with proper analysis or design models that simulate actual site condition. In many cases raft foundation induces excessive settlement which is not acceptable due to serviceability requirements. Placing a number of piles in suitable manner under the raft reduces such settlement and restricts settlement to an acceptable range. In addition to settlement control, the bearing capacity of the whole system also improves. The conventional design methods used for pile groups may lead to a higher number of piles under the pile cap in some critical cases. Using piled raft, this number can be reduced where pile is mainly used as settlement reducer.

2 FORMULATION OF PROBLEM

2.1 *Idealisation*

Idealisation is the first step of solving any complex problem. Idealisation has been done in such way that it will well simulate the practical problem making the problem easy to solve. In the present study, the raft is idealised as a thick plate freely resting on soil, which is idealised as semi-infinite, isotropic and homogeneous elastic continuum. Pile is represented as spring, the stiffness of which is evaluated using simplified approach by considering it as a shaft in the elastic half space.

2.2 *Solution procedure*

Discretisation of plate-half space interface is done using two-dimensional isoparametric quadrilateral elements, in which the plate is discretised into eight noded isoparametric plate-bending finite elements based on Mindlin's plate bending theory, which allows transverse shear deformation. Each node of the plate bending element has three degrees of freedom, namely vertical displacement and two orthogonal rotations. To maintain a node to node correspondence at the plate soil interface the half-space also discretised into eight-noded isoparametric elements.

The stiffness matrix of the soil is obtained by boundary element method inverting the soil flexibility matrix. Suitable transformation is carried out to maintain the compatibility at the raft-soil interface, to apply coupling of matrices obtained from two different approaches FEM and BEM. The stiffness of pile located below the raft is determined using elastic theory which is applied for the determination of the settlement of pile where pile is considered as a shaft in the half-space. Combined stiffness of the pile raft system is obtained by coupling the stiffness of the soil and raft obtained from different approaches to which the stiffness contribution of piles are added at the specific pile locations. A computer programme in FORTRAN language is then developed to calculate the natural frequency.

2.3 *Determination of stiffness matrix of soil*

The raft soil-soil interface is divided into a number of interconnected two dimensional isoparametric quadrilateral quadratic elements and the assumed quadratic shape functions are converted into natural co-ordinate system ξ_1 and ξ_2. In order to determine unit outward normal at any point on the boundary of the element two vectors tangential to the local co-ordinate ξ_1 and ξ_2 are defined.

A numerical integration is then carried out on the kernel to find the flexibility matrix of the soil from the developed relation between traction vector and settlement.

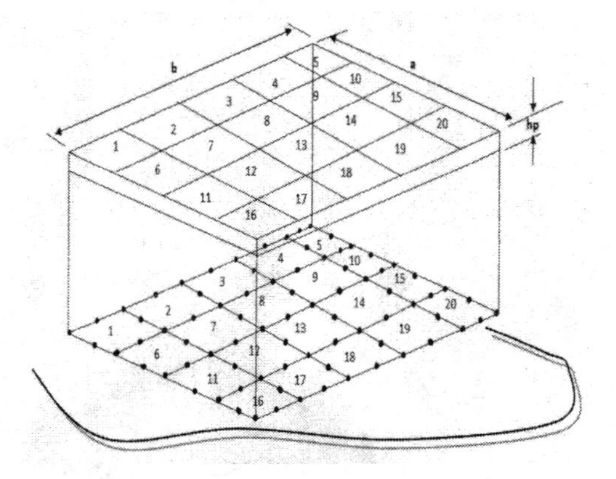

Figure 1. Plate on elastic continuum—discretisation of soil foundation system.

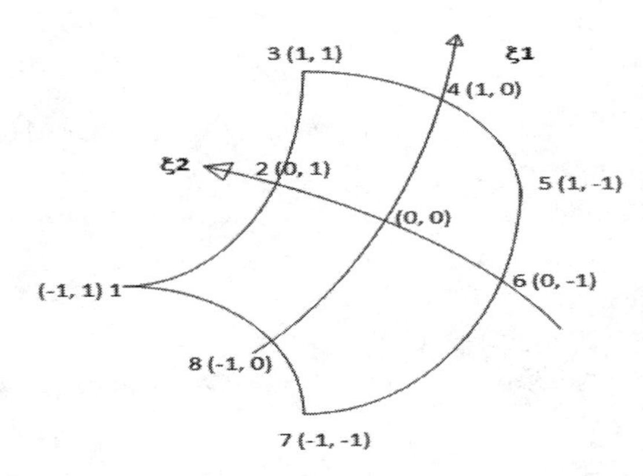

Figure 2. Two-dimensional isoparametric quadrilateral quadratic element.

2.4 *Determination of pile stiffness*

Stiffness of pile is calculated using simplified approach for floating pile in which pile settlement is expressed as

$$\rho = \frac{PI}{E_s d} \tag{1}$$

So pile stiffness will be;

$$K_{fp} \frac{P}{\rho} = \frac{E_s d}{I} \tag{2}$$

where,

ρ = Settlement of pile head,
P = Applied axial load,
E_s = Modulus of elasticity of soil.

$$I = I_o x R_k x R_h x R_v \tag{3}$$

I_O = Settlement influence factor for incompressible pile in semi-infinite mass,
R_k = Compressibility correction factor for pile,
R_h = Correction factor for finite depth of layer on a rigid base,
R_v = Correction factor for soil Poisson's ratio (v_s), d = total depth of soil layer.

2.5 *Determination of plate stiffness matrix*

Stiffness matrix of plate is derived after discretising the plate into isoparametric quadrilateral quadratic plate bending element in which transverse shear deformation is taken into account by including these deformation to the rotational degrees of freedom at each node i.e

$$\theta_y = -\frac{\partial w}{\partial x} + \gamma_{xz} \text{ and } \theta_x = -\frac{\partial w}{\partial y} + \gamma_{yz},$$

where γ_{xz} and γ_{yz} are the transverse shear strain.

2.6 *Coupling of FE and BE method*

Soil flexibility matrix determined by using boundary element method (BEM) is an asymmetric matrix by nature. But all the stiffness matrices determined using FEM is symmetric in nature. There are various methods available to convert BE equivalent asymmetric stiffness matrix to symmetric matrix. In this study a simple method described by Brebbia and Georgiou (1979) is adopted in which the condition for which the square of the error caused by forcing the asymmetric matrix to be symmetric is made minimum to get the converted symmetric matrix i.e

$$\frac{\partial(e^2)}{\partial Kij(new)} = 2K_{ij}^{(new)} - K_{ij}^{(old)} - K_{ji}^{(old)} \quad (4)$$

Which yields $K_{ij}^{(new)} = \frac{1}{2}\left[K_{ij}^{(new)} + K_{ji}^{(old)}\right];$ (5)

Where $e =$

$$\frac{1}{2}\left[\left(K_{ij}^{(new)} - K_{ij}^{(old)}\right) + \left(K_{ji}^{(new)} - K_{ji}^{(old)}\right)\right]; \quad (6)$$

Different terms of force unknowns are also used i.e nodal forces for FEM and nodal traction for BEM. Hence coupling of BE and FE method is a very difficult task which is done by imposing different interface conditions at the interface of plate and soil interface and satisfying compatibility and equilibrium condition. Equilibrium of force on the interface is satisfied by making the summation of all the forces (i.e. traction) on a node equal to the externally applied force to that node which is expressed as follows

$$(t_x)^{BE} = -(t_x)^{FE}; \quad (7)$$
$$(t_y)^{BE} = -(t_y)^{FE}; \quad (8)$$
$$(t_z)^{BE} = -(t_z)^{FE}; \quad (9)$$

Here the subscripts (x, y, z) denotes the direction of forces.

Combined stiffness of pile-raft-soil is then obtained by summing up the stiffness matrix of plate-soil system with the stiffness of pile at respective pile locations. The final stiffness matrix can be expressed as

$$[K_{pr}] = [[K_p] + [K_{fp}] + [[M][A^{-1}]]] \quad (10)$$

Here,
$[K_{pr}]$ = combined stiffness matrix of pile, raft, soil
$[K_p]$ = stiffness matrix of raft/plate,
$[K_{fp}]$ = stiffness matrix of floating pile and
$[A]$ = flexibility matrix of soil.

2.7 *Formulation for free vibration analysis*

It is not possible to apply Mindlin's classical solution to dynamic problem in the BE framework as it doesn't contain any inertia term associated with the dynamics of half-space. Hence the free vibration analysis is conducted within the framework of stiffness formulation and natural frequency of the system is determined using general equation of motion i.e.

$$[M]\{\ddot{u}\} + [K_{ps}]\{u\} = 0; \quad (11)$$

By putting $\{u\} = \{\bar{u}\}\sin\omega t$ and $\{\ddot{u}\} = -\omega^2\{\bar{u}\}\sin\omega t = -\omega^2\{u\}$ in the equation (11) we get.

$$\left(\left[K_{ps}\right] - \lambda[M]\right)\{\bar{u}\} = 0 \quad (12)$$

where $\lambda=\omega^2$, The above equation is called generalized eigen problem. The equation has only trivial solution when $\rho = \frac{Pl}{E_s d}$ For non-trivial solution we have to determine the eigen values λ that satisfy

$$\det([K_{ps}] - \lambda[M]) = 0; \quad (13)$$

For each eigen value λ_i, there is an eigen vector $\{u_i\}$,which is sometimes called normal mode. The lowest non-zero ω_i is called natural frequency.

This eigen value problem is solved by Generalized Jacobi Method which is an iterative method used to solve eigen value problem. As the large number of degrees of freedom associated with the problem it is preferred to use the diagonal terms of the consistent mass matrices to avoid any problem which may occur due to large relative ratio of the element of the matrix. The HRZ scheme (Hinton et al., 1976) is an effective method which has been applied to diagonalise the mass matrix.

3 RESULT

For the purpose of this study i.e effect of various pile, raft and soil parameters on the natural frequency of pile-raft, a model of raft size 30 m × 15 m is taken with nine piles at a specified spacing as shown in figure below. Effect of all variables on the natural frequency of the pile raft is studied one by one keeping other variables constant.

For the analysis raft is discretized into 72 elements. 12 elements along the longer direction of the raft and 6 elements along the shorter direction of the raft. Number of degrees of freedom at each node is 3. Degree of freedom in vertical direction at node 1 is taken as 1.

Initially a convergence study is made for checking the accuracy of the solution algorithm and computer code based on results obtained from various formulas and method. For circular foundation on elastic half-space natural frequency is obtained from the stiffness and soil parameter [Timoshenko and Goodier (1951), Sung(1953)]. For this convergence study mass density of the raft material is taken as 2.5 Kn-s2/m4.This is done on raft without pile. After getting satisfactory result same programme is used adding the stiffness of pile at specified node at specified direction.

A parametric study is carried out on the natural frequency of raft. The results obtained are presented with various figures for easy comprehension.

The analysis of natural frequency is carried out for different parameters. It has been found that modulus of elasticity and poisson's ratio of soil and raft material has no significant effect on the natural frequency of pile- raft but raft thickness, pile numbers, diameter of piles and spacing of pile affect the natural frequency significantly. Hence an optimum choice of these parameter is necessary for safe and serviceable design of pile-raft foundation. With the increase of raft thickness natural frequency of the system decreases whereas with the increase of pile length, diameter and spacing (along both X and Y direction) natural frequency increases sharply. All the figures of variation of natural frequency with respect to various parameter is shown in the following figure.

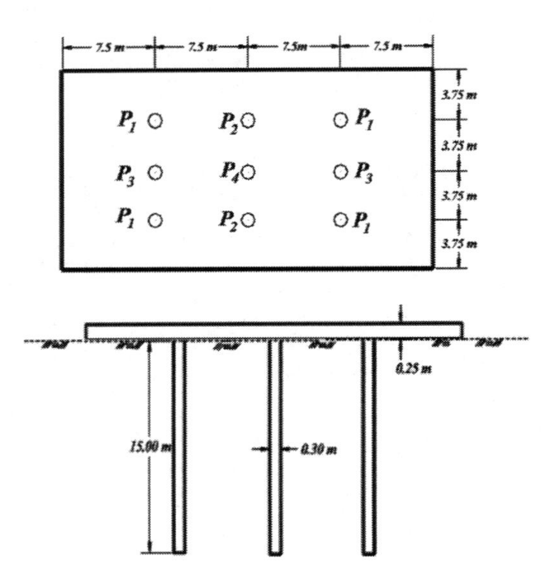

Figure 3. Plan and elevation of the raft and pile position.

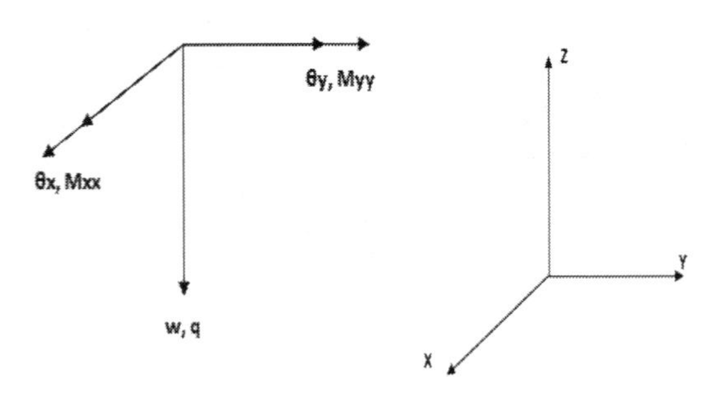

Figure 4. Degrees of freedom and stress resultants for the plate bending finite element.

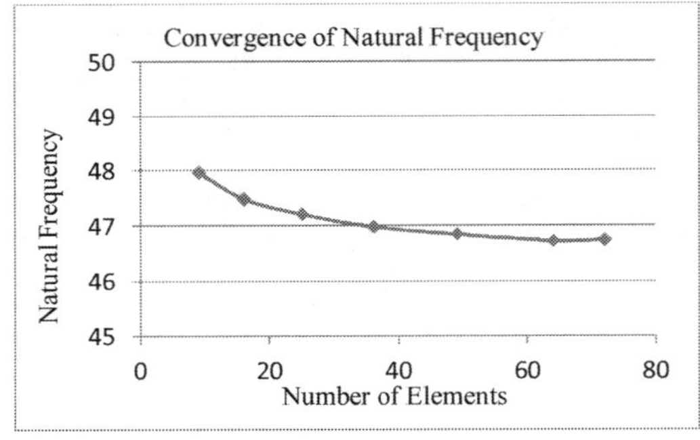

Figure 5. Natural frequency vs number of elements.

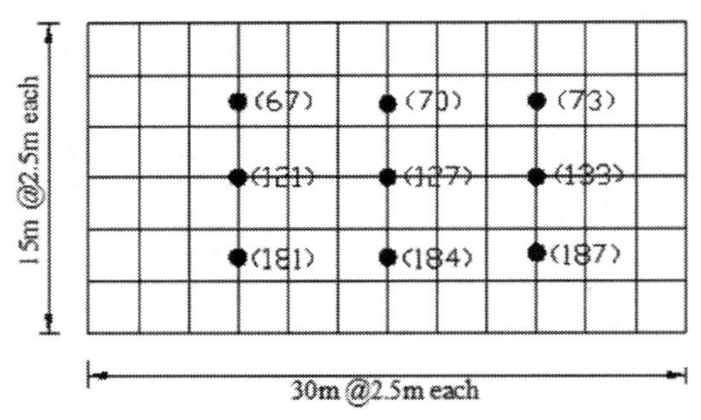

Figure 6. Plan view of pile-raft with pile position.

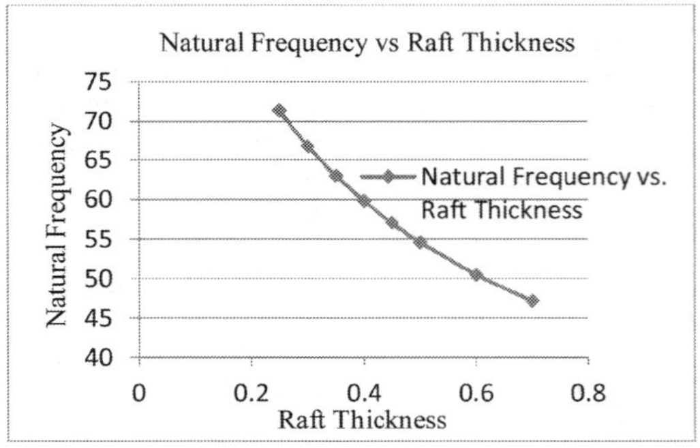

Figure 7. Natural frequency vs. raft thickness.

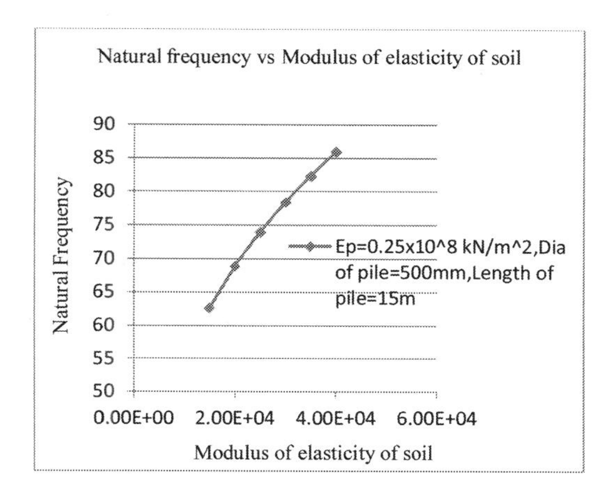

Figure 8. Natural frequency vs modulus of elasticity of soil.

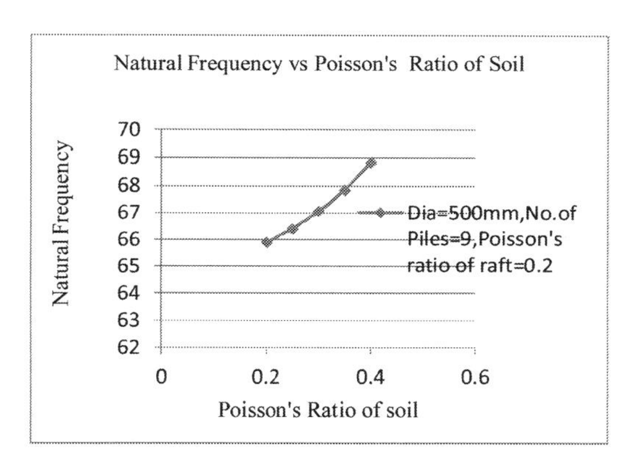

Figure 9. Natural frequency vs poisson's ratio of soil.

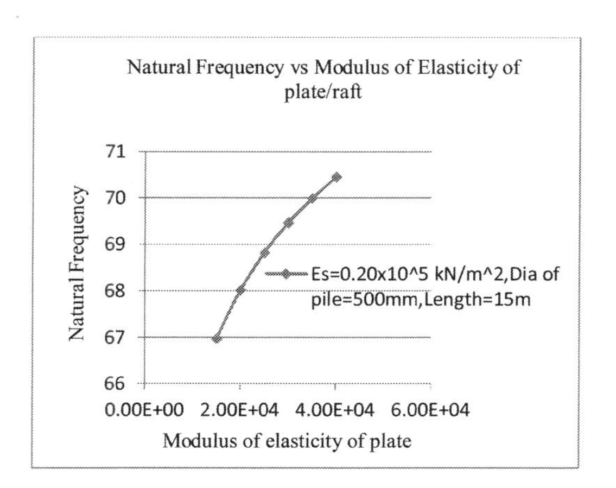

Figure 10. Natural frequency vs modulus of elasticity of pile/raft.

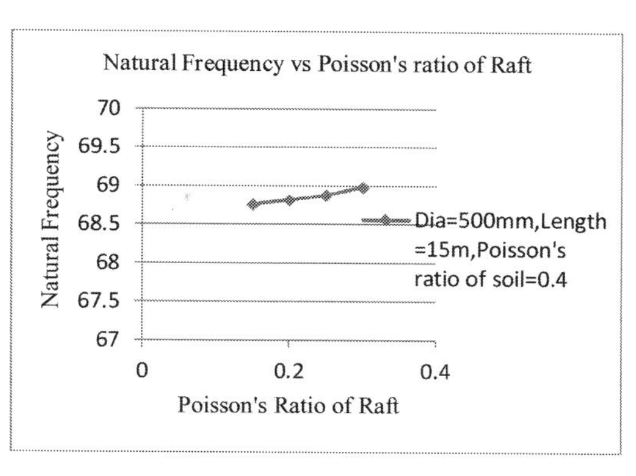

Figure 11. Natural frequency vs poisson's ratio of raft.

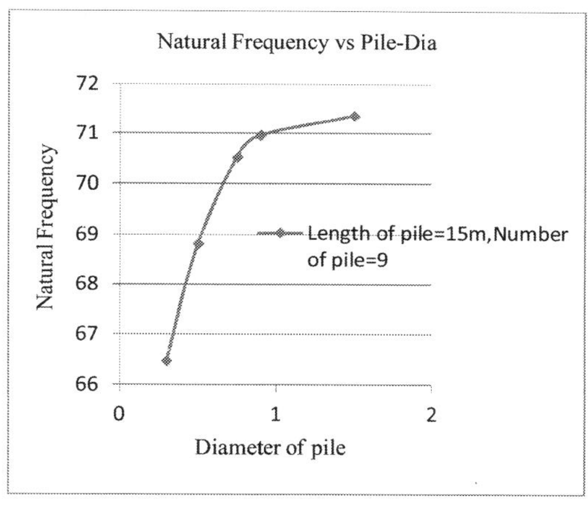

Figure 12. Natural frequency vs pile-dia.

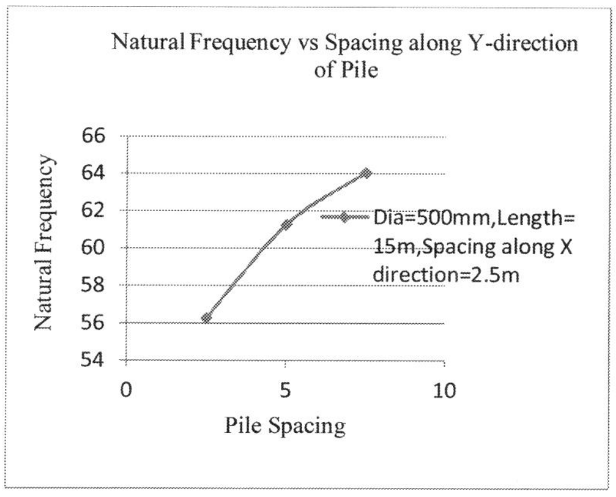

Figure 13. Natural frequency vs spacing along Y-direction of pile.

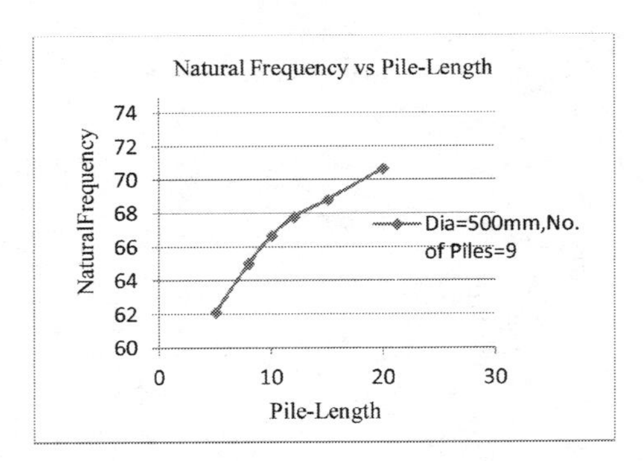

Figure 14. Natural frequency vs pile-length.

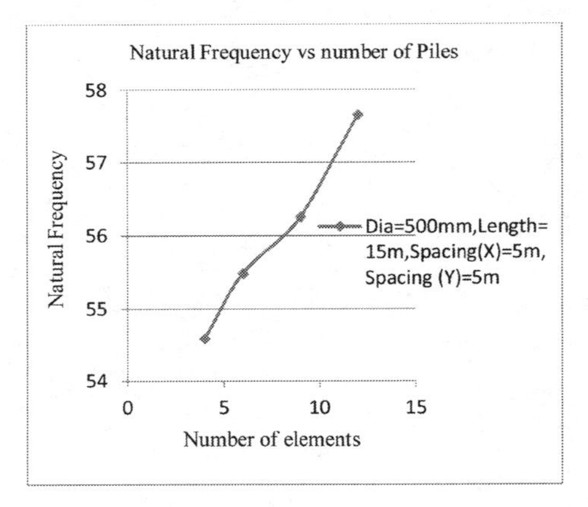

Figure 15. Natural frequency vs number of piles.

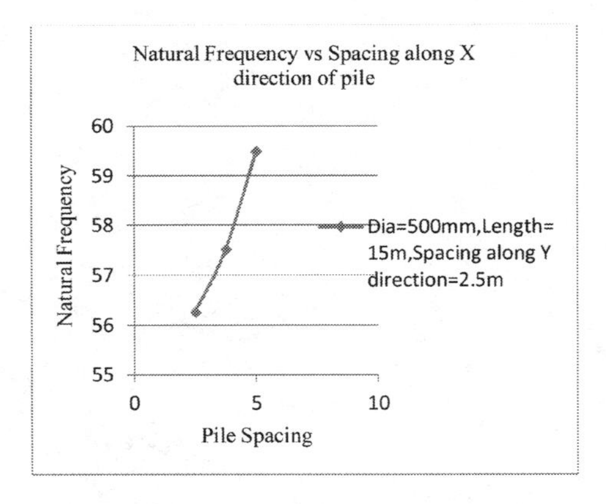

Figure 16. Natural frequency vs spacing along X-direction of pile.

5 CONCLUSION

In this present paper free vibration analysis of pile-raft foundation is carried out to find out the effect of various parameters on the natural frequency. A FORTRAN code is developed and validated for this purpose, the result of which shows a fast and good convergence. The effect of the physical property of pile, raft and soil are observed thoroughly. It has been found that with the increase of raft thickness, diameter of pile and length of the pile the system becomes more rigid and natural frequency also increases in a smooth pattern. Natural frequency increases with the increase of number of piles up to a certain number of piles depending on the size of raft and pile, after that the line becomes flat that indicates no significant increase in natural frequency. Poisson's ratio of raft, pile and soil has no significant effect on the natural frequency of pile-raft system. Hence in the analysis it can be ignored. In the present study compatibility is satisfied only in vertical direction. Thus the present study can be extended for carrying out dynamic analysis of pile-raft foundation under transient and earthquake excitation. Further study is needed considering multi soil layer and adhesion property between pile and soil to simulate the practical problem more accurately.

REFERENCES

Bhattacharya, B. 1977. Free vibration of plates on Vlasov's foundation, *Journal of Sound and Vibration* 54 (3): 464–467.

Brebia, C.A. and Georgiou. 1979. Combination of boundary and finite element for electrostatics. *Applied Math. Modeling* 3: 212–220.

Davis, E.H. and Poulos, H.G. 1972. The Analysis of Piled Raft Systems. *Aust. Geomechs. J.* G2: 21–27.

Feng, Y.T. and Owen, D.R. 1996. Iterative solution of coupled FE/BE discretisations for plate-foundation interaction problems. *International Journal for Numerical Methods in Engineering* 39: 1889–1901.

Hinton, E. and Campbell, J.S. 1974. Local and global smoothing of discontinuous finite element functions using a least square method *International Journal for Numerical Methods in Engineering* 8 (3): 461–480.

Mandal, J.J. 1998. A Coupled FE – BE Approach for Response Analysis of Raft Foundations. *Dissertation for the degree of Doctor of Philosophy, Department of Civil Engineering, Indian Institute of Technology, Kharagpur*.

Mandal, J.J. and Ghosh, D.P. 1999. Prediction of Elastic Settlement of Rectangular Raft Foundation – A Coupled FE-BE Approach. *International Journal for Numerical and Analytical Methods in Geomechanics* 23: 263–273

Mendonca, A.V. and Piava, J.B. 2003. An Elastostatic FEM/BEM Analysis of Vertically Loaded Raft and Piled Raft foundation. *Engineering Analysis with Boundary Elements* 27: 919–933.

Russo, G. 1998. Numerical Analysis of Piled Rafts. *Int. Jl. for Numerical and Analytical Methods in Geomechanics* 22 (6): 477–493.

Roy Tapabrata. 2014. Prediction of Settlement of Rectangular Piled-Raft Foundation. *M.Tech Thesis, N.I.T.T.T.R, Kolkata.*

Zienkiewicz, O.C., Kelly, D.W., Bettess.P. 1977. The coupling of finite element method and boundary solution procedure. *International Journal for Numerical Methods in Engineering* 11: 355–375.

Numerical Methods in Geotechnical Engineering IX – Cardoso et al. (Eds)
 ISBN 978-1-138-33198-3

An iterative sequential Monte Carlo filter for Bayesian calibration of DEM models

H. Cheng, S. Luding & V. Magnanimo
Multi Scale Mechanics, Faculty of Engineering Technology, University of Twente, Enschede, The Netherlands

T. Shuku
Graduate School of Environmental and Life Science, Okayama University, Okayama, Japan

K. Thoeni
Centre for Geotechnical and Materials Modelling, The University of Newcastle, Callaghan, Australia

P. Tempone
Division of Exploration and Production, Eni SpA, Milano, Lombardy, Italy

ABSTRACT: The nonlinear history-dependent macroscopic behavior of granular materials is rooted in the micromechanics at contacts and irreversible rearrangements of the microstructure. This paper presents an iterative sequential Monte Carlo filter to infer micromechanical parameters for DEM modeling of granular materials from macroscopic measurements. To demonstrate the performance of the new Bayesian filter, the stress–strain behavior of fine glass beads under oedometric compression is considered. The parameter sets are initially sampled uniformly in parameter space and then resampled around highly probable subspaces, which shrink towards optimal solutions iteratively. The proposed calibration approach is fast, efficient and automated, because it uses the posterior distribution after a completed iteration as the proposal distribution for the succeeding iteration, and thereby allocating computational power to more probable simulation runs. The Bayesian filter can also serve as a powerful tool for uncertainty quantification and propagation across various scales in multiscale simulation of granular materials.

1 INTRODUCTION

The Discrete Element Method (DEM) captures the collective behavior of a granular material by tracking the kinematics of the constituent grains (Cundall and Strack 1979). By just a few micromechanical parameters, DEM can provide comprehensive cross-scale insights (Cheng et al. 2016, Cheng et al. 2017) that are difficult to obtain in either state-of-the-art experiments or sophisticated continuum models. Nevertheless, fast and automated parameter estimation is still lacking for DEM models of granular materials against the time—or history-dependent macroscopic behavior measured in experiments. A successful calibration method should provide two key ingredients for the DEM models, namely, particle configuration and micromechanical parameters, conditioned to the granular material and experimentally measured macroscopic behavior. Each takes significant computational effort. This work aims to bring together the two ingredients for DEM modeling of granular materials within an iterative Bayesian framework.

Among the existing optimization approaches, the design of experiments (DOE) methods are efficient in searching for potential optimal solutions in the high-dimensional parameter space, with a manageable number of DEM model runs (Hanley et al. 2011, Rackl and Hanley 2017). The DOE methods are efficient when the prior knowledge of the micro–macro transitions is available. The optimizations aim to calibrate the micromechanical parameters against characteristic properties like Young's modulus, rather than time—or history-dependent macroscopic behaviors. This is very likely to hinder the predictive capacity of DEM models. Because local rather than global optimal solutions to the parameter estimation problem might be identified and adopted in the model.

The sequential data assimilation techniques can be applied to overcome the above-mentioned difficulties in the "inverse-modeling calibration" problems (Nakano et al. 2007). In our previous work (Cheng et al. 2017, Cheng et al. 2018), a novel probabilistic approach is developed using the sequential quasi-Monte Carlo (MC) to search optimal parameters in five-dimensional parameter space. However, the probabilistic approach requires a large number of model runs, which becomes very computational costly when dealing with high-dimensional parameter space. To improve the efficiency, a new iterative Bayesian filtering framework is proposed. In each iteration, the sequential MC is employed to estimate the probability density function (PDF) of the micromechanical parameters, conditioned to the stress–strain response given. The sequential MC filter uses the recursive formula of Bayesian filtering and the importance sampling scheme, which can jointly consider the effects of loading history on the elastoplastic behavior of granular materials (Shuku et al. 2012). The basic idea of the iterative Bayesian filter is to search for global optimal solutions in the parameter space with various resolutions of the posterior PDF.

A few new Bayesian calibration methods were proposed recently. Hadjidoukas (Hadjidoukas et al. 2014, Hadjidoukas et al. 2015) employed the Transitional Markov Chain Monte Carlo algorithm to identify optimal parameters and quantify the associated uncertainties for DEM simulations of silo discharge. A Bayesian model selection framework was also proposed for investigating the robustness of the contact laws, conditioned to the experimental measurements given. In the present work, the micromechanical parameters for DEM modeling of glass beads under cyclic oedometric compression are calibrated by the iterative Bayesian filter. The particle configuration is obtained from 3D feature-based image analysis. Together with the iterative Bayesian filtering framework, a fast, efficient and automated calibration procedure for image-based DEM modeling of dry granular materials is proposed. To the authors' knowledge, this work is the first attempt to closely bridge DEM simulation and experiments at both grain—and macro-scales, through image analysis and iterative Bayesian calibration.

2 AN ITERATIVE BAYESIAN FILTERING FRAMEWORK FOR FAST AND AUTOMATED PARAMETER ESTIMATION

In this section, we describe a nonlinear/non-Gaussian Bayesian filtering framework which iteratively estimates the posterior PDF of the micromechanical parameters of a DEM model, conditioned to the experimental measurements given. Either quasi-random sequences or Gaussian mixtures are used to create ensembles for the recursive Bayesian filtering (see Section 2.3). Over the consecutive iterations, the parameter space is explored from coarse to fine scales (see Section 2.4), enabling fast, efficient and automated identifications of the global optima. Fig. 1 shows the schematic workflow of the iterative Bayesian filter.

2.1 *Nonlinear non-Gaussian state space model*

Having its geometrical representation rooted in the 3DXRCT images, the dynamical model, in our case the DEM model, relies on force–displacement contact laws defined between interacting grains, to predict the macroscopic stress–strain behavior of the representative glass bead packing (termed as the physical system). Given the physical system and the dynamical model, the macroscopic measurements and numerical predictions resulting from the DEM model can be described in a nonlinear and non-Gaussian state space model (Kitagawa 1996):

$$\mathbf{x}_t = \mathbb{F}(\mathbf{x}_{t-1}) + \nu_t \tag{1}$$

$$\mathbf{y}_t = \mathbb{H}(\mathbf{x}_t) + \omega_t \tag{2}$$

where $\mathbf{x}_t$ and $\mathbf{y}_t$ are the state and measurement vectors at assimilation step t; each consists of three independent variables, namely, porosity n, mean stress p and deviatoric stress q. ν_t and ω_t are the stochastic model error and measurement error. Assuming zero means for both error terms, ν_t and ω_t are set to $N(0,\Sigma_t^m)$ and $N(0,\Sigma_t^d)$ where Σ_t^m and Σ_t^d are the covariance matrices accounting for modeling and measurement uncertainties. Note that for simplicity the system noise is neglected. $\mathbb{F}$ is the operator that represents the Lagrangian

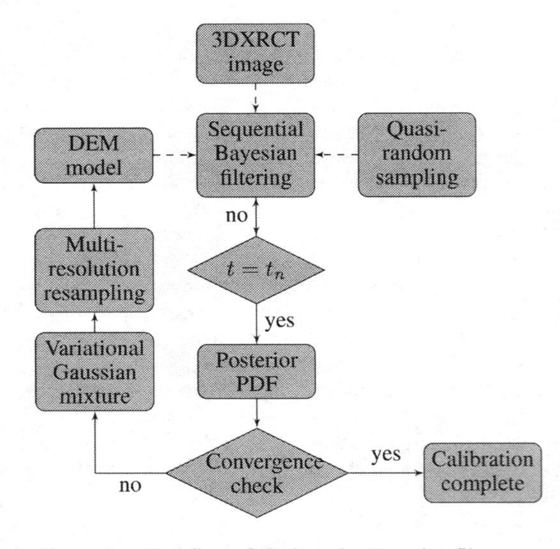

Figure 1. Workflow of the iterative Bayesian filter.

dynamical model of the granular material. The macroscopic mechanical behavior predicted by the model evolves over assimilation steps, depending on the entire loading history experienced. The measurement operator $\mathbb{H}$ is reduced to an identity matrix of size three for the sake of simplicity.

2.2 *Dynamical model*

The dynamical behavior of Eq. 1 is described by the DEM model which has four parameters to be identified, i.e., the contact-level Young's modulus E_c, interparticle friction angle μ, rolling stiffness k_m and rolling friction η_m. The geometrical representation of the DEM model is based on the particle morphology and configuration characterized by 3D feature-based image analysis. The DEM models granular materials as packings of solid particles with simplified geometries and vanishingly small interparticle overlaps. The kinematics of each particle are tracked within the explicit time integration scheme, based upon the net force resulting from the interparticle forces. Because the fine glass beads are mostly spherical, the simplified Hertz-Mindlin contact law is used to account for the nonlinearity of normal and tangential stiffnesses. To mimic the effect of surface roughness, rolling/twisting stiffness is considered in calculating the kinematics of the particles. Both interparticle tangential forces and contact moments are bounded by Coulomb type yield criteria. Over the loading history, the macroscopic state variables n_t, p_t and q_t are extracted and applied in Eq. 1.

2.3 *Sequential Monte Carlo as a standard Bayesian filter*

The solution lies in iteratively estimating and refining the posterior probability distribution of the state vector, i.e., the filtered PDF $p(\mathbf{x}_t \mid \mathbf{y}_{1:t})$, conditioned to the past measurements $\mathbf{y}_{1:t-1} = \{\mathbf{y}_1, \mathbf{y}_2, ..., y_{t-1}\}$ and the current measurement $\mathbf{y}_t$. Each iteration within the iterative Bayesian filtering framework is a complete process of the sequential MC, which recursively performs (i) one-step-ahead forecast and (ii) update at each assimilation step t. The state PDF that is propagated sequentially over t to obtain the forecast PDF $p(\mathbf{x}_t \mid \mathbf{y}_{1:t-1})$ can be expressed by the Chapman-Kolmogorov equation as $p(\mathbf{x}_t \mid \mathbf{y}_{1:t-1}) = \int_{-\infty}^{\infty} p(\mathbf{x}_t \mid \mathbf{x}_{t-1}) p(\mathbf{x}_{t-1} \mid \mathbf{y}_{1:t-1}) d\mathbf{x}_{t-1}$. While the forecasting tracks the evolution of the dynamical model from assimilation step $t-1$ to t $p(\mathbf{x}_t \mid \mathbf{x}_{t-1})$ directly in the propagation of the state PDF, the updating corrects the forecast PDF using the current measurement data $\mathbf{y}_t$ through the Bayes' theorem, resulting in the posterior PDF:

$$p(\mathbf{x}_t \mid \mathbf{y}_{1:t}) = \frac{p(\mathbf{y}_t \mid \mathbf{x}_t) p(\mathbf{x}_t \mid \mathbf{y}_{1:t-1})}{p(\mathbf{y}_t \mid \mathbf{y}_{1:t-1})} \tag{3}$$

where $p(\mathbf{y}_t \mid \mathbf{x}_t)$ is the likelihood, and the denominator $p(\mathbf{y}_t \mid \mathbf{y}_{1:t-1})$ is the evidence which involves an integral $\int p(\mathbf{y}_t \mid \mathbf{x}_t) p(\mathbf{x}_t \mid \mathbf{y}_{1:t-1}) d\mathbf{x}_t$.

A simple way of approximating the posterior, forecast and evidence PDFs for the nonlinear, non-Gaussian state space model (Eqs. 1 and 2) is through discretizing the state space with a set of deterministic model runs (referred as an ensemble), using parameters either initially sampled by quasi-random sequences or resampled from updated proposal distributions (see Section 2.4.1 and Section 2.4.2).

Taking advantage of the forecast ensemble $\{\mathbf{x}_{t|t-1}^{(1)}, \mathbf{x}_{t|t-1}^{(2)}, ..., \mathbf{x}_{t|t-1}^{(N_p)}\}$ and associated weights $\{w_{t-1}^{(1)}, w_{t-1}^{(2)}, ..., w_{t-1}^{(N_p)}\}$, the forecast PDF $p(\mathbf{x}_t \mid \mathbf{y}_{1:t-1})$ after the prediction update (Eq. 4, referred to as time update in (Chen 2003)) and the filtered posterior PDF $p(\mathbf{x}_t \mid \mathbf{y}_{1:t})$ after the measurement update (Eq. 3) become:

$$p(\mathbf{x}_t \mid \mathbf{y}_{1:t}) \approx \sum_{i=1}^{N_p} \tilde{w}_t^{(i)} w_{t-1}^{(i)} \delta(\mathbf{x}_t - \mathbf{x}_{t|t-1}^{(i)}) \tag{4}$$

where N_p is the number of realizations, $\delta(\cdot)$ is the Dirac delta function, and the superscript (i) indicates the ith realization of the forecast ensemble $\mathbf{x}_{t|t-1}^{(i)}$ and the associated weights $w_{t-1}^{(i)}$ and $w_t^{(i)}$ before and after the measurement update related by $w_t^{(i)} = \tilde{w}_t^{(i)} w_{t-1}^{(i)}$. Noted that each weight is no less than zero and the summation of them must be one, i.e., $0 \le w_{t-1}^{(i)} \le 1$ and $\sum_{i=1}^{N_p} w_{t-1}^{(i)} = 1$. The weights are updated through $\tilde{w}_t^{(i)}$, which is defined as

$$\tilde{w}_t^{(i)} = \frac{p(\mathbf{y}_t \mid \mathbf{x}_{t|t-1}^{(i)})}{\sum_j p(\mathbf{y}_t \mid \mathbf{x}_{t|t-1}^{(j)}) w_{t-1}^{(j)}} \tag{5}$$

Assuming that the measurement system is linear and the error in Eqs. 1 and 2 are independent, $p(\mathbf{y}_t \mid \mathbf{x}_{t|t-1})$ follows the multi-variable normal distribution:

$$p(\mathbf{y}_t \mid \mathbf{x}_{t|t-1}^{(i)}) = \frac{1}{(2\pi)^{m/2} \mid \Sigma_t^d \mid} \times \exp\{-\frac{[\mathbf{y}_t - \mathbf{H}_t(\mathbf{x}_{t|t-1}^{(i)})]^T {\Sigma_t^d}^{-1} [\mathbf{y}_t - \mathbf{H}_t(\mathbf{x}_{t|t-1}^{(i)})]}{2}\} \tag{6}$$

where m is the number of the independent variables in the state vector $\mathbf{x}_t, \Sigma_t^d$ is a predetermined covariance matrix for the measurement error, and $\mathbf{H}_t$ is the matrix form of the measurement operator $\mathbb{H}$.

In the iterative Bayesian filtering framework, the sequential MC filter is employed as the standard algorithm for each iteration of the Bayesian filtering. However, two different (re)sampling algorithms are implemented for the first iteration and the rest of the iterations. Our goal is to be able to explore the entire parameter space with multiple levels of resolution, such that global optimal parameter sets can identified in a fast, efficient and automated manner.

2.4 *(Re)sampling algorithms for iterative Bayesian filtering*

2.4.1 *Initial quasi-random sampling*

To ensure that no global optimal solutions are missed, quasi-random sampling is preferred in the 1st iteration, to search for potential solutions in the parameter space at a coarse scale. In fact, replacing the pseudo-random sequences in the original sequential MC filter with deterministic quasi-random (also known as low discrepancy) sequences, leads to the sequential quasi-MC (Halton 1994). It has recently been proved in (Gerber and Chopin 2015) that the sequential quasi-MC produces more accurate estimates with smaller errors than its original counterpart, and the error rate of the former is smaller than that of the latter. Nevertheless, using more homogeneous low-discrepancy sequences does not alleviate the *weight degeneracy* problem, namely, only few of the weights will because nonzero after a few assimilation steps. This is disadvantageous because enormous computational effort is wasted on the realizations that produce negligible weights. A possible way to alleviate the weight degeneracy problem is to perform resampling using more sensible proposal distributions.

2.4.2 *Resampling using updated proposal distributions*

In this work, the ensemble for the current kth iteration $(k > 1)$ of the Bayesian filtering is re-initialized using parameters resampled from the proposal distribution that is updated from the previous $(k-1)$ th iteration. The posterior PDF $p^{k-1}(\mathbf{x}_{t_n} \mid \mathbf{y}_{1:t_n})$, obtained at the final assimilation step t_n of the previous iteration, is employed as the proposal distribution $p^k(\mathbf{x}_0)$ to perform the current resampling. This is advantageous in three aspects:

1. Unknown proposal distributions can be iteratively updated from the *discrete* posterior probabilities and approximated by mixture models (e.g. Gaussian), conditioned to the same experimental measurements.
2. The iterative resampling algorithm keeps allocating realizations in the parameter subspaces where the posterior probabilities are more probable to be higher than elsewhere, and thus allows a coarse-to-fine search in the parameter space.
3. Resampling from an updated proposal distribution only takes place before each sequential MC, which makes it possible to prevent the weight degeneracy problem, while keeping the dynamical balance intact within the sequential MC.

Choosing an appropriate proposal distribution is pivotal to the efficiency of any sequential MC filters. Here the complex-shaped posterior probabilities calculated by Eq. 4 are approximated by a mixture of Gaussian distributions. The number of the components, and their means and covariance matrices are estimated by variational Bayesian inference using the Dirichlet process (DP) as a nonparametric prior (Blei and Jordan 2004). The basic idea is to treat the estimation of a posterior PDF from given *discrete* samples as an optimization problem. The variational inference method applied here is an extension of the expectation-maximization, which iterates for maximizing the lower bound on the likelihood. The DP mixture is approximated in the stick-breaking representation: a truncated distribution with a fixed maximum number of components. Unlike the Gaussian mixture model which depends on the number of mixture components provided as the input, the DP Gaussian mixture model tends to automatically choose the smallest number of components that can effectively represent the complex-shaped distribution. Interested readers are referred to (Blei and Jordan 2004) for the theoretical background and technical details.

3 APPLICATION TO DEM MODELING OF GLASS BEADS UNDER OEDOMETRIC COMPRESSION

From the **a priori** particle size distribution and packing configuration, deterministic DEM simulations of glass beads under cyclic oedometric compression (i.e., model runs) are managed and tracked by the (re)sampling (Section 2.4) and forecast-updating (Section 2.3) algorithms. The **a posteriori** micromechanical parameters, E_c, μ, k_m and η_m and their uncertainties are evaluated by iteratively applying the sequential MC filter to the dynamical model, conditioned to the loading path-dependent experimental measurements. Note that although the geometrical representation of the DEM model is facilitated by importing the particle configuration directly from the 3DXRCT images, the DEM packing needs to undergo a computationally less expensive "dynamic packing relaxation" stage, in which the particle positions are adjusted locally in order to recover the periodicity of the representative DEM packing.

3.1 *Identified micromechanical parameters*

As the oedometric compression proceeds simultaneously in all DEM model runs that belong to the kth iteration, the posterior PDF $p^k(\mathbf{x}_t \mid \mathbf{y}_{1:t})$ is approximated by recursively applying the sequential MC filter to the state vector, represented by the ensemble[1] $\mathbf{x}_t^{(i)} = \{n_t^{(i)}, p_t^{(i)}, q_t^{(i)}\}, i = 1,\ldots,N_p$ at assimilation step t, and the measurement vector $\mathbf{y}_{1:t}$ available until t. To track the evolutions of the optimal parameters from the posterior PDF in the four-dimensional parameter space, the weighted mean values at assimilation step t are calculated by

$$\bar{\mathbf{\Phi}}_t = \sum_{i=1}^{N_p} w_t^{(i)} \mathbf{\Phi}^{(i)} \tag{7}$$

where $\mathbf{\Phi}^{(i)}$ contains the four micromechanical parameters, i.e., $E_c^{(i)}, \mu^{(i)}, k_m^{(i)}$ and $\eta_m^{(i)}$, associated with the (i)th realization, and $\bar{\mathbf{\Phi}}_t$ contains the weighted means $\bar{E}_c(t), \bar{\mu}(t), \bar{k}_m(t)$ and $\bar{\eta}_m(t)$ for the four parameters updated until the assimilation step t. Note that the weighted means $\bar{\mathbf{\Phi}}_t$ are termed as *identified micromechanical parameters* instead of the specific parameter sets that have high posterior probabilities.

In this work, the parameter estimation for DEM modeling of the glass beads (d = 40–80 μm) under cyclic oedometric compression is chosen as an example to demonstrate the performance of the iterative Bayesian filter. Four iterations of nonlinear non-Gaussian Bayesian filtering are conducted, which eventually lead to accurate identification of the optimal parameters and the associated posterior PDF in the four-dimensional parameter space. In each iteration, Eq. 7 is applied to the ensemble and the associated weights estimated at every assimilation step. 100 DEM model runs (N_p = 100) are performed during each iteration to update the ensemble in accordance with the measurements. The corresponding parameter sets needed for the model runs are (re)sampled differently depending on the initial or updated proposal distribution (see Section 2.4). The weighted-averaging scheme gives the respective evolutions of the identified parameters against axial strain ε_a during the four iterations of the Bayesian filtering in Fig. 2. The number of DEM model runs stays the same over iterations for the sake of consistency. Note that the evolutions of the identified parameters are plotted as functions of axial strains ε_a, which is used to control the quasi-static loading in the simulations and corresponds to assimilation steps t.

[1]The iteration superscript k is omitted for simplicity. However, one should note that the equations applies to the state vector obtained in the kth iteration.

Based on the feasible ranges of the micromechanical parameters obtained via sensitivity analyses (see Table 1), a four-dimensional Halton sequence (De Rainville et al. 2012) is generated, and employed as the initial quasi-random MC samples for the first Bayesian filtering. During the first sequential MC (red line), it is clear in Figs. 2a–2d that all identified parameters E_c, μ, k_m *and* η_m and keep adjusting until the beginning of the unloading-loading cycles. Interestingly, the evolutions of the identified parameters do not change much during the unloading-loading cycles, except for μ. This is in line with the fact the interparticle friction plays a significant role in the plastic deformation of granular materials. The data assimilation steps during the unloading-loading cycles allow the posterior PDF to undergo more prediction and measurement updates (Eq. 4) which are relevant to the elastoplastic behavior of the glass beads.

Similarly, the evolution of k_m during the unloading-loading cycle of the second iteration (green line in Fig. 2c) suggests that the identified parameters can adjust themselves as more experimental measurements become available through the sequential MC. It has been known that k_m is another key parameter that controls the shear strength of granular materials (Cheng et al. 2017). Nevertheless, the adjustment of k_m during the unloading-loading cycles in the first iteration only occurs locally (red line in Fig. 2c), unlike that in the second iteration. The reason for this is that only potential optimal solutions are investigated, which means the resolution in the first iteration is too coarse to identify any optimal values for k_m. Because the DP Gaussian mixture model is capable of constructing a smooth and continuous proposal distribution over separate sample points that have high importance weights, potential optimal solutions can be efficiently searched near these samples in the next iterations of the Bayesian filtering. Using the proposal distribution roughly estimated in the first iteration, the sequential MC can identify more probable values for k_m with better resolution in the second iteration. The refinement of the other identified parameters during the loading history can also observed in Figs. 2a, 2b and 2d. Note that the "optimal" parameters identified at the end of the first iteration are almost identical to the identified parameters at the beginning of

Table 1. Initial guess of parameter ranges for the first iteration of the Bayesian filtering.

1–5	E_c (GPa)	μ	k_m (10^{-3} N·mm)	η_m
Φ_{max}	200	0.5	0	0.1
Φ_{min}	100	0.3	10	0.5

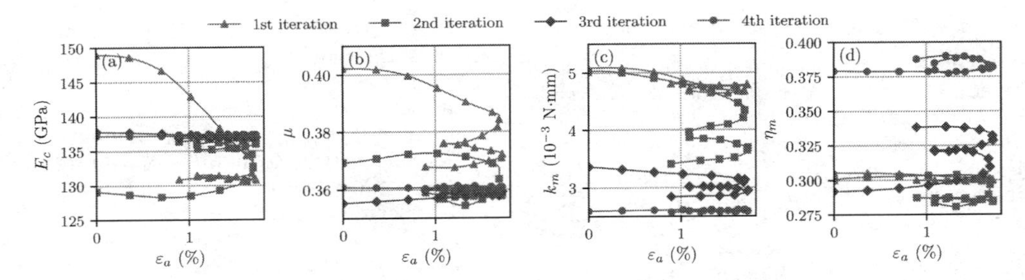

Figure 2. Evolution of identified parameters over the history of axial strain ε_a during different iteration of the filtering.

the second iteration (red and green lines in Fig. 2). This means that the prior knowledge updated from the previous iteration has been successfully passed to the following iteration without spoiling the weighted averages, through the resampling step enabled by the DP Gaussian mixture (see Section 2.4.2). Nevertheless, the parameter spaces investigated in these two iterations are different.

The transfer of the previously identified parameters to the following Bayesian filtering can also be found among other iterations. While the readjustment of E_c and μ becomes hardly noticeable in the third and fourth iterations (blue and purple lines in Figs. 2a and 2b), the identified values for k_m and η_m are still changing, particularly during the unloading-loading cycles in the third iteration (blue lines in Figs. 2c and 2d). In the fourth iteration, only the rolling friction is readjusted locally during the unloading-loading cycles. The fast and slow convergences in different plots of Fig. 2 indicate that E_c and μ, which have clear physical meanings, are easier to calibrate, whereas the non-physical parameters k_m and η_m requires more iterations to find optimal values. In this work, we choose to stop the iterative parameter estimation after a total number of four iterations, because of the marginal difference between the posterior PDFs estimated at the beginning and the end of the last iteration (see Section 3.3).

3.2 *Numerical predictions versus experimental data*

To understand how the quality of the identified parameters is improved over iterations, the numerical predictions of the macroscopic variables, n, p and q are compared with the experimental data in Fig. 3. The DEM simulation results therein are reproduced using the parameter sets associated to the top four highest posterior probabilities, which are estimated by the third sequential MC. The stress–strain relationships predicted using the four parameter sets identified after the first iteration, cannot give good agreement for both the $\varepsilon_a - q/p$ and $p - n$ responses. As the sequential MC iterates with the ensemble being updated by the new proposal distribution, the agreement between the numerical prediction and experimental data becomes increasingly high as shown in Fig. 3. After the three iterations, the numerical results predicted by the DEM model lie perfectly on top of the experimental measurements, which proves that the posterior PDF estimated after the third iteration is accurate enough.

3.3 *Evolution of posterior probability distribution over iterations*

Fig. 4 compares two snapshots of the posterior PDF estimated at the beginning and the end of the first iteration of the Bayesian filtering. While the diagonal panels indicate the marginals of the joint

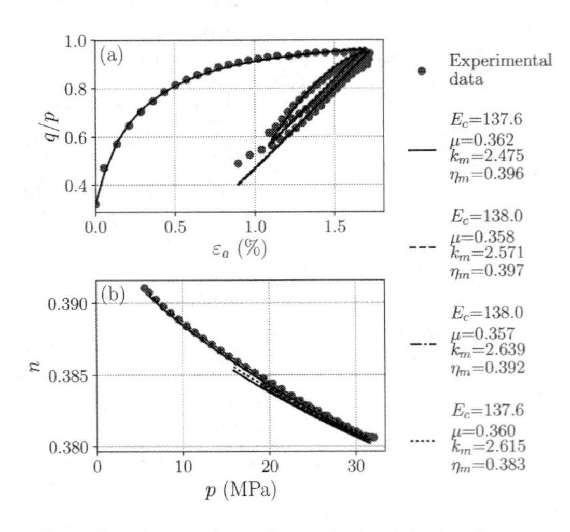

Figure 3. Comparison of experimental data and numerical results predicted using the parameter sets associated to the top four highest posterior probabilities at the third iteration.

posterior PDF, the above—and below-diagonal panels present the 2D projections of the posterior PDF at the beginning (blue) and the end (red) of the filtering. The projections in the above—and below-diagonal 2D parameter spaces are colored by the associated posterior probability densities. By discretizing the parameter space with the four-dimensional low-discrepancy sequence, the parameter space is explored uniformly at the coarse scale. After the prediction and measurement updates are sequentially performed over the oedometric loading path, the 2D projections of the posterior PDF obtained at the end of the first iteration appear to be multimodal. Each mode of the 2D distributions suggests a highly probable parameter subspace in which a global optimal solution can be potentially located. The resampling step for the second iteration is performed via the four-dimensional multi-variable proposal distribution, constructed with the posterior PDF shown in the below-diagonal panels of Fig. 4. Note the above-diagonal panels in Figs. 5–6 show the posterior PDFs already processed through one-step Bayesian filtering, and thus may look slightly different from their proposal distributions.

From the proposal distribution roughly estimated at the coarse scale, more samples are put into the highly probable parameter subspaces, as illustrated by the posterior PDF in the above-diagonal panels of Fig. 5. Note that the 2D projections of the posterior distribution in the above-diagonal panels of Fig. 5 are less bumpy than those in the below-diagonal panels of Fig. 5. This again confirms that the DP Gaussian mixture model allows the resampling to take place between the highly probable parameter subspaces, which merges the separated modes in the below-diagonal panels of Fig. 4 into a complex-shaped smoother distribution as shown in the above-diagonal panels of Fig. 5. Although the two posterior PDFs are estimated with two different ensembles at the two temporally-separated assimilation steps, the distributions are linked with the same weighted means $\bar{\Phi}$, as suggested by the red and green lines in Fig. 2. As the sequential MC proceeds, a better description of the posterior PDF in the highly probable parameter subspace is obtained. The resulting posterior PDF is again multimodal, which is due to the discrete nature of the sequential MC filter. Comparing the posterior distributions obtained at the beginning (blue) and the end (red) of the second iteration in Fig. 5, it can be seen that the modes are not shifted as much as in the first iteration. The primary mode of the proposal distribution is simply refined into multiple modes that are located within the same parameter subspace. The readjustment of weights is also present in the evolutions of the identified parameters (green lines in Fig. 2). From the evolutions of the identified parameters and posterior PDF in the first and second iterations, it can be concluded that while the initial sequential quasi-MC tells which parameter subspaces to look into, the subsequent sequential MC plays a key role in identifying the optimal solutions.

Following the same resampling strategy, the ensemble for the third sequential MC filter is re-initialized using a smooth and simple multimodal proposal distribution. Note that the parameter

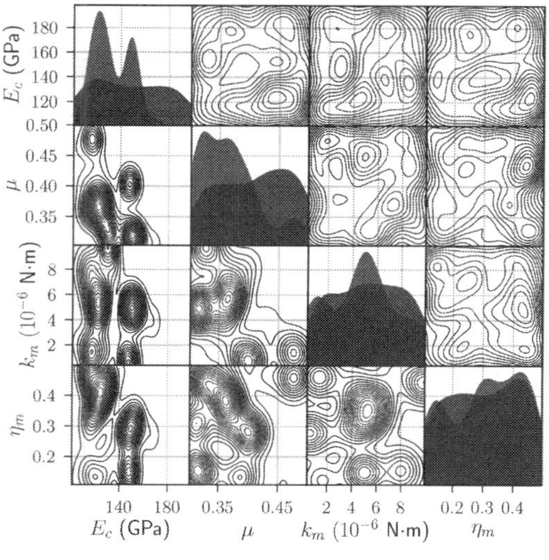

Figure 4. Posterior PDF estimated at the beginning (blue) and the end (red) of the first iteration of the Bayesian filtering.

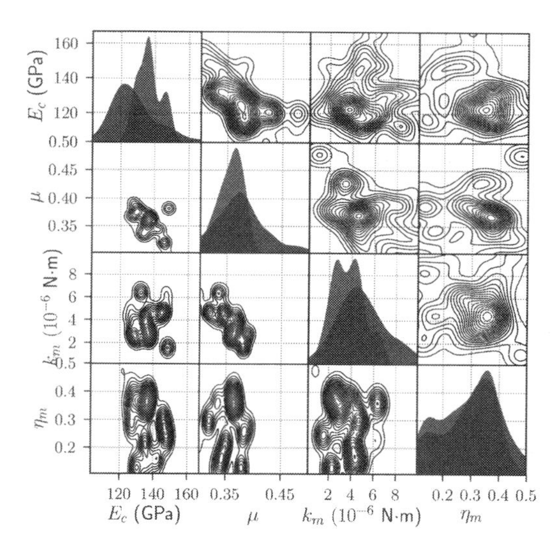

Figure 5. Posterior PDF estimated at the beginning (blue) and the end (red) of the second iteration of the Bayesian filtering.

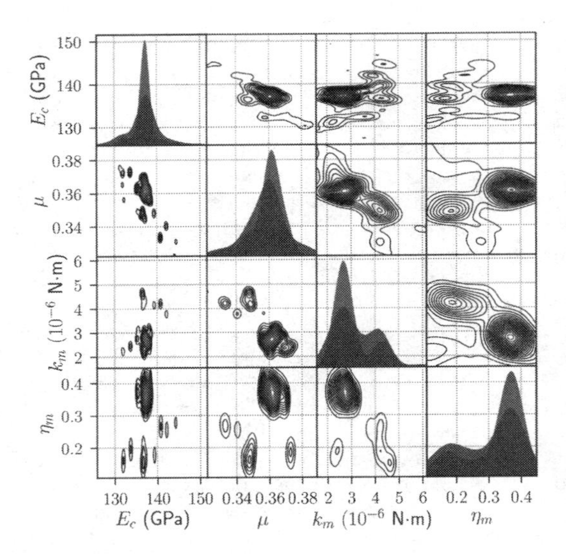

Figure 6. Posterior PDF estimated at the beginning (blue) and the end (red) of the third iteration of the Bayesian filtering.

subspace explored in the third iteration (Fig. 6) is much smaller than the initially sampled parameter space in the first iteration (Fig. 4). At this fine scale, the posterior PDF remains to be multimodal consistently, throughout the entire sequence of the sequential MC in the third iteration. Similar to the evolution of the modes in the second iteration, the primary and secondary modes at various assimilation steps lie in the same parameter subspace, as shown in the above—and below-diagonal panels of Fig. 5. While the importance weights close to the primary mode are growing, low importance weights are assigned to the realizations near the secondary modes. The subtle changes of the posterior PDF during the third iteration are caused by the readjustment of weights associated to the two non-physical parameters k_m and η_m, as can be observed from the shrinking 2D projections except for those on the E_c–μ plane. The readjustment, particularly during the unloading-loading cycles, can also be understood from the evolutions of k_m and η_m (blue lines in Fig. 2c and d). It appears that the non-physical parameters k_m and η_m are more difficult to calibrate than E_c and μ which have clearer physical meanings.

4 CONCLUDING REMARKS

A fast, efficient and automated calibration procedure for DEM modeling of dense granular materials is proposed, which aims at fast packing generation and parameter estimation, through 3D feature-based image processing and iterative approximation of the posterior PDF of the parameters, conditioned to a given loading history. The previously developed probabilistic calibration framework has been extended to enable a coarse-to-fine search for the optimal solutions in large parameter space, by introducing resampling steps between the iteratively performed nonlinear non-Gaussian Bayesian filtering. Within each iteration, the sequential Monte Carlo filter is applied to the ensemble of randomized numerical predictions and the measurements, obtained sequentially from the beginning until the end of the loading history. The resampling takes advantage of the Dirichlet process mixture models to connect highly probable parameter subspaces with resampled parameter sets. Over iterations, the resampling scheme allows the filter to zoom into these parameter subspaces to characterize the posterior PDF with greater detail. The convergence of the posterior PDF and its weighted means are adopted as the stopping criterion of the iterative assimilation process.

The major advantage of implementing the iterative version of the sequential Monte Carlo filter is that the prior knowledge, which is difficult to acquire or assume, can now be learned through iterations of Bayesian filtering and resampling. The computational power is thus allocated on the DEM model runs conducted with the parameter sets that potentially lead to high posterior probability. The efficiency of the data-driven Bayesian calibration is thereby improved, as the sequential Monte Carlo repeats with consistent weighted means of the posterior PDF. It enables fast, efficient and automated calibration of micromechanical parameters, which can be handled by a Python script with the open-source DEM package YADE. The current implementation is also easily parallelizable, because of the "off-line" estimation of the posterior PDF. Although the "off-line" processing is simple to implement, it would save even more computation cost by incorporating the resampling and convergence check during the sequential Monte Carlo, which would need an "on-line" processing framework for simulation control, data handling, etc. In addition to the implementation of "on-line" processing, future work will involve the uncertainty quantification and propagation across the grain scale and the macroscale in multiscale simulations of granular materials.

ACKNOWLEDGMENTS

This work was financially supported by Eni S.p.A.

REFERENCES

Blei, D.M. & M.I. Jordan (2004). Variational inference for Dirichlet process mixture models Mean field variational inference. *Bayesian Anal.* (1), 1–9.

Chen, Z.H.E. (2003). Bayesian filtering: from Kalman filters to particle Filters, and beyond. *Statistics (Ber). 182*(1), 1–69.

Cheng, H., T. Shuku, K. Thoeni, & H. Yamamoto (2017). Calibration of micromechanical parameters for DEM simulations by using the particle filter. *EPJ Web Conf. 140*, 12011.

Cheng, H., T. Shuku, K. Thoeni, & H. Yamamoto (2018). Probabilistic calibration of discrete element simulations using the sequential quasi-Monte Carlo filter. *Granul. Matter 20*(1), 11.

Cheng, H., H. Yamamoto, & K. Thoeni (2016). Numerical study on stress states and fabric anisotropies in soilbags using the DEM. *Computers and Geotechnics 76,* 170–183.

Cheng, H., H. Yamamoto, K. Thoeni, & Y. Wu (2017). An analytical solution for geotextile-wrapped soil based on insights from DEM analysis. *Geotext. Geomembranes.*

Cundall, P.A. & O.D.L. Strack (1979). A discrete numerical model for granular assemblies. *Géotechnique 29*(1), 47–65.

De Rainville, F.-M., C. Gagné, O. Teytaud, & D. Laurendeau (2012). Evolutionary optimization of lowdiscrepancy sequences. *ACM Trans. Model. Comput. Simul. 22*(2), 9:1–9:25.

Gerber, M. & N. Chopin (2015). Sequential quasi Monte Carlo. *J.R. Stat. Soc. Ser. B Stat. Methodol. 77*(3), 509–579.

Hadjidoukas, P., P. Angelikopoulos, C. Papadimitriou, & P. Koumoutsakos (2015). _4U: A high performance computing framework for Bayesian uncertainty quantification of complex models. *J. Comput. Phys. 284*, 1–21.

Hadjidoukas, P., P. Angelikopoulos, D. Rossinelli, D. Alexeev, C. Papadimitriou, & P. Koumoutsakos (2014). Bayesian uncertainty quantification and propagation for discrete element simulations of granular materials. *Computer Methods in Applied Mechanics and Engineering 282*, 218–238.

Halton, J.H. (1994). Sequential monte carlo techniques for the solution of linear systems. *J. Sci. Comput. 9*(2), 213–257.

Hanley, K.J., C. O'Sullivan, J.C. Oliveira, K. Cronin, & E.P. Byrne (2011). Application of Taguchi methods to DEM calibration of bonded agglomerates. *Powder Technology 210*(3), 230–240.

Kitagawa, G. (1996). Monte Carlo filter and smoother for non-Gaussian nonlinear state space models. *Journal of Computational and Graphical Statistics 5*(1), 1–25.

Nakano, S., G. Ueno, & T. Higuchi (2007). Merging particle filter for sequential data assimilation. Nonlinear Processes in Geophysics 14, 395–408.

Rackl, M. & K.J. Hanley (2017). A methodical calibration procedure for discrete element models. *Powder Technol. 307*, 73–83.

Shuku, T., A. Murakami, S.-i. Nishimura, K. Fujisawa, & K. Nakamura (2012). Parameter identification for camclay model in partial loading model tests using the particle filter. *Soils and Foundations 52*(2), 279–298.

Numerical Methods in Geotechnical Engineering IX – Cardoso et al. (Eds)
© 2018 Taylor & Francis Group, London, ISBN 978-1-138-33198-3

Particle-based modelling of cortical meshes for soil retaining applications

F. Gabrieli & A. Pol
Department of ICEA, University of Padova, Italy

K. Thoeni
Centre for Geotechnical Science and Engineering, University of Newcastle, Callaghan, NSW, Australia

N. Mazzon
Maccaferri Innovation Center, M.I.C., Bolzano—Bozen (BZ), Italy

ABSTRACT: Metallic cortical wire meshes are extensively used to protect slopes and infrastructures in mountain regions. These structures are constituted by periodic patterns of steel wires and have a ductile non-linear behaviour. The installed mesh panels generally experience non-trivial boundary conditions which combined with their high deformability and the chance of local ruptures make these structures difficult to be modelled as a continuum. The Discrete Element Method (DEM) is particularly well suited for modelling the soil-mesh interaction problem, taking advantage of its potentials in handling large deformation, contact-detachment and interlocking between particles and the mesh. The mesh is modelled with two approaches and a puncture test on a single mesh panel is used for calibration. After that a simplified soil-mesh problem with a single mesh panel is considered and compared with the results of a punch-mesh problem. The force-displacement behaviour experienced by anchors and edges of the panels are also discussed.

1 INTRODUCTION

Cortical wire meshes are common protection structures used for the safety of infrastructures, buildings, and people near natural or artificial cliffs and slopes. Differently from the passive net barriers which mostly work as "energy dampers", this kind of meshes actively wrap the cliff surface, hence they prevent the detachment and the slipping of fragments and rocky blocks along the slope. For these reasons, such kind of deformable structures act simultaneously as passive and active structures and they can be classified in between both categories (Flum et al. 2004, Bertolo et al. 2006). Their use was recently extended for the retaining of banks of granular slopes and in soil nailing applications. In some of these problems the meshes are also coupled with geotextiles to prevent the finer particles to pass through or with shotcrete to increase the stiffness of the external face.

In conventional applications, these meshes are draped from the top of the slope and anchored in some points to the underneath firm soil/rock layer by means of rod bars and plates (Ferraiolo & Giacchetti 2005).

Their numerical modelling is of interest, not only for the improvement of the design of such in-situ civil engineering applications but also for research and development of new products. In this context, the use of a reliable numerical model may help in finding the best geometrical and physical properties of the mesh and of the wires that maximize their mechanical performances.

In engineering practice, the wire meshes are considered as membranes for revetment. They are then modelled as deformable plate elements. Recently, more refined finite element numerical models able to describe the meshes as ordered structures of trusses (supporting only tensile stresses), beams or other elements have been developed (Cazzani & Frenez 2002, Castro-Fresno et al. 2008). Nevertheless, their coupling with the retained materials (soil, gravel or rocks) actually represents a big challenge.

Other numerical methods such as the Discrete Element Method (DEM) appears more suitable for this purpose. The mesh can be described wire by wire as remote interactions or cylindrical bodies (Albaba et al. 2017, Gabrieli et al. 2017) while the granular material behind the mesh can be reproduced with particles of different shape and size. The handling of large deformations, detachments, local failures as well as the use of different shaped bodies (e.g. spheres, cylinders, polyhedra) are some of the key potentials of this method.

In this work, two different DEM approaches for the modelling of the mesh are considered focusing on the analysis of a simplified soil-mesh interaction problem. An experimental punch test was used for the calibration of the mesh model.

2 DEM MODELLING OF WIRE MESHES

There are at least three approaches in the literature suitable for the discrete element modelling of wire meshes. The first one can be titled "cell-based" (*CB*) as it identifies the periodic elementary cell that rules the behaviour of the mesh and it expresses its relationships with the neighbouring cells (Nicot et al. 2001). The second method is called "node-wire-based" (*NWB*) and it substitutes the nodes of the mesh (intersection points between the wires) with spherical particles having fictitious remote interactions superimposed to the wires (Bertrand et al. 2008, Thoeni et al. 2013). The third method can be called "cylinder-wire-based" (*CWB*) as it replaces the wires with series of cylindrical bodies having a custom tensile behaviour and physical contact interaction with any other bodies along their lateral surface (Effeindzourou et al. 2016). The use of the latter method was proved to be effective with the partial drawback of a moderate higher computational cost and simulation time (Gabrieli et al. 2017). In this work, the last two approaches (i.e. both *NWB* and *CWB* methods) will be compared with reference to the hexagonal double-twisted steel wire mesh with wire diameter equal to 2.7 mm.

The first approach considers the replacement of nodes with spheres having a radius equal to 4 times the wire diameter and the density scaled in order to keep the same aerial mass of the wire mesh (see Fig. 1a). Single wires and double-twisted wires are represented by distinct remote interactions having custom tensile mechanical behaviour. It should be noted that in this case only the nodes of the mesh can interact with the other external bodies (e.g. spheres, walls) while remote interactions are practically invisible to any other bodies. In the second approach, wires are described as pieces in longitudinal direction deformable cylinders of a diameter of 2.7 mm and these cylinders are connected at the nodes (see Fig. 1b). Bending and torsion of cylinders are neglected and only the tensile properties are considered for direct comparison with the first approach. The same tensile mechanical behaviour as used for the node-based approach has been applied here. Clearly, different contact detection and contact mechanics are implemented in this approach (Effeindzourou et al., 2016) taking into

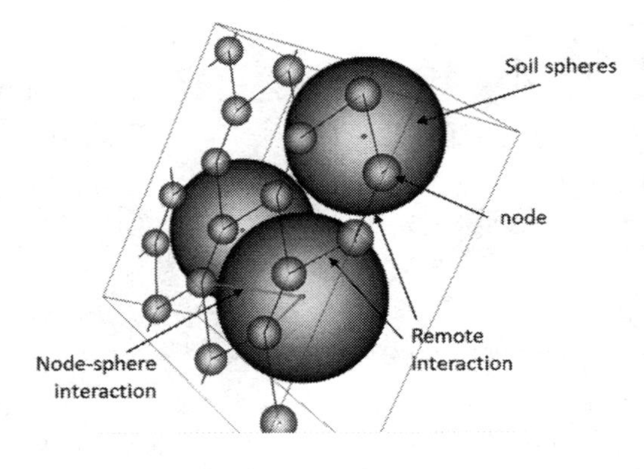

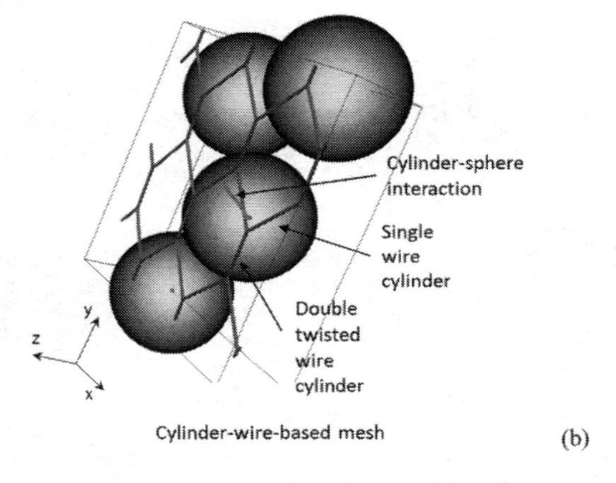

Figure 1. (a) Node-wire-based (*NWB*) and (b) cylinder-wire-based (*CWB*) representations of the soil-mesh interaction problem.

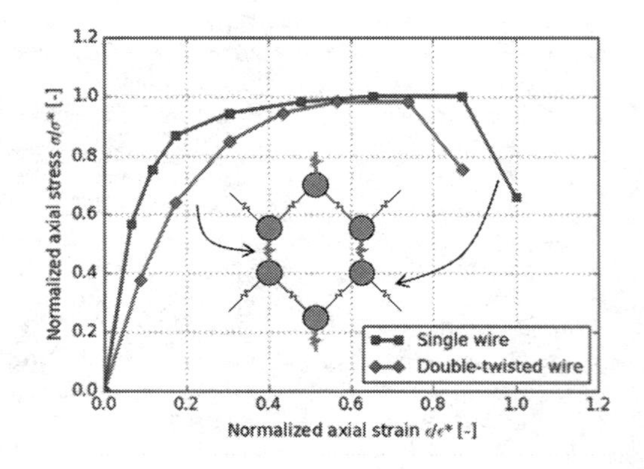

Figure 2. Tensile behavior of the two wire types of the hexagonal double-twisted steel wire mesh with wire diameter 2.7 mm.

account the chance of cylinder-sphere, cylinder-plane and cylinder-cylinder contacts.

A different tensile mechanical behaviour has been attributed to single wire and double-twisted wire elements (see Fig. 2). The tensile constitutive models have been extracted from experimental tensile tests and implemented as piecewise functions (Thoeni et al. 2013). Experimentally, wires exhibit a moderate stiff response up to a tensile strain equal to 4% for the single and equal to 7% for the double-twisted wires and then a ductile-plastic behaviour with an important tensile elongation before failure.

The different behaviour of single wire and double-twisted wire is also reported by Bertrand et al. (2008) and it is ascribable to the different industrial building processes.

3 PUNCH TEST

Punch tests are simple laboratory tests used to compare mesh performance in terms of force-displacement curve of the mesh panel for quasi-static out-of-plane loading conditions. The method is described in UNI 11437-2012. In the experimental test, the mesh is fixed at the edges of a steel frame and loaded by lifting a punching element, in the mesh out-of-plane direction, until the mesh panel fails. The punching element is a cup-shape disk with a diameter of 1 m, smoothed at the edges and with a constant curvature radius equal to 1.2 m. The mesh panel sample is squared and each side has a nominal dimension of $3 \pm 0.2 \times 3$ m. The displacement and force experienced by the punching element are continuously recorded during the test. In Figure 3, a scheme of the punch test geometry taken from the initial instant and the failure instants of the numerical test is reported.

For calibration and validation of the numerical model of the double-twisted hexagonal wire mesh, the data of an experimental punch test are used. A comparison between numerical and experimental results is reported in Figure 4. The results obtained

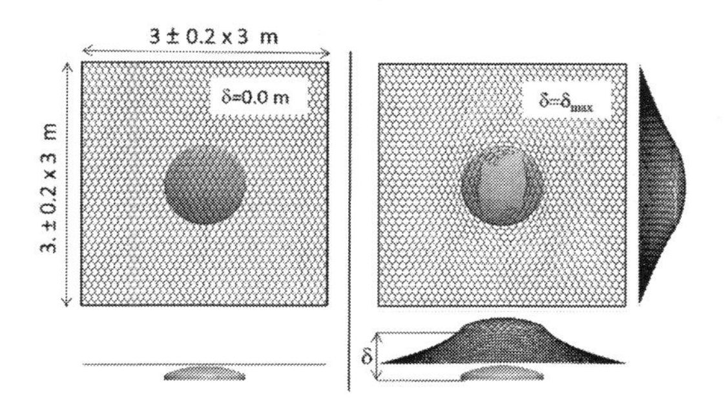

Figure 3. Basic layout of the punch test problem.

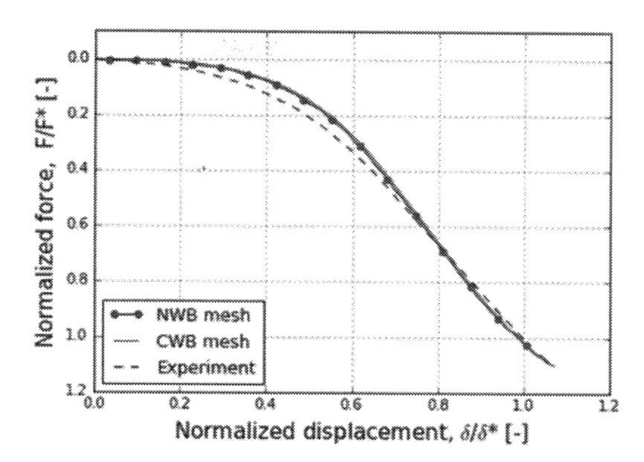

Figure 4. Force-displacement curve of the experimental punch test and the numerical predictions with the node-wire-based (*NWB*) and the cylinder-wire-based (*CWB*) modelling approaches.

with the numerical simulations are in very good agreement with the laboratory data.

With regards to the two DEM approaches, both node-wire-based and cylinder-wire-based methods report the same behaviour with a negligible difference probably caused by some slight differences in contact mechanics (position and direction of contacts). More detailed information about the above mentioned test can be found in Gabrieli et al. (2017).

4 DEM MODELLING OF THE SOIL

The soil considered for the numerical study of the soil-mesh interaction problem has a mean grain size diameter of $d_{50} = 0.12$ m. This grain size has been chosen so that it is larger than the maximum opening size of the mesh (*mos* = 0.08 m). Hence, particles cannot easily fall through the mesh. The other physical, geometrical and mechanical parameters of the soil are reported in Table 1.

The corresponding macroscopic peak and residual friction angles, have been derived from numerical cubical triaxial tests performed at different confinement stresses (σ'_3 = 100, 150, 200, 250, 300 kPa) on periodic granular samples having an initial porosity equal to $n = 0.405$. In Figure 5, the ultimate stress loci and the loading paths are depicted in the mean vs. deviatoric stress plane (p'-q).

Preliminary analyses have shown that the influence of the soil friction angle is not very important in this application. However, further in-depth analysis has to be planned for investigating the effect of all the soil parameters.

Table 1. Micromechanical parameters of the soil.

Diameter	d_{50}	0.12	m
Solid density	ρ_s	2900	kg/m^3
Young modulus	E	100	MPa
Contact friction coeff.	μ	0.61	–
Poisson's ratio	ν	0.30	–

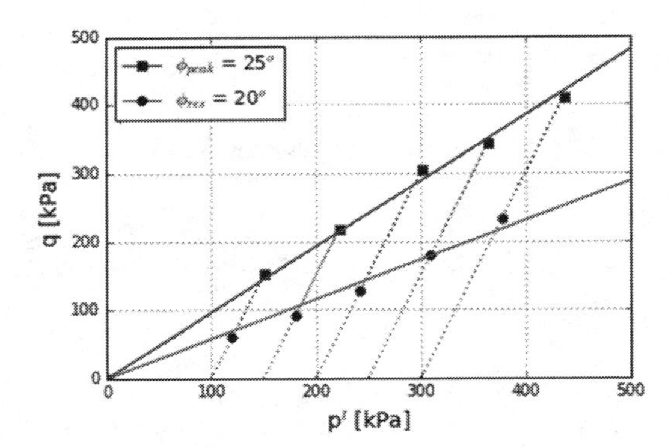

Figure 5. Results of periodic triaxial tests on discrete element samples of soil.

5 SOIL-MESH RETAINING PROBLEM

A simplified soil-mesh retaining problem has been modelled in a custom box with a vertical opening at the bottom with the aim of investigating a more realistic behaviour of a cortical mesh.

Initially, a very loose soil sample was generated with a random placement of spheres. Then particles are let free to settle under gravity in a frictionless closed prismatic box until the final stable configuration is obtained. The final soil sample fills a column of size $2.88 \times 2 \times 11$ m^3 generating a vertical stress of about 192 kPa at the base of the box. After the deposition and the stabilization of the granular sample, a mesh panel was created and vertically arranged in adherence to the soil column at the bottom, on the xy-plane. A small portion of the wall which has the same size and position of the mesh panel is used as temporal retaining wall for the column (the dark blue wall in Fig. 6a,b). The overall process of the sample generation is depicted in Figure 6.

The wire mesh type used for this test is the same calibrated with the punch test. The only differences are the boundary conditions

The mesh panel is 2.88×3 m^2 and it is anchored at its corners. Note that boundary conditions in node-wire-based and cylinder-wire-based discrete element representation of meshes are implemented as displacement constraints at nodes

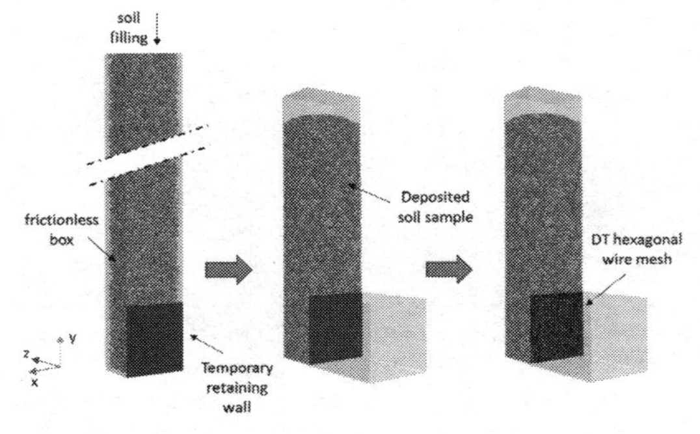

Figure 6. Sketch of the soil-mesh sample generation: (a) random placement of spheres; (b) gravitational deposition; (c) removal of the temporary wall and positioning of the wire mesh panel.

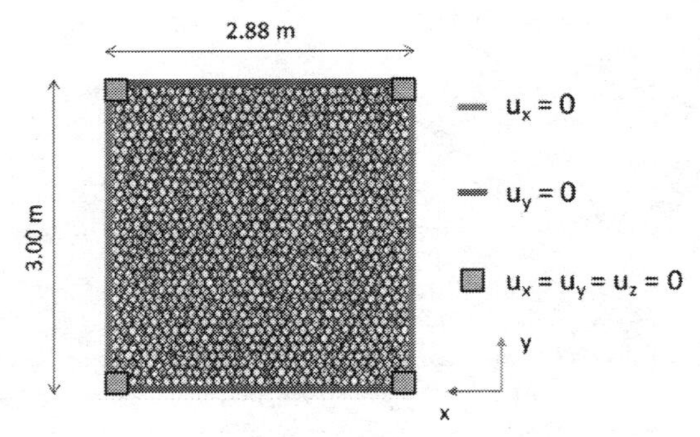

Figure 7. Front view of the soil-mesh problem and relevant boundary conditions.

while rotations and bending of wires are not constrained. Each plate anchor, which has a dimension of 0.16×0.16 m^2, is therefore schematized by blocking node displacements. In Figure 7, a scheme of the geometrical configuration of the mesh panel with relevant boundary conditions is reported. The other boundary conditions, reported in Figure 7, applied to the edges of the panel aim to model it as a sort of periodic panel and then taking into account the influence of the neighbouring panels to describe an idealized in-situ condition.

The test begins when the small vertical retaining wall is removed and the soil behind starts to push on the mesh panel till failure. Throughout this phase, displacement and forces on the boundaries as well as the stress and strain paths of each wire are monitored. By integrating each axial component of the nodal forces on the anchors and on the edges (i.e. the constraint reactions), the total force supported by the panel is derived. The absence of residual dynamic forces is also checked during the numerical test.

5.1 *Punch-mesh problem*

The same sample of mesh panel with boundary conditions identical to the latter configuration was also tested to punch with a standard punching element but vertically aligned. As in the soil-mesh problem the panel lies on the xy-plane and it is subjected to gravity. Clearly such conditions are not reproducible in laboratory but they can easily be implemented in a numerical model. Hence, it is possible to directly compare the two cases: mesh punched by soil and mesh punched by a rigid massive body. These analyses allow relevant information of the influence of the loading conditions on the mesh system mechanical response to be obtained. The force-displacement curve of the entire mesh panel as well as of each component (i.e. wires, edges, corner plates) can be compared to highlight the peculiarity of the soil-mesh interaction problem.

6 RESULTS

In this paragraph the distinction between the two approaches (NWB, CWB) will not be used anymore because of the similarity of their results. Figures 8–9 show the results of tensile forces along three edges (AB, AC, CD) and on two corner anchors (A, C) are presented for both the soil-mesh and the punch-mesh problem. For symmetry considerations along the y-axis, the results of the anchors B and D have not been reported being equal to anchors A and C. The same considerations are valid for the edge AC which behaves like the edge BD.

Note that all results are given as an absolute value and they are normalized over the maximum force and displacement values of the regular experimental punch test for direct comparison of the magnitudes. Nevertheless, their real positive or negative values (i.e. their real directions) can be easily deduced. For example, a normalized y-force value of 0.1 for one anchor means that it is supporting 10% of the maximum force of the panel evaluated with the regular punch test. The direction of the pulling force is always oriented towards the centre of the panel.

In both numerical models, the x-forces and the y-forces are zero along the x and y edges respectively as consequence of the boundary conditions applied at that nodes which mimic the existence of two axes of symmetry. Also z-forces are zero along all the edges because of the boundary conditions. Conversely to the edges, in both cases (i.e. punch-mesh and soil-mesh problem), the corner anchors experience skew forces with all the three spatial components which are not-negligible.

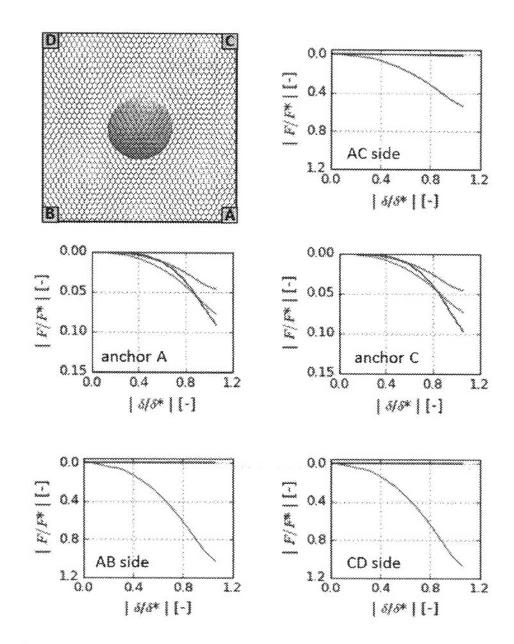

Figure 8. Front view of the panel in the punch-mesh problem. In all the figures, forces along x,y,z directions are represented with red, green and blue lines respectively (xyz = RGB).

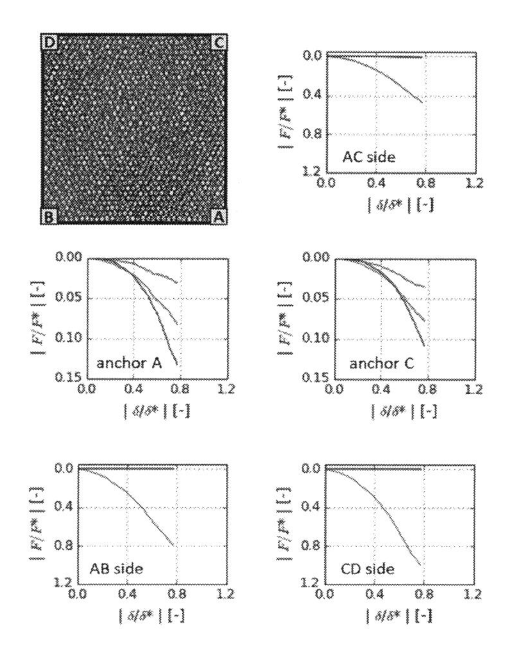

Figure 9. Front view of the panel in the soil-mesh problem. In all the figures, forces along x,y,z directions are represented with red, green and blue lines respectively (xyz = RGB).

Out-of-plane forces (i.e. z-forces, the blue curves) are only supported by the anchors, whereas the forces along x and y directions are supported also by the panel's sides (i.e. the neighbour panels). Summing all force components, the zones close to the anchors are the most stressed.

For the horizontal edges the tensile y-forces (i.e. the green curves) are slightly higher at the contact with the hypothetical upper panel (CD side) than those at the bottom edge (AB). This result can be attributed to the combined effect of the non-perfect symmetry of the mesh along these sides which increases the tensile forces of the upper edge by about 4% (for both the punch-mesh and soil-mesh problem) and for about 23% to the tangential thrust acting at the interface (only in the soil-mesh problem). This is reasonably due to the frictional and interlocking effects between soil and mesh that lead to the rise of vertical tangential forces along the internal surface of the panel. For reaction, this contributes to increase the tensile y-force experienced by the CD side. On the other hand, in the case of the punch-mesh problem, the tensile force along the y-direction measured for the anchor A is slightly greater ($\approx$6%) than the one of the anchor C. This is again due to the non-perfect geometrical symmetry of the mesh panel with respect to the x-axis.

Finally, Figures 10–11 show the overall tensile forces on the mesh panel for each distinct spatial component both for the punch-mesh and for the soil-mesh problem respectively. These forces are calculated as a sum, component by component, of the forces experienced by anchors and edges of the panel.

It is possible to note that the resulting forces along the x direction (the red curves) are approximatively

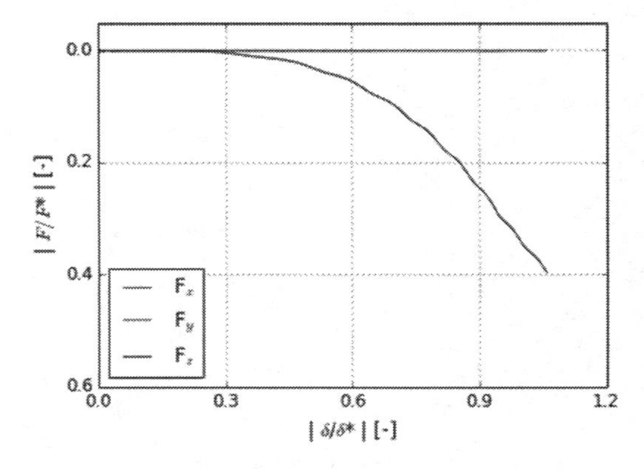

Figure 10. Results of the overall behaviour of the mesh panel for the punch-mesh problem in terms of force-displacement.

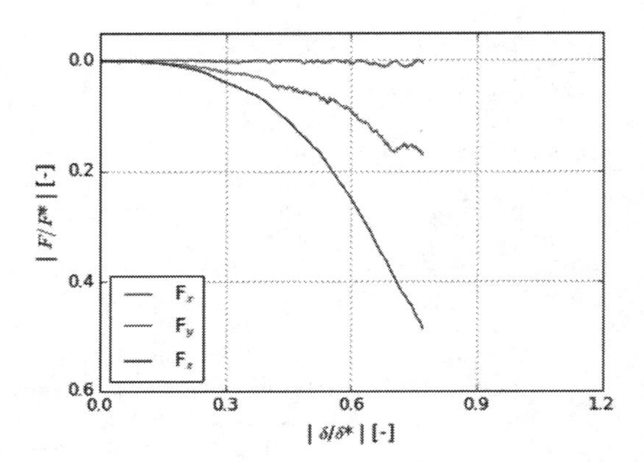

Figure 11. Results of the overall behaviour of the mesh panel for the soil-mesh problem in terms of force-displacement.

zero in both cases. Hence, the symmetry condition along this direction is met. On the other hand, in the y direction there is a small non-zero overall force caused by the weight of the mesh. However, its value is negligible if compared with the magnitude of the other force component; therefore, in the punch-mesh problem also the overall force in y direction can be considered approximatively zero. Instead, in the case of the soil-mesh problem, this y-force is further amplified and has a non-negligible value (approximatively $F_z/3$) due to the obliquity of the resultant earth pressure. Therefore, interlocking and frictional effects between the soil particles and the mesh play an important role in this kind of problem. It is also interesting to note that the inclination of the resultant force ($\approx$ 19°) has an order of magnitude that is comparable to the residual friction angle of the soil measured in the triaxial tests.

In Table 2, the total force values acting on the mesh panel and the displacements at failure for the studied cases, normalized with the standard UNI punch test results, are reported. The total force is computed as the Euclidean norm of the force component vector.

Comparing the overall response in terms of force-displacement curve in the two analysed configurations, the mesh panel behaves in a more rigid way for the soil-mesh problem. Such a difference in the mechanical response could be preliminary ascribed to the dissimilarity of the loading conditions. In fact, the punching element represent a local load, whereas the soil acts like a distributed load. A distributed load yields to a more uniform distribution of the stress field on the mesh and therefore to a better mechanical and stiffer response of the mesh panel.

Table 2. Total force and displacement values for the different tested configurations.

	F/F*	δ/δ^*
	[-]	[-]
Punch-mesh	0.40	1.06
Soil-mesh	0.52	0.77

7 CONCLUSIONS

In this work, it was shown that the mechanical response of a double-twisted cortical wire mesh can be effectively modelled with the Discrete Element Method. Both the node-wire-based and the cylinder-wire-based approaches used to model the hexagonal double-twisted cortical mesh resulted comparable using the same tensile properties for the wires: small differences have been reported probably mainly related to the different contact mechanics specific to the considered test.

With regards to the soil-mesh interaction problem a simplified but instructive model was developed. This allows the potentialities of the discrete element approach to be highlighted: handling large deformations, complex contact mechanics and separation events between the mesh and the soil behind, are only some of them. The specific mechanical behaviour of the mesh in this context (soil-mesh problem) was compared with a sort of "in-situ" punch test. The force-displacement curves of the whole panel and of each boundary component (edges and anchors) revealed a stiffer response of the mesh in the soil-mesh problem. This can be easily attributed to the different mechanisms of load distribution on the mesh panel. The rigid punch concentrated the stresses and strains just behind it, mostly yielding the wires which are in that zone. The soil instead distributes the tensile stresses at the boundaries along few wires.

Moreover, the soil-mesh problem underlines also the important contribution of the frictional and the interlocking effects of the soil particles on the mesh which should not be neglected. Such numerical approach can be further extended for the study of different soil-structure interaction aspects focusing on the resulting forces, on their magnitude and direction. These future results could be then exploited for an in-depth knowledge of those aspects and for improved considerations at design level.

The numerical model and the information derived from the simulations are of great interest for the design of this kind of meshes in field applications. Despite the simplifications adopted for the soil model (which was represented by large and moderately soft spheres), the problem revealed very promising results which can be used to enhance the system. Different grain shapes and sizes as well as different loading conditions and different anchorage specifications can be investigated in the future. A systematic study of the soil-mesh problem by changing the mechanical properties of the soil and the boundary constraints is under way.

REFERENCES

Albaba, A. et al. 2017. DEM modeling of a flexible barrier impacted by a dry granular flow. *Rock Mechanics and Rock Engineering* 50(11): 3029–3048.

Bertolo, P. et al. 2004. Metodologia per prove in vera grandezza sui sistemi di protezione corticale dei versanti. *Geoingegneria Ambientale e Mineraria* 44: 5–12.

Bertrand, D. et al. 2008. Discrete element method (DEM) numerical modeling of double-twisted hexagonal mesh. *Canadian Geotechnical Journal* 45(8): 1104–1117

Castro-Fresno, D. et al. 2008. Evaluation of the resistant capacity of cable nets using the finite element method and experimental validation. *Engineering Geology* 100: 1–10.

Cazzani, A. & Frenez, T. 2002. Dynamic finite element analysis of interceptive devices for falling rocks. *International Journal of Rock Mechanics and Mining Sciences* 39(3): 303–321.

Effeindzourou, A. et al. 2016. Modelling of deformable structures in the general framework of the discrete element method. *Geotextiles and Geomembranes* 44(2): 143–156.

Ferraiolo, F & Giacchetti, G. 2005. Rivestimenti corticali: alcune considerazioni sull'applicazione delle reti di protezione in parete rocciosa. *Proceedings of the Conference: Bonifica dei versanti rocciosi per la protezione del territorio, Trento, Italy, 11–12 March 2004:* 147–176.

Flum, D.et al. 2004. Dimensionamento di sistemi di consolidamento flessibili superficiali costituiti da reti in acciaio ad alta resistenza in combinazione a elementi di ancoraggio in barra. *Proceedings of the Conference: Bonifica dei versanti rocciosi per la protezione del territorio, Trento, Italy, 11–12 March 2004*: 461–470.

Gabrieli, F. et al. 2017. Comparison of two DEM strategies for modelling cortical meshes. *V International Conference on Particle-based Methods – Fundamentals and Applications, Particles 2017*: 489–496.

Nicot, F. et al. 2001. Design of rockfall restraining nets from a discrete element modelling. *Rock Mechanics and Rock Engineering* 34: 99–118.

Suiker, A.S.J. & Fleck, N.A. 2004. Frictional collapse of granular assemblies. *Journal of Applied Mechanics* 71(3): 350–358.

Thoeni, K et al. 2013. Discrete modelling of hexagonal wire meshes with a stochastically distorted contact model. *Computers and Geotechnics* 49: 158–169.

Numerical Methods in Geotechnical Engineering IX – Cardoso et al. (Eds)
© 2018 Taylor & Francis Group, London, ISBN 978-1-138-33198-3

Usability of piezocone test for finite element modelling of long-term deformations in soft soils

M. D'Ignazio, N. Sivasithamparam & H.P. Jostad
Norwegian Geotechnical Institute (NGI), Oslo, Norway

ABSTRACT: Numerical modelling of long-term behaviour of embankments on soft soils can require a significant amount of laboratory tests in order to derive all the necessary input parameters. Laboratory test results are strongly dependent on sample quality. Consequently, the quality of a finite element analysis is also dependent on the quality of the retrieved samples. The intent of this paper is then to evaluate the applicability of CPTu test results to determine preconsolidation stress for settlement calculations in soft soils when limited number of laboratory tests is available. The anisotropic Creep-SCLAY1S model is used for this purpose. The Creep-SCLAY1S model can capture the anisotropic visco-plastic behaviour and loss of soil structure after yielding. Settlement records from Murro test embankment are exploited to evaluate the practical implications of the proposed procedure.

1 INTRODUCTION

Accurate numerical modelling of long-term deformations in soft sensitive clays requires a significant effort in terms of laboratory testing needed to tune advanced constitutive models (e.g. Kelly et al. 2018). In this perspective, the performance of a numerical simulation will not only depend on the model capability of capturing the complex soil behaviour, but also on the quality of the retrieved soil samples that are used to determine the input parameters (e.g. Grimstad et al. 2016; D'Ignazio et al. 2017).

Behaviour of soft clays is rather complex. Soft clays are anisotropic, both in terms of particle orientation (intrinsic anisotropy) and stress-strain response (stress-induced anisotropy). For sensitive clays, soil structure is gradually lost when the soil is loaded beyond the yield stress or preconsolidation stress (p'_c). Moreover, the consolidation behaviour is not only governed by dissipation of excess pore pressure and consequent volume change, but also by viscous or creep deformations.

The material model Creep-SCLAY1S (Karstunen et al. 2013; Sivasithamparam et al. 2015), where "S" stands for "structure", is able to model these features. The Creep-SCLAY1S model is an extension of the anisotropic Creep-SCLAY1 model (Sivasithamparam et al. 2015) that includes, in addition to anisotropy and creep behaviour, soil destructuration after yielding.

Sivasithamparam et al. (2015) proved the validity of the Creep-SCLAY1 model through the back-calculation of the settlement of Murro test embankment (Karstunen et al. 2005). Murro test embankment is an instrumented test embankment built on a soft sensitive clay deposit located in Western Finland. However, inaccurate modelling of p'_c, might have influenced the results presented by Sivasithamparam et al. (2015). Indeed, Murro clay has been often referred to as a normally consolidated clay (e.g. Karstunen et al. 2005; Karstunen and Yin 2010, Koskinen 2014). Such a definition may possibly be the consequence of underestimation of the inferred p'_c values due to sample quality (D'Ignazio 2016).

The preconsolidation stress is generally determined from constant-rate-of-strain (CRS) or incrementally loaded (IL) oedometer tests. The interpreted value of p'_c corresponds to the yield stress of the soil in one-dimension oedometer compression (Fig. 1) and it gives indication of the overconsolidation of the soil through the overconsolidation ratio OCR ($= p'_c/p'_0$, where p'_0 is the in situ effective vertical stress). Several studies reported how p'_c is affected by sample quality (e.g. Lunne et al. 2006; Lunne et al. 2008; Karlsrud and Hernandez-Martinez 2013). For instance, Lunne et al. (2008) suggested that p'_c from constant-rate-of-strain (CRS) oedometer tests on conventional 54 mm piston samples can be up to 20–30% lower than from block samples.

The preconsolidation stress p'_c is also dependent on the strain rate used in the oedometer test. It increases with increasing strain rate (e.g. Vaid et al. 1979; Graham et al. 1983; Leroueil et al. 1985; Länsivaara 1999). As suggested by e.g. Leroueil et al. (1985), stress-strain behaviour from CRS tests may differ quite significantly from that observed in 24h IL tests. In principle, p'_c in CRS tests (p'^{CRS}_c) will be larger than in IL (p'^{IL}_c) tests because of the higher strain rates in CRS tests. D'Ignazio et al. (2016) proposed $p'^{CRS}_c/p'^{IL}_c = 1.27$. For Finnish clays, Kolisoja et al. (1989) and Hoikkala (1991) observed $p'^{CRS}_c/p'^{IL}_c = 1.16$–$1.30$. For Norwegian clays, Karlsrud and Hernandez-Martinez (2013) suggested $p'^{CRS}_c/p'^{IL}_c = 1.10$–$1.18$ based on block samples.

The scope of this study is to simulate the settlement behaviour of Murro test embankment using the anisotropic Creep-SCLAY1S model. The soil model is implemented as a user-defined model in the Finite Element software PLAXIS 2D (Plaxis 2016). Model parameters are selected according to previous studies on Murro test embankment, except for p'_c, which is based on piezocone (CPTu) and CRS test results on high-quality 132 mm tube samples of Murro clay. A well-established CPTu based correlation is calibrated against the limited number of CRS data and used to extrapolate *OCR* or p'_c in the layers were no measurements were available. The effects of soil destructuration and strain rate correction on p'_c on the calculated settlement are evaluated and discussed. The sensitivity of displacements is checked only with respect to the modelled *OCR*. Sensitivity of calculation results with respect to other soil parameters than *OCR* was beyond the scope of this study. Finally, some recommendations for practice are suggested based on the outcomes of this study.

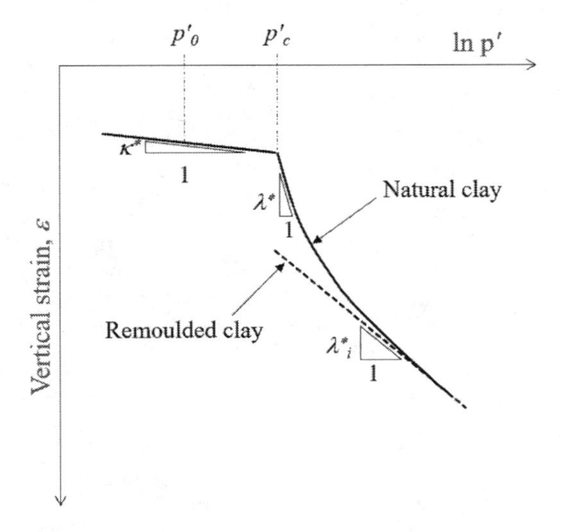

Figure 1. Stress-strain behaviour of a natural clay in one-dimensional oedometer compression test (after Karstunen and Yin 2010).

2 BACKGROUND

2.1 *Murro test embankment*

An instrumented test embankment was built in Murro, Western Finland, in 1993. The purpose was to collect experimental data on long-term settlement for the design of Highway 18 between the cities of Jyväskylä and Vaasa.

Murro test embankment is 2 m high, 10 m wide (top) and 30 m long with a gradient of 1:2 (Karstunen and Yin 2010). The body of the embankment consisted of crushed rock with unit weight γ = 19.6 kN/m^3. The embankment was built on a 23 m thick low organic sulphide silty clay deposit. The thickness of the dry crust layer at the site was 1.6 m. The ground water table was detected at 0.8 m below the ground level.

The extensive field instrumentation consisted of settlement plates, inclinometers, one extensometer and numerous pore pressure probes (Karstunen and Yin 2010). In addition, Tampere University of Technology (TUT) collected CPTu and strength data in 2013 (D'Ignazio 2016).

The available experimental data has been used in several studies on the modelling of strength and deformation behaviour of Murro clay (e.g. Karstunen et al. 2005; Karstunen and Yin 2010; Koskinen 2014; Sivasithamparan et al. 2015; D'Ignazio and Länsivaara 2016; D'Ignazio 2016).

Murro clay is a sulphide clay typical of the coastal areas around the Gulf of Bothnia, with organic content ranging between 2–4% (Koskinen 2014) and silt content of 70–74% (D'Ignazio 2016).

Figure 2 summarizes the main properties of Murro clay. Murro clay is a low to medium sensitive clay with sensitivity (S_t) of 2–10. The natural water content (w) increases from 65% near the ground surface, up to 100% at 5 m depth. After that point, w decreases with depth from 100% to

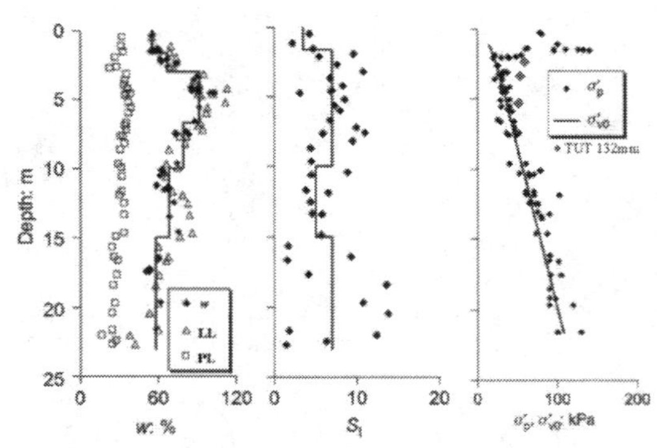

Figure 2. Characteristics of Murro clay (after Karstunen and Yin 2010).

about 60% below 15 m depth. The liquid limit (LL) varies between 55% and 120%, following the same trend observed for *w*.

Oedometer tests on 50 mm diameter piston samples suggested Murro clay to be nearly normally consolidated (Karstunen and Yin 2010). CRS tests on 132 mm tube samples of Murro clay performed by TUT indicate that the clay is slightly overconsolidated in the top 6 m below the embankment (Fig. 2), with OCR of 1.3–3.0. As reported by Di Buò et al. (2016), the 132 mm TUT tube sampler generally gives higher sample quality than the conventional 50 mm piston sampler. Therefore, sample disturbance might be the reason for the lower p'_c values observed by Karstunen and Yin (2010).

2.2 *Overconsolidation of Murro clay*

The study by Sivasithamparam et al. (2015) addressed the effect of anisotropy and secondary compression on the settlement of Murro test embankment. Soil structure was not modelled in that study. In their analysis, Murro clay was modelled as normally consolidated with pre-overburden stress (POP = $p'_c - p'_0$) of 2 kPa for the layers below the dry crust. According to D'Ignazio (2016), such hypothesis may seem conservative, as OCR was observed to vary between 1.3 and 3 in the top 5 m, with OCR 1.1–1.3 below 5 m depth. This was based on CPTu and CRS tests on 132 mm tube samples.

D'Ignazio (2016) discussed how p'_c from CRS tests should be corrected to be usable in settlement calculations. This is to account for the different strain rate between the CRS test (fast) and the IL 24h test (slow). In practice, p'_c from IL tests is normally used as a reference preconsolidation stress in settlement calculations.

As proposed by Jostad et al. (2018), the ratio p'^{CRS}_c/p'^{IL}_c can be also derived theoretically from Equation 1, which is based on the difference in strain rates in CRS and IL tests (ε_{CRS}, ε_{IL}) and the ratio $\mu^*/(\lambda_i^*-\kappa^*)$, where λ_i^* is the modified intrinsic compression index (Fig. 1), κ^* the modified swelling index (Fig. 1) and μ^* the modified creep index. Values for these indices are reported in Table 2. The calculated average p'^{CRS}_c/p'^{IL}_c from Equation 1 is about 1.15 for Murro clay for $\varepsilon_{CRS} = 0.144\ \text{day}^{-1}$ and ε_{IL} equal to μ^*/τ in Table 2, where $\tau = 1$ day is the reference time. This calculated value is in agreement with what suggested by Karlsrud and Hernandez-Martinez (2013) for Norwegian clays.

$$\frac{p'^{CRS}_c}{p'^{IL}_c} = \left(\frac{\varepsilon_{CRS}}{\varepsilon_{IL}}\right)^{\frac{\mu^*}{\lambda_i^*-\kappa^*}} \tag{1}$$

In this study, the p'_c profile at Murro test site is established from CPTu using Equation 2 (Mayne 1986):

$$OCR = k\left(\frac{q_t - p_0}{p'_0}\right) \tag{2}$$

Table 1. OCR used in Finite Element calculations.

Layer	OCR CRS	OCR $p'^{CRS}_c/p'^{IL}_c = 1.15$	OCR $p'^{CRS}_c/p'^{IL}_c = 1.27$
Layer 1–1	*	*	*
Layer 1–2	*	*	*
Layer 2	2.60	2.26	2.05
Layer 3	1.60	1.39	1.26
Layer 4	1.35	1.17	1.06
Layer 5	1.35	1.17	1.06
Layer 6	1.10	1.00	1.00

*POP = 20 kPa (Sivasithamparam et al. 2015).

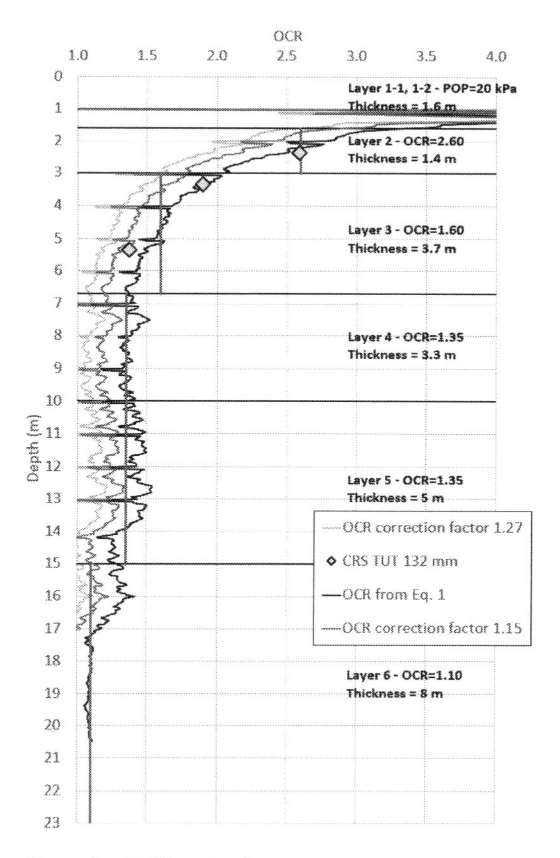

Figure 3. OCR vs depth.

where q_t = corrected cone tip resistance; p_0 = total vertical overburden stress; p'_0 = effective vertical overburden stress; and k = empirical constant generally varying between 0.2 and 0.5 (Lunne et al. 1997). For Murro clay, $k = 0.25$ seems to fit reasonably well the CRS tests, as shown in Figure 3.

The OCR profile from Equation 2 shown in Figure 3 is reduced by 1.15, calculated from Equation 1, and by 1.27, as suggested by D'Ignazio et al. (2016). OCR values for each layer are summarized in Table 1. Soil layering is based on Sivasithamparam et al. (2015) and shown in Figures 3 and 5. Piezocone measurements in the dry crust layer (layers 1–1 and 1–2 in Fig. 3) may not be accurate because of possible partial saturation. Therefore, OCR of layers 1–1 and 1–2 is modelled as suggested by Sivasithamparam et al. (2015) (Table 1).

2.3 *Creep-SCLAY1S*

The Creep-SCLAY1S formulation of anisotropy and degradation of structure (destructuration) is based on the S-CLAY1S model (Karstunen et al., 2005). The model assumes that there is no purely elastic domain and it uses a viscoplastic multiplier for creep strain rate.

The model comprises three reference surfaces as shown in Figure 4. The Normal Consolidation Surface (NCS) defines the boundary surface for normally consolidated state, the Current Stress Surface (CSS) represents the current state of the soil, and the Intrinsic Compression Surface (ICS) models the effect of degradation of soil structure.

The creep strain rate is formulated based on the concept of a constant rate of a viscoplastic multiplier $\dot{\Lambda}$ as

$$\dot{\Lambda} = \frac{\mu^*}{\tau}\left(\frac{1}{OCR^*}\right)^{\beta}\left(\frac{M^2 - \alpha^2 K_0^{nc}}{M^2 - \eta^2 K_0^{nc}}\right) \quad (3)$$

where

$$\mu^* = \frac{C_\alpha}{ln10(1+e_0)} \text{ and } \beta = \frac{\lambda_i^* - \kappa^*}{\mu^*}$$

M is the stress ratio at critical state, the actual stress ratio $\eta = q/p'$ and $\alpha_{K_o}{}^{nc}$ define the inclination of the ellipses in the normally consolidated state, μ^* is referred to as modified creep index and τ is the reference time (generally $\tau = 1$ day). OCR^* is ratio between mean preconsolidation stress and equivalent stress as a measure on the isotropic axis of the distance between the current stress and the mean preconsolidation stress. C_α is the secondary compression index and λ^* and κ^* are the modified compression and swelling indices, respectively (Fig. 1).

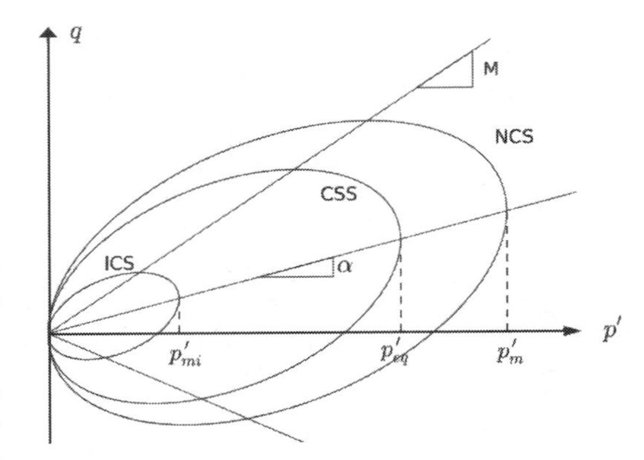

Figure 4. Distorted and rotated ellipse in q-p' plane representing the Current State Surface (CSS), Normal Consolidation Surface (NCS) and Intrinsic Compression Surface (ICS) of the Creep-SCLAY1S model (Adapted from Amavasai et al., 2018).

The destructuration that describes degradation of bonding due to plastic straining is formulated in such a way that both plastic volumetric and shear strains tend to decrease the value of the bonding parameter χ towards a target value of zero (Karstunen et al. 2005) as

$$d\chi = -\xi\chi\left(\left|d\varepsilon_v^p\right| + \xi_d\left|d\varepsilon_d^p\right|\right) \quad (4)$$

where ξ and ξ_d are additional soil constants. The parameter ξ controls the absolute rate of destructuration, while the parameter ξ_d controls the relative effectiveness of plastic deviatoric strains ($\varepsilon^p{}_d$) and plastic volumetric strains ($\varepsilon^p{}_v$) in destroying the inter-particle bonding (Karstunen et al. 2005).

The model mathematical formulation, description, parameter determination and implementation into the PLAXIS FE code can be found in Sivasithamparam et al. (2015), Amavasai et al. (2017), Gras et al. (2017a) and Gras et al. (2017b).

3 NUMERICAL ANALYSES OF MURRO TEST EMBANKMENT

3.1 *Finite element model and parameters*

The calculation phases have been modelled in plane-strain finite element analysis using PLAXIS 2D 2016 version including the "updated mesh" option since large deformations are expected. The Finite Element plane-strain model of Murro test embankment is shown in Figure 5. Only half of the embankment is considered in the analysis because of the symmetry of the problem. The lateral boundary is set 30 m from the symmetry axis.

Table 2. Creep-SCLAY1S model parameters for Murro test embankment (from Karstunen et al. 2012).

Layer*	M_c/M_e	e_0	α_0	λ_i^*	κ^*	μ^*/τ
Layer 1–1	1.4	0.63	0.070	0.0042	8.96E-4	1.6/1.04
Layer 1–2	1.4	0.63	0.070	0.0042	8.96E-4	1.6/1.04
Layer 2	1.8	0.63	0.064	0.0107	2.33E-3	1.6/1.04
Layer 3	2.4	0.63	0.072	0.0106	1.92E-3	1.6/1.04
Layer 4	2.1	0.63	0.066	0.0097	1.52E-3	1.6/1.04
Layer 5	1.8	0.63	0.076	0.0121	1.49E-3	1.6/1.04
Layer 6	1.5	0.63	0.059	0.0016	7.30E-4	1.6/1.04

*Subscripts c and e stand for "compression"; and "extension", respectively.

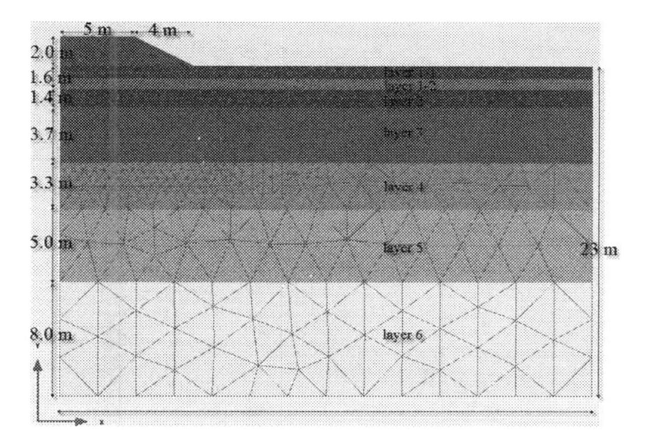

Figure 5. Finite Element mesh and layering.

A 2578 15-noded triangular element mesh was used. The mesh is refined in the area below the embankment where the majority of the deformations are expected to occur. Roller conditions are imposed at the side boundaries, while fully fixity is applied at the base. The groundwater table is set at 0.8 m depth. The initial stresses are generated from a K_0 procedure, since layers are horizontal. The embankment construction phase is modelled as a 2-day consolidation phase. The long-term settlement is calculated after a consolidation phase of 4000 days.

The layering shown in Figure 5 is based on Sivasithamparam et al. (2015). The embankment is modelled using the elasto-plastic Mohr-Coulomb model, as reported in Sivasithamparam et al. (2015). The soft clay and the dry crust are modelled using the Creep-SCLAY1S model described in Section 2.3. The input parameters for each soil unit given in Table 2 are based on Karstunen et al. (2012). The destructuration related parameters for Murro clay are based on Karstunen et al. (2005). The input OCR values are according to Table 1.

3.2 *Calculation results*

Figure 6 presents a comparison between calculated and measured settlement at the embankment centreline. The results in Figure 6 demonstrate the importance of an accurate modelling of *OCR* and soil destructuration. Increase in the *OCR* correction factor results in increase of settlement. The *OCR* correction factors used in the calculations (1.15–1.27) seem to provide a relatively good prediction of the embankment settlement, especially for $t > 1000$ days.

A degree of consolidation of approximately 90% is reached after ca. 2000 days, as also discussed by D'Ignazio and Länsivaara (2016). Therefore, for $t > 2000$ days, the settlement is mainly caused by creep strains. For $t > 2000$ days, the curve for *OCR* correction factor of 1.27 seems the most representative of the field measurements (Fig. 6). When using the *OCR* profile from CRS results (without any correction) as input to the calculation, the Creep-SCLAY1S model under predicts the settlement as shown in Figure 6.

The model over predicts settlement right after the construction of the embankment, possibly because of development of significant creep strains. Moreover, the model may underestimate the undrained shear strength of the soil. This is also reflected in the modelled horizontal displacement under the embankment crest (Fig. 7a) and vertical strain at the centreline (Fig. 8a). Overall, the model gives rather good settlement predictions at 3200 days for the corrected *OCR* profiles, while it over estimates horizontal displacements at shallow depths as shown in Figure 7b. The reason for the over prediction of displacements at shallow depth can be explained by Figure 8b as most of the strains seem concentrated in layers 2 and 3.

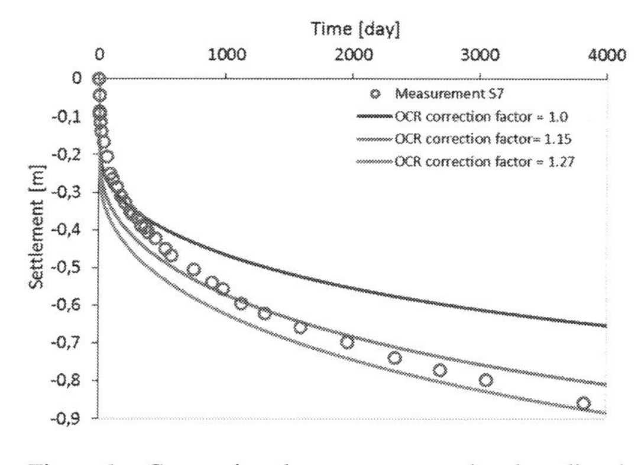

Figure 6. Comparison between measured and predicted settlements at the centreline of the embankment.

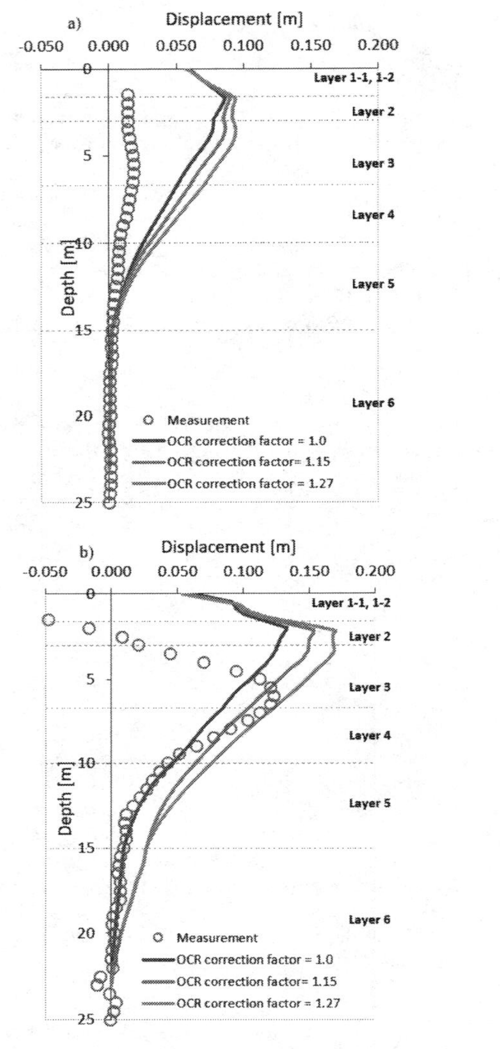

Figure 7. Comparison of horizontal displacements under crest. (a) after construction (b) after 3200 days.

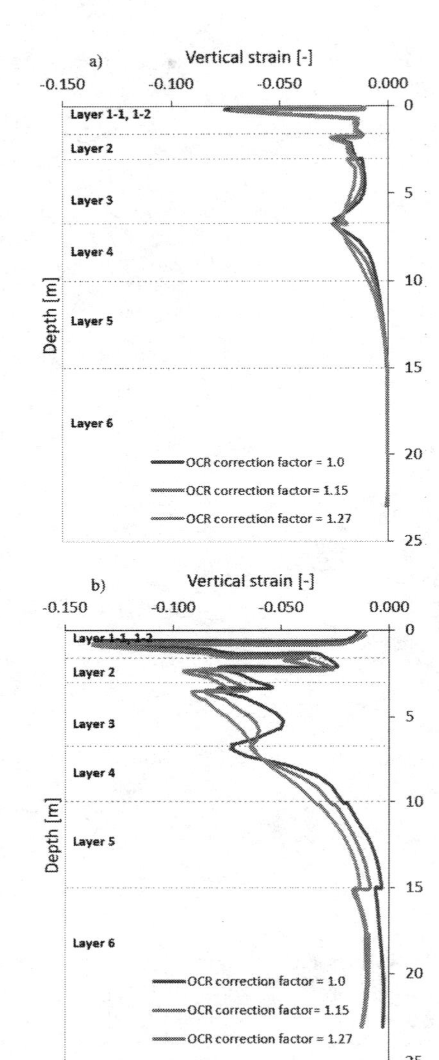

Figure 8. Comparison of vertical strains at the centre of embankment (a) after construction (b) after 3200 days.

When soil destructuration is not modelled, settlement at t = 4000 days is approximately 20% lower for the range of OCR considered.

4 DISCUSSION

Sivasithamparam et al. (2015) modelled Murro test embankment using Creep-SCLAY1 model, which does not account for bonding and destructuration. It was found that the model gives a good prediction of displacements compared to the measurements. However, they had assumed the soil layers 2–6 to be normally consolidated. The CPTu results presented in section 3 indicated the subsoil layers to be slightly over-consolidated. For Murro clay, the overconsolidation is possibly the result of ageing process that may have led to changes in the soil structure and, hence, caused an apparent preconsolidation stress. Furthermore, Murro clay is structured. The intact soil structure is lost after yielding with consequent loss of stiffness. Figure 1 illustrates the idealized one-dimensional compression behaviour of structured clays. By introducing bonding and destructuration of bonding in the Creep-SCLAY1S model, the model was able to predict fairly well the field measurements, especially in terms of long-term settlement.

The preconsolidation stress p'_c was shown to have a great impact on the FE results. Therefore, in absence of sufficient site-specific CRS or IL data, a proper calibration of the CPTu tests is crucial to achieve a reliable prediction of displacements. In this study, the CPTu-based correlation for *OCR* proposed by Mayne (1986) was used. High-quality oedometer data was available only for layers 2 and 3. Then, p'_c in the deeper layers was extrapolated by assuming a constant *k* factor in Equation 2, resulting from the fitting to the tests in the top 5–6 m. This assumption needs to be verified for layers 4 to 6 from high-quality oedometer tests. For instance, when the reference *OCR* from CRS is corrected for strain rate effects (Fig. 2), *OCR* values less than 1 are obtained from Equation 2 in layer 6. For this layer, $OCR = 1$ was used in the calculations (Table 2). This may suggest that *k* is not constant within a given site and may vary with e.g., soil properties. To the Authors' knowledge, there is no study that fully explains the variability of *k* or discusses the dependency of *k* on basic clay properties (i.e., plasticity, water content etc..). Needless to say, *k* will also depend on the strain rate used in the test because of the rate dependency of p'_c.

It is important to remark how, in this study, the sensitivity of displacements at Murro test embankment was checked only with respect to *OCR*. Checking the sensitivity of deformations with respect to other soil parameters than *OCR* was beyond the scope of this study. Other input parameters are expected to affect the calculation results, such as soil stiffness or permeability, among others. As for p'_c, these parameters are dependent on sample quality (e.g. Lunne et al. 2006; Lunne et al. 2008; Grimstad et al. 2016) and may have a notable impact on the deformation behaviour of the embankment (e.g. Kelly et al. 2018). Accurate tuning of model parameters for each layer from laboratory testing on good samples or back calculations is likely to result in improved predictions of both vertical and horizontal displacements (e.g. Jostad et al. 2018).

5 CONCLUSIONS

In this study, the long-term settlement of the benchmark Murro test embankment was modelled using the finite element anisotropic time-dependent Creep-SCLAY1S model. Focus was given to the modelling of preconsolidation stress p'_c (or *OCR*) when CPTu and a limited number of high-quality tests are available. Furthermore, the loss of soil structure after yielding was incorporated in the calculations for a more realistic description of the behaviour of Murro clay.

In previous studies, presumably because of sample disturbance, Murro clay was modelled as normally consolidated clay. The calibrated CPTu measurements suggested Murro clay to be slightly overconsolidated, especially in the top 5 m below the embankment.

The finite element results demonstrated the practical implications of correcting (reducing) p'_c from CRS tests to account for the strain rate difference between laboratory and field. Correction factors for p'_c were established both theoretically, considering the different strain rates between CRS and IL tests, and from literature. The use of factors in the range of 1.15–1.27 resulted in accurate prediction of settlement. However, horizontal displacements were slightly over predicted.

REFERENCES

Amavasai, A., Gras, J.P., Sivasithamparam, N., Karstunen, M., & Dijkstra, J. 2017. Towards consistent numerical analyses of embankments on soft soils. *European Journal of Environmental and Civil Engineering*, 1–19.

Amavasai, A., Sivasithamparam, N., Dijkstra, J. & Karstunen, M. 2018. Consistent Class A & C predictions of the Ballina test embankment. *Computers and Geotechnics* 93, pp. 75–86.

Di Buò B., D'Ignazio M., Selänpää J. & Länsivaara T. 2016. Preliminary results from a study aiming to improve ground investigation data. In *Proceedings of the 17th Nordic Geotechnical Meeting*: 187–197.

D'Ignazio, M. 2016. Undrained shear strength of Finnish clays for stability analyses of embankments. PhD thesis, Tampere University of Technology, Finland.

D'Ignazio, M., Phoon, K.K., Tan, S.A., & Länsivaara, T.T. 2016. Correlations for undrained shear strength of Finnish soft clays. *Canadian Geotechnical Journal* 53(10) 1628–1645.

D'Ignazio, M., & Länsivaara, T. 2016. Strength increase below an old test embankment in Finland. In *Proceedings of the Nordic Geotechnical Meeting (NGM) 2016, Reykjavik, Iceland.* Vol 1, 357–366.

D'Ignazio, M., Jostad, H.P., Länsivaara, T., Lehtonen, V., Mansikkamäki, J., & Meehan, C. 2017. Effects of Sample Disturbance in the Determination of Soil Parameters for Advanced Finite Element Modelling of Sensitive Clays. In *Landslides in Sensitive Clays* (pp. 145–154). Springer International Publishing.

Graham, J., Crooks, J., & Bell, A. 1983. Time effects on the stress-strain behaviour of natural soft clays. *Géotechnique* 33(3): 327–340.

Gras, J.P., Sivasithamparam, N., Karstunen, M. & Dijkstra, J. 2017. Permissible range of model parameters for natural fine-grained materials. *Acta Geotechnica* 1–12. DOI:https://doi.org/10.1007/s1144

Gras, J.P., Sivasithamparam, N., Karstunen, M. & Dijkstra, J. 2017. Strategy for consistent model parameter calibration for soft soils using multi-objective optimisation. *Computers and Geotechnics* 90, 164–175.

Grimstad, G., Haji Ashrafi, M.A., Degago, S.A., Emdal, A., & Nordal, S. 2016. Discussion of 'Soil creep effects on ground lateral deformation and pore water pressure under embankments'. *Geomechanics and Geoengineering 11*(1), 86–93.

Hoikkala, S. 1991. Continuous and incremental loading oedometer tests. M.Sc. thesis, Helsinki University of Technology, Espoo, Finland. [In Finnish.].

Jostad, H.P., Palmieri, F., Andresen, L., & Boylan, N. 2018. Numerical prediction and back-calculation of time-dependent behaviour of Ballina test embankment. *Computers and Geotechnics 93*, 123–132.

Karlsrud, K., & Hernandez-Martinez, F.G. 2013. Strength and deformation properties of Norwegian clays from laboratory tests on high-quality block samples 1. *Canadian Geotechnical Journal* 50(12):1273–1293.

Karstunen, M., Krenn, H., Wheeler, S.J., Koskinen, M., & Zentar, R. 2005. Effect of anisotropy and destructuration on the behaviour of Murro test embankment. *International Journal of Geomechanics* 5(2): 87–97.

Karstunen, M., & Yin, Z.Y. 2010. Modelling time-dependent behaviour of Murro test embankment. *Géotechnique* 60(10): 735–749.

Karstunen, M., Rezania, M., Sivasithamparam, N. & Yin, Z. 2012. Comparison of Anisotropic Rate-Dependent Models for Modelling Consolidation of Soft Clays. *International Journal Geomechnics* 15(5): A4014003.

Karstunen, M., Sivasithamparam, N., Brinkgreve, R.B.J., & Bonnier, P.G. 2013. Modelling rate-dependent behaviour of structured clays. In *International Conference on Installation Effects in Geotechnical Engineering (ICIEGE). Rotterdam, Netherlands,* pp. 43–50.

Kelly, R.B., Sloan, S.W., Pineda, J.A., Kouretzis, G., & Huang, J. 2018. Outcomes of the Newcastle symposium for the prediction of embankment behaviour on soft soil. *Computers and Geotechnics 93*, 9–41

Kolisoja, P., Sahi, K., & Hartikainen, J. 1989. An automatic triaxial-oedometer device. In *Proceedings of the 12th International Conference on Soil Mechanics and Foundation Engineering* pp. 61–64.

Koskinen, M. 2014. Plastic anisotropy and destructuration of soft Finnish clays. PhD thesis, Aalto University. Helsinki.

Länsivaara, T. 1999. A study of the mechanical behaviour of soft clay. PhD thesis, Norwegian University of Science and Technology, Trondheim.

Leroueil, S, Kabbaj, M., Tavenas, F., & Bouchard, R. 1985. Stress–strain–strain rate relation for the compressibility of sensitive natural clays. *Géotechnique* 35(2): 159–180.

Lunne, T., Robertson, P.K., & Powell, J.J.M. 1997. Cone penetration testing in geotechnical practice. Spon Press.

Lunne, T., Berre, T., Andersen, K.H., Strandvik, S., & Sjursen, M. 2006. Effects of sample disturbance and consolidation procedures on measured shear strength of soft marine Norwegian clays. *Canadian Geotechnical Journal* 43(7): 726–750.

Lunne, T., Berre, T., Andersen, K.H., Sjursen, M., Mortensen, N. 2008. Effects of sample disturbance on consolidation behaviour of soft marine Norwegian clays. In *Third International Conference on Site Characterization ISC: Geotechnical and Geophysical Site Characterization*, 3, pp 1471–1479.

Mayne, P.W. 1986. CPT indexing of in situ OCR in clays. In *Use of In Situ Tests in Geotechnical Engineering* (pp. 780–793). ASCE.

Plaxis, 2016. Plaxis user's manual. Plaxis, Delft, Netherlands.

Sivasithamparam, N., Karstunen, M., & Bonnier, P. 2015. Modelling creep behaviour of anisotropic soft soils. *Computers and Geotechnics* 69, 46–57.

Vaid, Y.P., Robertson, P.K., & Campanella, R.G. 1979. Strain rate behaviour of Saint-Jean-Vianney clay. *Canadian Geotechnical Journal* 16(1): 34–42.

Numerical Methods in Geotechnical Engineering IX – Cardoso et al. (Eds)
ISBN 978-1-138-33198-3

Study of the validity of a rectangular strip track/soil coupling in railway semi-analytical prediction models

D. Ghangale & J. Romeu
Acoustical and Mechanical Engineering Laboratory (LEAM), Universitat Politcnica de Catalunya, Spain

R. Arcos
Institute of Sound and Vibration Research (ISVR), University of Southampton, UK

B. Noori
AV Ingenieros, Spain

A. Clot
Department of Engineering, University of Cambridge, UK

J. Cayero
Automatic Control Department (ESAII), Universitat Politcnica de Catalunya, Spain

ABSTRACT: Semi-analytical prediction models for railway-induced vibration problems consider a two-dimensional track coupled with a three-dimensional soil. The semi-analytical models found in the literature assume a simplified model of force distribution for representing track/soil interactions, based on a rectangular strip load that acts on the soil surface. Because an accurate modelling of the track/soil interface forces is a key factor to obtain reliable prediction results of the vibration levels on the ground or on the surrounding buildings, this paper is intended to study the validity of considering a rectangular strip track/soil coupling load. The results associated to consider a rectangular strip track/soil coupling load are compared with the ones obtained from a two-and-a-half dimensional coupled finite element-boundary element model of the track/soil system, which is considered an accurate numerical model. For the present study, the soil is assumed as a homogeneous halfspace. Some proposals to enhance the accuracy of the rectangular strip modelling are also presented.

1 INTRODUCTION

Ground-borne vibrations induced by railway infrastructures are a source of annoyance to the inhabitants of the nearby buildings and can even cause structural damages. To ensure the accomplishment of the laws and regulations about this issue, it is necessary to predict the railway-induced ground-borne vibration levels that a new infrastructure will induce accurately. Several approaches are available in literature to model the vibration induced in the ground due to railway infrastructures. These approaches are of three different kinds: empirical, semi-analytical or numerical. Empirical models are based on large amount of experimental measurements performed on specific sites, giving efficient but potentially inaccurate predictions. Semi-analytical approaches, models the railway infrastructure using simple elastodynamic elements, like beams, plates, springs, dampers and halfspaces. Vibration predictions using these models requires performing numerical calculations, usually integral anti-transforms, in order to obtain system response. Usually, numerical models are based on the Finite Element Method (FEM) or on a combination of the former with the Boundary Element Method (BEM). Thus, the computational and engineering costs associated to semi-analytical models are significantly lower than the costs associated to numerical ones. However, numerical models are more accurate in predicting the behaviour of the system.

For the study of the railway-induced vibrations in buildings, railway infrastructures are commonly considered longitudinally invariant structures. This longitudinal invariance is exploited by using two-and-a-half-dimensional (2.5D) domain formulations. In semi-analytical approaches, one of the most well-established examples of this 2.5D modelling is the model of Sheng et al. (2004,

1999a, 1999b). As proposed by other existing semi-analytical approaches, this model assumes that the force distribution for representing track/soil interactions is a rectangular strip load, uniformly distributed along the perpendicular direction to the track, which acts on the soil surface. This assumption was discovered to be invalid for some particular frequencies in which the wavelength of the Rayleigh waves is close to the width of the track (Ntotsios et al. 2015). For arriving to this conclusion, Ntotsios et al. (Ntotsios et al. 2015) used an extension of the Sheng's semi-analytical model that accounts for the non-uniform distribution of tractions along the width of the track.

In the framework of numerical approaches, coupled FEM-BEM numerical models based on 2.5D formulations are popular numerical methods used for modelling soil-structure vibrations (Alves Costa et al. 2012, François et al. 2010, Galvín et al. 2010, Ozdemir et al. 2013, Romero et al. 2017, Sheng et al. 2005, Sheng et al. 2006). In a coupled FEM-BEM methodology, commonly a sub-domain decomposition is used to model the soil using Boundary Elements (BE) and the structure using Finite Elements (FE). Sheng et al. (2006, 2005) presented a coupled FEM-BEM methodology based on a 2.5D formulation to study soil-structure interactions. In their method, the structure is modelled with 2.5D FE and the soil is modelled by 2.5D BE. The BEM formulation used by Sheng et al. is derived based on the reciprocity theorem. This formulation employs the full-space fundamental solution for computing BE matrices. As full-space solutions are used, the method requires meshing of the layer interfaces when the soil is modelled as a layered halfspace. Mesh truncation is also used to model infinite surfaces and interfaces. The problem of singular fundamental solutions is solved by analytically computing the singular integrals. Alternatively, François et al. (François et al. 2010) proposed a 2.5D BEM formulation which avoids the computation of singular integrals. It is based on the substraction of static fundamental solution from dynamic fundamental solutions. This results in a global regularization scheme that makes the analytical computation of singular terms unnecessary. This method makes use of 2.5D Green's functions of layered halfspace (Schevenels et al. 2009) because of which meshing layer interfaces and use of mesh truncation is avoided. The coupled FEM-BEM presented in this paper is based on this methodology.

In the present paper, the ground response obtained by both semi-analytical and numerical previously selected models is compared in order to study the track/soil coupling assumption of the semi-analytical model. This comparison is done in order to perform a detailed examination of the track/soil coupling assumption of the semi-analytical model under study. The study has been performed in the framework of an at-grade ballasted track.

2 SEMI-ANALYTICAL METHOD

In this section, the model developed by Sheng et al. (2004, 1999a, 1999b) is summarized. As assumed by most of the other existing semi-analytical approaches, this model considers a two-dimensional (2D) track coupled with a three-dimensional (3D) soil. The 2D assumption is based on the consideration of the track symmetric loading, which is a result of considering the unevenness profiles of the wheel and the rail fully correlated between the two different rails and the two wheels of the same axle. As previously stated, the track/soil coupling tractions considered in this model are uniformly distributed along the perpendicular direction of the track, being this assumption the one that wants to be studied in the present paper.

The model is outlined in Fig. 1. In this model, the rails are modelled as a single Euler-Bernoulli beam, the sleepers are considered to be as a distributed mass, the fasteners are assumed to be a distributed massless spring, the ballast is considered to be a distributed vertical massive spring and the subgrade is assumed to be an homogeneous halfspace. In this model, $z_r = z_r(x,t), z_s = z_s(x,t)$ and $z_g = z_g(x, y_c, t)$, represent the rail vertical displacement, the sleepers vertical displacement and the vertical displacement of the ground surface at $y = y_c$, position where the coupling between the 2D track is coupled with the 3D soil. The model also accounts for the 3D motion of the ground surface in terms of $\boldsymbol{u}_g = \boldsymbol{u}_g(x, y, t)$. The mechanical and geometrical parameters of the rail model are its material Young's modulus E_r, its second moment of inertia I_r, its structural damping coefficient η_r, its material density $_r$, and its cross-sectional area S_r. The sleepers and the ballast distributed masses are m_s and m_b, respectively. Fasteners and ballast stiffnesses are defined by k_f and k_b, respectively. Viscous damping is considered for the ballast and the fasteners: c_f and c_b respectively.

This model works in the 2.5D domain, where the coupling of the equations of motion of the different subsystems is done. For the computation of

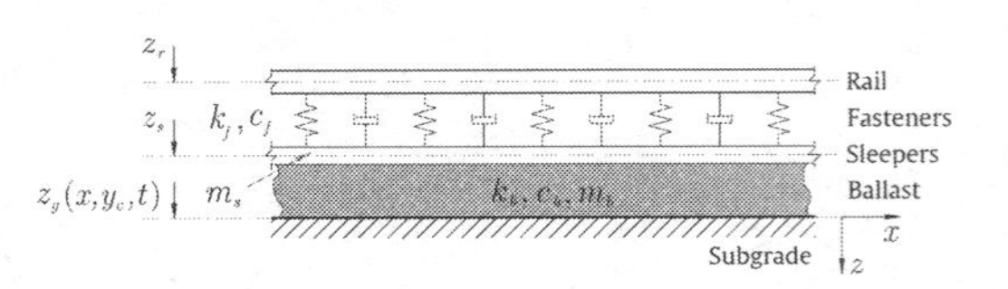

Figure 1. Scheme of semi-analytical model.

the 2.5D Green's functions associated to the surface of the homogeneous halfspace model of the subgrade, it is used the methodology presented in (Arcos et al. 2014) is used, which allows to reduce significantly the computational cost associated to this approach when a significant amount of discrete values of frequency are considered.

3 NUMERICAL METHOD

In this section we briefly summarize the numerical 2.5D FEM-BEM model used in this study. The approach is mainly based on the work of François et al. (François et al. 2010).

3.1 *FEM-BEM model*

To represent the at-grade ballasted track with the numerical method proposed, the system is decomposed in two domains i.e. unbounded semi-infinite homogeneous soil and the bounded structure. The domain associated to the track structure is modelled using FEM and the one associated with the soil using BEM. In order to avoid the meshing of the ground surface, the Green's functions of a homogeneous halfspace are the ones used in the proposed BEM. Thus, only the boundary interface common to soil and structure needs to be discretized. Since BEM satisfies Sommerfeld radiation condition, the infinite extent of soil is automatically taken into consideration. Continuity of tractions and displacements is enforced at soil-structure interface. This ensures the coupling between soil and structure. The coupling between FEM and BEM is done in finite element sense. A global stiffness matrix is constructed to obtain the response of the system. The equation for solving system modelled with the coupled 2.5D FEM-BEM method is given by

$$\left[\boldsymbol{K}_{pp} - ik_x\boldsymbol{K}^1_{pp} + k_x^2\boldsymbol{K}^2_{pp} - \omega \boldsymbol{M} + \boldsymbol{K}^s_{pp} \right] \bar{\boldsymbol{U}}_p = \bar{\boldsymbol{F}}_p, \quad (1)$$

where, k_x is the wavenumber, ω is the angular frequency, $\mathbf{K}_{pp}$, $\mathbf{K}^1_{pp}$ and $\mathbf{K}^2_{pp}$ are the stiffness matrices related to the FEM domain, $\mathbf{M}$ the mass matrix of the structure and $\bar{\mathbf{K}}^s_{pp}$ is the stiffness matrix provided by the BEM modelling. The stiffness and mass matrices of the structure are independent of the wavenumber and the frequency while the stiffness matrix of the soil and is a function of them. Moreover, $\bar{\boldsymbol{F}}_p$ represents the vector of external forces and $\bar{\boldsymbol{U}}_p$ is the vector of displacements, both defined in the 2.5D domain. In general, the bar notation denotes variables in wavenumber-frequency domain, vectors are denoted by upper case bold italic alphabets and matrices and tensors are represented by upper case upright bold letters.

In order to obtain the stiffness matrix of the BEM domain, tractions acting on the FEM-BEM interface are needed. The tractions acting on the FEM-BEM interface can be obtained by solving the boundary integral equation. The matrix form of the boundary integral equation is obtained following an iso-parametric nodal collocation approach and is given by

$$\left[\bar{\boldsymbol{T}} + \boldsymbol{I} \right] \bar{\boldsymbol{U}}_s = \bar{\boldsymbol{U}}\bar{\boldsymbol{T}}_s, \quad (2)$$

where, $\bar{\mathbf{T}}, \bar{\mathbf{U}}$ are traction and displacement boundary element system matrices, respectively, $\bar{\boldsymbol{U}}_s$ and $\bar{\boldsymbol{T}}_s$ are nodal displacements and tractions on the boundary, respectively. $\mathbf{I}$ represents the identity matrix. The tractions obtained by solving the above equation need to be converted to nodal forces in order to couple the FE and BE domains. A transformation matrix $\mathbf{T}_q$ is constructed with this objective. This matrix is independent of wavenumber and frequency and is given by

$$\bar{\mathbf{T}}_q = \int \mathrm{N}_s^T \mathrm{N}_s dS, \quad (3)$$

where N_s are the boundary element shape functions. The dynamic soil stiffness matrix can then be obtained as:

$$\mathbf{K}^s_{pp} = \bar{\mathbf{T}}_q \bar{\mathbf{U}}^{-1} \left[\bar{\mathbf{T}} + \mathbf{I} \right]. \quad (4)$$

Knowing the boundary tractions and displacements the radiated wave field at the evaluation points is computed using the following equation:

$$\bar{\boldsymbol{U}}_r = \bar{\boldsymbol{U}}_r \bar{\boldsymbol{T}}_s - \bar{\boldsymbol{T}}_r \bar{\boldsymbol{U}}_s, \quad (5)$$

where, $\bar{\boldsymbol{U}}_r$ is the vector of displacements of the radiated wave field at the evaluation points.

3.2 *Wavenumber sampling*

The sampling strategy used to sample the k_x vector is as follows. For each frequency, a specific set of wavenumbers are needed to describe accurately the Green's function at this frequency in the wavenumber domain. The sampling of the wavenumber is accomplished using the expression:

$$k_x = \kappa_x k_x^{\lim}. \quad (6)$$

The non-dimensional wavenumber κ_x is constructed using dense equispaced distribution from 0 to 1 and less dense logarithmic distribution up to 100. At first $k_x^{\lim}$ is set to $2k_\beta$, where

$$k_{\beta} = \frac{\omega}{\beta}, \tag{7}$$

where β is the wave speed associated to S-waves. In order to compute the k_x vector a pre-sampling procedure is made. In pre-sampling, procedure only the case of minimum and maximum frequency is computed, and for these cases $k_x^{\lim}$ is computed. Then, $k_x^{\lim}$ is extended to other values of the frequency assuming that $k_x^{\lim}$ linearly depends on the frequency.

3.3 *Green's function computation*

The Green's function used in this paper are computed using EDT toolbox (Schevenels et al. 2009). In the case of the elastodynamic problem associated to a halfspace, the Green's functions for stress and displacement depend only on the vertical position of the source, as the soil is invariant in the horizontal direction. The Green's functions are computed for a set of source locations for unique evaluation points and for all wavenumbers. Most of the times, the number of evaluation points used for computation can be reduced because of symmetry of the evaluation points grid in horizontal direction exists. Thus, by multiplying the Green's functions for positive horizontal evaluation points with a suitable matrix of symmetry, the Green's functions the associated negative horizontal evaluation points can be obtained. At first Green's functions for a given frequency for all source locations, for all evaluation points and for all wavenumbers are computed. A mapping scheme is then introduced which returns the Green's functions for one source location and its associated evaluation points for all wavenumbers from the globally computed Green's functions.

4 RESULTS AND DISSCUSION

The model of geometry of the 2.5D FEM-BEM of the studied system is shown in the Fig. 2. The geometry consists of a ballasted railway track resting on a homogeneous ground. The railway track is composed of the rails, the sleepers, the fasteners and the ballast. The fasteners consist of a top elastomer which is in contact with the rails, below which is a top plate, a base plate and another elastomer sandwiched between the top plate and the base plate. The materials associated to the rails, the fasteners, the sleepers and the ballast in FEM-BEM simulations are described with the Young's modulus E, Poisson's ratio v, density ρ and damping ratios D_p and D_s.

Linear triangular elements are used in finite element meshing, while linear line elements are used in boundary element meshing. The boundary element and finite element meshes match at FEM-BEM. Point collocation is used for obtaining

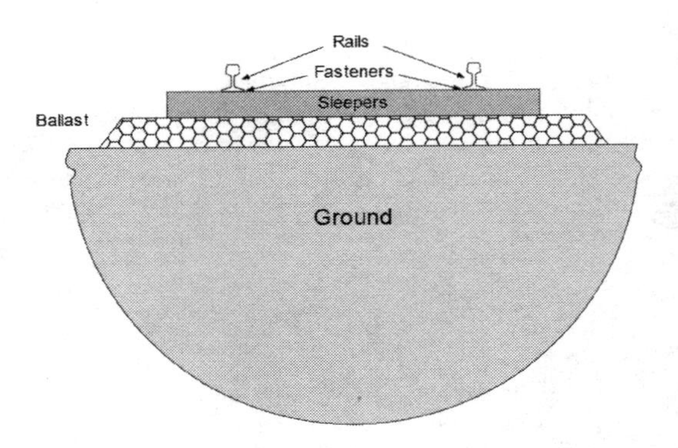

Figure 2. Geometry of the problem.

boundary element matrices. For computing the integrals appearing in the boundary element method, a 6-points Gaussian quadrature is used.

For both the numerical and the semi-analytical models, the system is loaded with vertically harmonic point load acting in downward direction and it is placed on top of each of the rails for the case of the 2.5D FEM-BEM model, and the rail, for the case of the semi-analytical model. An evaluator is also placed on each of the rails, which is used for computing the direct receptances. The ground response is computed on the points located at a distance of 2.5 m, 5.5 m, 10.5 m, 20.5 m from the center of the track (middle point between the to rails in the case of the 2.5D FEM-BEM model), on the ground surface. The response of the system is computed for frequencies upto 80 Hz. An inverse fast Fourier transform is used to obtain the results of the system in space-frequency domain.

At first the rail response is computed for the case of the stiff soil using FEM-BEM. The material properties of the track used in the FEM-BEM simulation are summarised in Table 1, the properties of stiff soil are given in Table 2. Using material properties and considering simple geometrical configuration for the fasteners, the sleeper and the ballast, equivalent stiffness and mass for the track are obtained. These stiffnesses of the track subsystems are then used to compute the rail response using semi-analytical method for the stiff soil. The parameters are then tuned again, so that the rail response for both method match at zero frequency. Fig. 3 shows the rail receptance obtained by both methods, the rail response is matched at zero frequency to obtain equivalent track parameters. The track parameters so obtained are given in Table 3.

After obtaining the equivalent parameters of the track, the response of the system for the soft soil with parameters defined in Table 2 is computed. Fig. 4 shows the rail receptance obtained by both methods for the case of the soft soil, asumming the same track/soil coupling load width for

Table 1. Properties of track used in numerical simulation.

	E [MPa]	ρ [kg/m³]	ν	D_p	D_s
Rail	$207 \cdot 10^3$	7850	0.30	0.001	0.001
Elastomer	1.81	1200	0.45	0.050	0.050
Top Plate	$207 \cdot 10^3$	7850	0.30	0.001	0.001
Elatomer	10	1200	0.35	0.050	0.050
Bottom Plate	$207 \cdot 10^3$	7850	0.30	0.001	0.001
Sleeper	$31 \cdot 10^3$	2500	0.20	0.001	0.001
Ballast	97	1900	0.12	0.030	0.030

Table 2. Properties of soil used in numerical simulation.

	E [MPa]	ρ [kg/m³]	ν	D_p	D_s
Soft	20	1950	0.30	0.040	0.030
Hard	5000	2500	0.30	0.025	0.025

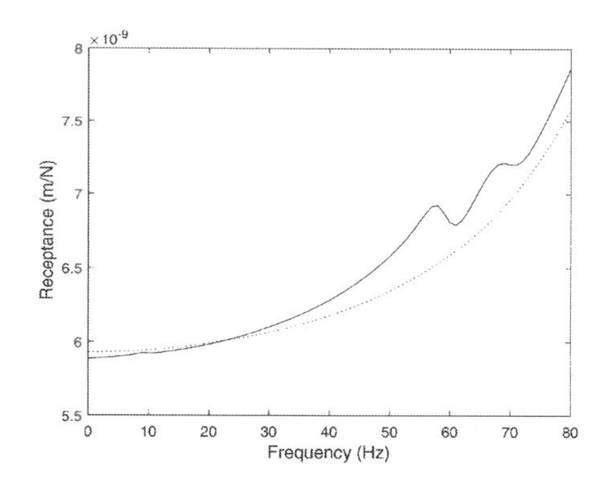

Figure 3. Rail receptance for the case of the stiff soil to obtain equivalent parameters for semi-analytical model for track/soil coupling load width of 2 m, where the solid line represents the solution obtained using the 2.5D FEM-BEM model and the dotted line the solution obtained using the semi-analytical method.

Table 3. Equivalent parameters of track for semi-analytical method.

Parameter	Units	Value
S_r	[m²]	$6.930 \cdot 10^{-3}$
I_r	[m⁴]	$23.50 \cdot 10^{-6}$
E_r	[GPa]	207
ρ_r	[kg/m³]	7850
η_r	[–]	0.001
k_F	[(N/m)/m]	$1.640 \cdot 10^8$
η_F	[–]	0.10
c_F	[(Ns/m)/m]	0
m_s	[kg/m]	1380
k_B	[(N/m)/m]	$9.120 \cdot 10^8$
η_B	[–]	0.060
c_B	[(Ns/m)/m]	0
m_B	[kg/m]	1810

semi-analytical method as was in case of the stiff soil.

The ground response at all evaluator positions, asumming the same track/soil coupling load width for semi-analytical method as was in case of the stiff soil, is also obtained. Fig. 6 shows the real part of vertical ground response, while Fig. 7 shows the imaginary part of vertical ground response.

Then the track/soil coupling load width is increased, for the case of the soft soil, so that the response of the rails are more similar. Fig. 5 shows the rail receptance obtained by both methods for the case of the soft soil, with a larger coupling load width between the track and soil for semi-analytical method.

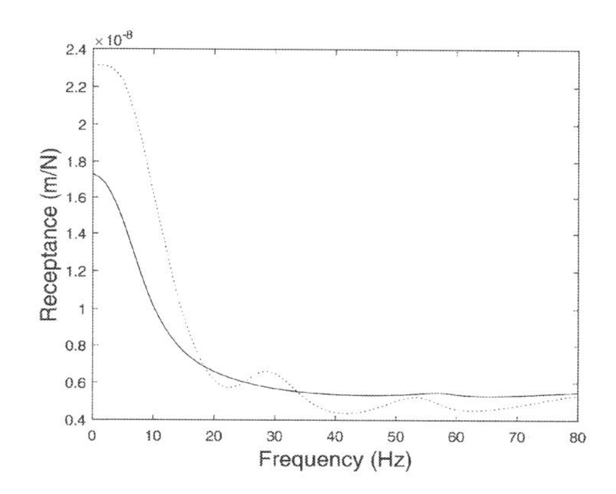

Figure 4. Rail receptance for the case of the soft soil for track/soil coupling load width of 2 m, where the solid line represents the solution obtained using the 2.5D FEM-BEM model and the dotted line the solution obtained using the semi-analytical method.

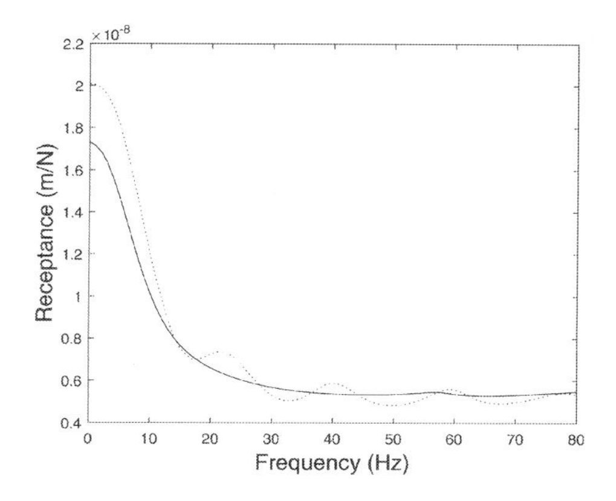

Figure 5. Rail receptance for the case of the soft soil for track/soil coupling load width of 2.8 m, where the solid line represents the solution obtained using the 2.5D FEM-BEM model and the dotted line the solution obtained using the semi-analytical method.

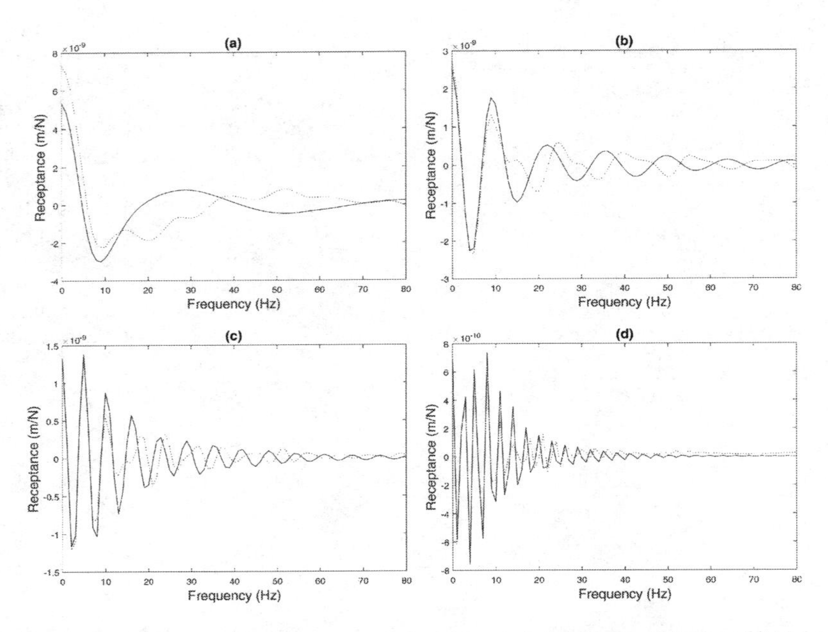

Figure 6. Real part of vertical ground receptances on evaluator located at: (a) 2.5 m; (b) 5.5 m; (c) 10.5 m; (d) 20.5 m, for the case of the soft soil for track/soil coupling load width of 2 m, where the solid line represents the solution obtained using the 2.5D FEM-BEM model and the dotted line the solution obtained using the semi-analytical method.

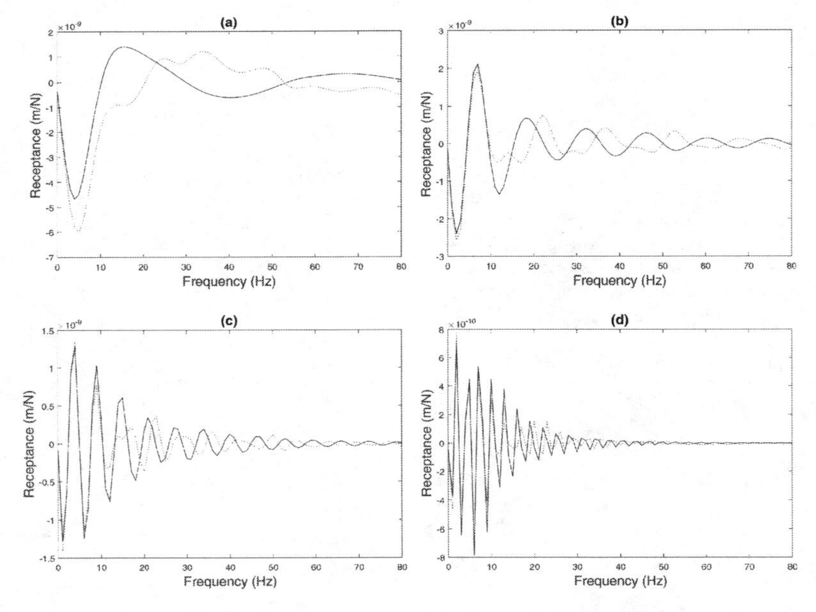

Figure 7. Imaginary part of vertical ground receptances on evaluator located at: (a) 2.5 m; (b) 5.5 m; (c) 10.5 m; (d) 20.5 m, for the case of the soft soil for track/soil coupling load width of 2 m, where the solid line represents the solution obtained using the 2.5D FEM-BEM model and the dotted line the solution obtained using the semi-analytical method.

The ground response at all evaluator positions, for this new track/soil coupling load width is shown in Fig. 8 and Fig. 9. Fig. 8 shows the real part of vertical ground response, while Fig. 9 shows the imaginary part of vertical ground response.

Despite having approximately the same receptance for the case of the stiff soil at zero frequency, from the Fig. 3 it can be said that the semi-analytical method and FEM-BEM approximate the response differently. For the case of the

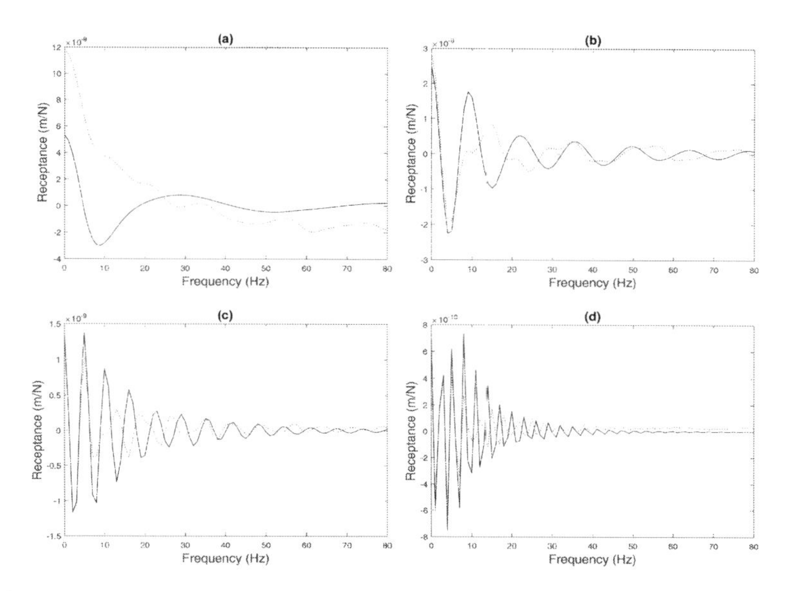

Figure 8. Real part of vertical ground receptances on evaluator located at: (a) 2.5 m; (b) 5.5 m; (c) 10.5 m; (d) 20.5 m, for the case of the soft soil for track/soil coupling load width of 2.8 m, where the solid line represents the solution obtained using the 2.5D FEM-BEM model and the dotted line the solution obtained using the semi-analytical method.

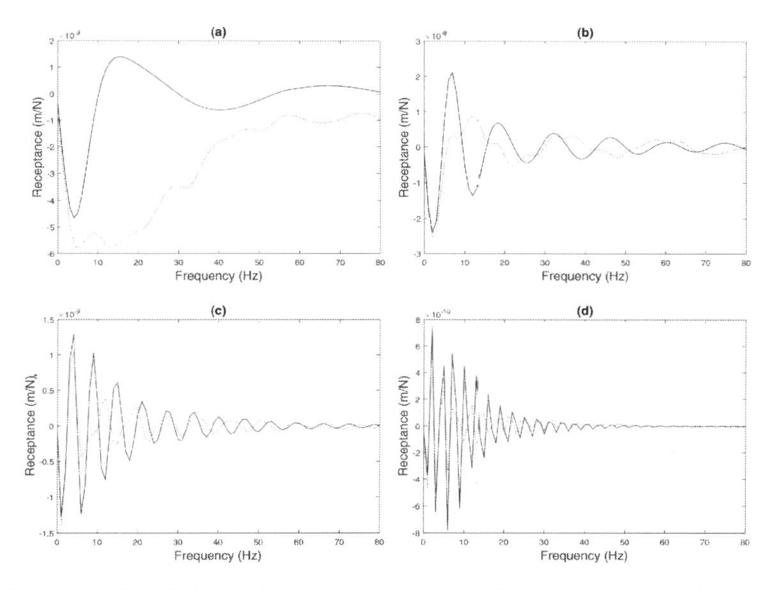

Figure 9. Imaginary part of vertical ground receptances on evaluator located at: (a) 2.5 m; (b) 5.5 m; (c) 10.5 m; (d) 20.5 m, for the case of the soft soil for track/soil coupling load width of 2.8 m where the solid line represents the solution obtained using the 2.5D FEM-BEM model and the dotted line the solution obtained using the semi-analytical method.

stiff soil, it can be said that the semi-analytical method is lagging the FEM-BEM. For the soft soil from Fig. 4 and Fig. 5 the semi-analytical method leads FEM-BEM. Increasing the track/soil coupling load width improves the track receptances and makes the response of semi-analytical method closer to FEM-BEM. Also, it can be said from the figures that in the case of the soft soil the semi-analytical method is estimating some vibration modes which are not shown in FEM-BEM, this behaviour is exactly opposite to the one showed by both methods in case of the stiff soil.

From Fig. 6 and Fig. 7 it can be said that as one moves away from the track the two methods seem to have better agreement. The two methods also seems to have good agreement for all ground evaluators when the frequencies are small. However from Fig. 8 and Fig. 9 increasing the track/soil coupling width results in a large mismatch between the two methods for ground response especially for first ground evaluator. This mismatch is because the first ground evaluator is within the track/soil coupling. The mismatch for the rest of the evaluators the distance between the track and evaluator positions is reduced for semi-analytical method. Also, the mismatch in the ground response could be due to the modes approximated by the semi-analytical method.

The modes that are present in the semi-analytical method for the case of the soft soil could be due to not accurately tuning of the parameters, as simplifying assumption of a complex geometry are made to get the equivalent mass and stiffness. The tuning of the parameters and also the system's response could also be affected by the simplifying assumption made on the coupling load between track and subgrade as it evident from Fig. 4, Fig. 5, Fig. 6, Fig. 7, Fig. 8 and Fig. 9.

5 CONCLUSIONS

This paper presents a preliminary study of the validity of considering a rectangular strip track/soil coupling load. The semi-analytical method is compared against a 2.5D FEM-BEM model for homogeneous soil. Equivalent parameters of the semi-analytical method required for comparing both methods were obtained by matching rail response of both the methods at zero frequency using a very stiff soil as subgrade. From the parameters obtained, the semi-analytical method was compared against FEM-BEM for the case for the soft soil.

Comparing the results obtained by simulation it can be concluded that a better approximation of the coupling load needs to be made in order to get accurate response of the system. The parameters required for testing the semi-analytical model also have an important role in making comparison with FEM-BEM and as such the obtaining of equivalent parameters for semi-analytical model from the parameters used in FEM-BEM, needs to be studied in more detail.

ACKNOWLEDGEMENT

This work has been supported by the Ministerio de Economía, y Competitividad (ES) under the research grant BES-2015-071453, for the project "Soluciones innovadoras para el aislamiento de edificios a las vibraciones producidas por infraestructuras ferroviarias soterradas".

REFERENCES

Alves Costa, P., R. Calçada, & A. Silva Cardoso (2012, jan). Track-ground vibrations induced by railway traffic: In-situ measurements and validation of a 2.5D FEM-BEM model. *Soil Dynamics and Earthquake Engineering 32*(1), 111–128.

Arcos, R., A. Clot, J. Romeu, & S.R. Martin (2014, 02). Fast computation of an infinite, longitudinally-varying and harmonic strip load acting on a viscoelastic half-space. *43*, 5867.

François, S., M. Schevenels, P. Galvín, G. Lombaert, & G. Degrande (2010, apr). A 2.5D coupled FE-BE methodology for the dynamic interaction between longitudinally invariant structures and a layered half-space. *Computer Methods in Applied Mechanics and Engineering 199*(23–24), 1536–1548.

Galvín, P., S. François, M. Schevenels, E. Bongini, G. Degrande, & G. Lombaert (2010, dec). A 2.5D coupled FE-BE model for the prediction of railway induced vibrations. *Soil Dynamics and Earthquake Engineering 30*(12), 1500–1512.

Ntotsios, E., D. Thompson, & M. Hussein (2015). Effect of rail unevenness correlation on the prediction of groundborne vibration from railways. *Journal of Sound and Vibration 402*(May), 1237–1242.

Ozdemir, Z., P. Coulier, M.A. Lak, S. François, G. Lombaert, & G. Degrande (2013). Numerical evaluation of the dynamic response of pipelines to vibrations induced by the operation of a pavement breaker. *Soil Dynamics and Earthquake Engineering 44*, 153–167.

Romero, A., P. Galvín, J. António, J. Domínguez, & A. Tadeu (2017). Modelling of acoustic and elastic wave propagation from underground structures using a 2.5D BEMFEM approach. *Engineering Analysis with Boundary Elements 76*(July 2016), 26–39.

Schevenels, M., S. François, & G. Degrande (2009). EDT: An ElastoDynamics Toolbox for MATLAB. *Computers and Geosciences 35*(8), 1752–1754.

Sheng, X., C. Jones, & D. Thompson (2005). Modelling ground vibration from railways using wavenumber finite and boundary-element methods. *Proceedings of the Royal Society A: Mathematical, Physical and Engineering Sciences 461*(2059), 2043–2070.

Sheng, X., C.J.C. Jones, & M. Petyt (1999a). Ground vibration generated by a harmonic load acting on a railway track. *Journal of Sound and Vibration 225*(1), 3–28.

Sheng, X., C.J.C. Jones, & M. Petyt (1999b, nov). Ground vibration generated by load moving along railway track. *Journal of Sound and Vibration 228*(1), 129–156.

Sheng, X., C.J.C. Jones, & D.J. Thompson (2004, may). A theoretical model for ground vibration from trains generated by vertical track irregularities. *Journal of Sound and Vibration 272*(3–5), 937–965.

Sheng, X., C.J.C. Jones, & D.J. Thompson (2006, jun). Prediction of ground vibration from trains using the wavenumber finite and boundary element methods. *Journal of Sound and Vibration 293*(3–5), 575–586.

Numerical Methods in Geotechnical Engineering IX – Cardoso et al. (Eds)
© 2018 Taylor & Francis Group, London, ISBN 978-1-138-33198-3

DEM modelling of dynamic penetration in granular material

N. Zhang, M. Arroyo & A. Gens
Polytechnic University of Catalonia (UPC), Barcelona, Spain

M. Ciantia
University of Dundee, Dundee, UK

ABSTRACT: Dynamic penetration has been widely employed in some site investigation procedures and for the installation of driven piles. A 3-dimensional discrete element method model has been developed to simulate dynamic rod penetration into a calibration chamber. The chamber has been filled with a scaled analogue of Fontainebleau sand. A novel method for simulating the physical rod as boundary condition is presented. It is shown that the method leads to a good agreement in static penetration comparisons. A parametric study of involved variables shows that penetration increases with force magnitude and impact time but tip resistance is quite insensitive to them. The tests are then extended to six conditions that combine two density levels and three confining stresses. It is found that increasing the confining stress does increase tip resistance, whereas it reduces penetration distance. As expected, penetration in dense sand results in larger penetration resistances and smaller penetration distances.

1 INTRODUCTION

Dynamic probing involve driving a device into the soil by striking it with a hammer. It has been employed for several widely used site investigation procedures such as Standard Penetration Test (SPT), Becker Penetration Test (BPT) and Dynamic Cone Penetration Test (DCPT), as well as for the installation of driven piles. It is therefore important to understand properly the dynamic response of the interaction between the driven device and the soil. Various approaches, like theoretical solutions (Smith 1960; Baligh 1985; Houlsby 1988), laboratory tests (Jardine et al. 2013; Yang et al. 2010) and numerical simulations (Chow 1981; Kiousis et al. 1988; Lobo-Guerrero & Vallejo 2007; Vallejo & Lobo-Guerrero 2005) have been proposed to address this topic. Discrete Element Method (DEM) has been successfully applied to many geotechnical problems, but not in tackling dynamic penetration problems in sands yet. Hence, our goal is to develop a DEM modeling procedure to simulate dynamic penetration processes.

The DEM code PFC3D (Itasca, 2016) has been used to construct the numerical models. In this paper, after briefly introducing the employed methodology of modelling dynamic penetration, a series of simulations of a rod driven into 3-dimensional samples of Fontainebleau sand are presented. Particular attention has been paid to the evolution of rod tip resistance under various force inputs and ground conditions.

2 METHODOLOGY

2.1 *Numerical approach*

In DEM, two independent approaches can be used to mimic rod driving. One is that a rod may be represented by a large amount of particles and then the hammer is dropped on the rod head. However, this approach leads to a high computational cost. Alternatively, the rod can be ideally represented by rigid wall. The hammer impact is then simulated by prescribing a time-dependent force on the rod, i.e. the force-time signal generated by the hammer impact is required as input. Schmertmann & Palacios (1979) proposed that the magnitude of the impact force can be calculated directly from the hammer velocity and hammer and rod properties. Thereby, it can be computed that the maximum force magnitude is 251 kN for the case of a 63.5 kg

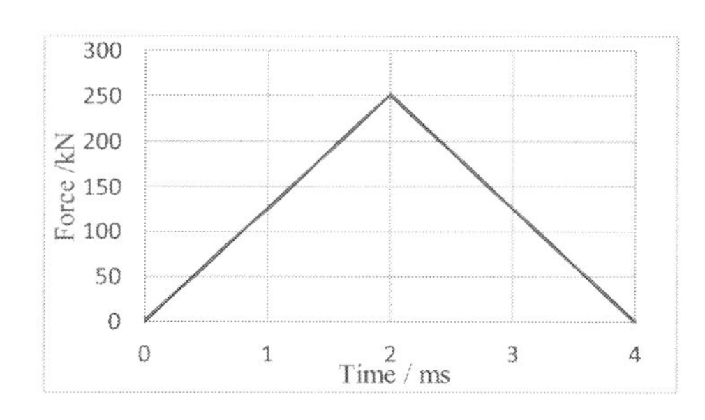

Figure 1. Computed impact force-time signal.

hammer falling on the rod head over a fixed fall distance of 0.76 m, corresponding to SPT practice (ASTM D1586-11, 2011). The impact lasts 4 ms. Figure 1 shows the computed force-time signal.

2.2 *Model construction*

The construction of 3-dimensional calibration chamber (CC) models was carried out by adapting the procedure described by Arroyo et al. (2011) and Ciantia et al. (2016). The chamber was filled with an analogue of Fontainebleau sand, which was previously tested by Ciantia et al. (2015, 2018). A scaling factor of 79 was applied in order to achieve a manageable number of particles. Isotropic compression to 5 kPa in which the inter-particle friction might be reduced was used to obtain - by trial and error - the targeted porosity. After equilibration, isotropic compression to a higher stress level was performed by using wall servo-control. The flat-ended rod is then created by using frictional rigid walls. During the penetration stage, boundary conditions were slightly changed by fixing the bottom wall. The model details can be seen in Figure 2 and Table 1.

2.3 *Simulation procedures*

The specimens were generated by combining two density levels, namely *dense* (n = 0.385) and *loose* (n = 0.44) and three confining stress levels (100 kPa, 200 kPa and 400 kPa). After the equilibration stage, the rod was firstly penetrated into the sample at a constant rate of 40 cm/s until the tip reached the depth of 15 cm. The purpose is to prevent the influence of chamber boundaries on dynamic penetration resistance. The rod was then slightly pulled out of the sample in order to eliminate the residual resistance. At this point, the sample is in an appropriate state for launching dynamic penetration. A series of single impact tests characterized by different input forces were performed on the dense specimen compressed to 100 kPa. Subsequently, dynamic penetration was applied to the other 5 specimens, prescribing the same force-time signal. All the simulations employed a local non-viscous damping of 0.05.

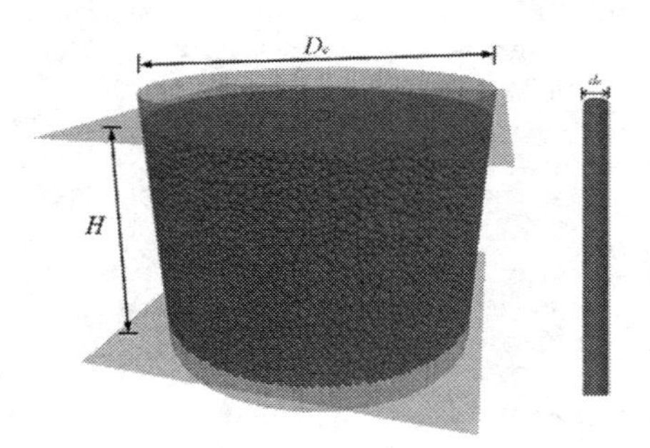

Figure 2. View of DEM model of calibration chamber and rod.

Table 1. Parameters of calibration chamber model.

Variable (unit)	Symbol	DEM
Chamber diameter (mm)	D_c	760
Rod diameter (mm)	d_c	72.1
Chamber height (mm)	H	500
Scaling factor	–	79
Particle mean size in DEM (mm)	D_{50}	21*79
Chamber/rod diameter ratio	$D_c / d_c = R_d$	15.0
Rod/particle ratio	$d_c / D_{50} = n_p$	3.06

3 RESULTS

3.1 *Verification of the proposed methodology*

To verify the proposed methodology for simulating dynamic penetration in sands, an impact was first applied to the rod head in the dense sample compressed to 100 kPa. Tip resistance, which can represent the reaction of the sand to the penetration of the rod, was obtained from the resultant of the force acting on the tip divided by the cross-sectional area of the tip. The evolution of tip resistance during dynamic penetration was recorded (Fig. 3). It shows that the penetration resistance curve produces results that follow the expected trend. The rod was driven to a permanent penetration of 0.02 m. It shows that the force applied on the rod head is large enough to drive the rod into the sand.

Recently, it has been shown empirically that, in granular materials, the dynamic penetration corresponds well with static cone penetration resistance (Schnaid et al., 2017). It is therefore attractive to try to verify this proposal by 3D numerical DEM models. A static penetration test has been conducted in the same specimen mentioned above.

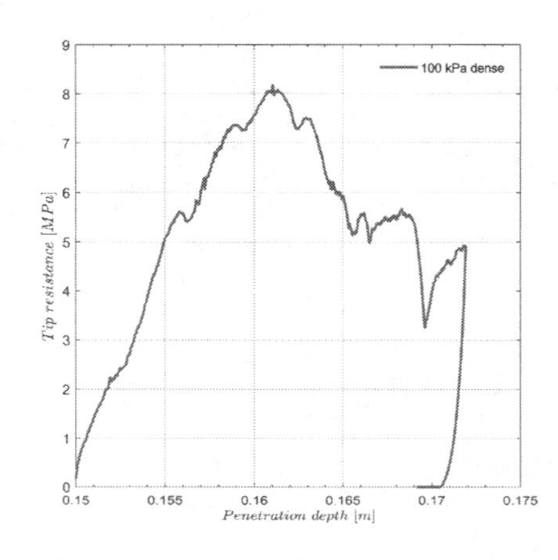

Figure 3. Penetration tip resistance vs penetration depth.

The final penetration was interrupted at the depth of 0.25 m. A direct comparison between profiles of static penetration and dynamic penetration is shown in Figure 4. It can be seen that both values show a good agreement.

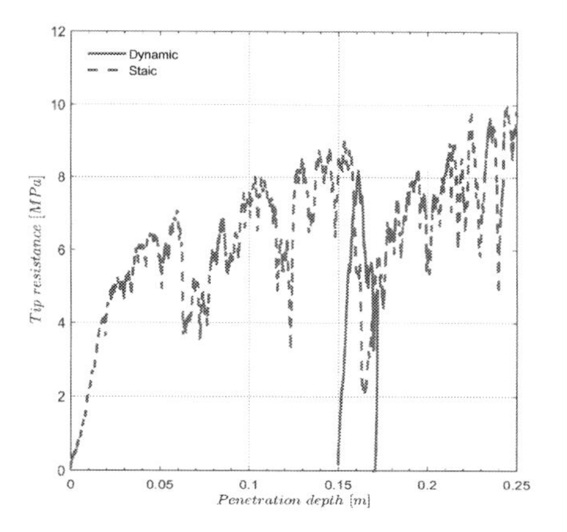

Figure 4. Comparison between static and dynamic penetration resistance.

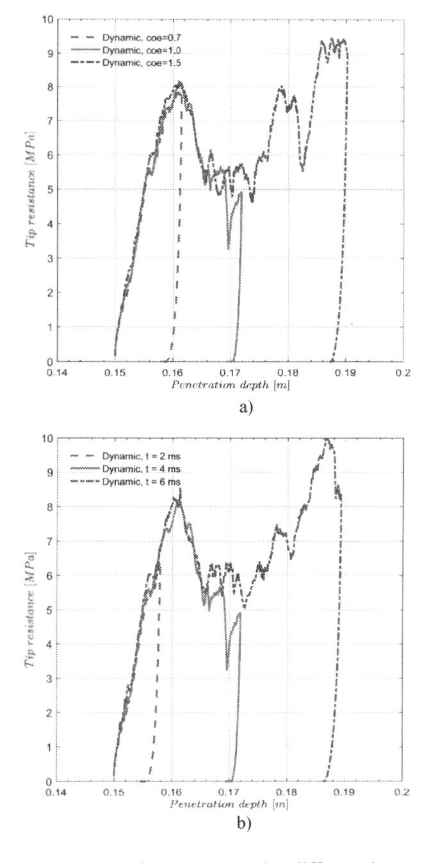

Figure 5. Penetration curves under different impact signals: a) different forces; b) different impact time.

3.2 *Effect of different force-time inputs*

In the previous section, the soundness of proposed approach has been highlighted, so it is of interest to perform a parametric study of some of the variables involved in dynamic penetration. As different hammer and rod configurations affect the force-time input signals, it is interesting to investigate the effect of various force inputs and different impact time on the dynamic penetration response. To this end, two series of tests were conducted: in one, the force used before is multiplied by 0.7 and 1.5 and the time is kept constant at 4 ms whereas in the other, an impact time of 2 and 6 ms is used applying the same force magnitude.

The corresponding penetration curves are presented in Figure 5. It can be noted that all the curves exhibit a similar evolution until a sharp reduction of the resistance occurs. Penetration increases with the magnitude of the force and the impact time but tip resistance is quite insensitive to them.

3.3 *Influence of ground conditions*

There are a great number of ground conditions that are known to influence the dynamic tip resistance

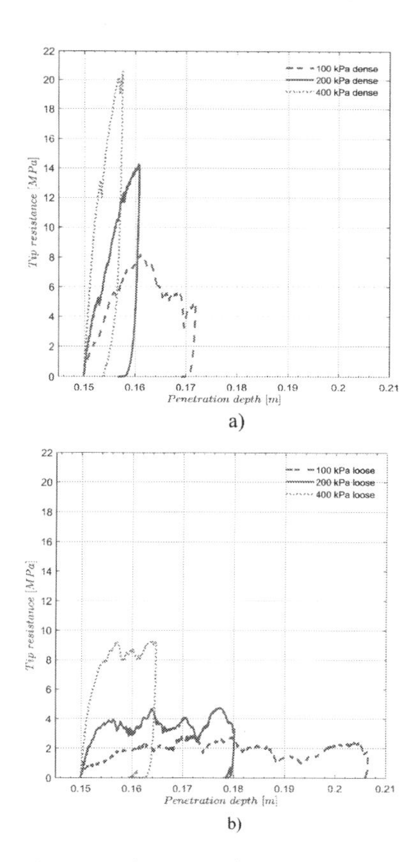

Figure 6. Dynamic penetration curves in: a) dense sand, b) loose sand.

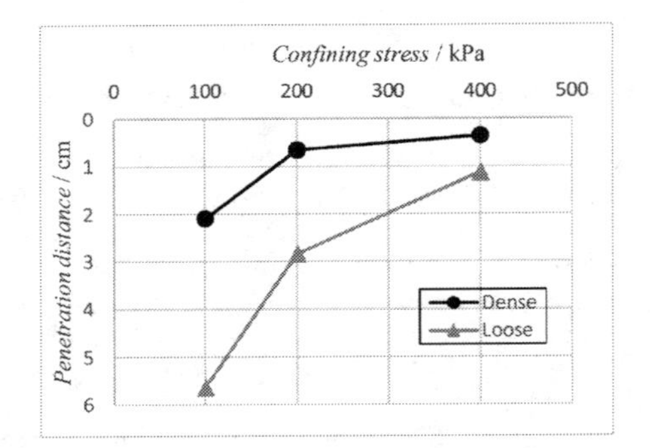

Figure 7. Penetration distance versus confining stress for dense and loose samples.

in sand. Here, the effects of two important factors (confining stress and porosity) are examined. Figure 6 shows the dynamic penetration curves obtained from dense samples and loose samples at three different confining stresses. The penetration distance for each sample is plotted in Figure 7. It is apparent that increasing the confining stress increases tip resistance, while it reduces penetration distance, for both dense and loose materials. Under the same confining stress, the dense material naturally results in a much higher tip resistance and a smaller penetration distance.

4 CONCLUSIONS

A 3-dimensional DEM model has been created to simulate rod dynamic penetration into calibration chambers filled with a scaled analogue of Fontainebleau sand. It has been shown that the proposed approach for simulating the physical rod as a boundary condition performs well in comparisons with static penetration tests. The parametric study of variables shows that penetration increases with force magnitude and impact time but tip resistance is quite insensitive to them. Increasing the confining stress increases tip resistance whereas it reduces penetration distance. As expected, dynamic probing in dense sand results in larger penetration resistances and smaller penetration distances compared to loose sand.

ACKNOWLEDGEMENTS

The support of EU through 645665 – GEO-RAMP – MSCA-RISE and China Scholarship Council is gratefully acknowledged.

REFERENCES

Arroyo, M., Butlanska, J., Gens, A., Calvetti, F., & Jamiolkowski. M., 2011. "Cone Penetration Tests in a Virtual Calibration Chamber." *Géotechnique* 61(6): 525–531.

Baligh, M.M. 1985. Fundamentals of deep penetration: I, soil shearing and point resistance II, pore pressures, Research Reports R85-9, R85-10, Dept.of.Civil Engineering, MIT. Teh C.I.

Chow, Y. K. 1981. "Dynamic Behaviour of Piles." University of Manchester.

Ciantia, M. O., Aroyo, M., Calvetti, F., & Gens, A., 2015. "An Approach to Enhance Efficiency of DEM Modelling of Soils with Crushable Grains." *Geótechnique* 65(2):91–110.

Ciantia, M. O., Arroyo, M., Butlanska, J., & Gens, A., 2016. "DEM Modelling of Cone Penetration Tests in a Double-Porosity Crushable Granular Material." *Computers and Geotechnics* 73:109–127.

Ciantia, M., Arroyo, M., O'Sullivan, C., Gens, A. & Liu, T. (2018) Grading evolution and critical state in a discrete numerical model of Fontainebleau sand, *Géotechnique*.

Houlsby G.T. 1988. Analysis of the cone penetration test by the strain path method, *Proc. 6th ICONMIG*, Innsbruck, pp.397-402, 1988.

Itasca Consulting Group. 2016. "PFC3D: Itasca Consulting Group, Inc. 2014. PFC — Particle Flow Code, Ver. 5.0 Minneapolis, USA."

Jardine, R. J., Yang Z. X., Foray, P., & Zhu, B. T. 2013. "Interpretation,of Stress Measurements Made around Closed-Ended Displacement Piles in Sand." *Géotechnique* 63(8):613–627.

Kiousis, P.D., Voyiadjis, G.Z. & Tumay M.T. 1988. A large strain theory and its application in the analysis of the cone penetration mechanism, *Int. J. Numer. Anal. Methods Geomech. 12*:1, 45-60.

Lobo-Guerrero, S. & Vallejo, L. E., 2007. "Influence of Pile Shape and Pile Interaction on the Crushable Behavior of Granular Materials around Driven Piles: DEM Analyses." *Granular Matter* 9(3–4):241–250.

Schmertmann, J. H. & Palacios, A., 1979. "Energy Dynamics of SPT." *Journal of the Geotechnical Engineering Division* 105(GT8):909–926.

Schnaid, F., D. Lourenço, & E. Odebrecht. 2017. "Interpretation of Static and Dynamic Penetration Tests in Coarse-Grained Soils." *Géotechnique Letters* 7(2):1–6.

Smith, E. A. 1960. "Pile-Driving Analysis by the Wave Equation." *Journal of the Engineering Mechanics Division* 86(EM 4):35–61.

Vallejo, L. E. & Lobo-Guerrero, S., 2005. "DEM Analysis of Crushing around Driven Piles in Granular Materials." *Géotechnique* 55(8):617–623.

Yang, Z. X., Jardine, R. J., Zhu, B. T., Foray, P., & Tsuha, C. H. C., 2010. "Sand Grain Crushing and Interface Shearing during Displacement Pile Installation in Sand." *Géotechnique* 60(6):469–482.

Numerical Methods in Geotechnical Engineering IX – Cardoso et al. (Eds)
ISBN 978-1-138-33198-3

A methodology for the 3D analysis of foundations for marine structures

P. Mira
Laboratorio de Geotecnia, CEDEX, Madrid, Spain

J.A. Fernández-Merodo
Instituto Geológico y Minero, Madrid, Spain

M. Pastor, D. Manzanal, M.M. Stickle & A. Yagüe
Universidad Politécnica de Madrid

I. Rodríguez
Formerly Puertos del Estado, Madrid, Spain

J.D. López
Puertos del Estado, Madrid, Spain

A. Tomás, G. Barajas & J. López-Lara
Environmental Hydraulics Institute, Universidad de Cantabria, Santander, Spain

ABSTRACT: A methodology for the 3D analysis of marine foundations is presented. The response in displacements, stresses and pore water pressures is obtained from a finite element coupled formulation implemented in the GeHoMadrid model. Loads due to wave action on the foundation are obtained from the IHFOAM model that solves the three-dimensional Reynolds Averaged Navier-Stokes (RANS) equations using a finite volume discretization and the volume of fluid (VOF) method. Additionally, the methodology includes a Generalized Plasticity based constitutive model for granular materials capable of representing liquefaction fenomena of sands subjected to cyclic loading, such as those frequently appearing in the problems studied. This methodology is applied to the study of the response of a caisson breakwater foundation.

1 INTRODUCTION

The development of an accurate numerical model to represent the behavior of foundations in a marine environment taking into account the different and complex hydraulic and mechanical interactions involved has been for many years a subject of great interest for the ocean engineering community. This interest is currently even greater due to the important development in recent years of offshore wind energy.

The purpose of this paper is to present a rational approach for the 3D analysis of this problem, based on the weak coupling of two models, one for each of the computational subdomains involved.

The first computational subdomain or fluid subdomain includes the ocean water. To represent the behavior of ocean water a numerical model based on the Reynolds Averaged Navier Stokes equations (RANS) is used. These equations are discretized using a Finite Volume technique combined with the Volume of Fluid (VoF) method to account for free interfaces.

The second subdomain includes the remaining components of the problem such as saturated granular media associated with the seabed or rockfill and "dry" solid media such as the upper foundation elements made of concrete or other solid materials (e.g. breakwater caissons, wind turbine towers, etc). To model the behavior of saturated granular media and "dry" media the Biot Equations are used and subsequently discretized using a coupled Finite Element formulation based on the displacements and pore pressure fields. "Dry" solid elements can be regarded as a particular case of saturated granular media, with zero pore fluid pressure.

The present work includes the description of the different ingredients of the model and an application to the analysis of a breakwater caisson foundation.

Section 2 presents the main features of the model associated with the first computational subdomain, the ocean water. Section 3 includes the mathematical and numerical models for saturated granular media: the Biot equations in three dimensions and a mixed formulation u-pw with enhanced deformations [Simo and Rifai, 1990], [Mira et al, 2003]. Section 3 also presents the principles main features of the Pastor-Zienkiewicz constitutive model of for loose sands, a basic ingredient of the analysis of the problem under study, if the phenomenon of seafloor liquefaction is to be adequately represented. Section 4 presents the application of the model to the case of a breakwater caisson foundation subjected to the action of a storm, along with the most representative graphical results.

2 FLUID MODEL

IHFOAM (www.ihfoam.ihcantabria.com) is a model developed in IHCantabria oriented to the three-dimensional study of the flow-structure interaction, that is, to the study of fluid action with structures, such as the effect of waves on maritime structures. The model has been widely validated and its results have been contrasted with laboratory measurements. This makes it possible to use the IHFOAM model to solve real problems, ensuring the reliability of the results.

IHFOAM is a three-dimensional numerical model of recent creation, and therefore in continuous development. It is based on OpenFOAM, a very advanced and robust multiphysics library that currently has a large presence in the industry.

The model developed is in the state of the art of hydraulic engineering, allowing the simulation of biphase flow, even inside porous media, such as breakwaters, and therefore providing it with the ability to simulate all types of structures, both hydraulic and of costs. This includes not only the static structures, but also the calculation of floating structures. Its capabilities cover the simulation both in laboratory scale and prototype scale, in which today we have been able to calculate domains in the order of 1 km^2.

Among its distinctive features, the application of self-developed boundary conditions in IHCantabria that allow the generation of all types of waves (regular, irregular, solitary...), linked to an active absorption system as well as any stationary flow condition or transient. This aspect makes it optimal for the simulation of hydraulic processes in flows with free lamina.

IHFOAM is an advance compared to other commercial models since it allows the generation of flow in the contours, which allows to shorten the calculation domains and reduce computational times. Another one of the advances is without a doubt the calculation of flow in porous media, without which it would be reduced to the simulation of impermeable structures.

IHFOAM allows simulating gravity waves (Higuera et al, 2013a,b), porous media flows (Higuera et al, 2014a,b) and wave interation with vegetation (Maza et al., 2015). It also incorporates a set of algorithms to generate and absorb waves at the boundaries without the use of relaxation zones, speeding-up the simulations and ensuring a correct representation of the wave-induced hydrodynamics in the numerical domain.

Additionally, dynamic meshing allows the calculation of floating structures that interact with fluids, responding to stresses based on the geometry and properties of the solid. All are defined by the effort made by IHCantabria to develop a numerical model as realistic as possible and validated in a very broad spectrum of laboratory cases.

IHFOAM solves the Navier-Stokes equations averaged by Reynolds in three dimensions for two incompressible phases by finite volumes. These equations are conservation of mass and conservation of the amount of movement, and are shown below:

$$\nabla \cdot U = 0$$

$$\frac{\nabla \rho U}{\partial t} + \nabla \cdot (\rho U U) - \nabla \cdot (\mu_{\text{eff}} \nabla U) = -\nabla p^* - g \cdot X \nabla \rho + \nabla U \nabla \mu_{\text{eff}} + \sigma \kappa \nabla \alpha$$

In these equations U is the velocity vector of the fluid; it is the density; it is the dynamic pressure; it is the acceleration vector of gravity; It is the vector position. The last term on the right side of the second equation takes into account the effect of surface tension: it is the surface tension coefficient; it is the curvature of the free surface; is the parameter of the VOF function, which will be explained below. Finally, it is the effective dynamic viscosity, that is, it takes into account both the dynamic molecular viscosity and the turbulent effects:

$$\mu_{\text{eff}} = \mu + \rho \nu_{\text{turb}}$$

The elements of the momentum conservation equation have a certain disposition. The elements of the left member are used to assemble the coefficient matrix, while the rest are calculated explicitly, and are part of the independent term of the equation.

The modeling of the free surface is carried out by means of the VOF (Volume Of Fluid) technique, which allows the characterization of very complex wave configurations in a simple way. This function is controlled by a parameter, called, which controls what type of fluid is present in the cells. If it takes the value 1 it indicates that the cell is full of water, whereas if it takes the value 0, the cell contains air. In case of intermediate values, the cell will contain a mixture of air and water. Using the VOF function it is straightforward to calculate the properties of the fluid in each cell, as the average averaged. For example, the density is calculated as follows:

$$\rho = \alpha\rho_{\mathbf{waguaer}} + (1-\alpha)\rho_{\mathrm{air}}$$

The VOF function controls the movement of the phases within the numerical domain. Its variation is controlled by an additional equation, in the same way as a classical advection equation:

$$\frac{\partial \alpha}{\partial t} + \nabla \cdot \boldsymbol{U}\alpha + \nabla \cdot U_c \alpha(1-\alpha) = 0$$

The only new term is, a compression speed that only affects the interface between both fluids. The last term allows an interface to be kept as narrow as possible, since the equation without this term is diffusive. Thanks to this, it helps to comply with one of the restrictions of using the VOF technique, which is to work on a non-diffuse interface. Other restrictions when applying the VOF technique include that the advection equation is conservative, which is fulfilled, and that the solution is always bounded between 0 and 1. This last condition is fulfilled due to the applied resolution method, called MULES (Multidimensional Universal Limiter for Explicit Solution), which applies limiting factors when solving the advection equation to ensure obtaining a bounded solution between 0 and 1.

In case of having porous media (areas with loose materials such as breakwaters or gabions), IHFOAM solves the so-called VARANS (Volume-Averaged Reynolds-Averaged Navier-Stokes) equations, which averaged these areas volumetrically.

It is noteworthy that IHFOAM also has a large number of turbulence models, including the k-ε and k-ω SST, which work both outside and inside porous media.

In summary, IHFOAM has tools that allow obtaining and interpreting completely three-dimensional results of pressures, velocities and turbulent variables. Likewise, the advanced set of boundary conditions that have been developed exclusively for the model, make it unique for the simulation of flows in channels and with mixed conditions of imposed flows and tidal or wave conditions. This makes it possible to calculate the flow demands on hydraulic and coastal structures, to evaluate their operation and functionality, or to obtain the spatial and temporal variation of any relevant variable in the field of hydraulic engineering.

In short, IHFOAM has tools that allow obtaining and interpreting completely three-dimensional results of pressures, velocities and turbulent variables. This makes it possible to calculate the waves' demands on maritime structures, to evaluate their operation and functionality, the hydrodynamic conditions in the swash zone or any other relevant variable in the field of coastal or hydraulic engineering.

3 SATURATED GRANULAR MEDIA MODEL

Traditionally problems in Geotechnics and Soil Mechanics have been solved assuming either that the soil is totally dry or that the soil is totally saturated. The concept of effective stress was proposed by [Terzaghi, 1936] and the first mathematical model that described the coupling between the solid skeleton and the interstitial fluid was proposed by [Biot, 1941,1955]. It was a model for linear elastic materials using as field variables the displacements of the solid matrix **u** and the displacements of the fluid phase relative to the solid matrix **w**. [Ghaboussi and Wilson, 1972] were the first to propose a numerical model using the finite element technique to solve the Biot equations. Later [Zienkiewicz and Bettess, 1982] reformulated the model more simply as a function of the displacements of the solid phase **u** and pressure of the fluid phase pore pressure p_w. This formulation assumes that the effect of high frequencies in the model response is negligible. Due to this assumption its range of application is smaller than Ghaboussi and Wilson's, although this does not affect the quality of results obtained in typical civil engineering applications as the present one. Subsequent versions extended the theory to nonlinear materials and large deformations, highlighting the work of Professor Zienkiewicz and his team at the University of Swansea [Zienkiewicz et al, 1980, 1990a, 1990b, 1999]. It is this mathematical model that is used in the present work.

This type of mixed problems is similar to others found in solid mechanics and fluid mechanics and can result in numerical instabilities unless certain requirements are met. There are two classical approaches to the problem of numerical instability in coupled formulations. The first is usually referred to as the Zienkiewicz-Taylor patch test for mixed formulations. As a consequence of this test, it is concluded that the degree of interpolation of the displacement field must be greater than that corresponding to the pressure field. This was the path adopted in the classical formulations of the Zienkiewicz group. Mathematically speaking, this is a necessary condition for stability. The second approach, mathematically more complex and demanding, is generally known as the Babuska-Brezzi inf-sup condition and constitutes a sufficient condition for stability.

However, it is possible to obtain stable formulations by avoiding the requirement of the degree of interpolation through the so-called stabilization techniques. These techniques were initially applied in the context of fluid mechanics and later extended to the mechanics of solids.

In this work, he presents the 3D version of a formulation in which stabilization is achieved through an approach based on the Simo-Rifai [Simo and Rifai, 1990], [Mira et al, 2003].

3.1 *Field equations*

Soils are porous materials constituted by a solid matrix traversed by a network of pores. The pores are saturated with water below the water table and contain air and water above the water table, a condition known as semi-saturation or partial saturation. The mechanical and hydraulic behavior of a soil is governed by the interaction of solid matrix and interstitial fluids.

Changes in boundary conditions (F, u, p, etc) produce fluid migrations from one area to another and changes in charge distribution between solid and fluids.

The model used, original of Biot contains the following ingredients:

a. Momentum balance

$$\mathbf{S}(\sigma' - \mathbf{m}p) + \rho\mathbf{b} = \mathbf{0}$$

b. Mass balance

$$\mathbf{m}^T\mathbf{S}\dot{\mathbf{u}} - \nabla^T\mathbf{k}\nabla p + \frac{\dot{p}}{Q^*} + \nabla^T\mathbf{k}\rho_w\mathbf{b} = 0$$

c. Constitutive relationships:

$$d\boldsymbol{\sigma}' = \mathbf{D}^{ep} d\boldsymbol{\varepsilon}$$

d. Kinematic compatibility conditions:

$$d\boldsymbol{\varepsilon} = \mathbf{S}\mathbf{u}$$

where:

$$\mathbf{S}^T = \begin{bmatrix} \frac{\partial}{\partial x} & 0 & 0 & \frac{\partial}{\partial y} & 0 & \frac{\partial}{\partial z} \\ 0 & \frac{\partial}{\partial y} & 0 & \frac{\partial}{\partial x} & \frac{\partial}{\partial z} & 0 \\ 0 & 0 & \frac{\partial}{\partial z} & 0 & \frac{\partial}{\partial y} & \frac{\partial}{\partial x} \end{bmatrix}$$

σ' = Effective stress vector.
$\mathbf{m}^T = [1\ 1\ 1\ 0\ 0\ 0]$
p = Pore pressure
ρ = Solid-water mixture density
$\mathbf{u}$ = Displacement vector
$\mathbf{b}$ = Gravity direccional vector

$$\nabla^T = \begin{bmatrix} \frac{\partial}{\partial x} & \frac{\partial}{\partial y} & \frac{\partial}{\partial z} \end{bmatrix}$$

Q^* = Compressibility coefficient for solid-fluid mixture.
k = Water permeability
ρ_w = Water density

3.2 *Numerical model*

The previously stated equations are used as a basis for the numerical model. An additional equation is introduced to account for the enhanced dtrain field [Simo and Rifai, 1990]:

$$\varepsilon = \nabla^s\mathbf{u} + \tilde{\varepsilon} \; where: \tilde{\varepsilon} = \mathbf{G}(\xi)\boldsymbol{\alpha}_e$$

$$and: \mathbf{G}(\xi) = \frac{j_0}{j(\xi)}\mathbf{F}_0^{-T}\mathbf{E}(\xi)$$

$$\mathrm{E}(\xi) = \begin{bmatrix} \xi & 0 & 0 & 0 & \xi\eta & 0 & 0 \\ 0 & \eta & 0 & 0 & 0 & \xi\eta & 0 \\ 0 & 0 & \xi & \eta & 0 & 0 & \xi\eta \end{bmatrix}$$

Discretizing in space with finite element techniques by enforcing the satisfaction of these equations in a weak fashion along with the orthogonality of total stresses and enhanced strains the following three equations are obtained:

$$\int_\Omega \mathbf{B}^T \boldsymbol{\sigma}' \, d\Omega - \mathbf{Q}_u\bar{\mathbf{p}} = \mathbf{f}_u$$

$$\int_\Omega \mathbf{G}^T \boldsymbol{\sigma}' \, d\Omega - \mathbf{Q}_\alpha\bar{\mathbf{p}} = 0$$

$$\mathbf{Q}_u^T\,\bar{\mathbf{u}} + \mathbf{Q}_\alpha^T\,\bar{\alpha} + \mathbf{H}\bar{\mathbf{p}} + \mathbf{C}\,\dot{\bar{\mathbf{p}}} = \mathbf{f}_p$$

Discretizing in the time domain with the following Newmark type expansions:

$$\bar{\mathbf{u}}_{n+1} = \bar{\mathbf{u}}_n + \Delta t\,\dot{\bar{\mathbf{u}}}_n^- + \beta\Delta t\,\Delta\dot{\bar{\mathbf{u}}}_n$$

$$\bar{\mathbf{p}}_{n+1} = \bar{\mathbf{p}}_n + \Delta t\,\dot{\bar{\mathbf{p}}}_n + \theta\Delta t\,\Delta\dot{\bar{\mathbf{p}}}_n$$

$$\bar{\boldsymbol{\alpha}}_{n+1} = \bar{\boldsymbol{\alpha}}_n + \Delta t\,\dot{\bar{\boldsymbol{\alpha}}}_n + \beta\Delta t\,\Delta\dot{\bar{\boldsymbol{\alpha}}}_n$$

and substituting into the space-discretized equations a new set of non linear equations is obtained. The non linear problem is then stated and solved in a Newton-Raphson format. In the process, the enhanced strain field $\boldsymbol{\alpha}$ is eliminated by static condensation. The problem in its final configuration is coupled and formulated in terms of **u** and p_w as global fields and $\boldsymbol{\alpha}$ as internal field.

This formulation is implemented in the finite element code GeHoMadrid for geotechnical applications.

3.3 *Constitutive model*

The choice of the constitutive model is a key aspect to obtain a good approximation. The constitutive equation must be able to reproduce the behavior of the soil under cyclic loading as well as phenomena such as liquefaction. The models most frequently used in the standard calculations carried out in geotechnical engineering can lead to errors, which can be important in the case of granular soils.

The theory of generalized plasticity was introduced by Zienkiewicz and Mroz in 1984 and developed by Pastor, Zienkiewicz, Leung and Chan to reproduce the behavior of soils under cyclic loading. These models are described in the text of [Zienkiewicz et al, 1999], as well as in the references [Pastor et al, 1985] [Zienkiewicz and Mroz, 1984], [Pastor and Zienkiewicz, 1986] and [Pastor et al, 1990]. Although Generalized Plasticity does not require the explicit definition of a yield surface, contrarily to Classical Plasticity it allows for plastic deformation even in unloading conditions. In a Generalized plasticity framework deformation and stress increments are related in the following manner:

$$d\boldsymbol{\varepsilon} = \left(\mathbf{D}^e\right)^{-1} ; d\sigma + \frac{1}{H_{L/U}} \mathbf{n}_{gL/U} \otimes \mathbf{n} : d\sigma$$

Or:

$$d\sigma = \left(\mathbf{D}^e + \frac{\mathbf{D}^e \mathbf{n}_{gL/U} \mathbf{n}^T \mathbf{D}^e}{H_{L/U} + \mathbf{n}^T \mathbf{D}^e \mathbf{n}_{gL/U}} \right) d\varepsilon$$

The model used in this work was proposed by Pastor, Zienkiewicz, Leung and Chan [Pastor et al, 1985,1990] to reproduce the most important aspects of the behavior of sands under monotonic and cyclic load. It requires the calibration of set of 12 constitutive parameters. In order to perform this calibration it is necessary to carry aut a series of standard triaxial tests: drained and undrained monotonic tests and undrained cyclic tests [Zienkiewicz et al., 1999].

4 APPLICATION TO A 3D FOUNDATION

4.1 *Geometry*

The foundation under study is a caisson type breakwater. The concrete caisson is 56.50 m long, 27.00 m tall and 20.85 m wide. It is supported by a series of porous media layers which in turn are supported by the seabed. The cross section is represented in Figure 1.

The model under study corresponds to an intermediate construction stage while the cross section shown in Figure 1 represents the final configuration of the foundation. The 3D representation of the model including the concrete caisson and the different porous transition elements between the caisson and the seabed are schematized in Figure 2

4.2 *Finite volume model*

The IHFOAM model used to obtain fluid pressure histories in wet contour consists of 7 million elements. However, in order to facilitate the coupling with the finite element model, the results have been presented on a mesh with a lower resolution than the calculation but which correctly represents the simulated hydrodynamics. In the Figure a mesh of 0.3 million elements is presented; in the upper image a general view of the mesh is represented, and in the lower one two cuts of it.

The simulation has a duration of 150 seconds with a time step of output of results of 0.1 s, which can be modified. The format of the results provided to CEDEX by IH Cantabria has consisted of a series of 1500 text files as many as outgoing time stations corresponding to a simulation period

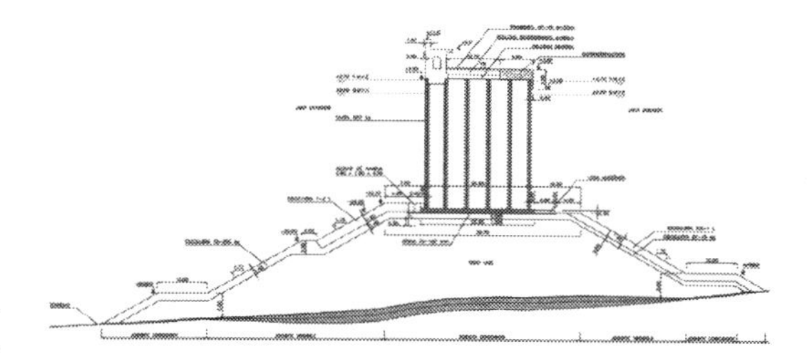

Figure 1. Cross section of the caisson breakwater.

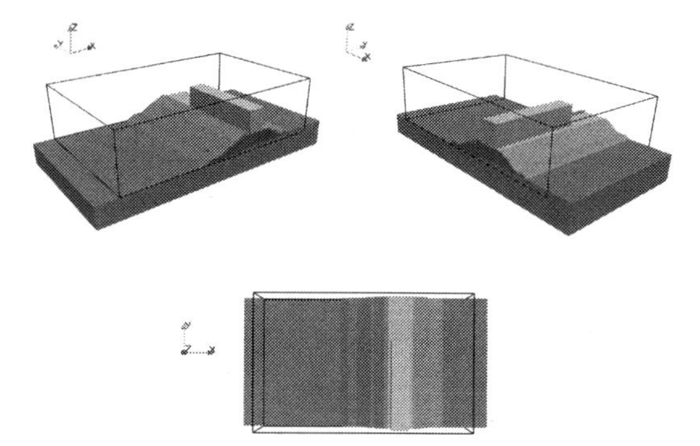

Figure 2. 3D Geometry of the model.

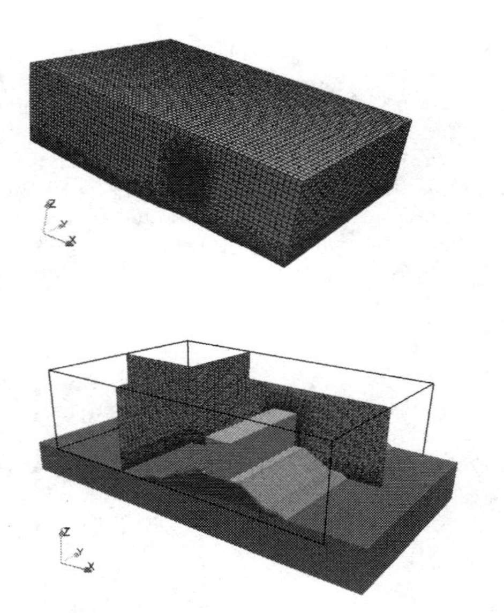

Figure 3. IHFOAM mesh.

of 150 seconds with a resolution or increase in output time of 0.1 seconds. Therefore, in each of these text files is the total interstitial pressure of each point of the wet contour. Each point is identified by its coordinates so that in each line of that text file there are four values: three spatial coordinates and the value of the total fluid pressure associated with the point of those coordinates and in the time associated with the file. The wet contour has been assumed fixed in this calculation.

4.3 *Finite element with u-pw formulation*

The product of the above calculation is a series of fluid pressure histories in a cloud of points that define the interface between the finite volume subdomain and the saturated granular media subdomain. The calculation domain in the previous phase is the volume of water contained in a 285×165 m space confined by the following elements:

- Seabed
- Rockfill transition elements
- Concrete caisson

In the present phase the calculation domain will include the aforementioned solid elements (Caisson, rockfill and seabed)

The interface between both subdomains will be defined by a cloud of points. Each of the points of this cloud is defined by its spatial coordinates. The coupled u-pw finite element model must be compatible with this interface. The histories of fluid pressures obtained with the previous model will be introduced in the finite element model as prescribed fluid pressures and external pressure forces in the interface.

To illustrate the magnitude of the storm, Figures 4a and 4b show the interstitial pressure histories in excess of the hydrostatic pressure at the two ends of the open sea contour of the model.

The calculations have four steps. The first three represent the construction process of the foundation. The fourth represents the storm whose impact on the foundation and seabed is modeled. These four steps are:

a. Application of gravity + hydrostatic pore pressure conditions on the seabed. For this phase, an approximate geostatic distribution of pore pressures and effective stresses are estimated by the distance of each point to the surface of the seabed. This distribution has subsequently been imposed on the mesh of 8 node **u**-p_w brick elements and the corresponding enhanced modes (B8P8SRC in abbreviated notation).
b. Activation of the mesh associated with the rockfill elements and application of the corresponding gravity forces.
c. Activation of the mesh associated with the concrete caisson and application of the corresponding gravity forces.
d. Application of the pore pressure histories representative of the storm.

The main objective of this model is to capture the response of the seabed. The concrete caisson and the rockfill transition elements have been assumed to behave linear elastically. The Pastor-Zienkiewicz model capable of representing the effects of the liquefaction in loose sands has been used only for the seabed.

Soil will be more prone to liquefy the closer to the seabed surface and therefore less confined. The

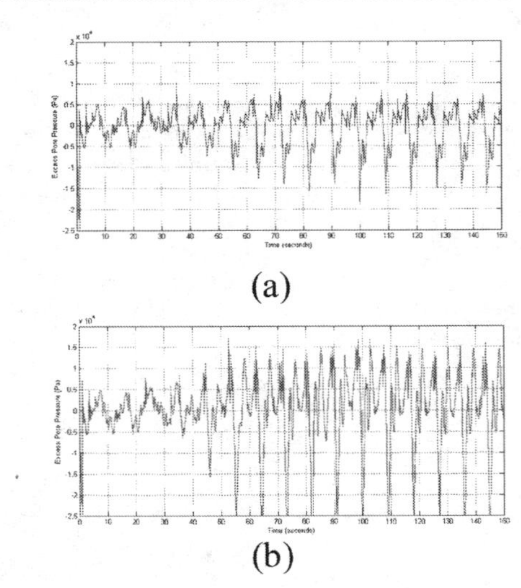

Figure 4. Pore pressure time histories in excess over hydrostatic pressure at the two ends of the open sea boundary of the model.

advent of liquefaction accentuates the non-linearity of the problem and the convergence difficulties. In this sense, a key factor for the difficulties of convergence of the problem is the proximity of the element integration points closest to the seabed surface.

The high degree of non-linearity of the problem and the problems of convergence that this implies make it important to control the size of the time step. An excessively large time step will have, in general, immediate repercussions on the speed of convergence. The calculations have been carried out with an explicit integration scheme and a strict control of the time step size through a maximum time step size restriction of 0.01 seconds and an additional adaptation reducing the time step even more when the number of iterations in the previous step exceeds five.

4.4 *Results*

The most representative results are shown in this section. This results focus on the seabed response. Contour graphs for different variables are shown. The graphs correspond to the end of the computation, in other words the 150 seconds time station. The variables represented in these graphs are:

a. Liquefaction coefficient or quotient between final effective pressure and initial effective pressure. The complete liquefaction would be reached when said coefficient takes the value of zero, that is, the effective pressure is completely canceled out.
b. Vertical effective tension
c. Pore pressure

The most representative graphs of the phenomenon are perhaps those associated with the liquefaction coefficient. As shown in Figure 6 this coefficient reaches its lowest value in those areas that satisfy the following conditions:

a. Lower confinement pressure, usually ssociated with proximity to the seabed surface.
b. Highest exposure to cyclic loading

Close to the seabed surface the liquefaction coefficient reaches a minimum value of 0.25 after 150 seconds.

Table 1. Pastor-Zienkiewicz parameter set for seabed.

K_{evo} (kpa)	G_{eso} (kpa)	M_g	α_g	M_f	α_f	H_0	β_0	β_1	γ
5000	3000	1.6	0.45	1.6	0.45	150	4.2	0.2	1

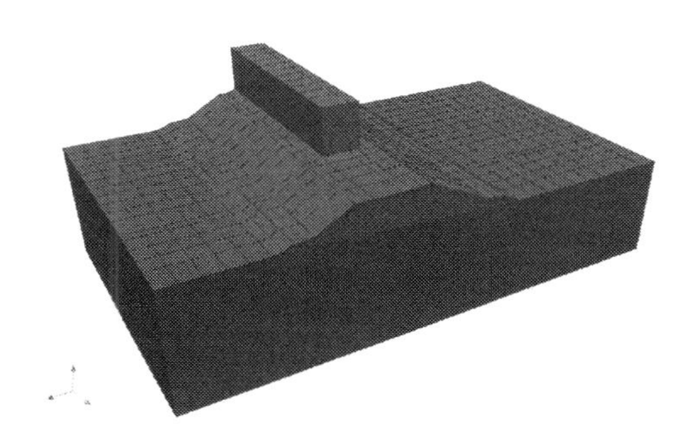

Figure 5. Model mesh.

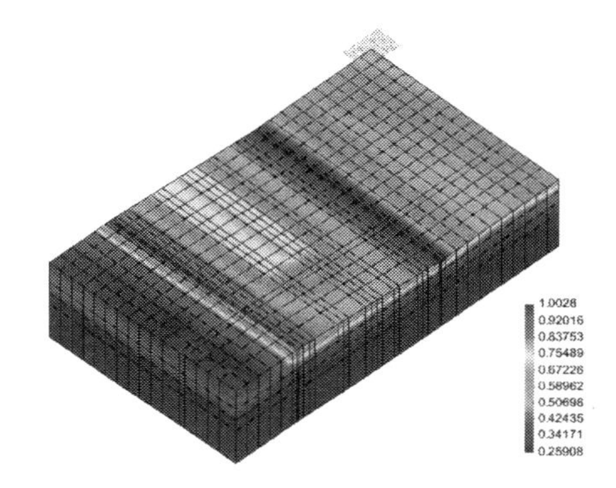

Figure 6. Liquefaction coefficient for t = 150.0 seconds.

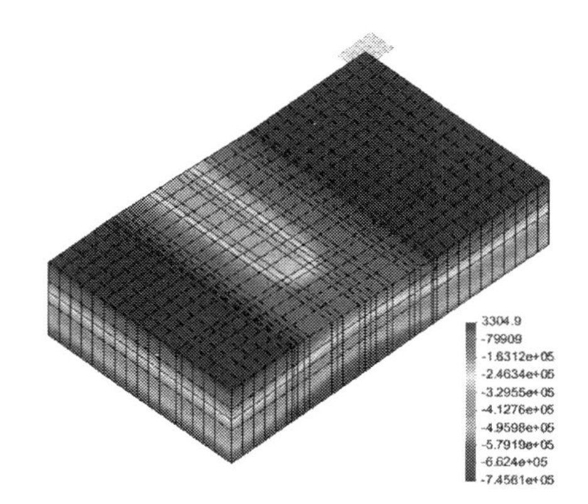

Figure 7. Vertical effective stresses for t = 150.0 seconds.

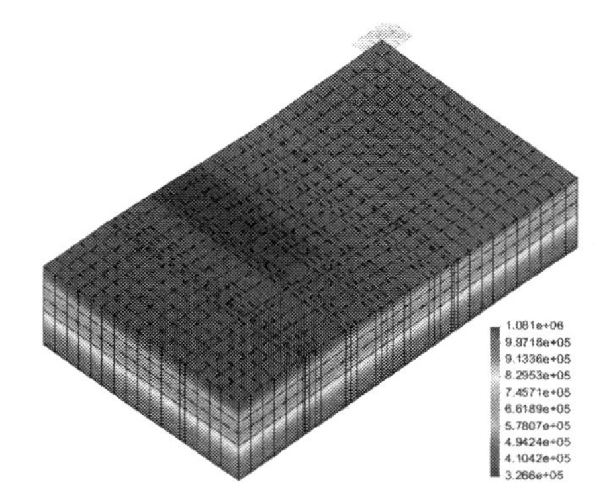

Figure 8. Pore pressures for t = 150.0 seconds.

5 CONCLUDING REMARKS

This work presents an application to a three-dimensional case of an analysis model of marine foundations in interaction with the surrounding fluid under the effects of a storm.

The mathematical model includes in the first place the Biot equations for saturated geomaterials, formulated in u displacements and pw interstitial pressures. The numerical model used is a coupled u-pw finite element formulation that includes enhanced deformations following the work by Simo&Rifai, a formulation implemented in the GeHoMadrid program.

For the marine fluid, the Navier-Stokes equations are used. Thenumerical model consist of. a formulation in Finite Volume with two incompressible phases and VoF (Volume of Fluid) function for the water-air interface, all implemented in the IHFOAM code of the University of Cantabria based on the multiphysics numerical library OpenFOAM.

The coupling between the two parts of the model is of weak type. The IHFOAM code taking a temporary record as input data performs a calculation with the restriction of fixed wet contours (seafloor, rockfill bench and concrete dam) that gives rise to a history of fluid pressures introduced below in GeHoMadrid as hydraulic contour loads and conditions.

To represent the constitutive behavior of the sea floor, the Pastor-Zienkiewicz model of Generalized Plasticity has been used. It is a model capable of representing the behavior of granular soils in general, including specifically important phenomena such as liquefaction of loose sands.

The application consisted of modelling the response under storm action of a three-dimensional model of a concrete caisson type breakwater anchored on rockfill. The section of modeled marine floor is 285 m long and 165 m wide.

As expected, the response shows how the most superficial areas of the seabed are closer to the state of liquefaction, a phenomenon observed in reality in marine foundations subjected to storms.

REFERENCES

Biot, M.A. 1941. General theory of three-dimensional consolidation. J. Appl. Phys., 12, 155–164.

Biot, M.A. 1955. Theory of elasticity and consolidation for a porous anisotropic solid. J.Appl. Phys., 26, 182–185

Castro, G. 1969. "Liquefaction of Sand". PhD Thesis, Division of Engineering and Applied Physics, Harvard University.

Chan A.H.C. 1988. A Unified Finite Element Solution to Static and Dynamic Problems in Geomechanics, University of Wales, Swansea, Department of Civil Engineering

Ghaboussi, J. y Wilson, E.L. 1972. Variational formulation of dynamics of-fluide-saturated porous elastic solids. J. Engng. Mech. Div. ASCE, 98, 947–963.

Higuera, P., Lara, J.L., Losada, I.J., 2013. Realistic wave generation and active wave absorption for Navier–Stokes models: application to OpenFOAM. Coast. Eng. 71, 102–118.

Higuera, P., Lara, J.L., Losada, I.J. 2013 Simulating coastal engineering processes with OpenFOAM. Coast. Eng. 71, 119–134.

Higuera, P., Lara, J.L., Losada, I.J. 2014a Three-dimensional interaction of waves and porous coastal structures using OpenFOAM. Part I: formulation and validation. Coast. Eng. 83, 243–258.

Higuera, P., Lara, J.L., Losada, I.J. 2014b. Three-dimensional interaction of waves and porous coastal structures using OpenFOAM. Part II: Applications. Coast. Eng. 83, 259–270.

Mira, P., Pastor, M., Li T., Liu, X. 2003. A new enhanced strain element with equal order interpolation for soil consolidation problems. Computer Methods in Applied Mechanics and Engineering. 192, 4257–4277

Pastor, M., Zienkiewicz, O.C. & Leung, K.H. 1985. "Simple Model for Transient Soil Loading in Earthquake Análisis II: Non-Associative Models for Sands", International Journal for Numerical and Analytical Methods in Geomechanics, 9, 477–498.

Pastor, M. y Zienkiewicz, O.C. 1986. "A Generalized Plasticity, Hierarchical Model for Sand under Monotonic and Cyclic Loading", 2nd International Symposium on Numerical Models in Geomechanic, Ghent, 131–149.

Pastor M., Zienkiewicz O.C. Y Chan A.H.C. 1990 « Generalized plasticity and the modelling of soils behaviour », Int. J. Num. Anal. Meth. Geomech, n°14, p. 151–190.

Pastor, M. y Quecedo, M. y Zienkiewicz, O.C. 1996, A mixed displacement-pressure formulation for numerical analysis of plastic failure, Comp. and Structures}

Pastor, M. y Li, T. y Liu, X. y Zienkiewicz, O.C. 1999. Stabilized low order finite elements for failure and localization problems in undrained soils and foundations, Comp.Meth.Appl.Mech.Eng., v 174, 219–234

Simo J.C. y Rifai M.S. 1990. A class of mixed assumed strain methods and the method of incompatible modes. International Journal for Numerical Methods in Engineering.Volume 29, Issue 8, pages 1595–1638, June 1990

Terzaghi, K. 1936. "The shear resistance of saturated soil", 1st International Conference of Soil Mechanics and Foundations Engineering, Cambridge, 54–56.

Zienkiewicz, O.C., Chang, C.T. y Bettess, P. 1980. Drained, undrained, consolidating dynamic behaviour assumptions in soils. Geotechnique 30, 385–395.

Zienkiewicz O.C. Y Mroz Z. 1984. Uniform formulation of constitutive equations for clays and sands. *Mechanics of Engineering Materials*, C.S. Desai Y R.H. Gallagher (éds.), Wiley, New York, Etats-Unis, p. 415–449,.

Zienkiewicz, O.C. y Chan, A.H.C. y Pastor, M. y Paul, D.K. y Shiomi, T., 1990a. Static and dynamic behaviour of soils: a rational approach to quantitative solutions, Part I: Fully saturated problems, Proc. Royal Society London, v 429, pp 285–309.

Zienkiewicz, O.C., Xie, Y.M., Schrefler, B.A., Ledesma, A., Bicanic, N. 1990b. Static and dynamic behaviour of soils: a rational approach to quantitative solutions. II. Semi-saturated problems. Proc. R. Soc. Lond. A 429, 311–321.

Zienkiewicz, O.C. Chan, A.H.C., Pastor, M., Schrefler, B.A. & Shiomi, T. 1999. "Computational Geomechanics". John Wiley & Sons.

Numerical Methods in Geotechnical Engineering IX – Cardoso et al. (Eds)
© 2018 Taylor & Francis Group, London, ISBN 978-1-138-33198-3

Axisymmetric formulation of the material point method for geotechnical engineering applications

V. Galavi, F.S. Tehrani, M. Martinelli, A.S. Elkadi & D. Luger
Geo-Engineering Department, Deltares, Delft, The Netherlands

ABSTRACT: The Material Point Method (MPM) is an effective method for large deformation analysis. However, compared to the Updated Lagrangian finite element method, extra calculation steps introduced in MPM make the method computationally more expensive. For axisymmetric large deformation problems, such as the cone penetration testing and pile installation, the 3-dimensional formulation of the method can be simplified to a less expensive, but physically valid, 2-dimensional axisymmetric form. The performance of the 2D-Axisymmetric formulation will be demonstrated through the numerical simulation of three axisymmetric problems such as a shallow foundation, soil column collapse and cone penetration testing.

1 INTRODUCTION

Many geotechnical engineering problems involve large deformations, such as e.g. cone penetration testing, pile installation and sinkhole development around damaged pipelines and landslides. The Updated Lagrangian Finite Element Method (UL-FEM), which is typically used in solving these types of problems, can be problematic due to mesh distortion. Some attempts have been made in the past to overcome this limitation by introducing more advanced techniques. One of these techniques is the Material Point Method (MPM), which was introduced by Sulsky *et al.* (1994). MPM can be considered as an extension of the UL-FEM. The method uses Lagrangian material points (or mesh) that is defined over the material of the body and a background (Eulerian) mesh that is defined on the computational domain. The Lagrangian material points carry all computational data including stresses and state variables. In the beginning of each calculation step, the computational data are projected from the Lagrangian material points to the background mesh. The equations are solved on the background mesh similar to the UL-FEM. Mesh distortion, associated with the Lagrangian method, is prevented by mapping the computational data from the background mesh to the material points at the end of each computational step and subsequently resetting the mesh before the next step. In this way, the method benefits from the advantages of both Lagrangian and Eulerian methods, while being free from the disadvantages of those methods, namely the mesh distortion in the Lagrangian method and the numerical dissipation in the Eulerian method.

Extra calculation steps introduced in MPM, compared to the UL-FEM, make the method computationally expensive. Furthermore, in order to properly describe the mechanical behaviour of soils, highly nonlinear material models are needed which makes the method more expensive.

The numerical simulation of large deformation problems, such as the cone penetration testing and pile installation can be carried out in a less expensive way, if the effect of rotational symmetry of the pile and the surrounding soil is taken into consideration in the formulation. This results in a new formulation of the method which is based on a two dimensional axisymmetric formulation developed by Sulsky & Schreyer (1996). This formulation of MPM can significantly reduce number of elements and material points compared to a three dimensional formulation. The validity of the method will be demonstrated through the numerical simulation of three axisymmetric large deformation problems such as a shallow foundation, a soil column collapse and cone penetration testing.

2 2D AXISYMMETRIC FORMULATION

The detailed formulation of the 3D material point method is presented in Al-Kafaji (2013). In this section, the formulation of a 2-dimensinal axisymmetric material point method together with the overall procedure for one calculation step is described.

2.1 *Strains in an axisymmetric geometry*

Gens & Potts (1984) described three examples in which three dimensional geotechnical engineering problems can be expressed by a two dimensional axisymmetric formulation, namely (a) a pile under torsional load; (b) a pile under vertical load; (c) hollow cylinder samples subject to torsion. This section covers the second case where the piles are subjected to a vertical load. In this case, the strain tensor has four non-zero components as follows ε_{rr} (radial strain), ε_{zz} (vertical strain), ε_{rz} (shear strain) and $\varepsilon_{\theta\theta}$ (circumferential strain) in the cylindrical coordinate system. This means that there is one more non-zero strain component in the 2D axisymmetric compared to the 2D plane strain, namely $\varepsilon_{\theta\theta}$. The displacements in the cylindrical coordinate system are denoted as u_r and u_z. The strain components are related to the displacements through the following relationships:

$$\varepsilon_{rr} = \frac{\partial u_r}{\partial r} \tag{1}$$

$$\varepsilon_{zz} = \frac{\partial u_z}{\partial z} \tag{2}$$

$$\gamma_{rz} = 2\varepsilon_{rz} = \left(\frac{\partial u_r}{\partial z} + \frac{\partial u_z}{\partial r}\right) \tag{3}$$

$$\varepsilon_{\theta\theta} = \frac{u_r}{r} \tag{4}$$

where r is the radial distance between the material point and the axis of symmetry. It should be noted that the mechanical sign convention is used in this study, which means that tensile strains are positive.

In the finite element method, strains are related to the displacements as:

$$\boldsymbol{\varepsilon} = \mathbf{B}_i \mathbf{u}_i = \begin{bmatrix} \partial N_i(x_p)/\partial r & 0 \\ 0 & \partial N_i(x_p)/\partial z \\ \partial N_i(x_p)/\partial z & \partial N_i(x_p)/\partial r \\ N_i(x_p)/r & 0 \end{bmatrix} \begin{bmatrix} u_{r,i} & u_{z,i} \end{bmatrix} \tag{5}$$

where $\mathbf{u}_i$ is a vector containing nodal displacements of node i. The matrix $\mathbf{N}$ contains shape functions at the local location (x_p) of integration points (material points) and the matrix $\mathbf{B}$ contains gradient of shape functions.

By expanding Eq.(5), the strain components in the 2D axisymmetric formulation can be expressed as:

$$(\varepsilon_{rr})_p = \sum_{i=1}^{N_n} u_{r,i} \frac{\partial N_i(x_p)}{\partial r} \tag{6}$$

$$(\varepsilon_{zz})_p = \sum_{i=1}^{N_n} u_{z,i} \frac{\partial N_i(x_p)}{\partial z} \tag{7}$$

$$(\gamma_{rz})_p = \sum_{i=1}^{N_n} \left(u_{z,i} \frac{\partial N_i(x_p)}{\partial r} + u_{r,i} \frac{\partial N_i(x_p)}{\partial z} \right) \tag{8}$$

$$(\varepsilon_{\theta\theta})_p = \sum_{i=1}^{N_n} u_{r,i} \frac{N_i(x_p)}{r_p} \tag{9}$$

in which N_n is the number of nodes of an element.

2.2 *Internal forces in an axisymmetric geometry*

The main difference between the 2D axisymmetric and the 2D plane strain formulations is in the matrix $\mathbf{B}$ (Eq.(5)), which is used to calculate internal forces and strains. The following equation is used to get the internal force:

$$f_i^{int} = \sum_{p=1}^{N_p} B_i^T(x_p)\sigma_p \Omega_p \tag{10}$$

where N_p is the total number of material points (integration points) within an element. Ω_p is the integration weight (volume) of material point p and σ_p is the stress tensor of material point p. The components of the internal force in the 2D axisymmetric formulation are written as

$$f_{r,i}^{int} = \sum_{p=1}^{N_p} \left\{ \left((\sigma_{rr})_p \frac{\partial N_i(x_p)}{\partial r} + (\sigma_{rz})_p \frac{\partial N_i(x_p)}{\partial z} + (\sigma_{\theta\theta})_p \frac{N_i(x_p)}{r} \right) \Omega_p \right\} \tag{11}$$

$$f_{z,i}^{int} = \sum_{p=1}^{N_p} \left\{ \left((\sigma_{zz})_p \frac{\partial N_i(x_p)}{\partial z} + (\sigma_{rz})_p \frac{\partial N_i(x_p)}{\partial r} \right) \Omega_p \right\} \tag{12}$$

2.3 *Overall procedure for one calculation step in 2D axisymmetric MPM*

The solution procedure for one calculation step of the 2D axisymmetric MPM in drained condition is summarised here:

1. Lumped mass matrix at nodes are calculated from the mass of material points:

$$m_i = \sum_{p=1}^{N_p} m_p N_i(x_p) \tag{13}$$

where m_p is the mass of material point p. The initial mass of the material point is calculated from

$$m_p = \rho_p \Omega_p \quad (14)$$

and

$$\Omega_p = r_p A_p \quad (15)$$

where ρ_p, A_p, r_p and Ω_p are the initial density, area, radius and volume of material point p, respectively. The initial area of the material point, A_p, is set based on the Gauss quadrature or standard quadrature which can be considered as a fraction of the element area. The volume of a material point is therefore calculated as the area of a material point multiplied by its radius r_p (Eq.(15)).

2. Momentum at nodes is calculated from the momentum at the material points by a mass-weighted mapping as

$$m_i v_i = \sum_{p=1}^{N_p} m_p v_p N_i(x_p) \quad (16)$$

where v_i and v_p are the velocities at nodes and material points, respectively.

3. Acceleration at nodes are calculated by solving the following discrete system of equations

$$m_i a_i = f_i^{ext} - f_i^{int} \quad (17)$$

where f_i^{int} and f_i^{ext} are the internal and external force vectors. The external force vector consists of body forces and traction forces. The internal force is calculated from stresses in the material points by Eq. (10). The components of the internal force vector are given by Eq. (11) and Eq. (12).

4. Acceleration at nodes are used to update the velocities of material points using the shape functions:

$$v_p^{t+\Delta t} = v_p^t + \sum_{i=1}^{N_n} \Delta t N_i\left(x_p\right) a_i^t \quad (18)$$

5. The nodal velocities are then calculated from the updated material point velocities by solving Eq.(16).
6. The nodal incremental displacements can be calculated by integrating the nodal velocities as

$$\Delta u_i^{t+\Delta t} = \Delta t v_i^{t+\Delta t} \quad (19)$$

7. The strains and stresses are calculated at material points as

$$\Delta \varepsilon_p^{t+\Delta t} = B(x_p) \Delta u_e^{t+\Delta t} \quad (20)$$

The subscript e stands for element which means that the nodal incremental displacements of the element are used. The calculated strains can then be used for calculation of stresses using a constitutive model. The components of the strain vector in the 2D axisymmetric formulation are given by Eqs. (6) to (9).

8. The volume and density of particles are updated using the volumetric strain $\Delta \varepsilon_{vol}$:

$$\Omega_p^{t+\Delta t} = \left(1 + \Delta \varepsilon_{vol,p}^{t+\Delta t}\right) \Omega_p^t \quad (21)$$

$$\rho_p^{t+\Delta t} = \frac{\rho_p^t}{\left(1 + \Delta \varepsilon_{vol,p}^{t+\Delta t}\right)} \quad (22)$$

9. Displacements and positions of material points are updated according to:

$$u_p^{t+\Delta t} = u_p^t + \sum_{i=1}^{N_n} N_i(x_p) \Delta u_i^{t+\Delta t} \quad (23)$$

$$x_p^{t+\Delta t} = x_p^t + \sum_{i=1}^{N_n} N_i(x_p) \Delta u_i^{t+\Delta t} \quad (24)$$

10. At this step, the mesh is reset and a space search is performed to find new positions of material points in the background mesh.

A similar procedure can be developed for the two-phase formulation (Van Esch *et al.* 2011) where generation and dissipation of excess pore water pressure can be simulated. For the sake of brevity, the overall procedure of two-phase formulation is not reported in this paper.

3 EXAMPLES

Three examples are considered here to verify the implementation of the two dimensional axisymmetric formulation of MPM, namely a circular surface load on an elastic ground, a circular soil column collapse and the cone penetration testing (CPT).

3.1 *Circular surface load on an elastic ground*

First, an analytical solution for circular surface load on an elastic ground is given and then the solution is compared with results of MPM and FEM. The commercial finite element software, PLAXIS 2D (www.plaxis.nl), is used for the finite element analysis.

3.1.1 *Analytical solution*

According to Boussinesq's equations, the vertical stress at a point (r, z) under a point load can be found from

$$\sigma_{zz} = \frac{3q}{2\pi z^2} \frac{1}{\left[1 + \frac{r}{z}^2\right]^{5/2}} \tag{25}$$

By integrating it over a circular area with the radius of R, the vertical stress below the centre point at any z can be found from

$$\sigma_{zz} = q\left[1 - \frac{z^3}{\left(R^2 + z^2\right)^{3/2}}\right] \tag{26}$$

Similarly, the horizontal radial stress is derived as

$$\sigma_{rr} = \frac{q}{2}\left[1 + 2\nu - \frac{2(1+\nu)z}{\left(R^2 + z^2\right)^{1/2}} + \frac{z^3}{\left(R^2 + z^2\right)^{3/2}}\right] \tag{27}$$

where ν is the Poisson's ratio. For the centre point right below the circular load, the vertical and horizontal radial stresses can be found by setting z to zero in Eq. (26) and Eq. (27), *i.e.*:

$$\text{For } z = 0: \sigma_{zz} = q; \quad \sigma_{rr} = \frac{q}{2}[1 + 2\nu] \tag{28}$$

The vertical strain ε_z, due to the triaxial load under the centre point is given by

$$\varepsilon_{zz} = \frac{1}{E}\left(\sigma_{zz} - 2\nu\sigma_{rr}\right) \tag{29}$$

in which E is the elastic modulus. By substituting σ_{zz} and σ_{rr} from Eq. (26) and Eq. (27) into Eq. (29) and integrating from $z = z$ to $z = \perp$, the elastic deformation D under the centre point at any z can be found. Thus:

$$D = \frac{q}{E}\left[\left(2 - 2\nu^2\right)\left(R^2 + z^2\right)^{1/2} - \frac{\left(1+\nu\right)z^2}{\left(R^2 + z^2\right)^{1/2}} + \left(\nu + 2\nu^2 - 1\right)z\right] \tag{30}$$

For $z = 0$, total displacement below the surface load can be found:

$$\text{For } z = 0: D = \frac{qR}{E}\left(2 - 2\nu^2\right) \tag{31}$$

3.1.2 *Numerical simulations*

The circular footing is simulated using the UL-FEM in the commercial finite element software PLAXIS and the developed two dimensional axisymmetric MPM. A surface pressure of -1000 *kPa* with a radius of 1 *m* is applied on an elastic medium with a stiffness of $E = 10^4$ *kPa* and Poisson's ratio of 0.3. From the analytical solutions (Eq. (28) and Eq.(31)), the vertical and horizontal stresses and the vertical settlement at the centre point below the surface load are $\sigma_{zz} = -1000$ *kPa*, $\sigma_{rr} = -800$ *kPa* and $D = 0.182$ *m*. It should be noted that the analytical solutions are based on the small deformation assumption.

The simulated vertical and horizontal stresses for both models are depicted in Figure 1 to

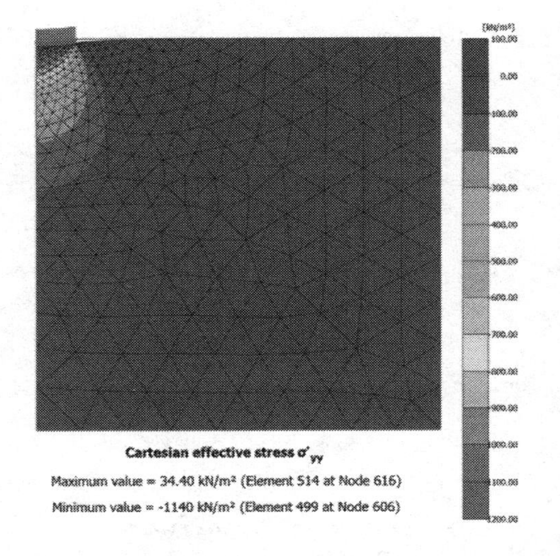

Figure 1. Vertical stress field obtained from PLAXIS 2D (UL-FEM).

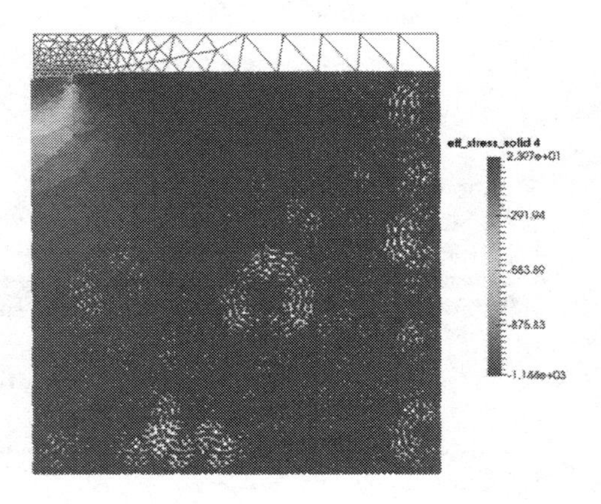

Figure 2. Vertical stress field obtained from 2D axisymmetric MPM.

4. The vertical stresses at the centre of the footing, obtained from PLAXIS 2D and MPM code, are (σ_{zz} = –1140 kPa, σ_{rr} = –937 kPa) and (σ_{zz} = –1144 kPa, σ_{rr} = –832 kPa) respectively, which are in good agreement with the analytical solution. The vertical displacement at the centre of the footing (see Figures 5 & 6) is predicted as 17.16 cm and 17.46 cm by the PLAXIS 2D and the MPM code, respectively.

By comparing the results of the numerical simulations with the analytical solutions, it can be concluded that the two dimensional axisymmetric formulation, developed for the material point method, is capable of predicting the stresses and deformation of the circular footing problem.

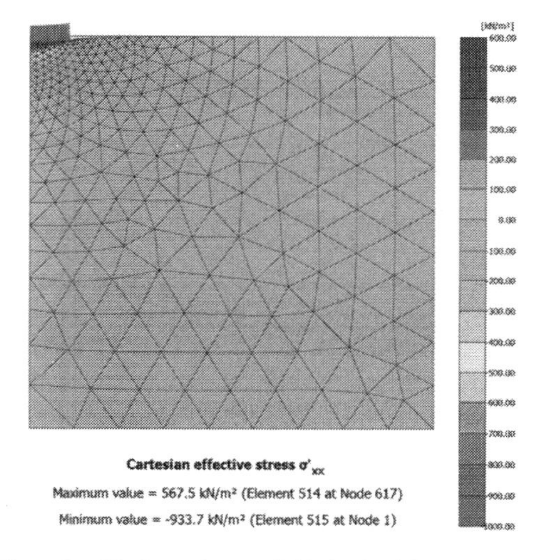

Figure 3. Horizontal stress field obtained from PLAXIS 2D (UL-FEM).

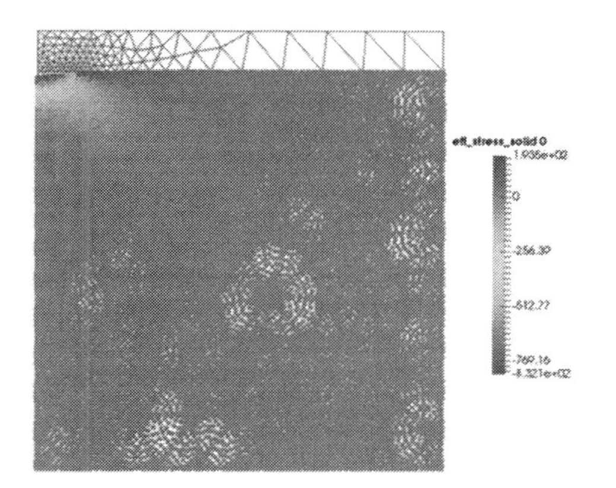

Figure 4. Horizontal stress field obtained from 2D axisymmetric MPM.

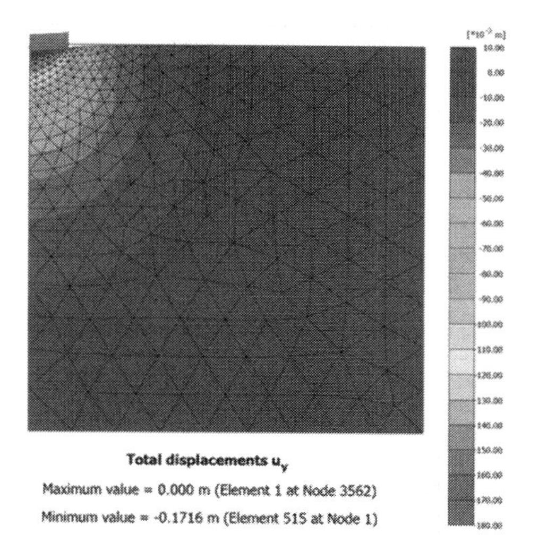

Figure 5. Vertical displacement field obtained from PLAXIS 2D (UL-FEM).

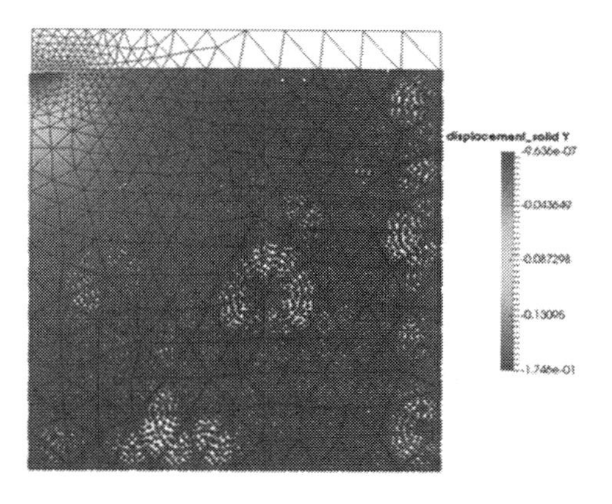

Figure 6. Vertical displacement field obtained from 2D axisymmetric MPM.

3.2 *Soil column collapse in axisymmetric condition*

In this example, the two-dimensional axisymmetric MPM is compared against the three-dimensional MPM (Al-Kafaji, 2013) in simulating a soil column collapse problem in the rotational symmetric condition.

Figure 7 shows the geometry of the soil column. The 3D element and material point discretisation are shown in Figure 8. The mesh consists of 24460 tetrahedral elements and 5230 material points. Figure 9 shows the 2D mesh together with the material point discretisation. The mesh consists of 540 triangle elements and 312 material points.

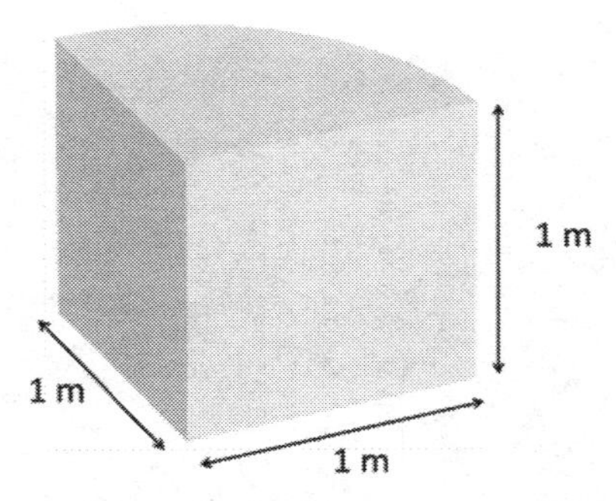

Figure 7. Geometry of the soil column.

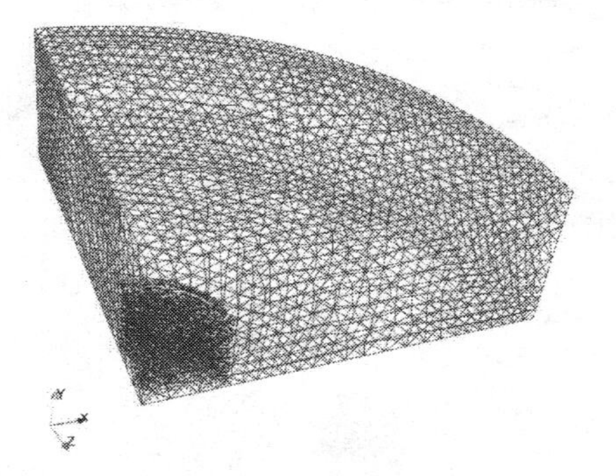

Figure 8. Element and material point discretisation for the 3D model.

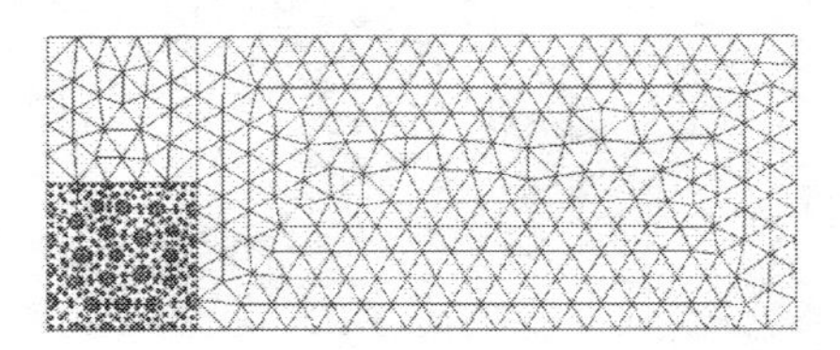

Figure 9. Element and material point discretisation for the 2D model.

The Mohr-Coulomb material model is used to describe the mechanical behaviour of the soil. The material parameters are listed in Table 1.

Figure 10 shows results of the three dimensional model versus the two dimensional model at different steps. It has been observed that the results obtained from the three dimensional model are the same as the ones obtained from the two dimensional axi-symmetric model. It can be concluded that both models show a similar failure pattern. The total time needed to simulate the 3D problem using one core is measured as 28 *s*, while the simulation of the 2D axisymmetric case, on the same machine, takes only 0.47 *s* which renders an almost 60 times faster calculation. This is due to the fact that the two-dimensional axisymmetric

Table 1. Material properties used for the soil column collapse.

E [kPa]	ν[–]	c [–]	φ [*degrees*]	ψ [*degrees*]
1000	0.2	0	30	0

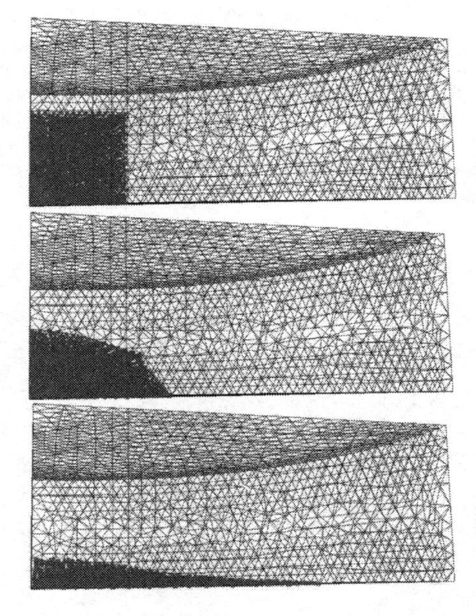

Figure 10. Soil column collapse; 3D model (red dots) versus 2D axisymmetric model (blue dots).

formulation needs much less number of elements and material points for the simulation.

3.3 *Cone Penetration Testing (CPT)*

The cone penetration test (CPT) is one of the most used methods for site characterization. Due to its simplicity, many design methods have been developed based on the cone resistance correlations. Numerical simulation of the CPT has always been a challenging task due to the large deformations involved in the problem. Different methods have been used in the past for numerical simulation of the CPT, such as the arbitrary Lagrangian-Eulerian finite element technique (*e.g.* Tolooiyan and Gavin, 2011), Particle Finite Element Method (PFEM) (*e.g.* Monforte *et al.*, 2017), the cavity expansion method (*e.g.* Randolph *et al.*, 1994) and the discrete element method (*e.g.* Arroyo *et al.*, 2011).

Tehrani & Galavi (2018) studied the effectiveness of the numerical spherical cavity expansion method and MPM in simulating the cone penetration testing in sand. The same problem as studied in Tehrani & Galavi (2018) is presented in more detail here. The geometry of the test is shown in

Figure 11. The cone penetrator has a diameter of $d_c = 0.036$ *m* and is initially embedded 0.36 *m* ($= 10d_c$) below the soil surface.

A constant vertical velocity of 0.02 *m/s* is applied to all degrees of freedom of the cone penetrator in order to model it as a rigid body. A frictional contact formulation is considered at the boundary between the cone penetrator and the soil. The algorithm was developed by Bardenhagen *et al.* (2000) in order to properly model the interaction between two bodies. It allows frictional sliding and separation while it prevents interpenetration. The frication angle between the soil and the steel cone is assumed to be 20 degrees which is equal to 2/3 of the constant volume friction angle of the soil (*i.e.* $\varphi'_c = 30°$). To increase the accuracy in the contact formulation, the moving mesh approach (Al-Kafaji, 2013) is used to ensure that the contact nodes of the two bodies are always at the same elevation. Therefore, the top part of the background mesh is moving together with the cone penetrator while the bottom part is being compressed.

For the numerical simulations, Baskarp sand with three different relative densities (D_R), namely 30%, 50% and 90% is considered. The initial stress in the ground is set based on the unit weight of the soil and a surcharge. The unit weight of the soil is kept constant and equal to 16 kN/m^3 for all the relative densities. Although this is not physically correct, but it ensures that the initial stresses are the same for all cases, which makes it more convenient to study the effect of different strength and stiffness properties of the soil on final results. A thin layer with higher density is placed on the ground surface in order to apply a uniform surcharge of 25 *kPa* which results in an additional horizontal effective stress of ~12.5 *kPa* in the model. According to Salgado & Prezzi (2007), the cone resistances q_c can be found from:

$$q_c = 1.64 p_A \exp(A)(\sigma'_h / p_A)^{(0.841-0.47D_R)} \quad (32)$$

$$A = 0.1041\phi_c + (2.64 - 0.02\phi_c)D_R \quad (33)$$

where σ'_h is the horizontal effective stress and p_A is the reference stress and equal to 100 *kPa*. In the formulas above D_R takes a value between 0 and 1.

Using Eq.(32), the cone resistance q_c is obtained for a reference point 1 *m* below the surcharge layer as 2.26 *MPa*, 3.95 *MPa* and 12.1 *MPa* for relative densities of 30%, 50% and 90%, respectively.

The Mohr-Coulomb model is used to describe the mechanical behaviour of the soil. The stiffness parameters were determined using the equations given by Lunne & Christophersen (1983) for normally consolidated and uncemented predominantly silica sands:

$$E_{oed} = 4q_c \quad (34)$$

$$E = E_{oed}\frac{(1-2\nu)(1+\nu)}{1-\nu} \quad (35)$$

The peak friction angle φ'_p is determined using the correlations proposed by Bolton (1986) for triaxial condition. The dilation angle ψ, is calculated

Table 2. Material properties of the soil in the CPT.

D_R [%]	E [kPa]	ν [-]	c' [kPa]	φ'_p [°]	ψ [°]
30	7894	0.2	1	32.8	2.8
50	14000	0.2	1	36.5	6.5
90	43659	0.2	1	43.4	13.4

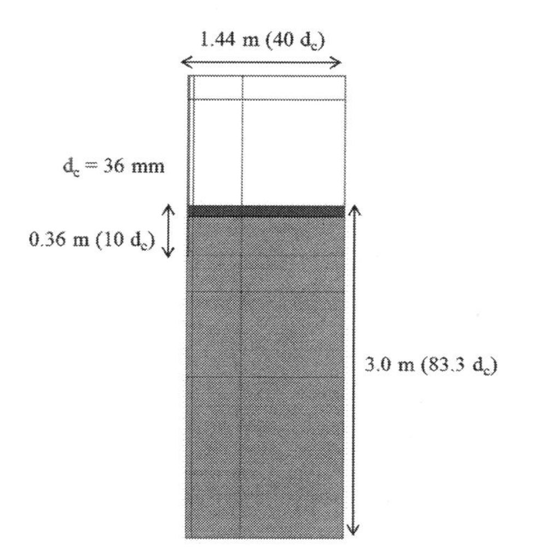

Figure 11. Geometry of the CPT model.

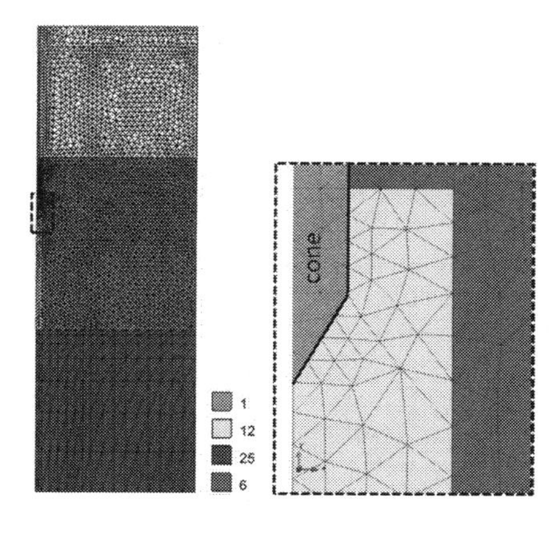

Figure 12. Mesh and material point discretisation for the CPT model.

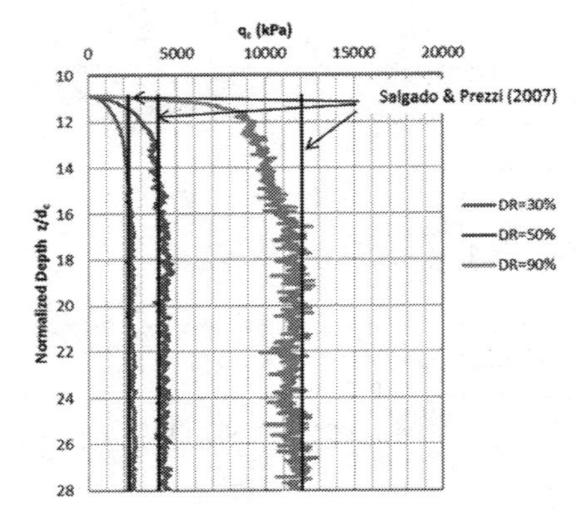

Figure 13. Averaged stresses at the cone (q_c)) obtained from the MPM analyses.

as the difference between the peak and the constant volume friction angles.

Table 2 summarizes the material parameters of the sand.

Figure 12 shows the mesh discretisation in the model. In total 5950 linear triangle elements together with 28782 material points are defined. The elements are refined in the vicinity of the cone penetrator in order to increase the accuracy.

The cone resistance q_c obtained from MPM simulations are plotted in Figure 13 against the q_c values from Salgado & Prezzi (2007) (Eq (32)). It can be seen that there is a good agreement between the numerical and the correlation results.

4 CONCLUSIONS

In this paper the formulation of a two dimensional axisymmetric MPM is presented. MPM is an extension of the Updated-Lagrangian FEM in which a Lagrangian description of material is defined over an Eulerian background mesh. The method is free from the disadvantage of the Lagrangian FEM, *i.e.* element distortion, by transferring the calculation information from the background mesh to the material points at the end of each calculation step. In the beginning of each step, the mesh is reset and the stored data in the material points is mapped to the mesh. Mapping data back and forth between the material points and the background mesh, and the space search to find the location of material points in the background mesh, makes the method computationally more expensive compared to the UL-FEM. For practical geotechnical applications with an axisymmetric geometry, such as CPT and onshore and off-shore piles, the developed 2D axisymmetric code can significantly reduce the computational costs by reducing the number of elements, material points and degrees of freedoms. The capability of the developed code was demonstrated by successful numerical simulations of three axisymmetric problems, namely a shallow foundation, soil column collapse and the cone penetration testing.

ACKNOWLEDGMENT

This research has been made possible within JIP-SIMON project with a grant from the Dutch TKI-Wind op Zee program (Topsector Energiesubsidie van het Ministerie van Economische Zaken) as well as financial support from the project partners, Allnamics, Boskalis, Cape Holland, Deltares, IHC IQIP, Innogy and van Oord.

REFERENCES

Al-Kafaji, I. 2013. Formulation of a dynamic material point method (MPM) for geomechanical problems, PhD thesis, Universität Stuttgart.

Bardenhagen, S.G, Brackbill, J.U. & Sulsky, D. 2000. The material-point method for granular materials. Computer Methods in Applied Mechanics and Engineering, 187 (3–4):529–541.

Gens, A. & Potts, D.M. (1984). Formulation of quasi-axisymmetric boundary value problems for finite element analysis. Engineering Computations, Vol. 1 Issue: 2, pp.144–150.

Lune, P.K. Robertson, J.J.M. Powell., 1997. «Cone Penetration Testing in Geotechnical Practice". Blackie Academic & Profesional, London.

Monforte, L., Carbonell, J.M., Arroyo, M., Gens, A. (2017). Performance of mixed formulations for the particle finite element method in soil mechanics problems. Computational Particle Mechanics, 4(3), pp. 269–284.

Randolph, M. F., Dolwin, R., & Beck, R. (1994). Design of driven piles in sand. Geotechnique, 44(3), 427–448.

Salgado, R. & Prezzi, M. (2007). Computation of cavity expansion pressure and penetration resistance in sands. Int. J. Geomech. 7, No. 4, pp. 251–265.

Sulsky D, Chen Z, Schreyer, H.L. (1994). A particle method for history-dependent materials. Computer Methods in Applied Mechanics and Engineering, 118:179–96.

Sulsky, D. & Schreyer, H. L.1996. Axisymmetric form of the material point method with applications to upsetting and Taylor impact problems. Computer Methods in Applied Mechanics and Engineering, Journal of Applied Mechanics, Vol. 139, pp. 409–429.

Tehrani, F.S. & Galavi, V. (2018). Comparison of Cavity Expansion and Material Point Method for Simulation of Cone Penetration in Sand. Submitted to CPT18, Delft, the Netherlands.

Tolooiyan, A., Gavin, K. (2011). Modelling the cone penetration test in sand using cavity expansion and arbitrary Lagrangian Eulerian finite element methods. Computer and Geotechnics. 38(4), pp. 482–490.

Van Esch, J., Stolle, D. & Jassim I. (2011). Finite element method for coupled dynamic flow-deformation simulation. In 2nd International Symposium on Computational Geomechanics (ComGeo II), Cavtat-Dubrovnik, Croatia, April.

Numerical Methods in Geotechnical Engineering IX – Cardoso et al. (Eds)
© 2018 Taylor & Francis Group, London, ISBN 978-1-138-33198-3

Finite element modeling of innovative energy geo-structure behaviour

F. Ronchi, D. Salciarini & C. Tamagnini
Department of Civil and Environmental Engineering, University of Perugia, Perugia, Italy

ABSTRACT: The paper presents the results of a series of 3-Dimensional (3D), Finite Element (FE) analyses, conducted to investigate the effects of the heating cycles on innovative energy foundations of small diameter (Energy Micro-Piles, EMP) installed in fine-grained soils. This study has been performed during the design stage of a full-scale prototype EMP currently under development at the University of Perugia for the exploitation of low-enthalpy geothermal energy in the retrofitting of existing buildings. The FE simulation program has been focused mainly on the evaluation of the heat flux variations of the prototype during functioning of the system, by varying thermal properties of the soil, the inlet fluid rate, and the primary circuit pipe material. The results of the numerical analyses showed a major influence given by the thermal properties of the soil surrounding the micro-pile and a limited effect provided by the material used for the circulation pipes.

1 INTRODUCTION

As defined by Brandl (2006) and Laloui & Di Donna (2013) *Thermally–active ground structures* or *energy geostructures* represent recent applications developed for exploiting the Low Enthalpy Geothermal Energy (LEGE) beside the more common *Borehole Heat Exchangers* (BHEs). The key feature of these technologies is the possibility of exploiting the structures in direct contact with the soil like heat exchangers through the embedment of the *primary circuit* pipes of a GSHP system within the concrete. This permits withdrawing heat from the soil in the winter season and releasing heat into the soil in the summer season so as to guarantee the conditioning of the building with an important reduction of the CO_2 production compared to standard systems.

Examples of energy geostructures are foundations, retaining walls, bulkheads or tunnels. This paper presents another promising technical solution for exploiting LEGE, which belongs to the class of the energy foundations: the energy micro-piles (EMPs). They represent a different perspective in the field of LEGE use in civil engineering, mainlyrelated to the retrofitting of existing buildings, where the enhancement of the thermal features can be obtained together with the improvement of the structural set up.

Although similar to the more common EPs, the limited dimensions—in terms of length and diameter, *i.e.* surface in contact with soil—make the EMP behavior different, both from the mechanical point of view (axial load supported mainly by lateral resistance with almost null end bearing capacity) and thermal point of view (generally, lower energy efficiency for the single element due to the short thermal circuit but potential of having a large number of elements in the foundation).

In this framework, two full-scale prototypes of EMPs have been installed at the Engineering Campus of the University of Perugia so as to conduct *in-situ* tests aimed at evaluating their response to the temperature variations induced by the GSHP system functioning. The final objective of the research is to better understand the possible large-scale applicability of EMPs as sustainable energy systems. Currently, only one of the two prototypes has been studied by means of both experimental investigations and numerical Finite Element (FE) analyses. The FE model has been first calibrated using the results obtained form a Thermal Response Test (TRT), and then used to make a prediction of the long-term behavior of the prototype in operating conditions of the GSHP functioning during the summer regime. Sensitivity analyses have been conducted by varying some key functioning parameters in view of an optimization of the long-term behavior of the EMP technology. The results of this study are presented in this paper in terms of the temporal evolution of the specific heat flux generated by the EMP. The complete description of the experimental *in-situ* testing setup with the details of the prototype installation, the instrumentation used, the methodology and the results obtained form the TRT are presented in Ronchi et al. (2018).

1.1 *Full-scale prototype of the EMP*

The considered EMP—bored and cast in place—is 12.0 m long and 0.18 m in diameter. To ensure the conduction of a series of thermal testing, the prototype has been equipped with high-density polyethylene (HDPE) absorber pipes, for which the inlet and outlet sections have been thermally insulated at the heads to limit the influence of the climatic daily variations on the heat exchange process.

The key feature of the EMP prototype is the particular shape of the tip—called, in what follows, *energy tip*, and is characterized by a steel tank having a maximum capacity of about 6 liters. This has been properly connected to the inlet and outlet points of the primary circuit pipe so as to complete the U-loop shape circulating system of the heat-carrier fluid. Figure 1 shows a schematic view of the energy tip of EMP prototype that has been investigated in the present work.

The tip has been designed with the aim of enhancing the heat exchange of the prototype with the surrounding soil so as to increase its energy performance. This is ensured by the presence of water in the tank that remains in thermal equilibrium conditions with the soil during the system functioning. This geometry has been thought to balance the shorter length of the EMP with respect to the more common EPs.

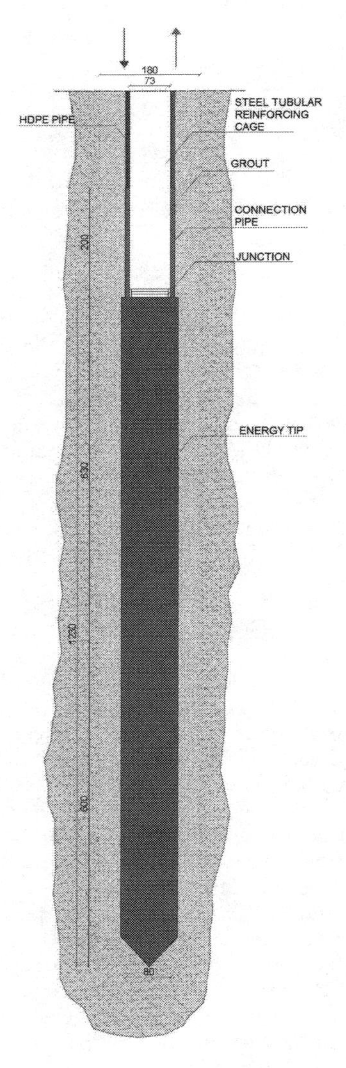

Figure 1. Scketch of the bottom portion of EMP prototype.

2 THE THERMAL EXCHANGE PROCESS

The mathematical formulation of the heat exchange process taking place in the energy system (pipe–pile–soil) accounts for the following phenomena: (i) the convective heat exchange between the heat carrier fluid and the wall of the exchanger pipes; (ii) the heat conduction within the concrete of the EMP; (iii) the heat conduction into the soil. [t]

The heat transfer through the pipe wall, the concrete and the soil has been described by the classical equation of heat conduction in transient regime, with no internal heat generation (Lewis & Schrefler 1998):

$$\rho C_p \frac{\partial T}{\partial t} = -\nabla \cdot \mathbf{q} \tag{1}$$

where ρ [kg/m³] is the material density, C_p [J/(kg·K)] the material specific heat at constant pressure, $\mathbf{q}$ [W/m²] the vector of the heat flux transferred by conduction and T [C] the temperature. Adopting the Fourier's law to define the heat flux vector $\mathbf{q}$, (1) can be reformulated as:

$$\rho C_p \frac{\partial T}{\partial t} = \nabla \cdot (\lambda \nabla T) \tag{2}$$

in which λ [W/(m·K)] is the thermal conductivity of the material (a scalar parameter in the isotropic hypothesis).

The non–isothermal flow in the pipe channels has been reduced to a 1D problem by modeling them as linear elements. The related governing equations have been fully coupled to those describing the heat transfer phenomena in the surrounding solid domain.

In particular, the momentum and mass conservation equations describing the non–isothermal flow in the pipes can be written as the following equations (3) and (4), respectively (Barnard et al. 1966):

$$\rho_f \frac{\partial \mathbf{u}}{\partial t} = -\nabla p - f_D \frac{\rho_f}{2d_h} \mathbf{u} |\mathbf{u}| + \mathbf{F} \quad (3)$$

$$\frac{\partial A_{pi}\rho_f}{\partial t} + \nabla(A_{pi}\rho_f \mathbf{u}) = 0 \quad (4)$$

where ρ_f [kg/m³] is the fluid density, $\mathbf{u}$ [m/s] is the cross section averaged fluid velocity along the tangent of the center line of the pipe, p [N/m²] is the pressure, f_D [-] is the Darcy friction factor, d_h [m] is the mean hydraulic diameter and A_{pi} [m²] is the cross section area of the pipe. The meaning of each parameter is described in Ronchi et al. (2018).

The energy balance for an incompressible fluid flowing in a pipe is described by the following equation (Lurie & Sinaiski 2008):

$$\rho_f A_{pi} C_{pf} \frac{\partial T}{\partial t} + \rho_f A_{pi} C_{pf} \mathbf{u} \cdot \nabla T = \nabla A_{pi} \lambda_f \nabla T + \\ + f_D \frac{\rho_f A_{pi}}{2d_h} |\mathbf{u}|^3 + q'_{\text{wall}} \quad (5)$$

where C_{pf} [J/(kg·K)] is the specific heat at constant pressure of the fluid, T [C] is the temperature, λ_f [W/(m·K)] is the fluid thermal conductivity, and the other terms have been already defined. The second term on the right hand side of (5) corresponds to the heat dissipated due to the internal friction in the fluid, while q'_{wall} [W/m] is a source term that accounts for the radial heat transferring from the pipes to the surrounding domain.

The coupled system of partial differential equations given by (2), (3), (4) and (5), can be (numerically) integrated given the appropriate initial and thermal boundary conditions for the particular problem considered. To this end, the FE method has been used, as detailed subsequently.

3 MODELING THE THERMAL BEHAVIOR OF THE ENERGY MICRO-PILE PROTOTYPE

3.1 *Geometry of the FE model*

The EMP prototype has been modeled in the FE code *Comsol Multiphysics* (Comsol 2011), within a cylindrical volume of homogeneous silt soil of 10 m in diameter and 30 m in depth (Fig. 2), whose thermo-mechanical properties have been shown in Table 1.

As described previously, simulation of 3D flow and heat transfer inside the cooling channels are computationally expensive. An efficient possibility

Table 1. Material thermo–physical properties.

	C_p	λ	ρ
	[J/(kg·K)]	[W/(m·K)]	[kg/m³]
Silt	800	1.34 (from TRT)	1835
Grout	880	1.60	2300
Steel	475	45	7850
Copper	385	302	7850
HDPE	2000	0.60	950
Water	4186	0.60	1000

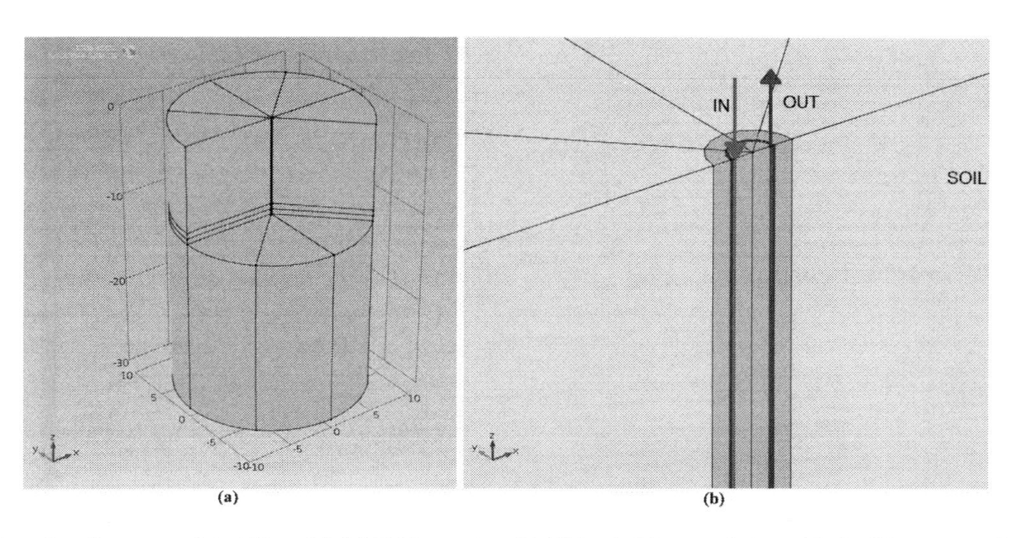

Figure 2. Geometry of the FE model: (a) EMP prototype highlighted; (b) zoomed view with the 1D representation of the circulation pipes.

is to model the flow and heat transfer in the cooling channels with 1D pipe flow equations, and modeling the surrounding domain in 3D (see *i.e.*, Ozudogru et al. 2014 and Ozudogru et al. 2015).

3.2 *Discretization*

The spatial discretization adopted for the entire domain is shown in the zoomed view of Figure 3. A total of 34860 hexahedral elements having quadratic interpolation for the temperature field have been used for discretizing the solid domains except for the inner part of the EMP—occupied by the grout—where 21530 tetrahedral elements (with the same characteristics) have been considered. The pipes have been discretized with 3810 linear elements having quadratic interpolation for the pressure and temperature fields and linear interpolation for the tangential velocity field. The pile–soil interface has been considered as perfectly rough (full adhesion) and no specific soil–pile interface elements have been introduced. The numerical integration in time

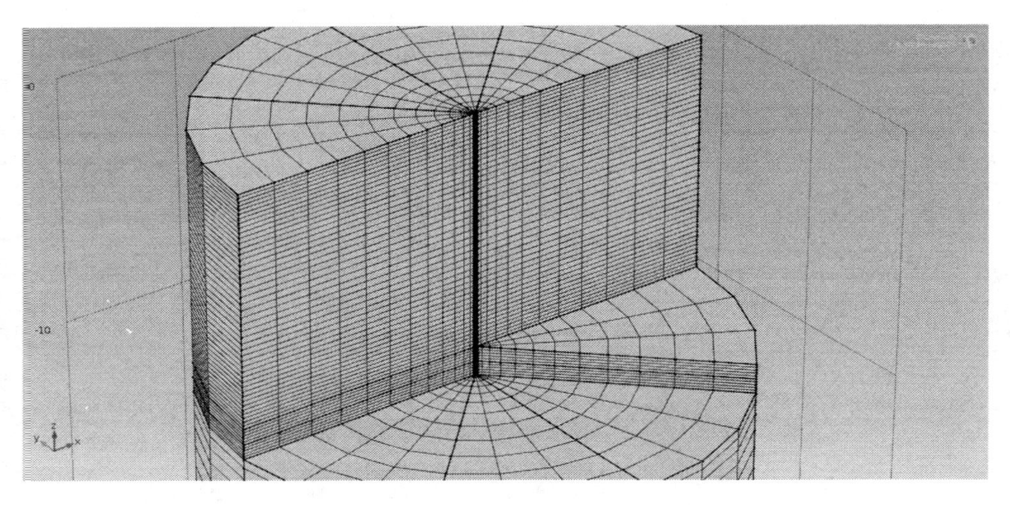

Figure 3. Spatial discretization used in the FE model of the system.

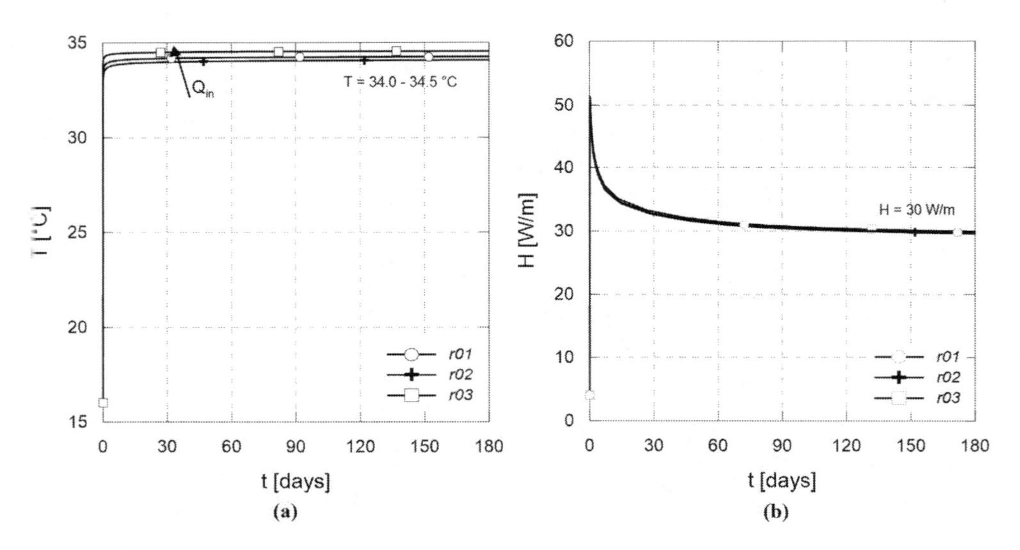

Figure 4. Comparisons between the simulations with different values of the inlet heat carrier fluid flow rate in terms of the evolution with time of: (a) the outlet heat carrier fluid temperature; (b) the specific heat flux.

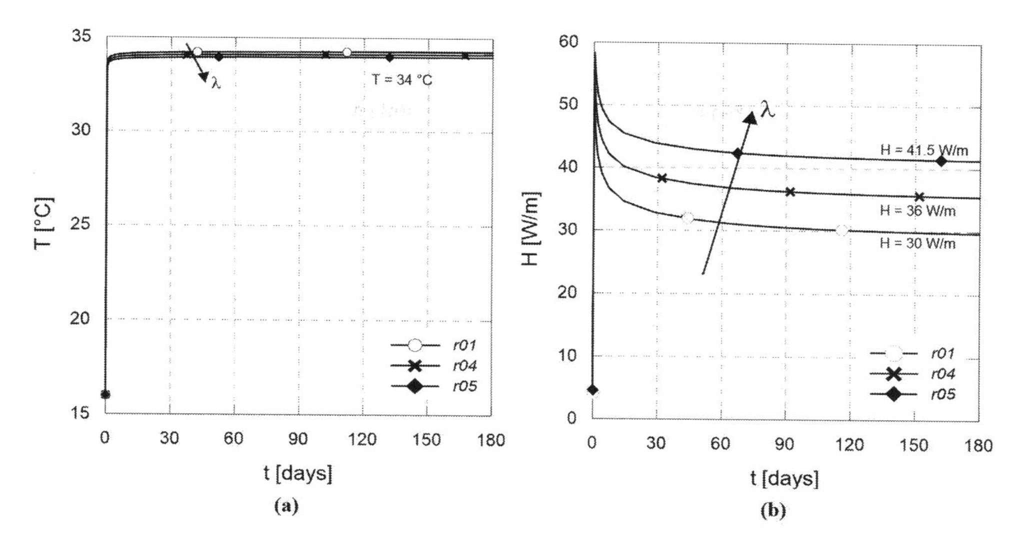

Figure 5. Comparisons between the simulations with different values of the soil thermal conductivities in terms of the evolution with time of: (a) the outlet heat carrier fluid temperature; (b) the specific heat flux.

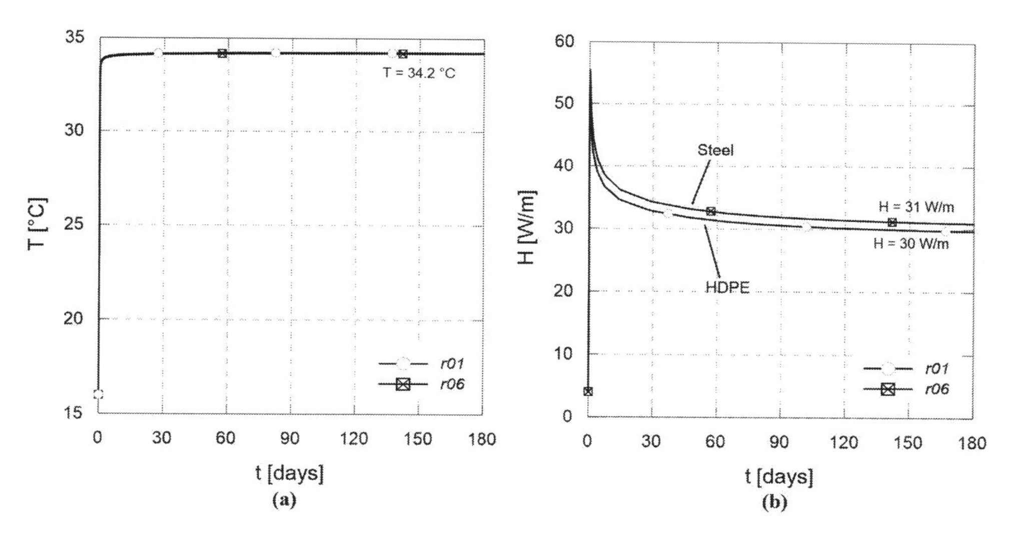

Figure 6. Comparisons between the simulations with different values of the material of the circulation pipe in terms of the evolution with time of: (a) the outlet heat carrier fluid temperature; (b) the specific heat flux.

of the resulting semidiscrete system of ordinary differential equations has been carried out using a fifth–order implicit backward differentiation formula algorithm (Ascher & Petzold 1998).

3.3 *Limit conditions and simulation program*

The solid domains have been considered initially at a constant and almost uniform temperature with depth, equal to the one registered during the *in-situ* testing, *i.e.* of about 16°C (average value), neglecting some minor oscillations at the surface level. The heat carrier fluid has been initially assumed in stationary condition with velocity vector equal to zero, pressure equal to 1.1 bar (circuit pressure), and temperature equal to that of the surrounding soil.

As for the boundary conditions, a constant temperature of 16°C has been assigned to the lower and lateral surfaces of the soil domain, assuming the latters to be sufficiently far from the pile so as

Table 2. Simulation program.

Analysis	Q_{in}	λ	
Id Code	[kg/s]	[W/(m(K)]	2*Material
*r*01	0.10	1.34	HDPE
*r*02	0.08	1.34	HDPE
*r*03	0.17	1.34	HDPE
*r*04	0.10	1.85	HDPE
*r*05	0.10	2.5	HDPE
*r*06	0.10	1.34	Steel

to be not influenced by the heat exchange process induced by the pile itself. The same condition has been considered to the upper surface at the ground level.

At the inlet point of the primary circuit a variation of the heat-carrier fluid temperature has been applied. By starting from the initial equilibrium value of 16°C, the temperature has been increased linearly in one day up to the maximum value of 35°C and then kept constant for the total duration of the analysis of 6 months. At the outlet of the primary circuit has been maintained a pressure equal to 1.1 bar. These assumptions have allowed to simulate the system functioning and to make a prediction of the prototype behavior in the summer regime (cooling of the building).

The possibility of guaranteeing a technological optimization of the newly developed EMP in energetic terms has been evaluated by carrying on various simulations with different values of the following parameters:

1. flow rate of the heat carrier fluid at the inlet point of the primary circuit (*i.e.* the velocity of the fluid itself) (Q_{in});
2. soil thermal conductivity (λ);
3. material of the circulation pipe.

Table 2 shows the analyses conducted.

The simulation with identification code $r01$ has been considered as the starting point for the parametric study (reference analysis). Then, with respect to the parameters characterizing simulation $r01$, Q_{in} has been changed in $r02$ and $r03$, *mbda* in $r04$ and $r05$, the pipe material in $r06$.

4 RESULTS

This section summarizes the results obtained from the FE simulations listed in Table 2. For each simulation of the sensitivity analyses, the history of the heat-carrier fluid temperature at the outlet point of the circulation pipe and the specific heat flux H generated by the EMP have been evaluated and compared to those obtained from the reference one. The specific heat flux has been determined by the following equation:

$$H = \frac{1}{L}\int_S \mathbf{q} \cdot \mathbf{n} da \tag{6}$$

where s and L are, respectively, the lateral surface and the length of the EMP, $\mathbf{q}$ is the heat flux vector, and $\mathbf{n}$ is the normal vector to the micro-pile surface. Further trends of the temperature have been investigated by Ronchi et al. (2018).

As can be noticed in Figure 2(a), Figure 2(a) and Figure 2(a), the outlet heat carrier fluid temperature is almost the same in all the cases, apart from slight fluctuations of a few tenths of Celsius degrees. In general, T_{out} increases rapidly at the beginning of the heat conduction process and then remains almost constant until the end of the analysis, where values around 34°C are reached.

In terms of the specific heat flux generated by the prototype, major influence is given by the variability of the soil thermal conductivity. As shown in Figure 2(b) and Figure 2(b), passing from λ of 1.34 W/(m·K) to 2.5 W/(m·K) leads to an increase of H of about 40%. This allows to obtain final values of 41.5 W/m, which are fully comparable with those of traditional energy piles. Conversely, the use of steel pipes instead of HDPE, does not allow increasing significantly the thermal efficiency of the system. In fact, H is equal to 31 W/m—at $t = 180$ days—in $r06$ simulation (Fig. 2(b)) instead of 30 W/m in $r01$ simulation. In general, the specific heat flux tends to increase rapidly during the first time stations of the system functioning, when the pile/soil temperature gradients are greater, and decreases monotonically in time until reaching the final steady–state conditions.

5 CONCLUSIONS

In this paper a series of 3D FE analyses of a prototype of energy micro-pile has been conducted. With the aim of analyzing the performance of this innovative technology for the exploitation of LEGE in the field of the renovation of existing buildings, the numerical study has been carried out varying some key operating conditions of the GSHP system functioning for simulating real operating conditions in the summer regime.

The obtained results have allowed to verify that although the energy efficiency of this technology is lower than that of energy piles, it is encouraging in view of its application in large scale air–conditioning installations if a high number of elements within the foundation is provided.

The analyses conducted by changing the inlet fluid flow rate, Q_{in}, the soil thermal conductivity, λ, and the material of the heat carrier pipe, show that the specific heat flux generated by the EMP varies between a minimum of 30 W/m to a maximum of 42 W/m, at the final time station $t = 180$ days.

In particular, the results have shown:

- the important role played by the soil, confirming that the proposed technology can provide different values of the thermal efficiency depending on the soil conductivity that characterize the installation site. This suggests that a geotechnical investigation, with the evaluation of the thermal properties of the soils in which the geothermal system is realized, is fundamental for a first evaluation of the thermal efficiency achievable;
- the performance of the system cannot be optimized so much by using circulation pipes made of steel instead of HDPE;
- the flow rate of the inlet fluid has shown a smaller influence on the response of the system; despite that, it needs to be calibrated in a suitable manner according to the characteristics of the GSHP system.

REFERENCES

Ascher, U.M. & L.R. Petzold (1998). *Computer methods for ordinary differential equations and differential-algebraic equations*, Volume 61. Siam.

Barnard, A.,W. Hunt,W. Timlake, & E. Varley (1966). A theory of fluid flow in compliant tubes. *Biophysical Journal 6*(6), 717–724.

Brandl, H. (2006). Energy foundations and other thermo-active ground structures. *Géotechnique 56*(2), 81–122.

Comsol, A. (2011). Comsol multiphysics user's guide. *Version: September.*

Laloui, L. & A. Di Donna (2013). *Energy geostructures: innovation in underground engineering*. John Wiley & Sons, New Jersey, NY, USA.

Lewis, R.W. & B.A. Schrefler (1998). The finite element method in the deformation and consolidation of porous media *(2nd edn)*. John Wiley and Sons Inc., New Jersey, NY, USA.

Lurie, M.V. & E. Sinaiski (2008). *Modeling of oil product and gas pipeline transportation*. Wiley Online Library.

Ozudogru, T., C. Olgun, & A. Senol (2014). 3d numerical modeling of vertical geothermal heat exchangers. *Geothermics 51*, 312–324.

Ozudogru, T.Y., O. Ghasemi-Fare, C.G. Olgun, & P. Basu (2015). Numerical modeling of vertical geothermal heat exchangers using finite difference and finite element techniques. *Geotechnical and Geological Engineering 33*(2), 291–306.

Ronchi, F., D. Salciarini, N. Cavalagli, & C. Tamagnini (2018). Thermal response prediction of an energy micro-pile prototype. *Geomechanics for Energy and the Environment*.

Numerical Methods in Geotechnical Engineering IX – Cardoso et al. (Eds)

 ISBN 978-1-138-33198-3

Effects of sandy soils permeability variation on the pore pressure accumulation due to cyclic and dynamic loading

H. Bayraktaroglu & H.E. Taşan
Department of Civil Engineering, Middle East Technical University, Ankara, Turkey

ABSTRACT: Cyclic and dynamic behavior of saturated sands is significantly influenced by their porosity and permeability. During loading, unloading and reloading compaction, loosening and re-compaction of sands take place, which lead to change in pore volume and permeability. In this study, the behavior of water saturated sands subjected to cyclic and dynamic loading is analyzed numerically. For the finite element analyses, a three-dimensional fully coupled two-phase finite element is developed and implemented on the basis of a two-phase model to consider the pore water pressure development in saturated sands. In addition, a hypoplastic constitutive model is used to describe the material behavior of sandy soils. The porosity-permeability variation is taken into account by Kozeny-Carman relationship. Comparing with experimental test results documented in the literature, the influence of porosity-permeability variation on the strain and pore pressure accumulation is investigated. The necessity of the consideration of porosity-permeability variation for realistic modeling of the cyclic and dynamic behavior of saturated sandy soils is assessed.

1 INTRODUCTION

The behavior of soils under cyclic and dynamic loading is recognized as one of the most challenging fields of soil mechanics. Understanding the development process of accumulation is critical for a proper description of the soil behavior under cyclic and dynamic loading.

Saturated soils subjected to cyclic and dynamic loading tend to build-up of pore pressure, shear strength degradation and softening (Martin et al. 1975, Andersen 2009, Cary & Zapata 2016). The accumulation of irreversible strains in the soil due to cyclic and dynamic loading leads to displacements which have to be considered in the design of structures. The strain and pore pressure accumulations in soil are key issues on the design of offshore foundations (Wichtmann et al. 2010, Cuéllar 2011, Taşan et al. 2011, Triantafyllidis 2016).

Numerical analysis of pore pressure development in saturated soils is a challenging task, due to the inherently strong coupling effects inside the two-phase physical system, and lack of reliable constitutive models that capture the mechanical behaviors of soil realistically under cyclic loading (Niemunis et al. 2005, Tang & Hededal 2014).

It is acknowledged that the porosity of the soil varies with large deformation and depending on the change in the porosity it is assumed that permeability of the soil also changes (Di & Sato 2003). For a sufficiently accurate description of the behavior of saturated soils under cyclic and dynamic loading is considered necessary to take into account the porosity-permeability variation (Taşan 2016).

The objective of the present paper is to investigate the effects of porosity-permeability variation on the strain and pore pressure accumulation. For this purpose, a sophisticated numerical model based on the three-dimensional fully coupled two-phase finite element was developed. A hypoplastic constitutive material model was used to describe stress-strain behavior and strength of the sandy soil subjected to cyclic and dynamic loadings. The Kozeny-Carman relationship was implemented to describe the variation of porosity-permeability.

The effects of the porosity-permeability variation were investigated under cyclic and dynamic loading conditions. Starting with a simple soil column example and subsequent simulation of a centrifuge test conducted to study a dyke under earthquake loading, the development of excess pore pressure and soil displacements with and without considering the porosity-permeability variation was determined and compared.

Before beginning with the presentation of the numerical analysis results, the used two-phase model, hypoplastic constitutive material law and the Kozeny-Carman relationship will be shortly introduced.

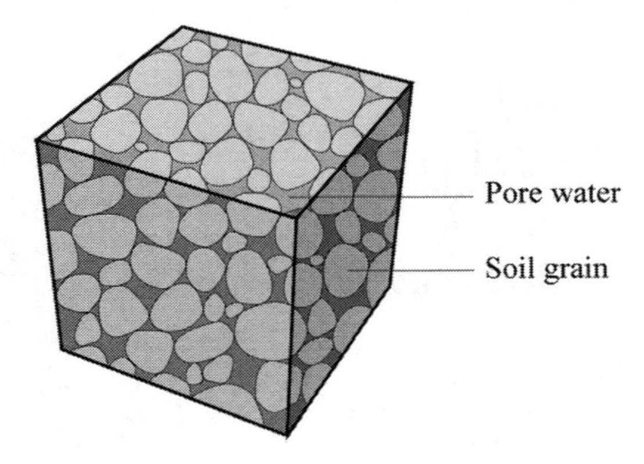

Figure 1. Two-phase model (Taşan et al. 2011).

2 TWO-PHASE MODEL

The two-phase model which is originally formulated by Biot (1941), and then further modifications and improvements performed by Zienkiewicz et al. (1980) and Zienkiewicz et al. (1982), called as u-p-model, was used to investigate the mechanical interaction between the soil grains and the pore water in fully saturated soils.

As its name implies, the two-phase model consists of a solid phase, skeleton and a fluid phase which fully occupies the pores in the skeleton (Fig. 1).

The governing variables of the system in the model are the absolute displacements **u** of the solid phase and the pore water pressure p. The u-p-model include the equation for balance of momentum for the mixture which can be expressed with considering Terzaghi's principle of effective stress principle as

$$\mathbf{L}^T(\boldsymbol{\sigma}' - \mathbf{m}p) + \rho\mathbf{b} = \rho\ddot{\mathbf{u}} + \varsigma\dot{\mathbf{u}} \tag{1}$$

where,

$\boldsymbol{\sigma}'$: Effective stress vector

ρ: Density of mixture

$\mathbf{b}$: Body forces

ζ: Damping ratio

Divergence operator **L** and **m** are given as,

$$\mathbf{L}^T = \begin{bmatrix} \partial/\partial x & 0 & 0 & \partial/\partial y & 0 & \partial/\partial z \\ 0 & \partial/\partial y & 0 & \partial/\partial x & \partial/\partial z & 0 \\ 0 & 0 & \partial/\partial z & 0 & \partial/\partial y & \partial/\partial x \end{bmatrix} \tag{2}$$

$$\mathbf{m}^T = [1,1,1,0,0,0] \tag{3}$$

The combined equation for the mass balance and linear momentum of the fluid in the pores is given as,

$$\mathbf{m}^T\mathbf{L}\dot{\mathbf{u}} - \nabla^T\left(\frac{\mathbf{K_p}}{\eta_w}\nabla p\right) + \frac{\dot{\mathbf{p}}}{Q^*} + \nabla^T\frac{\mathbf{K_p}}{\eta_w}\rho_w\mathbf{b} = 0 \tag{4}$$

where,

$\mathbf{K_p}$: Permeability matrix

ρ_w: Density of water

η_w: Dynamic viscosity of the water

Q^* in Equation 4 is the coupled volumetric stiffness of solid grains and fluid and expressed as

$$\frac{1}{Q^*} = \frac{n}{K_w} + \frac{1-n}{K_s} \tag{5}$$

In Equation 5, n is the porosity of the soil, K_s is the bulk modulus of solid grains and K_w is that of the pore water.

For isotropic problems, the permeability matrix can be written as $\mathbf{K_p} = k_p\mathbf{I}$ with unit matrix **I**. The hydraulic conductivity k_d can be derived from the permeability k_p by

$$k_d = \frac{\rho_w g}{\eta_w}k_p \tag{6}$$

where,

g is gravitational acceleration.

For the formulation of u-p-model, it was assumed that those inertial components that are related to the relative movement of grains and pore water are negligible. The validity of this simplification is given for most geotechnical problems in connection with saturated soils (Zienkiewicz et al. 1999).

In order to convert the partial differential Equations 1 and 4 to ordinary differential equations, standard Galerkin techniques are used. Thereby, spatial approximations for the displacement and pore water pressure are constructed as

$$\begin{aligned} \mathbf{u} &= \mathbf{N_u}\bar{\mathbf{u}} \\ p &= \mathbf{N_p}\bar{\mathbf{p}} \end{aligned} \tag{7}$$

where,

$\mathbf{N_u}$, $\mathbf{N_p}$ are the shape functions and $\bar{\mathbf{u}}, \bar{\mathbf{p}}$ are the corresponding vectors of unknowns. The resulting local coupled equation system of a finite element can be represented as follows:

$$\mathbf{M}\ddot{\bar{\mathbf{u}}} + \mathbf{C}\dot{\bar{\mathbf{u}}} + \mathbf{K}\bar{\mathbf{u}} - \mathbf{Q}\bar{\mathbf{p}} = \mathbf{f_u} \tag{8}$$

and

$$\mathbf{Q}^T\dot{\bar{\mathbf{u}}} + \mathbf{S}\dot{\bar{\mathbf{p}}} + \mathbf{H}\bar{\mathbf{p}} = \mathbf{f_p} \tag{9}$$

with,

$$\mathbf{M} = \int_{\Omega} \mathbf{N}_u^T \rho \mathbf{N}_u d\Omega, \mathbf{C} = \int_{\Omega} \mathbf{N}_u^T \varsigma \mathbf{N}_u d\Omega,$$

$$\mathbf{K} = \int_{\Omega} \mathbf{B}^T \mathbf{D}_t \mathbf{B} d\Omega, \mathbf{Q} = \int_{\Omega} \mathbf{B}^T \mathbf{m} \mathbf{N}_p d\Omega,$$

$$\mathbf{H} = \int_{\Omega} \mathbf{B}_p^T \mathbf{k}_p \mathbf{B}_p d\Omega,$$

$$\mathbf{f}_u = \int_{\Omega} \mathbf{N}_u^T \rho \mathbf{b} d\Omega + \int_{\Omega} \mathbf{N}_u^T \tilde{\sigma} d\Gamma,$$

$$\mathbf{S} = \int_{\Omega} \mathbf{N}_p^T \frac{1}{Q^*} \mathbf{N}_p d\Omega,$$

$$\mathbf{f}_p = \int_{\Omega} \mathbf{B}_p^T \mathbf{k}_p \rho_w \mathbf{b} d\Omega - \int_{\Gamma} \mathbf{N}_p^T \tilde{q} d\Gamma$$

where,

$\mathbf{B} = \mathbf{L}\mathbf{N}_u$ and $\mathbf{B}_p = \nabla \mathbf{N}_p$

$\mathbf{D}_t$ is the matrix of tangential modulus which can be determined from a nonlinear stress-strain relationship. $\tilde{\sigma}$ and $\tilde{q}$ are the prescribed tractions and water flux at the boundaries Γ of the domain Ω.

Time integration of the Equations 8 and 9 is performed by applying Newmark method (Bathe 1996). The detailed discretization of the Equations 1 and 4 are available in Taşan (2011).

Based on the u-p-model, a three-dimensional continuum element, called as u20p8, is implemented. Here, the displacement field is approximated using triquadratic interpolation functions and the pressure field is approximated using trilinear interpolation functions. Twenty nodes for the displacement and eight nodes for the pore pressure (u20p8) have been used for this element type as given in Figure 2. The choice of a higher order of displacement field is required to ensure the stability of the elements (Zienkiewicz 1986).

The numerical implementation of u-p-model and its verification are described in Taşan (2011).

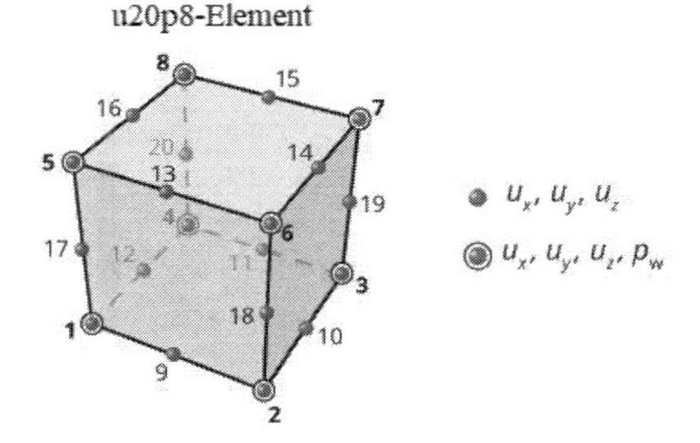

Figure 2. There-dimensional coupled two-phase element, u20p8.

3 HYPOPLASTIC CONSTITUTIVE MODEL

Deformation behavior of the cohesionless soils can be described realistically by non-linear hypoplastic constitutive model. Unlike the elasto-plasticity models, the hypoplastic model does not distinguish between the elastic and plastic deformation and it does not contain any yield surface, plastic potential, flow and hardening rule. The model is suitable to consider the effects of barotropy, pycnotropy, dilatancy and material softening which is observable during shearing of dense material.

A single tensorial equation is used to formulate the hypoplastic constitutive model. For the characterization of soil state, the granular effective stress and void ratio are required.

The first hypoplastic model was developed by Kolymbas (1988) and then further modifications and improvements took place (Bauer 1996, Wu 1992, von Wolffersdorf 1996). The basic hypoplastic model needs a total of eight material parameters those determinations are physically well founded (Herle 1997).

The hypoplastic constitutive model was able to predict the mechanical behavior of the granular materials under steady loading but not under repeated loading. The model is expanded by Niemunis & Herle (1997) to model realistically the accumulation effects and the hysteretic material behavior under cyclic loading. For this purpose, an additional state variable, which is the intergranular strain, is introduced to taking into account the influence of changing direction of deformation on the mechanical behavior of soil. Therefore, five more material parameters are added to the model.

In this study, the hypoplastic constitutive model with intergranular strain proposed by Niemunis & Herle (1997) is used.

The suitability of the material model using the u20p8 element was confirmed in Taşan et al. (2010).

4 KOZENY-CARMAN (KC) EQUATION

The Kozeny-Carman equation is a well-known semi-empirical, semi-theoretical formula developed for predicting the permeability of porous media. The theory of the model is simply based on the direct relationship between media properties and flow resistance in pore channels (Kozeny 1927, Carman 1938).

According to the KC relationship, the soil hydraulic conductivity k_d is defined as

$$k_d = C \frac{e^3}{(1+e)} \tag{10}$$

where e is the void ratio of the medium and C is a constant which depends on factors such as fluid viscosity, grain size distribution, shape of grains. The constant C can be determined experimentally.

The KC relationship provides satisfactory results for the evaluation of hydraulic conductivity of sandy soils (Lambe, & Whitman 1969, Di & Sato 2003).

Due to load dependent change of void ratio, the hydraulic conductivity of soil is updated for each element based on Equation 10 after each time step.

5 CYCLIC LOADING CONDITIONS

To investigate the effects of the porosity-permeability variation on the cyclic behavior of sand, first a simple soil column according to Figure 3 with a dimension of $0.1 \times 0.1 \times 10$ m was modeled using the coupled two-phase finite element u20p8.

The boundary conditions, which are imposed on the mesh, are fixing of nodes at the bottom of the mesh against displacement in all directions and at the edges of the model over the depth z in both lateral directions. A free drainage condition was defined at the top surface of the soil column and all other surfaces assumed to be impermeable. The pressure at the top surface of the column is assumed to be atmospheric.

The Hochstetten Sand is used and its hypoplastic material parameters according to Niemunis and Herle (1997) are given in Table 1. A submerged unit weight $\gamma' = 9.32$ kN/m^3, initial void ratio $e_0 = 0.695$ and hydraulic conductivity $k_d = 10^{-4}$ m/s are representing the sandy soil conditions.

The bulk modulus of solid grains and pore water are taken into account as $K_s = 2 \times 10^9$ kPa and $K_w = 2 \times 10^6$ kPa respectively.

Prior to the first phase of the simulation, the stresses after the initial loading must be defined for the sand to determine the required state variables of the hypoplastic model. Therefore, as first a calculation under gravity loading was performed. Subsequently, one-way cyclic loading according to Figure 3 was applied. Analyses including a cyclic loading with a maximum value $q_{max} = 200$ MPa and a period of $T = 6$ s were performed each for the case of with and without considering the KC relationship. In order to achieve a comparability of the results, for the analyses with considering the KC relationship, the constant C of KC in Equation 10 was determined so that at the beginning of the cyclic loading both analysis starts with the same permeability values. Accordingly, considering the value of the initial void ratio e_0 and hydraulic conductivity k_d the constant C is obtained to 5.05×10^{-4} m/s.

The time histories of excess pore pressure in soil at $z = 5$ m are shown in Figure 4. The maximum value of excess pore water pressure is determined in both analyses at the first cycle of cyclic loading. For the analysis case with KC, the compaction of soil within a loading cycle causes simultaneously the reduction of its permeability whereby the drainage condition is affected in an unfavorable manner. Hence the excess pore pressure values are significantly higher than those for the case without KC.

An accumulation of excess pore pressure with increasing cyclic load number is not determined for both cases. After first cyclic loading, a decrease of amplitudes with increasing time is calculated for the case with KC up to $t = 100$ s and the case without KC up to $t = 25$ s. The pattern of excess pore pressure does not change then from these points of times. The amplitude of excess pore pressure for the case with KC is noticeably higher than those without KC at each specific time (Fig. 4).

The time histories of settlements at $z = 5$ m, which is shown in Figure 5, indicate that the increase of settlements for the case of without KC is considerably higher than those for the case of with KC. For both a stabilization of settlements with increasing number of cycles is obtained.

6 DYNAMIC LOADING CONDITIONS

In this part the effects of the porosity-permeability variation are investigated based on the observations

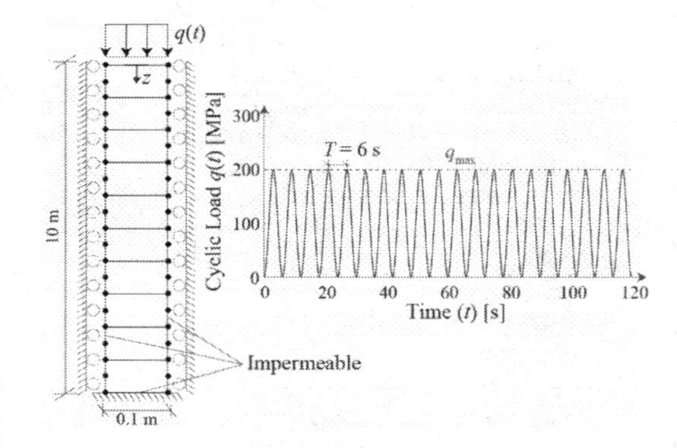

Figure 3. Soil column and applied cyclic load.

Table 1. Soil parameters for Hochstetten Sand (Niemunis & Herle 1997).

φ_c [°]	h_s [MPa]	n	e_{d0}	e_{c0}	e_{i0}	α	β
33.0	1000	0.25	0.55	0.95	1.05	0.25	1.0

R	m_R	m_T	β_r	χ
1·10–4	5.0	2.0	0.5	6.0

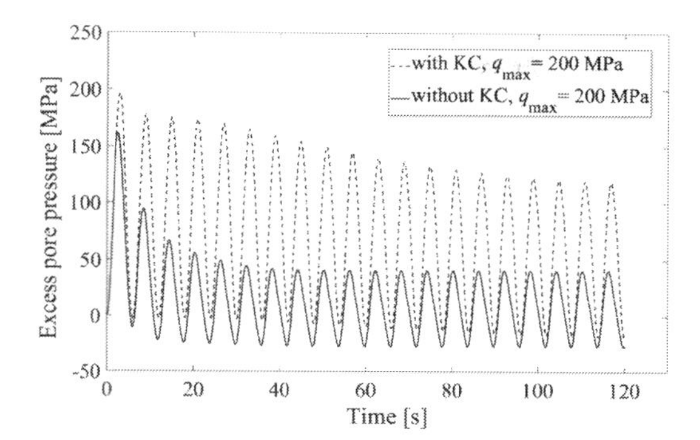

Figure 4. Time histories of excess pore pressure at 5 m depth.

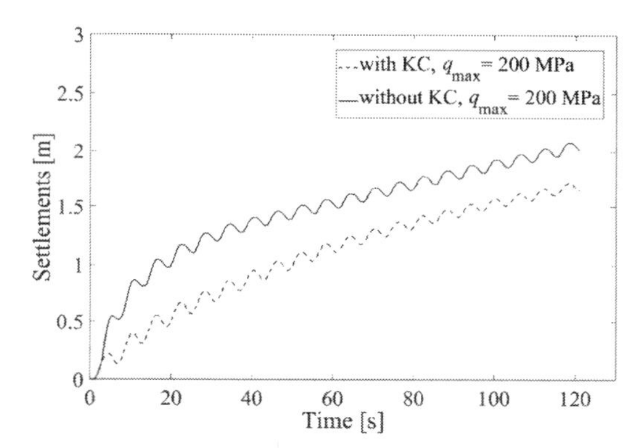

Figure 5. Time histories of settlements at 5 m depth.

from physical modeling, using data from centrifuge test that is presented in C-Core (2014). The experiment is performed to investigate the phenomena of soil liquefaction due to the earthquake loading.

The model configuration is given in Figure 6. The soil layers in the test have a length of 51.6 m and a width of 20.0 m. A drainage layer with a thickness of 1.4 m formed by highly permeable coarse sand was placed on the bottom of the model. A 5 m dense sand layer is overlaid by a loose sand layer with a thickness of 10.6 m at the toe. The water level is 1.0 m at the crest and 8.0 m at the toe above the soil surface level.

The FE model consisting of the coupled two-phase finite element u20p8 is shown in Figure 7. The displacements were fixed horizontally on vertical borders of the model and vertically on the bottom of the domain. The friction between the sand and the container was reduced in the test by lining the inside of the container with stain-less steel sheets and hence, this friction was neglected in the numerical model.

Non-permeable conditions are assumed on all boundaries, except boundaries on model surface, where pore pressures due to existing water level were considered.

In the centrifuge tests Fraser River sand with a specific gravity of 2.71 is used and the relative densities are 40% for loose and 80% for dense sand.

The initial void ratio of the Fraser River sand was calculated based on C-Core (2014) as $e_0 = 0.812$ for loose sand and $e_0 = 0.68$ for dense sand. Initial hydraulic conductivity corresponding to the 40% relative density was given as $k_d = 4.3 \times 10^{-4}$ m/s and the constant in KC relationship (Eq. 10) is determined as $C = 1.455 \times 10^{-4}$ m/s so that the initial permeability of both finite element and experimental model will be equal to each other.

The behavior of drainage layer consisting of coarse sand is assumed as linear elastic with a modulus of elasticity of $E = 1.10^5$ kN/m^2 and poisson's ratio $\nu = 0.3$ due to missing soil data. The hydraulic conductivity of coarse sand is set according to C-Core (2014) as $k_d = 4.3 \times 10^{-2}$ m/s. An effect from the porosity-permeability dependence is not considered for the drainage layer in the analyses.

The bulk modulus of solid grains and pore water are considered as $K_s = 1 \times 10^9$ and $K_w = 2 \times 10^6$ kPa.

The hypoplastic parameters of Fraser River sand are obtained in Holler (2006) and given in Table 2.

Again, in order to determine the required state variables of the hypoplastic model, calculation under gravitational loading was performed and then a dynamic earthquake load given in Figure 8 was applied to the system. Pore pressure accumulations results were compared with the experimental data obtained using pore pressure transducers (P).

Table 2. Soil parameters for Fraser River sand according to Holler (2006).

φ_c [°]	h_s [MPa]	n	e_{d0}	e_{c0}	e_{i0}	α	β
35.0	1600	0.39	0.62	0.94	1.08	0.2	1.0

R	m_R	m_T	β_r	χ
1·10–4	2.5	9.0	0.25	9.0

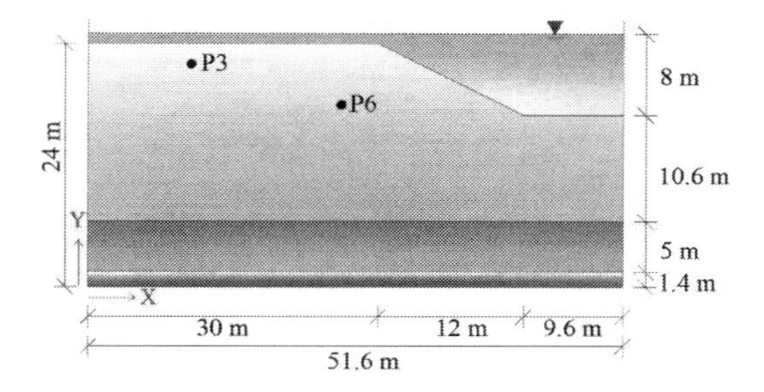

Figure 6. Model configuration.

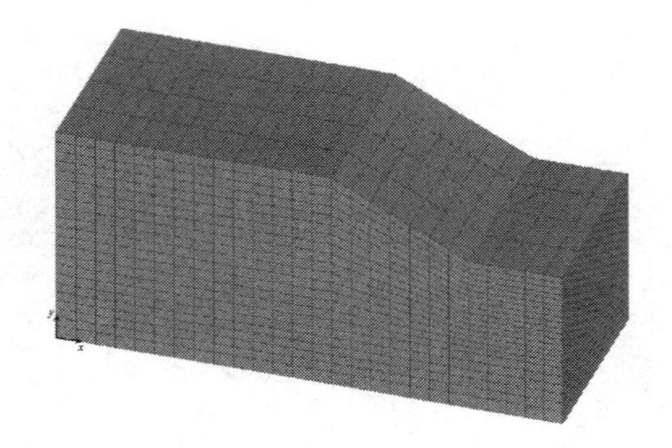

Figure 7. Mesh of 3D FE Model.

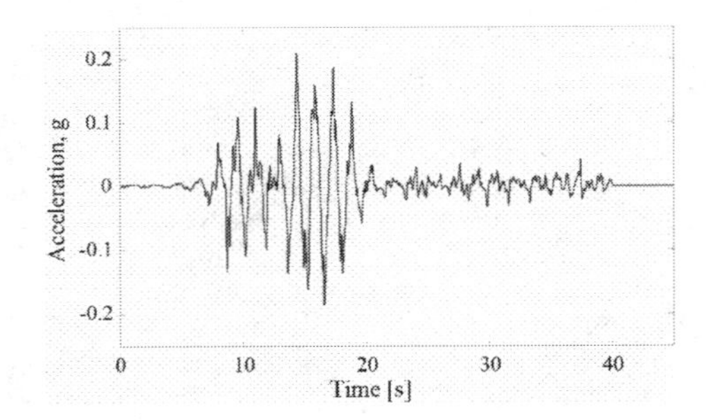

Figure 8. Horizontal earthquake input.

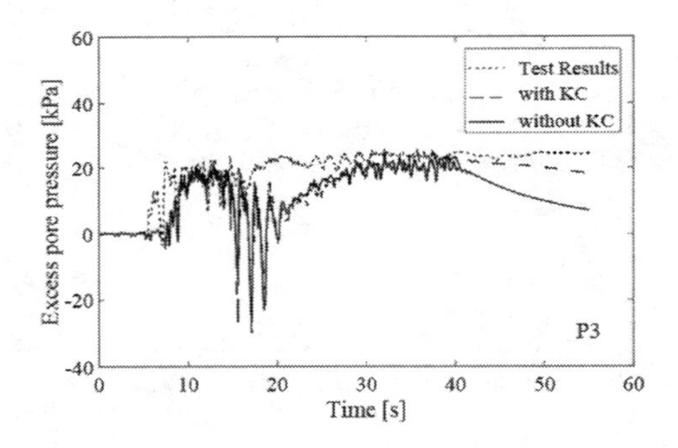

Figure 9. Time histories of excess pore pressures at P3.

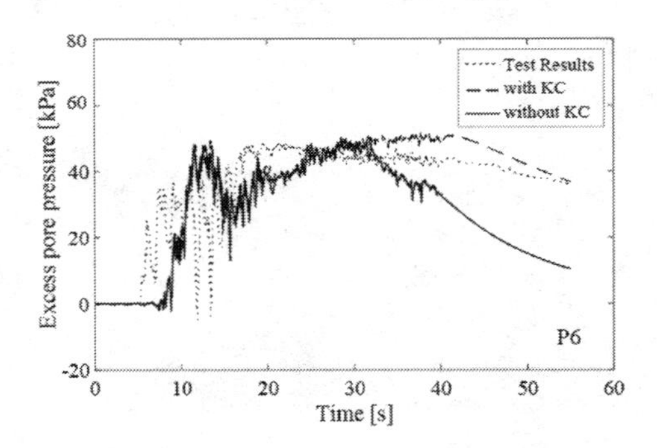

Figure 10. Time histories of excess pore pressures at P6.

In Figure 9 and 10 the measured and calculated excess pore pressure generations are shown at locations P3 with $x = 10$ m and $y = 22$ m and at P6 with $x = 24.5$ m and $y = 18$ m according to the Figure 6 resulting from during and after the dynamic earthquake loading procedure. Comparison of the experimental test results with FE results indicates that the simulation with KC yields more satisfactory results compared with the case without KC.

The main difference between two simulations is that without KC the pore pressure generation starts to decrease rapidly after the completion of dynamic earthquake loading process. However, the simulation case with KC allows the more accurate consideration of pore pressure dissipation phase based on the compared experimental results.

7 CONCLUSION

The effects of the porosity-permeability variation based on Kozeny-Carman relationship were investigated numerically under cyclic and dynamic loading conditions. For the case of dynamic loading conditions the results are compared with centrifuge results provided by C-Core (2004).

The results of finite element simulations for investigated cases provide the main following conclusions:

Cyclic loading condition:

- During the loading phase, higher excess pore pressure generations are determined for the analysis with KC compared to the simulation without KC.
- The stabilization of pore pressure amplitudes occurs for the analysis without KC earlier compared to with KC.
- The settlements with increasing number of loading cycles for the case of without KC is considerably higher than those for the case of with KC at each specific time.
- An accumulation of pore pressure is not obtained for investigated numerical example.

Dynamic loading condition:

- For the analysis case without KC, the dissipation of excess pore pressures after completing earthquake loading takes rapidly place compared with experimental results provided by C-Core (2004).
- A better agreement of simulation results with test results for the case with KC were achieved.

For modeling of the cyclic and dynamic behavior of saturated sandy soils, consideration of the porosity-permeability variation is necessary.

In order to more realistically describe the behavior of saturated sands subjected to cyclic and dynamic loads considering the porosity-permeability variation further numerical and extensive experimental investigations are planned.

ACKNOWLEDGEMENTS

The work of 2nd author presented within this paper is the result of a research project TA 1235/2-1, which is funded by the German Research Foundation. Its support is gratefully acknowledged.

REFERENCES

Andersen, K.H. 2009. *Bearing capacity under cyclic loading – offshore, along the coast, and on land.* The 21st Bjerrum Lecture presented in Oslo, 23 November 2007. Can Geotech J, 46(5): 513–535.

Bathe, K.-J. 1996. *Finite element procedures.* New Jersey: Prentice Hall.

Bauer, E. 1996. *Calibration of a comprehensive hypoplastic model for granular materials.* Soils and Foundations, 36(1): 13–26.

Biot, M.A. 1941. *General theory of three-dimensional consolidation.* Journal of Applied Physics, 12(2), 155–164.

Carman, P.C. 1937. *Fluid flow through granular beds. Transactions, Institution of Chemical Engineers,* London. 15: 150–166.

Cary, C.E. & Zapata, C.E. 2016. *Pore Water Pressure Response of Soil Subjected to Dynamic Loading under Saturated and Unsaturated Conditions.* Int J Geomech, 16(6): D4016001-9.

C-CORE. *Earthquake Induced Damage Mitigation from Soil Liquefaction. Data report – Centrifuge Test CT2.* Contract Report Prepared for University of British Columbia. C-CORE Report R-04-027-145, July 2004.

Cuéllar, P. 2011. *Pile Foundations for Offshore Wind Turbines: Numerical and Experimental Investigations on the Behaviour under Short-Term and Long-Term Cyclic Loading,* Doctoral Thesis, TU Berlin.

Di, Y. & Sato, T. 2003. *Liquefaction analysis of saturated soils taking into account variation in porosity and permeability with large deformation.* Computers and Geotechnics, 30(7):623–635.

Herle, I. 1997. *Hypoplastizität und Granulometrie einfacher Korngerüste. Veröffentlichung des Institutes für Bodenmechanik und Felsmechanik der Universität Fridericana in Karlsruhe.* Heft 142.

Kolymbas, D. 1988. *Eine konstitutive Theorie für Böden und andere körnige Stoffe. Veröffentlichung des Institutes für Bodenmechanik und Felsmechanik der Universität Fridericana in Karlsruhe,* Heft 109.

Kozeny, J. 1927. *Über kapillare Leitung des Wassers im Boden. Wien,* Akad. Wiss. 136(2a): 271–306.

Lambe, T.W., & Whitman, R.V. 1969. *Soil mechanics.* John Wiley & Sons, New York.

Martin, G.R., Finn, W.D.L.& Seed, H.B. 1975. *Fundamentals of liquefaction under cyclic loading. J Geotech Eng Div.* 101(5): 423–438.

Niemunis, A., Wichtmann, T. & Triantafyllidis Th. 2005. *A high-cycle accumulation model for sand.* Computers and Geotechnics, 32(4): 245–263.

Niemunis, A. & Herle, I. 1998. *Hypoplastic model for cohesionless soils with elastic strain range, Mechanics of Cohesion-Fractional Materials.* 2(4): 279–299.

Taşan, H.E., Rackwitz F. & Savidis S. 2010. *Behaviour of Cyclic Laterally Loaded Large Diameter Monopiles in Saturated Sand.* 7th European Conference on Numerical Methods in Geotechnical Engineering, NUMGE, Trondheim, Norway, 889–894.

Taşan, H.E. 2011. *Zur Dimensionierung der Monopile-Gründungen von Offshore-Windenergieanlagen. Dissertation.* Veröffentlichungen des Grundbauinstitutes der Technischen Universität Berlin.

Taşan, H.E. 2016. *Zum Tragverhalten von Suction-Buckets unter zyklisch lateralen Einwirkungen. Bauingenieur.* 91(12): 496–505.

Triantafyllidis, Th., Wichtmann, T., Chrisopoulos, S. & Zachert, H. 2016. *Prediction of long-term deformations of offshore wind power plant foundations using engineer-oriented models based on HCA. 26th International Ocean and Polar Engineering Conference (ISOPE-2016),* Rhodos.

Wichtmann, T., Niemunis, A., Triantafyllidis, Th. 2010. *Application of a high-cycle accumulation model for the prediction of permanent deformations of the foundations of offshore wind power plants.* International Symposium: Frontiers in Offshore Geotechnics, Perth, Australia.

von Wolffersdorff, P.-A. 1996. *A hypoplastic relation for granular materials with a predefined limit state surface.* Mechanics of Cohesive-Frictional Materials. 1(3):251–271.

Wu, W. 1992. *Hypoplastizität als mathematisches Modell zum mechanischen Verhalten granularer Stoffe.* Veröff. Inst. für Bodenmech. u. Felsmech. der Universität Fridericiana in Karlsruhe, Heft 129.

Zienkiewicz, O.C., Chang C.T. & Bettess P. 1980. *Drained, undrained, consolidating dynamic behaviour assumptions in soils.* Géotechnique. 30:385–95.

Zienkiewicz, O.C. & Bettess, P. 1982. *Soils and other saturated media under transient, dynamic conditions: general formulation and validity of various simplifying assumptions.* In: Pande GN, Zienkiewicz OC, editors. Soil mechanics – transient and cyclic loads. 1–16.

Zienkiewicz, O.C. & Shiomi, T. 1984. *Dynamic behaviour of saturated porous media; the generalized Biot formulation and its numerical solution.* International Journal of Numerical Analytical Methods in Geomechanics 8(1):71–96.

Zienkiewicz, O.C. 1986. *The patch test for mixed formulations.* International Journal for Numerical Methods in Engineering, 23(10), 1873–1883.

Zienkiewicz C., Chan A.H.C., Pastor M., Schrefler B.A. & Shiomi T. 1999. *Computational geomechanics with special reference to earthquake engineering.* John Wiley & Sons.

Numerical Methods in Geotechnical Engineering IX – Cardoso et al. (Eds)
© 2018 Taylor & Francis Group, London, ISBN 978-1-138-33198-3

Numerical study of pile setup for displacement piles in cohesive soils

Y.X. Lim, S.A. Tan & K.K. Phoon
National University of Singapore, Singapore

ABSTRACT: Pile setup refers to the gain of pile resistance over time. Its accurate prediction can increase efficiency of foundation design. The contribution of pile setup for displacement piles is mainly from shaft resistance rather than from end bearing resistance. This paper performs a numerical investigation of pile setup caused by increased in shaft resistance along clayey soils using the modified Cam-clay soil model. The press-replace method is used to simulate the initial undrained pile installation for a circular closed-ended flat-tip pile. Results from stress path analyses show how the mobilised undrained shear strengths of soil adjacent to pile shaft increases over time. Radial effective stresses also change during consolidation and load tests. The numerical results are then compared to the available field data by either considering the ratio of increase in undrained shear strength, or radial effective stress, or different combinations of both.

1 INTRODUCTION

1.1 *Background to pile setup*

Pile setup is gain in pile resistance over time. Accurate prediction of pile setup can lower costs and time of designed foundations. Several empirical equations have been proposed to describe soil setup with time and a summary has been provided by Haque et al. (2017). Amongst them, one of the most popular relationships is from Skov & Denver (1988):

$$\frac{R_t}{R_{to}} = 1 + A\left(\log\frac{t}{t_{to}}\right) \tag{1}$$

where R_t = total pile resistance and R_{to} = total pile resistance at reference time t_o, A = parameter to describe increase in pile resistance.

Pile setup of displacement piles are mainly due to increase in shaft resistance rather than end bearing resistance (e.g. Haque et al. 2017, Ksaibati & Ng 2017, Haque et al. 2014). To quantify the effects from the shaft a similar form of the equation can be used:

$$\frac{f_s}{f_{so}} = 1 + A\left(\log\frac{t}{t_{to}}\right) \tag{2}$$

where f_s = shaft resistance and f_{so} = shaft resistance at reference time t_o. Haque et al. (2017) and Bullock et al. 2005 reported results from long piles by distinguishing the contributions between clays and other soil types. Haque et al. (2017) used t_o = 15 min = 0.010 day and found that A for clay layers varies from 0.2 to 0.53 with an average value of 0.36 for piles tested to about 10 days after end of driving (EOD). Bullock et al. 2005 measured the performance of their piles over a much longer period and used t_o = 1 day instead. See Figure 1. Data from Bullock et al. 2005 is labelled at the depths where clay layers are encountered. Also presented in Fig. 1 is shaft resistance measurements collected by Konrad & Roy (1987) for a short pile in clay by using t_o = 1 day. Details of the above mentioned cases are summarised in Table 1.

1.2 *Limitations of theoretical solutions*

As illustrated in Figure 1, the increase of pile shaft setup in clay can be quite different. Developing theoretical solutions will enable better interpretation of field data that is subjected to large variabilities like differences in pile types, methods of installation and soil stress history etc. However robust and comprehensive analytical solutions for piles are

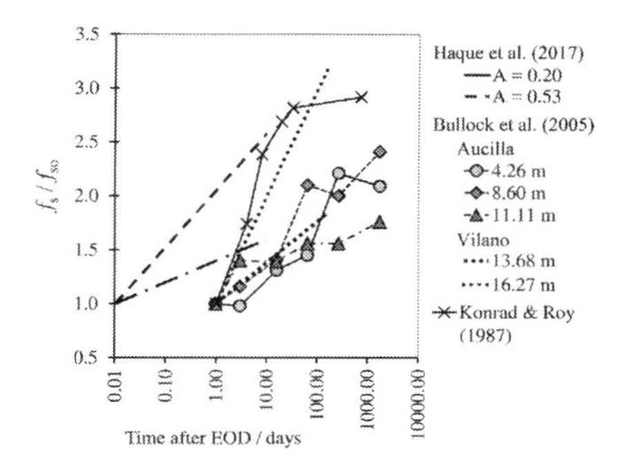

Figure 1. Pile setup from shaft resistance increase in clay.

Table 1. Summary of pile setup data used in Figure 1.

Site	Type	Instal-lation	Width	Length*
			m	m
Louisiana, USA 4 different sites	Prestressed concrete square piles	Driven	0.41–0.76	37–61
Florida, USA 1)Aucilla River Bridge 2)Vilano Bridge West	Prestressed concrete square piles	Driven	0.457	1) 19 2) 18
St Alban, Canada	Circular closed-ended steel pipe	Jacked	0.22	6

* Embedded length

difficult due to complicated large deformation in soils caused by the pile installation processes.

Numerical attempts to understand pile setup mostly employ, partly or in full, cavity expansion methods through radial expansion of a long cylinder to generate initial stresses along the shaft (e.g. Rezania et al. 2017, Rosti et al. 2016, Basu et al. 2014). However this form of loading is quite different for actual driven or jacked in piles, where soil experiences different stress loadings as well as stress rotations caused by advancement of pile toe.

1.3 *Aim of study*

The press-replace method (PRM), originated by Andersen et al. (2004) to model suction anchors and its implementation refined by Engin et al. (2015) for piles, was shown to produce stresses around a circular close-ended flat-tip pile in sand that is comparable to that obtained from material point method (Tehrani et al. 2016). Sivasithamparam et al. (2014) used PRM to model undrained pile installation with modified Cam-clay and S-CLAY1 soil models.

This paper explores the possibility of using the finite-element numerical approach, the press-replace method, to investigate pile setup due to shaft resistance increase along cohesive soils. A pile of arbitrary dimensions is considered and the modified Cam-clay soil (MCC) model is used. The undrained pile installation process is first simulated using PRM. This is followed by a consolidation phase where excess pore pressures are allowed to dissipate with time. Pile load tests (LT) are then numerically performed at different durations measured from the end of installation to track the increase in undrained soil strength.

2 NUMERICAL SETUP

2.1 *Press-replace method for modelling pile installation*

A pile of arbitrarily selected dimensions is chosen to be modelled here as the purpose it not to replicate any specific pile but to do a general comparison to the empirical data shown in Figure 1. The numerical pile has a diameter, D, of 1.00 m and total length of 10.0 m.

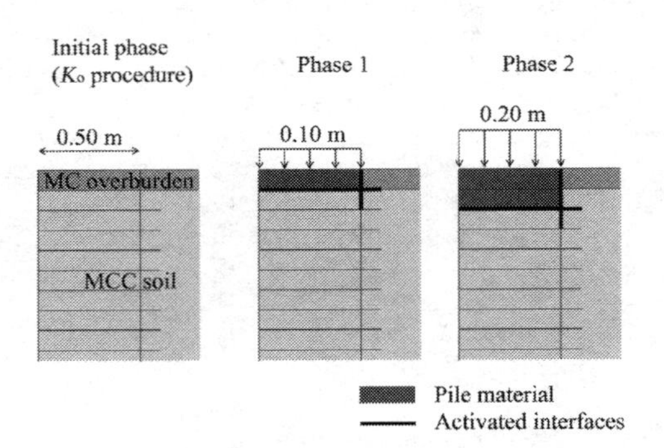

Figure 2. Press-replace method for pile installation.

To create uniform stresses within the modified Cam-clay soil model, a zero unit weight is given. Initial stresses are then generated using a thin but heavy overburden Mohr-Coulomb (MC) soil layer in the "initial phase" with a "K_o procedure". The water table is placed below the entire soil mass and the weight of overburden layer is chosen such that a uniform effective vertical stress, σ_{vo}', of 100 kPa is defined. Initial horizontal effective stresses are then determined from K_o input, found using the relationship proposed by Mayne & Kulhawy (1982), $K_o = (1-\sin\phi')OCR^{\sin\phi'}$ which is 0.61 when $\phi' = 30°$ and initial OCR = 1.5.

The technique of PRM adopted here is as described in (Engin et al. 2015) and performed using software PLAXIS 2D 2017 in an axisymmetric setup. Slice spacing, interface extensions lengths and incremental line displacements are 0.1D, and hence 100 press-replace phases are needed to complete the entire pile penetration to 10.0 m depth.

Details of the "initial phase" and first two PRM phases are illustrated in Figure 2. Interaction between the pile and soil is controlled by "interfaces" (van Langen 1991) and more information regarding interfaces can be found from the PLAXIS manual.

2.2 *Finite-element mesh and input parameters*

Triangular elements make up the 20 m by 20 m mesh in this study, see Figure 3. In total 4860 elements 41573 nodes are generated with a minimum mesh quality of 0.5517 (maximum mesh quality is 1 for an equilateral element). The top, bottom and sides of mesh are permeable. Pile material is "linear elastic" and it stiffness is assumed to be like steel with $E' = 200$ MPa and Poisson's ratio, $v' = 0.10$. Both horizontal and vertical permeability have values of 1.0×10^{-9} m s^{-1}. The MCC soil parameters are presented in Table 2, where λ = virgin recompression index, κ = recompression index, e_0 = void ratio and M is the slope of the critical state line on q–p 'plane. q is deviator stress and p' is mean effective stress. M is related to ϕ' in triaxial compression through the relationship $M = 6 \sin\phi'/(3\text{–}\sin\phi')$ (Wroth 1984). The M value chosen here corresponds to $\phi' = 30°$.

Interface strength during installation is taken as the undrained shear strength, s_u, of this soil in undrained triaxial compression. Its value can be determined using an equation from Chang et al. (1999) for K_0 consolidated specimen which works out to be 42 kPa.

During consolidation and numerical pile LT, if this interface cohesion value is not increased, the shaft will still fail at 42 kPa despite the gain in soil strength due to dissipation of excess pore pressures. Hence the interface cohesion value is increased such that it is twice the value used during installation, at 84 kPa. This is later checked to be sufficiently higher than the undrained shear strength reached along the shaft.

Pile LTs are simulated by increasing the prescribed displacement at the top of the pile for an additional 0.050 m. The duration of the load test is ignored here as the calculations are done as "plastic" where soil behaves as undrained. This pile movement has been verified to adequately mobilise shaft deviator stresses to a maximum unchanging value at failure, q_f.

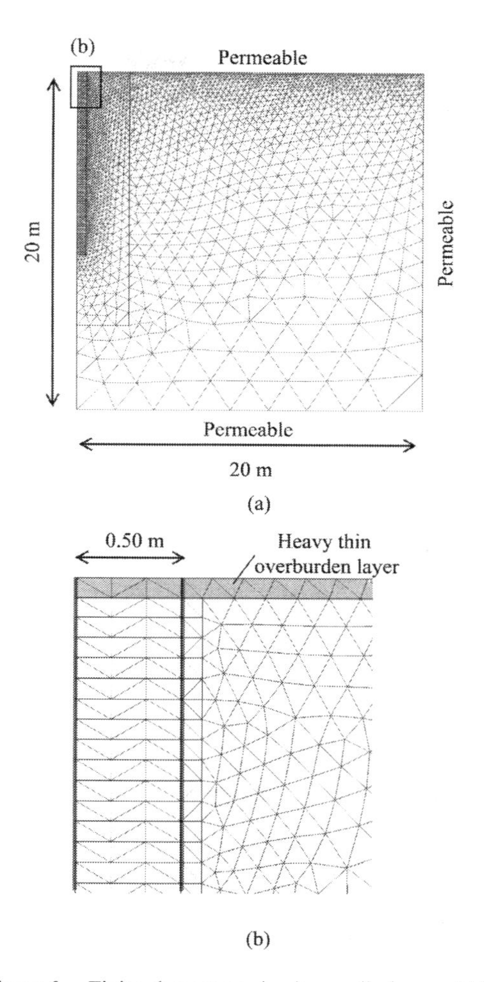

Figure 3. Finite-element mesh; a) overall view and b) enlarged portion at upper left-hand corner of mesh.

Table 2. Modified Cam-clay input parameters used in study.

λ	κ	e_0	v'	M
0.30	0.060	2.0	0.20	1.200

3 RESULTS OF NUMERICAL STUDY

3.1 *Identification of stress point for analyses*

The numerical pile penetrates from the surface to a depth of 10D but where along this pile shaft are the stresses most representative of shaft response? Assuming the steady-state is reached after 6D, five points are selected from 6D to 10D along the pile shaft to identify locations where shaft responses are uniform, see Figure 4a. These stress points reach different q_f during the various numerical LTs conducted after end of installation, see Figure 4b. The trends of stress points K, L and M are similar, while stress points nearer the pile toe, N and O, indicate increasing influence from the toe. This means that K, L and M are representative of shaft response and subsequent results are reported using data from stress point L.

3.2 *Pile setup from increase in shaft resistance*

The stress paths of stress point L over installation, consolidation and the pile load tests are shown in Figure 5. During installation, the initial stress path is vertically upwards when the pile toe starts from the surface and pushed to about 3D depth, reflecting the initial elastic response of stress point

L due to overconsolidation. Cap hardening occurs from $3D$ to approximately $5D$, shown as the slight arc before reaching critical state line (CSL). CSL is reached when pile toe reaches depth of $5D$. When pile toe reaches the level of stress point L at $7D$, it creates a slight and temporary decrease in q due to stress rotation as the pile toe advances. By $8D$, the soil state is back on the CSL and remains there until end of pile installation at $10D$. During 1000 days dissipation of excess pore pressures, the stress paths predominantly moves to the right showing gain in mean effective stress. Subsequent pile load tests show how the stress paths curve upwards like a soil with OCR = 1 towards the CSL to reach higher q at failure, reflecting the gain in soil strength due to consolidation.

From Figure 5, it is clear that undrained soil strength increases with consolidation. However how can this information be compared to field data shown in Figure 1? From traditional pile design methods involving α and β methods, it is implied that the amount of shear strength developed in the soil during shearing and the actual shaft resistance experienced by the pile are not the same. More specifically α methods suggest that f_s is related to s_u, and β methods use the effective stress concepts where greater effective radial stresses, σ_r', will lead to higher f_s.

In a numerical study like this, both the changes in q_f (and hence s_u) and σ_r' during pile LTs are known. The notation σ_r' is used here to denote radial effective stress at the start of pile LT and σ_{rf}' is used to denote radial effective stress corresponding to q_f at the end of pile LT. The following relationships use either the concepts from either α or β methods, or combinations of both to plot the numerical results against field data in Figure 6:

$$\frac{f_s}{f_{so(1day)}} = \frac{q_f}{q_{f(1day)}} \tag{3}$$

$$\frac{f_s}{f_{so(1day)}} = \frac{\sigma'_{rf}}{\sigma'_{rf(1day)}} \tag{4}$$

$$\frac{f_s}{f_{so(1day)}} = \frac{q_f}{q_{f(1day)}} \times \frac{\sigma'_{rf}}{\sigma'_{rf(1day)}} \tag{5}$$

$$\frac{f_s}{f_{so(1day)}} = \frac{\sigma'_r}{\sigma'_{r(1day)}} \tag{6}$$

$$\frac{f_s}{f_{so(1day)}} = \frac{q_f}{q_{f(1day)}} \times \frac{\sigma'_r}{\sigma'_{r(1day)}} \tag{7}$$

The field data in Figure 1 are also presented in Figure 6 as light grey plots in the background for easy reference. For this comparison, the field data from (Haque et al. 2017) is omitted as their testing program was over a short period of time. Similar to the data from Bullock et al. (2005), a reference time of 1.0 day is used for the plots.

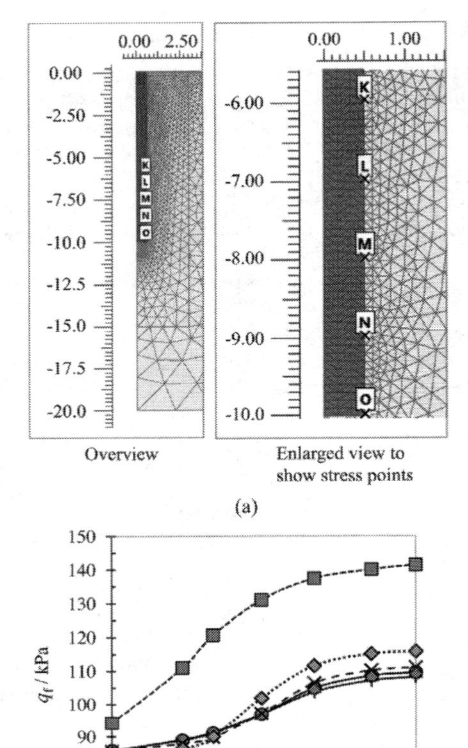

Figure 4. Stress points along shaft; a) locations and b) failure deviator stress reached during numerical pile load test.

Equations (3) and (4) produce identical curves and they show $f_{s(1000\ day)}/f_{so(1\ day)}$ rising to about 1.3. The combination of both equations (3) and (4), i.e. equation (5) compares well to the lower bound of the field data, reaching ratio of 1.6, and is in the opinion of the authors to be the most logical relationship to use amongst the various options. Equation (6) is similar to the β method which considers an initial state rather than the end state of a pile load test. The difference however is that β methods commonly relates to σ_{ro}' (undisturbed in situ soil state) whereas the numerical results here uses σ_r'; σ_r' is greater than σ_{ro}' as a result of

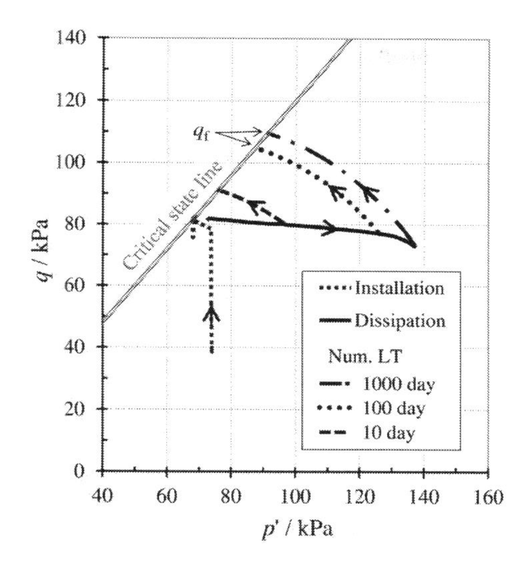

Figure 5. Stress paths of stress point L along pile shaft.

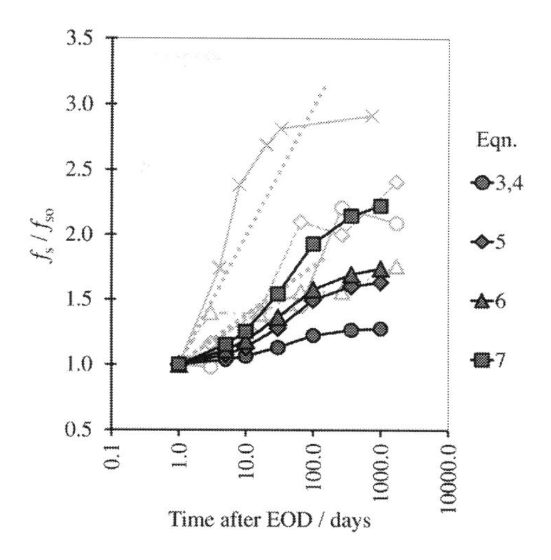

Figure 6. Pile setup at shaft; comparison of numerical results with Figure 1's field data (grey).

pile installation induced stresses and subsequent dissipation. The maximum f_s/f_{so} ratio reached is about 1.7 at 1000th day LT. Equation (7) is the combination of equation (3) and (6) and it produces the greatest pile setup at 2.2.

4 DISCUSSION

The numerical calculations show that PRM can produce stress changes around a pile shaft such that subsequent consolidation will cause soil strength to increase. Depending on how the results are interpreted, the amount of pile shaft setup either compares well or underestimates field data. If PRM really underestimates pile setup, it is due to the technique of PRM being unable to produce accurate radial stresses along the shaft? Hence the next question is; can the radial stresses along level of stress point L be compared to field data or theoretical solutions to prove the performance of PRM? One way to do so is to check the magnitude of excess pore pressures, Δu, generated during installation. However to validate this is difficult as there are limited experimental or field data available for closed-ended flat-tip circular pile sections; sufficiently long for end effects to be negligible, with field instrumentations also placed at those levels that are representative of only shaft responses.

In an attempt to address the above question an additional simulation is done to compare the variation of excess pore pressures with radial distance from shaft, with the field data at St Alban clay (Roy et al. 1981) and theoretical strain path method solutions from Whittle (1993). Although it

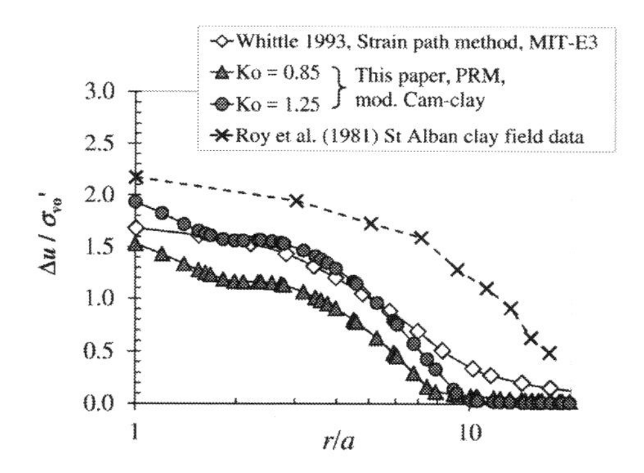

Figure 7. Normalized excess pore pressures at the end of pile installation.

is a poor comparison, some reference is still better than nothing at all. Many differences and uncertainties are present here; MCC soil model cannot represent the highly sensitive St Alban clay, MCC model and parameters are different from Whittle's (1993) MIT-E3 soil model, assumptions have to be made on the amount of shaft roughness to be modelled, the placement of instruments at St Alban pile may be influenced by tip stresses, and the strain path method is an approximate theoretical solution to pile penetration problem.

For this simulation, the setup described earlier and the soil parameters shown in Table 2 are kept with exception of the following changes; two seperate K_0 values are investigated, 0.85 from total load cell measurements and 1.25 back-calculated from pile load tests (Konrad & Roy 1987), initial OCR of soil set as 2.0 to compare with same value used by Whittle (1993) and St Alban clay also has OCR about 2.2 (Roy et al. 1981), the shaft roughness is assumed to be 75% of undrained shear strength of soil at undrained triaxial compression.

Comparisons of normalized results are shown in Figure 7. Δu are normalized with σ_{vo}' and radial distance from pile centre, r, is normalized with pile radius a. Results from PRM are after 0.1 day consolidation to account for time of installation. Both theoretical solutions (strain path and PRM) underestimate the normalized excess pore pressures when compared to the field data from St Alban clay. The PRM solution with $K_o = 1.25$ compare reasonably closely to Whittle (1993) until distance of $6a$.

5 CONCLUSIONS

Press-replace method is used to simulate pile installation in a setup similar to calibration chamber where initial effective stresses are uniform. A modified Cam-clay soil model is used to represent the soil. After installation excess pore pressures are allowed to dissipate. Numerical pile load tests are then performed at different points in time during consolidation.

At distance $2D$ above pile toe and $6D$ below surface shaft stresses are nearly uniform. Stress paths diagrams illustrate how soil adjacent to the shaft of pile gains undrained shear strength with time. Depending on the interpretation of results PRM can predict ratio of $f_{s(1000\ day)}/f_{so(1\ day)}$, between 1.3 to 2.2.

PRM has demonstrated potential of investigating pile setup due to shaft resistance increase in clays but further validation is required.

REFERENCES

Andersen, K.H., Andresen, L., Jostad, H.P., and Clukey, E.C. (2004). "Effect of skirt-tip geometry on set-up outside suction anchors in soft clay." *ASME 2004 23rd International Conference on Offshore Mechanics and Arctic Engineering*, June 20–25, 2004, Vancouver, British Columbia, Canada, 1035–1044.

Basu, P., Prezzi, M., Salgado, R., and Chakraborty, T. (2014). "Shaft Resistance and Setup Factors for Piles Jacked in Clay." *Journal of Geotechnical and Geoenvironmental Engineering*, American Society of Civil Engineers, 140(3), 4013026.

Bullock, P.J., Schmertmann, J.H., McVay, M.C., and Townsend, F.C. (2005). "Side Shear Setup. II: Results From Florida Test Piles." *Journal of Geotechnical and Geoenvironmental Engineering*, American Society of Civil Engineers, 131(3), 301–310.

Chang, M.-F., Teh, C.I., and Cao, L. (1999). "Critical state strength parameters of saturated clays from the modified Cam clay model." *Canadian Geotechnical Journal*, NRC Research Press, 36(5), 876–890.

Engin, H.K., Brinkgreve, R.B.J., and van Tol, A.F. (2015). "Simplified numerical modelling of pile penetration - the Press-Replace technique." *International Journal for Numerical and Analytical Methods in Geomechanics*, 39(15), 1713–1734.

Haque, M., Abu-Farsakh, M., Chen, Q., and Zhang, Z. (2014). "Case Study on Instrumenting and Testing Full-Scale Test Piles for Evaluating Setup Phenomenon." *Transportation Research Record: Journal of the Transportation Research Board*, Transportation Research Board, 2462, 37–47.

Haque, M.N., Abu-Farsakh, M.Y., Tsai, C., and Zhang, Z. (2017). "Load-Testing Program to Evaluate Pile-Setup Behavior for Individual Soil Layers and Correlation of Setup with Soil Properties." *Journal of Geotechnical and Geoenvironmental Engineering*, American Society of Civil Engineers, 143(4), 4016109.

Konrad, J.-M., and Roy, M. (1987). "Bearing capacity of friction piles in marine clay." *Géotechnique*, 37(2), 163–175.

Ksaibati, R., and Ng, K. (2017). "Medium-Scale Experimental Study of Pile Setup BT - Medium-Scale Experimental Study of Pile Setup." *Geotechnical Testing Journal*, 40(2), 269–283.

van Langen, H. (1991). "Numerical Analysis of Soil-Structure Interaction." Ph.D. thesis, Geo- Engineering Section, Delft University of Technology, The Netherlands.

Mayne, P.W., and Kulhawy, F.H. (1982). "Ko-OCR Relationships in Soil." *Journal of the Geotechnical Engineering Division*, 108(6), 851–872.

Rezania, M., Mousavi Nezhad, M., Zanganeh, H., Castro, J., and Sivasithamparam, N. (2017). "Modeling Pile Setup in Natural Clay Deposit Considering Soil Anisotropy, Structure, and Creep Effects: Case Study." *International Journal of Geomechanics*, American Society of Civil Engineers, 17(3), 4016075.

Rosti, F., Abu-Farsakh, M., and Jung, J. (2016). "Development of Analytical Models to Estimate Pile Setup in Cohesive Soils Based on FE Numerical Analyses." *Geotechnical and Geological Engineering*, 34(4), 1119–1134.

Roy, M., Blanchet, R., Tavenas, F., and La Rochelle, P. (1981). "Behaviour of sensitive clay during pile driving." *Canadian Geotechnical Journal*, 18(1), 67–85.

Sivasithamparam, N., Engin, H.K., and Castro, J. (2014). "Numerical modelling of pile jacking in a soft clay." *Computer Methods and Recent Advances in Geomechanics*, CRC Press, London, 985–990.

Skov, R., and Denver, H. (1988). "Time-dependence of bearing capacity of piles." *Proc. Third International Conference on the Application of Stress-Wave Theory to Piles*, Ottawa, 879–888.

Tehrani, F.S., Nguyen, P., Brinkgreve, R.B.J., and van Tol, A.F. (2016). "Comparison of Press-Replace Method and Material Point Method for analysis of jacked piles." *Computers and Geotechnics*, 78, 38–53.

Whittle, A.J. (1993). "Assessment of an Effective Stress Analysis for Predicting the Performance of Driven Piles in Clays." *Offshore Site Investigation and Foundation Behaviour*, 28, 607–643.

Wroth, C.P. (1984). "The interpretation of in situ soil tests." *Géotechnique*, 34(4), 449–489.

Numerical Methods in Geotechnical Engineering IX – Cardoso et al. (Eds)
© 2018 Taylor & Francis Group, London, ISBN 978-1-138-33198-3

A relook into numerical simulations of the pressuremeter test for the calibration of advanced soil models

Q.J. Ong & S.A. Tan
National University of Singapore, Singapore

ABSTRACT: In the current state of practice for the design of geotechnical projects, the extensive reliance on empirical correlations to obtain design parameters sometimes fails to capture realistic soil behaviour, and in other cases produces severely conservative design parameters. Proper calibration of numerical models with appropriately applied soil constitutive models is essential to obtaining reasonable numerical predictions of the performance of projects. This paper aims to provide some insights on the calibration process of advanced soil constitutive models for use in numerical analyses, in particular the Hardening Soil model, by proposing a sequential approach of calibration with the use of triaxial tests, oedometer tests and the pressuremeter test. Modelling techniques for the simulation of the pressuremeter test are briefly discussed, followed by some investigation into the influence of boundary conditions, loading conditions and the effects of borehole disturbance.

1 INTRODUCTION

For numerical analyses in geotechnical design, it is impractical to apply constitutive models that capture the whole range of soil behavior, due to the numerous parameters required. Hence the engineer should select suitable constitutive models that are able to reproduce the various characteristics that the soil is expected to exhibit in his project. The capabilities and limitations of the constitutive model used must be understood reasonably well by the engineer in order to avoid design errors.

It is good practice to first calibrate the constitutive model of choice with soil element laboratory tests and then verify that the calibration is reasonable by simulation of field tests. In this study, drained triaxial test and oedometer test data are first interpreted to obtain a first cut calibration of the numerical model, then the numerical model is verified against the pressuremeter test.

Samples used in laboratory tests generally experience unloading of stresses when they are extracted from the ground, and are sometimes subjected to disturbances during transportation and handling. In comparison, the pressuremeter test does not require extraction of soil samples and is a useful in-situ test as it provides a direct measurement of the stress and strain response of the soil. Results from finite element simulations of pressuremeter tests can agree reasonably well with those conducted in-situ and it is shown that it can be used to verify parameters obtained from laboratory tests.

This paper will discuss the process of calibrating geotechnical design parameters from conventional laboratory tests, followed by some investigation on the influence of boundary conditions, loading conditions and effects of borehole disturbance for simulations of the pressuremeter test. Finite element code PLAXIS (PLAXIS 2D 2017.0, 2017) was used for the simulation of the tests in this study.

2 HARDENING SOIL MODEL

The linearly elastic perfectly plastic Mohr-Coulomb model is widely used in the analyses of geotechnical projects due to it being easy to understand and use, or simply because the analyst does not know how to calibrate advanced constitutive models. While the Mohr-Coulomb failure criterion is safe for design, the bilinear soil model does not replicate the stress paths of real soils and fails to produce correct strains during soil deformation. As a result of these inadequacies, ground deformations predicted by the Mohr-Coulomb model may be grossly incorrect.

Where the Mohr-Coulomb model is lacking, advanced constitutive models can capture the various characteristics of soil behavior required for a safe yet economical design. A noteworthy advanced constitutive model suitable to capture the behavior of soils reasonably well in geotechnical projects involving unloading and reloading, is the Hardening Soil model.

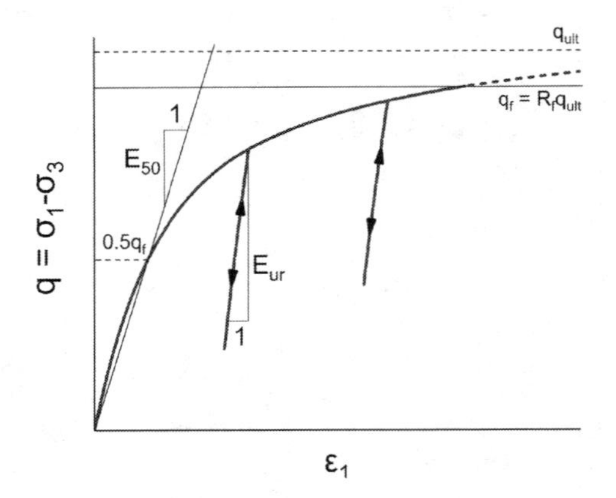

Figure 1. Hyperbolic stress-strain response of the Hardening Soil model in primary loading, and the definition of E_{50} and E_{ur}.

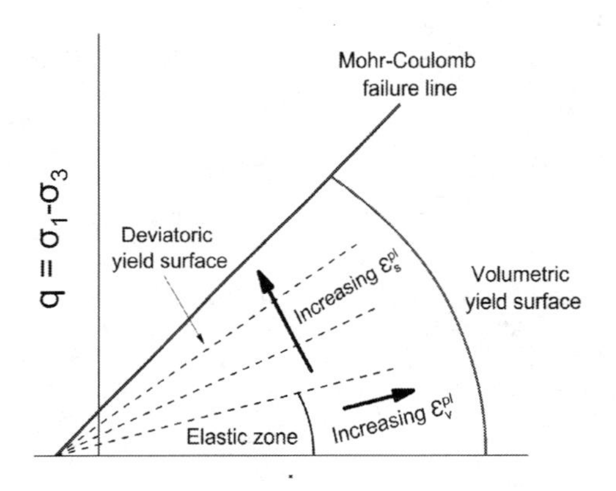

Figure 2. Yield surfaces and the double hardening mechanism of the Hardening Soil model.

The advanced constitutive model features non-linear stress dependent stiffness, distinct stiffness for primary loading in shear and compression as well as unloading/reloading, and memory of pre-consolidation stresses (Schanz et al., 1999). Figure 1 describes the hyperbolic stress-strain relationship of the Hardening Soil model in primary loading for a standard drained triaxial test, where q represents the deviatoric stress, ε_1 represents the axial strain and R_f represents the failure ratio. The Hardening Soil model can also distinguish between the volumetric hardening and shear hardening modes of plasticity, shown in Figure 2, where p' is the mean effective stress. Further development of the constitutive model has led to the Hardening Soil model with small strain stiffness which captures the small strain behavior of soils through strain dependency.

2.1 *Calibration through element tests*

In order to use numerical models effectively, they have to be first calibrated against soil element tests. As the Hardening Soil model was formulated from the hyperbolic relationship between the vertical strain and deviatoric stress in primary triaxial loading, calibrating the model from the standard drained triaxial test is ideal. The primary loading stiffness E_{50} can be obtained by interpreting the secant modulus of the hyperbolic stress-strain curve at mobilization of 50% of maximum shear strength q_f, as shown in Figure 1. As the Hardening Soil model utilizes the Mohr-Coulomb failure criterion, the traditional approach of Mohr envelopes may be used to obtain the effective strength parameters of the soil. The relationship between E_{50} and E_{50}^{ref} is described by Equation 1, where the reference pressure p^{ref} determines the reference stiffness modulus, and the amount of stress dependency is given by the power m. p^{ref} can be taken as the default value of 100 kPa or calibrated against the in-situ effective confining stress σ_3', where either value will give the right soil behavior provided that the calibration against the laboratory and field tests were done correctly. Equation 1 can be linearized to obtain Equation 2. With three different confining stresses conducted in a triaxial test, three sets of stiffness can be obtained, thus a straight line can be plotted with Equation 2. The stress dependency power may be directly obtained from the gradient, and the reference stiffness modulus may be interpreted from the intercept. Similar steps can be performed for the unloading/reloading stiffness E_{ur}^{ref}. A summary of the procedure is shown in Figure 3.

$$E_{50} = E_{50}^{ref}\left(\frac{c'\cos\phi' + \sigma_3'\sin\phi'}{c'\cos\phi' + p^{ref}\sin\phi'}\right)^m \tag{1}$$

$$\ln\left(E_{50}\right) = \ln\left(E_{50}^{ref}\right) + m\ln\left(\frac{c'\cos\phi' + \sigma_3'\sin\phi'}{c'\cos\phi' + p^{ref}\sin\phi'}\right) \tag{2}$$

$$\ln\left(A\right) = \ln\left(\frac{c'\cos\phi' + \sigma_3'\sin\phi'}{c'\cos\phi' + p^{ref}\sin\phi'}\right) \tag{3}$$

While test results from the drained triaxial test can be used to characterize the shear hardening response, test results from the oedometer test can be used to characterize the volumetric hardening response of soils. The confined stiffness E_{oed}^{ref} from the oedometer test is input independently from the drained triaxial stiffness, where the p^{ref} in this case is the primary loading stress σ_1', as opposed to using σ_3' for the triaxial stiffness. E_{oed}^{ref} is obtained from the tangent modulus as shown in Figure 4.

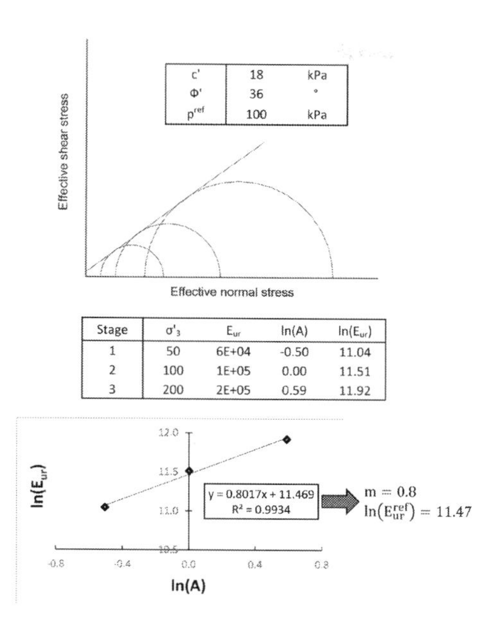

Stage	σ'₃	E_{ur}	ln(A)	ln(E_{ur})
1	50	6E+04	-0.50	11.04
2	100	1E+05	0.00	11.51
3	200	2E+05	0.59	11.92

Figure 3. Example of obtaining reference unloading-reloading stiffness and stress dependency by graphing Equation 2.

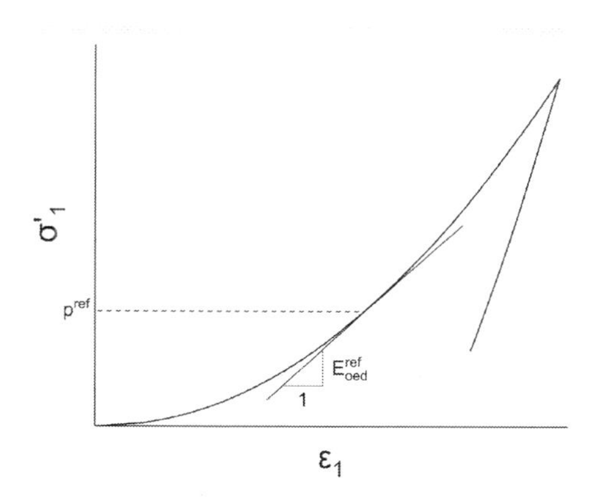

Figure 4. Definition of reference stiffness E_{oed}^{ref}.

Alternatively, E_{oed}^{ref} can be calculated from the compression index c_c and initial void ratio e_0, also obtained from the oedometer test, Equation 4.

$$c_c = \frac{2.3(1+e_0)p^{ref}}{E_{oed}^{ref}} \tag{4}$$

Capturing the actual in-situ soil stress history is critical for accurate prediction of settlements in geotechnical problems, where the pre-consolidation stress p'_c can be used to characterize it. p'_c may be obtained from the oedometer test with several approaches such as Casagrande's empirical method, the bi-logarithmic method by Butterfield (1979) or the incremental work method by Becker et al. (1987). Onitsuka et al. (1995) presented that out of the various bi-logarithmic methods, the ln (1+e) – log P' approach held the most theoretical weight as it can be verified with the work approach mathematically, and so produced the best results. In this bi-logarithmic approach, the pre-yield lines and post-yield lines are clearly defined, and the intersection of the two lines represents the yield stress where there is a change from small compressibility to large compressibility. The yield stress may be interpreted as the cause of critical state for structure collapse of clay and is synonymous with the pre-consolidation stress of soils.

Grozic et al. (2003) and Boone (2010) have reviewed the various methods of obtaining p'_c. Grozic et al. concluded that while the work method by Becker et al. and the bi-logarithmic method by Onitsuka et al. should be used to determine the p'_c of soils, Casagrande's method can also be used as a verification of results due to the wide experience from using the traditional method. Boone noted that Casagrande's method was susceptible to scaling issues, leaving some ambiguity in the interpretation of results. Boone also proposed an alternative mathematical and graphical approach using the recompression index, maximum compression index, in-situ vertical effective stress and void ratio at the in-situ stress.

Test curves with "rounded" shapes may not be an indication of sample disturbance, but rather a characteristic of the soil. In such cases, less ambiguous methods should be used in the interpretation of p'_c. An example of applying the method of Onitsuka et al. is shown in Figure 5. Since only two lines are drawn, there is less room for uncertainty in the value of p'_c.

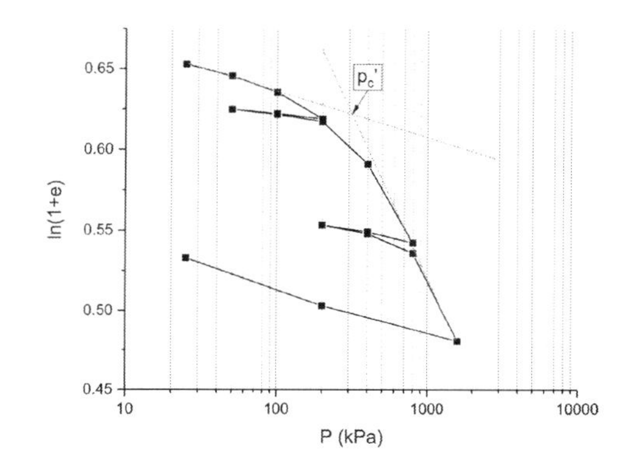

Figure 5. Onitsuka's bi-logarithmic method on a "rounded" test curve.

2.2 *Unloading and reloading behavior of soils*

In geotechnical projects such as excavations or tunneling, the dominant behavior of soils is unloading and reloading. In these cases, it is advantageous to specify laboratory and field tests to have unloading and reloading cycles. It is generally recommended to conduct more than one unloading and reloading cycle in both laboratory and in-situ tests. The preliminary idea is that the unloading-reloading stiffness is not a simple constant but a function of both stress and strain. As pointed out by Briaud (2013) and Goh et al. (2012), the unload-reload response in the pressuremeter test depends on the strain over which the unloading and reloading is performed, Figure 10.

An oedometer test conducted in stiff sandy soil is shown in Figure 6. With a logarithmic scale used, it may seem that the unloading-reloading stiffness is almost similar. However, when plotted in an undistorted scale, Figure 7, the increase in stiffness with increasing strain is very apparent.

Advanced constitutive models with the ability to capture the strain dependency of stiffness, such as the Hardening Soil model with small strain stiffness should be used when hysteretic behavior of soils is expected. Figure 8 illustrates a numerical simulation of a pressuremeter test with repeated unloading and reloading cycles with the Hardening Soil model with small strain overlay.

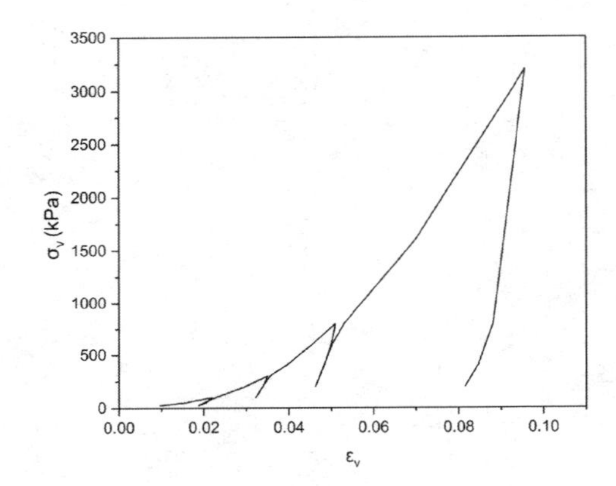

Figure 7. Same oedometer test plotted in non-distorted scale of vertical stress against vertical strain.

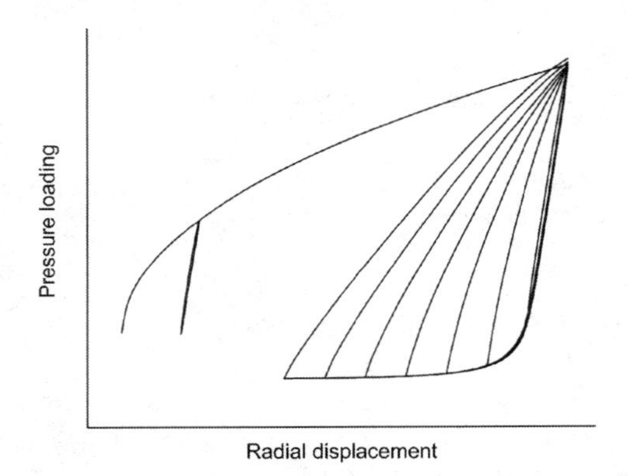

Figure 8. Numerical simulation of a pressuremeter test with repeated unloading and reloading cycles, using the Hardening Soil model with small strain stiffness.

3 PRESSUREMETER TEST

One of the more suitable field tests that can be used to verify element test results is the pressuremeter test. The pressuremeter test can be described as a cylindrical device with the ability to apply a uniform pressure to test pocket walls by a flexible membrane. Measurements of the pressure applied and increment in volume or radial displacement are made during the expansion of the pressuremeter to characterize the soil. Since it is conducted in-situ, the pressuremeter test takes into account all environmental factors, in-situ stresses, soil stress history, drainage conditions, and furthermore there is minimal soil disturbance in stable borehole conditions. In addition, the pressuremeter test has well-defined boundary conditions which permit a more rigorous theoretical analysis.

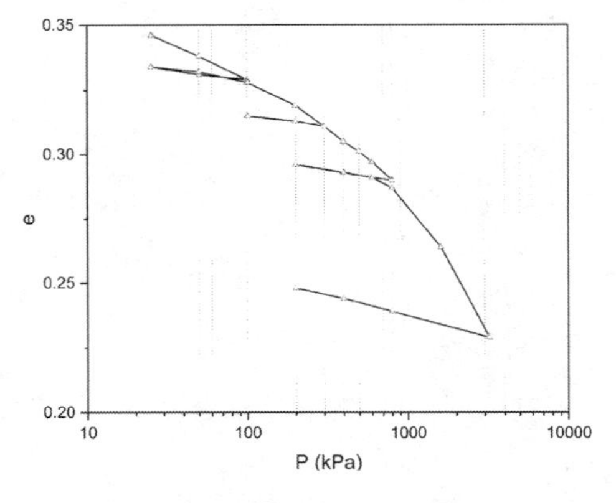

Figure 6. Typical oedometer test results plotted in e-logP.

There are mainly three types of pressuremeters, namely the self-boring pressuremeter, pre-bored pressuremeter and the push-in pressuremeter. The loading curve shown in Figure 9 describes the various stages that the typical pre-bored pressuremeter test undergoes. The initial cavity expands until first contact with the cavity wall where the horizontal at rest pressure is reached, followed by pseudo elastic expansion and then finally large plastic deformation.

The pressuremeter is a well-established site investigation tool with mechanics that was initially interpreted by empirical methods. Subsequently, fundamental interpretations obtained by simplifying the expansion problem into axisymmetrical and plane strain conditions allowed for application of the cavity expansion theory.

Rather than behaving as a compressive process, the expansion of the cavity in the pressuremeter test behaves completely as a shearing process (Mair & Wood, 2013). Wroth (1984) noted that the difference between the in-situ test and conventional laboratory tests is that the fields of stress and strain do not remain homogenous in the expansion test. This leads to high gradients of excess porewater pressure in the radial direction, and subsequently partial consolidation will occur. To overcome this, most pressuremeter tests are conducted at a much higher strain rate than conventional laboratory tests. Implications are such that since simple theoretical models are unlikely to fully capture the behavior of the pressuremeter cavity expansion such as partial consolidation or account for strain rate, there is a mismatch in solutions. Carter et al. (1979) suggested that a full numerical analysis rather than a closed form solution may be the only means of obtaining a solution in the pressuremeter cavity expansion due to two main reasons. The first of which is that soil is a multi-phase material, and secondly the prediction of behavior both during and after the test is of interest. In addition, closed form solutions may not be able to capture the complexity of advanced constitutive models.

Various studies on the modelling of the pressuremeter test have been conducted previously. Biarez et al. (1998) used PLAXIS with the "Advanced Mohr Coulomb" model to show that parameters obtained from the triaxial test or pressuremeter test can be in good agreement. Schanz et al. (1999) validated the Hardening Soil model with loose Hostun sand by back analyzing an experimental study of a pressuremeter test in a calibration chamber at the IMG in Grenoble. The unloading and reloading behavior of soils in these tests were not investigated however. Figure 10 shows that the unloading-reloading stiffness is not a simple linear constant. Thus, there is a need to make further use of numerical models to adequately characterize the pressuremeter test.

Sound understanding of the theoretical mechanics of the cylindrical cavity expansion theory is needed for an accurate interpretation as well as simulation of the in-situ test. For calibration of numerical models, an approach must be derived from the measured relationship between the pressure and volume of the pressuremeter, that describes the stress and strain relationship of the soil with some numerical parameters. The simulation of the pressuremeter test allows for a back calculation of these parameters.

This section aims to discuss modelling techniques with things to look out for and potential pitfalls in simulations of the boundary value problem.

3.1 *Influence of boundary conditions*

Numerical simulations of pressuremeter tests allow for a finite length, unlike theoretical solutions such as a perfect cylindrical cavity expansion in a semi-infinite half space. While simulations of pressuremeter tests will allow for calibration of design parameters, care should be taken when applying boundary conditions to the finite element model.

What has not been investigated in great detail is the influence of boundary conditions in the simulation of pressuremeter tests. The extent of yielding which happens in the field test may be different

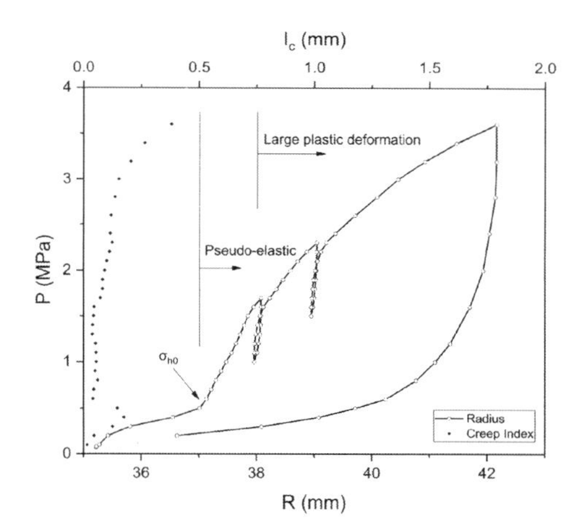

Figure 9. Typical pressuremeter test curve.

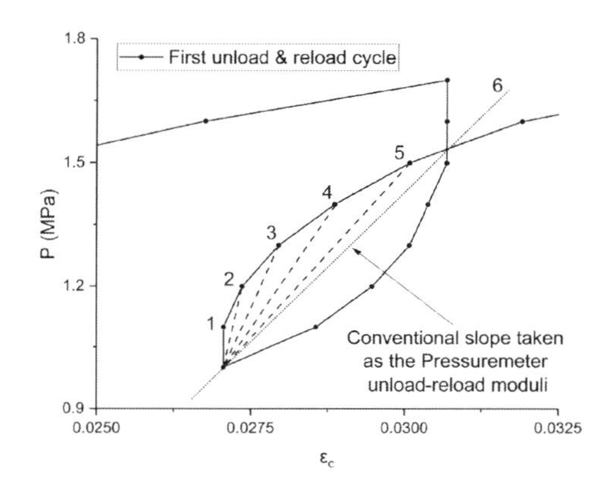

Figure 10. Pressuremeter unload-reload moduli as a function of strain level.

from what is observed in the calibration chamber because of the boundary conditions applied. These boundary conditions will always have an influence on the problem and may affect the calibration results.

A full borehole shown in Figure 11 was used to compare with a calibration chamber type of environment, where only the soil of interest is modelled and the soil above is represented by an overburden pressure, as illustrated in Figure 12. Both models utilize an axisymmetric setup. The results turn out to be in direct agreement with each other. It is good practice to always perform a quick check on the yielding profile of the soil to judge whether the numerical model is truly representative of in-situ conditions, Figure 13. In general, with the typical length over diameter ratio of the pressuremeter, the calibration chamber model should work.

A full numerical model of the borehole has certain advantages as compared to the calibration chamber type of model. Firstly, the borehole model allows for a direct input of initial stress conditions into the various soil layers. With the calibration chamber model, additional computation phases of loading and unloading to generate the initial stress state are required in order to match the in-situ soil stress history. Secondly, parameters obtained from laboratory tests can first be input into the numerical model of the borehole, and with the use of command lines, placement of the pressuremeter can be adjusted quickly to match the field test position. After calibration of the constitutive model, the borehole can be extended to form the base of the finite element model for design simulations of the geotechnical project, rather than creating another model.

Furthermore, if the soil is anisotropic in nature, the calibration chamber type of environment as the modelling approach of analysis may not be a reasonable representation of in-situ conditions due to the assumed boundary conditions. Anisotropic soil and the presence of discontinuities in the ground violate the principles for a true cylindrical cavity expansion. A full 3D numerical analysis with constitutive models of the ability to capture anisotropy needs to be applied in order to replicate the behavior of the cavity expansion in that case.

3.2 *Loading conditions*

The expansion of a pressuremeter may be controlled by applied pressure or applied strain rate. Ideally, stress controlled and strained controlled pressuremeter tests should produce similar results.

The application of either radial line loads or prescribed line displacements can be used to model both stress controlled test and strain controlled test with the correct boundary conditions applied. A hypothetical pressuremeter can be modelled with

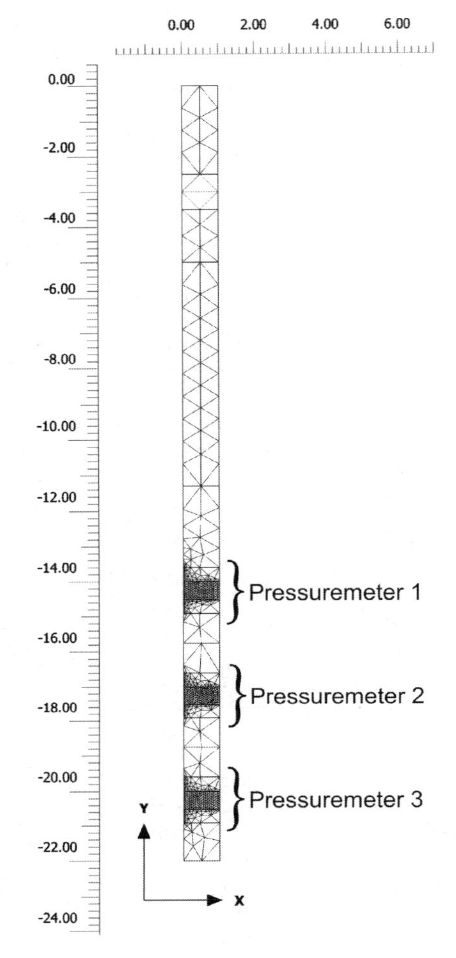

Figure 11. Typical finite element model of a borehole with 3 pressuremeter tests.

the use of prescribed displacement to reproduce a perfect cylindrical expansion. However, the actual expansion of a pressuremeter resembles that of an ellipse. A pressure loading will be more representative of actual conditions. If pressure loading is to be simulated in the finite element model, a free flexible boundary should not be used but a geogrid element should be applied to reduce the exaggeration of the pressure redistribution. Geogrid elements can only sustain tensile stresses and are more representative of the stiff rubber balloon of the pressuremeter sonde, compared with other types of elements.

It was found that the main differences in behavior between pressure loading and prescribed displacement are: firstly the displacement profile of the stress controlled simulation of the pressuremeter expansion resembles more an ellipse, while the strain controlled simulation resembles more a

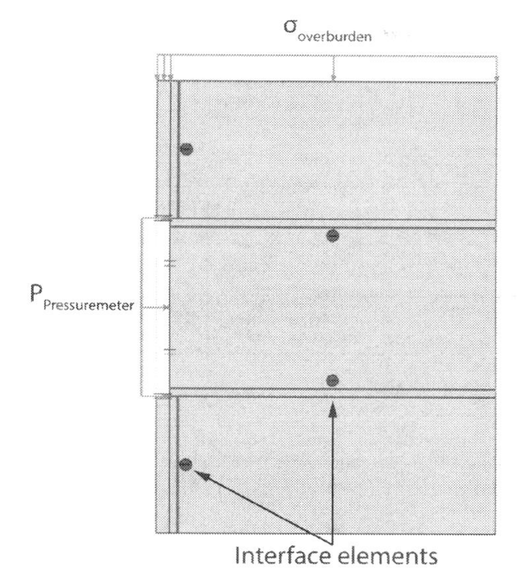

Figure 12. Finite element model of a calibration chamber environment for simulations of the pressuremeter expansion.

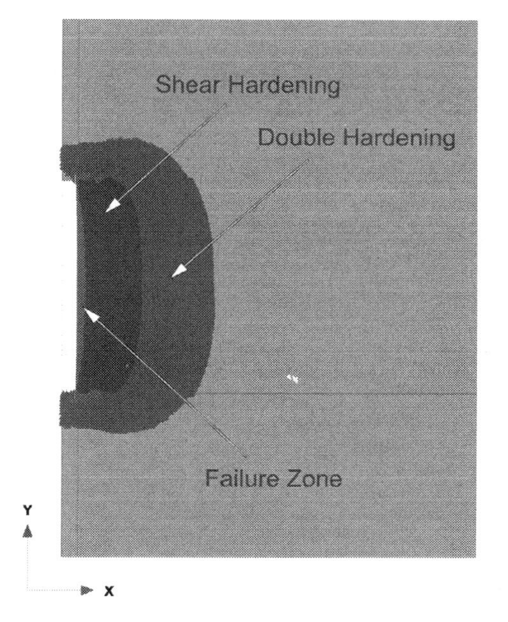

Figure 13. Yielding profile during a pressuremeter cavity expansion.

perfect cylindrical cavity, Figure 14; secondly stress concentrations occur along the boundary when prescribed displacements were used, Figure 15. Interface elements are modelled to allow for slippage between the elements at the top and base of the pressuremeter, to reduce the stress concentrations at the corners, Figure 12.

3.3 *Effect of borehole disturbance*

Apart from numerically controlled boundary and loading conditions, the actual field conditions can be vastly different from the numerical model. The most commonly employed pressuremeter test in Singapore is the pre-bored pressuremeter test, and so the performance of this pressuremeter variant will be discussed in this section.

As its name suggests, the pre-bored type of pressuremeter test must be installed in pre-drilled test pockets. Unfortunately, this results in almost inevitable soil disturbance in the initial stages due to the reduction of in-situ stresses, Figure 16. In some cases where sandy soils were encountered, the test pockets collapsed completely, Figure 17.

When compared to a numerically simulated perfect borehole, it is interesting to note that even though the initial stress-strain response of the cavity was severely affected by the disturbed soil, the unloading and reloading behavior is observed to be relatively unaffected by the disturbance. Furthermore, with repeated unloading and reloading cycles conducted on the soil, the imperfect borehole does not affect the unloading-reloading stiffness very much.

Unlike a real pressuremeter test conducted in-situ, disturbance effects cannot be captured within a computer model due to the uncertainty of the actual conditions. Figure 16 shows that the pressuremeter test curve can only be fitted by the numerical simulation when sufficient contact with the cavity wall is established.

Perhaps the correct way to conduct a pressuremeter test is to first load, then repeat the unloading and reloading cycle at different strain levels, until a reliable unloading-reloading stiffness is observed, somewhat illustrated by Figure 17. Then a back-analysis can be performed to determine the corresponding E_{50}, as the value cannot be measured in the field accurately due to the initial seating problems of the pre-bored pressuremeter test.

A great number of published work on pressuremeter tests focuses on the interpretation of the test results idealized as cylindrical expansions, or comparison to other in-situ or laboratory tests. It

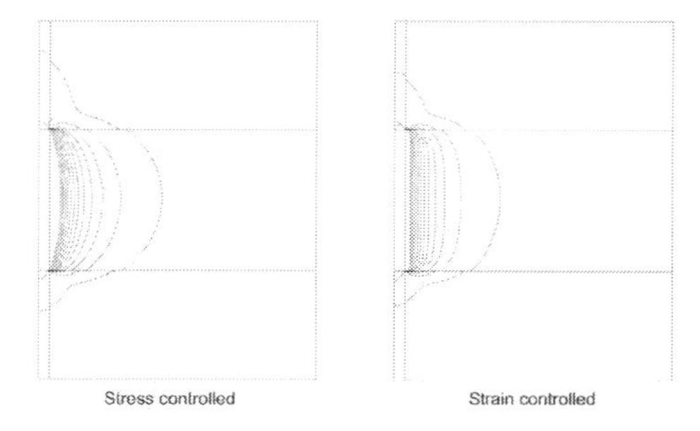

Figure 14. Comparison of displacement profile.

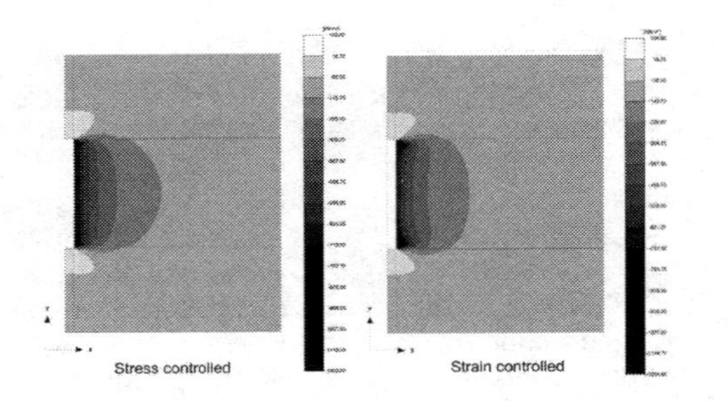

Figure 15. Comparison of horizontal stress distribution.

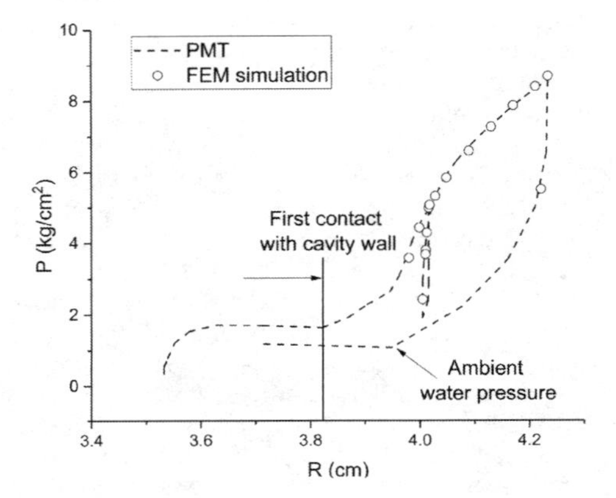

Figure 16. Fitted pressuremeter curve by calibration.

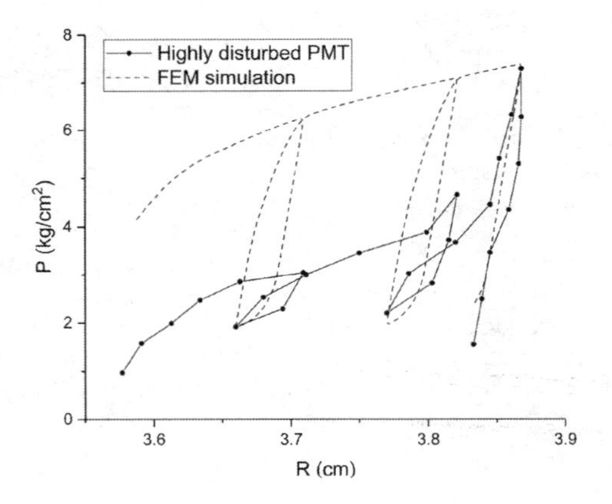

Figure 17. Highly disturbed pressuremeter test.

should however be noted that another important facet of the site investigation test to be considered is the effect of installation techniques. Clearly, it is difficult to account for installation effects of the pressuremeter test. Clarke and Gambin (1998) listed down various effects of installation for consideration such as softening of the ground adjacent to the probe during the pre-boring or the probe alignment not being vertical, both of which consequently create a non-cylindrical pocket. Such disturbance effects also make the reference strain datum of the cavity uncertain and difficult to determine.

4 CONCLUDING REMARKS

The calibration of the Hardening Soil constitutive model has been discussed. Simulations of the pressuremeter test can be used to verify geotechnical parameters obtained from soil element laboratory tests, however certain considerations must be taken into account when modelling the pressuremeter test.

REFERENCES

Becker, D.E., Crooks, J.H.A., Been, K., & Jefferies, M.G. (1987). Work as a criterion for determining in situ and yield stresses in clays. *Canadian Geotechnical Journal, 24*(4), 549–564.

Biarez, J., Gambin, M., Gomes-Corriea, A., Falvigny, E., & Branque, D. (1998). Using pressumeter to obtain parameters to elastoplastic models for sands. In *Proc of the first international conference on site Characterization, ISC* (Vol. 98, pp. 19–22).

Boone, S.J. (2010). A critical reappraisal of "preconsolidation pressure" interpretations using the oedometer test. *Canadian Geotechnical Journal, 47*(3), 281–296.

Briaud, J.L. (2013). Ménard Lecture: The pressuremeter Test: Expanding its use. In *Proceedings of the 18th International Conference on Soil Mechanics and Geotechnical Engineering, Paris, France* (pp. 107–126).

Butterfield, R. (1979). A natural compression law for soils (an advance on e-log p'). *Géotechnique, 29*(4).

Carter, J.P., Randolph, M.F., & Wroth, C.P. (1979). Stress and pore pressure changes in clay during and after the expansion of a cylindrical cavity. *International Journal for Numerical and Analytical Methods in Geomechanics, 3*(4), 305–322.

Clarke, B.G., & Gambin, M.P. (1998). Pressuremeter testing in onshore ground investigations. A report by the ISSMGE Committee 16. In *International Conference on Site Characterization, Atlanta.*

Goh, K., Jeyatharan, K., & Wen, D. (2012). Understanding the stiffness of soils in Singapore from pressuremeter testing. *Geotechnical Engineering Journal of the SEAGS & AGSSEA*, 43(4), 21–29.

Grozic, J.L., Lunne, T., & Pande, S. (2003). An oedometer test study on the preconsolidation stress of glaciomarine clays. *Canadian Geotechnical Journal, 40*(5), 857–872.

Mair, R.J., & Wood, D.M. (2013). *Pressuremeter testing: methods and interpretation.* Elsevier.

Onitsuka, K., Hong, Z., Hara, Y., & Yoshitake, S. (1995). Interpretation of oedometer test data for natural clays. *Soils and Foundations, 35*(3), 61–70.

PLAXIS 2D (Version 2017.0). (2017). [Software]. Available from www.plaxis.nl

Schanz, T., Vermeer, P.A., & Bonnier, P.G. (1999), "The hardening soil model: formulation and verification." *Beyond 2000 in computational geotechnics*, 281–296.

Wroth, C.P. (1984). Interpretation of in situ soil tests. *Geotechnique, 34*(4), 449–489.

Numerical Methods in Geotechnical Engineering IX – Cardoso et al. (Eds)
© 2018 Taylor & Francis Group, London, ISBN 978-1-138-33198-3

Direct infinite element for soil structure interaction in time domain

Y. Bakhtaoui
National Center Of Studies And Integrated Research On Building, Algiers, Algeria
Faculty of Civil Engineering, USTHB University, Algiers, Algeria

A. Chelghoum
Faculty of Civil Engineering, USTHB University, Algiers, Algeria

ABSTRACT: In dynamic soil structure interaction modelling, one of the main difficult comes from treating the far field effects by finite elements. This paper presents the formulation and implementation of a direct infinite element for simulating far-field problems in time domain. The procedure is based on infinite element method using the direct scheme with viscous boundary in layer form absorbing. Under dynamic loading with a non-vanishing time average, correlation of the obtained results shows good performance for the new procedure using explicit as well as implicit integration schemes in time domain.

1 INTRODUCTION

Soil-structure interaction problem leads usually to differential equations resolution where the definition field is not measurable. In this case specific boundary conditions must be imposed in order to obtain a solution. Some analytical solutions proposed by Boussinesq, Mindlin and Cerruti (Donida et al 1988) cannot be generalized; therefore numerical approaches coupling specifics boundaries become unavoidable.

This necessity becomes imperative for dynamic problems where the lateral and inferior borders of the model become reflective returning the waves to the structure modifying its response. Over the last few decades several studies have been carried out in order to establish numerical procedures based on absorbing boundaries or finite elements to avoid interferences on the solution.

Widely used, absorbing boundaries classified in local and non-local forms are a good compromise for solving dynamic problems. The non-local absorbing boundaries are accurate only in the frequency domain but with a high calculation cost. These non-local absorbing boundaries are too complicated to be practical and cannot be used for non-linear problems. On the other hand, local absorbing boundaries coupled with finite elements can be applied for nonlinear analysis in both frequency and time domains but will require a high number of elements to reach acceptable accuracy.

Infinite element method has been introduced initially to solve problems of fluid and propagation of waves at infinity for 2-D systems by Ungless & Anderson (1973), Bettess & Zienkiewicz (1975–1977), Bettess (1977–1980), then the formulation was adapted for static problems by Medina (1981), and Marques & Owen (1984). According to the decay type, several forms of infinite elements have been developed by Zienkiewicz et al. (1983), Curnier (1983), Bettess & Bettess (1984), Chen & Poulos (1993), Viladkar (1990). They allow the simulation of various types of static problems whereas in dynamic the existence of several types of waves complicates the use of this approach. Indeed the three-dimensional medium can convey body waves (expansion wave P and shear wave S) and surface waves (Rayleigh waves). The existence of multiple types of waves invalidates the direct application of elastostatic infinite elements.

To circumvent these limitations, Chow & Smith (1981), Medina & Penzien (1982), and Medina & Taylor (1983), have introduced elastodynamic infinite elements where elastostatic infinite elements are coupled with terms representing the dynamic part whose characteristics must meet the condition of radiation wave. Later Yun et al (1995–2007) developed more efficient infinite elements in dynamic layered media. Zhao & Valliappan (1993), and Park & Watanabe (2004) developed them in three dimensions. Recently, Seo et al (2007), and Kazakov (2010) developed infinite elements using wave functions containing various wave components; Bagheripour et al (2010) used them in the structure-ground-interaction by numerical representation of wavelet theory. Using the same approach, Düzgün & Budak (2011) evaluated the effects of topographical and geotechnical irregularities on the dynamic response of 2-D systems. Kazakov (2012) developed decay and mapped elastodynamic infinite elements based on modified Bessel shape functions to solve soil-structure interaction problems. Based on the frequency domain, most of these formulations were cloning method for non-linear dynamic problems involving the inverse of the Fast Fourier Transform to obtain the results of the time domain.

Dynamic infinite element was first presented by Haggblad and Nordgren (1987) to solve nonlinear soil-structure interaction problems in time domain. This approach consists of adding an absorbing layer on the infinite element in damping form.

This concept was recently tested by Su & Wang and Edip et al (2013) on mapped infinite elements in dynamic soil-structure interaction.

In this paper a new formulation and implementation of a procedure to simulate far-field problems in time-domain is presented using direct infinite element and an adjacent absorbing layer obtained through a modification of the Haggblad concept. From the viscous formulation, wave impedances are added on the interface finite element located between the mesh and the infinite element to form an absorbing interface ensuring dissipation, while elastic recovery is ensured by an exponential direct infinite element having a classical decay type.

Using original software based on FORTRAN language, the validation is provided by using examples from published papers (Modaressi & Benzenati 1992, Shridar & Chandrasekaran 1995, Nour & Alam 2002, Lehmann 2005).

2 GENERAL FORMULATION

To simulate far-field effects, the present procedure proposes to combine an infinite element with an absorber layer represented by a finite element. From a specific decay function, the direct infinite element is built. The absorbing layer is expressed in the form of a concentrated matrix on the interface finite element from the viscous boundary.

2.1 *Formulation of infinite elements*

According to Bettess's formulation (1977, 1980, 1984) the direct infinite element's interpolation functions are obtained from a standard parent finite element shape function combined with decay functions. In 2D isoparametric representation, the geometry x, y, and displacements u, v, of the element can be formulated by:

$$x = \sum_{i=1}^{n} N^i x_i, \text{and } z = \sum_{i=1}^{n} N^i z_i \tag{1}$$

$$u = \sum_{i=1}^{n} H^i u_i, \text{ and } v = \sum_{i=1}^{n} H^i v_i \tag{2}$$

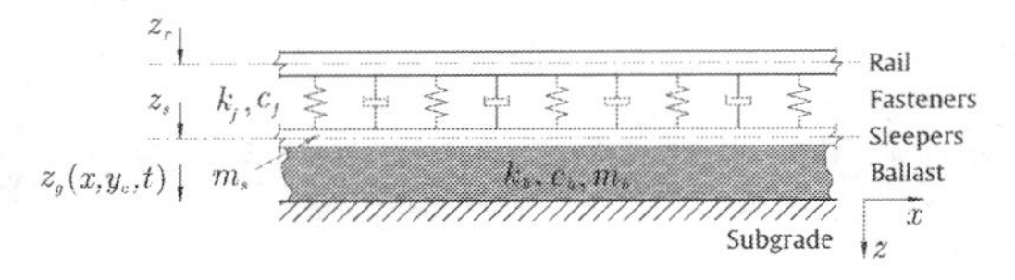

Figure 1. Local and global representation of an infinite element.

x_i and z_i are the global coordinates at node i, while the displacement field is expressed by the global nodal displacements u_i and v_i. n is the number of nodes for each infinite element.

From shape functions of a linear three noded finite element using Lagrange polynomials defined for i = 1 to n-1, and by application on a four two-dimensional element (according the convention of Fig. 1), the coordinate functions $N^i(\xi,\eta)$ are obtained in form

$$[N] = -\xi \begin{bmatrix} \frac{(1-\eta)}{2}; -\xi\frac{(1+\eta)}{2}; \frac{(1+\xi)(1+\eta)}{2}; \\ \frac{(1+\xi)(1-\eta)}{2} \end{bmatrix} \tag{3}$$

The displacement functions $H^i(\xi,\eta)$ in direct formulation are formulated by the relation:

$$H^i(\xi,\eta) = f^i(\xi).N^i(\xi,\eta) \tag{4}$$

With $f^i(\xi)$, the Decay-functions in ξ, fixed in exponential form (Bettess 1977, 1980, 1984) by

$$f^1 = e^{(-1-\xi)/L} \tag{5}$$

$$f^2 = e^{-\xi/L} \tag{6}$$

The displacements functions can be expressed by:

$$[H] = \begin{bmatrix} -\xi\frac{1-\eta}{2}e^{\frac{-1-\xi}{L}} - \xi\frac{1+\eta}{2}e^{\frac{-1-\xi}{L}} (1+\xi)\frac{1+\eta}{2}e^{\frac{-\xi}{L}} \\ (1+\xi)\frac{1+\eta}{2}e^{\frac{-\xi}{L}} (1+\xi)\frac{1-\eta}{2}e^{\frac{-\xi}{L}} \end{bmatrix} \tag{7}$$

The local coordinates η and ξ are defined such as $\eta \in [-1,+1]$ and $\xi \in [-1,+\infty]$.

Using specific numerical integration schemes, this approach must satisfy the following conditions:

- Radiation energy at infinity
- Completeness and consistency
- Finite values at infinity

That imposes to verify:

$$\begin{cases} f^i(\xi \to +\infty) = 0 \\ H^i(\xi_i, \eta_i) = 1 \\ \xi \in [-1,0] \Rightarrow \sum_{i=1}^{4} H^i \cong 1 \\ \xi \to +\infty \Rightarrow \sum_{i=1}^{4} H^i = 0 \end{cases} \quad (8)$$

The severity of the decay is determined by the exponential decay length L. It is possible to set L to unity, or as the distance between the nodes. The value of the exponential decay length can be obtained from the behaviour of the exterior solution for example its Green's function (Bettess 1980).

2.2 *Stiffness and mass matrices for the direct infinite elements*

In the same way as for the finite elements, the stiffness matrix [K^e] for an element is given by the virtual work theorem

$$[K^e] = \int_{-1}^{+1}\int_{-1}^{+\infty} [B]^T [D][B] \det[J].t.d\eta d\xi \quad (9)$$

[B]: deformation matrix.

[D]: material matrix.

Computation of [K^e] is carried out using specific numerical integration schemes of Gauss-Laguerre (Pissanetzky 1983) quadrature.

Referring to the approach proposed by Hinton et al. (1976), the evaluation of the mass to be allocated to each node of the infinite element is based on the concept of consistent mass matrix [Mc]

$$[M_c] = \int_{-1}^{+1}\int_{-1}^{+\infty} [H]^T \rho.[H].\det[J].t.d\eta.d\xi \quad (10)$$

Using total mass of the infinite element given by

$$M_{tot} = \int_V \rho dv = \int_{-1}^{+1}\int_{-1}^{+\infty} \rho \det[J].t.d\eta d\xi \quad (11)$$

The distribution of the mass m_i on the nodes of the infinite element is obtained by:

$$m_i = \frac{M_{ii}}{trace[Mc]}.M_{tot} \quad (12)$$

where trace [M$_c$] reflecting the total mass extracted from the consistent mass matrix, and M_{ii} are the diagonal term.

Using [H] as the matrix of displacements shape functions, the quadrature Gauss-Laguerre (Pissanetzky 1983) is used for numerical integration.

2.3 *Implementation of the viscous boundary*

Widely used for the simulation of soil structure interaction (Connolly et al 2013 and Fattah et al 2015), the viscous boundary was developed and implemented with success by Lysmer and Kuhlemeyer 1969 to avoid reflection of inducing waves in the model.

Using the formulation of the viscous boundaries, the procedure developed by Haggblad and Nordgren (1987), Su and Wang (2013) and Edip et al (2013) applied an absorbing layer to a mapped infinite element. In the proposed approach the scheme is repeated by taking into account the following changes.

1. A direct infinite element is used
2. The absorbing layer is applied on the finite element serving as interface between the mesh and the infinite element.

Hence, the absorbing interface is modelled by an absorbing finite element ensuring the wave dissipation while the elastic recovery is ensured by direct infinite elements. On the interface, the wave is transmitted from finite to the infinite element so the equality of impedances must be insured.

The viscous boundary formulation starts from using incident wave equation, defining normal and shear stresses for each interface node. In the two-dimensional plane strain model (Fig. 2), the stresses on the nodes of the border are expressed by:

$$\begin{aligned} \sigma &= -a\rho.V_p.\dot{u}_n \\ \tau &= -b\rho.V_s.\dot{u}_t \end{aligned} \quad (13)$$

where Vp and Vs are the velocities of P (compression) and S (Shear) waves, respectively. To maximize the absorption rate of P and S wave of the viscous boundary (White et al 1977):

a = b = 1.

Written in following nodal form:

$$\begin{bmatrix} \sigma \\ \tau \end{bmatrix} = [D^*][\dot{U}_N] = \begin{bmatrix} a\rho V_P & 0 \\ 0 & b\rho V_S \end{bmatrix} \begin{bmatrix} \dot{u}_n \\ \dot{v}_s \end{bmatrix} \quad (14)$$

The normal and tangential velocities are represented by the vector $[\dot{U}_N]$, while the equivalent nodal damping are represented by the matrix [D*]

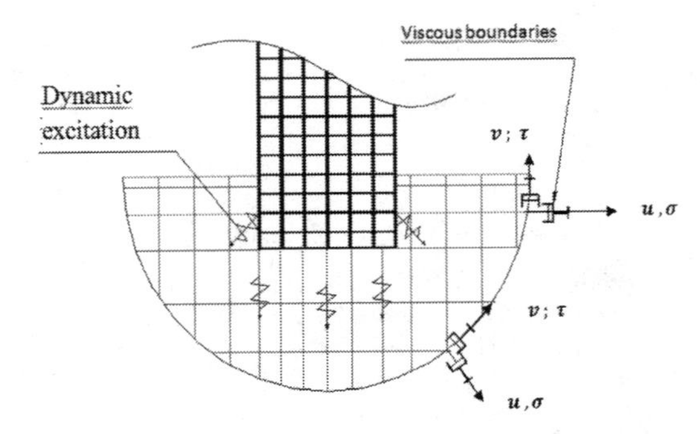

Figure 2. Planar viscous boundary.

$$[D^*] = \rho\begin{bmatrix} aV_p & 0 \\ 0 & bV_s \end{bmatrix} \tag{15}$$

The consistent damping matrix (representing the absorbing layer) is given by:

$$[C] = \int_S [H]^T [P]^T [D^*][P][H] ds \tag{16}$$

Where P is the projection matrix relating normal and tangential velocities to the global Cartesian velocities components while [H] is the matrix of displacement interpolation functions of interface finite element. The damping elements are calculated using a 2 points Gauss numerical integration. The evaluation of the damping allocated to each node and on each side of the finite element can use Hinton's Method in the explicit procedure.

Hence, each type of dissipative system can implicitly combined in the formulation of the global equations of the system as:

$$[M]\{\ddot{U}\} + [C]\{\dot{U}\} + [K]\{U\} = \{Q\} \tag{17}$$

3 APPLICATIONS

The validation of the proposed procedure under dynamic loading is carried out by using original software developed in FORTRAN language through numerical simulations on typical engineering problems.

Using two-dimensional finite elements for the modelling of the near-field media in the proposed software, a step by step integration algorithm with implicit procedure have been implemented. The domain is discretized by 4 nodes isoparametric finite elements while the masses and rigidities are evaluated using Gaussian numerical integration.

The performance of this system is proved on the base of the following parameters such as:

- Boundary conditions
- Type of elasticity
- Dynamic loading

3.1 *First application: one dimensional case*

From works of Modaressi & Benzenati (1992) and Nour and Alam, the application was used for testing respectively an absorbing boundary from Paraxial approximation and a modification of generalized Smith boundary procedure.

The solution concerns the evaluation of soil column's responses subjected to a unit heavy-side type loading on the top. The same data and geometry are considered to carry out the correlation of the results.

To evaluate the efficiency of the present procedure, three soil models are used (Fig. 3). Chosen as the reference model with a selected time window small enough to prevent the reflected wave disturbing the response, the first model is a 300 m semi-infinite height soil column with a rigid boundary at the base. The second model is a 100 m height soil column tested alternately by original Haggblad's procedure, the proposed procedure, and the fixed boundary conditions. Finally, the third model is a 130 m height soil column devoted to evaluating the variation of responses according to the location of the numerical boundary proposed.

The vertical displacement (Fig. 4) is calculated at three points respectively 20, 50 and 80 m from the top and performance of the procedure is illustrated on Figures 4–5.

Data:
Elastic modulus: $E = 4.108 \times 10^8$ Pa
Poisson ratio: 0.3
Mass density of the soil: 1620 kg/m3

Until the 0.25 seconds good agreement is observed with the reference solution at 20 and 50 m. From 0.25 seconds discrepancies appear at 50 m while at 80 m the discrepancies start from the 20th second. The evolution of response (Figure 4) shows the solution given by rigid boundary moving away from the reference solution whereas the solutions obtained by the present procedure and by Haggblad's method remain stable with same shapes.

Starting from the 0.30 sec all solutions obtained by absorbing infinite elements approaches seem distorted. This behavior can be justified by the change of the location of boundaries on the third model (Fig. 5) where a real improvement in the response can be seen compared to the reference solution. The new solutions obtained by the procedure display good efficiency at 20 m up to 0.35 secs and at 50 m, up to 0.30 secs. On the other hand, the discrepancies start from the 0.25th secs at 80 m.

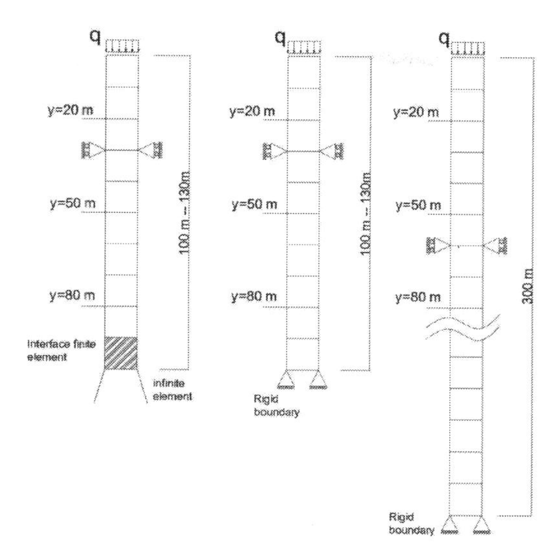

Figure 3. Soil column models used.

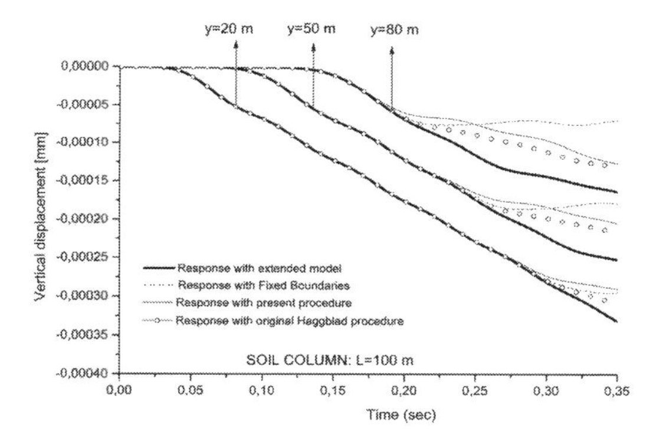

Figure 4. Vertical displacements with 100 m height soil.

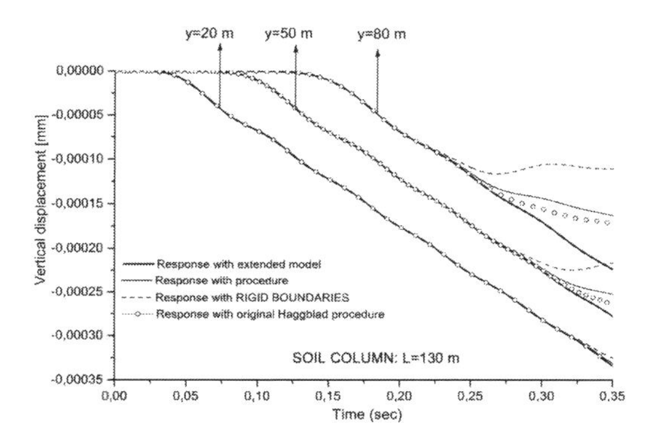

Figure 5. Vertical displacements with 130 m height soil.

Comparing with the solution obtained by the rigid boundary in both models, this improvement can be related to the absorbing layer formed by the viscous boundary. Indeed, several studies have highlighted the influence of the location of viscous boundary on the solution. Its performance is known to deteriorate when approaching the source of perturbation (Kellezi 2001). It is noted that there are no significant differences between the proposed procedure and Haggblad's approach.

3.2 *Axisymmetric case under loads having non-vanishing time average*

Based on works of Simons & Randolph 1986, Shridar & Chandrasekaran 1995, and Nour & Alam 2002, this application consists of evaluating the response of a soil profile with a flexible base under a vertical load having a non-vanishing time average. This dynamic load could represent the effect of a sudden load on an elastic infinite medium, such as the effect of a pile driving, or response of foundation for a transmission line tower when the cables on one side snap. Applied in form of a vertical disc with unity intensity at t = 0, then the loading is maintained constant thereafter.

The static solution is given by expression (Timoshenko and Godier 1951)

$$\delta = \frac{2qr(1-\upsilon^2)}{E} = 0.75 \tag{18}$$

To check the efficiency of the procedure, the reference model used is obtained by extending the mesh with rigid boundary conditions located at a sufficient distance, so that the wave after hitting the wall of reflection will not have time to disrupt the solution. For the extended model, a 30×30 mesh is used (Fig. 6a).

For boundary testing, the referential model is a 9×9 mesh (Fig. 6b) on which rigid boundary conditions, viscous boundaries, and infinite boundaries are imposed on the side and bottom borders (Fig. 6c).

The size of the finite element is taken equal to unity in order to meet the criteria of wave simulation (transmission of high frequencies *fmax*).

$$L_e = \frac{Vs}{(8-20).f\max} \tag{19}$$

$$T\min = \frac{Le}{Vm}(\text{Minimal period}) \tag{20}$$

Propagation velocities are chosen as Vp = 1.732 m/s and Vs = 1 m/s.

With: $Vm = (V_P + V_S)/2$ (21)

And: Le = $\min(\Delta x, \Delta y)$

Taking constant, the time step is expressed by

$$\delta t \leq \alpha \frac{T\min}{\pi} \tag{22}$$

And complementary Data:

$A = 0.25$, E = 2.5, Density = 1, Thickness = 1.

The obtained solutions presented concern vertical displacement at point A.

Compared with the large model, the discrepancy with the rigid model (on mesh 9X9) appeared clearly from the 12th second (Fig. 7) where the solution becomes distorted, confirming the inefficiency of this boundary type in dynamic problems.

The response given by the infinite elements without absorbing layer (Fig. 9) oscillates around the static solution, but discrepancies start from 22 secs. The response obtained by viscous boundaries (Fig. 8) exhibits until 25 seconds an excellent behaviour. After that, the solution deviates asymptotically from the true solution. The new solution shows good agreement with the reference solution providing an excellent behaviour until 50 seconds (figs 7–9).

3.3 *Plane strain under ricker wavelet case*

Carried out by Von Estorff & Prabucki (1990), and Lehmann (2005), to test their formulations,

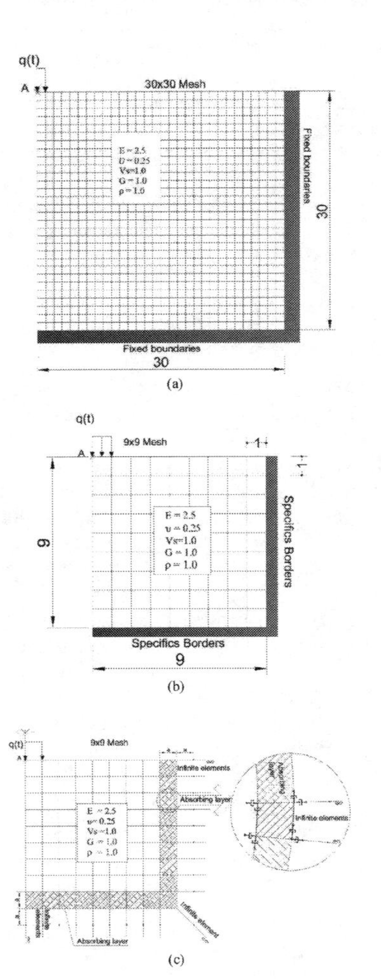

Figure 6. (a) The large model, (b) The used model (c) model of method proposed.

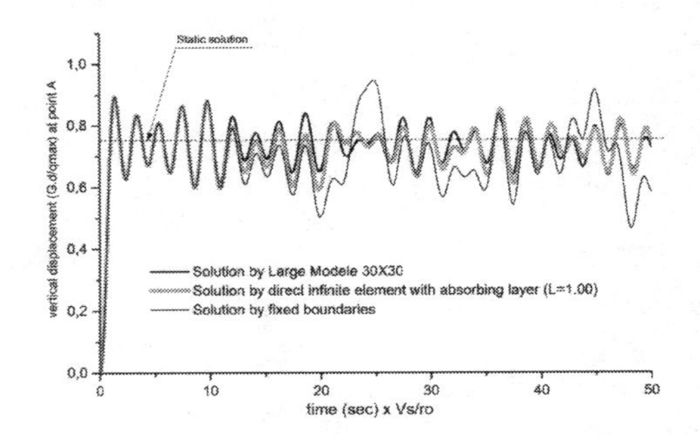

Figure 7. Comparison the response of the procedure with fixed boundaries.

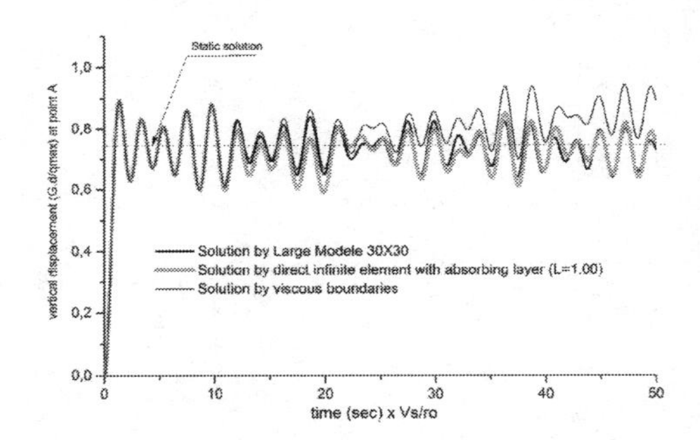

Figure 8. Comparison the response of the procedure with viscous boundaries.

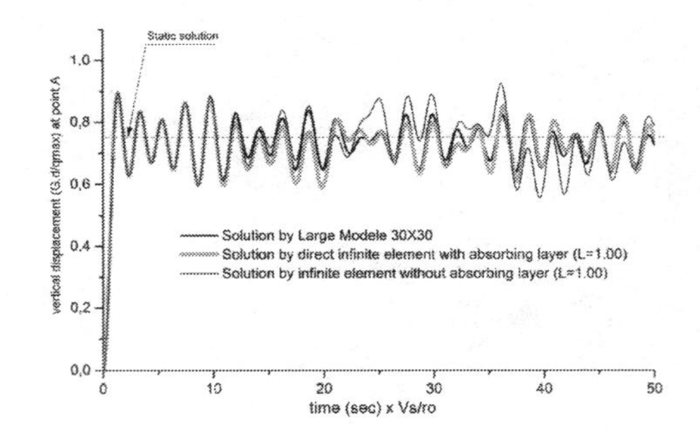

Figure 9. Comparison the response of the procedure with infinite element without absorbing layer.

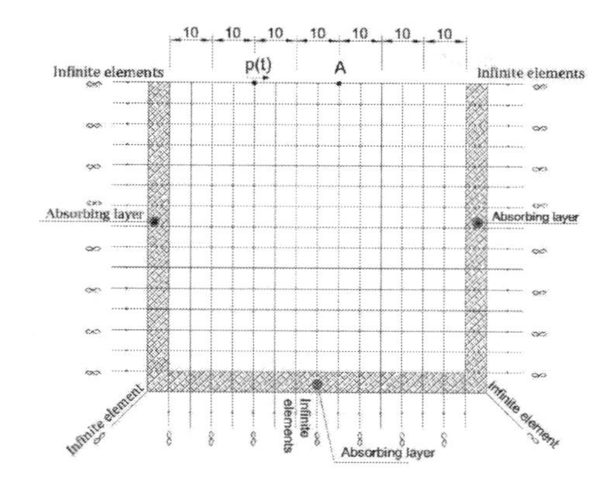

Figure 10. Model of the elastic half-space.

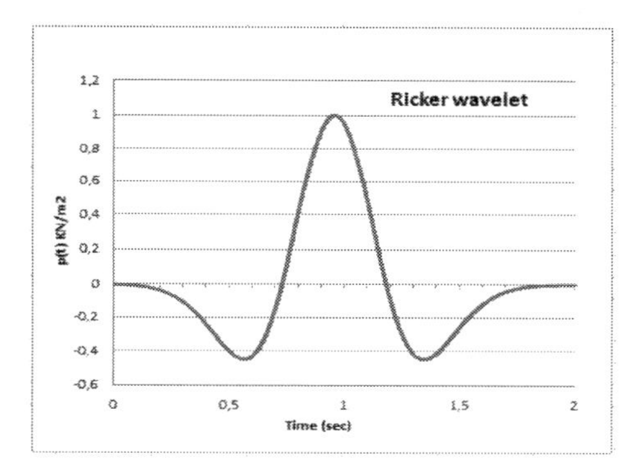

Figure 11. Load function p(t) (Ricker wavelet).

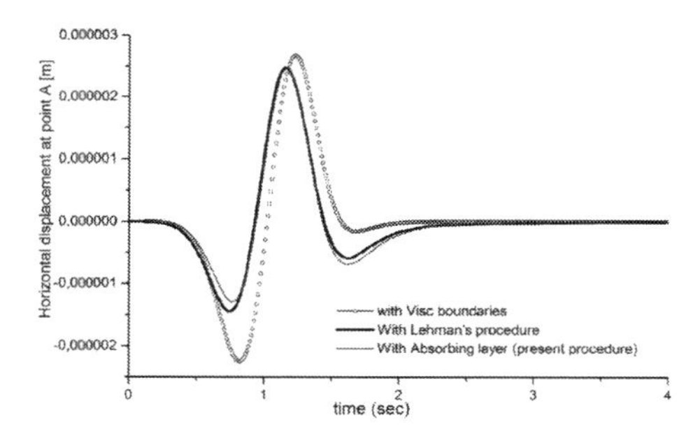

Figure 12. Comparison the responses with Boundary viscous coupled with the infinite element.

this example consists of evaluating the response of an infinite half-space loaded by a time-dependent horizontal load. Figure 10, shows the loading and an observation point (A) on the surface of the elastic half-space.

Using the same finite element, the elastic region (plane strain) has:

A Young's modulus $E = 2.66.10^5$ kN/m^2,

A Poisson's ratio $\nu = 0.33$,

And a mass density $\rho = 2.0 \times 10^3$ kg/m^3.

Figure 11, shows the time-dependent horizontal loading p(t) called Ricker-wavelet, where p(t) is given by:

$$p(t) = (1 - 2\tau^2)\exp^{-\tau^2} \quad (23)$$

$$where: \tau = t\pi - 3 \quad (24)$$

And the time-step used: $\Delta t = 0.002$ is used.

Compared to the solution given by Lehman's procedure, the horizontal displacement of point A illustrated in Figure 12, shows the efficiency of the proposed procedure (infinite element + finite element absorbing layer) while the solution given by uncoupled viscous boundary tends to diverge.

4 CONCLUSIONS

Starting from a direct infinite element approach, a new formulation and implementation of a dissipative interface in absorbing layer form to simulate far-field problems in time-domain are presented. In this approach, wave impedances are added to the adjacent finite element on the interface: "mesh-infinite element". Therefore, in the new model, the interface is modelled by finite elements ensuring the wave dissipation while the infinite elements ensure the elastic recovery. The procedure is easy to implement, and the number of elements and nodes is reduced. Compared to other methods such as viscous boundaries, paraxial methods, extrapolation algorithm, and boundary method, the procedure shows good performances for dynamic cases and gives good accuracy for usual dynamic geotechnical problems in the time domain.

REFERENCES

Bagheripour MH., Rahgozar R. & Malekinejad M. 2010. Efficient analysis of SSI problems using infinite elements and wavelet theory. *Geomechanics and Engineering, Vol. 2, No. 4:229–252.*

Bettess P. 1977. Infinite elements. *Int. Jour. Num. Meth. Eng., 11: 53–64.*

Bettess P. & Zienkiewicz OC. 1977. Diffraction And Refraction Of Surface Waves Using Finite And Infinite Elements. *International journal for numerical methods in engineering, vol. 11: 1271–1290.*

Bettess P. 1980. More on infinite elements. International journal for numerical methods in engineering, vol. 15:1613–1626

Bettess P. & Bettess JA. 1984. Infinite elements for static problems. *Engineering Computations, Vol. 1:4–16*

Chen L. & Poulos HG. 1993. Analysis Of Pile-Soil Interaction Under Lateral Loading Using Infinite And Finite Elements. *Computers and Geotechnics 15:189–220*

Chow YK. & Smith IM. 1981. Static And Periodic Infinite Solid Elements. *International Journal For Numerical Methods In Engineering, Vol. 17: 503–526*

Connolly D, Giannopoulos A & Forde MC. 2013. Numerical modelling of ground borne vibrations from high speed rail lines on embankments. *Elsevier Soil Dynamics and Earthquake Engineering, 46: 13–19*

Curnier A. 1983. A static infinite element. *Int. J. Num. Meth. Eng., 19:1479–1488.*

Donida G., Bruschi R. & Bernetti R. 1988. Infinite elements in problems of geomechanics. *Comp. and Struct. Vol 29, N°1:pp 63-67.*

Edip, K., Garevski M., Butenweg C., Sesov V., Cvetanovska J., & Gjorgiev I. 2013. Numerical Simulation of Geotechnical Problems by Coupled Finite and Infinite Elements. *Journal of Civil Engineering and Architecture, ISSN 1934–7359, USA, Jan. 2013, Vol. 7, No. 1 (No. 62): 68–77*

Fattah MY, Hamoo MJ & Dawood SH. 2015. Dynamic response of a lined tunnel with transmitting boundaries. *Earthquakes and Structures, 8,(1): 275–304*

Haggblad B. & Nordgren G. 1987. Modelling nonlinear soil-structure interaction using interface elements, elastic-plastic soil elements and absorbing infinite elements. *Comp. And Struct., Vol. 26 N°1/2: 307-324.*

Hinton E, Rock & T, Zienkiewicz OC. 1976. A note on mass lumping and related process in the finite element method. *Jour. Earth. Eng. Struc. Dyn., 4: 245–249.*

Kazakov KS. 2010. Elastodynamic infinite elements with united shape functions for soil–structure interaction. *Elsevier Finite Elements in Analysis and Design, Vol. 46: 936–942*

Kazakov K.S. 2012. Elastodynamic infinite elements based on modified Bessel shape functions, applicable in the finite element method. *Structural Engineering and Mechanics, Vol. 42, No. 3: 353–362*

Kellezi L. & Takemiya H. 2001. An Effective Local Absorbing Boundary for 3D FEM Time Domain Analyses. *International Conferences on Recent Advances in Geotechnical Earthquake Engineering and Soil Dynamics.*

Lehmann L. 2005 An effective finite element approach for soil-structure analysis in the time-domain, *Structural Engineering and Mechanics, Vol. 21, No. 4: 437–450.*

Lysmer J. & Kuhlemeyer RL. 1969. Finite dynamic model for infinite media. *Jour. Eng. Mech.: 859–877.*

Marques JMMC. & Owen DRJ. 1984a. Infinite elements in quasi-static materially Nonlinear problems. *Pergamon Press Ltd Comp.and Structures, Vol. 18. No. 4: 739–751.*

Medina F. 1981. An Axisymmetric Infinite Element. *International Journal For Numerical Methods In Engineering, Vol. 17:1177–1185*

Medina F. & Penzien J. 1982. Infinite elements for elastodynamics. *Earthquake Engineering And Structural Dynamics, Vol. 10: 699–709.*

Medina F. & Taylor RL. 1983. finite element techniques for problems of unbounded domains. *International Journal For Numerical Methods in Engineering, Vol. 19: 1209–1226.*

Modaressi H. & Benzenati I. 1992. An absorbing boundary element for dynamic analysis of two-phase media. *10th World conference on earthquake engineering 1992, 19–24 July Madrid, Spain p. 1158–62.* Rotterdam: Balkema.

Nour A & Alam, MJ. 2002. Use of generalized Smith boundary for loads having non-vanishing time average. *Elsevier Computers and Geotechnics, 29: 235–255.*

Park KL, Watanabe E. & Utsunomiya T. 2004. Development of 3D elastodynamic Infinite Elements For Soil-Structure Interaction Problems. *International Journal of Structural Stability and Dynamics Vol.04: 423–441.*

Pissanetzky S. 1983. An infinite element an a formula for numerical quadrature over an infinite interval. *Int. Jour. Num. Meth. Eng. 19: 913–927.*

Seo CG., Yun CB. & Kim JM. 2007. Three-dimensional frequency-dependent infinite elements for soil–structure interaction, *Engineering structures N°29: 3106-3120 Elsevier*

Simons HA & Randolph MF. 1986. Short communication comparison of transmitting boundaries in dynamic finite element analysis using explicit time integration. *Int. Jour. Num. and An. Met.in Geom. Vol 10: 329–342.*

Shridar DS. & Chandrasekaran VS. 1995. Use of transmitting boundary for loads having non-vanishing time average. *Comp. and Structure, Vol 54, N°3: 547-550*

Su J. & Wang Y. 2013. Equivalent dynamic infinite element for soil–structure interaction. *Elsevier Finite Elements in Analysis and Design, 63: 1–7*

Timoshenko S., Goodier JN. Theory of elasticity. *Mc Graw Hill (1951).*

Ungless RL. 1973. An infinite finite element. *MASc Thesis, University of British Columbia 1973.*

Viladkar MN, Godbole PN & Noorzaei J. 1990. Some new three dimensional infinite elements. *Comp. and Structures, Vol 34, N°3: 455-467.*

White W, Valliapan S & Lee IK. 1977. Unified boundary for finite dynamic models. *Jour. Eng. Mech. Div. ASCE103: 949–964.*

Yun CB, Kim JM & Hyun CH. 1995. Axisymmetric elastodynamic infinite elements for multi-layered half-space, *Int. J. Numer. Meth. Eng., 38: 3723–3743.*

Yun CB., CHANG SH. & SEO CG. 2007. Dynamic Infinite Elements For Soil-Structure Interaction Analysis In A Layered Soil Medium. *International Journal of Structural Stability and Dynamics, Vol. 7, No. 4: 693–713.*

Zhao C. & Valliappan S. 1993. A Dynamic Infinite Element For Wave Problems Three-Dimensional Infinite-Domain. *International Journal For Numerical Methods In Engineering, Vol. 36: 2567–2580.*

Zienkiewicz OC. & Bettess P. 1975. Infinite elements in the study of fluid structure interaction problems. *Proc. 2nd Int. symp. On comp. Methods Appl. Sci. Versailles.*

Zienkiewicz OC., Emson C. & Bettess P. 1983. A novel boundary infinite element. *International Journal for Numerical Methods in Engineering, Vol. 19: 393–404.*

Numerical Methods in Geotechnical Engineering IX – Cardoso et al. (Eds)
© 2018 Taylor & Francis Group, London, ISBN 978-1-138-33198-3

A robust numerical technique for analysis of coupled problems in elasto-plastic porous media

O. Ghaffaripour & A. Khoshghalb
School of Civil and Environmental Engineering, UNSW Sydney, Kensington, NSW, Australia

ABSTRACT: A robust numerical technique is presented for solution of coupled problems in saturated porous media with elasto-plastic behavior. The proposed method, called edge-based smoothed point interpolation method, is developed by application of the strain smoothing technique and the generalized smoothed Galerkin method to the governing equations of two phase porous media. Point interpolation shape functions are adopted for approximation of the independent variables in the problem domain. A novel approach for the evaluation of the coupling matrix of the porous media is adopted. Integration of the elasto-plastic constitutive model is performed using the substepping method, and the modified Newton-Raphson approach is utilized to address the global nonlinearities of the problem. The performance of the presented formulation is evaluated by studying two numerical examples and comparing the results of the analysis with analytical solutions and those obtained from the conventional finite element method.

1 INTRODUCTION

The finite element method (FEM) has been the major technique in dealing with elasto-plastic behavior of materials in a numerical point of view (Belytschko et al., 2000; Smith et al., 2013). In the recent years, many meshfree methods (MMs) have been developed to overcome deficiencies associated with the FEM. In the 2000s, the point interpolation was proposed by (Liu and Gu, 2001b). Despite their simplicity and many advantages over other MMs (Wang et al., 2001; Khoshghalb and Khalili, 2010; Khoshghalb and Khalili, 2013; Khoshghalb and Khalili, 2015), PIMs are not theoretically reliable. therefore, a robust category of MMs based on PIMs were formulated by Liu and Zhang (2013). In this approach, the generalised gradient smoothing technique is applied to the PIM resulting in a new class of MMs known as the Smoothed PIM (SPIM), and the problem associated with the incompatibility of the approximation function in PIM is circumvented by adopting a constructed, rather than a compatible, strain field and therefore removing the need for calculation of the derivation of the shape functions.

In this work, an edge-based SPIM (ESPIM) is presented for coupled hydromechanical analyses of elastoplastic porous media. The problem domain is discretized using triangular background elements. Edge-based smoothing domains are then constructed using these elements. The displacement field is created adopting the polynomial PIMs. The smoothed strain field is constructed by applying the smoothing operation technique over the smoothing domains and a simple node selection scheme is used to adopt support nodes for shape function. The conventional Gaussian points inside the smoothing domains are also used for the calculation of the compressibility and coupling matrices (Ghaffaripour et al., 2017). For stress integration, a substepping scheme (Sloan, 1987) is utilised, and finally the nonlinear system of equations is solved at each time step using the modified Newton-Raphson iteration scheme. The numerical model is then verified using a consolidation problem.

2 GOVERNING EQUATIONS

The general equations governing flow and deformation in a saturated deforming porous medium based on the equilibrium of the soil water mixture and the combination of the mass balance equation for the fluid phase with Darcy's law for the fluid flow in porous media are expressed as follows (Lewis and Schrefler, 1999)

$$\partial^{\mathrm{T}}\sigma + \mathbf{F} = \mathbf{0} \tag{1}$$

$$\mathrm{div}\left[\frac{\mathbf{k}_{\mathrm{f}}}{\mu_{\mathrm{f}}}(\nabla p_{\mathrm{f}} + \rho_{\mathrm{f}}\mathbf{g})\right] - a_f \frac{dp_{\mathrm{f}}}{dt} + \mathrm{div}(\frac{d\mathbf{u}}{dt}) = 0 \tag{2}$$

where ∂ is the differentiation matrix, ∇ is the gradient operator matrix defined as $\nabla = \partial^{\mathrm{T}}\delta$, with $\delta = [1 \;\; 1 \;\; 0]^{\mathrm{T}}$, and div stands for the divergence operator. In these equations, bold imprints denote vectors and matrices. σ is the total stress matrix and $\sigma = \sigma' + \eta p_f \delta$ and $d\sigma' = \mathbf{D}^{ep} d\varepsilon$, where $\mathbf{D}^{ep}$ is the tangent constitutive matrix, $\varepsilon = \partial \mathbf{u}$ is the Cauchy small strain vector, p_f is the fluid pressure and $\eta = 1 - \frac{c_s}{c}$; $\mathbf{u}$ is the soil displacement vector; $\mathbf{F}$ is the vector of body force per unit volume; $\mathbf{g}$ is the gravity acceleration vector; $\mathbf{k}_f$ indicates the intrinsic permeability; μ_f is the dynamic viscosity of the fluid phase; ρ_f is the density of the fluid; and ρ is the porous medium density. $a_f = n(c_f - c_s) + \eta c_s$, where, n is the porosity and c_f, c_s and c are the compressibility of the fluid phase, compressibility of the solid grains, and solid drained compressibility, respectively.

3 SHAPE FUNCTION CREATION

Polynomial point interpolation method (PIM) (Liu and Gu, 2001a) is considered for determination of the nodal shape functions in which the field function f is approximated at the point of interest $\mathbf{x} = [x \quad y]$ as

$$f(\mathbf{x}) = \sum_{i=1}^{p} P_i(\mathbf{x}) a_i = \mathbf{P}^{\mathrm{T}}(\mathbf{x})\mathbf{a} \tag{3}$$

where $P_i(\mathbf{x})$ are the polynomial basis functions obtained from the Pascal's triangle of monomials for 2D problems, $\mathbf{a}$ is a coefficient vector and p is the number of supporting nodes for the point of interest (Liu and Gu, 2001a).

4 EDGE-BASED SMOOTHED POINT INTERPOLATION APPROACH

In the current model, the generalised smoothed Galerkin (GS-Galerkin) weak formulation is employed to discretise the system of equations governing the problem. For this purpose, smoothing domains are constructed on top of the triangular background mesh and a constant smoothed strain is assigned to each smoothing domain. Construction of the smoothing domains is performed based on the edges of the triangular background cells. For taking the coupling effect of solid and fluid phases into account, conventional Gauss points are considered inside the triangular background cells other than Gauss points on the boundaries of the smoothing domains that are normally used for constructing the strain-displacement matrix in SPIMs (See Figure 1). To secure non-singularity of the moment matrix, a Tr3 node selection scheme is employed in which three nodes of the cell hosting

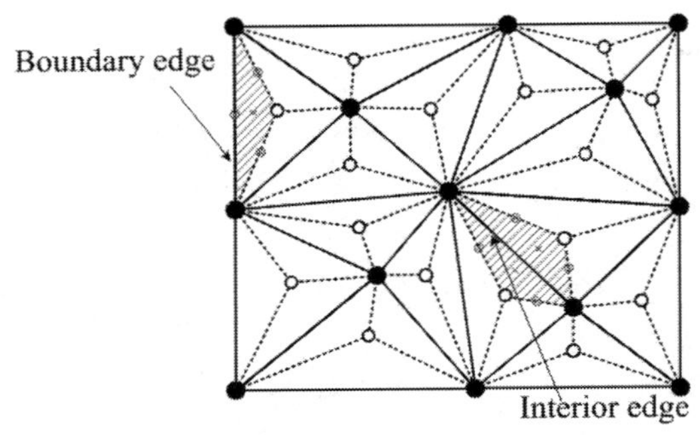

Figure 1. The schematic depiction of domain triangularisation, smoothing domains and Gauss points.

the Gauss point of interest are selected as the supporting nodes.

5 CONSTRUCTION OF SMOOTHED STRAIN FIELD

The compatibility condition is not necessarily satisfied in the global domain in using PIM shape functions (Liu and Gu, 2005). Therefore, the generalised smoothed Galerkin (GS-Galerkin) weak formulation, which works for both compatible and incompatible displacement fields is employed to discretise the system of equations in the current smoothed FEM model and utilising the Divergence theorem, the integration of the gradient of the field function over the smoothing domains is transformed into the integration of the field function itself over the boundary of the smoothing domains (Liu and Zhang, 2008). Hence, derivation of an incompatible displacement field resulting from PIM shape functions is avoided in the formulation. This leads to the definition of the smoothed strain over the smoothing domains. The smoothed strain can be defined for both compatible and incompatible displacement fields as follows

$$\hat{\varepsilon}_k = \frac{1}{A_k^{SD}} \int_{\Gamma_k^{SD}} \mathbf{L}_n \mathbf{u}(\mathbf{x}) d\Gamma \tag{4}$$

where $\widehat{\varepsilon}_k$ is the constant smoothed strain over the k th smoothing domain with the boundary Γ_k^{SD}, A_k^{SD} is the k th smoothing domain area, and $\mathbf{L}_{\mathrm{n}}$ is the matrix of the unit outward normal vector. The smoothed strain can be stated in terms of nodal displacements for each smoothing domain

$$\widehat{\varepsilon}_k = \sum_{i=1}^{q} \begin{bmatrix} \widehat{b}_{1_i} & 0 \\ 0 & \widehat{b}_{2_i} \\ \widehat{b}_{2_i} & \widehat{b}_{1_i} \end{bmatrix} \begin{Bmatrix} u_i \\ v_i \end{Bmatrix} = \widehat{\mathbf{B}}_1 \bar{\mathbf{u}} \tag{5}$$

where the strain displacement matrix $\widehat{\mathbf{B}}_1$ is

$$\widehat{\mathbf{B}}_1 = \begin{bmatrix} \widehat{b}_{1_1} & 0 & \widehat{b}_{2_1} & 0 & & \widehat{b}_{q_1} & 0 \\ 0 & \widehat{b}_{1_2} & 0 & \widehat{b}_{2_2} & \cdots & 0 & \widehat{b}_{q_2} \\ \widehat{b}_{1_2} & \widehat{b}_{1_1} & \widehat{b}_{2_2} & \widehat{b}_{2_1} & & \widehat{b}_{q_2} & \widehat{b}_{q_1} \end{bmatrix} \tag{6}$$

and the nodal displacement vector $\bar{\mathbf{u}}$ is

$$\bar{\mathbf{u}} = \begin{bmatrix} u_1 & v_2 & u_1 & v_2 & \cdots & u_q & v_q \end{bmatrix}^{\mathrm{T}} \tag{7}$$

u_i and v_i are components of nodal displacements in x and y directions. Now, using the Gaussian integration scheme it can be written that

$$\widehat{b}_{i_l} = \frac{1}{2A_k^{\mathrm{SD}}} \sum_{m=1}^{n_{\mathrm{seg}}} \left(L_m^k w_{\mathrm{G}}^m \varphi_i(\mathbf{x}_{\mathrm{G}}^m) n_l^m \right), \quad l = x, y \tag{8}$$

with q being the total number of supporting nodes of all the Gauss points on the boundaries of the k th smoothing domain. n_{seg} is the number of line segments of the boundary Γ_k^{SD}, n_l^m is the component of the unit outward normal vector to the mth segment of Γ_k^{SD} in x or y direction, $\mathbf{x}_{\mathrm{G}}^m$ is coordinate of the Gauss point on the mth segment of Γ_k^{SD}, L_m^k is the length of the mth segment of Γ_k^{SD}, and w_{G}^m is the corresponding Gauss integration weight. $\varphi_i(\mathbf{x}_{\mathrm{G}}^m)$ is the shape function value at note i at the point of interest $\mathbf{x}_{\mathrm{G}}^m$.

6 SPATIAL DISCRETISATION

The governing equations (1) and (2) are discretised adopting the GS-Galerkin method are as follows

$$\int_{\Omega_k^{\mathrm{SD}}} \widehat{\mathbf{B}}_1^{\mathrm{T}} \sigma' \, d\Omega + \eta \mathbf{Q} \mathbf{P} = \mathbf{F}_{\mathrm{u}} \tag{9}$$

$$\eta \mathbf{Q}^{\mathrm{T}} \dot{\mathbf{U}} - \mathbf{H}\mathbf{P} - a_{\mathrm{f}} \mathbf{S} \dot{\mathbf{P}} = \mathbf{F}_{\mathrm{p}} \tag{10}$$

where $\mathbf{U}$ is the vector of nodal displacements, $\mathbf{P}$ is the nodal pore fluid pressure vector, $\mathbf{F}_{\mathrm{u}}$ is the vector of nodal forces, $\mathbf{F}_{\mathrm{p}}$ is the vector of nodal fluxes, and $\mathbf{Q}$, $\mathbf{S}$ and $\mathbf{H}$ are the global property matrices of the system.

Adopting the smoothing operation, the global tangent stiffness matrix is obtained by assembling the local stiffness matrices as

$$\mathbf{K} = \sum_{k=1}^{n_{\mathrm{SD}}} \mathbf{K}_k^{\mathrm{SD}} = \sum_{k=1}^{n_{\mathrm{SD}}} \left(\int_{\Omega_k^{\mathrm{SD}}} \widehat{\mathbf{B}}_1^{\mathrm{T}} \mathbf{D}^{\mathrm{ep}} \widehat{\mathbf{B}}_1 d\Omega \right) = \sum_{k=1}^{n_{\mathrm{SD}}} A_k^{\mathrm{SD}} \widehat{\mathbf{B}}_1^{\mathrm{T}} \mathbf{D}^{\mathrm{e}} \widehat{\mathbf{B}}_1 \tag{11}$$

In the same manner, the global permeability matrix is evaluated as follows

$$\mathbf{H} = \sum_{k=1}^{n_{\mathrm{SD}}} \mathbf{H}_k^{\mathrm{SD}} = \sum_{k=1}^{n_{\mathrm{SD}}} \left(\int_{\Omega_k^{\mathrm{SD}}} \widehat{\mathbf{B}}_2^{\mathrm{T}} \frac{\mathbf{k}_{\mathrm{f}}}{\mu_{\mathrm{f}}} \widehat{\mathbf{B}}_2 d\Omega \right) = \sum_{k=1}^{n_{\mathrm{SD}}} \frac{A_k^{\mathrm{SD}}}{\mu_{\mathrm{f}}} \widehat{\mathbf{B}}_2^{\mathrm{T}} \mathbf{k}_{\mathrm{f}} \widehat{\mathbf{B}}_2 \tag{12}$$

where $\widehat{\mathbf{B}}_2$ is defined as

$$\widehat{\mathbf{B}}_2 = \begin{bmatrix} \widehat{b}_{1_1} & \widehat{b}_{1_2} & \cdots & \widehat{b}_{1_q} \\ \widehat{b}_{2_1} & \widehat{b}_{2_2} & & \widehat{b}_{2_q} \end{bmatrix} \tag{13}$$

To obtain the coupling matrix $\mathbf{Q}$, again the assembly process for all the smoothing domains along with the smoothing operation is required as follows

$$\mathbf{Q} = \sum_{k=1}^{n_{\mathrm{SD}}} \mathbf{Q}_k^{\mathrm{SD}} = \sum_{k=1}^{n_{\mathrm{SD}}} \left(\widehat{\mathbf{B}}_3^{\mathrm{T}} \int_{\Omega_k^{\mathrm{SD}}} \mathbf{\Phi}^{\mathrm{pq}} d\Omega \right) \tag{14}$$

in which $\partial^{\mathrm{T}} c$ is the shape function matrix for the pore fluid pressure at each point of interest, which includes all the supporting nodes of the Gauss points along the boundary of the smoothing domain of interest, and $\widehat{\mathbf{B}}_3 = \begin{bmatrix} \widehat{b}_{1_1} & \widehat{b}_{2_1} & \widehat{b}_{1_2} & \widehat{b}_{2_2} & \cdots & \widehat{b}_{1_q} & \widehat{b}_{2_q} \end{bmatrix}$.

The term $\int_{\Omega_k^{\mathrm{SD}}} \mathbf{\Phi}^{\mathrm{pq}} d\Omega$ in Eq. (14) has to be evaluated over each smoothing domain by dividing each of the interior smoothing domains into two triangles and using the standard Gauss integration method for the triangular areas, considering one Gauss point per triangle as illustrated in Figure 1. Therefore, the coupling matrix can be obtained as follows

$$\mathbf{Q} = \sum_{k=1}^{n_{\mathrm{SD}}} \left(2\widehat{\mathbf{B}}_3^{\mathrm{T}} \sum_{i=1}^{n_{\mathrm{tr}}} A_i^{tr} w_i \mathbf{\Phi}_i^{\mathrm{pq}} \right) \tag{15}$$

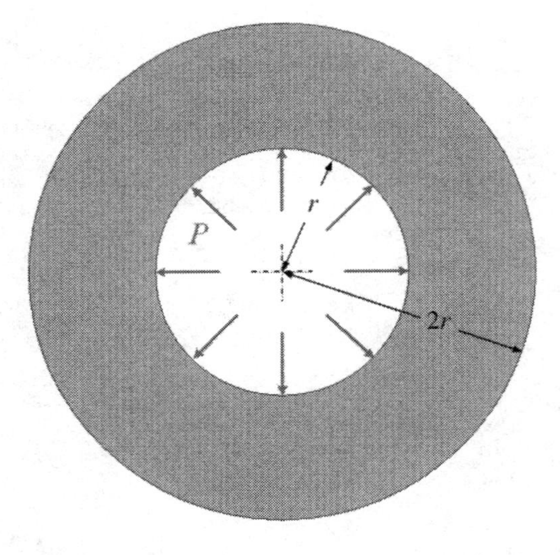

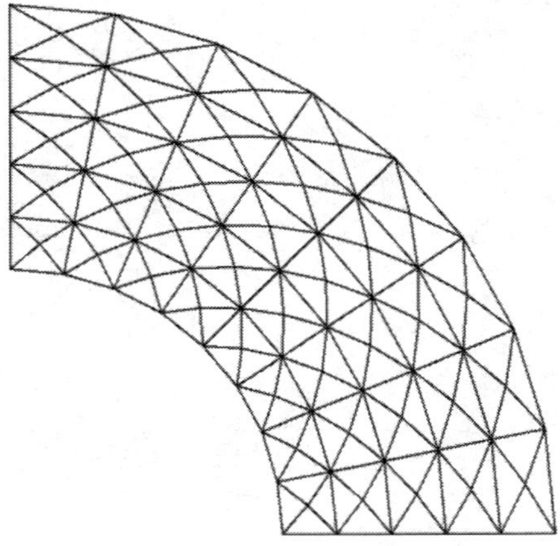

Figure 2. Schematic view of the problem domain and the assumed background mesh

where A_i^{tr} is the area of the triangle hosting the Gauss point of interest, and w_i is the weight corresponding to that Gauss point. Φ_i^{pq} is the fluid shape function matrix calculated at the Gauss point of interest. Note that n_{tr} equals 2 for interior smoothing domains and 1 for boundary smoothing domains.

The compressibility matrix can be obtained without using the smoothing operation as

$$\mathbf{S} = \sum_{k=1}^{n_{SD}} \int_{\Omega_k^{SD}} \mathbf{\Phi}^{pq\mathrm{T}} \mathbf{\Phi}^{pq} d\Omega \tag{16}$$

and finally the right hand side matrices in Equations. (9) and (10) are evaluated as

$$\mathbf{F}_{\mathrm{u}} = \sum_{k=1}^{n_{SD}} \left(\int_{\Omega_k^{SD}} \mathbf{\Phi}^{uq\mathrm{T}} \mathbf{F} d\Omega + \int_{\Gamma_k^{SD}} \mathbf{\Phi}^{uq\mathrm{T}} \mathbf{T} d\Gamma \right) \tag{17}$$

$$\mathbf{F}_{\mathrm{p}} = \sum_{k=1}^{n_{SD}} \int_{\Gamma_k^{SD}} \mathbf{\Phi}^{pq\mathrm{T}} q_{\mathrm{f}} d\Omega \tag{18}$$

where q_{f} is the imposed fluid flux across the boundaries, and $\mathbf{T}$ is the boundary traction.

7 TIME DISCRETISATION

Using a novel three-point time marching approach (Khoshghalb et al., 2011), equations (9) and (10) are discretised in time as follows

$$\int_{\Omega} \widehat{\mathbf{B}}_1^{\mathrm{T}} (\boldsymbol{\sigma}')^{t+\alpha\Delta t} + \eta \mathbf{Q} \mathbf{P}_{\mathrm{f}}^{t+\alpha\Delta t} = \mathbf{F}_{\mathrm{u}}^{t+\alpha\Delta t} \tag{19}$$

$$\begin{aligned} &\eta \mathbf{Q}^{\mathrm{T}} \left(A\mathbf{U}^{t+\alpha\Delta t} - B\mathbf{U}^{t} + C\mathbf{U}^{t-\Delta t} \right) - \Delta t \mathbf{H} \mathbf{P}_{\mathrm{f}}^{t+\alpha\Delta t} \\ &- a_{\mathrm{f}} \mathbf{S} \left(A\mathbf{P}_{\mathrm{f}}^{t+\alpha\Delta t} - B\mathbf{P}_{\mathrm{f}}^{t} + C\mathbf{P}_{\mathrm{f}}^{t-\Delta t} \right) = \Delta t \mathbf{F}_{\mathrm{p}}^{t+\alpha\Delta t} \end{aligned} \tag{20}$$

where α is the growth factor in the time discretisation scheme. The detailed time marching approach is available in (Khoshghalb et al., 2011).

8 NONLINEAR ALGORITHM

Equations. (19) and (20) are solved at each time step using the modified Newton-Raphson iterative method. To do so, the vectors of the nodal displacement and pore fluid pressure at iteration i of the current time step $t+\alpha\Delta t$ are improved at iteration $i+1$ as

$$\begin{Bmatrix} \mathbf{U}_{i+1}^{t+\alpha\Delta t} \\ \mathbf{P}_{i+1}^{t+\alpha\Delta t} \end{Bmatrix} = \begin{Bmatrix} \mathbf{U}_{i}^{t+\alpha\Delta t} \\ \mathbf{P}_{i}^{t+\alpha\Delta t} \end{Bmatrix} + \begin{Bmatrix} d\mathbf{U}_{i+1}^{t+\alpha\Delta t} \\ d\mathbf{P}_{i+1}^{t+\alpha\Delta t} \end{Bmatrix} \tag{21}$$

where the improvement vector is obtained using the Jacobian matrix $\mathbf{J}$ as follows

$$\begin{Bmatrix} d\mathbf{U}_{i+1}^{t+\alpha\Delta t} \\ d\mathbf{P}_{i+1}^{t+\alpha\Delta t} \end{Bmatrix} = -\mathbf{J}_i^{-1} \begin{Bmatrix} \left(\mathbf{\Psi}_{\mathrm{u}}\right)_i^{t+\alpha\Delta t} \\ \left(\mathbf{\Psi}_{\mathrm{p}}\right)_i^{t+\alpha\Delta t} \end{Bmatrix} \tag{22}$$

in which $\left(\mathbf{\Psi}_{\mathrm{u}}\right)_i^{t+\alpha\Delta t}$ and $\left(\mathbf{\Psi}_{\mathrm{p}}\right)_i^{t+\alpha\Delta t}$ are the residual vectors. More details regarding the nonlinear algorithm can be found in another research paper (Ghaffaripour et al., 2017).

9 NUMERICAL EXAMPLE

In order to verify the proposed model, internal pressurisation of a thick-walled cylinder is

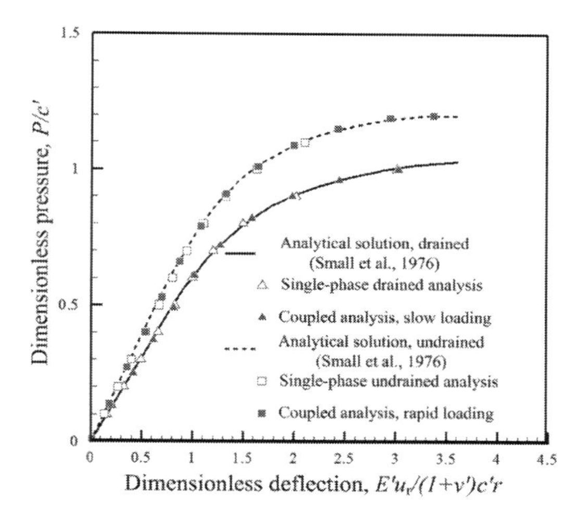

Figure 3. Dimensionless internal pressure versus dimensionless deflection of the inner radius.

considered in both drained and undrained conditions. Analytical solutions are available for this problem (Small et al., 1976). The pressure P is applied to the cylinder internally as illustrated in Figure 2. In this figure, also can be found the background mesh used in the numerical analyses. Only one quarter of the cylinder is simulated due to symmetrical nature of the problem.

The cylinder is assumed to initially be at a zero stress state and the drained and undrained properties of the solid skeleton for the adopted non-associative Mohr-Coulomb constitutive law are as follows: $E'/c' = 200$, $v' = 0$, $\phi' = 30^\circ$, $\psi' = 0$, $E_u = \dfrac{3E'}{2(1+v')}$, $v_u = 0.49$, $\varphi_u = 0$ and $c_u = \dfrac{2c'\sqrt{N_\varphi}}{1+N_\varphi}$, where $N_\varphi = \dfrac{1+\sin\phi'}{1-\sin\phi'}$.

The coupled behaviour of material is examined through two types of analyses. First, a single-phase analysis is carried out considering the drained or undrained properties of the material. Second, two extreme loading conditions are used along with the two-phase analyses: a fast loading rate to simulate the undrained behaviour of the material, and a slow loading rate to account for the drained response. More details regarding the loading rates can be found in Ghaffaripour et al. (2017).

The numerical results from the mesh free model for both single-phase and coupled analyses are compared with the analytical solutions in Figure 3. In this figure the vertical axis illustrates the dimensionless pressure P/c', while the horizontal axis represents the dimensionless deflection of the cylinder inner radius $\frac{E'u_r}{(1+v')c'r}$, where u_r is the deflection of the inner radius. From the figure, it is visible that in both drained and undrained analyses, the numerical results of the proposed method show perfectly match the analytical solutions.

10 SUMMARY

In this work, the application of the edge-based smoothed point interpolation method to coupled hydromechanical problems in porous media was evaluated. The capability of the formulation was examined through the internal pressurization of a thick cylinder. From the comparison between the numerical results of the proposed method and the analytical solutions, it can be concluded that, for flow-deformation problems in elastic-perfectly plastic materials, the proposed method offers very accurate solutions.

REFERENCES

Belytschko, T., Liu, W. K. & Moran, B. 2000. Finite elements for nonlinear continua and structures. *J. Wiley*.

Ghaffaripour, O., Khoshghalb, A. & Khalili, N. 2017. An edge-based smoothed point interpolation method for elasto-plastic coupled hydro-mechanical analysis of saturated porous media. *Computers and Geotechnics*, 82(99–109.

Khoshghalb, A. & Khalili, N. 2010. A stable meshfree method for fully coupled flow-deformation analysis of saturated porous media. *Computers and Geotechnics*, 37(6), pp 789–795.

Khoshghalb, A. & Khalili, N. 2013. A meshfree method for fully coupled analysis of flow and deformation in unsaturated porous media. *International Journal for Numerical and Analytical Methods in Geomechanics*, 37(7), pp 716–743.

Khoshghalb, A. & Khalili, N. 2015. An alternative approach for quasi-static large deformation analysis of saturated porous media using meshfree method. *International Journal for Numerical and Analytical Methods in Geomechanics*.

Khoshghalb, A., Khalili, N. & Selvadurai, A. 2011. A three-point time discretization technique for parabolic partial differential equations. *International Journal for Numerical and Analytical Methods in Geomechanics*, 35(3), pp 406–418.

Lewis, R. & Schrefler, B. 1999. The finite element method in the static and dynamic deformation and consolidation of porous media. *Meccanica*, 34(3), pp 231–232.

Liu, G.-R. & Gu, Y.-T. 2005. An introduction to meshfree methods and their programming: Springer.

Liu, G.-R. & Gu, Y. 2001a. A point interpolation method for two-dimensional solids. *International Journal for Numerical Methods in Engineering*, 50(4), pp 937–951.

Liu, G.-R. & Zhang, G.-Y. 2013. Smoothed point interpolation methods: G space theory and weakened weak forms: World Scientific.

Liu, G. & Gu, Y. 2001b. A local radial point interpolation method (LRPIM) for free vibration analyses of 2-D solids. *Journal of Sound and vibration,* 246(1), pp 29–46.
Liu, G. & Zhang, G. 2008. Edge-based smoothed point interpolation methods. *International Journal of Computational Methods,* 5(04), pp 621–646.
Sloan, S. W. 1987. Substepping schemes for the numerical integration of elastoplastic stress–strain relations. *International Journal for Numerical Methods in Engineering,* 24(5), pp 893–911.
Small, J. C., Booker, J. & Davis, E. H. 1976. Elasto-plastic consolidation of soil. *International Journal of Solids and Structures,* 12(6), pp 431–448.
Smith, I., Griffiths, D. & Margetts, L. 2013. *Programming the Finite Element Method*: John Wiley & Sons.
Wang, J., Liu, G. & Wu, Y. 2001. A point interpolation method for simulating dissipation process of consolidation. *Computer Methods in Applied Mechanics and Engineering,* 190(45), pp 5907–5922.

Numerical Methods in Geotechnical Engineering IX – Cardoso et al. (Eds)
© 2018 Taylor & Francis Group, London, ISBN 978-1-138-33198-3

Soil-structure interaction in coupled models

Belén Martínez-Bacas & Davor Simic
Geotechnical Division, Ferrovial Agromán S.A.

Marta Pérez-Escacho & Carlos J. Bajo-Pavía
Bridge and Civil Structures Division, Ferrovial Agromán S.A.

ABSTRACT: Nowadays, in order to improve interaction some new applications have been recently released. Those are able to connect the geotechnical software to the structural software, sharing loads and movements through an interface tool. The main value of this tool is to analyze the interaction soil-structure on the fly.

This paper presents this coupling methodology analyzing the foundation of a 52 m diameter and 50 m height storage structure with shallow foundation over underlying stiff clayey sand layers. The importance of the decision making about the mechanical behavior of soils has been highlighted when comparing final displacement of the structure using different soil models: non-linear plastic Hardening soils, linear elastic-plastic Mohr-Coulomb and the traditional linear elastic springs. This fact has in turn influence about the efficiency of the solution as well as assessing a model as close to the actual situation as possible where the use of springs has resulted to be conservative.

1 BACKGROUND AND CURRENT SITUATION OF DESIGN PROCESSES

In the last decades, the technical advances in materials, procedures and computerized support have oriented the engineering world towards specialization.

The current methods available for treating the soil-structure interaction, usually present the following characteristics:

- Separate work teams, geotechnics vs structure, where the ultimate goal in both cases is the parameterization and isolated response of their elements to the respective actions.
- Independent calculation processes, where the collaboration of the other team is required in an iterative process, and the interaction between soil and structure is observed as an imposed condition: load, settlement, and support.
- Processes that require a high degree of coordination, where the team management is basic to avoid extra costs in materials and/or time, due to the influence of transitional parameters.

With the objective to perform an optimum global design there are some companies with multidisciplinary working teams whose increased synergy provides solutions with a better cost/benefit ratio.

The XXI century offers a wide range of growing powerful technologies and applications. New tools, that allow for increasingly complex models, in capacity and speed, emerge and each working area integrates them achieving an improvement of their competence in the project.

Furthermore, the convergence between the technological progress and the way of working with global teams, has resulted in the integration of the technological improvement into the coordinated formats.

The BIM methodology is transcending naturally, generating platforms where there is a single combined solution and all teams regardless of their mission and physical location intervene in the process. The results increase its solidity, and the possibility of failure is minimized, with which more and more administrations demand this type of global 3D-4D-5D models. However, optimization of the process is not always accomplished, and sometimes the results are relegated to the non-interference between elements.

2 AIM OF THE ANALYSIS

The Construction Technical Office from Ferrovial Company considers that being at the forefront of the ongoing improvement processes, where the new technological opportunities are integrated with its own experience and way of working as a team, is a competitive advantage in all stages of a project: tender, design, execution and maintenance.

On this regard, the possibility of applying the BIM methodology is considered not only as a sharing platform, but also a tool to obtain optimal integrated solutions. Thus, the standard connections of the BIM platforms should be improved.

The current technologies allow for a more complex approach to the reality, empowering high-performance work-stations, both in individual and network capacity.

Habitually, the interaction between soil and structural responses are obtained using tools and auxiliary parameters looking for the best compatibility. This methodology generates uncertainties that often force to oversize the solutions, or leave the structures in a medium controlled risk situation.

The exploration of new processes and computational tools for soil-structure interaction have been advised as strategic in the current scenario and therefore they are the subject of the present analysis.

The scope of this paper is to analyze and review the influence in the decision making of a new compatibility tool generated to connect a FEM software for geotechnical design and a well-known structural software. These computer software are commonly used by the areas of Geotechnics and Structures, but thanks to the new tool, both software are forced to work together in an integrated way looking for the optimum solution and avoiding the requirement of processing any auxiliary correlation element.

The methodology of the soil-structure interaction that is being applied in each case is not always unique, and depends on the type of soil, the structural element involved, the geographical location of the project, etc.

On this regard, a preliminary differentiation characterized by the function of the element present in the interaction should be considered.

- Foundation of buildings and bridges
- Walls, Slurry walls and Piled walls:
- Tunnels and shafts

Thus, the soil response should be studied to choose a suitable mechanical model that simulates the real situation at service and construction stage in order to withstand the actions that the structure requests at each time. Moreover, the structural forces depend directly on the conditions and relationships of stiffness of the ground for both, short and long term, and due to this issue the forces are conditioned by the differential settlements and the horizontal movements.

This paper is focused on the first group, foundation of buildings and bridges.

In this cases, the interaction between the structural behavior and the supporting soil has been driven through the geotechnical calculation of the modulus of subgrade reaction, whose mission is the characterization of the soil in the structural model by linear-elastic springs. However, this parameter is always an approximation of the soil response, and it is provided with a large range of potential variation by means of a sensibility analysis. On the other hand, the subgrade reaction cannot represent the non-linear elastoplastic soil behavior.

As stated above, to overcome these problems, a coupling program has been developed to connect the geotechnical software with the structural one. This is an innovating tool programmed to carry out on the fly iterative analysis to resolve coupled substructure and superstructure. Thus, the soil can be modelled with the mechanical behavior that better adapts to the field situation. So it is possible to check the effect of the soil model on the superstructure forces and obtain a suitable solution without any geotechnical simplification.

This tool requires an iterative coupled analysis between the geotechnical and structural models for a particular load case and a particular construction stage.

A full example is developed in the following sections.

3 COUPLING METHODOLOGY FOR A SHALLOW FOUNDATION PROJECT

3.1 *Project definition*

This paper presents the analyses of the behavior of a large storage structure with shallow foundation over underlying stiff to very stiff clayey sand layers.

3.2 *Coupling tool steps*

The first step is to develop both, the structural model, considering the suitable restrains on the base of the structure (Figure 2) and the geotechnical model, including soil layers, base slab, interfaces, etc. (Figure 3)

On the process, the tool creates onto the foundation plate the connection point loads in the same position where the structural restrains are located. The coupling tool connects automatically both software, being its interface the foundation slab.

The equilibrium equations are solved by this tool, knowing the boundary conditions and actions of the system. The displacements calculated by the geotechnical software are transferred into the structural one for each connected node. Next, the structural program evaluates the reactions due to the imposed deflections, and updates them into the geotechnical model in an iterative process.

The iterations stop when the difference between displacements of the connected nodes is less than a given target value. For this paper 1 mm has been considered.

Figure 1. Storage structure to be designed.

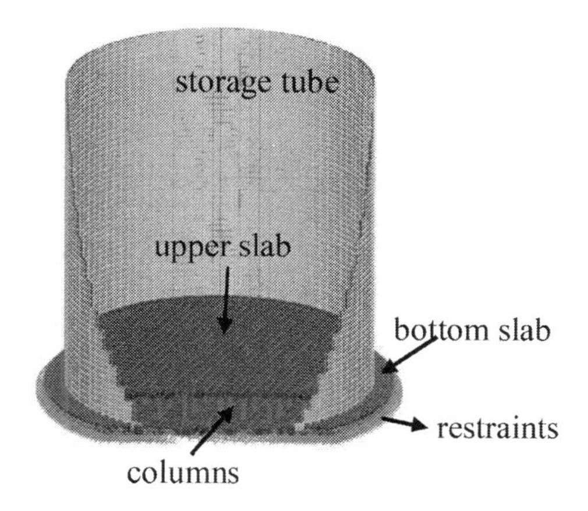

Figure 2. Structural model developed in Sap2000.

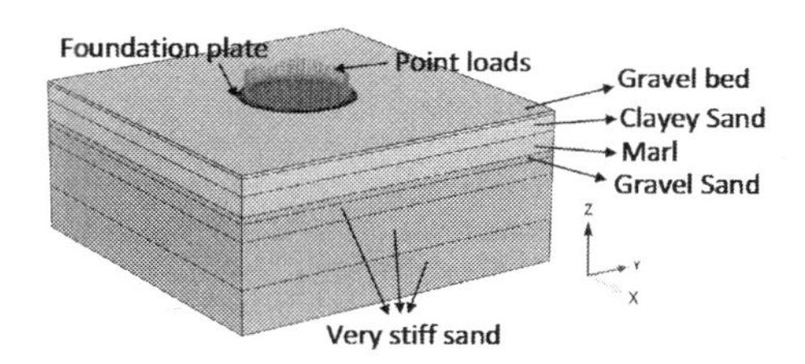

Figure 3. Geotechnical model developed in Plaxis3D.

4 SHALLOW FOUNDATION DATA

4.1 *Structural model*

The structure consists of a cylindrical deposit, with a diameter of 52 m and a height of 50 m. This structure can be separated in two main parts: the

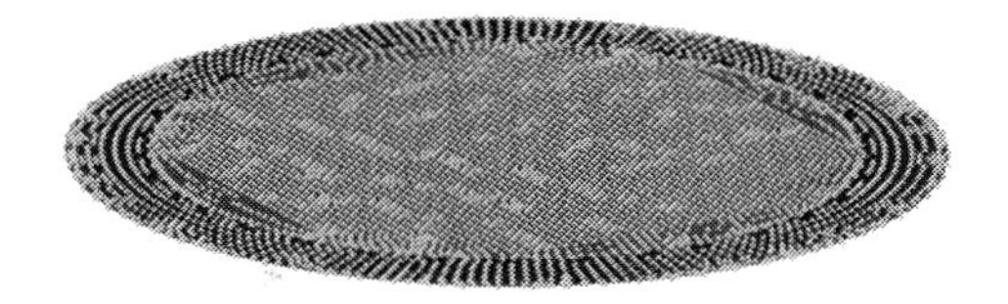

Figure 4. Geotechnical foundation model (plate+point loads).

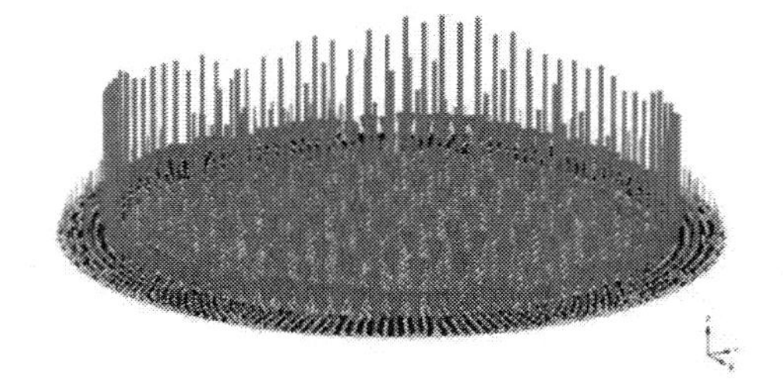

Figure 5. Loads transferred by structural program.

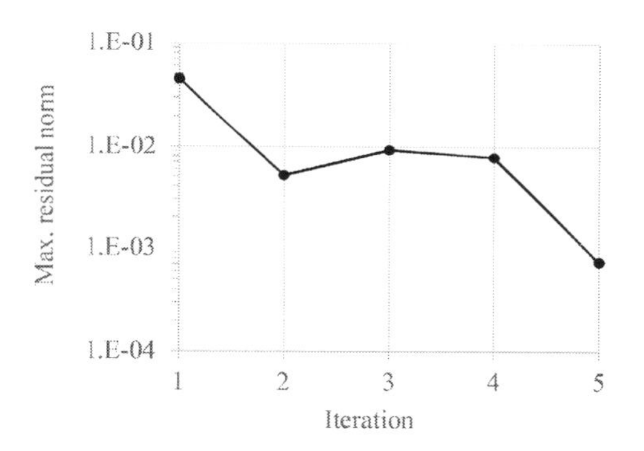

Figure 6. Residual norm versus number of calculus iterations.

storage body and the foundation. The tank walls are 0.42 m thick, while the foundation is formed by two separated thick slabs connected by columns. The upper slab has 1 m thickness, it supports directly the storing loads. The bottom slab has 1.1 m thickness and it is in direct contact with the soil. Finally, the columns between slabs are 4 m height, 1 m diameter and are separated 4 m (Figure 2). This system of columns makes both slabs work as a high-depth footing.

The simulation of the structure has been carried out in Sap2000 by means of shell elements simulating walls and slabs. Beam elements have been used to perform the connecting columns.

Regarding soil connection, a set of vertically restrained supports has been assigned to the bottom slab. They are responsible for transferring the structural information, reactions, to the geotechnical foundation model and also for receiving the imposed settlements. In this case, it has been observed that the more transfer points, the better for the compatibility.

The loads considered into the structural model, apart from the imposed deflections coming from Plaxis, are the following: self-weight of the structure, entirely made in concrete, 25 kN/m^3; permanent loads, 4.5 kN/m^2; vertical surcharge due to the weight of the stored material, acting as a distributed load on the top slab, 301.4 kN/m^2; vertical frictional load produced on the interface between the stored material and the vertical walls, acting as a linear load distributed along the wall perimeter, 2341.7 kN/m.

In this particular case, the bottom slab was decided to be shared in both models, fully connected, so as to ensure that the imposed deflections coming from the geotechnical model are properly understood by the structural model and reversed. That means the stiffness of the foundation member composed by both slabs and the connecting set of columns is much higher than the ground response, and therefore, the results highly depend on the accuracy of the load distribution but scarcely on variations of the foundation stiffness.

4.2 *Soil parameters*

The storage building is founded on four soil layers very overconsolidated, as the pressurometer tests in the area showed, as well as the gravels composed predominantly of quarzt. The soil layers named from top to bottom (figure 3):

Gravel bed: 1.40 m thickness
Clayey sand: 9 m
Marl: 9 m
Gravel-sand:3 m
Very stiff sand: 79 m

In the present paper it was analyzed the influence of the non-linear behavior of the soils over the structure forces. Therefore, two soil mechanical models have been chosen: Mohr Coulomb and Hardening Soil. The parameters are presented in Table 1 and Table 2 respectively.

The Mohr-Coulomb model (MC) is the well-known linear elastic perfectly plastic model. This model uses a constant stiffness for every soil layer.

The Hardening Soil model (HS) is an advanced model with non-linear elastoplastic behavior that allows for shear hardening plasticity as well as compression hardening to simulate compaction of soil under primary load. This model is suitable for both sand and gravel as well as clays and silts.

The calculations for different soil models were performed under drained conditions with the water level at 5 m depth.

In addition, Figure 7 presents the modulus of subgrade reaction (k), calculated in Plaxis using Mohr-Coulomb model for all soil layers (Table 1) and a plate foundation equal to the bottom slab of the tank structure (Table 3). This auxiliary parameter, k, is usually used instead the coupling tool.

Table 1. Mohr-Coulomb parameters.

Parameters	Gravel Bed	Clayey sand	Marl	Gravel-sand	Very stiff sand
γ_{sat} (kN/m^3)	21	20	21	21	21
E'(MPa)	100	103	280	144	347
ν	0.33	0.3	0.3	0.3	0.3
k_0	1	3.7	1.23	1.13	0.47
φ'(°)	–	35	25	32	32
c' (kPa)	0	10	15	5	10

Table 2. Hardening soil parameters.

Parameters	Clayey sand	Marl	Gravel-sand	Very stiff sand
E_{50}^{ref}(MPa)	103	280	144	347
E_{oed}^{ref} MPa)	103	280	144	347
E_{ur}^{ref}(MPa)	309	840	432	1041
ν_{ur}	0.2	0.2	0.2	0.2
m	0.05	0.05	0.05	0.05
φ'(°)	35	25	32	32
c' (kPa)	10	15	5	10
p^{ref} (kPa)	100	150	195	100
k_0^{NC}	0.43	0.57	0.47	0.47
k_0	3.7	1.23	1.13	0.47
OCR	10	3	4	1

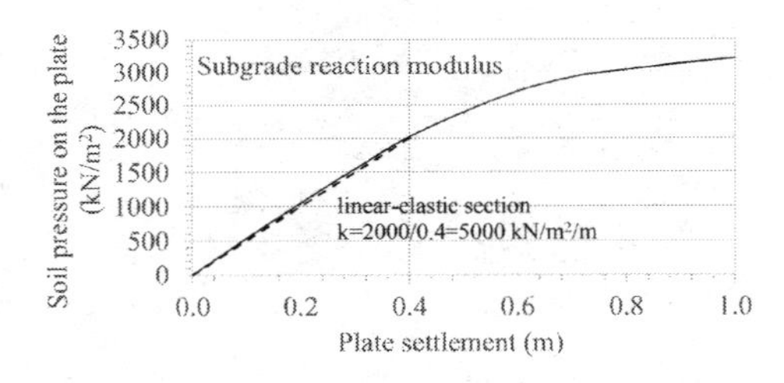

Figure 7. Subgrade reaction modulus for Mohr-Coulomb model.

4.3 *Geotechnical model*

The geotechnical model consists of soil layers, plate foundation, interface between plate and soil and point loads on the plate. The dimensions are 160 m long x160 m width x70 m depth. The number of zones is 92630. The size of the mesh is medium. The plate properties are shown in Table 3.

4.4 *Iterative process*

Figure 8 shows the structural joint reactions versus the geotechnical point loads in a cross section plate between columns and a cross section plate along the central column line for the Hardening Soil model. These figures shows a good correlation considering only 5 iterations with 1 mm of maximum residual norm (Figure 6). However, it is detected that this target value applied under the columns, as a single imposed deflection in a Sap2000 model, is indicative that more iterations are required. Therefore, a more accurate target is recommended.

5 SHALLOW FOUNDATION RESULTS

5.1 *Settlements of the slab foundation*

The main results provided by the coupling analysis at the end of the iterative process settlements, soil pressure and bending moment on bottom slab are summarized in the graphics bellow.

Figure 9 and 10 show that Plaxis3D and Sap2000 provide the same deflection results, meaning that it depends only on the soil model.

On this regard, Mohr Coulomb models show higher settlements than the Hardening Soil models. However, the largest settlement is coming from

Table 3. Plate parameters.

	E (Mpa)	v	Diameter (m)	Thickness (m)
Plate	32000	0.2	52	1.1

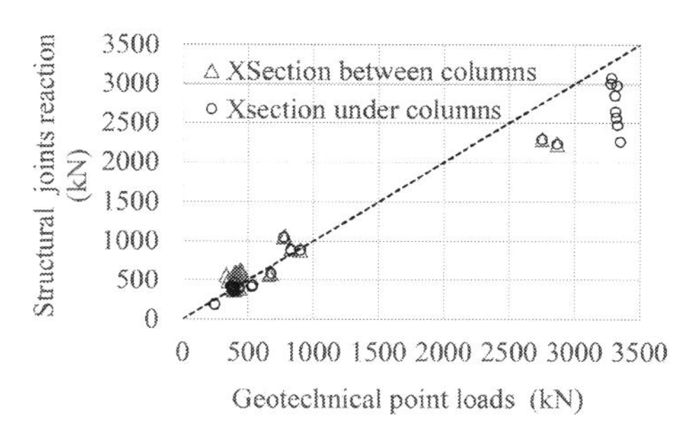

Figure 8. Structural joint reactions-geotechnical point loads.

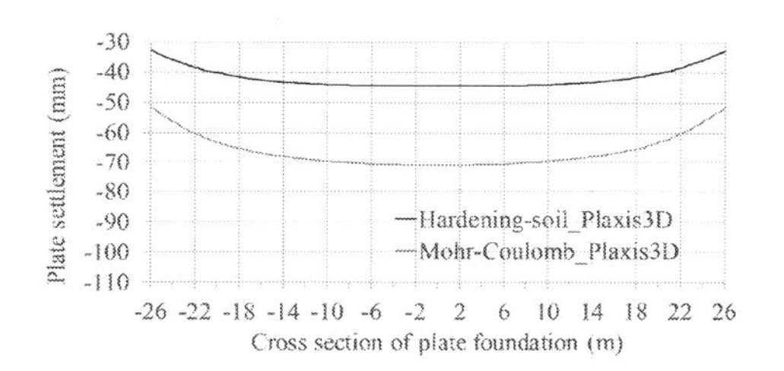

Figure 9. Settlement of the slab foundation on Plaxis3D.

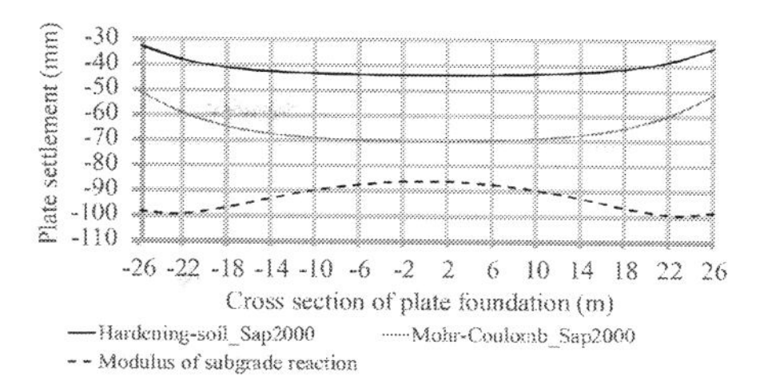

Figure 10. Settlement of the slab foundation on Sap2000.

a model with linear-elastic springs defined by the modulus of subgrade reaction, *k,* in Figure 7. This auxiliary parameter was introduced as a constant and uniform value, so *k* does not consider neither the geometry of the problem nor pressure. However, the Hardening soil model considers both the geometry and the pressure, becoming stiffer when the soil received more pressure. On the other hand, Mohr-Coulomb model considers only the geometry. These behaviors can be also observed in the soil pressure on the plate shown in Figure 10.

5.2 *Soil pressure on the plate*

Figure 10 and Figure 11 pointed out some important issues:

First the calculus with Sap2000 with a constant and uniform subgrade modulus implied not to take into account the geometry of both soil and structure. Thus, the relation between the displacement and the pressure is lineal. However, when the Hardening and the Mohr-Coulomb models have been applied, the geometry is considered and the soil surrounding the foundation contributes opening its principal stress isolines and increasing the shear stress, as Figure 12 illustrates.

Additionally, the stiffness of Hardening soil model depends on the stress state of the soil, so the area that receives more pressure is going to hardening inducing less settlement and more soil pressure, as Figure 11 shown. The edges of the slab foundations received larger load due to the frictional load than the middle (see section 4.1).

Next, in projects where the foundations are as presented in this paper, the differences between the real behavior of the soil and the uniform subgrade modulus can be compensated using subgrades modulus variables along the plate. Figure 9 and Figure 11 provided the displacement and pressure for the Hardening soil models, being possible to calibrate the subgrade modulus for each plate zones.

From the structural point of view to carry out the analysis with all load combinations and with linear superposition it is necessary the substitution of the real non-linear behavior of the soil by a suitable subgrade modulus calibrated with a coupling tool.

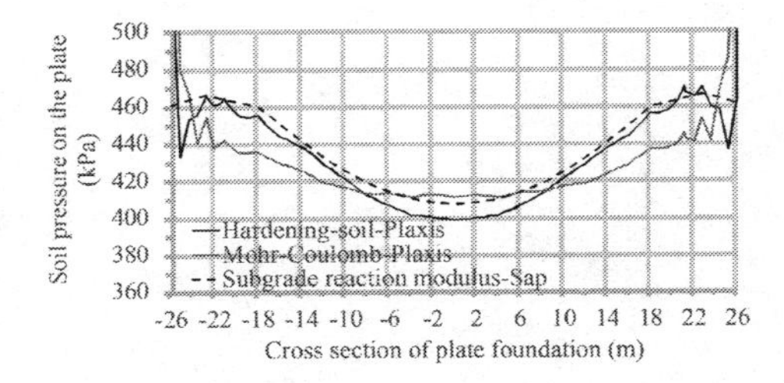

Figure 11. Soil pressures on the plate.

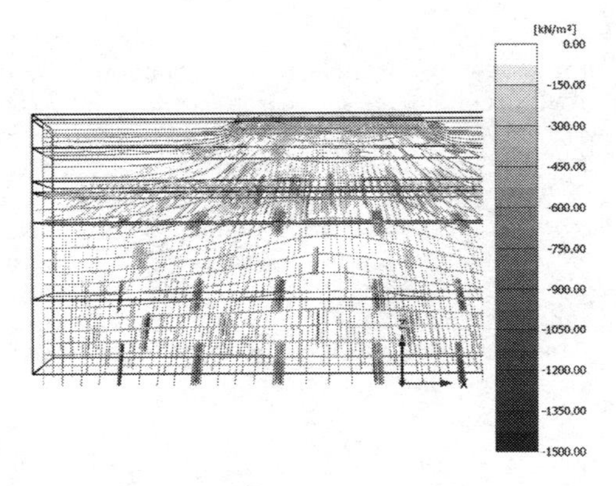

Figure 12. Principal stress directions ($\sigma 1$) and isolines of Hardening Soil model.

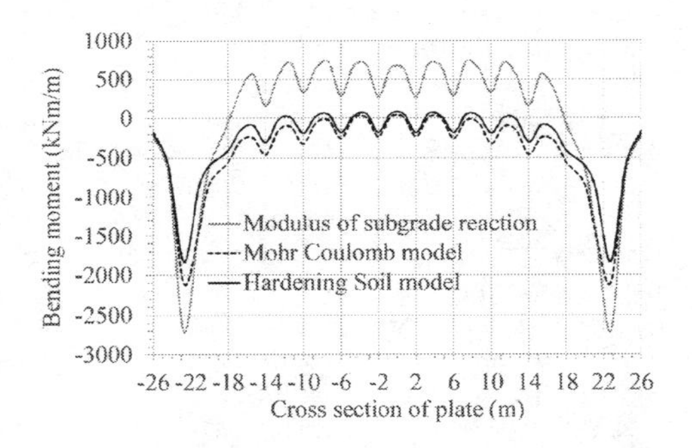

Figure 13. Bending moments in Sap2000 models.

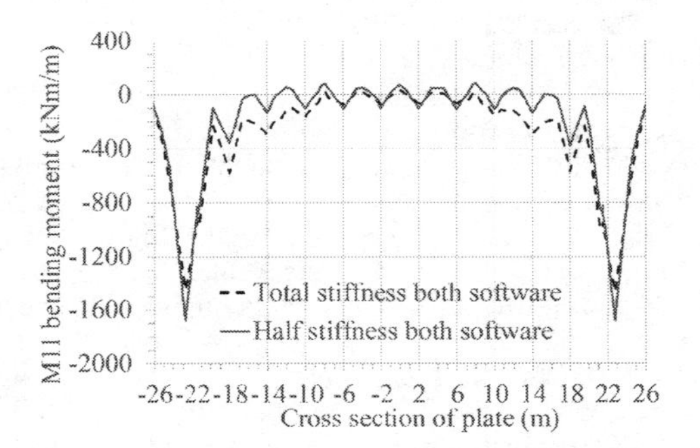

Figure 14. Bending moments in Plaxis3D models cross section plate between columns.

Finally once all load combinations are done with the structural software, the extreme or particular load cases can be verified by means of an explicit analysis with the real soil behavior using a coupling tool.

5.3 *Bending moments of the plate foundation*

Figure 13 compares the bending moment obtained in Sap2000 between different soil models analyzed with the coupling tool and linear-elastic calculation using the subgrade modulus (k). The bending moments were considered in a plate band between columns. The model with subgrade modulus presents the higher values. The successive peaks along the cross section belong to the influence area of the columns.

Next, in order to analyze the influence of the bottom slab stiffness, two cases have been calculated for the Hardening Soil model, one with total stiffness in both plates (E = 32 MPa), one plate is in Plaxis3D and one plate in Sap2000 model, which are coupled. The second case was calculated with half stiffness in both slabs (E = 16 MPa). Figure 14 shows the results from Plaxis output of the bending moments in a cross section plate between columns. Similar results are observed. This happens because the stiffness of the total foundation member, composed by bottom and upper slabs and the connecting set of columns, is much higher than the ground response.

6 CONCLUSIONS

This paper presents the analyses of the behavior of a 52 m in diameter and 50 m in height storage structure with shallow foundation over underlying stiff clayey sand layers, using two soil models: non-linear plastic Hardening soils vs linear elastic-plastic Mohr-Coulomb and the subgrade reaction modulus. On this regard, the process of applying a coupling tool has been introduced to connect the soil software to the structural sharing loads and movements.

The following observations can be indicated:

Regarding the particular case presented, it was developed a methodology more approximated to the real situation, taken into account non-linear models, without losing velocity and flexibility comparing with traditional elastic-linear calculus with load combinations.

Generally, the structural design is carried out separately, applying the subgrade reaction methods obtained by empirical methods. The structural software are very qualified for the structures but very limited for geotechnical problems and vice versa regarding to geotechnical software. The coupling tool between software allows to take advantage of the best of each one.

Reliability and probability analysis

Numerical Methods in Geotechnical Engineering IX – Cardoso et al. (Eds)
© 2018 Taylor & Francis Group, London, ISBN 978-1-138-33198-3

Impact of considering oriented rock variability on tunnel excavation

D. Ferreira
Rendel Limited, London, UK
Formerly University of Coimbra, Coimbra, Portugal

A.M.G. Pedro
ISISE, University of Coimbra, Coimbra, Portugal

P.A.L.F. Coelho & J. Almeida e Sousa
University of Coimbra, Coimbra, Portugal

D.M.G. Taborda
Imperial College London, London, UK

ABSTRACT: Soils and rocks are by nature heterogeneous materials whose properties are strongly dependent on the existent variability. In the case of rock massifs the influence of such variability is usually higher and more evident due to the tectonic movements that are associated with their formation. These processes often originate a preferential direction of the rock mass variability, which frequently does not correspond to the traditional horizontal layering observed in soils. In this paper, the impact of the orientation of rock variability is assessed by analysing the excavation of a circular tunnel in a rock mass characterised by a uniform and isotropic stress state. The variability is introduced in the analyses by considering multiple random fields, generated using the Local Average Subdivision (LAS) method. The impact of the orientation of variability is evaluated both in terms of displacements around the excavation and forces in the tunnel lining. The results show that both displacements and forces are affected by the orientation of the variability and that such effect depends on the parameters employed in the generation of the random fields, particularly on the ratio of anisotropy considered for the rock mass.

1 INTRODUCTION

Soils and rocks are materials that due to its formation usually present heterogeneous properties in space and in magnitude (Zhang, 2013). As a result these materials are often difficult to fully characterise and the adoption of a single deterministic value for a given mechanical property can cause severe consequences leading to damage and malfunctioning of structures (Breysse et al., 2005).

More recently the traditional deterministic approaches usually employed in the design of geotechnical structures have begun to be replaced by probabilistic analysis, where the variability of the soils and rocks can be taken into account by using methods, such as the Covariance Matrix Decomposition (*CMD*) (Le Ravalec et al., 2000), the Turning Bands Method (*TBM*) (Jones et al., 2002) or, probably the most employed in practice, the Local Average Subdivision (*LAS*) (Fenton & Griffiths, 2008; Vanmarcke, 2010). With these methods, it is possible to generate scalar multidimensional random fields (*RFs*) of a given property which can be integrated into a finite element program for numerical analysis. The ability of these methods to reproduce accurately the real soil and rock conditions has already been demonstrated successfully in different types of geotechnical applications, including slope stability (Hicks & Samy, 2002; Cho, 2007), foundations (Griffiths et al., 2002; Jimenez & Sitar, 2009), retaining walls (Fenton et al., 2005; Sert et al., 2016), tunnels (Cai, 2011; Pedro et al., 2017), among others.

However, these methods commonly rely on the fact that the horizontal and vertical directions are the principal directions of the field. This limitation can create a problem and limit the ability to generate *RFs*, particularly in rock masses, where, due to the tectonic movements, the variability is usually oriented in a preferential direction, often not coincident with the typical horizontal layering observed in soils, and concentrated around faults and bends.

In order to overcome the aforementioned issue, the *LAS* method was extended so that *RFs* having a user defined orientation, α, could be generated. The impact of this new feature when modelling geotechnical structures was assessed through a series of analyses performed in a well-defined application: the excavation of a circular tunnel in a rock mass. In order to isolate the effect of the introduction of oriented rock variability, symmetrical geometry and boundary conditions were adopted in the model, and a uniform isotropic initial stress state was assumed. The influence of the input parameters used in the generation of the *RF*s, namely the rotation of the field, the ratio of anisotropy, the standard deviation and the spatial correlation distance on the tunnel lining displacements and forces were evaluated through an extensive parametric study.

2 NUMERICAL MODELLING OF VARIABILITY

2.1 *Local Average Subdivision (LAS) method*

The *LAS* method, developed by Fenton & Vanmarcke (1990), relies on the assumption that most properties in engineering practice can be represented within the analysis domain by their average value. Based on this principle it is possible to generate multi-dimensional *RFs,* where a local average value of the property is determined while keeping the statistics initially defined consistent with the resolution desired for the field. Despite its complex formulation the *LAS* method only requires a reduced number of parameters, all with physical meaning (Fenton & Griffiths, 2008), to generate a *RF.* Another advantage of the method is the possibility of being coupled with the finite element method making it suitable for employment in numerical analysis.

A comprehensive explanation of the method and of its advantages and limitations can be found in Fenton & Griffiths (2008) and consequently only a brief description of its main features are described here. The method can be understood as having a top-down structure. The algorithm starts with the definition by the user of a suitable distribution for the variable property within the initial field. Usually, a Gaussian distribution, defined by its average value (μ) and standard deviation (σ), is adopted, although other types of statistical distribution can be selected (Fenton & Griffiths, 2008; Ferreira, 2016). In the next steps the initial field is progressively subdivided into 2^{in} subfields, with i being the number of dimensions (e.g. 2 for 2D) and n the increasing number of levels of the problem. These new subfields are characterised by an individual value that is conditioned to the initial average value of the adopted distribution but affected by a random noise. Within each level all values must have a correct standard deviation according to averaging theory and are also correlated through a covariance matrix obtained from a spatial correlation function. This function is a key point of the method since it establishes the existing relationship between the values assumed for a given property in two separate subfields.

The 2D spatial correlation function employed in the present work is the Gauss-Markov (Griffiths & Fenton, 2008; Zhang, 2013) defined by Equation 1, where θ_x and θ_y are the spatial correlation distances, $\tau_{ij,x}$ and $\tau_{ij,y}$ are the distances between the centres of the subfields i and j in the x and y directions, respectively, and σ is the standard deviation of the initial distribution.

$$\rho(\tau_{ij}) = \sigma^2 \cdot exp\left[-2\left(\frac{|\tau_{ijx}|}{\theta_x}+\frac{|\tau_{ijy}|}{\theta_y}\right)\right] \quad (1)$$

If the spatial correlation distances in both directions (θ_x and θ_y) assume the same value, then the correlation function and consequently the generated field is isotropic. However, if θ_x is different from θ_y an anisotropic *RF* can be generated. Depending on the adopted ratio of anisotropy (θ_x/θ_y), it is possible to generate almost vertically and horizontally layered *RFs* for small and high ratios of anisotropy, respectively.

2.2 *Implementation of the oriented variability*

Similar to other methods, *LAS* was developed for orthogonal fields having the horizontal (x) and vertical (y) directions as principal directions of the field and of the correlation function. However, it is possible to introduce oriented variability through the rotation of the main directions of correlation of the *RF* while keeping its boundaries vertical and horizontal. In the methodology proposed the rotation is introduced by an angle, α, measured between the principal directions of the field (x' and y') and the original coordinate system of the problem (x and y), being positive as defined in Figure 1(a) (clockwise starting from horizontal). In order to accommodate the rotation, an auxiliary field with adequate dimensions and having as main directions the rotated axis (x', y') is initially

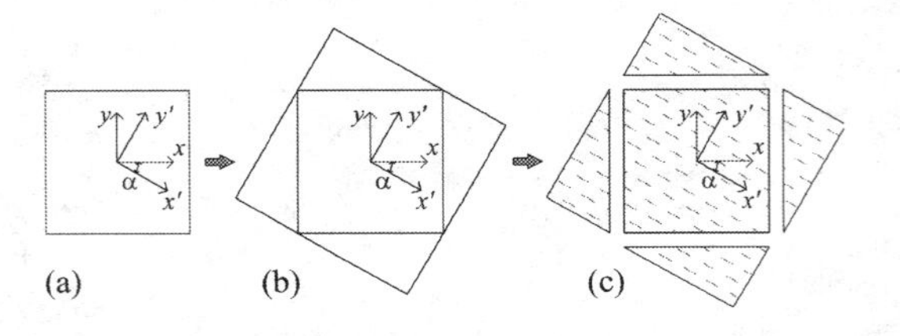

Figure 1. Rotated *RF* generation algorithm scheme.

created (Figure 1(b)). Based on these directions, and assuming that the correlation directions of the *RF* remain orthogonal, the *RF* is generated and then cropped to the size of the original field as displayed in Figure 1(c).

It should be noted that the rotation of the *RF* results in some loss of resolution of the field, though the correlation distances, average and standard deviation values remain unaltered. The loss of resolution occurs due to the cropping needed to reduce the generated *RF* to its original size. As previously mentioned, each level, n, of a 2D field generated by *LAS* has 2^{2n} subfields. As can be understood by a simple comparison of areas in Figure 1 the original field will only have $1/(\sin(2\alpha)+1)$ subfields of the auxiliary field. That implies that the number of subfields in the original field will vary between a minimum of half when $\alpha = 45°$ and the same number of subfields as the auxiliary field when α is equal to 0° or 90°. Consequently, it is recommended that at least one more level of discretization should be used for rotated *RFs* in order to have a number of subfields similar to that of a conventional *RF*.

Figure 2 presents an example of 4 *RFs*, generated by the developed software, UC2DRF, assuming a Normal distribution with $\mu = 100$, $\sigma = 10$, $\theta_x = 32$, $\theta_x/\theta_y = 8$ but with values for the rotation, α, of 0°, 90°, 45° and -45°. The figure clearly illustrates that the proposed methodology is capable of introducing oriented variability in the *RFs* allowing for a more adequate simulation of soil and rock massifs that were subjected to tectonic movements.

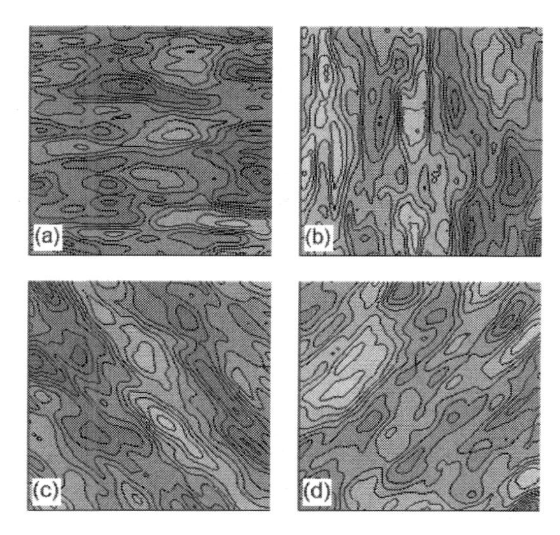

Figure 2. Examples of *RFs* generated for $\mu = 100$, $\sigma = 10$, $\theta_x = 32$, $\theta_x/\theta_y = 8$: (a) α=0°; (b) α=90°; (c) α=45°; (d) α=–45°.

3 NUMERICAL MODEL

3.1 *General considerations*

In order to evaluate the impact of considering oriented variability, a study of the excavation of a circular tunnel in a rock massif subjected to a uniform initial stress state was performed using the in-house finite element program UCGeoCode coupled with the *RF* module, UC2DRF. The 2D mesh used is presented in Figure 3 and consists of a 120 m by 120 m square, having at its centre an 8 m diameter tunnel. A total of 812 solid elements, each with 8 nodal points and 4 Gauss points for displacement and stress evaluation, respectively, were used to simulate the rock mass. The lining was simulated using 48 solid elements with 0.20 m thickness. The boundary conditions were set so that no displacements were possible at all boundaries of the mesh.

The stress relaxation method (Potts & Zdravković, 2001) was employed in the simulation of the tunnel excavation. In this method the 3D effects usually associated with tunnel construction, such as stress redistribution through arching and time delay with which the lining is installed (Moller, 2006), are simulated in two separate stages. In the first stage the solid elements of the tunnel are removed but only a percentage of the resulting stresses (ω) is applied in the contour of the excavation. In the second stage the lining is installed, and the remainder of the unbalanced stresses is applied ($1-\omega$) so that a final equilibrium state is achieved. Considering the dimensions of the tunnel, the stress state and the characteristics of the rock mass a value of 80% was adopted for the stress relaxation factor, ω. It should be noted that for simplicity of the interpretation of the results the presence of a water table was disregarded.

The generalised Hoek-Brown failure criterion (Hoek et al., 2002) was adopted for the rock mass (Equation 2).

$$\sigma_1' = \sigma_3' + \sigma_{ci} \cdot \left(m_b \cdot \frac{\sigma_3'}{\sigma_{ci}} + s \right)^a \tag{2}$$

The parameters of the model and of the stiffness of the rock mass were obtained using Equations 3 to 6, where GSI is the Geological Strength Index (Hoek et al., 1998); m_i is the constant of the intact rock; σ_{ci} and E_i are the compression strength and Young's modulus of the intact rock, respectively, and D is the disturbance factor. The values assumed for each parameter, together with the Poisson's ratio, ν, are presented in Table 1 and were derived by Pedro (2007) for a real rock mass. For the tunnel lining, a linear elastic behaviour with a

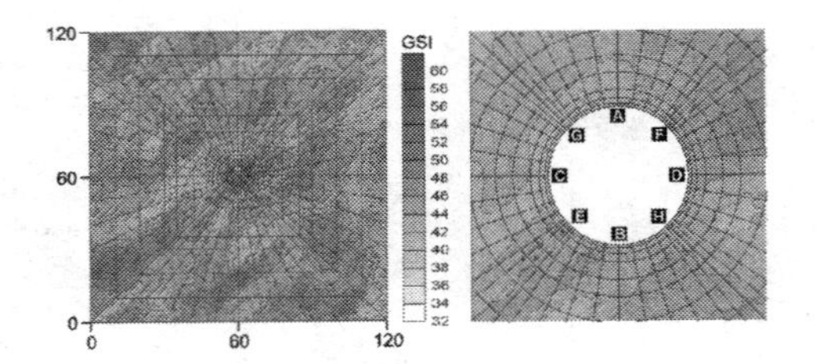

Figure 3. Finite element mesh and detail of the tunnel excavated.

Table 1. Rock mass parameters.

GSI	E_i (MPa)	σ_{ci} (MPa)	v	m_i	D
46	8100	27	0.25	9	0

Table 2. Deterministic analysis results.

Radial disp. (mm)	Radial convergence (mm)	Hoop force (kN)
2.5	5.0	338

Young's modulus of 20 GPa and a Poisson's ratio of 0.25 was adopted.

$$m_b = m_i \cdot e^{\left(\frac{GSI-100}{28-14D}\right)} \tag{3}$$

$$s = e^{\left(\frac{GSI-100}{9-3D}\right)} \tag{4}$$

$$a = \frac{1}{2} + \frac{1}{6} \cdot \left(e^{GSI/15} - e^{20/3} \right) \tag{5}$$

$$E_m = E_i \cdot \left(0.02 + \frac{1 - D/2}{1 + e^{\left(\frac{60+15D-GSI}{11}\right)}} \right) \tag{6}$$

3.2 *Deterministic analysis*

So that the influence of variability could be asserted, a reference deterministic analysis using the parameters defined in Table 1 was performed. As expected, given the symmetry of the problem and of the initial stress conditions, the calculation results in uniform radial displacements and convergences in the soil. Consequently, the forces and pressures acting in the lining are also uniform and a zero bending moment was determined. The results obtained in the deterministic analysis are summarised in Table 2.

4 RANDOM FIELD GENERATION

The variability in the *RFs* was introduced through the variation of the *GSI*, as this parameter is directly correlated with both strength and stiffness of the rock mass as can be seen from Equations 3 to 6. A Normal statistical distribution, as suggested by Cai (2011), with a mean value of 46 was adopted in all analyses. The remainder parameters of the constitutive model were also kept constant and with the values as defined in Table 1 in all analyses. The impact of oriented variability was assessed by using different values for the rotation, α, of the field: 0°, which corresponds to the analyses without rotation and 45° and -45°, which correspond to intermediate values.

The influence of the each of the other parameters required for the generation of the *RFs* was assessed by performing individual parametric studies. In the first of these, the ratio of anisotropy, θ_x/θ_y, was varied, assuming the values of 1/8, 1/4, 1/2, 1, 2, 4 and 8. In this set of analyses the standard deviation was kept constant with a value of 7 while the reference spatial correlation distance was assumed to be equal to the tunnel radius, 4 m. This implies that in all analysis with a θ_x/θ_y smaller or equal than 1 a θ_x of 4 m was adopted while for ratios higher than 1 it was the value of θ_y that was considered to be equal to 4 m.

In the second study, the influence of the standard deviation of the *GSI* was assessed by using values of 5, 7 and 9. In this study the ratios of anisotropy and the spatial correlation distances used were equal to those described in the first case.

Finally, in the last study, the influence of the spatial correlation distance was evaluated by adopting reference values of 4, 8 and 12 m. Moreover, in this study, the same ratios of anisotropy as in the first study were used, while a single standard deviation of 7 was employed in all analyses.

The introduction of variability in the numerical analysis is based on random processes implying that multiple calculations have to be performed to establish the pattern of the behaviour of the problem. Based on their experience in different applications, Fenton & Griffiths (2008) recommend a minimum of 500 realisations for each variable parameter. In the present work, a total of 1000 realisations were performed for every case. The amount of output data generated in each case cannot be treated as in a single deterministic analysis, requiring the use of statistical analysis to interpret and present the results. Given the nature and the objective of the problem, the interpretation focuses on the results of the hoop force acting on the lining and on the convergences (positive values correspond to movements towards the excavation), determined in the vertical (A-B), horizontal (C-D) and 45° diagonal (E-F and G-H) directions, as illustrated in Figure 3.

The 1000 values obtained for each case were then processed and fitted to a Normal and Lognormal

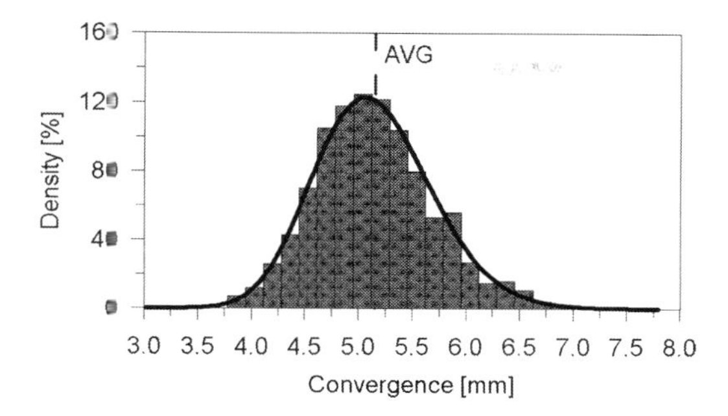

Figure 4. Example of a Lognormal distribution fit to the vertical convergence (A-B) results of an ordinary case study.

statistical distribution defined by two parameters: average value (AVG) and Coefficient of Variation (*COV*), which is the ratio between the standard deviation (*STD*) and the average value (AVG) of the distribution. Based on the quality of the fit determined for each result a Lognormal distribution was adopted for the convergences while for the forces in the lining a Normal distribution was selected.

Figure 4 presents an example of the application of the fitting process of a Lognormal distribution to the results of the vertical convergence (A-B) for one of the cases performed. Globally, it was observed that 1000 realisations were sufficient in all cases to adjust the parameters of both fitted distributions with high degree of confidence.

5 IMPACT OF ORIENTED VARIABILITY

5.1 *Influence of the ratio of anisotropy*

The effect of the ratio of anisotropy on the convergences of the tunnel is depicted in Figure 5. For clarity, the fitting parameters of the convergences determined in the vertical (A-B) and horizontal (C-D) directions are plotted in (a) and (c) and the results obtained in the two 45° diagonals (E-F and G-H) in (b) and (d). From the figure, it is possible to observe that the ratio of anisotropy has a similar influence on the *AVG*, regardless of the direction of the convergence. In all cases an increase of approximately 3% is observed in comparison with the deterministic value (dashed line). This difference can be explained by the plasticity observed in the analyses that have a smaller *GSI*, which tend to present higher displacements than those with high *GSI*, which are still within the elastic domain. This effect, that would not be observed if a linear elastic model had been employed, is only observed in terms of the magnitude of the convergences since the number of analyses with values above and below the deterministic value is statistically the same.

The results of the *COV* are at first sight more complex to interpret. The first aspect observed is that for all directions of the convergence the smallest *COV*, with about 11% of change, is obtained for the isotropic condition, i.e. $\theta_x/\theta_y = 1$. As expected, the rotation of the field has no particular influence in isotropic conditions, since the correlation function for these conditions remains circular ($\theta_x=\theta_y$) regardless of the direction of the correlation considered.

However, for anisotropic conditions, an increase in *COV* is observed, which can reach values of around 17% for the extreme cases considered (θ_x/θ_y = 1/8 and 8). This variation appears to be strongly dependent of the relation between the ratio of anisotropy, the rotation of the field and the direction of the convergence. For the vertical convergence (A-B) it is possible to verify that the highest *COV* is observed when the field is not rotated ($\alpha = 0°$) and when the ratio of anisotropy is 1/8, scenario that corresponds to having an almost vertical layered rock mass. A similar result occurs for the horizontal convergence (C-D) when the field is also not rotated but the ratio of anisotropy is such that the *RF* is almost horizontally layered ($\theta_x/\theta_y = 8$). For all other ratios of anisotropy considered, it is possible to observe some variation in the vertical and horizontal convergence regardless of the rotation of the field. The results obtained for the convergences determined in the 45° diagonals follow a similar pattern, i.e. the maximum *COV* in the G-H direction occurs for a rotation of the field of 45° with an anisotropy ratio of 8 and for a rotation of -45° with an anisotropy ratio of 1/8. As expected, in the other diagonal (E-F) the inverse is observed showing that the highest variability occurs when the rotation of the field has the same orientation given by the extreme case of the ratio of anisotropy. It is also interesting to note that the *COV* of the convergence in the direction of the rotation of the field appears to stabilize for ratios of anisotropy that produce an almost perpendicular layering.

Figure 6 depicts the variation of the parameters (*AVG* and *COV*) of the Normal distributions fitted to the hoop force acting along the lining for the cases where field rotations of 0° (Figure 6(a) and 6(c)), -45° and 45° (Figure 6(b) and 6(d)) were considered. The results are shown for three different ratios of anisotropy, 1/4, 1 and 4 and, similar to the convergences, it is possible to verify that anisotropic conditions (i.e. $\theta_x/\theta_y \neq 1$) result in higher variations (*COV*), which can reach up to 13% for the cases considered. The average values (*AVG*) of the hoop force in Figure 6 (a) and (b) show that this parameter is not very sensitive to the ratio of anisotropy and to the rotation of field with an almost uniform value obtained along the entire perimeter of the lining. Once again, the average value is about 3% higher than that obtained in the deterministic analysis, which, as mentioned before, can be justified by the plasticity observed in the analyses with smaller *GSI*.

The interpretation of the *COV* in Figure 6 (c) and (d) show that the isotropic *RF* ($\theta_x/\theta_y = 1$) has a slightly undulating pattern, higher in the rotated fields, with small peaks located at every 45° and averaging a value of about 9.5%. The COV for anisotropic conditions shows a similar behaviour but with much higher fluctuations, with the localisation of the peaks of the hoop force being dependent on the rotation of the field and on the ratio of anisotropy. When no field rotation is considered (α = 0°), ratios of anisotropy above 1 originate higher hoop forces located in the vertical direction (90° and 270°) and smaller in the horizontal direction (0° and 180°), while the opposite is observed for ratios below 1. A similar pattern is observed for the other rotations of the field considered, with the difference being the location of the maximum and minimum peaks of the hoop forces. In the case of α = 45° for ratios of anisotropy above 1 the maximum values occur at 135° and 315° and the minimum at 45° and 225°, while for ratios below one the locations are shifted by 90°, i.e. maximum at 45° and 225° and minimum at 135° and 315°. For α = -45° the exact opposite occurs. Based on these results it can be concluded that for ratios of anisotropy above 1 the minimum variation of the hoop force occurs for directions aligned with that of the preferential orientation of the field, while the maximum variation occurs for directions perpendicular to these. For ratios of anisotropy below 1 the opposite is verified.

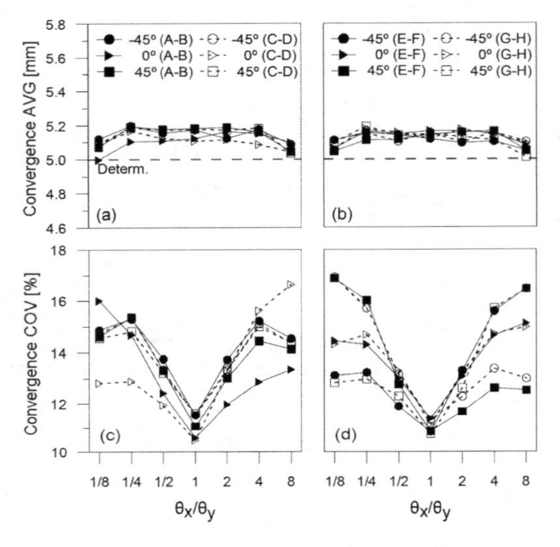

Figure 5. Effect of ratio of anisotropy in the convergences of the tunnel: (a) AVG A-B and C-D; (b) AVG E-F and G-H; (c) COV A-B and C-D; (d) COV E-F and G-H.

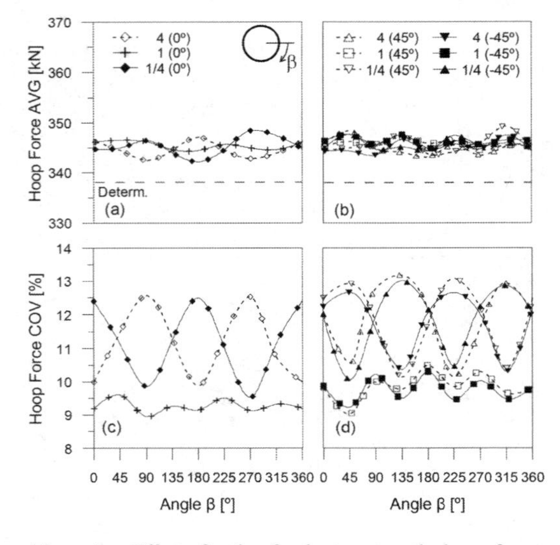

Figure 6. Effect of ratio of anisotropy on the hoop force: (a) AVG for α = 0°; (b) AVG for α = -45° and α = 45°; (c) COV for α = 0°; (d) COV for α = -45° and α = 45°.

5.2 *Influence of the standard deviation*

The influence of the standard deviation (σ) of the input variable, *GSI*, on the convergences and hoop forces acting on the lining is presented in Figures 7 and 8, respectively. Figure 7 presents the fitting parameters for the A-B and G-H convergences for different anisotropy ratios, two rotations of the *RF*, 0° and -45°, and for three values of the *GSI* standard deviation: 5, 7 and 9. The results clearly show that an increase of the standard deviation results in an increase of the variability, with both AVG and COV increasing, regardless of the anisotropy ratio, rotation of the *RF* or direction of the convergence considered. The observed increase in AVG values is justified by the increasing number of analyses where plasticity was achieved due to the smaller values of *GSI* generated. This contributes to a more uneven distribution of the values of convergence. However, it is possible to observe that for a standard deviation of 5 the AVG determined in all cases are very close to the deterministic value, which is a direct consequence of the *RF* behaving more 'elastically'. Although there are some differences, it appears that, for all directions of convergence considered, the rotation of the field has a limited effect on the value of AVG.

In terms of the COV, a similar behaviour can be observed. The increase of the standard deviation of the GSI originates an increase of the COV which appears to be independent of the rotation of the field. Naturally, and as mentioned before, the maximum variations are observed for anisotropic conditions and when the direction of the convergence and the preferential orientation of the field are aligned. Variations in the value of the convergence of more than 20% for a standard deviation of 9 are observed in such conditions. Even for a smaller standard deviation of 5 and isotropic conditions a minimum variation of about 8% in the convergence value is achieved.

The fitting parameters of the hoop force acting along the line for the isotropic condition, i.e. $\theta x/\theta y = 1$, and for an anisotropy ratio of 1/4 are depicted in Figure 8 for *RF* rotations of 0° and -45°. Similar to the observed response for the convergences, the results show that an increase of the standard deviation of GSI translates into higher variability both in the AVG and COV. In the case of isotropic conditions of the *RF* both AVG and COV values are almost constant along the lining presenting only some small fluctuations. In contrast, for anisotropic conditions it is possible to observe that both AVG and, in particular, COV values present considerable fluctuations that tend to increase with the increase of the standard deviation of the *GSI*. A variation in the hoop forces at the locations that are perpendicular with the direction of the rotation of the field of up to 15% is observed for the maximum value of the standard deviation considered. Another important aspect to highlight is the asymmetry of the hoop forces, which tend to increase with the standard deviation and that can result in additional problems in the lining design.

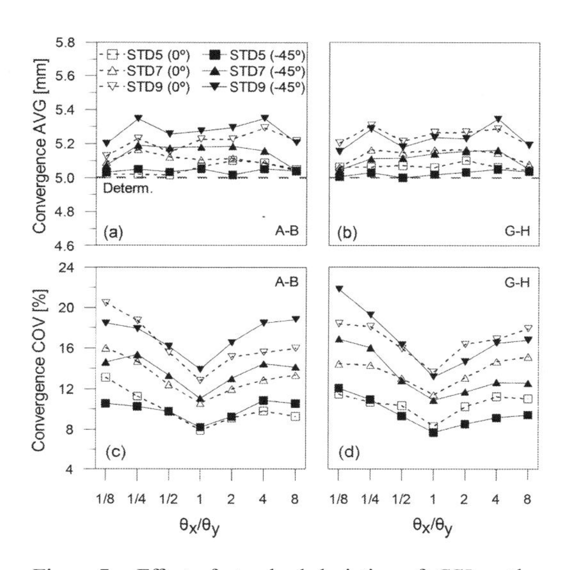

Figure 7. Effect of standard deviation of *GSI* on the convergences: (a) A-B AVG; (b) G-H AVG; (c) A-B COV; (d) G-H COV.

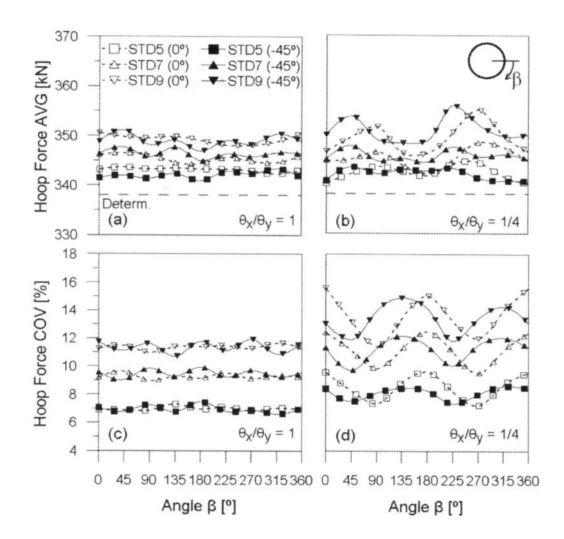

Figure 8. Effect of *GSI* standard deviation on the hoop force: (a) AVG for $\theta_x/\theta_y = 1$; (b) AVG for $\theta_x/\theta_y = 1/4$; (c); COV for $\theta_x/\theta_y = 1$; (d) COV for $\theta_x/\theta_y = 1/4$.

5.3 *Influence of the spatial correlation distance*

The influence of the spatial correlation distance was evaluated by adopting values for this quantity of 4, 8 and 12. Figure 9 presents the fitting parameters for the A-B and G-H convergences for different anisotropy ratios and two field rotations. The overall behaviour is very similar to that obtained for the variation of the standard deviation, with the variability increasing for higher values of the spatial correlation distance. This occurs since, with the increase of the spatial correlation distance, the areas in the rock mass that are correlated also increase, meaning that large areas of the rock mass with poor quality are more likely to occur. This effect can be particularly relevant depending on the geometry of the mesh. A spatial correlation distance of 4 m corresponds to only the tunnel radius, meaning that the probability of the area affected by the excavation of the tunnel being entirely located in a poor quality rock mass is small. With the increase of the value to 12 m that scenario becomes probable leading to a much higher COV that can reach up to 24%. Moreover, this study demonstrates that the AVG value remains largely unaffected by the spatial correlation distance and by the rotation of the field. In contrast, the COV

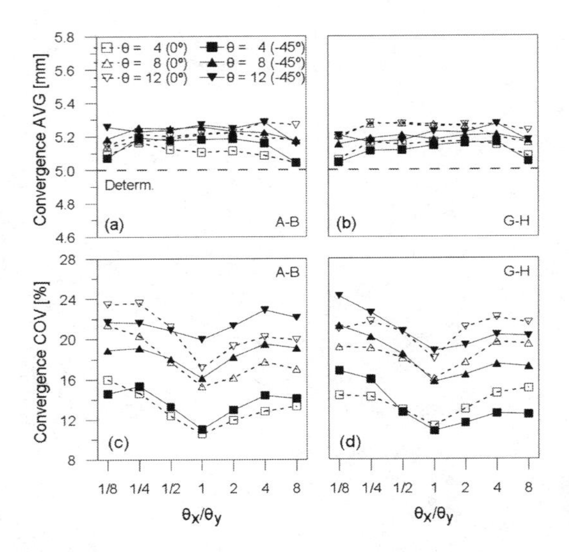

Figure 9. Effect of the spatial correlation distance on the convergences: (a) A-B AVG; (b) G-H AVG; (c) A-B COV; (d) G-H COV.

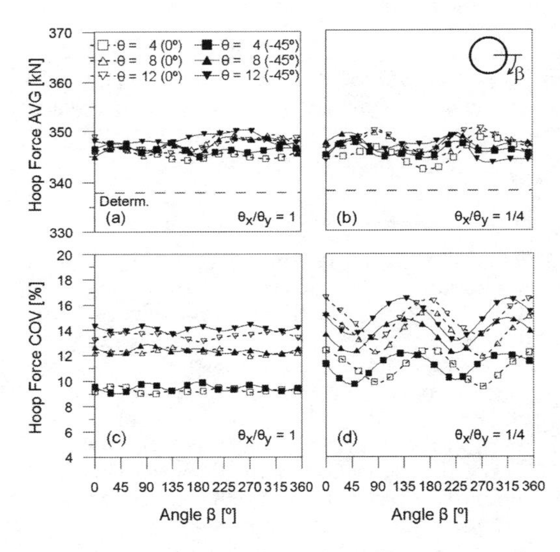

Figure 10. Effect of spatial correlation distance on the hoop forces: (a) AVG for $\theta_x/\theta_y = 1$; (b) AVG for $\theta_x/\theta_y = 1/4$; (c) COV for $\theta_x/\theta_y = 1$; (d) COV for $\theta_x/\theta_y = 1/4$.

values depend on the alignment between the preferential orientation of the field and the direction of the convergence, with the highest values observed when they are aligned and the ratios of anisotropy originate a *RF* layered in the same direction.

The hoop forces acting along the lining are also affected by the spatial correlation distance value, as can be seen in Figure 10. Although the AVG values remain approximately constant throughout the entire perimeter of the lining for all spatial correlation distances considered, the COV shows a reasonable variation with this parameter. Also in this study the COV values present almost no fluctuation for isotropic conditions, regardless of the rotation of the field. In anisotropic conditions it is observed, like in the previous studies, that the fluctuations of the COV values increase significantly and that the peaks occur at locations of the lining that are dependent of the rotation considered for the *RF*, with minimum values corresponding to that direction and maximum values to its perpendicular.

6 CONCLUSIONS

The demand for performing reliability analysis has led to the development of new numerical methods capable of incorporating aspects such as the existing variability in soils and rock massifs in the analyses. However, these methods tend to consider the variability as being isotropic or oriented in the directions of the field. In this paper an extension of the LAS method is proposed so that the direction of the variability can be controlled when generating *RFs*. The algorithm suggested is capable of simulating oriented variability in any direction, which is particularly useful when modelling rock massifs that were subjected to tectonic movements and that present variability that is not coincident with the typical horizontal layering observed on soils.

The impact of considering oriented variability was assessed through the modelling of the excavation of a circular tunnel in a rock mass. Multiple random fields were generated using the LAS method and the variability was introduced by assuming a Gaussian distribution of the GSI parameter. Three parametric studies were conducted in order to evaluate the influence of the parameters that control the generation of the *RFs*. Based on the obtained results the following conclusions can be drawn:

- With the introduction of variability it was possible to establish a reliable distribution of convergences and hoop forces in the lining and detect different behaviours that cannot be predicted by performing a simple deterministic analysis.
- The impact of considering oriented variability is mainly visible when anisotropic conditions are considered for the field. Its effect, both in convergences and displacements is maximum in the locations of the lining where the direction of the rotation of the RS is aligned with the layering of the rock mass induced by the ratio of anisotropy.

– As expected, an increase of the standard deviation of the GSI and of the spatial correlation distance originates higher variations in the convergences and hoop forces in the lining, which can reach an increase of up to 20% for the cases analysed.

The study performed demonstrates that using oriented variability in combination with other parameters allows for a more accurate simulation of the behaviour of rock massifs, while providing reliable information for performing a reliability-based design of geotechnical structures.

REFERENCES

Breysse, D., Niandou, H., Elachachi, S. & Houy, L. (2005) A generic approach to soil-structure interaction considering the effects of soil heterogeneity. *Geotechnique*, **55** (2), pp. 143–150.

Cai, M. (2011) Rock Mass Characterization and Rock Property Variability Considerations for Tunnel and Cavern Design. *Rock Mechanics and Rock Engineering*, **44** (4), pp. 379–399.

Cho, S. E. (2007) Effects of spatial variability of soil properties on slope stability. *Engineering Geology*, **92** pp. 97–109.

Fenton, G. & Griffiths, D. V. (2008) *Risk assessment in geotechnical engineering.* John Wiley & Sons. pp. 467.

Fenton, G. A. & Vanmarcke, E. H. (1990) Simulation of random fields via local average subdivision. *Journal of Engineering Mechanics-Asce*, **116** (8), pp. 1733–1749.

Fenton, G. A., Griffiths, D. V. & Williams, M. B. (2005) Reliability of traditional retaining wall design. *Geotechnique*, **55** (1), pp. 55–62.

Ferreira, D. (2016) *Modelling the Influence of Soil Variability on Geotechnical Structures.* MSc thesis. University of Coimbra, Coimbra.

Griffiths, D., Fenton, G. A. & Manoharan, N. (2002) Bearing capacity of rough rigid strip footing on cohesive soil: probabilistic study. *Journal of Geotechnical and Geoenvironmental Engineering*, **128** (9), pp. 743–755.

Griffiths, D. V. & Fenton, G. (2008) *Probabilistic methods in geotechnical engineering.* CISM courses and lectures. Springer. pp. 346.

Hicks, M. A. & Samy, K. (2002) Influence of heterogeneity on undrained clay slope stability. *Quarterly Journal of Engineering Geology and Hydrogeology*, **35** pp. 41–49.

Hoek, E., Marinos, P. & Benissi, M. (1998) Applicability of the Geological Strength Index (GSI) classification for very weak and sheared rock masses. The case of the Athens Schist Formation. *Bulletin of Engineering Geology and the Environment*, **57** (2), pp. 151–160.

Hoek, E., Carranza-Torres, C. & Corkum, B. (2002) Hoek-Brown failure criterion-2002 edition. *Proceedings of NARMS-Tac*, **1** pp. 267–273.

Jimenez, R. & Sitar, N. (2009) The importance of distribution types on finite element analyses of foundation settlement. *Computers and Geotechnics*, **36** (3), pp. 474–483.

Jones, A. L., Kramer, S. L. & Arduino, P. (2002) *Estimation of uncertainty in geotechnical properties for performance-based earthquake engineering.* Pacific Earthquake Engineering Research Center, College of Engineering, University of California.

Le Ravalec, M., Noetinger, B. & Hu, L. Y. (2000) The FFT moving average (FFT-MA) generator: An efficient numerical method for generating and conditioning Gaussian simulations. *Mathematical Geology*, **32** (6), pp. 701–723.

Moller, S. (2006) *Tunnel induced settlements and structural forces in linings.* PhD thesis. University of Stuttgard, Stuttgard.

Pedro, A. (2007) *Mato Forte tunnel—Back analysis of its Behaviour.* MSc Thesis. University of Coimbra, Coimbra, Portugal (in Portuguese).

Pedro, A. M. G., Ferreira, D., Coelho, P. A. L. F., Sousa, J. A. e. & Taborda, D. M. G. (2017) Modelling the influence of rock variability on geotechnical structures. In *Proceedings of the 19th International Conference on Soil Mechanics and Geotechnical Engineering, Seoul, South Korea.* pp. 1869–1872.

Potts, D. M. & Zdravković, L. (2001) *Finite element analysis in geotechnical engineering: application.* Thomas Telford. London.

Sert, S., Luo, Z., Xiao, J. H., Gong, W. P. & Juang, C. H. (2016) Probabilistic analysis of responses of cantilever wall-supported excavations in sands considering vertical spatial variability. *Computers and Geotechnics*, **75** pp. 182–191. (in English).

Vanmarcke, E. H. (2010) *Random fields: analysis and synthesis.* The MIT press. pp. 394.

Zhang, L. M. (2013) Characterizing geotechnical anisotropic spatial variations using random field theory. *Canadian Geotechnical Journal*, **50** (7), pp. 723–734.

Numerical Methods in Geotechnical Engineering IX – Cardoso et al. (Eds)
© 2018 Taylor & Francis Group, London, ISBN 978-1-138-33198-3

Conditional random field simulation for analysis of deep excavations in soft soils

C.J. Sainea-Vargas & M.C. Torres-Suárez
Facultad de Ingeniería, Universidad Nacional de Colombia, Bogotá, Colombia

G. Auvinet
Laboratorio de Geoinformática, Instituto de Ingeniería, Universidad Nacional Autónoma de México, Mexico

ABSTRACT: Deep excavations are commonly used in densely populated areas with the aim of cost and space optimization. Soft soil deposits present in Bogotá are inherently variable, so a combination of numerical and probabilistic approaches are suitable toassess the safety conditions of the excavations to prevent damages to neighboring buildings. In the analyses, the soil constitutive parameters can be treated as random fields to deal with uncertainty due to spatial variability. Geostatistical simulation techniques are useful to generate random fields of soil properties using available measures. Two issues related with random field simulations are the number of required realizations to obtain the final estimate and the horizontal scale of fluctuation because measurements in that direction are usually scarcer. Constitutive model Hardening Soil Small Strain was considered to represent soil behavior and parameters were obtained from laboratory and field test results, mostly SCPTu. Descriptive and spatial statistical analyses were performed on obtained sets of residuals after detrending. Sequential Gaussian co-simulations were used to obtain conditional random fields for selected constitutive parameters considering different number of realizations for each simulation and different anisotropy ratios for the scales of fluctuation. Random fields were mapped to numerical models and 3D finite element analyses were performed on an idealized deep excavation model obtaining the system response in terms of damage potential indexes.

1 INTRODUCTION

The assessment of safety conditions during construction for deep excavations in soft soils respect to serviceability limit states is a complex problem. Given the natural variability in lacustrine deposits as those present in Bogotá and the complex behavior of such soft soils, the use of numerical and probabilistic techniques is a suitable approach (Rodriguez 2006, Pineda, Estévez, & Daza 2014, Sainea-Vargas & Torres-Suárez 2017) Soil properties can be considered as random variables or random fields and system response as a random variable to evaluate safety conditions in terms of probabilities of undesired performance in reliability based analysis (Auvinet 2002, Baecher & Christian 2003, Sánchez 2010). Probability distribution and correlation structure of random fields for soil parameters can be studied using experimental measures such as given by CPT test results (Stefanou 2009, Hicks 2014).

There are different algorithms to generate random fields (Fenton & Griffiths 2008, Stefanou 2009). When modeling random fields using experimental data, it is convenient to constrain their values at the locations of actual measurements using geostatistical techniques such as kriging (Auvinet 2002, Lloret-Cabot, Hicks, & Van den Einjden 2012). However, as sampling becomes more spaced random fields based on kriging show a smoothing effect, so is preferable to use geostatistical simulation methods as the sequential Gaussian simulation (Olea 2009, Emery 2013) or sequential Gaussian co-simulation for correlated variables. Generated random fields depend on the spatial correlation so it is important to study the influence of the fluctuation scale, in particular in horizontal direction in which sampling is scarcer. As multiple equiprobable simulations are required to generate random fields for soil properties, assessing the number of simulations is also of interest.

In this paper, constitutive model Hardening Soil Small Strain (Schanz, Vermeer, & Bonnier 1999, Benz 2007) was considered to represent soft soil behavior. Descriptive and spatial statistical analyses were performed on selected soil constitutive parameters obtained from correlations with CPTu tests performed on Bogotá soft soils. Random fields were generated for the selected soil constitutive parameters using results of statistical analyses

Table 1. Descriptive and inferential analyses results for residuals.

Property	wE_0^{ref}	wE_{50}^{ref}
n	13247	13247
Mean μ_0 (kPa)	$-1.723x10^{-6}$	$-3.672x10^{-8}$
Standard deviation σ_0 (kPa)	11309.244	838.238
Skewness C_s	–0.164	0.225
Kurtosis C_k	3.249	3.865
pdf	IV Cauchy	IV Cauchy
Par1	16.9744	6.3507
Par2	7.21872	–1.70128
Location	13865.930	–409.922
Scale	61368.212	2578.4821
D	0.057	0.067
p-value	1	0.999

considering different values for the horizontal scales of fluctuation and number of realizations. Finally, random fields were mapped to numerical models and 3D finite element analyses were performed on an idealized deep excavation model obtaining the system response in terms of damage potential indexes.

2 METHODS

2.1 *Base information*

Base information was taken from a deep excavation project under construction in lacustrine soft soils of Bogotá (IGR 2014). In surface it was found a fill layer of 1.5 to 2 m, 38 to 42 m of silty-clay soils with soft consistency classified as CH, MH and OH, followed by intercalations of firm clays and dense sands, clayey silts and sand lenses up to 60 m, where gravels of sandstone were found. The soils of interest in this paper are those present from 2 to 40–44 m depth. The water level in the ground was found at 1.5 to 2 m depth. The results of the 8 SCPTu performed between 31 and 46 m depth indicate the presence of clays and silty clays with organic soils, mainly located in normalized behavior types 3 and 4. Obtained data of corrected tip resistance q_t were detrended, constitutive parameters were found from correlations (Mayne 2007, Obrzud & Truty 2014, Robertson & Cabal 2015) and the values found were fitted simulating available laboratory tests results; constitutive parameters to consider as random, E_0^{ref} and E_{50}^{ref}, were selected using principal components analysis (Sainea & Torres 2017).

Table 2. Results from spatial statistical analyses.

Transformed variable	Nugget	Sill (m)	θ_v (m)
wE_0^{ref}	0	1	3
wE_{50}^{ref}	0	1	5.2
$wE_0^{ref}.wE_{50}^{ref}$	0	0.52	7

2.2 *Descriptive and inferential statistics*

A third order polynomial trend was removed from the data of each SCPTu record. Descriptive and inferential statistical analyses were performed on residuals of parameters selected as random, wE_0^{ref} and wE_{50}^{ref}. Results in Table 1 are for the whole set of data of each random variable obtained from the 8 SCPTu records.

2.3 *Spatial statistical analyses*

Given natural variability in Bogotá soft soil deposits, it is important to characterize it for subsequent analyses. As the data had been detrended and belong to the same soft soil deposit, the assumption of statistic homogeneity or stationarity was considered (Phoon & Kulhawy 1999). The correlation length or scale of fluctuation θ is a measure of the variability in a random field, and it is defined as the distance within which soil properties are significantly correlated (Vanmarcke 1983, Fenton & Griffiths 2008). In sedimentary soils it is expected to find different scales of fluctuation in vertical θ_v and horizontal θ_h directions caused by the deposition process. There are several techniques to estimate the scales of fluctuation and fitting theoretical variogram models to sampling variograms is one of them (Uzielli, Lacasse, & Phoon 2007, Fenton & Griffiths 2008). Variogram $2\gamma(h)$ is a function commonly used in geostatistics to express the spatial structure of a data set, and there are several theoretical models available (Auvinet 2002, Baecher & Christian 2003, Uzielli, Lacasse, & Phoon 2007, Olea 2009, Emery 2013). $2\gamma(h)$ is the second order moment of increment $[Z(\mathbf{x}+\mathbf{h})-Z(\mathbf{x})]$:

$$2\gamma(h)=E\{[Z(\mathbf{x}+\mathbf{h})-Z(\mathbf{x})]^2\} \quad (1)$$

In this paper theoretical variogram models were fitted to empirically calculated sampling variograms and cross-variograms to find values of θ_v, using software SGeMS v2.5b (Remy, Boucher, & Wu 2009). As the sets of data were not univariate normal, they had to be transformed to normal scores for subsequent analyses (Olea 2009, Emery 2013). Table 2 summarizes results of statistical spatial analyses for transformed residuals. The order of magnitude of θ_h can be given by geological model on study site; for example for soft soils of similar origin of those considered in this paper, θ_h values reach several thousand meters (Juárez-Camarena, Auvinet-Guichard, & Méndez-Sánchez 2016).

2.4 *Sequential Gaussian co-simulation*

A regionalized variable z in geostatistics is a variable distributed in space having a correlation structure; it can not be known exhaustively but only by means of a limited set of data. Being d the dimension of workspace, $\mathbf{x} = (x_1,...,x_d)$ a vector of spatial coordinates and D the domain in space R^d, $z = \{z(\mathbf{x}), \mathbf{x} \in D\}$. As $\mathbf{x}$ runs over D a set of random variables $Z = \{Z(\mathbf{x}), \mathbf{x} \in D\}$ is obtained, constituting a random function or random field (Emery 2013). Interpolation methods as kriging produce realistic estimations depending on sampling interval, but as sampling becomes sparser the estimates tend to show an increasing smoothing effect; so, the generation of several equiprobable realizations using simulation techniques is a most common approach (Olea 2009). By means of geostatistical simulations, multiple artificial realizations Z_s of the random function Z can be obtained reflecting its statistical properties. Among all possible simulations or artificial realizations all those in which the simulated values coincide with experimental ones called conditional simulations are preferable Auvinet 2002, Remy, Boucher, & Wu 2009).

$$Z_s(\mathbf{u}_i) = E\{Z/z(\mathbf{x}_j), 2\gamma(h)\} \tag{2}$$

There are different methods of simulation of random fields (Fenton & Griffiths 2008, Olea 2009, Remy, Boucher, &Wu 2009, Emery 2013), and in this paper the sequential Gaussian simulation was chosen. This method requires multivariate probability density function of random function to simulate Z to be Gaussian, so data were converted to normal standard by anamorphosis. Having an original random function $Z = \{Z(\mathbf{x}), \mathbf{x} \in D\}$ with cumulative distribution function $F(z)$ to be transformed to another $Y = \{Y(\mathbf{x}), bfx \in D\}$ with cumulative distribution function Gaussian standard $G(y)$, a transformation is made such as $F(z) = G(y)$. Gaussian anamorphosis function $\varphi(y)$ relates Gaussian values with original data. Once having Y and its variogram, andhaving transformed available data on sites $\{\mathbf{x}_1,...,\mathbf{x}_n\}$, the sequential algorithm to obtain conditional random fields includes the following steps (Olea 2009, Emery 2013):

a. Define a random path to visit each node of the grid $\{\mathbf{u}_1,...,\mathbf{u}_m\}$.
b. For the next location to consider $\mathbf{u}_i$, use multi-Gaussian kriging using transformed experimental measures $\{Y(\mathbf{x}_1),...,Y(\mathbf{x}_n)\}$ and previously simulated values $\{Y(\mathbf{u}_1),...,Y(\mathbf{u}_{i-1})\}$. As a result an estimated value $Y^*(\mathbf{u}_i)$ and a standard deviation of estimation $\sigma^*(\mathbf{u}_i)$ are obtained.
c. Define a normal distribution of mean $Y^*(\mathbf{u}_i)$ and standard deviation $\sigma^*(\mathbf{u}_i)$. Draw at random a value $Y_s(\mathbf{u}_i)$ using that normal distribution. $Y_s(\mathbf{u}_i)$ is the simulated value at location $\mathbf{u}_i$.
d. Add $Y_s(\mathbf{u}_i)$ to the sample comprised by $\{Y(\mathbf{x}_1),...,Y(\mathbf{x}_n)\}$ and $\{Y(\mathbf{u}_1),...,Y(\mathbf{u}_{i-1})\}$.
e. Continue the previous process until reaching location $\mathbf{u}_m$.
f. Back transform all values to the original sampling space using Gaussian anamorphosis function $\varphi(y)$. To evaluate a given soil parameter, a sufficient number of realizations must be generated and average values in each point can serve as final estimates.

When considering two correlated variables, primary and secondary as in the case considered in this paper, sequential Gaussian co-simulation is needed (Remy, Boucher, & Wu 2009, Emery 2013). In this case the previously described steps change slightly:

a. Select the random path to follow.
b. The conditioning data consist of neighboring original data, previously simulated values and secondary data. Then a simple co-kriging is performed using those data and primary variable is simulated as a Gaussian vector with first two moments given by cokriging estimate and cokriging variance.
c. Simulated values are added to database to serve as conditioning data for next sites to simulate.
d. f variables are non-Gaussian they must be transformed at the beginning and back transformed at the end following the same procedure described earlier.

Software SGeMS v2.5b (Remy, Boucher, & Wu 2009) was used to generate random fields using sequential Gaussian co-simulation, obtaining final estimates in each cell as the averages of multiple equiprobable realizations. Simulations were conditioned to available data and search neighborhoods were limited to θ_h, θ_v. Examples of point-wise average values of simulated isotropic and anisotropic random fields are shown in Figure 1. Table 3 summarizes the anisotropy ratios a_r and considered values for θ_h.

2.5 *Numerical modeling*

Simulated random fields have to be discretized and mapped to finite element mesh to perform numerical analyses. A volume support was defined for the simulations performed of the random function. To define the size of the support, a vertical dimension of stochastic finite element of $\theta_v/4$ to $\theta_v/2$ were tried to maintain [?]. The size in horizontal directions were assumed to keep an adequate aspect ratio in the finite elements. Defined dimensions for each block in volume support and each finite continuum element to model soft soils were 1.5x4.0x4.0. Mapping was performed using centroid coordinates of each block in the random field and equivalent finite elements in the mesh.

Finite element model shown in Figure 2 was elaborated with ZSoil v2016software. The excavation is located in a corner and is surrounded by 10 storey buildings modeled as plates with distributed loads of 100 kPa. It is an idealized excavation of40 m × 64 m and 12 m depth, supported by a retaining system composed of 30 m depth diaphragm walls and concrete beams with $f'_c = 28$ MPa. 18 m length and 0.60 m diameter piles spaced every 4 m were used in the bottom of the excavation to prevent heave.The model extents to a 160 m × 184 m × 60 m block. The construction

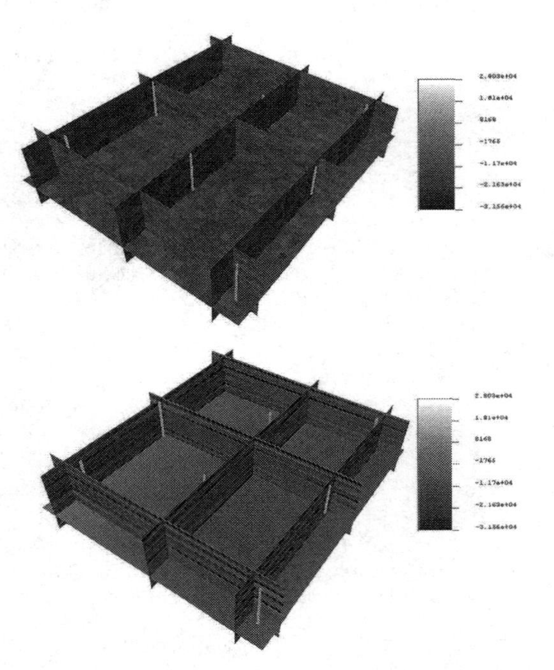

Figure 1. Simulated random fields for wE_0^{ref} obtained as point-wise average values from multiple equiprobable realizations.

Table 3. Anisotropy ratios a_r and θ_h values considered.

	θ_h	θ_h	θ_h
$a_r = \theta_h / \theta_v$	wE_0^{ref}	wE_{50}^{ref}	$wE_0^{ref}.wE_{50}^{ref}$
1	3	5.2	7
5	15	26	35
10	30	52	70
20	60	104	140
50	150	260	350
100	300	520	700
1000	3000	5200	7000

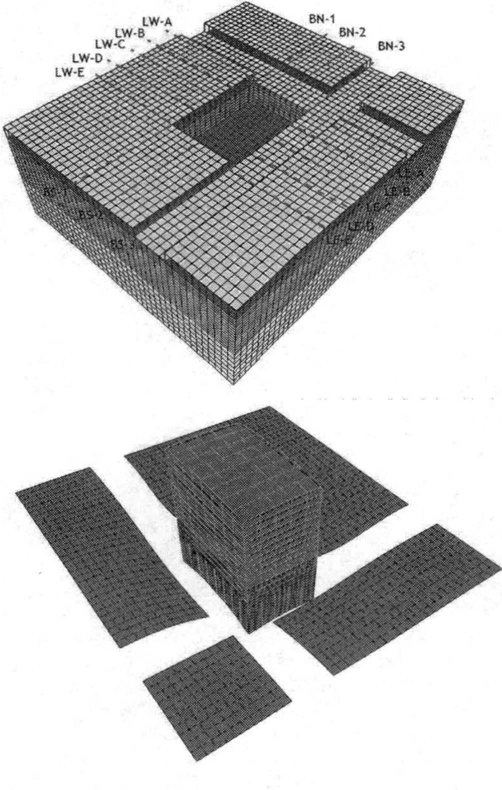

Figure 2. Finite element model of a deep excavation (up) and completed building and deformed retaining system after last construction stage (down).

stages are: a) Pile installation, b) Retaining wall construction, c) First level beams and plate (0.0 m), d) First excavation (0.0 m to −3.0 m), e) Second level beams and plate (−3.0 m), f) Second excavation

(−3.0 m to −6.0 m), g) Third level beams and plate (−6.0 m), h) Third excavation (−6.0 m to −9.0 m), i) Fourth level beams and plate (−9.0 m), j) Fourth excavation (−9.0 m to −12.0 m) k) Bottom slab and l) Framed building construction.

2.6 *Damage potential index (DPI)*

To define system response a serviceability limit state is considered, in this case possible damage in adjacent buildings to excavations due to sagging and hogging patterns of deformation. As a response variable it was chosen the damage potential index (DPI) (Schuster, Kung, Juang, & Hashash 2009). DPI is based on the maximum principal stress strain ε_p originated by excavation in adjacent buildings which combines angular distortion β, lateral deformation ε_l and the direction of crack formation measured from vertical plane θ_{max} (Son & Cording 2005). Node displacements obtained in numerical analyses on sections shown in Figure 2 were used to compute DPI values.

3 RESULTS

3.1 *Number of simulations*

When simulating a random field, the number of realizations must be large enough to obtain stable results, but increasing this number could lead to high computational effort in tridimensional models as the one considered in this paper. With the aim of looking for an adequate number of realizations N_{sim}, isotropic and anisotropic conditional random fields were simulated for wE_0^{ref} and

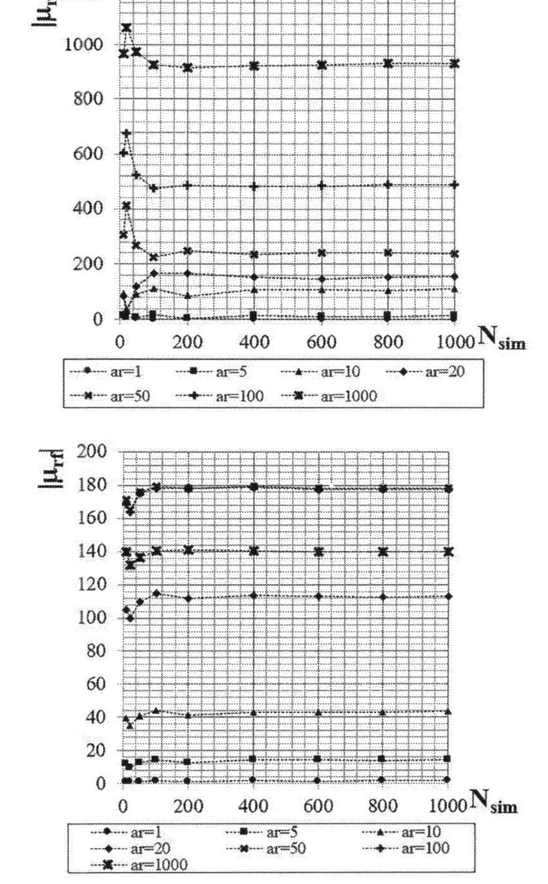

Figure 3. $|\mu_{rf}|$ for random fields of wE_0^{ref} (up) and wE_{50}^{ref} (down) for different N_{sim} and a_r values.

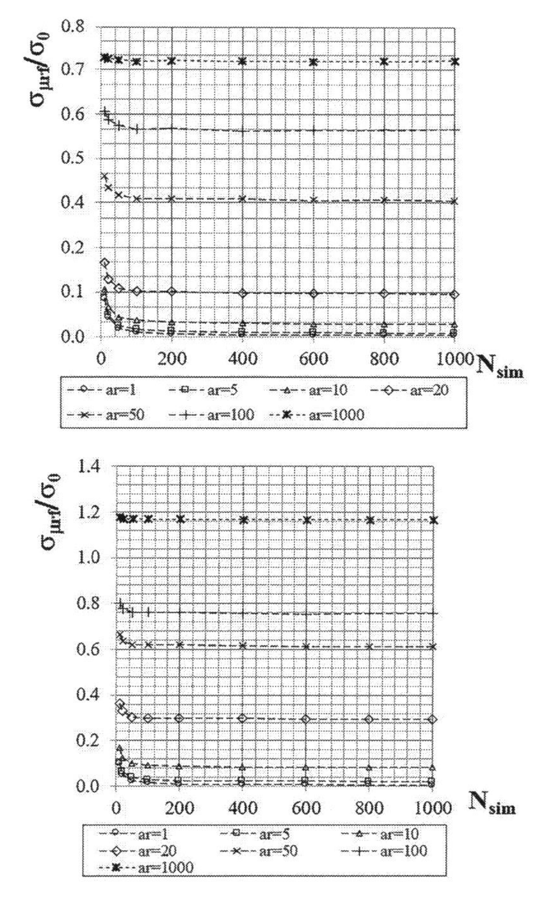

Figure 4. $\sigma_{\mu rf}/\sigma_0$ for random fields of wE_0^{ref} (up) and wE_{50}^{ref} (down) for different N_{sim} and a_r values.

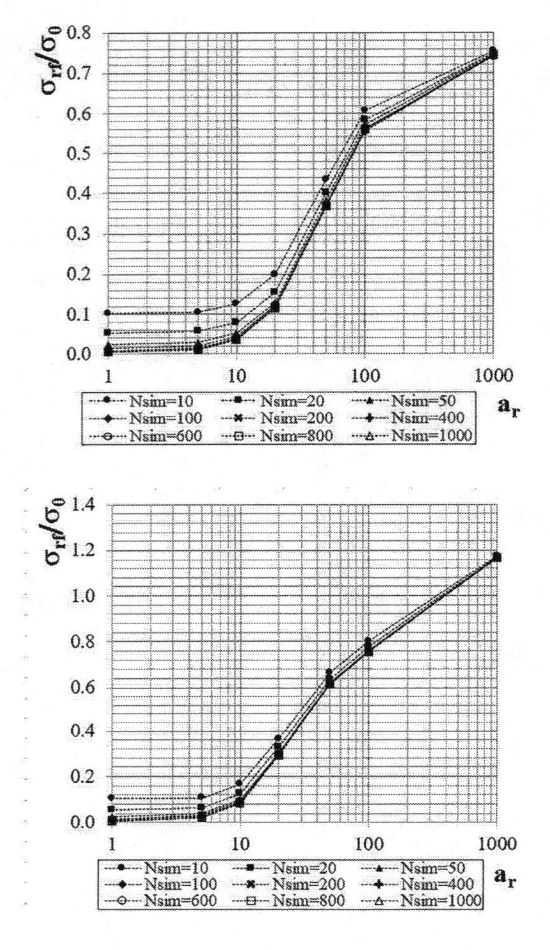

Figure 5. $\sigma_{\mu rf}/\sigma_0$ for random fields of wE_0^{ref} (up) and wE_{50}^{ref} (down) for different wE_{50}^{ref} and N_{sim} and a_r values

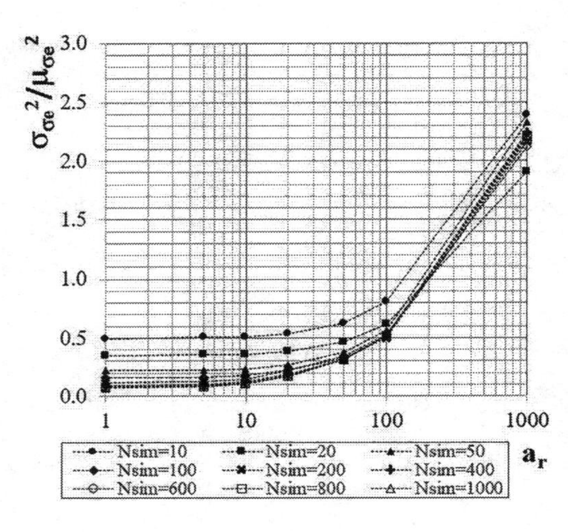

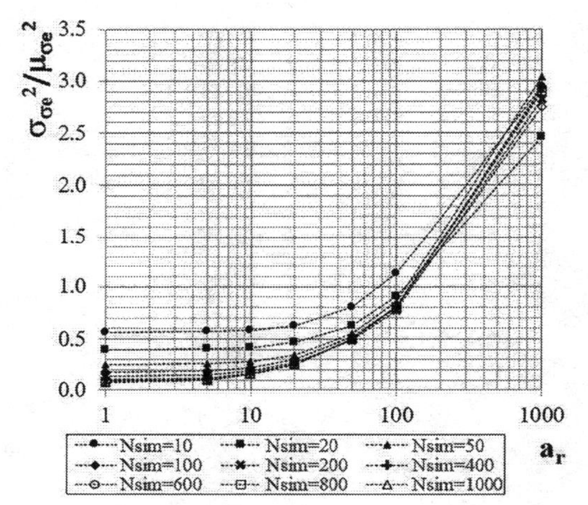

Figure 6. $\sigma_{\sigma^2_{\sigma e}}/\mu_{\sigma^2_{\sigma e}}$ of random fields of wE_{50}^{ref} (up) and wE_{50}^{ref} (down) for different N_{sim} and a_r values.

wE_{50}^{ref}, considering different N_{sim} values: 10, 20, 50, 100, 200, 400, 600, 800 and 1000. In each case and for each cell local conditional expectation estimates were obtained by point-wise average of simulated realizations μ_{rfi} and corresponding maps were generated. Mean μ_{rf} and standard deviation $\sigma_{\mu rf}$ values were obtained from histograms of μ_{rfi} including local estimates in all cells of the model. Absolute values of μ_{rf} are presented and $\sigma_{\mu rf}$ values were normalized using σ_0 as shown in Figure 3 and Figure 4. Also local conditional variance was computed for each cell $\sigma^2_{\sigma ei}$, and $\mu_{\sigma^2_{\sigma e}}$ and $\mu_{\sigma^2_{\sigma e}}$ values were obtained from histograms of $\sigma^2_{\sigma ei}$, including local estimates in all cells of the model; results of $\sigma^2_{\sigma ei}/\mu\sigma^2_{\sigma e}$ are presented in next section.

In Figure 3 and Figure 4 it can be seen that from 100 to 200 simulations data tend to stabilize so no more realizations of the random fields would be needed beyond that point. Despite of starting with $\mu_0 \approx 0$ for both ωE_0^{ref} and ωE_{50}^{ref}, the simulation with increasing a_r values induces a skew in simulated random fields. When compared to removed trend values the average induced skew is less than $0.01tE_0^{ref}$ for a_r = 1,5,10, is between 0.02 and $0.05tE_{50}^{ref}$ for ar = 20,50,100 and over $0.05tE_{50}^{ref}$ for a_r = 1000. Average induced skew is less than $0.02tE_{50}^{ref}$ for a_r = 1,5,10, is between 0.05 and $0.09tE_{50}^{ref}$ for a_r = 20,50 and close to $0.1tE_{50}^{ref}$ for a_r = 100,1000. On the other hand, results for $\sigma_{\mu rf}$ show simulated random fields tending to be more homogeneous or smoother as the number of realizations increases. For example upper random field of Figure 1 is for $N_{sim} = 10$, while the other is for $N_{sim} = 1000$. In maps of $\mu_{\sigma^2_{\sigma e}}$ as N_{sim} increases maps become smoother and in general show higher variances except near measurement locations where uncertainty is smaller than in other locations of the map.

3.2 *Anisotropy ratio*

As seen in Figure 3, in wE_{50}^{ref} random field, μ_{rf} values start to deviate from $|\mu_{rf}| = 0$ for a_r $ 10 and as a_r grows $|\mu_{rf}|$ increases. In simulated wE_{50}^{ref} random fields $|\mu_{rf}|$ tend to increase for $5 \leq a_r \leq 50$ and decrease for higher anisotropy ratios. This is the skew induced in simulated random fields as mentioned in previous section, which depends on considered a_r. Simulated anisotropic random fields for both variables wE_{50}^{ref} and wE_{50}^{ref} have different gradients of change in μ_{rf} for small $a_r \leq 5$, medium $5 < a_r \leq 50$ and high $a_r > 50$ anisotropy ratios.

In Figure 4 and Figure 5, $\sigma_{\mu rf}$ values for wE_{50}^{ref} random field are smaller than σ_0 with values between $0.01\sigma_0$ and $0.74\sigma_0$. On the other hand, $\sigma_{\mu rf}$ values for wE_{50}^{ref} random field are smaller than σ_0 for $a_r \leq 100$; and higher than σ_0 for $a_r = 1000$; in former case values are between $0.01\sigma_0$ and $0.8\sigma_0$, and around $1.17\sigma_0$ in the latter. In general $\sigma_{\mu rf}$ values increase for wE_0^{ref} and wE_{50}^{ref} random fields with different gradients for small $a_r \leq 10$, medium $10 < a_r \leq 100$ and large $a_r > 100$ values. Finally, $\sigma_{\sigma^2_{\sigma e}}/\mu_{\sigma^2_{\sigma e}}$ values in Figure 6 tend to increase with a_r with increasing gradients for small $a_r \leq 10$, medium $10 < a_r \leq 100$ and large $a_r > 100$ values respectively. μ_{rf}, $\sigma_{\mu rf}$ and $\sigma_{\sigma^2_{\sigma e}}/\mu_{\sigma^2_{\sigma e}}$ charts tend to concentrate for N_{sim} $ 100.

This results are related with the way in which the search neighborhoods were defined to simulate random fields, because they change with chosen θ_h. As θ_h grows there are more hard data available to simulate random field values at each location $\mathbf{u}_i$ in each step of the process. The change in behavior from small to medium a_r values seems to obey to the average spacing D among sampling locations used to estimate parameter values E_0^{ref} and E_{50}^{ref}. As $\theta_h > D$, the information of adjacent SCPTu tests is starting to being used simultaneously for a given location. All available SCPTu records are used in the same window as θ_h is of the order of the horizontal dimensions of the domain considered for the model;the change in behavior from medium to high a_r is related with that. As a result, estimated random fields tend to be markedly layered as a_r increases. Difference in behaviors for E_0^{ref} and E_{50}^{ref} can be explained by their θ_v values because $\theta_{vw}E_{50}^{ref}$ is about 1.7 times $\theta_{vw}E_{50}^{ref}$ and as said those values are related with search neighborhood size through a_r. Finally, growing pattern of $\sigma_{\sigma^2_{\sigma e}}/\mu_{\sigma^2_{\sigma e}}$ is originated by a limited sampling because as search windows are getting bigger, hard data to make estimations tend to be scarcer per area unit.

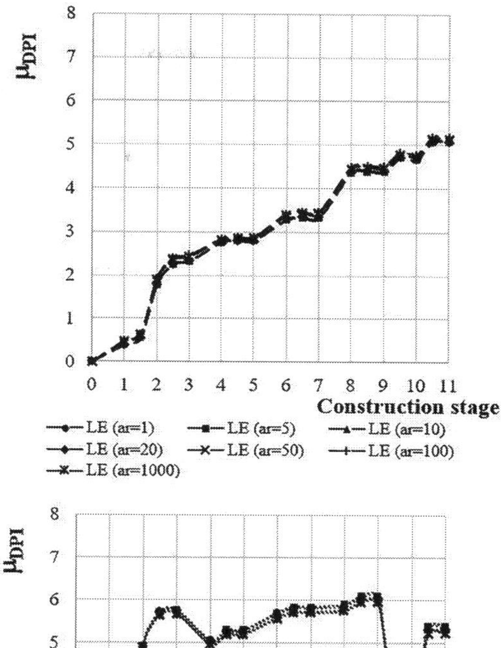

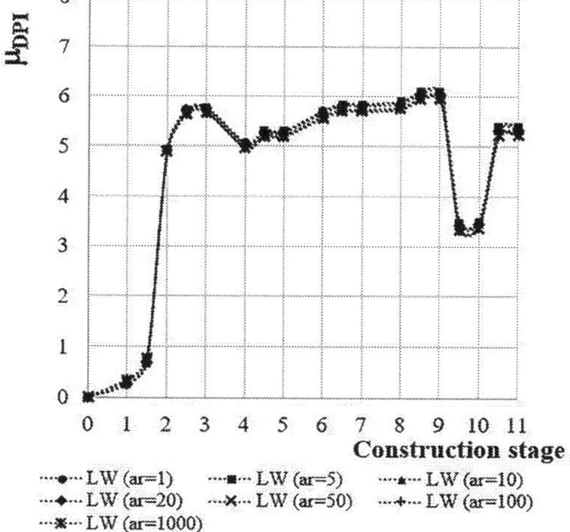

Figure 7. μ_{DPI} for LE (up) and LW (down) sides of excavation in each construction stage for different a_r values.

3.3 *DPI results for numerical models*

Until this point the response of the numerical models has not been considered in the analyses, so random fields obtained with $N_{sim} = 100$ were mapped to finite element models elaborated in ZSoil v2016software. One model was made for each one of those simulated random fields for

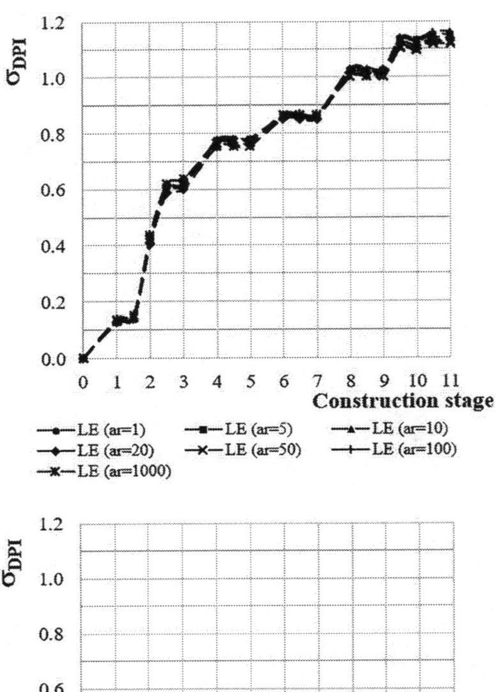

Figure 8. σ_{DPI} for LE (up) and LW (down) sides of excavation in each construction stage for different a_r values.

a_r = 1,5,10,20,50,100 and 1000 to perform deterministic analyses. Deep excavations in soft soils described previously were analyzed and displacements were obtained in sections shown in Figure 2. Obtained nodal displacements were used to calculate *DPI* values for hogging and sagging patterns of deformation for sections in each side of the excavation and in each construction stage described previously. As said deterministic analyses were performed, but expected μ_{DPI} and standard deviation σ_{DPI} values were computed using data from each side of the excavation; this with the aim of summarizing obtained results.

A limit from slight to moderate damage with DPI_{crit} = 25 for sagging and DPI_{crit} = 20 for hogging as defined by (Schuster, Kung, Juang, & Hashash 2009) was assumed as serviceability requirement. Both sets of *DPI* results were under assumed DPI_{crit}, being in this case the sagging patterns of deformation the ones that give greater values. Also, computed *DPI* values for long sides of the excavation, LE and LW, were similar the ones found for the short sides BN and BS. So, μ_{DPI} and σ_{DPI} results for LE and LW sides of excavations and for sagging patterns of deformation are analyzed in this section. In Figure 7 and Figure 8 computed μ_{DPI} and σ_{DPI} values are shown for $1 < a_r \leq 1000$ and for the different construction stages.

μ_{DPI} values in LE and LW sides of excavations have different patterns, but each set of data shows the same pattern. While in the former case μ_{DPI} increases until reaching its maximum value at final construction stage, in the latter case greatest μ_{DPI} values appear around stages 3 and 9 and decay around stage 10. In both LE and LW sides of excavation in models with random field simulated with the considered values a_r for all stages tend to be very close among them, with maximum values of $\mu_{DPI} = 5.2$ for stage 10 in LE and $\mu_{DPI} = 6.04$ for stages 9,10 in LW. In each set of data LE and LW there is also a similar trend for μ_{DPI} values. Close values are found in random fields simulated with all considered a_r values. In LE side it grows until a maximum of $\mu_{DPI} = 1.16$ is attained at stage 11, while for LW side the maximum is $\mu_{DPI} = 0.31$ at stage 2.

Despite of the differences in simulated random fields discussed earlier, μ_{DPI} and σ_{DPI} results tend to be the same for all considered μ_{DPI} values. This occurs because the results shown are for deterministic analyses performed on an average random field. There is a smoothing effect related with the obtained average of several realizations of the random field. This leads to the need of analyzing a whole set of models to perform probabilistic analyses following a sampling scheme such as given by Monte Carlo Simulations or Point Estimate Method (Auvinet 2002, Baecher & Christian 2003, Sánchez 2010). In each case a realization of the random field and not an average of realizations should be used.

4 CONCLUSIONS

Simulation of cross-correlated conditional random fields to represent soil constitutive parameter variability for its application in 3D FEM serviceability limit states analysis of deep excavations in soft soils is illustrated in this paper. The required number of realizations to simulate random fields, the influence of anisotropy ratio and response of the models in terms of potential damages to neighboring buildings are discussed.

The material presented here can serve as a general methodological guide to simulate and map conditional random fields for subsequent numerical and probabilistic analyses of deep excavations in soft soils. However, despite of potential damage index resultswere presented in statistical terms, they come from deterministic analyses as only one model was used to obtain results for each considered random field. Probabilistic analyses of deep excavations in soft soil for serviceability limit states using cross-correlated conditional random fields is a topic that remains to be treated in a posterior paper.

ACKNOWLEDGEMENTS

The research leading to these results is supported by Universidad Nacional de Colombia and Colciencias. First author acknowledges also support and guidance from Dr. Auvinet and researchers of Laboratorio de Geoinformática, Instituto de Ingeniería, UNAM.

REFERENCES

Auvinet, G. (2002). Uncertainty in geotechnical engineering. In *Sixteenth Nabor Carrillo Lecture - XXIth National Meeting of the Mexican Society for Soil Mechanics, Querétaro, México*, pp. 1–139.

Baecher, G. & Christian, J. (2003). *Reliability and statistics in geotechnical engineering.* Chichester West Sussex, England: John Wiley & Sons, Ltd.

Benz, T. (2007). *Small-Strain Stiffness of Soils and its Numerical Consequences.* Stuttgart, Germany: Institut f¨ur Geotechnik der Universit¨at Stuttgart.

Emery, X. (2013). *Geoestadística.* Santiago, Chile: Facultad de Ciencias F´ısicas y Matem´aticas Universidad de Chile.

Fenton, G. & Griffiths, D. (2008). *Risk assessment in geotechnical engineering.* Hoboken, New Jersey: John Wiley & Sons, Inc.

Hicks, M. (2014). Application of the random finite element method. In *ALERT Doctoral School 2014 Stochastic Analysis and Inverse Modelling*, Aussois, France, pp. 181–207.

IGR (2014). *Exploración para el proyecto Parque Colina Colpatria*. Bogotá, Colombia: IGR.

Juárez-Camarena, M., Auvinet-Guichard, G. & Méndez-Sánchez, E. (2016). Geotechnical zoning of Mexico Valley subsoil. *Ingeniería, Investigación y Tecnología 17(3)*, 297–308.

Lloret-Cabot, M., Hicks, M. & Van den Einjden, A. (2012). Investigation of the reduction in uncertainty due to soil variability when conditioning a random field using kriging. *Géotechnique Letters 2*, 123–127.

Mayne, P. (2007). *NCHRP Synthesis 368: Cone Penetration Testing A Synthesis of Highway Practice.* Washington, D.C.: NCHRP, TRB.

Obrzud, R. & Truty, A. (2014). *The Hardening Soil Model – A Practical Guidebook.* Lausanne, Switzerland: Zace Services Ltd.

Olea, R. (2009). *A Practical Primer on Geostatistics. Open-File Report 2009–1103.* Reston Virginia, USA: U.S. Geological Survey.

Phoon, K. K. & Kulhawy, F. H. (1999). Characterization of geotechnical variability. *Canadian Geotechnical Journal 36 (4)*, 612–624.

Pineda, J., Est´evez, L. & Daza, N. (2014). A case study of a deep excavation on a soft lacustrine clay of Bogot´a, Colombia, with emphasis between predicted and measured deformations. In *Geotechnical Aspects of Underground Construction in Soft Ground - Proceedings of the 8th International Conference of TC28 of the ISSMGE*, Seoul, Korea, pp. 327– 332.

Remy, N., Boucher, A. & Wu, J. (2009). *Applied Geostatistics with SGeMS: A User's Guide.* Cambridge, UK: Cambridge University Press.

Robertson, P. K. & Cabal, K. (2015). *Guide to Cone Penetration Testing for Geotechnical Engineering, 6th edition.* Signal Hill, California: Gregg Drilling & Testing, Inc.

Rodríguez, J. (2006). Case study of a deep excavation in soft soils with complex ground water conditions in Bogot´a. In *Geotechnical Aspects of Underground Construction in Soft Ground - Proceedings of the 5th International Conference of TC28 of the ISSMGE*, London, England, pp. 881–886.

Sainea, C. J. & Torres, M. C. (2017). A comparison of two approaches for parameter selection for numerical and probabilistic analyses of deep excavations in soft soils. In *The 15th International Conference of the International Association for Computer Methods and Advances in Geomechanics - 15thI-ACMAG*, Wuhan, China.

Sainea-Vargas, C. J. & Torres-Suárez, M. C. (2017). Numerical and probabilistic analyses of deep excavations in soft soils. *Electronic Journal of Geotechnical Engineering 22 (10)*, 3899–3924.

Sánchez, M. (2010). *Introduccion a la confiabilidad y evaluación de riesgos: Teoría y aplicaciones en Ingeniería (2a ed.).* Bogotá, Colombia: Universidad de Los Andes.

Schanz, T., Vermeer, P. & Bonnier, P. (1999). The hardening soil model: Formulation and verification. In *Beyond 2000 in Computational Geotechnics - 10 Years of PLAXIS.*, Rotterdam, Netherlands, pp. 1–16.

Schuster, M., Kung, G. T. C., Juang, C. H. & Hashash, Y. (2009). Simplified model for evaluating damage potential of buildings adjacent to a braced excavation. *Journal of Geotechnical and Geoenvironmental Engineering 135(12)*, 1823–1835.

Son, M. & Cording, E. (2005). Estimation of building damage due to excavation-induced ground movements. *Journal of Geotechnical and Geoenvironmental Engineering 131(2)*, 162–177.

Stefanou, G. (2009). The stochastic finite element method: Past, present and future. *Comput. Methods Appl. Math. Engrg. 198*, 1031–1051.

Uzielli, M., Lacasse, S. & Phoon, K. K. (2007). Soil variability analysis for geotechnical practice. *Characterization and engineering properties of natural soils 3*, 1653–1752.

Vanmarcke, E. (1983). *Random fields: Analysis and Synthesis.* Cambridge, Massachusetts: MIT Press.

Numerical Methods in Geotechnical Engineering IX – Cardoso et al. (Eds)
© 2018 Taylor & Francis Group, London, ISBN 978-1-138-33198-3

Stability assessment of the unsaturated slope under rainfall condition considering random rainfall patterns

Gaopeng Tang, Jinsong Huang, Daichao Sheng & Scott Sloan
Centre of Excellence for Geotechnical Science and Engineering, Faculty of Engineering and Built Environment, The University of Newcastle, Callaghan, NSW, Australia

ABSTRACT: Using a typical two-dimensional unsaturated slope, this paper investigates the effects of random rainfall patterns on the stability of unsaturated slope under rainfall condition. Rainfall information is presented in the form of Intensity-Frequency-Duration (IFD) curves. Random Rainfall Patterns (RRPs) are simulated based on Random Cascade Model (RCM) and Monte Carlo Method (MCM). The Conditional Failure Probability (CFP) of the unsaturated slope is investigated by considering the numerous generated RRPs. Meanwhile, the Annual Failure Probability (AFP) of the unsaturated slope is estimated considering also the occurrence frequencies of rainfall events. The results show that the slope stability is sensitive to rainfall patterns, and the RRPs can be considered in the determination of slope reliability.

1 INTRODUCTION

Rainfall-induced landslides are common in the tropical or sub-tropical region where the unsaturated soil is abundant (Rahimi et al., 2010). These landslides occur most commonly during or after rainfall period, and leading to casualties and significant economic losses (Afungang and Bateira, 2016). Hence, the need for unsaturated slope stability assessment under rainfall condition is pronounced.

IFD curves are generally derived to characterize rainfall information based on historical rainfall data (Green et al., 2014). However the rainfall patterns related to rainfall events cannot be reflected in these IFD curves. Owing to the difficulty in distinguishing rainfall pattern, most research studied the rainfall-induced slope using only one or several representative rainfall patterns. These rainfall patterns are generally linearly-distributed patterns which are artificially simplified from the actual rainfall patterns without considering the inherently stochastic nature of the rainfall patterns (White and Singham, 2012, Yuan et al., 2015). In reality, rainfall intensity randomly fluctuates over the rainfall duration, which may lead thousands of random rainfall patterns even when average intensity and duration of a rainfall event are the same(Peres and Cancelliere, 2014). In this regard, a rainfall simulation model is needed to generate a series of random rainfall patterns for covering as many natural rainfall patterns as possible. For a rainfall event extracted from IFD curve, the RCM, proposed by Menabde and Sivapalan (2000) can provide a simplified and rapid approach for reproducing the random rainfall patterns with very few statistical parameters.

In addition, landslide frequency analyses are essential in reliability-based design and assessment for engineering project because the probability of failure needs to be set in a reference timeframe (or design life) to be used in risk analyses (Rodríguez-Ochoa et al., 2015). The AFP, as a straightforward index, is generally introduced for decision makers to measure the landslide frequency (Phoon and Ching, 2014, AGS, 2007). Ohtsu et al. (2005) studied the AFP of rainfall-induced slope using uniform rainfall pattern. Sekiguchi et al. (2009) compared the AFP of slope under three typical rainfall patterns. However, such researches are still rather limited, and the AFP of the unsaturated slope is rarely investigated considering random rainfall patterns.

The primary objective of the present research is to investigate the stability of unsaturated slope under rainfall condition considering random rainfall patterns. IFD curves are introduced to characterize the information of rainfall events. Numerous RRPs are generated based on RCM and MCM. The CFP of unsaturated slopes is performed to quantize the stability states of the unsaturated slope considering the generated RRPs. Further, the AFP of the unsaturated slope is further estimated. Meanwhile, three widely used representative rainfall patterns are introduced to carry out comparative analysis.

2 THEORETICAL CONSIDERATIONS

2.1 *Intensity-Frequency-Duration (IFD) curves*

Intensity-Frequency-Duration (IFD) curves, derived from the statistical analysis of historical extreme precipitations, are used to describe the relationship between average rainfall intensities, rainfall durations, and the occurrence frequencies of rainfall events (or equivalently, return period T). These IFD curves are often required by geotechnical engineers for the design and analysis of water-related structures, such as levees and slopes. To derive these IFD curves, the generalized extreme value (GEV) distribution is often used (Coles et al., 2001). For a fixed rainfall duration, the relationship between the average rainfall intensity I_a and occurrence frequency [here, characterized by annual exceedance probability (AEP)] can be expressed using the inverse cumulative distribution function or quantile function of GEV distribution, as follows:

$$I_a(\lambda) = \mu + \frac{\rho}{\kappa}[(-\log(1-\lambda))^{-\kappa} - 1] \quad \xi \neq 0 \tag{1}$$

where, λ denotes the AEP; $I_a(\lambda)$ denotes the average rainfall intensity I_a at particular λ; and μ, ρ and κ are the location, scale and shape parameter of GEV distribution, respectively.

2.2 *Random and representative rainfall patterns*

Rainfall events represented by the same values of duration and cumulative depth, may correspond to thousands of random rainfall patterns owing to chaotic nature of rainfall process. The RCM is introduced here to generate a series of random rainfall patterns with statistical properties same to the observed data to represent the field conditions (Menabde and Sivapalan, 2000). For a certain rainfall event represented by rainfall duration D_r and average rainfall intensity I_a, the basic structure of the RCM used to simulate random rainfall pattern is illustrated in Figure 1.

As shown in Figure 1, in every step of disaggregation (or cascade step), the RCM distributes rainfall depth of an interval on two successive regular subintervals. The rainfall depth in one of two subintervals is calculated by multiplying the interval rainfall depth Q by the random cascade weight W. The rainfall depth Q_l^i for ith rainfall subinterval after lth disaggregation can be obtained as

$$Q_l^i = Q_0 \prod_{j=1}^{l} W_j^{f(i,j)} \quad \text{for } i = 1,2,...2^l \tag{2}$$

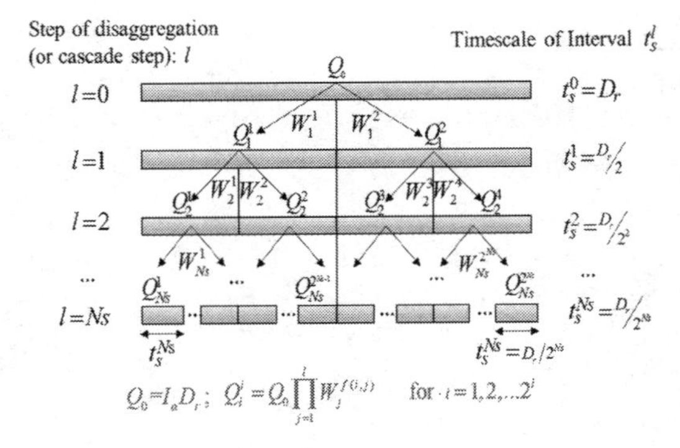

Figure 1. Schematic diagram of the RCM.

where, Q_0 denotes original cumulative rainfall depth, which equals to $I_a D_r$; the sequence of randomly generated weight $W_j^{f(i,j)}$ is steered at following jth disaggregation by the function $f(i,j)$, which is given by rounding up $i/2^{l-j}$ to the nearest integer. Obviously, the performance of the cascade-based rainfall disaggregation is governed by proper assumption regarding the cascade weight W. Following the existing literature (Menabde and Sivapalan, 2000), the probability distribution of W is assumed to follow the symmetrical *Beta* distribution with a sole parameter, ζ, governing the variance of W:

$$f(W) = \frac{1}{B(\zeta)} W^{\zeta-1}(1-W)^{\zeta-1} \tag{3}$$

where $B(\zeta)$ is the *Beta* function. The shape parameter ζ can be parameterized by the following scaling law:

$$\zeta(t_s^l) = \zeta_0 (t_s^l)^{-H_0} \tag{4}$$

where, t_s^l is the timescale of rainfall cell after lth disaggregation; and ζ_0 and H_0 are constants, which can be estimated from regional rainfall data (Menabde and Sivapalan, 2000). Meanwhile, to conserve rainfall depth exactly at each cascade step, the cascade weight W, as a random variable, should be constrained by the following condition:

$$\sum_{z=1}^{2} W_l^{(2(i-1)+z)} = 1 \quad \text{for } i = 1,2,...2^{l-1} \tag{5}$$

The abovementioned disaggregation is repeated Ns times until the rainfall duration D_r is divided into a series of basic rainfall intervals with desired timescale $t_s^{Ns} = D_r / 2^{Ns}$. The rainfall intensity of each basic rainfall interval can be obtained as $I_i = Q_{Ns}^i / t_s^{Ns}$, and then a realization of a random

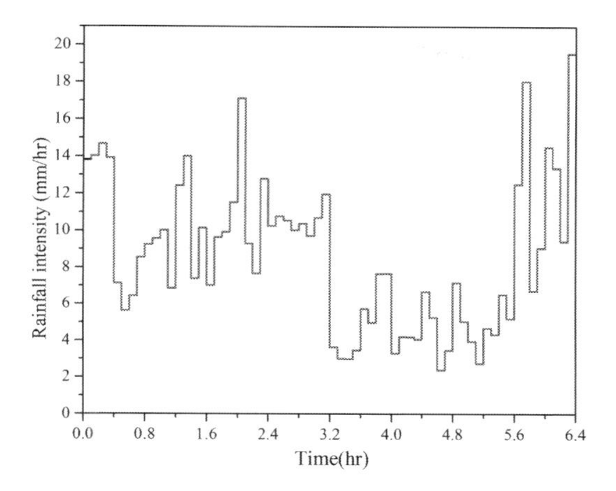

Figure 2. One realization of the random rainfall pattern.

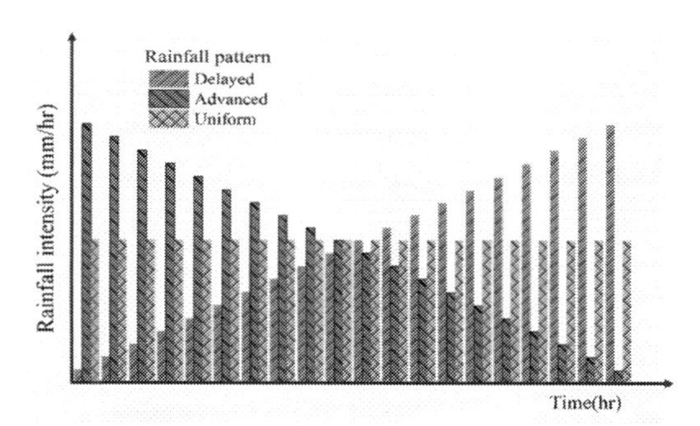

Figure 3. Three representative rainfall patterns.

rainfall pattern with time-varying rainfall intensity can be generated. One realization of random rainfall pattern with $N_s = 6$ and $t_s^{Ns} = 0.1\text{hr}$ is shown in Figure 2.

Three widely used representative rainfall patterns including uniform, advanced and delayed rainfall patterns, as shown in Figure 3, are adopted herein for the latter comparative analysis.

2.3 *Slope stability analysis under rainfall filtration*

The commercial software SEEP/W (Geo-slope International Ltd, 2010a) is adopted to carry out the unsaturated infiltration analysis. The water-flow governing partial differential equation is as follows (Fredlund and Rahardjo, 1993):

$$\frac{\partial}{\partial x}(k_{wx}\frac{\partial H_w}{\partial x}) + \frac{\partial}{\partial y}(k_{wy}\frac{\partial H_w}{\partial y}) + q = m_w \gamma_w \frac{\partial H_w}{\partial t} \quad (6)$$

where H_w is the total head; γ_w is the unit weight of water; t is the time; q denotes the boundary flux at the surface of slope; k_{wx} and k_{wy} donate the permeability coefficients in x- and y- directions (horizontal and vertical directions), respectively; and m_w is the slope of the soil-water characteristic curve (SWCC). Following the existing literature (Fredlund and Xing, 1994), the three-parameter SWCC model with correction factor $C(\psi)$ is used in the present study:

$$\theta_w = \theta_s C(\psi)\left\{\frac{1}{[\ln(e+(\psi/\alpha)^n)]^m}\right\} \quad (7)$$

where θ_w is the volumetric water content; θ_s is the saturated volumetric water content; e is the base of the natural logarithm; α, m and n are fitting parameters which are related to air-entry value, the residual water content and the maximum slope of SWCC, respectively; ψ is the matric suction; and $C(\psi)$ denotes the correction function According to the recommendation by Leong and Rahardjo (1997), $C(\psi)$ is taken to be equal to 1 in this study. The unsaturated permeability function, k_w, is derived based on the following function presented by Fredlund and Xing (1994):

$$k_w = k_s \Theta^p \quad (8)$$

where k_w is the permeability coefficient of the unsaturated soil; k_s is the saturated coefficient of permeability; Θ is a feature function related to SWCC, and the dimensionless parameter Θ can be expressed as (θ_w/θ_s); and p is the fitting parameter related to the slope of the permeability function.

2.3.1 *Slope stability analysis*

The factor of safety (*FS*) is introduced as a stability assessment index to reflect the stability state of the slope, herein. The stability analysis was performed in the present study by considering the unsaturated shear strength equation proposed by Fredlund and Rahardjo (1993), as follows:

$$\tau = c' + (\sigma_n - u_a)\tan\varphi' + (u_a - u_w)\tan\varphi^b \quad (9)$$

where τ denotes the shear strength of unsaturated soil; c' denotes the effective cohesion; φ' denotes the effective internal friction angle; φ^b denotes the friction angle with respect to matric suction; σ_n denotes the total stress; u_a denotes the pore-air pressure; u_w denotes the pore-water pressure; σ_n-u_a is the net normal stress, and u_a- u_w is the matric suction.

Under the conditions of rainfall infiltration, the pore water pressure distribution for a period of simulation time can be obtained using SEEP/W, and the corresponding results (e.g. pore water pressure) can be used as input data in slope stability analysis using SLOPE/W (Geo-slope International Ltd, 2010b), then the time-varying FS for unsaturated slopes under a certain rainfall pattern can be obtained. Here, the minimum value of FS (i.e. FS_{min}) during the whole rainfall duration is extracted and adopted to reflect the most instability state of the slope, which truly controls the safety of slope.

2.4 *Conditional failure probability under random rainfall patterns*

For a certain rainfall event $E(I_a, D_r)$ extracted from IFD curve, there may exist thousands of random rainfall patterns $\boldsymbol{X}$. The failure state of the slope under the rainfall event E is defined by the limit state function $g(\boldsymbol{X}| E)$, as follows:

$$g(\mathbf{X} \mid E) = FS_{min}(\mathbf{X} \mid E) - 1.0 \tag{10}$$

where $\boldsymbol{X}$ is the random rainfall patterns generated by RCM using the information of E extracted from IFD curves. The limit state function is customarily defined as

$$\begin{aligned} g(\mathbf{X} \mid E) &\geq 0 \quad \rightarrow \text{Safe} \\ g(\mathbf{X} \mid E) &< 0 \quad \rightarrow \text{Failure} \end{aligned} \tag{11}$$

The conditional (given that the corresponding rainfall event E extracted from IFD curve occurs) failure probability of unsaturated slope, $P(\mathrm{F}|E)$, is then defined as:

$$P(F \mid E) = \boldsymbol{Prob}[g(\mathbf{X} \mid E) < 0] \tag{12}$$

The $P(F|E)$ is estimated by application of the MCM and can be rewritten as $P(F|E) \approx N_f/N_{\text{total}}$, where N_f is the total number of failures, N_{total} is the total number of random simulations conducted in the MCM.

2.5 *Annual failure probability estimation*

Basically, AFP of the unsaturated slope, as an unconditional probability of landslide, can be obtained by integrating the joint probabilities of failure (the product of the conditional probabilities of failure and the corresponding probabilities of occurrence of rainfall events) over all different rainfall events $\boldsymbol{E}$. In the present work, the D_r is treated as a fixed value but the average rainfall intensity I_a is variable, thus the AFP can be estimated as follows:

$$AFP = \int_{all\ I_a} P(F|I_a) f(I_a) dI_a \tag{13}$$

where, $f(I_a)dI_a$ denotes the corresponding occurrence probability of rainfall even E with average rainfall intensity I_a; and $f(I_a)$ denotes the probability density function of the average rainfall intensity I_a. Notably, the $P(F|I_a)$ is essentially a continuous function with varying I_a, and this function is known as fragility curve (Schultz et al., 2010). However, the limit state function cannot be expressed explicitly due to the complexity of rainfall-induced landslide analysis, hence $P(F|I_a)$ estimated by MCM exists in the form of a set of discrete points. In this case, AEP is usually approximated by a discrete form (Phoon and Ching, 2014). An example of IFD curve and fragility curve combined to estimate an AFP is shown in Figure. 4.

As shown in Figure 4, IFD curves are usually discretized in several intervals in order to obtain a representative value (here, the average) and an occurrence probability value for each interval subsequently used to compute the AFP. As such, the calculation of AFP is given by the following sum, and in discrete form:

$$\text{AFP} = \sum_{j=1}^{N_{int}} P(F \mid I_{a,j}) v(I_{a,j}) \tag{14}$$

where, N_{int} is the number of intervals used to calculate the sum; The occurrence probability on the interval j, $v(I_{a,j})$, is computed as the difference between the AEP of lower and upper bound of the interval j (i.e. $v(I_{a,j}) = \lambda(I_{aj}^{L}) - \lambda(I_{aj}^{U})$; and $P(F|I_{a,j})$ denotes the CFP of the jth interval's representative value.

In the discretization of IFD curve, two special intervals, "*Negligible interval*" and "*Overwhelming interval*", as shown in Figure 4, should be primarily

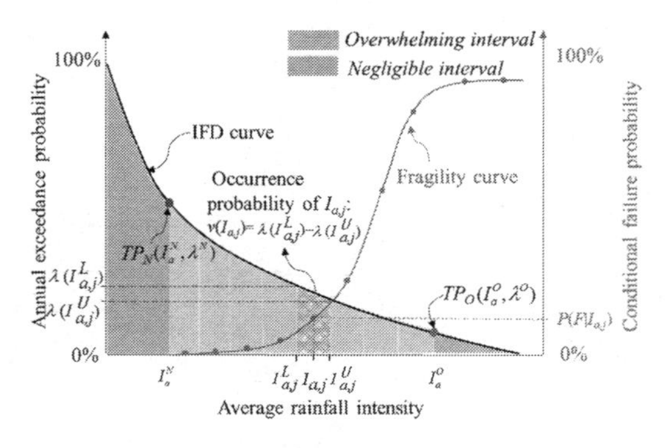

Figure 4. The combination of the IFD curve and fragility curve for estimating AFP.

considered for reducing computational burden. The *Negligible interval* can be established based on a upper threshold average rainfall intensity I_a^N, below which the CPF and the corresponding AFP of interval are negligible. The *Overwhelming interval* can be established based on a lower threshold average rainfall intensity I_a^O, beyond which the CPF is very close to 100% and corresponding AFP of interval is approximately equal to $\lambda(I_a^O)$. From their concept, the trial-and-error method is used to determine the two threshold since the two threshold average rainfall intensity are unknown in advance. Once the two specific intervals are determined, only the IFD curve between two threshold points (TP_N and TP_O) are needed to be discretized in the estimation of AFP. The number of intervals can be determined by increasing its value step by step until the value of AFP becomes relatively stabilized.

For the three representative rainfall patterns, the random rainfall patterns is not considered, so that:

$$P[F \mid I_{a,j}] = 0.0 \quad \text{for } I_{a,j} \le I_a^C$$
$$= 1.0 \quad \text{for } I_{a,j} > I_a^C \tag{15}$$

then equation (14) simplifies to:

$$\text{AFP} = P[I_{a,j} > I_a^C] \tag{16}$$

Combined with IFD curve, this is also expressed in terms of frequency of failure simply as a function of the frequency of exceeding the critical average rainfall intensity I_a^C: $\text{AFP} = \lambda(I_a^C)$. I_a^C is the threshold average rainfall intensity which makes the slope reach a critical instability state. This can be achieved by changing the value of I_a^C until the absolute value between its corresponding *FSmin* and 1 is smaller than a prescribed tolerance (e.g.0.001).

3 CASE STUDY

3.1 *Boundary and initial conditions of the unsaturated slope*

The typical unsaturated slope model selected from Rahardjo et al. (2007) is adopted in this study, as shown in Figure 5.

The slope angle, β is 45° and slope height, H_s, is 10 m. The boundary conditions of the unsaturated slope model, as can be seen from Figure. 5, are set as follows: (1) *ah*, *de* and *gf* are specified as the impermeable boundaries where zero nodal flux is applied to; (2) *ef* and *gh* are defined as total water head boundaries equal to the specified groundwater level, which are used to define the initial groundwater table; (3) *ab*, *bc* and *cd* are rainfall infiltration boundaries. To simulate the actual field conditions, ponding is not allowed to occur at the ground surface (*abcd*). The initial condition of pore water pressure distribution is provided by setting the initial water table position. The groundwater table is set to H_g = 5.0 m below the toe of slope at an inclination of 7° to the horizontal. The negative pore water pressure is proportional to the vertical distance from the groundwater table, but should be set to a limit for preventing the generation of unrealistic pore water pressures. Herein, the limited negative pore-water pressure is selected to be -50 kPa.

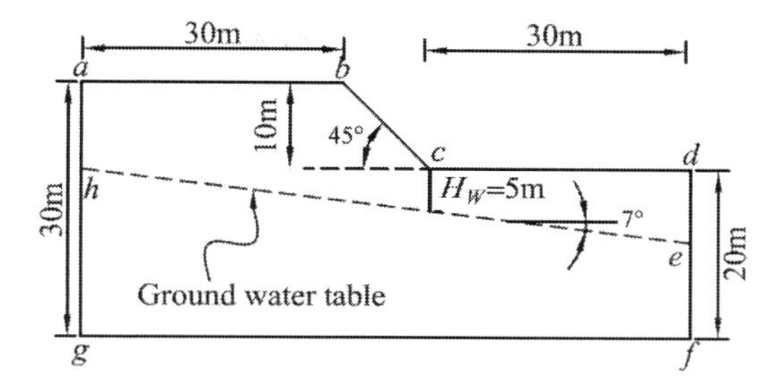

Figure 5. Geometry and boundary conditions of the unsaturated slope.

Table 1. Material and hydraulic properties of the sandy soil.

Parameter	Value
Unite weight γ(kN/m³)	20
Friction angle ϕ^b (°)	17.3
Effective friction angle ϕ'(°)	26
Effective cohesive c'(kPa)	0
Saturated volumetric water content θ_s(%)	45
Saturated permeability coefficient k_s (m/s)	2×10^{-6}
Fitting parameter related to air-entry pressure (kPa)	100
Fitting parameter related to residual water content m	1
Fitting parameter related to the slope of SWCC n	1
Fitting parameter related to the permeability function p	4

3.2 *Material properties of unsaturated soil*

The variations of the shear strength and unit weight are not considered in order to focus on the effect of random rainfall patterns on slope stability. The soil properties for this case are presented in Table 1.

The corresponding SWCC and permeability function (PF) for this sandy soil are displayed in Figure 6.

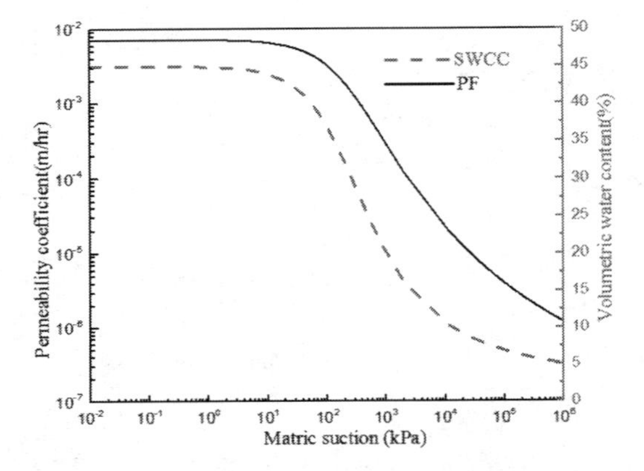

Figure 6. Soil-water characteristic curves and permeability function (PF) for the sandy soil.

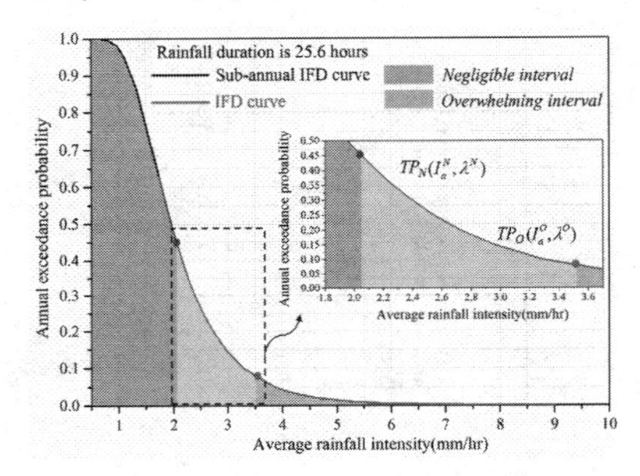

Figure 7. Average rainfall intensity versus annual exceedance probability for 25.6 hr rainfall duration.

3.3 *The generation of IFD curves and random rainfall patterns*

It can be seen from section 2.2 that the two statistical parameters, ζ_0 and H_0 [see, in equation (3)] determine the probability distribution of the cascade weight W. The weight W randomly generated in each step of disaggregation would further affect the rainfall patterns. The primary purpose of this paper is to study the effect of random rainfall patterns on the stability analysis of slope. Therefore, these parameters are set to the constants of ζ_0 = 12.27 and H_0 = 0.47 following Menabde and Sivapalan (2000), which are estimated based on the rainfall dataset with 6 min intervals from Melbourne, Australia. The number of cascade step, Ns, is set to 8 for generating random rainfall patterns with 25.6 hr rainfall duration and the t_s^{Ns} = 0.1 hr (6 min). To maintain consistency with the adopted statistical parameters, Melbourne is chosen as the research area to obtain the corresponding IFD curves, and the standard information of IFD curves under these rainfall durations can be extracted from the website of BOM (http://www.bom.gov.au/water/designRainfalls/ifd/index.shtml). The IFD curve for the exceedance probability of average rainfall intensities under 25.6 hr rainfall durations is shown in Figure 7.

It is worthwhile mentioning that the sub-annual IFD curves (the black curves) extracted from BOM, are also drawn here just for the completeness of IFD curve.

4 RESULTS AND DISCUSSION

The *Negligible and Overwhelming* intervals are represented by light blue and green regions, as shown in Figure 7. The light grey section between the two rainfall threshold points, TP_N and TP_O, is divided into 30 intervals, then the acceptable AFP can be obtained. Meanwhile, it can be seen from Figure 7 that the sub-annual IFD curves belong to negligible intervals, which has no effect on the estimation of AFP of unsaturated slope as mentioned above.

The relationship between the CFP of slope or the FS of slope and average rainfall intensity under random rainfall events and three representative rainfall events are determined to illustrate the effect of random rainfall patterns and three representative rainfall patterns on the stability of the unsaturated slope. The results are depicted in Figure 8.

As expected, the CDF of the unsaturated slope increases with the increase of rainfall intensity under random rainfall event. FS_{min} decreases with increasing rainfall intensity regardless of typical rainfall patterns. For three representative rainfall patterns, the slope is stable when the I_a is smaller than 2.52 mm/hr, and is unstable when the I_a is

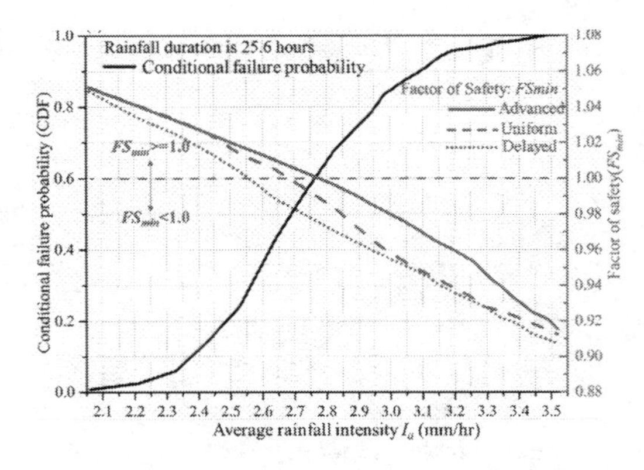

Figure 8. Variations of the CPF and FS of the slope under random rainfall patterns and three representative rainfall pattern.

Table 2. The AFP of unsaturated slope.

Representative rainfall patterns			Random rainfall patterns
Advanced	Uniform	Delayed	
0.202	0.223	0.294	0.214

greater than 2.78 mm/hr, regardless of rainfall types. However, when I_a is in the range of 2.52–2.78 mm/hr, the stability of slope is sensitive to the representative rainfall pattern. In this average rainfall intensity range, the slope is in the state of stable under advanced rainfall pattern, while is unstable in other representative rainfall pattern (e.g. delayed rainfall pattern). It means that estimating the stability of slope with a certain rainfall pattern cannot provide the most rational solution. Notably, when considering the random rainfall patterns, even if the slope is stable under three representative rainfall patterns from the FS_{min} perspective, the conditional failure probability of slope is relatively high. Namely, the slope is unstable under a lot of generated random patterns. Therefore, using only a single FS_{min} under a certain typical rainfall pattern cannot provide the most rational solution, and then the stability state of the unsaturated slope under rainfall event may not be adequately estimated.

According to the section 2.5, the AFPs of unsaturated slope under random and representative rainfall patterns are obtained and listed in Table 2.

As shown in Table 2, the AFP of the unsaturated slope for the delayed rainfall patterns to be the highest (i.e. least safe), which is much larger than that for random rainfall patterns, and the difference reaches to 37%; obviously, this will cause a conservative solution, which will lead to an uneconomical design; the AFP of the unsaturated slope for advanced rainfall patterns to be the lowset(i.e. the safest) is slightly smaller than that for random rainfall, and the difference is 6%. It indicates that using advanced rainfall will lead to an optimistic result and underestimate the risk of landslide, which will pose a potential danger in the design of slope. Relatively, the AFP of the unsaturated slope for the uniform rainfall patterns is the closest to that of the random rainfall patterns, which means that the uniform rainfall pattern can provide the most acceptable solution for this case to a certain extent. The main reason may be that the realizations of random rainfall patterns are closer to advanced and uniform rainfall patterns, rather than delayed rainfall pattern. It must also be mentioned that even though using uniform and advanced rainfall pattern can provide a relatively reliable solutions. Nevertheless they still cannot provide the accurate estimation of AFP, compared with using random rainfall patterns.

5 CONCLUSION

This paper studies the stability of unsaturated slope under rainfall conditions considering the random rainfall patterns. The CFP and AFP of the unsaturated slope are investigated. The results show that the slope stability is sensitive to rainfall patterns and the random rainfall patterns should be considered in the stability analysis of unsaturated slopes especially in the estimation of AFP.

REFERENCES

Afungang, R.N. & Bateira, C.V. 2016. Temporal probability analysis of landslides triggered by intense rainfall in the Bamenda Mountain Region, Cameroon. *Environmental Earth Sciences,* 75.

AGS 2007. Guideline for landslide susceptibility, hazard and risk zoning for land use planning. *Australian Geomechanics,* 42, 13–36.

Coles, S., Bawa, J., Trenner, L. & Dorazio, P. 2001. *An introduction to statistical modeling of extreme values*, Springer.

Fredlund, D.G. & Rahardjo, H. 1993. *Soil mechanics for unsaturated soils*, John Wiley & Sons.

Fredlund, D.G. & Xing, A. 1994. Equations for the soil-water characteristic curve. *Canadian geotechnical journal,* 31, 521–532.

Geo-Slope International Ltd 2010a. Seep/W User's Guide for Finite Element Seepage Analysis. 2010 ed. Calgary, Alberta, Canada: GEO-SLOPE International Ltd.

Geo-Slope International Ltd 2010b. Slope/W User's Guide for Slope Stability Analysis. Calgary, Alberta, Canada.: GEO-SLOPE International Ltd.

Green, J., Xuereb, K. & Jolly, C. Enhancing the new intensity-frequency-duration (IFD) design rainfalls-sub-annual ifds. Hydrology and Water Resources Symposium 2014, 2014. Engineers Australia, 605.

Leong, E.C. & Rahardjo, H. 1997. Review of soil-water characteristic curve equations. *Journal of geotechnical and geoenvironmental engineering,* 123, 1106–1117.

Menabde, M. & Sivapalan, M. 2000. Modeling of rainfall time series and extremes using bounded random cascades and levy-stable distributions. *Water Resources Research,* 36, 3293–3300.

Ohtsu, H., Janrungautai, S., Takahashi, K., Supawiwat, N. & Ohnishi, Y. 2005. Risk Evaluation Of Slope Failure Induced By Rainfall Using The Simplified Storage Tank Model. *Geotechnical Engineering,* 36, 157–164.

Peres, D.J. & Cancelliere, A. 2014. Derivation and evaluation of landslide-triggering thresholds by a Monte Carlo approach. *Hydrology and Earth System Sciences,* 18, 4913–4931.

Phoon, K.-K. & Ching, J. 2014. *Risk and reliability in geotechnical engineering*, CRC Press.

Rahardjo, H., Ong, T.H., Rezaur, R.B. & Leong, E.C. 2007. Factors Controlling Instability of Homogeneous Soil Slopes under Rainfall. *Journal of Geotechnical and Geoenvironmental Engineering,* 133, 1532–1543.

Rahimi, A., Rahardjo, H. & Leong, E.-C. 2010. Effect of antecedent rainfall patterns on rainfall-induced slope failure. *Journal of Geotechnical and Geoenvironmental Engineering,* 137, 483–491.

Rodríguez-Ochoa, R., Nadim, F., Cepeda, J.M., Hicks, M.A. & Liu, Z. 2015. Hazard analysis of seismic submarine slope instability. *Georisk: Assessment and Management of Risk for Engineered Systems and Geohazards,* 9, 128–147.

Schultz, M.T., Gouldby, B.P., Simm, J.D. & Wibowo, J.L. 2010. Beyond the factor of safety: Developing fragility curves to characterize system reliability. Engineer Research And Development Center Vicksburg MS Geotechnical And Structures Lab.

Sekiguchi, N., Ohtsu, H. & Yasuda, T. 2009. Study on road slope disaster prevention integrated management system. *Prediction and Simulation Methods for Geohazard Mitigation.* Kyoto, Japan: Taylor & Francis.

White, J.A. & Singham, D.I. 2012. Slope Stability Assessment using Stochastic Rainfall Simulation. *Procedia Computer Science,* 9, 699–706.

Yuan, J., Papaioannou, I. & Straub, D. 2015. Reliability Analysis of Infinite Slopes under Random Rainfall Events. *Geotechnical Safety and Risk V*, 439.

Numerical Methods in Geotechnical Engineering IX – Cardoso et al. (Eds)
© 2018 Taylor & Francis Group, London, ISBN 978-1-138-33198-3

Numerical evaluation of fragility curves for earthquake liquefaction induced settlements of a levee using Gaussian Processes

F. Lopez-Caballero
Laboratoire MSS-Mat CNRS UMR, CentraleSupélec Paris-Saclay University, France

ABSTRACT: The major cause of earthquake damage to an embankment is the liquefaction of the soil foundation that induces ground level deformations. It is well known that the liquefaction appears when the soil loses its shear strength due to the excess of pore water pressure. The aim of this paper is to assess numerically the effect of the liquefaction-induced settlement of the soil foundation on an embankment due to real earthquakes. For this purpose, a 2D Finite Element (FE) model of a levee founded on a layered soil/rock profile was considered. An elastoplastic multi-mechanism model was used to represent the soil behaviour. The crest settlement of the embankment was selected as the quantifiable damage variable of the study. Fragility functions were drawn to give the probability exceedance of some proposed damage levels as function of a seismic severity parameter. However, FE analysis can be expensive due to very large number of simulations needed for an accurate assessment of the system failure behaviour. This problem is addressed by building a Gaussian Process (GP) emulator to represent the output of the expensive FE model. A comparison with the FE reference results suggests that the proposed GP model works well and can be successfully used as a predictive tool to compute the induced damage on the levee.

1 INTRODUCTION

Earthquakes are the most natural phenomenon that cause damage to the soil and to the structures, in addition to other losses such as human and economic losses. Liquefaction phenomenon is considered as one of the most devastating and complex behaviours that affect the soil due to shakings. It is defined as the loss of the soil shear strength due to the excess of pore water pressure. The most affected structures by liquefaction foundation are the earth dams (Matsuo 1996, Ozutsumi et al. 2002, Unjoh et al. 2012, Okamura et al. 2013). Thus, to best design an embankment, its stability and performance should be taken into account.

There are several ways of failure of a levee: a disruption by a fault movement, a slope failure, a piping failure through cracks or a crest settlement. The crest settlement is the parameter that more easily quantifies the damage failure of an earth levee. According to Swaisgood (2003), the damages could be divided into four levels based on the crest settlement of the embankment and the peak ground acceleration of the input signal.

The following paper aims to assess numerically the effect of soil liquefaction-induced failure to a dam due to real earthquakes. It is based on the Performance Based Earthquake Engineering methodology (PBEE) developed by the federally funded earthquake engineering research center (Pacific Earthquake Engineering Research PEER) (Porter 2003). A deterministic study to quantify a failure way of a levee (crest settlement) and a probabilistic study to find the probability of exceedance of a certain level of performance, took place. In the later, the response to the earthquake is represented by an engineering demand parameter (*EDP*). Thus, fragility functions are drawn for this purpose. Hence, two stages of the PEER methodology were satisfied. In order to account for the natural hazards, in which an intensity measure (*IM*) parameter is identified, the input ground motions were used and chosen to be real motions to be consistent with the seismic parameter, magnitude, site to source distance, design and the duration of the earthquake. The finite element calculations were performed using the *GEFDyn* code (Aubry and Modaressi 1996). An elastoplastic multi-mechanism model was used to represent the soil behaviour.

Consequently, a database including a great number of ground motions is required to provide enough information to estimate in a reliable way the parameters defining the fragility curves (Baker and Cornell 2008, Saez et al. 2011). However, due to the high computational cost to perform the numerous non-linear dynamic calculations, it is no feasible to explore a large design space using the complex proposed FE model. In this context, fast-running models, also called surrogate models could be implemented by means of input-output data sets to approximate the response of the original FE model (Sacks et al. 1989, Dubourg et al. 2013, Stern et al. 2017). Hence, in this work a Gaussian process emulator (GP) was used as a surrogate model for the levee-foundation system, so as to reduce the computation time associated keeping an accurate prediction. The GP model was built using input model parameters that are relevant to represent the system response of the inelastic transient FE analysis. Once the GP model was trained and validated, it is applied to quantify the effect of soil liquefaction-induced failure on a levee subject to a large variety of earthquake events. In particular, the maximal induced crest settlement is computed and the corresponding fragility curves for a given damage threshold are estimated.

2 NUMERICAL MODEL DESCRIPTION

The geometry of the model, as shown in Figure 1, consists of an embankment of 9 m high composed of dry dense sand. The soil foundation is composed by a liquefiable loose sand of 4 m at the top of a saturated dense sand of 6 m. The bedrock at the bottom of the dense sand has a shear wave velocity V_s = 800 m/s. The water table is situated at 1 m below the base of the levee, which remains dry. The levee's inclination is a slope of 1:3 (V:H).

A 2D coupled finite element modelling with *GEFDyn* code (Aubry and Modaressi 1996) is carried out using a dynamic approach derived from the $\underline{u} - p_w$ version of the Biot's generalized consolidation theory (Zienkiewicz and Shiomi 1984). The FE model is composed of quadrilateral isoparametric elements with eight nodes for both solid displacements and fluid pressures. So the total number of nodes is 2331. The soil permeability for the loose sand is taken as $k_s = 10^{-4}$ m/s and is supposed to be isotropic. The FE analysis is performed in three consecutive steps: i) a computation of the initial *in-situ* stress state due to gravity loads; ii) a sequential level-by-level construction of the embankment; and iii) a seismic loading analysis in the time domain.

2.1 *Soil constitutive model*

The ECP's elastoplastic multi-mechanism model (Aubry et al. 1982, Hujeux 1985), commonly called Hujeux model is used to represent the soil behaviour. This model can take into account the soil behaviour in a large range of deformations. The model is written in terms of effective stress. The representation of all irreversible phenomena is made by four coupled elementary plastic mechanisms: three plane-strain deviatoric plastic deformation mechanisms in three orthogonal planes and an isotropic one. The model uses a Coulomb type failure criterion and the critical state concept. The evolution of hardening is based on the plastic strain (deviatoric and volumetric strain for the deviatoric mechanisms and volumetric strain for the isotropic one). To take into account the cyclic behaviour a kinematical hardening based on the state variables at the last load reversal is used. The soil behaviour is decomposed into pseudo-elastic, hysteretic and mobilized domains.

The obtained curves of cyclic stress ratio ($SR = \sigma_{v-cyc}/(2 \cdot p'_o)$, with σ_{v-cyc} the cyclic vertical stress applied in the cyclic loading) as a function of the number of loading cycles to produce liquefaction (N) and $G/G_{max} - \gamma$ curves are given in Figure 1. As qualitative comparison, the modelled test results are compared with the experimentally obtained curves given by Byrne et al. (2004) for Nevada sand at different densities (i.e. D_r = 40% and 60%) and with the reference curves given by Seed and Idriss (1971).

2.2 *Input earthquake motion*

The selection of input motions for geotechnical earthquake engineering problems is important as it is strongly related to the non-linear dynamic analyses. So as to obtain analytical fragility curves, it is necessary to analyse the embankment response to a wide range of ground motions. In addition, when dealing with surrogate models, it is required to have a representative set of data to train, to validate and to test the proposed meta-model. A total of 540 unscaled records were chosen from the Pacific Earthquake Engineering Research Center (PEER) database, the Center for Engineering Strong Motion Data and the Kiban Kyoshin strong-motion network (KIK-NET). The events range between 5.2 and 7.6 in magnitude and the recordings have site-to-source distances from 15 to 50 km and concern dense-to-firm soil conditions.

The database was split as follows: 95 signals concern the learning database (LDB), 50 ground-motions are used for the validation set (VDB) and the test database (TDB) is composed of 395 unscaled records. Concerning the response spectra of input earthquake motions, Figure 2 shows the mean and the response spectra curves (structural damping ξ = 5%) of all set of input motions.

The statistics on some input earthquake characteristics obtained for each database are summarized in Table 1. These earthquake characteristics are maximal outcropping acceleration ($a_{max\ out}$), Arias intensity (I_A), mean period (T_m) (Rathje et al. 1998), peak ground velocity (PGV), period of equivalent harmonic wave ($T_{V/A} = \alpha \cdot PGV/PHA$) and significant duration from 5% to 95% Arias intensity ($D_{5\text{-}95}$).

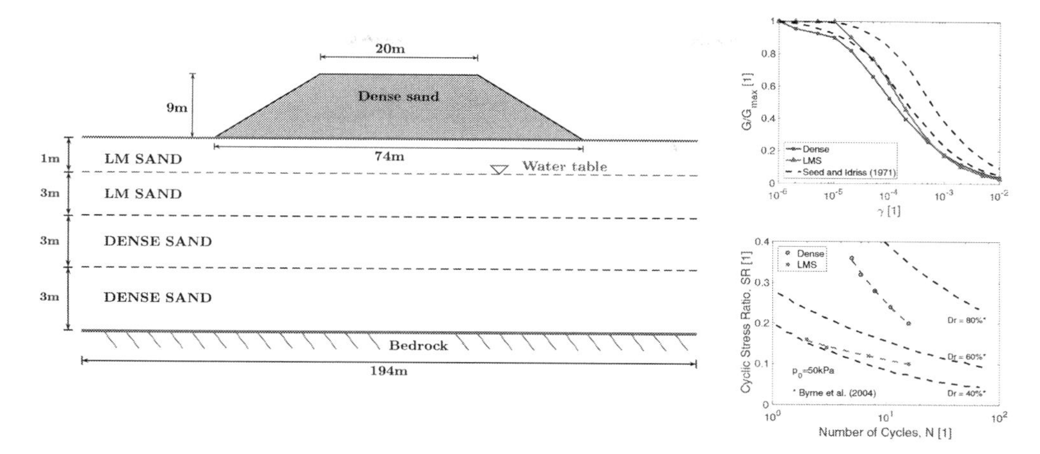

Figure 1. Geometry and behaviour of soils used in the numerical model.

Table 1. Statistics characteristics for the selected earthquakes in each database.

Parameter	LDB Range	VDB Range	TDB Range
$a_{max\ out}$ [g]	0.01–1.93	0.03–1.16	0.03–1.93
T_m [s]	0.12–1.69	0.17–1.69	0.17–1.81
$T_{V/A}$ [s]	0.09–1.91	0.13–1.42	0.13–1.32
I_A [m/s]	0.001–20.64	0.04–4.13	0.004–20.64
$D_{5\text{-}95}$ [s]	2.26–69.84	2.96–42.77	2.26–47.36
PGV [cm/s]	0.23–167.6	4.27–83.58	0.86–166.1

3 GAUSSIAN PROCESS EMULATOR

A meta-model is an analytical function used to provide rapid approximations of more expensive models (e.g. an analytical model or a finite element numerical model). In the Gaussian process (GP), the responses and input values are combined statistically to create functional relationships in a non-intrusive approach (i.e. the original model is considered as a black box). One of the advantages of Gaussian processes is that they are flexible enough to represent a wide variety of complex models using a limited number of parameters (Sacks et al. 1989).

Let us consider a non-linear computer model response, that could be represented by a multivariate function $\mathbf{y} = f(\mathbf{x})$; where $\mathbf{x}$ is a d-dimensional vector describing the input parameters of the model and $\mathbf{y}$ is a vector of n observed outputs. Usually, $f(\mathbf{x})$ is deterministic and it is also assumed that its evaluation is computationally expensive, thus, only limited function evaluations $\mathbf{y}_1 = f(\mathbf{x}_1),\ldots,\mathbf{y}_n = f(\mathbf{x}_n)$ are available. These evaluations are called experimental design (ED) and they are used as a database for learning the meta-model (i.e. the learning database LDB). The purpose of the meta-model is therefore to predict the response ($\mathbf{y} = f(\mathbf{x}^*)$) for a new data set where only the input $\mathbf{x}^*$ are known (i.e. the test database, TDB) (Sacks et al. 1989, DiazDelaO et al. 2013, Strong and Oakley 2014). Hence, it is possible to obtain a statistical approximation to the output of a numerical model after evaluating a small number n of design points if $f(\mathbf{x})$ is modelled as a Gaussian process (GP). A GP is a collection of random variables, which have a joint multivariate Gaussian distribution. The GP model will be separated in mean and covariance functions:

$$f(\mathbf{x}) = \mathbf{h}(\mathbf{x})^T \beta + Z(\mathbf{x}) \tag{1}$$

where $\mathbf{h}(\mathbf{x})^T \beta$ is the mean function (usually modelled as a generalized linear model and sometimes times assumed to be zero), $\mathbf{h}(\mathbf{x})$ is a vector of known functions and β is a vector of unknown coefficients. The function $Z(\cdot)$ is a Gaussian process with mean zero and covariance function $\mathrm{Cov}(Z(\mathbf{x}), Z(\mathbf{x}') \mid \sigma^2, \theta)$ between output points corresponding to input points $\mathbf{x}$ and $\mathbf{x}'$:

$$\mathrm{Cov}(Z(\mathbf{x}), Z(\mathbf{x}') \mid \sigma^2, \theta) = \sigma^2 \cdot c_\theta(\mathbf{x}, \mathbf{x}') \tag{2}$$

where σ^2 is the variance of Z, θ the range parameter and $c_\theta(,)$ its correlation function. The GP assumes that the correlation between $Z(\mathbf{x})$ and $Z(\mathbf{x}')$ is a function of the "*distance*" between $\mathbf{x}$ and $\mathbf{x}'$. The covariance can be any function having the property of generating a positive definite covariance matrix (Rasmussen and Williams 2006, Iooss et al. 2010). A wide variety of covariance functions could be used in the Gaussian process framework, however, for a sake of brevity, in this work only one correlation function was used, namely, exponential (equation 3).

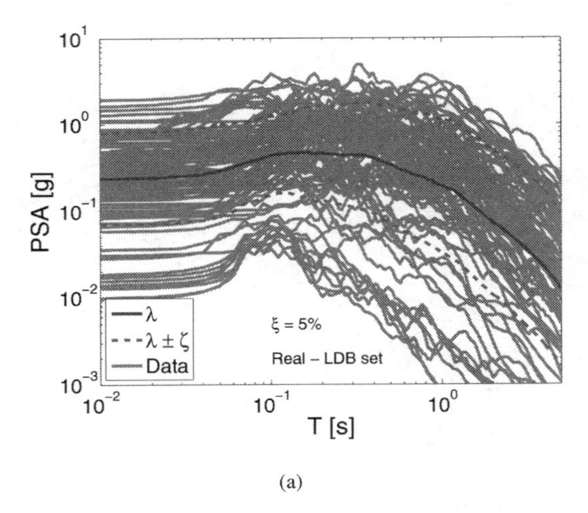

(a)

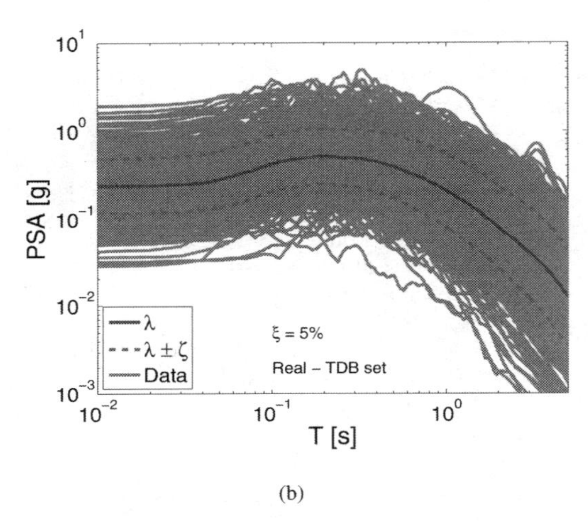

(b)

Figure 2. Response spectra of input earthquake motions (5% damped), a) LDB set; b) VDB and TDB set.

$$c_\theta(\mathbf{x},\mathbf{x}') = \exp\left\{ -\sum_{i=1}^{d} \frac{|x_i - x_i'|}{\theta_i} \right\} \tag{3}$$

for $\mathbf{x} = (x_1,\ldots,x_d)$. Finally, the hyperparameters involved in each covariance function are estimated by likelihood maximization.

4 NUMERICAL ANALYSIS USING FEM

For embankments placed in seismic zones, it has been shown that the widespread damage to such embankments occurred mainly due to the liquefaction of foundation soil, resulting in excessive settlements, lateral spreading and slope instability (Sharp and Adalier 2006, Oka et al. 2012, Okamura et al. 2013). Thus, in this study, the crest settlement is chosen to be the mode of failure because it is a quantifiable measurement.

Even if the earthquake loading applied to the soil-levee system is very complex, it is necessary to select few strong-motion intensity parameters that can accurately represent the levee behaviour. Swaisgood (2003) analysed a historical database on the performance of dams during earthquakes and found that the crest settlement is directly related to some input ground motion characteristics (i.e. the peak ground acceleration and magnitude). In addition, he proposes four damage levels according to the induced crest settlement. Following Swaisgood's proposition, in this work the obtained percentage crest settlement ($\delta u_{z,rel}/H$, where $u_{z,rel}$ is the crest settlement, H is the height of the dam plus the foundation which, is 19 m as seen in Figure 1) is compared to the peak ground acceleration at the outcropping bedrock ($a_{max\ out}$). To take into account all the signals in the LDB set, the crest settlement was calculated accordingly and it was drawn as function of $a_{max\ out}$ (Figure 3). It is interesting to note that, as expected, the calculated crest settlement increases when the acceleration at the outcrop increases.

5 GP MODEL IDENTIFICATION

One of the problem of calibrating or training a surrogate model (GPM) to observations from the numerical model (FEM) deals with finding input values such that the GPM outputs match the observed data as closely as possible. According to the previous works, several strong-motion intensity parameters have a great influence on the levee response, e.g. $a_{max\ out}$, $1/T_{V/A}$ and I_A among others (Koutsourelakis et al. 2002, Lopez-Caballero and Modaressi-Farahmand-Razavi 2013). Thus, it means that the proposed GPM will be a multiple-input single-output one. Other aspect concerns the correlation function defining the Gaussian process itself (i.e. found the unknown hyperparameters). As recalled before, only the exponential correlation function is tested in this section (equation 3). The hyperparameters for this model are estimated with the R-code packages for the Analysis of Computer Experiments developed by Roustant et al. (2012). Once the GPM was trained with the LDB set, the possible model is validated, hence, a comparison of all responses in terms of $\delta u_{z,rel}/H$ obtained with FEM and predicted by GPM using the VDB set will be done. Further, the selected GP model is tested on a database (TDB) that is similar in structure to the database which was used for train-

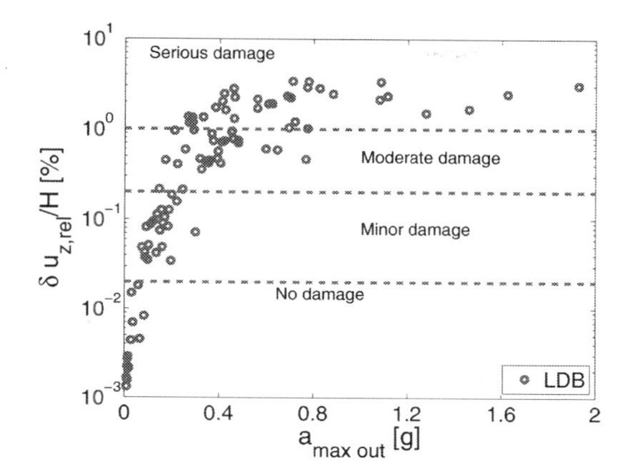

Figure 3. Scatter plot of crest settlement ratio of the FE model as a function of $a_{max\ out}$.

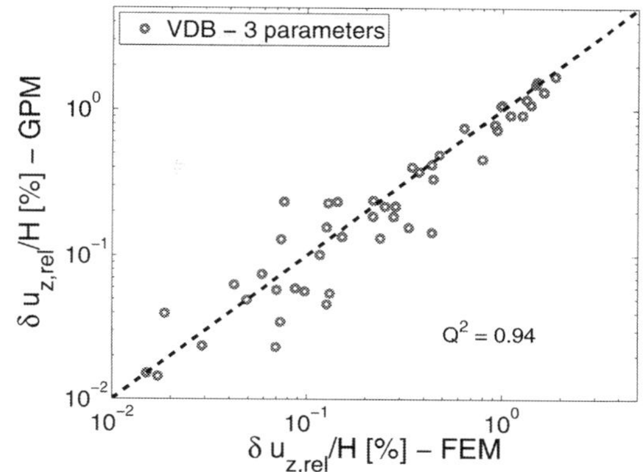

Figure 4. Comparison of $\delta u_{z,rel}/H$ values obtained with FEM and with GPM approaches. Case of VDB sets.

ing, but was not used to built the surrogate model (Table 1).

In order to assess how well the GP model has been trained (i.e. Validation phase) a comparison between the $\delta u_{z,rel}/H$ values obtained with the FEM and the mean predicted ones by the GPM using the VDB set is done (Figure 4). Thus, the relative error or discrepancy between the GPM predictions (y_i^{pred}) with the FEM computations (y_i) is calculated with the predictive squared correlation coefficient (Q^2):

$$Q^2\left(y_i, y_i^{pred}\right) = 1 - \frac{\sum_{i=1}^{N}(y_i^{pred} - y_i)^2}{\sum_{i=1}^{N}(y_i - \mu_y)^2} \tag{4}$$

where μ_y is the mean of the N observations (i.e. FEM computations). It ranges between 0 and 1.

Finally, the GPM is now used to simulate other earthquake scenarios (i.e. TDB with 445 signals). The mean $\delta u_{z,rel}/H$ values predicted by the GPM are shown in Figure 5. It can be noted that the GP model selected for this study shows a reasonable capability to reproduce the variation of $\delta u_{z,rel}/H$ as a function of $a_{max\ out}$.

6 EVALUATION OF VULNERABILITY USING GPM

In the context of the PBEE, the damage analysis, which is the third stage of this methodology, is a procedure to quantify the structural damage. It consists of setting fragility functions in order to find the conditional probability of the design to exceed a certain level of performance for a given seismic input motion parameter. Usually, fragility curves are constructed by using a single parameter

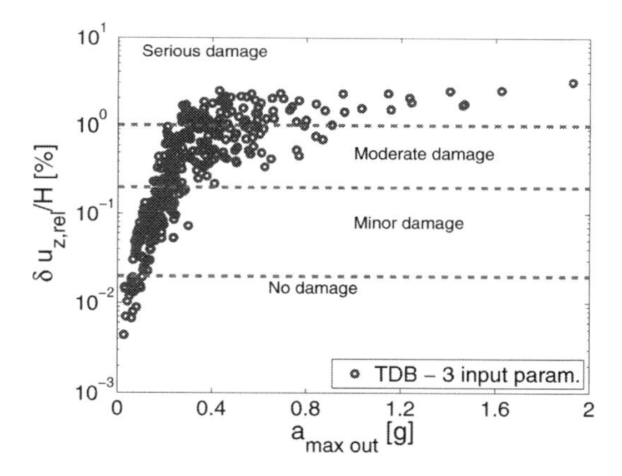

Figure 5. Mean predicted $\delta u_{z,rel}/H$ as a function of $a_{max\ out}$ obtained with the GPM approach. Case of TDB sets.

to relate the level of shaking to the expected damage (Koutsourelakis et al. 2002, Baker and Cornell 2008). So as to derive analytical fragility functions, it is necessary to define damage states in terms of some mechanical parameters (*EDP*) that can be directly obtained from the analysis (e.g. $\delta u_{z,rel}/H$). The damage states limits or the performance levels of the levee are those proposed by Swaisgood (2003). The three damage levels thresholds are superposed in Figures 3 and 5. They correspond to $\delta u_{z,rel}/H = 0.02$, 0.2 and 1.0%. In this work, the fragility curves are constructed following the methodology proposed by Shinozuka et al. (2000), i.e. the maximum likelihood method is used to compute numerical values of the estimators $\hat{\alpha}$ and $\hat{\beta}$ of Log-normal distribution.

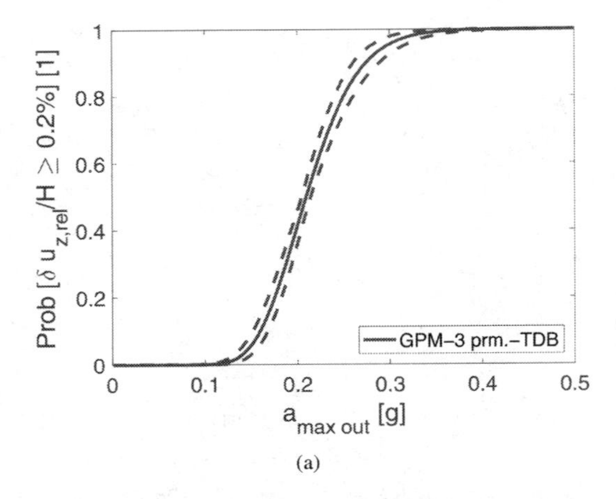

(a)

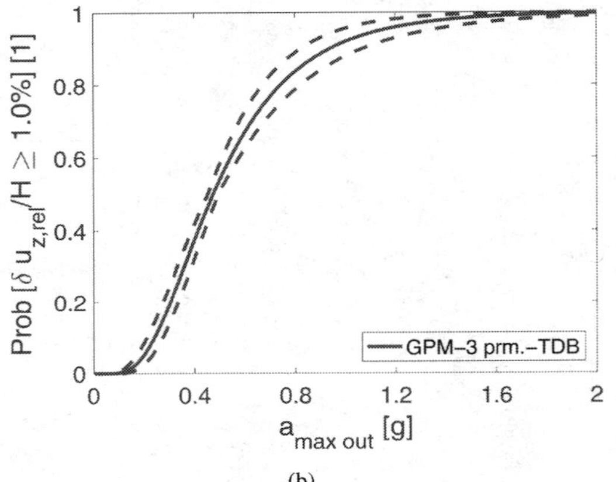

(b)

Figure 6. Computed fragility curves for two damage levels following GP model approach, a) minor to moderate and b) moderate to serious damages.

The obtained fragility curves for the third and fourth state damages (i.e. minor to moderate and moderate to serious damages) are shown in Figure 6. These curves are drawn as solid lines whereas the statistical confidence of the derived fragility curves are drawn as dashed lines (i.e. $\left\{\hat{\alpha}\hat{\beta}\right\} \pm \left\{\sigma_1\sigma_2\right\}$). This confidence is a function of the information provided by the size of motion database over the parameters $\hat{\alpha}$ and $\hat{\beta}$ describing the shape of each curve and it is computed via the Fisher information matrix (Saez et al. 2011). Figures 6(a) and 6(b) present fitted fragility functions obtained for two damage levels with respect to $a_{max\ out}$ using the proposed GPM and using 445 ground motions (i.e. VDB and TDB set).

The benefit of using a meta-model appears principally in the reduction of the statistical confidence of the derived fragility curves for both damage levels by increasing the size of tested motions (i.e. the obtained σ_1 and σ_2 values. This study confirms that the use of a well constructed surrogate models allows to obtain fragility curves with a reasonable accuracy and with a manageable computational effort.

7 CONCLUSIONS

A Finite Element analysis and a meta-model were used to investigate the soil liquefaction induced settlement and associated damage for an levee due to real earthquakes. Fragility functions were obtained for that purpose. The main conclusions drawn from this study are as follows:

- Seismic demand fragility evaluation is one of the basic elements in the framework of performance-based earthquake engineering (PBEE). For solving the absence of sufficient Finite Element responses to obtain fragility curves with a reasonable accuracy, a Gaussian process model was build to mimic the FEM and used to increase the number of levee model evaluations reducing the computational time.
- The predictive capability of the adopted GPM was assessed comparing the obtained levee settlements and induced damage levels with the ones simulated with the FEM. With respect to the case study considered, the GPM has shown a good capability of approximating the non-linear FEM response.
- Results reveal that a GPM with three inputs parameters (i.e. $1/T_{V/A}$, $a_{max\ out}$ and I_A) is able to describe the liquefaction induced settlement of a levee.
- Based on these analyses, it is concluded that the proposed Gaussian process model is accurate enough for practical purposes and represents an important economy in CPU consumption time. It is confirmed by the comparison between the fragility curves obtained by the two methods on the test data set.

ACKNOWLEDGEMENTS

The research reported in this paper has been supported in part by the SEISM Paris Saclay Research Institute.

REFERENCES

Aubry, D., J.-C. Hujeux, F. Lassoudière, & Y. Meimon (1982). A double memory model with multiple mechanisms for cyclic soil behaviour. In *International symposium on numerical models in geomechanics*, pp. 3–13. Balkema.

Aubry, D. & A. Modaressi (1996). GEFD*yn*. Manuel scientifique, Ecole Centrale Paris, LMSS-Mat.

Baker, J.W. & C.A. Cornell (2008). Uncertainty propagation in probabilistic seismic loss estimation. *Structural Safety 30*(3), 236–252.

Byrne, P.M., S.-S. Park, M. Beaty, M. Sharp, L. Gonzalez, & T. Abdoun (2004). Numerical modeling of liquefaction and comparison with centrifuge tests. *Canadian Geotechnical Journal 41*(2), 193–211.

DiazDelaO, F.A., S. Adhikari, E.I. Saavedra Flores, & M.I. Friswell (2013). Stochastic structural dynamic analysis using bayesian emulators. *Computers & Structures 120*(1), 24–32.

Dubourg, V., B. Sudret, & F. Deheeger (2013). Metamodel-based importance sampling for structural reliability analysis. *Probabilistic Engineering Mechanics 33*, 47–57.

Hujeux, J.-C. (1985). Une loi de comportement pour le chargement cyclique des sols. In *Génie Parasismique*, pp. 278–302. V. Davidovici, Presses ENPC, France.

Iooss, B., L. Boussouf, V. Feuillard, & A. Marrel (2010). Numerical studies of the metamodel fitting and validation processes. *International Journal On Advances in Systems and Measurements, IARIA 3*, 11–21.

Koutsourelakis, S., J.H. Prévost, & G. Deodatis (2002). Risk assessment of an interacting structure-soil system due to liquefaction. *Earthquake Engineering and Structural Dynamics 31*(4), 851–879.

Lopez-Caballero, F. & A. Modaressi-Farahmand-Razavi (2013). Numerical simulation of mitigation of liquefaction seismic risk by preloading and its effects on the performance of structures. *Soil Dynamics and Earthquake Engineering 49*(1), 27–38.

Matsuo, O. (1996). Damage to river dikes. *Soils and Foundations 36*(1), 235–240.

Oka, F., P. Tsai, S. Kimoto, & R. Kato (2012). Damage patterns of river embankments due to the 2011 off the pacific coast of tohoku earthquake and a numerical modeling of the deformation of river embankments with a clayey subsoil layer. *Soils and Foundations 52*(5), 890–909.

Okamura, M., S. Tamamura, & R. Yamamoto (2013). Seismic stability of embankments subjected to predeformation due to foundation consolidation. *Soils and Foundations 53*(1), 11–22.

Ozutsumi, O., S. Sawada, S. Iai, Y. Takeshima, W. Sugiyama, & T. Shimazu (2002). Effective stress analyses of liquefaction-induced deformation in river dikes. *Soil Dynamics and Earthquake Engineering 22*(9–12), 1075–1082.

Porter, K.A. (2003). An overview of PEER's Performance-Based Earthquake Engineering Methodology. In *Ninth International Conference on Applications of Statistics and Probability in Civil Engineering (ICASP9) July 6–9, San Francisco*.

Rasmussen, C.E. & C.K.I. Williams (2006). *Gaussian Processes for Machine Learning*. the MIT Press.

Rathje, E.M., N.A. Abrahamson, & J.D. Bray (1998). Simplified frequency content estimates of earthquake ground motions. *Journal of Geotechnical and Geoenvironmental Engineering 124*(2), 150–159.

Roustant, O., D. Ginsbourger, & Y. Deville (2012). DiceKriging, DiceOptim: Two R packages for the analysis of computer experiments by kriging-based metamodeling and optimization. *Journal of Statistical Software 51*(1), 1–55.

Sacks, J., W.J. Welch, T.J. Mitchell, & H.P. Wynn (1989). Design and analysis of computer experiments. *Statistical Science 4*(4), 409–423.

Saez, E., F. Lopez-Caballero, & A. Modaressi-Farahmand-Razavi (2011). Effect of the inelastic dynamic soil-structure interaction on the seismic vulnerability assessment. *Structural Safety 33*(1), 51–63.

Seed, H.B. & I.M. Idriss (1971). Simplified procedure for evaluating soil liquefaction potential. *Journal of Soil Mechanics and Foundations Division – ASCE 97*(SM9), 1249–1273.

Sharp, M.K. & K. Adalier (2006). Seismic response of earth dam with varying depth of liquefiable foundation layer. *Soil Dynamics and Earthquake Engineering 26*(11), 1028–1037.

Shinozuka, M., Q. Feng, J. Lee, & T. Naganuma (2000). Statistical analysis of fragility curves. *Journal of Engineering Mechanics - ASCE 126*(12), 1224–1231.

Stern, R., J. Song, & D. Work (2017). Accelerated Monte Carlo system reliability analysis through machine-learning-based surrogate models of network connectivity. *Reliability Engineering & System Safety 164*(1), 1–9.

Strong, M. & J.E. Oakley (2014). When is a model good enough? deriving the expected value of model improvement via specifying internal model discrepancies. *SIAM/ASA Journal on Uncertainty Quantification 2*(1), 106–125.

Swaisgood, J.R. (2003). Embankment dam deformation caused by earthquakes. In *Pacific Conference on Earthquake Engineering*, pp. 014.

Unjoh, S., M. Kaneko, S. Kataoka, K. Nagaya, & K. Matsuoka (2012). Effect of earthquake ground motions on soil liquefaction. *Soils and Foundations 52*(5), 830–841.

Zienkiewicz, O.C. & T. Shiomi (1984). Dynamic behaviour of saturated porous media; the generalised biot formulation and its numerical solution. *International Journal for Numerical and Analytical Methods in Geomechanics 8*, 71–96.

Numerical Methods in Geotechnical Engineering IX – Cardoso et al. (Eds)
© 2018 Taylor & Francis Group, London, ISBN 978-1-138-33198-3

Reliability analysis of constant total stress foundations subjected to water table fluctuations

José A. Alonso-Pollán & Luis M. Muñoz
Technical Directorare. Departments of Ground Engineering and Structures, Dragados, S.A., Madrid, Spain

R. Jimenez
Escuela de Caminos, Technical University of Madrid, Spain

ABSTRACT: Constant total stress foundations are a special class of geotechnical problems in which the total load transmitted by the foundation is assumed constant, and in which water table fluctuations (below the ground surface) change effective stress, hence leading to possible settlements (when the water table descends) or heave of the foundation (when it ascends).

We address a particular case of this problem, in which a building with a diaphragm wall foundation is planned at a site in which a high permeability gravel layer, that is hydraulically connected to a nearby river, overlays a fine-grained stratum with lower permeability. We analyze how transient water table ascensions (associated, for instance, to the design flood for a given return period) affect the effective stress at the base slab of the building, with particular attention to the number of days that such high water levels must be kept at higher-than-normal elevations so as to produce likely foundation problems. The geotechnical and flow problems are solved using a finite element code and, to be able to assess the influence of existing uncertainties on the computed results, we conduct a reliability analysis of the problem using the simplified procedure recommended by the Spanish Recommendations for the Design of Maritime and Harbour works. Finally, we conduct a sensitivity analysis to assess the influence of the main geometric variables associated to the diaphragm wall design on the computed results.

1 INTRODUCTION

Constant total stress foundations are a special class of geotechnical problems in which the total load transmitted by the foundation is assumed constant, and in which water table fluctuations (below the ground surface) produce changes of effective stress due to changes of the pore water pressure; hence leading to possible settlements (when the water table descends) or heave of the foundation (when it ascends), or even to possible problems due to the loss of equilibrium of the structure due to uplift by water pressure ("buoyancy" failure mode).

We address a particular case of this problem, that generated from our practice, in which a building with several basement levels and a diaphragm wall foundation is planned at a site in which a high permeability granular layer, that is hydraulically connected to a nearby river, overlays a fine-grained stratum with lower permeability. More specifically, we analyze how transient water table ascensions (associated, for instance, to the design flood for a given return period) affect the effective stress at the base slab of the building, hence affecting the possibility of occurrence of a "buoyancy" failure mode; and we also analyze the influence of the hydrological regime of the river or, in other words, the number of days that such high water levels have to be kept at higher-than-normal elevations so as to produce likely foundation problems. Computations are conducted using a finite-element formulation that accounts for the coupled effects of the deformability of the ground (that controls the amount of heave) and its permeability (that, together with deformability, controls the velocity of the process).

Furthermore, given the uncertainties associated to the geotechnical parameters involved, we tackle this problem using a probabilistic (or reliability-based) methodology, in which uncertainties are quantitatively characterized and incorporated into the analysis. And, finally, we also assess the influence of other design parameters—such as the depth of embedment of the base slab within the impervious layer; or the depth of penetration of the wall toe below the slab—on the computed results.

2 GROUND MOVEMENTS DUE TO PHREATIC OSCILLATIONS

2.1 *Introduction*

The influence of water table (or phreatic) oscillations on the behaviour of geotechnical designs is a classical problem that has been analyzed by several authors in the past (Terzaghi, Peck, & Mesri 1996, Terzaghi 1943, Puertos del Estado 2005, Venkatramaiah 1995). The reason is that ground water fluctuations are common in most sites (due for instance, to seasonal changes of weather conditions; or to changes of river or sea levels).

Therefore, pore pressures also change due to such phreatic fluctuations, hence producing changes of effective stresses within the ground. For instance, an elevation of the phreatic surface under constant external load increases the pore water pressure much more than it increases the soil's total stress at any depth level, as the increase of soil's weight (from its apparent total unit weight to its saturated one) is much less than the water's unit weight, hence significantly reducing the effective stress of the soil and increasing its volume (heave). (One exception is when water level varies above the ground surface, as fluctuations of the water level above the ground surface do not change the effective stress within a soil mass.)

It is also important that such changes of pore pressures and effective stresses are not immediate in low permeability soils. They take some time to develop (*i.e.*,stresses suffer transient changes before reaching their final, hydrostatic, values), as the capability of the soil to deform in response to the effective stress changes is limited by the rate at which water can enter (or leave) the soil mass or, in other words, by the permeability of the soil. (The geometry of the problem is of course also important.) Therefore, geometry, deformability and permeability are important factors that control the velocity of the consolidation process that develops in response to ground water level fluctuations below the ground surface.

2.2 *Governing equations*

Accepting the common assumptions that water and soil particles are incompressible, the process of consolidation under compression (or of heave under unloading) of a soil mass is controlled by one fundamental idea (Lambe & Whitman 1995, Jiménez Salas & de Justo Alpañés 1976, ?): the net flow of water in a soil element must be equal to the change of volume of water within the soil element under consideration. If the resulting continuity equation is combined with the equations of equilibrium and compatibility, a finite element (FE) formulation of the process can be developed.

Here, we employ the FE implementation of coupled consolidation available in program RS2 (Phase2 9.0), a 2D finite element program for soil and rock applications (Rocscience 2017). Phase2 implements Biot's equations of coupled consolidation, in which the soil skeleton "is treated as a porous elastic solid with laminar pore fluid coupled with it"; the governing equation is (Rocscience 2017):

$$\frac{K'}{\gamma w}\left[k_x\frac{\partial^2 u_w}{\partial x^2}+k_y\frac{\partial^2 u_w}{\partial y^2}+k_z\frac{\partial^2 u_w}{\partial z^2}\right]=\frac{\partial u_w}{\partial t}-\frac{\partial p}{\partial t} \quad (1)$$

where K' is the bulk modulus of the soil, γ_w is the unit weight of water, k_x, k_y, k_z are the soil permeabilities in different directions, u_w is the pore water pressure, and p is the mean total stress. (t denotes time.)

Considering also the equilibrium equation in 2D, together with the constitutive laws for the solid skeleton and for fluid flow (elastic behaviour and generalized Darcy's law, respectively), as well as the compatibility equations of deformability; and using $\mathbf{r}$ to denote the column vector of nodal displacements, and $\mathbf{u_w}$ to denote the column vector of pore water pressures at the nodes —*i.e.*, $r=\{u_1,v_1,\ldots,u_4,v_4\}$ and $\mathbf{u_w}=\{u_{w1},\ldots,u_{w4}\}^T$ for 4-noded elements—, the following equations are obtained (Rocscience 2017):

$$\mathbf{KMr}+\mathbf{C}u_w=\mathbf{f} \quad (2)$$

$$\mathbf{C}^T\frac{d\mathbf{r}}{dt}-\mathbf{KP}u_w=0 \quad (3)$$

where **KM** and **KP** are the elastic solid and fluid matrices, **f** is the external loading vector, and **C** is a rectangular coupling matrix. (For additional details, see Rocscience (2017) and Smith & Griffiths (1997).)

3 RELIABILITY ANALYSIS

To estimate the probability of failure of a geotechnical design (or, equivalently, its reliability index) we employ a simple method indicated in the Spanish Recommendations for the Design of Maritime and Harbour works, or ROM (Puertos del Estado 2005). It associates the probability of failure to the usual methods to assess the performance of such design using global or partial factors of safety; such "Level I" methods can be considered as "ordered sensitivity analyses" similar to first-order second-moment (or FOSM) methods, and are therefore simpler than more advanced reliability methods—such as FORM, SORM, or Monte Carlo methods—that are referred to as Level II and Level III methods in Puertos del Estado (2001).

The method starts with the computation of a "centred" safety factor, F^*, computed for average values; it then repeats the computations changing the value of one input variable of interest at a time (out of the set of input variables whose variability is to be considered; in this case the permeability and stiffness of the ground) and keeping the other variables constant, to obtain the safety factor corresponding to such modified situation. Usually, the change is made equal to the standard deviation of the corresponding variable and, in the interest of precision, it is generally preferred to make such change both in favourable and unfavourable directions. A sensitivity coefficient, ν, can then be obtained for each input variable whose variations are of interest as:

$$\nu = \frac{F^+ - F^-}{2F^*}, \tag{4}$$

where F^+ is the safety factor associated to the change of input parameter in the favorable direction (and F^- refers to one associated to the unfavourable direction). In addition, when we aim to compute the reliability of a design under an "accidental" combination of loads, it is also possible to consider the sensitivity to changes on the extreme loading variable, ν_A. (In this case, it is associated to the rise of the phreatic surface.) It is computed as:

$$\nu_a = \frac{\Delta F_a}{F^*}, \tag{5}$$

where ΔF_a is the change of F due to the change of the loading.

Once the sensitivity coefficients to changes on the loads and input variables are available (*i.e.*, in our case, to changes on the water table level, and on the soil stiffness and permeability), the global sensitivity is computed as:

$$\nu_F = (\nu_a^2 + \nu_k^2 + \nu_E^2)^{1/2}. \tag{6}$$

And, once F^* and ν_F are obtained, the reliability index, β, for such situation can be computed computing the number of standard deviations that the "centred safety factor" needs to be moved so that failure (theoretically) occurs; *i.e.*, so that $F = 1$. Assuming a lognormal distribution of, F the solution is:

$$\beta = \frac{\ln F^*}{\xi} - \frac{1}{2}\xi, \tag{7}$$

with given as:

$$\xi = \sqrt{(\ln(1 + \nu_F^2))} \tag{8}$$

The probability of failure is uniquely associated to the reliability index by the following expression:

$$P_f = \Phi(-\beta), \tag{9}$$

where $\Phi(\cdot)$ is the cumulative density function (CDF) of the standard normal distribution.

4 APPLICATION EXAMPLE

4.1 *Introduction*

Next, we apply the methodologies described to analyze the evolution of pore pressures under the base slab of a 10-story building (with 7 stories above ground and three levels of underground basements) which is constructed at a site with a highly permeable gravel layer overlying a much more impervious clay formation. The gravel layer is hydraulically connected to a nearby river in which, despite the regulation effects provided by some dams constructed upstream, significant increases of water levels are expected during the life span of the structure. (See Section 4.2 for details.)

The stratigraphy of the site can be simplified as follows:

Layer 1 (Gravels): With a thickness of 12 m, and with the following geotechnical parameters: apparent unit weight $\gamma = 20.0 kN/m^3$; saturated unit weight $\gamma = 23.0 kN/m^3$; cohesion $c' = 5$ kPa; friction angle $\phi' = 35°$; unloading Young's modulus $Eu,r = 150$ MPa; coefficient of (isotropic) permeability, $k = 10^{-1}$cm/s.

Layer 2 (Plastic clay with fine sand seams): Down to indefinite depth from below 12 m. Its geotechnical parameters are: $\gamma = 23.0 kN/m^3$; $c' = 10$ kPa; $\phi' = 29°$; $Eu,r = 150$ MPaMPa; $k = 10^{-5}$cm/s.

The building is founded on a base slab with dimensions, 100 × 45m², and with a thickness of 1.50m. The excavation needed to construct the three levels of basements is constructed with the aid of diaphragm walls of 0.6 m thickness, and with 24 m depth. The depth of excavation (*i.e.*, the depth of the bottom of the base slab) is 14 m. Note that this means that the base slab has an embedment of $d = 2$m within the clays; and that the embedment of the toe of the diaphragm walls below the excavation bottom is defined by $t = 10$m. (In Section 4.5 we conduct sensitivity analyses to assess the influence of d and t on the computed results.)

The phreatic surface is normally located 5 m below the ground surface (*i.e.*,within the gravel layer), although it can increase during flooding events, as it is hydraulically connected to the river. However, given the embedment of the diaphragm wall into the low permeability clays, it is expected that the associated increase of pore water pressure at the bottom of the base slab needs some time to develop. This means that the problem is also controlled by the duration of the flooding events, a topic that is discussed in the next section.

4.2 *Discussion of water table elevations with limited duration*

The site analyzed herein is located within the fluvial environment of a city in Northeast Spain, which is controlled by River Ter—the longest river in the area, and the one with the largest flow rate—and affected by some additional tributary streams. This section aims to estimate the increase of water levels during flooding events in the area, and it mainly focuses on the analysis of river Ter, as it is the main river affecting the area and the one for which more data are available.

The flow rate of river Ter in the area of the study site can be established at around 250m^3/s for its maximum ordinary flooding event; and at around 2700m^3/s for its flooding event associated to a return period of 500 years (or T500 event) (ACA 2015). Such increase of the flow rate is associated to river levels that can rise up to 6,5–7,5 m for the T500 flood or, in other words, to increases of the river levels of up to 4.0–5.0 m. This is shown in Figure 1, where the river limits for the maximum ordinary flood are shown, together with the river limits associated to the T500 flood event. Similarly, Figure 1 also shows the associated flooded areas, as obtained from geomorphological-historical studies that determine, based on the analysis of the ground morphology and of historical information, the frequency of flooding events and the city areas which are liable to future flooding (Díez, Laín, & Llorente 2006). Finally, in relation to the duration of flooding events, there is available historical information suggesting that previous flooding events in the area can last up to five days for events associated to return periods lower than 500 years (Ayala et al. 1988).

4.3 *Numerical simulations*

Figure 2 shows the mesh of the Phase2 numerical model employed to simulate the stratigraphy at the site and the foundation of the building, considering the construction sequence of the diaphragm walls and of the base slab: three levels of anchors are installed, which are later substituted by the support provided by the structure—shown with dashed black lines—and by the base slab—shown

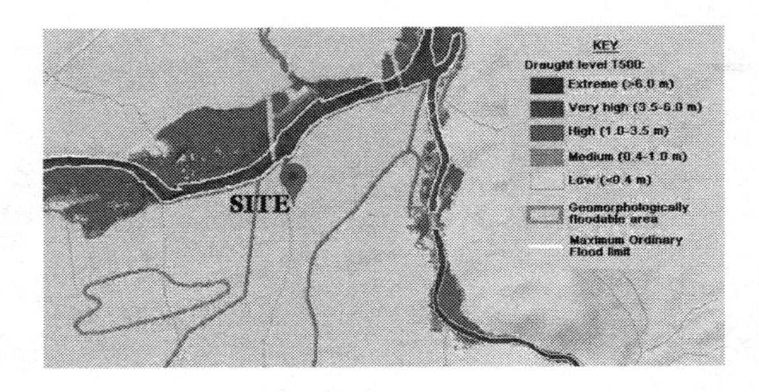

Figure 1. Flood levels for river Ter at the study site associated to the T500 flood event.

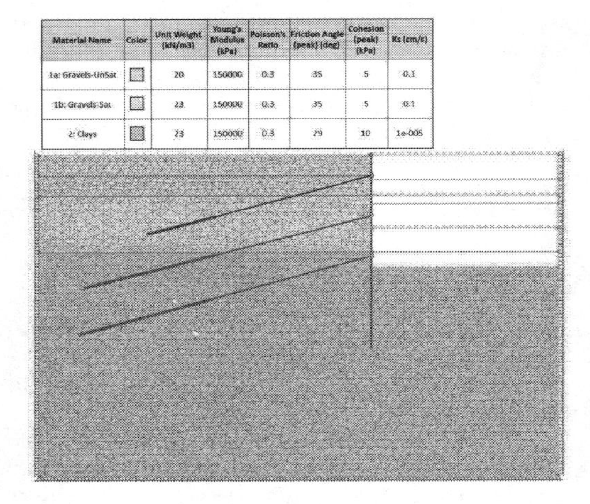

Material Name	Color	Unit Weight (kN/m3)	Young's Modulus (kPa)	Poisson's Ratio	Friction Angle (peak) (deg)	Cohesion (peak) (kPa)	Ks (cm/s)
1a: Gravels-UnSat		20	150000	0.3	35	5	0.1
1b: Gravels-Sat		23	150000	0.3	35	5	0.1
2: Clays		23	150000	0.3	29	10	1e-005

Figure 2. Set-up of the numerical model simulating the stratigraphy and construction sequence.

in orange at the base of the excavation—. The set of input geotechnical parameters employed for that specific analysis are also listed in the Table included within Figure 2.

Once the construction of the building has been simulated and the "long term" equilibrium situation has been achieved (*i.e.*,after all excess pore pressures due to the construction process have been dissipated), an increase of the phreatic surface is imposed, so as to simulate the increase of the water level in the river. (Remember that the gravel layer is assumed to be hydraulically connected to the river.) Following the available information summarized in Section 4.2, such elevation of the water level within the river is assumed to reach up to 5 m (*i.e.*,up to the ground surface) during a two-day period: *i.e.*, the water level rises 2,5 m per day during two days. Then, the water table is maintained at such high level, and analyses are conducted to estimate the distribution of pore water pressures below the base slab at different times after the flooding: times of $t = \{1,3,5,7,14\}$ days are considered, and the "long-term" situation is also simulated. (Note

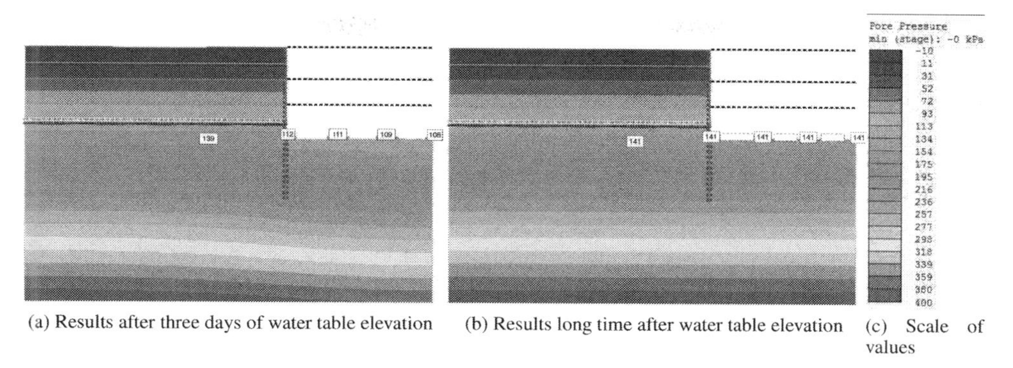

(a) Results after three days of water table elevation (b) Results long time after water table elevation (c) Scale of values

Figure 3. Examples of results computed with the transient coupled finite element model.

Table 1. Summary of finite element analyses conducted.

Coment.	Case	d [m]	t [m]	E_{clay} [MPa]	k_{clay} [cm/s]	Phreatic elev. [m]
"Ref. case"	1.1	2	10	150	1E-05	5,00
	1.2	2	10	125	1E-05	5,00
Reliability	1.3	2	10	175	1E-05	5,00
analyses	1.4	2	10	150	5E-05	5,00
(ROM)	1.5	2	10	150	5E-06	5,00
	1.6	2	10	150	1E-05	5,65
Sensitivity	2.1	0	10	150	1E-05	5,00
to t	2.2	2	10	150	1E-05	5,00
	2.3	4	10	150	1E-05	5,00
Sensitivity	3.1	2	0	150	1E-05	5,00
to d	3.2	2	5	150	1E-05	5,00
	3.3	2	10	150	1E-05	5,00
	3.4	2	15	150	1E-05	5,00

that, as discussed in Section 4.2, assuming such long periods of elevated water tables may not be realistic, but it allows us to (i) assess the influence of the time during which the water table maintains its elevated position; and (ii) verify the correctness of the numerical results, as the "long-term" solution should converge to the hydrostatic distribution of pore water pressures. Computations are conducted with the Phase2 finite element program, using Biot's coupled formulation described in Section 2.2, and considering the transient water flow that develops due to the low permeability of the clay layer; see Section 2.1.) Since pore water pressures are the main concern for this analysis, an elastic behaviour for the soil mass is assumed.

Figure 3 shows an illustration of the results obtained: Figure 3(a) illustrates the distribution of pore pressures at an intermediate time after elevation of the phreatic surface; and Figure 3(b) illustrates the equilibrium long-term solution (with hydrostatic pore pressures).

Table 2. Random Variables: Mean value and standard deviation (or shift).

Variable	Parameters	
	Mean value	Standard deviation (or interval)
Young's modulus—Clay		
E_{clay} [MPa]	150	25
	(as per pressurement data)	
Permeability—Clay		
k_{clay} [cm/s]	1E-05	[1E-05/2, 2 × 1E-05] (reasonable interval)
Extreme water level	On surface	13% (as per ROM 0.5/05)

Table 1 summarizes the input parameters—illustrating different geotechnical properties and different geometric definitions for the diaphragm walls—employed for all the analyses conducted. Cases labeled as 1.*i* are required for the reliability analyses following the ROM procedure (see Section 3); whereas the 2.*i* and 3.*i* analyses are employed, respectively, to assess the sensitivity of results to (i) the depth of embedment of the base slab within the clay; and to (ii) the depth of embedment of the wall toe below the base slab.

4.4 *Results of the reliability analysis*

To compute the reliability index (or the probability of failure) of a given design using the procedure explained in Section 3, we need to characterize the uncertainty associated to the random variables of interest (in this case, the stiffness and permeability of the clay layer). To that end, we employ their standard deviation or their coefficients of variation, as summarized in Table 2.

In addition, we need to define a limit state function that separates "safe" from "unsafe" states. In this case, our analyses suggest that a "buoyancy"

Table 3. Summary of reliability results for different times.

		Time [days]				
		1	3	5	7	14
Sensitivity Coef.						
E_{clay}	v_E	2,4E-03	6,1E-04	2,9E-03	1,5E-03	3,0E-03
k_{clay}	v_k	1,3E-01	1,3E-01	1,2E-01	1,0E-01	6,1E-02
Extreme loading	v_a	5,0E-02	4,7E-02	4,5E-02	4,3E-02	4,2E-02
Joint sensitivity	v_F	1,4E-01	1,4E-01	1,3E-01	1,1E-01	7,0E-02
Reliability index,	β	2,03	1,23	0,86	0,56	0,35
Failure probability,	P_f	2,1E-02	1,1E-01	2,0E-01	2,9E-01	3,6E-01

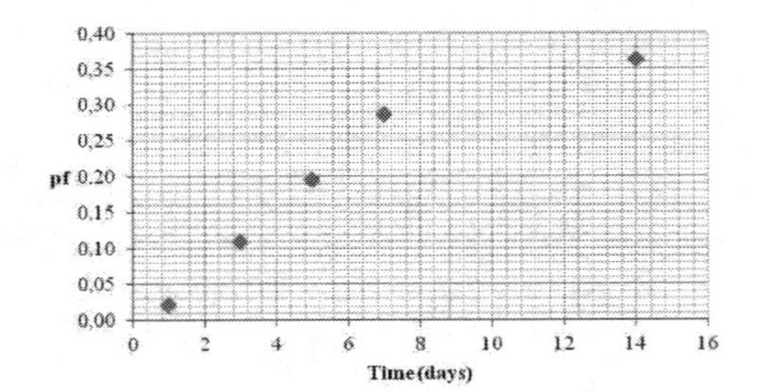

Figure 4. Evolution of reliability results as a function of time.

failure mode is critical when a rise of the phreatic surface occurs (*i.e.*,the "buoyancy" failure mode occurs before that the base slab fails due to flexural stresses imposed by the increased water pressure); for that reason, the safety factor is defined as:

$$F = \frac{R_{cp}}{U}, \tag{10}$$

where R_{cp} is the resultant force due to the permament load transmitted by the structure; and where U is the resultant force due to the pore water pressures developed at the base of the slab.

We have conducted several reliability analyses to assess the reliability as a function of time; *i.e.*, assuming that the flood produced by the river is maintained in its high-level position for a long enough time. (Obviously, and as expected, the reliability decreases with such time, as in the long term the water pore pressures tend towards their hydrostatic values, hence producing and ultimate limit state associate to "buoyancy" due to excessive pore pressures.) The computed results (for t = {1,3,5,7,14} days) are presented in Figure 4 and listed in Table 3.

4.5 *Results of the sensitivity analyses*

Additional finite element computations were conducted to assess the sensitivity of results—as expressed by the average pore water pressure below the base slab—to changes on the geometric parameters that define the diaphragm wall foundation: *i.e.*,to changes on the depth of embedment of the base slab within the clay layer; and to changes on the depth of embedment of the wall toe below the base slab. Results are summarized, respectively, in Figures 5(a) and 5(b).

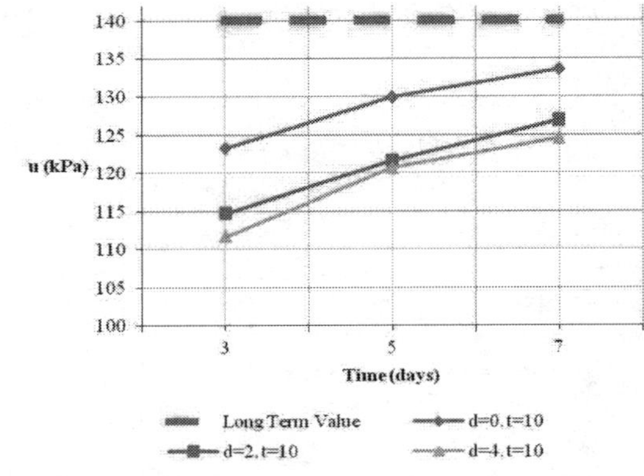

(a) Sensitivity to embedment of base slab

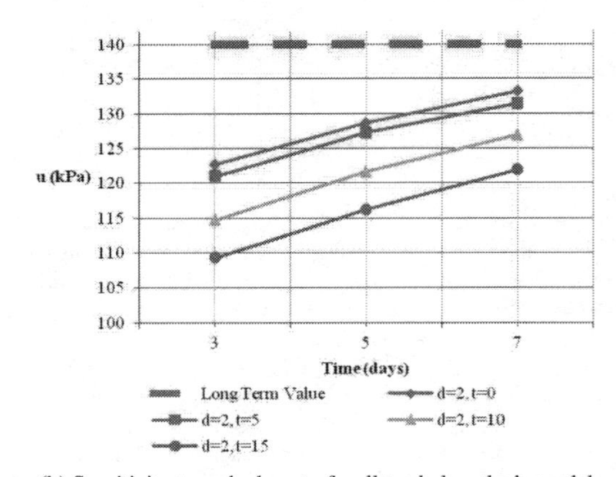

(b) Sensitivity to embedment of wall toe below the base slab

Figure 5. Results of the sensitivity analysis.

5 CONCLUDING REMARKS

This paper addresses the problem of design of a deep foundation with diaphragm walls and a base slab below three levels of basements. The geotechnical profile consists of a layer of gravel at the surface that is hydraulically connected to a nearby river, which overlays a much more impervious clay formation, in which the diaphragm wall and the base slab are embedded. The analysis considers the possible occurrence, as an extraordinary event, of a significant rise of the river water level due to floodings following periods of intense rain; and a transient finite element model is constructed to compute the increasing pore water pressures below the base slab that result from such water level increase at the river.

Based on the results of such analyses, we can make the following remarks in relation to the design of the diaphragm wall and of the base slab:

- It seems adequate to consider a hydrostatic distribution of water pressures at the exterior side of the diaphragm wall, as the increase of water pressures in that region is quite fast.
- The increase of water pressures below the base slab takes more time to develop and, as illustrated by the sensitivity analyses conducted, it is affected by (i) the depth of embedment of the base slab within the clay (in other words, on the position of the clay-gravel contact, if a constant depth of the base of the slab is considered); and (ii) on the depth of embedment of the toe of the diaphragm wall within the clay.
- For a deterministic analysis based on "mean" or "characteristic" values, it may seem plausible to consider a limited increase of the pore water pressure of the base slab; such value could be computed using a finite element analysis with transient flow, and results would depend on (i) the geometry of the problem, (ii) the geotechnical parameters considered (stiffness and permeability), (iii) the assumed rise of the river level during the flood, and (iv) the assumed duration of the flood.
- However, considering the existing uncertainties (on river elevations during floods, as well as on geotechnical parameters) could have a significant effect on the computed results. In this case, for instance, results suggest that the reliability is mainly controlled by the permeability of the clay, with ν_k sensitivies that are very similar to the overal sensitivity, given by ν_F(ν_k is larger than ν_E by several orders of magnitude; and larger than ν_a by about one order of magnitude). This difference is more significant for smaller times (*i.e.*,shortly after the water level increase of the river), or when it is more relevant for our problem.

In that sense, results suggest that, in this case, a value of $k_{clay} \approx 10^{-5}\,\text{cm/s}$ might be the upper bound so that the reliability is admissible during "early times" after a flood event. It is also expected that the reliability would significantly increase for lower average clay permeability values $\left(\text{of, say,} k_{clay} \le 10^{-6}\,\text{cm/s}\right)$; or if the uncertainty of permeability values is reduced with respect to those employed in Table 2.

ACKNOWLEDGMENTS

This article has benefited from comments and feedback from the consulting companies that collaborated with Dragados for the analysis of the project that motivated this study; special thanks are due to Mr. Luis Ortuño and Mr. Tomás Murillo (Uriel y Asociados, S.A.). The research was also funded, in part, by the Spanish Ministry of Economy and Competitiveness [grant number BIA 2015–69152-R]. Their support is gratefully acknowledged.

REFERENCES

ACA (2015). *Planificació de l'Espai Fluvial de les conques del Baix Ter.* Agencia Catalana del Agua.

Ayala, F.J. et al. (1988). *Catálogo Nacional de Riesgos Geológicos.* Instituto Geológico y Minero de España.

Díez, A., L. Laín, & M. Llorente (2006). *Mapas de peligrosidad de avenidas e inundaciones: métodos, experiencias y aplicación.* Madrid: Instituto Geológico y Minero de España.

Jiménez Salas, J.A. & J.L. de Justo Alpañés (1976). Propiedades de los suelos y de las rocas. In *Geotecnia y Cimientos* (2 ed.), Volume I. Madrid: Editorial Rueda.

Lambe, T.W. & R.V. Whitman (1995). *Mecánica de Suelos.* México, D.F.: Editorial Limusa.

Puertos del Estado (Ed.) (2001). *ROM 0.0. General Procedure and Requirements for Design of Maritime and Harbour Structures (Part I).* Madrid: Ministerio de Fomento.

Puertos del Estado (Ed.) (2005). *ROM 0.5–05. Geotechnical recommendations for the Design of Maritime and Harbour works.* Madrid: Ministerio de Fomento.

Rocscience (2017). Phase2 v.9.0. theory manual. theory: Coupled consolidation.

Smith, I. & D. Griffiths (1997). *Programming the Finite Element Method* (3rd ed.). West Sussex, England: John Wiley and Sons Ltd.

Terzaghi, K. (1943). *Theoretical soil mechanics.* New York: John Wiley and Sons.

Terzaghi, K., R.B. Peck, & G. Mesri (1996). *Soil Mechanics in Engineering Practice* (3rd ed.). New York: John Wiley and Sons, Inc.

Venkatramaiah, C. (1995). *Geotechnical engineering.* New Age International.

Numerical Methods in Geotechnical Engineering IX – Cardoso et al. (Eds)
 ISBN 978-1-138-33198-3

A practical case study of slope stability analysis using the random finite element method

T. de Gast, A.P. van den Eijnden, P.J. Vardon & M.A. Hicks
Section of Geo-Engineering, Faculty of Civil Engineering and Geosciences, Delft University of Technology, Delft, The Netherlands

ABSTRACT: The Random Finite Element Method (RFEM) has been shown in many theoretical publications to offer advantages in the quantification of the probability of failure. However, it has rarely been applied in real situations (geometry, material properties, soil layers) and seldom, if at all, to a well instrumented geotechnical failure. This paper reports a case study of a full-scale controlled dyke failure, where the heterogeneity was previously measured via CPTs (Cone Penetration Tests), and the dyke itself was highly instrumented. This offers the opportunity to compare and apply various techniques previously developed (e.g. random field conditioning) with field data, rather than to computer generated data. The RFEM analyses presented are compared with deterministic analyses, demonstrating the relative performance of the methods.

1 INTRODUCTION

A large number of numerical benchmark tests have shown the veracity of the Random Finite Element Method (RFEM) in simulating the probalility of failure. However, limited data are available of field tests that can be compared with an RFEM analysis. In particular, the spatial variability (heterogeneity) of material properties has seldom been investigated in field tests. In this paper, the instrumented failure of a dyke, coupled with a site characterisation focused on identifying the site's material variability and heterogeneity in the vertical and horizontal planes, is used to compare deterministic FEM and RFEM analyses.

Several sources of uncertainty may be identified in dyke stability analysis, the two main ones being: (1) natural variability, either temporal or spatial; (2) knowledge uncertainty, relating to model uncertainties and corresponding uncertainties in material properties. In this paper, model uncertainties have not been taken into account, and the boundaries between soil layers have been taken as deterministic. The uncertainties in natural variability have been modelled using RFEM analyses, with the variation in the subsurface within layers being modelled using unconditional random fields or random fields conditioned on local CPT measurements.

In this paper, four analyses are compared for the particular conditions recorded at failure: (1) Unconditional RFEM (UC-RFEM), where the point and spatial statistics are based on local CPTu data; (2) Conditional RFEM (C-RFEM), with random fields conditioned to actual CPTu measurements; (3) FEM using the average value per layer; (4) FEM using the average value minus one standard deviation per layer. Using these approaches, a range of responses corresponding to the moment of failure are computed and compared.

2 FAILURE TEST

In the dyke failure test, over the period of a month the dyke was saturated with water, and soil in front of the toe was excavated in steps and replaced by water, effectively increasing the height of the dyke. In the final stage, the water in the excavation was removed and the dyke failed under its own weight. This, in combination with an extensive site investigation and laboratory testing programme, provided detailed information that was used to investigate the dyke failure both deterministically and stochastically. In the days before the controlled failure, the slopes of the ditch were steepened to 1:1 and the ditch was excavated to the bottom of the peat layer at 2.5 m depth.

In Figure 1(a), the data recorded during the pumping period until failure are shown at the centre cross-section of the excavation. The figure indicates the differential displacements (bold arrows) and excess pore water pressures (vertical arrows). During the experiment, the main failure occurred just south of the centre. Large differential displacements were measured in the toe, below the peat and organic clay boundary, and in the organic clay layer. Measured differential displacements below the crest of the

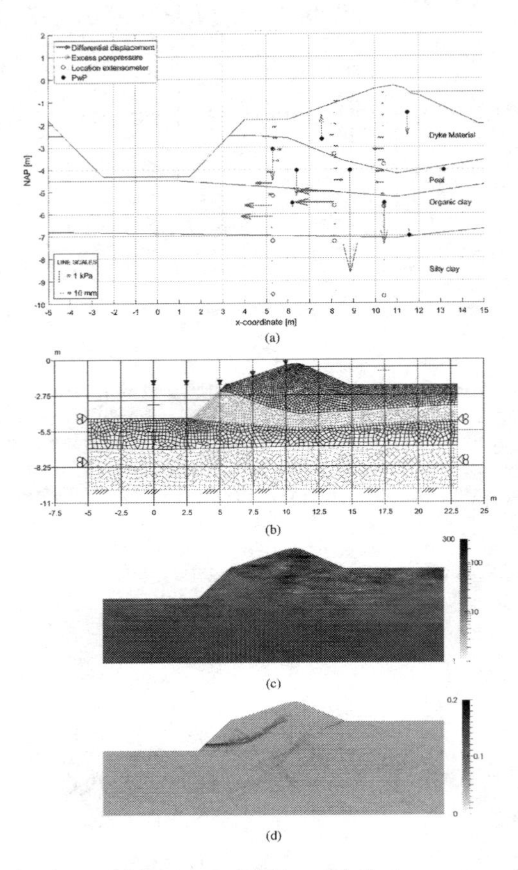

Figure 1. Analysis of the embankment: (a) measured differential displacements and excess pore pressures at the moment of failure; (b) finite element mesh used for analysis, with triangles indicating CPT locations, water level in ditch at NAP –3.15 m and waterlevel upstream at NAP –0.6 m; (c) example of an RFEM analysis, with shear strength illustrated on a logarimic scale in kPa (d) the calculated shearbands as contours of deviatoric strain.

dyke were distributed near the boundary between the peat and dyke material. The failure occurred between a drawdown of 1.5 m and 2.0 m; the precise water level at failure could not be determined, but was estimated to be 1.6 m, i.e. NAP –3.5 m.

3 MATERIAL PARAMETERS

A dataset of CPT and laboratory test data had been collected to investigate and determine the vertical and horizontal heterogeneity. This dataset was taken over an area of 50 × 15 m² in the immediate vicinity of the dyke failure, and included 100 CPTu tests that were obtained over a two week period. De Gast et al. (2017) evaluated and showed that the vertical heterogeneity under the dyke was influenced by compression of the material.

From the data at the cross-section of the failure location, both the vertical and horizontal scales of fluctuation were determined using the CPTs. The average horizontal interval of the CPTs was 2.5 m (perpendicular to the dyke). Due to the deposition history, it is generally assumed that the horizontal scale of fluctuation is larger than the vertical scale of fluctuation, so that the spacing of data in the horizontal plane was anticipated to be acceptable. The scale of fluctuation was estimated using the method elaborated in Gast et al. (2017).

The shear strength s_u in kPa was determined using

$$s_u = \frac{q_t - \sigma_v}{N_{kt}} \tag{1}$$

Where q_t is the total cone resistance (kPa), σ_v is the total vertical stress (kPa) and N_{kt} is an empirical correction factor (–). Values for N_{kt} suggested by Robertson (2009) range from 10–20 and, by comparing the CPT and laboratory data (20 consolidated undrained triaxial tests and four direct simple shear tests), the N_{kt} values for the different materials were refined. The values of N_{kt} for the four materials are: (1) dyke material, $N_{kt} = 20$; (2) peat, $N_{kt} = 15$; (3) organic clay, $N_{kt} = 10$; and (4) silty clay, $N_{kt} = 10$.

Table 1 presents the results of the analysis of the CPT data. These comprise the mean shear strength, the standard deviation of shear strength, and estimates of the vertical and horizontal scales of fluctuation, θ_v and θ_h, respectively. Note that the mean and standard deviation of the shear strength were obtained from the dataset before de-trending.

4 RFEM VS FEM

To account for the spatial variability of the soil parameters, FEM has been combined with random field theory within a stochastic (Monte Carlo) process. This involves multiple simulations (i.e. realisations) of the same problem, a procedure often referred to as RFEM. In each realisation of an RFEM analysis, a random field of material properties is generated, based on the point and spatial statistics of the material properties. The method has proven to be an efficient approach for conducting stochastic slope stability analyses (e.g. Hicks & Samy Hicks & Samy 2002).

Spatial variation has been modelled by random fields generated using covariance matrix decomposition, with local averaging for unstructured meshes (van den Eijnden & Hicks 2017). This method starts by generating a field with a standard normal distribution, in which the spatial variation of property values is related to a correlation function incorporating the scales of fluctuation. The standard normal field is then transformed to the appropriate distribution based on the mean and standard deviation of the variable being modelled.

In this paper, only the undrained shear strength is spatially random, while other parameters are assumed to be constant. In the first RFEM analysis the spatially random undrained shear strength has been generated only from the input statics (μ, σ, θ_v, θ_h); in the other RFEM analysis the uncertainty has been reduced by conditioning the spatially random undrained shear strength to CPT data (Li et al. 2016). Both the conditional and unconditional RFEM analyses assume a lognormal strength distribution. Moreover, for each analysis four different material layers, with each layer having its own random field, were discretised using a mesh of 4250 eight-node elements, with each element using 2 × 2 Gaussian integration points. Nearer to the top of the mesh (where most of the failure mechanism was expected) a finer mesh was used. On each side of the dyke a load was applied representing the waterload at the time of failure; on the right side, the waterlevel was at NAP –0.6 m and on the left side (the excavation) the waterlevel was at NAP –3.5 m, i.e. equal to a drawdown of 1.6 m. Figure 1(b) shows the mesh used in the analyses, indicating the waterlevels at both sides of the dyke. Figure 1(c) shows one of the RFEM realisations, indicating the shear strength, and Figure 1(d) indicates the calculated shear strains at failure of the same realisation.

Figure 2 presents the results of the analyses in terms of safety factor (SF). Specifically, the solid curve is the unconditional RFEM analysis and the broken curve is the conditional RFEM analysis, with each curve based on the results of 400 realisations; the dotted line with crosses is the FEM analysis using the mean strength value and the solid line with circles is the FEM analysis using a strength estimate one standard deviation below the mean.

The unconditional RFEM analysis shows the largest range of solutions for SF, from 0.47 to 1.15, with an average SF of 0.83. The conditional RFEM analysis shows a reduced range of SF, from 0.76 to 1.09, with a higher average SF of 0.93. The deterministic FEM analysis based on mean strength values gives SF = 1.16 and the FEM analysis using strengths of one standard deviation lower than the mean gives SF = 0.73.

Table 1. Material parameters based on CPT data at the failure location.

	$S_{u,\mu}$ [kPa]	$S_{u,\sigma}$ [kPa]	θ_v [m]	θ_h [m]
Dyke material	19.5	16.4	0.4	2.13
Peat	10.4	5.7	0.76	2.84
Organic clay	14.9	3.9	0.76	2.84
Silty clay	22.2	3.8	0.26	2.1

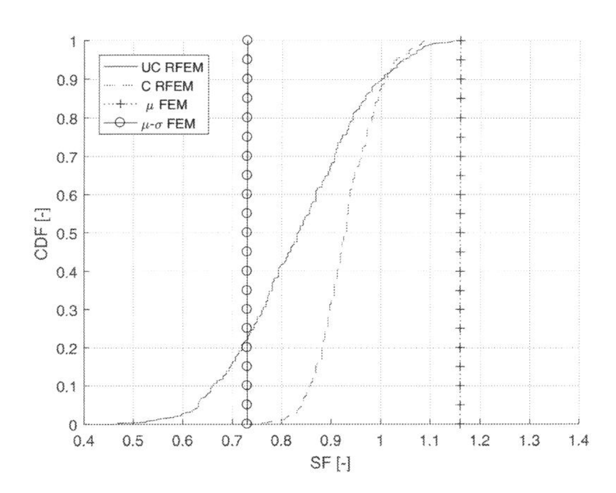

Figure 2. Analyses results for safety factor (SF).

5 DISCUSSION

As previously discussed, the analyses are undertaken for the conditions when the dyke failed; therefore, SF would be expected to be approximately 1. In the deterministic FEM analysis the SF using the mean shear strength is 1.16, overestimating the safety of the dyke. Comparing this result to the RFEM analyses, the mean SFs predicted (at CDF = 0.5) are 0.83 and 0.93 for the unconditional and conditional RFEM analyses, respectively. This means that, for this example, the conditional RFEM analysis has led to a calculated SF almost indistinguishable from reality and therefore gives confidence in using this method to calculate the slope reliability.

Both the conditional and unconditional RFEM analyses compute lower SF values than the deterministic FEM result based on the mean. This is as expected, due to the failure passing through weaker zones of the materials, as has been previously reported by Hicks & Samy (2002) and Hicks & Spencer (2010). The distribution of SF calculated by the conditional RFEM is significantly narrower than for the unconditional RFEM, also as expected, due to the variation in the spatial distribution of the material parameters being smaller (while the variation in the point statistics is the same).

Note that at the 95% confidence level, the unconditional RFEM analysis has an SF of 0.63, whereas for the conditional RFEM analysis it is 0.84. This has a significant implication for the assessment of dykes, as it could make the difference between a dyke being assessed as reliably safe or requiring costly improvement.

Further work is needed to investigate the impact of this work on 3D failures; so far only 2D stochastic analyses have been carried out. The methodology to incorporate CPT data in 3D slope stability assessments, including the possible impact on slope reliability, has been theoretically investigated by Li et al. (2016). Moving from 2D to 3D simulations has been investigated deterministically as part of this research (details not presented here), and from 2D to 3D incorporating spatial variability by Li et al. (2015) and Varkey et al. (2017). In the deterministic, case the SF increases by ≈15% when moving from 2D to 3D, due to the impact of the sides of the failure surface; however, in cases where the spatial variability has been incorporated, it may even be possible for the SF to decreases by several percent, depending on the scale of fluctuation in the longitudinal direction of the dyke (relative to the dyke length).

Additional information from measurements can also be incorporated into analyses, further reducing the variation in the calculated SF, especially on the hydro-mechanical behaviour and impact of a variable phreatic surface, as demonstrated by Vardon et al. (2016). However, it is noted that with all additional measurements and analyses, there are financial and time implications. Therefore, a cost benefit judgement must be made.

6 CONCLUSION

A real dyke failure has been induced and monitored, with the conditions of the dyke at failure being used in a series of comparative numerical analyses using deterministic FEM and RFEM approaches. The observed and calculated failure modes are similar, although the computed safety factors differ significantly depending on the adopted approach (deterministic FEM versus RFEM) and on the relative use of data (i.e. conditional versus unconditional analysis). By incorporating spatial variability the confidence in the stability can be calculated, and by incorporating additional measurements the confidence can be increased. This leads to a higher calculated reliability and more efficient design.

ACKNOWLEDGEMENTS

This research was supported under project number 13864 by the Dutch Technology Foundation STW, which is part of the Netherlands Organisation for Scientific Research (NWO), and which is partly funded by the Ministry of Economic Affairs.

REFERENCES

de Gast, T., P.J. Vardon, & M.A. Hicks (2017). Estimating spatial correlations under man-made structures on soft soils. *Geo-risk 2017 GSP 284*, 382–389.

Hicks, M.A. & K. Samy (2002). Influence of heterogeneity on undrained clay slope stability. *Quarterly Journal of Engineering Geology and Hydrogeology 35*(1), 41–49.

Hicks, M.A. &W.A. Spencer (2010). Influence of heterogeneity on the reliability and failure of a long 3D slope. *Computers and Geotechnics 37*(7–8), 948–955.

Li, Y.J., M.A. Hicks, & J.D. Nuttall (2015). Comparative analyses of slope reliability in 3D. *Engineering Geology 196*, 12–23.

Li, Y.J., M.A. Hicks, & P.J. Vardon (2016). Uncertainty reduction and sampling efficiency in slope designs using 3D conditional random fields. *Computers and Geotechnics 79*, 159–172.

Robertson, P.K. (2009). Interpretation of cone penetration tests: a unified approach. Canadian *Geotechnical Journal 46*(11), 1337–1355.

van den Eijnden, A.P. & M.A. Hicks (2017). Efficient subset simulation for evaluating the modes of improbable slope failure. *Computers and Geotechnics 88*, 267–208.

Vardon, P.J., K. Liu, & M.A. Hicks (2016). Reduction of slope stability uncertainty based on hydraulic measurement via inverse analysis. *Georisk: Assessment and Management of Risk for Engineered Systems and Geohazards 10*(3), 223–240.

Varkey, D., M.A. Hicks, & P.J. Vardon (2017). Influence of spatial variability of shear strength parameters on 3D slope reliability and comparison with an analytical method. *Georisk 2017 GSP284*, 400–409.

Numerical Methods in Geotechnical Engineering IX – Cardoso et al. (Eds)
© 2018 Taylor & Francis Group, London, ISBN 978-1-138-33198-3

Modes of improbable slope failure in spatially variable cohesive soils

A.P. van den Eijnden & M.A. Hicks
Section of Geo-Engineering, Faculty of Civil Engineering and Geosciences, Delft University of Technology, Delft, The Netherlands

ABSTRACT: Accounting for spatial variability in probabilistic slope stability analysis using the random finite element method (RFEM) typically leads to a distribution of calculated factors of safety as well as a distribution of resulting depths of the sliding body. Factor of safety (or its corresponding probability of failure) and sliding depth are weakly correlated when looking at the results of classical Monte Carlo simulations, but the mean (i.e. the expected) depth of the sliding surface as a function of the factor of safety shows a clear trend. This trend becomes more prominent when looking at the weak tail of the factor of safety distribution in more detail. In this paper, subset simulation is combined with RFEM to address the weak tail in an efficient manner and to simulate slope failure events at low probability levels, thereby demonstrating the tendency to shallow modes of failure in such cases. The mechanism behind the tendency to shallow modes of failure is then further investigated by simplification of the problem to limit equilibrium. Slide-specific variance reduction through spatial averaging is demonstrated as the mechanism behind the tendency to a shallow sliding surface as the mode of improbable slope failure.

1 INTRODUCTION

Reliability analysis of slopes in heterogeneous soils is generally based on the strength reduction method (in the case of cohesive soils), c-ϕ reduction (for cohesive-frictional soils) or increased gravitational loading (in the general case of complex material behaviour with material models that have no direct strength parameters). The reduction factor that needs to be applied to trigger slope instability is then used as the factor of safety for the structure and is, generally speaking, the ratio between the resisting and driving forces. This approach, as initially used for the analysis of slopes in homogeneous soils, can be successfully applied in reliability analysis within a probabilistic framework. However, care is needed when it is applied to spatially variable soils, as strength reduction is applied homogeneously over the soil domain and thus represents uncertainty in the mean strength.

For addressing uncertainty in the spatial variability of strength parameters in finite element methods, the random finite element method (RFEM) is often used (e.g. Griffiths & Fenton (2004), Hicks & Spencer (2010)), which employs random field theory (Vanmarcke, 1983) to characterise the spatial variability. In this method, strength reduction is applied by factoring down the shear strength over the full domain by a factor SRF and the minimum value of SRF to bring the slope to failure is found iteratively. This factor is proportional to F_μ, which is the factor required to bring a homogeneous slope, with a strength equal to the same statistical mean strength μ_c, to failure. The ratio between F_μ and SRF defines the realisation-specific factor $F = F_\mu / SRF$, which signifies the additional factor in the mean strength that is required to compensate for the influence of spatial variation in that realisation. The cumulative distribution of the values of F in a Monte Carlo simulation gives the reliability curve for the slope. In this context, F is referred to as the factor of safety based on the mean.

When studying the mode of failure (here characterised by the maximum depth d of the sliding surface below the toe of the slope), a correlation can be found between F and d, which becomes more prominent when the simulation moves towards the (weaker) tail of the probability distribution. This was first demonstrated by van den Eijnden & Hicks (2017), who used subset simulation to efficiently generate realisations at failure probability levels $p_f < 10^{-6}$ and studied the probability-dependent mean sliding depth of a slope in a heterogeneous cohesive soil. Indeed, the mean sliding depth tends to be shallower for slopes with a lower probability of failure. This research was extended in van den Eijnden & Hicks (2018) for a range of slope angles and coefficients of variation, confirming the tendency towards shallow modes of failure even in slopes with only a gentle inclination.

Variance reduction through spatial averaging of the shear strength along the sliding surface is considered as the mechanism behind the tendency towards shallow slope failures in heterogeneous soils. This paper investigates this hypothesis by applying a probabilistic limit equilibrium analysis to the slope stability problem. These new results are then used to compare with the previous findings from the RFEM analysis, thereby providing a more comprehensive basis for the explanation of the tendency towards shallow slope failure for improbable slope failure events in heterogeneous soils.

2 IMPROBABLE SLOPE FAILURE AND RFEM

A slope of height $H = 5$ m resting on a foundation layer of 5 m thickness is modelled using RFEM with boundary conditions as specified in Figure 1. The spatial variability of shear strength is characterised by an exponential correlation function, incorporating the correlation length characterised by the vertical and horizontal scales of fluctuation θ_y = 1.25 m and θ_x = 8.0 m A lognormal distribution is used as the distribution function for the point statistics.

A simulation example is given here for an α = 45° slope with a coefficient of variation (the ratio between the mean shear strength μ_c and standard deviation σ_c) of $CoV = 0.25$ (note that the absolute value of μ_c is irrelevant due to the definition of F). Figure 2 shows the computed probability of failure p_f as a function of the factor of safety based on the mean, obtained from a Monte Carlo analysis applying shear strength reduction in 500 realisations (continuous line).

While the failure probability can be estimated accurately for safety factors close to the median of F (i.e. F 1.08), the estimation of p_f becomes unreliable towards the tail of the distribution (i.e. at small probabilities). To address the tail of the distribution efficiently, subset simulation is employed on the same problem and subsets of realisations are generated with subsequently higher factors of safety based on the mean. For background on subset simulation, see Au & Wang (2014), whereas for details on the subset simulation procedure applied here, see van den Eijnden & Hicks (2017). The calculated p_f based on subset simulation is included in Figure 2, with the crosses indicating results for subsequent subsets. Three subsets, A, B and C, are identified for further investigation.

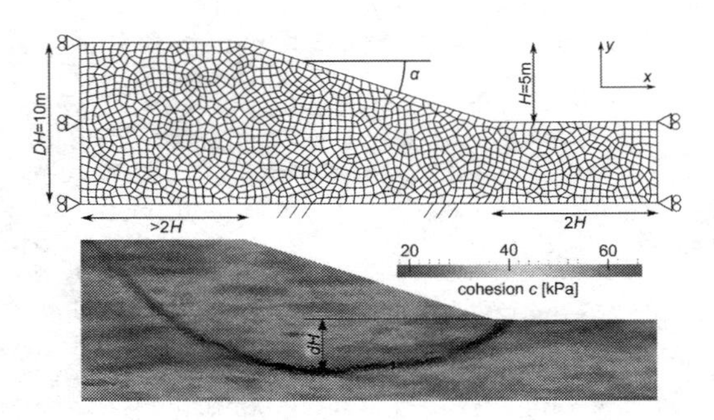

Figure 1. Details of boundary value problem. Above: slope mesh with $D = 2$ and $\alpha = 18.4°$. Below: typical realisation with spatially varying cohesion, for μ_c = 32 kPa and CoV = 0.25 (large deviatoric strain contours shown in black, indicating the sliding surface.

Figure 3 shows a selection of the sliding surfaces that are found in subsets A, B and C, as well as for the Monte Carlo simulation and the analysis of a homogeneous slope. Note that the realisations belonging to a subset correspond to the range before the corresponding factor of safety (e.g. subset A corresponds to $F < 1.06$, subset B to $1.27 < F < 1.31$). The sliding surfaces for MCS show a wide range of depths, with most surfaces being close to the foundation layer. Subset A, with realisations corresponding to $p_f > 75\%$, shows mainly deep sliding surfaces, whereas subset B, with $p_f = 0.001$ at the end of the domain of the MCS results, shows a higher concentration of shallow sliding surfaces, indicating the tendency to move from deep to shallow modes of failure with increasing F. Subset C, at $p_f = 10^{-6}$, clearly demonstrates that at very small levels of probability the mode of failure is shallow.

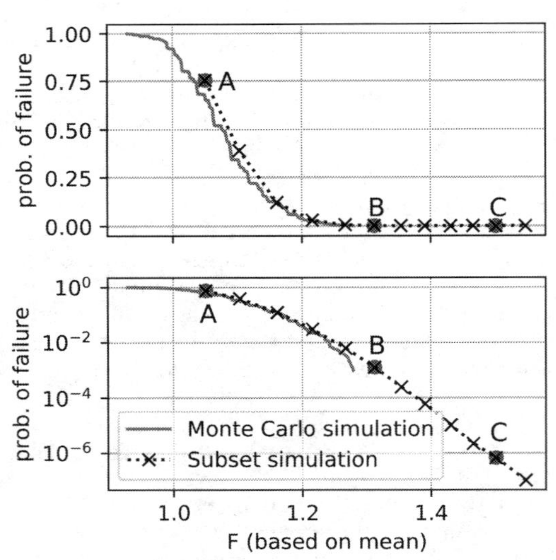

Figure 2. Probability of failure as a function of the factor of safety based on the mean, for $\alpha = 45°$, $Cov = 0.25$, and computed from 500 MCS and by subset simulation.

3 PROBABILITY-DEPENDENT FAILURE MODES IN RFEM

Following the same procedure as outlined above, a rigorous simulation campaign was performed on slopes with a range of slope angles and *CoV* ranging from 0.0 to 0.4 (van den Eijnden & Hicks, 2018). This resulted in a map of the expected failure probability-dependent sliding depth, of which contour lines can be projected on the *CoV*-plane (Figure 4). The expectation, being the probability-dependent mean of the simulation results, is used to characterise the mode of failure. The homogeneous solution ($CoV = 0.0$) is indicated by the thicker continuous line, switching from deep to shallow mode of failure roughly between 55° and 65°, much like the predicted solution of Taylor (1948) for cohesive soils (53.6° and 60.5° for the transition from deep failure to shallow toe failure respectively). The dotted lines present the probability-dependent mean sliding depth for levels of probability from $p_f = 0.5$ (i.e. the median) to $p_f = 10^{-12}$.

For all slope angles α and levels of *CoV*, the tendency towards shallow modes of failure is expressed as a migration towards $d = 0$ of the probability-dependent mean sliding depth with decreasing failure probability. Moreover, the results show that with higher *CoV*, the tendency is more pronounced and the difference in expected sliding depth is already apparent at moderate levels of probability (i.e. 0.1–0.01).

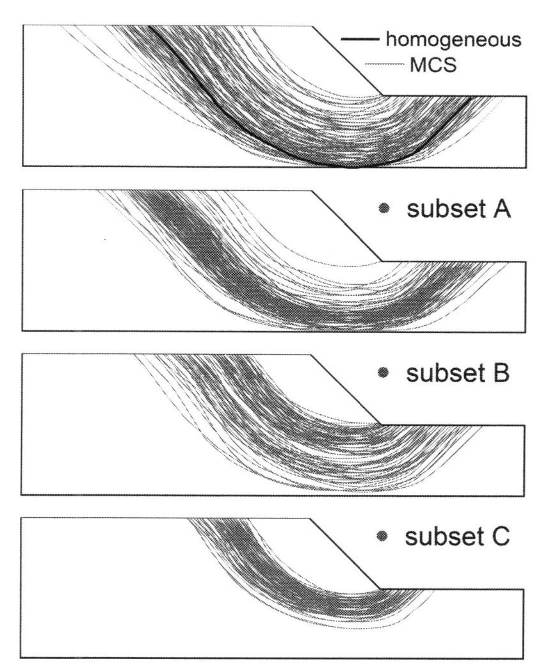

Figure 3. Selections of 100 sliding surfaces obtained by Monte Carlo simulation and subset simulation for subsets A, B and C (Figure 2) at p_f equal to 0.75, 0.001 and 10^{-6} respectively.

4 PROBABILISTIC LEM MODEL CONCEPT

To further examine the effect of probability-dependent modes of slope failure, and to verify the mode-dependent variance reduction through averaging of spatially variable shear strength as the mechanism behind the tendency towards shallow modes of failure at low levels of probability, the problem is simplified to an analysis using the limit equilibrium method (LEM) based on Bishop circles in cohesive soils (Bishop, 1955).

The assumption that the sliding surface is circular greatly simplifies the analysis, although this prevents a rigorous comparison with RFEM. A probabilistic limit equilibrium approach is employed and the analysis is performed based on the reliability index of sliding circles with constrained depth. The concept of this method is similar to the probabilistic slope stability analyses of Low and co-workers (2007, 2012, 2014) and Cho (2007), but reformulated for purely cohesive soils with random field theory accounting for the spatial variability. This reformulation leads to a direct expression of the reliability index β, bypassing the need to identify a design point in parameter space.

4.1 *Formulation of the probabilistic LEM*

For a certain slip circle S the factor of safety F_S is given by:

$$F_S = \frac{C_S}{W_S} \tag{1}$$

where C_S is the total resistance against sliding through the mobilised shear strength along the sliding surface S and W_S is the total driving force for sliding defined as:

$$W_S = \int_{s \in S} \rho_{clay} g h \sin\gamma \frac{dx}{ds} ds \tag{2}$$

with h, γ and x depending on s and representing respectively the local depth, angle of inclination and x-coordinate of the sliding surface. The total resisting force C_S is obtained as the product of the average shear strength $\bar{c}_S$ along the sliding surface S and its length L_S:

$$C_S = \int_{s\in S} c(s)ds = \bar{c}_S L_S \quad (3)$$

In the case of spatial variability of the shear strength with first order stationarity, C_S is probabilistic with mean and variance:

$$\mu_{C_S} = \mathrm{E}[C_S] = \mu_c L_S \quad (4)$$

$$\sigma^2_{C_S} = \sigma^2_c \Gamma_S L^2_S \quad (5)$$

with Γ_S being the variance reduction factor due to averaging along the sliding surface accounting for the spatial correlation function $\rho(s, t)$ between any two points, s and t, on the sliding surface S:

$$\Gamma_S = \frac{1}{L^2_S} \int_{s\in S} \int_{t\in S} \rho(s,t)\, ds dt \quad (6)$$

This expression can be approximated by discretisation of the sliding circle into N sections of equal arc length $l \ll \theta_x, \theta_y$ and correlation matrix $\mathbf{R}$ with components $R_{ij} = \rho(s_i, s_j)$ as the correlation between the shear strength at any two positions, s_i and s_j, on the sliding circle S:

$$\Gamma_S \approx \frac{1}{N^2} \sum_{i=1}^{N} \sum_{i=1}^{N} R_{ij} \quad (7)$$

For stationary fields with a lognormal distribution, random fields are usually generated through a transformation from Gaussian (i.e. standard normal) fields; $\boldsymbol{Z}_c = \exp(\boldsymbol{X})$, with $\boldsymbol{X} \sim \mathcal{N}(\mu_X, \sigma^2_X, \mathbf{R}_X)$. Note thatthe normal transform of the random field along the sliding surface is defined as $\boldsymbol{X} = \mu_X + \sigma_X \mathbf{R}^{1/2}_X \xi$, with ξ being a vector of uncorrelated standard normal random numbers. The correlation matrix $\mathbf{R}_X$ corresponds to the standard normal transform of the field $\boldsymbol{Z}_c$ and the correlation matrix $\boldsymbol{R}$ of the lognormal field is defined as:

$$R_{ij} = \sigma_c^{-2} e^{2\mu_X + \sigma^2_X} \left(e^{\sigma^2_X R_{X_{ij}}} - 1 \right) \quad (8)$$

$$\mu_X = \ln \left(\frac{\mu_c}{\sqrt{1 + CoV^2}} \right) \quad (9)$$

$$\sigma_X = \sqrt{\ln(1 + CoV^2)} \quad (10)$$

4.2 *Reliability index*

The probabilistic representation of F_S outlined above, accounting for the effect of spatial variability characterised by random fields, can be used in reliability analyses (e.g. FORM, SORM). The reliability index β_S for a specific circle S wasdefined by Hassofer & Lind (1974) and characterises the reliability as the distance between the origin and the point for the corresponding parameter set in uncorrelated standard normal parameter space. Here, only a single parameter is to be considered, this being the mobilised resistance C_S along the sliding surface given by Equation (3). To account for the non-Gaussian shape of the distribution of F_S and to keep track of its corresponding probability, the generalised reliability index (after Breitung (1984) and Ditlevsen (1981)) is used, and expressed here as:

$$\beta_S = -\Phi^{-1}\left(\mathrm{P}[C_S < W_S, \mu_c L_S, \sigma^2_{C_S}] \right) \quad (11)$$

where $\Phi(.)$ is the standard normal cumulative distribution function and P[.] is the probability function of C_S, which, in the following, is assumed to remain lognormal after averaging out along the sliding circle and to be characterised adequately by the mean and variance (see Appendix). The most likely sliding circle is the one with the lowest reliability index and minimisation of β_S with respect to the probabilistic parameters involved in the slope reliability analysis leads to the system reliability index β for the slope.

4.3 *Iterative minimisation of* β_S

For a given slope geometry, a sliding circle is defined by the coordinates of its centre x_S, y_S and its radius R_S. By varying the coordinates and radius, minimisation of β_S leads to the most likely sliding surface under the given conditions (i.e. μ_c, CoV, slope geometry). For efficient minimisation, different strategies can be followed (e.g. Hasofer & Lind (1999), Zolfaghari et al. (2005)). As the given problem here is simple, the minimisation of β_S against x_S, y_S and R_S is done by a full Newton method using numerical differentiation, which proves accurate and highly efficient for this application. In addition, the mode of failure is constrained to either a fixed depth or a sliding circle passing through the toe of the slope, which makes R_S directly dependent on y_S (and x_S in the case of toe failure). Note that the reliability index represents a single stochastic variable (C_S) and therefore expresses the exact probability of failure. This probability of failure, however, is slide-specific and non-failure along this specific circle does not exclude failure along other circles, which, in combination with correlation between slides, prevents accurate assessment of the total probability of failure. As such, the probability corresponding to the most likely sliding circle can be considered as a

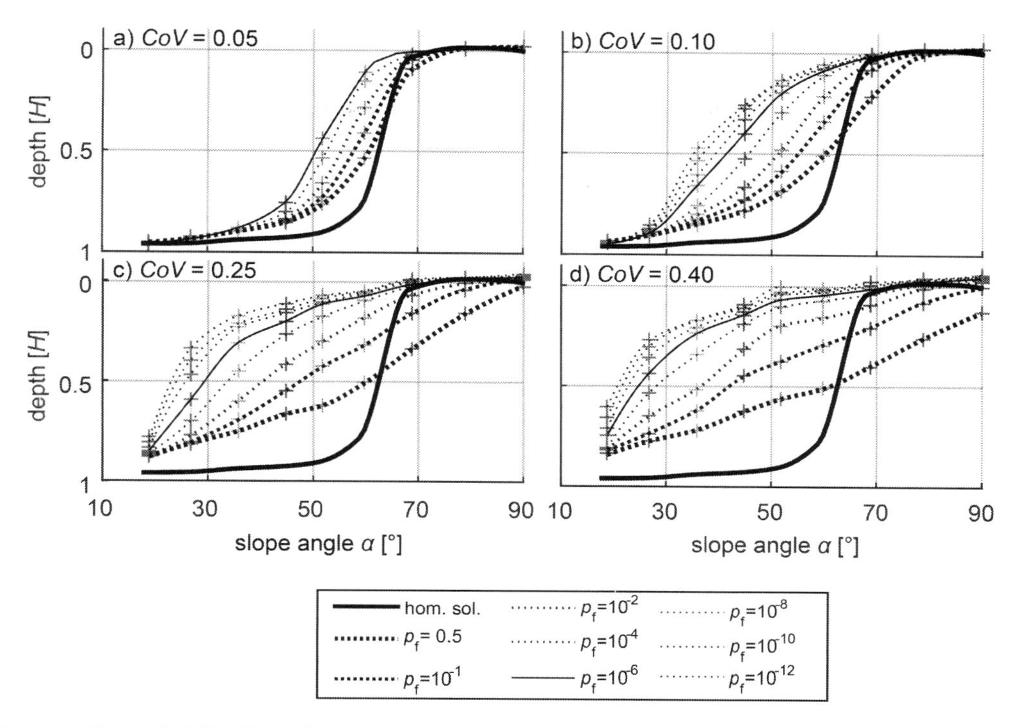

Figure 4. Expected sliding depth d below slope toe as a function of slope angle α, for different probabilities of failure p_f and coefficients of variation CoV. The calculated sliding depth for a homogeneous slope is given as a reference.

Table 1. Calculation example for slope failures constrained to toe failure and deep failure (d = 1) for different mean strength values μ_c and a lognormal distribution. / separates values for toe and deep failure respectively; bold text indicates critical values of F_S and β.

μ_c kPa	W kN / kN	L_S m / m	Γ –/–	μ_{C_S} kN / kN	σ_{C_S} kN / kN	F_S –/–	β –/–
17.00	201 / 524	11.8 / 29.6	0.225 / 0.103	200 / 502	23.7 / 40.2	1.00 / **0.96**	–0.08 / **–0.57**
18.00	200 / 523	11.7 / 29.5	0.225 / 0.103	211 / 531	25.1 / 42.5	1.06 / **1.02**	0.41 / **0.15**
19.00	200 / 522	11.7 / 29.4	0.225 / 0.103	223 / 559	26.4 / 44.8	1.12 / **1.07**	0.86 / **0.82**
20.00	199 / 521	11.7 / 29.4	0.225 / 0.103	234 / 587	27.8 / 47.1	1.17 / **1.13**	**1.30** / 1.46
22.00	199 / 519	11.7 / 29.3	0.225 / 0.103	257 / 644	30.5 / 51.6	1.29 / **1.24**	**2.11** / 2.65
25.00	198 / 516	11.6 / 29.1	0.225 / 0.103	291 / 728	34.5 / 58.5	1.47 / **1.41**	**3.19** / 4.25
32.00	197 / 512	11.6 / 28.9	0.225 / 0.103	370 / 924	43.9 / 74.3	1.88 / **1.81**	**5.27** / 7.33

lower-bound solution for the total probability of failure, under the assumption that the method of slices is an adequate physical model and that the probability function P[.] is known.

In comparing β_S with the classical interpretation of the reliability index of the design point (Hasofer & Lind (1974)), β_S represents here the reliability index related to the design point in the N-dimensional sampling space ofthe vector $\boldsymbol{\xi}$, used to generate the random field $\boldsymbol{Z}_c$ as the array vector of all shear strength values in the N sections along the sliding surface S. The sliding circle that minimises β_S for the slope geometry, provides the slope reliability index β. In addition to the potential for failure along that circle, β gives a first order estimation of the system reliability. However, due to the high-dimensional context, the first order estimation of the system reliability might be inaccurate

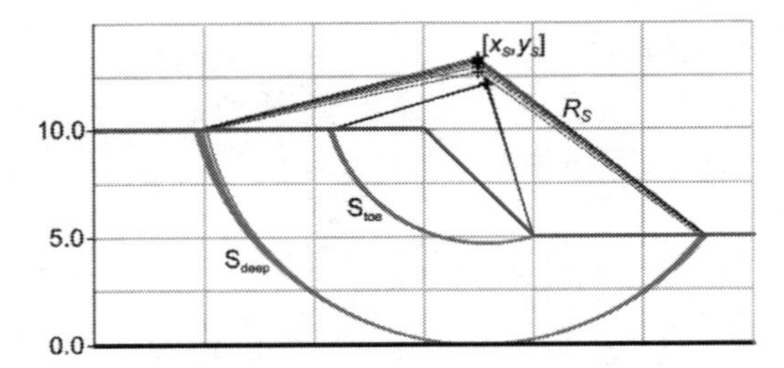

Figure 5. Bishop circles with constrained reliabibity indices β_{toe} and β_{deep} for a series of mean strengths μ_c (see Table 1).

and should, at best, only be used in relative comparison and qualitative interpretation.

5 PROBABILISTIC LEM SIMULATIONS

A series of probabilistic LEM simulations is performed for various mean strengths and sliding circles constrained to either deep (i.e. d = 1) or toe failure. Figure 5 gives the corresponding sliding circles, passing either through the foundation or the toe of the slope. Note that all sliding circles that are constrained to deep failure have their centre of rotation $[x_S, y_S]$ exactly above the middle of the slope. Toe failures have a centre more to the right.

Table 1 gives the computational data for determining the factor of safety and the reliability index for the mean strength values considered in Figure 5. Due to the centre and size of the constrained sliding circles being hardly affected by a change in mean strength parameters, the driving force W, the sliding length L and the variance reduction factor Γ, show negligible variation with mean strength μ_c. Hence, the mean μ_{C_S} and standard deviation σ_{C_S} of the mobilised resistance show a nearly linear relation with the point mean strength μ_c. For μ_c close to 17.73 kPa, which is when the factor of safety based on the mean F = 1.0 in the finite element analysis, the driving force W and mobilised resistance C_S show little difference. Hence, calculation of the factor of safety F_S and reliability index β using Equations (1) and (11) leads to values of F_S close to unity and β around 0. At these levels of μ_c, deep sliding circles show the lowest safety factors as well as the lowest reliability indices, which indicates that deep modes of failure are most likely to be expected.

With W being practically insensitive to variations in μ_c, higher values for the mean strength (i.e., giving larger factors of safety based on the mean) correspond to a bigger difference between W and μ_{C_S}. The linear relation between μ_{C_S} and μ_c results in an approximately linear increase in F_S with μ_c, thereby predicting a lower total factor of safety for deep sliding circles. This corresponds to analyses of homogeneous slopes, in which the effect of spatial variability is not accounted for.

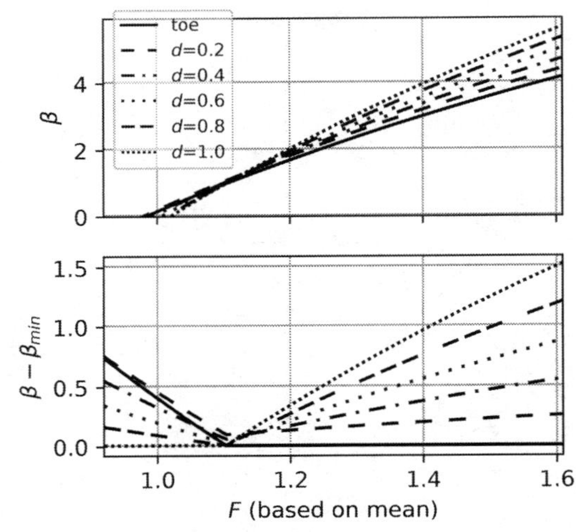

Figure 6. Above: reliability index β for different sliding depths d. Below: relative difference between depth-dependent reliability indices with respect to the smallest of the 6 computed reliability indices. Slope angle α = –45°, CoV = 0.25, θ_x = 8.0 m and θ_y = 1.25 m.

Unlike the factor of safety, the reliability index β_S does account for the effect of spatial variability through the variance reduction factor that is present in the standard deviation. Since the variance reduction factor is smaller for large sliding surfaces, the mobilised resistances of the corresponding deep sliding circles have a lower coefficient of variation and are thus more standard deviations away from the driving force W than toe circles for the same factor of safety. Therefore, the reliability index increases faster with increasing mean strength for deep sliding circles. Indeed, with a large enough mean strength, β_{toe} becomes smaller than β_{deep}, indicating a switch from a predominantly deep mode of failure to a shallow mode of failure with increasing mean strength. This effect is directly visible in the calculated indices given in Table 1, with toe circles showing the lowest reliability indices for $\mu_c > 19$ kPa. The difference between β_{toe} and β_{deep} at μ_c = 32 kPa, which can be expressed as a relative difference in the probability of occurrence, $P_{toe} / P_{deep} = \Phi(-5.27)/\Phi(-7.33) = 6.0 \times 10^5$, emphasises the significant change in the likelihood of finding shallow modes of failure with decreasing total probability of failure.

Figure 6 shows the calculated reliability index as a function of the factor of safety based on the mean F for different imposed sliding depths d, as well as the

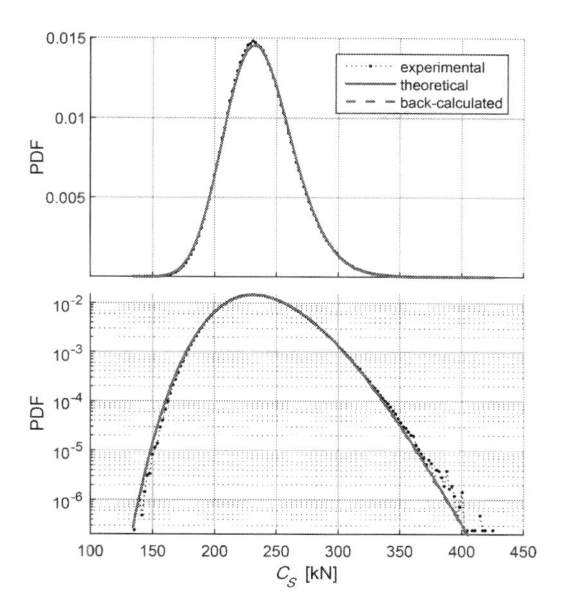

Figure 7. Distributions of average strength along the toe sliding circle at μ_c = 20.0 kPa, determined by Monte Carlo simulations ('experimental'), the theoretical lognormal distribution as presented in this paper ('theoretical') and the back-calculated lognormal distribution ('back-calculated').

relative difference in reliability index with respect to the smallest value of the corresponding reliability indices. Indeed, slopes close to F = 1.0 show the highest reliability against shallow modes of failure and, although the difference is small here, predominantly deep modes of failure can be expected. Looking at the difference between depth-specific reliability indices and the overall reliability index clearly shows the evolution of mode of failure with F. Starting from a deep-failure dominated mode at F = 1.0, a point exists where the depth-specific indices are almost identical(F = 1.09), after which toe failure becomes the dominant mode, gaining in importance towards higher values of F.

6 DISCUSSION

The probabilistic limit equilibrium method as outlined above is able to provide the most likely sliding circle for spatially variable cohesive soils (i.e. characterised by random fields), and gives an exact probability of occurrence for the sliding circle under consideration. However, linking it to a total probability of slope failure is difficult, as other sliding surfaces have similar probabilities of occurrence (albeit consistently smaller) and correlations between these alternative circles is difficult to systematically and correctly account for. A rigorous and accurate evaluation of the probability of failure by means of probabilistic limit equilibrium methods is therefore unlikely without relying on Monte Carlo simulations (Cho, 2013).

Although the method presented above is easily extended towards non-circular slip surfaces or even cohesive-frictional soils, limit equilibrium methods fail to account for complex material behaviour as only mobilised shear strength is accounted for. Nevertheless, the computational simplicity of the probabilistic limit equilibrium methods make them attractive for qualitative parameter studies of slope stability problems involving spatially variable soils, upon which a detailed RFEM simulation campaign can be formulated for obtaining quantitative and detailed results.

7 CONCLUSIONS

Modes of improbable slope failure, being slope failures due to the spatial variability of strength in predominantly stable slopes at low levels of failure probability, have been demonstrated based on finite element computations in a framework combining the random finite element method with subset simulation. The mechanism influencing the tendency to shallow modes of failure at small levels of probability is considered to be sliding surface specific variance reduction arising from the averaging of spatially varying mobilised shear strength. This mechanism can be qualitatively demonstrated by limit equilibrium methods, which corroborates the results of more advanced finite element computations.

ACKNOWLEDGEMENTS

This research is supported under project number 13864 by the Dutch Technology Foundation STW, which is part of the Netherlands Organisation for Scientific Research (NWO). This work was carried out on the Dutch national e-infrastructure with the support ofSURF Cooperative.

REFERENCES

Au, J.K. & Y. Wang (2014). *Engineering risk assessment with subset simulation*. John Wiley & Sons.

Beaulieu, N.C. & Q. Xie (2004). An optimal lognormal approximation to lognormal sum distributions. *IEEE Transactions on Vehicular Technology 53*(2), 479–489.

Bishop, A.W. (1955). The use of the slip circle in the stability analysis of slopes. *Geotechnique 5*(1), 7–17.

Breitung, K. (1984). Asymptotic approximations for multinormal integrals. *Journal of Engineering Mechanics 110*(3), 357–366.

Cho, S.E. (2007). Effects of spatial variability of soil properties on slope stability. *Engineering Geology 92*(3), 97–109.
Cho, S.E. (2013). First-order reliability analysis of slope considering multiple failure modes. *Engineering Geology 154*, 98–105.
Ditlevsen, O. (1981). *Uncertainty modeling with applications to multidimensional civil engineering systems*. McGraw-Hill International Book Co.
van den Eijnden, A.P. & M.A. Hicks (2017). Efficient subset simulation for evaluating the modes of improbable slope failure. *Computers and Geotechnics 88*, 267–280.
van den Eijnden, A.P. & M.A. Hicks (2018). Probability-dependent failure modes of slopes and cuts in heterogeneous cohesive soils. *Submitted for publication*.
Griffiths, D.V. & G.A. Fenton (2004). Probabilistic slope stability analysis by finite elements. *Journal of Geotechnical and Geoenvironmental Engineering 130*(5), 507–518.
Hasofer, A.M. & N.C. Lind (1974). Exact and invariant secondmoment code format. *Journal of the Engineering Mechanics division 100*(1), 111–121.
Hassan, A.M. & T.F. Wolff (1999). Search algorithm for minimum reliability index of earth slopes. *Journal of Geotechnical and Geoenvironmental Engineering 125*(4), 301–308.
Hicks, M.A. & W.A. Spencer (2010). Influence of heterogeneity on the reliability and failure of a long 3D slope. *Computers and Geotechnics 37*(7), 948–955.
Ji, J., H.J. Liao, & B.K. Low (2012). Modeling 2-D spatial variation in slope reliability analysis using interpolated autocor-relations. *Computers and Geotechnics 40*, 135–146.
Low, B.K. (2014). FORM, SORM, and spatial modeling in geotechnical engineering. *Structural Safety 49*, 56–64.
Low, B.K., S. Lacasse, & F. Nadim (2007). Slope reliability analysis accounting for spatial variation. *Georisk: Assessment and Management of Risk for Engineered Systems and Geohazards 1*(4), 177–189.
Taylor, D. (1948). *Fundamentals of soil mechanics*. Chapman and Hall Limited.
Vanmarcke, E. (1983). *Random fields: analysis and synthesis*. MIT Press.
Zolfaghari, A.R., A.C. Heath, & P.F. McCombie (2005). Simple genetic algorithm search for critical non-circular failure sur-face in slope stability analysis. *Computers and Geotechnics 32*(3), 139–152.

APPENDIX

The distribution of the sum of (correlated) lognormal random numbers does not have a general analytical solution based on the first two moments only. Depending on the distribution parameters (mean and covariance function), different approximate expressions have been proposed (see, for example, Beaulieu & Xie [?]). In the context of this paper, the approximation by a lognormal distribution based on the mean and standard deviation is used. To validate this approximation, the distribution of C_S was determined for the case of toe failure at $\mu_c = 20.0\,\text{kPa}$ over 10^6 random fields, based on the summed resistances generated along the slip circle.

Figure 6 shows the probability density function obtained from this simulation (denoted as 'experimental'). The theoretical lognormal distribution of C_S based on the mean and variance according to Equations (4) and (5) is alsoshown ('theoretical'). In the context of this paper, the match between the experimental distribution and the theoretical expression validates the assumption of a lognormal distribution for C_S. In addition, a lognormal distribution based on the back-calculated mean and variance of the simulated values of C_S is given ('back-calculated') to demonstrate that the mean and variance of C_S are correctly defined by Equations (4) and (5) respectively.

Numerical Methods in Geotechnical Engineering IX – Cardoso et al. (Eds)
© 2018 Taylor & Francis Group, London, ISBN 978-1-138-33198-3

3D slope stability analysis with spatially variable and cross-correlated shear strength parameters

D. Varkey, M.A. Hicks & P.J. Vardon
Section of Geo-Engineering, Faculty of Civil Engineering and Geosciences, Delft University of Technology, Delft, The Netherlands

ABSTRACT: The paper investigates the stability of slopes with spatially variable and cross-correlated shear strength parameters in 3D. The influence of various cross-correlation coefficients between these parameters on the probability of 3D slope failure has been considered for different levels of anisotropy of the heterogeneity in the shear strength. Specifically, 3D random fields of cohesion and friction angle were generated using the Local Average Subdivision method, and these were correlated with eachother by various degrees. The fields were then linked to finite element analyses within a Monte Carlo framework. The results indicate that a positive cross-correlation between the parameters reduces the slope reliability, whereas a negative cross-correlation between the parameters increases the reliability.

1 INTRODUCTION

The inherent nature of soil is to be spatially variable (Phoon & Kulhawy 1999). The uncertainty in the spatial variability of parameters arises due to a combination of various geologic, environmental and physio-chemical processes. However, quantification of this heterogeneity is not a trivial task and demands extensive field and laboratory tests (Jaksa et al. 1999, de Gast et al. 2017). There can also be other types of uncertainties, such as geometric uncertainty in the form of uncertain soil layer boundaries, or epistemic uncertainties associated with sampling, modeling, and so on. The uncertainty in the spatial variability of the shear strength parameters alone has been considered in this paper.

Conventionally, the stability of slopes is calculated deterministically, i.e., by ignoring the spatial variability in heterogeneity within soil layer(s) and considering the entire slope to be made up of a single or multiple homogeneous layers. The outcome of such an analysis is a single factor of safety (FS), which gives no information about the reliability. Ignoring the heterogeneity within the soil has been shown to have a significant influence on computations of FS (Hicks & Samy 2002, Hicks 2007, Cho 2007, among others) and also on the failure mechanisms (Hicks & Spencer 2010). Various reliability-based methods have been developed to include heterogeneity; for example, the first order second moment method, first order reliability method, point estimate method, stochastic response surface methods and the random finite element method (RFEM) (Fenton & Griffiths 2008). The outcome of RFEM is a range of possible responses of the structure. Research has also been done to efficiently use the available data to condition random fields, for improving the confidence in results (Lloret-Cabot et al. 2012, Li et al. 2016).

Soils generally exhibit spatial variability in a range of parameters. These parameters, in addition to being correlated over certain lengths, may also be correlated to each other. The influence of cross-correlation between effective cohesion (c') and effective friction angle (φ') on bearing capacity predictions has been investigated by Cherubini (2000) and Fenton & Griffiths (2003). The influence of this cross-correlation on the reliability of slopes (Le 2014, Javankhoshdel & Bathurst 2016, among others), as well as different methods for constructing the bivariate distributions (Tang et al. 2015) and their influence on the reliability of retaining walls (Li et al. 2015) have also been investigated. Griffiths et al. (2009a) identified critical values of the coefficients of variation of the shear strength parameters, beyond which ignoring the spatial variability gives unconservative results with or without cross-correlation between them.

Research has also been done on the reliability analysis of slopes in 2D to understand the influence of various levels of heterogeneity in the mechanical and hydraulic parameters (Arnold & Hicks 2011), and on making use of inverse analysis techniques to reduce the uncertainty in hydraulic conductivity by using pore pressure measurements (Vardon et al. 2016). All these studies are based on the assumption that the mechanical and hydraulic parameters are correlated over an infinite distance in the third dimension. Although this is generally not the case, only a limited amount of research has been done regarding full 3D probabilistic analysis, possibly due to the large computational requirements.

Vanmarcke (1977) pioneered 3D reliability assessments of slopes by assuming the governing soil parameter to be the spatial average of the randomly varying parameter over a predefined surface. In contrast, 3D RFEM does not make any assumption regarding equivalent soil parameter or failure mechanism, although it requires a large computational effort to carry out multiple realisations. Spencer (2007), Griffiths et al. (2009b), Hicks & Spencer (2010) and Li et al. (2015) used 3D RFEM to investigate the influence of anisotropy of the heterogeneity in undrained shear strength and slope length in the third dimension on the estimation of failure probability. Hicks & Spencer (2010) grouped the failure modes into three different categories based on the anisotropy of the heterogeneity in shear strength relative to the slope dimensions. Strategies for quantification of the failure consequences have also been developed (Hicks et al. 2008, Huang et al. 2013, Hicks et al. 2014).

This paper considers the spatial variability and cross-correlation between shear strength parameters (c' and φ') for an idealised long slope. The random fields of the parameters were generated using the 3D Local Average Subdivision method, and linked with the finite element model within a Monte Carlo framework. Different values of anisotropy of the heterogeneity in the shear strength were considered. The influence of different cross-correlation coefficients between these parameters on the probability of failure of a 3D slope has been investigated.

2 RANDOM FINITE ELEMENT METHOD (RFEM)

The mathematical representation of the spatial variability of soil parameters can be made in the form of a random field. This can be univariate or multivariate, depending whether the field value at a point in space is a random variable or a random vector.The field is said to be stationary if the mean (μ) and variance (σ^2) of the random variables are constant and the autocorrelation coefficient (ρ) is only dependent on the separation between the points (t,t') under consideration.The correlation structure for the random variable (X) between these points is given as:

$$\rho(X_t, X_{t'}) = \frac{E[(X_t - E[X_t])(X_{t'} - E[X_{t'}])]}{\sigma_t \sigma_{t'}} \quad (1)$$

where X_t and $X_{t'}$ are the respective values of X at t and t', and $E[X_t]$, $E[X_{t'}]$ and σ_t, $\sigma_{t'}$ are the expectations and standard deviations of X, respectively. For a stationary random field, $E[X_t] = E[X_{t'}] = \mu$ and $\sigma_t \sigma_{t'} = \sigma^2$.

In the context of finite element analysis, the mechanical response of a system is approximated by the spatial discretization of the geometry. RFEM combines random fields with finite elements and hence discretization of the random fields is required, as carried out in this paper by Local Average Subdivision (LAS) [?]. In this method, a local integral process is obtained by integrating X over a moving window (T), such that the new process has the same average as X and is smoother than X, i.e. with a reduced variance to account for local averaging. The variance reduction ($\Gamma(T)$) is dependent on the correlation function ($\rho(\tau)$) for a stationary process, and is given in 1D as:

$$\Gamma(T) = \frac{2}{T}\int_0^T \rho(\tau)d\tau - \frac{2}{T^2}\int_0^T \tau\rho(\tau)d\tau \quad (2)$$

where τ is the lag distance. A 3D separable Gauss Markov correlation structure is used in this paper, with the correlation in the vertical (z) direction separated from the two horizontal (x and y) directions. The 3D covariance function ($\beta = \sigma^2\rho$) is:

$$\beta(\tau_x, \tau_y, \tau_z) = \sigma^2 exp(-\frac{2\tau_z}{\theta_z} - \sqrt{(\frac{2\tau_x}{\theta_x})^2 + (\frac{2\tau_y}{\theta_y})^2}) \quad (3)$$

where θ_x, θ_y and θ_z are the scales of fluctuation and τ_x, τ_y and τ_z are the lag distances in the respective directions.

The separation of the vertical correlation structure from the two horizontal directions was done to model the long-term depositional characteristic in the soil. It is assumed that the horizontal layers were deposited at the same instant, whereas the vertical deposition occurs over time.

LAS is a top-down recursive approach, which begins with generating a random number (from a standard Gaussian distribution) which is assigned as the initial global mean for the entire domain. Proceeding downwards, the domain is subdivided into equal halvesin each direction, i.e. each cell is divided into 2^3 cells at each subdivision level in 3D LAS. In the subdivision process, the global average is preserved by the top-down approach, whereas the variance of the local average reduces and tends towards the target variance as the number of subdivision levels increases. In this paper, the minimum required subdivision level is determined in order to have a variance reduction value not less than 0.8, i.e. as given by the scale of fluctuation of the process being at least four times the averaging window in the last sub-division level (Li 2017).

2.1 *Cross-correlation between variables*

The generated random fields, for say n parameters, can be correlated to each other by using the correlation matrix (R) given in Equation 4, with $\rho_{X_iX_j}$ being the correlation coefficient between the randomly varying parameters, X_i and X_j, at the same point in space.

$$R=\begin{bmatrix} 1 & \rho_{X_1X_2} & .. & \rho_{X_1X_n} \\ \rho_{X_2X_1} & 1 & .. & \rho_{X_2X_n} \\ .. & .. & .. & .. \\ \rho_{X_nX_1} & \rho_{X_nX_2} & .. & 1 \end{bmatrix} \tag{4}$$

The generated n univariate Gaussian random variables are cross-correlated by using the Cholesky decomposition (LL^T) of R, and using the following matrix transformation to generate the cross-correlated fields (ζ_i):

$$\begin{pmatrix} \zeta_1 \\ .. \\ .. \\ \zeta_n \end{pmatrix} = L \begin{pmatrix} X_1 \\ .. \\ .. \\ X_n \end{pmatrix} \tag{5}$$

where L is the lower triangular matrix. The above transformation requires that the individual random fields are stationary, and the Cholesky decomposition fails if R has negative eigenvalues. Figure 1 shows the random variables for two parameters, X_1 and X_2, correlated to each other with different values of $\rho_{X_1X_2}$.

Figure 2 shows the covariance structure obtained in the field of X_2 cross-correlated to X_1 for different values of $\rho_{X_1X_2}$. The generated fields coincide with the exact covariance structure (Eq. 3) for an isotropic field with $\theta = 1m$. Hence, cross-correlation does not affect the covariance structure within the fields.

Note that for generating anisotropic random fields, the authors have first generated isotropic random fields for each uncertain parameter by LAS using $\theta = \theta_x = \theta_y = \theta_z$ in Equation 3, followed by cross-correlating the fields using Equation 5. This field was then post-processed by squashing and/or stretching in the respective directions to generate

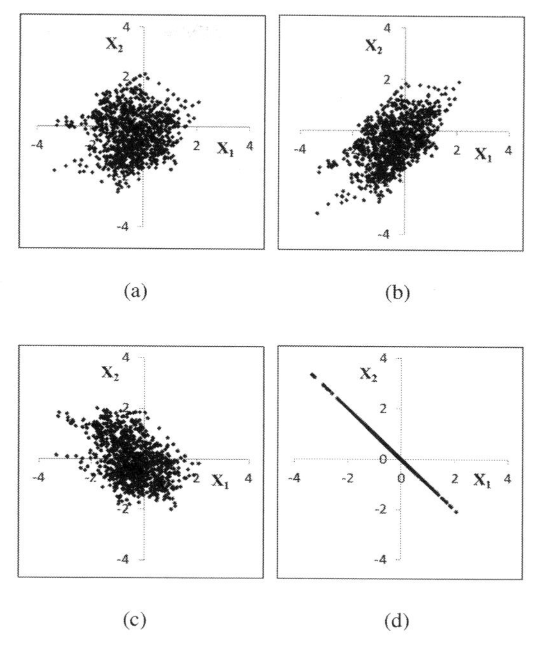

Figure 1. Cross-correlated fields in standard normal space; (a) $\rho_{X_1X_2}=0$, (b) $\rho_{X_1X_2}=0.5$, (c) $\rho_{X_1X_2}=-0.5$, (d) $\rho_{X_1X_2}=-1$.

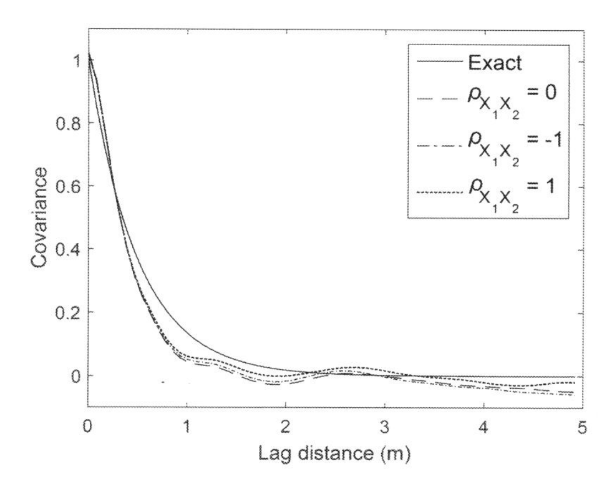

Figure 2. Covariance structure obtained in a cross-correlated isotropic 3D field with $\theta = 1m$ and domain side length of $5m$, for different cross-correlation coefficients ($\rho_{X_1X_2}$).

the required level of anisotropy ($\xi = \theta_h/\theta_v$); see Hicks & Samy and (2002) and Hicks & Spencer (2010) for details. The cross-correlated random fields corresponding to each parameter were then transformed into their physical space using the point statistics and type of parameter distribution.

3 PROBLEM DESCRIPTION

A 50 m long slope, with the cross-sectional geometry shown in Figure 3, was analysed by RFEM. Different values of the cross-correlation coefficient ($\rho_{c'\varphi'}$) between the shear strength parameters (c' and φ') were considered. The parameters of the model are summarised in Table 1. The slope was meshed with a total of 4000 20-node regular hexahedral elements with a $2 \times 2 \times 2$ Gaussian integration scheme. The elements were of size $1m \times 1m$ in plan and $0.5m$ in depth. The boundary conditions applied to the model were: fixed along the base, rollers on the side face, and rollers on the vertical end-faces allowing only vertical movement; see Spencer (2007) for an explanation of these boundary conditions. The random field variables corresponding to each uncertain parameter, after post-processing, were assigned to the Gauss points within each element. A linear elastic, perfectly plastic Mohr-Coulomb model was used to define the stress-strain conditions within the problem domain. In each realisation, the in-situ stresses were generated by applying gravity loading in a single step, and the slope was checked for stability under its own weight using the strength-reduction method. A total of 500 realisations were carried out for each set of statistics of the parameters, and a distribution of the FS was determined.

A wide range of values for the cross-correlation coefficient between c' and φ' have been reported in the literature. The different values of $\rho_{c'\varphi'}$ are attributed to different soil types, sampling techniques and testing rates used. The results for different values of $\rho_{c'\varphi'}$ are summarised in the next section.

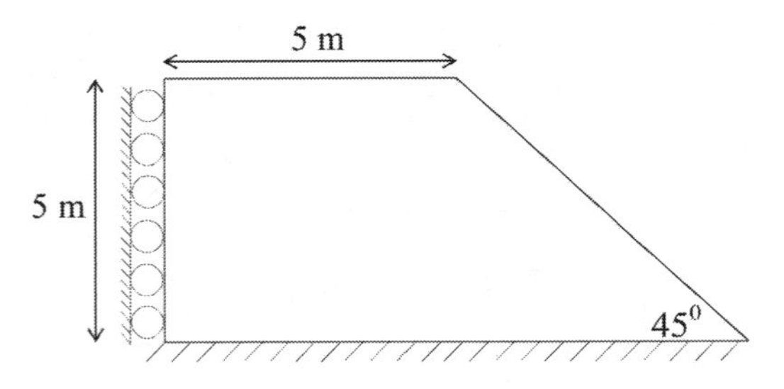

Figure 3. Sketch of the cross-sectional geometry.

Table 1. List of parameter values.

Parameter	Mean	Standard deviation
Cohesion	10 KPa	2
Friction angle	25°	5°
Dilation angle	0°	–
Young's modulus	1×10^5 kPa	–
Poisson's ratio	0.3	–
Unit weight	20^3	–

4 RESULTS

In this section, the response of the structure, in terms of FS distributions obtained from 500 realisations of the problem by RFEM, are presented. For simplicity, the same value of ξ was used to generate the random fields for c' and $\varphi'^{m}e$. The vertical scale of fluctuation θ_v was fixed to 1 m in all the analyses. A 2D deterministic analysis of the slope for the mean values given in Table 1 gave a FS of 1.4. Figure 4 plots the FS obtained in each realisation using perfectly positive and perfectly negative

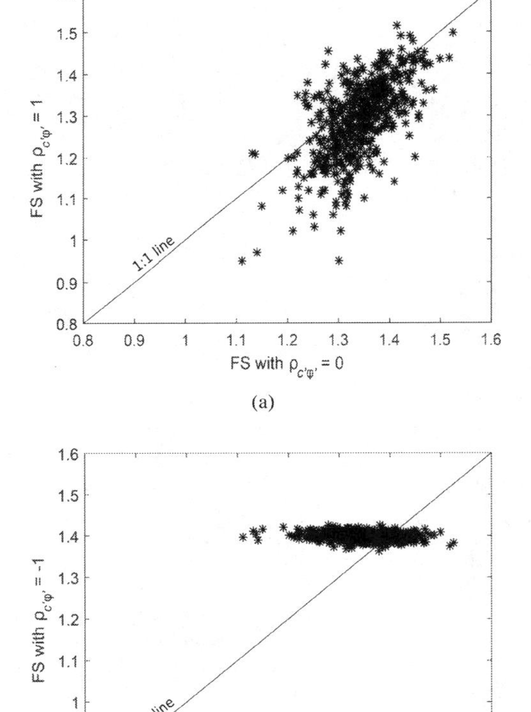

Figure 4. FS obtained with (a) $\rho_{c'\varphi'} = 1$ and (b) $\rho_{c'\varphi'} = -1$ against $\rho_{c'\varphi'} = 0$ for θ_h = 12 m.

cross-correlated c' - φ' fields, against the FS obtained from uncorrelated c' - φ' fields for θ_h = 12 m. Extreme values of $\rho_{c'\varphi'}$ compared to values reported in literature have been chosen, to highlight the differences between the solutions. For positively cross-correlated fields of the shear strength parameters, the weak zones (and the strong zones) of the shear strength are exaggerated compared to uncorrelated fields, making it easier to seek out the failure path. Hence, the positive cross-correlation decreases (or increases) the safety factor for each realisationand increases the range of possible solutions. In contrast, a negative cross-correlation between the shear strength parameters reduces the range of possible solutions.

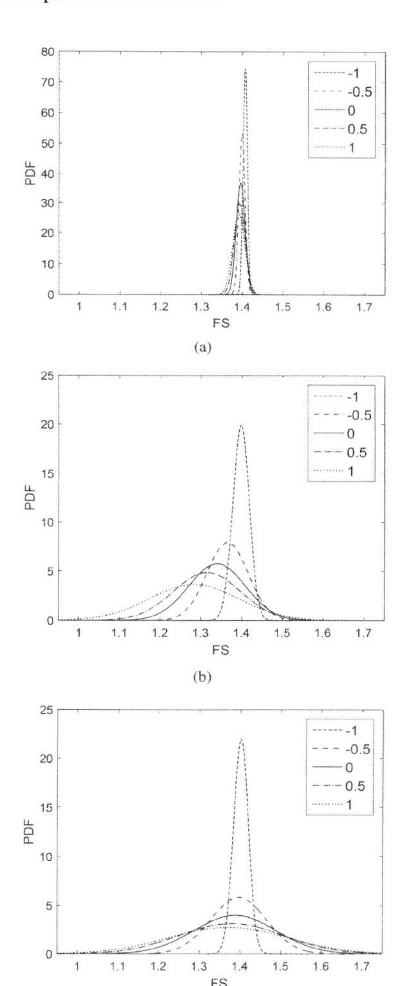

Figure 5. Probability density functions of FS for different values of $\rho_{c'\phi'}$; (a) $\xi = 1$, (b) $\xi = 12$, (c) $\xi = 2000$.

Figure 5 compares the distributions of FS at different values of ξ, for different $\rho_{c'\phi'}$. The different values of ξ considered in Figure 5(a-c) are similar to deterministic, 3D stochastic and 2D stochastic solutions, respectively, as in Hicks & Spencer (2010) and Varkey et al. (2017). In Figure 5(b), for the case of ξ = 12, i.e., for a value of θ_h lying between the slope height and half of the slope length, there is the possibility of discrete weak zones generated within each realisation (Spencer 2007, Varkey et al. 2017). This results in the mean FS being lower than 1.4, which is also the case for other values of θ_h lying in this range (not shown in Fig. 5). For positive values of $\rho_{c'\phi'}$, the failure propagates through even weaker zones and the mean FS reduces further below 1.4. In contrast, for negatively cross-correlated fields of c' and ϕ', the average of the mobilised shear strength over all the realisations increases. This results in the mean FS tending towards the deterministic FS for $\rho_{c'\phi'} = -1$. Also, the range of possible solutions decreases considerably compared to the uncorrelated and positively cross-correlated fields, and the variance of FS therefore reduces considerably.

For the case of a very large θ_h relative to the slope length (Fig. 5(c)), there is a wide range of possible solutions for uncorrelated fields and an even wider range for positively correlated fields. This wide range is due to the relative locations of very extensive weak zones through which the failure propagates.

For very small scales of fluctuation relative to the slope height, as in Figure 5(a), extreme averaging takes place and thus there is a negligible difference between the responses with different values of $\rho_{c'\phi'}$.

Figure 6 shows the reliability obtained at different values of F for slopes with $\rho_{c'\phi'} = -0.5, 0$ and 0.5, for the range of ξ values considered in Figure 5. Here, F is defined as the factor of safety based only on the mean shear strength. The reliability at each F for a given set of input statistics is calculated as:

$$\text{Reliability} = 1 - \frac{N_f}{N} \tag{6}$$

where N is the total number of realisations and N_f is the number of realisations in which the slope fails at a value less than or equal to F.

A negative cross-correlation between c' and ϕ' increases the reliability, whereas a positive cross-correlation decreases the reliability of the structure for all values of ξ considered.

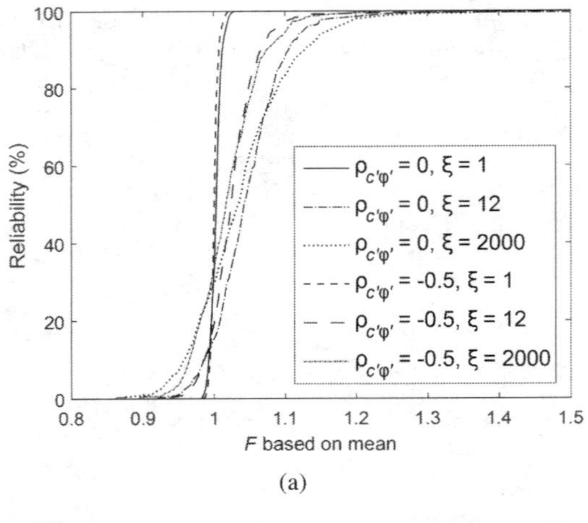

(a)

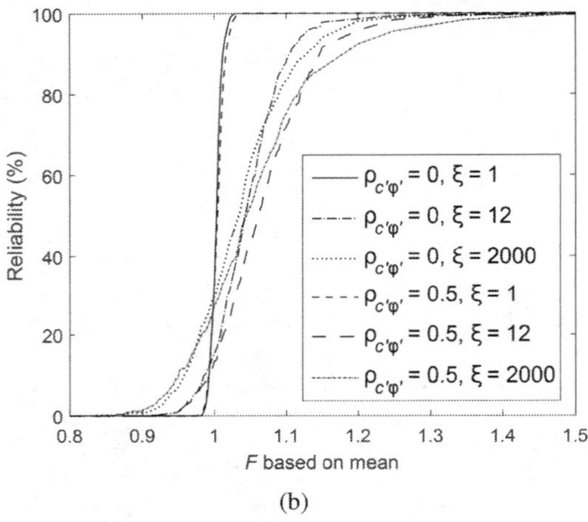

(b)

Figure 6. Reliability of the slope for various values of anisotropy of the heterogeneity (ξ) in the shear strength; (a) $-0.5 \le \rho_{c'\phi'} \le 0$, (b) $0 \le \rho_{c'\phi'} \le 0.5$.

5 CONCLUSIONS

An idealised 50 m long slope has been analysed by RFEM for various degrees of cross-correlation between the shear strength parameters (c' and ϕ'). It has been shown that assuming a positive cross-correlation between the parameters reduces the reliability, whereas a negative cross-correlation between the parameters increases the reliability of the slope. At intermediate and very large horizontal scales of fluctuation of c' and ϕ', assuming a perfectly negative cross-correlation considerably reduces the range of possible outcomes and makes the mean safety factor tend towards the plane strain safety factor based on the mean values alone. Hence, caution is needed when assigning cross-correlation coefficients between the shear strength parameters in an analysis.

ACKNOWLEDGEMENT

This work is part of the research programme Reliable Dykes with project number 13864 which is financed by the Netherlands Organisation for Scientific Research (NWO), and was carried out on the Dutch National e-infrastructure with the support of SURF Foundation.

REFERENCES

Arnold, P. & M.A. Hicks (2011). A stochastic approach to rainfall-induced slope failure. In *Proc. 3rd Int. Symp. Safety and Risk*, Munich, pp. 107–115.

Cherubini, C. (2000). Reliability evaluation of shallow foundation bearing capacity on c', φ' soils. *Canadian Geotech. J. 37*, 264–269.

Cho, S.E. (2007). Effects of spatial variability of soil properties on slope stability. *Engng. Geology 92*(3), 97–109.

de Gast, T., P.J. Vardon, & M.A. Hicks (2017). Estimating spatial correlations under man-made structures on soft soils. In *Proc. 6th Int. Symp. Geotech. Safety and Risk*, Colorado, USA, pp. 382–389.

Fenton, G.A. & D.V. Griffiths (2003). Bearing-capacity prediction of spatially random c-φ soils. *Canadian Geotech. J. 40*, 54–65.

Fenton, G.A. & D.V. Griffiths (2008). *Risk Assessment in Geotechnical Engineering.* John Wiley & Sons.

Fenton, G.A. & E.H. Vanmarcke (1990). Simulation of random fields via local average subdivision. *J. Engng. Mech. 116*(8), 1733–1749.

Griffiths, D.V., J. Huang, & G.A. Fenton (2009a). Influence of spatial variability on slope reliability using 2D random fields. *J. Geotech. Geoenviron. Engng. 135*(10), 1367–1378.

Griffiths, D.V., J. Huang, & G.A. Fenton (2009b). On the reliability of earth slopes in three dimensions. *Proc. R. Soc. London A: Math., Phys. Engng. Sc. 465*, 3145–3164.

Hicks, M.A. (2007). *Risk and Variability in Geotechnical Engineering.* Thomas Telford.

Hicks, M.A., J. Chen, & W.A. Spencer (2008). Influence of spatial variability on 3D slope failures. In *Proc. 6th Int. Conf. Computer Simulation Risk Analysis and Hazard Mitigation*, Kefalonia, pp. 335–342.

Hicks, M.A., J.D. Nuttall, & J. Chen (2014). Influence of heterogeneity on 3D slope reliability and failure consequence. *Comp. Geotech. 61*, 198–208.

Hicks, M.A. & K. Samy (2002). Influence of heterogeneity on undrained clay slope stability. *Quart. J. Engng. Geology and Hydrogeology 35*(1), 41–49.

Hicks, M.A. & W.A. Spencer (2010). Influence of heterogeneity on the reliability and failure of a long 3D slope. *Comp. Geotech. 37*(7), 948–955.

Huang, J., A.V. Lyamin, D.V. Griffiths, K. Krabbenhoft, & S. Sloan (2013). Quantitative risk assessment of landslide by limit analysis and random fields. *Comp. Geotech. 53*, 60–67.

Jaksa, M., W. Kaggwa, & P. Brooker (1999). Experimental evaluation of the scale of fluctuation of a stiff clay. In *Proc. 8th Int. Conf. Appl. Statistics and Probability in Civil Engng.*, Sydney, pp. 415–422.

Javankhoshdel, S. & R.J. Bathurst (2016). Influence of crosscorrelation between soil parameters on probability of failure of simple cohesive and c-φ slopes. *Canadian Geotech. J. 53*(5), 839–853.

Le, T.M.H. (2014). Reliability of heterogeneous slopes with cross-correlated shear strength parameters. *Georisk: Assessment and Management of Risk for Engineered Systems and Geohazards 8*(4), 250–257.

Li, D.Q., L. Zhang, X.-S. Tang, W. Zhou, J.-H. Li, C.-B. Zhou, & K.-K. Phoon (2015). Bivariate distribution of shear strength parameters using copulas and its impact on geotechnical system reliability. *Comp. Geotech. 68*, 184–195.

Li, Y. (2017). *Reliability of long heterogeneous slopes in 3D*. Ph.D. thesis, Delft University of Technology, The Netherlands. Li, Y., M.A. Hicks, & J.D. Nuttall (2015). Comparative analyses of slope reliability in 3D. Engng. Geology 196, 12–23.

Li, Y., M.A. Hicks, & P.J. Vardon (2016). Uncertainty reduction and sampling efficiency in slope designs using 3D conditional random fields. *Comp. Geotech. 79*, 159–172.

Lloret-Cabot, M., M.A. Hicks, & A.P. van den Eijnden (2012). Investigation of the reduction in uncertainty due to soil variability when conditioning a random field using Kriging. *Géotechnique Letters 2*(3), 123–127.

Phoon, K.-K. & F.H. Kulhawy (1999). Characterization of geotechnical variability. *Canadian Geotech. J. 36*(4), 612–624.

Spencer, W.A. (2007). *Parallel stochastic and finite element modelling of clay slope stability in 3D*. Ph.D. thesis, University of Manchester, UK.

Tang, X.-S., D.Q. Li, C.-B. Zhou, & K.-K. Phoon (2015). Copula-based approaches for evaluating slope reliability under incomplete probability information. *Struct. Safety 52*, 90–99.

Vanmarcke, E.H. (1977). Reliability of earth slopes. J. *Geotech. Engng Div. 103*(11), 1247–1265.

Vardon, P.J., K. Liu, & M.A. Hicks (2016). Reduction of slope stability uncertainty based on hydraulic measurement via inverse analysis. *Georisk: Assessment and Management of Risk for Engineered Systems and Geohazards 10*(3), 223–240.

Varkey, D., M.A. Hicks, & P.J. Vardon (2017). Influence of spatial variability of shear strength parameters on 3D slope reliability and comparison of analysis methods. In *Proc. 6th Int. Symp. Geotech. Safety and Risk*, Colorado, USA, pp. 400–409.

Numerical Methods in Geotechnical Engineering IX – Cardoso et al. (Eds)
 ISBN 978-1-138-33198-3

Variability in offshore soils and effects on probabilistic bearing capacity

E.A. Oguz, N. Huvaj & C.E. Uyeturk
Civil Engineering Department, Middle East Technical University, Ankara, Turkey

ABSTRACT: Risk assessment in geotechnical engineering requires consideration of uncertainties and variabilities in soil parameters. These uncertainties result from measurement errors, transformation uncertainties and inherent soil variability. In this study, the variability of soil properties in Turkish offshore sea beds are investigated based on nine CPT measurements in 40–64 m water depths. Statistical evaluation of the CPT tip resistance and sleeve friction data is performed and vertical spatial correlation lengths are reported for clays and sands, separately. The effects of the spatial correlation length and coefficient of variation on the probability of failure and bearing capacity of a shallow strip foundation on c-ϕ soil (with cross-correlation between shear strength parameters) are investigated. The bearing capacity analyses are performed by two-dimensional random finite element method using plane strain condition and Von Mises constitutive model for the soil. Shear strength parameters, stiffness and dilation angle are assumed to be lognormally distributed random variables. The results of this study indicate that the variability of soil parameters, in terms of coefficient of variation and vertical spatial correlation length, has a significant effect on the bearing capacity and probability of failure.

1 INTRODUCTION

Uncertainty and variability in soil properties are increasingly gaining more importance and popularity in the literature as well as in geotechnical engineering practice. Uncertainties in soil parameters result from measurement errors (equipment errors, human factor), transformation uncertainties and inherent soil variability (Phoon & Kulhawy 1999a). In reliability based analyses, all uncertainties should be addressed to achieve a reliable design. By utilizing Random Finite Element Method (RFEM), a combination of finite element methodology and random field theory, these uncertainties can be integrated with the conventional methods. The abovementioned uncertainties of the parameters can be defined via the mean (μ) and standard deviation (σ) (or coefficient of variation, COV) with a statistical distribution at a point. In addition, the spatially-varying property of soil (the heterogeneity) can be represented by a spatial correlation length (SCL), which is also called the scale of fluctuation, and correlation function. In the reliability-based design of offshore foundations, higher reliability indexes and lower probability of failure can be achieved with the use of spatial correlation length approach (Liu et al. 2015, Jha 2016, Chiasson et al. 1995, Akkaya & Vanmarcke 2003, Firouzianbandpey et al. 2014).

Zhang et al. (2016) state that information on the spatial variability of the marine seabed soils is relatively limited as compared to the onshore soils. Although there are some studies on this topic (e.g. Lacasse et al. 2007, Nadim 2015, Lacasse & Nadim 2007), there is no study for quantifying the SCL of offshore sea bed soils in Turkey. In this study, the variability of soil properties in Turkish offshore sea beds is investigated based on nine cone penetration test (CPT) measurements to determine the vertical SCL. The effects of the SCL and COV of soil properties on the probability of failure and bearing capacity of shallow strip foundation on c-ϕ soil (with cross-correlation between shear strength parameters; cohesion and friction angle) are investigated. Bearing capacity of shallow strip foundation is analyzed by utilizing the random finite element-based RBEAR2D software (Fenton & Griffiths 2008) using Monte Carlo simulations. The effect of the number of simulations is investigated and the analyses are conducted for isotropic and anisotropic cases, where SCL in the vertical and horizontal directions are taken equally for isotropic case and a ratio of 10 (Baecher & Christian 2003) is used for anisotropic case. The results of the study indicate that the SCL and COV of soil properties have important effects on the bearing capacity of shallow foundations.

Figure 1 demonstrates the soil variability by the deviation from the trend (Phoon & Kulhawy 1999a) and the soil property can be written as:

$$Y(x) = T(x) + \varepsilon_r(x) \tag{1}$$

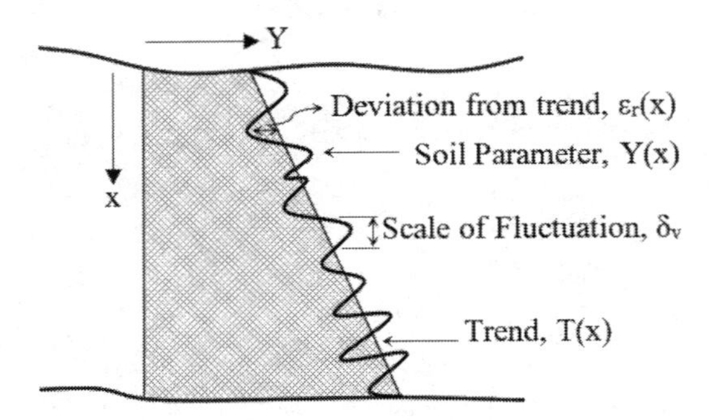

Figure 1. Illustration of the variability of soil properties: trend and deviations from the trend with depth.

The fluctuations, the deviations from the trend, are considered as statistically homogenous with depth, which means that the mean and the standard deviation are constant and the correlation between two measurements is only related to the distance of separation. Therefore the measurements should be detrended and then the condition can be satisfied. Likewise, Vanmarcke (1977) and Jaska et al. (1999) state that treating statistically homogeneous requires standardizing the data (zero mean and unit standard deviation) or detrending (zero mean).

2 RANDOM FINITE ELEMENT METHOD

In the RFEM a random field of soil properties are generated by utilizing statistical properties, a correlation function and then model is matched with finite element meshes. Statistical properties are the mean, standard deviation with a distribution model (such as normal, lognormal distributions) and SCL with a correlation function (Fenton & Griffiths 2008, Elachachi et al. 2012, Luo et al. 2014, Jha 2016). In this method, probabilistic analyses are conducted by performing a large number of Monte Carlo simulations.

2.1 *Spatial correlation length*

Spatial correlation length, which is also called the scale of fluctuation, is the distance over which the soil parameters are correlated to each other. Akkaya & Vanmarcke (2003) also defines SCL as the distance where the two points in that distance tend to be on the same side (above or below) of the trend. The study of Vanmarcke (1977) is one of the earliest research on the evaluation of SCL and provides a spatial averaging process to evaluate the scale of fluctuation where the soil parameter is averaged through distance and the standard deviation of the averages decreases as the distance of averaging increases. This decrease is defined by reduction function as follows:

$$\Gamma_u(\Delta z) = \frac{\tilde{u}_{\Delta z}}{\bar{u}} \tag{2}$$

where $\tilde{u}_{\Delta z}$ is the standard deviation of spatially averaged parameters while $\bar{u}$ is the standard deviation of the data. The square of reduction function is called as variance function and as the averaging interval increases, the variance function becomes inversely proportional to the interval (Vanmarcke 1977). The above-mentioned relationship brings us to the scale of fluctuation, δ_u, as follow:

$$\Gamma_u^2(\Delta z) = \frac{\delta_u}{\Delta z} \tag{3}$$

Table 1. Correlation functions and corresponding SCL (Vanmarcke 1977).

Function Name	Correlation Function	SCL
Exponential	$e^{-(\Delta z/a)}$	$2a$
Squared Exponential	$e^{-(\Delta z/b)^2}$	$\sqrt{\pi}b$
Cosine Exponential	$e^{-(\Delta z/c)}\cos(\Delta z/c)$	c
Sec. Order Autoregressive	$e^{-(\Delta z/d)}\,[1+(\Delta z/d)]$	$4d$

Vanmarcke (1977) also states that the scale of fluctuation may be evaluated by correlation functions used to fit the correlation coefficients and provide four different correlation functions and corresponding scale of fluctuations (Table 1).

The autocorrelation structure of soil parameters can be defined by autocovariance function, c_k, (Eq. 4) and autocorrelation coefficient, ρ_k (Eq. 5).

$$c_k = Cov(X_i, X_{i+k}) = E\left[\left(X_i - \bar{\bar{X}}\right)\left(X_{i+k} - \bar{\bar{X}}\right)\right] \tag{4}$$

$$\rho_k = \frac{c_k}{c_0} = \frac{\sum_{i=1}^{N-k}\left(X_i - \bar{\bar{X}}\right)\left(X_{i+k} - \bar{\bar{X}}\right)}{\sum_{i=1}^{N}\left(X_i - \bar{\bar{X}}\right)^2} \tag{5}$$

where X_i and X are the actual and trend (mean) values at point i.

Exponential (Marcov) (Firouzianbandpey et al. 2014, Akkaya and Vanmarcke, 2003, DeGroot 1996, Zhang et al. 2016, Peng et al., 2017) and squared exponential autocorrelation (Gaussian) functions are widely used in the literature. Therefore, exponential and squared exponential

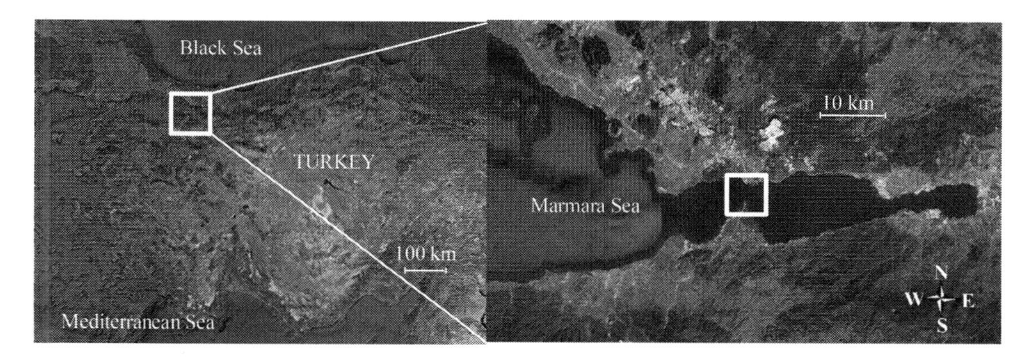

Figure 2. Location of the CPT borings.

correlation functions are employed in this study and corresponding SCL's are reported.

3 DESCRIPTION OF THE SITE

The project site is located in Gulf of Izmit in the Sea of Marmara, on the northwestern part of Turkey (Fig. 2). Total of nine CPT soundings, arranged in a rectangular grid pattern (240 m × 200 m separation distances in the plan) are analyzed. The average length of the CPT soundings and average water depths are 40.4 m (39.5 m–42.5 m) and 51.5 m (40.4 m–64.2 m), respectively. The cone tip resistance and sleeve friction measurements are taken at each sounding with a 0.02 m vertical spacing.

Two CPT profiles (tip resistance and friction ratio) are given in Figure 3 as an example. The classification of the soil profile is made by Robertson's soil behaviour type chart (Robertson, 2010): the soil profile at nine CPT soundings include clays/clay-silt mixtures and sands/silty sands according to the soil behavior types. Therefore, soils types are broadly grouped into two; "Clays" and "Sands", and all statistical analyses are conducted for these two types of soil groups separately.

Undrained shear strength, c_u, and relative density, Dr, are estimated through depth by using Eq. 6 and Eq. 7.

$$c_u = \frac{q_t - \sigma_{v0}}{N_k} \tag{6}$$

where q_t is the measured cone tip resistance, σ_{v0} is the total in situ vertical stress and N_k is the constant that has a range from 14 to 20.

$$Dr = \left(\frac{1}{0.0296}\right) \ln \left[q_t / \left[2.494\left(\sigma'_{v0}\left(\frac{1+2K_0}{3*100}\right)\right)^{0.46}\right]\right] \tag{7}$$

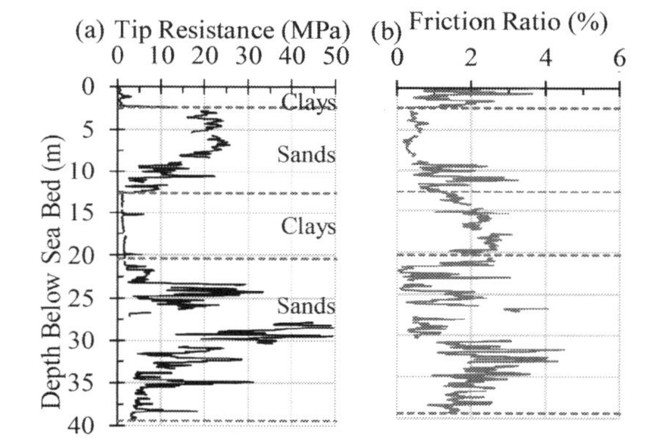

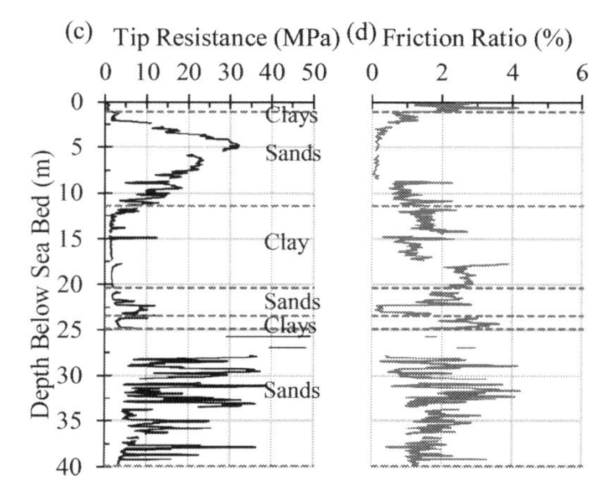

Figure 3. Two representative CPT profiles and soil layers; (a)-(b) and (c)-(d) are tip resistance and friction ratio of soundings 1 and 2.

where σ'_{v0} is the effective overburden pressure and K_0 is the at-rest earth pressure coefficient. In Figure 4, undrained shear strength and relative density profiles are provided for two CPT

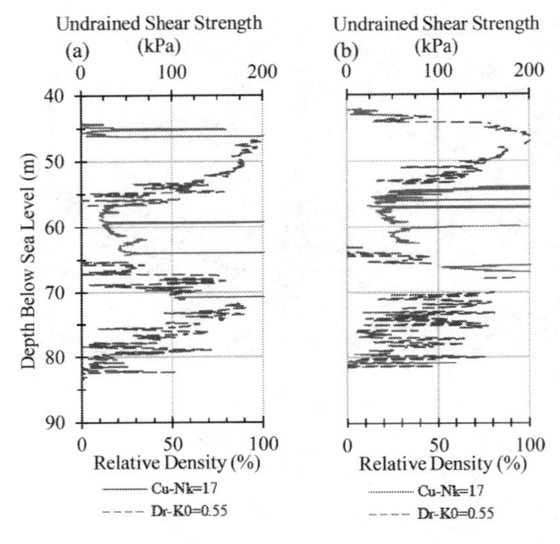

Figure 4. Two representative undrained shear strength and relative density profiles; (a) Sounding-1 (b) Sounding-2, where N_k and K_0 are taken as 17 and 0.55, respectively.

soundings. N_k value is taken as 17. The results of the nine CPT soundings indicate that the average c_u is 30 kPa (for N_k = 20)– 43 kPa (for N_k = 14) in the upper 10 m and 42 kPa (for N_k = 20)–60 kPa (for N_k = 14) for deeper Clays. Additionally, the relative density of the upper Sand layers is mostly medium-dense to dense, whereas the lower sands are in loose to medium-dense state.

4 SPATIAL CORRELATION LENGTH

Spatial correlation length is calculated for 9 CPT soundings, using cone tip resistance and sleeve friction, for Clays and Sands separately, using the exponential and squared exponential correlation functions as described in Table 1, and the results are given in Table 2. As can be seen in Table 2, all of the SCL values are within 0.04–0.49 m range.

The SCL values do not change significantly for exponential and squared exponential correlation functions. However, the squared exponential function gives slightly higher SCL values. In both Sands and Clays, SCL based on tip resistance is slightly larger than SCL based on sleeve friction. Likewise, Liu & Chen (2010) reported that tip resistance gives larger correlation length than sleeve friction. Sands have larger SCL values based on tip resistance and lower SCL values based on sleeve friction when compared to the Clays. In the study of Akkaya & Vanmarcke (2003) the similar result was obtained.

Table 2. Summary of Results.

		SCL (m)	
Soil type	Cor. function	Tip resistance	Sleeve friction
Clays	Exponential	0.16 (0.08–0.40)	0.14 (0.06–0.36)
	Sqr. exponential	0.16 (0.08–0.49)	0.15 (0.07–0.44)
Sands	Exponential	0.16 (0.04–0.42)	0.13 (0.04–0.37)
	Sqr. exponential	0.17 (0.04–0.49)	0.14 (0.04–0.42)

In the literature, the vertical SCL values for Clays have been reported as 0.7–1.1 m (Phoon et al. 1995), 0.2–0.5 m (Phoon & Kulhawy 1999a, 1999b), 0.19–0.72 m (Cafaro & Cherubini 2002), 0.26–3.14 m (Akkaya & Vanmarcke 2003), 0.13–0.32 m (Shuwang & Linping 2015) based on CPT tip resistance. In addition, the values for Sands have been reported as 0.1–2.2 m (Phoon et al. 1995), 0.61–3.72 m (Akkaya & Vanmarcke 2003). The values reported in this study are within the ranges reported in the literature, however towards the lower bound.

5 THE EFFECT OF SPATIAL CORRELATION LENGTH ON BEARING CAPACITY OF A SHALLOW FOUNDATION

The effects of SCL and COV of soil parameters on bearing capacity of the shallow foundation are investigated by using RBEAR2D software (Fenton and Griffiths, 2008) using random finite element method. A 4-m wide rigid strip footing located on the ground surface (Figure 5) and a general, c- ϕ soil is considered in the analyses where a negative correlation coefficient of –0.5 is employed between c and ϕ. 5 levels of COV, between 5% and 40%, for the soil parameters are utilized in the analyses. The model parameters are given in Table 3. In an analysis, same COV levels are applied to all random variables. Both isotropic and anisotropic analyses are performed where the ratio of horizontal SCL (δ_h) to the vertical (δ_v) is taken as 10 (Baecher and Christian, 2003). Soil is modeled with elasto-plastic Von Mises constitutive soil model. Von Mises model may not be the most appropriate model to capture soil behavior, however it was used for simplicity, as the purpose of this study was not to model the deformations to the highest accuracy but to observe the general effects of spatial correlation length on bearing capacity.

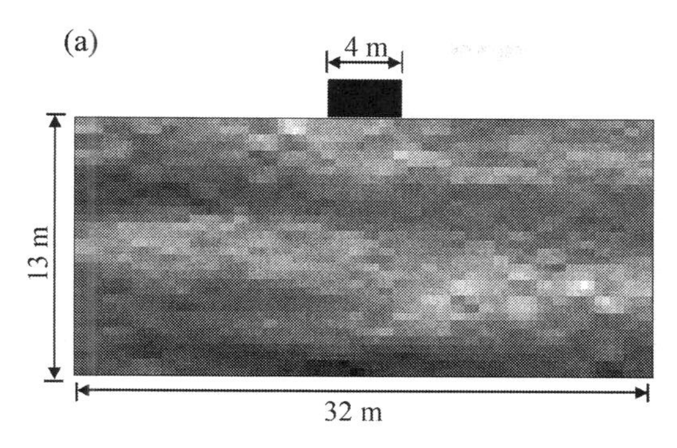

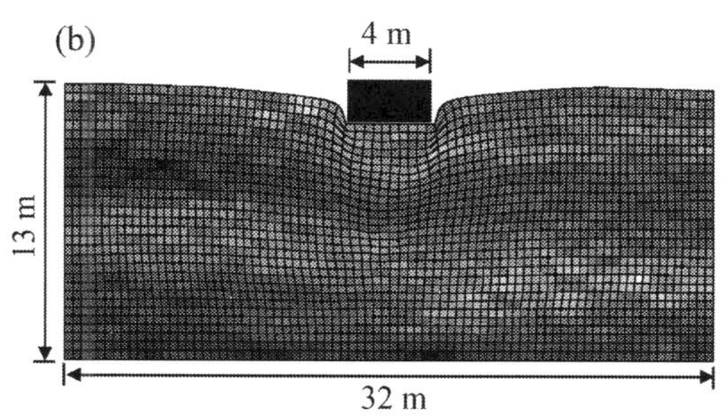

Figure 5. Representation of model with random field of cohesion (SCL_v = 10 m and SCL_h = 100 m); (a) Geometry of the model, (b) deformed mesh.

Table 3. Parameters used in the model.

Soil parameters	Statistical distribution	Mean value
Cohesion (kPa)	Lognormal	10
Friction angle (deg.)	Lognormal	35
Dilation angle (deg.)	Lognormal	5
Elastic modulus (kPa)	Lognormal	20000
Poisson's Ratio	Deterministic	0.25
Random field parameters		
Spatial Correlation Length (m)	0.025–0.05–0.1–0.25–0.5–1–2–3–4–5–6–7–8–9–10–15–20	
COV (%)	0–5–13.75–22.5–31.25–40	
Correlation coefficient ($c-\phi$)	–0.5	

Effect of the number of Monte-Carlo simulations is investigated by performing analyses with simulation numbers of 1 to 10000. The analyses are performed for COV of 22.5% and SCL was taken as 1, 2 and 3 m in both directions (isotropic). Figure 6 indicates that the mean bearing capacity converges to a stable point at about 2000simulation numbers regardless of the SCL value. Therefore, in all analyses, the number of simulations is taken as 2000. In the literature,

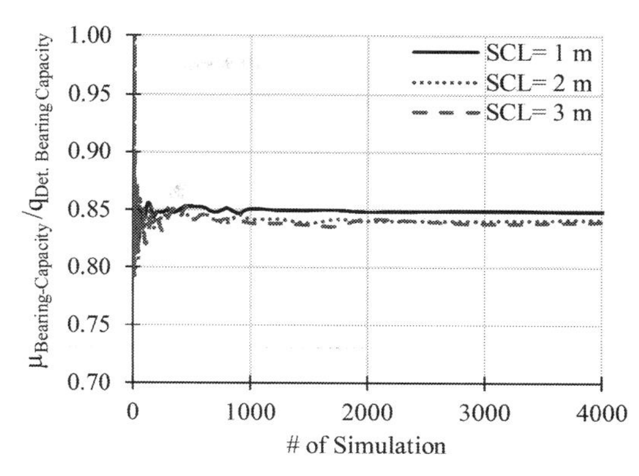

Figure 6. Effect of number of Monte Carlo simulations on the mean bearing capacity for COV = 22.5%.

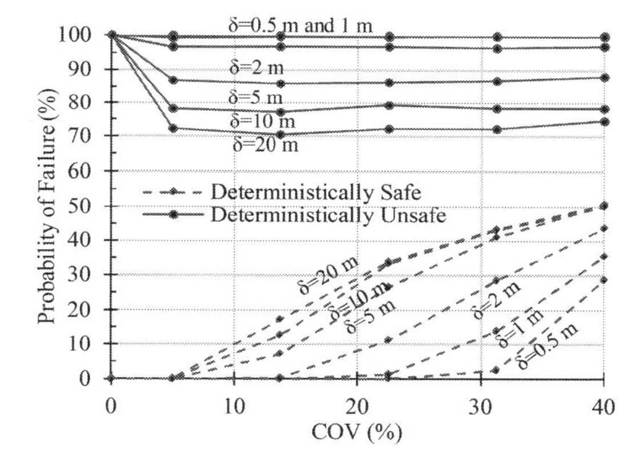

Figure 7. Effect of spatial correlation length (δ) with different $COV_{soil\ parameters}$ levels on probability of failure at deterministically safe and unsafe conditions.

The effect of SCL on the probability of failure is illustrated in Figure 7 for the foundation under an applied load that is greater and less than the deterministic bearing capacity of the soil, which are called "deterministically safe" and "deterministically unsafe" cases, respectively. For deterministically safe case, increasing SCL increases the probability of failure because having larger weaker zones under the foundation becomes possible with increasing SCL and larger weaker zones initiate the bearing failure. On the contrary, increasing SCL decreases the probability of failure for the deterministically unsafe case. The reason is that, relatively stronger zones are formed by increasing SCL and larger loads can be carried by the foundation. Sarma et al. (2014) investigated the effects of spatial variability of soil parameters on soil slopes and obtained similar results that increasing SCL increase the probability of failure for deterministically safe case. Likewise, Jiang et al. (2014) and Le et al. (2014) reported similar results.

The mean value of bearing capacity of 2000 Monte Carlo simulations decreases with increasing variability of soil parameters (Fig. 8). The average bearing capacity can decrease by about 25% (110 kPa) of its deterministic value (450 kPa). Additionally, the results show that for a given COV, increasing SCL from 0.025 m to 3.0 m can

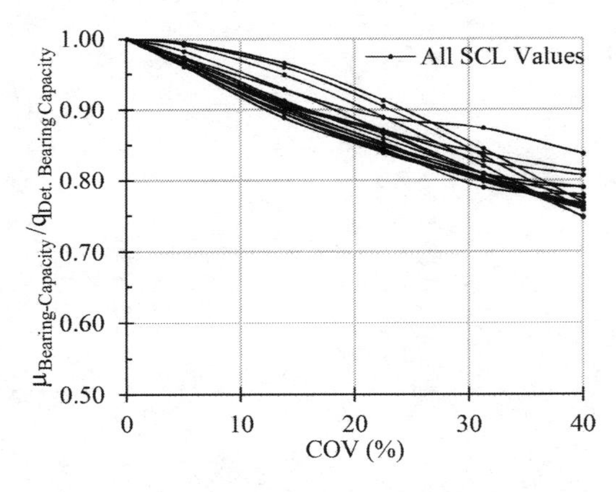

Figure 8. Effect of spatial correlation length with different $COV_{soil\ parameters}$ levels on mean bearing capacity.

Figure 9. Probability density function of δ = 20 m & 1 m for $COV_{soil\ parameters}$ of 5% and 40%.

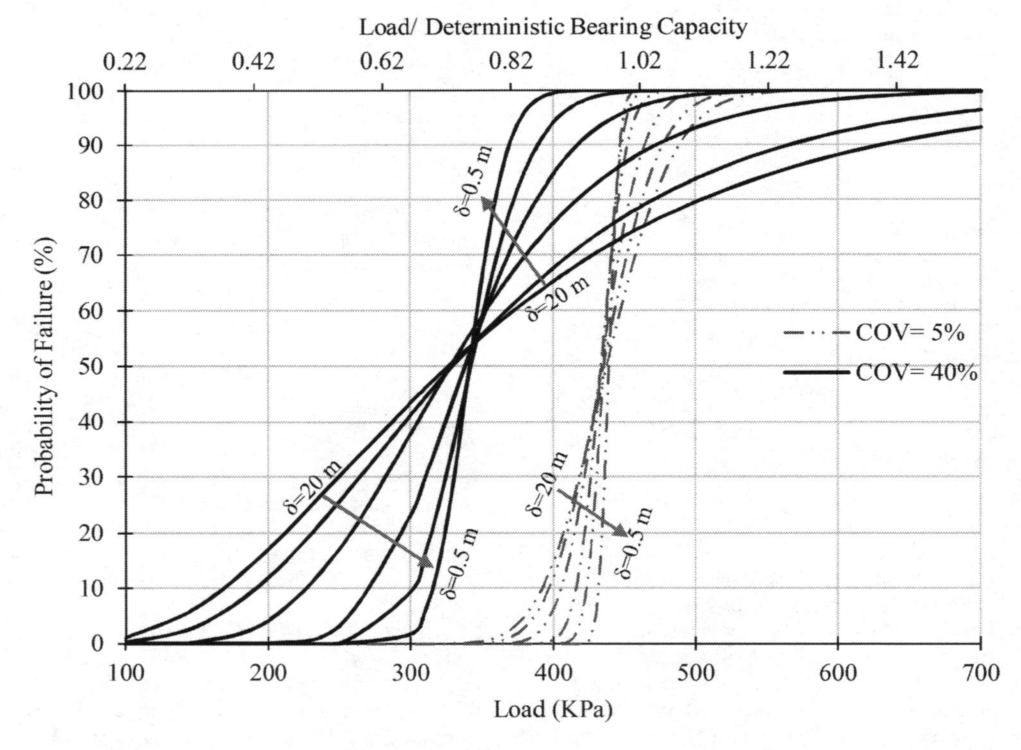

Figure 10. Effect of spatial correlation length (δ) with different COV soil parameters levels on probability of failure under different load.

decrease the mean bearing capacity as much as 11% (50 kPa) of its deterministic bearing capacity.

In Figure 9, when SCL increases for the same $COV_{parameter}$, the standard deviation of evaluated bearing capacity increases and a wider range of results are obtained. Likewise, when SCL is kept constant and $COV_{parameter}$ increases from 5% to 40%, the standard deviation increases. That is, either increasing SCL or $COV_{parameter}$ significantly increases the standard deviation of evaluated bearing capacity.

The effect of SCL and COV of soil parameters can be seen in Figure 10. It is seen that for a given loading, the probability of failure increases as COV of soil parameters increases. For the same COV level for soil parameters, increasing the SCL may increase or decrease the probability of failure (Fig. 10) according to the safety level, which is illustrated in Figure 7. The inverse effect of SCL on probability of failure for different cases are also obtained in the study of Griffiths and Fenton (2001) and Fenton and Griffiths (2000).

6 CONCLUSIONS

In this study, nine CPT soundings are analyzed; undrained shear strength and relative density profiles are obtained for clays and sands, respectively. Soil types are grouped into two broader categories: Clays and Sands, and spatial correlation length is calculated for each soil type separately. The Clays are found in soft to medium stiff state (c_u = 41–59 kPa with a standard deviation of 27–38 kPa) and Sands are found in loose to dense state with an average D_r of about 50%.

Spatial correlation lengths obtained in this study are within the range of values reported in the literature. Based on the tip resistance, SCL is found as 0.16 m for Clays and 0.16–0.17 m for Sands. Based on the sleeve friction, the values are 0.14–0.15 m for Clays and 0.13–0.14 for Sands. It is seen that SCL of Sands based on the tip resistance is slightly higher than SCL of Clays based on tip resistance. Likewise, SCL of Clays based on sleeve friction is larger than SCL of Sands based on sleeve friction. It is also seen that SCL has no depth dependent behavior (i.e. both upper and lower clay layers have similar SCL).

The effect of SCL and coefficient of variation of soil parameters are investigated by using RBEAR2D, random finite element method. All analyses are conducted with 2000 Monte-Carlo simulations. The effect of SCL on the probability of failure, P_f, changes according to the safety level. Increasing SCL increases the P_f of deterministically safe cases, while it decreases the P_f of deterministically unsafe cases. Mean bearing capacity decreases with increasing COV of soil parameters. In addition, the mean bearing capacity may decrease by about 11% of its deterministic bearing capacity with the increase of SCL. The standard deviation of 2000 bearing capacity simulations increases by increasing SCL values and therefore the probability of failure is affected significantly.

Similar conclusions are reached about the effect of SCL and COV on the probability of failure in anisotropic SCL condition where the ratio of SCL in the horizontal to vertical directions is 10 (Oguz 2017).

The results of this study add to the database of SCL of offshore sea bottom soils based on CPT. Also, it emphasizes that the SCL and COV of soil parameters have a significant influence on the probabilistic bearing capacity evaluations.

REFERENCES

Akkaya, A., & E.H. Vanmarcke. 2003. Estimation of Spatial Correlation of Soil Parameters Based on Data from the Texas A&M University NGES. Probabilistic Site Characterization at the National Geotechnical Experimentation Sites, 29–40. doi:10.1061/9780784406694.ch03.

Baecher GB & Christian JT. *Reliability and Statistics in Geotechnical Engineering*. Wiley: New York, 2003.

Cafaro, F. & C. Cherubini. 2002. Large Sample Spacing in Evaluation of Vertical Strength Variability of Clayey Soil. Journal of Geotechnical and Geoenvironmental Engineering 128 (7): 558–68. doi:10.1061/(ASCE)1090-0241(2002)128:7(558).

Chiasson, P., J. Lafleur, M. Soulié & K.T. Law. 1995. Characterizing Spatial Variability of a Clay by Geostatistics. Canadian Geotechnical Journal 32 (1): 1–10. doi:10.1139/t95-001.

DeGroot, D.J. 1996. Analyzing spatial variability of in-situ soil properties. In C.D. Shackleford, P.P. Nelson and M.J.S. Roth (eds.), Uncertainty in the Geologic Environment: From Theory to Practice, Geotechnical Special Publication No. 58: 210–238. New York: ASCE.

Elachachi, S.M., Breysse, D., & Denis, A. 2012. The effects of soil spatial variability on the reliability of rigid buried pipes. Computers and Geotechnics, 43, 61–71.

Fenton, G.A., & Griffiths, D.V. 2008. *Risk Assessment In Geotechnical Engineering*. Hoboken, NJ: John Wiley & Sons.

Firouzianbandpey, S., D.V. Griffiths, L.B. Ibsen & L.V. Andersen. 2014. Spatial Correlation Length of Normalized Cone Data in Sand : Case Study in the North of Denmark. Canadian Geotechnical Journal 857 (July 2013): 844–57. doi:10.1139/cgj-2013-0294.

Jaksa, M.B., W.S. Kaggwa & P.I. Brooker. 1999. Experimental evaluation of the scale of fluctuation of a stiff clay, In Proc. 8th Int. Conf. on the Application of Statistics and Probability, 1:415–422, December 1999, Sydney, AA Balkema, Rotterdam.

Jha, S.K. 2016. Reliability-Based Analysis of Bearing Capacity of Strip Footings Considering

Anisotropic Correlation of Spatially Varying Undrained Shear Strength. International Journal of Geomechanics, 6016003. doi:10.1061/(ASCE) GM.1943-5622.0000638.

Lacasse, S., & Nadim, F. 2007. Probabilistic geotechnical analyses for offshore facilities. Georisk, 1(1), 21–42.

Lacasse, S., Guttormsen, T., Nadim, F., Rahim, A. & Lunne, T. 2007. Use of statistical methods for selecting design soil parameters. In Offshore Site Investigation And Geotechnics, Confronting New Challenges and Sharing Knowledge. Society of Underwater Technology.

Liu, C.-N., and C.-H. Chen. 2010. Estimating Spatial Correlation Structures Based on CPT Data. Georisk: Assessment and Management of Risk for Engineered Systems and Geohazards 4 (2): 99–108. doi:10.1080/17499511003630504.

Liu, Z., S. Lacasse, F. Nadim, M. Vanneste, and G. Yetginer. 2015. Accounting for the Spatial Variability of Soil Properties in the Reliability-Based Design of Offshore Piles. In Frontiers in Offshore Geotechnics III, 978-1.

Luo, Z., Wang, L., Khoshnevisan S. & Juang C.H. Effect of spatial variability on the reliability-based design of drilled shafts. Proceedings of the Geo-Congress 2014,GSP (Geotechnical Special Publication);234, 23–26 Feb 2014, Atlanta, Georgia, 3274–3282.

Nadim, F.. 2015. Accounting for Uncertainty and Variability in Geotechnical Characterization of Offshore Sites. doi:10.3233/978-1-61499-580-7-23.

Oguz, E.A. 2017. Spatial Probabilistic Evaluation of Sea Bottom Soil Properties and Its Effect on Foundation Design. Middle East Technical University. M.S. Thesis. Ankara, Turkey.

Peng, X.Y., L.L. Zhang, D.S. Jeng, L.H. Chen, C.C. Liao, & H.Q. Yang. 2017. Effects of Cross-Correlated Multiple Spatially Random Soil Properties on Wave-Induced Oscillatory Seabed Response. Applied Ocean Research 62. Elsevier B.V.: 57–69.

Phoon, K.-K. & F.H. Kulhawy. 1999a. Characterization of Geotechnical Variability. Canadian Geotechnical Journal 36 (4): 612–24. doi:10.1139/t99-038.

Phoon, K.-K. & F.H. Kulhawy. 1999b. Evaluation of Geotechnical Property Variability. Canadian Geotechnical Journal 36 (4): 625–39. doi:10.1139/t99-039.

Phoon, K.-K., F.H. Kulhawy, & M.D. Grigoriu.1995. Reliability-based design of foundations for transmission line structures. Electric Power Research Institute, Palo Alto, Calif., Report TR-105000.

Robertson, P.K., Campanella, R.G., Gillespie, D. & Greig, J. 1986. Use of piezometer cone data. Use of In-Situ Tests in Geotechnical Engineering (GSP 6), ASCE, Reston, VA: 1263–1280.

Shuwang, Y., & G. Linping. 2015. Calculation of Scale of Fluctuation and Variance Reduction Function. Transactions of Tianjin University 21 (1): 41–49.

Vanmarcke, E.H. 1977. Probabilistic Modelling of Soil Profiles. Journal Of The Geotechnical Engineering Division, ASCE, 1227–46.

Zhang, L.L., Y. Cheng, J.H. Li, X.L. Zhou, D.S. Jeng, & X.Y. Peng. 2016. Wave-Induced Oscillatory Response in a Randomly Heterogeneous Porous Seabed. Ocean Engineering 111: 116–27. doi:10.1016/j.oceaneng.2015.10.016.

Numerical Methods in Geotechnical Engineering IX – Cardoso et al. (Eds)
 ISBN 978-1-138-33198-3

Stochastic study of stability of unsaturated heterogeneous slopes destabilised by rainfall

T.M.H. Le
Norwegian Geotechnical Institute, Oslo, Norway
Klima2050—Centre for Research-based Innovation for Risk Reduction through Climate Adaptation of Building and Infrastructure, Trondheim, Norway

ABSTRACT: Failure of unsaturated soils due to rainfall infiltration is a threat in many parts of the world with temperate or hot climate. Understanding of unsaturated slope stability during rainfall infiltration is an important but complex problem. The shear strength of unsaturated soil depends on soil suction and positive pore water pressure which are governed by the water flow. In addition, the soil porosity can vary spatially quite significantly which generates atypical preferential paths for the water flow. This can lead to unusual suction reduction and/or build-up of positive pore pressure over the domain. This study nummerically investigates the stability of an unsaturated slope with spatially varying porosity. The random finite element method and the shear strength reduction technique are employed to probabilistically estimate the change in factor of safety during and after a rainfall event. Through Monte Carlo simulation, the study shows that the heterogeneity of porosity can alter the failure mechanism and the probability of failure of the unsaturated slope.

1 INTRODUCTION

Understanding of unsaturated slope stability during rainfall infiltration is of critical importance in many parts of the world (e.g. southern Europe, Brazil, HongKong) where the hot climate give rise to extensive areas with unsaturated soils. On one hand, the unsaturated state actually improves slope stability in dry or partial saturation condition because the existence of suction (i.e. negative pore pressure) raises the soil strength compared with its saturated state. On the other hand, the diminish of the unsaturated condition due to, for example, wetting process by rainfall can lead to a reduction of the soil shear strength because of a reduction in suction and, possibly, a rise in positive pore water pressure. Both these effects directly contribute to destabilising unsaturated slopes. If porosity varies considerably in the soil mass, an extra level of complexity is added because the wetting front no longer migrates in a smooth and regular pattern over time as in a domain with uniform porosity. Instead, water follows atypical preferential paths causing uneven reduction of suction and, possibly, build-up of positive pore pressure in the soil masses. A soil element experiencing large suction loss might reach the failure condition earlier or under lower stresses than other neighboring elements experiencing a limited suction reduction. The slip surface then preferably passes through these failed elements despite having to follow an unusual shape. Heterogeneity of porosity can therefore alter the failure mechanism of an unsaturated slope subjected to rainfall.

Stability of a slope can normally be assessed using an analytical solution (for simple infinite slopes) or the traditional Limit Equilibrium Method (LEM) and/or the more advanced Finite Element Method (FEM). The FEM is often employed to deal with complex slope problems such as, for example, those having unusual geometries, subjected to complicated loading sequences, undergoing progressive failure or showing heterogeneous material properties. The problem considered in this study is rather complex and highly non-linear, hence the use of the FEM is necessary.

When it comes to probabilistic investigation of slope stability, most of studies reported in literature assumed that the slope is fully saturated (or fully dried). In many cases, the investigations focused on the spatial variability of shear strength, including undrained shear strength, or effective friction angle (ϕ') and/or effective cohesion (c'). Many past stochastic studies employ the Monte-Carlo approach due to its conceptual simplicity and its capability of handling complicated geometries and variability patterns without requiring over-simplified assumptions (e.g. (Griffiths and Fenton, 2004, Hicks and Onisiphorou, 2005, Griffiths and Marquez, 2007, Griffiths et al., 2009)). Some studies demonstrated that the influence of soil heterogeneity on stability depend on various factors such as, for example, the slope characteristics (e.g. slope angle) or the combination of input parameters (e.g. mean and standard deviation of the random soil property). Several studies have integrated the random field theory and the finite element method to model soil spatial variability. The method is often referred to as the "random finite element method" (Hicks and Samy, 2002b, Hicks and Samy, 2002a, Griffiths and Fenton, 2004, Fenton and Griffiths, 2005, Hicks and Onisiphorou, 2005, Hicks and Spencer, 2010, Le et al., 2015).

Past studies on unsaturated slope stability are almost exclusively limited to homogeneous soils. Griffiths and Lu (2005) and Lu and Godt (2008) devised a formula of suction stress which takes into account both soil characteristics and infiltration rate. The suction stress was then used to analytically predict the stability of an unsaturated slope in a steady seepage condition. A partially saturated slopes can also be analysed by the LEM in an uncoupled manner. This usually involves an independent flow analysis to obtain the pore water pressure distribution assuming rigid soil behaviour under given hydraulic boundary conditions. This pore water pressure distribution is then used as the groundwater condition to input into a subsequent limit equilibrium analysis (Fredlund and Rahardjo, 1993).

This study investigates the stability of an unsaturated slope with spatially varying porosity using the random finite element method. The aim is to demonstrate the use of stochastic methods in investigating the risk associated with unsaturated soil heterogeneity at different points in time during and after a rainfall event. Particularly, the effects of soil variability (quantified by different values of the Coefficient of Variation, *COV*) on the factor of safety of the slope are examined.

2 METHODS

2.1 *Slope geometry and boundary condition*

This study investigates a slope is 10 m height, having an angle of 2:1 (or 26.6°) and resting on a base of 20 m thickness (Fig. 1). The initial stress distribution is in equilibrium with gravity (9.81 m2/s) while the initial pore water pressure distribution is in hydrostatic equilibrium with the initial water table position at 5 m below the ground surface. The hydrostatic condition results in a maximum initial suction of approximately 150 kPa at the crest (AB) which falls in the low end of the prevalent surface suction range of semiarid or arid environments such as in Australia (Cameron et al., 2006).

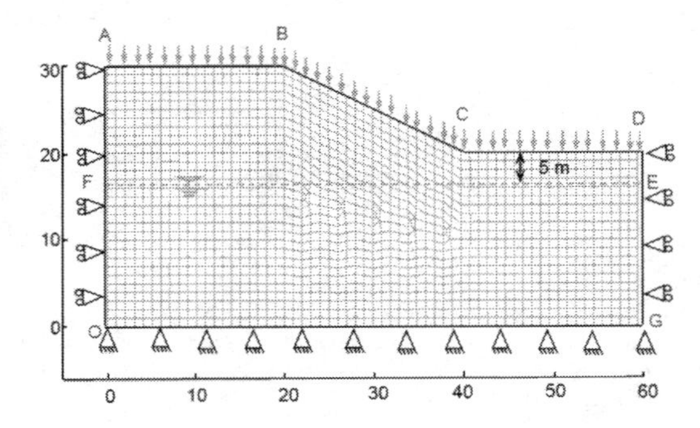

Figure 1. Slope dimensions, geometry and boundary conditions (scale in metre).

The slope is subjected to a rainfall event which extends continuously over 10 days after which the rainfall is terminated but the simulation was continued for another 355 days. The stability of the slope is assessed at different times both during and after the rainfall as explained in detail in subsequent sections. Rainfall is assumed to infiltrate into the slope through boundaries AB, BC and CD over the ten days. These boundaries (AB, BC and CD) then become impermeable once the rainfall has stopped (i.e. day 10 onward). This means that there is no subsequent rainfall infiltration or evaporation at the ground surface, after the initial 10 days of rainfall. Boundaries DG, GO and OA are assumed to be impermeable both during and after the rainfall. The factor of safety are estimated at 0, 0.5, 5, 10, 15, 20, 100, 365 days from the starting time.

2.2 *Hydraulic and mechanical models*

The van Genuchten (1980) and van Genuchten and Nielsen (1985) water retention curve and permeability function are selected because they can represent the soil hydraulic behaviour in a simple and numerically stable way. Details of the selected hydraulic constitutive relationships can be found in Le et al. (2015).

$$S_e = \left(1 + \left(\frac{s}{s_e}\right)^{\frac{1}{1-m}}\right)^{-m} \tag{1}$$

$$S_e = S_{eo}\exp(\eta(\phi_o - \phi)) \tag{2}$$

$$k_s = k_{so}\frac{\phi^3}{(1-\phi)^2}\frac{(1-\phi_o)^2}{\phi_o^3} \tag{3}$$

$$k_r = \sqrt{S_e}(1-(1-S_e^{1/m})^m)^2 \tag{4}$$

$$\mathbf{q} = -k_s k_r \nabla \left(\frac{u_w}{\rho_w g} + z \right) \tag{5}$$

The soil water retention curve (SWRC) van Genuchten (1980) is given by equation 1, which relates the effective degree of saturation (S_e) to suction s=-p_w through the air entry suction parameter s_e. In equation 2, the parameter s_e is in turn related to the porosity (ϕ) through parameter η which controls the rate at which s_e deviates from its reference value s_{eo} when ϕ deviates from its reference value ϕ_o (Rodríguez et al., 2007, Zandarín et al., 2009). Similarly, equation 3 describes the variation of the saturated permeability (k_s) from its reference value k_{so} when ϕ deviates from its reference value ϕ_o, as proposed by (Kozeny, 1927). Equation 4 describes the van Genuchten and Nielsen (1985) permeability curve linking the relative permeability k_r to the effective degree of saturation S_e through the parameter m, which can be geometrically interpreted as the curve gradient. The unsaturated permeability k_u is the product of the saturated and relative permeability (i.e. $k_u = k_s k_r$). Finally, the unsaturated flow **q** is calculated using the generalized Darcy's law (equation 5). The symbols u_w, ρ_w, g and z indicate the pore water pressure, the water density, the gravitational acceleration and the elevation coordinate, respectively.

The values of hydraulic parameters for the water retention curve and the permeability function are shown in Table 1. The values of the constitutive parameters are selected to be at around the middle of the likely range for each parameter in order to avoid overly large or small input parameters and, hence, possible unrepresentative or unrealistic results. For example, an extremely low permeability (i.e. very small k_{so}) would result in the rainfall having no effect on slope stability because water would not have sufficient time to infiltrate into the unsaturated soil domain. The rainfall intensity is assumed to be at 43.2 mm/day.

Table 1. Values of hydraulic and mechanical soil parameters.

Hydraulic			Mechanical		
Symbols	Units	Values	Symbols	Units	Values
m		0.2	E	kPa × 10^3	100
η		5	v		0.3
ϕ_o		0.333	ϕ'	°	20
k_{so}	m/s	10^{-5}	c'	kPa	5
s_{eo}	kPa	20	ϕ^b	°	18
S_s		1	ρ_s	kg/m^3	2700
S_r		0.001			

A linear elastic model with an extended Mohr-Coulomb (MC) failure criterion for unsaturated soils is selected to model mechanical behaviour (Fredlund et al., 1978). The values adopted for the mechanical parameters are also shown in Table 1. These values are kept constant throughout the study. The typical property values of a clay are selected for the strength parameters (c', ϕ' and ϕ^b) and the elastic parameter, i.e. Young modulus (E) and Poisson ratio (v).The nonlinear equations associated with flow and mechanical problems are solved in a fully coupled manner using the New-Raphson method (Olivella et al., 1996).

The Factor of Safety (FoS) in the FEM is defined by relating the actual soil strength to the failure strength. In particular, the FoS is estimated in the FEM by inducing slope failure through a numerical process called shear strength reduction. This involves incrementally and continuously reducing the values of the shear strength parameters until instability is initiated. For unsaturated soil, Equation (1) is adopted in this study to conduct the shear strength reduction process which involves incremental reduction of tanϕ', c' and tanϕ^b (Le, 2011, Le et al., 2015).

$$FoS = \frac{c'_{actual}}{c'_{fail}} = \frac{\tan\phi'_{actual}}{\tan\phi'_{fail}} = \frac{\tan\phi^b{}_{actual}}{\tan\phi^b{}_{fail}} \tag{6}$$

2.3 *Random variable and random finite element method*

Porosity is probably one of the most easily observed soil parameters which exhibit spatial variability. The variability characteristics of porosity have been investigated in a few studies (e.g. (Le et al., 2013a, Phoon and Kulhawy, 1999)). In unsaturated soils, porosity variability leads to heterogeneity in the water retention behaviour and permeability of the soil (Le et al., 2015, Le et al., 2012, Le et al., 2013b). During a rainfall event, infiltrated water is thus likely to follow unusual preferential flow paths in a porosity heterogeneous slope. Therefore, those soil elements subjected to an earlier or larger infiltration will loose the cohesive component of strength contributed by suction earlier than other soil elements or by a more substantial amount. This non-uniform strength reduction causes variation in the overall factor of safety.

In this study, the heterogeneity of porosity in each realization is created by generating first a random field of void ratio (e). Note that each "realization" refers to a deterministic distribution of porosity generated during Monte Carlo simulation. The e random field is modelled by a log-normal distribution which ensures that the generated random values are always positive. The values of porosity ϕ calculated using $\phi = e/(1+e)$ are therefore always bounded between 0 and 1 as required. The values of mean $\mu(e)$ and correlation length $\theta(e)$ are kept constant at 0.5 and 8 m, respectively. A range of coefficient of variation, COV_e, between 0.1 and 1.6 are considered. This results in the corresponding values of $\mu(\phi)$ and $COV\phi$ as presented in Table 2. As the COV_e increases at a constant $\mu(e)$, the $\mu(\phi)$ decreases while the $COV\phi$ increases. The largest value of $COV\phi$ are larger than the upper bound of the usual $COV\phi$ range which can be found in, for example, Phoon and Kulhawy (1999) and Le et al. (2013a). The selection of such large values are to demonstrate the influence of porosity more clearly and to consider the extreme situation where such a large variability in porosity is possible (e.g. in fill materials).

The same mesh of 1515 quadrilateral (squares or parallelograms) elements is employed consistently in every realization to eliminate undesired variations associated with mesh discretisations (Fig. 1). This mesh was shown to produce a reliable estimate of the factor of safety of homogeneous slope, which remained essentially unchanged for finer discretisations (Le et al., 2015). For each realisation, a random field with a grid of square cells is first generated with one random value allocated to each cell. A mapping process is then performed which involves superimposing the random field on the mesh. The superimposition allows to search and assign each finite element the random value from the cell having centroid closest to the centroid of the element. Each square element, and the vast majority of parallelogram elements, has an area of 1 m^2, with a few exceptions in the centre of the mesh. The relatively uniform sizes and similar shapes of the mesh elements ensure marginal discrepancy (caused the mapping procedure) between the variability characteristics of the random finite element mesh and the underlying random field. An exemplary comparison between the statistics of the random variable mapped onto the mesh and the theoretical statistics is shown in Figure 2. The figure demonstrates that the probability density functions of both random field e and the mapped values of e on the finite element mesh are reasonably consistent with the theoretical probability

Table 2. Parameters of porosity (ϕ) random field corresponding to different input values of COV_e and $\mu(e) = 0.5$.

COV_e	0.1	0.2	0.4	0.8	1.6
$\mu(\phi)$	0.333	0.330	0.323	0.300	0.256
COV_ϕ	0.067	0.132	0.254	0.456	0.720

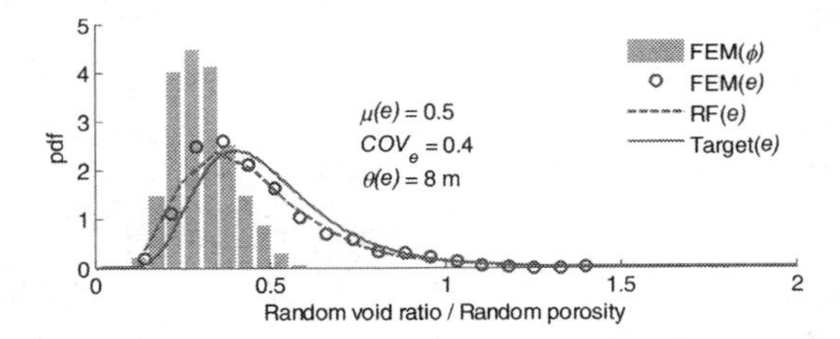

Figure 2. Histogram of a sample realization of random porosity mapped onto the mesh (FEM(ϕ)) and the pdf of associated void ratio (FEM(e)), of the underlying random void ratio field (RF(e)) and of the theoretical distribution generated from input parameters (Target(e)).

distribution function (pdf) calculated from the input $\mu(e)$ and COV_e. The consistency however tends to deteriorate with increasing input COV_e. The corresponding random values of ϕ follows a pdf that appeared to be skewed-left and confined between zero and 1.

3 RESULTS AND DISCUSSIONS

3.1 *Statistical characterisations*

The main outputs of the Monte Carlo analysis in this study are the *FoS*. In order to be considered reliable, the statistics of the *FoS* should be stable with the number of realizations (i.e. does not change with more realizations). The convergences of the $\mu(FoS)$ and $\sigma(FoS)$ are assessed by plotting the evolution of mean with increasing number of realizations as, for example, shown in Figure 3. Individual data points and 95% confidence interval are also represented on these figures to indicate the range of variability, hence fluctuation of $\sigma(FoS)$. Both the $\mu(FoS)$ and $\sigma(FoS)$ converge rather quickly after around 60 realizations. The 95% confidence interval also becomes relatively narrow after 60 realizations. A difference of approximately 0.06 is observed between the lower and upper bounds at this point indicating the stabilization of $\sigma(FoS)$. The *FoS* for individual realizations however spread over a reasonably wide range from just above 1 to just below 2 which is caused by the variation in suction distribution between individual realizations. Note that Figure 3 shows variation of data corresponds to 5 days of continuous rainfall. Inspection of the results show that the *FoS* at this time (5 days) tends to be most varied. For other times, the convergence can be achieved with a lower number of realizations than 60.

Various techniques including visual observation of histograms, probability plots and chi-squared goodness of fit tests are used to find the most suitable pdf to represent the distribution of the *FoS*

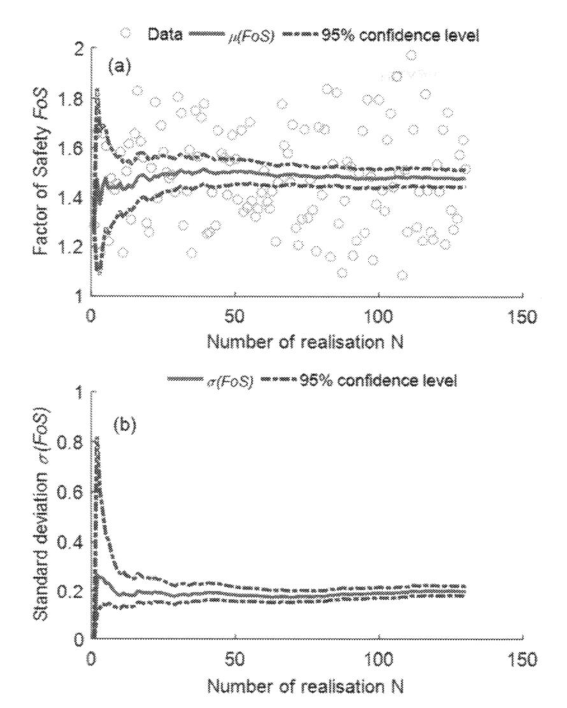

Figure 3. Convergence of *FoS* and its standard deviation with the number of realizations (5 days, $\mu(e) = 0.5$, $COV_e = 0.8$, $\theta(e) = 8$ m).

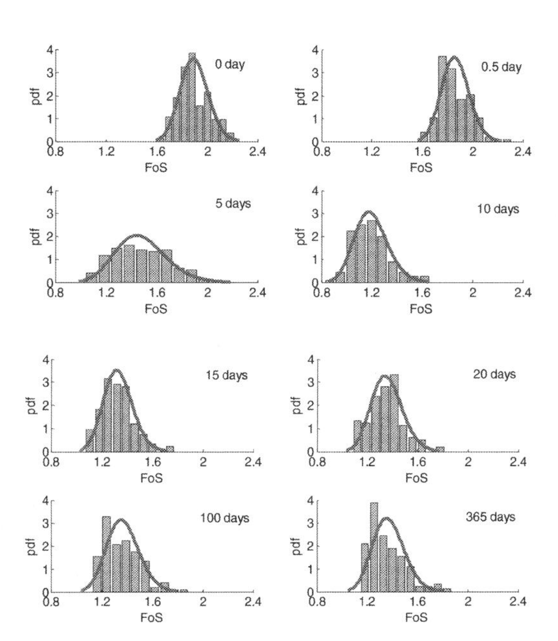

Figure 4. Histograms (bars) with the fitted log-normal distribution function (continuous line) of *FoS* at different times during the rainfall ($\mu(e) = 0.5$, $COV_e = 0.8$, $\theta(e) = 8$ m).

data. It is found that the histograms of the *FoS* can be described reasonably well by the log-normal distribution function. As an example, the *FoS* histograms with their fitted log-normal pdfs at different times are shown in Figure 4 for different times.

3.2 *Failure mechanism during and after rainfall*

A sample realization of random porosity is illustrated in Figure 5. The porosity values were calculated from the associated random field of e ($COV_e = 0.8$ and $\underline{\theta(e)} = 8$ m) and mapped onto the finite element mesh (Fig. 5). The porosity value of the elements ranges between 0.05 to 0.75 which is a slightly large range compared with the typical range of porosity variation within a single domain of this size. The inclusion of such a large variation

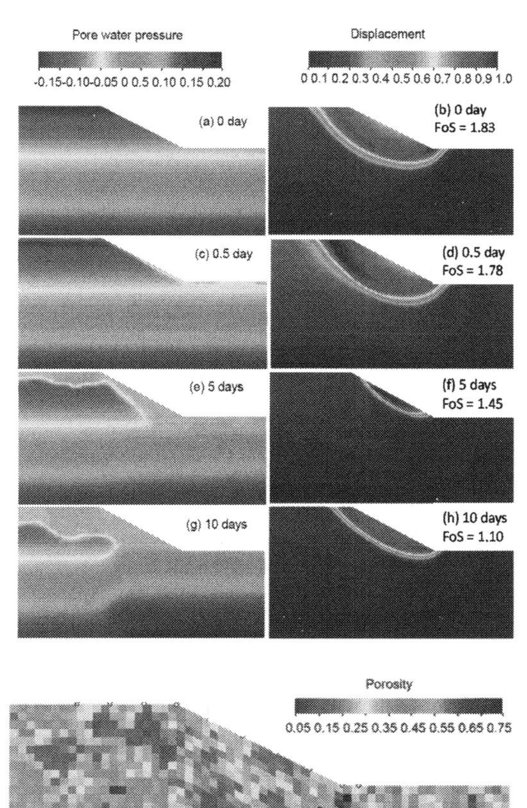

Figure 5. A typical realization of porosity distribution (transformed from the random void ratio field with μ(e) = 0.5, COV_e = 0.8, θ(e) = 8 m).

ranges is however considered reasonable in this study to highlight the effects of heterogeneity.

The changes in failure mechanism during and after a rainfall event due to the changes in suction and pore water pressure distribution are shown in Figure 6 (corresponding to porosity distribution shown in Fig. 5). The pore water pressure distributions are highly irregular at intermediate times from 5 to 20 days (Fig. 6). The *FoS* are lower while the sliding masses are also smaller (Fig. 6) at these times compared with other times. Conversely, the beginning (0–0.5 day) and the end (100–365 days) of the simulation period display less variability in the pore water pressure distribution and higher *FoS*. The area and depth of the failure masses at these times (0–0.5 or 100–365 days) are however significantly larger than at intermediate times (5–20 days). The rise of the water table at the end of the simulation compared to the initial condition causes not only a drop in the *FoS* but also a reduction in the sliding mass size.

3.3 *Influence of porosity variability on Factor of Safety*

The variation of the $\mu(FoS)$ and COV_{FoS} at various COV_e show very similar patterns over time (Fig. 7). The $\mu(FoS)$ decreases over the rainfall (day 0 to 10). The lowest $\mu(FoS)$ occurs unanimously at 10 days just before then rainfall stops. From day 10 to day 365, infiltration is no longer occurring leading to the recovering of the $\mu(FoS)$ (Fig. 7a). The assumption of impermeable vertical and bottom boundary after the rainfall leading to accumulation of water in the soil domain. Consequently, the *FoS* long after the rainfall has stopped (i.e. 365 days) is lower than the initial *FoS* (i.e. 0 day). The COV_{FoS} is almost the same after half a day of rainfall, then increases from day 0.5 to day 5, peak at day 5 for most cases (except $COV_e = 0.1$). The COV_{FoS} then decrease from day 5 to day 20 and fluctuate within a small range after 20 days (Fig. 7b).

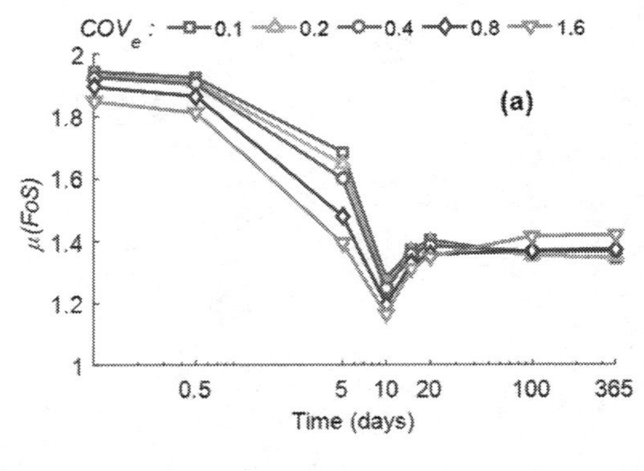

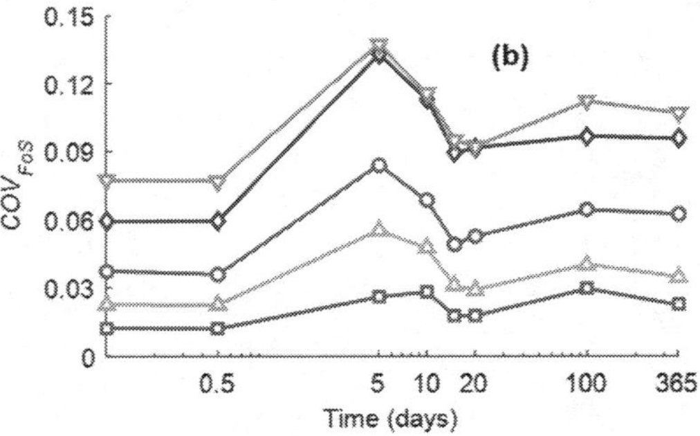

Figure 7. Variation of (a) $\mu(FoS)$ and (b) COV_{FoS} of the *FoS* over time at various COV_e ($\mu(e) = 0.5$, $\theta(e) = 8$ m).

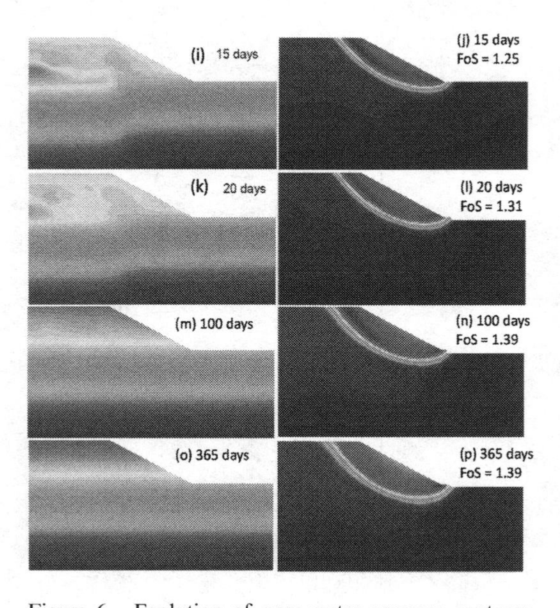

Figure 6. Evolution of pore water pressure contours and slip surfaces at different times during the rainfall (corresponding to the porosity distribution in Figure 5, $\mu(e) = 0.5$, $COV_e = 0.8$, $\theta(e) = 8$ m).

The porosity heterogeneity within the soil domain increase with increasing COV_e leading to the increasing variability in both stresses distribution (induced by soil overburden weight and degree of saturation heterogeneity) and pore water pressure distribution (induced by permeability heterogeneity). The former leads to the differences in $\mu(FoS)$ and COV_{FoS} between different input COV_e at the initial time step because there is no infiltration so far at this point. The latter exacerbates these differences during the rainfall as can be seen, for example, by the wider distances between data points at 5 days (Fig. 7).

As the degree of variability becomes higher (i.e. COV_e increases), the lowest values of the *FoS*

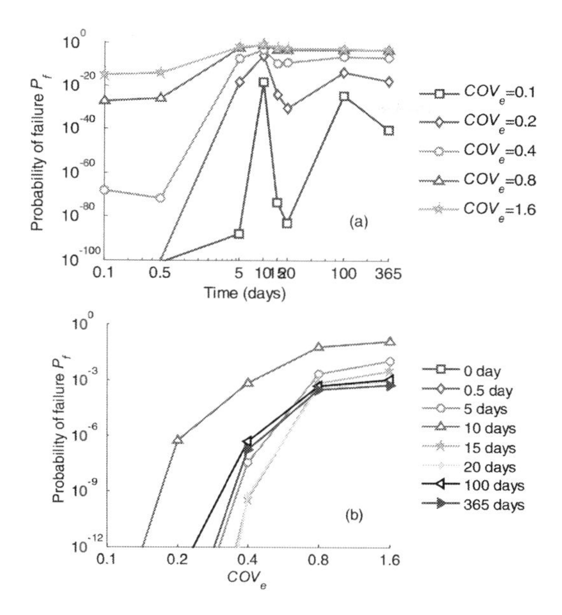

Figure 8. Probability of failure (P_f) against (a) Time (b) Coefficient of variation of void ratio (COV_e).

tend to become smaller because the "weakest" slip surface can have lower strength. This leads to the slight decreases in $\mu(FoS)$ (except at 100 and 365 days) and the significantly increases in COV_{FoS} over the whole simulation period.

The value of probability of failure (P_f) calculated from the above $\mu(FoS)$ and COV_{FoS} are presented against time and COV_e, assuming a log-normal distribution function governing the FoS data (Fig. 8). The results consistently show the increasing risk of failure with increasing COV_e (Fig. 8). For all the values of COV_e investigated, the P_f generally increases over the rainfall and peaks at 10 days then decreases from 10 days onward (Fig.8a). Some fluctuations at extremely small P_f values causing deviations from this general trend occur early in the rainfall (0.5 days) or close to the end of the simulation period (100 and 365 days). These fluctuations are likely due to the sensitivity of the very small values of P_f to the $\mu(FoS)$ and COV_{FoS}. Note that the vertical axis in Figure 8b has been scaled to limit between 1 and 10^{-12} hence some data points representing $P_f < 10^{-12}$ are not presented in this figure.

4 CONCLUSIONS

This study has demonstrated the use of numerical methods and probabilistic analysis to evaluate the factor of safety and probability of failure of a heterogeneous unsaturated slope subjected to rainfall infiltration. The followings can be concluded:

- The combination of the finite element method and the random field theory can be powerful and is capable of dealing complex problem which is highly variable in time and space.
- By using Monte-Carlo analysis, variability of porosity has been shown to significantly influence the mean and coefficient of variation of the factor of safety through its influence on the unsaturated flow.
- Increasing variability of porosity leads to lower mean value but higher coefficient of variation of the factor of safety. Both of these effects lead to higher probability of failure. Over time, the risk of slope keeps dropping and reaches the lowest at the end of the rainfall, then recovers after the rainfall stops.
- The study highlights that both time and space are important factors to be taken into account in dealing with wetting of unsaturated heterogeneous slopes.

ACKNOWLEDGEMENT

The authors gratefully acknowledge the financial support by the Research Council of Norway and several partners through the research Centre SFI Klima 2050 (www.klima2050.no).

REFERENCES

Cameron, D. A., Jaksa, M. B., Wayne, P. & O'Malley, A. (2006) Influence of trees on expansive soils in southern Australia. IN Al-Rawas, A. A. & Goosen, M. F. A. (Eds.) *Expansive soils: recent advances in characterization and treatment.* London, UK, Taylor & Francis.

Fenton, G. A. & Griffiths, D. V. (2005) A slope stability reliability model. *Proceedings of the K.Y. Lo Symposium.* London, Ontario.

Fredlund, D. & Rahardjo, H. (1993) *Soil mechanics for unsaturated soils,* New York, John Wiley & Sons.

Fredlund, D. G., Morgenstern, N. R. & Widger, R. A. (1978) The shear strength of an unsaturated soil. *Canadian Geotechnical Journal,* 15, 313–321.

Griffiths, D. V. & Fenton, G. A. (2004) Probabilistic Slope Stability Analysis by Finite Elements. *Journal of Geotechnical and Geoenvironmental Engineering,* 130, 507–518.

Griffiths, D. V., Huang, J. S. & Fenton, G. A. (2009) Influence of Spatial Variability on Slope Reliability Using 2-D Random Fields. *Journal of Geotechnical and Geoenvironmental Engineering,* 135, 1367–1378.

Griffiths, D. V. & Lu, N. (2005) Unsaturated slope stability analysis with steady infiltration or evaporation using elasto-plastic finite elements. *International Journal for Numerical and Analytical Methods in Geomechanics,* 29, 249–267.

Griffiths, D. V. & Marquez, R. M. (2007) Three-dimensional slope stability analysis by elasto-plastic finite elements. *Géotechnique,* 57, 537–546.

Hicks, M. A. & Onisiphorou, C. (2005) Stochastic evaluation of static liquefaction in a predominantly dilative sand fill. *Géotechnique,* 55, 123–133.

Hicks, M. A. & Samy, K. (2002a) Influence of anisotropic spatial variability on slope reliability. *Proc. 8th Int. Symp. Num. Models Geomech.* Rome, Italy.

Hicks, M. A. & Samy, K. (2002b) Influence of heterogeneity on undrained clay slope stability. *Quarterly Journal of Engineering Geology and Hydrogeology,* 35, 41–49.

Hicks, M. A. & Spencer, W. A. (2010) Influence of heterogeneity on the reliability and failure of a long 3D slope. *Computer and Geotechnics,* 37, 948–955.

Kozeny, J. (1927) Über kapillare Leitung des Wassers im Boden. *Akad. Wiss. Wien,* 136, 271–306.

Le, T. M. H. (2011) Stochastic Modelling of Slopes and Foundations on Heterogeneous Unsaturated Soils. *School of Engineering.* Glasgow, UK, The University of Glasgow.

Le, T. M. H., Eiksund, G. & Strøm, P. J. (2013a) Statistical characterisation of soil porosity. IN Deodatis, G., Ellingwood, B. & Frangopol, D. (Eds.) *Proceeding of the 11th In International Conference on Structural Safety & Reliability.* Columbia University, New York, USA, CRC Press/Balkema.

Le, T. M. H., Gallipoli, D., Sánchez, M. & Wheeler, S. (2015) Stability and failure mass of unsaturated heterogeneous slopes. *Canadian Geotechnical Journal,* 52, 1747–1761.

Le, T. M. H., Gallipoli, D., Sanchez, M. & Wheeler, S. J. (2012) Stochastic analysis of unsaturated seepage through randomly heterogeneous earth embankments. *International Journal for Numerical and Analytical Methods in Geomechanics,* 36, 1056–1076.

Le, T. M. H., Gallipoli, D., Sanchez, M. & Wheeler, S. J. (2013b) Rainfall-induced differential settlements of foundations on heterogeneous unsaturated soils. *Geotechnique.*

Lu, N. & Godt, J. (2008) Infinite slope stability under steady unsaturated seepage conditions. *Water Resources Research,* 44.

Olivella, S., Gens, A., Carrera, J. & Alonso, E. (1996) Numerical formulation for a simulator (CODE-BRIGHT) for the coupled analysis of saline media. *Engineering Computations,* 13, 87–112.

Phoon, K. & Kulhawy, F. (1999) Characterization of geotechnical variability. *Canadian Geotechnical Journal,* 612–624.

Rodríguez, R., Sánchez, M., Lloret, A. & Ledesma, A. (2007) Experimental and numerical analysis of a mining waste desiccation. *Canadian Geotechnical Journal,* 44, 644–658.

van Genuchten, M. T. (1980) A closed form equation for predicting the hydraulic conductivity of unsaturated soils. *Soil Science Society of America Journal,* 44, 892–898.

van Genuchten, M. T. & Nielsen, D. R. (1985) On describing and predicting the hydraulic properties of unsaturated soils. *Annales Geophysicae,* 3, 615–627.

Zandarín, M. T., Oldecop, L. A., Rodríguez, R. & Zabala, F. (2009) The role of capillary water in the stability of tailing dams. *Engineering Geology,* 105, 108–118.

Large deformation – large strain analysis

Numerical Methods in Geotechnical Engineering IX – Cardoso et al. (Eds)
© 2018 Taylor & Francis Group, London, ISBN 978-1-138-33198-3

Optimizing the MPM model of a reduced scale granular flow by inverse analysis

M. Calvello, P. Ghasemi & S. Cuomo
Department of Civil Engineering, University of Salerno, Italy

M. Martinelli
Deltares, Delft, The Netherlands

ABSTRACT: MPM is a mesh-free method designed for large deformation problems. This paper presents the calibration by inverse analysis of a MPM model for a dry sand flow experiment conducted in a miniature flume. The simulations were carried out using the Anura3D code and adopting a Mohr-Coulomb frictional constitutive law for the propagating soil. An inverse analysis algorithm was used to evaluate the sensitivity and optimize some of the input parameters of the model. Four sets of observations have been considered to define the error functions, representing the soil thickness profiles along the slope at four distinct times. The results are presented in relation to: the ability of the model to simulate the behaviour of the sand flow; the sensitivity of the results to changes in the parameter values; the effectiveness of the calibration procedure; the type and number of observations used.

1 INTRODUCTION

Modelling the propagation stage of landslides is a relevant issue for urban planning and landslide risk mitigation (Fell et al., 2008) and the knowledge of the rheological properties of the propagating soil, i.e. the relationships between shear stress and shear strain rate, is a key ingredient of any physically-based model.

Rheology is difficult to assess experimentally because in-situ investigations provide information on soil characteristics before failure and laboratory equipment can hardly reproduce in-situ conditions for the propagation stage. Therefore, rheological properties are usually estimated through the back-analysis of case histories and best-fitting values are computed through trial-and-error procedures based on expert judgment and subjective evaluations.

In this paper an inverse analysis procedure is tested for a flume test of a dry granular flow, which can be considered a benchmark case for many phenomena occurring in the Alps. Previous contributions about similar flume tests and real landslides were proposed by Cascini et al. (2014), Cuomo et al. (2013, 2015) and Calvello et al. (2017) using Smooth Particle Hydrodynamics (SPH) models. Herein, numerical modelling is performed using the Material Point Method (MPM).

2 MATERIALS AND METHODS

2.1 *Experimental test*

Small scale experiments with flows of dry sand were carried out by Denlinger and Iverson (2001). The material propagated through a rectangular flume with a bed surface, inclined 31.4°, joined to a horizontal runout surface by a curved section with a 10 cm radius of curvature (Figure 1).

A vertical gate spanning the entire flume width (20 cm) was positioned in the uppermost part of

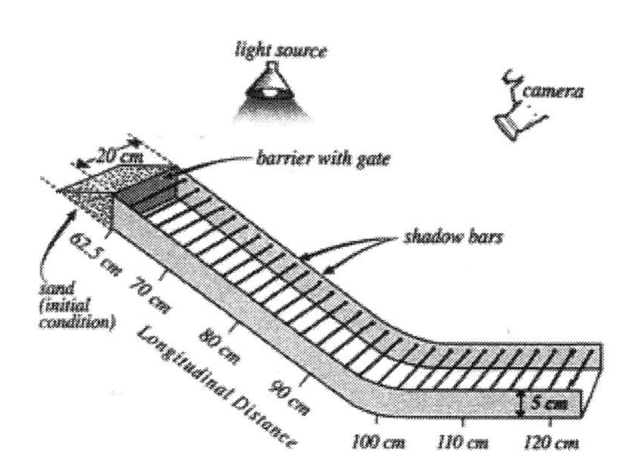

Figure 1. Schematic of the flume used for propagation tests of granular flows (Denlinger and Iverson, 2001).

the slope. About 290 cm^3 of loosely packed, well-sorted and well-rounded dry sand was placed behind the gate ensuring a horizontal soil surface.

The bulk density of the soil was approximated as 1600 kg/m^3. The flume bed was surfaced with Formica. The static bed frictional angles measured using tipping table tests of sand sliding across the Formica was reported as $29° \pm 1.4°$. The internal friction angle of the sand was reported equal to $40° \pm 1°$.

The experiment started by suddenly opening the entire gate. The flow accelerated, elongated, and thinned rapidly after the gate opened. A non-invasive optical shadowing technique was used to measure the soil thickness during the flow and at the end of the experiment.

When the leading edge of the flow reached the break in slope located 37.5 cm downslope from the gate, the sand that first reached the depositional area was only slightly pushed forward by subsequently arriving sand.

In the experiment considered herein, the sand deposition was complete 1.5 s after the flow release. Sand thicknesses normal to the flume bed are reported by Denlinger and Iverson (2001) at 0.32 s, 0.53 s, 0.93 s and 1.5 s after the gate released (Figure 2).

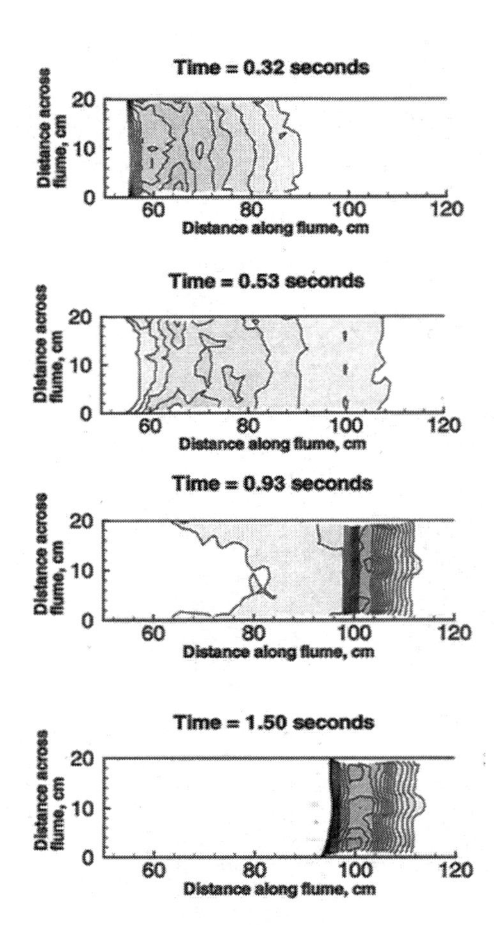

Figure 2. Overview of the experimental results achieved in the flume test considered herein (Denlinger and Iverson, 2001).

2.2 *Material Point Method (MPM)*

MPM is considered to be a variant of FEM, which utilises two discretization structures: the material points and the background mesh. All information (e.g. body forces, stresses, state variables,..) are stored in the material points and move through the background mesh during the simulation. The mesh is used only to solve the balance equations and to transfer information to the material points.

Since its adaptation for solid mechanic problems (Sulsky et al., 1994), a wide range of slope stability applications have used this method, such as: landslide and debris flow (Shin, 2009), wave breaking on a dike (Kafaji, 2013), levee failure propagation (Bandara and Soga, 2015), rainfall induced flow slide (Wang et al., 2016) and earthquake induced slope failure (Abe et al., 2017).

Here in, the MPM software ANURA 3D (http://www.mpm-dredge.eu/) has been used to model the flume test previously described. In order to simulate the contact between the sand and the basal surface, the contact formulation proposed by Bardenhagen et al. (2001) was employed. This algorithm is a predictor-corrector scheme, in which the velocity is predicted from the solution of each body separately and then corrected using the velocity of the two bodies following a Coulomb friction law (Kafaji, 2013). Thus, sliding of two adjacent surfaces depends on a parameter defined as frictional coefficient.

2.3 *Inverse analysis*

Inverse analysis works in the same way as a non-automated calibration approach: parameter values and other aspects of the model are adjusted until the model's computed results match the observations made for the behaviour of the system. Herein, model calibration by inverse analysis is conducted using UCODE (Poeter and Hill 1998), a computer code designed to allow inverse modelling posed as a parameter estimation problem.

In UCODE the parameters are optimized by minimizing, using a modified Gauss-Newton method, the following weighted least-squares objective function $S(\underline{b})$:

$$S(\underline{b}) = \left[\underline{y} - \underline{y}'(\underline{b})^T \right] \underline{\omega} \left[\underline{y} - \underline{y}'(\underline{b}) \right] = \underline{e}^T \underline{\omega}\, \underline{e} \tag{1}$$

where: $\underline{b}$ is the vector of the parameters being estimated; $\underline{y}$ is the vector of the observations being matched by the regression; $\underline{y}'(\underline{b})$ is the vector of the corresponding computed values; ω is the weight matrix, being the weight of every observation taken as the inverse of its error variance; $\underline{e}$ is the vector of residuals.

The objective function should be seen as a measure of the ability of the models to correctly represent the physical process. The following two convergence criteria are used at any given iteration: i) maximum parameter change lower than a user-defined percentage of the value of the parameter at the previous iteration; ii) objective function changes lower than a user-defined amount for three consecutive iterations.

After the model is optimized, the final set of input parameters is used to run the numerical model one last time and produce the final "updated" results. More details on the inverse analysis procedure adopted herein can be found in Calvello (2014, 2017).

2.4 *Observations*

The longitudinal cross-sections of the propagating soil have been drawn, at different experimental times, using the contour lines presented in Figure 2. Figure 3 shows, as an example, the cross section highlighting the base of the apparatus and the position of the soil at the end of the test, corresponding to an experimental time equal to 1.5 s.

The adopted coordinate system employs a vertical axis starting at the level of the rightmost horizontal surface and a horizontal axis with the same longitudinal distances reported in Figure 2. The soil surface is discretized by means of 18 points, almost equally spaced along the horizontal axis.

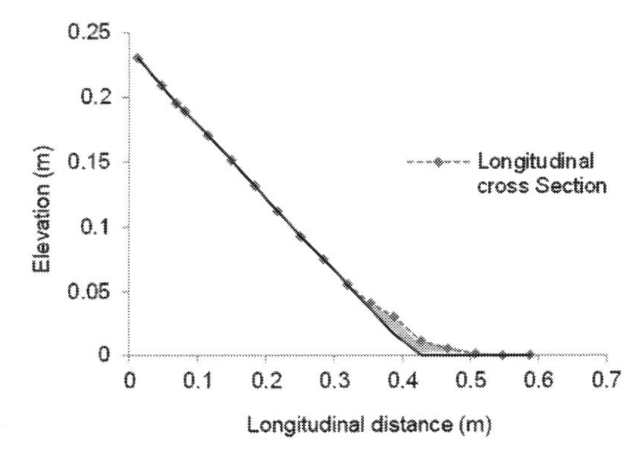

Figure 3. Observations, at time equal to 1.5 s, along the longitudinal cross-section of the flume used to calibrate the MPM model.

The observations used in the inverse analysis of the MPM model are the values of the elevation of these points. When, at any given experimental time, the soil is not present at these locations, the elevation of the base of the flume is used.

3 CASE STUDY

3.1 *MPM model*

The MPM model was created adopting the Anura3D MPM code. The domain was discretized by 12'555 elements (Figure 4). The material points representing the soil are initially positioned in a relative small area located in the uppermost portion of the mesh comprising 440 elements. Each one of these elements initially contains 4 material points. The experimental gate was simulated by applying horizontal fixities at the right boundary of the soil domain. To initialize the soil stresses, a quasi-static calculation was carried out at the beginning of the simulation. Subsequently, the horizontal fixities were removed and the soil was allowed to propagate downwards along the slope.

An elastic perfectly plastic constitutive law is used to simulate the behavior of the soil. The constitutive is model based on the Mohr-Coulomb failure criterion and adopts 5 input parameters: stiffness modulus (E); Poisson's ratio (ν); cohesion (c); friction angle (ϕ); and dilatancy angle (ψ). In addition to these, two other parameters are also needed to define the initial conditions of the soil: porosity (n); and specific gravity of the soil grains (G_s). The contact with the base of the experimental apparatus was simulated adopting a frictional law with a single input parameter: the contact coefficient (μ).

The values of the input parameters of the initial numerical simulation were determined considering the values of the sand properties reported by Delinger and Iverson (2001) and the results of a numerical simulation already performed, for the same case study, by Ceccato and Simonini (2016). They are equal to: E = 1000 kPa, $\nu = 0.3$, $c = 0$, $\phi = 40°$, $\psi = 0$, $n = 0.4$, $G_s = 2.65$.

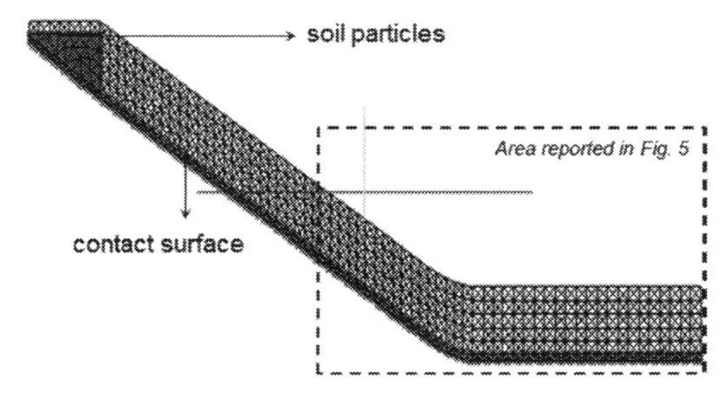

Figure 4. Scheme of computational domain.

Considering the above conditions, the time needed to run one model simulation is approximatively equal to 60'. The comparison between the experimental observations and the results of the initial MPM simulation is reported in Figure 5 considering the position of the soil at the end of the test, i.e. experimental time equal to 1.5 s. The numerical results of the MPM model are "stored", at the end of each time step, at the location of the material points, which are of course moving within the mesh during the simulation. On the contrary, the 18 points used as observations for a given experimental time (Fig. 3) are fixed in space. A purposefully defined numerical algorithm is herein used to extract the values of the elevations of the MPM material points corresponding to the adopted observations. Buffer zones having a width equal to D_b are defined at the location of each observation, i.e. longitudinal distance X_i, as follows:

$$X_{i\_min} = X_i - D_b \tag{2}$$

$$X_{i\_max} = X_i + D_b \tag{3}$$

where: X_{i_min} is initial longitudinal distance of the buffer zone for the i-th observation; X_{i_max} is final longitudinal distance of the buffer zone for the i-th observation.

As depicted in Figure 5, the numerical value to compare to the elevation of the i-th observation, at any given experimental time, is equal to the maximum elevation of all the material points falling within the corresponding buffer zone at that time.

3.2 *Sensitivity analysis*

The relative importance of the input parameters being simultaneously estimated by the adopted inverse analysis algorithm can be defined using: statistics representative of the sensitivity of the predictions to changes in parameters values; and statistics derived from the variance-covariance matrix.

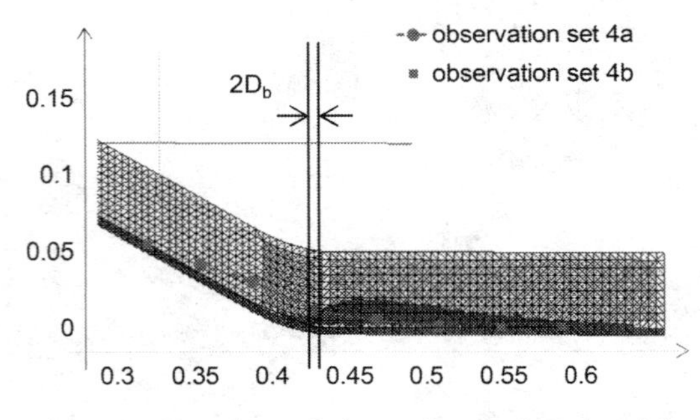

Figure 5. Experimental observations and results of the initial MPM simulation at the end of the test.

Among the statistics able to evaluate the sensitivity of the predictions to parameters changes, the composite scaled sensitivities, css_j, are herein used:

$$css_j = \left[\sum_{j=1}^{ND} \left(\left(\frac{\partial y_i'}{\partial b_j} \right) b_j \omega_{ii}^{1/2} \right)^2 \Bigg|_{\underline{b}} / ND \right]^{1/2} \tag{4}$$

where: y'_i is the ith simulated value; X_{ij} is the sensitivity of the ith simulated value with respect to the jth parameter; b_j is the jth estimated parameter; ω_{ij} is the weight of the ith observation, ND is the number of observations.

Multiple runs of the MPM model are required to compute the sensitivity matrix (X_{ij}). To this aim, a perturbation method is used. Every input parameter b_j is independently perturbed by a fractional amount to compute the results' response to its change. To this aim, all the available elevation data (72 observations) have been considered. Table 1 reports the values of the composite scaled sensitivities for 7 of the 8 input parameters of the MPM model. Parameter c is not considered in the sensitivity analysis as its value is always assumed to be zero. The values of css were computed considering the input parameters values of the initial numerical simulation. The perturbations were always set to 1% of the parameter values. The results of the sensitivity analysis clearly indicate that the input parameter whose changes have the highest impact on the model results is the contact coefficient, μ. Not surprisingly, among the 5 parameters used in the constitutive law adopted to simulate the soil behaviour, the highest composite scaled sensitivity value refers to the friction angle, ϕ.

The results of the sensitivity analysis also indicate that the input parameters are not highly correlated among themselves. To this aim, the parameter statistics to look at are the correlation coefficients, $cor(i,j)$. They are derived from the computed parameter covariances as follows:

Table 1. Composite scaled sensitivities of the model input parameters.

Parameter	Value	Perturbation	css
E	1000 kPa	0.01	5.59
v	0.3	0.01	6.89
c	0	None	–
ϕ	40°	0.01	9.58
ψ	0.1	0.01	4.28
n	0.4	0.01	5.21
G_s	2.65	0.01	5.80
cc	0.40	0.01	29.43

$$cor(i,j) = cov(i,j) / \left(\sqrt{var(i)}\sqrt{var(j)}\right) \quad (5)$$

where: $cor(i,j)$ is the correlation coefficient between parameter i and parameter j; $cov(i,j)$ equal the off-diagonal elements of the variance-covariance matrix; $var(i)$ and $var(j)$ refer to the diagonal elements of the variance-covariance matrix.

The highest correlation coefficient values are reported for the correlation between parameters E and ϕ (–0.60), between parameters n and ψ (-0.55) and between parameters n and ψ (0.47). The absolute values are always significantly lower than 0.8, thus indicating that only mild correlations exist between some of the input parameters.

Based on the results of the sensitivity analysis, it has been chosen to calibrate only the two input parameters to which the model results are most sensitive, i.e. the contact coefficient between the soil and the base of the apparatus and the soil friction angle.

3.3 *Calibrated MPM simulation*

The values of two model parameters, ϕ and μ, are calibrated adopting the regression analysis algorithm described in section 2.3. To this aim, many different sets of observations have been used, considering that elevation data of the propagating soil are available at four experimental stages corresponding to the following times: 0.32 s, 0.53 s, 0.93 s and 1.5 s (Fig. 2). For each experimental time, the 18 elevation points adopted to describe the soil surface profile along a longitudinal cross-section have been further subdivided in two classes, in relation to whether the observations refer to areas with or without soil. This distinction allows to differentiate between experimental data carrying information of the absolute value of the soil depth at a given location (i.e. observation sets a) and data only reporting the "absence" of soil at that location (i.e. observation sets b).

Six inverse analyses have been performed considering the following observation sets: all the available elevation data (72 observations), type-a elevation data (36 observations), and single stage elevation data from each one of the four experimental stages considered (18 observations). Contrarily to what one may have expected, the best results have not been obtained when all the observations are used. Indeed, the largest objective function reduction refers to the regression analysis conducted using only the soil elevation values at the end of the propagation, i.e. observation sets 4a and 4b (Figure 6).

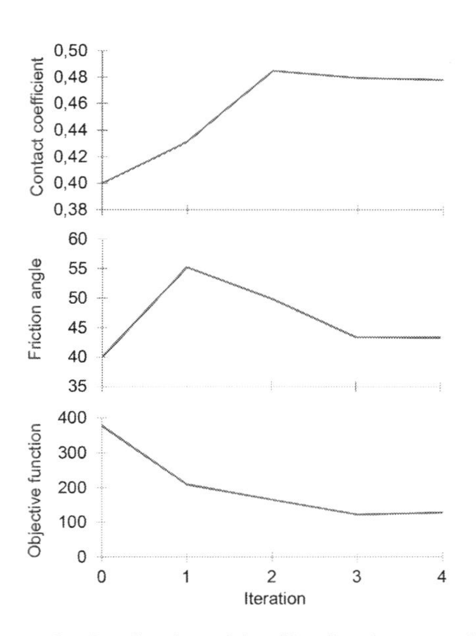

Figure 6. Result of model calibration by regression using observations at time t = 1.5 s.

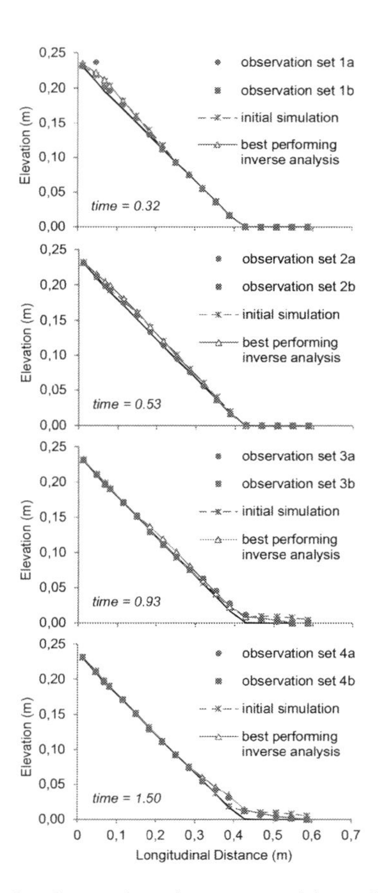

Figure 7. Comparison between model results and experimental observations.

In particular, the best fit between computed and experimental data is obtained, in this case, by increasing the values of both the parameters being calibrated: from 40° to 43.43 for parameter ϕ, and from 0.40 to 0.48 for parameter μ. Figure 7 shows the comparison between the observations and the results of the model for all the four experimental stages. The latter are reported for both the initial simulation and the calibrated one, i.e. best performing inverse analysis. The improvement of the model in reproducing the final cross section of the soil mass (i.e. observation time 1.5 s) is manifest.

The calibrated model is almost perfectly reproducing the position of the deposited soil mass and only slightly overpredicting the soil deposition heights. On the contrary the initial simulation shows a soil mass propagating few centimetres below the tip of the experimental observations as well as an almost total absence of soil along the final part of the slope. Moreover, the calibrated model is also adequately, although not optimally, reproducing the evolution of the soil during propagation (i.e. observation times 0.32 s, 0.53 s, 0.93 s). It means that the information carried by the final position of the soil mass allows the calibration of the important model parameters (ϕ and μ) that ensures a good simulation of the time-dependent behaviour of the propagating mass.

4 CONCLUSIONS

The paper presented the calibration by inverse analysis of the MPM model of a debris flow experiment conducted in a small-scale flume. The results of the optimized model demonstrated the ability of a MPM schematization of the test to simulate, in space and time, the behaviour of a dry sand flow. The case study also highlighted the effectiveness of the adopted calibration procedure to detect the input parameter values producing the best fit between experimental observations and simulated soil heights. The role of the observations adopted in the optimization algorithm was also investigated. It showed that the deposited soil heights at the end of the propagation are sufficient to adequately calibrate the numerical model. This should be considered very good news because, differently to what happens for laboratory experiment, it is very difficult to get in-situ observations of debris flows related to the propagating soil mass.

REFERENCES

Abe, K., Nakamura, S., Nakamura, H., & Shiomi, K. 2017. Numerical study on dynamic behavior of slope models including weak layers from deformation to failure using material point method. *Soils and Foundations*, *57*(2): 155–175.

Bandara, S., & Soga, K. 2015. Coupling of soil deformation and pore fluid flow using material point method. *Computers and geotechnics*, *63*: 199–214.

Bardenhagen, S. G., Guilkey, J. E., Roessig, K. M., Brackbill, J. U., Witzel, W. M., & Foster, J. C. 2001. An improved contact algorithm for the material point method and application to stress propagation in granular material. *CMES: Computer Modeling in Engineering & Sciences*, *2*(4): 509–522.

Calvello, M. 2014. Calibration of soil constitutive laws by inverse analysis. In Hicks M.A., Jommi, C. (eds), *ALERT Doctoral School 2014 Stochastic Analysis and Inverse Modelling*: 239–262.

Calvello, M. (2017). From the observational method to "observational modelling" of geotechnical engineering boundary value problems. Geotechnical Special Publication No. 286, Geotechnical Safety and Reliability: Honoring Wilson H. Tang, ASCE, p. 101–117.

Calvello, M., Cuomo, S., & Ghasemi, P. 2017. The role of observations in the inverse analysis of landslide propagation. *Computers and Geotechnics*, *92*: 11–21.

Cascini L., Cuomo S., Pastor M., Sorbino G., & Piciullo L. 2014. SPH run-out modelling of channelized landslides of the flow type. *Geomorphology*, 214, 502–513.

Ceccato, F., & Simonini, P. 2016. Study of landslide run-out and impact on protection structures with the Material Point Method. In *INTERPRAEVENT 2016-Conference Proceedings*. Lecerne, Switzerland

Cuomo, S., Calvello, M., & Villari, V. 2015. Inverse analysis for rheology calibration in SPH analysis of landslide run-out. In Giorgio, L. et al (eds). *Engineering Geology for Society and Territory-Volume 2* 1635–1639: Springer, Cham.

Cuomo, S., Pastor, M., Vitale, S., & Cascini, L. 2013. Improvement of irregular DTM for SPH modelling of flow-like landslides. In E. Oñate, D.R.J. Owen, D. Peric and B. Suárez (eds). *Proceedings of the XII International Conference on Computational Plasticity. Fundamentals and Applications (COMPLAS XII)*: 3–5. Barcelona: Spain.

Denlinger, R. P., & Iverson, R. M. 2001. Flow of variably fluidized granular masses across three-dimensional terrain: 2. Numerical predictions and experimental tests. *Journal of Geophysical Research: Solid Earth, 106*(B1): 553–566.

Fell, R., Corominas, J., Bonnard, C., Cascini, L., Leroi, E., & Savage, W. Z. 2008. Guidelines for landslide susceptibility, hazard and risk zoning for land-use planning. *Engineering Geology*, *102*(3): 99–111.

Kafaji, I. K. A. 2013. Formulation of a dynamic material point method (MPM) for geomechanical problems. PhD Thesis, University of Stuttgart.

Poeter, E. P., & Hill, M. C. 1998. Documentation of UCODE, a computer code for universal inverse modeling. USGS Report 98–4080, 116 p.

Shin, W. K. 2009. Numerical simulation of landslides and debris flows using an enhanced material point method Ph.D. thesis, University of Washington

Sulsky, D., Chen, Z., & Schreyer, H. L. 1994. A particle method for history-dependent materials. *Computer methods in applied mechanics and engineering*, *118*(1–2): 179–196.

Numerical Methods in Geotechnical Engineering IX – Cardoso et al. (Eds)
 ISBN 978-1-138-33198-3

Modelling rockfall dynamics using (convex) non-smooth mechanics

G. Lu, A. Caviezel, M. Christen, Y. Bühler & P. Bartelt
WSL Institute for Snow and Avalanche Research SLF, Davos Dorf, Switzerland

ABSTRACT: An accurate and efficient modelling of rockfall dynamics has great significance in natural hazards prevention e.g. predicting rock trajectories and run-out zones on a complex terrain. To this end, a novel rockfall simulation module, employing non-smooth mechanics coupled with hard contact laws, has been developed for the software package RAMMS (**RA**pid **M**ass **M**ovement Simulation) at the WSL Institute for Snow and Avalanche Research SLF. In this paper, the major features of the RAMMS::ROCKFALL module will be first summarized. Subsequently, the principle of the modelling strategy (non-smooth mechanics coupled with hard contact laws) will be dissected and compared with the so-called discrete element model (DEM) method. Finally, a RAMMS::ROCKFALL simulation will be performed demonstrating the effectiveness of the software in capturing rockfall dynamics.

1 INTRODUCTION

1.1 *Characteristics of rockfall*

Rockfall, which can be simply defined as downward movement of detached rock fragments, occurs nowadays frequently in mountainous regions (Dorren 2003). Typically, the mass and the volume of the falling rocks fall into the following ranges: 1–300,000 kg and 10^{-3}–10^{2} m^3. If the volume of the falling rocks increases further, e.g. to the range of 10^{2}–10^{5} m^3, rockfall can be also termed as mass fall (Crosta et al. 2015).

Compared to avalanche, i.e. downward movement of massive amount of small particles (e.g. the total volume can be as high as 10^{7} m^3, while the particle size is 0.1–0.3 m), one very significant characteristic of a rockfall process is that the interactions among the falling rocks can be neglected. Moreover, rockfall is typically observed as a single-phase (solid) flow, whereas an avalanche of granular materials can be either single-phase (solid) or two-phase (solid-liquid). A falling rock, normally accompanied by discrete contacts and impacts between the body of rock and the surface of terrain, exhibits rich physical, dynamical phenomena such as free falling, rolling, bouncing and sliding (Fig. 1). Hence, rockfall is usually regarded as a mechanically non-smooth system. In contrast, an avalanche, a system of particles characterized by granular shearing and frictional sliding, is governed by smooth, continual particle interactions.

Modelling rockfall dynamics, despite its challenging nature, has attracted great interests among the geotechnical engineering community. Because it has significance in such as: (1) predicting rock trajectories and run-out zones on three-dimensional terrain for creating hazard maps, (2) designing of protective measures (e.g. rockfall dams, sheds, nets) at optimal location, (3) studying the role of mountain forests in rockfall hazards

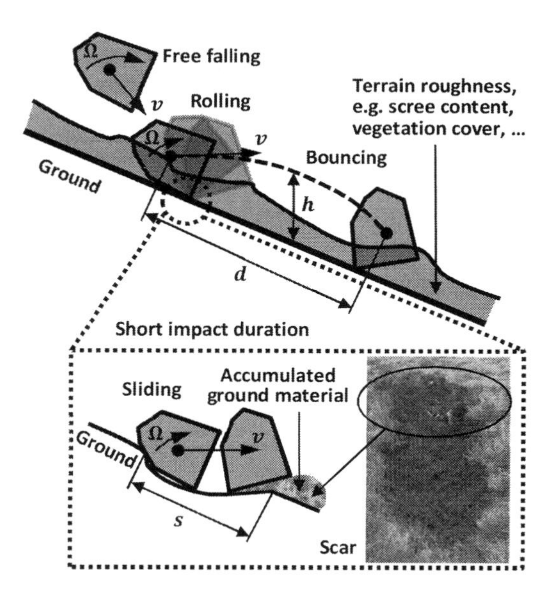

Figure 1. Scheme of different modes of rock motion: free falling, rolling, bouncing and sliding. Note that the sliding motion of rock might accumulate ground material and generate scars on the surface of terrain. The symbols represent the following parameters: Ω - Angular velocity; v - Translational velocity; h - Jump height; d - Jump distance; s - Slippage distance.

mitigation, and (4) assessing the safety of public utilities such as roads, buildings, and mining sites.

In principle, rockfall dynamics obeys the classical Newton's law. Nonetheless, it is demanding in reality to precisely model a rockfall process due to its inherent uncertainties (or stochasticity) (Li & Lan 2015): (1) it is very challenging to identify the rockfall starting regions, the properties of rock (e.g. material, volume, mass, and shape), and the initial release conditions (e.g. rock velocity and orientation), (2) it is extremely difficult to accurately determine the terrain parameters such as terrain topography, mechanical parameters of ground, vegetation and forest cover, and scree distribution, (3) energy dissipation of rock occurs only in short-duration impacts and depends strongly on the rock's shape and orientation at impact, and the properties of slope (which may even change during the rock collision), and (4) rock fragmentation might occur due to intensive impacts, adding additional difficulty to the prediction of rock trajectories and run-out zones.

1.2 *Rockfall modelling methods*

Two fundamental issues are significant in a rockfall process modelling (Leine et al. 2014): How to represent the shape of a rock and how to model its interaction with terrain?

Geometrically, the shape of a rock can be represented by using point-mass, sphere, polyhedron, or discrete element complex shape (Fig. 2a). The point-mass model simplifies rock's geometry into a point; thus no rock shape (moment of inertia) is captured. Therefore, the energy partition between the linear and rotational velocities of a rock is not considered. The sphere model is a natural improvement to the geometric description of a rock shape without introducing much more difficulties to rock-terrain contact detection. However, the centrosymmetric shape of a sphere will lead to an excessive run-out length, and a lack of gyroscopic forces. The polyhedron model is based on the fact that in nature rocks typically have sharp vertices and edges. Hence, gyroscopic effects due to the rotation of non-spherical rocks can be fully considered. Although contact detection between a polyhedron and terrain is in general computationally more expensive, it is nowadays acceptable thanks to the leap of computer powers. Finally, the discrete element complex shape model 'assembles' small spheres to construct a non-spherical rock geometry. This method has an evident advantage since the detection of rock contacts is intrinsically the same as that in the sphere model. Nonetheless, a lot of small spheres are required in order to reproduce the sharp boundaries of a rock, resulting in a tremendous increase of computational effort in the contact detection. Moreover, engineers will need to build the rock shape, which can be inconvenient from a practical point of view. Due to all the above considerations, RAMMS::ROCKFALL has implemented the polyhedron model to represent non-spherical rocks.

Rock-terrain interaction can be modelled on different kinematic levels, namely displacement, velocity, and acceleration (Fig. 2b). The traditional impact rebound model, usually applied to a point-mass, treats rock-terrain interaction on velocity and displacement levels. The pre—and post-impact velocities are linked via the so-called apparent restitution coefficient; thus rock-terrain contact forces are not available. The well-established soft sphere model is frequently applied in the discrete element model (DEM) community. Here, a local (i.e. at the contact point) mechanics model, such as a spring and dash-pot element, is introduced to account for the contact forces and the energy loss of rock during an impact. In order to avoid using a very stiff differential equation to describe the dynamics of a rock at contact, researchers normally use a rather small value for the spring stiffness to ensure that the time integration of motion equation is stable (smooth). This strategy speeds up the simulation but causes the spring stiffness to be a numerical rather than a physical parameter. Alternatively, the hard contact model does not require local, mechanical elements to model contacts and is thus

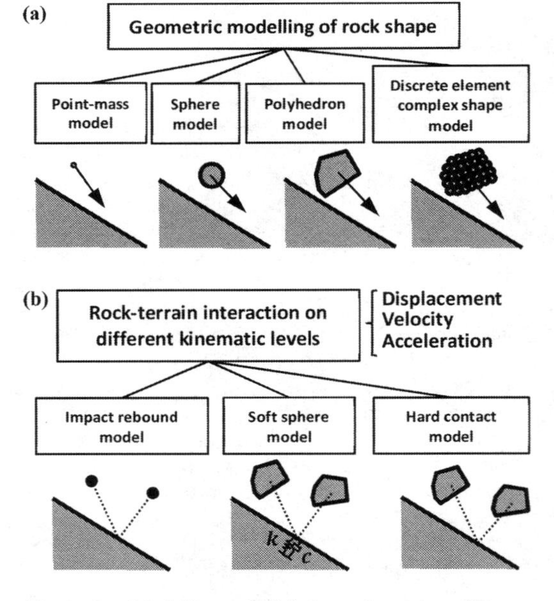

Figure 2. Modelling rockfall dynamics using different methods: (a) geometric representation of rock shape, and (b) rock-terrain interaction on different kinematic levels.

less harsh than the soft sphere model in parameter configuration. This model resolves the state of rocks on velocity and displacement levels (not on acceleration level) through calculating time integration of the external forces which are acting on the rocks. At the impact moment, (non-smooth) velocity jump occurs at the rock's boundary points, and the pre—and post-impact velocities at each contact point are linked via the restitution coefficient. In RAMMS::ROCKFALL we have employed the hard contact model to simulate rock-terrain interactions. This method will be compared with DEM simulations in detail in the following section.

2 MODELLING ROCKFALL DYNAMICS

2.1 *Key features of RAMMS::ROCKFALL*

The implementation of the RAMMS::ROCKFALL module was addressed elsewhere explicitly (Leine et al. 2014, Christen et al. 2012) and will not be repeated here except the core features of the software.

There are three main units for a three-dimensional rockfall simulation: rock, terrain, and contact. The rock of polyhedron shape is obtained through the input point cloud, and the terrain is constructed using the digital elevation model (Bühler et al. 2012). Two coordinate frames are necessary to describe the rock motion, i.e. the inertial frame I (global, fixed at the origin O of terrain) and the rock eigenframe K (local, fixed at the mass center S of rock and extended along its principal axes of moment of inertia). It is more convenient to describe the rotational motion of rock in K since the rock's inertia tensor is timely invariant in this frame. Figure 3 illustrates a numerical simulation environment in RAMMS::ROCKFALL.

To establish the equation of rock motion, the so-called generalized coordinates are introduced:

$$q := \begin{pmatrix} r_{OS}^{I} \\ p_{IK} \end{pmatrix} \in \mathbb{R}^7 \tag{1}$$

where p_{IK} is the quaternion satisfying $|p_{IK}| = 1$ and captures the instant, spatial orientation of a rock. The time derivative of q is the so-called generalized velocities:

$$u := \begin{pmatrix} v_{S}^{I} \\ \Omega^{K} \end{pmatrix} \in \mathbb{R}^6 \tag{2}$$

where $v_S{}^I$ is the translational velocity and Ω^K is the rotational velocity. The time evolution of $v_S{}^I$ is:

$$m\dot{v}_S^I = F_g^I + F_d^I \tag{3}$$

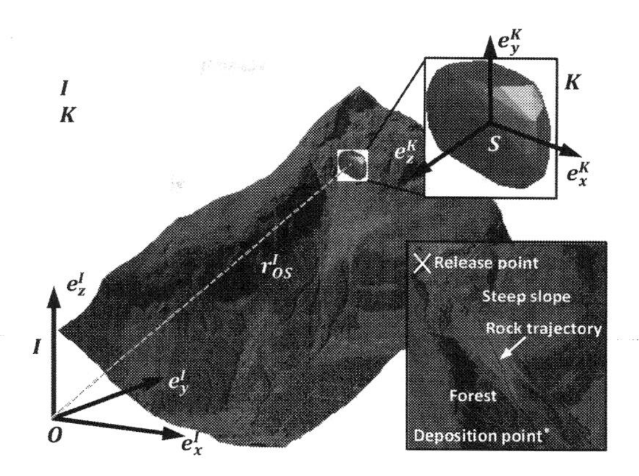

Figure 3. RAMM::ROCKFALL simulation: three-dimensional (3D) and two-dimensional (2D) numerical environment obtained at Evolène, Switzerland. The translational motion of rock is recorded in the inertial frame I, and the rotational motion of rock is described in the rock eigenframe K.

where m is the rock mass, $F_g{}^I$ is the gravitational force, and $F_d{}^I$ is the external damping force which acts at the rock's mass center. Correspondingly, the time evolution of Ω^K is given by Euler's equation:

$$\theta_S^K \dot{\Omega}^K + \Omega^K \times \theta_S^K \Omega^K = M^K \tag{4}$$

where $\theta_S{}^K$ is the rock's inertia tensor represented in the eigenframe K, and M^K is the external torque generated by forces acting on the rock's boundary (in case of no rock-terrain contact, $M^K = 0$). Hence, the equation of rock motion can be written as:

$$\begin{cases} M\dot{u} - h(q,u,t) = W(q)\lambda \\ M = \begin{bmatrix} mI_{3\times3} & 0_{3\times3} \\ 0_{3\times3} & \theta_S^K \end{bmatrix} \\ h(q,u,t) = \left(F_g^I + F_d^I \quad -\Omega^K \times \theta_S^K \Omega^K\right)^{\mathrm{T}} \end{cases} \tag{5}$$

where $W(q)$ is the matrix of generalized force directions, and λ is the vector containing all the contact forces. Note, $W(q)$ transfers the contact forces λ acting on the boundary of a rock to the forcing (force and torque) at the mass center S, producing additionally translational and rotational motion for the rock.

To relate gap G, i.e. the penetration of rock into terrain along the direction $e_z{}^I$ (Fig. 4), with the associated normal contact force λ_N, the Signorini condition is applied, which says:

$$0 \le \lambda_N \perp G \ge 0 \tag{6}$$

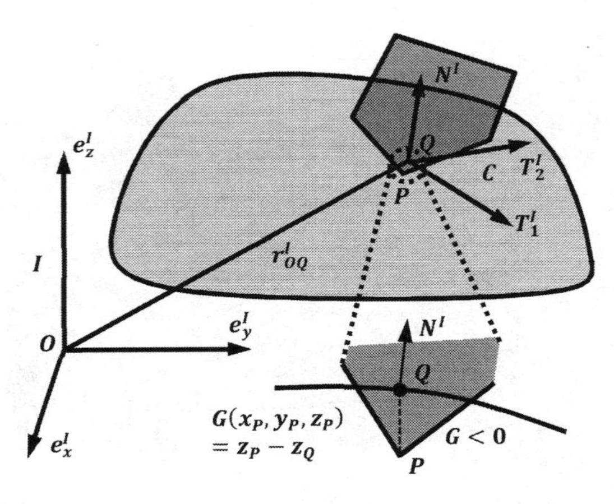

Figure 4. Gap function G and contact frame C for a rock-terrain contact. The contact frame C (i.e. normal direction N^I and tangential directions T_1^I and T_2^I) is represented in the inertia frame I, and is originated at the contact point Q, i.e. the projection of the rock's boundary point P along e_z^I on the surface of terrain.

This means for a closed contact $G = 0$, we have $\lambda_N \geq 0$; for a case of no contact $G > 0$, we have $\lambda_N = 0$. In principle, the hard contact model allows no penetration of rock into terrain; whereas in numerical simulations we use $G < 0$ for the contact detection.

In contact tangential directions, the spatial Coulomb's friction law is applied, which says:

$$-v_T = \begin{cases} \{0\} & \text{if } \lambda_T < \mu\lambda_N \text{ sticking mode} \\ \mathbb{R}_0^+ \lambda_T & \text{if } \lambda_T = \mu\lambda_N \text{ slipping mode} \end{cases} \tag{7}$$

where μ is the friction coefficient. The system switches from sticking to slipping mode if the frictional force λ_T increases and reaches the sliding frictional force $\mu\lambda_N$. Note that λ_T is acting in the counter-direction of the tangential velocity v_T.

In mathematics, it has been shown by using convex analysis that both the normal and the tangential contact force laws satisfy a form of the so-called normal cone inclusion, which is equivalent to an implicit proximal point function (Leine et al. 2014). In the time-stepping scheme, we need to discretize both the equation of rock motion (explicit, semi-implicit or implicit) and the inclusion problem on velocity level, and utilize iterative methods to solve for the impulse (or Percussion) P added to the rock during impact/contact. P is understood as the impulse due to impact, plus the time integration of non-impulsive contact forces. The solution set which P belongs to is a subset of $\mathbb{R}^n$ and is thus convex. This ensures a unique solution to the proximal point formulation. In physics, care must be taken when restitution coefficient is appointed for each contact point. If the same restitution coefficient is used for simultaneous, multiple, frictionless impacts, no additional energy will be pumped into the system (Möller 2011).

In the RAMMS::ROCKFALL module, a special friction model has been implemented to account for the change of friction coefficient due to scarring. When a rock slides on the surface of terrain, it may accumulate ground material and generates plastic deformation to the terrain (Fig. 1). This will potentially increase the friction coefficient and force the rock to roll and jump again (imagine that a bicycle is running downwards along an inclined plane. If one suddenly breaks, the bicycle tends to rotate around its front wheel and jump from ground). The model of friction coefficient reads:

$$\mu(s) = \mu_{\min} + \frac{2}{\pi}(\mu_{\max} - \mu_{\min})\arctan(\kappa s) \tag{8}$$

where $\mu_{\max}$ and $\mu_{\min}$ provides the upper and the lower boundary for the friction coefficient μ, respectively, and κ controls how fast μ increases with increasing s. This friction model is enabled when at least one contact point is active between rock and terrain; otherwise s decays following:

$$\dot{s} = -\beta s \tag{9}$$

2.2 *Hard contact model vs. discrete element model*

Nowadays DEM has a wide application in simulating dense granular flows since the kinematic information of all levels (displacement, velocity, and acceleration) as well as the transient or static contact forces acting on particles can be readily obtained from simulation (Lu et al. 2016). The sphere model is most frequently used; however various non-spherical particle models are also available (Lu et al. 2015). As aforementioned, one significant distinction between DEM and hard contact model is that, one needs to incorporate locally mechanical elements in the DEM to model the rebound and the energy dissipation of objects for a contact problem. In this study we have tested three simple cases aiming to clarify the pros and cons of these two particle modelling methods.

Case I: In the first test case we simply simulate free falling of a sphere towards a horizontal plane (one dimensional). The sphere contacts with ground with a single contact point. No friction but only the normal contact force in the vertical direction is considered. Table 1 lists the main mechanical parameters used in the simulations.

Table 1. Mechanical parameters used for simulating free falling of a sphere.

Simulation method	Hard contact model	DEM
Sphere radius R	0.1 m	
Density ρ	2700 kg/m^3	
Restitution coefficient e	0.6	
Spring stiffness k_N	NA	2×10^7 N/m
Damping coefficient η_N	NA	0.1605
Gravitational acceleration g	10 m/s^2	
Release height H	0.5 m	
Initial velocity v_N	0 m/s	
Simulation time t	2 s	

In the DEM simulation, a linear spring and dash-pot element is used to model contact mechanics, and the contact force in the normal direction (along the vertical direction) is calculated as:

$$F_N = \max\left(0, k_N\delta_N - 2\eta_N\sqrt{mk_N}v_N\right) \quad (10)$$

where δ_N is the overlap between the sphere and ground in the normal direction. Here, negative values of F_N must be prevented since we do not intend to model attractive forces between the contacting objects. For a single contact point, the damping coefficient η_N is linked to the restitution coefficient e_N by

$$e_N = \exp\left(\eta_N\pi / \sqrt{1-\eta_N^2}\right) \quad (11)$$

In order to observe more pronounced rebound phenomenon after the sphere-ground impact, e_N is set as 0.6. In addition, more effort is required in choosing k_N. It is normally required by DEM that k_N should be sufficiently large such that the maximum δ_N during a contact is less than 1% of the sphere diameter. However, k_N cannot be too large since it directly affects the time step dt through affecting the duration of collision t_{col}, which can be estimated assuming the contact as a damped harmonic motion (nonetheless this tends to overestimate t_{col} since a contact is not a damped harmonic motion):

$$t_{col} = \pi / \sqrt{\frac{k_N}{m}\left(1-\eta_N^2\right)} \quad (12)$$

It can be seen that t_{col} decreases with increasing k_N. We ensure that the whole contact process is resolved by at least 25 time steps (Cleary et al. 2008).

Figure 5 plots the height of sphere center H as a function of simulation time t. In Figure 5a, the results are obtained using the hard contact model with different orders of time steps. It can be seen that a time step of 10^{-3} s is sufficient to reproduce the rebound behavior of the sphere, i.e. the bouncing amplitude gradually decreases until the sphere rests on the surface of ground. As shown by the inset of Figure 5a, the impact/contact is resolved in one single time step by the hard contact model. The change of the sphere's momentum upon impact/contact is accounted for by Percussion P. However, the values of contact forces cannot be retrieved from the hard contact model simulation.

In our DEM framework, k_N is set to be 2×10^7 N/m, resulting in a rather small time step 2.5×10^{-5} s (1/40 of the time step 10^{-3} s used in the hard contact model). Figure 5b shows that the DEM simulation can reproduce very well the height of sphere center. In addition, one advantage of the DEM simulation is that the contact force (under

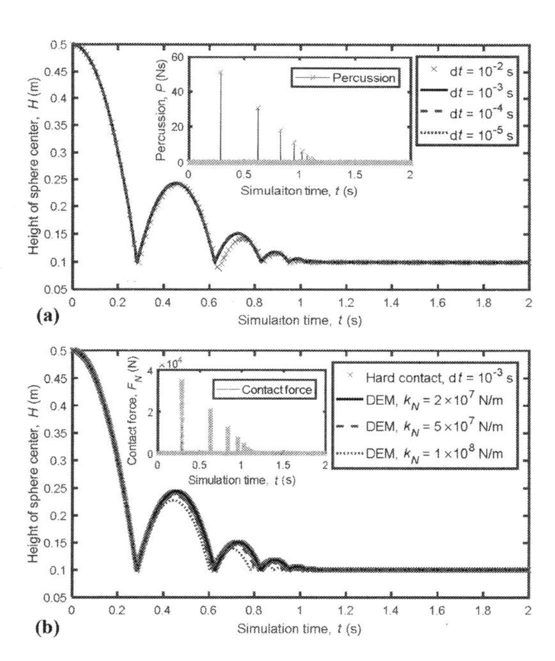

Figure 5. Plots of height of sphere center H as a function of simulation time t: (a) Free falling of a sphere simulated using the hard contact model. The time step dt is varied from 10^{-2} to 10^{-5} s. The inset plots Percussion P exerted to the sphere for the simulation with a time step of 10^{-3} s. (b) Comparison of the trajectories of free falling spheres simulated using the hard contact model (dt = 10^{-3} s) and the DEM with different k_N. The inset shows the normal contact force F_N obtained for the sphere-ground contact (DEM, $k_N = 2 \times 10^7$ N/m).

the chosen k_N) is available for any time step during t_{col}, as shown by the inset of Figure 5b. However, since k_N is a numerical parameter, the DEM simulation result seems to be sensitive to the choice of k_N, e.g. the rebound height of sphere decreases with increasing k_N. It is suggested that one uses an even smaller dt for a larger k_N to more accurately capture the contact dynamics. Unfortunately, a larger k_N will, in turn, increase the total computational time.

Case II: In the second test case the particle shape is changed into cubic, i.e. we simulate free falling of a cube towards a horizontal plane (one dimensional). The cube, which is volume-equivalent to the sphere used in Case I, has a length of edge about 0.161 m. The other mechanical parameters (Tab. 1) are kept unchanged. Before releasing, one face of the cube is parallel to the ground. Thus, the four lower boundary points of the cube will impact with ground at the same time and have the same contact condition. In the hard contact model simulation ($dt = 10^{-3}$ s), we calculate Percussion P generated at all of the four lower boundary points, and update the height of the cube's mass center accordingly. In the DEM simulations, we model the cube-ground contact either at a single contact point which is located at the center of the cube's lower face, or at multiple contact points which are located at the four lower boundary points.

It can be seen from Figure 6 that the DEM simulation with the single contact point model can correctly calculate the rebound behavior of cube, which agrees very well with the result obtained from the hard contact model simulation. However, the DEM simulation with the multiple contact points model predicts a significantly lower rebound height of cube comparing to the other simulations. It seems that an excessive damping force has been added to the cube. Note, in the DEM the mechanics element (spring and dash-pot) is local, functioning separately at each boundary point of the cube. The contact force calculated at one boundary point does not depend on the values of contact forces calculated at other boundary points. Therefore, overdamping can be expected for the multiple contact points model if the same damping coefficient is applied as for the single contact point model. This overdamping phenomenon has also been demonstrated in DEM simulations of particles constructed by 'assembling' small spheres (Kodam et al. 2009).

The multiple contact points model is very suitable for RAMMS::ROCKFALL considering that irregular rock shapes are represented by polyhedron with in general hundreds of facets. Indeed, modelling contacts at multiple points is critical for rockfall simulation since the impulses generated at the rock boundary will influence the jump height and the spin significantly. To this end the hard contact model has an advantage over the DEM.

Case III: In the third test case we incorporate frictional force (using our slippage based friction model) and simulate a three-dimensional cube (the same as in Case II) sliding and rolling downwards along an inclined plane. It is intended to examine whether both simulation methods can produce similar results. Here, instead of tilting the plane we 'tilt' the direction of gravitational force, which simplifies the system configuration. Initially the cube is put on the surface of plane, with one face being parallel to the horizontal. Subsequently, the gravity is tilted by the given angle, generating a driving force which enables the cube to slide along the direction of one of its edge. Though the simulations are performed in three dimensions, one can expect that the motion of cube is mainly two-dimensional, occurring along the inclination of the plane (Fig. 7e).

To calculate the frictional force in DEM, we have introduced another spring and dash-pot element in the contact tangential directions, which are parallel to the plane. The magnitude of the frictional force is governed by Coulomb's law, namely:

$$F_T = \min\left(\mu k_N \delta_N,\ k_T \delta_T - 2\delta_T \sqrt{m k_T} v_T \right) \tag{13}$$

Where k_T is the spring stiffness in the tangential direction, and η_T is the damping coefficient in the tangential direction, which is linked to the restitution coefficient e_T using the same equation as Eqn. (11). δ_T is the tangential elongation applied to the spring, which equals to time integration of the tangential velocity at the contact point $\delta_T = \int v_T\, dt$. The magnitude of the frictional force is bounded by $\mu k_N \delta_N$. In this test case we simulate multiple contact points for both the hard contact model and the DEM. Hence, e_N and e_T are set to zero to prevent the overdamping effect in the DEM, resulting in values of one for the damping coefficients η_N and η_T. Table 2 lists the other modelling parameters used in Case III.

The simulations are performed under MATLAB R2016b environment on a desktop computer running Windows 10. Approximately, it requires 15 s to perform the hard contact model simulation but 634 s to complete the DEM simulation (634 s / 15 s ≈40). This discrepancy in computational efficiency agrees with the relation of dt used by the two numerical methods.

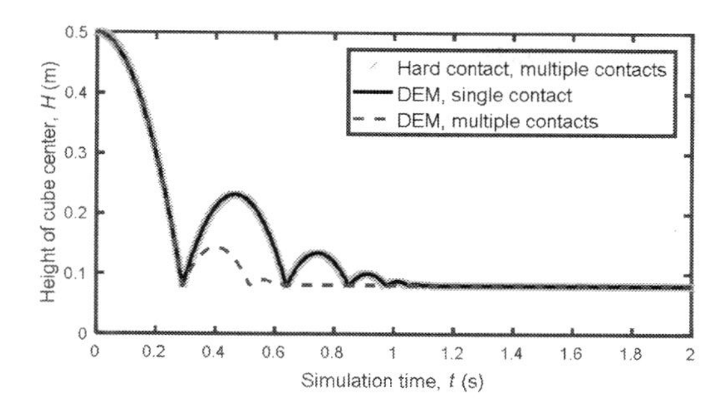

Figure 6. Plot of height of cube center H as a function of simulation time t. Simulations are performed using the hard contact model with multiple contact points, the DEM with single contact point, and the DEM with multiple contact points.

Table 2. Mechanical parameters used for simulating sliding and rolling of a cube downwards along an inclined plane.

Simulation	Hard contact	DEM
method model		
Spring stiffness k_T	NA	2×10^7 N/m
Plane inclined angle α	50°	
Minimum friction coefficient μ_{min}	0.4	
Maximum friction coefficient μ_{max}	1.2	
Friction model parameter β	200.0	
Friction model parameter κ	4.0	
Time step dt	1×10^{-3} s	2.5×10^{-5} s
Simulation time t	3 s	

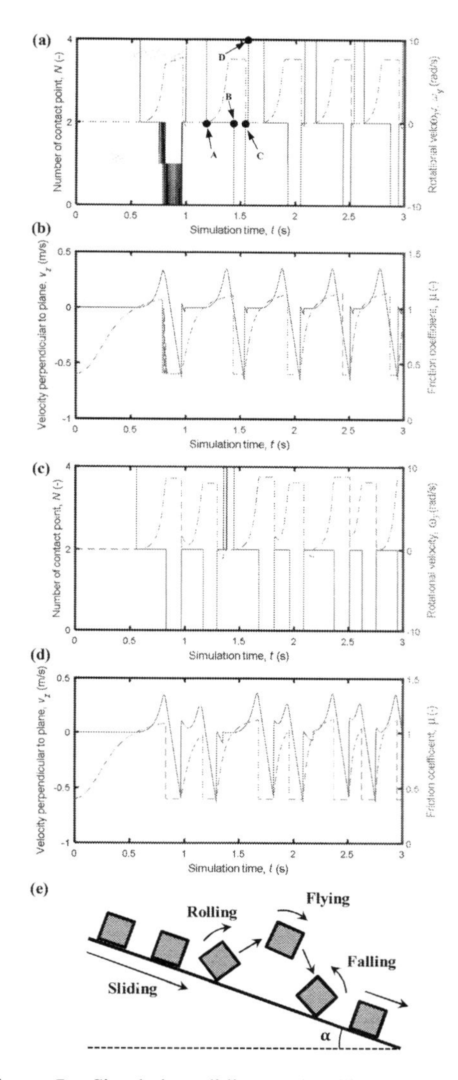

Figure 7. Simulating sliding and rolling of a cube downwards along an inclined plane using (a, b) the hard contact model and (c, d) the DEM: (a, c) plot of number of contact point N (black, solid) and rotational velocity ω_y (gray, dashed) as a function of simulation time t; (b, d) plot of velocity perpendicular to plane v_z (black, solid) and friction coefficient μ (gray, dashed) as a function of simulation time t. The letters **A** to **D** in (a) mark the critical transition points between the different modes of cube motion shown in (e): sliding, rolling, flying, and falling.

Figure 7 shows the results obtained in both simulations, i.e. number of contact point N, rotational velocity ω_y, velocity perpendicular to plane v_z, and friction coefficient μ as a function of simulation time t.

From the hard contact model simulation (Figs 7a, b): one can clearly distinguish the different modes of cube motion after the initial ~1 s transition section. The cube slides with four boundary points contacting with the plane, and μ increases during this period. At a certain moment (e.g. point **A**, $\mu > 1$), the cube starts to roll with two boundary points contacting with the plane. μ keeps on increasing until the cube completely detaches (flies) from the plane (e.g. point **B**), and then μ goes back to μ_{min}. Afterwards the cube falls and touches the plane with two boundary points (e.g. point **C**), causing a counter-directional, rotational velocity to the cube and forcing it go into the sliding mode again (e.g. point **D**). Note, there is no Percussion generated to the cube in the latter part of the rolling motion (section **AB**), though the cube is contacting with the plane. These different modes of motion can be also identified from the cube's velocity perpendicular to plane v_z.

From the DEM simulation (Figs 7c, d): one cannot observe sliding motion of the cube after the transition phase. Here, the cube switches directly from a flying mode into a rolling mode. If the magnitude of ω_y at the end of the flying mode is relatively small, the subsequent contact forces can lead to a negative ω_y for the cube. In the rolling mode there are always contact forces acting on the cube as long as at least one contact is closed. Overall, the 'periodic' motion of the cube looks less regular compared to that obtained from the hard contact model simulation. The same observations can be made using the cube's velocity perpendicular to plane v_z.

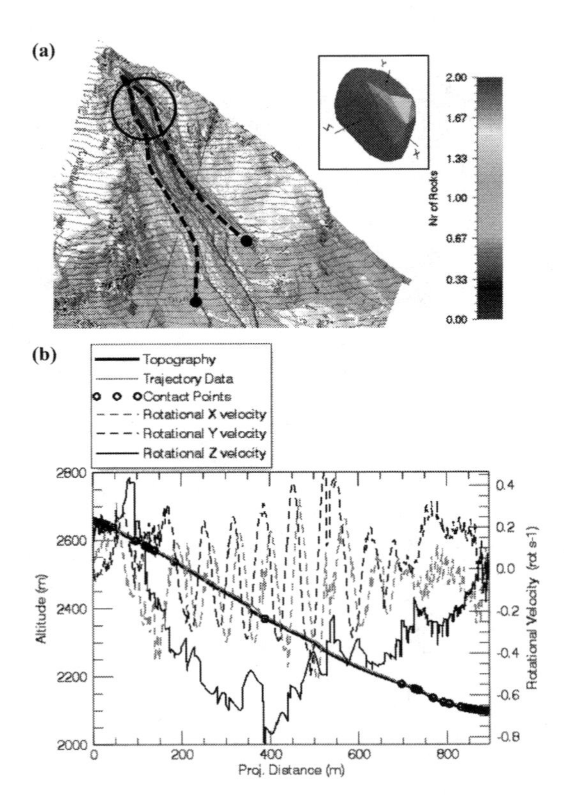

Figure 8. A simulation using RAMMS::ROCKFALL with the terrain data obtained at Evolène, Switzerland: (a) Trajectories of rocks observed in the simulation. The black circle marks the locally higher topography near the release point. (b) Rotational velocity of one selected rock around its three principal axes of moment of inertia. The axis Z is apparently dominating, resulting in the wheeled motion of rock around this axis.

3 A ROCKFALL SIMULATION USING RAMMS

A simulation example will be provided in this section, demonstrating that RAMMS::ROCKFALL can effectively model rockfall dynamics.

The simulation is performed with the terrain data obtained at Evolène, Switzerland (Fig. 3). One 200 m^3 flat-shaped rock (8.95/8.02/5.09 m) is released from an altitude of about 2650 m. For the terrain, μ_{min} and μ_{max} is set as 0.4 and 2.0, respectively. In total 200 simulations are performed with different initial rock orientations. It turns out that the mean jump height of these rocks is 4.78 m, and the mean rock velocity is 11.82 m/s.

Figure 8a shows the rock trajectories. Although the rock is released from the same point, the trajectories can bifurcate from each other, leading to two dominating deposition areas. This is mainly due to the relatively higher topography highlighted by the black circle, which divides the falling rocks into two streams (see the dashed lines in Fig. 8a). The numerical observation agrees well with the historical record. Another important feature of the falling flat-shaped rocks is that they can generate a wheeled motion. Due to the rock's dominating moment of inertia around the axis Z, the gyroscopic effect comes into play and forces the rock to rotate like a wheel. RAMMS::ROCKFALL captures this particular motion very well (Fig. 8b).

4 CONCLUSIONS

The major features of RAMMS::ROCKFALL have been summarized. This is a novel software module incorporating non-smooth mechanics coupled with the hard contact model, which allows us to study the influence of rock shape on rockfall dynamics.

In particular, two widely used particle modelling methods, i.e. the hard contact model and the DEM, have been compared using three test cases simulating a free falling sphere, a free falling cube, and a sliding and rolling cube along an inclined plane. We have highlighted the unique advantages of the hard contact model in rockfall simulation. Firstly, this method does not require additionally local, mechanical elements to calculate the impulses generated by impact/contact; thus the simulation result is not as sensitive as the DEM in choosing numerical parameters. Secondly, the hard contact model does not have the overdamping problem existing in the DEM simulation; thus it is more suitable for modelling rock contacts at multiple points (which is significant in rockfall dynamics modelling). Finally, the hard contact model captures the modes of rock motion more efficiently compared to the DEM simulation.

A simulation example has been provided using RAMMS::ROCKFALL, showing the effectiveness of the software in rockfall dynamics modelling. For the next step, the software will be systematically calibrated using our field test data (Caviezel et al. 2017, Niklaus et al. 2017). It is hoped that the endeavor will lead RAMMS::ROCKFALL to successful simulations of rockfall events (with arbitrary convex rock geometries) on arbitrary slopes.

REFERENCES

Bühler, Y., Marty, M. & Ginzler, C. 2012. High resolution DEM generation in high-Alpine terrain using airborne remote sensing techniques. *Transactions in GIS* 16(5): 635–647.

Caviezel, A., Schaffner, M., Cavigelli, L., Niklaus P., Bühler, Y., Bartelt, P., Magno, M. & Benini, L. 2017. Design and evaluation of a low-power sensor device for induced rockfall experiments. *IEEE Transactions on Instrumentation and Measurement* PP(99): 1–13.

Christen, M., Bühler, Y., Bartelt, P., Leine, R., Glover, J., Schweizer, A., Graf, C., McArdell, B.W., Gerber, W., Deubelbeiss, Y., Feistl, T. & Volkwein, A. 2012. Integral hazard management using a unified software environment: Numerical simulation tool 'RAMMS' for gravitational natural hazards. In G. Koboltschnig et al. (eds), *12th Congress INTERPRAEVENT 2012; Proc., Grenoble, France, 23–26 April 2012*. International Research Society.

Cleary, P.W., Sinnott, M.D. & Morrison, R.D. 2008. DEM prediction of particle flows in grinding processes. *International Journal for Numerical Methods in Fluids* 58(3): 319–353.

Crosta, G.B., Agliardi, F., Frattini, P. & Lari, S. 2015. Key issues in rock fall modelling, hazard and risk assessment for rockfall protection. In G. Lollino et al. (eds), *Engineering Geology for Society and Territory—Volume 2*: 43–58. Cham: Springer.

Dorren, L.K.A. 2003. A review of rockfall mechanics and modelling approaches. *Progress in Physical Geography* 27(1): 69–87.

Kodam, M., Bharadwaj, R., Curtis, J., Hancock, B. & Wassgren, C. 2009. Force model considerations for glued-sphere discrete element method simulations. *Chemical Engineering Science* 64(15): 3466–3475.

Leine, R.I., Schweizer, A., Christen, M., Glover, J., Bartelt, P. & Gerber, W. 2014. Simulation of rockfall trajectories with consideration of rock shape. *Multibody System Dynamics* 32(2): 241–271.

Li, L.P. & Lan, H.X. 2015. Probabilistic modeling of rockfall trajectories: a review. *Bulletin of Engineering Geology and the Environment* 74(4): 1163–1176.

Lu, G., Hidalgo, R.C., Third, J.R. & Müller, C.R. 2016. Ordering and stress transmission in packings of straight and curved spherocylinders. *Granular Matter* 18: 34.

Lu, G., Third, J.R. & Müller, C.R. 2015. Discrete element models for non-spherical particle systems: From theoretical developments to applications. *Chemical Engineering Science* 127: 425–465.

Möller, M. 2011. Consistent integrators for non-smooth dynamical system. *PhD thesis, ETH Zürich*: No. 19715.

Niklaus, P., Birchler, T., Aebi, T., Schaffner, M., Cavigelli, L., Caviezel, A., Magno, M. & Benini, L. 2017. StoneNode: A low-power sensor device for induced rockfall experiments. *IEEE Sensors Applications Symposium (SAS), Glassboro, NJ, USA, 13–15 March 2017*. IEEE.

Numerical Methods in Geotechnical Engineering IX – Cardoso et al. (Eds)
© 2018 Taylor & Francis Group, London, ISBN 978-1-138-33198-3

On the use of the material point method to model problems involving large rotational deformation

L. Wang, W.M. Coombs & C.E. Augarde
Department of Engineering, Durham University, Durham, UK

M. Brown, J. Knappett, A. Brennan & C. Davidson
School of Science & Engineering, University of Dundee, Dundee, UK

D. Richards & A. Blake
Faculty of Engineering and the Environment, University of Southampton, Southampton, UK

ABSTRACT: The Material Point Method (MPM) is a quasi Eulerian-Lagrangian approach to solve solid mechanics problems involving large deformations. In order to improve the stability of the MPM, several extensions have been proposed in the last decade. In these extensions, the sudden change of stiffness when a point crossing an element boundary in the standard MPM is avoided by replacing a material point with a deformable particle domain. The latest extensions are Convected Particle Domain Interpolation approaches, primarily including the CPDI1 and recently published CPDI2. We have unified the standard MPM and CPDI approaches into one implicit computational framework, and here investigate their ability to model problems involving large rotational deformation, which is essential in the installation of screw pile foundations. It was found that the CPDI2 approach can produce erroneous results due to particle domain distortion, while the CPDI1 approach and standard MPM can predict more physically realistic mechanical responses.

1 INTRODUCTION

The Material Point Method (MPM) is a numerical method used to simulate massive deformation of solids combining advantages of both Eulerian and Lagrangian approaches for solving solid mechanics problems. In the MPM a body is described by a number of Lagrangian material points, at which state variables are stored and tracked. Computation for an incremental loading is then carried out on a background computational mesh. As demonstrated in Figure 1 for a simple shear problem, the total deformation and other state variables are stored at the material points, while the background mesh is extended with the incremental displacment, thus avoiding the mesh distortion seen with the standard FEM for large deformation problems. In fact, the background mesh can be any mesh at the beginning of each load step. Because of this attraction the MPM has been applied to several large deformation problems particularly in the area of geotechnical engineering, e.g. (Ceccato et al. 2016, among others).

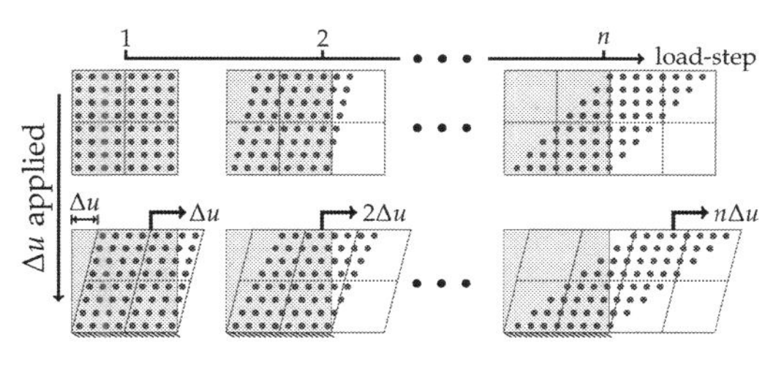

Figure 1. Demonstration of the standard MPM under simple shear: a high quality mesh (red) is used in the computation of equilibrium for each loading increment, while the material points (blue) track the total deformation.

As part of wider study, we are developing a program based on the MPM to model the soil response to the installation of a screw pile, with the engineering application of providing a computer-aided design tool for engineers to optimise pile design for offshore wind turbine foundations (Wang et al. 2017). During installation of a screw pile, the pile is pushed and rotated into the ground. Our program focuses on computation of the vertical force and torque applied on the pile for this installation. Modelling large torsion is essential for this purpose.

The method of mapping the state variables back and forth between the material points and background mesh nodes in the MPM is a crucial step. For example, in a static elasticity analysis, the stresses at material points are used to work out the nodal internal force; and a material point's position is updated by adding incremental displacements interpolated from the nodal increment displacements. In the *standard MPM (sMPM)*, a material point only relates to its *parent element*, (i.e. the backgound element in which it is currently located). However, the sMPM has an inherent instability when a material point crosses an element boundary. This instability is due to the sudden transfer of stiffness between elements, this can result in some elements having very little stiffness, or some internal elements losing all stiffness. Therefore, several extensions to the sMPM have been proposed, each of which replace the material point with a deformable particle domain. The most notable of these extensions are the *Generalized Interpolation Material Point (GIMP)* (Bardenhagen and Kober 2004), *Convected Particle Domain Interpolation (CPDI1)* (Sadeghirad et al. 2011), and *Second-order CPDI (CPDI2)* (Sadeghirad et al. 2013). Assigning a particle domain to each material point enables the influence domain of a material point to cover more than one element. For a 2D problem, the particle domain is a rectangle in GIMP, a parallelogram in CPDI1 and a quadrilateral in CPDI2. Therefore, the CPDI2 approach has particle domains for completely covering a general body, while are gaps or overlaps with the other two approaches. Previously, these methods have not been examined in terms of their ability to model problems involving large rotational deformation, and this is the focus of this paper.

Applying boundary conditions is another challenge when using the MPM, because the mesh is independent of the probem domain (unlike in the FEM) and the domain boundary generally is not included in the computational mesh. This is probably why numerical examples in many published papers consider problems in which body forces only are present. With a regular grid as the computational mesh, Dirichlet boundary conditions can be applied using the implicit boundary method, e.g. Cortis et al. (2017). However, in our project the geometry of a screw pile is more complex than allowed for in that paper. In this framework, the *moving mesh concept* (Beuth 2012), which simply modifies the computational mesh such that the boundary of domain is explicitly included in the mesh, is adapted as the rotating mesh associated with the rotation, as detailed in Wang et al. (2017). In order to generate mesh including body boundary, an unstructured mesh is used. With the mesh including the domain boundary at the beginning of each load-step, boundary conditions are applied on these mesh nodes, straightforwardly as in the FEM.

In this paper, the sMPM and CPDI approaches are unified into one computational framework. This framework solves a quasi-static problem for the deformation of elasto-plastic material. The finite stain theory is used to characterise the deformation. The plastic yield condition is governed by the von-Mises criterion. An implicit stress return algorithm is employed for finding the stress state after material is yielded, such that the yield condition is accurately enforced, with more details included in Coombs (2011). An implicit scheme is also employed for solving the system of equations. This allows large load-steps and increases in stability and accuracy, as shown in Charlton et al. (2017).

This unified computational framework is verified and investigated for modelling large torsion. The performance of these methods is compared in Section 3 and conclusions are drawn in Section 4.

2 COMPUTATIONAL FRAMEWORK

This computational framework solves implicitly the weak form of the equilibrium equtions for the quasi-static finite deformation problem of an elasto-plastic material. The formulation is largely based on Charlton et al. (2017) but uses different particle domains and basis functions, which are detailed below.

Compared to the FEM, the MPM requires mapping between material points and the computational mesh. However, the computation inside each load-step, e.g. each column in Figure 1, is the same as in the FEM. Therefore, we present a unified computational framework for integrating the standard material point methods and the CPDI approaches as follows.

i. Set up problems: generate computational mesh, specify boundary conditions and material parameters, and generate
   - IF (sMPM): material point coordinates
   - IF (CPDI1): material point coordinates and two vectors, (s^0, t^0), of the parallelogram particle domain for each material point
   - IF (CPDI2): corners and connectivity for the particle domains

and material point volumes.

ii. A load-step starts, incremental boundary conditions are specified.

iii. Find influence elements of material points in the mesh:
   - IF (sMPM): FOREACH material point, p, find its parent element

- IF (CPDI1): FOREACH material point, p, to compute the coordinates of four corners of its particle domain as

$$x_p+[-1\ 1\ 1-1]\frac{s_p}{2}+[-1-1\ 1\ 1]\frac{t_p}{2}$$

and find parents elements for these corners
- IF (CPDI2): FOREACH corner to find its parent element

iv. Compute the *basis functions* S_{ip} and their *spatial gradients* ∇S_{ip} for node i and material point p: FOREACH material point apply the following:
- IF (sMPM)

$$S_{ip}=S_i(x_p),\qquad \nabla S_{ip}=\nabla S_i(x_p),$$

where $S_i(")$ are the standard FEM basis functions and x_p are the coordinates of material point p.
- IF (CPDI1)

$$\begin{aligned}S_{ip}&=\frac{1}{4}\left[S_i(x_p^1)+S_i(x_p^2)+S_i(x_p^3)+S_i(x_p^4)\right],\\ \nabla S_{ip}&=\frac{1}{2V_p}\left\{\left(S_i(x_p^1)-S_i(x_p^3)\right)\begin{bmatrix}s_y-t_y\\ t_x-s_x\end{bmatrix}\right.\\ &\quad\left.+\left(S_i(x_p^2)-S_i(x_p^4)\right)\begin{bmatrix}s_y+t_y\\ -s_x-t_x\end{bmatrix}\right\},\end{aligned}$$

where the superscript indicates the index of particle domain corners.
- IF (CPDI2)

$$\begin{aligned}S_{ip}=&\frac{1}{24v_p}\left[(6v_p-a-b)S_i(x_p^1)\right.\\ &+(6v_p-a+b)S_i(x_p^2)+(6v_p+a+b)S_i(x_p^3)\\ &\left.+(6v_p+a-b)S_i(x_p^4)\right],\\ \nabla S_{ip}=&\frac{1}{2V_p}\left\{S_i(x_p^1)\begin{bmatrix}y_p^2-y_p^4\\ x_p^4-x_p^2\end{bmatrix}+S_i(x_p^2)\begin{bmatrix}y_p^3-y_p^1\\ x_p^1-x_p^3\end{bmatrix}\right.\\ &\left.+S_i(x_p^3)\begin{bmatrix}y_p^4-y_p^2\\ x_p^2-x_p^4\end{bmatrix}+S_i(x_p^4)\begin{bmatrix}y_p^1-y_p^3\\ x_p^3-x_p^1\end{bmatrix}\right\},\end{aligned}$$

where

$$\begin{aligned}a&=(x_p^4-x_p^1)(y_p^2-y_p^3)-(x_p^2-x_p^3)(y_p^4-y_p^1),\\ b&=(x_p^3-x_p^4)(y_p^1-y_p^2)-(x_p^1-x_p^2)(y_p^3-y_p^4).\end{aligned}$$

For more detail on these basis functions see (Sadeghirad et al. 2013).

v. The nodal internal force can then be obtained from the stresses at the material point using

$$f_i^{int}=-\sum_p\nabla S_{ip}\sigma_p v_p,$$

where σ_p is the Cauchy stress and v_p the volume of the material point p. The system of equations for the nodal force equilibrium can then be formed

$$f^{int}-\left(f^{rea}+f^{ext}\right)=0, \tag{1}$$

where the superscripts *rea* and *ext* indicate nodal reactions and external forces.

vi. Solve (1) with the Newton-Raphson (NR) iterative solver to obtain the nodal incremental displacement.

vii. Map nodal incremental displacements onto the material points, and added onto the total displacement field as

$$x_p=x_p+\Delta u_p=x_p+\sum_i S_{ip}\Delta u_i.$$

for the sMPM and CPDI1 approach. For the CPDI2 approach, the corners of particle domains are updated in the same manner. Other state variables, e.g. deformation gradient **F** and Cauchy stress, are also updated. In addition, the particle volume are updated:
- IF (sMPM): $v_p=v_p^0\det\boldsymbol{F}_p$, where superscript 0 indicates a value in the initial configuration
- IF (CPDI1): $s_p=\boldsymbol{F}_p s_p^0, t_p=\boldsymbol{F}_p t_p^0$, and $v_p=|s_p\times t_p|$
- IF (CPDI2): v_p is the volume of particle domain with updated corners

viii. The computational mesh is simply modified by the moving mesh concept and go back to (ii) for simulation of next load-step.

Employing an unstructured mesh, the cost in step (iii) for finding the position in a mesh for a point is increased, because it cannot be determined by computing the offset as with a structured mesh. Instead, a searching routine has be to used by looping over elements for a point. However, the independence of the job of checking whether a point is inside of the an element for different points and different elements enables us to parallelise these loops and speed-up computation in this framework. The searching algorithm can be further speeded up by reducing the searching region with a assistant bucket data structure, e.g. a multilevel octree bucket in Nie et al. (2012).

3 NUMERICAL EXAMPLES

The computational framework is first validated through benchmark problems. The performance of all the methods are then investigated for a problem involving large rotational deformation.

3.1 *Validation*

The simple stretch of a square block is simulated by the FEM, sMPM and CPDI approaches under the plane-strain condition. The block has dimension 2×2 with the elasto-plastic material, specified by the von-Mises constitutive law, with the yield criterion

$$\rho = \sqrt{2J_2} / \rho_c - 1,$$

where J_2 is the second invariant of the deviatoric part of the Kirchhoff stress tensor. The material parameters are Young's modulus $E = 1000$, Poisson's ratio $\nu = 0$ and yield strength $\rho_c = 400$. All variables have compatible units.

For the FEM case, the block was discretised by four bi-linear quadrilateral elements. For other cases, the same mesh was used with four material points per element. Figure 2 shows the geometry, mesh and boundary conditions for the CPDI2 approach in both initial and deformed configurations. Roller boundary conditions were applied on the bottom and left sides, with a horizontal displacement of 0.2 per load-step applied on the right. The simulation was run up to 20 steps. Due to the moving right boundary and displacement applied there, the moving mesh concept described above was adopted such that the computed incremental horizontal displacement was applied to the mesh at the end of each load-step.

Quadratic convergence rate of solver for the global equilibrium is expected, since the nonlinear system of equations are solved by the NR iteration. The convergence of the CPDI2 approach is shown in Figure 3 by plotting the residual, i.e. L_2 norm of the left hand side of (1), against the NR iteration steps in each load-step. Particularly, the

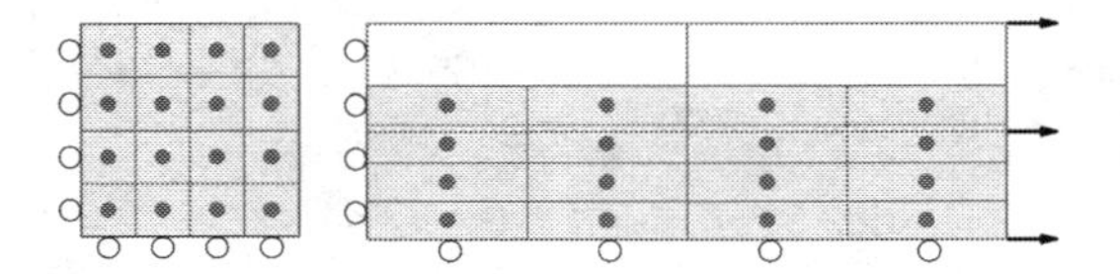

Figure 2. Left: Initial geometry (grey), computation mesh (red), material points and particle domains (blue) with roller boundary conditions in the CPDI2 approach. Right: deformed geometry (grey) subject to uni-axial stretch and plane-strain condition.

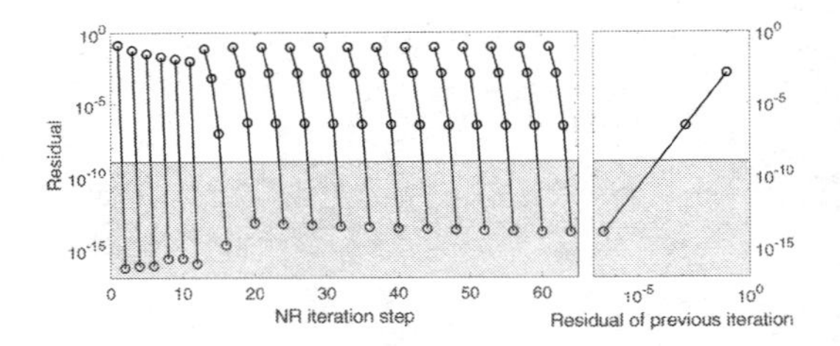

Figure 3. Left: The residual against the NR iteration step for each load-step, in which markers along a curve indicates NR iterations in that load-step, in the simulation with the sMPM. Right: The residual in the (n+1)-th iteration against that in the n-th iteration for the last load-step. The slope of two segments are 1.9951 and 1.9965, showing the convergence rate is quadratic.

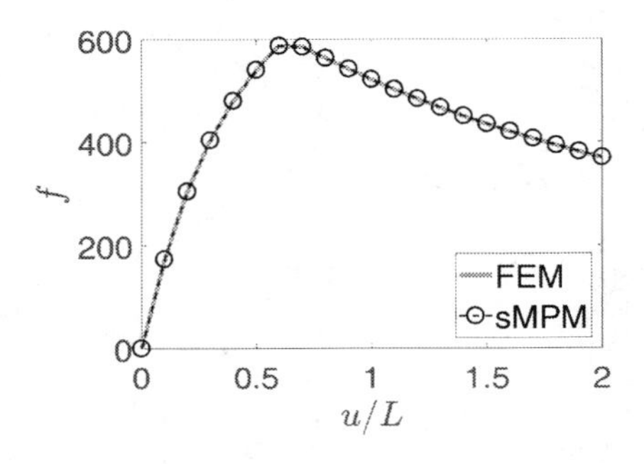

Figure 4. The reaction force on the right end of the block. The results from the CPDI approaches are the same with those from the sMPM.

residual of an iteration against that of the previous iteration shows that the average convergence rate is 1.996. The sMPM and CPDI1 approach also have this correct quadratic convergence rate.

The x-component of reaction force on the right end from the sMPM and FEM are plotted in Figure 4, where the markers show the load-steps. The force increases non-linearly in the first six steps due to the large deformation mechanics, i.e. the geometric nonlinear finite strain measure used in the computational framework. In the seventh step, the material yields and so this reaction force starts to decrease gradually. The results from the sMPM and CPDI approaches are the same with the FEM, as this FEM simulation is not affected by mesh distortion. This agreement validates the computational framework and computer codes of the sMPM and CPDI approaches.

3.2 *Doughnut twist*

To test the formulations performances for large torsion problems, a challenging problem, used involving the azimuthal shear of a confined annular domain, is termed the Doughnut Twist problem here. The geometry, boundary conditions and material parameters of this problem are shown in Figure 5.

Two groups of simulation were carried out with circular and elliptical holes, respectively. Through these simulations, we are investigating:

1. the variation in response among the sMPM, CPDI approaches and FEM; and
2. the capacity of each method for modelling this large rotation.

In order to simulate this problem, the moving mesh was adopted as the rigid rotating mesh fixed to the inner circle or ellipse.

3.2.1 *Variation in response*

For question 1), the results from different methods when $\alpha = 10°$, i.e. small torsion, are first compared. In this case the deformed FE mesh suffers only minor distortion, thus the FE results are still reliable as a control. With the FEM, fine (160 × 1152 elements) and coarse (10 × 72 elements) meshes were used, in which the number before the sign × indicates the number of elements in the radial direction and the other is in the circumferential direction. The agreement of the results between fine and coarse meshes (Figure 6) suggests the coarse mesh is good enough for modelling this deformation. Therefore, this coarse mesh was used as computational mesh in the simulation with the sMPM and CPDI approaches. In all simulations, four material points per element were used. All simulations were run with the rotation increment $\Delta\alpha = 5°$ in the clockwise direction up to $\alpha = 10°$ in two load-steps.

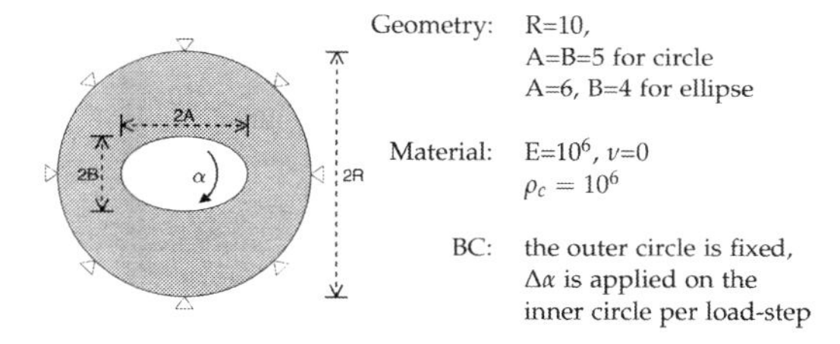

Figure 5. The geometry, material parameters and boundary conditions (BC) for the simulation of doughnut twist.

The variations in response predicted by these methods are demonstrated by comparing the displacements along a radius (Figure 6) and the reaction force (Figure 7) on the inner circle during this simulation. The reaction force f is computed as the summation of the reaction force magnitudes at all nodes along the inner circle. The mesh nodes along the radius at the central angle $\theta = 0°$ were selected as sampling points, indicated by markers in Figure 6. From the FEM simulation, these nodal displacement were used directly. However, these sampling points have to be added as the material points in the sMPM and CPDI1 approach, in order to obtain comparable total displacements. These extra material points were also assigned a tiny square domain with edge length 0.001, for computing the initial volumes in the sMPM and for specifying particle domains in the CPDI1 approach. In contrast, the CPDI2 approach did not need these material points. Instead the displacements at corners of some particle domains were used. That is because the corners of some particle domains coincide with

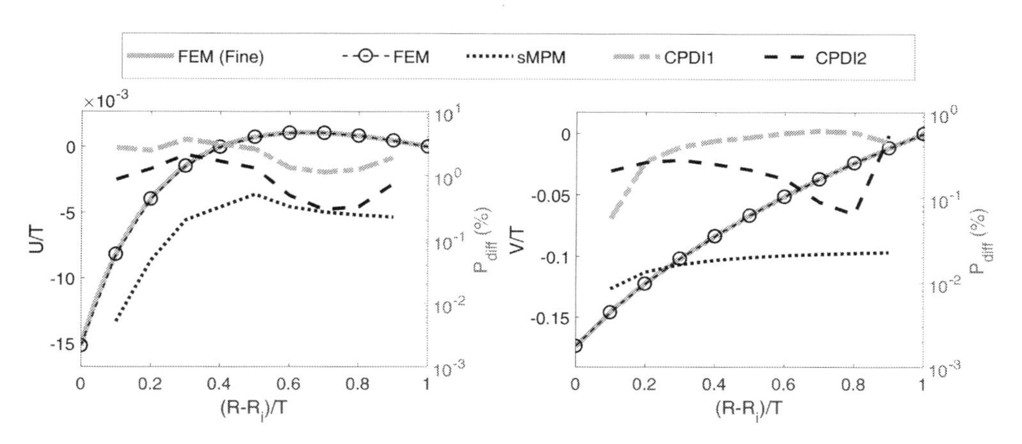

Figure 6. The two components of displacement across the wall of doughnut when $\alpha = 10°$. The displacements U, V are non-dimensionalised by the thickness of doughnut T. The FE results with both fine (160 × 1152 elements) and coarse (10 × 72 elements) meshes are plotted. The difference in the sMPM and CPDI approaches results to the FE results are shown on the right hand axes.

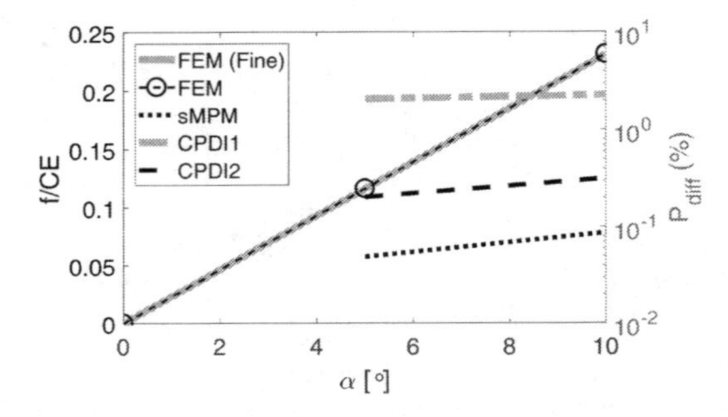

Figure 7. The normalised reaction force in FE simulation with fine and coarse mesh, and the difference in the sMPM and CPDI approaches results to the FE results. The force f is summation of reaction at all nodes along the inner circle, and C is circumference of the inner circle and E is the Young's modulus.

the sampling points in the initial configuration and these domains follow the deformation.

The difference in the displacement from the sMPM and CPDI approaches were computed with respect to the FEM results as

$$P_{diff} = \frac{u^m - u^{FE}}{u^{FE}}, \tag{2}$$

where u^m is the displacement component from the sMPM and CPDI approaches and the superscript m indicates a method. The comparison in Figure 6 shows the CPDI1 approach to differ the most from the FEM result (<%5), followed by the CPDI2 approach(<1%), while the sMPM is the most similar to the FEM (<0.05%). The difference in CPDI1 result and the FEM result is because the CPDI1 approach uses a parallelogram for a particle domain, and there are consequently overlaps and/or gaps among particle domains for covering this problem domain. The difference between the collection of these parallelogram particle domains and the actual deformed domain results in the difference in the stiffness matrix for simulation. In contrast, quadrilateral particle domains are employed in the CPDI2 approach, so avoiding this difference. The observation, that the results from the sMPM are closer to those from the FEM than from the CPDI2 approach, is because of the mapping of displacement information between material point and mesh nodes, which needs to be done in two steps in the CPDI2 approach but only one in the sMPM. For example, updating of material points, positions at the end of a load-step needs: 1) interpolation of nodal displacements onto the corners of particle domains; and 2) computation of incremental displacements at material points based on the corners of their own particle domains. In the sMPM, the incremental displacement at material points is simply interpolated from the nodal incremental displacement.

3.2.2 *Capacity of methods for large torsion*

For question 2), all methods were applied to simulate the doughnut twist problem up to $\alpha = 80°$. The magnitude of reaction force along the inner circle against the boundary condition α is plotted in Figure 8. These reaction forces are almost the same in the elastic region when $\alpha < 25°$, but the FEM results are very different to the sMPM and the CPDI approaches in the plastic region when $\alpha = 30°$.

When the material yields, the reaction force is expected to decrease, as observed in the results of the sMPM and CPDI approaches. The reaction force in the FEM is erroneously increasing due to errors brought via mesh distortion. Around $\alpha = 35°$, a larger peak was observed in the sMPM than in the CPDI approaches. The results of the sMPM are closer to those of the FEM than the CPDI approaches before $\alpha = 40°$ as demonstrated in the previous section. The drop following the peak appears to be due to the stress relaxation when the inner material yields while the external material is still elastic, as shown in Figure 9. The incremental displacements are shown for $\alpha = 30°$, 45° and 60°. The first layer mesh nodes are rotated in the opposite direction–counter-clockwise to the boundary condition which is applied clock-wise on the inner circle when $\alpha = 45°$, showing the stress relaxation.

Both the sMPM and CPDI1 approaches predict the same constant reaction force for $\alpha > 50°$, but the CPDI2 approach has a slight increase. This erroneous increase is because of the distortion

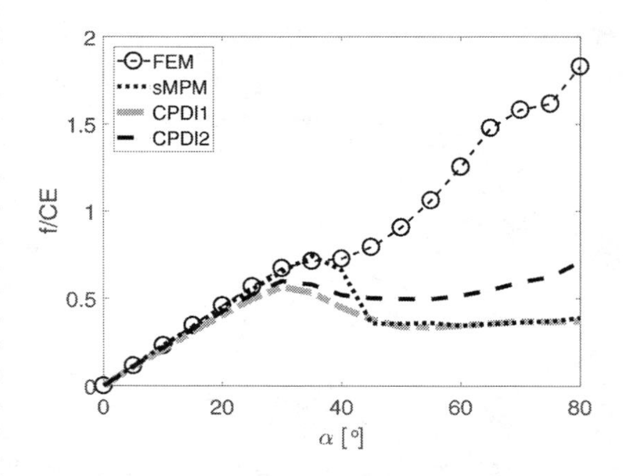

Figure 8. The normalised reaction force in the simulation with the same mesh (5 × 18 elements) in the FEM, sMPM and CPDI approaches.

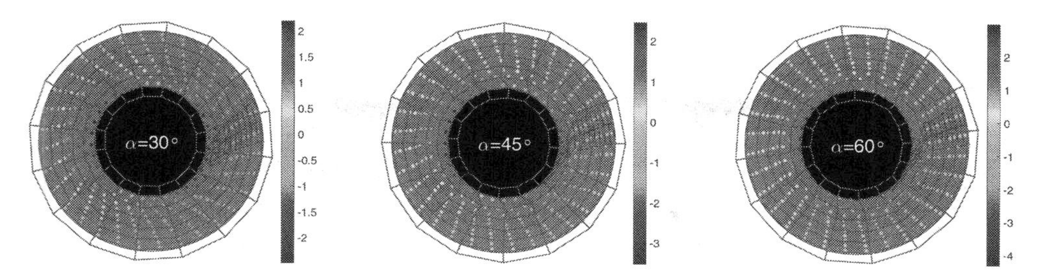

Figure 9. Deformed mesh at end of the load-step for $\alpha = 30°$, $45°$ and $60°$, with color on material points showing the total vertical displacement in the CPDI1 simulation. The blue lines show the particle domains.

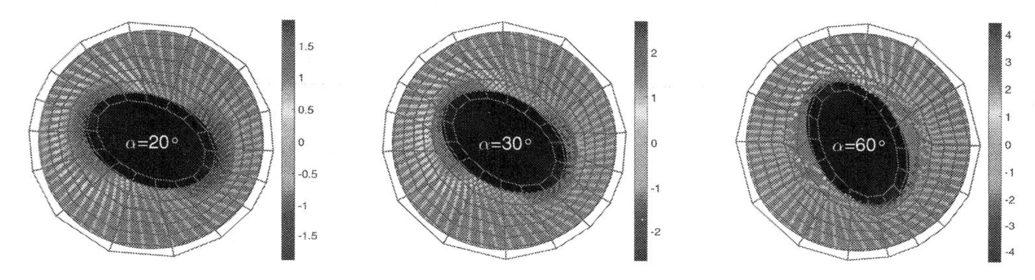

Figure 10. Deformed mesh at end of the load-step for $\alpha = 20°$, $30°$ and $60°$, with color on material points showing the total vertical displacement in the sMPM simulation.

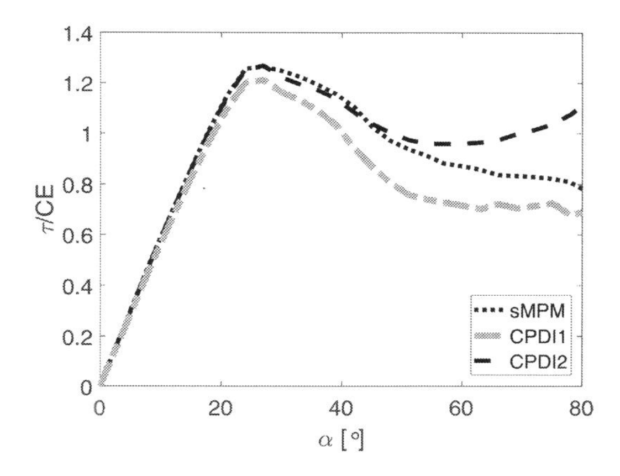

Figure 11. Normalised torque versus rotation α for the elliptical hole, where C is the circumference of the ellipse and E is the Young's modulus.

of particle domains. Recall that in the CPDI2 approach the particle domains exactly follow the deformation as in a FE mesh, but the distortion causes more serious error in the FEM than the CPDI2 approach.

The performance of these methods was also investigated through the simulation of doughnut twist with an elliptical hole. In these simulations, the mesh at the ends of the long axis were locally refined because the deformation gradient and stress tends to concentrate in these positions. The deformed mesh with incremental displacement and material points in the load-steps for $\alpha = 20°$, $30°$ and $60°$ are shown in Figure 10. Due to the stress concentration at ends of long axis of the ellipse, the material yields earlier than for a circular hole. As the ellipse further rotates in the plastic material, two regions with fewer material points are created after the ends of long axis passing, e.g. see Figure 10 when $\alpha = 60°$. The magnitude of torque against α in Figure 11 shows that the CPDI2 approach predicts erroneous increase in the plastic region again. Both the CPDI1 approach and the sMPM predict a more physically realistic response.

4 CONCLUSIONS

This paper has presented a common computational framework for the sMPM and its extensions, i.e. the CPDI approaches, which is first validated and then applied to problems involving large rotational deformation. In the published papers, the CPDI2 approach is the latest extension of the sMPM. Using the CPDI approaches can increase the stability of the computation, but as shown here can lead to erroneous results if the particle domains are distorted. As observed in the simulations of the doughnut twist problem, the increase in the torque predicted using the CPDI2 approach does not seem physically correct. The CPDI1 approach appears to have less negative effect on the results. In contrast, the sMPM predicts physically reasonable response, and stable reaction force and torque after the material is yielded. In the simulations for the doughnut with the circular hole, both the sMPM and the CPDI1 approach predict the same asymptotic reaction force, but for an elliptical hole they predict different values. Without referring to a reliable control, we cannot tell which is the closest to the exact solution. But both the sMPM and the CPDI1 approach predict what appear to be more physically reasonable responses than the CPDI2 approach. In practice, increasing the number of material points per element could improve the stability of the sMPM. In conclusion, the CPDI2 approach appears not to be suitable for simulating these large rotational problems.

ACKNOWLEDGEMENTS

We are grateful for the support by the UK Engineering and Physical Sciences Research Council grant (No. EP/N006054/1).

REFERENCES

Bardenhagen, S. & E. Kober (2004). The generalized interpolation material point method. *Computer Modeling in Engineering and Sciences 5*(6), 477–496.

Beuth, L. (2012). *Formulation and application of a quasi-static material point method*. Ph. D. thesis, University of Struttgart, Germany.

Ceccato, F., L. Beuth, P.A. Vermeer, & P. Simonini (2016). Two-phase material point method applied to the study of cone penetration. *Computers and Geotechnics 80*, 440–452.

Charlton, T., W. Coombs, & C. Augarde (2017). iGIMP: An implicit generalised interpolation material point method for large deformations. *Computers & Structures 190*, 108–125.

Coombs, W. (2011). *Finite deformation of particulate geomaterials: frictional and anisotropic Critical State elasto-plasticity*. Ph. D. thesis, Durham University, UK.

Cortis, M., W. Coombs, C. Augarde, M. Brown, A. Brennan, & S. Robinson (2017). Imposition of essential boundary conditions in the material point method. *International Journal for Numerical Methods in Engineering 133*(1), 130–152.

Nie, Y., Y. Li, & L. Wang (2012). Parallel node-based local tetrahedral mesh generation. *Computer Modeling in Engineering & Sciences(CMES) 83*(6), 575–597.

Sadeghirad, A., R.M. Brannon, & J. Burghardt (2011). A convected particle domain interpolation technique to extend applicability of the material point method for problems involving massive deformations. *International Journal for Numerical Methods in Engineering 86*(12), 1435–1456.

Sadeghirad, A., R.M. Brannon, & J. Guilkey (2013). Second-order convected particle domain interpolation (CPDI2) with enrichment for weak discontinuities at material interfaces. *International Journal for Numerical Methods in Engineering 95*(11), 928–952.

Wang, L., W.M. Coombs, C.E. Augarde, M. Brown, J. Knappett, A. Brennan, D. Richards, & A. Blake (2017). Modelling screwpile installation using the MPM. *Procedia Engineering 175*, 124–132.

Numerical Methods in Geotechnical Engineering IX – Cardoso et al. (Eds)
 ISBN 978-1-138-33198-3

Issues with the material point method for geotechnical modelling, and how to address them

C.E. Augarde, Y. Bing, T.J. Charlton, W.M. Coombs & M. Cortis
Department of Engineering, Durham University, UK

M.J.Z. Brown, A. Brennan & S. Robinson
Civil Engineering, University of Dundee, UK

ABSTRACT: The Material Point Method (MPM) for solid mechanics was first proposed by Sulsky and co-workers in the 1990s. Since then it has been developing a growing band of followers not least because of its ability to handle large deformation problems with ease. This feature has more recently come to the notice of geotechnical researchers who have plenty of problems to solve involving large deformations. It is clear from recent publications, however, that many geotechnical researchers have found difficulties with the use of the MPM in a number of areas. In this paper we visit three of these problem areas and highlight solutions we have developed. It is to be hoped that this can remove some of the roadblocks to the use and further development of the MPM for geotechnical problems in future.

1 INTRODUCTION

The problems that have to be solved by geotechnical engineers are wide-ranging, covering tunnelling, foundations, slopes and many other areas, however there are common features of these problems that provide challenges to numerical modellers in particular. Principally these are material and geometric non-linearity. The former has been recognised as crucial for accurate modelling from before the development of computational geotechnics, in the recognition that plasticity is a vital part of any constitutivemodel for soil. Indeed material non-linearity is likely to feature in the majority of the papers at this conference. Geometric non-linearity has received less attention to date largely because material non-linearity is crucial to a wider range of geotechnical problems than geometric non-linearity. When using this term, we mean the ability to model large deformations and the use of strain definitions which are no longer linear with displacement, as opposed to infinitesimal strain measures for standard analyses.

The finite element method (FEM) remains the method of choice for most geotechnical numerical modelling, and with good reason. However, there are issues with its use for the class of problems mentioned above, i.e. in the modelling of large deformation problems. If large deformations are to be modelled then any mesh-based method (the FEM included) will require an update of the mesh during a stepped non-linear solve to avoid the inaccuracies associated with distorted elements. Any change of mesh will then require a mapping of state variables from the old to the new mesh. While both of these actions bring potential errors the FEM has been successfully adapted for large deformation problems via modifications such as the Arbitrary Lagrangian Eulerian (ALE) method and there are other examples which do much the same thing, such as the Coupled Eulerian-Lagrangian (CEL) method; indeed some of these techniques are available in commercial software such as ABAQUS, and have been used for geotechnical problems (e.g. shortciteNKim2015). Having said this, there is a school of thought that says adherence to mesh-based methods places a restriction of the development of numerical modelling, and there are many examples of mesh-free methods being developed, although most incur a greater computational cost to the FEM for standard problems at present (Heaney et al. 2010).

In this paper we are concerned with a relative newcomer to computational geotechnics, called the Material Point Method (MPM) which seems to offer advantages over the standard FEM for large deformation problems, without some of the complexities of rival approaches mentioned above. The MPM is of key and current interest in geotechnics having featured in Alonso's 2017 Rankine Lecture, and its use in geotechnics having been the feature of a major conference in the same year (Anura3D MPM Research Community 2017). In this paper we highlight a number of issues that geotechnical modellers might face when using the MPM and present recent solutions developed by the authors.

2 THE MATERIAL POINT METHOD

The MPM for solid mechanics was developed from an earlier method for fluids (the FLIP method) by Sulsky and co-workers (Sulsky et al. 1994). It is usually described in an explicit form where it is used for time-stepping analyses of problems with inertiaand acceleration. It is however equally possible to formulate the MPM in implicit form (Guilkey and Weiss 2003) for quasi-static problems (usually of most interest in geotechnics) where to deal with non-linearity a total applied action is split into a number of substeps, in which each requires the solution of a linear system involving a stiffness matrix and an unknown vector of displacements.

The method is often referred to as an Eulerian-Lagrangian method (Muller & Vargas 2014) but this is not really correct; there are aspects of the method that make it look that way. In fact the calculations are just Lagrangian. In the MPM, a problem domain is defined with a set of material points (MPs, sometimes also referred to as particles). These MPs carry all the information relating to that location in the problem domain throughout a calculation, i.e. total displacement, strain, stress and if necessary other state variables required by a constitutive model. All calculations are however carried out on a finite element mesh (often referred to as a background grid) to which data are mapped back and forth from the material points. The feature in the MPM that ismost attractive to those wishing to model large deformation is that the deformation of the problem domain is represented at the material points only and the background grid can be discarded after each time step. This means never having to calculate on a distorted grid, and also means that a regular structured grid can be reused each time, avoiding the overhead linked to unstructured mesh generation. A secondary but important aspect of the MPM, from the implementation point of view, is that much finite element technology can be seamlessly transferred to the MPM, especially items such as constitutive models and basis functions, reducing the overhead of code development and, perhaps, providing some confidence in the use of the method.

An important problem dealt with in the literature with the standard MPM is grid-crossing instability, which is a numerical artefact linked to a deformation pattern in which a material point leaves one grid element and enters another. In the standard MPM, where domain volume (or mass) is concentrated at a material point, this leads to a sudden change in stiffness of a grid element which causes non-physical numerical effects. This problem has been tackled by developing variants of the standard MPM where each material point carries a spatially-defined and potentially deforming volume (or mass) with it. Methods include the Generalised Interpolation MPM (Bardenhagen and Kober 2004, Charlton et al. 2017) and CPDI methods (Sadeghirad et al. 2011, Nguyen et al. 2017). While the sevariants effectively reduce grid-crossing instabilities they lead to additional complexities in calculations (especially the CPDI methods) and all suffer from the same issues which are covered in this paper.

3 THE MPM FOR GEOTECHNICS

It is clear there is keen and current interest in the use of the MPM in geotechnics as evidenced by an increasing number of papers describing its use for various geotechnical problems, for instance soil-structure interaction (Ma et al. 2014) and slope stability (Zabala and Alonso 2011), and a recent conference was devoted to MPM and geotechnics (Anura3D MPM Research Community 2017). In addition there are a number of survey papers, e.g. Sołowski and Sloan (2015). The MPM has also been developed to model coupled problems important in geotechnics, for example Jassim et al. (2013). The following sections cover three issues affecting use of the MPM or its variants for geotechnical analysis. The discussion relates primarily to our experiences with implicit MPM codes for elastic and elasto-plastic statics problems rather than explicit MPM with dynamics.

3.1 *Essential boundary conditions*

Essential (or Dirichlet) boundary conditions (BCs) are necessary to fully define a boundary value problem. They do not (usually) enter the weak form description of the problem but are subsidiary conditions that have to be incorporated. In the standard FEM, essential BCs are usually imposed directly, i.e. those degrees of freedom with essential BCs are treated differently, or the stiffness matrix is amended, as described in standard texts, e.g. Potts and Zdravković (1997). This is possible due to the Kronecker delta property of standard FE shape functions which delivers an *interpolation* of nodal unknowns. In contrast most weak-form based meshless methods, such as the element-free Galerkin method (Belytschko et al. 1994) have basis functions, often derived from moving least squares, that do not possess the Kronecker delta property and hence lead to *approximations* rather than interpolations. Consequently essential BCs

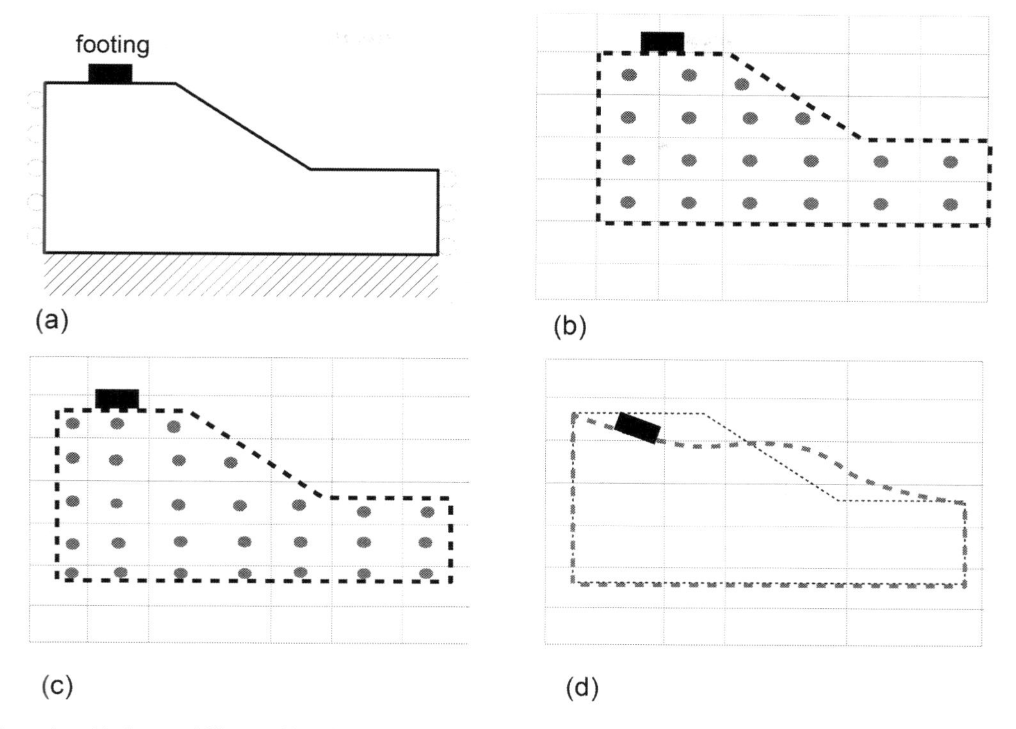

Figure 1. (a) slope stability problem geometry; (b) MPM solution requiring collinear grid and domain boundaries; (c) ideal MPM model with no requirement of collinear gird and domain geometry; (d) deformed domain. Red dots are the material points in each case.

have to be imposed indirectly (Fernández-Méndez and Huerta 2004).

The MPM (in all of its different flavours as mentioned above) has a problem with essential BCs of a slightly different nature. Consider Fig. 1 which shows a typical 2D plane strain discretisation of a slope stability problem, in which the vertical boundaries are subject to an essential boundary condition of zero displacement normal to the boundary while the bottom horizontal boundary is subject to an essential BC of full fixity or zero normal displacement. Providing the essential boundaries align exactly with the background grid (Fig. 1(b)), these conditions can be applied in the MPM directly just as in the standard FEM. All calculations are carried out on the background grid and therefore as the domain boundary coincides with edges of elements in the background grid for these three boundaries, there is exact coincidence with imposing essential BCs on the grid. However, if the problem domain boundary does not coincide with edges of elements in the background grid, an essential boundary condition cannot be applied this way. The example so far has concentrated on essential BCs which act as fixities in a boundary value problem, however geotechnical problems often require non-homogeneous essential BCs such as for the modelling of the footing shown at the top of the slope in Fig. 1. In this case, while it might be possible to arrange for the initial grid to align with the starting position of the footing, subsequent loading steps (in this case increments of applied displacement to the footing) would require grids to follow the predicted movements exactly, thus losing a key advantage of the MPM in that a regular grid, not connected to the domain geometry, can be used. While we now present a solution to this problem, in fact, Fig. 1 will be used to illustrate the other problems with the MPM on which this paper focusses in later sections. The issue outlined above is linked to a large body of emerging literature in computational solid mechanics on non-matching mesh methods which, as the name suggests, are numerical methods where solutions are calculated without matching the discretisation mesh to the problem domain, examples include immersed FE methods and the Finite Cell method (Ramos et al. 2015, Schillinger et al. 2012). In the case of the MPM, we have developed a new method for the imposition of essential BCs which removes any need for alignment of mesh edges and problem domain boundaries. It is based on Kumar and co-workers' *implicit boundary approach* (Kumar et al. 2008) and can be viewed as a form of penalty method of applying the essential BCs. In the method, essential boundaries are defined by signed distance functions ϕ_j where $\phi_j < 0$ indicates the exterior and $\phi_j > 0$ the interior of the problem domain for the jth boundary. Essential boundary (or Dirichlet) functions $d_j(\phi_j)$ are then defined for each boundary, which are equal to zero on the boundary and rise to unity a small distance δ on the domain side of the boundary. d_j are often simple discontinuous quadratic functions in ϕ_j. An example is shown in Fig. 2 where the Dirichlet function is given by

$$d = \begin{cases} 0, & \phi < 0 \\ 1 - \left(1 - \dfrac{\phi}{\delta}\right)^2 & 0 \le \phi \le \delta. \\ 1 & \phi > \delta \end{cases} \tag{1}$$

At points where more than one essential boundary is active, i.e. at a domain corner, then the product of the $d(\phi_j)$ forms the essential boundary function and in general we write the net essential boundary at a point as

$$D_k = \prod_j d(\phi_j), \tag{2}$$

Where k refers to the component of displacement defined at that boundary. D_k are then the components of the diagonal matrix $[D]$ used to redefine the trial functions for the grid elements as

$$\{u'\} = [D]\{u\} + \{u^a\} \tag{3}$$

where $[D] = diag(D_1, \ldots, D_{n_d})$ and n_d is the dimensionality. In Eqn 3, $\{u\}$ is the standard approximation for displacement in a finite element while $\{u^a\}$ is the essential boundary condition (i.e. zero for a fixed degree of freedom or non-zero for a prescribed displacement). Within the narrow band adjacent to the implicit boundary, the first term in Eqn 3 will be suppressed via the matrix $[D]$, enforcing the essential boundary condition in the second term. Substituting these trial functions and some suitable test functions into the standard weak form for equilibrium leads to expressions for the stiffness matrices of elements containing essential boundary conditions. Fig. 3 shows a grid element (cell) cut by an essential boundary. For this element the net stiffness matrix would be

$$[k^E] = [K_1] + ([K_2] + [K_2]^T) + [K_3], \tag{4}$$

where $[K_1]$ is the standard finite element stiffness matrix for the part of the element occupied by the problem domain and is obtained through the summation of the MP contributions, while $[K_2]$ and $[K_3]$ contain Dirichlet functions and their derivatives. The additional stiffness matrices $[K_2]$ and $[K_3]$ are effectively penalty terms imposing the essential BC crossing the element and their components are calculated by numerical integration in the bandwidth δ as shown in Fig. 3. It is straightforward to implement essential BCs which cut an element at an angle with respect to the element global coordinates; a transformation matrix is applied to components of the matrices forming $[K_2]$ and $[K_3]$. An example of the use of the approach is given in Fig. 4 where a rigid footing penetrates a unit distance into an elastic square domain of side 3 units. The roller sides and base are imposed as implicit homogeneous essential BCs while the footing is an implicit non-homogeneous essential BC. Full details of this approach for implementation of essential boundaries in the MPM are given in Cortis et al. (2017).

3.2 *Tracking domain boundaries*

Fig. 1(d) serves to illustrate another problem with the MPM for geotechnical problems, that of tracking evolving boundaries to a problem domain and application of traction (Neumann) boundary conditions. The former is particularly troublesome given that geotechnical engineers would like to use the MPM for problems in which this can be a crucial output, e.g. a key prediction for long runout landslides is the final surface profile of the disturbed material (e.g. Wang et al. (2016)). To date, researchers wishing to track free surfaces in an MPM analysis have been limited to determining the location of the edges of material point individual volumes which exist in GIMP and CPDI methods but not in the

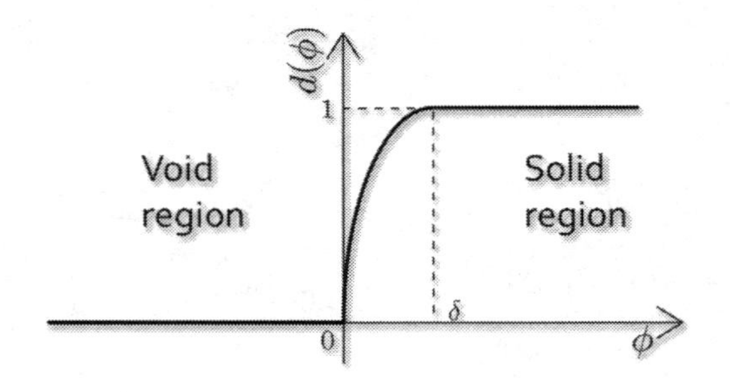

Figure 2. A Dirichlet function in 1D.

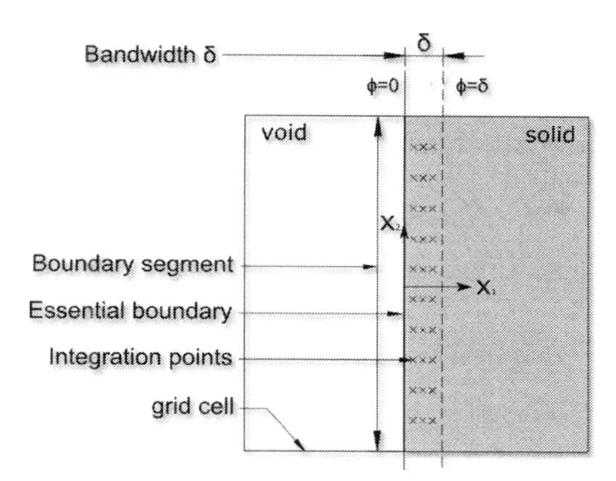

Figure 3. A single grid element (cell) crossed by an essential boundary condition.

standard MPM. The best that these can deliver are piecewise linear edges (CPDI2) or stepped approximations (GIMP and CPDI1). As regards the application of traction boundary conditions this has been attempted in the past by placing loads at material points, which is clearly an inaccurate representation, particularly when wishing to exploit the capabilities of the MPM for large deformation problems.

In recent research undertaken at Durham, Bing (2017) describes a new method which can deliver both accurate tracking of domain boundaries and application of surface tractions in any variant of the MPM for arbitrary large deformations. A local cubic B-spline interpolation is used to represent the boundary based on a set of defined boundary material points. These can be additional material points (with near-zero volumes) placed on the boundaries or an outer layer of standard material points in the physical domain. Spline segments are fitted between adjacent sampling points to calculate the B-spline control points and a suitable knot vector determined completing the description of the B-spline curve. Although fitting a curve globally to the boundary (Piegl & Tiller 1997) would result in higher continuity, it would not be capable of reproducing sharp corners which are important features in many geotechnical problem domains. Local fitting, as used here, constructs curves in a piecewise fashion so that only local data are used at each step, so a fluctuation in data only affects the curve locally. As regards spline order, a local cubic interpolation has a simpler formulation than the local quadratic interpolation (Piegl & Tiller 1997) and no special cases or angle calculations are needed.

Once an accurate boundary representation is in place we can consider how to apply boundary conditions. Essential boundary conditions on B-spline boundaries can be imposed with the implicit boundary method described above, with a few minor alterations. The application of tractions to a B-spline defined boundary is also straightforward with the only complexity arising in the integration required, as will be outlined now. In the standard FEM, nodal forces $\{f^t\}$ consistent with a surface traction $\{t\}$on a surface $d\Omega$ are obtained as

$$\{f^t\} = \int_{\partial\Omega} [M]^T \{t\} d\Omega, \tag{5}$$

where $[M]$ contains the standard finite element shape functions. So for the case of the B-spline boundaries we have to consider how to carry out this integration. A pth-degree B-spline curve can be integrated numerically by using $(p-1)$th order Gauss quadrature. However, the local coordinate of 1D Gauss quadrature has a range of $[-1, 1]$, whereas, the local coordinate of a B-spline curve has positive values only, therefore mapping between these two systems is required to complete the integration (in addition to the standard map between the local and global coordinates, or physical space for isoparametric FEs). To allow for this, an additional space, called the *parent domain* is introduced over which quadrature takes place (Hughes, Cottrell, & Bazilevs 2005). Fig. 5 shows an illustration of the three spaces: the physical space, $\{x\}$, the parametric space, ξ, and the parent domain, $\tilde{\xi}$. In the physical space, the boundary geometry is defined in global coordinates. The parametric space contains the knots (local coordinates) which run along the curve and the parent domain is simply a local system where $\tilde{\xi} \in [-1,1]$ on which numerical integration is performed. To carry out the integration, the B-spline curve segment is pulled back from the physical space to the parametric space, i.e. the local coordinates (ξ_j and ξ_{j+1}) of the start and the end point of the segment are identified by using their global coordinates. A linear transformation between the parent domain, $\tilde{\xi} \in [-1,1]$, and the parametric space, $[\xi_j, \xi_{j+1}]$ maps the locations of Gauss points between these two spaces. In order that the integration can be completed, a Jacobian mapping is required between the parent and physical spaces between the Gauss quadrature lengths, which are defined over the local coordinates $\tilde{\xi} \in [-1,1]$, and the physical space. Due to the use of a two local coordinate systems, the Jacobian contains two components

$$\{J_B\} = \left\{\frac{dC}{d\tilde{\xi}}\right\} = \left\{\frac{dC}{d\xi}\right\}\frac{d\xi}{d\tilde{\xi}} \tag{6}$$

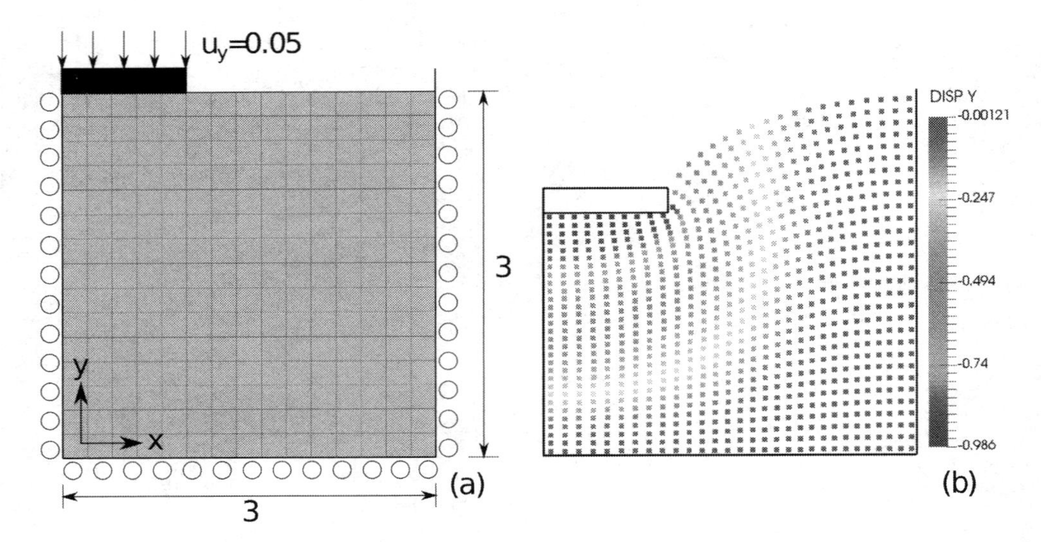

Figure 4. Rigid footing penetrating into an elastic domain: (a) problem definition, (b) deformation prediction using the MPM. (From Cortis et al. (2017)).

where C is the B-spline curve defining a segment in the physical boundary. Applying Gauss quadrature to (5), we obtain

$$\{f^t\} \cong \sum_{i=1}^{n_{gp}} [M_i]^T \{t\}_i \| \{J_B\}_i \| w_i, \tag{7}$$

where n_{gp} is the number of Gauss points used to integrate over the segment within the background grid element, w_i is the weight associated with Gauss point i, $\|(\cdot)\|$ denotes the L2 norm of $(\cdot)$ and in this case the L2 norm of the boundary Jacobian, $\{J_B\}$, which maps the length of the boundary between the parent and physical spaces.

To illustrate the use of this method Fig. 6 shows an elastic cantilever beam of length 10 m and depth of 2 m is subjected to a constant pressure of 1500 Pa applied along the top boundary; a traction which remains perpendicular to the top surface of the cantilever throughout the analysis. A background grid with 1.5 m by 1.5 m elements is used, and the problem domain is discretised using 896 uniformly distributed standard material points. The outer layer of the material points are identified as the problem boundaries which are approximated using B-splines. The initial discretisation and the final deformed cantilever beam are shown in Fig. 7a. The advantage of this approach is that the boundaries can be tracked after each load step without plotting out all the material points (see Fig. 7b) and the deformed shape appears to be successfully captured by the B-spline approximation. Further examples and full details of this approach

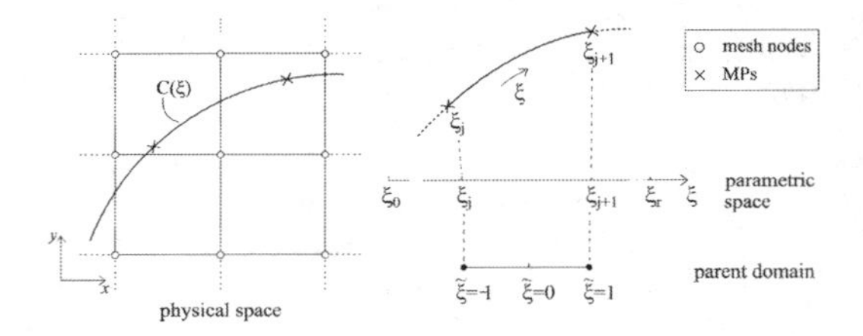

Figure 5. Physical and parametric spaces for numerical integration of boundary conditions.

to accurately track domain boundaries and apply traction boundary conditions in the MPM are given in Bing et al. (2018) and Bing (2017).

3.3 *Volumetric locking*

In the MPM the material points are integration points for the calculation of grid element stiffness, and since they are allowed to convect through the grid, they cannot provide the accuracy of an equivalent number of properly placed Gauss points. Consequently it is normal to use many more material points per grid element than the number required for accurate numerical integration. Combining this with the fact that the grid is usually comprised of low order elements, e.g. bilinear quadrilaterals in 2D, means that the method is susceptible to volumetric locking (resulting in over-stiff behaviour) when modelling near-incompressible materials. This volumetric locking is caused by excessive constraints placed on an element's deformation. That is, the constitutive

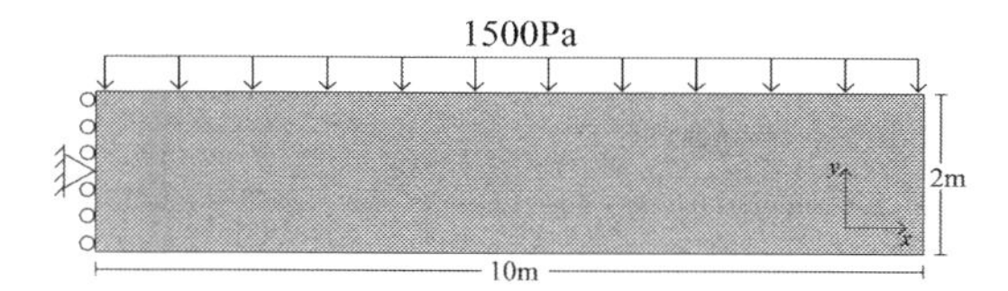

Figure 6. Cantilever beam geometry.

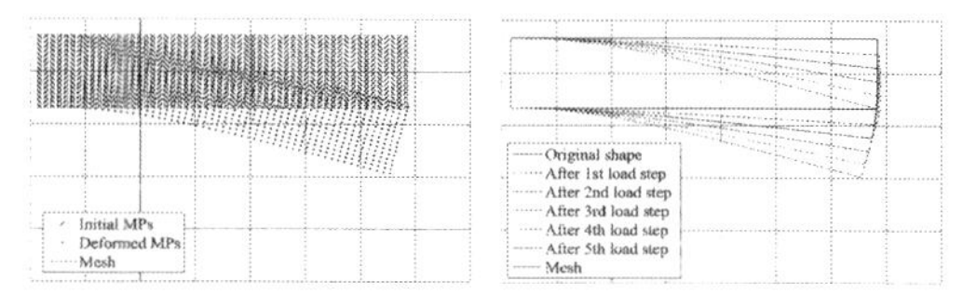

(a) Final deformation of the material points. (b) Boundary visualisation after each load step.

Figure 7. Cantilever beam deformation.

model will require near-isochoric behaviour at the integration (or material) point's location within the element and each of these points places a constraint on the deformation of the element. A common technique to avoid volumetric locking in finite element methods is to use higher order elements with reduced Gaussian integration. However, this is not viable in MPMs since it is not known how many material points will be in any given element at a given load step. In the context of finite deformation solid mechanics, a number of formulations have been proposed to overcome volumetric locking in finite elements. (A review can be found in de Souza Neto et al. (2008)). The issue of volumetric locking in MPMs has received little attention to date with the notable exception of Mast et al. (2012) who investigated the issue of kinematic (volumetric and shear) locking in the standard MPM and developed a complex multi-field variational principle based approach which introduces independent approximations for the volumetric and the deviatoric components of the strain and stress fields.

For the MPM we have instead adopted the $\bar{F}$ approach hitherto applied to the standard FEM by de Souza Neto et al. (1996) for the following reasons: (i) unlike mixed approaches it does not introduce any additional unknowns into the linear system, (ii) it is simple to implement within existing finite element codes (and therefore also the MPMs), (iii) the approach can be used with any constitutive model and (iv) it does not introduce any additional *tuning* parameters into the code. In the $\bar{F}$ approach applied to standard FEM, the volumetric and deviatoric components of the deformation gradient are sampled at different locations. The deformation gradient becomes

$$\bar{F}_{ij} = \left(\frac{det\,(F^0_{ij})}{det\,(F_{ij})} \right)^{1/n_D} F_{ij}, \tag{8}$$

where n_D is the number of physical dimensions and F^0_{ij} is the deformation gradient obtained from the deformation field at the centre of the element. Therefore the volumetric component of the deformation gradient for all of the Gauss points within an element is obtained from a single point, thus relaxing the volumetric constraint on the element when the material behaviour is near incompressible.

For the MPM, where we have large deformations, we adopt the incremental equivalent of (8), giving the $\bar{F}$ deformation gradient increment as

$$\Delta\bar{F}_{ij} = \left(\frac{det(\Delta F^0_{ij})}{det(\Delta F_{ij})} \right)^{1/n_D} \Delta F_{ij}, \tag{9}$$

where ΔF^0_{ij} is the volumetric component of the deformation gradient increment. It is straightforward to modify the standard material point method by replacing ΔF_{ij} with $\Delta\bar{F}_{ij}$ in the finite deformation formulation. This is because the shape functions are directly adopted from the finite element basis. However, it is more appropriate to use the geometric centre of the material points located within a given finite element rather than the centre of the element. This is due to two key reasons:

1. when a single material point is used to integrate the background grid cell the $\bar{F}$ deformation gradient, (9), equals the standard deformation gradient; and
2. when a background grid cell is only partially filled with material points the volumetric behaviour is centred on the physical region.

To demonstrate the performance of the MPM with the $\bar{F}$ approach results are presented for the analysis of a smooth square rigid footing bearing onto a 3D weightless elasto-plastic domain. Due to symmetry only a quarter of the physical problem is modelled and the footing has a half width of 0.5 m and the simulated domain is 5 m in length in each direction. The same material properties were adopted as (de Souza Neto et al. 2008) for their plane strain analysis of a rigid footing. The smooth footing was displaced vertically (z-direction) by 0.002 m over 200 loadsteps and roller boundary conditions were imposed on the sides and the base of the domain. All of the boundary conditions were imposed using the implicit boundary method discussed above. A relatively coarse regular background grid of tri-linear hexahedral elements with $h = 0.2$ m was used to analyse the problem and the physical domain was discretised using 8

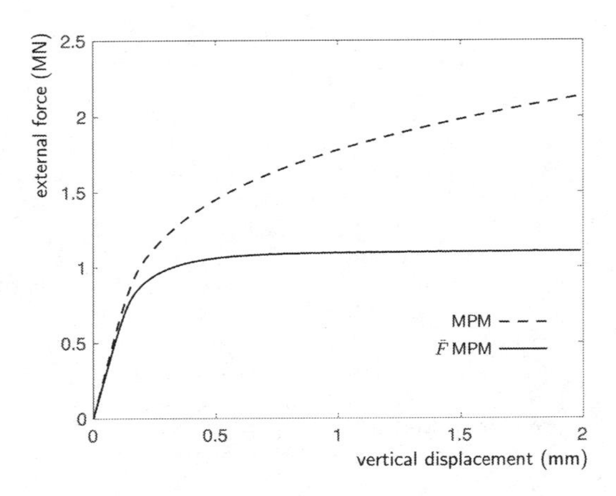

Figure 8. 3D footing: force displacement response for standard and $\bar{F}$ MPMs. (From Coombs et al. (2018)).

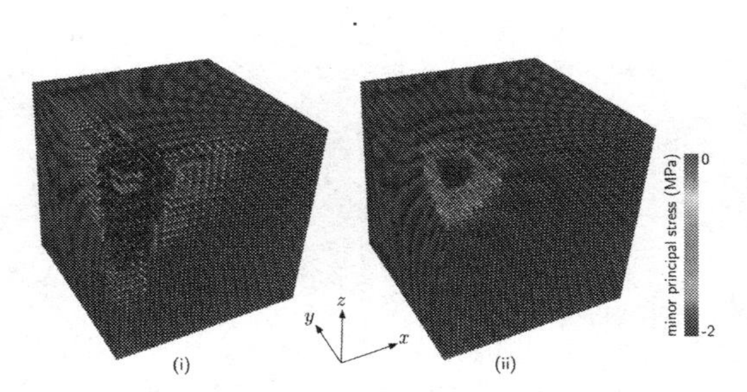

Figure 9. 3D footing: minor principal stress for (i) standard and (ii) $\bar{F}$ MPMs. (From Coombs et al. (2018)).

standard material points per background grid cell (125,000 material points in total). The force versus displacement response for the standard and $\bar{F}$ material point methods are shown in Fig. 8. The standard formulation locks and predicts an overstiff response whereas the $\bar{F}$ formulation reaches a limit load, as expected for this type of analysis. Due to the small imposed displacement, material points do not cross between background grid cells and both formulations give a smooth response. The minor principal (most compressive) stress distribution at the end of the analysis for the two formulations are shown in Fig. 9. The standard material point formulation contains spurious stress oscillations caused by volumetric locking. In particular, the column of material points underneath the footing oscillate between tensile and compressive stress states. The $\bar{F}$ formulation stress distribution shown in Fig. 9 (ii) demonstrates the correct compressive region underneath the footing, as shown by the blue-shaded particles. Full details of this approach to deal with locking in the MPM and other variants such as the GIMP method are given in Coombs et al. (2018).

4 CONCLUSIONS

In this paper we have attempted to summarise a number of issues with the use of the MPM that may afflict geotechnical analyses more than other applications for which it may be used. The literature to date shows that while these challenges have often been evident, many researchers have avoided tackling them head-on, especially in the area of imposition of essential boundary conditions. We have presented four issues but have also presented solutions developed in our research group at Durham University. It is to be hoped that these solutions will provide useful to others attempting geotechnical modelling with the MPM in future.

ACKNOWLEDGEMENTS

Some of the work described in this paper has been undertaken for the UK EPSRC funded project *Seabed ploughing: modelling for infrastructure installation* (EP/M000362/1 & EP/M000397/1). The thirds author was supported by an EPSRC DTA Studentship during PhD study (grant ref EP/K502832/1) which has contributed to parts of this research.

REFERENCES

Anura3D MPM Research Community (2017). *First International Conference on the Material Point Method for Modelling Large Deformation and SoilWaterStructure Interaction*. http://mpm2017.eu/home.

Bardenhagen, S. & E. Kober (2004). The generalized interpolation material point method. *CMES-Computer Modeling in Engineering and Sciences 5*(6), 477–495.

Belytschko, T., Y.Y. Lu, & L. Gu (1994). Element-free Galerkin methods. *International Journal for Numerical Methods in Engineering 37*, 229–256.

Bing, Y. (2017). B-spline based boundary method for the material point method. MScR thesis, Durham University, UK.

Bing, Y., M. Cortis, T. Charlton, W. Coombs, & C. Augarde (2018). B-spline based boundary conditions in the material point method. *Computer Methods in Applied Mechanics and Engineering*. Under review.

Charlton, T.J., W.M. Coombs, & C.E. Augarde (2017). iGIMP: An implicit generalised interpolation material point method for large deformations. *Computers & Structures 190*, 108–125.

Coombs, W., T. Charlton, M. Cortis, & C. Augarde (2018). Overcoming volumetric locking in material point

methods. *Computer Methods in Applied Mechanics and Engineering*. Accepted for publication.
Cortis, M., W.M. Coombs, C.E. Augarde, M.J. Brown, A. Brennan, & S. Robinson (2017). Imposition of essential boundary conditions in the material point method. *International Journal for Numerical Methods in Engineering*.
de Souza Neto, E., D. Perić, M. Dutko, & D.R.J. Owen (1996). Design of simple low order finite elements for large strain analysis of nearly incompressible solids. *International Journal of Soilids and Structures 33*, 3277–3296.
de Souza Neto, E., D. Perić, & D. Owen (2008). *Computational methods for plasticity: Theory and applications*. John Wiley & Sons Ltd.
Fernández-Méndez, S. & A. Huerta (2004). Imposing essential boundary conditions in mesh-free methods. *Computer Methods in Applied Mechanics and Engineering 193*, 1257–1275.
Guilkey, J. & J. Weiss (2003). Implicit time integration for the material point method: Quantitative and algorithmic comparisons with the finite element method. *International Journal for Numerical Methods in Engineering 57*, 1323–1338.
Heaney, C., C. Augarde, A. Deeks, W. Coombs, & R. Crouch (2010). Advances in meshless methods with application to geotechnics. In *Proc. NUMGE Trondheim*, pp. 239–244.
Hughes, T., J. Cottrell, & Y. Bazilevs (2005). Isogeometric analysis: Cad, finite elements, nurbs, exact geometry and mesh refinement. *Computer Methods in Applied Mechanics and Engineering 194*(39), 4135–4195.
Jassim, I., D. Stolle, & P. Vermeer (2013). Two-phase dynamic analysis by material point method. *International Journal for Numerical and Analytical Methods in Geomechanics 37*(15), 2502–2522.
Kim, Y., M. Hossain, D. Wang, & M. Randolph (2015). Numerical investigation of dynamic installation of torpedo anchors in clay. *Ocean Engineering 108*(Supplement C), 820–832.
Kumar, A.V., S. Padmanabhan, & R. Burla (2008). Implicit boundary method for finite element analysis using nonconforming mesh or grid. *International Journal for Numerical Methods in Engineering 74*(9), 1421–1447.
Ma, J., D. Wang, & M. Randolph (2014). A new contact algorithm in the material point method for geotechnical simulations. *International Journal for Numerical and Analytical Methods in Geomechanics 38*(11), 1197–1210.
Mast, C.M., P. Mackenzie-Helnwein, P. Arduino, G.R. Miller, & W. Shin (2012). Mitigating kinematic locking in the material point method. *Journal of Computational Physics 231*(16), 5351–5373.
Muller, A.L. & E.A. Vargas (2014). The material point method for analysis of closure mechanisms in openings and impact in saturated porous media. In *48th U.S. Rock Mechanics/Geomechanics Symposium, 1–4 June, Minneapolis, Minnesota*. American Rock Mechanics Association.
Nguyen, V.P., C.T. Nguyen, T. Rabczuk, & S. Natarajan (2017). On a family of convected particle domain interpolations in the material point method. *Finite Elements in Analysis and Design 126*(Supplement C), 50–64.
Piegl, L. & W. Tiller (1997). *The NURBS Book (2 Nd Ed.)*. New York, NY, USA: Springer-Verlag New York, Inc.
Potts, D. & L. Zdravković (1997). *Finite element analysis in geotechnical engineering: Theory*. Thomas Telford.
Ramos, A., A. Aragn, S. Soghrati, P. Geubelle, & J.-F. Molinari (2015). A new formulation for imposing Dirichlet boundary conditions on non-matching meshes. *International Journal for Numerical Methods in Engineering 103*(6), 430–444.
Sadeghirad, A., R.M. Brannon, & J. Burghardt (2011). A convected particle domain interpolation technique to extend applicability of the material point method for problems involving massive deformations. *International Journal for Numerical Methods in Engineering 86*(12), 1435–1456.
Schillinger, D., M. Ruess, N. Zander, Y. Bazilevs, A. Duster, & E. Rank (2012). Small and large deformation analysis with the p- and b-spline versions of the finite cell method. *Computational Mechanics 50*(4), 445–478.
Sołowski, W.T. & S.W. Sloan (2015). Evaluation of material point method for use in geotechnics. *International Journal for Numerical and Analytical Methods in Geomechanics 39*(7), 685–701.
Sulsky, D., Z. Chen, & H. Schreyer (1994). A particle method for history-dependent materials. *Computer Methods in Applied Mechanics and Engineering 118*, 179–196.
Wang, B., M.A. Hicks, & P.J. Vardon (2016). Slope failure analysis using the random material point method. *Géotechnique Letters 6*(2), 113–118.
Zabala, F. & E. Alonso (2011). Progressive failure of Aznalcóllar dam using the material point method. *Géotechnique 61*(9), 795–808.

Numerical Methods in Geotechnical Engineering IX – Cardoso et al. (Eds)
© 2018 Taylor & Francis Group, London, ISBN 978-1-138-33198-3

LDFEM analysis of FDP auger installation in cohesive soil

J. Konkol & L. Bałachowski
Faculty of Civil and Environmental Engineering, Gdańsk University of Technology, Gdańsk, Poland

J. Linowiecki
Menard Polska Sp. z o.o., Warszawa, Poland

ABSTRACT: This paper deals with Large Deformation Finite Element (LDFE) preliminary modelling of Full Displacement Pile (FDP) installation in cohesive soil deposit located in Jazowa, Poland. The detailed FDP auger geometry is applied and the drilling process is modelled with full 3D Coupled Eulerian-Lagrangian (CEL) formulation. The total stress approach and elastic-perfectly plastic model with rate-dependent Mises plasticity is used. The interaction between auger and surrounding soil is controlled by adhesive contact algorithm. The radial total stress distributions along FDP auger in selected distances from the pile axis are investigated. The reaction force and rotational moment which act on the auger are determined and the results of numerical analysis are compared with data registered by drilling rig. The aim of this paper is to clarify the installation effects which take place during FDP pile installation in soft cohesive soil.

1 INTRODUCTION

Full Displacement Piles (FDPs) are widespread and commonly used technology characterized by relatively high pile capacities. The idea of FDPs construction is to move soil outward the pile with minimum amount of material unearthed. The FDP installation consists of few steps. Firstly, the drilling tool is inserted into the subsoil by pushing and rotation which introduces large deformations and disturbance of the soil structure. Then, after reaching the installation depth, the concrete is pumped with simultaneous extraction of the auger. Finally, the reinforcement cage is pushed into the concrete pile. The above mentioned steps result in many crucial elements during pile construction that influence the pile bearing capacity and long-term performance.

Numerical modelling of complex FDP technology introduces significant problems. In recent years, the jacked piles, the easiest to be analyzed, have been successfully implemented in numerical models involving large deformation problems (e.g., Mabsout & Tassoulas, 1994; Van Den Berg, 1994; Mahutka et al., 2006; Dijkstra et al., 2011; Henke & Grabe, 2009). The numerical modelling of jacked piles has shown the importance of some factors which directly influence the stress state around the pile. For instance, Mahutka et al. (2006), have shown the zones of soil compaction and dilation as well as radial stress change around the jacked piles. It has been also shown that the radial stress changes around the pile influence directly the pile shaft capacity (e.g., Sheng et al., 2005; Mahutka et al., 2006; Konkol & Bałachowski, 2017). The numerical modelling of drilling is much more difficult due to complex auger geometry and loading conditions. Moreover, the extraction of the auger and concreting phase is also problematic to solve. Until now the common practice was to simplify the drilling auger to the jacked pile as the similar effects in terms of radial stress changes were assumed. This simplification has been used previously by authors (Konkol & Bałachowski, 2017) as well as others (e.g., Larisch, 2014). However, advances in computational method allows for more detailed and advanced analysis as it has been shown by Pucker and Grabe (2012) who have modelled full displacement auger drilling.

The possibility of numerical capture of drilling tool behavior may be very attractive in pile industry. Currently, the new augers are designed via experience and technological knowledge without detailed insight considering the changes they can produce in the soil. The numerical verification of augers performance and, consequently, the limitation of possible design in preliminary field testing can be economically and qualitatively desired. Numerical analysis of auger drilling may help to optimize the pitch, tilt and size of auger's flights. Further, the numerical studies can be useful in auger designation due to specific soil conditions.

This paper presents the preliminary numerical modelling of FDP auger drilling in cohesive soil deposit in Jazowa, Poland. The Coupled Eulerian-Lagrangian formulation is used to deal with large deformation problem and the results of numerical studies are compared with in situ measurements in order to validate numerical model. All appeared discrepancies are commented in details. The range of influence due to auger drilling is determined and the stress state changes in the vicinity of the auger are briefly described. The aim of this paper is to recognize of the significant factors which influence the drilling tool resistance such as mobilized undrained shear strength of the soil and the possible failure mechanism. Finally, the conclusions concerning the conducted numerical tests are drawn and a wide range of possible further improvements is presented.

2 JAZOWA TESTING SITE

Jazowa is located in Vistula Marshlands, 50 km southeast of Gdansk, Poland and lies within the S7 highway, currently under construction. The testing site is supervised by consortium formed by Gdańsk University of Technology and Menard Polska under the cooperation within the National Centre of Research and Development grant. At the four research plots totally 69 FDPs, 40 cm in diameter with different embedded lengths have been drilled in accordance to the Controlled Modulus Column (CMC) technology. The CMC's use specifically designed auger connected with the displacement component which prevent the upward movement of soil. The field investigation include CPTu, DMT and FV tests supplemented by a wide scope of laboratory testing. For the purpose of this paper the first pile drilled at the research plot 2 has been assumed as the reference one to get rid off the influence of neighbouring pile drillings. However, it should be noticed that the column spacing was 2 m and all piles drilled at the Jazowa site have similar drilling logs and this issue will be also highlighted in the present study.

The Jazowa site soil profile with nearby CPTu soundings and drilling log of reference pile is shown in Figure 1. The column drilling will be modelled in the reference depth range chosen from 7 to 14 meters depth below ground level. This will provide a uniform and homogenous cohesive soil layer for the purpose of numerical model. However, the drilling log shows that the sufficient comparison between numerical studies and field registered data can be made beginning from 10 m depth. The drilling data for 7 to 10 m depth range are strongly influenced by sand interlayer, see Figure 1e.

The numerical analysis is assumed to be conducted in accordance to total stress approach. The undrained shear strength profiles assessed from

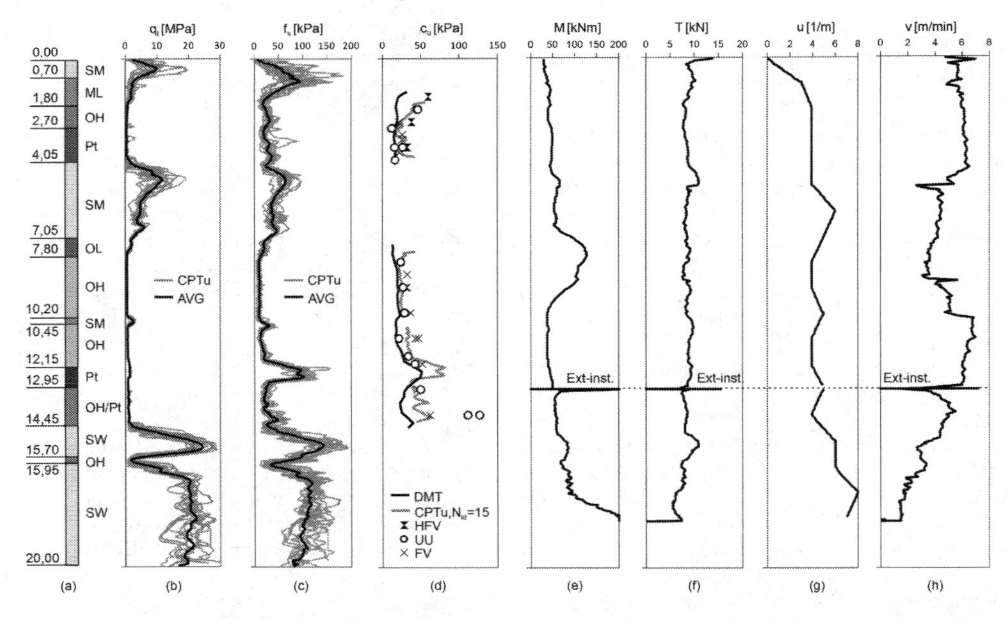

Figure 1. Jazowa soil profile (a); CPTu soundings results (b)-(c); Undrained shear strength profile (d); Drilling log for reference CMC pile drilling (e)-(h); Note: q_t = corrected cone resistance; f_s = sleeve friction; c_u = undrained shear strength; M = rotational moment; T = pushing force; u = rotation per 1 m; v = penetration velocity; Ext-inst = extension mounting.

cone penetration tests (CPTu), dilatometer tests (DMT), field electrical vane (FV) tests (corrected values), hand field vane tests (HFV) and unconsolidated undrained (UU) triaxial compression tests are presented in Figure 1d. Generally, a moderate scatter of data was achieved, which hampered the accurate determination of c_u. Laboratory UU triaxial tests on reconstituted Jazowa clayey mud have also shown the rate dependency of the material and this observation has been confirmed with rate CPTu tests at the site. Finally, the last laboratory testing of Jazowa clayey mud has been concerned on interface behaviour between soil and smooth steel. As a result, the interface tests on Jazowa clayey mud—smooth steel suggest the adhesion factor α of 0.36 (for the shear rate of 0.033%/s).

3 NUMERICAL MODELLING OF FULL DISPLACEMENT AUGER DRILLING

3.1 *General consideration*

The numerical calculation of full displacement auger drilling is conducted with Coupled Eulerian Lagrangian (CEL) formulation (e.g., Noh, 1963; Benson, 1992) and total stress analysis using Abaqus software suite. The concept of CEL in drilling process is to use Eulerian domain to describe the soil and Lagrangian domain to define the pushed and rotating auger. Eulerian domain allows for arbitrary large deformation and the interaction with auger is modelled with the general contact algorithm (Dassault Systèmes, 2014) where penalty pressures are applied within the Eulerian mesh to push soil outward the auger. On the other hand, the penalty forces are applied on the auger described with Lagrangian manner to simulate the resistance of the soil. The auger is drilled with prescribed boundary conditions in terms of penetration and rotational velocities which are assumed in respect to the averaged drilling log data for 7–15 m depth and they are equal to 0.074 m/s and 2.083 rad/s, respectively.

The basic problem in numerical modelling of full displacement auger drilling is the complex geometry of the drilling tool. The step-by-step simplification of the auger is presented in Figure 2. The auger's bit have been replaced with conical one and the drilling tool surfaces have been smoothed, see Figure 2b. The simplification is mainly concentrated on displacement component of CMC auger. However, the auger's flights have been also simplified and smoothed. The drilling tool have been modelled as a discrete rigid body meshed with 6046 elements, as can be seen in Figure 2c. The soil domain is cylindrical, 13 meters height and 6 meters wide, Eulerian body, discretized with 35100 elements. The assembled numerical model is presented in Figure 3. The boundary conditions are applied to support the in-situ conditions in agreement with total stress analysis. Three parts of the soil domain can be specified as it is marked in Figure 3. The upper 3 meters of soil domain, corresponding to 7–10 meters depth in the field is the transition zone to allow the auger to be fully drilled in the soil. The target modelling depth to be analysed is designated as 10 to 15 meters depth. The last part of soil domain is the boundary zone to exclude the boundary influence. Authors initially tried to differentiate the soil layers in terms

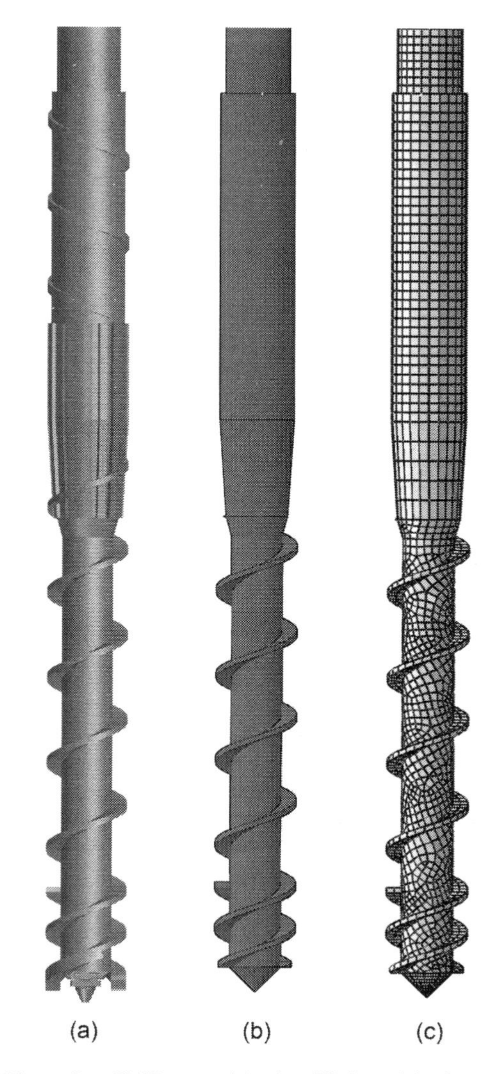

Figure 2. CMC auger (a), simplified model of auger (b), meshed auger part (c).

of constitutive modelling to provide the c_u varying with depth. However, many contact errors have been encountered during computational process when auger was crossing first two soil layers. Consequently, one constitutive model have been applied to the whole soil domain with constant reference undrained shear strength. As only the target drilling zone is important in numerical investigation the authors consider that presented simplification is reasonable due to calculation success and generally constant value of undrained shear strength of soil that can be assumed between 8 and 12 meters depth, see Figure 1d. Thus, the final drilling depth has been chosen as 12 m and the numerical modelling presented in this paper will be treated as a preliminary study.

3.2 *Rate dependency of Jazowa clayey mud*

As the Jazowa clayey mud is rate dependent an elastic-perfectly plastic model with rate dependent Mises plasticity was used to describe the constitutive behaviour of soil. The undrained shear strength is assumed after Kulhawy and Mayne (1990) proposition:

$$c_u = c_{u0} \times \left(1 + 0.1 \times \log(\dot{\varepsilon})\right) \quad (1)$$

where c_u = actual undrained shear strength; c_{u0} = reference undrained shear strength corresponding to strain rate of 1%/h; $\dot{\varepsilon}$ = strain rate.

For the purpose of preliminary modelling the average value of reference undrained shear strength equal to 25.0 kPa was applied due to modelling assumptions explained in section 3.1. The shear modulus is assumed as 2000 kPa and this value is adequate for moderate and large strain shearing scenario according to unconsolidated undrained triaxial compression tests conducted in laboratory. The rate dependent Mises plasticity with isotropic liner elasticity has been implemented in Abaqus software as a user subroutine VUMAT and the appropriate parameters are summarized in Table 1. The contact between auger and soil is modelled with Coulomb model with friction coefficient equal to 0.1 and limit shear stress equal to 9.0 kPa. The limit shear stress is estimated from sleeve friction of CPTu probing with standard penetration rate for the depth range of 8–12 m. Laboratory direct shear interface testing also suggests the values close to 10.0 kPa as the limit shear stress. Further, the rate CPTu probings at Jazowa testing site show lack of rate influence on sleeve friction which supports the auger drilling condition at the interface presented above. Consequently, the applied friction model with limit shear stress at the interface equal to 9.0 kPa corresponds to adhesive contact algorithm without need for explicit adhesion coefficient specification.

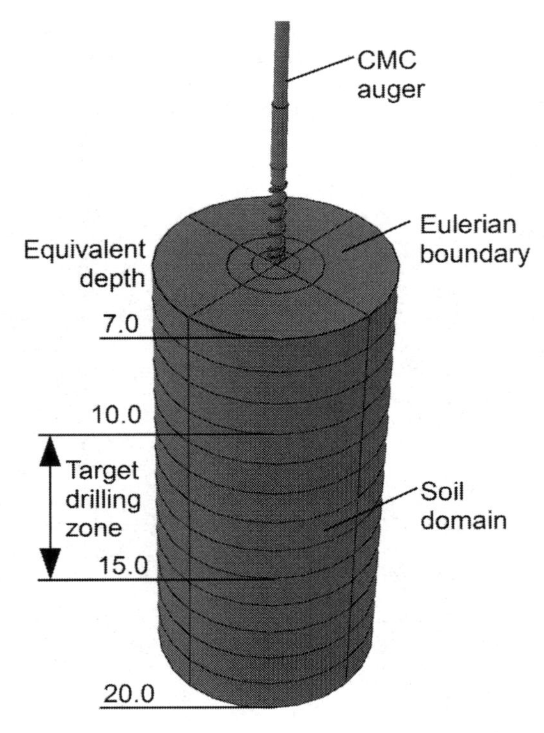

Figure 3. Numerical model of CMC auger drilling problem.

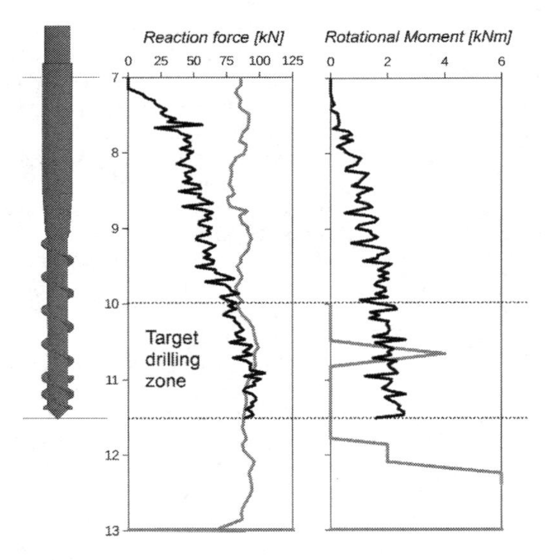

Figure 4. Reaction force and rotational moment of CMC auger—numerical results versus field registered data.

Table 1. Jazowa clayey mud parameters used in user subroutine VUMAT.

Parameter	Symbol	Value	Unit
Soil saturated density	ρ_{sr}	1.66	g/cm^3
Shear modulus	G	2000	kPa
Undrained Poisson's ratio	υ_u	0.49	-
Reference undrained shear strength	c_u	25	kPa
Reference strain rate	$\dot{\varepsilon}$	1.0	%/h

4 ANALYSIS OF NUMERICAL RESULTS

The auger drilling has been successfully modelled up to the depth of 11.5 m where the calculations were terminated due to contact modelling error. The mobilized undrained shear strength around the auger due to rate-dependency was ranging between 30 and 35 kPa which is corresponding to 1.2 and 1.4 of the reference undrained shear strength obtained at the shearing rate of 1%/h, respectively. The values of mobilized c_u are generally consistent with FV tests results performed at the site. However, the precise detection of zones of c_u increase was impossible due to high variation of strains rates induced by boundary conditions and drilling process. The other results are discussed in details below.

4.1 *Field measurements versus numerical ones*

The numerical results of axial reaction force and rotational moment mobilized during drilling are compared with machine registered data in Figure 4. To make sufficient comparison the total axial force from the drilling log is used as it depends mainly on the auger bit behaviour. However, the rotational moments may be influenced by the upper soil layers. In this case, the increment of rotational moment, beginning from 10 meter is considered for comparison. The comparison range is chosen from 10 to 11.5 m depth due to assumptions presented above in this paper.

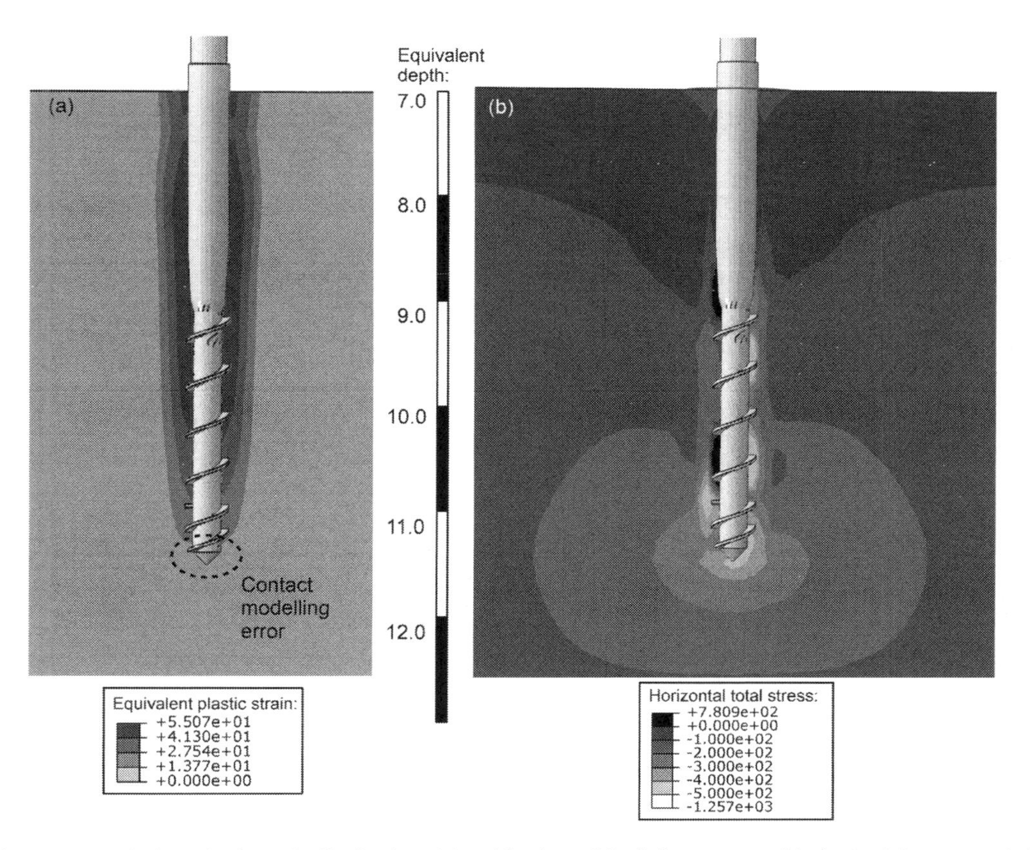

Figure 5. Equivalent plastic strain distributions (a) and horizontal (radial) stress map (b) obtained from numerical study.

As one can see, a very good fit has been obtained in terms of reaction pushing force. The stabilization of the axial pushing force is obtained after passing 10.5 m. Thus, the calculated force can be treated as a reasonable estimate. However, the analysis of the rotational moment output was rather confusing. The total rotational moment can be divided into contact and frictional part. It was found that contact component of rotational moment provides highly overestimated results which is probably induced by auger surface discretization and general contact algorithm performance. Similar overestimation of rotational moment was also obtained for simple, cylindrical pile which was pushed and rotated. Consequently, the rotational moment distribution due to contact pressure is assumed as highly non-realistic. Thus, as a representative distribution of rotational moment only the friction-induced part the rotational is considered in Figure 4. The rotational moment due to friction reaches constant and very low value equal to approx. 2.5 kNm. Similar, very low values of rotational moment increase has been registered in the field, see Figure 4. Generally, the numerical model reflects the registered drilling data in terms of pushing force and rotational moment and it encounters for further development.

4.2 *Zone affected by auger drilling*

The equivalent plastic strains developed around the auger at the end of penetration are presented in Figure 5a. As one can see, the influence of pile driving in clayey muds is limited to a zone approximately 1.5D wide on each side. This observation is in agreement with drillings data, where almost the same response of the drilling rig have been achieved for all piles at the site. Further, the CPTu and DMT testing one week after CMC piles construction in the distance of 2D from the pile axis reveals no significant changes in soft soil. Consequently, the numerically obtained affected zone seems to be reasonable for Jazowa site. It is worth to notice that similar narrow-ranged and equal to ~2D influence of CMC drilling has been directly measured in the field by Suleiman et al. (2015) in soft sandy silt (c_u = 6 kPa). Finally, Figure 5a shows also the contact error which induces the calculation termination at the depth of 11.5 m. As one can see, no plastic strains are developed under the auger tip. Thus, the contact modelling needs to be improved in the further models of CMC auger drilling.

The horizontal total stresses distribution around the auger at the end of drilling is presented in Figures 5b and 6. The horizontal stress map,

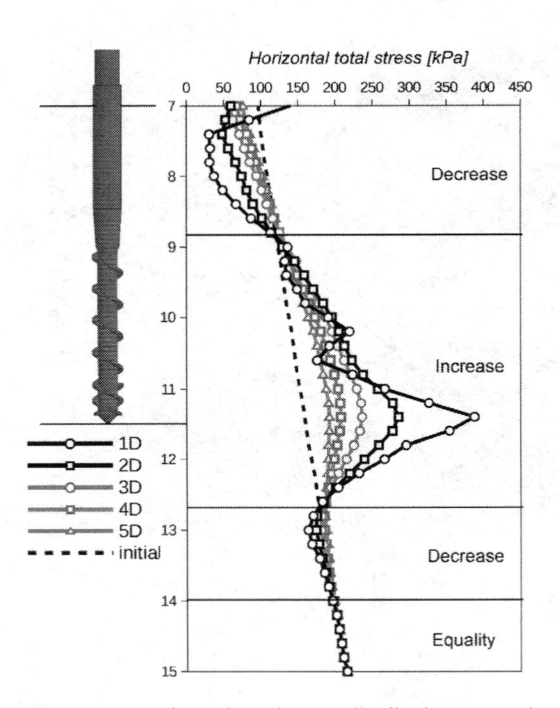

Figure 6. Horizontal total stress distributions around the auger.

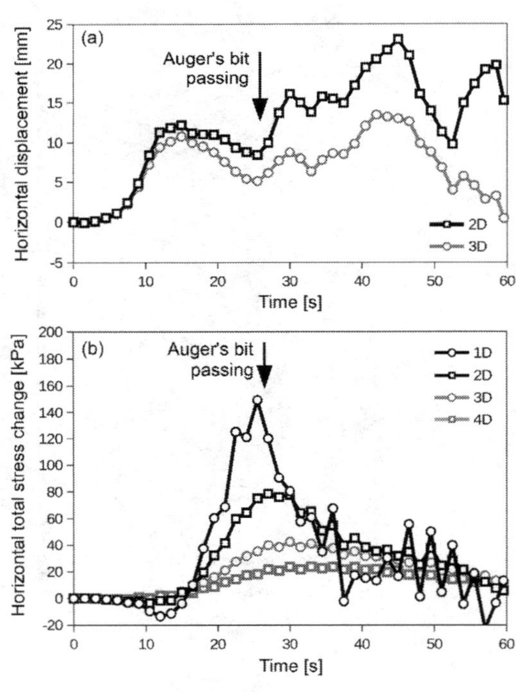

Figure 7. Horizontal soil displacement (a) and horizontal total stress change (b) for the depth of 90 m.

presented in Figure 5b, shows a significant stress change around the augers flights as well as zones of compression and extension in the surrounding soil. The highest horizontal stress changes are observed in the vicinity of the drilling tool which corresponds to obtained equivalent plastic strain distribution. The horizontal total stresses distributions around the auger at different distances from the pile shaft are presented in Figure 6. The results are consistent with numerical modelling of FDP auger drilling presented by Pucker & Grabe (2012). Three areas of soil response can be recognized, i.e., the zone of increased horizontal total stresses between approx. 9 and 13 m, the zone of decreased stresses between 7 m and approx. 9 m as well as between approx. 13 and 14 m and stress equality zone below 14 m depth. This phenomenon can be summarized as a decrease-increase-decrease path in accordance to horizontal total stress. Around the auger's flights and in some distance below the auger bit, the significant stress increase can be observed, see also Figure 5b. The impact of this phenomenon becomes more smooth and decreases outward the auger. On the other hand, in the distance of 3D below the auger as well as just above the flights a decrease in radial total stresses can be noticed. Although the total stress approach allows for valuable insight into the process of CMC auger drilling, for more detailed analysis the effective stress modelling of CMC auger is required.

The changes in lateral displacement and horizontal total stress at the depth equal to 9.0 m are shown in Figure 7. The lateral displacements of the soil body are shown for the distances of 2D and 3D from the pile axis due to acceptable modelling conditions of large and moderate strain. As on can see in Figure 7a, significant lateral soil movement has been achieved with characteristic drop when auger's bit is passing 9.0 m depth. Similar characteristic drop has been observed in the field measurement by Suleiman et al. (2015). The change in horizontal total stress is shown in Figure 7b with the highest value obtained when the auger's bit is passing the measurement point. In terms of horizontal total stresses the results are also quantitatively consistent with Suleiman et al. (2015) research.

4.3 *Viscous effects influence*

The influence of rate dependency of Jazowa clayey mud is shown on the basis of horizontal stress change. The numerical model of pile drilling with rate independent Mises material with $c_{u0} = 25$ kPa is considered for comparison purposes. The comparison will be carried out for the depth of 11.5 m which corresponds to auger's bit position, see Figure 6. The results are summarized in the

Table 2. Example of viscous effects influence: horizontal total stress at the depth of 11.5 m.

Distance from the pile axis	Rate independent model	Rate dependent model	Ratio
1D	310.16	387.86	1.25
2D	266.95	285.17	1.07
3D	234.71	236.49	1.01
4D	211.91	207.62	0.98
5D	197.53	192.27	1.01

Table 2. As one can see, the influence of viscous effects is limited to the distance of approx. 2–2.5D and it results in significant increase in horizontal total stresses equal to ratio of 1.25 in the distance 1D from the pile axis.

5 CONCLUSIONS

The numerical study presented in this paper confirms that LDFE methods open a new possibilities in geotechnical investigations, although many improvements have to be done in the future. In this paper authors conducted successful auger drilling using CEL formulation through the 4.5 meters depth of soft soil deposit. The soil parameters used in the calculation were determined in comprehensive field and laboratory investigation including rate effects. The soil-steel contact characteristics were determined based on sleeve friction from CPTU tests and interface direct shear tests. The adhesive contact algorithm and rate dependent constitutive soil model was used.

The numerical model returns satisfactory distribution of pushing force and rotational moment distributions. It was also found that the zone influenced by pile drilling in soft soil is approximately 1.5D wide, which is supported by pile drilling logs and CPTu and DMT probings performed one week after piles construction as well as other research on CMC drillings (Suleiman et al., 2015). The investigation of horizontal total stresses has shown the zones of decreased and increased values due to drilling. The horizontal total stress change and lateral soil movement is also quantitatively consistent with Suleiman et al. (2015) research. However, more detailed analysis in accordance to the effective stress approach is required for more complex comparisons.

The presented numerical model has failed when layered soil have been used. Numerical model also shows sensitivity to the complex auger surface which led to calculation termination during general contact algorithm execution. These issues have to be improved in the future modelling. The second branch in future improvement consists of application of effective stress approach and more appropriate constitutive modelling of soil. However, despite many imperfections mentioned above the LDFE modelling in terms of CEL formulation is valuable investigation tool in geotechnical engineering, which allowed for qualitative capture of soil behavior around the FDP auger.

ACKNOWLEDGEMENTS

The research is supported by the National Centre for Research and Development grant PBS3/B2/18/2015. The calculations were carried out at the Academic Computer Centre in Gdańsk (CI TASK).

REFERENCES

Benson, D.J. 1992. Computational methods in Lagrangian and Eulerian hydrocodes. *Computer methods in Applied mechanics and Engineering*, 99(2): 235–394.

Dassault Systèmes. 2014. Abaqus 6.14 Documentation.

Dijkstra, J., Broere, W. & Heeres, O.M. 2011. Numerical simulation of pile installation. *Computers and Geotechnics*, 38(5): 612–622.

Henke, S. & Grabe, J. 2009. Numerical modeling of pile installation. In M. Hamza et al. (eds.), *The Academia and Practice of Geotechnical Engineering: 1321–1324, Proceedings of the 17th International Conference on Soil Mechanics and Geotechnical Engineering, Alexandria, Egypt, 5–9 October 2009.* Amsterdam, The Netherlands: IOS Press.

Konkol, J. & Bałachowski, L. 2017. Influence of installation effects on pile bearing capacity in cohesive soils—large deformation analysis via finite element method. *Studia Geotechnica et Mechanica*, 39(1): 15–26.

Kulhawy, F.H. & Mayne, P.W. 1990. *Manual on estimating soil properties for foundation design*. Palo Alto, California, USA: Electric Power Research Institute.

Larisch, M. 2014. *Behaviour of stiff, fine-grained soil during the installation of screw auger displacement piles*. PhD Thesis. Queensland: University of Queensland.

Mabsout, M.E. & Tassoulas, J.L. 1994. A finite element model for the simulation of pile driving. *International Journal for Numerical Methods in Engineering*, 37(2): 257–278.

Mahutka, K., König, F. & Grabe, J. 2006. Numerical modelling of pile jacking, driving and vibratory driving. In T. Triantafyllidis (eds.), *Numerical Simulation of Construction Processes in Geotechnical Engineering for Urban Environment, Proceedings of International Conference on Numerical Simulation of Construction Processes in Geotechnical Engineering for Urban Environment, Bochum, Germany, 23–24* March 2006. Boca Raton, USA: CRC Press.

Noh, W.F. 1963. *CEL: a time-dependent, two-space dimensional, coupled Eulerian-Lagrangian code*. Lawrence Radiation Lab. Livermore: University of California.

Pucker, T. & Grabe, J. 2012. Numerical simulation of the installation process of full displacement piles. *Computers and Geotechnics*, 45: 93–106.

Sheng, D., Eigenbrod, K.D. & Wriggers, P. 2005. Finite element analysis of pile installation using large-slip frictional contact. *Computers and Geotechnics*, 32(1): 17–26.

Suleiman, M.T., Ni, L., Davis, C., Lin, H. & Xiao S. 2015. Installation effects of controlled modulus column ground improvement piles on surrounding soil. *Journal of Geotechnical and Geoenvironmental Engineering*, 142(1): 04015059

Van Den Berg, P. 1994. *Analysis of soil penetration*. PhD Thesis. Delft: Delft University of Technology.

Numerical Methods in Geotechnical Engineering IX – Cardoso et al. (Eds)
 ISBN 978-1-138-33198-3

Large deformation finite element analyses for the assessment of CPT behaviour at shallow depths in NC and OC sands

H.K. Engin, H.D.V. Khoa, H.P. Jostad & D.A. Kort
Norwegian Geotechnical Institute, Oslo, Norway

R. Bøgelund Pedersen & L. Krogh
Ørsted Wind Power, Copenhagen, Denmark

ABSTRACT: The assessment of shallow CPT results is of particular importance for shallow foundations for offshore structures such as suction caissons, Gravity Based Structures (GBS), etc. Estimation of the relative density is crucial for the assessment of strength and stiffness parameters, not the least, reconstitution of sand for laboratory testing at in-situ conditions. This study aims to provide insight into shallow CPT behaviour, i.e. clarifying the development of failure mechanisms by utilizing Finite Element (FE) analysis adopting idealized soil profiles to understand the observed CPT behaviour at shallow depths. Hence, the work focuses on reproducing the stress and density dependent CPT resistance and by that correlating unique features of measurements to particular soil states. The Arbitrary Lagrangian-Eulerian (ALE) and Coupled Eulerian-Lagrangian (CEL) methods are employed to better capture the large deformation mechanism of the soil. The sand behaviour (i.e. state dependent strength, stiffness, and deformation behaviour) is modelled using the hypoplasticity.

1 INTRODUCTION

1.1 *Motivation*

For marine site investigations for offshore structures, CPT is often the preferred in situ testing method to obtain shear strength parameters to be used for geotechnical design. Interpretation of offshore CPT data in sand at shallow depth is often done by using interpretations methods derived from calibration chamber tests. These methods are empirical methods taking primarily the stress state of the sand into account to derive the relative density, D_r. The established methods cover sands that are normally consolidated and structure-free, which is not necessarily typical for an offshore site. The empirical methods, e.g. Jamiolkowski et al. (2003), Baldi et al. (1986), normally assume that the results are measured in full flow conditions meaning at a depth where the influence of the surface is not felt by the tool (i.e. $\sigma_v' > 50kPa$. Emerson et al. (2008) developed an empirical method that accounted for the effect of the soil surface. However, this latter method needs further evaluation.

NGI, GEO and Ørsted Wind Power formed a Joint Industry Project (JIP) in order to develop a method for interpretation of CPTs focusing on sands with *overconsolidated* (OC) behaviour at shallow depth, considering the typical North Sea conditions. This JIP focused on three legs: an extended field testing and sampling program at an onshore site, a comprehensive laboratory testing program and advanced finite element analyses to study the soil behaviour and different failure mechanisms involved in the cone penetration testing. This paper focuses on the numerical part of the work.

1.2 *Computational methods used for analysing CPT*

In the last few decades, numerous computational methods including mesh-based methods (finite element, boundary element, material point, etc.) and mesh-free methods (discrete element, smoothed-particle hydrodynamics, etc.) have been extensively developed and successfully applied in simulating cone penetration testing. This section briefly reviews some relevant numerical methods proposed in literature.

Van den Berg (1994) used the ALE FE method to analyse the cone penetration test in both clays and sands. Although chamber tests have been performed and documented by Van den Berg (1994), no direct comparison of numerical results with measured values of cone penetration outputs was done. Therefore, no correlation of cone factors and soil properties has been given for analysis or comparison.

Walker (2007) applied the Abaqus ALE method to simulate the deep penetration of a cone penetrometer in both clay and sand materials. The calculated results agree well with others available in the relevant literatures as well as with the cavity expansion solutions. The good agreement supports the efficiency of the ALE method in predicting the CPT results

Yu and Mitchell (1998) presented a comprehensive review of different methods used for analysing cone resistance. Yu et al. (2000) published a novel FE procedure for the analysis of steady state cone penetrating into granular materials. The comparison with the strain path method (Baligh, 1985; Teh & Houlsby, 1991) shows that the steady state FE method is more accurate, robust, and efficient. However, the method neglects transient deformation of the soil around the cone.

Huang and Ma (1994) used the discrete element method (DEM) to simulate cone penetration focused. Their focus was on the effect of soil-penetrometer interface friction. Tannant and Wang (2002) applied the DEM to model penetration of a wedge into oil sands. Iqbal (2004) used a 2D discrete element model to simulate a cone penetrometer in a coarse grained soil with rigid single sized particles. His calculated results agrees reasonably well with the experimental and other numerical methods. Although the DEM can be very helpful in understanding the micro level mechanisms, it is not yet at a practical level due to extensive computational effort required, and lacking adequate validation against experimental data compared to continuum based numerical methods (e.g. FEM).

The aforementioned approaches have had limited success in practice in modelling the cone penetration test because of either due to the lack of robust constitutive models, geometric nonlinearity and relevant validation against experimental testing or the significant computation effort required.

The objective of this study is to use both the Arbitrary Lagrangian-Eulerian (ALE) and the Coupled Eulerian-Lagrangian (CEL) methods, which are available in the FE program Abaqus, in combination with the advanced hypoplasticity constitutive model proposed by Masin (2011) to simulate the entire penetration process of a CPT cone in sand. The FE-model and the material parameters have been calibrated against laboratory tests, in-situ tests and general soil mechanics correlations.

2 LARGE DEFORMATION FINITE ELEMENT ANALYSES OF CPT AT SHALLOW DEPTH

This section briefly describes the CEL and ALE FE-models as well as the hypoplasticity constitutive model used to evaluate the deformation behaviour of the soil and the cone resistance during cone penetration in sand. Comparison between the CEL and ALE FE analyses of the cone penetration is presented. A series of different FE analyses to validate the FE-models as well as to examine the expected effects of relative density D_r, coefficient of lateral earth pressure K_0 and overburden stress OCR on the CPT resistance are also summarised in this part.

2.1 *FE-model*

A standard cone geometry with diameter D_{cone} = 0.036 m and tapered tip (60 degrees) penetrating in to different granular soils is modelled using both the ALE and CEL methods available in Abaqus/Explicit. An axisymmetric FE-model presented in Figure 1a is used in analyses where the ALE method is applied while a three-dimensional FE-model shown in Figure 1b is used in analyses with the CEL. This is because the Eulerian method in Abaqus only supports the hexahedral elements. In both FE-models, the displacements at the bottom are fully fixed and they are constrained in lateral direction at the side boundaries.

For the CEL FE-model, owing to axisymmetry, only a 30-degree slice of the full 3D problem is modelled in order to reduce computational time without causing excessive skewing of the elements close to the axis of symmetry. Thus, the symmetric boundary conditions are in addition imposed on the two planes of symmetry by constraining the two in-plane rotations.

At the symmetry-axis roller conditions are used in the CEL FE-model, while a zipper-technique (Cudmani & Sturm, 2006) is employed in the ALE FE-model. This technique is necessary to improve contact between cone and surrounding soil during large penetration.

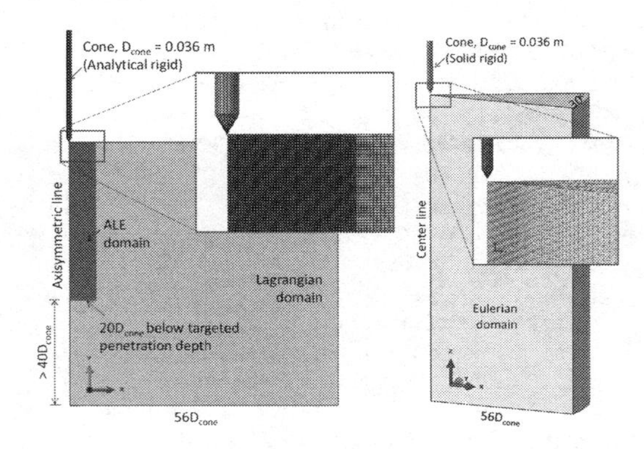

Figure 1. Overview of a) ALE, b) CEL FE-model of cone penetrometer.

Table 1. Hypoplasticity with intergranular strain model input parameters (after Masin, 2011 and Niemunis, 2003).

Parameter	Unit	Description
ϕ_c	°	Critical state friction angle
h_s & n	kPa	Granular hardness and exponent controlling the shape of limiting void ratio curves (normal compression and critical state lines)
e_{i0}, e_{c0}, e_{d0}	-	Reference void ratios specifying limiting void ratio curves
α	-	Controls the dependency of peak friction angle on relative density
β	-	Controls the dependency of soil stiffness on relative density
e_0	-	Initial void ratio at zero stress level
m_R & m_T	-	Multipliers for monotonic stiffness, which correspond to 90° and 180°reversal of the strain path.
R	-	Small strain range
χ & β_R	-	Control the evolution of intergranular strain

In the ALE FE-models the cone is modelled by analytical rigid elements. The soil domain is discretised with Lagrangian elements except for the soil within the targeted depth and a radius of 0.1 m from the axisymmetric line where the ALE elements are applied in order to avoid any mesh distortion problems. In the CEL FE-model, the cone is modelled by Lagrangian rigid solid elements whereas the soil is discretised with Eulerian elements.

If not otherwise specified, a friction coefficient of 0.38 ($\alpha_f = 0.38$) is assumed at the interface between the cone and the soil and all analyses are conducted with a constant penetration rate of 0.1 m/s.

In order to represent the essential characteristics of sand (e.g. stress-dependency, hardening-softening, dilatancy and contractancy) and to improve numerical stability, especially in large deformation problems (e.g. Henke & Grabe, 2006), the modified version of von Wolffersdorff (1996)'s hypoplastic model (Masin, 2010) was employed in the numerical analyses. Hypoplasticity provides flexibility and modelling capabilities in terms of capturing hardening and softening behaviour, as well as void ratio and stress dependent stiffness.

The hypoplasticity model employed in this study has 9 basic (von Wolffersdorff, 1996) and 5 additional input parameters (Niemunis and Herle, 1997), which are summarized in Table 3.1.

Detailed information on the determination of the hypoplastic model parameters can be found in literature (e.g. Herle and Gudehus, 1999; Rondon et al, 2007; Anaraki, 2008; Masin, 2011). The parameters representing the material properties are independent of the density state. This is based on the assumption that the size and shape of the particles do not change. At high stress and strain levels, this assumption may not hold as the particles may be rounded or even crushed. If high levels of stresses or crushable material will be modelled, for a realistic representation, the model should be modified so that the change in the geometry of particles is considered (e.g. Rohe, 2010, Engin et. al, 2014).

2.2 *Sensitivity study*

The effect of discretisation is investigated using a CEL FE-model. A total of three mesh densities with different element sizes around the cone tip of $0.06D_{cone}$ (fine mesh), $0.11D_{cone}$ (medium mesh) and $0.17D_{cone}$ (coarse mesh) are constructed and simulated. The calculated cone tip resistances indicated that the mesh density of $0.11D_{cone}$ below the cone tip is considered sufficiently fine to satisfy both the accuracy and the computational efficiency of the simulations. This CEL result agrees reasonably well with the ALE result published by Andresen&Khoa (2013).

In most real installation problems, the rate of penetration is so slow (i.e. < 0.02 m/s) that dynamic effects can be disregarded and the problem can be regarded as being static. Solving a static problem with Abaqus/Explicit requires choosing a penetration velocity which does not introduce any inertia effects of significance for the solution. Four different cone penetration rates are simulated using the CEL FE-model. The calculated cone tip resistances plotted in Figure 2 show that the rate of 0.1 m/s gives almost the same result as the rate of 0.05 m/s while the two other cases with higher rate result in remarkably higher cone resistance compared to the cases with lower the penetration rate. Hence, the penetration rate of 0.1 m/s is adopted to optimize the calculation time.

To evaluate the effect of cone-soil interface roughness, three cases of a fully smooth cone, a frictional cone with $\alpha_f = 0.38$ and a cone with frictional tip and smooth shaft are analysed. Figure 3 shows that the two latter cases having rough tip yield almost the same result up to a penetration of about 0.6, which indicates that the shaft roughness does not affect the tip resistance particularly at shallow depths (which is in agreement with experience from field tests). Penetrating

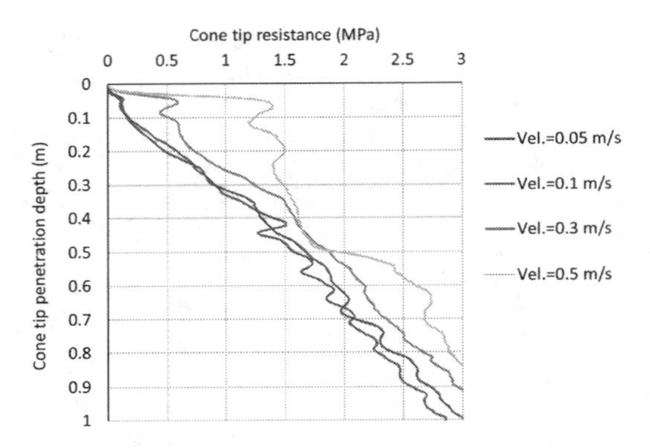

Figure 2. Effect of cone penetration rate with respect to cone tip resistance.

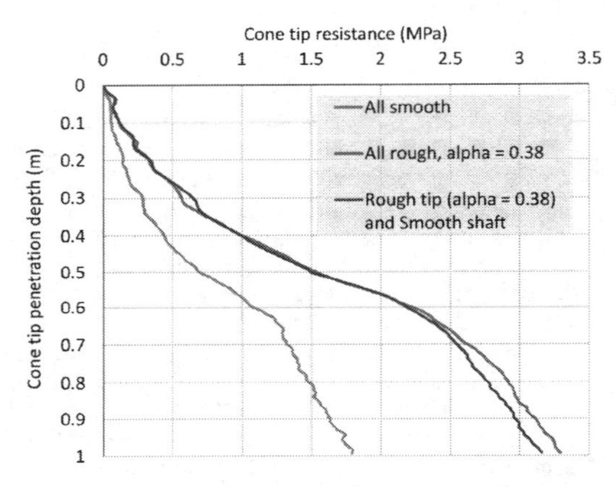

Figure 3. Effect of modelling of cone roughness with respect to cone tip resistance.

the cone further, the influence of the shaft roughness on the tip resistance becomes more and more pronounced, however, the difference between these two cases are negligible (i.e. < 3–4%) up to 1.0 m penetration depth.

2.3 *ALE versus CEL*

Figure 4 shows a comparison of the cone tip resistance and shear stresses calculated using the ALE or CEL FE analyses, respectively. It can be seen that the tip resistance below about 0.55 m penetration depth is less with the CEL method then with the ALE method. One of the main reasons is that the CEL is less accurate in capturing the contact interface behaviour. This issue is confirmed by the shaded contour plot of the calculated Tresca stresses at 1.0 m penetration depth as shown in Figure 4.

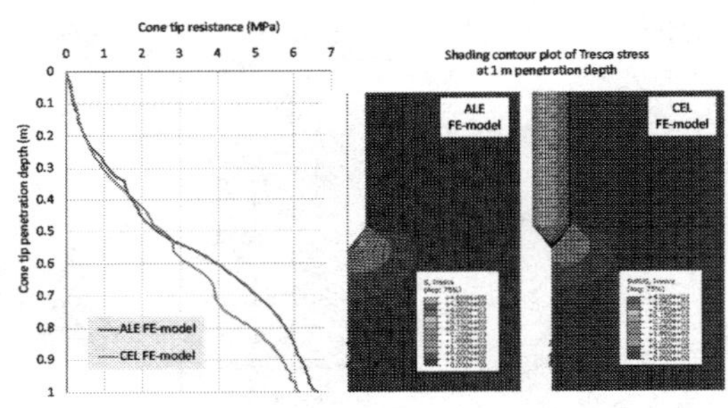

Figure 4. Comparison between ALE and CEL FE analyses of cone penetration.

The following are two other important observations from the comparison:

- the axisymmetric ALE FE-model runs significantly faster than the 3D-slice CEL FE-model;
- there is no mesh deformation in the CEL FE-analysis while the mesh is moderately deformed, especially around the cone tip, in the ALE FE-analysis. In some cases where the penetration object has a complex geometry, the ALE FE-mesh may become severely deformed and distorted, which can result in an inaccurate solution or event in a numerical divergence of the analysis.

2.4 *FE analyses of cone penetration*

In order to investigate the cone penetration response in a NC and OC sands, $D_r = 0.65$ and $D_r = 0.85$ are considered and several FE analyses are performed in which both CEL and ALE methods are employed.

Ticino sand is considered in the analyses in order to be able to compare the numerical results with results found in literature, which are mostly calibration chamber (CC) tests performed using Ticino sand (e.g. Baldi et al. (1986, Jamiolkowski et al., 2003). The hypoplastic model parameters were based on literature (e.g. Rondon et al, 2007, Cudmani & Sturm, 2006).

It should be noted that the finite element analysis (FEA) is compared with tests reported in literature utilizing calibration chamber tests. Typically, stress states are applied in these tests, which correspond to depths larger than 5 m. Since no data are available for z < 5 m (less than 50 kPa), the comparison is made by extrapolating the results from the literature, which is out of bounds of those correlations.

The penetration starts from seabed level. FEAs starting from an embedded level of 5 m are also included as a side check of the behaviour observed in the field (e.g. cone resistance builds up to the same level as the CPT starting from seabed level).

Table 2. Basic hypoplastic model parameters determined for Ticino sand (Cudmani & Sturm, 2006).

ϕ_c	h_s	n	e_{d0}	e_{c0}	e_{i0}	α	β
°	MPa	-	-	-	-	-	-
31	250	0.68	0.5762	0.9261	1.11	0.11	1

Table 3. Intergranular strain parameters of the hypoplastic model parameters determined for Ticino sand (Cudmani & Sturm, 2006).

m_R	m_T	R_{max}	β_R	χ
-	-	-	-	-
5	2	0.0001	1	1

The following main cases have been considered for both $D_r = 0.65$ and $D_r = 0.85$, representing medium dense and dense sand cases:

- $K_0^{nc} = 0.415$, penetrating from 0 to 6 m.
- $K_0^{nc} = 0.415$, penetrating from 0 to 1 m with:
  - $q_{overburden} = 20\ kPa\ (\sim 2m\ below\ seabed)$
  - $q_{overburden} = 50\ kPa\ (\sim 5m\ below\ seabed)$

  to allow comparison with calibration chamber test data.
- $K_0^{oc} = 0.86$, penetrating from 0 m to 6 m.
- $K_0^{oc} = 1.0$, penetrating from 0 m to 6 m.

In addition, effect of varying K_0 profile considering past overburden pressure, hence varying OCR with depth is also investigated. The K_0 can be estimated using Kulhawy & Mayne (1990).

In this study, K_0 is limited to the passive earth pressure ($K_0 \leq K_p$). The effects of varying K_0 profile and $K_0 > 1$ are evaluated by using the ALE FE-model since there are some numerical issues in the test runs with the CEL FE-model.

Table 2 and Table 3 summarize the hypoplastic material model parameters for the Ticino sand is used in the FEAs. The initial void ratio has to be updated to simulate the relevant density as:

$$e_0 = e_{c0} - D_r \cdot (e_{c0} - e_{d0}) \quad (2)$$

Hence, $e_0 = 0.699$ was used for $D_r = 0.65$, and $e_0 = 0.629$ was used for $D_r = 0.85$

3 FE RESULTS & DISCUSSIONS

3.1 *Cone penetration resistance*

This part summarizes the FEA results in comparison with the empirical correlations given by Jamiolkowski et al. (2003), Emerson et al. (2008), and Senders (2010). The OC cases (in this study considered as increased K_0) predicted by these empirical correlations (developed for NC sands) are plotted merely for comparison.

Figure 5 plots the CEL FE results for the $D_r = 0.65$ and $D_r = 0.85$ cases with different K_0 (i.e. OCR) values. The analyses are carried out until 2 m max penetration level, which is equivalent to $z/D_{cone} \approx 55$. All the cone resistances, q_c, are normalized by the maximum cone resistance of all FEAs, i.e. $q_{c,ref}$. It is observed from the same figure that the effect of increased K_0 is only evident after approximately $z/D_{cone} \approx 20-25$ of penetration. Very similar behaviour for the two high K_0 cases ($K_0 = 0.86$ *and* $K_0 = 1.0$) are obtained. The effect of density on the cone tip resistance, which is evident from the empirical correlations, seems to be captured reasonably well.

Moreover, for the FEA results of the medium dense cases ($D_r = 0.65$) given in Figure 5, the shallow failure mechanisms seem to prevail up to $z/D_{cone} \approx 14$. From that level to $z/D_{cone} \approx 22$ for NC case and ≈ 28 for OC case, a transition of shallow to deep failure mechanism is evident. After this level, deep failure mechanisms seem to prevail. Regarding the dense cases ($D_r = 0.85$), the shallow failure mechanisms seem to prevail up to $z/D_{cone} \approx 17$ for NC and ≈ 25 for OC cases. From those levels to $z/D_{cone} \approx 24$ for NC case and ≈ 31 for OC case, a transition of shallow to deep failure mechanism is evident. Below this penetration depth, deep failure mechanisms seem to prevail.

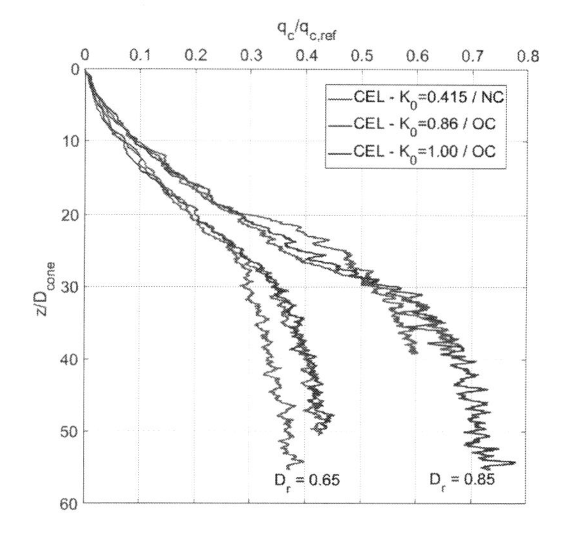

Figure 5. Effects of OCR and D_r on cone penetration resistance calculated from CEL FEAs.

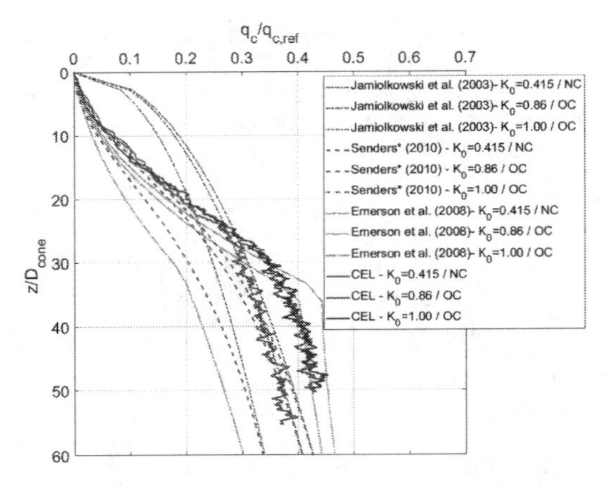

Figure 6. Comparison of CEL results with empirical predictions for case of $D_r = 0.65$.

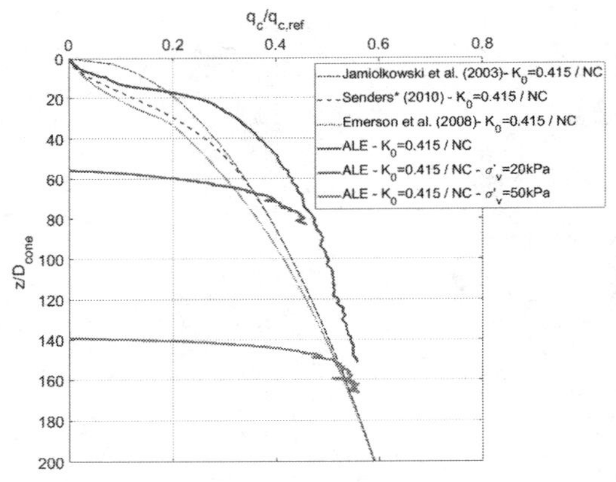

Figure 7. Comparison of ALE results with empirical predictions case of $D_r = 0.65$.

Figure 6 compares the CEL FE results against the empirical correlations for the medium dense case ($D_r = 0.65$). It is seen that none of the empirical methods fit the normalised cone resistance curves, $q_c / q_{c,ref} - z / D_{cone}$, obtained by CEL for the NC case. The overall cone resistance curves for the OC cases calculated by the CEL simulations seem to match reasonably with the predictions by Emerson et al. (2008) up to $z / D_{cone} \approx 31$. Jamiolkowski et al. (2003) match the rest of the curves below that penetration depth (i.e. $z / D_{cone} \approx 31$). It is also noted from the figure that Senders*(2010) confirms the simulation results quite well up to about $z / D_{cone} \approx 17$.

Due to the fact that fine CEL FEAs require both enormous computational resources and significant computation time, a series of ALE FEAs were performed to improve the understanding of the soil behaviour and the cone resistance at penetration depth deeper than $z / D_{cone} \approx 60$.

Figure 7 plots the calculated results of the cone penetrating to $z / D_{cone} \approx 170$ for the NC medium dense case (i.e. $K_0 = 0.415$ and $D_r = 0.65$). The empirical correlations by Jamiolkowski et al. (2003), Emerson et al. (2008), and Senders* (2010) are also illustrated in the same figure. It is seen that the calculated results at deep penetration levels converge to the well-established empirically picked values. Furthermore, it is interesting to observe that the penetration cases having different overburden pressures as initial condition with the one commencing at seabed level converges to same level after $z / D_{cone} \approx 28$ or less. Both results validate the field behaviour where the stationary cone resistances match the latter curve in about $z / D_{cone} \approx 28$ penetration. The results, hence, confirm the model consistency.

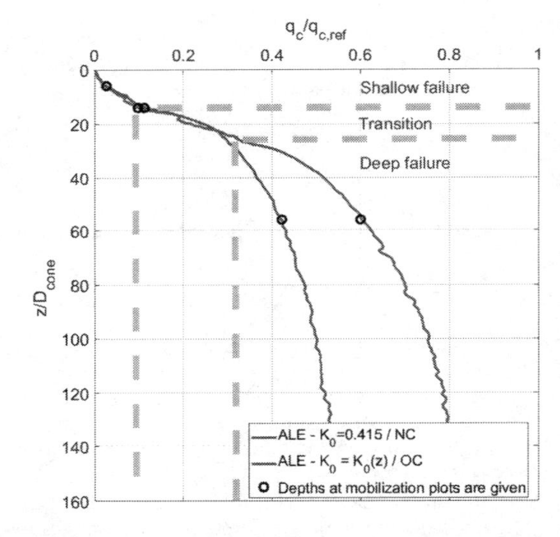

Figure 8. Depths at which the failure mechanisms of the selected cases are investigated for $D_r = 0.65$ (approximate zones of failure mechanisms presented for $K_0^{NC} = 0.415$ and $K_0^{OC} = K_0(z)$.

All of the cases (both of the densities considered and the K_0 variations) the depths at which deep penetration mechanisms prevail are shallower than Emerson et al. (2008) and Senders (2010) and seem to increase with increasing K_0.

3.2 *Failure mechanisms*

This section presents the ALE FE results to shed a light on the failure mechanisms developing during the penetration of a cone from shallow to deeper depths.

The velocity fields give an indication of the failure surfaces, and hence are presented in the following figures at the selected depths, as indicated in Figure 8 for the dense sand case ($D_r = 0.65$). Figure 8 indicates that the shallow failure mechanisms prevail up to $z/D_{cone} \approx 14$. From that level to $z/D_{cone} \approx 22$ for NC case ($K_0^{NC} = 0.415$) and ≈ 28 for OC case ($K_0^{OC} = K_0(z)$), a transition of shallow to deep failure mechanism seem to occur. After the transition depths, deep failure mechanisms fully develop. Insight of the different failure mechanisms is given in Figures 9–11, which plot the shaded contours of the velocity fields calculated at the three selected penetration depths: $z/D_{cone} \approx 6{,}14$ and 56 , respec-tively.

Figure 9 displays shallow failure mechanisms for both of the NC and OC cases. The mobilized zone has a radial extent of $\approx 5D_{cone}$ from shaft of the CPT probe and a depth of $\approx 1.5D_{cone}$ and $\approx D_{cone}$ from tip of the cone for the NC and OC cases, respectively. Increasing K_0 seems to force the failure towards sides as the confinement towards the tip increases for the OC case.

Figure 10 shows combined type of failures. It is observed that deep failure mechanism has been clearly developing. The radial extent of the failure

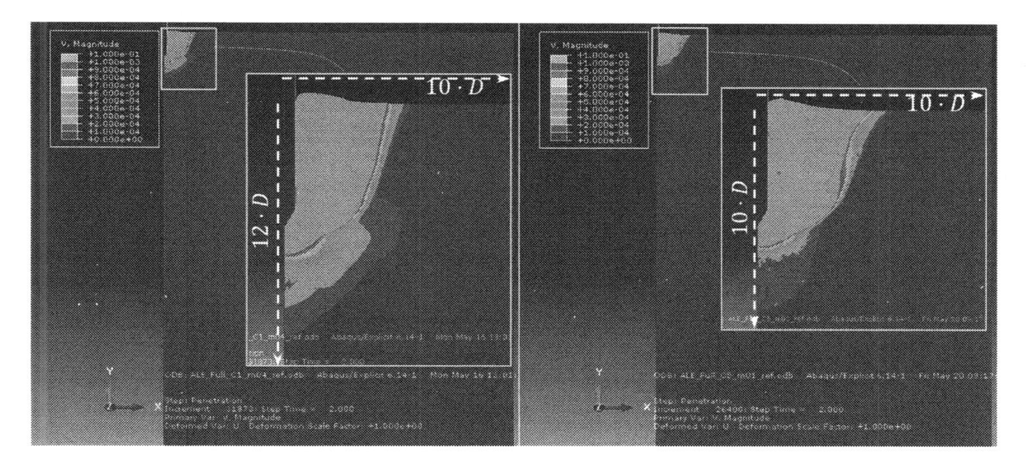

Figure 9. Scaled velocities as indication of failure surfaces at z/D$_{cone}$ = 6 for $K_0^{NC} = 0.415$ (left) and $K_0^{OC} = K_0(z)$ (right) for $D_r = 0.65$.

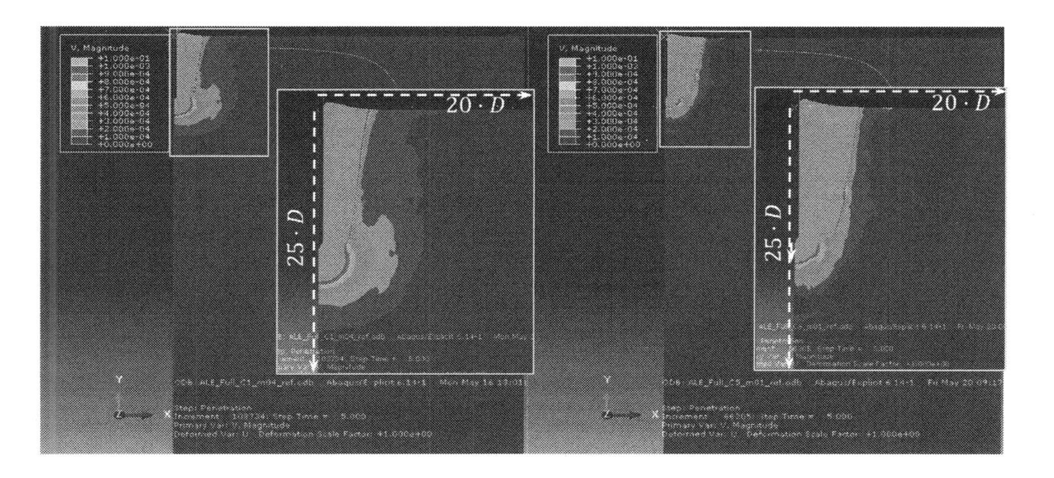

Figure 10. Scaled velocities as an indication of the failure surfaces at z/D$_{cone}$ = 14 for $K_0^{NC} = 0.415$ (left) and $K_0^{OC} = K_0(z)$ (right) for $D_r = 0.65$.

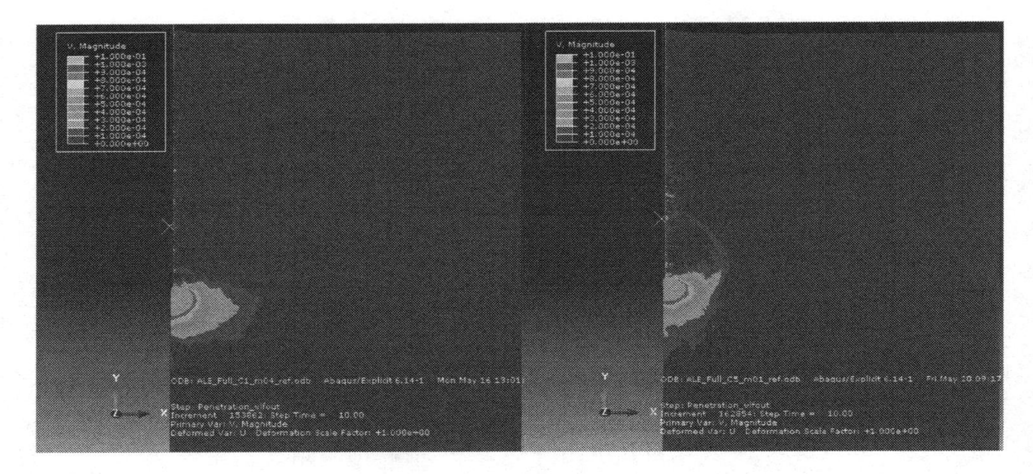

Figure 11. Scaled velocities as an indication of the failure surfaces at z/D$_{cone}$ = 56 for $K_0^{NC} = 0.415$ (left) and $K_0^{OC} = K_0(z)$ (right) for $D_r = 0.65$

surface stabilizes at around $\approx 4.5\text{D}_{cone}$ from shaft of the CPT probe for the NC case (based on the intermediate analyses results, which are omitted in this paper due to limited space). The OC case have the extent of the radial failure envelope up to $\approx 6\text{D}_{cone}$. The depth of failure zone extends to $\approx 2.5\text{D}_{cone}$ and $\approx 2\text{D}_{cone}$ from tip of the cone for the NC and OC cases, respectively. Increasing K_0 results in an effect of pushing the failure towards sides as the confinement towards the tip increases for the OC case.

At very deep penetration depth, $z/\text{D}_{cone} \approx 56$, the plots of the calculated velocities given in Figure 11, clearly indicate deep failure mechanisms for both NC and OC cases. It is also observed that the failure envelope depth of the OC case is smaller than the NC case, which is believed to be due to difference in confinement levels.

In general, similar observations can be made for the dense cases (the results of which are omitted in this paper due to limited space). The denser cases mobilize the deep failure mechanism at slightly shallower depths.

4 CONCLUSIONS

This paper summarizes the numerical analyses performed as a part of the Joint Industry Project (JIP) - Interpretation of CPT in shallow depth (2015). Objective of this study was to investigate the initial density and stress on the CPT behaviour at shallow penetration depths. The following conclusions can be made:

- The CPT cone resistances of (NC) Ticino sand in CC sand can be reproduced with the FE technique (i.c. CEL and ALE) and the soil model (i.c. hypoplasticity) employed in the numerical simulations. Hence, the global CPT behaviour in sand can be captured by the numerical analyses.
- Field behaviour, i.e. q_c obtained for pre-embedded case (e.g. initial overburden pressure) penetration from surface level converging in about 1 m of penetration, can be reproduced.
- Both aforementioned validations confirms the reliability of the numerical method employed for the CPT analyses at least for the NC sand.

Figure 10. Scaled velocities as an indication of the failure surfaces at z/D$_{cone}$ = 56 for $K_0^{NC} = 0.415$ (left) and $K_0^{OC} = K_0(z)$ (right) for $D_r = 0.65$

- In the view of computational efficiency and easiness of imposing complex K_0 profiles, one may recommend using ALE instead of CEL method for the CPT simulations.
- The same type of evolution of failure mechanisms, i.e. shallow to deep penetration observed by Senders* (2010), resulting in a concave to convex $q_c - z$ curve, namely, a concave part (shallow failure), linear to convex part (transition zone) and convex part (deep failure) can be obtained.
- It was challenging to model the OC effect in the FEAs (especially using CEL method) imposing an initial stress considering an increased K_0, as it had to be limited to 1.0 for CEL and K_p for the ALE analyses. The FEAs have shown that using (slightly) higher K_0 increased q_c values for depths larger than $\sim 20 - 25\text{D}_{cone}$. The same analyses showed no significant effect of K_0 in a depth less than approximately $\sim 20 - 25\text{D}_{cone}$. In view of the field experience, it is recommended that the OC and the corresponding soil behaviour should be modelled focusing on the

history and coupled behaviour (i.e. pore pressure effects, which could be significant especially for saturated OC sands, e.g. typical North Sea soil conditions). This way, the FEAs would provide better insight to the mechanisms prevailing the response during CPT.

ACKNOWLEDGEMENTS

The work done under this Joint Industry Project (JIP) - Interpretation of CPT in shallow depth during 2015 and 2017 was carried out with the project partners Ørsted Wind Power and GEO. The financial and technical support of all partners is highly appreciated.

Support of Dr. Gökhan Saygılı (UT Tyler) with running the CEL models on the supercomputers as well as pre-processing (e.g. scripting for supercomputer runs) is also gratefully acknowledged.

REFERENCES

Anaraki, K.E. (2008) Hypoplasticity investigated parameter determination and numerical simulation. Master's thesis, Delft University of Technology, The Netherlands, 2008.

Andresen, L. & Khoa, H.D.V. (2013) LDFE analysis of installation effects for offshore anchors and foundations, Proceedings of the International Conference on Installation Effects in Geotechnical Engineering, Rotterdam: 162–168.

Baldi, G., Bruzzi, D., Superbo, S., Battaglio, M. and Jamiolkowski, M. (1986) Interpolation of CPTs and CPTUs; 2nd part: drained penetration of sands. Proc. of the Fourth International Geotechnical Seminar, Singapore, 143–56.

Baligh, M. (1985) Strain path method, ASCE Journal of Geotechnical Engineering, 111(9), 1108–1136.

Cudmani, R. and Sturm, H. (2006) An investigation of the tip resistance in granular and soft soils during static, alternating and dynamic penetration. In H. Gonin, A. Holeyman, and F. Rocher-Lacoste, editors, TransVib 2006: International Symposium on vibratory pile driving and deep soil compaction, pp. 221–231.

Henke, S., & Grabe, J. (2006). Simulation of pile driving by 3-dimensional finite-element analysis. In Proceedings of 17th European Young Geotechnical Engineers' Conference. V. Szavits-Nossan, Croatian Geotechnical Society.

Herle, I. and Gudehus, G. (1999) Determination of parameters of a hypoplastic constitutive model from properties of grain assemblies. Mechanics of Cohesive-frictional Materials, 4(5):461–486.

Huang, A.B. & Ma, M.Y. (1994) An analytical study of cone penetration tests in granular material, Canadian Geotechnical Journal 31(1), 91–103.

Iqbal M.S. (2004) Discrete Element Modelling of Cone Penetration Testing in Coarse Grain Soils. M.S. thesis, Civil and Environmental Engineering, University of Alberta, Edmonton.

Jamiolkowski, M., Lo Presti, D.C.F. and Manassero, M. (2003)

Evaluation of Relative Density and Shear Strength of sands from Cone Penetration Test (CPT) and Flat Dilatometer (DMT). ASCE Geotechnical Special Publication No. 119.

Kulhawy, F.H. and Mayne, P.H. (1990) Manual on estimating soil properties for foundation design, Palo Alto, California, Report Report, EL-6800, pp. 2–38.

Masin, D. (2010) Plaxis implementation of hypoplasticity. Technical Report, Plaxis bv.

Masin, D. (2011) PhD course on hypoplasticity, Faculty of Science, Charles University in Prague.

Niemunis, A. (2003) Extended Hypoplasticity Models for Soils, Ruhr University Bochum, Germany. ISSN: 1439–9342. http://www.pg.gda.pl/~aniem/pap-zips/Hab-19–12–2002.pdf

Niemunis, A., & Herle, I. (1997). Hypoplastic model for cohesionless soils with elastic strain range. Mechanics of Cohesive-frictional Materials, 2(4), 279–299.

Ørsted Wind Power, GEO & NGI (2015). Interpretation of CPT in shallow depth, Joint Industry Project (JIP).

Rohe, A. (2010) On the modelling of grain crushing in hypoplasticity, Technical Report, TUDelft, NL.

Rondon, H.A., Wichtmann, T., Triantafyllidis, Th. And Lizcano, A. (2007) Hypoplastic material constants for a well-graded granular material for base and subbase layers of flexible pavements, Acta Geotechnica (2007) 2:113–126, DOI 10.1007/s11440–007–0030–3.

Senders, M. (2010) Cone resistance profiles for laboratory tests in sand. Proc. 2nd Int. Conf. on Cone Penetration Testing, Huntington Beach, CA, paper no. 2–08.

Tannant D.D., and Wang C. (2002) PFC Model of Wedge Penetration into Oil Sands. Discrete Element Methods: Numerical Modelling of Discontinua. Proceedings of the Third International Conference, Santa Fe, September 2002, Reston, Virginia, ASCE, 311–316.

Teh, C.I. & Houlsby, G.T. (1991) An analytical study of the cone penetration test in clay, Géotechnique, 41(1), 17–34.

van den Berg, P. (1994) Analysis of soil penetration, PhD thesis, Delft University of Technology.

von Wolffersdorff, P.A. (1996) A hypoplastic relation for granular material with a predefined limit state surface. Mechanics of cohesive-fractional materials 1: 251–271.

Walker, J. (2007) Adaptive finite element analysis of the cone penetration test in layered clay, PhD thesis, University of Nottingham

Yu, H.S., and Mitchell, J.K. (1998). Analysis of cone resistance: a review of methods. Journal of Geotech Geoenv Eng ASCE. 124(2), p.140–9.

Yu, H. S, Herrmann L. R and Boulanger R.W. (2000). Analysis of steady cone penetration in clay. Journal of Geotech Geoenv Eng ASCE.126(7), p. 594–605

Numerical Methods in Geotechnical Engineering IX – Cardoso et al. (Eds)
© 2018 Taylor & Francis Group, London, ISBN 978-1-138-33198-3

Pipe-seabed interaction under lateral motion

H. Sabetamal, J.P. Carter & S.W. Sloan
ARC Centre of Excellence for Geotechnical Science and Engineering, The University of Newcastle. Newcastle, NSW, Australia

ABSTRACT: Controlled on-bottom lateral buckling of partially embedded pipelines is a novel and cost-effective solution to relieve the axial compressive stresses resulting from cycles of thermal expansion and contraction of operating pipelines. To investigate the mechanisms involved in pipe-soil interaction under large-amplitude cyclic movements, large deformation finite element analyses have been conducted. The dynamic coupled simulations not only capture the complex changes in seabed geometry during lateral buckling, but also consider likely episodes of undrained or partially drained behaviour interspersed with periods of soil consolidation.

1 INTRODUCTION

Deep-water pipelines are used to transport products from offshore oil and gas reserves to shore or field-processing facilities. They are usually installed by laying them from a vessel where the motion of the lay vessel and any hydrodynamic action on the hanging span will cause the pipe to move dynamically. The laying procedure is relatively quick and therefore imposes an undrained loading condition on the seabed soil so that excess pore pressures are generated. Once laid, the vertical pipe-soil load is reduced to the submerged pipe weight and dissipation of the excess pore pressures with time takes place leading to some consolidation settlements. The consolidation of soil around the pipe results in an increase in the bearing capacity of the soil immediately surrounding the pipeline. In operation, axial compressive stresses are generated due to thermal expansion in the pipe, which may lead to lateral buckling. To relieve the axial stresses, buckles are usually allowed to form; they are engineered such that no excessive bending of the pipe takes place. The changing operating conditions and shutdowns of the pipeline result in cycles of expansion and contraction that produce cyclic lateral movements of the partially embedded pipes (Bruton et al. 2008).

Design of on-bottom pipelines with regard to lateral buckling requires assessment of the pipe-soil interaction forces that result from lateral motion and requires accounting for the associated changes in seabed geometry and strength. The main uncertainty in the buckling design is the soil resistance faced by the pipe during lateral movement. The plasticity solutions for the vertical collapse load of a shallowly embedded pipeline may closely predict the deformation mechanisms during pipe penetration, but they do not allow for large deformation effects, such as heave, to be incorporated. In addition, including the dynamic effects involved in the lay process is very difficult. Furthermore, the generation of excess pore pressures around a partially embedded pipe, together with their subsequent dissipation, may also have a significant effect on the vertical penetration and horizontal breakout resistance of the pipe. The steady lateral response of the soil–pipe system also involves the growth of a soil berm, which is created as the pipe moves laterally from the initial embedment position. Pore-pressure dissipations can occur during lateral sweeping and for the period of start-up and shutdown events. This gives rise to the reconsolidation of the disturbed soil within the berm that can significantly increase the berm resistance.

A few experimental investigations have been performed at large scale and at reduced scale in a centrifuge (e.g., Cheuk et al. 2007, Burton et al. 2008, Dingle et al. 2008, Cardoso & Silveira 2010), which have led to some empirical expressions for the lateral breakout resistance and the subsequent steady residual resistance. However, these approaches are subject to significant scatter and uncertainty and do not provide detailed information on the mechanisms of soil movement during large-amplitude lateral pipe displacements. Therefore, a better understanding of the soil–pipe response requires robust and realistic numerical modelling in order to address the different phenomena involved in the pipe-soil response.

Previous numerical studies have only explored the undrained response of a pipe-soil system under large movements, considering only the first cycle of lateral movement (Wang et al. 2010, Chatterjee et al. 2012, Dutta et al. 2014). These studies typically involve a total stress analysis with a Tresca soil model, essentially ignoring dynamic effects. Moreover, the pipe–soil interface is assumed to be fully smooth or fully rough in order to avoid numerical difficulties arising from frictional forces developed at the soil–pipe interface. Therefore, it is necessary to consider the whole of life behaviour of a pipe element sweeping repeatedly within a buckle and to capture the consolidation effects within the large lateral deformation framework. The present study considers simulation of such problems in a dynamic effective stress analysis approach. The analysis results provide some insights into the mechanism of pipe-soil response under large amplitude lateral motions as well as the generation and dissipation of excess pore pressures during the whole event.

2 NUMERICAL PROCEDURE

A computational procedure for the analysis of coupled geotechnical problems involving finite deformation, inertial effects and changing boundary conditions has been developed by Sabetamal (2015). The procedure involves new finite element (FE) algorithms, which have been formulated and implemented into SNAC, a bespoke FE code developed by the geomechanics group at the University of Newcastle, Australia. A comprehensive overview of the computational scheme and its validation have been provided recently (Sabetamal et al. 2016a). A summary of this procedure is presented below.

2.1 *Governing equations*

The field equations for a two-phase saturated porous medium have been derived from the theory of mixtures, extended by the concept of volume fractions, which models the dynamic advection of fluids through a fully saturated porous solid matrix (Morland 1972). A numerical solution of the governing differential equations was then obtained by the FE method. A **U-P-V** formulation was used to describe both incompressible and compressible fluids, in which the resulting mixed formulation predicts all field variables, including the displacement of the solid soil skeleton, **U**, the pore-fluid pressure, **P**, and the Darcy velocity of the pore fluid **V**. The resulting approximate FE equation system governing the behaviour of the soil-water mixture may be written in matrix form as

$$\begin{bmatrix} \mathbf{M}_{ss} & \mathbf{0} & \mathbf{M}_{sr} \\ 0 & 0 & 0 \\ \mathbf{M}_{rs} & \mathbf{0} & \mathbf{M}_{rr} \end{bmatrix} \begin{bmatrix} \ddot{\mathbf{U}} \\ \ddot{\mathbf{P}} \\ \dot{\mathbf{V}}_r \end{bmatrix} + \begin{bmatrix} \mathbf{C}_s & 0 & 0 \\ \mathbf{C}_{ps} & -\mathbf{C}_{pr} & \mathbf{C}_{pp} \\ 0 & \mathbf{C}_{rr} & 0 \end{bmatrix} \begin{bmatrix} \dot{\mathbf{U}} \\ \dot{\mathbf{P}} \\ \mathbf{V}_r \end{bmatrix} + \begin{bmatrix} \mathbf{K}_\sigma & \mathbf{K}_{sp} & 0 \\ 0 & \mathbf{K}_{rp} & 0 \\ 0 & \mathbf{K}_{pp} & 0 \end{bmatrix} \begin{bmatrix} \mathbf{U} \\ \mathbf{P} \\ \mathbf{0} \end{bmatrix} = \begin{bmatrix} \mathbf{F}^s \\ \mathbf{F}^p \\ \mathbf{F}^r \end{bmatrix} \quad (1)$$

where $\mathbf{M}_{ss}$, $\mathbf{M}_{rr}$, $\mathbf{M}_{rs} = [\mathbf{M}_{rr}]^T$ and $\mathbf{C}_{\alpha\beta}$ are the solid mass, fluid mass, coupled fluid mass and damping matrices, respectively. $\mathbf{K}_\sigma$ and $\mathbf{K}_{pp}$ are, respectively, the stiffness and compressibility matrices, while $\mathbf{K}_{\alpha\beta}$ represents coupling matrices and $\mathbf{F}_s$, $\mathbf{F}_p$ and $\mathbf{F}_r$ are the vectors of external nodal forces (Sabetamal et al 2016b). Chang & Hulbert's (1993) generalised-α algorithm was employed to integrate the governing equations of the two-phase saturated porous media in the time domain. Moreover, the cone energy-absorbing boundary of Kellezi (2000), which consists of dashpots and springs, was adopted to absorb the outgoing bulk waves and to avoid spurious wave reflections at the mesh boundaries.

2.2 *Large deformation and mesh refinement*

The Updated Lagrangian (UL) approach with Jaumann's objective stress rate was utilised to consider the effects of finite deformation while an Arbitrary Lagrangian-Eulerian (ALE) approach was also incorporated in order to avoid excessive mesh distortions. The ALE convective term in the time integration of the constitutive equations is handled by the Van Leer advection algorithm (van Leer 1977), which is an integration point-based convection scheme possessing second-order accuracy. The use of high order schemes in the solution of the advection equations and the subsequent remap of stresses and history dependent material properties is significantly important in order to alleviate numerical instability and numerical diffusion throughout the advection process (Sabetamal et al. 2017).

2.3 *Interface modelling*

The interaction between the soil and a pipe is modelled utilising a frictional contact algorithm. The formulation of the contact kinematics and constraints is based on the so-called mortar segment-to-segment approach (Fischer & Wriggers 2005), which allows the interpolation functions of the contact elements to be of order *n*. This facilitates the incorporation of smooth and continuous surfaces at a contact interface utilising high order elements, and thus circumvents the occurrence of numerical oscillations in the predicted response (Sabetamal et al. 2014). The contact constraints arising from the requirement for continuity of the contact traction and the fluid flow across the contact interface are enforced using a penalty approach, which is regularised with an augmented Lagrangian method (Sabetaml et al. 2016b). Therefore, for a node that comes in contact with a corresponding contacting pair, contact contributions arising from constraining solid displacement, Darcy's velocity and pore pressure are added to the tangent stiffness matrix and the residual vector during the global Newton iterations. In a frictional contact element, two conditions, stick and slip, are distinguished on the basis of the level of interface frictional force compared with the Coulomb frictional force. The formulation of frictional contact has been developed in terms of effective forces. To differentiate between the stick and slip cases, the concept of a moving friction cone (Wriggers & Haraldsson 2003) has been used, which is a relatively efficient methodology in deriving the contact kinematics.

In the coupled consolidation-contact algorithm developed here, free-draining conditions are automatically adopted for the nodes that lose connection with their possible contacting pair, whereas impermeable or semi-impermeable conditions are adjusted for soil nodes that come in contact with another surface. This feature is particularly important for a pipe-seabed interaction problem where large lateral motions impose variable drainage conditions at the seabed.

3 NUMERICAL SIMULATION

It is assumed that an elastic pipe, 0.8 m in diameter, was first laid rapidly onto deformable seabed soil by applying a uniform pressure load to it, and the soil was then allowed to consolidate. When the consolidation process was completed and all generated excess pore pressures were dissipated, the vertical force was reduced to a specific value, equivalent to the submerged weight of the pipe. Finally, the pipe was swept back and forth across the model (at a constant velocity) while the vertical load remained constant and the pipe was free to move up or down.

A 2D finite element mesh containing 3,240 plane strain triangular elements and 6,695 nodal points was employed for the simulation of this problem, as depicted in Figure 1. The elastic pipe was modelled using 24 quadratic triangular elements, producing a smooth continuous surface. The Modified Cam Clay (MCC) constitutive model was adopted to represent the soil in this study using the model parameters listed in Table 1. Further, the coefficient of friction at the pipe–soil interface, μ, was taken as 0.1. The analysis started by establishing the initial stresses in the soil due to its self-weight and an effective overburden pressure of 75 kPa. Application of this overburden pressure may not be entirely realistic, but was adopted to ensure stability of the numerical solution, avoiding the case of a normally consolidated soil profile with zero undrained shear strength at the surface of the seabed.

Subsequently, energy-absorbing boundaries that comprised springs and dashpots were applied to the boundary nodes to absorb the energy of the impacting stress waves and to eliminate possible wave reflections. The initial stresses in the seabed were generated assuming the soil to be 1D (K_0) consolidated with $K_0 = 0.67$, and the water table was assumed to be located at the soil surface. During this stage of the analysis, the horizontal components of the solid movement and fluid flow were prevented on the side boundaries. The bottom boundary was also fixed against vertical and horizontal solid displacements, whereas fluid flow tangential to the boundary only was allowed. Drainage was allowed at the top soil surface through the nodal points not in contact with the impermeable pipe surface. Finally, the pipe was laid on the soil by applying a uniform pressure $q = 4.75\ s_{u0}$ to it over the dimensionless time increment of $\Delta T_v = 1.4 \times 10^{-6}$, where $s_{u0} = 38.8$ kPa is the undrained shear strength at the pipe invert, obtained from the assumed MCC parameters for K_0-consolidated soil.

The dimensionless time T_v is defined as

$$T_v = \frac{c_v t}{D^2} \tag{2}$$

where t represents the actual time and c_v denotes the coefficient of consolidation given by:

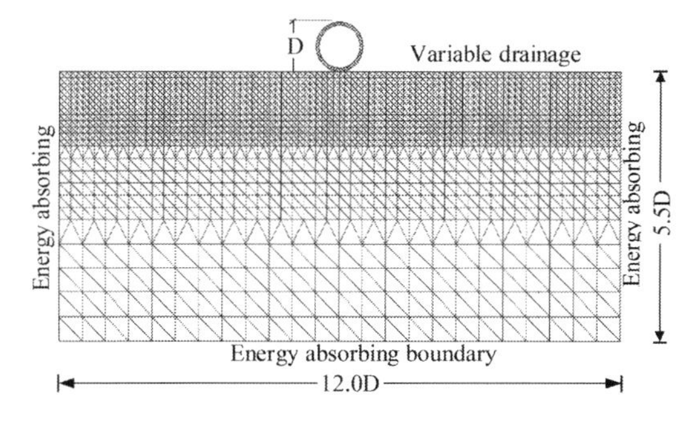

Figure 1. FE model for pipe–soil interaction under lateral movement.

Table 1. MCC material parameters.

Parameter	Value
Friction angle	$\phi' = 25°$
Slope of normally consolidated line*	$\lambda = 0.205$
Slope of unloading-reloading*	$\kappa = 0.044$
Initial void ratio	$e_0 = 2.14$
Over consolidation ratio	OCR = 1
Poisson's ratio	$\nu = 0.3$
Saturated bulk unit weight	$\gamma_{sat} = 15$ kN/m^3
Unit weight of water	$\gamma_w = 10$ kN/m^3
Permeability of soil	$k = 10^{-8}$ m/s

* in e-ln(p') space, where $p($ is mean effective stress

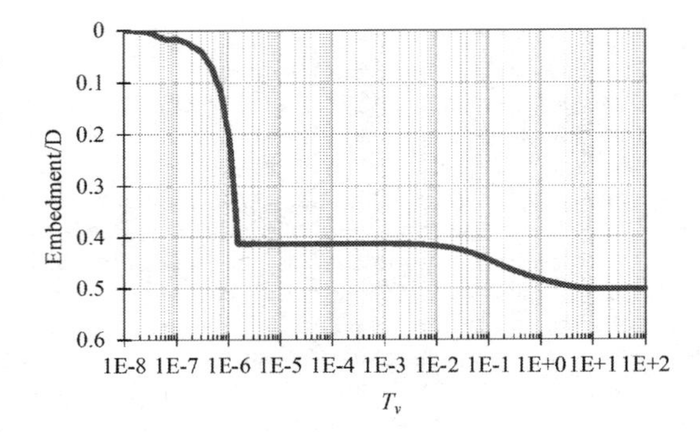

Figure 2. Normalised embedment versus time factor.

$$c_v = \frac{k}{m_v \gamma_w} \quad (3)$$

where k is the soil permeability, m_v is the volume compressibility and γ_w denotes the unit weight of water. The virgin compressibility in the Cam Clay model can be expressed as

$$m_v = \frac{\lambda}{(1+e_0)\,p'_0} \quad (4)$$

where p'_0 denotes the initial mean effective stress at the pipe invert.

3.1 *Vertical penetration*

Accurate prediction of the pipeline embedment is considerably important for the assessment of its subsequent lateral behaviour, and embedment is the most important parameter that affects the lateral response. Figure 2 depicts the variation of the normalised settlement of the pipe during the loading process and subsequent consolidation stage. According to Figure 2, the pipe embedment at the end of the loading ($T_v = 1.4 \times 10^{-6}$) is 0.35D; however, the pipe penetration continues to increase up to $T_v = 1.54 \times 10^{-6}$, when it comes to rest at an embedment depth of 0.41D. The increased settlement after the installation is because of inertial effects. It is noted that for a static analysis the pipe settlement due to installation should not increase after the vertical loading process is completed. Sabetamal et al. (2014) showed that the excess pore pressure at the pipe invert is not significantly affected by the inertia forces, whereas, the embedment depth increases markedly because of the dynamic forces. Therefore, a dynamic approach is, indeed, necessary for coupled problems of pipe–seabed interaction involving very fast loading for which static coupled solutions might not be appropriate. This conclusion is in agreement with practical observations of the as-laid pipeline embedment, which is typically much greater than would be expected from the static pipeline weight. Although the increased penetration could be attributed to different mechanisms, three major mechanisms here include a stress concentration at the pipe invert (touchdown point), induced inertia forces and subsequent potential for partial liquefaction of the soil under the pipeline. Other possible causes of the increased embedment in practice are an additional vertical force near the touchdown point resulting from catenary effects, as well as remoulding or displacement of the soil resulting from small-amplitude cyclic movements of the pipeline throughout the laying process, which may cause significant self-burial. However, the last two phenomena were not considered in the analysis described here.

Existing methods of estimating pipeline embedment are based on correlations from model tests (e.g., Verley & Lund 1995), theoretical and numerical analyses using limit plasticity (Murff et al. 1989, Randolph & White 2008) or the FE method (Aubeny et al. 2005, Merifield et al. 2009, Chatterjee et al. 2013). These works do not take inertial effects into account when deriving the relation for vertical penetration, whereas the analysis described here shows that dynamic effects considerably increase the vertical penetration and these effects should not be necessarily ignored.

Figure 3 depicts the contours of the excess pore-water pressure at different times during the embedment process. It is evident from these contour plots that the maximum pore pressure is observed at a point below the pipe invert diminishing towards the free surface along the pipe wall. As the pipe embeds deeper into the soil, the pore-pressure contours grows in size and shape. The extent of the compressive pore pressure contours indicates the amount of soil undergoing compression and shearing because of the pipe loading.

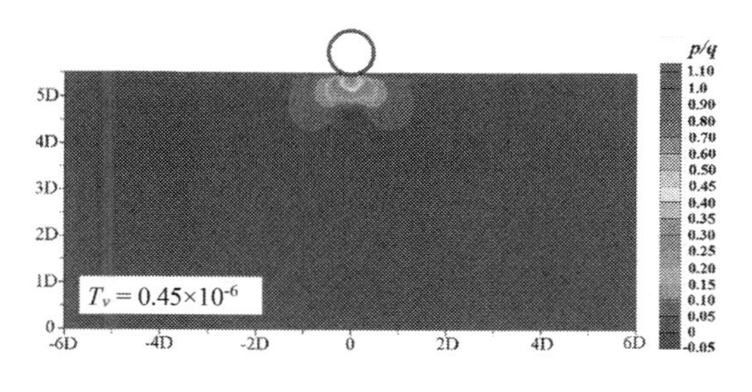

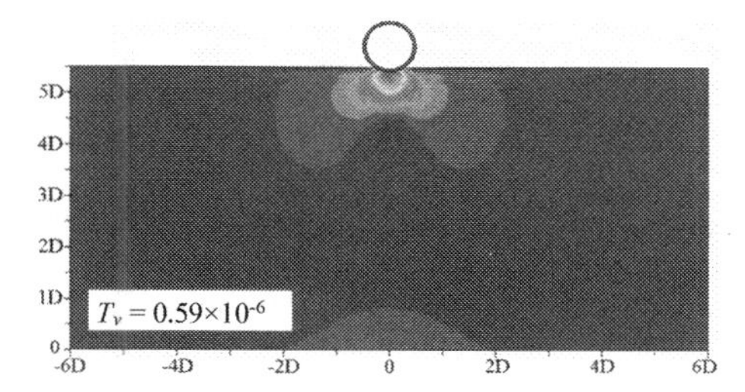

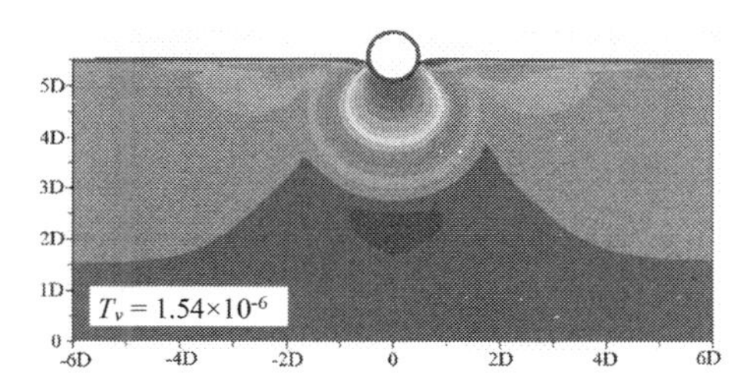

Figure 3. Excess pore-water pressure contour plots during the loading stage.

The shape of the contours is a reflection of the curved surface geometry of the pipe. A small zone of suction can be seen advancing downwards as the pipe is embedded deeper. This is due to the development of a plastic expansion (softening) region beneath the pipe beyond the compression zone. At the end of loading, pore water pressures larger than $0.2q$ are observed in a vertical interval up to around 3D under the pipe invert. The displaced soil during pipe embedment causes the gradual formation of soil heave at the two sides of the pipe, leading to pipe–soil contact over more of the pipe perimeter when compared with a whished-in-place pipe. The enlarged contact surface increases the resistance to both axial and lateral movement. Further, the level of thermal insulation provided by the soil would increase because of the reduced exposure of the pipe to the free water circulating around the pipe. The excess pore pressures at and around the heave may affect pipe stability and, in particular, its lateral buckling, as elevated pore pressures could soften heaved soil, thereby possibly

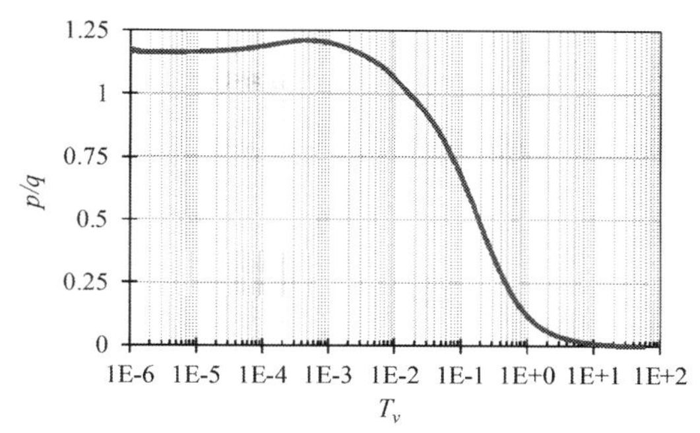

Figure 4. Dissipation of excess pore pressure at the pipe invert.

providing additional resistance to a buckling pipe. However, the time delay for the dissipation of pore pressures around the heave zone would be relatively small compared to the pipe invert, as it is closer to the free surface.

Once the loading process is completed, the analysis proceeded to the consolidation stage. The axial and lateral resistance of the pipeline are significantly affected by the degree of consolidation following installation. The dissipation of compressive excess pore water pressures results in an increase in the shear strength of the soil near the pipe and alters the strength distribution around it, leading to an increase in the breakout resistance. Figure 4 depicts the time history of the normalised excess pore-water pressure at the pipe invert, indicating that the developed excess pore-water pressure had entirely dissipated when $T_v = 10$. During the soil consolidation, the excess pore pressures around the pipe wall dissipate with time, considerably influenced by the horizontal length from the pipe invert as pore water essentially moves away from the pipe invert towards the free surface. This trend of pore pressure dissipation has been shown in Sabetamal (2015) through visualising the direction of the pore-water flow during the consolidation stage.

3.2 *Lateral movement*

After the pipe was fully embedded and the consolidation process had finished, the vertical force was reduced to $q' = 2.7\ s_{u0}$, which is 57 per cent of the force required for embedment, corresponding to an overloading ratio of $q/q' = 1.76$. This specific value of overloading was chosen in order to attain a relatively steady embedment following the lateral displacement. Depending on whether the pipe is heavy or light, relative to the vertical bearing capacity at the current embedment, lateral breakout will be accompanied by either downward

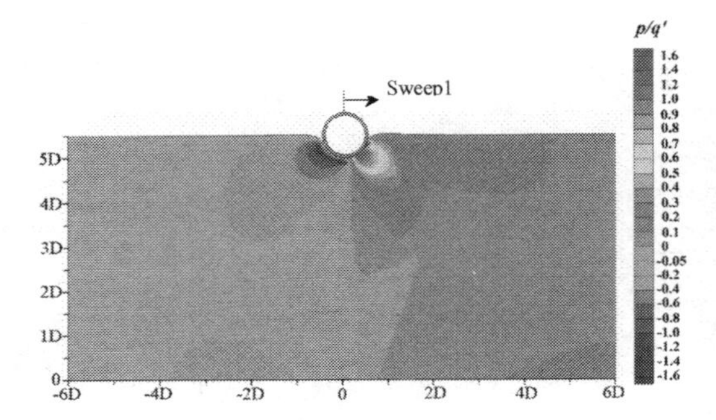

Figure 5. Excess pore-pressure contours during lateral movement.

or upward movement. Nevertheless, the unloading was applied during a period of $\Delta T_v = 1.4 \times 10^{-6}$, which was the same rate as adopted in the initial loading stage. This relatively rapid unloading process resulted in some negative pore pressure (suction) in the soil around the pipe surface.

The pipe was then moved horizontally while keeping the vertical load constant and allowing the pipe to freely move up or down. The rate of lateral displacement was 0.625D/s, and it was applied in a rightwards direction for a period of 2 s (sweep1). This rate of movement was sufficiently high to invoke dynamic effects in the analysis and effectively impose undrained conditions within the soil around the pipe. It is notable that analyses with different rates of movement, as well as static solutions to the problem, could be obtained, but such comprehensive parametric analyses are not the focus of this study.

According to Figure 5, suction pore pressures are generated behind the pipe as soon as it is displaced laterally to the right, whereas compressive pore pressures are developed ahead of the pipe. It is notable that the suction behind the pipe generates a tensile force. This force also potentially resists the lateral movement of the pipe, but it was not considered in the analysis through the contact algorithm. Its inclusion is a relatively trivial task, but it was not followed in any detail in this study.

When the pipe movement at sweep1 was finished (the pipe invert had been displaced 1.25D horizontally in the rightwards direction), the generated active berm was deposited (Fig. 6a) and the direction of pipe movement was reversed to pass its initial position during sweep2. The backwards movement was continued in the leftwards direction (sweep3) to the same amount (–1.25D) and then reversed again to the first place (sweep4). The deformed FE mesh at the extremities of sweep 1 and sweep 3 are depicted in Figure 6. As it can be

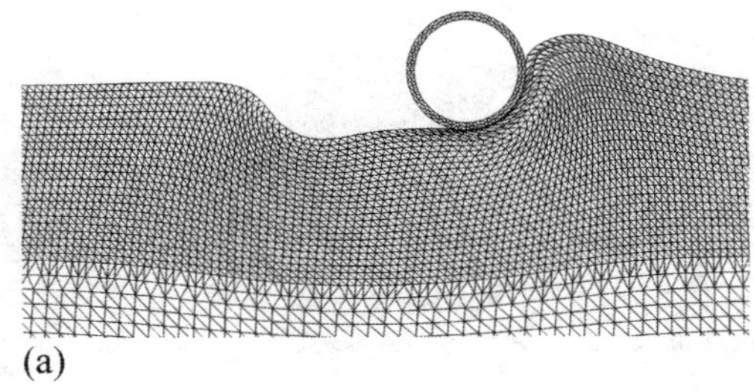

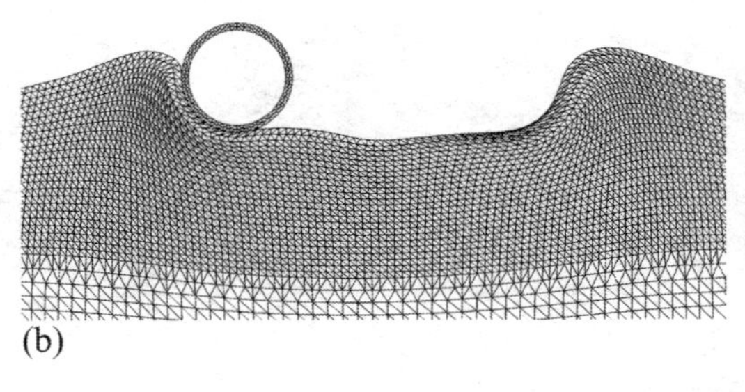

Figure 6. Deformed FE mesh at the extremity of (a) sweep1; (b) sweep3.

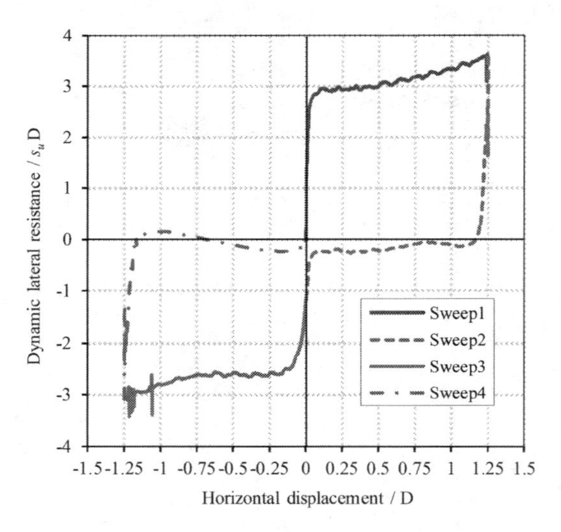

Figure 7. Dynamic lateral resistance: 1st, 2nd, 3rd and 4th sweeps.

seen the quality of the deformed mesh is reasonably well preserved.

Figure 7 depicts the lateral dynamic load-displacement response of the pipe over the four sweeps. For sweep1, the initial breakout of the pipe occurs at a lateral displacement of ~0.1D, followed by a steady residual resistance up to the horizontal

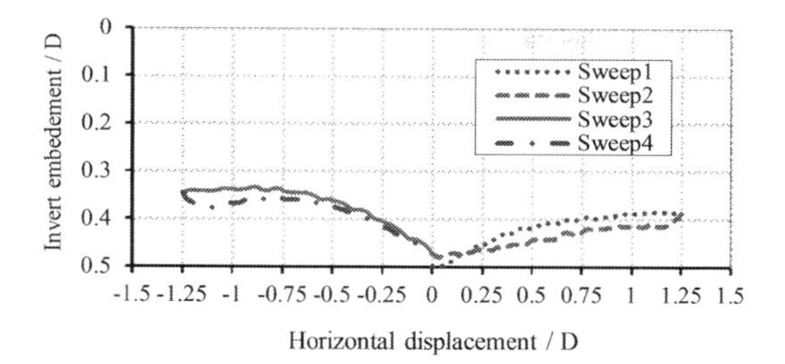

Figure 8. Pipe invert trajectory during lateral movement.

displacement of 0.60D, during which the pipe has an upward trajectory, as depicted Figure 8. This upward movement of pipe is typical of light pipes. From this point on, the pipe approached a relatively steady elevation, and the lateral resistance increased gradually because of the steady growth of the active berm ahead of the pipe (Fig. 6a). Upon reversal of the movement direction (sweep2), the lateral resistance was released (see Fig. 7).

Throughout sweep2, the pipe slid backwards over the created trench and slightly deepened surface profile of the trench, as depicted in Figure 8. When the pipe approached its initial location, it faced the pre-existing soil berm created during the embedment process. Accordingly, a steep increase in soil resistance was then experienced by the pipe (Fig. 7). The pipe continued its backwards movement (sweep3) and pushed the berm ahead, in which the second breakout occurred at ~-0.20D lateral movement (Fig. 7). A similar behaviour was then repeated as observed for sweep1, such that a steady residual resistance was followed after a breakout while the pipe moved slightly upwards, and then the lateral resistance increased gradually because of the steady growth of the berm ahead of the pipe. However, in sweep3, the embedment of the pipe invert decreased more than that observed for sweep1 (see Fig. 8). This suggests that the left berm mobilised more resistance compared to the right berm initially faced by the pipe in sweep1. Accordingly, the upward movement of the pipe was larger compared to sweep1, during which the pipe pushed a smaller volume of soil and ultimately led to a decreased breakout and residual lateral resistance. The higher passive resistance of the left berm resulted from the existence of suction pore pressures generated during sweeps1–2. At the extremity of sweep3, the generated active berm was deposited and the pipe reversed to travel the created trench through sweep4, during which the lateral resistance was released (Fig. 7).

4 CONCLUSIONS

A large deformation finite element methodology combined with the Modified Cam Clay (MCC) plasticity soil model was developed for this study to explore the coupled consolidation behaviour of pipe-soil interaction throughout the full process of vertical pipe penetration followed by soil consolidation and large-amplitude lateral pipe displacement. Accordingly, the whole of life behaviour of a pipe element sweeping repeatedly within a buckle was simulated, while capturing the excess pore pressure generations and likely interspersed consolidation effects within the large lateral deformation framework. The embedment of the as-laid pipes was seen to be affected markedly by dynamic forces which therefore should be included in any empirical relations describing pipe behaviour.

According to the analysis results, three key stages involved in the force-displacement response of the system including (1) initial breakout of the pipe; (2) suction release, which occurs at the back face and beneath of pipe and can cause a sudden drop in soil resistance upon separation of the pipe from the soil behind it; (3) resistance increase associated with the growth of an active berm ahead of the pipe; and (4) additional passive resistance because of suction forces within a pre-existing dormant berm.

REFERENCES

Aubeny, C.P, Shi. H, Murff, J.D. 2005. Collapse load for cylinder embedded in trench in cohesive soil. *Int J Geomech* 5(4): 320–325.

Brutton, D.A.S, White, D.J, Carr, M, and Cheuk, C.Y. 2008. Pipe-soil interaction during lateral buckling and pipeline walking—the safebuck JIP. *Proc Offshore Technology Conf*, Houston, TEX, OTC 19589.

Cardoso, C.O, & Silvira, R.M.S. 2010. Pipe-soil interaction behavior for pipelines under large displacements on clay soils: a model for lateral residual friction factor. *Proc Offshore Technology Conf*, Houston, TEX, OTC 20767.

Chatterjee, S, White, D.J, Randolph, M.F. 2012. Numerical simulations of pipe-soil interaction during large lateral movements on clay. *Géotechnique* 62(8): 693–705.

Chatterjee, S, White, D.J, Randolph, M.F. 2013. Coupled consolidation analysis of pipe-soil interactions. *Can Geotech J* 50: 609–619.

Cheuk, C.Y, White, D.J, Bolton, M.D. 2007. Large-scale modelling of soil–pipe interaction during large amplitude cyclic movements of partially embedded pipelines. *Can Geotech J*, 44, 977–996.

Chung, J & Hulbert, G.M. 1993. A time integration algorithm for structural dynamics with improved numerical dissipation: the generalized-α method. *J Appl Mech* 60: 371–375.

Dingle, H.R.C, White, D.J, Gaudin, C. 2008. Mechanisms of pipe embedment and lateral breakout on soft clay. *Can Geotech J* 45: 636–652.

Dutta, S, Hawlader, B, Philips, R. 2014. Finite element modeling of partially embedded pipelines in clay seabed using coupled Eulerian-Lagrangian method. *Can Geotech J* 52: 58–72.

Fischer, K.A, and Wriggers, P. 2005. Frictionless 2D contact formulations for finite deformations based on the mortar method. *Comput Mech* 36: 226–244.

Kellezi, L. 2000. Local transmitting boundaries for transient elastic analysis. *Soil Dyn Earthq Eng* 19(7): 533–547.

Merifield, R.D, White, D.J, Randolph, M.F. 2009. The ultimate undrained resistance of partially embedded pipelines. *Géotechnique* 58(6): 461–470.

Morland, L.W. 1972. A simple constitutive theory for a fluid-saturated porous solid. *J Geoph Res*, 77: 890–900.

Murff, J.D, Wanger, D.A, Randolph, M.F. 2009. Pipe penetration in cohesive soil. *Géotechnique* 39(2): 213–229.

Randolph, M.F & White, D.J. 2008. Upper-bound yield envelopes for pipelines at shallow embedment in clay. *Géotechnique* 58(4): 297–301.

Sabetamal, H. 2015. Finite element algorithms for dynamic analysis of geotechnical problems. PH.D. Thesis, University of Newcastle, Newcastle, Australia.

Sabetamal, H, Carter, J.P, Nazem, M, Sloan, S.W. 2016a. Coupled analysis of dynamically penetrating anchors. *Comput Geotech* 77: 26–44.

Sabetamal, H, Nazem, M, Sloan, S.W, Carter, J.P. 2016b. Frictionless contact formulation for dynamic analysis of nonlinear saturated porous media based on the mortar method. *Int J Numer Methods Engng* 40(1): 25–61.

Sabetamal, H, Carter, J.P, Sloan, S.W. 2017. Pore pressure response to dynamically installed penetrometers. *submitted to Int J Geomech*.

Sabetamal, H, Nazem, M, Carter, J.P. Sloan, S.W. 2014. Large deformation dynamic analysis of saturated porous media with applications to penetration problems. *Comput Geotech*, 55: 117–13.

Sabetamal, H, Nazem, M, Sloan, S.W, Carter, J.P. 2014. Numerical modelling of offshore pipe-seabed interaction problems. *Proc 14th Int Association for Computer Methods and Recent Advances in Geomechanics* 2014. IACMAG, 655–660.

van Leer, B. 1977.Towards the ultimate conservative difference scheme IV. A new approach to numerical convection. *J. Comput. Phys.* 23: 276–299.

Verley, R. & Lund, K.M. 1995. A soil resistance model for pipelines placed on clay soils. *Proc 14th Int Offshore Mech Arctic Eng Conf*, OMAE 5: 225–232.

Wang, D, White, D.J, Randolph, M.F. 2010. Large deformation finite element analysis of pipe penetration and large-amplitude lateral displacement. *Can Geotech J* 47(8): 842–856.

Wriggers, P. & Haraldsson, A. 2003. A simple formulation for two-dimensional contact problems using a moving friction cone. *Commun Numer Meth Eng* 19: 285–295.

Numerical Methods in Geotechnical Engineering IX – Cardoso et al. (Eds)
© 2018 Taylor & Francis Group, London, ISBN 978-1-138-33198-3

Validation of Coupled Eulerian-Lagrangian (CEL) method by means of large scale foundation testing

D. Heinrich, T. Quiroz & A. Schenk
Fraunhofer Institute for Wind Energy Systems IWES, Germany

ABSTRACT: Numerical investigations with finite element models are a powerful option used in combination with geotechnical model tests to clarify the load-deformation behavior of foundations. A crucial issue is the correct representation of the installation effects and the load history in the numerical model. A recent common practice is the numerical implementation of special methods capable to deal with large soil deformations. In this paper the challenges emerging from performing numerical simulations with the Coupled Eulerian-Lagrangian (CEL) method in an elasto-plastic model will be presented for jacked piles and circular footings. The results of the numerical simulations will be compared with experimental evidence obtained from large scale model tests carried out in saturated non-cohesive soil. The analysis includes the evaluation of the potentials and limitations of the CEL method to represent the installation process, the stiffness and the load bearing capacity of shallow and piled foundations.

1 INTRODUCTION

The accurate prediction of the load-displacement behavior of a foundation system is nowadays still one of the central subjects in geotechnics. With the latest development of more powerful computers, finite-element-based calculations are widely used to solve geotechnical problems involving large soil deformations. Currently a number of methods that enable large strain modelling are used in combination with traditional finite element formulations with remarkable efficiency and accuracy (Wang et al., 2015). One of the most promising methods for geotechnical applications is the Coupled Eulerian-Lagrangian (CEL) method.

The applicability of the CEL method for a selection of diverse geotechnical problems was outlined by Qui et al. (2009). In this work it was shown that the singularities derived from large soil movement in a conventional finite element calculation (i.e. implicit) can be avoided through the CEL implementation and the inclusion of a void volume in the model to allow the unrestricted material movement during the soil deformation phase. Gütz et al. (2013) explored the robustness of the CEL method on the basis of numerical investigations to simulate the installation process of spud cans for offshore jack-up platforms.

The potentials of the CEL method to simulate the installation of piled foundations were addressed in Henke (2013) and Moorman et al. (2015) with special focus on thin-walled structures such as steel open ended piles. These studies revealed that the soil plug occurrence inside the pile, as a result of the increase of the soil stresses during pile penetration, is possible to recreate correctly with this method.

For an accurate estimation of the load-bearing behavior of soils subjected to large deformations, calibrated finite element models are needed. Relevant aspects required for the selection of the simulation technique are advanced knowledge on constitutive soil modeling and the definition of material parameters. The material parameters for soil modeling are defined by means of standardized soil laboratory tests. Advanced soil models (i.e. Hardening Soil with small strain stiffness or Hypoplasticity) are not broadly implemented in commercial programs and require the definition of more material parameters.

The computational calculations performed with calibrated finite element models can be validated by comparing the numerical results against field measurements of prototypes or by involving experimental results from investigations with physical models under fully controlled environment. In the geotechnical pit of the Test Center for Support Structures (TTH) of the Leibniz Universität Hannover, large scale experimental investigations with foundation systems are performed by Fraunhofer IWES with focus on model validation.

One of the main advantages of this test facility is the possibility to recreate uniform and controllable soil test conditions, ensuring repeatability of the experiments and opening the possibility to validate numerical models.

In this work the applicability of the CEL method will be evaluated for two case studies involving large soil deformations. The considered geotechnical systems are: 1) the jacked installation of a steel circular open ended pile; 2) the complete failure mechanism of an axially loaded rigid circular footing. Both systems rest on saturated dense sand. The load-deformation behavior of the soil will be idealized using an elasto-plastic formulation based on the Mohr-Coulomb failure criterion.

The analysis will be carried out on the basis of the comparison between numerical and experimental results. The information obtained from the comparison will be used to evaluate the potentials and limitations of the CEL implementation and to confirm the model assumptions.

Figure 1. Geotechnical test pit after sand preparation at the TTH test facility.

2 LARGE SCALE MODEL TESTS

2.1 *Geotechnical test pit*

The large scale model tests presented in this paper were carried out in the geotechnical facilities of the TTH. The large dimensions of the geotechnical test pit with a length of 14 m, a width of 9 m and a depth of 10 m offer ideal features (i.e. homogenous soil and advanced actuator systems) for large scale testing (Figure 1).

To prepare the soil sample in the geotechnical pit a volume of around 1200 m^3 uniformly graded medium silica sand ("Rohsand 3152") was distributed in 30 cm layers across the pit. The sand layers were successively compacted at optimal compaction moisture with electro-dynamic vibratory plates. After the compaction of each layer representative soil samples were extracted and analyzed in the laboratory to estimate the soil density. The physical properties of the model sand are listed in Table 1.

The mean value of the relative density of the compacted sand reached a value of $D_R = 0.76$. Finally, the prepared sand was uniformly saturated by filling water very slowly (30 cm/day) from the bottom of the pit through a drainage system.

Table 1. Physical properties of the model sand "Rohsand 3152".

Property	Symbol	Unit	Value
Maximum void ratio	e_{max}	-	0.83
Minimum void ratio	e_{min}	-	0.44
Specific gravity	G_s	-	2.65
Coeff. of uniformity	C_u	-	1.97
Coeff. of curvature	C_c	-	0.98
Grain diameter to 60% Passing material	d_{60}	mm	0.407

Figure 2. Experimental setup of the pushed pile at the end of pile installation.

Further information regarding the soil preparation method used in the geotechnical pit of TTH is described in Spill et al. (2017) and Schmoor et al. (2017).

2.2 *Jacked pile installation*

For the jacked installation a steel open ended model pile (steel grade S355) was considered as test specimen. The circular open ended pile with a diameter of D = 0.29 m, a wall thickness of t = 0.01 m and a pile embedded length of h = 4.0 m was pushed continuously in to the dense saturated sand of the pit with a constant penetration velocity of v = 0.003 m/s.

During the jacked installation the required axial force and the installed pile length were recorded using sensors and a data collection system. The experimental setup at final installation depth is shown in Figure 2. It consists of a large steel loading frame and a hydraulic cylinder connected to the horizontal beam for the vertical load application. In correspondence to the designated maximum embedded length of h = 4 meters, a pushing force V of 1150 kN was recorded.

2.3 *Rigid circular footing under axial loading*

A circular steel plate with a diameter D = 0.65 m and a plate thickness of t = 0.055 m was utilized to simulate the load-settlement behavior of a rigid circular footing under axial loading. The test infrastructure required for the physical experiment was nearly identical to that used for the pile test described in Section 2.2. In this case the bottom of the foundation was resting on the top of the sand surface before the pressing force was applied. To ensure perpendicularity between contact surfaces during the initial phase of the load application, the sand surface was prepared rigorously.

The load bearing capacity test was carried out using a hydraulic actuator, which was coupled to the circular footing and fixed to the loading frame. The actuator pushed the footing into the saturated sand with a constant displacement velocity of v = 0.01 mm/s idealizing a "quasi static" loading. As part of the experiment an unloading and re-loading phase was induced (see Figure 7). During the test the maximal force reached a value of V = 83 kN and occurred at a settlement of u = 18 mm. The pushing force was recorded with a load cell integrated in the cylinder. The footing settlement was registered with the internal positioning sensor of the cylinder and an external inductive position transducer (LVDT).

3 NUMERICAL MODEL

3.1 *CEL method*

The CEL method, implemented numerically in the commercial ABAQUS/Explicit (Simulia), is used widely in geotechnics to simulate geotechnical systems exposed to large soil deformations. The main feature of this method is its mesh-based description, which combines the advantages of the Lagrangian and Eulerian finite element formulations.

In the Lagrangian approach the material movement is described as function of the time and the material coordinates, resulting in a material-based mesh. In this case the mesh nodes are displaced when the material undergoes deformations. Large deformations can cause mesh distortions and consequently lead to convergence problems during the simulation.

The Eulerian modelling approach formulates the material movement as function of the time and the space coordinates. Since the Eulerian mesh is fixed in space and the material can "flow" through the mesh, the mesh integrity remains intact at large material deformations. In the CEL method the structure, which should not experience high deformations, will be declared as Lagrangian rigid solid and the body suffering the higher deformations as Eulerian. Consequently, for geotechnical applications the soil is discretized through Eulerian and the structure through Lagrangian elements. The contact between soil and structure is given by the general contact definition, which enforces contact between all the areas interacting with the surrounding geometry of the Lagrangian mesh.

One of the features of the CEL method is the utilization of the explicit solution algorithm, which postulates the existence of an explicit solution only for discrete time steps Δt and not for arbitrary times t. Thus, the stability of the solver is a critical issue to be considered. Additionally, the control over the explicit solution is more complicated, because the convergence has to be ensured. Consequently, the stability and accuracy of the calculation depend on the selected time step Δt.

To ensure a stable solution the selected time step should not exceed the critical time step $\Delta t_{crit.}$ By reducing the element deformation as a wave propagation problem, the critical time step can be formulated by Equation 1 as:

$$\Delta t_{crit} = \frac{Le}{cd} \tag{1}$$

where L_e is the smallest element length and c_d is the dilatory wave propagation velocity. From Equation 1 it is clear that the number of time steps and the calculation time may increase considerably for thin-walled structures.

3.2 *Soil constitutive model and material parameter*

In this study, the well-established Mohr-Coulomb failure criterion was adopted. The main advantages of this model over more complex ones are the reduced number of material parameters required and the direct computational implementation of these parameters obtained from standard soil laboratory test.

The elasto-plastic Mohr-Coulomb formulation is included as predefined soil model in the finite

Table 2. Soil parameters for the numerical simulations.

		Jacked pile	Circular footing
Symbol	Unit	Value	Value
E	MN/m²	36	25.5
v	-	0.3	0.3
γ_s'	kN/m³	10.16	10.16
c'	kN/m²	1.5	1.5
φ'	°	33.0*[2]	36.0*[1]
ψ	°	φ' - 30°	φ' - 30°
K_0	-	1 - sinφ'	1 - sinφ'

*[1] Peak friction angle
*[2] Critical state friction angle

element program ABAQUS/Explicit. In this soil model the strain-stress dependency behaves linear within the elastic range and it is defined by two parameters from the elasticity theory; named the Young's modulus E and the Poisson's ratio v. The failure criterion is defined by the friction angle ϕ' and the cohesion c. The flow rule is described by the dilatancy angle ψ.

The soil input parameters for the two finite element models presented in this study (jacked pile installation and rigid circular footing under axial load) are listed in Table 2:

The values of the Young's modulus E used in the numerical simulations were determined by oedometer test in accordance with the German standard DIN 18135 (2012). For the numerical implementation the initial soil stress level within the assumed influence zone below the foundation (four times the foundation diameter) was averaged to select the Young's modulus from the oedometer reloading curve. Since the soil model used in the numerical model does not support the stress-dependency of the Young's modulus, its initial value remained constant during the computations.

The effective friction angle φ' of the model sand "Rohsand 3152" was determined by means of direct shear tests according to DIN 18137–3 (2002) and drained triaxial tests according to DIN 18137–2 (2002). From the direct shear tests a critical friction angle of $\varphi'_{cr} = 33°$ was derived. The peak friction angle φ'_p was obtained from the interpretation of drained triaxial tests and it reached a mean value of $\varphi'_p = 36°$.

4 FE-MODEL JACKED PILE

4.1 *CEL Model*

Based on the geometry of the reference jacked pile described in Section 2.2, the boundary conditions of the geotechnical pit of the TTH were reproduced in a numerical model created with the finite element program ABAQUS/Explicit. The geometry of the numerical model is illustrated in Figure 3. The soil is modeled with a depth of 10 m. The CEL formulation supports the use of 3D elements. Consequently, even though the treated geotechnical system is axisymmetric, a wedge corresponding to one eighth of the entire geometry was simulated. By doing this the number of elements can be reduced and the calculation time efficiency enhanced.

To avoid boundary effects influencing the computations, the boundary conditions of the cylindrical-slice-shaped model were defined at 5 meters from pile center. The boundary conditions set on the most external soil elements restrict the material movement inside the limits of the model geometry.

Above the soil an additional free material layer with a height of 2 m was incorporated to enable upwards soil movement during the pile penetration process. The soil and the free material layer were defined as Eulerian instances and discretized using three dimensional elements EC3D8R. This element type is defined by eight nodes and a reduced integration scheme. The entire Eulerian volume is assembled with a total of 204,576 elements.

The material parameters of the soil are listed in Table 2. Since the pile deformations during the pushing process are not relevant for this study, the steel pile was modeled as rigid body.

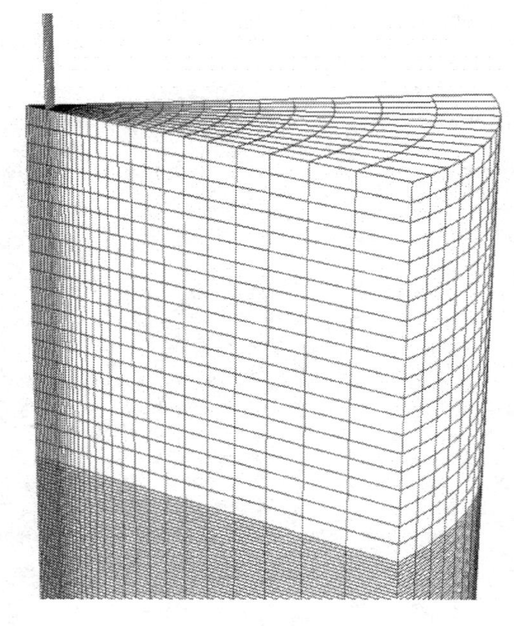

Figure 3. Segmented view of the CEL finite element model for the simulation of pile installation.

The jacking process begins after the pile tip is positioned and placed in contact with the soil surface. As already mentioned, the contact between steel and sand is given by the general contact algorithm and the Coulomb friction law is used as frictional condition. The wall friction angle δ between the steel open ended pile and the sand can be estimated from the following equation:

$$\delta = \frac{2}{3}\phi' \tag{2}$$

4.2 *Numerical results and discussion: Jacked Pile*

The results of the numerical implementation of the CEL method are presented as a load-penetration curve and compared with the results of the large scale experiment in Figure 4. The comparison delivers a good agreement between experimental and numerical curves.

In the pile segment corresponding to an installation depth h of 1.25 m to 1.75 m, the numerical curve underestimates the penetration force by 10–20%. The underestimation becomes practically negligible at an installation depth within 3.25 m and 4.00 m, where the values deviate between 3 to 9% from the experimental curve.

The load-settlement behavior of jacked piles is influenced by the soil stress distribution markedly. Especially for open ended piles, the pile geometry and the soil stresses influence the soil plug formation. Figure 5 illustrates the distribution of the soil stresses inside and outside the pile after the simulation of the jacked installation. The contour plot displays the stress influence zones at pile tip in the vertical and horizontal direction, showing higher stresses inside compared to the outside of the pile. The high stress concentration inside the pile at the pile tip is the evidence of soil plug development.

The soil plug affects the penetration resistance of the pile due to the increase of the pile tip area and as consequence of this the pile behaves as a close ended pile. The formation of a soil plug was observed both in the numerical simulation and measured by the large scale experiment. Besides, in Figure 5 the formation of a pressure bulb below the pile tip can be recognized. This observation infers once more the existence of a soil plug at the final installation depth.

At this depth the slender open ended pile behaves in the numerical model nearly as a full-displacement pile. From the comparisons it can be affirmed that the implementation of CEL method in the program Abaqus/Explicit, using Mohr-Coulomb elasto-plasticity allows the construction of the load-penetration curve of the reference jacked pile with good fidelity. The selection of the effective critical friction angle φ'_{cr}, obtained from direct shear tests, was decisive to improve the accuracy of the simulation.

In the drained direct shear and triaxial tests of the model sand in dense condition, it was observed that at high soil stresses, after reaching a peak value, the shear strength decreases causing large shear strains. In analogy to this, very high stresses and continuous soil shearing are expected during the jacked pile installation. This could explain why the residual shear strength is governing the selection of the effective critical friction angle (φ'_{cr}) for this geotechnical problem. Overall this analysis reveals that the pile load-penetration behavior of the reference situation can be simulated using the CEL method. Besides, with the utilization of the elasto-plastic Mohr-Coulomb soil model an acceptable qualitative and quantitative compatibility of numerical results and experimental evidence was achieved.

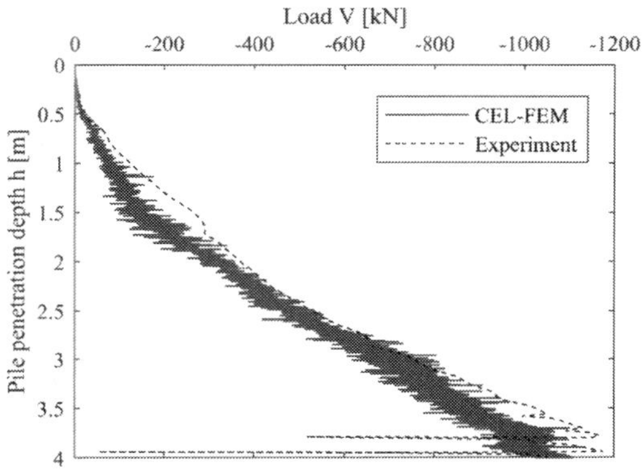

Figure 4. Comparison of the load-penetration curves of the jacked pile.

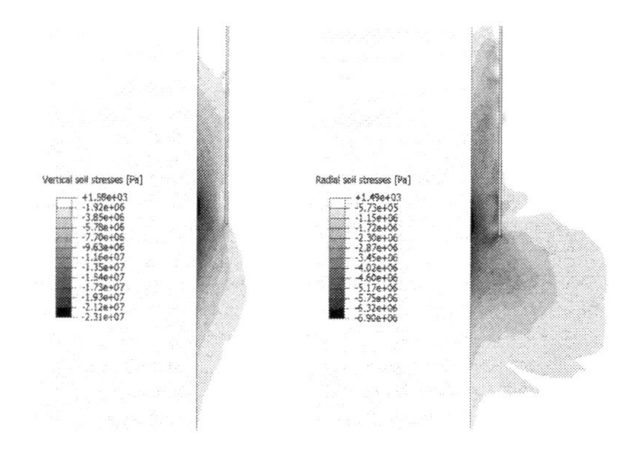

Figure 5. Contour plot of the vertical effective stresses (left) and radial effective stresses (right) at failure.

5 FE-MODEL RIGID CIRCULAR FOOTING

5.1 *CEL-Model*

A finite element model of a rigid circular footing under axial loading resting on sand was generated in the program ABAQUS/Explicit recreating the test and boundary conditions of the large scale experiment described in Section 2.3. For the reasons stated earlier in Section 4.1 even though the geotechnical system is axisymmetric the soil geometry was idealized as a half cylinder with a diameter of 6 m and a height of 6 m.

Directly under the footing, a near-surface soil layer with a thickness of 2.5 m was discretized with a mean element size of 30 mm (see Figure 6). Underneath this layer and for the rest of the model depth, the element size was doubled. A free material layer with a thickness of 0.5 m. was placed above the upper sand surface to provide an empty volume to be filled with the material "flowing" adjacently to the footing during the soil deformations.

The complete Euler-Body was discretized through 214,248 EC3D8R elements. The rigid circular footing was idealized in the model as a rigid body. The interface condition between the surface of the sand in contact with the footing and the bottom of the footing was defined as frictionless. The selection of the Mohr-Coulomb elasto-plastic soil model and its parameters for the numerical simulation was discussed in Section 3.2.

5.2 *Numerical results and discussion: Circular footing*

The result of the numerical simulation, displayed as a load-settlement curve, is presented in Figure 7 and presents a rather good agreement with the experimental data. It is remarkable how the numerical results fit the stiffness at initial strains and the values of load bearing capacity with high precision.

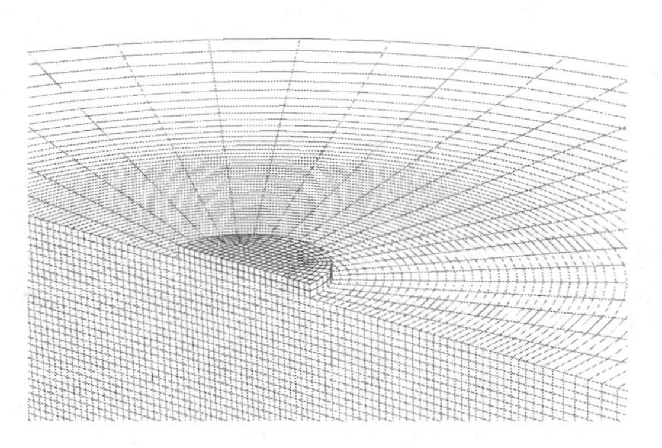

Figure 6. Segmented view of the CEL-FE model for the simulation of the circular footing.

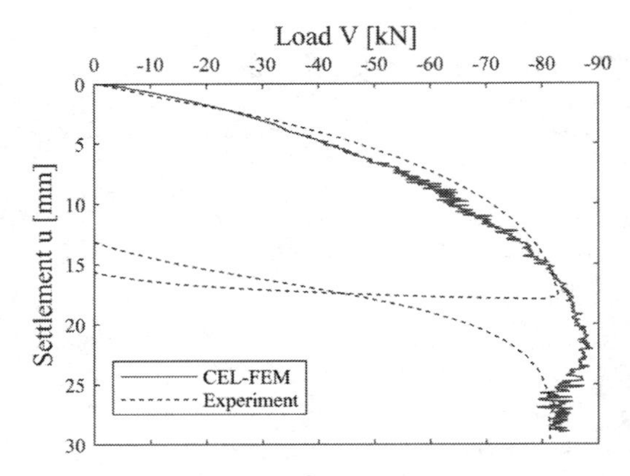

Figure 7. Comparison of load-settlement curves of the rigid circular footing.

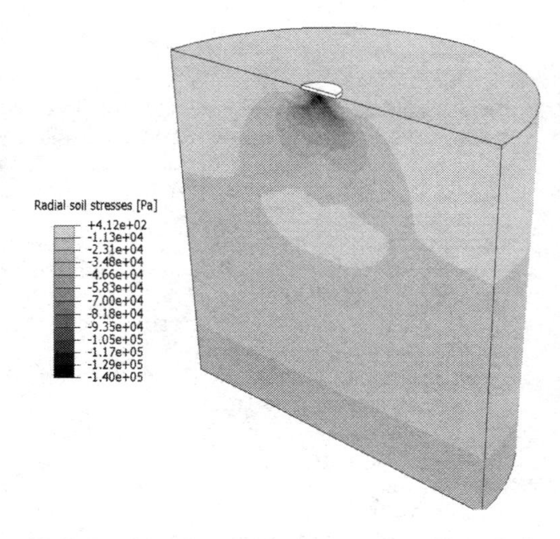

Figure 8. Model overview and contour plot of the radial soil stresses at failure.

The peak load of $V = 88$ kN occurs in the numerical curve at a settlement of around $u = 21$ mm.

Even though the experimental curve presents an unloading-reloading branch, it is evident that both experimental and numerical load-settlement curves follow the same behavioral pattern, characterized by a steady value of the load bearing capacity at a displacement of 28 mm, which is equivalent to a relative settlement of 4.3% of the plate diameter.

At this displacement the load bearing capacity estimated with the CEL method is with 83 kN equivalent to the value obtained from the large scale experiment. The contour plot of the effective radial soil stresses occurring at the failure settlement of 28 mm is displayed in Figure 8 and shows that the induced soil stresses are not affected by the geometry of the model, confirming in turn the choice taken regarding the model size.

6 CONCLUSIONS AND OUTLOOK

In this study, the versatility of the CEL method in the finite element program ABAQUS/Explicit for the simulation of two selected geotechnical problems involving large deformations was evaluated. To enable the evaluation, the computational results were compared with experimental evidence obtained from a large scale model test of a rigid circular footing and a jacked installation of a large scale steel open ended model pile, both resting on dense saturated silica sand.

The simulation technique presented in this work and the consequent methodology used to select the soil input parameters could be validated by comparing the numerical and the experimental load displacement curves of the reference situations. It seems that the selection of the peak soil friction angle obtained from drained triaxial tests is an adequate practice for modeling geotechnical problems involving dilatant soil behavior at very low initial stresses (shallow foundations).

When the potential for soil dilatancy is restricted (i.e. due to large soil confining pressure) and soil failures occur in a continuous manner (installation of jacked piles) the critical friction angle seems to be appropriate. Future comparison and validation works should include the validation of the CEL method for complementary model scales and additional foundation systems as reference.

ACKNOWLEDGEMENTS

The research works related to the jacked installation of a steel open ended pile were realized with the financial support of the German Federal Ministry for Economic Affairs and Energy (BMWi) within the scope of the national research project "Sealence", FKZ 0325912C. The investigation works concerning the rigid circular footing were realized with the financial support of Fraunhofer IWES and the technical assistance of the staff of the Test Center of Support Structures in Hannover.

REFERENCES

DIN 18137–2 2002. Soil, investigation and testing. – Determination of shear strength—Part 2: Triaxial test. German Institute for Standardization. 2002.

DIN 18137–3 2002. Soil, investigation and testing—Determination of shear strength—Part 3: Direct shear test. German Institute for Standardization. 2002.

DIN 18135 2012. Soil, investigation and testing—Oedometer consolidation test. German Institute for Standardization. 2012.

Gütz, P., Abdel-Rahman, K., Achmus, M. & Peralta P. 2013. Numerical modeling of spudcan footing penetration in sand. *SIMULIA Customer Conference,* Vienna.

Henke, S. 2013. *Untersuchungen zur Pfropfenbildung infolge der Installation offener Profile in granularen Böden,* Hamburg: Veröffenltichungen des Instituts für Geotchnik und Baubetrieb (29).

Moormann, C., Labenski, J. & Aschrafi, J. 2015. Simulation of soil plug effects in open steel open ended piles considering the complex soil-structure-interaction during installation. *40th Annual Conference on Deep Foundations,* Oakland, California.

Qiu, G., Henke, S. & Grabe, J. 2009. Applications of coupled eulerian-lagrangian method to geotechnical problems with large deformations. *SIMULIA Customer Conference,* London.

Schmoor, K.A., Achmus, M., Foglia, A. & Wefer, M. 2017. Reliability of design approaches for axially loaded offshore piles and its consequences with respect to the North Sea. *15th International Conference of the International Association for Computer Methods and Advances in Geomechanics.*

Spill S., Kohlmeier M., Wefer M., Dührkop J. & Maretzki S. 2017. Design of Large-scale Tests Investigating the Lateral Load-bearing Behavior of Monopiles. In proceedings of ISOPE 2017.

Wang, D., Bienen, B., Nazem, M., Tian, Y., Zheng, J., Pucker, T., & Randolph, M.F. 2015. Large deformation finite element analyses in geotechnical engineering. Computer and Geotechnics 65, 104–114.

Numerical Methods in Geotechnical Engineering IX – Cardoso et al. (Eds)
© 2018 Taylor & Francis Group, London, ISBN 978-1-138-33198-3

A coupled constitutive model for modelling small strain behaviour of soils

S. Seyedan & W.T. Sołowski
Department of Civil Engineering, Aalto University, Finland

ABSTRACT: Laboratory experiments on soils have shown that the deviatoric stress—deviatoric strain relationship in soils is highly non-linear, especially in the small strain range. That led to several formulations quantifying that nonlinearity. However, the constitutive models which aim to replicate the small strain nonlinearity are often complex and rarely used in geotechnical engineering practice. The goal of this study is to offer a simple way for updating the existing constitutive models to take into account small strain stiffness changes. The study use an existing small-strain relationship to derive a yield surface. When the yield surface is introduced to an existing soil model, it enhances the model with the nonlinear deviatoric stress—deviatoric strain relationship in the small strain range. Subsequently, the paper gives an example of such a model enhancement by combining the new yield surface with the Modified Cam Clay constitutive model. The simulations with the upgraded constitutive model of the undrained triaxial tests on London Clay replicate the experiments in clearly better than the base model.

1 INTRODUCTION

Many studies (e.g. Hardin & Drnevich (1972), Jardine et al. (1984), Jardine et al. (1986), Burland (1989), Simpson (1992) and Benz (2007)) have investigated nonlinear shear stress—shear strain behaviour in different types of soils. The evidence gathered since 1970's led to numerous formulation giving the variation of shear modulus with strains (e.g. Correia *et al.* (2001), Darendeli (2001), Oztoprak & Bolton (2013)). Although the highly nonlinear shear stress-strain behaviour in soil in the small strain range is now well documented, geotechnical engineering practice tends to ignore it as discussed in Doherty and Muir Wood (2013). Soil constitutive models popular in engineering practice typically employ linear or simple non-linear elasticity and fail to describe the nonlinear shear stress-shear strain behaviour at the small strain range (e.g. Atkinson (1993)). Furthermore, the advanced constitutive models which can describe nonlinear behaviour of soils in the small strain range require special soil investigation and extra parameters (Oztoprak & Bolton (2013)) which prevents engineers to apply them in routine practice. Therefore, the need for simple constitutive models, yet capable of capturing nonlinear shear stress—shear strain behaviour in the small strain range is evident.

This research shows how to introduce a yield surface, enhancing existing elastic and elasto-plastic constitutive models so they can capture the nonlinear shear stress—shear strain behaviour of soils at small strain. The study uses existing model giving the secant shear modulus variation in small strain range to derive an elasto-plastic yield surface leading to replication of the shear modulus changes of the base model. The obtained yield surface can be in principle combined with any elasto-plastic model which uses constant shear modulus. As an illustration, the paper combines the new yield surface with the Modified Cam Clay model. Finally, the study shows that the combined model can replicate laboratory tests on a clayey soil in the small strain range.

2 NEW YIELD SURFACE

Many researchers gave formulas for nonlinear shear stress-shear strain behaviour of soils in the small strain range. Hardin & Drnevich (1972) proposed one of the first formulation in the hyperbolic form:

$$G = \frac{G_0}{1 + \dfrac{\gamma}{\gamma_r}} \tag{1}$$

where G is the secant shear modulus, G_0 is the maximum shear modules at the very small strain, γ is the shear strain and γ_r is the reference shear strain equal to the maximum shear stress divided to G_0. This formulation requires calibration of γ_r, which can be difficult (Oztoprak & Bolton (2013)).

Therefore, Darendeli (2001) suggested a similar hyperbolic formulation for sand

$$G = \frac{G_0}{1 + \left(\frac{\gamma}{\gamma_r}\right)^a} \tag{2}$$

where a is a curvature parameter and γ_r is the reference shear strain at which the shear modulus reduces to half of its maximum value, which simplifies the model calibration

Around the same time Correia et al. (2001) and Santos et al. (2001 & 2003) introduced a formulation to predict secant shear modulus in the small strain range for both clays and sandy soils

$$G = \frac{G_0}{1 + \left(0.385 \frac{\gamma}{\gamma_r}\right)} \tag{3}$$

where γ_r is the reference shear strain and is equal to shear strain in which shear modulus reduces to 70% of its maximum value. This equation is only applicable in small strain region.

This paper uses Eq. (3) for finding a yield surface which would lead to small strain behaviour replicating (3). That formulation has been chosen due to its simplicity, as well as its applicability for both clayey and sandy soils. Nevertheless, other formulations could be adopted in a similar manner.

The shear strain in equations 1–3 and the deviatoric shear strain are linked:

$$\gamma = \frac{3}{2}\varepsilon_q \tag{4}$$

Therefore, (3) may be recast into the deviatoric stress space:

$$G = \frac{G_0}{1 + 0.385 \frac{\varepsilon_q}{\varepsilon_{qr}}} \tag{5}$$

where the deviatoric strain and the reference deviatoric strains (ε_q, $\varepsilon_{q\,r}$) replaced the shear strain and the reference shear strain (γ, γ_r).

Elastic (recoverable) and plastic (irrecoverable) strains are the two components of the deviatoric strain in (5). However, change of stress is only due to elastic part of the strain. Therefore:

$$q = 3\,G\,\varepsilon_q = 3\,G^e\,\varepsilon_q^e \tag{6}$$

In (6) q is the deviatoric stress, G^e is the elastic shear modules and ε_q^e is the elastic deviatoric strain. G^e is not constant and changes as shear strain evolves. Investigations have shown that G^e is equal to the maximum modulus G_0 when shear strain is smaller than a threshold (Oztoprak & Bolton (2013)). As shear strain evolves G^e decreases. Therefore, considering elastic formulation with varying G^e during shearing results in more accurate predictions. On the other hand, using varying G^e prevents easy coupling of the new yield surface with constitutive models that use constant G^e, e.g. the Modified Cam Clay model with a constant shear modulus. The goal of this study is to propose an easy upgrade path for the constitutive models widely used in geotechnical practice. As most of these models use constant G^e, representing soil behaviour at medium strains, this study also assumes such a value of constant G^e.

Under assumption of constant elastic modulus Eq. (6) allows for calculation of plastic deviatoric strain:

$$\varepsilon_q^p = \varepsilon_q - \varepsilon_q^e = \frac{q}{3}\left(\frac{1}{G} - \frac{1}{G^e}\right) \tag{7}$$

where ε_q^p is the plastic deviatoric strain. Unfortunately, such a formulation leads to a negative plastic strain when G has a higher value than G^e, which happens at very small stress. Once G becomes smaller than G^e, ε_q^p becomes positive. The negative plastic strain is not physically meaningful, but is necessary mathematically, as it offsets the overestimation of ε_q^e, so the overall deviatoric strain remains correct.

In order to find a yield surface describing small strain nonlinearity, an equation which relates the shear strain to the shear plastic strain is necessary. Combining (3), (6) and (7) provides a function for ε_q^p based on ε_q. The inverse of this function defines ε_q based on ε_q^p:

$$\varepsilon_q = f\left(\varepsilon_q p\right) = \frac{-1 + d + b\varepsilon_q p - \sqrt{\left(1 - d - b\varepsilon_q p\right)^2 + 4b\varepsilon_q p}}{2b} \tag{8}$$

In the equation above, b is equal to $0.385/\varepsilon_{q\,r}$ and d is equal to the ratio of G_0 and G^e, that is $d = G_0/G^e$.

Formulation (8) allows for calculation of the deviatoric strain based on the plastic strain:

$$q = q_0 = \frac{3\,G_0}{\frac{1}{f(\varepsilon_q^p)} + \mathrm{b}} \tag{9}$$

where q_0 is small strain hardening parameter, which tracks the deviatoric stress history. Eq. (9) helps to specify the yield surface for shearing in small strain shearing, which is:

$$F_1 = q_0 - q = 0 \tag{10}$$

The study assumes an associated flow rule. This flow rule and independency of yield surface from mean pressure results in plastic volumetric strain to be zero during small strain shearing.

In the introduced yield surface, q_0 is the hardening parameter and is only dependent to ε_q^p. Therefore, the hardening rule should relate changes of q_0 to variations of ε_q^p. Differentiation of (10) based on ε_q^p results in such a relation. Equation 11 shows the calculated hardening rule.

$$\delta q_0 = \frac{3}{2}\frac{G_0}{\left[1+b\,f(\varepsilon_q^p)\right]^2}\left[1-\frac{\left(1+d+b\,\varepsilon_q^p\right)}{\sqrt{(1-d-b\,\varepsilon_q^p)^2+4\,b\,\varepsilon_q^p}}\right]\delta\varepsilon_q^p \tag{11}$$

Application of the consistency condition ($dF_1 = 0$) leads to calculation of plastic deviatoric strain changes. Equation 12 shows the formulation for those changes in strain.

$$\delta\varepsilon_q^p = \frac{2}{3}\frac{\left[1+b\,f(\varepsilon_q^p)\right]^2}{G_0}\left[1-\frac{\left(1+d+b\,\varepsilon_q^p\right)}{\sqrt{(1-d-b\,\varepsilon_q^p)^2+4\,b\,\varepsilon_q^p}}\right]^{-1}\delta q \tag{12}$$

The introduced yield surface defines small strain behaviour in soils using equation 3. Figure 1 compares variation of secant shear modulus calculated by Eq. (3) and the new yield surface assuming G_0, G^e and γ_r to be 30 MPa, 1 MPa and 0.015%, respectively. Upon examination of figure 1, it is clear that the new yield surface exactly replicates the small strain behaviour defined by equation 3.

Coupling this yield surface with existing constitutive model allows them to describe this aspect of soil behaviour more realistically. This paper uses the Modified Cam Clay model with a constant shear modulus for coupling the yield surfaces. Other constitutive models with a constant shear modulus may add the derived extra yield surface to get the ability to predict small shear strain nonlinearity.

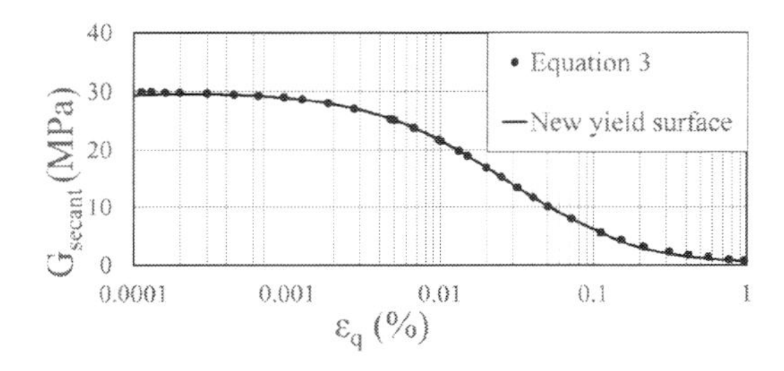

Figure 1. Verification of the new yield surface.

3 MODIFIED CAM CLAY

Roscoe & Schofield (1963) introduced Cam Clay constitutive model for soils. Few years later, Roscoe & Burland (1968) presented the Modified Cam Clay model. The Modified Cam Clay popularity is due to its simplicity—the model yield surface uses only a single shape parameter M. It also can predict soil behaviour qualitatively well and is easy to enhance to include more advanced aspects of soil behaviour. For all those reasons, the model is widely used for simple numerical modelling of soils. This paper also used the Modified Cam Clay model as a base model to present the new yield surface enhancement.

Equation 13 shows the yield and plastic potential surfaces of Modified Cam Clay.

$$F_2 = q^2 - M^2\left[p(p_0 - p)\right] = 0 \tag{13}$$

In this equation, p is the mean pressure, p_0 is the compressive hardening parameter of the model and M is the slope of the critical state line. This yield equation represents a family of eclipses passing through (0, 0) and (p_0, 0) point of stress plane (p:q).

The hardening rule for Modified Cam Clay constitutive model is:

$$\delta p_0 = \frac{\upsilon}{\lambda-\kappa}\, p_0\ \delta\varepsilon_p^p \tag{14}$$

where ν is the specific volume of soil, λ and κ are the slopes of normal compression and unloading-reloading lines in compression plane (lnp: ν).

Modified Cam Clay hardening is not affected by any change in the plastic deviatoric strain. Therefore, during the shearing at small strain, which leads to plastic deviatoric strain, p_0 and the Modified Cam Clay yield surface remain unchanged. The Modified Cam Clay model uses nonlinear elasticity with constant shear modulus. Equation 15 shows the formulation for calculation of elastic strain changes

$$\begin{bmatrix}\delta\varepsilon_p^e\\ \delta\varepsilon_q^e\end{bmatrix} = \begin{bmatrix}\dfrac{\kappa}{\upsilon p} & 0\\ 0 & \dfrac{1}{3G}\end{bmatrix}\begin{bmatrix}\delta p\\ \delta q\end{bmatrix} \tag{15}$$

and Equation 16 shows the relations for calculation of plastic strain changes

$$\begin{bmatrix} \delta\varepsilon_p^{\,p} \\ \delta\varepsilon_q^{\,p} \end{bmatrix} = \frac{\lambda-\kappa}{\upsilon p\,(M^2+\eta^2)} \begin{bmatrix} M^2-\eta^2 & 2\eta \\ 2\eta & \frac{4\eta^2}{M^2-\eta^2} \end{bmatrix} \begin{bmatrix} \delta p \\ \delta q \end{bmatrix} \tag{16}$$

where η is a variable equal to q/p.

4 COUPLING YIELD SURFACES

This study shows how to enhance the existing constitutive models for soils with the capability of replicating soil behaviour at small shear strains. Therefore, the coupling procedure has to keep the base constitutive model (Modified Cam Clay here) intact. Furthermore, any change in the soil behaviour from the small strain shearing to shearing in a higher strain range should be a smooth transition. Finally, the enhanced model should ensure a consistent and realistic unloading-reloading at any shear strain.

As such, the combined model have to use the elastic law of the Modified Cam Clay with the constant G^e. This elastic rule controls the soil behaviour until the stress state reaches one or both yield surfaces. Afterwards, further loading leads to elasto-plastic deformations.

Elasto-plastic behaviour on the new additional yield surface produces deviatoric elastic and deviatoric plastic strains (in case loading has also the volumetric component, change of mean stress is predicted according to the elastic rules). During loading, the deviatoric plastic strains is negative and moves the new yield surface upward. Figure 2 shows a schematic of yield surface changes during small strain loading. Furthermore, no change in the volumetric plastic strain means that Modified Cam Clay yield surface is unaffected.

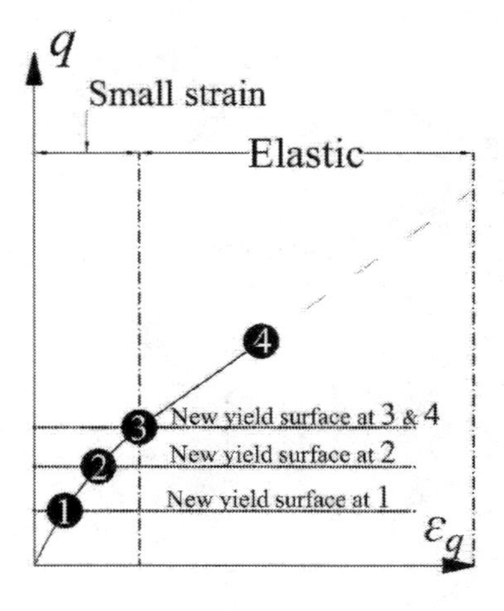

Figure 2. Schematic of small strain loading.

The elasto-plastic behaviour on the Modified Cam Clay yield surface produces deviatoric elastic, volumetric elastic, deviatoric plastic and volumetric plastic strains. The produced deviatoric plastic strain is positive and moves the new yield surface downwards until it reaches 0. Afterwards, changes of deviatoric plastic strain create no change in the deviatoric hardening parameter of the new surface. Figure 3 shows a schematic of new yield surface changes during loading on the Modified Cam Clay yield surface.

Changes in the volumetric plastic strain affect only the Modified Cam Clay yield surface. Consequently, both the Modified Cam Clay and the new yield surface can change when loading happen on the Modified Cam Clay yield surface. Should the stress state satisfy both the Modified Cam Clay and the new yield surface equations, the positive plastic deviatoric strain obtained from the Modified Cam Clay will lead to stress state moving away from the new yield surface, hence yielding will occur only on the Modified Cam Clay yield surface. Therefore, the plastic behaviour of the Modified Cam Clay model remains intact on all the possible stress paths.

Shearing behaviour of soils changes between small strain and higher strain range in a smooth manner. The procedure enforces such a smooth change by checking the tangent shear modulus based on the new yield surface. When the tangent shear modulus reaches the value of the elastic shear modulus, the coupling prevents any further change of q_0. That leads to smooth change of behaviour to that of Modified Cam Clay (Figure 4).

The unloading can be elasto-plastic or purely elastic. That depends on the stress state at the beginning of unloading. If the unloading starts or reaches the new yield surface, the elasto-plastic deformation occur. Such an unloading results in positive deviatoric strain change, moves the new yield surface downward and reverses the small strain loading. The unloading is purely elastic if the previous loading on the Modified Cam Clay surface increased the deviatoric plastic strain above zero, moving the new yield surface to q = 0. Figure 4 shows a schematic of unloading from

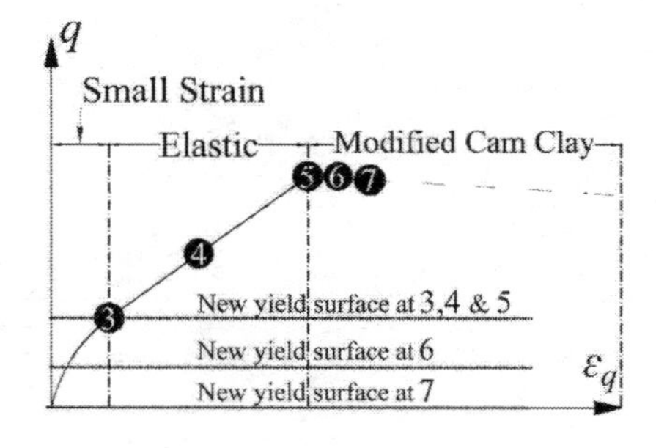

Figure 3. Changes of new yield surface during loading on the Modified Cam Clay.

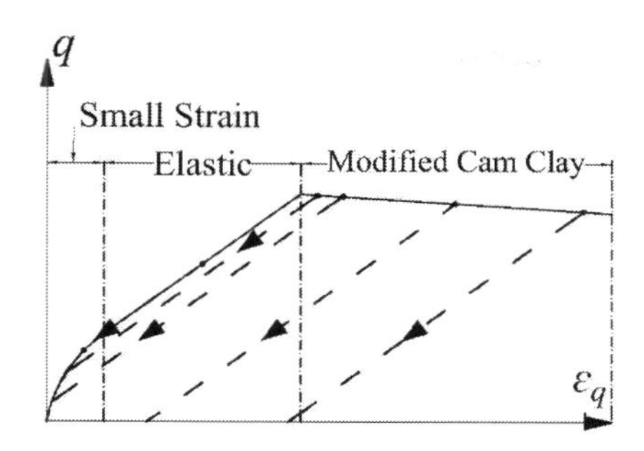

Figure 4. Schematic of unloading from different points.

different points in the deviatoric strain-deviatoric stress space. In this figure, the continuous line represents the loading of the soil while the dashed lines shows different cases of unloading.

5 REPLICATION OF THE TRIAXIAL TESTS

Jardine et al. (1984) performed two undrained triaxial tests on intact overconsolidated London Clay samples instrumented with an axial displacement gauge measuring the internal axial strains during the tests. The presented simulations use parameters of London clay based on investigations of Gasparre (2005) and Viladesau (2004). In addition, equations and data on London Clay from Viggiani & Atkinson (1995) allowed for estimation of the initial shear moduli of London Clay. The new model driver has been implemented with explicit stress integration with error control and NICE technique (Sołowski & Sloan (2016)).

Figure 5 shows the variation of the secant shear modulus in triaxial tests. Laboratory results in figure 5 allows for estimation of $\varepsilon_{q\,r}$ and G^e of the coupled model. These parameters are chosen to be 0.035% and 2.5 MPa. Using these two parameters, the model predicts variation of secant shear modulus in undrained test as shown in figure 5. Table 1 summarizes all the parameters used in the modelling of laboratory tests.

Figure 6 shows the variation of deviatoric stress in the triaxial tests and compares them with the predictions of the Modified Cam Clay and the coupled model. Upon examining the figure, it is clear that during shearing at small strain, the coupled model replicate the nonlinear shear stress—shear strain behavior well. When tangent shear modulus in the coupled model reduces to the elastic modulus, the behavior is no different to that of Modified Cam Clay. However, the improvement in small strain range leads to overall improved predictions, as demonstrated in figure 6.

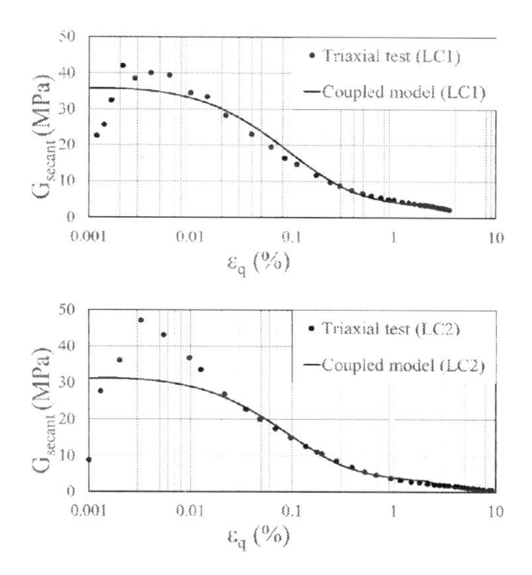

Figure 5. Variation of secant shear modulus based on data from Jardine et al. (1984).

Table 1. Parameters of London Clay in the simulations.

	M	λ	k	G_0 (MPa)	$\varepsilon_{q\,r}$ (%)	G_e (MPa)
LC1				36.82		
	0.55	0.182	0.091		0.035	2.5
LC2				32.21		

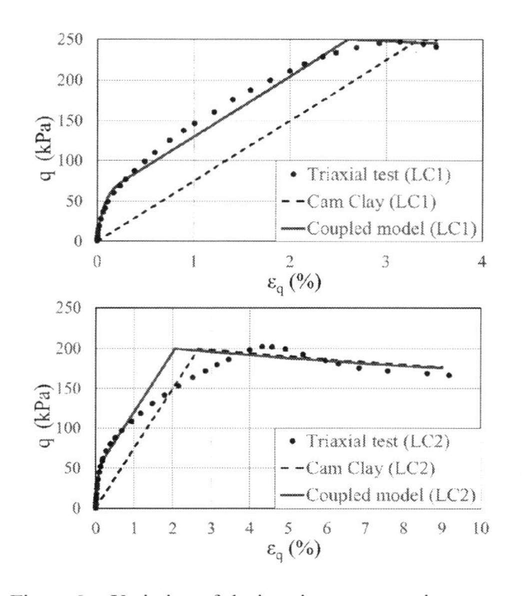

Figure 6. Variation of deviatoric stress—strain.

6 CONCLUSION

This study addresses the need of geotechnical engineering practice for simple constitutive models capable of capturing realistic nonlinear shear stress—shear strain behaviour in the small strain range. First, the paper derives a yield surface describing the nonlinear behaviour in small strain region based on formulation given in Correia et al. (2001) and Santos et al. (2001 & 2003). Then, it enhances a widely used constitutive model, Modified Cam Clay, with the derived yield surface. Finally it illustrates the capabilities of the proposed approach by comparing results of models with laboratory tests on a clayey soil.

The introduced procedure upgrades the capability of base model and provides better description of shearing in small strain region while not affecting the base model behaviour at large strains. On the other hand, this procedure only models the deviatoric stress—deviatoric strain non-linearity at small strains and the elastic volumetric behaviour of soil is unchanged. That may be revisited in the future research.

The shown enhancement of the model via an extra yield surface can be applied to great many other elasto-plastic models. The procedure typically does not require any changes in the original model driver as the yield loci are independent. Therefore, it may be used to enhance constitutive models without the need of altering the source code associated with the original model. The only requirement is the assumption of a linear shear modulus in the base model.

The small strain nonlinearity as shown may be implemented into THMC framework (Abed & Sołowski (2017)) to predict the behaviour of swelling clay barriers more accurately. It will also be incorporated into the MPM code where the more accurate replication of small strain effects may lead to improvements in prediction of e.g. initiation of landslides, as well as forces acting on landslides barriers (Seyedan & Sołowski (2017)).

REFERENCES

Abed, A.A., & Sołowski, W.T. (2017). A study on how to couple thermo-hydro-mechanical behaviour of unsaturated soils: Physical equations, numerical implementation and examples. *Computers and Geotechnics*, *92*, 132–155. https://doi.org/10.1016/j.compgeo.2017.07.021.

Atkinson, J.H. (1993). A note on modelling small strain stiffness in Cam clay. In *Predictive Soil Mechanics* (pp. 111–120). London.

Benz, T. (2007). *Small-Strain Stiffness of Soils and its Numerical Consequences.* Universität Stuttgart.

Burland, J.B. (1989). Ninth Laurits Bjerrum Memorial Lecture: "Small is beautiful"—the stiffness of soils at small strains. *Canadian Geotechnical Journal*, *26*(4), 499–516. https://doi.org/10.1139/t89–064

Correia, A., Santos, J., Barros, J., & Niyama, S. (2001). An approach to predict shear modulus of soils in the range of 10–6 to 10–2 strain levels. In *International conferences on recent advances in Geotechnical Earthquake Engineering and soil dynamics*.

Darendeli, M.B. (2001). *Development of a new family of normalized modulus reduction and material damping curves.* The University of Texas at Austin.

Doherty, J.P., & Muir Wood, D. (2013). An extended Mohr–Coulomb (EMC) model for predicting the settlement of shallow foundations on sand. *Géotechnique*, *63*(8), 661–673. https://doi.org/10.1680/geot.12.P.008

Gasparre, A. (2005). *Advanced laboratory characterization of London clay. Civil Engineering.*

Hardin, B.O., & Drnevich, V.P. (1972). Shear modulus and damping in soils: measurement and parameter effects. *Journal of Soil Mechanics & Foundations Div*, *98*(sm6).

Jardine, R.J., Potts, D.M., Fourie, A.B., & Burland, J.B. (1986). Studies of the influence of non-linear stress–strain characteristics in soil–structure interaction. *Géotechnique*, *36*(3), 377–396. https://doi.org/10.1680/geot.1986.36.3.377

Jardine, R.J., Symes, M.J., & Burland, J.B. (1984). The measurement of soil stiffness triaxial apparatus. *Géotechnique*, *34*(3), 323–340. https://doi.org/10.1680/geot.1985.35.3.378

Oztoprak, S., & Bolton, M.D. (2013). Stiffness of sands through a laboratory test database. *Géotechnique*, *63*(1), 54–70. https://doi.org/10.1680/geot.10.P.078

Roscoe, K.., & Burland, J.B. (1968). On the generalized stress-strain behaviour of wet clay.

Roscoe, K.., & Schofield, A.N. (1963). Mechanical behaviour of an idealized "wet" clay. In *Proc. 3rd Eur. Conf. Soil Mech. Wiesbaden, 1963* (Vol. 1, pp. 47–54).

Santos, J.A. Dos, & Correia, A.G. (2001). Reference threshold shear strain of soil. Its application to obtain an unique strain-dependent shear modulus curve for soil. In *XV International Conference on Soil Mechanics and Geotechnical Engineering* (pp. 267–270). Istanbul, Turkey.

Santos, J.A. dos, Correia, A.G., Modaressi, A., Lopez-Caballero, F., & Gomes, R.C. (2003). Validation of an elastoplastic model to predict secant shear modulus of natural soils by experimental results. In *3rd International Symposium on Deformation Characteristics of Geomaterials: Deformation Characteristics of Geomaterials* (pp. 1057–1061). Lyon France: Swets & Zeitlinger.

Seyedan, S., & Sołowski, W.T. (2017). Estimation of Granular Flow Impact Force on Rigid Wall Using Material Point Method. In *V International Conference on Particle-based Methods.*

Simpson, B. (1992). Retaining structures: displacement and design. *Géotechnique*, *42*(4), 541–576. https://doi.org/10.1680/geot.1992.42.4.541

Sołowski, W.T., & Sloan, S.W. (2016). Explicit stress integration with streamlined drift reduction. *Advances in Engineering Software*, *99*, 189–198. https://doi.org/10.1016/j.advengsoft.2016.05.011

Viggiani, G., & Atkinson, J.H. (1995). Stiffness of fine-grained soil at very small strains. *Géotechnique*, *45*(2), 249–265. https://doi.org/10.1680/geot.1995.45.2.249

Viladesau Franquesa, E. (2004). *The influence of swelling on the behaviour of London Clay*. Universitat Politècnica de Catalunya.

Numerical Methods in Geotechnical Engineering IX – Cardoso et al. (Eds)
© 2018 Taylor & Francis Group, London, ISBN 978-1-138-33198-3

Three-dimensional analysis of penetration problems using G-PFEM

L. Monforte, M. Arroyo & A. Gens
Universitat Politècnica de Catalunya—BarcelonaTech, Barcelona, Spain

J.M. Carbonell
Centre Internacional de Mètodes Numèrics en Enginyeria, CIMNE, Barcelona, Spain

ABSTRACT: Numerical simulation of large displacement problems in geomechanics has attracted the interest of many researchers over the past decades. In many circumstances these problems can be treated as two dimensional: plane stress or axisymmetric. However, some problems—such as the penetration of a square foundation or the dilatometer test (DMT)- do not enjoy any of these conditions. This contribution presents the extension to three dimensions of a numerical code for the simulation of large displacement fluid-saturated porous media at large strains. The proposal relies, on one hand, on the Particle Finite Element Method, known for its capability to tackle large deformations and rapid changing boundaries, and, on the other hand, on constitutive descriptions well established in current geotechnical analyses. The performance of the method is assessed in several benchmark examples, ranging from the insertion of a rigid square footing to a rough ball penetrometer.

1 INTRODUCTION

Numerical simulation of large displacement problems is relevant for many geomechanical problems. Discrete element models offer one alternative (Ciantia et al. 2016) but most efforts are still centered on continuum-based methods, like the finite element method (FEM). FEM can deal with the nonlinearities that arise from the simulation of large strains problems. Lagrangian formulations are well suited for path dependent material models, of the kind frequently applied in geotechnics. However, if a Lagrangian formulation is employed the mesh may experience severe distortion, leading to numerical inaccuracies and even rendering the computation impossible. To overcome the mesh distortion problem, several methods using the particle concept have been proposed: Material Point Methods (Wang et al. 2016, Iaconeta et al. 2017) Galerkin Mesh Free Methods (Navas et al. 2016, Iaconeta et al. 2017) or the Smoothed Particle Hydrodynamics (Blanc and Pastor 2013).

In this work, the Particle Finite Element method (PFEM) is employed. The method is characterized by a particle discretization of the domain: every time step a finite element mesh—whose nodes are the particles—is build using a Delaunays tessellation and the solution is evaluated using a well shaped, low order finite element mesh (Oñate et al. 2004, Carbonell et al. 2013).

Two-dimensional idealizations (plane stress and axisymmetric conditions) are computationally advantageous and fairly realistic for many important cases (e.g. CPTu). However, a large number of problems do not enjoy any symmetry; for instance, the penetration of a square foundation or the DMT. For this kind of problem three-dimensional models are required (Yu et al. 2008).

This work presents the extension to deal with three-dimensional problems of a PFEM implementation, sometimes referred to as G-PFEM (Geotechnical-PFEM), based on the Kratos framework (Dadvand et al. 2010). The work is structured as follows: first, the numerical method is outlined—including the basics of PFEM and the balance equations-; afterwards, the proposed approach is assessed against a benchmark example—the penetration of a footing near a vertical cut—and, finally, the three-dimensional penetration of a ball penetrometer subject to an anisotropic initial stress state is presented.

2 NUMERICAL MODEL

This section outlines the numerical procedures used in this work. First the Particle Finite Element method (PFEM) is briefly reviewed; then, the balance equations are highlighted. Finally, the constitutive equations are described.

2.1 *Particle finite element method*

In PFEM, the continuum is modelled using an Updated Lagrangian formulation; that is, a Lagrangian description of the motion is used where all variables and their derivatives are referred to the deformed configuration. In this implementation only low order elements are used—linear triangles in 2D and linear tetrahedron in 3D. Nodes discretizing the analysis domain are treated as material particles whose motion is tracked during the solution.

Periodically, the FE mesh is re-triangulated to alleviate distortion problems. In addition, h-adaptive techniques are employed to obtain a better discretization in areas of the domain with large plastic deformation (Rodriguez et al. 2016).

A typical solution algorithm involves the following steps (Oñate et al. 2004):

1. Discretize the domain with a Finite Element mesh. Define the shape and movement of the rigid structure,
2. Identify the external boundaries. Search the nodes that are in contact with the rigid structure,
3. Compute some time-steps of the (hydro)-mechanical problem,
4. Construct a new mesh. This step may include re-triangulation of the domain, introduce new particles in and adaptive fashion and interpolate the state variables between the previous mesh and the new one,
5. Go back to step 2. and repeat the solution process for the next time-steps.

2.2 *Balance equations*

The soil-water mixture is modeled as a two-phase continuum employing a finite deformation formulation. The equations of balance of linear momentum and mass balance for the mixture are written following the movement of the solid skeleton. The unknown fields are the solid skeleton displacements and the fluid pressure $(\mathbf{u}-p_w\ \mathbf{formulation})$. Inertial effects are neglected and the Updated Lagrangian form of the governing equations reads (Larsson and Larsson 2002):

$$\begin{cases} \nabla \cdot \sigma' + \nabla p_w + \mathbf{b} = 0 & \text{in } \Omega_t \times (0,T) \\ \dfrac{\varphi}{K_w} \dot{p}_w + \nabla \cdot \mathbf{v} + \nabla \cdot \mathbf{v}^d = 0 & \text{in } \Omega_t \times (0,T) \end{cases} \quad (1)$$

This system is completed with the appropriate initial and boundary conditions. $\sigma' = \hat{\sigma}'(\mathbf{F}, V)$ stands for the effective Cauchy stress that is computed with the appropriate constitutive equation for path dependent materials, $\mathbf{F}$ stands for the deformation gradient whereas V stands for the set of internal variables. p_w is the water pressure, $\mathbf{b}$ are the external volumetric loads, $\mathbf{u}$ is the solid skeleton velocity, $\dot{p}_w$ is the material time derivative of the water pressure (computed following the solid motion), $\mathbf{v}^d$ is Darcys velocity and K_w is the water compressibility.

The system of equations is non-linear geometrically since balance equations are imposed in the new (unknown) configuration. In the present implementation both equations are solved in a monolithic approach and an implicit time-marching scheme is employed. The space is discretized with linear triangles in 2D and linear tetrahedron in 3D; the same shape functions are used for both displacements and water pressure.

Incompressibility produce volumetric locking in low order finite elements (Sun et al. 2013). In hydro-mechanical simulations there are two sources of incompressible behavior: nearly undrained conditions and the constitutive model: for instance, Critical State soil models predicts zero volume change at the Critical State line.

To alleviate volumetric locking a mixed form is used:

$$\begin{cases} \nabla \cdot \breve{\sigma}' + \nabla p_w + \mathbf{b} = 0 & \text{in } \Omega_t \times (0,T) \\ J - \theta = 0 & \text{in } \Omega_t \times (0,T) \\ -1K_w \dot{p}_w + \nabla \cdot \mathbf{v} + \nabla \cdot \mathbf{v}^d = 0 & \text{in } \Omega_t \times (0,T) \end{cases} \quad (2)$$

where $J = \det \mathbf{F}$ is the determinant of the deformation gradient, θ is the volumetric deformation and $\breve{\sigma}' = \hat{\sigma}'(\breve{\mathbf{F}}, V)$ is the effective Cauchy stress evaluated with an assumed deformation gradient, $\breve{\mathbf{F}}$ that is defined as:

$$\breve{\mathbf{F}} = \left(\tfrac{\theta}{J}\right)^{1/3} \mathbf{F} \quad (3)$$

that is, the deviatoric part of the deformation gradient is preserved whereas the volumetric part is replaced with that of the variable θ. Although the Cauchy stress tensor depends on displacement and Jacobian, the usual strain-driven format of the local problem remains.

After obtaining the weak form of Equation (2), linear shape functions are used to discretize the three unknown fields, $\mathbf{u}-\theta-p_w$. To further alleviate volumetric locking, the mass balance equation and the Jacobian balance equation is stabilized with the PPP technique (Bochev et al. 2006). More details are given in (Monforte et al. 2017).

As detailed elsewhere (Monforte et al. 2017), the code may also be used for total stress analysis, with appropriately defined stabilized formulations.

2.3 *Constitutive equations*

In the literature, two main families of schemes have been proposed for the analysis of large deformation elasto-plastic problems (Simo and Hughes 1998). The first one is based on the use of hypoelastic rate models and an additive decomposition of the spatial rate of deformation in an elastic and plastic part. This scheme may be regarded as an extension of the usual small strains algorithms; care is required to use stress rates that fulfill objective transformation and frame indifference. In the second family, deformation itself is decomposed multiplicatively into an elastic and plastic part; with the former being defined by an hyperelastic model. By its construction, the scheme fulfills inherently the objectivity requirements (Simo and Hughes 1998).

Although an hypoplastic scheme has also been implemented for comparison, all the results presented below were obtained with the hyperelastic scheme that G-PFEM uses as default.

Within this constitutive framework, the deformation gradient is assumed to split multiplicative into and elastic and plastic part. That is, an intermediate configuration of irreversible (plastic) deformation is introduced, relative to which the elastic response of the material is characterized. As a consequence, the definition of the constitutive problem reads (Simo 1998):

$$\begin{cases} \mathbf{F} = \mathbf{F}^e \cdot \mathbf{F}^p \\ \tau = \frac{\partial W(\varepsilon^e)}{\partial \varepsilon^e} \\ \mathbf{l}^p = \dot{\gamma} \frac{\partial g(\tau, h)}{\partial \tau} \\ f(\tau, h) \leq 0 \\ h = h(\varepsilon^p) \end{cases} \tag{4}$$

where $\varepsilon = 0.5\ln(\mathbf{F}^e \cdot \mathbf{F}^{eT})$ is the elastic Hencky strain and $W(\varepsilon^e)$ is the stored-energy function.

Stress integration is performed with an explicit scheme with adaptive sub-stepping and correction for the yield surface drift. Further details may be found in (Monforte et al. 2015).

3 NUMERICAL ANALYSES

In this section several analyses are presented to illustrate the performance of the method. The first one consist of the two-dimensional, total stress analysis of the penetration of a footing into a weightless Tresca soil and serves to demonstrated the benefits of the use of mixed formulations to deal with incompressibility. Then, results of two three-dimensional problems are presented: the penetration of a footing into the soil and penetration of a ball.

3.1 *Footing near a vertical cut*

The first computational analysis consist of total stress penetration of a strip footing that is located near a vertical cut. This is a classic example problem analyzed by Pastor et al. (1999). The domain consists of a square whose edges are five times the width of the footing; all the displacements are restricted at the bottom of the domain whereas the horizontal displacements are restricted to zero in the right boundary.

The soil is assumed weightless and characterized by a shear modulus, G = 100 kPa, a Poissons ratio, ν = 0.49, and an Undrained shear strength, S_u = 1 kPa. A vertical velocity is applied on the top of the footing, idealized as an elastic material of shear modulus two orders of magnitude higher than that of the soil. Remeshing is disabled in a first computation, to assess the effect of the mixed formulation in isolation. The mesh is depicted in Figure 2.

Figure 1 presents the curve normalized settlement vs normalized soil resistance. The first thing to note is that in the curve obtained by using the primal formulation the normalized resistance increase continuously. On the other hand, by using the mixed stabilized formulation after a normalized penetration of $z/B = 0.02$ the value of the resistance remains almost constant. In general higher resistances are found at every displacement when using the primal formulation. Finally, it is noted that the normalized resistance obtained with the mixed-stabilized formulation is slightly higher than 2, which is in agreement of the results presented by Pastor et al. (1999).

The behavior of the primal formulation is a consequence of the severe volumetric locking affecting low order finite elements. Furthermore, as shown in Figure 2, the results obtained using the primal formulation present high amplitude spatial oscillations on the total mean stress field whereas a much smoother distribution is found when using the mixed-stabilized formulation. Additionally, using the primal formulation, localization takes place in a shear plane whose orientation is influenced by the preferential mesh orientation. On the other hand the localization plane obtained by

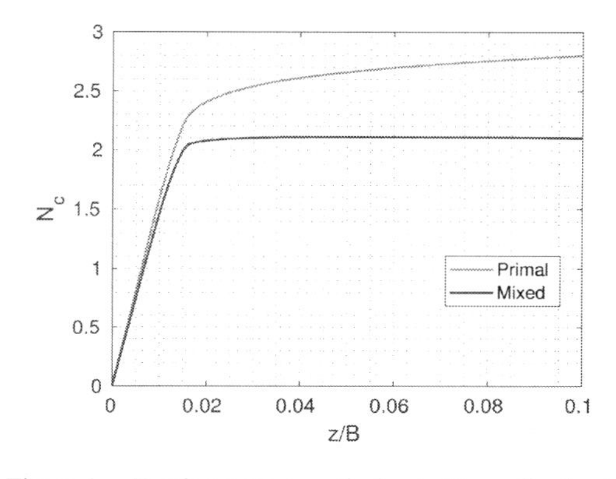

Figure 1. Footing near a vertical cut. Normalized settlement vs normalized resistance using the primal and mixed formulation.

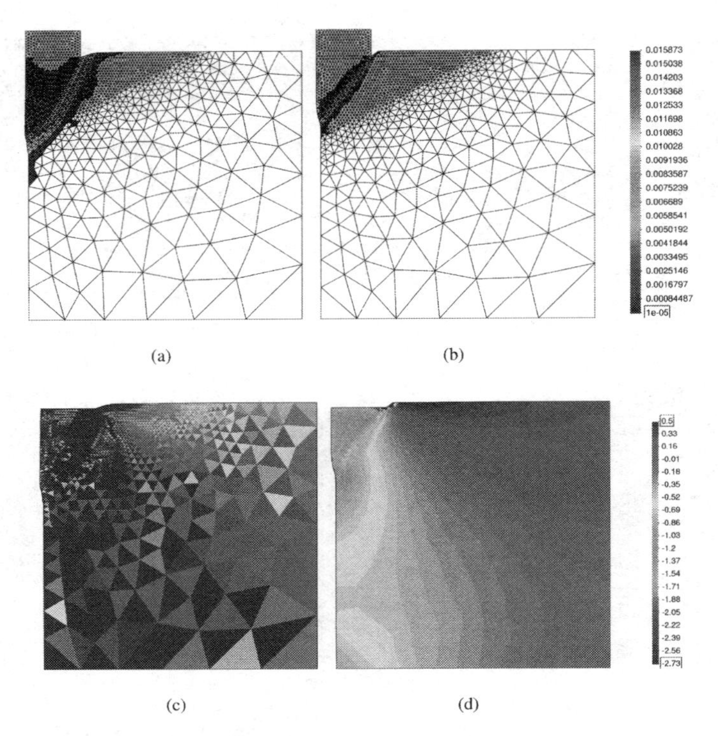

Figure 2. Footing near a vertical cut. Incremental plastic shear strain, (a) and (b), and total mean stress, (c) and (d), using the primal formulation, (a) and (c), and the mixed-stabilized formulation, (b) and (d).

the stabilized-mixed formulation is very similar to that obtained by Pastor et al. (1999).

To extend the analysis the effect of the Rigidity index and the Poisson's ratio is now examined. The model is slightly changed in that, instead of applying the vertical movement to a rigid footing, vertical displacement is directly imposed on the soil. Large vertical displacements up to 1 width of the footing are now simulated; as a result the remeshing algorithms of PFEM enter into play. Only the mixed-stabilized formulation is used, since its benefits has been illustrated in the previous numerical analysis.

Figure 3(a) shows the normalized resistance curves for Poissons ratios ranging from 0.45 to 0.499 and a fixed $I_r = 100$. All the simulations share the same undrained shear strength, $S_u = 1$ kPa; the Shear modulus has been modified to maintain the same rigidity index. For this particular problem the Poisson's ratio does not seem to play a prominent role since almost the same resistance is obtained irrespectively of the Poisson's ratio. The curves present slight oscillations that are caused by the introduction of new nodes in the vicinity of the nodes with prescribed displacement.

The, the effect of the Rigidity Index is assessed for a Poisson ratio of $\nu = 0.49$ in Figure 3(b). Similarly to the rigid strip footing on a soil layer (Monforte et al. 2017), the effect of the rigidity index is very pronounced at the beginning of the loading, but less important as penetration progresses. In all the cases a failure mechanism such as that presented in Figure 2(b) takes place and the drastic change of slope of the penetration curve corresponds to the moment when the failure mechanism is completely formed. Figure 3(c) illustrates the failed final state.

3.2 *Rectangular footing on poroelastic media*

In the second example a rectangular footing is pushed into a porous saturated soil. The soil is assumed to behave as a linear elastic material, $E = 500\,kPa$ and $\nu = 0.3$ with a permeability $K = 10^{-5}$ m/day. The footing Young modulus is two orders of magnitude larger than that of the soil.

To investigate the three dimensional effects, several footing shapes are studied, all of them with a height of 0.5 m and length of 1 m but widths to lengths ratios variable between 1 and 4. A surface load of 200 kPa is applied on top of the footing, ramped up over a period of one day. Drainage is only allowed through the free surface. Due to the symmetry of the problem, only a quarter of the geometry is simulated, see Figure 4(c).

Figure 4 presents the evolution of the footing settlement and the water pressure at a point

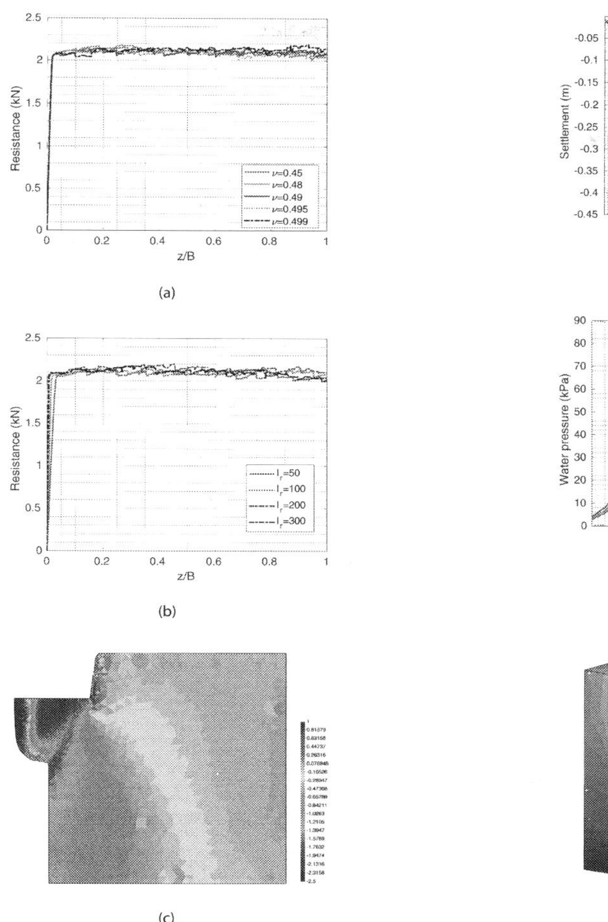

Figure 3. Footing near a vertical cut. Resistance in terms of the normalized penetration. Effect of the Poisson ratio for a Rigidity Index, $I_r = 100$, (a). Effect of the Rigidity index for $\nu = 0.49$, (b). Vertical total stress for $\nu = 0.499$ and $I_r = 100$ after a penetration of $z/B = 1$, (c).

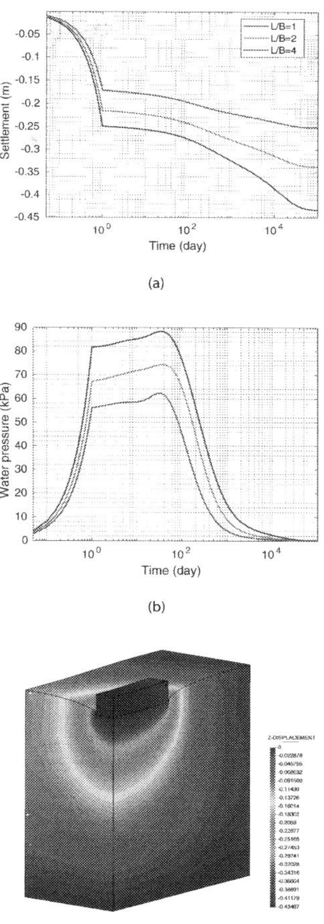

Figure 4. Evolution, over time, of the settlement of the Footing, (a), and water pressure at a depth equal to one width, (b), for several ratios of the length and width. Vertical displacement for the case L/B=4 after the consolidation phase, (c).

located at a depth equal to one width over time. As the ratio B/L increases, the settlement and the water pressure also increase. All the curves that depict the water pressure evolution show a marked Mandel-Cryer effect.

These results have been compared with the analytical solution developed by Brown (1978) for rectangular, stiff rafts resting in an homogeneous isotropic half-space. As shown in Table 1, systematically lower settlements are obtained by the numerical solution, between 20% to 40%. Several are the causes of this mismatch: in the numerical model it has been considered that the raft is flexible and that the interface is completely rough whereas the analytical solution assumes a completely stiff raft whose interface is smooth; additionally, in the numerical simulation the domain is finite, whereas a half-space is used in the reference solution. Altough it seems a crude discrepancy, a similar order mismatch between numerical simulations and analytical solutions has been observed in axisymmetric conditions (Wang et al. 2015).

3.3 *Ball penetrometer in anisotropically stressed clay*

In this final example, the displacement of an initially embedded ball penetrometer is studied. The sphere is initially whished in placed in the middle of a cubical domain with side equal to 20 times the diameter of the sphere.

Table 1. Settlement at the center of the footing after the loading phase (S_l) and the consolidation phases (S_c). Comparison with the solution of Brown (1978).

	S_l (m)	S_c (m)		
L/B	PFEM	Ref	PFEM	Ref
1	0.172	0.226	0.252	0.316
2	0.215	0.317	0.338	0.444
4	0.25	0.416	0.433	0.582

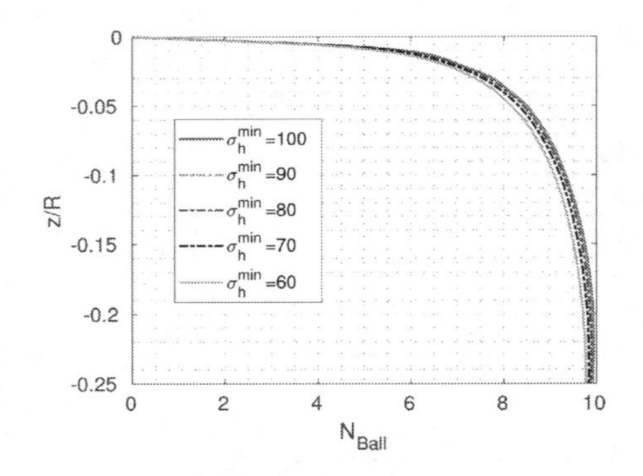

Figure 5. Ball penetrometer. Effect of the horizontal stresses on the on the ball resistance. Curves are labeled in terms of the minimal horizontal stress.

The soil is described by an hyperelastic Modified Cam Clay model; all the cases share the same set of constitutive parameters: $\kappa^* = 0.01, \lambda^* = 0.1, G = 10^4\,kPa$ and initial preconsolidation pressure of $p_c = 150\,\text{kPa}$. The yield surface shape in the deviatoric plane is described using the fully convex formulation proposed by Panteghini and Lagioia (2014). Permeability was set at a relatively low value of $K = 10^{-8}\,\text{m/s}$, so undrained penetration is expected. A smooth interface between the clay and the ball is considered.

All simulations are carried under constant vertical effective stress of 100 kPa. Horizontal stresses are adjusted, increasing one and decreasing the orthogonal one by the same amount, with the purpose of maintaining constant the mean effective stress at $p_0' = 100\text{kPa}$. It is this anisotropy of horizontal stress what makes the problem three-dimensional.

Figure 5 shows the normalized penetration vs resistance curve for a set of simulations in which the minimum horizontal effective stress varies between 60 and 100 kPa. Results are presented as curves of resistance factor N_{ball} against normalized displacement. The normalized resistance factor is obtained as the ratio of mobilized resistance (i.e. the total vertical force acting on the ball divided by the pro-

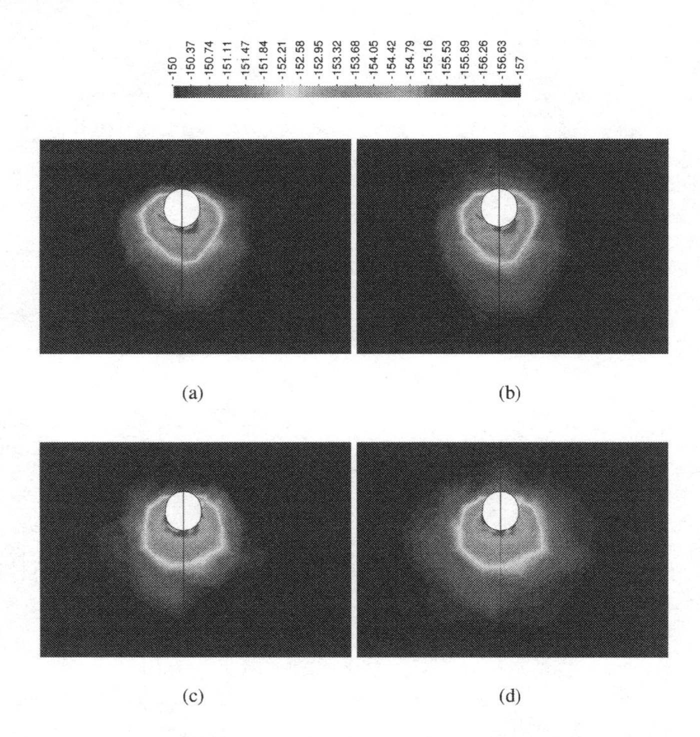

Figure 6. Ball penetrometer. Preconsolidation contour plots for minimal horizontal stress of $\sigma_{h0}^{\boxtimes} = 90\,kPa$, (a) and (c), and $\sigma_{h0}^{\boxtimes} = 60\,kPa$, (b) and (d). On top results on the plane normal to the minimal horizontal stress whereas, on the bottom, normal to the maximum horizontal stress.

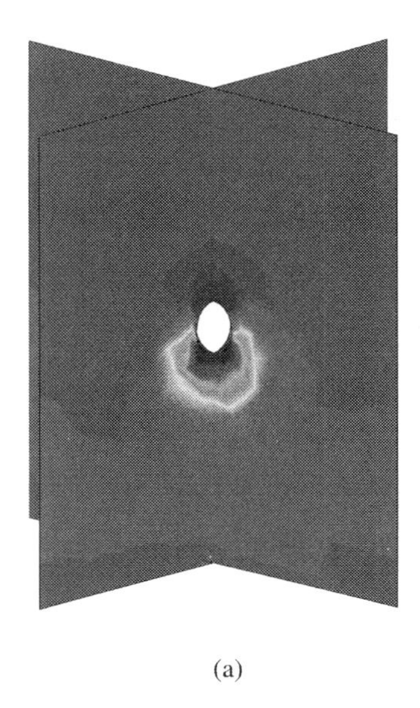

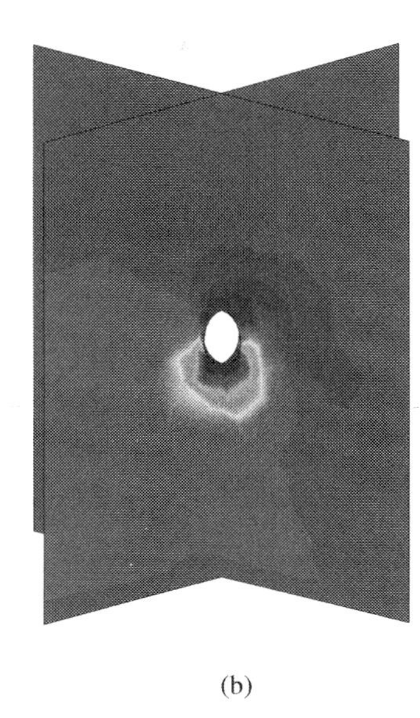

(a) (b)

Figure 7. Ball penetrometer. Water pressure contours along the two two planes for a minimum horizontal stress of $\sigma'_{h0} = 90\,kPa$, (a), and $\sigma'_{h0} = 60$, (b).

jected area) to the triaxial compression undrained shear strength, which, for MCC is given by

$$S_u = \frac{1}{2}\sigma'_{v0}\,M\left(\frac{R}{2}\right)^{\Lambda} \tag{5}$$

Using the input data the undrained shear strength is $S_u = 38.59\,\text{kPa}$; which implies a rigidity index $I_r = G/Su = 259$. The resistance factor approximates a limit value close to $N_{ball} = 9.93$, with a very small effect (less than 3%) of the anisotropic horizontal stress state. The limit value obtained is slightly below the reference value $N_{ball} = 10.43$, predicted by Einav and Randolph (2005) using a combination of upper bound and strain path methods. Two reasons may explain this discrepancy. The first is the different material model employed (a Von Mises model was employed by Einav and Randolph). The second is that only relatively shallow penetrations have been computed and the expected full flow mechanism has not been yet formed.

Despite the small difference observed in the resistance factor, the anisotropic initial stress state has consequences. For instance (Figure 6) the shape of the plastic zone is highly affected by the horizontal stress anisotropy: it is observed that the plastic zone is much more extended in the plane normal to the smallest horizontal stress and more reduced in the plane normal to the largest horizontal stress. The pore pressure field also exhibit a three-dimensional effect: as the stress anisotropy increases, larger water pressures are developed in the plane normal to the maximum horizontal stress and smaller excess water pressures are found at the plane normal to the minimum horizontal stress; in fact, in some simulations a decrease of the water pressure is found in large regions behind the ball (Figure 7).

4 CONCLUSIONS

In this work, a numerical framework for the analysis of saturated porous media undergoing large deformations based on the Particle Finite Element Method has been presented.

By means of several numerical analysis the performance and accuracy of the proposal have been analyzed. First, the benefits of the use of stabilized mixed formulations to deal with incompressibility constraints have been illustrated in a two-dimensional problem.

Preliminary results of the three-dimensional of the Ball penetrometer under anisotropic initial stress states, maintaining the same initial effective pressure, suggest that mobilized ball resistance is not heavily influenced by initial stress anisotropy; however, significant differences on the shape of the plastic region have been observed and the penetration mechanism was not yet fully developed. More research on this topic is currently ongoing.

The developed numerical scheme appears to be a promising tool for the three-dimensional simulation of penetration problems in geomechanics.

ACKNOWLEDGEMENTS

The financial support of the Ministry of Education of Spain through research grant BIA2014–59467-R is gratefully appreciated.

REFERENCES

Blanc, T. & M. Pastor (2013). A stabilized Smoothed Particle Hydrodynamics, Taylor-Galerkin algorithm for soil dynamics problems. *International Journal for Numerical and Analytical Methods in Geomechanics 37*(1), 1–30.

Bochev, P.B., C.R. Dohrmann, & M.D. Gunzburger (2006). Stabilization of low-order mixed finite elements for the Stokes equations. *SIAM Journal on Numerical Analysis 44*(1), 82–101.

Brown, P.T. (1978). Stiff rectangular rafts subject to concentrated loads. *Australian Geomechanics Journal*, 40–49.

Carbonell, J.M., E. Oñate, & B. Suárez (2013). Modelling of tunnelling processes and rock cutting tool wear with the particle finite element method. *Computational Mechanics 52*(3), 607–629.

Ciantia, M.O., M. Arroyo, J. Butlanska, & A. Gens (2016). DEM modelling of cone penetration tests in a double-porosity crushable granular material. *Computers and Geotechnics 73*, 109–127.

Dadvand, P., R. Rossi, & E. Oñate (2010, Sep). An objectoriented environment for developing finite element codes for multi-disciplinary applications. *Archives of Computational Methods in Engineering 17*(3), 253–297.

Einav, I. & M.F. Randolph (2005). Combining upper bound and strain path methods for evaluating penetration resistance. *International Journal for Numerical Methods in Engineering 63*(14), 1991–2016.

Iaconeta, I., A. Larese, R. Rossi, & Z. Guo (2017). Comparison of a material point method and a Galerkin meshfree method for the simulation of cohesive-frictional materials. *Materials 10*(10), 1150.

Larsson, J. & R. Larsson (2002). Non-linear analysis of nearly saturated porous media: Theoretical and numerical formulation. *Computer Methods in Applied Mechanics and Engineering 191*(36), 3885–3907.

Monforte, L., M. Arroyo, J.M. Carbonell, & A. Gens (2017). Numerical simulation of undrained insertion problems in geotechnical engineering with the particle finite element method (PFEM). *Computers and Geotechnics 82*, 144–156.

Monforte, L., M. Arroyo, A. Gens, & J.M. Carbonell (2015). Explicit finite deformation stress integration of the elastoplastic constitutive equations. *Computer Methods and Recent Advances in Geomechanics, IACMAG 2014*, 267–272.

Monforte, L., J.M. Carbonell, M. Arroyo, & A. Gens (2017). Performance of mixed formulations for the particle finite element method in soil mechanics problems. *Computational Particle Mechanics 4*(3), 269–284.

Navas, P., C.Y. Rena, S. López-Querol, & B. Li (2016). Dynamic consolidation problems in saturated soils solved through u–w formulation in a LME meshfree framework. *Computers and Geotechnics 79*, 55–72.

Oñate, E., S.R. Idelsohn, F. Del Pin, & R. Aubry (2004). The particle finite element methodan overview. *International Journal of Computational Methods 1*(2), 267–307.

Panteghini, A. & R. Lagioia (2014). A fully convex reformulation of the original Matsuoka–Nakai failure criterion and its implicit numerically efficient integration algorithm. *International Journal for Numerical and Analytical Methods in Geomechanics 38*(6), 593–614.

Pastor, M., T. Li, X. Liu, & O. Zienkiewicz (1999). Stabilized low-order finite elements for failure and localization problems in undrained soils and foundations. *Computer Methods in Applied Mechanics and Engineering 174*(1–2), 219–234.

Rodriguez, J.M., J.M. Carbonell, J.C. Cante, & J. Oliver (2016). The particle finite element method (PFEM) in thermo-mechanical problems. *International Journal for Numerical Methods in Engineering 107*(9), 733–785.

Simo, J.C. (1998). Numerical analysis and simulation of plasticity. *Handbook of numerical analysis 6,* 183–499.

Simo, J.C. & T.J.R. Hughes (1998). *Computational inelasticity.* Springer Science & Business Media.

Sun, W.C., J.T. Ostien, & A.G. Salinger (2013). A stabilized assumed deformation gradient finite element formulation for strongly coupled poromechanical simulations at finite strain. *International Journal for Numerical and Analytical Methods in Geomechanics 37*(16), 2755–2788.

Wang, B., M.A. Hicks, & P.J. Vardon (2016). Slope failure analysis using the random material point method. *Géotechnique Letters 6*(2), 113–118.

Wang, D., B. Bienen, M. Nazem, Y. Tian, J. Zheng, T. Pucker, & M.F. Randolph (2015). Large deformation finite element analyses in geotechnical engineering. *Computers and Geotechnics 65*, 104–114.

Yu, L., J. Liu, X. Kong, & Y. Hu (2008). Three-dimensional RITSS large displacement finite element method for penetration of foundations into soil. *Computers and Geotechnics 35*(3), 372–382.

Numerical Methods in Geotechnical Engineering IX – Cardoso et al. (Eds)
ISBN 978-1-138-33198-3

The use of MPM to estimate the behaviour of rigid structures during landslides

L. González Acosta, I. Pantev, P.J. Vardon & M.A. Hicks
Section of Geo-Engineering, Faculty of Civil Engineering and Geosciences, Delft University of Technology, Delft, The Netherlands

ABSTRACT: In geotechnical engineering, proper design of retaining structures is of great importance, since failure of these structures can lead to catastrophic consequences. Nowadays, the finite element method is seen as a reliable numerical technique to analyze soil behaviour and is widely used to assess the interaction between soil and rigid structures. However, a disadvantage of this method is the difficulty of simulating contact between separate bodies. Because of this, the event of a slope failing and colliding with a rigid body cannot be analyzed, so that the additional forces acting against the rigid body caused by the motion of the ground are neglected. With the recent development of the Material Point Method (MPM), this limitation has been overcome and problems involving large deformation and multiple bodies in contact can be analyzed. In this paper, the effect of a landslide colliding with a rigid wall has been studied, and multiple initial conditions have been considered in order to identify the critical case.

1 INTRODUCTION

During the early years of geotechnical engineering, the only methods available to assess the stability of rigid retaining structures (retaining walls, sheet piles, etc.) were those related with limit equilibrium. The biggest disadvantage of such methods is their inability to take account of the complexity of the problem; for example, with respect to the geometry, construction stages, material behavior, variability of soil properties, and irregular loading conditions.

Years later, with the implementation of numerical techniques such as the finite element method (FEM) and the finite difference method (FDM), taking account of complexities in geotechnical analysis has become more feasible. Using these techniques, it is now possible to compute the stability and interaction between the soil and rigid structures, returning important information for the design process. Moreover, in the particular case of retaining structures, FEM has proven to be appropriate for solving this type of problem under both static (Hosseinzadeh & Joosse, 2015) and dynamic (Gazetas et al. 2016) loading conditions. However, because the connectivity between the mesh and domain is essential for most standard FEM analyses, there is a limit to the range of problems that can be solved with the method.

During the last 20 years, new numerical techniques such as mesh free and meshless methods have been developed, to eliminate the need of using a continuum mesh to simulate the material. Among these techniques, the material point method (Sulsky et al. 1994, 1995) has been proven to solve problems with satisfactory results. Since, in MPM, the material is not attached to the mesh, large deformations can be simulated, enabling the analysis of complex geotechnical problems such as a progressive slope failures (Wang et al. 2016) and landslides (Soga et al. 2016). Moreover, due to the development of a contact detection algorithm (Bardenhagen et al. 2000, 2001), the interaction between separate bodies is feasible, so that the simulation of a landslide impacting on a retaining wall has become possible.

In the first part of this paper, the background of MPM and the contact detection algorithm are outlined, followed by the analysis of an elastic body bouncing on a rigid surface to demonstrate the efficiency and accuracy of the formulation. Finally, the analysis of a slope colliding with a rigid body is performed, including a parametric study to find the critical conditions.

2 THEORETICAL FORMULATION

2.1 *MPM background*

A great advantage of MPM over other mesh free and meshless techniques is its significant overlap with FEM, albeit with two key differences. The first difference is that material properties are attached to the material points and not to the elements;

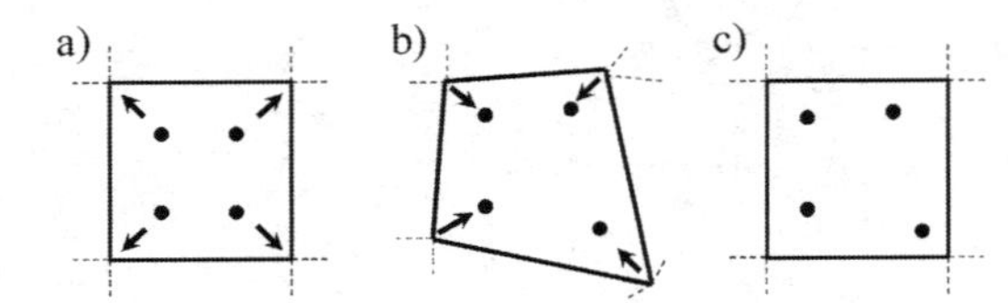

Figure 1. MPM solution steps: a) mapping phase, b) solution of momentum balance at the nodes, and c) convection phase.

the second difference is that, after the equations of motion have been solved at the nodes and the mesh is distorted, the material points adopt their new positions and the mesh is reset to its original position. Figure 1 illustrates the MPM solution procedure. Firstly, particle variables such as mass, momentum, velocity and stresses are mapped to the nodes; next the momentum equation is solved for the velocities and accelerations at the nodes; then, the updated solution is transferred to the material points by mapping the nodal values using shape functions; and finally, the mesh is reset.

Since MPM shares similarities with the FEM mechanics framework (Chen et al. 2015), the weak form of the momentum conservation is given as

$$\int_{\Omega} \rho \mathbf{a} \cdot \delta u dV + \int_{\Omega} \rho \boldsymbol{\sigma} \nabla \cdot \delta u dV = \int_{\Omega} \rho \mathbf{b} \cdot \delta u dV + \int_{\Gamma} \rho \boldsymbol{\tau}^{s} \cdot \delta u dA \tag{1}$$

where ρ is the material density, $\mathbf{a}$ the acceleration, δu the virtual displacement, V the body volume, $\boldsymbol{\sigma}$ the Cauchy stress, $\mathbf{b}$ the body forces, $\boldsymbol{\tau}^s$ the prescribed tractions, Γ the traction boundary, and Ω the solution domain.

Note that since the integration of internal and external forces, as well as kinematic variables, is performed considering the new positions of the material points, the oscillation of stresses, velocities, accelerations, and many other quantities is inevitable, reducing the accuracy of the results. However, an explanation of these matters is beyond the scope of this work.

2.2 *Contact detection*

Contact detection was developed in MPM to allow interaction between bodies (e.g. collision, penetration, sliding, and adhesion). To detect contact, the velocity field of a single body is compared to the velocity field accounting for all bodies in the system. Due to the discrete domain of the shape functions, these will only be different at the nodes where contact occurs, i.e. at those nodes with contributions from more than one body. The contact at the nodes is computed as

$$\mathbf{v}^{cm} - \mathbf{v}^{i} \neq 0 \tag{2}$$

where $\mathbf{v}^{cm}$ is the velocity field accounting for all bodies, and $\mathbf{v}^{i}$ is the body velocity field.

After contact is established, further behaviour, such as approach or departure, is computed by using the surface normal direction at every contact node, i.e.

$$\left(\mathbf{v}^{i} - \mathbf{v}^{cm}\right) \cdot \mathbf{n}^{i} > 0 \tag{3}$$

where

$$\mathbf{n}^{i} = \sum_{p=1}^{nmp} G_{ip} m_{p} \tag{4}$$

In which $\mathbf{n}^{i}$ is the body surface normal, nmp is the number of material points in the support elements, G_{ip} is the gradient of the shape function with respect to the position of the material point, and m_p is the material point mass.

Figure 2 gives an illustration of the contact nodes detected after the total velocity at the nodes diverges from the single body velocity contribution.

3 BENCHMARK PROBLEM

3.1 *Bouncing of a block*

To demonstrate the accuracy and performance of MPM with the contact algorithm, a benchmark

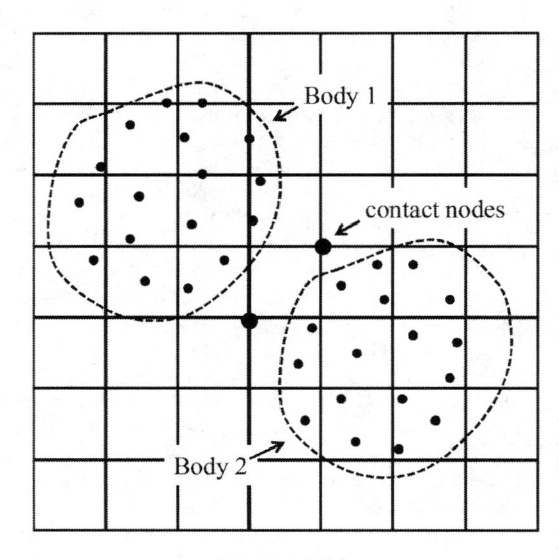

Figure 2. Contact detection nodes after mapping velocities on the background mesh.

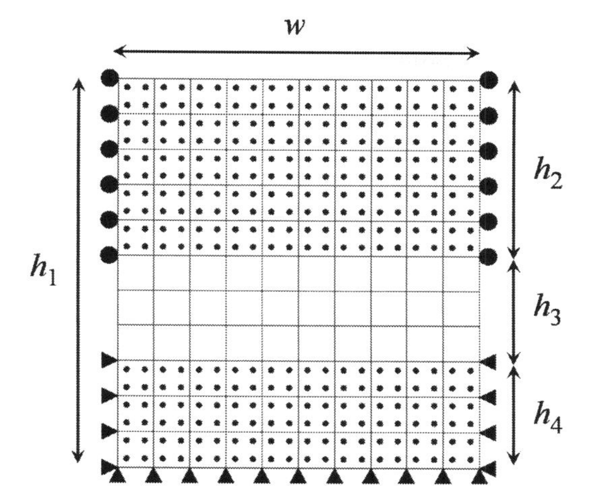

Figure 3. Collision between an elastic block and a rigid body.

problem has been analysed. Specifically, an elastic body is allowed to free fall, so that it collides with a second body acting as a rigid surface. The falling body is a block made up of 200 material points and it is dropped from a height of $h_3 = 0.30$ m relative to the second body representing the infinite rigid surface, which is composed of 120 material points. Figure 3 shows a sketch of the problem. As can be seen, the background mesh is made up of square elements of dimensions $\Delta x = \Delta y = 0.10$ m, and the material points are initially evenly spaced at $\xi = \pm 0.5$ and $\eta = \pm 0.5$ in terms of local coordinates. The total height of the domain is $h_1 = 1.1$ m, and the falling body and the rigid body have a height of $h_2 = 0.5$ m and $h_4 = 0.30$ m respectively. The width of the domain is $w = 1.0$ m.

The block and the rigid surface are considered as perfectly elastic bodies, both with an elastic modulus and Poisson's ratio of $E = 500$ kN/m^2 and $\nu = 0.49$ respectively. The reason for choosing such a high Poisson's ratio is to avoid high compression and deformation of the bodies during contact, thereby avoiding stress oscillation problems caused by the material points crossing element boundaries.

The boundary conditions for this problem are as follows. For the rigid surface, the nodes at the right, left and bottom are fixed in both directions to prevent any displacement. In contrast, the nodes at the right and left boundary for the falling block are fixed just in the horizontal direction, in order to allow free fall and avoid horizontal deformation. To consider the surface body as a rigid body with infinite stiffness, the displacements, velocities, and accelerations are erased after each step to prevent any distortion and the effects this can cause to the body representing the falling block. Finally, the initial velocity of the falling block is zero, and gravity is the only external force acting on it, whereas the body representing the rigid surface is not affected by gravity and is only affected by the forces developed during the contact.

3.2 *Results*

Figure 4 shows the analytical velocity at time t of a body bouncing over a surface, and the average velocity of all material points from the falling block bouncing over a second body considering two element sizes, $\Delta x = 0.1$ m and $\Delta x = 0.05$ m. As can be seen, the average velocity of both solutions matches the analytical one perfectly during free fall, but neither reaches the analytical maximum velocity. After two bounces, the maximum velocity reached reduces substantially. It is also evident that when using a smaller mesh, the solution is closer to the analytical solution, showing that the contact and body behavior depend on the mesh size. The main reason why the velocity of the free falling block does not reach the maximum analytical velocity, is that contact detection happens if the material points of both bodies are in neighboring elements, leading to an early bounce. The solution converges to the analytical one as the element size reduces.

In Figure 5, the average vertical position h of the falling body is plotted against time and compared with the analytical vertical position for a perfectly elastic body bouncing on a rigid surface. As before, the comparison is made using two different cell sizes. The position after bouncing for the analytical solution was computed using the analytical velocity after bouncing, leading to the perfect bounce that reaches the initial position.

As was seen in the previous figure, the computed results match the analytical solution closely during free fall, but the bounce occurs earlier. Also, after bouncing the block is unable to reach

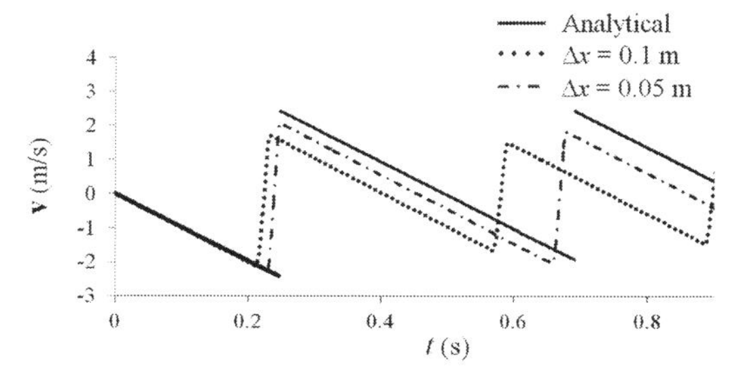

Figure 4. Analytical and experimental average velocities of material points of the falling body before and after bouncing.

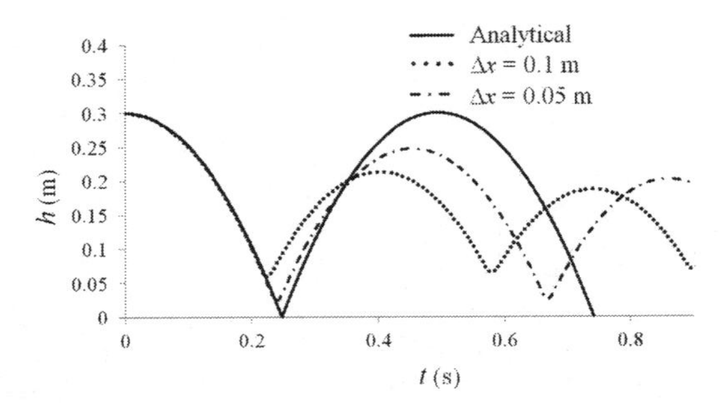

Figure 5. Analytical and experimental average vertical positions of the bouncing body with time.

the initial position, and this becomes more evident after the second bounce. As mentioned before, the reason why the body does not reach the rigid surface ($h = 0$ m), is because contact takes place at a distance of 1 element. Again, with the reduction of element size, the contact distance is smaller, leading to a solution closer to the analytical one. Finally, it is clear that the maximum altitude reached by the bouncing body is lower than the analytical solution because of energy losses. These are due to (i) the resetting of the rigid body, and (ii) stress oscillation in MPM.

The above analysis proves that the contact algorithm works properly and that interaction between bodies can be simulated with suitable results using MPM. For more information regarding the mathematical background of the contact algorithm, as well as benchmark problems to validate its use with MPM, the reader is directed to Pantev (2016).

4 GEOTECHNICAL APPLICATION

4.1 *Collision of a landslide with rigid structure*

To illustrate the application of the combined MPM and contact algorithm in geotechnical problems, the behavior of a rigid body when a landslide collides with it was analized. Figure 6 shows a sketch of the problem, which consists of 2 bodies. Body 1 (B_1) is a low strength block of soil that, after failure, collides with body 2 (B_2) representing a rigid wall. Both bodies are constructed using background meshes of size $\Delta x = \Delta y = 0.1$ m, and each element is initially filled with 4 and 9 material points for body 1 and body 2 respectively. The height of B_1 is $h_1 = 2.0$ m and its length is $l_1 = 3.0$ m, whereas the height of body 2 is $h_2 = 1.0$ m and its length is $l_1 = 1.0$ m. During this investigation, the distance (d) between the bodies is varied to find the distance at which the impact on B_2 is highest. Note that gravity only acts on B_1, and that the bottom nodes

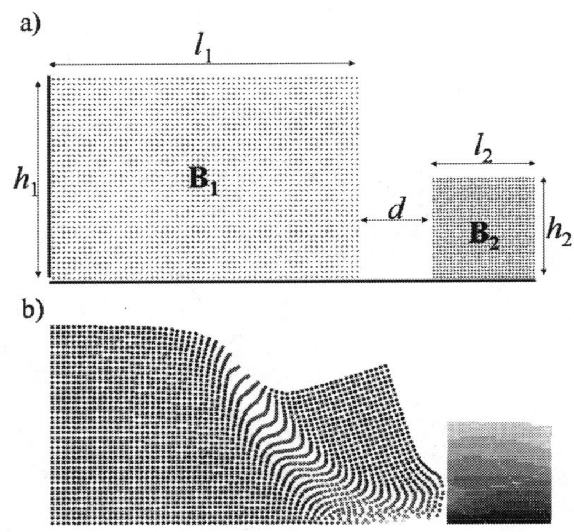

Figure 6. a) Sketch of the problem, and b) final configuration including displacement contours in rigid body.

of B_2 are fixed in both directions to avoid a potential slide or roll-over of the body after impact.

The soil making up body 1 has a peak cohesion of $c_p = 6$ kPa, a residual cohesion of $c_r = 1.5$ kPa, a softening modulus of $s_m = -15$ kPa, an elastic modulus of $E = 1000$ kPa, and a Poisson's ratio of $\nu = 0.30$. Body 2 is purely elastic, with an elastic modulus of $E = 2000$ kPa and a Poisson's ratio of $\nu = 0.49$. The friction coefficient acting between the soil and the bottom boundary, and between the soil and the rigid body, is $\mu = 1.0$, so that after contact the material sticks to the rigid body. To quiantify the reaction in body 2, the mean value of the deviatoric stress (q) was computed based on the 9 points in the base element, in which the reaction is bigger. The stresses in body 1 and body 2 are computed using CMPM as in Gonzalez et al. (2017), rather than using classical MPM, in order to have more accurate results, especially for body 2 which develops high stresses due to the incompressibility caused by the high Poisson's ratio.

4.2 *Results*

Figure 7 shows that the critical distance (i.e. the distance at which the reaction in body 2 is highest) is $d_c = 0.5$ m, whereas for $d = 0.2$ and 1.5 m the reactions are similar. The reason for similar values at $d = 0.2$ and 1.5 m is because at these values the velocity of the landslide is similar and small; in the first case, it is because failure has not fully developed in body 1, so that the velocity of the body has not increased significantly, whereas in the second case the body has completely failed but the velocity developed has started to decrease because the travel distance of the slide is longer. Note that the collision times for the different analyses are different because the distance d is different; hence, the measure of the deviatoric stress is computed relative to the time of the collision (t_c) and not before.

Figure 8 illustrates the configuration of the landslide at the critical condition, as well as the collision at the maximum distance (in which the effect of the impact is smaller).

After the critical distance was found, another set of analyses were performed for $d = d_c$, in order to investigate the influence of the properties of body 1 on the stresses built up in body 2. The property that is varied is the residual cohesion, with values of c_r = 1.5, 2.0, 2.5 and 3.0 kPa. As in the previous analysis, the change in the residual cohesion affects the behavior of body 1, as well as causing the contact to occur at different times for each analysis. As before, the time from the moment of collision was used to plot the results.

Figure 9 shows that the highest deviatoric stresses are generated when the residual cohesion of body 1 is the smallest, and that, with an increase of the residual cohesion, the impact on body 2 decreases. Clearly, if the residual cohesion is small, the forces resisting the displacement of body 1 are smaller, leading to a more sudden failure that will cause the development of higher velocities and thereby a bigger impact with the rigid body.

Finally, the mesh size effect has been analyzed, using the critical distance d_c and the critical residual cohesion of c_r = 1.5 kPa. Specifically, four analyses are performed considering mesh sizes of Δx = 1/10, 1/8, 1/6, and 1/4 m.

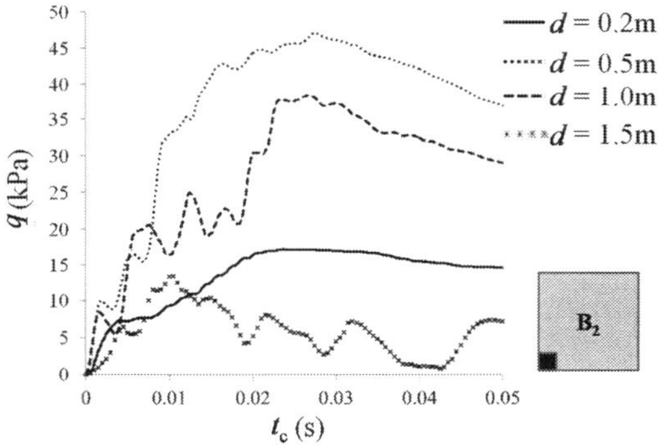

Figure 7. Mean deviatoric stress evolution with time at the base of body 2 as a function of initial distance between the bodies.

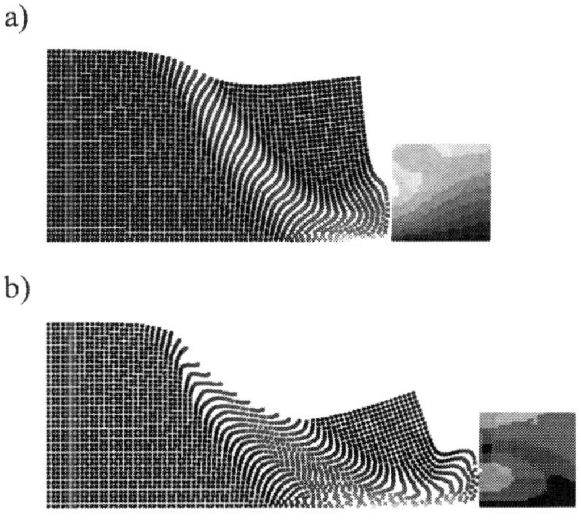

a)

b)

Figure 8. Landslide collision a) at the critical distance, and b) at the maximum distance.

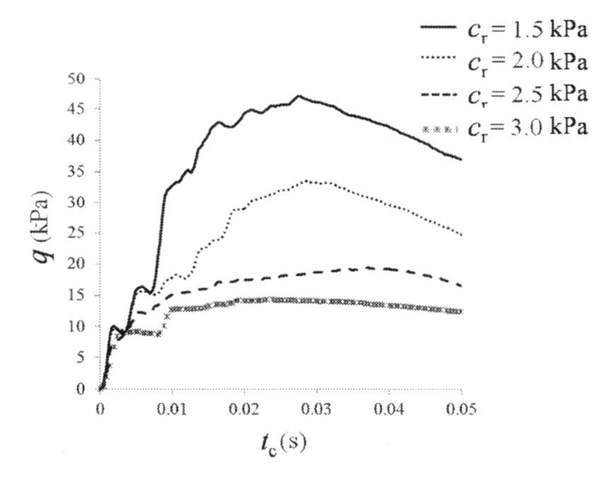

Figure 9. Deviatoric stress in body 2 considering different residual cohesion values for body 1.

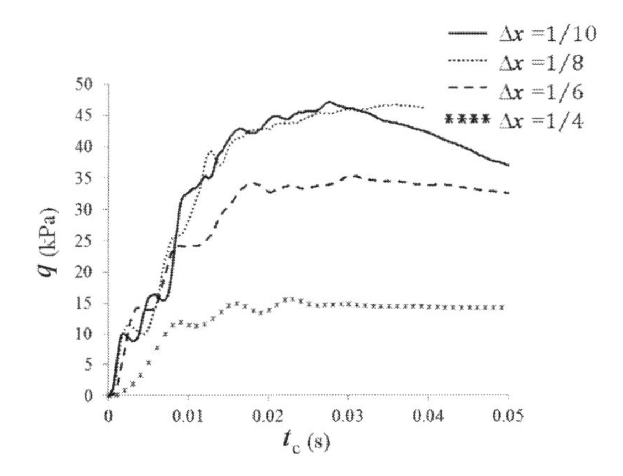

Figure 10. Deviatoric stress in body 2 for different mesh sizes.

As can be seen in Figure 10, the maximum deviatoric stress is reached when the mesh size is smallest, and this is thought to be mainly a consequence of no strain regularization induced in the current verison of the algorithm. Hence, for a smaller element size, the shear band in B_1 is narrower and the slope failure more sudden, causing an increase in the velocity of the slide and a greater impact on B_2.

5 CONCLUSIONS

This paper has investigated the performance of MPM with a contact algorithm to analyse the behaviour of a rigid structure during impact from a landslide, considering the variation of the mechanical properties of the sliding material and the initial conditions of the problem.

The results have shown that the contact algorithm in MPM is suitable for determining if bodies are interacting or not. Also, the transfer of information between bodies in contact has been demonstrated, although improvements are needed in the computation of the normal direction at the boundary of the bodies, since this can severely impact the accuracy of the analysis. The oscillations in the solution caused by material points moving between elements during contact is also a significant problem that causes an irregular redistribution of the contact forces, especially if contact is maintained over a longer time.

The results are an indication that MPM is an appropriate tool for analysing geotechnical problems. Considering a broad combination of initial conditions and a range of mechanical properties, it is possible to obtain a wide range of results in order to detect the most critical scenario.

REFERENCES

Bardenhagen, S.G., Brackbill, J.U., & Sulsky, D. 2000. The material-point method for granular materials. *Computer Methods in Applied Mechanics and Engineering 187(3–4)*: 529–541.

Bardenhagen, S.G., Guilkey, J.E., Roessig, K.M., Brackbill J.U., & Foster J.C. 2001. An improved contact algorithm for the material point method and application to stress propagation in granular material. *Computer Modeling in Engineering and Sciences 2(4)*: 509–522.

Chen, Z.P., Qiu, X.M., Zhang, X., & Lian, Y.P. 2015. Improved coupling of finite element method with material point method based on a particle-to-surface contact algorithm. *Computer Methods in Applied Mechanics and Engineering 293*: 1–19.

Gazetas, G., Garini, E., & Zafeirakos, A. 2016. Seismic analysis of tall anchored sheet-pile walls. *Soil Dynamics and Earthquake Engineering 91*: 209–221.

Gonzalez Acosta, L., Vardon, P.J., & Hicks, M.A. 2017. Composite material point method (CMPM) to improve stress recovery for quasi-static problems. *Procedia Engineering 175*: 324–331.

Hosseinzadeh, S., & Joosse, J.F. 2015. Design optimization of retaining walls in narrow trenches using both analytical and numerical methods. *Computers and Geotechnics 69*: 338–351.

Pantev, I. 2016. Contact modelling in the material point method. *MSc thesis, Delft University of Technology,* 2016.

Soga, K., Alonso, E., Yerro, A., Kumar, K., & Bandara, S. 2016. Trends in large-deformation analysis of landslide mass movements with particular emphasis on the material point method. *Géotechnique 66(3)*: 248–273.

Sulsky, D., Chen, A., & Schreyer H.L. 1994. A particle method for history dependent materials. *Computer Methods in Applied Mechanics & Engineering 118(1–2)*: 179–196.

Sulsky, D., Zhou, S., & Schreyer H.L. 1995. Application of a particle-in-cell method to solid mechanics. *Computer Physics Communications 87(1)*: 236–252.

Wang, B., Vardon, P.J., & Hicks, M.A. 2016. Investigation of retrogressive and progressive slope failure mechanisms using the material point method. *Computers and Geotechnics 78*: 88–98.

Numerical Methods in Geotechnical Engineering IX – Cardoso et al. (Eds)
 ISBN 978-1-138-33198-3

Assessment of dike safety within the framework of large deformation analysis with the material point method

B. Zuada Coelho, A. Rohe & A. Aboufirass
Department of Geo-engineering, Deltares, Delft, The Netherlands

J.D. Nuttall
Department of Numerical Simulation Software, Deltares, Delft, The Netherlands

M. Bolognin
Geo-Engineerging Section, Delft University of Technology, Delft, The Netherlands

ABSTRACT: Dike infrastructure is of vital importance for the safety against flooding. The standard methodologies for the assessment of dike safety for macrostability are based on limit equilibrium methods, which result in a safety factor against shear failure. The more advanced alternative consists of using finite element method to compute the safety factor against shear failure. However, these approaches do not take into account the capacity of the dike to retain water, but are only concerned with the mechanical equilibrium of the dike's initial composition. With the recent advancements in the modelling of large deformations within geotechnical engineering, e.g. by means of the material point method, the post failure behaviour of the dike can be predicted. The material point method is a mesh-free method that has been developed to address the problem of large deformation on a continuum level. The material point method offers the possibility to redefine the concept of factor of safety against shear failure. The initial shear failure of a dike does not necessarily lead to the loss of the dike's capability to retain water. In reality, after the initial shear failure the mass of soil will move and reach a new equilibrium position. This paper, after a brief description of the material point method, presents the analysis of a progressive dike failure, where the post failure behaviour is examined and a proposal is made to redefine the concept of factor of safety.

1 INTRODUCTION

Slope stability of dikes is an important issue in geotechnical engineering as dikes are often the first line of defence against water flooding. Traditionally, slope stability analysis is carried out using limit equilibrium methods (LEM) or the finite element method (FEM). Limit equilibrium methods have the advantage of being computational efficient, but have many drawbacks when compared to other numerical methods. The main limitation of limit equilibrium methods regards the definition of safety, as it strongly depends on the assumptions made for the slip surface (Griffiths and Lane 1999, Li et al. 2009). Different limit equilibrium methods are distinguished by the assumptions for the slip surface. A detailed review of limit equilibrium methods is reported by Duncan (1996).

The finite element method is a more advanced methodology that has significant advantages for the analysis of slope stability. The shape and location of the slip surface are not predefined, and are automatically found. The analysis provides insight into the initiation of failure and failure mechanism. The finite element method also allows the use of complex models for the stress-strain behaviour of the soil (Farias and Naylor 1998, Griffiths and Lane 1999, Li et al. 2009).

In limit equilibrium methods, failure is defined when the shear strength of the soil is equal to the shear strength required to guarantee the slope equilibrium. In the finite element method, failure is traditionally defined when there is a non-convergence of the solution (Griffiths and Lane 1999). Slope failure and numerical non-convergence occur simultaneously, and are accompanied by a major increase in the nodal displacements within the mesh. This leads to problems related to mesh distortion that prevent the use of the finite element method for post-failure analysis of slopes. This means that, although the finite element method is an appropriate tool to analyse slopes up to the moment of initial failure, the behaviour following this initial failure cannot be correctly described.

In order to overcome the limitations of the finite element method, mesh-free methods that address the problem of large deformation on a continuum level have been developed, such as the material point method (MPM). In the material point method the continuum material is represented by a set of Lagrangian points (material points) that move through an Eulerian background mesh. The material points contain all the properties of the continuum, such as mass, stress, strain and material parameters. Therefore, the material point method can be seen as a combination of both Lagrangian and Eulerian formulations. The problems related to mesh distortion under large deformations are circumvented, as well as the diffusion associated with the convective terms of the Eulerian approach (Sulsky et al. 1994, Sulsky et al. 1995).

The use of Lagrangian material points conserves mass and allows the use of complex history dependent stress-strain material models. The discrete equations for momentum balance are obtained on the background grid similar to the finite element method with an updated Lagrangian formulation. The material point method has been successfully used for geomechanics problems involving slope stability (e.g. Zabala and Alonso 2011, Soga et al. 2016, Fern et al. 2017).

In this paper, the material point method will be used to study the stability of a dike. After an initial validation of the method against literature results, the material point method will be applied to a typical Dutch dike profile, and the differences with the standard methodologies will be highlighted and discussed.

2 VERIFICATION

The material point method analyses have been performed with the software Anura3D (2017). The validation has been performed by simulating a slope previously presented by Hicks and Wong (1988). The cross section of the 3D slope geometry and mesh for the material point method analysis are presented in Figure 1. The slope is assumed to be fully saturated with water, and the analysis is undrained. The system of equations is solved explicitly in the time domain. The domain was discretised using low-order tetrahedral elements (in total 1086 active elements), with initially 4 material points in each element. The displacements are constrained in both vertical and horizontal directions for the bottom boundary, and the displacements are constrained in the horizontal direction along the vertical boundaries. The slope material has been modelled as a Mohr-Coulomb material (see Table 1).

The slope factor of safety was assessed by increasing the gravity multiplier, while keeping all material parameters constant. The factor of safety is defined as the ratio between the gravity multiplier at the failure state and the initial gravity multiplier, i.e. 1g (Swan and Seo 1999, Li et al. 2009, Zheng et al. 2006)

Figure 2 shows the vertical displacement for point A (see location in Figure 1), located at the edge of the slope crest, obtained by finite element method (Hicks and Wong 1988), material point method and limit equilibrium method. The limit equilibrium analysis were performed by using Bishop's method within the D-GeoStability software (Deltares 2016).

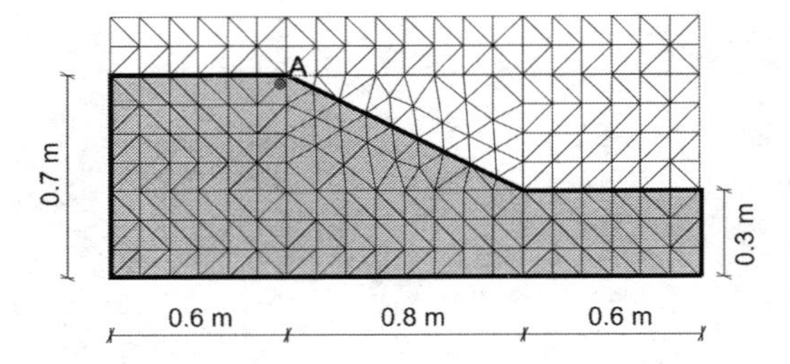

Figure 1. Cross section of 3D slope geometry and mesh (based on Hicks andWong 1988).

Table 1. Mohr-Coulomb material parameters for the slope (based on Hicks andWong 1988).

Parameter	Unit	Value
Volumetric weight	kN/m^3	20
Young modulus	kPa	1000
Poisson ratio	–	0.3075
Poisson ratio undrained	–	0.499
Cohesion	kPa	0.6
Friction angle	o	30
Dilantancy angle	o	0

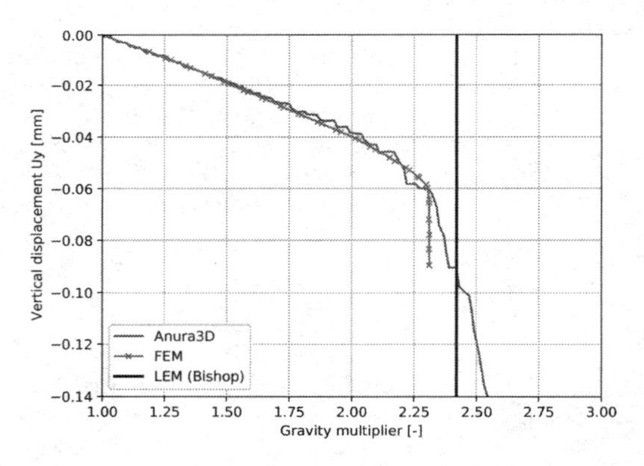

Figure 2. Comparison between FEM (Hicks & Wong 1988) MPM and LEM results for the vertical displacement of the slope crest.

It follows that the results provided by the material point method calculation are in agreement with the results from the finite element analysis, up to the moment of failure (gravity multiplier of ≈ 2.3), which illustrates the correctness of the material point method solution for small strain analysis. The finite element computation has the limitation of not providing any insight into the dike behaviour after the initial failure, while the material point method is able to simulate beyond this point. The limit equilibrium method provides a result that is also in agreement with the finite element analysis. This shows that when the objective is to estimate the initial factor of safety of slopes the limit equilibrium method provides a good solution with fewer computational resources.

Following the traditional design methodology it is assumed that the slope factor of safety is 2.3. However, for this factor of safety the vertical displacement of point A is only 0.08 mm (0.002% of the slope height), which means that the slope still fulfils its function of retaining high water levels. This will be further discussed in the next section.

3 DIKE ANALYSIS AND SAFETY

3.1 *Problem description*

In order to illustrate the benefits of using the material point method for slope stability analysis, a generic dike section has been studied. The dike geometry corresponds to a typical Dutch dike section, with non-symmetric inclination of the slopes (outer slope = 1/4; inner slope = 1/5). The dike was considered fully saturated with the water level placed at the height of the crest. The cross section of the 3D dike geometry and mesh for the material point method analysis are presented in Figure 3.

The system of equations is solved explicitly in the time domain. The domain was discretised using low-order tetrahedral elements (in total 7017 active elements), with initially 4 material points in each element. The displacements are constrained in both vertical and horizontal directions for the bottom boundary, and the displacements are constrained for the horizontal direction along the vertical boundaries. The dike material was considered to be uniform and homogeneous, and has been modelled as a Mohr-Coulomb material (properties available in Table 2). The mesh is refined in domains where higher shear strains are expected.

The application of distributed loads in the material point method are not straight forward, as the material points can move through the mesh. To overcome this issue, the water loading was modelled by representing the water reservoir as an elastic material, with the properties presented in Table 2. Note that the water is not infiltrating into the dike.

Figure 4 shows the vertical effective stress and pore water pressure, after the initialisation of gravity loading (gravity multiplier 1g). It follows that the proposed methodology is appropriate to model the water loading.

Similar to the slope presented in the previous section, the dike safety was assessed by increasing the gravity multiplier. However, as the aim of the analysis is to analyse dike deformation, a problem arises as the dike deformation incorporates both the deformation due to shearing and the deformation due to the elastic deformation caused by the gravity increase. In order to be able to distinguish between the two, an additional one-dimensional soil column subjected to gravity increase was computed with thematerial point method, and its results subtracted from the dike results. The results from the one-dimensional soil column correspond entirely to the deformation caused by the gravity increase,

Table 2. Mohr-Coulomb material parameters for the dike and water.

Material	Parameter	Unit	Value
Dike	Volumetric weight	kN/m³	20
	Young modulus	kPa	7500
	Poisson ratio	–	0.33
	Poisson ratio undrained	–	0.499
	Cohesion	kPa	1
	Friction angle	o	35
	Dilantancy angle	o	0
Water	Volumetric weight	kN/m³	10
	Young modulus	kPa	150000
	Poisson ratio	–	0
	Poisson ratio undrained	–	0.499

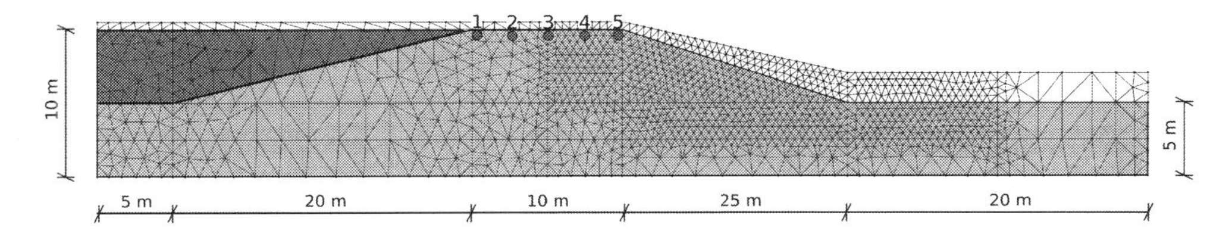

Figure 3. Cross section of dike geometry and mesh.

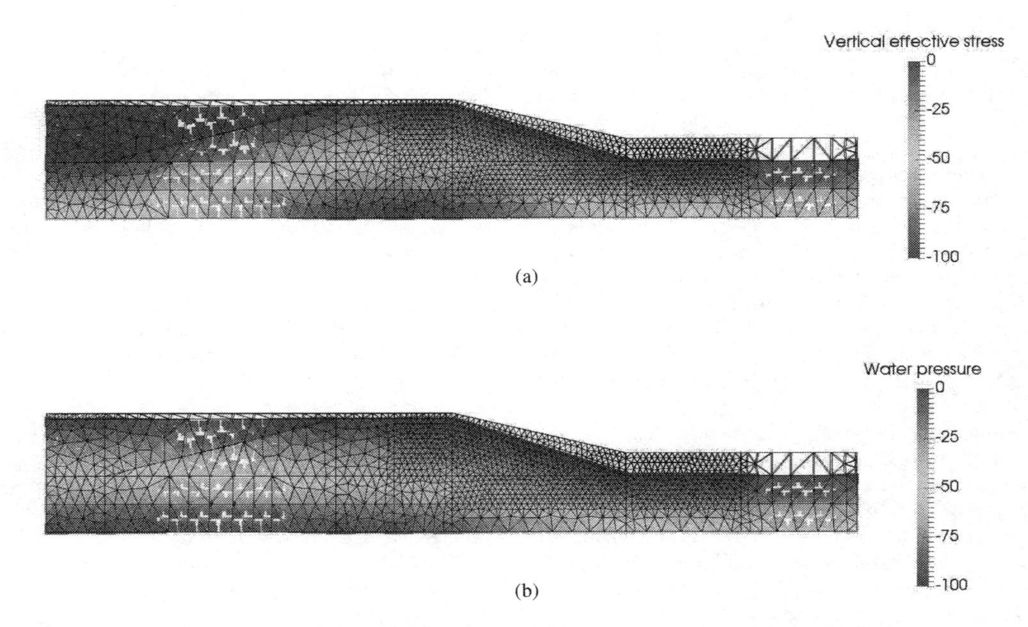

Figure 4. Initialisation stage: (a) vertical effective stress and (b) water pressure at gravity multiplier 1g.

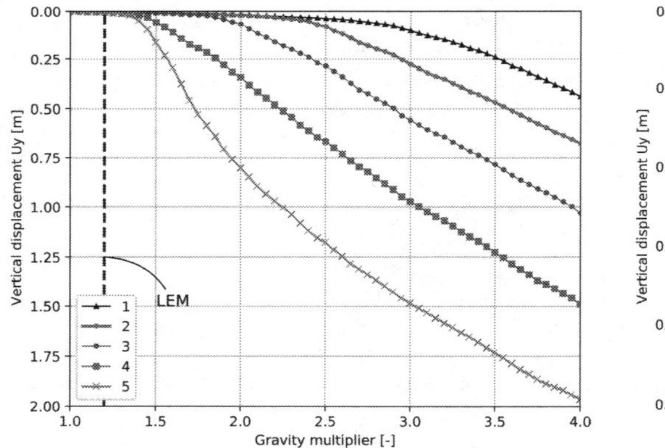

Figure 5. Vertical displacement for several points along the dike crest.

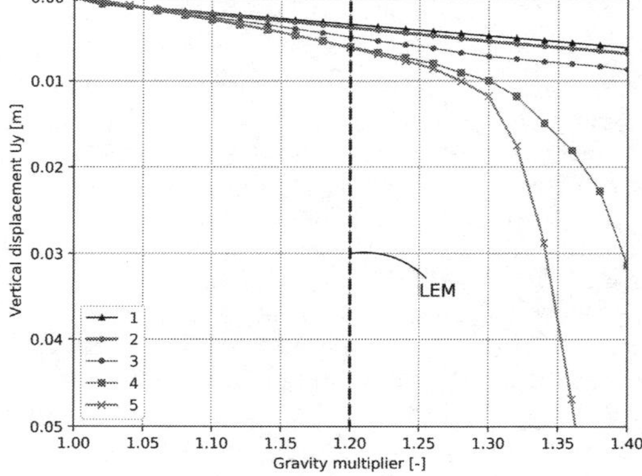

Figure 6. Vertical displacement for several points along the dike crest for low gravity multipliers.

as no shearing occurs. Therefore, the results presented in the following section are corrected, and merely concern the shear deformation.

3.2 *Results*

Figure 5 shows the vertical displacement of several points located along the dike crest, together with the factor of safety obtained by means of the limit equilibrium method. The location of the points are indicated in Figure 3. The dike crest exhibits higher displacements at points towards the inner slope. Point 1, which is located close to the outer slope, and next to the water loading has a smaller vertical deformation.

For a gravity multiplier of 1.2, which corresponds to the factor of safety according to the limit equilibrium analysis, the crest deformation is very small. This is further illustrated in Figure 6, which presents the vertical crest displacement for low values of the gravity multiplier. For a gravity multiplier of 1.2 the maximum dike crest deformation occurs at point 5 and is smaller than 7 mm. This means that, if this dike was analysed following the standard methods, the dike would be considered to fail with a maximum

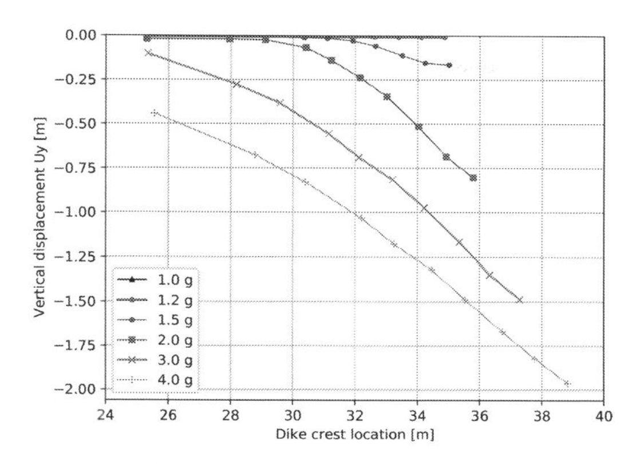

Figure 7. Dike crest profile for different values of gravity multiplier.

displacement of 7 mm. However, for such a small deformation it can be expected that the dike still fulfils its primary function of retaining water.

Figure 7 presents the dike crest profile at different levels of the gravity multiplier. For a gravity multiplier of 1.2 the dike crest exhibits no visible displacement. At a gravity multiplier of 1.5 the dike crest starts to move, but only for the points closer to the inner slope. Only at a gravity multiplier of 3 a clear movement of the entire dike crest is identified. From the figure it is also found that the dike crest not only settles, but also elongates.

The main advantage of the material point method for the analysis of dike safety is related to the possibility of analysing beyond the initial failure. As the gravity multiplier increases, the dike continuously

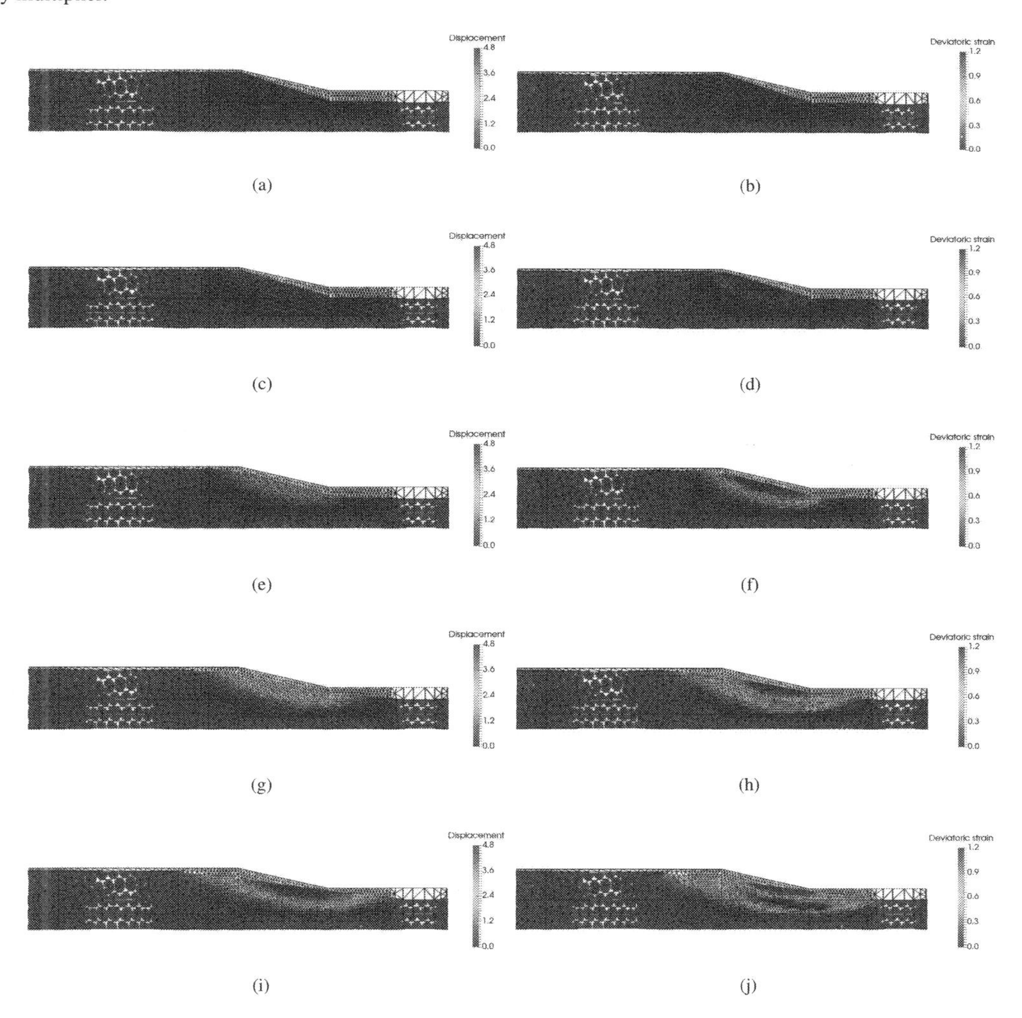

Figure 8. Displacement and deviatoric strain fields for different values of gravity: (a) and (b) 1.2g, (c) and (d) 1.5g, (e) and (f) 2g, (g) and (h) 3g, (i) and (j) 4g.

deforms and reaches new equilibrium positions. This is because failure is progressive and not abrupt as it is assumed by the limit equilibrium methods.

From a safety point of view, if a maximum allowable displacement of the dike crest is defined, a new factor of safety can be established based on this displacement criterium. For example, for this particular dike, if it would be acceptable to have a maximum crest displacement of 25 cm, the factor of safety would be larger than 1.5, instead of 1.2 as defined from the limit equilibrium method.

The displacement and shear strain fields of the dike are presented in Figure 8, at different gravity multipliers. The dike is found to exhibit a classic macro-stability failure by movement of the inner slope. Up to a gravity multiplier of 2 no significant displacement or shear bands occur. For gravity multipliers of 2 and higher it is clear the formation of the shear band and its effects on the dike displacement are clearly visible.

3.3 *Influence of Young modulus*

The standard methodology of dike safety assessment based on limit equilibrium methods, does not take into account the dike deformation, hence the

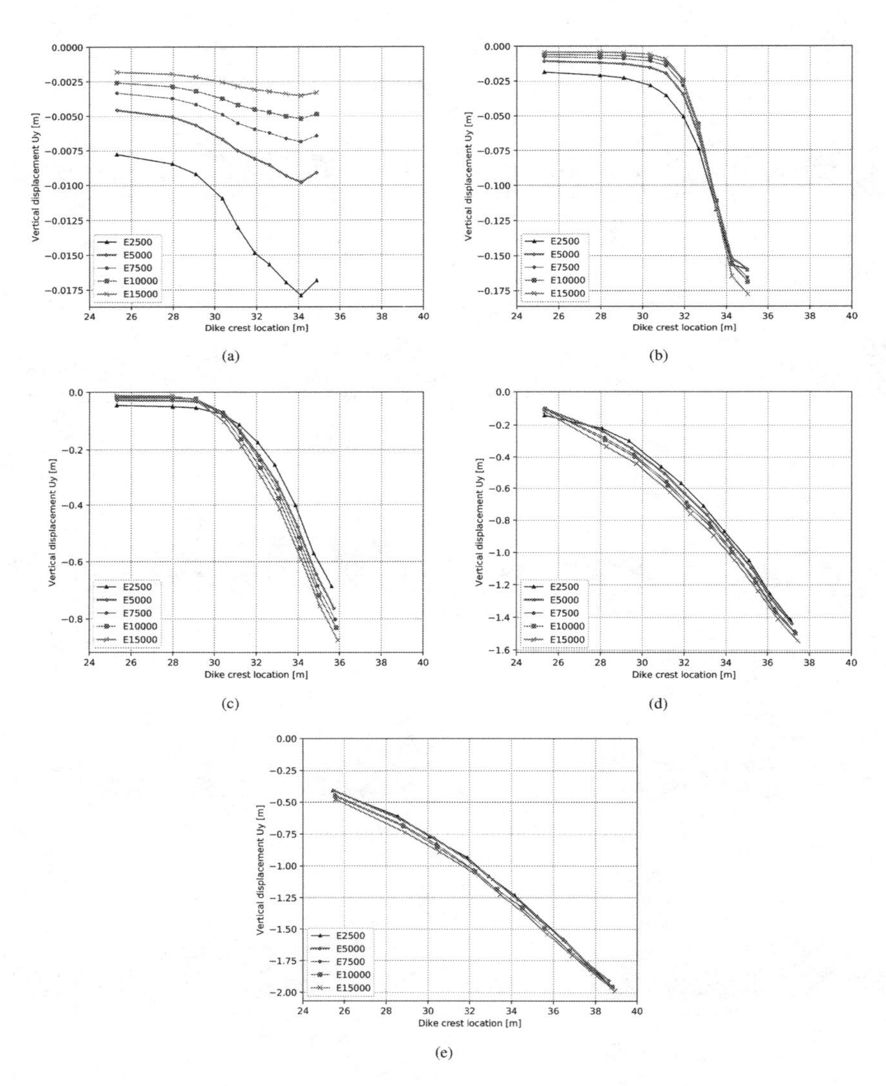

Figure 9. Effect of the Young modulus on the crest profile for different values of gravity: (a) 1.2g, (b) 1.5g, (c) 2g, (d) 3g and (e) 4g.

stiffness has no influence on the dike safety. However, when going towards an assessment based on displacement criteria, the stiffness is of importance, as it is fundamental to correctly model the displacement field. Therefore an analysis on the Young modulus is performed in the range between 2500 and 15000 kPa.

Figure 9 presents the results of the dike crest displacement for several cases. It is clear the effect of the Young modulus on the dike crest displacement results. For low gravity multipliers (Figure 9a and 9b), the deformation pattern is the same for all Young moduli, while the lower the Young modulus the larger the crest deformation is.

As the gravity multiplier increases the deformation pattern changes. At a gravity multiplier of 2 (Figure 9c) the displacement at the outer slope is larger for the lower Young modulus, however at the inner slope, the displacement is larger for the larger Young moduli. An inflection point is found halfway along the dike crest, where the influence of the Young modulus on the displacement is inverted. This is likely caused by stress redistributions caused by the large deformations, and illustrates the complexity of large deformation analysis. For the higher gravity multipliers (3.3d, and 3.3e), this effect is enhanced, whereby higher crest displacements occur for higher Young modulus.

These results illustrate the importance of correctly estimating the Young modulus when performing more advanced computations by means of the material point method. Also, it shows that as the design shifts towards displacement criteria, more soil investigation is needed in order to be able to correctly parametrise the material models.

4 CONCLUSIONS

This paper presented the validation of the Anura3D material point method software, and dike safety analysis based on displacement criteria.

The standard methodology for the assessment of dike safety is based on limit equilibrium methods, which do not take into account a deformation criterion. Based on material point method calculations it has been shown that, for the factor of safety corresponding to the limit equilibrium method, the dike crest displacements are very small. This means, that the factor of safety does not imply that the dike fails its requirement of retaining water, but just that an initial failure took place. After this initial failure the dike stabilises in a new equilibrium position.

The recent developments of the material point method enable to look beyond this initial failure. From the analysed dike section, it has been shown that the dike continues deforming with increasing gravity, and that the dike failure is progressive. Profiles for the dike crest deformation were computed, which can be used to assess the dike safety, by combining it with criteria for the maximum allowed displacement of the dike crest. Moving towards a displacement based safety assessment of dikes can have a positive impact for existing dikes, as it will provide a better prediction of their safety, as well as for dikes to be constructed, as it will allow for more economical designs.

The analyses presented in this paper can be considered as a first step in a transition towards a safety assessment based on displacement criteria. Future work concerns the definition of the maximum allowed crest displacement and this must be established within the probabilistic framework of dike safety. The transition towards displacement criteria is likely to require more and different soil investigation, in order to determine the stiffness parameters reliably.

REFERENCES

Anura3D (2017). Anura3D MPM Research Community. www.Anura3D.com.

Deltares (2016). *Slope stability software for soft soil engineering*. D-Geo Stability. v16.1.1.

Duncan, J.M. (1996). State of the art: Limit equilibrium and finite-element analysis of slopes. *Journal of Geotechnical Engineering 122*(7), 577–596.

Farias, M. & D. Naylor (1998). Safety analysis using finite elements. *Computers and Geotechnics 22*(2), 165–181.

Fern, E.J., D.A. de Lange, C. Zwanenburg, J.A.M. Teunissen, A. Rohe, & K. Soga (2017). Experimental and numerical investigations of dyke failures involving soft materials. *Engineering Geology 219*(Supplement C), 130–139.

Griffiths, D.V. & P.A. Lane (1999). Slope stability analysis by finite elements. *Géotechnique 49*(3), 387–403.

Hicks,M. &W.Wong (1988). Static liquefaction of loose slopes. In *Numerical methods in Geomechanics*, pp. 1361–1367. Balkema.

Li, L., C. Tang, W. Zhu, & Z. Liang (2009). Numerical analysis of slope stability based on the gravity increase method. *Computers and Geotechnics 36*(7), 1246–1258.

Soga, K., E. Alonso, A. Yerro, K. Kumar, & S. Bandara (2016). Trends in large-deformation analysis of landslide mass movements with particular emphasis on the material point method. *Gotechnique 66*(3), 248–273.

Sulsky, D., Z. Chen,& H. Schreyer (1994).A particlemethod for history-dependent materials. *Computer Methods in Applied Mechanics and Engineering 118*(1), 179–196.

Sulsky, D., S.-J. Zhou, & H.L. Schreyer (1995). Application of a particle-in-cell method to solid mechanics. *Computer Physics Communications 87*(1), 236–252.

Swan, C.C. & Y.-K. Seo (1999). Limit state analysis of earthen slopes using dual continuum/FEMapproaches. *International Journal for Numerical and Analytical Methods in Geomechanics 23*(12), 1359–1371.

Zabala, F. &E. Alonso (2011). Progressive failure of Aznalc´ollar dam using the material point method. *Géotechnique 61*(9), 795–808.

Zheng, H., L. Tham, & D. Liu (2006). On two definitions of the factor of safety commonly used in the finite element slope stability analysis. *Computers and Geotechnics 33*(3), 188–195.

Numerical Methods in Geotechnical Engineering IX – Cardoso et al. (Eds)
© 2018 Taylor & Francis Group, London, ISBN 978-1-138-33198-3

Numerical simulation of pile installation in saturated soil using CPDI

C. Moormann, S. Gowda & S. Giridharan
Institute for Geotechnical Engineering, University of Stuttgart, Germany

ABSTRACT: In diverse applications including pile driving, excavations, foundations, prediction of landslides and slope failure, large deformation and material movement is observed in the soil medium. In order to simulate these processes, pure Lagrangian based numerical methods, popularly known as the Finite Element Method pose certain challenges. This can be overcome by coupling the Lagrangian method with the Eulerian approach. Material Point Method (MPM) is one such implementation capable of simulating large deformations in soil. Certain numerical difficulties in MPM such as grid-crossing error, incapability for extension etc., are alleviated in the Convected Particle Domain Interpolation (CPDI) approach, an improvement bought about to MPM.

CPDI is further extended to a two-phase model for simulating saturated soil. Soil and water phases are represented by a single particle, based on their respective porosities. The third phase—air/void is neglected in the present work. A multi-velocity formulation, i.e. solid-water velocity formulation (v-w formulation) is adopted to precisely capture the saturated soil behavior. This includes a full set of equation for soil and water; which can capture both stress and pressure waves. Effective stress concept is followed by accounting pore pressure effects in the deformation of soil. A few examples are presented to validate the two-phase CPDI implementation. The method is then evaluated by means of a dynamic application, a pile installation simulation in saturated sand. A 2D axisymmetric model is considered together, with a penalty contact method for evaluating the interaction between soil and pile. The pile is hammered into the sand by means of pulse loading applied on head of the pile. A Hypoplastic constitutive model is chosen for the sand which allows the study of soil parameters such as stress, void ratio, friction angle, pore pressure etc., during the installation process. The effect of saturated soil on pile installation is studied and compared with the results from the installation in dry soil.

1 INTRODUCTION

1.1 *Overview*

Simulating large deformation problems has been of particular interest in the field of geotechnical engineering. Notably, studying numerical simulation of pile installation has captivated the interest of research community for quite a while. Various methods have been adopted to simulate and study the behaviour of sand during the installation of pile. (Dijkstra et al. 2011) presented two modelling approaches to simulate the installation of piles in sand; a fixed pile approach which models the flow of soil around the pile, and the moving pile approach which models the penetration of the pile from the surface level. Coupled Eulerian-Lagrangian (CEL) approach has also shown to be robust in simulating geomechanical problems that involve large deformation (Qiu et al. 2011). CEL's popularity in application to geotechnical problems has further been bolstered due to it's availability in popular commercial packages like ABAQUS. Pile installation has proven to be challenging in terms of both testing and execution. Before proceeding to installation on-site, predicting an efficient installation method is essential, and requires a study of various parameters like soil behaviour, force required, type of loading etc., and happens to be some of the minimum parameters that are to be known beforehand. Performing scaled model tests helps in making certain decisions about various parameters, but fails to replicate the real behaviour fully. A need therefore, emerges to develop a numerical technique to test real scale models easily, within a certain degree of accuracy. This can then be used to predict ranges for parameters, thereby reducing the number of experimental investigations required.

Pile installation simulation in sand involves handling large deformation in soil during the penetration process. This renders the conventionally used numerical simulation methods inadequate. Finite Element Method (FEM) has been one of the widely used methods for numerical simulation of physical processes. Due to heavy mesh distortion caused during simulation of large deformation problems, and its subsequent limitations posed owing to

FEM's inability in handling large deformations, many alternate methods have been proposed to handle problems like pile installation, slope failure, excavation etc. One such method that has been widely employed for large deformation problems is called as Material Point Method (MPM). MPM finds its roots in Particle-in-Cell (PIC) method, wherein the material is represented by Lagrangian mass points, referred to as particles, moving through a computational grid (Sulsky et al. 1994). This procedure proved highly successful in tracking contact discontinuities and in modelling highly distorted flows. MPM thus belongs to the family of meshless methods, and is a particle-based method that combines aspects of Eulerian and Lagrangian procedures to achieve an efficient method to simulate large deformation (Bardenhagen et al. 2000). In MPM, the continuum is represented by a cloud of particles known as material points, and carry the history-dependent information associated with the state parameters. Information stored within the particle is updated by solving the momentum equation on the background computational mesh. The deformation of material is represented by tracking and updating the location of material points, and by taking into account the momentum balance without the convection term.

MPM has known to exhibit certain numerical shortcomings which cause oscillations in the stress states during severe deformation. To overcome this, several variations of MPM have been proposed. The new algorithm, referred to as the Convected Particle Domain Interpolation (CPDI) has been developed to improve the accuracy and efficiency of the MPM for problems where large tensile deformations and rotations are anticipated. CPDI has shown to be robust in the application of large deformation problems, in particular, applications to geotechnical engineering (Hamad 2016). The objective of this paper is to present an overview of the CPDI method, it's extension to two-phase formulation, examples validating the robustness of said method, and present an application of the two-phase CPDI method to a simulation of pile installation in saturated sand.

1.2 *Convected particle domain interpolation*

Early versions of MPM assumed a collocated distribution for material properties across the particle domain. This distribution causes numerical oscillations when particles cross element boundaries. To alleviate this shortcoming, a version of MPM, called Generalised Interpolation Material Point (GIMP) method has been proposed, in which the concentrated mass of the material point is distributed over a finite sub-domain, which can be thought of as the Dirac delta function in MPM being replaced with a smoother interpolation function (Bardenhagen & Kober 2004). GIMP has shown to exhibit instability during regimes of large distortion due to its inability to follow particle deformation, and during regimes of extension. A further enhancement, called CPDI, has been proposed (Sadeghirad et al. 2011), wherein the domain of a single particle is allowed to deform as a parallelogram, and is constantly updated using the deformation gradient, which is evaluated at the particle locations. The standard grid basis functions are replaced with a set of alternative functions that accounts for the fact that the particle is spread over finite domain defined by the vertices of a parallelogram. CPDI has since been extended for allowing higher-order continuity, in which the particles are tracked as quadrilaterals, rather than parallelogram, and has been named as CPDI2 (Sadeghirad et al. 2013). In this work, the CPDI approach has been used for all the results presented.

2 TWO-PHASE FORMULATION

Saturated soil is treated using the so-called v-w formulation in this present work. A physical based mapping of water momentum between particles and grid points when considering the material point method is provided. Keeping in mind that simulating a porous material in itself, introduces certain complexities to the mechanical behaviour of the materials, and consequently, to the numerical simulation. Ultimately, the capability of the v-w formulation to precisely capture the physical response of saturated soil under dynamic loading is attributed to the fact that in this formulation, all acceleration terms are considered. Reading from a physical view point, pressure should be interpolated one order lower than that of velocity for consistency between variations in pressure and stress. A v-w-formulation automatically ensures the consistency between pressure and stress. Coupling of solid and fluid phases is ultimately achieved by making use of this formulation. Highlighting the total unknowns in the v-w-formulation, there are in total four that are to solved for: water pressure (p), effective stress(σ'), and the velocities of solid (v) and water phase (v_w), respectively. A more detailed explanation of the parameters involved in the formulation is presented in the work of Jassim et al. (2013). Nodes serves as the position at which the velocities are calculated as the primary variables which are then interpolated to the particles, while pressure and stress are calculated at the particles.

The formulation considers only one particle for both the phases, i.e., the single CPDI particle accommodates both water and solid mass through the porosity definition. Following equations define the masses based porosity value (n),

$$m_w = \rho_w V, \tag{1}$$

$$m_d = \rho_d V, \tag{2}$$

$$m_{sat} = \rho_{sat} V, \tag{3}$$

where m_w, m_d and m_{sat} are the mass of water, dry soil and mixture, respectively. V is elemental volume, ρ_w is the density of water, and the dry density of soil, ρ_d is given by the relation

$$\rho_d = (1-n)\rho_s, \tag{4}$$

where ρ_s is the solid density. Density of saturated sand, ρ_{sat} is given by the relation

$$\rho_{sat} = \rho_d + n\rho_w. \tag{5}$$

From these relations, the internal (f_w^{int}, f^{int}), the external forces (f_w^{ext}, f^{ext}), the gravity (f_w^g, f^g) and the drag forces (f^{drag}) are calculated for water and mixture separately. The governing equations are then solved to obtain velocities.

To begin with, the velocity of water, v_w is calculated using the relations

$$m_w a_w = f_w^{ext} + f_w^g - f_w^{int} - f^{drag}, \tag{6}$$

$$v_w = a_w * dt. \tag{7}$$

The velocity of mixture v is then calculated using the equations

$$m_s\, a = -m_w' a_w + f^{ext} + f^g - f_w^{int}, \tag{8}$$

$$v = a\, dt. \tag{9}$$

Further these velocities are used to calculate the strain increments $\delta\varepsilon$ and excess pore pressure δp from the relations

$$\delta\varepsilon = B v\, dt, \tag{10}$$

$$\delta p = dt \frac{K_w}{n}[(1-n)Bv + nBv_w], \tag{11}$$

where B is the strain displacement matrix, and dt is the time increment.

The stresses at the particles are evaluated from the constitutive model, and the excess pore pressure is accumulated to obtain the total pore pressure p. From the effective stress concept, total stress is obtained from the relation

$$\sigma = \sigma' + p. \tag{12}$$

These are used to calculate the internal forces of the mixture and water. The multi-velocity concept effectively considers the effect of water and solid in a two-phase medium.

2.1 *Validation cases*

In this section, two examples are presented, along with its analytical solution, to demonstrate the capability of the two-phase CPDI method to capture the behaviour of continuum response, that is: geostatic stress build-up in a column of saturated sand under gravity load, and the well known solution to one-dimensional consolidation problem presented by Terzaghi.

2.1.1 *Geostatic stresses*

To validate the two-phase formulation, a column of soil, assumed to be in fully saturated, is subjected to a gravity load. The soil is assumed to be elastic and with the material properties being Young's Modulus, E = 10,000 kPa, Poisson's ratio, v = 0.33, solid density, ρ_s = 2,143 kg/m^3, porosity, n = 0.3 and permeability, k = 0.001 m/s. The column is modelled with a width of 0.06 m and height of 1 m. The model is meshed to contain a total of 600 particles. During the solution stage, gravity is applied to the model instantaneously in order to achieve the final geostatic stresses. Oscillation in stresses are observed throughout the model initially, due to the sudden application of the gravity field. Later, due to the drag force generated, the relative velocities between the soil and water acts to dissipate the oscillation, and eventually, equilibrium is achieved. Effective stresses and pore pressure distribution in the column is shown in Figure 1.

The values obtained from the simulation are in good agreement with the analytical values for this problem set-up. The pore pressure, p can be calculated using the relation,

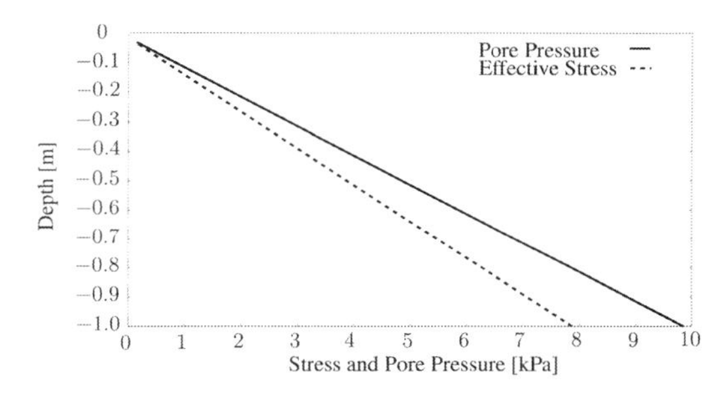

Figure 1. Effective stress and Pore pressure in a fully saturated soil column under gravity.

$$p = \rho_w g h, \tag{13}$$

where ρ_w is the density of water, g is the acceleration due to gravity, which is assumed to be 10 m/s^2 for ease of calculation, and h being the total height of the column. The effective stress is calculated using the relation,

$$\sigma' = [(1-n)\rho_s + n\rho_w - \rho_w] g h. \tag{14}$$

The total stress σ is given as a sum of the effective stress σ' and the pore pressure, p, presented by Equation 12. We can, from the above relations, confirm that the analytically calculated values and the values obtained from the simulation are in good agreement with each other.

2.1.2 *One-dimensional consolidation*

Terzaghi's problem of one-dimension consolidation is presented to validate the two-phase formulation. For this simulation, an elastic soil layer of thickness 1 [m] is considered. The material properties for this problem is considered similar to the values used in the simulation of geostatic stresses, i.e., Young's Modulus, E = 10,000 kPa, Poisson's ratio, v = 0.33, solid density, ρ_s = 2,143 kg/m^3, porosity, n = 0.4 and permeability, k = 0.001 m/s. In this case, however, no gravity field is considered acting on the body, and only traction loads are applied on the soil column. Initially, traction loads are applied on both the soil and water components. This leads to build-up of pore pressure in the column, owing to the high bulk modulus of the water. After having achieved a state of equilibrium, and after having achieved a constant value for the pore pressure in the column, the traction load on the water component is removed. This effectively allows the water to seep out of the soil column, and realise the build-up of effective stresses, and dissipation of pore pressure.

The pore pressure dissipation at different time intervals obtained from the numerical simulation can be seen in Figure 2. The results are in good agreement with the analytical solution for the given setup.

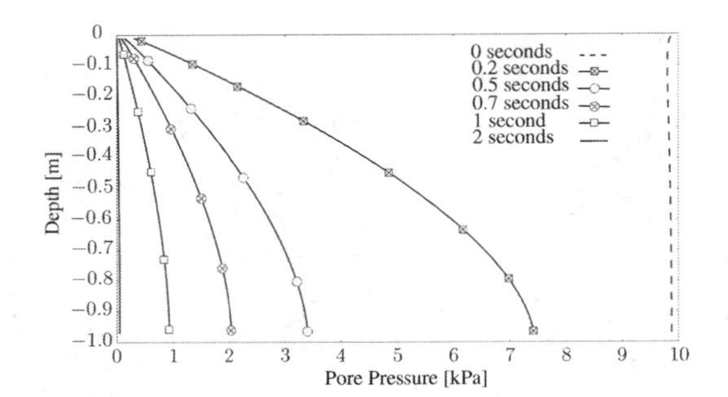

Figure 2. Pore pressure at different intervals of time during consolidation in a soil column.

3 CONTACT FORMULATION

Standard MPM formulation provides an automatic no-slip condition. Contact is detected when the material points of two different entities contribute to the same computation mesh grid node. In this case, consequently, corrections to account for contact is triggered before the actual contact ever having taken place. Furthermore, a lack of smoothing function contributes to oscillation in contact stresses. In the present implementation, a penalty function method, that is often implemented in FE analysis is introduced within the framework of CPDI. Here, the contact forces are assumed to be proportional to the residual of the impenetrability constraints and the surface stiffness. The surface of the continuum is discretised separately from the volume discretisation. By setting a finite amount of mass to the interface, the surface nodes are able to follow the deformation of the continuum. Upon the equation of motion, the entities interact in accordance to the penalty function. Frictional forces are traced back as external contact forces acting on the boundary. Early works relating to contact in MPM focussed on relaxing the no-slip and no-separation condition (Bardenhagen et al. 2000), and was improved in the further works of Bardenhagen et al. (2001). Hamad et al. (2017) proposes an improved contact algorithm, in which the penalty function method is used to evaluate the contact forces between interacting bodies. In this method, the surface of the continuum is discretised separately from the volume discretisation.

In the penalty function method, if a Γ_c exists where contact violation takes place, the potential energy is penalised proportional to the amount of constraint violation, by using a penalty function method P, and is expressed as,

$$P = \frac{1}{2}\omega_n \int_{\Gamma_c} g_n^2 \, d\Gamma_c + \frac{1}{2}\omega_t \int_{\Gamma_c} g_t^2 \, d\Gamma_c, \tag{15}$$

where, ω is the penalty parameter, g is the gap function, and subscripts n and t refer to the normal and tangential directions, respectively. In this method, the interface of the continuum is discretised using two-node linear segments. Normal and tangential stiffness is assigned to these elements, which here, are the penalty parameters in the formulation. A distinction between master and slave is made in the formulation, although purely for numerical convenience. For a detailed formulation of the algorithm, the readers are directed towards fhe work of Hamad et al. (2017).

4 CONSTITUTIVE MODEL

Hypoplastic constitutive soil model, first proposed by von Wolffersdorff (1996), and its subsequent extension to small strain stiffness proposed by Niemunis & Herle (1997) has been used in this work. The hypoplastic constitutive equation can be written as

$$\mathring{\sigma} = \boldsymbol{G}(\sigma, e, \varepsilon) \tag{16}$$

where, $\mathring{\sigma}$ the Jaumann rate of stress, e the current void ratio, and ε the strain rate tensor. The Jaumann stress rate tensor is given as $\dot{\sigma} + \sigma\dot{\omega} - \dot{\omega}\sigma$, where, $\dot{\sigma}$ is the Cauchy stress rate, and $\dot{\omega}$ is the rotation rate tensor. It is observed here that the constitutive model is applied in the rate form, and uses the strain rate tensor $\dot{\varepsilon}$ along with the definition of the Jaumann stress rate to ensure material objectivity.

Due to the drawback of over predicting displacements with the use of classical hypoplastic sand model, the concept of intergranular strain tensor is introduced that accommodates the elastic deformation of the intergranular layer. If applied strain is within the limit of the intergranular strain, the small strain stiffness is activated. Else, the model reverts back to the original model. For the present study, the values for the parameters of the constitutive model are chosen for that of Karlsruhe sand. The material parameters are tabulated in Table 1.

5 PILE INSTALLATION

To demonstrate the efficacy of applying the two-phase CPDI for predicting pile installation effects, the simulation of installation of solid pile into a hypoplastic sand is carried out assuming undrained soil conditions. The behaviour of soil domain is observed in this dynamic process. In the previous works, a simplified form of the two-phase formulation was used to simulate saturated sand. This involved calculating the hydrostatic pore pressure separately, and computing the excess pore pressure part as a product of the bulk modulus of water, and the volumetric strain. In this work however, two-phase model as described in Section 2 to represent the fully saturated soil is adopted.

Table 1. Hypoplastic model properties for Karlsruhe Sand.

$\phi_c(°)$	p_t	$h_s(MPa)$	$n(-)$	$e_{d0}(-)$
33	–	1600	0.19	0.44
$e_{c0}(-)$	$e_{i0}(-)$	$\alpha(-)$	$\beta(-)$	$m_R(-)$
0.85	1.0	0.25	1	5
$m_T(-)$	$R(-)$	$\beta_r(-)$	$\chi(-)$	e_0
2	$1e^{-4}$	0.5	2	–

5.1 *Initial conditions*

An axisymmetric soil domain is modelled with the dimensions of height 10 m and width 5 m, as shown in Figure 3. The pile is assumed to be solid for this simulation. The domain is meshed with 10,500 particles for the soil body and the pile, and an additional 2000 linear elements for the boundary elements. The hypoplastic constitutive model is considered with an initial void ratio of $e_0 = 0.65$, which corresponds to medium dense Sand. A porosity of $n = 0.4$ and the solid density of $\rho_s = 2{,}600$ kg/m^3. The solid pile is assigned to be linear elastic material, with a length of 10 m, and radius of 0.15 m. The properties of the pile are as follows: Young's modulus $E = 10^7$ kPa, Poisson's ratio $\nu = 0.33$ and density $\rho = 3000$ kg/m^3. Four-node quadrilateral discretisation is adopted for the computational mesh. For the material points, however, a non-regular discretisation is chosen. This is done in order to refine the mesh near tip of the pile. The bottom part of the domain is assumed to be fully fixed and roller boundary conditions are applied on the outer boundaries. A Coulomb type friction is considered between the pile and soil, with friction coefficient μ chosen for all the computations as 0.2.

The installation simulation is carried out in two stages. Initially, gravity stresses are built up linearly over a period of one second, after which dynamic loading phase is started. In the loading phase, a pulse load is applied on top of the pile to reproduce the hammering effect. Five blows are applied per second with the maximum traction of 1000 kN.

5.2 *Simulation*

A total of 9 seconds of pile installation is simulated, wherein the first second is alloted to the gravity loading. Two simulations with similar dimensions and values for soil parameters are carried out, assuming different soil-water conditions. While one simulation assumes pile installation in saturated soil, the other assumes installation in dry soil. This has been carried out to compare and highlight the differences in behaviour in both cases.

5.3 *Parameter study*

5.3.1 *Installation depth*

After initial penetration of the pile into sand due to gravity, the pile is hammered into sand by activating the dynamic load condition. In the present simulation, hammering of the pile has been car-

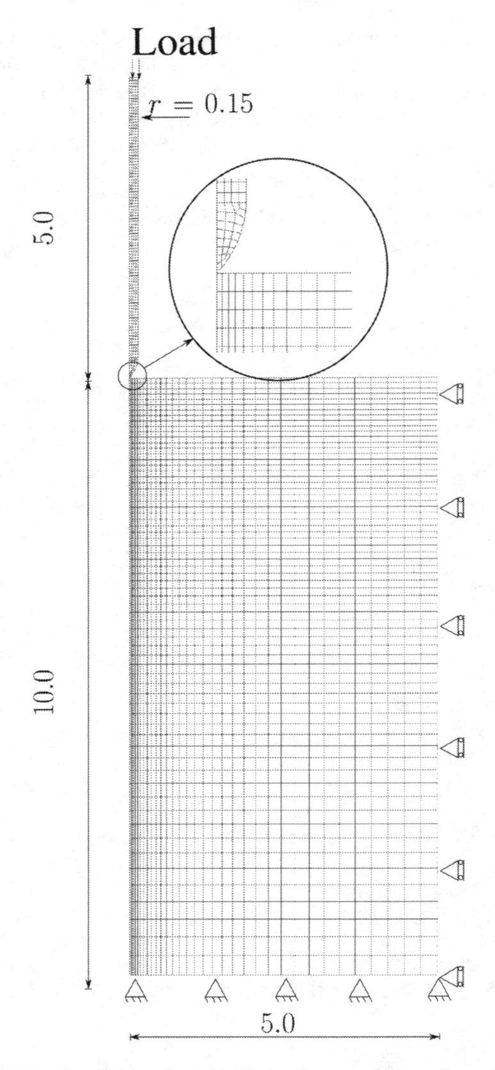

Figure 3. Dimensions and boundary conditions for soil and pile system.

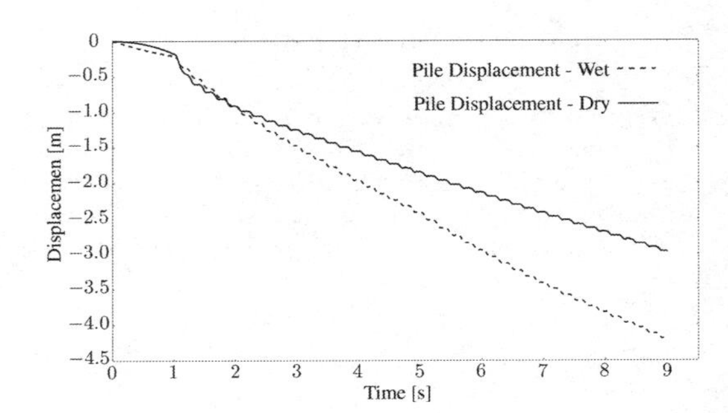

Figure 4. Pile penetration curve in dry and wet sand for 9s of simulation.

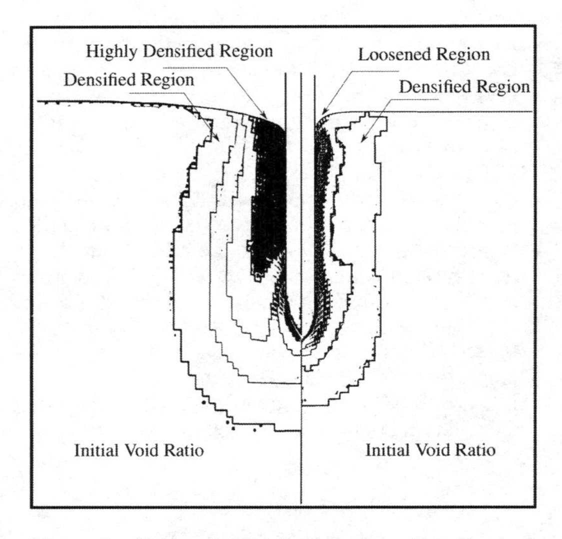

Figure 5. Void ratio distribution along the pile shaft after 2m penetration in dry (left) and wet (right) sand.

ried out for 9 seconds with frequency of 5 blows per second. At the end of 9 seconds, it is observed that the pile has reached a final depth of about 4 m in the case of saturated sand, and a final depth of about 3 m in the case of dry sand. The differences in depth of penetration for both these cases is shown in Figure 4. It is evident that more penetration is achieved with the same loading condition, and the same soil parameters in the case of the saturated sand, than in the case of dry sand.

5.3.2 *Void ratio*

The void ratio is yet another key parameter which aids us to understand the differences between saturated and dry sand simulations. In dry sand, at 2 m of pile penetration, it is observed that the void ratio has decreased around the pile shaft, as shown in Figure 5.

This implies that the sand has compacted, offering more resistance to the pile during further penetration. The compaction zone is approximately 4 times the diameter of the pile, from the pile shaft. In case of wet sand however, the sand has not been compacted to the extent as was in the case of dry sand. A minuscule reduction of void ratio is observed between the zone spanning 0.5 to 4 times the pile diameter, but an increase in void ratio is observed in the regions adjacent to the pile shaft. A small dilation region is observed around the pile shaft. Upon further penetration, the compaction and dilation regions around the pile in wet sand is distinctly observed, see Figure 6.

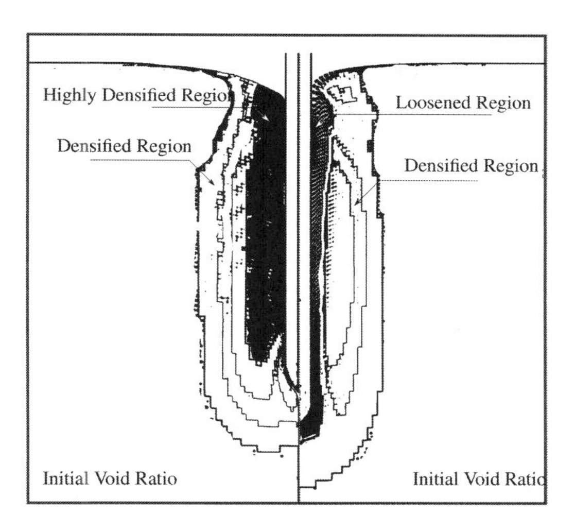

Figure 6. Void ratio distribution after 40 blows in dry (left) and wet (right) sand.

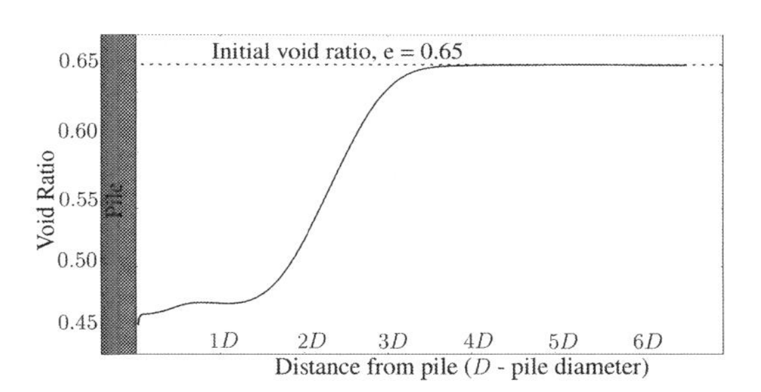

Figure 7. Void ratio variation along a radial line at the depth of 2.5m in dry sand.

In dry sand, however, the compaction zone becomes more exacerbated, with no distinctive dilation region obervable. This effect is better visualised by plotting the changes in void ratio along a radial line at a depth of 2.5 m. In the case of dry sand, the value of void ratio is lesser than the initial value of 0.65 near the pile shaft, and gradually reaches the initial value at a distance 4 times the pile diameter, see Figure 7.

In the case of wet sand, in the region adjacent to the pile shaft (about 0.5 times the pile diameter), a value of void ratio higher than the initial value is observed, indicating loosening in the regions adjacent to the pile. Between the regions of 0.5 to 4 times the pile diameter, a lower value of void ratio is observed indicating compaction, finally reaching the initial void ratio beyond this region, as shown in Figure 8. A similar observation can be seen in the work of Henke & Grabe (2006).

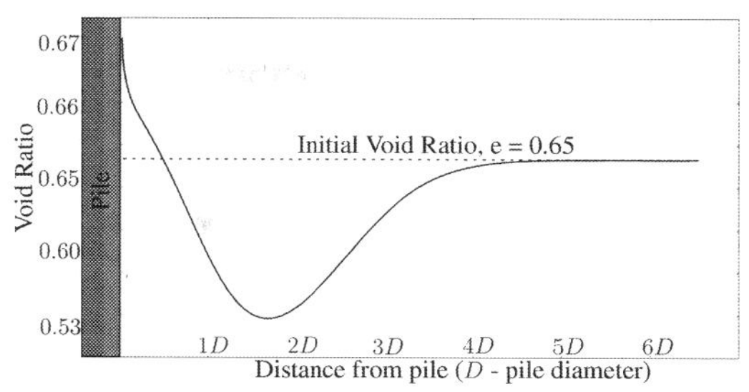

Figure 8. Void ratio variation along a radial line at the depth of 2.5 m in wet sand.

Figure 9. Increase in water pressure near the pile tip during the pile installation in wet sand.

5.3.3 *Water pressure*

Pore pressure build up during the installation process is observed in saturated sand. During a hammer blow, the load is transferred to the water segment of the model, which increases the water pressure, thereby reducing the effective stress. Water pressure distribution and the increase in pressure during a blow near the pile tip is shown in Figure 9.

6 CONCLUSIONS

The simulations performed using CPDI demonstrates the ability to test real scale applications involving large deformation during dynamic processes. Specifically, the installation of pile can be efficiently simulated and the soil behaviour can be studied.

CPDI provides a better interpolation which eliminates non-physical oscillation caused due to grid crossing error. CPDI achieves this without the need for higher order shape functions. With the implementation of two-phase model in CPDI framework, simulating fully saturated soil is achieved. As the two-phase formulation takes into account both water and soil velocities, many assumptions previously made in the simplified two-phase formulation has been avoided. This allows to replicate behaviour close to reality. Void ratio distribution in case of saturated sand captures the dilation and compaction regions distinctly, which simplified two-phase model failed to. Additionally, the water pressure change, stress states and other parameters shows a marked improvement, allowing a more accurate gauging of the soil behaviour. Finally, it is observed the reduction in effective pressure in saturated sand leads to faster penetration than in dry sand. This paves the way to carry out more dynamic applications such as liquefaction simulation and seismic analysis where it is expected for the effective stress in the soil becomes nearly zero.

REFERENCES

Bardenhagen, S., J. Brackbill, & D. Sulsky (2000). The material–point method for granular materials. *Computer Methods in Applied Mechanics and Engineering 187*(3–4), 529–541.

Bardenhagen, S., J. Guilkey, K. Roessig, J. Brackbill, W. Witzel, & J. Foster (2001). An improved contact algorithm for the material point method and application to stress propagation in granular material. *CMES: Computer Modeling in Engineering & Sciences 2*(4), 509–522.

Bardenhagen, S. & E. Kober (2004). The generalized interpolation material point method. *CMES: Computer Modeling in Engineering & Sciences 5*(6), 477–495.

Dijkstra, J., W. Broere, & O. Heeres (2011). Numerical simulation of pile installation. *Computers and Geotechnics 38*(5), 612–622.

Hamad, F. (2016). Formulation of the axisymmetric CPDI with application to pile driving in sand. *Computers and Geotechnics 74*, 141–150.

Hamad, F., S. Giridharan, & C. Moormann (2017). A penalty function method for modelling frictional contact in material point method. *Procedia Engineering 175*, 116–123.

Henke, S. & J. Grabe (2006). Simulation of pile driving by 3-dimensional Finite-Element analysis. In *Proceedings of the 17th European Young Geotechnical Engineers Conference, Zagreb*, pp. 215–233.

Jassim, I., D. Stolle, & P. Vermeer (2013). Two–phase dynamic analysis by material point method. *International Journal for Numerical and Analytical Methods in Geomechanics 37*(15), 2502–2522.

Niemunis, A. & I. Herle (1997). Hypoplastic model for cohesionless soils with elastic strain range. *Mechanics of Cohesive-frictional Materials 2*(4), 279–299.

Qiu, G., S. Henke, & J. Grabe (2011). Application of a coupled Eulerian–Lagrangian approach on geomechanical problems involving large deformations. *Computers and Geotechnics 38*(1), 30–39.

Sadeghirad, A., R. Brannon, & J. Burghardt (2011). A convected particle domain interpolation technique to extend applicability of the material point method for problems involving massive deformations. *International Journal for Numerical Methods in Engineering 86*(12), 1435–1456.

Sadeghirad, A., R. Brannon, & J. Guilkey (2013). Second-order convected particle domain interpolation (CPDI2) with enrichment for weak discontinuities at material interfaces. *International Journal for Numerical Methods in Engineering 95*(11), 928–952.

Sulsky, D., Z. Chen, & H.L. Schreyer (1994). A particle method for history–dependent materials. *Computer Methods in Applied Mechanics and Engineering 118*(1), 179–196.

vonWolffersdorff, P.-A. (1996). A hypoplastic relation for granular materials with a predefined limit state surface. *Mechanics of Cohesive-frictional Materials 1*(3), 251–271.

Numerical Methods in Geotechnical Engineering IX – Cardoso et al. (Eds)
© 2018 Taylor & Francis Group, London, ISBN 978-1-138-33198-3

Multi-material arbitrary Lagrangian-Eulerian and coupled Eulerian-Lagrangian methods for large deformation geotechnical problems

M. Bakroon, R. Daryaei, D. Aubram & F. Rackwitz
Chair of Soil Mechanics and Geotechnical Engineering, Technische Universität Berlin, Berlin, Germany

ABSTRACT: The performance of two advanced numerical methods using Multi-Material Arbitrary Lagrangian-Eulerian (MMALE) and Coupled Eulerian-Lagrangian (CEL) formulations is studied. The evaluation is based on two large deformation benchmark cases which classical pure Lagrangian methods cannot model. MMALE is an enhanced version of CEL in which the computational mesh can be rezoned in an arbitrary way so that mesh nodes are concentrated in areas of interest. This form of solution adaptivity provides more data in regions undergoing large deformations compared to the fixed mesh in CEL methods. MMALE has gained popularity in the field of fluid dynamics. In this study, the applicability of MMALE to geomechanical problems is investigated with regard to accuracy and robustness. Two geomechanical problems, pipeline displacement and sand column collapse, are analyzed for this purpose. It can be concluded that MMALE handles such large deformation problems more efficiently than CEL.

1 INTRODUCTION

In computational geomechanics, the numerical simulation of soil-structure-interaction problems where the soil material undergoes large deformations has become an active area of research (Aubram et al. 2017; Wang et al. 2015). Classical finite element methods (FEM) based on a Lagrangian formulation suffer from mesh elements distortion which may deteriorate accuracy or even stop the solution at early stages of the calculation. Novel techniques and advanced numerical methods have been developed to resolve these issues associated with the Lagrangian approach. These methods have been proven a powerful and reasonably accurate alternative to experimental and analytical solutions. The Arbitrary Lagrangian-Eulerian (ALE) and the Coupled Eulerian-Lagrangian (CEL) are two of such methods.

CEL has been extensively evaluated in the context of geotechnical problems. For example, a study done by (Wang et al. 2015) compared the performance of CEL with other numerical methods for large deformation geotechnical problems. Another study thoroughly evaluated the CEL compatibility with a complex soil material model and the results were compared to an experimental test (Bakroon et al. 2017a).

On the other hand, application of ALE to geotechnical problems is rather new and mostly uses a simplified mesh formulation (so-called simplified or single-material ALE). Moreover, current ALE research is focused on rather technical aspects of the method (Barlow et al. 2016; Aubram 2013; Aubram et al. 2017), hence further studies are required concerned with its application in geotechnical engineering. A recent work done by (Bakroon et al. 2017b) compares simplified ALE with the classical FEM using a benchmark case where an analytical solution is available. They conclude that ALE provides more accurate and stable results when applied to large deformation geotechnical problems. A more advanced variant of ALE, called Multi-Material ALE (MMALE), is studied in this paper, and two example applications are used to thoroughly evaluate its performance.

2 METHODS

Both ALE and CEL calculations are based on the operator-split scheme (Benson 1992); see Figure 1. First, a Lagrangian step is performed, where the computational mesh deforms with soil particles. In the second step, a new mesh is generated which is called the remeshing resp. rezoning step. The difference of the CEL method and the ALE method arises in this step (Figure 2). In CEL, the rezoned mesh is simply the original computational mesh while in ALE a new distinct mesh is generated. Finally, the solution from the old mesh is transferred into the new mesh which is called the remapping/advection step. This step is comparable to the solution in a classical Eulerian method.

ALE methods can be subdivided into two main groups, single-material/simplified ALE (SALE) and Multi-Material ALE (MMALE). The SALE method handles one material per element while MMALE can consider multiple materials in one single element. The MMALE is considered more powerful since the performance of SALE is strongly influenced by material interfaces. In other words, the remeshing step is constrained by material boundaries, which is not the case in MMALE chosen here.

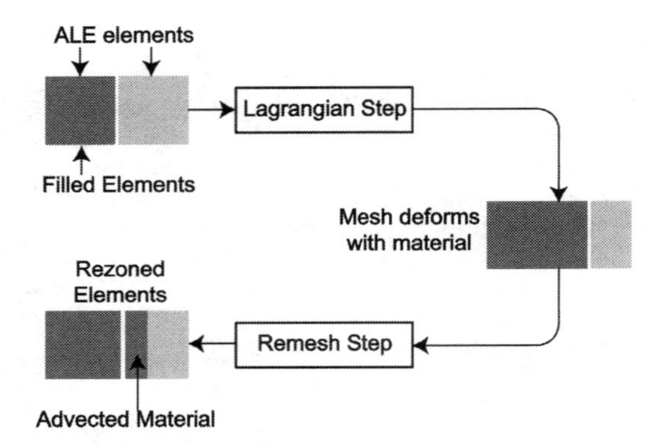

Figure 1. Diagram of the operator-split scheme applied in CEL and MMALE methods.

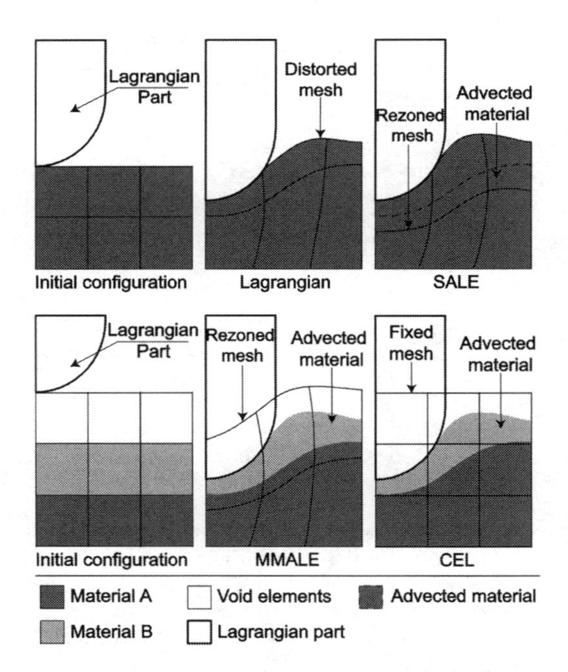

Figure 2. Schematic diagram of SALE, CEL, and MMALE approaches comparing the effect of mesh rezoning and advection steps of the solution (Bakroon et al. 2017b).

In this study, the MMALE method is evaluated against CEL method via two benchmark problems: pipeline displacement and sand column collapse. Each problem tackles specific aspects of the numerical approaches. The problems are checked against analytical and experimental results as well.

3 NUMERICAL EXAMPLE 1: PIPELINE DISPLACEMENT

3.1 *Background*

Pipelines are one of the key components in offshore industrial projects. Pipes are initially placed on the seabed. After installation, the pipe penetrates the soil due to its own weight. Moreover, the varying thermal effects of the pipe induces a lateral force resulting e.g. in a lateral movement. Calculating combined horizontal and vertical resistance of the soil against pipe movement can lead to a more optimized and safe design. There is a large amount of literature concerned with various aspects of embedded pipeline behavior in seabed in the field of theoretical, physical, and numerical modeling (Merifield et al. 2009; Wang et al. 2015).

The vertical and horizontal resistance force is usually calculated based on bearing capacity theory for a shallow embedded footing (Skempton 1951). The equations are modified to take the problem conditions into account such as soil heaving, buoyancy, shape of pipe etc.

The schematic view of the problem is illustrated in Figure 3. After a pipe with diameter D penetrates to a depth w, the soil starts to heave to a width of B_{heave} with the height of H_{heave}. This increases the lateral resistance of the soil which can be taken into consideration to reach an optimum design.

The analytical equations calculating the horizontal and vertical resistance forces are presented in Equations 1 and 2 (cf. (Merifield et al. 2009)). These equations consist of two terms. The first term is attributed to the undrained soil strength while the second term considers the self-weight effects of the soil.

$$\frac{V}{D} = N_{cV}\, s_u + N_{swV}\, \gamma' w \tag{1}$$

$$\frac{H}{D} = N_{cH}\, s_u + N_{swH}\, \gamma' w \tag{2}$$

$$N_{cV} = a\widehat{w}^{b} = 5.3\widehat{w}^{0.25} \tag{3}$$

$$N_{cH} = c\widehat{w}^{d} = 2.7\widehat{w}^{0.64} \tag{4}$$

$$N_{swV} = \frac{1}{2\hat{w}}\left(1+\frac{1}{\lambda}\right)\times \left[\frac{\sin^{-1}\left(\sqrt{4\hat{w}(1-\hat{w})}\right)}{2} - (1-2\hat{w})\sqrt{\hat{w}(1-\hat{w})}\right] \quad (5)$$

$$N_{swH} = \frac{\hat{w}}{2} + \left(\frac{4}{\lambda}\right)\times\left[\frac{\sin^{-1}\left(\sqrt{4\hat{w}(1-\hat{w})}\right)}{2\sqrt{\hat{w}(1-\hat{w})}} - (1-2\hat{w})\right] \quad (6)$$

where V = vertical resistance force, H = horizontal resistance force; $\hat{w}$ = normalized penetration depth ($\hat{w} = w/D$); λ = parameter approximating the amount of heaving; γ' = submerged unit weight of soil; D = pipe diameter; N_{cV} and N_{cH} = coefficient for vertical and horizontal strength, respectively; and N_{swV} and N_{swH} = coefficient for vertical and horizontal strength for self-weight term, respectively (see Figure 3).

As suggested by (Merifield et al. 2009) the values of 3 and 1.6 are used for λ in vertical and horizontal force, respectively, as well as the corresponding coefficients a, b, c, and d in Equations 3 and 4.

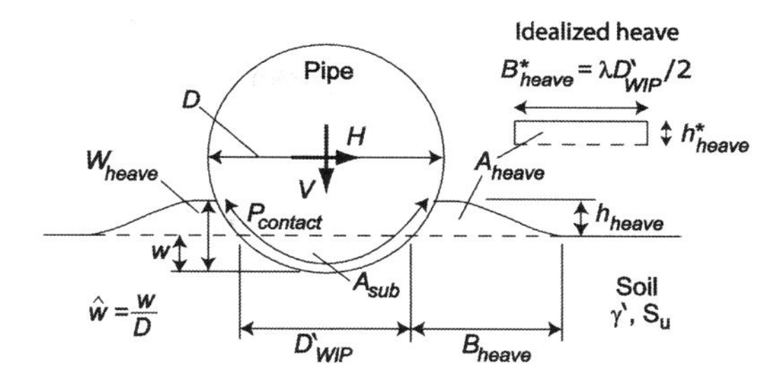

Figure 3. Schematic view of the pipeline displacement problem (Merifield et al. 2009).

3.2 *Numerical model*

In this problem, a pipe is placed above the soil which represents the seabed. The soil is considered to be fully saturated with average shear strength, su = 1.5 kPa in overall depth. The submerged unit weight of the soil is $\gamma' = 6$ kN/m^3. The Young's modulus E of the soil is calculated by E/su = 500. The elastoplastic material model with Tresca yield criterion was employed. Due to significant strength difference between pipe and soil, the pipe was considered as a rigid part. No friction was considered between pipe and soil (smooth surface).

The pipe is moved in vertical direction until depth of 1.0D to simulate the embedment of the pipe. The velocity rate of 0.01 m/sec was assigned to ensure quasi-static loading conditions. The vertical resistance force was calculated and compared to analytical equations.

To compare the horizontal resistance of the pipe with analytical solution, ten models were developed, where the pipe was displaced horizontally at different embedment depths from 0.1D to 1.0D with intervals of 0.1D. No vertical displacement was allowed at this stage.

MMALE and CEL methods are used for numerical simulation. The mesh configuration of the model is shown in Figure 4. The minimum element size was 0.04 m which increased at the boundaries to 0.15 m resulting in total number of 10,900 elements. A void layer of 1 m height was defined above the soil layer to allow soil heaving simulation. The model was considered as a 2D problem, however, 3D solid elements were used since no 2D Eulerian elements are available. The thickness in normal direction is considered as one element.

3.3 *Results and discussion*

A Lagrangian model was first adopted. The quality of mesh elements reduced drastically after significant penetration as shown in Figure 5. Hence, the results were considered unreliable. This emphasizes on a need for more advanced models for this problem.

Subsequently, MMALE and CEL methods were used for the simulation. Both methods converged to solution after 1.0D penetration. Vertical and horizontal resistance forces of CEL and MMALE are checked against analytical equations in Figure 6 and Figure 7, respectively. The vertical resistance force for both CEL and MMALE are in a good agreement with analytical equations. In Figure 6, it is shown that both methods provide acceptable results. The CEL method gives stiffer behavior than MMALE. After about 0.2 m penetration, both results from the MMALE and CEL with coarse mesh, start experiencing oscillation, while the MMALE with fine-mesh is smoother. It can be argued that due to coarser mesh, less coupling nodes area available which causes the oscillation. Nevertheless, this oscillation does not cause significant errors.

For calculated horizontal force in Figure 7, both methods give higher values at low penetration depths, but lower values at higher depths in comparison to analytical equation. This can be attributed to complex mechanism of heaving and its effect on the resistance force mentioned in (Merifield et al. 2009).

By using the same model configuration, MMALE captures a better mesh resolution for areas of interest than CEL. This argument is supported by Figure 8 where CEL and MMALE interfaces are compared together using the initial mesh element size. MMALE provides a smoother interface than CEL. Hence, it is possible to achieve

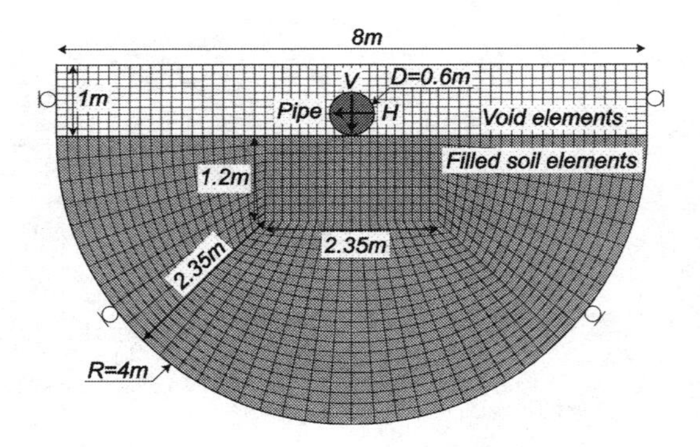

Figure 4. Mesh configuration of the pipeline displacement model.

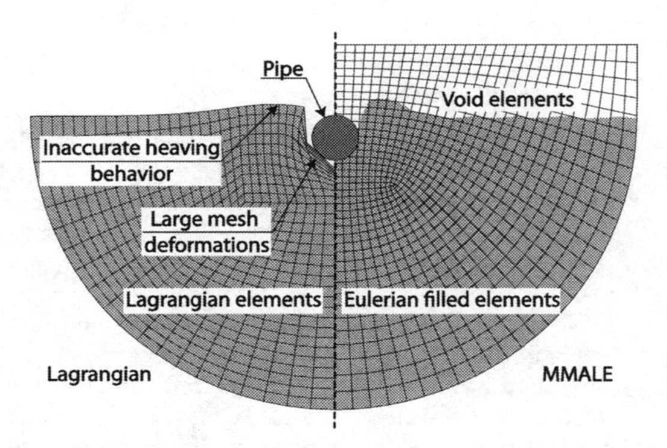

Figure 5. Mesh distortion using Lagrangian (left) and MMALE (right) methods.

an acceptable accuracy with increasing the element size in MMALE.

The velocity vectors of soil after application of vertical displacement are shown in Figure 9a and Figure 9b. The velocity vectors of the horizontal displacement after reaching $0.5D$ penetration are shown in Figure 9c and Figure 9d. The arrow at center of the pipe shows its movement direction.

The velocity field shows clearly which part of the soil regime undergoes significant movement. This movement is due to the shear band mechanism appearing due to excessive pipe movement and soil softening.

In Figure 9a and Figure 9b, the soil flow regime is distinguished by dense arrows. This is similar to failure mechanism of a strip footing in general soil mechanics theory. In Figure 9c and Figure 9d, a new shear zone is developed. At both displacement modes the velocity field is uniform which is a criterion for stability of the numerical methods.

A more realistic model has been developed to account for simultaneous displacement of pipe in horizontal and vertical direction. Similar to previ-

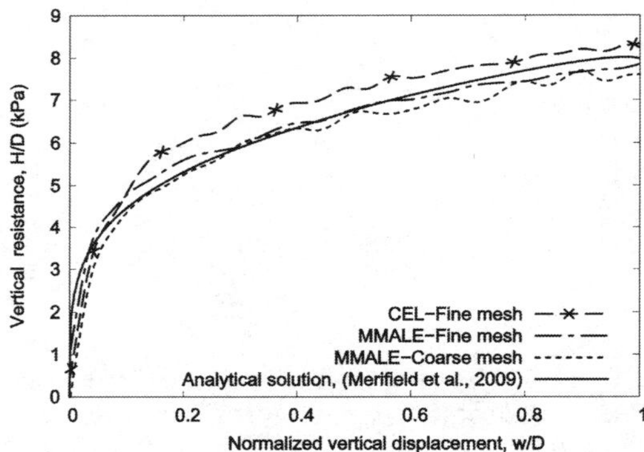

Figure 6. Comparison of vertical resistance obtained from different numerical methods and analytical results.

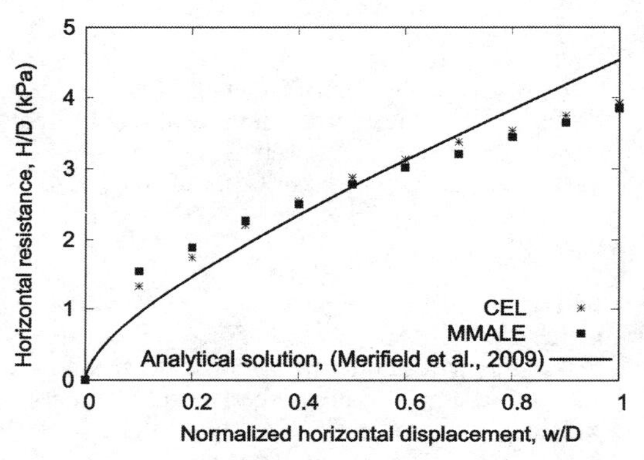

Figure 7. Comparison of horizontal resistance obtained from different numerical methods and analytical result.

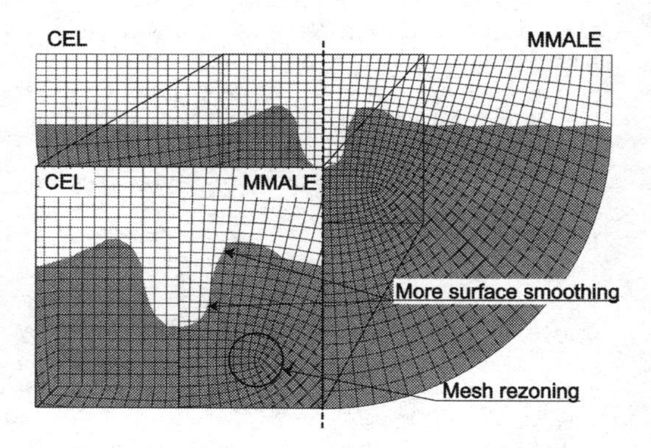

Figure 8. Comparison of MMALE and CEL interface reconstruction.

ous model, the pipe is initially placed above the soil. Then the pipe moves vertically with the rate of 0.01 m/s into the soil until the depth of $0.5D$ for

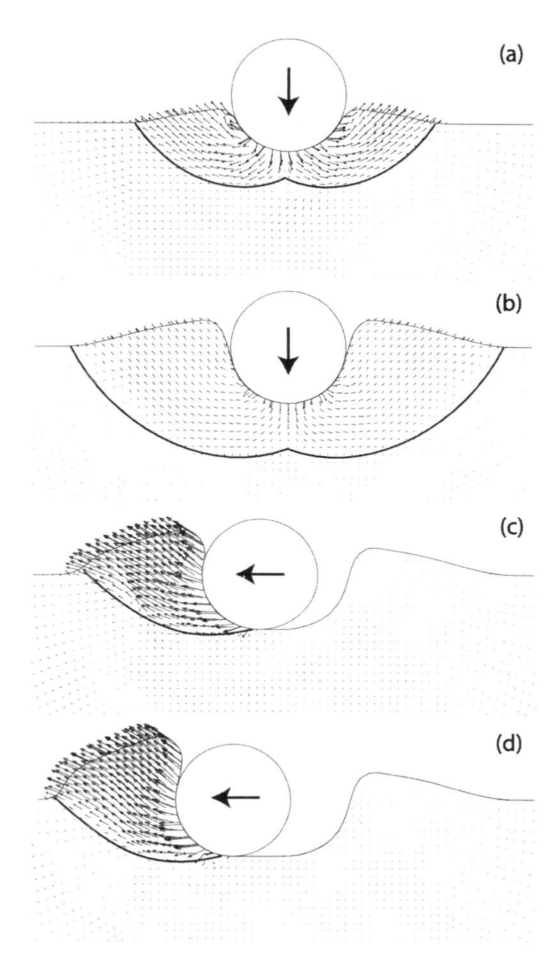

Figure 9. Velocity vectors of sand movement during vertical pipe displacement of a) $0.25D$ b) $0.5D$ and horizontal pipe displacement of c) $0.25D$ and e) $0.5D$.

Table 1. Calculation time comparison for MMALE and CEL for pipeline displacement problem.

Numerical method	Calculation time h:min:sec
CEL	02:38:43
MMALE	01:42:14

simulation of partial embedment. Subsequently, the horizontal displacement was applied with the same rate of 0.01 m/s. During horizontal movement, 60% of the obtained maximum vertical force in the last phase was maintained to model the self-weight of the pipe and its containing fluid. Vertical movement is allowed during this stage.

The model was solved with both CEL and MMALE. To reduce the number of irrelevant affecting variables, the simulations were conducted under same configurations and conditions on a conventional personal computer with 4-core CPU with 3.2 GHz.

Figure 10 shows the final deformed shape for both CEL and MMALE. The mesh in MMALE model is significantly concentrated around the pipe, which is also the interested area of study. In addition, due to mesh concentration, more nodes are available which enhances coupling with Lagrangian elements leading to more accurate results. Besides, more Eulerian elements at coupling interface reduce the possibility of leakage. In Figure 11 the velocity field of the soil is shown. The arrow in the middle of the pipe shows the pipe movement direction. Compared to Figure 9, more soil volume is displaced. Results obtained from the model agrees well with similar tests available in the literature (Dutta et al. 2015).

Figure 12 shows a comparison of horizontal force for MMALE and CEL. Both results converge to a similar value with negligible differences.

Furthermore, the calculation time for both CEL and MMALE is considered as a comparison crite-

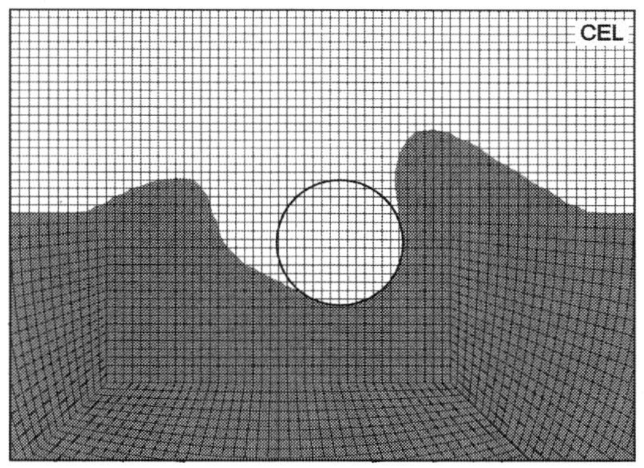

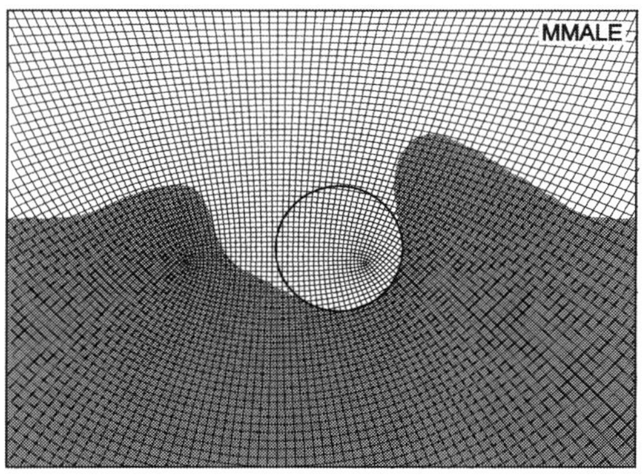

Figure 10. Final deformed shape of soil and mesh using MMALE and CEL methods, element concentration is clearly observed in MMALE.

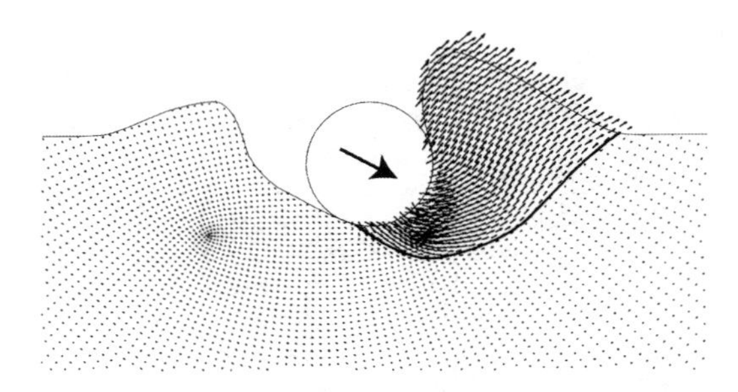

Figure 11. Final velocity vectors of sand movement after enforcement of both vertical and horizontal pipe displacement.

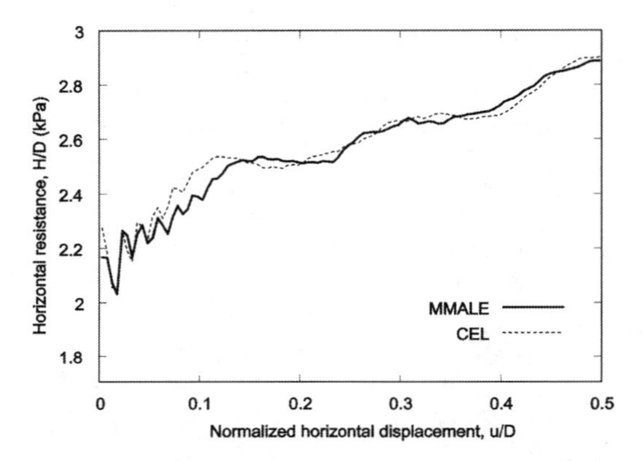

Figure 12. Pipeline response during penetration and lateral displacement.

rion. As illustrated in Table 1, MMALE was about 35% faster than CEL. Although MMALE has one more step in calculation process (e.g. remeshing step), which increases calculation cost in comparison to CEL, it is not necessary to perform it at each calculation step. Hence, the remeshing and remapping step can be performed after several Lagrangian steps without affecting the results, which leads to less computation time.

4 NUMERICAL EXAMPLE 2: SAND COLUMN COLLAPSE

4.1 *Background*

Collapse of sand column has been extensively studied as an experimental test and benchmark or numerical methods verification. Conventionally, a sand specimen is deposited inside a container. As shown in Figure 13, at least a side of the container is released abruptly which allows the sand to flow on the surface. Then, the corresponding parameters such as run-out distance, slope angle, etc. are studied; see (Lube et al. 2005) for further information.

4.2 *Numerical model*

(Lube et al. 2005) carried out several experiments on two dimensional sand columns. The results of this experiment are used as a benchmark case for numerical assessment of MMALE and CEL. The evaluation parameters are the run-out distance and sand column height.

The configuration of the numerical model which is obtained from the experiment is shown in Figure 14. The column of sand is at rest until one side of the container is removed to let the soil flow by its own weight. The initial width and height of the soil column is $d_i = 0.0905$ m and $h_i = 0.635$ m leading to height to width aspect ratio $a = 7$. In the experiment, the depth of the soil in direction normal to flow is 0.2 m. It was reported that in this direction no relative difference in run-out distance was observed. Therefore, it is possible to model the experiment in two dimensions. However, due to lack of 2D Eulerian elements in the commercial hydrocodes used, a 3D model with depth of 1 m was developed consisting of hexahedral elements with 1-point integration. The Mohr-Coulomb material model is used in the present study, and the surface friction angle is assumed equal to the internal friction angle of the sand. Material properties are summarized in Table 2 based on a research by (Solowski & Sloan 2013). It should be noticed that the density was assumed by the authors as an average value of sand density (Table 2). In reasonable range of sand density, the effect of this parameter was observed to be negligible.

For both MMALE and CEL, a void region should be defined to allow the soil to flow in this region after the collapse has started. The gravity acceleration is taken as 9.806 m/s^2. The total calculation time of the problem is 2 seconds. Rigid parts are employed to model the container and the flowing surface. The container is assumed smooth and frictionless. The gate is released after in-situ stresses are initialized.

4.3 *Results and discussion*

Again, the problem was first modeled using the classical Lagrangian approach. In Figure 15, the deformed shape of sand clearly shows the inability of the method for such large material deformations. At about 30% of the calculation time, the mesh quality is significantly reduced. Consequently, the time step size decreased drastically. Even if the termination time was reached, the resulting mesh size would have made the results unreliable

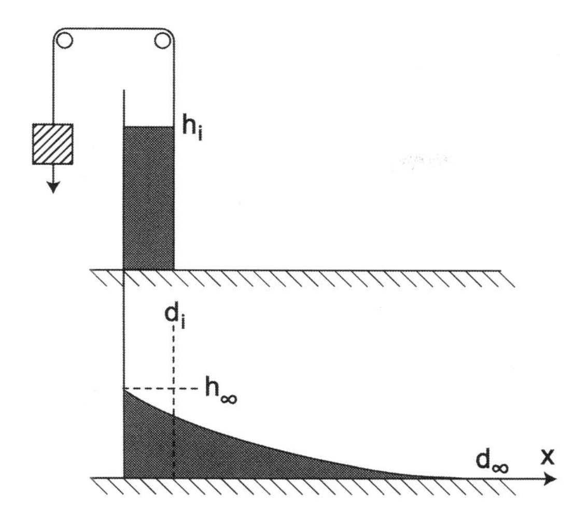

Figure 13. Schematic view of the sand column problem.

Table 2. Mohr-Coulomb properties for the sand column collapse model (Solowski & Sloan 2013).

Parameter	Value
Density (kg/m³)	1,600
Friction Angle (°)	31
Dilatancy angle (°)	1
Cohesion (kPa)	0.01
Poisson ratio	0.3
Elastic Modulus (kPa)	840

Table 3. Calculation time comparison for MMALE and CEL for sand column collapse problem.

Numerical method	Calculation time h:min:sec
CEL	00:05:13
MMALE	00:03:24

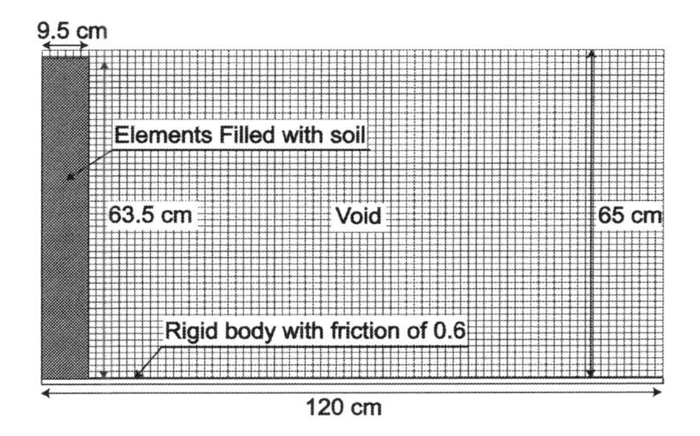

Figure 14. Initial configuration of the sand column collapse model.

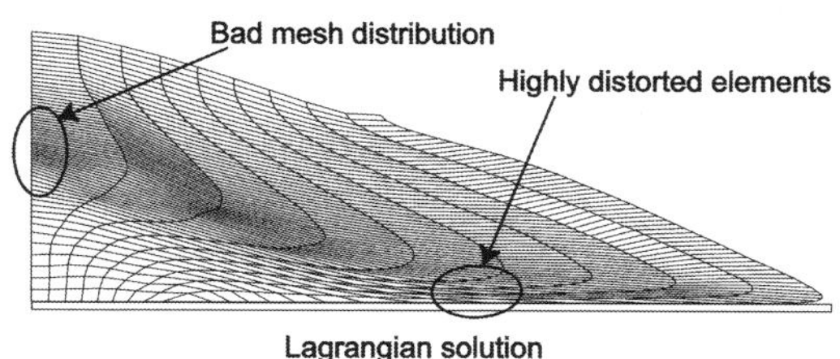

Figure 15. Mesh deformation at an intermediate stage of sand column collapse using a classical Lagrangian method.

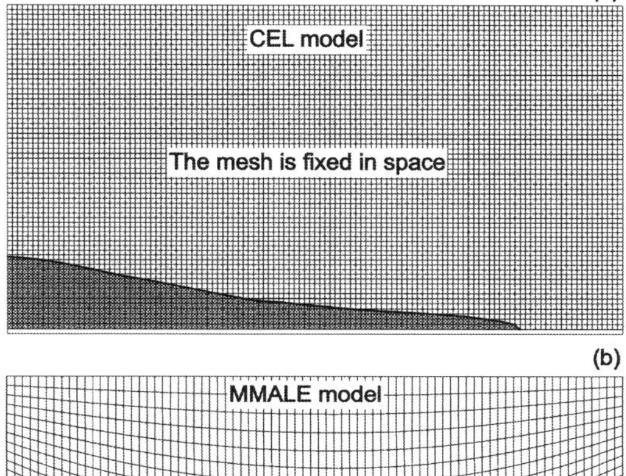

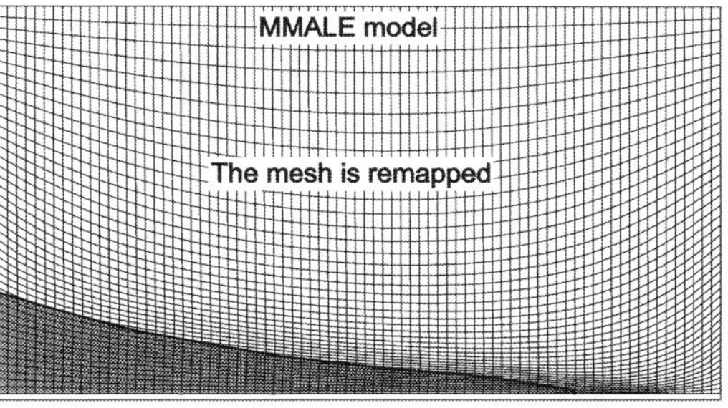

Figure 16. CEL mesh (above) and MMALE mesh (below) at the end of the calculation.

due to excessive mesh distortion. In contrast to the Lagrangian method, both CEL and MMALE simulations reached a converged solution. This is due to the implemented advection technique which enables the calculation of sand motion independent of mesh deformation. In Figure 16, the final soil shape is shown. Mesh element size was initially taken as 15 mm for both MMALE and CEL. Despite convergence, the initial CEL model results in a poorly resolved free surface of the collapsing sand column. Hence, the mesh element size was refined to 7.5 mm. The sand column shape after flow was also evaluated in terms of its measured run-out distance and height. This is shown in Figure 17 for different times.

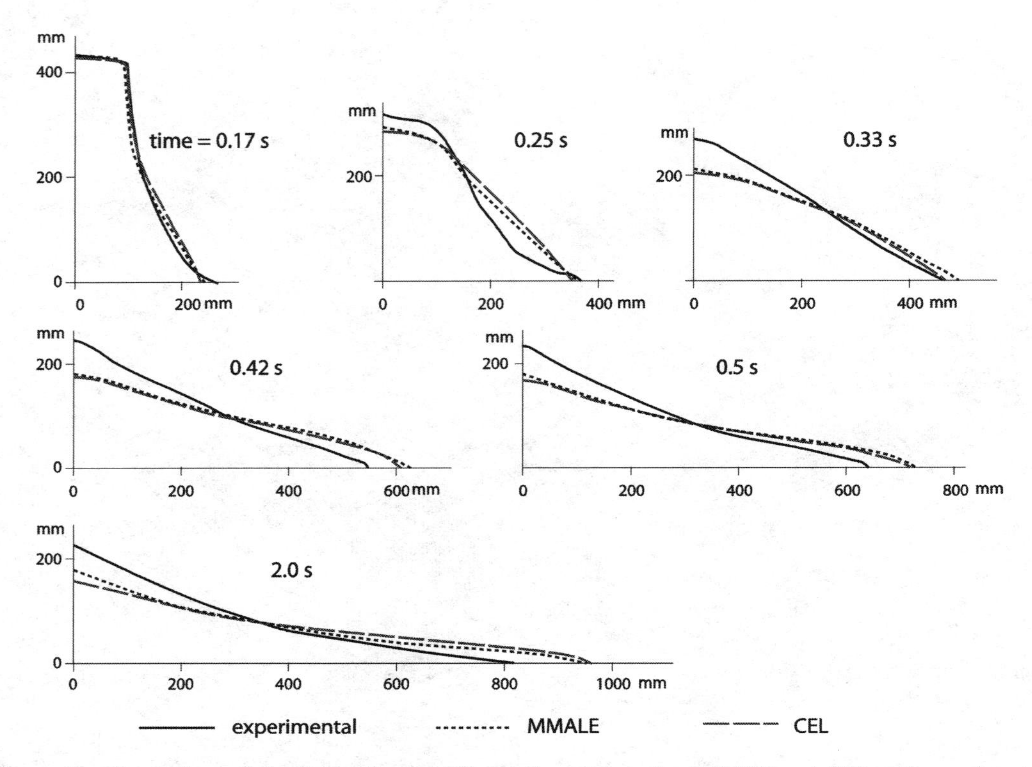

Figure 17. Comparison of the free surface at different time stations for MMALE and CEL with experimental results.

Owing to the remeshing feature in MMALE, a mesh density at the free surface comparable to that of the fixed CEL mesh could be reached using a coarser initial mesh. Clearly, the mesh can be adapted to material deformations during remeshing, which renders MMALE computationally less expensive than CEL at comparable accuracy. The computation time for CEL and MMALE using the mesh of Figure 16 are summarized in Table 3.

5 CONCLUSION

In this study, the performance of two numerical analysis approaches tailored for large deformation problems, CEL and MMALE, was evaluated using two example applications. These examples cannot be solved using classical Lagrangian methods since the latter stop at early stages or provide unacceptable results. For the first problem addressing lateral pipeline displacement an analytical solution is available. On the other hand, for the second problem of sand column collapse, experimental measurement is available. Therefore, it was possible to thoroughly investigate both methods and compare their results. Both methods provided comparable results within acceptable calculation time which proves their efficiency and robustness. One of the major differences between MMALE and CEL lies in the remeshing resp. rezoning step. In CEL the mesh is rezoned to its original configuration, while in MMALE the mesh is rezoned to an arbitrary mesh, including the Eulerian (fixed) or Lagrangian mesh as limit cases. The utilized rezoning technique in MMALE has several advantages. At the same mesh size, MMALE interface resolution is generally higher in comparison to CEL. Moreover, an MMALE mesh provides a natural form of solution adaptivity, meaning that mesh density and resolution is increased in areas of interest. On the other hand, it is possible to use coarser meshes in MMALE simulations than in CEL simulations at comparable accuracy in order to reduce calculation times. Additionally, for problems with structural (Lagrangian) parts, MMALE provides more coupling nodes which increases the robustness of the model and decreases the problem of material leakage in CEL methods. The findings of this study highlight that the Multi-Material

Arbitrary Lagrangian-Eulerian method is suitable for simulation of large deformations, and it can be considered as a promising tool for modelling more complex geotechnical problems.

REFERENCES

Aubram, D. 2013. Arbitrary Lagrangian-Eulerian Method for Penetration into Sand at Finite Deformation: Shaker Verlag, Aachen. Germany.

Aubram, D., Rackwitz, F. & Savidis, S.A. 2017. Contribution to the non-Lagrangian formulation of geotechnical and geomechanical processes. In Theodoros Triantafyllidis (ed.); *Holistic Simulation of Geotechnical Installation Processes,* Springer International Publishing, Cham, pp. 53–100.

Bakroon, M., Aubram, D. & Rackwitz, F. 2017a. Geotechnical large deformation numerical analysis using implicit and explicit integration. In Huriye Bilsel (ed.); *3rd International Conference on New Advances in Civil Engineering*, pp. 26–36.

Bakroon, M., Daryaei, R., Aubram, D. & Rackwitz, F. 2017b. Arbitrary Lagrangian-Eulerian Finite Element Formulations Applied to Geotechnical Problems. In Jürgen Grabe (ed.); *Workshop on Numerical Methods in Geotechnics.* Veröffentlichungen des Institutes Geotechnik und Baubetrieb (41), Breit schuh & Kock GmbH, Hamburg, Germany, pp. 33–44.

Barlow, A.J., Maire, P.-H., Rider, W.J., Rieben, R.N. & Shashkov, M.J. 2016. Arbitrary Lagrangian–Eulerian methods for modeling high-speed compressible multimaterial flows. *Journal of Computational Physics* (322), pp. 603–665.

Benson, D. 1992. Computational methods in Lagrangian and Eulerian hydrocodes. *Computer Methods in Applied Mechanics and Engineering* (99), pp. 235–394.

Dutta, S., Hawlader, B. & Phillips, R. 2015. Finite element modeling of partially embedded pipelines in clay seabed using Coupled Eulerian–Lagrangian method. *Canadian Geotechnical Journal* (52), pp. 58–72.

Lube, G., Huppert, H.E., Sparks, R.S.J. & Freundt, A. 2005. Collapses of two-dimensional granular columns. *Physical review. E, Statistical, nonlinear, and soft matter physics* 4 Pt 1 (72), p. 41301.

Merifield, R.S., White, D.J. & Randolph, M.F. 2009. Effect of Surface Heave on Response of Partially Embedded Pipelines on Clay. *Journal of Geotechnical and Geoenvironmental Engineering* (135), pp. 819–829.

Skempton, A.W. 1951. The Bearing Capacity of Clays. *Proc. Building Research Congress*, pp. 180–189.

Solowski, W.T. & Sloan, S.W. 2013. Modelling of sand column collapse with material point method. *Proceeding of the Third International Symposium on Computational Geomechanics (ComGeo III)* (553), pp. 698–705.

Wang, D., Bienen, B., Nazem, M., Tian, Y., Zheng, J., Pucker, T. & Randolph, M.F. 2015. Large deformation finite element analyses in geotechnical engineering. *Computers and Geotechnics* (65), pp. 104–114.

Artificial intelligence and neural networks

Numerical Methods in Geotechnical Engineering IX – Cardoso et al. (Eds)
© 2018 Taylor & Francis Group, London, ISBN 978-1-138-33198-3

Use of artificial neural networks to analyse tunnelling-induced ground movements obtained from geotechnical centrifuge testing

A. Franza
Department of Engineering, University of Cambridge, Cambridge, UK

P.G. Benardos & A.M. Marshall
Faculty of Engineering, University of Nottingham, Nottingham, UK

ABSTRACT: In geomechanics, centrifuge modelling and digital image analysis enable the acquisition of large amounts of high-quality data related to ground movements. In this paper, modern intelligent methods based on a feedforward Artificial Neural Network (ANN) architecture are applied to study tunnelling-induced ground displacements. Soil displacement data obtained from a geotechnical centrifuge test are used to investigate the capabilities of ANNs in this context. Because this work represents a feasibility study, the centrifuge dataset is limited to a single test. The trial-and-error process is used to identify three architectures of varying complexity that achieve a good level of performance. Predictions are evaluated both statistically (R^2) and qualitatively (analysing the shape of vertical and horizontal displacement profiles). Results show the applicability of modern intelligent analysis methods for analysing centrifuge datasets and highlight certain strengths and deficiencies of feedforward ANN architectures compared to empirical methods.

1 INTRODUCTION

In geomechanics, centrifuge modelling and digital image analysis enable the acquisition of large amounts of high-quality data related to ground movements. These data are used for qualitative studies of soil behaviour and soil-structure/foundation interaction mechanisms as well as for validation of numerical and analytical models. In addition, because of the complexity of soil behaviour and the preference for closed form solutions, researchers frequently adopt empirical methods, based on standard regression analyses such as curve-fitting, to express the experimental outcomes. Despite their straightforward use in practice, empirical methods may be limited in scope when complex geotechnical problems are studied.

An alternative approach to empirical regression analyses is the use of modern intelligent methods such as artificial neural networks (ANNs), which have the potential to perform better than user-defined curve-fitting techniques because of their ability to capture complex and non-linear relationships between the variables involved in the physical problem.

The aim of this paper is to carry out a feasibility study to assess the performance of ANNs with respect to a centrifuge experimental dataset consisting of a two-dimensional spatial displacement field of ground movements obtained using an image based measurement technique (White et al. 2003). In particular, the problem of plane-strain tunnelling-induced ground movements above a shallow tunnel excavated in dry dense sand is considered. In this paper, the centrifuge dataset is limited to a single test carried out by Marshall et al. (2012).

By combining feedforward ANNs and centrifuge modelling, this work addresses with a novel approach the need of engineers to estimate both horizontal and vertical displacements induced by underground excavations

1.1 *Tunnelling-induced movements*

Engineers are interested in the prediction of soil movements induced by the excavation of tunnels in urban areas in order to assess the effects of tunnelling on existing structures and infrastructure. Tunnelling-induced ground movements in greenfield conditions (i.e. when there are no other structures near the new tunnel) are often studied as a reference term for soil-structure interaction analyses. In this work, the steady-state condition developing transversely and far away from the tunnel heading is considered. Figure 1 displays the geometry of the problem and a qualitative distribution of surface and subsurface transverse ground movements above the tunnel in both the horizontal and vertical directions. Empirical methods based on Gaussian and modified Gaussian curves are used for the analysis of tunnelling-induced settlement troughs measured during experiments and field monitoring because of their efficiency and simplicity. Horizontal displacement profiles are often related to these empirical settlement curves by adopting a simplifying assumption on the orientation of the ground movement vectors (Mair et al. 1993, Mair and Taylor 1997, Marshall et al. 2012). Although past studies have provided a satisfactory empirical framework for tunnelling in clays, limited guidance is available in the case of sandy soils; in the latter scenario, tunnelling results in a more complex displacement mechanism that is affected by volumetric strains that occur within the soil (Marshall et al. 2012). In tunnelling, the area of the settlement trough is generally related to the tunnel volume loss, $V_{l,t}$, which is the ratio between the tunnel ground loss (i.e. the over-excavated ground at the tunnel periphery) and the notional final area of the tunnel cross-section. Therefore, in the case of dry sands, the main variables of the tunnelling problem are the tunnel volume loss, $V_{l,t}$, the tunnel to axis depth, z_t, the tunnel diameter, D, the soil relative density, I_d, and the spatial coordinates, x and z.

1.2 *Feedforward neural network*

The main characteristic of feedforward ANNs is that they are able to learn from a given set of training data. During the training process, they identify complex and non-linear patterns between the input and target data. This allows for a generalisation capacity that allows ANNs to be used as prediction tools (Benardos and Vosniakos 2007).

In previous research, feedforward ANNs have been successfully adopted to carry out complementary analysis of tunnelling problems using data obtained with time-consuming numerical modelling or field monitoring. The capabilities of ANNs for the estimation of tunnelling-induced settlements have been illustrated by several studies (Ahangari et al. 2015, Benardos and Kaliampakos 2004, Santos and Celestino 2008, Suwansawat and Einstein 2006). However, the use of ANNs for horizontal movements has received little attention in the literature, and there have been no attempts to adopt them to analyse the outcomes of centrifuge testing combined with image based measurement techniques.

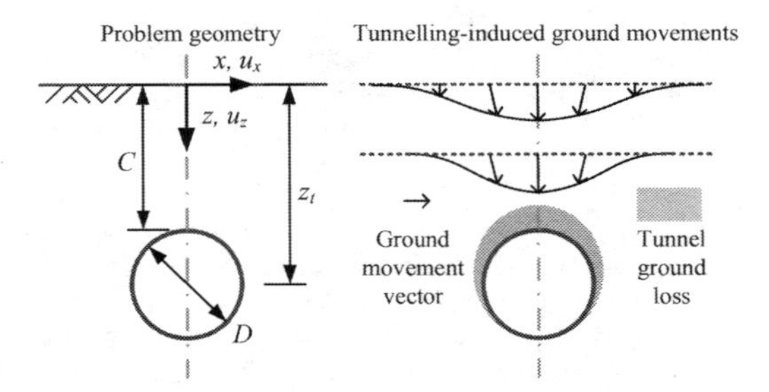

Figure 1. Excavation of a shallow tunnel in greenfield conditions: problem geometry (left) and resulting ground movements (right).

2 ANN DEVELOPMENT AND RESULTS

2.1 *Overview*

In general, feedforward ANNs are models consisting of interconnected neurons arranged in a network with a layered topology. The first layer allocates the network input (corresponding to independent variables), the final layer produces the model outputs. One or more hidden (intermediate) layers are also present with a given number of neurons defined by the ANN architecture. Each neuron is connected to all the nodes in the previous and following layers. Each link between neurons consists of a weight (that multiplies the signal) and a bias. Each neuron at the hidden and output layers adds up the signals received from the previous layer and applies an activation function to the signal.

The training phase of this model is achieved, for a given series of input values, by adjusting the weights and biases of the connections to obtain a satisfactory match between the model output at the final layer and the target data (i.e. the actual data).Note that the training phase should avoid over-fitting of the training data to achieve a good generalisation performance so that the ANN can satisfactorily respond to a generic input within the domain of training.

The combination of weights and layered connections allows description of highly non-linear relationships between inputs and outputs without requiring the definition of deterministic relationships, as in the case of empirical methods. Note that ANNs architecture should not be too complex in order to prevent over-fitting and poor generalisation; on the other hand, simplistic networks may not fully capture the non-linear relationships between input and output layers.

2.2 *Model development*

In this paper, the neural network toolbox of Matlab was used. The topology of an ANN implemented in this paper consisting of a single hidden layer is presented in Figure 2.

Centrifuge data (both inputs and targets) are pre-processed with [i] a normalisation (to allow future work considering a wider dataset consisting of a series of tests with tunnels of different diameters) and [ii] a sampling (to achieve a regular distribution of data within the input domain). The adopted normalised input variables are z_t/D, I_d,

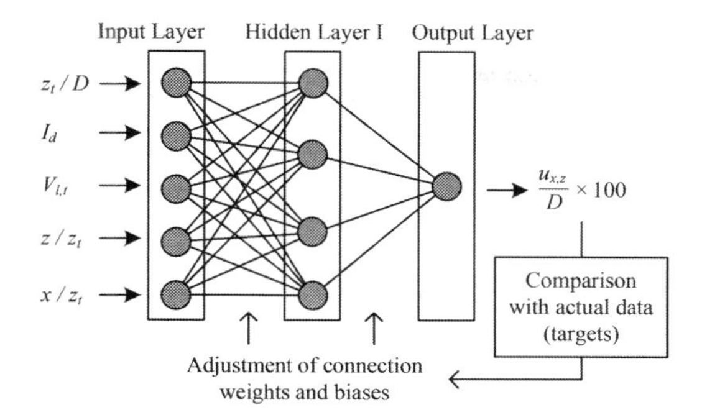

Figure 2. An example of a feedforward ANN with one hidden layer with 4 neurons.

$V_{l,t}$, x/z_t and z/z_t, whereas the adopted normalised targets are $u_x/D \times 100$ and $u_z/D \times 100$. The input data are sampled from the centrifuge dataset corresponding to a regular grid with $V_{l,t} = [1, 1.5, 2, 2.5, 3]$, $z/z_t = [0, 0.1, 0.2, 0.3, 0.4, 0.5]$; $x/z_t = [-2, -1.8, \ldots, 1.8, 2]$. Note that, in this paper, the dataset is limited to one centrifuge test; therefore $z_t/D = 1.7$ (i.e. $C/D = 1.3$) and $I_d = 0.9$.

Different activation functions were used in the input, hidden, and output layers. The identity function, the continuous non-linear tan-sigmoid transfer function, and the linear transfer function were used, respectively, in the input layer, the hidden layer, and the output layer. Prior to training, weight and bias values were initialised to a starting point using the Nguyen-Widrow method (Nguyen and Widrow 1990). During training, the Levenberg-Marquardt backpropagation algorithm (Hagan and Menhaj 1994) along with early stopping was adopted to adjust the connection weights and biases whereas the MSE (the mean squared error) was used as a performance function.

For learning and validation purposes, the preprocessed input data were randomly divided into training, validation, and testing subsets with a respective ratio of 0.7, 0.15, and 0.15. The training subset was used for training the ANN, whereas the validation subset was used for the early stopping feature of the Levenberg-Marquardt algorithm (to avoid overfitting). At the conclusion of the training phase, the testing subset was used to assess the generalisation performance of the model by comparing model outputs corresponding to the testing inputs to the target data, which are unknown to the ANN.

2.3 *Statistical evaluation of ANN performance*

In this section, a series of architectures with varying complexity are developed and their performance evaluated to determine the best architecture using a trial-and-error method. At this stage, the criterion used to statistically quantify the ANNs performance (both training and generalisation) is the coefficient of determination R^2. Firstly, several test runs were performed to identify a suitable group of training, validation, and generalisation subsets. Subsequently, six ANNs with varying complexity were trained using the subset partition identified at the previous stage. Table 1 reports the obtained coefficients of determination of the trained ANNs. A value of R^2 closer to unity represents a better fit of the model outputs with the actual targets. Results in Table 1 display that the architectures with 3 and 4 hidden neurons achieved a good balance between performance and number of neurons. The criterion used for this selection is based on the observation that, when the increase of hidden neurons isnot associated with an increase of performance, the ANN reliability may be compromised and the unnecessary added complexity may result in data overfitting. For the selected architecture (with 4 hidden neurons or $5 \times 4 \times 1$), the regression of the generalisation performance is plotted in Figure 3. The predictions of the selected architecture are compared with centrifuge displacement profiles for a qualitative assessment later in the paper. The architecture with 4 hidden neurons was chosen for the later analysis.

2.4 *Comparison of experimental and model displacement profiles*

A qualitative assessment of the ANNs outputs is carried out by comparing the model predictions of the optimum architecture (using 4 hidden neurons) with centrifuge data of vertical and horizontal displacements to illustrate strengths and deficiencies of feedforward ANN architectures and to provide guidance for future applications. Figure 4 and 5 compare normalised vertical and horizontal ground movements, respectively, at three values of volume

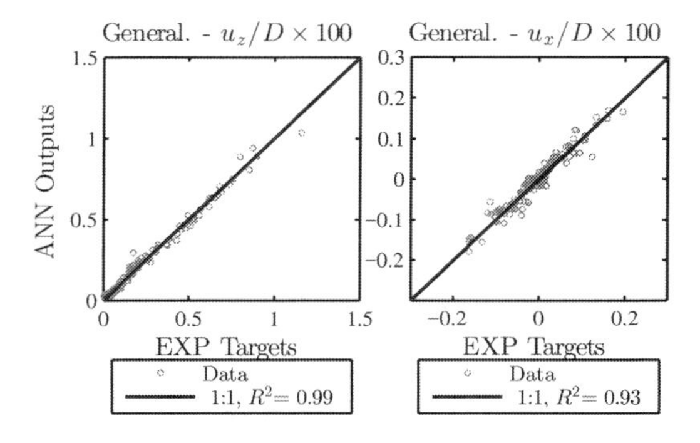

Figure 3. Predicted versus measured vertical (left) and horizontal (right) displacements using the architecture with 4 hidden neurons.

Table 1. Performance of ANNs.

R^2	Horizontal movements		Vertical movements	
Hidden neurons	Training response	Testing general.	Training response	Testing general.
2	0.66	0.73	0.96	0.95
3	0.93	0.91	0.98	0.98
4	0.94	0.93	0.99	0.99
5	0.95	0.94	0.99	0.99
10	0.96	0.95	1.00	0.99
25	0.99	0.99	1.00	0.99

loss ($V_{l,t} \approx 1, 2, 3\%$) and relative depth ($z_t/z = 0, 0.25, 0.5$). Solid lines are used for the ANNs output whereas markers are adopted for the entire centrifuge dataset (rather than the data used for training the model). The following observations can be made.

- Overall, there is good agreement between model predictions and experimental measurements, which was expected because of R^2 values reported in Table 1. Results confirm that simple architectures are able to describe the complex displacement mechanisms induced by tunnelling.
- It is interesting to highlight that ANN is able to fully describe the variability of the horizontal displacement field with both the spatial coordinates and tunnel volume loss, which has not been demonstrated by previous research (see Figure 5).
- Despite good performance at $V_{l,t} \approx 2, 3\%$, Figure 4 shows that the ANN performs less well for prediction of settlements at low volume loss ($V_{l,t} \approx 1\%$), especially near the surface where there is an inconsistency in the trough shape. Although these differences in settlements between model outputs and centrifuge targets at $V_{l,t} \approx 1\%$ may not be statistically significant if the fit is measured with MSE or R^2, the achievement of a consistent shape is essential to promote the use of ANNs in engineering practice. In general, results in Figure 4 indicate that engineers should also check the consistency and acceptability of the shape of the displacement mechanism (not only the magnitude) comparing ANN predictions with measured displacement profiles.
- Both horizontal and vertical displacement profiles are characterised by the lack of symmetry with respect to the tunnel centreline (i.e. z-axis). The use of ANN trained on the entire spatial domain with experimental measurements of displacement does not guarantee the achievement of a perfectly symmetric solution with respect to the tunnel axis. ANNs provide an asymmetric output with respect to the tunnel centreline, as shown in both Figures 4 and 5, because of i) the lack of constraints in the ANN learning, ii) the scatter present in the experimental input dataset, and iii) the variability of tunnel modelling experiments that, generally, do not provide perfectly symmetric results. It should be noted that, since the training was based on MSE, the ANN is likely to provide a better performance at the higher values of tunnel volume loss of the input domain. The use of normalised displacement fields such as $u/(DV_{l,t})$ in future applications may overcome this issue. However, preliminary analyses using this alternative approach suggest that the normalised displacement pattern may have a more complex distribution because of the effects of volumetric strains; therefore, further studies are needed to overcome this issue.

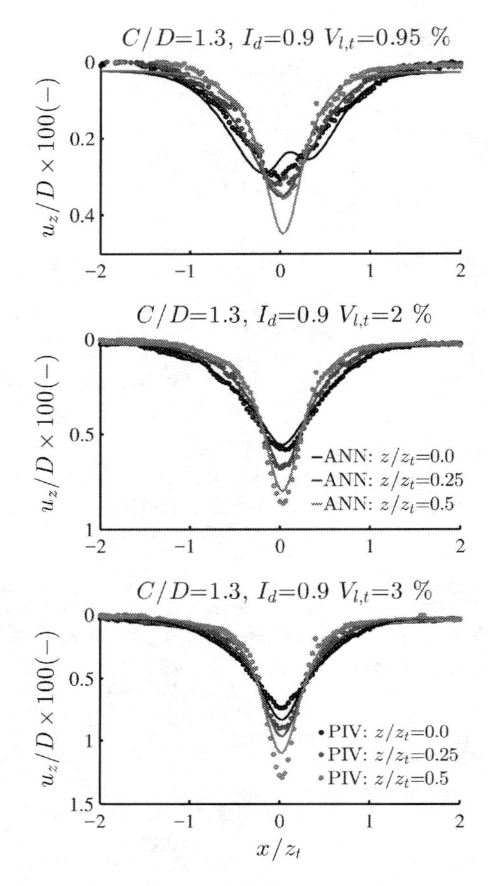

Figure 4. Centrifuge and model vertical displacement profiles.

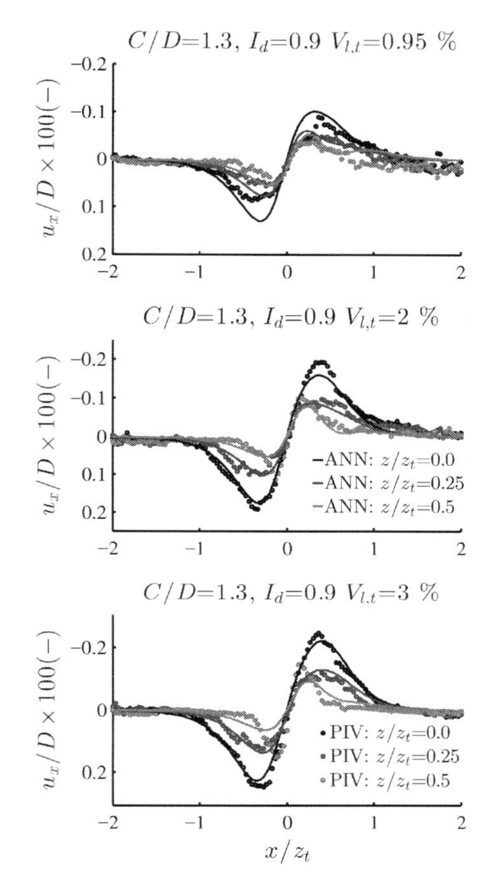

Figure 5. Centrifuge and model horizontal displacement profiles.

3 CONCLUSIONS

Feedforward ANNs are a powerful tool to describe the complex and non-linear displacement fields induced by tunnelling. This study illustrated that, after normalisation of centrifuge data and its sampling into a regular input domain, modern intelligent methods such as ANNs are able to describe the variability of the displacement field measured with an image based techniqueduring centrifuge testing. In general, they have the potential to provide insights into complex centrifuge datasets of ground movements.

ANNs have the advantage over empirical methods in that they do not require the preliminary definition of relationships relating inputs and outputs. However, ANNs may provide ground movement distributions with a shape that is not consistent with the original data and/or present a degree of asymmetry. This issue should be addressed in future works.

In general, this work highlighted that, to validate ANNs as a design tool, the assessment of model performance should not be limited to a statistical regression (for instance using R^2). Although a statistical regression of the generalisation performance is useful to define the best architecture, the model prediction should be evaluated over a continuous and representative portion of boundaries of the training domain to ensure that the distribution (both shape and magnitude) of displacements is physically acceptable.

Further work is required to extend this application to a wider centrifuge dataset including tests with varying tunnel cover-to-diameter ratio and soil relative density.

REFERENCES

Ahangari, K., S.R. Moeinossadat, & D. Behnia (2015). Estimation of tunnelling-induced settlement by modern intelligent methods. *Soils and Foundations 55*(4), 737–748.

Benardos, A.G. & D.C. Kaliampakos (2004). Modelling TBM performance with artificial neural networks. *Tunnelling and Underground Space Technology 19*(6), 597–605.

Benardos, P.G. & G.-C. Vosniakos (2007). Optimizing feedforward artificial neural network architecture. *Engineering Applications of Artificial Intelligence 20*(3), 365–382.

Hagan, M.T. & M.B. Menhaj (1994). Training feedforward networks with the Marquardt algorithm. *IEEE transactions on Neural Networks 5*(6), 989–993.

Mair, R.J. & R.N. Taylor (1997). Theme lecture: Bored tunneling in the urban environment. In *14th International conference on soil mechanics and foundation engineering*, Hamburg, pp. 2353–2385. Balkema.

Mair, R.J., R.N. Taylor, & A. Bracegirdle (1993). Subsurface settlement profiles above tunnels in clay. *Géotechnique 43*(2), 315–320.

Marshall, A.M., R. Farrell, A. Klar, & R. Mair (2012). Tunnels in sands: the effect of size, depth and volume loss on greenfield displacements. *Géotechnique 62*(5), 385–399.

Nguyen, D. & B. Widrow (1990). Improving the learning speed of 2-layer neural networks by choosing initial values of the adaptive weights. In *Proceedings of the International Joint Conference on Neural Networks*, pp. 21–26. IEEE.

Santos, O.J. & T.B. Celestino (2008, sep). Artificial neural networks analysis of S˜ao Paulo subway tunnel settlement data. *Tunnelling and Underground Space Technology 23*(5), 481–491.

Suwansawat, S. & H.H. Einstein (2006, mar). Artificial neural networks for predicting the maximum surface settlement caused by EPB shield tunneling. *Tunnelling and Underground Space Technology 21*(2), 133–150.

White, D., W. Take, & M. Bolton (2003, jan). Soil deformation measurement using particle image velocimetry (PIV) and photogrammetry. *Géotechnique 53*(7), 619–631.

Numerical Methods in Geotechnical Engineering IX – Cardoso et al. (Eds)
© 2018 Taylor & Francis Group, London, ISBN 978-1-138-33198-3

Construction of bedrock topography from airborne-EM data by artificial neural network

A.K. Lysdahl, L. Andresen & M. Vöge
Norwegian Geotechnical Institute, Oslo, Norway

ABSTRACT: Airborne Electromagnetics (AEM) has been used frequently in the past years to investigate ground properties for planning and optimization of large road or railroad projects in Norway. Converting electric properties from the AEM survey into useful geotechnical information requires a thorough interpretation of large amounts of data. A supervised multi-layer perceptron neural network has been used to aid interpretation of depth to bedrock from AEM-data. A subset of manual interpreted AEM data points serves as training data, and the remaining interpretation points as well as ground thruthed drillings serve as validation data. The python module scikit-learn is used to build the neural network, which serves as a non-linear operator that maps the AEM models onto bedrock depth and completes the interpretation effectively. In addition, interpretation uncertainty is likewise calculated. Results are compared to those obtained by a linear function approximator and to ground thruthed validation data. In case of inhomogeneous geologic models, the neural network clearly proves superior in terms of consistency and mismatch error. With this approach only a fraction of the AEM data has to be interpreted manually, which makes the interpretation process much more efficient. The methodology gives reliable predictions of depths to bedrock and reduce the manual time and cost involved in AEM data interpretation.

1 INTRODUCTION

Airborne Electromagnetics (or AEM) is a fast and non-intrusive geophysical technique that measures the electrical properties of the subsurface. Developed in the 50's, the method was originally applied in mineral exploration. There has been a rapid technology development the last decades (Sørensen and Auken 2004), and todays AEM instruments provide high enough resolution and penetration depth to be used for geotechnical investigations. Since 2013, NGI has carried out nine large surveys for about 360 km of road- or railroad construction projects in Norway. Successful application of the AEM data gives a better understanding of the subsurface stratigraphy and to some extent geotechnical properties at an early stage in the planning. This reduces the amount of geotechnical boreholes necessary at a later stage, allows for smarter positioning of boreholes and finally provides a basis for geologic and geotechnical ground models that are spatial continuous (as opposed to sparsely distributed boreholes).

The data output from an AEM survey is typically a 3D model of the ground's electrical resistivity (the inverse of the electrical conductivity), obtained by inversion of the measured voltage soundings spaced evenly along flightlines:

$$\mathbf{M} = \begin{bmatrix} x_1 & y_1 & z_1 & r_1^1 & r_1^2 & \cdots & r_1^L \\ & & \vdots & & & \ddots & \vdots \\ x_N & y_N & z_N & r_N^1 & r_N^2 & \cdots & r_N^L \end{bmatrix} \quad (1)$$

Each row in the resistivity model **M** contains one vertical AEM sounding (with coordinates x_n, y_n, topography z_n and resistivity values $r_n^1 \dots r_n^L$ for N soundings with L layers). The model's spatial variation (both 1st and 2nd spatial derivative of resistivity) is normally constrained within certain limits in order to avoid overfitting and unrealistic, meaningless resistivity models. The resistivity model therefore becomes intrinsically smooth with no sharp boundaries.

To make use of the AEM data in a geotechnical context, the resistivity values have to be translated into geotechnical parameters, like depth to bedrock, which is discussed here. This is a demanding task which require knowledge not only about the local geology but also about geophysical inversion and how different materials are imaged by electromagnetic induction, resolution thereof etc. Consequently, the interpreter uses information from many different sources when interpreting bedrock depth from the AEM data, and performing a fully automatic transformation based on one or two AEM data attributes (like resistivity threshold or gradient search) will in general not produce satisfying results (Anschütz et al. 2016). A complete manual interpretation process may therefore seem unavoidable.

There is, however, potential for automation and the objective of this work is to develop a semi-automatic approach. In such an approach, the interpreter's knowledge is combined with an advanced interpolation method that learns the "interpretation rule", which is subsequently applied to the whole dataset, making repetitive and exhausting manual work obsolete. Such an approach would also make interpretations more objective.

The simplest way to implement such a method is to assume a linear relation between the AEM resistivity data M and the interpretation d, such that:

$$\mathrm{d}_{\mathrm{train}} = \mathrm{M}_{\mathrm{train}}\mathrm{w} \quad \mathrm{M}_{\mathrm{train}} \in \mathrm{M} \tag{2}$$

where w is the linear operator mapping M on d. It can be seen as the weights given to each column in M (each AEM sounding attribute) in order to calculate d_{train}. In the case of bedrock depth interpretation, d_{train} is a vector containing the manual interpreted picks of bedrock depth, and the rows in M_{train} are the corresponding AEM soundings at the pick locations. Then, M_{train} can simply be inverted in order to calculate w, and a prediction can be made for all AEM soundings by:

$$\mathrm{d}_{\mathrm{pred}} = \mathrm{M}\left(\mathrm{M}_{\mathrm{train}}{}^{-1}\mathrm{d}_{\mathrm{train}}\right) \tag{3}$$

However, as the interpretation of depths is performed manually, there will be a certain degree of randomness related to the vector d_{train}, and hence it can be seen as a stochastic variable with a probability density function (PDF) for each depth value, so that a number of different solutions for w are possible to a certain set d_{train}. In 2015, Guldbrandsen et al. proposed a method called "Localized Smart Interpretation" (LSI) where the PDF of d_{train} and hence the PDF on w is assumed Gaussian. The covariance of w is calculated iteratively by using a Tikhonov regularization scheme to find the best possible prediction (by calculating the cross validation error), whereupon the final w is found and predictions for all AEM soundings is made:

$$\mathrm{w} = \left(\mathrm{M}_{train}^{T}\mathrm{C}_{\mathrm{d}}^{-1}\mathrm{M}_{\mathrm{train}} + \lambda^{2}\mathrm{I}\right)^{-1}\mathrm{M}_{train}^{T}\mathrm{C}_{\mathrm{d}}^{-1}\mathrm{d}_{\mathrm{train}} \tag{4}$$

Here, C_d is the standard deviation of d_{train} and λ the regularization parameter.

The LSI method is fast and can be used to aid the interpreter during work, because a new prediction can quickly be performed each time the d_{train} is updated. However, the relation between d and M is still assumed to be linear, which means that the same w (or the same "interpretation rule") eventually is used everywhere, regardless of the nature of the particular area. In cases where for example the geologic setting changes drastically within the AEM model extent, w should rather dependent on location. The problem then becomes non-linear.

2 METHOD

This section describes how an artificial neural network is used as a non-linear operator between the training data d_{train} and the AEM distribution M. In this example, we illustrate the methodology by interpreting bedrock depth.

An Artificial Neural Network (ANN) is trained on d_{train} and M_{train}, and then used to predict d_{pred} on all AEM soundings. The network is a so-called Multi-layer Perceptron (MPL) regressor and is implemented in the python package "scikit-learn" (Pedregosa et al. 2011). The deterministic linear relation (equation 2) can be seen as a MLP regressor without hidden layers. In order to achieve non-linear regression, one hidden layer with a non-linear activation function g is added to the network. This results in the non-linear regression function

$$\mathrm{d}(\mathrm{M}) = \mathrm{W}_2\ \mathrm{g}\left(\mathrm{W}_1{}^{\mathrm{T}}\mathrm{M} + \mathrm{b}_1\right) + \mathrm{b}_2 \tag{5}$$

where W_1 and W_2 are weight matrixes defining the regression in addition to g, and b_1 and b_2 are bias constants which are determined during the backpropagation training process. The cost (or misfit) function is a square error loss function:

$$\mathrm{Loss} = \frac{1}{2}\left\|\mathrm{d}_{\mathrm{pred}} - \mathrm{d}_{\mathrm{train}}\right\|_2^2 + \frac{\alpha}{2}\left\|\mathrm{W}_k\right\|_2^2 \tag{6}$$

In addition to predicting the error, the loss function also measures the variation within the weight parameters of network layer k in terms of the square Frobenius L_2 norm of W_k ($W_{k2}^2 = \sum_i \sum_j |w_{ij}|^2$). The purpose of this is to penalize complex models, i.e. regressions where the elements in W varies too much). The regularization can be adjusted by the user through the coefficient α. The loss is backpropagated through each hidden layer and used to update the weights of each layer, where upon a new prediction is made. The iteration scheme used here is the L-BFGS solver, implemented in the package *scipy* (Jones et al 2001).

For comparison of results, the LSI method has also been implemented setting $\mathrm{C}_{\mathrm{d}} = \mathrm{I}$ and using a Golden section search (Kiefer 1953) iterative algorithm.

3 DATA

Two different AEM datasets have been chosen to illustrate and compare the methods. The first one is a small survey (27 kilometers of flightlines covering 2.8 square kilometers) flown in the middle of a broad glacier-eroded valley in Eiker, Norway. The sediments are mostly marine clay with a relatively high electric conductivity and the bedrock is resistive limestones, resembling a homogeneous geologic situation (although depth to bedrock varies). Seven geotechnical boreholes have been drilled in the area, confirming the bedrock depth at these points.

The second survey is much larger (167 line-km) and covers a landscape of various geologic origin in Ringerike, Norway. In the north, Pre-Cambrian bedrock (gneiss) is covered by thick ground moraine, some leached marine clay and thick ridges of river sediments. Further south, bedrock drops to greater depths, and the imaged subsurface is dominated by marine clay and a top layer of river transported sediments like sand and gravel. Covered by river sediments bedrock is again shallower further south-east, but consists now of Cambro-Silurian sedimentary rocks, namely layered mudstones/shales and limestones. Then, bedrock comes to the surface and the landscape is dominated by hills and clay-filled valleys. Lastly, a fresh-water lake covers the very southern part. A total of 431 geotechnical boreholes confirm depth to bedrock in the area.

Both surveys were flown using a SkyTEM 304 system with flightline spacing of about 100 m and sounding spacing of about 30 m. Data was inverted using the Aarhus Workbench software resulting in a 25 layer spatially constrained smooth resistivity model.

4 RESULTS

Interpretations of bedrock depth d_{train} was available from previous projects, consisting of 27 training points for the Eiker survey and 1240 training points for the Ringerike survey. These manual picks are created in the following way: The interpreter has looked at the resistivity data, available boreholes and map information and placed an interpretation point with uncertainty bars in such a way that boreholes lie within the uncertainty range, and that the predicted bedrock model is as consistent and meaningful as possible. As a consequence, the training points do not always match the borehole depths exactly. A few such points are placed on every flightline. For the Eiker survey, the training points were used as a basis for the methodology presented here. For the Ringerike survey, which was more challenging, training points had to be picked much denser, and a standard spline interpolation scheme was used to construct the bedrock model. However, because of its complexity, it is well suited to demonstrate the methods and is therefore included in the following.

The performance of the methods is evaluated by selecting a subset of the available interpretation data as training data and the remaining points as validation data. In order to investigate the effect of the amount of training data, the size of the training subset is varied (4-23 for Eiker, 10–500 for Ringerike), constituting one evaluation sequence.

The error in the prediction is somewhat dependent on the choice of subset, especially for small subsets and complex AEM models. Some training points are designed to fit a certain geologic feature, some to fit another. Therefore, the list containing training data is shuffled prior to the evaluation sequence. Then, many such sequences are performed with different stacking of the training data, and the results are averaged.

The average absolute error defined as distance in meters between predicted bedrock depths and validation depths is calculated for each method, serving as a measure for the methods technical performance. Also, the mismatch between predicted bedrock depths and borehole depths are calculated, serving as a measure for the geotechnical effectiveness of the methodology. Here, the closest prediction point to a certain borehole is used, if there is any within a radius of 50 m from the borehole.

When considering the results, we should remember that AEM resolution is decreasing with depth. One should therefore expect the mismatch to be larger at great depths and smaller at shallow depths.

Validation point mismatch are plotted in figure 1 and figure 2 for the Eiker and Ringerike survey, respectively. Borehole mismatch are similarly plotted in figure 3 and 4. Figure 5 shows an exemplary section of the Ringerike AEM model interpreted with both the LSI (upper pane) and ANN (lower pane) methods.

Run times of the methods are very similar, indeed, the golden search algorithm implemented in the LSI may be more time consuming than the L-BFGS solver. The training takes only a couple of seconds for around 20 training data and up to a minute for 500 training data.

5 DISCUSSION

Considering the Eiker dataset, the evaluation proves both methods quite successful, only a few training points are needed to get the average

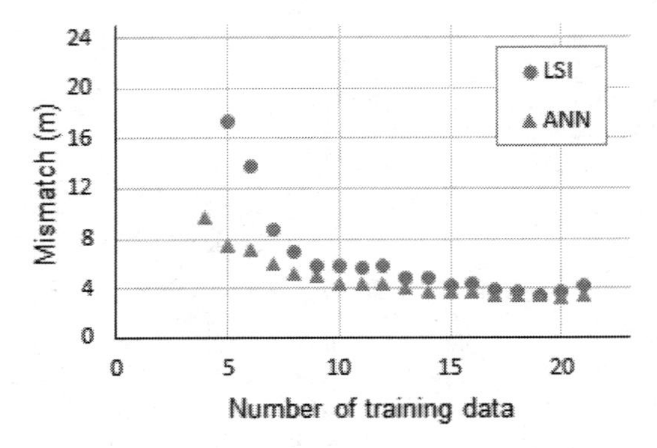

Figure 1. Average mismatch, validation points, Eiker.

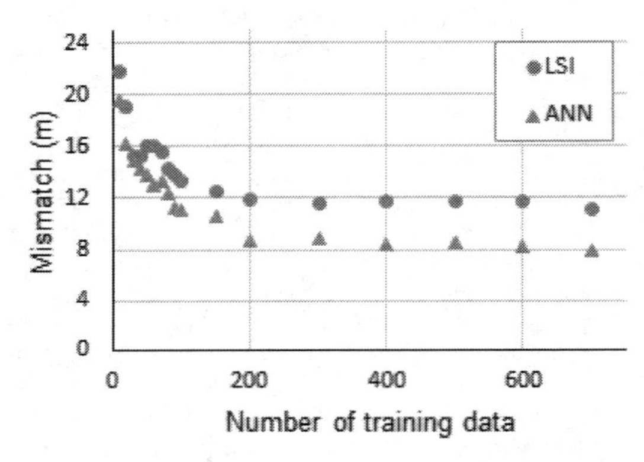

Figure 2. Average mismatch, validation points, Ringerike.

mismatch below 5 m, converging towards 3,5 for all training points. For most training points, prediction mismatch is not even this large (only a meter or less), but a few training points are poorly fitted, contributing to the average values displayed in the figure. Also, bedrock is as deep as 70 m at places, and better than 10 m uncertainty cannot be expected from AEM data at these depths. The rapid convergence and low mismatch witnesses a strong resistivity contrast around the bedrock interface which correlates well with the interpretation picks. Furthermore, there is little difference in performance between the two methods (the AN network being slightly better), which makes good sense considering the homogeneous geologic condition.

Although of good quality, the Ringerike dataset is more challenging, given its larger size and complexity.

Here, the average mismatch decreases with number of training points and reaches a minimum of around 8 m at about 200 training points. With more training points, the fit improves only little. Further training data works then only as adjustments of the prediction, not as reinforcement, meaning that there is not enough information in the AEM model to support learning of the given interpretation. Again, it is stressed that some training points contribute more to this mismatch than others. This can be seen f.ex. at around 5500 m in figure 5 (upper pane, LSI), where the prediction fails to match the deepest training points. The non-linear ANN performs better than the linear LSI method, reducing mismatch by around 30% consistently (referring to the same corresponding spot in figure 5, lower pane). It is obviously able to adjust the mapping operator according to the different geologic regions.

When interpretation is made, drillings are plotted onto the AEM data sections for the interpreter's information. It is not necessarily a good idea to blindly use the borehole bedrock depth as training data, because 3D variations, smoothness, moraine transitions or lack of penetration depth etc. might cause such a training to be less consistent. It is rather advised to use boreholes as a guideline and

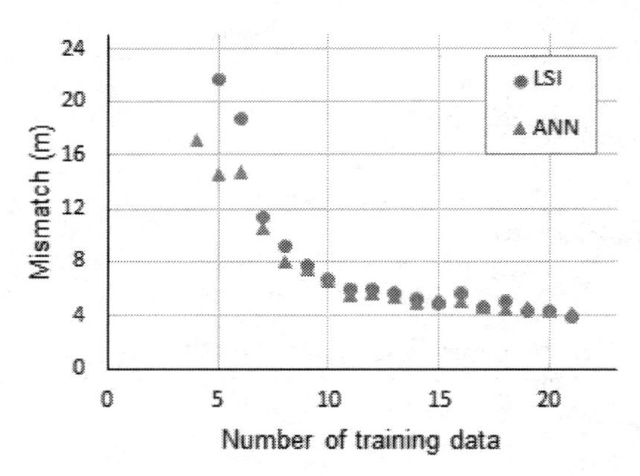

Figure 3. Average mismatch, boreholes, Eiker.

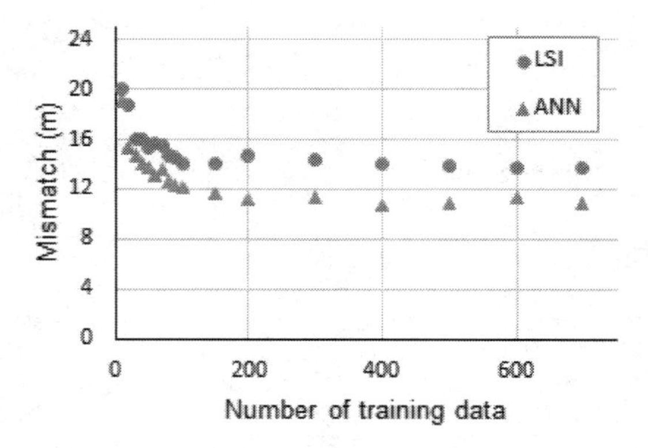

Figure 4. Average mismatch, boreholes, Ringerike.

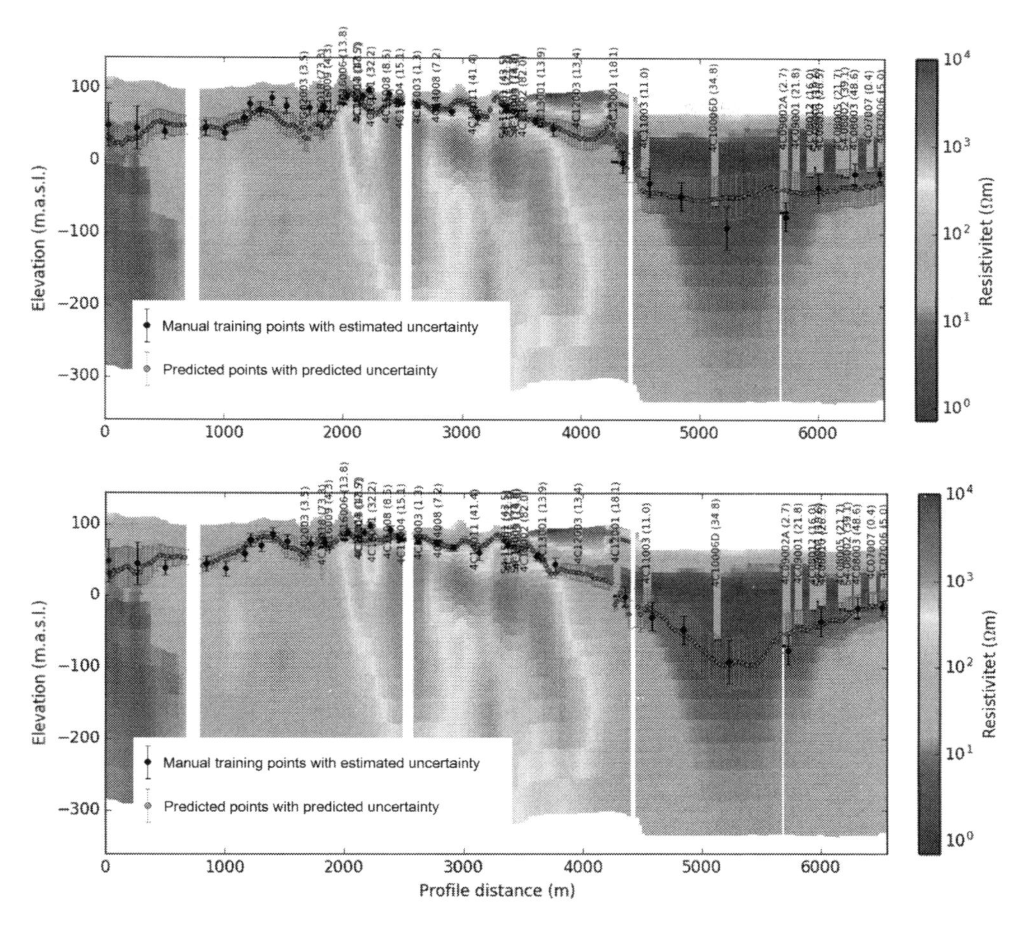

Figure 5. Exemplary section of AEM data from Ringerike, interpreted with LSI (upper pane) and ANN (lower pane).

try to locate correlating features in the AEM data. Nevertheless, comparison with drillings are a valuable measure, and similar to the validation misfit, the mismatch between predicted depth and borehole depths has been calculated for varying training data sizes.

Considering figure 3 (the Eiker survey), borehole mismatch decays gradually with the number of training data. Considering that the validation mismatch curve was relatively flat, this result indicates that new training data are successfully fitted while still preserving a good fit for the former training data, in other words; the operator tackles adjustment without losing consistency.

For the Ringerike survey (figure 4), borehole mismatch follows the same trend as the validation mismatch, with slightly larger values. This is as expected considering that not all boreholes fit the AEM features equally well.

In geotechnical projects, a prediction error of 4–8 m is significant when tunnels and rock cuts are planned. The interpretation uncertainty, measured in meters, is assigned and predicted in the same way as the bedrock depth value, in such a way that borehole depths lie within the range of uncertainty (see Figure 5, lower pane). As stated above, AEM resolution is limited to around 1–2 m at the subsurface and is around 10 m at 70–80 m depth, depending on materials and geologic complexity. Moreover, geophysical inversion of AEM raw data adds non-uniqueness to the solution, as there are several ways of regularizing the resistivity models.

To make the method even better, especially in complex geologic situations, different regularization schemes and solvers should be tested, as well as different AEM inversion schemes, ways of creating training datasets smarter and lastly inclusion of geologic information in the network learning step.

6 CONCLUSION

The study presents and evaluates an artificial neural network approach for semi-automatic interpretation on airborne-EM data, in both simple and more complex geological settings. Prediction mismatch lies between 4–8 m (averaged for all depths), however, borehole depths are within the uncertainty range of the prediction. The number of training data is shown to improve the prediction up to a certain limit where further training points only barely reduces the mismatch. The performance is compared with a linear function approximator algorithm, and shows similar behavior in case of the simple geology, however in case of the more complex AEM models, the non-linear neural network-derived operator greatly improves the prediction, both in terms of validation error and borehole mismatch.

REFERENCES

Anschütx, H., Vöge, M., Lysdahl, A.K., Bazin, S., Sauvin, G., Pfaffhuber, A.A. & Berggren, A.-L. 1980. From manual to automatic AEM bedrock mapping. *Journal of Environmental & Engineering Geophysics* 22 (1): 35–49.

Gulbrandsen, M.L., Bach, T., Cordua, K.S. & Hansen, T.M. 2015. Localized Smart Interpretation – a data driven semi-automatic geological modelling method. *Expanded Abstracts: ASEG-PESA*, 24th International Geophysical Conference and Exhibition, Perth, Australia.

Jones E, Oliphant E & Peterson P. 2001. SciPy - Open Source Scientific Tools for Python. http://www.scipy.org/ (Online; accessed 2017-12-21).

Kiefer, J. 1953. Sequential minimax search for a maximum. *Proceedings of the American Mathematical Society* 4 (3): 502–506.

Pedregosa et al. 2011. Scikit-learn: Machine Learning in Python. *JMLR* 12: 2825–2830.

Sørensen, K.I. & Auken, E. 2004. SkyTEM – A new high-resolution helicopter transient electromagnetic system. *Exploration Geophysics* 35: 191–199.

Numerical Methods in Geotechnical Engineering IX – Cardoso et al. (Eds)
 ISBN 978-1-138-33198-3

Artificial neural networks in the analysis of compressibility of marine clays of Grande Vitória, ES, Brazil

A.G. Oliveira Filho & K.V. Bicalho
Department of Civil Engineering at the Federal University of Espírito Santo, Brazil

W.H. Hisatugu
Department of Computing and Electronic at the Federal University of Espírito Santo, Brazil

C. Romanel
Department of Civil Engineering at the Pontifical Catholic University of Rio de Janeiro, Brazil

ABSTRACT: The compression index C_C is used to calculate the consolidation settlement of foundation on clayey soils. Several empirical relationships linking the compressibility parameters of clayey soils to their index properties have been published in the literature. This paper evaluates some of the empirical equations, which determine the Cc from laboratory index properties of soft marine clay deposits from Grande Vitória, ES (GV-ES), Southeast Brazil. Regression analyses are carried out to suggest simple correlations using both single and multiple soil parameters for estimating C_C. This paper also illustrates the potential of using Artificial Neural Networks (ANN) in the analysis of compressibility of marine clays from GV-ES. Actual laboratory test data are used in training the neural network. The results indicate that the ANN model is able to predict the compressibility of marine clays from GV-ES. It was concluded that ANN is a good alternative to the empirical equations.

1 INTRODUCTION

Projects of civil engineering structures on soil profiles with soft clays may require the determination of parameters of soil compressibility, such as the compression index (C_C), which is used to calculate the significance of the consolidation settlement of soft clayey soils. The conventional laboratory oedometer test is utilized to obtain the C_C values. However, oedometer test is a test of long duration and requires sufficiently undisturbed soil samples. Thus, several empirical relationships linking the compressibility parameters of clayey soils to their index properties, such as liquid limit (W_L), bulk density (G_S), plasticity index (PI), natural water content (W_n) and initial void ratio (e_0), have been published in the literature for different soft fine-grained soils.

By using the regression analysis, empirical equations for predicting C_C are proposed by several publicatins (Terzaghi & Peck 1967, Azzouz et al. 1976, Oh & Chai 2006, Ozer et al. 2008, McCabe et al. 2014) for different soft clays. Some studies have suggested different correlations for Brazilian marine clays (Castello & Polido 1986, Futai et al. 2008, Bicalho et al. 2014, Coutinho & Bello 2014, Baroni & Almeida 2017). As the previous published equations have an empirical nature and are highly discrepant, it is important to estimate their reliability.

Since the work of Rumelhart (1986) about the backpropagation algorithm, the use of neural networks obtained more relevance in several publications in geotechnical engineering. Studies of mechanical strength of sands and hydraulic conductivity in clays may be found in Goh (1995). In the geotechnical approach, the work of Isik (2009) about the applications of Artificial Neural Networks (ANNs) in the prediction of soil expansion index (C_S) can be also highlighted. Najjar et al. (1996) predicted the value of optimum water content and the maximum dry density of the soil for compaction projects; (Nawari et al. 1999, Nejad et al. 2009, Benali & Bouzid 2013) used neural network in predictions of consolidating settlement. Diminsky (2000) has been publishing studies with ANNs in prediction of pile loading capacity, besides modeling of the tension-deformation behavior of the soil and prediction of the geotechnical characteristics of the subsurface. Concerning the application of ANN in studies of prediction of the clay compression index, it can be mentioned the researches of (Kolay et al. 2008, Ozer et al. 2008, Kalantary & Kordnaeij 2012, Kurnaz et al. 2016).

This study evaluates some of the empirical equations, which predict the compression index Cc from laboratory index properties of soft marine clay deposits from Grande Vitória, ES, Southeast Brazil. Regression analyses are carried out to suggest simple correlations using both single and multiple soil parameters for estimating Cc. Soil parameters include natural water content, initial void ratio, liquid limit, bulk density and plasticity index.

Considering the potential applications of ANNs in predicting the compression index values for soft clayey soil from the learning of existing patterns among the physical properties of the soil. This paper also illustrates the use of artificial neural networks (ANNs) in the analysis of compressibility of marine clays from Grande Vitória, ES, GV-ES, Brazil. Actual laboratory test data are used in training the neural network.

2 THE TESTS SITE

The metropolitan region of GV-ES, is located on the coast of the state of Espírito Santo, southeast Brazil, is composed by the capital called Vitória and six other cities, adding more than half the population of the state, and where there is a strong dynamic of urbanization.

In the Grande Vitória's geology it can be found great relevance the presence of sedimentary associations in the flattened fluvial-marine region dating from the Tertiary and Quaternary, formed by process of accumulation of sediments of fluvial-marine origin, under the influence of the climate and the sea-level variation, according to Martin et al. (1997). Figure 1 shows the location of the test site.

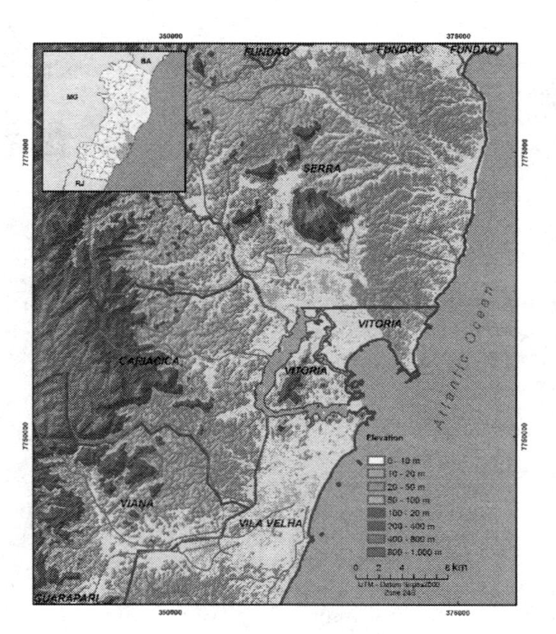

Figure 1. Location of the test site: Grande Vitória, ES, Brazil.

3 DATABASE COMPILATION

3.1 *Evaluation of the input data used in the research*

The laboratory oedometer test results and the corresponding index properties of 56 soil samples of marine clays obtained in different locations of the GV-ES were investigated in this research.

The soil samples of marine clay have been collected in relation to the recommendations of the Brazilian Standard (NBR), according to ABNT (1997). The oedometer tests have been carried out on soil samples taken from Shelby tube, at different depths.

The index soil properties of the investigated clays are: liquid limit (W_L), bulk density (G_S), plasticity index (PI), natural water content (W_n) and initial void ratio, besides the compression index (C_C). The statistical description of the input experimental data is presented in Table 1.

In general, the studied marine clays from GV-ES have a compression index ranging from 0.10 to 1.53 with a mean value of 0.68 indicating high compressibility clays. The W_L and PI data show that the investigated marine clays are classified as silty clays with high to very high plasticity (mean W_L greater than 73.21%) with liquid limit varying between 32.80% and 142.00%. The plasticity index values range from 12.60% to 99.70% with an average of 44.59%. It is observed a large variability in the initial void ratio values of the studied clays, with values between 0.64 and 6.20, and average of 1.76.

An evaluation by linear regression of the influences of clays GV-ES index properties on their compressibility, shows that the liquid limit presents the highest correlation with the compression index (R^2 = 0.50), followed by natural water content (R^2 = 0.460, plasticity index (R^2 = 0.44) and initial

Table 1. Statistical description of the input data.

	W_L %	PI %	e_0	W_n %	G_S	C_C
Minimum	32.80	12.60	0.64	19.23	2.22	0.10
Maximum	142.00	99.70	6.20	128.00	2.91	1.53
Average	73.21	44.59	1.76	62.94	2.69	0.68
SD	25.32	18.78	0.88	20.39	0.11	0.37

void ratio index ($R^2 = 0.36$). And, the bulk density presents a weak correlation with C_C ($R^2 = 0.03$).

3.2 *Predictions of C_C using W_n*

One of the commonly used correlations in geotechnics to estimate C_C of a soft soil is their W_n, which normally has a strong correlation with the compression index. In this context, Table 2 presents several proposals to estimate C_C from W_n and the corresponding RMSE values. The equations used have all presented a low value of RMSE, despite the coefficient of determination (R^2) has been approximately 0.46 and the maximum RMSE is 0.044.

3.3 *Predictions of C_C using W_L*

The liquid limit (W_L) is a parameter with high correlation with C_C. For those correlations, the R^2 results close to 0.5 showed a higher correlation between the estimated and the laboratory C_C. The RMSE values have also been low, less than 0.055 (Table 3).

3.4 *Predictions of C_C using e_0*

From de variables commonly used in estimating the compression index, e_0 has the lowest C_C compared to W_L and W_n. Table 4 shows some correlations to estimate the C_C with initial e_0. R^2 is close to 0.36 and the maximum RMSE is 0.060.

Table 2. Some previous published empirical compression index correlation using W_n.

Reference	Correlation	RMSE
Azzouz et al. (1976)	$C_C = 0.0100\ W_n - 0.05$	0.039
Castellho & Polido (1986)	$C_C = 0.014\ W_n - 0.17$	0.036
Futai et al. (2008)	$C_C = 0.013\ W_n$	0.040
McCabe et al. (2014)	$C_C = 0.014(W_n - 22.7)$	0.042
Bicalho et al. (2014)	$C_C = 0.0128\ W_n - 0.0951$	0.036
Coutinho & Bello (2014)	$C_C = 0.0070\ W_n + 0.4010$	0.044
Baroni & Almeida (2017)	$C_C = 0.011\ W_n$	0.036

Table 3. Some previous published empirical compression index correlation using W_L.

Reference	Correlation	RMSE
Azzouz et al. (1976)	$C_C = 0.006\ W_L - 0.054$	0.054
Terzaghi & Peck (1967)	$C_C = 0.009(W_L - 10)$	0.038
Castello & Polido (1986)	$C_C = 0.01(W_L - 8)$	0.035
Bicalho et al. (2014)	$C_C = 0.011(W_L - 8.3)$	0.035
Baroni & Almeida (2017)	$C_C = 0.125\ W_L$	0.047

Table 4. Some previous published empirical compression index correlation using e_0.

Reference	Correlation	RMSE
Azzouz et al. (1976)	$C_C = 0.400e_0 - 0.100$	0.044
Castello & Polido (1986)	$C_C = 0.228e_0 + 0.22$	0.040
Bicalho et al. (2014)	$C_C = 0.4961(e_0 - 0.1163)$	0.049
Baroni & Almeida (2017)	$C_C = 0.5284e_0$	0.060

3.5 *Correlations between soil compression index and multi soil's parameters*

Oh & Chai (2006) propose the following empirical correlation to calculate C_C from the combination of e_0, PI and W_L:

$$C_C = 0.5393e_0 - 0.0074PI + 0.0094W_L - 0.1248 \quad (1)$$

Oh & Chai (2006) correlation showed a modest $R^2 = 0.306$ and RMSE = 0.0616. Those results suggest that the correlation with one variable is more efficient for the analysis of the clays of GV- ES.

4 ARTIFICIAL NEURAL NETWORK

4.1 *Artificial neural network fundaments*

Artificial Neural Networks (ANNs) are characterized as artificial intelligence (AI) techniques that are inspired in the structure of the human brain, in order to simulate their functioning in computational systems in a simplified way. The main ANN skill lies on its ability to generalize and learn from errors. Artificial neural are distinguished by performing three essential operations: learning and storing knowledge; apply the knowledge acquired in solving proposed problems; as well as acquire new knowledge from constant learning (Khanna, 1990).

The artificial neuron is the basic processing element of an artificial neural network and it is formed by the following parts: a set of input connections (X_j); synaptic weights (W_{kj}), where k is the number of the neurons and j corresponds to the input stimulus; *bias*, where (b_k), is a weighting parameter that can increase or decrease the value of the linear combination of inputs of the neuron activation function $f(.)$. Figure 2 presents a simplified model of an artificial neuron, where (u_k) represents the linear combination of input signals, and (y_k) corresponds to the output value of the neuron.

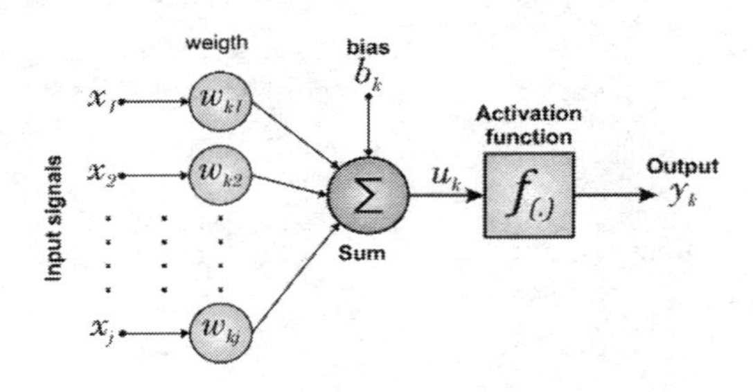

Figure 2. Example of an artificial neuron model (adapted from Haykin 2001).

In Figure 2 it is possible to observe in (w_{kj}) that the bigger the weight is (w), the greater will be the contribution of the inputs (k) to the sum of the weights. Thus, the input weighting process represents the learning rate acquired by an ANN. Hence, the weights are adjusted as the input data set is presented to the network. The supervised learning process in an ANN is based on the adjustment of the synaptic weight, therefore, the output value is approximated to the expected value.

The activation function $f\,(.)$ corresponds to the normalized amplitude of the neuron, which aims to limit the input signals in a given range, usually among [0;1] and [−1;1]. The functions commonly used in geotechnical studies are log-sigmoid, tan-sigmoid and linear.

The ANN architecture is the way the networks present the arrangement of their neurons. The structure can be in one layer or in several layers. The layers between the input and output neurons are referred to as the hidden layer, or middle layer. The number of hidden layers is connected to the degree of complexity of the problem to be solved. Thus, determining the appropriate configuration of the ANN is a great challenge for the application of the networks by the researchers.

Generally, it is possible to identify mainly three architecture classes of an ANN: feedforward single-layer network, multi-layer feedforward networks with hidden layers, and neural networks with recurrence.

As noted, the number of hidden layers and their neurons is not a trivial task, being defined by means of the experience acquired in solution of similar problems through trial and error. Although a single hidden neuron layer, operating a sigmoidal activation function is sufficient to model a wide range of practical problems, Flood & Kartam (1994) suggest, based on the behavior of the network activation, it is possible to note the potential gain to develop solutions that are close to the training pattern, however, the same authors indicate that a large number of neurons make the network operations slower and more likely to lead to the wrong solutions.

Another relevant point in relation to the hidden layers is the definition of the number of neurons that each layer should have. Caudil (1988) suggests that the hidden layer must be provided by a number neurons governed by the expression ($2i + 1$), where (i) represents the number of input variables. Nawari et al. (1999) who point out that best result can be found by starting the network with a small number of hidden neurons and gradually increasing the number.

In this hand, it is possible to note a good approach to solve the topology problem of a network is the adoption of different configurations of the number of layers and hidden neurons, in order to improve the network performance and accuracy of the modeled results, as observed by Najjar et al. (1996), besides the use of different activation functions to better delimit the contour of the desired solution.

4.2 *Proposed artificial neural network*

In an ANN, the data to be analyzed are generally, divided into three groups: training, validation and testing. In the present study, 10% of the samples were separated for a cross-validation test, they were samples that were only presented to the network after the training stage. Of the remaining samples, 70% of the total of 50 were selected for training, 15% for validation and 15% for testing. All samples were randomly selected.

In order to carry out the training of the networks, Matlab2012software has been used through toolbox nntool. Multi-layer perceptron neural networks have been used and trained with the Levenberg-Marquardt (LM) algorithm, which is an optimization of the backpropagation algorithm.

The determination of the best network architecture has been obtained through trial and error, several networks with one and two hidden layers have been trained, varying the number of neurons of each intermediate layer. The different networks have been trained with different input combinations and six entries adopted: E1 (W_L, PI, e_0, W_n), E2 (W_L, PI, e_0, $W_{n,}$ G_S), E3 (W_L, e_0, $W_{n,}$ G_S) E4 (W_L, PI, $e_{0,}$ G_S), E5 (W_L, PI, e_0) and E6 (W_L, W_n).

For improving the network performance, the set of inputs have been normalized to the interval [0;1], which is within the domain of the activation functions used.

The performance of the proposed ANN model has been evaluated by coefficient of determination (R^2) and the root mean square error (RMSE). Those statistical indexes have been used by Onyejekwe et al. (2014) for assessment of empirical equations for the compression index of fine-soils. The coefficient R^2 and RMSE are defined as:

$$R^2 = \sum_{i=1}^{n} (X_{i,m} - \bar{X}_{i,m})(X_{i,p} - \bar{X}_{i,p}) \quad (2)$$

$$RMSE = \sqrt{\frac{1}{n}\sum_{i=1}^{n} \left(X_{i,m} - X_{i,p}\right)^2} \quad (3)$$

where $X_{i,m}$ = input value, $X_{i,p}$ = estimated output value, $\bar{X}_{i,m}$ = mean input values, $\bar{X}_{i,p}$ = mean estimated output values and n = number of variables.

In general, the coefficient of determination has the objective of evaluating the relationship between two variables, from "n" observations of those variables, indicating how much the independent variable can be explained by the fixed variable, the close to "1" the better the fit of the correlation. The RMSE is the squared root of the square mean of the difference between the estimated and input values, which consequently attributes greater weight to the larger errors. Values close to "0" indicate better model performance.

Several ANNs has been trained and Table 5 presents a summary of the networks with the best statistical achieved results ($R^2 > 0.684$ and RMSE < 0.200), especially the networks E2_NN6 and E5_NN13, where H is the number of hidden layers and N is the number of neurons in each hidden layers. The E2_NN6 network was trained with 5 inputs (W_L, PI, e_0, W_n, G_S), with a single hidden layer with 18 neurons. Log-sigmoid and linear activation functions have been used in the hidden and output layers, respectively. For the training phase, the value of 0.660 has been obtained for R^2 and 0.220 for RMSE. For the testing, $R^2 = 0.920$ and RMSE = 0.100 have been determined. The E5_NN13 network was modeled considering 3 entries (W_L, PI, e_0), with 2 hidden layers, 14 and 8 neurons in each layer, respectively. In the training phase, the performance resulted in $R^2 = 0.647$ and RMSE = 0.212 have been obtained. In the testing phase it has been obtained $R^2 = 0.898$ and RMSE = 0.199.

Table 5. Neural networks results for C_C estimated by ANNs with the best statistical achieved results.

ANN		H	N	N	Training R^2	RMSE	R^2	RMSE
E1	NN13	1	18	10	0.750	0.180	0.930	0.200
E2	NN6	1	18	–	0.660	0.220	0.920	0.100
E3	NN8	2	8	5	0.700	0.210	0.800	0.200
E4	NN10	2	12	7	0.560	0.236	0.684	0.098
E5	NN13	2	14	8	0.647	0.212	0.898	0.199
E6	NN18	2	20	11	0.710	0.197	0.937	0.137

The networks highlighted in Table 5 shows that the great part of the ANN have a high coefficient of determination (R^2), both in the training and testing phases; although the RMSE values are relatively high.

Figures (3) and (4) present a comparison curve between the estimated and laboratory values of the compression index in the training phase of the neural network. It is possible to notice in the graphs a dispersion of the values of (C_C) for different zones. The same characteristic is observed in the best fit of the curves. The neural network E2_NN6 has presented an absolute mean error of 0.162 and a standard deviation of 0.145. The E5_NN13 network has presented similar values, mean absolute error of 0.165 and standard deviation of 0.132.

The generalization capacity of the networks is highlighted in Figures (5) and (6), where it is possible to focus the best performance of the E2_NN6 and E5_NN13 networks, which present estimated values close to the reference curve of the measured C_C values in the laboratory. The E2_NN6 network presented lower values of absolute average error and standard deviation (0.083 and 0.062), respectively, compared to the second network (0.149 and 0.147).

In addition, it is possible to observe a tendency of the ANNs in overestimating the compression index value; in the higher part of the networks the estimated values are close or higher than the mean measured values both in the training and testing phases (see Figure 5).

4.3 *Evaluation of other artificial neural networks*

Although the best training results have been found with the use of log-sigmoid activation functions in the hidden layer and linear functions in the output layer, other architectures have also been evaluated with the use of log-sigmoid and tan-sigmoid functions both in the hidden and output layers. The results have not been satisfactory, indicating that for C_C analysis of clays, best convergence of the network have been found with the use of a linear function in the output layer. The same conclusion was obtained in the researches of (Kalantary & Kordnaeij 2012, Kurnaz et al. 2016).

Figure (7) shows a comparison curve between the predicted and laboratory values of the compression index in the test phase of the modified E2_NN6 network. In this neural network it has been used the log-sigmoid function in the hidden layer and tang-sigmoid function in the output layer. It is possible to observe a low correlation value between the estimated and laboratory values.

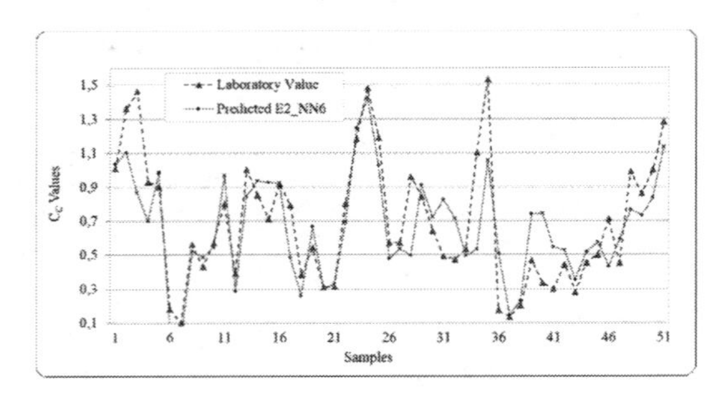

Figure 3. Training data set for C_C for network E2_NN6.

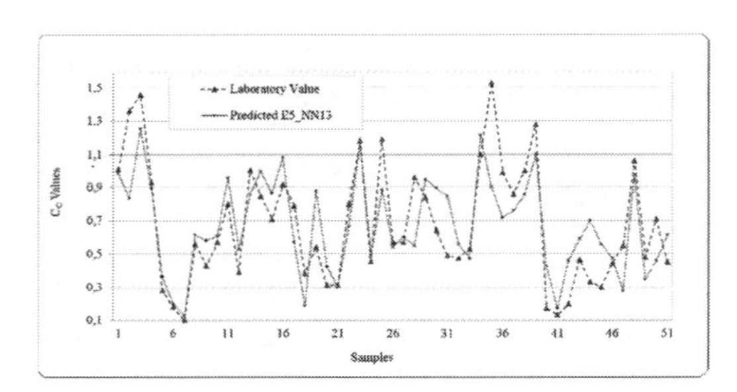

Figure 4. Training data se for C_C for network E5_NN13.

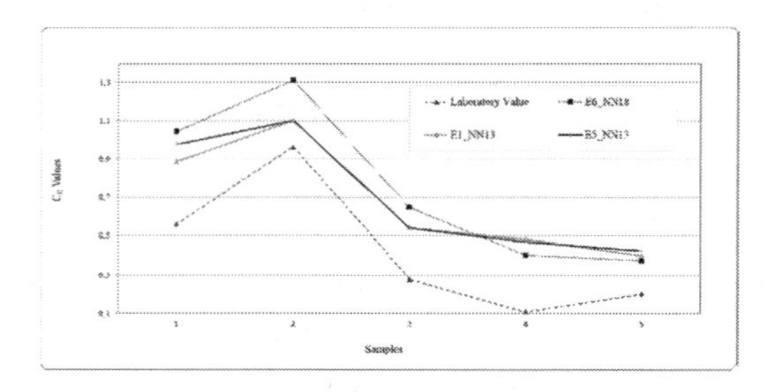

Figure 5. Testing data set for C_C for network E6, E1 and E5.

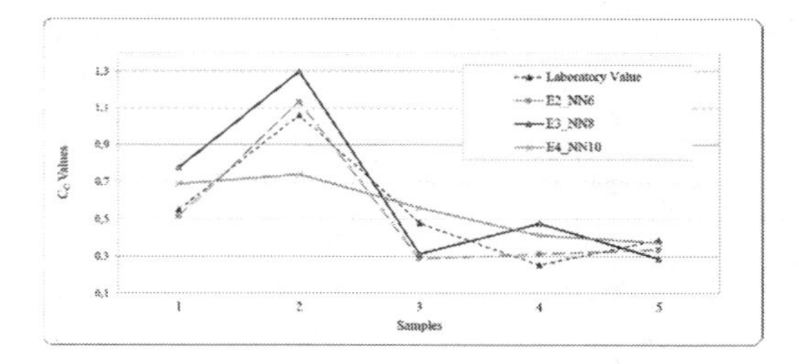

Figure 6. Testing data set for C_C for network E2, E3 and E4.

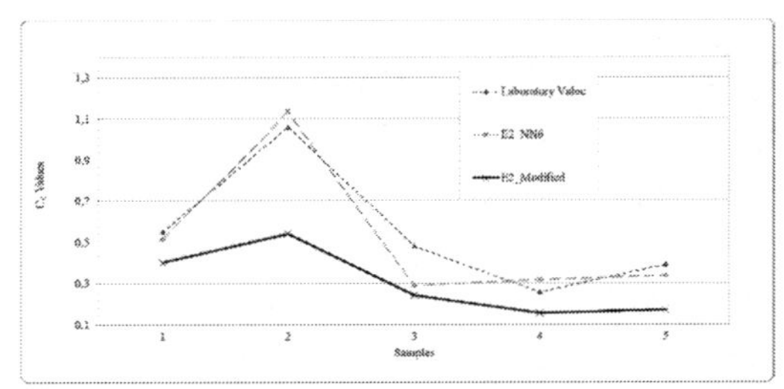

Figure 7. Testing data set for C_C for network E2 modified.

5 CONCLUSIONS

In this research, empirical correlations and artificial neural networks have been applied to estimate the compression index value from the analysis of soil physical parameters. It has been investigated a database composed of 56 samples of soft marine clays from different surveys in Grande Vitória, ES. Several ANN have been modeled and trained, what has made possible to highlight two networks with best statistical results: E2_NN6, composed of 5 inputs (W_L – PI – G_S – e_0 – W_n) and one hidden layer with 18 neurons; and E5_NN13, composed of 3 inputs (W_L – PI – e_0) and two hidden layers with 14 and 8 neurons, respectively. The Levemberg-Marquardt (LM) training algorithm has been applied, and the log-sigmoid and linear activation functions have been used in the hidden layer and output layer, respectively. The first network presented $R^2 = 0.660$ and RMSE = 0.220, during the training, and $R^2 = 0.920$ and RMSE = 0.100 during testing stage. The second network presented $R^2 = 0.647$ and RMSE = 0.212 during training, and $R^2 = 0.898$ and RMSE = 0.199 during the tests. The achieved results suggest that ANNs with one or two layers are able to converge to good prediction of C_C.

Evaluating the correlations used for the calculation of C_C, it is possible to highlight that the best performance occurs with the W_L-C_C correlation, with RMSE below 0.05, and R^2 close to 0.50.

Comparing the results obtained with the ANNs and the correlations, it is possible to emphasize that the C_C values estimated by the networks present a higher value of R^2, however it presents the most unfavorable value of RMSE.

The preliminary achieved results of this study indicate the potential of the application of the neural networks in predicting the compression index values for soft marine clays from GV-ES, from the learning of existing patterns among the physical properties of the investigated soils.

ACKNOWLEDGEMENTS

The second author is grateful to CNPq and FAPES for financial support.

REFERENCES

ABNT (Associação Brasileira de Normas Técnicas) 1997. NBR 9820. 1997. *Assessment of undisturbed low consistency solid samples from boreholes – Procedure. ABNT*, Rio de Janeiro, Brazil (in Portuguese).

Azzouz, A. S. Krizek, R. J. & Corotis, R. B. 1976. Regression analysis of soil compressibility. Tokio. *Soils Foundations, 16 (2)*: 19–29.

Baroni, M. & Almeida, M. S. S. 2017. Compressibility and stress history of very soft organic clays. *Proceeding of Institution of Civil Engineers – Geotechnical Engineering:* 148–160. Thomas Telford Ltd.

Benali, A. Nechnech & Bouzid, Ammar. 2013. Principal Component Analysis and Neural Networks for Predicting the Pile Capacity Using SPT, *International Journal of Engineering and Technology vol. 5, no. 1*: 162–169.

Bicalho, K. V. Bastiti, V. G. & Ximenes, R. B. 2014. Compressibility of marine clays, Grande Vitória, ES, Brasil. Geological Communications, 101 Special III, 1093–1095.

Castello, R. R & Polido, U. F. 1986. Some characteristics of consolidation of marine clays form Vitória, ES. In. *VIII Brazilian congress of soil mechanics and foundation engineering (1986)*: 149–159. Rio Grande do Sul, Porto Alegre.

Caudil, M. 1988. Neural networks primer, Part III. *AI Expert, 3(6)*: 53–59.

Coutinho, R Q. & Bello, M. I. M. C. V. 2014. Geotechnical characterization of Suape soft clays, Brazil. *Soils and Rocks, 37 (3)*: 257–276.

Diminsky, A.S. (2000). Analyze of geotechnical problems through neural networks. Thesis (Dr. in Civil Engineering) – Pontifical Catholic University of Rio de Janeiro, Rio de Janeiro.

Flood, I. & Kartam, N. 1994. Neural networks in civil engineering I: Principles and understanding. *Journal of Computing in Civil Engineering, 8(2)*: 131–148.

Futai, M. M. Almeida, M. S. S. Lacerda, W. A. & Marques, M. E. S. 2008. Laboratory behavior of Rio de Janeiro soft clays. Index and compression properties– part 1. *Soils and Rocks* 31 (2): 69–75.

Goh, A. T. C. 1995. Modeling soil correlations using neural networks. *Journal of Computing in Civil Engineering, ASCE, 9(4)*: 275–278.

Haykin, S. 2001. *Neural Networks. A Comprehensive Foundation. Second Edition, Pearson Education*, McMaster University, Hamilton, Ontario, Canada.

Isik, N. S. 2009. Estimation of swell index of fine grained soils using regression equations and artificial neural networks. *Scientific Research and Essays, Vol. 4 (10)*: 1047–1056.

Kalantary, F. & Kordnaeij, A. 2012. Prediction of compression index using artificial neural network. *Scientific Research and Essays, Vol. 7 (31)*: 2835–2848.

Khanna, T. 1990. *Foundations of neural networks*. United States of America: Addison-wesley Publishing Company.

Kolay, P. K. Rosmina, A. B. & Ling, N. W. 2008. Prediction of compression index for tropical soil by using artificial neural network. *International Association for Computer Methods and Advances in Geomechanics (IACMAG)*. 1–6 October, India.

Kurnaz, T. F. Dagdeviren, U. Yildiz, M. & Ozkan, O. 2016. Prediction of compressibility parameters of the soils using artificial neural network. *SpringPlus. 5:1801*. p. 1–11.

Martin, L. Suguio, K. Flexor, J. & Domingues, J. M. L. 1996. *Geology of the Quaternary of the coastal of north Rio de Janeiro and Espírito Santo*. Belo Horizonte: CPRM.

McCabe, B. A. Sheil, B. B. Long, M. M. Buggy, F. J & Farell, E. R. 2014. Empirical correlations for the compression index of Irish soft soils. *Proceeding of Institution of Civil Engineers – Geotechnical Engineering:* 510–517. Thomas Telford Ltd.

Najjar, Y. M. Basheer, I. A. & Naouss, W. A. 1996. On the identification of compaction characteristics by neuronets. *Computers and Geotechnics, 18(3)*: 167–187.

Nawari, N. O. Liang, R. & Nusairat, J. 1999. Artificial intelligence techniques for the design and analysis of deep foundations. *Electronic Journal of Geotechnical Engineering.*

Nejad, F. P. Jaksa, Mark B. Kakhi, M. & McCabe, Bryan A. 2009. Prediction of pile settlement using artificial neural networks based on standard penetration test data. *Computers and Geotechnics, 36(7)*: 1125–1133.

Oh, E. Y. N. & Chai, G. W. K. 2006. Characterization of marine clay for road embankment design in coastal area. *Proceeding of Sixteenth International Offshore and Polar Engineering Conference, San Francisco, USA, Vol.* 2: 560–563.

Onyejekwe, S. Kang, X. & GE, L 2014. Assessment of empirical equations for the compression index of fine-grained soils in Missouri. *Bulletin of Engineering Geology and the Environment,* [s.l.], v. 74, n. 3, p.705–716, 28 ago. 2014. Springer Nature.

Ozer, M. Isik, N. S. & Orhan, M. 2008. Statistical and neural network assessment of the compression index of clay-bearing soils. *Bull Eng Geol Environ (2008) 67*:537–545.

Rumelhart, D. E. Hilton, G. E. & Williams, R. J. 1986. Learning representation by back-propagation erros. *Nature, 323*: 533–536.

Terzaghi, K. & Peck, R. B. 1967. *Soil mechanics in engineering practice.* 2nd edn. Wiley, New York, NY, USA.

Numerical Methods in Geotechnical Engineering IX – Cardoso et al. (Eds)
 ISBN 978-1-138-33198-3

Rock and soil cutting slopes stability condition identification based on soft computing algorithms

J. Tinoco
ISISE—Institute for Sustainability and Innovation in Structural Engineering/ALGORITMI Research Center, School of Engineering, University of Minho, Guimares, Portugal

A. Gomes Correia
ISISE—Institute for Sustainability and Innovation in Structural Engineering, School of Engineering, University of Minho, Guimares, Portugal

P. Cortez
ALGORITMI Research Center/Department of Information Systems School of Engineering, University of Minho, Guimares, Portugal

D. Toll
School of Engineering and Computing Sciences, University of Durham, Durham, UK

ABSTRACT: This study aims to develop a tool able to help decision makers to find the best strategie for slopes management tasks. It is known that one of the main challenges nowadays for every developed or countries undergoing development is to keep operational under all conditions their tranpostations infrastructure. However, considering the network extension and increased budget constraints such chalenge is even more difficult to accomplish. In the framework of transportations networks, particularly for railway, slopes are perhaps the element for which their failure can have a strongest impact at several levels. Therefore, it is important to develop tools able to help minimizing this situation. Aiming to achieve this goal, we take advantage of the high flexible learning capabilities of Artificial Neural Networks (ANNs) and Support Vector Machines (SVMs), which have been used in the past to model complex nonlinear mappings. Both data mining algorithms were applied in the development of a classification tool able to identify the stability condition of a rock and soil cutting slopes, keeping in mind the use of information usually collected during routine inspections activities(visual information) to feed them. For that, two different strategies were followed: nominal classification and regression. Moreover, to overcome the problem of imbalanced data, three training sampling approaches were explored: no resampling, SMOTE (Synthetic Minority Over-sampling Technique) and Oversampling. The achieved results are presented and discussed, comparing the performance of both algorithms (ANN and SVM) according to each modeling strategy as well as the effect of the sampling approaches. Also, a comparisson between both types of slopes is presented and discussed. An input-sensitivity analysis was applied allowing to measure the relative inlfuence of each model attribute.

1 INTRODUCTION

In the framework of a transportation network, one of the biggest challenges today is to keep it operational under all conditions, mainly if we take into account its extension and the increased budget limitation for maintenance and repair tasks. Indeed, this is one of the main concerns of every developed or countries undergoing development that have invested and keep investing to build a safe and functional transportation network. Thus, taken into account the key importance of the transportation system in modern societies, it is fundamental to develop new tools able to help in its management.

In the framework of transportations networks, in particular for a railway, slopes are perhaps the element for which their failure can have the strongest impact at several levels. Therefore, it is important to develop ways to identify potential problems before they result in failures. Over time, several efforts have been made toward the development of a system to detect slope failures. However, most of the systems were developed for natural slopes, presenting some constraints when applied to engineered (human-made) slopes. In addition, they have limited applicability as most of them were developed based on particular case studies or using small databases. Furthermore, another aspect that can limit its applicability is related with the information required to feed them, such as data taken from complex tests or from expensive monitoring systems. Pourkhosravani & Kalantari (2011) summarized in their work some of the current methods for slope failure detection, which were grouped into Limit Equilibrium (LE) methods, Numerical Analysis methods, Artificial Neural Networks and Limit Analysis methods. There are also approaches based on finite elements methods (Suchomel et al. 2010), reliability analysis (Husein Malkawi, Hassan, & Abdulla 2000), as well as some methods making use of data mining (DM) algorithms (Cheng & Hoang 2014, Ahangar-Asr, Faramarzi, & Javadi 2010, Yao, Tham, & Dai 2008). More recently, a new flexible statistical system was proposed by Pinheiro, Sanches, Miranda, Neves, Tinoco, Ferreira, & Gomes Correia (2015), based on the assessment of different factors that affect the behaviour of a given slope. By weighting the different factors, a final indicator of the slope stability condition is calculated.

In summary, most of the approaches so far proposed share the main limitations, which are related with its applicability domain or dependency on information that is difficult to obtain. In fact, the assessment of the stability condition of given slope is a multi-variable problem characterized by a high dimensionality.

Artificial Neural Networks (ANNs) and Support Vector Machines (SVMs) are two of the most well known Data Mining (DM) algorithms, which have been applied with success in different knowledge domains, such as web search, spam filters, recommender systems, and fraud detection (Domingos 2012). Also in civil engineering field, several application can be found. For example, artificial neural networks and support vector machines were applied in the study of physical and mechanical properties of jet grouting columns (Tinoco, Gomes Correia, & Cortez 2014, Tinoco, Gomes Correia, & Cortez 2016). Indeed, the high learning capabilities of these algorithms give them the ability to model complex nonlinear mappings. Thus, in this work we take advantage of ANNs and SVMs capabilities and fit them to a large database of rock and soil cutting slopes in order to predict the stability condition of a given slope according to a pre-defined classification scale based on four levels (classes). One of the underlying premises of this work is to identify the real stability condition of a given slope based on information that can be in a someway easily obtained during visual routine inspections. For that, more than fifty variables related with data collected during routine inspections as well as geometric, geological and geographic data were used to feed the models. This type of visual information is sufficient from the point of view of the network management, allowing the identification of critical zones for which more detailed information can then be obtained in order to perform more detailed stability analysis, which is out of the scope of this study. In summary, our proposal will allow to identify the stability condition level of a given rock or soil cutting slope based on visual information that, in most of the cases, can be easily obtained during routine inspections. Such novel approach is intended to support railway network management companies to allocate the available funds in the priority assets according to its stability condition.

2 METHODOLOGY

2.1 *Data characterization*

To fit the proposed models for stability condition identification, from this point referred to as EHC (Earthwork Hazard Category (Power, Mian, Spink, Abbott, & Edwards 2016)), of rock and soil cutting slopes two database were used respectively. Hence, two databases were compiled containing information collected during routine inspections and complemented with geometric, geological and geographic data of each slope. Both databases were gathered by Network Rail workers and are concerned with the railway network of the UK. For each slope a class of the EHC system was defined by the Network Rail Engineers based on their experience/algorithm (Power Mian, Spink, Abbott, & Edwards 2016), which will be assumed as a proxy for the real stability condition of the slope for year 2015. The EHC system comprises 4 classes ("A", "B", "C" and "D") where "A" represents a good stability condition and "D" a bad stability condition. In other words, the expected probability of failure is higher for class D and lower for class "A".

Both databases contain a significant number of records. The rock slopes database comprises 5945 records, while the soil cutting slopes database is bigger, having 10928 records available. Figure 1 depicts the distribution of EHC classes for each database. From this analysis, it is possible to observe a high asymmetric distribution (imbalanced data), in particular for the rock cutting slopes database. In fact, more than 86% of the rock slopes are classified as "A" Although this type of asymmetric distribution, where most of the slopes present a low probability of failure (class "A"), is normal and desirable from the safety point of view and slope network management, it can represent an important challenge for data-driven models learning, as detailed in next section. The proposed models for EHC identification of rock and soil cutting slopes were fed with more than fifty variables normally collected during

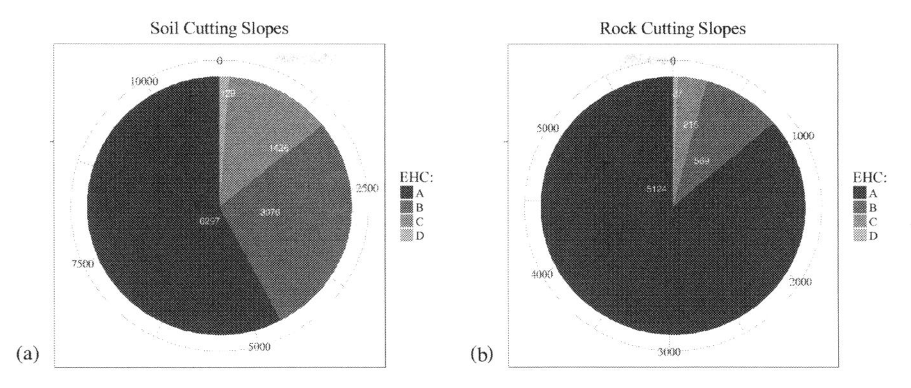

Figure 1. Rock and soil cutting slopes data distribution by EHC classes: (a) Soil cutting slopes; (b) Rock cutting slopes.

routine inspections and complemented with geometric, geographic and geological information. To be precise, 65 variables were used in the rock slopes study and 51 variables in soil cutting slopes. Since the number of analysed variables is high (65/51), just a few examples of the variables used to feed the models are enumerated: height, slope angle, presence of rock outcrops, animal activity, presence of boulders, ground cover, rock type, dangerous trees, number of root balls, rock strength, etc.

2.2 *Modelling*

In this work we applied two of the most well known DM algorithms, namely ANNs and SVMs to model EHC prediction of rock and soil cutting slopes. These two algorithms are not new, but are supported in a strong background. Indeed, they have been applied in the past with high success in different knowledge domains including in civil engineering (Chou, Yang, & Lin 2016, Gomes Correia, Cortez, Tinoco, & Marques 2013). There are also some examples of ANNs and SVMs applications in slope stability analysis (Wang, Xu, & Xu 2005, Cheng, Roy, & Chen 2012).

ANN are learning machines that were initially inspired in functioning of the human brain (Kenig, Ben-David, Omer, & Sadeh 2001). The information is processed using iteration among several neurons. This technique is capable of modelling complex non-linear mappings and is robust in exploration of data with noise. In this study we adopt the multilayer perceptron that contains only feed forward connections, with one hidden layer containing H processing units. Because the network's performance is sensitive to H (a trade-off between fitting accuracy and generalisation capability), we adopt a grid search of $\{0, 2, 4, 6, 8\}$ under an internal (i.e. applied over training data) three fold cross validation during the learning phase to find the best H value. Under this grid search, the H value that produced the lowest MAE (Mean Absolute Error) was selected, and then the ANN was retrained with all of the training data. The neural function of the hidden nodes was set to the popular logistic function $1/(1 + e^{-x})$.

SVMs were initially proposed for classification tasks (Cortes & Vapnik 1995). Then it became possible to apply SVMs to regression tasks after the introduction of the ϵ-insensitive loss function (Smola & Schölkopf 2004). The main purpose of the SVM is to transform input data into a high-dimensional feature space using non-linear mapping. The SVM then finds the best linear separating hyperplane, related to a set of support vector points, in the feature space. This transformation depends on a kernel function. In this work the popular Gaussian kernel was adopted. In this context, its performance is affected by three parameters: γ, the parameter of the kernel; C, a penalty parameter; and ϵ (only for regression), the width of an ϵ-insensitive zone (Safarzadegan Gilan, Bahrami Jovein, & Ramezanianpour 2012). The heuristics proposed by (Cherkassky & Ma 2004) were used to define the first two parameter values, $C = 3$ (for a standardised output) and $\epsilon = \hat{\sigma} / \sqrt{N}$, where $\hat{\sigma} = 1.5 / N \cdot \sum_{i=1}^{N} (y_i - \hat{y}_i)^2$ y_i is the measured value, $\hat{y}_i$ is the value predicted by a 3-nearest neighbour algorithm and N is the number of examples. A grid search of $2^{\{-1, -3, -7, -9\}}$ was adopted to optimise the kernel parameter γ, under the same internal three-fold cross-validation scheme adopted for ANN.

As a first attempt, EHC prediction of rock and soil cutting slopes was approached following a nominal classification strategy. Then, aiming to improve the models performance, the problem was also addressed following a regression strategy, adopting a regression scale where A = 1, B = 2, C = 4, D = 10, which was that leading to the best performance.

In addition, in order to minimize the effect of the imbalanced data (see Figure 1), Oversampling

(Ling & Li 1998) and SMOTE (Synthetic Minority Over-sampling Technique) (Chawla, Bowyer, Hall, & Kegelmeyer 2002) approaches were applied over the training data before fitting the models. When approaching imbalanced classification tasks, where there is at least one target class label with a smaller number of training samples when compared with other target class labels, the simple use of a soft computing training algorithm will lead to data-driven models with better prediction accuracies for the majority classes and worst classification accuracies for the minority classes. Thus, techniques that adjust the training data in order to balance the output class labels, such as Oversampling and SMOTE, are commonly used with imbalanced datasets. In particular, Oversampling is a simple technique that randomly adds samples (with repetition) of the minority classes to the training data, such that the final training set is balanced. SMOTE is a more sophisticated technique that creates "new data" by looking at nearest neighbours to establish a neighbourhood and then sampling from within that neighbourhood. It operates on the assumptions that the original data is similar because of proximity. More recently, Torgo, Branco, Ribeiro, & Pfahringer (2015) adapted the SMOTE method for regression tasks. We note that the different sampling approaches were applied only to training data, used to fit the data-driven models, and the test data (as provided by the 5-fold procedure) was kept without any change.

For models evaluation and comparison, we calculated three classification metrics: recall, precision and $F_{1\text{-score}}$ (Hastie, Tibshirani, & Friedman 2009). The recall measures the ratio of how many cases of a certain class were properly captured by the model. In other words, the recall of a certain class is given by *TruePositives*/(*TruePositives* + *FalseNegatives*). On the other hand, the precision measures the correctness of the model when it predicts a certain class. More specifically, the precision of a certain class is given by *TruePositives*/(*TruePositives* + *FalseNegatives*). The $F_{1\text{-score}}$ was also calculated, which represent a trade-off between the recall and precision of a class. The $F_{1\text{-score}}$ correspond to the harmonic mean of precision and recall, according to the following expression:

$$F_{1-score}s = 2 \cdot \frac{precision \cdot recall}{precision + recall} \quad (1)$$

For all three metrics, the higher the value, the better are the predictions, ranging from 0% to 100%.

The generalization capacity of the models was accessed through a 5-fold cross-validation approach under 20 runs (Hastie, Tibshirani, & Friedman 2009). This means that each modelling setup is trained 5 × 20 = 100 times. Also, the three prediction metrics are always computed on test unseen data (as provided by the 5-fold validation procedure).

All experiments were conducted under the R statistical environment (Team 2009). ANN and SVM algorithms were trained using the rminer package (Cortez 2010), which facilitates its implementation, as well as different validation approaches such as the cross-validation adopted in this work.

3 RESULTS AND DISCUSSION

Following are presented and discussed the achieved performance in EHC prediction of rock and soil cutting slopes through the application of soft computing techniques, comparing both soft computing algorithms (ANN and SVM) performance for each one of the two implemented strategies (nominal classification and regression) as well as for the three training sampling approaches explored: Normal (no resampling), OVERed (Oversampling) and SMOTEd (SMOTE). A particular emphases is also given to the comparison between soil and rock cutting slopes studies.

Tables 1 and 2 give an overview of all models performance in soil and rock cutting slopes respectively, based on recall, precision and $F_{1\text{-score}}$.

Concerning to soil cutting slopes study, a very promising performance is observed, namely according to a nominal classification strategy, which learned better EHC prediction than following a regression strategy as shown in Figure 2a. For example, soil cutting slopes of class "A" can be correctly identified, particularly by ANN model, with or without re-sampling. Also for classes "B" and "C" a promising performance is observed, with an $F_{1\text{-score}}$ around 55%, in particular by the ANN algorithm. Concerning the class "D", although an $F_{1\text{-score}}$ lower than 36% was achieved, the obtained value for recall metric around 57% shows a good performance for class "D" prediction according to ANN algorithm. Following a regression strategy, and by comparison with the nominal classification strategy, the main differences are related with the effect of the sampling approaches, which is not so relevant, particularly for the minority classes. Moreover, analysing Figure 3a that shows the relation between observed and predicted EHC values according to the best fit, we can see that the models performance is very promising. Indeed, according to a nominal classification strategy with SMOTE re-sampling, ANN algorithm is able to predict correctly around 57% of soil cutting slopes of class "D", which represents a very promising performance if we take into account that this is the minority class. For class "C", although the accuracy is

Table 1. Metrics in EHC prediction of rock slopes (best values in bold).

	Model	Approach	AUS	Recall				Precision				$F_{1\text{-score}}$			
				A	B	C	D	A	B	C	D	A	B	C	D
Classification	ANN	Normal	**0.46**	96.23	52.95	20.40	3.65	94.66	49.06	39.22	13.71	**95.44**	50.93	26.84	5.77
		SMOTEd	0.37	88.10	67.60	36.58	17.3	98.50	38.36	26.14	10.89	93.01	48.95	30.49	**13.37**
		OVERed	0.44	90.21	67.96	39.58	**12.84**	98.01	41.27	33.47	12.70	93.95	**51.35**	36.27	12.77
	SVM	Normal	0.33	97.39	39.79	6.44	0.41	91.63	48.57	42.95	18.75	94.42	43.74	11.20	0.80
		SMOTEd	0.29	85.53	82.64	2.07	1.49	97.24	33.08	34.36	17.19	91.01	47.25	3.90	2.74
		OVERed	0.13	**99.78**	7.14	0.00	0.00	86.95	**62.83**	NA	0.00	92.92	12.82	NA	NA
Regression	ANN	Normal	0.43	93.7	48.3	41.77	3.38	95.01	41.38	40.19	30.49	94.35	44.57	**40.96**	6.09
		SMOTEd	0.35	85.97	68.37	**45.84**	4.32	98.07	33.85	32.95	**35.56**	91.62	45.28	38.34	7.70
	SVM	Normal	0.34	96.32	49.83	0.30	0.00	92.56	46.33	**54.17**	NA	94.40	48.02	0.60	NA
		SMOTEd	0.16	77.13	**93.15**	11.12	0.00	**99.40**	27.61	48.33	NA	86.86	42.59	18.08	NA

Table 2. Metrics in EHC prediction of soil cutting slopes (best values in bold).

	Model	Approach	AUS	Recall				Precision				$F_{1\text{-score}}$			
				A	B	C	D	A	B	C	D	A	B	C	D
Classification	ANN	Normal	–0.05	90.36	64.01	45.61	14.53	87.23	60.36	59.21	42.57	88.77	62.13	51.53	21.67
		SMOTEd	–0.08	80.87	66.59	46.07	**56.78**	91.68	54.49	51.48	21.63	85.94	59.94	48.62	31.33
		OVERed	–0.04	82.05	58.75	**63.77**	38.41	91.13	55.02	49.77	33.71	86.35	56.82	**55.91**	**35.91**
	SVM	Normal	–0.12	90.33	66.82	34.11	2.25	86.85	**58.34**	57.71	22.31	**88.56**	62.29	42.88	4.09
		SMOTEd	–0.27	73.65	79.27	24.96	24.88	91.50	47.90	53.53	30.81	81.61	59.72	34.05	27.53
		OVERed	–1.35	**94.79**	24.74	1.54	1.32	63.25	52.35	62.98	62.96	75.87	33.60	3.01	2.59
Regression	ANN	Normal	–0.05	87.41	64.47	47.94	25.62	87.74	57.88	59.2	44.87	87.57	61.00	52.98	32.62
		SMOTEd	–0.03	85.34	68.68	48.53	23.64	89.32	57.00	60.23	54.08	87.28	62.30	53.75	32.90
	SVM	Normal	–0.16	83.66	82.02	15.7	0.00	91.07	52.89	60.00	NA	87.21	**64.31**	24.89	NA
		SMOTEd	–0.27	66.30	**85.38**	33.77	0.62	**93.43**	45.81	**66.37**	**66.67**	77.56	59.63	44.76	1.23

lower (around 40%), when not predicted as "C" they are classified as belonging to the closest class, that is, "B" or "D". This type of misclassification is also observed for classes "A", "B" and "D", which can be interpreted as an advantage. Concerning to classes "A" and "B", the ANN model was able to identify it very accurately.

Relating to rock cutting slopes study, the achieved performance is somewhat lower, either following a nominal classification or regression strategies. Although a very high performance is observed in class "A" identification ($F_{1\text{-score}}$ higher than 95%), for class "C" and particularly for class "D", all models evidence difficulties in predicting these classes correctly. In fact, and using $F_{1\text{-score}}$ as reference, the best performance in identification of slopes of class "D" is lower than 14% (see Figure 2b) which was achieved by the ANN algorithm following a nominal classification strategy with SMOTE re-sampling. From Figure 3b analysis, which plots the relation between observed and predicted EHC values based on ANN algorithm following a nominal classification with oversampling (best fit), it is clear the model difficulties in correctly predicting class "C" and particularly class "D", for which the expected probability of failure is higher. As shown, only around 12% of rock cutting slopes classified as "D" are correctly identified, which represents a low performance, namely when compared with soil cutting slopes study. Overall, these results show that the methodology applied for EHC prediction of rock cutting slopes needs future development in order to overcome this gap.

Comparing the achieved results of soil and rock cutting slopes studies, the proposed models for soil cutting slopes are more effective, namely in the identification of classes "C" and "D" for which the probability of failure is higher (see Figure 2). A possible explanation for the lower performance, namely in classes "C" and "D" identification of rock cutting slopes could be related with the EHC classes being assumed as representative of the real stability condition of each slope. Indeed, analysing the number of slope failures by EHC class for rock slopes there are some indications that the classification attributed to each rock slope could lack some accuracy as reported in the work of Power, Mian, Spink, Abbott, & Edwards (2016), that used the same source of information.

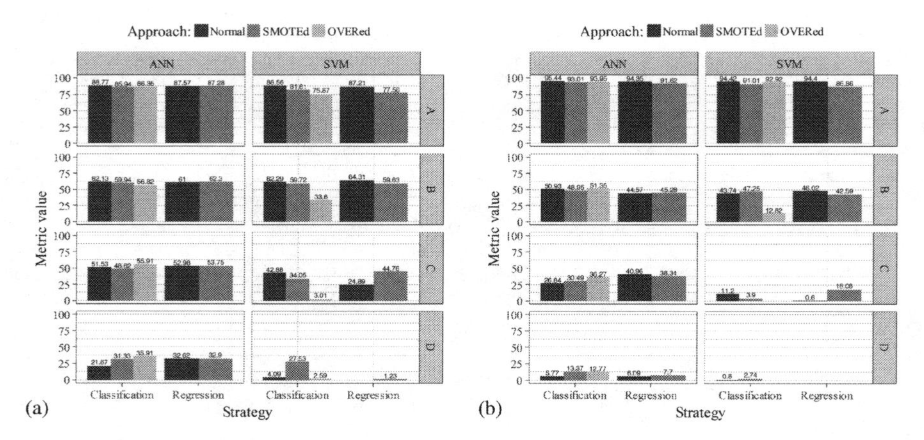

Figure 2. Nominal classification and Regression strategies performance comparison based on F1-score: (a) Soil cutting slopes study; (b) Rock cutting slopes study.

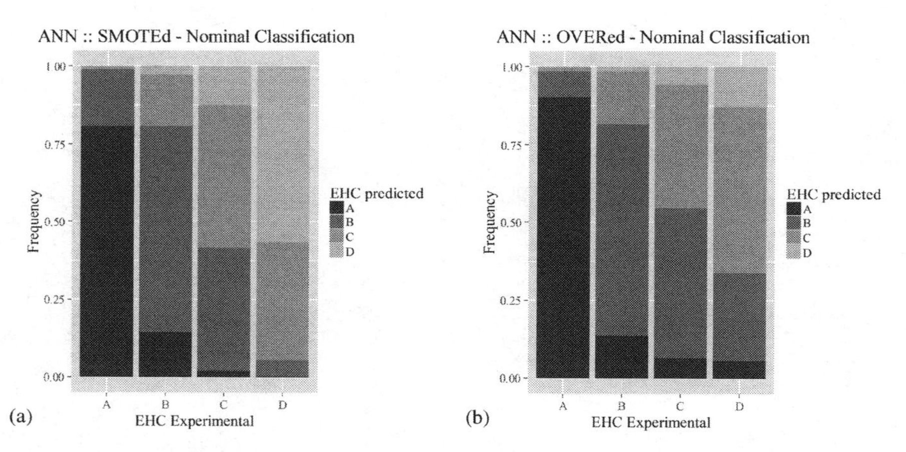

Figure 3. ANN models performance comparison according to a nominal classification strategy in EHC prediction of: (a) Soil cutting slopes following an SMOTEd approach; (b) Rock cutting following a OVERed approach.

It would be expected that most of the failures would occur in slopes of classes "C" and mainly "D". However, for rock slopes such behaviour is not observed as reported on Power, Mian, Spink, Abbott, & Edwards (2016). In fact, the number of failures for each EHC class is almost constant from "A" to "D", particularly when compared with soil cuttings. For example, the number of failures observed in rock cutting slopes of class "C" is only twice higher when compared to class "A". This observation shows that the defined classes for rock slopes have a poor correlation with actual failures.

These results show that a deeper data analysis is required, particularly in the study of rock cutting slopes. For example, the number of variables taken as model attributes might be too high and may be influencing the generalization performance of the models. Aiming to check if a better generalization could be achieved using the most relevant inputs, we performed additional experimentation using a fast feature selection method that is based on a sensitive analysis (Cortez & Embrechts 2013), which allows to measure the relative importance of each input of a classification or regression method. Taken as reference the two models that achieved the overall best performance in EHC prediction of soil and rock cutting slopes (see Figure 3), we applied the sensitivity analysis to measure the relevance of each input variable in EHC prediction. Figure 4 shows the relative importance of the 20 most relevant variables in both soil and rock cutting slopes studies. Following these results, all models were re-trained (including both strategies and the three re-sampling approaches) considering only the 12 and 16 most relevant variable in soil and rock cutting slopes studies respectively. Using

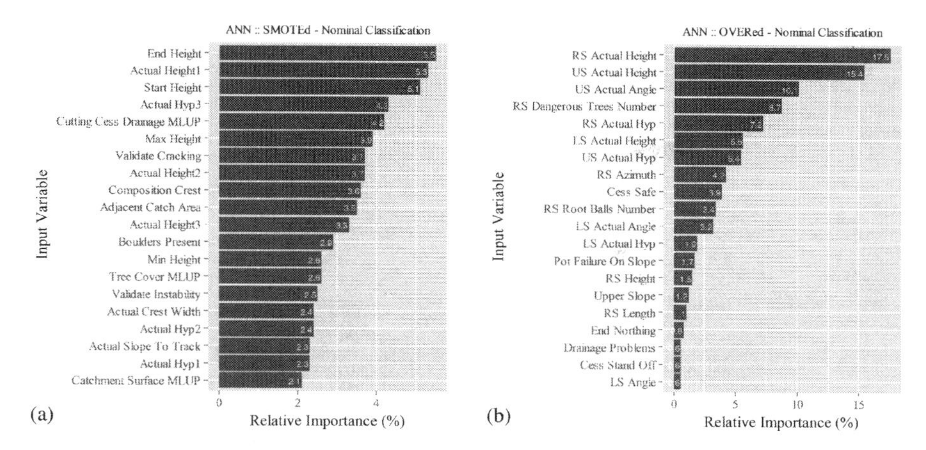

Figure 4. Relative importance bar plot of the 20 most relevant variables according to ANN models based on a nominal classification strategy in EHC prediction of: (a) Soil cutting following a SMOTEd approach; (b) Rock slopes following an OVERed approach.

Table 3. Difference between $F_{1\text{-score}}$ values of the full input model with a feature selection model that included the most relevant inputs according to a sensitivity analysis procedure.

			Soil cutting slopes				Rock cutting slopes			
	Model	Approach	A	B	C	D	A	B	C	D
Classification	ANN	Normal	7.11	16.00	18.81	15.09	1.53	13.9	19.27	3.72
		SMOTEd	9.15	12.25	23.60	19.04	2.38	7.51	3.15	5.26
		OVERed	8.82	20.31	16.31	21.00	3.28	12.96	10.31	6.30
	SVM	Normal	7.44	15.26	28.56	3.49	0.90	13.34	7.31	NA
		SMOTEd	6.23	13.78	–1.12	13.17	90.87	29.78	NA	NA
		OVERed	–0.49	–17.01	–8.51	–2.21	0.91	–25.02	NA	NA
Regression	ANN	Normal	10.00	9.40	17.25	25.82	1.23	–2.20	14.46	5.81
		SMOTEd	10.38	10.90	17.44	28.29	1.24	1.47	9.66	NA
	SVM	Normal	6.49	14.74	1.69	NA	0.74	3.65	0.18	NA
		SMOTEd	0.72	11.79	17.27	NA	–1.70	–0.56	2.34	NA

$F_{1\text{-score}}$ as comparison metric, Table 3 shows the difference between the full models (with 51/65 inputs for soil and rock slopes respectively) and feature selection ones (with 12/16 most relevant inputs). The results from Table 3 show that the feature selection tends to present a lower performance, with lower $F_{1\text{-score}}$ values. Thus, in the light of the achieved results, and as a future works, we intend to apply a more sophisticated feature selection method aiming to improve models performance. For instance, by using a multi-objective evolutionary computation method that simultaneously maximizes prediction performance and minimizes the number of inputs used.

As a final observation, and considering the overall performance of all models, we would like to underline the potential of soft computing algorithms, namely the ANNs, in EHC prediction of soil and rock cutting slopes.

ACKNOWLEDGEMENTS

This work was supported by FCT - "Fundação para a Ciência e a Tecnologia", within ISISE, project UID/ECI/04029/2013 as well Project Scope: UID/CEC/00319/2013 and through the post-doctoral Grant fellowship with reference SFRH/BPD/94792/2013. This work was also partly financed by FEDER funds through the Competitivity Factors Operational Programme—COMPETE and by national funds through FCT within the scope of the project POCI-01-0145-FEDER-007633. This work has been also supported by COMPETE: POCI-01–0145-FEDER-007043. A special thanks goes to Network Rail that kindly made available the data (basic earthworks examination data and the Earthworks Hazard Condition scores) used in this work.

REFERENCES

Ahangar-Asr, A., A. Faramarzi, & A.A. Javadi (2010). A new approach for prediction of the stability of soil and rock slopes. *Engineering Computations 27*(7), 878–893.

Chawla, N.V., K.W. Bowyer, L.O. Hall, & W.P. Kegelmeyer (2002). Smote: synthetic minority over-sampling technique. *Journal of artificial intelligence research 32*, 321–357.

Cheng, M.-Y. & N.-D. Hoang (2014). Slope collapse prediction using bayesian framework with k-nearest neighbor density estimation: Case study in taiwan. *Journal of Computing in Civil Engineering 30*(1), 04014116.

Cheng, M.-Y., A.F. Roy, & K.-L. Chen (2012). Evolutionary risk preference inference model using fuzzy support vector machine for road slope collapse prediction. *Expert Systems with Applications 39*(2), 1737–1746.

Cherkassky, V. & Y. Ma (2004). Practical selection of svm parameters and noise estimation for svm regression. *Neural Networks 17*(1), 113–126.

Chou, J.-S., K.-H. Yang, & J.-Y. Lin (2016). Peak shear strength of discrete fiber-reinforced soils computed by machine learning and metaensemble methods. *Journal of Computing in Civil Engineering*, 04016036.

Cortes, C. & V. Vapnik (1995). Support vector networks. *Machine Learning 20*(3), 273–297.

Cortez, P. (2010). Data mining with neural networks and support vector machines using the r/rminer tool. In P. Perner (Ed.), *Advances in Data Mining: Applications and Theoretical Aspects, 10th Industrial Conference on Data Mining*, Berlin, Germany, pp. 572–583. LNAI 6171, Springer.

Cortez, P. & M. Embrechts (2013). Using sensitivity analysis and visualization techniques to open black box data mining models. *Information Sciences 225*, 1–17.

Domingos, P. (2012). A few useful things to know about machine learning. *Communications of the ACM 55*(10), 78–87.

Gomes Correia, A., P. Cortez, J. Tinoco, & R. Marques (2013, Mai). Artificial intelligence applications in transportation geotechnics. *Geotechnical and Geological Engineering 31*(3), 861–879. doi:10.1007/s10706–012–9585–3.

Hastie, T., R. Tibshirani, & J. Friedman (2009). *The Elements of Statistical Learning: Data Mining, Inference, and Prediction* (Second Edition ed.). Springer-Verlag New York.

Husein Malkawi, A.I., W.F. Hassan, & F.A. Abdulla (2000). Uncertainty and reliability analysis applied to slope stability. *Structural Safety 22*(2), 161–187.

Kenig, S., A. Ben-David, M. Omer, & A. Sadeh (2001). Control of properties in injection molding by neural networks. *Engineering Applications of Artificial Intelligence 14*(6), 819–823.

Ling, C.X. & C. Li (1998). Data mining for direct marketing: Problems and solutions. In *KDD*, Volume 98, pp. 73–79.

Pinheiro, M., S. Sanches, T. Miranda, A. Neves, J. Tinoco, A. Ferreira, & A. Gomes Correia (2015). A new empirical system for rock slope stability analysis in exploitation stage. *International Journal of Rock Mechanics and Mining Sciences 76*, 182–191. http://dx.doi.org/10.1016/j.ijrmms.2015.03.015.

Pourkhosravani, A. & B. Kalantari (2011). A review of current methods for slope stability evaluation. *Electronic Journal of Geotechnical Engineering 16*.

Power, C., J. Mian, T. Spink, S. Abbott, & M. Edwards (2016). Development of an evidence-based geotechnical asset management policy for network rail, great britain. *Procedia Engineering 143*, 726–733.

Safarzadegan Gilan, S., H. Bahrami Jovein, & A. Ramezanianpour (2012). Hybrid support vector regression–particle swarm optimization for prediction of compressive strength and rcpt of concretes containing metakaolin. *Construction and Building Materials 34*, 321–329.

Smola, A. & B. Sch¨olkopf (2004). A tutorial on support vector regression. *Statistics and Computing 14*(3), 199–222.

Suchomel, R. et al. (2010). Comparison of different probabilistic methods for predicting stability of a slope in spatially variable c-φ soil. *Computers and Geotechnics 37*(1), 132–140.

Team, R. (2009). R: A language and environment for statistical computing. R Foundation for Statistical Computing, Viena, Austria.Web site: http://www.r-project.org/.

Tinoco, J., A. Gomes Correia, & P. Cortez (2014, January). Support vector machines applied to uniaxial compressive strength prediction of jet grouting columns. *Computers and Geotechnics 55*, 132–140. http://dx.doi.org/10.1016/j.compgeo.2013.08.010.

Tinoco, J., A. Gomes Correia, & P. Cortez (2016, June). Jet grouting column diameter prediction based on a datadriven approach. *European Journal of Environmental and Civil Engineering 0*(0), 1–22.

Torgo, L., P. Branco, R. Ribeiro, & B. Pfahringer (2015). Resampling strategies for regression. *Expert Systems 32*(3), 465–476.

Wang, H., W. Xu, & R. Xu (2005). Slope stability evaluation using back propagation neural networks. *Engineering Geology 80*(3), 302–315.

Yao, X., L. Tham, & F. Dai (2008). Landslide susceptibility mapping based on support vector machine: a case study on natural slopes of hong kong, china. *Geomorphology 101*(4), 572–582.

Numerical Methods in Geotechnical Engineering IX – Cardoso et al. (Eds)
 ISBN 978-1-138-33198-3

Estimating spatial correlations from CPT data using neural networks and random fields

J.D. Nuttall
Department of Numerical Simulation Software, Deltares, Delft, The Netherlands

ABSTRACT: Vertical and horizontal scales of fluctuation are measures of spatial correlation and variability in soils, and as such are extensively used in the modelling of soils in reliability methods, such as the Random Finite Element Method (RFEM). These parameters are conventionally estimated from site surveys, commonly using Cone Penetration Test (CPT) data. By fitting theoretical correlation functions to the site data, the horizontal and vertical scales of fluctuation can be estimated. Presented is a new approach that trains a Convolution Neural Network (CNN) with pseudo CPT data taken from generated Random Field (RF) data with known scales of fluctuation. Once trained the network can predict these measures of spatial variability from real CPT data. This paper presents the results of a study training a network to predict vertical scales of fluctuation.

1 INTRODUCTION

Site investigation and surveying is an expensive and time consuming process. Methodologies which improve accuracy and thus reduce the intensity of site investigation could potentially reduce associated costs.

Typically in stochastic finite element analysis the spatial variation of soil is modelled, by incorporating random fields into the model using spatial statistics recovered from CPT data, including θ_v and θ_h, the spatial correlation coefficients in the vertical and horizontal directions respectively. These statistics are typically recovered by fitting an autocorrelation function to the CPT data (Vanmarcke 1977), or by using the Rice method, derived from signal theory methods (Rice 1944, Pieczynska-Kozlowska 2015, Phoon & Kulhawy 1999, Vanmarcke 1977).

This paper will introduce a new methodology for the prediction of θ_v which utilizes the Convolution Neural Network (CNN) techniques, trained using simulated CPT data from random fields. This methodology will then be compared with the typical approaches mentioned in literature. (Pieczynska-Kozlowska 2015, Phoon & Kulhawy 1999)

2 TRADITIONAL METHODOLOGIES

The two traditional approaches will be described and used as comparison to the methodology developed.

2.1 *Autocorrelation method*

The fitting of a suitable autocorrelation function (e.g. Gaussian or Markov) was suggested by (Vanmarcke 1977). This methodology fits the correlation, ρ, and lag, τ, within the CPT data (typically cone tip resistance) with a suitable correlation function. Table 1 shows the commonly used correlation functions to model soil variation.

The correlation functions given in Table 1 are also typically used in the generation of random fields when modelling spatial variability in soils. In this paper, the generation of random fields, is limited to those generated by using a Markov correlation function.

Firstly any depth trend is removed from the data and then the normalized correlation function, $\hat{\rho}$, is estimated from the data, using (de Gast, Vardon, & Hicks 2017):

$$\hat{\rho} = \frac{\hat{\gamma}(\tau)}{\hat{\gamma}(0)} \tag{1}$$

Table 1. Typical correlation functions for a single variable (Pieczynska-Kozlowska 2015).

Correlation model	Function
Gaussian	$\rho(\tau) = e^{-\pi(\frac{\lvert\tau\rvert}{\theta})^2}$
Markov	$\rho(\tau) = e^{\frac{-2\lvert\tau\rvert}{\theta}}$

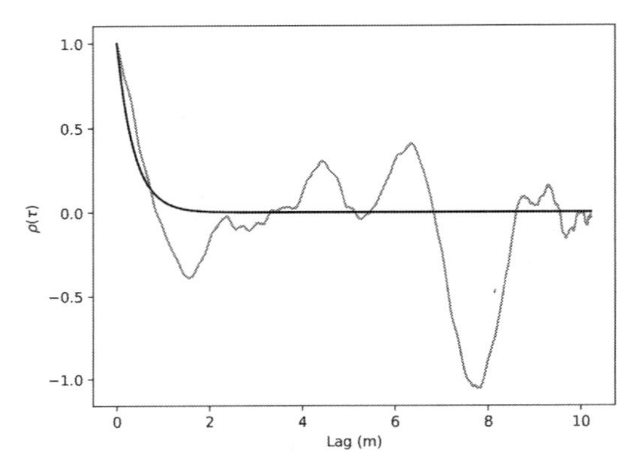

Figure 1. An illustration of the prediction of spatial variation by the Autocorrelation fitting method. (Lines: Gray - CPT data, Black - Fitted Function).

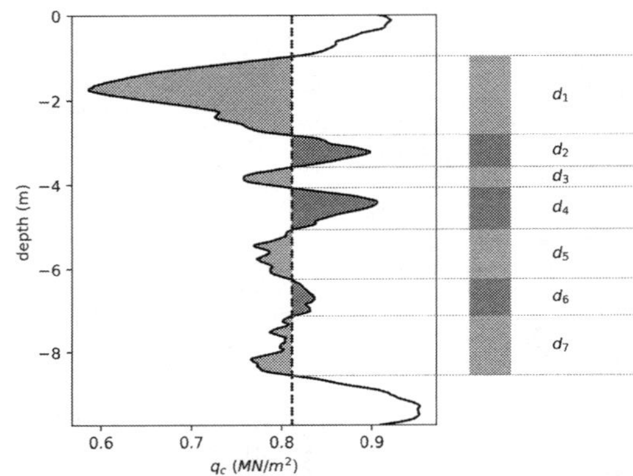

Figure 2. An illustration of the prediction of spatial variation using Rice's method, where q_c is the corrected tip resistance, as adjusted from the CPT data.

where,

$$\hat{\gamma}(\tau)=\frac{1}{(t-1)}\sum_{j=1}^{t}(x_j-\hat{\mu}_{j+\Delta\tau})(x_j-\hat{\mu}_{j+\Delta\tau}) \tag{2}$$

with

- $\hat{\gamma}$ is the covariance,
- x_j is the value in the dataset,
- $\hat{\mu}_j$ is the mean of the dataset,
- τ is the lag distance
- j is the observation number
- k is the number of observations
- t is the number of pairs of data at lag $\Delta\tau$

The correlation function generated from the CPT data is then fitted to the required correlation function in Table 1, thus predicting θ, as illustrated in Figure 1. i.e. the value of θ providing the best fit of the CPT data.

In this study the fitting was undertaken within a Python based code, using the fitting functionality of the NumPy library.

2.2 *Rice method*

The Rice Method (Rice 1944), is based on Signal Theory, and has been adapted for the interpretation of CPT data with respect to θ_v (Pieczynska-Kozlowska 2015), it is similar to the method discussed in Phoon & Kulhawy 1999, Vanmarcke 1977. The method is simply to de-trend the CPT data and then take the average distance between mean crossing of the de-trended data. This is best explained in Figure 2.

In Figure 2, the average crossing distance, $\bar{d}$, is taken as,

$$\bar{d}=\frac{1}{n}\sum_{i=1}^{n}d_i \tag{3}$$

It had been previously suggested by Vanmarcke (1977), that this average value, $\bar{d}$, should be adjusted to give a better approximation

$$\theta_v \approx 0.8\times\bar{d} \tag{4}$$

However, the adaptation of the method by Pieczynska-Kozlowska (2015), proposes that this relation is based on the correlation model selected, as shown in Table 2.

3 NEURAL NETWORK METHODOLOGY

The new methodology trains a convolution neural network (CNN) using simulated CPT data (cone tip resistance, q_t), generated from 1D random fields generated by the Local Average Subdivision (LAS) method (Vanmarcke 1977), as previously stated using a Markov correlation function (Table 1). Furthermore, as the 1D random fields generated can be considered trendless, the cone tip resistance requires no correction.

It can be accurately described that this methodology is an inverse parameter determination of the random field generation parameters.

3.1 *Convolution Neural Network*

Although beyond the scope of this paper, a brief explanation of a Convolution Neural Network is offered.

A Neural Network is a collection of neurons, which mimic those found in the brain. This collection of neurons, receives a signal, or number which causes the neuron to fire, or not, passing on it a further signal. The network is trained with known

Table 2. Rice equations for scale of fluctuation (Pieczynska-Kozlowska 2015).

Correlation model	Scale of Flucuation θ
Gaussian	$\approx \bar{d}\sqrt{\frac{2}{\pi}} = 0.798\bar{d}$
Markov	$\approx \frac{2\bar{d}}{\pi} = 0.637\bar{d}$

input and output data, adjusting the weighting of the connections between the neurons.

A convolution neural network (CNN) is a network designed predominantly to recognise the features within an image. It is this ability that is utilized in the prediction of spatial variability properties, as this variation can be considered in a visual manner. Figure 3 shows the basic steps incorporated into the CNN developed and trained for this study.

The CNN is adapted for this study firstly by adapting the input from an image to a column of sampled data from the CPT, i.e. replacing the standard image in Figure 3 ("5") with an image of a 1D random field. This gray scale image visualizes the tip resistance where the random field cell value corresponds to the depth of gray in the image. Secondly it is adapted by including a final layer beyond the categorization layer which accumulates the results into a single value.

3.2 *CPT data simulation*

The CPT data was simulated from random fields generated using known statistical properties, thus both the input, i.e. the CPT values, and the output data, in this case the vertical scale of fluctuation, θ_v, was known to enable the training of the neural network. Furthermore the data was therefore trend-less and other measures such as the Mean, μ_{q_t}, and standard deviation, σ_{q_t}, were controllable.

In this study the neural network was trained using 200 epochs of 40000 realisations (8,000,000 different realisations). An epoch is typically the number of times over a training set the neural network will iterate in the training phase, or the number of training sets available.

The CPTs were modelled as 1D LAS random fields with 1024 cells, each cell measuring 0.01 m, thus each CPT had a depth of 10.24 m. Each field was generated using a mean tip resistance of $\mu_{q_t} = 4.1\ \text{MN/m}^2$ and $\sigma_{q_t} = 0.38\ \text{MN/m}^2$, however these statistics are irrelevant, as the simulated CPT values are normalized, thus removing both the mean and standard deviation from the data. The vertical scale of fluctuation, θ_v, is varied between 0.1 and 3.0 m. These statistics are based on the statistical properties of sand and clay, provided by the literature review undertaken in Phoon (1999).

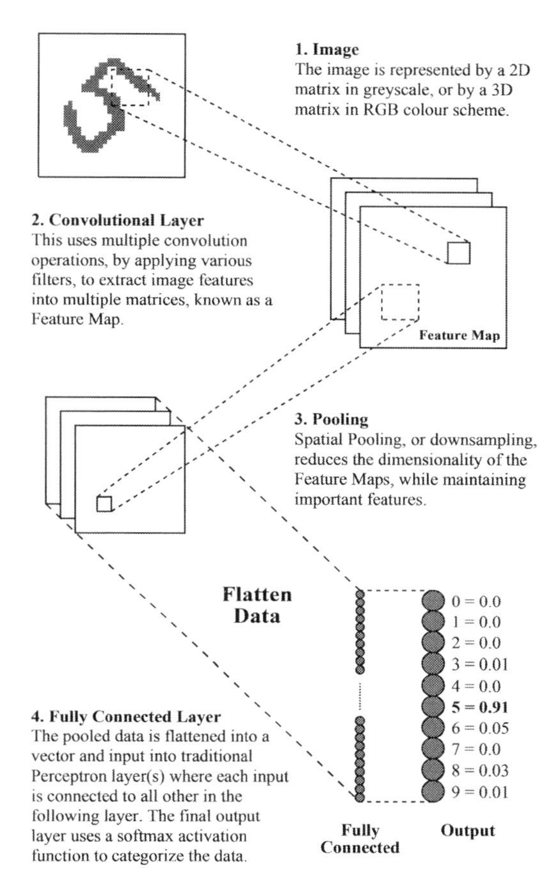

Figure 3. An illustration of an example Convolution Neural Network (CNN) used to predict image categories.

The CNN was implemented using a Python script using the Keras libraries used to generate Neural Networks.

4 RESULTS

Following the training of the data, the described methods were used to predict θ_v in the range 0.1–3.0 m, generated using the same 1D Random Field generator described in Section 3.2, however previously unseen by the trained CNN.

Figure 4 shows a selection of the results based upon the generation of 100 realisation predictions.

The results show that the range of the predictions is generally relatively more accurate and narrower for the CNN method than the other methods. While the autocorrelation fitting method, provides

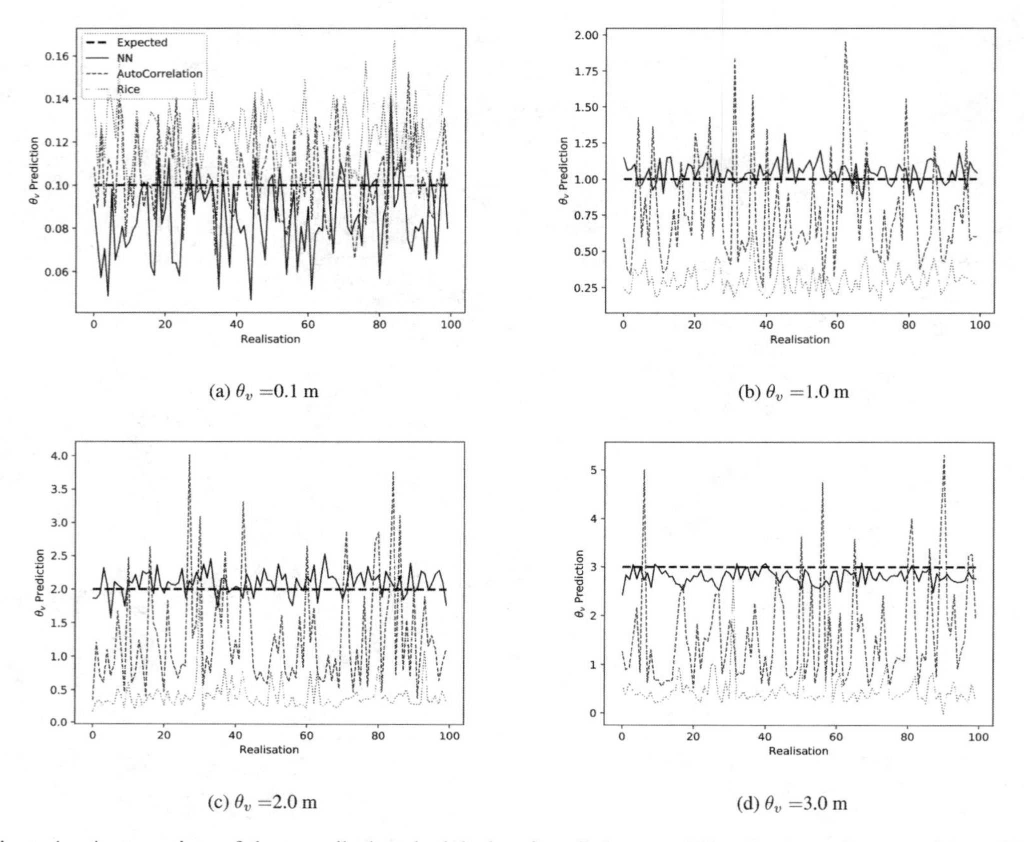

(a) θ_v =0.1 m (b) θ_v =1.0 m

(c) θ_v =2.0 m (d) θ_v =3.0 m

Figure 4. A comparison of the prescribed methodologies of prediction over 100 realisations of generated faux CPT data.

a more accurate prediction at very low θ_v values, the CNN prediction method is more accurate and consistent where $\theta_v > 0.2$ m.

Figure 5 shows the prediction ranges of the prescribed methods over the range of θ_v property ranges (0.1 m $\leq \theta_v \leq$ 3.0 m).

It is clear that the CNN methodology provides a more precise and consistent prediction of θ_v in the properties range. These prediction capabilities indicate that the CNN method would require less data when predicting the properties of a full CPT site investigation.

The accuracy of the two other methodologies are dependent on having significantly more CPT data, i.e. from other CPTs in a full site investigation to predict the spatial variation.

The Rice method, initially showed poor accuracy, however after smoothing the data by using moving averages, some improvement was found. However the Author had difficulty in applying the method consistently, resulting in significant error within the prediction.

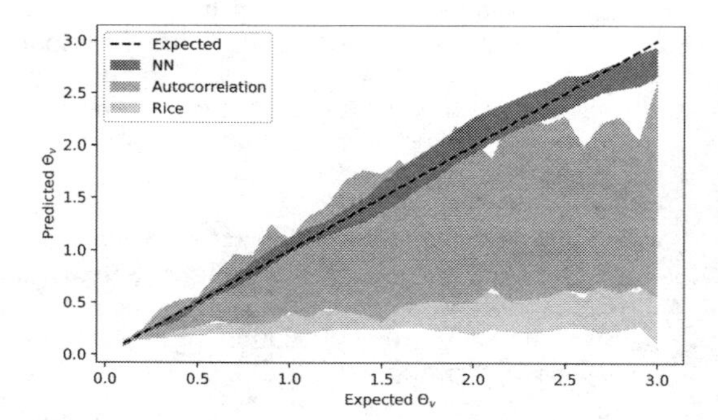

Figure 5. Comparison of the prediction ranges from the prescribed prediction methodologies.

The autocorrelation fitting method, appears to be more consistent at lower values of θ, where the amount of data at these lower lag τ values is significantly greater.

The relative accuracy and range of the new approach at higher values of θ_v are likely due to the values at either end of the spectrum acting as bounds, as the network was not trained with values above or below, therefore as the θ_v values approach 3.0 m the predictions are bounded below. This may be remedied by training the networks over a greater range than operationally required, e.g. $0.01 \text{ m} \le \theta_v \le 4.0 \text{ m}$.

Furthermore the methodology was tested with different CPT sampling rates and depths from those over which it was trained. It must be noted that the CNN method will only work with CPT data of a consistent size, i.e. 1024 cells, thus the input and prediction results must be adjusted accordingly. The results were found to be equally as accurate, after adjusting the resulting prediction by a factor equivalent to the ratio of the old to new sampling rate or distance.

5 CONCLUSION

The results have shown that the new method of prediction in most circumstances, i.e. $\theta_v \ge 0.2$ m is superior to those previously used, in similar situations. However it should be noted that this comparison has been conducted with simulated data and not with real CPT data, as such it can not be concluded that the new methodology is superior without further in-depth testing with real data, which would include noise. However the initial indications from this proof of concept is that this methodology has potential and should be investigated further.

The Authors opinion is that the methodology is superior over more statistical based methods due to the fact that the measurement of the spatial variable is based upon the identification of features within the CPT profiles, similar to the features that are identified by CNNs categorizing images such as letters, numbers, or objects within pictures. In this way the CNN methodology relies more on the quality of data to identify the feature, rather than the quantity of data, which is shown by the narrow range of predictions compared with those from more traditional methods. However the Author recognises that more work should be done in applying the alternative methods, before solid conclusions can be asserted.

Within this paper there is an assumption that a 1D LAS lognormal random field using a Markov correlation function, is a suitable model for the subsurface soil, although this may be the case, this new methodology should allow for the use of many differing models, and can be trained to be predict the necessary values from them.

6 FUTURE WORK

Firstly the new approach should be investigated with real CPT data, therefore testing the new methodology with data that is noisy. Unfortunately the issue with using real CPT data is that the spatial correlation values are unknown and can only be compared with measurements from other methodologies. An intermediate analysis could be to incorporate noise into the simulated CPT data, to see whether the CNN approach still recovers the necessary statistics.

The future of this work is to expand the investigation into 2D and 3D, investigating the use of the CNN approach in identifying properties in full site surveys, where each CPT will have varying spatial statistics based upon where the CPT is taken.

Further to this the properties identified can be extended in 2D and 3D to include the horizontal scale of fluctuation θ_h, a property which is difficult to obtain with limited data, rather than thousands of CPT sampling points which are used to estimate the vertical scale, the horizontal scale is restricted to the number of CPTs taken.

REFERENCES

de Gast, T., P.J. Vardon, & M.A. Hicks (2017). Estimating spatial correlations under man-made structures on soft soils. In *Geo-Risk 2017*.

Phoon, K.-K. & F.H. Kulhawy (1999). Characterization of geotechnical variability. *Canadian Geotechnical Journal 36*(4), 612–624.

Pieczynska-Kozlowska, J. (2015). Comparison between two methods for estimating the vertical scale of fluctuation for modeling random geotechnical problems. *Studia Geotechnica et Mechanica 37*, 95–103.

Rice, S.O. (1944). Mathematical analysis of random noise. *The Bell System Technical Journal 23*(3), 282–332.

Vanmarcke, E. (1977). Probabilistic modeling of soil profiles. *ASCE Journal of the Geotechnical Engineering 103*, 1227.

Numerical Methods in Geotechnical Engineering IX – Cardoso et al. (Eds)
© 2018 Taylor & Francis Group, London, ISBN 978-1-138-33198-3

Human-driven machine-automation of engineering research

M.D.L. Millen, A. Viana Da Fonseca & X. Romão
Department of Civil Engineering, Faculty of Engineering of University of Porto, Portugal

ABSTRACT: This paper presents a framework for efficiently producing engineering research in a global collaborative effort in a rigorous scientific manner. The proposed framework reduces subjective analysis, automates several mundane research tasks and provides a suitable formal structure for efficient information sharing and collaboration. The implementation of the framework involves multiple research groups setting up different web-servers that can perform the steps of the scientific method and automatically determine the quality and value of new research by directly communicating between servers via public and private Application Programming Interfaces (APIs) using a set of object-oriented protocols. The automation of many mundane research tasks (e.g. data manipulation), would allow researchers to focus more on the novel aspects of their research efforts. The increased clarity around the quality and value of research would allow the research efforts of individuals and available research funding to be better disbursed. The paper discusses the major aspects of the scientific method, object-orientated programming, the application of the proposed research framework for experimental/analytical/numerical engineering research, some of the potential benefits and drawbacks, as well as the current state of implementation.

1 INTRODUCTION

A major issue with the current research process in civil engineering is that it requires an enormous amount of effort to stay up-to-date with the latest research in a given field. From this issue stem several others including: redundant research (because the author and reviewer were not up-to-date), poorly founded research (the author did not understand the necessary existing literature that supported the new findings), weak research (the new hypotheses provides weaker conclusions than previous research), to name but a few issues. These issues make it difficult to perform further research based on the current global understanding, therefore researchers typically build off only their own research and the research of a select few. There is definitely more research being done than in the past, but we cannot 'stand on the shoulders of giants' as efficiently as Isaac Newton.

Historically this has happened before. Mathematical research in the 19th century still relied on human intuition, and in some cases, had inconsistencies and lacked formal proofs that underpinned major branches of mathematics. There was a movement, "Hilbert's program", to rigorously rebuild mathematics from its foundations using a set of axioms to re-prove and formalise old theorems (Hilbert 1902). The rebuilding and formalisation of mathematics allowed mathematicians to more easily collaborate and to develop more advanced and consistent theories. Another case is the work of René Descartes (1596–1650), who wanted to remove all doubt from science and philosophy by completely rebuilding it from nothing, the first truth: "I think, therefore I am". Currently, many fields of science are suffering from poorly organised global research and non-reproducible results, which has prompted new initiatives such as the 'Reproducibility Project' in Psychology (Poldrack and Poline 2015). While in other fields of research they have fully formalised the research process and use robots and machine learning to automate scientific discovery for some narrow research problems such as drug development (Sparkes et al. 2010). Engineering research does not need to be as extreme as Descartes or move completely to robotics, but a greater focus on consistency is required now to allow research to happen efficiently and effectively at a global level.

In addition to the poor consideration of consistency, many research fields have changed considerably in the last century, moving from the study of individual mechanisms and phenomena to system-based effects (Foster 2006). The study of complex systems typically requires large simulations with large amounts of data and some knowledge of computational techniques. Djorgovski (2005) argues that applied computer science in research fulfils the role of applied mathematics in the 17th to 19th centuries, providing a formal framework for exploring science. The increased complexity of system-based effects research makes it harder to quickly repeat an experiment/simulation or manually check the underlying calculations. To improve the repeatability of new research, one solution would be to present new research in a machine-readable format to allow everyone to re-run the exact same simulation or have a computer cross-check experimental results against other existing results.

Fortunately in the field of civil engineering research, there are several initiatives that are greatly improving the situation by open-sourcing their findings and providing machine-readable output from their research (e.g. software or databases). OpenSees (McKenna 2011), an open-source software for structural and geotechnical engineering, allows research to be embedded into OpenSees subroutines, which enables others to easily validate those findings and use them in their own research. The use of exactly the same subroutines across various research projects also provides some level of consistency in that the underlying assumptions of the numerical simulations are the same.

The open-source object-orientated programming language Python, has also grown a large user base of engineering researchers, where standard libraries now exist for site-response analysis, ground motion analysis and unit analysis, among other libraries. This effort has recently been supported with the allocation of Digital Object Identifiers (DOIs) for software packages through Zenodo, a CERN and OpenAire initiative. The DOIs support having multiple versions of a software which "enables users to update the records files after they have been made public and researchers to easily cite either specific versions of a record or to cite, via a top-level DOI, all the versions of a record" - Zenodo (2018). In support of the direct implementation of scientific research into software, frameworks for the evaluation of the quality and predictive capabilities of scientific software have been developed for particular situations (e.g. Oberkampf et al. (2003), Bradley (2011)).

The creation of online experimental and field data repositories (e.g. The Europe SERIES database (University of Patras 2018) or the US DesignSafe-CI database (Rathje et al. 2017)) also enables a reviewer to more easily validate research findings, by downloading the raw data and comparing the results against existing theoretical work or other experimental work. Other databases such as the NGA2 strong ground motion database (Ancheta et al. 2013) and the European strong ground motion database (Akkar et al. 2013), also facilitate the validation of experimental and numerical research in earthquake engineering, where all researchers have the ground motions readily available, making it easier to re-construct the experiment/simulations of existing literature.

However, these advancements are still not nearly enough. The effort required to replicate a numerical study of another researcher often takes months and typically requires requesting additional information not supplied in the original publication. The replication of experimental results or the comparison of two analytical expressions can be equally difficult.

Müller (1958) conceived a solution to these problems for research in the physical sciences: "It would seem eminently feasible, however, to punch program cards for tens of thousands of cards for as many empirical or fundamental equations and the data to which they apply. In a suitable computer center or agency these could be interrogated when necessary and the detailed data sent to a subscriber by teletype or more leisurely by mail." To address these problems in engineering research, this paper presents a framework for implementing the empirical and fundamental equations of engineering research in computer code, similar to Müller's concepts. However, using modern technology thus avoiding the need for teletype and mail!

The framework outlines a process to rapidly improve the validation time of existing research and allow researchers to stay up-to-date with new developments, as well as increasing the quality and value of new research. The major aspects of object-orientated programming and the scientific method are discussed and the proposed research framework is presented along with the current application of the framework to research assessing the earthquake performance of buildings on liquefiable soils.

2 THE SCIENTIFIC METHOD

Engineering research is first and foremost a scientific pursuit. It is therefore underpinned by the scientific method, a procedure to develop new knowledge by evaluating the predictive capabilities of hypotheses against measurements from experiments. The exact steps vary for different scientific pursuits and it is an iterative procedure, however, the process must develop a falsifiable hypothesis (i.e. there could conceivably be experiment that could prove the hypothesis to be false), and the experiment should test whether the hypothesis is false. Some engineering research can have a different focus than pure science, such as research that is for the practical improvement of a product. Although this research can still follow the scientific method, it is typically better defined by product development methodologies.

In Figure 1 the main steps of the scientific method are outlined and can be described as:

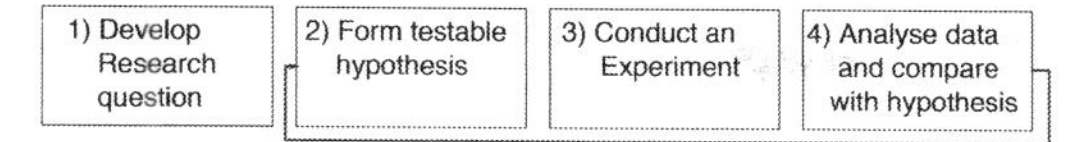

Figure 1. The steps of the scientific method.

1. Develop a research question. This step is typically driven by curiosity about why something works or from more practical demands.
2. Develop a hypothesis. This is the formulation of a concept that could be used to provide a prediction for the the research question.
3. Conduct an experiment. Collection of data that could prove the hypothesis is incorrect.
4. Analyse data and compare with hypothesis. In this step the comparison between experimental results and the prediction attempts to nullify the proposed hypothesis.

High quality blind-predictions of experimental tests are a clear application of the scientific method in engineering research and provide an objective quantification of the true predictive abilities of current hypotheses. However, given the current difficultly of fully understanding someone else's research, often this process gets performed by a single person or research team, where hypotheses are presented and experiments are performed in a closed group without a comparison to existing hypotheses and data from other groups. An accepted scientific theory should be able to sufficiently explain all prior experimental observations and the best current hypothesis should be the hypothesis that best explains all of the prior experimental observations.

3 OVERVIEW OF OBJECT-ORIENTED PROGRAMMING

Engineering research involves the understanding of physical *objects* and therefore it is well suited to the paradigms of object-orientated programming (OOP). Conceptually OOP allows the representation of physical and conceptual objects and their parameters as numerical objects (i.e. the software can store in memory a building and its parameters (e.g. height, width, floor weights, etc)). The alternative to OOP is typically procedural programming, where a software is written based on functions and subroutines and therefore the parameters do not have additional relationships defined outside of what is performed within the function/subroutine.

The additional structure required for OOP can sometimes be a hindrance, but the structure can also be used to enforce compatibility between different parts of a software and greatly reduce the amount of repeated source code. Figure 2

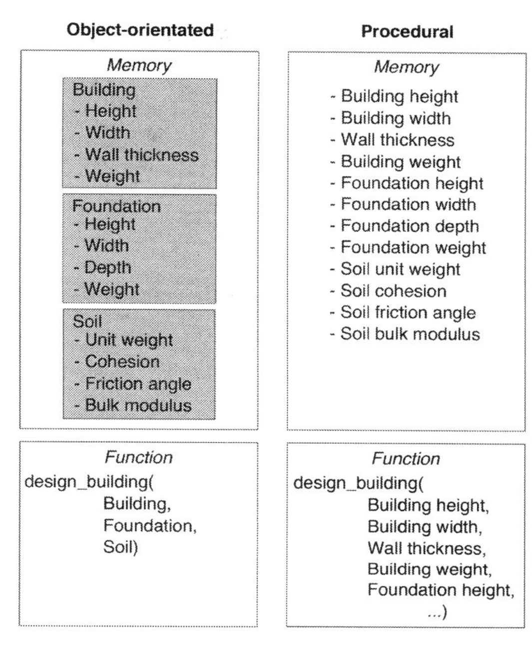

Figure 2. Conceptual comparison between object-orientated programming and procedural programming.

conceptually shows the difference between the two approaches, where in OOP approach the additional structure of the parameters is stored in memory and can be passed directly into the function to design a building, while the procedural approach requires explicitly passing each parameter to the function and then re-establishing the relationships between parameters inside of the function. The OOP approach is particularly well suited to engineering because parameters typically have many interconnected relationships.

When multiple people use the same code it is important that they have the same definition of the parameters and the relationships between them (e.g. is the foundation height measured to the base of the concrete or the lowest layer of the reinforcing steel?). The structure that is inherent when using OOP means that the parameter definitions are tied to the objects and therefore you just need to know if you are using the same object (in Figure 2, this reduces the chance of inconsistent parameters from twelve to three). The structure of OOP is therefore a key aspect to improve collaboration directly at the calculation level.

An integral part of research is the development of new concepts and the adaption of existing concepts for new needs, and therefore flexibility in a research workflow is very important. OOP provides a toolset for extending and adapting existing

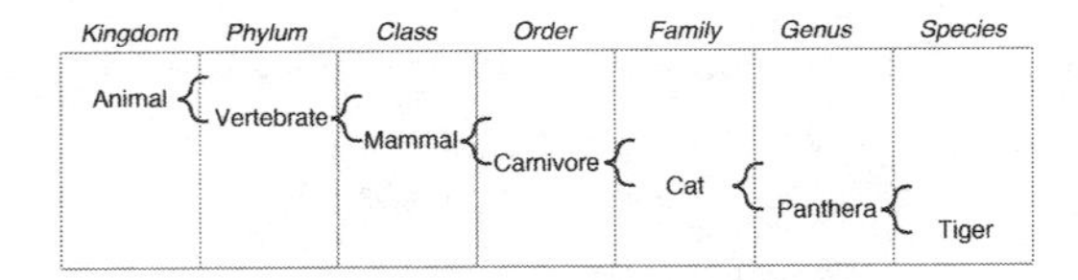

Figure 3. Abstraction applied to the taxonomic ranks of the animal kingdom.

concepts through a technique called abstraction. Typically in OOP, an object is defined by inheriting parameters from a parent object and then adding the unique parameters specific to that object, while another object would also inherit from the same parent object but have a different set of unique parameters. The process of abstraction is defining which parameters should belong to the parent object and what should be unique. Abstraction is not just used in OOP. In fact it used in many engineering processes and scientific concepts such as the taxonomic ranks in biology. The Tiger species in Figure 3 can be abstracted to the genus Panthera, which includes lions, leopards and jaguars, which all share the same flattish skull among other attributes. The Panthera genus can be abstracted to the family, Cat, which includes all cats, and can be abstracted further to the order, Carnivore, class, Mammal, phylum, Vertebrate, kingdom, Animal. At each abstraction there are a set of common attributes that are shared, and this structure has allowed researchers to easily communicate the attributes of new biological forms without having to explain all of the features but by simply stating what genus it is from and defining the unique characteristics at the species level.

For engineering research a series of base objects (e.g. soil, foundations, beams and columns) could therefore be extended to cover the unique cases of new research. When objects inherit from the same parent they have parameters that are common and, in some cases, can be used interchangeably by other functions and objects. It is this attribute of OOP that allows research to be developed with consistency.

4 A HUMAN-DRIVEN MACHINE-AUTOMATED RESEARCH PROCESS

The framework proposes implementing all of the existing engineering hypotheses and experimental and field data into computer readable form in a way that allows it to be automatically crossed checked for consistency, similar to the concepts of Müller (1958). The experimental data and hypotheses could exist on a distributed set of web-servers ("Müller machines"), that would expose application programming interfaces (APIs) to receive input parameters as objects and return objects or values that represent results of an experiment or the prediction from a hypothesis. A research question could be defined on a web-server using objects and a set of criteria to evaluate the accuracy (see Section 4.4), thus initiating a automated research workflow. The web-server would then check available APIs to see whether they can provide experimental data or predictions.

A path finding algorithm could also be employed to formulate a prediction by connecting the inputs and outputs of multiple hypothesis to provide the required outputs from the given inputs. Figure 4 shows how multiple hypotheses could be pulled together to provide a prediction for building damage in terms of foundation settlement, column damage and beam damage based on a set of inputs (Earthquake fault, Building location, Soil profile, Foundation, Building). Each hypothesis shown in Figure 4, could be formulated by many smaller element—or mechanism-level hypotheses and each hypothesis could be validated at the element level and at the whole system level against available experimental data sets. There could potentially be many different combinations of hypotheses that could provide a prediction. In the example in Figure 4 the hypotheses could be considering the ground shaking as a time-series or as a spectral quantity. Different set of hypotheses could be used that consider the influence of the building's torsional response or ignoring the influence of the site response or soil-foundation-structure interaction. All possible combinations that take the required

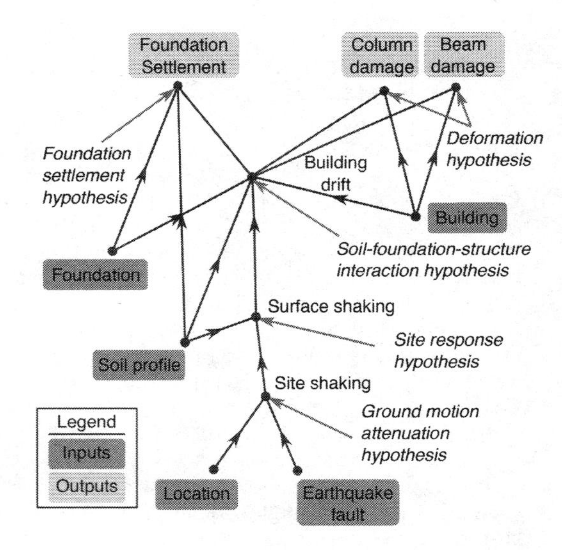

Figure 4. A prediction of building damage parameters from objects.

inputs and produces an estimation could be scored (based measures of level of uncertainty, bias, physical insight, etc) to provide a scorecard of the current knowledge for answering a given question. Limitations could also be imposed on the range and use of particular hypotheses to avoid excessive extrapolation of trends. The research question could remain an active question, such that as more experimental data becomes available and new hypotheses are developed, they are automatically assessed against the existing experimental data and hypotheses.

In the following sections the main aspects of the framework are outlined, the application to analytical and experimental research is explained and the role of scoring functions is explored. The final section discusses the current implementation of the framework for research on the earthquake performance of buildings on liquefiable soil, as part of the LIQUEFACT H2020 research project.

4.1 *Main aspects*

The framework is intended to improve the quality and value of new research in an efficient and easy to understand manner. To achieve this, the following philosophies have been adopted:

1. Follow the scientific method. Question ® Hypothesis ® Experiment ® Comparison
2. Extendable. The project allows the development and connection of new yet-to-be-conceived research
3. Convention over Configuration. The benefits of a convention (e.g. standard format for presenting experimental data) out-weigh the benefits of infinite configurability
4. Repeatable research. Clear documentation of new research such that it could be repeated by an equally equiped and skilled researcher
5. Human readable. The reasoning behind a hypothesis must be interpretable by a human
6. Open-source research. Research should be provided open-source to allow others to use and validate it

Note that this list does not include common research goals of accuracy, physical insight, simplicity of description or application. However, these are important attributes of high quality engineering research and they will be dealt with in Section 4.4 on scoring functions, where the quality of research is discussed. Also note that this framework does not require the use of artificial intelligence or neural networks. In fact obscure processes such as neural networks often struggle to fulfil the main steps of the scientific method of having repeatable results and a clear line of reasoning. While they can be helpful for finding trends and formulating a hypothesis, similar to the role of human intuition, if they are applied on their own for predictive purposes, they can be considered a completely different approach to developing knowledge.

4.2 *Application for experimental, field and long numerical simulation data*

Experimental and field data is often collected to evaluate a hypothesis and is a key part of the scientific method. More recently numerical simulations are being used to evaluate hypotheses, in the same role as experimental data. Therefore, numerical simulations can fulfil two roles in the scientific method: a pseudo experiment (i.e. can a simplified method (hypothesis) capture the behaviour of the numerical simulations (experiment)), as well as being a hypothesis (can the numerical model simulate the experimental behaviour). In the proposed framework short open-sourced numerical procedures can be considered as hypotheses, which could be used to provide a prediction for new experiments, while long numerical simulations using commercial software cannot easily be applied to new data sets and therefore only fulfil the role of a pseudo experiment.

One of the key parts of experimental research is the comparison of a data set with existing hypotheses and other data sets. This comparison not only provides a level of validation of the data set but it provides useful insights into unique intricacies that were recorded in the new or existing data sets. This comparison is essentially data processing and typically involves a large amount of mundane manual work. This framework aids the comparison, where standard input and output files are written for each experiment, which define the measured input and output objects using a language agnostic format such as json or yaml (Figure 5 - step 1). The new data set is then uploaded to an online database and the standard input and output files would be submitted to a web-server, where an API would automatically be built for the dataset. The web-server could then compare the experimental results against all existing data sets and against all existing relationships (Figure 5 - step 2). The web-server could easily produce the required plots of the comparisons and build the appropriate references, ready for journal publication.

4.3 *Application for analytical and empirical research*

Analytical research is referred to here as the development of new hypotheses based on the adaption of existing hypotheses using mathematical reasoning. Empirical research is referred to here as the development of new hypotheses based on the fitting of a curve through a data set using statistical regression. Typically new hypotheses in engineering are developed through a combination of both analytical and empirical techniques.

A new hypothesis is typically developed to answer a specific research question (e.g. estimating liquefaction triggering (Boulanger and Idriss 2014)). The new hypothesis is then compared to existing less complex situations (e.g. elastic solutions) and experimental/field data or numerical simulations that are independent of the data set that was used to originally develop the hypothesis. This comparison process is an important part of validating a new hypothesis but it can take an enormous amount of effort to re-interpret independent data sets, especially if they are in different formats.

The framework aids this validation, where the new hypothesis can be written into an OOP language in which the inputs and outputs are compatible with a standard set of objects. The new equation is then submitted to a web-server where an API is generated for it to receive the input objects and output the predictions. The web-server then automatically compares the new hypothesis against all existing data sets and against all exist hypotheses (step 2 in Figure 5).

This process would provide an independent and objective 'score' of the validity and value of new research, as well as highlight inconsistencies between hypotheses. Completely independent validation is a key role of the reviewer for publication, and therefore the review process for publication could be greatly reduced.

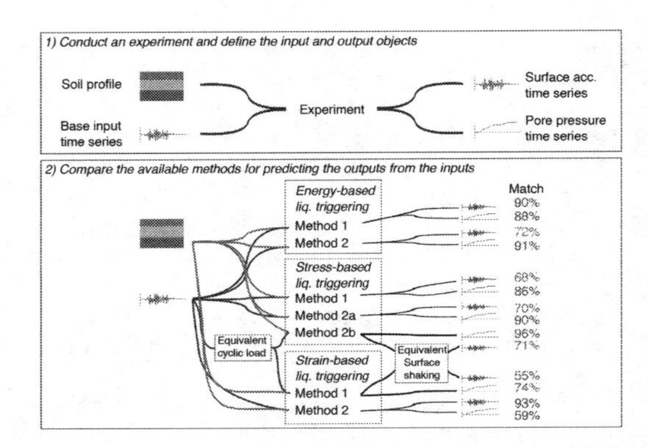

Figure 5. Conceptual application of the engineering consistency project for liquefaction triggering analysis.

4.4 *Scoring functions*

The framework could provide an independent evaluation of the quality and value of research. However, this requires an objective quantification of quality and value.

Research quality is largely focused on accuracy. For experimental work, accuracy is difficult to completely establish and typically relies on an honest account from the experimental researcher or a repeated study. For new hypotheses the accuracy can be directly measured through validation and verification procedures that quantify uncertainty and error for given validation metrics. Oberkampf et al. (2003) outlines the key steps in the validation and verification procedure for assessing computational physics software, highlighting that the specification and use of validation metrics is the most important part of the validation process. Validation metrics are the measures of the response that can be used to evaluate the difference between the experimental result and the prediction from the simulation. The choice of metrics is not a trivial step, and is typically highly dependent on the end-use of the prediction, where metrics should be used that allow the minimisation of the uncertainty for the end-use. Therefore for a given set of experiments, and given set of analytical/empirical/numerical predictions, the accuracy would be dependent on the end-use. As an example, for the seismic response of a soil deposit, the end-use case may be the seismic response of a building and therefore the energy of the surface motion in the frequencies closest to the natural frequencies of the building are the most important metrics, while if the end-use was the estimation of design loads on a buried pipeline, then the soil strains and displacements at the depth of the pipe are of greatest interest.

Bradley (2011) outlines a framework for the validation of constitutive models for modelling the seismic response of a soil deposit, suggesting the use of a vector of engineering demand parameters for the validation metrics (e.g. peak ground acceleration, displacement and Arias intensity). Generally engineering demand parameters offer a useful tool for validation metrics as they tend to be quantities that are of interest for a wide-variety of end-use applications and they are a single quantity. However, some measures are two dimensional (e.g. pore-pressure build up versus time) or even three-dimensional (inter-storey drift versus height versus time). The error and uncertainty associated with two-dimensional (and three-dimensional) parameters have additional considerations compared to a single value due to their inter-dependence. Simply computing the difference in the two dimensions (e.g. x-offset and y-offset in Figure 6) may fail to capture the accuracy of the prediction which may be offset by scale or slope. Further work is required to develop standard validation metrics. However, for initial implementation purposes the researcher who asks the initial question (Step 1 of the scientific method) would also define the validation metrics.

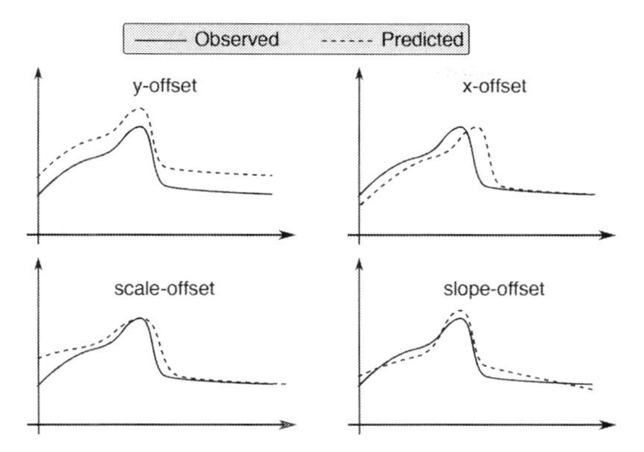

Figure 6. Several types of errors for 2 dimensional data.

The measure of research value for field studies and experimental work can be directly quantified by the reduction in the uncertainty of existing hypotheses (or of the null hypothesis) especially for previously untested ranges of input parameters. Also repeating an existing experiment can offer immense value by reducing the uncertainty around the experimental results. The measure of research value for a new hypothesis, especially in engineering research, should be concerned with attributes such as physical insight, simplicity in application and sufficient generality as outlined by (Wolf and Deeks 2004), as well as considering end use applications. The attributes could be quantified by:

- Physical insight—Difficult to quantify as it is concerned with the ability to communicate the inner workings of the physical problem, but could be evaluated by assessing the number of parameters that have a physical meaning.
- Simplicity in application—Quantified by the number of input parameters or number of calculations or ease of determining the inputs.
- Generality of application—Quantified by the extent of the domain of inputs where the hypothesis could provide a prediction (linked to a level of sufficient accuracy)
- End use application—Could be quantified based on the number of predictions that it can be part of for a given set of research questions.

By quantifying the quality and value of all research the framework could quantify the current state of understanding and it would continuously update with the addition of new hypotheses and data sets.

4.5 *Implementation*

The implementation of the framework would provide a flexible system that allows research teams to have full control of their research process and the publication of their data and hypotheses. However, it requires some level of understanding of OOP and web-servers as well as the basic concepts of the scientific method. To achieve ease the implementation, the web-server and database architecture could be designed as in Figure 7. A user develops new hypotheses on their local computer, and then can submit the hypothesis to a private server that copies the current state of the public web-server and performs the comparison of the new hypothesis against all of the publicly available data sets and all existing hypotheses. The user can then publish the new hypothesis to the central web-server where it becomes part of the existing hypotheses available for others to use and forms part of the current state of knowledge. Experimental data could be held privately in a private database (in fact in the same database as the public data, just with a private database address), to allow researchers to run comparisons prior to publishing the address and making the data public.

The central web-server allows anyone to ask a research question, which would involve setting the validation metrics, and then the available data sets would be queried and available hypotheses would be compiled and compared to provide an answer to the question.

4.6 *Benefits*

The presented framework offers many benefits to the researcher and the research community:

- Repeatable research. The comparisons would be performed independently by the server, therefore they can be requested by any end-user.
- Updatable. A researcher can update their hypothesis after publication by incrementing a version number, and each version would be stored but they could select which one should be used as the default.
- Quantitative importance. The framework could directly calculate the improvement that a new analytical expression provides to the estimation of an output compared to existing literature. It could also quantify how much a new experimental data set improves validation of existing theories. This would reduce the amount of redundant research and could even highlight to end-users where experiments are needed for validation and where existing hypotheses are weak.
- Automation of recurrent work. The structure applied to research data allows tools to be built that can automate the generation of research outputs such as tables, figures, references, methodologies and many other recurring aspects of research. Potentially the only aspects that

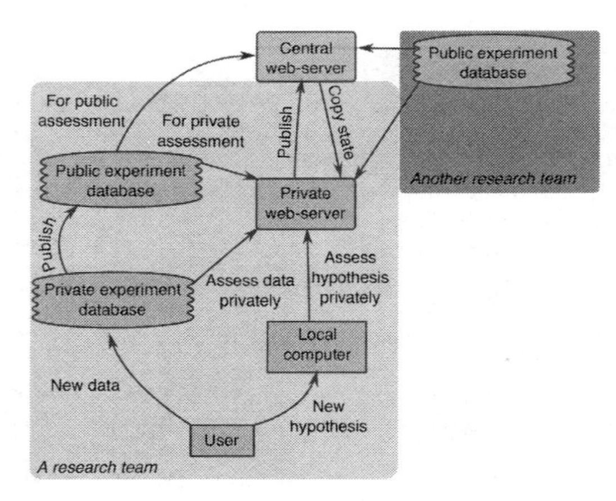

Figure 7. Proposed server and database architecture for the research framework.

cannot be automated are the unique aspects of the research.

- More citations. The paper references (DOIs) to individual hypotheses or data sets can be embedded into the source code or objects that they generate. Thus, when someone publishes a comparison using the framework, the references could automatically be included in the new paper.
- Connect to industry. Research could easily connect to object-oriented software platforms for engineering and architecture (e.g. Building Information Modelling, BIM).

4.7 *Drawbacks*

The presented framework potential contains several drawbacks to the researcher and the research community:

- Exposure to criticism. The exact implementation of a set of equations and the experimental data sets are made publicly available and therefore could more easily be scrutinised
- Effort/Learning. There is an effort involved in learning enough about OOP to implement and make use of the framework
- Excessive references. Because the framework would test for consistency against all available research, there would then be a large amount of references for each new publication. However, as more research is produced in a particular field, it could be expected that a proper literature review would result in increasing amounts of references. The approach to referencing work might need to change to allow this, where research outputs could be built using a blockchain that clearly stated the exact inputs, equations and data sets that were used to generate the output.
- Gaming of the system. Researchers may focus only on research that would result in a high score (quality and value), while this is the intended purpose of quantifying the quality and value of research, it could misalign research if the scoring failed to recognise the importance of different types of research.

4.8 *Current state*

The goal of the project is large and open-ended, because new theories/hypotheses are continuously being developed. The first step is to focus on the area of engineering research related to the response of buildings on liquefiable soil as part of the LIQUEFACT project. The major hypotheses around liquefaction triggering, building settlement, site response analysis, building response, soil-foundation-structure interaction, are being implemented into Python using a standard set of objects (models for Soil, Foundation, Building, TimeSeries, etc.). The current version of these implementations are publicly available at https://github.com/eng-tools.

5 CONCLUSION

This paper presents a more rigorous approach to the application of the scientific method to engineering research. The presented approach takes advantage of web-server technology and object-oriented programming to provide a consistent, structured and continuously updating state of knowledge. The key use cases as well as the potential advantages and drawbacks of the procedure are highlighted to emphasise that the new approach could rapidly speed up the development of new high quality and valuable research. The imposed structure and consistency of the approach, as well as the ability to automatically update the state of knowledge provides the necessary foundation to efficiently develop research at a global level both now and into the future.

ACKNOWLEDGEMENTS

This paper was produced as part of the LIQUEFACT project (Assessment and mitigation of liquefaction potential across Europe: a holistic approach to protect structures / infrastructures for improved resilience to earthquake-induced liquefaction disasters) has received funding from the European Union's Horizon 2020 research and innovation programme under grant agreement No GAP–700748.

REFERENCES

Akkar, S., M.A. Sandıkkaya, M. S¸ enyurt, A. Azari Sisi, B. ö. Ay, P. Traversa, J. Douglas, F. Cotton, L. Luzi, B. Hernandez, and S. Godey (2013). Reference database for seismic ground-motion in europe (resorce). *Bulletin of Earthquake Engineering 12*(1), 311–339.

Ancheta, T.D., R. Darragh, J.P. Stewart, S. Emel, W. Silva, B. Chiou, K.E.Wooddell, R.W. Graves, A.R. Kottke, D.M. Boore, T. Kishida, and J. Donahue (2013). Peer nga-west2 database. Technical Report 03.

Boulanger, R.W. and I.M. Idriss (2014). Cpt and spt based liquefaction triggering procedures. Technical report.

Bradley, B.A. (2011). A framework for validation of seismic response analyses using seismometer array recordings. *Soil Dynamics and Earthquake Engineering 31*(3), 512–520.

Djorgovski, S.G. (2005). Virtual astronomy, information technology, and the new scientific methodology. *arXiv.org,* 125–132.

Foster, I. (2006, mar). 2020 computing: A two-way street to science's future. *nature.com 440*(7083), 419–419.

Hilbert, D. (1902). Mathematical problems. *Bull. Amer. Math. Soc.*, 429–437.

McKenna, F. (2011). Opensees: a framework for earthquake engineering simulation. *ieeexplore.ieee.org 13*(4), 58–66.

Müller, R.H. (1958). Computer center for basic physical science data proposed. *Analytical Chemistry 30*(8), 55 A–55 A.

Oberkampf, W.L., C. Hirsch, and T.G. Trucano (2003). Verification, validation, and predictive capability in computational engineering and physics. Technical report, Sandia National Laboratories (SNL), Albuquerque, NM, and Livermore, CA.

Poldrack, R.A. and J.-B. Poline (2015). The publication and reproducibility challenges of shared data. *Trends in Cognitive Sciences 19*(2), 59–61.

Rathje, E.M., C. Dawson, J.E. Padgett, J.-P. Pinelli, D. Stanzione, A. Adair, P. Arduino, S.J. Brandenberg, T. Cockerill, C. Dey, M. Esteva, F.L. Haan Jr., M. Hanlon, A. Kareem, L. Lowes, S. Mock, and G. Mosqueda (2017). Designsafe: New cyberinfrastructure for natural hazards engineering. *Natural Hazards Review 18*(3), 06017001.

Sparkes, A., W. Aubrey, E. Byrne, A. Clare, M.N. Khan, M. Liakata, M. Markham, J. Rowland, L.N. Soldatova, K.E. Whelan, M. Young, and R.D. King (2010). Towards robot scientists for autonomous scientific discovery. *Automated Experimentation 2*(1), 1–11.

University of Patras (2018). Series database. http://www.dap.series.upatras.gr/default.aspx. Accessed: 2018–01–20.

Wolf, J.P. and A.J. Deeks (2004). *Foundation Vibration Analysis: A Strength of Materials Approach* (1 ed.). Butterworth-Heinemann.

Zenodo (2018). Zenodo now supports doi versioning! http://blog.zenodo.org/2017/05/30/doi-versioninglaunched/.Accessed: 2018–01–20.

Ground flow, thermal and coupled analysis

Numerical Methods in Geotechnical Engineering IX – Cardoso et al. (Eds)
 ISBN 978-1-138-33198-3

Heat transfer process in a thermo-active diaphragm wall from monitoring data and numerical modelling

D. Sterpi, A. Angelotti, O. Habibzadeh Bigdarvish & D. Jalili
Politecnico di Milano, Milan, Italy

ABSTRACT: The thermo-active diaphragm walls represent a solution that permits the exploitation of near-surface geothermal energy for the thermal conditioning of buildings and infrastructures. Preliminary studies showed that some time-dependent thermal conditions, such as the thermal input at the inlet of the heat exchanger pipes and the temperature at the excavation side, play a crucial role in the overall energy performance. An on-site monitoring programme allowed to enhance the understanding of the heat transfer process occurring in the real scale structure in the different heating/cooling operating modes. The research was then focused on the numerical model calibration based on the monitoring data. The results proved the effectiveness of the numerical model in predicting the heat transfer process and the energy performance, but also the high sensitivity of the analysis with respect to parameters of uncertain definition.

1 INTRODUCTION

The thermo-active geostructures, for the near surface geothermal energy exploitation, nowadays represent an effective solution towards the growth in the use of renewable and green energy sources (Laloui & Di Donna 2013, Bourne-Webb et al. 2016b). For instance, a reinforced concrete diaphragm wall, conventionally used to support an excavation, can host heat exchanger pipes to harvest or disperse heat, for the thermal conditioning of building and infrastructures (Amis et al. 2010, Soga et al. 2015).

Serving two functions, the design of thermo-active geostructures should be based on both concepts of optimal energetic performance and structural functionality and safety. As to the structural aspect, these geostructures are subjected to combined mechanical and thermal loads, the latter induced by the heat transfer occurring during the seasonal geothermal system operations. Consequently, the need arises for a better understanding of the heat transfer process and its effects, in order to adopt proper design criteria. It is worth noting that methods conventionally adopted for the heat transfer analysis in borehole heat exchangers or in thermo-active piles do not apply to diaphragm walls, due to their specific features, and purposely conceived methods should be considered in this case (Kürten et al. 2015, Sun et al. 2013).

While extensive studies on thermo-active piles have already been presented in the literature, the behaviour of diaphragm walls is less investigated. Nevertheless, from monitored cases and numerical investigations on the energy performance and on the geotechnical and structural response (Bourne-Webb et al. 2016a, Di Donna et al. 2016, Sterpi et al. 2017, Xia et al. 2012), it can be inferred that the thermal boundary conditions and the heating/cooling input, related to the building energy demand, play a crucial role in the overall behaviour.

The thermal boundary conditions are represented by the time-dependent heat flux at the ground surface and within the basement adjacent to the excavation. While the former depends on climatic conditions, the latter depends on the use of the basement space, the control of its temperature and ventilation, and the presence of connections with the outside climate. The heating/cooling input can be estimated from the analysis of the building physics, that includes the natural heat fluxes between the building, the atmosphere and the soil.

The complexity and the uncertainties in the definition of these variables, combined with the sensitivity of the numerical analysis with respect to them, undermine the prediction potential of the numerical approach.

This contribution aims at addressing some of these issues, focusing firstly on the interpretation of monitoring data from a real case, and secondarily on the calibration of a three-dimensional finite element model, devised to predict the temperature variations within the wall and the soil mass and the energy performance.

2 MODELLING OF THE CASE STUDY

2.1 *The thermo-active geostructure*

A zero-energy residential building was recently built in Northern Italy, with a six storeys elevation and a three floors basement (Bertani 2016). The excavation, on an almost squared floor plan of 40.m side, reaches 10.8 m depth and is supported by two pairs of facing diaphragm walls, with a height of 15.2 m and thickness of 0.5 (Fig. 1). The entire perimeter comprises 66 single wall panels, each 2.4 m wide and equipped with two exchanger pipes, fixed to the reinforcing steel cage on the soil side and covering 0.80 m in width. Each loop has six vertical segments, finally resulting in a total length of about 90 m (Fig. 2).

Previous analyses showed the influence of the pipe loop on the energy performance of the diaphragm wall (Sterpi et al. 2014).

The total surface of the perimeter wall, available for the heat transfer, is equal to about 2400 m^2. However, 70% out of it represents the portion which is not fully embedded in the ground but exposed to the excavation on one face. This value highlights the major influence that the thermal condition within the basement might have on the energy performance.

The geothermal system was equipped with a comprehensive sensor system (Todeschini 2016), for a continuous inspection of its operation (thermal power, fluid temperatures at the pipes inlet/outlet, mass flow rate, etc) and for monitoring the temperature variations at various positions (Fig. 1). A series of temperature sensors was fixed to the upper anchor (i.e. the plane x = 0 in Figure 2) and another series was embedded in the wall, into empty tubes fastened to the steel cage, at both faces of the wall, between the inlet and outlet positions (plane x = 0.8 m).

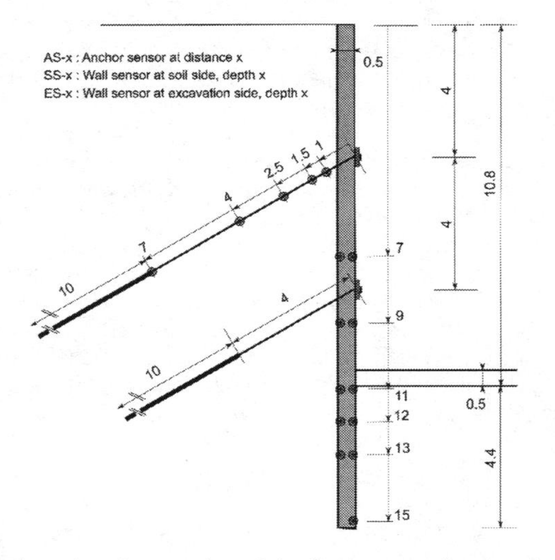

Figure 1. Cross section of the diaphragm wall and positions of the temperature sensors (dots): along the upper anchor (AS), at the soil side (SS) and at the excavation side of the wall (ES) (unit m).

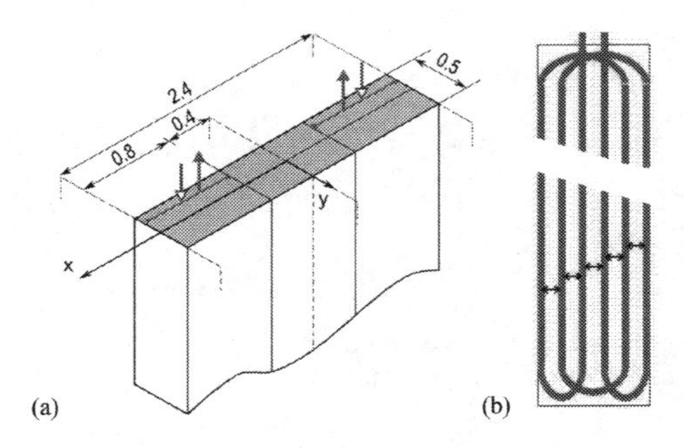

Figure 2. Sketches of (a) the wall single panel with the pipes inlet/outlet locations; (b) the single pipe loop (unit m).

In the following, only the first year of operation is considered, i.e. within the short-term transient period. Besides natural variations between different years, that depend on climatic conditions, a gradual thermal drift is expected to occur within the soil mass, if there is no balance between absorbed and injected heat in respectively winter and summer periods. The system tends to energy equilibrium and long-term characteristic efficiency after some years of operation (Sterpi et al. 2017).

Selecting the most relevant monitoring data, for the sake of briefness, figures 3 shows that, for a given depth above the floor slab (-7 m) the temperature measured at the soil side (SS7) follows the same trend than the one measured at the excavation side (ES7), but slightly damped. This damping is due to the mitigating effect that the soil mass has on the left side of the wall, while the right side is directly influenced by the temperature variations of the basement. Consequently, the temperature gradient along the transversal horizontal axis, and the associated heat flux, have a positive direction that changes depending on the season being considered.

On the contrary, the temperatures at depths below the floor slab (-13 m), shown in figure 4, are no longer influenced by the basement but rather by the presence of the heat exchanger, circulating hot fluid in summer and cold fluid in winter. In fact, at the soil side (SS13) the temperature is lower in winter time and higher in summer time with respect to the excavation side (ES13), and the two values approximately coincide during the idle periods. At these depths the whole soil mass, fully surrounding the wall, positively contributes to the heat transfer with the geothermal system, regardless the season.

The temperature variations along the upper anchor are shown in figure 5. These temperature fluctuations, naturally delayed and damped with increasing depth, could help in the assessment of the soil thermal properties. However, they are affected not only by the soil properties, but also by other parameters: the temperature variations of

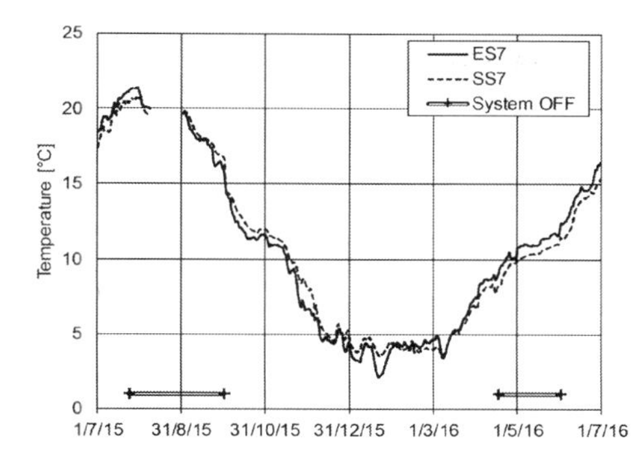

Figure 3. Monitored temperatures at adjacent sensors above the floor slab (-7 m), at excavation side ES7 and soil side SS7. Idle periods of the system are also indicated as "System OFF".

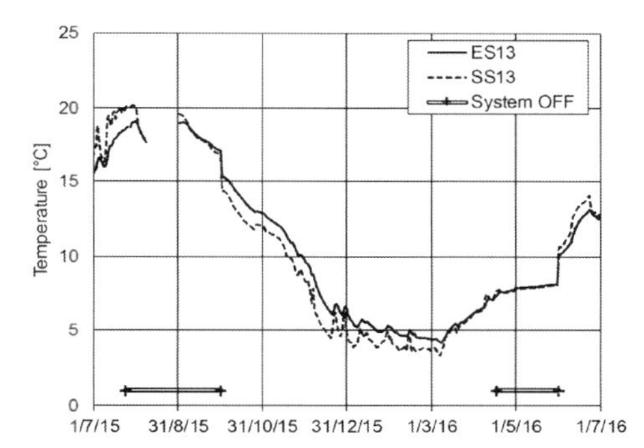

Figure 4. Monitored temperatures at adjacent sensors below the floor slab (-13 m), at excavation side ES13 and soil side SS13.

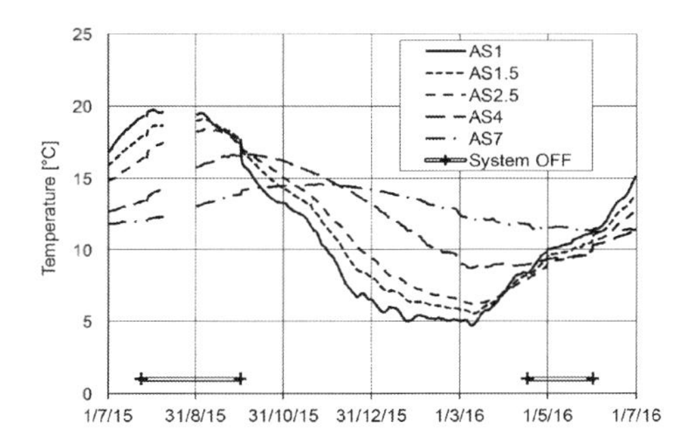

Figure 5. Monitored temperatures along the upper anchor.

the heat carrier fluid, the thermal properties of the steel anchor that eases the heat transfer at depth, and finally the combined effect of seasonal temperatures at ground surface and excavation side.

In fact, even the deepest position (AS7) turns out to lay within the zone of influence of both

Table 1. Physical and thermal properties.

Material	Density	Thermal Conductivity	Specific Heat
	kg/m³	J/(s m K)	J/(kg K)
Heat carrier fluid	1000	0.57	4186
Soil grains	2750	3.6	820
Saturated soil	1930	2.2	1642
Reinforced concrete	2500	2.6	880

boundaries, according to the analytical solution of semi-infinite space subjected to sinusoidal temperature variation at the surface (e.g. Hillel 2003).

Therefore, the properties of the soil, regarded to as a saturated material of given porosity, were estimated from appropriate assumptions on the properties of the single phases (Rees et al. 2000), and are listed in Table 1.

2.2 *The numerical modelling*

The most accurate finite element analysis requires a 3D modelling since the pipe loop is not uniformly arranged within the wall and a thermal gradient arises in the longitudinal direction. It is also possible to recognize symmetry planes, so to reduce the 3D domain to a thin slice of the entire structure, of 1.2 m width (Fig. 2).

The thermal analysis is based on convection and conduction heat transfer processes, respectively in the pipe and in soil and diaphragm wall. The time-dependency is introduced by the heat transfer phenomenon and by the thermal boundary conditions. These consist of an initial temperature field $T(\underline{x},0)$ and of prescribed values $T(\underline{x}_B,t)$ at given boundaries $\underline{x}_B$: a constant value at the ground base, and time-dependent values at ground surface, basement surfaces, and inlet of the exchanger pipe. The latter must be completed with the input on the fluid flow rate, to correctly assign the heat flux. The lateral sides are assumed as adiabatic boundaries, as well as the front and back faces of the 3D domain, being symmetry planes. If the fluid velocity at the pipe inlet is set to zero, the analysis with these same thermal conditions provides the initial temperature field $T(\underline{x},0)$. This assumption entails that the excavation and construction time is long enough to let the soil mass reach a steady state initial condition before activating the geothermal system. This steady state condition is reached after a 5/6 years analysis.

The Environmental Protection Agency (ARPA-Lombardia, Italy) provided the ground surface air temperatures at the site, from which the 2010–2015 monthly averaged cyclic variation was computed, to be considered as boundary condition (Fig. 6). At the base of the domain, the constant

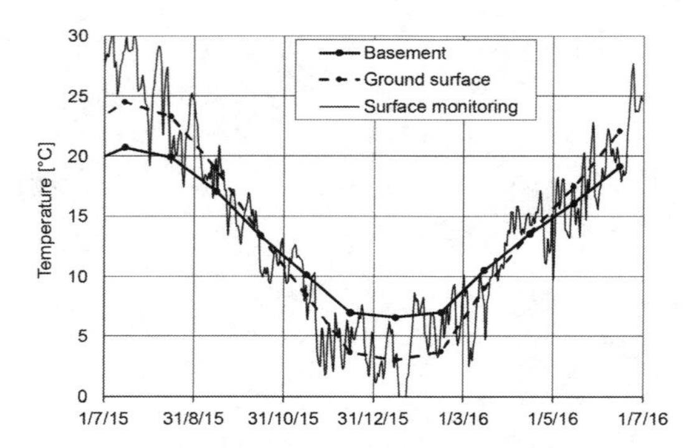

Figure 6. Monitored temperatures at the ground surface and temperature boundary conditions assumed in the analyses.

value of 13.4°C was assigned, equal to the yearly average of the surface temperatures, to have a thermally stable system.

Concerning the condition at the basement, an unheated parking lot, the hypothesis was introduced that the temperature yearly fluctuation is characterized by the same average value as the ground surface, but damped amplitude. The damping coefficient, equal to 0.66, was calibrated based on the monitoring data during the brief period of inactive geothermal system (summer 2015), when the temperature at the excavation side (sensors ES7, ES9) is likely equal to the temperature of the basement without time delay.

As to the heat flux at the pipe inlet, both fluid temperature and flow rate are established from the system monitoring. Again, monthly averaged values are calculated, to simplify the analysis. This assumption leads to inaccuracies in the idle periods, when the operation is not continuous, but intermittent, due to milder external climate. In these months, a shorter time interval should be adopted to calculate the average input values (Faizal et al. 2016).

Finally, the thermal power is calculated, knowing the fluid specific heat and the fluid flow rate, from the temperature difference between pipe inlet and outlet. A decrease in fluid temperature is assumed as positive; consequently, a positive thermal power is calculated during the cooling mode operation.

3 RESULTS AND DISCUSSION

3.1 *Temperature field*

The results of the numerical analysis are discussed first with reference to the temperature variations induced by the heat transfer process. It is worth noting that the analysis concerns a transient period and it might be not representative of the steady state long-term conditions (Coletto & Sterpi 2016).

Figure 7 shows the temperatures at two positions along the anchor: AS1.5, close to the excavation, and AS7, far from equally the ground surface and the excavation.

The analysis is able to predict the general trend but with inaccuracies, that are large or small depending on the distance from the boundaries.

The largest inaccuracies are shown in locations within the wall, for instance at the soil side SS7 (Fig. 8) and SS11 (Fig. 9), and generally consists in a tendency to overestimate the values of temperature.

Moreover, the sharp variations measured at the end of idle periods, when the system starts the phases of heating (temperature drop on October 1st) or cooling (temperature gain on June 1st), are not correctly reproduced in the analyses, unless in these periods a refined calculation of the heat flux input data is worked out.

Therefore, the prediction potential of the numerical analysis, for the temperature field, seems to be affected by the high sensitivity to input data such as the temperature variations at the boundaries and the heat flux. A minor sensitivity was observed

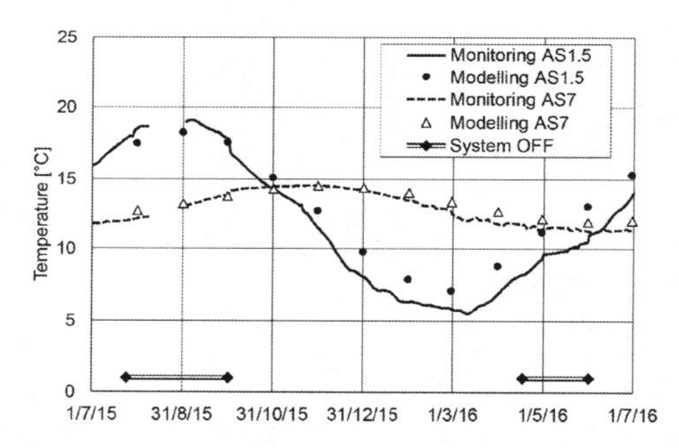

Figure 7. Monitored temperatures and numerical results at sensors AS1.5 and AS7 along the anchor (cf. figure 1).

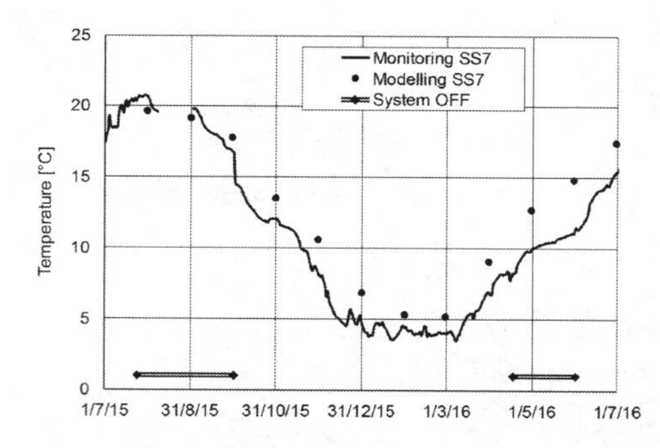

Figure 8. Monitored temperatures and numerical results at sensor SS7 within the wall at the soil side (cf. figure 1).

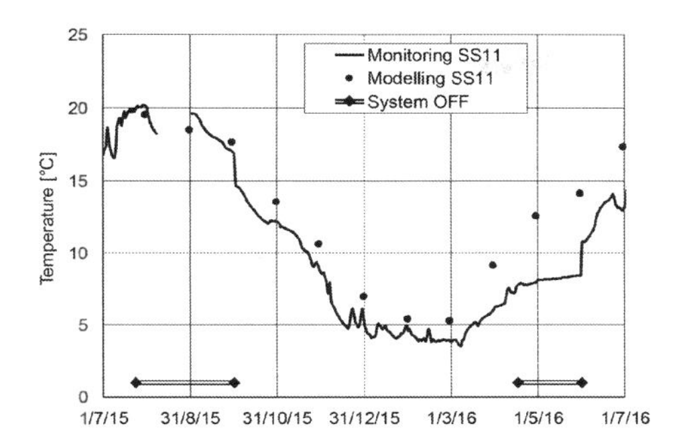

Figure 9. Monitored temperatures and numerical results at sensor SS11 within the wall at the soil side (cf. figure 1).

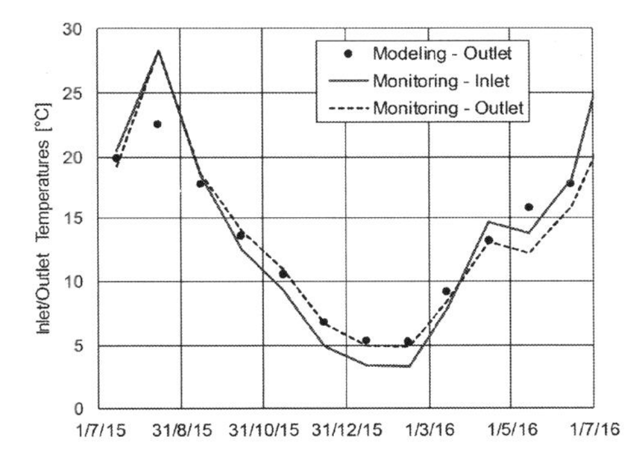

Figure 10. Pipe inlet and outlet temperatures from monitoring data (solid and dashed lines) and from numerical modelling (dots).

to variations of the thermal properties, if chosen within the range of likely values, given the soil nature and conditions (density and water content).

3.2 *Energy performance*

The fluid temperature at the pipe outlet contributes to the thermal power calculation. Figure 10 reports a comparison between monitored and computed values, together with the values of inlet temperature, that are worked out from monitoring data and used as input data of the numerical model. In two periods only, a major difference between monitored and computed values is observed, namely in August '15 and May '16, both falling within an idle period. The discrepancy could be due to the difficulty to work out the monthly average value of the heat flux in case of intermittent operations.

The comparison between monitored and computed thermal power is shown in figure 11, with reference to one single loop. An estimate of the thermal power of the entire diaphragm wall can

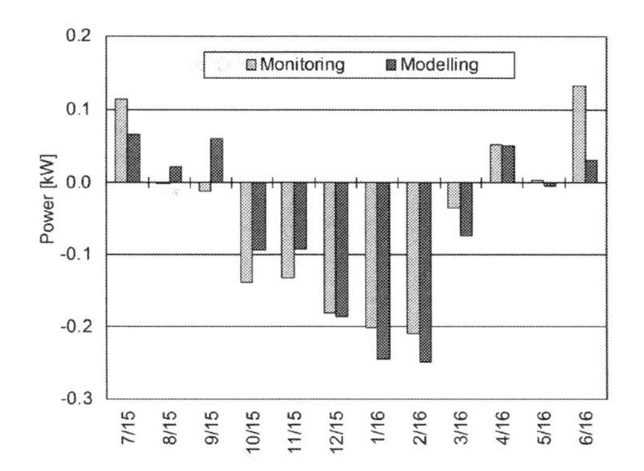

Figure 11. Thermal power from monitoring data and numerical modelling.

be obtained by multiplying by the total number of loops (132 loops). The observed discrepancies are due to both inaccuracies in the numerical computation of fluid outlet temperatures (Fig. 10) and difficulties in averaging the monitored fluid flow over monthly periods.

Given the testing conditions of the first months and the transient nature of this phase, a balance between injected and extracted thermal energy on a yearly basis is not expected.

4 CONCLUSIONS

Combined monitoring programme and numerical analyses allowed to get insights into the heat transfer process characterizing a thermo-active diaphragm wall, with a limitation to the initial transient period (first year of operation).

The continuous inspection of the system operation and the measurement of temperature variations within the wall and at depth highlighted the role of the thermal boundary conditions and the factors that influence the magnitude and the direction of the heat fluxes across the wall.

The differential in temperatures across the wall width (soil side and excavation side) at different depths (fully embedded or laterally exposed positions), confirmed that the basement space always represents a negative contribution to the building heating/cooling potential. In fact, while the surrounding soil mass always acts as a heat source in heating periods and a heat sink in cooling periods, the basement space draws heat from the circulating fluid in the heating operating mode and transfers heat to it in the cooling mode. The overall energy performance could be anyway satisfactory, especially if the heating and cooling of the basement space is within the purposes of the geothermal system.

The monitoring data could not provide an information about the soil thermal properties, because influenced by factors that make the interpretation uncertain. In absence of suited thermal response tests carried out at the specific site, the estimation of the soil thermal properties still must rely on information from the hydro-geological site survey and thermal properties of soil/rock classes from literature databases (Beier et al. 2011, Vieira et al. 2017).

The numerical analysis could be an effective predictive tool, in terms of temperature variations and energy potential, only when based on an accurate calibration of the boundary thermal conditions and the heat flux input.

The identification of these parameters is affected by many factors. For the excavation side, a hypothesis is required about the temperature fluctuations, based on the actual use of the basement space. The heat flux at the pipe inlet depends on the building energy demand and is subjected to abrupt variations, especially in periods of low energy demand and intermittent system operation. These abrupt variations lead to inaccuracies if the average values, to be used as input data, are worked out on long time periods.

In spite of the approximations introduced in the numerical modelling and the inaccuracies in the prediction of the temperature field, the results in terms of pipe outlet temperature and thermal power can be considered rather satisfactory in the periods of stable and continuous operation.

As a conclusion, for thermo-active diaphragm walls the thermal conditions are crucial to determine the heat transfer process and the consequent energy performance. The behaviour can be predicted by numerical analyses but the analysis is very sensitive to parameters of uncertain definition.

ACKNOWLEDGEMENTS

The authors acknowledge the contributions by Ing. G. Bertani (Ingg. Bertani e Baselli & C. SpA, Italy) and Ing. L. Todeschini (Tecnoel srl, Italy). D. Sterpi acknowledges the support of COST Action TU1405 GABI (Geothermal Applications for Building and Infrastructures).

REFERENCES

Amis, T., Robinson, C.A.W. & Wong. S. 2010. Integrating geothermal loops into the diaphragm walls of the Knights-bridge Palace Hotel project. In *Geotechnical Challenges in Urban Regeneration; Proc. 11th DFI/EFFC Int. Conf.*, London.

Beier, R.A., Smith, M.D. & Spitler, J.D. 2011. Reference data sets for vertical borehole ground heat exchanger models and thermal response test analysis. *Geothermics* 40:78–85

Bertani, G. 2016. Personal communication.

Bourne–Webb, P.J., Bodas Freitas, T.M. & da Costa Gonçalves, R.A. 2016a. Thermal and mechanical aspects of the response of embedded retaining walls used as shallow geothermal heat exchangers. *Energ. Buildings* 125:130–141.

Bourne–Webb, P.J., Burlon, S., Javed, S., Kürten, S. & Loveridge, F. 2016b. Analysis and design methods for energy geostructures. *Renew. Sust. Energ. Rev.* 65:402–419.

Coletto, A. & Sterpi, D. 2016. Structural and geotechnical effects of thermal loads in energy walls. *Procedia Eng.* 158:224–229.

Di Donna, A., Cecinato, F., Loveridge, F. & Barla, M. 2016. Energy performance of diaphragm walls used as heat exchangers. *P.I. Civil Eng.-Geotech. 1*70:232–245.

Faizal, M., Bouazza, A., Singh, R.M. 2016. An experimental investigation of the influence of intermittent and continuous operating modes on the thermal behaviour of a full scale geothermal energy pile. *Geomech. Energ. Environ.* 8:8–29.

Hillel, D. 2003. *Introduction to environmental soil physics.* Elsevier.

Kürten, S., Mottaghy, D., Ziegler, M. 2015. A new model for the description of the heat transfer for plane thermo-active geotechnical systems based on thermal resistances. *Acta Geotech. 1*0:219–229.

Laloui, L. & Di Donna, A. 2013. *Energy Geostructures.* ISTE and John Wiley & Sons.

Rees, S.W., Adjali, M.H., Zhou, Z., Davies, M. & Thomas, H.R. 2000. Ground heat transfer effects on the thermal performance of earth-contact structures. *Renew. Sust. Energ. Rev.* 4:213–265.

Soga, K., Rui, Y. & Nicholson, D. 2015. Behaviour of a thermal wall installed in the Tottenham Court Road station box. *Proc. Crossrail Conference*, Crossrail Ltd and Federation of Piling Specialists, City Hall, London, 112–119.

Sterpi, D., Angelotti, A., Corti, D., Ramus, M. 2014. Numerical analysis of heat transfer in thermo-active diaphragm walls. *Proc. 8th NUMGE Conf.*, London: Taylor & Francis Group; Vol.2, 1043–1048.

Sterpi, D., Coletto, A. & Mauri, L. 2017. Investigation on the behaviour of a thermo-active diaphragm wall by thermo-mechanical analyses. *Geomech. Energ. Environ.* 9:1–20.

Sun, M., Xia, C. & Zhang, G. 2013. Heat transfer model and design method for geothermal heat exchange tubes in diaphragm walls. *Energ. Buildings* 61:250–259.

Todeschini, L. 2016. Personal communication.

Vieira, A. et al. 2017. Characterisation of ground thermal and thermo-mechanical behaviour for shallow geothermal energy applications. *Energies* 10(12):2044

Xia, C., Sun, M., Zhang, G., Xiao, S. & Zou, Y. 2012. Experimental study on geothermal heat exchangers buried in diaphragm walls. *Energy Build.* 52:50–55

Numerical Methods in Geotechnical Engineering IX – Cardoso et al. (Eds)
© 2018 Taylor & Francis Group, London, ISBN 978-1-138-33198-3

Stress dependency of the thermal conductivity of a regular arrangement of spheres in a vacuum

J.R. Maranha & A. Vieira
LNEC (National Laboratory for Civil Engineering), Lisbon, Portugal

ABSTRACT: In this paper the thermal conductivity of two regular arrangements of spheres in a vacuum is computed, using the FEM, as a function of the isotropic stress. This represents a lower limit for the thermal conductivity of a dry soil. The stress dependency found was in agreement with the model of Sakatani et al. (2017).

1 INTRODUCTION

The thermal conductivity of soils has a primordial influence in the performance of thermoactive geostructure systems (Vieira & Maranha, 2017). The study of the micro-structural factors influencing soil thermal conductivity in the macro-scale is relevant to formulate expressions for the aggregate thermal conductivity of the soil as a function of such variables as the void ratio, the stress, the mineral composition of the solid particles, the degree of saturation and granulometry among others. In this work, the thermal conductivity of a regular arrangement of spheres with the same radius in a vacuum is analysed. It represents a simplified version of a dry soil neglecting the air thermal conductivity. The heat conduction performed by the solid particles alone gives a lower limit for the soil thermal conductivity. It is also of interest to model the thermal conductivity of soils in planets, planetary satellites and asteroids with no atmosphere, as occurs on the Moon. In this case there is also heat transfer between the particles due to radiation, which will not be considered in this work.

According to Hertz elastic contact theory, heat conduction between spheres is only possible if there is a finite contact area, which needs that the stress acting on the spheres be non-zero. In the absence of stress only heat transfer by radiation is possible.

2 SPHERE ARRANGEMENT THERMAL CONDUCTIVITY

The analysis of the thermal conductivity of a regular spatial arrangement of spheres can be reduced to the analysis of a unit module, which by successive mirror reflections across its bounding planes can reproduce the given spatial arrangement. Symmetry dictates that the unit module boundaries parallel to the heat flow direction are adiabatic (no heat can flow across them) and those perpendicular to it have constant temperature. The thermal conductivity of a medium composed of any number of equal unit modules will be the same as that of a single unit module.

Due to the complex geometry of the solid fraction of the unit module, the time rate of heat (power) $\dot{Q}_m$ flowing through it for a unit temperature difference, ΔT, unit conductivity of the solid material, λ, and unit particle radius, R, is evaluated by means of the Finite Element Method (FEM) in steady state conditions. Because it is a linear problem, the heat flow rate $\dot{Q}_s$ in the solid fraction of a module with other values of R, λ and ΔT is given by

$$\dot{Q}_s = \lambda R \Delta T \dot{Q}_m \tag{1}$$

The values of $\dot{Q}_m$ are evaluated for different values of the δ/R ratio, defining the relative shortening of the radius of the sphere at the centre of the circular contact area. To the increasing values of δ/R will correspond increasing values of contact area, of the contact force, F, and of the associated isotropic stress acting on the arrangement of spheres.

The relative homogenised thermal conductivity, that is, including both solid particles and voids, of a module of arbitrary size, can be obtained from the heat flow computed for a unit module with the FEM, $\dot{Q}_m(\delta/R)$, by

$$\frac{\lambda_h}{\lambda} = \frac{RL_m}{A_m} \dot{Q}_m \left(\frac{\delta}{R} \right) \tag{2}$$

where λ_h is the homogenised thermal conductivity, i.e. the conductivity of the arrangement of spheres, L_m is the length of the module in the direction of the heat flow and A_m is the area of the cross section of the module perpendicular to the flow direction. Because both L_m and A_m are proportional to R and R^2 respectively, the thermal conductivity is independent of the particle radius.

Two sphere arrangements, cubic and face centred cubic (FCC), with different densities are analysed.

2.1 *Cubic packing*

The cubic arrangement of spheres is relatively loose. The module used is a half cube containing a half sphere. The length and cross section of the module are

$$L_m = R - \delta \tag{3}$$

and

$$A_m = 4(R - \delta)^2. \tag{4}$$

The relative thermal conductivity is obtained from the FEM computed heat flow rate for the unit module as

$$\frac{\lambda_h}{\lambda} = \left[4\left(1 - \frac{\delta}{R}\right)\right]^{-1} \dot{Q}_m\left(\frac{\delta}{R}\right). \tag{5}$$

2.2 *FCC packing*

The FCC arrangement of spheres is the densest possible for an arrangement of equal spheres. The module used is an equilateral triangular prism containing two half spherical wedges, each 1/12 of the volume of one sphere, in contact (see Figure 1) and is shown integrated in a larger scale assemblage in Figure 2. The length and cross section of the module are

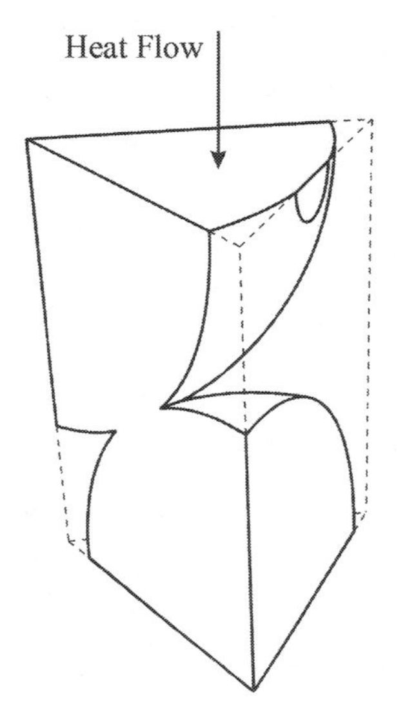

Figure 1. Module for the FCC sphere arrangement.

$$L_m = \frac{2\sqrt{2}}{\sqrt{3}}(R - \delta) \tag{6}$$

and

$$A_m = \frac{1}{\sqrt{3}}(R - \delta)^2. \tag{7}$$

In this case, the relative thermal conductivity is obtained from the FEM computed heat flow rate for the unit module as

$$\frac{\lambda_h}{\lambda} = 2\sqrt{2}\left(1 - \frac{\delta}{R}\right)^{-1} \dot{Q}_m\left(\frac{\delta}{R}\right). \tag{8}$$

The temperature results for $\delta/R=10^{-4}$ are shown in Figure 3 for the FCC arrangement. The temperature is practically constant in almost the entire volume of each sphere with a very steep gradient in the immediate vicinity of the contact area. This is even more pronounced in the case of the smaller contact areas ($\delta/R=10^{-5}$ and 10^{-6}).

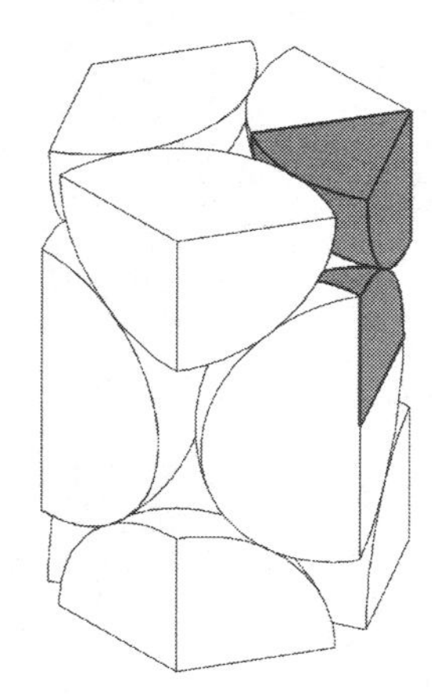

Figure 2. Larger scale assemblage showing one module for the FCC arrangement.

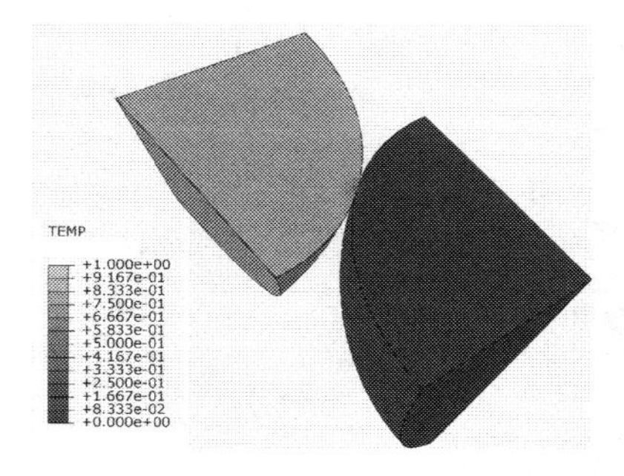

Figure 3. Computed temperature distribution in the FCC module for $\delta/R=10^{-4}$.

3 STRESS INFLUENCE

Due to intrinsic symmetries of both arrangements of spheres, equal magnitude forces at every contact between spheres imply an isotropic stress state. This, combined with the Hertz theory for contact between elastic spheres, allows the determination of the ratio δ/R as a function of the isotropic stress, p, acting on the arrangement of spheres. It is then possible to determine the effect of p on the homogenised thermal conductivity.

In the case of the cubic packing there are 6 contacts per particle and the isotropic stress is obtained as

$$p=\frac{2}{3}E^{*}\left(\frac{\delta}{R}\right)^{\frac{3}{2}} \tag{9}$$

where

$$E^{*}=\frac{E}{2(1-\nu)^{2}}. \tag{10}$$

E and ν are respectively the Young's modulus and the Poisson's ratio of the spherical particles. In the case of the FCC arrangement there are 12 contacts per particle and the resulting isotropic stress is

$$p=\frac{4\sqrt{2}}{3}E^{*}\left(\frac{\delta}{R}\right)^{\frac{3}{2}}. \tag{11}$$

The stress of the FCC arrangement is $2\sqrt{2}$ times the cubic one for the same contact area (defined by δ/R). This is expected due to the higher density of the FCC packing.

The values of the heat flow rate in the unit module (R = 1, λ = 1 and ΔT = 1), $\dot{Q}_m$, were computed with the FEM for the following values of the ratio δ/R: 10^{-6}; 10^{-5}; 10^{-4}; 10^{-3}; 10^{-2} and 0.1. In the case of the cubic packing module, the solid part is a truncated half sphere (obtained by removing a spherical cap) and an axisymmetric solution using a 2D mesh was computed. The finest mesh corresponded to the smallest contact area and comprised 38422 linear triangular elements with minimum length 5×10^{-5}. In the case of the FCC packing a 3D mesh was used and the finest mesh (δ/R = 10^{-6}) had 517,007 linear tetrahedral elements with minimum length 10^{-4}.

The relative thermal conductivity of both arrangements as a function of δ/R, determined using the values of the unit module's heat flow rates computed by the FEM, are presented in Figure 4. The relative thermal conductivity is the ratio of the homogenised conductivity to the solid material conductivity. The denser FCC packing has higher thermal conductivity.

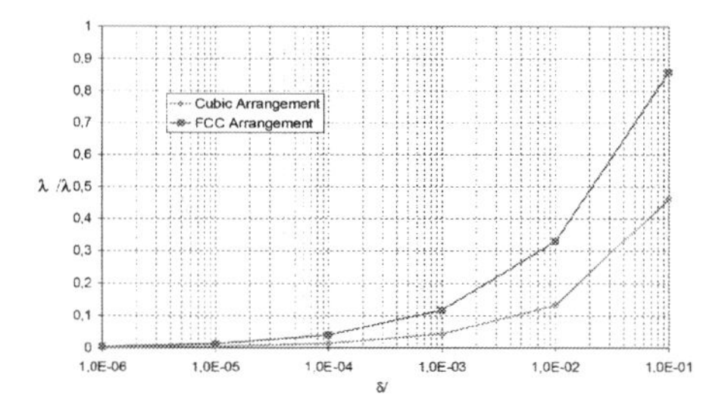

Figure 4. Variation of thermal conductivity of the cubic and the FCC arrangements of spheres relative to the solid thermal conductivity with the radial contraction at the centre of contact area.

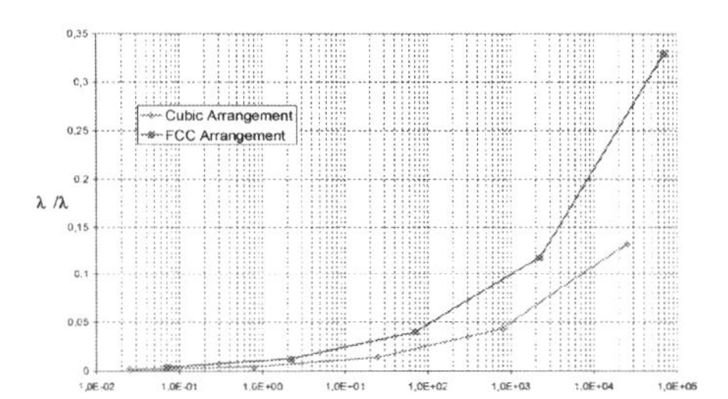

Figure 5. Relative thermal conductivity of the cubic and FCC arrangements of quartz spheres versus isotropic stress.

The conductivity as function of the stress, p, was evaluated for spherical particles made of quartz (E = 72.69 GPa and ν = 0.167) and is represented in Figure 5 for both arrangements. The denser FCC arrangement has higher thermal conductivity for the same stress.

4 COMPARISON WITH OTHER EXPRESSIONS

Many different expressions have been presented for the thermal conductivity of soils. For example, one of the expressions used is the geometric mean (Rees et al., 2000) which for a two-phase soil (dry or saturated) becomes $\lambda_h = \lambda^{1-n}\lambda_v^n$, where n is the porosity and λ_v is the thermal conductivity of the material filling the voids. In the case of a vacuum, $\lambda_v = 0$ resulting in $\lambda_h = 0$, which would be correct only if the stress were zero. Of the various expressions presented in Dong et al. (2015) none considers the effect of the stress. Sakatani et al. (2017) present an analytical model for the solid conductivity of general arrangements of equal sized spheres in a vacuum dependent on the stress as

$$\frac{\lambda_h}{\lambda}=\frac{4\sqrt{2}}{\pi^2}n_s C\sqrt{\frac{\delta}{R}}, \tag{12}$$

where n_s is the solid fraction and C is the coordination number (number of contacts per particle). A comparison between this expression and the results computed from our model is shown in Figure 6.

The values used for the solid fractions of both arrangements are

$$n_s^{cubic} = \frac{\pi}{12}\left[2 - 3\left(\frac{\delta}{R}\right)^2\left(3 - \frac{\delta}{R}\right)\right]\left(1 - \frac{\delta}{R}\right)^{-3} \quad (13)$$

and

$$n_s^{FCC} = \frac{\pi}{\sqrt{2}}\left[\frac{1}{3} - \left(\frac{\delta}{R}\right)^2\left(3 - \frac{\delta}{R}\right)\right]\left(1 - \frac{\delta}{R}\right)^{-3}. \quad (14)$$

Also, the values of the coordination numbers are $C^{cubic} = 6$ and $C^{FCC} = 12$. The ratio between the FEM solution and that of Sakatani et al. (2017) is presented in Figure 7. For most of the δ/R range, the more precise FEM solution gives values of the relative thermal conductivity between 20% and

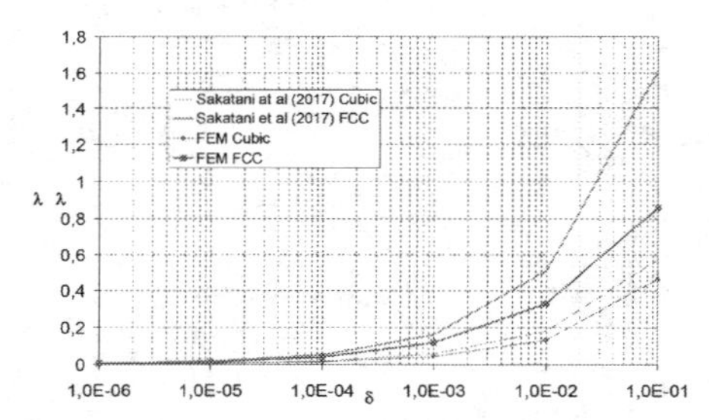

Figure 6. Comparison between the results computed with FEM and those of the expression of Sakatani et al. (2017) for Cubic and FCC arrangements and different values of the radial shortening ratio at contact.

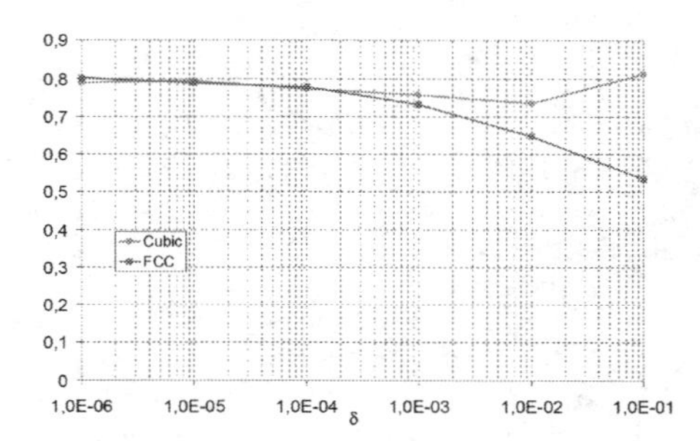

Figure 7. Ratio between the relative thermal conductivity computed with FEM and that of the expression of Sakatani et al. (2017) for cubic and FCC arrangements and different values of the radial shortening ratio at contact.

30% inferior to those of the expression of Sakatani et al. (2017).

The expression $\lambda_h / \lambda = a\sqrt[3]{p}$, meaning that the relative thermal conductivity is proportional to the cubic root of the isotropic stress for a given value of the porosity, gives a good fit to the computed results of the model presented in this work, with a = 0.0047 kPa^3 for the cubic arrangement and a = 0.0097 kPa^3 for the FCC packing. This is also the dependency on the isotropic stress in the model of Sakatani et al. (2017), but with different values of a.

5 CONCLUSIONS

The thermal conductivity of the solid skeleton of a soil in a vacuum, ignoring heat transfer by radiation, gives a lower limit to the thermal conductivity of a dry soil. Precise numerical computations using the FEM revealed an important dependence of the thermal conductivity of regular arrangements of spheres on the applied isotropic stress. This dependence is similar to that given by the model of Sakatani et al. (2017), which was in agreement with experiments performed on glass spheres in vacuum. It suggests that the stress might have a significant influence on the thermal conductivity of dry or almost dry soils.

ACKNOWLEDGEMENTS

The authors acknowledge the support provided by the FCT (Portuguese Foundation for Science and Technology), under Project Success (Sustainability of shallow geothermal systems. Applied studies to southern Europe climates. PTDC/ECM-GEO/0728 /2014).

REFERENCES

Dong, Y., McCartney, J.S. & Lu, N. 2015. Critical Review of Thermal Conductivity Models for Unsaturated Soils. Geotech. Geol. Eng., 33:207–221, DOI 10.1007/s10706–015–9843–2.

Rees, S.W., Adjali, M.H., Zhou, Z., Davies, M. & Thomas, H.R. 2000. Ground heat transfer effects on the thermal performance of earth-contact structures. Renewable and Sustainable Energy Reviews, 4: 213–265.

Sakatani, N., Ogawa, K., Iijima, Y., Arakawa, M., Honda, R. & Tanaka, S. 2017. Thermal conductivity model for powdered materials under vacuum based on experimental studies. AIP ADVANCES **7**, 015310.

Vieira, A. & Maranha, J. 2017. Thermoplastic analysis of a thermoactive pile in a normally consolidated clay. International Journal of Geomechanics, Vol. 17, Issue 1.

Numerical Methods in Geotechnical Engineering IX – Cardoso et al. (Eds)
© 2018 Taylor & Francis Group, London, ISBN 978-1-138-33198-3

Factors affecting the thermo-mechanical response of a retaining wall under non-isothermal conditions

E. Sailer, D.M.G. Taborda, L. Zdravković & D.M. Potts
Imperial College London, UK

ABSTRACT: Sustainable and low-carbon thermal energy can be extracted from the ground by means of heat exchangers buried within it. In the last decades, Ground Source Energy Systems (GSES) have been incorporated within geotechnical structures such as foundation piles, tunnel linings and retaining walls. To date, limited field data regarding thermo-active walls exist and therefore there is considerable uncertainty regarding their mechanical response under non-isothermal conditions. This paper investigates the short and long term behaviour of a hypothetical thermo-active diaphragm wall installed in London by performing fully coupled Thermo-Hydro-Mechanical (THM) Finite Element (FE) analyses using the Imperial College Finite Element Program (ICFEP). A first set of analyses shows the influence of different types of analyses on the wall's response, whereas a parametric study highlights the impact of varying ground properties on the simulated behaviour. The results demonstrate the importance of assessing the thermal load and estimating accurately the soil's thermal and hydraulic parameters as these factors may affect the serviceability and stability of thermo-active retaining walls.

1 INTRODUCTION

Shallow geothermal energy, i.e. the energy stored within the ground at shallow depths (up to 300 m—Banks 2012), is widely recognised as being an efficient, sustainable and low cost energy source for space heating and cooling.

Since the '80 s, closed loops have been incorporated within geotechnical structures, such as foundation piles, tunnels and retaining walls (Adam & Markiewicz 2009), thus creating double-purpose structures able to provide both stability and energy to buildings. Concerns regarding the stability of these structures, which may be subjected to considerable thermal loads, led to numerous field and numerical studies in the last decade, the focus of these being mainly thermo-active piles. Indeed, thermo-active retaining walls have been the subject of limited studies and many uncertainties remain regarding their safe design.

Despite the lack of field data, it is clear that retaining walls will behave in a different manner when compared to pile foundations and that the behaviour will be governed by the geometry of the wall, the type of connection with, and the stiffness of, the permanent internal structure (such as slabs) and the environmental loads applied at the exposed face of the wall (e.g. temperature, wind speed, etc.). Various numerical studies have been performed to investigate the thermo-mechanical behaviour of thermo-active walls. Soga et al. (2014) simulated the thermo-active wall installed at the Tottenham Court Road Station (London) and concluded that heating/cooling cycles may cause a delay in the dissipation of the pore water pressures and a permanent deformation of the soil mass. However, the effects on the structural forces seem to be of limited significance. Bourne-Webb et al. (2016) investigated the behaviour of the wall of a cut and cover tunnel, analysing the influence of the soil's coefficient of thermal expansion and the boundary condition at the exposed face of the wall in terms of wall horizontal movements and bending moments. The obtained results seem to indicate little effects of the thermal load on the performance of the wall. Sterpi et al. (2017) analysed the behaviour of a basement wall installed within granular material and observed that thermal loads do not greatly affect earth pressures but may have a significant effect on structural forces and displacements. In addition, the three-dimensional nature of the problem was shown to be important for thermo-active walls, as a considerable thermal gradient may develop across the width of a panel. It should be noted, however, that few studies present fully coupled transient thermo-hydro-mechanical (THM) analyses and that additional research is required in terms of soil-structure interaction.

This paper presents a numerical study of the THM behaviour of a hypothetical thermo-active diaphragm wall located in central London, with the aim of analysing the effect of the soil's properties—coefficient of thermal expansion, soil's thermal conductivity and permeability—on the thermo-mechanical response of the wall in terms of forces and displacements. However, prior to simulating the effects of thermal loads on the behaviour of the retaining wall, it is important to demonstrate the ability to accurately simulate soil-structure interaction phenomena of the employed modelling procedure. This assessment was carried out by reproducing a case study of an 18 m deep and 800 mm wide retaining wall constructed in London (Wood & Perrin 1984), for which extensive field data gathered during several construction stages exist. The analyses are carried out using the Imperial College Finite Element Code (ICFEP, Potts & Zdravkovic 1999), which is capable of performing fully coupled THM simulations. The THM finite element formulation, its implementation and validation are described in Cui et al. (2017).

2 NUMERICAL ANALYSIS

2.1 *General aspects*

Thermo-active retaining walls behave differently compared to piles given their different geometry. The analysis of the thermo-mechanical response of pile foundations is typically quantified by the restrained strain, i.e. the part of the thermally induced axial strain that is impeded from developing due to the restrictions imposed to the structure, either by the soil or the end restraints, and which causes thermally-induced mechanical loads to develop within the structures (Amatya et al. 2012 and Bourne-Webb et al. 2013). Unlike thermo-active piles, retaining walls are not fully surrounded, and therefore restricted, by soil. In fact, the part of the wall where the greatest restriction takes place is along the embedded section. Furthermore, the restrained strain for thermo-active walls is also highly dependent on the type of connection with internal structures, such as slabs.

The excess pore water pressures generated under undrained conditions, as a consequence of a change in temperature, are related to the difference between the coefficients of thermal expansion of water and soil skeleton. These thermally-induced pore water pressures are evaluated through the following equation (Cui et al. 2017):

$$\Delta u = \frac{3n(\alpha_w - \alpha_s)\Delta T}{\dfrac{n}{K_f} + \dfrac{1}{K_s}} \tag{1}$$

where Δu is the change in pore water pressure, ΔT is the change in temperature, n is the porosity, α_w and α_s are the coefficients of thermal expansion of pore water and soil skeleton, respectively, while K_f and K_s are the bulk moduli of pore fluid and soil skeleton, respectively. This equation implies that, for example, a positive change in temperature will lead to compressive (positive) pore water pressure, since $\alpha_w > \alpha_s$ (Campanella & Mitchell 1968). It should be noted that constant values for α_w and α_s are employed in all the analyses reported in this paper.

Note that, in the remainder of this text, the adopted sign convention is such that positive values refer to tensile axial forces, upwards vertical movements and compressive pore water pressures.

2.2 *Analysed problem*

An 18 m deep and 800 mm thick diaphragm wall retains an 11 m deep excavation for an underground parking, comprising three levels, of a six-storey building located in London (Wood & Perrin 1984). The excavation is supported by three temporary prop levels and three permanent concrete slab levels, each 350 mm thick, while the base slab is 1.5 m thick.

The ground profile, illustrated in Figure 1, consists of 4.8 m of Made Ground (MG), 2.0 m of Terrace Gravel Deposits (TGD), approximately 40.0 m of London Clay (LC), 12 m of Lambeth Group Clay (LGC) and 7 m of Thanet Sand (TS) overlying Chalk (CH). The groundwater table was located at 4 m below ground level and the pore water pressure was assumed to vary hydrostatically within the drained materials (i.e. Made ground, Terrace Gravel Deposits and Thanet Sand). In terms of the lower aquifer, the water table was assumed to be at the top of the Thanet Sand. Within the London Clay and Lambeth Group Clay, the pore water pressure profile was underdrained, as shown in Figure 1(a). The K_0 profile is displayed in Figure 1(b). The initial temperature

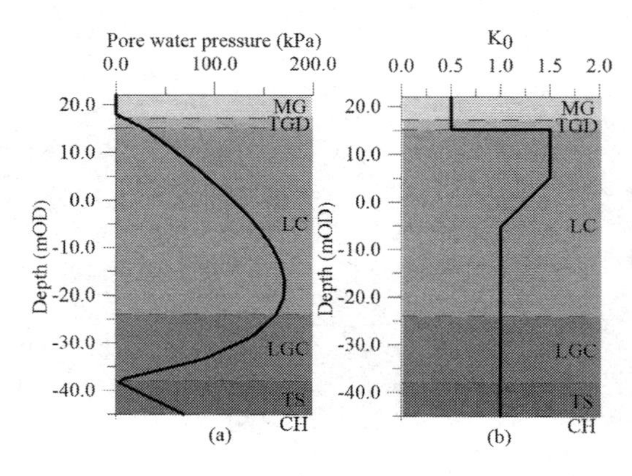

Figure 1. Initial ground conditions (a) pore water pressure and (b) K_0 profiles.

of the ground was assumed to be 13°C (Loveridge et al. 2013).

2.3 *Modelling procedure*

Plane strain FE analyses were performed using ICFEP. The excavation and construction of the permanent structures was simulated by performing coupled hydro-mechanical (HM) analyses. The subsequent hypothetical usage of the studied retaining wall as a heat exchanger was simulated using fully coupled THM analyses.

The finite element mesh, extending in the horizontal direction from half-width of the excavation to 100 m beyond the wall, is composed of 2755 eight-noded quadrilateral elements. Displacement and temperature degrees of freedom are associated to each node of the FE mesh, while for undrained materials, a pore pressure degree of freedom exists only at the corner nodes (see Cui et al. (2016) for studies on the performance of such hybrid elements in THM analyses).

Concrete was modelled as a linear-elastic material with a Young's modulus of 30 GPa, as indicated in Wood & Perrin (1984). The Made Ground was modelled as linear elasto-plastic with a Mohr-Coulomb failure criterion, whereas the remaining soil layers were modelled as non-linear elasto-plastic with a Mohr-Coulomb failure criterion and the IC.G3S non-linear elastic stiffness model (Taborda et al. 2016) to better predict the pre-yield soil response. The undrained materials (London Clay and Lambeth Group Clay) were modelled with a non-linear permeability model, according to which this property varies with mean effective stress p':

$$k = k_0 e^{-Bp'} \tag{2}$$

where k_0 and B are model parameters. The mechanical soil properties were adopted from Gawecka et al. (2017), while the hydraulic and thermal parameters are listed in Table 1. Note that, to facilitate the interpretation of the analyses' results, it was assumed that the thermal properties for all soil types were identical.

The domain was restrained from moving in the vertical direction along the bottom boundary and in the horizontal direction along the lateral and bottom boundaries. Hydraulic boundary conditions consisted of no water flow across the lateral boundaries, while the Made Ground, the Terrace Gravels and Thanet Sand were considered to be free-draining materials (i.e. pore water pressures were assumed to remain constant along the interfaces between these materials and the layers of London Clay and Lambeth Group Clay). The wall was "wished in place" and full friction was assumed at the soil-wall interface. The connection between the diaphragm wall and the internal structures was simulated as a pin connection, i.e. only axial forces can be transferred. Along the interface between the underside of the base slab and the soil, a pore water pressure of 0 kPa was prescribed, simulating the existence of a drainage system, as reported by Wood & Perrin (1984). The adopted modelling procedure and the assumed hydro-mechanical properties were shown to be adequate by the excellent agreement obtained between the predicted wall movements and the monitoring data published in Wood & Perrin (1984). These results are not presented here for brevity.

Table 1. Material properties for reference analyses (HM and THM).

Concrete	
Young's Modulus, E (GPa)	30.0
Linear coefficient of thermal expansion, α_c (m/m/K)	1.0×10^{-5}
Poisson's ratio, μ (−)	0.3
Soil	
Linear coefficient of thermal expansion of soil, α_s (m/m/K)	1.0×10^{-5}
Linear coefficient of thermal expansion of pore water, α_w (m/m/K)	6.9×10^{-5}
Permeability model parameter k_0 (m/s)	1.0×10^{-10}
Permeability model parameter B (−)	0.0023
Volumetric heat capacity of mixture, ρc_p (kJ/m^3 K)	3000
Thermal conductivity of mixture, λ (W/mK)	2.0

Before applying the thermal load, full dissipation of excess pore water pressures was allowed through the consolidation process in order to facilitate the interpretation of the results of the subsequent thermal analyses and the stiffness of all soils was reset to its maximum value (see Gawecka et al. (2017) for additional details on this procedure). A prescribed temperature change of 15°C over 10 days (i.e. 1.5°C/day) was applied to all elements of the diaphragm wall and the final temperature was kept constant for 10 years. Although this is unlikely to represent a realistic operation mode for a GSES, its use in the analysis provided valuable insight into transient phenomena taking place within the soil. It should also be noted that prescribing the temperature over the whole wall structure is a simplified approach that neglects factors such as pipe arrangement, advection within the pipes, non-uniform temperature distribution along the pipe and the heat conduction through the concrete.

Table 2. Parameters for parametric study.

Parameter	Reference analysis	Analysed range	
α_s (m/m/K)	1.0×10^{-5}	$0.5\,\alpha_s$	$2\,\alpha_s$
λ (W/mK)	2.0	$0.5\,\lambda$	$2\,\lambda$
k_0 (m/s)	1.0×10^{-10}	$0.1\,k_0$	$10\,k_0$

A further thermal boundary condition was imposed as constant temperature along the ground surface, while no heat transfer was allowed between the ground and the building through the base slab. Moreover, all other boundaries of the domain were modelled as insulated boundaries, i.e. no heat flux was allowed across these boundaries, an assumption which was shown not to affect the results as these were placed sufficiently far away from the modelled structure.

A first set of analyses aimed at identifying the influence of the type of analysis on the simulated thermo-mechanical behaviour of the diaphragm wall. For this purpose, two analyses with reference parameters (Table 1) were performed, namely a HM analysis simulating only the thermal expansion of the wall (i.e. no heat transfer is simulated within the soil) and a THM analysis, in which thermal energy is transferred through the soil. Subsequent analyses were carried out in order to investigate the effects of ground thermal conductivity, the soils' coefficient of thermal expansion and hydraulic permeability on the behaviour of the wall. These parameters were varied within ranges considered realistic for soils and are listed in Table 2. For each analysis, only one parameter at a time was varied, keeping the others equal to the ones of the reference analysis.

3 RESULTS OF THE ANALYSES

The behaviour of the wall is analysed at different time instants in order to characterise the transient effects on wall axial forces, vertical wall and ground movements and pore water pressures. Since further analyses have shown that, in the present case, the temperature seems to have a limited effect on bending moments and horizontal wall movements, these results are not presented herein.

3.1 *Effect of transient thermal behaviour*

The two analyses, HM and THM, are compared in order to provide insight into the fundamental aspects of the behaviour of a thermo-active retaining wall and to assess the transient nature of such a problem. It has been observed that the behaviour is heavily controlled by the development and subsequent dissipation of thermally-induced pore water pressures.

The changes in pore water pressures are illustrated in Figure 2 for the two analyses after 10 days and 1 year from the start of heating. In the short term, as seen in Figure 2 (a), suctions (maximum of −23.3 kPa) are generated in the HM analysis along the soil-wall interface, as a consequence of shearing along the embedded part of the wall due to its expansion. Concurrently, compressive pore water pressures – of a maximum value of 26.5 kPa—develop at the toe where the wall, upon expansion, pushes against the soil. For the THM analysis (Fig. 2 (b)), where the soil expands due to temperature changes, only compressive excess pore water pressures develop, since the linear coefficient of thermal expansion of water is greater than that of soil (see Eq. 1). The maximum compressive pore water pressures (95.0 kPa) are generated close to the toe of the wall.

In the long term, in the HM analysis (Fig. 2 (c)), the thermally-induced excess pore water pressures equilibrate (after 1 year they are negligible), while for the THM analysis (Fig. 2 (d)) this process is slower, since the magnitude of pore water pressures due to heating is larger in this case and the propagation of heat through the soil mass leads to a continuous generation of additional compressive pore water pressures.

Figure 3 (a) shows the thermally induced axial forces along the depth of the wall for different time instants. Initially, as the wall is heated and the concrete expands, compressive axial forces develop

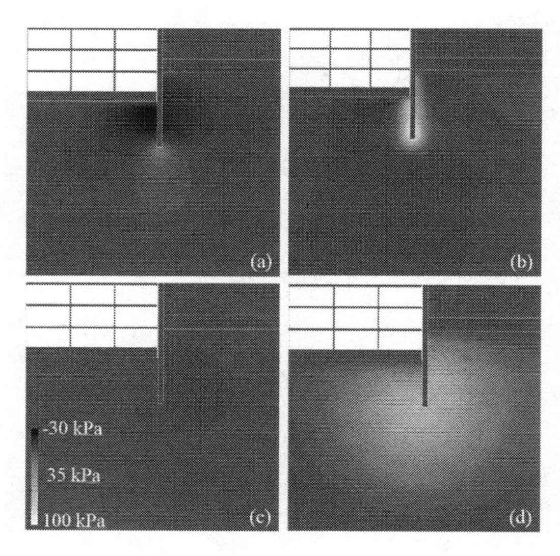

Figure 2. Changes in pore water pressures after 10 days of heating for the (a) HM and (b) THM analyses and after 1 year for the (c) HM and (d) THM analyses.

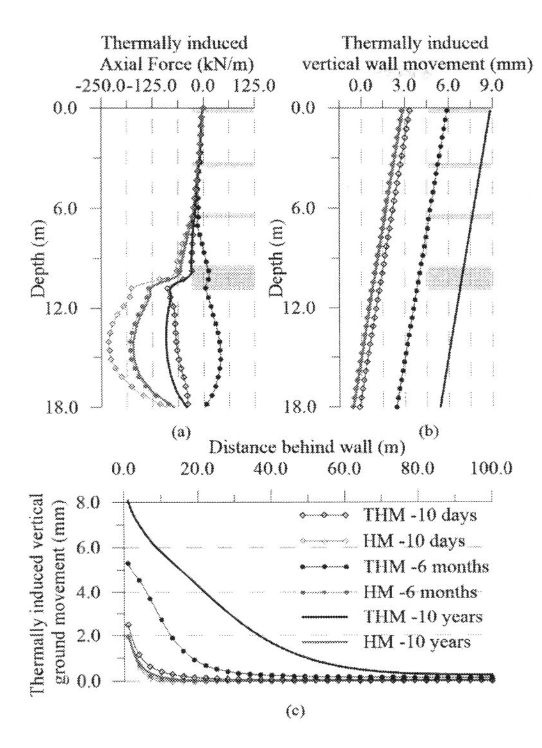

Figure 3. Influence of type of analysis on thermally induced (a) axial force, (b) vertical wall movements and (c) vertical ground surface movements.

as the soil impedes part of the wall expansion. However, it is observed that the soil restrains the wall to a much lesser extent than what has been observed in piles (see Gawecka et al. (2017) for an example of the thermo-mechanical behaviour of this type of structures). The simulated restriction is clearly larger within the embedded part of the wall, where it is fully surrounded by soil and, therefore, where larger changes in axial force are registered (maximum increase of compressive forces of -230 kN/m and -89.0 kN/m for the HM and THM analysis, respectively). Conversely, the upper part of the wall experiences only a small increase in compressive axial forces (maximum of –30 kN/m). It can be noted that the increase in compressive axial force is much larger in the analysis where the soil is not thermally active (HM analysis), since in this case the absence of thermal expansion of the ground results in a higher restriction to the wall expansion. Furthermore, in the THM analysis, the development of compressive excess pore water pressures and consequent reduction in effective stress level leads to soil swelling, which induces additional tension within the wall reducing the compressive axial force. With time, the thermally induced axial forces develop differently in the two analyses. This can be clearly observed in Figure 4, which displays the development with time of axial forces in the wall at a depth of 14.0 m below ground level, where, on average, the greatest changes in axial forces have been observed to take place. In the HM analysis, the value of compressive axial force reduces as a result of the dissipation of excess pore water pressure generated during the heating of the wall. Indeed, since the thermally induced excess pore water pressures are tensile along the shaft, their dissipation will lead to expansion of the soil during equilibration, inducing tension to the wall, hence reducing the compressive axial force. It can also be noted that this process takes place mainly within the first 6 months, after which the changes in axial force are minor as dissipation is almost completed. In the THM analysis, during the first 6 months, the compressive axial force reduces by 118.0 kN/m (a reduction of 175% in respect to the initial increase, leading to tensile forces at the end of this period), due to the thermal expansion of the soil as the heat front propagates, which reduces the restriction of the structure inducing tensile forces in the wall. As the soil mass around the wall reaches thermal steady state, the mechanical effects of consolidation prevail over the thermal ones. In fact, the tensile axial force reduces as a consequence of the dissipation of the compressive excess pore water pressures, which lead to a compression of the soil and hence the wall. As a result, after 10 years the axial force reached a final value which is compressive and 36% higher than the one due to initial heating of the wall (i.e. after 10 days).

The vertical movements of the wall and those of the retained ground are shown in Figures 3 (b) and (c), respectively. As expected, no transient behaviour is observed in the HM analysis. Conversely, the THM analysis displays an increase in vertical wall and ground movements as the soil expands when heated. It should be noted that the increase in vertical wall movements with time, i.e. after the

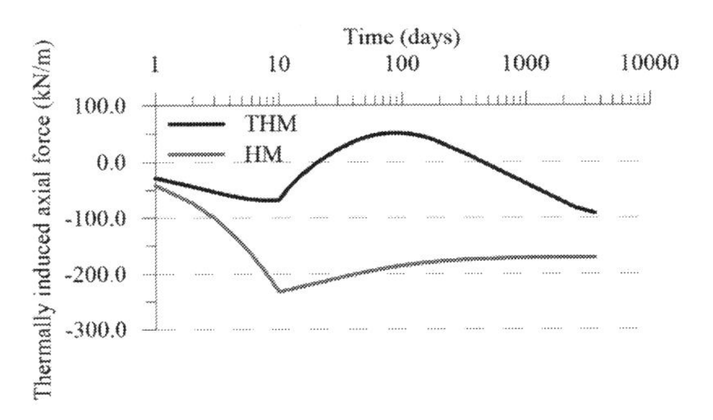

Figure 4. Influence of type of analysis on development with time of thermally induced axial force at depth of 14.0 m.

temperature of the wall stops changing, is due to the expansion of the soil, which pushes the wall upwards, as can be concluded from the vertical displaced shapes in Figure 3 (b) being parallel. Furthermore, it should be observed that the vertical expansion of the wall, of 3.45 mm, is close to the value of free expansion (3.51 mm), which explains the limited increase in axial force. However, this will reduce as more restrictions are applied to the wall, e.g. full moment connections to the slabs and the presence of a top restraint. Regarding the long-term vertical ground movements due to temperature changes behind the wall, these are quite significant (8 mm) and extend for a considerable distance, meaning that there could be an impact on the serviceability of neighbouring structures.

3.2 *Parametric study*

To provide further insight into the factors controlling the behaviour of a thermo-active retaining wall, additional THM analyses were carried out where the coefficient of thermal expansion, the thermal conductivity and the permeability were varied according to the parameters shown in Table 2.

3.2.1 *Effect of coefficient of thermal expansion*

In the previous analysis, the coefficient of thermal expansion of soil and concrete were equal, meaning that the structure and the surrounding soil were expanding equally for the same temperature change. When the soil expands less than the concrete ($0.5\alpha_s$), the former applies a larger restriction to the structure which leads to an initial increase in compressive axial force of –76.0 kN/m (12.5% more than the reference THM analysis), as shown in Figure 5(a). Conversely, when the soil expands more than the concrete, the soil induces tension in the structure hence reducing the compressive axial force (maximum value of –52.0 kN/m, i.e. 23.0% less than the reference THM analysis). With time, the axial force develops as described in the previous section, where the case with higher coefficient of thermal expansion shows a higher rate of reduction due to greater soil expansion. After reaching thermal steady state around the wall (i.e. when consolidation effects start becoming more dominant), the differences between the three analyses remain approximately constant. This suggests that the generated excess pore water pressures are similar for the three cases (indeed, the differences in pore water pressures with respect to the reference THM analysis are limited to ± 5%).

The vertical wall movements, depicted in Figure 5 (b), are very similar in the short term for all the three analyses, as the heat front has not yet propagated within the soil, meaning that limited thermal expansion has taken place. With time, as can be expected, larger movements are calculated for the analysis with larger coefficient of thermal expansion.

3.2.2 *Effect of thermal conductivity*

The thermal conductivity has a major impact on the temperature distribution within the soil. As a result, this property will also affect the overall expansion of the soil due to heating. As shown in Figure 6 (a), with a lower conductivity ($0.5\ \lambda$), the

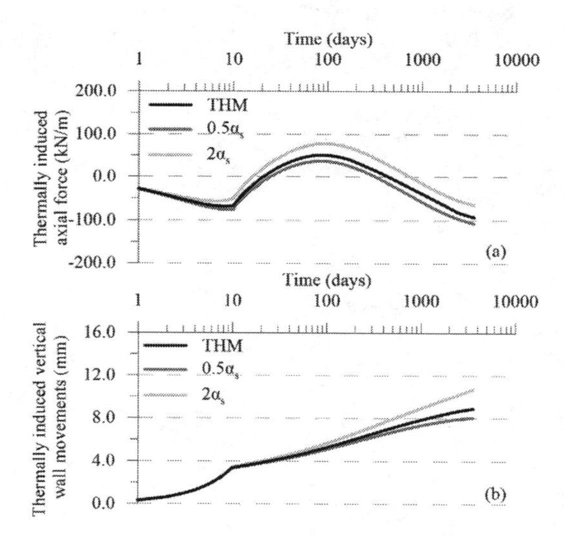

Figure 5. Influence of coefficient of thermal expansion on thermally induced (a) axial force at depth of 14.0 m and (b) vertical wall movements at the top of wall.

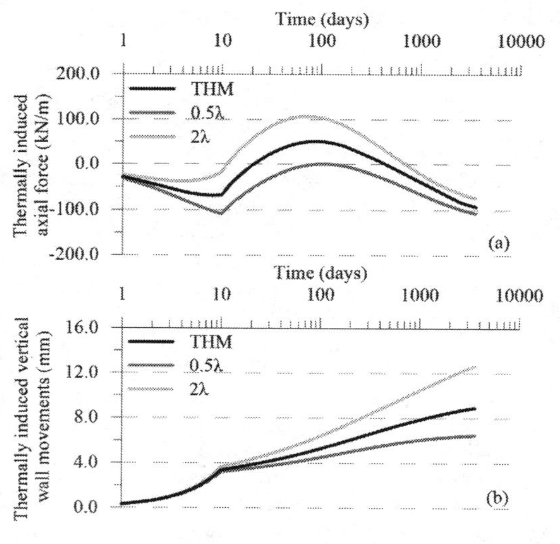

Figure 6. Influence of thermal conductivity on thermally induced (a) axial force at depth of 14.0 m and (b) vertical wall movements at the top of wall.

structure is subjected to a higher initial compressive force since the soil is cooler and therefore expands less and restricts the structure more. The opposite is true for a higher thermal conductivity of the soil. It should also be noted that the effect is more pronounced than the one seen for the coefficient of thermal expansion: at the end of the initial heating phase, the axial forces for low and high thermal conductivity are respectively of -109 kN/m and -17.5 kN/m (i.e. 61% more and 74% less than the reference THM analysis). Figure 6 (a) demonstrates also that the thermal conductivity affects the development of axial force with time. It can be observed that the peak in reduction of compressive axial force (which is due to thermal soil expansion) is reached earlier for a higher conductivity. This is because thermal steady state is reached sooner for higher values of thermal conductivity, meaning that consolidation effects become dominant earlier in time. Furthermore, the subsequent increase in compressive axial force due to consolidation takes place at a higher rate in the case with higher thermal conductivity, indicating that close to the wall dissipation of excess pore water pressures is faster, as thermally-induced pore water pressures stop occurring sooner. However, the overall larger temperatures in the soil at given time instants mean that higher initial excess pore water pressures are observed for a larger thermal conductivity (15% higher than the references analysis and 30% higher than the analysis with lower thermal conductivity), increasing the potential for higher settlements due to consolidation.

In the long term, vertical wall movements, which are displayed in Figure 6 (b), are larger with a higher thermal conductivity since higher temperatures in the soil lead to larger expansions. Moreover, the higher rate of heat transfer simulated for higher thermal conductivity results in a larger rate of expansion with time.

3.2.3 *Effect of permeability*

With the thermal properties being the same, this set of analyses investigates the influence of the hydraulic behaviour of the soil (i.e. generation and dissipation of pore water pressures) on the evolution of wall forces and displacements.

The thermally-induced pore water pressure distributions are displayed in Figure 7. Larger compressive excess pore water pressures are experienced for low permeability soil, both along the shaft and beneath the toe of the wall (maximum value of 123.0 kPa, 32% higher than that in the reference analysis), while considerably lower values are predicted when a high permeability material is simulated, due to the faster dissipation (maximum value of 50.0 kPa, 54% less than in the reference analysis). With time, the excess pore water pressures equilibrate faster in the high permeability analysis, not only because of the higher permeability, but also because the thermally-induced pore pressures to dissipate are considerably lower.

Figure 8 (a) shows that both analyses display slightly higher initial compressive axial forces with respect to the reference analysis. In particular, a lower permeability induces a slight increase of 6.2 kN/m (9%) due to a stiffer soil response (behaves in a more undrained manner); with a higher permeability the force increases by 24.0 kN/m (35%) since the compressive pore water pressures are smaller and therefore the soil swells less, applying less tension. However, the reduction with time of the compressive axial force and subsequent compression due to dissipation is more pronounced for lower permeability values: the maximum reduction reaches tensile forces of 177.5 kN/m for 0.1 k_0, while for 10 k_0 the wall forces

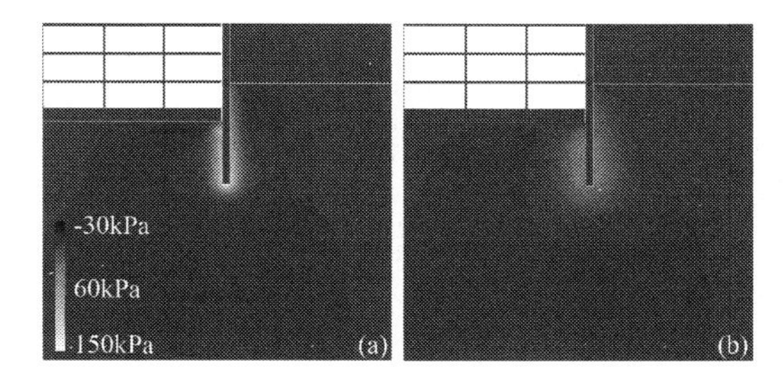

Figure 7. Thermally induced pore water pressures after 10 days—influence of permeability (a) 0.1 *k* and (b) 10 *k*.

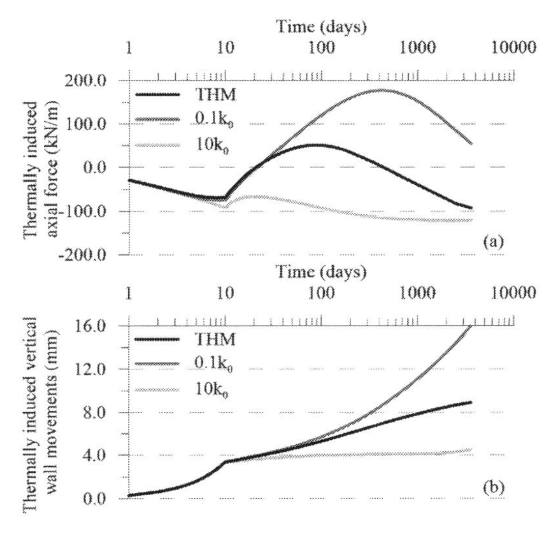

Figure 8. Influence of permeability on thermally induced (a) axial force at depth of 14.0 m and (b) vertical wall movements at the top of wall.

remain compressive. Moreover, it can be observed that not only the magnitude of axial force is influenced by the permeability, but also its transient behaviour. As seen earlier, the change in axial force is due to a combination of effects related to excess pore water pressure dissipation and reduction in restriction provided by the soil through thermal soil expansion. Clearly, in this study the latter is equal for all three analyses, since the thermal parameters are the same. Therefore, the observed differences between the three analyses confirm that the dissipation of thermally-induced excess pore water pressures has a major role in the response of the structure. Indeed, the high permeability soil experiences limited changes in axial force with time, whereas the low permeability soil requires a greater amount of time to reach the peak in reduction of compressive axial force because pore water pressures are dissipated at a slower pace. In addition, the larger amount of excess pore water pressures generated during the initial heating stages lead to greater transient effects. This is confirmed by the analysis of the wall displacements which are depicted in Figure 8 (b). These transient effects are more pronounced in the case of less permeable soil since slow drainage impedes the dissipation of excess pore water pressures and thermal phenomena prevail over consolidation (which would lead the soil to settle).

4 CONCLUSION

In this paper, the thermo-mechanical behaviour of a retaining wall installed in London Clay subjected to a heating load was investigated by performing a series of fully coupled thermo-hydro-mechanical finite element analyses. The complexity of a retaining wall problem is related to its geometry, the connection to structural components and the environmental loads at the exposed face of the wall.

In order to facilitate the interpretation of the results, pore water pressures due to excavation were allowed to dissipate before applying the thermal load, while pinned connections were assumed between the wall and the internal structures. Furthermore, the thermal load was applied as constant temperature over the whole cross-section of the wall for a period of ten years to provide insight into transient effects.

It was shown that the behaviour of a thermo-active retaining wall is strictly related to the development of thermally-induced excess pore water pressures, which are compressive when the soil is heated, due to the coefficient of thermal expansion of water being larger than that of the soil structure. In a coupled THM analysis, initial compressive axial forces develop due to heating as the wall expands and the soil restricts part of the expansion. With time, the axial forces decrease for a period of time, which can in some cases be substantial, before a final increase in the long term is observed. The initial reduction in axial forces with time is a consequence of soil expansion which reduces restriction, whereas the following increase is due to the dissipation of compressive excess pore water pressures which induce settling of the soil and consequent compression of the wall. These mechanical effects of consolidation become important once thermal steady state is reached. Although the values of thermally-induced axial forces appear to be relatively low, these may be unconservative as different connections between the wall and the permanent structure may lead to larger forces. Therefore, when designing thermo-active walls, the magnitude of the thermally-induced forces should be assessed using an accurate representation of its connections to the internal structure.

The effect of the transient phenomena was investigated by comparing the results of simulations with and without heat transfer in the soil. The initial changes in axial forces due to the thermal load are greater in the latter case as the soil imposes a larger restriction to the structure. Since the soil was not thermally active, the generated excess pore water pressures were very different when compared to those obtained in the fully coupled THM analysis. In fact, these were tensile along the shaft and of a much lower magnitude, which induced a decrease of the axial force with time, displaying therefore a very different transient behaviour from that observed in the THM analysis. It was also shown that the long-term movements and changes in forces are greater in the case where the soil was thermally-active, as it is able to expand as its temperature increases.

The performed parametric studies demonstrate the effects of the soil's coefficient of thermal expansion (α_s), thermal conductivity (λ) and permeability (k_0). It was found that a lower thermal conductivity or coefficient of thermal expansion lead to larger compressive axial forces both in the short and long terms, although it should be noted that the effects of α_s are significantly less pronounced. Conversely, higher thermal conductivity or coefficient of thermal expansion cause larger vertical wall movements in the long term. In addition, the permeability, as expected, influences the dissipation of excess pore water pressures, with a lower value of this property leading to higher excess pore water pressures. Lower permeability soils are also subjected to greater transient effects, with noticeably greater changes in axial forces registered compared to a soil with a larger value of permeability. Moreover, the displacements are larger for a less permeable soil as the dissipation of excess pore water pressure is delayed, enhancing the thermal effects.

According to these results, it is evident that adequate estimates of the hydraulic and thermal parameters of the soil are important to guarantee the safe and efficient operation of thermo-active walls.

REFERENCES

Adam, D. & Markiewicz, R. 2009. Energy from earth-coupled structures, foundations, tunnels and sewers. *Géotechnique* 59(3): 229–236.

Amatya, B., Soga, K., Bourne-Webb, P.J., Amls, T. & Laloui, L. 2012. Thermo-mechanical behaviour of energy piles. *Geotechnique* 62 (6): 503–19.

Banks, D. 2012. *An Introduction to Thermogeology: Ground Source Heating and Cooling.* 2nd Edition. Chichester: Wiley-Blackwell.

Bourne-Webb, P.J., Soga, K. & Amatya, B. 2013. A framework for understanding energy pile behaviour. *Proceedings of the Institution of Civil Engineers—Geotechnical Engineering* 166(2): 170–177.

Bourne-Webb, P.J., Bodas Freitas, T.M. & Da Costa Gonçalves, R.A. 2016. Thermal and mechanical aspects of the response of embedded retaining walls used as shallow geothermal heat exchangers. *Energy and Buildings* 125: 130–141.

Campanella, R.G. & Mitchell, J.K. 1968. Influence of temperature variations on soil behaviour. *ASCE Journal Soil Mechanics and Foundation Engineering Division* 4 (3): 709–734.

Cui, W., Potts, D.M., Zdravković, L., Gawecka, K.A., Taborda, D.M.G. 2017. An alternative coupled thermo-hydro-mechanical finite element formulation for fully saturated soils, *Computers and Geotechnics*, In press.

Cui, W., Gawecka, K.A., Taborda D.M.G., Potts, D.M., & Zdravković, L. 2016. Time-step constraints in transient coupled finite element analysis. *International Journal for Numerical Methods in Engineering* 106(12), 953–971.

Gawecka, K.A., Taborda D.M.G., Potts, D.M., Cui, W., Zdravkovic, L. & Haji Kasri, M. 2017. Numerical modelling of thermo-active piles in London Clay. *Proceedings of the Institution of Civil Engineers—Geotechnical Engineering*, 170(3): 1–19.

Loveridge, F., Holmes, G., Powrie, W., & Roberts, T. 2013. Thermal response testing through the Chalk aquifer in London, UK. *Proceedings of the Institution of Civil Engineers. Geotechnical Engineering* 166(2): 197–210.

Potts, D.M. & Zdravković, L. (1999) *Finite element analysis in geotechnical engineering. theory*. London: Thomas Telford.

Schroeder F.C., Potts D.M. and Addenbrooke T.I. 2004. The influence of pile group loading on existing tunnels. *Géotechnique* 54 (6): 351–362.

Soga, K., Qi, H., Rui, Y. & Nicholson, D. 2014. Some considerations for designing GSHP coupled geotechnical structures based on a case study. *7th International Congress on Environmental Geotechnics.* Melbourne.

Sterpi, D., Coletto, A., & Mauri, L. 2017. Investigation on the behaviour of a thermo-active diaphragm wall by thermo-mechanical analyses. *Geomechanics for Energy and the Environment* 9: 1–20.

Taborda, D.M.G, Potts, D.M. & Zdravković, L. 2016. On the assessment of energy dissipated through hysteresis in finite element analysis. *Computers & Geotechnics* 71: 180–194.

Wood, L.A. & Perrin, A.J. 1984. Observations of a strutted diaphragm wall in London clay: A preliminary assessment. *Géotechnique* 34(4): 563–579.

Numerical Methods in Geotechnical Engineering IX – Cardoso et al. (Eds)
© 2018 Taylor & Francis Group, London, ISBN 978-1-138-33198-3

Energy efficiency evaluation in thermoactive geostuctures: A case study

J. Sequeira
Mota-Engil México, México

A. Vieira
National Laboratory of Civil Engineering, Lisbon, Portugal

R. Cardoso
Lisbon Technical University, Lisbon, Portugal

ABSTRACT: A numerical study was performed for evaluating the geothermal potential of foundation piles used as heat exchangers with the ground, based in a case-study of a building in Aveiro University Campus. The influence of the soil thermal properties, the location of the water table level and the presence of the building's slab on the thermal performance of the pile were investigated regarding the efficiency of the system. Climate data records concerning temperature from different sources in the city of Aveiro were considered. As expected, when the water table is located closest to the surface, the GSHP performs better than if considered at lowest point. Overall, considering a slab as a temperature boundary between soil and building hinders the efficiency of the thermo-active pile, reversing the direction of the heat flux in the soil, if a long term analysis is considered (5 years).

1 INTRODUCTION

Ground source heat pumps (GSHP) are a mature technology that allows the transport of thermal energy from a heat source to a heat sink by means of an electrical compressor. This technology may reach point-of-use efficiency of 500% and over and uses a renewable source of energy, reducing greenhouse emissions by more than 66% and using less than 75% of electricity when compared with conventional HVAC systems (*e.g.* Brandl 2006; Rees, 2016).

In this work, an actual GSHP system is analyzed regarding its thermal interactions with the surrounding soil and with the building associated with it. The first goal to be met consists on investigating and outlining the different aspects that affect the performance of an actual ground source heat pump (GSHP) system working since 2013, in an academic building located in Aveiro, Northwest of Portugal. The building is equipped with a shallow geothermal heat pump system (GSHP) designed to support most part of the energy demands for the building's acclimatization.

The work has been carried out for a preliminary assessment of the thermal behaviour of the mentioned shallow geothermal heat pump system. This analysis is performed under an efficiency perspective, by evaluating how well can the soil transfer heat in different conditions and what is the thermal response of that soil under the thermal load imposed by the GSHP. The soil conductivity was obtained by means of empirical relations and through laboratory characterization tests. The modeling of a single pile (0.60 m diameter and 10 m length) was performed using FLAC (Itasca, 2011) and the main results are presented in (Sequeira, 2017).

2 SHALLOW GEOTHERMAL ENERGY

2.1 *The different uses for geothermal energy*

Geothermal energy can be sorted into three different categories based on the temperature level of the source: (i) high enthalpy resources, used mainly for electricity generation; (ii) medium enthalpy resources, used mostly for direct heating; (iii) low enthalpy resources, used for indirect heating (*e.g.* Willliams *et al.*, 2011).

Low enthalpy resources may also be called shallow geothermal resources, because they use the thermal energy existing in the shallow layers of the foundation soil. The concept behind shallow geothermal energy utilization states that one may harness or dump thermal energy from and to the soil, respectively. In the Winter, thermal energy would be gathered from shallow soil and transported to a building in order to heat the colder air. The opposite occurs in the Summer season, where soil works as a deposit for the thermal energy retrieved from a building's usable space. Such processes are associated with indirect heating and cooling, which requires the support of certain machines called heat pumps. Heat pumps use the same principle as a refrigerator, moving heat from one place to another, with the help of a little electrical energy input.

2.2 *Heat transfer in soils*

There are three main modes of heat transfer in soil (*e.g.* Rees *et al.* 2000): convection, conduction and radiation. Conduction heat results from the random spread of heated particles that occurs when a higher energy region contacts with a lower energy region or, in other words, heat diffusion. Convective heat transfer combines diffusion with the transport of heated particles by the motion of a fluid, which is what defines advection. Radiation is related to the heat transfer across vacuum or a transparent medium by propagation of electromagnetic waves.

Thermal properties are what determines the behaviour of materials when subject to any given thermal action. The most important thermal properties are thermal conductivity () and heat capacity (C). The heat equation can be written as Equation 1, assuming identical thermal conductivity in the three directions of space, where T is temperature, ρ is density and ∇^2 is the Laplacian operator.

$$\nabla^2 T = \frac{\rho C}{\lambda}\frac{\partial T}{\partial t} \tag{1}$$

Soils are heterogeneous materials, which comprise three different phases: solid, liquid (mainly water) and gas (mainly air, assumed dried). Therefore the concept of effective thermal properties λ_{eff} and C_{eff} may be used to characterize the soil's combined thermal conductivity and heat capacity, respectively, among the individual properties of the multiphase constituents. These properties may be determined by empirical methods, laboratory tests and field tests (*e.g.* Vieira *et al.*, 2017).

2.3 *Shallow geothermal heat transfer systems*

There are two main approaches to effectively use low enthalpy thermal energy. One of them gathers heat from a heat wasting source and stores it underground in an underground thermal energy store (UTES) for further use. The second method utilizes the already mentioned ground source heat pumps (GSHP) to transfer heat from a source to a usable space. Although both systems are similar in many concepts, for the purpose of the case study heat transfer systems refer to the applications using GSHP.

A ground source heat pump system can be decomposed in three global components: primary circuit, heat pumps and secondary circuit. The primary circuit establishes thermal trades with the soil, using of lengths of heat exchanger pipes installed in the soil. The secondary circuit transports the usable heat energy to the end user. Both these circuits are connected by the heat pump, which transports heat from one circuit to another, with the help of a small electrical input (*e.g.* Brandl, 2006).

The primary circuit is always installed by means of a geotechnical application, including grouted boreholes, horizontal trenches, ground water wells, energy foundations. Energy Foundations are the denomination given to structural foundations equipped with heat exchanger pipes integrated into its steel mesh. The pipework involved in the mentioned solutions are usually made from polyethylene due to its flexibility and durability. Diameters range from 25 mm to 60 mm. This work focuses on a case study of a building that includes, in its design, more than 80 energy piles (Lapa, 2014).

3 CASE STUDY

3.1 *Building overview*

The building evaluated in this investigation is designated by CICFANO (*Complexo Interdisciplinar de Ciências Físicas Aplicadas à Nanotecnologia e Oceanografia*), and was built in 2012. The GSHP was integrated in the project since the construction of the building, but it has only been functional since mid-2013. The building is founded on 110 piles, 85 of which being equipped with heat exchanger pipes (Figure 1). The piles are 10 meters long and diameters vary between 0.4 m and 0.6 m.

The acclimatization of the building results of a combination of techniques, including a GSHP and a Biothermal Heat Pump System (BHPS) as heat sources. The geothermal system is controlled in real-time by a software, and the heat pump equipment works with a COP of 4.1.

In order to study the thermal behaviour of the soil underneath the CICFANO building, when subject to the thermal loads of the GHPS, three elements must be assessed: climate; energy demands of the building and soil thermal characterization.

3.2 *Climate analysis and energy demand*

For the climate analysis, temperature data was retrieved from two sources (Campus of Aveiro

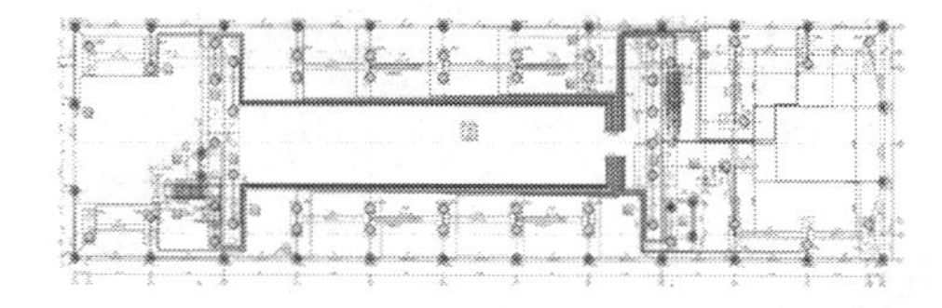

Figure 1. CICFANO's building foundation and hydraulic network schematics (no scale).

University weather station and https:// www. worldweatheronline.com). The best analytical fit for the different temperature data was a sinusoidal function (R^2>0.98) (Sequeira, 2017). After comparing the curves obtained for both sources and for different time frames, the chosen function was:

$$T = 15.45 + 4.52 * \sin\left(t * 2\pi / 31536000 + 2.36\right) \quad (2)$$

This function is used in the numerical simulation to represent the climate action at the surface of the soil. This specific curve represents temperatures felt during the 2013/2014 season (year of the inauguration of the GSHP from the CICFANO).

This assessment of energy demand focuses on two aspects: what are the building requirements and what trades are expected from the thermo-active piles. In accordance with the data provided by the University of Aveiro, the building's requirements are: 96.1 MWh/year for heating and 62.3 MWh/year for cooling. Due to the limited depth of the energy piles, the system is only expected to satisfy 75% and 65% of heating and cooling demands, respectively. As a simplification, it is assumed that building only requires heating if temperatures drop below the average temperature and, similarly, it only requires cooling if temperatures rise higher than that value (Sequeira, 2017). It is assumed, also as a simplification, that the thermal load on a single pile is applied uninterruptedly over 365 days of cooling/heating. The resulting step function of the heat load applied on a single pile during one year is in Figure 2.

3.3 *Soil analysis*

The soil from the CICFANO's foundations can be divided in an upper sand layer, about 5 meters thick and a stiff lower silty clay layer with at least 15 meters of thickness. Disturbed samples from both layers were analyzed in the laboratory regarding their grain size distribution. Classification and results for that analysis can be seen in Table 1.

The thermal conductivity of the upper sand layer was measured by Cruz (2017). For the dry sand, a dry effective thermal conductivity of 0.44 W/mK was obtained (void ratio $e_0 = 0.334$ and degree of saturation $S_R = 0\%$).

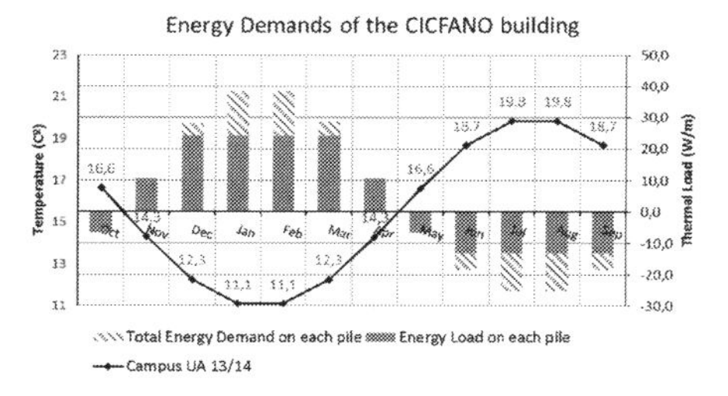

Figure 2. Energy demands of the CICFANO building.

Using that value as reference, investigations were performed on the lower clayey soil, in order to establish a relation to estimate the conductivity in fully saturated conditions. Table 2 shows the results obtained from the analysis of an undisturbed sample from the lower layer, regarding its void ratio, porosity and degree of saturation. Table 3 shows the results obtained from the mineralogical evaluation performed using the reference intensity ratio (RIR) method on the X-Ray diffraction test.

According to the mineralogical evaluation, both layers will have similar thermal conductivity of the solid phase, *s*. Since sample from the lower layer displays a porosity of 32.6%, effective thermal conductivity may be calculated empirically estimated to be λ_{eff} = 1.96 $W.m^{-1}.K^{-1}$considering saturated state and a geometric average arrangement among the constituents (Rees *et al.*, 2000). As a summary, the geotechnical model used in the numerical simulation is presented in Figure 3.

Table 1. Foundation soil identification.

SP: Poorly Graduated Sand				ML-CL: Silty Clay with sand			
Cc	Cu	% Gravel	% Sand	LL	PL	PI	% fines*
0.82	3.75	27	71	22.9	17.8	5.1	75

*Particles with diameter D 0.075mm

Table 2. Foundation soil characteristics.

w (%)	*Gs*	γ_{dry} (KN/m^3)	*Sr* (%)	e_0
18.2	2.70	18.2	100	0.483

Table 3. Mineralogical composition of the foundation soil.

Crystalline Composites (%)					
Layer	Quartz	Microcline	Abite	Microcline	Orthoclase
Upper	93	1	1	4	93
Lower	92	1	1	3	93

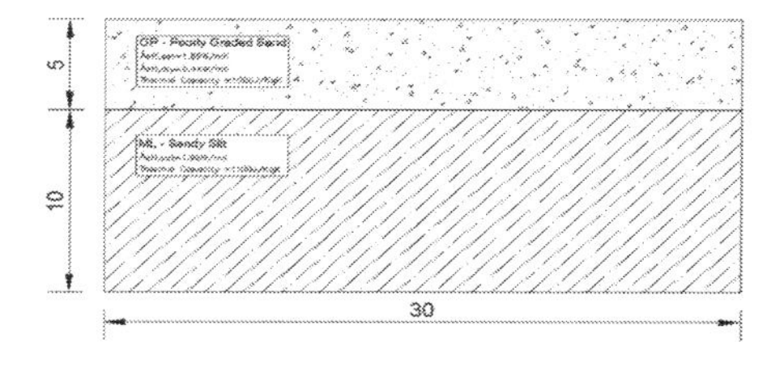

Figure 3. Geotechnical model of the foundation soil of the CICFANO building (dimensions in meters).

4 NUMERICAL MODELLING

4.1 *Mesh used and profiles analyzed*

The model used for the analysis performed is presented in figure 4. The grid for the model of the pile was defined by Sequeira (2017). It was found that the grid should have at least 40 meters of width. The inputs and boundary conditions adopted depend on the uses of the model: verification and simulation of the case study. They will be explained latter.

Figure 5 presents the location of the different profiles analyzed in order to investigate heat conduction through the soil, as well as the points which results are analyzed in the next sections.

4.2 *Preliminary numerical verifications*

Prior to the development of the case study model using this numeric tool, a few summary verification problems are solved in order to demonstrate the reliability of upcoming results. The goal is to compare the results obtained using FLAC with well described analytical solutions. A summary of the verifications is presented in Table 4. They consist in defining two models (infinite line and infinite half space), in which temperature was applied and thermal conduction was computed. The numerical results were compared with the analytical results (in Table 4).

Numerical outputs are very close to analytical values (see Figure 6), with differences lower than 0.05°C. Both transient and steady state conduction have been verified for two different kinds of heat source: a temperature boundary and a heat flux boundary.

4.3 *Numerical modeling of a thermo-active pile*

An evaluation of the thermal behaviour of the CICFANO GSHP system was performed,

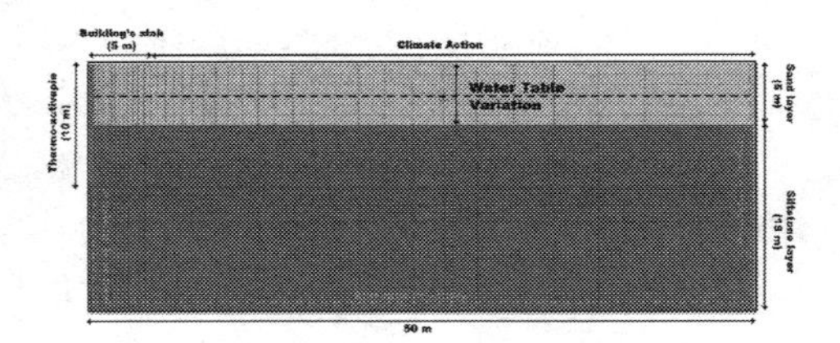

Figure 4. Model used in the numerical simulation with the relevant boundaries, inputs and dimensions.

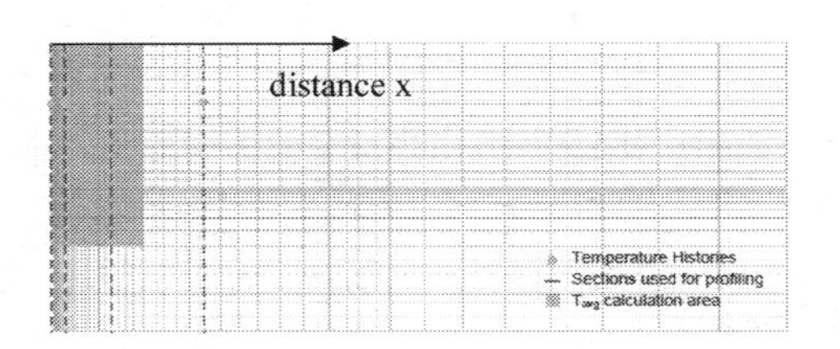

Figure 5. Detail of the different measurements performed.

considering an efficiency perspective. Three cases were considered, the first two based on two realistic positions of the water table, and the third focused on the influence of the building's bottom slab as a boundary condition.

Case 1 considers the sand layer fully saturated, which corresponds to winter case. Case 2 considers the sand layer completely dry, which corresponds to summer case. Case 3 is based on the summer case model (Case 1), and is performed taking in consideration the existence of a slab on the surface of the model, surrounding the head of the pile (assumed to be in the corner of the building). This slab is a realistic boundary condition modelled as an applied temperature (constant indoor temperature). Table 5 resumes the 3 cases all carried out under soil surface temperatures cyclic variation dictated by Eq. 2 and a step function pile thermal load as shown in Figure 1.

These models are defined considering axisymmetric conditions. Isotropic conduction was the chosen thermal model for all simulations. Mechanical stress-strain implications induced by temperature changes are not computed. All the input parameters are considered to be temperature independent. The simulated pile is 10 meters long and has radius of 0.60 mm, according to the

Table 4. Verifications done to validate the model.

Model	Verification	Analytical solution (cited by sequeira, 2017)	Grid
Infinite line source	Steady state conduction	Nowacki (1962)	20 × 20 and 40 × 40
	Transient state conduction	Nowacki (1962)	40 × 40
Infinite half space	Steady state conduction	Fourier Law of conduction	40 × 40
	Transient state conduction	Wealthy (1974)	40 × 40

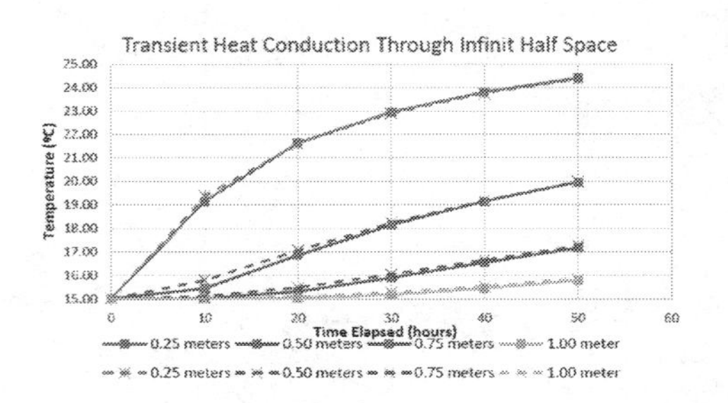

Figure 6. Transient heat conduction through infinite half space, measured in 4 points during the first 50 hours of the simulation (dashed lines are FLAC estimations), points in Figure 5.

structural design. The size of the simulated space was defined based on pile length. Hence, 5 times the length of the pile in terms of horizontal dimension and 2 times in terms of the vertical dimensions.

The simulation performed computes temperature variations in the soil mass. However, different kinds of temperature data can be extracted from models such as the one shown in Figures 7 and 8 (evolution in depth for different time intervals, for cases 1 and 2 and for cases 2 and 3):

- Vertical temperature profiles: Values of the temperature registered on every grid point located at a certain distance x from the axis.
- Extreme temperature profiles: For the total duration of the simulation, these profiles specifically register the highest and lowest values of temperature computed on each grid point located at a certain distance x from the axis.
- Temperature histories: For a certain grid point with coordinates (x,y), a history registers the values of temperature computed on each time step.
- Temperature average (T_{avg}): Variable formulated to determine the average temperature felt among a certain number of grid points of the model.

When modeling the soil as fully saturated or completely dried, the only difference is in the values of the effective thermal conductivities adopted for the layer of sand. For a simulation lasting 365 days, the vertical profiles with the maximum and minimum temperatures are presented in Figure 7.

The inclusion of a slab above the pile axis is a simple attempt to replicate the existence of the CICFANO building on top of the foundation soil. In the numerical studies it was assumed that the floor slab showcases a constant temperature equal to the indoor room temperature of 22ºC. The simulation was performed only considering the dry soil conductivity, which is considered the worst scenario.

Temperatures obtained for case 2 (dry sand) are more extreme than the ones obtained for case 1 (saturated sand). However, average temperatures throughout elapsed time are not affected by that fact. The influence on the thermo-active pile on the soil temperatures is becomes clear by examining the differences in the extreme temperature profiles, between the reference profile (no pile installed) and the remaining profiles (Figure 8).

Table 5. Summary of the analysis carried out.

	Water table	*Slab boundary*	λ (W/m.K)
Case 1	At surface	Non existing	Sand: 1.89 Silt: 1.96
Case 2	At 5 m depth	Non existing	Sand: 0.44 Silt: 1.96
Case 3	At 5 m depth	Constant indoor temperature	

The introduction of the slab as a boundary condition (case 3) produces globally higher temperatures than cases 1 and 2. Furthermore, minimum temperatures in the soil are radically increased from cases 1 and 2. In case 3, temperatures are never lower than their initial value in some areas (always increasing).

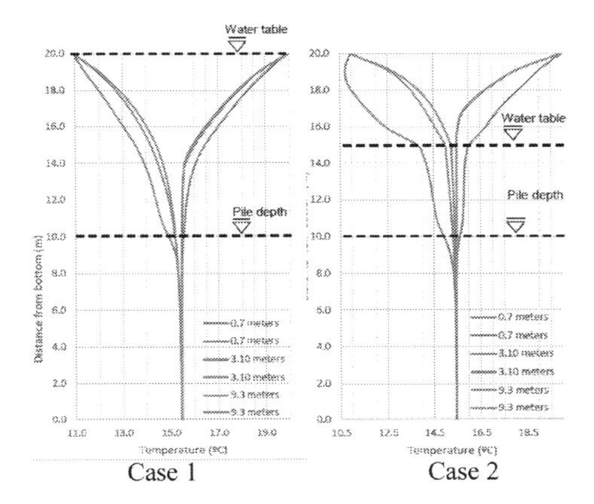

Figure 7. Maximum and minimum temperatures computed for case 1 (left) and case 2 (right).

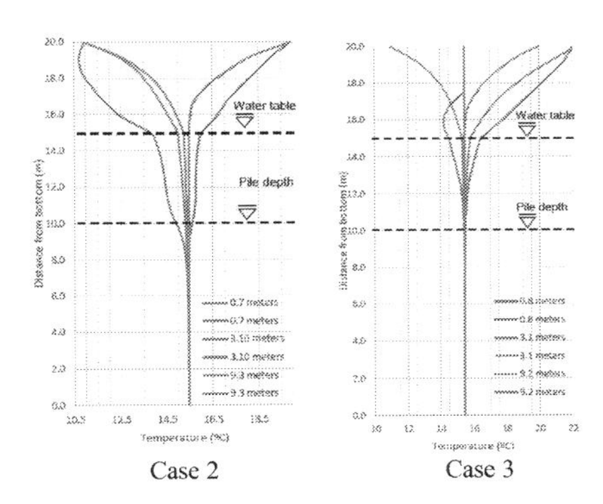

Figure 8. Maximum and minimum temperatures computed for case 2 (left) and case 3 (right).

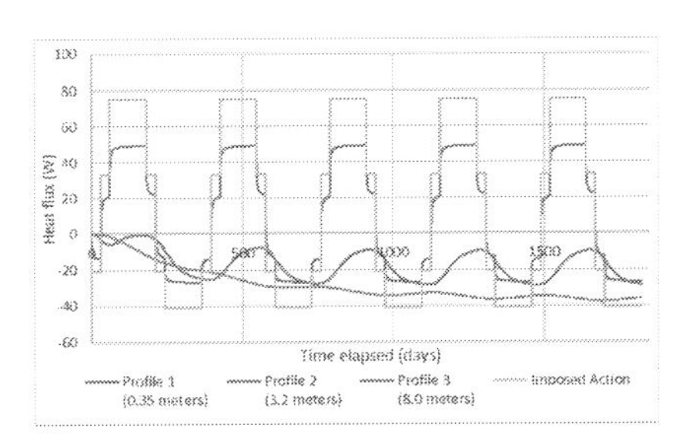

Figure 9. Thermal power held by the soil mass in 3 different areas for case 3, over 5 years.

From the data acquired in the previous simulations, it is possible to elaborate a calculation of the heat flux passing through the soil, perpendicularly to the pile axis (horizontal flux). Since this analysis is purely conductive, the heat flux on each point was calculated by using Fourier's law of conduction. After calculating the heat flow for each point of a profile was calculated, they were integrated in the cylindrical area respective to each profile (area identified in Figure 5), becoming a measurement of the energy per second (Watt) transferred by the soil at a certain distance from the source.

Overall, case 1 (saturated) transfers heat better than case 2 (dry). In fact, the flux wave through the soil travels faster and more efficiently in the saturated sand due to the high thermal conductivity of water when compared with that of dry air. In case 3, heat flux reverses its natural direction right after the first year, when compared with cases 1 and 2. This is a result of the presence of the top slab at the same time as the pile. Such analysis can be seen in Figure 9.

5 CONCLUSIONS

The study carried out to evaluate the energy efficiency of a single thermo-active pile installed in a building located at Aveiro University Campus (CICFANO) was described. For the two soil layers of the foundation a similar constituent arrangement was assumed and identical thermal conductivities were estimated for both in fully saturated conditions. Therefore, the different thermal parameters used depended mostly on the hydrological conditions assumed for the soils.

The numerical results showed that the saturated scenario produces slightly higher temperatures in the soil over time than the dry one. The dry soil seems to be incapable of transferring heat as efficiently as the saturated soil. A reduction in the amount of heat transferred by the dry soil has a real impact on the building's acclimatization process.

Among the three cases studied, temperatures either stabilize after the first year (case 2), or they rise at an average rate of 0.05°C per year, over a time lapse of 5 years (case 1). Including the presence of the slab into the simulation causes the system to become overheated. In fact, after 3 years, the pile is attempting to absorb heat from the soil, and instead, heat is moving away from the pile.

Overall, using a conduction only analysis and assuming no water advection within the soil mass, the GSHP system seems to be sustainable over a 5-year span for cases 1 and 2. However, assuming that the building's slab is at a constant indoor temperature, the system will not be able to respond efficiently, at least in the long therm. Nevertheless, the GSHP was analyzed as a separate part of the CICFANO building. In reality, the complementary biological source heat pump as proven effective in supplying thermal energy when the GSHP cannot. Therefore, the building's acclimatization using renewable energy sources is always secured.

The study of peak thermal loads on the thermo-active and the system's "down time" was not studied in this work. This is important because certain hydro-geological conditions may allow a typical amount of heat to pass through the soil, but a peak load might not have the same response.

Including a coupled advection-conduction model in this simulation, which would allow heat to be transported by the moving water inside the sandy layer. This would be a step towards accuracy, since the CICFANO building is close to the river and the sea. Therefore, a strong underground water percolation is expected. A thermo-hydro-mechanical coupled analysis provides a complete understanding of the system and this will be done in the future.

ACKNOWLEDGMENT

Acknowledgement is due to FCT, project Success PTCDT/ECM-GEO/0728/2014, for the funding.

REFERENCES

Brandl, H. (2006). Energy foundations and other thermo-active ground structures. *Geotechnique*, *56*, 81–122.

Cruz, J.M.Q (2017). *Experimental determination of the parameters necessary to modelling thermo-active structures*. MSc Thesis, Instituto Superior Técnico, University of Lisbon, Portugal (in Portuguese).

Itasca (2011). FLAC Online Manual, version 5.0, 3058 pp.

Lapa, J.M. (2014). *Casos de aplicação da geotermia superficial em sistemas de estruturas termoactivas em edifícios escolares da Universidade de Aveiro.* 2° Seminar of the Portuguese Platform of Geotthemal Energy, Lisbon (in Portuguese).

Rees, S. (2016). *Advances in Ground Source Heat Pump Systems*; Woodhead Publishing: Sawston, UK, 2016; ISBN 978-0-08-100311-4.

Rees, S.; Adjali, M.; Zhou, Z.; Davies, M.; Thomas, H. (2000). Ground heat transfer effects on the thermal performance of earth-contact structures. *Renew. Sustain. Energy Ver.*, *4*, 213–265.

Sequeira, J. (2017), *Energy Efficiency Evaluation in Thermo-Active Structures.* MSc Thesis, Instituto Superior Técnico, University of Lisbon, Portugal.

Vieira, A.; Alberdi-Pagola, M.; Christoudolides, P. et al. (2017). *Characterisation of Ground Thermal and Thermo-Mechanical Behaviour for Shallow Geothermal Energy Applications.* Energies, 10(12), 2044; doi:10.3390/en10122044

Williams, C. F.; Reed, M. J.; Anderson, A. F. (2011). *Updating the Classification of Geothermal Resources.* Proceedings of Thirty-Sixth Workshop on Geothermal Reservoir Engineering, Stanford.

Numerical Methods in Geotechnical Engineering IX – Cardoso et al. (Eds)
© 2018 Taylor & Francis Group, London, ISBN 978-1-138-33198-3

Hydro-mechanical modelling of an unsaturated seal structure

D.F. Ruiz, J. Vaunat, A. Gens & M.A. Mánica
Universitat Politecnica de Catalunya (UPC), Barcelona, Spain

ABSTRACT: The paper contains a coupled numerical analysis of the hydration of a potential sealing system for isolating shafts, ramps and connecting galleries in a deep geological repository for high level nuclear waste. The modelling adopted for the different components of the seal are described first together with the basic features of the numerical model. Subsequent the results of the analyses, concerning both mechanical and hydraulic parameters, are presented and discussed. Finally, it is shown that the use of a double structure model for the behaviour of the bentonite affects the evolution of the mechanical response of the seal but it does not modify significantly the basic hydration process.

1 INTRODUCTION

The design of a deep geological repository for high-level nuclear waste must include sealing systems to ensure the isolation of the connecting galleries, ramps and shafts of the facility after the end of the operational phase.

The preliminary configuration of the seal system analyzed in the work reported here consists of a central core built with unsaturated swelling bentonite placed between two support concrete plugs. The plugs are in contact with the filling of the gallery, made with recompacted crushed host rock. (Figure 1). As reported in Gens (2003), the performance of these systems relies on achieving low permeability values in the bentonite and surrounding rock (particularly in the Excavation Damaged Zone, EDZ) and low transmissivity values along possible discontinuities (contacts and potential gaps). The issues concerning seal behaviour are similar to those arising in the design of engineered barriers, with the important difference that no high temperatures are present; therefore, the hydro-mechanical coupling governs the response of system.

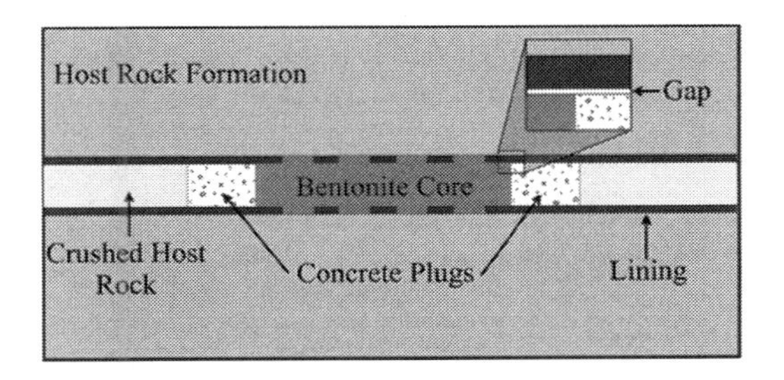

Figure 1. Configuration of the seal structure.

The performance of the sealing system is affected by all phases of the repository:

- The excavation creates an EDZ around the openings.
- The ventilation applied during the operational phase that may influence the adjacent rock.
- The installation conditions such dry density and compaction water content of the sealing material, presence of gaps and mode of eventual filling.
- The long-term behaviour after water pressure equilibration as long-term creep deformation of the host rock will compress the sealing system.

2 MODELLING COMPONENTS OF THE SEAL SYSTEM

2.1 *Host rock*

Although the seal system is applicable to any type of host rock, this study is focused on the Callovo-Oxfordian (COx) claystone. This argillaceous formation located at a depth of about 500 m in the site of Meuse/Haute Marne (MHM) is considered as a potential host rock for high level radioactive nuclear waste repository in France (ANDRA, 2012). Important features of the mechanical behaviour of the material are brittleness and anisotropy (Gens, 2011a). In addition, COx claystone has a low permeability that is also anisotropic. The higher values of permeability are often associated with flow parallel to bending planes. Damage process due to excavation has been observed to increase permeability by several orders of magnitude.

The intact host Formation (COx) and EDZ are simulated by a phenomenological model through a visco-elastoplastic stress-strain relationship based on a smoothed Mohr-Coulomb yield criterion

(Mánica et al. 2017). The following features are incorporated:

- Non-linear hardening/softening law.
- Anisotropic failure criterion (Mánica et al; 2016).
- Creep deformations using a modified Lemaitre's Law in which long-term deformations are function of the deviatoric stress (See Equations 1, 2 and 3).

$$\dot{\varepsilon}_{ij}^{vp} = \frac{2}{3}\frac{\dot{\varepsilon}_{vp}}{q}S_{ij} \tag{1}$$

$$\dot{\varepsilon}_{vp} = \gamma(q-\sigma_s)^n(1-\varepsilon_{vp})^m \tag{2}$$

$$q = \left(\frac{3}{2}s:s\right)^{1/2} \tag{3}$$

where $\dot{\varepsilon}_{ij}^{vp}$ = visco-plastic strain tensor; q = devitoric stress; σ_s = Stress threshold from which visco-plastic strains are activated; γ = viscosity parameter; m,n = fitting parameters; and ε_{vp} = state variable of time dependent response.

Hydraulic phenomena are modelled by the van Genuchten retention curve, the variation of the intrinsic permeability Ki with porosity and the variation of the relative hydraulic conductivity with degree of saturation Sr (Olivella et al. 1996).

This argillaceous material is being intensely studied by means of laboratory and in situ tests al both small and large field scales (Seyedi et al. 2017). Figure 2 and Figure 3 show two types of laboratory tests performed at sample scale. A triaxial test under confined pressure of 12 MPa, a value close to the in situ stress state and creep tests at the same confined pressure and different deviatoric stress levels. The dotted lines correspond to the numerical modelling of the tests. The constitutive model, that incorporates features such as non-linear hardening, softening and creep deformations, is able to provide a good approximation to the experimental results.

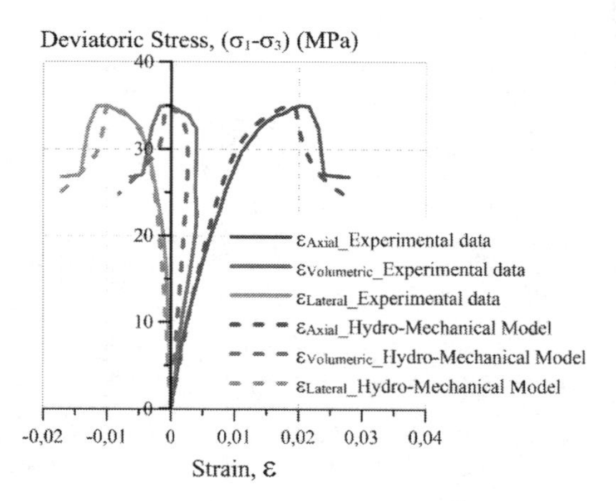

Figure 2. Numerical modelling of triaxial test—Experimental data from Seyedi et al. (2017).

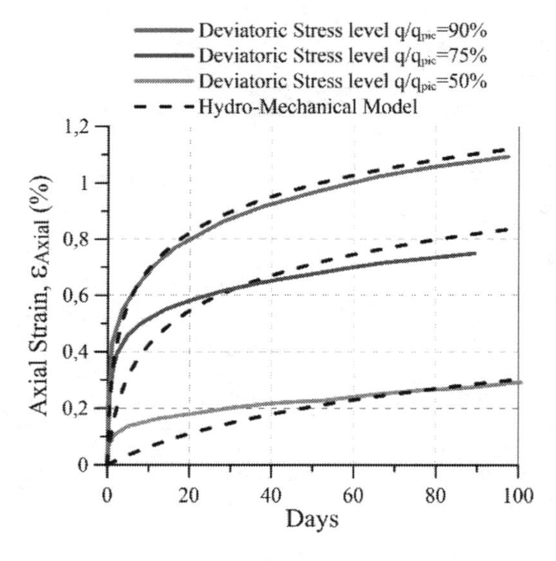

Figure 3. Numerical modelling of creep test—Experimental data from Seyedi et al. (2017).

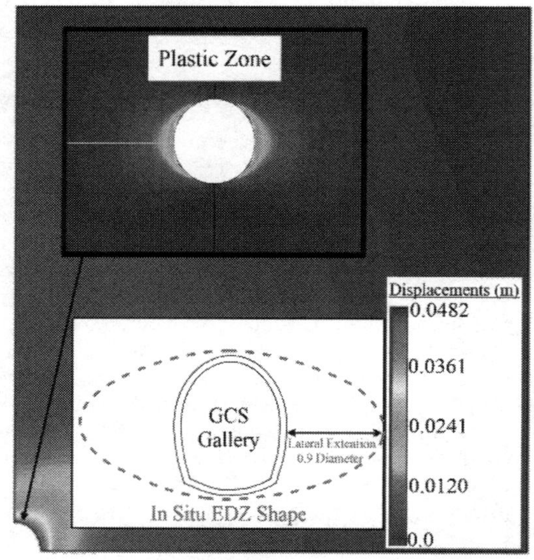

Figure 4. Modelling of excavation of a gallery aligned with the major principal horizontal stress. Excavation Damaged Zone (EDZ) shape.

The formation presents anisotropic stress state and the shape of the EDZ depends on the orientation of the excavation. The analysis of a case of a gallery in the underground laboratory where the orientation ensures that the cross-section in situ stress is nearly isotropic is shown in Figure 4 and Figure 5. In the modelling, the shape of the EDZ is represented by contours of the plastic multiplier. Both the EDZ and the anisotropic and long-term convergence measurements are well captured. This response incorporates also consolidation hydro-mechanical phenomena that together with creep deformations results in the development of strain with time.

2.2 *Contacts*

The installation of seal structure involves contacts between its different components. Therefore, modelling of the contact effects is required in order to account for the real performance of the structure.

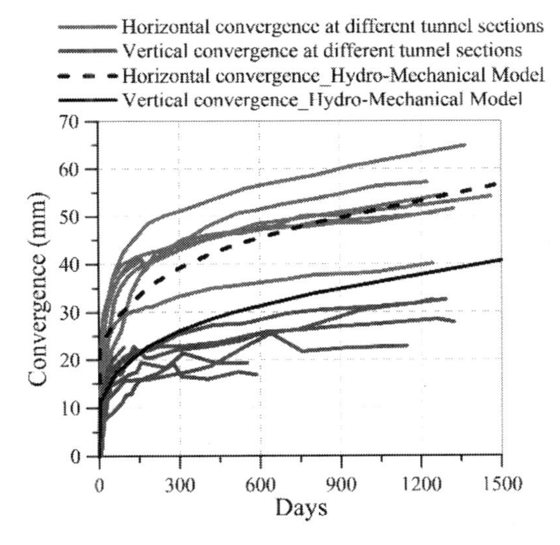

Figure 5. Modelling of excavation of a gallery aligned with the major principal horizontal stress. Convergence comparison of computed results and observations—In situ data from Seyedi et al. (2017).

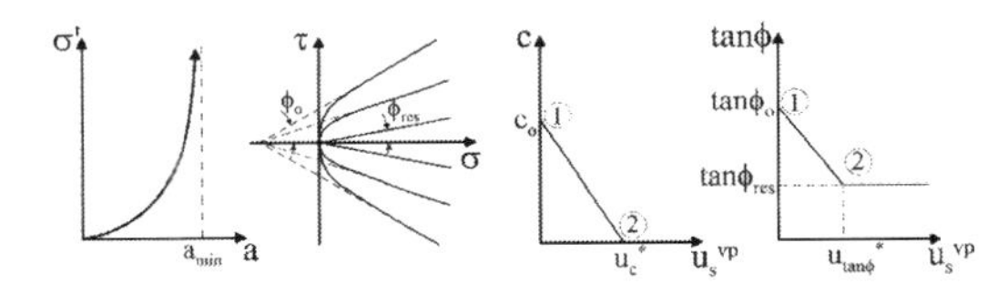

Figure 6. Constitutive law of the contact element. From left to right: elastic law, yield surface and softening law for cohesion and friction angle.

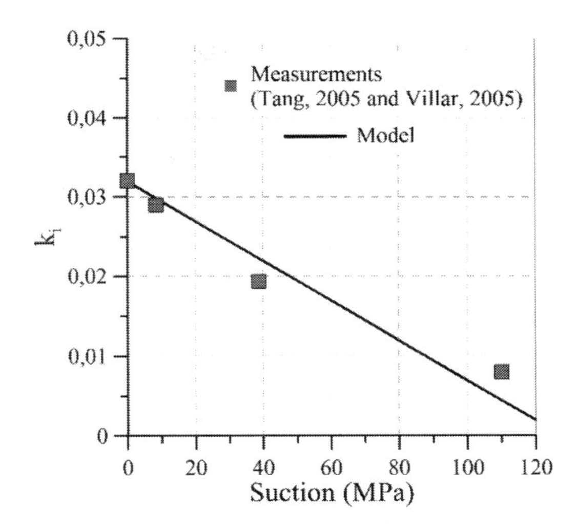

Figure 7. Variation of MX-80 bentonite elastic stiffness with suction (from Zandarin et al. 2011).

Stress distributions and the stability of the concrete plugs as supporting elements are only simulated realistically if contact effects are considered in an appropriate manner.

A joint element formulated by Zandarín (2010) has been implemented in the finite element code CODE-BRIGHT (Olivella et al. 1996). Tangent shear stiffness is considered constant but the normal stiffness depends on the aperture of the contact. Viscoplastic behaviour is generated with a hyperbolic Mohr-Coulomb yield criterion together with a Perzyna approach. After peak, a softening law function of relative displacement is adopted (Figure 6).

Transversal fluxes across the joint element depend on the permeability of the porous media that bound the contact whereas the longitudinal fluxes are function of the joint aperture and roughness profile that in the case of artificial contacts is considered negligible. Perfect mobility of the water inside the contact has been assumed.

2.3 *Bentonite core*

The main part of the seal structure is the expansive core composed, initially, of unsaturated bentonite. The goal of this core is generate a strong swelling in order to recover a stress state similar to the initial one and to achieve a very low permeability throughout. The design swelling pressure is 7 MPa.

Bentonite powder, compacted blocks and mixture of powder and pellets are being considered as potential buffer materials. A common feature of these materials is a double structure in which microporosity inside clay aggregates coexists with macroporosity. The modelling approach could be based on the definition of different constitutive models of increasing complexity.

Here a single structure approach is first adopted. With this approach, it is possible to generate a progressive swelling upon hydration. A modified version of the elastic law of the Barcelona Basic Model (Alonso et al. 1990) has been used for this purpose. Equations 4 and 5 describe the dependence of the elastic parameters on mean stress and suction.

$$k_i = k_{io}(1+\alpha_{is}s) \quad (4)$$

$$k_s = k_{so}(1+\alpha_{sp}\ln P/P_r)\exp(\alpha_{ss}s) \quad (5)$$

where k_i = elastic parameter related to net stress changes; k_{io} = initial elastic parameter related to net stress changes; k_s = elastic parameter related to suction changes; k_{so} = initial elastic parameter related to suction changes; $\alpha_{is}, \alpha_{ss}, \alpha_{sp}$ = fitting parameters; s = suction; and P = mean net stress.

The different parameters have been calibrated to obtain the desired design swelling pressure. Some values can be derived from measurements reported in the literature such as the variation of the elastic stiffness with suction (Figure 7).

A more advanced model capable of reproducing the transient collapse of the macrostructure requires a double structure model approach. This collapse has been often observed in laboratory tests (Figure 8). A modified version of the double structure model presented in Gens et al. (2011b) has been implemented and used in a complementary analysis that is reported at the end of the paper.

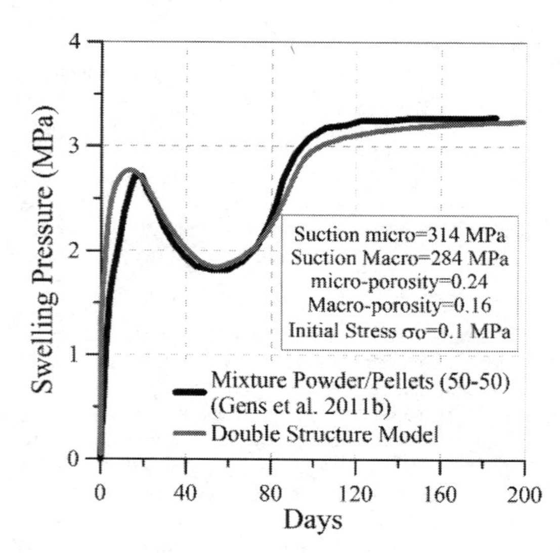

Figure 8. Laboratory oedometer swelling pressure tests and its reproduction with a double structure model.

3 NUMERICAL MODELLING OF THE HYDRATION OF A SEAL STRUCTURE

The hydro-mechanical modelling of the seal structure has been performed using the finite element code CODE-BRIGHT (DIT-UPC, 2002) solving the field equations related to mass balance of solid and water and the balance of momentum. The mass balance for water can be established following the scheme illustrated in the Figure 9.

$$\frac{\partial}{\partial t}(1-\varphi)\rho_s + \nabla\cdot(j^s) = f^s \quad (6)$$

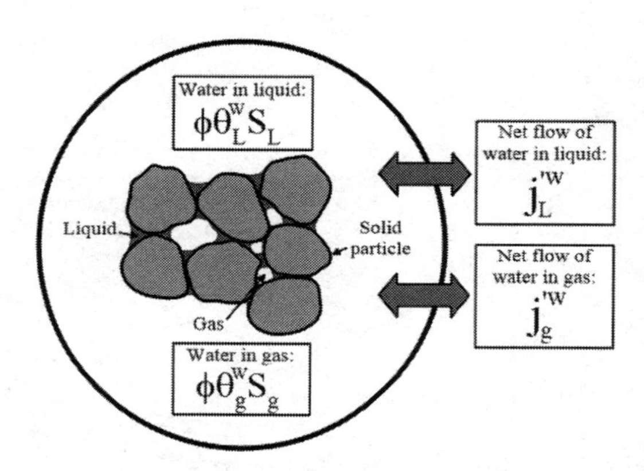

Figure 9. Scheme to establish the equation for mass balance of water in a single structure porous medium (Gens, 2010).

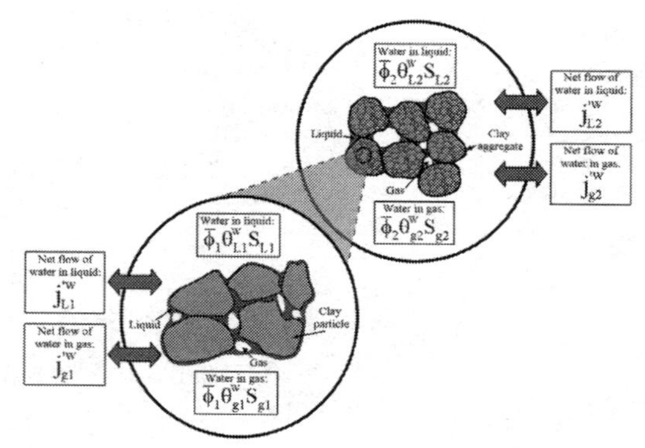

Figure 10. Scheme to establish the equation for mass balance of water in a double structure porous medium (modified from Gens, 2010).

$$\frac{\partial}{\partial t}(\theta_L^w S_L \varphi - \theta_g^w S_g \varphi)\rho_s + \nabla \cdot (j_l^w + j_g^w) = f^w \tag{7}$$

$$\frac{\partial}{\partial t}(\theta_L^a S_L \varphi - \theta_g^a S_g \varphi)\rho_s + \nabla \cdot (j_l^a + j_g^a) = f^a \tag{8}$$

$$\nabla \cdot \sigma + b = 0 \tag{9}$$

where ϕ = porosity; ρ_s = bulk density of the solid; j^s, j_l^w, j_g^w = total mass flux of solid and water; f^s, f^w = source or sink of solid and water mass; θ_l^w, θ_l^g = mass fraction in phases; S_l^w, S_l^g = degree of saturation of liquid and gas phases; σ = total stress tensor; and b = body forces vector.

The above equation describe porous media with a single structure like host rock formation, contact element, concrete and the bentonite core modelled with a modified Barcelona Basic Model (BBM).

If a double structure approach is used to model properly the structural evolution of the bentonite core during the hydration process, the field equations necessary to solve are:

- The balance of momentum (Equation 9)
- The mass balance of solid (Equation 6).
- The mass balance of water for the medium macro (Equation 10) and the medium micro (Equation 11).
- The mass balance of air for the medium macro (Equation 12) and the medium micro (Equation 13).

An extended scheme of water balance for double-structure porous media is shown in the Figure 10.

$$\bar{\phi}_2 \frac{D(\theta_{L2}^w S_{L2} + \theta_{g2}^w S_{g2})}{Dt} + (\theta_{L2}^w S_{L2} + \theta_{g2}^w S_{g2}) \left(\frac{d\bar{\varepsilon}_{v2}}{dt}\right) \nabla \cdot (\mathbf{j}_{L2}^w + \mathbf{j}_{g2}^w) = -\Gamma^w \tag{10}$$

$$\bar{\phi}_1 \frac{D(\theta_{L1}^w S_{L1} + \theta_{g1}^w S_{g1})}{Dt} - (\theta_{L1}^w S_{L1} + \theta_{g1}^w S_{g1}) \left(\frac{d\bar{\varepsilon}_{v1}}{dt}\right) = \Gamma^w - (\theta_{L1}^w S_{L1} + \theta_{g1}^w S_{g1})(1-\phi)\frac{d\rho_s}{\rho_s} \tag{11}$$

$$\bar{\phi}_2 \frac{D(\theta_{L2}^a S_{L2} + \theta_{g2}^a S_{g2})}{Dt} + (\theta_{L2}^a S_{L2} + \theta_{g2}^a S_{g2}) \left(\frac{d\bar{\varepsilon}_{v2}}{dt}\right) + \nabla \cdot (\mathbf{j}_{L2}^a + \mathbf{j}_{g2}^a) = -\Gamma^a \tag{12}$$

$$\bar{\phi}_1 \frac{D(\theta_{L1}^a S_{L1} + \theta_{g1}^a S_{g1})}{Dt} - (\theta_{L1}^a S_{L1} + \theta_{g1}^a S_{g1}) \left(\frac{d\bar{\varepsilon}_{v1}}{dt}\right) = \Gamma^a - (\theta_{L1}^a S_{L1} + \theta_{g1}^a S_{g1})(1-\phi)\frac{d\rho_s}{\rho_s} \tag{13}$$

where the subscripts 1 and 2 are referred to the medium micro and macro respectively; Γ = mass flux per unit total volume with respect to the solid skeleton going from the macro to medium micro; $\bar{\varphi}_2$ = volumetric volume fraction for macrostructure; and $\bar{\varphi}_1$ = Volumetric volume fraction for microstructure It is important remark that the volume fractions are different to the macro and micro porosities, ϕ_2 and ϕ_1, because they are referred to the total volume of the porous medium.

Special assumption is made for the water fluxes in the medium micro. The water in the medium micro has a lower mobility and this is accounted for in the balance equation themselves, by assuming that total mass flux of water is null. This simplification allows treating the water pressure micro as a history value to be integrated in the gauss point level during the stress updating, instead to treat it as nodal variable.

The modelling considers an axisymmetric geometry around the axis of the gallery (Figure 11). The model domain has a width of 60 m and a length of 100.5 m. The outside diameter of the gallery is 8.72 m and a 0.7 m-thick concrete lining is provided for support. Finally, the sealing system includes a 40.5 m long swelling core (made of unsaturated bentonite), two 5 m long concrete plugs and a fill of disaggregated and recompacted COx (considered as a non-expansive material) (see Figures 1 and 9). With this approach, it is not possible to reproduce the anisotropy of the EDZ but the other features are retained in order to capture the short—and long-term response.

The stages of the modelling include the excavation of the gallery, the installation of the lining six months later, 100 years of ventilation as an extreme case during the operational time of the repository, the sealing system installation and finally the hydration process from the host rock until that the initial water pressure (4.85 MPa) is recovered.

Artificial stress concentration appears close to the contact between the concrete support plugs

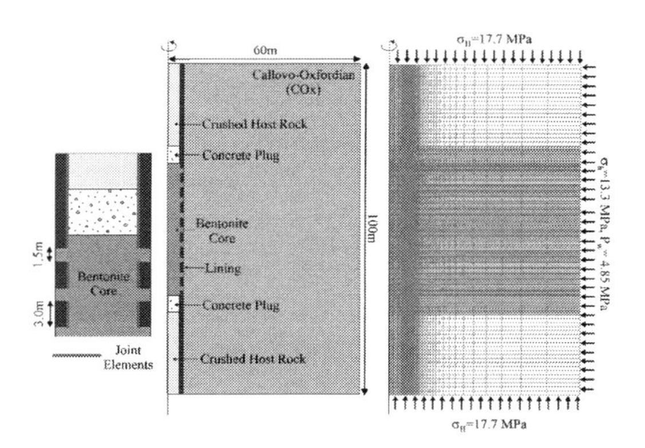

Figure 11. Seal scheme and model geometry.

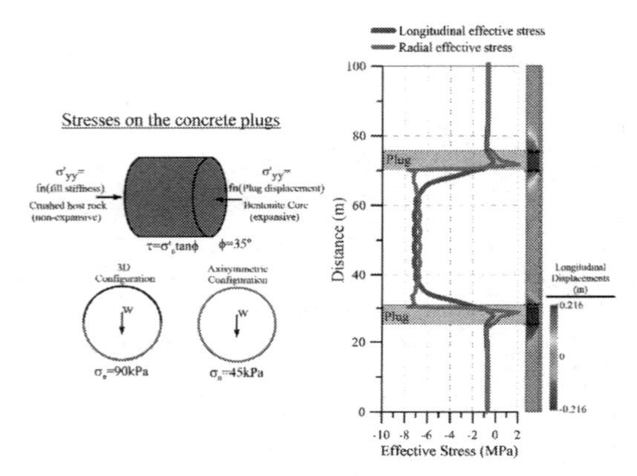

Figure 12. Distribution of longitudinal and radial stresses and contours of radial displacements (Negative values mean compression).

plugs. However, a central part of the core maintains the high swelling pressure and the radial pressure is constant along the gallery due to the lining (Figure 12).

The evolution of the degree of saturation and the water pressure during the hydration process is shown in Figure 13. It can be observed that the evolution is affected by the different phases of the repository. In this analysis, the total saturation of the seal is reached after 1200 years and the initial water pressure (4.85 MPa) is recovered after 2500 years.

The swelling pressure increases during the hydration processes, in accordance with the single structure approach, until it reaches the design swelling pressure value (Figure 14). As indicated in the same Figure, the joint is closed at stresses lower than the final state of the bentonite core, because

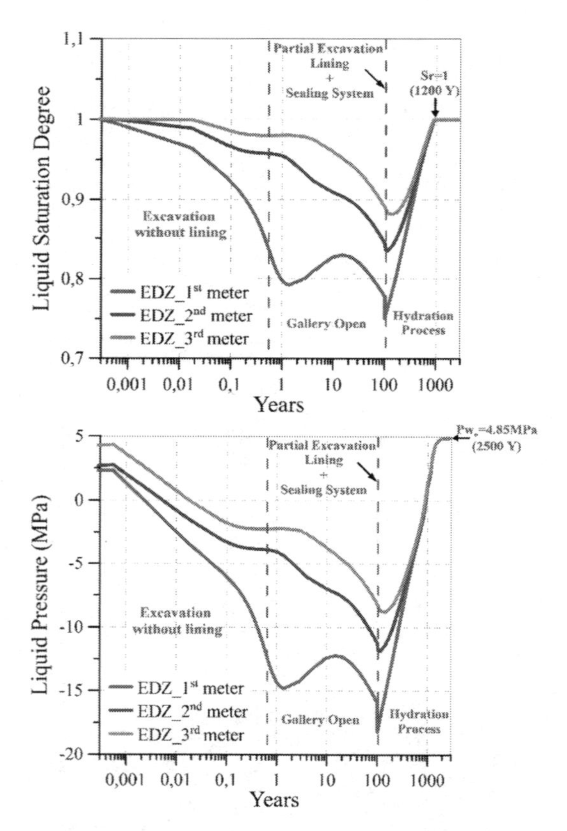

Figure 13. Hydraulic response during seal hydration.

and the lining. This effect is due to the fact that the contact is considered non-sliding. A longitudinal displacement of the plugs of 0.20 m is computed and the longitudinal stresses vanish near the

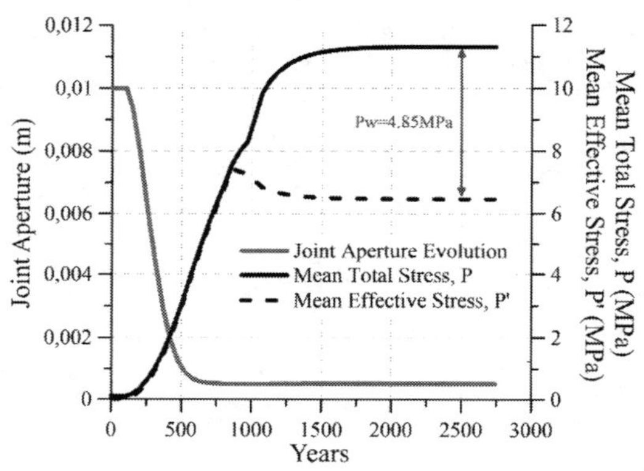

Figure 14. Swelling pressure of the bentonite and joint aperture evolution.

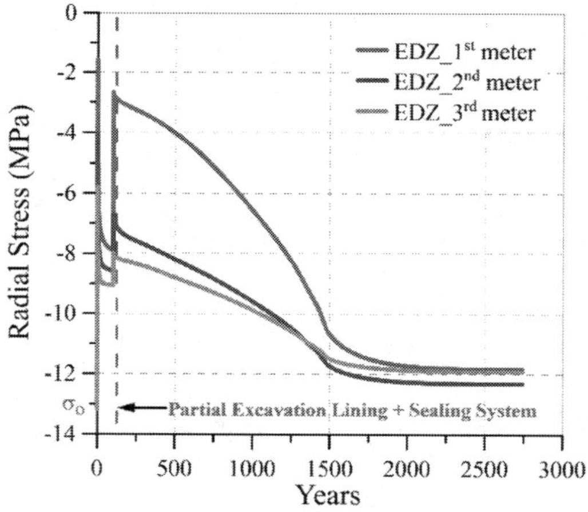

Figure 15. Recompaction of the EDZ after seal installation. (Negative values mean compression).

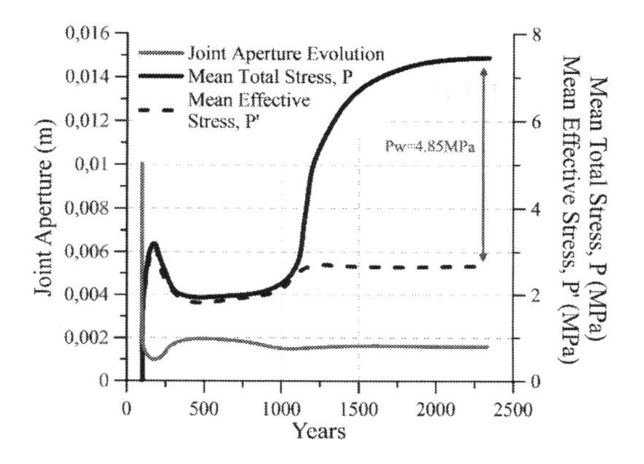

Figure 16. Swelling pressure of the bentonite and join aperture evolution with a double structure approach.

of its low normal stiffness. Finally the creep deformations generate a long-term recompaction of the EDZ increasing the radial stresses (see Figure 15).

Finally an additional analysis has been performed using the double structure model mentioned above. A swelling pressure of about to 3 MPa has been adopted in this case. It can be seen that the transient macrostructural collapse occurs also at this large scale generating a long period when the gaps re-open before the eventual swelling finally closes them. The hydrostatic pore pressure (4.85 MPa) is reached at 2500 years, a time similar to that of the single-structure analyses indicating that the overall hydraulic evolution depends basically on the properties of the host rock (Figure 16).

4 CONCLUDING REMARKS

The hydration of a sealing system for a repository for high level nuclear waste has been simulated using a coupled hydro-mechanical approach. The behaviour of the seal structure is influenced by all phases of the repository, including both the construction, ventilation and the operational periods. It has been shown that, by incorporating in the analysis special features such the constitutive modelling of short—and long-term behaviour of the host rock, a specific contact formulation and the expansive behaviour of the unsaturated bentonite core, a reasonable response of the system is obtained. It has also been shown that the use of a double structure model for the behaviour of the bentonite affects the evolution of the mechanical response of the seal but it does not modify significantly the basic hydration process.

REFERENCES

Alonso, E.E. Gens, A. Josa, A. (1990). A constitutive model for partially saturated soils. Géotechnique 40, No. 3. 405–430.

ANDRA (2012). Référentiel du comprotement des formations sur le site de Meuse/Haute Marne. Report D.RP.AMFS.1 2.0024. Châtenay-Malabry: ANDRA.

DIT-UPC (2002). CODE_BRIGHT, a 3-D program for Thermo-Hydro-Mechanical analysis in geological media: User'sguide, CIMNE, Barcelona.

Gens, A. (2003). The role of geotechnical engineering for nuclear utilisation. (Special Lecture). Proc. 13th Eur. Conf. Soil Mech. Geotech. Engng, Prague: IOS Press, p. 25–67.

Gens, A. (2010). Soil–environment interactions in geotechnical engineering. Géotechnique, 60(1): 3–74.

Gens, A. (2011a). On the HM behaviour of argillaceous hard soils-weak rocks. "Proc 15th Eur. Conf. Soil Mech. Geotech. Engng: geotechnics of hard soils, weak rocks". Atenes: IOS Press, p. 71–118.

Gens, A., Valleján, B., Sánchez, M., Imbert, C., Villar, M.V., Van Geet, M. (2011b) Hydromechanical behaviour of a heterogeneous compacted soil: experimental observations and modelling. Géotechnique., 61(5):367–86.

Mánica, M., Gens, A., Vaunat, J., Ruiz, D. F. (2016). A cross-anisotropic formulation for elasto-plastic models. Géotechnique Letters 6, 1–7

Mánica, M., Gens, A., Vaunat, J., Ruiz, D.F. (2017). A time-dependent anisotropic model for argillaceous rocks, Application to an underground excavation in Callovo-Oxfordian claystone. Computer & Geotechnics., 85, 341–350.

Olivella, S., Gens, A., Carrera, J., Alonso, E.E. (1996). Numerical formulation for a simulator (CODE_BRIGHT) for the coupled analysis of saline media. Engineering Computations., 13(7), 87–112.

Seyedi, D.M., Armand, G., Noiret, A. (2017). "Transverse Action" – A model benchmark exercise for numerical analysis of the Callovo-Oxfordian claystone hydromechanical response to excavation operations. Computer & Geotechnics., 85, 287–305.

Zandarin, M.T. (2010). "Thermo-Hydro-Mechanical analysis of joints—A theoretical and experimental study". PhD Thesis. Universitat Politécnica de Catalunya (UPC).

Zandarin, M.T., Gens, A., Olivella, S., Alonso, E.E. (2011). Thermo-hydro-mechanical model of the Canister Retrieval Test. Physics and Chemistry of the Earth., 36, 1806–1806.

Numerical Methods in Geotechnical Engineering IX – Cardoso et al. (Eds)
© 2018 Taylor & Francis Group, London, ISBN 978-1-138-33198-3

Finite element modelling of excess pore fluid pressure around a heat source buried in saturated soils

W. Cui, A. Tsiampousi, D.M. Potts, K.A. Gawecka, L. Zdravković & D.M.G. Taborda
Imperial College London, London, UK

ABSTRACT: Soils may be exposed to significant temperature variation in many geotechnical engineering problems. When a thermal load is applied to the soil surrounding a geotechnical structure, a change in pore fluid pressure may be observed due to the fact that the thermal expansion coefficients of the pore fluid and the soil particles are different. To model the behaviour of this thermally-induced excess pore fluid pressure, a robust coupled Thermo-Hydro-Mechanical (THM) finite element formulation is presented and employed in this study. Subsequently, numerical analyses are carried out using the proposed formulation to model the coupled consolidation and heat transfer around a cylindrical heat source buried in saturated soils. The predicted temperature and excess pore fluid pressure are compared to the existing approximate analytical solutions available in the literature. Finally, a centrifuge test involving heating of a cylinder buried in clay is simulated and numerical predictions are compared to experimental measurements.

1 INTRODUCTION

Significant temperature variations in soils may be observed in many geotechnical engineering problems, such as those involving thermo-active piles or the disposal of radioactive waste. A temperature rise in soils is usually accompanied by an increase in pore fluid pressure due to the different thermal expansion coefficients of the pore fluid and the soil particles. If there is insufficient drainage, this thermally induced excess pore fluid pressure may become significant and may even result in the thermal failure of soil (Gens 2010).

A number of laboratory (e.g. Britto et al. 1989, Savvidou & Britto 1995, Lima et al. 2010, Mohajerani et al. 2012) and in situ (Gens et al. 2007, Francois et al. 2009) tests have been conducted to investigate the time-dependent behaviour of thermally induced pore fluid pressures. To model this behaviour, where both transient heat transfer and consolidation are involved, appropriate numerical tools capable of simulating the fully coupled thermo-hydro-mechanical (THM) response of soils are required. Booker & Savvidou (1985) presented the governing equations for a transient coupled THM problem in soils and subsequently derived a closed form solution for the thermally induced pore fluid pressure around a point heat source. Subsequently, an approximate solution was derived for a cylindrical heat source by integrating the point source solutions. Since then, various coupled THM finite element (FE) formulations for soils have been presented (Lewis et al. 1986, Britto et al. 1992, Vaziri 1996) and comparisons made between the numerical predictions and the theoretical solutions given by Booker & Savvidou (1985). Furthermore, these coupled THM FE formulations were also applied to simulate the behaviour of transient heat transfer and consolidation in triaxial and centrifuge heating tests (Britto et al. 1989, Savvidou & Britto 1995, Vaziri, 1996).

The study presented here employs an alternative FE formulation which is capable of modelling the coupled THM behaviour of soils and has been implemented into the bespoke Imperial College Finite Element Program (ICFEP, Potts & Zdravkovic 1999). The hydraulic equation is derived following a new approach which employs only the principle of mass conservation of the pore fluid, while mass conservation laws of both the pore fluid and the solid particle were commonly used to formulate the same hydraulic equation in the literature (e.g. Olivella et al. 1996, Lewis & Schrefler 1998, Francois et al. 2009). An advantage of the presented approach is that pore fluid pressure induced by the difference in thermal expansion coefficients of the pore fluid and the soil particles can be naturally accounted for. The coupled THM facilities are subsequently employed to simulate the heat transfer and consolidation around a cylindrical heat source and the obtained results are verified against the approximate analytical solutions of Booker & Savvidou (1985). Finally, the centrifuge heating test reported by Britto et al. (1989) is modelled and numerical predictions are compared against the experimental measurements.

2 NUMERICAL FORMULATION FOR A COUPLED THM PROBLEM

2.1 *Hydraulic governing formulation*

For a fully saturated soil, applying the principle of mass conservation for the fluid phase leads to the following expression:

$$\frac{\partial(n\rho_f dV)}{\partial t}+\left[\nabla\cdot(\rho_f\{v_f\})-\rho_f Q^f\right]dV=0 \tag{1}$$

where ρ_f is the density of the pore fluid, $\{v_f\}$ represents the vector of the seepage velocity, $\nabla\cdot$ is the symbol of divergence which is defined as $\nabla\Theta = \partial\Theta/\partial x+\partial\Theta/\partial y+\partial\Theta/\partial z$, dV is an infinitesimal volume of the soil, n is porosity, Q^f represents any pore fluid sources and/or sinks, and t is time. Substituting $n = e/(1+e)$, where e is the void ratio, and $dV = (1+e)dV_s$, where dV_s is the infinitesimal volume of the soil particles, into Equation (1) yields:

$$\frac{\partial(e\rho_f dV_s)}{\partial t}+\left[\nabla\cdot(\rho_f\{v_f\})-\rho_f Q^f\right](1+e)dV_s=0 \tag{2}$$

Under isothermal conditions, dV_s is generally assumed to be constant, regardless of the change in effective stresses. However, under non-isothermal conditions, dV_s is temperature dependent and can be written as:

$$dV_s=(1+\varepsilon_{vT})dV_{s0} \tag{3}$$

where dV_{s0} is the initial infinitesimal volume of the soil particles and ε_{vT} is the thermal volumetric strain of the soil particle, which is generally assumed to be equal to that of the soil skeleton (Campanella & Mitchell 1968). It is noted that dV_{s0} is assumed to be constant here, which differs from the approach of Thomas et al. (2009) who assume that dV_s is constant for a coupled THM problem. Substituting Equation (3) into Equation (1) and eliminating dV_{s0} leads to:

$$\frac{\partial[e\rho_f(1+\varepsilon_{vT})]}{\partial t}\frac{1}{(1+e)(1+\varepsilon_{vT})}+[\nabla\cdot(\rho_f\{v_f\})-\rho_f Q^f]=0 \tag{4}$$

The pore fluid density can be expressed by a function of temperature, T, and pore pressure, p_f, as (Fernandez, 1972):

$$\rho_f=\rho_{f0}\exp[-\frac{1}{K_f}(p_f-p_{f0})-3\alpha_{T,f}(T-T_0)] \tag{5}$$

where ρ_{f0} is the reference pore fluid density under the corresponding reference pore pressure p_{f0} and reference temperature T_0. K_f is the bulk modulus of the pore fluid and $\alpha_{T,f}$ is the linear thermal expansion coefficient of the pore fluid. Differentiating Equation (5) with respect to time yields:

$$\frac{\partial\rho_f}{\partial t}=\rho_f(-\frac{1}{K_f}\frac{\partial p_f}{\partial t}-3\alpha_{T,f}\frac{\partial T}{\partial t}) \tag{6}$$

Noting that $\Delta\varepsilon_{vT}=3\alpha_T\Delta T$ where α_T is the linear thermal expansion coefficient of the soil skeleton, substituting Equation (6) into Equation (4) yields:

$$\begin{aligned}&\frac{\rho_f}{1+e}\frac{\partial e}{\partial t}+n\rho_f(-\frac{1}{K_f}\frac{\partial p_f}{\partial t}-3\alpha_{T,f}\frac{\partial T}{\partial t})\\&+\rho_f\frac{3n\alpha_T}{(1+\varepsilon_{vT})}\frac{\partial T}{\partial t}+[\nabla\cdot(\rho_f\{v_f\})-\rho_f Q^f]=0\end{aligned} \tag{7}$$

Assuming that the changes in void ratio are only a result of the mechanical volumetric strain (i.e. $(\varepsilon_v-\varepsilon_{vT})$) and ignoring the effect of pore fluid buoyancy, $\{v_f\}^T\nabla(\rho_f)$, Equation (7) can be further derived as:

$$\begin{aligned}&\rho_f\frac{\partial(\varepsilon_v-\varepsilon_{vT})}{\partial t}-\rho_f\frac{n}{K_f}\frac{\partial p_f}{\partial t}-\rho_f(\frac{3n\alpha_T}{1+\varepsilon_{vT}}-3\alpha_{T,f})\frac{\partial T}{\partial t}\\&+\rho_f\nabla\cdot\{v_f\}-\rho_f Q^f=0\end{aligned} \tag{8}$$

Assuming that $1+\varepsilon_{vT}\approx 1$ and eliminating ρ_f lead to:

$$\begin{aligned}&\nabla\cdot\{v_f\}-\frac{n}{K_f}\frac{\partial p_f}{\partial t}+3n(\alpha_T-\alpha_{T,f})\frac{\partial T}{\partial t}-Q^f\\&=-\frac{\partial(\varepsilon_v-\varepsilon_{vT})}{\partial t}\end{aligned} \tag{9}$$

In Equation (9), the first two terms on the left-hand side represent the flow of pore fluid into and out of the soil element and the changes in the volume of the pore fluid due to its compressibility, respectively, while the third term denotes the excess pore fluid generated by the difference in thermal expansion coefficients between the soil particles and the pore fluid. It should be noted that Equation (9) is the same as that obtained by Lewis & Schrefler (1998), who followed a different approach which combines the mass balance equation for the solid phase with the mass balance equation for the fluid phase. Also, it is shown by Cui et al. (2017) that adopting the principle of volume conservation of the pore fluid leads to the same hydraulic governing equation as Equation (9). Equation (1) was adopted by Thomas & He (1997) as the starting point, however, the third term on the left-hand side of Equation (9) (i.e. $3n(\alpha_T-\alpha_{T,f})\partial T/\partial t$) was missing in their final form of the hydraulic governing equation due to the fact that dV_s and ρ_f are assumed to be constant in the derivation process.

Adopting the generalised Darcy's law leads to the expression of the seepage velocity $\{v_f\}$ in Equation (9) as:

$$\{v_f\} = [k_f]\left(\frac{\{\nabla p_f\}}{\gamma_f} + \{i_G\}\right) \tag{10}$$

where $[k_f]$ is the permeability matrix (or hydraulic conductivity) of the soil, $\{\nabla p_f\}$ represents the gradient of pore fluid pressure, the vector $\{i_G\}^T = \{i_{Gx}\ i_{Gy}\ i_{Gz}\}$ is the unit vector parallel, but in the opposite direction, to gravity.

2.2 *Thermal governing formulation*

Adopting the law of energy conservation gives the governing equation of heat transfer in a fully saturated soil as:

$$\frac{\partial\{[n\rho_f C_{pf} + (1-n)\rho_s C_{ps}](T - T_r)dV\}}{\partial t} + \left[\nabla\cdot[\rho_f C_{pf}\{v_f\}(T - T_r)] - \nabla\cdot([k_T]\{\nabla T\}) - Q^T\right]dV = 0 \tag{11}$$

where C_{pf} and C_{ps} are the specific heat capacities of the pore fluid and soil particles respectively, ρ_s is the density of the soil particle, T_r is a reference temperature, $[k_T]$ is the thermal conductivity matrix and Q^T represents any heat sources and/or sinks. The first term in Equation (11) denotes the heat content of the soil per unit volume, while the second term expresses the heat flux per unit volume including both heat conduction and heat advection. Applying the principle of mass conservation for each phase and following a similar procedure to that for the hydraulic equation, yields:

$$[n\rho_f C_{pf} + (1-n)\rho_s C_{ps}]\frac{\partial T}{\partial t} + \rho_f C_{pf}(T - T_r)\frac{\partial e}{\partial T} + \nabla\cdot[\rho_f C_{pf}\{v_f\}(T - T_r)] - \nabla\cdot([k_T]\{\nabla T\}) = Q^T \tag{12}$$

2.3 *Mechanical governing formulation*

Under non-isothermal conditions, the incremental total strain $\{\Delta\varepsilon\}$ can be expressed as the sum of the incremental strain due to stress change (mechanical strain), $\{\Delta\varepsilon_\sigma\}$, and the incremental strain due to temperature change (thermal strain), $\{\Delta\varepsilon_T\}$:

$$\{\Delta\varepsilon\} = \{\Delta\varepsilon_\sigma\} + \{\Delta\varepsilon_T\} \tag{13}$$

where $\{\Delta\varepsilon_T\}^T = \{\alpha_T\Delta T\ \alpha_T\Delta T\ \alpha_T\Delta T\ 0\ 0\ 0\}$. Applying the principle of effective stress, the total stress can therefore be given as:

$$\{\Delta\sigma\} = [D'](\{\Delta\varepsilon\} - \{\Delta\varepsilon_T\}) + \{\Delta\sigma_f\} \tag{14}$$

where $\{\Delta\sigma_f\}^T = \{\Delta p_f\ \Delta p_f\ \Delta p_f\ 0\ 0\ 0\}$ and $[D']$ is the effective constitutive matrix which depends on the adopted constitutive relations (e.g. linear elastic, non-linear, elasto-plastic).

2.4 *Finite element formulation*

Applying the standard finite element discretisation to Equations (9), (12) and (14) (Cui et al. 2017) and the time marching method (Potts & Zdravković 1999), the coupled THM finite element formulation for fully saturated soils can be derived as:

$$\begin{bmatrix} [K_G] & [L_G] & -[M_G] \\ [L_G]^T & -\beta_1\Delta t[\Phi_G] - [S_G] & -[Z_G] \\ [Y_G] & -\beta_2\Delta t[\Omega_G] & \alpha_1\Delta t[\Gamma_G] + [X_G] \end{bmatrix} \begin{Bmatrix} \{\Delta d\}_{nG} \\ \{\Delta p_f\}_{nG} \\ \{\Delta\theta\}_{nG} \end{Bmatrix} = \begin{Bmatrix} \{\Delta R_G\} \\ \{\Delta F_G\} \\ \{\Delta H_G\} \end{Bmatrix} \tag{15}$$

where α_1, β_1 and β_2 are time integration parameters, the values of which should be between 0.5 to 1.0 to ensure the stability of the marching process. The matrices in Equation (15) are the same as those detailed in Cui et al. (2017), where the hydraulic formulation was derived using the law of volume conservation.

These fully coupled THM facilities have been implemented into the bespoke FE software ICFEP (Potts & Zdravković 1999, 2001), which is employed to carry out all the FE analyses presented in this work.

3 VALIDATION EXERCISE

3.1 *Numerical modelling*

To demonstrate and verify the performance of the proposed THM formulation in simulating thermally-induced pore fluid pressures, a series of axisymmetric benchmark analyses, representing the example of elastic consolidation around a cylindrical heat source proposed by Booker & Savvidou (1985), has been performed.

The same material properties as those from Lewis et al. (1986) were employed (see Table 1), ensuring the same conditions adopted in the numerical example illustrated by Booker & Savvidou (1985). The adopted FE mesh is shown in Figure 1, employing 8-noded quadrilateral elements, with displacement, pore fluid pressure and temperature degrees of freedom at all element

Table 1. Material properties for the validation exercise.

Young's modulus, E (Pa)	6.0×10^3
Poisson ratio, ν (−)	0.4
Permeability, k_f (m/s)	3.92×10^{-5}
Initial void ratio, e_o (−)	1.0
$\rho_s C_{ps}$, $\rho_f C_{pf}$ $(kJ/m^3\ K)$	167.2
Conductivity k_T $(kJ/m\ s\ K)$	4.3
α_T $(m/m\ K)$	3.0×10^{-7}
α_{Tf} $(m/m\ K)$	2.1×10^{-6}

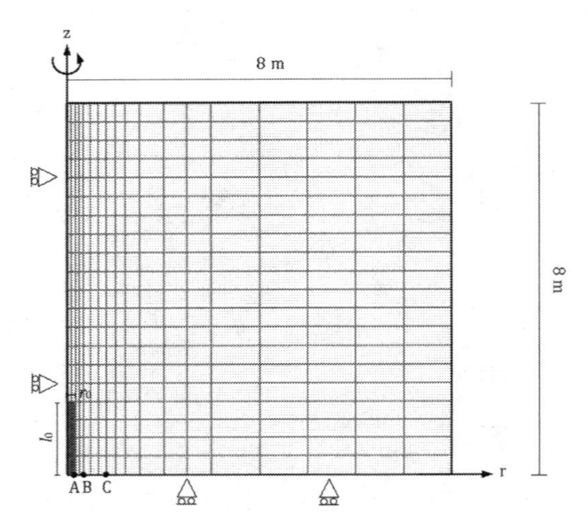

Figure 1. Finite element mesh for modelling consolidation around a cylindrical heat source ($l_0/r_0 = 10.0$).

nodes. A domain of 8 m × 8 m, representing the upper quadrant of the problem, was used and was shown to be sufficiently large such that the heat front did not reach the boundary during the analysis. The cylindrical heat source has a length of $2l_0$ and a diameter of $2r_0$. As part of the study, the value of l_0 was varied. For convenience, a value of $r_0 = 0.16\ m$ is adopted to ensure that the simulation time in the analysis, t, is the same as the time term $\tilde{t}$ ($\tilde{t} = \{[n\rho_f C_{pf} + (1-n)\ \rho_s C_{ps}]\ r_0^2 t\}/k_T$) used in the analytical solutions by Booker & Savvidou (1985). All the boundaries were assumed to be impermeable and insulated and a total constant heat input of 1000 W was prescribed over the elements representing the cylindrical heat source. The pore fluid pressure was assumed to be initially hydrostatic and a time-step of 0.1 sec was used in the analysis.

3.2 *Numerical results*

The changes in temperature and pore fluid pressures at three different points on the plane $z = 0$, i.e. A (r_0, 0), B ($2r_0$, 0) and C ($5r_0$, 0), were monitored throughout the analysis. To compare the numerical results to the exiting solutions presented by Booker & Savvidou (1985), the predicted temperature change, ΔT, was normalised with respect to the final temperature change obtained at the point A (i.e the maximum value in the mesh), ΔT_A. The predicted pore fluid pressure change was normalised by the change of pore fluid pressure at the point A assuming that the soil was impermeable, $\Delta p_{f,N}$. The expression of $\Delta p_{f,N}$ was given by Booker & Savvidou (1985) as:

$$\Delta p_{f,N} = \frac{E}{1-2\nu}\left\{\frac{1-\nu}{1+\nu}[3n\alpha_{T,f} + 3(1-n)\alpha_T] - \alpha_{T,f}\right\} \quad (16)$$

As shown in Figures 2 and 3, very good agreement was found in both temperature and pore fluid pressure changes between analytical solutions and numerical predictions with a ratio of $l_0/r_0 = 10.0$ in this study. Conversely, the numerical results obtained by Britto et al. (1992) with their

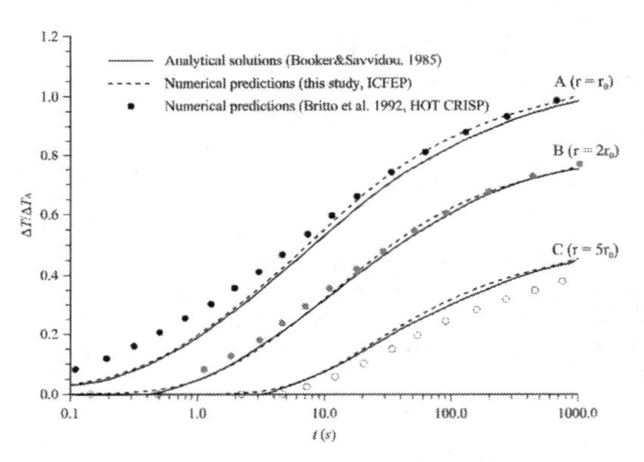

Figure 2. Variation of normalised temperature with time ($l_0/r_0 = 10.0$ for numerical results in this study).

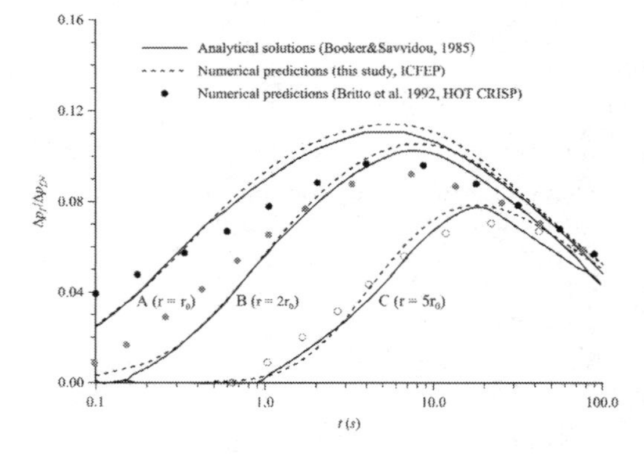

Figure 3. Variation of normalised pore fluid pressure with time ($l_0/r_0 = 10.0$ for numerical results in this study).

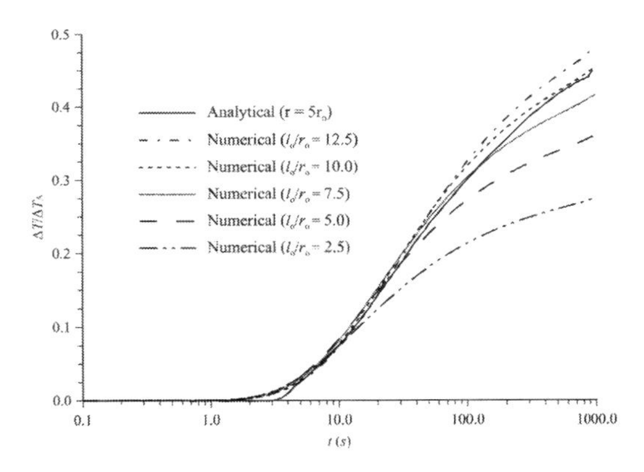

Figure 4. Variation of normalised temperature with time for different l_0/r_0 at point C.

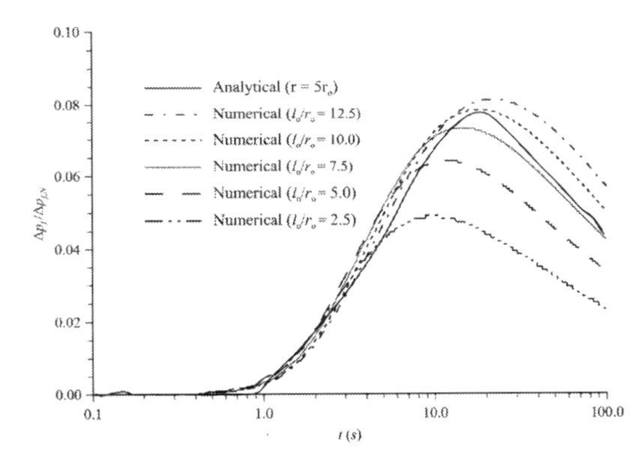

Figure 5. Variation of normalised pore fluid pressure with time for different l_0/r_0 at point C.

FE program HOT CRISP (value of l_0/r_0 was not given) differed substantially from the analytical solution. However, it should be noted that the size of the heat source, i.e. the ratio of l_0/r_0, was found to significantly affect the numerical results. As shown in Figures 4 and 5, good agreement between numerical and analytical results was found only for values l_0/r_0 between 7.5 and 10.0 at point C ($5r_0$, 0). The same conclusion also applies to the variations of temperature and pore fluid pressure changes with time at points A (r_0, 0) and B ($2r_0$, 0).

4 MODELLING OF THERMALLY INDUCED PORE FLUID PRESSURES IN A CENTRIFUGE TEST

4.1 *Experimental setup*

To investigate the behaviour of coupled heat flow and consolidation around a nuclear waste canister, a series of 100 *g* centrifuge tests were carried out by Maddocks & Savvidou (1984), where a 6 *mm* in diameter and 60 *mm* long model cylinder was buried in a steel tub containing fully saturated Kaolin clay. Two 5 *mm* thick sand layers were placed in the clay at some distance below and above the canister (see Figure 6) to help accelerate the consolidation process. A constant power supply was applied to the model canister after its installation. The temperature and pore fluid pressure changes in the clay surrounding the canister were monitored by transducers (i.e. thermocouples on the surface of the canister and pore pressure transducers adapted to also measure temperature at the location shown in Figure 6).

A representative centrifuge test (CS5), with all the experimental results detailed in Britto et al. (1989), was chosen for the numerical case study using the coupled THM formulation presented above. In this test the Kaolin clay was normally consolidated and a constant power of 13.9 *W* was supplied to heat the canister.

4.2 *Numerical modelling*

A prototype axisymmetric finite element analysis was performed to model the centrifuge test CS5. It should be noted the scaling law from Kutter (1992), was adopted for the numerical modelling. Therefore, a finite element mesh, which is 52.5 *m* × 35.0 *m* in dimension, as shown in Figure 6, was used in the analysis.

Hydraulic and thermal material properties of Kaolin clay, which are the same as those reported by Britto et al. (1989), were employed in this analysis (see Table 2). The modified cam-clay model, with all the parameters measured and presented by Maddocks & Savvidou (1984), was adopted to model the mechanical behaviour of the normally consolidated Kaolin clay (see Table 3). It should be noted that plastic volumetric strains were only

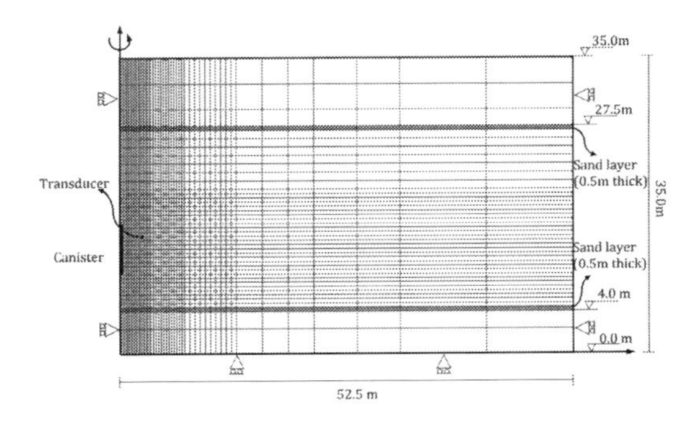

Figure 6. Finite element mesh for modelling the centrifuge test.

Table 2. Thermal and hydraulic material properties for modelling the centrifuge test.

Setting	Kaolin clay	Sand
α_{Tf} $(m/(m\ K))$	6.7×10^{-5}	6.7×10^{-5}
α_T $(m/(m\ K))$	8.3×10^{-6}	8.3×10^{-6}
ρ_f (kg/m^3)	1000	1000
ρ_s (kg/m^3)	2610	2650
C_{pf} $(kJ/(kg\ K))$	4.2	4.2
C_{ps} $(kJ/(kg\ K))$	0.94	0.83
k_T $(kJ/(s\ m\ K))$	1.5×10^{-3}	2.0×10^{-3}
k_f (m/s)	2.5×10^{-9}	2.5×10^{-5}

Table 3. Mechanical material properties of Kaolin clay for modelling the centrifuge test.

Slope of the compression line, λ	0.25
Slope of the swelling line, κ	0.05
Angle of friction, ϕ	23°
Poisson's ratio, v	0.25
Specific volume at 1 kPa, υ	3.58

observed in the numerical analysis in very small zones of the clay below and above the canister due to the thermal expansion of the canister, while stress paths in other parts of the clay were found to lie within the yield locus due to the significant rise in the excess pore fluid pressure. As the sand layers were extremely thin and had negligible mechanical effect in the analysis, a simple linear elastic model with a Young's module of $E = 3.0 \times 10^5\ kPa$ and a Poisson's ratio of $v = 0.25$ was employed for modelling the mechanical behaviour of the sand, while typical values of thermal and hydraulic properties of a sand, listed in Table 2, were adopted.

All boundaries in the mesh were assumed to be smooth. As shown in Figure 6, no lateral displacement was allowed at both the axis of symmetry and the right vertical boundary, while a no vertical movement boundary condition is prescribed at the bottom boundary. The two vertical and top horizontal boundaries were impermeable and water could only leave the mesh through the bottom boundary where a zero change in the pore fluid pressure was prescribed. A constant scaled heat flux of 0.819 kW/m^3 applied over the volume, equivalent to the total power supply of 13.6 W in the centrifuge test, was injected into the canister throughout the analysis and all boundaries of the mesh were assumed to be thermally insulated.

The same initial stress profile as that presented in Britto et al. (1989) was used for this study. A hydrostatic initial pore fluid pressure profile over the depth of the mesh with a zero pore fluid pressure at the top boundary, as well as an initial temperature of 20°C, was adopted. The initial vertical effective stresses was determined as $\sigma_v' = (\gamma_{sat} - \gamma_w)h$, where h represents the depth from the top boundary of the mesh, and γ_w and γ_{sat} are the specific unit weights of water and saturated soil respectively. A value of $\gamma_{sat} = 16.7\ kN/m^3$ (no scaling is needed) for the Kaolin clay was deduced from the in situ initial stress profile provided by Britto et al. (1989). The initial horizontal effective stress profile was obtained as $\sigma_h' = K_0\sigma_v'$ and a value of $K_0 = 0.69$, as suggested by Britto et al. (1989), was adopted.

8-noded quadrilateral elements were employed in the numerical analysis, where each node has displacement degree of freedom and only the corner nodes have temperature and pore fluid pressure degrees of freedom. The variation of temperature was monitored at the surface of the canister where a thermocouple was placed in the test, while variations in both temperature and excess pore fluid pressure were monitored at the position where the pore pressure and temperature transducer was placed (see Figure 6).

4.3 *Numerical results*

Figure 7 compares temperature evolutions with time between numerical and experimental results at both the canister surface and the transducer (its location is shown in Figure 6). A good match was observed at the canister surface which implies that the scaling law applied to the numerical modelling is appropriate. A slightly higher numerical prediction was found at the transducer, which may be due to the fact that the boundaries of the centrifuge apparatus were not completely insulated from the environment and some heat was lost through those boundaries.

Due to the lack of experimental data at the canister surface in the literature, Figure 8 compares

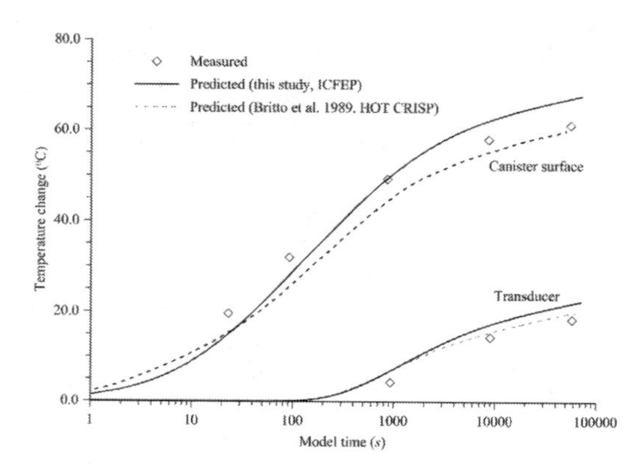

Figure 7. Comparison of temperature evolution with time between numerical predictions and experimental measurements for centrifuge test.

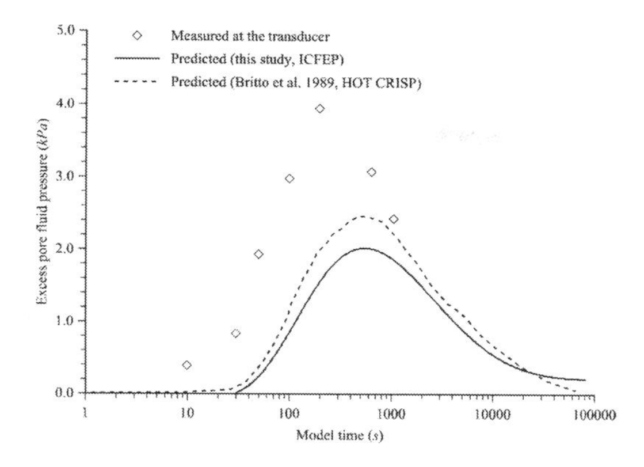

Figure 8. Comparison of pore fluid pressure evolution with time at the transducer between numerical predictions and experimental measurements for centrifuge test.

only the predicted and measured evolutions of thermally induced excess pore fluid pressure with time at the transducer. Although a similar response was predicted by both ICFEP used in this study and HOT CRISP used by Britto et al. (1989), both underestimated the peak values. This is thought to be caused by the simplification of using constant values of the thermal expansion coefficient of water and soil permeability in the numerical analysis, as both properties are known to vary significantly with temperature (Delage et al. 2000, Cengel & Ghajar 2011). Further research is needed to account for the non-linear temperature-dependent behaviour of these soil properties.

5 CONCLUSIONS

This paper briefly presents a numerical formulation necessary for modelling the transient behaviour of thermally induced pore fluid pressures. A new approach for deriving the hydraulic equation for a coupled THM problem is detailed. While other approaches found in the literature apply the mass conservation laws of both the pore fluid and the solid particles, the current formulation is derived by considering only the principle of mass conservation of the pore fluid phase. Despite this, the same hydraulic formulation capable of accounting for the excess pore fluid flow induced by the difference between thermal expansion coefficients of the pore fluid and the solid particles, is obtained.

In the first study presented in this paper, an existing analytical solution to the cylindrical heat source problem was used to verify the adopted THM formulation. An excellent match was found between the analytical solution and numerical predictions demonstrating the capabilities of the coupled THM finite element formulation in the simulation of thermally induced pore fluid pressures.

The second study involved simulation of an existing centrifuge heating tests. Although the predicted temperature evolution is similar to that measured in the experiment, the peak excess pore pressure was underestimated compared to the experimental measurements. Further study is necessary to investigate the effect of accounting for the non-linear temperature dependency of soil properties, such as the thermal expansion coefficient of the pore fluid and soil permeability.

ACKNOWLEDGEMENTS

The research presented in this paper was funded by the Geotechnical Consulting Group (GCG) in the UK.

REFERENCES

Britto, A.M., Savvidou, C., Maddocks, D.V., Gunn, M.J., & Booker, J.R. 1989. Numerical and centrifuge modelling of coupled heat flow and consolidation around hot cylinders buried in clay. *Géotechnique* 39(1): 13–25.

Britto, A.M., Savvidou, C., Gunn, M.J. & Booker, J.R. 1992. Finite element analysis of the coupled heat flow and consolidation around hot buried objects. *Soils and Foundations* 32(1): 13–25.

Booker, J.R., & Savvidou, C. 1985. Consolidation around a point heat source. *International Journal for Numerical and Analytical Methods in Geomechanics* 9(2): 173–184.

Campanella, R.G. & Mitchell, J.K. 1968. Influence of temperature variations on soil behaviour. *ASCE Journal Soil Mechanics and Foundation Engineering Division* 4(3): 709–734.

Cengel, Y.A., & Ghajar, A.J. 2011. *Heat and Mass Transfer: Fundamentals and Applications*. 4th ed. New York: McGraw-Hill.

Cui W., Potts D.M., Zdravković L., Gawecka K.A. & Taborda D.M.G. 2017. An alternative coupled thermo-hydro-mechanical finite element formulation for fully saturated soils. *Computers and Geotechnics*. (ISSN: 0266-352X)

Delage, P., Sultan, N., & Cui, Y.J. 2000. On the thermal consolidation of Boom clay. *Canadian Geotechnical Journal* 37(2): 343–354.

Fernandez, R.T. 1972. Natural convection from cylinders buried in porous media. PhD thesis, University of California.

François, B., Laloui, L., & Laurent, C. 2009. Thermo-hydro-mechanical simulation of ATLAS in situ large scale test in Boom Clay. *Computers and Geotechnics* 36(4): 626–640.

Gens, A., Vaunat, J., Garitte, B., & Wileveau, Y. 2007. In situ behaviour of a stiff layered clay subject to thermal loading: observations and interpretation. *Géotechnique* 57(2): 207–228.

Gens, A. 2010. Soil-environment interactions in geotechnical engineering. *Géotechnique* 60 (1): 3–74.

Kutter, B.L. 1992. Dynamic centrifuge modeling of geotechnical structures. *Transportation Research Record 1336, Transportation Research Board*, Washington, DC.

Lewis, R.W. & Schrefler, B.A. 1998. *The finite element method in the static and dynamic deformation and consolidation of porous media*. 2nd ed. Wiley.

Lima, A., Romero, E., Gens, A., Muñoz, J., & Li, X.L. 2010. Heating pulse tests under constant volume on Boom clay. *Journal of Rock Mechanics and Geotechnical Engineering* 2(2): 124–128.

Maddocks, D.V., & Savvidou, C. 1984. The effects of heat transfer from a hot penetrator installed in the ocean bed. In: *Proceedings of the Symposium on Application of Centrifuge Modelling to Geotechnical Design*, Universiiy of Manchester.

Mohajerani, M., Delage, P., Sulem, J., Monfared, M., Tang, A.M., & Gatmiri, B. 2012. A laboratory investigation of thermally induced pore pressures in the Callovo-Oxfordian Claystone. *International Journal of Rock Mechanics and Mining Sciences* 52: 112–121.

Olivella, S., Gens, A., Carrera, J. & Alonso, E.E. 1996. Numerical formulation for a simulator (CODE_BRIGHT) for the coupled analysis of saline media. *Engineering Computations* 13(7): 87–112.

Potts, D.M. & Zdravković, L. 1999. *Finite Element Analysis in Geotechnical Engineering: Theory*. London: Thomas Telford.

Potts, D.M. & Zdravković, L. 2001. *Finite Element Analysis in Geotechnical Engineering*: Application. London: Thomas Telford.

Savvidou, C., & Britto, A.M. 1995. Numerical and experimental investigation of thermally induced effects in saturated clay. *Soils and foundations* 35(1): 37–44.

Thomas, H.R. & He, Y. 1997. A coupled heat–moisture transfer theory for deformable unsaturated soil and its algorithmic implementation. *International Journal for Numerical Methods in Engineering* 40(18): 3421–3441.

Thomas, H.R., Cleall, P., Li, Y.C., Harris, C. & Kern-Luetschg, M. 2009. Modelling of cryogenic processes in permafrost and seasonally frozen soils. *Géotechnique* 59(3): 173–184.

Vaziri, H.H. 1996. Theory and application of a fully coupled thermo-hydro-mechanical finite element model. *Computers and structures* 61(1): 131–146.

Numerical Methods in Geotechnical Engineering IX – Cardoso et al. (Eds)
© 2018 Taylor & Francis Group, London, ISBN 978-1-138-33198-3

Numerical investigation of the effects of thermal loading on the mechanical behaviour of energy piles in sand

I. Kamas & E. Comodromos
University of Thessaly, Volos, Greece

D. Skordas & K. Georgiadis
Aristotle University of Thessaloniki, Thessaloniki, Greece

ABSTRACT: Integrating geothermal loops into conventional structural piles constitutes an environmentally friendly way for the heating and cooling of a building. Energy piles take advantage of the shallow geothermal energy in order to satisfy the thermal needs of the building. However, the exploitation of geothermal energy creates additional thermal loads. Thermomechanical finite element analyses are conducted in order to examine the effect of the additional thermal loads on the mechanical behaviour of single energy piles in sand. Both mechanical and thermal loads are applied. The paper summarizes the effect of the thermal loads on the axial stress response and the settlement of thermoactive piles. The effect of pile head restraints (free or fixed pile head), the duration of thermal loading and the coefficient of thermal expansion are investigated in the parametric analyses. Several observations made in previous studies are confirmed here and important factors that control the behaviour of energy piles are quantified for the case of single axially loaded piles in sand.

1 INTRODUCTION

Ground source heat pumps (GSHPs) that exploit shallow geothermal energy are routinely employed in many countries to improve the energy performance of buildings. Shallow geothermal energy is used mainly to cover heating and cooling needs and constitutes an extremely economic renewable energy source.

The technology of burying heat exchanger pipes in boreholes is the most commonly used worldwide. Over the last few decades, however, building foundation elements such as shallow and deep foundations, and retaining walls, are increasingly used in order to harvest geothermal energy (Brandl 2006). These structures are called energy geostructures. The use of concrete structural elements for the exploitation of geothermal energy is particularly appealing because these elements are already required for example to retain an excavation or to transfer the loads imposed by the superstructure to the ground and because of the concrete's high thermal conductivity that results in more efficient heat transfer. Optimal design and installation of such systems can lead to a reduction of up to 50% of a building's carbon dioxide emissions (Laloui et al. 2006). Moreover, energy geostructures have lower installation costs, compared to geothermal boreholes.

A single energy pile can transfer 15–120 W/m depending on its size, the construction details, the type of the surrounding soil and the system's operating mode (Bourne-Webb 2013).

At first, the technology of using structural piles as ground heat exchangers was implemented in Austria and Switzerland (Brandl 2006; Markiewicz & Adam 2009). Ever since, the technology has been applied in many other European countries, the USA and Japan. Although there is a number of high profile buildings (e.g. as the One New Change Building, the Lambeth College, the Dock Midfield E at Zurich airport and many others) that are examples of successful use of structural foundation elements as renewable sources of energy, there are unresolved issues related to their performance and safety that need to be investigated. In recent years, lots of research has been conducted in order to resolve issues regarding the complex issue of thermal and thermo-mechanical behaviour of energy geostructures. Moreover, issues that concern the thermal performance of such systems have been investigated. These investigations include full-scale (Laloui et al. 2006; Bourne-Webb et al. 2009; Murphy et al. 2015) and small-scale (Goode & McCartney 2014; Ng et al. 2015; Nguyen & Tang 2017) experiments, numerical investigations that examine the thermo-mechanical performance of single energy piles and energy pile-groups with the load—transfer approach (t-z function) (Péron et al. 2011; Knellwolf et al. 2011; Suryatriyastuti et al. 2014) and finite element studies (Laloui et al. 2006; Bodas Freitas et al. 2013; Tsetoulidis et al. 2016; Jeong et al. 2014; Salciarini et al. 2013; Gawecka et al. 2017; Loria & Laloui 2017a; Bourne-Webb et al. 2015).

According to Laloui et al. (2006) and Bourne-Webb et al. (2009), a thermoactive pile, when heated or cooled, expands or contracts, respectively, relative to the surrounding soil. The lateral shear stress and the reaction of the head and base of the pile act as restrictions to the thermal dilation or contraction of the pile. Consequently, during heating, compressive axial stresses develop, while during cooling the corresponding thermal stresses become tensile, decreasing the total axial stress distribution along the pile (Amatya et al. 2012). However, there are some studies in the literature (Bodas Freitas et al. 2013; Bourne-Webb et al. 2015) that show that heating of the pile can also lead to the development of tensile stresses.

The thermal expansion or contraction of a pile depends on the stiffness of the soil and the superstructure (Bourne-Webb 2013). Goode & McCartney (2014) studied the thermo-mechanical behaviour of a fixed and a free—head pile, in sandy soil, by conducting centrifuge experiments. They observed an increase of about 50% of the axial stresses of a fixed—head pile compared to those of an identical free—head pile when heating loads were applied. Accordingly, in the free—head pile, thermal strains developed, and thus the vertical displacements were greater as the pile expanded freely.

Ng et al. (2015) investigated the effect of thermal loading on the bearing capacity of an energy pile in medium dense sand. They observed an increase of 13% and 30% for a temperature increase of 15°C and 30 respectively. Finite element analyses of energy piles in clay presented by Tsetoulidis et al. (2016) showed that heating or cooling of a pile in clay causes insignificant changes to its bearing capacity (approximately 3% for cooling and 1.3% for heating).

Rotta Loria et al. (2015) & Tsetoulidis et al. (2016) studied the effect of thermal loading on the pile axial load distribution of a free—head pile working under different axial loads. They observed that the magnitude of the working load applied prior to the thermal cycles affected the side shear resistance mobilization of the pile and consequently the axial stress distribution along the pile.

The additional axial stresses developed due to the cooling—heating operation also depend on the magnitude of the thermal loading and the thermal properties of the soil. The ground temperature is directly affected by the thermal loading imposed on the pile. These thermal changes can cause irreversible volumetric strains in sandy (Agar et al. 1986) and clayey (Cekerevac & Laloui 2004) soils. As a result, pile heave or settlement may be induced, changing the lateral friction along the pile (Di Donna et al. 2016), while at the same time the end—bearing resistance of the pile may be altered (Amatya et al. 2012). Bodas Freitas et al. (2013) studied the influence of the coefficient of thermal expansion (a_{soil}) of a clayey soil on the thermo-mechanical behavior of a single energy pile, for steady state. They observed that a_{soil} can significantly affect the thermo-mechanical response of the energy pile.

Although a lot of research has been done on the thermo-mechanical behavior of single energy piles, there are still aspects of their behavior, especially in the case of piles in sand, that have not yet been thoroughly examined. The purpose of this study is to investigate the effects of: (i) the axial pile head load and pile head restraint, (ii) the duration of cooling/heating, and (iii) the coefficient of thermal expansion of the soil (a_{soil}), on the thermo-mechanical performance of an energy pile in sand.

2 FINITE ELEMENT ANALYSIS

2.1 *Numerical model*

The numerical analyses presented in this paper were conducted with the finite element software ABAQUS. An axisymmetric model of a circular pile with a length of $L_p = 30$ m and a diameter of $D_p = 1$ m in uniform fully saturated sandy soil was considered. The calculations are based on the following assumptions: i) uniform temperature along the thermo-active pile, ii) thermally insulated superstructure, hence no thermal flux between the superstructure and the soil-pile system was allowed.

The sandy soil behavior is modelled as linear elastic—perfectly plastic with a Mohr-Coulomb failure criterion. The mechanical material parameters selected for the ground were: angle of shearing resistance $\phi = 30°$, Young's modulus $E_s = 30$ MPa, dilation angle $\psi = 0.1°$, Poisson's ratio $v_s = 0.3$, bulk unit weight $\gamma_s = 20$ kN/m^3. The concrete pile is modelled as linear elastic with Young's modulus $E_p = 2.9$ GPa, Poisson's ratio $v_s = 0.1$ and unit weight $\gamma = 25$ kN/m^3.

The values of the ground thermal parameters were selected based on existing literature (Jeong et al. 2014; Salciarini et al. 2014; Rotta Loria & Laloui 2017b). According to the above studies, the thermal expansion coefficient a_{soil} of a sandy soil ranges between $10–30 \cdot 10^{-6}$ m/m/°C. In order to examine the effect of a_{soil} on the thermo-mechanical behavior of the pile, values ranging between 0-30.10^{-6} m/m/°C were considered. Typical thermal properties selected for the materials of the ground and the pile are shown in Table 1.

The finite element mesh used for the single pile analysis is shown in Figure 1. The length and the diameter of the mesh are equal to 60 m ($2 \cdot L_p$). The finite element mesh consists of 10846 nodes and 3347 elements. 8-node biquadratic axisymmetric quadrilateral elements are used for both the pile and the soil. For the pile, the approximate element size is equal to 0.2 m. For the soil, a denser mesh is used in a cylindrical region around the pile with a diameter of about 1.5 m, with element sizes ranging between 0.1 m to 0.2 m. The element size increases as the distance from the pile increases and reaches between 5 m to 10 m at the mesh right and bottom boundaries.

The vertical and horizontal displacements are tied at the bottom boundary, while only the normal displacements are tied at the vertical boundaries. The ground water table is at the ground surface and the initial lateral stresses are calculated with a coefficient of earth pressure at rest $K_o = 1$. The mechanical pile head load is imposed as an equivalent prescribed displacement for fixed-head piles and as a uniformly distributed load for the free—head piles. As a consequence, the axial force at the pile head varies during thermal loading in the case of fixed-head piles and remains constant in the case of free-head piles. A constant temperature field of 15°C is prescribed to the whole mesh at the beginning of each analysis and the same temperature is prescribed as a boundary condition at the bottom and right boundaries throughout the analyses. The axis of symmetry is assumed adiabatic. For the soil—pile interface, frictional behaviour is considered, with an angle of shearing resistance equal to the angle of shearing resistance of the soil.

Table 1. Thermal properties of the soil and the concrete.

	Sand	Concrete
a: m/m/°C ($\cdot 10^{-6}$)	0, 10[1], 20[2], 30[3]	10
λ: W/m · °C	2.5	1.8
c: J/kg/°C	1200	1000

[1]Jeong et al. 2014; [2]Rotta Loria & Laloui 2017b; [3]Salciarini et al. 2014.

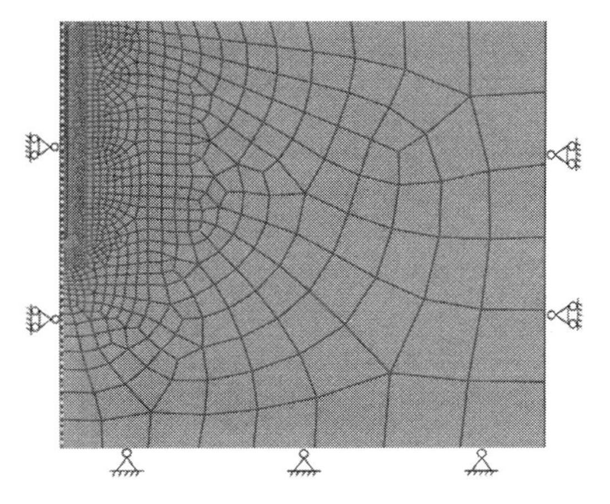

Figure 1. Finite Element mesh.

2.2 *Steps of the thermal and thermo-mechanical analysis*

The thermal loading imposed to the pile is considered in a separate heat transfer analysis. Thermal flow finite element analyses are conducted in both transient and steady state conditions. The same finite element mesh with the mechanical analyses is used, with 8 – node quadratic axisymmetric heat transfer quadrilateral elements. Temperature fields calculated with the heat transfer analysis are then imposed on the same time steps of the mechanical FE analysis in order to calculate the mechanical response to the temperature changes. The pile is initially cooled with $\Delta T = -15$°C and then heated with $\Delta T = +30$°C, under constant pile head conditions (settlement for the fixed-head piles or axial load for the free-head piles).

3 RESULTS

3.1 *Overview*

The effect of the pile—head restraint for different initial pile head loads on the axial force distribution in the piles is first examined. Then, the effect of the duration of thermal loading and of the coefficient of thermal expansion on the axial stress and the vertical displacement distributions along the pile is investigated, considering different thermal loads under constant initial working load. The working load was taken equal to half the bearing capacity of the pile (Factor of Safety 2). The bearing capacity of the pile was initially calculated analytically equal to Q_{ult} = 11900 kN. Then, the bearing capacity of the pile was calculated using ABAQUS by applying increasing mechanical load to the pile—head until failure and a bearing capacity of 12000 kN was computed, which is in excellent agreement with the analytically calculated theoretical value. Thus, the bearing capacity was estimated equal to Q_{ult} = 12000 kN and the working load equal to $Q_{ult}/2$ = 6000 kN.

3.2 *Effect of pile head restraint*

Three different initial pile—head axial loads are considered: $Q_1 = 0$ kN (only self—weight); $Q_2 = 2000$ kN ($1/6 \cdot Q_{ult}$); $Q_3 = 6000$ kN ($1/2. Q_{ult}$). As noted above, the mechanical load is imposed as an equivalent prescribed displacement load in the case of fixed—head piles. The thermal load applied is the same in all cases: cooling from an initial temperature of 15°C to 0°C, followed by heating to 30°C. In the analyses presented below $a_{soil} = 10 \cdot 10^{-6}$ m/m/°C and thermal load duration t = 6 months.

3.2.1 *Fixed pile*

Figure 2 shows the initial (Mechanical) distribution and the distribution after 6 months of cooling (0°C) and further 6 months of heating (30°C) of the axial forces along a fixed—head energy pile, for different initial axial pile head loads. It is evident that cooling and heating significantly affects the distribution of axial forces along the pile. Differences of about 1000 kN in the axial forces can be observed for both cases (i.e. cooling and heating) between the mechanically only loaded pile and the thermo-mechanically loaded pile cases regardless of the magnitude of the mechanical load. The stiffness of the superstructure prevents the vertical movement of the pile. During cooling, the superstructure does not allow the pile to contract resulting in the development of tensile thermal stresses. Consequently, a decrease of the overall compressive forces is observed along the entire length of the pile. On the contrary, when the pile is heated, the superstructure prevents the thermal expansion of the pile. As a result, compressive thermal stresses are mobilised and an increase of the compressive forces along the pile occurs. It can also be observed in Fig. 2 that the axial force variations with depth after cooling or heating have the same shape as the initial (mechanical) variations. The magnitude of the initial applied pile head load does not seem to significantly affect pile response to heating or cooling.

3.2.2 *Free pile*

Contrary to the fixed—head pile case, the free—head pile is not restricted to contract or expand. As a result, the axial stresses induced by the thermal loads are less significant. However, a more complex behaviour is noticed especially during cooling. In Figure 3, it can be seen that heating the pile causes an increase of the compressive axial forces along the pile due to the pile thermal expansion. However, as illustrated in the same figure, the magnitude of the applied mechanical load significantly influences the behavior of the pile when cooled. For the case of zero axial loading, cooling the pile results in a decrease of the axial forces. However, when the initial applied pile head load is increased a reversal in behavior takes place, which is already evident from the intermediate axial load of 2000 kN, where an increase of the compressive forces occurs in the upper 7 m of the pile. For greater mechanical loads, an increase of the compressive axial loads in the greater part of the pile is observed, which is caused by the reduction in the interface shear strength due to cooling. This unexpected behavior is discussed in detail in Tsetoulidis et al. (2016), who report a similar response for the case of a free—head pile in clay. The above behavior is observed for all values of a_{soil} and becomes more pronounced as a_{soil} increases.

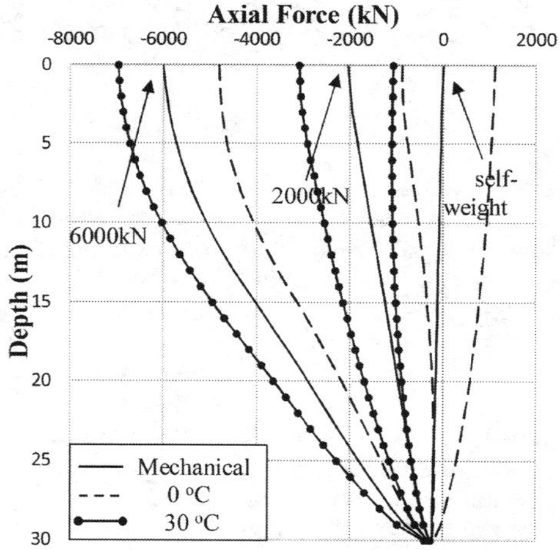

Figure 2. Axial force distributions with depth for fixed-head pile.

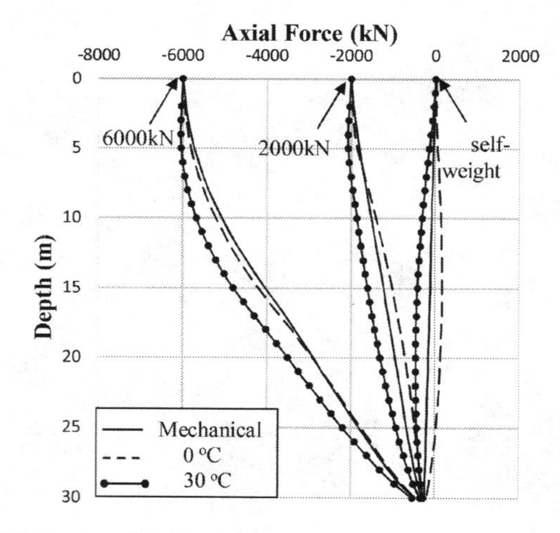

Figure 3. Axial load distributions with depth for free-head pile.

3.3 *Effect of the duration of the thermal loading*

Although normal operation mode of typical heating—cooling systems (e.g. radiators, air-conditions) ranges between several hours per day and some months, in environments with extreme weather conditions, numerical investigations already published concerning the effect of the thermal loads on the thermo-mechanical response of energy piles, predominantly consider steady state conditions. For this reason, this investigation examines different durations of the operation of the cooling—heating cycles on the mechanical behavior of an energy pile in sand.

Four durations are examined (1 day; 1 month; 6 months; steady state). Initially, mechanical load equal to 6000 kN is applied, followed by cooling to 0°C and heating to 30°C, under constant pile head conditions (displacement or load for fixed- or free-head piles, respectively). Figure 4 shows the developed axial stresses along the fixed—head piles, and Figure 5 shows the variation of axial displacement along the free-head piles. In the analyses presented below $a_{soil} = 30 \cdot 10^{-6}$ m/m/°C.

3.3.1 *Fixed pile*

For both the fixed- and the free- head pile the duration of the thermal loading affects the response of the energy pile and the longer the duration of the thermal load, the more significant is the influence. As seen in Figure 4, for durations ranging between 1 day and 6 months, the decrease or increase of the axial stresses along the pile, when the pile is cooled or heated respectively, is similar, ranging between the maximum values of 1450 kPa and 1870 kPa in the pile head when the pile is cooled and 1000 kPa and 1330 kPa when heated. This indicates that the after the initial instantaneous stress change, the rate of change is very slow. At steady state conditions the stress changes are much greater: a maximum decrease of 3100 kPa and increase of 2000 kPa during cooling or heating, respectively, are computed. However, steady state conditions are unlikely in practice and based on Fig. 4 such an assumption is over conservative.

3.3.2 *Free pile*

Figure 5 shows the axial displacement variation of the pile with depth for the free—head pile. It is evident that the effect of the duration of the thermal loading is more significant when heating loads are applied. When the pile is cooled, contraction occurs, that results in an increase of the settlement of the pile head. The increase in axial downward displacement reduces with depth, and becomes almost zero (and even slightly negative) at the pile base for the case of cooling of up to 6 months. During cooling the whole pile moves downward and only for very short-term thermal loading (i.e. 1 day and 1 month) the base of the pile heaves negligibly. The maximum settlement occurs, as expected, at the pile head. For the case of cooling under steady state conditions, the maximum increase was 36%, while for a duration of 1 day the corresponding increase was 25%.

During heating, a more complex behavior was observed. Heating the pile for a period ranging between 1 day and 6 months caused the pile to expand in a similar way. The upper part of the pile heaved and the rest of the pile settled. However, for steady state conditions the whole pile heaved due

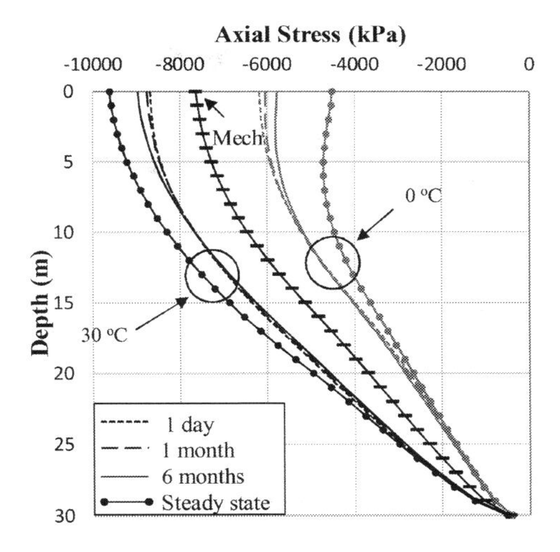

Figure 4. Axial stress distributions with depth of fixed-head pile for different durations of thermal loading.

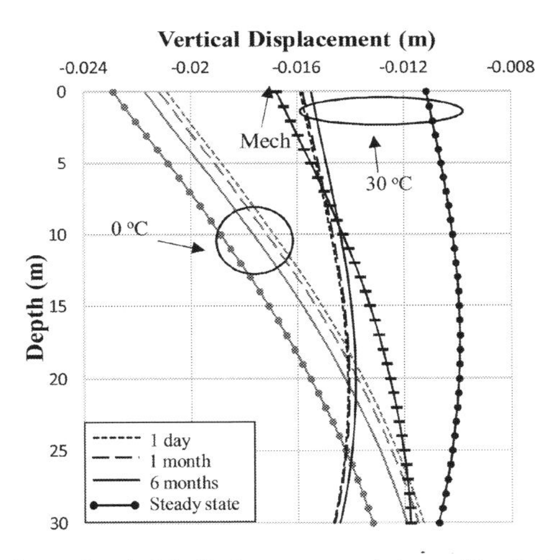

Figure 5. Axial displacement variation with depth of free-head pile for different durations of the thermal loading.

to heating. This response of the pile can be attributed to the higher value of the coefficient of thermal expansion of the ground relative to the pile. More specifically, when the pile is heated, the soil surrounding the pile is also heated. At steady state conditions, when a large volume of soil has heated up, the soil expands more than the pile causing an upward movement of the pile. As a result, although the lower part of the pile moves downward during the first six months of heating (with most of this movement taking place in the first day), at steady state the upward pressure imposed by the expanding soil moves the whole pile upward.

3.4 *Effect of the coefficient of thermal expansion of the ground a_{soil}*

The effect of the coefficient of thermal expansion of the soil (α_{soil}) on the thermo-mechanical response of the pile is investigated. Four values of the coefficient of thermal expansion representative for sandy soils are examined: 0; $10 \cdot 10^{-6}$ m/m/°C; $20 \cdot 10^{-6}$ m/m/°C; $30 \cdot 10^{-6}$ m/m/°C, (See Table 1). Analysis were conducted for both fixed and free—head piles. For all the cases examined, the thermal loading steps described above were applied: mechanical loading followed by cooling to 0°C and heating to 30°C. All the analyses were performed for four durations of thermal loading (1 day; 1 month; 6 months; Steady state). However, significant effects of α_{soil} were only observed for steady state conditions.

3.4.1 *Fixed pile*

Figure 6 presents the distribution of axial stresses along a fixed-head pile and Figure 7 presents the

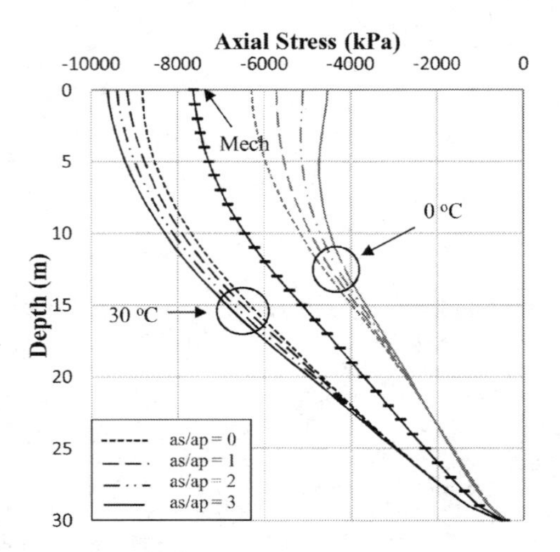

Figure 6. Axial stress distributions with depth of fixed-head pile for different coefficients of thermal expansion of the ground.

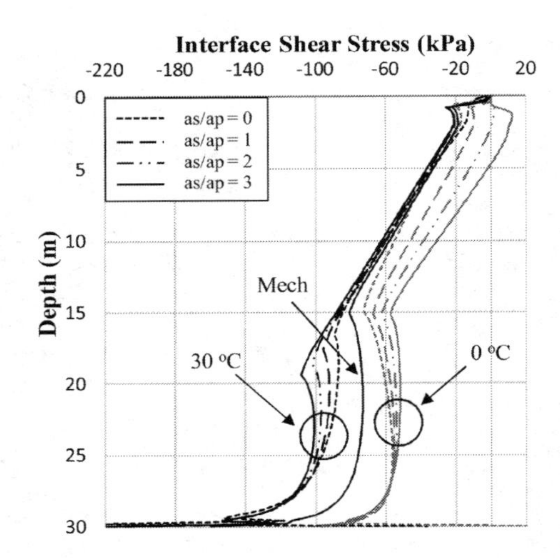

Figure 7. Interface shear stress distributions with depth of fixed-head pile for different coefficients of thermal expansion of the ground.

mobilized interface shear stresses, for different values of α_{soil}. The effect of α_{soil} was found to be insignificant for thermal loading of up to 6 months. For this reason, only the results for steady state conditions (cooling and heating) are shown in Figures 6 and 7.

It is obvious that the greater the soil's coefficient of thermal expansion the greater are the changes of the axial stresses along the pile. It is observed that the effect of the a_{soil} is more evident in the case of cooling. During cooling, the maximum axial stress change calculated along the pile is + 40%, while during heating the corresponding change is - 26%. This can be attributed to the fact that when the pile is cooled, the interface shear stresses decrease along the whole pile for all the values of the (a_{soil}) examined, with a greater decrease occurring for increasing values of a_{soil} (Figure 7).

The decrease of the shear stresses is associated with the relative movement of the pile and the soil during cooling. The restrained pile head cannot settle further, while the soil is free to contract. This vertical movement of the ground induces shear stresses opposite to the shear stresses mobilized due to the mechanical load, resulting in a decrease of the total shear stresses along the whole depth. For increasing values of a_{soil} more significant decrease of the interface shear stresses occur. As seen in Figure 7, for the case of $a_s/a_p = 3$, the total interface shear stresses become positive in the first 5 m.

In contrast, when the pile is heated an increase of the interface shear stresses is noticed only on the lower half of the pile which is also more pronounced for increasing a_{soil}. In this case, owing to the head fixity, the pile cannot expand upward during heating, while the ground expands freely. In this case, relative movements develop at the interface and the mobilized shear stresses are in the same direction with the shear stresses due to mechanical load. The increase in these shear stresses is limited, however, by the shear strength of the pile-soil interface, which is mobilized in the upper part of the pile. Thus, an increase of the total interface shear stresses occurs only in the lower half of the pile, resulting in smaller change of the total axial stresses along the pile than in the case of cooling.

3.4.2 *Free pile*

Figures 8 presents the axial displacement variation with depth for the case of free—head pile. As in the case of fixed-head piles discussed above, the effect of a_{soil} was found to be insignificant for heating/cooling of up to 6 months and only at steady state conditions were significant effects observed. Therefore, only the results for steady state heating and cooling are presented in Figure 8.

As seen in this figure, axial displacements are clearly affected by the value of a_{soil} at steady state conditions. During cooling, the whole pile settles, apart from the case of thermally inert soil where the lower 5 m of the pile heave. It is evident that as a_{soil} increases the pile settlement increases uniformly along the length of the pile (i.e. for $a_s/a_p = 3$ the pile—head settlement increases by about 34%, while for $a_s/a_p = 1$ the corresponding increase is 26%), the displacement vs depth curve translates to the left and the neutral point moves downward. This behavior can be attributed to the greater contraction of the soil for greater values of a_{soil}. The effect of the a_{soil} is more pronounced when the pile is heated. For thermally inert soil, only the first 10 m of the pile heave while for increasing values

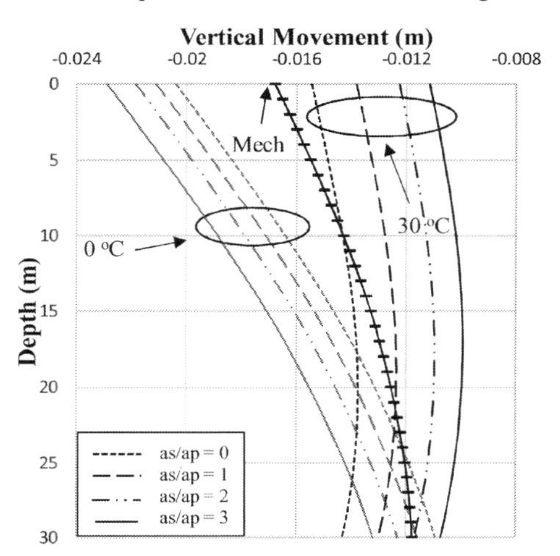

Figure 8. Axial displacement variation with depth of free-head pile for different coefficients of thermal expansion of the ground.

of a_{soil}, greater part of the pile heaves and the neutral point moves downward. For the case of $a_s/a_p = 3$, however, the whole pile heaves when heated. This is due to the fact that the soil, in this case, expands more compared to the pile.

4 CONCLUSIONS

The effect of thermal loading on the mechanical behavior of a single energy pile, in sandy soil, was investigated, with the use of finite element analysis. At first, the effect of pile head restraint under different pile head working loads was examined. Then, the effect of the duration of the thermal cycles and of the coefficient of thermal expansion of the soil on the mechanical behaviour of both fixed and free—head piles was investigated. The results of this study can be summarized as follows:

a. The magnitude of the pile head load (which is constant in the case of free-head piles but varies during heating/cooling for fixed-head piles) does not affect significantly the change in axial forces during heating or cooling in the case of fixed – head piles. It does, however, affect the response of free-head piles, especially during cooling.
b. As expected, the effect of heating or cooling on the axial pile forces is much greater in fixed-head than in free-head piles.
c. Thermally induced axial stress (and consequently also axial force) changes in fixed-head piles increase with time, however the rate of increase is very slow. After an initial almost instantaneous stress change with the application of the thermal load, the axial stresses do not change much over the first six months of constant heating or cooling.
d. Cooling of a free-head pile was found to cause downward movement of almost the entire pile from the first day. The downward movement increases with the duration of cooling and the neutral point moves well below the base of the pile.
e. During heating, and after the initial thermally induced axial displacement of a free-head pile, where the upper part of the pile moves upward and the rest of the pile moves downward, only small changes are observed over the studied 6-month heating period and the neutral point remains close to the pile head. However, at steady state conditions, the whole of the pile moves upward and the neutral point moves well below the base of the pile, as in the case of cooling.
f. The assumption of steady state conditions leads to overly conservative estimation of the thermally induced axial stress changes and axial displacements.

g. The effect of the value of the coefficient of thermal expansion of the ground (a_{soil}) on pile behavior is insignificant for cooling or heating durations of up to 6 months. It becomes important if steady state conditions are assumed.
h. In the case of steady state conditions, the value of a_{soil} affects both the axial stresses in fixed-head piles and the vertical displacements of free head piles. For fixed—head piles, the effect is greater in the case of cooling, while for free-head piles it is greater in the case of heating. In the latter case the value of a_{soil} not only controls the magnitude of the thermally induced displacements but also profoundly affects the position of the neutral point.

REFERENCES

Agar, J., Morgenstern N. & Scott, J., 1986. Thermal expansion and pore pressure generation in oil sands., pp. 327–333.

Amatya, B.L., Soga K., Bourne-Webb P.J., Amis T. & Laloui L., 2012. Thermo-mechanical behaviour of energy piles. *Géotechnique*, 62(6), pp. 503–519.

Bodas Freitas, T.M., Cruz Silva, F. & Bourne-Webb, P.J., 2013. The response of energy foundations under thermo-mechanical loading. *18th Intl Conf Soil Mechanics and Geotechnical Engineering, Paris, Intl. Society for Soil Mechanics & Geotechnical Engineering*, pp. 3347–3350.

Bourne-Webb, P.J., 2013. An overview of observed thermal and thermo-mechanical response of piled energy foundations. *Proc. Eur. Geothermal Conference, Pisa, Italy*, pp. 1–8.

Bourne-Webb, P.J. et al., 2009. Energy pile test at Lambeth College, London: geotechnical and thermodynamic aspects of pile response to heat cycles. *Géotechnique*, 59(3), pp. 237–248.

Bourne-Webb, P.J., Bodas Freitas, T.M. & Assunção, R.M.F., 2015. Soil – pile thermal interactions in energy foundations. *Géotechnique*, 66, pp. 1–5.

Brandl, H., 2006. Energy foundations and other thermo-active ground structures. *Géotechnique*, 56(2), pp. 81–122.

Cekerevac, C. & Laloui, L., 2004. Experimental study of thermal effects on the mechanical behaviour of a clay. *International Journal for Numerical and Analytical Methods in Geomechanics*, 28(3), pp. 209–228.

Di Donna, A., Rotta Loria, A.F. & Laloui, L., 2016. Numerical study of the response of a group of energy piles under different combinations of thermo-mechanical loads. *Computers and Geotechnics*, 72(March 2016), pp. 126–142.

Gawecka, K.A. et al., 2017. Numerical modelling of t hermo-active piles in London Clay. *Proceedings of the Institution of Civil Engineers - Geotechnical Engineering*, 170(3), pp. 201–219.

Goode, J.C. & McCartney, J.S., 2014. Evaluation of Head Restraint Effects on Energy Foundations. *Proceedings of GeoCongress 2014 (GSP 234), ASCE, Reston, VA, USA*, pp. 2685–2694.

Jeong, S., Hyunsung L., Joon K.L. & Junghwan K., 2014. Thermally induced mechanical response of energy piles in axially loaded pile groups. *Applied Thermal Engineering*, 71, pp. 608–615.

Knellwolf, C., Peron, H. & Laloui, L., 2011. Geotechnical Analysis of Heat Exchanger Piles. *Journal of Geotechnical and Geoenvironmental Engineering*, 137(10), pp. 890–902.

Laloui, L., Nuth, M. & Vulliet, L., 2006. Experimental and numerical investigations of the behaviour of a heat exchanger pile. *International Journal for Numerical and Analytical Methods in Geomechanics*, 30(8), pp. 763–781.

Loria, A.F.R., Gunawan A., Shi, C., Laloui, L., Ng, C.W.W., 2015. Numerical modelling of energy piles in saturated sand subjected to thermo-mechanical loads. *Geomechanics for Energy and the Environment*, 1, pp. 1–15.

Loria, A.F.R. & Laloui, L., 2017a. Displacement interaction among energy piles bearing on stiff soil strata. *Computers and Geotechnics*, 90, pp. 144–154.

Loria, A.F.R. & Laloui, L., 2017b. The equivalent pier method for energy pile groups. *Géotechnique*, 67, pp. 691–702.

Markiewicz, R. & Adam, D., 2009. Energy from earth-coupled structures, foundations, tunnels and sewers. *Géotechnique*, 59(3), pp. 229–236.

Murphy, K.D., McCartney, J.S. & Henry, K.S., 2015. Evaluation of thermo-mechanical and thermal behavior of full-scale energy foundations. *Acta Geotechnica*, 10(2), pp. 179–195.

Ng, C.W.W. Shi C., Gunawan A., Laloui L. & Liu H.L., 2015. Centrifuge modelling of heating effects on energy pile performance in saturated sand. *Canadian Geotechnical Journal*, 52(8), pp. 1045–1057.

Nguyen, V.T. & Tang, A.M., 2017. Long-term thermo-mechanical behavior of energy pile in dry sand. *Acta Geotechnica*, 12(4), pp. 729–737.

Péron, H., Knellwolf, C. & Laloui, L., 2011. A method for the geotechnical design of heat exchanger piles. *Proceedings of the American Society of Civil Engineers, Geo-Frontiers,*, 397(48), pp. 470–479.

Salciarini, D., Ronchi F., Cattoni E. & Tamagnini C., 2014. Thermomechanical Effects Induced by Energy Piles Operation in a Small Piled Raft. *International journal of Geomechanics*, 15(2), pp. 1–14.

Suryatriyastuti, M.E., Mroueh, H. & Burlon, S., 2014. A load transfer approach for studying the cyclic behavior of thermo-active piles. *Computers and Geotechnics*, 55, pp. 378–391.

Tsetoulidis, C., Naskos, A. & Georgiadis, K., 2016. Numerical investigation of the mechanical behaviour of single energy piles and energy pile groups. *Proceedings of the 1st International Conference on Energy Geotechnics, Kiel, Germany*, pp. 569–575.

Yavari, N.Ã. et al., 2016. Mechanical behaviour of a small-scale energy pile in saturated clay. *Géotechnique*, 66, pp. 1–10.

Numerical Methods in Geotechnical Engineering IX – Cardoso et al. (Eds)
 ISBN 978-1-138-33198-3

Finite-element modelling of thermo-mechanical soil-structure interaction in a thermo-active cement column buried in London Clay

Y. Ouyang
Cementation Skanska, UK
Formerly: University of Cambridge, UK

L. Pelecanos
University of Bath, UK
Formerly: University of Cambridge, UK

K. Soga
University of California, Berkeley, USA
Formerly: University of Cambridge, UK

ABSTRACT: Geothermal energy attracted considerable attention over the past years, as it has been considered as a green, renewable and sustainable energy source. Several attempts were made to use existing infrastructure to extract it with minimum amount of additional construction cost, such as foundation piles. This, however, raises issues regarding the thermal degradation of material properties and hence structural integrity. This paper presents a practical approach to modeling thermo-mechanical soil-structure interaction using load-transfer analysis and then shown an application to a recent thermal response test of a cement column buried in London Clay. The proposed model was able to reproduce the observed field behavior including the formation of thermal tensile cracks and it is therefore considered appropriate for future modeling of such thermo-mechanical soil-structure interaction problems.

1 INTRODUCTION

Geothermal energy attracted considerable attention over the past years, as it has been considered as a green, renewable and sustainable energy source (Brandl, 2006, Loveridge & Powrie, 2013). Several attempts were made to use existing infrastructure to extract it with minimum amount of additional construction cost, such as foundation piles, tunnels, and boreholes.

This, however, raises issues regarding the effects of the temperature-induced mechanical reaction from infrastructure (Ng et al, 2014, Pasten & Santamarina, 2014), such as expansion and contraction, which may potentially, for e.g. in the case of energy piles, (a) degrade the soil-structure interaction at the pile interface and (b) affect the structural integrity of the pile (e.g. cracking).

A number of relevant field tests were carried out (Bourne-Webb et al., 2009, 2013; Amatya et al., 2012; Ouyang, 2014; Ouyang et al., 2018) to investigate actual the field behaviour of energy piles. In parallel, a significant number of numerical models were developed, mainly based on the load-transfer approach (Knellwolf et al., 2011; Ouyang et al., 2011) and the solid finite-element (FE) approach using relevant constitutive models (Dupray et al., 2014; Ozudogru et al., 2014; Gawecka et al., 2016, Rotta Loria & Laloui, 2016).

This paper focuses on the modelling the mechanical response of an instrumented cement column in London Clay from the extended Thermal Response Test (TRT). This is a new test with new monitoring field data The aim is to understand the impact of thermal loading on soil-structure interactions, and also investigate if and how cracking can be formed in piles. A finite-element model is developed which couples thermal and mechanical loads in assessing deformations. The column is modelled with beam elements and the surrounding soil with nonlinear springs, which depend on the material properties of the soil. Thermal loads are considered as field variables that induce additional mechanical response, such as expansion and contraction. Moreover, the model is able to consider variations in the column properties (such as axial rigidity) to accommodate localised weaknesses due to the inclusion of plastic spacers within the cement. Nevertheless, the unique feature of this model is the ability to model cracks during the cooling phases when localised strains become tensile. A detailed description of the latter modelling feature will be included which is essential in modelling cracking during the cooling phase to capture the thermal impact on the structural integrity of heat-exchanger structures such as energy piles.

2 FINITE ELEMENT MODEL

2.1 *Load-transfer analysis of pile-soil interaction*

The adopted model follows the load-transfer approach in which the cement column is modeled as an 1D beam discretised with two-noded linear beam elements and the surrounding soil with 1D vertical nonlinear springs that follow a hyperbolic-type load-transfer (t-z) curve (Knellwolf et al., 2011; Ouyang et al., 2011). The springs follow a nonlinear relationship according to a hyperbolic model for the backbone (monotonic) loading (Bond et al., 1991; Bica et al., 2014). Unloading-reloading hysteresis is also considered using the Masing rules, as shown in Figure 1.

Local equilibrium is satisfied at each node in the column and global equilibrium is satisfied for the overall column according to:

$$[K_c + K_s]\{u\} = \{F\} - \alpha_c EA\{\Delta T\} \tag{1}$$

where, $[K_c]$ and $[K_s]$ are the stiffness matrices for column and soil, $\{u\}$ is the vector of DOFs (vertical axial displacements) and $\{F\}$ is the applied load vector. The thermal expansion/contraction component is added in the equilibrium where α_c, E, A are the thermal expansion coefficient, Young's modulus and cross-sectional area of the column respectively, whereas {ΔT} is the vector of temperature difference along the column depth. The local column element stiffness matrix is given by:

$$K_{c-EL} = \frac{EA}{L_{EL}}\begin{bmatrix} 1 & -1 \\ -1 & 1 \end{bmatrix} \tag{2}$$

where, L_{EL} is the length of the column beam element.

2.2 *Thermo-mechanical load-transfer model*

The nonlinear soil springs (t-z and q-z) are defined by the hyperbolic type load-transfer curve model (Pelecanos & Soga, 2017a, b; 2018; Pelecanos et al., 2018) given by:

$$t = \frac{k_m z}{\left[1 + \left(\frac{k_m}{t_m} z\right)^{hd}\right]^{\frac{1}{d}}} \tag{3}$$

where t is the shaft friction, z is the nodal displacement, and k_m, t_m, d, h are the model parameters related to maximum soil stiffness, maximum shaft friction, stiffness degradation and hardening respectively. The first two parameters can be obtained by relevant geotechnical engineering relations, whereas the latter two are purely model parameters that are obtained by curve fitting of load tests. In the absence of relevant load tests, the paper adopts h = 1 and d = 1 to represent a standard hyperbolic model suggested by Pyke

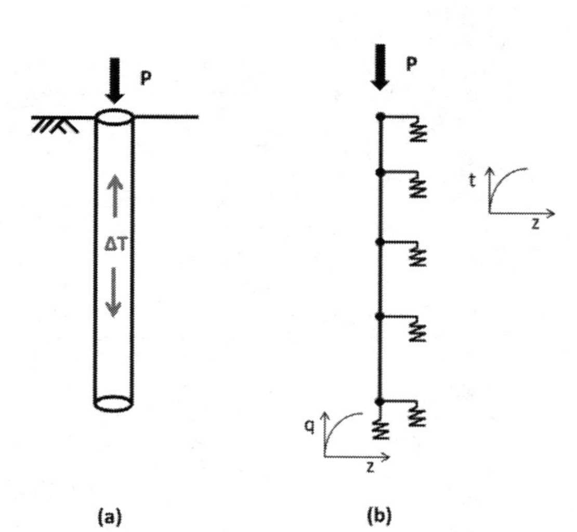

Figure 1. Load-transfer analysis of buried cement column.

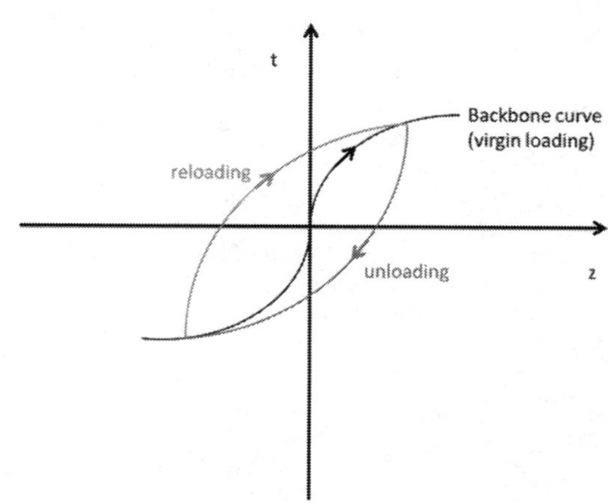

Figure 2. Nonlinear load-transfer model: backbone curve and unloading-reloading Masing rules.

(1979) with only two stiffness and ultimate friction parameters and that is close to other similar t-z curves (Frank & Zhao, 1982; Abchir et al., 2016; Seo et al., 2017a,b).

The stiffness of the soil spring for the column shaft, k_m, can be obtained using the Randolph & Wroth (1978) relationship, given by:

$$k_m = \frac{2\pi G L_{EL}}{\ln\left(2.5\ L_{tot}\frac{1-v}{r}\right)} \tag{4}$$

where G is the shear modulus of the soil, r is the column radius, v is the soil's Poisson ratio and L_{tot} is the total length of the column. The stiffness of the soil spring for the column base, k_{mb}, can be obtained by:

$$k_{mb} = \frac{4rG}{1-v} \tag{5}$$

The maximum shaft friction, t_m, and base capacity, q_m, depend on the yield parameters of the soil (S_u, c, ϕ) depending on the soil (Tresca or Mohr-Coulomb) and are given by:

$$t_m = a\ S_u \tag{6}$$

$$t_m = \beta\ \sigma'_{v0} = K_0 \tan\delta\ \sigma'_{v0} \tag{7}$$

$$q_m = N_c\ S_u \tag{8}$$

$$q_m = N_q\ \sigma'_{v0} \tag{9}$$

Both mechanical and thermal actions can be considered in this FE model. Boundary conditions and loads can be specified by the force and temperature load vectors, where the solution provides vertical displacements, and subsequently strains and stresses can be obtained as post-processing from the analysis.

3 APPLICATION IN LONDON

3.1 *Structural geometry & site description*

The cement column was installed at the basement of an existing building in London. The concrete slab under the basement was drilled and the column was buried in London Clay, as shown in Figure 1.

The column was instrumented with distributed Fibre Optic sensors using the BOTDR (Brillouin Optical Time-Domain Reflectometry) approach, and with conventional point vibrating-wire strain gauges (VWSGs). The BOTDR technique provides spatially-continuous (i.e. distributed) strain data along the entire length of an optical fibre (Soga, 2014; Soga et al. 2015). A detailed description of the theory of distributed FO strain sensing and its applications in civil and geotechnical infrastructure may be found by Kechavarzi et al. (2016, 2019), while examples of application to piles are obtained from Ouyang et al. (2015, 2018a, b) and Pelecanos et al. (2016, 2017b, 2018). More examples on monitoring soil-structure interaction may be found by Acikgoz et al. (2016, 2017), Di Murro (2016, 2019), Pelecanos et al. (2017a), Schwamb et al. (2014), Seo et al. (2017a,b) and Soga et al. (2015, 2017).

Because of the lack of significant reinforcement (as if it was a pile), the monitoring instruments were installed on an auxiliary steel bar which was run at the middle of the column. In addition, in order for this steel bar to be straight and be kept at the middle of the column, some auxiliary spacers were also installed. These spacers were made of plastic (shown in grey bars in Figure 3) were installed to keep the instrumentation system in place. However, the inclusion of such objects in the thin column has resulted in significant compromise of the column's axial rigidity and therefore resulted in localized cracking.

3.2 *Thermal-response test*

The thermal response test comprised of a number of heating and cooling cycles, as shown in Figure 4. More specifically, the test started at T = 17°C and

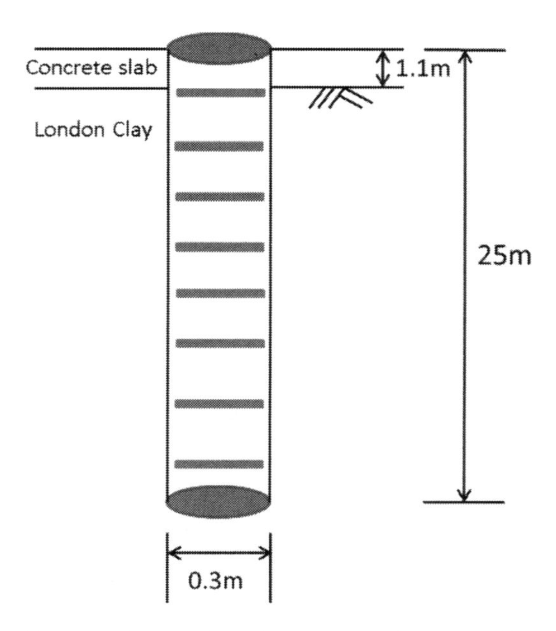

Figure 3. Description of the cement column buried in London Clay.

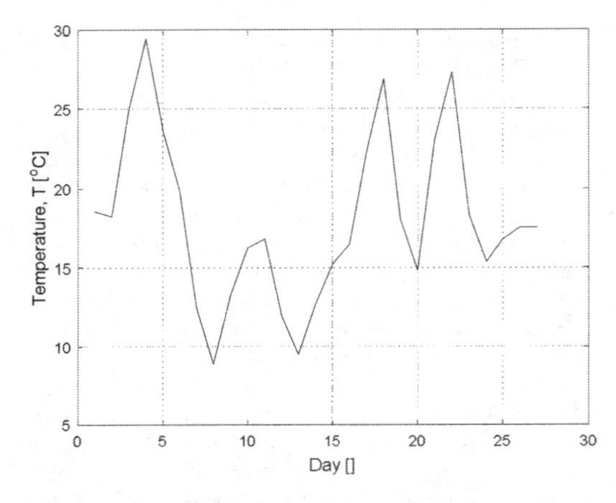

Figure 4. Time-history of the TRT including the heating and cooling cycles.

followed a series of heating and cooling cycles reaching a peak temperature of T = 29°C and low temperature of T = 8°C.

3.3 *Finite-element analysis*

The numerical load-transfer model described above was employed and the model parameters were defined based on the local SI information listed in Table 1. Then, the thermo-mechanical behaviour of the column was simulated during the TRT test duration. It is assumed that the soil is under undrained conditions.

The current FE model allows prescribing a variable EA along the column to accommodate the development of the (observed) spikes at six positions. These changes in column properties are expected to compromise the axial stiffness of the column to account for the "localised necking" along the column by reducing EA in these six locations.

The numerical results are presented together with the field measurements in Figure 5 to describe the behaviour of the column during the heating and cooling stages from. Figure 9 (a) simulates the column response to the heating phase, whereas Figure 9 (b) shows the response of the column during the cooling phase. Both figures clearly show the effect of tensile cracks generated in the first cooling stage on the pile behaviour in the subsequent cooling and heating stages with the corresponding field measurements. Both numerical results and field measurements agree well.

It is shown that in both heating and cooling stages, the mechanical strain gets to zero at the top of the column which is because of the free boundary condition. Also, in the case of heating, there is some small (non-zero) strain at the bottom of the column which is because of the base resistance of the column which reacts to the heating induced thermal expansion. Finally, the real observed strains at the middle of the pile seem to get close to zero which is because of the confining effect of

Table 1. Model parameters for the t-z FE analysis of the thermal column test.

	Depth	k_m	t_m	d	h
Layer	m	kN/m^3	kPa	–	–
1	0–4	5350	50	2	1
2	4-end	5350	120	2	1

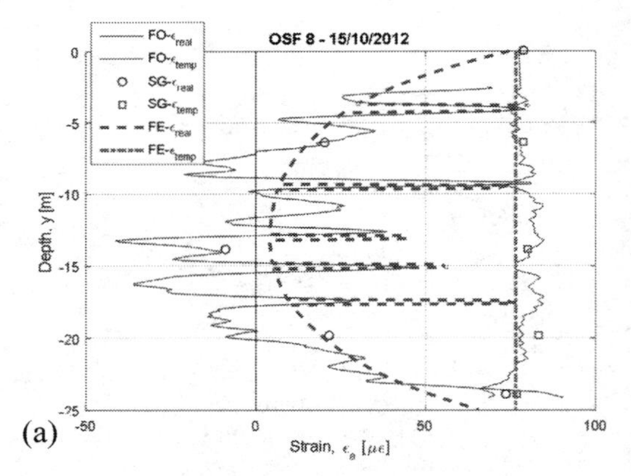

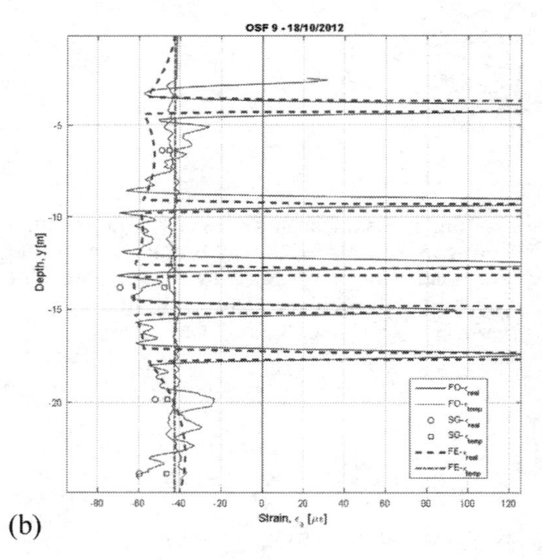

Figure 5. FE analysis of the TRT test: (a) heating phase, (b) cooling phase.

the surrounding soil and that the column does not expand freely at the middle.

4 CONCLUSIONS

This paper presents an approach to practical numerical modeling of thermo-mechanical response of soil-structure interaction. A model based on the load-transfer finite-element approach is presented using nonlinear t-z curves and it is applied in a recent field case study of a thermal response test of a slim cement column buried in London Clay.

The model is capable of considering the axial thermal expansion contraction of the slim cement column along with the hysteretic cyclic behavior of the surrounding soil due to the repeated heating-cooling cycles. Particular attention is paid on the development of a number of discrete cracks in the cement column which were observed during the thermo-mechanical field test and which can be reproduced by the numerical model.

Finally, it is shown that this efficient approach can be applied to the analysis of similar thermo-mechanical soil-structure interaction problems such as thermo-active buried structures, energy piles, heat-exchanger tunnels etc.

ACKNOWLEDGEMENTS

The Authors would like to acknowledge the financial and other support provided by Cementation Skanska, Arup, Concept, Geosense, Gecco, Canary Wharf Contractors (CWCL), the University of Cambridge and the Cambridge Centre for Smart Infrastructure and Construction (CSIC).

REFERENCES

Abchir, Z., Burlon, S., Frank, R., Habert, J. and Legrand, S., 2015. t–z curves for piles from pressuremeter test results. Géotechnique, 66(2), pp.137–148.

Acikgoz, M.S., Pelecanos, L., Giardina, G., Aitken, J. and Soga, K., 2017. Distributed sensing of a masonry vault due to nearby piling. Structural Control and Health Monitoring, 24 (3), e1872.

Acikgoz, M.S., Pelecanos, L., Giardina, G., Aitken, J. and Soga, K., 2016. Field monitoring of piling effects on a nearby masonry vault using distributed sensing. In: International Conference on Smart Infrastructure and Construction, 2016-06-25.

Amatya, B.L., Soga, K., Bourne-Webb, P.J., Amis, T., & Laloui, L. (2012). Thermo-mechanical behaviour of energy piles. Géotechnique, 62(6), 503–519.

Bica, A.V.D. et al., 2014. Instrumentation and axial load testing of displacement piles. Proceedings of the ICE - Geotechnical Engineering, 167(3), pp.238–52.

Bond, A.J., Jardine, R.J. & Dalton, J.C.P., 1991. The design and performance of the Imperial College Instrumented Pile. Geotechnical Testing Journal, 14(4), pp.413–24.

Bourne-Webb, P.J., Amatya, B., Soga, K., Amis, T., Davidson, C., & Payne, P. (2009). Energy pile test at Lambeth College, London: geotechnical and thermodynamic aspects of pile response to heat cycles. Geotechnique, 59(3), 237–248.

Bourne-Webb, P.J., Amatya, B. and Soga, K., 2013. A framework for understanding energy pile behaviour. Proceedings of the Institution of Civil Engineers-Geotechnical Engineering, 166(2), pp.170–177.

Brandl, H. (2006). Energy foundations and other thermo-active ground structures. Geotechnique, 56(2), 81–122.

Di Murro, V., Pelecanos, L., Soga, K., Kechavarzi, C. and Morton, R.F., 2016. Distributed fibre optic long-term monitoring of concrete-lined tunnel section TT10 at CERN. In: International Conference on Smart Infrastructure and Construction, 2016-06-25.

Di Murro, V, Pelecanos, L, Soga, K, Kechavarzi, C, Morton, RF, Scibile, L, 2019, Long-term deformation monitoring of CERN concrete-lined tunnels using distributed fibre-optic sensing., Geotechnical Engineering Journal of the SEAGS & AGSSEA. 50 (1) (Accepted).

Dupray, F., Laloui, L. and Kazangba, A., 2014. Numerical analysis of seasonal heat storage in an energy pile foundation. Computers and Geotechnics, 55, pp.67–77.

Frank, R. & Zhao, S.R., 1982. Estimation par les parametres pressiometriques de l'enforcement sous charge axiale de pieux fores dans des sols fins. Bull. Liaison Lab. Ponts Chaussees, 119, pp.17–24.

Gawecka, K.A., Taborda, D.M., Potts, D.M., Cui, W., Zdravković, L. and Haji Kasri, M.S., 2016. Numerical modelling of thermo-active piles in London Clay. Proceedings of the Institution of Civil Engineers-Geotechnical Engineering, pp.1–19.

Kechavarzi, C., Soga, K., de Battista, N., Pelecanos, L., Elshafie, M. and Mair, R., 2016. Distributed optic fibre sensing for monitoring civil infrastructure:A practical guide. Thomas Telford.

Kechavarzi, C., Pelecanos, L., de Battista, N., Soga, K. 2019. Distributed fiber optic sensing for monitoring reinforced concrete piles. Geotechnical Engineering Journal of the SEAGS & AGSSEA. 50 (1) (Accepted)

Knellwolf, C., Peron, H. and Laloui, L., 2011. Geotechnical analysis of heat exchanger piles. Journal of Geotechnical and Geoenvironmental Engineering, 137(10), pp.890–902.

Loveridge, F.A. and Powrie, W., 2013. Pile heat exchangers: thermal behaviour and interactions. Proceedings of the ICE-Geotechnical Engineering, 166(2), pp.178–196.

Ng, C.W.W., Shi, C., Gunawan, A., Laloui, L. and Liu, H.L., 2014. Centrifuge modelling of heating effects on energy pile performance in saturated sand. Canadian Geotechnical Journal, 52(8), pp.1045–1057.

Ouyang, Y., Soga, K. and Leung, Y.F., 2011. Numerical back-analysis of energy pile test at Lambeth College, London. In Geo-Frontiers 2011: Advances in Geotechnical Engineering (pp. 440–449).

Ouyang, Y., Broadbent, K., Bell, A., Pelecanos, L. and Soga, K., 2015. The use of fibre optic instrumentation to monitor the O-Cell load test on a single working pile in London. In: XVI European Conference on Soil Mechanics and Geotechnical Engineering, 2015-09-13–2015-09-17.

Ouyang, Y. (2014) Geotechnical behaviour of energy piles. PhD Thesis. University of Cambridge, UK.

Ouyang, Y., Pelecanos, L., Soga, K., Nicholson, D.P. (2018a) Thermo-mechanical response test of a cement grouted column embedded in London Clay. Geotechnique (Under review).

Ouyang, Y., Pelecanos, L., & Soga, K. (2018b). The field investigation of tension cracks on a slim cement grouted column during the thermal response test. Paper presented at DFI-EFFC International Conference on Deep Foundations and Ground Improvement: Urbanization and Infrastructure Development-Future Challenges, Rome, Italy.

Ozudogru, T.Y., Olgun, C.G. and Senol, A., 2014. 3D numerical modeling of vertical geothermal heat exchangers. Geothermics, 51, pp.312–324.

Pasten, C. and Santamarina, J.C., 2014. Thermally induced long-term displacement of thermoactive piles. Journal of Geotechnical and Geoenvironmental Engineering, 140(5), p.06014003.

Pelecanos, L., Skarlatos, D. and Pantazis, G., 2017a. Dam performance and safety in tropical climates – recent developments on field monitoring and computational analysis. 8th International Conference on Structural Engineering and Construction Management (8th ICSECM), Kandy, Sri Lanka.

Pelecanos, L., Soga, K., Chunge, M.P.M., Ouyang, Y., Kwan, V., Kechavarzi, C. and Nicholson, D., 2017b. Distributed fibre-optic monitoring of an Osterberg-cell pile test in London. Geotechnique Letters, 7 (2), pp. 1–9.

Pelecanos, L., Soga, K., Elshafie, M., de Battista, N., Kechavarzi, C., Gue, C.Y., Ouyang, Y. and Seo, H., 2018. Distributed Fibre Optic Sensing of Axially Loaded Bored Piles. Journal of Geotechnical and Geoenvironmental Engineering. 144 (3) 04017122.

Pelecanos, L., Soga, K., Hardy, S., Blair, A. and Carter, K., 2016. Distributed fibre optic monitoring of tension piles under a basement excavation at the V&A museum in London. In: International Conference on Smart Infrastructure and Construction, 2016-06-25.

Pelecanos, L., & Soga, K. 2017a. Innovative Structural Health Monitoring Of Foundation Piles Using Distributed Fibre-Optic Sensing. 8th International Conference on Structural Engineering and Construction Management 2017, Kandy, Sri Lanka.

Pelecanos, L., & Soga, K. 2017b. The use of distributed fibre-optic strain data to develop finite element models for foundation piles. 6th International Forum on Opto-electronic Sensor-based Monitoring in Geo-engineering, Nanjing, China.

Pelecanos, L. & Soga, K. 2018, Using distributed strain data to develop load-transfer curves for axially-loaded piles. Journal of Geotechnical and Geoenvironmental Engineering. (Under Review).

Rotta Loria, A.F., & Laloui, L. (2016). The interaction factor method for energy pile groups. Computers and Geotechnics, 80, 121–137.

Randolph, M.F. and Wroth, C.P., 1978. Analysis of deformation of vertically loaded piles. Journal of Geotechnical and Geoenvironmental Engineering, 104(ASCE 14262).

Seo, H, Pelecanos, L, Kwon, Y-S & Lee, I-M 2017a, 'Net load-displacement estimation in soil-nailing pullout tests' Proceedings of the Institution of Civil Engineers - Geotechnical engineering, 170 (6) 534–547.

Seo, H & Pelecanos, L 2017b, 'Load Transfer In Soil Anchors – Finite Element Analysis Of Pull-Out Tests'. 8th International Conference on Structural Engineering and Construction Management 2017, Kandy, Sri Lanka.

Soga, K., Kechavarzi, C., Pelecanos, L., de Battista, N., Williamson, M., Gue, C.Y., Di Murro, V. and Elshafie, M., 2017. Distributed fibre optic strain sensing for monitoring underground structures - Tunnels Case Studies. In: Pamukcu, S. and Cheng, L., eds. Underground Sensing. Elsevier.

Soga, K., Kwan, V., Pelecanos, L., Rui, Y., Schwamb, T., Seo, H. and Wilcock, M., 2015. The Role of Distributed Sensing in Understanding the Engineering Performance of Geotechnical Structures.

Earthquake engineering, soil dynamics and soil-structure interaction

Numerical Methods in Geotechnical Engineering IX – Cardoso et al. (Eds)
© 2018 Taylor & Francis Group, London, ISBN 978-1-138-33198-3

Numerical noise effects and filtering in liquefiable site response analyses

Y.Z. Tsiapas & G.D. Bouckovalas
Geotechnical Department, School of Civil Engineering, National Technical University of Athens, Athens, Greece

ABSTRACT: Numerical noise, in the form of high frequency acceleration spikes, may become a major obstacle to the prediction of peak seismic ground acceleration and short period spectral accelerations. This problem is examined herein for the special case of liquefiable sites, where noise effects become more pronounced. It is shown that the acceleration spikes are not related to the numerical solution algorithm and to problem-specific modelling assumptions, but are associated to the abrupt changes in soil stiffness which occur during loading reversals, after the onset of soil liquefaction. The use of Rayleigh damping may moderate noise effects only in cases of low cyclic shear strain amplitude, as in the case of dry soil profiles. For liquefiable sites, where cyclic strains become large, numerical noise may be effectively eliminated by filtering separately the pre- and the post-liquefaction segments of the predicted acceleration time-histories, using different low-pass filters.

1 INTRODUCTION

The nonlinear simulation of the free-field seismic response of liquefiable sites, using effective stress numerical (Finite Difference or Finite Element) algorithms and advanced plasticity constitutive soil models becomes increasingly popular today for research and design purposes. This is because, when properly calibrated, such numerical analyses may capture realistically the nonlinear hysteretic soil response as well as the earthquake-induced excess pore pressure buildup and the associated soil softening. The credibility of this approach has been demonstrated in a number of older and recent studies (e.g. Arulanandan et al. 1994, Byrne et al. 2004, Andrianopoulos et al. 2010, Taiebat et al. 2010, Ziotopoulou et al. 2012, Shahir et al. 2012, Bouckovalas et al. 2016, 2017, Tsiapas 2017), which simulate the recorded seismic motion of liquefiable sites, in field case studies and centrifuge experiments. In addition, following the drastically increased capacity of office-oriented computational tools, the required computational cost of such analyses has dropped to readily affordable levels even for practicing engineers.

Nevertheless, there is one feature of the aforementioned nonlinear elasto-plastic numerical analyses that should be treated with some wariness: the apparent sensitivity of acceleration estimates due to the presence of high frequency spikes. The result of such spikes is to overrate predicted peak seismic ground acceleration and spectral accelerations at the short period range, leading thus to overly conservative estimation of the pseudo-static seismic loads to superstructures. Note that this problem is not unique for geotechnical earthquake engineering applications, but it has been previously identified in studies related to the numerical analysis of the nonlinear (particularly frictional and self-centering) structural systems per se (e.g. Kelly 1982, Dolce & Cardone 2003). Some of these studies have attributed the high frequency acceleration spikes to numerical modelling choices, such as the choice of time-step (e.g. Rodriguez et al. 2006, Wiebe & Christopoulos 2009), but the majority has concluded that they are most likely calculated upon abrupt stiffness change, e.g. near nonlinear elements with nearly rigid-plastic behavior and near degrees of freedom with small masses (Rodriguez et al. 2002, Tremblay et al. 2008, Wiebe & Christopoulos 2010). The latter effect is definitely related to the response of liquefied soil elements, where the stiffness change may become abrupt due to intense dilation and unloading-reloading. However, as discussed in later sections, it appears that, when dealing with a liquefied site response, numerical modelling choices are more significant and overshadow the abrupt change in stiffness observed in the actual soil element response.

The aim of the present study was first to highlight the problem of high frequency spikes, hereafter considered as "numerical noise", in nonlinear numerical analyses of liquefiable sites and to evaluate its significance for the prediction of the seismic response of the liquefied ground and the associated pseudo-static design loads applied to the superstructure. In the sequel, it was examined whether the observed noise is a result of the applied numerical analysis algorithm, including the constitutive model that is used to simulate the seismic soil response. Finally, various methods were applied in order to reduce the noise to acceptable levels, starting with modelling refinements (e.g. mesh discretization, time-step, boundary conditions) and proceeding with various noise filtering techniques. It is thus concluded that an efficient method to eliminate noise from predicted seismic ground motions is by using different low-pass filters for the pre- and the post-liquefaction segments of predicted acceleration time-histories.

2 DEVELOPMENT OF NUMERICAL NOISE

2.1 *Dry soil conditions*

To get acquainted with the general problem of numerical noise, a sensitivity analysis was initially conducted for dry soil conditions. These conditions are simpler to simulate than the saturated ones which prevail in liquefiable soil deposits and consequently the probability of modelling inaccuracies, which may affect noise, is significantly reduced. An additional and perhaps more important advantage is that the results can be compared either with analytical solutions from wave propagation theory or with results of simpler linear equivalent analysis (SHAKE-type) which are well established in practice and considered noise-free. In the sequel, the numerical investigation was extended to saturated soils and liquefiable soil condition.

For this purpose, fully-coupled elasto-plastic nonlinear dynamic numerical analyses were performed in the Finite Difference Code FLAC (Itasca 2011), while the critical state plasticity constitutive model NTUA-SAND (Andrianopoulos et al. 2010, Karamitros 2010), calibrated against laboratory test results for saturated fine Nevada sand (Arulmoli et al. 1992), was employed in order to simulate the cyclic response of the dry sand layer. The baseline case for the sensitivity analyses consists of a 10 m-deep uniform sand layer with relative density D_r = 75%, which is simulated in the Finite Difference analyses with a single column of elements with constant (width × height) dimensions equal to 1.0 m × 0.5 m, as shown in Figure 1a. Free-field lateral boundaries were simulated with the "tied-node" technique, which imposes the same (horizontal and vertical) boundary displacements at grid-points of the same elevation. The base of the soil column was shaken with the 12-cycle harmonic excitation of Figure 1d, with maximum acceleration a_{max} = 0.20 g and frequency f_{exc} = 5 Hz (Fig. 1d). For comparison purposes, equivalent linear analyses (Schnabel et al. 1972) were also conducted for the same soil profile and harmonic base excitation.

Figures 1b and 1c show the horizontal acceleration time-histories at the soil surface and at mid-depth computed with FLAC for dry soil conditions. It is observed that high frequency spikes are superimposed to the otherwise harmonic shape of both time-histories, indicating the existence of numerical noise in the conducted analysis. Figure 2 compares the elastic response spectrum computed with FLAC at the soil surface to the pertinent spectrum from the equivalent linear analysis. The response spectrum of the base excitation, which is common for both types of analysis, is also depicted in the same figure. It is observed that the numerical simulation with FLAC captures adequately the

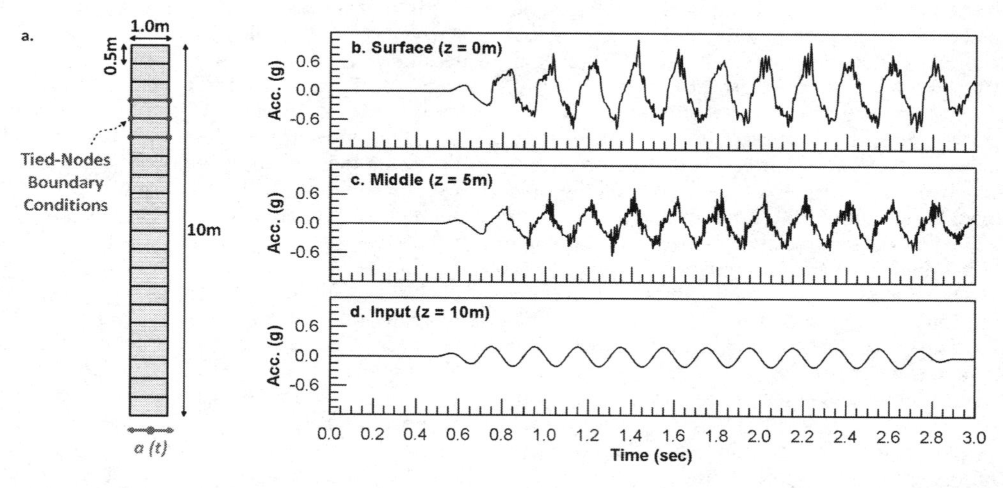

Figure 1. (a) Finite different mesh and (b) - (d) horizontal acceleration time-histories at different depths for dry soil conditions.

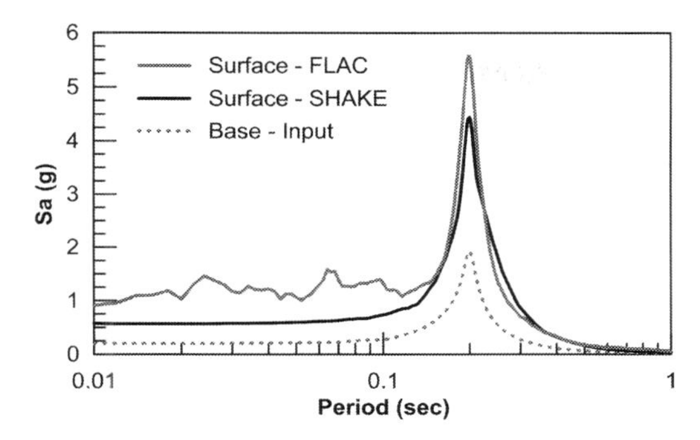

Figure 2. Comparison of the response spectrum at the surface predicted with FLAC with the pertinent spectrum predicted using the equivalent linear method for dry soil conditions.

seismic response predicted from the equivalent linear analysis, for periods $T > 0.15$ s. The small differences that are observed around resonant period $T \approx 0.20$ s are of secondary importance as they are attributed to unavoidable minor differences in the damping ratios used in the two analysis methods. On the contrary, in the short period range (i.e. for $T < 0.15$ s), spectral and peak ground acceleration values predicted with FLAC are overrated by about 100% compared to the values obtained from the equivalent linear analyses. This is the result of the aforementioned high frequency noise and should be considered with skepticism as it leads to unintentional overestimation of the pseudo-static seismic loads which are applied to superstructure design

To investigate whether the observed numerical noise is due to defects of the constitutive soil model (i.e. NTUA-SAND) or constitutes a more general problem of the numerical solution algorithm, the NTUA-SAND model was replaced in the FLAC analyses with the simpler elasto-plastic Mohr-Coulomb model. A uniform value of shear wave velocity equal to $V_S = 250$ m/s was considered for the sand column, which corresponds to the average V_S value with depth for the baseline case, the Poisson's ratio was set to $v = 0.33$, while the friction angle was equal to $\phi = 36°$. Two different cases were examined: the first with zero cohesion ($c = 0$) and the second with very high cohesion ($c = 100$ kPa) so that there would be no yielding at any zone of the numerical model and hence the response would be linear elastic. The acceleration time-histories at the soil surface from these two analyses are compared in Figures 3a and 3b. It is now observed that numerical noise is also developed in Mohr-Coulomb model, but only when yielding occurs, i.e. for the $c = 0$ analysis in Figure 3a. This is a clear indication that the numerical noise is not related to the accuracy of the applied constitutive soil model, but it is a more general problem of the numerical solution algorithm which arises upon yielding and rapid stiffness changes, either during loading or during a subsequent load reversal.

This finding is also confirmed, when the NTUA-SAND model is employed, but the development of plastic strains is restrained. To show this, the baseline case was modified with two different approaches: *(i)* by increasing considerably (i.e by 1000-times) the value of plastic modulus constant of NTUA-SAND, which leads to a very large value of plastic hardening modulus and *(ii)* by reducing the amplitude of the input acceleration to $a_{max} = 0.03$ g. The pertinent acceleration time-histories at the soil surface, presented in Figures 4a and 4b, show indeed that the numerical noise in both analyses has been practically eliminated.

As mentioned in the introduction, the correlation between numerical noise and plastic strains development in nonlinear numerical analyses has been also demonstrated in a number of previous studies regarding the dynamic response

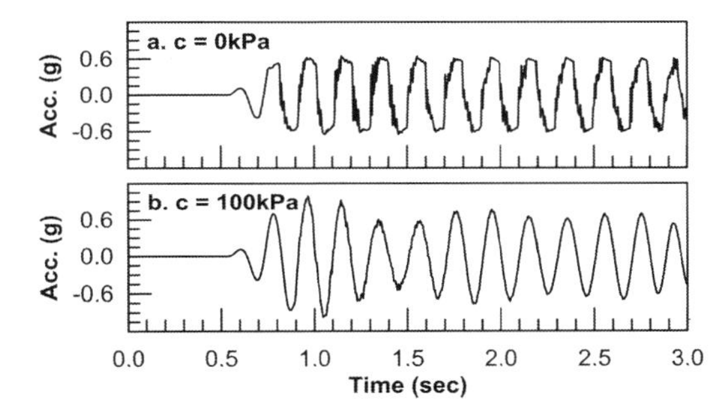

Figure 3. Comparison of horizontal acceleration time-histories at the soil surface using Mohr-Coulomb with (a) $c = 0$ and (b) c = 100 kPa.

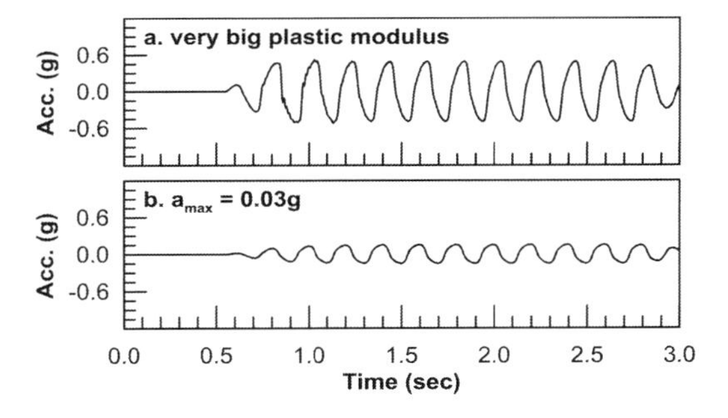

Figure 4. Comparison of horizontal acceleration time-histories at the soil surface using NTUA-SAND considering (a) very big plastic hardening modulus and (b) very low input acceleration.

of structural systems. For instance, Wiebe & Christopoulos (2010) developed a closed-form mathematical model to characterize the response of two degree-of-freedom systems undergoing abrupt changes in stiffness (e.g. plastic loading followed by elastic unloading). The formula that was developed "*showed that very large accelerations at high frequencies are likely to be calculated near nonlinear springs with nearly rigid-plastic behavior and near degrees of freedom with small masses*". Both these conditions result in rapid stiffness increases at high velocities, as in the examples presented earlier for nonlinear elasto-plastic soil conditions.

Based on an extensive sensitivity analysis that was performed as part of the present study (Tsiapas 2017), it was found that the aforementioned noise effects could not be eliminated by simple refinements of the numerical modeling, such as increasing the width of the mesh by adding more element columns, using finer mesh discretization, changing the method of input motion application or using different types of free-field boundary conditions. The only refinement that provided encouraging results for the case of dry soil conditions was to add a small amount of Rayleigh damping, as suggested by the mother company of FLAC (personal communication). Namely, Rayleigh damping with minimum damping ratio ξ_{min} = 2% anchored at the predominant frequency of the ground response, was added at the baseline analysis for dry soil with FLAC, shown in Figure 1a. The horizontal acceleration time-histories and the respective response spectra at the soil surface, with and without Rayleigh damping, are compared in Figure 5. It is observed that the problem of numerical noise has been effectively reduced, as the acceleration spikes of the baseline case have been almost disappeared. In addition, the response spectrum in the lower period range (i.e. $T < 0.15$ s) has been smoothed and approached the response spectrum from the equivalent linear analyses, which is considered noise-free.

2.2 *Saturated sand conditions*

The above findings, regarding the development of numerical noise in nonlinear seismic response analyses for dry soil profiles were also observed in the sensitivity analyses that were performed for saturated soil conditions. In fact, the high frequency acceleration spikes were intensified after the onset of liquefaction in the sand layer. It has been also observed that, despite its success for dry soil conditions, the use of Rayleigh damping proved insufficient for the elimination of numerical noise when the numerical analyses are repeated for saturated soil conditions.

This is shown in Figure 6 which compares the acceleration time-histories and the corresponding elastic response spectra at the soil surface, with and without Rayleigh damping, obtained for the baseline case from a typical FLAC analysis for saturated soil conditions. At a first glance, it is observed that numerical noise persists in the predicted seismic ground motions despite the use of Rayleigh damping. However, a closer look at the numerical predictions in this figure reveals that the use of Rayleigh damping works sufficiently well until the onset of liquefaction, which occurs approximately at t = 0.8 s, as the high frequency spikes have been effectively eliminated from the acceleration time-history until that point. However, it has a minor effect on predicted seismic ground motions after the onset of liquefaction. This is evidenced by a visual comparison of the acceleration time-histories in Figures 6a and 6b, and also from the direct comparison of the associated elastic response spectra with and without Rayleigh damping, shown with continuous (black and red) lines in Figure 6c, and the noise-free spectrum shown with dotted line in the same figure. It is noted that the noise-free spectrum was approximately estimated, for comparison purposes, using the noise filtering technique that

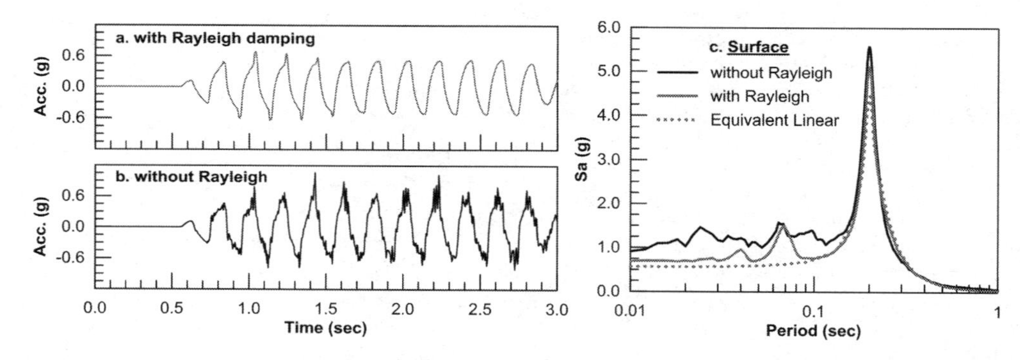

Figure 5. Comparison of acceleration time-histories and elastic response spectra at the soil surface with and without adding Rayleigh damping for dry soil conditions.

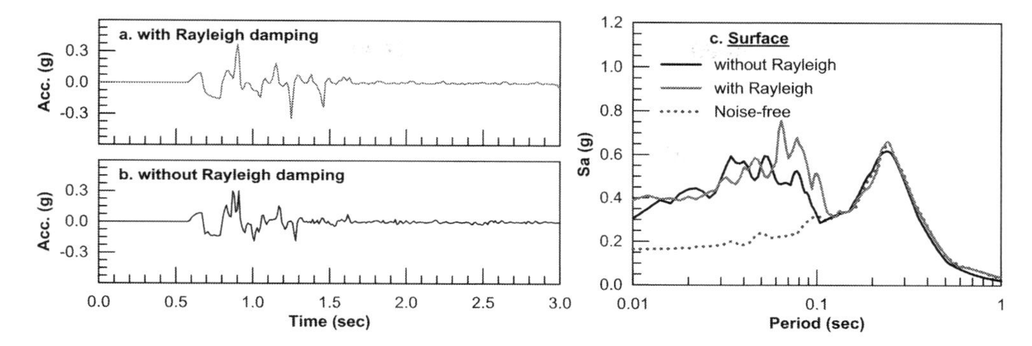

Figure 6. Comparison of acceleration time-histories and elastic response spectra at the soil surface with and without adding Rayleigh damping for undrained soil conditions.

is described in the following section. The significantly different effect of Rayleigh damping for dry and liquefied soil conditions is attributed to the considerably larger plastic strains which develop in the latter case.

3 SELECTIVE NOISE FILTERING

The previous analyses have shown that the numerical predictions of the liquefied ground response are inevitably accompanied by numerical noise, which is not related to deficiencies in the analysis algorithm or inadequate modelling assumptions. Hence, it was next attempted to overcome this problem explicitly, i.e. by intervening directly to the noise-contaminated numerical predictions through low-pass filtering of the high frequency acceleration spikes. In the development of an efficient filtering technique, it was taken into account that the onset of liquefaction in the soil triggers two basic changes in the predicted seismic ground response (e.g. Fig. 6): (a) the numerical noise is significantly intensified while (b) the frequency content of the ground motion is drastically reduced due to liquefaction-induced soil softening. As a result, it was realized that filtering cannot be unique for the entire time-history of predicted accelerations, but different low pass filters should be used for the pre- and the post-liquefaction segments. To satisfy this requirement, a selective filtering technique was developed, referred hereafter as "*Selective Filtering Method*" (or SFM).

In particular, the characteristic low-pass filtering frequencies that are used to remove the noise from the post-liquefaction segment of the predicted seismic motion are much lower relative to that for the pre-liquefaction segment of the motion. Note that, if the low frequency post-liquefaction filters were also applied to the pre-liquefaction segment of the seismic motion, they would interfere with the much shorter significant frequency content of this segment and would drastically distort the entire seismic ground motion. In the opposite case, i.e. if the pre-liquefaction high frequency filters were also applied to the post-liquefaction segment of the seismic motion, there would be practically no effect to the noise-affected frequency content of that segment.

In more detail, the steps required to apply the selective filtering method in practice are the following:

i. Determine the onset time of liquefaction at the predicted seismic ground motion, $t_{L,gr}$. This may be approximately achieved by visual comparison of the wave-forms at the soil surface and at the base of the soil column, or with other more elaborate techniques such as the use of time-frequency spectra (Kramer et al 2015).
ii. Apply a typical low-pass filter with maximum frequency $f_{filter} \geq 15$ Hz to the pre-liquefaction segment of the predicted acceleration time-history.
iii. Estimate the natural soil period of the liquefied ground $T_{soil,L}$ from the normalized (surface-to-base) response spectra that correspond to the post-liquefaction segment.
iv. Apply a low-pass filter with $f_{filter} \geq 3/T_{soil,L}$ to the post-liquefaction segment.

The efficiency of SFM is initially evaluated for the undrained analysis of Figure 6a, in which the base excitation is harmonic. The noise-clean acceleration time-history at the soil surface, as obtained with the aforementioned technique, is shown in Figure 7a, on top of the noise-affected time-history of Figure 6a, while the corresponding comparison in terms of response spectra is depicted in Figure 7c. It is observed that the SFM can effectively eliminate numerical noise, as noise-affected spectral accelerations at $T < 0.1$ s have been significantly decreased.

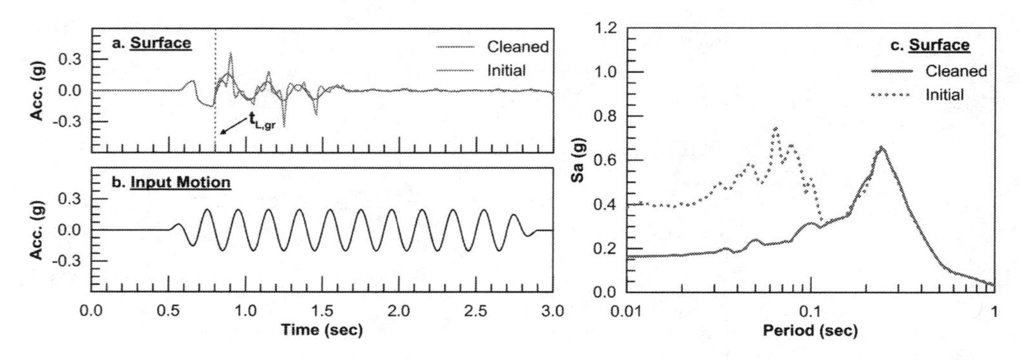

Figure 7. Evaluation of the selective filtering method in terms of horizontal acceleration time-histories and response spectra for harmonic seismic excitations.

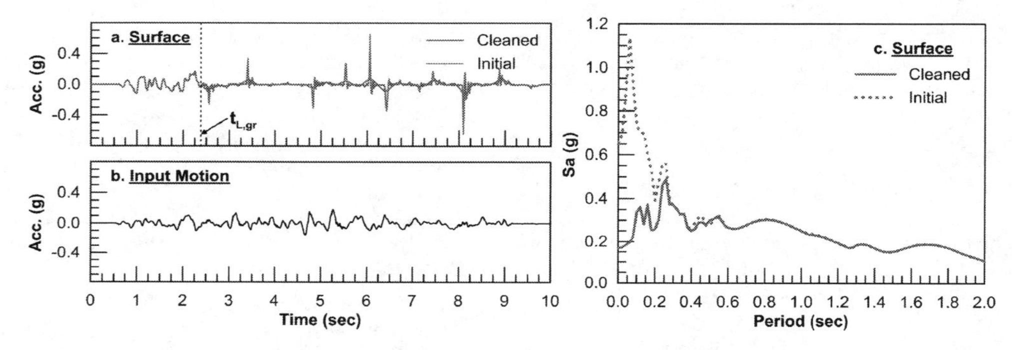

Figure 8. Evaluation of the selective filtering method in terms of horizontal acceleration time-histories and response spectra for actual seismic excitations.

The evaluation of SFM for the case of real input seismic excitations is presented in Figure 8. In particular, Figure 8a compares the predicted noise-clean and noise-affected acceleration time-histories at the ground surface, when the seismic recording at Anderson Dam from the Kobe (1995), $M_w = 6.9$, earthquake (Fig. 8b) is applied to the soil column of the baseline case. In addition, Figure 8c compares the noise-clean and the noise-affected elastic response spectra predicted from the same analyses. These comparisons suggest that the SFM application to analyses with real seismic excitations is as effective as in the previous case of harmonic seismic motion.

As mentioned in the introduction, it is reasonable to question whether the high frequency spikes in the predicted seismic ground response are due to the actual response of liquefied soil elements (i.e. abrupt stiffness changes due to dilation and/or unloading—reloading) and should not be eliminated, or due approximations in the numerical simulation of the seismic site response. It is our opinion that both above causes are valid, but the former is much less significant that the latter and it

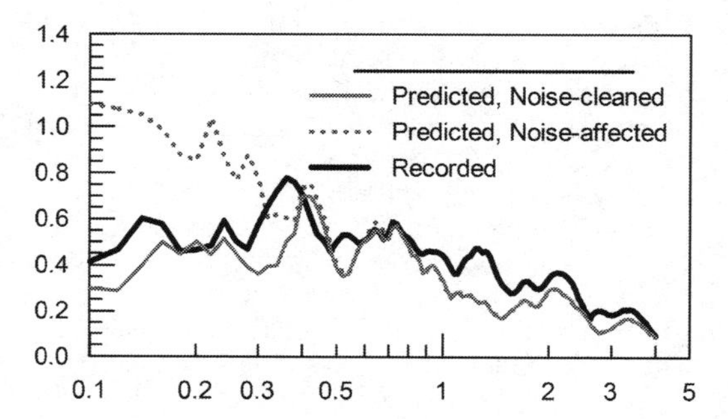

Figure 9. Evaluation of the selective filtering method in terms of response spectra for the WLA site during the Superstition Hills (1987) earthquake.

is probably physically damped out in field recordings. To show this, Figure 9 presents the elastic response spectrum of the recorded motion at the surface of the "Wildlife Liquefaction Array" (WLA) during the Superstition Hills (1987) earthquake, of $M_w = 6.6$ (Holzer et al. 1989), and compares it to the numerically predicted response spectra (Tsiapas 2017) prior and after the application of SFM. Observe that the high frequency spikes are present only in the numerically predicted spectrum prior to selective filtering indicating that they are mainly attributed to numerical causes.

4 CONCLUDING REMARKS

Numerical noise, in the form of high frequency spikes in predicted acceleration time-histories, is a major obstacle for the accurate prediction of seismic ground response, as it may unrealistically increase the peak seismic ground acceleration and the associated spectral accelerations at the short period range. Based on the results of the present study, it is concluded that:

a. Nonlinear numerical analyses of seismic ground response are inevitably accompanied by numerical noise, which is not related to deficiencies in the numerical analysis algorithm or to inadequate modeling assumptions.
b. Numerical noise becomes more intense upon soil liquefaction, due to abrupt changes in soil stiffness within each loading-unloading cycle, from very soft during loading to stiff during unloading.
c. The use of Rayleigh damping can moderate noise effects only in cases with limited development of shear strains, as is the case of dry soil conditions or weak seismic loading of liquefiable sites where excess pore pressures remain low. However, its use becomes insufficient in cases of extensive liquefaction where plastic strains become high.
d. For such cases, a selective filtering method (SFM) has been developed and evaluated in this study, where the pre-liquefaction and the post-liquefaction segments of the seismic ground motion are filtered separately, the first with a high frequency and the second with a low frequency filter, and then superimposed.

It should be acknowledged that the proposed noise-filtering methodology does not offer the advantages of a robust mathematical solution that could be incorporated into the numerical analysis algorithm. However, it is simple and can be indiscriminately applied to various numerical analysis and soil modelling methodologies in order to obtain realistic estimates of the liquefied ground response. It is also recognized that the efficiency of SFM needs to be further evaluated against actual recordings of liquefied ground response obtained from carefully instrumented model experiments and seismic arrays. An additional benefit that should be expected from such comparisons is to clarify two practical issues that need further study: *(i)* the extent to which high frequency acceleration spikes represent a real characteristic of the physical system response, and *(ii)* whether these spikes are relevant or not to superstructure damage.

ACKOWLEDGEMENTS

The Authors would like to thank C & M Engineering S.A. for partly funding this research.

REFERENCES

Andrianopoulos, K.I., Papadimitriou, A.G. & Bouckovalas, G.D. 2010. Bounding surface plasticity model for the seismic liquefaction analysis of geostructures. *Soil Dynamics and Earthquake Engineering*, 30(10), 895–911.

Arulmoli, K, Muraleetharan, K.K., Hossain, M.M., Fruth, L.S. 1992. *VELACS: verification of liquefaction analyses by centrifuge studies; Laboratory Testing Program – Soil Data Report*. Research Report, Earth Technology Corporation.

Arulanandan, K., Dobry, R., Elgamal, A.W., Ko, H.Y., Kutter, B.L., Prevost, J., Riemer, M.F., Schofield, A.N., Scott, R.F., Seed, R.B., Whitman, R.V. & Zeng, X. 1994. Interlaboratory studies to evaluate the repeatability of dynamic centrifuge model tests. *ASTM Special Technical Publication*, ASTM, 400–422.

Bouckovalas, G.D., Tsiapas, Y.Z., Zontanou, V.A. & Kalogeraki, C.G. 2017. Equivalent Linear Computation of Response Spectra for Liquefiable Sites: The Spectral Envelope Method. *Journal of Geotechnical and Geoenvironmental Engineering*, 143(4), 4016115.

Bouckovalas, G.D., Tsiapas, Y.Z., Theocharis, A.I. & Chaloulos, Y.K. 2016. Ground response at liquefied sites: seismic isolation or amplification? *Soil Dynamics and Earthquake Engineering*. 91, 329–339.

Byrne, P.M., Park, S.-S., Beaty, M., Sharp, M., Gonzalez, L. & Abdoun, T. 2004. Numerical modeling of liquefaction and comparison with centrifuge tests. *Canadian Geotechnical Journal*, 41(2), 193–211.

Dolce, M. & Cardone, D. 2003. Seismic protection of light secondary systems through different base isolation systems. *Journal of Earthquake Engineering*, 7(2), 223.

Holzer, T.L., Youd, T.L., & Hanks, T.C. 1989. Dynamics of liquefaction during the 1987 Superstition Hills, California, earthquake. *Science*, 244(4900), 56–59.

Itasca. 2011. FLAC version 7.0. *Itasca Consulting Group Inc.*

Karamitros, D.K. 2010. *Development of a numerical algorithm for the dynamic elastoplastic analysis of geotechnical structures in two and three dimensions*. PhD Thesis, Dept of Civil Engineering, NTUA, Athens.

Kelly J.M. 1992. The influence of base isolation on the seismic response of light secondary equipment. *Research Report UCB/EERC-81/17*, University of California, Berkeley, USA.

Kramer, S.L., Sideras, S.S. & Greenfield, M.W. 2015. The timing of liquefaction and its utility in liquefaction hazard evaluation. *6th International Conference on Earthquake Geotechnical Engineering*, Christchurch, New Zealand, 1–4 November.
Rodriguez, M.E. Restrepo, J.I. & Blandón, J.J. 2006. Shaking Table Tests of a Four-Story Miniature Steel Building- Model Validation. *Earthquake Spectra*, 22(3), 755–780.
Rodriguez, M.E., Restrepo, J.I., & Carr, A.J. 2002. Earthquake-induced floor horizontal accelerations in buildings. *Earthquake Engineering & Structural Dynamics*, 31(3), 693–718.
Schnabel, P.B., Lysmer, J. & Seed, H.B. 1972. SHAKE: A computer program for earthquake response analysis of horizontally layered sites. *Rep. No. UCB/EERC-72/12*, Earthquake Engineering Research Center, University of California at Berkeley.
Shahir, H., Mohammadi-Haji, B. & Ghassemi, A. 2014. Employing a variable permeability model in numerical simulation of saturated sand behavior under earthquake loading. *Computers and Geotechnics*, 55, 211–223.
Taiebat, M., Jeremić, B., Dafalias, Y.F., Kaynia, A.M. & Cheng, Z. 2010. Propagation of seismic waves through liquefied soils. *Soil Dynamics and Earthquake Engineering*, 30(4), 236–257.
Tremblay, R., Lacerte, M. & Christopoulos, C. 2008. Seismic Response of Multistory Buildings with Self-Centering Energy Dissipative Steel Braces. *Journal of Structural Engineering*, 134(1), 108–120.
Tsiapas, Y. 2017. *Seismic response analysis of liquefiable ground with computational methods*. PhD Thesis, Dept. of Civil Engineering, NTUA, Athens.
Wiebe, L. & Christopoulos, C. 2009. Mitigation of Higher Mode Effects in Base-Rocking Systems by Using Multiple Rocking Sections. *Journal of Earthquake Engineering*, 13(sup1), 83–108.
Wiebe, L. & Christopoulos, C. 2010. Characterizing acceleration spikes due to stiffness changes in nonlinear systems. *Earthquake Engineering and Structural Dynamics*, 39(14).
Ziotopoulou, K., Boulanger, R.W. & Kramer, S.L. 2012. Site response analysis of liquefying sites. *GeoCongress 2012*, American Society of Civil Engineers, Reston, VA, 1799–1808.

Numerical Methods in Geotechnical Engineering IX – Cardoso et al. (Eds)
© 2018 Taylor & Francis Group, London, ISBN 978-1-138-33198-3

Influence of water table fluctuation on soil-structure interaction

M.P. Santisi d'Avila
Université Côte d'Azur, CNRS, LJAD, Nice, France

L. Lenti
Université Paris Est, IFSTTAR, Paris, France

S. Martino
Dipartimento di Scienze della Terra, CERI Research Centre, Università Roma La Sapienza, Roma, Italy

ABSTRACT: This study is focused on the influence of seasonal variation of the water table depth in the seismic response of a multilayered soil with a building at the surface, taking into account the Soil-Structure Interaction (SSI). The effect of SSI on the seismic response of the building, having frame structure and rigid shallow foundation, is estimated using a one-directional propagation of three-component seismic waves (1D-3C). The model of 1D-3C wave propagation, in a nonlinear multilayered soil profile, is assembled with a multi-story multi-span frame model in a finite element scheme, solved directly. The case study is the area of San Carlo village in Emilia Romagna region (Italy), struck by the 20 May 2012 M_w 5.9 Emilia earthquake. Results are provided in terms of ground motion at the building base, structural deformation, profiles with depth of maximum excess pore water pressure, strains, stresses and motion.

1 INTRODUCTION

This research aims at studying the impact of the water table depth on the seismically induced effects in a multilayered soil profile affected by a strong earthquake. A finite element model allows the simulation of the response to seismic loading of a saturated soil column.

The propagation of three-component seismic waves, in a multilayered soil basin, is simulated using SWAP_3C code (Santisi et al. 2012, 2013, Santisi & Semblat 2014) that has been verified and validated during the Prenolin benchmark (Regnier et al. 2016) in the case of dry soil having nonlinear behavior.

The soil-structure interaction (SSI) modeling technique adopted in this research has been proposed by Santisi d'Avila and Lopez Caballero (2018) in the case of total stress analysis. A frame structure is assembled to a vertical seismic wave propagation model, in a finite element scheme, under the assumption of rigid shallow foundation and reduced rocking effects. Fares et al. (2018) compare the one-directional three-component (1D-3C) wave propagation approach for SSI analyses with a model where a three-dimensional (3D) soil domain with lateral periodic boundary conditions is used and show an excellent fit, in the case of vertical propagation of three-component seismic waves in a horizontally multilayered soil.

The Iwan's elasto-plastic model (Iwan 1967, Joyner 1975, Joyner & Chen 1975) is adopted to represent the 3D nonlinear behavior of soil. Its main feature is the faithfully reproduction of nonlinear and hysteretic behavior of soils under cyclic loadings, with the minimum number of parameters characterizing the soil properties. The model is calibrated using the elastic moduli in shear and compression and the shear modulus decay curve employed to deduce the size of the yield surface.

The correction to the shear modulus proposed by Iai et al. (1990a,b) is employed for saturated cohesionless soil layers. Liquefaction front parameters are calibrated by a trial-and-error procedure to best reproduce the curves obtained by cyclic consolidated undrained triaxial tests, that are the deviatoric strain amplitude and normalized excess pore water pressure vs the number of cyclic loading.

During the 20 May 2012 M_w 5.9 Emilia earthquake, in Italy, liquefaction phenomena have been observed (Chini et al. 2015; Emergeo Working Group 2013) in several sites (including the village of San Carlo) and inventoried in the CEDIT catalogue (Fortunato et al. 2012, Martino et al. 2014).

In this research, the stratigraphy and related geotechnical parameters of a soil profile in San Carlo village (Emilia Romagna, Italy) are used to investigate the influence of the seasonal fluctuation of water table depth on the seismic response of the analyzed soil profile. The soil-structure interaction with a building at the surface is also taken into account.

First, the seismic response of the soil-structure assembly to the 20 May 2012 M_w 5.9 Emilia earthquake is estimated in terms of total stresses, under the assumption of dry soil, to show how this simplification cannot be considered as suitable in the case of expected liquefaction phenomena. Then, it is discussed the effect of a building at the surface, having fundamental frequency higher than the frequency associated to the soil column. Finally, the results obtained using a corrected constitutive model for saturated soil are shown for the average position of the water table depth and its expected variation.

2 1D-3C WAVE PROPAGATION MODEL

2.1 *Spatial discretization*

The soil basin is assumed as horizontally layered and is modeled as a 1D soil profile (Fig. 1), considering its very large lateral extension. The multilayered soil is assumed infinitely extended along the horizontal directions x and y and, consequently, no strain variation is considered in these directions. A three-component seismic wave propagates vertically in z-direction from the top of the underlying elastic bedrock to the free surface. The soil is assumed to be a continuous medium, with nonlinear constitutive behavior.

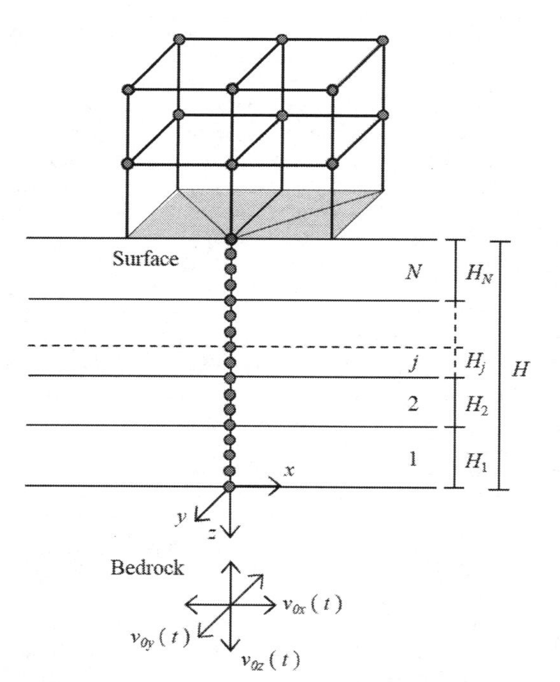

Figure 1. Spatial discretization of a horizontally layered soil with a frame structure at the top, loaded by a three-component seismic motion applied at the soil-bedrock interface in terms of incident velocity.

The soil profile is discretized, using a finite element scheme, into quadratic line elements having three translational degrees of freedom per node. The finite element model applied in the present research is completely described in Santisi d'Avila et al. (2012, 2013).

2.2 *Boundary condition*

The system of horizontal soil layers is bounded at the bottom by a semi-infinite elastic medium representing the seismic bedrock. The absorbing boundary condition proposed by Joyner & Chen (1975) is applied at the soil-bedrock interface. The three-component velocity time history at the bedrock level is obtained by deconvolution of the seismic signals recorded at the outcropping bedrock. The obtained movement at the soil-bedrock interface is composed of the incident and reflected waves. At this regard, the interested reader can refer to Santisi d'Avila et al. (2012) for more details.

2.3 *Time discretization*

The implicit dynamic process is solved step-by-step by the Newmark algorithm. The two parameters $\beta = 0.3025$ and $\gamma = 0.6$ guarantee an unconditional numerical stability of the time integration scheme (Hughes 1987).

Moreover, the nonlinearity of soil demands the linearization of the constitutive relation within each time step. The discrete dynamic equilibrium equation does not require an iterative solving, at each time step, to correct the tangent stiffness matrix, if a small fixed time step $dt = 10^{-4}$ s is selected. Gravity load is imposed as static initial condition in terms of strain and stress.

2.4 *Constitutive model for saturated soil*

The 3D elasto-plastic model for soils used in the presented finite element scheme is inspired from that suggested by Iwan (1967) and applied by Joyner (1975) and Joyner and Chen (1975) in a finite difference formulation, in terms of total stresses. The adopted model for dry soils (total stresses) satisfies the so-called Masing's criteria (Kramer 1996) and does not depend on the number of loading cycles. According to Joyner

(1975), the tangent constitutive matrix is deduced from the actual strain level and the strain and stress values at the previous time step. Then, the knowledge of this matrix allows calculating the stress increment. Consequently, the stress level depends on the strain increment and strain history but not on the strain rate. Therefore, this rheological model has no viscous damping. The energy dissipation process is purely hysteretic and does not depend on the frequency. The failure curve $\Delta\tau(\gamma)=G(\gamma)\Delta\gamma$, where $G(\gamma)$ is the shear modulus decay curve versus shear strain γ is needed to characterize the soil behavior. The main feature of Iwan's model is that the mechanical parameters to calibrate the rheological model are easily obtained from laboratory dynamic tests on soil samples.

The applied constitutive model does not depend on the selected initial loading curve. In the present study, normalized shear modulus decay curves are provided by laboratory tests, as resonant column, and fitted by the function $G(\gamma)/G_0 = 1/(1+|\gamma/\gamma_{r0}|)$, where γ_{r0} is a reference shear strain corresponding to an actual tangent shear modulus $G(\gamma)$ equivalent to 50% of the elastic shear modulus G_0. This model provides a hyperbolic stress-strain curve, having asymptotic shear stress $\tau_0 = G_0\gamma_{r0}$ in the case of simple shear. The Poisson ratio is assumed constant during the time history, in this study. It means that the normalized compressional modulus decay curve E/E_0 corresponds to the shear modulus decay curve G/G_0.

A correction of mechanical properties is applied for saturated soil layers, according to Iai's model (Iai et al. 1990a,b). Iai's rheological model for saturated soils allows attaining larger strains with proper accuracy. Seven parameters have to be fixed to calibrate Iai's correction of shear modulus for saturated soils. They are the shear friction angle ϕ', the phase transformation angle ϕ'_p, the parameter c_1 that corrects the elastic shear work and the four parameters S_1, w_1, p_1 and p_2 that influence the relationship between the liquefaction front parameter S_0 and the normalized shear work w. The shear friction angle ϕ' and the phase transformation angle ϕ'_p are obtained from static consolidated undrained triaxial tests. According to Iai et al., 1990a,b, parameters c_1, S_1, w_1, p_1 and p_2 are obtained numerically, by try-and-error procedure, to best fit the curves provided by cyclic consolidated undrained triaxial (CTX) tests. Three curves have to be reproduced: the cyclic deviatoric stress, the deviatoric strain amplitude and the normalized excess pore water pressure $\Delta u/p'_0$ vs the number of cyclic loading N, where $\Delta u = p'_0 - p'$ is the excess pore water pressure, p'_0 and p' are the initial and actual average effective stress.

2.5 *Building model*

According to the 1D-3C propagation approach for SSI analyses, a building is rigidly connected to the soil, under the assumption of rigid shallow foundation and negligible rocking effects (Fig. 1). As in actual design practice, the 3C seismic motion is the same at the base of all columns of the building. Rotational degrees of freedom of nodes at the base of columns are blocked. Live and dead loads are imposed on the beams in terms of mass per unit length.

The 3D frame structure is modeled using Timoshenko beam elements having six degrees of freedom per node. A linear constitutive behavior is assumed for the structure. The damping due to non-structural elements is taken into account by the damping matrix that is assumed as mass and stiffness proportional, according to Rayleigh's approach.

3 THE ANALYZED SOIL COLUMN

The definition of the analyzed soil column, used for the 1D numerical model, is derived from the engineering-geological model obtained for the San Carlo village (Emilia Romagna, Italy), at about 17 km from the epicenter of the 2012 Emilia earthquake, based on direct geotechnical investigations (boreholes and in situ geotechnical tests) and geological surveys (Romeo et al. 2015). The stratigraphy and the mechanical features identified for each layer are presented in Table 1. Two sandy-silt liquefiable soil levels are interlayered with two silty-clayey levels (not susceptible to liquefaction) which rest above a sandy-gravel level at about 20 m b.g.l.. The front liquefaction parameters are also reported (Table 2) for noncohesive soil layers, subjected to possible liquefaction phenomena. The first trial values $S_1 = 0.005$ and $c_1 = 1$ are suitable for the analyzed soils. The variation of the elastic shear modulus with the depth z is taken into account for the liquefiable soil layers.

A ground motion recorded at an outcropping bedrock close to this soil basin is not available. Consequently, a synthetic motion produced by the Italian ENEA Agency is used as outcropping reference motion. The three-components of the reference outcropping motion are halved and applied as incident wave at the base of the analyzed soil columns. The reference incident motion applied at the soil-bedrock interface is shown in Fig. 2, in terms of acceleration and Fourier spectrum. The peak acceleration is 2.54 m/s^2 in North-South direction (named x in the model), 1.51 m/s^2 in East-West direction (named y) and 0.33 m/s^2 in Up-Down direction (named z). All input and output signals

Table 1. Stratigraphy and geotechnical parameters of the C1 soil profile.

	H-z	t	ρ	v_s	v_p	γ_r	Soil
Soil layer	m	m	kg/m³	m/s	m/s	‰	sample
1 sandy silt	9.7	9.7	1750	185	1234	0.39	S1
2 silty clay	12.0	2.3	1800	180	465	0.48	
3 sandy silt	14.0	2.0	1850	180	402	0.39	S3
4 silty clay	21.5	7.5	1800	200	632	0.48	
5 sandy gravel	27.5	6.0	1850	275	710	0.39	S3
6 silty clay	33.0	5.5	1900	280	723	0.49	
7 sandy gravel	73.0	40.0	1975	385	994	100	
8 silty clay	113	40.0	2125	595	1536	100	
bedrock	> 113		2200	700	1807		

Table 2. Liquefaction parameters obtained by numerical calibration of CTX tests.

Soil	ϕ'	ϕ_p	p_1	p_2	w_1
sample	°	°			
S1	29	25	0.4	0.6	10
S3	32	26	0.4	0.6	12

are filtered using a 4-pole Butterworth bandpass filter in the frequency range $0.1-15\,\text{Hz}$. The predominant frequency of horizontal motion is $0.7\,\text{Hz}$ (Fig. 2 bottom).

The soil column has a fundamental frequency equal to $1.6\,\text{Hz}$. The 3-floor 1-span building at the surface of the analyzed soil profile has the first and second natural frequencies equal to $3.1\,\text{Hz}$. The frame structure has equivalent inertia distribution in the two orthogonal directions x and y. The number of stories of the frame structure modifies the building fundamental frequency and, consequently, the SSI effect. On the other hand, the enlargement of floor area, by increasing both mass and stiffness, has a little influence on the building fundamental frequency and it is not captured using a 1D-3C wave propagation approach for SSI. For this reason, a building having only one span is considered in this analysis. The interstory height of the analyzed building is 3.2 m and the span length is 5 m in both horizontal directions. The beam elements have rectangular cross-section of dimensions $30\times 60\,\text{cm}$. The adopted elastic modulus in compression, Poisson's ratio and density, for the structure, are $E=31220\,\text{N/mm}^2$, $\nu=0.2$ and $\rho=2500\,\text{kg/m}^3$, respectively. It is assumed the damping ratio $\zeta=5\%$, as in structural design of typical reinforced concrete buildings. The dead load of $800\,\text{kg/m}^2$ is converted in consistent mass of beams.

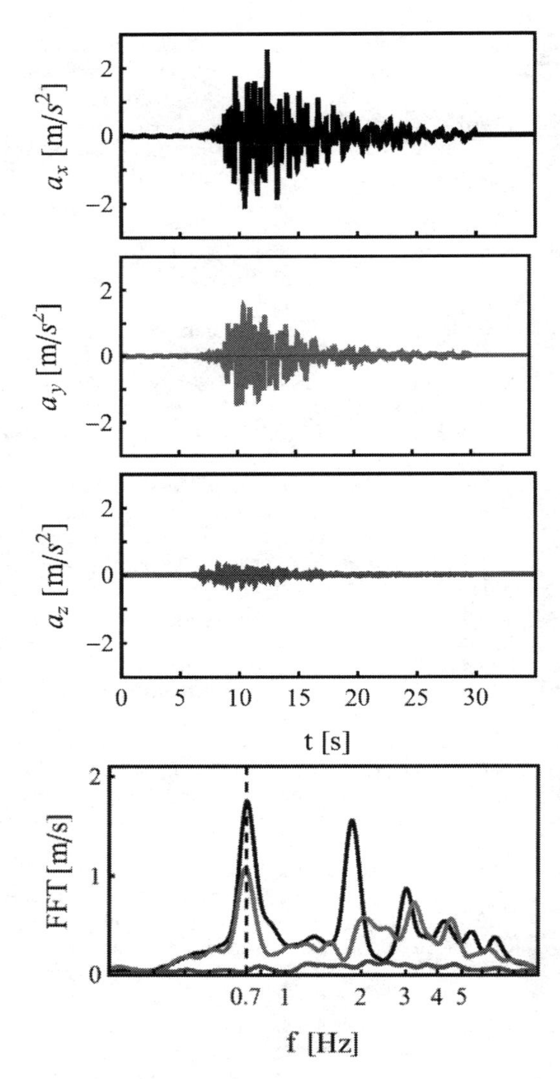

Figure 2. (Top) Three components of the incident motion applied at the soil-bedrock interface in terms of acceleration. The peak acceleration is 2.54 m/s^2 in x-direction, 1.51 m/s^2 in y-direction and 0.33 m/s^2 in z-direction. (Bottom) Fourier spectrum of the three components of motion.

4 INFLUENCE OF WATER TABLE DEPTH

The assumed water table depth is $z_w=5.8\,\text{m}$, according to the available technical reports. The difference between the assumption of dry soil and the condition of saturated soil in the seismic response of the analyzed soil profile is firstly investigated. Moreover, a water table variation of $\pm 1\,\text{m}$ is considered and its influence in the soil column response to the seismic loading is analyzed.

Profiles with depth of the peak acceleration and velocity, shear strain and stress for the soil profile are represented in Figs 3 and 4, in the cases of dry soil and saturated soil with variable water table position. Strains are increased in liquefiable layers, compared with the dry soil assumption.

In liquefiable soil layers, the shear modulus is reduced during the process and the reference shear strain γ_{r0} is numerically corrected when the liquefaction front parameter S_0 is lower than 0.4, according to Iai's model (1990a,b). The minimum values attained by the shear modulus during the process, at each depth, and the maximum reference shear strain are shown in Fig. 5, for the analyzed soil profile. Fig. 5 shows also the profile with depth of excess pore water pressure. The time history of excess pore water pressure, at a given depth in a liquefiable soil layer, is shown in Fig. 6.

Hysteresis loops in non-liquefiable soil layers follow the first loading stress-strain curve as maximum shear stress (Fig. 7 top). The shape of loops in liquefiable soil layers is influenced by the reduction of shear modulus during the process (Fig. 7 bottom). In Fig. 7 (bottom), the backbone curve for dry soil (dashed line) is different in the case of saturated soil (continuous line), at the same depth, because the adopted elastic shear modulus and the reference shear strain have different values. These mechanical properties are assumed constant in the soil layer when a dry soil model is used and are considered variable with depth, into each liquefiable soil layer, if the saturated soil model is adopted.

Observing Figs 3–7, the considered variation of the water table position equal to $\pm 1\,\text{m}$, due to seasonal fluctuation, is not influent in the seismic response of the analyzed soil profiles. On the contrary, the assumption of dry soil in the model, in a stratigraphy where there are liquefiable soil layers, totally modifies the seismic response, neglecting the reduction of soil stiffness and the increase of ground motion. According to Figs 3–4, the dry soil assumption underestimates the peak values of motion.

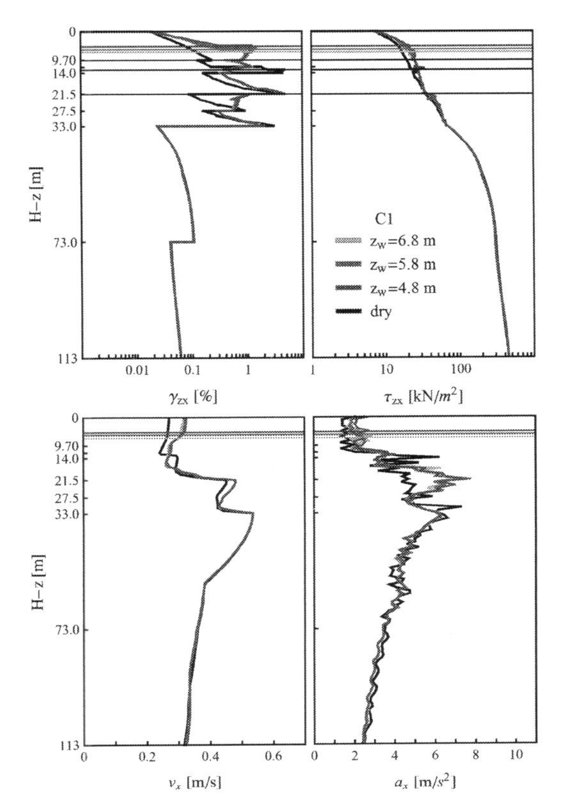

Figure 3. Profiles with depth of shear strain and stress, horizontal velocity and acceleration for different water table depth and for dry soil in the C1 soil column (seismic response in x-direction).

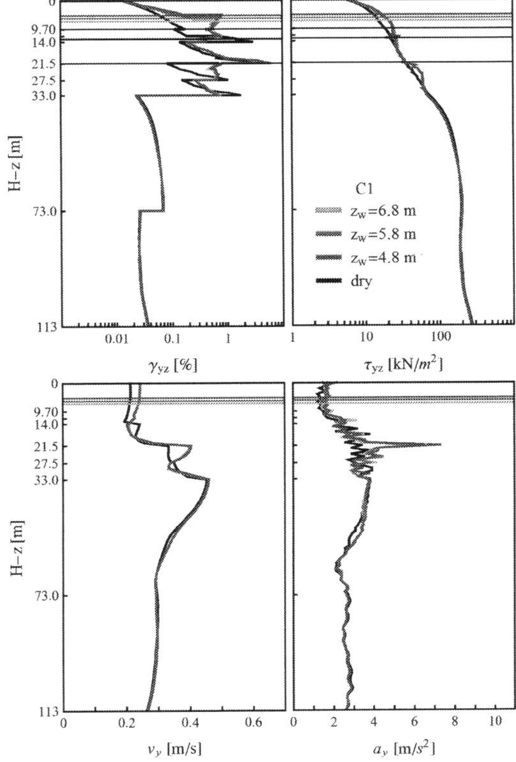

Figure 4. Profiles with depth of shear strain and stress, horizontal velocity and acceleration for different water table depth and for dry soil in the C1 soil column (seismic response in y-direction).

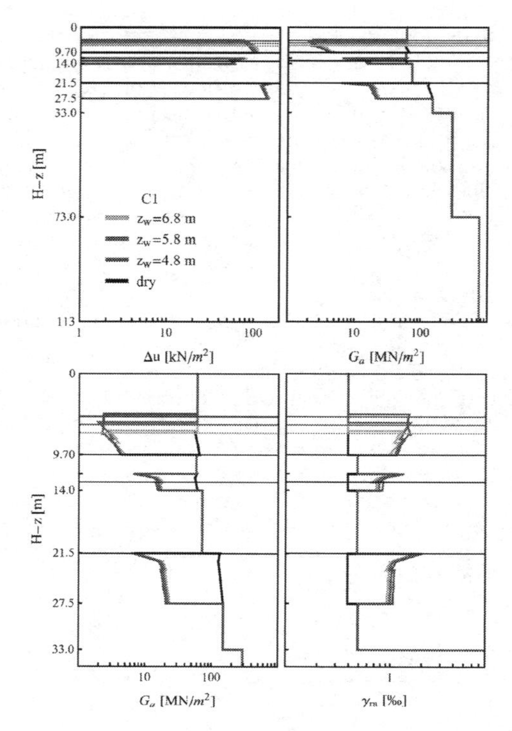

Figure 5. (Top) Profiles with depth of excess pore water pressure and shear modulus, for different water table depth and for dry soil, in the C1 soil profile. (Bottom) Profiles with depth of shear modulus and reference shear strain in the top layers.

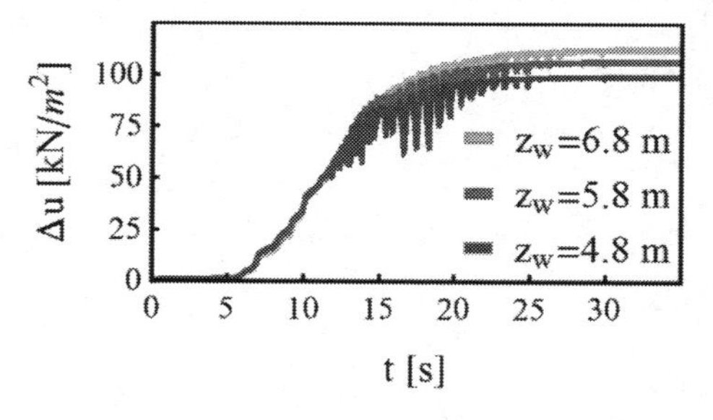

Figure 6. Excess pore water pressure time history, for different water table depth at 9.7 m in C1 soil profile.

Fig. 8 shows that the variability in the ground motion and structural deformation time history with the water table position is negligible.

In the analyzed case study, the frequency content of the seismic motion (predominant frequency equal to 0.7Hz) does not particularly excite the analyzed building (first and second natural frequency equal to 3.1Hz). The SSI effect in not evident in this case study and the difference between the free-field and the SSI solutions is negligible.

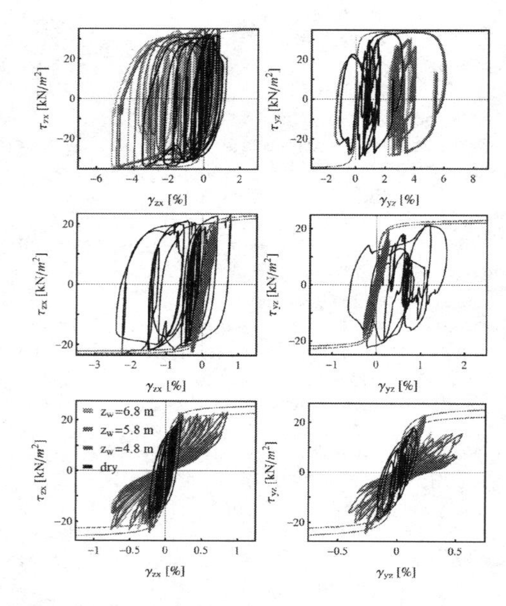

Figure 7. Hysteresis loops in C1 soil profile for different water table depth and for dry soil: (top) at 21.5 m in a non liquefiable soil layer; (middle) at 13 m in a liquefiable soil layer; (bottom) at 9.7 m in a liquefiable soil layer.

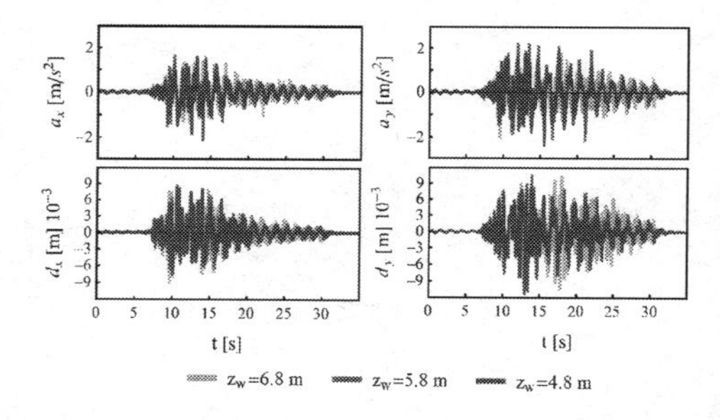

Figure 8. Acceleration (top) and displacement (bottom) time history in x- and y-direction at the ground surface, for different water table depth.

5 CONCLUSIONS

During the 20 May 2012 M_w 5.9 Emilia earthquake, diffuse liquefaction effects have been observed. A one-directional propagation model of a three-component seismic wave (1D-3C approach), in a finite element scheme, is used to investigate the seismic response of a soil profile derived from the stratigraphy of San Carlo village (Emilia Romagna, Italy) and soil-structure interaction. A corrected record of the 20 May 2012 M_w 5.9 Emilia earthquake is applied as reference input motion at the soil-bedrock interface.

The 3D nonlinear soil behavior is described by the shear modulus decay curve and the assumption of constant Poisson's ratio during strain increase, using the Iwan's elasto-plastic model. A correction of the shear modulus is applied for saturated soil layers, according to Iai's model.

The seismic response of the analyzed soil profile having a frame structure at its surface is investigated for different hypotheses of the water table depth. Profiles with depth of maximum shear strain and stress, peak of the horizontal motion and excess pore water pressure, hysteresis loops in liquefiable soils and the time histories of the ground motion and the excess pore water pressure are obtained.

The impact of saturated soil modeling in the numerical seismic response of a soil profile, compared with a dry soil assumption, is observed and shows that the dry soil assumption is a simplification that underestimates the peak values of the ground motion.

The influence in the numerical seismic response and in the stress-strain effects due to the variability of the water table depth, caused by seasonal fluctuation, is investigated and it appears negligible in the analyzed case study.

A parametric analysis should be performed in further studies, using different input motions having predominant frequency close to the natural frequencies of soil and structure. Moreover, other columns that can be extracted from the available high resolution engineering-geological model of the San Carlo village.

ACKNOWLEDGMENTS

Seismograms and soil stratigraphic setting used in this study were obtained in the framework of the Project S_2-2012 by INGV-DPC 2012–2013 – UR4 titled: "Validation of Seismic Hazard through observed data; Constraining OBservations into Seismic hazard (COBAS)" (scientific responsible: Laura Peruzza; UR4 co-ordinator R.W. Romeo).

The Authors wish to thank the Servizio Geologico della Regione Emilia technical office for the availability of technical data and reports.

REFERENCES

Chini, M., M. Albano, M. Saroli, L. Pulvirenti, M. Moro, C. Bignami, E. Falcucci, S. Gori, G. Modoni, N. Pierdicca & S. Stramondo 2015. Coseismic liquefaction phenomenon analysis by COSMO-SkyMed: 2012 Emilia (Italy) earthquake. *Int. J. Appl. Earth Obs. Geoinf.*, 39: 1–14.

Emergeo Working Group 2012. Liquefaction phenomena associated with the Emilia earthquake sequence of May–June 2012 (Northern Italy). *Nat. Hazards Earth Syst. Sci.*, 13: 935–947.

Fares, R., M.P. Santisi d'Avila & A. Deschamps 2017. Advantages and detriments of 1-Directional 3-Component wave propagation approach for Soil-Structure Interaction modeling, *Procedia Eng.*, 199: 2426–2432.

Fortunato, C., S. Martino, A. Prestininzi, R.W. Romeo, A. Fantini & P. Sanandrea 2012. New release of the Italian catalogue of earthquake-induced ground failures (CEDIT). *Italian J Eng Geol Environ*, www.ceri.uniroma1.it/index_cedit.html.

Hughes, T.J.R. 1987. *The finite element method - Linear static and dynamic finite element analysis*, Prentice Hall, Englewood Cliffs: New Jersey.

Iai, S., Y. Matsunaga & T. Kameoka 1990a. Parameter identification for a cyclic mobility mode, Report of the Port and harbour Research Institute, 29(4): 27–56.

Iai, S., Y. Matsunaga & T. Kameoka 1990b. Strain space plasticity model for cyclic mobility, Report of the Port and harbour Research Institute, 29(4): 57–83.

Iwan, W.D. 1967. On a class of models for the yielding behavior of continuous and composite systems, *J. Appl. Mech.*, 34: 612–617.

Joyner, W. 1975. A method for calculating nonlinear seismic response in two dimensions, *Bull. Seism. Soc. Am.*, 65(5): 1337–1357.

Joyner, W.B. & A.T.F. Chen 1975. Calculation of nonlinear ground response in earthquakes, *Bull. Seism. Soc. Am.*, 65(5): 1315–1336.

Kramer, S.L. 1996. *Geotechnical earthquake engineering*, Prentice Hall: New Jersey.

Martino, S., A. Prestininzi & R.W. Romeo 2014. Earthquake-induced ground failures in Italy from a reviewed database. *Nat. Hazards Earth Syst. Sci.*, 14: 799–1814.

Regnier, J. et al. 2016. International benchmark on numerical simulations for 1D non-linear site response (PRENOLIN): verification phase based on canonical cases, *Bull. Seismic Soc. Am.*, 106(5): 2112–2135.

Romeo, R.W., S. Amoroso, J. Facciorusso, L. Lenti, C. Madiai, S. Martino, P. Monaco, D. Rinaldis & F. Totani 2015. Soil liquefaction during the Emilia, 2012seismic sequence: investigation and analysis, In G. Lollino et al. (eds), *Engineering Geology for Society and Territory*, 5: 1107–1110.

Santisi d'Avila, M.P., L. Lenti & J.F. Semblat 2012. Modeling strong seismic ground motion: 3D loading path vs wavefield polarization, *Geophys. J. Int.*, 190: 1607–1624.

Santisi d'Avila, M.P. & F. Lopez Caballero 2018. Analysis of nonlinear Soil-Structure Interaction effects: 3D frame structure and 1-Directional propagation of a 3-Component seismic wave. *Comput. Struct.*, in press.

Santisi d'Avila, M.P., J.F. Semblat & L. Lenti 2013. Strong Ground Motion in the 2011 Tohoku Earthquake: a 1Directional - 3Component Modeling, *Bull. Seism. Soc. Am.*, Special issue on the 2011 Tohoku Earthquake, 103(2b): 1394–1410.

Santisi d'Avila, M.P. & Semblat J.F. 2014. Nonlinear seismic response for the 2011 Tohoku earthquake: borehole records versus 1Directional - 3Component propagation models, *Geophys. J. Int.*, 197: 566–580.

Numerical Methods in Geotechnical Engineering IX – Cardoso et al. (Eds)
© 2018 Taylor & Francis Group, London, ISBN 978-1-138-33198-3

Numerical simulation on the ground response in saturated sand

M. Morigi & G.M.B. Viggiani
Dipartimento di Ingegneria Civile ed Informatica, Università di Roma "Tor Vergata", Roma, Italy

R. Conti
Università di Roma "Niccolò Cusano", Roma, Italy

C. Tamagnini
Dipartimento di Ingegneria Civile e Ambientale, Università di Perugia, Perugia, Italy

ABSTRACT: Positive excess pore water pressures may develop in saturated fine sands during the strong motion phase of the earthquake, inducing a reduction of the state of effective stress, and hence of the strength and of the stiffness of the soil, which, in extreme cases, may lead to liquefaction. A reliable numerical prediction of seismically induced excess pore water pressures requires the development of a dynamic fully-coupled formulation capable of reproducing solid-fluid interaction together with the adoption of advanced constitutive models. In this work, the propagation of shear waves in layered and homogeneous saturated sand is examined numerically. The paper describes an original fully coupled dynamic u-p formulation, which was implemented in the finite element code *FEAP 8.4* (Taylor 2013) together with the advanced constitutive model SANISAND (Dafalias and Manzari 2004), which was adopted to describe the response of the solid skeleton. The field equations are formulated in a fully saturated condition.

1 INTRODUCTION

It is well known in the literature that earthquake induced liquefaction phenomena can occur in saturated loose-to-medium dense sand deposits, leading to a complete or partial loss of the shear strength of the soil (Jeremić et al. 2008). Well documented case studies, where recorded accelerations were available at different depths within the soil deposit, have demonstrated that liquefaction phenomena can induce either amplification or attenuation of the transmitted accelerations (Zeghal and Elgamal 1994; Iai et al. 1995; Matasovic and Vucetic 1996; Cubrinovski et al. 1996; Bonilla et al. 2005). In all these cases it was observed that, when liquefaction occurs, driven by a progressive build up of pore pressures due to cyclic loading, the waveform of the accelerations within the liquefied layer starts to change with respect to the underlying non-liquefied layers. Specifically, if the sand layer exhibits a dilatant behaviour, recorded accelerations are usually amplified, showing distinct spikes concurrently with sharp drops in the time history of Δu. This correlation was observed also in dynamic centrifuge tests carried out on reduced-scale models and numerical analyses (Jeremić et al. 2008; Elgamal et al. 2005; Elgamal et al. 2002; Zeghal et al. 1999). On the contrary, when the sand exhibits a pure contractant behaviour, accelerations are progressively filtered by the liquefied layer.

A theoretical explanation was recently suggested by Kokusho (2014), showing that the attenuation of surface accelerations is due to a reduction of the energy transmitted from the underlying non-liquefied layer to the overlying liquefied one, due to a substantial reduction of the shear wave velocity of the latter layer. Always using a simple visco-elastic model as reference, Bouckovalas et al. (2016) emphasized that accelerations can be either amplified or deamplified by the liquefied layer depending on the dimensionless ratio H/λ, where H is the depth of the liquefied layer and $\lambda = f/V_s$ is the wavelength of the transmitted signal.

This paper presents the main results of a finite element numerical study on the effects of liquefaction induced phenomena on the dynamic response of saturated sand deposits. The main goal of the work is to explore the influence of the frequency content of the input earthquake on the response of the liquefied layer, particularly when the dilatant behaviour of the soil is activated. To this end, two distinct cases will be examined, consisting of an ideal homogeneous deposit of loose sand ($D_r = 8\%$) and medium dense sand ($D_r = 43\%$) respectively, and different input earthquakes will be applied.

2 NUMERICAL TOOL

The numerical work presented in this paper was carried out using the finite element code *FEAP 8.4*, developed at the University of California, Berkeley (Taylor 2013). Within this framework, a 2D-plane strain finite element was implemented to solve the coupled hydro-mechanical problem for a fully saturated porous medium. Moreover, an advanced constitutive model was implemented, capable of reproducing the complex behaviour of sands under cyclic loading.

2.1 *FE formulation of the hydro-mechanical coupled problem*

The fully coupled dynamic field equations were solved using the u-p formulation proposed in (Zienkiewicz et al. 1980), introducing the assumption of isothermal process and neglecting the compressibility of grains.

By indicating with **d** the displacement of the mixture and with u the pore fluid pressure, the system of equations governing the problem can be written in weak form, as:

$$\begin{cases} \int_\Omega \left[\sigma_{ij,j} + \rho b_i - \rho \ddot{d}_i\right] d_i^* d\Omega = 0 \\ \int_\Omega \left[\dot{\varepsilon}_{ii} + \frac{k_i}{\rho_f g}\left[-u_{,i} + \rho_f b_i - \rho_f \ddot{d}_i\right]_{,i} + \frac{\dot{u}n}{K_f}\right] \cdot \\ \cdot u^* d\Omega = 0 \end{cases} \tag{1}$$

where $\mathbf{d}^*$ and u^* are the virtual displacement and virtual pore fluid pressure respectively, σ is the Cauchy stress vector, $\dot{\varepsilon}$ is the strain rate vector, ρ is the mass density of the mixture, ρ_f is the mass density of the fluid phase, **k** is the permeability vector, g is the gravity acceleration, **b** is the body load for unity of mass, n is the porosity and K_f is the bulk modulus of the fluid phase.

The boundary conditions for the problem at hand are as follows:

$$\begin{cases} \mathbf{d} = \overline{\mathbf{d}} & \text{on}\,\partial\Omega_d \\ \sigma \cdot \hat{\mathbf{n}} = \overline{\mathbf{t}} & \text{on}\,\partial\Omega_\sigma \\ u = \overline{u} & \text{on}\,\partial\Omega_u \\ \hat{\mathbf{n}} \cdot \frac{\mathbf{k}}{\rho_f g}\left[-\nabla u + \rho_f g - \rho_f \ddot{\mathbf{d}}\right] = \overline{q} & \text{on}\,\partial\Omega_q \end{cases} \tag{2}$$

2.1.1 *Approximation in space and time*

Displacement and pore pressure fields can be written as:

$$\begin{cases} d_i(x,y,t) \approx \sum_{j=1}^{8} N_{ij}^d(x,y)\hat{d}_j(t) \\ u(x,y,t) \approx \sum_{i=1}^{4} N_i^u(x,y)\hat{u}_i(t) \end{cases} \tag{3}$$

where $\mathbf{N}_j^i$ are the shape functions for the j-th node and the i-th unknown field, while $\hat{(\cdot)}$ are the nodal unknown. As shown in Figure 1, two different orders of approximation are adopted for the displacement and the pore pressure field. Specifically, bi-quadratic shape functions are used to approximate the displacement field, while bi-linear shape functions are adopted for the pore pressure field.

Substitution of (3) and (2) in (1) leads to the following system of algebraic equations:

$$\begin{cases} \mathbf{M}\ddot{\hat{\mathbf{d}}} - \mathbf{Q}\hat{u} + \mathbf{f}_{int}(\hat{\mathbf{d}}) - \mathbf{f}_{ext} = 0 \\ \mathbf{M}^*\ddot{\hat{\mathbf{d}}} - \mathbf{Q}^T\dot{\hat{\mathbf{d}}} + \mathbf{S}\dot{\hat{u}} + \mathbf{H}\hat{u} - \mathbf{q}_{ext} = 0 \end{cases} \tag{4}$$

where:

$$\mathbf{M} = \int_\Omega \rho \mathbf{N}^{d^T}\mathbf{N}^d dV \quad \mathbf{Q} = \int_\Omega \mathbf{B}^T\mathbf{m}\mathbf{N}^u dV$$

$$\mathbf{f}_{int} = \int_\Omega \mathbf{B}^T\sigma'(\hat{\mathbf{d}})dV \quad \mathbf{H} = \int_\Omega \mathbf{E}^T\frac{\mathbf{k}}{\rho_f g}\mathbf{E}dV$$

$$\mathbf{f}_{ext} = \int_\Omega \rho \mathbf{N}^{d^T}\mathbf{b}dV + \int_{\partial\Omega_t} \mathbf{N}^{d^T}\overline{\mathbf{t}}dA$$

$$\mathbf{S} = \int_\Omega \frac{n}{K_f}\mathbf{N}^{u^T}\mathbf{N}^u dV \quad \mathbf{M}^* = \int_\Omega \mathbf{E}^T\frac{\mathbf{k}}{g}\mathbf{N}^d dV$$

$$\mathbf{q}_{ext} = \int_\Omega \mathbf{E}^T\frac{\mathbf{k}}{g}\mathbf{b}dV - \int_{\partial\Omega_q} \mathbf{N}^{u^T}\overline{q}dA$$

In the above expressions, $\mathbf{m} = (1\,1\,1\,0)^T$, **B** is the strain-displacement transformation matrix, $\mathbf{E} = \nabla\mathbf{N}^u$ and Ω is the problem domain.

Equations (4) are solved by numerical integration in time domain. Two different schemes are adopted to solve the system, *i.e.*: a Generalised Newmark algorithm (GN22) for the displacement

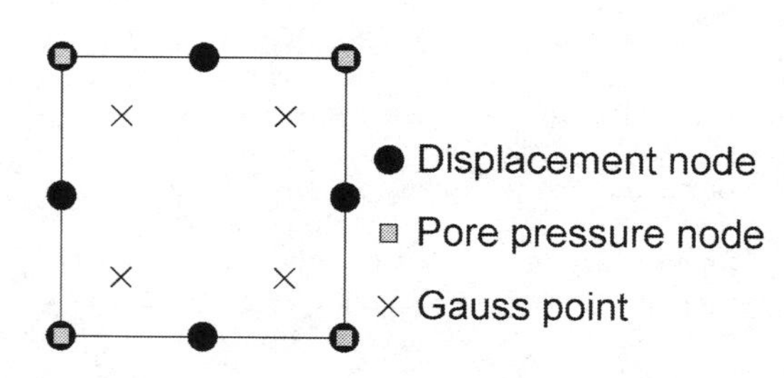

Figure 1. Layout of the 2D finite element implemented in *FEAP 8.4*.

field and a Single Step algorithm (SS11) for the pore pressure field (Zienkiewicz and Taylor 2000):

$$\begin{cases} d_{n+1} = d_n + \Delta t \dot{d}_n + \dfrac{\Delta t^2}{2}\left[(1-\beta_2)\ddot{d}_n + \beta_2 \ddot{d}_{n+1}\right] \\ \dot{d}_{n+1} = \dot{d}_n + \Delta t(1-\beta_1)\ddot{d}_n + \Delta t \beta_1 \ddot{d}_{n+1} \\ u_{n+1} = u_n + (1-\theta)\Delta t \dot{u}_n + \theta \Delta t \dot{u}_{n+1} \end{cases} \quad (5)$$

where $\beta_1(=0.5), \beta_2(=0.5)$ and $\theta(=1.0)$ are integration parameters of the algorithm. It is worth mentioning that, using these values for β_1, β_2 and θ, the GN22 and SS11 algorithms correspond to the constant average acceleration method and the backward Euler method respectively.

2.2 *Constitutive model for the soil*

The constitutive model SANISAND (Dafalias and Manzari 2004) was implemented in *FEAP*. The model was developed within the framework provided by the critical state soil mechanics and

Table 1. Summary of the constitutive equations for the SANISAND model.

Equation descriptions	Equations	Constants
Critical state line	$e_c = e_0 - \lambda_c\left(\frac{p'_c}{p_{at}}\right)^{\xi}$	e_0, λ_c, ξ
Elastic deviatoric strain increment	$d\mathbf{e}^e = d\mathbf{s}/2G$	–
Small strain shear modulus	$G = G_0 p_{at}\left[(2.97-e)^2/(1+e)\right]\left(p'/p_{at}\right)^{0.5}$	G_0
Elastic volumetric strain increment	$d\varepsilon_v^e = dp'/K$	–
Small strain bulk modulus	$K = 2(1+\nu)G/[3(1-2\nu)]$	ν
Yield surface	$f = [(\mathbf{s}-\alpha p'):(\mathbf{s}-\alpha p')]^{0.5} - \sqrt{2/3}\,p'm$	–
–	$\mathbf{n} = (\mathbf{s}/p' - \alpha)/(\sqrt{2/3}m)$	m
Plastic deviatoric strain increment	$d\mathbf{e}^p = \langle L \rangle \mathbf{R}'$	–
Plastic deviatoric strain direction	$\mathbf{R}' = B\mathbf{n} - C(\mathbf{n}^2 - 1/3\mathbf{I})$	–
–	$B = 1 + \frac{3}{2}\frac{1-c}{c}g(\theta,c)\cos(3\theta)$	–
–	$C = 3\sqrt{\frac{3}{2}}\frac{1-c}{c}g(\theta,c)$	–
Plastic modulus	$K_p = (2/3)p'h(\alpha_\theta^b - \alpha):\mathbf{n}$	–
Boundary surface	$\alpha_\theta^b = \sqrt{2/3}[g(\theta,c)M\exp(-n^b\psi) - m]\mathbf{n}$	M, c, n^b
–	$h = b_0/(\alpha - \alpha_{in}):\mathbf{n}$	–
–	$b_0 = G_0 h_0(1-c_h e)(p_{at}/p')^{0.5}$	h_0, c_h
Plastic volumetric strain increment	$d\varepsilon_v^p = \langle L \rangle D$	–
Dilatancy	$D = A_d(\alpha_\theta^d - \alpha):\mathbf{n}$	–
Dilatancy surface	$\alpha_\theta^d = \sqrt{2/3}[g(\theta,c)M\exp(n^d\psi) - m]\mathbf{n}$	n^d
–	$A_d = A_0(1 + \sqrt{3/2}\langle \mathbf{z}:\mathbf{n} \rangle)$	A_0
Fabric-dilatancy tensor evolution	$d\mathbf{z} = -c_z\langle -d\varepsilon_v^p \rangle(\sqrt{2/3}z_{max}\mathbf{n} + \mathbf{z})$	c_z, z_{max}
Back-stress ratio tensor evolution	$d\alpha = \langle L \rangle (2/3)h(\alpha_\theta^b - \alpha)$	–

the bounding surface plasticity, in order to reproduce the main features of the behaviour of sands under cyclic loading. The constitutive equations of the model are briefly recalled in Table 1, while the reader is referred to Dafalias and Manzari (2004) for a thorough discussion on their derivation.

The constitutive rate equations were integrated using an explicit method, together with an automatuc error control and an adaptive substepping procedure (Sloan et al. 2001). Specifically, the Runge-Kutta method with an accuracy of *2nd* and *3rd* order (RK23) was used to solve the equations.

Table 2 reports the physical and mechanical parameters adoped in this work for the fully saturated sand medium, while all the model constants, together with the RK23 tolerances, are shown in Table 3.

3 GROUND RESPONSE ANALYSIS

The problem under investigation refers to the one-dimensional vertical propagation of SH shear

Table 2. Physical and mechanical parameters of the fully saturated sand medium.

Parameter	Symbol	Value
Mass density of the mixture	ρ	1.85 t/m^3
Mass density of the fluid phase	ρ_f	1.0 t/m^3
Porosity	n	0.5
Permeability	**k**	(0.0; 0.0005)m/s
Fluid bulk modulus	K_f	$2.2 \cdot 10^6$ kN/m^2

Table 3. Constitutive model constants and integration tolerances.

Constant	Symbols	Value
Elasticity	G_0	125
	ν	0.05
Critical State	M	1.25
	c	0.712
	e_0	0.934
	λ_c	0.019
	ξ	0.7
Dilatancy surface	n^d	2.1
	A_0	0.704
Plastic modulus	c_h	0.934
	n^b	1.25
	h_0	7.05
Fabric-dilatancy tensor	c_z	600
	z_{max}	2
Yield surface	m	0.01
Tolerance RK23	–	10^{-4}
Tolerance drift	–	10^{-5} kPa

waves within an ideal homogeneous soil deposit of saturated sand overlying a rigid bedrock. The sand layer, with a total depth of 10 m, is characterised by two different values of the initial relative density, *i.e.*: $D_r = 8\%$ (loose sand) and $D_r = 43\%$ (medium dense sand).

Figure 2 shows the adopted mesh, consisting of 20 elements with dimensions of 0.5 m × 0.5 m, together with the two profiles assumed for the initial shear wave velocity and void ratio. To highlight the different behaviour expected for the two soil deposits under investigation, Figure 2 also shows the corresponding profile of the critical void ratio, thus anticipating a potential contractant and dilatant behaviour for the loose and medium dense sand respectively.

An initial hydrostatic distribution is assumed for the pore pressure, with the water table located at $z = 0\,\text{m}$, and an initial geostatic stress state is applied. Hydraulic boundary conditions consist of fixed pore water pressure at the top of the sand layer $(u = 0\,\text{kPa})$ and null flux along the lateral sides of the soil column. Moreover, standard periodic constraints are applied to the nodes on the lateral boundaries of the mesh, *i.e.* they are free to move in both vertical and horizontal direction, while tied to one-another in order to enforce the same displacements of the two boundaries. The seismic input is applied to the bottom nodes of the mesh, in terms of a displacement time history in the horizontal direction.

Two different input signals were considered in this study, both recorded during real earthquakes and characterized by substantially different frequency contents, namely the Friuli (Italy)

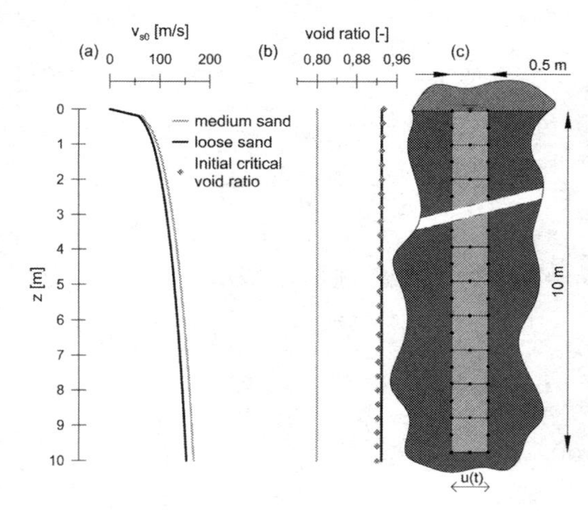

Figure 2. Profiles of the shear wave velocity (a), the initial and critical void ratios (b), together with the layout of the problem (c).

earthquake of 1976, and the Kobe (Japan) earthquake of 1995. Figure 3 shows the acceleration time histories (a, b) and the amplitude Fourier spectra (c) of the two inputs, while Table 4 summarizes the corresponding values of peak acceleration, *PGA*, dominant and mean frequency, f_d and f_m, Arias intensity, I_a, and strong motion duration, T_{5-95}. Both recordings were scaled to a maximum acceleration of 0.2 g. The frequency content of the Friuli earthquake is mostly comprised in the range $[1-4]\text{Hz}$, while the energy of the Kobe earthquake spread over a wider range of approximately $[0.5-8]\text{Hz}$.

As a guide for the interpretation of possible amplification/attenuation phenomena within the two sand deposits, the initial value of their fundamental frequency was estimated as:

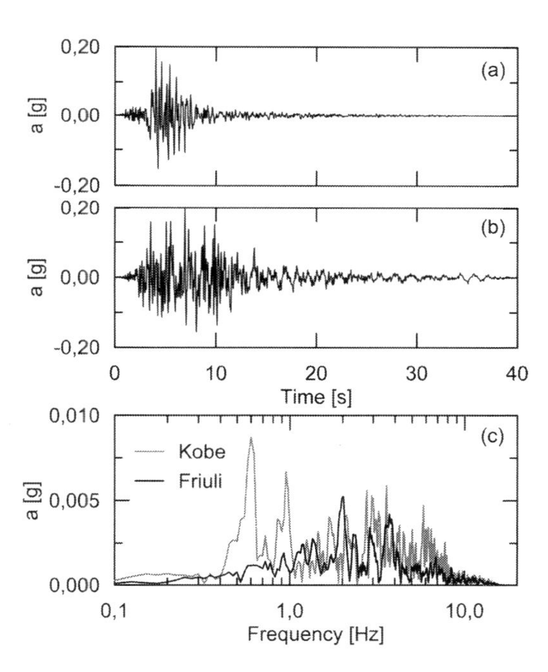

Figure 3. Acceleration time histories of the Friuli (a) and the Kobe (b) earthquake, together with the corresponding amplitude Fourier spectra (c).

Table 4. Ground motion parameters of the applied earthquakes.

	PGA	f_d	f_m	I_a	T_{5-95}
Eq.	[g]	[Hz]	[Hz]	[m/s]	[s]
Friuli	0.2	2.00	3.40	0.28	4.25
Kobe	0.2	0.60	3.76	0.61	12.87

$$f_{0,ini} = \frac{\overline{V}_{s0}}{4H} \approx 3.5\text{Hz} \tag{6}$$

where $\overline{V}_{s0}$ is the mean value of the small strain shear wave velocity within the sand layer.

3.1 *Results*

Figure 4 shows the time histories and the Fourier amplitude spectra of the accelerations computed at the bottom (bedrock) and at the top (free-field) of the soil layer, for both the loose sand $(D_r = 8\%)$ and medium dense sand $(D_r = 43\%)$ deposits, and for both the Friuli and the Kobe earthquakes. In all four cases, two distinct waveforms can be recognised in the free-field acceleration time histories, reflecting the occurrence of liquefaction phenomena within the saturated sand deposit during the earthquakes. Specifically, during the first $3-5\text{s}$ (first waveform) the free-field signal is similar to the bedrock one, thus suggesting that no appreciable amplification or attenuation phenomena occur within the sand deposit during the first instants of the earthquakes. A completely different trend is observed in the subsequent stages (second waveform), during which the behaviour exhibited by the two sand deposits starts to deviate. On the one hand, free-field accelerations are completely de-amplified by the loose sand deposit (Figure 4a,b), thus suggesting that attenuation phenomena do not depend on the frequency content of the earthquake in this case. On the other hand, moving to the medium dense sand case (Figure 4c,d), the observed behaviour strictly depends on the applied input signal. Specifically, free-field accelerations are attenuated during the Friuli event, as for the loose sand deposit, while amplifications occur during the Kobe event, where large isolated spikes can be recognised in the free-field acceleration time history.

Looking at the amplitude Fourier spectra (Figure 4e-h), it is apparent that the loose sand deposit filters approximately all the significant frequencies of the applied earthquakes, while the medium dense one tends to amplify frequencies below $f_{0,ini}$, while filtering the high-frequency components of the input signals. However, as the standard Fourier transform applies to the whole signal, the amplitude Fourier spectra in Figure 4 must be taken as representative of the mean frequency content of the free-field signal along the whole duration of the earthquake. In other words, this representation does not allow to discern between the two distinct waveforms just identified in the time domain.

A better understanding of the problem at hand can be obtained using the wavelet transform,

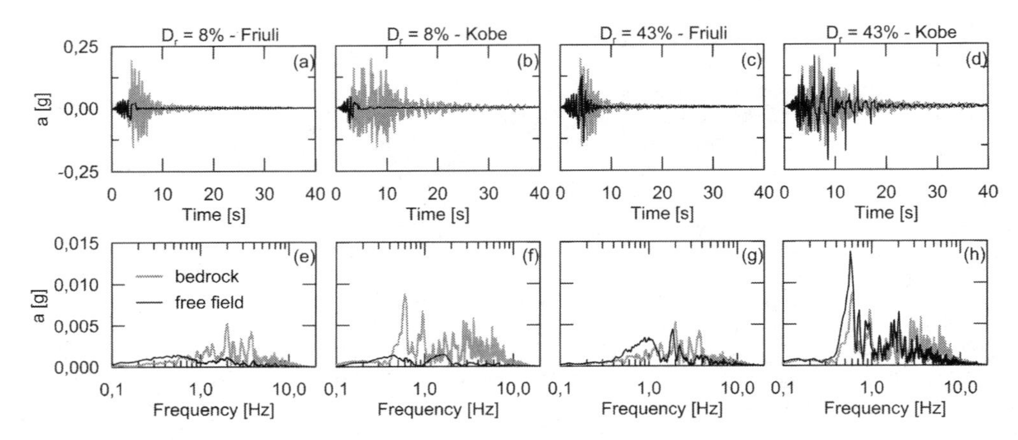

Figure 4. Time histories (a, b, c, d) and Fourier amplitude spectra (e, f, g, h) of the accelerations computed at the bottom (bedrock) and at the top (free field) of the soil layer, for both the loose sand (D_r = 8%) and medium dense sand (D_r = 43%) deposit.

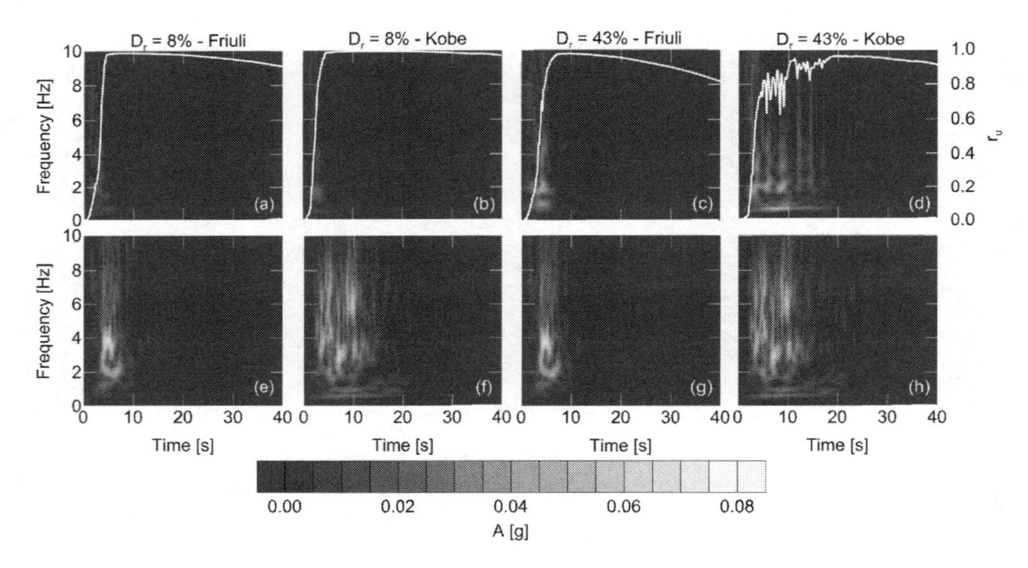

Figure 5. Time-frequency spectra of the free field accelerations together with the $\overline{r}_u$ parameter for both loose (a-b) and medium dense (c-d) sand with the input signal (e-f-g-h).

providing knowledge of how the frequency content of the signal varies in time. Specifically, Figure 5 shows the contours of the amplitudes of the wavelet transform, applied to both free-field (Figure 5a-d) and bedrock accelerations (Figure 5e-h), for the four numerical analyses investigated. Moreover, (Figure 5a-d) also shows (white line) the evolution of the normalised parameter $\overline{r}_u$, defined as:

$$\overline{r}_u = \frac{1}{H}\int_0^H \Delta u(z,t) / \sigma'_{v0}(z) \qquad (7)$$

taken as an average indicator of the evolution of the excess pore pressures within the whole deposit during the applied earthquakes. As far as the loose sand layer is concerned, $\overline{r}_u$ exhibits a monotonic increase during the first stage of the earthquakes, and the free-field accelerations almost nullify in the whole frequency range as soon as $\overline{r}_u \approx 1$. In these time instants, the saturated sand deposit looses its shear resistance and the horizontal accelerations cannot be propagated further, irrespective of the frequency content of the input signal. On the contrary, the medium dense sand layer shows

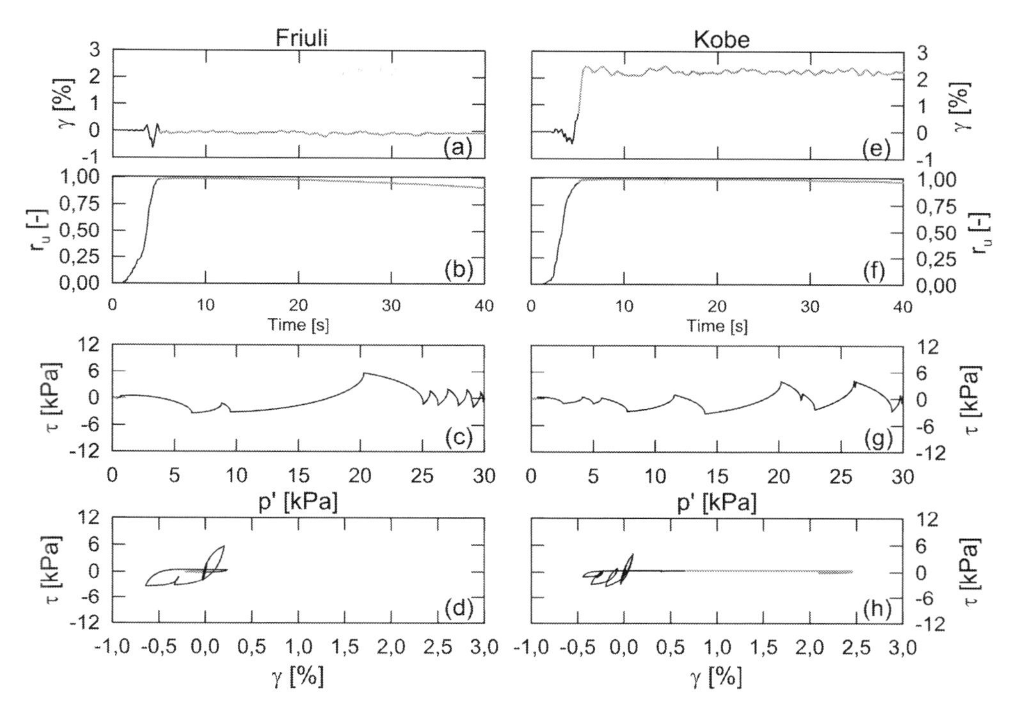

Figure 6. Loose sand deposit (D_r = 8%): time histories of the shear strain at mid depth of the soil deposit (a-e) and of $\overline{r}_u$ (b-f); stress path in the $\tau: p'$ plane (c-g) and stress-strain response in the $\tau: \gamma$ plane (d-h) during the Friuli (left) and the Kobe (right) earthquake.

a different behaviour, depending on the applied earthquake. When looking at the Friuli earthquake, $\overline{r}_u$ increases again monotonically, even though at a lower rate with respect to the previous two cases, eventually leading to a sharp reduction in the transmitted acceleration. However, when the Kobe earthquake is applied $\overline{r}_u$ increases, but without reaching unity, showing instead an oscillating trend clearly related to the contractant-dilatant behaviour of the saturated sand. Concurrently, the high-frequency components of the input signal are de-amplified, while a significant amplification occurs at low frequencies, particularly for $f < 1\text{Hz}$.

3.2 *Interpretation*

The results of the four numerical analyses just outlined indicate that the frequency content of the input signal can play a crucial role in the dynamic response of a saturated medium dense sand deposit. On the contrary, attenuation phenomena occurring in a saturated loose sand deposit are essentially independent of the frequency content of the earthquake.

A possible interpretation of these results can be obtained looking at the mechanical behaviour exhibited by the sand as a result of the imposed dynamic excitation. To this end, Figure 6 shows, for the loose sand deposit and for both the Friuli (left) and the Kobe (right) earthquake, the time histories of the shear strain (a,e) and of the $\overline{r}_u$ parameter (b,f), together with the computed stress path in the $\tau: p'$ plane (c,g) and stress-strain response in the $\tau: \gamma$ plane (d,h). Both stress and strain quantities were computed at mid depth of the soil deposit ($z = 5\text{m}$). Moreover, Figure 7 shows the same results for the medium dense sand deposit. For the sake of clarity, different colors were used for the first time instants of the applied earthquakes (first waveform: black lines) and the subsequent stage (second waveform: gray lines).

With reference to the first waveform, the following observations apply to all the four cases: (i) small shear deformations develop within the soil layer; (ii) $\overline{r}_u$ increases monotonically; (iii) the mean effective stress, p', gradually reduces; and (iv) the secant shear stiffness mobilised during the applied loading cycles progressively reduces. Looking at the second waveform, the loose and the medium

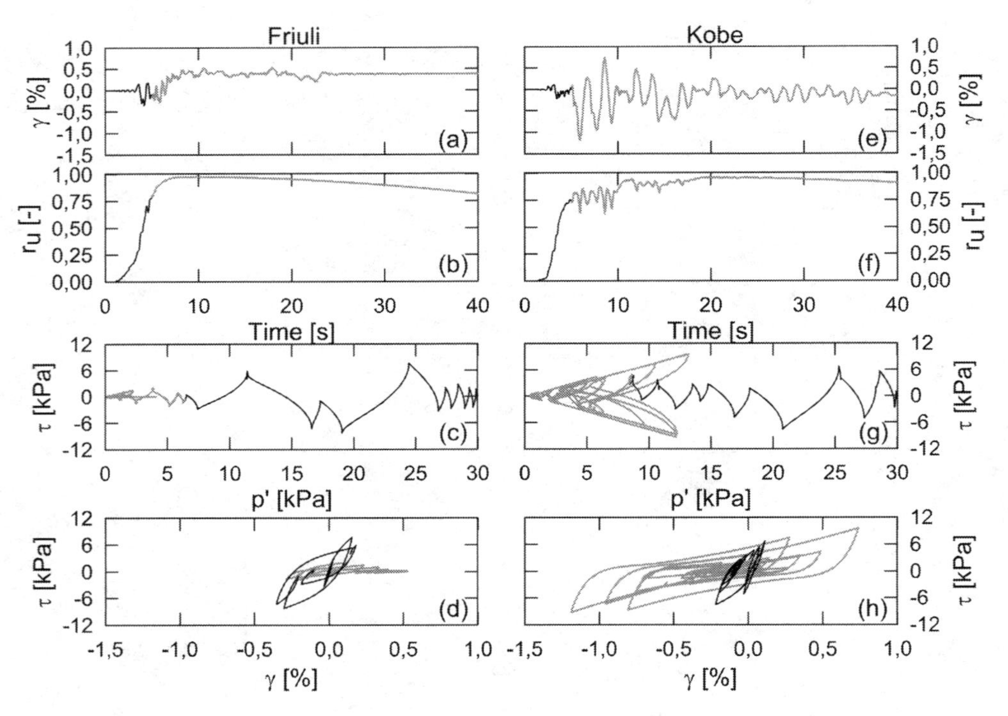

Figure 7. Medium dense sand deposit (D_r = 43%): time histories of the shear strain at mid depth of the soil deposit (a-e) and of $\overline{r}_u$ (b-f); stress path in the $\tau : p'$ plane (c-g) and stress-strain response in the $\tau : \gamma$ plane (d-h) during the Friuli (left) and the Kobe (right) earthquake.

dense saturated sand deposits exhibit a different behaviour. In the first case (Figure 6) only positive Δu can develop during the earthquake, progressively increasing after each loading cycle imposed by the dynamic excitation. These positive Δu induce a progressive reduction of the shear strength and stiffness of the sand, eventually leading to a fully liquefaction condition when p' approaches to zero. In the second case, instead, both positive and negative Δu can develop, thus inducing a stabilization of the response of the material in both the $\tau : p'$ and the $\tau : \gamma$ planes. However, the possibility to develop negative Δu is strictly related to the dilatant behaviour of the sand and, also, to the amplitude of the imposed deformation. In other words, relatively small shear strains cannot trigger the dilatant behaviour of the sand and, then, only positive Δu arise, thus leading to a progressive reduction of p'. Instead, large shear strains can induce also negative Δu, thus inducing a stabilization of the response of the soil. Indeed, this latter behaviour is what observed when the Kobe earthquake is applied to the medium dense sand deposit (Figure 7).

In the light of the mechanical behaviour mentioned above, the different trends observed for the medium dense sand deposit are clearly related to the frequency content of the two earthquakes applied. As a matter of fact, the progressive reduction of p' during the first seconds induces a concurrent reduction of the first natural frequency of the soil deposit, which, then, tends to amplify lower frequencies, while filtering the larger ones. As a result, the behaviour observed during the subsequent stage (second waveform) depends on the low frequency content of the applied signal. For the Kobe earthquake, low frequencies in the input signal are highly amplified by the sand layer, thus inducing large shear strains and negative Δu. For the Friuli earthquake, instead, the low-frequency components of the input are small and no amplification phenomena can occur within the sand layer, the resulting shear strains being too small to trigger the dilatant behaviour of the sand.

4 CONCLUSION

This paper addressed the issue of the ground response analysis with reference to an ideal homogeneous soil deposit of saturated sand overlying a rigid bedrock, characterised by two different values of the initial relative density ($D_r = 8\%$ e $D_r = 43\%$). To this end, a 2D-plane strain finite element was implemented in *FEAP* to solve the coupled hydro-mechanical problem for a fully saturated porous medium, together with the constitutive model SANISAND, proposed by (Dafalias and Manzari 2004) to reproduce the cyclic behaviour of sand.

Two different input earthquakes were considered in this work: the Kobe earthquake, characterised by a significant frequency content even below the initial fundamental frequency of the soil deposit, $f_{0,ini}$, and the Friuli earthquake, whose energy content is shifted towards relatively higher frequencies.

Numerical results showed that the seismic response of the loose sand deposit does not depend on the frequency content of the input signal, as the soil always reaches a condition of complete liquefaction induced by a progressive accumulation of positive Δu.

A different behaviour is exhibited by the medium dense sand deposit. In this case, if the input earthquake has a significant frequency content below $f_{0,ini}$, amplification phenomena can occur within the soil layer, mobilising large strains and activating the dilatant behaviour of the sand, as observed also by Elgamal et al. (2005). A possible explanation is the progressive reduction of the natural frequency of the soil deposit, which occurs during the initial stage of the earthquake as a consequence of the concurrent accumulation of positive Δu. The dilatant behaviour of the sand can induce large spikes in the free field accelerations, as recorded during the Wildfile Refuge and Kushiro-Oki earthquakes. On the other hand, no amplification phenomena can occur if the input earthquake has a poor energy content below $f_{0,ini}$. As a consequence, the dilatant behaviour of the sand cannot be triggered and the progressive development of positive Δu eventually leads to a condition of full liquefaction, as in the case of the loose sand deposit.

REFERENCES

Bonilla, L.F., R.J. Archuleta, & D. Lavallée (2005). Hysteretic and dilatant behavior of cohesionless soils and their effects on nonlinear site response: Field data observations and modeling. *Bulletin of the Seismological Society of America 95*(6), 2373–2395.

Bouckovalas, G.D., Y.Z. Tsiapas, A.I. Theocharis, & Y.K. Chaloulos (2016). Ground response at liquefied sites: seismic isolation or amplification? *Soil Dynamics and Earthquake Engineering 91*, 329–339.

Cubrinovski, M., K. Ishihara, & F. Tanizawa (1996, June 23–28). Numerical simulation of the kobe port island liquefaction. In Pergamon (Ed.), *11 WCEE: Eleventh World Conference on Earthquake Engineering, June 23–28, 1996*, Number 330, Acapulco, Mexico.

Dafalias, Y.F. & M.T. Manzari (2004). Simple plasticity sand model accounting for fabric change effects. *Journal of Engineering Mechanics 130*(6), 622–634.

Elgamal, A., Z. Yang, T. Lai, B.L. Kutter, & D.W.Wilson (2005). Dynamic response of saturated dense sand in laminated centrifuge container. *Journal of Geotechnical and Geoenvironmental Engineering 131*(5), 598–609.

Elgamal, A., Z. Yang, & E. Parra (2002). Computational modeling of cyclic mobility and post-liquefaction site response. *Soil Dynamics and Earthquake Engineering 22*(4), 259–271.

Iai, S., T. Morita, T. Kameoka, Y. Matsunaga, & K. Abiko (1995). Response of a dense sand deposit during 1993 kushiro-oki eaqrthquake. *Soils and Foundations 35*(1), 115–131.

Jeremić, B., Z. Cheng, M. Taiebat, & Y. Dafalias (2008). Numerical simulation of fully saturated porous materials. *International Journal for Numerical and Analytical Method in Geomechanics 32*, 1635–1660.

Kokusho, T. (2014). Seimic base-isolation mechanism in liquefied sand in terms of energy. *Soil Dynamics and Earthquake Engineering 63*, 92–97.

Matasovic, J. & M. Vucetic (1996, June 23–28). Analisys of seismic records from the wildlife liquefaction site. In Pergamon (Ed.), *11 WCEE: Eleventh World Conference on Earthquake Engineering, June 23–28, 1996*, Number 209, Acapulco, Mexico.

Sloan, S.W., A.J. Abbo, & D. Sheng (2001). Refined explicit integration of elastoplastic models with automatic error control. *Engineering Computations 18*, 121–154.

Taylor, R.L. (2013). Feap - - a finite element analysis program – version 8.4 user manual. Technical report, University of California at Berkeley, Berkeley, California.

Zeghal, M. & A.W. Elgamal (1994). Analysis of site liquefaction using earthquake records. *Journal of Geotechnical Engineering 120*(6), 996–1017.

Zeghal, M., A.W. Elgamal, X. Zeng, & K. Arulmoli (1999). Mechanism of liquefaction response in sandsilt dynamic centrifuge tests. *Soil Dynamics and Earthquake Engineering 18*(1), 71–85.

Zienkiewicz, O.C., C.T. Chang, & P. Bettess (1980). Drained, undrained, consolidating and dynamic behavior assumptions in soils. *Géotechnique 30*(4), 385–395.

Zienkiewicz, O.C. & R.L. Taylor (2000). *The finite element method - Volume 1 – The basis*. Oxford: Butterworth-Heinemann.

Numerical Methods in Geotechnical Engineering IX – Cardoso et al. (Eds)
© 2018 Taylor & Francis Group, London, ISBN 978-1-138-33198-3

Simulating the seismic response of laterally spreading ground after its passive stabilization against liquefaction

A.G. Papadimitriou & Y.K. Chaloulos
Department of Geotechnical Engineering, National Technical University of Athens, Athens, Greece

G.I. Agapoulaki
Department of Civil Engineering, University of Thessaly, Volos, Greece

K.I. Andrianopoulos
Atkins, Member of the SNC-Lavalin Group, Leeds, UK

Y.F. Dafalias
Department of Mechanics, National Technical University of Athens, Athens, Greece
Department of Civil and Environmental Engineering, University of California Davis, Davis CA, USA

ABSTRACT: This paper compares numerical simulation results to recordings from two (2) sets of dynamic centrifuge tests that emphasize on the lateral spreading response of a saturated sand layer during shaking, before and after passive stabilization against liquefaction with colloidal silica. The first corresponds to the seismic response of a gently sloping sand layer, while the second to the seismic response of a pile group within a gently sloping sand layer. The former test shows significantly reduced horizontal displacements and settlements of the stabilized layer, while the latter test depicts smaller horizontal pile displacements, but also lower pile bending moments. These beneficial effects of stabilization are well reproduced by the simulations, which were obtained by employing the finite difference method and a constitutive model for sands named NTUA-SAND. The stabilization is simulated by reducing the pore fluid bulk modulus in the coupled analyses, thus simulating the seemingly increased compressibility of colloidal silica in the sand pores.

1 INTRODUCTION

Passive (site) stabilization is a new method of ground improvement (Gallagher 2000) against liquefaction, which has competitive advantages for use at developed sites (under structures, around lifelines). It concerns the low-pressure injection of colloidal silica into the sand pores under a structure, or around lifelines. Colloidal silica is an aqueous dispersion of silica particles, which initially has very low viscosity and gels abruptly after well-controlled time. The gelled colloidal silica in the sand pores alters the mechanical response of the sand, making it less vulnerable to plastic strain accumulation and strength degradation related to liquefaction. The insitu injection of colloidal silica may be performed via injection and extraction wells on either side of the site (see Fig 1).

So far, the experimental study of this new technique has focused on the increase of cyclic soil resistance (e.g. Gallagher and Mitchell, 2002), the colloidal silica rheology (e.g. Gallagher, 2000) and its injectability in sand volumes (e.g. Gallagher & Lin 2009). The modeling of the response of this new geomaterial (stabilized sand) has attracted little interest in the literature until now, since only Kodaka et al. (2005) has proposed a dedicated constitutive model. From another point of view, Andrianopoulos et al. (2016) proposed the simulation of the effect of stabilization in fully coupled analyses, by introducing approximate properties of colloidal silica for the pore fluid component, instead of the default values for water. Their proposal was based on simulations of element tests, but mainly on the simulation of a dynamic centrifuge test studying the seismic response of a horizontal stabilized layer.

This paper adopts the proposal of Andrianopoulos et al. (2016) and investigates its accuracy via comparisons with other, more complicated, dynamic centrifuge tests. Hence, section 2 provides information with respect to the employed numerical methodology, while section 3 verifies its accuracy against recordings of a centrifuge test involving lateral spreading of a gently sloping sand layer. Section 4 does the same, but for a pile group located within such a layer, and the paper ends in section 5 with conclusions regarding the reliability of the employed numerical methodology, as well as the effectiveness of this novel ground improvement method.

2 NUMERICAL METHODOLOGY

The mechanical response of this new geomaterial named stabilized sand has been only macroscopically investigated in the literature. In this perspective, Andrianopoulos et al. (2016) showed that existing constitutive models for sands can be used for the phenomenological simulation of the response of stabilized sands. They propose that the effect of stabilization can be incorporated in fully coupled analyses by introducing the properties of colloidal silica for the pore fluid component, instead of the default values for water. In its simplest form, this can be achieved via a decreased value for the pore fluid (bulk) modulus K, following the observations of Towhata (2007) who performed unconfined compression tests on gelled colloidal silica samples and found that this material has seemingly increased volume compressibility in comparison to water. Based on their analyses, a good estimate for K in passively stabilized media is a value of 500 to 1000 times lower than $K_w = 2 \times 10^6$ kPa, the bulk modulus of (de-aired) water.

In this paper, we adopt this proposal for fully coupled dynamic analyses using the finite difference method. For the cyclic response of sands, the well-established bounding surface plasticity model NTUA-SAND (Andrianopoulos et al., 2010a, 2010b) is employed. In all analyses presented herein a value of K that is 975 times smaller than K_w is used for the stabilized sands (i.e. essentially the upper-bound of the proposed range), whereas

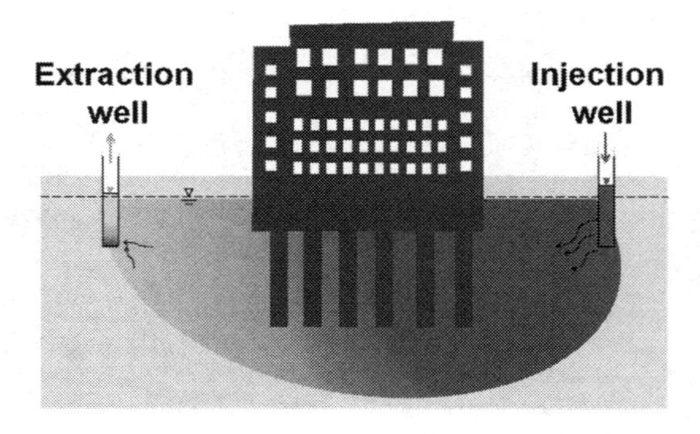

Figure 1. Concept of passive stabilization under an existing structure.

K_w is assigned to the fluid modulus of the saturated untreated soils. The 2 sections that follow present the details of the simulations for the 2 centrifuge test series, as well as the comparisons of simulations to recordings.

3 LATERAL SPREADING OF A GENTLY SLOPING LAYER

The accuracy of the numerical simulation of stabilized sand response in boundary value problems is first evaluated via a dynamic centrifuge test on a gently sloping stabilized sand layer (Conlee et al. 2012). The centrifuge model consists of two symmetrical soil profiles with a 3° inclination towards the centerline of the model (Fig. 2), which are divided by a latex membrane. The soil profiles consist of 3 layers (in prototype scale): a 0.75 m bottom layer of dense Monterey sand, a 4.8 m layer (maximum thickness at the lateral boundaries) of liquefiable Nevada sand (relative density D_r = 40%) and a top 1 m layer of soft silty clay (Yolo loam), which forms a central channel. As shown in Fig. 2, the model is 24.75 m long and its mesh discretization is quite dense (0.5 m × 0.5 m elements, on average).

The sand layers of the left part of the model are stabilized with Ludox®-SM colloidal silica at a concentration per weight of CS(%) = 9, while the respective layers of the right part of the model remain untreated, i.e. with CS(%) = 0. The model is subjected to a base acceleration consisting of 20 sinusoidal cycles of 0.10 g peak acceleration at a period T = 0.5 sec. Nodes at the lateral boundaries that are at the same height are tied to one another in order to have the same horizontal and vertical displacement, thus simulating the use of a laminar box in the centrifuge test. The values of model constants of NTUA-SAND used for Nevada sand are as proposed by Andrianopoulos et al. (2010a, 2010b) for both sides of the centrifuge model, and are also used for the Monterey sand layers. The difference is only the void ratio, which is equal to 0.49 for dense Monterey sand, as compared to 0.72 for the medium-dense Nevada sand. For the Yolo loam layers, the Mohr-Coulomb model is used, with parameters that are compatible with the soft nature of this soil (Kamai and Boulanger, 2013). Given the plane strain geometry, the analyses were performed with FLAC v7.0 (Itasca Consulting Group Inc., 2011).

The first comparison of the numerical simulation results with the recordings is presented in Fig. 3, in terms of acceleration time histories in the mid height of the stabilized and the untreated layer (see locations of accelerometers ACC-T4, ACC-U28 in Fig. 2). The grey lines depict the recordings,

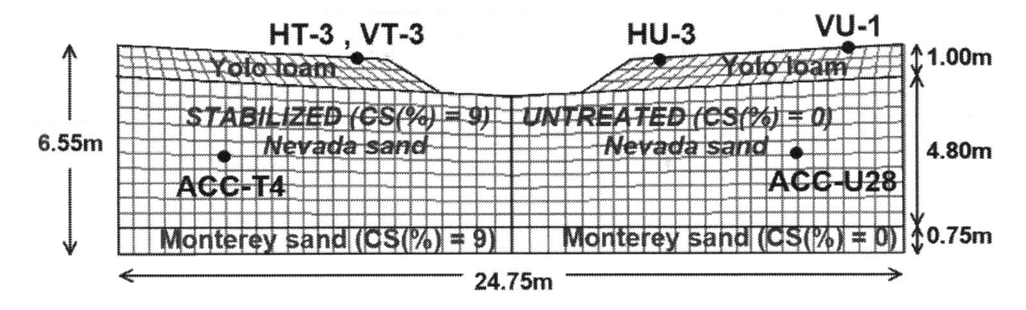

Figure 2. Model geometry, soil types, location of recorders and grid for the numerical simulation of the dynamic centrifuge test of Conlee et al. (2012).

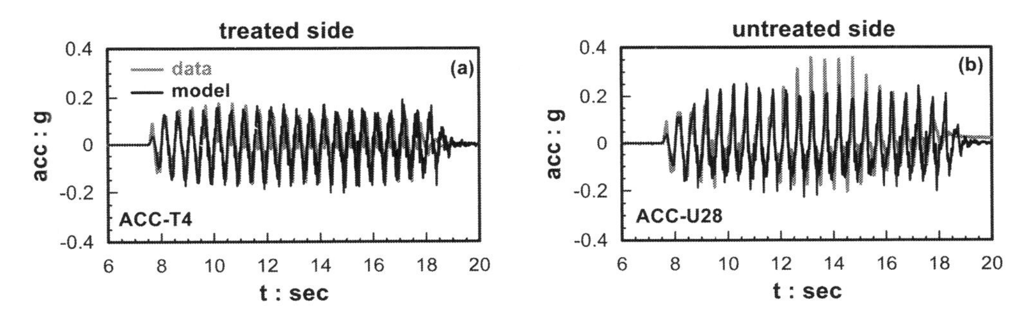

Figure 3. Comparison of computed versus recorded time histories of acceleration at mid depth of the treated (a) and untreated (b) side of the dynamic centrifuge model of Conlee et al. (2012).

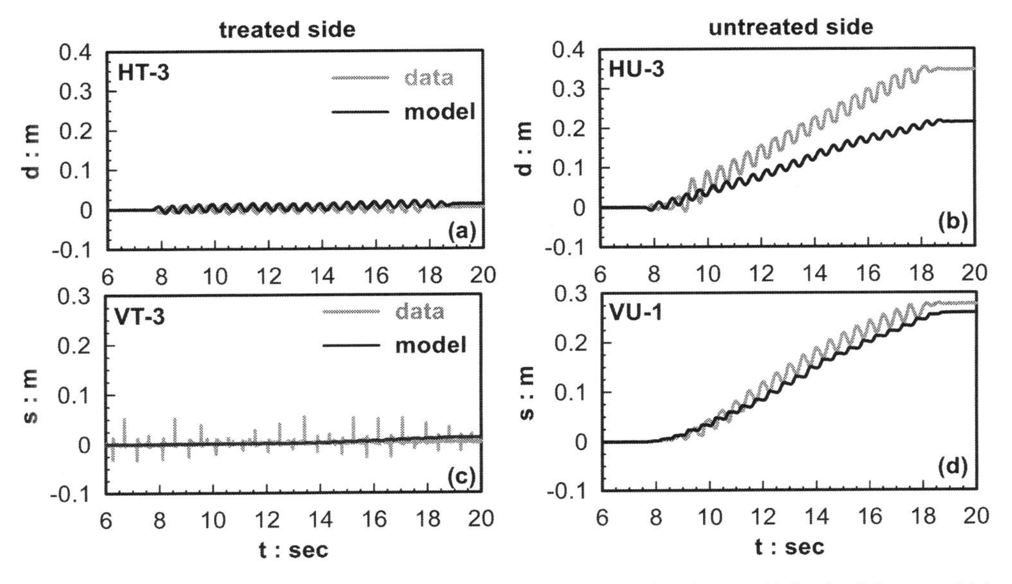

Figure 4. Comparison of computed versus recorded time histories of acceleration at mid depth of the treated (a) and untreated (b) side of the dynamic centrifuge model of Conlee et al. (2012).

while the thicker black lines the simulations. This figure shows that both acceleration time histories are simulated quite well. It also shows that the simulation captures that the acceleration of the untreated side is more intense at this depth.

The comparison of the numerical results with the recordings is continued in Fig. 4 in terms of time histories of horizontal displacements (HT-3, HU-3; denoted by d) and settlements (VT-3, VU-1; denoted by s) of the ground surface, for the stabilized side (Figs 4a, 4c on the left) and the untreated side (Figs 4b, 4d on the right). Again, the exact location of the LVDTs may be found in Fig. 2. This figure illustrates that the displacement time histories from the analysis (black lines) are in good agreement with the recordings (gray lines), and show significantly reduced (practically zero) horizontal displacements and settlements for the stabilized side. The agreement between recordings and simulations is less perfect for the untreated side, but still the huge benefit from passive stabilization becomes evident, thus underlining that the effects of this novel ground improvement method can be simulated successfully.

4 PILE GROUP WITHIN LATERALLY SPREADING SAND

Here, the accuracy of the numerical simulation of stabilized sand response is evaluated via a series of dynamic centrifuge tests on a pile group within gently sloping stabilized sand (Pamuk et al., 2007). The centrifuge model geometry is three dimensional and the soil profile (in prototype scale) is 16 m wide, 35.5 m long and 10 m tall, while in its center lies a 2 × 2 pile group with a pile cap, as shown in Fig. 5. It has a uniform 2° inclination along its longer side, which creates the conditions for lateral spreading if left untreated. The soil profile consists of 3 layers:

- The lower 2 m comprise a lightly cemented non-liquefiable Nevada sand, where the pile group is founded.
- The middle 6 m correspond to the liquefiable Nevada sand layer, with $D_r = 40\%$.
- The upper 2 m consist of a a lightly cemented non-liquefiable Nevada sand layer, where the pile cap lies.

The 4 piles (and the pile cap) were made of metal, had a diameter of D = 0.6 m and where at center-to-center distance of 3D = 1.8 m. They did not carry any load other than their own weight (and the weight of the pile cap). Given the symmety of the 3D model, the fully coupled analysis was performed using FLAC^{3D} (Itasca Consulting Group Inc., 2012), by discretizing half of the centrifuge model (as shown in Fig. 5). The upper and lower 2 m of lightly cemented non-liquefiable Nevada sand layers were simulated with an elastic model, with the lower layer being significantly stiffer than the upper one in order to account for the effect of much higher confining pressure. The intermediate 6 m of liquefiable sand were simulated with the NTUA-SAND model, with the values of model constants as proposed by Andrianopoulos et al. (2010a, 2010b) for Nevada sand. The piles (and the pile cap) were considered elastic, while the properties of the piles were chosen so as to comply

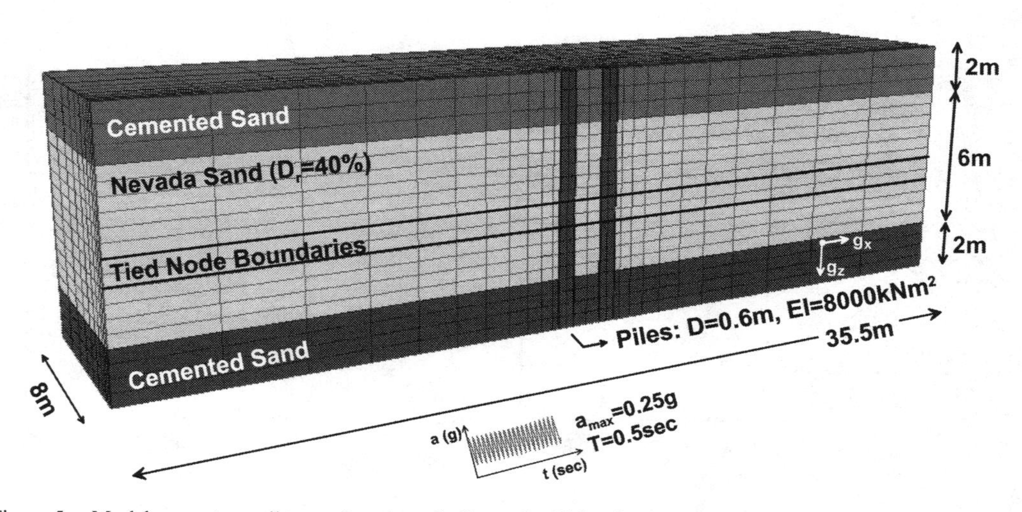

Figure 5. Model geometry, soil types, location of piles and grid for the numerical simulation of the dynamic centrifuge test of Pamuk et al. (2007).

with the value of EI = 8000 kNm^2 mentioned by Pamuk et al. (2007). The soil-pile interface angle was assumed equal to 36°.

The model is subjected to a base acceleration consisting of 32 sinusoidal cycles of 0.20 g peak acceleration at a period T = 0.5 sec. Nodes at the lateral boundaries that are at the same height are tied to one another in order to have the same horizontal and vertical displacement, thus simulating the use of a laminar box in the centrifuge test. The mild inclination of the laminar box was taken into account by rotating slightly the vector of acceleration of gravity. Two (2) centrifuge tests were executed, one with the sand profile left untreated (CS(%) = 0) and a second one with the sand profile being stabilized with Ludox®-SM colloidal silica at a concentration per weight of CS(%) = 5. Correspondingly, two (2) simulations were performed, one with the soil layer left untreated (i.e. by considering water as the pore fluid, $K = K_w$) and another where the soil layer was passively stabilized. Following the rationale of Andrianopoulos et al. (2016), this second analysis was run by employing NTUA-SAND with the model constants for Nevada sand, but with a reduced pore fluid modulus $K = K_w/975$, i.e. as was performed in the experiment of Conlee et al. (2012) above.

In the experiment of Conlee et al. (2012), the same test studied both the response of the untreated soil, as well as that of the stabilized one, and Figures 3 and 4 showed the accuracy in the simulation of both. In the study of Pamuk et al. (2007), two (2) separate tests were run, and hence here the validation of the test for the untreated sand which undergoes lateral spreading is shown first in Figure 6. Particularly, Fig. 6a shows the time history of the horizontal displacement d of the pile cap on the basis of the experiment (thin line) and the simulation (thick line). Then, Figs. 6b and 6c compare recordings to simulations for the horizontal diplacements d with depth z of the soil profile at the free field (far from the pile group) at 2 distinct time instances: t = 6 sec (in the first part of the shaking) and t = 17 sec (at the end of shaking). Finally, Figs. 6d and 6e compare recordings to simulations for the (upslope) pile bending moments M at the foregoing time instances.

Based on Figs. 6b and 6c it is deduced that the soil undegoes severe lateral spreading, with displacements that accumulate during shaking and reach approximately 70 cm at the end of shaking. These horizontal displacements are due to the middle 6 m-thick layer of the soil profile that liquefies, since the top and bottom 2 m do not contribute to the soil's overall displacement. These characteristics of the soil response are typical for lateral spreading (e.g. Valsamis et al. 2010) and are satisfactorily simulated by the model, despite the overprediction of the liquefied sand layer displacements at the later stages of the shaking.

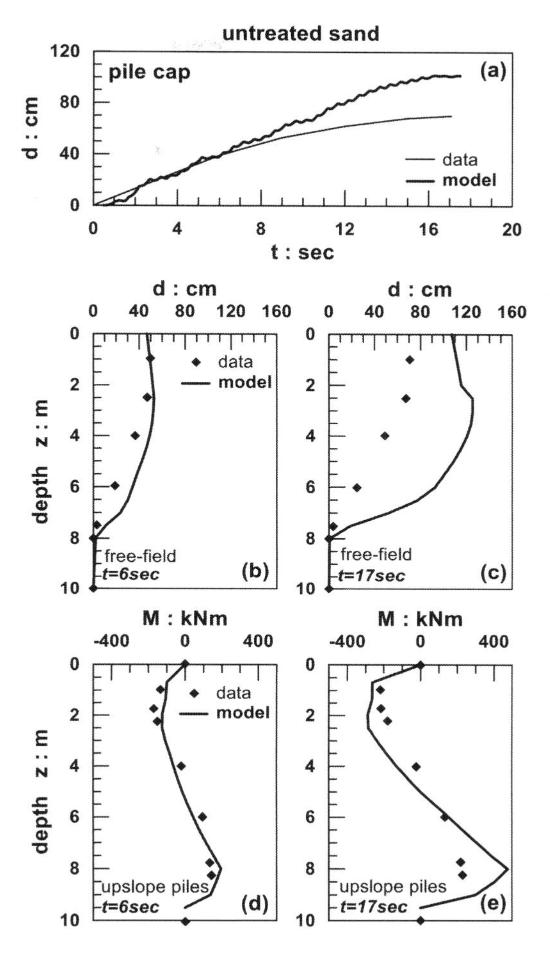

Figure 6. Comparison of recordings to simulations in terms of pile cap displacements, free-field soil displacements with depth and bending pile moments with depth for the dynamic centrifuge test of Pamuk et al. (2007) with the untreated sand layer.

The foregoing significant free field soil displacements drag the pile group towards the downslope direction, leading to significant pile cap displacements, which are somewhat overpredicted in the later stages of the shaking (see Fig. 6a). In turn, these pile cap displacements lead to significant bending of the piles, since their tips remain in place. As expected, the peak bending moments M increase with time and appear at the intefaces of the liquefiable layer with the top and bottom layers, and this is satisfactorily captured by the model predictions.

Fig. 7 presents a comparison of model predictions to recordings for the second centrifuge test, where the soil profile is stabilized with colloidal silica. To ease the interpretation and enable a direct comparison with the previous test, the format of Fig. 7 is identical to Fig. 6, with the exception of the scales of displacements and bending moments in the 2 figures.

Based on Figs. 7b and 7c it is deduced that the soil undegoes very small displacements that reach just 3 cm at the end of shaking. Again, it is mostly the middle 6 m that contribute to the overall soil displacement, although the relative contribution of the various layers is not clear as the stabilized layer seems to behave similarly to the lightly cemented top and bottom layers. As an effect, the pile cap does not undergo significant displacements (Fig. 7a), and hence pile bending moments are insignificant as well. These significant benefits of passive stabilization in reducing displacements and pile bending moments are very satisfactorily captured by the similation, especially until t = 12 sec. In the later stages of shaking (t = 12–17 sec), soil and pile displacements are somewhat over-predicted, and so are the induced pile bending moments.

However, overall, the simulation of this quite complicated centrifuge test is very satisfactory. In closure note that an over-prediction of displacements in the later stages of shaking is observed in both Figs. 6 and 7, which means that the relative effect of stabilization is satisfactorily simulated throughout the shaking. This problem in accuracy is attributed to the simulation of the response of top and bottom layers, which is necessarily simplistic given the limited related information included in Pamuk et al. (2007).

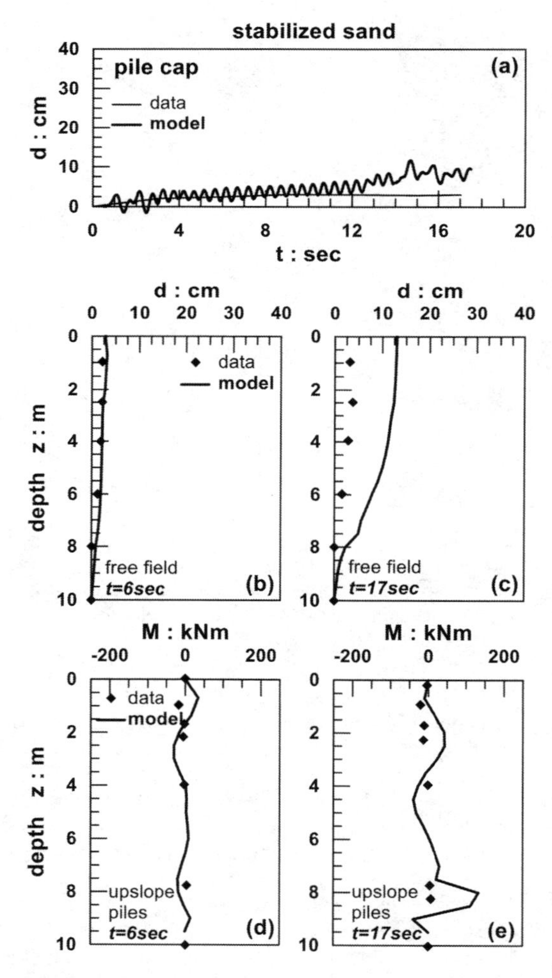

Figure 7. Comparison of recordings to simulations in terms of pile cap displacements, free-field soil displacements with depth and bending pile moments with depth for the dynamic centrifuge test of Pamuk et al. (2007) with the stabilized sand layer.

5 CONCLUSIONS

The following conclusions are drawn from this study:

- A decrease of the pore fluid bulk modulus K can constitute a simple framework for the fully coupled analysis of dynamic boundary value problems involving stabilized sands. This framework is generic and can be used with any constitutive model that captures the liquefaction response for sands.
- Passive (site) stabilization can effectively mitigate settlements and lateral spreading displacements of otherwise liquefiable soils. As a result, this ground improvement method effectively reduces horizontal displacements and related bending moments of piles within a laterally spreading layer.

ACKNOWLEDGMENTS

This research has been co-financed by the European Union (European Social Fund ESF) and Greek national funds through the Operational Program "Education and Lifelong Learning" of the National Strategic Reference Framework (NSRF)–Research Funding Program: Thales. Investing in knowledge society through the European Social Fund.

Partial funding from C&M Engineering S.A. is acknowledged.

Partial funding from the European Research Council under the European Union's Seventh Framework Program FP7-ERC-IDEAS Advanced Grant Agreement no. 290963 (SOMEF) is acknowledged.

The last author acknowledges sabbatical leave of absence from the CEE Department of the University of California, Davis, that allowed him to participate in the research.

REFERENCES

Andrianopoulos, K.I., Agapoulaki, G.I., and Papadimitriou, A.G. (2016). "Simulation of seismic response of passively stabilised sand." *Geotechnical Research, ICE Publishing*, 3(2), 40–53.

Andrianopoulos, K.I., Papadimitriou, A.G., and Bouckovalas, G.D. (2010). "Bounding surface plasticity model for the seismic liquefaction analysis of geostructures." *Soil Dynamics and Earthquake Engineering*, 30(10), 895–911.

Andrianopoulos, K.I., Papadimitriou, A.G., and Bouckovalas, G.D. (2010), "Explicit integration of bounding surface model for the analysis of earthquake soil liquefaction", *International Journal for Numerical and Analytical Methods in Geomechanics*, 34(15): 1586–1614.

Conlee, C.T., Gallagher, P.M., Boulanger, R.W., and Kamai, R. (2012). "Centrifuge Modeling for Liquefaction Mitigation Using Colloidal Silica Stabilizer." *Journal of Geotechnical and Geoenvironmental Engineering*, ASCE, 138(11), 1334–1345.

Gallagher, P.M. (2000). "Passive site remediation for mitigation of liquefaction risk." *PhD Thesis*, Virginia Polytechnic Institute and State Univ., Blacksburg Va.

Gallagher, P.M., and Lin, Y. (2009). "Colloidal Silica Transport through Liquefiable Porous Media." *Journal of Geotechnical and Geoenvironmental Engineering*, ASCE, 135(11), 1702–1712.

Gallagher, P.M., and Mitchell, J.K. (2002). "Influence of colloidal silica grout on liquefaction potential and cyclic undrained behavior of loose sand." *Soil Dynamics and Earthquake Engineering*, 22(9–12), 1017–1026.

Itasca Consulting Group, Inc. (2011). *"FLAC - Fast Langrangian Analysis of Continua, Ver. 7.0."* Minneapolis: Itasca.

Itasca Consulting Group, Inc. (2012). "FLAC^{3D} - Fast Langrangian Analysis of Continua in three dimensions, Ver. 5.0." Minneapolis: Itasca.

Kamai, R., and Boulanger, R.W. (2013). "Simulations of a Centrifuge Test with Lateral Spreading and Void Redistribution Effects." *Journal of Geotechnical and Geoenvironmental Engineering*, ASCE, 139(8), 1250–1261

Kodaka, T., Oka, F., Ohno, Y., Takyu, T., and Yamasaki, N. (2005). "Modelling of cyclic 678 deformation and strength characteristics of silica treated sand." In *Proceedings, 1st Japan – US Workshop on Testing, Modelling, and Simulation* (GSP 143)

Pamuk A., Gallagher P.M., and Zimmie T.F. (2007). "Remediation of piled foundations against lateral spreading by passive site stabilization technique", *Soil Dynamics and Earthquake Engineering*, 27(9), 864–874.

Towhata, I. (2007). "Developments of soil improvement technologies for mitigation of liquefaction risk." *Geotechnical, Geological and Earthquake Engineering*, 6, 355–383.

Valsamis, A., Bouckovalas, G.D., and Papadimitriou, A.G. (2010), "Parametric investigation of lateral spreading of gently sloping ground", *Soil Dynamics and Earthquake Engineering*, 30(6): 490–508, June

Numerical Methods in Geotechnical Engineering IX – Cardoso et al. (Eds)
© 2018 Taylor & Francis Group, London, ISBN 978-1-138-33198-3

Artificial neural networks for the evaluation of impedance functions of inclined pile groups

A. Franza & M.J. DeJong
Department of Engineering, University of Cambridge, Cambridge, UK

M. Morici
SAAD, University of Camerino, Ascoli Piceno, Italy

S. Carbonari
DICEA, Universitá Politecnica delle Marche, Ancona, Italy

F. Dezi
DESD, University of San Marino, San Marino

ABSTRACT: This paper presents the application of Artificial Neural Networks (ANNs) to the impedance functions of inclined pile groups. Firstly, the dataset required for the training and testing of the ANNs is obtained through a numerical FE model. A parametric investigation is carried out for the frequency-dependent impedance functions of 2 × 2 pile groups rigidly connected at the head and embedded in homogeneous soil deposits; soil-foundation systems with different geometric characteristic and properties are investigated (e.g. inclination angles, pile-soil stiffness ratios, pile length-to-diameter ratios). Subsequently, several ANNs consisting of two hidden layers are trained, tested, and validated to optimise the generalisation performance of the model. Results show that ANNs are able to capture the trends of the impedance functions of inclined pile groups on the basis of the adopted dimensionless parameters. In addition, the complexity of the ANN models achieving good performance confirms the need for advanced regression models.

1 INTRODUCTION

Inclined piles may be used to resist large lateral loads under static conditions. Recent research indicated that inclined piles have potential for good seismic performances and that previous failures during earthquakes were mostly due to an improper design (Carbonari et al. 2017, Giannakou et al. 2010, Padrón et al. 2010). Therefore, engineers should take into account the opportunity of using inclined pile in seismic design. Although several numerical models have been developed to investigate the dynamic response of inclined piles and to perform seismic soil-foundation-structure interaction analysis, their use is limited in the practice. In addition, there is a lack of time-efficient tools that can be implemented within inertial soil-structure interaction analyses to bypass the numerical modelling of the dynamic behaviour of the foundation. In previous research, Artificial Neural Networks (ANNs) have been successfully adopted to dynamic soil-structure interaction problems (Ahmad et al. 2007, Badreddine and Goudjil 2012, Deng et al. 2017, Farfani et al. 2015, Pala et al. 2008). In particular, Ahmad et al. (2007) successfully estimated kinematic soil pile interaction response parameters of vertical isolated piled. ANNs can learn from a given set of training dataset identifying complex and non-linear patterns without requiring the definition of deterministic relationships, as in the case of curve-fitting using user-defined relationships. ANNs have general suitable potentials for their use as a prediction tool Benardos and Vosniakos (2007). ANNs have been applied by researchers as a tool for the development of predictive models on various geotechnical problems but, to the Author knowledge, they have never been used to predict the dynamic behaviour of pile groups. The aim of this paper is to develop artificial neural networks (ANNs) of different complexity that can directly generate the impendence functions of inclined pile groups. The dataset required for the training and testing of the ANNs is obtained using the numerical model of Dezi et al. (2016).

2 STUDIED CONFIGURATIONS

The impedance functions of 2 × 2 pile groups rigidly connected by a cap at the head, embedded in homogeneous soil deposits, and characterised by two symmetry axes are considered. The configuration is sketched in Figure 1. In the framework of the sub-structure approach to solve soil-structure interaction problems, foundation impedances are the force-displacement relationships that characterise the compliant restraints to be used in the inertial analysis of the superstructure. Thus, impedances should be referred to the soil-superstructure interface. For the applications of the present paper, impedance functions are evaluated at the centroid of the pile cap, at the level of the piles head, where the master node of the rigid foundation cap is located.

In the parametric study, geometric and mechanical properties of case studies are provided in non-dimensional form. In particular, the following non-dimensional parameters govern the dynamic response of the soil-foundation systems: dimensionless frequency $a_0 = \omega d/V_s$, inclination angle θ, pile-soil Young's modulus ratios E_p/E_s, pile spacing-to-diameter ratio s/d, and pile length-to-diameter ratio L/d, pile-to-soil density ratio ρ_p/ρ_s, where ω is the frequency, V_s is the soil shear wave velocity. The range of investigate input parameters is reported in Table 1, which covers most of the practical scenarios. Particularly, the parametric study was carried out for a_0 = [0: 0.05: 1], θ = [0; 5; 10; 15; 20; 25; 30], E_p/E_s = [50; 75; 100; 200; 500; 1000; 3000; 10000], s/d = [2; 3; 4; 5; 6; 7], L/d = [8: 2: 30], and ρ_p/ρ_s = [1.5; 1.6; 1.7; 1.8; 1.9; 2.0]. In addition, the soil Poisson's ratio ν_s = 0.4, the soil hysteretic damping ratio ξ_s = 0.05, and the pile Poisson's ratio ν_p = 0.2.

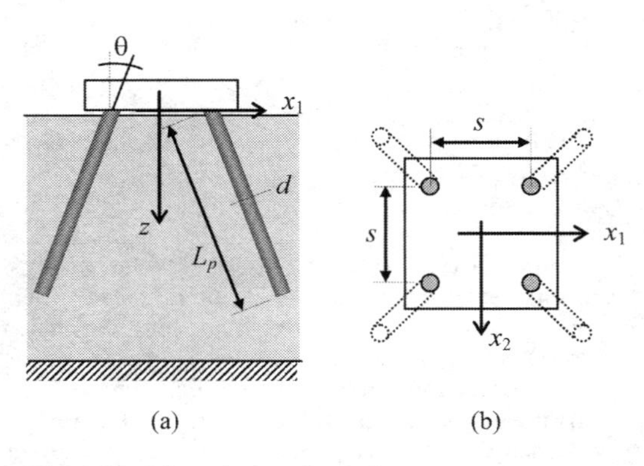

Figure 1. Pile group configuration.

Table 1. Ranges of investigated input parameters.

	$\omega d/V_s$	θ	s/d	L/d	E_p/E_s	ρ_p/ρ_s
min	0	0	2	8	50	1.5
max	1	30	7	30	10000	2.0

3 FINITE ELEMENT MODEL

The dataset used to train and evaluate the performance of ANNs is obtained through a numerical model developed by Dezi et al. (2016), which solves the kinematic interaction analysis of inclined pile groups in the frequency domain.

Analytically, the piles and soil are modelled by means of Euler-Bernoulli beams and independent horizontal infinite layers, respectively. The model accounts for the pile-soil-pile interaction and the radiation problem through the use of elastodynamic Greens functions, available in the technical literature (for more details refer to Dezi et al. (2016)). The constraint of the rigid cap is analytically imposed through a geometric matrix that defines the relationships between displacements of pile heads and the cap master node. The solution of the problem is achieved numerically by means of the finite element method in the displacement-based approach. Figure 2 schematically depicts the soil-foundation model and the reference system to which impedances will be referred in the sequel. The dynamics of the soil-foundation system is governed by the following complex linear equation system:

$$\begin{bmatrix} \mathbf{Z}_{FF} & \mathbf{Z}_{FE} \\ \mathbf{Z}_{EF} & \mathbf{Z}_{EE} \end{bmatrix} \begin{bmatrix} \mathbf{d}_F \\ \mathbf{d}_E \end{bmatrix} = \begin{bmatrix} \mathbf{f}_F \\ \mathbf{f}_E \end{bmatrix} \tag{1}$$

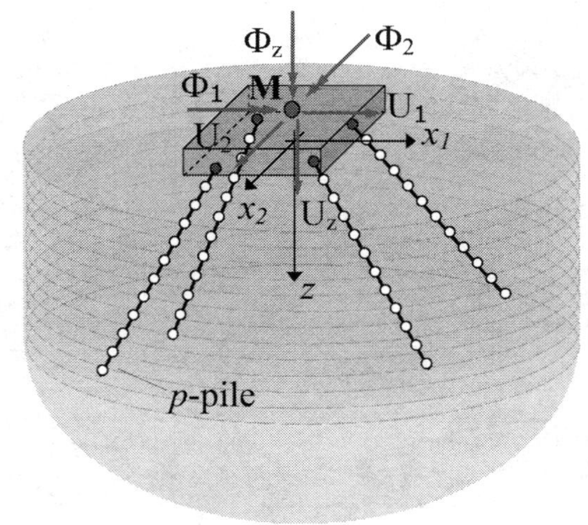

Figure 2. Pile group model (Dezi et al. 2016).

$$\begin{bmatrix} \mathbf{f}_F \\ \mathbf{f}_E \end{bmatrix} = \mathbf{A}^T\mathbf{f} \tag{2}$$

$$\begin{bmatrix} \mathbf{Z}_{FF} & \mathbf{Z}_{FE} \\ \mathbf{Z}_{EF} & \mathbf{Z}_{EE} \end{bmatrix} = \mathbf{A}^T\left(\mathbf{K}_p - \omega^2\mathbf{M}\mathbf{K}_s\right)\mathbf{A} \tag{3}$$

where $\mathbf{K}_P$ is the global stiffness matrix of the piles, $\mathbf{M}$ is the mass matrix of the piles, $\mathbf{K}_S$ is the global impedance matrix of soil, $\mathbf{A}$ is geometric matrix imposing the rigid connection at the pile heads, $\mathbf{d}$ is the displacement vector of the pile group, $\mathbf{f}$ is the the vector of external loads due to the free-field motion, $\mathbf{d}_F$ and $\mathbf{d}_E$ are the displacement vectors of the cap (the master node has six generalised components) and of the embedded piles, respectively.

By manipulating Equation (1), the complex-valued foundation impedance matrix ($\mathbf{J}$) and the foundation input motion ($\mathbf{d}_F$), namely the actual displacements of the master node as a consequence of seismic wave propagation in the soil (free-field motion), are obtained:

$$\mathbf{J}(\omega) = \left(\mathbf{Z}_{FF} - \mathbf{Z}_{FE}\mathbf{Z}_{EE}^{-1}\mathbf{Z}_{EF}\right) \tag{4}$$

$$\mathbf{d}_F(\omega) = \mathbf{J}^{-1}\left[\mathbf{f}_F - \mathbf{Z}_{FE}\mathbf{Z}_{EE}^{-1}\mathbf{f}_E\right] \tag{5}$$

Because of the symmetry of the considered foundation configuration and the selected position of the master node, the non-null components of the impedance matrix, expressed in dimensionless form, are $J_z/(E_sd)$, $J_x/(E_sd)$, Jr/(E_sd^3), and $J_{xr}/(E_sd^2)$, where subscripts z and x indicate the vertical and horizontal directions, respectively, while r indicates the rotational degree of freedom. It is worth noting that the impedance matrix includes a coupled roto-translational term (J_{xr}).

4 ARTIFICIAL NEURAL NETWORK MODELS

4.1 *Overview*

In general, feedforward ANNs are models consisting of interconnected neurons arranged in a network with a layered topology. The first layer allocates the network input (corresponding to independent variables), the final layer produces the model outputs. One or more hidden (intermediate) layers are also present with a given number of neurons defined by the ANN architecture. Each neuron is connected to all the nodes in the previous and following layers. Each link between neurons consists of a weight (that multiplies the signal) and a bias. Each neuron at the hidden and output layers adds up the signals received from the previous layer and applies an activation function to the signal. The combination of weights and layered connections allows description of highly non-linear relationships between inputs and outputs. The training phase of this model can be achieved, for a given series of input values, with a back-propagation technique that consists of adjusting the weights and biases of the connections to obtain a satisfactory match between the model output at the final layer and the target data (i.e. the actual data). As an example, Figure 3 displays an architecture with 4 input parameters $P_{i,i}$, two hidden layers with 5 neurons, and an output layer with 3 neurons for the output parameters $P_{o,i}$ (i.e. $4 \times 5 \times 5 \times 3$).

4.2 *Model development*

The topology of an ANN implemented in this paper consists of two hidden layers. Input and output parameters are expressed in the dimensionless form to reduce their number and, thus, reduce both the size of the required FEM parametric study and the complexity of the ANN's architecture. Six input parameters $P_{i,i}$ (a_0, θ, s/d, L/d, Log$[E_p/E_s]$, ρ_p/ρ_s) and eight output parameters $P_{o,i}$ (both real and imaginary parts of the non-null impedance components $J_z/(E_s,d)$, $J_x/(E_s,d)$, $J_r/(E_s,d^3)$, and $J_{xr}/(E_s,d^2)$) are considered. Rather than defining eight ANNs (one for each output parameter with $6 \times n \times n \times 1$ architectures, where n is the number of hidden neurons), in this work a single architecture ($6 \times n \times n \times 8$) is adopted to describe the dependence of the entire impedance matrix on the dimensionless input parameters.

Different activation functions were used in the input, hidden, and output layers. The identity function, the continuous non-linear tan-sigmoid transfer function, and the linear transfer function were used, respectively, in the input layer, the hidden layer, and the output layer. Prior to training, weight and bias values were initialised to a starting

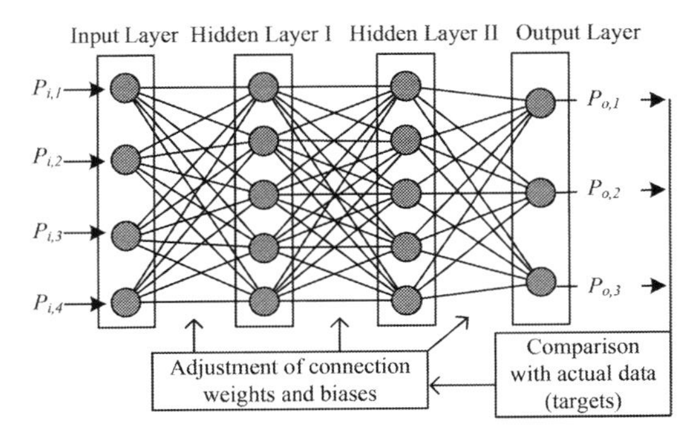

Figure 3. Example of a feedforward ANN architecture.

point using the Nguyen-Widrow method (Nguyen and Widrow 1990). The training phase was carried out avoiding the over-fitting of the training data to achieve a good generalisation performance. During training, the Levenberg-Marquardt backpropagation algorithm (Hagan and Menhaj 1994) along with early stopping was adopted to adjust the connection weights and biases, whereas the MSE (the mean squared error) was used as a performance function. Note that scaled conjugate gradient backpropagation and resilient backpropagation algorithms were also tested; despite a lower computational time than the Levenberg-Marquardt algorithm, the performance of the resulting ANNs was not satisfactory.

For learning and validation purposes, the pre-processed input data were randomly divided into training, validation, and testing subsets with a respective ratio of 0.7, 0.15, and 0.15. The training subset was used for training the ANN, whereas the validation subset was used for the early stopping feature of the Levenberg-Marquardt algorithm (to avoid overfitting). At the conclusion of the training phase, the testing subset was used to assess the generalisation performance of the model by comparing model outputs corresponding to the testing inputs to the target data, which are unknown to the ANN. Note that identical subset partitions (for the training, testing, and validation) were used for all ANNs.

4.3 *Evaluation of training and generalisation performance*

Firstly, the performance of architectures with varying complexity is evaluated in terms of coefficient of determination (R^2). R^2 associated with both response (training) and generalisation (testing) of six architectures with two hidden layers are displayed in Figure 4. R^2 close to unity is associated with a good fit between the model outputs and the targets given by the FEM model.

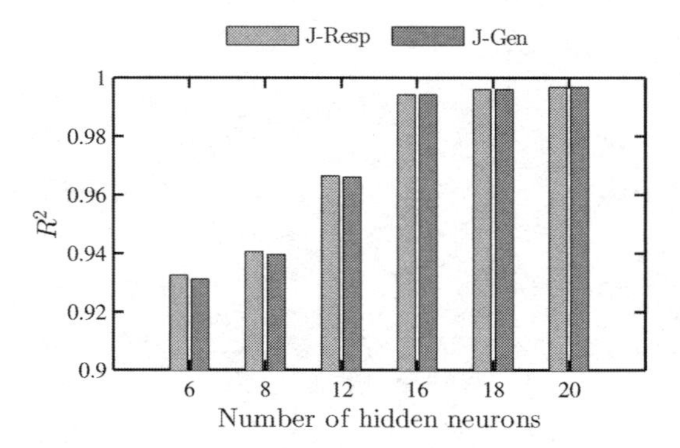

Figure 4. Performance of ANNs.

Figure 4 shows that the ability of the ANNs to match the training dataset improves with the number of neurons. In addition, the generalisation and training performance is comparable, which indicates that the complexity of the architectures is adequate and does not result in the over-fitting of the training dataset. Therefore, the architecture with 20 hidden neurons (($6 \times 20 \times 20 \times 8$)) is selected and its predictions evaluated in the following.

The target data ($J_z/(E_s,d)$, $J_x/(E_s,d)$, $J_r/(E_s,d^3)$, and $J_{xr}/(E_s,d^2)$) used for the training and testing (which are, respectively, 70% and 15% of the entire dataset) are plotted against the predictions of the chosen ANN in Figure 5 to display the response and generalisation performance. Interestingly, despite $R^2 \approx 1$, the ANN predictions do not perfectly match the target FEM results. Therefore, it is interesting to analyse the performance of the chosen ANN with respect to each non-null component of the impedance matrix, as displayed by Figure 6. In Figure 6, the relationship between predictions and target data is reported in terms of generalisation capability for the testing sub-set. Results show that a low degree of error between predictions and targets was achieved by the ANN for the vertical impedance values (J_z) and roto-translational impedance (J_{xr}), whereas the prediction of translational (J_x) and rotational (J_r) impedance components is better.

4.4 *Prediction of impedance trends with dimensionless frequency*

Although the previous section displayed that the statistical evaluation of the ANN's performance is useful to identify an appropriate architecture, the coefficient of determination only describe the magnitude of the impedance and it provides no information on the impedance trend with the dimensionless frequency a_0. In order to be

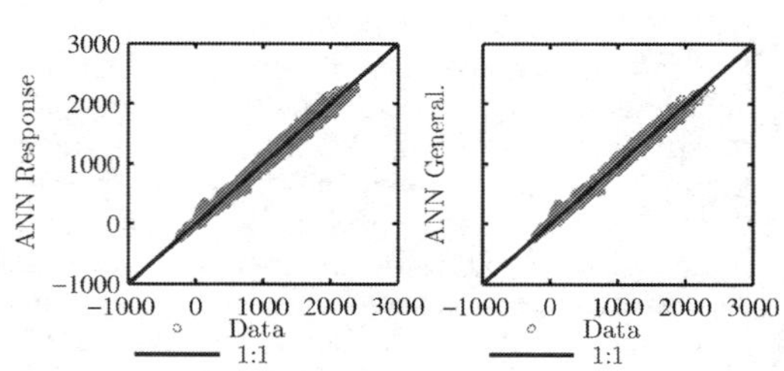

Figure 5. Predicted versus testing (left) and validation (right) impedance for the architecture $6 \times 20 \times 20 \times 8$.

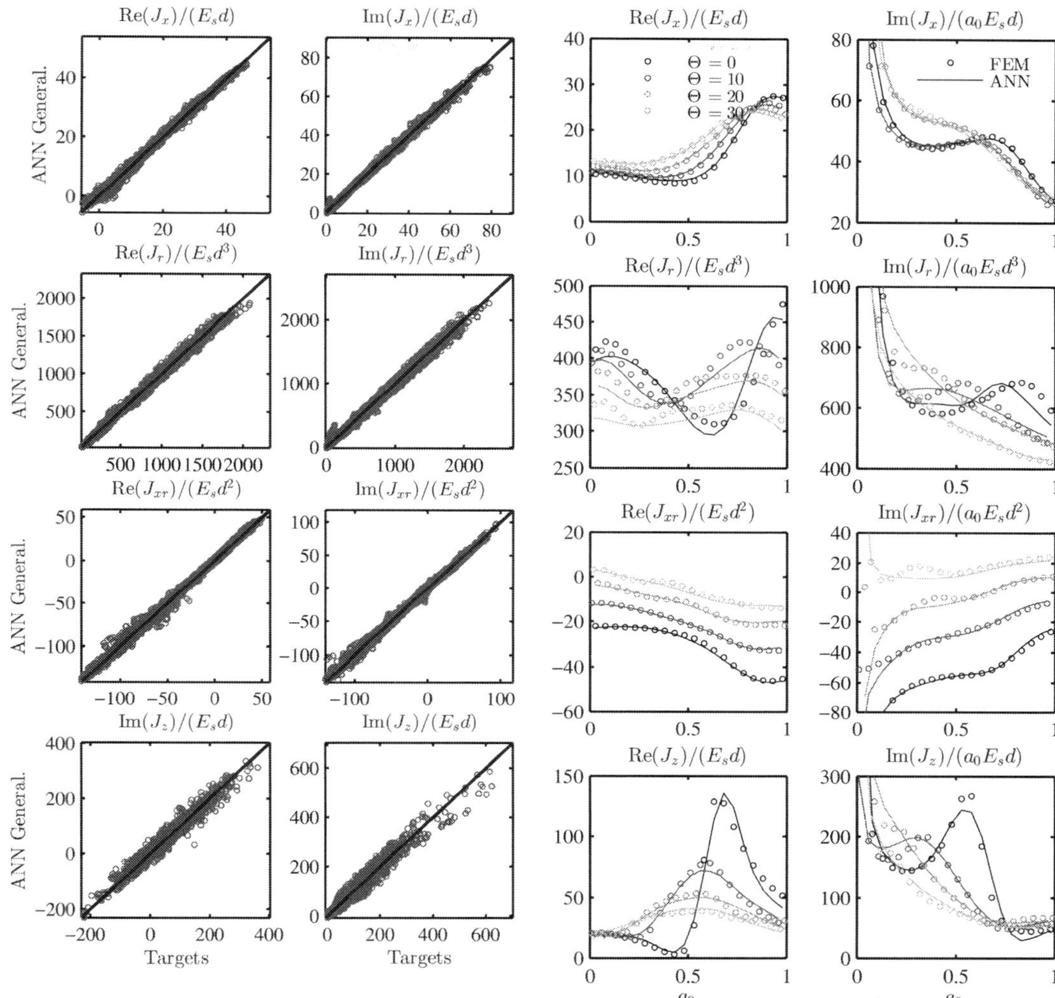

Figure 6. Impedance components: generalisation performance of the ANN (6 × 20 × 20 × 8).

Figure 7. Non-null components of impedance matrix for configurations $\theta = 0$, 10, 20, 30, $E_p/E_s = 10^3$, $s/d = 5$, $L/d = 14$, and $\rho_p/\rho_s = 1.5$.

applicable by practitioners, the predictions of the ANNs need to satisfactorily capture the variation of the impedance **J** with a_0.

For brevity, only results corresponding to pile rake angles $\theta = 0$; 10; 20; 30 are displayed in Figure 7 for $E_p/E_s = 10^3$, $L/d = 14$, $\rho_p/\rho_s = 1.5$ and $s/d = 5$. The agreement between the FEM target data (lines) and the ANN predictions (markers) is satisfactory. Despite minor scatter (e.g. $\mathrm{Re}(J_r)/(E_s d^3)$ for low a_0 values in Figure 7), the ANN is able to capture the full complexity of the impedance dependency on the frequency. Also note that the scatter between the ANN's predictions and the FEM model data is within the variation resulting from modifying the approach/assumptions of the 3D numerical model (Dezi et al. 2016).

5 CONCLUSIONS

This paper displayed that feedforward ANNs can be trained to efficiently predict the frequency dependent impedance functions of deep foundations with vertical and inclined piles, which were obtained from a parametric study using a Winkler-based FEM model. The chosen ANN was able to predict the frequency-dependent impedance considering the effects of all main dimensionless geometrical and mechanical parameters governing the dynamic kinematic problem. In particular, accurate trends of the impedance components with frequency were obtained; this achievement was important to allow for the use of ANNs in the engineering practice. However, it should be noted that, to achieve the desired accuracy, ANNs having multiple output parameters (equal to the components of the impedance matrix) require complex architectures; coefficient of determinations R^2 for both the ANN generalisation and response performance should be close to unity.

This work was limited to a single pile group configuration (2 × 2) and a unique stratigraphy (homogeneous deposit). Future works will deal with a greater number of possible scenarios and the estimation of both the foundation input motion and the impedance values.

REFERENCES

Ahmad, I., M. Hesham El Naggar, & A.N. Khan (2007). Artificial neural network application to estimate kinematic soil pile interaction response parameters. *Soil Dynamics and Earthquake Engineering 27*(9), 892–905.

Badreddine, S. & K. Goudjil (2012). Prediction of Dynamic Impedances Functions Using an Artificial Neural Network (ANN). In *Applied Mechanics and Materials*, Volume 170–173, pp. 3588–3593.

Benardos, P.G. & G.-C. Vosniakos (2007). Optimizing feedforward artificial neural network architecture. *Engineering Applications of Artificial Intelligence 20*(3), 365–382.

Carbonari, S., M. Morici, F. Dezi, F. Gara, & G. Leoni (2017). Soil-structure interaction effects in single bridge piers founded on inclined pile groups. *Soil Dynamics and Earthquake Engineering 92*, 52–67.

Deng, H., M. Gu, & X. Jin (2017). Research on the dynamic properties of piled structures using the neural networks and the support vector machines. In *IABSE Symposium Vancouver 2017: Engineering the Future*, Volume 109, pp. 2141–2146.

Dezi, F., S. Carbonari, & M. Morici (2016). A numerical model for the dynamic analysis of inclined pile groups. *Earthquake Engineering & Structural Dynamics 45*(1), 45–68.

Farfani, H.A., F. Behnamfar, & A. Fathollahi (2015). Dynamic analysis of soil-structure interaction using the neural networks and the support vector machines. *Expert Systems with Applications 42*(22), 8971–8981.

Giannakou, A., N. Gerolymos, G. Gazetas, T. Tazoh, & I. Anastasopoulos (2010). Seismic Behavior of Batter Piles: Elastic Response. *Journal of Geotechnical and Geoenvironmental Engineering 136*(9), 1187–1199.

Hagan, M.T. & M.B. Menhaj (1994). Training feedforward networks with the Marquardt algorithm. *IEEE transactions on Neural Networks 5*(6), 989–993.

Nguyen, D. & B. Widrow (1990). Improving the learning speed of 2-layer neural networks by choosing initial values of the adaptive weights. In *Proceedings of the International Joint Conference on Neural Networks*, pp. 21–26. IEEE.

Padrón, L.A., J.J. Aznárez, O. Maeso, & A. Santana (2010). Dynamic stiffness of deep foundations with inclined piles. *Earthquake Engineering & Structural Dynamics 39*(12), 1343–1367.

Pala, M., N. Caglar, M. Elmas, A. Cevik, & M. Saribiyik (2008). Dynamic soil–structure interaction analysis of buildings by neural networks. *Construction and Building Materials 22*(3), 330–342.

Numerical Methods in Geotechnical Engineering IX – Cardoso et al. (Eds)
© 2018 Taylor & Francis Group, London, ISBN 978-1-138-33198-3

3-D source-to-site numerical investigation on the earthquake ground motion coherency in heterogeneous soil deposits

F. Gatti, S. Touhami & F. Lopez-Caballero
Laboratoire Mécanique des Sols, Structures et Matériaux (MSSMat) UMR CNRS, CentraleSupélec, France

D. Pitilakis
Department of Civil Engineering, Aristotle University of Thessaloniki, Greece

ABSTRACT: The effect of soft sediments on the dispersion of the earthquake ground motion is numerically investigated herein, to assess the effects of (i) the 3-D geological interfaces and (ii) of the spatial fluctuations of the mechanical properties. A source-to-site computational model is built-up, for a complex applicative case, configured as a 3-D soft basin embedded in outcropping bedrock. The transient wave-field is accurately computed by means of SEM3D, based on the spectral element method in elastodynamics and vectorized over large parallel supercomputers, for efficient scalability. Broad-band (0–7 Hz) earthquake simulations at regional scale (tens of kilometers) is performed, including the irregular edges of the basin. The spatial fluctuation of the shear modulus is integrated into the model as a multi-variate stationary random field, sampled at the computational nodes. The effect of soil heterogeneity is compared to the homogeneous counterpart, to assess the influence of the basin on the ground motion coherency.

1 INTRODUCTION

1.1 *A holistic approach to earthquake simulation*

In the past two decades, the seismic hazard analysis and vulnerability assessment took progressively advantage of the ever-increasing computational power available (Paolucci et al. 2014). This outstanding technological and numerical progress seemingly broke through the evergreen and most stringent bottleneck in computational seismology: the impossibility to solve the complete source-to-site seismic wave propagation problem in a single-step analysis. All the ingredients (i.e. source, path and siteeffects) can nowadays naturally be convolved in one-step all-embracing analysis, capable of predicting a realistic seismic wave-field and to explain the observed time-histories in sedimentary deposits (local scale, i.e. approximatively 1–10 km of characteristic length) and/or at the continental scale (i.e. 100 km or greater, De Martin 2011). Once shattered these computational barriers, a new *holistic* philosophy took place, driven by the deterministic modelling of the physics underlying each aspect of the earthquake phenomenon, for more accurate sensitivity analyses and uncertainty quantification of models and related parameters.

In spite of the inherent complexity and the huge dimensions of those computational models, their power is essentially embodied by the higher broad-band accuracy they provide (i.e. up to 4–5 Hz, De Martin 2011), gradually bridging the gap between low-frequency source models obtained via wave-form inversion techniques and the structural modal frequencies (i.e. up to 20 Hz). This achievement is paving the way to fully couple the large scale seismological models for the region of interest, with local engineering models for geotechnical, site-effect and structural analyses, in the next few years.

1.2 *Broad-band source-to-site simulation*

A major challenge related to the high-fidelity earthquake numerical simulation is represented by the accuracy of the predicted wave-motion at the frequency of interest. Unfortunately, due to the well-known spatial aliasing of the computational grid, major computational efforts are required to enlarge the wave-field frequency bandwidth propagated with accuracy, since finer meshes are required. Moreover, the simulation of realistic ground shaking scenarios in a broad-band frequency range (BB2S2S) requires are liable estimation of several different parameters, related to the source mechanism, to the geological configuration and to the mechanical property of the soil layers and crustal rocks. The great impedance contrast between soft sedimentary layers and crustal bedrock entail the need for smaller time steps, i.e. increasing the overall CPU-time required to simulate realistic time-histories (e.g. of approximately 30 s), containing the P- and S- wave strong phases, as long as the *coda*-waves. Finally, due to the enormous extension of those regional scale scenarios, the degree of uncertainty associated to the whole earthquake process (from fault to site) is extremely high.

At this point, it appears necessary to build up a multi-tool HPC-platform (High Performance Computing) capable to tackle the following issues:

1. to mesh the domain of interest, its geological conformation (bedrock to sediment geological surfaces), the surface topography and the bathymetry (if present)
2. to represent the material rheology (i.e., elastic, viscous-elastic, non-linear hysteretic)
3. to describe the natural heterogeneity of the Earth's crust and soil properties, at different scales (i.e., regional geology, local basin-type structures and heterogeneity of granular materials)

To this end, a plethora of extremely efficient numerical methods, for instance, the Finite Difference Method (FDM, Graves 1996), the Finite Element Method (FEM, Taborda and Bielak 2011) and the Spectral Element Method (Faccioli et al. 1997, Komatitsch and Vilotte 1998, i.e. the two seminal papers) has been progressively applied to the seismological problem. Several numerical benchmarks have been recently performed during the past years, with the aim at testing the accuracy and the efficiency of different simulation methods when handling complex domain geometries and material interfaces (Chaljub et al. 2015).

1.3 *The Mygdonian basin test case (greece)*

The Mygdonian basin is located 30 km east-northeast of Thessaloniki, northeastern Greece (see Fig. 1a), at the center of the sedimentary basin between the Volvi and Lagada lakes, in the epicentral area of the magnitude 6.5 event that occurred in 1978 and damaged the city of Thessaloniki.

The Mygdonian basin represents an experimental site of several different investigation campaigns since a decade ago (Pitilakis et al. 2009, Pitilakis et al. 2013, among others). Therefore it has been instrumented with surface and borehole accelerometers, and the 3-D elevation and geological models are available (see Figs. 1b-1c, Manakou et al. 2010). Maufroy et al. 2015 performed verification/validation tests at the Mygdonian basin by running 3-D earthquake simulations between 0.7–4 Hz, addressing the description of the material heterogeneities, of the attenuation model, the approximation of the free surface, and of the absorbing boundaries. The predictions well reproduce some, but not all, features of the actual site effect, due to the uncertainty on the source parameters (location, hypocentral depth, and focal mechanism), on the description of the geological medium (damping, internal sediment layering structure, and shape of the sediment-basement interface). Another reference work is represented by the numerical benchmark presented in Chaljub et al. 2015, where the Mygdonian basin was chosen as reference site to compare different wave-propagation codes for large scale simulations.

Based upon the works of Chaljub et al. 2015 and Maufroy et al. 2015 (EUROSEISTEST Verification and Validation Project, E2VP), in this paper, the Mygdonian basin (Greece) is chosen as a suitable case study to test the performances of the mentioned BBS2S HPC-platform, constructing a numerical model of the earthquake scenario (valid in the 0.15–7 Hz frequency bandwidth) and focusing on the influence of the 3-D geology and of the soil heterogeneity on the earthquake ground motion spatial variability.

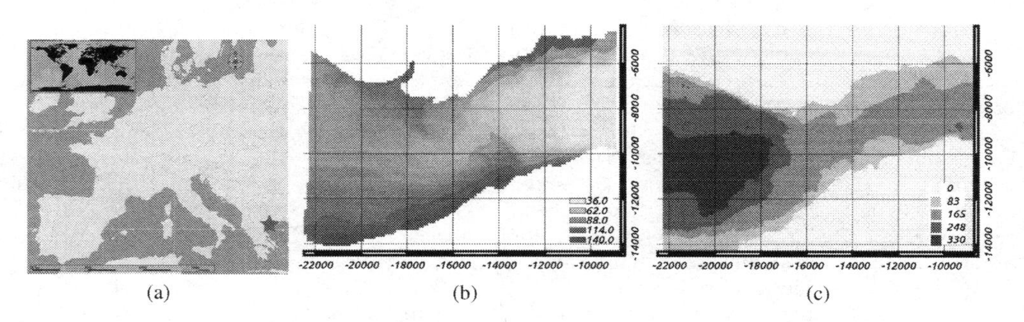

Figure 1. (a) Geo-referenced map of Europe with Mygdonian basin location. (b) Georeferenced elevation contour of the Mygdonian basin (Hatt Projection, Bessel ellipsoid). (c) Georeferenced depth contour of the sedimentary basin (all the layers), in Hatt coordinates.

2 A HPC-PLATFORM FOR BBS2S SIMULATIONS

2.1 *An efficient 3-D wave-propagation solver*

In the following section, a description of the BBS2S HPC platform employed for the earthquake simulations presented in this paper is provided. The main tool is represented by SEM3D, a numerical code tailored to solve the 3-D wave propagation problem in large solid/fluid computational domains. SEM3D is based on the Spectral Element Method (SEM, based on the pioneering works in fluid dynamics performed by Patera 1984, Korczak and Patera 1986, Maday et al. 1987, Mayday et al. 1989) and it has been developed upon the code RegSEM (Cupillard et al. 2012, Festa and Vilotte 2005), thanks to the synergistic effort of three research teams, at CentraleSupélec, Institut de Physique du Globe de Paris (IPGP) and the Comissariat d'Energie Atomique (CEA), respectively. The SEM is a high-order version of the Finite Element Method, well known for its accuracy in solving the elasto-dynamic problem in highly heterogeneous visco-elastic media (Seriani 1998, Komatitsch and Tromp 1999 and for its straight-forward extension to parallel implementation on supercomputer architectures (Göddeke et al. 2014). The SEM bares on a non-isoparametric piecewise polynomial approximation of the wave-field, featuring orthogonal Lagrangian polynomials, sampled at the Gauss-Lobatto-Legendre (GLL) quadrature points. The original core of the SEM3D software allowed to solve the wave propagation problem in any velocity model, including anisotropy, intrinsic attenuation and Newtonian fluid-structure interaction. Among the advantages of SEM3D, two main aspects must be stressed, namely (1) its efficient and cost-effective massively parallel implementation (by Message Passing Interface, MPI) on large supercomputers and (2) its ability to accurately take into account 3-D discontinuities such as the sediment-rock interface. Due to the properties of the semi-discretized SEM formulation (with diagonal mass matrix), coupled with an explicit yet energy preserving time-marching scheme (i.e. the *leap-frog* one Simo et al. 1992, Festa and Vilotte 2005), a fast and effective solving scheme is obtained, that reduces to a multi-DOF system of algebraic equations of motion for each node. The latter is solved by means of an efficient and highly scalable parallel solver that allows a cascade solution at each time-step, with limited communication time (see the scalability curves in Fig. 2). Moreover, the code makes use of a library called *HexMesh* (https://github.com/jcamata/HexMesh.git), that implements an efficient linear 27-tree finite element mesh generation scheme (de Abreu Corrêa, Camata, de Carvalho Paludo, Aubry Cottereau, & Coutinho 2015) and based on the previous works of (Camata & Coutinho 2013)) and it is capable to generate large computational grids (i.e. ≈ 100 km) by extruding the Digital Elevation Model (DEM) provided and progressively top-down coarsen it, so to obtain a non-structured grid. However, in this study, a flat topography was considered, as first approach, focusing on the complex geology instead.

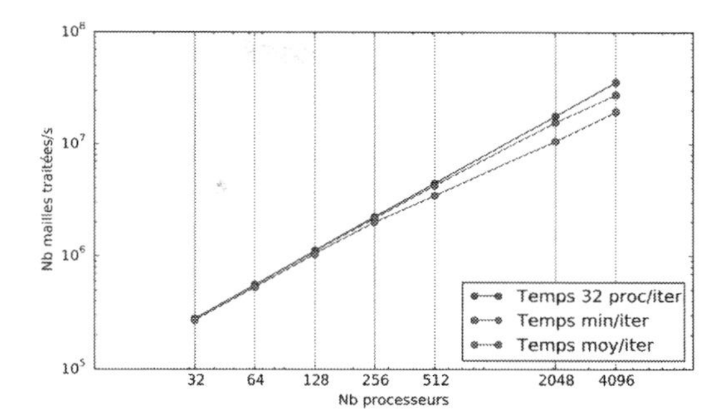

Figure 2. Scalability curves (number of MPI processes vs. number of mesh cells) for SEM3D in terms of time per iteration for 32 MPI processes (blue curve), minimum time per iteration (green curve) and average time per iteration.

2.2 *Modelling soil heterogeneity by random fields*

The SEM is well known for the high number of integration points in each element, compared to the FEM. This feature render the SEM naturally prone to model the heterogeneity of soil material, which is among the main issues responsible of wave dispersion and scattering. A common strategy to model heterogeneous media consists in considering the generic mechanical property $U(\underline{x})$ (usually a scalar quantity, such as the elastic shear modulus $G(\underline{x})$ as one (non-)stationary uni-variate random field, featured at least by its average $\mu_U(\underline{x})$ and its auto-covariance model $\mathcal{C}_{UU}(\underline{x})$. The latter is then point-sampled at N_p integration GLL points $\underline{x}_i^{GLL}$ in the computational domain Ω, at which both mass matrix and internal force vectors are computed in the SEM semi-discretized equations. To generate random samples, an on-source library ScaRL was developed linked to SEM3D (Paludo et al. 2015). Scarl provides realizations $U^k(\underline{x})$ of the random field by initially generating a Gaussian one $\Theta(\underline{x}_{i=1,N_p}^{GLL})$:

$$\Theta\underline{x} = \sum_{n\leq N_\phi} \sqrt{2S_{\Theta\Theta}(\underline{k}_n)\,\|\underline{\Delta k}_n\|}\cos \underline{k}_n.\underline{x} + \phi_n \tag{1}$$

whose Cumulated Distribution Function (CDF) is then mapped point-wise to the desired first-order marginal density (e.g. log-normal) via the non-linear Rosenblatt transform Rosenblatt 1952. Following this approach, the wave-number domain spanned by $\underline{k}_n$ is discretized over a regular grid of size $\underline{N} = N_x, N_y, N_z^T$, indexed by $\underline{n} = n_x, n_y, n_z^T$. $\mathcal{S}_{\Theta\Theta}(\underline{k}_n)$ represents the power spectral density of the random field, whereas $\|\underline{\Delta k}_n\|$ is the unit volume in the spectral domain and the random variables ϕ_n are the independent elements of a $\underline{N}$-dimensional random variable with uniform density over [0, 2π]. The Fast Fourier Transformcan be used to bring the complexity of this generation method to $\mathcal{O}\left(N_\phi \log N_\phi\right)$ (Shinozuka and Deodatis 1991), although an uniform grid is required and since the GLL points are not uniformly distributed in space, the random field must then be interpolated to the N_p GLL nodes. However, this spectral approach to random field discretization was adopted so to avoid the computationally expensive Cholesky factorization of $\mathcal{C}_{UU}\left(\underline{x}^{GLL}_{i=1,N_p}\right)$, which scales as $\mathcal{O}\left(N_p^3\right)$ in the general case. However, modelling soil heterogeneity over large scale domains implies a major computational effort: despite its characteristic dimension L ($\approx 10^4$ m) the computational grid must resolve both the wavelength and the heterogeneity correlation length $\ell_{c\theta}$ ($\approx 10^2$ m). ScaRL overcomes this problem (that in the end turns into scalability issue), by generating realizations U^k over the entire domain as superposition of I smaller independent realizations $\Theta^k|_{\Omega_i}$ supported on overlapping subdomains Ω_i of Ω (Paludo et al. 2015):

$$\Theta^k(\underline{x}) = \sum_{i\in I} \sqrt{\psi_i(\underline{x})}\Theta^k|_{\Omega_i}(\underline{x}). \quad (2)$$

where the set of functions $\psi_i x$ forms a partition of unity of Ω (i.e. $\sum_{i\in I}\psi_i(\underline{x}) = 1$ for any $\underline{x} \in \Omega$), supported by the set of subdomains Ω_i. Using this approach, the complexity becomes $\mathcal{O}\left(n_p \log\left(n_p\right)\right)$ where $n_p = N_\phi / P$ and P is the number of processors. Essentially, this means that the scheme is $\mathcal{O}(1)$ when a constant number of GLL nodes per processor is considered. The overlapping Eq. (2) involves an approximation that does not alter the average and variance of the resulting field $\Theta(\underline{x})$. The influence on the correlation structure depends on the overlap, relative to the correlation length (Paludo et al. 2015).

3 CONSTRUCTION OF THE GEOLOGICAL MODEL

3.1 *Geological model of the Mygdonian basin*

Manakou et al. 2010 provided a detailed description of the geometry and of the dynamic properties of the soil layers within the Mygdonian basin, obtained by inverting data from microtremor array measurements, seismic refraction profiles, boreholes, and geotechnical investigations. S-wave velocity profiles were inverted from the phase-velocity dispersion curves obtained at 27 different sites placed in the whole valley and then by interpolation with previously available data. The proposed 3-D model describes the geometry and shear-wave velocities of the Mygdonian and pre-Mygdonian sedimentary systems, and the top bedrock surface (Tab. 1). In this paper, the pre-Mygdonian system was solely considered as first approach, surrounded by the outcropping bedrock with an average shear-wave velocity of 650 m/s (see Tab. 2). Fig. 3 depicts the SEM3D representation of the Mygdonian basin, along with two cross sections of the basin (A-A' and B-B' in Fig.s3b and 3c).

3.2 *Introducing heterogeneity*

In this paper, the fluctuation of the soil properties within the sedimentary basins, observed *in-situ*, is reproduced numerically. ScaRL is employed, coupled with SEM3D, so to obtain a spatially variable field for the elastic shear modulus $\mu(\underline{x})$ for the first 500 m depth. As portrayed in Fig. 2, the elastic shear modulus $\mu(\underline{x})$ is represented by a non-stationary random field all over the uppermost part of the Earth's crust. $\mu(\underline{x})$ was generated by ScaRL, characterized by an anisotropic Von Karman Correlation structure ($\ell_{C-x} = \ell_{C-y} = 90$ m, $\ell_{C-z} = 40$ m), over a domain of $\approx$13 km × 11 km × 0.4 km. Its standard coefficient of variation $CV = \mu_p / \sigma_p$ is equal to 10%,

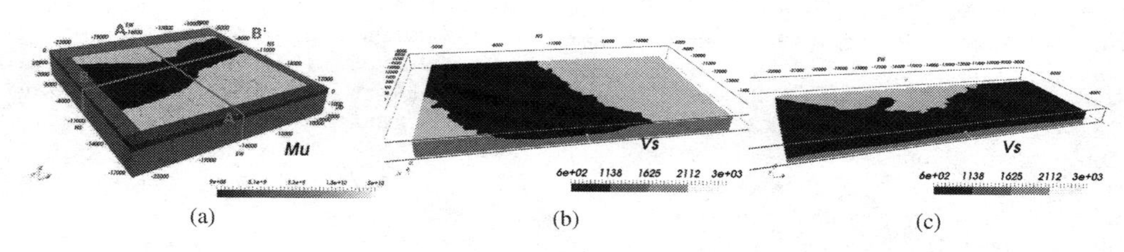

(a) (b) (c)

Figure 3. (a) Layout of the SEM3D model of the Mygdonian basin. Two representative cross-sections are depicted: (b) A-A' (NS) and (c) B-B' (EW).

Table 1. Weighted average V_S velocities with plus or minus one standard deviation for the two sedimentary units and the bedrock of the Mygdonian basin (Manakou et al 2010).

Geological units	V_S
(sedimentary layers)	[m/s]
A+B (Mygdonian system)	185–257
C+D (Mygdonian system)	337–426
A+B+C+D (whole Mygdonian system)	291–379
E+F (pre–Mygdonian system)	532–668
Bedrock	> 1000

Table 2. Geological properties of the SEM3D model of the Mygdonian basin.

	V_S	V_P	rho
	[m/s]	[m/s]	[kg/m^3]
E+F	650	2500	2200
Bedrock	2600	4500	2600

with a first order marginal that is assumed to follow a log-normal pdf-distribution.

4 MAJOR FINDINGS AND DISCUSSION

4.1 *Numerical model*

The numerical grid employed in this analysis was designed with a structure texture, featured by average element sizes of approximately 90 m × 90 m × 40 m (resulting into $5.5 \cdot 10^{+07}$ hexahedral elements and $5.8 \cdot 10^{+07}$ GLL), to grant the propagation of a seismic wave at 7 Hz (with 7 *GLL* points per minimum wave-length). The model was enclosed by non-reflecting Perfectly Matched Layers (PML) and no surface topography was introduced. This mesh was used to perform an extensive comparison between three geological models of the Mygdonian basin area is presented, namely (1) horizontally bi-layered half-space (LAY), (i.e. classical Green's function model), (2) homogeneous (HOM) 3-D basin (EF sediments, see Table 2) embedded into outcropping bedrock (see Fig. 3a) and (3) heterogeneous (HET) 3-D basin (EF sediments, see Table 2) embedded into outcropping bedrock (see Fig. 4). To integrate the complex 3-D geology into the computational grid, a *not-honouring* approach was employed, i.e. the geological interface was not directly meshed, but the heterogeneous mechanical properties were interpolated over the structured computational grid. No viscous dissipation was introduced at this stage.

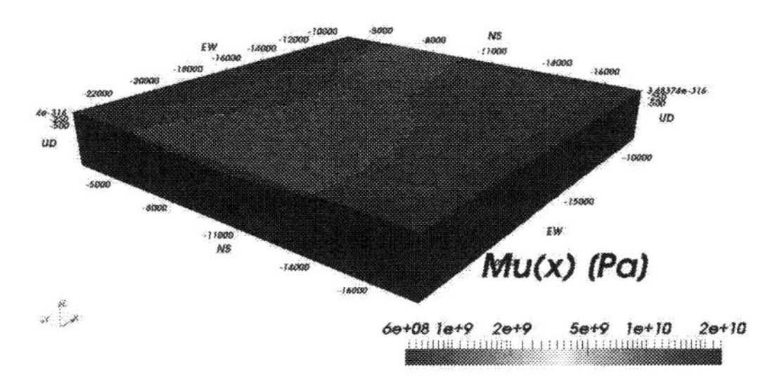

Figure 4. Example of the heterogeneous shear modulus field $\mu(\underline{x}) = \sqrt{\mu(\underline{x}) / \rho(\underline{x})}$.

A point-wise double-couple seismic source ($\phi_S = 22.5°$, $\lambda = 0°$, $\delta = 90°$) was employed at this stage,located southwards, at 5 km depth (in the bedrock stratum), to discount the effect of the extended fault rupture from the wave-field at surface (Chaljub et al. 2015).

The computations were performed on the supercomputer cluster *Fusion*, held by the Université Paris Saclay. The LAY model took approximately 310 h CPU-time to run 30 seconds of simulation on 216 MPI cores. HOM and HET took instead approximately 1487 h CPU-time.

4.2 *Influence of the 3-D basin structure on the wave-motion*

The effects of complex 3-D basin-like geological structures have been vastly observed and studied in the past (see Faccioli and Vanini 2003 and Smerzini et al. 2017, for instance).

The present case study confirms the complexity introduced by the complex interaction between the incident seismic wave-motion and the 3-D interfaces, when the former impinges the latter while travelling towards the surface. Fig. 5 shows the comparison between synthetic velocigrams obtained for three different geological models (Fig.s 5a-5c). The strong phase arrivals are delayed in the LAY case (black traces), as expected. Moreover, both in the HOM (blue traces) and HET (red traces) cases, late wave-forms appear (some sort of *coda-waves*), due to the wave-motion interaction with the basin structure. Smaller differences are encountered between the two latest cases, highlighting the moderate contribution of the soil unit mass heterogeneity in this case dominated by the 3-D basin effect. However, in Fig.s 5d-5f, the corresponding acceleration Fourier's spectra are plotted: those spectra better show the actual different response obtained by either considering the heterogeneity or not. It is worth to mention that the vertical components have amplitudes comparable to the horizontal ones, suggesting that some surface waves have been

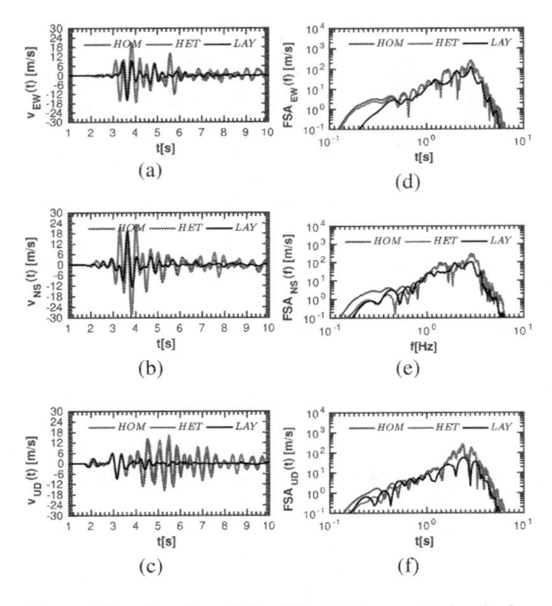

Figure 5. On the left: Velocigrams obtained for HOM (blue), HET (red) and LAY (black) respectively at the intersection between sections A-A' and B-B' (EW = -16100, NS = -10400, see Fig. 3.1). On the right: the corresponding Fourier's spectra. The time-histories and Fourier's spectra (referring to synthetics filtered at 7 Hz) correspond to EW direction (a-d), NS direction (b-e) and UD direction (c-f) respectively.

generated, due to both the presence of the basin and the added soil heterogeneity.

Fig. 6 portrays the interferograms obtained for section A-A' (polarization EW) and B-B' (polarization NS). From the interferogram, it is noticeable that a complex wave-motion is generated by the multi-pathing effect taking place within the basin's borders due to the high impedance contrast with

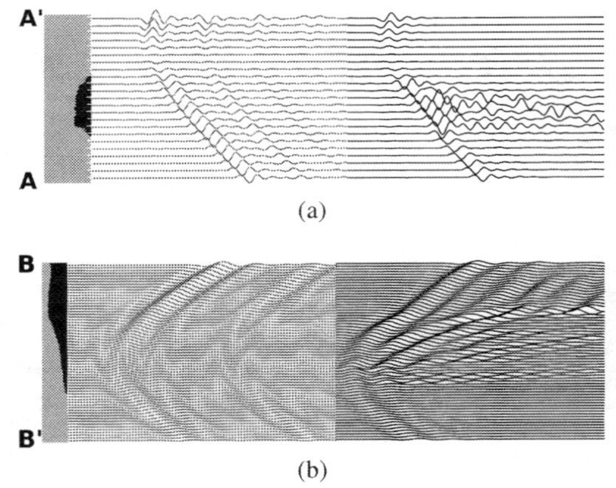

Figure 6. Interferograms for section A-A' (a) and section B-B' (b). Red velocigrams refer to the layered test case; black velocigrams refer to the homogeneous basin case.

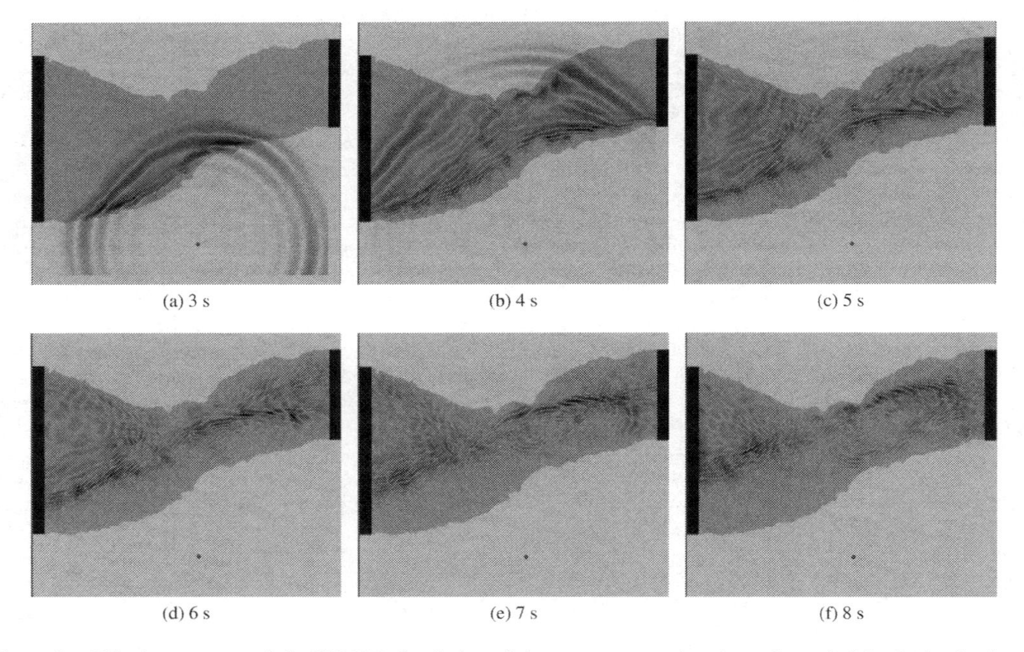

Figure 7. Velocity contours of the SEM3D simulation of the wave propagation throughout theMygdonian basin.

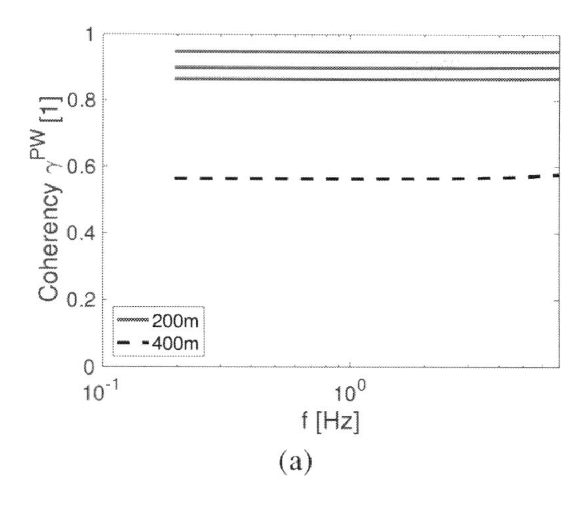

(a)

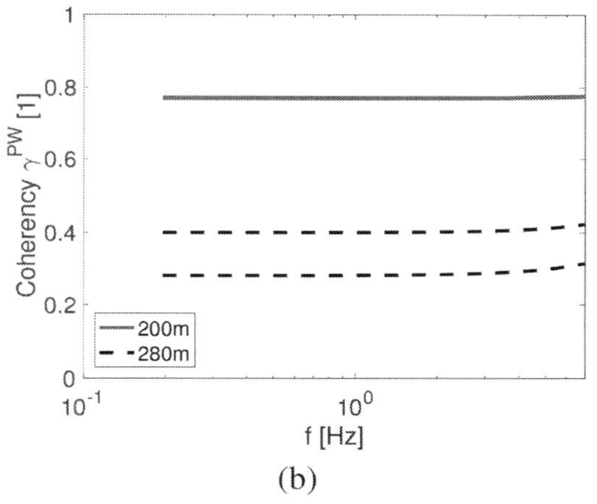

(b)

Figure 8. Plane-wave coherency curves obtained for the HET case at three stations: (a) aligned along the A-A' section and (b) aligned along the 45° diagonal

respect to the surrounding outcrop. Long *coda-waves* can be noticed within the softer sedimentary area, that seem trapped within it. The latter remark is even more evident when one looks at the velocity contours showed in Fig. 7, where wavelets seem trapped within the basin's boundaries.

4.3 *Analysis of wave motion coherency*

Finally, the wave motion coherency of the synthetic wave motion was analyzed. For the sake of brevity, just the HET case is considered herein. Fig. 8a shows some plane wave coherency curves γ^{PW} calculated along the section A-A' (at surface) at a lag-distance of 200 m and 400 m. The wave passage effect was removed. A coherency drop can be clearly observed, although the coherency curves appear to be rather flat, most probably due to the short duration of the *coda*-waves and to the limited frequency range. It is interesting to notice that the same coherency drop can be observed even at closer lag-distances, for instance when considering a EW-NS diagonal line across A-A' and B-B', as depicted in Fig. 8b. The coherency drop appears at 280 m, indicating the fact that the basin and the soil heterogeneity effectively scatter the incident wave motion propagate by the source.

5 CONCLUSIONS

In this paper, a 3-D source-to-site earthquake scenario is presented. The selected case study is the well documented Mygdonian basin, in Greece. A 3-D numerical model (of approximately 10 km of characteristic size) of the shallow sedimentary basin is constructed, by means of a SEM model. The heterogeneous model of the Earth's crust at the regional scale (the soft soil deposit within the basin valley) is integrated by heterogeneous soil properties at the mesoscopic scale, obtained by generating a random field representing one spatial realization of the unit mass distribution. The 3-D analyses are accurate up to 7 Hz.

The effect of the basin is studied in terms of time histories recorded at the surface and in terms of wave motion coherency. The basin scatters the wave motion propagated from the point-wise source, trapping the radiated energy due to the great basin-crust great impedance contrast. The soil heterogeneity, on the other hand, acts at a smaller scale, inducing local scattering which however is poorly visible in this frequency range. Both the basin configuration and soil heterogeneity, however, cause a coherency drop in the wave motion at surface.

As future developments, the softer basin layers will be included in the analysis, along with an increased frequency range to be able to follow the coherency drop broad-band and specify the role of the soil heterogeneity at higher frequencies.

ACKNOWLEDGMENTS

This work has been developed in the framework of the IMPEC project (On the broadband synthetic sIgnals enhanceMent for 3D Physic based numerical analysis, the EUROSEISTEST Case study), financed by the Seismological and Earthquake Engineering Research Infrastructure Alliance for Europe and it was also partly supported by the SEISM Paris Saclay Research Institute.

REFERENCES

Camata, J.J. & A.L.G.A. Coutinho (2013). Parallel implementation and performance analysis of a linear octree finite element mesh generation scheme. *Concurrency and Computation: Practice and Experience* (6), 826–842.

Chaljub, E., E. Maufroy, P. Moczo, J. Kristek, F. Hollender, P.-Y. Bard, E. Priolo, P. Klin, F. de Martin, Z. Zhang, W. Zhang, & X. Chen (2015). 3-D numerical simulations of earthquake ground motion in sedimentary basins: testing accuracy through stringent models. *Geophysical Journal International* (1), 90–111.

Cupillard, P., E. Delavaud, G. Burgos, G. Festa, J.-P. Vilotte, Y. Capdeville, & J.-P. Montagner (2012). RegSEM: a versatile code based on the spectral element method to compute seismic wave propagation at the regional scale. *Geophysical Journal International* (3), 1203–1220.

de Abreu Corrêa, L., J.J. Camata, L. de Carvalho Paludo, L. Aubry, R. Cottereau, & A.L.G.A. Coutinho (2015). Wave propagation in highly heterogeneous media: scalability of the mesh and random properties generator. *Submitted for publication in Computers & Geosciences*.

De Martin, F. (2011, December). Verification of a Spectral-Element Method Code for the Southern California Earthquake Center LOH.3 Viscoelastic Case. *Bulletin of the Seismological Society of America* (6), 2855–2865.

Faccioli, E., F. Maggio, R. Paolucci, & A. Quarteroni (1997). 2D and 3D elastic wave propagation by a pseudo-spectral domain decomposition method. *Journal of Seismology* (3), 237–251.

Faccioli, E. & M. Vanini (2003). Complex seismic site effects in sediment-filled valleys and implications on design spectra. *Progress in Structural Engineering and Materials* (4), 223–238.

Festa, G. & J.-P. Vilotte (2005, June). The Newmark scheme as velocity-stress time-staggering: an efficient PML implementation for spectral element simulations of elastodynamics. *Geophysical Journal International* (3), 789–812.

Göddeke, D., D. Komatitsch, & M. Möller (2014). Finite and Spectral Element Methods on Unstructured Grids for Flow and Wave Propagation Methods, Chapter 9, pp. 183–206. Springer.

Graves, R.W. (1996). Simulating seismic wave propagation in 3D elastic media using staggered-grid finite differences. *Bulletin of the Seismological Society of America* (4), 1091–1106.

Komatitsch, D. & J. Tromp (1999). Introduction to the spectral element method for three-dimensional seismic wave propagation. *Geophysical journal international* (3), 806–822.

Komatitsch, D. & J.-P. Vilotte (1998). The Spectral Element Method: An Efficient Tool to Simulate the Seismic Response of 2D and 3D Geological Structures. *Bulletin of the Seismological Society of America* (2), 368–392.

Korczak, K.Z. & A.T. Patera (1986). An isoparametric spectral element method for solution of the Navier-Stokes equations in complex geometry. *Journal of Computational Physics* (2), 361–382.

Maday, Y., A.T. Patera, & E.M. Ronquist (1987). A well-posed optimal spectral element approximation for the Stokes problem.

Manakou, M., D. Raptakis, F. Chavez-Garcia, P. Apostolidis, & K. Pitilakis (2010). 3D soil structure of the Mygdonian basin for site response analysis. *Soil Dynamics and Earthquake Engineering* (11), 1198–1211.

Maufroy, E., E. Chaljub, F. Hollender, J. Kristek, P. Moczo, P. Klin, E. Priolo, A. Iwaki, T. Iwata, V. Etienne, F. De Martin, N.P. Theodoulidis, M. Manakou, C. Guyonnet-Benaize, K. Pitilakis, & P.-Y. Bard (2015, June). Earthquake Ground Motion in the Mygdonian Basin, Greece: The E2VP Verification and Validation of 3D Numerical Simulation up to 4 Hz. *Bulletin of the Seismological Society of America* (3), 1398–1418.

Mayday, Y., A. Patera, & E. Rønquist (1989). Optimal Legendre spectral element methods for the multi-dimensional Stokes problem. *SIAM J. Num. Anal.*

Paludo, L., V. Bouvier, R. Cottereau, & D. Clouteau (2015). Efficient Parallel Generation of Random Field of Mechanical Properties for Geophysical Application. In *th International Conference on Earthquake Geotechnical Engineering*.

Paolucci, R., I. Mazzieri, C. Smerzini, & M. Stupazzini (2014). Physics -Based Earthquake Ground Shaking Scenarios in Large Urban Areas. In A. Ansal (Ed.), *Perspectives on European Earthquake Engineering and Seismology*, Volume 34 of *Geotechnical, Geological and Earthquake Engineering*, pp. 331–359. Springer.

Patera, A. (1984). A spectral element method for fluid dynamics: Laminar flow in a channel expansion. *Journal of Computational Physics* (3), 468–488.

Pitilakis, K., G. Manos, D. Raptakis, A. Anastasiadis, K. Makra, & M. Manakou (2009). The EUROSEISTEST experimental test site in Greece. In *EGU General Assembly Conference Abstracts*, Volume 11, pp. 13729.

Pitilakis, K., Z. Roumelioti, D. Raptakis, M. Manakou, K. Liakakis, A. Anastasiadis, & D. Pitilakis (2013). The EUROSEISTEST Strong-Motion Database and Web Portal. *Seismological Research Letters* (5), 796–804.

Rosenblatt, M. (1952). Remarks on a multivariate transformation. *The annals of mathematical statistics* (3), 470–472.

Seriani, G. (1998). 3-D large-scale wave propagation modeling by spectral element method on Cray T3E multiprocessor. *Computer Methods in Applied Mechanics and Engineering* (1), 235–247.

Shinozuka, M. & G. Deodatis (1991). Simulation of stochastic processes by spectral representation. *Applied Mechanics Reviews* (4), 191–204.

Simo, J., N. Tarnow, & K. Wong (1992). Exact energy-momentum conserving algorithms and symplectic schemes for nonlinear dynamics. *Computer Methods in Applied Mechanics and Engineering* (1), 63–116.

Smerzini, C., K. Pitilakis, & K. Hashemi (2017, Mar). Evaluation of earthquake ground motion and site effects in the Thessaloniki urban area by 3D finite-fault numerical simulations. *Bulletin of Earthquake Engineering* (3), 787–812.

Taborda, R. & J. Bielak (2011). Large-scale earthquake simulation: computational seismology and complex engineering systems. *Computing in Science & Engineering* (4), 14–27.

Numerical Methods in Geotechnical Engineering IX – Cardoso et al. (Eds)
© 2018 Taylor & Francis Group, London, ISBN 978-1-138-33198-3

Analysis of observed liquefaction during the 2016 Kumamoto earthquake

Bashar Ismael & Domenico Lombardi
University of Manchester, Manchester, UK

ABSTRACT: This paper evaluates the reliability of different approaches used to determine the factor of safety against liquefaction, namely Eurocode 8 (1998) and Idriss and Boulanger (2008), at three real sites struck by the 2016 Kumamoto earthquake (Japan). The computed factors of safety are compared with results from a series of numerical analyses performed in the object-oriented software framework OpenSees. The results show that the different methods yield similar factors of safety, which are consistent with the numerical results and field observations. Additionally, the numerical results confirm the high liquefiability of the abovementioned sites in which liquefaction is predicted to be triggered on the surface and/or at some depth below the ground surface, based on the deposit nature and the characteristics of the applied ground motion.

1 INTRODUCTION

Liquefaction phenomenon could be considered as the main source of failure in areas stricken by an earthquake, especially if the ground is formed from loose or medium dense sand. In general, liquefaction is described as the transformation from a solid state to the liquid one. From a soil mechanics standpoint, the liquefaction phenomenon represents the condition in which most of the soil's stiffness and strength are lost via the application of cyclic loading or any rapid loading. In saturated sand, pore water pressure generated during the application of seismic load causes a decrease in the effective stress according to the effective stress principle. In an extreme case, the effective stress becomes zero, and the soil grains lose contact with each other. At this stage, the soil grains are floating in the pore water without any confinement support from the surrounding soil.

Infrastructures located in seismic regions are highly susceptible to liquefaction-induced damage, especially if the ground water is shallow. To minimise the damage caused by liquefaction, there is a need to predict liquefaction vulnerability for the sites under investigation. The ground deformations as a result of liquefaction should also be predicted so that the structures could be designed with a reasonable factor of safety against liquefaction-induced damage.

Several factors could be correlated to liquefaction vulnerability. One is the ground geology where previous earthquakes confirm that the majority of the liquefied soil was found to be fine-grained sand and sandy silt (Huang & Jiang, 2010). The grain size distribution of the soil deposit could also be considered as a boundary to distinguish between liquefiable and non-liquefiable soil, as suggested by Tsuchida, (1970). The last author introduced a grain size distribution curve, in which two boundaries are proposed for liquefiable and highly liquefiable soil. Furthermore, liquefaction amenability could be anticipated by other factors such as the shape and hardness of the particles, besides the percentage of fines in the sample (Verdugo, 1989). In-situ tests could be used to predict liquefaction vulnerability, such as Standard Penetration Test (SPT), Cone Penetration Test (CPT) and in-situ shear wave velocity, by using empirical equations such as the codes of practice (e.g. Eurocode 8 (1998)) and simplified procedures, such as the one suggested by Idriss & Boulanger (2008(. Numerical modelling is a versatile tool which has been used recently to estimate liquefaction taking into account several factors that could not be considered in the empirical equations. Generally, liquefaction vulnerability could be assessed using the ratio of the excess pore water pressure to the effective overburden stress, r_u in a soil deposit. Several constitutive models have been used recently to predict the aforementioned ratio (r_u) and its correlated deformation for a specific site. However, this task could be considered challenging in cases involving liquefaction because of the continuous change in the soil condition and state during liquefaction, where the effective pressure keeps varying as a result of the mobilisation/dissipation of excess pore water pressure. Additionally, the non-uniformity of the load (i.e.

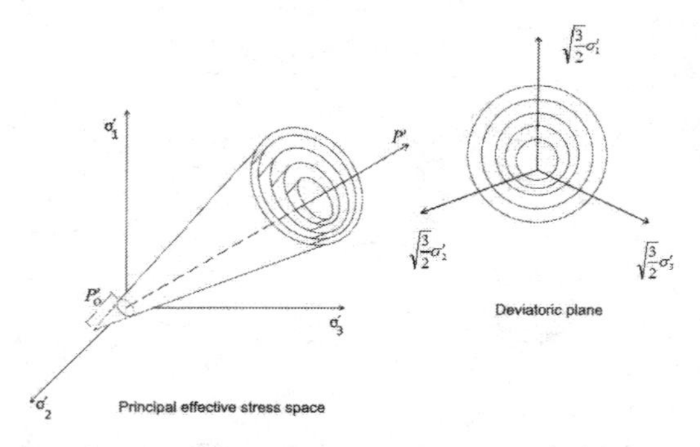

Figure 1. Conical yield surface in principal stress space and deviatoric plane (redrawn from Yang et al., 2003).

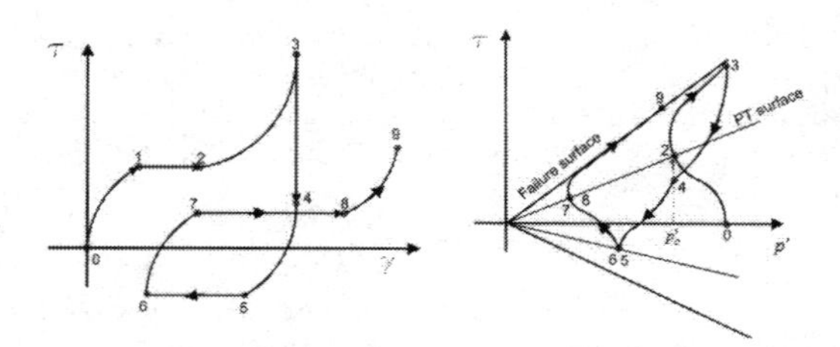

Figure 2. Shear stress-strain and effective stress path under undrained shear loading conditions (redrawn from (Elgamal, 2007)).

earthquake motion) could complicate the analysis of liquefaction-related problems.

The main aims of the presented paper are to evaluate the reliability of different approaches used to determine the factor of safety against liquefaction at three real sites struck by the 2016 Kumamoto earthquake and to investigate the parameters governing liquefaction onset in a real field.

2 SOIL CONSTITUTIVE MODEL

The saturated soil is modelled as a two-phase material which is known as a $u-p$ formulation, in which u represents the soil skeleton displacement and p refers to the pore water pressure. Using the aforementioned formula, the interaction between solid soil skeleton and pore fluid could be accounted for enabling the liquefaction phenomenon to be studied.

The material that could be used for modelling liquefaction is called PressuredependMultiYield02. Figure 1 presents the multi-yield surface plasticity model introduced by Prevost, (1985) used to simulate the saturated sand response under seismic and/or cyclic loading. The main distinct characteristic of the presented model is the ability to capture the cyclic plastic shear-strain accumulation (Fig. 2).

Table 1. Input parameters required by the constitutive model.

Parameter	Dr = 30%	Dr = 40%	Dr = 50%	Dr = 60%	Dr = 75%
Saturated mass density(ton/m3)	1.7	1.8	1.9	2.0	2.1
Reference shear modulus (MPa)	60	90	100	110	130
Reference Bulk Modulus (MPa)	160	220	233	240	260
Friction Angle (Degree)	31	32	33.5	35	36.5
Phase Transformation Angle (Degree)	31	26	25.5	26	26
Peak shearstrain	0.1				
Reference pressure (kPa)	101				
Pressure dependent coefficient	0.5				
Contrac1	0.087	0.067	0.045	0.028	0.013
Contrac3	0.18	0.23	0.15	0.05	0.0
dilat1	0.0	0.06	0.06	0.1	0.3
dilat3	0.0	0.27	0.15	0.05	0.0
Void ratio	0.85	0.77	0.7	0.65	0.55

Table 1 presents the input parameters required by the constitutive law to conduct the Finite Element (FE) analysis (Yang & Elgamal, 2008). It is worth mentioning that when using the aforementioned material different parameters are required for different relative densities/void ratios because it is not formulated following the critical state soil mechanics framework.

3 FACTOR OF SAFETY AGAINST LIQUEFACTION

Two approaches, namely Eurocode 8 (1998) and Idriss & Boulanger (2008), have been used to evaluate liquefaction vulnerability for three sites where liquefaction has been observed.

Boreholes data and SPT count numbers for the sites under investigation were obtained from the electronic national data hub for Japan organised by the Geospatial Institute of Japan publically available via http://maps.gsi.go.jp/#14/32.753375/130.8059861/&base = std&ls = std&disp = 1&vs = c1 j0l0u0f0&d = v. The blowcount number N_{SPT} obtained from the SPT test has been normalised to a reference overburden stress of 100 kPa and the ratio of the impact energy to theoretical free-fall energy (0.6) is taken as equal to 1.2. In both the EC8 and Idriss and Boulanger approaches the maximum peak ground acceleration recorded at

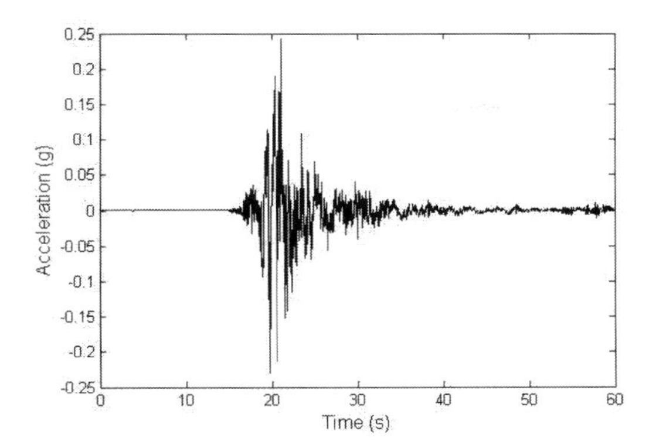

Figure 3. Maximum acceleration time history recorded during the main shock (borehole).

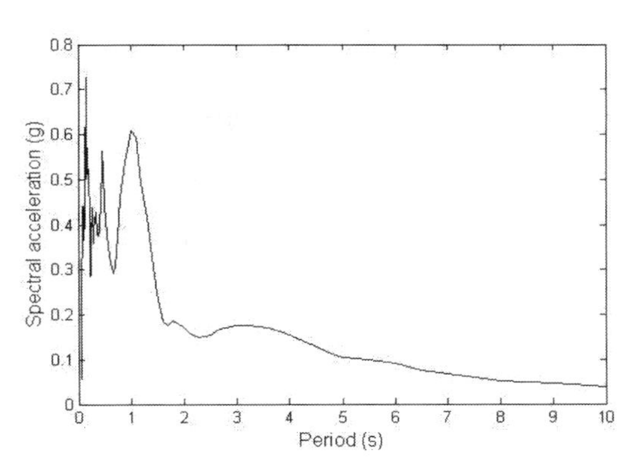

Figure 4. Response spectra at 5% damping ratio.

the surface (1.2 g) is considered in the liquefaction vulnerability assessment.

4 NUMERICAL MODEL FOR LIQUEFACTION PREDICTION

To track the generated r_u during the ground shaking, an 8-node brick element with both displacements and pore water pressure degrees of freedom is used. Using the aforementioned element type, the interaction between solid soil skeleton and pore fluid could be accounted for enabling the liquefaction phenomenon to be studied. Additionally, the equal Degrees Of Freedom (DOF) command is recommended to obtain the same lateral deformation for nodes that share the same vertical location simulating a 1D analysis. The nodes above the water table are fixed in the pore pressure DOF to represent the open drainage condition. Similarly, the nodes at the base of the soil column have to be fixed against vertical displacement to account for the bedrock layers that exist below the modelled soil profiles. After defining the model and running the analysis, the output of this analysis is visualised using post-process software (i.e. GiD software).

5 GROUND MOTION

The ground motion characteristics were obtained from KiK-NET and K-NET operated by the National Research Institute for Earth Science and Disaster Resilience (NIED) in Japan. The maximum ground motion for the largest mainshock registered at the borehole for KMMH16 station located in Mashiki town (32.7967°N, 130.8199°E) (Fig. 3) was used in the analysis.

Figure 4 presents the response spectra at 5% damping ratio, as recommended by most of the current codes such as Eurocode 8 (1998).

6 ANALYSIS OF OBSERVED LIQUEFACTION DURING THE KUMAMOTO EARTHQUAKE

The calculated factor of safety (from the two abovementioned approaches) for Akitsu River at different depths is introduced in Figure 5 beside the borehole log and the corresponding SPT count number. The relative density is determined based on the corrected N_{SPT} value, which has been normalised to a reference overburden stress of 100 kPa, and the ratio of the impact energy to theoretical free-fall energy (0.6) is taken as equal to 1.2. At Akitsu River, the factor of safety is below unity from 2 m below the ground surface to up to 8 m.

The two approaches show approximately a similar trend comparable with numerical analyses performed in the object-oriented software framework OpenSees. The numerical analysis shows that full liquefaction (i.e. r_u $ 0.8) is mobilised for the total depth from 3.5 m up to 5.5 m, as can be seen in Figure 5. The decrease in the r_u value with further depth could be correlated to the increase of the overburden pressure with depth. For that reason, a greater intensity of dynamic disturbance is required to overcome the internal pressure between sand particles at higher depth (i.e. at a depth higher than 6 m). Accordingly, even loose soil can hardly liquefy at deeper elevation. After the onset of liquefaction, which may continue for several seconds, minutes or even hours, the process of solidification starts. Consequently, the generated excess pore water pressure starts to decrease as a result of the water squeeze out of the soil deposit (i.e. sand boils). The occurrence of liquefaction

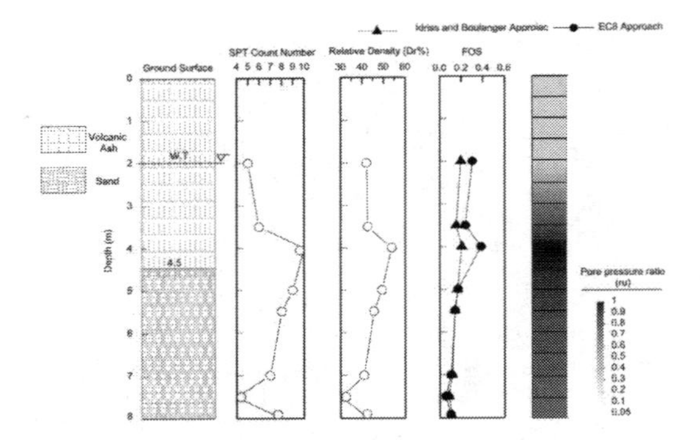

Figure 5. Borehole log, SPT count number, liquefaction assessment using two approaches and maximum r_u generated at Akitsu-river, as calculated by OpenSees software.

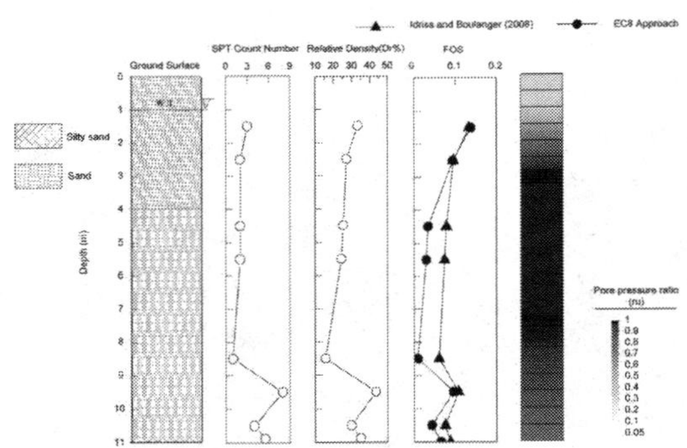

Figure 7. Borehole log, SPT count number, liquefaction assessment using two approaches and maximum r_u generated between Shirakawa and Midorikawa rivers, as calculated by OpenSees software.

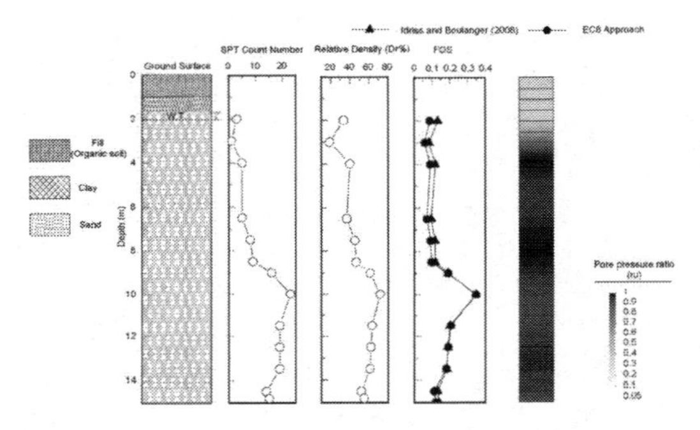

Figure 6. Borehole log, SPT count number, liquefaction assessment using two approaches and maximum ru generated at Kameizu Lake, as calculated by OpenSees software.

in limited zones (i.e. at 3.5 m up to 5.5 m depths) confirms the fact that liquefaction may occur at the surface and/or at some depth below the surface where the required conditions for liquefaction triggering are available (i.e. soil state and the triggering motion).

The factor of safety is found to be less than 1 up to 15 m depth at Kameizu Lake except for the surficial layer, where the soil is comprised of non-liquefiable organic soil and clay, as can be seen in Figure 6. The numerical model shows that full liquefaction initiates in the confined zone below 3 m (Fig. 6). It is suggested that the surficial non-liquefiable layer with a limited permeability prevents the dissipation of the generated r_u leading the ground to spread laterally, as confirmed in the site investigation near the examined area (Goda et al., 2016).

The liquefaction assessment for the rectangular belt between Shirakawa and Midorikawa rivers shows that the soil contains silty sand and sand (Fig. 7) is susceptible to liquefaction up to 11 m depth. The numerical analysis shows that full liquefaction develops in the sandy/silty sand layers from 2.5–9 m depth, as depicted in Figure 7. The presence of sand and silty sand explains the high liquefiability of the examined area, where previous earthquakes confirm that the majority of the liquefied soil was found to be fine-grained sand and sandy silt (Huang & Jiang, 2010). It is worth mentioning that, when the surficial layer liquefies, the confining pressure applied on the lower layer reduces, leading to liquefaction propagation from the top down. Otherwise, if there is a decrease in the density with depth, then liquefaction may occur at a depth below the surface where the soil is loose enough to trigger liquefaction. Several cyclic triaxial tests conducted by Seed & Lee (1966) under undrained condition proved that liquefaction resistance reduces simultaneously with the reduction of the applied confining pressure. The assessment results in the current study were proved by the field survey, where clear signs of liquefaction were apparent at all the studied sites.

Finally, the initiation of liquefaction in a confined zone (i.e. level ground area) would have a minor effect on infrastructure foundations related to the depth of the liquefied layer, especially if there are no signs of upward flow.

7 CONCLUSION

An assessment of liquefaction vulnerability for the three visited sites is conducted using the EuroCode 8 (1998) approach and Idriss and Boulanger (2008) method. The high liquefaction probability is apparent for all sites under investigation, especially for sandy layers. For that reason, special care must be taken to stabilise the soft soil before infrastructure construction. The numerical analysis for the aforementioned sites shows that liquefaction could occur at the surface and/or at some depth below the ground surface, based on the soil state and the induced ground motion. The minor damage for structures at the sites under investigation may be correlated to the existence of liquefiable soil in a confined area. Finally, liquefaction is a complex phenomenon and the simplified approaches may be considered conservative because of the limitation to consider all factors that may affect liquefaction triggering for a specific site.

REFERENCES

Eurocode 8. 1998. Design of structures for earthquake resistance part 5: Foundations, retaining structures and geotechnical aspects (English). European Committee de Normalisation.

Elgamal, A. 2007. Nonlinear modeling of large-scale ground-foundation-structure seismic response. ISET J. Earthquake Technology, 44(2), 325–339.

Goda, K., Campbell, G., Hulme, L., Ismael, B., Ke, L., Marsh, R., Sammonds, P., So, E., Okumura, Y. & Kishi, N. 2016. The 2016 Kumamoto earthquakes: cascading geological hazards and compounding risks. Frontiers in Built Environment, 2, 19.

Huang, Y. & Jiang, X. 2010. Field-observed phenomena of seismic liquefaction and subsidence during the 2008 Wenchuan earthquake in China. Natural Hazards, 54(3), 839–850.

Idriss, I. & Boulanger, R.W. 2008. Soil liquefaction during earthquakes: Earthquake engineering research institute.

Prevost, J.H. 1985. A simple plasticity theory for frictional cohesionless soils. International Journal of Soil Dynamics and Earthquake Engineering, 4(1), 9–17.

Seed, H.B. & Lee, K.L. 1966. Liquefaction of saturated sands during cyclic loading. Journal of the Soil Mechanics and Foundations Division, 92(6), 105–134.

Tsuchida, H. 1970. Prediction and countermeasure against the liquefaction in sand deposits. Abstract of the seminar in the Port and Harbor Research Institute, 1970. 3.1–3.33.

Verdugo, R. 1989. Effect of fine content on the steady state of deformation on sandy soils. Master thesis, University of Tokyo.

Yang, Z., Elgamal, A. & Parra, E. 2003. Computational model for cyclic mobility and associated shear deformation. Journal of Geotechnical and Geoenvironmental Engineering, 129(12), 1119–1127.

Yang, Z., Lu, J. & Elgamal, A. 2008. OpenSees soil models and solid-fluid fully coupled elements. User's Manual. Ver, 1.

Numerical Methods in Geotechnical Engineering IX – Cardoso et al. (Eds)
 ISBN 978-1-138-33198-3

Potential mechanism for recurrent mid-span failure of pile supported river bridges in liquefied soil

Piyush Mohanty
University of Surrey, UK
Scientist, CSIR-Central Building Research Institute, Roorkee, India

Subhamoy Bhattacharya
Chair, Geomechanics Group, University of Surrey, UK

ABSTRACT: The soil structure interaction in case of liquefied soil is not completely understood. As a result, it has been recurrently observed that the middle piers of the pile supported bridges fail in case of liquefaction while the abutments and its adjacent piers remain stable. The current paper proposes a mechanism behind such collapses. It is well-known that the natural periods of the piers increases as the soil liquefies. Due to the inherent river bed profile, the natural period of the central pier increases more as compared to the other ones. Correspondingly, the displacement demand for the central piers increases more. Hence, if the seating length is inadequate, the spans may fall off the piers due to this differential displacement demand. A case study of failure of Showa Bridge has been provided to corroborate this mechanism. It has been observed that the displacement demand for the central piers increased by more than 200% for the central piers, while for the piers close to the abutment it increased by just 50–60%.

1 INTRODUCTION

Bridges are essentially lifeline structures. Therefore, the design of bridges needs to be earthquake resilient so that life and property losses can be minimised and the traffic movement can be resumed in minimum time at the event of any disaster. But, sometimes the liquefaction due to an earthquake at a bridge site poses a special threat to its foundation. Especially the river bridges are quite vulnerable as these are basically founded on alluvial plains with high groundwater table, which is quite susceptible to liquefaction. These bridges on highly liquefaction-susceptible soils such as loose to medium dense sands are often built on pile foundations, since it is required to transfer the axial load to the lower strata of higher bearing capacity. Nonetheless, the phenomenon associated with soil structure interaction in such a case still seems uncertain and inadequately assessed. Therefore, the liquefaction continues to be a cause of failure for many bridges in the past earthquakes; e.g. Showa Bridge in Niigata earthquake (1964), Rio-Viscaya and Rio Estrella Bridge in Telire-Limon earthquake (1991), Miaoziping bridge in Wenchuan earthquake (2008), Juan Pablo II Bridge in Maule earthquake (2010) etc. Some of these bridges failures are shown in Figures 1, 2 and 3. It can be easily observed that the piers close to the mid-span are more damaged as compared to the ones close to the abutments.

It has been well understood that the pile may fail because of excessive bending due to lateral spreading of soil (Boulanger et al. 1999; Brandenberg et al. 2007; Chang GS and Kutter 1989; Varun 2010). This kind of failure is known as bending failure. Sometimes the shear force exerted on the pile due to the inertial load may exceed its design shear capacity (Tang and Ling 2014). This kind of failure is known as the shear failure of piles. Further, the pile may fail due to the buckling owing to the large unsupported length due to liquefaction. This is known as buckling failure. Sometimes it may happen that a significant amount of skin friction resistance may get wiped out because of the liquefaction. Hence, the pile settles further to mobilise more skin friction and end bearing resistance (Armstrong et al. 2014; Fellenius et al. 2008). It induces differential settlement, which is primarily known as settlement failure. But all these failure mechanisms cannot explain the reasons for the midspan failure of pile supported bridges. The recent study carried out by Mohanty et al. (2017), Mohanty and Bhattacharya (2018) found that the piers may fail because of the effects related to the elongation of their natural period due to liquefaction. The present paper highlights this failure mechanism in the light of mid-span failure of pile supported bridges.

Figure 1. Collapse of Rokko Bridge during 2011 Tohoku (Japan) earthquake.

Figure 2. Collapse of Miaoziping Bridge during 2008 Wenchuan (China) earthquake.

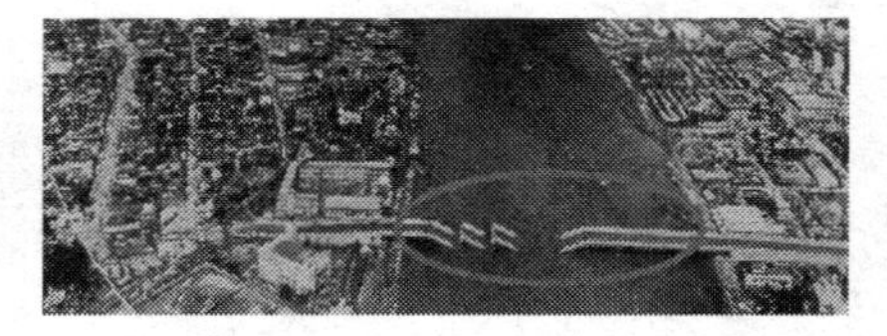

Figure 3. Collapse of Showa Bridge during 1964 Niigata (Japan) earthquake.

2 FAILURE OF PILES DUE TO EFFECTS RELATED TO CHANGE IN NATURAL PERIOD DUE TO LIQUEFACTION

Recently, the study carried out by Mohanty et al. (2017) found that the piers of the pile supported bridges may fail because of the effects related to the elongation of natural period of piers due to the liquefaction. It has been observed that as the liquefaction sets in, the unsupported length of the pile increases as compared to that of before liquefaction. This effect in turn increases the natural period of the piers.

Due to the natural riverbed profile, water depth increases as we move from abutments towards the center of the river channel. Hence, the unsupported length of the mid piers in the event of liquefaction is higher as compared to ones near the abutments. So, the mid-span piers will have higher elongation in their natural periods. Subsequently, the displacement demand for these central pier increases more as compared to the adjacent ones and it may dislodge the deck if sufficient seating length is not provided.

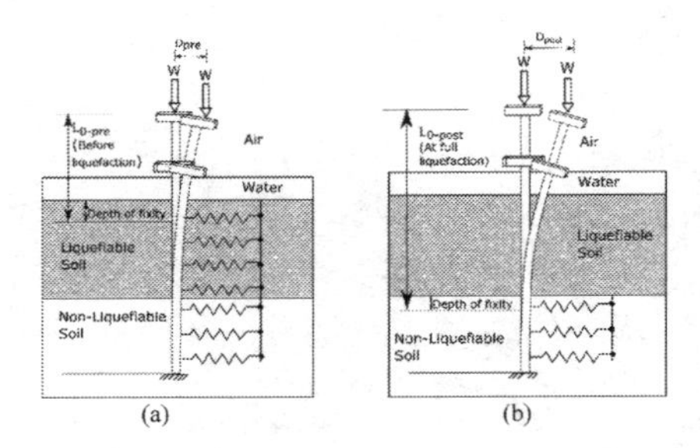

Figure 4. Illustration of a single pile of a bridge for (a) Pre-liquefaction Stage; (b) Post-liquefaction stage.

The equivalent static analysis (single-mode method) has been used for carrying out the analysis. Figure 4 shows mathematical idealization of the problem which is inspired from the Showa and Rokko Bridge configuration (See Figure 1 and 3) and the assumptions used in the analysis are listed below:

(a)Each pier is considered separately and the interactions due to the adjacent piers are neglected; (b) The same section property is used throughout for pile and pier; (c) Piles are flexible and hence, there will be only flexure and no rigid body rotation; (d) The effect of group action on the individual pile is ignored. (e) Pile is axially stable i.e. enough base capacity to resist bearing failure; (f) Piles are laterally unsupported in liquefiable zone and as a result there are no Winkler springs (p-y springs) in the liquefiable part.

The next section of the paper deals with the methodology adopted to carry out the calculations to estimate the natural period of the piers before and after liquefaction and their subsequent displacement demand.

I. Calculate unsupported length of the pile ($L_{0\text{-}pre}$, $L_{0\text{-}post}$):

Before liquefaction only the portion of the piers, which is above the ground, remains laterally unsupported. The depth of fixity is further added to the length of this portion to obtain unsupported length of the pile before liquefaction ($L_{0\text{-}pre}$) (see Figure 4(a)). For the case of full liquefaction, the depth of liquefaction can be determined from the ground profile obtained by carrying out field investigations or standard SPT tests and with the application of simplified methodology depicted in different codes of practice (Idriss and Boulanger 2008). Corresponding depth of fixity is considered in addition to find out the total unsupported length of the pile ($L_{0\text{-}post}$). It can be estimated following Bhattacharya and Goda (2013) and is schematically shown in Figure 4(b).

II. Calculate natural period of vibration (T_{pre}, T_{post}):
Following the work of Lombardi and Bhattacharya (2014), the natural period for the pier is estimated based on idealization depicted in Figure 4. It can be observed that for the pre-liquefaction stage, the stiffness offered by the both the layers contributes to the natural period of the pile-pier system. The equation (1) can be used to find out the natural period for such a case, where $K_{e\text{-}pre}$ is the stiffness of the equivalent pile–pier system before liquefaction and M_e is the equivalent mass lumped at the top of the pile-pier system shown in Figure 4(a).

$$T_{pre} = 2\pi\sqrt{\frac{M_e}{K_{e-pre}}} \quad (1)$$

The complete pile-pier system can be idealized as a fixed cantilever with an unsupported length $L_{0\text{-}pre}$. So the stiffness before the liquefaction becomes

$$K_{e-pre} = \frac{3EI}{L_{0-pre}^3} \quad (2)$$

where EI is the flexural stiffness of the pile.

Conversely, the pile is laterally supported by only the underlying non-liquefiable soil layer after liquefaction. Hence, for the post liquefaction case, the natural period can be estimated using equivalent post liquefaction stiffness $K_{e\text{-}post}$ and equivalent lumped mass M_e for the pile-pier system shown in Figure 4(b). So the natural period becomes

$$T_{post} = 2\pi\sqrt{\frac{M_e}{K_{e-post}}} \quad (3)$$

Where, the complete pile-pier system can be idealized as a cantilever with an unsupported length $L_{0\text{-}post}$. Similarly, the stiffness of the combined pile-pier system at full liquefaction:

$$K_{e-post} = \frac{3EI}{L_{0-post}^3} \quad (4)$$

Besides these formulation, based on Indian Road Congress(IRC) 78 (2014), the period of the pier can be obtained using equation (5) where the pier-pile is idealized as a single cantilever beam carrying the superstructure mass, resting on a foundation.

$$T = 2.0\sqrt{\frac{D}{1000F}} \quad (5)$$

where D = Appropriate dead load of the superstructure and live load in kN

F = Horizontal force in kN required to be applied at the top of the bearings for the earthquake in the longitudinal direction for one mm horizontal deflection at the top of the pier/ abutment

However, Eurocode 8: Part 2 (2011) recommends that the period of the pier can be evaluated by Flexible Deck Model as shown by Equation 2.

$$T = 2\pi\sqrt{\frac{\sum M_i d_i^2}{g\sum M_i d_i}} \quad (6)$$

Mi = mass at the ith nodal point of the pier as when discretized

di = displacement in the direction under examination when the structure is acted upon by force gMi acting at all nodal points in the horizontal direction considered before liquefaction.

The estimation of natural period by the Indian Road Congress (IRC-78, 2014) tends to give a lower natural period as compared to the formulation given by Lombardi and Bhattacharya(2014). In addition, the formula proposed by the Eurocode 8: Part 2(2011) requires the estimation of displacement at different nodal points of the structure due to the inertial force in the horizontal direction, which is usually unknown. Hence, the formulation given by Lombardi and Bhattacharya(2014) has been used in the current analysis.

III. Calculate peak acceleration $\left(A_{pre}, A_{post}\right)$:
The peak acceleration before the liquefaction A_{pre} is determined from T_{pre} using the design response spectra given in various regional codes;e.g.JRA(2002) and Indian Road Congress(IRC-78,2014). As the design response spectra are determined by taking the maximum values of the response of the single degree of freedom system, the applicability of it for the current study is quite pertinent. Similarly, the peak acceleration at full liquefaction A_{post} is determined using T_{post} and the response spectrum provided in the codes of practice as mentioned already.

IV. Calculate peak displacement $\left(D_{pre}, D_{post}\right)$:
The peak displacement D_{pre} is determined from A_{pre} and T_{pre} by using the following relationship.

$$D_{pre} = \left[\frac{T_{pre}}{2\pi}\right]^2 .A_{pre} \quad (7)$$

Similarly, the peak displacement at full liquefaction D_{post} is determined from A_{post} and T_{post} by using the following relationship.

$$D_{post} = \left[\frac{T_{post}}{2\pi}\right]^2 .A_{post} \quad (8)$$

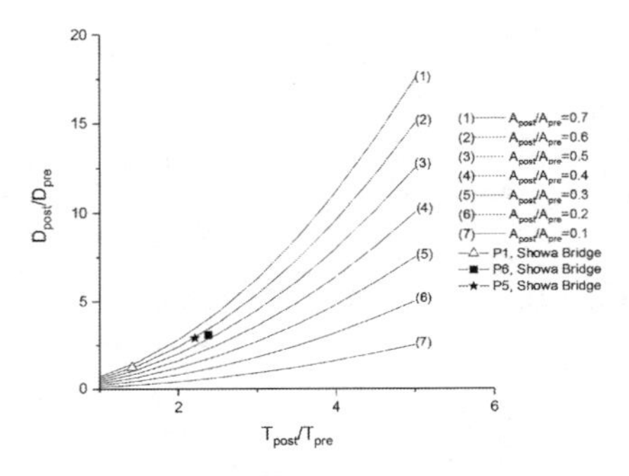

Figure 5. Variation of peak displacement with the increase in natural period.

Furthermore, equations (7) and (8) can be combined to compare the enhanced spectral displacement at full liquefaction and is shown in equation (9). It may be noted that the change in spectral displacement is a function of two parameters: (a) T_{post}/T_{pre}; (b) A_{post}/A_{pre}. However, based on equation (9), it is quite clear that elongation of natural period has a greater influence on the lateral displacement relative to the reduction in spectral acceleration.

$$\frac{D_{post}}{D_{pre}} = \left[\frac{T_{post}}{T_{pre}}\right]^2 \frac{A_{post}}{A_{pre}} \tag{9}$$

Figure 5 shows a graphical representation of Equation (9) for different values of A_{post}/A_{pre} and T_{post}/T_{pre}. It can be easily noticed that as T_{post}/T_{pre} will be higher for mid-span piers relative to the piers adjacent to the abutments, the correlation illustrated in Figure 5 may explain higher lateral displacement demand.

The failure of Showa Bridge in case of Niigata earthquake (1964) is taken as an example to apply the formulation.

3 CASE STUDY OF SHOWA BRIDGE FAILURE DURING NIIGATA EARTHQUAKE (1964)

The Niigata earthquake occurred on 14th June, 1964 and registered a moment magnitude of 7.6. The Showa Bridge, spanning across the Shinano River, was located some 55 km from the epicentre and it collapsed because of the earthquake. The total length of the bridge was about 307 m. The bridge had 12 composite girders and its breadth was about 24 m. The case study is very well documented in and the details of the bridge and the earthquake can be obtained from Bhattacharya et al. (2014).

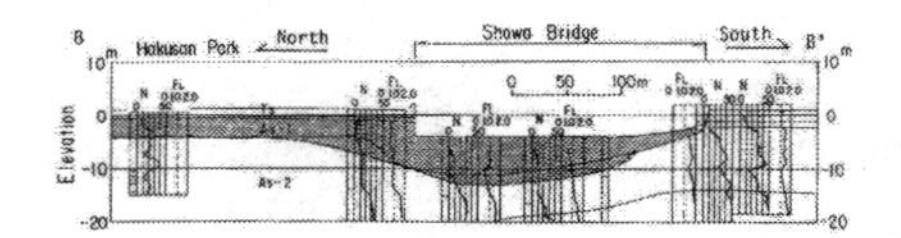

Figure 6. Soil liquefaction profile (in grey) (Hamada and O'Rourke 1992).

3.1 *Liquefaction profile*

Hamada and O'Rourke (1992) estimated the ground liquefaction profile and it has been shown in Figure 6. The soil at the site liquefied to a maximum depth of about 10 m below the riverbed, and this depth decreased towards the abutments (Bhattacharya et al. 2014).

3.2 *Foundation and structural details of the bridge*

The foundation of each supporting pier was a single row of 9 tubular steel piles connected laterally by a pile cap. Each pile was 25 m long with outer diameter (D) of 0.609 m. The wall thickness of the upper 12 m of the pile was 16 mm and the thickness for the bottom 13 m was 9 mm. The material of the Showa Bridge piles, as per the Japanese standard JIS-A: 5525 (JSA, 2004) was assumed to be SKK490 grade steel pipe with the yield strength (σ_y) and ultimate strength (σ_u) of 315 MPa and 490 MPa respectively.

A dead load of 6662 kN from the girder and the slab was being transferred to pier, which in turn was shared by the 9 piles (Bhattacharya 2003). The design live loads are ignored as there was no significant traffic on the bridge during its failure.

3.3 *Estimation of pertinent dynamic design parameters of Showa Bridge*

3.3.1 *Unsupported length of the pile (L_{0-pre}, L_{0-post})*

The unsupported length of the pile in pre-liquefaction and post liquefaction stage is determined with the appropriate depth of fixity as per Bhattacharya and Goda (2013). The unsupported length for the piles is estimated as per the data available from the available literatures (Bhattacharya et al. 2014) and is presented in Table 1.

3.3.2 *Natural period (T_{pre}, T_{post})*

For the estimation of natural period before and after the liquefaction at the Showa Bridge, the geometry of the individual pile foundation, length of the pile above the ground surface and depth of liquefied soil have all been taken from available literatures (Bhattacharya et al. 2014; Dash et al. 2010). Using the equation (1) and (3) the natural period of different piers have been calculated for both the extreme condition and presented in Table 1.

Table 1. Analysis for the Showa Bridge.

Pier No	H_{air} (m)	H_{water} (m)	H_{liq} (m)	$L_{0\text{-}pre}$ (m)	T_{pre} (sec)	D_{pre} (m)	$L_{0\text{-}post}$ (m)	T_{post} (sec)	D_{post} (m)	% Increase in D	Remarks*
P1	6	0	5	9	1.60	0.95	13.4	2.91	1.17	23.15	NC
P2	6	2.5	5	11.5	2.31	1.2	15.9	3.77	1.76	46.66	NC
P3	6	3	6.5	12	2.47	1.3	17.9	4.50	2.8	**115.3**	C
P4	6	3	8	12	2.47	1.3	19.4	5.08	3.3	**153.8**	C
P5	6	3	9	12	2.47	1.3	20.4	5.48	3.78	**190.7**	C
P6	6	3	10	12	2.47	1.3	21.4	5.88	4.0.	**207.6**	C
P7	6	4	4.5	13	2.78	1.6	16.9	4.13	2.5	56.25	NC
P8	6	4.5	1	13.5	2.95	1.6	13.9	3.08	2.2	37.5	NC
P9	6	5	1	14	3.11	1.79	14.4	3.25	2.4	34.07	NC
P10	6	2	0.5	11	2.17	1.09	11	2.14	1.5	37.61	NC
P11	6	0	0.5	9	1.60	0.63	9.4	1.57	0.675	7.142	NC

*H_{air}: Mean height of each pier in air, H_{water}:Mean height of water column at each pile, H_{liq}:Mean depth of liquefaction; NC: Piers didn't collapse after full liquefaction; C: Piers collapsed after full liquefaction;

3.3.3 *Peak acceleration and displacement* *($S_{a\text{-}pre}$, $S_{a\text{-}post}$) and ($S_{d\text{-}pre}$, $S_{d\text{-}post}$)*

The response spectrum of Type I (Level 2) earthquake as prescribed by Japanese code of practice (JRA 2002) is used for the analysis.

The peak acceleration for different piers are estimated from the expressions proposed in Table 6.4.1 ***(Standard Values of the Design Horizontal Seismic Coefficient for Level 2 Earthquake Ground Motion (Type I), khc0) of DESIGN SPECIFICATIONS FOR HIGHWAY BRIDGES*** of Japanese codes of practice (JRA 2002). This ground motion corresponds to an inter-plate earthquake with less probability of occurrence during the bridge service life but strong enough to cause critical damage. This response spectrum was considered to take the worst-case scenario into account while doing a generalised design. T_{pre}, T_{post} are determined for different piers, the same is used to estimate the peak accelerations. These acceleration values are used to find out the peak displacements at the head of each of piers using equations (7) and (8). All these values are mentioned in the Table 1.

4 RESULTS AND DISCUSSIONS

The following points may be noted based on the Table 1:

1. It can be observed that the natural period of different piers increases due to the liquefaction. For soil with greater depth of liquefiable soil, the margin of increase is even higher. For instance; the natural period of the P6 becomes almost 5.8 seconds at full liquefaction, whereas before the liquefaction it was only 2.47 seconds.
2. Secondly, the peak displacement prior to liquefaction ($S_{d\text{-}pre}$) for the different piers can range from 0.9 m to 1.7 m. This value increases as the liquefaction sets in. For example, the peak displacement at full liquefaction ($S_{d\text{-}post}$) for pier no P6 increases up to a value of 4 m.
3. Thirdly, when the degree of increase in the peak displacement of the pile for the different piers before and after liquefaction are compared, it can be seen that for pier no P3, P4, P5, P6, the margin of increase is much higher than that of the other piers. The displacements of the aforementioned piles increased by more than 100% to about 200%, where as for other piles the margin of increase was at most or less than 50%. It may be noted that as the natural period of the pile increases to a higher value due to the liquefaction, the equivalent static force acting at the pile top reduces due to reduction in the value of the peak acceleration. Nonetheless, the lateral displacement due to this effect becomes so high that it may have unseated the deck, which may have caused the failure.
4. It is of interest to note here that this resulting higher displacement may trigger yielding of material of the pile, which can further enhance the final displacement of the members. Sometimes it may happen that this higher resulting displacement may induce geometric nonlinearity (P-Δ) effect in the structure. However, this is beyond the scope of present work and will be carried out in the future course of action.

5 CONCLUSION

A review of various river bridge collapses in liquefiable soils from different earthquakes was carried out where a repeated observation of midspan collapse was noted as given in the Mohanty et al.

(2017). The collapse is caused due to excessive bending of the pier and its foundation leading to fall of the decks. It was also noted that foundations close to the abutments were relatively stable as compared to that of near the midspans despite large lateral spreading being observed. This paper provides an explanation for such observations through well-established analytical calculations.

The mechanism of failure is based on differential elongation of natural period of different piers supporting a river bridge due to subsurface liquefaction. Due to the riverbed profile (i.e. water depth variation along the river width) the increase in natural period for the central piers is more as compared to the adjacent once. Correspondingly, the displacement demand on the central pier also increases as soil progressively liquefies further promoting differential pier-cap displacement. The well-known example of bridge failure (Showa Bridge) is taken to validate the proposed mechanism of failure. It was noted that the displacement demand for central piers increased by more than 100% due to seismic liquefaction owing to the enhanced flexibility of the bridge piers. Hence, the collapse of the spans occurred due to the insufficient seating length. This is in contrast to the piers close to the abutments where the increase in corresponding displacement demand is in the order of 30–50%. It is of interest to note here that this resulting higher displacement may induce geometric nonlinearity (P-Δ) effect in the structure which may further worsen the geometry. However, the calculation concerning this geometric nonlinearity effect has been kept outside the scope of this present work and will be taken in the future course of action.

River bridges in seismic areas are lifeline structures and they must operate even after an earthquake. As codes of practice do not explicitly mention this proposed mechanism, it may have been overlooked in many designs and there remains a risk of such failures. One of the solutions is to increase the stiffness of the foundations for middle piers so that there is limited increase in natural period owing to liquefaction. Future work needs to develop a method to identify the existing unsafe bridge and methods to retrofit them as well as propose a new design methodology for the new upcoming bridges.

REFERENCES

Armstrong, R.J., Boulanger, R.W., and Beaty, M.H. (2014). "Equivalent Static Analysis of Piled Bridge Abutments Affected by Earthquake-Induced Liquefaction." *Journal of Geotechnical and Geoenvironmental Engineering*, 140(8), 1–10.

Bhattacharya, S. (2003). "Pile instability during earthquake liquefaction." University of Cambridge, UK.

Bhattacharya, S., and Goda, K. (2013). "Probabilistic buckling analysis of axially loaded piles in liquefiable soils." *Soil Dynamics and Earthquake Engineering*, 45, 13–24.

Bhattacharya, S., Tokimatsu, K., Goda, K., Sarkar, R., Shadlou, M., and Rouholamin, M. (2014). "Collapse of Showa Bridge during 1964 Niigata earthquake: A quantitative reappraisal on the failure mechanisms." *Soil Dynamics and Earthquake Engineering*, Elsevier, 65, 55–71.

Boulanger, R.W., Curras, C.J., Kutter, B.L., Wilson, D.W., and Abghari, A. (1999). "Seismic Soil-Pile-Structure Interaction Experiments and Analyses." *Journal of Geotechnical and Geoenvironmental Engineering*, 125(9), 750–759.

Brandenberg, S.J., Boulanger, R.W., Kutter, B.L., and Chang, D. (2007). "Static Pushover Analyses of Pile Groups in Liquefied and Laterally Spreading Ground in Centrifuge Tests." *Journal of Geotechnical and Geoenvironmental Engineering*, 133(9), 1055–1066.

Chang GS, and Kutter, B.L. (1989). "Centrifugal modeling of soil-pile-structure interaction." Engineering Geology and Geotechnical Engineering: Proceedings of the 25th Symposium, Reno, Nevada. Rotterdam, the Netherlands, 327–336.

Dash, S.R., Bhattacharya, S., and Blakeborough, A. (2010). "Bending-buckling interaction as a failure mechanism of piles in liquefiable soils." *Soil Dynamics and Earthquake Engineering*, Elsevier, 30(1–2), 32–39.

Eurocode 8: Part 2. (2011). Eurocode 8: Design of structures for earthquake resistance - Part 2: Bridges. European Standard.

Fellenius, B.H., Asce, M., Siegel, T.C., and Asce, M. (2008). "Pile Drag Load and Downdrag in a Liquefaction Event." *Journal of Geotechnical and Geoenvironmental Engineering*, 134(9), 1412–1416.

Hamada, M., and O'Rourke, T.D. (1992). Case studies of liquefaction and lifeline performance during past earthquakes, Vol. 1. National Center for Earthquake Engineering Research, Technical Report NCEER-92–0001.

Idriss, I.M., and Boulanger, R.W. (2008). "Soil liquefaction during earthquakes." *Earthquake Engineering Research Institute*, 136(6), 755.

Indian Road Congress(IRC) 78–2014. (2014). "Standard specifications and code of practice for road bridges."

IRC:6. (2014). *Standard specifications And Code of Practice For Road Bridge. IRC:6–2014*, Indian Road Congress.

JRA. (2002). *Seismic Design Specifications for Highway Bridges*. Japan.

Lombardi, D., and Bhattacharya, S. (2014). "Modal analysis of pile-supported structures during seismic liquefaction." *Earthquake Engineering and Structural Dynamics*, 43(1), 119–138.

Mohanty, P., and Bhattacharya, S. (2018). "Reasons for mid-span failure of pile supported bridges in case of subsurface liquefaction." *Proceedings of Fifth Geo-China International Conference, ASCE*, HangZhou, China.

Mohanty, P., Dutta, S.C., and Bhattacharya, S. (2017). "Proposed mechanism for mid-span failure of pile supported river bridges during seismic liquefaction." *Soil Dynamics and Earthquake Engineering*, 102, 41–45.

Tang, L., and Ling, X. (2014). "Response of a RC pile group in liquefiable soil: A shake-table investigation." *Soil Dynamics and Earthquake Engineering*, 67, 301–315.

Varun, V. (2010). "A non-linear dynamic macroelement for soil structure interaction analyses of piles in liquefiable soils." Georgia Institute of Technology.

Numerical Methods in Geotechnical Engineering IX – Cardoso et al. (Eds)
© 2018 Taylor & Francis Group, London, ISBN 978-1-138-33198-3

Effect of earthquake characteristics on permanent displacement of a cantilever retaining wall

Junied Bakr & Syed Mohd Ahmad
School of Mechanical, Aerospace and Civil Engineering, University of Manchester, UK

ABSTRACT: One of the most significant components of performance-based methods in the seismic design of retaining walls is the accurate estimation of the anticipated permanent displacement. Little attention has been paid to predict the seismic permanent displacement of a cantilever-type retaining wall. Finite element method is used in the current study to predict the seismic permanent displacement of a cantilever retaining wall considering many realistic aspects associated with the real seismic behaviour of the wall-soil system. This study mainly focuses on the evaluation of the effect of the earthquake characteristics and seismic earth pressure on the seismic permanent displacement. The results show that the Newmark sliding block method overestimates the seismic permanent displacement. The most critical scenario, causing maximum permanent displacement, is the one when the ground motion having maximum amplitude but a minimum frequency content. The seismic earth pressure has a low impact on the permanent displacement.

1 INTRODUCTION

For proper seismic design of different types of retaining walls, there has been an increased push in using the performance-based design methods. One of the very important components of the performance-based design process is an accurate prediction of the permanent displacement of the retaining walls. Richards & Elms (1979) proposed the first analysis method to compute the permanent displacement of a gravity type retaining wall. Newmark sliding block method, which was derived to evaluate the seismic stability of slopes (Newmark (1965)), has also been widely used to compute the accumulated permanent displacement in a time history profile of a gravity retaining wall. This method was based on the introducing the concept of yield acceleration in which the gravity retaining wall will accumulate permanent displacement when the earthquake acceleration time history exceeds the yield acceleration level. Following that, extensive efforts have been made to predict the seismic permanent displacement of a gravity-type retaining wall. However, little attention has been paid to compute the seismic permanent displacement of a cantilever retaining wall. The same Newmark sliding block method has still been used to estimate the permanent displacement of a cantilever retaining wall although the basic difference that the cantilever retaining has more complicated geometry than a gravity retaining wall, and it maintains its stability from the weight of backfill soil above footing slab in addition to its self-weight. In general, the Newmark sliding block method has assumed that the seismic earth pressure force, which already computed by using force-based methods like M-O method (Mononobe & Matsuo (1929)), as a part of total driving force causing the permanent displacement; however, the recent experimental methods like Nakamura (2006) and (Jo et al. 2014) have shown that the force-based methods are quite conservative of a gravity and cantilever retaining wall respectively. On the other hand, the real seismic response of a cantilever retaining wall covers by many fundamental factors like the material properties of the wall, backfill and foundation soil, the interface between the wall and backfill soil as well as between the footing slab and foundation soil, boundary condition effect, amplification of acceleration response and phase difference issue. So, it is difficult for the simplified Newmark sliding block method to account all abovementioned problem associated with the real response of a cantilever retaining wall. Some researcher like (Jo et al. 2014), (Candia et al. 2016), and (Jo et al. 2017) conducted a series of centrifuge tests to estimate the seismic earth pressure behind the stem of the wall in order to provide a safe seismic structural design of the stem of the wall; however, few researchers like (Green et al. 2008) and (Kloukinas et al. 2015) carried out numerical study and shaking table test respectively to investigate the permanent displacement of a cantilever type retaining wall.

Despite the fact that the permanent displacement is a focal point in the performance-based methods, the proper design of a cantilever retaining wall should not ignore considering other aspects that directly affect the amplitude of permanent displacement like earthquake characteristics and seismic earth pressure. Hence, little emphasis has been given to the effect of earthquake characteristics and seismic earth pressure on the permanent displacement of the cantilever retaining wall. So, finite element method is proposed in the current study to account all abovementioned aspects in order to allow for more realistic estimation of seismic permanent displacement of a cantilever-type retaining wall and then study the effect of earthquake characteristics and the seismic earth pressure on the permanent displacement of a cantilever retaining wall.

2 PROBLEM DESCRIPTION

As shown in Figure 1 cantilever retaining wall, with a footing slab width L and total stem height H, is constructed to provide lateral support to a horizontal backfill layer. For a proper seismic design of the cantilever-type retaining wall, it should be ensured that retaining wall does not fail by excessive permanent displacement. For a real treatment of the permanent displacement of a cantilever-type retaining wall, it is required to consider the following parameters (Fig. 1): 1) Total increment seismic earth pressure force ($\Delta P_{ae} = P_{ae} - P_a$), computed along the vertical line passing through the heel; where, P_{ae} = total seismic earth pressure force; P_a = total static earth pressure force; 2) Total seismic inertia force of the retaining wall (F_W), which includes the total seismic inertia forces of the stem and footing slab; 3) Total seismic inertia forces of backfill soil above the footing slab (F_S); 4) Total friction resistance force between the footing slab and foundation layer (F_F).

The cantilever-type retaining wall maintains its stability from the weight of backfill soil above the footing slab in addition to its self-weight. So, the seismic earth pressure is assumed to be developed along the vertical line extended from the heel up to the backfill soil surface (Fig. 1). The results obtained from current finite element

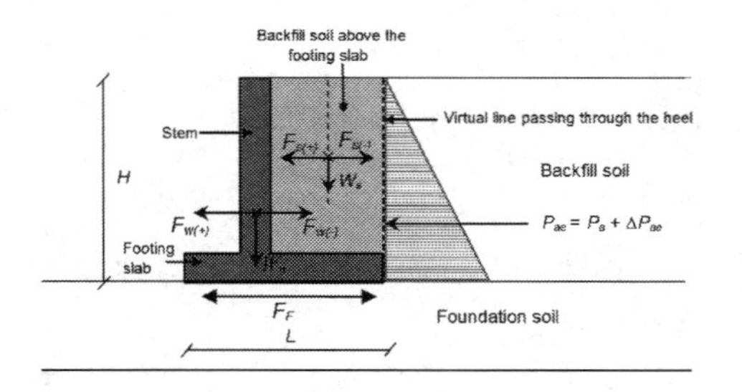

Figure 1. Cantilever retaining wall profile.

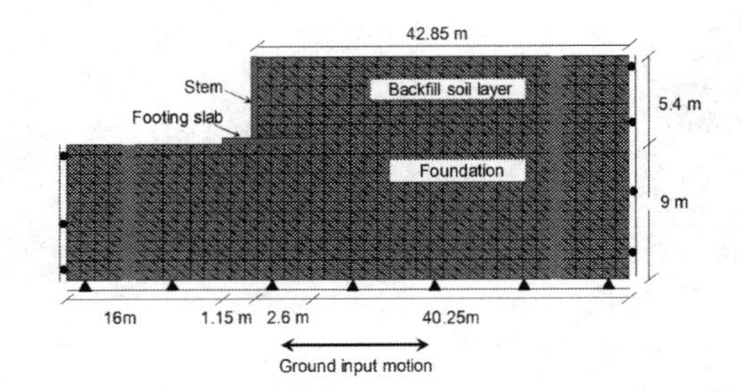

Figure 2. Finite element model of the wall-soil system.

analysis will be compared with traditional methods like Newmark sliding block method. After that, a variety of ground motions is used in order to investigate the effect of earthquake characteristics like amplitude and frequency content on the permanent displacement of a cantilever retaining wall. The effect of seismic earth pressure on the permanent displacement of a cantilever-type retaining wall is also investigated

3 FINITE ELEMENT METHOD

A finite element (FE) model has been developed in current study by using the PLAXIS 2D software (Brinkgreve et al. 2016) as shown in Figure 2 in order to investigate the seismic performance of a cantilever-type retaining wall.

As shown in Figure 2, the height of the retaining wall is 5.4 m and sits on a 9 m thick foundation soil. The stem member is assumed to have a fixed connection with the footing slab. The backfill soil and the foundation soil are modelled using 6-noded triangular elements of the PLAXIS 2D library (Brinkgreve et al. 2016), while the cantilever-type retaining wall was modelled using plate elements. The maximum height of the element is limited by 20% of minimum wavelength as recommended by Kuhlemeyer & Lysmer (1973).

The interaction between the cantilever-type retaining wall and backfill soil as well as between the footing slab and foundation layer has been modelled by using the 6-noded interface elements, available in the PLAXIS 2D library (Brinkgreve et al. 2016). The absorbing boundaries are applied to the vertical boundaries of the finite element model in order to reduce the effect of seismic wave reflection in the finite element domain and to increase the accuracy of the analysis.

4 MATERIAL BEHAVIOUR

The backfill soil and foundation layer are simulated by using hardening soil with small strain model, which is available in PLAXIS 2D (Brinkgreve et

Table 1. The parameters of soil and retaining wall used to run finite element model.

Parameter	Symbol	Unit	Value
Soil			
Relative Density	D_r	%	78%
Unit weight	γ	kN/m³	14.23
Effective friction angle of the soil	ϕ'	°	40
Reference stiffness modulus at 50% of ultimate soil strength	E_{50}^{ref}	MPa	46.8
Reference secant modulus of oedometer test	E_{oed}^{ref}	MPa	46.8
Reference stiffness modulus of unloading reloading	E_{ur}^{ref}	MPa	140.4
Dilatancy angle of the soil	ψ	°	10
Poisson's ratio for unloading-reloading	v_{ur}	-	0.2
Stress-level dependency of the stiffness of the soil	y	-	0.5
Initial shear modulus	G_o^{ref}	MPa	113
Reference shear strain at 70% of G_o^{ref}	$\gamma_{0.7}$	-	0.0002
Reference confining pressure	p^{ref}	kN/m²	100
Damping ratio	ξ	%	3
Failure ratio	R_f	-	0.9
Retaining wall			
Modulus of elasticity	E	MPa	68000
Moment of inertia	I	m⁴	0.00089
Poisson's ratio	v	-	0.334
Unit weight	γ	kN/m³	26.6
Damping ratio	ξ	%	3

al., 2016) library. However, the cantilever retaining wall is simulated by using a linear viscoelastic constitutive model. Table 1 shows the parameters of hardening soil with small strain model as well as the material parameters of a cantilever retaining wall.

5 GROUND SEISMIC MOTION

Real acceleration-time history of the 1952 Kern County earthquake used in the current study to simulate ground seismic motion and it is applied at the nodes of the base boundary of finite element model. The peak ground acceleration (PGA) of this ground seismic motion is 0.24g (see Fig. 3a). A fast Fourier transform analysis is conducted for the earthquake time history in order to obtain the dominant frequency of ground input motion. A shown in Figure 3b, the applied ground seismic motion

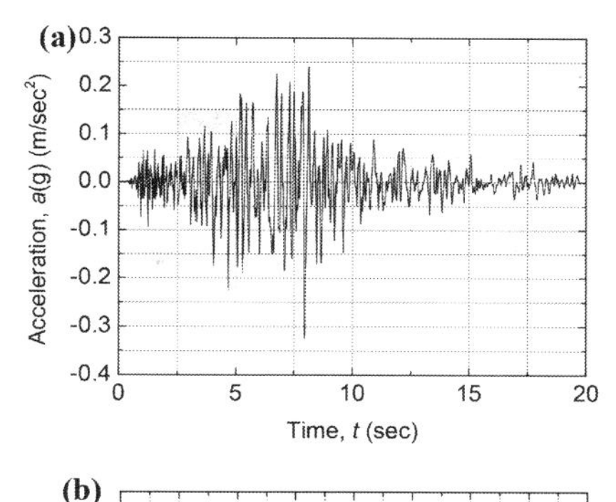

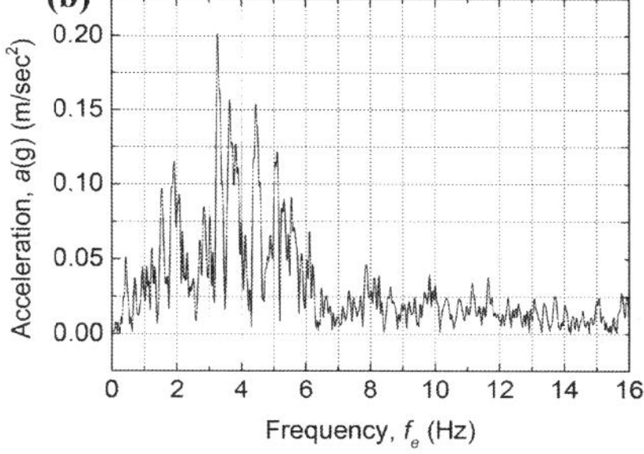

Figure 3. a) Real acceleration-time history of the 1952 Kern County earthquake; b) Frequency domain representation of the 1952 Kern County earthquake.

has three dominant frequencies (f_e) of (0.4 / 1.9 / 3.33 Hz). The uniform sinusoidal ground motions are also used in the current study to investigate the effect of earthquake characteristics on the permanent displacement of the wall, and they are simulated by three groups according to their frequency content; group 1: f_e = 0.5Hz, group 2: f_e = 2Hz, and group 3: f_e = 4Hz. For each group, the uniform sinusoidal ground seismic motion is scaled by three maximum amplitudes of 0.2g, 0.4g, and 0.6g.

6 SEISMIC PERMANENT DISPLACEMENT

Two permanent displacement profiles were predicted from the current finite element analysis in order to understand the seismic deformation mechanism of the wall-soil system. In the first instance, the permanent displacement was calculated between the footing slab of the retaining wall and and a point located in the foundation layer at a depth of 0.5m; while in the second instance, the permanent displacement was computed between the centre of gravity of the backfill soil above footing slab and a point located in the foundation layer at a depth of 0.5m below the footing slab.

It can be noted from Figure 4 that the normalised permanent displacement between the retaining wall and the foundation soil (*d/H*) attains a

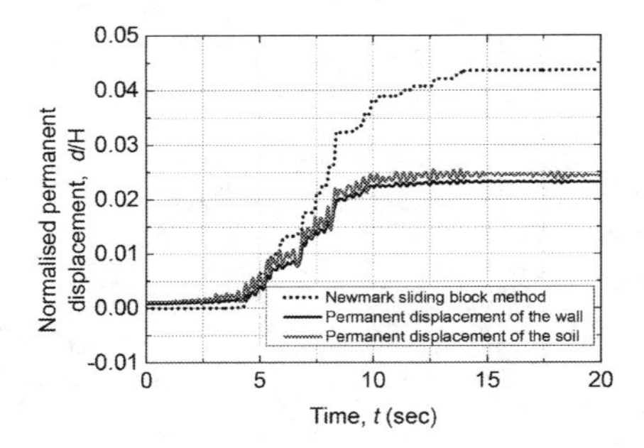

Figure 4. Seismic permanent displacement predicted by current finite element analysis of the wall-soil system.

maximum value of about 0.023 for the time duration 5–10sec. It is also observed that the normalised permanent displacement between the backfill soil and the foundation soil (*d*/*H*) also achieves its maximum value of about 0.024 for the time duration 5–10 sec. After 10 sec of the time of ground motion, it is observed that there is no change in the permanent displacement of both the wall and backfill soil until the end of the analysis. Thus, from Figure 4 it can be said that the retaining wall and backfill soil slide together away from the backfill soil, and almost move as a single entity. It can be noted from the Figure 4 that the wall-soil system is also accumulated permanent displacement towards the backfill soil, but their amplitudes are smaller than the amplitude of accumulated displacement of the wall-soil system away from the backfill soil. The permanent displacement of the wall-soil system is also computed by using Newmark sliding block method. The acceleration predicted in current finite element analysis at mid-height of the wall-soil system is used in Newmark method to compute the yield acceleration. The friction angle between the footing slab and foundation layer is assumed 20°, and the same was used in current finite element analysis. In conjunction the weight of the retaining wall and backfill soil above footing slab with the friction angle between the footing slab and foundation layer as well as the total seismic earth pressure force computed along the vertical line passing through the heel by using M-O method, the maximum yield acceleration (*N*.g) is found equal to 0.18g. The comparison between the results obtained from current finite element analysis and Newmark sliding block method as shown in Figure 4 shows that the normalised permanent displacement computed by Newmark method (d/H = 0.046) is remarkably larger than that predicted by finite element method. This is because the Newmark sliding block method did not take into account the problems associated with real seismic behaviour of wall-soil system like the realistic representation of seismic earth pressure force.

7 EFFECT OF EARTHQUAKE CHARACTERISTICS ON THE PERMANENT DISPLACEMENT

To investigate the effect of earthquake characteristics on the permanent displacement of the wall-soil system, a variety of ground motions are applied at the base of finite element model. Figures 5a, 6a, and 7a show three groups of ground motions applied at the base of the finite element model with frequency content 0.5Hz, 2Hz, and 4Hz respectively. It can also be noted that the amplitude of ground motion in each group is simulated by 0.2g, 0.4g, and 0.6g. Figure 5b, 6b, and 7b show the permanent displacement of the cantilever retaining wall for three groups of applied ground motion described above respectively. Figure 5b, 6b, and 7b that that as the amplitude of the input motion increases from 0.2g to 0.6g the permanent displacement of the retaining wall increases, while with an increase in the frequency content of the input motion from f = 0.5Hz to f = 4 Hz, the permanent displacement reduces. It is also interesting to note that the retaining wall slides by about d/H = 0.038 at an ground motion amplitude of 0.6g and a frequency content of 4 Hz (Fig. 7b) while it slides by about d/H = 0.05 for a ground motion amplitude of 0.4g and a frequency content of 2 Hz (Fig. 6b). This suggests that the frequency content of the input motion is a more dominating factor than its amplitude which contributes to the permanent displacement of the retaining wall. From the above discussion, it can be very safely

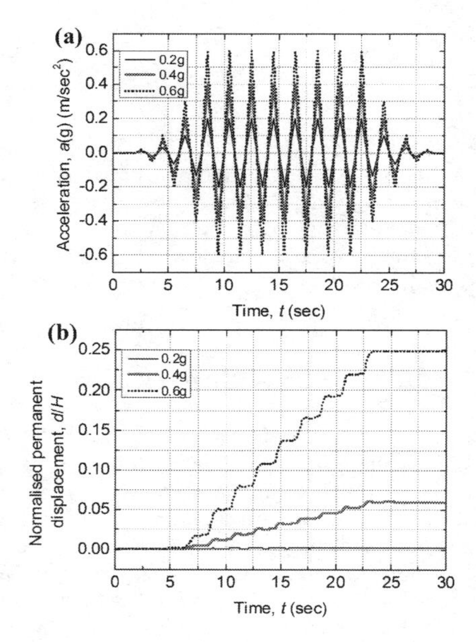

Figure 5. Group of f_e = 0.5Hz; a) Ground input motions b) Permanent displacement.

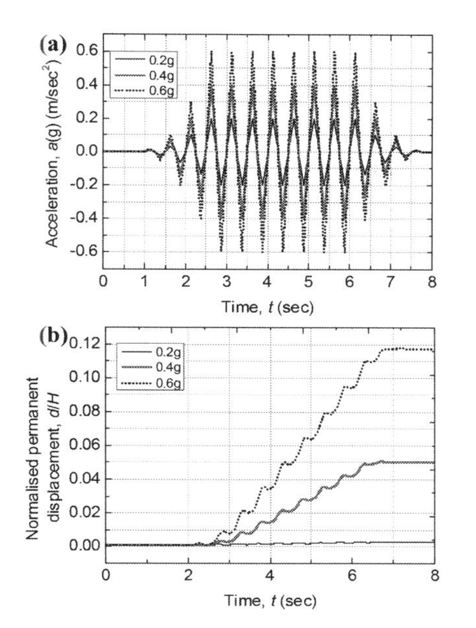

Figure 6. Group of $f_e = 2$; a) Ground input motions b) Permanent displacement.

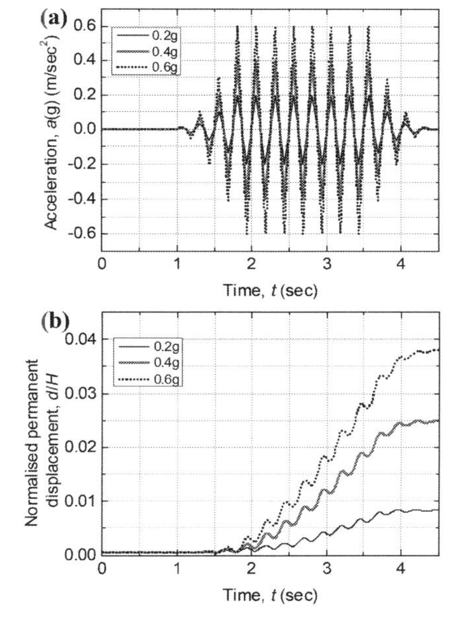

Figure 7. Group of f_e = 4Hz; a) Ground input motions b) Permanent displacement.

argued that a low frequency content of the seismic ground motion and maximum amplitude cause a critical case scenario. The results also show that the accumulated permanent displacement of the retaining wall is highly sensitive to the number of acceleration cycles (duration of the input motion) in which the retaining wall is still accumulated permanent displacement away from the backfill soil with increasing of the number of acceleration cycle.

8 TOTAL INCREMENT OF SEISMIC EARTH PRESSURE FORCE

The total seismic earth pressure force P_{ae} has been calculated along virtual line passing through the heel. The total increment of seismic earth pressure force ΔP_{ae} is estimated by subtracting the total seismic earth pressure force P_{ae} from the total static earth pressure force P_a. Figure 9 shows the total increment of seismic earth pressure force, estimated at the virtual line passing through the heel (ΔP_{ae}) between the time 5 sec -10 sec when the maximum permanent displacement of the wall-soil system has been accumulated. It can noted from the Figure 8 that when the ground acceleration applied towards the backfill soil (for example at time 5.5 sec, 7.1 sec, and 9 sec (Fig. 3a) the total increment of seismic earth pressure force is close to zero, and the total seismic earth pressure force is close to the static earth pressure force. However, when

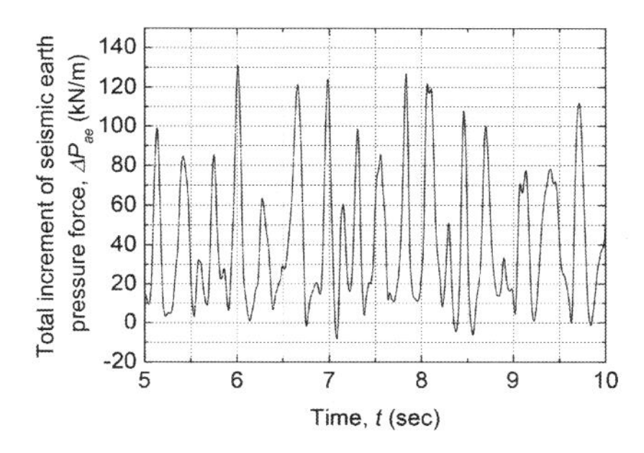

Figure 8. Total increment of seismic earth pressure force predicted by finite element analysis.

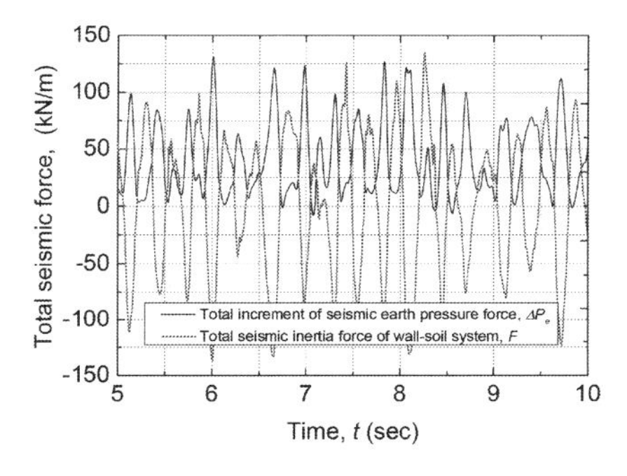

Figure 9. Total increment of seismic earth pressure force and total seismic inertia force predicted by finite element analysis.

the ground acceleration changes its direction away from the backfill soil (for example at time 6 sec, 7 sec, and 7.5 sec (Fig. 3a), the maximum increment of seismic earth pressure force is developed along the virtual line passing through the heel.

To more realistic treatment of the contribution of seismic forces causing the permanent displacement of the wall, Figure 9 shows the total seismic inertia force of the wall-soil system $F = F_W + F_S$, and it is combined with the total increment of seismic earth pressure force ΔP_{ae} between the time 5sec to 10 sec. Reading the Figure 9 with Figure 3a and 5 together, it can be noted that at the time when the acceleration of ground motion is applied towards the backfill soil (for example at time 5.5 sec, 7.1 sec, and 9 sec), the wall-soil system is accumulated maximum permanent displacement away from the backfill soil and the total inertia force of the wall-soil system is acting away from the backfill soil while the total increment of seismic earth pressure force has the minimum value. However, when the ground acceleration is applied away from the backfill soil (for example at time 6 sec, 7 sec, and 7.5 sec), the wall-soil system is accumulated permanent displacement towards the backfill soil, and the seismic inertia force of wall system is acting towards the backfill soil, and the wall-soil system is resisted by the maximum seismic earth pressure force. Thus, it can be observed that the total seismic earth pressure force has low impact in the accumulated permanent displacement away from the backfill soil, while it has a remarkable contribution to resist the accumulated permanent displacement towards the backfill soil.

9 CONCLUSION

The main aim of this paper is to use performance-based method for predicting the seismic permanent displacement of a cantilever-type retaining wall. Emphasis has been given to investigate the effect of earthquake characteristics and the seismic earth pressure on the earthquake-induced permanent displacement of the cantilever retaining wall. Finite element method is proposed in the current study to consider more real aspects of the seismic behaviour of the wall-soil system. The soil behaviour is simulated in the current numerical analysis by using hardening soil with small strain model. The results of the current study show that the Newmark sliding block method overestimated the permanent displacement of the cantilever retaining wall. A variety of ground input motions were applied at the base of the finite element model with different amplitude and frequency content to investigate the effect of earthquake characteristics on the permanent displacement of the cantilever retaining wall. The results of the parametric study shows that the critical scenario for the design of the cantilever retaining wall, predicting maximum permanent displacement, is when the ground motion is subjected to a ground motion having maximum amplitude but a minimum frequency content. The parametric study also shows that the permanent displacement is highly sensitive to the duration of the earthquake. The result also show that the seismic earth pressure force has a low impact in the accumulated permanent displacement away from the backfill soil, while the same has significant effect to resist the accumulated permanent displacement towards the backfill soil.

REFERENCES

Brinkgreve, R., Engin, E. & Swolfs, W. 2016. PLAXIS 2016. *PLAXIS bv, The Netherlands*.

Candia, G., Mikola, R.G. & Sitar, N. 2016. Seismic response of retaining walls with cohesive backfill: Centrifuge model studies. *Soil Dynamics and Earthquake Engineering* 90: 411–419.

Green, R.A., Olgun, C.G. & Cameron, W.I. 2008. Response and modelling of cantilever retaining walls subjected to seismic motions. *Computer-Aided Civil and Infrastructure Engineering* 23(4): 309–322.

Jo, S.-B., Ha, J.-G., Lee, J.-S. & Kim, D.-S. 2017. Evaluation of the seismic earth pressure for inverted T-shape stiff retaining wall in cohesionless soils via dynamic centrifuge. *Soil Dynamics and Earthquake Engineering* 92: 345–357.

Jo, S.-B., Ha, J.-G., Yoo, M., Choo, Y.W. & Kim, D.-S. 2014. Seismic behaviour of an inverted T-shape flexible retaining wall via dynamic centrifuge tests. *Bulletin of earthquake engineering* 12(2): 961–980.

Kloukinas, P., di Santolo, A.S., Penna, A., Dietz, M., Evangelista, A., Simonelli, A.L., Taylor, C. & Mylonakis, G. 2015. Investigation of seismic response of cantilever retaining walls: Limit analysis vs shaking table testing. *Soil Dynamics and Earthquake Engineering* 77: 432–445.

Kuhlemeyer, R.L. & Lysmer, J. 1973. Finite element method accuracy for wave propagation problems. *Journal of Soil Mechanics & Foundations Div* 99: 421–7.

Mononobe, N. & Matsuo, M. 1929. On the determination of earth pressures during earthquakes. *Proceedings, World Engineering Congress* 9: 179–187.

Nakamura, S. 2006. Reexamination of Mononobe-Okabe Theory Of Gravity Retaining Walls Using Centrifuge Model Tests. *Soils and Foundations* 46(2): 135–146.

Newmark. 1965. Effect of earthquakes on dams and embankments. *Geotechnique* 15: 139–159.

Richards JR, R. & Elms, D.G. 1979. Seismic Behaviour of Gravity Retaining Walls. *Journal of Geotechnical and Geoenvironmental. Dn., ASCE* 105: 449–464.

Rock mechanics

Numerical Methods in Geotechnical Engineering IX – Cardoso et al. (Eds)
© 2018 Taylor & Francis Group, London, ISBN 978-1-138-33198-3

Arching of granular flow under loading in silos

P. To & N. Sivakugan
College of Science and Engineering, James Cook University, Australia

ABSTRACT: Sublevel scheme is a major type of underground mining. At each sublevel, mine is blasted to form a silo, so that ore can be extracted from the bottom outlet of the silo. Due to the extraction, ground surface will have some significant subsidence, which can influence nearby infrastructure. To avoid this impact, a filling material might be stopped from upper sublevel during the extraction. This additional load creates spatial constraints for the granular flow and might stimulate the arching of this granular material. The research employs Discrete Element Method to simulate granular flow in silos under general vertical loadings. Particles are generated by Voronoi tessellation at a given porosity. The sample is shaken lightly so that particles can achieve a more random position and neutral contact. After that, they are released under gravity to form a granular flow. A parameter study is undertaken to find the condition of arching. A force is assigned to a top plane to simulate a top load, which increase stresses in force chains. The research draws out the influence of the geometry and top loads on arching probability in a silo.

1 INTRODUCTION

Mining is one of major contributors to economy of not only Australia but also many other countries. Because valuable minerals are often placed at a certain depth under a thick layer accumulated over millions years of residual soils and sedimentary rocks, underground mining is inevitable in many cases. Among many types of subsurface mining, sub-level scheme is a big category, included many frequent types such as sub-level caving (Jian 1991), sub-level stopping (Dowd & Elvan 1987), and sub-level shrinkage (Gu, Guo, Li, & Wang 2012). In a sub-level scheme of mining, the massive mine is divided to many levels. The ore is blasted to form huge silos and, then, extracted from outlet points on each level. After the extraction, huge voids like silos are left. They can be back filled during or after the extraction to avoid collapse (Sivakugan, Rankine, Rankine, & Rankine 2006, Kratzsch 2012).

This paper focuses on arching during the extraction. If the opening at the outlet points is too small, blasted ore particles may interlock to form a stable arch, which supports the ore mass and interrupts the extraction. This arching effect is of common interest to not only mining but also many other industries such as agriculture, chemistry, and pharmacy. In general, the shapes and dimensions of the moving particles and bottom outlet are of vital importance for arching and granular flow pattern (Oldal, Keppler, Csizmadia, & Fenyvesi 2012, Duran 2012, Schulze 2006). Besides, the stress state inside ore mass sometimes plays a decisive role. High loads may compact particles and make the density exceed the critical value of jamming density (Ogarko & Luding 2012). When particles are closer to each other, they have more potential to form force chains and interlock themselves (Hidalgo, Lozano, Zuriguel, & Garcimartín 2013).

Experimental studies are often undertaken in laboratory at a small scale (Terzaghi 1936, Zuriguel, Garcimartín, Maza, Pugnaloni, & Pastor 2005, Ting, Shukla, & Sivakugan 2010). The velocity fields are captured by Particle Image Velocimetry (Slominski, Niedostatkiewicz, & Tejchman 2007) while force chain analyses are facilitated by Photoelasticity (Tang & Behringer 2011). However, the tiny size of particles is a challenge for an accurate analysis in 3D. This paper employs numerical simulation to study the influence of (i) the contracting angle at the bottom outlet and the size ratio of particles to the opening size, and (ii) top loads.

When particle sizes are insignificant in comparison with the geometry of silos, stress effect on granular flows can be studied by using Finite Different Method (Li, Aubertin, Simon, Bussière, & Belem 2003). However, Discrete Element Method (DEM), where each ore particle is presented by an individual numerical object (Cundall & Strack 1979), is preferable when the particle sizes are not negligible. However, most of the current DEM studies on arching use spheres or blocks of spheres to take advantages of their mathematical simplicity (Sakaguchi, Ozaki, & Igarashi 1993, Zuriguel, Janda, Garcimartín, Lozano, Arévalo, & Maza 2011, Weinhart, Labra, Luding, & Ooi 2016). This may not be acceptable for the simulations of bulky ore particles with significant shape effects (To & Scheuermann 2014). Recently, Alonso-Marroquín, Ramírez-Gómez, González-Montellano, Balaam, Hanaor, Flores-Johnson, Gan, Chen, & Shen (2013) proposed to use sphero-polyhedra (Pournin & Liebling 2005, Galindo-Torres, Pedroso, Williams, & Li 2012) to investigate arching in 2D silos.

This paper extends the use of sphero-polyhedra to the simulation of silo in 3D. The outlet of a rectangular silo is altered to have different opening sizes and angles. To investigate the influence of load, a significant force is assigned to the top plane. The simulation results show that the applied force does not significantly influence the arching of the granular flow if the silo is high enough. Besides, some recommendations for rectangular silos are proposed. These may help the mining industry in explosion design at some level.

2 DISCRETE ELEMENT SIMULATION

The simulation is created in C++ with help from Mechsys (MechSys 2017), a multi-physics library for simulation in open source systems. Two sets of simulations are employed to investigate the arching of particles at the bottom outlet. The first set investigates the impact of outlet's geometry on the arching of particles. Meanwhile, the second set focuses on the influence of top loads. In both sets, the generated ore volume is of 4 m * 4 m * 5 m with the porosity of 0.4 (Figure 1). The mean particle size d is approximately 0.2 m. The algorithm chart of the simulation is shown in Figure 2, and a brief description is as below.

- Step 1: All necessary parameters, such as geometrical dimensions and geotechnical coefficients, are read from a text file.
- Step 2: After that, a set of points are generated based on the given dimensions of the silo and particle size distribution of ores. Basing on the set, a 3D Voronoi tessellation is generated to acquire a pack of angular particles. A number of particles will be removed so that the pack can obtain a given porosity (Figure 1).
- Step 3: A silo is created to contain all particles. All boundary planes are fixed and not able to move. Then, the pack is shaken slightly to avoid bonding contacts between particles and achieve a random distribution.
- Step 4 (optional): To investigate the impact of top loads, a significant force is assigned to the top plane. The vertical restraint of this plane is removed.
- Step 5: The bottom outlet is open and particles are released under gravity.

Although the required number of particles in step 2 can be calculated by dividing the given

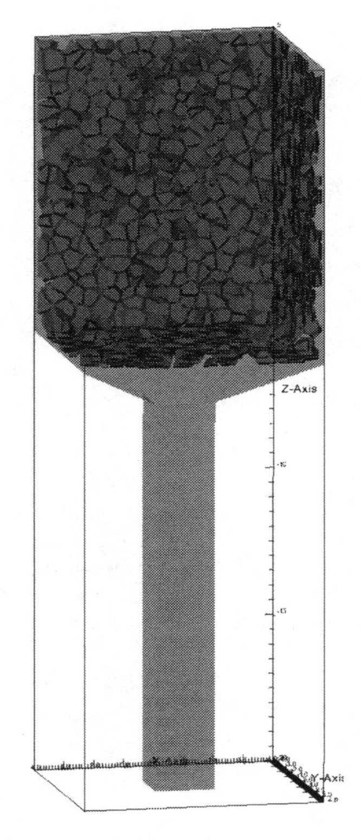

Figure 1. Silo setup.

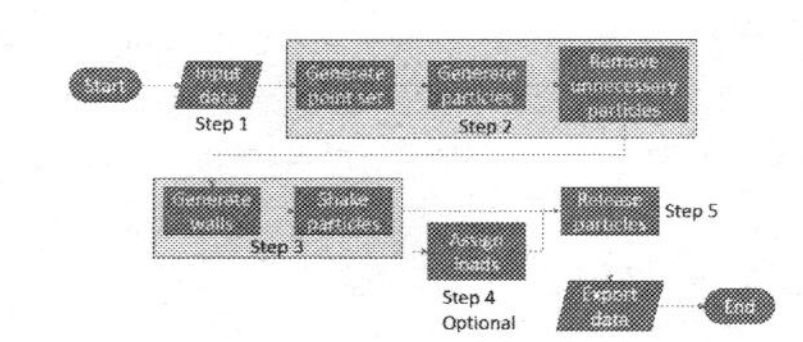

Figure 2. Algorithm chart of the simulation.

particle size distribution into equal intervals (To, Scheuermann, & Galindo-Torres 2015), this paper employs narrowly graded soils where particles are of a similar size to reduce the computational burden. Each particle is eroded slightly (Galindo-Torres, Pedroso, Williams, & Li 2012) to have a roundness radius R of vertices and edges at approximately 0.1 of the mean size d (Figure 3). This tiny erosion does not significantly change the volume of particles, but reduces loads of computational steps and avoid complicated systems of equations while finding contact force between two adjacent polyhedra (Figure 4).

The normal contact force $\vec{F}_n$ is calculated by the normal elastic stiffness K_n and overlapping distance δ (Galindo-Torres, Scheuermann, Mühlhaus, & Williams 2015, To, Torres, & Scheuermann 2015):

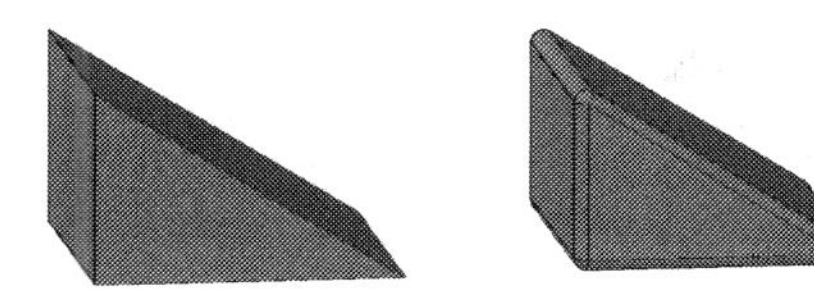

Figure 3. Sphero-polyhedron before (left) and after (right) erosion.

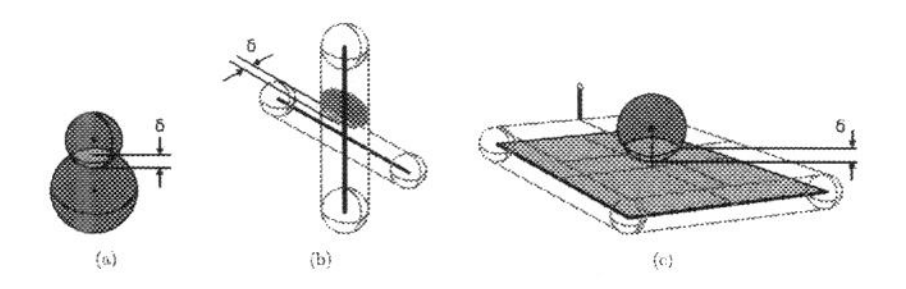

Figure 4. Three of typical contacts between sphero-polyhedra particles after Galindo-Torres et. al. (2012): a) vertex-vertex; b) edge-edge; c) vertex-face.

$$\vec{F}_n = K_n \delta \vec{n} \tag{1}$$

where: $\vec{n}$ = normal direction.

The tangential contact force during the collision, $\vec{F}_t$, is bounded by the Coulomb limit (Galindo-Torres, Pedroso, Williams, & Li 2012) and is estimated by:

$$\vec{F}_t = min(K_t \delta_t, \mu \vec{F}_n)\vec{t} \tag{2}$$

where: δ_t = tangential displacement (Luding 2008); μ = friction coefficient; and $\vec{t} = \vec{v}_t / v_t$ with $\vec{v}_t$ = tangential velocity.

Arching occurs when particles are bonded together to form a stable structure. To model bonding effect, the normal and tangential contact forces are calculated by a simplified Euler's Beam, based on the area of common face (Galindo-Torres, Pedroso, Williams, & Li 2012, Pöschel & Schwager 2005):

$$\vec{F}_n = B_n A R(\varepsilon_n)\vec{n} \qquad \vec{F}_t = B_t A(\varepsilon_t)\vec{n} \tag{3}$$

where: B_n, B_t = normal and tangential elastic moduli of the material, A = bonding face area, $R(\hat{I}_n)$ = ramp function depending on the normal strain,

$\varepsilon_n, \varepsilon_t$ = normal and tangential strains.

In many DEM simulations, time steps are defined as (O' Sullivan 2011, D' Addetta 2004):

$$dt \le \sqrt{\frac{m}{K_n}} \tag{4}$$

where: m = mass of the smallest particle; K_n = maximum normal stiffness.

However, this estimation seems to be conservative for a granular flow simulation because it was initially defined for a dense sphere packing, where particles move over an insignificant distance. When the travelling distance is remarkable, the time step selection depends on the maximum kinetic energy E_k and the maximum allowed overlapping distance (To, Galindo-Torres, & Scheuermann 2016). If a tiny particle moves too fast, it may go through the silo's boundary within a time interval smaller than the defined time step. Therefore, the collision could not be detected. Nevertheless, the prediction of the maximum kinetic energy is complicated because it changes after every collision. In this current study, the time step is input manually after several attempts and defined approximately 0.1 the traditional time step.

After an equal time interval d_{tout}, a video frame with particles' data is recorded for later analyses. Other DEM simulation parameters are given in Table 1.

Table 1. DEM simulation's parameters.

Parameter	Value	Unit
Normal stiffness, K_n	$3 \cdot 10^7$	N/m
Tangential stiffness, K_t	$1.7 \cdot 10^7$	N/m
Normal viscous coefficient, G_n	$1.6 \cdot 10^4$	s^{-1}
Tangential viscous coefficient, G_t	$0.8 \cdot 10^4$	s^{-1}
Time step, dt	10^{-5}	s
Intermediate output step, dt_{out}	0.1	s
Total time of simulation, t_f	30	s
Limit kintetic energy, K	10^{-7}	Nm
Rolling stiffness coefficient, β	0.12	
Plastic moment coefficient, ν	1.0	

3 INFLUENCE OF GEOMETRY

Granular flow patterns are often classified as mass or funnel flows (Jenike 1970). However, these patterns usually do not change if the geometry remains the same. To distinguish the granular flows influenced by different parameters, a new classification of flow pattern is proposed. The granular flow is considered as a free flow if there is a continuous chain/block of movement from the bottom to the top of the ore mass (Figure 5b). All particles in this chain have a remarkable velocity so that they can travel more than half of the mean size d per second ($\approx$ 100 mm/s). That means the particles cannot be assumed to be at the same position at the particle scale. If this fast moving block spread over the whole height of the ore mass (Figure 5d), it is considered as a partial flow. When arching occurs (Figure 5e), all particles stay still.

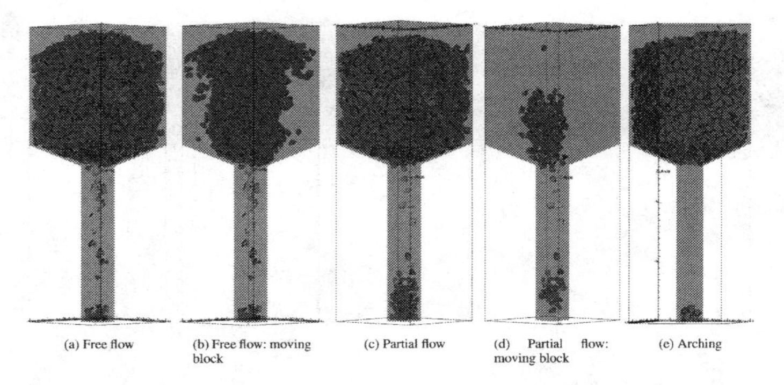

(a) Free flow (b) Free flow: moving block (c) Partial flow (d) Partial flow: moving block (e) Arching

Figure 5. Flow pattern. Particles in red (or lighter in the grey scale) are moving slowly. These particles are removed in picture b) and d) to have a clear view of the moving block.

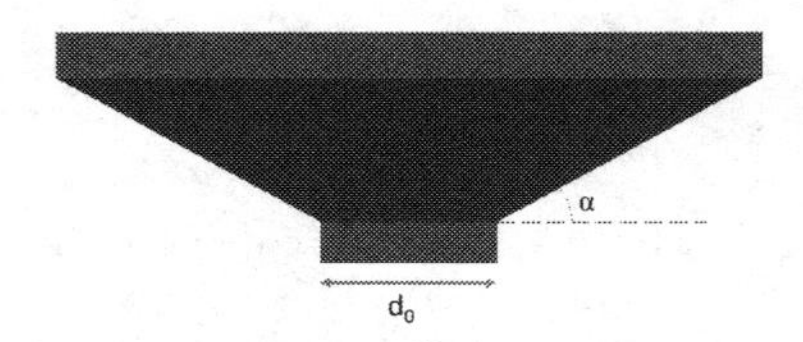

Figure 6. Size and angle of the opening outlet.

Note that the moving block is located at the middle and limited by some shear faces. However, unlike the simulation of spheres by DEM and other methods (Kafaji 2013, González-Montellano, Ramirez, Gallego, & Ayuga 2011), the shear faces are not smooth, obvious, and stable due to irregular shapes of particles.

The size d_0 and angle α of the opening outlet (Figure 6) are alternated to investigate the influence of geometry on arching of particles. d_o varies from $1.8d$ to $6.6d$, while α runs from 30° to 75° by step of 15°. These are frequent values for silos in mining industries. The simulation of each configuration have been undertaken by an MPI cluster 64 i7 @ 3.2 GHz for more than 20 times to obtain a probability of arching. The results are shown in Figure 7. Note that, the geometrical configurations with probability over 40% mostly have a partial flow pattern if there is no arching.

It is obvious from the graph that when $d_o > 4d$, the probability does not depends on α, and particles usually go through. When $d_o < 3d$, configurations with α = 45° has a more gradual change in probability of arching. The moving block in simulation with α = 30° is limited by shear faces with inclination of approximately 70° (Figure 3b and 3d). Therefore, the flow patterns where α = 30°, 60°, and 75° are actually similar. Note that, although silos with big α often have a mass flow pattern (Jenike 1970) the velocity field of angular particles are not homogeneous like of spheres. Top particles stay almost still while bottom particles have been extracted at a high rate (Figure 8). When $d_o < 2d$,

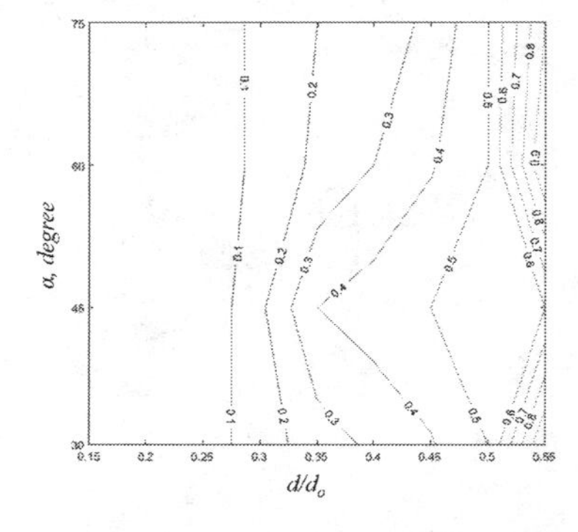

Figure 7. Probability of arching for varied sizes and angles of the opening outlet.

(a) All particles (b) Moving block

Figure 8. Partial flow with α = 45°, $d/d_o = 0.4$, and no loading.

the arching probability jumps up, and most of the flows stop within few seconds after releasing.

4 INFLUENCE OF LOADINGS

To investigate the influence of loading, The simulation set employ does not randomly change the particle packing as in the section above, but uses just one particle packing with exactly the same particle distribution. A force is assigned to the top plane to obtain a surcharge varied from 0.25 MPa to 2.5 MPa. This simulates the stress from back filling mass with the height from 10 to 100 m respectively. The simulation was undertaken with $\alpha = 30°$ and d/d_o runs from 0.35 to 0.55 (Table 2). The simulation results show that the application of small surcharge does not provide any additional constraints to the flow. In contrast, it facilitates the free flow by adding forces to the top particles, which barely move.

When the load increases, top particles are almost frozen and the flow returns to a partial flow pattern. The main reason is that particles interlock/bond together under big stresses. This increases friction and produces additional spatial constraints. This way, most of the load is carried by the wall and still particles around the moving block. Therefore the load does not have any direct impact on the moving block and granular flow at the bottom outlet. This leads to a statement that the loads from back filling materials may not have a significant impact on the ability of ore to be extracted.

As the particles around the moving block are compressed by the top load, they cannot give space to the moving block. Hence, particles in the block have a trend to move inside the block, i.e. toward the central line. This way, the angle of the shear faces are almost vertical (Figure 9b).

If the ore mass is not high enough to form a stable interlocking layer, the applied force acting directly on moving particles near the outlet. This leads to a simulation error because particles move too fast and exceed the maximum allowed overlapping distance. Thereby, the particle collisions are unreal and no more reflect particle behaviours.

Table 2. Granular flow under varied loads.

	d/d_o		
Loads	0.35	0.45	0.55
0.00 MPa	Free	Partial	Arching
0.25 MPa	Free	Free	Arching
0.50 MPa	Free	Free	Arching
0.75 MPa	Free	Partial	Arching
1.00 MPa	Free	Partial	Arching
1.75 MPa	Partial	Partial	Arching
2.50 MPa	Partial	Partial	Arching

Figure 9. Partial flow with α=30°, $d/d_o = 0.45$, and surcharge = 1MPa.

5 CONCLUSION

This study conducted two sets of simulation to investigate the influence of geometry and loadings on the granular flow of angular particles in rectangle silos.

Based on the results, the ratio d/d_o seems to be the decisive factor for the flow pattern and arching probability in most of the studied cases, except when $\alpha = 45°$. The block of fast moving particles is often limited by shear faces at the inclination of approximately 70°. When $\alpha = 45°$, the flow has a lower arching probability. However, the flow pattern in this situation is often partial. This may be good for mining thanks to a low discharge velocity.

Although top loads may provide any additional constraints to the granular flow, they do not have direct impact on the outlet. In contrasts, a small surcharge may enhance the movement at some level. When the stress is big, it compacts the particles surrounding the moving block. However, moving particles tend to travel inward to the block's central line. Therefore, the flow is not clogged and the shear faces are almost vertical.

Nevertheless, all simulations employ narrow graded particle size distribution. In a future study, particles of mixed shapes and sizes will be investigated.

ACKNOWLEDGEMENT

The author would like to express deep gratitude to Mr Patrick Wilson at Carpentaria Gold, who inspired the simulation. A special thank to Mr Jack Nicolosi at James Cook Univeristy, who worked under supervision of the first author. Although his results are not included in this paper, they sparked some ideas for this paper and future research.

REFERENCES

Alonso-Marroquín, F., Á. Ramírez-Gómez, C. González-Montellano, N. Balaam, D.A. Hanaor, E. Flores-Johnson, Y. Gan, S. Chen, & L. Shen (2013). Experimental and numerical determination of mechanical properties of polygonal wood particles and their flow analysis in silos. *Granular Matter 15*(6), 811–826.

Cundall, P.A. & O.D. Strack (1979). A discrete numerical model for granular assemblies. *Geotechnique 29*(1), 47–65.

D' Addetta, G.A. (2004). *Discrete models for cohesive frictional materials.*

Dowd, P. & L. Elvan (1987). Dynamic programming applied to grade control in sub-level open stopping. *Trans. IMM 96*, A171–A178.

Duran, J. (2012). *Sands, powders, and grains: an introduction to the physics of granular materials.* Springer Science & Business Media.

Galindo-Torres, S., D. Pedroso, D. Williams, & L. Li (2012). Breaking processes in three-dimensional bonded granular materials with general shapes. *Computer Physics Communications* 183(2), 266–277.

Galindo-Torres, S., A. Scheuermann, H. M¨uhlhaus, & D. Williams (2015). A micro-mechanical approach for the study of contact erosion. *Acta Geotechnica 10*(3), 357–368.

González-Montellano, C., A. Ramirez, E. Gallego, & F. Ayuga (2011). Validation and experimental calibration of 3d discrete element models for the simulation of the discharge flow in silos. *Chemical Engineering Science 66*(21), 5116–5126.

Gu, C.-H., Z.-Y. Guo, Q.-Y. Li, & W.-H. Wang (2012). Application of sublevel shrinkage caving method in fine-crushed ore-body. *Mining and Metallurgical Engineering 32*(5), 26–29.

Hidalgo, R., C. Lozano, I. Zuriguel, & A. Garcimartín (2013). Force analysis of clogging arches in a silo. *Granular Matter 15*(6), 841–848.

Jenike, A.W. (1970). *Storage and flow of solids.* University of Utah Salt Lake City.

Jian,W. (1991). Theory and practice of sub-level caving method in china [j]. *Journl of China Coal Society 3*, 000.

Kafaji, I.K. a. (2013). *Formulation of a dynamic material point method (MPM) for geomechanical problems.*

Kratzsch, H. (2012). *Mining subsidence engineering.* Springer Science & Business Media.

Li, L., M. Aubertin, R. Simon, B. Bussière, & T. Belem (2003). Modeling arching effects in narrow backfilled stopes with flac. In *Proceedings of the 3rd international symposium on FLAC & FLAC 3D numerical modelling in Geomechanics, Ontario, Canada*, pp. 211–219.

Luding, S. (2008). Cohesive, frictional powders: contact models for tension. *Granular matter 10*(4), 235–246.

MechSys (2017). Mechsys website.

O' Sullivan, C. (2011). *Particulate Discrete Element Modelling: A Geomechanics Perspective. Applied Geotechnics.* Spon Press/Taylor & Francis.

Ogarko, V. & S. Luding (2012). Equation of state and jamming density for equivalent bi-and polydisperse, smooth, hard sphere systems. *The Journal of chemical physics 136*(12), 124508.

Oldal, I., I. Keppler, B. Csizmadia, & L. Fenyvesi (2012). Outflow properties of silos: The effect of arching. *Advanced Powder Technology 23*(3), 290–297.

Pöschel, T. & T. Schwager (2005). *Computational granular dynamics: models and algorithms.* Springer Science & Business Media.

Pournin, L. & T.M. Liebling (2005). A generalization of distinct element method to tridimensional particles with complex shapes. In Powders and Grains 2005, Volume 2, pp. 1375–1378. AA Balkema Publishers.

Sakaguchi, H., E. Ozaki, & T. Igarashi (1993). Plugging of the flow of granular materials during the discharge from a silo. *International Journal of Modern Physics B 7*(09n10), 1949–1963.

Schulze, D. (2006). Storage of powders and bulk solids in silos. *Dietmar Schulze. com.*

Sivakugan, N., R. Rankine, K. Rankine, & K. Rankine (2006). Geotechnical considerations in mine backfilling in australia. *Journal of Cleaner Production 14*(12), 1168–1175.

Slominski, C., M. Niedostatkiewicz, & J. Tejchman (2007). Application of particle image velocimetry (piv) for deformation measurement during granular silo flow. *Powder Technology 173*(1), 1–18.

Tang, J. & R. Behringer (2011). How granular materials jam in a hopper. *Chaos: An Interdisciplinary Journal of Nonlinear Science 21*(4), 041107.

Terzaghi, K. (1936). Stress distribution in dry and in saturated sand above a yielding trap-door.

Ting, C.H., S.K. Shukla, & N. Sivakugan (2010). Arching in soils applied to inclined mine stopes. *International Journal of Geomechanics 11*(1), 29–35.

To, H. & A. Scheuermann (2014). Separation of grain size distribution for application of self-filtration criteria in suffusion assessment. In *Proc., 7th Int. Conf. on Scour and Erosion, ICSE*, pp. 121–128.

To, H.D., S.A. Galindo-Torres, & A. Scheuermann (2016). Sequential sphere packing by trilateration equations. *Granular Matter 18*(3), 1–14.

To, H.D., A. Scheuermann, & S.A. Galindo-Torres (2015). Probability of transportation of loose particles in suffusion assessment by self-filtration criteria. *Journal of Geotechnical and Geoenvironmental Engineering 142*(2), 04015078.

To, H.D., S.A.G. Torres, & A. Scheuermann (2015). Primary fabric fraction analysis of granular soils. *Acta Geotechnica 10*(3), 375–387.

Weinhart, T., C. Labra, S. Luding, & J.Y. Ooi (2016). Influence of coarse-graining parameters on the analysis of dem simulations of silo flow. *Powder technology 293*, 138–148.

Zuriguel, I., A. Garcimartín, D. Maza, L.A. Pugnaloni, & J. Pastor (2005). Jamming during the discharge of granular matter from a silo. *Physical Review E 71*(5), 051303.

Zuriguel, I., A. Janda, A. Garcimartín, C. Lozano, R. Arévalo, & D. Maza (2011). Silo clogging reduction by the presence of an obstacle. *Physical review letters 107*(27), 278001.

Numerical Methods in Geotechnical Engineering IX – Cardoso et al. (Eds)
© 2018 Taylor & Francis Group, London, ISBN 978-1-138-33198-3

Boundary stress distribution in silos filled with granular material

P. To & N. Sivakugan
College of Science and Engineering, James Cook University, Australia

ABSTRACT: Silo, a very popular structure in powder and mining industry, is a vertical container with an open outlet at the bottom and an optional inlet at the top. The design of silo requires a deep understanding of stress distribution at boundaries in both static and dynamic condition. Prior numerical studies use Finite Element Method or Finite Difference Method which shows an increase of vertical stress before it is leveled out by friction at a shallow depth. This is understandable for continuous media because the settlement caused by vertical stress must stop at some level. Nevertheless, experimental studies show that the vertical stress of the granular and porous media still increases with a constant rate even at a much greater depth. Although the Discrete Element Method (DEM) can simulate granular materials, it has some difficulties in determination of stress distribution because it is based on contact force, not the stress. This paper employs DEM with sphero-polyhedra shapes to simulate the behaviour of granular materials in silos. The stress distribution is calculated as average values. This requires a significant number of particles. Therefore, the paper focuses on narrowly graded materials. Some correlation with experimental data has been found.

1 INTRODUCTION

Silo, a common technological term in many industries, refers to a vertical container with a contracting outlet at the bottom. In underground mining, huge empty silos are what remain when ore bodies have already been extracted. These large voids may collapse and lead to ground subsidence, which influences mining activities and nearby infrastructure. Therefore, they are often closed and backfilled with waste rocks or tailings. These are called stopes. The closure requires some good understanding of vertical stress inside the stopes, which the outlet must support. The failure can cause not only an interruption of production but also fatal accidents for people working in a close proximity.

The failure rate of silos is more frequent than the collapse rate of some other industrial structures, including buildings (Theimer 1969, Dogangun, Karaca, Durmus, & Sezen 2009). The main reasons consist of, but not limited to, explosion and bursting (Li 1994, Mason & Lechaudel 1998), filling and discharging scheme (Piskoty, Michel, & Zgraggen 2005), earthquakes (Bechtoula & Ousalem 2005), thermal ratcheting (Carson 2001), and large axial stress (Bozozuk 1976). This paper focuses on vertical stress acting at the bottom of the silo.

To estimate the stress, traditional approaches often use a plane strain problem, which has been used to evaluate the stress at the bottom of trenches (Marston 1930), bins or hoppers. (Drescher et al. 1991) Li & Aubertin (2008) proposed an analytical solution for the distribution of vertical stresses and validated it with Fast Lagrangian Analysis of Continua - FLAC (Itasca Consulting Group 2017). Later, an analytical study showed that the increase of vertical stress should stop at a certain depth (Widisinghe & Sivakugan 2015). However, an experimental study showed that the hypothesis may be conservative. (Sivakugan, Widisinghe, & Wang 2013)

A possible reason is that most of the current numerical simulations for stress states consider contained materials as a continuous medium, which actually is a particulate one. The assumption is acceptable at some level if the size of particles is significantly smaller than the size of silos. However, a soil is a discrete medium, so a simulation with Discrete Element Method (DEM) may be required (To, Scheuermann, & Galindo-Torres 2015).

This study employs Voronoi tessellation and sphero-polyhedra (Pournin & Liebling 2005, Galindo-Torres, Pedroso, Williams, & Li 2012) to simulate angular particles. Uniform spheres are used to simulate round particles. The load acting on the bottom plane is recorded and used to calculate average the vertical stress. Material properties are defined experimentally and imported to numerical model, which is built from Mechsys, a C++ multi-physics library (MechSys 2017).

Due to the limit of computational ability, the simulation investigates only narrow-graded soils.

A respective laboratory setup is also built to have some comparison. Some good agreement between numerical and experimental results is observed.

2 DISCRETE ELEMENT SIMULATION

This section discusses DEM simulation employed in this study and some background information. Most of DEM numerical simulations start from a particle packing procedure (To, Galindo-Torres, & Scheuermann 2016). In this study two types of granules are studied: angular (or cube-like) and spherical. To some extent, spheres can represent fine non-cohesive soil particles, while angular particles can simulate gravels and massive ores. Although this assumption is very common in DEM simulation, a microscopic comparison between a sand and an artificial sand showed that sand particles, especially crushed sands, can be considered as angular at some level (To & Scheuermann 2014).

To generate a pack of angular particles, Voronoi tessellation in 3D is employed. A set of points is built in accordance to the given number of particles and particle sizes. These points are considered as tessellation centres of particles. The middle planes between the points are the boundary faces of particles. For an easy understanding, a 2D illustration is also demonstrated in Figure 1.

To facilitate the contact detection in DEM, the model employs sphero-polyhedra (Galindo-Torres, Pedroso, Williams, & Li 2012). All particles generated above are eroded/smoothed by a small amount to have a roundness radius of edges at approximately 0.1 of the size of the smallest generated particles (Figure 2). Given that the volume does not change significantly, this simplifies the determination of contact forces because colliding contacts between particles become typical contacts among spheres, cylinder, and planes. Otherwise, the estimation of the overlapping distance between two adjacent particles would be a complicated system of equations.

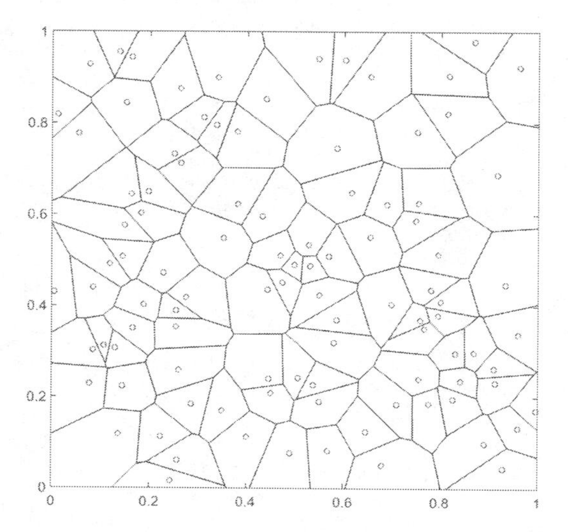

Figure 1. Voronoi tessellation.

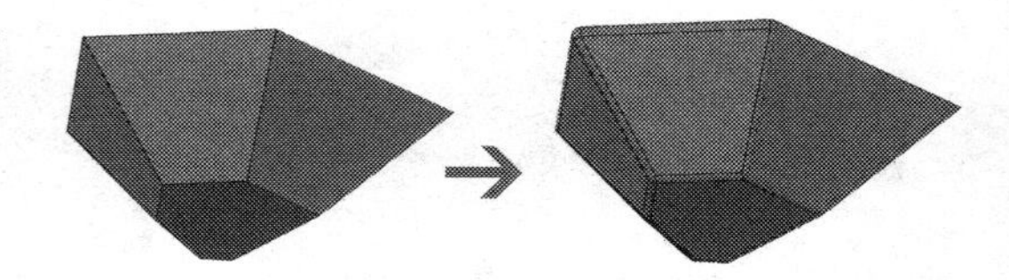

Figure 2. Formulation of a sphero-polyhedron.

The contact forces are defined by material properties and the overlapping distance. The normal contact force $\vec{F}_n$ in each contact is estimated by (Galindo-Torres, Scheuermann, Mühlhaus, & Williams 2015, To, Torres, & Scheuermann 2015):

$$\vec{F}_n = K_n \delta \vec{n} \tag{1}$$

where: K_n = normal stiffness; δ = overlap distance; $\vec{n}$ = normal direction.

The tangential contact force during the collision, $\vec{F}_t$, is bounded by the Coulomb limit (Galindo-Torres, Pedroso, Williams, & Li 2012) and is estimated by:

$$\vec{F}_t = min(K_t \delta_t, \mu \vec{F}_n)\vec{t} \tag{2}$$

where: K_t = tangential stiffness; δ_t = tangential displacement (Luding 2008); μ = friction coefficient; and $\vec{t} = \vec{v}_t / v_t$ with $\vec{v}_t$ = tangential velocity.

Because the tangential force depends on $\vec{t}$, which is defined by velocity $\vec{v}_t$, the friction diminishes when particles are static. Therefore, particles are released under gravity for an insignificant distance to observe the maximum friction before it plunges. The viscous force, $\vec{F}_v$, is added to dissipate the energy and simulate the inelastic collisions (Galindo-Torres, Scheuermann, Mühlhaus, & Williams 2015):

$$\vec{F}_v = G_n m_e v_n \vec{n} + G_t m_e v_t \vec{t} \tag{3}$$

where: G_n, G_t = normal and tangential dissipation constants respectively, m_e = effective mass of the colliding particle pair, v_n normal velocity.

To model bonding, the normal and tangential contact forces are calculated by a simplified Euler's Beam, based on the area of common face (Pöschel & Schwager 2005):

$$\vec{F}_n = B_n A R(\varepsilon_n)\vec{n} \qquad \vec{F}_t = B_t A(\varepsilon_t)\vec{n} \tag{4}$$

where: B_n, B_t = normal and tangential elastic moduli of the material, A = shared face area, $R(\varepsilon_n)$ = ramp function depending on the normal strain, ε_n,ε_t = normal and tangential strains.

There is an alternative model using overlapping area between dilated spheres (Potyondy 2015). However, this model use a bunch of spheres to simulate a plain surfaces. The environmental viscous force, $\vec{F}_{vv}$, acting on each particle is calculated, if needed, by the energy dissipation parameter, G_v:

$$\vec{F}_{vv} = -G_v m\vec{v} \tag{5}$$

where: m = mass of particle, $\vec{v}$ = velocity.

To simulate the roughness of particles, MechSys (2017) employed a rolling stiffness, β, which is calculated from tangential stiffness (Belheine, Plassiard, Donzé, Darve, & Seridi 2009). This method avoids the rolling of spherical particles with high angular velocity at one position. In many DEM simulations, time steps are defined as (O' Sullivan 2011, D' Addetta 2004):

$$dt \le \sqrt{\frac{m}{K_n}} \tag{6}$$

where: m = mass of the smallest particle; K_n = maximum normal stiffness.

However, this traditional estimation seems to be conservative. A practice in 2D simulation showed that this critical time increment should be deducted by a factor of 0.8 (O' Sullivan & Bray 2004). A big time step may lead to a big overlapping distance. Note that, if the overlapping distance is larger than the eroded distance described above, the simulation halts because of unrealistic collision or compression. For 3D simulations, time steps can be smaller (To, Torres, & Scheuermann 2015). Moreover, the time step in a simulation with sphero-polyhedra may be deducted further due to tiny spheres representing vertices. In this current study, the time step is input manually after several attempts. A rough estimation of the deducting factor is approximately 0.25. Other DEM simulation's paramater are introduced in Table 1

A given porosity of the numerical sample can be achieved by eliminating randomly a respective number of particles (Figure 3 a). The initial sample is stiff because most of the particle contacts are face to face. Therefore, it is shaken slightly before the simulation to obtain a more discrete distribution.

Apart from the packing of angular particles, two samples consisting of spheres are generated by mathematical algorithms (Figure 3 b, c). Although there are several approaches to pack spheres, including an authorial sequential algorithm (To, Galindo-Torres, & Scheuermann 2016), their application may not make a significant difference because spheres are of just one size. To simulate a smooth wall, spherical particles are generated by a cubic grid, so that no significant horizontal contact force will occur (Figure 3 c).

Note that, material parameters for numerical samples are imported from laboratory results (Table 2). The simulations are undertaken by some desktop computers i7@3.4GHz.

Figure 3. Numerical samples: (a) frictional wall and angular particles, (b) smooth wall and spherical particle, and (c) frictional wall and spherical particles.

Table 1. DEM simulation's parameters.

Parameter	Value	Unit
Time step, dt	10^{-5}	s
Intermediate output step, dt_{out}	0.1	s
Total time of simulation, t_f	3	s
Limit kintetic energy, K	10^{-7}	Nm
Rolling stiffness coefficient, β	0.12	
Plastic moment coefficient, v	1.0	

Table 2. Material parameters.

Parameter	Value	Unit
Young Modulus of particle, E	30	GPa
Cohesion, c	0	kPa
Internal Frictional angle, φ	34	°
Young Modulus of wall, E_{wall}	3.2	GPa
Wall cohesion, c_{wall}	0	kPa
Wall Frictional angle, φ_{wall}	12	°

3 BENCHMARKING MODELS

This section describes three different approaches to have benchmarks for the DEM model: experimental, analytical and numerical models. Because they have already been developed in prior studies, the models are not described in detail here.

3.1 *Laboratory model*

In order to have an experimental benchmark with DEM simulation, a specified silo has been built in laboratory from 12 rectangular sections of 150 mm × 150 mm × 75 mm, filled by a fine sand (Figure 4). To investigate the vertical stress a certain depth, the respective number of sections will be used. This model actually has been developed in a prior study with some extended application for vertical stress at the bottom of trenches (Sivakugan, Widisinghe, & Wang 2013). The sand and the sections are supported by a hanging hook at the top and a plane at the bottom. The load acting on the bottom plane is measured by a scale. A tiny gap between the bottom section and the plane to ensure that the measured load is solely the sum of vertical stresses at the bottom boundary. Meanwhile, the sum of tangential stresses at the vertical boundary is the total load acting on the hanging hook. The laboratory tests use fine sand, material parameters of which are represented in Table 2.

Figure 4. Laboratory setup.

3.2 *Analytical model*

Assuming soils as an elastic solid medium, shear load at the lateral boundary of a small layer can be calculated as (Sivakugan, Widisinghe, & Wang 2013):

$$dS = (K\sigma_z tan\delta + c_a)Pdz \tag{7}$$

where K = ratio of horizontal to vertical stresses, σ_z = vertical stress, δ, c_a = frictional angle and adhesion of the wall, P = lateral perimeter of the cross section, and dz = layer's thickness. This shear load causes the deduction from the overburden stress to the vertical stress. Solving the integral equation over the height of silo, Sivakugan, Widisinghe, & Wang (2013) proposed to calculate the vertical stress in a square cross section silo with no surcharge by:

$$\sigma_z = \frac{\gamma L}{4Ktan\delta}(1 - e^{-Ktan\delta\frac{4}{L}z}) \tag{8}$$

where: L = width of the square.

Note that, the estimation of K depends on stress state of soil (Mesri & Hayat 1993) and interaction with walls (Krynine 1945). In this benchmark, the active earth pressure coefficient (Rankine 1857) and a simplified coefficient for soil at rest (Jaky 1944) are employed:

$$K_a = \frac{1 - sin\phi}{1 + sin\phi} \qquad K_o = 1 - sin\phi \tag{9}$$

where: φ = frictional angle of the soil, K_a = active earth pressure coefficient, and K_o = earth pressure coefficient for soil at rest.

3.3 *Finite element model*

Soil model by Finite Element Method (FEM) and Finite Difference Method (FDM) has been used extensively to study stress state of soils for decades. However, its application on porous media is still in improving progress. This paper uses results of a prior FDM model in *FLAC^{3D}* conducted by Sivakugan, Widisinghe, & Wang (2013).

4 RESULTS & DISCUSSION

To normalise the influence of the sample size and soil density, the results are plotted in the graph of $\sigma_z/\gamma L$ versus a depth ratio z/L (Figure 5). The studied range of depth varies from 0 to 6L for analytical approaches, and from L to 6L for DEM simulation.

The experimental result shows an almost linear relationship. This trend may stop at a certain

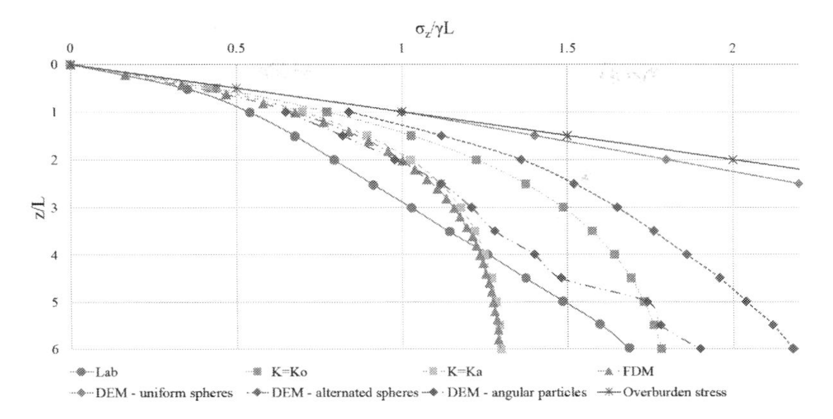

Figure 5. Results' comparison.

depth, which is not observed within the studied range. Meanwhile, results of FEM and analytical method using K_a show a good coherence to each other. It is logical as the analytical method tries to solve the integral of stress over a solid and continuous volume as FEM does. Note that, the analytical analysis employs a rough wall friction angle, where $\delta \approx \phi$. However, both methods underestimate the vertical stress and showed that it goes close to a vertical asymptote at a shallow depth.

The application of K_o with $\delta = \phi/2$ overestimates σ_z. However, It shows the same trend with application of K_a. Hence, the selection of K may not influence significantly the depth, where σ_z stops increasing (Figure 5).

The DEM simulation with a cubic arrangement of spheres (Figure 3b) shows a linear increase of σ_z with depth. Because the spheres are positioned by the mathematical algorithm, they can be stabilised by themselves and have almost no interaction with the walls. This simulates the behaviours of absolutely smooth walls, where σ_z is equal to the overburden stress.

Because the ideal situation above may not happen in the real world, an alternative arrangement has been used (Figure 3c). This way, spheres interact well with the walls. Hence, σ_z is reduced significantly and the result comes closer to the experimental result. Nevertheless, it still overestimates the vertical stress although the rates are similar. One possible reason for this difference is the assumption that soil particles are spherical, but they are not. This way friction is underestimated due to tiny contact areas, which are neglected in equation 2. Therefore, an increase of the number of particles may not help.

Among all employed approaches, DEM simulation with angular particles delivers the closest distribution of σ_z to the experimental result. In addition, it does not seem to meet an asymptote within the studied range of depth. The graph even does not signify any possible vertical asymptote. On the one hand, a smaller σ_z in comparison with the simulation of spheres is caused by the inclusion of bonding interaction and arching of particles, if occurs. It raises up a concern that the current approach for the tangential force between spheres may be inadequate for soil particles. On the other hand, a larger σ_z in comparison with the analytical and FEM analyses reflects the discrete medium of soil. Not the whole of the area of the vertical wall is in contact with particles to produce friction.

Note that, the trend of the simulation with angular particles is not so clear because the variation of σ_z is not smooth as the results of the other approaches. Some obvious reason may be the position, orientation, and arrangement of angular particles. A feasible solution is an increase of the sample size so that the particle size is insignificant in comparison with the sample size. This may be time consuming and exceeds the computational capacity of a normal desktop. A use of high performance computers (HPC) is proposed for the future studies.

5 CONCLUSION

In conclusion, this paper studies the vertical stress at the bottom boundary of silo and trenches by experimental, analytical and numerical (FEM and DEM) approaches. Although FEM and analytical models show a good coherence to each other, both of them underestimate the vertical stress. Therefore, the application of FEM and the analytical method should be scrutinised carefully as it may leads to some unforeseen collapse.

Simulations by DEM with simple spheres may not suitable for non-cohesive soils because the friction with wall may not be simulated well due to the lack of bonding interaction. This may be improved if the model introduce new way to calculate friction of spheres.

Meanwhile, the numerical simulation with angular particles by DEM show some agreement with the experimental data. Nevertheless, further simulations with more particles may be required.

ACKNOWLEDGEMENT

Authors would like to express deep gratitude to Dr Daniel Cocks at James Cook University for his help in setting up an Ubuntu Cluster for the simulation. A special thank to Dr Galindo Torres at The University of Queensland for his help with MechSys, the C++ multi-physics library used for the simulation.

REFERENCES

Bechtoula, H. & H. Ousalem (2005). The 21 may 2003 zemmouri (algeria) earthquake: Damages and disaster responses. *Journal of Advanced Concrete Technology 3*(1), 161–174.

Belheine, N., J.-P. Plassiard, F.-V. Donzé, F. Darve, & A. Seridi (2009). Numerical simulation of drained triaxial test using 3d discrete element modeling. *Computers and Geotechnics 36*(1), 320–331.

Bozozuk, M. (1976). Cbd-177 tower silo foundations. *Canadian Building Digest*.

Carson, J.W. (2001). Silo failures: Case histories and lessons learned. *Handbook of Powder Technology 10*, 153–166.

D' Addetta, G.A. (2004). *Discrete models for cohesive frictional materials*.

Dogangun, A., Z. Karaca, A. Durmus, & H. Sezen (2009). Cause of damage and failures in silo structures. *Journal of performance of constructed facilities 23*(2), 65–71.

Drescher, A. et al. (1991). *Analytical methods in bin-load analysis*. Elsevier.

Galindo-Torres, S., D. Pedroso, D. Williams, & L. Li (2012). Breaking processes in three-dimensional bonded granular materials with general shapes. *Computer Physics Communications 183*(2), 266–277.

Galindo-Torres, S., A. Scheuermann, H. Mühlhaus, & D. Williams (2015). A micro-mechanical approach for the study of contact erosion. *Acta Geotechnica 10*(3), 357–368.

Itasca Consulting Group, I. (2017). Flac software.

Jaky, J. (1944). The coefficient of earth pressure at rest. *Journal of the Society of Hungarian Architects and engineers*, 355–388.

Krynine, D. (1945). Discussion of stability and stiffness of cellular cofferdams by karl terzaghi. *Transactions, ASCE 110*, 1175–1178.

Li, H. (1994). Analysis of steel silo structures on discrete supports.

Li, L. & M. Aubertin (2008). An improved analytical solution to estimate the stress state in subvertical backfilled stopes. *Canadian Geotechnical Journal 45*(10), 1487–1496.

Luding, S. (2008). Cohesive, frictional powders: contact models for tension. *Granular matter 10*(4), 235–246.

Marston, A. (1930). The theory of external loads on closed conduits in the light of the latest experiments. In *Highway research board proceedings*, Volume 9.

Mason, F. & J. Lechaudel (1998). Explosion of a grain silo in blaye france. *Summary Rep*.

MechSys (2017). Mechsys website.

Mesri, G. & T. Hayat (1993). The coefficient of earth pressure at rest. *Canadian Geotechnical Journal 30*(4), 647–666.

O' Sullivan, C. (2011). *Particulate Discrete Element Modelling: A Geomechanics Perspective. Applied Geotechnics*. Spon Press/Taylor & Francis.

O' Sullivan, C. & J.D. Bray (2004). Selecting a suitable time step for discrete element simulations that use the central difference time integration scheme. *Engineering Computations 21*(2/3/4), 278–303.

Piskoty, G., S. Michel, & M. Zgraggen (2005). Bursting of a corn silo–an interdisciplinary failure analysis. *Engineering Failure Analysis 12*(6), 915–929.

Pöschel, T. & T. Schwager (2005). *Computational granular dynamics: models and algorithms*. Springer Science & Business Media.

Potyondy, D.O. (2015). The bonded-particle model as a tool for rock mechanics research and application: current trends and future directions. *Geosystem Engineering 18*(1), 1–28.

Pournin, L. & T.M. Liebling (2005). A generalization of distinct element method to tridimensional particles with complex shapes. In *Powders and Grains 2005*, Volume 2, pp. 1375–1378. AA Balkema Publishers.

Rankine, W.M. (1857). On the stability of loose earth. *Philosophical transactions of the Royal Society of London*, 9–27.

Sivakugan, N., S. Widisinghe, & V.Z. Wang (2013). Vertical stress determination within backfilled mine stopes. *International Journal of Geomechanics 14*(5), 06014011.

Theimer, O. (1969). Failures of reinforced concrete grain silos. *Journal of Engineering for Industry 91*(2), 460–476.

To, H. & A. Scheuermann (2014). Separation of grain size distribution for application of self-filtration criteria in suffusion assessment. In *Proc., 7th Int. Conf. on Scour and Erosion, ICSE*, pp. 121–128.

To, H.D., S.A. Galindo-Torres, & A. Scheuermann (2016). Sequential sphere packing by trilateration equations. *Granular Matter 18*(3), 1–14.

To, H.D., A. Scheuermann, & S.A. Galindo-Torres (2015). Probability of transportation of loose particles in suffusion assessment by self-filtration criteria. *Journal of Geotechnical and Geoenvironmental Engineering 142*(2), 04015078.

To, H.D., S.A.G. Torres, & A. Scheuermann (2015). Primary fabric fraction analysis of granular soils. *Acta Geotechnica 10*(3), 375–387.

Widisinghe, S. & N. Sivakugan (2015). Vertical stress isobars for silos and square backfilled mine stopes. *International Journal of Geomechanics 16*(2), 06015003.

Numerical Methods in Geotechnical Engineering IX – Cardoso et al. (Eds)

Numerical simulation of the advance of a deep tunnel using a damage plasticity model for rock mass

M. Schreter, M. Neuner & G. Hofstetter
Unit for Strength of Materials and Structural Analysis, Institute of Basic Sciences in Engineering Sciences, University of Innsbruck, Austria

ABSTRACT: Analyzing the complex mechanical problem of the advance of a deep tunnel by means of numerical methods requires appropriate constitutive models of the interacting materials to obtain reliable estimates of displacements and stresses of the support structure. A recently proposed damage plasticity model for rock mass is employed in finite element simulations of a deep tunnel, driven by a drill, blast, and secure procedure according to the New Austrian Tunneling Method. The model describes the nonlinear three-dimensional mechanical behavior of intact rock and is extended to model quasi-homogeneous, quasi-isotropic rock mass based on the geological strength index and the disturbance factor. In contrast to linear-elastic perfectly-plastic models, which are commonly used in practical applications, irreversible strains in the pre-peak region and degradation of stiffness and strength are represented. The numerical results based on the damage plasticity model for rock mass are compared with numerical results on the basis of a perfectly-plastic rock model and with available in-situ measurement data. Consideration of strain softening in the damage plasticity model for rock mass allows the prediction of the formation of shear bands emanating from the tunnel surface due to excavation. The obtained localized deformation zones indicate the transition of the rock mass from a quasi-continuum to a quasi-discontinuum, which can be interpreted as a precursor to failure of the rock mass.

1 INTRODUCTION

One construction principle within the New Austrian Tunneling Method (Rabcewicz 1964) is to mobilize the strength of the surrounding rock as an active load-bearing part by allowing controlled deformation of the ground. In tunnels with high overburden, during the excavation process the high prevailing geostatic stresses are strongly rearranged leading to structural changes involving potential damage in the rock mass.

For the construction of a deep tunnel, reliable estimates of displacements and loads acting on the support structure are of major interest. To this end, predictions of the mechanical behavior of the tunnel structure based on the finite element method are of great value. Therein, constitutive models for interacting materials consisting of rock or soil and the securing measures play a major role. For rock mass, commonly employed constitutive models are linear-elastic perfectly-plastic models (Negro & De Queiroz 2000), using Hoek-Brown (Hoek & Brown 1980) or Mohr-Coulomb type failure criteria with non-associated flow rules. However, in these models several simplifying assumptions in the mechanical behavior are inherent in comparison with experimentally observed behavior of rock.

To account for the nonlinear material behavior in the pre-peak and post-peak region of the stress-strain curve, a damage plasticity model for intact rock was proposed in (Unteregger et al. 2015) and was extended to model rock mass in (Unteregger 2015), denoted as RDP (Rock Damage Plasticity) model. It represents irreversible deformations, strain hardening and strain softening including degradation of stiffness and strength. In particular, softening material behavior may have a significant impact on the stability of the tunnel profile during tunnel advance, as pointed out in (Schreter et al. 2017) for tunnel excavation without securing measures.

In the present contribution, the influence of softening in the constitutive relations of rock mass is assessed based on two-dimensional finite element simulations of deep tunnel advance. To this end, the mechanical response of the tunnel structure is analyzed on the basis of a perfectly-plastic rock model in comparison with the advanced RDP model. Since this contribution focuses primarily on influences of the rock model, for sake of simplicity, a linear-elastic shotcrete model is employed.

2 CONSTITUTIVE MODELS

In the following, the investigated constitutive models for rock mass consisting of a linear-elastic perfectly-plastic model, serving as representative for commonly employed rock models, and a damage plasticity model, which can be regarded as an extension of the perfectly-plastic rock model, are described briefly.

2.1 *Linear-elastic perfectly-plastic model for rock mass*

A linear-elastic perfectly-plastic model, denoted as RP model, is employed as a reference model for commonly employed rock models in engineering practice. A Hoek-Brown yield criterion (Hoek and Brown 1980, Hoek et al. 2002) in the smooth version proposed by Menétrey & Willam (1995) is used to delimit the elastic domain. The evolution of plastic strains is described by a non-associated flow rule with a parabolic shape in the meridian plane, as proposed in (Unteregger et al. 2015).

The intact rock parameters of the RP model are the Young's modulus, E, Poisson's ν ratio, friction parameter m_0 and dilatancy parameter m_{g1}. In addition, for modeling rock mass the geological strength index GSI and the disturbance factor D, as proposed in (Hoek 1994, Hoek et al. 2002), are required.

2.2 *Damage plasticity model for rock mass*

To account for the highly nonlinear behavior of rock in both pre-peak and post-peak regime of the stress-strain curves, the damage plasticity model for intact rock proposed in (Unteregger et al. 2015) is employed. It was extended in (Unteregger 2015) to describe the behavior of quasi-homogeneous quasi-isotropic rock mass based on empirical down-scaling factors for rock mass proposed in (Hoek et al. 2002, Hoek and Diederichs 2006), which are related to the geological strength index GSI and the disturbance factor D, and is henceforth denoted as RDP model.

The RDP model combines linear elasticity, non-associated plasticity, nonlinear isotropic hardening and nonlinear isotropic softening. It is formulated within the theory of plasticity in conjunction with the theory of continuum damage mechanics, and its stress-strain relation is expressed as

$$\sigma = (1-\omega)\mathbb{C} : (\varepsilon - \varepsilon^p). \tag{1}$$

Therein, σ denotes the nominal Cauchy stress tensor, $\mathbb{C}$ the fourth order elastic stiffness tensor, ε the total strain tensor and ε^p the plastic strain tensor. The nominal stress tensor σ (force per total area) is related to the effective stress tensor $\bar{\sigma}$ (force per undamaged area) by the scalar isotropic damage variable ω, ranging from 0 (undamaged material) to 1 (fully damaged material), by $\sigma = (1-\omega)\bar{\sigma}$.

During isotropic hardening, the yield surface is evolving and finally attains the Hoek-Brown yield criterion in the smooth version. Thus, the predicted peak stress is corresponding to the perfectly-plastic RP model. In contrast to the latter, irreversible strains in the pre-peak region of the stress-strain curves, nonlinear response for predominantly hydrostatic compression and degradation of strength and stiffness are represented.

Since strain softening of the stress-strain curve leads to strain localization at structural level, its description in terms of a constitutive model without regularization would yield mesh-dependent results in finite element simulations. For the present model, the softening behavior is regularized based on the crack band approach (Bažant & Oh 1983), assuming that strain localization occurs in a zone attributed to an integration point. In this simplified approach, a mesh-adjusted softening modulus depending on the specific mode I fracture energy and a characteristic length of a finite element is calculated.

The additional parameters to be determined for the RDP model to those of the RP model consist of the uniaxial compressive yield stress f_{cy}, the model parameters for hardening A_h and C_h, the specific mode I fracture energy G_{fI}, and the model parameter for softening A_s.

3 SIMULATIONS OF DEEP TUNNEL ADVANCE

Two-dimensional finite element simulations of the advance of a deep tunnel are presented. The model is derived from a stretch of the Brenner Base Tunnel, located in Innsbruck quartz phyllite rock mass. It is characterized by a circular profile with full face excavation employing a drill, blast, and secure procedure according to the New Austrian Tunneling Method. In (Schreter et al. 2017), this stretch was part of similar numerical investigations, however, without considering a support structure in conjunction with softening rock mass. The numerical results obtained from the simulations employing the (non-softening) RP model are compared with the respective results employing the (softening) RDP model, both in combination with a linear-elastic shotcrete model.

3.1 *Model description*

An equivalent two-dimensional plane-strain model according to the convergence confinement method (Carranza-Torres and Fairhurst 2000), to

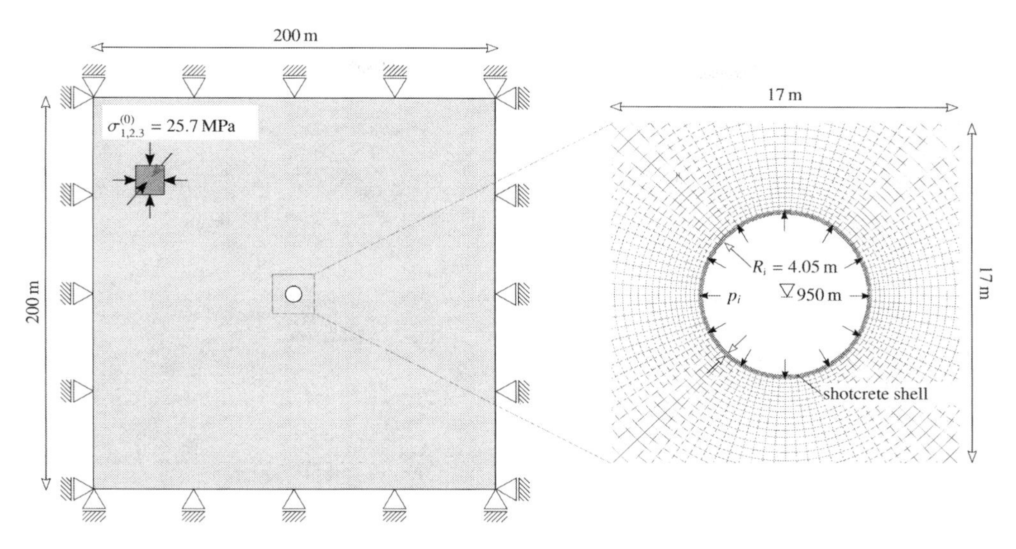

Figure 1. 2D finite element model of the IBVP – (left) full model and (right) detail center view after installation of the shotcrete shell.

account for the influence of the three-dimensional excavation process, is employed.

The initial boundary value problem (IBVP) with its geometry, initial conditions and boundary conditions is shown in Figure 1. The excavation area of the tunnel has a diameter of 8.5 m and is characterized by an overburden of 950 m. The initial geostatic stresses are assumed as hydrostatic with $\sigma^{(0)}_{11} = \sigma^{(0)}_{22} = \sigma^{(0)}_{33} = -25.7\,MPa$. The installed securing measures consist of a 0.2 m thick shotcrete shell.

The modeled domain of rock mass of 200 m × 200 m is discretized by 8-node fully integrated quadrilateral continuum elements. The shotcrete shell is modeled by the same finite element type employing 4 finite elements across the thickness.

The parameters for the RP model and for the RDP model for Innsbruck quartz phyllite were determined from a series of triaxial compression tests on-small scale intact rock specimens taken from the tunnel site (Schreter et al. 2017). The empirical down-scaling factors for the transition from intact rock to rock mass, the geological strength index *GSI* and the disturbance factor *D*, are specified according to the geological survey. The parameters for the RP model are summarized in Table 1. The additional parameters for the RDP model are listed in Table 2.

For the shotcrete shell, a time-independent, artificially low Young's modulus of 7000 MPa is assumed to account for time-dependent nonlinear material behavior in an approximate manner, as proposed by Pöttler (1990).

Table 1. Material parameters of Innsbruck quartz phyllite for the RP model.

E (MPa)	ν (–)	f_{cu} (MPa)	m_0 (–)	m_{g1} (–)	GSI (–)	D (–)
56670	0.2	41.6	12.0	9.9	45	0

Table 2. Material parameters of Innsbruck quartz phyllite for the RDP model in addition to the parameters of Table 1.

f_{cy} (MPa)	A_h (–)	C_h (–)	G_{f1} (N/mm)	A_s (–)
29.5	0.0045	8.8	0.15	40.0

Symmetry with respect to the vertical axis is not exploited since due to expected strain localization, potential unsymmetric failure modes arise. In addition, motivated by the actual non-uniform distribution of material properties in the rock mass, slightly weakened elements are introduced to trigger strain localization into the formation of shear bands. They are randomly distributed at a small share in the vicinity of the tunnel and a single configuration is investigated in this contribution. In (Schreter et al. 2017), it can be seen that their influence on the ground response is negligible.

The excavation and securing process is simulated following the convergence confinement method and is structured as follows: (i) application of the initial geostatic stress state together with the equivalent internal pressure $p_i^{(0)}$ acting on the rock-shotcrete interface; (ii) release of the internal pressure according to the initial stress release ratio λ_0 by $p_i = (1-\lambda_0)p_i^{(0)}$; (iii) installation of the securing measures consisting of the shotcrete shell; (iv) release of the remaining internal pressure.

For the present example, the initial stress release ratio is specified as λ_0 = 90%. The corresponding pre-displacements at the tunnel surface computed from the ground response curves based on the RP model and the RDP model, employing the parameters listed in Tables 1 and 2, are in the range of 18 to 25 mm (cf. Figure 2). For comparison, at the construction site, preplaced measurement devices recorded pre-displacements in the range of 11 mm to 43 mm (Neuner et al. 2017, Schreter et al. 2017). Since the RDP model represents softening behavior, at initiation of softening strains start localizing into narrow zones in the vicinity of the tunnel, resulting in a scatter of displacements along the tunnel perimeter (depicted by the black ground response band in Figure 2). Furthermore, in the simulation employing the RDP model beyond the normalized internal pressure of 5% loss of

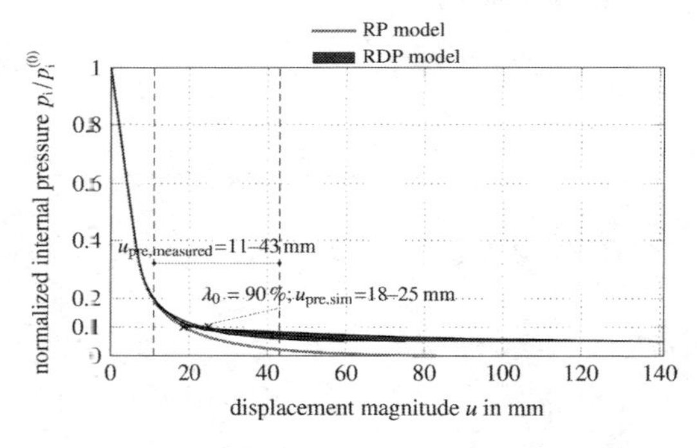

Figure 2. Ground response curve computed on the basis of the RP model (gray) and ground response band based on the RDP model (black).

equilibrium is observed since no support structure is considered.

3.2 *Results*

The influence of softening of rock mass on the mechanical behavior of the tunnel structure is investigated by comparing (i) the linear-elastic perfectly-plastic (non-softening) RP model and (ii) the (softening) RDP model, both in combination with the linear-elastic material model for the shotcrete shell. The predicted mechanical response of the tunnel structure is assessed by comparing the evolution of the displacements at the rock-shotcrete interface as well as the evolution of the circumferential stress obtained in the shotcrete shell.

The predicted evolution of the magnitudes of the displacement vectors at the rock-shotcrete interface in terms of the normalized internal pressure are shown in Figure 3, represented by the mean, minimum and maximum values along the tunnel perimeter. The displacements are composed of the pre-displacements of the unsupported rock mass according to the initial stress release of 90% and the further displacements after installation of the shotcrete shell.

For both rock models, the internal pressure can be fully released. Furthermore, it can be seen that the evolution of displacements is slowed done noticeably after installation of the support structure. Comparing the displacement magnitudes predicted by the two numerical models, it can be seen that the RDP model predicts increasingly larger displacements from the onset of softening behavior already during the initial stress release. At the same time, the strains start localizing into narrow zones of the rock mass leading to the formation of shear bands.

Comparing the increase of displacement magnitudes after installation of the support structure (cf. Figure 3, right) predicted by the two

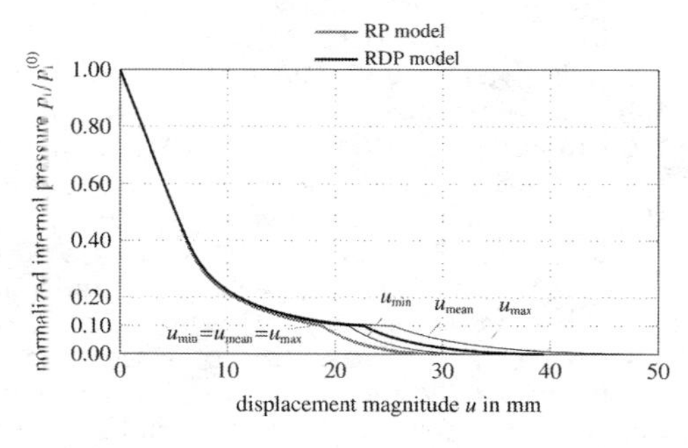

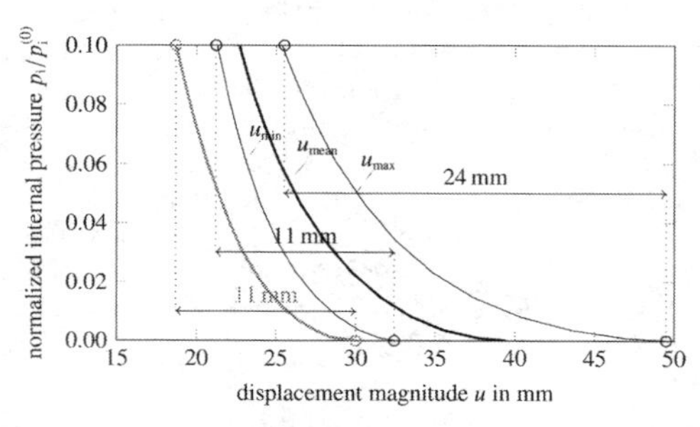

Figure 3. Magnitudes of displacement vectors at the rock-shotcrete interface in terms of the normalized internal pressure predicted by the RP model (gray) and the RDP model (black, minimum, mean and maximum values): (left) total view and (right) detailed view after installation of the shotcrete shell.

numerical models, substantial differences can be recognized: employing the RP model for the rock mass additional displacements of 11 mm are predicted in contrast to the RDP model, where additional displacements between 11 mm and 24 mm along the tunnel perimeter are obtained. The latter values allow for a comparison with the geodetic measurements, reported in (Neuner et al. 2017, Schreter et al. 2017): At the respective tunnel site displacement magnitudes ranging from 14 mm to 38 mm were recorded two weeks after installation of the support structure. By comparing the numerically predicted displacements with the measured displacements, it can be concluded that the displacements obtained from the RDP model are in better agreement with the in-situ measurement data. However, underestimation of the maximum value of displacement measurements is partially attributed to the assumed linear-elastic material behavior of shotcrete. Accordingly, the overestimated stiffness of the shotcrete shell, as discussed also in (Neuner et al. 2017), could be remedied by employing an advanced shotcrete model, which considers time-dependent effects like evolution of stiffness and strength, creep, and shrinkage.

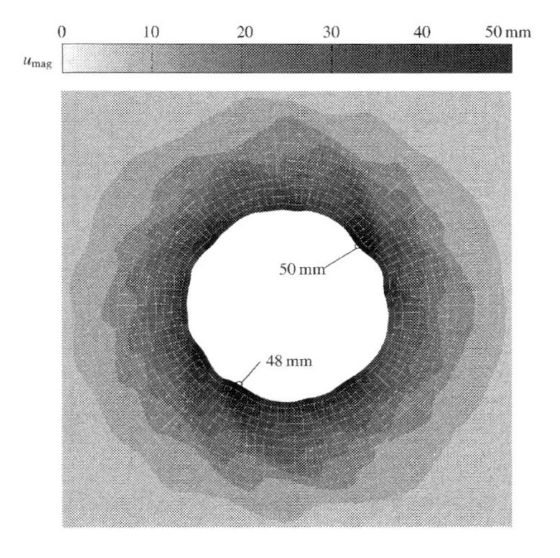

Figure 4. Distribution of the displacement magnitude after full release of the internal pressure (magnification factor of 20).

Figure 4 illustrates the deformation of the tunnel structure predicted by the RDP model after full release of the internal pressure. In two specific regions notable larger displacements are obtained, indicated by displacement magnitudes of 48 mm and 50 mm. These regions are affected by the formation of shear bands resulting in movement of blocks of rock mass towards the tunnel center.

The distribution of the damage variable in the rock mass predicted by the RDP model is illustrated in Figure 5 for two different levels of the normalized internal pressure. Since the onset of damage is already observed during initial stress release, immediately before installation of the shotcrete shell diffuse damaged zones are formed around the tunnel (cf. Figure 5a). Upon further release of the internal pressure, the formation of distinct shear bands can be seen (cf. Figure 5b).

Figure 6 depicts the evolution of the circumferential stress predicted in the shotcrete shell. According to the numerical simulation employing the RDP model, this refers to a zone where the adjacent rock mass is highly damaged. The influ-

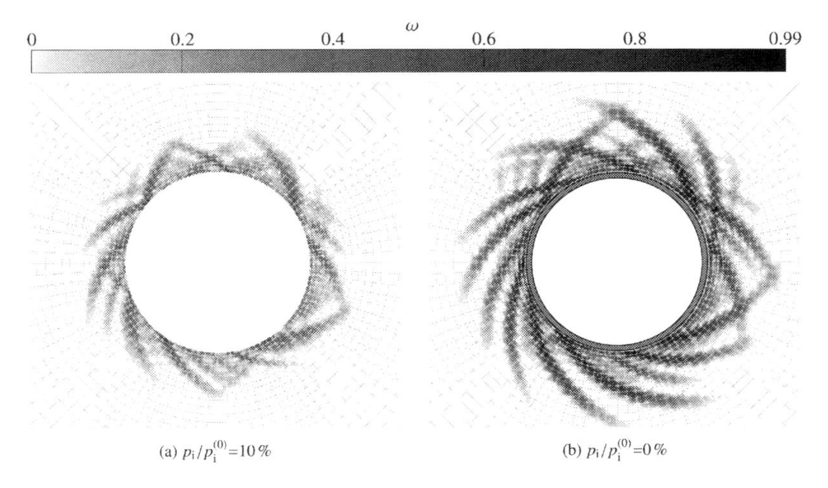

Figure 5. Distribution of the damage variable ω in the rock mass (a) immediately before installation of the shotcrete shell and (b) after full release of the internal pressure; the shotcrete shell is indicated in solid gray color.

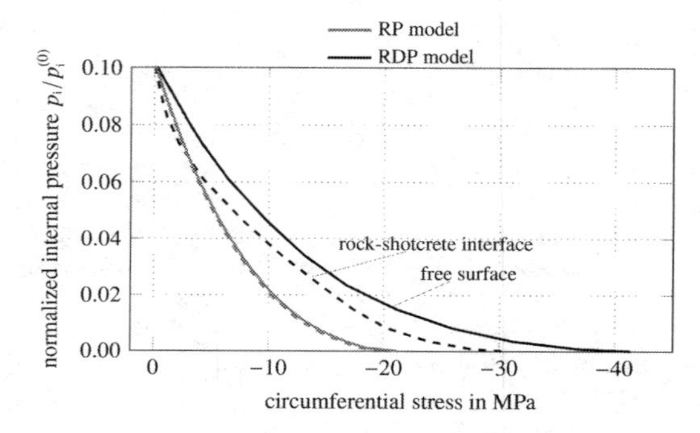

Figure 6. Evolution of the circumferential stress in the shotcrete shell predicted by the RP model (gray) and by the RDP model (black).

ence of prevailing bending moments in the shotcrete shell is analyzed based on a comparison of the circumferential stress obtained at the free surface and at the rock-shotcrete interface.

In case of the RP model, the circumferential stress obtained at the free surface and at the rock-shotcrete interface coincide, indicating the absence of bending moments. In contrast, in case of the RDP model due to the localized deformation non-uniformly distributed displacements are imposed on the shotcrete shell. As a result, the stress difference between the free surface and the rock-shotcrete implies a prevailing bending moment. Clearly, the temporal reduction of bending stresses due to creep, as discussed in, e.g., (Neuner et al. 2017), cannot be captured by employing the linear-elastic shotcrete model.

4 CONCLUSIONS

A damage plasticity model for rock mass was employed in two-dimensional finite element simulations of deep tunnel advance considering securing measures consisting of a shotcrete shell. The influence of softening was assessed based on a comparison with the mechanical response predicted by a linear-elastic perfectly-plastic rock model. In addition, the numerically obtained displacements were compared to in-situ displacement measurements.

Based on this numerical study, it was demonstrated that considering softening material behavior of rock has a considerable impact on the predicted mechanical behavior of the tunnel structure. In contrast to assumed perfectly-plastic material behavior of the rock mass, softening material behavior allows to capture the transition from initially quasi-continuous rock mass to quasi-discontinuous rock mass and, thus, potential failure modes can be captured.

Furthermore, since the shotcrete shell was simplified considered by a linear-elastic model, a more advanced shotcrete model could be employed accounting for time-dependent effects.

ACKNOWLEDGEMENTS

Financial support of the science fund provided by the Land Tyrol is gratefully acknowledged.

REFERENCES

Bažant, Z.P. & B.H. Oh (1983). Crack band theory for fracture of concrete. *Mater. Struct. 16*(3), 155–177.

Carranza-Torres, C. & C. Fairhurst (2000). Application of the convergence-confinement method of tunnel design to rock masses that satisfy the hoek-brown failure criterion. *Tunn. Undergr. Sp. Tech. 15*(2), 187–213.

Hoek, E. (1994). Strength of rock and rock masses. *ISRM News Journal 2*(2), 4–16.

Hoek, E. & E.T. Brown (1980). Empirical strength criterion for rock masses. *J. Geotech. Eng. Div., ASCE 6*(GT9), 1013–1035.

Hoek, E., C. Carranza-Torres,&B. Corkum (2002). Hoek-brown failure criterion – 2002 edition. *Proceedings of NARMSTAC 1*, 267–273.

Hoek, E. & M.S. Diederichs (2006). Empirical estimation of rock mass modulus. *Int. J. Rock. Mech. Mining Sci. 43*(2), 203–215.

Menétrey, P. & K. Willam (1995). Triaxial failure criterion for concrete and its generalization. *ACI Struct. J. 92*(3), 311–318.

Negro, A. & P. De Queiroz (2000). Prediction and performance: A review of numerical analyses for tunnels. In *Geotechnical Aspects of Underground Construction in Soft Ground. Rotterdam*, pp. 409–418. Kusakabe O., Fujita K., Miyazaki Y., editors. Balkema.

Neuner, M., M. Schreter, D. Unteregger, & G. Hofstetter (2017). Influence of the Constitutive Model for Shotcrete on the Predicted Structural Behavior of the Shotcrete Shell of a Deep Tunnel. *Materials 10*(6), 577.

Pöttler, R. (1990). Time-dependent rock-shotcrete interaction—a numerical shortcut. *Comput. Geotech. 9*, 149–169.

Rabcewicz, L. (1964). The new austrian tunnelling method. *Water Power 16*(11), 453–456.

Schreter, M., M. Neuner, D. Unteregger, G. Hofstetter, C. Reinhold, T. Cordes, & K. Bergmeister (2017). Application of a damage plasticity model for rock mass to the numerical simulation of tunneling. In *Proceedings of the 4. International Conference on Computational Methods in Tunneling and Subsurface Engineering (EURO:TUN 2017)*, pp. 549–556. G. Hofstetter et al.; Innsbruck University, Austria.

Unteregger, D. (2015). *Advanced constitutive modeling of intact rock and rock mass.* Ph. D. thesis, Innsbruck University.

Unteregger, D., B. Fuchs, & G. Hofstetter (2015). A damage plasticity model for di_erent types of intact rock. *Int. J. Rock. Mech. Mining Sci. 80*, 402–411.

Numerical Methods in Geotechnical Engineering IX – Cardoso et al. (Eds)
© 2018 Taylor & Francis Group, London, ISBN 978-1-138-33198-3

Experimental validation of numerical rockfall trajectory models

A. Caviezel, Y. Bühler, G. Lu, M. Christen & P. Bartelt
WSL Institute for Snow and Avalanche Research SLF, Davos Dorf, Switzerland

ABSTRACT: Numerical simulation tools have become a key tool for practitioners in the field of natural hazard decision making and risk analysis. Over the past several years different numerical rockfall codes have been developed and applied for geotechnical risk assessment. The reliability of the simulation output depends on well calibrated input parameters for any given set of initial and boundary conditions. At present, calibration of rockfall models is performed using empirical back-calculations of specific case studies. Direct comparison to experimental data, however, would be preferred. A validation procedure is presented for rockfall simulation codes based on induced rockfall experimental data. The presented routine is a start point for facilitating and securing consistent (re-)calibration of physics based rockfall trajectory models. This is of key importance for guaranteeing the physical consistency of the numerical simulation kernel upon model enhancements or changes.

1 INTRODUCTION

Rockfalls pose a threat to infrastructure such as roads, railway lines and houses. They are therefore a major concern to residents of many mountain communities. It is now expected that the frequency of rockfall hazard will increase due to extreme precipitation events coupled with extreme thermal forcing, leading to the increased destabilization of many rock slopes. The ever growing spatial requirements of alpine settlements in addition to the decreasing risk tolerance of societies is causing a demand for more detailed rockfall hazard mitigation strategies. As municipalities are confronted with the problem to quantify hazard potential, the need for more accurate predictions of rockfall risk is clearly growing.

The planning of mitigation measures relies on different data sources such as field observations and measurements. Increasingly, numerical simulation tools are used by engineers to supplement field observations. The combination of field observations and numerical modelling facilitates a scenario-based risk assessment. This development is being supported by an enormous leap in geographic mapping technology by means of satellite remote sensing and, equally, by manned and unmanned aerial vehicles. These technologies provide practitioners with digital terrain models of unprecedented quality and quantity. Alongside with the ever growing computational power available at low cost, these developments have paved the way for civil engineers to move from simple one-dimensional simulation models to a full three-dimensional approaches.

Different numerical tools have been investigated, developed and implemented over the past few years to model rockfall trajectories (Jones et al. 2000, Agliardi and Crosta 2003, Lan et al. 2010, Dorren 2010, Bourrier et al. 2012, Leine et al. 2014). The aim of all methods is to assist in risk assessment and planning of rockfall mitigation strategies. Each method, however, is based on a physical model that ultimately depends on parameters governing rock-ground interaction. As most of the models are calibrated via empirical back-calculation using specific case studies, there is little inter-comparison between different methods, and therefore modelling parameters. The value of modelling parameters is restricted to one particular modelling approach. A more favorable path would be to calibrate rockfall simulation codes via real-world in-field measurement data. Data sets of high quality field measurements of rockfalls, however, are scarce and difficult to obtain.

Here, data sets of an experimental rockfall campaign consisting of small (masses less than 100 kg), irregular shaped rocks are presented. It is demonstrated how the measured data can be used to calibrate model parameters of the rockfall simulation code. The analysis includes statistical measures of runout length and lateral dispersion as well as kinematic information such as translational and angular velocities. A well calibrated model is the base for investigations of rockfall energies and important parameters used in mitigation planning such as jump heights and lengths.

2 EXPERIMENTAL METHODS AND RESULTS

2.1 *Experimental methodology*

The data from the *Small Rock Experimental Campaign* (SREC) presented here have been obtained at a test site close to the village Tschamut in Canton Grisons (Switzerland). It is a typical alpine meadow slope interspersed with rocks providing a slope of larger than 30° with a maximal inclination of 42°. The location has been chosen due to its high accessibility and its uniform runout area. The full SREC contains two series, the first one conducted during wet conditions, the latter conducted during frozen ground conditions, both subsequently labeled as RS_w and RS_f respectively. Each series contains repeated manual releases of differently shaped boulders with weights ranging from 30–80 kg. The rocks have been digitized in order to perform RAMMS::ROCKFALL simulations (Leine et al. 2014) with the exact specific shape. As the numerical routines perform numerous checks on rock faces and edges, rock shape point clouds should not exceed 100 points to prevent unnecessary computation slow downs.

Figure 1 shows the Sneed and Folk (1958) rock shape classification scheme for the used rock ensemble. The input, denoted in Table 1, is based on the digitized axis lengths from the convex hull algorithm of the RAMMS::ROCKFALL rock builder. The rocks range from platy rocks (Rock 1,4,8) over an elongated form (Rock 3) to a perfectly compact equant shape boulder (EOTA). This rock represents the norm rock of the European Organization for Technical Assessment used

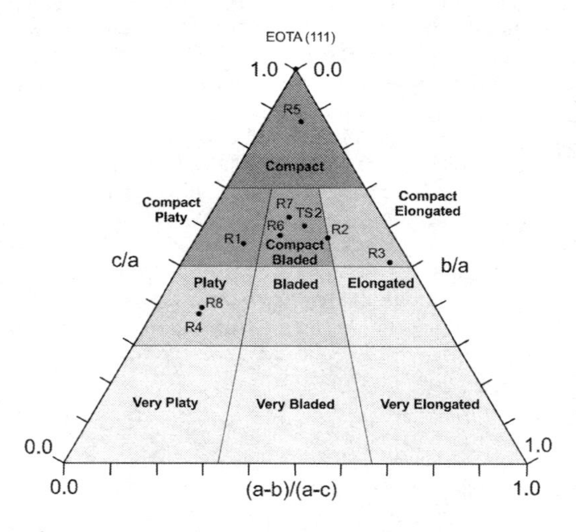

Figure 1. Sneed and Folk classification diagram (elongation and flatness) of the SREC rock ensemble.

Table 1. Details of the experimental rock ensemble. The rock set consists of eight granite rocks (Rock 1-8) used in RS_w and two additional rocks in RS_f (the heaviest test rocks are labeled TS2 and EOTA).

Rock Nr	Dimensions (m)	Weight (kg)	Volume (m³)
1	0.36/0.32/0.20	30.5	0.011
2	0.41/0.30/0.24	34.8	0.013
3	0.49/0.25/0.27	42.4	0.016
4	0.43/0.48/0.18	34.8	0.012
5	0.37/0.34/0.32	42.6	0.016
6	0.45/0.37/0.26	41.1	0.016
7	0.33/0.40/0.25	29.9	0.011
8	0.52/0.58/0.22	52.2	0.019
9 (TS2)	0.50/0.39/0.30	78.4	0.030
10 (EOTA)	0.30/0.30/0.30	44.0	0.016

in standardized rock fence testing procedures in official European Technical Approval Guidelines (ETAG027 2013). The chosen set of rocks illustrates the importance of rock shape and its influence on rockfall dynamics (Glover 2015). Thus, the SREC provides a cost-efficient opportunity to screen rock shape dependencies in real world conditions.

The majority of the experimental rocks were instrumented with a low-power sensor node, in-situ tracking accelerations and rotations. These so-called *StoneNodes v1.0* are described in detail in Caviezel et al. (2017) and Niklaus et al. (2017). Main components are a ST H3LIS331DL tri-axial accelerometer featuring a measurement range of 400 *g* and an InvenSense ITG-3701 3-axis gyroscope recording up to 4'000 degrees/sec at a detection rate of 500 Hz. This work focuses on the entire data ensemble. A detailed investigation of a single trajectory data stream together with a quality control of the data is presented in Gerber and Caviezel (2018). Measurements of the deposition points are performed using a high precision Trimble GPS hand-held data collector from the Geo Series with a two digit centimeter accuracy in favourable local weather, and satellite availability. A high-resolution digital elevation model (DEM) of the test site was constructed via aerial remote sensing using an Ascending Technologies Falcon 8 Octocopter equipped with a Sony Alpha NEX-7 camera. The photogrammetric work flow with the described acquisition and post-processing is described in Bühler et al. (2012) and Bühler et al. (2017) (and references therein). The UAV measurements provide a DEM resolution of 5 cm. Normally, DEM resolution is reduced to 1 or 2 m in post-analysis to adopt the generally available DEM resolution used in practice. This will likely change in the near future with the growing market of affordable unmanned aerial mapping systems. The simulations shown here have been performed with a 2 m resolution grid.

2.2 *Parameters of interest in rockfall events*

Figure 2 displays a contour plot of the experimental site alongside with the fundamental parameters of interest in a rockfall event or experiment. A rock—irrespective whether belonging to a natural rockfall or an artificially triggered event—originates from a source, which can be a point, line or area. The rock then describes a trajectory path on its way to the deposition point consisting of simple parabolic flight sections interrupted by short contact phases. While those contact phases are most often reduced to a single interaction point in time by models using restitution coefficients, the process itself exhibits overwhelming complexity. An in-depth treatment of impacts recorded during these experiments is presented in Gerber and Caviezel (2018).

The free flight sections exhibit each its maximal jump heights and lengths as shown in the lower inset of Fig. 2. Be aware that the specifications of jump heights depend whether they are given plum normal or surface normal with respect to the DEM surface cell at the given point. The outermost deposition points mark the maximal runout distance L for a given site, the maximal distance within the trajectory envelope displays the maximal lateral dispersion W. Here, W is defined as the sum of the maximal right- and left-lateral normal distances with respect to the cardinal heading of the deposition ensemble of any given rock. The cardinal heading is given by the ray connecting the release point with the mean deposition point of the deposition points—also denoted as center of mass (COM)—as indicated in Fig. 7 (a). The mean travelling distance d_{COM} is defined as the length of

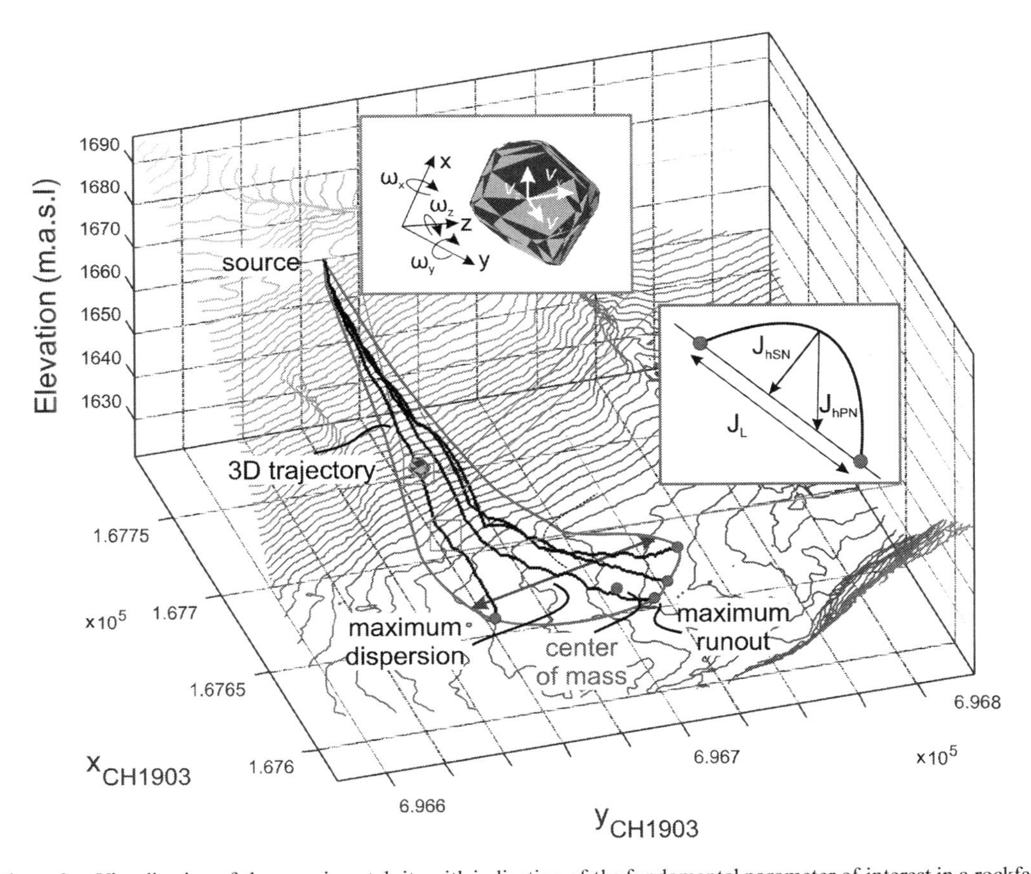

Figure 2. Visualization of the experimental site with indication of the fundamental parameter of interest in a rockfall event. Indicated are the maximal runout distances L and the maximal lateral spread W. The upper inset shows the eigenframe of the rock with its kinematic vectors defined in this reference frame. Inset two shows the jump heights and lengths and the difference between surface normal and plum normal jump height definition.

the cardinal heading ray from release point to the COM.

The rock position is fully described via its position vector $\mathbf{r} = (x,y,z)$ given in the DEM's reference coordinate system and rotation quaternion $\boldsymbol{q} = s + ui + v\cdot + w\ell$ with s,u,v,w scalars and i, j, ℓ three imaginary numbers. The temporal evolution of the position quaternion defines the rotational velocities $(\omega_x,\omega_y,\omega_z)$ given in the rock's eigenframe as sketched in the upper inset of Fig. 2. To facilitate comparison between experimental measurements and simulated results, kinematic vectors such as translational velocities (v_x,v_y,v_z) are extracted in the DEM's coordinate frame, whereas accelerations and angular rotations are calculated in the rock's eigenframe (this is what the in-situ sensors are measuring).

3 RESULTS AND ANALYSIS

3.1 *Experimental results*

The SREC data set comprises 48 deposition points recorded in wet conditions RS_w and 57 deposition points in frozen conditions RS_f. Figure 3 visualizes both data sets. While for RS_w only Rock 1-8 have been used, the RS_f series is extended with the largest rock from an earlier experiment (TS2) and an EOTA concrete block. Figure 3 shows the calculated COM within an ellipse representing the standard deviation of deposition points (SDE). In an ideal environment, these are two key parameters to quantify the deposition pattern. Obviously, terrain features such as couloirs, ridges, etc., but also the rock shape have a strong influence on these ideal parameters. For strongly channelized terrain, the main axis orientation of the SDE might become questionable. However, At this site the runout area is considered to be uniform enough such that changes in deposition pattern are solely caused by the rock shape.

All available sensors have been deployed such that 21 of the 48 runs in RS_w and 26 out of 57 runs in RS_f have been recorded with a StoneNode v1.0. As a typical data stream we chose an EOTA rock run recorded during RS_f, the according deposition point is marked with an arrow in the right panel of Fig. 3. The data stream is plotted in Fig. 4. While the left panel shows the accelerations spikes happening at each impact and recording zero g during free flight, the middle panel shows the angular velocities in all three axis. The resultants of both parameters are shown in the right panel of Fig. 4. This sequence has been chosen as it is also the sequence treated in great detail in Gerber and Caviezel (2018). The EOTA block represents the only perfectly symmetric shape in the rock ensemble. The rotational data shows, that the initiated symmetry breaking caused by the sensor hosting hole is sufficient to establish a predominant rotation axis. The gyrodynamical stabilization of a rigid body with rotations around the axis with either the largest or smallest moment of inertia is discussed in Ashbaugh et al. (1991). A comprehensive display of sensor data for different rock shapes is provided in Caviezel et al. (2017).

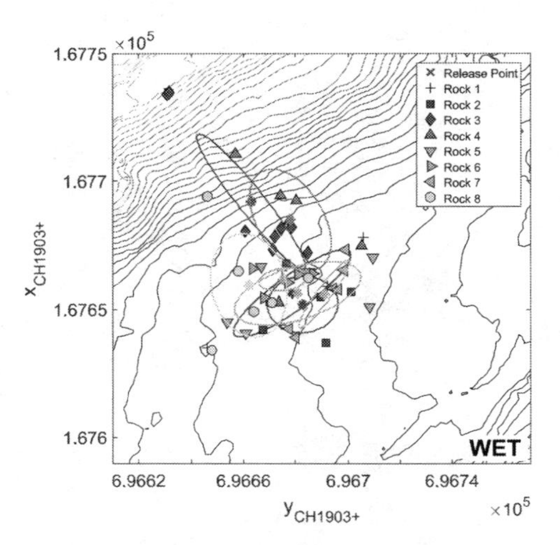

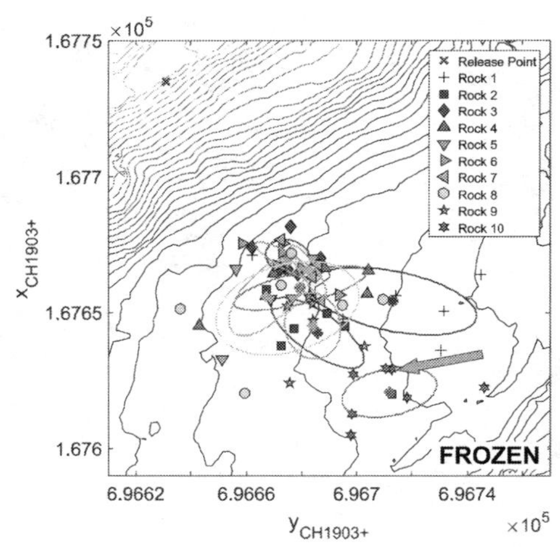

Figure 3. Deposition points for both experimental series, wet conditions RS_w shown in the left panel, frozen conditions RS_f shown in the right panel. For each rock the standard deviation ellipse with the center of mass of the individual rock deposition are plotted. The EOTA deposition point used for the sensor data stream plotting is marked with an arrow.

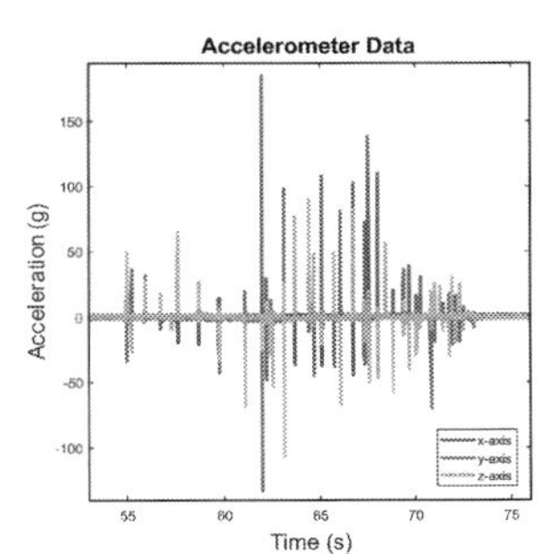

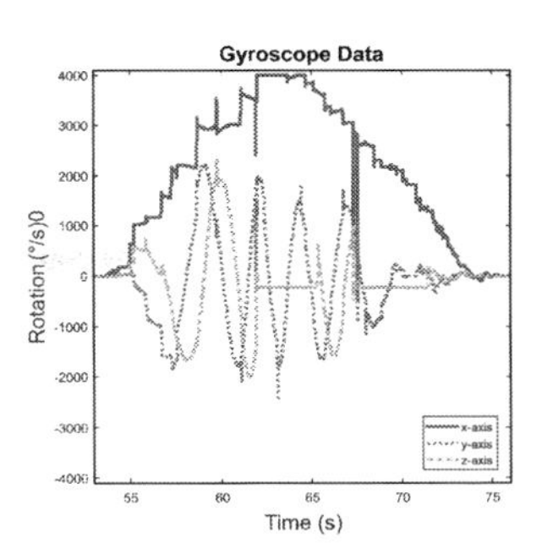

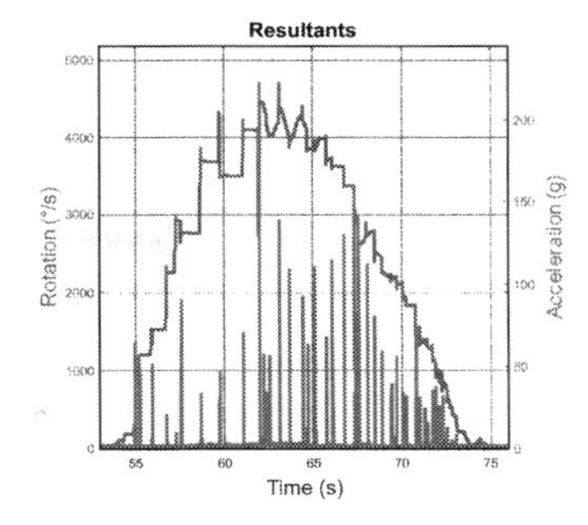

Figure 4. A typical experimental data stream from the StoneNode v1.0. Here, a run of the EOTA rock is shown. The left panel shows the acting accelerations during each contact and the middle panel the angular velocities in all three axis. The right panel features the resultants for both sensor streams.

3.2 *Simulations and calibration*

3.2.1 *The RAMMS software modules*

The RAMMS::ROCKFALL simulation tool is a numerical software package applying non-smooth mechanics coupled with hard contact laws to the rockfall problem. The tool has been developed at the WSL Institute for Snow and Avalanche Research SLF (Leine et al. 2014). Together with the modules for avalanche and debris flow simulations, the RAMMS software provides a unified software environment for integral hazard management of snow avalanches, debris flows and rockfalls (Christen and Bühler 2012). A detailed description of the non-smooth mechanics and numerical solution methods can be found in Leine et al. (2014). An interesting comparison of non-smooth mechanical approaches and discrete element type methods can be found in Lu et al. (2018).

The first versions of the RAMMS::ROCKFALL model employed a frictional sliding rebound interaction parameterized by

$$\mu(s) = \mu_{min} + \frac{2}{\pi}(\mu_{max} - \mu_{min})\arctan(\kappa s) \quad (1)$$

where the effective friction $\mu(s)$ evolves from a smaller μ_{min} to a higher μ_{max} during the contact time governed by a time scale κ. This equation governs the rolling, and above all, stick-slip motion occurring at every impact (Leine et al. 2014, Lu et al. 2018). The slipping distance s describes the travelling distance of the rock's center of mass during the contact phase. The temporal evolution of the slipping distance decays to zero when the rock leaves the terrain; that is, when no closed contact between rock and DEM is detected anymore:

$$\dot{s} = -\beta s. \quad (2)$$

The parameter β thus controls how fast the friction is released while exiting from the ground scar. Large β means, that friction is immediately removed as the rock moves away from the ground. Conversely, small β indicates acting sliding friction even after the rock is no longer in contact with the ground. The parameter β is linked to the penetration depth of the rock into the ground. Larger penetration depths (softer materials) are associated with smaller β values (Bartelt et al.). Note, that this contact treatment is fully deterministic by calculation of acting forces and torques. No stochastic modelling processes are introduced within the rock-ground contact mechanism. This is in strong contrast to most other rockfall simulation software where impact directions are often stochastically distributed to obtain the needed lateral spread in runout behavior. In RAMMS::ROCKFALL, ground interaction, and therefore lateral dispersion, is governed entirely by the randomized start positions or irregularly shaped rocks. The calculated trajectory is determined by momentum conservation and the non-smooth model governing the contact/rebound dynamics using hard contact laws.

The above friction law governs the scarring behavior when the rock is fully in contact with the hard rebound layer. The contact time, energy dissipation, torques and exit angles thus are governed by the parameters defined in Eqn. 1. The current RAMMS::ROCKFALL user version contains eight ground parameter default values ranging from *Extra Soft* to *Extra Hard* with the addition of *Snow*. An extract of those can be found in Table 2 (Bartelt et al.). Note, that RAMMS::ROCKFALL sets the rebound coefficient to zero and the large change for the μ coefficients for snowy conditions.

3.2.2 *Calibration methodology*

Despite of the benefits that numerical models are providing to geotechnical and hazard mitigation applications, they feature a major drawback: Plausible final results may well be obtained via implausible or nonphysical model parameter settings. The presented verification via experimental data is one approach to strengthen the chosen model settings. New developments to the RAMMS::ROCKFALL code kernel are intended to be calibrated to any newly available experimental data. This is performed via a parameter sweep over the model parameters, including the slippage parameters from Table 2. Although the measured data set has strong limitations in statistical significance for individual rock sequences, the full SREC serves as an ideal testing ground for the development of calibration routines as the simulation domain is rather small and data handling for large parameter sweeps become more feasible.

The experimental results and case studies involving large scree fields identified the need of the expansion of the current friction model, especially to include for effects from surface roughness. The current model comprises solely the hard contact layer. Figure 5 visualizes the extended view of friction layers which can occur during a rockfall impact: With decreasing distance to the ground the rock travels through a 1) possible canopy layer including discrete tree stems and lower lying vegetation, 2) a roughness layer created by scree fields, general terrain variation etc., and finally 3) a scarring layer with the 4) hard contact layer whose friction law is given by Eqn. 1.

Table 2. Extract of default terrain parameters in the current RAMMS::ROCKFALL user version.

Terrain	μ_{min}	μ_{max}	β	κ	ε	Drag
Extra Soft	0.20	2	50	1	0	0.9
Soft	0.25	2	150	1.25	0	0.8
⋮	⋮	⋮	⋮	⋮	⋮	⋮
Hard	0.55	2	185	3	0	0.4
Extra Hard	0.80	2	200	4	0	0.3
Snow	0.1	0.35	150	2	0	0.7

The test site contains no forest and barely any low lying vegetation. Additionally, the overall homogeneity of the slope requires no modelling of a roughness layer. Thus, the parameters governing layer 1) and 2) are not necessary when performing a parameter sweep. The parametrization of the scarring layer is done via a minimal scarring height h_s and an acting drag force within this scarring layer s_d. The translational drag force D_t thus is defined by

$$D_t = s_d \cdot v\|v\|. \tag{3}$$

This drag acts only within the scarring layer height defined by h_s. The scarring layer drag exhibits a mass dependency, that is larger rocks create deeper scars. Treating the small masses of this experimental campaign has the advantage that the mass dependency is of no further concern. Only one rock mass category is considered. Consequently, only s_d and h_s are included in the sweep.

3.3 *Simulation results and analysis*

Figure 6 shows the RAMMS::ROCKFALL user interface after the completion of a parameter sweep. The default representation mode is the display of the 95% quantile value of the parameter values of interest, in this case the velocity. The simulation domain of a sweep basically covers the simulation area from a frictionless to a highly

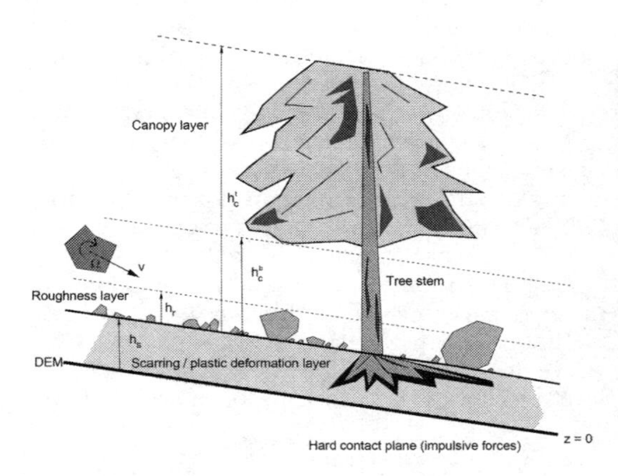

Figure 5. The four friction layers in rockfall: 1) the forest layer including both the canopy drag layer and discrete tree stems, 2) the roughness layer, 3) the scarring or plastic deformation layer, 4) the hard contact or rebound layer.

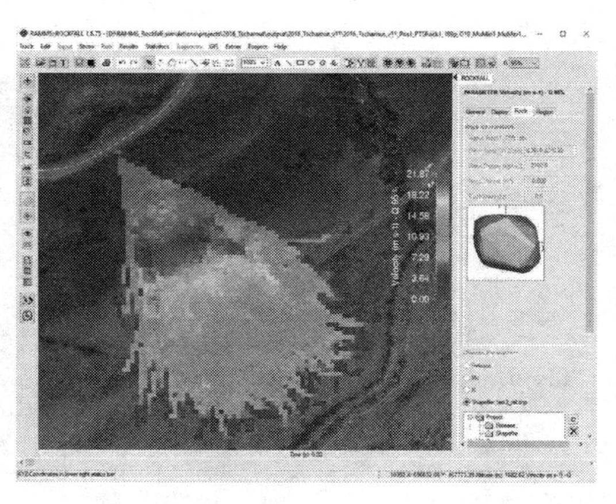

Figure 6. RAMMS::ROCKFALL user interface after a completed simulation sweep. The default mode is the statistics mode representation where every touched grid cell displays the 95% quantile value of the chosen parameter—in this case velocity.

adhesive environment. The user interface allows the display of much information in the statistics mode as well as more detailed visualizations and animations in the trajectory mode (Christen and Bühler 2012). The search for a best fitting parameter set, however, is externally scripted. The simulation results are searched with respect to

- the difference of mean deposition point between experiment and simulation,
- the lateral spread *W* normal to the cardinal heading,
- the spread in global *X* and *Y* coordinates,
- the mean rotation in the principal axis of rotation,
- the agreement in runtime duration for a single trajectory and thus mean velocities.

The foremost reduction to the parameter sets is given by eliminating all of whose COM deposition is outside a 10% radius with respect to d_{COM} and fail to possess a lateral spread of 90% compared to the experimental lateral spread. The relative boundary conditions incorporate the need for larger deviations when runout distances become larger. The pronounced stabilization around a favorable axis in rotation allows for a quite rigorous condition on the mean rotation. Table 3 shows an extract of the measured angular velocities. The lower rotational speeds for more elongated forms is striking. For platy rocks, data sets with mean angular velocities below 3000 degree/s in the principal axis have been discarded. The maximal angular velocity is used as a feasibility check, as the simulated results should not exceed by more than 25%. The temporal differentiation of the gyroscope data stream yields exact information on temporal impact locations and run time duration. Simulation results should exhibit similar run time duration, such that unwanted effects like slowly creeping and tumbling rocks are eliminated from the results.

Figure 7 shows deposition point evolution for the different parameter sets for the EOTA boulder. Panel (a) shows the best fit parameter evaluated via the above procedure. The cardinal heading is denoted as dashed line. The fit corresponds to a

Table 3. Extract of rotational sensor data for the SREC: Mean rotational speeds $\overline{\omega}_{x/y/z}$ for Rock 1 and 3 for both data sets, added with the EOTA data for RS_f. The maximal resultant rotational velocity ω_{tot}^{max} together with the resultant mean angular velocity $\overline{\omega_{tot}}$.

Rock Nr	$\overline{\omega_x}$	$\overline{\omega_y}$	$\overline{\omega_z}$	ω_{tot}^{max}	$\overline{\omega_{tot}}$
1 (RS_w)	3518	2086	1622	3923	1608
3 (RS_w)	1993	857	1312	2050	662
1 (RS_f)	3940	2600	2162	4259	1890
3 (RS_f)	2890	2147	1692	3191	1151
EOTA (RS_f)	2652	3677	3587	5313	2196

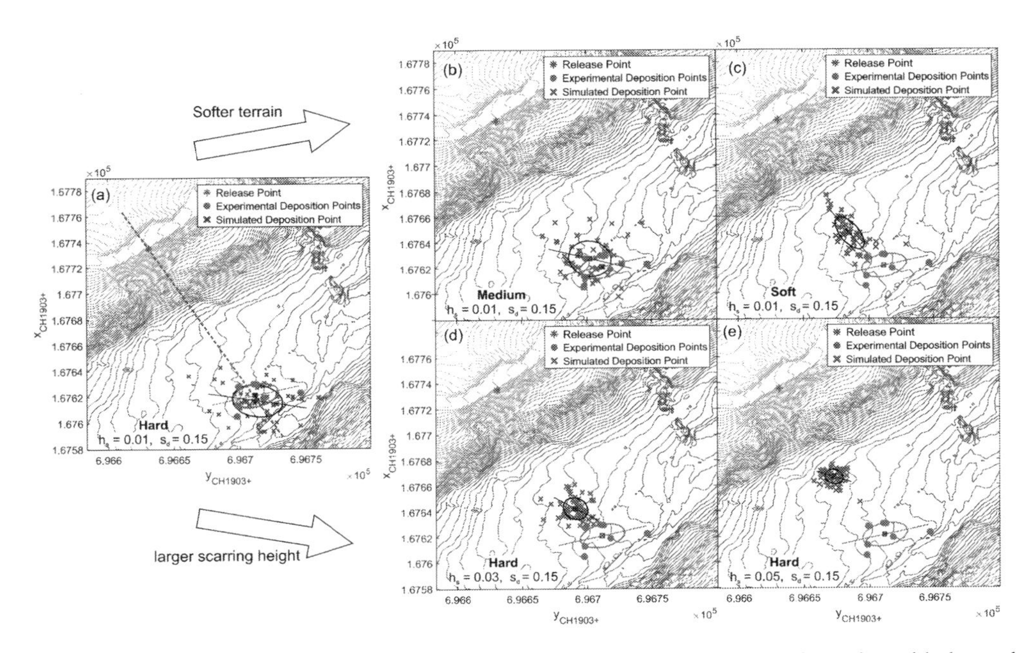

Figure 7. Simulation results for the EOTA boulder. Panel (a) showing the best fit deposition points with the marked cardinal heading, the ray connecting release point with center of mass of the deposition ensemble. Panel (b) and (c) show the effects of softer terrain, panel (d) and (e) the influence of the scarring height.

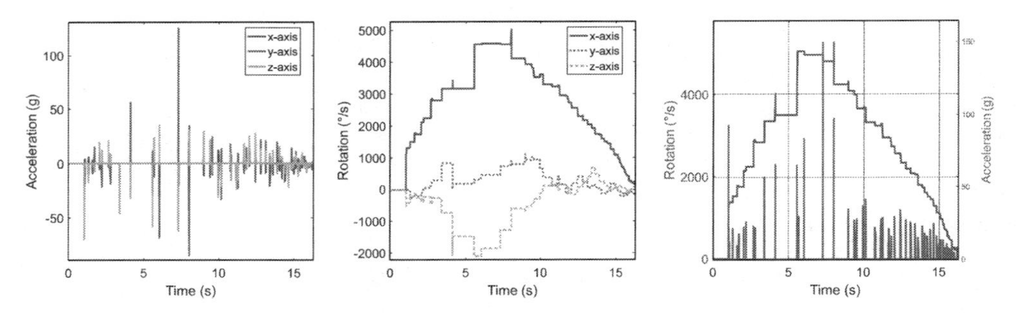

Figure 8. Simulated accelerations and angular velocities for the EOTA rock with the best fit parameters obtained via the described procedure. Qualitative and quantitative resemblances are striking.

Hard terrain from the RAMMS::ROCKFALL terrain classification with very small scarring depth value of $h_s = 0.01$ and scarring drag $s_d = 0.15$. Being a reasonable fit for frozen terrain it also strengthens the preceding slippage parameter calibration from the existing user version. The upper panels (b)-(c) of Fig. 7 visualize the effect on deposition pattern when iterating from *Hard* over *Medium* to *Soft* parameter settings for the friction law according to the RAMMS::ROCKFALL terrain classification. Note that the drag is kept constant. The lower panels (d) and (e) demonstrate the effect of larger scarring heights with the same acting drag. It becomes clear that for such small rocks energy dissipation and thus shorter and narrower deposition pattern can be achieved via different friction mechanisms. Figure 8 visualizes simulated output obtained with the best fit parameter setting as used in Fig. 7 (a) equivalently put into graphs as the experimental data in Fig. 4. Quantitatively and qualitatively similar data patterns are recognizable. While the real EOTA rock exhibits wobbling around the y and z axis, the rotations in the simulated rock remain more stable due to its perfect symmetric shape. All other rock forms exhibit stabilization around their principal axis of inertia as also manifested in the experimental results (Caviezel et al. 2017).

4 CONCLUSIONS

Full-scale field tests with instrumented rocks are the starting point to establish an experimental foundation for the calibration of numerical rockfall models. Here we demonstrate how experimental results can be used to calibrate a newly introduced constitutive law both on the statistical and kinematic levels. Although only two in-situ experimental data streams are available, the congruence between experiment and simulation bolsters the confidence in other simulated measures such as jump heights and lengths.

The small spatial dimensions of the field test site is advantageous from both the experimental as well numerical modelling standpoints. The small size of the site simplifies the experimental logistics. It also facilitates numerical back-calculations and data handling. Conversely, it poses one of the major limitations in the presented study. Although rock shape and terrain consistence (e.g. wet vs. frozen) can lead to differences in observed runout and lateral dispersion, the traveled distances are rather small. It is therefore hardly possible to statistically separate experimental results from different measurement campaigns. The rocks remain in a low energy and velocity regime. Additionally, the limited data sets for identical rock shapes poses a further constraint. Thus, the provided calibration routine remains in the qualitative regime.

In future experiments will be performed at a site with a more extreme vertical drop in altitude and larger surface roughness. Here, differences in individual trajectories would help segregate different parameter sets. The SREC data set also demonstrates the need for higher replication rates with the same rock shape in order to gain a statistically more significant data set. This goal should be tackled with the construction of artificial rocks made from reinforced concrete such that full control over shape and mass is given together allowing a maximal experimental repeatability. Additionally, the experimental data stream should be enriched with data on jump heights and velocities. The experiments should be carried out with larger masses, preferably greater than 1000 kg. First tests with artificial EOTA boulders with masses of 200 kg and 800 kg at a new alpine test site have been successfully performed to address the problem of size in rockfall modelling.

The primary conclusion of our experimental work is that mass and size scalability is simply not given in rockfall engineering. This makes the rockfall problem far from trivial, especially the problem of how to validate numerical models. Expanding the experimental range to the feasibility limits is of urgent importance to provide data as close as possible to real case studies. Setting up induced rockfall experiments with the purpose of validating numerical models should be a future priority in rockfall research.

REFERENCES

Agliardi, F. & G. Crosta (2003). High resolution three-dimensional numerical modeling of rockfalls. *International Journal of Rock Mechanics and Mining Sciences 40*(4), 455–471.

Ashbaugh, M.S., C.C. Chiocone, & R.H. Cushman (1991, Jan). The twisting tennis racket. *Journal of Dynamics and Differential Equations 3*(1), 67–85.

Bartelt, P., C. Bieler, Y. Bühler, M. Christen, M. Chirsten, L. Dreier, W. Gerber, J. Glover, & M. Schneider. RAMMS::ROCKFALL User Manual. http://ramms.slf.ch/ramms/downloads/RAMMS_ROCK_Manual.pdf, accessed: 01/03/2018.

Bourrier, F., F. Berger, P. Tardif, L. Dorren, & O. Hungr (2012). Rockfall rebound: comparison of field experiments and alternative modeling approaches. *Earth Surface Processes and Landforms 37*(6), 656–665.

Bühler, Y., M.S Adams, A. Stoffel. & R. Boesch (2017). Photogrammertic reconstruction of homogenous snow surfaces in alpine terrain applying near-infrared uas imagery. *International Journal of Remote Sensing 38*(8–10), 3135–3158.

Bühler, Y., M. Marty, & C. Ginzler (2012). High resolution dem generation in high-alpine terrain using airborne remote sensing techniques. *Transactions in GIS 16*(5), 635–647.

Caviezel, A., M. Schaffner, L. Cavigelli, P. Niklaus, Y. Bühler, P. Bartelt, M. Magno, & L. Benini (2017). Design and evaluation of a low-power sensor device for induced rockfall experiments. *IEEE Transactions on Instrumentations and Measurement PP*(99), 1–13.

Christen, M. & Y. Bühler (2012). Integral hazard management using a unified software environment numerical simulation tools ramms. *Proc. Congress Interpraevent*, 7786.

Christen, M., J. Kowalski, & P. Bartelt (2010). RAMMS: Numerical simulation of dense snow avalanches in three-dimensional terrain. *COLD REGIONS SCIENCE AND TECHNOLOGY 63*(1–2), 1–14.

Dorren, L.K.A. (2010). Rockfor3d revealed description of complete 3d rockfall model,. *Tech. rep., EcorisQ*,.

ETAG027 (2013). Guidline for European technical approval of falling rock protection kits. https: // www.eota.eu/en-GB/content/etages-used-ad-ead/26/, accessed: 01/03/2018.

Gerber, W. & A. Caviezel (2018). Measurement and analysis of ground contacts during rockfall events. In preparation.

Glover, J. (2015). *Rock-shape and its role in rockfall dynamics*. Ph. D. thesis, Durham University.

Jones, C.L., J.D. Higgins. & R.D. Andrew (2000, March). Mi66 colorado rockfall simulation program, version 4.0.

Lan, H., C.D. Martain, & C. Lim (2007). Rockfall analyst: A {GIS} extension for three-dimensional and spatially distributed rockfall hazard modeling. *Computers & Geo-sciences 33*(2), 262–279.

Leine, R., A. Schweizer, M. Christen, J. Glover, P. Bartelt, & W. Gerber (2014). Simulation of rockfall trajectories with consideration of rock shape. *Multibody System Dynamics 32*(2), 241–271.

Lu, G., A. Caviezel, M. Christen, Y. Bühler, & P. Bartelt (2018). Modelling rockfall dynamics using (convex) non-smooth mechanics. Submitted to NUMGE 2018.

Masuya, H., K. Amanuma, Y. Nishikawa, T. Tsuji (2009). Basic rockfall simulation with consideration of vegetation and application to protection measure. *Natural Hazards and Earth System Sciences 9*(6), 1835–1843.

Niklaus, P., T. Birchler, T. Aebi, M. Schaffner, L. Cavigelli, A. Caviezel, M. Magno, & L. Benini (2017, March). Stonenode: A low-power sensor device for induced rockfall experiments. In *2017 IEEE Sensors Applications Symposium (SAS)*, pp 1–6.

Sneed, E.D. & R.L. Folk (1958). Pebbles in the lower Colorado river, texas a study in particle morphogenesis. *The Journal of Geology 66*(2), 114–150.

Numerical Methods in Geotechnical Engineering IX – Cardoso et al. (Eds)
 ISBN 978-1-138-33198-3

Providing perfect numerical simulations of flexible rockfall protection systems

A. Volkwein
Mountain Hydrology and Mass Movements, Swiss Federal Research Institute WSL, Birmensdorf, Switzerland

ABSTRACT: Although numerical simulations of flexible rockfall protection fences are common today, evaluating the reliability of such simulations remains a challenge. This contribution provides insight into and discusses certain finesses regarding the simulation of flexible rockfall protection fences. It identifies the bottlenecks of suitable simulation calibration and describes how to handle numerical effects which can cause non-realistic simulation results. The article not only gives an overview of the existing approaches used to simulate flexible barrier systems, but also introduces some detail on how the simulations work, which also serves as a form of evaluation manual for people confronted with simulation results.

1 INTRODUCTION

Flexible rockfall protection systems are an efficient and effective way to protect buildings and infrastructure from this considerable natural hazard. Rockfall protection systems usually consist of steel nets which are mounted in the field using steel ropes and posts that in turn are anchored to the ground. Special energy absorbing elements between the steel ropes and the anchors provide high deformation capabilities and act as a form of load limiting device.

For the usage of such flexible barriers there exist corresponding instructions. For example, the New Zealands design considerations for passive rockfall protection structures (NZL 2016) inform customers how to select suitable barriers. The product manuals of the barrier's manufacturers usually provide instructions for installation in the field. The barriers themselves are pre-designed by the manufacturers. I.e., the client places an order regarding the necessary barrier length and height, as well as the expected rockfall energy.

Flexible rockfall protection barriers have recently reached a stunning level of performance. It is now possible to retain up to 10,000 kJ of energy. Compared to the typical energy capacities of rockfall barriers of the mid-eighties this is an improvement by a factor of 50 within about 30 years.

Usually, the reliable performance of a flexible barrier was previously evaluated through type testing according to official standards. The first guidelines of this kind (Gerber 2001) were published in 2001 in Switzerland. In 2008, the guideline ETAG 027 (EOTA 2008) became valid. Barriers certified according to this European guideline obtained a European Technical Approval (ETA) and could be CE marked. Flexible rockfall protection kits are treated as construction elements and are subjected to the European Civil Product Directions (CPD 1989). In 2013, an update of the CPD was issued termed the Civil Product Regulations (CPR 2013). Based on these regulations the existing ETAG 027 is now being converted into a so-called European Assessment Document (EAD) which forms the basis for future European Technical Assessments, abbreviated to ETAs. The ETAG 027 covers rockfall barriers with expected maximum rockfall energies at impact ranging from 100 kJ to more than 4500 kJ. For barriers with a maximum impact energy below 100 kJ a different EAD (2016) serves as a reference for ETAs. A list of assessed rockfall fences is available on the EOTA website (http://valideta.eota.eu/pages/valideta and http://issuedeta.eota.eu/pages/issuedeta, last visited 19.02.2018).

However, such standards only define standard load cases. They usually define a block of a certain shape and mass impacting the barrier with a certain speed into the centre of the middle field of the barrier. It is not clear what happens if different load cases were to occur. For example, the block could enter the barrier eccentrically or it might first hit a post or rope instead of the net. Moreover variation in block shape, or an usual block rotation is not considered. NZL (2016) describes these deficits:

"While ETAG 027 is useful for comparing fence systems developed by different manufacturers, the procedure has some limitations related to cost and technology that the designer must consider including:

- the system is not tested for impacts other than in the centre of the fence panel; in practice the falling rock may impact anywhere on the fence, including posts and anchor cables.

- rotational effects are not considered, as the system is tested with a non-rotating block.
- the test does not account for the *bullet effect* in which a smaller block traveling at a higher velocity (same energy rating) could potentially punch through the net."

The deficits listed above could be covered by systematically testing rockfall barriers for all of the various load cases which could occur. However—apart from the fact that it is almost impossible to cover all natural variants—such a test series would be too costly to implement. Therefore, computational numerical simulation might provide a practical alternative to study the performance of flexible barriers regarding non-standardized rockfall load cases as e.g. shown in Wendeler et al. (2016) or Wienberg et al. (2008).

Numerical simulations are also very efficient when studying the effect of structural changes. For example, the braking distance of the block might increase if the barrier height is increased. However, in such a case, the maximum loads exerted on the anchors may be reduced due to the increased braking distance, the braking forces would be distributed over a longer time interval and therefore lowered. Furthermore protection systems which function in a different way, for example the so-called attenuating systems as described in Glover et al. (2012) can be simulated. For such systems the purpose of the barrier is not to brake the impacting block, but to guide it between a net (hanging along a slope) and the slope down into the valley.

Additionally through the use of numerical simulations the flexible barriers can be checked against load cases completely different to rockfalls, such as debris flows (Wendeler 2008) or shallow landslides (von Boetticher 2013).

Despite the fact that numerical simulations form an efficient and effective alternative to field tests problems still remain in their delivery of reliable results. The user must be sure of the validity of the simulation results and must have full confidence in the simulations in order to use them. The user must also be able to confirm the predicted braking distances or anchor loads produced by the simulations.

When referring to the "user" different people have to be included in this definition. First there is the programmer of the software or the input models used for the simulations. Usually, these people are rather qualified because they know many details of the simulations. Secondly, there are the engineers who have to apply or convert the simulation results to the real barrier. They should be able to interpret the simulations. Finally, there is the client who has purchased or ordered a flexible barrier system and should be able to evaluate the reliability of the simulations.

It is quite simple to produce attractive figures and animations that show how the block is stopped by the barrier. Such animations can be accurately physically modelled using graphical software common in the gaming industry. However, the big benefit of good simulations is that not only are the simulation results delivered but additional information on the validity, reliability and precision of the simulation is also provided. Only with this information can full confidence in the simulations be gained.

2 BASIC NUMERICAL PROCEDURES

As a result of deflections and deformations of flexible rockfall protection barriers they undergo highly geometrical non-linearities, non-linear material and element behaviours alongside continuous structural invariances. The latter usually occurs when for example the net slides along the support ropes or—after sliding a certain distance—is prevented from sliding further and blocked by a post head. Often the structure of the net consists of loosely connected net elements and their connection nodes transfer tensile loads only as well as the steel ropes.

The best numerical scheme to cover these irregularities are—and this is common to all published numerical approaches—explicit time step algorithms. Here, the model is divided into many small mass points and their movements are calculated over time: if a force acts on the node it is converted into a nodal acceleration. Integration over time converts the accelerations into velocities and displacements. The explicit time step algorithms do not search for a dynamic acceleration-force-equilibrium for each time step (as would be the case for an implicit time step procedure) but a rough equilibrium is assumed and a possible imbalance is automatically corrected by the algorithm: a force disequilibrium on a single mass point results in an acceleration of the point and hence a corresponding movement over time. This movement is usually visible until an equilibrium is found. The disadvantage of the explicit algorithms is the need for very small time steps in the magnitude of 1–100 μs (microseconds).

The application of the above approaches varies from discrete elements methods (DEM) to finite element methods (FEM) but also particle based methods (DPM). The approaches use either specially developed, open-source or commercial codes. An overview is provided in the following section.

3 EXISTING SIMULATION APPROACHES

Since the early nineties different approaches to numerically simulate flexible rockfall protection barriers have been used. Mustoe & Huttelmaier (1993) use the Discrete Element Method not to model today's typical steel fence-like structures, but columns consisting of old car tires hanging in a row along a top support rope.

Nicot et al. (2001) first set up a simulation capable of simulating a full flexible rockfall fence considering all of the different components. In their simulation the main focus was on modelling the main net, which consisted of steel rings where each ring is connected to six neighbouring rings. The mechanical model of the net rings consisted of truss elements, one for each connection between neighbouring rings. The truss elements went from one ring centre to the neighbouring ring centre. The model mesh is similar in appearance to a triangular mesh.

Cazzani et al. (2002) modelled a different net type, a cable net, through a grid of truss elements using Finite Element software ABAQUS (2014). They produced visualizations of the dynamic reaction of single net panels and full barriers. In addition to the deformations of the barrier they focussed on the distribution of speeds of the single mass points by providing isotach figures.

Volkwein (2004) formulated a specially programmed discrete element code for flexible barriers which have a primary mesh consisting of steel rings, but where each ring only has four surrounding neighbouring rings. Furthermore, this software presented the first approach capable of mapping the sliding effects of the net along the support ropes, i.e. the so-called curtain effect, on an element base without classical and computationally expensive master-slave contacts using for example the Penalty approach. This software was later enhanced by von Boetticher (2013) to also model chain-link wire meshes. Grassl (2002) formulated a different approach for the same type of net rings.

Bertrand et al. (2012) set up a DEM to model full-scale barriers with a main net consisting of clipped steel ropes.

The basic numerical code ANSYS (2005) was used by Castro-Fresno et al. (2008) whereas Gentili et al. (2012) used ABAQUS to formulate simulations of full-scale flexible barriers. Dhakal et al. (2011) applied the explicit code LS-DYNA (2017) for modelling. Thoeni et al. (2011) implemented an approach into the particle code YADE (2017) allowing the simulation of both typical rockfall stopping fences but also drapery systems which guide the block between the mesh and the hill slope down into the valley. Effeindzourou et al. (2017)) used the same approach. Such a particle code requires a completely different approach to model barrier components compared to the ones presented in the earlier paragraphs. Latorre et al. (2017) also used this approach using the code DEMpack (2017).

Ghoussoub et al. (2014) present new approaches to model a net and sliding effects along the support ropes. Here the nets are modelled as a membrane.

The simulations of Moon et al. (2014) use the Finite Element Code LS-DYNA (2017) to model a barrier.

Escallon (2016) set up a fully detailed model of flexible rockfall protection fences using ABAQUS/explicit. Here, every shackle and other structural detail were fully modelled in 3D. The simulation results are exceptional, however the simulation time is rather long. Coulibaly et al. re-visited mechanical models of ropes and net rings to formulate mechanically consistent models (Coulibaly et al. 2017, Coulibaly et al. 2017a).

3.1 *Comparing numerical simulations*

The previous section described numerous different simulation approaches and solutions. It would be beneficial if these simulations were comparable. If so, one could evaluate the qualities of single simulations and choose the most reliable one. However such easy comparison is hindered by the fact that the single approaches were each calibrated for only one type of protection barrier. Furthermore, the single approaches are often not able to simulate different barrier types.

All of the approaches presented above have been published—without exception—by academic institutions. This—in principle—should guarantee a certain level of detail of the published simulations, i.e. a traceability or repeatability. However, academics usually collaborate with different manufacturers of flexible protection barriers. All simulation approaches are the result of common projects between the industry and the academic world and each project usually focusses on the products of the corresponding manufacturer only. Even if it was of interest of the academic institutions to also model products from different manufacturers, the contracts between academic and industrial partners often do not allow such extensions.

Therefore, each simulation approach has to be evaluated individually. Cross-references are not available and cannot be carried out.

4 EVALUATION OF NUMERICAL SIMULATIONS

Reliably mapping a real barrier with a numerical simulation is a challenge. However the challenge does not lie in the simulation itself, but rather in judging whether a simulation is realistic enough to be reliable. But how can this be tested?

Initially all numerical details of a simulation should be available. For example information on: how the model was set up, which boundary conditions were used and which element types and which material laws were implemented. Having these details provides a level of retraceability that—if necessary—allows the reproduction of a simulation.

Secondly, no accurate simulation can exist without calibration. The load-deformation-characteristics of each individual element of a rockfall protection system have to be adjusted. Different approaches exist for calibrating individual elements. For example if typical material properties are known, stress-strain-relationships can be formulated. However this approach has its limits; when the loading of an element reaches the domain of plastic deformation, the material properties change significantly. Furthermore, care has to be taken regarding stability issues. For example, as long as a steel post does not fail due to high stresses its buckling load might already have been reached.

Another approach uses experimental component tests to set up the numerical properties of a simulation. This procedure is quite common for e.g. the barrier's mesh. The testing procedures mostly consist of quasi-static component tests. An advantage of this approach is that the load limit of individual components can be incorporated into the element properties. Usually the material or element properties are non-linear. Care should however be taken that such tests reflect the loading of individual components as it typically would occur in the complete protection barrier. For example, rope elements are usually validated only for tension. However, their sensitiveness to shear loads requires further consideration of the bending radii of a rope in the field.

Not only is the mechanical load application relevant, but also the loading rate: not every component reacts the same when loaded quasi-statically or dynamically (experiences shown in Studer (2001), Zünd et al. (2006) and Giacomini et al. (2008)). For example, all frictional processes most probably show a dynamic characteristic differing from to the quasi-static one. Therefore, information on the loading rate dependency of individual components is essential. The frictional processes which have been considered, along with their contribution, should also be documented.

Frictional processes not only happen within the individual components of a barrier but mainly between the different components. For example, when the net slides along the support ropes (curtain effect), the support ropes slide along the post heads and bottoms. The friction between impacting block and net surface is essential for the performance of the net at the impact location. For all of these effects assumptions have to be made which correctly describe the type of friction and corresponding friction coefficients. All these assumptions should be reported.

Finally, whether and how a numerical simulation has been calibrated and validated must be clearly stated. Usually, the calibration process is carried out in order to adapt a simulation to certain results, i.e. simulation parameters are adjusted. Validation however uses fixed simulation parameters for a different simulation and documents the performance of this simulation in comparison with the results of a corresponding field experiment. Hence, validation is a way of testing the reliability of a simulation.

5 EXAMPLES FOR NUMERICAL DETAILS

5.1 *Impact between a block and the mesh*

When a block impacts the net the contact area between net and block is subjected to the greatest load. Usually, the impact of the block applies forces which act to stretch single mesh openings. This stretching process is an enormous load for the mesh and is often decisive for the successful performance of the barrier.

There are some numerical possibilities to adjust the maximum net load imposed by the block:

Friction: If there is a sticking or sliding friction defined in the contact between the block and the net, the net load can be reduced significantly. Table 1 shows an example of how the degree of utilization of the net changes depending on Coulomb friction coefficients. Without friction the net's capacity is almost reached, whereas a friction coefficient of 0.3 leaves a 36% residual capacity. Yet, a change in the friction coefficient from 0.3 to 0.5 only influences the degree of utilization by 3%. Therefore, a proper simulation should explain both the contact conditions used in the simulation

Table 1. Influence of Coulomb friction between net and block on the degree of utilization of the net.

Friction coefficient	Degree of utilization of net
0.0	96%
0.1	85%
0.3	64%
0.5	61%
0.7	52%
1.0	46%

and an example of the sensitiveness of the barrier's performance regarding friction.

Contact forces: Today's testing methods in the field often provide the measurement of the block's deceleration during impact. Such data serve excellently to a comparison with corresponding simulations. If the simulated heavy block comes into contact with the—in comparison very light weight—net nodes the latter might still impose a very high acceleration to the block. Such an acceleration peak only has a very short duration and does not strongly influence the block's trajectory. However, if the maximum acceleration of the block is to be evaluated these peaks do not reflect reality. The scatter of acceleration peaks as shown in Figure 1a therefore has to be treated. One approach would be to use the running mean or median when displaying the acceleration over time. Volkwein (2004) used a different approach by numerically applying a 0.1–1 *mm* thick elastic layer around the block which dampens the acceleration peaks efficiently, as shown in Figure 1b.

For the interpretation of simulation results we therefore recommend not only using given maximum values, but we insist on using plots over time to additionally evaluate the given maximum value.

5.2 *Component loads*

Another reason not to solely rely on single values but rather to evaluate plots over time is visualized in Figure 2a: the graph shows the measured and the simulated rope force of a lateral suspension rope of a barrier. Although the maxima of both curves are rather similar their shapes differ a lot. A continuous load curve was obtained from field tests, whereas the simulation revealed a form of vibration in the structure. Here, it should be questioned whether the numerical model was set up correctly. The reason for such a different numerical behaviour should be evaluated and provided with the simulation results. For example, some numerical models react with vibration as soon as the impact load reduces. This effect can clearly be seen in Figure 2b where the load in the retaining ropes

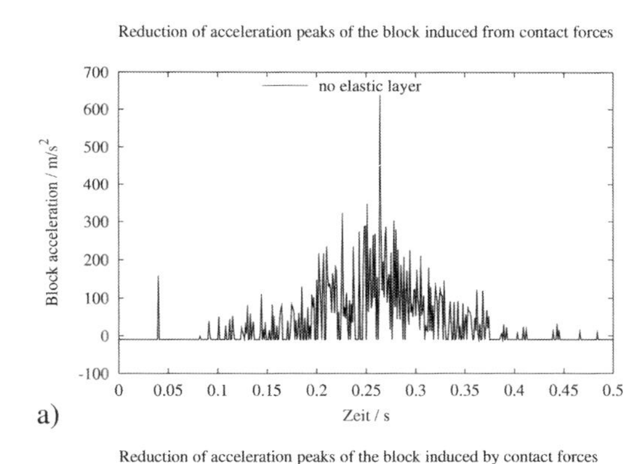

Reduction of acceleration peaks of the block induced by contact forces

elastic layer 0.1mm
elastic layer 1mm

Zeit / s

b)

Figure 1. Block acceleration over time: (a) direct collision of lightweight net nodes with a heavy rigid block and (b) damping of the unrealistic short timed acceleration peaks by applying a numerical elastic layer around the block. Legend:

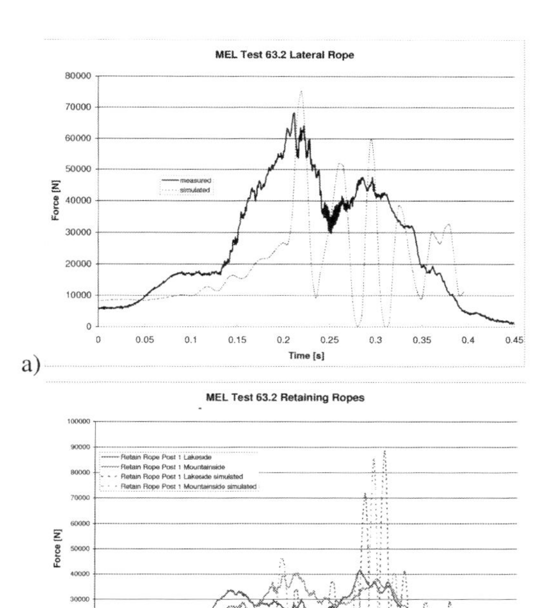

Figure 2. Vibrations in simulated barrier ropes after overall load reduction: (a) lateral suspension rope, (b) retain ropes. The vibrations are so strong that their maxima might be much higher than during the loading process of the impacting block.

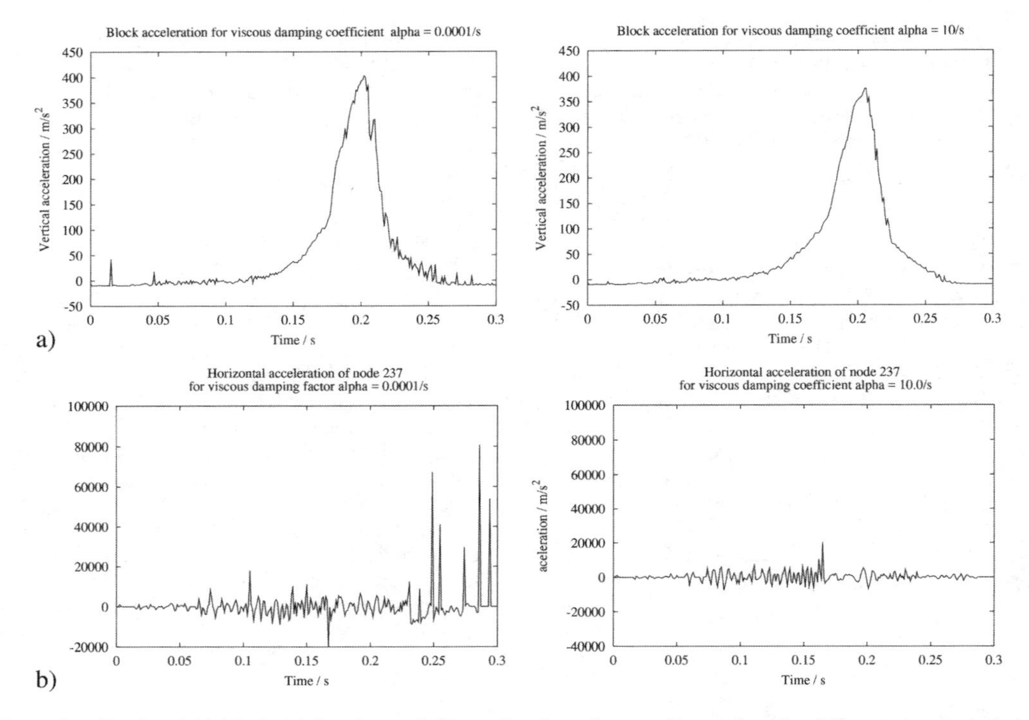

Figure 3. Simulated (a) block deceleration and (b) acceleration of net node over time for different viscous damping coefficients.

reaches high maxima although the main impact has already occurred.

5.3 *Numerical stability*

The vibrations mentioned above can be numerically handled by introducing for example a viscous damping for all mass points. The overall influence of such a numerical damping should be documented because if the damping coefficient is too high, too much energy is absorbed by the virtual damping. Figure 3 shows—for example—a reduction in the maximum block deceleration by about 10% for different viscous damping factors.

5.4 *Material laws*

The material laws which have been implemented should be stated. For example, rope elements could be simulated as linear-elastic components. This is fully sufficient in most cases. Only in cases where the energy absorbing capabilities of ropes are mapped should the material law reflect physical rope behaviour. Figure 4 shows that a non-prestretched rope has a higher deformation capacity for the first loading and stiffens only after some loading cycles.

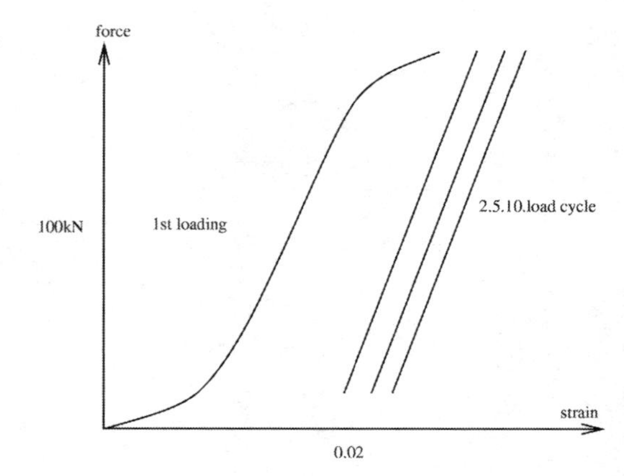

Figure 4. Schematic force-strain diagram of a non-prestretched steel rope.

6 CONCLUSIONS

Numerical simulations are an elegant way to prove the performance of flexible rockfall protection barriers. Even if a barrier has been full-scale type tested, for e.g. CE certification, the simulation allows the consideration of special local demands or structural changes. It is important that any client who orders a barrier is not blinded by attractive simulations. He or she should be able to judge whether the simulations results given are reliable. A few aspects on how to achieve such a judgment are given in this contribution.

To summarize we recommend that good simulation always deliver information on how the simulation procedures have been calibrated and how they have been validated. Knowledge of the ultimate loads the barrier components and the barrier as a whole can withstand are also vitally important. Only with this information can residual capacities be calculated.

The input files used for simulations should be available or reconstructable. Furthermore, information should be provided about who carried out the simulation, only the work of qualified modellers should be accepted.

Simulation results should not consist of singular numbers only, rather plots over time should be provided as they make evaluation of the reliability of the simulation much easier.

REFERENCES

ABAQUS (2014) 6.14 Documentation. Dassault Systems Simulia Corporation.

ANSYS (2005) User's Manual: Procedures, commands and elements. Vols. I, II and III. Swanson Analysis System.

Bertrand, D., Trad, A., Limam, A. et al. (2012) Full-Scale Dynamic Analysis of an Innovative Rockfall Fence Under Impact Using the Discrete Element Method: from the Local Scale to the Structure Scale. Rock Mech Rock Eng 45: 885. https://doi.org/10.1007/s00603-012-0222-5

von Boetticher, A. (2013) Flexible Hangmurenbarrieren: Eine numerische Modellierung des Tragwerks, der Hangmure und der Fluid-Struktur-Interaktion. Shaker.

Castro-Fresno, D., del Coz Diaz, J.J., López, L.A. and Nieto, P.G. (2008). Evaluation of the resistant capacity of cable nets using the finite element method and experimental validation. Engineering geology, 100(1), 1–10.

Cazzani, A., Mongiovi, L., Frenez, T. (2002) Dynamic finite element analysis of interceptive devices for falling rocks. International Journal of Rock Mechanics and Mining Sciences, 39(3), 303–321.

Cerema (2016) Falling rock protection kits with MEL less than 100 kJ. EOTA 15-34-0089-01.06.

Coulibaly, J.B., Chanut, M.A., Lambert, S., Nicot, F. (2017) Sliding cable modeling: An attempt at a unified formulation. International Journal of Solids and Structures.

Coulibaly, J.B., Chanut, M.A., Lambert, S., Nicot, F. (2017) Nonlinear Discrete Mechanical Model of Steel Rings. Journal of Engineering Mechanics, 143(9), 04017087.

DEMpack - Discrete/finite element simulation software. http://www.cimne.com/dem

Dhakal, S., Bhandary, N.P., Yatabe, R., and Kinoshita, N. (2011). Experimental, numerical and analytical modelling of a newly developed rockfall protective cable-net structure. Natural Hazards and Earth System Sciences, 11(12), 3197–3212.

Effeindzourou, A., Thoeni, K., Giacomini, A., Wendeler, C. (2017) Efficient discrete modelling of composite structures for rockfall protection. Computers and Geotechnics, 87, 99–114.

EOTA (2008) ETAG 027, Guideline for European Technical Approval of falling rock protection kits. EU, Brussels, Belgium.

Escallon, J.P. (2015) Simulation of flexible steel wire-net rock-fall barriers via finite element model updating. PhD thesis, ETH Zurich. https://doi.org/10.3929/ethz-a-010484303.

European Council (1989) Construction Products Directive CPD. Council Directive 89/106/EEC). EU, Brussels, Belgium.

European Council (2011) Construction Products Regulation CPR. Regulation (EU) No 305/2011. EU, Brussels, Belgium.

Gentilini, C., Govoni, L., de Miranda, S., Gottardi, G., Ubertini, F. (2012) Three-dimensional numerical modelling of falling rock protection barriers. Computers and Geotechnics, 44, 58–72.

Gerber, W. (2001) Guideline for the approval of rockfall protection kits. Technical report, Bundesamt für Umwelt, Wald und Landschaft (BUWAL), Eidgenössische Forschungsanstalt WSL, Bern.

Ghoussoub, L., Douthe, C., Sab, K. (2014) Analysis of the mechanical behaviour of soft rockfall barriers. RocExs 2014 - 5th Interdisciplinary Workshop on Rockfall Protection, May 2014, Italy. 4 p.

Giacomini, A., Giani, G.P., Migliazza, M. (2008) Quasi-static and dynamic response of energy dissipators for rockfall protection. Australian Centre for Geomechanics. ISBN:9780980418552.

Glover, J.; Denk, M.; Bourrier, F.; Volkwein, A.; Gerber, W. (2012) Measuring the kinetic energy dissipation effects of rock fall attenuating systems with video analysis. In: Koboltschnig, G.; Hübl, J.; Braun, J. (eds) Proc. 12th Congress INTERPRAEVENT, 23–26 April 2012 Grenoble - France.Vol. 1, pp 151–160.

Grassl, H.G. (2002) Experimentelle und numerische Modellierung des dynamischen Trag- und Verformungsverhaltens von hochflexiblen Schutzsystemen gegen Steinschlag. Doctoral dissertation, Swiss Federal Inst. of Technology Zurich, Switzerland.

Latorre, S., Celigueta, M.Á., Irazábal, J., Salazar, F. and Oñate, E. (2017) Design and validation of rockfall protection systems by numerical modeling with discrete elements. Proc. 6th Interdisciplinary Workshop on Rockfall Protection (RocExs), Barcelona.

LS-DYNA (2017) Livermore Software Technology Corporation, Livermore, California.

Moon, T., Oh, J., Mun, B. (2014) Practical design of rockfall catch fence at urban area from a numerical analysis approach. Engineering Geology, 172, 41–56.

Mustoe, C.G.W. and Huttelmaier, H.P. (1993) Dynamic simulation of rockfall fence by the discrete element method. Microcomputers in Civil engineering, 8:423–437.

Nicot, F., Cambou, B., Mazzoleni, G. (2001) Design of rockfall restraining nets from a discrete element modelling. Rock Mechanics and Rock Engineering, 34(2), 99–118.

NZL (2016) Rockfall: Design considerations for passive protection structures. Ministry of Business, Innovation & Employment MBIE, New Zealand. ISBN: 978-0-947493-61-3 (print) 978-0-947493-62-0 (online).
Studer, C. (2001) Simulation eines Bremsrings im Steinschlagschutzsystem. Diploma thesis,ETH Zurich.
Thoeni K, Lambert C, Giacomini A, Sloan S.W. (2011) Discrete modelling of a rockfall protective system. In: Onate E, Owen D, editors. Particle-based methods II: Fundamentals and applications. II International conference on particle-based methods. CIMNE International Center for Numerical Methods in Enginering; p. 24–32.
Volkwein, A. (2004) Numerische Simulation von flexiblen Steinschlagschutzsystemen (No. 289). vdf Hochschulverlag AG.
Wendeler, C. (2008) Murgangrückhalt in Wildbächen - Grundlagen zu Planung und Berechnung von flexiblen Barrieren, Ph.D. thesis, ETH Zurich. *English version: Wendeler, C. (2016) Debris flow protection systems for mountain torrents - basic principles for planning and calculation of flexible barriers., WSL report.*
Wendeler, C., Volkwein, A., Biedermann, B. (2013) A flexible rockfall gallery in the high-energy zone as an alternative to concrete galleries or tunnels - design and application. Zeitschrift für Wildbach-, Lawinen-, Erosions- und Steinschlagschutz 77, 171: 90–97.
Wienberg, N., Weber, H., Toniolo, M. (2008). Testing of flexible barriers – behind the guideline. In Interdisciplinary workshop on rockfall protection, edited by Volkwein, A., Labiouse, V., and Schellenberg, K (pp. 114–116).
Yade - Open Source Discrete Element Method. https://yade-dem.org (last visited: 7.12.2017).
Zünd, T., W. Gerber, M. Sennhauser und A. Müller (2006) Dynamische Tests von Bremsringen. Techn. Ber., Eidg. Forschungsanstalt WSL and Fatzer AG.

Numerical Methods in Geotechnical Engineering IX – Cardoso et al. (Eds)
© 2018 Taylor & Francis Group, London, ISBN 978-1-138-33198-3

Time-lapse crosshole seismic tomography for characterisation and treatment evaluation of the Ribeiradio dam rock mass foundation

M.J. Coelho & R. Mota
LNEC—Laboratório Nacional de Engenharia Civil IP, Lisboa, Portugal

A. Morgado
COBA—Consultores de Engenharia e Ambiente SA, Lisboa, Portugal

J. Neves
EDP—Gestão da Produção de Energia SA, Porto, Portugal

ABSTRACT: Ribeiradio dam, located on Vouga River, Portugal, is the main dam of the multipurpose hydro- scheme of Ribeiradio-Ermida. It is a concrete dam with a gravity profile, with a maximum height of 83 m and a crest length of 265 m. Ribeiradio dam is founded on a granitic rock mass, predominantly constituted by medium to coarse grained granite. Initially in the foundation rock mass there were highly weathered and fractured rock zones until significant depths, especially in the upper elevations of the left bank and at intermediate elevations on the right bank. In the dam's foundation, 34 vertical boreholes were drilled to obtain seismic velocity tomographies for various sections between boreholes. These tests were performed before (Phase 1) and after (Phase 2) the foundation treatment to improve the rock mass's mechanical properties. In Phase 1, the seismic tomographies contributed to the geological-geotechnical zoning of the dam's foundation. With the results of Phase 2, it was possible to evaluate the effectiveness of the rock mass treatment.

1 INTRODUCTION

The Ribeiradio dam at the Vouga River is located about 4 km east from Sever de Vouga, Aveiro (NW Portugal). This is the main dam of the multipurpose hydro-scheme of Ribeiradio-Ermida, whose aims are power generation, water supply for urban, industrial and agricultural uses, and flood control. It is a concrete dam with a gravity profile, with a maximum height of 83 m and a crest length of 265 m, according to a 240 m radius circular alignment, amounting to a total concrete volume of 300,000 m^3.

In geological terms, the foundation of the Ribeiradio dam lies on a NNW-SSE elongated intrusion of medium to coarse grained granite. The foundation rock mass is quite heterogeneous presenting variable mechanical characteristics namely in the upper elevations of the left and right banks, where highly weathered and fractured rock until significant depths occurs.

Crosshole seismic tomography is a geophysical technique that can provide a high-resolution imaging for the subsurface, in terms of seismic P-wave velocity (V_P). Consequently, this is an especially adequate technique to characterize, to zone and delineate weak areas on rock mass foundations for large structures, in construction or already constructed, as demonstrated in several studies, for e.g. Cartmell et al. 1997, Demanet & Jongmans 1997, Coelho 2000, Plasencia et al. 2000, Coelho et al. 2004; Oliveira & Coelho 2004, Barton 2007, Coelho 2007, Lehmann 2007, LNEC 2015a, b, Butchibabu et al. 2017, etc.

Seismic wave velocities are directly related to the quality and strength of the rock masses (mechanical properties) and so they can also be used to evaluate the effectiveness of foundations treatment, when, for e.g., time-lapse seismic tomographies are carried out before and after the treatment (Coelho et al. 2004, LNEC 2015a, b, and LNEC 2017). Generally the rock mass consolidation treatment consists in grout injections to fill joints that usually are open or filled with soft erodible materials (previously removed before the injection). If this filling is effective, there is a strength increase and usually the seismic wave velocities also increase, especially in the highly fractured zones.

Using 34 vertical boreholes drilled in Ribeiradio dam's foundation, specifically for seismic crosshole tests, it was possible to obtain V_P seismic tomographies for various sections and profiles between boreholes. The seismic tests and corresponding tomographies were carried out before (Phase 1) and after (Phase 2) the dam's foundation global treatment (consolidation and grout curtain). Phase 1 seismic tomographies evidenced low and high velocity zones in the rock mass, agreeing, in general, with the previous geological-geotechnical zoning of the dam's foundation. Phase 2 time-lapse seismic tomographies, and consequent analysis of variation of seismic velocity in the rock mass between Phase 1 and Phase 2, allowed assessing the effectiveness of the foundation treatment.

2 GEOLOGICAL AND GEOTECHNICAL SETTING

At the dam site, the Vouga River runs ENE-WSW through a V shaped narrow valley with steep side slopes and a 30 m wide thalweg. A cross-section of the valley in this area shows that both banks are nearly symmetric, the left and right banks with average slopes of 30° and 35°, respectively.

The evaluation of the geological and geotechnical conditions at Ribeiradio's dam site was carried out through a comprehensive investigation program, in order to study the rock mass foundation characteristics.

Geological and geotechnical investigation works included detailed surface geological mapping, mechanical and geophysical site investigations (38 refraction seismic profiles) and in situ and laboratory tests in order to characterize the complex geological and geotechnical site framework.

Surface geological mapping showed the dam is placed on a NNW-SSE elongated anisotropic hercynian granite outcrop, roughly 100 m wide, which has neighbouring contacts with other geological units: injection migmatites, micaschists and quartzites.

Site exploration results also showed that the weathering pattern is more intense on the valley's upper levels and especially penetrative above level (90) on the left bank. All the formations are intensely tectonized and jointed until significant depths and ten different joint sets were identified on each bank, mostly subvertical. Main faults, usually corresponding to zones where jointing of the rock mass is more intense and/or where fillings are only a few centimetres thick, were grouped in six sets, also mainly subvertical.

By the end of the excavation stage, the results of structural mapping works had revealed 366 faults grouped in six main sets, preferably orientated ENE-WSW—which constraint the river alignment on this stretch.

Regarding the rock mass permeability, Lugeon tests showed pronounced absorptions up to a depth of 20–30 m. At lower levels the tests revealed a progressive reduction in water absorption as the quality of the rock mass increased, generally associated to less intense jointing. The impervious boundary was reached at depths of 35–45 m on both banks and 20–25 m at the valley's lower levels.

3 FOUNDATION TREATMENT

The design of the foundation surface considered the results from geotechnical zoning presented in Figure 1.

To limit the excavation depth to values compatible with the height and type of dam, most of the dam foundation surface was set on GZ2 rock mass, showing RMR values between 40 and 60 (fair quality). This design conception lead to estimated excavations depths of 5–7 m on the central zone of the valley, 8–10 m along intermediate levels of both banks and 12–15 m above level (90) on both banks (see Figure 1).

In spite of the envisaged excavation, the remaining rock mass at low and intermediate levels of the valley only showed fair conditions and, on the upper levels, the geotechnical conditions were even poorer. This situation called for a systematic consolidation treatment to improve the dam's foundation operational behaviour (see Figure 2), reducing deformability and increasing strength conditions by filling rock mass discontinuities (open or filled with soft, erodible materials). This treatment considered 20–25 m deep grout holes, in order to reach the limit between GZ2 and GZ1 horizons. The efficiency of this treatment greatly benefited from the several joint sets intersections and consequent intercommunication, which promoted a significant increase of the areas subjected to cement grouting, mainly in the rock mass's superficial levels.

The consolidation holes were organized in specific treatment sections composed by upstream and downstream fans of inclined boreholes, executed from the main drainage gallery as well as

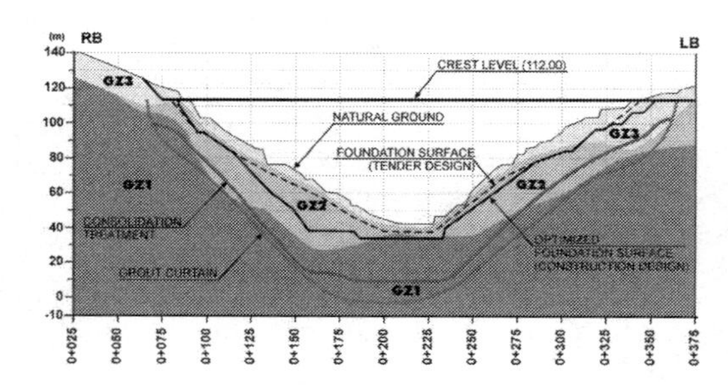

Figure 1. Ribeiradio dam: geotechnical zoning, foundation surface and ground treatment.

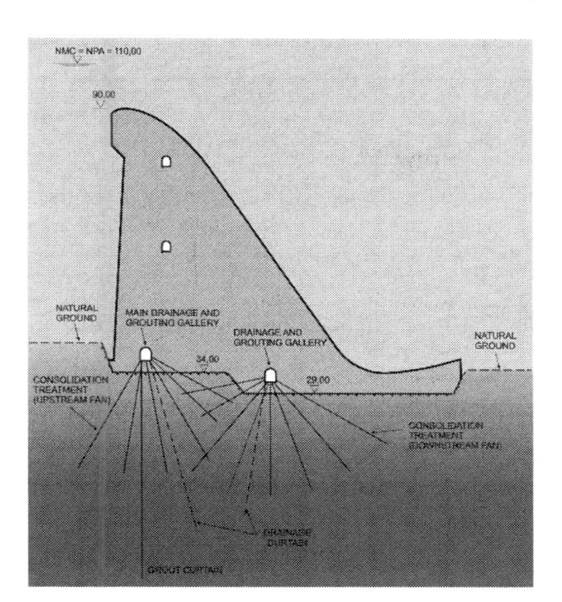

Figure 2. Ribeiradio dam: foundation treatment.

the downstream drainage gallery (see Figure 2), which allowed injecting systematically with greater density and efficiency the weathered and fractured upper part of the dam foundation. Total drilling amounted to 19,400 m, and total cement absorption reached 706 $\times 10^3$ kg.

Owing to the fact that the water tests undertaken inside the boreholes showed occurrences of easy flow at great depths on both banks, an impermeabilization treatment was considered necessary, resorting to a single row grout curtain executed from upstream the gallery incorporated in the dam. The grout curtain reached 20 to 35 m below the foundation surface (see Figure 2), with total drilling amounting to 12,600 m and total cement absorption reaching 266 $\times 10^3$ kg.

Two drainage curtains were also foreseen from both drainage galleries available to allow controlling residual seepage below the foundation after the consolidation and waterproofing treatments (see Figure 2), but mainly to reduce installation of uplift stresses at the base of the dam.

The effectiveness of the global treatment of the rock mass foundation was evaluated recurring to: a) water absorption tests carried out systematically inside injection holes; b) real time analysis of the injected cement (in each stage); c) crosshole seismic tomographies (indirect method) and d) dam monitoring. It will be addressed in the next sections that treatment was proven to be effective and that the results from the indirect method (c) were, in general, in good agreement with those from the other methods.

4 CROSSHOLE SEISMIC TOMOGRAPHY

4.1 *Method guidelines*

Crosshole seismic tomography is a wide used geophysical method to study the elastic properties of materials in the crosshole section, for a large variety of scales and applications, including the engineering site characterisation and the foundation evaluation (Jackson & McCann 1997, Coelho 2000, Barton 2007, Lehmann 2007, LNEC 2015a, b and Butchibabu et al. 2017).

In the crosshole seismic tomography here considered, P-wave velocity (V_P) distribution is reconstructed from the travel time ray paths, corresponding to several crosshole seismic measures as illustrated in Figure 3, assuming a purely bidimensional wave propagation on the crosshole section.

In data acquisition, for each seismic source activation along the source borehole, there is a multichannel seismic record from the receivers array along the receiver borehole.

In this study all P-wave measured travel times for a unique crosshole section or for a set of adjacent crosshole sections (profile), were jointly inverted into a V_P matrix assigned to a cells grid which covers (discretizes) the area with seismic ray paths, using a numerical method. The inversion was performed by a SIRT (simultaneous iterative reconstruction technique) type algorithm implemented by Pessoa (1990). This algorithm uses straight ray paths (direct waves) assumption, constant velocity for each grid cell

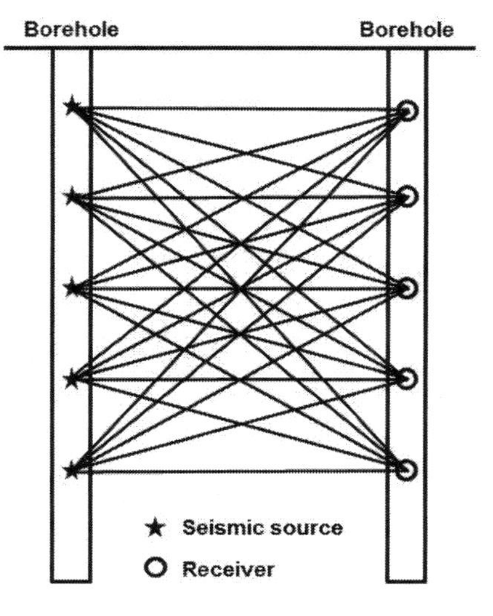

Figure 3. Data acquisition layout for crosshole seismic tomography.

and assumes isotropic materials. It starts with an initial velocity matrix (initial model) for which are computed both the travel times for all the straight ray paths and the differences between the calculated and the measured travel times. These time differences, called residual times, are used to improve the velocity model in the next iterations until the next residual times will be minimized.

4.2 *Data acquisition and processing*

In Ribeiradio dam's foundation there were carried out 41 crosshole seismic tomography tests using 34 vertical boreholes along the drainage and visit galleries (see Figure 4). These boreholes were drilled specifically for the seismic tests.

The seismic data was acquired in two different times for the 41 crosshole sections: the first time or Phase 1, before any foundation treatment, and the second time or Phase 2, after the global foundation treatment already described.

Test boreholes have lengths (depths) between 27 and 47 m, 17 of them having more than 40 m length. They were steel cased, sealed at the bottom and filled with water, in order to carry out the seismic crosshole tests.

For each tested crosshole section, spacing between consecutive source and receiver positions along the boreholes was 1 m, which generated a huge quantity of gathered data, especially for the longer crosshole sections (several of them with more than 1000 ray paths). Tested crosshole sections have distances between boreholes from about 10 to 17 m.

In Phase 1 two seismic sources were used, a sparker and electrical detonation caps, while for Phase 2 only electrical detonation caps were used. The receivers used were hydrophones molded in two multi-channel cables, each having 24 hydrophones with 1 m spacing.

Seismic data (P-wave travel times) from some adjacent sections were jointly processed and inverted for a unique tomographic planned profile. So, for each phase, two tomographies were calculated for the upstream alignment (see Figure 5a, b), three tomographies for the downstream alignment (one of them at Figure 6a, b), and nine tomographies for the radial (transversal) sections or profiles with two adjacent sections (one of them at Figure 7a, b).

For the tomographic inversion all sections or planned profiles were discretized by grids with 1 m side square cells. The initial model (velocity matrix) was a uniform velocity model which value was equal to straight rays mean velocity. The final inverted models were chosen to have low enough mean residual error (ratio between the mean of absolute values of residual times and the mean of measured times, in percent value) and, if possible, to avoid the increment of velocity artefacts. Artefacts due to data errors and limitations and due to the inversion algorithm itself are often present in seismic tomography, causing anomalous high or low velocity zones (Pessoa 1990, Demanet & Jongmans 1997, Coelho 2000, LNEC 2015a, b).

Considering all the tomographies obtained for both Phases 1 and 2, the mean residual errors were less or equal than 10%, being, in general, lower in Phase 2 than in Phase 1, for the same sections or profiles. This reduction of the mean residual errors in Phase 2 was expected due to either the higher data quality in relation to Phase 1, either the greater homogeneity of the rock mass in the last phase.

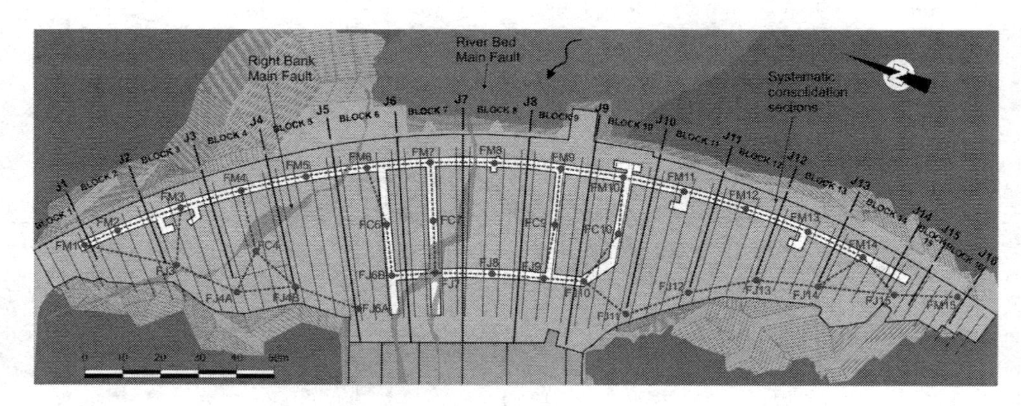

Figure 4. Ribeiradio dam plan with: blocks, joints, drainage and grouting galleries, test boreholes (blue dots) and crosshole tomography sections (blue dashed line), and consolidation treatment sections; main geological faults are also superimposed on this plan.

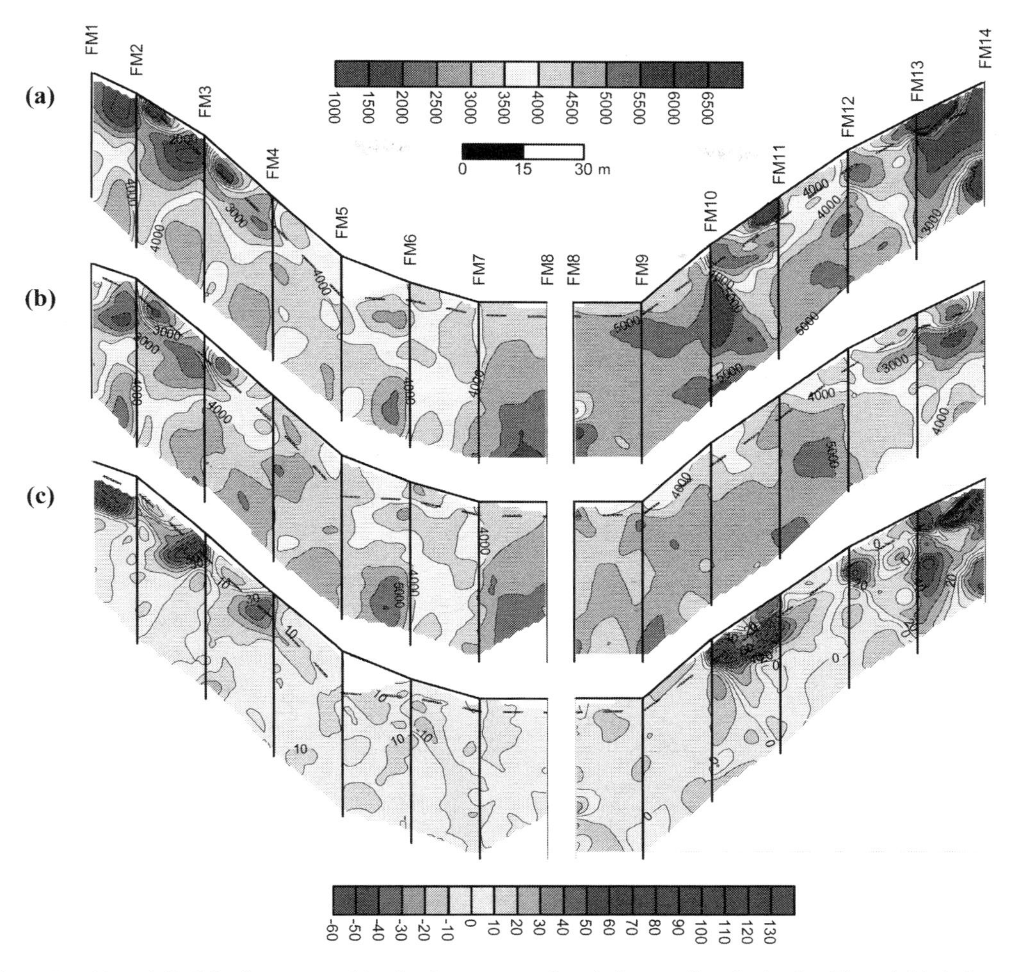

Figure 5. (a) and (b) Seismic tomographies for the upstream foundation profile, obtained at Phase 1 (a), before the treatment, and at Phase 2 (b), after the treatment; (c) V_P percent variation at Phase 2 in relation to Phase 1. Black dashed line represents the limit of the concrete layer, over the rock mass. The FM1 borehole top is at level (95).

4.3 *Results*

From all the seismic tomographies achieved for the dam's foundation those presented on Figures 5–7 were chosen as examples. In their parts (a) and (b) are respectively displayed the V_P seismic tomographies for Phases 1 and 2. The (c) part presents the V_P percent variation between Phase 2 and Phase 1 (obtained from the ratio between Phases 2 and 1 V_P difference, and Phase 1 V_P).

In Phase 1 seismic tomographies revealed a relatively heterogeneous rock mass with velocity ranging between 1500 to 5500 m/s, as can be observed in Figures 5a, 6a and 7a. Typically V_P takes values about 1500–2500 m/s in both margins' upper levels, agreeing with the poorer geotechnical zones; velocities of about 4500–5500 m/s were found on the central zone of the valley and on the left bank lower levels, coinciding with higher quality rock mass. Tomographies evidenced the lower and higher quality rock mass zones, which agreed, in general, with the previous geological-geotechnical zoning of the dam's foundation, having allowed its complement and detailing.

The geological main faults depicted at Figure 4 were remarkably evidenced in Phase 1 tomographies as weak zones, with relatively low velocity until greater depth, in particular in FM5-FM6-FM7 crosshole sections (Figure 5a left), in FJ3-FJ4 A-FJ4B crosshole sections (Figure 6a) and in FJ6B-FC6-FM6 profile (Figure 7a). According with (Morgado et al., in prep. and Morgado et al. 2017), the Right Bank main fault (see Figure 4) dips 40–60° towards the river, while the River Bed

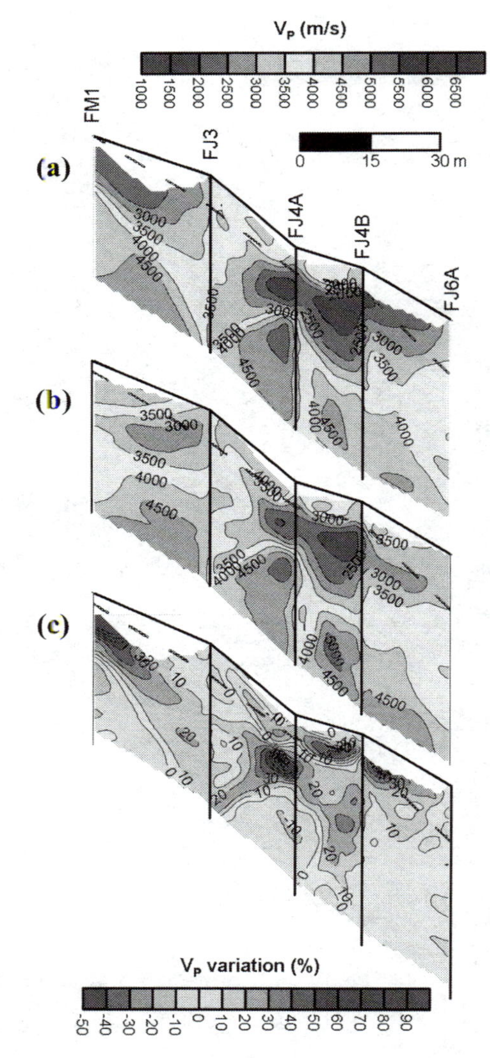

Figure 6. (a) and (b) Seismic tomographies for the FM1-FJ3-FJ4 A-FJ4B-FJ6 A downstream profile, obtained at Phase 1 (a) and at Phase 2 (b); (c) V_P percent variation at Phase 2 in relation to Phase 1.

main fault (see Figure 4) dips 40–60° towards the right bank. In this way these two main faults cross each other in the proximity of FM6-FM7 cross-hole section and, probably, also in the vicinity of both FJ3-FJ4 A-FJ4B-FJ6 A and FJ6B-FC6-FM6 crosshole sections.

Some tomographies, especially those from Phase 1, present located anomalies that may be artefacts. Examples of this are the excessive high velocities (≥ 5500 m/s) in the concrete layer, where usually V_P takes values around 4000 m/s, or near very low velocity zones (some of them also

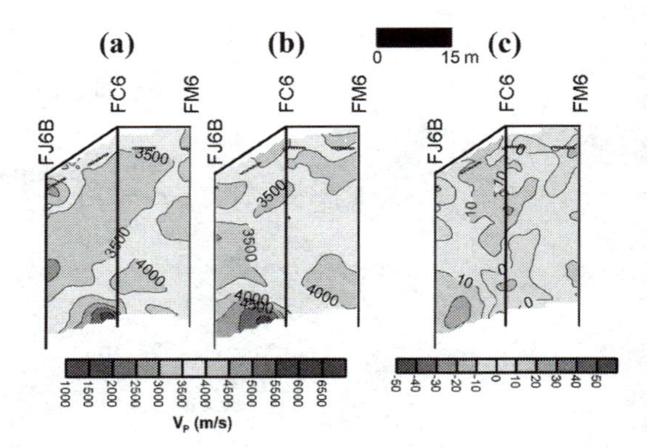

Figure 7. (a) and (b) Seismic tomographies for the FJ6B-FC6-FM6 radial profile, obtained at Phase 1 (a) and at Phase 2 (b); (c) V_P percent variation at Phase 2 in relation to Phase 1.

existing in the concrete layer), as can be observed in tomographies of Figure 5a. In Phase 2, either the velocity contrasts between superficial and deep materials, either the rock mass heterogeneity, have diminished; also higher quality seismic data, with higher S/N ratio, were achieved. Consequently, there were less velocity artefacts in Phase 2 tomographies.

Phase 2 tomographies showed, in general, a significant V_P increase (> 20%) on the superficial zones of both banks' upper levels, where the rock mass showed lower quality, given its weathering degree and intense jointing, and consequently where V_P took the lowest values (see Figure 5b, c and Figure 6b, c).

The higher velocity increase (>20%) in the surficial zones of the rock mass, at upper levels of the right and left banks, occurred in the zones which, before the treatment, had higher alteration and fracturing and where V_P took the lowest values.

Velocity increases higher than 10% were also noticeable on Phase 2 tomographies in the above mentioned fault zones, for both FM1-FJ3-FJ4 A-FJ4B-FJ6 A longitudinal downstream profile (Figure 6b, c), and FJ6B-FC6-FM6 radial profile (Figure 7b, c).

In central zone of the valley and in the left bank lower levels, as expected, there were no significant variations of V_P on Phase 2 due to the higher quality of the massif. There were even some small velocity reductions that, however, are irrelevant since the initial V_P values were high enough and might even been raised by the Phase 1 tomographic processing.

Other negative V_P variations were also observed in some zones where strong velocities contrast occurred or in some border zones, but are generally

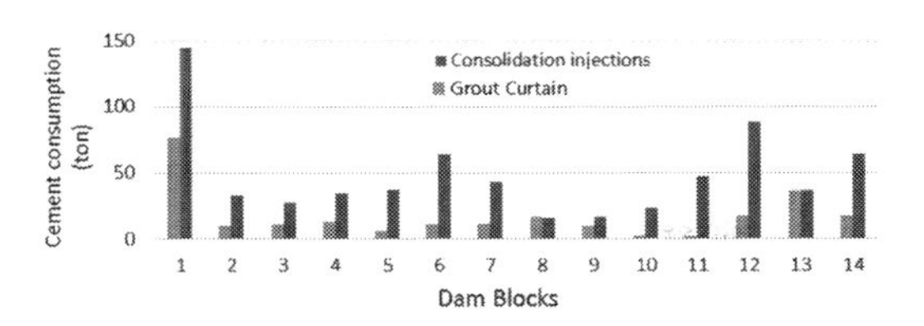

Figure 8. Cement consumptions during consolidation injections and grout curtain in Ribeiradio dam foundation.

of small extent and result possibly from data errors and from data processing method itself.

5 DISCUSSION AND CONCLUSIONS

Time-lapse seismic tomographies provided a qualitative assessment regarding the effectiveness of Ribeiradio's dam foundation rock mass treatment, which included a consolidation treatment and a grout curtain.

Figure 8 displays the cement consumption for each dam block (see blocks distribution at Figure 4), until Block 14, during consolidation injections and grout curtain. Higher cement consumption occurred in blocks where time-lapse tomographies revealed significant velocity increase, as can be seen in the examples showed in Figure 5, Figure 6 and Figure 7.

After consolidation and grout curtain treatments, the water absorption measurements (based on Lugeon tests) evolved from total loss (especially in GZ2), before any treatment, to values between 0 and 2 Lugeon units in the grout curtain zone. This substantial permeability reduction proofed the effectiveness of the rock mass injection treatment.

Dam monitoring considers the measurements of the main loads and of the representative parameters of the structural responses. Their combined analysis, that allows assessing the dam's behaviour over time, didn't reveal any anomalous situation during the first reservoir filling (LNEC 2015c), what seems to corroborate the effectiveness of the rock mass foundation treatment.

In summary, this case study illustrates that seismic tomography proves to be a valid complementary tool that allows evaluating the improvement of rock mass mechanical properties after its treatment. Moreover it provides a continuous image of the rock mass dam foundation in opposition to discrete measurements obtained from other methods.

This method may also be used as a management tool to monitor the rock mass behaviour along the dam's working lifetime. Their results can additionally be used to calibrate the mechanical parameters for the modelling of the dam's structural behaviour.

ACKNOWLEDGEMENTS

The authors are thankful to EDP—Gestão da Produção de Energia SA, for the permission to publish this work and for providing all the necessary data for it.

REFERENCES

Barton, N. 2007. Rock quality, seismic velocity, attenuation and anisotropy. Leiden:Taylor & Francis / Balkema.

Butchibabu, B., Sandeep, N., Sivaram, Y.V., Jha, P.C., Khan, P.K. 2017. Bridge pier foundation evaluation using cross-hole seismic tomographic imaging. *Journal of Applied Geophysics* 144: 104–114.

Cartmell, S.J., Conn, P.J., Pugh, T.D. 1997. An example of the use of crosshole tomography in dam wall foundation studies. In D.M. McCann et al. (eds.), *Modern geophysics in engineering geology*: 141–151. London: The Geological Society.

Coelho, A.G. 2007. The added value of geology in site investigation. XIV European Conference on Soils Mechanics and Geotechnical Engineering, Madrid, September 2007: 203–216.

Coelho, M.J. 2000. *Crosshole seismic tomography for geotechnical investigation*. (in Portuguese). Monography (Trabalho de síntese), Public examination for LNEC Research Assistant (Provas para Assistente de Investigação do LNEC). Lisboa: LNEC.

Coelho, M.J., Salgado, F.M., Fialho-Rodrigues, L. 2004. The role of crosshole seismic tomography for site characterization and grout injection evaluation on Carmo convent foundations. In A. Viana da Fonseca & P.W. Mayne (eds.), *ISC'2, 2nd International Conference on Site Characterization, Porto, September 2004, Vol. 1*: 443–449. Rotterdam: Millpress.

Demanet, D. & Jongmans, D. 1997. Seismic tomography survey under La Gileppe dam. In D.M. McCann et al. (eds.), *Modern geophysics in engineering geology*: 175–182. London: The Geological Society.

Jackson, P.D. & McCann, D.M. 1997. Cross-hole seismic tomography for engineering site investigation. In D.M. McCann et al. (eds.), *Modern geophysics in engineering geology*: 247–264. London: The Geological Society.

Lehmann, B. 2007. Seismic traveltime tomography for engineering and exploration applications. Houten: EAGE.

LNEC. 2015a. Crosshole seismic tomography at the Ribeiradio dam foundation, Phase 1 (before treatment). (in Portuguese). Report LNEC 12/2015. Lisboa: LNEC.

LNEC. 2015b. Crosshole seismic tomography at the Ribeiradio dam foundation, Phase 2 (after treatment). (in Portuguese). Report LNEC 394/2015. Lisboa: LNEC.

LNEC. 2015c. *Ribeiradio dam – Numerical simulation of the observed structural behaviour*. (in Portuguese). Technical report - Inspection visit to P3 level of Ribeiradio dam (Annex 2). Lisboa: LNEC.

LNEC. 2017. Crosshole seismic tomographies at the foundation of Baixo Sabor dam, after the first reservoir filling. (in Portuguese). Report LNEC 209/2017. Lisboa: LNEC.

Morgado, A., Oliveira R., Costa C., Monteiro G., Lima C., Queralt M. 2017. Ribeiradio and Ermida dams foundation treatment – Design, execution and effectiveness control. *The International Journal on Hydropower and Dams. Seville, 9–11 October 2017.*

Morgado, A., Monteiro, G., Oliveira, R., Queralt, M., Cacilhas, F., Lima, C. in prep. Ribeiradio dam foundation treatment – design, effectiveness control and performance. *Submitted to the 26th Congress on Large Dams*. Vienna: ICOLD 2018.

Oliveira, M. & Coelho, M.J. 2004. Decomposed rock mass characterization with crosshole seismic tomography at the Heroísmo station site (Porto). In A. Viana da Fonseca & P.W. Mayne (eds.), ISC'2, 2nd International Conference on Site Characterization, Porto, September 2004, Vol. 1: 531–538. Rotterdam: Millpress.

Pessoa, J.M.N.C. 1990. *Application of tomographic techniques to crosshole seismic exploration*. (in Portuguese). Monography (Trabalho de síntese), Public examination for Teacher Assistant (Provas de aptidão pedagógica e capacidade científica). Aveiro: Universidade de Aveiro.

Plasencia, N., Coelho, M.J., Lima, C., Fialho-Rodrigues, L. 2000. The role of seismic tomography in the Venda Nova II powerhouse rock mass characterization. (in Portuguese). *VII Congresso Nacional de Geotecnia, Porto, Abril 2000, Vol. I*: 279–285.

Numerical Methods in Geotechnical Engineering IX – Cardoso et al. (Eds)
 ISBN 978-1-138-33198-3

Opening effect on mechanical behaviour of rock brick

Y.L. Gui
School of Engineering, Newcastle University, UK

ABSTRACT: Openings in rock play a significant role in the performance of rock related structures. The well-established knowledge in this area can contribute to the engineering practices, for example, underground space design, planning and optimisation in Civil and Mining Engineering and wellbore stability in Drilling Engineering, among others (Gui et al. 2017). In this paper, a numerical study on the effect of non-banded openings, i.e., circular opening, on the rock mechanical behaviour is performed using UDEC. It is revealed that the proposed simulation method can reproduce reasonably the crack initiation and propagation, and predict well the change of the mechanical behaviour due to the openings.

1 INTRODUCTION

Rock is heterogeneous and normally contains defects. The defects in the rock (faults, joints, beddings, and openings) can propagate, interact and coalescence under internal and external stresses, leading to damage or failure of the rock. The influence of these defects particularly affects in two aspects: (1) rock mechanics properties (e.g., compressive strength and stiffness) decrease seriously due to existing defects; (2) defects may propagate or coalesce with other defects under loading, leading to a further degradation in stiffness and strength (Cao et al. 2015). Therefore, the study of the defects effect on the rock mechanics and rock engineering is meaningful.

The main objective in the paper is to numerically study the influence of openings on the stiffness, strength and fracture pattern of rock. The numerical simulations are compared with the experimental work. All numerical specimens described in the paper are cut from the specimen of the uniaxial compressive test to keep the numerical model consistent. Two patterns of opening (i.e., single opening and multi-openings) are investigated using single set of parameters. The results have demonstrated that opening in rock can significantly influence rock mechanical behaviours, including stiffness, strength and fracture pattern under compression.

2 METHOD

The material used in the study is Hong Kong granite with grain size ranging from 0.3 to 3 mm (Lin et al.2015). All the rock models are discretised using Voronoi tessellations with average size of 0.7 mm using UDEC. Voronoi tessellation has proved to be non-mesh bias and allows the crack to initiate and propagate its pattern to maximally achievable (Gui et al. 2016a, c). Voronoi grain size effect on the mechanical behaviour of rock can be found in the work of (Gui et al. 2016b). The contacts between adjacent grains are treated using an elastic-brittle fracture model and the grain itself is elastic. Cracking can only occur at the contact once its tensile strength or shear strength is exceeded. Material parameters used in the simulations are calibrated by modelling uniaxial compression test in Wong et al. (2006) and are listed in Table 1. In the laboratory uniaxial compression test, a cylindrical specimen with dimensions of 100 mm height and 50 mm diameter as suggested by ISRM (International Society for Rock Mechanics) was tested. The crack initiation and propagation simulations of specimens with openings (Figure 1) are performed based on the specimens obtained by cutting the numerical model of uniaxial compression simulation to avoid the potential mesh variation induced discrepancy, and keep all numerical models consistent (Gui et al. 2017).

The boundary condition used in the simulations was that the bottom is fixed along the vertical direction while a constant loading rate is applied on the top boundary. The loading rate in the laboratory test was 0.003 mm/s. Theoretically, the loading rate used in the simulation should be the same as the one used in the experiment. However, the low loading rate adopted in experiment will result in large number of computational cycles so that produces the target deformation in the simulations. Therefore, the selected loading rate in the simulations is raised and a mechanical local damping ratio of 0.8 is used to eliminate the dynamic effect (Kazerani and Zhao 2012). Considering the

Table 1. Summary of the model parameters.

ρ (kg/m3)	2601	k_s (Pa/m)	4×10^{13}
E (GPa)	80	σ_t (MPa)	10
υ	0.34	c (MPa)	20
kn(Pa/m)	8×10^{13}	ϕ (°)	40

Note: ρ-density; E-Young's modulus; υ-Poisson's ratio; k_n-normal stiffness; k_s-shear stiffness; σ_t-tensile strength; c- cohesion; ϕ-friction angle.

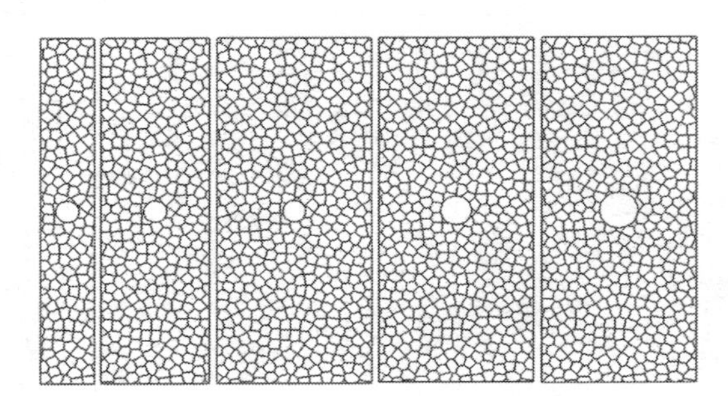

Figure 1. Numerical models for the specimens with single hole.

efficiency and accuracy, the loading rate used in the simulations is 0.01 m/s. Plane-strain analysis was used for all the simulations.

3 RESULTS AND DISCUSSIONS

3.1 *Width effect in single hole specimen*

The numerical models of the five specimens presented in Wong et al. (2006) are presented in Figure 1. The height of the five specimens is 24 mm, while the width of the five specimens is 3.5, 7, 10, 10, and 10 mm, respectively, as listed in Table 2. The radius of the centre holes in the five specimens is 0.75, 0.75, 0.75, 1.0 and 1.25 mm, respectively. The Young's modulus and compressive strength obtained from the simulations are also presented in Table 2. It can be seen that for the specimens with the same hole size (0.75 mm), the Young's modulus and compressive strength are both increased with increasing the specimen's width, from 28.65 GPa and 71.87 MPa to 31.54 GPa and 124.01 MPa, respectively, which agrees well with the experiments. It is considered that increasing the width of the specimen will increase the cross-sectional area, thereby increasing the compressive strength and Young's modulus.

The cracks initiate from the north and south poles of the holes and then propagate vertically since the maximum tensile stress is produced at the north and south poles. In addition to the propagation of the cracks starting from the south and the north poles, there are other cracks initiating randomly across the whole specimen, especially when the loading stress closely reaches to the peak stress. The random cracks primarily initiate and propagate along vertical direction. Accompanying the cracks growth is volumetric expansion and vertical slabbing of the rock specimen. It is also noted that enlarging the width of specimen can increase the crack initiation stress for the three specimens. The cracks propagate nearly vertically (Gui et al. 2017).

Table 2. Summary of single hole specimens and their simulation results.

H	24	24	24	24	24
W	3.5	7	10	10	10
N	1	1	1	1	1
R	0.75	0.75	0.75	1	1.25
n	0.0210	0.0105	0.0074	0.0131	0.0204
E	28.65	31.15	31.54	30.97	30.52
σc	71.87	108.66	124.01	112.15	91.46

Note: H-height (mm); W-width (mm); R-hole radius (mm); N-number of holes; n-opening ratio; E-Young's modulus (GPa); σ_c-compressive strength (MPa)

3.2 *Hole size effect in single hole specimen*

For the specimens with identical width (i.e., 10 mm), the hole radius is varied from 0.75 to 1.25 mm, as shown in the right three models in Figure 1. As indicated in Table 2, the Young's modulus and compressive strength are both decreased from 31.54 GPa and 124.01 MPa for specimen with hole radius of 0.75 mm to 30.52 GPa and 91.46 MPa for specimen with hole radius of 1.25 mm. This also can be explained by the effective cross-section area which decreases if increasing the hole radius, thereby decreasing the bearing capacity (i.e., compressive strength) as well as global stiffness (i.e., Young's modulus).

It is also found that larger hole radius induces lower crack initiation stress. The first cracks also always start from the north and south poles of the hole and propagate nearly vertically. Volumetric expansion and vertical slabbing are also occurred in the three specimens.

3.3 *Hole alignment effect in multi-hole specimen*

The effect of the multi-holes on the mechanical behaviour of rock is investigated based on the laboratory testing (Lin et al. 2015). In the laboratory testing, the specimen dimension was

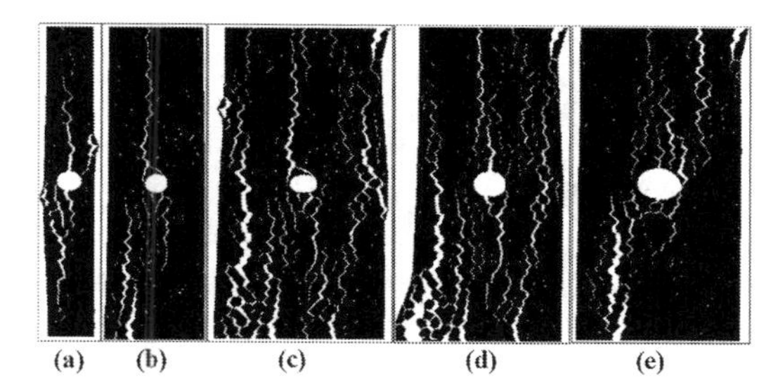

Figure 2. The final failure pattern of the specimen with single hole.

15 mm × 30 mm, and four hole configurations were adopted: vertical, horizontal, diagonal and random alignments. The numerical models are shown in Figure 3. In Figure 3, H, V, D and R represent horizontal, vertical, diagonal and random alignment, respectively. The radius of the holes are the same (i.e., 0.75 mm) and the rock bridge length (distance between two neighbouring holes measured from the hole boundary) is 1.5 times the hole radius (i.e., 1.25 mm) except D3 for which the bridge length is 1 mm due to the size limitation in the experimental specimen preparation.

The simulated Young's modulus and compressive strength together with the experimental results are listed in Table 3. It is obvious that the hole alignment has significant impact to the mechanical behaviour of the rock, i.e., the Young's modulus and compressive strength. More specifically, for the horizontal alignments (i.e., H1, H2 and H3), the simulated Young's modulus and compressive strength are both reduced if increasing the number of holes: from 30.38 GPa and 112.28 MPa for the specimen with 3 holes (H1) to 26.67 GPa and 88.95 MPa for the specimen with 15 holes (H3), respectively. To demonstrate the correctness of the UDEC simulations, the experimental compressive strengths (both average value and value range) are listed in Table 3. It is shown that the modelled compressive strengths are generally in good agreement with the experimental compressive strength range. However, the simulated compressive strength of H2 is intermediate among the three specimens, which is the lowest in the experimental average results. The observations of the experimental results can be due to the heterogeneity among the rock specimens in terms of mineralogy contents, spatial distribution etc. The size of the experimental specimens is very small (30 mm × 15 mm) and its mineral grain size ranges from 0.3 to 3 mm (Lin et al. 2015), which is inevitable to induce the heterogeneity in the specimens; however, this heterogeneity can be avoided in numerical simulation through using the same set of parameters.

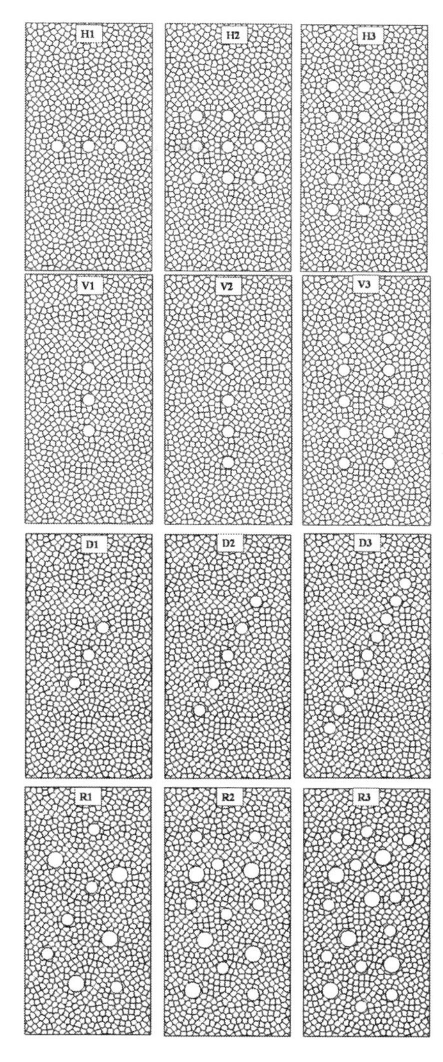

Figure 3. Numerical models for the specimens with multi-holes.

It is noted that the row number is varied but the number of holes in each row is fixed in the specimens with horizontal alignment (i.e., H1 to H3). The vertical alignment both changes the number of holes in one column (V1 and V2) and the number of holes column (V2, V3). From the simulations, it is found that increasing the number of holes can slightly decrease the Young's modulus but increase the compressive strength (e.g., V1 and V2 in Table 3) for a single column of holes. However, both the Young's modulus and compressive strength decline with increase of the number of column with same number of holes in each column (e.g., V2 and V3 in Table 3). The simulated

Table 3. Summary of single hole specimens and their simulation results.

	H1	H2	H3
H	30	30	30
W	15	15	15
N	3	9	15
R	0.75	0.75	0.75
n	0.0118	0.0354	0.0589
E	31.20	29.35	27.62
σc *	108.2	73.2	84.3
σc **	77.8–138.5	62.5–94	65.6–103.0
σc	112.28	107.47	88.95
	V1	V2	V3
H	30	30	30
W	15	15	15
N	3	5	10
R	0.75	0.75	0.75
n	0.0118	0.0196	0.0393
E	31.55	31.04	29.43
σc *	111.5	98.7	74.6
σc **	72.3–161.8	68.0–142.7	67.9–81.3
σc	116.31	124.46	100
	D1	D2	D3
H	30	30	30
W	15	15	15
N	3	5	9
R	0.75	0.75	0.75
n	0.0118	0.0196	0.0354
E	31.19	30.65	27.40
σc *	68.4	54.0	24.6
σc **	66.7–70.0	46.0–62.0	20.6–28.6
σc	84.75	53.60	44.76
	R1	R2	R3
H	30	30	30
W	15	15	15
N	9	13	17
R	0.75, 1.0	0.75, 1.0	0.75, 1.0
n	0.0475	0.0663	0.0850
E	28.14	26.72	25.58
σc *	57.8	55.6	42.8
σc **	48.8–62.4	46.2–69.4	36.7–48.8
σc	70.54	70.32	52.27

Note: *H*-height (mm); *W*-width (mm); *R*-hole radius (mm); *N*-number of holes; *n*-opening ratio; *E*-Young's modulus (GPa); σ_c -UDEC simulated compressive strength (MPa); (*) -average experimental compressive strength; (**)-experimental compressive strength range.

compressive strengths are also in good agreement with the experimental value ranges. Since the failure in the uniaxial compression test of brittle plate with vertically aligned holes are primarily in the form of splitting (mode-I failure), the vertical alignment of the holes can decrease the accumulation/concentration of the tensile stress on the south and north poles of the holes, therefore smaller tensile stress will occur if more holes are aligned vertically.

For diagonal alignment, the simulated Young's modulus and compressive strength both decrease with the number of holes, from 31.19 GPa and 84.75 MPa for 3 holes (D1) to 27.40 GPa and 44.76 MPa for 9 holes (D3), respectively, which agrees with the trend observed in the experiment. It is found that the specimens with diagonal patterns show large ductile phenomenon, especially for D2 and D3. This is because of the shear band along the diagonal orientation and the loading on the top boundary make the shear band under biaxial compression condition. The stress along and perpendicular to the shear band is approximated as

$$\begin{cases} \sigma_n = \sigma_1 / \sqrt{5} \\ \sigma_s = 2\sigma_n = 2\sigma_1 / \sqrt{5} \end{cases} \quad (1)$$

where σ_1, σ_n and σ_s are the applied compressive stress on the top boundary of the specimen, the stress perpendicular to the shear band, and the sliding stress along the shear band, respectively. Both the two stresses on the shear band increase with the applied stress on the top boundary. However, the increasing rate of the sliding stress (σ_s) is higher than the normal stress (σ_n) on the shear band. Once the anti-friction capacity is overwhelmed by the sliding stress, the shear deformation is accelerated. Thus, there is no obvious volumetric expansion and vertical slabbing compared with the specimens with horizontally and vertically aligned holes (Figure 4), due to the dominant shear failure along the shear band in the specimens with diagonal holes.

The fracturing progress of the horizontal and vertical alignment is now discussed. Overall, pre-peak cracks are more random compared with single-hole specimens. The cracks are more initiated and propagated when the compressive stress is close to the compressive strength. More holes induce lower stress of crack initiation. Crack coalescence is mainly formed between two neighbouring holes along the vertical direction. In addition, the failure of the specimens is mainly splitting failure. Volumetric expansion and vertical slabbing are also observed. Distinct from the regular alignments (vertical and horizontal), the cracks clearly start from the holes in both the diagonal and random alignments. In diagonal alignment, the cracks propagate initially along vertical direction, but they turn to diagonal direction to its adjacent holes, therefore the primary propagation direction

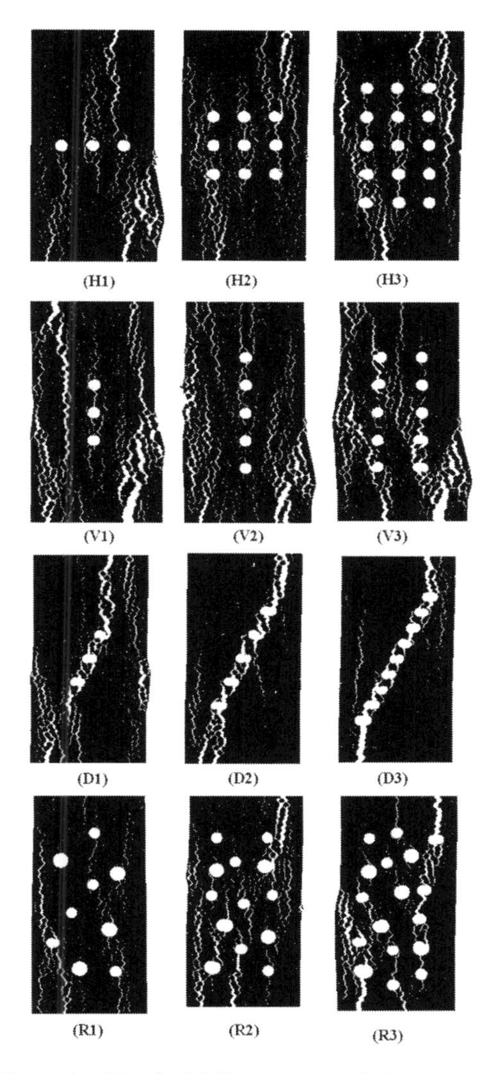

Figure 4. The final failure pattern of the specimens with multi-holes.

is inclined, resulting in a clear inclined shear band. In random alignments, the cracks start from both large and small holes, and coalescence is mainly formed along two nearest holes to the top or bottom of each other.

4 CONCLUSIONS

Effect of openings on rock mechanical behaviour is investigated using UDEC. In the simulation, all the rock specimens are discretized using Voronoi tessellation grains. The models with single opening and multi-opening are simulated. The simulation results demonstrate that the opening in rock can significantly influence the stiffness, strength and fracture pattern under uniaxial compression. It is also demonstrated that the proposed method using UDEC is appropriate and applicable by comparing the simulation results with the experimental results.

REFERENCES

Cao, P., Liu, T.Y., Pu, C.Z. & Lin, H. 2015. Crack propagation and coalescence of brittle rock-like specimen with pre-existing cracks in compression. *Engineering Geology* 187: 113–121.

Gui, Y.L., Bui, H.H., Kodikara, J., Zhang, Q.B., Zhao, J. & Rabczuk, T. 2016a. Modelling the dynamic failure of brittle rocks using a hybrid continuum-discrete element method with a mixed-mode cohesive fracture model. *International Journal of Impact Engineering*. 87: 146–155.

Gui, Y.L., Zhao, Z.Y., Ji, J., Wang, X.M., Zhou, K.P. & Ma, S.Q. 2016b. The grain effect of intact rock modelling using DEM with Voronoi grains. *Geotechnique Letters* 6(2): 136–143.

Gui, Y.L., Zhao, Z.Y., Kodikara, J., Bui, H.H., Yang, S.Q. 2016c. Numerical modelling of laboratory soil desiccation cracking using UDEC with a mix-mode cohesive fracture model. *Engineering Geology* 202:14–23.

Gui, Y.L., Zhao, Z.Y., Zhang, C. & Ma, S.Q. 2017. Numerical investigation of the opening effect on the mechanical behaviours in rock under uniaxial loading using hybrid continuum-discrete element method. *Computers and Geotechnics*. 90:55–72.

Itasca Consulting Group Inc. 2008. UDEC manual (Version 5.0), USA.

Kazerani, T. & Zhao, J. 2012. A discrete element model for predicting shear strength and degradation of rock joint by using compressive and tensile test data. *Rock Mechanics and Rock Engineering* 45: 695–709.

Lin, P., Wong, R.H.C. & Tang, C.A. 2015. Experimental study of coalescence mechanisms and failure under uniaxial compression of granite containing multiple holes. *International Journal of Rock Mechanics and Mining Sciences* 77: 313–327.

Wong, R.H.C., Lin, P. & Tang, C.A. 2006. Experimental and numerical study on splitting failure of brittle solids containing single pore under uniaxial compression. *Mechanics of Materials*.38: 142–159.

Author index